Small Arms Of The World

A BASIC MANUAL OF SMALL ARMS

a basic manual of small arms

STACKPOLE BOOKS HARRISBURG, PENNSYLVANIA

SMALL ARMS of the WORLD

the classic by W. H. B. Smith

TENTH EDITION COMPLETELY REVISED BY

Joseph E. Smith, ARMY MATERIEL COMMAND

The Stackpole Company
Cameron and Kelker Streets, Harrisburg, Pa., 17105

Tenth Edition
Printed in the U.S.A.

LC72-90881 SBN8117-1566-3

Preface

This the tenth edition of "Small Arms of the World" clearly indicates that new small arms continue to appear all over the globe. Of particular interest are new developments in the Peoples Republic of China, (mainland China). In the past, that country had mostly utilized Soviet designed material or slight modifications thereof. But a definite new tack in design has appeared. The Soviets, for their part, continue to produce modification upon modification of basic designs; and the Czechs have produced some brand new designs.

In the Western world most of the activity in the development of military weapons appears to be in the realm of commercial firms, with West Germany leading the list. Much development in both East and West seems to be aimed at the "third World" export market since the weapons developed are frequently offered in a multiplicity of calibers.

You may find minor variations in the weapons characteristics listed here, in other books, or even when checked against a sample weapon. Weights will frequently vary by as much as one-half pound among weapons of the same type having wooden stocks, because of the varying density and humidity content of the woods used. Lengths may vary because of the number of decimal places used in conversion from metric to English measurements. Data given in various publications, even official handbooks or manuals, does not always make clear whether the weights given are of the weapon with sling, loaded or unloaded, with cleaning kit, etc.

SMALL ARMS OF THE WORLD is as factual as possible, and we appreciate any correction of factual errors within the limitations noted above. I try to answer personally all letters of inquiry or comment regarding the contents of this book, but sometimes the press of business makes the immediate answer of correspondence impossible; I am, however, grateful for all suggestions. I have tried to be objective in this book and have limited my comments to particular advantages or disadvantages of various systems. The opinions expressed are entirely my own.

Charlottesville, Virginia

Joseph E. Smith

Acknowledgments

Many individuals, organizations and business firms have contributed to this and previous editions of this book. The list includes those who have assisted by encouragement as well as those who have contributed data and photographs.

UNITED STATES: These individuals and organizations of the U.S. Army Materiel Command: General Henry A. Miley, Mr. Nat Cancilla, and Mr. Tom Cosgrove, and Dr. Fred H. Carten, Mr. Bill Bonkemeyer and Mr. Ralph Palese formerly of that organization; the U.S. Army Foreign Science and Technology Center and Col. Robert A. J. Dyer, Dr. John Ord, Mr. M. A. Shapiro, Mr. Hal Johnson, Mr. Craig Burden, Mrs. Barbara Rebich and Mr. Donald Rhode of that organization also Mr. Jerry Reen, and Mr. Phil Valentini formerly of that organization. Mr. Dick Maguire and Mr. Lou Artioli of the U.S. Army Weapons Command; Mr. Tom Wallace of the Springfield Armory Museum Inc., Mr. Howard Johnson, of Aberdeen Proving Ground and Mr. Karl Kempf, Mr. Robert Farris and Col. G. B. Jarrett (USAR Retd), formerly of that installation, Mr. Ludwig Olson of the National Rifle Association, Mr. Donald Bady of New York City, Mr. Val Forgett of Service Armament Inc., Mr. C. Goins of the Smithsonian, Washington, D.C., Interarms, Alexandria, Va., especially Mr. Sam Cummings, President and Mr. Richard Winter his director at Alexandria, Mr. Tom Nelson, President of Replica Models Inc., Alexandria, Va., the Mars Equipment Co. of Chicago, Ill., the Armalite Co. of Costa Mesa, California, Colts of Hartford, Conn. Smith and Wesson of Springfield, Mass., Cadillac Gage Co. of Detroit Mich., and Mr. Eugene Stoner designer of the AR 10 and 15, and the Stoner 63 Series of weapons, New England Armament Corp of Boston, Mass., Century Arms of St. Albans, Vt., my daughter Kathleen Smith who assisted with typing and my wife and the rest of my family who retained their patience during the gestation period of the revision.

ARGENTINA: Mr. I.J. Osacar of Buenos Aires and Armas & Equipos, R. L. of Cordoba.

BELGIUM: Fabrique Nationale of Herstal and Messrs. R. Laloux and Vervier of that firm, and the Musee Royal De L' Armee et D' Histoire Militaire.

CANADA: Mr. V. "Jack" Kroma of Willowdale, Ontario.

DENMARK: Warrant Officer V.G.B. Christensen, G. Larsen, Tojhus Museum and Dansk Industri Syndicat of Copenhagen.

FINLAND: Mr. P. Janhunen, Helsinki.

FRANCE: Messrs. G. Demaison and M. Roy, and MAB, Bayonne.

GREAT BRITAIN: Mr. J. R. Ward and Mr. Herbert Woodend Royal Small Arms Factory, Enfield Lock, Major F. A. W. Hobart (Retd) Royal Military College of Science, Shrivenham, and BSA Guns Ltd. of Shirley, Solihull, Warwickshire.

WEST GERMANY: Mr. Hans Lockhoven of Cologne, Mr. O. Moraiewitz of W. Berlin, Heckler and Koch A. G. of Oberndorf and Carl Walther of ULm A/D.

ITALY: Armi P. Beretta of Gardone V. T., Brescia, and Luigi Franchi of Brescia.

NETHERLANDS: Mr. H. L. Visser

SPAIN: Star Bonifacia Echeverria S. A. Eibar

SWITZERLAND: Mr. Fred Datig of Luzurn, SIG of Neuhausen am Rheinfalls.

URUGUAY: Mr. J. A. Nin, Montivideo

To the numerous others who in some small way helped make this edition complete and factual, I extend my thanks.

Contents

Part One: Historical

Part Two: Current Weapons

Contents

Part One: Historical

Part Two: Current Weapons

Contents

Contents

PART 1

Historical

1 Origins of Gunpowder and Firearms

Before entering upon a history of small arms, it is appropriate to consider the history of that substance—75 percent potassium nitrate (saltpeter), 15 percent charcoal, 10 percent sulphur—which provided the propelling power for all these weapons until the late 19th century. Gunpowder was certainly known as a chemical compound with explosive properties for a long time before its capacity to project items with lethal energy was realized. In other words, the grenade preceded the gun, apparently by a considerable period of time.

Written records of the earliest use of firearms, and of the development of gunpowder as a firearms propellant, are far from complete.

HISTORY OF GUNPOWDER

PARTINGTON'S STUDIES

Recent research by Professor J.K. Partington, a British chemist, indicate that the Chinese were using saltpeter-based compositions by 1000 A.D., long before they were known in the West. Professor Partington's book, A HISTORY OF GREEK FIRE AND GUNPOWDER, utilizes the original Chinese, Sanskrit, and Arabic texts and transliterates them into modern terminology. The fact that Professor Partington is a chemist lends more authenticity to his work than that of many of his predecessors in this field. Unfortunately much previous writing on the origin of gunpowder has been overly flavored with nationalism, the British favoring Bacon, the Germans favoring Berthold Schwartz, etc. Partington rather effectively proves that the "Greek Fire" which was used extensively in the West did not contain saltpeter. This is also borne out in the "Alexiad," the history of the Roman Emperor Alexius I who ruled from 1081 to 1118, written contemporaneously by the Emperor's daughter Anna. This work is the chief source of reliable data on the First Crusade and the Byzantine Empire. It lists numerous formulas for "Greek Fire" but none are detonating substances. All evidence indicates that gunpowder was not known in the West until the 13th Century.

EARLY RECORDS OF USE OF GUNPOWDER

LIBER IGNIUM, the work of Marcus Graecus (Mark the Greek), indicates a knowledge of both saltpeter and gunpowder. Recent research indicates that this manuscript is not a Latin translation of a Greek work as originally thought, but a Latin translation of an Arabic work produced about the time of Roger Bacon and Albertus Magnus.

Bacon, an English monk, refers to gunpowder in two works; the OPUS MAJUS of 1268, and DE SECRETIS OPERIBUS ARTIS ET NATURAE which may have been written about 1248. At roughly the same time another manuscript DE MIRABILIBUS MUNDI appeared which had similar formulae to those appearing in LIBER IGNIUM. This work is attributed to Albert Magnus (1193 - 1280).

The German monk Berthold Schwartz may actually have been the figment of someone's imagination. He was credited by some with the invention of gunpowder at Freiberg during the 1300s, but gunpowder was known in the West at least a century earlier and references to Schwartz (Black Berthold), did not appear until the end of the fifteenth century.

The only thing that is sure about the invention of gunpowder, a substance which for good and evil has had as much influence on the history of man as any other substance or item yet known, is that no one knows who invented it or who first realized its deadly capability. Many learned men have produced works on the origins of gunpowder, but none have ever been able to prove definitely that a specific individual invented gunpowder at a specific time and a specific place. While application of this explosive substance to the projection of objects cannot be traced to a particular individual, information on its origin is a bit more definitive.

FIREARMS

DEFINITION

Firearms may be defined as tubes used with a charge of gunpowder (or like explosive substance) to hurl projectiles. In their earliest forms as true projectile instruments, these hurled the same stone or metal balls, or variations of the spears and arrows, which were then in use for launching from bows, slings, ballistae, and the like.

Flame Throwing Tubes

It is impossible to establish just when and where firearms as defined above were introduced. Tubes were used long before the advent of gunpowder to direct inflammable mixtures, much as the modern flame thrower does. "The Alexiad" specifies such use, for instance.

"Roman Candle" Tubes

The first use of gunpowder to launch objects without thought of penetration may well have been in the Far East. Both the Tartars and the Arabs are known to have made very early use of a variety of "Roman Candle" tubes to start fires. These instruments were commonly hollow tubes of wood or bamboo. They were tightly wrapped around with hide or hemp or wire for strength. They were loaded from the muzzle with alternate charges of powder and an incendiary ball, often of tallow, though some seem to have been cloth saturated with crude petroleum. They were ignited at the

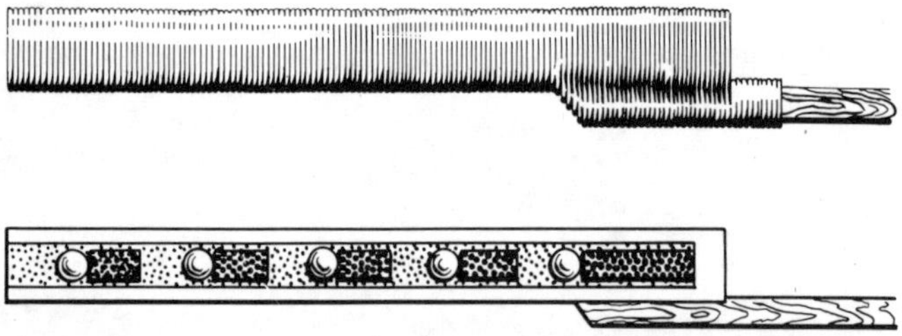

"Roman Candle" Tube. From a thirteenth century Saracen Manuscript.

muzzle, and as the fire worked around each ball it touched off the powder to launch the ball from the tube. The German "Zeitschrift fur historische Waffenkunde," a former learned society which went to great lengths to research ancient arms and armor, found mention in old Chinese annals authentically dated 1259 of this use of powder.

THE EARLIEST RECORDS

Early Saracen records have been translated to show that stone-throwing cannon were used in 1247 in the defense of Seville. However, the general evidence indicates that these were actually mechanically operated projectors—not gunpowder types.

German writers have made various claims that the town of Amberg had a cannon in the year 1301. This and similar claims have not been documented. The reports of many writers and researchers of the 19th Century indicating Flanders as the source of the origin of the first projectile gunpowder arms have been seriously questioned by later writers who had far better research facilities, notably Sir Charles Oman and Oscar Guttman.

The Manuscript of Walter de Millimete

The first contemporaneous illustration which can be fully authenticated showing an unquestionable gunpowder cannon is found in the Millimete Manuscript at Oxford. The illustrations on the manuscript show very fine illuminations of cannon. One vignette shows a soldier in armor firing a bottle-shaped cannon at a fortress gate at close range. The gun itself is on a four-legged mount. From its muzzle projects a huge spear-like "bolt" of the pattern known as a "garrot" or "carreau" such as is shown in early illustrations dealing with mechanically launched projectors like the espringale. Smaller bolts of this type are shown at a later date being used in hand arms. Such "bolts" often had brass "feathers" to stabilize flight, patterned of course after the arrow, though many seem to have been used to impart rotation to the flying missile much in the manner that rifling does.

It must be noted that this manuscript is a dedicatory address which Walter de Millimete gave to King Edward III upon accession to the throne in 1327. The text makes no mention whatever of the illustrated items. However, since we know that Edward III was among the earliest monarchs to employ cannon in battle, and since the authenticity of the manuscript is beyond question, the historical value is self-evident.

Edward III

Shortly after he was crowned, Edward III led an army against the Scots. It is said that he had with him a group of cannoneers from Hainault, indicating that his guns came from Flanders. The weapons were apparently of the type shown in the Millimete manuscript, often spoken of as "pot-de-fer." Records of Ghent for the year 1313 indicate such weapons, while later records refer to their use at Metz in 1324. It is apparent therefore that Edward III used cannon, but did not develop them.

John Barbour's THE METRICAL LIFE OF ROBERT THE BRUCE is the most often quoted source on this Scotch campaign, speaking of cannon as "crayks of war." The facts are in all probability accurate in general, though the careful historian must note that Barbour wrote late in life when he was Archdeacon of Aberdeen. He was only seven years old in 1327.

Many historians state that Edward III used two or three cannon at the battle of Crecy. This, too, may be true.

Other Records

It is impossible to establish whether the handgun or the artillery cannon was first used as a powder weapon. It is quite possible that the development may have been simultaneous. The construction and the loading and firing principles were the same for all types in the early stages of arms development. The ignition was the same.

In general the only source of recording knowledge in those days

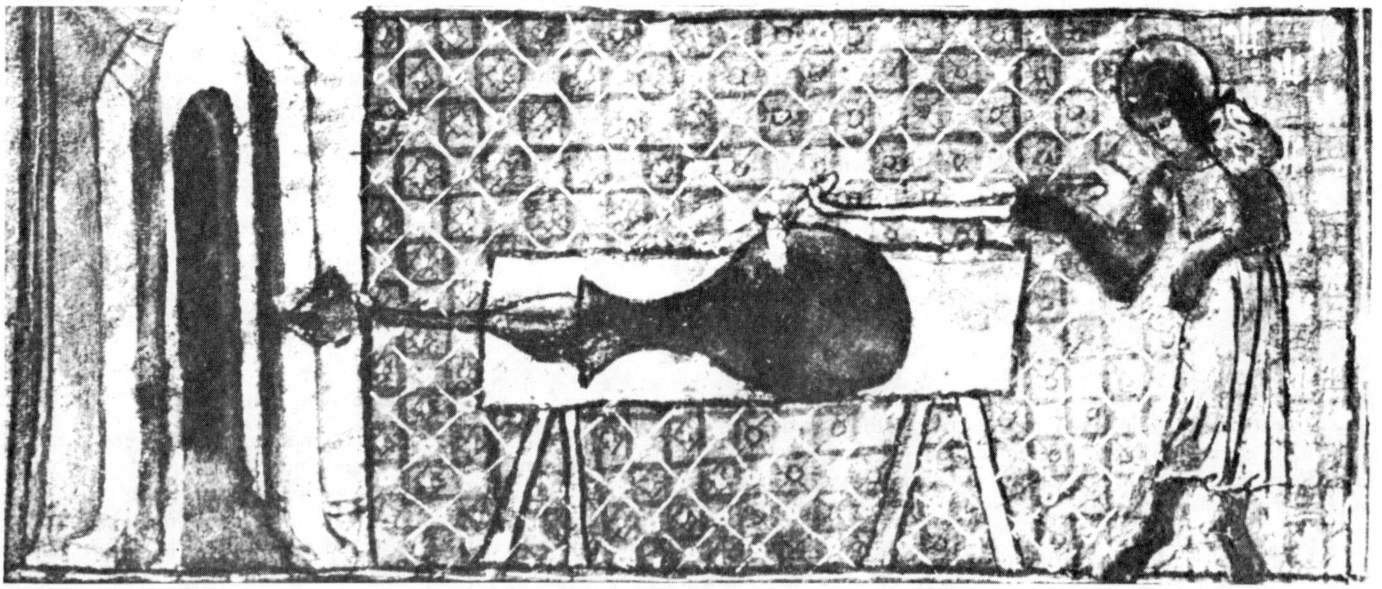

The oldest verifiable illustration of a weapon using gunpowder for hurling projectiles. An illustration from the de Millimete Mss. executed in the year 1326 A.D.

rested with the religious groups, more particularly with those in Europe, since few outside church circles could read or write. Moreover, as the era of record keeping developed early in the 14th Century, a new terminology built around the new weapons developed in various countries. The French records speak of "quenon" or "cannon," for instance. The Lowlanders described them most often as "vogheleer." Among the Italians they were recorded as "bombardes." The German records still extant list them as "Buchsen."

There are very few firearms made during the early part of the 14th Century whose authenticity can be even vaguely established. However, art in the form of tapestries, paintings, church frescoes and illustrative drawings was developed to a very high degree about this time, particularly in Italy and France. These art forms together with more detailed written records have enabled arms historians to correlate the terms in various countries, and thus to provide a pattern of continuity to show the development of firearms from this point on.

For the record, however, it must here be set forth that in the last century there were in existence in Italy two handguns which historians of great ability believed to be the oldest then extant. The first of these was thoroughly discussed by German General Köhler in his comprehensive work on early arms ENTWICKLUNG DES KRIEGSWESENS, ETC. It was of bronze, ornamented with a Greek cross and oak leaves, bore the number 1322 (believed to indicate the date of manufacture) and the letters PPF. Köhler described it in his book published in 1887, though it had been described earlier (in 1847) by Count d' Arco who owned it at one time and also by the reliable Major Angelucci in his DOCUMENTI INEDITI published in Turin in 1869. This gun was stolen from the Monastery of St. Orsola at Mantua in 1849.

The second "sclopos" said to date from the early part of the 14th Century was a wrought iron piece having a very roughly finished barrel and shaft attachment which was understood to have been excavated from the ruins of ancient Monte Vermini castle in 1841. Assuming it to be genuine, this piece would date from at least 1341.

While these particular weapons are lost to historians, from this period on Italian records from unimpeachable sources verify the existence of their types; and from these records and those of other European countries we can establish a chronology of development accurate in every respect except that exact dates cannot also be provided. Some of these can be tied down to a given year, but in the main all but the official Annals and Chronicles can be assigned only to a period which may vary as much as 25 years. Of course, even in our own times it is often impossible to exceed that record on quite recent developments.

2

Evolution of Firearms

Written records of the early development of firearms are quite incomplete. Who it was that first fashioned a metallic tube through which to propel a projectile at a distant or near target, using an explosive substance as a propellant, and just when this was done, we do not precisely know.

The term "lock" was used with early firearms to indicate the device used to fire the weapon. Discussed below are cannon locks, matchlocks, wheel locks, flintlocks (including the Snaphaunce and Miquelet), and percussion locks.

The development of percussion arms and of repeating arms is examined, as well as that of the transitions arms of the period, and of metallic cartridges.

THE CANNON LOCK

The earliest firearms are classed as "cannon locks." These arms, whether in the form of hand or artillery types, were fired by holding a lighted coal or hot iron against powder placed over a touchhole at the rear of the barrel section. The flame flashed through the hole into the powder charge, which had been loaded down the muzzle of the piece.

Cannon locks were commonly made by casting brass, bronze, or similar alloy. Some were made of wrought iron. Many had straight bores, though others had clearly defined chambers. The hand types were usually attached to staffs or pikes. The heavier types used various forms of mounts, ranging from cradles to carts.

Granulation was not known at this period. Gunpowder was actually a true powder and required considerable wadding between charge and projectile to allow a buildup of gas pressure. The low saltpeter content of the mixtures also was a factor in slowing down combustion. The wads used were usually soft wood. Gas leakage around the wad naturally limited velocity and power.

Illustrations of the projectiles used appear in profuse detail in many tapestries and manuscript illuminations. Balls of stone, iron, brass, and even lead were used. Bolts—called variously "quarrels" and "garros a feu"—were also employed. These bolts all developed from the crossbow types of missiles. They were made in a huge range of sizes for handgun and siege gun work.

It is with these elementary firearms that our chronology of development begins.

When we remember that Italy was the best educated and most scientifically advanced country at this period, it is not strange that we find the best records in Italian archives. However, toward the middle of the century some authentic English records appear, most of the important ones indicating actual manufacture of the arms in the Lowlands (specifically Belgian areas which are still a source of export arms). French records at this period also become more specific and authoritative. Toward the close of the century the finest records available are German manuscripts which begin to chronicle the developments of that ever scientific and sometimes warlike Teuton mentality.

EARLY ITALIAN RECORDS

The most reliable sources of original material of the earliest days which the author has been able to weed out of literally thousands of manuscripts, drawings, records, and tapestries studied are listed below, sometimes by date.

1324. ARCHIVO DE FLORENCE, REG. 23, DE RIFORMAGIONI, PAGE 65. This record definitely establishes that various firearms were in general use in Florence at that time. It is obvious that the arms mentioned must have been in use for some time before this date, since such matters usually are not chronicled until well after all experimental stages have been passed.

1331. CHRONICLES OF CIVIDALE. This was an Italian town in Venezia. Its CHRONICON EXTENSE dated three years later also makes definite mention of handguns (sclopetus).

1340. The famous artist Paolo del Maestro Neri began work in this year on a series of frescoes in a church near Sienna, Italy. The receipt signed by the artist upon receiving payment for completing the work is dated 1343. It is in the Library at Sienna today.

The Neri frescoes can still be seen, though they are badly peeled. They are the most important records in the entire history of early firearms. These frescoes show clearly both land and naval warfare of the period. Besiegers are shown firing cannon against

Cannon locks in military use. From the Italian Neri Frescoes dated 1343.

Wheel mounted cannon in siege. Defenders in tower are firing hand cannon locks. From Neri Frescoes.

Cannon lock drawing. From the English Burney Manuscript Number 169, Folio 127, dated 1469.

a castle in one panel. The defenders are shown using hand cannon and bows in defense.

1364. CHRONICLES OF PERUGIA. These are important as showing the extent of firearms in use at this period. They cite an order placed for "500 bombarde." The record shows that the specifications required that these handarms must penetrate armor!

The Frescoes show guns of various types which are in all essentials identical with those shown in German manuscripts produced near the end of the 14th Century. By showing in the same panels the use of both siege and hand firearms, they establish historically that such developments were either parallel or very closely related.

EARLY ENGLISH RECORDS

1347. Among the English records of early days only those of Thomas de Roldeston, Keeper of the King's Privy Wardrobe, are truly specific. He served during the period of Edward III. Items cover work on guns, purchase of gunpowder at 18 pence per pound, and substantial purchases of sulphur and saltpeter.

In other sources mention is made of the "ribauldequin" used by Edward III. This is one of the earliest types of multi-fire weapons which were the precursors of the machine gun. It consisted of several iron barrels arranged to fire simultaneously, and its earliest recorded use is in 1339.

1386. The first recorded use of the term "handgun" is found in English records. The Chamberlain is recorded as having been sent three such by one Ralph Hutton.

EARLY GERMAN RECORDS

The earliest true scientific approach to recording the progress of all firearms is found in German records. It is unlikely that the original gun as such was of German origin. There can be no question, however, that the most productive of the succeeding development work and its recording were done by the Germans.

An instance in point is the CODEX GERMANICUS 600. This is a manuscript in the Munich Museum. Some German authorities have dated it as far back as 1345. It is more likely that the date is that set by less prejudiced researchers—about 1390.

This manuscript translates "Directions for Preparation of Gunpowder. How to Load Guns and Discharge Them." It is extensive, detailed, and scientific.

Many of the finest examples of German art, manuscript illumination, and tapestries dealing with early firearms were destroyed or looted during the period of the two world wars in this century; so reference will be made here only generally to certain records which were unquestionably authenticated, or of which copies exist in accepted museums or libraries.

1389. The Vestaburg Inventory listed firearms, among them being "Handbuchsen" or hand firearms. Of course, this is again direct evidence that such arms were in use for a considerable time before the date of the inventory, or at least had gone through the long experimental stages required in all scientific and mechanical developments.

1399. The TANNENBERGER BÜCHSE is an actual handgun excavated in 1849 at the site of a once powerful fortress. Vesta Tannenberg was a notorious robber stronghold in the late 14th Century. It was stormed in the year 1399 and every effort was made to destroy it utterly, quite as in the case of Carthage. For centuries tales about its very existence were considered legendary, so utterly was it obliterated.

The gun itself is of cast bronze. Its design and construction is so superior to most authenticated arms of the period that scientists

Gunpowder manufacture. From Codex Germanicus 600. Dated reliably 1390.

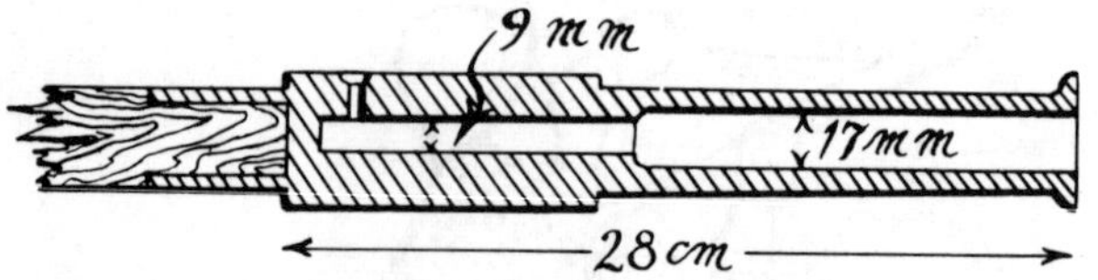

The Tannenberger Büchse in Section.

for a time related it to a much later date, but archeologists finally established the date satisfactorily.

Among other early German records meeting the qualifications given above may be mentioned the following: The GEMEINER REGENSBURGER CHRONIK. There is, for instance, an entry by a Ratisbon gunsmith noting the delivery of handguns set in wood stocks, and weighing about 11 pounds each. In 1381 in the CHRONICLES OF AUGSBURG are listed firearms purchased by the City Council to ward off expected attacks. CHRONICLES OF NUREMBERG, MOHRINGEN, AND KAUFBEUREN of this period all list firearms.

From 1396 to 1405 considerable research was recorded, notably the KRIEGSBUCH BELLIFORTIS by Konrad Kyeser. This manuscript, Codex Ms. phil 63 at Göttingen University covers in detail much of the transition period when the hand-fired "cannon lock" system was giving way to the next evolutionary step—the matchlock.

This manuscript and others, CODEX 719 AT NUREMBERG and CODEX 734 AT MUNICH, follow in detail the evolution of firearms at the time, particularly the shoulder-fired arms. Stocks are discussed in detail, and the one-piece military stock is shown. Drilling methods were improved to provide tighter gas locks and better accuracy and range. Barrels, flash pans to protect the priming from wind and rain, recoil blocks—these and many other technical aspects are covered in detail.

The manuscripts and illuminations of this period show in great detail such items as multi-barrel guns which were the forerunner of the Gatling guns. Two- and three-barreled arms are shown, both over-under and side-by-side types. There are guns to fire single balls, others on the shotgun principle to fire a charge of shot. It is safe to say that every type of firearm in existence today was to some extent visualized and attempted in Germany in this period. The one thing that prevented great advances then was the lack of a suitable cartridge—a development which did not occur until five centuries later.

As the 14th Century drew to a close, primitive forms of a new firing mechanism began to appear, apparently developed in Burgundy. No advance in scientific method ever immediately wipes out the old methods, and so the cannon lock continued to be used far into the 15th Century. All early dates, therefore, must be taken as general except where otherwise specified with reference to new developments.

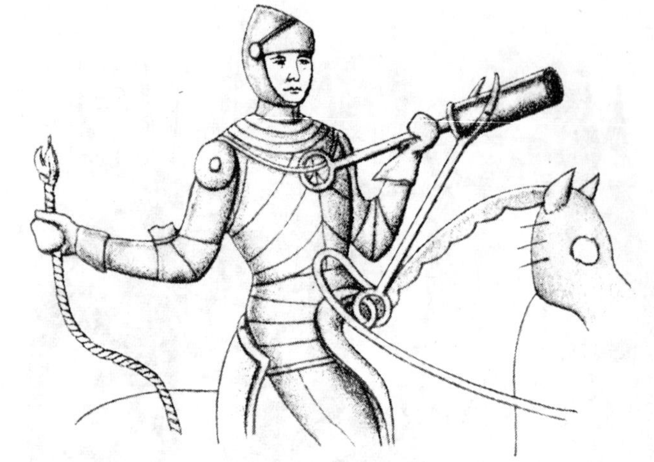

From contemporary fourteenth century drawings. Upper: Firing the cavalry Petronel cannon lock. Lower: Firing the infantry hand cannon.

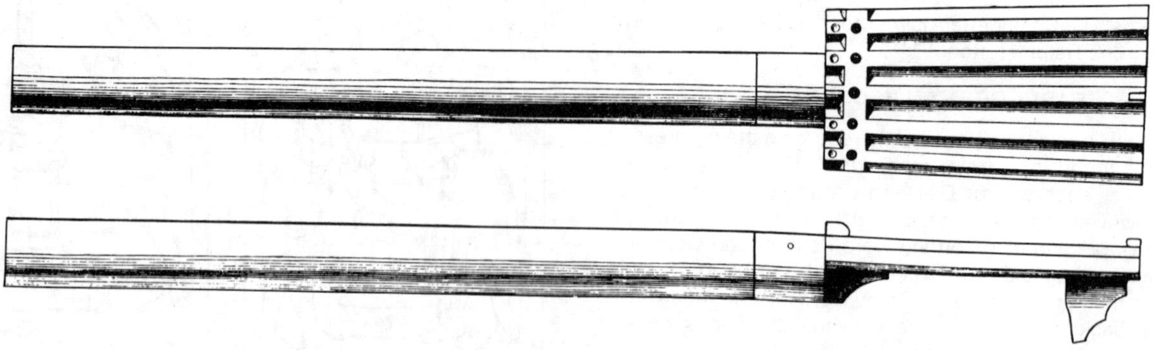

Upper: Five shot "Lade Büchse" with bronze barrels. Early German cannon lock. Lower: Early German Hakenbüchse for wall firing.

THE MATCHLOCK

Since the earliest times the ignition system in firearms has been called the "lock." The cannon lock, as we have seen, required the shooter to touch a lighted coal or hot iron directly to the priming charge in order to fire.

The "matchlock" was originally a very simple "C"-shaped piece of metal pivoted to the side of the stock. It was split to grip a "slow match"—a cord generally made of hemp fiber treated with saltpeter and other chemicals to make it slow burning. It was commonly twisted and then held in shape by thread wrapped around it. The "match" burned at a rate of about 3 to 5 inches an hour.

This simple development revolutionized the use of firearms in its day. Up to the time of its introduction, sights were not used, since one hand had to be kept free to apply the burning coal. Field use under such conditions was all but impossible in hand arms. Only artillery was really practical.

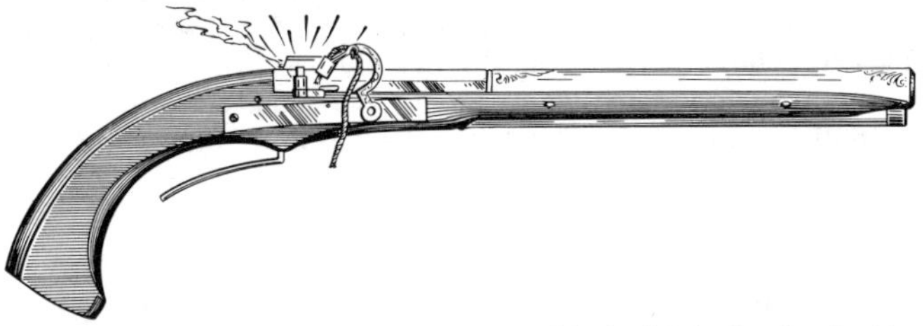

Original drawing from specimen early matchlock pistol, showing ignition system.

With the matchlock as initially introduced, the tip of the burning cord was held in the split iron holder, and could be pushed with the finger to bring the burning end into contact with the priming powder to fire the weapon. This made it possible to aim hand arms.

This revolutionary device was immediately seized upon by the soldiery of all nations, and development speedily brought it to its peak of possible perfection. The first positive record is found in the CODEX 3069 AT VIENNA authentically dated 1411. Erlangen and also Vienna records of the period also detail features of the matchlock; but it took some 50 years of development to produce a good combination of trigger, hammer, and pan mechanism. The "C" form of holder was replaced with a bent iron piece resembling an "S," and known as the "serpentine." This form was used in 1471 by Edward IV at Ravenspur. English Yeomen of the Guard were armed with it in 1485. It was the weapon which turned the tide of the battle of Pavia against the French when used by the Spaniards. Swiss models of the era still in existence carry both ramrods for loading and cleaning needles for clearing the priming touchhole.

From Codex Icon. 222, Munich. Harquebussier of Maximilian I, with early matchlock.

The early developments of this lock included the addition of a sort of hammer spring. A button trigger was supplied. This feature, of course, speeded the ignition and made sighting still easier. At a later date the more formal trigger type was introduced.

THE MERZ MATCHLOCK

Martin Merz was the most celebrated firearms authority of his time. A truly revolutionary form of the matchlock was developed by him. It is shown in detail in the CODEX GERMANICUS 599 published in 1475.

This arm might have revolutionized warfare had its significance been grasped by the rulers of the day. It featured a lock plate covering a series of levers and springs which allowed use of a short sear. It brought the burning match forward away from the eye of the shooter, making aiming still easier. Like the Swiss matchlocks previously mentioned (which seem to have been developed from Merz' ideas) it carried both a cleaning needle and a ramrod. The importance of this latter item, which is meaningless to us in this day, cannot be overrated. It was not until 1698 that the iron ramrod below the barrel for ready field use in reloading was introduced. Prince Leopold 1 of Anhalt Dessau armed his forces with it. It was credited with playing a major part in providing the firepower which won the battle of Mollowitz in 1730! In addition to these other features, the Merz Matchlock had both front and rear sights, allowing a maximum of accurate aiming in those days when most arms did not have even a muzzle sight.

Had Merz been an unknown, the failure of the military to pick up his ideas would have been understandable; but in the light of his standing it is one of the unsolved riddles of firearms history that they were so long overlooked.

VARIETIES OF MATCHLOCKS

The earliest form of the matchlock, as we have seen, was merely a pivoted iron holder for the slowmatch which was pushed by the firer manually.

The Germans seem to have been the first to develop the next form already mentioned—the "button lock." The cock or serpentine (equivalent to a hammer) was drawn back and held by a sear under spring pressure. Pushing a button released the cock and brought

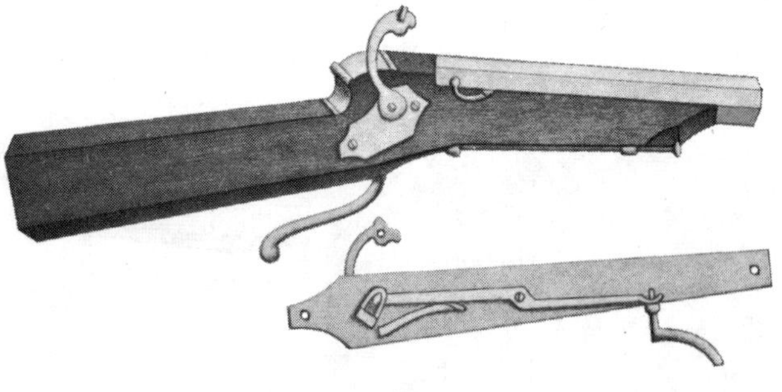

Merz matchlock and plate. From Codex Germanicus 599, published in 1475.

Early use of matchlock in war. From Codex Germanicus 734.

the lighted end of the match down on the priming. This was the standard military form used on all small arms until about 1520. It is shown in all the important arms research works of the time, such as the Arsenal Book of the Emperor Maximilian.

The next development was the most successful. It is used even to this day in some areas of Africa and Asia. This was the "pressure lock." Cornelius Johnson, an English gunsmith, manufactured locks of this pattern as early as 1521 in the Tower of London, but available evidence again indicates the originals were probably German. This type in all its varieties consists essentially of a cock (equivalent of a hammer), a tumbler, sear, two flat springs, and a hammer. Basically, it is the thumbcocked "single action" system as shooters know it today. Pulling the trigger released the tumbler. It revolved 90 degrees to bring the match into contact with the priming.

The final form of the matchlock was the "snap lock" which was in use in Europe in the 1570's and later. It was called by the Germans "light snapping lock" or in some forms "tinder lock." The names derived from the fact that instead of a long dangling slow match as in the regular matchlock forms, this pattern held a small tube in the jaws of the cock. In this tube was a piece of tinder or a short section of match. The advantage (if any) was that the soldier did not expose himself by showing a light until he was ready to go into action, at which time he ignited the match or tinder. One of the disadvantages of the matchlock, of course, was that the enemy could see the men approaching with lighted slow-matches. Captain John Smith, of Pocahontas and tobacco fame at a later date, in early life fought the Turks as a mercenary in Europe. He tells of an instance where he frightened off a night attack by having a few soldiers carry lines between them with pieces of slow match burning at intervals of a yard. The enemy thought a huge force of matchlock soldiers was approaching them. The snap lock was intended as an answer to this sort of maneuver. This form of lock was probably the forerunner of the "snaphaunce"—a flintlock arm of much later date.

DEVELOPMENTS DURING THE MATCHLOCK PERIOD

Rifling

Most of the important principles of firearms which have since been developed were at least attempted in the matchlock period. Rifling was introduced, for instance, though again no one can say precisely where or by whom. One manuscript at Leipzig dated 1498 gives an account of a system of cutting grooves in barrels as developed by Caspar Köllner of Vienna. The grooves were straight, as described, however. This would indicate that Köllner was interested primarily in rapid loading and easy cleaning, rather than in giving the bullet rotational flight to increase accuracy, as is the purpose of true rifling. Rifling has also been ascribed to August Cotter who worked at Nuremberg from 1500 to about 1520. Again, this statement is not subject to proof. However, about this time twist rifling does appear. An Italian Inventory for the year 1476, for example, lists "firearm with spiral grooved barrels." At the British Woolwich Arsenal Museum are a group of arms and barrels most of which have spiral grooves, though a few are cut straight. Danish records indicate that the military use of the rifled barrel was introduced first in that country by Christian IV. He lived 1577 to 1648. It is noteworthy that one of these Danish rifles dated 1611 is still in the Woolwich Arsenal.

Sights

Yet another development of the matchlock period was a very thorough use of sights. A manuscript at Erlangen University dated 1500 gives detailed descriptions. Some German documents indicate that shooting was done at 200 yards, a distance which would require both rifling and fairly good sight equipment.

Interchangeable Barrels

The system of interchangeable barrels was also introduced at this early time. They are widely pictured in manuscripts of the early 16th Century. Some were dropped into the stock and secured at the forward end by a projection passing through a mortise. Others were secured by tang screws. These permitted use of different caliber barrels or of shot barrels at the option of the shooter much as ours do today.

Breechloaders

Breechloaders were commonly attempted. Due to the lack of a good gas seal at the breech none were particularly successful, of course. Many of these operated on the lifting block principle used by our forces even at the beginning of the Spanish-American War in 1898—the basic difference being only that we had brass cartridge cases which would expand on firing and prevent gas from blasting back through the breech. The famous British breechloader made in 1866 to convert the muzzle loading Enfields of that day to cartridge breechloaders was developed by an American, Jacob Snider. The British paid him for it. Actually the conversion was

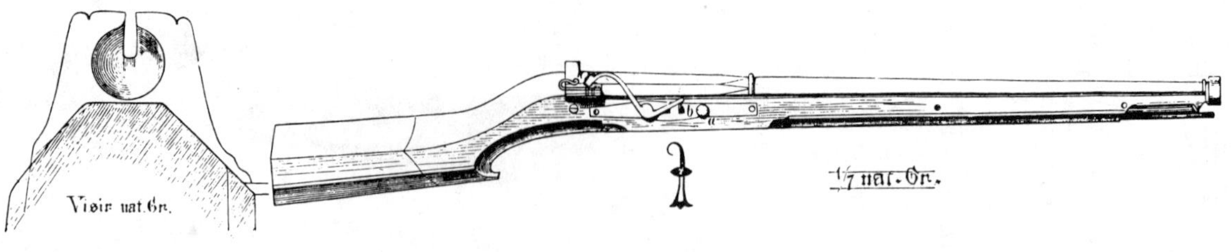

Early button lock with receiver sight and cleaning rod from Basle, 1500
A.D. Gun has a bronze barrel.

Examples of historical sources: Upper: German tapestry 1584 A.D. Lower: Drawing by von Solms 1550 to 1560.

almost a duplicate of one introduced in England by Henry VIII in 1537 except for the cartridge used. A sample of the 1537 conversion is in the Tower Museum!

Chamber Pieces

"Chamber pieces" were also used. These were the forerunners of cartridges. They were commonly quite heavy and of steel. They were loaded with powder, wad and ball and were inserted at the breech of the arm—pistol, rifle and even repeater. The idea of course was to furnish a method of quick reloading. Since all firing at that period depended upon use of flame or sparks, the ignition system also doomed the breech-loading system of that day to failure. Every nation in Europe at that period did some work with breech-loading and ample records are available to show how widespread was the recognition of the value of the system.

Repeating Arms

Repeating arms were also attempted. Many were cylinder arms much like the later Colt revolvers. Others had successive charges rammed down the same barrel, with the charges separated by heavy wads. Individual touch holes on the side of the piece were primed and fired independently. However, as in the case of the Roman Candle, fire would get around the wads and fire successive charges. Often this would blow up the arm. In the face of this development work of the early 16th Century, it is interesting to find that during our own Civil War experimental rifles on this pattern were attempted. These were fired by percussion caps, but again they failed because of fire getting behind the protective wads.

Multi-barreled Guns

Every type of multi-barrel gun of course was also tried. Records show clear drawings of two-, three-, and four-barrel guns arranged quite like our modern types. Others were built on the "pepperbox" principle used in America in the early 1800 s, several barrels being grouped around a central axis.

Volley fire field guns—early forms of modern machine arms—were also common. Leonardo Da Vinci designed a wide variety of these. Some were designed to be breechloaded. In general these were multi-barreled arms with barrels side by side. Only the lack of a satisfactory ignition system prevented the development of every arm we know today except our automatic designs.

The Matchlock Handgun

With the arrival of the handgun with matchlock the infantry battalions became "the strength of battle." No more were the heavily armored knights the deciding factor in the field. In the history of England, Warwick actually brought about a revolution in military art when he introduced the Burgundian mercenaries at the second battle of St. Albans. It is true that in that engagement the matchlock did little damage, a heavy rain handicapping the gunners by wetting their priming and their matches. But the start was made that day. The efficiency of the handgun with all its draw-

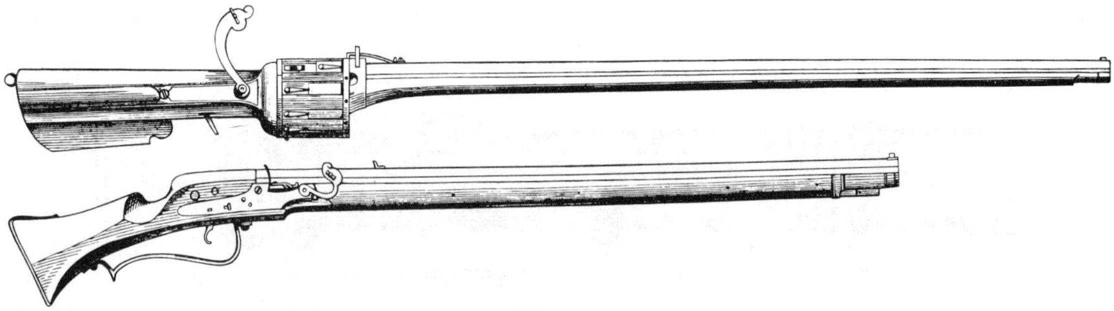

Early matchlock revolving rifle and breechloader, German.

backs grew rapidly until by the time of Bannockburn the infantry became such a power that knights did not again engage in combat from horseback until the time of Henry VIII when the introduction of the Wheel lock made shooting possible by mounted men.

The Knight so famed in history as the Flower of Chivalry, "without fear and above reproach" whose name, ironically enough, in our time is the trademark of a Belgian arms firm (H. Pieper)—the Chevalier Bayard—so feared and detested the new firearms that despite his knightly honor he summarily hanged every Spanish handgun man who fell prisoner to him! Yet when it came his time to die, it was at the hand of a detested Spanish gunner who shot him from his steed.

The Arquebus

Throughout the entire early history of firearms we find special interests and the ultra-conservative—quite as in our own time—trying to discredit and to laugh off that which they most feared. In Fourquevaus' "Instructions for the Warres" translated from the French by Paul Ive in 1589 is the following passage: "The Harquebuse hath been invented within these few years, and is verie good—." He continues on to say that not many are yet expert in its use so that "in a skirmish wherein tenne thousand Harquebussados are shot, there dieth not so many as one man, for the Harquebusiers content themselves with making of a noyse, and so shoot at all adventure."

Quite a different sidelight, however, is that we find in the "Vitae Illustrorum Virorum" of Paulus Jovius, quoting Farbrizio Colonna who was with the great Captain Gonsalvo de Cordoba when the Spanish arquebus men were placed in shallow trenches to await the charge of the Flower of France on a bitter April day in 1503 at the bloody and decisive engagement of Cerignola: "Neither courage of the troops," he says, "nor the steadfastness of the general won the day; but a little ditch—and a parapet of earth—and the arquebus!"

Even in far off Asia Minor the arquebus was cancelling out the armor and the speed of the mounted man, as evidenced in contemporary illustrated manuscripts of the Persians dealing with the battle with the Turks at Tchaldrian on the 23rd of August, 1514, a manuscript still in existence at Oxford.

In the "Kriegswesen" of Max Jänns we find an inventory of Innsbruck for 1515 listing "469 best quality and 662 smaller arquebuses of brass; 1125 large and 665 small ones of iron;" still another index of the extent of arms development.

This then, was the picture when the wheel lock, the next evolutionary step came.

THE WHEEL LOCK

The military and sporting needs for an ignition system which would not depend on lighted match or live coals led directly to the development of the "wheel lock." The principle is simple. We use it today in our cigarette lighters. Spin a serrated steel wheel against a flint to give off a shower of sparks. The sparks ignite the priming powder to flash fire down through the touchhole into the powder charge in the firing chamber.

Again history is not clear on just when, where and by whom this

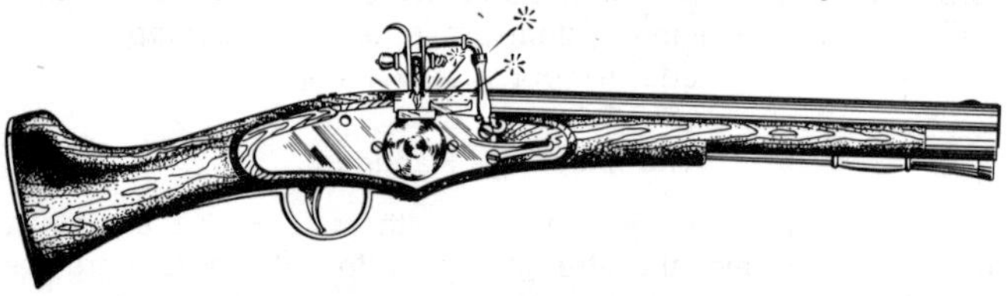

Original drawing of an early wheel lock showing ignition principle.

system was invented. Various accounts of early historians credit Johann Kiefuss as inventing it at Nuremberg in 1517. Others cite the same inventor but place him in Vienna. In any event, an earlier form than the true wheel lock existed in the form of the so-called "Monchbüchse" of Dresden. This was a simple tube fitted with a rasp and piece of flint mounted near the priming hole. This of course was manually operated.

The true wheel lock principle involved use of a steel wheel with knurled or grooved edge mounted in a frame and connected with a chain and spring. The outer side of the wheel had a projection at its axis. A wrench fitted over this projection. It was used to wind the chain and compress the spring accordingly. The cock held a piece of flint or pyrites. In preparation for firing the head of the cock was pulled down until the flint rested against the edge of the wheel. Pulling the trigger released the wheel. As it spun around it showered down the firing sparks.

The heavy trigger pull required led directly to the development of the set-trigger pretty much as we use it today. By pushing one arm of the trigger when the weapon was cocked, the piece was pre-set so that a touch on the second trigger would fire it with a minimum of risk of shifting the pistol off the target.

Literally thousands of variations of the mechanism were developed. Such systems were expensive to make, and their use was generally confined to that of the wealthy and some of the military.

Again we find that the system was used in all European countries. Da Vinci described it in detail. Collardo writing in Italy in 1586 ascribed its development to the Germans. Henry the VIII used it in England.

Because of the firing system employed, much attention was now given to various types of safeties. There were trigger safeties, hammer and sear safeties, manual and automatic safeties. Every form of device now in use to prevent accidental firing in manually operated arms can be found, in at least elementary form, in the early wheel locks.

Rifling also received increased attention because the firing system and the set-trigger and sight refinements permitted greater

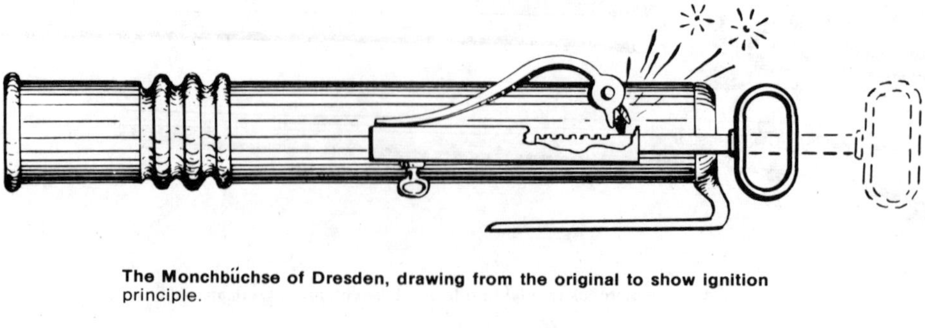

The Monchbüchse of Dresden, drawing from the original to show ignition principle.

and greater accuracy. During this period experimental work was carried on covering use of every number of grooves from 2 to 34. Rate of twist of the rifling was also subject for experimentation. And even the shallow groove rifling currently being advertised as "micro-groove" was considered, some examples still existing in specimens at Woolwich and Zürich.

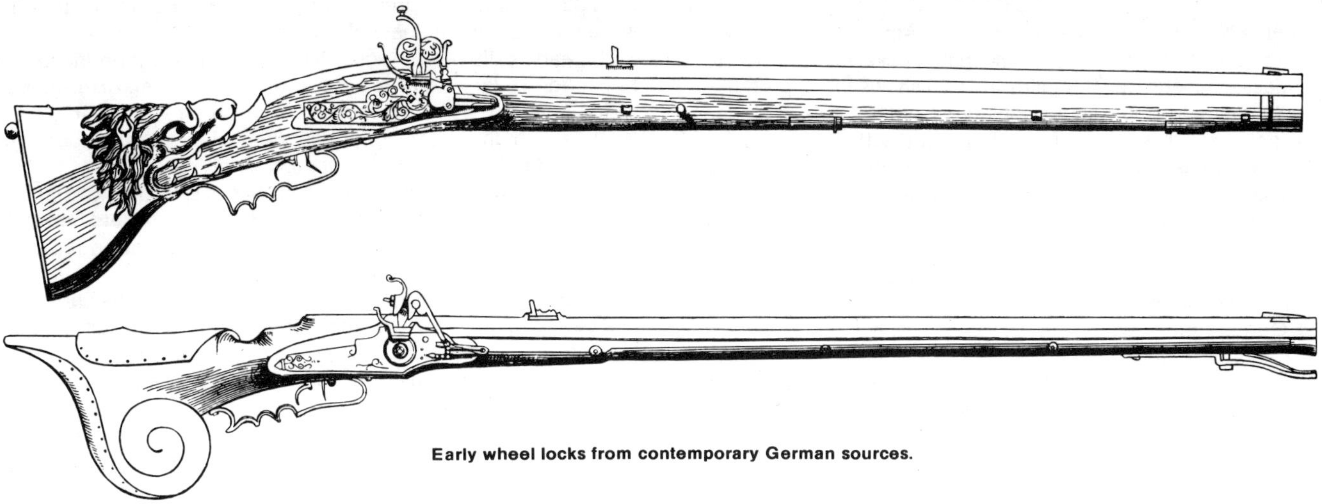

Early wheel locks from contemporary German sources.

THE SNAPHAUNCE

In place of the expensive wheel mechanism, there appeared about 1525 a simple form of mechanical fire striking lock called the "snaphaunce." It is really little more than an adaptation of the old matchlock "light snapping lock" already described. In the jaws of the cock a piece of pyrites was fastened instead of a tinder or match holder. A steel point or anvil was provided near the touch-hole. Pulling the trigger released the cock. Its spring forced it ahead. The flint striking a glancing blow at the steel anvil showered sparks into the priming powder. Just as simple as that! Yet this was one of the great developments in the history of firearms.

Its name seems to have derived from the Dutch "schnapphan"—meaning snapcock. Some writers ascribe the name to the Dutch translation "chicken thieves." They theorize that it was developed by thieves who wished to poach for game without the tell-tale lighted match. The more probable explanation is that it merely indicated the snapping of the cock against the anvil to give sparks. Such derivations, of course, indicate Flemish origin for the system. However, it is referred to in many early German records as "der Spanische schnappschloss," and many authorities have claimed Spain as the place of origin. It was in general use throughout Europe not very long after the introduction of the wheel lock, so that even France and Italy lay some claim to its invention.

In any event, the new system was recognized everywhere as a great forward step. Experimentation to improve it started promptly. However, it was a full 100 years before a perfected form of the system appeared to mark a still greater advance in firearms technology.

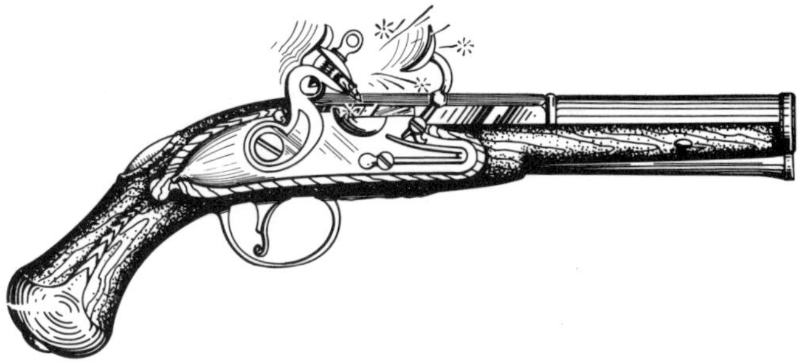

Original drawing from an early snaphaunce showing ignition principle.

THE FLINTLOCK

The true "Flintlock" is basically the snaphaunce as described plus a hinged steel right-angle pan cover hinged over the priming pan. As the cock falls the flint strikes this hinged piece (called the "batterie" or "frizzen"). The blow throws the hinged cover back, exposing the priming powder in the pan below. At the same time the flint scraping along the steel showers sparks down into the priming powder to fire the arm.

DEVELOPMENT OF THE FLINTLOCK

This development is credited to the French. They successfully combined the principles of the snaphaunce with a rather similar system known as the Spanish "miquelet." Again we find that all European nations experimented with and finally adopted the system. It was officially adopted by the British in 1690. This arm was popularized in history and legend as "Brown Bess" from the color of its barrel plus the erroneous idea that it was adopted by Queen Elizabeth, a queen who had been dead 87 years at the time of its adoption! This anecdote gives some ideas of the reliability of much that is popularly accepted in the story of firearms.

One model flintlock manufactured by the French at Charleville was officially adopted by them in 1746 and with minor changes was the official pattern until 1842. A slight variant known as the 1763 Model was the basic arm of our forces under Washington

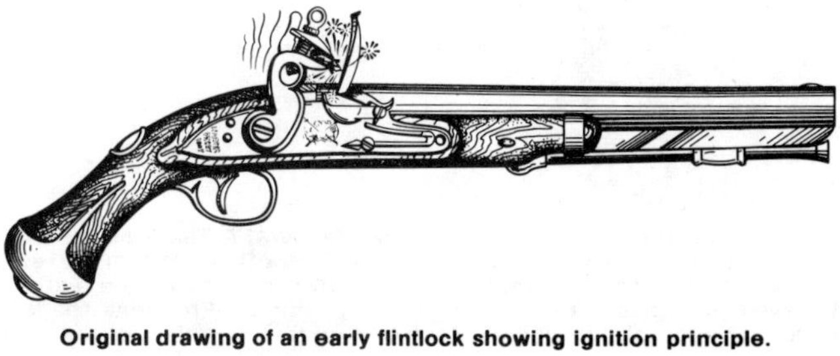

Original drawing of an early flintlock showing ignition principle.

during the Revolutionary War. Prussia, curiously, did not adopt the system until 1808.

The Flintlock in America

Very little is known about early firearms in America. We do know that the earliest settlers brought some matchlocks; and one record of the Massachusetts Bay Company in 1628 shows the importation of snaphaunces for their paid troops. Very little else is recorded of the earliest days, but it is certain that with the need for protection from Indians as well as for game shooting, considerable work must have been done by local gunsmiths.

The first scientific approach to arms making was done by the German settlers in Pennsylvania. They had a heritage of mechanical and gun skills to draw upon which the Puritan and Pilgrim settlers did not.

The German and Swiss settlers brought with them the finest arms of their countries. Generally these were large bore, heavy arms of general European pattern. The barrels were quite short as a rule. However, the gunmakers quickly realized the need for modifications required for American usage and changes came fast. Barrels were lengthened to provide better powder combustion and better sighting arrangements. Useless weight was removed by stock design. Full stocks of American maple and walnut were fashioned to protect the barrel in forest use. Calibers were changed in the interest of accuracy and range. By the year 1700 flourishing small gun businesses were producing gems of the art. In 1732 the noted Heinrich Brothers and Peter Leman and others were turning out the finest rifled arms available anywhere.

The early calibers averaged .54 but as time went on these were reduced for varying types of hunting so that the original balls weighing about half an ounce were greatly reduced in weight. Sights of a high order were provided on all these in various styles from bead to blade fronts and special notch rears.

But the biggest development of all was in speed of loading. The European types of the period required that the bullets be started down the barrel with a mallet, then punched down on top of the charge with a ramrod. The Pennsylvania gunmakers introduced the idea of wrapping the bullets in linen or buckskin patches which had been soaked in tallow. This system permitted easier and faster loading, and at the same time gave a degree of gas check hitherto unknown. This in turn gave better accuracy and longer range.

These early rifles are famous as "Kentuckies," taking the name from the fact that the vast territory between the Cumberlands and the Mississippi River was then called Kentucky; and most of the settlers and hunters who ranged that area were outfitted in Pennsylvania. Tales of the accuracy of these rifles have been grossly

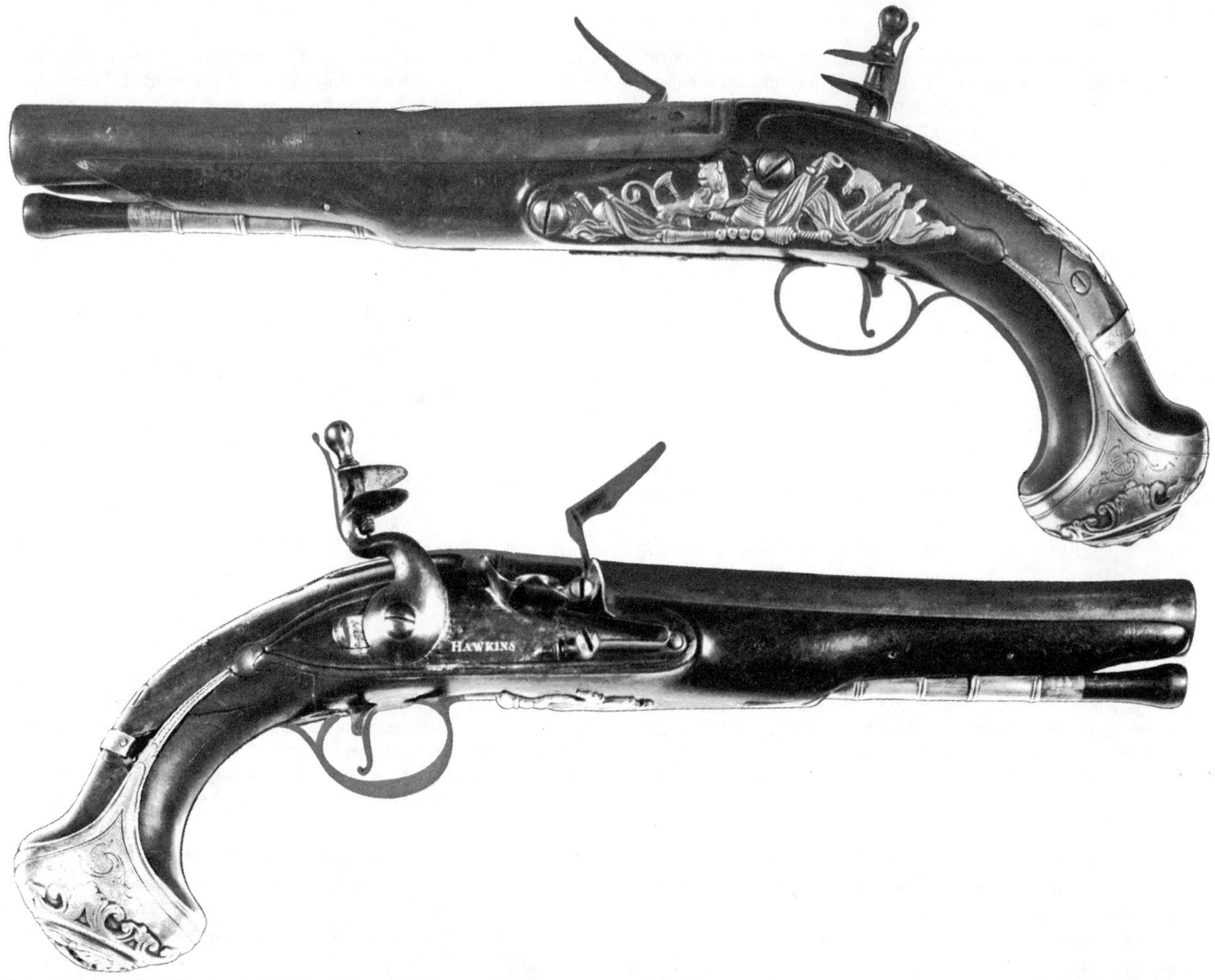

A Pair of George Washington's Pistols, Caliber .67. These pistols were purchased at the sale of the effects of Batholomew Dandridge, Jr. in 1804. He was Washington's private secretary after the Revolution and it is believed the pistols were presented to him by our first President. The lock plates of the pistols are signed "Hawkins" and their barrels bear the mark of R. Wilson, both London gunsmiths. Their silver inlay has British proofmarks for 30th May 1748-30th May 1749, and the stocks are walnut. After passing through various hands they were acquired along with documents by Mr. Clendenin Ryan of New York who presented them to the West Point Museum where they now are.

overdone, of course. But the fact remains that they were the greatest precision arms of their day—and far ahead of any European designs for years to come.

Our earliest official military arms are those of the Committee of Safety set up at the time of the Revolution. Committees in the 13 colonies authorized their adoption. The arms themselves were made by over 200 gunsmiths in Massachusetts, Rhode Island, Maryland, and Pennsylvania. For the most part they were smoothbore muskets of flintlock pattern. They weighed about 10 pounds average, calibers varying from .72 to .80.

The American Revolution compelled the military forces of the world to adopt rifled arms to replace the traditional smoothbores. In spite of all the European experimentation with rifling from the early days of the matchlock, it was not until the American settlers armed with their highly accurate muzzle-loading rifles had taken drastic toll of trained British troops that the British army accepted

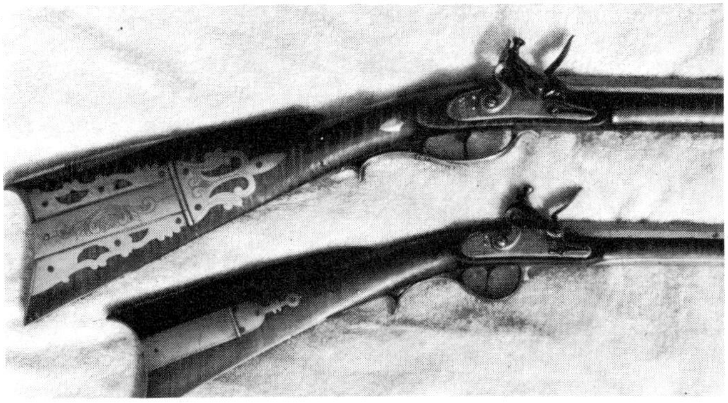

A representative American Kentucky flintlock rifle.

the value of rifling. British Captain Hanger was in command of one of the German Jäger Corps, trained riflemen employed by the British, in a desperate effort to counter the long-range accuracy and fighting ability of the colonists. In his writings he expressed the greatest respect for both the deadliness and the manufacture of the American rifles.

Into the scene at this time came one of the most remarkable persons of that day, a British officer and gentleman whose name and fame are but little known, Major Patrick Ferguson, 2nd Battalion, 71st Regiment of Highlanders. Soldier, inventor, marksman extraordinary, inventor of the first breech-loading rifle of English manufacture used in war by the British forces, Ferguson for one brief instant held the very life of the American Nation in his hands, when at Germantown on the field of battle General George Washington accompanied by a solitary French aide came under the sights of his rifle. Ferguson, one of the deadliest riflemen of his time, had been assigned to pick off the American general. When he saw him, he refused to believe that that casually dressed, fearless man who merely turned and stared at his call to identify himself could be the American Commander-in-Chief. On Ferguson's inability to realize that Washington could be so confident or reckless as to go about without pomp or bodyguard, on that one failure to grasp Washington's stature, the general's life was spared. Ferguson held his fire.

In British patent specification No. 1139 of the year 1776 we find the then Captain Ferguson's description of his breech-loading flintlock. He does not claim the invention is new. He states that previous attempts of the sort have failed, that his is "an improvement in breech-loading firearms."

The specification also sets forth an elevating rear sight. A single turn of the trigger-guard lever drops the attached 3 to 12 thread vertical screw plug which passes through the breech end of the barrel. As the quick-thread screw goes down, the ball is dropped into the chamber through the opening in the top of the breech, the powder poured in (or paper cartridge inserted) and the trigger guard swung back to elevate the breech plug. The pan is primed, the hammer cocked, and the weapon is ready.

A "History of the War in America between Great Britain and her Colonies" published in Dublin in 1785 says of Ferguson: "He was perhaps the best marksman living...It has been reported that George Washington owed his life at the Battle of Germantown to the gentleman's total ignorance of his person, as he had him sufficiently within reach, and view during that action for the purpose." The story originally was told by Ferguson himself to a Tory, a member of the New York De Peyster family who joined with Ferguson in his mission of organizing in America those forces which sided with the British Army in its efforts to stamp out the Revolution.

Ferguson headed the Tory forces who were finally trapped on King's Mountain on October 7, 1780 in the battle which will ever be a tribute to the skill and marksmanship of those early Americans

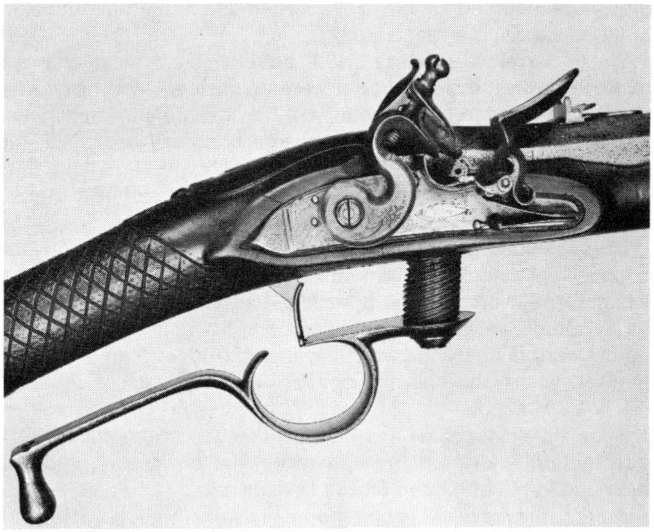

Ferguson rifle with action open ready for loading through breech.

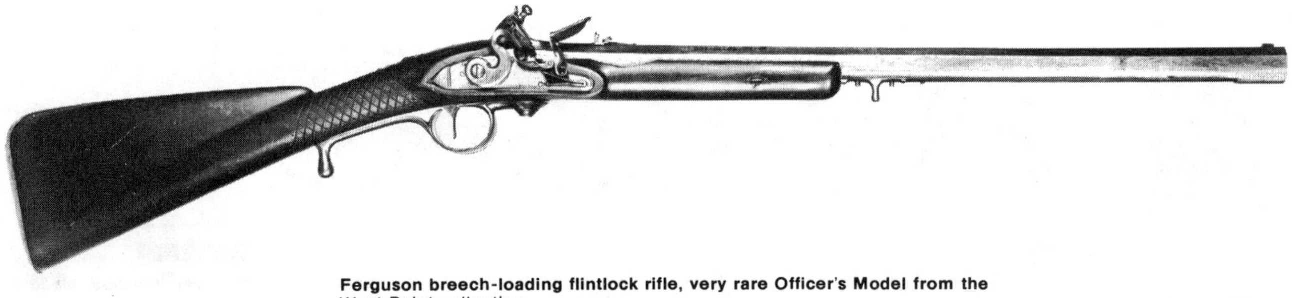

Ferguson breech-loading flintlock rifle, very rare Officer's Model from the West Point collection.

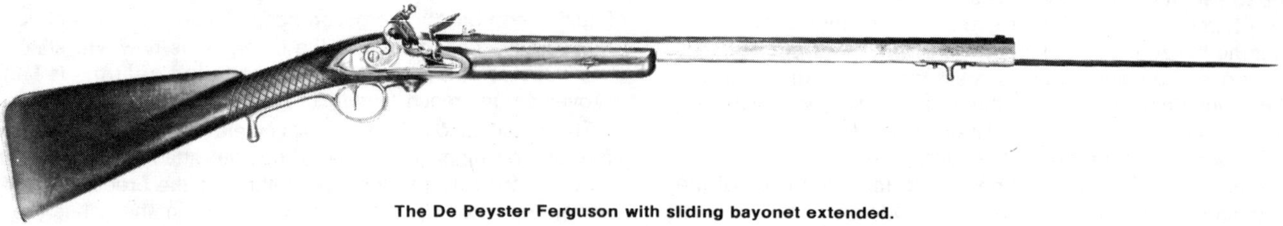

The De Peyster Ferguson with sliding bayonet extended.

who fought for freedom. When a sniper finally killed Ferguson, the living Tories surrendered after some 400 out of the contingent of 1100 were casualties. The Ferguson rifles seized in that engagement were highly prized, but today we know of the existence of only three of them. Ironically enough, one specimen now in the Smithsonian Institute was donated by a descendant of the De Peyster who received it from Ferguson—he was somewhere else when the King's Mountain engagement took place.

Other Developments of the Flintlock

As was to be expected, every form of rifle which had been experimented with in earlier days was again tried with the flintlock system. In addition, many relatively new ideas were tried out, and some of these proved successful at a later date when improved ignition was developed.

Of the earlier forms which were repeated as flintlocks we find the familiar two- three- and four- and more barreled patterns. Again they were mounted side by side, over-under, and around an axis in all conceivable arrangements.

Single-barrel "repeaters" with several separated charges rammed down the single barrel were again tried. And of course they again failed. All the ingenuity of the inventors using sliding-rack arrangements for individual charge firing still could not get by the lack of a gas seal between the barrel charges; and firing one often set off the whole firing chain with disastrous effects to gun and shooter.

Breechloaders of a hundred types were tried. Again some were supplied with steel chambers which were individually loaded and which were inserted in the breech just as we do fixed cartridges today. Others were to be loaded directly from powder horn. Still others were to be loaded with the paper cartridges then becoming familiar, where the ball and powder were individually wrapped ready for shooting.

Special Italian models had devices attached to the cock which automatically primed the arm—when the devices worked! All previous forms of hinged blocks were used.

Among the newer designs, however, were breechblocks and also barrels hinged and operated by trigger-guard levers. Some had barrels which slid forward or pivoted to the side when a lever was operated. Some of these patterns paved the way for the modern hinged shotgun.

Revolving barrel systems were again tried, as were revolving cylinder systems. Volley guns firing a number of barrels simultaneously and other multi-barreled types firing in rapid succession were again developed. And at least one unusual repeating type found its way to America. This was the fantastic "Cookson," an arm far ahead of its time.

The Cookson Repeating Flintlock. There is a tantalizing entry in Pepys Diary, written by the amazing Samuel at his house in Seething Lane on July 3, 1662, an entry never elaborated and one on which no other records or descriptions have ever been unearthed. It reads: "Dined with the officers of the Ordnance, where Sir Compton, Mr. O'Neile, and other great persons were. After dinner was brought to Sir W. Compton a gun to discharge seven times, the best of all devices that I ever saw and very serviceable and not a bauble; for it is much approved of and many made thereof."

What was this repeating gun? We do not know positively, but it may have been one of the "Cookson" guns, a specimen of which is now in the Smithsonian. It may have been a somewhat similar arm developed about that time by an unknown gunsmith in Vienna, samples of whose works are known. Or it may have been one of the guns made by a famous Italian gunmaker, "Antonio Constantine"—the first repeating arm of which we have record which had a buttstock magazine (charge and priming powder chambers were loaded through the butt by raising the butt plate). In any event it belongs to the type called the "Cookson" because the outstanding specimen known to be in existence bears that name engraved on it.

While the name "Cookson" is English, this gun is believed to be of Lorenzoni type designed late in the sixteenth century.

This arm was evidently brought to Maryland by early English colonists. It was one of the guns confiscated after a search for arms by the Provost Marshal in Baltimore in 1863 and remained in United States Government custody until the close of the Civil War. Unclaimed by 1867, it was then thrown in the scrap heap. Purchased by an individual for a small sum of money, it was turned

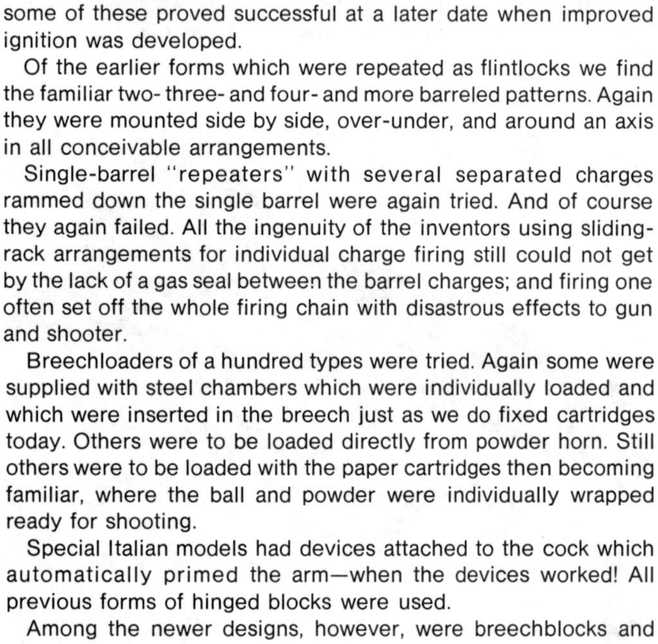

Italian "over-under" double flintlock. Barrels are hand rotated for firing. Gun in Birmingham Museum.

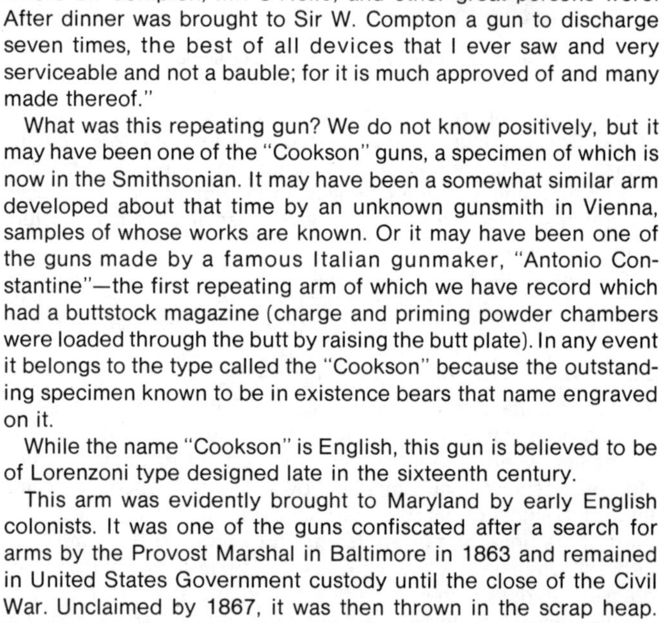

Italian breechloader with removable steel chamber "chamber piece." May 1694 A.D. by Aqua Fresca a-Borgia.

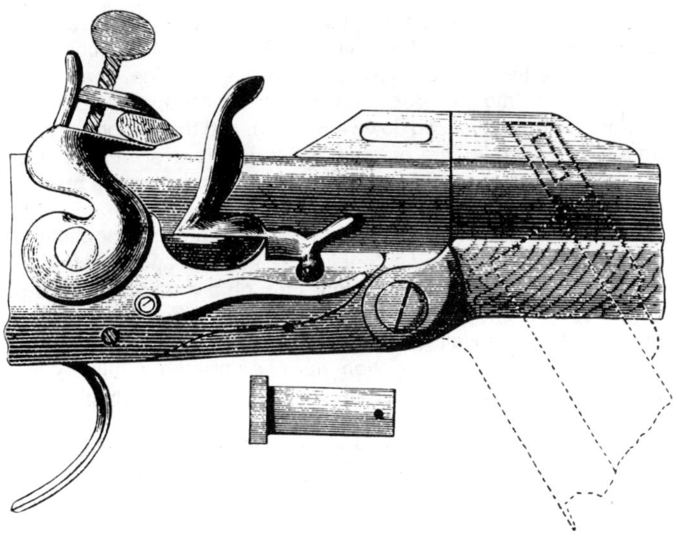

Early French flintlock breechloader with hinge barrel and steel "chamber piece" for loading. Cross bolt locked as in modern double shotguns.

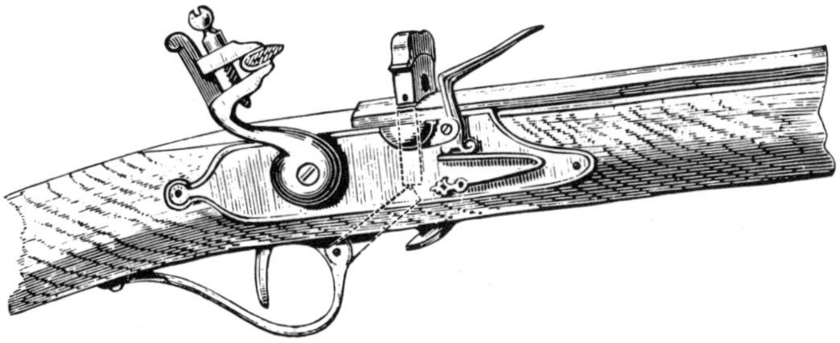

Early German Theiss flintlock breechloader. Lever operated. Block lock.

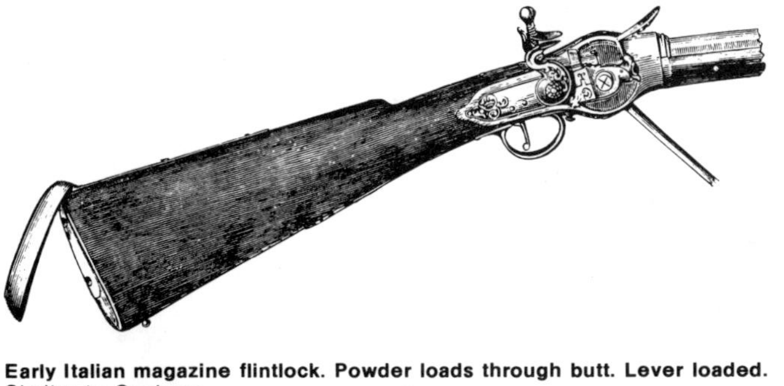

Early Italian magazine flintlock. Powder loads through butt. Lever loaded. Similar to Cookson.

over to a gunsmith for checking. For the first time the unusual nature of the weapon was then realized.

Its front sight is like a Turkish crescent. The metal work is all heavily engraved with flags, drums, piles of cannon balls, cannon being fired, stacks of muskets, boarding pikes, and the like. On top of the barrel is stamped "John Cookson, Fecit." (Made by John Cookson.) This is a magazine flintlock gun of smooth bore. It fires spherical bullets weighing 260 grains and a charge of about 125 grains of black powder. The arm had a capacity of 10 rounds, with a magazine fitted into the lock carrying a similar number of priming charges. (Other known specimens have smaller magazines.)

There is a hinged flap on the left side of the receiver which is opened to permit bullets to be put into one compartment and powder into an adjoining one. Cylindrical passages lead from the compartments to the central chamber in the receiver in which is located a solid cylindrical block with its axis from right to left.

This cylinder forms a recoil block. It has two radial cavities large enough to hold a ball and a charge, respectively, and is so located that when revolved the cavities will be opposite the passages from the magazines.

At this point the ball drops into the first cavity and the powder into the second. By turning the cylinder to the front the passages are closed and the ball and charge are brought in front of the rear part of the bore. (Since this is a gravity loader, the muzzle must be held down.) The bullet drops in and the block remains with the charge in line with the bore. The powder cavity, which has a diaphragm to prevent the bullet from dropping into it, is connected by a vent through the axis of the cylinder to the pan.

Turning the lever on the left side revolves the cylindrical breechblock and also cocks the hammer and closes the pan. The lock and trigger are of ordinary design; the sear, sear spring, and mainspring are the same as those used on the last models of flintlock guns. The highest mechanical skill is indicated in the manufacture of this weapon, the tolerances being astonishingly close.

This gun was far in advance of its time. Had it been provided with a wad to serve as a gas check, it might have established repeating rifles two hundred years earlier than their time.

Once the magazine is loaded with bullets and powder, no counting or measuring is necessary, as the charges are automatically measured and the loading is quite accurate.

With a magazine fully charged, 10 shots could be fired very nearly as rapidly as with the modern metallic cartridge repeater!

EARLY AMERICAN-MADE REPEATING RIFLES

The Jennings and the North

About the period of 1815 single-barreled arms were experimented with in America which were merely flintlock versions of Emperor Maximilian's early repeaters, notably the "superimposed loading" system. In this dangerous system, already alluded to, a charge of powder and ball was rammed down the muzzle, a heavy wad pounded down, and then another load placed in on top of the wad. Some of these arms were provided with separate locks which would fire the charges in succession—the shooter fervently hoped!—starting from the one nearest the muzzle. Others like that of L. Jennings of New York (his later percussion rifle was one of the first reasonably successful repeating arms and was a forerunner of the present Winchester line) were fitted with a sliding lock which was pushed forward on a track until it was opposite the first flash hole. It was moved progressively back as each charge was fired. A single trigger and one sear only were needed in this design. In 1824 and 1825 repeaters on this system were made by that master gunmaker, Simeon North of Middletown, Connecticut. Sooner or later, of course, flame from charge one managed to bypass the wad and reach the other charges—a situation which speedily discouraged use of this system. Hope lives long in the human breast, however, and research is not utilized as it might be; and so in 1862 we find Springfield Armory experimenting with this superposed load system in the later percussion days under Lindsey's patents!

American Breech-loading Flintlocks

The Hall. The American Hall Rifle, patented in 1811 by Colonel John H. Hall of North Yarmouth, Maine and William Thornton, is generally accepted as the first reasonably successful breechloading arm to be extensively used. Originally a flintlock, it was adapted to the percussion and rimfire systems in a later day.

Hall developed his rifle without being aware that similar systems were being experimented with in Europe. His was the only one to achieve a measure of success. It had an outside hammer, of course, and was fitted with a special chamber pivoted at its rear

Typical American Hall rifle. Action of typical American Hall breech-loading flintlock. Breech is open ready for loading from forward end. Block snapped down to lock.

Collier flintlock action. This was made as a shotgun, rifle, and as a pistol.

and in a box receiver behind the barrel. Pushing down a short lever below the fore-end elevated the front end of this chamber, which had radial grooved locking blocks on either side. The charge was inserted into the front end of this tipped-up chamber, the chamber was then locked down into place, and the weapon primed and fired. Naturally, the gas seal at the juncture of the chamber and barrel was not good, but the arm was a step forward in breech-loading construction.

The Hall was officially adopted in 1816 as an experimental arm, and was used to a limited extent in the Seminole and the Mexican wars, in .54 caliber. The arm went through a wide succession of improvements and changes in lockwork and caliber. The engrossing history of the Hall rifles is still to be written, for while a vast amount of data concerning them has been printed in military and sporting records, the data has not yet been satisfactorily gathered together.

The Jenks. The one other American flintlock breechloader to receive official attention was the Jenks. Raising and pushing back a lever on top of the grip exposed the chamber for loading. (This arm is more commonly found as a percussion lock; but as a flintlock of .69 caliber, it was tested against the Hall.)

The Collier Revolving Flintlock. Soon after Artemus Wheeler received a patent for a revolving flintlock in the U.S. (1818), one

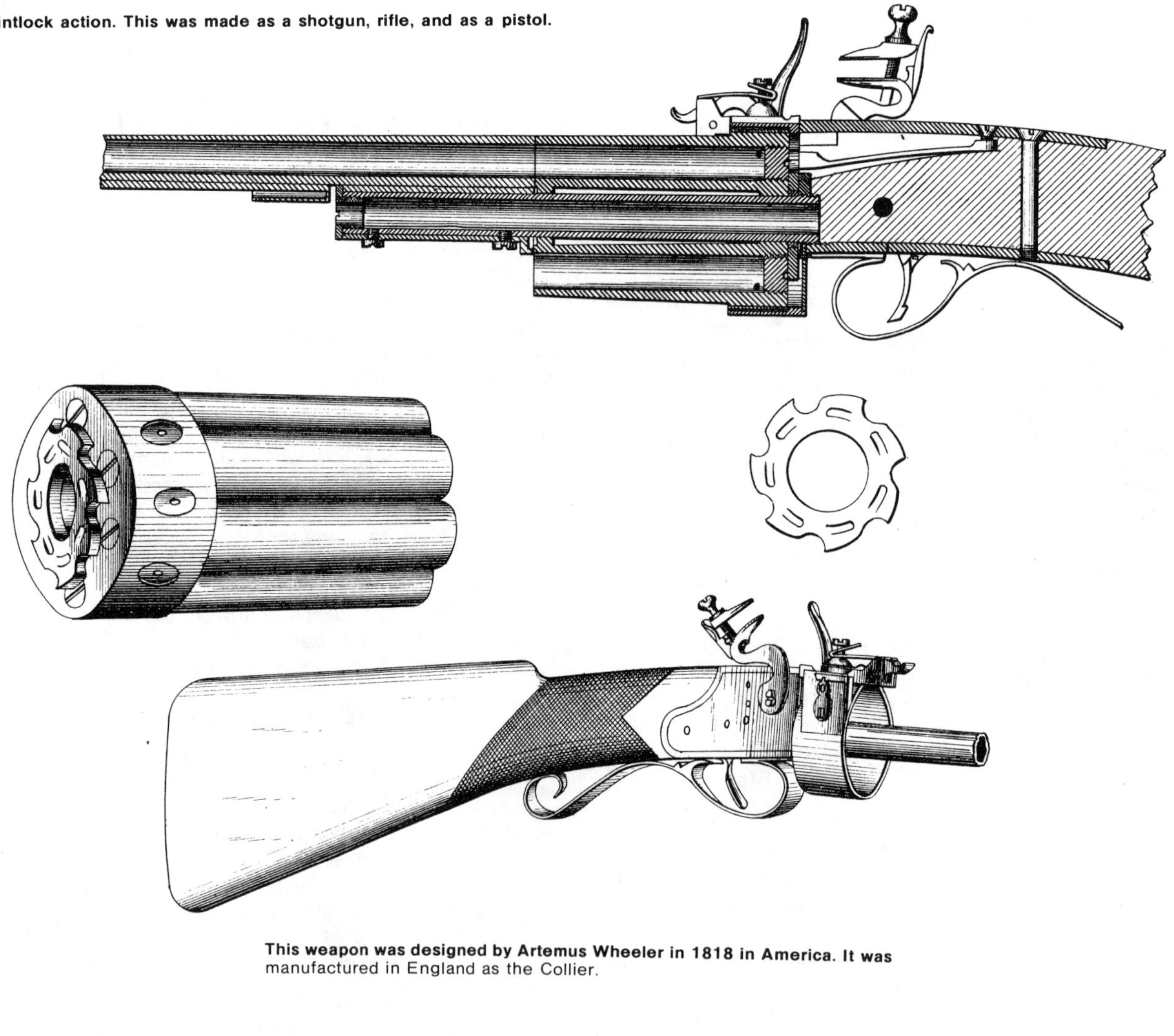

This weapon was designed by Artemus Wheeler in 1818 in America. It was manufactured in England as the Collier.

Elisha Collier of Boston appeared in London with the same idea and took out a patent for the same weapon in England. Another Bostonian named Coolidge obtained a patent in France at about the same time. The details of how the remarkable coincidences occurred is lost in the mists of time, but it would appear that Mr. Wheeler spoke to too many people and trusted too many in a day before international patent agreements were in effect—not that it couldn't, hasn't, and doesn't occur all too frequently in these days as well!

Collier had pistols, rifles, and shotguns made to the Wheeler revolving design in England and some of these weapons are still extant.

In the original Collier, a tube in the butt carried spare priming powder. Priming for about 10 charges was carried in a box which formed the frizzen. Each movement of the frizzen automatically dropped the correct amount of the priming in the flash hole.

A revolving cylinder with four, five or eight chambers was loaded from the muzzle end. These chambers were chamfered (cut back) at the muzzle. The cylinder was held forward by a spring which thrust the chamber in line with the barrel directly ahead and over the rear end of the barrel to form a reasonably tight gas seal. When the weapon was fired it was necessary to press a release stud to free the cylinder, which was then drawn back from its position over the barrel until it could be turned to bring the next chamber in line. The cylinder was then moved forward so that the end of the next chamber fitted over the barrel breech. The hammer or cock carrying the flint had to be thumb cocked for each shot.

The Close of the Flintlock Era

The flintlock marked the end of a period of purely mechanical changes to furnish ignition for firearms use.

From direct application of a lighted coal to the priming, development passed on as we have seen to the mechanical application of lighted slow match, then to the still mechanical system of showering sparks into the priming by wheel and by flint.

The next step, percussion, brings us to the stage of chemistry where a new priming system of firing by detonation eliminates the need for direct fire or sparks.

THE PERCUSSION LOCK

Percussion firing requires use of an explosive which will detonate when struck a sharp blow. Gunpowder of an early day could not be exploded this way, though certain types made today can be.

In its initial form priming powder was fulminate of mercury. A small quantity was placed in the gun pan over the touchhole after the gunpowder charge, the wad and the bullet had been rammed down the muzzle. When the hammer fell it struck this fulminate and the crushing blow exploded it. Fire from the explosion passed down the touchhole to set off the propelling charge of gunpowder below.

FIRST USE OF PERCUSSION

French scientists had recognized the existence of the principle as early as 1703, but no application of it was made until a Scottish minister, Rev. Alexander John Forsyth, began experimenting with it in 1793.

We will probably never know how old the discovery of detonation really is. We do know that Forsyth was definitely among the first to employ it. We also know that he worked the principle out himself without any knowledge that others were working along the same line at about the same period.

Let us at this point consider another documented instance of how a principle can be "lost" for centuries, this time in relation to percussion. In the astonishing diary of Samuel Pepys we find the following: "November 11, 1662. At noon to the coffee house, where, with Dr. Allen, some good discourse about Physics and Chymistry. And among other things I telling him what Dribble the German Doctor do offer of an instrument to sink ships; he tells me that which is more strange, that something made of gold, which they call in chymistry, Aurum Fulminans, a grain, I think he said, of this put into a silver spoon and fired will give a blow like a musquette and strike a hole through a silver spoon downwards without the least force upward."

No British records have yet been unearthed which give any further clue to this matter; but it is obvious that the principle of detonation lay fallow for well over a century at least.

Rev. Forsyth lived and died a Presbyterian minister in the village of Belhelvie. He spent his spare time hunting, repairing guns, and working on chemistry experiments. He himself made the first known gunlock to use detonating powder, and showed it to friends in 1805. He was induced by Lord Moira to travel to London where in the Tower he was given facilities for chemistry experimentation.

In 1812 Forsyth in partnership with James Watts, inventor of the steam engine, invented several types of percussion locks.

One curious sidelight on official thinking can be found in Forsyth's story. His friend Lord Moira was replaced by a politician who saw no sense in Forsyth's work; and in April of 1807 Forsyth was ordered out of the Tower. Three months later, on July 4th, Forsyth was granted a British patent for his invention, which was soon adopted by the British Army. Politics being politics, the Government did not make any payment to the minister until 23 years later, at which time they granted him 200 pounds. This niggardly treatment created much public furore, and an award of an additional 800 pounds was made in 1843. Unfortunately, Forsyth died three months before the grant was made.

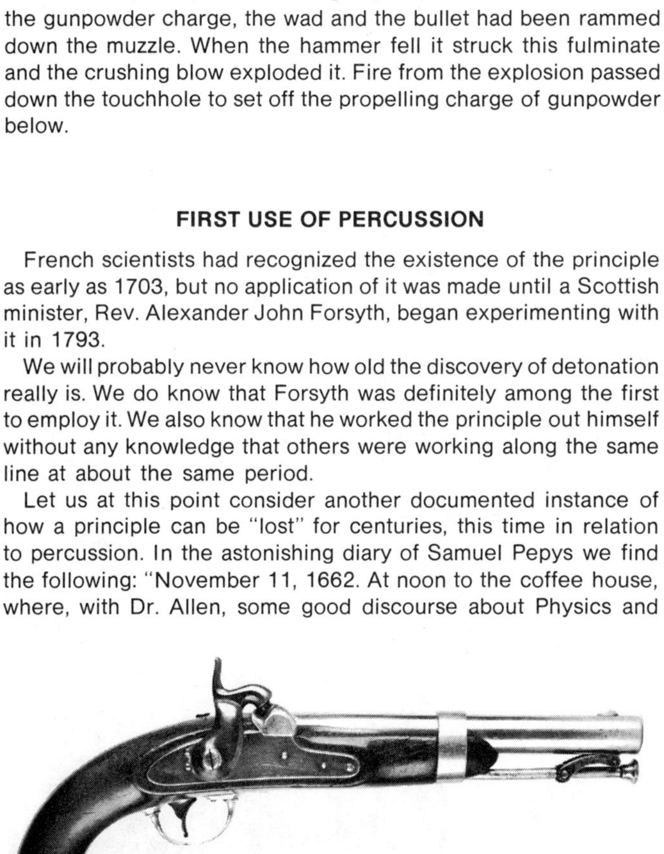

U.S. Army Aston muzzle-loading smoothbore percussion pistol. Caliber .54 Model of 1842. Representative percussion lock type.

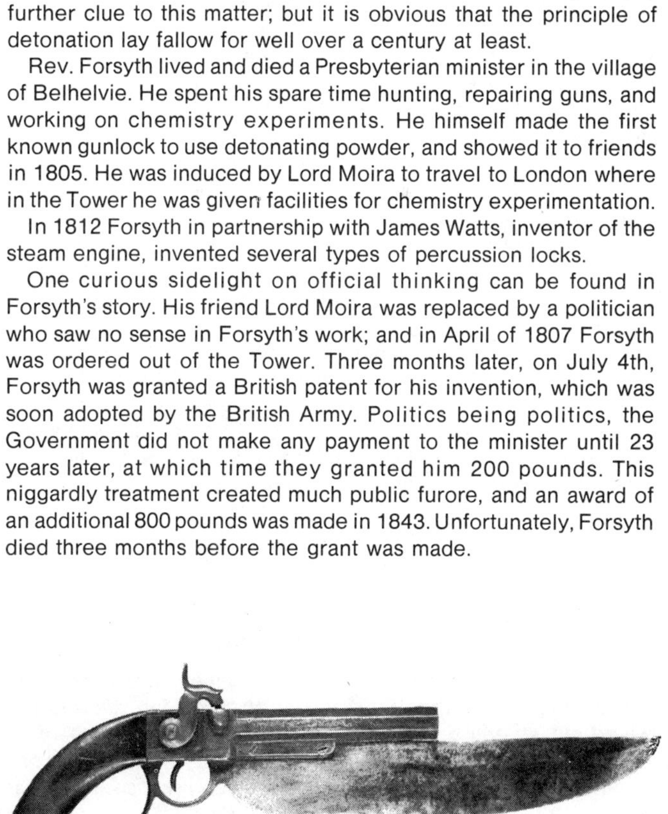

The famous U.S. Elgin Navy cutlass pistol, percussion type.

Forsyth originally used the power directly in the pan. However, he soon developed a system of "caps" by placing the powder between two pieces of paper. His next move was to make a roll of these "caps" to speed up the process of priming arms. These devices, of course, are the same basically as those used in toy cap pistols to this very day, but in their time they were the wonders of the firearms world.

OTHER PERCUSSION DEVELOPMENTS

A French gunsmith named Pauly used a paper fulminating device in 1808. It was fired when a needle pierced it. This was doubtless the forerunner of the ammunition of the "needle gun" later used effectively by the Germans, as well as of the American "volcanic" system which initiated the true lever action repeating rifle.

In 1816 the famous London gunmaker "Old Joe" Manton filed specification No. 3895 with the patent office covering the first attempt at a "tube lock."

Manton's true "tube lock," however, was not filed until 1818, under specification No. 4285. The name comes from the small copper tube carrying the fulminate which was inserted in the vent, its upper end resting against an anvil to form a lower support as the hammer crushed the tube to fire it.

William Westley Richards of Birmingham introduced one of the most popular arms of the time in his "detonating gun" specification No. 4611 in 1821. This simple arm has a cock which strikes into the flash pan, which has a pivoted lid operated by a spring. As the hammer falls it forces against a point of a pivoted lever whose other end operates the pan cover. This arm could be primed with powdered fulminate, with paper caps, or with fulminate balls of pellets which were also called "pills." In America this was usually called the "pill lock."

We do not know positively who developed the copper cap to hold fulminate. This cap, of course, led eventually to modern cartridge development. The best evidence available indicates it was first produced by Joshua Shaw, an Englishman. He was working in Philadelphia, Pennsylvania when it first was brought to public notice in 1814.

Some British authorities ascribe the development to Joseph Egg, the noted gunmaker; and there is no question that he at least introduced it into England. A very wordy sportsman of the day, one Colonel Hawker, made some claim to having worked with Joe Manton in such an invention, but the facts are hazy at best. A large number of lesser lights also laid claim to it—after the fact, of course.

In 1824 the Berenger patent was granted covering percussion pellets fed from a magazine. A year later Joe Manton patented a revolving magazine primer. In 1834 Baron Heurteloup's continuous primer, a long tube of metal holding the fulminate which was fed forward and suitably cut by an edge on the hammer, was patented. Egg's self-acting primer magazine was introduced in 1836. In 1845 Dr. Edward Maynard was granted a U.S. patent for his "tape primer," consisting of priming pellets embedded in strips of paper or linen so they could be fed automatically as the hammer was cocked.

Rifling in the Percussion Era

While a brigade of the British 95th Regiment was armed with Baker flintlock rifles in 1800, and at Waterloo the rifle brigade wiped out several brigades of French artillery, little military attention was given to rifling until 1838. The early British Baker had a rifling of one-quarter turn in the 30-inch barrel; the caliber was .615.

In the year 1789 one J. Wilkinson in English specification No. 1694 described a barrel rifled with two spiral grooves, the missiles to have belts or wings to fit the grooves. In 1836 a Board at Woolwich selected the two-groove Brunswich rifle for service. The caliber was about .704, the rifling made one turn in 30 inches (the length of the barrel), and the bullet was a 555 grain belted ball, the ball corresponding to the width and depth of the grooves. (A few Lancaster rifles, two-grooved but firing conical balls with side ribs to fit the rifling, were issued during the Kaffir War from 1846 to 1852, the Lancaster being the first conical ball used in military service.) This form of rifling fouled up quickly, and its design was considered impractical, though in the metallic cartridge era it was again tried.

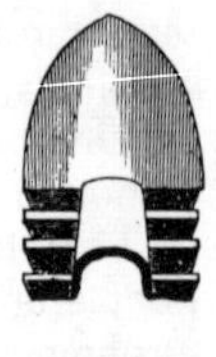

The Minié Ball.

The Minié rifling next introduced had four grooves and fired a conical ball. As introduced in British service in 1851 it was of .702 caliber. It too, proved unsatisfactory, though some were used in the Crimean war. It was replaced in 1853 by the first famous Enfield, a three-groove rifle of .577 caliber, with a pitch of one turn in 78 inches. This rifle was made in the Royal Small Arms Factory at Enfield which was equipped with special American-made machinery during this period.

Bullet Development

Since the bullet in a muzzle-loading rifle must be smaller than the bore to permit it to be rammed down effectively, much thought was given to a bullet which would enter easily but would then expand to fill the grooves and utilize the full force of the powder gases formed behind it. The problem was to design a ball which could be dropped down the barrel, then satisfactorily expanded. France did considerable experimentation on this problem. One of the first solutions was that of the Delvigne bullet. The powder rested in a chamber smaller than the barrel bore. Thus the round ball, resting against the shoulders of the powder chamber, expanded when rammed down. Delvigne next introduced a hollow base bullet which was to be expanded by the exploding powder. The Thouvenin system, famous throughout Europe at the time, consisted of a stem, a sort of anvil projecting out of the breeching into the barrel. The gunpowder charge was distributed around this stem, while the cylindro-conical bullet dropped down the barrel after the powder, was pounded down until it expanded and took the rifling. This was famous as the system of the "carabine a tige"—or carbine with stem. This was followed by the famous Minié ball, designed by Captain Minié of the French Army, a conical ball with an iron cup in its base. The cup was thrust ahead and expanded the lead bullet against the rifling when fired.

Charles Claude Etienne Minié received 20,000 pounds from the British government for this little device, just twelve years after W. Greener by his own story offered it to the British Army only to have it rejected as "useless and chimerical." Greener always insisted that Minié had learned of the bullet through seeing a French translation of an early Greener book; and after a series of

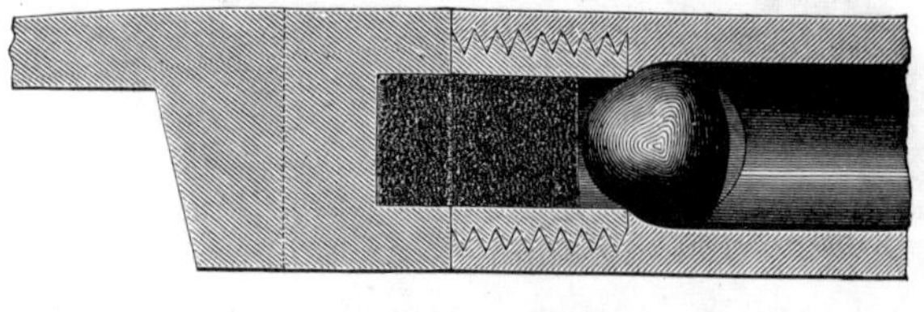

The Delvigne System.

suits the British Government finally paid Greener 1000 pounds. However, since a Captain Norton suggested a similar idea 13 years before Greener, the facts in the case remain in doubt.

In any event, the English version differed from the original French Minié. When in 1852 a new rifle was decided upon, the Pritchett bullet suggested by Metford (inventor of the Metford rifling) was used. First having only a hollow base, the bullet finally was made with a heavy plug fitted to the base, the whole being heavily lubricated, and was of .55 caliber to fit the .577 bore. These were the bullets which legend has it were a factor in the Indian Mutiny, the native soldiers because of caste distinctions refusing to bite off the ends of the paper cartridges which were heavily smeared with animal fat. As a result of this mutiny, which took two years to put down, the British Crown took over rule of India from the East India Company.

Because of poor manufacturing tolerances, in both bullets and arms, Joseph Whitworth, a pioneer in the field of scientific measurements, was called in to experiment. He suggested a hexagonal bullet and rifling; and while some experimental arms and ammunition were made, his ideas on these subjects were not employed.

The Minié ball played a tremendous role in American history. It was the ball generally used in all the muzzle loaders used by both the North and the South during our Civil War; it was responsible in very large measure for the catastrophic loss of life in that terrible conflict.

Percussion Breech-Loading Firearms

The most successful of all percussion types of course, were those using the well-known "percussion cap." In this system the priming is held in a small, hollow metal cup. The cap is placed over a projecting hollow nipple at the breech of the arm. The falling hammer crushes the metal cap to detonate the priming and fire the barrel charge.

As in all earlier developments, every conceivable type of single and multi-shot system was attempted using these caps. Also used were many kinds of percussion systems, only a few of which can be described herein.

The most important feature from an arms development standpoint, however, was the intensive work on breechloading systems at this period. These mechanical experiments laid the direct groundwork for most modern arms and ammunition. And, as usual, most of them were the result of private development by individuals and manufacturing firms—not by government subsidized or controlled groups.

Indeed, the conservatism of the military groups throughout the world was such that in 1859 in his noted book RIFLES AND RIFLE PRACTICE we find Lieut. C. M. Wilcox of the U. S. Army writing: "In Prussia part of the infantry is armed with the needle breechloading rifle; in Sweden and Norway the breechloader is partly introduced; and in France the Cent Guards are so armed. With the above exceptions, no breech-loading rifles are in the hands of European troops...the future will determine whether or not the breech-loading arm is to be more generally introduced into service, or to be abandoned."

While the Dreyse Needle gun was adopted in Prussia in 1841, four years after its development, other government groups spent more time on telling what was wrong with it than on how to improve the principle.

Breech-Loading Systems of the Era

At this period four systems of breech construction were in use, though literally hundreds of variations were attempted. First in importance was the BOLT SYSTEM. This was generally called the "slide system," and was introduced by Nicholas Dreyse in his Zündnadelgewehr or Needle gun. When the bolt was withdrawn, a paper cartridge was inserted. A primer was attached to the base of the bullet, behind it was fastened the bag with the powder. A long needle inside the bolt functioned as the modern striker. When the trigger was pulled, the needle was driven ahead by its spring, its point passed through the powder and hit the primer ahead. The theory was that this placement of the primer would give more complete combustion of the charge. The tremendous importance of this invention, completely overlooked at the time, was the employment of the turn-bolt principle. With the coming of metallic cartridges, the turn bolt was to prove the simplest and perhaps the strongest breech-loading system ever developed. In its early form, however, it could not seal the breech properly, with the paper cartridges used; and the needle corroded rapidly and was subject to breakage. There is one undated specimen of bolt pattern gun in Paris in the Musee d'Artillerie. So far as we know, however, the Dreyse is the first successful application to firearms of the simple system of the age-old turning bolt as used on doors.

In 1854 Charles W. Lancaster patented English specification No. 2089, a bolt action breechloader using a special paper cartridge

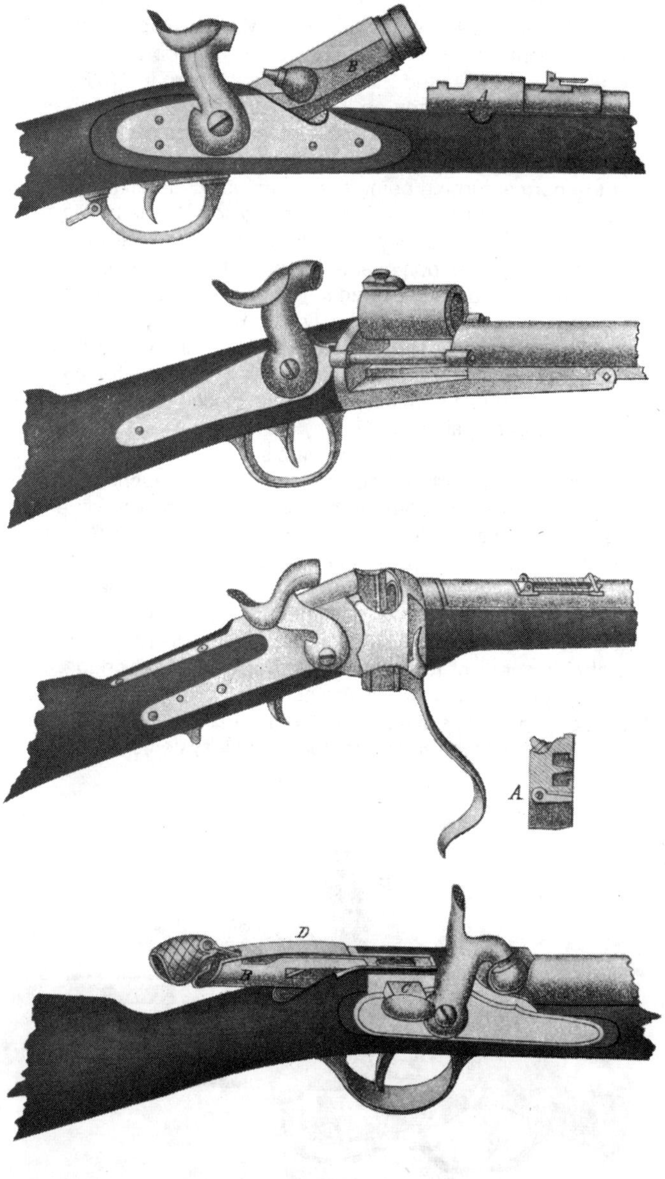

Representative specimens of percussion breechloaders. There were literally thousands of types, variations, alterations, and conversions.

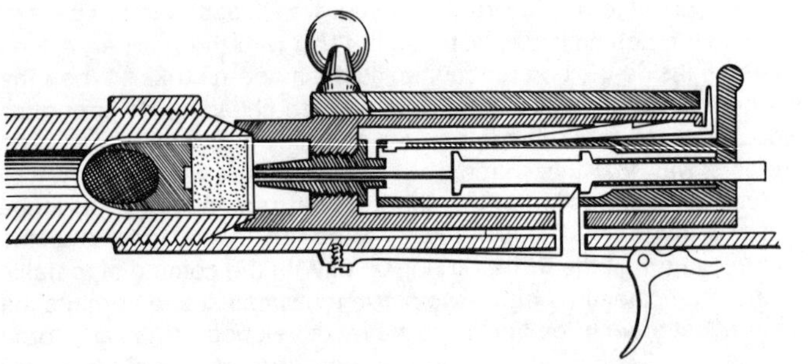

Original drawings showing Dreyse needle gun mechanism and cartridge.

fired by an ordinary lock and percussion cap placed over an outside nipple; but again the gas seal was not satisfactory.

As evidence of the fact that the true bolt principle was not grasped, we have the fact that the early American Greene carbine, an arm in which a forward trigger was pulled to unlock the barrel, permitting it to be twisted to the left and then slid forward for loading, was listed with the Dreyse as a "slide system." This quite unusual arm, invented by Col. J. Durrel Greene of the United States of America, was equipped with a Maynard tape primer. It is not to be confused with the later Greene oval-bored bolt-action rifle tested by the United States Ordnance Board in 1872. In this later arm, a release button on the tang is pressed to free the bolt, which can then be turned up to the left and pulled back. A bullet—not a cartridge—is inserted in the breech. The bolt is pushed ahead past the normal turning point, thereby causing an auxiliary rod to force the bullet ahead in the chamber and seat it. The bolt is again withdrawn. A paper cartridge having another bullet at the rear is next inserted and the bolt pushed forward. Turning the bolt handle down revolves two front locking lugs into place—probably the first time this now universal front lug system was employed. Thus the rear bullet serves as a gas seal, and after firing this rear bullet is forced ahead by the bolt to become the projectile for the next round! This arm has a ring-type hammer placed below the barrel. It is fired by percussion cap.

The dropping block American Sharp's was classed together with the hinged frame early breechloaders of the Lefaucheux as "the Hinge System" by noted writers of this period. The other two classifications were listed as the "screw" or "trap-door" systems and the "revolver system" as used by Colt.

The Terry (1852) was a bolt-action arm with two symmetrical lugs at the rear of the bolt. The Westley Richards, generally listed as "the best of the capping breechloaders," gained its reputation from the effectiveness of its gas seal and the felt wad cartridge used. This lifting hinged block arm was not mechanically in a class with the American Sharp's of the period, which marked the beginning of the long line of lever action dropping block systems.

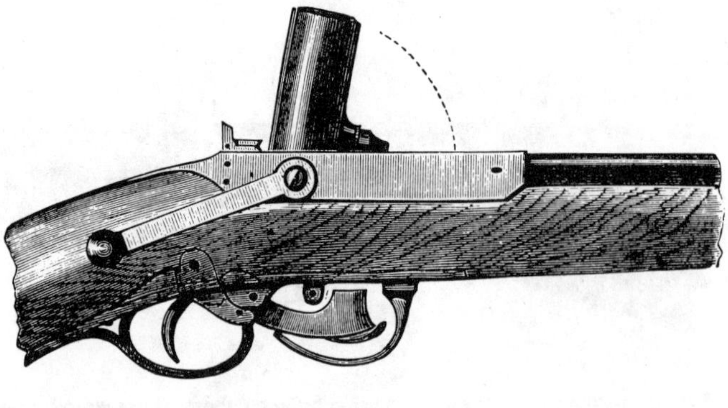

Norwegian 1842 carbine. Percussion breech-loading system.

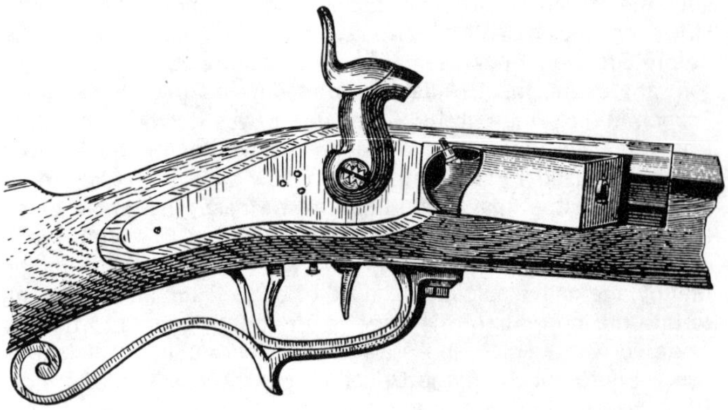

Swiss 1851 d'Abezz percussion breech-loading system.

In 1866 the French introduced the Chassepot, a rifle based on the Dreyse. The forward end of the bolt was fitted with the de Bang obturator of India rubber which expanded in the chamber behind the paper cartridge and was quite effective until hardened by the heat of firing. The Chassepot cartridge carried its primer in the head of the paper cartridge where the needle pierced and fired it.

The Norwegian military arm, 1842 model, had a chamber which was swung up for loading in the manner of the old American Hall. A side lever operated the block. The block had a nipple on its underside and was fired by an under hammer striking the cap on the nipple.

One unusual design was the Swiss d'Abezz, in which turning a guard around the trigger guard withdrew the neck of the chamber block from the barrel, then swung the block on a pivot to the right so it could be loaded from its front end. This arm had the standard nipple and a side hammer.

AMERICAN PERCUSSION ARMS

The official U.S. muskets and rifles of the period are standard muzzle—loading weapons adapted to the percussion system. The 1841 rifle, often erroneously called the "1842," used in hunting and in the Indian and Mexican Wars and even in the Civil War, was noted throughout the world for its accuracy. Variously known as the "Yaeger"—a corruption of the old German Jäger, it is also known as the "Harper's Ferry Rifle" because many of them were made there and also as the "Mississippi." This arm used standard percussion caps. The new 1855 model employed the Maynard Tape primer.

With the coming of the percussion system, many Kentucky flintlocks were altered to "pill locks," the pan being replaced by an iron bowl and the cock shaped to form a hammer with a projection which would hit down into the bowl to discharge the pill.

The Kentucky rifle now evolved into the "Plains Rifle" which played so important a part in American history. Gunsmiths such as Hawkins and Dimick in St. Louis adapted the Kentucky type to the needs of the settlers moving West, and the changes they originated caught on generally. The "Plains Rifle" weighed from 6.5 to 10.5 lbs. as a rule. The calibers ranged from .26 to .40. Barrels measured from 26 to 38 inches—occasionally longer. Set triggers were the general rule. Front bar locks were employed (the Forsyth system of having the lock mechanism behind the hammer was tried out but was not popular because it seriously weakened the stock, which in Plains service had to stand up under considerable punishment).

The Sharps Rifles

Christian Sharps patented in 1848 a mechanism which is unquestionably the most important American breechloader of this time. It introduced the principle of the successful "drop block" actions. When World War II began over forty actions of this pattern

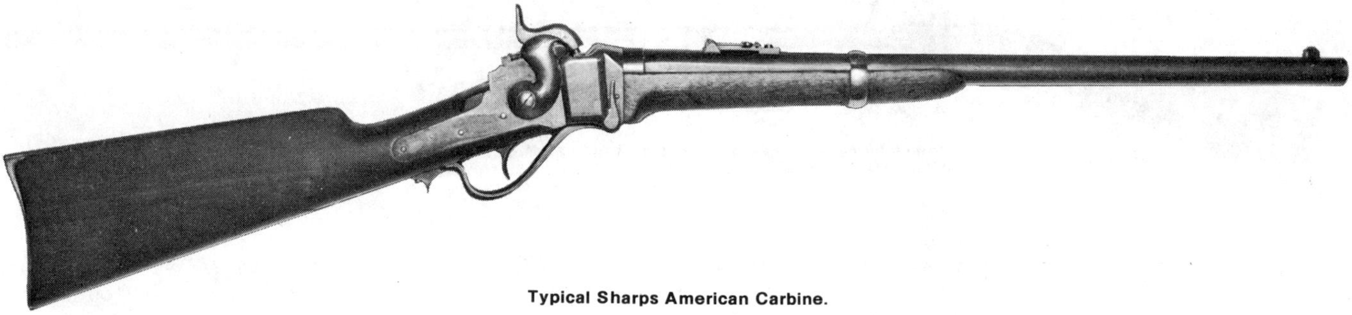

Typical Sharps American Carbine.

were still made in Europe for target and civilian use. Indeed a variety of the single-shot, drop-block rifle was used in 1953 by Indian Forces guarding Korean prisoners under United Nations authority!

Sharps designed it in caplock days. It was originally loaded with a paper cartridge. The design was merely modified as required to make it one of the great rifles of the early metallic cartridge era. It offered a fairly good sealing of the breech even with paper cartridges. Its action is simple. Lowering the hinged trigger-guard lever drops the breechblock down in mortises in the frame. When paper cartridges were used, the end of the cartridge was automatically sheared off as the breech was closed by reversed trigger-guard action. This allowed the percussion primer to flash directly into the charge and speeded up loading as well. Later a blade was added to the breechblock to give still better cutting of the cartridge. Some types were designed to be usable as muzzle-loaders if the breech fouled too badly! These patterns had ramrods.

The Sharps rifle and carbine (the latter being merely a short rifle, of course) were widely used during the Civil War. When adapted to metallic cartridges, these arms were the famous buffalo guns of our early Western meat and buffalo-robe hunters. The arm passed through a tremendous range of alterations and modifications as time went on, none of which are important to this work and must therefore be passed over.

Other American Breechloaders

The first Joslyn carbine patented in 1855 had a ring above the small of the stock which when lifted raised the breechblock for loading. The Starr, of which over 25,000 were purchased during the Civil War, had a special two-piece breechblock, one part falling back to expose the breech. Literally hundreds of other types were produced experimentally or in some quantity.

Arms which require some passing mention are the Burnside, the Gallagher, and the Maynard. The first, invented by Union General A.F. Burnside, used a special brass cartridge of conical type. A hole at the rear permitted the cap to flash through into the charge; 55,567 of these rifles and 21,819,200 cartridges were made in the original model. The Gallagher, invented by Mahlon J. Gallagher, had a trigger-guard lever operated breeching system which swung the barrel ahead and down for loading with a paper cartridge; 22,728 Gallagher rifles with 8,294,023 cartridges were purchased by the Government. The Maynard, like the Burnside and the Smith, approached being a transition arm; for while it was fired by percussion cap, it used a freak brass cartridge case. The case had a rimmed base with a hole in the center to permit the cap to flash through. The barrel was hinged down for loading by pushing the trigger-guard lever. The Maynard mechanical system was later employed when true metallic cartridges were introduced. Its inventor was a Washington dentist. Earlier he had invented the common tape primer or cap roll as already mentioned.

Gilbert Smith's carbine is worthy of mention here, as records show that 30,062 were purchased by the Union Government during the Civil War, and the rifle was well received in Europe. It hinged at the breech somewhat in the manner of the common shotgun. The release mechanism was a small forward trigger within the trigger guard. It fired a freak India rubber cartridge with base perforated to permit a common cap to be used on a vent to fire the charge. Some were later altered to use the weird Crispin metallic cartridge. Over 13,000,000 Smith cartridges were produced during the Civil War.

REPEATING ARMS OF THE PERCUSSION PERIOD

The United States inventors, spurred on by the need for improved firearms in the opening of the West as well as for war, repeated in this period every design heretofore experimented with to give many shots with a single loading and in some respects went far beyond anything ever done up to that time.

Arms with two loads in the same barrel, as we have seen, were again tried. Legend has it that the Lindsay arm of this type made at Springfield Armory was designed by the inventor as a "surprise" weapon to be used against Indian tactics. The story goes that Lindsay's soldier brother was killed in an Indian attack when the red men charged his unit, drew their fire, and then overran them before they could load again. Lindsay's idea was to let the enemy see you fire an apparently single-shot rifle, then be ready to give him the second "surprise" load as he charged. Most of the surprises in this system, however, usually came to the shooter, who was

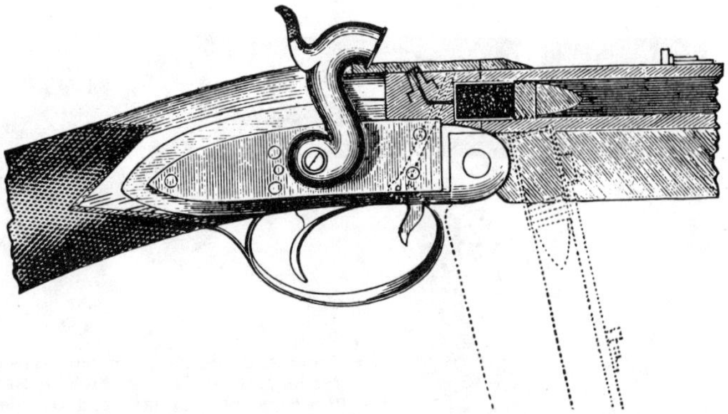

Action of Gilbert Smith's American carbine with hinged frame.

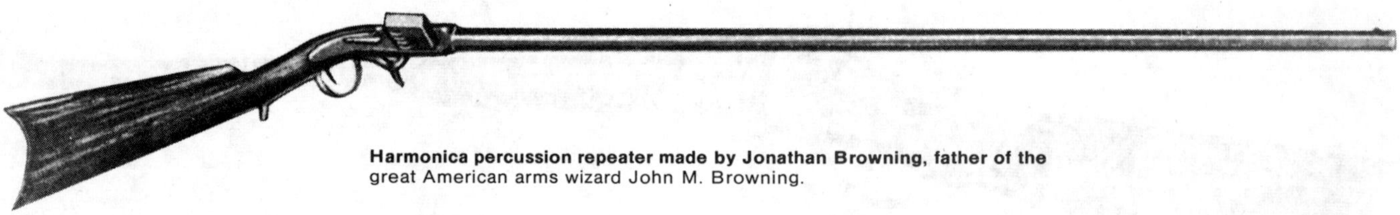

Harmonica percussion repeater made by Jonathan Browning, father of the great American arms wizard John M. Browning.

fortunate indeed when only one load went off at the first trigger pull.

Multi-barreled rifled mounted like the conventional pepperbox also appeared, the barrels commonly being turned by hand to fire in succession from a single lock. The "Sweet" rifle was just one of this type.

"Harmonica" repeaters on the Jarrẽ system were also made. One used by General Sam Houston and now in the Smithsonian has a rectangular feed block which is loaded from the forward end (there are five chambers), and is then inserted in the breech opening on the side. The chamber block, actuated by the hammer, slides across the breech horizontally, bringing each chamber successively in line for firing down the barrel. The "Kendall" of this system has an underhammer. When this is half-cocked, pressing the spring catch on the receiver permits the block to be slid across. John Browning's father made rifles of this type.

A weird collection of revolving cylinder types appeared, quite distinct from the familiar Colt type. These fell into two general classes. First was the "turret" type. The "Cochran" was of this design. The turret type cylinder on top of the receiver was loaded with paper cartridges thrust in from the muzzle openings. There were usually eight chambers. At all times when the arm was loaded, some chambers were pointing their bullets at the firer—and when fire flashed from a firing chamber to an adjoining one, anything could happen.

The "radial" type, to which the "Porter" belonged, operated on the common water-wheel principle, carrying the charges around in succession to line up with the bore as the trigger-guard lever was operated. Again, however, the shooter had loaded chambers pointing right at him when firing, and again the ammunition in use could not be prevented from flashing to adjoining chambers.

In 1852 North & Skinner issued a cylinder type resembling the older Collier, in which the cylinder was drawn back off the barrel breech and turned to slip the chamfered mouth of the next chamber over the barrel mouth for firing. Dr. Le Mat of New Orleans evolved this cylinder arm which was later made in pin fire; and in the same year Alexander Hall also produced a cylinder rifle; while from Europe in this period came the Adams, the Deane, and a host of others inspired by or frankly imitating the already famous Colt.

The Colt Revolving Rifle

While no arm without fixed ammunition has ever made a satisfactory breechloader or repeating arm, Sam Colt's first products came closest to doing so.

The first Colt Revolving Rifle—merely the Colt percussion revolver of the period with a shoulder stock and long barrel—was produced in 1836. Improvements were added in 1840, 1842, and 1855, and the arm was sold and used throughout the known world. The Colt Repeating Rifle Model 1855 with side hammer of Root patent is famous as being the first repeater bought and used by the U.S., though not many were used. These were .56 caliber arms with five-shot cylinders. In their day they were the wonder of the age, tales of their accuracy being told and re-told until they achieved the heights of the ridiculous. In Cleveland's HINTS TO RIFLEMEN, published in 1864, we find them credited with accuracy at 680 yards! A more reliable yet still glowing account is that of Captain Hans Busk of the Victoria Rifles who in 1860 wrote his book THE RIFLE AND HOW TO USE IT, the best work available on the military weapons of that period. He praised the Colt as being the best military arm produced up to that time.

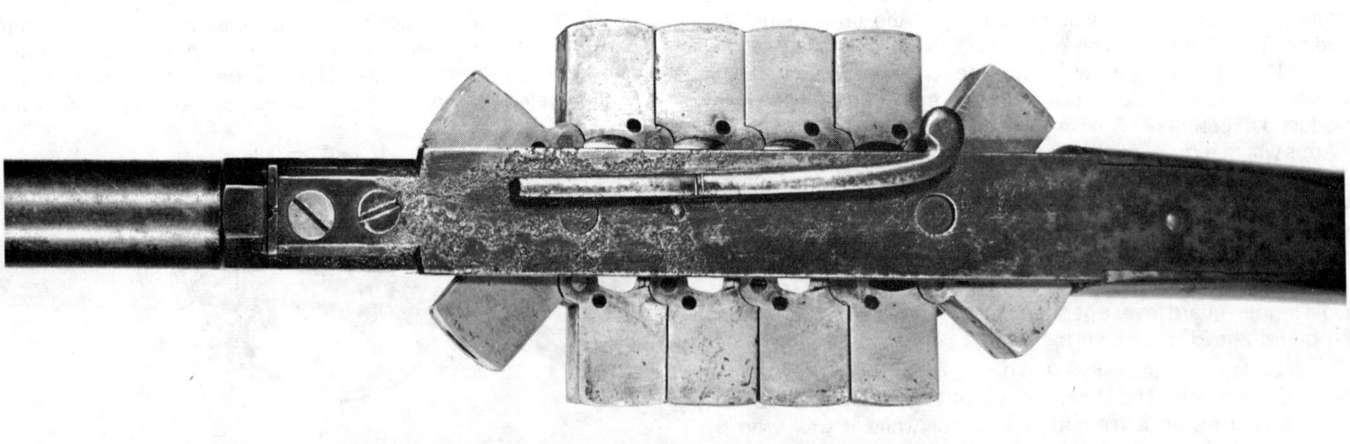

Bennett breech-loading percussion rifle, caliber .40. Original inventor's model patented in 1838 by E. A. Bennett. The breech has twelve individual blocks which were fastened together and rotated horizontally instead of up and down as in the Colt and other revolvers and revolving rifles.

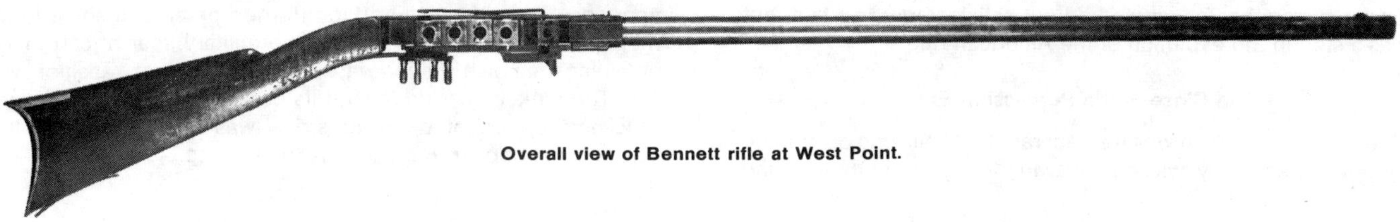

Overall view of Bennett rifle at West Point.

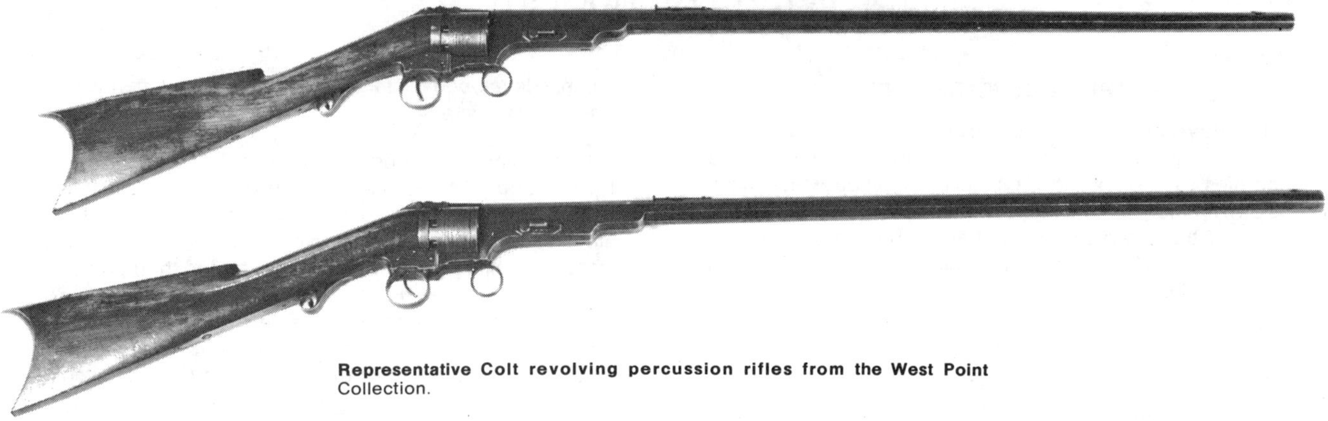

Representative Colt revolving percussion rifles from the West Point Collection.

Captain Busk's account of the general accuracy and shooting ability of the English soldiers of the day may be gathered from the following extracts: "A patrolling party at the Cape," he says, "used 80,000 cartridges in 1851 to kill or disable 25 savages." This figures out about 3,200 rounds per man. By our modern standards this is most efficient! He tells us that "At the Crimea the French used 25,000,000 cartridges without stopping 25,000 Russians." Of the American rifleman he had high regard, quoting their record at Churubusco where they killed or wounded 800 Mexicans with an average cartridge expenditure of only 125 rounds!

In service use the Colt early received a bad reputation as a rifle. There was so much trouble encountered with the chamber being fired flashing to adjoining chambers that it was the practice of shooters to turn the attached rammer down, then to hold it for forward support. This way they escaped burns on the left hand and wrist from fire spitting at the barrel and chamber juncture or from

Repeating rifle with a revolving cylinder made by Jonathan Browning.

multi-chamber firing. Even in metallic cartridge days the revolving cylinder system has never worked out satisfactorily in rifles, but it was a step in the evolution of the American rifle.

The Close of the Percussion Era

The percussion era in which a separate tape, pill, cap or priming charge was necessary, was now drawing to a close both here and abroad. Freak cartridges of all orders would still be tried, but all of importance would feature self-contained priming in some form. The popular, and more particularly the military, fear of cartridges containing their own ignition was waning. A period of transition was here. The metallic cartridge with its perfect breech seal was not yet to be fully accepted, but its arrival was foreshadowed in the freaks now to appear in a short passing phase.

THE PERIOD OF TRANSITION

CARTRIDGE DEVELOPMENT

With the development of the percussion cap, inventors—particularly in England, France, and the United States—turned their energies to attempts to produce ammunition which would carry its own ignition, as well as arms to use such ammunition. The importance of the complete gas seal at the breech was not fully recognized until after thorough war tests had been made of the Dreyse Zündnadelgewehr.

Pauly and His Inventions

Jean Samuel Pauly of Geneva invented a breechloader in Paris, working out the principle between 1808 and 1812. This was the first attempt at the needle gun, and the self-consuming cartridge was discharged when the needle pierced a paper detonating cap attached to the cartridge. Out of this first crude design Dreyse probably developed his famous needle gun, but that was not until some 30 years later.

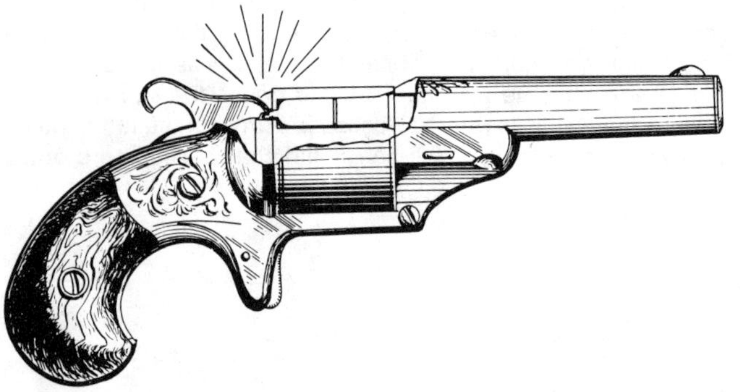

Moore Tit Fire. One of the scores of unsuccessful evolutionary metal cartridge types. This was an example of the American period of transition. Original drawing to show firing system.

At this point it might be well to consider Pauly's developments of 1814 and 1816, since they are generally confused with the inventions of more practical and successful designers. His English patent specification of 1814 (No. 3833) primarily concerns the discharge of black powder by heat generated through compressed air. The patent does not apply to later successful percussion systems. Since standard forms of black powder explode at temperatures around 600 degrees F., Pauly, Newmarch, and others developed experimental mechanisms to discharge the powder charge by obtaining such temperatures through piston arrangements which would produce detonation. A secondary part of this specification describes a hinged-lever breechloader of lifting block design.

Pauly's specification of 1816 is of historic interest, on the other hand, because it clearly recognizes and states the principle of effective breech sealing. It says, "A plug is so placed in the gun as to come between the charge of gunpowder and the movable breeching in all cases, and is formed of lead, copper, or such other ductile materials as will give way to the explosive force of the charge." Here clearly stated for the first time we find official record of the principle of the ductile expanding case which was to be introduced by other Frenchmen, Lefaucheux in 1836 and Houiller in 1847, though even those developments were foreshadowed by the introduction in 1835 of the French Flobert cap. This cap, however, was evolved directly from the familiar copper detonating cap; it did not use a propelling charge. The priming mixture itself was sufficient to drive the light, small gallery or "saloon" bullet out. The Flobert cap, which is practically our "BB" cap of today, was a cap evolution in theory and design, rather than a breech sealing cartridge.

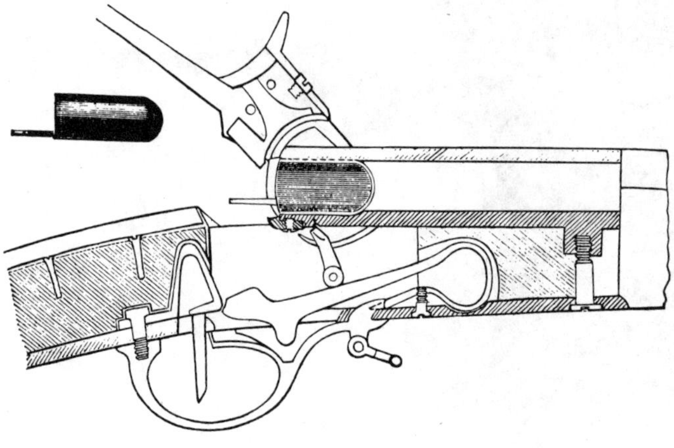

The French Demondion transitional breechloader.

Demondion Lifting Block Breechloader

Meanwhile in 1831 Augustus Demondion of Paris patented (No. 6137) an unusual rifle and cartridge. This rifle was manufactured by the noted gunsmith, Robert. It was a most important transition design as it marked the introduction of the system of the lifting breechblock hinged at the front. This was the forerunner of literally dozens of lifting block designs throughout the world. The Demondion is also noteworthy as introducing a weird cartridge which later provided the germ of the idea for the successful pin fire cartridge. This Demondion cartridge had a flat percussion tube in the shape of a tail projecting from the lower base of the cartridge. When the rifle was opened for loading, lifting the block with its lever automatically compressed a powerful mainspring below the breech. The cartridge was inserted with tail to the bottom. When the trigger was pulled, the flat mainspring snapped up and crushed the detonating tail against the underside of the breechlock. The cartridge case being self-consuming, no extractor was employed—but the breech was not fully sealed against gas escape. However, this design paved the way for the first true gas-tight design which was to revolutionize arms manufacture.

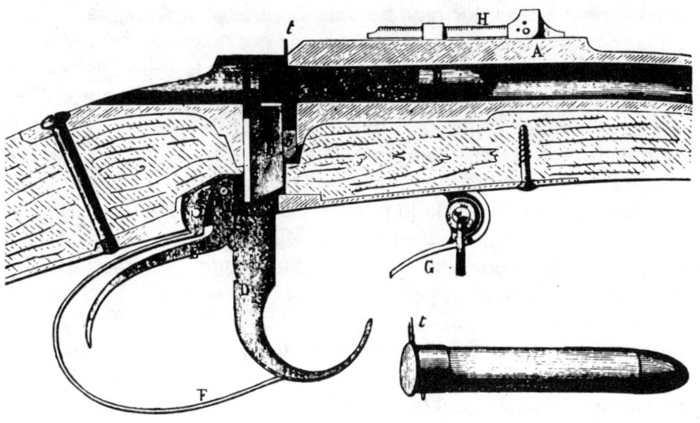

Action of the French Mousquetoon Des Cent Gardes.

Lefaucheux Paper Cartridge

For all practical purposes, the appearance in 1836 in France of the Lefaucheux paper cartridge with metal expanding base, cardboard body and projecting brass pin, marks the beginning of the true breech seal. Enthusiasts still argue as to whether this pin fired cartridge or that of Houiller in 1847 marked the "true" pin fire. The matter is entirely academic, since both have projecting pins which when hit by the hammer crush the detonator within the case head and fire the charge. In 1836 Lefaucheux introduced both this cartridge and a hinged-frame gun to handle it. This gun is essentially the grandfather of all our modern double-barreled shotguns. The Lefaucheux system was not adapted to military needs, either in its early form using a non-sealing cartridge or its improved form with gas seal case.

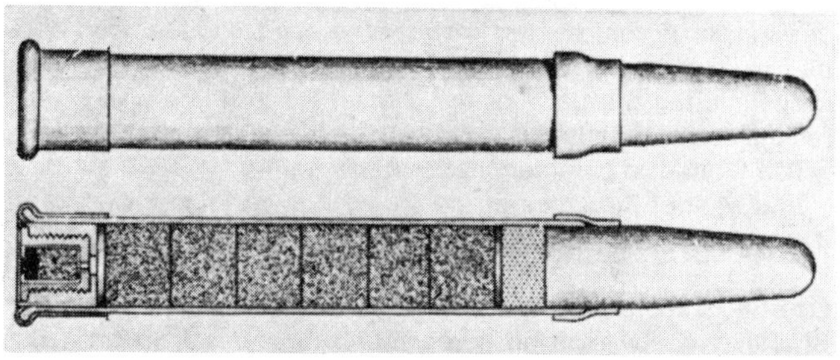

French Mitrailleuse cartridge and section.

Houiller Patents

Houiller's patents ranging from 1847 to 1850 touch on pin, rim, and center-fire ignitions; the outstanding feature of the specifications, however, being the use of full metal cases.

Mousquetoon des Cent Gardes Rifle and Pinfire Cartridge

About 1854 the French "Mousquetoon des Cent Gardes" came into use. This strange arm used a cartridge of modified pinfire type with a small pin projecting from its metal on the bottom to fire it, while a longer pin projected from the top to extract it. The breechblock was lowered to load the chamber. When the trigger was pulled the breechblock was driven up by the mainspring, and as it reached its top position a projection at the bottom of the block hit the pin and fired the cartridge!

The Needle Fire Chassepot Rifle

This, then, was the general state of development in France when the Government adopted the Chassepot Rifle already mentioned. It is said that Pauly offered his original needle gun to the great soldier Napoleon Bonaparte, who refused it. In the state of its development at that time, and considering the general conservatism of mankind whether dealing with breechloaders or atomic energy, that is not difficult to understand. By 1830 when Dreyse introduced his first needle gun many of the objections had been worked out. (As a matter of record, it must be pointed out that the gun and cartridge patented in England in 1831, No. 6196 by Moser were patented for Dreyse, English patent office regulations of the time making it wise to have patents handled by English agents.)

When the Dreyse was introduced into the Prussian service it was a "military secret" of the first order. Like most "military secrets" it was a secret only to those naive branches of the military who never seem to be aware of what has been done in their line—those artless individuals with which every country is regularly afflicted, and who strangely enough seem to be nearly always in a position to make policy while submerging the real experts who are present in any army.

And so it was that foreign observers of European powers watched with dread the success of the Dreyse gun in the Schleswig-Holstein war in 1864 and the Austrian War of 1866 where the Prussians again wrought terrible havoc by pitting this inefficient breechloader against the still more inefficient muzzle-loaders of their enemies.

In far-off America by this time both the Spencer and the Henry repeating rifles with their metallic cartridges were in use. Breech-loading arms of nineteen types were accepted even by the hidebound military and banking-type minds which worried about the heavy ammunition expenditure! These must be dealt with in subsequent chapters, but they require mention here because of their bearing on reactionary and pinch-penny psychologies which have cost so many nations so much throughout the course of history.

In his brilliant treatise, THE PAST AND FUTURE OF ARTILLERY, the French ruler of that time, Napoleon III, wrote: "Inventions that are before their age remain useless until the stock of general knowledge comes up to their level....Not only does routine scrupulously preserve, like some sacred deposit, the errors of antiquity but it actually opposes, might and main, the most legitimate and the most evident improvements."

Perhaps, as some humorist would have it, the Emperor did not read his own book. Perhaps, with the French Maginot-line mentality of an earlier day, he was so obsessed with the terrible efficiency of the Germans that he could not conceive of any other nation developing instruments of war superior to theirs. Whatever the cause, in the face of the great strides his own inventors had made, in the face of the ghastly efficiency of the metallic cartridge demonstrated in the tragic American Civil War, in the face of the ultra-conservative British War office testing some 50 systems for converting their old Enfield muzzle-loaders to breechloaders and accepting the Snider with its gas-sealing cartridge—in the face of all these developments he adopted a modified needle gun! Oh, it was a great stride forward! It used an obturator of rubber to seal the breech more effectively than the German. (The rubber hardened and became worse than useless after a little firing; but that didn't turn up until later.) The cartridge was reduced to .434 caliber and fired a conical bullet of 380 grains to an extreme range of 1800 yards—but what of its actual efficiency? Consider what a contemporary writer says: "It is said that the Frenchmen spit on their cartridges, force their fingers into the breech action, and give every possible sign that after a few shots, the Chassepot gets so foul they do not know how to treat it. There is a difficulty in getting the cartridge into the chamber when the rifle is foul; if force is used it becomes dangerous—the cartridge being soft, the percussion cap is compressed between the bullet and the point of the bolt, and has been known to explode in the act of loading, in many cases injuring the hand of the shooter." The rifling was given a left instead of a right twist to offset somewhat the bad trigger pull which tended to pull the muzzle to the right!

And what of the value of the long-range cartridge? Apparently it wasn't very accurate beyond 100 yards in actual battle use. Yet the Chassepot was admittedly an improvement on the Dreyse. On the Prussian side, consider the report of our General Phil Sheridan who was an observer at the Meuse and Sedan in that fateful War of 1870: "The cavalry," he says, speaking of a charge of the French Hussars, "behaved most gallantly, were sheltered from fire till the last moment, were skillfully and bravely led, and the ground they charged over was not more than 400 yards; yet the result was their destruction as a military body without any effect whatever." The Prussian infantry in this engagement held the fire of their needle guns until the cavalry was within 150 yards, a case of disciplined men recognizing the limitations of their arms.

The Franco-Prussian War converted every nation to the idea of metallic cartridge breechloaders overnight—as time passes in military evolution, that is. In the United States the Dreyse had been imitated and sold by P. Klein in New York City, some 500 having been made by George Foster at Taunton, Mass. These were never popular however. In England, Needham, Sears, Schlesinger, and others developed needle type arms, while in France the Comte de Chateauvillier patented a needle gun with a lock resembling the earlier Demondion. This Chateauvillier is interesting because it employed a form of toggle-joint lock which, like the American Jennings, pointed the way to the successful toggle-locking system of automatic days.

Needham Repeating Rifle

Of all these foreign transition arms only one form of the Needham warrants attention at this point. When one remembers that the original W. Greener was one of the greatest gun designers and writers of all time, his enthusiasm for the Needham Repeating Rifle can be explained away only on the ground of excessive British patriotism. This weird specimen utilized a tube below the barrel for a magazine, plus a revolving cylinder. Special cartridges were used which were in two pieces, the base being rimless and tapering to the rear; a collared cap at the forward end fitting the exterior of the cap and projecting inside the case proper. When the percussion cap in the base of the cartridge was detonated, and the powder charge ignited, the collar about the case was supposed to drive ahead and expand to seal the joint between cylinder and barrel. When the magazine was loaded a spiral spring forced the cartridges back. Operating a finger lever trigger guard carried the cylinder through a one-quarter revolution, partly raised and then dropped the hammer to eject the fired case forward out of a chamber, lined up a chamber to receive a loaded cartridge, lined up the chamber with the barrel, and then cocked the hammer! Remember that metallic cartridge repeaters were, as Greener puts it, "flooding the market" at that time, then consider his appraisal: "This, the only English repeating rifle, is of novel construction; the idea, most happily conceived, combines the principles of the ejecting hammerless gun and the revolver with a tubular magazine gun...The system is to be greatly commended. It is decidedly original, of sound principle, and unlike any of the numerous magazine arms now flooding the market." This closing remark referred to such arms as the Spencer, Winchester, and Vetterli! The Needham, needless to say, was a complete failure.

The fantastic English Needham repeating rifle and cartridge.

While the Needham was the only one manufactured, several transition English designs in repeaters were patented. The Restall had a tube magazine below barrel, used special freak cartridges and was operated by a sliding trigger guard. The Maberly carried special cartridges in an endless chain device in the buttstock. None of the European transition arms are of fundamental importance, however.

AMERICAN TRANSITION ARMS

American arms design during all phases of this period was the soundest in the world. Germany had the best military machine, France the best chemists and cartridge experimenters, England the best writers and public relations men. But the U.S.A. was most prolific in both design and manufacture of practical arms. The need for arms for frontier service, of course, had much to do with this.

The Germans were hampered psychologically by the fact that they were actually winning wars with the needle gun and had no time to consider change. The French were handicapped by their national psychology and by studying Prussia to the exclusion of all other countries. English inventors, faced with the fact that the War Office considered any cartridge with self-ignition dangerous and impracticable, and any breechloader as a useless expender of large quantities of ammunition, spent their energies on a hundred useless designs to make a more efficient cap-fired muzzle-loader!

But in America there was a civilian hunting, sporting, and protective market to absorb really good designs. As a result, while every conceivable system of breech locking appeared, in every major instance the arm was fired by an elementary tube, cap, or pill. No effort was wasted on the abortive groups of cartridges used in Europe which sought to combine the ignition with the charge, but without considering the necessity for an effective gas seal. In short, the American line of design, with only one truly notable exception, moved directly from percussion lock to metallic cartridge design.

The Jennings Repeating Rifle

Walter Hunt of New York City patented a lever operated "Volitional Repeater" in 1849, but it was an impractical weapon employing a special rocket-type cartridge with separately fed primer and was operated by two levers. Lewis Jennings improved on Hunt's basic idea.

Without question the Jennings was the most important firearms development of this era. Its system led directly to the transitional Volcanic which in its turn produced the rimfire metallic cartridge Henry rifle, which in turn became the first famous Winchester rifle.

The story of Windsor, Vermont, that great cradle of firearms development, manufacture, and invention, has never been written. High among its achievements ranks the manufacture by Robbins and Lawrence of the first reasonably successful approach to the solution of the problem of the repeating rifle, the repeater of Lewis Jennings, patented Dec. 25, 1849 (Pat. No. 6,973).

Some samples of the Jennings in collections have the Maynard tape primer. A tube below the barrel carries twenty conical bullets with hollow bases. The propelling charge is in the hollow bullet itself. The charges are loaded by gravity into the breech, no springs being used as in later cartridge types. The muzzle is raised to allow the bullets to slide back until the first one drops onto a carrier. This carrier is operated by a ratchet through the operation of a ring trigger. So far as we know, this is the first successful use of the familiar below-barrel magazine tube.

Much more important, however, is the breech locking system. This Jennings is the direct source of the mechanical principle of locking of some of the most successful of all automatic arms later produced. It was brought to a higher state of efficiency in its direct successor—the Volcanic pistol and rifle.

Modification of Jennings by Maxim. The Volcanic system of locking was modified in the Henry cartridge rifle and in the first rifle to bear the Winchester name. Hiram Maxim, one of the greatest arms inventors of all time, adapted the early Winchester rifle to an automatic loader by using a false butt plate and a system of springs. More important, however, he obtained from this pattern the idea for his later machine gun locking system.

This toggle system of locking the breech was perfected by Maxim and used on the Maxim Machine guns. With only slight modifications it became the locking system of the Vickers Machine Gun, the toggle being inverted.

Modification of Jennings by Borchardt. Another inventor who drew on Jennings toggle-lock system was Hugo Borchardt. He was employed at Winchester for a time. He developed one of the earliest successful autoloading pistols. When he could not interest American makers in it, Borchardt took his design to Germany. It was manufactured there for a time. The Borchardt was developed into the world-famous Luger pistol—the toggle buckling up to unlock. Many experimental and a few successful recent machine guns use the principle, one being the Swiss Fürrer.

Historical Note on the Jennings. Jennings has received scant attention from firearms historians, though as we have seen he was one of the great pioneers of design. A note on the operation of his arms based on contemporary reports and drawings may therefore be of interest.

In a copy of the International Magazine published in 1852, one W. M. Ferris contributed a description of the Jennings rifle. A magazine on top of the breech was filled with percussion pills or priming powder, he says. A tubular magazine below the barrel was then loaded with hollow cartridges containing the propelling powder. These cartridges were merely bullets with cavities in them to hold the powder in the rear section. The magazine tube held 24 cartridges.

The forefinger of the firing hand was passed through the ring which formed the lever end below the receiver. The thumb was placed on the hammer. The muzzle of the rifle was then elevated sufficiently to let the cartridge nearest the breech slip by gravity onto the carrier. The lever was then pushed forward. Raising the hammer moved the breech pin back and raised the carrier, placing the cartridge in line with the barrel. Pulling the lever back forced the breech pin forward to push the cartridge into the firing chamber. This same motion took a percussion priming from the magazine by action of a special priming rack. This rack revolved the pinion forming the bottom of the magazine, and also threw up a toggle behind the breech pin securely locking it for the moment of discharge. An upward pressure of the finger within the ring released the hammer. The hammer pin struck the priming and flashed fire into the powder within the ball cavity. Releasing the finger pressure permitted the ring to take its normal rest position ready for the next shot.

Early reports give the tube capacity as 24 bullets and claim a rate of fire of twenty shots per minute.

Christian Buchel also of New York City patented a repeater also in 1849 intended to use a special papercase cartridge resembling the needle type and operated by sliding the magazine tube. This, too, was an impractical arm.

The Jennings of the same year was actually manufactured as a repeater. Some are found in collections today as single-shot arms, the gravity tubes having been dispensed with when found

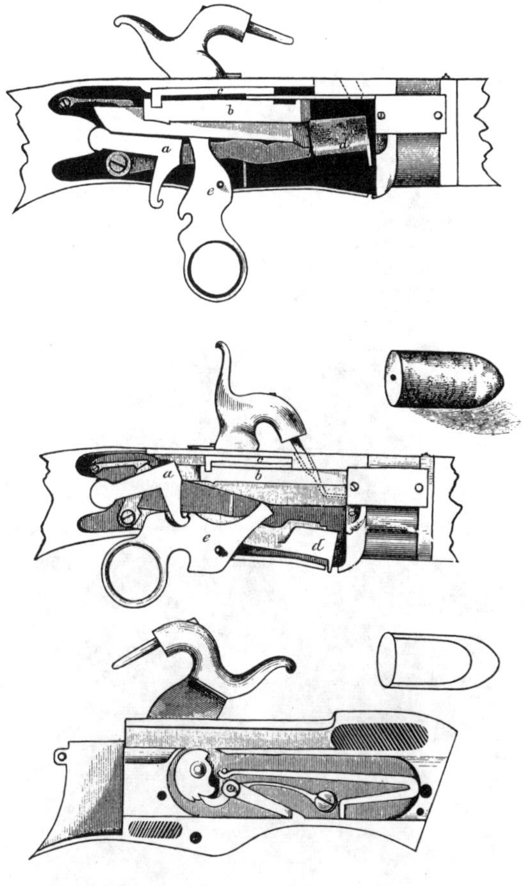

Action of original Jennings Repeating Rifle.

impractical under hard field conditions. Horace Smith improved the original patent with a reservoir for the percussion pills.

The Volcanic Rifle

However, the really important improvement in design came with the introduction of the positive double-toggle lock joint (pat. 10535) of Smith and Wesson filed on Feb. 14, 1854. Tyler Henry, an outstanding mechanic employed at the Robbins and Lawrence plant at Windsor, Vermont, had helped build the original Jennings. He was engaged to combine the Jennings and the Smith patents to produce the first Volcanic.

The Volcanic cartridge was new here, though several resembling it had been attempted in Europe. It consisted of a hollow base conical bullet having a charge of black powder in the hollow. The powder was kept in place by a cardboard disc which contained the primer. Since this cartridge had no gas seal, it was foredoomed to failure. This cartridge design in those days permitted only a low-power charge, and gas escape at the breech was impossible to prevent.

Originally introduced as the "Smith & Wesson" and manufactured at Norwich, Conn., in 1854 (both as a rifle and as a pistol), the arm was taken over by a corporation formed in July 1855 and renamed the "Volcanic." In 1856 the company moved to New Haven, and Smith and Wesson left to found their pistol company. When the company was thrown into bankruptcy in 1857, Oliver F. Winchester, one of the original stockholders, bought up the assets and formed the New Haven Arms Company.

Winchester at that time was neither the "successful manufacturer in the textile field" nor yet the "shirt manufacturer" that some have made him out. He was actually a prosperous haberdasher. True, he knew little about arms. But he had in him the native strains of shrewdness, foresight, and courage which brought fame and fortune to so many New Englanders of the period. Oliver Winchester had to a supreme degree the characteristics which helped to mold America in those early days. Himself a tireless worker, he drove all those around him. He had the sound executive ability to hire and trust experts whose services he needed—a tremendous ability in itself; and he did not hesitate to back his judgment with his own cash—a characteristic that distinguished him from many of his critics.

In the October 9th issue for the year 1858 of "Frank Leslie's Illustrated Newspaper" is a long and glowing account of the Volcanic Repeating Rifle which has been the basis for hundreds of interesting if rather foolish stories about the arm. As the forerunner of the Winchester, the Volcanic is an epochal development. As a practical arm it was a distinguished failure.

Leslie's said: "It combines every quality requisite in such a weapon...thirty shots can be fired in less than one minute...the balls may be soaked in water with perfect impunity, and can be kept any length of time. They cannot be exploded by contact with flame...."

"The squad of police sent down to Staten Island during the Quarantine riots," the article continues, "were armed with these weapons to the extent of some eighty or ninety rifles and forty pistols. We should judge that for police purposes and as cavalry carbines these Volcanic firearms can have no competitor, and the pistols are undoubtedly superior to anything yet produced. Their waterproof qualities render them especially desirable for the use of travelers and in the Navy."

B. Tyler Henry, one of the great American developers of arms and ammunition. The letter H appearing on the head of every Winchester rimfire cartridge case is in honor of his work in rimfire cartridge development.

Oliver P. Winchester, founder of the great arms and ammunition combine which today is part of the Olin-Mathieson Chemical Corporation.

The closing blurb, however, is an index to the unreliability of much of the published literature of the time: "A Good Shot will hit a Quarter of a Dollar at Eighty Yards with one of the Navy pistols." Actually, of course, it is quite a problem to hit a flour barrel full of quarters—or full of anything else—at eighty yards with this cartridge. But the publicity sold weapons, helped to establish Oliver Winchester, and has been a great boon to writers ever since its original publication.

As a matter of further record, the Volcanic vanished from the American scene within two years of the publication of this glowing article! An actual machine rest test of a Volcanic with specially loaded cartridges (no original ones could be found which would fire) placed 5 shots just inside a 21-inch circle at 80 yards when fired from a practically unused collector's specimen.

With the coming of World War II, the "wheel of design," like so many other things in history, completed a turn. Experiments conducted by the Japanese on Volcanic-like ammunition with the charge contained in a hollow in the ball, met with some success in properly sealed automatic cannon. In the United States freak cases with metal heads and weak plastic sides permitted us to develop the outstanding piece of portable artillery, the Recoilless Cannon—some of the powder gases escaping through the case walls to a Venturi tube where they blow back over the shoulder of the firer and counterbalance the recoil. In both the U.S. and Germany considerable headway was made on plastic cartridge cases; while some German developments in self-consuming cases used with modern breechlock designs met with a certain measure of success. ...In short, the historical period of Firearms Transition repeated itself at its new place on the spiral of development; but for practical purposes, the metallic cartridge is still master of the ammunition world.

METALLIC CARTRIDGES

As we have seen, breechloaders were known even in the 14th Century. However, the successful breechloader became a possibility only when the first metallic cartridges containing both built-in ignition and expanding case were introduced.

Some soft metal cases were used in early European muzzle-loaders. The rear end was removed and the case containing the powder and ball was rammed down the muzzle. Similar types were tried with the early American Maynard. In all such cases the ignition was from a separate cap. The metal case itself was blown out the barrel with the bullet. Such systems were devised solely to speed up muzzle loading.

The modern breechloader requires three things of its ammunition. First, it must be a "fixed" charge. That is to say, each must contain case, primer, propelling charge, and bullet. Second, it must have a ductile metal case which when fired will expand against the walls of the firearm chamber to prevent any gas blowing back along the sides of the case; and which will contract automatically after the gas pressure drops so that extraction will be easy. Third,

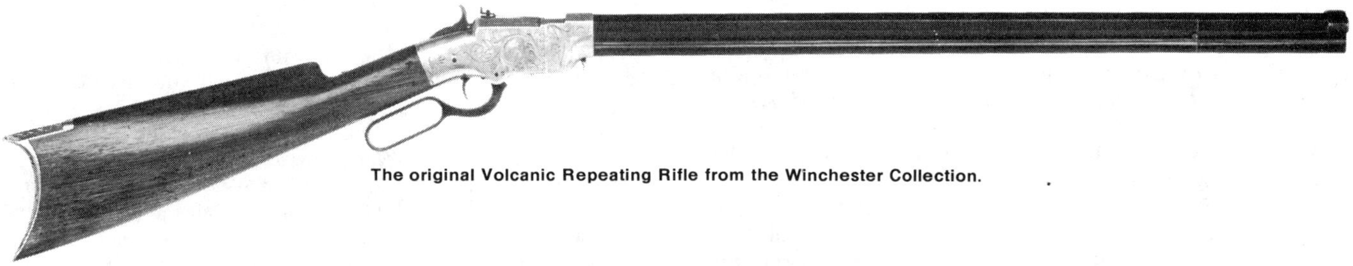

The original Volcanic Repeating Rifle from the Winchester Collection.

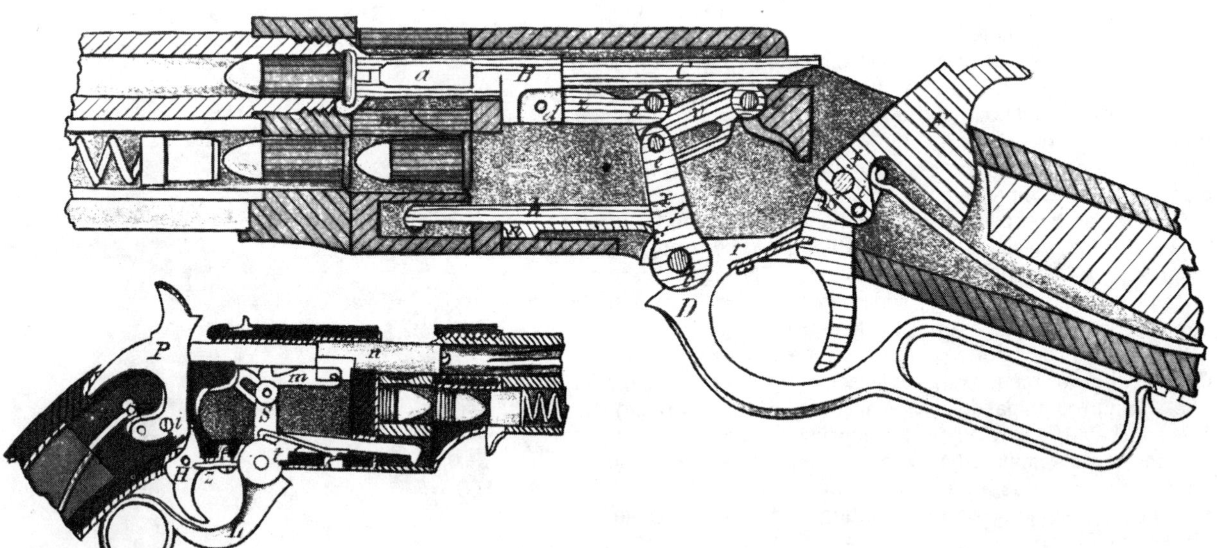

Contemporary drawings showing evolution from Volcanic (below) with its freak cartridge to Henry (above) with rimfire cartridge.

its primer must have built-in a mechanical anvil in center-fires or a folded head in rimfires which will allow the firing pin blow to crush and detonate the priming charge.

The varieties of such fixed ammunition are discussed separately in the order of their development.

THE PINFIRE CARTRIDGE

The first successful breech-loading cartridge was the Pinfire.

Mention has already been made of Pauly's experiments in France on ammunition. He tried sealing the breech with ductile metal discs.

The honor of introducing the first truly successful breech-loading cartridge and gun, however, belongs to Pauly's successor, E. Lefaucheux, a famous Parisian gunsmith. This was in 1836. Here again a certain qualification is necessary because of improvements made in the Lefaucheux case design by Houiller, another Frenchman, in 1847.

The cartridges themselves are known as "Pinfires." In its earliest form the Pinfire introduced by Lefaucheux resembled a modern shotshell. It was a paper case which contained shot and powder. It was fitted with a metal head which contained the priming and which would expand as a gas seal at the instant of firing. A pin projected from the side of the case head and led down into the priming mixture. The gas seal was not too good. Houiller's improvement gave a good gas seal. Note must be made that seven years before Houllier's development yet another Parisian gunmaker, Bastin LePage, introduced a freak cartridge in which the copper cap and anvil projected from the base of the case. The design gave extraction trouble. It must not be confused with the successful Lefaucheux and Houiller developments.

Lefaucheux introduced his new cartridge and a hinged frame shotgun to use it in 1836. The gun itself is the parent of most of our common double-barrel shotguns, though its breeching mechanism was crude and not too efficient.

In Pinfire arms (rifles and revolvers and pistols as well as shotguns are made on the principle) a notch is provided in the arm itself to position the pin when the cartridge is chambered. As the hammer falls it drives the pin down into a reversed cap which is seated in a cup-shaped base within the case head, and which contains the priming mixture.

When Lefaucheux showed his double shotgun at the London Exhibition in 1851, the cartridges shown with it incorporated the Houiller improvements. All Pinfires are very commonly referred to as "the Lefaucheux system," though Houiller played a considerable part in the ultimate design.

The Lefaucheux design was an instantaneous success in sporting use, both as a cartridge and as a gun design. The military, understandably, did not rate the principle very highly. The hinged frame principle of breech loading, of course, is not suited to military needs, primarily because so much shooting and reloading must be done from prone positions where the drop-barrel system is impractical. Except as substitute standard arms purchased in times of stress where any arms available had to be used, this system has never been in official military use. The pinfire cartridge principle was used in military service only on such freaks as the Cent Gardes Musketoon (or "Treuille de Beaulieu" rifle, to give it its true name, which derives from the name of the French Artillery Captain who was responsible for its adoption); and the impractical rifle-shotgun combination of Dr. Alexandre Le Mat of New Orleans. This arm, having a revolving cylinder which fired through an upper rifle barrel while mounted on a lower shotgun barrel, was used by the Confederacy as a percussion arm. After the Civil War, the French and Belgian plants manufacturing it produced the weapon in pinfire design.

The pinfire system has been widely used in shotguns, revolvers, and pistols in Europe, though very sparingly in rifles. These arms are of little interest and less importance to us in this study. The cartridge itself, whether of metal-and-cardboard shotgun type or of metallic pistol and rifle type, is also of little importance in the United States. No arms are made here for such cartridges.

However, the principle of the gas-tight cartridge, which completely revolutionized firearms design and the art of war, found its first successful expression and acceptance in the pinfire; a fact which assures the system a high place in the history of arms and ammunition evolution.

Why the principle of the gas-tight cartridge was not recognized by the military as it was by sporting gun manufacturers at that 1851 Exhibition is one of the mysteries of firearms. As late as 1866, as we have seen, the French adopted the impractical Chassepot needle cartridge. In 1867 the Russians adopted the Karl modification, also a bolt action needle system. In 1867 the Italians transformed their Carcano muzzle-loaders to a bolt action needle system also!

And what about the United States Army? Well, we too, had observers and exhibitors at the London Exhibition in 1851. Yet at the outbreak of the Civil War the Union had on hand as arms for an infantry war some 530,000 muskets which were essentially only modified Revolutionary War design! They were fitted with rifled barrels and percussion locks, but were still muzzle-loaders. In the initial year of that terrible war the Union armories turned out about 10,000 such muzzle-loaders. In the second year production was stepped up to 200,000. In the next year the machine potential of the industrialized North swung into high gear and produced about 500,000 more. Still more muzzle-loaders were purchased from abroad, ranging from some 428,000 Enfield .577's made in England to 60,000 Austrian tube lock alterations purchased from the Belgian firm of Herman Boker and Company. The North alone bought 726,705 European muzzle-loaders in this period as shown by official War Department records.

When we add in any estimate for the arms of the Confederacy—the vast bulk of which were also muzzle-loaders—it becomes apparent that in an advanced day of breech-loading systems our own government was also caught short with muzzle-loaders. The official Union tally lists 4,022,130 muzzle-loading muskets and 90,000,000 rounds of lead bullets issued during the war!

The inefficiency of the muzzle-loader in battle conditions is highlighted in an official War Department report quoted by many writers from Napoleon III on: The Battle of Gettysburg occurred July 1, 2 and 3, 1863. Of 37,574 muskets salvaged from the field and sent to Washington for inspection, one musket had 23 rounds in the barrel; 6,000 had from 3 to 10 rounds each in the barrel; 12,000 had 2 rounds in the barrel; 6,000 had one load in the barrel. The remainder were unloaded. Allowing 13,574 fired and 6,000

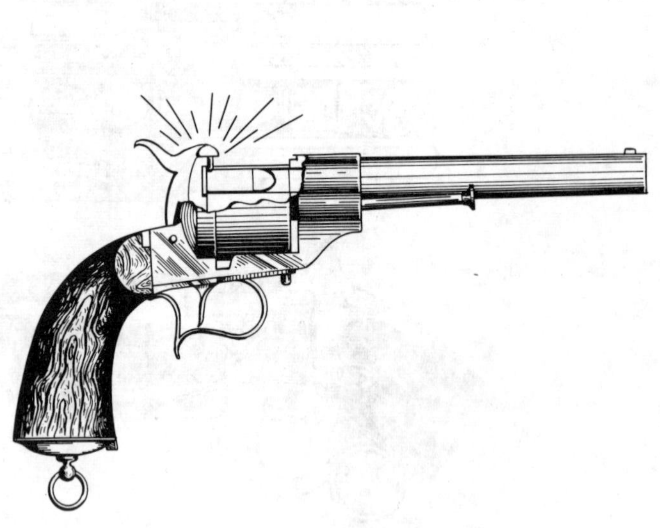

Pin fire principle—the Lefaucheux. Original drawing to show method of ignition.

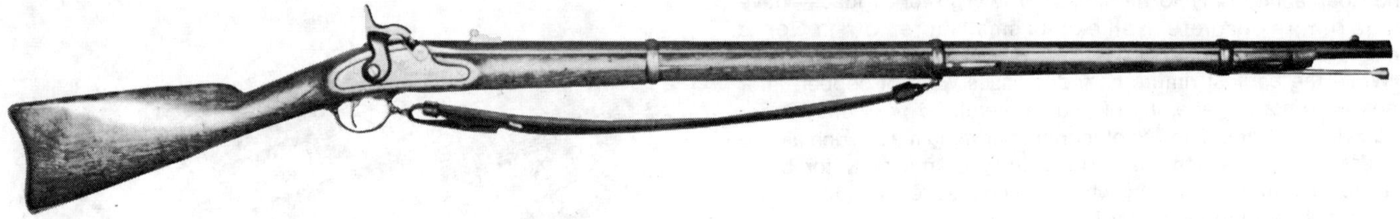

Typical representative American Civil War muzzle-loader.

correctly loaded, we get a very graphic picture of what happened under stress. Some men inserted paper cartridges incorrectly so the capfire couldn't reach the powder. Some tried to fire without caps. Others plugged the barrels in many ways. None of these things could occur with breechloaders, of course. Thus from one-third to one-half of the men engaged in this great battle on both sides are shown to have been practically useless as riflemen! And all because they used muzzle-loaders!

THE RIMFIRE CARTRIDGE

The next successful stage was the development of the Rimfire cartridge and arms to use it. In this design, still in common use in some calibers, the casehead has a projecting rim. The priming is spun into the space between the folds of this rim. Unlike the Pinfire which must be deliberately positioned so that the hammer will hit the pin directly, the Rimfire is loaded without positioning. The underside of the projection commonly rests on the face of the breech which thus serves as an anvil. When the firing hammer or pin hits the top rim of the case, it crushes the metal and detonates the priming.

Some of Houiller's French patents bear on a rimfire principle. So far as we know, however, he did nothing to develop the principle.

Flobert Cartridge

The noted French gunmaker Flobert took the common percussion cap of the time and re-shaped it to give it a rim. He loaded this with a small lead projectile. From 1835 to 1847 Flobert produced pistols and parlor rifles to use this ammunition. Note that such cartridges did not carry a projectile charge of powder. The priming fulminate itself generated enough power for the type of indoor (or "salon") shooting for which Flobert developed firearms. The present day "BB Cap" (meaning bulleted-breech cap) is almost identical with Flobert's cartridge. He too showed his arms and ammunition at the 1851 London Exhibition in the Crystal Palace.

Smith & Wesson

There is no question that the first truly practical Rimfire cartridge was the Smith & Wesson .22 produced in 1857. From his own research there is no question in the author's mind that the founders of the great Smith & Wesson Arms Company developed this cartridge directly from samples of Flobert's cap.

In 1851 Horace Smith, Daniel B. Wesson, and B. Tyler Henry were all intimately associated with Robbins & Lawrence of Windsor, Vermont. This firm entered a series of rifles and interchangeable parts at the London Exhibition. Representatives of the firm secured samples of the Flobert caps and returned with them to America.

Smith & Wesson undertook the development of a low cost method of drawing the cartridge cases. This was their great contribution. And it was indeed a great one in its day. The original .22 Smith & Wesson cartridge is practically identical with the common .22 Rimfire cartridge we know today.

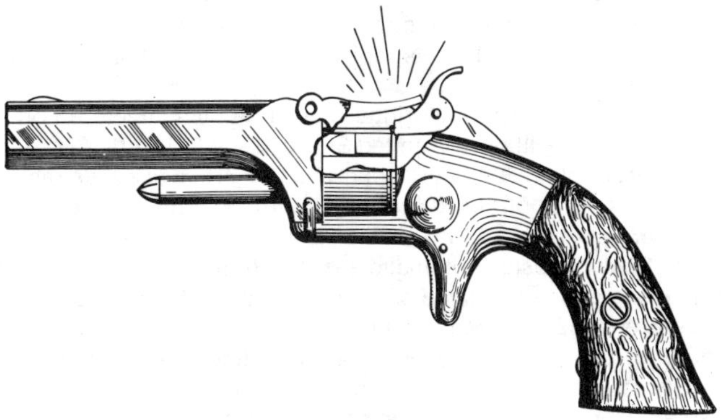

Rimfire principle—the first Smith & Wesson metallic cartridge, revolver, caliber .22. Original drawing to show principle of ignition.

Other Rimfire Developments

B. Tyler Henry redesigned the old Volcanic rifle and chambered it for a rimfire cartridge of .44 caliber. The rifle patent was granted in 1860. It is noteworthy that Henry at no time ever laid claim to having designed the cartridge. However, Winchester, for whom he worked and later re-designed the .44 Henry to become the famous .44 Winchester of frontier days, stamped an "H" on the head of the copper cartridge case. To this day the tradition is maintained. Every Winchester-made Rimfire cartridge case bears the letter "H" on its head.

The rimfire was the cartridge of the early successful American repeaters used during the Civil War to a limited extent: the rifles of Spencer, Ball, and Henry. It was the cartridge of the early single-shots, too: Peabodys, Remingtons, Ballards, later-day Sharps, of the Joslyns, converted Starrs, Cochrans, Whitneys, Millers, Needhams, Robert-Springfield conversions, Allen & Wheelocks, Ball & Williams—and of scores of others whose records may be found in the files of the Ordnance Corps.

In their time these rifles and carbines with the rimfire ammunition made history not only in America but throughout the world. In Europe the Peabodys and the Remingtons in particular were widely adopted by the military and were a big factor in foreign design. The rimfire cartridge itself was adopted for use in the Austrian Wänzl in 1867, the Albini-Braendlin made in England in 1867, the Belgian Terssen of 1868, the Swiss Vetterli repeater of 1869, and the Italian Vetterli-Vitali repeater of 1869.

The faults of the rimfire cartridge—the uneven distribution of priming around the rim, case weakness caused by rim crimping, and adaptability only to medium and low-powered loads—have made rimfire obsolete in all except small-bore, low-powered designs.

While the earliest rimfire cartridge cases were of copper, later ones were made generally of gilding metal, 95 per cent copper and 5 per cent zinc. The use of mercuric priming made such a case necessary because the natural affinity of mercury for brass resulted in defective priming when attempts were made to use the rimfire system in the stronger brass cases.

With the development of non-mercuric priming in recent years, much rimfire ammunition has used strong brass cases, thereby allowing higher pressure in high speed ammunition. Steel cases, too, have become practicable as mechanics and metallurgy have developed.

Originally loaded with black powder, by 1905 these cartridges were also being loaded with a blend of black and smokeless of which the "Lesmok" cartridges were typical. While this mixture gave better ignition and accuracy, it left a heavy residue which fouled the small-bore barrels rapidly. The common load today is smokeless powder.

In recent years rimfire cartridges have been manufactured only in the following calibers for United States use: .22 BB and CB caps for indoor shooting; .22 Short, Long, and Long Rifle in a wide range of trademarks for plinking, gallery shooting, and for precision small-bore shooting; .22 magnum, .22 W.R.F. for target and small game; .22 Winchester Automatic and .22 Remington Autoloading for plinking and small game shooting; .25 Short Stevens and .25 Stevens and .32 Short and Long to meet the demands from large groups of owners of older weapons handling these cartridges; .41 Short to meet a large demand for cartridges for the long obsolete Derringer type pistols; and .41 Swiss to supply the cartridge demand from farm areas where huge numbers of obsolete Swiss Vetterli rifles, originally sold by Bannerman of New York City early in the century, are still in use.

THE CENTER-FIRE CARTRIDGE

The development of the center-fire cartridge permitted the design of the really high-power firearm. In its common form it is the familiar rimmed or rimless case (though it may be semi-rimmed or belted also), in which the primer is an individual element seated in a depression in the center of the case head. This design is much more positive in its ignition than any rimfire can possibly be and it also lends itself to reloading which the crushed rim of the rimfire design does not.

Again the evolution of the design is such that it is not possible to ascribe it definitely to any one source. For all practical purposes, however, the French Pottet case patented in France in 1857 is the direct ancestor of all center-fire cases. This design closely resembles our modern shotgun cartridge case, and its specifications are so clearly set forth in the patent that there is no question about its originality.

Many European writers in the past have credited the English gunmaker Charles Lancaster with the development. Actually the Lancaster case was merely an attempt to overcome the shortcomings of the earlier pinfire design. True, it was center-fire in action. But it bears no such resemblance to the modern center-fire, as does the Pottet. Lancaster's development was an expedient only. Pottet's development was a new principle of ignition.

The first widely used center-fire cartridge case was patented in England in 1861. Patent is in the name of F.E. Schneider of Paris. This case was originally sold by the English gunmaker, G.H. Daw. In 1866 Daw sued W.T. Eley for making a similar cartridge. The suit was thrown out of court, and soon thereafter many varieties of center-fire modifications were developed in England.

British observers of the Danish-German wars were highly impressed with the field performance of the Dreyse Needle Breechloader. Based on their reports the Russel Committee was set up in 1864 in England to investigate the possibilities of Breechloading systems. While the Henry and Spencer repeaters were already in use in the United States at the time, no thought was given by the English to such arms. Military thinking everywhere still favored the single-shot. And hard economic facts also had to be faced—even as today.

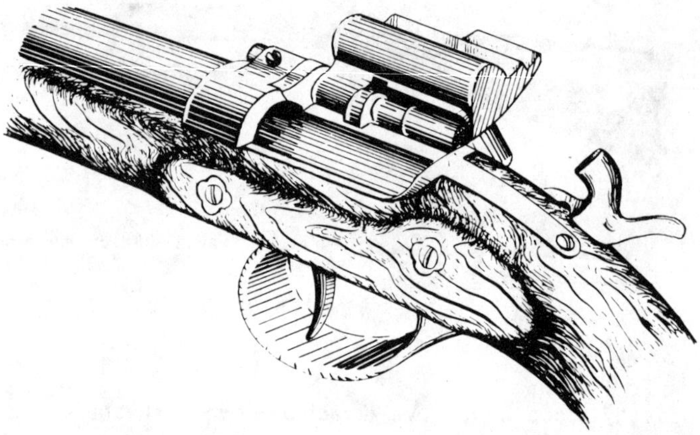

The Snider conversion mechanism. Original drawing to show rifle open ready for loading.

The Snider Rifle Cartridge

The British tests not only covered rifle alteration and design. They also extended to ammunition; and in that connection historical note may be taken here of the Snider Rifle and its cartridges.

In answer to a Government advertisement in August of 1864, about fifty systems were offered to meet the requirements stipulated. Chief of these was that the system should be suitable for converting the Enfield muzzle-loaders to breechloaders, and that proper ammunition should be available. This, it must be understood, was strictly an expedient. England realized her predicament in a breech-loading infantry world at that time and sought a quick makeshift arm. Most of the systems presented were designed to utilize the available Enfield lockwork, naturally.

In 1867 a system offered by Jacob Snider, an American, was officially adopted. The conversion consisted of cutting away the breech of the arm to fit a breechblock hinged upon the right-hand side of the barrel. A firing pin and spring were mounted in the breechlock. The original Enfield lockwork was employed. A spring catch locked the breechblock when in place. Turning the breechblock up on its hinge exposed the chamber for direct insertion of a cartridge. The block was then locked down into place, the hammer cocked, and the trigger pulled. When the block was turned up on its hinge, it could then be pulled back a short distance, compressing a spring around the hinge pin. A "draw cartridge" or claw extractor was attached to the breechblock and the rearward pull drew the empty case out of the chamber. Turning the rifle over dropped the case out.

We have already noted the resemblance of this system to that of the old breechloaders of Henry VIII in the Tower. However, it is doubtful that Snider ever heard of those venerable relics. On the other hand, Snider may have gotten his inspiration from the little known American "Hubbell" caplock, patented by W.W. Hubbell (U.S. No. 3649) on July 1, 1844. This arm was well received in some military circles on the Continent. It was hinged to open from the right, and used a cartridge fired by an external cap. The principle of the Snider follows it very closely. Incidentally, U.S. officials rejected the Snider before it was offered in England.

The cartridge originally used in the Snider was not successful. This cartridge had a metal base, center-fire priming, and a paste-board body. The accuracy was poor. Colonel Boxer of the Royal Laboratory was assigned to iron out the cartridge troubles. He produced and patented an improved form of the original Pottet case. The walls of the famous Boxer cartridge were made of coiled brass, the bullet had a hollow taper at its base and a taper plug first of wood and later of clay to expand the bullet to fill the grooves, as well as a hollow in the front end which gave better weight distribution. With the new cartridge the accuracy exceeded that of the highly prized Enfield muzzle-loader.

These conversions were known as "Snider-Enfield." They could be loaded and fired at a rate as high as 18 shots per minute and had a life as high as 70,000 rounds. For some years the Snider system was an official one in the Montenegrin, Serbian, and Turkish armies. It was, however, merely a conversion introduced to rearm the British infantry quickly and cheaply at a time when the new breechloaders were upsetting nations and vanquishing armies. It soon gave way to a rifle of entirely new design—again based on an American invention, the Martini-Henry. This arm was evolved directly from the already famous American Peabody, the original "falling block" breech-locking design.

The Boxer Primers and Cases

Colonel Boxer of the British Army was highly praised for the development of his coiled brass case with its iron base head and separate cap chamber. The introduction of that expanding case making a perfect gas seal started the trend of military change to center-fire metallic cartridge breechloaders in Europe. In the United States sporting breechloaders and repeaters of superior design were already in use but handled rimfire cartridges. The cost of the Boxer cases of thin coiled brass was quite low. This doubtless had much to do with the fact that this type of case was continued in service by the British long after the superior American Berdan drawn-brass case had been put into use. On the heels of the Egyptian campaign of 1885, however, complaints poured in at such a rate that, except for target use, the British replaced the Boxer coiled type with one of drawn brass. Today the Boxer case is obsolete, of course.

The Boxer primer, on the other hand, was such an outstanding achievement that today it is the standard American primer form. This primer is merely a simple metal cup containing the priming mixture and having its own anvil (a pointed metal piece) crimped in so the point is imbedded in the priming. The single unit is retained in the pocket in the head of the cartridge case. The firing pin hitting the base of the primer drives it in against the point of the anvil. The friction ignites the priming which flashes through the vent in the case and in turn ignites the propellant.

This elementary device has been improved upon only to the extent of providing advanced priming mixtures and in improved primer sealing and waterproofing. The mechanical principle itself remains the same.

American Case and Primer Designs—The Berdan

Most military cartridges made in the United States in the 1860 period were folded head types. The cases resembled the rimfire externally, though ignition was actually center-fire. A priming cup was inserted down the mouth of the case. Indentations in the case just below the cup seat secured it in place. Priming was held in a dip at the center of the cup. Priming was crushed between this cup and the case head when the firing pin struck. The system was at once expensive and unsatisfactory. It was difficult to reload. Firing was very often uncertain. Still, this remained the generally used type in center-fire until well into the 1880's.

In 1870 Colonel Hiram Berdan of the U.S. Ordnance Department developed a new primer to be used with a brass case he also designed. The case was a big advance technically, as was the system for rapid and low cost drawing of the brass which Berdan worked out.

The Berdan primer, though still widely used in Europe, is not made in the United States. It is inferior in many ways to the Boxer which, ironically enough, is the standard in this country and little used elsewhere.

The Berdan differs from the Boxer primer in that while it, too, retains the priming mixture, it lacks the integral anvil feature. In lieu of this important feature, the anvil is formed by part of the primer pocket in the case. Two small holes in the case allow the priming flash to reach the propelling charge. This style of primer was used in the United States only about 20 years.

The Boxer uses a comparatively large SINGLE flash hole. This in turn allows easy punching out of fired primers, so that cases using this primer are easy to reload. Also, in addition to being simpler to manufacture than the Berdan, the Boxer primer with its relatively large central hold allows use of higher intensity primers than the Berdan; and less set back and primer swelling is likely to be found.

3 Single-Shot Metallic Cartridge Arms and Lock Systems

The single-shot breechloader in practically all its varying forms has at one time or another been used as a rifle, pistol, and shotgun. Some are still made in each of these classifications. A somewhat detailed study of the origins and development of one covers the others for all practical purposes.

A modern single-shot firearm is one in which the action must be opened manually at least for the first shot. The cartridge must be inserted manually into the chamber or onto a loading trough in the breech. The action must then be closed by hand.

The range of variations is tremendous. This work covers all the successful ones, but lack of space prevents inclusion of more than a fraction of the experimental ones with which the writer has come in contact.

PURPOSE OF SINGLE-SHOT STUDY

This record of single-shot arms is, of course, primarily intended for historical purposes. It may also serve another purpose—that of providing a ready reference work for designers and inventors who so often spend years of intense work on such matters only to find after the veil of developmental secrecy has been raised that their big "secret" has been the subject of much past experimentation.

In the course of business contact the author is constantly exposed to "secret developmental work" for which our government is paying on a cost-plus basis which is merely duplication of early experiments. This condition arises from lack of knowledge on the part of those placing contracts and those accepting them. Lack of research facilities accounts for some such wastage. This chapter, therefore, seeks to combine and simplify many of the basic elements of design; and to present them so they will be of practical value.

Some may question what value a single-shot study, for instance, has in this day of machine weapons. The answer is simple. There is absolutely no element of locking principle or operating or firing principle used in the most modern automatic arms which is not reflected in single-shot design; and these principles are constantly the subject of costly experimentation.

Even electrical ignition encountered in modern experiments in an endeavor to speed up fire rates was the subject of design studies in the 1880's! All the basic features of tomorrow's designs are to be found in yesterday's patterns. The creator of today may save himself time and energy by studying what the great men of the past did and did not succeed in accomplishing.

One more note must be made here on the reason for a single-shot study. The history of the world was affected to a tremendous degree by the arms developments of America during the period of the 1860's and 70's. Few American histories have given more than passing notice to these. This record seeks to spread some little of this historical lore into popular channels—lore which heretofore has been largely confined to the arms enthusiast.

THE DROPPING BLOCK LOCKING SYSTEM

THE SHARPS DESIGN

This is the earliest and one of the strongest and most successful of all breech-locking systems. Variants of it are in use today in both small arms and artillery. Christian Sharps, as we have seen, introduced the system during the percussion period. Its success with the metallic cartridge was immediate, of course, since it gave great breech strength and a perfect gas seal. Thousands of his early arms were converted to handle rimfire and center-fire cartridges.

All dropping block systems in which the breechblock slides vertically (or nearly vertically) in mortises in the receiver walls are merely modifications of the Sharps. In the years between 1875 and 1880 this immensely strong single-shot action was made by Sharps in heavy calibers and heavy weights for buffalo shooting. Professional hunters using telescopes frequently staked out and did sniper shooting into herds at distances up to 1000 yards! These were the principal rifles which cleared the plains of big game. Indeed the very name "sharpshooter" derives from the astonishing accuracy and rapidity of fire the old percussion Sharps produced in the Civil War. Col. Hiram Berdan, later the designer of the Berdan primer and drawn-brass case, led a regiment armed with Sharps which did much to save the Union Army from being wiped out in the battles in the Potomac Valley. Many of his troops were equipped with telescope Sharps rifles. "Berdan's Sharpshooters" have gone down in the annals of history for their deadly accuracy. The Union forces used nearly 100,000 Sharps' arms in the Civil War.

The Sharps in all calibers maintained their high accuracy ratings from the days when Christian Sharps himself dominated policy down to the closing days of the firm's history when the incredible P.T. Barnum, then Mayor of Bridgeport, Conn. joined with other businessmen to build a plant to transfer the Sharps manufacturing from Hartford to Bridgeport.

ROD AND GUN for Sept. 30, 1876 carries an interesting account of the International Match of 1876 in which Sharps rifles captured the matches against 40 competitors using the arms of 6 manufacturers. In the following year the Sharps at the International Match averaged 420 points per man against 414 by the best competitor; while in the 1878 N.R.A. Fall Meeting at Creedmoor, Sharps took 12 out of 16 matches!

The original Sharps' design used the familiar external large

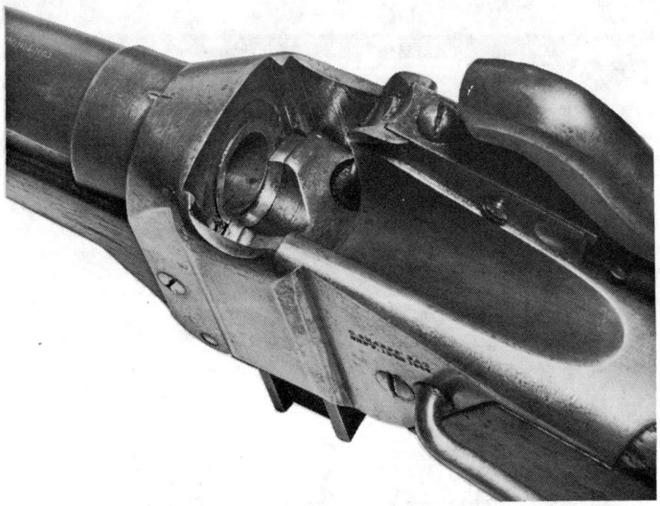

Typical Sharps carbine and action. Forerunner of the Browning-Winchester and other famous single-shot rifle locks.

hammer reminiscent of the military muzzle-loaders of the day. In 1870 the Sharps-Borchardt modification was tested by U.S. Ordnance. This model concealed the hammer within the frame in what has become popularly known as the "hammerless" design. Actually there is a hammer in this model (though some other arms designs to be discussed later are truly hammerless). Drawing down the trigger-guard lever lowers the breechblock, cocks the hammer, and extracts and ejects the empty shell from the firing chamber. The safety is automatically applied as the action is closed by drawing back on the lever. A release behind the trigger must be pulled to prepare for firing. With the original external hammer, of course, one could tell at a glance if the arm were cocked. The 1878 Model of the Sharps-Borchardt was made in too many styles, weights, and calibers for listing here. The action is still in great demand by cartridge experimenters and shooters as a basis for building special rifles.

In 1881 a French repeating rifle was built on the dropping block system. It was not considered a military success because a lever design does not lend itself to rapid fire when shooting prone, as so much military shooting is, in theory at least.

OTHER DROPPING BLOCK DESIGNS

The Sharps was followed by a long line of fine variants, chief of which were the American Ballards, the Sims, and the line of Stevens "Ideals." Also using a rather minor modification of Sharps' principle was the original Winchester Single-Shot .22 Rifle. This arm was developed by the late great John M. Browning. Its purchase and manufacture by Winchester started Browning on his career as the most prolific and successful of all small arms designers.

The basic design is still used by literally hundreds of small gunsmiths and gun plants in Europe today, though its use is confined in general to small-bore, high-priced, precision match rifles.

THE FALLING BLOCK SYSTEM

PEABODY FALLING BLOCK

The next great original breech-locking system was that of Henry O. Peabody of Boston, Massachusetts. His basic patent was granted July 22, 1862.

The Peabody principle is that of hinging the breech-block at the rear and above the line of the barrel bore. It is unlocked by pulling the trigger-guard lever down and forward. The front end of the block is drawn below the line of the chamber, permitting rapid loading. The first model used the familiar external military hammer of the day and was produced in .45 caliber rimfire. Its block system prevented action jamming by expanded case heads. It was a great advance over anything in the market at the time of its introduction.

This arm was tested at Watertown Arsenal in 1862. It is a commentary on U.S. governmental thinking of the day that while this superior metallic cartridge arm was available then, it was adopted only by some militia units. In Europe, however, its value was recognized almost from the first.

It was 1865 before the Examining Board at Springfield Armory got around to testing it seriously. Its performance alongside 64 competitors was so remarkable that it was recommended for purchase. The Civil War had ended by then and funds were not available for a change over to a new model, however superior to the old.

Just as in the instance of Britain with her Snider conversion, American energies were turned to trying expedients to convert our enormous stock of muzzle-loaders to breechloaders. This wasteful system still plagues every nation after a war. In the face of the most revolutionary advances, most efforts to completely modernize equipment fail because of the mistaken ideas of the uninitiated. In the determination to "save" money, both money and efficiency are invariably sacrificed by the winners.

It might be pointed out in this connection that the tremendous German equipment advances encountered in War II stemmed largely from the fact that as a defeated and disarmed country at the end of War I, Germany was able to start manufacture and design for War II equipment without the winner's psychological handicap. This process again is being repeated. West Germany has been able to start from scratch—using the best that experience has established, but developing the new without the hindrance of

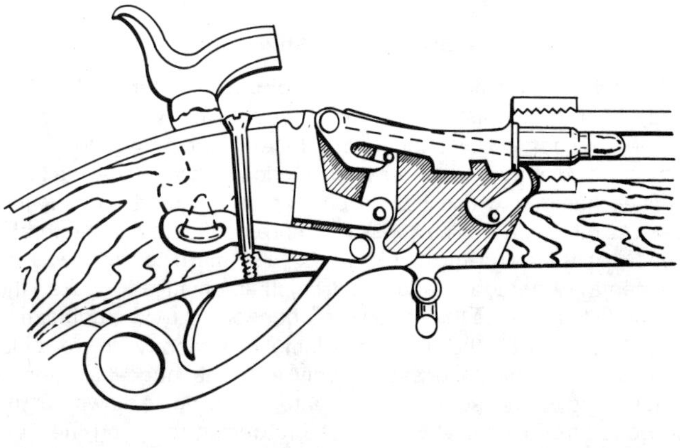

Peabody falling block mechanism. Original drawing to show principle of operation.

The famous Browning-developed Winchester single-shot rifle.

stocks on hand and old tools and plant equipment. History never changes.

Fortunately for the Peabody company, Canada in 1885 placed an order for 3000 new rifles, the caliber to be .50-60 rimfire (the ".50" indicated the caliber; the "60" indicated the powder charge in grains). Thus encouraged, the manufacturers turned to other world markets. In the next few years the Peabody was purchased by the military in such diverse areas as Austria, Bavaria, Denmark, France, Mexico, Rumania, and Switzerland. In many of these countries the Peabodys played an important historical part; and in many others they influenced the target arms design to a point where some manufacture still persisted into the 1960s.

Switzerland, that staunch little nation of riflemen, recognized the potentialities of the Peabody. In 1867 Captain Michel of their Ordnance Corps came to Providence to supervise an order for 15,000 rifles in caliber .41 rimfire. Hard on the heels of this order came one from Denmark. Rumania bought 15,000 in rimfire and then another 10,000 in .45 center-fire. Canada and Mexico totaled another 24,000. France, caught up in the War of 1870, managed to get delivery of 39,000 Peabodys to help reduce her military handicap, but had to take them in the .43 Spanish caliber. Austria and Bavaria both purchased and tested them. Reports on all hands spoke glowingly of their rapidity of fire—17 shots a minute aimed and 35 shots a minute off-hand being recorded in U.S. Government tests in 1866. As to accuracy, these arms were the marvels of their time. The Royal Danish Legation and Consulate General at New York reported to the manufacturers that a Royal Testing Commission reported "good hitting shots were obtained at 2400 feet." The Austrian Imperial Breechloading Arms Commission reported fine accuracy at 600 yards. At 700 yards range (the practical maximum then, and not greatly exceeded today) the center of impact of the .43 caliber was 40.79 inches; the mean deviation only 23.6 inches.

The Martini-Henry Modification

The Swiss, unhampered by military dogmatism, turned the Peabody over to a master mechanic, Frederich von Martini of Frauenfeld, for modification. In the United States it was felt that a self-cocking army weapon was not practical. If it didn't have a huge outside hammer which had to be manually hauled back, it just wasn't a military arm. The Swiss had more confidence in the intelligence of their riflemen. Martini's first modification was an elementary leverage arrangement which automatically cocked the Peabody hammer as the action was opened. His next modification completely buried the name and fame of Peabody, the original designer. Martini dispensed entirely with the huge hammer of flintlock days and substituted an internal firing lock within the breechblock. The result was a pleasing design retaining the Peabody breech lock but which introduced the now famous Martini firing mechanism. This is operated by a lever to the rear of the trigger guard.

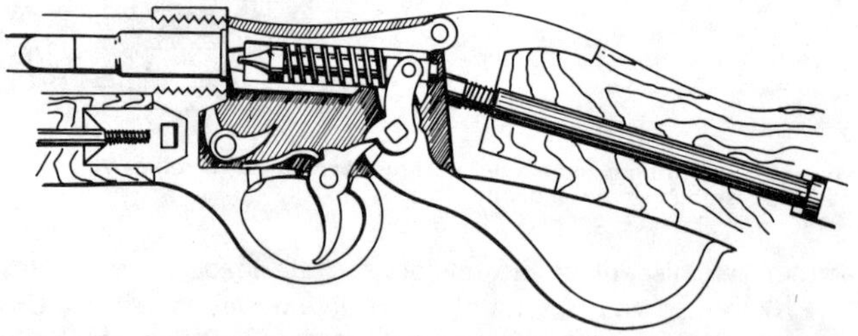

Martini-Henry mechanism. Original drawing to show variation and improvement over original Peabody system.

Martini submitted this design to England for tests. It was slightly modified at Enfield and fitted with a barrel rifled on the English Henry system. The Henry bore is polygonal. Lands are formed at the angles. They provide additional bearings for the projectile as it travels down the barrel. The resulting arm was named the "Martini-Henry." It was officially adopted in April 1871 as the British service rifle in caliber .45 using a Boxer rolled cartridge. During War II this rifle, altered to the current British caliber .303, was used by the Home Guard services in Britain. As a small-bore precision rifle it is still made by the Birmingham Small Arms Company, in slightly modified form.

The Providence Tool Company made and delivered to the Ottoman Government 600,000 Peabody rifles in .45 caliber during the year 1873. The chamber was smaller for the Turkish cartridge than for the British. These rifles are still encountered in the Balkans. They are clearly stamped "Peabody & Martini Patent. Man'f'd by Providence Tool Co., Prov. R.I. U.S.A." These have the Peabody breechblock and the Martini firing system, of course, but the firing system is somewhat modified from the original along the Enfield alteration line.

The British adopted their version under the name "Martini-Henry" and the name has tended to obscure Peabody's fame completely as years have gone by. Even in America this incorrect designation is the accepted one.

The Werder Modification

In Bavaria the Peabody lock was modified along quite different lines than in other places. Nevertheless, the basic "falling block" system was retained. The modification there was by J. L. Werder, noted director of a Nuremberg arms plant. He studied Peabody tests which a Bavarian Arms Commission had made over a period of a year, tests which proved very favorable except that they pointed up the military need for an arm which could be prone fired without shifting the position of the shooter's body. Werder accordingly modified the system to replace the under-lever release. In its place he used a special lever placed inside the trigger guard ahead of the firing trigger. This lever looks like a reversed trigger. It is forward of the firing finger position.

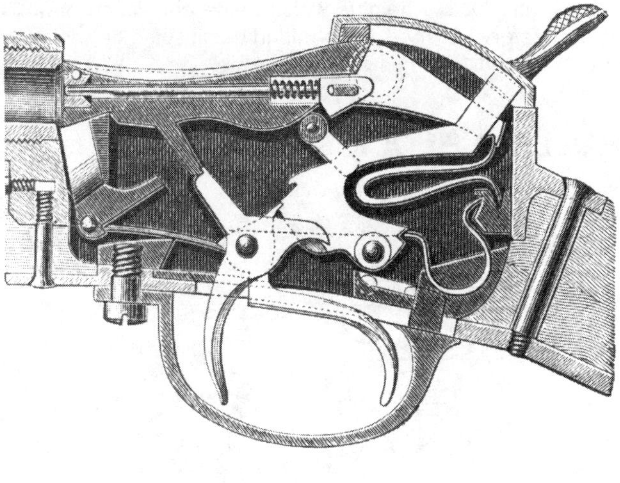

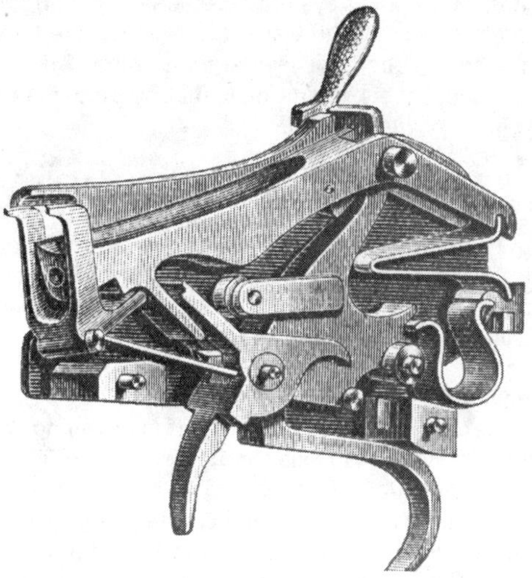

Upper: Bavarian lightning (Werder) lock. Left: Action closed ready to fire. Right: Action open ready to load. Lower: The Bavarian Werder Rifle. Caliber 11.15mm.

The shooter pushes his firing finger ahead against this lever to unlock the arm. The lever releases a locking prop within the mechanism. This lowers the front face of the breechblock and exposes the chamber for loading. A cartridge is slid down the shaped slot in the face of the block and into the chamber with ease. On the right side of the arm there is a cocking piece shaped rather like an old time hammer. This is drawn back by the thumb of the firing hand. This action cocks the arm, and at the same time elevates the breechblock and locks it at its forward end. The rapidity of fire of this type of weapon, whether rifle or pistol, is astonishing when used by an experienced operator. A forward thrust of the firing finger opens the breech and expels the empty case. A cartridge is thrust down the concave face of the block into the breech with ease. A backward flip of the thumb of the firing hand locks and cocks the arm. Here again the breech-locking system is but a variant of the Peabody. Again Henry Peabody received no credit. In passing it might be noted that the Werder "Bavarian Lightning" pistol and rifle were both used to some extent in Eastern Europe during World War I.

Other Variations of the Peabody

There are a score or more of other variants of the Peabody lock, usually combined with some modification of the Borchardt or Martini firing mechanism. Today's Belgian Francotte-Martini arms form one type. The British Martini is another. In England one still encounters old Swinburns, Stahls, and Westley-Richards; while the Continent harbors any number of modifications made up by small gunmakers who appropriated Peabody's breechblock.

All but a few of the precision single-shot rifles made in Europe and Britain in recent years fall in this general classification. In fact, the only commercially successful dropping block action which doesn't, and which can lay any claim to originality, is the German Aydt. The Aydt pivots its block below the line of the chamber.

Historically the principle of Peabody is also of interest as the breeching system of the early Spencer American repeating shotgun with which we will deal later. When Winchester introduced its first great "pump" shotgun invented by John M. Browning, the New York used-military equipment firm of Francis Bannerman sued to restrain the manufacture. Bannerman had bought out the defunct Spencer firm for a song. The suit was lost by Bannerman, but it is strange that no one on either side seems to have recognized the derivation of the Spencer lock.

This then is a small part of the story of the Peabody lock. The Peabody-Martini lighted up the scene of history in a blaze of glory in the Russo-Turkish War of 1877-78, when in the course of a few hours of battle its use by the Turks slaughtered the Russians at long range as they approached en masse. And in that same encounter at Plevna the American name Winchester forged to the front in the world of military thought, completing the destruction the long-range Peabody-Martini started on the fateful day. The Winchester repeater in the hands of the Turks completed the massacre at close range which the Peabody began at long range. The Winchester's terrific firepower that day sealed the doom of all military single-shot rifles, though it was long before its lesson was effectively grasped in our own country. That story too belongs in another place in these pages, but first we must consider the remaining important single-shot locking designs.

As to Peabody himself, in all the long history of firearms few inventors have played a greater part in shaping military events than did the invention of this obscure New Englander. It is a strange irony of circumstance that his name is so little known outside the circles of early cartridge collectors, while those of the men who utilized and modified his system have achieved world renown. It is unfortunate that we have little space here to devote to the man himself and his background; for like so many other Americans of his time, the products of his genius led to machine and manufacturing developments which started the United States on its way to the productive heights which made our nation great. The Peabody story deserves to be told in detail some day as a thrilling chapter in American business enterprise.

THE REMINGTON ROLLING BLOCK SYSTEM

The rolling block action is still used to some extent in shotgun, pistol, and rifle types in Europe. The basic design principle is simple and can be made strong enough to handle almost any conceivable load.

The mechanism for locking the breech consists of two parts. First is the breechblock itself. This is a heavy rolling member mounted in the receiver on an axis pin. It contains the firing pin in a hole in its center face; and is machined with a thumb-extension to allow it to be rolled back away from the breech when its locking support is withdrawn. The second unit is the firing hammer. It is a rotating member pivoted on its axis pin behind the breechblock. The lock is simplicity itself: the breechblock is spring-supported against the face of the breech when ready to fire; as the hammer falls when the trigger is pulled, a projecting under surface on the hammer rolls firmly into place below and behind the breechblock, supporting it immovably. This locking surface reaches its full support position before the striking surface on the hammer can mechanically reach the firing pin.

In its day this was the most widely-used army rifle lock design in the world. Its effective military life, like that of the Peabody, was short, however. It, too, came in during the Civil War when procurement officials had to concentrate on manufacturing the muzzle-loaders we were tooled up to make, rather than many of the superior breechloaders which were new at the time, untested and untried. And it, like all other single-shot rifles, became obsolete militarily on the day of the Russo-Turkish battle of Plevna.

The history of the Remington long arm has never been written. The story of Eliphalet Remington's hand-forged flintlock barrel, of his success as a maker of custom guns in the early days of the republic, of the rise of the firm under Eliphalet Jr. and the manufacture of Harper's Ferry Rifles and Jenks Carbines during the Civil War, all form part of the vast canvas of American firearms history which we cannot deal with here.

However, the Remington Rolling Block Lock is an integral part of the story of modern American design, the more so since it, too, deals with an American inventor whose name is almost unknown in the annals of arms. While Remington marketed the mechanism, he did not invent it.

THE GEIGER PATENT

The inventor was Leonard Geiger of Colburn, Vermont. The first true patent for the rolling block lock with the hammer hung to the rear of the axis of the breechblock and supporting the block immovably is that granted to Leonard Geiger (U.S. No. 37501) January 27, 1863.

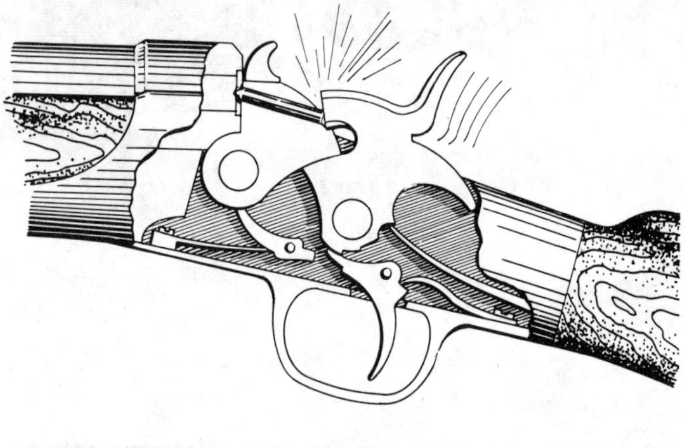

Remington rolling block system. Original drawing to show firing and breech locking mechanisms.

THE RIDER PATENT

Joseph Rider's (or Ryder's) original patent clearly shows the hammer forward of the breechblock axis. The Remingtons purchased Rider's patent and also gave him a job in their Ilion, N. Y., factory. When Remington bought Geiger's patent, Rider and some other employees made minor changes which resulted in the original Remington "split breech" design, one of the simplest and strongest breech actions ever devised; and one ironically enough almost universally called the "Remington-Rider."

The "breechpiece" which closed against the head of the cartridge was combined with the extractor, while the hammer also was shaped to function as a breech bolt. The "breechpiece" in this first design was split to permitt the firing pin in the hammer nose to strike through at the cartridge case head, resulting in a comparatively weak piece.

When the breechblock was made solid, being pierced only to receive a firing pin mounted within it, the design was an immediate military success. Cocking the hammer rolls the under locking surface out from its support of the breechblock. The breechblock is then pulled back by a projection on its right side to expose the chamber for loading. When the piece is closed, a flat spring holds it firmly against the head of the cartridge. When the trigger is pulled and the hammer falls, its shaped under-surface travels below and behind the breechpiece locking it firmly as the hammer

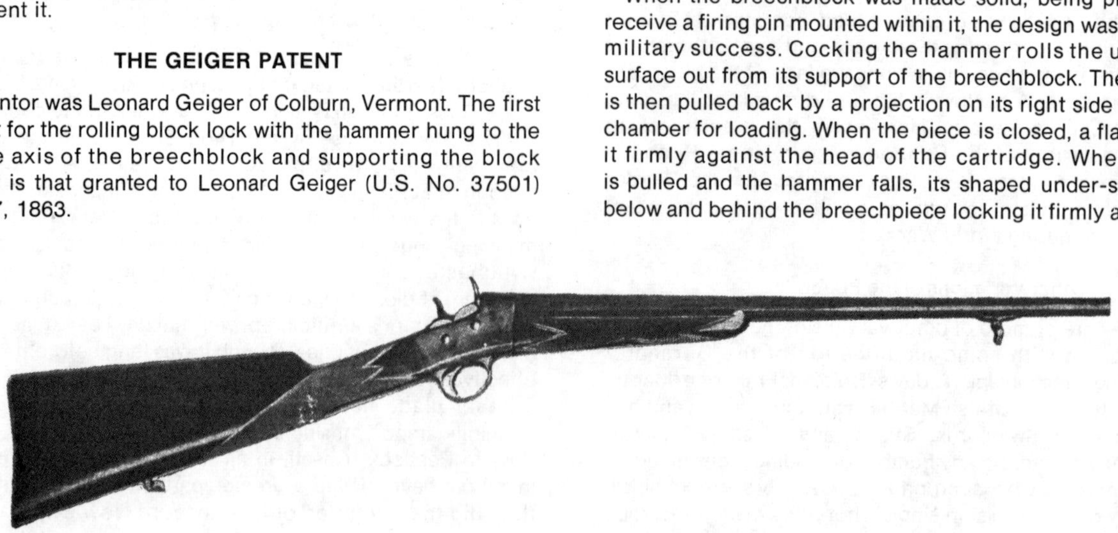

Typical Remington rolling block carbine.

hits the firing pin. This perfected design was completed in 1866, but production of the rifles did not get under way until the following year when the U.S. Navy ordered 12,000.

This rifle was adopted by Denmark in 1867, by Sweden and Norway in 1867, by Spain in 1871, by Egypt in 1870, and by Argentina in 1879. It was also widely used in China, Austria, Italy, and in several South American countries, and was the subject of military experimentation everywhere. In all, over 1,000,000 rolling block rifles and carbines were manufactured and sold!

The rolling block system was continued in low priced lines into the twentieth century but is no longer manufactured by Remington. Shotguns on this design are still made in Sweden, where the design is quite popular. A form of rolling block actuated by a trigger guard lever was made for some years, notably by Hopkins & Allen in America. The American "Buck" action lever repeater also used a form of rolling block.

THE REMINGTON-HEPBURN

For a time Remington manufactured, under the patent of Lewis L. Hepburn (Oct. 7, 1879), a dropping block action which is often erroneously referred to as a modification or "improvement" of the rolling block. This Remington-Hepburn as first issued had a right side lever operated by the thumb which through compound leverage unlocked the breechblock and lowered it in receiver mortises for loading. A later modification of this type, the Remington-Hepburn No. 3 Schuetzen Match Rifle, was operated by an under lever, the action being designed to thrust home a cartridge only partly inserted in the chamber as the block moved up and ahead.

Remington single-shot rifles are currently manufactured only in modern bolt action designs.

THE WHITNEY DESIGN

The Whitney Arms Company of New Haven, Conn., submitted a rolling block action to our Ordnance Board of 1865. It was not accepted for service use. The system was patented by Captain T.T.S. Laidley (No. 54,743) and closely resembled the Remington, differing radically only in the shape of the extractor and in the way in which the hammer supported the breechpiece. This arm was not successful. (It is not to be confused with the later Whitney "Phoenix" line of rifles. In these the breechblock could be hinged up from the side when the hammer was half cocked. As in the case of the Snider and other side lifting breechblock actions, the Phoenix block, by passing across the head of the cartridge case instead of to its rear as in the rolling block, encountered extraction trouble if a primer blew out or a weak case was encountered.) Eli Whitney, founder of the original company, was one of the first arms manufacturers for our Government. He instituted interchangeability of parts in 1798. (The idea had been previously suggested in France, but nothing was done about it.) The cost proved prohibitive in those early days and the principle was temporarily abandoned.

OTHER AMERICAN SINGLE-SHOT DESIGNS

While some other American designs were of importance in their day, none approach the original ones already described, and none are truly of American inception.

THE CAM LOCK

At the U.S. Trials for Breech-Loading Carbines held at Springfield Armory in 1865, there were 65 designs submitted. Five designs were selected for alteration of the muzzle-loaders on hand—the same approach originally made by the British. The one finally selected was suggested by E.S. Allin, Master Armorer of Springfield Armory. After a score of modifications this became the official U.S. Rifle with which we armed all our Indian fighters, including Custer, whose command was massacred because the Indians opposing him had repeating rifles. This "cam lock" as it was officially known, was our first-line infantry arm until the introduction in limited quantity of the Krag-Jorgensen magazine rifle in 1892. This old black-powder cam lock was still in use by U.S. forces at the battle of San Juan Hill in the Spanish-American War. This arm is the familiar rifle (or carbine) with top hinged breechblock locked by a cam piece at the right rear. The huge hammer is pulled back to full cock and the cam latch is pushed to unlock and then to lift up the breechblock. This action extracts the empty case and leaves the chamber exposed ready for loading. This form of action can be traced back to the French Demondion freak rifle of 1831; and its resemblance to a score of other foreign and American designs is too pronounced to escape notice. It, at least, served the purpose of arming our troops with cheap breechloaders; and we were fortunate that its first trial in war was against a foe so inferior in numbers that the quality of the weapon did not cost us too heavily in lives because of our conservatism and our penury.

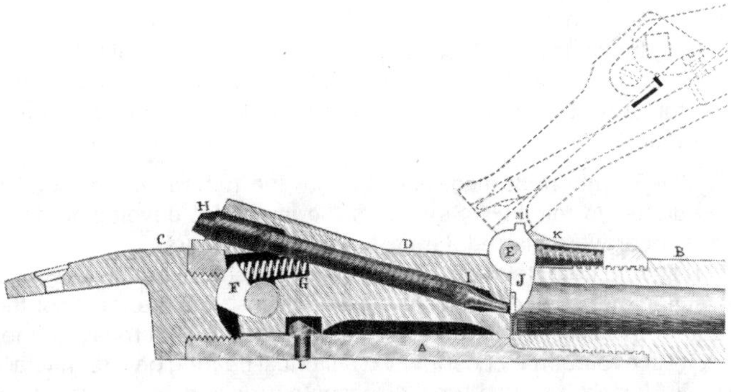

The U.S. cam lock mechanism.

General Berdan claimed the 1866 Springfield used his single-joint breechblock and he instituted suit. The suit dragged on long after his death, but in 1895 a judgment of $95,000 was finally allowed to his widow by the Government. The final form of this single-shot was the 1888 Model, Cal. .45-70.

THE MAYNARD HINGED FRAME

Of all the other types developed, none has persisted but the hinged frame evolved from the Maynard, and that only to a minor degree. The Maynard, of course, was not the first hinged-frame rifle: that system was used in Italy as early as 1670; and was used with pinfire cartridges by Lefaucheux in France, as we have seen.

Stevens, Hamilton, Wesson, Wurfflein and others made low-priced rifles on this system—a trigger guard or other lever being pushed to unlock the barrel from the standing breech and tilt it up for loading. In Europe this system is still widely used for high-power game rifle design, though not for military purposes. In the United States its modified use is confined largely to shotguns and to target pistols.

VARIOUS OTHER TYPES OF SINGLE-SHOT

The sliding barrel design in which pressure on the trigger-guard lever slides the barrel forward away from the standing breech for loading (as in the Sharps & Hankins); the side-swinging barrel of the Frank Wesson Pocket Rifle (he also made hinged frame types); the swingout breechblock of the Quackenbush; the Phoenix type of turn-out breech piece; and the Berdans were perhaps the outstanding designs. None have anything but historical importance today.

The Berdan Rifles

The Berdans are of interest because they were developed by General (Civil War Colonel) Hiram S. Berdan. Berdan resigned his commission in January, 1864, to undertake the construction of a new rifle. A test was made of one conversion, the cal. .58 Berdan, on November 12, 1864.

The Berdan I was a curious composite resembling the Albini block (swinging up on a forward hinge) and the Chassepot firing mechanism. A thumbpiece projected from the rear of the receiver and was drawn back to cock the arm. A spiral mainspring drove the locking bolt against the striker in the center of the breechblock, thereby assuring that the breechblock would be positively locked before the cartridge fired. In 1868-70 the Russians purchased 30,000 of the Berdan I pattern. The Russians developed an improved form, the Berdan II, using the Russian .42 cal. center-fire cartridge. It was manufactured at Tula Arsenal in Russia. It has a circular thumbpiece at the rear of the bolt. This Russian cartridge is of military interest as it was the first center-fire bottlenecked cartridge with outside Berdan primer. This development, of course, marks the beginning of the era of high-power small-bore rifles with great range and accuracy and relatively low trajectory.

When in 1872 a Board was established by Act of Congress to give further consideration to breech-loading designs, 108 designs were submitted! It is noteworthy as an index of the way the United States has forged ahead on arms development at the time that only nine of the new designs were foreign. Since none of these types are of further evolutionary interest, they cannot be included in this work.

NOTEWORTHY HISTORIC AND FREAK AMERICAN DESIGNS

The Quackenbush Block System

This featured a breechblock pivoted below the chamber. The block was pushed to the right on its pivot to expose the cartridge chamber for loading and extraction. The striker was housed within the breechblock. A cocking handle on the right side was drawn back to cock the arm. Gas sealing was not efficient. Nevertheless, the arm was so simple to manufacture that it was copied by European makers for export even in recent years. In general, its use has been confined to low-power cartridges, since breech wear sets in rapidly.

The 1870 Joslyn-Tomes

In this design pulling back the hammer also retracted the breechblock and ejected the empty case. When a cartridge was placed in the receiver, the breechpiece was pushed forward against it forcing it into the chamber. Pulling the trigger fired the cartridge.

The Laidley-Chick

This design, a modification of the Remington Rolling Block, differed in the way the breechpiece was backed by the hammer. It had a three-click hammer and a safety. Legend has it that the name derived from the fact that the proud inventor, Captain Laidley, usually referred to it affectionately as "my chick."

The Jenks 1868

A top hammer when drawn back rotated and lowered a breechblock to expose the breech in this design, which was used in the Civil War.

The Cochran of 1865

This design is of interest as a modification of the Hall principle which was also attempted in Europe. Pushing down the trigger-guard lever raises the front of the breechblock (which is pivoted at the rear as in the Hall). When a cartridge is inserted in the chamber, the lever draws the breechblock down to back it up. This arm has an external hammer and a firing pin in the breechblock. Cochran also invented the rather impractical revolving rifles already mentioned.

The Joslyn, 1862

Several types were made. Withdrawing a lock pin on the right permitted the breechblock to be turned to the left on a hinge. An inclined face on the block forced the cartridge home. This early design had a gas vent in the top of the block, a noteworthy item even today. Another unusual feature was a plane on the face of the block which acted to extract the cartridge. The Joslyn is of historical interest because 11,261 were bought by the North during the Civil War; and because the design was copied in Europe.

Miscellaneous

Other early metallic cartridge breechloaders of some importance included: The Ballard lever operated block rifle, 1509 bought by the Union in the Civil War; Ball magazine rifle, 1002 purchased same period; Cosmopolitan (also called Union), 9342 purchased; Gallagher hinged frame, 22,728; Gibbs sliding and tilting barrel, 1052 delivered; Lindner, 892; Merril locked by a lifting top strap, 14,495; Smith hinged barrel, converted from special rubber cartridge to metallic design; Warner side-swinging block with hand extractor, 4,001; and Wesson with barrel tilting when released by special trigger.

Multi-Barrel Rifles

Rifles with more than one barrel have never been well received in the United States. Their manufacture here has never been of military or commercial importance. When encountered they usually represent merely a shotgun action in double or treble design fitted with rifle barrels. Of course, there was a rash of such arms developed by individuals from time to time, but none were produced in quantity.

The one arm of the period which might merit attention is the G. R. Remington 3-barrel rifle. The breechlock was a modified rolling block, the block carrying 3 firing pins. Essentially this was a "volley gun," all 3 barrels firing simultaneously. The caliber was customarily .32 rimfire. These constituted a sort of long-range shotgun.

NOTE: Important magazine rifles of the period will be considered later in this text. Several of the important developments of this order were started during the Civil War years.

It should be noted that the foregoing constitute but a very small fraction of the list of early American designs. It is the intent of the author to cover in this work only those designs which have either a direct evolutionary bearing or a historical bearing on later military arms design and use; or which contain design elements which have still not been completely exploited.

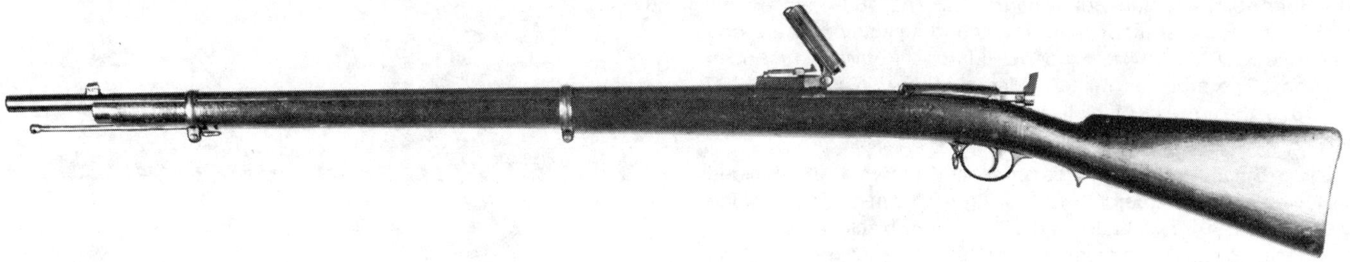

Berdan I, Russian rifle as made by Colt for the Russian Czarist Government.

THE EARLY METALLIC CARTRIDGE RIFLE IN GREAT BRITAIN

Great Britain's contributions to small-arms design, while great indeed in the development of the sporting double-barreled shotgun, have been particularly significant in the field of the rifle only in the evolution of rifling. Very little original rifle design has been done by the British; and even their modifications have not been particularly noteworthy.

The original machinery used at Enfield Armory was designed and manufactured in the United States, as Britain was not equipped with either facilities or designers to produce the required machine tools. In 1854 Lord Hardinge assigned Joseph Whitworth, a pioneer in the field of exact measurements in England, to experiment on the manufacture and design of rifle barrels. English gunsmiths and writers generally glossed over Whitworth's findings and recommendations as being merely a rehash of what was already known. Few realized that while Whitworth knew nothing about firearms, and was therefore ridiculed, he was a scientific observer whose findings couldn't be pushed airily aside by responsible authorities. Whitworth's rifle was not adopted. His recommendations for reduction in caliber were not given attention until some years later. However, his methods started the British in 1860 on a system of interchangeable parts, a mighty step forward in arms manufacture in a nation which had always prided itself on the indispensability of its hand fitting of parts.

THE BRITISH METFORD AND THE ENFIELD RIFLING SYSTEMS

British developments in the field of rifling did much to advance the tempo of military small-arms design in the United States and Europe. The Metford was the only really important rifling design for black powder loads. The Enfield rifling is the world standard form of rifling for smokeless powder cartridges. The Danish Krag and the Japanese Arisaka rifles still used in World War II the rifling designed by the noted British consulting engineer Metford in 1865.

The Lee Metford Mark I system had a bore diameter of .303 inches, land width of .023 inches, mean rifling depth of .0045 inches, and a left twist of 1 turn in 10 inches. There were 7 grooves. (The later Danish Krag uses 6 grooves, the Japanese rifle 4.)

Before the introduction of the Metford form, rifling generally was deep or polygonal. Used with black powder, the grooves fouled rapidly. The basic principle of the Metford rifling is very shallow grooving with grooves and lands slightly rounded. This system minimized both fouling and bullet and bore deformation to such a degree that it quickly outmoded all previous forms of rifling. The grooves being shaped to the segment of a circle, Metford is often called "segmental" rifling. The absence of sharp angles in the rifling provides a minimum hold for fouling.

Enfield rifling is today the form used in practically every military rifle made, with the exceptions already noted. Indeed, except for the very limited use in special rifle designs of the famous American Pope rifling and the modified English Lancaster oval-bore rifling used by the American Newton, it is in general use throughout the world in sporting rifles as well.

The Metford rifling was the perfect one for black powder use. But with the introduction of the new smokeless "cordite" propellant, the British found that erosion set in very fast, and the shallow Metford rifling washed out rapidly. (NOTE: With the new powders developed during and since World War II, shallow rifling is again practicable. As this is written experimentation is being conducted here and abroad in both military and sporting circles with so-called "micro-groove" rifling.)

British experimentation resulted in the Enfield form of rifling for smokeless powder use. The basic principle of this Enfield form is deeper grooves concentric with the bore, the grooves and lands both having square corners. As originally used by British forces, the Enfield rifling had a bore diameter of .303 inches, land width of .0956 inches, mean rifling depth of .0065 inches, and a twist of 1 turn to the left in 10 inches. It was rifled with 5 grooves. (In most military usage today 4 to 6 grooves are common. During World War II some barrels were rifled with only 2 grooves in the interest of economy and speed of manufacture; and for practical military use they are quite satisfactory.) Enfield rifling is commonly referred to as "concentric" rifling.

THE BRITISH MARTINI-HENRY SINGLE-SHOT RIFLE

We have already touched upon the early British conversions such as the Snider in our consideration of metallic ammunition development.

The Snider was merely a temporary expedient to permit rapid change over to breechloading. In 1866 a special committee was convened to consider new breechloaders. About 120 different rifles and 49 types of cartridges were submitted to the committee, representing domestic and foreign designs. All were rejected. In the following year the committee sat to examine further into the matter. This time they separated the matters of caliber and barrel types from the action designs. Barrels submitted were tested with a Henry action. Actions submitted for tests were required to be adaptable for barrels meeting given requirements as to caliber, type of cartridge, and so forth. Forty-five additional action designs were submitted which were also tested by these scientific methods. Bolt actions were glossed over because some early forms occasionally fired accidentally on the closing thrust of the bolt, a defect comparatively simple to overcome.

As a result of this detailed testing, the committee decided on the Henry barrel, using Boxer cartridges. The 480-grain bullets had waxed wads for lubrication. The action selected was the one developed by the Swiss, Frederich Martini, from the American Peabody. The arm was named "Martini-Henry." Tested in 1869 and 1870, the arm was officially adopted in April, 1871. The caliber was .45.

A new Small Arms committee, convened first in 1883, decided on lowering the caliber to .402, using a drawn-brass cartridge case with a 380-grain bullet and 85 grains of black powder. This new design employed a modified form of Metford 7-groove rifling and the arm was designated the "Enfield-Martini." While many were made, they were not officially issued at the time. (In .303 caliber this type was used by Home Guards to some extent in World War II.)

By 1866 the continental powers had awakened to the value of the small-bore rifle with smokeless powder, and Great Britain began experiments with the new rifle and cartridge designed in Switzerland by Colonel Rubin.

Before passing on to a consideration of the British adoption of the Lee Magazine Rifle, we must pause to examine the few worthwhile single-shot designs, since these designs marked practically the end of rifle design in England. Because of economic and geographical factors, there was neither need nor market for enough sporting rifles to warrant more than a passing interest in their development.

Consciously or not, the Small Arms committee was unquestionably influenced by the success of the falling block locking system on the continent when they set up the rules for tests. The design selected required that the cartridge be thrust fully home into the chamber before the action could be closed. In field use, when troops are under tension, such a system leads invariably to serious jamming of the rifle as the soldier tries to close the breech without first fully chambering the round. In addition, the extraction depended on a single pushing movement of the extractor which was insufficient to free swollen cases. As a military arm for use against the then modern armies, this design was a failure when it was adopted. However, in the hands of native troops for basically guerrilla-type fighting in Africa and Asia it still serves a purpose.

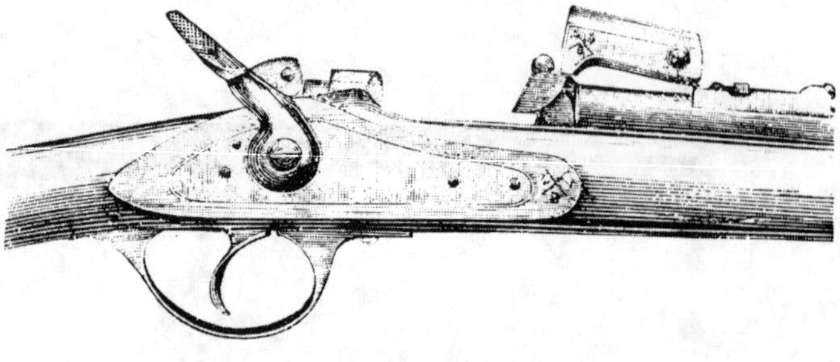

Albini-Braendlin rimfire, caliber .60.

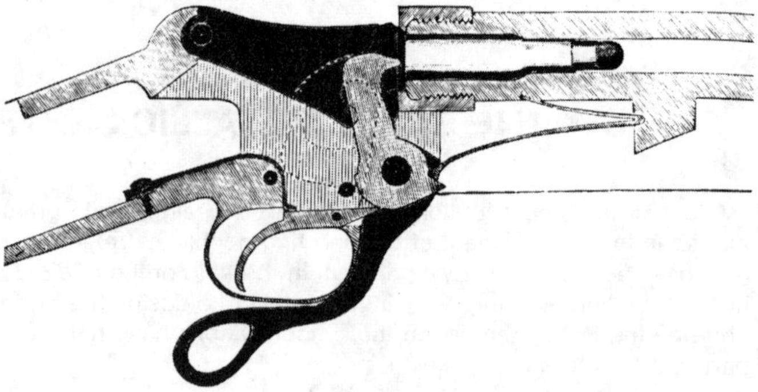

English Westley-Richards falling block system.

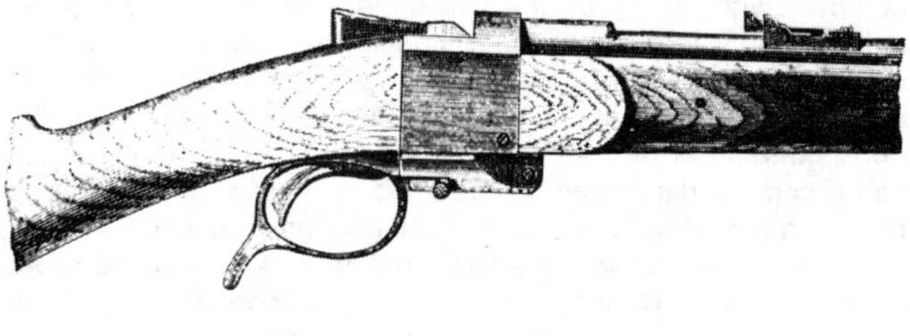

English Westley-Richards sliding block system.

THE SOPER SINGLE-SHOT

One of the most widely publicized rifles of its day was the British Soper. In 1866 the inventor offered an arm which was rejected as being too complicated. The day after the conclusion of the 1868 trials at which the Martini-Henry was accepted, the new Soper arrived—too late for entrance. British papers and magazines of the day tell remarkable stories of the rapidity of fire of this arm; some of the stories fail to stand up in the light of modern research. The arm had a side-hinged swinging breechblock on the general principle of the earlier American Joslyn. It was quite simple and efficient in design, an exceptionally fine feature being the fact that the hammer was low in the tang and when cocked, the opened breech could be cleaned from the rear. When the arm was fired, a thumb lever on the right side was pushed down. This cocked the internal hammer (whose upper end emerged through a hole in the tang when firing to hit the pin in the breechblock). This action also forced the striker back, raised the hinged block and ejected the empty case. The extractor worked off a lever operating from the hammer while being cocked, affording a form of primary extraction prior to the ejecting motion. In a rapid-fire contest at Wimbledon in 1870 the Soper delivered 770 shots against 306 by a team of Sniders. However, in spite of tales of its "firing 60 times in one minute" as recorded in THE FIELD in April, 1870, documented tests show it firing 68 shots in 2 minutes against 62 for the Martini-Henry—hardly a superiority to raise all the tales it did.

OTHER INTERESTING BRITISH SINGLE-SHOT DESIGNS

The Albini-Braendlin

This .60 caliber rimfire was a rifle designed by Colonel Albini of the Italian Army, manufactured in Birmingham, England, and actually used in Belgium. The Belgians used this system in 1867 to convert breechloaders. The action operated like the familiar American Springfield Single Shot: when the hammer was cocked the breechblock could be swung up over the barrel on its hinge. The design was a common one, in this case apparently stemming from the familiar percussion lock alteration patented by W. Mont Storm in 1857 in England, and extensively used in Europe, more particularly in Russia and England.

The British Henry

This arm was a single shot made by the designer of the Henry barrel. At first it was little more than an imitation of the American Sharps dropping block. Using the general design of the American arm, Henry made the operating lever a separate lever mounted around the trigger guard. In later models he copied from the Martini system the idea of enclosing the firing mechamism in the breechblock itself and used a Martini-type extractor. This arm did have the good feature of having a block which dropped and exposed the barrel for cleaning from the breech end.

The Field

The Field Sliding Block. This is a dropping block action operated by a thumb lever on the right side of the receiver. Pushing it forward acts through a system of levers to cock the hammer, lower the block, and extract the case. The compound extractor system is unusually good for a block rifle. This arm very closely resembles our Hepburn.

The Field Hinge Block. This was merely another variant of the American Peabody. The block hinged at the rear falls at the front to expose the chamber when the lever is operated.

The Money-Walker

About 1868 this strange design appeared. In effect, it was the old Peabody idea again with a different opening system. While the block was hinged at the rear to drop at the front end to expose the chamber, the operating lever was a prolongation of the breechblock itself. It lay above the small of the stock. When lifted, it pivoted the block, dropping the front end. This design was copied on the continent.

The Swinburn

This was a modified Peabody-Martini. It too, had a falling block hinged at the upper rear and operated by a lever below the small of the stock. Its "unusual" feature was to replace the spiral spring striker with an internal hammer and a V-spring. This arm also had an exposed right side lever to permit the hammer to be half-cocked and recocked. At this time spiral springs were in bad repute in England.

The Westley Richards

This arm was made in two lock designs.

Falling Block. The original Westley Richards "falling block" was on the Peabody pattern with an internal hammer and a block hinged at the upper rear. Operating a lever around the trigger guard cocked the hammer within the block and lowered the block. The extraction system was better than in most block types, being effected by direct powerful leverage. The hammer had the firing pin as part of its face. A V-mainspring was positioned below the chamber. This was one of the simplest designs of this type, but had to be cleaned from the muzzle.

A later model of this design used a horizontal striker operated by a V-spring behind it, making a much more compact action.

Sliding Block

The Westley Richards "sliding block" was designed by two employees of the W-R firm named Anson and Deeley. These two designed the famous Anson & Deeley double-barrel shotgun action, the best low-priced action of that pattern ever designed, and still the basis for most of the world's low-priced shotguns. Their military design, however, was not acceptable, coming as it did at the close of the single-shot era in military usage. A vertical dropping block on the Sharps idea is lowered and raised by action of a trigger-guard lever. The firing mechanism is concealed within the breechblock. As the block is lowered, a tumbler is cocked and the extractor operated. One good feature of the action is that it is designed to permit cleaning the barrel from the breech end.

MODERN BRITISH SINGLE-SHOT RIFLES

None are intended, of course, for military use. The B.S.A. and Vickers models are merely modifications of the old Martini-Henry action and are used generally for precision small-bore shooting. Birmingham Small Arms also makes a line of .22 single-shot bolt-action rifles on the order of the familiar American types.

European arms design differs radically in its approach from that of the United States. The comparative results offer an interesting study in psychology and economic systems.

From the very beginning of rifle manufacture in America, the emphasis has been very largely centered on sporting arms to equip the individual pioneer, hunter, sportsman, or target shooter. Except for the direct efforts of Springfield Armory as the prime designer for our armed forces, rifle development has been primarily in the interest of the free individual—the average citizen. From this fact stems the tremendous line of arms and ammunition types evolved in our country. While American arms played great roles in battles throughout the world, they were still primarily commercial sporting types used for more grim purposes. It is only since our involvement in World Wars I and II that specific manufacturing attention has been turned to military design as a practical necessity.

4

Early European Single-Shot Rifles

EUROPEAN PREOCCUPATION WITH MILITARY ARMS

In Europe practically all rifle design since the earliest times has settled around just one consideration—what type will kill the most men at the greatest distances. Hence, the sporting designs of European countries have been largely borrowed from successful American types, or have evolved from military weapons. Thus pump and semiautomatic arms of a nonmilitary character sold abroad have been in practically every important instance either a product of the Belgium F.N. factory manufacturing weapons designed by our own John M. Browning, cheap Belgian or Spanish imitations of Winchester types originally designed by Browning, or Browning designs made under license in Germany. Only in target single-shots has any originality been shown.

As a natural outgrowth of a situation where military thought and fears constantly haunt the horizon, where few men are free to possess rifles except when handed to them by the government together with a uniform and a demand for subserviency, early European designers understandably had little to offer beyond modifications of types already known.

When we come to the period of military magazine and automatic arms the picture is quite different as we shall see in due course. But in the opening of the metallic cartridge era, little was developed beyond that which has already been discussed with one notable exception. That exception of course is the true turn bolt action which marked the opening of a new arms era throughout the world; and it is credited almost entirely to the German, Peter Paul Mauser.

Of all European nations, only Switzerland and Turkey followed very closely the event of the American Civil War and kept well abreast of our developments. It was natural, of course, that the Swiss, a nation of riflemen, would see and understand the importance of the arrival of the breech-sealing metallic cartridge and the repeating rifle principles. On the other hand, it is a distinct tribute to the Turkish leaders of the time that they, expecting a war with Russia, should have had the progressiveness to see that the best European military minds were so blinded by European military might that they were not sufficiently attentive to the amazing arms development in America. The Turks, not having manufacturing facilities, turned to the United States first for Peabodys, then for Peabody-Martinis, then for Winchester repeaters; arms which they knew were not primarily military, but which would establish superiority until European military design woke up. The Swiss, after experimenting with the Peabody, focused attention on the development of a true military design—a bolt action, the Vetterli.

Since all modern military magazine rifles were evolved directly from the German Dreyse and the French Chassepot Needle guns, the story of European design is essentially that of the bolt action. However, since the value of the bolt system was not fully recognized until after the Franco-Prussian War of 1870, a wide variety of other systems were tried before the metallic cartridge bolt action design was militarily perfected.

It is with those variants that we shall deal first because of their historical value.

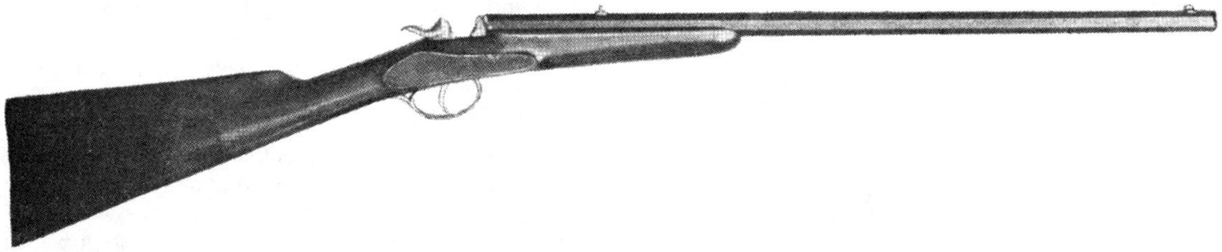

French Flobert system.

EARLY EUROPEAN SINGLE-SHOT DESIGNS

THE FRENCH FLOBERT

The very earliest form, interestingly enough, was the French Flobert which appeared about 1845. It handled the first successful forerunner of the rimfire, the primed copper case and lead ball combination we know today as the "BB" or bulleted-breech cap. In that original Flobert the thumb-actuated breech piece was not backed up by a hammer lock as in our own Remington rolling block of later date. The French had the germ of one of the most successful locking systems later developed, but the military was too occupied with making a better needle gun to realize the prize within their reach. As has already been pointed out, Flobert's arms and ammunition were exhibited and listed as patented at the London Exhibition in 1851. Yet the French rifle of 1870 was a paper cartridge, needle fire!

According to Colonel Rudolf Schmidt of the Swiss Army, a noted authority writing first in 1877, the original Flobert action appeared in 1845. The Swiss experimented with rimfires at a very early date, probably about 1847, but the first practical and successful metallic rimfire was the American Smith & Wesson.

THE SWISS MILBANK-AMSLER CONVERSION

As early as 1865 the Swiss designers were experimenting with the Milbank-Amsler conversion. This system is like our later Springfield 1873. The hammer is cocked and the forward hinged breechblock turned up over the top of the barrel to expose the chamber. A perfected form of this design was adopted by the Swiss in 1867 and used until actually replaced by the bolt type Vetterli. (The Vetterli was officially adopted several years before it was actually put in service.)

BELGIAN DESIGNS

In 1868 a strange design somewhat resembling the American Cochran of that period appeared in Belgium. Known as the "Reilly-Comblain," it had a side lever which served to elevate the front of the breechblock to give access to the chamber. Like the percussion American Hall, the breechblock was hinged at the rear. Since the R. C. was a metallic cartridge breechloader, the solid block was pierced only for the striker. The better known "Comblain," the dropping block design, did not appear until about 1870. It and the British-made Albini-Braendlin already described were in general Belgian use for some years, the Comblain being more generally used by the Civil Guards. Both of these last mentioned rifles in .43 caliber were actually used to some extent even in World War I! The Terssen of 1868 was still another hinged-block Belgian conversion on the order of the Albini and the American Springfield 1873. The Comblain drop-block model was adopted quite early by Brazil, Chile and Peru.

The Chabot of 1865 and the Austrian Wänzl of 1867 were also conversions based on the principle of the lifting hinge block.

EUROPEAN MODIFICATIONS OF U. S. SYSTEMS

The Swiss "Schmidt & Jung" rifle of 1865 was on the principle of the earlier American Joslyn, although one gathers from Colonel Schmidt that he thought his breechblock hinging to the left was quite original.

In 1867 the original Remington rolling block swept over Europe. It is an interesting commentary on the business ability of the heads of the Remington concern at that time, that they were so well covered by patents—and by foreign representatives!—that no attempt was made to infringe on their basic system. Without exception every other American type was copied, modified, or otherwise altered by foreign designers.

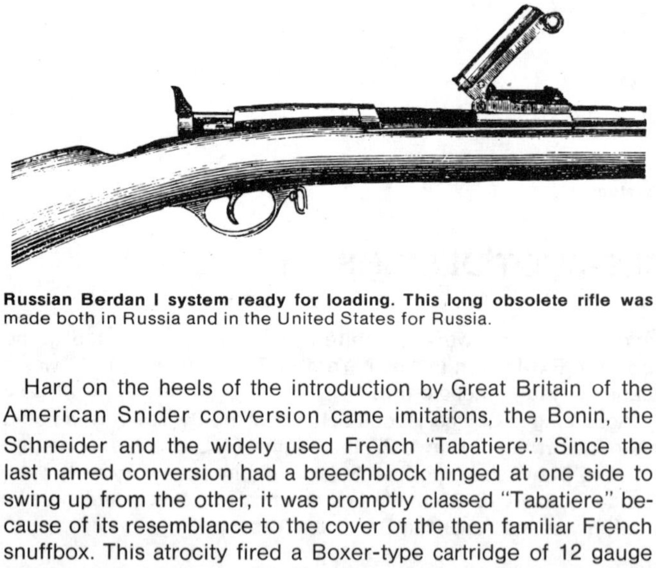

Russian Berdan I system ready for loading. This long obsolete rifle was made both in Russia and in the United States for Russia.

Hard on the heels of the introduction by Great Britain of the American Snider conversion came imitations, the Bonin, the Schneider and the widely used French "Tabatiere." Since the last named conversion had a breechblock hinged at one side to swing up from the other, it was promptly classed "Tabatiere" because of its resemblance to the cover of the then familiar French snuffbox. This atrocity fired a Boxer-type cartridge of 12 gauge whose conical bullet was hollow from base nearly to point, the hollow being plugged with papiermache to expand the bullet to take the grooves. The receiver of the British Snider was cut away to permit insertion of the breechblock, yet to leave plenty of breech metal behind it for the shooter's protection. The receiver of the Tabatiere, on the other hand, was cut away at the rear, making it easier to insert a cartridge and to clean the barrel. This weakened the breech, however. Defective cartridges made the design unsafe, and the French probably were just as well off that they entered the War of 1870 with needle guns. The Russian Krnka conversion of 1869 is another variant of the Snider, with all the defects of the Tabatiere. An iron breechblock was fitted into a bronze receiver well, and was hinged on the left. This Bohemian design automatically extracted the case as the breechblock when lifted acted on an extracting lever.

In 1867 Spain adopted the American Berdan lifting hinged block as a conversion unit for its muzzle-loaders, and in the same year Russia adopted the striker-fired hinged block American "Berdan I" already described. (The Berdan II adopted in 1871 by Russia must not be confused with the Berdan-type lifting hinged block design. The name "Berdan II" generally used in Europe applied to the turning bolt action similar to that of the Mauser, and was made at the Russian Tula Arsenal.)

The Bavarian Werder, adopted in 1869, has already been described. This modification of the American Peabody was one of the noteworthy developments of its day, since it was not merely a conversion of an existing muzzle-loader, but was actually an entirely new rifle.

In 1871 the Swiss "Kaestli," a striker-fired modification of the American Sharps dropping block system appeared. The French "Mesnier" dropping block (followed by a modified rolling block form which is little known) also appeared about this time.

The year 1874 saw the adoption by Greece of a modified dropping block rifle, the "Mylonas," which used a centrally hung hammer and was operated by a lever positioned ahead of the trigger guard.

In 1868 the Werndl rifle was introduced in Austria. This arm had a block hinged on the left to be turned up away from the breech when the hammer was cocked. A thumbpiece jutted out on the right side. This was the last single-shot rifle officially used by the Austrians.

The Belgian Martini-Francotte, a hammerless modification of the Peabody noteworthy because the entire lockwork unit can be removed without tools for cleaning, was later distributed by Greener, and is sometimes called the "Greener-Martini." This Francotte lock is still in extensive use on European single-shot sporting rifles.

The early Krag-Petersson single-shot was also a modification of the Peabody system, the block being hinged at the rear to drop at the front for loading. This arm appeared in the early 1870's and was the basis for a later magazine rifle. About this time also appeared the French "Duprez," a hinged block operated by a side lever, a freak design of no particular importance, except that it illustrates the paucity of French design at that period.

In 1885 the last military design on the falling block principle appeared. This was the Guedes-Castro adopted by Portugal. The arm was made in 8mm caliber at the Steyr works in Austria. It was a direct modification of the British Martini-Henry and not, as it is often listed, a bolt-locked rifle.

The only specimens of rolling block locks to appear at all in European design emerged after the passing of the value of the Remington system. In 1881 Nagant in Belgium brought out a hammer rifle with a rolling block lock operated by a lever below the trigger guard; and in the following year the Mesnier hammerless, its action operated by a trigger-guard lever, made a brief appearance.

With this short survey of the outstanding modifications of American systems, we turn now to the bolt action military design which Europe pioneered and perfected, and to the magazine arm, as it evolved.

5

Evolution of U.S. Military Bolt Actions

The great part played by the United States in the use, evolution, and development of the early bolt action has been given little attention in the written histories of firearms. It is true that all modern bolt actions stem from the German Dreyse Needle gun. It is also true that the direct descendant of that arm, the German Mauser, is the parent of most of its perfect forms.

However, American bolt-locking systems for metallic cartridges were manufactured and used first in the U.S. Civil War. The North issued 1001 Palmer carbines using .52 caliber rimfire cartridges. These arms were patented by William Palmer (41,017) December 22, 1863—a good four years earlier than the original Norris-Mauser. True, the two arms have a resemblance only in their method of opening and closing the breech; still, the Palmer bolt breech locking design is a first evolutionary step. This carbine had screw threads at a rear section of the bolt which locked into corresponding receiver cuts when the bolt was pushed forward and the handle turned down. Since the Army at that time didn't consider a weapon a weapon unless it had a huge thumb cocking lock (or hammer) on the right side, Palmer provided one. The falling hammer hit the cartridge on the rim. This arm had a spring extractor on top of the bolt, an automatic ejector in the form of a pivoted lever in the side of the receiver, and had the receiver side cut away to make ejection positive. The Palmer, made by E. G. Lamson at Windsor, Vermont, was the source of design factors used in several later bolt actions in this country, though it seems to have been little known in Europe.

EARLY BOLT ACTION DEVELOPMENTS

The first use of the bolt forward locking lug principle was, as we have seen, incorporated by Lt. Col. J. Durrel Green, U.S.A., in his percussion arm of 1857, though that principle was not used for metallic cartridges until the French (Lebel) introduced it, in 1886.

The first bolt action arm to use center-fire cartridges, and also the first to have the firing pin directly propelled within the bolt, was the Ward-Burton Single-Shot, first patented Dec. 20, 1859 (No. 26,475 to Bethel Burton). Springfield Armory manufactured 313 carbines and 1000 rifles in caliber .50-70 C.F. and the arms were field tested in 1872 trials. Some were later made commercially in other calibers. The Ward-Burton bolt was unlocked in standard fashion by turning the bolt handle up. This disengaged lock threads at the rear of the bolt body from corresponding receiver cuts (as in the Palmer). The cartridge could be dropped into the magazine well, as in the best European modern single shot practice. Pushing the bolt home chambered the cartridge and cocked the piece. Ejection was automatic on retracting the bolt. This weapon was fitted with a special bolt lock which had to be released by the right thumb before the handle could be lifted, a device which could also be used as a safety. Soldiers long used to old side hammers soon gave the bolt design a bad name when they encountered accidental fire on bolt closing. The comparatively minor design changes necessary to overcome this defect were passed over at the time. (The British, too, were influenced against bolt actions because of possible accidental fire until Paul Mauser evolved his simple cam-cocking system.)

In 1872 Ward-Burton offered a rifle of this hammerless bolt design with a magazine tube below the barrel. It, too, was rejected by the Ordnance Board. This was the first American bolt action magazine rifle using the by then familiar Winchester magazine tube. Only the Swiss Vetterli and the Austrian Früwirth in this design were ahead of the Ward-Burton, both having appropriated the Winchester magazine idea before 1870 as a result of experiments with the Winchester 1866.

Also tested and rejected at the 1872 U. S. tests was a bolt action designed by G. Merrill, samples of which may be encountered with the stamp of the Brown Mfg. Co. of Newburyport, Massachusetts. This arm had an exposed hammer behind the bolt. When a right side lever was actuated to unlock the bolt and the action was opened, the bolt cocked and rode over the hammer. This system also appeared in later designs. The Merrill caliber was .58 C.F. Merrill alterations on Enfields bought from England during the Civil War (North and South each purchased some 400,000 of these muzzle-loading Enfields!) were used by various state militia groups to some extent.

All told the Ordnance Board viewed 12 bolt action types in the 1872 tests out of a total of 108 designs—including 10 repeaters—but decided to retain a modified Springfield Single-Shot, the 1873 Model, with its hinged block lock.

The 1878 Ordnance Board tested 29 rifles of magazine type. Eight were bolt actions. The Hotchkiss was officially adopted as our first official bolt-action magazine arm. The inventor, B. B. Hotchkiss, was an American. At the time he was living in France; and it was there that his model rifles were made. (We shall encounter Hotchkiss again under the history of machine guns.)

The Hotchkiss rifle came to the attention of Winchester when it was shown at the Philadelphia Centennial Exposition in 1876. Arrangements were made for manufacture by Winchester. First U. S. patents were issued in 1860. Modifications of the rifle continued through 1878.

The magazine of the Hotchkiss rifle was a variant of the earlier Spencer, a spring-fed tube in the buttstock. New features were magazine cutoffs allowing the soldier to hold his magazine in reserve, and a good manual safety. Winchester supplied 2500 actions to the Navy and 1000 to the Army. Barrels and furniture were supplied by Springfield Armory, thus allowing utilization of Springfield single-shot bands, sights, and bayonets in the interest of economy.

In 1882 Congress appropriated $50,000 to develop and test magazine rifle designs—the small sum being an indication of how hard it was in those days to convince people of the need for improved military arms in the United States.

THE LEE INVENTIONS

James Paris Lee was born in Scotland and educated in Canada. However, all his arms design work was done in the United States after he was granted citizenship here. He is another of the inventing genius group to whom due credit has never been extended.

The name "Lee" in firearms circles immediately conjures up the British Lee-Enfield and Lee-Metford rifles, the primary small arms of the British forces from 1888. Lee's development of those actions and of the straight pull Lee rifle are well-known. His other achievements need some passing notice, however.

Lee patented in 1879 the arm known as the Lee U. S. Navy Rifle, caliber .45- 70 Government. This rifle actually marked the introduction of the magazine system which revolutionized military rifle design—the centrally positioned box magazine in which cartridges are stored directly below the bolt. It seems simple today, but its development was a stroke of design genius at the time.

Lee further developed a magazine design which had been patented by three Englishmen; Walker, Money, and Little in 1867. This box magazine overcame many of the troubles which had been encountered with tubular magazines.

But what of the inventor himself? James Lee was a man early dogged by troubles. In July, 1862, he patented a rimfire rifle, caliber .36 as first issued, which was manufactured by the Lee Arms Company, Milwaukee, Wisconsin. This was a single-shot weapon with an exposed hammer. With the hammer at half-cock the barrel could be swung out to the right for loading, a barrel lug being provided for hand extraction. Lee lived at Stevens Point, Wisconsin, at the time, and it was there he worked out the patterns of many of his later designs. The Milwaukee plant was probably the only Middle Western works of any importance during the Civil War. Lee began manufacture on a War Department order for 1000 carbines early in 1865, having sunk most of his available capital in the Milwaukee plant. He placed an order with Remington at Ilion, N. Y., for the manufacture of the barrels. Assuming the Government order was for .44 rimfire long, since the specifications did not state exactly, he found that the Ordnance Department had not so intended. The Government refused to accept his finished carbines. This unfortunate venture cost Lee his capital. The resulting suits were a further strain on Lee.

In 1874 Congress appropriated $10,000 for the manufacture and testing of a new Lee Single-Shot—a curious design so impractical that one wonders if some conscious-stricken official was trying to make amends to Lee by providing the money to manufacture 134 of them. Having learned the lesson of tying up his own money in governmental experimentation, this time Lee went to Springfield on salary to supervise manufacture. The arm was a failure and Lee turned his abilities to the design of a new bolt rifle and a new magazine.

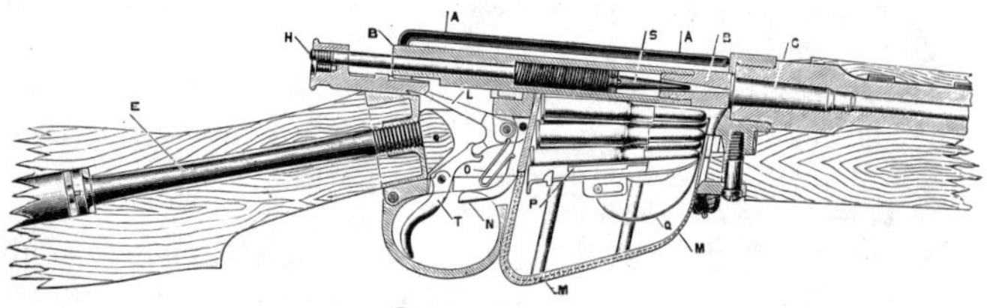

Mechanism of the Lee-Speed Magazine Rifle Mark II.

Lee submitted his Navy Magazine rifle to the Equipment Board in July, 1879. He arranged for manufacture at Bridgeport by a new concern consisting of the group then operating as the Sharps Rifle Co. The group suddenly decided there was no money in the arms business and being unable to see enough future in it to warrant putting up more capital, turned the manufacture over to Remington at Ilion, N. Y. Lee went to work at Remington. Remington, which had been in danger of losing all its world markets now that magazine rifles were obviously here to stay, took a new lease on life. In the years that followed much of Remington's success was due to Lee and his designs. (One of the men who sold Lee short by failing to back up his developments in the original Lee Arms Company was, curiously enough, old P. T. Barnum, a man whose business acumen in most matters was quite outstanding.)

The 1879 Lee was designed without a cutoff. The magazine was intended to be removed from the action when the rifle was used as a single-shot. Aside from the U. S. Navy caliber, the model was also made for use in China and in Spain in caliber .43 Spanish. It was a Civil Guards weapon for many years.

Lee was granted patents in England, Belgium, and Russia. He sued both Mannlicher and Mauser for alleged infringements when they introduced their first box magazine rifles. Every student of design or invention, of course, is aware of the phenomenon of "parallel development." When a need is recognized for any advance, geniuses in various areas seem to work out the same approach often in complete secrecy and without any way of physically knowing about other research in the field.

A study of the Mauser and Mannlicher records shows that such was the case with Lee and his magazines. The original work records of the German and the Austrian inventors show clearly that they were both experimenting with box designs before any word of Lee's magazine could have reached them. However, it is quite possible that both men benefited by later studies of the Lee work, and may have perfected their own types after seeing Lee's. It is a fact that the Lee was openly tested in both England and the United States well before the commercial introduction of the initial Mannlicher or Mauser box types.

In 1885 Great Britain tested the Lee action with the American Bethel-Burton side-feed magazine. By then, however, other European nations had beaten the British to the adoption of magazine rifles. Those in use at that time in France, Germany, and Russia used tube magazines. Hence, as usual, the British authorities also leaned towards the tube pattern. Lee made further improvements on his box magazine. The British learned the Germans were concerned with a box design about this time. This again swayed official ideas. In 1887 the British tested both the Lee-Burton and Lee actions with Metford rifling in a new caliber .402. Meanwhile Switzerland had reduced its caliber, and after experiments with a Swiss Rubin of caliber 7.5mm, the British hit on the idea of combining the Lee bolt and magazine, the Metford rifling, and a new .303 cartridge originally using compressed black powder but later changed to a cordite load. (The change was made in 1892.) In December, 1888, the final form of the rifle was decided, the name Lee-Metford Magazine Rifle Mark I assigned, and Britain was now among the nations using magazine arms. The further story of these Lees must be told in the history of Great Britain, but this was the first of the series with which the British Tommies forged and protected the Empire from that day on. Meanwhile, the official U. S. arm was still the Single-Shot Springfield cam lock!

Remington made the Lee as a sporting rifle in several calibers, following the design with the modified Remington-Lee Small-Bore Military Rifle, an arm chambered for the 6mm U.S. Navy, the 7mm Mauser (30,000 of which were delivered to Cuba), the 7.65 mm Mauser and the .30-40 U.S. Krag cartridges. Two thousand in

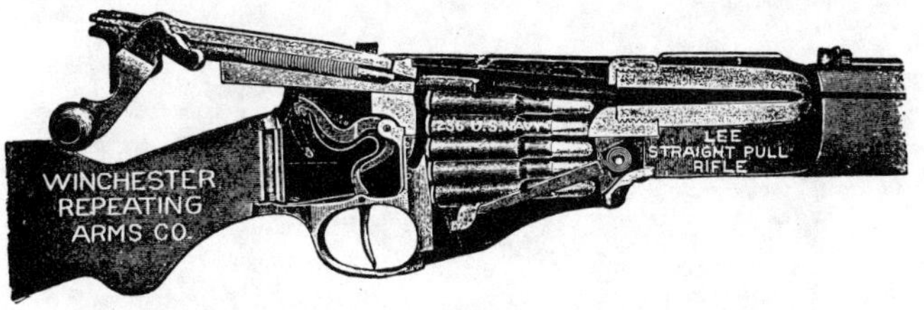

Navy Lee straight pull 1895 model. Caliber 6mm (.236).

the last caliber were used by the Michigan Militia, but the design was not adopted by the Federal Government.

In 1895 a Naval Board adopted the Lee Straight Pull in caliber 6mm (.236) the smallest military caliber ever adopted, although some as low as 5mm were experimented with abroad. Pulling the bolt straight back unlocked the arm, ejected the empty case, and cocked the action. A forward thrust chambered a cartridge and locked the bolt. The locking design was decidedly original. Winchester was given a contract for manufacture and 10,000 were delivered. This was our first clip-loaded official rifle. Though excellent in theory, the rifle did not work out well as a military design.

OTHER EARLY AMERICAN MILITARY MAGAZINE RIFLES

Many new American designs followed fast on the heels of the Lee 1879. For historical value, we shall here touch upon the outstanding ones.

Major W. R. Livermore and Captain A. H. Russell of the U. S. Ordnance Department made extensive researches into magazine construction about the same time that Lee introduced his noted magazines. What is often called the "Russell-Livermore" rifle, an arm which received considerable attention in European arms circles, is actually an altered Lee-Speed rifle fitted with a special feed case system. The case was a piece of bent sheet steel designed to carry enough cartridges to fill the magazine, permitting magazine loading with a single motion, the case being discarded after loading.

The "Russell" magazine rifle invented by the same A. H. Russell of Ft Union, N. M., was a different development, an unsuccessful straight-pull bolt action. The lock was ingenious. It consisted of a crosspiece within the bolt body of slightly greater diameter than the bolt, provided with cam-shaped wings extending into receiver seatings. The initial pull on the bolt acted to revolve the crosspiece out of locking engagement, and the continuing pull extracted and ejected the case. This arm was made with both a bottom magazine feed, Lee style, and a special side or lateral magazine.

In 1882 the Ordnance Board also recommended tests of a rifle offered by Gen. J. N. Reece, and known as the Chaffee-Reece. It had a tube magazine in the buttstock. The loading point was a trap in the butt plate which could be opened only when the bolt was in open position. A magazine cutoff was provided. Setting a thumbpiece at half-cock set the safety and locked the bolt. One thousand were ordered. They were the first U. S. magazine rifles made at Springfield Armory. Like the Hotchkiss, it was designed to take the old single-shot sights, bayonet, and fittings. It was soon discarded as too complicated and unreliable for field service.

The 1882 Board also arranged for testing of an improved Hotchkiss bolt action, the Model 1883. This arm was made by Winchester. It is of interest here only because its tests definitely decided U.S. design opinion against the buttstock-tube magazine for large cartridges. From that time on this tube system was confined to small-bore arms. Winchester made the design as a sporting arm also.

The third design tested was the Lee Magazine Rifle 1882, usually called the Remington-Lee. While the arm could be used as a single-shot bolt action without magazine, the 5-shot magazine itself actually constituted part of the rifle mechanism. As originally designed, several Lee magazines were intended to be issued to the soldier, who was to carry them loaded. They were costly to make and were not intended to be expendable. This magazine system was never favored by the U.S. military, and its later success was entirely in the sporting field where the individual owner is likely to be more careful of the fragile magazines than is the typical soldier.

In passing it must be noted that with the coming of World War II, we designed our carbine to use detachable magazines; for by this time the manufacturing methods had improved, costs had been lowered, and the question of cost on the field of battle has now been subordinated to increased firepower such as is possible by rapidly replaceable magazines. Once again we see Lee's original ideas succeeding.

It is interesting to note in passing that the Lee was the first choice of the 1882 Board out of a total of 40 rifles submitted. These 40 represented the designs of only 13 inventors, however, and many models were merely modifications of a basic pattern. Of the remaining systems considered by the Board, only the Marlin Lever Action and the Remington-Keene Bolt Action were later produced commercially.

The Remington-Keene was a tube-under-barrel rifle originally patented by J. Keene in 1874. It had a cocking piece shaped like a common hammer attached to the extreme end of the bolt. The Keene was tested by both Army and Navy in the Government .45 caliber, and is of interest as the only full rifle ever used by the U.S. Army with below-barrel tube magazine, other types having been carbines. An outstanding feature of the design was that it could be magazine-loaded with bolt closed through the bottom of the action. The magazine was provided with a cutoff.

When the 1882 Board concluded its tests, the War Department settled back confident that the Springfield Single-Shot was still the proper arm. Fortunately, we had no occasion to need an Army for a number of years. Meanwhile we were nearing the time when smokeless powder would be commonly used and a new design would be demanded.

THE KRAG-JORGENSEN

By 1890 every important military group in the world was equipped with magazine rifles except the United States. Our

U.S. single-shot black powder rifle, principal arm in Spanish-American War.

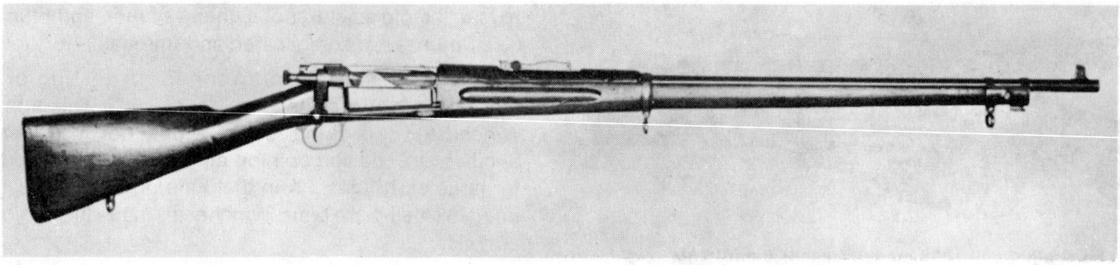

Representative Krag-Jorgensen rifle.

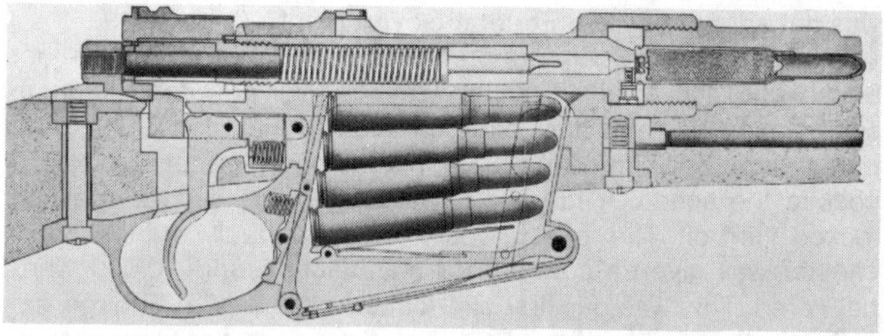

M-91 Mauser.

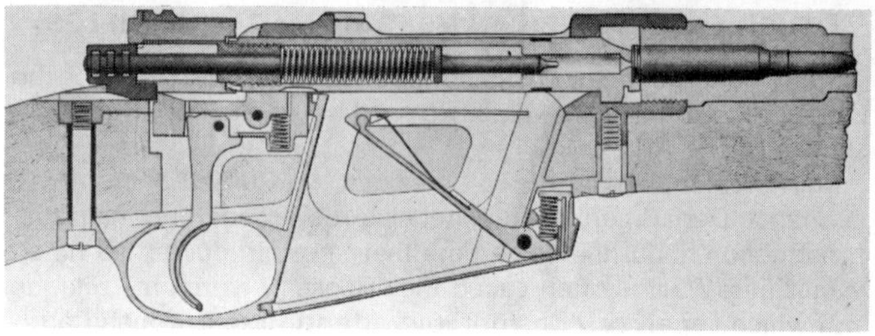

M-92 Mauser (also called M92/93).

Ordnance Board convened to consider magazine systems. Fifty-three systems were presented, both domestic and foreign arms. The Board eventually selected the Norwegian Krag-Jorgensen system around which to build our new rifle. It passed through five modifications before it was accepted officially. The adoption of this system is a sterling example in history of how important timing is in military decisions, and of how utterly inadequate theoretical tests may be. At a time when the simple Mauser bolt was already known, when the superior locking features of the dual front lug systems was evident, when the simplicity of both the Mauser and the Mannlicher types of clip-loading devices had been field proved, we nevertheless adopted a bolt with a single locking lug and a cumbrous side box loader magazine which had to be loaded with single cartridges! A variety of clips were tried, but the Krag system did not allow a sensible clip to be developed. In all fairness it must be admitted that the ordnance authorities in Great Britain and Germany seem to have been afflicted with the same virus as our own at that period. England had adopted the American Lee with its rear locking system, a system that from conservatism and from financial considerations they did not change over to any degree until the production of the FN after World War II, although experimentation was begun in 1913 and early 1914. Germany herself, meanwhile, had adopted the hodge-podge known as the "Commission, Mannlicher Loading System," the Gewehr 1888, flying in the face of Paul Mauser, her premier designer who at that time had a superior design in being. Austria adopted a Mannlicher with a wedge rear lock. Instead of selecting the best design the military all sought something different.

While smokeless powder had come into its own by 1892, making obsolete the single-shot black-powder rifle for first line military purposes, the flanged (rimmed) cartridge case was still in accepted use. For that matter, it still was used by both Great Britain and Russia in World War II, despite the superiority of the rimless.

However, by 1892 foreign use and practice had established the desirability of calibers of about .30 inches for maximum military efficiency. These facts doubtless influenced the development and adoption of the .30-40 Krag cartridge decided upon by our ordnance authorities. The original 1892 rifle was not particularly satisfactory in accuracy or usage, even with the cartridge of its time; but by 1896 Springfield Armory had remedied design and material defects to such a great degree that the Model 1898 (which mechanically differed only in minor details such as rear sights) established a record as one of the sturdiest rifles ever made. In all some 400,000 were manufactured. The 1896-1898 models in general sporting use today are excellent arms in their class, reliable and accurate within the limits of the ballistics of the cartridge used, though they cannot of course compare with the Mauser action for strength.

The design could not be made strong enough to be used with the powerful Government loads later developed, a factor which many experts contended should have been considered at the time of the official adoption of the Krag. In any event, at the outbreak of the Spanish-American War only the Regular Army (which numbered only 27,000 total) was armed with Krags. State militia units had a variety of rifles of their own purchase, mostly single-shot arms. As a result we actually entered the War of 1898 armed chiefly with old .45-70 black-power arms.

Spain by this time had armed her Cuban forces with a variety of modern small arms. The principal one was the 7mm Mauser rimless made by Loewe, a precision cliploader. The Battle of San Juan Hill showed that the Krag just wasn't good enough—not to mention the single-shot Springfield. Our attack force of 15,000 was aided by Cuban insurgent groups. The garrison was only 700 men—but they had modern arms. The hill was taken, but not until some 1400 casualties were inflicted upon us!

THE "SPRINGFIELD"

The long series of U.S. Government bolt-action rifles listed as the Model 1903 and its modifications stem directly from the

Spanish 7mm smokeless Mauser magazine rifle used by Spain in War of 1898.

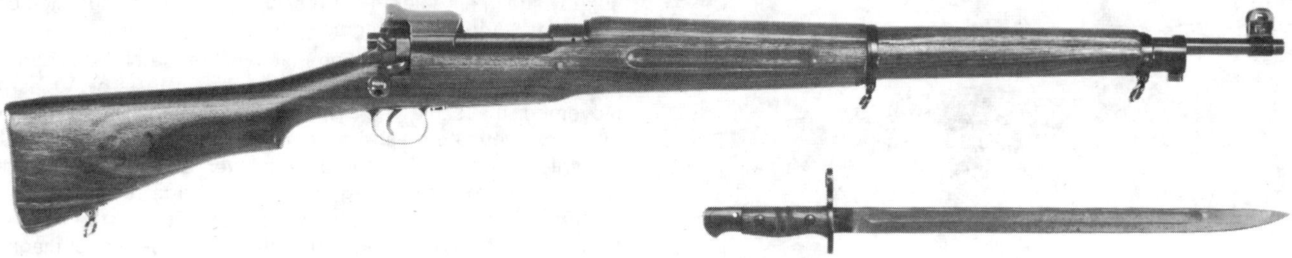

U.S. Model 1917 (Enfield) Rifle.

German Mauser. In 1900 and 1901 our Ordnance Department conducted intensive tests with captured German-made Spanish Mausers, and made studies of the new German Mauser designs then in use in Germany as the official "Gewehr 98."

The first of the new rifles was produced at Springfield Armory in 1900. It was chambered for a new .30 cartridge which was not very efficient. Actually it was little better than the Krag it replaced. An improved rifle was issued to our troops in 1904 and 1905 under the designation "Model 1903." From its point of design and manufacture it was popularly known as the "Springfield." The name has stuck through the years, even though the rifle has been made at other arsenals and manufacturing plants.

We paid Mauser $200,000 for manufacturing rights. The design was altered somewhat. A magazine cutoff was combined with the bolt release; a cocking piece was provided which could be pulled back for manual recocking in case of a misfire; and the excellent German one-piece firing pin was replaced with a two-piece design. Other minor alterations were made, resulting in probably the finest military bolt action produced. Its chamber support of the cartridge case is not as efficient as the German; but this a technical factor which normally affects only individuals who dangerously overload used cartridge cases for experimental shooting.

The design and variations of the Springfield are dealt with in detail in the chapter on United States small arms. All use a variant of the U.S. 1906 Model ammunition. These are based on a cartridge developed by the United States in 1906 after experimenting with the German 1905 pointed "Spitzer" bullet. Springfields are often described as "caliber .30-06," indicating the Model 1903 rifle chambered and adapted for the 1906 cartridge. The sights of the original 1903 were also changed and the cleaning rod bayonet was eliminated.

THE "ENFIELD"

The "U.S. Model 1917" rifle was an emergency design only. Yet in two wars it served its purposes well. It was a modification of a British design and was first produced in 1917. The name "Enfield" is popularly applied to it from the British arsenal where it was developed.

The outbreak of World War I found the British experimenting with a new Mauser-pattern .276 rifle originally intended to replace the "Lee-Enfield" designs. British manufacturing facilities were not capable of handling additional production, so U.S. arms makers were asked to manufacture the new design, but chambered and altered for the British standard .303 cartridge to conform to regular issue.

When the U.S. entered the war in 1917, there was a shortage of Springfields. Since both Winchester and Remington were tooled by then for the British Pattern '14 (Enfield), the United States Ordnance Department had those companies undertake alterations of the design to handle the standard U.S. rifle cartridge.

This rifle has been issued to allies throughout the world. It is commonly referred to as the "American Enfield" to distinguish it from the earlier version chambered for the British cartridge. It is discussed in detail in a later chapter on current weapons of Great Britain. While heavy and rather clumsy, it is one of the strongest military rifles ever built. It bears the mark of the manufacturer also, which may be Winchester, Remington, or Eddystone; and these markings at times cause confusion to foreign writers.

With this short summary of American bolt-action development, we now revert to an earlier day for the American story of repeating and magazine rifle development.

REPEATING AND MAGAZINE RIFLE DEVELOPMENTS

While the terms are very often used interchangeably, the word "repeater" in general use ordinarily indicates a tube-cartridge reservoir; while the term "magazine" ordinarily indicates a military type arm with cartridges held in a box below the bolt.

THE SPENCER—THE WORLD'S FIRST SUCCESSFUL REPEATER

The Spencer (1860) was the first successful repeater. Of course, its development was made possible only by the introduction of the metallic cartridge case. It had a spring-fed magazine tube in the buttstock. Firing was by an external hammer. Operation was by trigger-guard lever. A semi-circular breechblock operated by the lever had both a falling (unlocking) motion and a rotating (feeding) motion.

As an item of genuine American history as well as firearms history, the Spencer story deserves to be told.

The story, part of another little-known saga of American arms inventors, began on a day early in the Civil War when a youth of 20, harried and self-conscious, haunted the corridors of the bustling War Department in Washington. He had had to run the customary gantlet of military red tape even to gain entrance to the sacred confines. To the busy ordnance officers he was just another "young man with a gun"; just another crazy inventor come to take up their valuable time with senseless talk of a revolutionary weapon. Every armory and every capable manufacturer in the North was turning all his efforts to producing more of those fine muzzle-loading percussion rifles. Yet here was an adolescent talking gibberish about not only a metallic cartridge breechloader, but actually one which would fire repeatedly after being loaded with several cartridges! How could the war ever be won with all these interruptions of precious routine?

Dispirited, discouraged, ready to quit, Spencer complained bitterly to the one man he found who would listen to him: a humble doorkeeper at the War Department Building, not an arms "expert,"

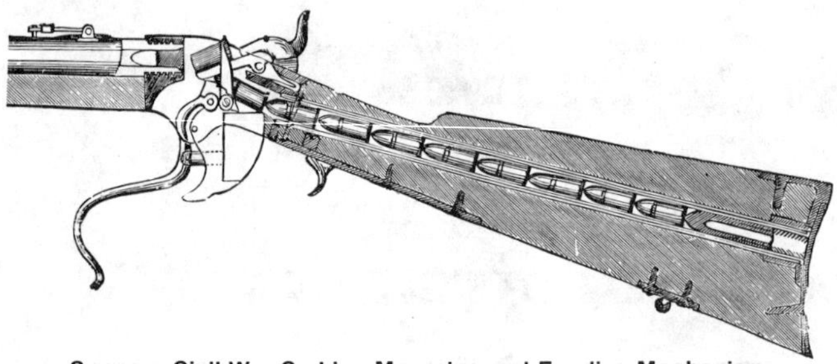
Spencer Civil War Carbine Magazine and Feeding Mechanism.

—just a simple man who sensed in the manner of the boy something the experts had not taken time to observe. That plain American doorman examined the weapon the boy carried. He watched the dummy cartridges inserted in the tube in the hollow butt, watched the lever of the crude model thrown down to feed a cartridge into the chamber, studied the action as young Spencer brought the lever back to seat the cartridge and seal the breech. Another forward movement of the trigger-guard lever and the dummy cartridge in the chamber was ejected and a second one fed; a closing movement of the lever and the breech was sealed again! Unbelievingly the humble man watched the stream of dummies ejected as Christopher Spencer worked the lever. Here was a development which might actually make killing so fast and terrifying that it might shorten the fratricidal war!

"You come back here after I am through for the day and I will take you to a man who will examine your gun." Those were the words the doorman spoke as Christopher M. Spencer recalled them in a later day. Heartened but doubting, the youngster left, alternating between hope and fear as the day dragged on.

Later that day the two made their way to the great White Mansion next door to the grim War Department Building. The man was confident that the great humble personage in the White House would have the time no Army man could spare; the boy was tense and unbelieving at the prospect of seeing the Commander-in-Chief himself.

Why President Lincoln had such confidence in the simple door tender we do not know. That he did is evidenced by his actions, for he saw them both in his chambers. His plaid shawl about his shoulders, Abe Lincoln sat gaunt and drawn, attentively watching and listening as the boy demonstrated the action with dummy cartridges. Slowly he rose. "It works all right," Spencer recalled the President saying, "but the proof lies in the shooting. Let's go out and shoot it."

As they walked down the grounds of the White House, Lincoln absently noted that the pocket of his coat was torn. "This is a nice dress for the Chief Magistrate to appear in public," he said, as he asked for a pin. While the tall gray man pinned his torn pocket, young Spencer picked a weathered shingle from the ground and stood it against a nearby tree.

With nervous fingers he dropped seven cartridges into the hollow in the butt, pushed forward and locked the feeder spring. Carefully now he swung the carbine towards the target. Quickly he thrust the lever down, catching the glint of the copper cartridge case as the spring drove the first cartridge into the chamber when the breechpin left its catch. Quickly he pulled the lever back, seeing the breechpin thrust the cartridge home as it was rocked forward behind the head of the case. He thumb-cocked the bulky side hammer expertly, aimed fleetingly, fired. Again he fired. Again, and again. The gun functioned perfectly!

The boyish inventor reloaded the weapon and handed it to President Lincoln. Gravely the great man aimed and fired; levered, cocked the hammer, and again aimed and fired. Again.

Together the two walked down to inspect the riddled shingle. Noting that Spencer had made the better pattern, Old Abe grinned slightly and in his solemn drawl said, "When I was your age I could do better." And the testing continued.

Out of that strange shooting match came the Navy Department test of June 1861 and the War Department tests of August and November that same year. The first delivery of the new Spencer .52 caliber repeaters was made on December 31, 1862. By June of the following year 7500 had been delivered. Many of these rifles were issued to the Michigan Cavalry Brigade, and a French Observer, Colonel LeCompte, wrote that at the Battle of Gettysburg, "The Michigan Brigade, armed with seven-shooter magazine guns, the Spencer breechloader, caused a great loss to the enemy, who on this account thought themselves opposed by a very much superior force."

From October 3, 1863 to August 31, 1865 new Spencer .52 caliber carbines to the number of 61,685 were delivered. Of these the Comte de Paris in a contemporary history of our Civil War wrote, "The Spencer carbine is an excellent arm whose use has spread more and more in the Federal Army...Wonderful examples are given of individual defenses due to the rapidity of fire of this arm. Many of the Federal regiments which used it were most effective."

Had the French studied the reports of their field observers on the use of the Spencer, the story of the Franco-Prussian War of 1870 might have had a different ending. Instead the overly conservative French entered that war with the hopeless paper-cartridge needle gun, the Chassepot, while reports on the marvelous new Spencer repeater moldered in their files.

During the Civil War, the North spent nearly $3,000,000 for Spencers—the arms could not be manufactured rapidly enough even to approximate the demand. The armory in the Chickering piano factory at Tremont and Camden Streets in Boston closed its doors in the fall of 1869. The war ended, its financial backers saw no further market for their arms; and that more far-seeing Yankee, Oliver Winchester, promptly bought up surplus Spencer stocks at auction and sold them from 1869 to 1872. With one move he thereby eliminated his only important competitor and managed a good business deal on a basis of hard Yankee cash.

It was only after the Winchester 1866 raised havoc at Plevna in 1877 that Europe awoke to the value of the repeater. By that time the Spencer Repeating Rifle Company was a part of history. It will remain a part of an important history as long as free men believe in individual initiative, and as long as free Americans maintain an interest in firearms.

The Spencer, then, stands as the first successful repeating arm in history, a record attributable to the strange combination of the genius of an American youth who at the age of 19 solved mechanical problems which were too much for the combined military design brains of the world in that day; of a humble and intelligent door tender who had the vision to see what the experts could not and the character and confidence to do something about his convictions; and of a truly great President without whose active support this new invention might have been hopelessly sidetracked in the vast wastes of departmental red tape.

THE HENRY REPEATING RIFLE

As we have seen, B. Tyler Henry was granted a patent for a lever-action repeating rifle using metallic cartridges, a direct evolution of the earlier Volcanic Rifle, on October 16, 1860 (No. 30446). While this was only a few months later than Christopher Spencer's patent of March 6, 1860 (No. 27393), the Henry did not receive the prompt acceptance accorded the Spencer in actual service. During the Civil War the Government purchased only 1731 Henry rifles, though approximately 10,000 more were bought privately and used by State regiments. In the march through Georgia two regiments of the command of Major General Dodge carried Henrys.

Inexplicable though it seems today, the Army at that time failed to realize the potentiality of this weapon; for every endeavor by Oliver Winchester to get Government orders or assistance to enlarge his plant facilities failed. The Spencer could be loaded in a fraction of the time required for the Henry, that is true. And it is also true that the Henry was primarily a sporting rifle. Nevertheless, once loaded it was ready for 12 rapid shots—13 if the chamber was also loaded—and it could spew bullets at the rate of some 25 per minute, factors which should have impressed muzzle-loader enthusiasts, but didn't.

The Henry magazine was below the barrel. To load, the magazine spring was first drawn up into the muzzle end of the tube. This muzzle section could then be pivoted to permit cartridges to be inserted headfirst down the tube. The muzzle section was then pivoted back in place, thus allowing the compressed spring to force the cartridges back toward the breech.

The breechblock was toggle locked. Pulling the trigger-guard lever down first disjointed the toggle to unlock it. From then on the toggle drew the breechblock back in a straight line. The rear of the breechblock rode down and thus cocked the hammer. A carrier block containing the first cartridge from the magazine tube now came in line with the breechblock. Drawing the lever back drove the cartridge into the chamber, dropped the block to receive the next cartridge, and on final movement straightened out

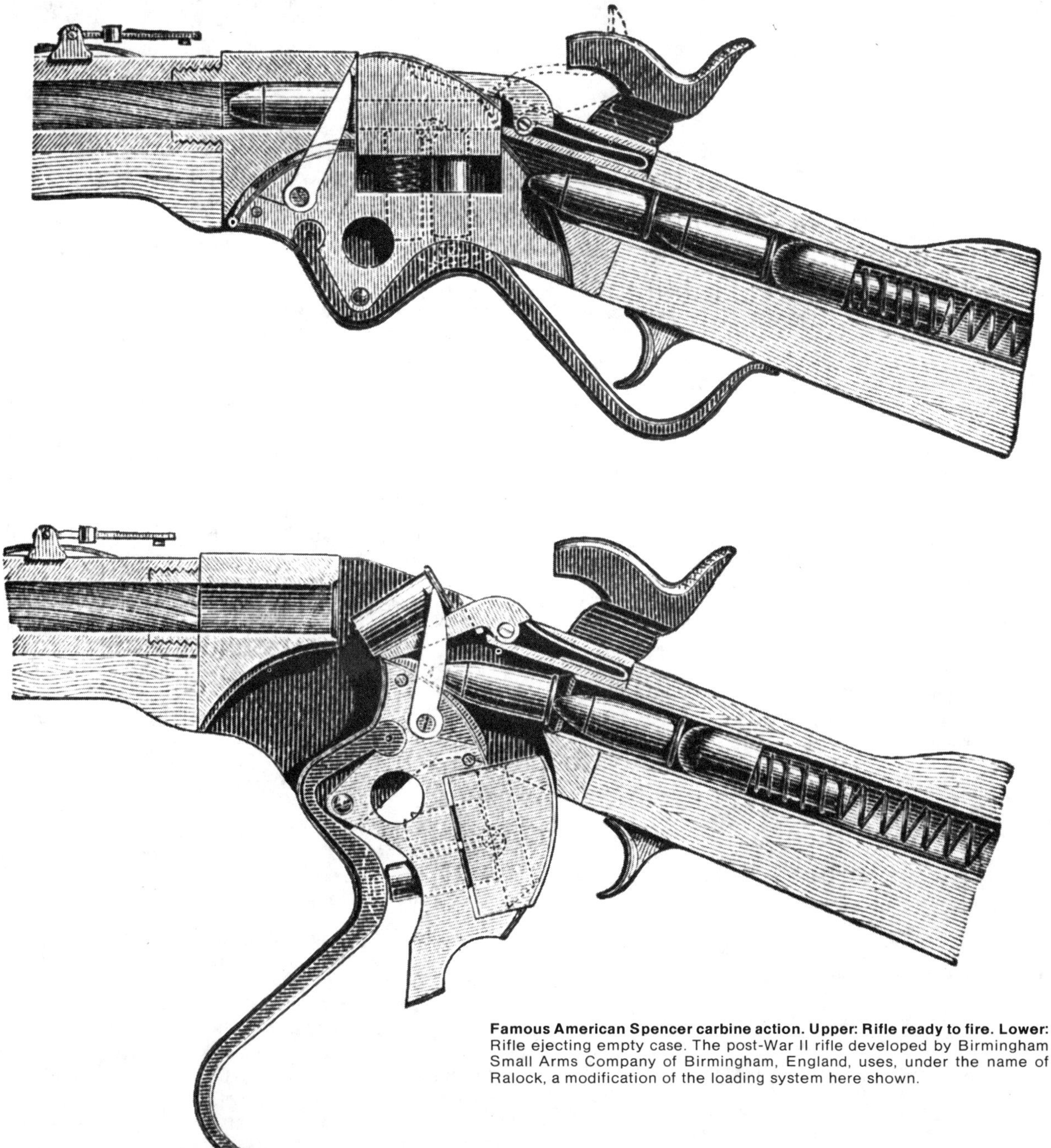

Famous American Spencer carbine action. Upper: Rifle ready to fire. Lower: Rifle ejecting empty case. The post-War II rifle developed by Birmingham Small Arms Company of Birmingham, England, uses, under the name of Ralock, a modification of the loading system here shown.

the joint in the toggle to lock the breech firmly. The trigger was now in position to actuate the sear and thereby allow the hammer to fall on the firing pin.

A few Henrys were made in caliber .44-25-216 (meaning .44 caliber with 25 grains of black powder and a 216-grain bullet). They were stamped "Henry's Patent Oct. 16, 1860. Manufact'd by the New Haven Arms Company, New Haven, Conn." These rifles used .44 rimfire cartridges in which the bullet caliber was .44. The barrel however was bored .42 and rifled with six .005 inch deep grooves with a "gain twist" rifling.

In 1860 the Henry Repeating Rifle and Carbine Model 1860 was introduced. This had the "King improvement"—the spring-loaded gate in the breech wall which allowed cartridges to be inserted from the breech end with ease.

The manufacturers, The New Haven Arms Company, dissolved in 1866.

WINCHESTER LEVER-ACTION MILITARY RIFLES THE 1866 MODEL

The Winchester Repeating Arms Company was formed in 1866 at New Haven, Conn. It was merely a reorganization of the New Haven Arms Company. Its early records and catalogs show the Henry Rifle with the King loading gate being listed as the Winchester Model 1866. One Winchester catalog states, "The latest improvements consist in an entire change in the magazine and arrangement for filling it." Actually as hair-splitting collectors will point out, it was a very minor modification of the last Henry.

This 1866 Winchester is still another American invention which helped to revolutionize warfare in its day, yet has been passed by with rather scant attention in this country. Not normally rated a military arm, the lever action Winchester 1866 was responsible for two great Turkish victories over the Russians at Plevna. That story, seldom told and little known, deserves a place in this historical outline.

Plevna—The Last Days of the Single-Shot Military Rifle

Little known today, and even less remembered, the 30th of July, 1877, cast shadows which even now are there for anyone with eyes to see. The great Russian army massed before the Turkish defenses of Plevna on the morning of that day stood in their might prepared to sacrifice themselves to the full, determined to storm the trenches ahead of them regardless of cost. They would not be halted; their front ranks would inevitably fall before the defense of the courageous Turks. They were prepared for that. So thought General Krudener and his staff as they studied the quiet fields ahead on that eventful morning. In their trenches the Turks, barely half as numerous as the Russians, waited grimly, almost eagerly, for the beginning of the assault.

The Russian Guards soberly checked their Berdan rifles, their bottle-necked .42 caliber cartridges. Other units swung up the side locking blocks of their .63 caliber Krnka breechloaders, confident that the alteration of the Bohemian Sylvester Krnka had provided them with the speed of loading necessary to match the American Peabody-Martini rifles in the trenches ahead of them.

Bugles sounded. Officers shouted. The men roared. In massed formations the long lines advanced stolidly, inexorably. The solid lines marched on until, suddenly, a cloud of smoke rose from the black-powder rifles in those trenches far ahead. Strangely and with terrifying accuracy, a plunging hail of lead ripped into those massed Russian ranks before the reports of the rifles reached them. The Russian staff stared aghast through their glasses as ranks thinned out before they were able to fire a shot. They claimed later that the slaughter began at a distance of two kilometers, some 2200 yards! Other observers on the Russian side claimed even more fantastic killing ranges for the Peabody-Martinis in the hands of the Turks—as much as 3000 yards! Perhaps the observers were too excited to measure correctly. Perhaps they were seeking to alibi their terrible mistake in ordering massed men to certain death. The true answer we can never know. But we do know that the American rifles had shown deadly efficiency at 700 yards; and it is possible that massed plunging fire at 1000 yards may have decimated the advancing Russians. Riflemen of today armed with Springfields or Garands could not hit consistently at the ranges claimed in 1877 for those old Peabody-Martinis. Of one thing, however, we are sure. At a range considered far beyond that of a rifle of those days, the Peabody-Martini began chopping down the numerical superiority on which the Russian staff had counted for victory.

With the stoic calm of the true Slav, the Russian advance continued to ranges of 500 yards, 400, 300—200. In the Turkish trenches there was a momentary hush, a pause. The Turks laid down their single-shot Peabody-Martinis. And the Russians charged madly ahead.

Russian Intelligence had duly reported the delivery of 30,000 Winchester repeaters to the Turkish cavalry. Those were Tyler Henry's lever action tube loaders with the new King patent side-loading gate added, the first arm to bear the name "Winchester." The caliber was .44 Rimfire Turkish. True, the arm had shown its merit in wild, far-away America; but what use would it be in real military combat on European fields? Little if any, they surmised.

Russian Intelligence had not learned that the cavalry had been disbanded and that their Winchesters had been issued to the defenders in the trenches. They had not known the intensive drill the defenders had been given in using the arms. They could not conceive what was about to happen ... And so the charge went on.

At 100 yards the storm broke. All down the line a hail of rapid-fire lead burst from the muzzles of 30,000 Winchesters. "Each Turk," wrote General Todleben, the new Russian commander, to General Brialmont in a letter dated January 18, 1878, "carried 100 cartridges, and had a box containing 500 placed beside him. A few expert marksmen were employed to pick off the officers ... the Turks did not even attempt to sight, but, hidden behind the trenches, loaded and fired as rapidly as they could ... The most heroic endeavors of our troops were without effect, and divisions of over 10,000 men were reduced to an effective strength of between 4,000 and 5,000." But the Russians were stubborn. Instead they repeated the attack on Plevna with the same tactics—and the same results—on September 11, 1877! In all they lost 30,000 men in the useless assaults!

Those Winchesters broke the back of the Russian attacks in their closing phases, after the long-range Peabody-Martinis had whittled the advancing ranks down with harrowing fire. The "quick loaders" fastened to the sides of the Berdans and the Krnkas to hold cartridges ready for insertion in the breech were no possible answer to the true repeating arm.

A hush settled over the chancelleries of Europe that summer of 1877. Every European nation now set itself to re-arm with repeaters as rapidly as possible. The Turks turned first to Winchester for another 140,000 repeaters, then to Germany for a long succession of advanced Mauser designs.

In this day of the nuclear bomb it is difficult to conceive the way the success of the Winchester then altered economic planning and military thinking around the globe. The hush that began over the field of Plevna and spread throughout Europe was more than just another episode in the duel between Turks and Russians. It was a hush that presaged a development of arms and a course of diplomacy leading inevitably to the years of strain that lay ahead.

Truly, the day of the single-shot military rifle was over.

WINCHESTER MODEL 1895—RUSSIAN

One other Winchester Lever Action has played a part in military usage—the Model 1895 Box Magazine. This rifle was designed for Winchester by the late great inventor John M. Browning, about whom we must have more to say when we deal with pistols and machine guns.

The Model 1895 differs radically from the other lever actions which Browning designed for Winchester and whose uses have been sporting. It differs primarily by having a built-in box magazine in the receiver directly below the line of the breechblock. The magazine was designed to hold military cartridges in single line one atop the other, the column being forced up by a spring actuated follower as in regular military bolt-action feeding.

The positioning of the magazine required the locking bolt (which secures the breechblock to the receiver when firing) to be placed to the rear of the receiver. Working out a forward locking bolt to be lever-actuated would have interfered with the simplicity and weight required. The action proved entirely strong enough for the normal military cartridges then in use (quite as does the British Lee-Enfield type with its rear-placed lugs).

During World War I, however, Russian purchasing commissions came to the United States to place orders for the manufacture of bolt-action rifles of their own pattern. One of these groups was impressed with the performance of the Model 1895 Winchester and a Russian Government order was placed for them. These rifles were chambered for the standard Russian rifle cartridge. A special alteration was made to allow these rifles to be loaded rapidly by inserting a standard Russian army cartridge clip in feed guides, then pressing the column of cartridges down into the magazine with a single thrust. This of course was merely an adaptation of the Mauser system of clip loading; but it is the only instance of its use in a lever action arm.

These Russian 1895's were used in both wars. Captured German intelligence bulletins reported them as in use on all fronts to some degree; probably, of course, indicating pressing shortages of small arms at the time of their use.

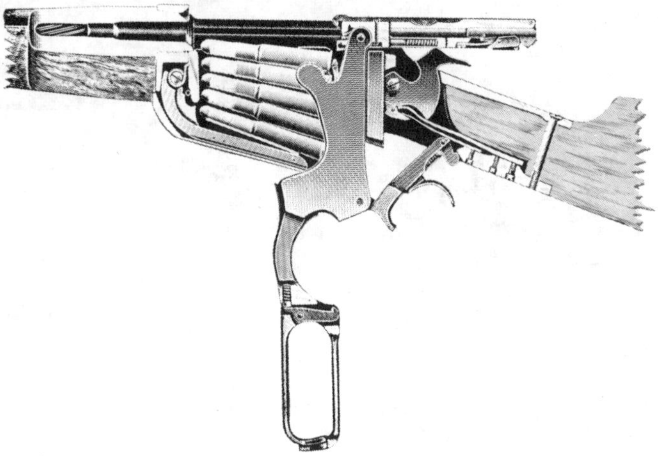

Action of the Winchester Model 1895 box magazine rifle ready for closing bolt stroke by operating the lever. This is the same as the Russian version except for the caliber and lack of clip slot loading element.

OTHER AMERICAN LEVER ACTIONS

At one time or another all United States sporting rifles have turned up in guerrilla use, somewhere on the world's frontiers. The Winchester '73, '86, and '94 have all been encountered at some time or other in the course of troubles in South America, Asia, Africa, and the Balkans. So have the old Burgess, the Colt Lightnings, the Marlins, Ballards, Savages, and even the Adirondacks and the Kennedys.

Of all these, however, only the Savage 1899 has an action strong enough for military cartridge use as we know it today; and it was the only one to receive any serious official attention. It was tested, but not adopted, by our Navy in its early development form in the year 1895.

THE SAVAGE 1899

This American magazine lever rifle merits some attention because of its magazine system and its breechlock. The magazine is a revolving box type housed below the bolt. The cartridges are separated at all times so there is no danger of deformation of the cartridges, or similar troubles. The magazine is a modification of the Spitalsky pattern which was perfected by Mannlicher and Schoenauer in Austria.

The breechlocking system is operated by lever, but can easily be altered to make a semiautomatic on the gas actuation principle. As the lever is drawn down and ahead, it first lowers the strong steel breechblock out of locking mortises in the receiver walls. Once clear of its locking seat, the breechblock is then drawn to the rear by continuing lever action to extract and eject. Rearward movement of the trigger-guard lever chambers a cartridge from the turning "spool" magazine; and then elevates the breechblock into locked position.

Experiments by Winchester and others on breech locks of this pattern have established that they are capable of sustained automatic use even with .50 caliber machine-gun cartridges.

Note on American Lever-Action Repeaters

The foregoing outline merely touches upon the successful systems or those which played an important historical role.

The author's files alone list several hundred attempts by inventors at designing lever-action repeaters for military use. It would be impossible to give space even to listing these abortive designs—though it is entirely possible that some of them may in time reappear. Inventors are forever turning up old and often discarded principles as something quite new!

The "Ball" for instance was patented in 1863 and 1002 were used by the Northern forces during the Civil War. This arm had a

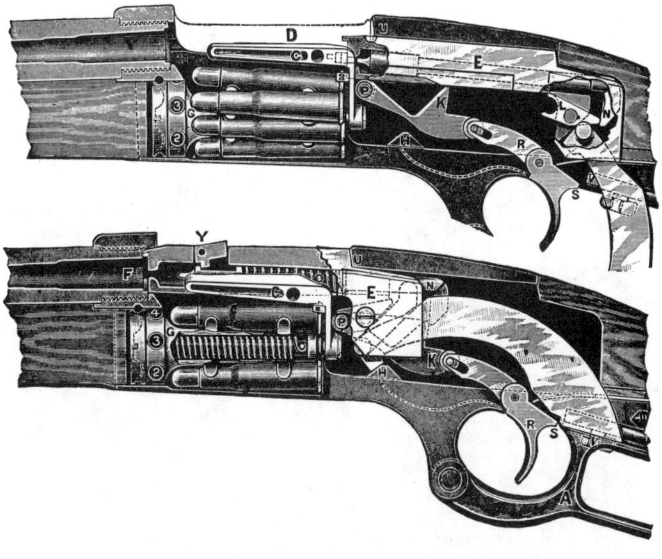

Savage 1899 Model Action.

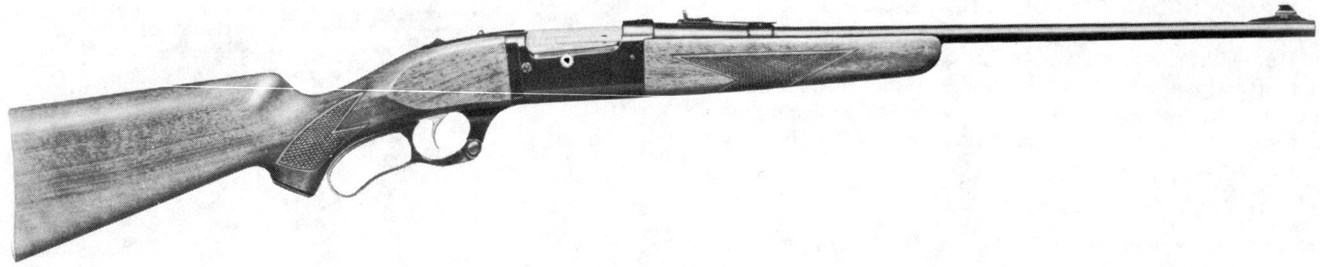

Savage Model 1899 Repeating Rifle.

below-barrel magazine tube and was loaded through the receiver. Its feeding, locking, and ejecting systems, however, were mechanical freaks, and the design failed.

One of the weirdest of all was the Evans Rimfire .44. It was patented in 1868 and was made to some extent by the Evans Repeating Rifle Co. at Mechanics Falls, Maine until 1880. This weird design had a buttstock magazine not unlike von Mannlicher's first design. The tube formed the small of the stock and extended through the buttstock to the butt plate. A four-fluted cylinder within this tube revolved one-quarter turn with each movement of the breechblock, the action being caused by a pawl. The magazine was loaded through the butt plate by dropping in a cartridge, then operating the lever. The magazine held from 26 to 34 cartridges. Loading was slow and had to be carefully done. A jam just about ruined the arm. No magazine spring was needed. These and similar oddities, however, had no part in the true evolution of the lever-action system; they were and are collectors curios only.

The Evans in its day was the object of study by some Russian designers. And, of course, the Archimedian screw feature eliminating the need for springs might some day lend itself to a buttstock magazine to hold, say a year's supply of .22's!

In any event, having covered in historical outline these lever developments, we must now turn to mention of the bolt-action development in Europe.

6

The Military Bolt Action in Europe

In any study of the development of military bolt-action rifles in Europe, an outstanding name is that of Peter Paul Mauser, a mechanical design genius of Oberndorf, Germany. His first rifle, the Model 71, while based on the Dreyse action, was truly basic, and one of the first successful metallic-cartridge, bolt-action rifles. It included automatic cam cocking, the principle of bolthead design, elastic extractor, primary extraction, ejector, manual safety, and improvement in the locking lug system. Almost every good original feature of the metallic cartridge turning bolt action design was the work of Peter Paul Mauser, who systematically developed his basic design. He received strong support from the power-hungry German government, which saw in the Mauser rifle a possible means of dominating Europe.

In Switzerland appeared the Vetterli, said to be the first successful metallic-cartridge, bolt-action rifle, which was designed by Frederic Vetterli. It used a copper rimfire cartridge and was in part based on the American Henry-Winchester rifle.

The most prolific inventor of firearms was the Austrian baron, Ferdinand Ritter von Mannlicher, who actually made and tested more than 150 types of weapons, most of them original creations. Many of Mannlicher's mechanical designs were too far in advance of the metals, machines, and cartridges of his day to be fully utilized. Austria adopted in 1886 a Mannlicher wedge-locked, straight-pull rifle with the Mannlicher magazine, using a 11mm black-powder cartridge. The Austrian 1888 Mannlicher rifle was like the 1886 pattern but, influenced by the new French Lebel, no longer used a black-powder cartridge but an 8mm smokeless.

In France the 1886 Lebel Rifle was the first bolt action to use smokeless powder. While the Lebel rifle may not have been outstanding in design, its use of smokeless powder created a world-wide stir as other countries began to redesign their firearms to emply the new smokeless powder.

Now, in the late sixties, we observe that the bolt-action rifle as a military weapon is being replaced with automatic and semiautomatic small arms by all countries which can afford to do so. In the future the bolt action will primarily be for sporting use. While their military value is waning, doubtless there will be bolt actions used somewhere in the world for many decades.

THE EARLY MAUSERS

As we have noted, one of the first successful metallic-cartridge, bolt-action rifles was designed by Peter Paul Mauser. One of his brothers, Wilhelm, was associated with him at the time in a business capacity. The mechanical brain, however, was that of Peter Paul, the youngest in a gunsmith's family of thirteen children.

Young Mauser graduated from elementary school in 1852, at the age of 14. He went to work as an apprentice in the Government Firearms Factory in Oberndorf. He was trained by his father and by four older brothers who also worked in the same factory. In 1859 he was drafted for military service. He was, of course, an expert on rifles. Military organizations being about the same then as now, it is perhaps not too surprising to find that he was assigned to the artillery.

During his military service he gave considerable attention to the Dreyse Needle gun (Zündnadelgewehr) which was then the pride of his army. Naturally, he also spent some thought on the subject of artillery.

Upon completion of his term of active service, he returned to Oberndorf where he soon developed a model of a breech-loading cannon. He had neither the financial, military, nor social station to command any attention in the artillery field, as he soon discovered. Thereupon he turned to active work on the idea of a bolt-action rifle. By 1865 he had worked out an action based on use of a metallic cartridge. An older brother, Franz, had emigrated to America and worked for Remington. Peter Paul had several sources of information on metallic cartridge developments aside from this one, however.

The unsuccessful attempts at bolt actions pointed the way for his invention, as step by step he overcame each drawback. First was his system of automatic cam cocking, whereby the firing pin was withdrawn at the opening movement of the bolt; this did away with the danger of firing accidentally as the bolt was closed, and also made way for easy extraction and fast operation of the weapon. Next was an effective breech seal and an ultra-strong lock which paved the way for the powerful military cartridges to come.

However, the Prussian authorities were satisfied with the Dreyse, so they rejected the new Mauser. Austria was interested, but had just been committed to the Wänzl design.

However, the Austrian ambassador was so impressed personally that he introduced Peter Paul and Wilhelm Mauser to an American, Samuel Norris of Springfield, Massachusetts, who was then the Remington representative in Europe.

Norris gave them a contract under which he would finance manufacture for them in Belgium, which was then as now a great small-arms manufacturing center. Norris agreed to take out patents in the United States and was to pay a royalty on each rifle made.

And that is the way in which the first Mauser came to be patented

Peter Paul Mauser, the great German arms inventor and manufacturer, 1838-1914.

—not in Germany but in the United States! The patent, 78603 of June 2, 1868, was granted for the "Norris-Mauser" in the names of Norris and the two Mauser brothers.

When the news broke, Remington quite naturally went after their Mr. Norris. Norris found himself unable to raise the capital he had promised. Thereupon he tried to interest the Prussian Government in the arm, hoping to raise money on a government contract. The Mauser contract was breached when he did not succeed.

However, Norris had done the Mausers quite a lot of good in an indirect way. The government called Wilhelm to Spandau for conferences. Out of these developed a contract which resulted in the first successful military bolt action—the Infanteriegewehr M. 71.

The success of the Mauser was immediate. Initial production was at Spandau in 1872 where the Prussian government made them and paid Mauser a royalty. The Mausers raised enough money to set up a plant, but their work was given a serious setback when the factory they opened in 1873 burned down a few weeks after the official opening. Württemberg gave the Mausers a contract of 100,000 rifles and sold them the government factory at Oberndorf. From there on development was rapid. By 1878 Peter Paul was intensively experimenting with repeating systems, having been influenced by the Turkish use of Winchester repeaters against the Russians at Plevna, which we have already discussed.

He developed a wide series of magazine systems adaptable to his M. 1871. Since all military eyes were still on the Plevna battles, it is perhaps not surprising that his first production repeater was a below-barrel tube system along the lines of the Winchester 1866. Thus in 1880 we find him adding a tube magazine to the 1871—even though in the following year Serbia was still single shot conscious to the extent of buying 100,000 Mausers of the type. These rifles were still encountered in the Balkans in World War II.

It was 1884 before the Prussian High Command accepted the Mauser tube repeater in caliber 11mm for general issue under the designation M. 71/84.

The German M. 1888 was NOT a Mauser development, but was the outgrowth of studies by an Army commission. The bolt was modified from the early Mauser, and the magazine system was taken from the Austrian Mannlicher.

The first truly modern Mauser came, of course, after the development of smokeless powder and the small-bore caliber. This was the Belgian Model of 1889, using the strip-in clip system. This was followed in 1890 by a modified version for Turkey; others for Argentina, Bolivia, Colombia, and Ecuador in 1891; by the Spanish model of 1893 which first successfully incorporated the staggered box magazine as used in our M. 1903; and by succeeding modifications culminating in the German Model 1898. (Note: Since these systems are dealt with in the later chapter on German World War II small arms, there is no need for further examination of them here.)

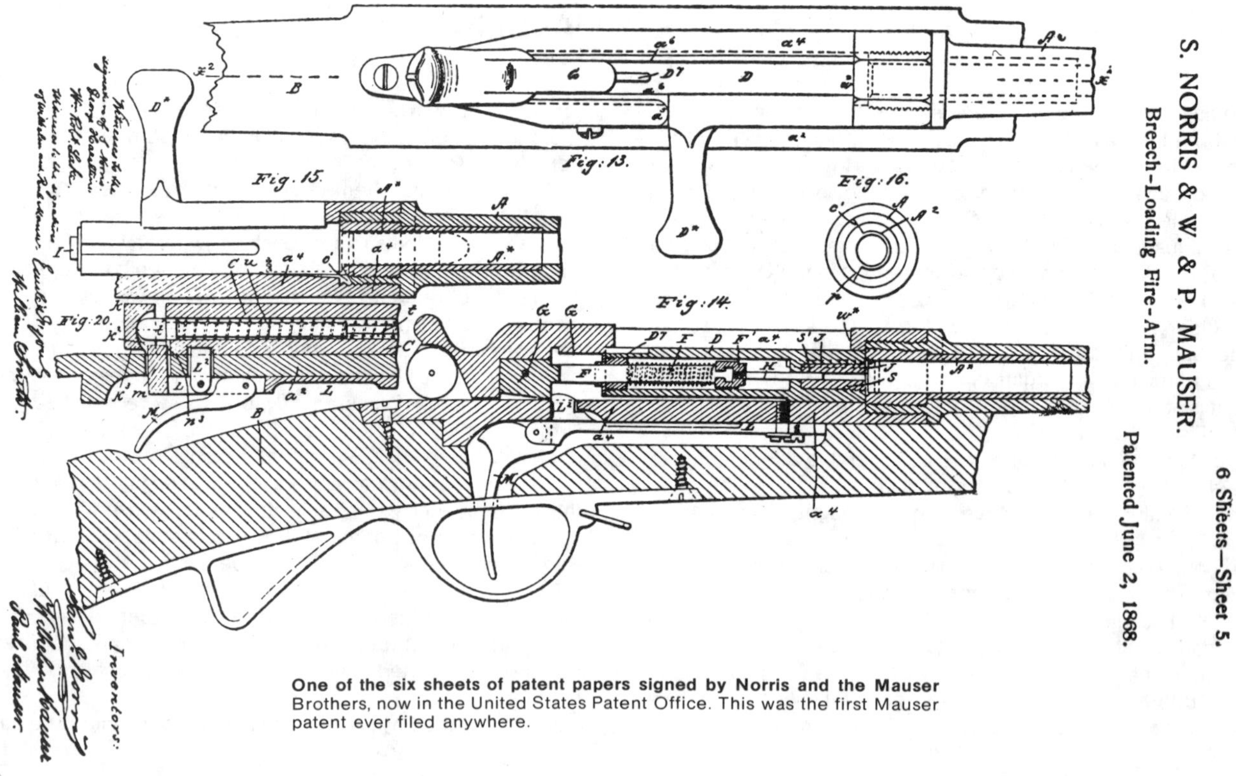

One of the six sheets of patent papers signed by Norris and the Mauser Brothers, now in the United States Patent Office. This was the first Mauser patent ever filed anywhere.

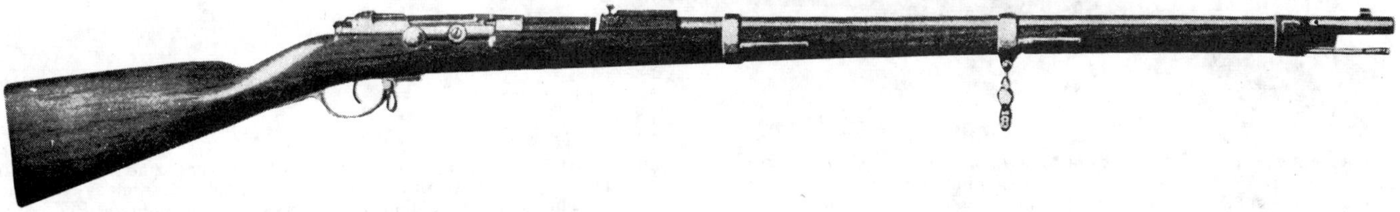

Model 71 German Infantry Rifle. Right side view with action closed showing original Mauser single shot metallic cartridge rifle adopted by the German Army in 1871 and actually introduced into service in the following year. This model used the removable bolt head and split bridge receiver later used on turn-bolt Mannlicher rifles. The German Gew. 1888 bolt and receiver were evolved from the M. 71, not from Mannlicher designs. Model 71 rifles, both original and converted types, were used in the Balkans even in World War II.

OTHER EARLY EUROPEAN BOLT ACTIONS

Having considered Mauser and his developments—some of them out of time context for purposes of continuity—we now turn to the story of bolt and repeater development in Europe generally.

SWISS VETTERLI

While Mauser invented the first successful military bolt action, his was not of course the first attempt in the field. The Swiss were actually first in the field to combine the turn bolt and the copper metallic cartridge.

Swiss ordnance men, and more particularly Frederic Vetterli, had studied closely the first American Volcanic and its successful metallic cartridge counterpart, the Henry repeater, as well as the Spencer. They had watched, too, the growing might and arrogance of the Prussian Junkers. They saw Schleswig-Holstein overrun with the aid of the terrible new Dreyse Needle gun; they also saw Austria fall before it, and saw France brace itself for the coming attack which was to be the disaster of 1870-71.

Grimly the Swiss set about to devise an arm which in case of dire peril might outshoot the dread German weapon. At this late date it is difficult to visualize the terror spread by the Dreyse, the first widely used European breechloader. But long before any major European power could think in terms of metallic cartridges, the Swiss actually had a repeater on the drawing boards, although by that time repeaters were an old story to the American sportsman. As their whole psychology revolved around defense of their precious homeland, the rifle-minded Swiss approached the design of their rifle from a far different viewpoint than either the Americans or the other Europeans. The American rifle being primarily a hunting or mounted-use arm, our inventors turned in the direction of the lever-action repeater. The typical European military mind outside of Germany thought in terms of better needle guns, or of conversions which would salvage the tremendous stocks of muzzle-loaders. German military thought itself was so oversold by the success of the Dreyse that Mauser had considerable trouble selling his first rifles! Characteristically, however, once the Prussians were sold, they went all out for further development right up to the day the United States introduced the Garand.

However, a few Swiss and Austrian designers saw the military value of the bolt repeater very early in the game. Since the vertical magazine system of Lee and Mauser and Mannlicher had not yet appeared, and since the successful American systems were tube repeaters, it is understandable that early European magazine design centered around tubes. Basically what the Swiss did was adapt the Henry-Winchester to their needs. They produced an improved rimfire .41 cartridge with copper case. They altered the King loading gate and the Henry cartridge carrier and applied them to turn-bolt actions which could be better handled prone than could the lever action.

The Federal Assembly of the Swiss Confederation meeting in solemn sessions in July and December 1866 officially approved the adoption of the repeating principle. Production of the Vetterli started in 1867, but it took two years to iron out manufacture. In 1871 and 1878 new models were produced at Waffenfabrik Bern. The rifle was used officially until 1889. As issued it had a 12-shot magazine. The caliber was 10.4mm. With 313-grain lead bullet, the rifle achieved a muzzle velocity of 1338 feet per second. Maximum range was given at about 3000 meters.

When the Swiss discarded these old Vetterlis they were sold to Bannerman in New York. He in turn sold them by mail throughout the United States. It is an eloquent testimony to the workmanship and materials of the Swiss that the Vetterli was still used in America to such an extent that cartridges were made for them in this country until just before World War II.

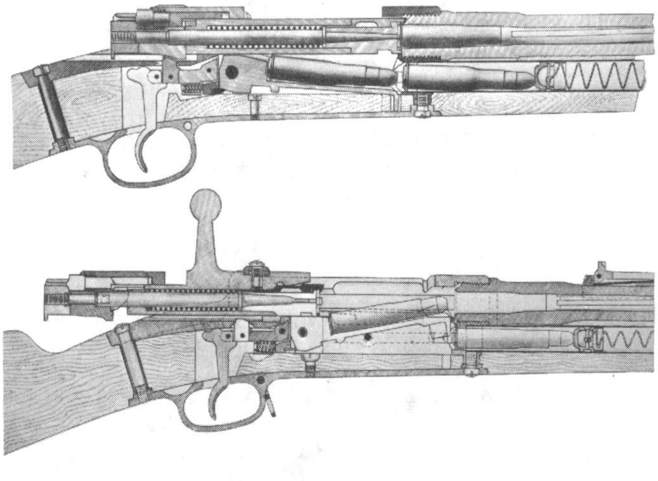

Drawing of the first Mauser Repeating Rifle, the Infanteriegewehr M.71/84. Note that this is a below-barrel tube loader.

SWISS RUBIN

The next development of interest in Switzerland was in 1883, when Major Rubin designed a straight-pull bolt system. A special revolving cam turned the locking lugs out of receiver engagement and then brought the bolt back in a straight line to eject when the operating handle was pulled to the rear. Lugs for locking were at the rear of the bolt. A special detachable vertical box magazine was provided. The design was improved through the years as the model 1911 and model 1931. The caliber is 7.5mm Swiss. The Swiss straight-pull has never been imitated elsewhere. (NOTE: Current designs are discussed later in the chapter on Switzerland.)

Model 89 Belgian Infantry Rifle. Right Side View With Action Closed. This is the standard rifle officially adopted by Belgium after experimenting with the Mauser Model 88 trial rifle. Note that in this weapon a barrel jacket is employed which covers the barrel almost to the muzzle. It was designed to protect the hands of the user from heat during firing and to protect the rather thin barrel from injury in field service. The magazine is considerably modified from the earlier design, and is adapted to rimless cartridges. Fabrique Nationale in Belgium made this model in quantity. Photograph from original Mauser design records.

AUSTRIAN FRÜWIRTH 1869

Ferdinand Früwirth's bolt-action tube repeater was in general use even before the Vetterli, though the design actually came later than the Swiss. Adopted as the official Austro-Hungarian rifle for Gendarmerie in 1869, it was generally issued in 1870. Its use was not extended to the military because when it was presented the Austrians had already been committed to manufacture the single-shot Werndl block action.

The caliber was 11mm. The bolt handle turned down very close to the stock. A cocking piece at the rear of the bolt resembled a hammer.

OTHER EARLY OFFICIAL BOLT ACTIONS

Dutch Beaumont

When Germany (Prussia) officially adopted the Mauser in 1871, all other European nations frantically stepped up rifle manufacture. Beaumont, a Dutch engineer, produced a weird single-shot bolt rifle using a center-fire 11mm cartridge. Coil springs were not in too good repute at the time due to tempering troubles. The Beaumont had a freak V-spring in the bolt handle which powered the striker. Holland adopted the design.

Swiss Vetterli in Italy

Italy adopted the Swiss Vetterli in single-shot pattern in 1871. The caliber of 10.4mm was kept, but the cartridge itself was a center-fire type. The rifle was made under Swiss license in Italian arsenals.

Russian Berdan II

In 1871 Russia and Bulgaria adopted the Russian bolt pattern known as the Berdan II, already mentioned. Caliber was .42, muzzle velocity about 1440 feet per second. To compensate for slow fire, the Krnka and other "quick loaders" were developed. These devices were designed to be attached to the side of the rifle. The theory was that as the soldier ejected the empty case he could readily pull a loaded cartridge out of the loader and feed it into the rifle. This "poor man's repeater" was a failure, of course.

French Developments

France recovered enough from the beating she had taken in 1870 so that by 1873 Chassepots were being altered to handle 11mm metallic cartridges. In 1874 the Gras conversion was adopted officially, work being done at Chatellerault and St. Etienne.

It is a commentary on French military thinking that although their ordnance groups had already done much experimentation with Winchester, Spencer, and other repeaters, they still followed German thinking on design and caliber. The French Navy in particular at this period did exhaustive and intelligent testing of repeater designs; but all that came of them were reports. The new cartridge ballistically was almost a duplicate of the German. Among the rifles tested by the French Navy was a very unusual design—a falling block action with tube magazine. Firing was by a hammer. The design was not too practical, but in it may be found elements of later automatic design by the Norwegians who developed it. This arm was known as the Krag-Petersson.

The French tested and adopted to a limited degree a Gras system in 1874. In 1878 they first adopted on limited scale a repeater, the 11mm bolt action Kropatschek. It was issued to French Marines. The design was later improved in Vienna, by Gasser. No real use was made of the alteration, however.

Russian Mosin Nagant

In 1883 Russian designers at Tula Arsenal experimented with the impractical American Evans repeater already touched upon, but wisely stopped in the experimental stages. It was not until the smokeless powder era (actually in 1891) that the Russians finally adopted a magazine rifle. This was the Mosin Nagant, a turn bolt action which has never been adapted to any other designs. (Note: This and its variations are described in the chapter on the USSR.)

Lee Magazine; Spitalsky Magazine

The year 1879 marked a great step forward in military rifle design. J. P. Lee introduced his vertical box magazine, which may have played a part in helping Mauser and Mannlicher to perfect their own designs in this line. Also in this year the first practical "spool" or revolving box magazine was introduced at Steyr Armory in Austria by Spitalsky. This was perfected later by Schoenauer and utilized by Von Mannlicher.

Serbian Mauser-Milovanovitch

Serbia in 1881 adopted a single-shot known as the Mauser-Milovanovitch. They could have had a repeater at the time, but national pride dictated otherwise because of single-shot modifications Milovanovitch had made in the Mauser action which could not readily be applied to a repeater.

Portuguese Guedes-Castro

In 1885 Portugal adopted a single-shot with a Martini dropping block action, the Guedes-Castro. Here was another instance where national pride in design overrode common sense in procurement. Even in 1904 when Portugal adopted a German-made Mauser she still had to clutter it up with "improvements" under the designation of "Mauser-Vergueiro." It and the current Mauser (98 pattern) are described in the chapter on Portugal, hence do not merit further attention here.

Austrian Schulhof

The Austrian Schulhof rifle was offered for use in 1882. This was a bolt action with thumb trigger and a most involved stock magazine of very large capacity. The side of the stock had a hinged plate which was lifted to expose several compartments for loading.

A combination of gravity and traveling rail operated by the bolt controlled the feed. Later Schulhofs used spool magazines of a more practical nature, but the designs were just too complicated to compete with the Mausers, Mannlichers, and Lees then available.

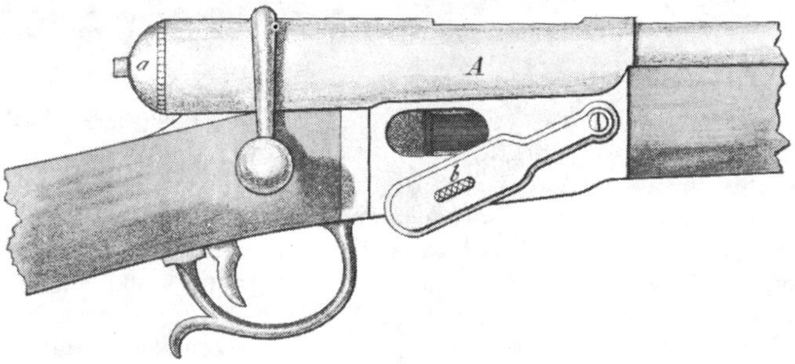

Vetterli Loading Gate, modified from American King type.

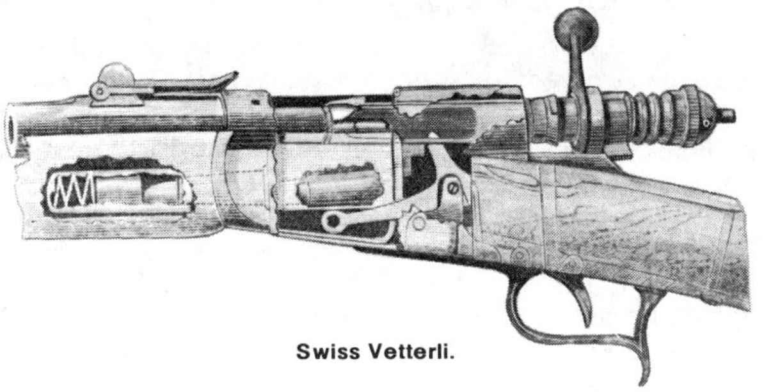

Swiss Vetterli.

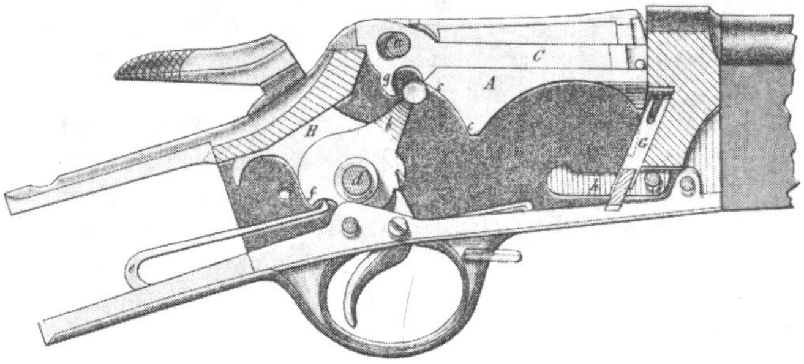

Sectional view of Krag-Petersson. Its locking and feeding features influenced later Danish designs.

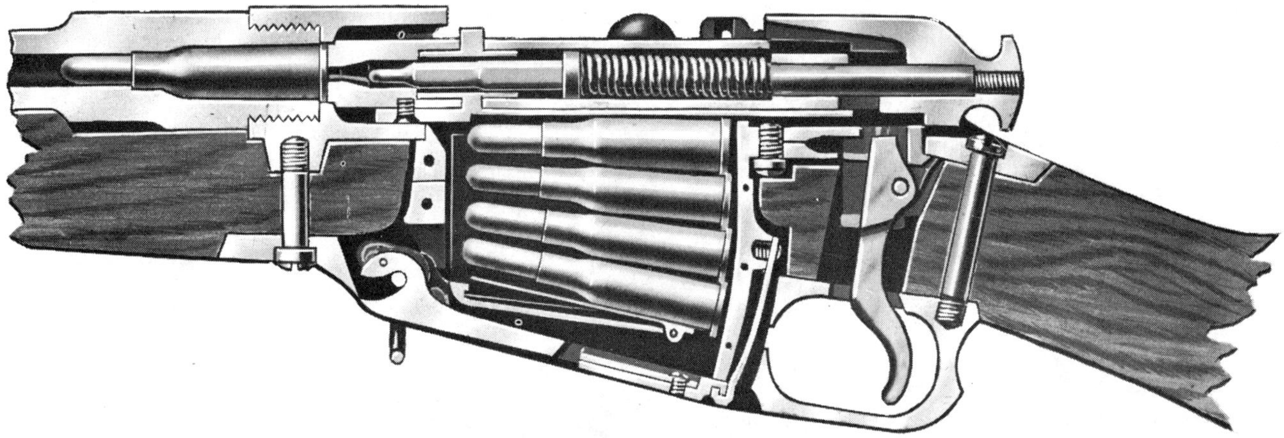

Sectionalized Russian Nagant Rifle. With very minor alterations this was the primary Russian military rifle of World War II. Also it was the basic rifle supplied to Chinese Communist forces in Korea.

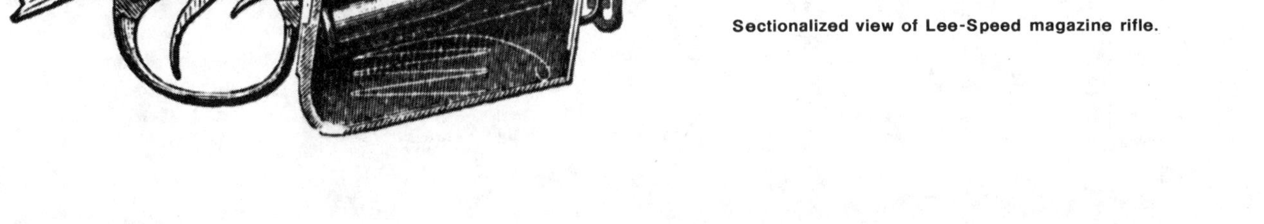

Sectionalized view of Lee-Speed magazine rifle.

THE EARLY AUSTRIAN MANNLICHERS

Von Mannlicher was one of the great firearms designers of all history. He was at all times in his thinking far ahead of the ammunition and metallurgy developments of his period. His automatic development work will be touched upon later. We can be concerned here only with his successful bolt action. All the important successful patterns are described in detail in the chapter on Austria; but historically some others require mention at this point in our historical outline. Von Mannlicher actually produced over 150 designs. Had the Austrian Government backed him up financially as the Germans did Mauser, the arms history of the world could have been far different.

Note that the dates are those of the actual introduction of the items mentioned, not the times of design, for Von Mannlicher was usually ahead of other designers.

In 1880 he produced a turn-bolt action with 3- and 4- tube magazines in the buttstock. These worked automatically on bolt operation, and allowed 15 to 20 large cartridges to be carried within the rifle. These were chambered for the Austrian M. 77 cartridge.

In the following year a new design of turn bolt was offered. It also featured a slant steel box magazine from below. The magazine was detachable.

In 1882 he introduced a simplified turn-bolt action with magazine tube below the barrel; and another one with box magazine positioned on the right top side of the receiver as in European machine-rifle systems such as the much later Bren.

In 1884 came a very unusual item—a straight pull with revolving lugs positioned to the rear of the receiver well, and with a left-side box magazine. This was the forerunner of the famous 1895 breech lock.

In 1885 he introduced his straight-pull, hinged-wedge lock pattern. Here for the first time appeared the "Mannlicher" clip system in which the cartridges are loaded in a special clip which is loaded into the magazine complete. When the last shot has been fired, the clip is dropped out the bottom of the rifle. Our Garand clip is a modification of this 1885 Mannlicher pattern.

His 1886 model was adopted by Austria—a straight-pull, wedge-locked breech system. It used the perfected form of his new clip for loading.

The next year he brought out his first models with revolving-box magazines. These used still another design of turn-bolt action. And in 1887 came the first "Mannlicher-Schoenauer" with turn-bolt lock, special spool magazine for single-shot or clip loading, and with cartridge release for rapid emptying of magazine.

The 1888 models included a new Austrian service rifle in 8mm caliber, straight pull with wedge lock; and the new German 7.9mm rimless cartridge rifle known as the Model 1888, already referred to. Here only the magazine and furniture were Mannlicher design (the breechblock having been modified from a Mauser by the German Commission); but the initial production of the arm itself was done at Steyr under Von Mannlicher.

1890 saw the introduction of an official Austrian carbine with straight-pull action and forward lug locking. In rapid succession then came developments and modifications in conjunction with other countries—Italy, France, Rumania, Switzerland, Greece and Holland.

The most widely known of all Von Mannlichers' military arms was the famous straight-pull action of 1895.

With this brief summary, then, we must pass on to other early European bolt developments.

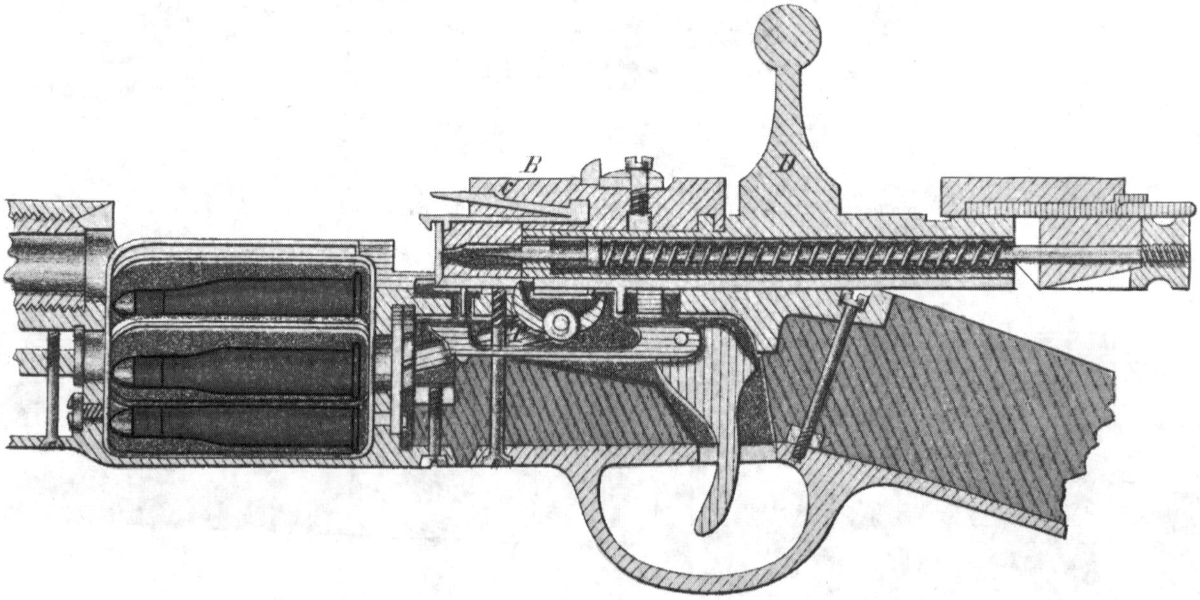

Spitalsky revolving box magazine perfected later by Von Mannlicher.

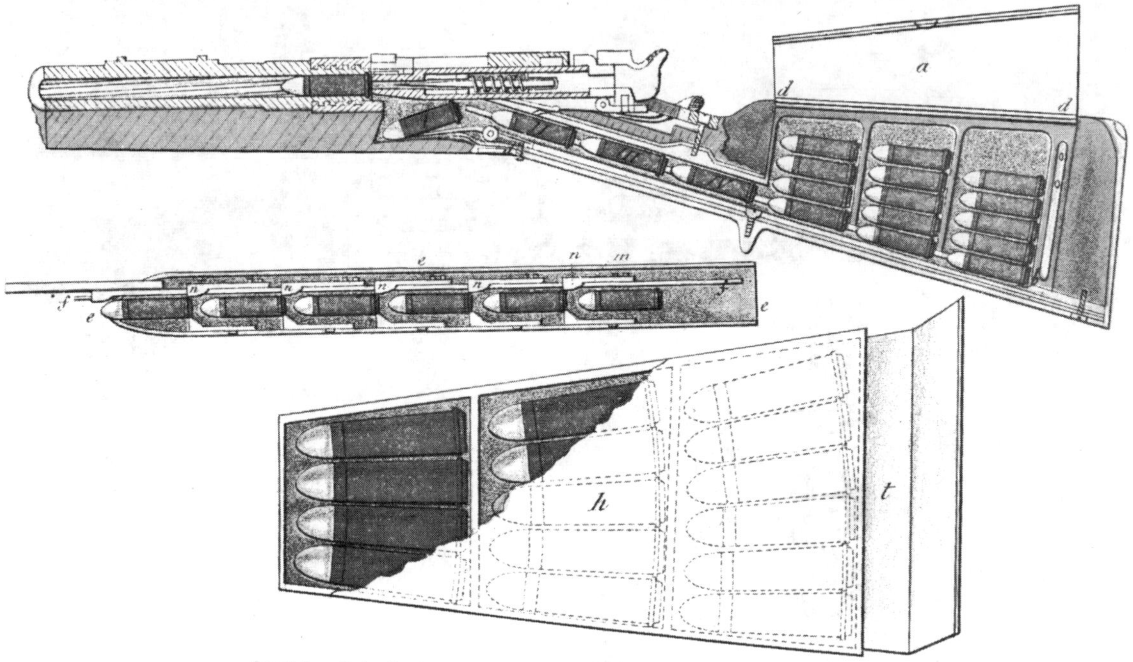

Austrian Schulhof, showing details of magazine system and cartridge transport.

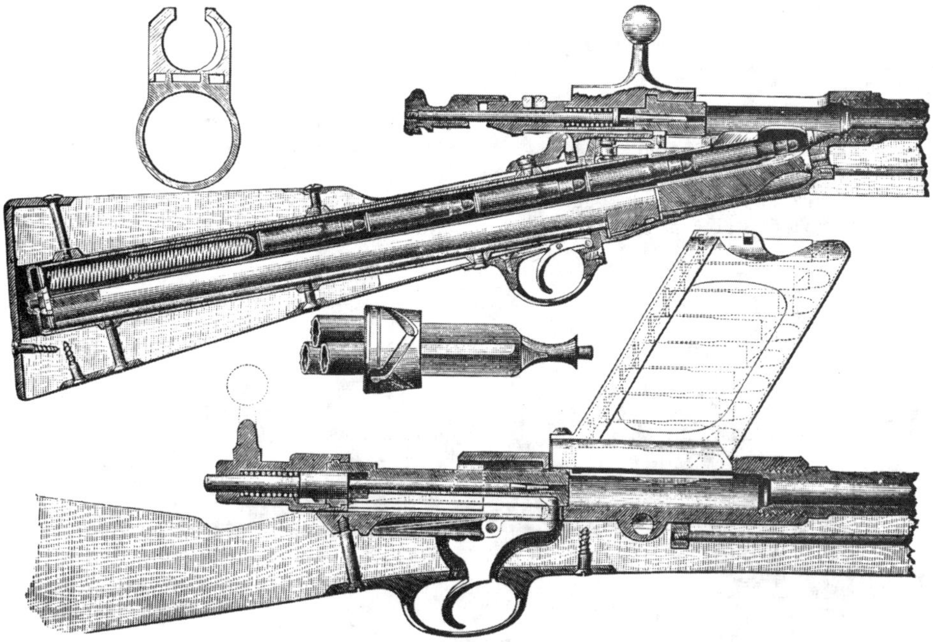

Early Mannlicher repeaters with tube and top box feeds sectionalized to show operation. Details of action and loading systems.

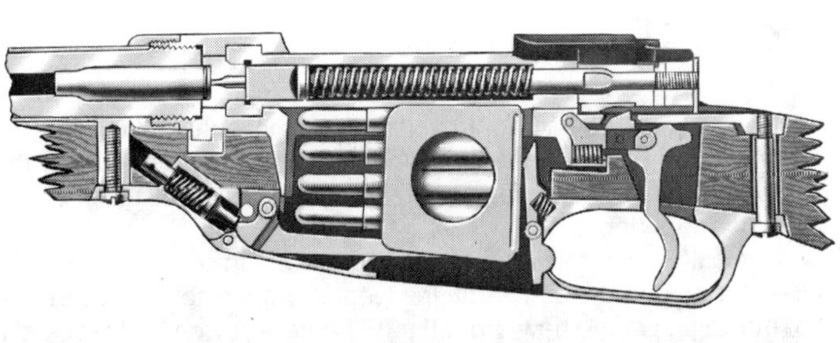

German Model 1888 rifle system. The packet loading system was Von Mannlicher's development. A variation of it is used in our M1 (Garand) rifle today.

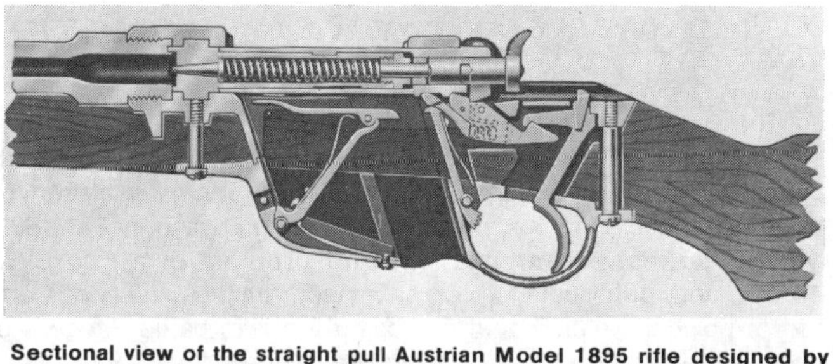

Sectional view of the straight pull Austrian Model 1895 rifle designed by Von Mannlicher. Variants of this locking system appear today in many successful machine weapons.

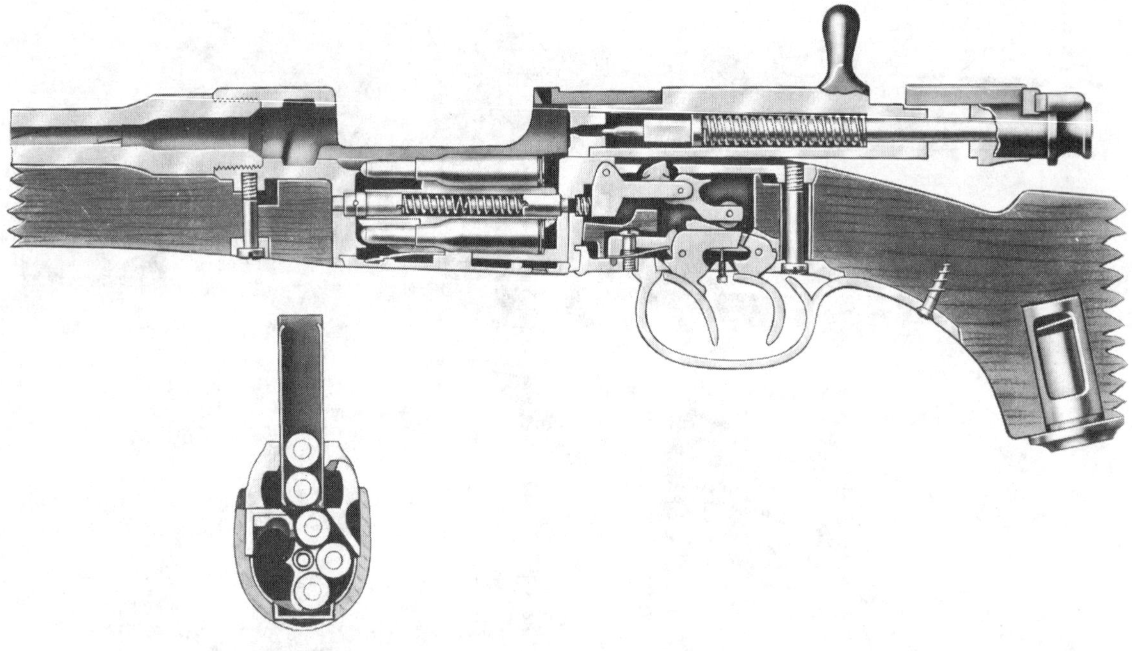

The Mannlicher-Schoenauer Rifle with turn bolt and spool magazine. Details of action and loading systems.

FRENCH 1886 LEBEL

In the year 1886 France produced one of the best-known rifles ever turned out for military use, one which is of prime historic importance as it was the first bolt rifle to handle the new smokeless powder. French chemists more than made up for the lack of ability in their arms designers. The new smokeless powder allowed the design of the first successful small-bore (8mm) cartridge for military use. This compelled a revolution in military arms design.

We must digress here for some historical notes on smokeless powder.

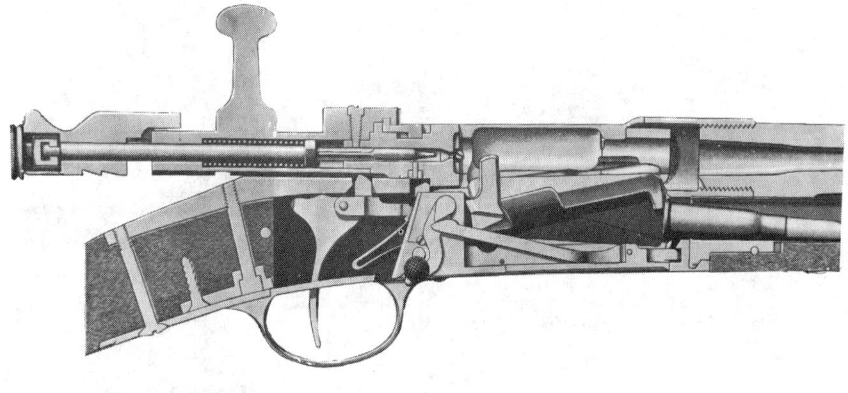

Section of French 1886 (Lebel). Tube repeating rifle showing operating system.

SMOKELESS POWDER

The first successful smokeless powder for rifles was that developed by M. Vieille in 1884 for the French Government. His successful experiments followed the trails originally blazed by earlier French chemists, notably Pelouze and Braconnet in 1832.

The earliest researches into the problem of a smokeless propellant would seem to have stemmed from the military demand for a powder which would overcome the twin battle dangers of black powder. Riflemen opening fire in the early days at once disclosed their position to the enemy, a serious matter indeed in the day of slow reloading. In volley fire—more than ever with the introduction of the successful repeating and magazine arms—the clouds of powder smoke actually hindered vision of the firers to a dangerous degree.

Ironically enough, when Vieille perfected smokeless powder it developed that the comparative freedom from smoke was but one of the values of the new propellant! The new powders possessed a chemical and a ballistic stability far exceeding those of the best black powders. The ballistic efficiency, weight for weight, was far greater, thus permitting higher velocities without excessive pressures. As a result of the increased velocity, smaller, longer, and lighter weight projectiles were necessary. The greater accuracy given by increased sectional densities of the longer bullets also changed tactics. Since the common lead bullet could not take the quick twist, jackets were introduced. A jacketed bullet was designed by a Swiss, Major Bode, in 1875. The improved copper-jacketed lead-core bullet, introduced by Major Rubin, Director of the Government Laboratory in Thun, Switzerland, in 1881, proved strong enough to stand the friction and torsion effects when a small-bore long bullet was driven down the rapid twist rifling. Rubin's original cartridge used compressed black-powder pellets. When the new smokeless powder was used instead, the results were startling. The longer range and the greater penetration of these new bullets propelled by smokeless powder completely altered military tactics.

The "invention" of smokeless powder was more in the nature of an evolution, actually. Its first reasonably effectual appearance was in 1846, when the German chemist Schönbein introduced his "cotton powder" (guncotton), although chemists in all European countries had been concerned with the problem for years. Schönbein's invention was not suitable as a propellant, but as an explosive it started an important chain of experiments. The substance was not well understood, however, and soon a terrific explosion which wrecked the plant of Hall and Son at Faversham, England was followed by similar disasters in other places. For several years both manufacture and experimentation was dropped out of sheer fear, except in Austria.

General von Lenk of the Austrian Artillery persisted in experiments with guncotton which, while unsuccessful, were of value in pointing the way for British chemists to do further experimentation. As a propellant it still proved useless, however.

Dr. Hartig was able to control the combustion rate to some extent, and in 1867 the German Captain Schultz invented the smokeless powder which in improved form was used in shotguns as "Schultz Powder." It was not successful in rifle use, however. "Collodin," Frederick Volkmann's smokeless powder patented in 1871, was the first real approach to a successful smokeless powder for rifles. The next step was the entirely controlled powder of Vielle in 1884, which was originally a mixture of nitrocellulose and picric acid. The picric acid was later dispensed with, the new powder being a mixture of nitrocellulose, ether, and alcohol. In 1847, the Italian chemist Sobrero working at Turin first produced nitroglycerin, out of which came Alfred Nobel's "dynamite" four years later. This development was not a propellant, but led the way for the mixture which Nobel patented as "Ballistite" which was a propellant. Out of Nobel's work the British Explosives Committee headed by Sir Frederick Abel developed a propulsive powder of good ballistic efficiency and excellent stability; this mixture of nitroglycerin and guncotton gelatinized by a solvent, and to which mineral jelly was added as a stabilizer and fouling preventive, was named "Cordite" because of the cord-like shape it takes when manufactured. In modified form it is still used in Great Britain.

These, then, were the basic propulsive powders with which the modern small-bore, high velocity rifles were first used. The further evolution is part of the story of Ballistics.

ADDED EUROPEAN MILITARY BOLT ACTIONS

FRANCE

France adopted in 1890 and in 1892 versions of the Berthier carbine, an individual turn-bolt breech-locking system with an adaptation of the Mannlicher clip-loading magazine system. Modifications were made from 1907 to 1916, barrel lengths varying and magazine capacities being either 3- or 5-shot. These arms, together with the Model 1886, all used the rimmed 8mm cartridge and were actually the basic rifles of the French forces during both World Wars!

In 1932 France introduced a quite new design with a turn-bolt breechlock, and a modified Mauser staggered box magazine. An improved model was introduced in 1936, caliber 7.5mm rimless.

(These are described in the chapter on France, hence we shall pass over them at this point.)

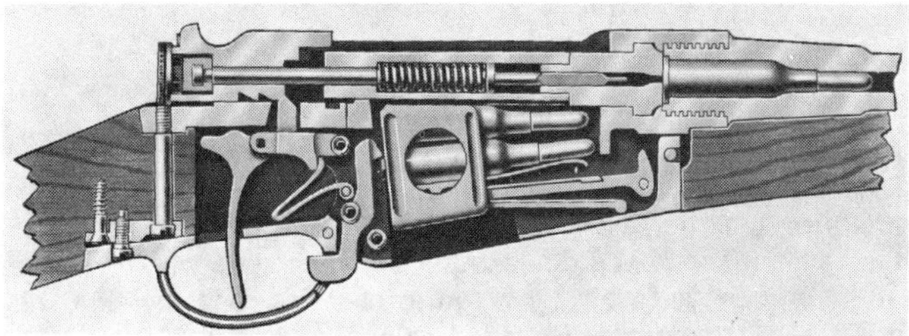

Section of representative French Berthier system carbine using Mannlicher type loading system.

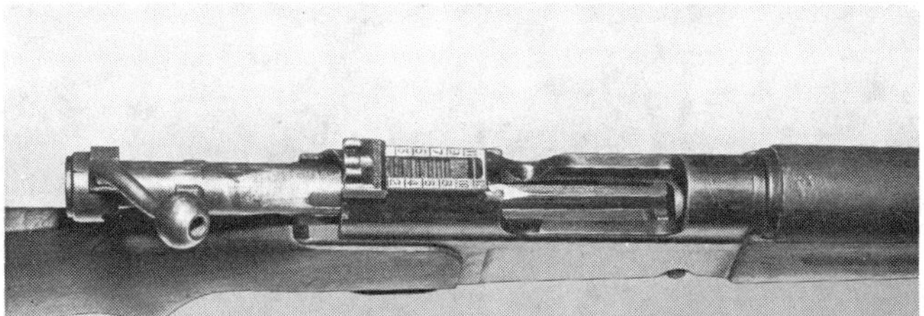

French 1936 Rifle, top action view.

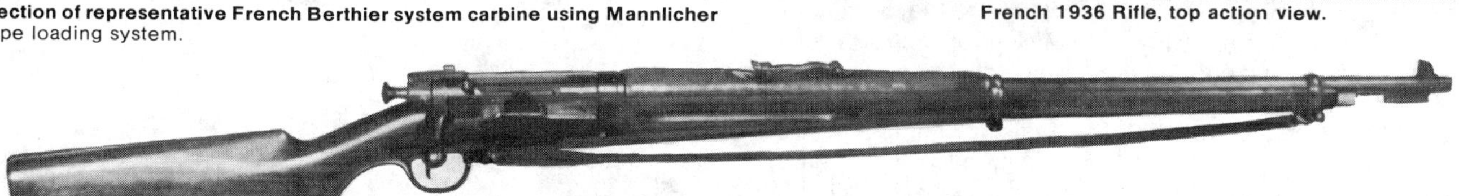

Norwegian Krag-Jorgensen, Caliber 6.5mm.

Representative Italian Carcano pattern rifle.

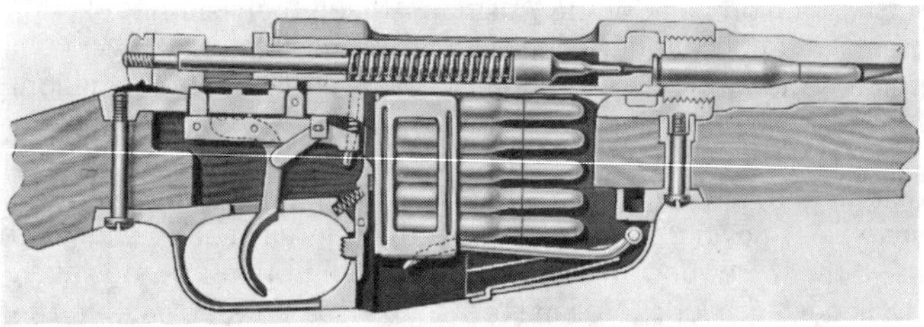

Representative Italian Carcano pattern rifle section.

NORWAY AND SWEDEN

Norway and Sweden both adopted the Jarmann in 1887, still another example of military authorities being swayed by national origin rather than by efficiency. This rifle was a below-barrel tube magazine design with a turn-bolt lock. Caliber was 10.5mm. Basically it was merely the Jarmann single-shot of 8 years earlier with a tube and lifter mechanism added. Norway, in 1894 adopted a version of the Krag-Jorgensen. Sweden in 1894 adopted the Mauser Carbine and in 1896 adopted the Mauser rifle.

ITALY

Italy adopted the Vetterli in 1871 as a breech lock system. However, they altered it to utilize a vertical fixed-box magazine in 1887. This magazine system used a special coil follower spring. This magazine is a variant of the Lee. Its inventor's name was incorporated with that of Vetterli, and the rifle was popularly known as the Vetterli-Vitali. The cartridge caliber of 10.4mm was retained, but the cartridge itself was center-fire. Magazine cutoff and manual safety were provided.

The Italians adopted the Carcano in 1891. This was a modified arm incorporating elements of the Mauser breech lock with the Mannlicher clip magazine. As the M38 in 1938 it was modified to handle a new 7.35mm cartridge instead of the old 6.5mm.

HOLLAND

Holland in 1888 adapted the Vitali pattern magazine to their old Beaumonts, the resulting hodge-podge being characterized as the M71/88. In 1895 Holland adopted a Mannlicher turn-bolt design (not the Austrian service straight pull) which was made under contract in Austria and Holland. The magazine system is the familiar Mannlicher clip type. The bolt, however, has forward lugs as in the Mauser system, but has a detachable bolt head in distinction to the Mauser which features the much safer and stronger solid head. The caliber is 6.5mm Dutch.

JAPAN

A closing word must be said for an Asiatic country which had produced military arms in the period before World War II—Japan. The first Japanese service rifle was the turn-bolt Murata of 11mm caliber. In 1887 this design was modified from a single-shot to a tube repeater and at the same time chambered for a new rimmed 8mm cartridge which was the cartridge of their China campaign in 1894. In 1897 a new rifle of Mauser pattern was designed by one Colonel Arisaka. It was the Japanese rifle in the war with Russia, and was subject to criticism as it could be incorrectly assembled so as to be dangerous when fired. In 1905 this 6.5mm caliber rifle was again redesigned. It is one of the strongest rifle actions in existence. It was the basic Japanese rifle in World War II. An altered version in 1939 increased the caliber to 7.7mm.

Netherlands (Dutch) Mannlicher turn-bolt rifle.

GREAT BRITAIN

The Lee-Metford already mentioned was adopted in 1888. The barrel had Metford rifling. The magazine was a Lee pattern holding 8 rimmed .303 caliber cartridges. It was listed in British practice as the Mark I. This was followed in 1892 by the Mark II, with double column 10-shot magazine and simplified bolt. In 1892 the Mark I was altered and was listed as Mark I*. The Mark II* of 1895 was the same as the Mark II except for a safety catch at the rear of the bolt.

The first of the long line of British Lee-Enfields—the real work horses of the British in both World Wars—was introduced in 1895. It was identical with the Lee-Metford Mark II* except that the rifling was the new Enfield pattern and the sights were changed. The L. E. Mark I*, just to confuse the soldier, was the Mark I without cleaning rod!

The succeeding Lee-Enfields and other British service designs will be found in detail in the chapter on Britain and the British Commonwealth. However, mention should be made of one colonial arm, the Canadian Ross. This was adopted in 1905 in caliber .303 to at least make a common ammunition supply with the mother country. It was an unusual straight-pull arm with many of the strength features of the Mauser design. Bolt lugs were revolved by cam action on straight pull and thrust to unlock and lock the lugs into receiver seats directly behind the cartridge head. It was modified in 1910. Short trench usage in World War I established that the principle was not adequate for war use where regular attention cannot always be given to cleaning, and the rifle was replaced by the British Short Magazine Lee-Enfield.

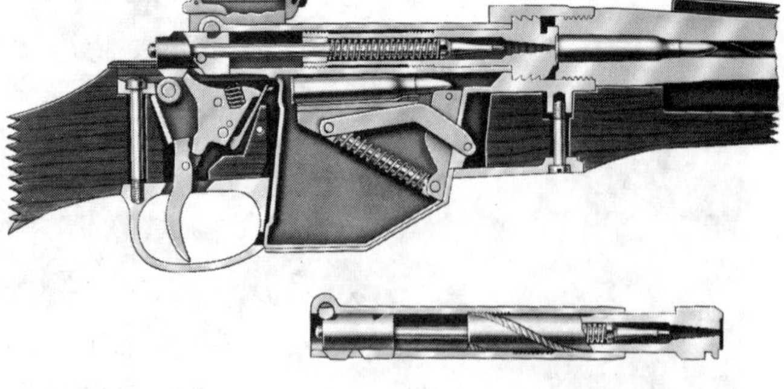

Representative Canadian Ross straight pull rifle and bolt in section to show operation.

7 Semiautomatic Rifle and Carbine Development

A semiautomatic rifle (often popularly but erroneously called an "automatic rifle) is actually a self-loading arm. In appearance, general design, and weight, it approximates the standard manually-operated type of rifle. Its magazine system in military usage is commonly only a variation of the box magazine found in manually-operated rifles. In commercial arms it may be box or tube magazine fed, however.

It must be hand loaded for the first shot. For succeeding shots, however, only an individual pull on the trigger is required. A very small part of the energy developed as the cartridge is fired is made to unlock the rifle, extract and eject the empty case, cock the firing mechanism, chamber a cartridge, and close and lock the breech.

Successful military systems have narrowed down the types of operation. Gas through a barrel hole is by far the most efficient.

The gas cylinder may be below the barrel as in the case of the U.S. M1 (Garand) or may be on top of the barrel as in the case of the Russian rifles. In any event, a small hole is commonly drilled in the barrel. As the bullet passes over it, gas under pressure is passed through the hole into the cylinder where it drives the operating piston. The piston is connected or in contact with the bolt or other breech-locking mechanism. The rearward thrust of the piston does all the work normally done manually in getting ready for the next shot. In some designs, no piston is used; the gas itself goes through a tube and acts directly against a bolt carrier. A disconnector mechanism is incorporated in this design so that the trigger must be deliberately released after each shot before it can be pulled again for the succeeding shot. This is necessary to prevent the gun from mounting out of control in full automatic fire. When, as in some instances, 20-shot detachable box magazines are supplied with such an arm, a trained man can fire as many as 60 aimed shots a minute.

A CARBINE is usually defined as a short rifle, generally one having a barrel length of less than 22 inches.

SEMIAUTOMATIC RIFLES

EARLY DEVELOPMENTS

Hiram Maxim

The American, Hiram Maxim, in his scientific experiments in 1881 and 1883 actually produced the first practical method of semiautomatic operation in a rifle.

While one of those early experiments was merely occupied with converting Winchester lever action repeating arms by linking up levers and springs with the recoil, these experiments pointed the way for Maxim's successful introduction of true automatic operation.

It was out of these experiments that Maxim in 1884 developed his original basic patent for locked-breech recoil operation. This recoil principle has proven the most efficient one for machine-gun use where high rate of fire and continuous volume is essential and where factors of weight and bulk are not the primary consideration.

With the sole exception of Johnson, the American, no semi-automatic rifle of high power has been produced in any quantity using recoil locked-breech operation. The Browning long-recoil hunting rifles all use medium-power cartridges. Because the desirable weight of the military rifle is preferably well under 10 pounds, and because military rifles require a minimum of exposed operating parts which may be affected by dirt or otherwise fouled in action, further application of recoil operation to military rifles is unlikely.

Maxim later developed one of the earliest gas-operated rifle designs which, like his recoil system, formed the basis for scores of related inventions. In the years between, designers throughout the world were widely influenced by all his automatic developments and he may be credited indirectly for scores of imitation designs at this period.

Von Mannlicher

In 1885 the Austrian, Von Mannlicher, following close on the heels of Maxim, produced a short-recoil rifle. It was a crude arm, but could be fired either semi- or full automatically. It incorporated possibly the first appearance of the pivoted accelerator for speeding up breechblock travel after the unlocking motion, the system found in the Browning machine gun. The breech-locking

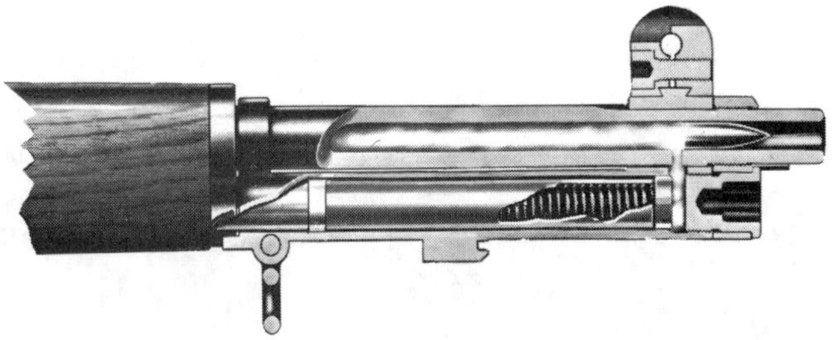

Garand Gas System. Gas passes through port in barrel after bullet passes over it. Expanding gas acts against piston and springs to operate weapon. Variants of this system represent the most successful form of actuation in light weight military rifle caliber arms.

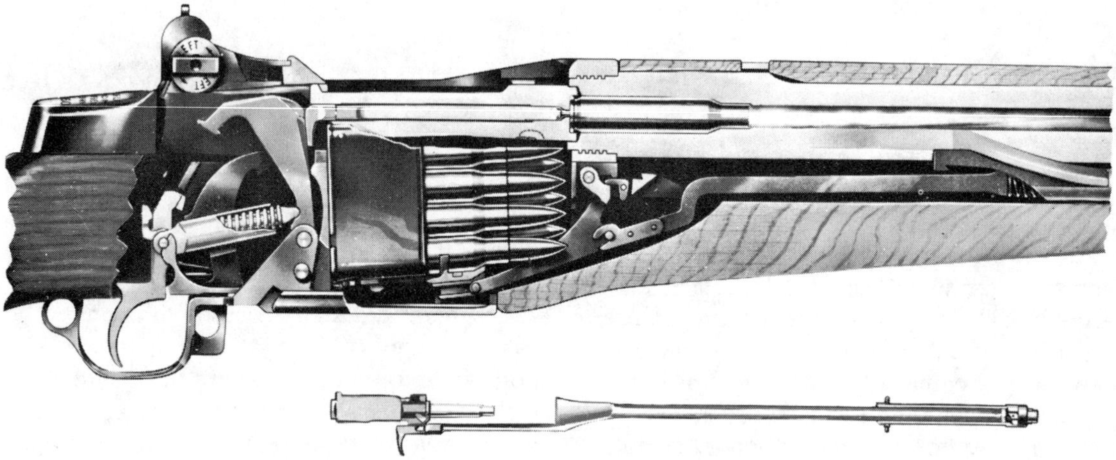

The U.S. M1 (Garand) in section. The trigger action requires release of finger pressure after firing each shot.

was effected by the camming up and down of a locking block from the floor of the receiver into the lower side of the breechblock. Cocking was on rearward motion by a lever which entered the breechblock and engaged a notch in the striker. The recoil spring was mounted in the receiver behind the breechblock. All these details are strikingly similar to those in the modern Browning machine gun. Since Browning did much of his design work in Belgium and Von Mannlicher's efforts were well known and carefully followed there, it seems quite likely that the genesis of many of the Browning principles actually came directly or indirectly from a study of this earliest of the Von Mannlicher automatic weapons.

Other Developments

Of the scores of developments in this field in the next few years, however, none was successful and few were worthy of even passing attention. The Schlund & Arthur British patent of 1885 deals with a long-recoil method of functioning, a form which has proved practical only in connection with shotgun and sporting rifle production. Most developments at this period were efforts to apply the automatic-loading principle to the military arms of the time. In England, for example, Needham & Paulson produced automatic unloading and cocking of the British Martini-Henry. Since this is a lever-action single-shot rifle, one which would be most difficult to operate from a magazine, the value of their efforts may be readily assessed.

The black-powder ammunition of this period precluded truly successful automatic development in any event. It was not until 1886, when the French Lebel rifle cartridge, using a rimless solid-drawn brass case, a cupro-nickel jacketed bullet, and a load of smokeless powder was introduced, that the semiautomatic rifle became really a practical design for experimentation.

Successful military weapon functioning requires that while the high point of pressure must pass quickly after firing, residual pressure in the barrel must still be sufficient to operate the action. Black-powder pressure dissipates too rapidly to meet this requirement.

Maxim's Gas Operated Rifle

In 1891 Hiram Maxim received British patent number 22859 which dealt specifically with a rifle, rather than with the machine guns to which most of his attention had been directed. This Maxim rifle was an application of the short-stroke gas piston principle of operation. By bleeding off gas through a port quite close to the chamber, Maxim utilized the expanding gas to operate a piston in the British Martini-Henry. Since his attention was basically given to his now successful machine gun, Maxim did not follow through on this gas rifle development, which, had it been applied to a magazine arm, might have been of tremendous importance.

Browning's First Automatic Rifle

John M. Browning's initial attempt at automatic rifle development followed pretty much along the line of Hiram Maxim except that he utilized gas instead of recoil to operate a lever-action. Variations of this muzzle-cap design, which does not require

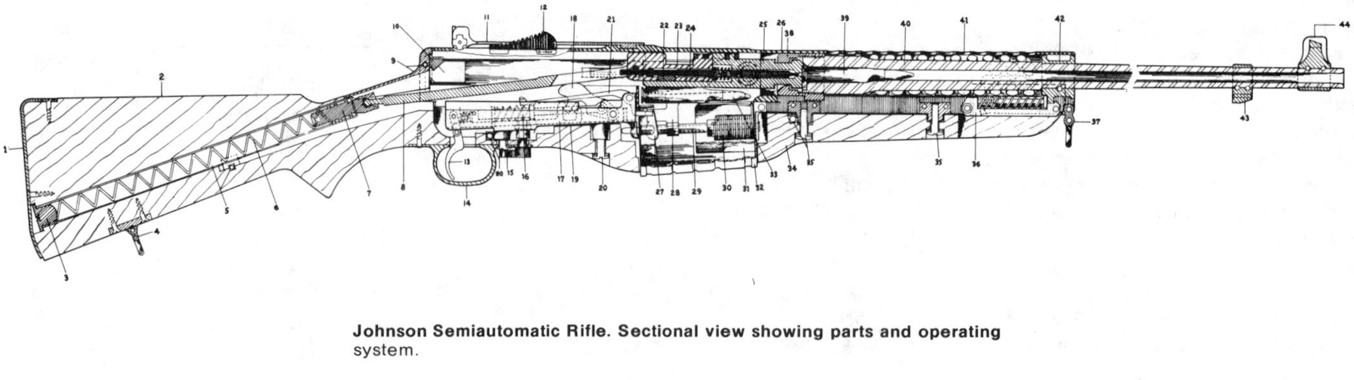

Johnson Semiautomatic Rifle. Sectional view showing parts and operating system.

Gas Operated Danish Bang rifle (circa 1927).

drilling the barrel, have been tried at various times since Browning's original experiment, but the mechanical complications of the system offset its advantages.

Browning's next model (see the chapter on the development of machine guns and automatic firearms) used gas taken off through a hole in the barrel to operate an under-barrel piston. This design is actually a true forerunner of the modern military semiautomatic rifle, though its piston operation differs.

Browning utilized this gas principle for the production of the Colt machine gun and at the time did nothing about it in connection with rifle development.

Von Mannlicher's Semiautomatic Rifles

In 1891 Von Mannlicher in Austria introduced two rifles which were modified versions of his 1885 design. These rifles had very many of the most modern design characteristics, including a standard Mannlicher box magazine system which was clip-loaded through the top of the open action, a hold-open device to keep the breech open when the last shot was fired, and a standard pattern cocking handle on the right side of the weapon. Lack of military interest in Austria prevented general manufacture of either rifle. The inventor was ahead of his time.

Von Mannlicher in 1893 again gave evidence of his protean design ability by introducing two new semiautomatic rifles. One of these was an adaption of his turn-bolt magazine rifle to semiautomatic action by adjusting the pitch of the normal locking lugs and their chamber seats. Actually the principle he utilized here was on the hesitation system, later found in the Villar Perosa and similar machine weapons. The opening action began immediately on the discharge of the cartridge, but the turning motion of the lugs in their helical grooves slowed the opening appreciably. However, even with the cartridges of the pressures of those days, breech opening was much too rapid for good operation.

The second rifle was an adaptation of the same lug principles to his straight-pull rifle. Again, the breech was hesitation operated and not fully locked against the discharge. These Von Mannlicher designs were actually the forerunner of the much discussed Blish principle used in the first American Thompson submachine guns at a much later date. They directly influenced the later machine-gun developments of the Austrian, Andrea Schwarzlose, also. By properly adjusting the angles of the receiver cuts, Von Mannlicher sought to produce a condition where at the moment of firing the lugs would jam in their seats, serving as an actual full lock. As the pressure fell rapidly, and the residual gas continued its thrust against the inside of the case head, he sought to produce a condition where the friction would lessen gradually, and the action would open without too much violence. While devices of this sort do serve to slow down the action somewhat, the function is still frequently too violent for application to the standard patterns of military rifle cartridges.

In 1894 Von Mannlicher came forward with two new designs, this time solid-breech actions with barrels blowing forward. The breech of the action being solid, as the weapon fired the movable barrel was thrust ahead in its guides and turned. Note that these arms were not true locked-breech weapons in the common sense of the term. The blow-forward systems slowed recoil and absorbed a great deal of energy as the heavy barrel went forward and was literally pulled off the fired case to allow ejection. The later Schwarzlose automatic pistols utilized a variation of part of this principle. The Schwarzlose barrel, however, did not turn.

(Also in 1894, the English Griffiths & Woodgate short-recoil rifle was introduced. This arm showed many promising possibilities, but lack of interest by the military halted further expensive development on it.)

In the following year Von Mannlicher introduced a gas-operated rifle with a true locked breech. This was a piston-operated rifle with a below-barrel gas cylinder. The bolt was cammed to the left on a hinge to open the breech by operation of the connected gas piston. A standard Mannlicher clip was used. This rifle utilized in elementary form many of the best operating principles found in our modern Garand military rifle.

The year 1900 saw the introduction of still another Von Mannlicher rifle. For the first time he utilized a revolving box magazine, loaded, however, from a strip-in clip. This was a locked-breech arm operated by gas taken off close to the breech and utilizing a short-stroke piston. A cylindrical bolt was actuated by a stud on the operating rod, rather a forerunner of the later Lewis machine-gun principle.

Italian Developments

The Italians had done considerable experimental work on semiautomatic weapons during the years after Maxim's initial work, but it was not until 1900 that a really worthwhile design was produced by them. Major Cei-Rigotti used a gas cylinder, with port in the barrel, and a tappet system for operating the bolt. This

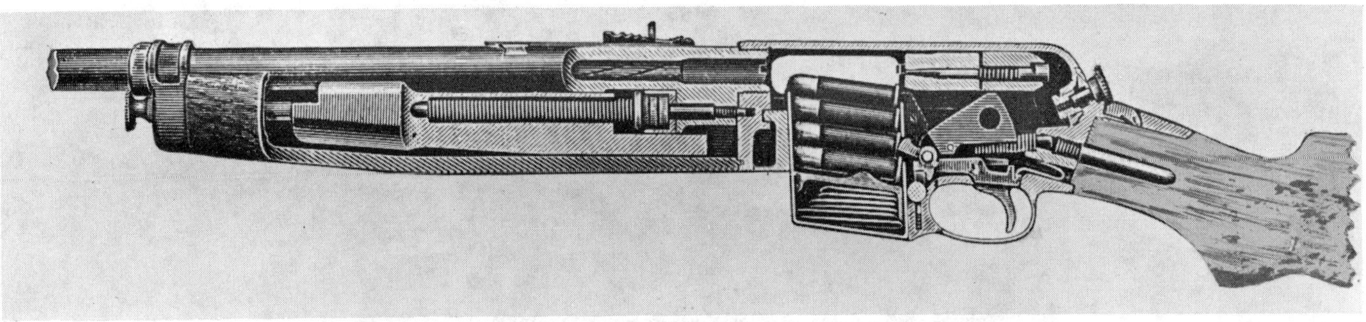

Winchester Semiauto Blowback, Cal. .401 designed by T.C. Johnson, Action open.

system was used in World War II by the Russians and the Germans in their successful semiautomatic rifles and is present today in the F.N. light auto rifle currently being used by many NATO nations.

Unlike the later adaptions of the tappet principle, however, this early Italian rifle hooked the system up to a rotating bolt with dual locking lugs at the forward end as in conventional practice. This rifle utilized the hold-open principle when the last shot was fired. Its box magazine was loaded from strip-in clips. Special magazines were provided holding as many as 50 cartridges. In one test in 1900, 300 rounds were fired in a model equipped with a full automatic switch. The 300 were actually fired in about one minute! The barrel heated so badly that it could not be used further.

Mauser's Semiautomatic Rifles

While Mauser had done considerable experimental work since the introduction by Maxim of the short-recoil principle, it was not until 1898 that his first production rifle of semiautomatic operation was generally exhibited. This was a short-recoil operation using the standard Mauser box magazine. The arm is of interest basically because the locking principle of the breech was a development of the early Swedish Friberg principles.

1902, four years later, saw the introduction of still another German Mauser recoil rifle for military cartridges. This however, was a long-recoil weapon which was on the principle later adapted by Browning to Remington and FN commercial autoloading rifles. In some respects this early Mauser followed the Roth Austrian design of 1899, but the locking mechanism was quite distinctive. Mauser sought to make a practical military arm by using a box magazine, a cylindrical bolt securely locked to the barrel receiver by two locking lugs, and utilizing a striker firing design. The only truly successful application of this long-recoil principle to rifles is that of John Browning in the commercial medium-power rifles already mentioned; as a military design it has been a failure.

Mauser also introduced in 1902 still another semiautomatic military rifle operated by gas trapped at the muzzle in a blast cone. The gas was backed up to operate a rod on the piston principle to function the action for unlocking. (In World War II the Mauser factory introduced the experimental G 41(M) utilizing a slight variation of this gas principle. The design was not successful. The tappet system developed at the Walther plant along the old Cei-Rigotti line was a very efficient low-cost arm late in World War II.)

Winchester Self-loading Rifles

In 1903 a series of self-loading rifles was developed at Winchester by Thomas C. Johnson, one of the best of the commercial designers. These were elementary blowback arms of quite ingenious design but have no true military application, though in larger calibers they have been extensively used for police work and some gendarmerie operations.

Swedish Rifle

Kjellman's development of the earlier Friberg principle was used in a rifle in 1904 in Sweden, in caliber 6.5mm Mauser. The operation principle was short recoil, the truly interesting feature being the locking mechanism which we have already mentioned.

Japanese Nambu Semiautomatic Rifle

Major Nambu of the Japanese Artillery also exhibited a semiautomatic rifle in 1904 which proved unsuccessful. However, Nambu turned his ideas to the development of machine guns along the same general lines. Actually these were variations of the French Hotchkiss machine-gun principle. The only reasonably successful Japanese self-loading rifle was a World War II imitation of our Garand with modifications.

Remington

The Browning-designed Remington rifle introduced in 1906 was a long-recoil sporting rifle of medium caliber, employing the turning-bolt principle and long-recoil action already mentioned as having been earlier conceived by Mauser. It had not enough military significance to warrant attention, though it has been used extensively for special police work.

Danish Semiautomatic

In 1911 the Danish Bang Rifle was tested by the United States Ordnance Department and showed considerable promise. This operated on the principle of the cone about the muzzle trapping the gas and pulling the cone ahead to draw forward an attached operating rod which operated the action. This, of course, was a variation of the original rifle experiments of both Maxim and Browning. (An improved form of the design was again offered in

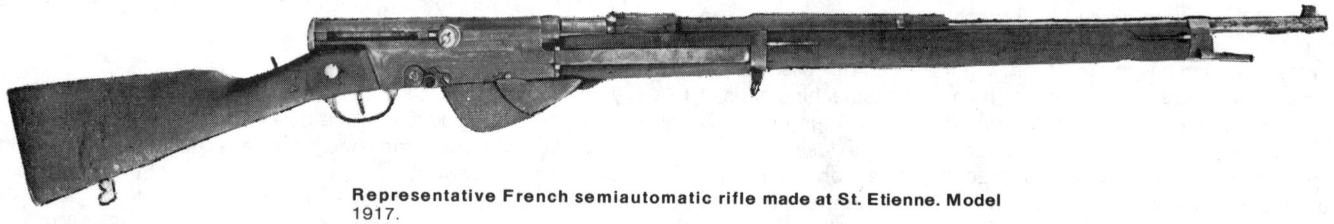

Representative French semiautomatic rifle made at St. Etienne. Model 1917.

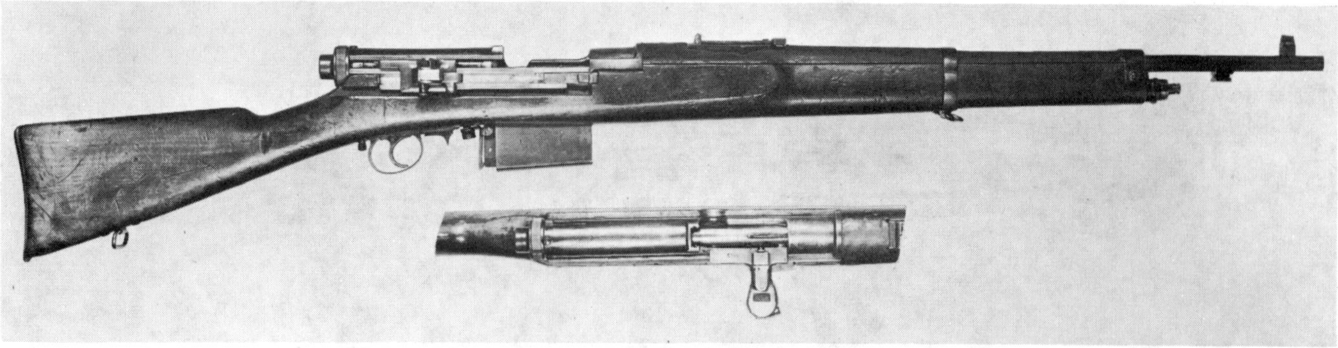

Mondragon Rifle. Side and top views. Made in Switzerland. Invented by Mexican General Mondragon. Used early in 1914 by German observers in airplanes before machine guns were mounted.

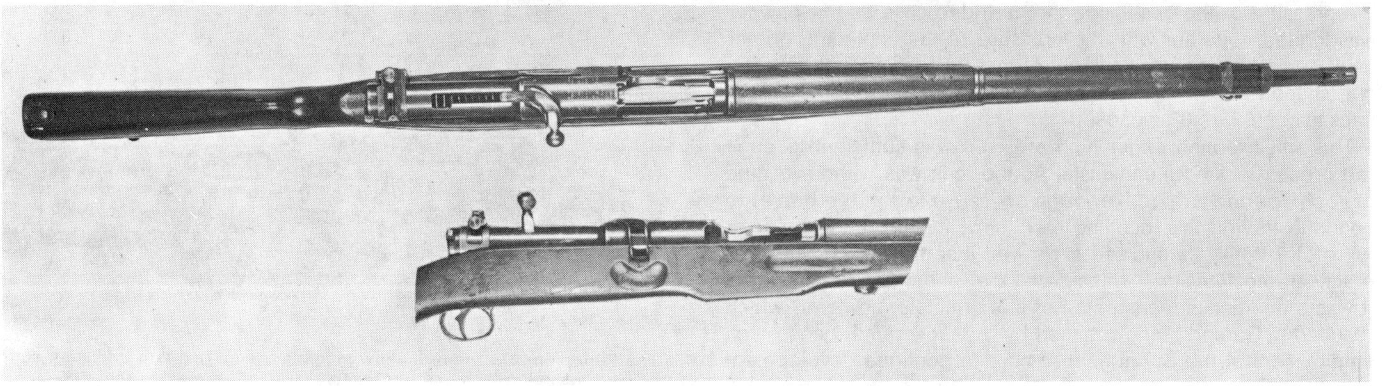

Thompson Semiautomatic Rifle, Caliber .30. Top view closed. Insert shows breech open.

1920 as modified by U. S. Captain James Hatcher. However, high production costs were envisioned and the design was passed over.)

French St. Etienne Rifles

In 1894 the French began their experiments with semiautomatic rifles, out of which came the later St. Etienne rifles. These designs, named after the arsenal of manufacture, were used to some extent by the French in 1917 and 1918 and were gas-operated, rotating-bolt rifles with box magazines. The operating rods were exposed on the right side of the forearm in these rifles, a condition which led to considerable field trouble. However, they represented the first actual use of semiautomatic rifles in general combat, though Mondragons and others had been tried earlier by the Germans for special duty use.

Mexican Mondragon

Passing mention must be made of this Mexican Mondragon. Invented by a Mexican Army officer, the rifle was produced in small quantity in Switzerland. In several instances early in World War I, observers in planes were armed with these rifles. The Mondragon had a box magazine, a gas port about three-quarters the distance to the muzzle, and used a 10-shot box magazine. As in the case of all Swiss manufactured arms, the workmanship was excellent. The Mondragon was in production at the beginning of the war in limited degree, and actually saw battle service earlier than did the French, but the French rifles mentioned were the first ones to receive actual tactical battle tests.

The Garand Semiautomatic Rifle

Besides the Bang rifle already mentioned, as altered by Colonel James L. Hatcher, 1920 saw several other semiautomatic designs tested by the United States Ordnance Department. The Thompson rifle using a variation of the Blish hesitation principle, did not prove satisfactory because of oiling requirements and because of jams and feeding trouble due to too rapid opening of the action. John C. Garand offered his semiautomatic rifle with primer-actuated mechanism at these trials.

In the Roth primer system, the primer was blown back in its seat against the head of the heavy striker. The striker, forced back inside the breechblock which housed it, was utilized to cam back locks to unlock the bolt from the receiver. The design was quite simple but the required special ammunition was not practical.

John Garand realized the impossibility of using other than standard cartridges. He therefore designed a primer-operated semiautomatic rifle action which would utilize a regular service rifle cartridge.

His firing striker had a cup-shaped face which would fit over the primer in the center of the cartridge case. When the rifle fired,

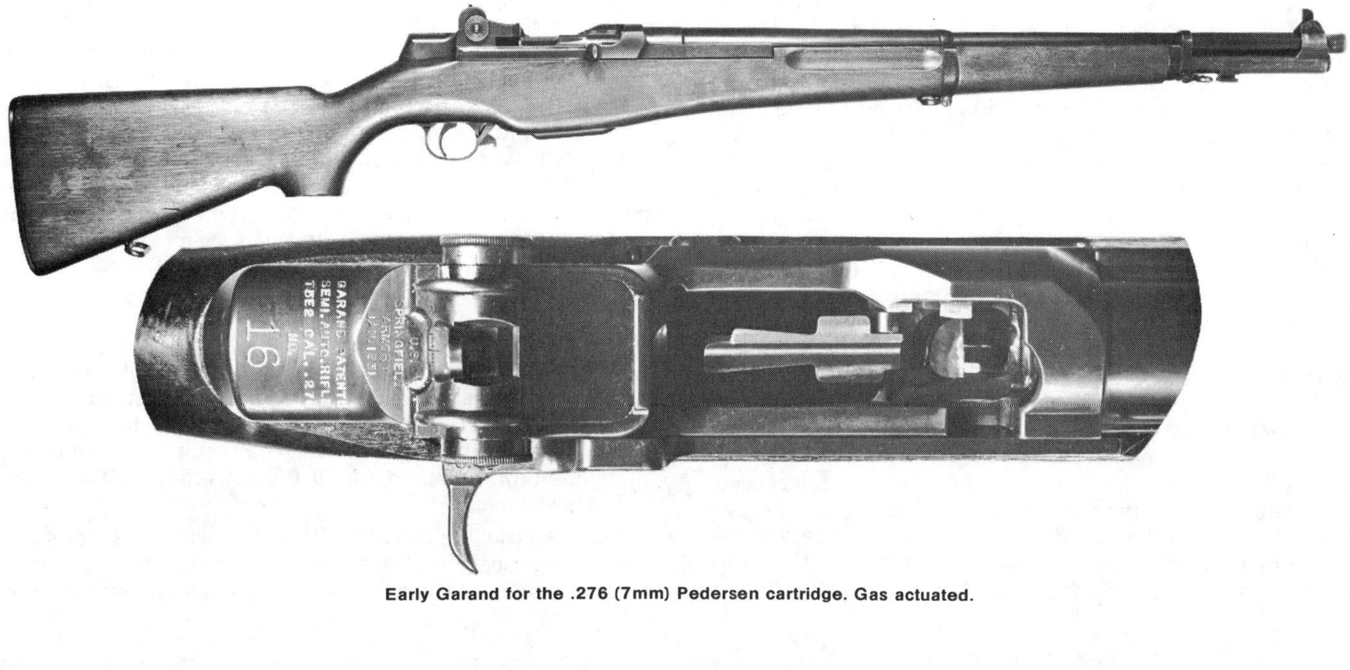

Early Garand for the .276 (7mm) Pedersen cartridge. Gas actuated.

the pressure of the gas inside the cartridge case as the powder was ignited, together with the pressure of the exploding primer itself, forced the primer back out of the case far enough into the cup shaped indent in the face of the striker to drive the striker back about .02 to .03 inches.

This small motion under high pressure was sufficient to open the breech-locking mechanism. As the lock was withdrawn, the breech opened with still enough pressure left in the barrel to continue driving the recoiling parts back for the length of their full stroke. While Garand's tests showed that the system itself was practical, the individual rifles developed structural weaknesses. It was an earlier demonstration of this principle, however, which resulted in Garand being retained by the Government and given employment at the Springfield Armory to continue development of automatic arms. He, of course, is the inventor of our famous M1 service rifle, which is gas-operated.

Future development will unquestionably take into consideration this primer actuation principle of Garand's. It has a tremendous number of advantages over other systems. Unlike the short-recoil operation, for example, it does not require a recoiling barrel with all the necessary complications brought about by utilization of floating parts. With reference to the gas-operated system, it offers simplification of manufacture, fewer parts, and substantially longer operating life without attention. Naturally it has disadvantages, too, but those may be overcome.

The design points out one way to arms simplification, very few of which are possible so long as our current system of ammunition is utilized.

The Pedersen Rifle

In 1923, another American, J. D. Pedersen, also was hired to do design work at Springfield. He was given unusual latitude, being permitted to develop a cartridge as well as a rifle. This is not standard procedure, rifles normally being designed around a specified cartridge. It is also not very practical, in view of the enormous investment every government has in equipment to produce the cartridges in general usage.

The rifle Pedersen finally produced was not a true locked-breech system. It was a hesitation blowback, actually. It utilized a toggle principle somewhat like the old Maxim, but did not utilize either a recoiling barrel or a gas actuation which would permit the toggle to completely lock the breech. Instead the toggle was designed to permit, as in standard blowback operation, immediate recoil of the moving parts at the moment of discharge. However, the joints were such that the arm was held closed longer than under normal blowback procedure, permitting an approximation of a locked-breech action by delaying the opening sufficiently to prevent burst case heads in most instances. The leverage was brilliantly worked out. As in all such systems, however, ruptured cases were encountered when the cartridges were not lubricated. Pedersen therefore developed a dry wax process for the cartridges to prevent such occurrences. This was impractical from a military standpoint and the design with its new .276 cartridge was not adopted. The military was sufficiently interested in the new cartridge, however, that it was approved by the Army as one to be eventually developed into our semiautomatic rifle cartridge, though time has proven this decision wrong also. All arms entered in our official 1929 tests were in this caliber with the sole exception of the Garand, which used our standard .30.

Fortunately, however, in 1932 a decision was made to retain the original .30-06 cartridge. In view of later developments, including the necessity for utilizing our huge capacity for manufacturing cartridges of this caliber and design, it is most fortunate that the change as indicated by the tentative acceptance of the Pedersen cartridge was not accomplished.

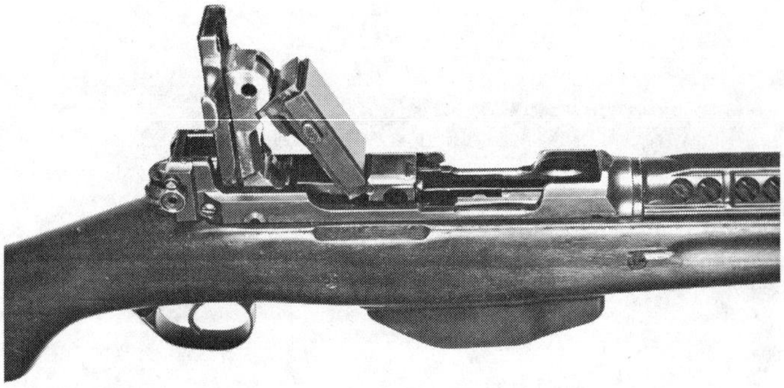

The Pedersen .276 open to show toggle system. This is a brilliantly conceived hesitation-lock system.

The Pedersen Device

Pedersen had much to do with developing a magnificent line of firearms for Remington. However, these, were commercial rifles and shotguns and a pistol. One item for which he is noted is the rather fantastic Pedersen Device which was one of the great United States "secrets" of World War I. In order to keep from the enemy any knowledge of this weapon, it was listed officially as, "Automatic pistol, caliber .30, Model 1918."

This device was a self-contained unit which could be placed in the receiver of a Springfield bolt-action rifle when the regulation bolt had been removed. It was locked in place by modifying the cutoff mechanism of the regular rifle to serve as a lock. Its barrel was the exact length, size, and shape of the regular rifle cartridge case, in effect being merely a chamber at its forward end. When the device was installed, a short pistol type .30 caliber cartridge would chamber properly in it. A 40-round double-line box magazine was provided which projected at a 45-degree angle to the barrel, the ejection port being cut in the rifle receiver to line up with the ejection port in the Pedersen Device. In practice a trained operator could remove the bolt from his Springfield and install the Pedersen Device in about 15 seconds. In actual field work it did not work that way. There is a considerable difference between what a trained operator will do and what an average infantryman under fire will do, as the course of all history including our own Civil War has shown. The idea was to provide these devices and ammunition to infantrymen going into battle. They were to use the regular ammunition and rifle for long-range work, but at close-up work were to install the device and thereby become automatic riflemen for in-fighting.

The device was made a "top secret" item; 65,000 were actually made, and Springfield rifles altered as required to handle them. All but a few were destroyed after the war. It was intended as one answer to the development of the submachine gun by the Germans in World War I. Like most makeshifts, it was good in theory but not in practice. The few devices which escaped destruction are collector's items today. The cartridges, too, were "top secret," though they were little more than modified .32 auto pistol cartridges.

The Garand Adopted

After 1929 tests at Aberdeen Proving Ground ruled out other experimental rifles introduced, the Garand was adopted. General Douglas MacArthur was then Chief of Staff and fortunately he disapproved of the .276 caliber. As a result our Garand was developed for the standard .30-06 cartridge. In 1936 it was officially adopted.

Julian S. Hatcher's BOOK OF THE GARAND deals thoroughly with the various experimental designs which were rejected in favor of the Garand, and the interested reader is referred to it.

The Johnson Semiautomatic Rifle

In 1936 the first working models of the Johnson semiautomatic military rifle were introduced by their inventor, Melvin M. Johnson. Twenty-three models of this short-recoil arm were made before production was started.

The Johnson was the first truly successful application of the short-recoil principle to a rifle. In 1940 it was put into competition with the gas-operated Garand. By this time over 50,000 Garands were actually in use, and the field testing and manufacturing modifications which had been made had eliminated most of the difficulties from the design. Johnson's almost untried rifle was pitted against this thoroughly tested and modified design in competition. It is not surprising, therefore, that the Johnson came off second best in tests. The Garand was put into production at Springfield and a contract for manufacture was also placed with Winchester. During the course of World War II the Garand performed beautifully and fully warranted the hopes which had been placed in it, except for the inevitable complaints about its weight.

The Johnson Machine Gun saw some service in the war but like the military shoulder rifle of the same caliber and same general design, it did not stand up too well in service, principally because it had not had the years of evolutionary development behind it that the other weapons had.

Winchester Rifle Tested by USMC

In 1940 a new gas-operated rifle was offered to the United States Marine Corps for tests by Winchester. This arm operated on a short-stroke piston of the type later utilized in the U. S. Carbine M1 (an arm which also was designed by Winchester). In this system gas is taken off near the breech and drives the piston back about .01 inches, the drive being passed on to the operating slide which in turn moves back to operate breech mechanism as in the Garand method.

The breechblock mechanism itself in this first Winchester rifle design resembled that of the old Savage Model 99. It was a block lifted at the rear in mortises to lock position. When in position it was held up by a wide steel surface behind the action opening giving tremendous strength; the block was first lowered and then carried back after being so unlocked to activate the mechanism. It required fluted chambers to ease extraction, a factor which also was found in Russian rifle design because of difficulty with extracting unlubricated cases where adequate primary extraction is not provided. This was the one real objection to the lock.

Russian Semiautomatic and Automatic Development

In 1916 the Russians began development of their semiautomatic rifle. One of their first products was the rifle developed by Simonov. It was issued in 1936. The arm in general is a semiautomatic military rifle though some designs have been encountered with full automatic features. It is a gas-operated arm of considerable effectiveness. The original wedge breechblock was very complicated. The design received a tryout by the Russians during the Spanish Civil War when they provided a quantity to the so-called Loyalists so battle efficiency could be observed.

In 1938, therefore, they introduced their new design, the first of a line of Tokarevs. A modification was made in 1940 and additional changes have since been made, none of them drastic, however.

The Czech Holek Rifle

The only foreign design of any note before 1936 was in Czechoslovakia. Here the Holek Rifle was under development, which was known as the ZH29, a tilting breechblock design on the gas piston system. These rifles were magnificently manufactured but had so much detailed hand work in them that general production was never attained.

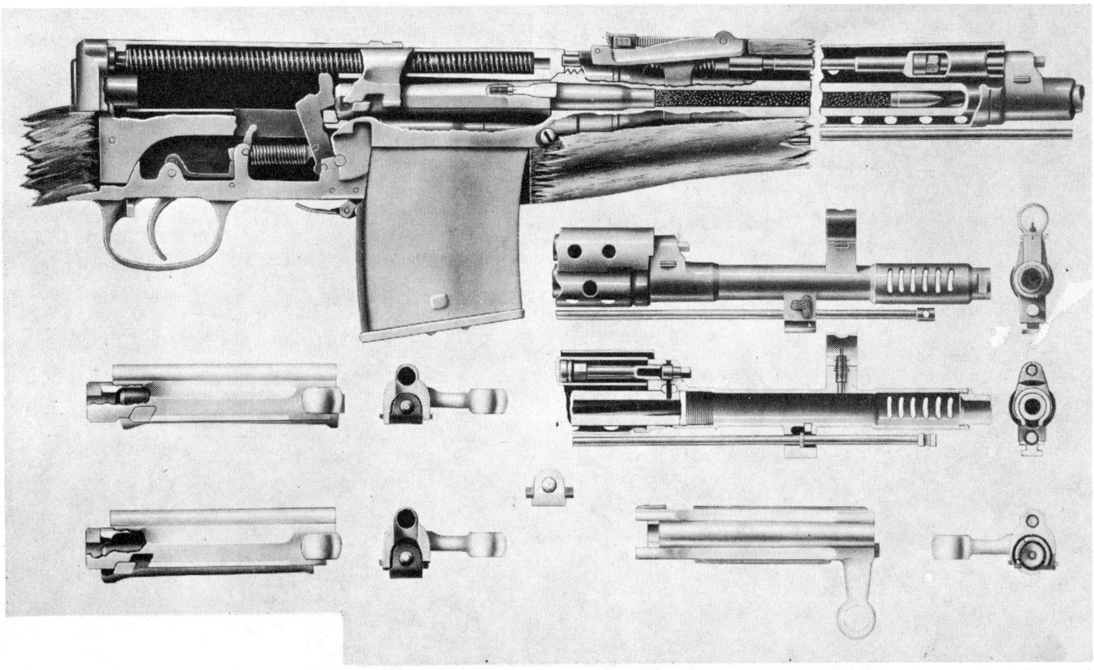

The Russian Tokarev Semiauto Rifle. Phantom view showing action locked at instant of firing. Details show the muzzle brake, gas operation and bolt locking systems.

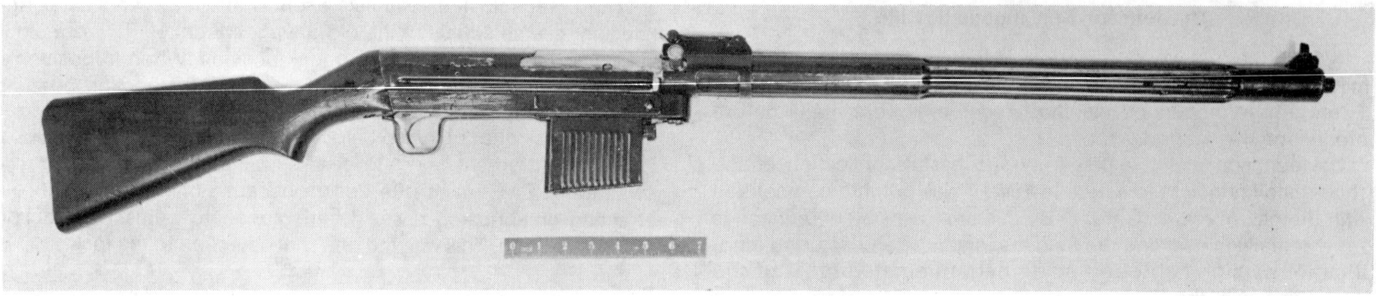

An early German Walther experimental gas-operated design.

German 41W. Designed by Walther. Gas Operated. Muzzle cap design.

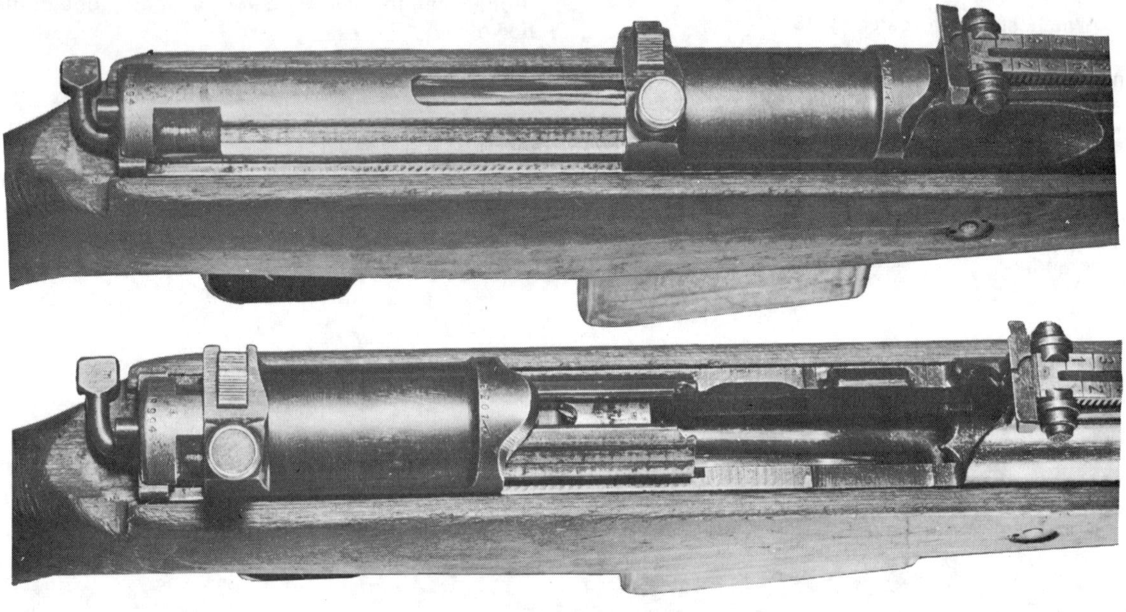

German Gewehr 41W. Breech closed and open.

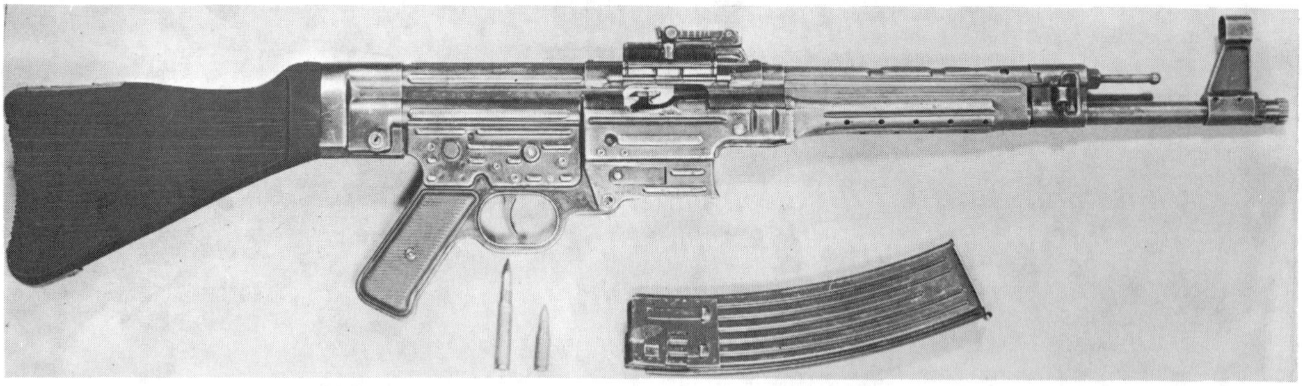

German MP 44. This arm was the first approach to the modern all-purpose military rifle. For comparison, cartridges shown are the standard German 7.92mm and the short form 7.92mm for the MP 44.

They were used, however, to some limited extent in Ethiopia. Most specimens seen come from the Middle East. This design, by the way, was one of those entered in the tests we conducted in 1929; it was passed over for the Garand.

German Semiautomatic and Automatic Rifles

During 1941 the Germans, brought to a realization of the desirability of a semiautomatic rifle by use of the Russian arms against them in the Eastern theater of war, produced their muzzle trap gas-operated rifles. These arms were in many respects similar to earlier Mauser experimental designs. Various types were issued, the 41M by Mauser and the 41W by Walther.

By 1943 Germany realized the necessity for an arm which could be manufactured at very low cost and under conditions utilizing high-speed production equipment. In a very short period they produced the crude appearing but highly efficient Gewehr 43. The designation K43 was later applied to a slightly modified form. The locking system was basically that of the early Mauser, which in its turn was an adaptation of the early Friberg.

Toward the end of the war Germany frantically produced a number of potentially semiautomatic and full automatic arms capable of tremendous future development. Noteworthy among these is the Fallschirmjaeger Gewehr which will be covered in the chapter on German World War II small arms.

Perhaps the outstanding German design was the MP 44. This is a locked-breech arm operated by gas. It uses a new short cartridge form intended to offset and go far beyond our own carbine experimentation. The magazine capacity is 30 cartridges. The arm is gas operated by piston having, of course, a locked breech. It is actually a machine or automatic carbine for special purpose use.

The Germans showed far more intelligence and imagination in developing a cartridge for this weapon than we did in our adoption of one for our own carbine. Their cartridge was merely a shortened version of their 7.92mm standard rifle cartridge. The case was bottlenecked but shorter than the standard case. The regular pointed bullet was retained, the bullet weight being reduced somewhat, however. Ballistically the cartridge was a tremendous improvement over the one that we developed.

CURRENT AUTOMATIC RIFLES

Since World War II a large number of automatic rifles and assault rifles have appeared throughout the world. These are dealt with in detail in Part II of this book under their country of origin.

CARBINES

A carbine has been defined as a short rifle, generally one having a barrel length of less than 22 inches.

Like most of the very old terms used in the history of firearms, the actual origin of the name cannot be positively established. The most reliable early records, including those in Gays' Memoires pour l'Artillerie published in 1548, indicate that the short form of shoulder-fired gun acquired its name from the fact that it was extensively used in Spain by cavalry groups then called "Carabins."

While some researchers indicate that the name originated from the fact that short shoulder arms were early used by Calabrian troops, and still others trace its derivation to its alleged use for repelling boarders on small ships called Carabs, the best documentation seems to support the theory of the Spanish cavalry origin.

In passing it might be mentioned that still another somewhat fanciful explanation for the derivation of the name carbine is that it stems from the Arabic "Karab," meaning a weapon.

One of the classic Wallhausen military drawings dated 1616. It shows a carabiner firing a wheel-lock carbine. The weapon is a short form for use from horseback. A wheel-lock pistol carried in its holster slung across the "pistallo" or pommel of the saddle is also a historic item, as it is an early piece of evidence indicating that the term "pistol" quite probably derives from the position in which mounted soldiers first carried it.

In military usage in the past for the most part, the carbine has been merely a form of the standard rifle. It features the same lock work but has a shorter barrel and modifications of the stock and forestock as required to lessen the weight. In general, in early times carbines were basically cavalry arms carried in a boot on the saddle.

Exceptions generally were in Italy and Spain where short lightweight rifled shoulder arms, using normally the 9mm Parabellum in Italy, or in Spain the 9mm Bayard or Parabellum cartridge, have been designated as carbines. These have appeared in turnbolt, full automatic and semiautomatic weapons. However, such arms have invariably been intended for basic police work and for border patrol service, not for military usage as such.

In any event, even in sporting arms both here and abroad the term carbine has been used to designate a rifled arm similar to the parent rifle, except that it is shorter and lighter by reason of lesser barrel length and modified stocks, et cetera.

U. S. CARBINES

The United States Carbine M1 is the first arm in general military use to deviate from the definition given above, being a light and short rifled weapon—but one which uses a cartridge of its own and has no parts interchangeable with the U.S. M1 (Garand) rifle.

In 1940 the Ordnance Department provided a set of specifications and called for designs to meet them. These were to be semiautomatic (and later also full automatic) carbines. In this instance an entirely new military cartridge was developed for the short lightweight weapon whose weight was specified originally at 5.5 pounds. Curiously the cartridge developed was almost identical with the original commercial .32 Winchester autoloading cartridge of 1905 except the bullet was shorter and lighter.

Guns made by Auto Ordnance, Harrington and Richardson, Hyde, Savage, Springfield Armory, Winchester, and Woodhull were submitted. The locked-breech Winchester with its short stroke action was selected, partly because of its highly efficient

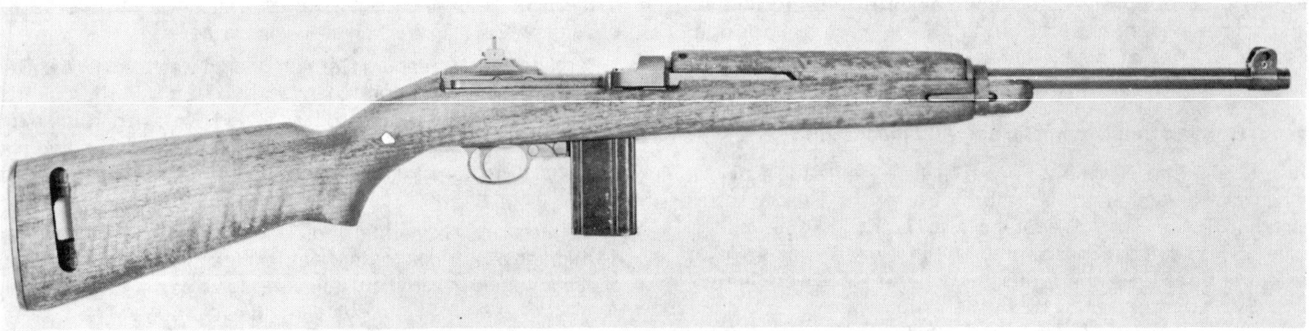

Representative U.S. MI carbine. Unlike earlier U.S. carbines, this arm does not use the same cartridge as the MI rifle, nor does it employ any parts from the rifle. Earlier U.S. carbines all used the standard rifle cartridge and most of the same parts, being modified only by shortening and decreasing weight.

lock action which resembled the Garand quite closely. The gas operating system, however, was the type used in the Winchester 1940 rifle.

When adopted, this carbine was cataloged as the U.S. Carbine, Caliber 30 M1. It has since gone through numerous modifications, including paratroop models and full automatic designs, most of which will be found in the chapter on the United States current small arms.

The U.S. Carbine, caliber .30 M1 is not to be confused with the U.S. Rifle .30 M1, the cartridge being shorter, lighter and very, very much less powerful. It is a special purpose arm not to be confused with rifle requirements.

The U.S. Carbine Caliber .30 M1 and its modifications represent one approach to the subject of increasing the efficiency of the very light weight, readily portable automatic weapon. German specialists investigated this subject thoroughly in 1933 and 1934, pointing out at that time the two ways of accomplishing it; either by enlarging the type of pistol ammunition in use or by decreasing the size of the standard rifle type ammunition as used in the light machine gun.

In its inception the U.S. carbine was intended to produce a compromise weapon having the general characteristics of a group of arms. Many hopefully expected it to actually replace the pistol, the submachine gun, and even perhaps the rifle. No truly serious military students ever had any illusions on this possibility, however.

Argentine 7.65mm Model 1909 Cavalry Carbine.

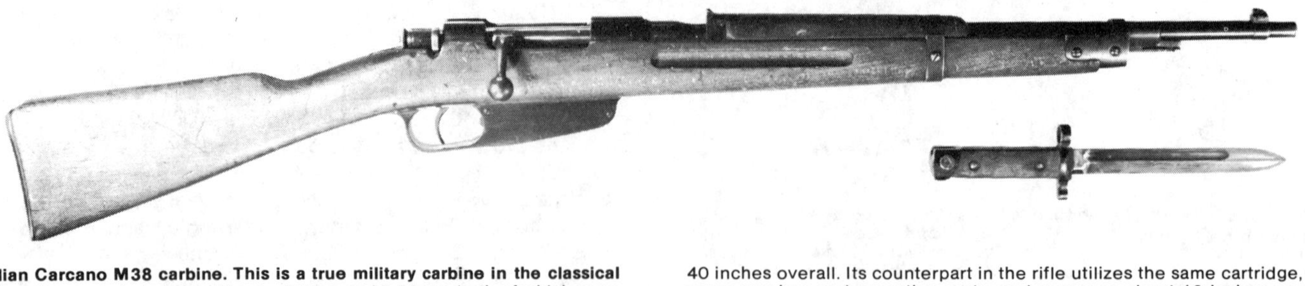

Italian Carcano M38 carbine. This is a true military carbine in the classical sense. It was made in two calibers—6.5mm and 7.35mm, both of which were official rifle calibers in World War II. This carbine has a 21-inch barrel and is 40 inches overall. Its counterpart in the rifle utilizes the same cartridge, the same receiver and operating parts, and averages about 10 inches greater barrel length and overall measurement.

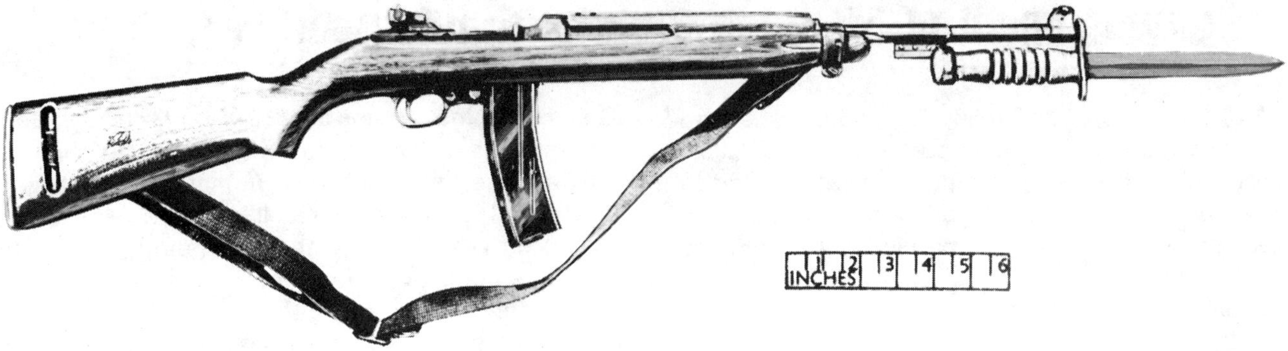

U.S. Carbine, caliber .30 M2. This is a modification of the semiautomatic Carbine M1.

Note on .30 M1 Carbine Cartridge

This cartridge was developed in 1940 by the Winchester Repeating Arms Company at New Haven at the direct request of United States Ordnance Department. The Company was furnished with specifications calling for a compact, rimless, cased cartridge with a bullet of 110 grains, developing a velocity in the neighborhood of 1800 feet per second from an 18-inch barrel.

The cartridge as originally produced by Winchester had an overall length of 1.67 inches with a case length of 1.29 inches. Originally head stamped .30 SL by Winchester, the cartridge later became our .30 Carbine M1.

It is a most unusual coincidence that the cartridge developed was actually one very close to the .32 Winchester Self-Loading Rifle cartridge developed by that company for use in its model 1905 Self-Loading rifle. This is a short, semi-rimmed cartridge case with straight side wall. The slightly altered version was issued to several developers to use in producing carbines for Government tests. Entered in the tests were two versions of carbines developed by John Garand. Among others Winchester had developed a 7 1/2 pound carbine or experimental rifle for this cartridge. It proved so successful that Winchester undertook the development of a 5-pound carbine to handle the newly accepted cartridge.

They developed the first handmade sample carbine in the record-breaking time of 14 days. The initial tests brought out several bugs as was inevitable in such a rush model. By working round the clock for 34 days, the Winchester organization rushed through the model which became the U.S. Carbine Caliber .30 M1.

Samples of this new Winchester were tested on September 15, 1941 against improved models from Springfield, and by Hyde and Reising, as well as a gas-operated rifle by R. J. Turner. The outstanding superiority of the Winchester design was clearly demonstrated.

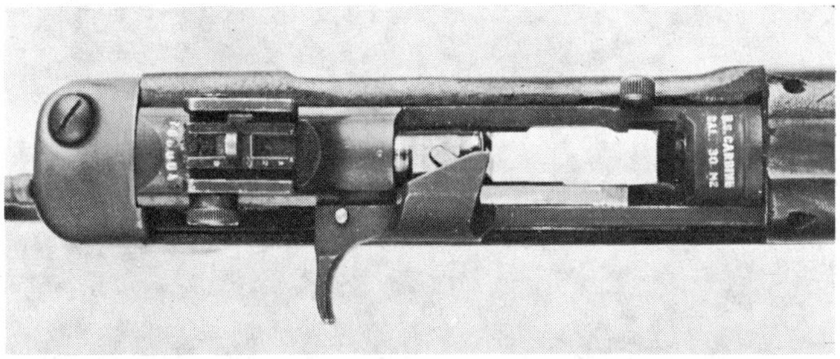

The M2 is a special purpose weapon. Shown without magazine. The full auto switch is seen at the receiver above the stock. Externally it resembles the Carbine MI. This specimen has the folding MIAI-type stock. Top view showing Selector (switch). Breech open.

BRITISH AND U.S. NOMENCLATURE

Great Britain has recently changed their terminology to designate as "submachine guns" in accordance with United States practice, all the various weapons heretofore listed by them as "machine carbines."

For all practical purposes, the term "carbine" will normally indicate, either in military or sporting usage, a rifled shoulder weapon with all the basic characteristics of the rifle version, except that the barrel will be shorter and the stocks and sights altered accordingly to reduce weight and make for a more compact weapon.

The tremendous number of "U.S. Carbines M1" and modifications now in use throughout the world will undoubtedly require the continuation of this designation, although the weapon itself actually falls into a new arms classification.

Development of Machine Guns and Automatic Firearms

8

The idea of using rapid fire, volley fire, and concentrated fire, particularly for defensive purposes in military practice, is almost as old as the history of gunpowder itself. Of course it was not until the development of the metallic cartridge that even the manually-operated machine type of arm was feasible, much less the fully automatic type with which we are basically concerned today.

Among many designers of automatic small arms, particularly machine guns, the names of three are especially outstanding: Gatling, Maxim, and Browning. The Gatling gun, the creation in 1862 of Dr. Richard J. Gatling, was the first successful mechanical machine gun. Hiram Maxim developed the first automatic functioning of a weapon in 1883, using the recoil force to operate it. Maxim was also the first to use the expanding gas of a fired cartridge to provide automatic operation. John M. Browning, a famous designer of automatic small arms, designed the first successful gas-operated machine gun. His "Browning Automatic Rifle" and water-cooled heavy Browning .30 caliber machine gun, were extensively used in World War 1.

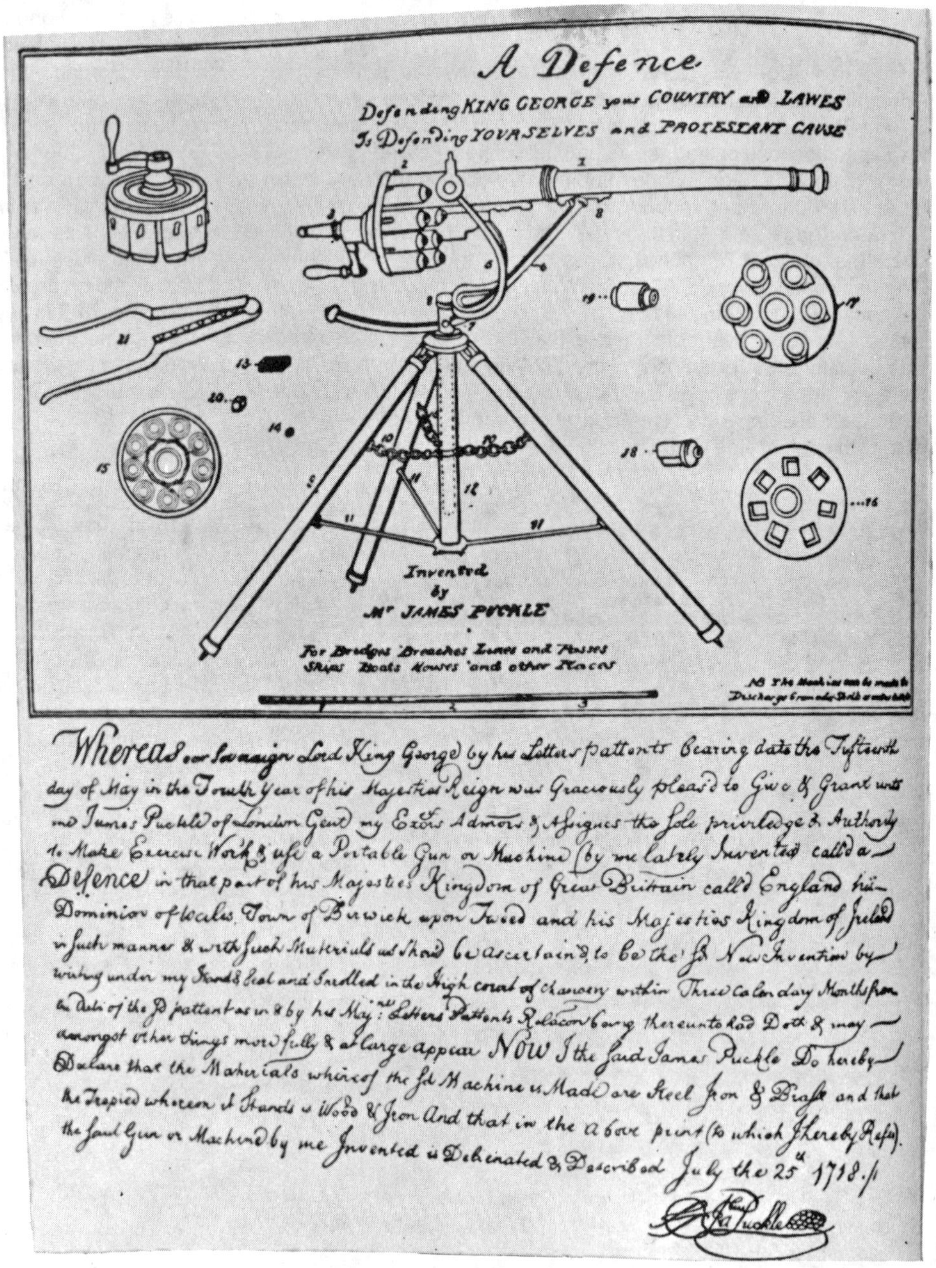

The famous Puckle design. So far as records show, it was never made in England. A version of this gun was actually built in Russia about 1850.

PRINCIPLES OF OPERATION OF AUTOMATIC AND SEMIAUTOMATIC SMALL ARMS

Automatic firearms are those in which forces of the discharge are utilized to reload, recock, and fire successive shots until the trigger is released, or until the ammunition supply is exhausted.

Semiautomatic arms are those in which the forces of the discharge are similarly used to reload and recock, but in which a deliberate release and pull of the trigger are required to fire the second and successive shots.

The discharge forces utilized may be merely those of the **recoil** occasioned by the discharge, as in the case of the so-called automatic pistol. Strictly speaking, of course, these arms are semiautomatic. It is necessary to deliberately release the trigger, and pull individually for each successive shot. If this were not done, the discharges would be so rapid and the recoil so great because of the light weight of the weapon that the arm would mount out of control. Recoil-actuated arms may be blowback (unlocked), hesitation (retarded opening), or positively locked designs.

A second principle of such operation is **utilization of the expanding gases,** either at the muzzle or as they are tapped off through a port in the barrel itself, to operate an unlocking and reloading and firing mechanism. Such actions are locked-breech designs (on all but freaks).

Still another type is that in which there is a **combination** of the use of the recoil forces and the use of the expanding gas. These are locked-breech designs.

CLASSIFICATION OF MACHINE GUNS

The general term "machine guns" is loosely applied to a wide variety of automatic arms. In the interest of clarity, the general characteristics of the various classes are now given.

DEFINITIONS

Light Machine Gun

This term customarily designates a weapon which can be fired full automatically and which may or may not be furnished with a device to permit semiautomatic or single-shot fire. The light machine gun normally fires the same cartridge as the military shoulder rifle of the using nation, whether bolt action or semiautomatic. The normal weight of this weapon is from 15 to 30 pounds. It is customarily fitted with a buttstock like a standard rifle and is intended normally to be fired from shoulder support while in prone position. The front end is commonly supported when firing by a mount, usually a folding two-legged steel design.

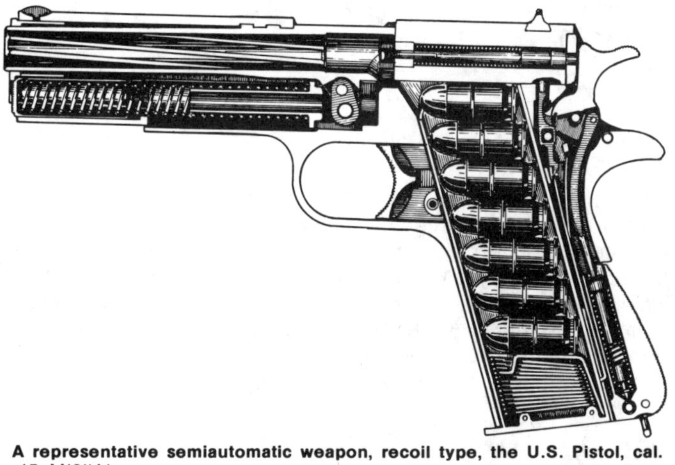

A representative semiautomatic weapon, recoil type, the U.S. Pistol, cal. .45, M1911A1.

The light machine gun is capable of use by a single man under emergency conditions. Normally, however, it is operated by at least a two-man crew, the gunner and an ammunition carrier who feeds the weapon. Furthermore, it must be recognized that this form of weapon uses ammunition at such a high rate under combat conditions that it really requires additional support in the form of reserve ammunition carriers or providers.

The feeding system may be any of the types found in other machine guns, though the most common types utilize detachable steel box magazines. Drum feeds are common in some European designs, notably German and Russian, but the belt feed is uncommon, normally being encountered only in designs where quick barrel change is a part of the design of the arm. This is because of the rapid overheating under continuous fire from a belt feed with its great ammunition carrying capacity in comparison to the box magazine patterns.

Operation may be either gas or recoil, and in some occasional experimental designs may even be blowback.

Automatic Rifle or Machine Rifle

This term is often used to designate an assault rifle or a light machine gun. It is a matter of national terminology. In general, the designation "automatic rifle" or "machine rifle" in the past was commonly employed with arms such as the Browning Automatic Rifle "BAR" or its European equivalents. The designation automatic comes from the fact that the arm can fire in full automatic operation by merely holding back the trigger with the selector in the proper position. The characteristics under this name are the same as those given under the designation "Light Machine Gun."

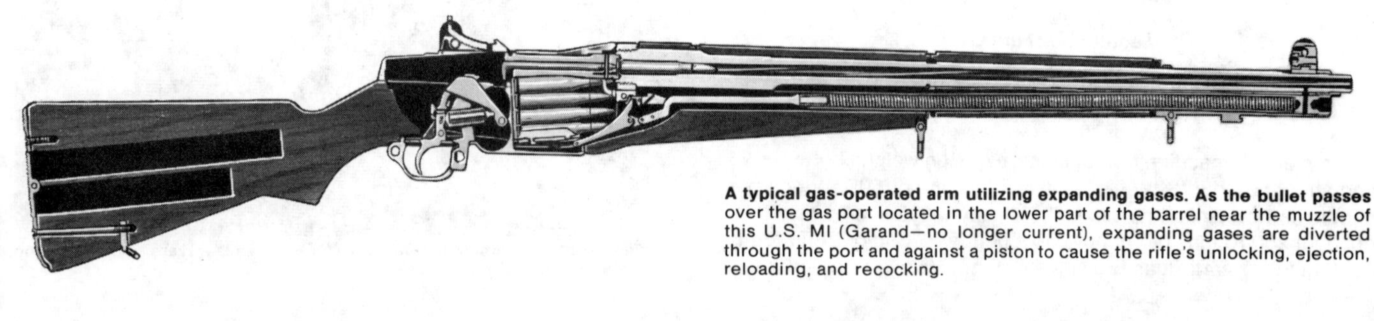

A typical gas-operated arm utilizing expanding gases. As the bullet passes over the gas port located in the lower part of the barrel near the muzzle of this U.S. M1 (Garand—no longer current), expanding gases are diverted through the port and against a piston to cause the rifle's unlocking, ejection, reloading, and recocking.

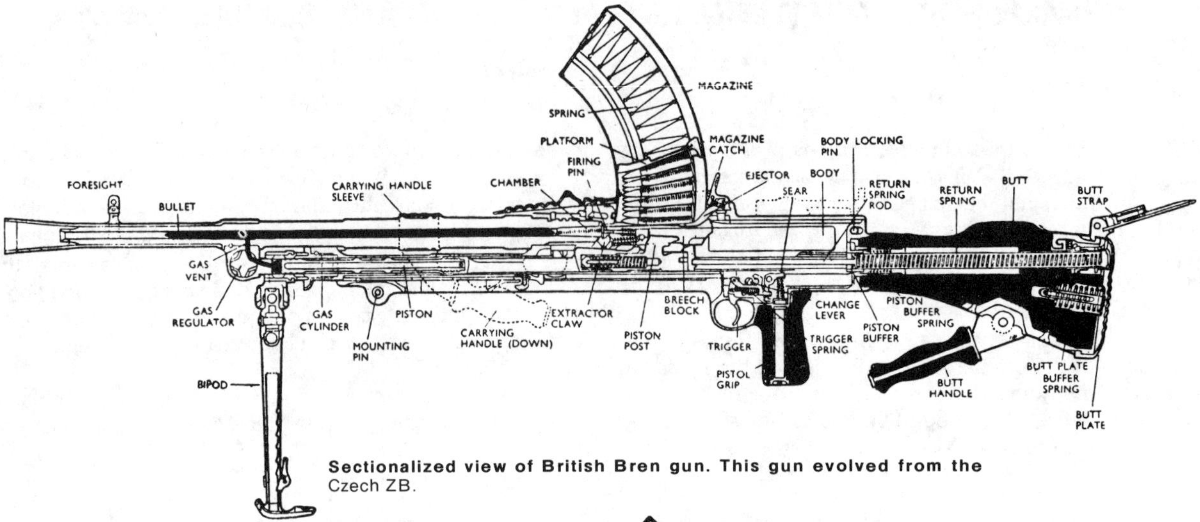

Sectionalized view of British Bren gun. This gun evolved from the Czech ZB.

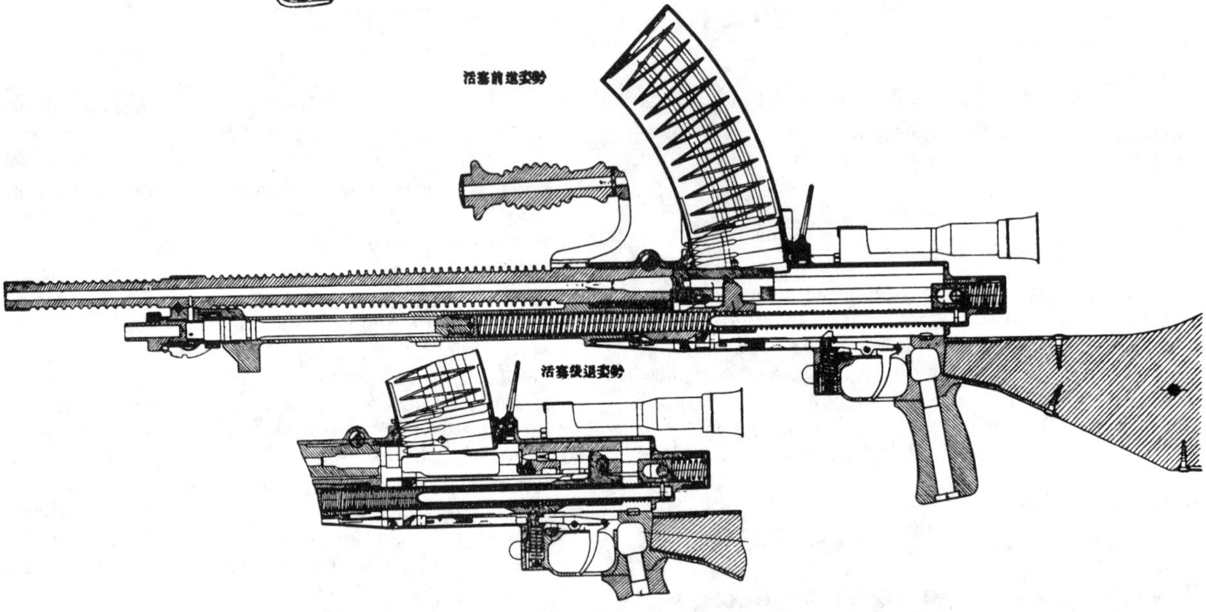

Section drawing of representative World War II Japanese light machine gun fed from top-mounted box.

At times the name is also applied to the post-War II forms of shoulder rifles having full-auto switches—the true "automatic rifles."

Weapons classed as machine rifles and light machine guns have the advantage of very high mobility. This together with their ease of handling by one man in an emergency make them the perfect weapon for frontline operations. Since the weight and bulk are low, these guns must be fired in bursts, and are not suitable for sustained firing except where extra barrels are readily available for interchange. Regardless of the firing rate, it is actually not practical to fire much more than 100 per minute from the light machine gun except in case of direst emergency.

Medium Machine Gun

The "medium machine gun" (commonly referred to merely as "machine gun") uses the same rifle cartridge as the light machine gun.

The medium machine gun of modern design weighs between 25 and about 60 pounds. While it has a firing rate of 500 rounds per minute or more, in actual usage it normally will deliver a maximum of about 250 rounds per minute from belt feed. Clip or drum feeds holding fewer cartridges provide substantially lower fire volumes.

Even when several interchangeable barrels are available, a gas-operated medium machine gun is seldom capable of sustained fire of more than 6,000 rounds per hour actually delivered. The water-cooled varieties on the other hand have been known to fire up to 15,000 rounds per hour, because of their more efficient but far bulkier and heavier cooling systems.

The medium machine gun is normally mounted on a folding

A representative light machine gun. The Russian Degtyarev in field action. Note drum-shape magazine.

American Johnson machine rifle of World War II. This, too, is a representative machine rifle. It is short-recoil operated. It feeds from a box magazine inserted in the left side of the receiver. The gun was designed and manufactured by Capt. Melvin M. Johnson, U.S.M.C.R.

Browning machine gun cal. .30. A typical medium machine gun using the standard rifle cartridge.

portable tripod. This arm also, of course, requires a gun crew to sustain it, or even to put it into action, since the minimum handling crew must be one man for the tripod, one the gun, and at least one for ammunition. Once the gun is fixed, of course, it can in emergency be operated by the firer alone.

Because of its weight, cooling system, and stable mounting, the medium machine gun can be used effectively at ranges three to four times greater than is possible with the light machine gun.

All the operating systems used in the light machine gun are also applicable to the medium machine gun.

General Purpose Machine Gun

Sometimes called a DUAL PURPOSE machine gun. These are weapons which are intended to be used on bipods as light machine guns and on tripods as medium (heavy rifle caliber) machine guns. They are normally belt-fed, but frequently use small ammunition boxes attached to the gun when used in the light machine gun role. The U.S. M60, the French M52, the FN type MAG, and the German MG42 fall in this category.

Heavy Machine Gun

In general the characteristics are those of the light machine gun except that the weight of all components is greatly increased, as is their size also. The term heavy machine gun is correctly applied today only to cartridges .50 caliber and larger. In general, these arms are merely scaled up models of the equivalent medium machine gun pattern, as in the case of our .50 caliber Browning. Such guns today play a considerable part in various forms of motorized equipment carrying machine guns and are used to some extent for low antiaircraft fire, often in multiples. Since the cartridges they use are too powerful for normal use in standard rifle pattern weapons, and since their mechanical and design characteristics are but minor modifications of the medium machine guns, no further coverage of them is required here.

French Hotchkiss cal. 13.2mm heavy machine gun.

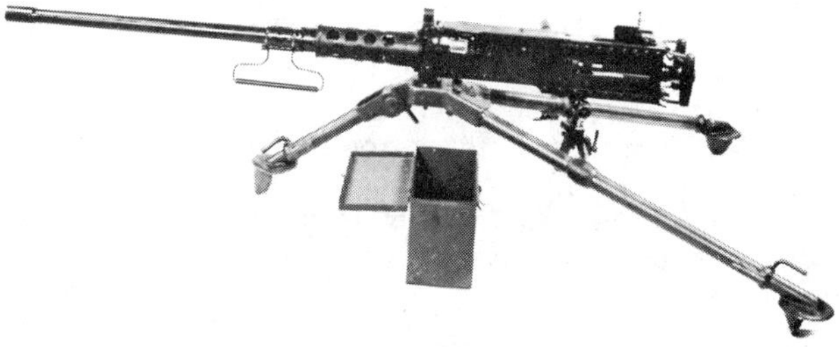

U.S. Browning machine gun cal. .50. A typical heavy machine gun using the .50 caliber cartridge which is too powerful for standard rifle use.

Polish Browning machine rifle cutaway. This gun is based on the Browning as made in Belgium. It differs in some design features from our BAR.

EARLY RAPID-FIRE CONCEPTS

EARLIEST DESIGNS

First Concepts

Early historical records are sketchy, sometimes referring to a concept of a rapid-firing weapon rather than to an actual weapon. There is, for example, one very noted contemporary record shown in Birch's History of the Royal Society published in England. The records of the Royal Society for March 2, 1663 quote Sir Robert Moray, F.R.S. as telling that body "there had come to Prince Rupert a rare mechanician who pretended to make a pistol shooting as fast as could be presented and yet could be stopped at pleasure; and wherein the motion of the fire and bullet within was made to charge the piece with powder and bullet, to prime it and bend the cock."

On the basis of this and similar fragmentary records, the British TEXT BOOK OF SMALL ARMS 1929 EDITION considers the automatic principle as having been an English invention. (In view of the fact that the records of the Royal Society show no further mention of the arm, such a claim is obviously rather farfetched.)

Among other fragmentary records of the day one might quote that unearthed by the Abbe J. Rouquette, found in the archives of the Province of Languedoc, France. A document there states that on the 21st of August, 1688, one Abraham Soyer was arrested. When arraigned before the Inspector of the Mission of the Cevennes, a pistol was found in his luggage. He stated he was taking the weapon to St. Etienne, which was an arsenal and arms development center at that time in France even as it is today. The description given indicates that the weapon was a breech-loading item using a crude cartridge which was loaded through the butt end of the stock quite in the manner of the common automatic pistol in use today! However, we find no further mention in French records of such a device until some 200 years later.

Still another mention which might possibly be descriptive of an automatic weapon is that found in the Pepys Diary for the 4th of March, 1664. Herein in mentioned a "new fashion gun brought by Lord Peterborough this morning, to shoot off often, one after another, without trouble or danger." Here again a search of the records of the period in which this was written does not disclose any further mention of such an arm.

One could quote equally vague and yet tantalizing reports from the Belgian, German, and Italian records of the early seventeenth century. In fact some of an even older date might be quoted. However, in no case are any of them sufficiently explicit to indicate that the arms actually were presented, but merely ideas about them were discussed.

For all practical purposes, therefore, we must face the fact that self-loading and self-cocking mechanisms as such were not capable of practical development until the perfected metallic cartridge of the twentieth century came into being.

In the United States patent records for the year 1863 we find patent number 2998 to one Regulus Pilon, which describes a system for recocking the hammer of a weapon by utilizing the barrel recoil. We find similarly in our records for the year 1866, patent number 1810 to W. Curtis for a proposed gas-operated system of firearms. An unsuccessful gas pistol, the Clair, appeared in France in 1863.

While items such as these are of interest to the true historical student, they have little bearing indeed on the actual development of firearms for military purposes. There is no indication that knowledge of such past developments were of any avail to the actual developers of our modern weapons.

The Organ Gun

Specific historical mention of rapid-fire small arms may properly begin with the "Orgue" or Organ gun and the Ribauld gun which appeared as early as the 14th Century.

In general they were a series of barrels or arms positioned together in a frame or on a wheeled carriage. These were intended to be moved in for a direct attack or to be placed in front of artillery gunners as a protection against cavalry charges. The Army of Burgundy as early as 1411 listed two thousand special carts having a number of barrels attached thereto for field use.

At the battle of Ravenna, Pedro Navarro, in command of the Spanish forces, positioned thirty of these carts, each mounted with a large number of arquebuses, in front of his infantry to protect them from the charge of the mounted French armored forces. They were most successful.

Leonardo Da Vinci in his life span (1452-1519) developed a large number of volley-type guns. Some of these fired in banks straight ahead, while others would fire fan-shaped volleys.

Da Vinci apparently even had provided some sort of rudimentary cartridge for loading this device. It is interesting to note that even during the period of our own Civil War very extensive use was made of this system of firing.

For that matter, as late as World Wars I and II, instances were known of arms being set to be fired mechanically in volleys, much on the order of these early Organ guns. Of course, most usage of this sort was from defensive positions.

Puckle's Patent

No history of machine weapons could be complete without at least passing mention of the British patent No. 418 issued on May 15, 1718, to James Puckle. This called for a revolver type of firearm on a mount. Its drawings are very specific. Nothing ever came of the weapon so far as we know. Many claims have been made for it as the progenitor of arms such as the Gatling gun. The revolving block intended to be used as shown by Puckle could hold seven or nine bullets. He specified that they should shoot round bullets in ordinary warfare against Christians; but square bullets when used against the infidel Turks! His drawing even showed his intention to provide cylinders for these types of projectiles.

Toward the close of the last century, qualified observers reported scores of Russian museum pieces at St. Petersburg, Tula Arsenal, and the Kremlin built on Orgue, Puckle, and other systems. Many were for cap-and-ball operation. None were actually very old, most being copies of early items from other lands. Thus far we have no verified records of actual use of these types.

In both France and England in the Eighteenth Century, volley arms were made. These were a group of barrels having one common touchhole. They were mounted in a single stock and fired by a single flint. As many as ten barrels appeared in these arms, which were intended for defense of fortifications. The famous London gunsmith Nock, who developed the Knox Form which is found on the barrel of the former British service rifle, was one of the manufacturers of this type of arm.

The Barnes Gun

Scores of attempts were made to develop multi-firing weapons in the day of the percussion cap. The only one of these worthy of any particular attention is that of C.C. Barnes of Lowell, Massachusetts. He was issued a patent on July 8, 1856 for a gun which suffered from the fact that it was too far ahead of the military thinking of its day and too far ahead of cartridge technology.

This was a crank-operated gun using a special linen cartridge.

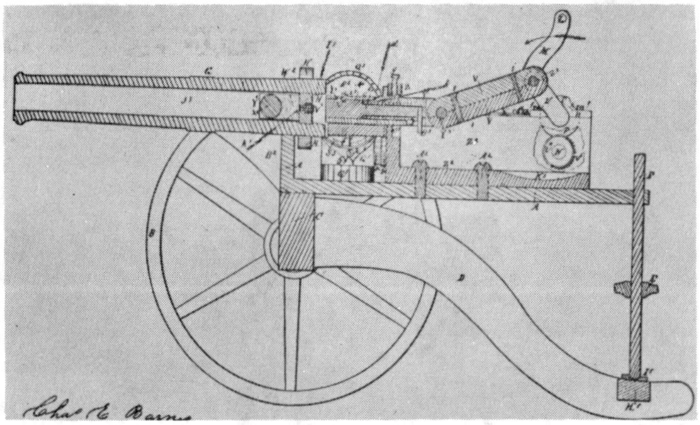

Barnes gun, 1856. From United States' Patent Office drawing. This gun suffered from being designed too early. It was not produced commercially.

Of course, the percussion cap had to be fitted separately over the firing nipple. This, however, was done mechanically after the weapon was locked.

The breech lock itself was a toggle joint arrangement. This, too, was in line with the final successful development in its field.

However, the metallic cartridge was on the verge of coming into its own about this time, and nothing much was done about the Barnes gun.

The Ripley Gun

With the introduction of the paper or linen cartridges, in which the powder for propelling and the ball itself were combined so they could be readily loaded down the muzzle end of a chamber, as was done with the original Colt percussion revolvers, several inventors turned their attention to developing machine arms to handle this new type of cartridge.

The most original one was that of Ezra Ripley of Troy, New York. He developed a gun idea in which a series of barrels were grouped about a common axis. A special breech block, made somewhat like a revolver cylinder of the day was loaded with paper cartridges. Of course, percussion caps were then placed over the nipples at the rear of each chamber, as in common revolver practice in that day. This cylinder was then placed over the breech end of the weapon, the holes lining up with the breech ends of the individual barrels. The breech was then locked into place. An operating crank was provided. When turned it fired each chamber in succession as rapidly as the action permitted. By having on hand a number of these cylinders already loaded, the Ripley might have been capable of use as an early form of the later successful Gatling gun. However, again because of rapid cartridge development, nothing came of this idea. It is of historical interest only, although it does have a bearing on the development of the later multi-barrel machine type arms.

Ripley machine gun drawings in the United States' Patent Office. This gun was designed in percussion days. The development of the metallic cartridge halted its production.

The Billinghurst Requa Battery Gun

During our Civil War, volley guns were used to a considerable extent. One of the best known of these came to be called "the Covered Bridge Gun." The name stemmed from the fact that most of the bridges in those days were covered or enclosed. Roof and sides were provided to protect the floor of the bridge and the supports from weather conditions. Because of this type of construction, the bridges could be easily protected by use of a volley type gun. The best known of these probably is the Requa Battery. This was a .58 caliber gun manufactured by Billinghurst at Rochester.

Twenty-five barrels were mounted flat side by side on a light platform. A sliding breech mechanism was loaded with individual steel chambers. The powder and ball were inserted exactly in the manner used in early days of the flint lock as shown in some of the exhibition pieces still to be found in the Tower of London.

The ignition system was interesting. When the breech was locked into place by its lever the openings or touch holes of the various cartridge chambers were exposed. A train of powder passed the hole in each of these chambers. A single ignition percussion cap ignited this train firing the twenty-five barrels. Only one hammer

Billinghurst Requa battery gun. This is the original working model which stood in the window of the Billinghurst shop in Rochester, New York. It demonstrates clearly the system of operation. This gun was percussion-fired.

Billinghurst Requa volley gun. The name stems from the fact that it fired all barrels as a single volley.

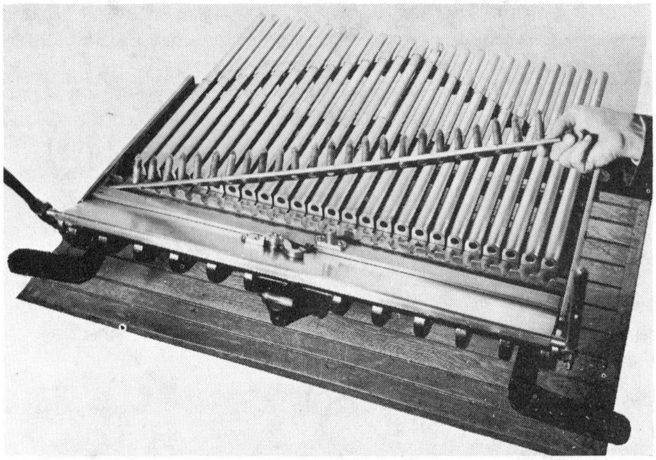

The breech mechanism is shown loaded ready to be inserted. The mount is a display mount used only at West Point. The gun has 25 barrels of .52 caliber. It is loaded with individual steel cartridges.

was used. It was cocked by hand and, when released by pulling an artillery type lanyard, fired all 25 barrels.

This, of course, was a very crude arm. Nevertheless, its barrels could be adjusted to height and width to cover the entrance to a bridge. A three-man crew normally operated the gun. Firing could be done at the rate of seven volleys a minute. The effective range was well over one thousand yards.

The gun's use was confined largely to bridges and fortifications where it could be enclosed, since any dampness which reached the powder train would prevent the firing of the arm. A confederate specimen of this type of gun was used at Charleston, South Carolina. It was of large caliber and weighed some 1300 pounds.

The Ager Coffee Mill Gun

Still another type of our Civil War period is the Ager Coffee Mill gun. Its name derives from the fact that it resembles a coffee grinding machine of that day. It, too, was a hand-cranked machine of .58 caliber. Unlike the Requa it only had one barrel. Cartridges of steel were loaded and fed into the hopper. Operation of the crank pushed the cartridge forward and held it in alignment as the wedge-lock was cammed forward. The barrel was stationary. A detailed report was made on this gun by Major Fosbery of the British Army, quite a noted authority of the day and one who had been employed by the government in India to report on machine arms for adoption by the Indian Army.

While tests indicated that the Ager was a serviceable and potentially terrible weapon, the military authorities of the day

Ager Coffee Mill machine gun, 1862. Caliber .58. This uses a short muzzle-loading chamber or charger capped with a percussion nipple. The loaded chambers were fed into a hopper at the top. As the crank was turned they were fed to firing position, after which the empty charger was ejected to be recovered and reloaded. Barrel, bullet, and charger into which the bullet and powder were loaded are seen below the gun. The mount and shield are those on a specimen at West Point. In use, the gun was normally mounted on a wheeled carriage.

refused to definitely commit themselves on recommending it. The matter of purchase was even carried to President Lincoln, who in turn refused to authorize purchase unless the military would specifically request it, since he did not pretend to be an authority in this field. The war passed without any action being taken.

Vandenberg Volley Gun

This weird item is a massive instrument which was based on the earlier Belgian Montigny Mitrailleuse. The Vandenberg had from 85 to 450 individual barrels! A special breech held cartridges in individual drilled chambers. This could be inserted and screwed into the rear of the barrel arrangement so that the cartridges came in a direct line with the individual barrels. A center charge was ignited by a percussion cap and the entire volley was simultaneously fired. The resulting recoil, of course, was terrific. The accuracy of the gun was surprisingly good. One British test with a 191-barrel type showed that ninety percent of the volley hit a six-

Vandenberg volley gun. 85 barrels. Caliber .50. Made by Robinson & Cutton of London, England. The breech is being opened to load.

The Vandenberg volley gun. Rear view showing mechanism ready for loading.

Confederate revolving cannon used in the Civil War. These arms were remarkable developments in the percussion period.

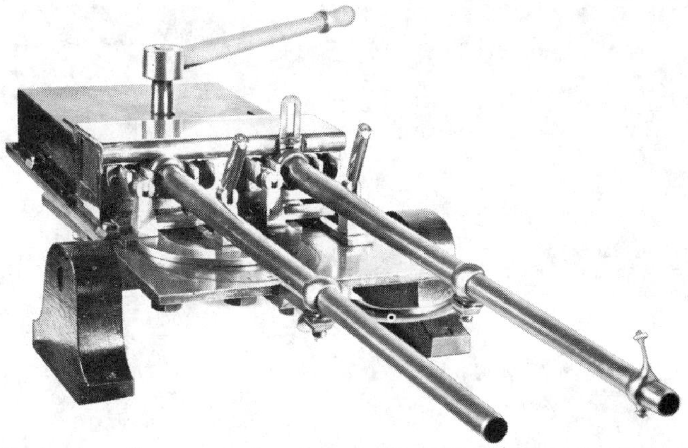

Claxton invention commonly called after its British makers, the Guthrie & Lee machine gun. Caliber .69. It has two barrels. Behind these barrels is a breechblock containing four removable cartridge chambers. There are also four loading troughs with loading pistons located near the rear ends of the barrels on each side. The gun is operated by a long arm which is moved back and forth for a limited travel. This is a percussion-fired gun.

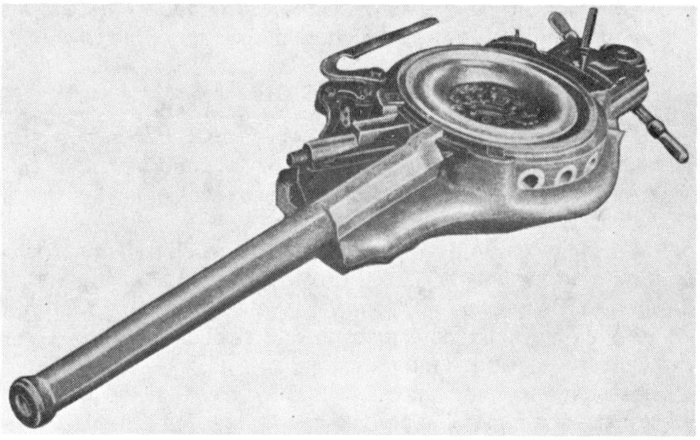

Confederate Gorgas gun on a variation of the Cochran principle.

foot square at one hundred yards. Its weight, of course, limited its value in land warfare. It was used to some extent experimentally by the British, who intended it for naval use.

Throughout the Civil War the Confederate States showed far more initiative and imagination in design then did the North. There were many other very unusual developments, several of which are to be seen at the United States Military Academy at West Point today. Notable among the types developed was one not unlike that of a common revolver.

MECHANICAL MACHINE GUNS

THE WILLIAMS MACHINE GUN

The first use made of the machine gun in actual battle of which we have any positive knowledge was that of the Confederate States of America during the Civil War. This gun was invented by Captain Williams of the Confederate Army who was from Covington, Kentucky. He developed a very unusual arm in the form of a one-pounder gun having a 1.57-inch bore. It was mounted on a howitzer-type limber to be drawn by a horse between shafts. The Confederate Bureau of Ordnance adopted the gun at the outbreak of the Civil War. It was considered a very serious secret weapon at the time.

This gun was operated by turning a crank. It had a single barrel. The crank pulled back and pushed forward the breech block. Self-consuming paper cartridges were dropped into the piece by assistants as the operator turned the crank. This gun could be fired at the rate of about 65 shots a minute. A battery of them proved particularly efficient in battle use on May 3, 1862, at Seven Pines, Virginia. Several of the guns under the control of Captain

Confederate Williams rapid-fire One-Pounder. Probably the first machine gun actually ever to see battle service. Several were used by the famous Pickett's Brigade.

William's Civil War Confederate rapid-fire gun. Gun open ready to load.

Williams himself were used by Pickett's Brigade. These arms received considerable use throughout the period of the Civil War.

THE GATLING GUN

The Gatling goes down in history as the first successful mechanical machine gun. It is to be differentiated from the later Maxim which was automatically operated by recoil of the individual cartridges as they were fired.

The Gatling falls into that classification of machine arms where operation by manual effort, usually by hand crank or lever, causes the gun to fire repeatedly, doing all the functions of loading, cocking, firing, extracting, ejecting and reloading. (Note: Later day Gatlings have been motor driven.)

These arms were so important in their day that the historian must make some mention of their inventor.

Richard J. Gatling was born in Hertford County, North Carolina, September 12, 1818. His most notable achievement was the gun which came to bear his name.

Gatling's father developed several machines in connection with cotton planting. His son assisted in the manufacture of some of these mechanical aids, and by himself patented a machine for planting rice. He adapted this planter to handling other grains and he traveled through various cities in Indiana, Ohio, and Missouri offering the invention to farmers.

After studying medicine for two years at La Porte, Indiana, he entered the Ohio Medical College where after a year he was given a degree. So far as is known he never practiced medicine. However, he always used the title and throughout his life was known as Dr. Gatling. At that time smallpox periodically swept the country. Legend has it that Gatling studied medicine merely to be able to protect his own family in case of illness. There is no documentation for this interesting story, however.

Gatling and one of his guns with modified Accles feed. Dr. Gatling, in his time, thought his weapon was so deadly that the terror it induced would prevent wars.

Gatling machine gun, caliber .58. One of the earliest models of the Gatling gun. Made in the rifle caliber of the period. The feed is missing.

His real vocation was mechanics and invention. In 1851 he developed the idea for a gun which he patented in 1862. He designed it to defend covered bridges and general installations. He seems to have studied both the Ager and the Ripley guns which we have already mentioned. His early model, which was a far cry from the later successful gun, was actually a combination of the fundamental system of the Ager coupled with an improvement of the multi-barrel system of the Ripley.

His early gun used paper cartridges and also could be adapted to handle steel chambers as in other models which we have discussed. The model, while good for its type, obviously could not be too efficient with the type of ammunition available. With the advent of the metallic cartridge he was able to build a practical model of a new revolving-barrel design.

After receiving some assurances that his arm would receive military attention if it could pass tests, he raised enough capital to start manufacture of six guns. Before the work was completed, however, the factory was destroyed by fire. The blueprints, all his data, and all the work were destroyed. After a period of reorganization, he raised enough capital to again attempt gun manufacture, this time at Cincinnati, Ohio. He produced twelve new guns. Gatling then hit on the idea of using new copper case cartridges instead of the steel cylinders or paper cartridges he had been using. He utilized the new metallic cartridges of rimfire pattern. Numerous mechanical changes were required in connection with the new ignition system, and this gun really should be listed as number two in his design series.

However, these early Gatlings operated on a principle analogous to the common revolver. There was a gap between the chambers holding the cartridges and the barrels themselves. Gas leakage, of course, was considerable because of the impossibility of keeping a tight joint.

These initial models had six barrels grouped around a central axis. The caliber was .58 rimfire.

Gatling was not a believer in nonsensical secrecy. Realizing that the success of his gun would hinge on general knowledge of its operation and how it would be made, he invited attention by publishing illustrated accounts giving full details of his design. European nations immediately were interested in the subject, but he received practically no support in this country.

He attempted to have the arm considered by the Chief of Ordnance in Washington, but was refused. Even in Civil War days we had our cloak-and-dagger mentalities. One reason later given for the refusal of the Ordnance Department to even test the Gatling was that confidential reports had been submitted indicating that Dr. Gatling was actually in sympathy with the South; and it was feared that he might have an ulterior motive in presenting the gun which might in some mysterious way be injurious to the Union!

There is little doubt that Gatling's sympathies were with the Confederacy. But official blindness in confusing Gatling's politics with his invention cost the Union a heavy toll.

However, the gun was brought to the attention of General Benjamin F. Butler, later to become infamous in Confederate history for his treatment of the women of New Orleans. General Butler was nevertheless one of the most remarkable people of his period. A self-made man, who during his lifetime amassed a huge fortune for those days, he later became Governor of Massachusetts. Butler early developed an antipathy toward professional militarists as an outgrowth of his inability, in spite of marked superior mental capacity, to get admitted to West Point. He made his way up the military ladder by sheer force of ability and drive.

Butler saw at once the terrific potential of the Gatling. He ordered twelve purchased at a price of $12,000 for the guns on carridges, together with twelve thousand rounds of ammunition. General Butler personally directed the use of these guns during the siege of Petersburg, Virginia, with telling effect. Official policy nevertheless was able to block further purchases.

In 1863 Gatling offered the guns to the French Royal Artillery. The offer was given considerable attention by French personnel but nothing was done about it. In 1870, seven years later, we find the French obsessed with the idea of a "secret weapon" based on the notoriously ineffective deReffye volley-type gun. The Gatling might have altered French history.

Gatling in 1865 entered into a contract with the Cooper Firearms Manufacturing Company of Frankford, Pennsylvania, to produce improved models of his Gatling gun. This new model did away with the separate chamber system and incorporated the chambers in the barrel breeches themselves.

This new model introduced a modified bolt and a cam arrangement for performing the functions of loading, firing, extracting, and ejecting. This basic gun is representative of all the successful Gatlings which followed it. While a loader places cartridges in the feed, the operator aims the gun and by turning a side crank fires it. A series of beveled gears turns the main shaft. The barrels, carrier, and cylinder all turn with it. Cartridges drop into the grooves of the carrier from the feed as the barrels are rotated by this action. The bolt, engaging in cam surfaces, is moved forward to push the round into the chamber. A cocking lug on the striker compresses the spring. When released at its high point, the striker is driven forward by the spring to hit the primer and discharge the cartridge.

Colt-made Gatling with Accles drum-feed.

Colt Gatling with feed removed to show direction of cartridge insertion. Gun is at West Point.

The bolt starts to the rear as the rotation is continued. The extractor hooks free the empty case and draw it to a point where the face of it hits the ejector. The empty case is thus thrown out of the gun through an opening in the housing.

Each barrel is fired as it reaches the lower right-hand position. The cycle of operation of each bolt and its barrel assembly is completed with one revolution, and fire continues as long as the crank is operated.

After a successful Army test in 1864, Gatling was granted a patent on his changes in May of 1865. Tests were run on a new design of one-inch caliber which had been recommended by the Chief of Ordnance. These were made at Frankford Arsenal at Philadelphia. All were highly successful.

In 1867 Gatling contracted with Colt's at Hartford to build 100 guns. These were delivered and all proved eminently satisfactory. The Colt's Company continued manufacture of the Gatling through the life of the gun's existence. The next development was the introduction of a new .50 caliber center-fire cartridge to replace the old .58 rimfire. This too was developed into an eminently successful gun. The United States adopted this 1865 model.

Shortly afterwards Gatling licensed Paget & Company at Vienna and W. G. Armstrong Company in England to manufacture the design. Both produced ten-barreled types to handle the cartridges used by the various governments for which they manufactured them. In addition, some one-inch models were produced with both 6-and 10-barrel types, the latter being of course extremely heavy and intended basically for fortification work. One celebrated test of the efficiency of the Gatling was that conducted at Carlsbad, in 1869. One hundred crack infantry troops with the Dreyse Needle gun, men who had been trained for volley fire, were pitted against the Gatling gun.

The range was 800 meters. The one hundred riflemen and the one Gatling of rifle caliber were to fire the same number of bullets. The soldiers succeeded in making 27% of their bullets hit the target, while the Gatling scored 88% hits. Strangely, this devastating demonstration of the Gatling potential still was not enough to awaken the Prussians in that day to the fact that their Needle gun was already obsolete.

We have already noted the Battle of Plevna in 1877, where American arms in the form of long-range rifles and rapid-firing Winchester 1866 rifles broke the back of the Russian assault on the Turkish position. Curiously enough, the Russians during that battle were equipped with some Gatling machine guns which they used, however, only for night firing and for use on bridge and approach covers. They had at hand the means to equalize the tremendous advantage the Turks had over them in the way of rapid repeating arms, but failed to recognize the potential of the Gatlings in their possession. The Turks, too, had a few Gatlings, and they made better use of theirs, though they did not understand all the implications of their use.

The British purchased Gatling guns but insisted that they handle the already outmoded British Boxer cartridge. This, of course, was not suited for machine gun use and the gun therefore was not too efficient. Usage was generally confined to the Navy.

The British used Gatling guns in the Zulu War of 1879. Their firepower proved of tremendous value in stopping charges of hordes of savages, but their liability to jam with the Boxer ammunition then used limited their value.

The gun was later improved by the addition of the Accles feed, a drum-type gravity feed much more positive than the vertical type previously used. The guns finally adopted by Russia were purchased from Colt's in Hartford. The head of the Russian Commission in 1871 was a General Gorloff. He had the guns chambered for the Russian infantry rifle cartridge. Four hundred guns were made and delivered. He insisted that each one have his name stamped in Russian.

Quite as today, it was just a question of time before the Russians forgot all about the inventor of the gun. The Gatling was later made in Russia's own arsenals, under the name "Gorloff." In the Russo-Turkish War the identical guns, except in different calibers of course, were used to some extent by both forces, as we have noted. The only difference was that the Turks used them under the American name of Gatling, the Russians called them (and still call them) Gorloff's.

From then on until the adoption of the Maxim, the British, too, used increasingly large numbers of Gatlings, even manufacturing them under license.

One ironic note in American history is always brought to mind by any historical discussion of Gatling. This is the famous Custer Massacre. In 1876, when Custer's command was wiped out at the Little Big Horn, his headquarters actually had in its possession four Gatling guns with a rate of fire of approximately 1,000 rounds per minute! These were the perfected types chambered for the .45-70 center-fire rifle cartridge, the cartridge with which Custer's troops were equipped. When Custer was surrounded by the Indians, his troops were armed with the single shot cam-locked Springfields (which were still in use, for that matter, at the time of the Spanish-American War). A few Indians had the terrific advantage of the firepower of Henry and Winchester rifles which they had bought from traders. Custer had, in the availability of these Gatlings, the solution to any such Indian tactics as those which resulted in the annihilation of his command. No use, unfortunately, was made of the Gatlings and another epic of history was written—on the wrong side of the ledger!

At the time of the Spanish-American War in 1898, Captain John H. Parker, recognizing the value of the Gatling, requested permission to organize a Gatling battery. His immediate superiors refused to permit it. He went directly, however, to his commanding officer and received permission to get both the equipment and the necessary skilled men to operate the guns. Parker made most effective use of these during the campaign in Cuba. He took up where General Benjamin Butler had been compelled to stop thirty-six years before at Petersburg, Virginia. To "Gatling" Parker goes much of the credit for developing the ideas which first led to our machine gun tactics.

As the years went on, Gatling worked constantly to improve his device. The introduction of the automatic Maxim, of course, doomed the mechanical principle on which the Gatling operated. However, Gatling was not to be dissuaded. He developed a strip-type feed and an electric motor built into the rear of the gun casing. The motor could be detached and the gun fired by a hand crank if electricity was not available. This and other attempts at power driving the gun gave a tremendous rate of fire, as much as 3,000 rounds per minute. A final development incorporated some actual gas operation. However, although his guns were chambered for the .30-40 Krag and even the .30-06 Springfield cartridge, they were never adopted by the U. S. armed forces. As a closing note it might be mentioned that in recent years special applications have again been made of the mechanical operation of the Gatling system, particularly in the aircraft gun field, such as the U. S. 20mm M61 aircraft cannon.

OTHER AMERICAN MECHANICALLY OPERATED MACHINE GUNS

The Gardner Machine Gun

Among the machine arms to appear in this general period must be mentioned the American Gardner. It had a certain degree of acceptance in Great Britain, though in spite of very successful tests in the United States we did nothing with it.

William Gardner was a Captain in the Northern army during the Civil War. He designed and built his original gun in 1874. He was unable to finance it. He turned manufacture over to Pratt & Whitney at Hartford, Connecticut, on royalty. Francis Pratt had been a master mechanic at the Colt's plant in Hartford and knew both design and production techniques. Within a year he actually was producing the gun in the Pratt & Whitney factory for military sale.

This organization was the parent firm of the great Pratt & Whitney group which later became noted for its aviation engine developments in the United States.

The Gardner gun was composed of two barrels placed parallel to each other, about 1 1/4 inches apart. They were housed in a single casing. Loading, firing, and ejection were all handled by turning a crank as in the Gatling system. In all crank-operated arms accuracy is not too good, partly because of torsion produced as the crank is turned. Yet the system was an efficient one in its era.

The Gardner was purchased in some quantity by the Royal Navy to supplement its Gatling guns. It was simpler, cheaper, and easier to transport. Some were used by British Navy land groups in the Sudan and Upper Nile campaigns in 1884 and 1885. Later attempts were made to convert the system to a belt-feed operation using smokeless powder cartridges, the modification being popularly known as the Robertson, after the British modifier. However, the approach to the recoil automatic operation of Maxim doomed all such hand-operated attempts at machine arms.

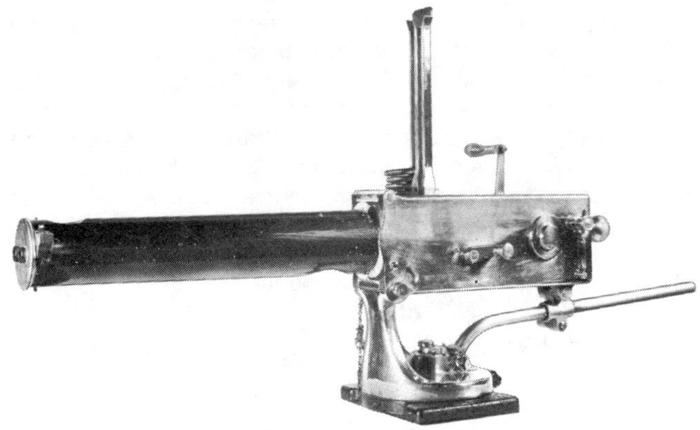

American Gardner machine gun, Model 1879, caliber .45.

The Lowell Machine Gun

Another gun externally resembling the Gatling, but otherwise quite different, was the Lowell, manufactured at the city of that name in Massachusetts. It was designed by De Witt C. Farrington in 1875.

Our Navy gave some attention to this unusual gun. The Russian Navy also became interested and purchased 20 of them. As an interesting aside, it must be noted that the State of California purchased three for protective use in its prisons. The city of Cincinnati, which had a very forward-looking police organization in those days, also purchased one for possible riot use. However, nothing of any true importance came from the development because its mechanical system had already been superseded by more developed design. In later years the Italian designer Alfredo Scotti utilized a variety of the Lowell's 3-barrel system in a gas-operated machine gun. (The principle is that of having three barrels so mounted that as one overheats the next can be brought immediately into position to sustain fire.)

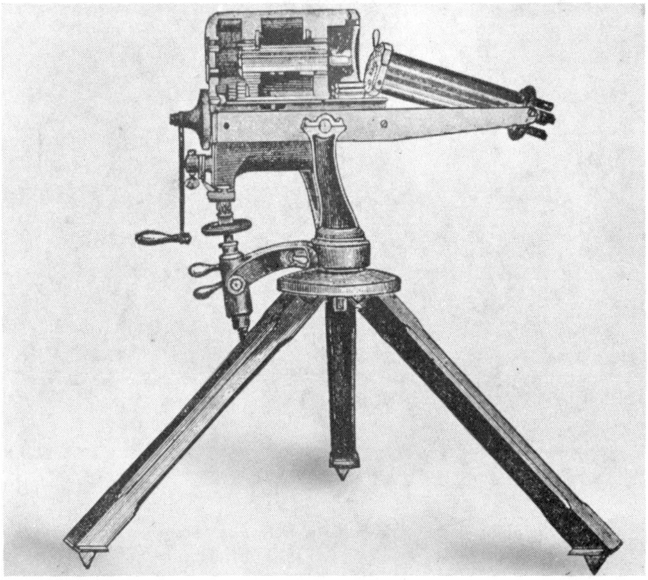

Lowell machine gun from a contemporary drawing. Barrels are hinged open. It was used to some extent and for developmental purposes by the Russian Navy.

Gardner machine gun, Model 1879, showing method of feed, breech open. Barrels are hinged open. This system had several meritorious features.

The Wilder Machine Gun

The Wilder gun developed in 1876 at Hillsborough, New Hampshire, by Elihu Wilder was another instance of a gun coming too late to be of practical value. Wilder arranged his barrels in a half-circle instead of in the full circle of the Gatling type. This permitted use of fewer locks and smaller parts, making for a far more reliable weapon. However, the inventor turned his attention to adapting the system to a 37mm gun which, had it been produced, might have been a tremendous competitor for the Hotchkiss. Wilder died before completing this model and the Wilder gun never was put into production.

McLean "Peace Makers"

Some passing mention must be made of the "McLean Peace Makers." These were a highly publicized group of promised weapons offered by a Scotch gentleman who was quite a medicine man in his day. None were ever made. Barnum probably took lessons from McLean.

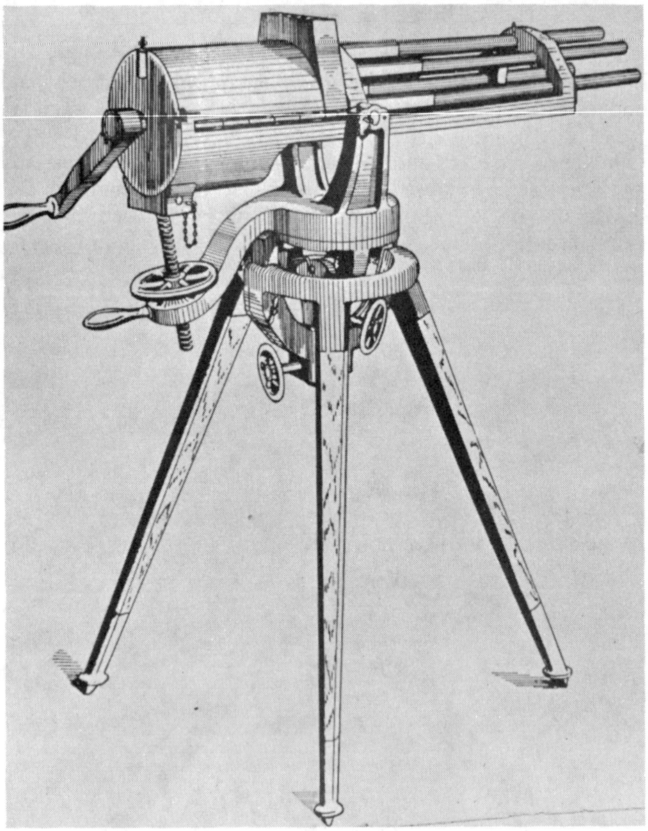

American Wilder machine gun. From contemporary drawing.

McLean made a fortune in the patent medicine game. And then he met one Myron Coloney of New Haven, Connecticut, a gentleman who considered himself an inventing genius. In fact he admitted it! These two men proceeded to issue extravagant pamphlets (and to offer stock for sale in ventures) describing an amazing range of alleged gun developments, none of which were ever actually developed, but all of which laid the groundwork for our modern Buck Rogers.

In a 200-page brochure these gentlemen offered such little items as a 48-shot repeating pistol—a 128-shot self-loading rifle—machine guns capable of firing 2,000 shots a minute and sweeping an area of 6 miles—and a host of other fantastic items.

Amazingly enough, out of this tremendous hodge-podge of impossible promises, McLean actually presented some drawings which did have an effect on the development of firearms, though he personally did nothing in this connection! One of his illustrations shows a cannon with a tubular feed. To produce the proper type of operation, the artist showed the cartridges having no rims and having an indentation (or cannelure) around the heads of the cartridge cases to permit extraction! This in a day when the rimmed case alone was known!

This depiction of a true rimless cartridge case, the development of which did so much to produce superior machine arms, was a direct influence on actual weapons and ammunition designers in the production of the rimless cartridge case as we use it today.

The Bailey Machine Gun

Fortune Bailey's machine gun of 1874 is worthy of passing comment, as it is part of the history of the great Winchester organization. A sample was made by the Winchester Arms Company to handle a .32 caliber rifle cartridge. This arm externally resembled the Gatling gun. However, it was a belt-feed operation of very unusual design. The amazing feature in the belt feeding was that the rounds were never removed from the belt as it passed through a conveyor and was fed through the gun for firing. Again, although this arm had a number of features which would have made it an extremely important weapon some 30 years earlier, it too was doomed by the recoil operation which at that time was getting worldwide attention.

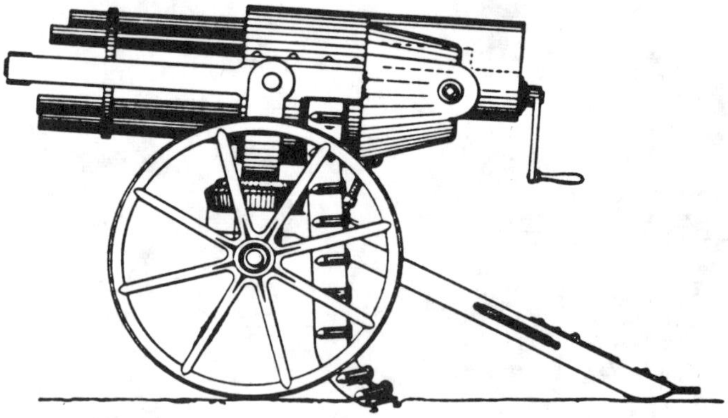

American Baily machine gun. Probably the first to use a belt feed. From contemporary drawing.

RECOIL OPERATION

THE STORY OF HIRAM MAXIM

Early History

It may truly be said that the real history of the machine gun and of automatic small arms began with the inventions of the American, Hiram S. Maxim.

Hiram Stevens Maxim was born February 5, 1840 at Sangersville, Maine, the youngest of a family of seven children. He is the unquestioned father of automatic firearms as we know them today. He was one of the great geniuses in mechanics of all times. A short account of his astonishing life has a place here in these pages.

At the age of six Maxim was enrolled in the local school at Sangersville. His unusual mechanical abilities were discernible very early in life. At the age of twelve he encountered a former sea captain who taught him to read latitude and longitude. Having no money to purchase a chronometer, with his own hands he built one which worked perfectly. As in the case of so many other successful self-educated men, a few books seemed to have had a very early and important development in shaping his life. His father, impressed with the interest young Maxim showed in the subject of astronomy, purchased for him a book on the subject which he studied avidly. Another of the formative books to which he was subjected was Comstock's Natural Philosophy, together with of course the inevitable and invaluable family Bible. The Maxim family was of French Huguenot descent. They had been driven out of France to England, and long after emigrated to Plymouth County, Massachusetts. Maxim himself tells us in his autobiography that the family emigrated "so they could worship God according to the dictates of their own conscience, and prevent others from doing the same." His wry sense of humor also was early developed! These characteristics stayed with him through his long and profitable span of life.

Maxim was apprenticed to a carriage maker at the age of fourteen. He was already noted for having built a boat and for being a natural mechanic. Working days of sixteen hours were quite common at that period, and from this very difficult early appren-

Hiram Stevens Maxim, 1840-1915. The father of the successful automatic system in firearms.

ticeship Maxim also developed many of the traits which stood him in such good stead later on.

This apprenticeship paid four dollars per month, not necessarily in cash. Maxim next turned his hand to manufacturing rakes for a few months. He returned to school and supplemented his income by selling the skins of animals he and his brother trapped. Hunting and trapping gave him a familiarity with arms and their mechanical principles as well as a rugged, healthy physique.

As an outgrowth of his apprenticeship, however, Maxim was able to get employment in the Flint Carriage Shop at Abbot Lower Village. This was a noted plant in its day, as it used machinery operated by water power. During this period young Maxim developed an ability at drawing. He spent considerable time designing parts and sketching various items he thought might improve the carriage business.

He spent four years at the Flint establishment. He left there while still a minor to operate a grist mill at Abbot, Maine. All grist mills, of course, were mice-infested and Maxim felt he must do something to get rid of the mice. Legend has it that one must merely develop a better mouse trap and the world will beat its way to one's door. Unfortunately the maxim did not work very well for young Hiram Maxim.

He actually did invent a better mouse trap, one operating through the use of coil springs, on a rather clocklike design. As the trap was tripped on catching one mouse, it would automatically reset itself for another. He tells us that his first experiment with it caught five mice in the first night. Nobody seemed to be unduly interested in his better mouse trap, however.

Maxim, as so often happened during his life, was ahead of his time even on mouse traps; years later this type of mouse trap was produced by dozens of small manufacturing plants for whom it earned large sums of money. Maxim, of course, never benefited.

The grist mill was successful but unfortunately payment was normally made in grain, not cash. When he couldn't buy clothes with grain, Maxim sold the business and moved to Dexter, Maine. There he obtained employment as a woodturner, an occupation in which he was engaged when the Civil War broke out. He joined a local home guard company but soon tired of "playing soldier," as he called it. Shortly thereafter Maxim decided the war could not last very long, and that if he enlisted he would soon have to hunt another job upon his discharge.

On the advice of a friend whose judgment he valued, Maxim left Maine and moved to Huntingdon, Canada. In later life this move was often brought up in the course of arguments he had with various business competitors. Many took occasion to point it out and to sneer that he had left the United States to dodge the draft. In all due justice to Maxim there is nothing in the record to indicate that he was so motivated. In fact two of his brothers did serve, and it was Lincoln's policy not to take more than two from any one family.

In Canada he worked at dozens of varying occupations ranging from painting signs to bartending and decorating sewing machines.

A rather strange incident resulted in his return to the United States. During the course of doing painting work, he was engaged to produce a new type of blackboard for a school house at St. Jean Chrisostome. He developed a paint which permitted the chalk to work well on a plain plaster wall, thereby saving the school hundreds of dollars.

He asked the huge price of six dollars for payment. The school board refused on the ground that although they were saving considerable money, Maxim himself had only used 40 or 50 cents worth of material!

From this experience he developed such a strain of bitterness against Canadians that he moved back to his old home in Maine. At this period he encountered still another book which had a great influence on his mental development. This was Ure's DICTIONARY OF ART, MINES AND MANUFACTURES. He tells us that he spent an entire winter reading and digesting this work, and that it marked a major step in his educational development. He recounts that the village girls teased and ridiculed him for spending time with a dictionary rather than with them. Curiously, he developed almost an antipathy for women as a result of this hazing, although a few years later he moved to Boston where he was married after an engagement of only a few months.

After traveling rather extensively through the West and South, working at one thing and another, he returned to Fitchburg, Massachusetts where he went to work for an uncle, Levi Stevens. Here he learned a great deal about casting methods and engineering generally. During this time his uncle had contracted to manufacture a number of automatic illuminating gas machines for the Drake Company of Boston. These were extremely important items in that day. Maxim made a study of the machines and was responsible for many improvements in them. As a result of this work the uncle determined to undertake the production of an improved version of the Drake machine. While it was in the course of preparation for manufacture, young Maxim developed an even better type! As a result the uncle took personal umbrage at his nephew and discharged him, apparently on the theory that Hiram should have had his second thoughts first.

Maxim went to Boston where he entered the employ of Oliver P. Drake whose basic business was instrument manufacture. He worked for him through the period of the closing days of the Civil War.

The Drake automatic illuminating gas machine on which he worked led directly to a number of other inventions of considerable importance in their day.

One outgrowth of the contacts made in the course of this work was the development by Maxim of the first successful automatic fire extinguishing system of which we have any knowledge. In response to a request from one of their Boston customers, Maxim developed an automatic sprinkler alarm which would shower water down on an area where the fire was burning and also sound an alarm in the fire house giving the exact location of the fire position.

The installation worked perfectly, but in spite of it Maxim was unable to interest other building owners in using it.

He did take out patents on the system. It is one of ironies of his business life that only after the patent had expired was the principle fully seized upon. Today practically all such sprinkler systems are based on the principles so early evolved by young Hiram Maxim.

His experience with trying to market the fire extinguisher system and the cost of patents led him directly to an attempt to sell his patentable ideas to other concerns rather than to invest in them himself.

At this point we see that he learned greatly from his sour business experiences. He proceeded to associate himself with Mr. A. T. Stewart, then one of the wealthiest men in the country. Together they formed the Maxim Gas Machine Company at 264 Broadway, New York City. From Stewart he received contracts to provide lighting systems for Stewart's mills and a hotel in New York City. By this association he was able to eliminate the problems of selling his ideas or his products. He proved the American truism that there is no substitute for capital.

About this time, however, electric lamps were becoming widely known and Maxim saw that the use of the illuminating gas system was on the wane. He turned therefore directly to the production of electric bulbs using carbon filaments. As a matter of record, the first electric lights to be used in New York City were those installed by Maxim's Company for the Equitable Insurance Company at 120 Broadway, the building wonder of its day.

As a result of this installation people from all parts of the country flooded into New York to see the new marvel. Out of this interest Maxim developed a rather amusing sideline which still produced money. Only direct current was used at the time. As people stood around watching the lighting equipment many had their watches magnetized and stopped. Maxim produced a machine to demagnetize watches and charged one dollar for the service! The business came to take a great deal of his time. He patented the machine he had developed and then offered it to watchmakers and jewelers. Until the introduction of the alternating current, these de-magnetizers were still big money makers for many jewelers.

His next venture was with the United States Electric Lighting Company where he received the then very large salary of $5,000.00 a year. He gave himself the title of Engineer, one to which he was very well entitled though no school had granted it to him and he had no formal education to warrant such usage.

He was sent to Europe on a trip to attend an exhibition dealing with the electrical industry then being held at Paris. This was in August, 1881.

When he returned home the electrical items on which he reported created such a stir that he was assigned the task of preparing literature and explanations covering all electrical appliances and data of which he had any knowledge. He searched both the French and the Belgian patent files thoroughly in preparation of this manuscript. The resulting data was of tremendous value to the United States Electric Lighting Company in later years when it was utilized to defend against law suits for infringement.

Maxim Turns To Arms Development

His next trip was to London where he went to reorganize the British subsidiary called the Maxim-Weston Company. While he and other American inventors were primarily interested in electricity, Maxim found that Europe was interested basically in arms. As a friend of his, also an electrical engineer, said according to Maxim "Hang your chemistry and electricity. If you want to make a pile of money, invent something that will enable these Europeans to cut each others throats with greater facility."

Use of Recoil

After a visit to Vienna where he again saw the European interest being exhibited in rapid-fire weapons, he turned his attention to the development of a new device in this field.

He tells us himself that his original idea for recoil operation of a weapon came as a result of some target shooting he did with an old Springfield .45-70 rifle years before.

The recoil of this monster left his shoulder black and blue and the thought came to him "Cannot this great force, at present merely an inconvenience, be harnessed to a useful purpose?"

Characteristically he filed the thought away. Characteristically, too, in due course he made use of it.

His examination of the European attempts at machine arms, all of which were operating on mechanical principles, convinced him that a single-barreled arm, recoil operated, would be the answer to military requirements.

Maxim drew heavily on the toggle lock of the Henry-Winchester rifle and on the general design of the American Gardner mechanically operated machine gun for his automatic design. But the basic principle of recoil operation was his. So was the basic principle of gas operation, often erroneously credited to John M. Browning. Browning was the first to use the gas principle successfully, however. Maxim concentrated on the successful recoil operation.

He turned his attention to the matter of making a standard lever action Winchester 1866 rifle operate through use of recoil. He attached a movable butt plate with a spring support on the stock of his Winchester. He arranged a series of jointed levers connecting it with the lever trigger guard.

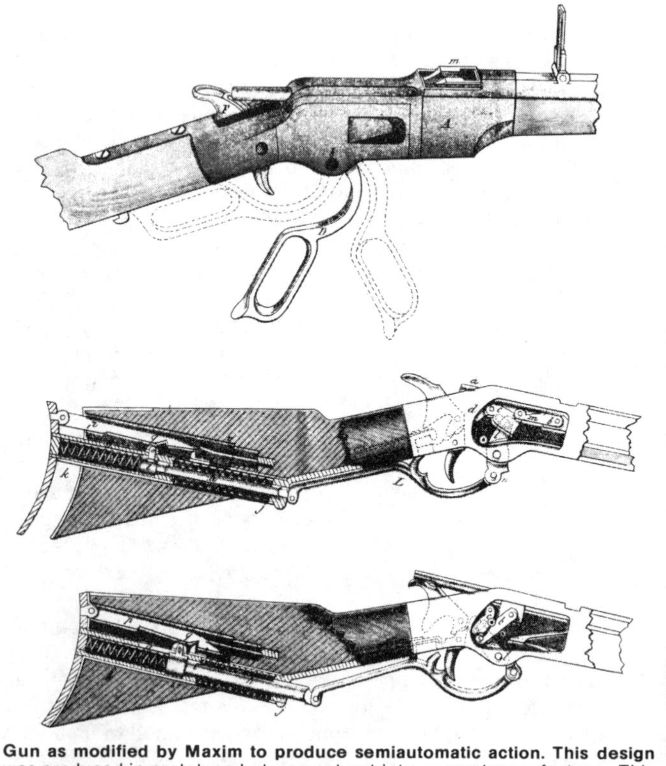

Gun as modified by Maxim to produce semiautomatic action. This design was produced in prototype but was not put into general manufacture. This weapon was not produced in quantity.

The recoil driving the piece back and compressing the spring between the weapon itself and the butt plate operated the levers to open the trigger guard lever. This action of course ejected the empty case and cocked the rifle. The reaction of the now compressed spring under the butt plate at the end of the recoil action forced the arm ahead, and working through the compound levers closed the action on the chambered cartridge ready for the next trigger pull.

This was the first truly successful automatic functioning of any firearm of which we have evidence.

This epochal development occurred in 1883. It achieved tremendous attention throughout Europe. It was applied to many Turkish Winchester .44's, specimens of which are occasionally encountered in arms collections abroad.

First Patent

In 1884 Maxim was granted his first basic patent. This covered a recoil operated lock breech system and was applied to machine guns. From it was evolved many of the important items which followed.

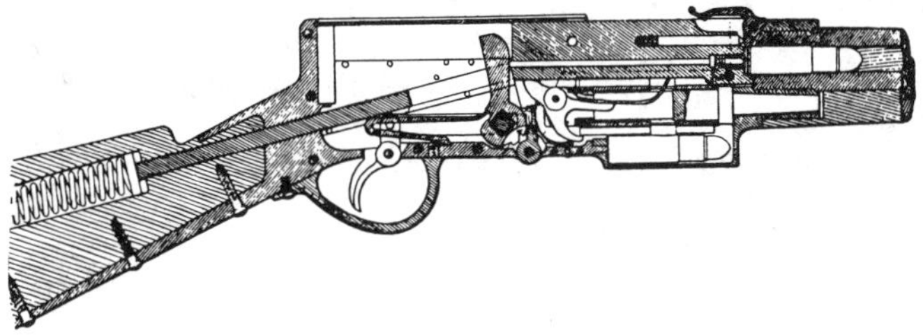

Early automatic rifle developed by Maxim. This weapon was not produced in quantity.

Automatic Rifles

One of his early developments was a fully automatic rifle loaded from a revolving magazine based on the Roper system which was then being tried in repeating shotguns in the United States. A study of the drawings of this arm are of interest. They show positioning of the recoil spring in the butt of the rifle, much as in the common practice in some of the world's finest lightweight automatic and semiautomatic arms rifles, shotguns, and machine weapons.

While Maxim invented several automatic rifles and automatic pistols in the course of his career, none of these was ever manufactured commercially. His entire successful developmental progress centered around the machine gun. It is to that field we must turn for most of our essential observations on his firearms developments, though his principles were exploited by others in lesser arms.

Machine-Gun Development

Maxim began experimental development of a machine gun at a small plant at 57 Hatton Gardens, London. Among the equipment he used was a new Browne and Sharpe milling machine recently imported from America, an item which itself had been developed primarily in relation to gun manufacture. He proceeded to develop all the tools for building the necessary parts, all the holding fixtures, jigs and gauges. The one thing he would not undertake to produce personally was the barrel; barrels were purchased.

The ability, genius, and hard work of Maxim at this point is difficult to imagine. If one remembers that he had no guideposts, since he was operating in an entirely new field, one gathers some little insight into the tremendous problems he faced and overcame despite warnings from "experts" that he could not possibly succeed. He operated in the face of such criticism and foreboding

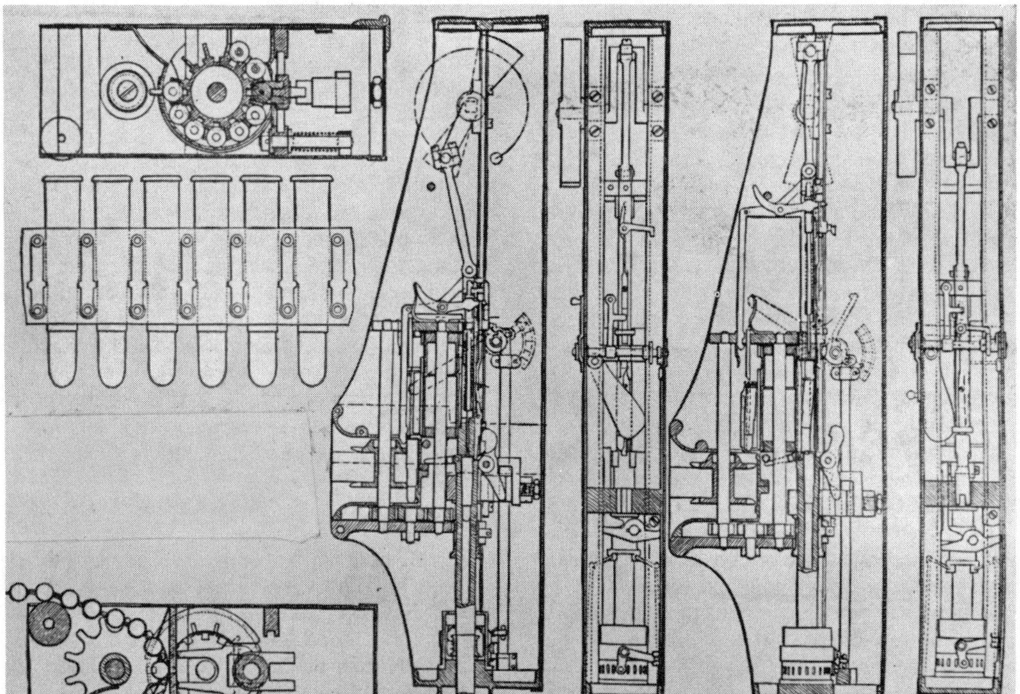

Drawings from Scientific American Magazine in 1884. The Maxim gun basic mechanism has changed very little through the years.

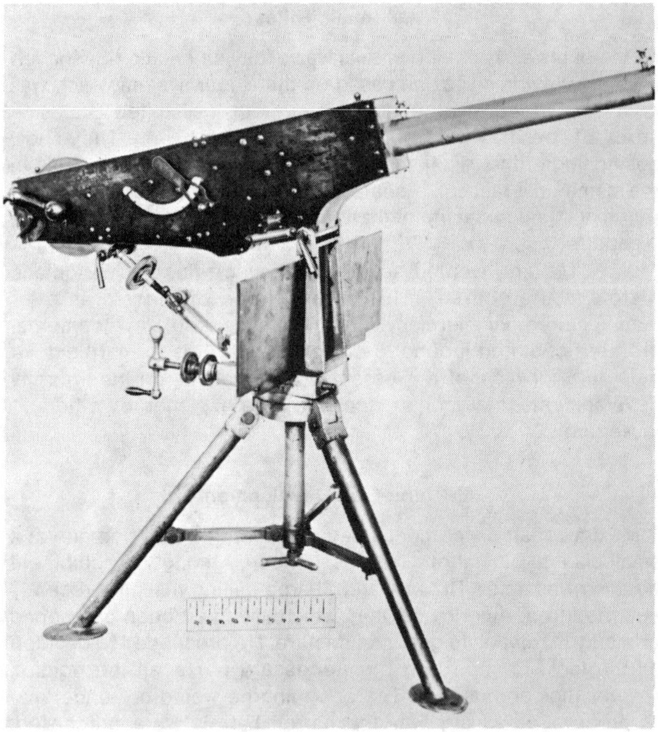

First model Maxim machine gun, 1884. The selector which allowed the weapon to be set on various rates of fire from one to 600 rounds per minute can be plainly seen on the right side of the receiver.

at all times during his life, constantly upsetting the experts in the field.

When his first experimental model was ready for testing, he used a .45 caliber British Gatling gun barrel. Placing six cartridges in the feedway, he discharged them all in about half a second.

From this he knew that his principle was sound and he proceeded to develop working drawings and to lay out methods for actual manufacture of both models and production arms.

While his experiments were made with an orthodox vertical feed, he soon realized that continuous fire with formal systems of this sort would not be practical in automatic arms. Thereupon he developed the cartridge belt feed which was the deciding factor in long firing records with machine arms.

Maxim by this period was well aware of the danger of demonstrating an arm before it was fully tested. He produced several handmade weapons before he released any press notices. Many of these were passed over with amusement by the British press which was generally skeptical of the statements that a single-barreled arm could possibly load and fire itself by recoil energy at a rate of six hundred rounds a minute as Maxim claimed.

However, his Royal Highness the Duke of Cambridge visited the shop to see a demonstration. He was very much impressed. As a result of this visit, important people in London generally traveled to Hatton Gardens to see demonstrations of this amazing new arm. Fashion followed the leader then as now.

On the advice of Lieutenant General Sir Andrew Clarke, Inspector General of Fortifications, Maxim proceeded to develop a system of parts so designed that they could be withdrawn from the mechanism without the use of tools. As a result he developed components which permitted change in case of breakage of practically any part in six seconds.

So sound was the original design that very little has been done to alter any basic features of the Maxim since the day of its introduction! Its modern version known as the Vickers differs very little except in styling and weight reduction and in the fact that the locking action of the toggle is inverted.

The principle of short-recoil operation itself is very simple. The bolt is locked to the barrel at the instant of firing. They recoil together customarily for a distance of about three-quarters of an inch. At this point the barrel travel is halted. The action is unlocked through movement of the toggle and the lock mechanism continues to the rear to extract, operate the feed mechanism, and compress the recoil spring for returning the action to battery and for firing the next round. The "toggle" system is like the human knee operation.

Maxim utilized a canvas belt based on that used by sportsmen to carry 333 cartridges. This belt was seven yards long. It had a clip device on the end to attach another loaded belt to give continuous sustained fire. Maxim's feed mechanism pulled the belt through the action automatically.

The first model was also equipped with an external rate control and firing arrangement which was later discontinued.

We must remember that Maxim was pioneering at this point, both in design and usage. The military itself had no conception of what could be done or even what was wanted in the field. Therefore, he had to develop the arm for use under practically any type of conceivable military situation. An indicator on the quadrant scale on the side of the gun gave the position for the selector at which the gun would fire at the rate of one round per minute. Moving the selector to the rear increased the rearward travel until a rate of six hundred rounds per minute could be achieved.

It was possible to fire in single shots or in bursts of ten, twenty, or one hundred per minute. The selector could also be set to maintain continuous fire at fast or slow speeds.

One of Maxim's ideas here was that should the gunner be killed, the gun would continue to function if so set until the cartridge belt had been exhausted. In actual practice this has never proven to be a desirable feature, so it is no longer encountered.

The rate of fire regulator was a simple hydraulic oil buffer system, modifications of which have been applied to many recoil systems since then.

Maxim also experimented with a drum type of magazine. It soon developed that the belt feed was far superior. He even adapted the belt feed to an automatic rifle intended to be fired from the shoulder, a device which did not prove practical because of belt-snarling.

Unexplored Principle

At this period he produced a rifle which was intended to operate on a principle which has never been fully exploited, nor fully explored. The case was corrugated to permit a slight elongation of it at the instant of firing. Such a system of course would permit the operation of an automatic arm without the utilization of a gas port or trap and without any movement of the barrel, yet permit the use of a simple lock and powerful cartridges.

The metallurgical knowledge of his day did not permit this system to operate properly. It is entirely possible, however, that in our present advanced stage of technology the principle might again be investigated for certain types of automatic weapons.

Maxim's Developments

In the period from 1883 to 1885 Maxim filed patent specifications on just about every conceivable system under which automatic fire might be produced. The results of his tests were followed avidly in Europe. Mauser, Mannlicher, and others proceeded to intensify their own research into the field. And some years later John M. Browning developed the first successful gas-operated machine gun on an earlier Maxim principle.

To Maxim must be credited not only the short recoil system found in so many of the world's finest automatic arms today, but also such developments as adjustable head space (which controls safe operation of the arm); the accelerator (to transfer energy from the halted barrel during recoil to the bolt assembly); T-slot extractor, such as we find in all our Browning machine guns; and the special ejector system. Most guns throughout development all over the world have utilized one of more of these principles. None have particularly been improved since Maxim introduced them.

FIRST MAXIM MACHINE GUN

The first Maxim was designed for a black powder cartridge of its day. In later years it has been built to handle practically every type and caliber of rifle cartridge made in the world as well as a great many types of artillery ammunition. The basic mechanism has remained unchanged through the years. Only cooling systems, methods of mounting, weight, and fire control systems have been altered as required by different cartridges, different usages intended, and improved metals.

The British Government gave Maxim his first order for machine guns in 1887. The order was for three guns for test, and though they passed all tests provided by the Government, the Maxim was not officially adopted.

In the same year 1887 the inventor made a trip to Switzerland where he entered his gun in competition with the Gatling, the Nordenfelt, and the Gardner, all being mechanically operated as true "machine guns."

Maxim's production guns were made by Albert Vickers, a steel manufacturer at Crayford, Kent. This relation between Maxim and Vickers continued to their mutual benefit and eventually evolved into the present "Vicker's gun" after the Maxim patents expired.

All Europe by then was awake to the desirability of a machine arm. Tests were conducted at Spezzia in Italy in competition against the Nordenfelt and also in Austria.

Archduke William, Field Marshal of the Austrian Army, was tremendously impressed with the demonstration. During the test British-made cartridges were used, of course. The Austrian officers desired to have a model tested with their own cartridge. In spite of the importance attached to the tests, Maxim still was unable to get a release from the Austrian government to carry a single Austrian service cartridge out of the country. The cartridge was classified as a top military secret by the Austrian brass of that day. In case this seems ridiculous we must point out that the same nonsensical military thinking continues to occur, particularly in our own country. Maxim made a mechanical drawing and took a sample of the brass which went into a case with him to England. The Birmingham Small Arms Company there produced cartridges for the arm. Not being conversant with the Austrian system of cartridge manufacture, they furnished ammunition with considerably less power than the Austrian.

In July 1888 the Austrian Commission conducted experiments with Maxim's guns in Austrian caliber and gave a very favorable report on them. Despite this, however, only a small number were purchased. The Austrians proceeded later to develop the Schwarzlose, an arm far inferior to the Maxim, which we shall discuss later.

Shortly afterwards Maxim tests were made in Germany. The Kaiser visited Spandau Arsenal where tests were conducted in his presence. The arm proved itself a terrific improvement over the Gatling which was operated by four men. The Kaiser was very much impressed by the one- or two-man Maxim operation. He specified that this be the gun that Germany adopt. It remained the basic machine arm of the German forces until the end of War I, and saw much use even in War II.

Later in 1887, Maxim took his guns to St. Petersburg for tests. The Russian military mentality was not as progressive as it is today. They absolutely refused to believe that the gun could operate at the rate of 600 rounds per minute, as Maxim had stated. Apparently the idea of automatic operation was completely unknown to them and they were unable to visualize it! Until the gun began firing they insisted that the handle could not be cranked back more than 200 times a minute, believing it had to be worked by hand. When the belt was placed in the gun and the entire belt load of 333 shots fired by recoil forces without stopping, the Russian staff for the first time realized the meaning of a true automatic weapon.

Despite the enthusiasm with which the military now accepted his product, the secret police proceeded to give Maxim a lot of trouble. He was summoned to Police Headquarters and called upon to give a dossier on himself. Talking with an English speaking official, he was very much disturbed to find that their principal interest seemed to be in his religion! He was advised that he could not remain in Russia unless he had a religion. Maxim said he had none. The official advised that Protestant would be the correct denomination. Whereupon Maxim tells us he said "Then put me down as a Protestant. I protest against this whole thing."

Note on Smokeless Powder

When Vieille developed smokeless powder in France in 1885 he accelerated the development of the automatic machine gun tremendously.

Up to that time Maxim, of course, had been compelled to use black powder ammunition in his gun. With black powder practically all the forces of the explosion are produced at the instant of ignition. Pressure drops very rapidly following its peak.

On the other hand, with smokeless powder ignition rapidly reaches its peak but there is a continuing expansion of gas in the bore following the achievement of maximum pressure. This is called "residual pressure." It continues to exert some rearward force for an instant after the bullet has left the barrel. By taking advantage of this continuing thrust after the pressure has dropped to safe limits, it is possible to fire submachine guns and low-power arms with unlocked breech safely; and also to use relatively lightweight locking mechanisms for the more powerful locked-breech machine guns.

The development had, then, a tremendous influence on the rapid development of machine weapons in other countries.

Early Rise of the Maxim Machine Guns

Maxim guns were first issued to the British forces in the year 1891. They very rapidly displaced the Gatling guns which had been used up to that time. The Maxim proved it's effectiveness at the battle of the Shanghai on the 25th of October 1893. A force of approximately 5000 Matabele warriors attacked a lager of wagons, in which were encamped 672 armed police of the Rhodesian Charter Company and their auxiliaries. The police were armed with Martini Henry rifles, 6 Maxims, one Nordenfeldt and one Gardner machine gun. They also had one Hotchkiss breech loader and one 7-pounder muzzle-loader. The Matabele were largely equipped with spears, but they did have Martini Henry rifles and assorted muzzle loaders of varying vintage. The Maxims and the Nordenfeldt fired about 3,000 rounds, the Gardner fired about 400, and the artillery fired about 35 rounds. At the end of the short action there were about 500 Matabele dead and twice that number wounded, in the area of the lager. The next effective use of the Maxim came soon thereafter in yet another British outpost, the Northwest frontier of Afghanistan. In the Chitral campaign there in 1895 the Maxims again stopped fanatical tribesmen attempting to rush the guns. When the British infantry finally charged the positions occupied by the natives they found nobody alive to oppose them.

The most effective use was made at Omdurman. This is the campaign in which the then "Leftenant" Winston Churchill figured in one of the last cavalry charges in British history. At that time, incidentally, Churchill carried one of the new German Mauser

Early Maxim. Marked on receiver is "Maxim Nordenfelt Maxim's Patent 1893. Guns and Ammunition Company, Ltd." Gun is at West Point; mount is for display only.

automatic pistols, the first time this weapon saw service. British casualties in this battle amounted to less than two percent of the relatively small forces involved. On the other hand over twenty thousand of the ferocious Dervishes were wiped out. Over three-quarters of those slain were officially credited to the machine gunners.

The gunners as yet were not familiar enough with the handling of the weapons to avoid considerable feeding trouble, a fact which makes even more remarkable the terrible effectiveness they showed. The heavy-wheeled mount used with the gun was unsuitable for such campaigns. It became the practice with the British to leave the mount to attract enemy attention and carry the gun to an emplacement. Up to this time, military men generally assumed that the gun could be used only for covering fire in an attack or as a defensive weapon protecting fortifications. It was some time later that the effectiveness of the machine gun as an attack weapon was forced on the attention of the military experts of the world.

The Maxim "Pom-pom"

In the Boer War the Maxim gun was first used on land by white men against other white men armed with the same terrible weapons. In this instance the Boers were also armed with the new Maxim 37mm "pom-pom."

A special gun was developed by Maxim at the request of the British for a gun to be used against light naval craft.

The British government did not promptly purchase the gun. Instead it was manufactured in England for sale to France. African natives in discussing the arm gave the sounds equivalent to "pom-pom" in attempting to tell what it sounded like. This name stuck to the gun through its continuing use.

The Boers made astonishing use of this new weapon. Since it used smokeless powder, they could use individual shots fired at British artillery positions to get the range without worrying the artillerymen particularly or disclosing position. Once they had the range zeroed in properly, the Boers would then cut loose an entire belt of 37mm cartridges into the area. This tactic was used time after time and was terribly efficient in destroying British artillery groups. The rate of fire of the pom-pom was about three hundred rounds per minute, although the belts normally held only twenty-five rounds because of the large size of the cartridges.

Unlike the British who had been responsible for its development, the French recognized the efficiency of this new Maxim gun and placed orders with the Maxim firm for them, primarily for use with the French navy and for resale.

USE OF MAXIM MACHINE GUNS IN RUSSO-JAPANESE WAR

It was during the Russo-Japanese war of 1904-5, however, that the automatic machine gun really came into its own. For the first time large numbers were used by each side under military conditions. They were grouped in batteries of from six to eight pieces in most engagements. Russia used the Maxim with a high-wheeled carriage for a mount. These proved to be unwieldy in action. They also exposed the gun crews to sharpshooter fire. Therefore the Russians substituted low mounts instead. These proved extremely efficient. They were equipped with shields, incidentally.

The Japanese were armed with a gas-operated gun, the French-made Hotchkiss with which we shall deal later. In the battles between the Russian and Japanese forces, the recoil-operated Maxim proved much the more efficient. German observers of this campaign noted the fact and the German army went into very serious examination and production of the Maxim gun. This was their really terrible infantry weapon at the opening of World War I.

The Japanese in a report after the war stated that at the battle of Hei-Kou-Tai cartridges jammed in the Hotchkisses once in about three hundred rounds as an average. This was considered quite dangerous. However, except for refining the gun the Japanese did relatively little about the design until well toward the end of World War II.

It is of historical interest that even the Russian Maxim guns in their newly adopted 7.62mm caliber were made by Vickers Sons & Maxim. The usage initially made in this caliber was at the Yalu River which has become known to every American as a result of our recent Korean troubles. It was here that the desirability of a low mount and a shield to protect the gunners was established by the Russians.

The Spandau (MG 08)

While the British army introduced the Maxim gun in 1891, it is noteworthy that no attention was given to the design in the United States. This was largely due to the fact that oceans at that time constituted a natural barrier and our military saw no need for rapid-firing guns of the type. Even though they had been very much impressed with its early tests, the Germans did not adopt the Maxim until 1899. However, from that point on they worked intensively on it, as already noted. The Maxim gun designated by them as Model 08 was actually produced at the German government armory at Spandau. This name is often given as generic for machine guns both of this type and others in some quarters in Europe today.

The Spandau was equipped with a very heavy sled mount. The mount was later altered and a lighter one produced at the Erfurt Arsenal. The caliber was the standard German 7.92mm Mauser. The German gun used a belt feed of 250-cartridge capacity.

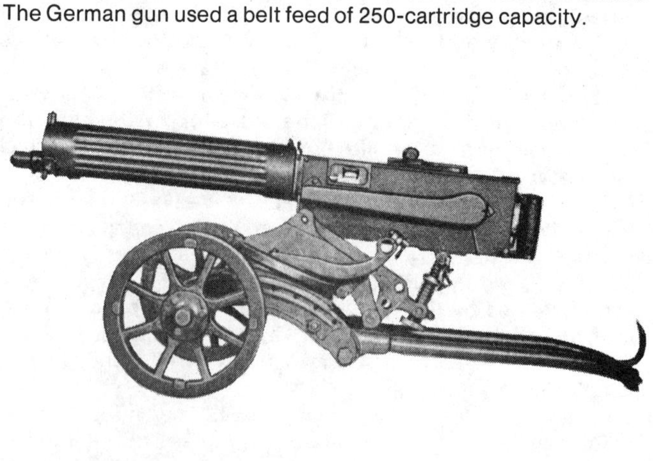

Maxim on wheel mount. With very slight modifications, this pattern was used by the Russians in both world wars. Also used by the Chinese Communists in the Korean War.

Maxim demonstrating the light weight of his L.M.G. (light machine gun) and mount. This was one of his favorite photographs.

Use by Russians

Russia continued, on the other hand, to purchase its guns from Vickers & Maxim. However, it adopted the Russian Sokolov mount. This mount uses two small wheels for field transport and can be equipped with a heavy steel shield to protect the operator. The gun is chambered for the standard Russian 7.62mm rifle cartridge.

Use by Balkan League

When the young Turks deposed Abdul Hamid in 1908, Turkey was so severely disrupted that Bulgaria was able to declare its independence. Other provinces were seized by Austria. A Balkan League was soon formed consisting of Greece, Rumania, Bulgaria, and Serbia. This group in its war against Turkey in 1912 seized all the European Turkish possessions except the city of Constantinople. The basic machine arm used by all these groups was the Maxim. Some even used the pom-pom. The machine guns made in the varying calibers of the different countries involved were in use during World Wars I and II.

Maxim Machine Guns in the United States

American forces during the Philippine insurrection seized some Maxim machine guns in the official British .303 caliber. These had been manufactured as early as 1895. Despite this we as a nation did nothing at that time to purchase any automatic machine guns.

The first official trial given to the Maxim in the United States was in 1888. None were bought. Our Navy did purchase a quantity of Colt 6mm guns which were used in the Spanish-American war.

Due in part to public apathy and Congressional lethargy, our military did little about machine arms until World War I was looming and we found ourselves hopelessly outdistanced by all other major nations.

In 1903 tests were made with the Vickers, the modified Maxim, so named after the Maxim patents ran out and some few changes had been made. The guns were then being built at Hartford by the Colt's Company. These guns were practically identical as to operating mechanism with the one Maxim had shown at the 1888 trials when he used black powder cartridges. The United States Army adopted the model as manufactured by Colt's to use our standard caliber .30 cartridges. It was designated the 1904 model Vickers. In 1916, as a result of tests made against seven competing machine guns, the Army Board recommended purchase of 4600 Vickers guns, M1915. At the time we did not have one machine gun in the Army which was actually of a type we would care to use on the European front.

Meanwhile the famous Browning gun had been developed. The heads of the Army were now so impressed with this model that they insisted that no official adoption be considered for any other machine gun. But in 1917 it was impossible to produce the gun for prompt use. We called on Colt's therefore to manufacture more Vickers with which to arm our forces. Colt's had been producing that machine gun for the Russians in 7.62mm. The necessary retooling to produce it for the United States caliber .30 cartridge hindered production and it was July, 1917, before initial deliveries were made. Altogether during the period from then to September 12, 1918, 12,125 of these guns were made. Over 7,000 American Vickers were shipped overseas.

There is a lesson here which we as a nation might learn if we would. It is shown by the figures of World War I, when the Germans at the outbreak of the war had 12,500 of the Spandau-made Maxims ready for service. Remember that we did not have a single acceptable machine gun ready for active military field use then. And at the end of the War we still had very few of our own at the front.

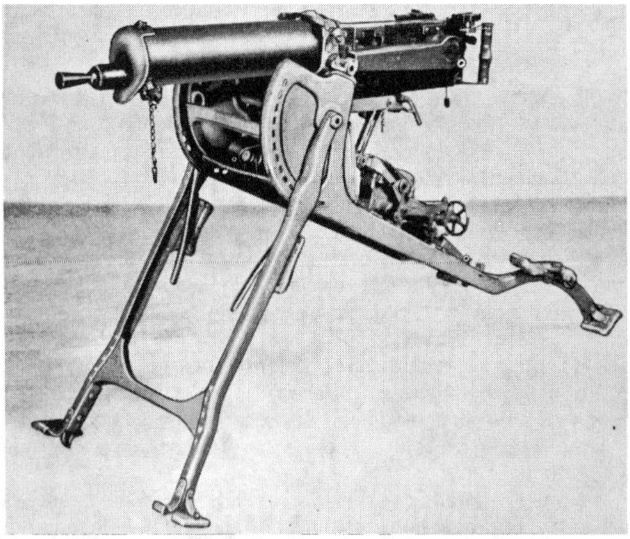

Spandau with folding sled mount. This was a prime German weapon in World War I and saw service in World War II also.

GAS OPERATION

THE AMERICAN DE KNIGHT MACHINE GUN

This gun was produced only in prototype. It is of interest basically because it represents another abortive attempt by Pratt & Whitney to enter the machine gun field—their first having been with the early mechanical Gardner, as we have noted.

The gun was invented by Victor De Knight of Washington, D. C. in 1898. It was a tripod-mounted gun for the .30-40 Krag service cartridge. Water cooled, it looked like a conventional recoil operated arm externally, but actually was a rather complicated gas-operated weapon. A pivoting block lock was used, actuated by a gas piston. Feed was by belt.

When we saw war coming in 1916, the De Knight gun was brought out of moth balls and subjected to test for possible adoption. It failed rather miserably. The original gun was made by Pratt & Whitney under arrangement with De Knight. It is doubtful, however, that it ever had a chance because of its involved design.

De Knight machine gun made by Pratt & Whitney.

THE AMERICAN MARLIN MACHINE GUNS

At the outbreak of World War I the Marlin-Rockwell Corporation at New Haven, Connecticut, received a contract to produce Colt-Browning Model 1914 machine guns to be used by infantry for training.

In view of the need for aerial machine guns, Marlin-Rockwell undertook to alter this infantry design. The pendulum type of operating lever below the barrel of the gas-operated Colt-Browning gun was impractical, of course, for aircraft work.

The modification was actually worked out by A. W. Swebilius, a Swedish-born American who began work at the age of 16 with Marlin and developed into one of the finest firearms designers in the country. Swedilius later organized the High Standard Arms Co. whose pistols and shotguns have achieved considerable commercial success.

He substituted a reciprocating piston below the barrel for the pendulum lever of the original Colt. The gun was lightened throughout and other modifications were made as required, including a cocking mechanism consisting of a handle on the right side of the gun itself.

It is a great tribute to the skill of Swebilius that in the course of a few weeks time he completed a major transformation in the original gun. The Marlin was used throughout the war and for several years thereafter as the principal synchronized machine gun of the

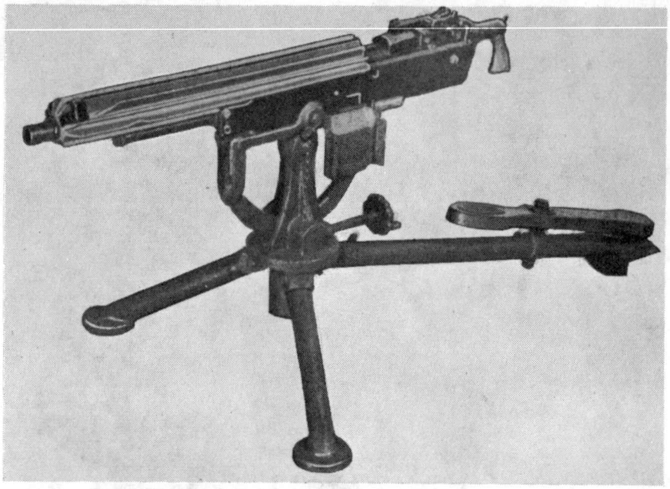

Marlin modification of the Colt-Browning machine gun for tank use. This gun was made primarily in aircraft types.

American Air Corps. It was also produced in a design for tank use and one was also designed for potential infantry service.

The Parabellum and the Vickers had been synchronized for aircraft work, a relatively simple matter since they were recoil operated. The Marlin was the first gas-operated weapon to be successfully synchronized.

As issued, the Marlin 1918 model was belt fed and gas operated. Cooling was by air. It weighed only 22 1/2 pounds and had a fire rate of 630 per minute.

The tank version could be used with a tripod mount. A longitudinal-finned cooling jacket was placed about the barrel. The grip section was copied from the Colt Model 95.

The Marlin ground gun patterned its barrel on the heavy design of the original Colt. The operating handle and the general design was that of the Swebilius modification.

THE AMERICAN LEWIS MACHINE GUNS

Since the various successful designs of the Lewis gun are discussed in detail in Part Two, we have place here only for its influence on the history of automatic weapons development.

First Airplane Machine Gun

Most noteworthy perhaps is the fact that the Lewis constitutes the first aerial machine gun, having been tested in June, 1912.

The test was made when the gun was taken up in a Wright pusher-type plane equipped with a crossbar on which the pilot and observer rested their feet during flight. The gun muzzle was rested across this bar while Captain Chandler fired the gun at the target as they flew over it.

This first experimental firing of a new type of machine gun from an airplane stirred the interest of the newspapers and magazines to great heights. Hundreds of writers recognized the import of this new development and it was given tremendous publicity.

The promoter of this gun was Isaac Newton Lewis, born on October 12, 1858 at New Salem, Pennsylvania. He graduated from the United States Military Academy at the age of 21 and was assigned to the Coast Artillery.

In 1910 the Automatic Arms Company of Buffalo, New York approached Lewis with a business proposition. They offered him stock in the Company if he would produce a machine gun for them. They owned a number of patents which they offered as a starting

Colonel I. N. Lewis, U.S.A., the developer (not inventor) of the famous gun which bears his name. The system was actually invented by Samuel McClean.

point. He accepted and his first prototype weapon was ready in 1911. The machine gun was the now well-known air-cooled, gas-operated Lewis with magazine mounted on top of the gun.

In August of 1914 two British airmen carried a Lewis gun aloft with them without authorization. At an altitude of 5,000 feet they emptied a drum at a German plane. They missed. This represented the first active use of the machine gun in aerial warfare.

When the pilots reported their adventure, instructions were immediately issued that planes were not to be so equipped because of the danger of inciting retaliation on the part of the enemy. Nevertheless, force of circumstances soon compelled the issuance of Lewis guns for airplanes. By September of 1914 British aircraft armed with the Lewis gun were in active service.

During World War I the Germans produced a terrifying psychological weapon in the famous Zeppelins.

They attempted raids on England, to counter which incendiary ammunition was developed at London by George Buckham. This was issued for tests with Lewis guns. Buckham's work was duly reported by agents to the German government, but fortunately somebody slipped up in evaluating the information they obtained. Twelve Zeppelins were destroyed in attacks over London, ten of them being officially shot down by Lewis guns firing the Buckham incendiary bullets. These bullets had flat-nosed jackets of cupronickel. The nose of each bullet held a charge of yellow phosphorus. This charge of eight grains was retained by a serrated plug of lead which was backed up by a larger base plug. A hole through the jacket near the juncture of the two plugs was filled with an alloy which would fuse at low temperature. When the bullet passed down the barrel the alloy melted and permitted the phosphorus to ignite. The yellow phosphorus, when it passed through the bags of the Zeppelins contacted the highly inflammable hydrogen gas therein. This exploded the gas, completely wrecking the dirigibles.

The Lewis gun, therefore, must be given full credit for destroying the power of that early German terror weapon, the Zeppelin, over England in the first World War.

Lewis Developments

It must be noted that Lewis was not the inventor of the gun. Actually Samuel Neal McClean was its developer and he assigned all patent rights to his machine gun to the Automatic Arms Company before Lewis joined the firm in 1910. The original design is generally known as the McClean-Lissak rifle, an unsuccessful pattern never produced.

Lewis' contributions aside from acting as a salesman were still very considerable. It was he who introduced the draft cooling system which is characteristic of the common variety of Lewis gun. He also developed the rate of fire regulator, the clock-type mainspring, and several other modifications.

The Lewis gas-operating system was actually not original in any true sense; its principles had been known for years. But it was the first truly successful portable arm of its class. We find its prototype in early Mannlicher automatic and semiautomatic rifle design and in the principle of the Mannlicher straight-pull rifle. An interesting observation is that the Germans utilized this same principle in a weapon developed just prior to the closing of World War II, the F.G.42; this was a highly efficient model designed for paratroop use.

Lewis Guns in Europe

Despite successful tests by the United States Ordnance Department, Lewis received no orders. Like so many technicians before

First aerial use of the machine gun. Infantry-type Lewis gun tested from plane in 1912. From news photograph.

him, he went to Europe with his guns in 1913. His first stop was at Liege in Belgium where he demonstrated the four guns he had manufactured for testing for the United States Army. Out of this grew several similar demonstrations for various foreign military groups. As a result a company was organized at Liege under the name of the Armes Automatiques Lewis. Manufacture was done at Liege in the opening stages. Later, however, the production was moved to the British Small Arms Company at Birmingham, England. The Lewis gun was revolutionary in its time. With either 47 or 96 cartridges, depending on the height of the magazine pan, and not being burdened with belts or jutting clips, it was possible for one man to operate the weapon and move rapidly with an infantry organization while carrying it.

When World War I began, the Birmingham Small Arms Company turned over most production facilities to making Lewis guns, since six could be manufactured in the same time as one Vickers.

Even the combined efforts of the British and Belgian factories, organizations which work together today even as they did in early World War I days, could not produce the number of guns desired.

Contracts were let to the Savage Arms Corporation at Utica, New York to manufacture them in the United States. Within two years the Savage organization was turning out four hundred Lewis guns a week.

During World War II, many Lewis guns were obtained by England from various friendly powers, including the United States. These were issued to the British Army after the defeat at Dunkirk where they lost so much of their equipment. And as more advanced arms later became available, these Lewis guns were issued to Home Guard units and to small ships in the fleet.

Lewis Guns in U. S.

A considerable controversy was stirred up in the United States in 1916 because of the Lewis gun. Colonel Lewis complained that our War Department was taking a negative attitude toward his gun. The Ordnance Department on the other hand claimed it was unable to get the Automatic Arms Company, manufacturers of the Lewis, to submit guns for testing. Regardless of the merits of the controversy, it is an established fact that some 40,000 Lewis guns had already been tested on the field of battle at a time when our Ordnance Department still could not make up its mind whether or not the gun should be adopted here.

It developed later that some unsatisfactory tests were the result of dimensional changes in the gun. These had been made without Lewis' knowledge because of the appalling blanket of secrecy which was then as now thrown over the most elementary developments in these fields. The United States Ordnance Department at one time contended that the gun could not satisfactorily handle the .30-06 United States Government cartridge because its velocity and chamber pressures were higher than those of the British .303 for which the Lewis was generally chambered. The Savage Arms Company then manufacturing the Lewis exploded this idea by building test guns chambered for our cartridge.

In very rigid U.S. tests the early Lewis guns did not stand up. This may have been due to faulty construction and manufacture or may have resulted from too rigid testing. Their battle service in British hands speaks for itself, though the .303 cartridge cannot compare with ours.

Astonishingly enough, although our combat Marine groups were equipped and trained with Lewis guns prior to landing in France during World War I, they were compelled to surrender them when embarking. They were not permitted to use the Lewis in service. Instead the fantastically inferior Chauchats were issued to them. This matter has never been satisfactorily explained. It indicates at once the confusion and the difficulties encountered by Colonel Lewis and by the Savage Arms Company in their dealings at the time.

Lewis guns have been officially used by the United States, Belgium, Great Britain, Portugal, Holland, Honduras, Nicaragua, France, Russia, Italy, and Japan. During World War II they were used throughout the Balkans and even by Russian troops, as indicated by specimens captured and reported by the German Intelligence Organizations at that time.

One closing note on Colonel Lewis is necessary. He sent checks for over $1,000,000 as donations to our Treasury Department. These represented his portion of royalties on Lewis guns which were purchased by the United States during and after World War I. He made a point of stating, "I will not accept one cent of royalty for a single Lewis gun purchased by the Government of my country."

An interesting sidelight, in view of the difficulties Lewis encountered in his earliest attempts to deal with the Ordnance Department, was that General Crozier, the Chief of Ordnance, even fought the acceptance of the first of these checks! Disparaging remarks were passed which embittered Lewis very much. Still, our Treasury kept Lewis' donations.

JOHN M. BROWNING AND HIS GUNS

The most famous of all American arms designers is the late John M. Browning. Born in Ogden, Utah in 1855, young Browning was the son of Jonathan Browning, a member of the Latter Day Saints, or as they are commonly known, the Mormons. His father moved to Ogden in 1852, where he set up a gun manufacturing shop.

The father, who was born in Tennessee, served an apprenticeship as a gunsmith and at the age of 21 was already operating his own business in that trade. While he worked at farming and blacksmithing, he also worked on the production of guns to order.

Among the items for which he is noted is an original repeating rifle of the so-called Harmonica type. The mechanism consists of a rectangular piece of iron chambered to take powder and balls, which resembles a harmonica. It was slipped through an opening in the breech from the left-hand side and operated by finger pressure on a small lever on the side of the rifle. As the magazine was inserted and the lever pressed, the action was jacked forward to form a tight seal between the individual chamber and the barrel. The arm, of course, was a cap-and-ball design since it was produced before the metallic cartridge era. It not only enabled a considerable number of shots to be fired rapidly from one loading, but furnished an opportunity to carry extra magazines fully loaded ready for sustained fire. In addition to this, Jonathan also produced a breech mechanism containing a six-chambered cylinder rather like that of the standard revolver of the time.

From all this it will be seen that young Browning had a family background to interest him in arms design. He went, however, far beyond the confines of the typical gunsmith. He became in his lifetime one of the most noted and legendary figures in the entire history of armaments.

Before he reached the age of 20 John Browning was actually building rifles himself, one of which he made for his brother Matthew. This was so good that even his father admitted it was the finest he had ever seen.

Jonathan had hauled a foot-operated lathe with him in his ox cart on his trip across the plains from Council Bluffs, Iowa. John served an apprenticeship of 10 years, his particular interest being in the operation of this lathe.

Young John's first design and patent was a single-shot rifle of the dropping block type. Lowering the trigger guard dropped the block to open the breech, extracted and ejected the empty case, and cocked the hammer. When the trigger guard was returned to battery position the hammer was below the line of sight, an obvious but quite noteworthy improvement in general rifle design. It was this rifle which first attracted the attention of the Winchester Arms Company of Hartford, Connecticut.

Upon the death of Jonathan, John Moses and his brother,

Matthew Sandefur Browning, took over the operation of the business. John devoted himself to production and Matthew, the better business man, turned his attention to handling the all-important matters of sales and money.

Matthew realized the pitfalls of building individual guns for individual cranks, a lesson which many fine gunsmiths find out the hard way even in this day and age. He interested his brother in the idea of producing a group of identical arms for sale. With this idea in mind, they actually built 600 of the drop block action rifles which John had invented and patented before offering any on the market. This simple piece of strategy probably marked the difference between their doing business and not doing business with Winchester.

The quantity produced enabled them to set up an assembly line of sorts, and by duplication of parts with common machine setups, allowed them to produce a gun which could be sold at a price well below that of any comparable weapon produced by the large gun manufacturers.

As a result of this investment of theirs in 600 rifles, the Winchester representative was impressed not only with the arm itself but with the danger that the Browning brothers might become a definite competitor in the West.

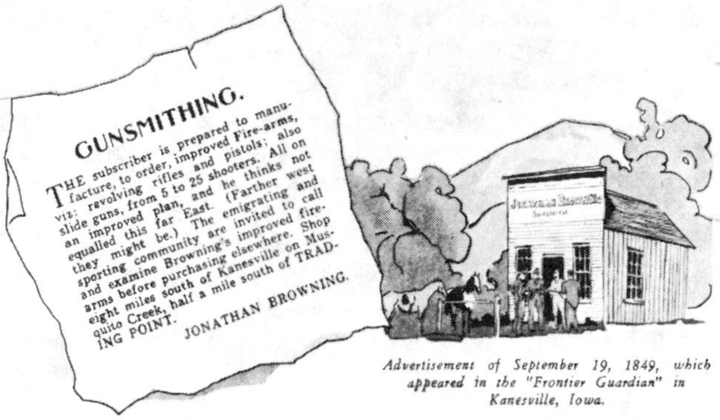

GUNSMITHING.

THE subscriber is prepared to manufacture, to order, improved Fire-arms, viz: revolving rifles and pistols; also slide guns, from 5 to 25 shooters. All on an improved plan, and he thinks not equalled this far East. (Farther west they might be.) The emigrating and sporting community are invited to call and examine Browning's improved fire-arms before purchasing elsewhere. Shop eight miles south of Kanesville on Musquito Creek, half a mile south of TRADING POINT.

JONATHAN BROWNING.

Advertisement of September 19, 1849, which appeared in the "Frontier Guardian" in Kanesville, Iowa.

John Moses Browning, 1855-1926. An original photograph of the great American arms inventor taken at the F.N. factory in Belgium in 1913. He is posed with his superb .22 autoloading rifle.

Winchester not only purchased the rights to manufacture the gun but bought out the entire stock of 600 rifles. The brothers opened a sporting goods store on the lower floor of the building they occupied. They sold standard arms of other manufacture as well as their own. The upper floor was turned into a workshop and a pattern-making shop where John did much of his early experimental design and development work.

He was only 26 when he designed and patented in 1884 the rifle which Winchester marketed two years later as their famous Model 1886.

This lever action, tube magazine arm was originally produced in .45 caliber. It remains today one of the sturdiest of all the medium-power hunting rifle designs ever produced.

Young Browning also designed the Winchester pump action rifle Model 90 in .22 caliber, the prototype of most of the familiar shooting gallery rifles. This was followed by the famous 92 and 94 Model Winchesters, the latter still among the most popular for deer hunting in the United States. After that came the pump shotgun, the Winchester Model 97, and the famous Model 95 box magazine rifle, specimens of which during World War I were purchased for use by Russia in their own 7.62mm caliber. The Model 1906 .22 Winchester repeater is also a Browning development.

The story of Browning's other Winchester developments and his problems with that company do not form a part of the historical section of this book, since we are concerned primarily with military weapons. However, we must note that as the result of an argument between him and Winchester, growing out of difficulties over an automatic shotgun which he offered them for production, Browning discontinued his contacts with Winchester and did not again enter the plant until World War I when he returned there to do work on machine arms in the interests of the United States Government.

The years in between were spent largely at the F.N. Works in Belgium, where his first automatic pistol was manufactured, and where his first automatic shotgun, the most widely imitated arm of its type, was first produced.

Out of the early model automatic pistols, most of which he designed at F.N., came the long line of automatic pistols known to us as Colt's, including the .25, .32, .380, .38, and .45 Government models. All were designed by Browning and were manufactured under license in the United States by Colt's. Several of the models were also made in slightly varying forms at F.N. for world sale.

According to folklore, John Browning in 1889 hit upon the idea of the possibility of automatic operation by utilization of gas. While out hunting he fired while walking through some bulrushes and noticed that the blast disturbed the rushes for some distance from the muzzle. The story goes on that Browning's mechanical mind realized that this represented a power source which could be utilized.

Since the principle of gas operation had been very widely exploited in Europe well before this time, and since Browning was practical enough that he was thoroughly familiar with patent office procedure, it is much more likely that his original ideas were based on knowledge of some of the European experimentation. The Browning in the bulrushes yarn is unlikely, even though his middle name was Moses.

Regardless of the genesis of the idea, the fact remains that Browning did in 1889 start experimental work by utilizing a muzzle attachment on his rifle to determine the effect of the blast. He placed a 4-inch square chunk of iron weighing some five pounds at a distance one inch forward of the muzzle of the rifle. A hole was drilled in the center of the block and the rifle was alined with the hole. A lanyard attached to the trigger fired the piece. The bullet passed through the drilled hole and the blast of gas following the flight of the bullet blew the iron block across the room. Browning proceeded next to produce a concave steel cap with a hole in the center which he secured over the muzzle of his rifle. By connecting this with a hinged arrangement to a spring-loaded operating lever he secured a condition where the blast blew the cap down, thereby pulling the loading trigger-lever forward. A spring returned the lever to lock position, thus preparing the arm for firing again merely by pulling the trigger. The next progression was tapping gas off through a hole in the underside of a magazine rifle and allowing the gas to actuate a piston to operate the action for unlocking, extraction, and reloading. This action was made fully automatic.

Out of such elementary experimentation in a little shop in Ogden, Utah, Browning evolved the basic design for what was to become the famous first successful gas-operated machine gun.

The gun was offered for production to Colt's in a letter from Browning on November 22, 1890. Prototype manufacture was undertaken, since the original gun was actually in .44 Winchester caliber, hardly a military arm even in that day.

In 1893 the United States Navy undertook some tests on the new Colt's gun. The design still had bugs. In 1895 guns were developed to handle both the .30-40 Krag cartridge used by the Army and the 6mm Lee cartridge then used by our Navy.

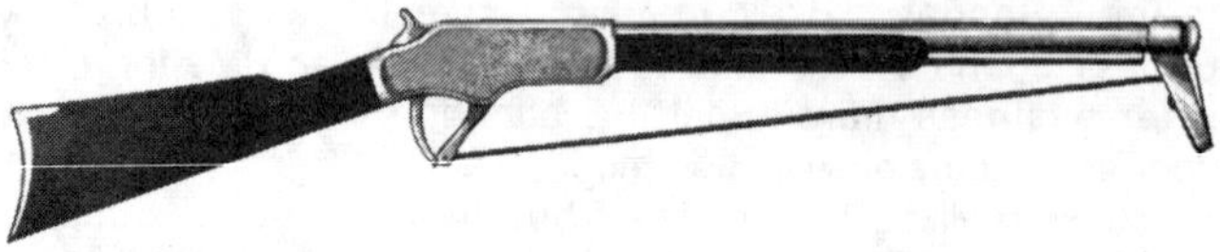

Browning's first experimental rifle, gas-actuated by muzzle blast.

Browning's second experimental gas-operated rifle. Actuation here was by gas passing through a barrel port and acting on a piston. This was one of the first approaches to the modern successful system found in nearly all lightweight automatic and semiautomatic weapons of military rifle caliber.

The first Browning gun shop, Ogden, Utah. The factory was on the second floor. From left to right are: Sam, George, John, Matt, and Ed Browning and a gunsmith whose name is unknown.

This first Colt's 95 Model was nicknamed the "potato digger," a name it still carries in some of the far corners of the earth where models of this gun are still in use. The name derives from the fact that there is a very heavy operating lever below the barrel which swings in a half arc. If the tripod is mounted too low, or the earth happens to be soft at the point of mounting, the gun tripod burrows in and the operating lever will churn up the ground.

A United States Navy Board was so impressed with the gun that in January, 1896, it ordered 50 guns from Colt's. At that time Colt's was also manufacturing the gun in various other calibers for Austrian, Mexican, and even German ammunition. While the Navy was farseeing enough that in 1898 it ordered 150 more of these Colt guns, the Army still remained adamant on the matter of retaining the crank-operated Gatling.

These Colt "potato diggers" in the hands of United States Marines were the decisive element that saved the lives of the inhabitants of the foreign legations at Pekin in the Boxer uprising.

The Navy Colt guns, originally in 6mm Lee caliber, were rechambered for the .30-40 rim Krag cartridge when a decision was made to standardize all service ammunition. These were later modified further to let them handle the .30-06 cartridge when that cartridge was introduced for general service use.

So fixed was the American military mind on the matter of mechanical gun operation as opposed to automatic that, as late as June 14, 1895, an Ordnance Board at Springfield Armory turned in a detailed report of a highly negative order on automatics! To quote directly from the closing of their opinion, "The Board is of the opinion that in its present form as shown by the tests made, this arm is not suitable for ordinary service and has no place in the land armament." This is the same gun which turned up in use even in War II in many places.

When World War I began we still had practically no machine arms available. Although no public information was released, in April, 1917 we had on hand 670 Benet-Mercies, 282 Maxims, and 158 Model 95 Colt machine guns, a total of some 1,100 outmoded pieces to begin a war of the scale in which we embarked!

For use as a training gun, Colt's was authorized to produce these "potato diggers" and several thousand were turned out by 1918. As we have already noted, the design was altered by the Marlin Rockwell Company to a piston operation rather than a lever type; and those constituted the important machine-gun equipment of our planes during the war and for some time thereafter.

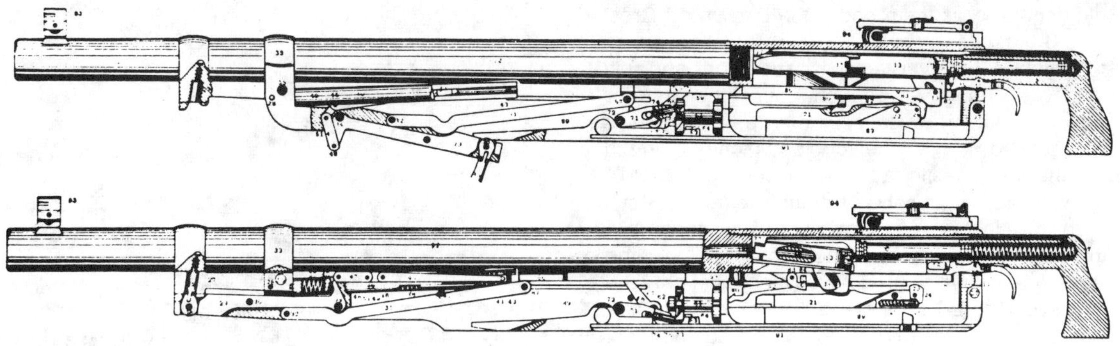

Section drawings of the famous 1895 Model Colt (Browning invented) gas-
operated machine gun, popularly called "The Potato Digger."

The Browning Recoil-Operated Guns

Browning's own experience with machine guns early convinced him that Maxim had been right on heavy arms for which the short-recoil principle was the ideal operating system. In 1900, therefore, he filed an application, which in 1901 was granted, covering a water-cooled short-recoil operated gun. The patent covers all the basic features of the World War II line of Browning Automatic Machine Guns which were the mainstay of our forces and those of our allies whom we supplied with machine arms.

Because the government would not cooperate financially toward the design or production, Browning put the item aside and concentrated on the production of designs for pistols, shotguns, and hunting arms for which there was a wide commercial market. On these he built a very large personal and family fortune.

Since there is a tendency to believe that this recoil-operated gun, which is our standard today, was not developed until the period of World War I, it should be pointed out that Browning presented in 1900 a written description of the operation of the gun he had then designed. This is in all essential features the same as the later Brownings which will be found in our discussion of the Browning arms in the chapter on U.S. arms.

As we have seen, Germany started World War I with 12,500 improved Maxim guns on hand and many more under construction. Against this we had 1,100 obsolete guns. Browning and other inventors were called upon to submit gun designs after the declaration of war. Such was the state of our military preparedness at the time.

Some time earlier Browning had delivered to Washington a sample of his heavy water-cooled machine gun and also of the light gas-operated machine rifle which was later known as the Browning Automatic Rifle (or BAR).

The Browning automatic rifle is covered thoroughly in the chapter on U.S. small arms, so only historical note may be made here. It was offered for tests and in due course public firing demonstrations of the light portable gun were made on February 27, 1917, outside the city limits of Washington, D.C. at Congress Heights.

Over 300 people were invited to attend the Washington tests of the BAR. Included in the list were representatives of the armed services of Belgium, France, Great Britain, and Italy, and for a refreshing change representatives of the press were also allowed to view the new secret weapon.

Even the breathless prose of those newspapermen hardly did justice to this new development of John Browning. At that time French military opinion was that a light automatic weapon was needed which could be used for "walking fire" by soldiers in the trench fighting then going on. This was a most impractical concept. In actual practice the arm proved tremendously useful in later days, but for quite different tactics. However, demonstrations of walking, semiautomatic and full automatic fire were given. A 20-shot magazine was emptied in two-and-one-half seconds. The BAR's 70 pieces were disassembled and assembled in 55 seconds—a marvelous record at that time.

While the design was excellent, the writers seemed to forget that we were already on the threshold of war and did not have the tools to permit it to be produced in volume. It was evident that numerous manufacturing facilities would be necessary to turn out the BAR as well as the heavy Browning in quantity, and arrangements were made for the Government to buy the rights to manufacture these Brownings from the Colt's Patent Firearms Company of Hartford, Connecticut. Colt's held an exclusive contract with Browning for the production of these arms and his automatic pistols in the United States; and since they were the best qualified they provided the drawings and the all-important gauges to the other companies assigned to manufacture.

Most of the early development of the BAR was done by Browning at the Colt's plant. Browning had not entered the Winchester plant in years, as the result of the early disagreement with them over production of his automatic shotgun which drove him to Europe to seek a manufacturer. Now, however, in the face of war involving his country, Browning pocketed his resentment. The great Winchester organization turned its full facilities to preparing and correcting working drawings. Many of the important refinements of the gun were developed there. Work actually began on the BAR at Winchester in February, 1918. As an inevitable result of the working pressure, some manufacturing bugs developed which resulted in parts not being completely interchangeable. After some delay in production, the necessary corrections were made and when the war ended Winchester was turning out these very difficult to manufacture weapons at the rate of 300 a day. This was a most remarkable performance which reflected great credit on the Winchester engineering and production groups.

In June, 1918, additional BAR production was commenced by the Marlin-Rockwell Corporation in a plant especially acquired for the purpose. When the war ended on November 11th of the same year, this producer was turning out 200 guns a day, also a remarkable achievement. Before the war ended some 52,000 BARs were delivered, only about 9,000 being made by the Colt's Company since they were simultaneously heavily engaged in other war production.

Our 79th Division landed in France in 1918, in July, armed with BARs, and their guns were put into action on the 13th of September. Historical note must be taken of the fact that Val Browning, son of the inventor and then a First Lieutenant in our Army, demonstrated the use of the arm against the Germans. Today he is the active head of the Browning family organization which still manufactures pistols and shotguns developed by his father at the

F.N. Plant in Belgium and sells them under the Browning Bros. name in the United States.

The French in particular were very enthusiastic about our new weapon, experience with their own Chauchat and other inferior French machine rifles having been very disillusioning. It is interesting to note that Browning's water-cooled machine gun did not receive nearly the enthusiastic reception accorded the BAR. There were many who felt that in spite of its fine features, it still was not superior to the Vickers and other guns of the time. As a matter of fact, the smooth feeding from the improved Browning belt and feed system was in itself so great an advance that only the prejudiced could have ignored its superiority. About the BAR there could be no question, since there was no true competition.

The Browning Machine Gun Model 1917

While our government seized upon the BAR for immediate adoption after its initial demonstration, and history since then has proven the judgment sound, action on the water-cooled Browning machine gun was not handled so expeditiously.

Under tests at Springfield Armory in May, 1917, the gun fired 20,000 rounds at a cyclic rate of 600 a minute without malfunction or breakage. This in itself was a truly remarkable performance. Browning personally ran another 20,000 cartridges through without failure of any of the component parts. A second gun was introduced to demonstrate that it was the design rather than the individual gun which was responsible for this sterling performance; and by securing together long belts loaded with ammunition, Browning succeeded in operating the gun continuously for 48 minutes and 12 seconds.

John M. Browning and his famous Model 1917 machine gun. It is used throughout the world.

World War I photo of the then Lt. Val Browning demonstrating his father's great BAR somewhere in France.

Here, too, for the first time demonstrations were made of dismounting and reassembling the gun blindfolded. This demonstration started the era of training which sought to enable gunners to become so proficient in handling their weapons that they could function almost subconsciously. Familiarity with the weapon is even more important than actually being able to make repairs.

Entirely aside from the reliability the Browning showed, the adoption was fully warranted by the fact that the design was suitable for high-speed production methods of that era. Colt's (who originally controlled the patents), Remington Arms-Union Metallic Cartridge Company, and New England Westinghouse were given contracts to produce this gun, which at the time was known as a "heavy" machine gun though today its classification is that of a "medium."

The performance of Westinghouse in this production was a tremendous tribute to the gun design and also to the flexibility and intelligence of the American industrial system. While in theory the gun companies with their knowledge of firearms should have had a manufacturing edge, the Westinghouse group produced a hand-made pilot model in 29 days from the time of the contract! A mere 34 days later they brought the first production gun off the assembly line. By the end of the war, Westinghouse was turning out 500 Browning Machine guns a day.

Remington produced 12,000 during the war as compared with Westinghouse's production of 30,150. The Colt's organization turned out only 600, though this figure is not truly representative of their capabilities since their facilities were largely engaged in preparing various gauges, tools, and mechanical drawings for other plants, as well as turning out British Vickers' guns.

As it happened, the Brownings reached Europe entirely too late to be given any extensive battle practice. Again Lieutenant Val Browning served as instructor to introduce this new machine gun of his father's design to our troops, but European observers were unable to evaluate the design because of its very limited war use. In very minor combat use in September by our 79th Division, which had a small detachment armed with them, four Brownings stood up very well indeed.

In a way, it was fortunate indeed that the Brownings did not get into extensive use in World War I, since some relatively minor weaknesses which did appear (and were not the result of the design necessarily) would have given the gun a bad name from which it might not have recovered.

One such weakness which may be cited is the bottom plate in the receiver. The original metal used for this part did not stand up under continuous fire. In view of the rapidity with which the gun had to be put into production, this metallurgical slip-up is understandable. When this inadequacy was recognized, however, immediate steps were taken to correct it by introducing a reinforcing "stirrup" to fit over the weak area on the outside of the receiver.

Over 25,000 original Brownings required this particular modification. Later production was altered, of course, to make this modification unnecessary.

During the period between World War I and II, manufacture and development of the Browning machine gun went ahead both in the United States Ordnance Department and at the F.N. Plant in Belgium. The guns made abroad were sold commercially throughout the world.

The Browning since 1918 has seen service everywhere in the world under all types of weather and battle conditions. The modifications made in it, while all relatively minor, have evolved a design which for the purposes intended is probably the equivalent of any produced since Browning developed it. No higher tribute can be paid Browning than recognition of that fact.

The Browning .50 Caliber Machine Gun

Colonel John Parker, in charge of the United States Army Machine Gun School at Gonducourt in France, was very much impressed with a new 11mm machine gun and cartridge the French introduced for particular use against observation balloons. He secured samples for shipment to the United States.

Colt's was given the task of altering eight Brownings to handle this new 11mm French cartridge. Aside from its use as a balloon gun with an incendiary bullet, the round was also used with armor piercing bullets against armored vehicles and defensive shields which were then appearing in use by the enemy.

The ammunition as designed by the French did not meet military needs as General Pershing saw them. He insisted that the bullet weight and muzzle velocity both be substantially greater. Pershing's attention to this seemingly minor matter proved to be of great importance.

The Winchester Repeating Arms Company at New Haven were by this time working with Browning on the development of a larger caliber machine gun cartridge than the standard rifle .30-06. They merely scaled up the standard .30-06 rifle cartridge to try and make it usable in both the heavy machine gun and this antitank gun, and in the course of doing so added a rim to the cartridge case.

Fortunately Pershing, who was following ammunition development very carefully, saw the new specifications and insisted that the case be redesigned to be rimless. His instructions were followed and a dummy cartridge of this pattern was sent to Browning at the Colt's Plant. Browning, meanwhile, had scaled up his .30 caliber machine gun, keeping all the mechanical features but adding weight, size, and strength to handle the larger cartridges with their increased pressures. A test gun was ready.

Browning moved the model from Colt's to Winchester, since the latter was making and developing ammunition as well as arms. The work developed so well that he stayed on at the Winchester Plant. Since Colt's had no ammunition manufacturing facilities, this move greatly accelerated the eventual development of the .50 caliber machine gun.

The first of the new .50 calibers was actually assembled on

The inventor posed firing an early model of his basic .50 caliber gun at the Colt plant at Hartford, Conn.

September 12, 1918, and Winchester proceeded to produce six duplicate handmade testing models. Although the new bullet developed was heavier than the minimum demand of General Pershing, the velocity was still low by several hundred feet per second.

During infantry tests it was found that even with the gun and tripod weighing 160 pounds, the increased energy engendered during firing the heavy rounds made it almost impossible to hold the gun on the target during full automatic fire. Other tests indicated that the bullet lacked sufficient penetration against improved enemy armor.

While the Browning design was still under way, our troops captured some German antitank rifles with a new variety of 13mm ammunition which fired an 800-grain bullet at a muzzle velocity well over 2,700 feet per second. This bullet was able to penetrate an inch of armor at 250 yards range. This was a fantastic performance at the time. Under normal conditions, allied military policy is to keep such developments top secret since any admission of enemy superiority is decried. The more practical German approach has been to seize immediately upon an advance regardless of its origin.

Fortunately, the normal system of military secrecy we throw up about such matters did not prevent some rounds reaching Winchester. The engineers there began immediate work on bringing their .50 caliber cartridge into the ballistic range of the German. After Winchester had worked out all the technical aspects of the new ammunition, development was transferred to the Government Arsenal at Frankford, Pennsylvania. The transfer was not effected until 1918. Much of the credit for our .50 caliber development belongs to Winchester and to their energy in taking advantage of the new German development so rapidly.

The new .50 caliber cartridge required, in addition to the scaling up of the .30 caliber parts, various other alterations. An oil buffer was added with the dual purpose of helping to absorb energy and providing a fire rate regulator. The pistol grip of the .30 caliber had to be replaced with a double spade grip as encountered generally in Maxim design, since the gun really needs both hands to be controlled. The gun is covered in detail in the chapter on U. S. weapons.

Note on Browning

The range of the development genius of John M. Browning is breath-taking. Unlike Maxim and Mannlicher, Browning did not develop any principles. He did adapt known principles to a vast

Outstanding productions of Browning's inventive genius.

range of firearms developments from the early Winchester single-shot .22 rifle through the entire field of medium and high-powered rifles (bolt actions excepted), shotguns, automatic pistols, and machine weapons, even to 37mm automatic cannon. From the standpoint of commercial success, no other designer even approaches him.

His commercial specialties were lever-action rifles such as the Winchester Model 94, .22 semiautomatic rifles such as the Remington Model 24, high-powered semiautomatic rifles such as the F. N. and the Remington Model 8, .22 pump rifles like the Model 06 Winchester Hammer Model, and the hammerless F. N. .22 caliber Trombone. His shotgun developments ranged from the early 97 Winchester Pump Gun with its exposed hammer, a shotgun used for military service in both wars in guard capacity, through the hammerless Stevens Model 520 pump gun to the world-famous Browning Automatic shotgun with its long recoil action, the most imitated automatic shotgun ever produced.

The Browning over-under (or superposed) shotgun is without peer in its field, one of the finest combinations of American genius and Belgian craftsmanship that the gun industry has produced. His range of automatic pistols starts with the Colt's, such as the .22 Woodsman, the .38 Pocket and Militaries, the .25, .32 and .380 Hammerless Pocket Models, and, of course, our own official .45 Automatic. His modification of our military pistol, the 9mm Belgian

John Moses Browning, the most successful arms inventor in history.

1935 High Power, has influenced automatic pistol design all over the world. In no instance have any of his commercial designs ever been a failure.

Few men in history have done more for the development of defensive weapons for their governments than John M. Browning. His patriotism may be summed up in the fact that he discussed payment with the War Department, through his representatives, only after his guns had been officially adopted by our government and their production had reached their greatest heights. His instructions to his fiscal agents then were to accept any price the government cared to set without hesitation or bargaining. He actually accepted royalties at a rate about one-tenth of that normally allowed.

In his development of military, sporting, and commercial arms, Browning achieved the heights of quality and reliability. The standards he set have been maintained to this very day.

THE EVOLUTION OF THE AUTOMATIC MACHINE GUN IN EUROPE

As we have seen, our own Hiram S. Maxim was the real father of automatic arms operation. He not only developed to the full the principle of recoil operation for automatic weapons systems but also did tremendous experimental work in the field of gas actuation, some of his systems being utilized by others.

We have space here only for passing mention by country of the various great foreign inventors and the arms for which they were particularly noted. All such arms as are in general use today, which means all those which were truly successful, will be found completely detailed in Part Two of this book. We shall content ourselves here with historical background on these developments as they emerged.

AUSTRIA

The Austrian Schwarzlose Machine Gun

This gun is remarkable for being the only machine gun on the retarded blowback system which met with any degree of success prior to 1950 and which was put into general military production.

It was invented by Andreas Wilhelm Schwarzlose of Charlottenberg, Germany in 1902 and was first produced by Steyr in Austria three years later. Schwarzlose is familiar to all students of the history of automatic arms as the developer of early pistol and rifle types. His production included an early short-recoil system and also several unusual "blow-forward" pistols, one of which was commercially manufactured and sold in the world market. These are touched upon in their proper place. At this point we are concerned with his successful machine gun primarily.

All gas and recoil-operated machine weapons involved high-cost machining and fitting in the early days of development. As a matter of fact, it was not until the Russian and German developments in World War II (notably the German MG42) that locked-breech systems were developed which were capable of fast production at low cost; and their manufacture was a matter of improved metals and manufacturing equipment even more than design.

Schwarzlose, therefore, turned his attention to a weapon having a fixed barrel, few moving parts, and rugged construction. His development, like all such attempts, had many limitations. But it was successful within those limits. The gun saw service in both world wars, and in several armies.

In this system the breech is at no time truly locked. When the gun fires, the rearward thrust of the exploding gases actually starts the action opening at the same instant it starts the bullet down the barrel. However, by using a very short barrel and a combination of very heavy recoil parts and springs, plus a system for developing a mechanical advantage in the mechanism, Schwarzlose produced a machine gun which permitted the use of military rifle cartridges without an impossibly heavy breech mechanism.

The gun was water cooled. Originally chambered for the 8mm Austrian military cartridge, it was modified at times for other calibers. The 1905 and its 1907 modification required lubrication of cartridge cases to prevent case ruptures caused by the action

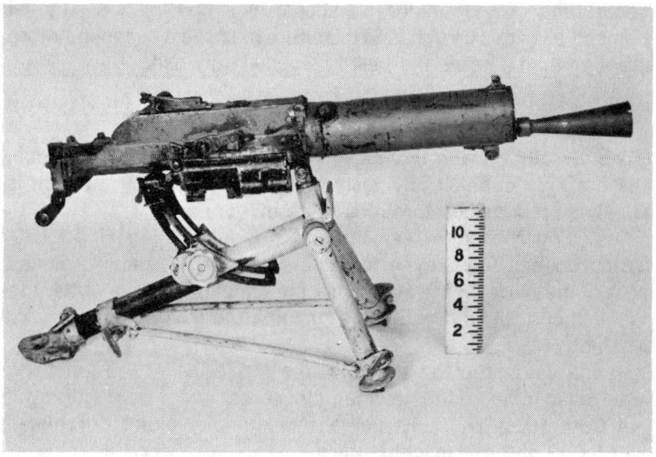

Representative Austrian Schwarzlose M.G. (machine gun).

opening so fast that it drew the case back while the bore pressure was still too great, and the case was still sealed by expansion to the barrel chamber walls. The designer built an oil pump into the gun to overcome this difficulty. On each stroke oil was squirted into the firing chamber to lubricate the incoming cartridge case. In the later modification in 1912 the pump was eliminated by considerable re-design.

A study of the Schwarzlose and the illustrations in the chapter on Austria will give a clear understanding of its operation. In general it may be stated that the operation required the use of a short barrel to allow fast dissipation of the gas pressure, unusually heavy and sturdy recoil spring and bolt to furnish the initial inertia, and a system of jointed levers which placed the bolt at a mechanical disadvantage in its action of compressing the huge recoil spring.

The mechanical disadvantage is produced through an elbow joint attached to the bolt and having another arm pivot-pinned to the heavy receiver. As the action starts to open, the elbow must move through an arc. The angle between this linkage and the operating crank is slight when in closed position. This results in much of the primary thrust being taken up by the receiver rather than the breech elements. The ingenuity in the parts design is truly remarkable, even the huge recoil spring being designed to serve the several purposes of recoil spring, buffer spring, and firing pin spring at the instant each function is required. The cartridge feed is equally unusual, having only two working parts. One, a roller, is acted upon by the moving belt to pull the belt into the gun, the cartridges passing through a wheel-section of the roller and being cammed back slightly before complete withdrawal—a form of initial extraction of the loaded round to be fed which is far better than the abrupt jerking motion customarily encountered in belt feeds. The only other part is a detent slide of relatively simple design.

The 1912 Model introduced a still heavier bolt and spring, plus a more angled linkage. This design, which of course is a form of toggle action, permitted the placing of the operating crank well off center, the effect being to control the pressure and make the receiver absorb even more shock. The bolt movement was still further slowed, as the initial thrust must first raise the off-center crank out of line. This additional disadvantage was sufficient to permit the designer to discard the lubricating pump of his earlier models.

Even with the relatively low-powered military cartridges used in the gun, so much flash was encountered at the muzzle that a large cone-shaped hider was required over the muzzle for night firing. The construction requires such heavy and sturdy parts that it did not have very much effect on other design development in portable weapons, though some cannon utilize varieties of hesitation or retarded blowback operation.

With the recent developments in permanent magnets and special metals, and in ammunition advances, new possibilities suggest themselves in this design, and the future may see it receiving more attention than it has in the past for special duty use.

The Austrian Solothurn Model 30

The Solothurn was technically Swiss, but since Rheinmetall was the only outlet it obviously was German. Hence the gun is usually so recognized throughout the military world.

Austria purchased this same model, also designating it Model 30, in 8mm caliber. The Steyr Arms factory in Austria actually furnished most of the parts for the Solothurn, manufacturing them under the noses of control authorities and shipping them to Switzerland for assembly.

Hungary adopted this model as the Model 31 in the following year. As an index of the volume produced, it might be pointed out that from 1930 to 1935 Austria and Hungary alone purchased over 5,000 of these machine guns.

The trigger system of the earlier model was retained. Its face has two curved depressions, pressure on the top one giving single-shot operation, and a pull on the lower one giving full automatic fire. This system allows the operator to fire single-shot or full automatic without any movement of the hand except the index finger and without shifting aim from the target.

Austrian Solothurn Model 30 made in Switzerland.

Although German army tests firing 100,000 rounds on this model produced very good results, other and more advanced designs were already on the way, and the Germans did not adopt it for general use although they did use many for drill.

The MG30 featured a special quick-change barrel system. Pressing a locking piece allows the shoulder stock to be turned 60 degrees to the left and pulled to the rear. The operating spring and guide remain in the rear section. The barrel with the attached bolt and other assembly are then shaken out the rear of the receiver. By disconnecting the barrel from the rest of the assembly, a cool barrel can be inserted.

To fire, a magazine is inserted with the bolt either open or closed. Assuming it to be closed, the operating handle is then pulled to the rear. The mechanism is held to the rear by the sear ready for discharge. When the trigger is pulled the bolt moves forward under the drive of the compressed operating spring. It picks up a cartridge from the magazine and pushes it into the chamber. The bolt (carried in a special carrier) drives the carrier, the barrel, and the barrel locking ring ahead under its spring thrust. The locking ring is turned under the influence of its lugs engaging in spiral grooves and locks the assembly. The firing pin is released as the action locks in its fully forward position.

As the arm fires and the mechanism recoils, when the unlocking position is reached the rollers on the locking ring engage the spiral grooves machined into the walls of the receiver.

The rearward movement turns the locking ring to free the bolt. The barrel, bolt extension, and locking ring are halted by a buffer and held in rearward position. The bolt and the mainspring guide continue rearward under the acceleration imparted by the spiral grooves. The extractor pulls back the empty case. As ejection takes place the continuing rearward action of the parts is finally halted. The firing-pin spring is compressed as the bolt starts into counter recoil position. At the end of the stroke and if the automatic trigger is being held, the action continues until the magazine is emptied.

In 1932 Rheinmetall introduced the MG15, an aircraft-type gun for the German forces.

This gun was only a modification of the MG30, a muzzle booster allowing an increase in rate of fire to as much as 1,000 per minute. Special drum magazines of the saddle type holding 75 rounds each were furnished for this gun. Each half of the drum is fired alternately, this preventing the center of gravity from being affected as the firing continues.

Later aircraft models designated as the MG17 incorporated solenoids for electrical firing of the guns. A heavy model, known as the MG131 used a 13mm cartridge with a velocity of 2,560 feet per second. This design also used the locking ring principle of the Solothurn.

The Skoda Machine Gun

Skoda is the world famous Czech center of development for all types of arms, both light and heavy, now unfortunately in Communist hands. Actually Skoda is a tremendous group of what we would call job shops, not fully integrated factories along the general American plan. As a metals center it also has no superior.

Soon after Maxim's gun appeared, an automatic machine gun was introduced by the Grand Duke Karl Salvator in conjunction with Colonel Von Dormus of the Austrian Army. The gun was patented in 1888. Manufacture began at a plant for arms manufacture, built in 1859 by the Count of Waldstein, which was acquired by Monsieur de Skoda in 1869. The works were at Pilsen, then in Austria-Hungary. This soon became one of the great armament plants of the world.

This first Skoda gun was a delayed blowback. It operated on the principle of being locked only by spring pressure and by the weights of the moving elements, though a retarding mechanism was provided which still further, through spring pressure and friction, slowed down the rapidity of opening so that the barrel pressure would be safe by the time the action had opened enough to permit extraction. This system has the advantage of simplified design and manufacture, but is not adaptable to truly high power ammunition.

It was officially adopted in Austria in 1893. In general it was intended for naval and fortification protection. It saw some service in the hands of an Austro-Hungarian detachment in defense of their legation at Pekin in 1900 at the time of the Boxer Rebellion.

This Skoda gun was completely redesigned and issued as the Austrian Model 1909, at which time the original hopper feed mechanism, which had not been satisfactory, was replaced with a belt feed.

In tests the original Skoda proved incapable of standing up alongside the competition to which it was subjected. The gun is interesting only because of the place and conditions of its production.

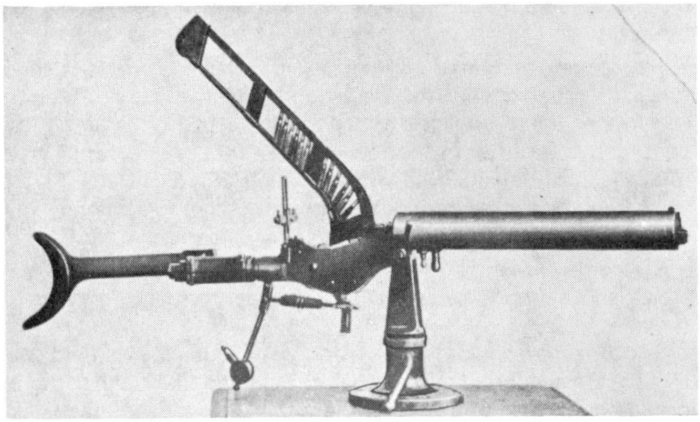

Upper: The Skoda of 1893. Lower: The Skoda of 1909 with shield.

BRITAIN

The English Beardmore-Farquhar Machine Guns

The Beardmore Farquhar was submitted for tests to the Royal Air Force in November, 1919. It also was intended for possible use as an infantry machine gun of the light pattern. Despite favorable tests it was not adopted. No further development has been done on the gun. However, because of its unusual nature, it requires some passing historical comment.

In general lines the gun resembled rather closely other patterns. Its outline for instance is not too dissimilar from that of the Russian Degtyarev. The mechanism, however, is quite different.

This arm is a strange combination of gas and spring functioning.

The forces of the explosion do not act directly on the unlocking mechanism of the bolt. Instead they compress and store up spring energy until the pressure within the bore has dropped to safe operating limits. At that point the bolt is unlocked by a very smooth action of the spring. This furnishes positive unlocking and yet does not give the sudden jerking effect found where a standard piston mechanism is employed.

The piston itself is in a tube below the barrel and connected to the bore by a drilled hole as in standard practice. The main spring is in the front end of a special spring tube having a sear device within it. A special spring for closing the bolt is housed about a central rod at the rear spring tube below the barrel.

The bolt remains locked until such time as the force required to unlock it is lower than the strength of the compressed mainspring, thereby preventing varying gas expansion factors from having any effect whatever on the operation. In other words, regardless of the gas pressure of varying cartridges, the operation of the functioning of the piece remains constant.

As the weapon is fired, the expanding gas in the cylinder forces the piston to the rear, thereby compressing the mainspring until it reaches a position where it is caught by its sear. This spring remains under compression until the resistance to turning the bolt head is so reduced that the strength of the spring overcomes it.

The bolt is operated for initial loading by a straight pull on the handle which passes through a slot in the left side of the gun. The bolt itself is composed of a non-rotating cocking piece and a head which rotates, the locking lugs being at the front of the bolt head and engaging in locking shoulders in the receiver directly to the rear of the chamber.

The feed mechanism is also distinctive. It is a two-layer rotary drum. Cartridges are under spring tension and are spring indexed and stopped as well. It is possible to unload the drum by pushing the two feed stops simultaneously, thereby releasing the full pressure of the springs to eject the cartridges.

The arm was so designed that the drum magazines which held 77 rounds could be replaced, if desired, with a 5-round magazine using infantry clips, thereby permitting it to be used as a shoulder type automatic rifle.

The gun was rated superior in many ways, especially in lighter recoil and less liability to jamming than the other guns then used. However, the design was not considered sufficiently advanced over the existing patterns to warrant the expenditure of producing it.

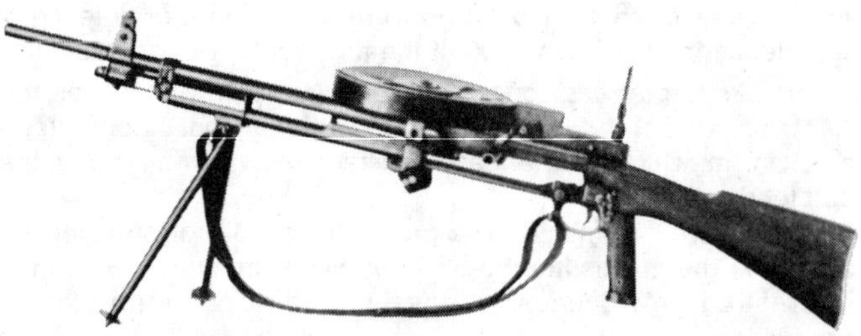

English Beardmore-Farquhar machine gun. Original in design but quite complicated.

The English B. S. A. Machine Gun

The B. S. A. is of interest only as a historical item today.

It was developed in 1924 by the Birmingham Small Arms Company with the intention of making it a .50 caliber gun, basically for use by airplane observers.

Externally it resembles the familiar Lewis gun. Its system is radically different in every respect, however. The ammunition was the same as that used in the .50 caliber Vickers aircraft gun. Besides the aircraft version, another water-cooled gun was provided for naval use.

The drum feed resembles the Lewis only externally. It carries 37 rounds of ammunition. This is much too small for practical use with the bursts involved.

Unlike the Lewis, which it resembles so closely in external line, this gun is technically a short-recoil action. However, it is a freak in its category. Instead of the normal travel to the rear of the barrel, bolt, and extension mechanism, this weird mechanism traversed a distance of 2 7/8 inches fully locked before the unlocking operation commenced. No explanation was given for this unnecessarily long stroke. It must be remembered that it has all the characteristics otherwise of a short-recoil action.

Because of the long locking stroke the rate of fire was very low, about 400 rounds per minute. This is much too low for aircraft purposes, but is ideal for a land gun.

Nothing was done with general manufacture of this gun in light of the advances and tactical developments after its introduction. It was not produced in quantity.

The British Besa Machine Gun

The ZB Plant was seized by the Germans early in World War II. It was operated under their control during the war as the Waffenwerke Brunn A. G.

The Germans maintained the working organization of ZB and used the arms produced there throughout the war. Currently the work is similarly progressing under Communist domination.

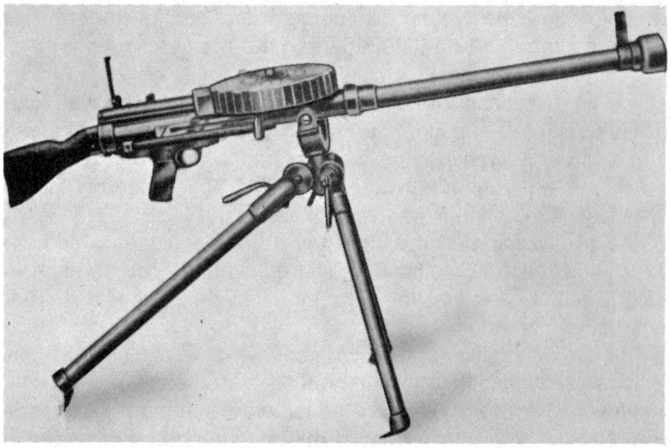

English Birmingham Small Arms machine gun.

The British Besa machine guns of World War II are also of Czech design. They stem from the 1937 gun of Vaclav Holek. The plant designation was ZB53 model 1937.

Officially this gun was known in Czech service as the Model 1937.

Its performance impressed the British to such an extent that again they secured licenses to manufacture this new product of the ZB Works.

The name Besa was applied to this gun. The B is for Brno. The E means Enfield. The SA are the initials of the Birmingham Small Arms Corporation which entered into the manufacture.

The introduction of this gun represented still another instance in British procedure in the line of accepting efficiency before national pride. Since the gun was chambered for the very efficient German 7.92mm rimless cartridge, the British continued use of this instead of their own .303, even in the face of possible ammunition complications. As a secondary advantage, however, it might be pointed out that captured German ammunition could be used in the gun.

The gun and modifications were known as Besa 7.92mm Mark I, II, III, and III*. Later a 15mm caliber was designed to utilize a high-velocity anti-tank cartridge. The new weapon was listed as the Besa 15mm Mark I and is a copy of the ZB 60 (Czech 15mm M1938 machine gun).

Again we find that in the interests of engineering design and manufacturing economy, the basic operating parts of the last models were kept. Alterations were made only as required for the new operation.

The cocking system of these guns is unique. The grip is pushed forward. A sear catches the extension of the gas piston. The grip and trigger guard are then pulled straight back, bringing all the operating parts to the rear and held by the sear. Grip is then returned to a normal position, care being taken to keep the finger away from the trigger during this motion as otherwise the gun will fire.

By the use of very heavy barrels, some flanged and others not, Holek succeeded in reducing the necessity for barrel cooling or replacement.

Because the shoulder stock of previous items in this field was removed (since the guns were basically intended for either tripod or tank mounting), the spring-loaded buffer system had to be housed within the receiver itself.

This buffer system permitted an increase or decrease of firing rate. By elevating the buffer system, the bolt was permitted a further rearward travel on its recoil stroke and the compressed operating spring alone served to return the parts to battery. This combination resulted in a slowing down of the action. The buffer of course functioned as an accelerator, though actually in effect it was not. Its function was to interfere with the length of the bolt stroke and by shortening it and compelling a faster recoil to move the gun back into battery at faster rate.

Thus it did not add energy to speed up the recoiling parts, but effected the same result by interfering with the length of travel and the rapidity of return.

Gas pressure could be increased if desired by turning a circular piece having two holes of different size. By rotating this member between the port in the barrel and the gas piston cylinder, control was obtained.

After the grip has been pushed forward to catch the piston assembly, pulled back to cock it and compress the operating spring, and then returned to position, the arm is ready for firing. A selector switch on the side of the handle permits single shot or full automatic fire.

Pressing the trigger releases the sear. The firing mechanism moves forward impelled by the powerful operating spring. A projection at the forward top of the bolt passes through the center of a cartridge link, pushing the cartridge ahead of it directly into the chamber.

As the bolt reaches firing position, its rear end is directly under the locking recess in the bolt extension. An extension on the gas piston is held to the rear by a beveled locking lug. It continues forward as the lug cams the rear of the bolt up into the locking recess and out of the piston extension path. The final travel of the piston extension serves to free the barrel holding catch. The barrel, bolt, piston, and extension move forward to firing position. The locking lug on the piston extension strikes the firing pin and fires the weapon. As in similar designs, such as the Swiss Furrer later discussed in this chapter, the recoil movement in this gas-operated arm starts slightly before the forward-moving parts contact the stationary receiver, thereby giving a cushioning effect and smoothing out the operation of the gun itself.

The gas port is about one-third of the way forward from the breech end. Gas striking the face of the piston drives it rearward in recoil. The back of the lug on the piston extension pulls the bolt down out of its locking recess.

The cartridge extractor is standard but the ejector is positioned in the belt guide. Ejection is through a slot in the piston extension and in the receiver.

A barrel recoil spring is mounted in the cover group to reduce the upward jump of the muzzle and increase accuracy.

The British Vickers-Berthier Machine Gun

The Vickers Company purchased the manufacturing rights to Berthier machine guns in 1925 and started limited production.

The light machine guns so produced were not particularly successful, although organizations such as the Latvian Army did adapt them for their rifle cartridge. Spain and South America bought limited quantities and the gun was also officially adopted by the Indian Army. It received relatively little service use during World War II.

This was a gas-piston-operated machine gun. It was modeled along lines somewhat like the Bren but was quite different in functioning. Its aircraft version appeared in 1928. It was outmoded because the turrets in the planes were fitted with belt-fed fixed-type weapons.

The bolt locking was by a tipping-up operation of the bolt itself. The magazine in the aircraft design resembled externally the Lewis but did not rotate. This gun has had no particular effect on armaments development, although at the time of its introduction it represented a step ahead in the small number of parts, the complete enclosure of those parts, and the ability to be dismounted and assembled without the use of tools.

CZECHOSLOVAKIA

The Czech ZB Machine Guns

Since these guns are dealt with in the chapter on Czechoslovakia, only historical references are required at this point. However, in view of the importance of the ZB Works, at this time when Czechoslovakia is under Communist domination, their designs cannot be too strongly emphasized.

A producing company set up at Brno (Brunn) was established in 1922. Today, of course, it is under Communist control. Little information about its operational heads is currently available, though the arms being manufactured there are well known because of worldwide distribution.

In addition, sporting rifles based on the Mauser pattern actions made at this plant were sold in the United States until quite recently. Sporting rifles using this action and automatic pistols of the pocket variety, which are used in police and military circles abroad, were also imported into the United States by both east and west coast importers in very large quantities.

The original company was named Ceskoslovenska Zbrojovka Akciova Spolecnost v Brno. Upon its formation in 1922, 75 per cent of the stock was retained by the Czech government and 20 per cent was held by the Skoda Works, the remaining 5 per cent being distributed among employees.

At this time very close liaison was developed with the French authorities. French technicians were supplied to aid in the development of the plant. In short order, however, they became subsidiary to the Czechs both in importance and ability.

In 1922 the ZB Works produced their first light machine gun, a Hotchkiss model of French design. It was modified, of course, but was still the French pattern.

By 1924 they had developed a prototype light machine gun based on the principles of the commonly known Berthier, the Hotchkiss, and the American and Belgian Browning Automatic Rifles.

When this gun was put into production it was introduced as the Brno ZB Model 1926 and was offered for world sale under that designation. The name probably was determined upon because of the difficulty foreigners would have in otherwise identifying it or pronouncing it.

The principal designers were two Czechs, Vaclav and Emanuel Holek, brothers. An Austrian engineer, Marek, and the Polish engineer, Podradsky, were also influential in design.

Vaclav Holek is a designer of outstanding ability. He is responsible for the production of a very remarkable semiautomatic rifle which was intended largely for sporting use but was curtailed by the beginning of the war. This gun is a masterpiece of simplified operation on the gas principle and is designed for high-speed production. It uses a large number of easily manufactured parts, including stampings and screw machine parts, and leans heavily on the utilization of torsion springs of very simple design.

This Holek rifle was manufactured for a number of Mauser type cartridges, notably the 7.9mm German military cartridge. It fires from a closed breech, utilizes an internal hammer, has an extremely simple takedown system, and is clip fed with a detachable box magazine. The cocking is rather unusual, being done by a sling strap attached to a swivel at the forward end of a sliding element over the gas tube.

Vaclav Holek is credited with the principal features of the combination he made in the design of the ZB light machine guns. It is worth noting that his remarkable ability was recognized and that he became a very wealthy man as a result of the international sales of the products of his genius which were manufactured in the ZB Works.

The ZB is one of the finest light machine guns ever developed. Because of the exceptional quality of the steel provided by the Skoda plant, spare barrels could be easily interchanged after rapid fire. Even red hot barrels of this steel can be plunged into cold water without injuring them. A quick change barrel feature was an important element in the design, since the gun could stand only about three complete magazines in full automatic fire before requiring cooling. The Japanese Type 97 tank gun is a slightly modified ZB26.

The British Bren Gun (Czech made)

This gun, which is discussed in the chapter on Great Britain, is of historical interest also, because of its evolution from the ZB arm, and because during the war many technicians from the Czech Works, who had contacted the British during the development of the Bren, left their native land and joined the military and engineering forces of Great Britain and the United States. When Great Britain instituted a series of tests for a new light machine gun in 1932, Madsen and the Vickers-Berthier were ranged alongside the ZB model for testing. The ZB out-performed competition to such an extent that even at the expense of losing

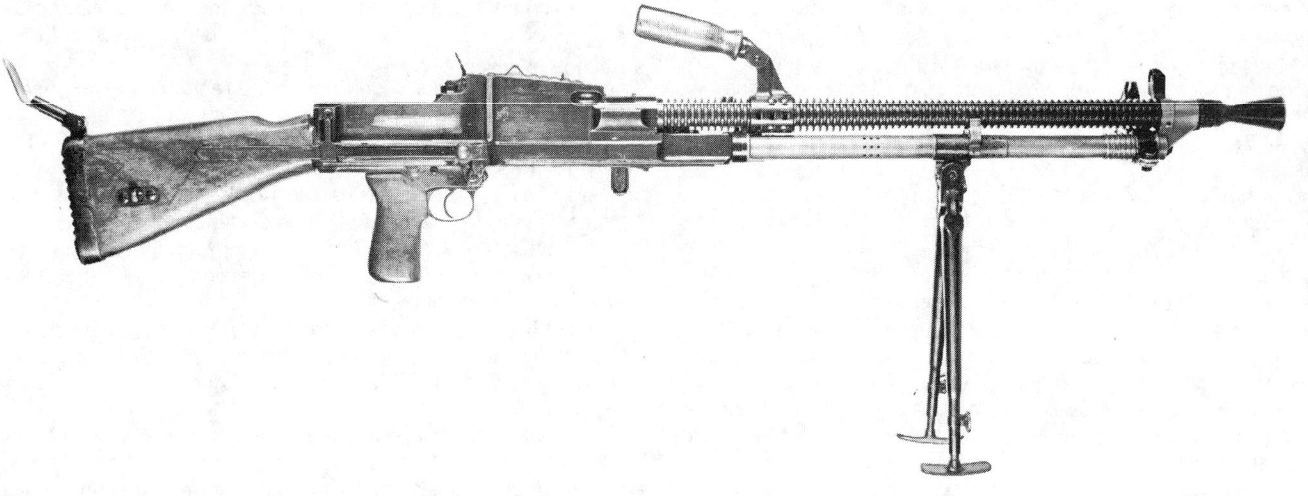

Representative Z.B. machine gun. Weapon captured by General George S. Patton in World War II. The right side of receiver is marked "Kulomet-26."

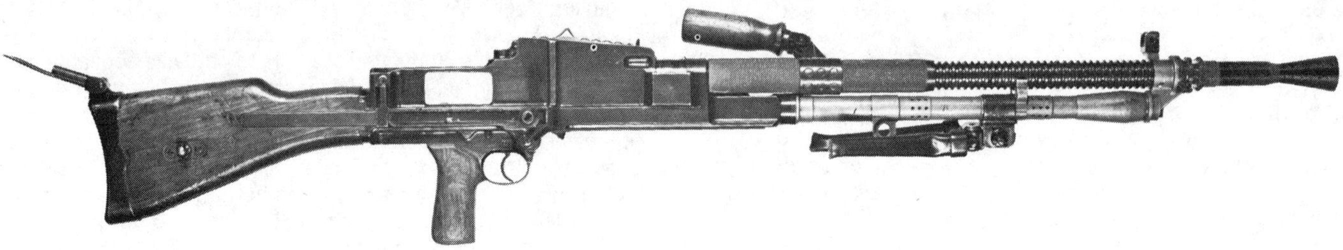

Representative Z.B. as made for Yugoslavia (ZB30J).

face the British decided to go outside the confines of the Island to obtain their new design. This took a considerable amount of moral courage on the part of the British military authorities. Practically all countries including our own are notorious for favoring national designs even when foreign ones are outstandingly superior.

The Czechs produced a model with alterations submitted by the British Government and labeled the gun the ZBG. The modifications, aside from rechambering the barrel for the British cartridge, were basically shortening the barrel and bringing the gas port nearer the breech to compensate for the shorter barrel. The Bren barrel does not move during firing. These modifications together with a newly designed and supported stock increased the rate of fire and gave a much smoother performance. Because of the recoil mechanism redesigning, the stock on the Bren as it is now known

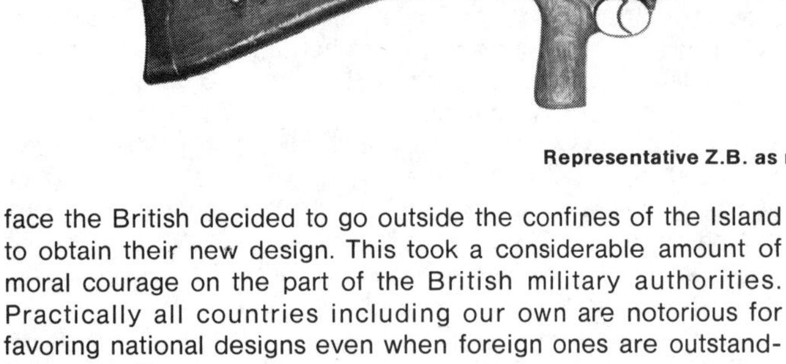

Representative Z.B. showing quick barrel change-lever. Magazine mounts on top of gun.

cannot be hinged.

The British Government was licensed by ZB to manufacture. In January, 1935, drawings were received from the ZB firm. By September, 1937, the Enfield Government Arms Plant produced its first guns. They were called the Bren. The first two letters were taken from Brunn (Brno) in Czechoslovakia. The last two letters were from Enfield, the British Arsenal.

In January, 1938, 200 guns were assembled. By the middle of the same year production had been stepped up to about 300 a week. That rate of production continued until September, 1939, when it was increased to about 400 a week.

The guns themselves continued to be made at the Royal Small Arms factory, but the B. S. A. and the Austin organization received contracts to produce spare magazines. Considerable initial trouble was encountered until it was found that the drawings had provided for magazines to hold rimless cartridges while the British ammunition of course was rimmed. In spite of this it was found that the magazine could be loaded with 29 instead of the customary 30 cartridges and still work well.

Upon the outbreak of World War II the Bren gun was manufactured in very large quantities in England and also in Canada for British service use.

The Canadian manufacturing plant also furnished large quantities of the gun to China in caliber 7.92mm, most of the military rifle equipment there at that time being of German Mauser type using the German service cartridge.

The Czech ZB50 Machine Gun

In 1932 the Czechs developed a new gun under the designation Model 50-1932. This gun is a short-recoil arm in caliber 7.92mm. This represents the first recoil-operated arm produced at ZB, although experimental work had preceded it.

The bolt assembly resembles that of the gas-operated models. However, the gas piston principle is eliminated. An accelerator transfers the energy at the time of recoil and at the instant of unlocking to transmit full force to the bolt, speeding it to the rear. This accelerator works very much like that of our own Browning machine gun.

Feed is by a non-disintegrating metal belt. The belt is expelled and the gun held open on the last shot. The design of the gun is largely credited to and patented by Anton Marek. A feature of his design is that only a loaded cartridge is needed to completely dismount the gun. The point of a bullet is used to depress a number of spring loaded detents, a system also utilized in many experimental designs by Holek.

A combination of a muzzle booster and the accelerator gave a firing rate of about 600 rounds per minute, the booster being also made to serve as a flash hider and forward barrel bearing. The influence of this design on later German developments can be seen.

A most unusual safety design is incorporated. It acts not only to lock the sear to prevent its release, but also lowers the bullet point of the cartridge in line with the bolt so that if by any breakage or freak action the sear might be slipped, the bolt going forward could not chamber the cartridge to fire it.

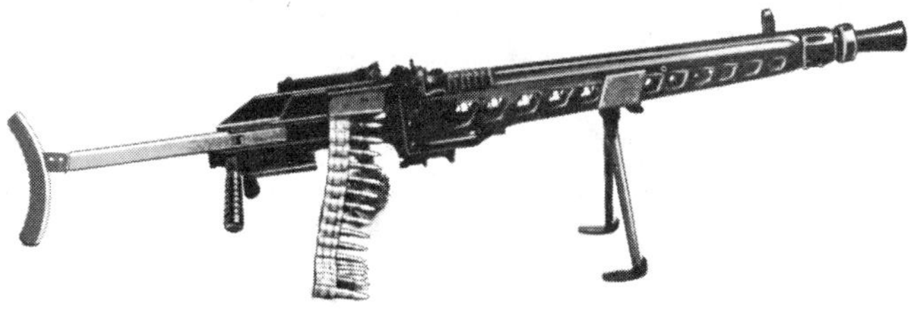

Czech Z.B. 50 machine gun.

DENMARK

The Danish Madsen Machine Guns

The machine guns marketed as the Madsens have been chambered for every conceivable military cartridge and been used in every corner of the world at one time or another since their introduction in 1902. Their widespread use stems from a variety of causes and reasons. First, the gun is a good design and is always at least reasonably well made. Its cost has always been relatively low. Also, it has had an aggressive worldwide sales group behind it. But perhaps most important of all, Madsen has stayed in the business of making machine guns and in emergencies all military groups have been able to turn to Madsen—not for what they wanted necessarily but for what Madsen had to offer in the line.

In the 1880s, a light machine gun was constructed by Mr. W. O. H. Madsen, who was then a captain of the artillery and later major general and minister of war, in cooperation with Mr. Rasmussen, technical foreman at the Arsenal. In Denmark, this weapon was called a recoil rifle, whereas to the rest of the world it became known by the name of the Madsen Machine Gun.

Both Captain Madsen and Mr. Rasmussen were employed by the Technical Services of the Danish Army, and therefore the utilization of the invention in Denmark was reserved to the Danish Defense Authorities.

In order to utilize the invention abroad, in which the constructors had great confidence, a syndicate was founded in 1896 by the name of "Dansk Rekyl-Riffel Syndikat" for utilization of the invention in all countries outside of Denmark.

In 1900, the syndicate was changed to a joint stock company which kept the same name.

In 1936, the name of the company was changed to "Dansk Industri Syndikat, Campagnie Madsen, A/S."

From a very modest start the production increased gradually as the weapon was improved. However, it was not until the end of World War I and the years thereafter that the boom occurred, especially after the factory was modernized. Thereby the quality of the weapon was improved and the capacity for competition greatly increased.

In spite of all improvements and modifications, the weapon, which in the production of the factory was called "The Standard Machine Gun," still operates on the recoil principle.

However, there is never any question about the Madsen to one who knows the elements of its design. It is the only "non-ramming" action used in machine guns. While all other patterns utilize the breechblock or its equivalent to drive or ram the cartridge out of the feed into the chamber, the Madsen system does not.

The breechblock is a variety of the original American Peabody and the Martini lock. The breechblock is pivoted at one end. On the opening action stroke it is swung up to expose the head of the cartridge case in the chamber for extraction and ejection. On its next action the block is dropped below the barrel line to permit a fresh cartridge to be chambered. The final motion brings the oscillating breechblock back up again into position to support the chambered cartridge as it is fired. A simple circular stud on the barrel extension operates in corresponding grooves in a switch plate secured to the receiver, to function feed and lock; all actions take place as the barrel extension recoils when the gun fires. The original feed was a top-mounted clip magazine; but improved belt feeds are also used on later models.

Experts for years have "proven" time and again that the Madsen is not a "sound" machine gun principle. Each round as it is rapidly chambered is somewhat distorted into an arc. Their theory can be proven by high speed photography. However, the gun works in spite of it; and in usage in every clime through the years it has built up a remarkable reputation for reliability.

In 1923 the Madsen syndicate altered its production considerably to bring it more into line with requirements of the time.

An aircraft type and tank version of the machine gun were produced. Weapons were made in a wide variety for almost any type or caliber. The general mechanism was not altered to any significant degree. The guns were made more streamlined in appearance and some were fitted with muzzle boosters to speed up the rate of fire.

Because of the tremendous spread of the Madsen arms throughout the world, these weapons may be said to be of more than passing interest.

Danish Bang Machine Gun

The Danish Bang of 1929 was a machine gun in prototype. It was based on the operational principle of the forward moving muzzle cone as mentioned in the chapter on semiautomatic rifles.

FINLAND

The Finnish Lahti (Suomi) Machine Gun

These very interesting arms are dealt with in the chapter on Finland. However, a note concerning them has a place here.

The designer was Aimo Johannes Lahti. Another of his noted products is the Lahti pistol. He served as Chief of Finland's Government Arsenal for a number of years. His inventions also cover rifles, submachine guns, and cannon. All of them were good; all of them were simple. His developments included both gas and short-recoil-operated arms.

The weapons themselves were first produced in 1926 at the

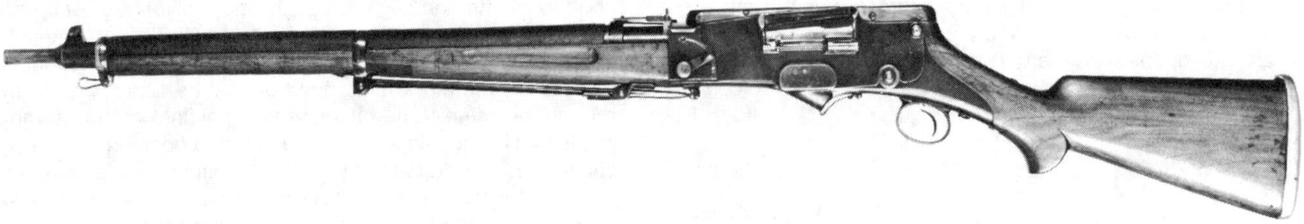

Model 1896 Danish semiautomatic rifle made by Madsen. This rifle did not receive the attention given to the full automatic machine gun version. It is relatively unknown.

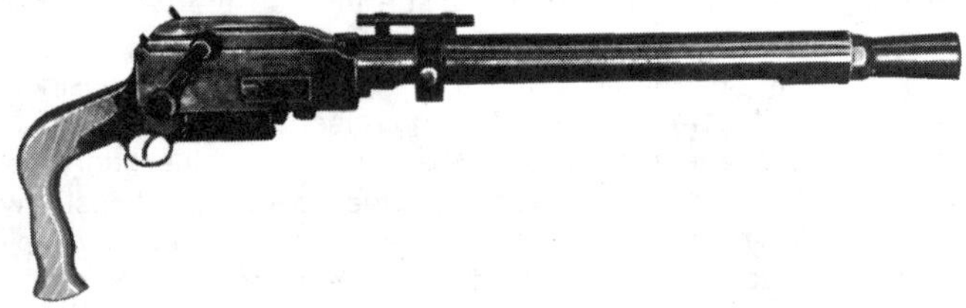

Representative Danish Madsen, tank pattern.

Valtion Kivaaritehdas which, translated, is State Rifle Factory. This is at Jyaskyla, Finland. The Lahti machine gun was a primary arm of the Finns in the defense of their country against the Russian invasion. Originally designed in 1926, it was modified in 1932. It is recoil-operated, air-cooled, and can be fired both full automatic and single-shot.

When the gun went into manufacture, the producers hoped to sell it to either the Germans or the British. As it happened it served a far better purpose in its part in defending Finland during the Russian attack.

FRANCE

The DeReffye Mitrailleuse

In this period when American inventors were introducing the wide range of new rapid-fire mechanisms mentioned—and hundreds for which we have no space here—European developers generally were blinded to anything but alterations of then-outmoded systems. The one most concentrated on by military and independent designers alike was the "Mitrailleuse."

While the design originated with Belgian Army Captain Fafchamps in 1851, the French gun was actually produced by a brilliant engineer of the day named Captain deReffye of France.

In 1867 Napoleon III was so impressed with versions of the Mitrailleuse that he placed it in secret manufacture at the French Arsenal at Meudon. And thus began another chapter in the long history of military misadventure based on undue secrecy.

The French press of the period really had a field day. The new "secret" weapon had such dread possibilities that none save the initiate might even receive a hint of its nature. Its transport was guarded, its presence covered at all times. The precautions taken in transporting our first atom bomb were amateurish alongside the measures taken by the French to hide the nature and identity of their secret weapon of 1870.

When the Prussians launched their attack in 1870, the French papers reported after each engagement that the new weapon had mowed down the enemy in fantastic numbers. The German artillery, in actual fact, was disposing of them every time they made an appearance. Had the French followed up the occasional successes they had when the guns were employed as infantry support, they might have given a far better account of themselves. Instead, they were considered and used by the French as artillery arms. Before the war ended, some Gatlings as well as other American machine arms reached France, but the secrecy surrounding the Mitrailleuse had blinded the French to the proper usage of such arms, and they played a very small part except for fortified defense use.

And what were these "secret" arms? The DeReffye is typical. It consists of 25 separate rifled barrels positioned inside an iron jacket. There is a loading plate provided with holes to receive cartridges to match up with the barrel positions. The plate when loaded is dropped into grooves in the breechblock. A hand crank is turned to drive the breechblock forward against the barrels, the plate providing the firing chambers. The firing mechanism is cocked by the closing of the breechblock, and the cartridge necks are forced into the barrel mouths. The gunner now releases the loading and locking crank and gives his attention to the right-hand crank which does the firing. As he turns the firing crank the 25 barrels are fired in succession, the speed depending upon the rate of turning. When the plate is empty it is withdrawn and replaced with a new loaded one.

This secret wonder mounted on an artillery-type carriage weighed with its limber well over two tons.

The French Hotchkiss Crank Operated Machine Gun

The original Hotchkiss machine gun, with which we will deal in passing, is the hand-operated machine type based somewhat on the Gatling. (We shall later consider the gas-operated automatic arms made by the Hotchkiss organization. These and modifications were in use in France and in Japan through both World Wars. All gas-operated Hotchkiss patterns were developed after the death of the man whose name they bear.)

Benjamin Berkley Hotchkiss was an American. He was born in Watertown, Connecticut, in 1826. He acquired his early knowledge of firearms while working at the Colt's Patent Firearms Company at Hartford. Many of the early improvements in the Colt revolver were accredited to Hotchkiss.

Like practically all great American arms inventors, he received little cooperation from the United States Government, but much attention from foreigners. In 1856 Hotchkiss sold a rifled field-piece to the Mexican government. Our Government did later adopt a percussion fuze which he developed for projectiles. While he did much small arms designing off and on throughout his life, it is in the field of light artillery that he is perhaps best known. Only his repeating rifle received any appreciable attention in his own country.

Hotchkiss was in charge of the City Arsenal at New York during the draft riots of 1863. He was given this appointment because of his knowledge of arms.

He went to France in 1867. His improved metallic cartridge case, which had received little attention here, was seized upon by the French to replace the combustible paper types used in the Chassepot rifle. However, the military failed to get production under way until too late for use in their dire need in 1870.

Hotchkiss revolving cannon. Caliber 1.5 inch. Developed in 1881. 5 barrels. The gun shown was made outside Paris. It is numbered 596. A shell is shown being loaded into the gun.

Hotchkiss organized a manufacturing company in France in 1875 at St. Denis, and proceeded to produce his new machine gun there.

This Hotchkiss has five rifled barrels of compressed steel. These are mounted parallel to each other about a central axis. They are held between two metal discs, and they rest in the frame which carries the trunnions. Turning a hand crank at the right side gives the necessary rotation for loading, firing, and extracting. While the gun externally resembles the Gatling in many ways, its design is completely original.

Hotchkiss began manufacture of a 37mm gun for both land and naval purposes. By 1884 the business had developed to such a large extent that additional works for production were necessary. Manufacture was begun by Armstrong in England.

The French plant sold over 10,000 cannon and 4,000,000 rounds of ammunition to the French Navy. During the period of its general use it was at one time or another employed by all the major nations in connection with naval operation. Details of its use will be found in the literature of Austria, Denmark, England, Germany, Holland, Italy, Russia, Turkey, and the United States.

Hotchkiss died in 1885, recognized and honored as one of the great inventors of his time. He did a great deal to push forward the development of repeating mechanical mechanisms. What is perhaps more important is that he trained the men under him to continue the operation of the plant after his demise. As a result, the name of Hotchkiss continues even to this day in connection with the gas-operated arms developed by his highly trained staff after his death.

It is France's misfortune that in recent times the effort, energy, and intelligence of the once-great Hotchkiss group has been allowed to diminish. Its history is considerable.

The Hotchkiss Automatic Machine Guns

When Benjamin Hotchkiss died in 1885, he left behind him a very capable engineering and production organization devoted to the development and manufacture of arms.

To this organization in 1893 went Austrian Army Captain, Baron Adolf von Odkolek with the new type of gas-operated machine gun which he had developed at Vienna.

The Hotchkiss Company by that time had expanded into a dual organization, though both were under French control. The original organization was changed after the death of Hotchkiss and in 1887 it became officially the Société Anonyme des Anciens Establissements, Hotchkiss et Cie. Its offices were in Paris, but manufacture continued at St. Denis. The English branch set up originally in 1884 with Armstrong at Elswick now became the Hotchkiss Ordnance Company Ltd. with offices in London.

Laurence V. Benét, an American who had been associated for years with Hotchkiss, was placed in charge of promotion and engineering—a rather unusual dual position. The fact that Benét's father was General S. V. Benét, then Chief of Ordnance in the United States, just might have had something to do with the appointment and the jobs. In any event it was he who arranged originally for his son to enter the Hotchkiss employ.

Mr. Benét had the faculty so important for the successful executive—he knew how to pick assistants and how to delegate authority to them. One such was Henri Mercié, a brilliant and tireless worker. Aside from the Benét-Mercié Machine Gun which was not to prove successful, these men played a very considerable role in the development and advancement of automatic arms.

Von Odkolek had the good fortune to present himself just when Benét was badly in need of a new product to sell his stockholders. And the Austrian definitely had something. Maxim had ruined the Hotchkiss world market with his superior automatic designs and

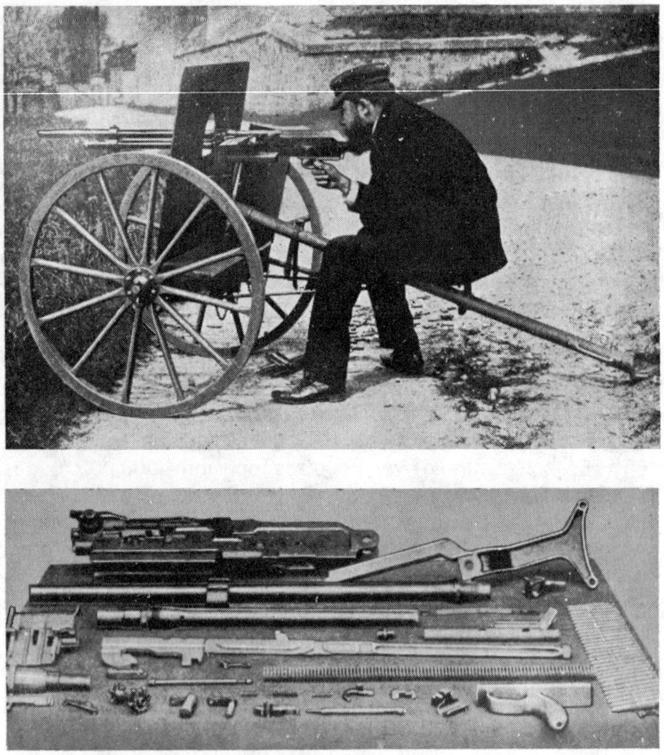

Upper: Laurence Benet, assistant to Hotchkiss, testing the original Hotchkiss. Lower: The first Hotchkiss field stripped to show simple parts.

he had the patents on recoil operation so sewed up that Hotchkiss couldn't find a way to compete. Von Odkolek provided the way. While his gun itself was not very good, being basically an early inventor's model, he had a new application of gas-powered operation. This was the now familiar gas piston housed in a cylinder below the barrel.

The Hotchkiss people drove a pretty hard bargain, refusing a royalty deal and paying instead a flat sum for outright purchase. The inventor in later years tried in many ways to get around his own patent, but he had done too good a job in the first place; and the Hotchkiss group through the years controlled the world market until the patents expired.

Benét and his associates re-worked von Odkolek's system until in 1895 they came up with the first version of the successful gas-piston-operated automatic arm. The gun was chambered for the 8mm Lebel cartridge and through the intervening years nothing basic has been done which could really improve it. The 1914 gun as detailed in the chapter on France is typical of all the variations, being itself a slight variation of the 1897 model.

The gun is locked-breech, clip-fed, gas-piston operated and air cooled. Gas is taken off to operate the piston through a port drilled in the barrel, the gas being exhausted as the piston passes back over an exhaust port.

The United States Navy was furnished a model for testing in 1896, the tests being unsuccessful because of lack of proper metal usage and tempering of parts. An American, Edward Parkhurst of Hartford, Conn., was engaged upon recommendation by our Navy to make design changes and manufacturing suggestions; and as a result of his work the 1897 Hotchkiss modification was produced.

Benét retained his American citizenship, leaving France to serve as an ensign during the Spanish-American War, but returning to Hotchkiss after his discharge.

While the French adopted the Hotchkiss gun in 1897 and placed a limited number in African desert service, it was not until the 1900 version with improved steel cooling fins was introduced that the gun really received serious military attention. Exhaustive tests were made by the United States at Springfield under which the gun stood up well, but in which the feed superiority of both the Maxim and Colt-Browning belt systems was established. The Japanese early purchased large quantities of Hotchkisses. In the Russo-Japanese War of 1904-5, the machine gun as such really came into its own when these Hotchkisses were pitted against the Russian Maxims in numbers. The German observers advised immediate concentration on machine guns as a result of their observations, and their government acted.

The Hotchkiss influence spread to other guns or more correctly to modifications of the Hotchkiss, for that is all that they are. We must now mention the more celebrated of these, since much confusion exists on the subject.

The French Puteaux Machine Gun

This gun takes its name from the fact that it was first produced at the French National Arsenal at Puteaux in 1905.

It is actually only a minor modification of the 1900 Hotchkiss, the various "features" developed by the French army designers being useless gingerbread for the most part. They built in, for instance, a firing regulator permitting firing cycle changes of from 8 to 650 per minute. This of course was merely a variation—and not a good one—of the regulator Hiram Maxim placed on his very first gun; and which he very soon discarded. Brass circular fins covered the barrel from breech to muzzle, and in desert campaigns these heated up faster than the standard pattern did. As a result of its desert failure, the Puteaux was withdrawn from field service and issued for use in fortifications. It is often referred to as the "Fortifications Model," though it was never actually intended as such, but just happened to end up there because in field use it was just too undependable for issue.

The French St. Etienne Machine Gun

This gun represents yet another unsuccessful attempt to improve the basic Hotchkiss. It takes its name from the fact that it was designed by officers and made at the French Arsenal at St. Etienne.

Just about the most stupid thing any design group can do is to make a change just for the sake of change—or for private empire building. The St. Etienne gun illustrates this technique.

The gun merely reverses the successful gas piston system of operation. The gas is utilized to drive the piston forward to unlock the bolt—instead of to the rear. This requires a gear rack and spring-loaded rod hooked up to the piston. The gear rack in turn is engaged with a spur gear which in turn is secured to an operating lever. This lever in forward position engages a cam slot in the bolt to lock it when firing.

As the gun fires, tapped-off gas blows the piston forward. This compresses the spring attached and also forces the spur gear to rotate. The operating lever turns with the gear through a half turn, thereby withdrawing the bolt and halting it in the rear horizontal position. The operating spring now reacts to drive the piston to the rear. This in turn brings the bolt back to firing position. The operating spring under the barrel is behind the piston and must be exposed at all times as otherwise heat would destroy its temper in very short order.

In War I complaints by soldiers resulted in its replacement in the field by the Hotchkiss. Its heavy cast brass receiver immediately identifies it.

The French Benét-Mercié Machine Rifle

About 1908 the Hotchkiss Company introduced a completely new design. It was the combined work of Benét and Mercié.

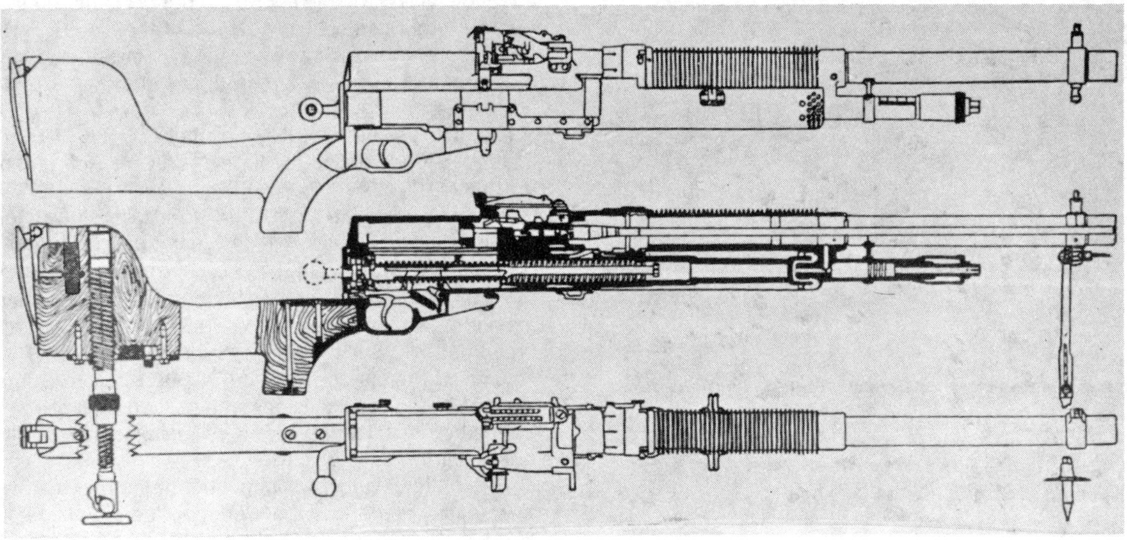

Sections of the Benet-Mercie 1909, modified from the original Hotchkiss.

Their idea for a light weight machine weapon was very good and most sound. Some of their design changes were excellent, too, resulting in a reduction of the number of parts to about 25. In Europe where it was adopted by the French in caliber 8mm Lebel in the year 1908 it was widely known as the "Hotchkiss Portative." The American terminology for the model we adopted in 1909 in our caliber .30-06 was "Benét-Mercié Machine Rifle"—from the inventors' names and the fact that it was light and had a shoulder stock.

It varied in two design ways from the Hotchkiss—both bad: In method of breech locking and in feeding the clips upside down. Cartridges were on the lower side of the clip during feeding. The locking was by a cylindrical fermature nut. A lug on the nut engaged the gas piston and locked the breechblock when firing. During operation the nut also did the unlocking after gas pressure had dropped.

The barrel in the gun could shift forward if a lock nut on the action worked loose; but the big troublemaker was the feed system. To fire the gun the feed clip had to be inserted upside down, and very carefully, in the right side of the receiver. The cocking handle was then pulled back to draw the operating units to the rear. The handle was then pushed forward as far as it would go, at which point it could be lined up with the letter "A" for full auto fire, or with the letter "R" for single shot or repetitive fire by turning it to the right.

This was the official United States machine arm until 1916. Fortunately we didn't send any overseas, keeping them at home for training purposes where they were at least heavy enough for the purpose, something that couldn't be said for the broomstick guns with which so many War II soldiers were trained.

In all fairness, the author must point out, however, that the United States purchased only 29 of these guns from Hotchkiss. Over the years Colts and Springfield Armory made only 670 more. In the face of the small orders, it is obvious that we could not hope to get either manufacturing or operational "bugs" out of any design. These facts must be considered in appraising our failure to improve the Benét-Mercié.

When Pancho Villa raided Columbus, New Mexico, in 1916 the town was "defended" with Benét-Mercié Machine Rifles. According to newspaper stories and some military historians, not one of them fired a shot. The gun squads had a perfectly good alibi—Villa attacked at night, and the feeding of their guns was so delicate and ticklish that they just couldn't get the guns started. So goes the story.

American newspapers lampooned the Benét-Mercié mercilessly, even to the point of suggesting that the rules of war be changed to prevent night fighting so we could use our machine arms. They thus became famous as the "daylight guns."

The Mexican guerrillas had some Colt-Brownings, as well as some early Hotchkiss guns they had captured from their own armed forces. These all worked very well when we managed to capture and test them.

What actually is the truth? Quite different. The raid was completely unexpected—but then so was Pearl Harbor. One of the most reliable and unbiased arms experts of all time was Major General Julian S. Hatcher, U.S.A., Ret. Throughout his long years of service and experience his reports always were noteworthy for accuracy and detail. In the Saturday Evening Post magazine issue of November 10, 1917 will be found an article by him dealing with these stories. His investigations showed that after the shock of surprise, the crews actually got their four Benét-Merciés into action and fired 20,000 rounds, despite some jams. Fire was often withheld because of darkness and lack of targets; but the guns did contribute positively to the defense of the town. The journalistic outbursts should have been leveled at those responsible for permitting a surprise attack—just as they should have at Pearl Harbor and Hickam Field.

The French Hotchkiss Model 1914

Although the Germans made no secret of their stock of Maxim guns, the approach of war in 1914 found the French desperately short of machine weapons. In their need the French called upon the Hotchkiss organization to produce in quantity, the Arsenal production and manufacturing methods being entirely inadequate. Hotchkiss performed miracles of production for that day. By 1916 they were really pouring guns out, and the quality and reliability astonished all observers—a real tribute to the design and the manufacturers. By 1917 the French troops were demanding Hotchkisses to replace the St. Etienne and other hybrids; and on July 15, 1918, it was the Hotchkisses which Gouraud's Fourth Army used to break up the all-out German attack which really folded up German offensive ability in War I. More than fifty per cent of the attack force was left on the field dead or wounded.

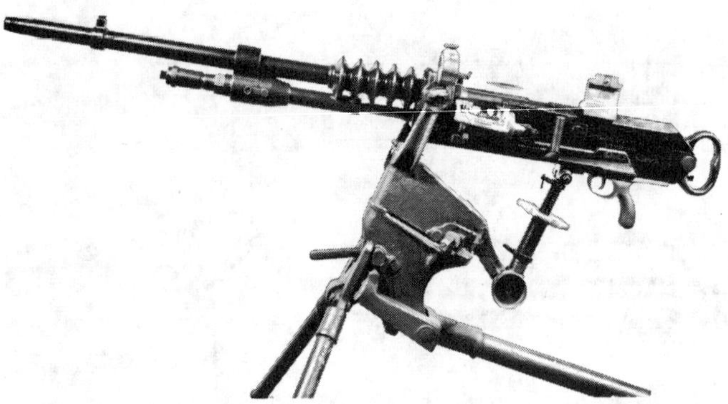

The French 1914 model Hotchkiss, caliber 8mm French. A famous gun of World War I.

American forces in France were armed with these 1914 Hotchkiss guns using the French 8mm cartridges.

The French Hotchkiss Balloon Gun

This gun, a typical Hotchkiss beefed up to handle an 11mm (.472 caliber) cartridge was introduced late in the war by Hotchkiss. It was originally intended as a long-range machine gun for use against artillery crews at ranges too great for standard rifle caliber guns to reach.

Its surprise use, however, came with the widespread introduction by the Germans of observation balloons for artillery spotting. These new Hotchkisses with their large size bullets were able to carry a heavy enough incendiary charge into the hydrogen-filled bags of the balloons to raise havoc. Firing at a rate of 400 to 500 per minute, they used not only conventional 20-shot clips but also a special metal belt holding 250 rounds.

While still in the prototype stage this gun was observed by our Colonel John Parker, who recognized in it features we could use in developing a heavy caliber arm of the type. He arranged for samples to be sent to the United States from France, and while the gun was not adopted because ballistically its ammunition did not meet our requirements, it really pointed the way for the eventual development of our famous .50 caliber cartridge and gun.

The French Chauchat Machine Gun

Like so many others we have considered, the Chauchat gun is known by the name of the man who pioneered it. Col. Chauchat was chairman of the French Commission which decided upon its adoption, and it was named in his honor. Many American soldiers swore by this gun, which was one of their arms in World War I. Many more swore at it as one of the clumsiest and balkiest pieces of equipment ever encountered, and they were by far the more numerous.

The gun is sometimes listed as the "C.S.R.G." This, too, is in the European tradition of "honoring" officials, for it represents the initials of the entire development group—Chauchat, Suterre, Ribeyrolle, and Gladiator! After World War I it was distributed to Belgium, Turkey, Rumania, and many other countries. In British circles the gun is often listed—incorrectly—as the "Chauchard."

It is ironic to recall that our own supply of automatic arms was almost nil when war broke out, and that we were forced first to equip our troops with this monstrosity in its 8mm French service caliber and to purchase some 37,000 of them from the French. Nine of our combat divisions were issued Chauchats in the United States before sailing, some in caliber 8mm, some altered for our .30-06 cartridge.

As a matter of record, it might be mentioned that the alterations were easily made, but they reduced the magazine capacity from 20 to 16 cartridges. As might have been expected when altering such poor equipment to handle far more powerful ammunition, the .30-06 models were even worse than the originals. Cases stuck

The famous World War I Hotchkiss balloon gun.

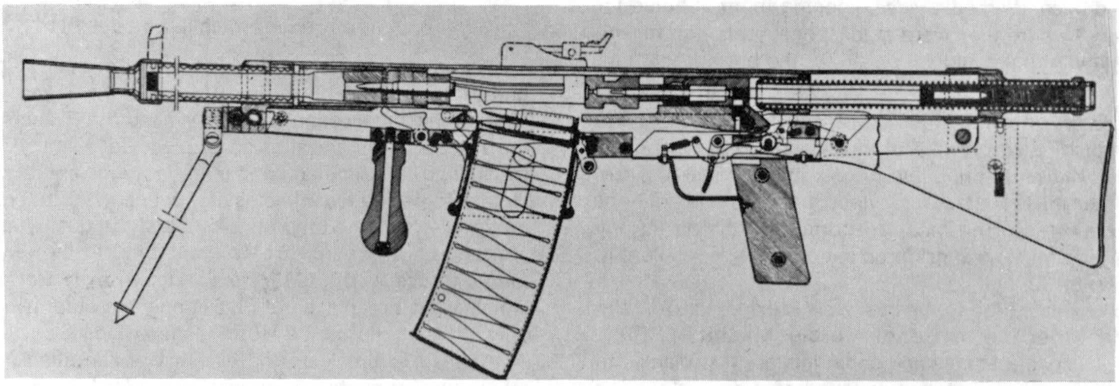

French Chauchat M.G. in section to show operation.

in the barrel after even moderate fire, parts broke—particularly springs—and the gun jammed under almost all field conditions. Still, the BARs were not in production and we were compelled to accept the philosophy of Chauchats being better than nothing at all. To make matters even worse, alterations and inspection were under French control, and little heed was paid to American suggestions and requests for improvements.

One French Chauchat training tactic was too much even for their own long-suffering troops. This was the "brilliant" idea of rear echelon experts that the way to use the Chauchat was to advance in line, each soldier firing a burst from the hip as he planted his left foot! Fortunately for the French, not much use was made of this and similar instructions, since it was difficult to keep the gun firing for more than two very short automatic bursts even in spite of its leisurely rate of operation.

The gun was a pretty accurate copy of a turn-of-the-century rifle designed in Hungary by R. Frommer, the father of the long recoil system of operation which works well in automatic shotguns, fairly well in pistols, and not well at all in machine rifles or machine guns.

Ths basic feature of this system is that the recoiling members, including of course the barrel, travel back the entire length of the cartridge case, fully locked. Compressed springs then assert themselves. The barrel is pushed ahead. Through cams acting in a non-recoiling sleeve, its movement causes a turning movement of the bolt head to unlock—the bolt tail being suitably held back at this point. As ejection occurs and a new round comes up for feeding, the bolt is released to chamber the cartridge as its spring drives it home. Cams turn the bolt head to lock the action. This is the basic Frommer action.

The gun could be fired only when the action was fully locked—when it could be fired. The quality and construction of the gun was undoubtedly the crudest ever to appear in any military arm. Despite a complete lack of normal tolerances, the gun parts still were not interchangeable. The arm was unique in that it utilized stampings, tubing, and lathe-turned parts throughout. The principle of manufacture was most sound. The trouble was in the crudity and carelessness of manufacture. The principle used in building the Chauchat was utilized by both the Germans and the Russians during World War II and produced some superb weapons, some very cheap ones—but all reliable because of care in manufacturing plus improved manufacturing methods. Even our own M3 submachine gun is built on this manufacturing principle, as are many other of today's best foreign arms in that category.

The French Berthier Machine Gun

This gun is usually listed as French because its inventor was a French Army Officer, Lt. (later General) Andre Berthier. His first patent was in Belgium in 1905. This was for a straight-pull rifle similar to the Mannlicher, but with an external gas cylinder housing a piston on the right side of the rifle where it could move the operating handle in straight line. Similar applications of this system were made at Steyr Armory before Berthier, yet he achieved considerable attention because of the design.

In 1908 he introduced the first model of his machine gun, the arm being made at the plant of Anciens Establissements Pieper at Herstal, Belgium, an organization set up in competition to the F.N. plant by one of its original founders. This is of interest since at the time there was great rivalry between the two firms, and John M. Browning was then a member of the F.N. combine as his heirs are today. Browning's later famous BAR has much in common with Berthier's first Belgian produced gun, particularly in the matter of its gas operation and its breechlock system.

The gun was light, well-made, and streamlined. It was designed for both infantry and cavalry use. A unique water-cooling system used a very tight barrel jacket with two compartments, water being circulated through them from rubber water bags squeezed by an

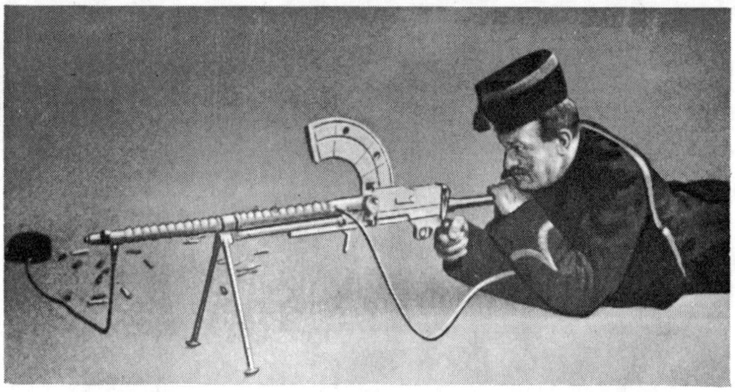

Model 1911 French-invented Berthier M.G. was manufactured in Belgium.

assistant during firing. This method did away with the bulky jacket design of standard guns of the type.

The cartridges were fed down through the top of the receiver from a sprng-operated magazine which had a capacity of 30. Firing rate was about 450 per minute. The breechblock was the prop-up type which is locked as its rear end is raised into a locking seat in the top of the receiver, allowing the actuating gas piston to continue ahead and its firing device to discharge the round in the chamber.

General Berthier visited the United States in 1916 in an attempt to set his gun up for mass production. After tests in June 1917 orders were placed by our Army for 5,000 and by the Navy for an additional 2,000. The guns were to be made at Norwich, Conn. by the arms firm of Hopkins & Allen—a manufacturer no longer in existence. The firm had set up a special division (The United States Machine Gun Company) for manufacture. They had financial troubles and the gun was not put into production.

With the end of the war, all U.S. interest in such developments also ended. The Berthier went the way of many similar arms. Some of its principles found their way into successful designs—e.g., the Vickers Berthier Mark III.

Laird-Menteyne Machine Gun

One of the final tests of off-trail machine arms was that conducted at Springfield Armory in the United States in September of 1913 when a test was made of the so-called Laird-Menteyne Automatic Machine Gun.

The arm was produced at the Coventry Ordnance Works in England. A representative of the company, of course, was with the gun to assist in the test.

The design was invented by two French mechanical engineers, one of whom was Paul Menteyne. Development work and original patents seem to have been done in 1909 but it was several years before the Coventry organization undertook development of a sample model.

It was an air-cooled, recoil-operated machine rifle to use the standard service cartridge. The arm had several items considerably in advance of its day, particularly in the matter of safety features. However, it failed to pass the rigid test to which it was subjected and was not put into production.

The French Darne Machine Gun

The original Darne guns were the invention of Regis and Pierre Darne, who operated a factory near St. Etienne. During World War I they produced small quantities of Lewis guns. The French Government gave them an order for their new gun and some were delivered in 1917. At the close of the war they continued their experimentation on this arm and eventually produced types which were used basically for aircraft work by French units and by some of the countries in the Balkans.

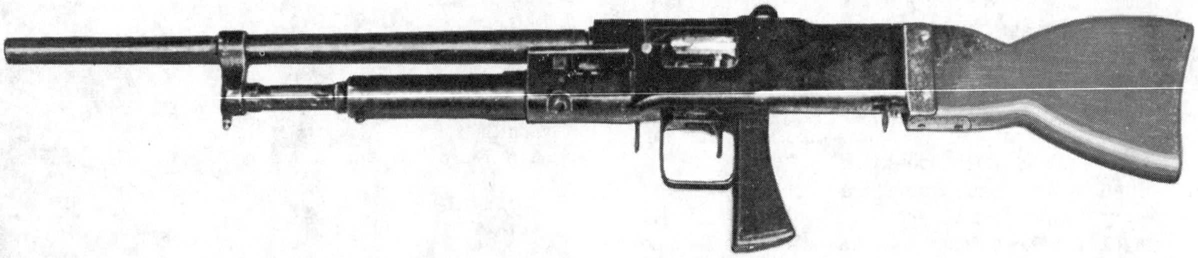

French Darne light machine gun. Probably the cheapest machine gun ever built. Good design but very bad manufacture.

The Darne is not, as had been suggested, just another type of Hotchkiss, though like that gun it is gas operated. Fundamentally however, it is quite a different arm. The feed system differs radically. Its unique design permits a very short bolt stroke. The functioning of the feed occurs during the course of the recoil. The Darne is capable of a very high rate of fire, as much as 1700 rounds a minute with the 7.5mm French rifle cartridge used in World War II.

In view of the high quality and extremely high selling price of the famous Darne shotguns in commercial use, it is difficult to reconcile the poor quality and workmanship of its very cheap machine gun.

The quality of the arm seems to have been deliberately sacrificed on the theory that the life of an airplane was short anyway and it was not worth while spending a lot of money on guns for it!

Darne machine guns were furnished to the government of France in 1931 at a price of approximately $28.00 in U.S. currency. In the period from 1918 to 1931, 11,000 went to Brazil, Italy, Serbia, and Spain as well as to France. The gun was also adopted by Lithuania after competitive tests in 1934, again on a price basis. In the following year a British commission visited the Darne plant. While the weapons passed French tests, they did not measure up to the British standards.

Darne produced a light and a heavy machine gun for infantry use as well as an antitank automatic gun of 11mm caliber at this time.

For all its terribly crude appearance, the Darne is potentially an efficient gun and probably the cheapest arm of its type ever produced anywhere. It is quite remarkable for not having one forged piece in its entire construction. It depends upon screw machine pieces and stampings for most of its parts. The action of course is gas operated. Primary extraction is provided and ejection is quite satisfactory. The bolt is quite heavy and is securely locked at the instant of firing by a shoulder on the gas piston which cams the rear of the bolt up into a locking recess in the receiver. This design, if better constructed, could easily be one of the finest weapons of its particular type. The answer to arms manufacture, for all practical purposes, should actually fall midway between the preposterously poor manufacturing methods, equipment, and materials of the Darne and that of the super finish and unnecessarily strong and costly designs with which we as a nation are afflicted. There is considerable room for thought and action in this connection.

The French Chatellerault Machine Gun

The Chatellerault was introduced in prototype in 1921. Its first official adoption was in 1924 after modifications.

Basically it is a combination of the American BAR and the earlier Belgian Berthier. Its name derives from the point of manufacture, Manufacture d'Armes de Chatellerault. This is a French government arsenal.

After the introduction of this new gun, which was prepared in utmost secrecy in common with such matters, the French newspapers proceeded to give ecstatic reports about it. Among other things it was stated that the soldier could fire 30 shots at one burst from the shoulder and that eventually every French soldier would carry this new weapon.

The Chatellerault design required the introduction of a new cartridge instead of the old rim case 8mm cartridge. The 7.5mm rimless now in use in the French service was developed to permit this gun to operate.

Before the gun was even in production the French govenrment offered it to Yugoslavia in 1925 at a very low price of 2,000 francs. A Yugoslav Military Commission went to France to check on the Chatellerault gun. The price by that time had risen since the French found it could not be produced within the original offering figure; but it was still far below that of other machine arms. The Yugoslav army needed machine arms badly and asked for demonstrations and competitive tests. Tests were so poor that Yugoslavia

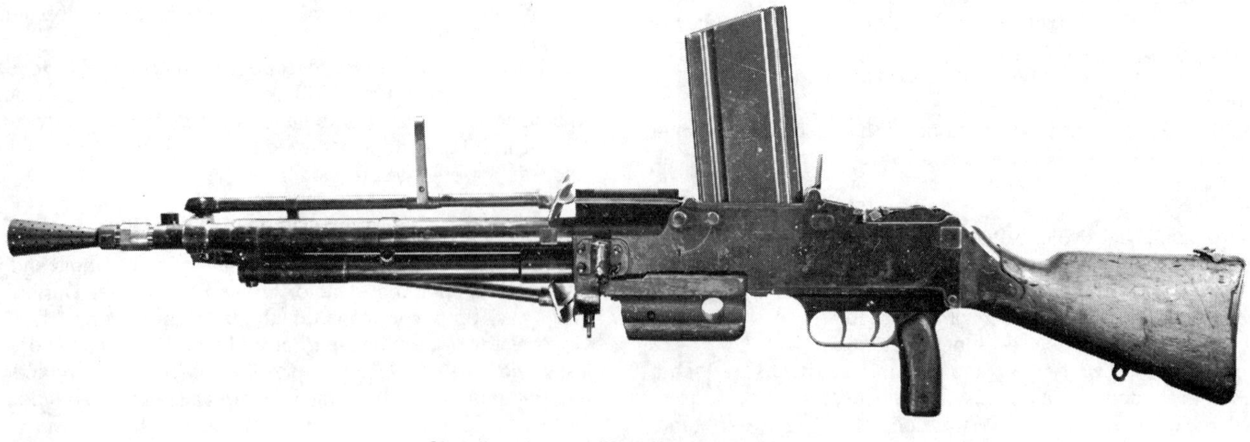

Chatellerault as modified. Used in Indo-China.

turned to Germany and Czechoslovakia instead of to France for its equipment.

Similar conditions arose when the gun was offered to Rumania. During the Rumanian tests at least one gun operator was seriously injured by an explosion in the receiver of one of the guns. The French immediately claimed sabotage. Tests were made to produce the same type of premature explosion in the Hotchkiss, which resulted in merely a swollen barrel. There was no such explosion as occured in the then inferior and lighter Charellerault.

For several years the French troops themselves were much afraid of the design after the early stories they had heard about it. They continued to demand the heavy Hotchkiss. Eventually, however, the manufacturing bugs, particularly those in relation to poor heat treatment and metals, were overcome.

Special models were made between 1934 and 1939 to adapt the 7.5mm arm for use as a fixed aircraft gun. Another design was intended for use as a tank, armored car, and fortress model. The so-called "fortress model" could be furnished with a drum-type magazine on either right or left side, ejection being through the bottom of the gun. The capacity of this weird magazine was 150 rounds.

This "fortress model" Chatellerault was another one of the great "secret weapons" of the French which tied in with their Maginot Line psychology. It was actually fitted with a device operated from the recoil and the counter recoiling motion of the bolt to inject water into the chamber end of the barrel for cooling. This action took place as the case was extracted and before the new round was chambered. This represented a tremendous military secret to the French. The theory was that in the Maginot Line the attack would be made directly from the front and any gun which could maintain long bursts of fire would unquestionably annihilate the Germans coming forward in mass formation! Like the forward-facing artillery of the Maginot Line, the guns did not get an opportunity to demonstrate themselves.

Chatellerault also furnished an aircraft version allegedly capable of firing 1600 rounds per minute. On actual tests, it proved to be much nearer 1300. The magazine was a monstrosity intended to hold 500 cartridges. The big "secret" in the design was allegedly a new type of spring having at least double the life of the customary type according to the production men at the Chatellerault Arsenal.

Actually the spring had already been used by the Russians. Aircraft machine guns on planes sent to Spain by Russia during the Civil War in 1937 utilized this type of spring. It consisted in coiling several smaller diameter springs of piano wire together to replace the single thicker coil. There is no question about the efficiency of the spring. The only question is about the secrecy with which its alleged development was surrounded.

GERMANY

The German Bergmann Machine Gun

While the Bergmann never achieved in Germany the success of the Maxim, it still was a machine gun to reckon with. In view of its many fine points, plus the fact that it was designed by a German, the noted inventor Theodor Bergmann of Gaggenau, it is noteworthy that the German Army passed over it as a first line gun and adopted instead the foreign Maxim. The German policy was always to try and create the best, but not to hesitate to put their own aside when a better product appeared. Most other nations could learn something militarily from this attitude.

Bergmann is noted for his production of automatic pistols, and also for the fact that he produced the first truly successful submachine gun as we shall see later.

His first machine gun was patented in 1900 and initial production at the Bergmann works in Suhl appeared two years later as the M. 1902. Modifications in 1903 related only to mounting and feed details and indeed others in 1910 were again little beyond mount modifications. As in the Maxim and the Hotchkiss, the design itself was so simple and correct that later mechanical improvements could only be minor ones. This is the true measure of any design genius.

The Bergmann utilized an exceptionally fine feed, the belt being a non-disintegrating type made of aluminum links in a day when only the canvass belt was in general usage. The feed extractor claw would engage, withdraw, and position cartridges inserted under conditions which would jam any other feed mechanism.

This simple gun was a water-cooled, short recoil-operated, belt-fed weapon with a quick-change barrel system unique in water-jacket types. Many of its general design features appeared in later machine guns, including our own Brownings. The back plate carried the grips and trigger mechanism and could be easily removed. The cover plate enclosed the feed system and was locked to the receiver, and when the cover was lifted the gunner had access to all the lockwork. The barrel could be removed without loss of water from the jacket by turning the muzzle down, pressing the bayonet-lock release catch on the receiver, and then pulling the barrel and its connected extension out the rear of the receiver, An assistant shoved a stopper in the barrel opening in the jacket as the barrel was withdrawn. The stopper was pushed out by the new barrel as it was inserted.

The parts recoiled locked for approximately one-half inch, after which the barrel and extension travel was halted, and the rising block was crammed down out of engagement with the breechlock, thus allowing that member to continue rearward travel to produce the action for extraction, ejection, feeding, and reloading.

The gun was very compact and had a straightline action of a very desirable type. Probably the long successful record of the Maxim dwarfed the Bergmann. Whatever the reason, though it saw battle service to some extent in both world wars, it never achieved large scale production.

M.G. Model 1910. Inventor Theodor Bergmann demonstrating his first type machine gun.

Bergmann water-cooled machine gun Model 1910.

The German Dreyse Machine Gun

The Dreyse is still another of the endless line of arms which have created confusion in military circles because of its name, much as in the case of the Madsen we have discussed.

In the first place, Johann Nikolaus Dreyse had been dead many years before the machine gun named in his honor was patented by Louis Schmeisser of Erfurt. That was in 1907. And it was five years later before the first model appeared. The gun was named in honor of the inventor of the needle gun by the heads of the factory where it was made—the factory originally founded by Dreyse, incidentally—the plant known as the Rheinische Metallwaren und Maschinenfabrik A.G. at Düsseldorf. This organization, as we shall see later, was the primary concern in setting up the small arms might of Hitler through its machinations with the Swiss Solothurn and other foreign arms makers.

The Dreyse 1912 and its 1918 modification rather resemble the Bergmann at first glance, and are often confused with that earlier gun. Actually they have little in common except appearance and the fact that both are recoil operated, water cooled, and belt fed.

Like the Bergmann, the Dreyse gave great consideration to feeding—one of the basic weaknesses in belt-fed guns. Its feed featured a three-clawed cartridge withdrawal unit, one of which was bound to pick up the cartridge regardless of how poorly it was positioned in the belt.

The 1918 Model replaced the tripod mount for most purposes with a bipod which allowed close-to-ground firing positions for the gunner. The receiver was equipped for telescopic sight. Unlike the Bergmann, the Dreyse incorporated an accelerator and buffer, giving a firing rate of about 600 per minute as against the other's 450.

The breechlock was a pivoting design pinned at its lower end to the barrel extension. During recoil its rear section rode up a ramp in the receiver, thus lowering the locking section out of engagement. Firing was by an internal hammer through a firing pin, the hammer utilizing the hammer-hook safety principle not unlike that on our modern Garand rifle.

At the close of World War I the victors sought to prevent fast German rearmament by prohibiting the manufacture of water-cooled machine guns—on the theory that only relatively heavy guns of that type constituted a real military threat.

The developers of arms such as the Dreyse found little difficulty in designing better and simpler light arms which could do just about everything that the heavy water-cooled types could—plus. And the makers of the Dreyse were also the manufacturers who arranged for parts manufacture outside Germany in preparation for the day of resurgence. Those who studied the Dreyse and its construction often foretold what its makers were capable of doing; but in a world of politics their warnings went unheeded until World War II broke and it was too late to take preventive action.

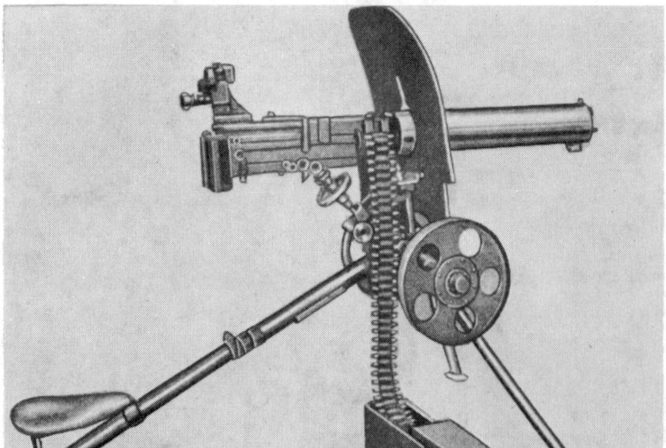

German Dreyse machine gun. The gun was named in honor of the inventor of the needle gun who was dead many years before this machine gun's production.

The German Maxim Model 08

These arms will be found generally discussed in the chapter on World War II materiel. As a historical note, however, it might be stated that the German Army had over 12,500 of them for immediate use at the outbreak of World War I and many more were in course of manufacture or on order. Only Germany was awake to the importance of the machine gun.

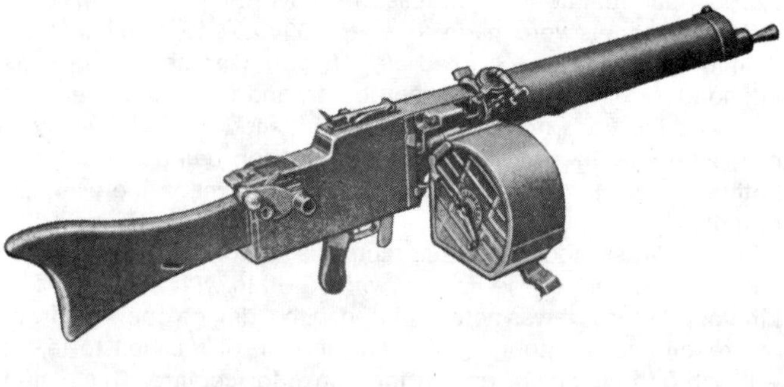

German Maxim light model 08/15. These guns appeared to some extent even in World War II.

These Maxims were also mounted as part of the equipment on Zeppelins to fight off attacks.

The German Parabellum Machine Gun

During World War I the Germans utilized the so-called Parabellum as their prime aircraft machine gun. Mention of it is made here only because a certain number of these guns equipped with special jackets and mounts were issued for infantry use or for use on mounts in Zeppelins. During World War II the Parabellum again made its appearance to a certain extent. The gun is basically a highly refined version of the German Maxim. It was developed at the Deutsche Waffen Und Munitions Fabriken at Berlin. This was the plant which produced the world-famous pistol known to us as the Luger. (In Germany it is known in military circles as either the Pistole 08 or the Parabellum, the latter name being merely Latin "for war," indicating a military type arm.)

In 1911 Karl Heinemann was assigned at the D.W.M. plant to work over and lighten the Maxim gun to handle the standard Mauser rifle cartridge. He produced the Parabellum, a name which was the telegraph code address of D.W.M.

It varies in action from the Maxim in that the locking toggle moves up as in the Luger instead of down as in the original Maxim. By other refinements Heinemann managed to reduce the weight to 22 pounds with a firing rate of about 700 rounds per minute.

The Parabellum was utilized by the German aircraft designer Fokker. He developed from an earlier Swiss patent a device for permitting safe fixed firing from an airplane through the movement of the propellers. This enabled the pilot to fire effectively while in flight.

Fokker was born in Batavia, Java. As a designer he offered his services to Great Britain before World War I broke out. He received absolutely no attention then from either the French or the British in this connection. The Germans however utilized his service with tremendous success. The British Secret Service during the war offered Fokker sums up to 2,000,000 pounds to try and inveigle him away from Germany!

The German Model MG13

When Hitler achieved control of Germany, action was taken immediately to rework the thousands of Dreyse Model 1918 water-cooled machine guns on hand. They were converted to more efficient air-cooled types. This work was done by Simson & Company of Suhl. The resulting modifications altered the appearance of the gun very much indeed.

During this period automatic weapons were given an identification number. The designation MG 13 was assigned to this particular modification. Until 1935 it remained the principal machine gun of the German Army.

The improvements over the old Dreyse consisted of eliminating the water jacket and furnishing a ventilated air-cooled barrel. A light-weight shoulder stock and a pistol grip were provided for prone shooting. Instead of the belt, a spring-loaded curved 25-shot magazine was provided. This could be loaded directly from the standard 5-shot army rifle clips.

The firing mechanism stayed open, held by a slide catch when the last shot had been fired. This, of course, notified the gunner and also permitted rapid reloading. A special system of cover operation was devised whereby the cover assembly was raised, allowing the back plate to be lowered. The two operations permitted immediate access for inspection of all operating parts.

The action also allowed a quick change of the heated barrel. This circumvented the League of Nations idea of preventing the Germans from developing arms which could have sustained fire, theretofore obtainable only by very heavy barrel versions or by water-cooled jacket patterns. A selector switch gave the operator his choice of single or automatic fire.

Later a muzzle booster and flash hider were incorporated in the design, as well as a saddle-type drum magazine to hold 75 cartridges.

When these guns were superseded by larger quantities of the perfected MG34, the Germans sold MG13's to Spain and Portugal. Spain kept the designation MG13. Portugal described them as "Dreyse M1938." Confusion sometimes results from this designation. The "1938" merely indicates the year in which these used guns were adopted by the Portuguese.

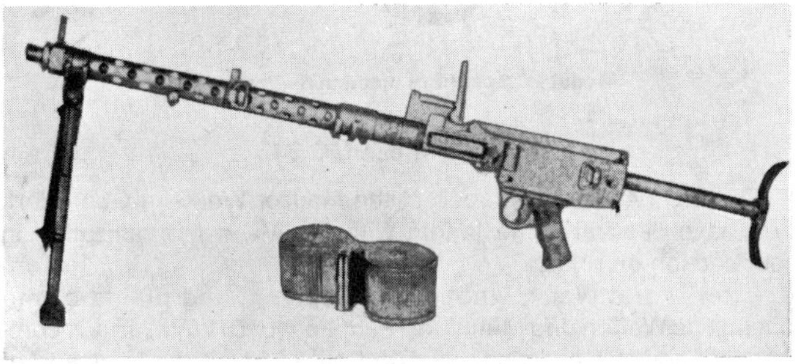

German MG13.

The German Secret Gast Machine Gun

Toward the end of World War I Germany was concentrating intensively on the development of a top secret machine gun. This was named after its developer, Carl Gast of Barmen, Germany. The principle curiously is closely allied to one patented in England in the year 1886 by Bethel Burton.

The Germans had hoped to produce this gun, capable of firing some 1,600 rounds per minute, as the answer to their aerial fighting problems. At the end of the war our Ordnance organization located drawings and details of the weapon and requested delivery to Springfield Armory of at least one gun with 4,000 rounds of test ammunition. When it finally did arrive, strangely enough it was unaccompanied by the drum magazines necessary to fire it. This again delayed testing at the Springfield Armory. It was not until 1923 that we were finally able to test this gun.

The Gast proved to be a dual-barreled affair with unusual drum magazines (spring fed) positioned on each side of the gun. The mechanism operated on the recoil locked-breech principle. These guns were provided with high-powered telescope sights for installation in aircraft. One of the most remarkable features of the design was the easy disassembly. Thumb pressure on the back plate latch permitted the gun to be field stripped within one minute.

Our tests (as usual) produced a report to the general effect that while the gun was mechanically good and operated reliably, it did not have sufficient advantage over the Browning gun to be of interest to us at the time.

German Gast machine gun. A secret development in World War I.

The German Rheinmetall-Borsig Machine Gun

The noted German Solothurn Machine Guns were made by the Rheinmetall-Borsig A. G., one of the largest munitions industries in Germany in World War II.

The parent of this organization was the original Rheinische Metallwaren und Maschinenfabrik A. G. founded in May, 1889.

It was set up by the firm of Horder Bergwerk of Westfalen in that year to manufacture new jacketed bullets for the German Army. The plant was at Dusseldorf in one of its suburbs. The company prospered in the armaments field. When World War I began it was second only to Krupp in the production of munitions. Among its products were the German version of the Maxim gun, the Model 08.

The plant was dismantled by order of the victors under the Versailles Treaty after World War I. However, the Germans managed to get some 23,000 tons of tools, dies, patents, and drawings out of Germany and into Holland where they were stored in warehouses at Rotterdam and at Delfzyl.

The Inter-Allied Control Commission allowed Rheinmetall to build, for the reconstituted small German army and navy which was authorized, all guns below 17cm bore diameter.

In effect this ruling concentrated all army weapon developments in this company. From 1925 to 1927 special facilities were instituted for development work.

After unsatisfactory arrangements with Holland, the German company in 1929 obtained control of the Waffenfabrik Solothurn A. G. in Switzerland. This plant was originally a watch manufacturing concern. It was purchased by a Swiss who had previously worked for the Deutsche Waffen und Munitionsfabrik A. G. in Germany.

He set up to manufacture weapons, but failed financially. The notorious Fritz Mandl of Hirtenberg, Austria then obtained control. Rheinmetall worked with Mandl in the development of the Solothurn Plant to bypass Allied restrictions on machine gun manufacture and development.

The German authorities had complete stock control of this works. Arrangements were made with armament works in Austria

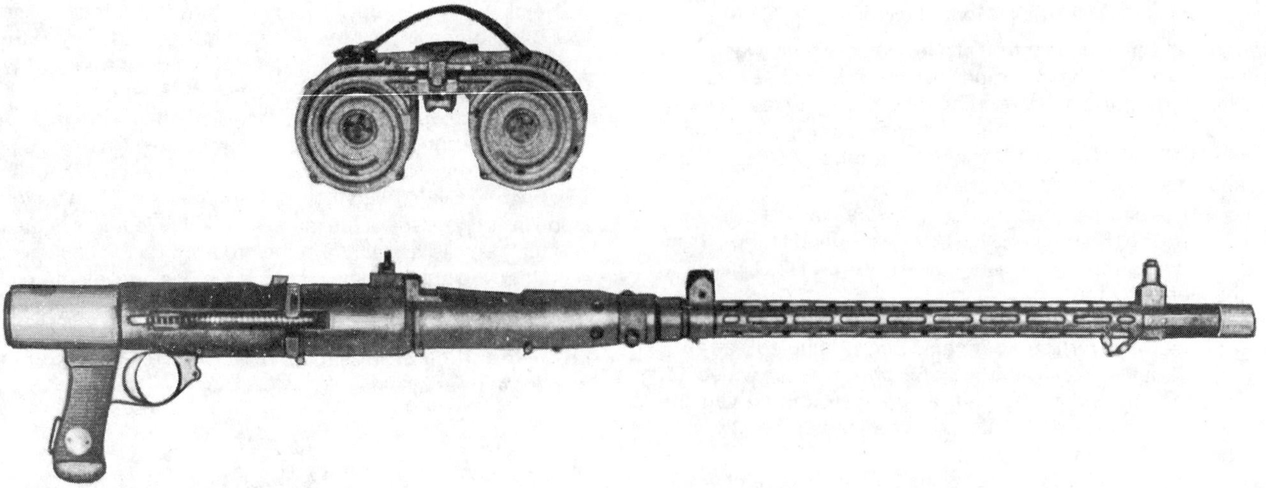

German R-B machine gun. Customarily an aircraft gun, this pattern was at times adapted for ground use. MG15.

and in Hungry to produce component parts for automatic arms.

Within a few months of the time Rheinmetall took over the Solothurn firm, weapons began appearing with the new name. The plant, of course, was used merely as an assembly point.

The Germans utilized this plan to sell throughout the world outmoded equipment as they developed new and improved forms for their own use.

This was common knowledge in arms circles at the time and it is incredible that government agencies and intelligence organizations either had no knowledge of it or failed to properly evaluate the information received. This permitted Germany a tremendous edge in rearmament which actually hastened World War II.

The Solothurn Model 29

The Model 1929 gun was offered for sale to military groups throughout the world just two months after Rheinmetall purchased the stock control.

The actual development of the gun is credited to Louis Stange, a German designer who worked originally in the Bergmann factory at Suhl under Schmeisser, who was one of the great developers of automatic arms systems.

Schmeisser, whose pistols and submachine guns we shall touch upon later, was also affiliated with Rheinmetall. It was no secret at any time that many of his patents were assigned to this company.

The first Solothurn was short recoil operated. Cooling was by air. The gun weighed only 17 pounds when chambered for the German 7.92mm cartridge.

This gun is extremely interesting in connection with current world design, as it pointed the way to the first really low-priced, high-production, dependable automatic weapons. A large number of the units were lathe turned or made by screw machines. Since this type of manufacture can be conducted by semi-skilled or at times even unskilled help, the guns had considerable acceptance in countries where mechanical skills are not highly developed. Replacement parts for the guns could be made in an emergency by almost any commercial metal turning shop.

Single shot or automatic fire is provided in the Solothurn.

The breech-locking action developed by Stange is most unusual. It is cylindrical in shape. A central locking ring rigidly holds the bolt and barrel together by six interrupted threads. Only in the fully-locked position is an obstruction removed from the path of firing pin. This prevents accidental discharge.

The magazine holds 25 cartridges. After the last shot is fired the bolt is held open.

A safety is provided to lock bolt and trigger together. This permits inserting a loaded magazine while the bolt is locked.

A refined version, the S-2-200 appeared in 1930. It was also called the MG 30, and is mentioned earlier in the text.

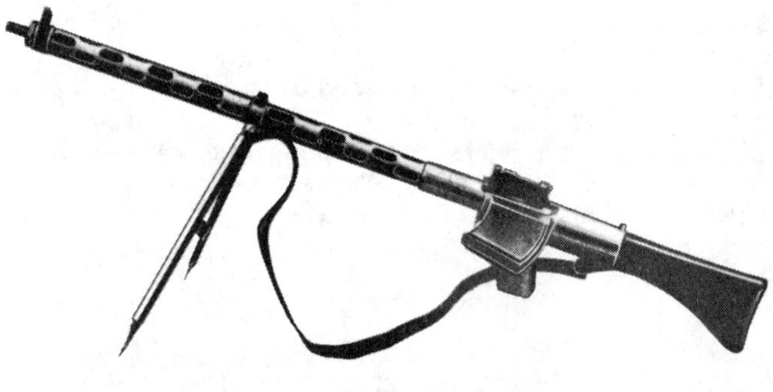

Model 29 Solothurn made in Switzerland.

The German MG 34

The MG 34 was developed at the Mauser Works in Oberndorf. We have dealt at some length with the Mauser organization in our section on rifles.

After World War I, when the equipment and plants of the Deutsche Waffen und Munitions-Fabriken (or D. W. M. as it is commonly known in America) were put out of operation by action of the Allies, the equipment, patents, and drawings covering the Luger "Pistole 08" were transferred to Mauser from D. W. M.

The work and models done by Heinemann in modifying the Maxim machine gun to the version we have discussed as the "Parabellum" were also transferred to Mauser.

No overt machine gun development was undertaken at the time, although considerable covert attention was given to the subject. It was not until 1934 when Hitler came into power and Germany began to rearm openly that active work was begun by Mauser on a new machine gun. The organization at that time was officially Mauser Werke, A. G.

A special all-purpose type machine gun to use the infantry ammunition was desired. It was specified that the new design must be adaptable to light and heavy machine gun use, as well as to tank and antiaircraft work. The general specifications to be met were

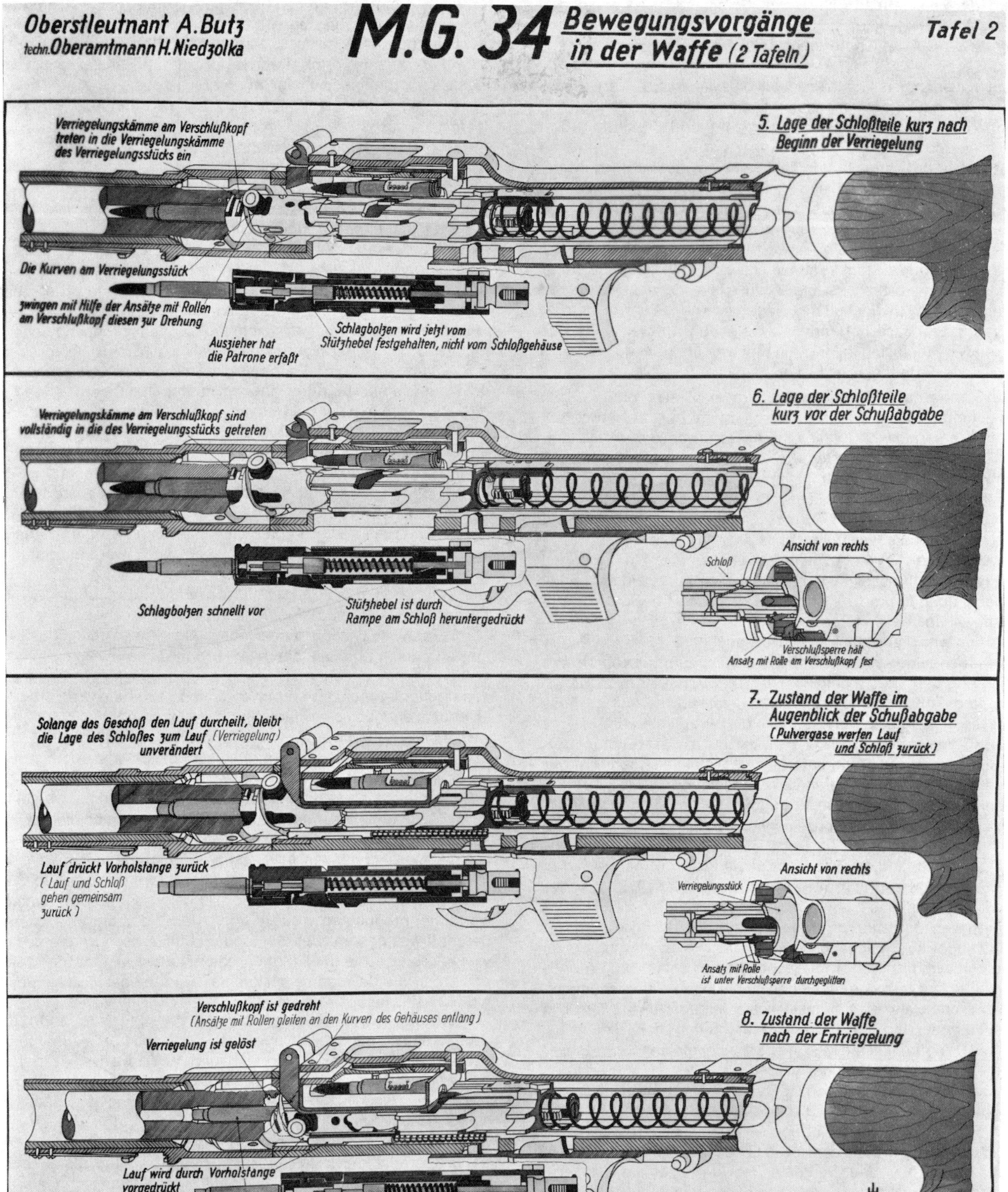

German Machine Gun 34. Sectional drawing showing operation as presented by German Ordnance.

provided by the German Military High Command. It was decided to begin work on the basic lock mechanism of Louis Stange of the Rheinmetall-Borsig Organization with which we have already dealt.

Paul Mauser himself had done work on a mechanism very similar to the breech lock now suggested for manufacture, although the version as produced by Stange went beyond the application Mauser had made of the principle. It might be stated in passing that all basic principles in this field were developed in the early periods by Maxim, Mauser, and Mannlicher. All designs since that time have been merely extensions of or improvements on the basic principles. Stange deserves considerable credit for furthering the original design however.

The locking ring system as already utilized on the Solothurn was not considered the best possible because its construction required the removal of all the locked units together with the barrel in order to effectuate a barrel change. It was sought to overcome this deficiency by replacing the locking ring with the rotating bolt head design, thus permitting removal of the barrel alone.

The Stange modification of the original Mauser rotating bolt head actuated by recoil utilized the energy of the barrel together with an accelerator to increase the travel of the bolt to the rear after unlocking. Most previous efforts utilized the residual chamber pressure to furnish this power. This modification among other things assisted greatly in developing reliability of functioning and increased rate of fire.

The muzzle brake was utilized also as both a flash hider and as a forward support for the recoiling barrel.

The rearward thrust of the expanding gases trapped in the brake, impinging upon the muzzle of the barrel, drove it back to operate the arm with increased force.

This gun was first put into actual production in 1936 in infantry form with shoulder stock and bipod mount. Elimination of the locking ring permitted a quick barrel removal system to be incorporated. This and all other general details of the gun are fully discussed in the chapter on German World War II materiel.

The MG 34-S and MG 34-41 are the same as the earlier models except for the styles of the barrel jackets. All these modifications differ from the initial model in being equipped for full automatic fire only; in having larger muzzles to permit greater surface for the brake-checked gas to bear upon; a larger buffer; elimination of firing pin nut; shorter barrel and incidental changes in the feeding method; and single trigger of standard design instead of the pivoted swinging dual-operation trigger. This trigger immediately identifies the modifications.

A further modification of the MG 34 was produced by Mauser in 1939 under the designation MG 81. This was an aircraft version of the gun with the rate of fire stepped up to about 1200 rounds per minute. This rate was developed by utilizing a different version of muzzle brake and exceptionally heavy buffer return spring. The feed was generally from flexible disintegrating metal link belts.

It must be noted that the MG 34 after its development by Mauser was also produced by Maget at Berlin, by Steyr-Daimler-Puch A. G. at Vienna, by Gustloff at Suhl; and, after the German occupation, at Waffenwerke-Brunn A. G. in Czechoslovakia. The distribution of manufacture of course left these plants, some of which are today in Russia control, in full possession of equipment, methods, know-how, and even personnel utilized by the Germans.

The MG 81 likewise was produced not only by Mauser but also by a large group including Krieghoff at Suhl and by the Waffenwerke Brunn A. G. It is obvious therefore that the designs are common knowledge to all anti-democratic experts as well as to our own.

The German MG 42 Machine Gun

This gun is one of the most remarkable machine weapons ever produced anywhere by anyone. Its design has influenced production methods and will continue to influence machine arm production in the years ahead.

While the gun is discussed in detail in the chapter on German World War II small arms, some historical notes are essential here because of the unusual background of the design.

As its name indicates, the MG 42 was first introduced in 1942. Its first appearance so far as American troops were concerned was at the terrible battle at Kasserine Pass, where its value was immediately recognized by all observers.

This gun is not an original invention. Rather it is one of the finest examples of what can be done by composite design and manufacture that the firearms industry has ever seen.

Without wasting unnecessary energy on trying to establish it as a purely developmental design of their own national genius, the German military proceeded to combine a group of elements of fine foreign guns to produce the result they wanted.

The quick barrel change system the Germans produced enabled maintenance of a terrific fire volume with a light machine arm.

By utilizing a dual feed pawl system wherein one round was positioned while the other one was being chambered, a feeding operation with metal belts was produced. This gave a smooth belt movement to prevent interference with the aim of the gunner by the customary violent jerking.

The muzzle device served to support the forward end of the barrel, to act as a booster by backing up expanding gases against the barrel muzzle to increase rearward motion during the recoiling action, to serve as a flash hider, and by slotting to serve as a muzzle brake to stabilize the arm when short bursts were fired at high rate.

The most remarkable aspect of the entire design of the MG 42, however, is its method of manufacture.

This is the utilization of the finest high speed techniques for mass producing metal stampings to form an arm which might be crude in appearance by gun standards, but which could be made at a fraction of the cost and in a fraction of the time required for an equivalent gun as produced by formal methods.

This part of the design was undertaken by a noted German industrialist, Doctor Grunow, a specialist in metal stamping work (or pressing, as it is commonly known abroad).

By working out a system of pressing, riveting, and spot welding, he was able to turn out the guns with heavy duty stamping equipment of the type normally utilized in automotive production.

A great deal of nonsense was poured out at the time by military intelligence circles and others attempting to prove that the Germans were in dire need of materials. Otherwise it was argued they would not produce so crude a gun. A considerable time passed before it was generally understood that the reason for the design was to produce the finest arms in the shortest time at the lowest cost; that this development stemmed from mechanical genius—not from shortages or desperation.

The gun was put into production in the plant where Doctor Grunow supervised the initial work, the Johannus Grossfuss Metall und Locierwarenfabrik at Dobeln in Saxony. Mauser Werke at Berlin also later manufactured it, as did Maget at Berlin. The gun was also produced at the Steyr Works in Vienna and at the Gustloff Works at Suhl. These items are mentioned again to establish that since many of the areas involved are in Russian hands, the methods are as fully known to the Russians as to United States technicians. There is nothing secret about the production techniques.

In 1943 an attempt was made in the United States to produce prototypes of this gun, the work being delegated to the Saginaw Steering Gear Division of General Motors, at Saginaw, Michigan.

The guns as produced in 1944 simply wouldn't work. The arm was modified and was again subjected to tests and again found wanting. The American-made version was listed as the Machine Gun Caliber .30 T-24. The reason for the failure of these American-made versions of the MG 42 to operate was really very simple. The United States .30 caliber cartridge is, as almost all shoot-

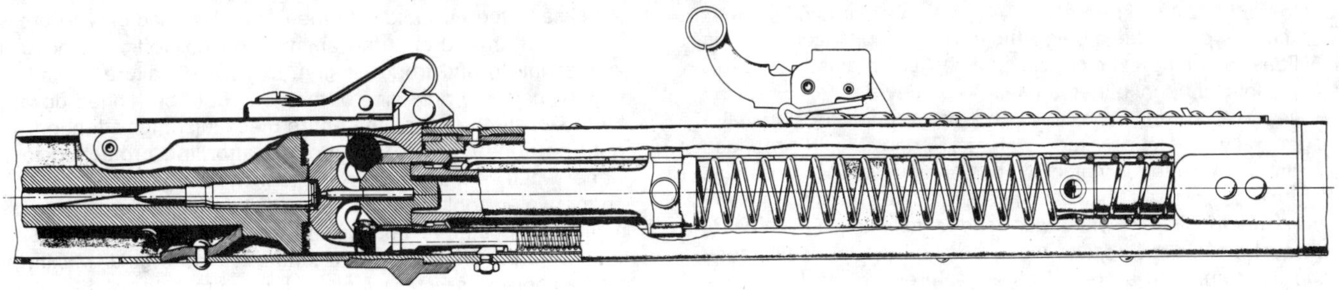

German MG42. Top phantom view showing gun locked and about to fire.

ers are aware, merely a somewhat lengthened version of the German 7.92mm cartridge, the common Mauser cartridge from which ours was originally developed. In the course of attempting to manufacture a gun to handle United States ammunition, our design and engineering groups overlooked entirely the elementary matter of the difference in the length of the cartridge cases.

As a result, the bolt face could not recoil far enough behind the ejection slot in the receiver to permit ejection, the receiver yoke interfering with the cartridge guide plate by about a quarter of an inch! On such oversights as this often rests the matter of savings of countless millions of dollars.

The German FG 42

This gun represents still another amazing German experiment in the field of machine arms improvisation.

The FG 42 was also designed on the principle of utilizing stampings and machine screw parts to the full, with the intention of issuance for service as a light-weight machine weapon for airborne troops. It was so utilized in its first appearance at the raid on Crete. The designation FG 42 which sometimes leads to confusion with the entirely different MG 42 derives from the name "Fallschirm-Jaeger Gewehr," the paratroop machine rifle.

The operating system of the mechanism was not taken from the Krieghoff sporting modification, which is underbarrel operated. It was a utilization of a form of the Lewis gas system.

A description of the gun and its operation including the gas system is to be found in the chapter on German World War II small arms.

The receiver is a stamping. The barrel is fixed, being permanently secured at front and rear end by swaging and by having the added security of locking pins. The muzzle brake is particularly efficient, and also serves as a flash hider. The gas selector, by affording access to different size port holes, allows the gunner to increase or slow down the action.

The trigger mechanism is sheet metal, welded, as is the unit containing the sear, the firing change lever, and safety latch system. In common with some arms like the American Johnson, it permitted firing from either an open or closed bolt, thus giving the gunner the opportunity for keeping the action open for cooling in case of full automatic fire or giving him the desirable feature of a non-slamming breech operation for accurate single shot firing.

The locking action is the basic Lewis gun system in which, after the bolt has completed its travel and chambered a cartridge, the still moving piston hits a cam in the bolt body, turning and rotating the bolt and its head so the locking lugs engage in their recesses to lock the weapon securely and get the obstruction out of the path so the firing pin on the gas piston can strike the primer and fire the cartridge.

On rearward action the gas piston has a travel of approximately one inch. During this time, of course, the bullet is well clear of the barrel before the unlocking action starts, through the cam rotation of the bolt. In addition to the operating spring, a heavy buffer spring is provided as required by the light weight design.

ITALY

The Italian Perino Machine Gun

The Perino is little more than a historical name in the field of automatic weapons. But for the stupid secrecy with which the Italian military surrounded the gun, Giuseppe Perino's 1901 invention might today be one of the world's outstanding arms developments.

In their determination to keep the gun a secret, the Italian army

German FG42. This is one of the most remarkable of all light automatic weapons. Shown with clip, cartridge, and detached bayonet.

went to the fantastic extreme of buying Maxims to equip its forces as a "cover-up" for developing the new Perino. Machiavelli in his wildest dreams never conceived the hare-brained ideas his descendants hit upon in their attempt to be devious.

While Italian designers have always been noted for unusual—and generally impractical—feed systems, Perino actually produced something new in feeds which still has an undeveloped potential. His gun was compact, had an unusual recoil operating system with a rather long stroke which was used to mechanical advantage, featured both air-cooled and water-cooled systems well ahead of the time, and had a host of other features which placed it, if perfected, actually ahead of the fabled Maxim.

In its early form the Perino feed was by metal trays of 25 cartridges fed through the gun from left to right. Trays could be started only when the bolt was forward, the mechanism locked a safety feature in itself. Pulling back the cocking handle opened the action and cocked the mechanism, meanwhile indexing a cartridge in line with the chamber.

On forward motion of the firing mechanism when the trigger is pulled to release it, the bolt pushes the cartridge out of the tray into the chamber, actually passing over the tray itself. Five 25-shot trays were placed in the ammunition feed box, and, as the gun fired, the trays fed from the bottom layer, this allowing the loader to replace loaded trays in the top of the box for continuous fire.

A most unusual arrangement withdraws each fired case from the grip of the extractor as it emerges from the chamber on recoil and snaps it down into its place in the loading tray under the bolt. This, of course, does away with the need for an ejection chute and for much of the involvement of belt feeds. As the 25-shot tray receives its 25th empty case, the next loaded tray follows it into the gun, the empty one being expelled to the right.

The breechlock is actuated by lugs thereon which engage receiver cams during the movements of the barrel, barrel extension, and bolt. The lock is a pivoted type. A unique linkage provides simple acceleration not only of the bolt in its rearward movement to the buffer, but also in its forward movement to chamber and lock. Part of this linkage is a formed lever which does away with the need for a separate driving spring behind the bolt, and also positions the bolt to compel breechlocking before the gun can fire.

A later feed was almost equally unique, consisting of a flexible strip of brass holding the cartridges coiled in an ammunition drum. The strips could be separated at 2-foot intervals.

Italian Perino machine gun. Never produced in quantity, yet one of the outstanding designs in its field.

In case of feed trouble with the Perino one had only to press a release button and take the ammunition box off to remove the troublesome round from the gun. In case of failure of any mechanical part, the right side of the gun could be hinged down instantaneously to expose the entire mechanism for attention. This could be done close to the ground, affording protection to the gunner while he made repairs without exposing himself.

In the water-cooled model, the recoiling barrel functioned as a small reciprocating pump, giving constant water circulation so that the gun in tests would fire long after the regular Maxim pattern had boiled its water supply. A built-in muzzle booster utilized the gas trapped after the bullet left the barrel to give added rearward thrust, thus giving a high cyclic rate in a gun whose normal recoil length was much too great to allow it with the formal arrangement. This long recoil, incidentally, afforded slow extraction to prevent ripping off case heads.

By the time the aura of secrecy wore off this gun, the Italians were too embroiled in other matters to do anything with it. Had this gun been put into production in the day of its development, it would have affected all later machine-gun design.

The Italian Revelli Machine Gun

In the chapter on Italy due attention is given to the Italian Model 1914 Revelli (or Fiat) machine gun in caliber 6.5mm.

The gun is mentioned here because of the unusual nature of this arm and of the background of its inventor.

It was developed in 1908 by Bethel Abiel Revelli of Rome, an officer in the Italian Army.

The design itself is a combination of short recoil and blowback. Locking by wedge is so slight that the gun is not a locked-breech weapon in the true sense of the word, but actually is a hesitation pattern blowback.

Models were manufactured at the Fiat Automobile Works at Turin, Italy and the name Fiat is often applied to the arm. It was tested both in the United States in 1911 and in Italy, of course. It utilized a 100-round magazine in the 1913 test.

One very unusual feature of the gun is the magazine system of so called "mousetrap" type. Theoretically it provides better flexibility than does the belt-fed mechanism. However, in actual practice as the mousetrap magazines are ejected, they are easily dented and rendered unserviceable.

All of Revelli's experimental and developmental work was of considerable value both to Fiat and the Italian government.

Italian Revelli (Safat) Machine Guns

The Fiat Company, the original producer of the Revelli gun, offered a lightweight machine gun in response to an Italian government request in 1926. In theory this was a new gun. In actual practice it turned out to be just another modification of the earlier Revelli.

The Italian government placed an order for 2,000. Fiat set up a new organization to handle this government contract under the name of Societe Anonina Fabrica Armi Torino (SAFAT). This organization actually manufactured the guns which were officially known as Fiat Model 1926. The nameplate however bore the initials of the technical manufacturer, "Safat."

Very much confusion often develops in considering Italian arms because of methods such as these. The Italian system has been to contract for machine guns based on identical specifications with various companies. Tests are then run to see which company produces the best gun. Often guns which are practically identical will be produced bearing various names, usually those of the individual manufacturers.

The Model 1928 did away with the retarded blowback system of the Revelli. It introduced a positive locking system developed by one of the Safat engineers named Mascarucci. This gun with

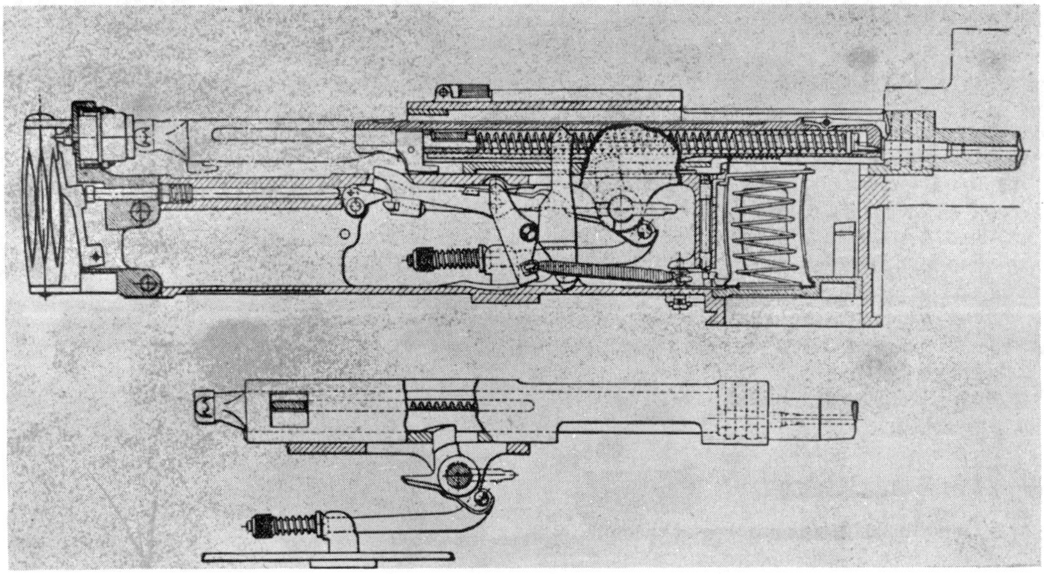

Italian Revelli M.G.

Italian Revelli (Fiat) machine gun.

tripod or shoulder piece weighs about 21 pounds. It can be taken apart in a few seconds without the use of tools. Feeding is from a special metal loader.

At the moment of firing a latch locks the bolt, barrel extension, and barrel together. After a travel of about 1/2-inch the link begins to elevate the bolt latch slowly. The recoiling parts then are suddenly released. This slow unlocking gives the effect of primary extraction, preventing ruptured cases. This original Fiat was chambered for the standard Italian 6.5mm rifle cartridge. A later model was designed for a 7.7mm cartridge. In 1935 Fiat produced a new model. This was merely a redesigned 1914 model Revelli. Strangely enough, they reverted to the retarded blowback system in preference to the locked bolt which had been successful.

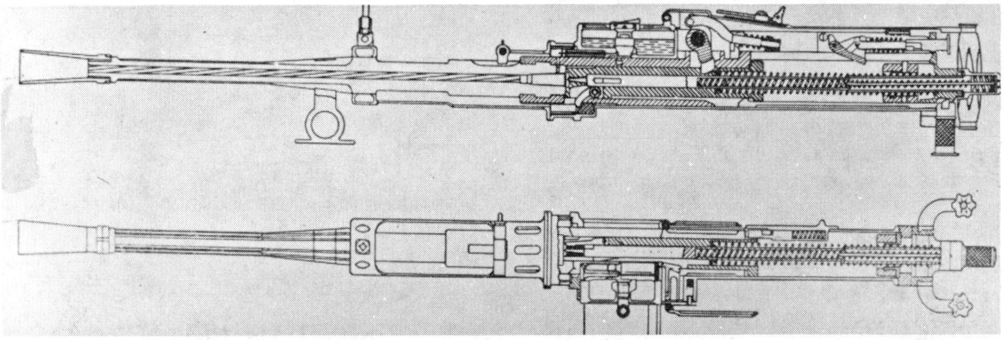

Section drawing of Revelli Model 1926, a hesitation-locked weapon.

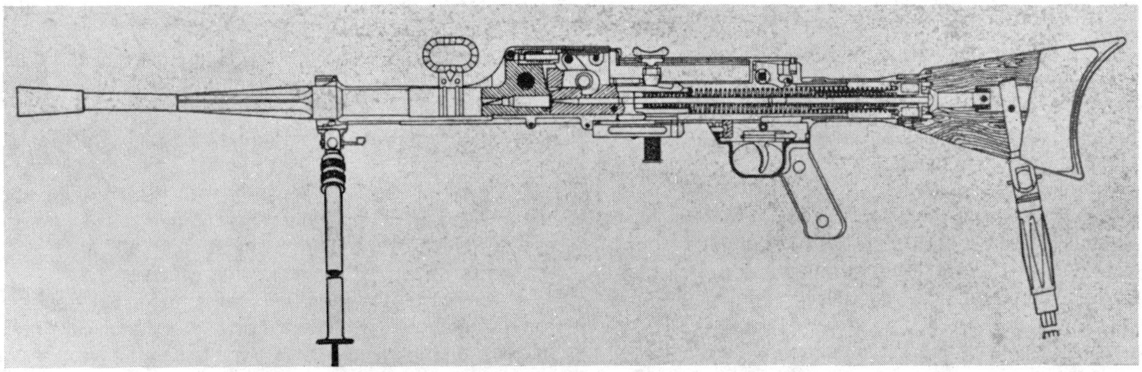

Italian Model 1928 with Mascarucci locked breech and quick-change barrel. This arm influenced many current designs.

Italian S.I.A. Machine Guns

The S.I.A. guns were actually built first in Italy in caliber 6.5mm Italian service. They are described in general in the chapter on Italy.

The name derives from the fact that the inventor, Giovanni Agnelli, transferred patent rights to the Societa Italiana Ansaldo at Turin. The designation S.I.A. comes from the initials of this organization. They were given a substantial order by the Italian government but their guns were not issued during World War I.

They were, however, introduced into service prior to the World War II period. These air-cooled guns are retarded blowback design and are provided with vertical feed top magazines. The heavy barrel and large aluminum fins extending to the flash hider at the muzzle give good cooling to the weapon.

The S.I.A. is of interest only historically today, far better designs being commonly available.

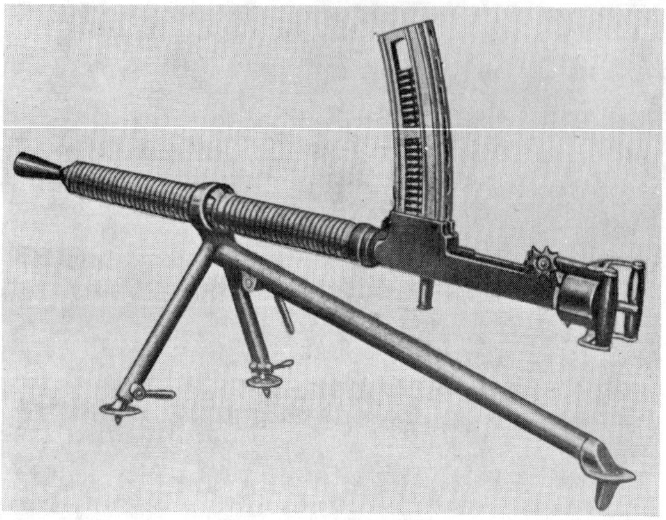

Italian S.I.A. used in both world wars. Initials indicate the manufacturer.

The Italian Brixia Machine Gun

The Brixia is an interesting Italian machine gun which was produced in 1920. Its manufacturers are producers of the unusual Brixia mortar.

It derives its name from the initials in Italian of its manufacturers, the Brescia Metallurgical Company. This organization had considerable experience in the manufacture of Fiat machine guns during World War I.

While the Brixia has never been produced in quantity, many of its design factors are such that it is deserving of some attention here for possible future design reference.

One of its unusual characteristics is a special rate-of-fire control system. The regulator is in the form of an eccentric bolt, whose cylindrical section is moved along an axial plane. The periphery of its other half has two grooves, placed at variable distances from the axis of rotation. Projections resting in the tops of the grooves act to either accelerate or slow down the recoil action. Pushing down the regulator button at the left of the gun, with the trigger depressed at different levels, controls the forward and rearward distance to which the bolt is permitted to travel. This regulates the speed of motion of the recoiling parts and hence the rate of fire.

The locking system is short recoil. The bolt, barrel, and bolt extension travel to the rear at the instant of explosion securely locked for a distance of some five millimeters. When the unlocking point is reached, the breechlock is flipped over backwards, thereby functioning as an accelerator.

The barrel extension being stopped by its buffer, the bolt which is now unlocked from the barrel, continues to the rear under the forces imparted to it. The empty case in the grip of the extractor in the bolt face strikes a projection on the receiver which forms the ejector. Ejection is to the upper left. The recoil of the parts is halted in standard fashion by a back plate buffer, at which time the recoil spring and buffer drive the operating parts forward for the next cycle. The bolt has a circular section. A rectangular projecting unit is attached to it. During the counter recoil movement, the projection is moved into a housing slot in the upper part of the receiver. Movable units in the upper section of the bolt projection act not only to unlock the bolt at the proper moment but are also the fire regulators already mentioned.

The gun was offered in both water-cooled and air-cooled models, the latter having an aircraft-type barrel with radiation flanges along its length. The feed is peculiar, consisting of rectangular boxes which are attached to the receiver. This is another instance of the involved mechanical feeding box to which Revelli originally gave so much attention.

One very unusual feature of the arm is the complete enclosure of all recoiling members. This not only prevents dust, sand, or other foreign matter from entering to interfere with the operation but also serves to protect the gunner in case of defective cartridge explosion.

Italian Brixia water-cooled machine gun. Very short recoil-operated.

The gun is mechanically designed to prevent firing until the breechlock is in closed position. This is accomplished by the mechanical positioning of the bolt in relation to the loading and the recoil mechanisms.

The Italian Breda Machine Guns

While the Breda guns are covered in the chapter on Italy, the background of their development is sufficient to merit historical mention at this point.

Societa Italiana Ernesto Breda at Brescia was originally a locomotive manufacturing plant. During World War I it produced the Model 1914 Revelli gun which we have already discussed, operating on sub-contract from the Fiat Company which held the Revelli patents.

Breda constructed a separate plant near the locomotive works to produce arms. Considerably enlarged, this works today is producing some of the world's finest automatic shotguns, which are commonly exported to the United States.

At the close of World War I Breda undertook development work on machine guns which resulted in the Model 1924 caliber 6.5mm, a gun weighing just under 20 pounds and having a rate of fire of about 500 per minute.

Its magazine system is entirely unique, consisting of a clip arrangement which can be pivoted out from the side of the gun for loading.

Another noteworthy advance at the time and still important today is the system of quick barrel exchange. Heavy interrupted threads secure the barrel extension and barrel. A flash hider was incorporated in the gun muzzle. The operation of all these Bredas is on the short-recoil principle. A locking ring engaging six cams in the receiver rotates the lock partially around the end of the barrel in closing. The locking lugs at the front end of the bolt engage the two pieces into position. The recoil stroke is about three-eighths of an inch.

Cartridges are loaded from a cardboard container directly into the magazine, which has been hinged out from the side of the receiver for loading. In closing the action, the last cartridge to be inserted in the magazine is picked up by the bolt.

The normal extraction on this arm is extremely violent. An oil apparatus is provided, which is operated by the recoil and counter recoil of the mechanism to squirt oil into the chamber between shots. This lubrication of the ammunition is necessary to prevent the violent breech opening from tearing the heads off cartridge cases while being extracted.

When finally issued, the first Breda was listed as a Model 1926. Much confusion at times is encountered in Italian arms designation because the Fiat produced in their "Safat" plant another model which was also designated Model 1926.

After some minor changes, Breda issued a modified model in 1928, the modifications being relatively small. Late in 1930 Breda absorbed Fiat. It assumed control over all machinery, patents, and equipment for the production of machine arms and automatic rifles. A new plant called the Breda-Fiat was opened at Piacenza.

The first product of the new plant was the 1930 model Breda. This appeared in the standard Italian 6.5mm caliber and also in 7mm and 7.92mm calibers for export, the latter, of course, being the standard German cartridge which is in common use in countries where Mauser rifles have been sold.

The Italian Breda Model 1937 Machine Gun

This model was an outgrowth of experimental work done in 1931. It is a gas-operated arm entirely unlike any of the preceding Bredas.

It too was noteworthy for still another freak type of feeding system. The feed is by tray, the cartridges being inserted from chargers holding 20 rounds, each of which is housed in a separate compartment.

In operating this model, which was made in an 8mm caliber, the gunner inserts a loaded 20-cartridge tray in the feed slot in the left side of the receiver. The cocking handle on the right side is pulled back to withdraw the gas piston and locking mechanism. This action also indexes the feed tray to bring a cartridge into feeding position.

There is no provision for single-shot fire. Pushing the trigger off the safe position and forcing the trigger in releases the sear from the gas piston assembly. The assembly, driven forward by the compressed spring, chambers the cartridge in its path.

The bolt stops directly behind the head of the cartridge case but the piston continues its forward movement, camming the breechlock up into recesses in the top of the receiver. As the movement continues a projection on the gas piston strikes the firing pin to discharge the cartridge in the chamber.

The gas port on the underside of the barrel is some two-thirds of the distance down from the breech end. The barrel weighs close to 10 pounds, being unusually heavy to permit firing large bursts.

As the gas passes through the orifice into the gas cylinder, pressure is exerted against the piston therein. The initial opening movement withdraws the firing pin and lowers the breechlock to free the bolt. Some initial extraction is provided. As the bolt is

A representative Italian Breda M.G. during quick barrel change.

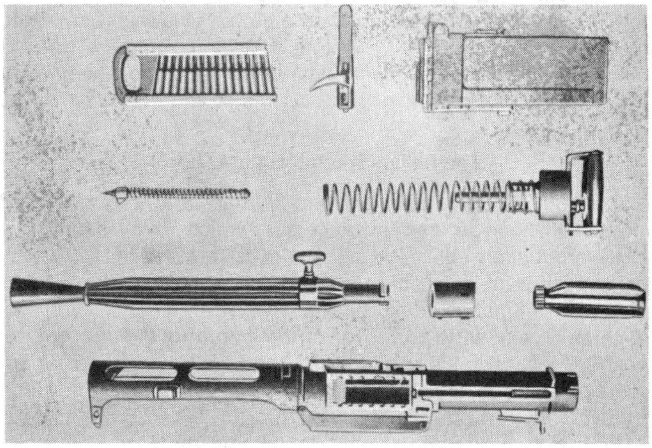

A representative Breda field-stripped to show simplicity of design.

unlocked it first moves the cartridge case back sufficiently to loosen it from its chamber.

From this point on the action is distinctive. As the cartridge case is extracted and drawn back by the extractor in the face of the bolt, its motion is halted by a dog on the receiver. Simultaneously a cam thrusts the empty case up into the space it originally occupied in the feed tray.

When the 20 cartridges have been discharged, the clip which contains all the empty cases is thrown out of the gun to the right.

The theory of this clip system is that it will make it easy to reclaim cartridge cases since they may be picked up in units of 20. In actual practice the clips are damaged so often that reuse of the clip or of the ammunition cases is seldom practical.

The Italian designers themselves came to this conclusion after exhaustive tests. As a result, in 1938 they issued a modified tank model which also saw considerable service during World War II. This design has an overhead clip-feed. Cartridge cases are ejected out the bottom of the receiver. The arm is furnished with a pistol-type grip. Beyond this, however, the mechanism and the principles are the same as with the 1937 model already described.

The Italian Breda-Safat Model 1935

These guns represent a combination of the old Mascarucci locking principle already discussed and design improvements developed by the engineers at the Breda Works.

Basically these were intended as aircraft machine guns. The lock was inverted to place it on the bottom. Muzzle boosters were furnished to increase the rapidity of action and hence rate of fire. Recoil-operated accelerators were also incorporated.

The guns were issued in 7.7mm caliber using the same cartridge as the British, the .303 rimmed. They were also made in 7.92mm German caliber and in 13.2mm for use against vehicles. These guns were designated Breda-Safat Model 1935.

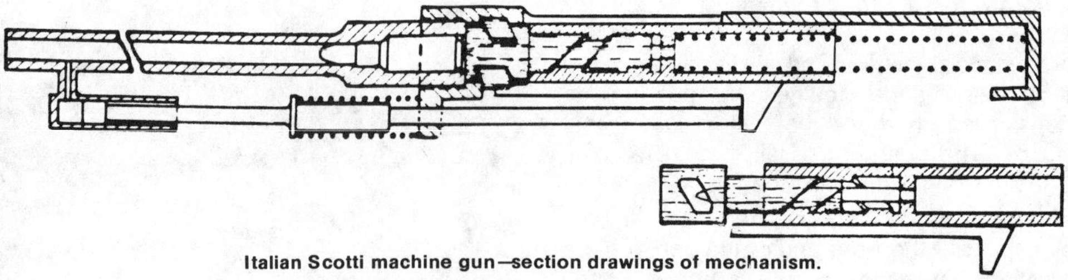

Italian Scotti machine gun—section drawings of mechanism.

A disintegrating metal link belt was provided for feed. It was so arranged that feeding on the aircraft model could be from either side. Ejection was through the bottom of the receiver. The chambers in the barrels of these guns were fluted. This was required to expedite extraction. The principle is that of allowing the gas expanding as the cartridge fires to "float" the empty case momentarily until pressure is lowered.

The Italian Scotti Machine Gun

Scotti, an Italian engineer in Brescia, Italy, introduced in 1928 the first of a long series of machine guns which he designed.

His production utilizes in all instances the delayed blowback system, with breech locking based on the earlier Mannlicher straight-line function of a turning bolt head.

While a very wide range of automatic and semiautomatic weapons including submachine guns and light machine guns appeared based on the Scotti patent, only his automatic cannon received any real attention.

His guns were manufactured largely by the Isotta-Fraschini Company, Italy's great automobile and aircraft engine plant. An organization he set up in Switzerland was eventually taken over by the great Swiss Oerlikon firm.

As we have indicated, his action was always based on the elementary principle of the gas port in the barrel tapping off gas to operate an under-barrel piston to unlock a rotating bolt which opens under residual gas pressure.

His use of feeding systems was considerably more varied, ranging all the way from the Perino system of metal clip loading, in which the empty case is reinserted automatically in the clip after being fired, to belt and drum feeds of all types.

The only really unusual gun produced under the Scotti patents was a machine gun utilizing three barrels. In operation (at least in theory) when one barrel was over-heated the gunner operated a mechanism to rotate a new barrel into position, thus giving continuous fire. The device was never accepted as truly functional, however.

Italian Sistar Machine Gun

The Italian Sistar patented in December, 1932, also was produced only in prototypes in caliber 6.5mm and 7.92mm. The Sistar used one of the original swing-out magazine clips to permit rapid loading. Operation was short recoil on the rising or lifting block-locking principle.

JAPAN

The Japanese Nambu Machine Guns

In the chapter on Japan all the basic Japanese machine arms are treated in detail. We have room here only for a historical note concerning their inventor and their development.

The first Japanese machine gun known as the Nambu was developed by Lieutenant (later General) Kijiro Nambu. The design was based strictly on the old French Hotchkiss which had been in use by the Japanese since the days of the Russo-Japanese War.

His first gun was a heavy machine gun produced in 1914. In 1922 he developed the type 11. In 1927 the general founded a company for the manufacture of firearms at Tokyo. Here he developed the Type 92 machine gun. This led to the arm which appeared in use against our forces during World War II known as the Type or Model 96.

Nambu's organization merged with that of the Chuo Kogyo Kaisha Company at Tokyo in 1937. In 1939 they introduced the Type 99 light machine gun.

The Nambu operating system was strictly that of the Hotchkiss, but the ejection principle of pivoting the cartridge out over the bolt body was a steal from the Lewis system. All essential details of these types are discussed in the chapters on Japan, where illustrations may also be found.

Typical Japanese heavy machine gun. Note Hotchkiss-type lines and clip.

Mexican 7mm 1934 Mendoza light machine gun.

MEXICO

The Mexican Mendoza Light Machine Gun

The Mendoza is described generally in the chapter on Mexico and calls for merely historical observation at this point.

It was developed by Senor Rafael Mendoza, then a foreman at the National Arms Factory at Mexico City. The design was started in 1920. The finished gun was not offered until 1932.

A very low number of parts (22), ease of barrel removal, and the relative simplicity of manufacturing requirements all rate the Mendoza as a very unusual light machine arm.

Some refinements and production changes were suggested by M. H. Thompson of New York City, a mechanical engineer with previous experience in production at the National Arms Factory.

The gun was adopted by the Mexican government in 1934 after the conclusion of successful tests.

It was issued to Mexican troops to replace the conglomeration of Colt, Hotchkiss, and Vickers guns with which they were then equipped. The standard Mexican 7mm rifle cartridges were used in the Mendoza.

Note on Mendoza

A new version of the Mendoza in caliber .30-06 is currently being produced in Mexico. This version is potentially one of the world's finest designs. The barrel is not quickly removable as in the 1932 type. However, provision is made for hinging down on the butt section for quick removal of all operating parts much in the manner incorporated in the Belgian F.N. Auto Rifle. See the chapter on Mexico for more detailed description.

SWEDEN

The Swedish Kjellman Machine Gun

The important operating principles of this unusual design were patented in 1870 by Swedish Army Lt. D. H. Friberg. Peter Paul Mauser in Germany did some work along similar lines at about this time also.

However, the ammunition then available did not lend itself to successful automatic action, and at the time nothing came of Friberg's brain child. With the development of advanced ammunition at the close of the century, however, the Friberg principle again received attention. Its principles were seized upon in 1907 by Rudolf Henrik Kjellman of Stockholm, Sweden. Variations of this intriguing locking system were experimented with by Mauser who developed several rifles and experimental automatic pistols using variations of it. In World War II variations of the system appeared both in the gas-operated Russian light machine guns and in the famous German MG 42 recoil-operated gun.

The feeding systems in all these variations differed distinctly from those employed by Kjellman and Friberg. But the breechlocks have similarities.

The original gun was water cooled and furnished with a tripod mount. It was belt loaded from the right-hand side. A later model with a bipod mount and with an arc type magazine positioned on top of the receiver was also introduced. However, although Sweden conducted various experiments on it, it was never officially adopted.

Perhaps the outstanding characteristic of the arm is the fact that the extremely short recoil was so utilized that it made for a breech-locking mechanism more secure than most of the other systems of its time. Spark photography showed the bullets traveling some 98 feet out of the rifle before the actual unlocking of the breech commenced.

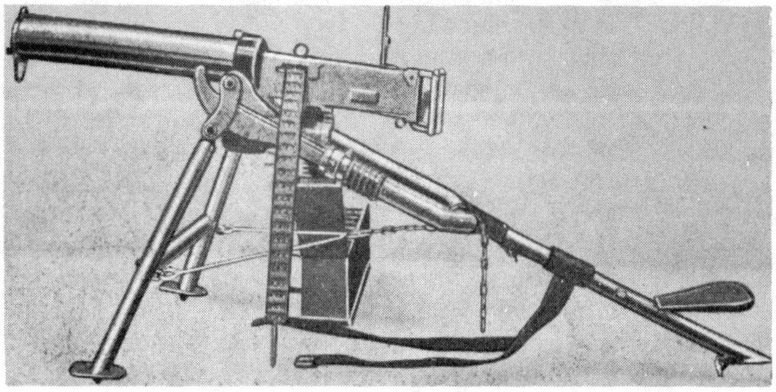

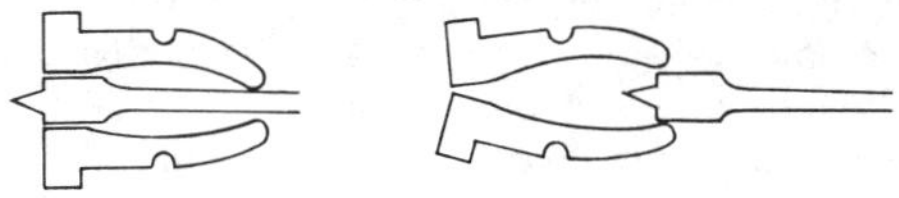

Swedish Kjellman heavy M.G. illustrating locking system. Firing striker in movement cams lugs into and out of engagement in locking recesses. Modification of this system is used in Russian Deg. L.M.G. (light machine gun).

SWITZERLAND

The Swiss Furrer Machine Gun

Swiss small arms in general have had very little direct influence on gun design, and even less throughout the world because they are normally very expensive to produce. The designs of Colonel Adolf Furrer are worthy of considerable attention both from a historical and design standpoint, nevertheless.

His prototype gun produced in 1925 was an 18-pound air-cooled ground machine gun with a shoulder stock. His principle of locking and operation has been applied to a wide variety of arms from rifles and machine arms through submachine guns and cannon. All have been mechanically successful. Few have seen any service outside of Switzerland.

The locking principle of the Furrer is the toggle joint system, basically as found in the Luger pistol design. As we have seen, this toggle action was developed long before Borchardt, the developer of the predecessor of the Luger, entered the field of firearms. It is also the basic locking principle of the various Maxim guns.

The Luger pattern of pistol was the official side arm of the Swiss army from early in the 20th Century until 1948. It is most likely that Colonel Furrer, who was Director of the Swiss Government's Small Arms Plant at Berne for some time, got the idea for his mechanism from that pistol. However, he went beyond the customary pistol locking procedure and developed a number of refinements.

The first Furrer gun was the Model 1925. This was an air-cooled, 18-pound light machine gun with shoulder stock. It was fed from a 30-shot clip magazine using the standard 7.5mm Swiss rifle cartridge.

Barrels were made by Hammerli of Lensburg, still one of the world's finest custom gun and barrel makers. They are producers of some of the finest precision pistols made and are today making auto target pistols based on the German Walther Olympia pattern.

Manufacture of the initial gun proceeded slowly while tests were made on its efficiency and also on methods of manufacture. In 1928 a sufficient number had been delivered to warrant full-scale adoption after it had survived all possible field and endurance tests. The Hammerli barrels, a feature of the weapon, are commonly considered to be accurate for at least 20,000 rounds.

Because of the light weight of the barrel and the relatively small surface allowed for heat radiation, the gun required cooling after about 400 shots had been fired from the full automatic position. Each gun carried a complement of 34 spare magazines. A flash hider, muzzle booster, and blast suppressor were incorporated in the gun construction. As customary in all Swiss firearms, the parts were magnificently machined but were so intricate that mass production was never possible.

A very unusual feature in the Furrer is the construction, which permits exceptional timing of the operations. The barrel and barrel extension are held in rear position after the unlocking stroke. The action, of course, is a typical short-recoil fully-locked system. Nevertheless, it incorporates an interesting operational feature not commonly found.

This feature is an arrangement whereby the gun is actually fired a small fraction of a second before the moving elements are stopped by the receiver. In this design, unlike the blowback, the barrel and bolt are securely locked, but firing actually occurs while the recoiling parts are moving forward under tension of the spring behind them. In effect this permits the gun to fire and start recoiling before the bolt metal strikes the receiver metal. This gives a buffing action which produces a very smooth firing cycle. It also cushions all the parts and adds greatly to their length of serviceable life.

The design was adapted for airplane use also. Guns of both flexible and fixed type were made. Belt feeds were arranged for these types, the belts normally being disintegrating. In these patterns longer barrels were employed on the aircraft guns than on the ground types.

The fire rate of the aircraft types was stepped up from the 450 per minute (normal for the infantry version) to about 1,200 rounds a minute through the use of a trap to utilize the expanding gases as the bullet leaves the bore. The backward thrust against the barrel muzzle accelerates the recoiling action. In addition, heavily spring-loaded buffers provide a faster return to step up firing rate.

The barrel and all the attached firing mechanism may be withdrawn from the rear of the receiver as an assembled unit. A new assembly can be inserted to continue operation during parts replacement in case any defects in operation are encountered. This is a rather farfetched provision which is not normally of value except possibly for fortification or defense work. It must be remembered that the energies of the Swiss have been concentrated for generations on just such employment, so the design makes sense in Switzerland, at least.

In operating the gun, a magazine is locked into position in the side of the receiver. The operating handle is pulled back to unlock and open the action. When the trigger is pressed, the sear is lowered allowing the compressed operating spring to thrust the firing mechanism forward.

When the bolt comes to the position of the feed a loaded round is driven forward from it into the firing chamber. On the last fraction of an inch of travel, the toggle locking joint is forced into line. As it locks, by cam action, it also cocks the piece. The sear holding the firing pin can rotate inside the frame of the bolt. It pivots and releases the firing pin which has its own spring. The pin is driven forward to fire.

As already noted, however, the action is still moving forward at the instant of firing, although all the elements of barrel, bolt, and barrel extension are rigidly locked at this point.

As the recoil starts, the barrel extension and bolt and barrel are still locked together during the first fraction of an inch of rearward travel. There are suitable guides for them, of course, in the receiver. The bolt is carried in a frame which is also traveling to the rear at this point. It connects by a link with the forward end of a pivoted member which is also a link. This rear link is mounted in the frame (so it can rotate on a pivot). The front end is attached to the barrel extension and the rear end connected by pivot pin to one of the support links.

As the unlocking point is reached, a stationary projection in the receiver breaks the joint of the toggle, allowing the bolt to open slowly to furnish primary extraction. As the unlocking motion is continued past the locked point, the further action draws the loosened cartridge case held by the extractor until the case hits the ejector in the receiver wall. The empty case is pivoted and ejected through the ejection port.

The barrel and barrel extension unlocked from the bolt remain in rear position. As the bolt completes its recoil stroke, stored-up operating spring energy drives it ahead to feed. The bolt compresses the firing pin spring to cock it only in the final act of locking the breech. This gun fires from a closed breech.

Detailed characteristics of weapon will be found in the chapter on Switzerland.

RUSSIAN FIREARM DEVELOPMENTS

The first reported machine gun manufacture in Russia was undertaken in 1905 at Tula Arsenal. This was purely the standard Maxim design. A number of the parts, including the water jacket, were made of bronze, a metal easily worked at the time but having great weight.

The 1905 model was redesigned in 1910 as an all-steel model, reducing the weight notably. The Navy continued use of the earlier models because of anti-corrosion factors.

Later modifications altered the smooth water jacket to a ribbed design modeled after the Vickers to give added strength. Various other minor alterations were made, particularly along the line of muzzle boosters, again following in the line of development of the Vickers.

In combat action in Korea, guns of this Maxim type were regularly encountered as first-line medium machine gun equipment. They have only one notable new feature, in the form of a special large-size cap on the top of the water jacket which allows rapid filling when the water tends to boil off. Earlier patterns are slow to fill.

The best known of the modern Russian designers is Degtyarev, born in 1880. To him is credited the light machine gun which is described in the chapter on the USSR, one of the most notable manufacturing and efficiency developments in the field of small arms.

He was engaged in work of experimental nature at Tula with the designer, Federov, on developing special ideas for light machine guns as far back as 1908.

The Degtyarev was adopted over various Maxim modifications because of the ease of manufacture and the simplicity of general design.

In addition to the infantry version, there is a tank adaptation utilizing a larger capacity drum. A modified version, the DPM, was produced to house the driving springs in a tube at the rear of the receiver, as overheating of the barrel affected the springs when housed as in normal infantry gun fashion below the barrel. This overheating, of course, was aggravated by the sustained rate of fire utilized in the tank modification.

In 1938 Degtyarev introduced his 12.7mm caliber gun. This is merely an oversized version of the original infantry model, with modifications. The drum feed is replaced by a belt feed of rugged and simple character to increase the volume firing potential. A 1939 7.62mm version introduced to utilize 250-round belts was otherwise very much in the regular Degtyarev tradition, except that it had a heavily flanged barrel to furnish greater radiation surfaces for more adequate cooling.

A recent development in the field in Russia is the Goryunov 1943 model. This arm also utilizes the standard 7.62mm rimmed Russian rifle cartridge. This gun is a subject for consideration from a design standpoint by military authorities throughout the world today.

9

Submachine Guns—Historical Development

A submachine gun is a lightweight weapon designed to use pistol ammunition and to be fired with two hands. It differs from the standard type of semiautomatic pistol (from which it was originally evolved) by the fact that it can fire cartridges full automatically as long as the trigger is held back. It may or may not, according to its design, incorporate a switch to permit the typical semiautomatic pistol type of action where an individual pull and release of the trigger is required for each shot fired. In general practice it fires from open bolt position.

The submachine gun as we know it was formerly called in Great Britain the "machine carbine"—now they conform to our terminology. It is known generally throughout Europe under various terms which translate "machine pistol." In most cases it is a very simple weapon operating without a breech-locking device of any formal kind, since the low-powered pistol cartridges do not require such locking mechanisms. More attention will be given to the technical phases of the submachine gun at a later point, but at the moment we are concerned basically with the history of its origin and its development, about which there has been a great deal of confusion.

ORIGIN

FIRST USE IN AIRCRAFT

Curiously enough, the submachine gun which has come to be a prime infantry arm, and also an invaluable adjunct to tank and transport groups, began as an aircraft weapon!

Early in 1914, before any general recognition had been made of the desirability of equipping aircraft with protective weapons, British and German observers in airplanes engaged in duels with pistols, rifles, and even shotguns. There is one recorded instance of a German plane being downed by shotgun fire from a British observer!

Wire Cage Attachment

About this time, and before the first Lewis gun was mounted in an airplane, there appeared a number of weird devices to attach to automatic pistols. The most common form used was that of a wire cage attached to the right side of a .45 automatic pistol. Used in conjunction with an over-sized magazine, it enabled the aviator to fire 20 rounds in standard pistol fashion, each ejected cartridge case being caught by the wire cage. This precaution was necessary to prevent the ejected cartridge case from striking the pilot or any sensitive part of the plane.

FIRST DESIGN—BY REVELLI IN ITALY

Villar Perosa

The credit for designing the first full automatic arm to fire pistol-type cartridges goes, however, to Bethel Abiel Revelli, about whom we have said considerable in the discussion of machine-gun design. This Italian was the inventor of the fully automatic arm which is commonly known as the Villar Perosa. The name derives from the company to which he originally assigned the patents he took out on April 8, 1914. This was the Villar Perosa Company of Pinerola, Italy. At a somewhat later date the weapon was put into production at the great Italian Fiat Factory, principally for intended infantry use. From this fact the gun is often designated as the "Fiat Model 15."

Revelli originally designed the arm, however, for aircraft use. It was substantially the same weapon as the later infantry version discussed in the chapter on Italy. It was extremely light, weighing only 14 pounds 4 ounces with dual loaded magazines in place, the magazines normally carrying 25 rounds each, although special ones for 50 were available. The cartridges were stacked in double rows. The gun actually consisted of two barrels each complete with a firing mechanism held rigidly together by a crossbar at the front

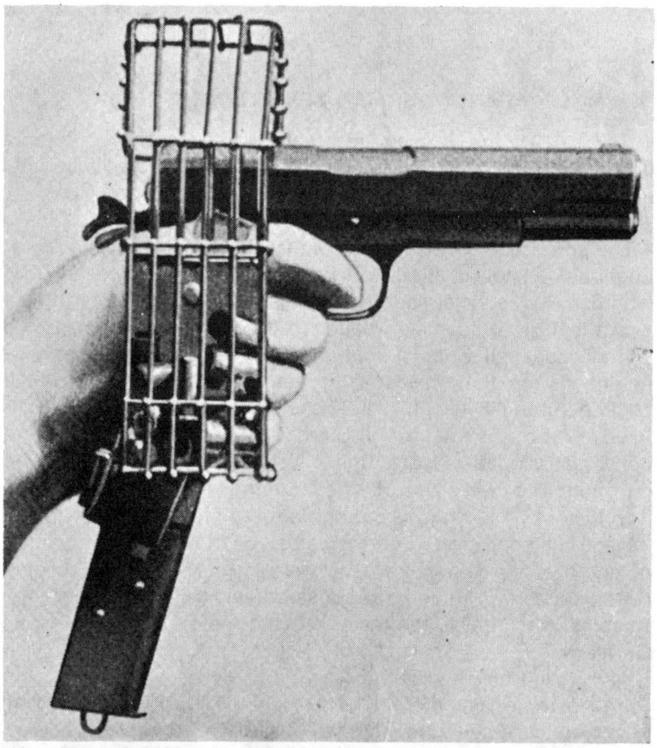

World War 1 cartridge cage designed for use by airmen in combat before the appearance of machine guns. The cage prevented ejected cartridge cases from injuring any delicate plane mechanism of that day. An oversize magazine was also provided.

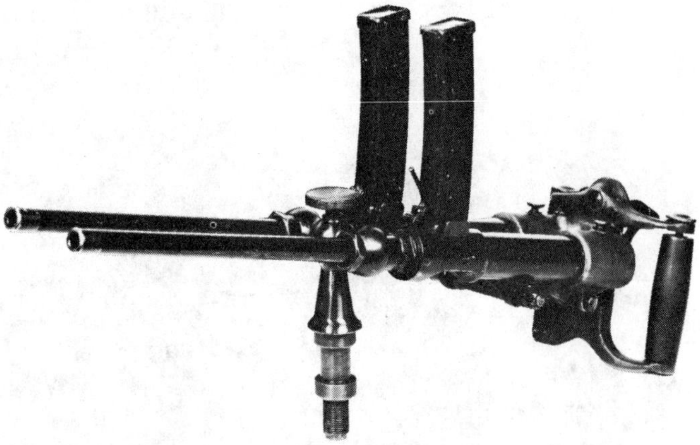

Italian 9mm Villar Perosa submachine gun of 1915.

end with a handle at the rear. The barrels were stationary. The action was retarded blowback or hesitation, each bolt being so designed that it was compelled to turn slightly to the right during its forward motion of chambering and firing.

At the instant of firing a camming action took place which delayed the bolt opening, adding greatly to the inertia required at the instant of the explosion to open the action. Each mechanism was provided with an individual trigger in the handle and each one could be fired individually by the thumb of one hand. Alternatively both could be fired simultaneously by jamming both triggers with the thumbs of both hands. Since the bolts and strikers weighed only about 10 ounces each and had a travel of less than 1 3/4 inches, the rate of fire of the weapon using the standard 9mm Parabellum pistol cartridge was extremely high. With the standard Italian service cartridge used at that time in the Glisenti pistol, neither the velocity nor the striking energy were comparable to the German 9mm pistol cartridge which, except for charge and bullet weight, was the same as the Italian. With the Italian service cartridge, each barrel of the Villar Perosa would fire at a rate of about 1500 rounds per minute.

The weapon soon proved unsuitable for aircraft use due to the under-powered cartridge it fired. The institution of the Lewis machine gun into the British air service, firing as it did a rifle type cartridge, sealed the doom of all low-powered pistol cartridge weapons for air use, even before the Villar Perosa was in general production.

Adaptation for Infantry Use

However, the Italians realized the potentiality of this full automatic weapon for infantry use long before any of the other powers did. As a result the gun, slightly modified, was put into production at Fiat. It was equipped with a bipod mount and also with a protective shield. This made it difficult to handle the arm except when in the prone position, but the Italians experimented with the use of straps secured around the crossbar and around the neck of the gunner in an attempt to fire the gun from the standing and kneeling positions also. A special holding device was developed to be attached to the belt buckle of the firer when attempting to shoot from the standing or kneeling positions. One of the most novel uses attempted in connection with it was mounting the gun on a bicycle. The rate of fire was so high, the dispersion so great, and servicing of the weapon so difficult that the gun was quickly outmoded for practical service use. Nevertheless, this model was the first actual use of a submachine gun in military service.

EARLY DEVELOPMENT

EARLY GERMAN DEVELOPMENTS

Early in 1917 the German Army realized the necessity for a rapid firing lightweight arm for close-quarters use. They tested automatic pistols, both Mauser and Luger types with lengthened barrels, shoulder stocks, and in some cases special magazines.

The principal German weapon of this sort was the Pistole 08, or Luger as we know it, of 1917. It was equipped with a special 7.87-inch barrel. A special snail-type magazine to hold 32 cartridges was a feature of this new weapon. Although this model required an individual pull for each shot, the fact that it was supported by two hands greatly increased the number of aimed shots which could be fired per minute. All 32 shots in the snail could be aim-fired in about 45 seconds. Basically the arms were issued to noncommissioned officers, primarily those attached to machine gun units. They were also utilized for defense work at close range.

In October 1917, the Germans and Austrians folded up the Italians in the terrific disaster of the battle of Caporetto. Very large numbers of the Villar Perosas were captured there. German intelligence recognized the possibilities in these arms, and samples were sent to the major arms development centers in Germany.

As a result of intensive studies, particularly at the Bergmann Plant, a rather clumsy but extremely efficient weapon was designed there by Hugo Schmeisser at the Suhl Plant of Bergmann. Schmeisser even before World War I was one of the primary German designers of small arms. His World War II designs are outstanding.

The German MP-18

The first weapon, often known as the Bergmann Muskete, was issued to the German forces late in 1918 under a designation MP18, meaning Maschinen Pistole 1918. While the Germans recognized that a more powerful cartridge than the standard pistol type would be preferable in such an arm, the ammunition industry was already overburdened in attempting to keep up with production. Hence the arm was designed to fire the standard German service pistol cartridge, the familiar Pistole 08. This 9mm cartridge used in conjunction with heavy recoiling parts obviated the necessity for a locked-breech mechanism.

Entirely aside from the ammunition factor, manufacturing facilities were available for producing the extra length barrels more suitable to this design and also the snail type magazines already used in the Luger or Pistole 08. For simplicity of action

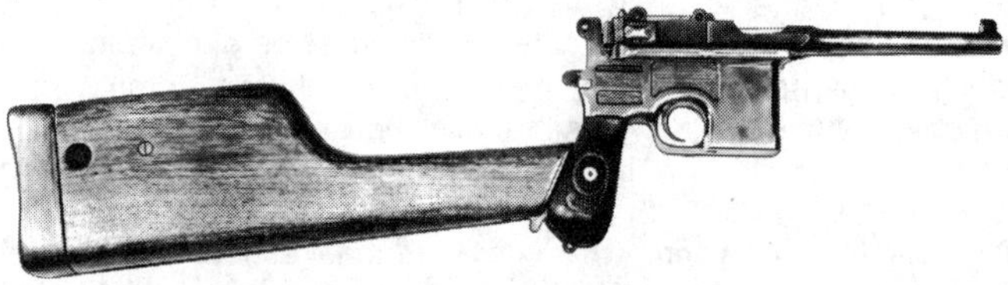

World War 1 Mauser pistol converted to caliber 9mm Parabellum and fitted with detachable shoulder stock and holster. Experimental models were full auto. After the war the design was also issued commercially with a full auto switch. This design is too light for practical true submachine gun use.

Early Mauser Pistol-Carbine, stock detached. Magazine built in. While this model was not originally intended to fire full auto, some were altered to allow selective fire. Like the Thompson, this pre-World War I Mauser was manufactured to use pistol cartridges. This weapon fires from a fully locked breech. Whether full or only semiauto, this arm is legally classed as a machine weapon in the U.S.

this initial design has never been surpassed. It fires only full automatic, but because of the weight of the recoiling parts, the cyclic rate is good for the purposes intended.

Since this weapon is treated extensively in the chapter on German World War II materiel, only general details are necessary at this point. It utilized a rifle type stock, fed from the side with a snail magazine, was fitted with a perforated jacket for cooling the barrel; and consisted only of the stationary barrel, the moving bolt driven forward by the recoil spring, and the simple trigger, sear, extractor and ejecting mechanism.

The design was so extremely simple that it was possible to issue the guns without any special training. As a matter of routine, all officers and noncommissioned officers and 10 per cent of all the men of the infantry companies were given instruction in its use. The original idea was to furnish a squad armed with six of the MP-18's to each company. Gunner and ammunition carrier were also to be armed with a standard rifle. A hand-drawn cartridge transport cart was assigned to a company.

Probably because of the terrific pressure the Germans were under in the summer and autumn of 1918 while events were turning steadily against them, very few official records were kept of the actual value of this new gun in service. However, the allies placed tremendous importance on its use. That is, all except the British. As a result, under the treaty of Versailles the 100,000-man army which was permitted Germany was denied the right to use these machine pistols or submachine guns. Some small use was authorized for police detachments. The manufacture of such weapons was ordered completely discontinued. The order was not particularly effective because German design went ahead on an underground basis and in other countries.

EARLY U. S. DEVELOPMENTS

Early in 1918 appeared the only World War I attempt by the U.S. at even an approximation of such an arm. This was the device which was listed as a top military secret and classed as the automatic pistol, Caliber .30 Model 18.

The Pedersen Device

This of course was the now familiar Pedersen conversion which we have discussed under rifles, a unit intended to replace the conventional turn bolt in an altered Springfield and allow the insertion of a special magazine to hold a small .30 caliber pistol type cartridge. In theory the Springfield rifle could thus be converted to a semiautomatic arm for close-up use. Since the user would have to carry two types of ammunition, two bolts, and a special magazine, and there was constant danger of losing equipment which had to be changed in the course of the heat of battle, the Pedersen was doomed to failure from the start.

Thompson Submachine Gun

The first American submachine gun was the now world-known

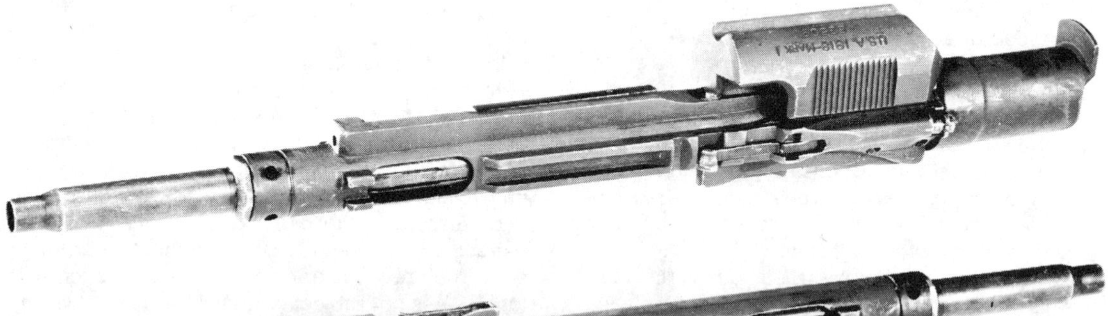

Pedersen device, caliber .30.

The famous Thompson submachine gun showing stock and magazine detached. A blowback gun with inertia lock.

Thompson. Designed by General John T. Thompson, and first placed on the market in 1921, it is ironic that the weapon was immediately seized upon not as a law enforcement gun but by the bootleggers in Chicago. The reputation it made there at that time was grim indeed. Police and prison units now saw the necessity for an arm of this type to counter the number which had found their way into lawless hands.

At this point the weapon received attention from the United States Ordnance Department. The Thompson, as originally made by Colt, had a removable buttstock. It was fitted with two pistol grips, one below the barrel for left-hand hold and the other conventionally behind the trigger. Designed to be fired from either hip or shoulder with both hands, it utilized the Cutts compensator to control the climb of the muzzle during automatic fire. Unlike the earlier Bergmann, it was equipped with a switch to permit either full automatic fire or to require a pull for each shot.

The cocking apparatus was originally a knob projecting through a slot in the top of the receiver. Drawing it back cocked the bolt mechanism ready to fire. The arm like most of its type fires from an open bolt position. The cartridge originally used was the .45 Colt auto pistol cartridge, but the weapon was furnished in other calibers including 9mm Luger and .38 Colt automatic; 50-shot and 100-shot circular drums were provided, but in actual practice the 100-shot drum because of its weight and feeding factors was not found too practical; 20- and 30-shot detachable clips were also issued. The original cyclic rate was of the order of 600 to 800 shots per minute. The original Thompson was equipped with the involved Thompson idea of a hesitation locking principle under which the factor of adhesion was utilized, in theory at least. In actual practice the gun would fire as well without the inertia wedge block as with it, though perhaps at a somewhat faster rate.

In later models of the arm as redesigned for the United States services the hesitation system was dispensed with entirely. The last of this famous arm is a straight blowback operation, the cocking handle now being placed on the right side of the arm in standard practice. It is far simpler and quite as efficient.

It might be noted in passing that the so-called "Blish principle" on which the original Thompson was intended to operate is very much like that which Revelli developed for the original Villar Perosa. These submachine guns, of course, are operated on the elementary blowblack principle. The theory of this method of retarding the breech opening is that the design is such that the rapidity of opening is controlled. In theory, because of the rapidity with which the barrel pressure reaches its maximum, the bolt is firmly locked by the adhesion of the inclined surfaces provided until the pressure has dropped to a safe value. Any successful application of such a principle depends very largely on the

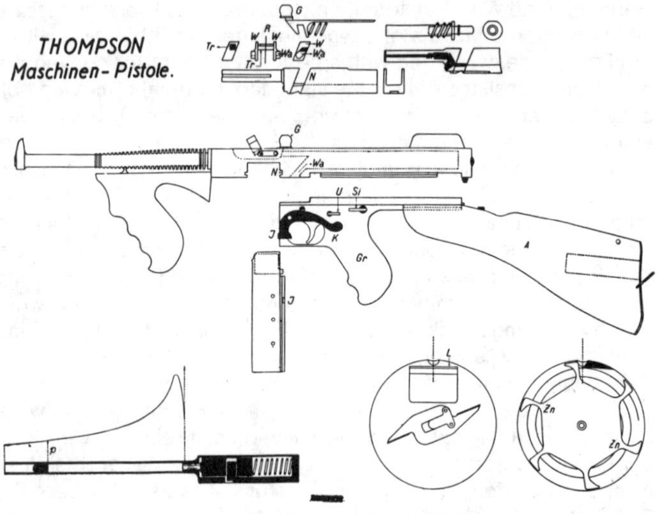

A German drawing showing locking detail of the Thompson. This gun received considerable attention from German military officials.

constant and equal lubrication of the surfaces of the sliding elements. The Thompson originally provided lubrication by an oil pad.

In theory at least, because of the very high friction resistance with normal gas pressure even when the lock is oiled, it is impossible for the bolt to slide freely in the receiver. As the projectile leaves the barrel, the gas pressure drops. The decrease of the friction resistance occasioned by the high pressure is immediately encountered. The lock is thereby freed. The residual gas pressure in the barrel pushes the bolt mechanism back to extract, eject, and cock the firing mechanism ready for the next shot. The breech oiler automatically keeps the bolt lubricated so long as it is not subjected to dust or dirt. As we have stated, in actual practice these systems operate as well without the frictional device installed as with it.

As originally listed in early catalogs by the Auto Ordnance Corporation, the Thompson sold for $200. While the price was high, the amount of machining and fitting involved entailed a great deal more work than in the later forms of the same gun. Since no other submachine gun was available in the world market until 1928, it achieved world-wide attention in all theaters where police troubles or minor military troubles were developing.

LATER WORLD DEVELOPMENTS

GERMANY

Meanwhile, although manufacture of submachine guns was prohibited in Germany by treaty, the development work was not interfered with. Hugo Schmeisser, then working with the Haenel firm in Suhl, acquired manufacturing rights for the MP-18 I which he had developed.

German MP Schmeisser 28-2

Under Schmeisser's guidance the arm was modernized and prepared again under the name of the MP Schmeisser 28-2. The design perfected, the Germans very easily sidestepped the Peace Treaty by transferring the manufacturing rights in name to the Pieper Firm in Herstal in Belgium. It was manufactured there for world distribution.

This new MP Schmeisser 28-2 was offered for the standard German 9mm Parabellum cartridge. (The word Parabellum from the Latin "for war" is the commercial designation for the pistol ammunition known in Germany as the Pistole 08.) The design was also made for the 7.63mm and the 9mm Mauser pistol ammunition which had been manufactured in Europe for years and which had received worldwide distribution through the Mauser firm. Thus the elements of the pre-war Mauser organization were able to offer a submachine gun to their former customers in calibers they desired.

The snail drum of the early design of Schmeisser was replaced with a box magazine carrying the ammunition in staggered column as in the familiar military rifle. Moreover, it was equipped with a switch to permit single or continuous fire. The switch was in a form of a button in the shaft above the trigger. Where the recoil spring of the first model had been provided with a guide, this was eliminated in the new model and instead the spring diameter was increased to utilize the interior dimension of the receiver tube.

Ironically, the fullest acceptance of this weapon in military circles as an official weapon was by the Belgian army in 1934. They listed it as the Mitraillette Model 34. (Mitraillette was a designation used in Belgium and in France for the submachine gun, indicating a small caliber full automatic arm).

German Bergmann MP Model 34

By 1932 the Bergmann Company then at Berlin had produced a very unusual modification and improvement of their original MP-18I. This weapon (Model 34) had been turned over to the Schultz & Larsen Manufacturing firm at Otterup in Sweden. This arm, slightly modified, was adopted by the Swedish army as its Model 39.

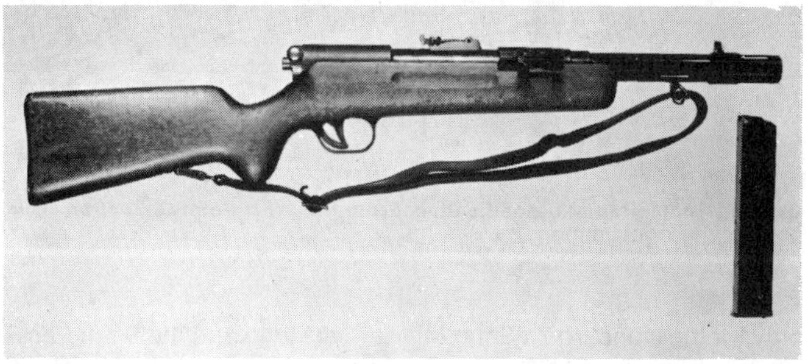

German Bergmann 9mm 1934 machine pistol or submachine gun.

While this 34 Model is blowback as in the case of the original Bergmann, it varies in many mechanical respects. The receiver is completely enclosed. The operating handle does not move during firing. The bolt may be withdrawn and dismounted without tools. A special firing pin device was incorporated, the firing pin protruding from the bolt to fire just before the end of the closing bolt stroke. The trigger system is a variation of one utilized in various Solothurn arms. The trigger is pivoted at its center, and a pull at the upper section fires a single shot, the breech staying open until the trigger is released and then pulled again, while the pull on the lower half gives continuous fire until the trigger is released. The barrel has a cooling jacket and is fitted with a compensator to counteract the muzzle climb and also a device to reduce recoil.

Originally issued as the MP Bergmann Model 34, the gun had obvious advantages over its earlier model. However, so much simplification was lost that the cost, naturally, increased considerably.

Erma Machine Pistol by Vollmer

Still another German submachine gun (or machine pistol) developed during this period was the Vollmer. Manufactured at the Erfurter Maschinenfabrik (The Erfurt Machine Works), it is also a blowblack weapon. Its recoil spring is positioned directly behind the bolt, enclosed in a telescoping tube to protect it from dirt, a system which was utilized during the World War II by the Germans in their submachine guns MP-38 and MP-40.

This arm was equipped with a right thumb switch to permit single or continuous fire and had a forward pistol grip to steady the arm as in the case of the early Thompsons. It was manufactured not

Erma machine pistol. Designed by Vollmer, this gun was made in both Germany and Spain. Its name derives from the manufacturer, Erfurter Maschinenfabrik.

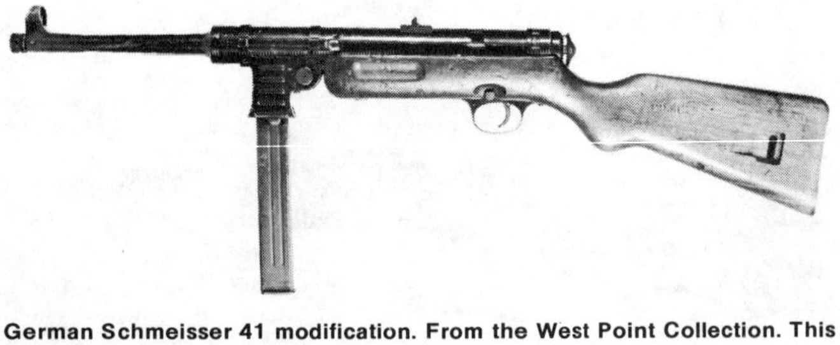

German Schmeisser 41 modification. From the West Point Collection. This pattern was not common.

only for the standard 7.63mm Mauser cartridge and the 9mm Parabellum, but also for the Luger 7.65mm cartridge and for the 9mm Bergmann, this last being a cartridge seldom seen except in Spain and in some areas in Scandinavia. It too was one of the submachine guns sent to Spain to the Franco forces during the Civil War there.

German MP-38

In 1938 the Germans introduced their MP-38. This was a standard simple blowback gun in 9mm caliber. It was a magnificent example of the utilization of the latest design and in the use of elementary turned parts. The operating spring, housed in telescopic tubes somewhat like the earlier Vollmer, prevented kinking of springs and gave complete protection from dirt or dust. This gun introduced the idea of the folding buttstock which could be folded so as not to interfere with hip firing. In 1940 a modification of this arm was introduced. The cyclic rate of fire of about 450 to 540 per minute is nearly ideal for submachine gun operation according to European theory, although American theory favors an even lower rate. In any event, a trained gunner can tap off single shots with this arm. In view of this, the Germans dispensed with the idea of a special single-shot trigger, thereby still further simplifying manufacture and reducing costs.

FINLAND

Finnish Lahti Suomi

The Tikkakoski factory in Finland began production of a machine pistol or submachine gun under the designation of Suomi. It was promptly adopted by the Army of Finland and some were bought by the Swedish, Swiss, and Danish armies. It is a well-made elementary blowback design. The recoil spring is positioned directly behind the bolt, the point of the firing pin protruding constantly. The arm fires as the bolt is still moving forward, providing an additional inertia factor to slow down the rapidity of opening. The operating handle is stationary during the firing period, an advantage in itself. The Suomi was fitted with a barrel which could be speedily changed without the use of tools. This too marked a step forward in design. Both box and drum magazines were provided.

This Suomi was developed by the Finnish designer Lahti. When the Russian attack opened on Finland in November of 1939, one of the weapons with which the Finns defended themselves was this submachine gun. Its deadliness and particularly its adaptability impressed the Russians tremendously.

In passing it may be noted that Lahti had produced his submachine gun in 9mm caliber. This, of course, represented German influence at the time, as well as the fact that Finland and Sweden also used the 9mm Parabellum cartridge.

USSR

While the United States was adopting the Thompson for special duty purpose, and England and France ignored the new weapon, the Russians proceeded in 1934 to develop a submachine gun. Operating on the theory that pistols were no longer adequate defense arms for the protection of heavy weapons by their crews, nor for the leaders of infantry units, Russian concentration on submachine gun development, while it came slowly, did eventually develop on a very wide scale.

Russian 7.62mm Submachine Guns

At the close of World War I the Russians' sidearm equipment had been a hodepodge of revolvers, mostly in 7.62mm Russian caliber with an intermingling of German Mauser pistols. The revolvers of course had rimmed cartridges which will not function effectively through box magazines of the type used in submachine guns. The Russian designers, therefore, turned their thoughts to a new automatic pistol cartridge. Since Mauser military model automatic pistols have been used extensively in Russia since early in the century, the 7.63mm bottlenecked high-speed cartridge used by this weapon had been given considerable attention by the Russian development groups.

The Russian designer Tokarev, using the Browning-Colt system as a model, worked out a very unusual development in low-cost production of that pattern and introduced in 1930 a new service Russian automatic pistol chambered for caliber 7.62mm. This actually used the slightly underpowered 7.63mm Mauser ammunition. Up to this point American experience had leaned toward the large caliber in submachine guns, as witness the fact that our pistols and the Thompson both use the .45 automatic pistol cartridge. The Germans and their supporters leaned toward the 9mm. The Russians felt that the 7.63mm caliber with its bottlenecked case, higher velocity, and greater range was more in line with their requirements. (Also, the same barrel-making equipment used for making their service 7.62mm rifles could be adapted to this pistol and submachine gun caliber.)

The Russian Model 34 and its modification, the Model 34/38, as well as their Model 40, which are very much a pattern, were not originally Russian in development. Their designs were evolved directly from the MP Schmeisser 28-2.

It will be noted that, in later modifications of this first Russian submachine gun and in most later types they have adopted, the Russians have adhered religiously to the principles set forth early by Hugo Schmeisser. They have sought to produce a weapon at an extremely low price, using the common machinery available in their country, utilizing to the full the application of stampings and metal turnings which are the simplest form of manufacture. While the resultant arms have all been crude in appearance, they have been very effective. Their low cost and rapid production possibilities have enabled the Russians to follow through on the original German idea of issuing them to all noncommissioned officers and to all special groups. In addition, also in line with the original theory, the guns have been kept so simple that the most ignorant and illiterate soldier can be taught to use them without extensive training.

Russian Model 1940 PPD Machine Pistol

As we have said, the early Russian submachine guns were basically little more than modifications of the German Schmeisser MP-28-2. These included the Model 1940 machine pistol equipped with a 71-round drum. This modification was ascribed to the Russian inventor V. Degtyarev. The usual designation is Model 1940: PPD.

Russian Model 1941: PPSH

In 1941 the Russians introduced a new one, the Model 1941: PPSH, the name stemming from the name of the designer Shpagin. This design differed radically from the earlier ones and is of true Russian origin. For the most part it is a series of simple stampings.

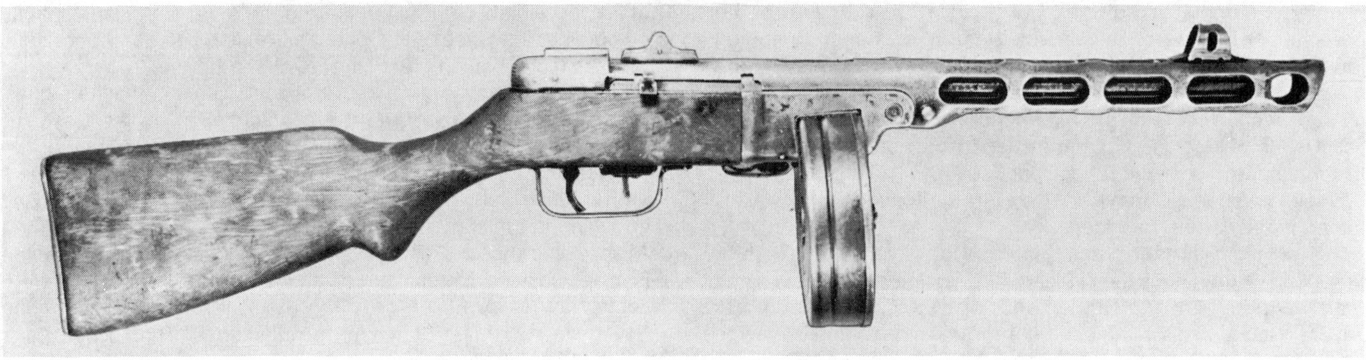

Russian Model PPSH 41. One of the cheapest yet most efficient submachine guns ever built, despite its crudity.

The receiver cover and the barrel jacket are a one-piece design. An extension of the barrel cover over the muzzle acts as a muzzle brake. The front end is designed to counteract muzzle climb. A switch provides for single and full automatic fire. The safety is a sliding design copied from that of the Model 34/38, and the drum magazine is similar except for the type of lip required.

Russian Models 1942 and 1943

In 1942 and again in 1943 new Soviet submachine guns appeared. The clumsy wooden one-piece buttstock was replaced with a folding steel stock modified along the lines of the German Model 38. These arms were designed for box type magazines only, the normal magazine capacity being 35 rounds.

SWITZERLAND

The Swiss MP Solothurn 34

In 1934 there also appeared the first of the Swiss-made Solothurn submachine guns (or machine pistols as they are called in Europe). As we have discussed in some length in our section on machine guns, the Solothurn Works was controlled by the great German Rheinmetall firm which had been authorized to produce rifles for the 100,000-man German army authorized by the Peace Treaty. Design work had continued, even though they were denied the right to manufacture machine pistols. Parts were manufactured in Denmark and also at the Steyr Plant in Austria and shipped to Switzerland to the Solothurn Works for assembly under German control.

The first issuance of the MP Solothurn 34 was in caliber 9mm Mauser. This is not to be confused with the 9mm Luger, the so-called Parabellum or official German Pistole 08 cartridge. It is a longer and much more powerful cartridge which was manufactured originally for the Mauser 9mm export model pistols, most of which were sold in South America. The original Solothurn idea was that this overpowered 9mm cartridge would be a step in the right direction in connection with future machine pistol development. They recognized the importance of a more powerful load for the submachine gun which, being heavier and intended to be used with two hands, was much more stable than the pistol.

The 9mm Solothurns were delivered to the Austrian Army where they were adopted as the MP-34. Since the Austrian police were basically equipped with still another type of 9mm automatic pistol, their Model 1912 or Steyr, a quantity of the guns were chambered for this cartridge for police use. Later, during the Spanish Civil War, considerable quantities of these weapons were sold by Solothurn to the Spanish Franco groups. In fact, it was the testing done in this Civil War with this caliber which led German observers to the firm conclusion that for machine pistols the 9mm Parabellum, whatever the theoretical considerations, in actual practice was as desirable as the 9mm Mauser patterns.

In the Solothurn the recoil spring is positioned in the rifle-like buttstock, being compressed by a rod hinged to the bolt. A change lever on the left side of the receiver permits single or automatic fire. The magazine is a double line box. A special loading attachment is provided. Since the arm was primarily designed for close quarters work, it was equipped to be fitted with a bayonet. Still another accessory tested in the Spanish Civil War was a container to be mounted over the ejection port. This was for tank and vehicle work where flying empty cartridge cases might be undesirable.

SPAIN

Spanish "Star" Submachine Gun

In 1935 the Spanish Echeverria firm at Eibar introduced a machine pistol. (Star is the trade name of this firm which manufactures pistols, submachine guns, and shotguns.) This initial Spanish machine pistol, like so many of the Spanish developments, was a modification of an American arm. An attempt was made to utilize the bolt system of the American Thompson. The gun was offered in various calibers including the standard 7.63mm and 9mm Parabellum and others.

It might be noted in passing that the various arms factories in Spain have for years issued copies and modifications of Mauser military pistols and of Colt automatic pistols. Star currently produces a pistol externally and in locking arrangement very much like the .45 Colt automatic. It can be fitted with an extension magazine. It also can accommodate a detachable buttstock. A firing switch permits the use of single-shot or full automatic fire. Even when used with a stock, however, this design is too light for practical machine pistol usage. Star also makes a German pattern submachine gun (Model Z-45).

HUNGARY

Hungarian M1939 Submachine Gun

In 1939 the Hungarian Army adopted officially their M1939 machine carbine in caliber 9mm Mauser. While it was never produced in quantity, it is an interesting example of the better type of retarded blowback or hesitation system.

In general lines it resembles a short rifle and is furnished with a box magazine which can be folded up below the barrel when not in use but can be swung into position quickly in case of necessity.

The breechblock has two connected parts, a pivoted lever being positioned in the front one. The upper end of this lever bears constantly against the rear section of the breechblock which is thrust forward by the recoil spring. The lower end of the lever rests in a slot in the lower section of the receiver.

As the cartridge is fired and the breechblock starts back, since the lower leg of the lever is positioned in the receiver, the bolt thrust against the upper half compels the lever to rotate. Consequently, the upper end of the lever thrusts the rear half of the breechblock backward. This causes the front section to be forced forward by reaction through the lever axis. By thus utilizing the opposing reaction which, of course, is equal to every action, the forward section of the block which supports the recoiling cartridge case is slowed down after the initial thrust, but when the rear section of the connected breechblock reaches its limit of rearward travel its transferred kinetic energy pulls the forward section back to completely open the breech.

This delaying system permits the use of a very much lighter breechblock than does the standard blowback arm using the same 9mm cartridge. On the other hand, of course, this system has several offsetting disadvantages. It is a much more complex design requiring more parts and more machining and fitting. However, this is one of the best of the delay systems and does offer experimenters a principle worthy of more intense study than it has received for application to other types of arms.

ITALY

The Italian Beretta Moschetto

A form of machine carbine, based originally on the Villar Perosa principle of the lightweight turning bolt to retard the blowback, is common in Italy. The magazine is the typical box magazine fed in through the top of the rifle. Ejection is through the bottom. This arm, while normally encountered as a single-shot or semiautomatic design, was also made to some extent as a full automatic. It uses the standard 9mm Italian service pistol cartridge and is commonly known as the Beretta Moschetto, after the factory manufacturer.

Italian Beretta MP-38/42

In 1938, however, the Beretta factory produced a true submachine gun of the very highest order. Heavily built and of rifle type (it weighed 10.3 pounds with a loaded 40-round magazine) it was equipped with barrel jacket for cooling, compensator, bayonet mount, dual triggers for single or full automatic fire, and was intended to be used with a heavily loaded 9mm cartridge popularly known as the M38. This gun was put into service in Africa. It was also issued to police in areas where smuggling was rampant. Because of the heavy weight of its recoiling parts, and the fine quality of the materials and the machining, this is one of the most efficient arms of its sort ever designed, although its weight and bulk are considered by many to take it out of the standard classification of the common machine pistol.

In 1942 technicians in the Italian factory proceeded to modify the Beretta. The new version was listed as the MP-38/42. The barrel was shortened and its forward end cut to furnish a built-in compensator. The cooling jacket was dispensed with in the interests of lessening weight.

BRITAIN

It is a matter of German record that when the German Armies entered Poland, Norway, and the Balkans they encountered very few submachine guns, most of those they did meet being early Bergmanns or American Thompsons.

Despite the fact that both Germany and Russia were testing several types of submachine guns of their manufacture or control in Spain during the Spanish Civil War, the British army failed to recognize the importance of this type of arm. Both the Germans and the Russians learned during the Spanish conflict the value of the machine pistol as a close combat arm for both attack and defense, with particular reference to its use in house to house and forest fighting. Its light weight and easy transportability gave it a terrific edge over even machine rifles in such close-quarters work. Its superiority over the pistol from the standpoint of rapidity of fire, possibility of firing bursts, and the extra range which could be obtained because of the longer barrel and two-hand operation, were all factors given great consideration.

The British Army was not then equipped with sub-machine guns. The immediate use in quantity of this design by the Germans compelled them to turn to America as a source of supply at the outbreak of war. Large quantities of Thompsons were purchased. These were both the 1928 commercial patterns with two hand grips and the military patterns with one grip and forestock. In the African campaigns trouble was shortly encountered with the involved locking mechanism on the Thompson. The locks were commonly removed. Large quantity orders were placed with Savage in the United States for manufacture of Thompsons. Over 1,000,000 were made.

Meanwhile the British, examining samples of the German MP-38, set out to design a low-cost submachine gun. The standard British side arm at that period was a .38 caliber revolver which, of course, used rimmed cartridges which would not feed properly through an automatic pistol magazine. This fact led to British developers to produce their submachine gun to handle the standard 9mm German pattern cartridge. This enabled them to utilize stocks of captured ammunition. Their own and United States ammunition factories had a table of experience on manufacture of the 9mm Parabellum cartridge, a fact which eliminated design and development work on a new type of cartridge.

British Sten

While they were greatly influenced by the German idea of utilization of stampings and machine turnings, the British did not have either the German machine equipment nor the manufacturing know-how to produce arms of a like quality. They developed, therefore, an ultra-cheap type of arm which could be manufactured in the most elementary machine shops. This weapon was really a master of design from the standpoint of low cost and rapid production. This is the famous Sten gun. The name derives apparently from the initials of the designers and the point of manufacture, Enfield Armory. Both the bolt system and the feed mechanism are a direct imitation of the Schmeisser MP-28.2. However, the firing pin is stationary in the bolt. The gun was designed to be easily dismounted and carried hidden by paratroopers and guerrillas. The arm originally met with considerable derision by German technicians, but its actual battle use very shortly established it as an arm capable of doing anything that the most expensive types of submachine guns could hope to do. The cost was brought down as low as the fantastic figure of $9.00 per gun during the height of the war and under intense manufacture in Canada!

This gun led the way to later manufacture in Australia of the quite unusual Austen or Australian Sten gun, which together with the Australian Owen, is completely described in the chapter on Australia.

FRENCH DEVELOPMENT

Like the British, the French paid little attention to submachine gun design in the period between World Wars I and II. The outbreak of war found them using a certain number of Thompsons in .45 caliber which had been purchased from the United States, and also a version of the Vollmer Erma produced in Germany and originally sold to the Communist groups in Spain. The French Army itself produced only the MAS 38 (the name comes from the point of design and manufacture, Manufacture d' Armes, St. Etienne). This gun will be found completely described in the chapter on France. An elementary blowback, it utilized a 7.65mm long pistol cartridge. Since the French were armed with a wide variety of revolvers and automatic pistols in varying calibers, they designed a special cartridge for this submachine gun. The cartridge is little more effective than the common American .32 automatic pistol cartridge. The cartridge is actually so close to the U. S. .30 Pedersen device cartridge of War I that it would seem that French designers merely "borrowed" the caliber.

This submachine gun was issued in considerable quantities to French Colonial troops in the recent Indo-China campaign. When the Germans stormed over France, they permitted the French police to use these weapons.

U.S. submachine gun. Cal. .45, serial #5. Experimental model of M3 sub- machine gun from the West Point collection.

UNITED STATES

U. S. M3 Submachine Gun

During the War the United States Ordnance Department developed its own stamped type of submachine gun based on the British Sten gun and the German Schmeisser. This gun is described in the U.S. chapter. The United States design, which was promptly dubbed the "grease gun" by our troops because of its crude appearance, is actually one of the most efficient submachine guns in the world today. It went through several modifications before finally getting all the bugs out, but the design was such as to lend itself to a maximum of highspeed production with stampings and screw machine equipment commonly available on a large scale in the United States. Neither as crude as the Sten nor as well-made and finished as the German, it still has all the elements of simplicity and ruggedness needed in this type of arm.

Other U. S. Designs

Another elementary blowback was made at Philadelphia by Sedgley, though this too was an expensive type of arm and with the coming of the M3 was completely outmoded. Our Ordnance M2 was also by-passed in test stages for the superior M3.

The U. D. developed at High Standard Arms for the Dutch in limited quantities was but an elementary blowback also.

The very unusual delayed blowback Reising gun which we also cover in the chapter on the United States did not stand up too well in use by United States Marines. This gun too was outmoded on a basis of cost, simplicity, and reliability by the M3. It is rather extensively used by law enforcement groups who purchased Reisings while the Thompson patterns were available only for military use.

THE ROLE OF THE SUBMACHINE GUN

The history of the submachine gun is small in relation to other arms types, as we have seen. It really began in 1915 in Italy with the Villar Perosa.

However, due to the intense development and the widespread usage both during World War II and in the various so-called police campaigns throughout the world since then, it is technically very far developed. Relatively few changes of any importance can be expected in this type of arm unless and until more powerful types of ammunition are available or adaptable.

While technical details of the various arms differ, of course, all use ammunition which was originally designed for one hand firing in pistols where effective ranges are necessarily short.

None handle ammunition which warrants usage against human targets at ranges of much more than 200 yards. While many are equipped with sights graduated up to 1,000 meters or more, such sights are useless except for possible plunging fire. Submachine guns have proven to be very essential arms equipment for both the Army and for certain types of police work, all, however, at close range.

For all practical purposes probably the 9mm Parabellum cartridge is the most suitable for this type of arm. In connection with standardization of equipment, it is quite likely that the .45 automatic cartridge in due course will give way to this smaller caliber. The advantage of greater bullet weight is more than offset by the greater weight of the ammunition and of the weapon itself.

Magazines for normal tactical purposes require a capacity of 25 to 30 rounds. Except for the possible development of the staggered 4-column box issued in Sweden and Finland, there is no practical way of increasing the magazine capacity much beyond this point without incurring disadvantages which will nullify the increased magazine capacity. The 70- to 100-round drum magazines proved unsatisfactory in the field wherever used and by whomever used. They are much more given to jams than are the simple straight line box types of magazine.

In practically all instances, firing from the open breech is and will continue to be the ideal method of operation of the submachine gun. Because of the heat generated in automatic fire the chamber must be left empty, as otherwise cartridges would "cook off" in a closed breech. There is little to be gained by increasing the barrel length, since the distance between sights is not particularly important in view of the relative inaccuracy of this type of arm and cartridge at distances over 200 yards. The short bullet travel permits simple design and construction of a blowback action, whereas the longer barrel would have the effect of increasing velocity somewhat but would introduce serious design factors with relation to the operation of the breech mechanism.

Cooling systems are unnecessary not only because of the relatively low power of the cartridges used and the relatively heavy weight of the barrels involved, but also because the amount of fire which can be put through a gun at any one time would hardly ever warrant the weight and complication of a cooling agent or section.

As we have mentioned, the Hungarian Model 1939 is an instance of a delaying device or hesitation device which might be the subject of some investigation, though its complicating factors with use of pistol ammunition more than offset the value of the lower weight breechblock which it permits. Madsen, as we have seen, tried to utilize a lighter breechblock by compelling it to carry with it the receiver cover, but here again the complication offsets the value of the weight saved, which in no case is likely to be more than six ounces. If more weight than that is saved, the cyclic rate of fire will be too high for proper submachine gun use.

Probably the one truly valuable field feature to be introduced in any new gun is that in the Swiss pattern in which the magazine may be swung from horizontal to vertical position. In all other types of submachine guns, either the magazine protrudes so far below the receiver that the gun cannot be fired from the prone position without exposing the shooter, or in side feeding it is clumsy and awkward in anything but prone shooting.

Much attention has been given in some quarters to the idea of using precision castings to simulate machined parts with an ease of manufacture which is expected to be comparable to that of stampings. In actual practice this just does not work. There is no process of manufacture which begins to approach stampings or pressings for low-cost production. On the other hand the design of an arm to utilize such procedures and to hold them to the tolerances found in the M3 submachine gun and in the German MP-40 is beyond the technical skill of any but the world's finest arms factories and metal working factories. As to precision castings, it is impossible in quantity at low cost to hold these to the types of tolerances which are commonly required in the firearms manufacturing business. They are excellent for developmental and prototype arms, but in production even the most formal methods of forging and machining permit a faster and far cheaper rate of manufacture.

Developments such as the so-called German MP-44 and the U.S. carbine M1 series constitute locked breech weapons utilizing cartridges of far higher power than can be handled in standard submachine guns. Such arms form an entirely new classification of weapons. Although they are often loosely placed in the category of submachine guns, and varieties of them may be used in full automatic fire, they actually constitute a new field weapon whose uses are being explored.

A study of the submachine guns in Part II covers pictorially every design which has been produced in any quantity.

Note on Design

It is generally agreed that the most efficient cyclic rate for a submachine gun is of the order of 450 rounds per minute. This is extremely difficult to achieve, particularly with a relatively light weight breechblock. Most designs have far higher rates which result in ammunition wastage through uncontrollable fire. While the heavy breechblock will reduce the rate materially, the weight of the moving block itself is such a sizable part of the entire weight of the arm that its movement creates vibration to a degree that interferes with accuracy. The M3 submachine gun minimizes this through use of a relatively heavy breechblock but by providing it with balanced recoil springs and by keeping the bolt completely guided on metal guides at all times during its forward and rearward travel.

Compensators are definitely valuable, particularly on the lighter variety of arms, because by deflecting some of the muzzle blast (upwards as a rule) they oppose the tendency of the barrel to climb during full automatic fire. The Beretta system of cutting slots in the upper section of the muzzle forward of the front sight is relatively effective, as the escaping gas produces a reaction equal to its own momentum.

The revolver is an arm intended to be fired with one hand ordinarily. It consists of a single fixed barrel positioned in a frame, and a revolving cylinder which contains several chambers for cartridges, one chamber being lined up and locked in line with the barrel at the moment of firing.

While the terms "revolver" and "pistol" are customarily used interchangeably in common lay practice, no firearm is actually a revolver except as described above.

From the earliest times attempts have been made to produce revolvers, however crude.

10 The Military Revolver —Historical Outline

EARLIEST REVOLVERS

THE COLT

The first actually successful revolver was, of course, that of our own Samuel Colt. However, it must be noted that his original patent was granted in England in 1835 and he did not seek out United States patents until later, the first being granted on Feburary 25, 1836.

Romantic legend has it that Samuel Colt conceived the idea of his revolving system while traveling on a schooner as a cabin boy and studying the operation of the spoked steering wheel. It is not strange that a person with the fantastic background and personality of Samuel Colt should be the subject of legends, though this one is considerably farfetched.

Samuel Colt. From a painting made at the height of his success.

There is in the Royal United Service Museum in Whitehall, England, a Snaphaunce revolver of the days of Charles the First of England. Its date of manufacture is before 1650. This is a brass-barreled giant of .500 caliber which weighs about 6 1/4 pounds. In every essential of cylinder operation it is so similar to Colt's original firearm that the researcher looking at it cannot help but wonder if Sam, one of the great opportunists of all times, did not see this model on one of his early trips to England. A ratchet is cut into the head of the cylinder and is set with six teeth. When the hammer is cocked, a metal hand attached to it pushes up against the ratchet to line a chamber up with the barrel. The familiar spring catch locks in a positioned notch in the cylinder to lock the action at this point.

Whatever the origin of his design, whether individually conceived or subconsciously borrowed, Samuel Colt is unquestionably the father of the successful revolver.

His original British patent covers the revolving of the cylinder by a hand attached to the hammer as the hammer is brought to full cock. It also specifies in great detail center-fire ignition produced by percussion caps firing through horizontal nipples separated by partitions to prevent accidental firing of adjoining chambers. It also deals in detail with the lockwork.

Colt's first model was made by an East Hartford gunsmith named Anson Chase. Sam lost the hard-earned dollars paid to Chase when the model blew up as fire flashed from one chamber to another. The failure was due to lack of the protective partitions between the chambers which he later specified in his patent. The famous "Anson Chase Model," Colt's number 2 design, is still to be seen at the Colt's Museum at Hartford. In weight, measurements, and general appearance it is a close relative of the Single Action Army Model of 1873 which was in manufacture until recent years, the famed "Peacemaker" of Wild West fame.

Colt's original weapons, of course, were developed in the percussion era. All the chambers were loaded with powder and ball (or with combustible cartridges which contained the powder and the ball) from the front end. Percussion caps were then positioned over the nipples on each of the chambers. These nipples had holes drilled in them to permit flash of the percussion fire into the powder charge inside the individual chambers. Partitions between the nipples confined the percussion flash and prevented accidental firing of adjoining chambers.

Colt's original patents were so well drawn that they prevented the actual manufacture here and abroad of any but freak revolvers

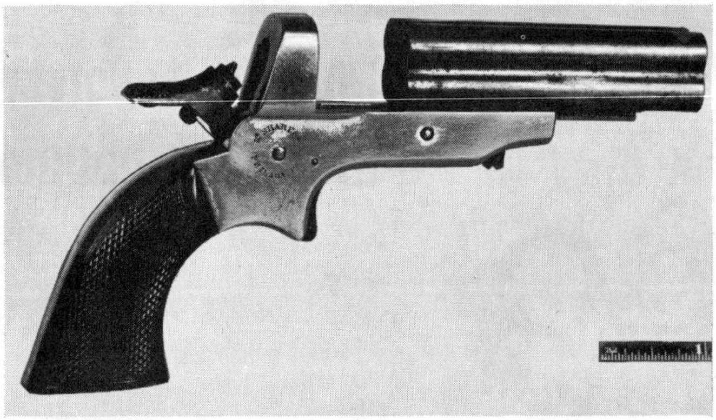

Early American Sharps 4-barreled pistol for metallic cartridges. Note that this and variant types which do not have revolving cylinders are not technically revolvers.

until 1850. He promptly squelched with law suits anyone attempting to infringe upon his idea.

Early Colts Made at Paterson

Colt's first revolvers were manufactured at Paterson, New Jersey. Today these are among the world's prime gun collection pieces. Very few are in existence. That original manufacturing company went bankrupt, but six years later Colt was back in production at Whitneyville, Connecticut. With the cooperation of one of the great assembly line production geniuses of all time, Eli Whitney, Colt proceeded to act upon suggestions from Captain Walker of the Texas Rangers to produce the heavy revolvers used at the outbreak of the Mexican War. From this business Colt managed to raise enough money to open his own factory at Hartford, still the best known revolver manufacturing plant in the world.

By 1850 Colt was exporting revolvers to England and elsewhere in Europe. In 1853 he opened a plant at London, using men he had brought from the United States as foremen to train the manufacturing groups there.

Colt Competitors

The English Deane-Adams revolver, patented in 1851, was the only arm which at that time offered any serious competition to Colt. This English revolver was the first true solid frame design. Since the barrel and frame were a single forging, this arm was a much stronger design than the Colt of the same period. Also, the English revolver was a double-action type in which a pull on the trigger turned the cylinder and cocked and dropped the hammer, a design favored by the British even in recent years. However, neither this nor other British revolvers such as the Lang & Witton and the Daw were able to compete commercially with the mass production Colt had initiated. His system of manufacture played quite as large a role in the introduction of his revolver as did the design itself.

The English Beaumont of 1855 furnished a true advance in lock work, permitting the revolver to be thumbcocked and fired by trigger pressure as in the Colt, or fired by direct double-action pull on the trigger as in the case of the Deane-Adams. The Beaumont action, in fact, was soon incorporated into the Adams revolver, the weapon then being officially adopted by the British as their service revolver.

The Colt cocking system was excellent for accuracy since it allowed a relatively light pull on the trigger to drop the hammer for firing without affecting the line of sight. However, it was relatively slow in operation in the hands of the average shooter. The Deane-Adams, on the other hand, with its double-action system allowed far greater rapidity of fire than the Colt, but gave relatively poor accuracy because of the long pull on the trigger and the additional force necessary to compress the main-spring and revolve the cylinder.

The value of the Beaumont double-action system becomes immediately apparent when we see that it lends itself either to thumb cocking for accurate shooting when time permits, or for straight pull through on the trigger when several shots must be fired in rapid succession or at close quarters.

None of the later percussion weapons, whether the Tranter, Webley, Kerr, Westley-Richards and similar English designs, or the Belgian Comblain or Amangeot, were of sufficient merit to affect Colt's markets.

When the Colt's patents expired, however, the United States was flooded with a tremendous variety of domestic percussion pattern revolvers, most of which were merely variations of the Colt. These included the Remington, which was a relatively standard design single-action percussion cap revolver. In the double-trigger Savage, the first trigger cocked the hammer and revolved the cylinder and the second trigger tripped the hammer to fire. The Pettingill had an enclosed hammer (so-called "hammerless") and quick removable cylinder design. The Savage double-action with a ring trigger and chamfered cylinder was reminiscent of the earlier Collier flintlock and the present day Russian Nagant gas seal revolver, the cylinder being moved forward as the weapon was cocked so that the individual chamber was actually thrust forward over the mouth of the breech of the barrel in an attempt to prevent gas escape at the juncture. (Many of these revolvers were later altered to fire metallic cartridges.)

All of these early patterns were familiar Civil War weapons, as were also a host of lesser types.

We tend to think of the military revolver as being a large caliber weapon, which in normal American usage it is. Even the early day Colts were of the .36 or .44 caliber pattern when used for regular issue Navy or Army service in the United States, though in Europe the caliber was often considerably smaller.

The Best-Known Colts

The most important revolver of our Civil War was the Colt 1860 Model. Over 200,000 of these .44 caliber 6-shooter percussion revolvers were made. Yet another Colt is probably the most famous revolver in history. This is the solid frame Single-Action Army revolver which was issued in 1873 for a new .45 Army caliber metallic cartridge, and which was a direct evolution from the percussion 1860 Model.

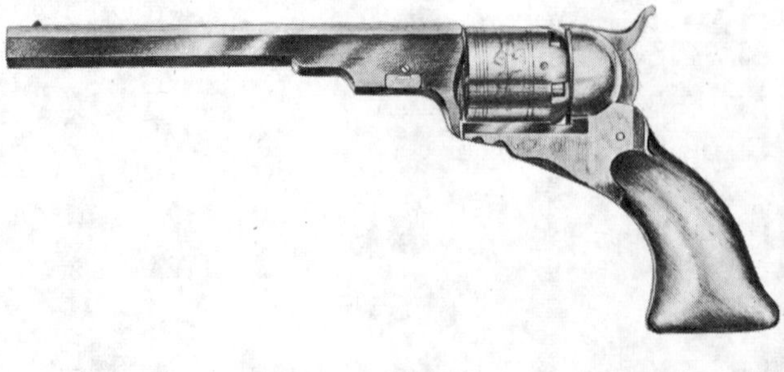

The Colt Texas Paterson. Caliber .40. Muzzle-loaded with powder and ball. Percussion-cap fired. Model 1836. Made at Paterson, New Jersey.

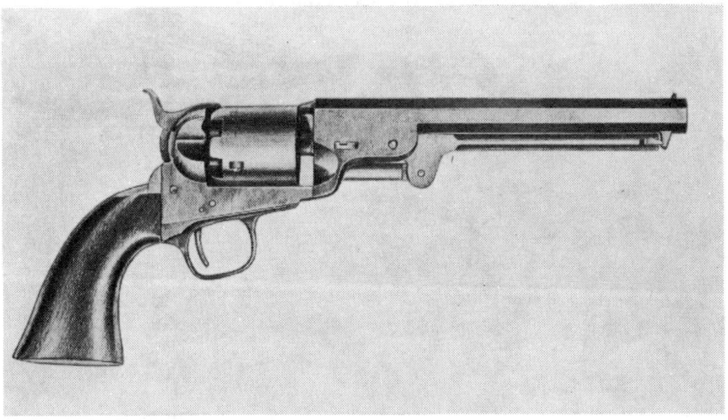

The Colt Navy Model 1851. Caliber .36.

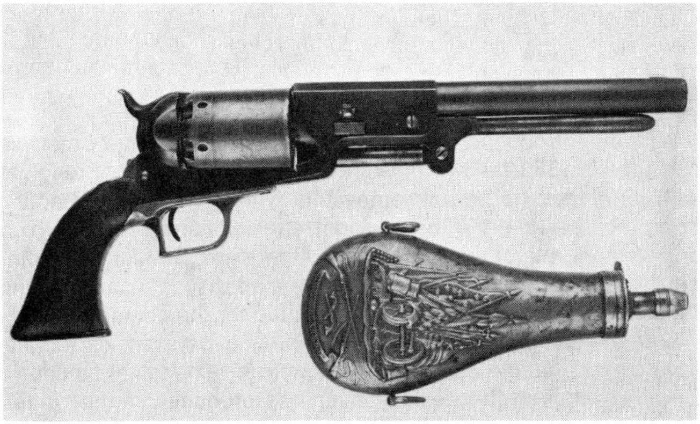

U.S. Percussion Revolver. Whitneyville-Walker Colt. Model of 1847. "C" Company No. 154. Caliber .44, six-shot, single-action.

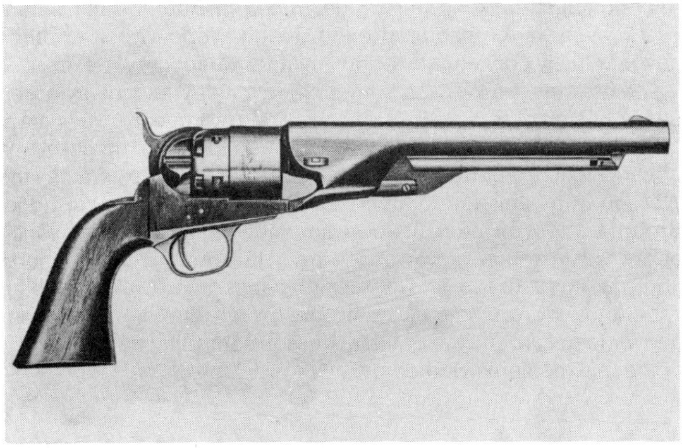

The Colt Army model of 1860. Caliber .44. Still muzzle-loaded and cap-fired, this was nevertheless the major revolver in our Civil War.

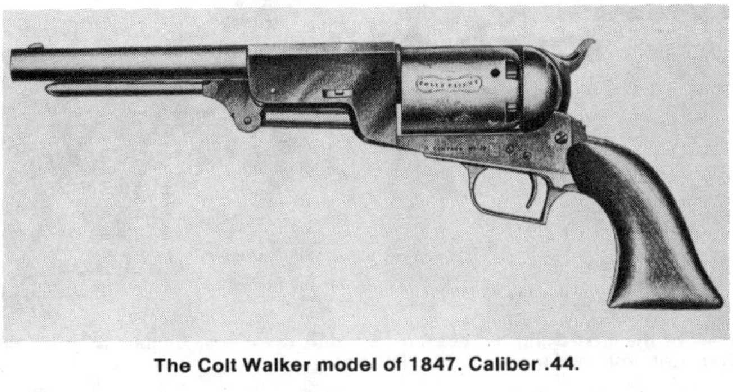

The Colt Walker model of 1847. Caliber .44.

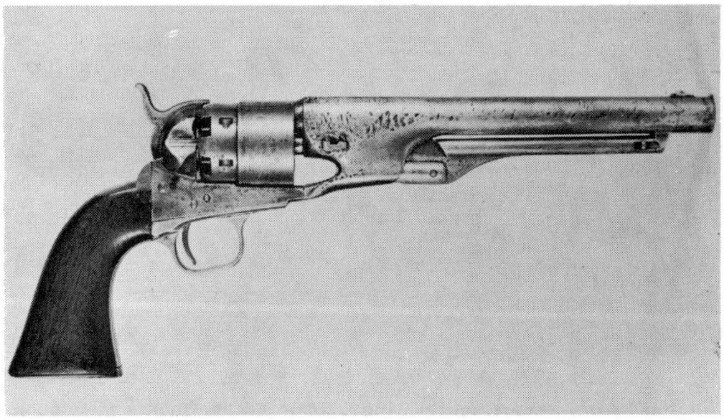

Colt Army 1860. Caliber .44, six-shot, single-action. The principal revolver of the Civil War. The first of the "Stream-lined" model Colt revolvers; 107,156 were furnished to the War Department between Jan. 4, 1861 and Nov. 10, 1863. It was known as the "New Model Army Revolver" at time of issue.

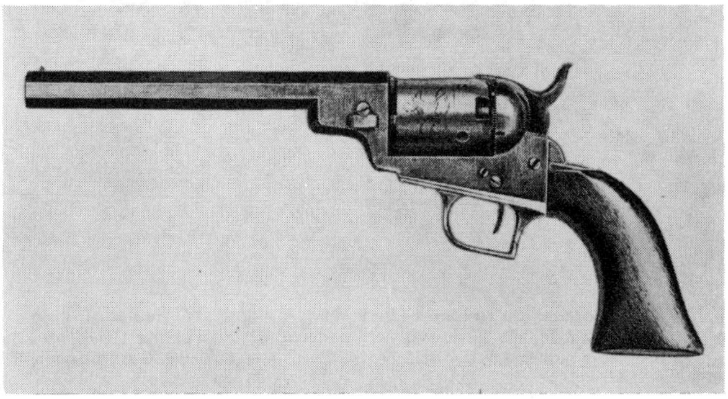

The Colt Wells-Fargo Model 1848. Caliber .31.

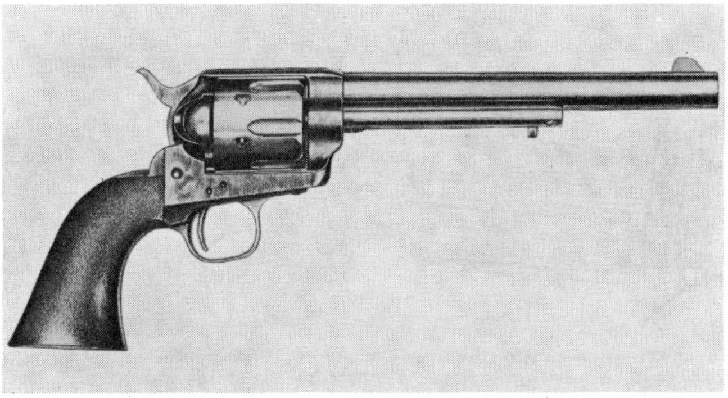

The Colt Peacemaker model of 1873. Caliber .45. The most famous revolver in history and legend, this is the revolver that "made all men equal" in the pioneer days of the West. Metallic cartridge pattern.

Remington Army, New Model (Civil War). Percussion revolver, caliber .44, six-shot single-action. 125,314 Remington revolvers were purchased by the U.S. Government during the Civil War—the most advanced design at the time, with the top strap—second only to Colt in number and popularity during the Civil War. The most accurate of the cap and ball revolvers.

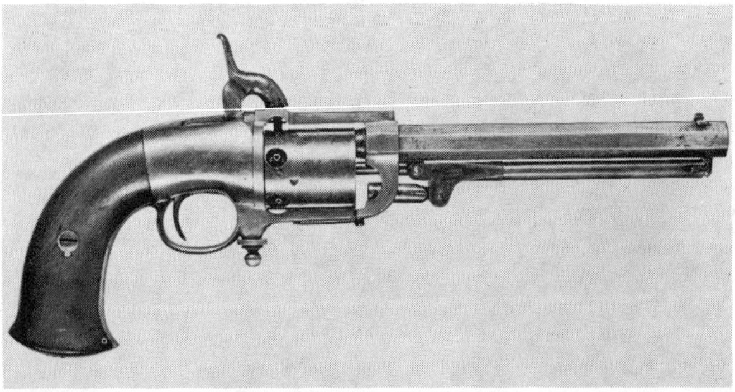

Butterfield Army percussion revolver. Caliber .44 five-shot, single-action. There is no record of Government purchase during the Civil War. This weapon used a mechanical feed for a disc primer fed from a magazine.

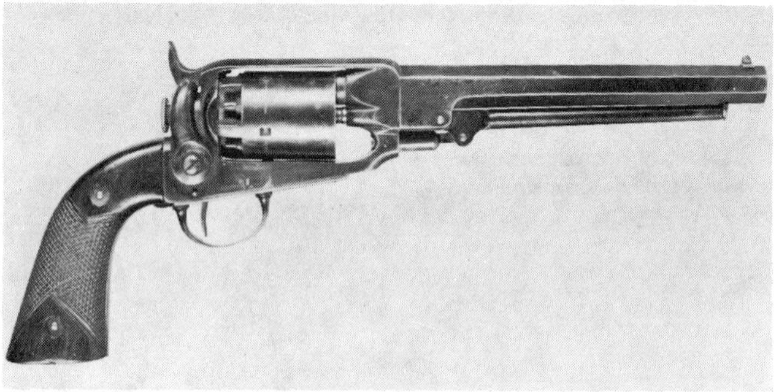

Joslyn Army percussion revolver (Civil War). Caliber .44, five-shot, single-action. A total of 1,100 Joslyn revolvers were purchased by the Government during the Civil War for Army and Navy. Of these, 875 were purchased in 1862.

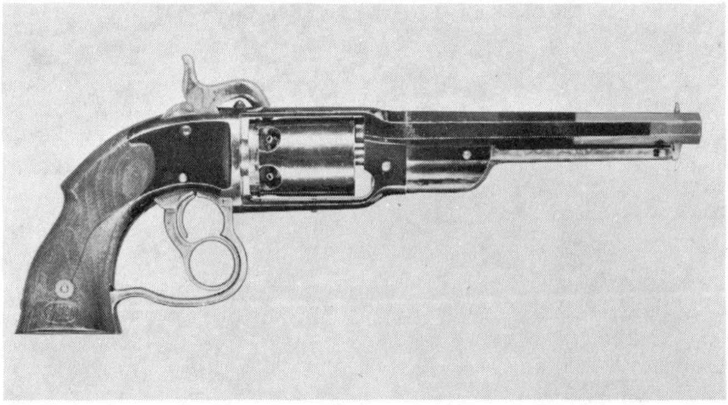

Savage Navy percussion revolver. Caliber .36, single-action. 11,284 Savage Navy revolvers were purchased by the Government during the Civil War. This is a transition from the single-action to the double-action revolver. The only U.S. revolver with the mouth of the chamber covering the breech of the barrel at the time of firing, giving a nearly gas-tight seal. This system of gas sealing was used in Russian revolvers. It is not efficient. Any American Colt or Smith & Wesson revolver gives better sealing because of close tolerance manufacture.

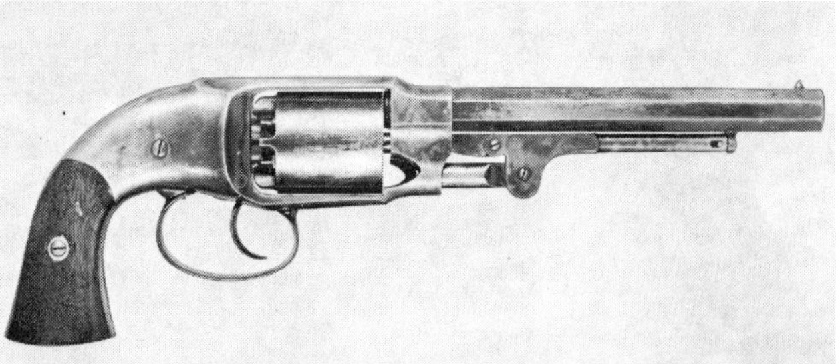

Pettingill Army percussion revolver (Civil War). The first and only hammer-less U.S. martial revolver. Caliber .44, six-shot, double-action. 2,001 were delivered to the Government on contract during the Civil War. It did not prove serviceable, since the mechanism was complicated and delicate, and could not stand rough usage under field conditions.

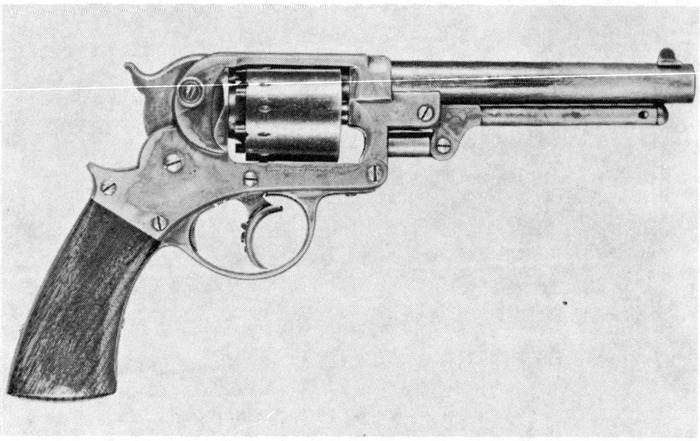

Starr Army double-action. Civil War. Caliber .44, six-shot. 47,952 Starr re-volvers were purchased by the Government during the Civil War.

SMITH & WESSON

With the introduction by Smith and Wesson of their .22 metallic cartridge in 1859, their astonishing little 7-shot .22 caliber revolver with its hinged-up barrel, removable cylinder, and single-action firing lock became the most sought-after sidearm of both Union and Confederate officers during the American Civil War. The combination of compactness, light weight, rapidity of loading, and reliability of the new cartridges accounted for the demand.

With the coming of the Lefaucheux pinfire system of cartridges, revolvers to handle such ammunition were extensively made in Europe, although the design never was produced in the United States except experimentally. This design, of course, requires that each chamber be so notched that the cartridge dropped into the chamber will have a resting place for its projecting pin in the cylinder face, where the hammer can strike it a downward blow. Although such designs were widely used in Europe until recent years, many being captured even during World War II, pinfires never actually constituted serious military weapons.

The first Smith & Wesson rimfire .22 revolver was actually based on the design of Rollin White of Lowell, Massachusetts. His patent of April 3, 1855 for a cylinder bored through to permit the insertion of cartridges from the breech end was purchased by Smith and Wesson who applied it to their new .22 caliber metallic cartridge. Smith & Wesson exercised a monopoly on the system, which permitted insertion of cartridges from the rear of the cylinder, until patents expired in 1869. During this time, Colt attempted various hybrid systems including the Thuer alteration in an endeavor to get around the Smith & Wesson controlled patents. None of these expedients worked out.

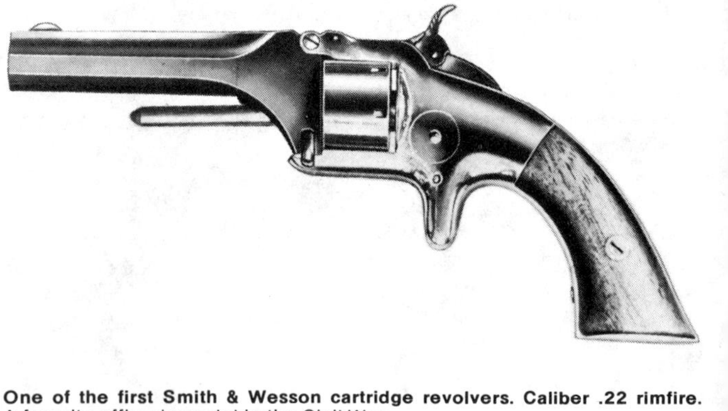

One of the first Smith & Wesson cartridge revolvers. Caliber .22 rimfire. A favorite officer's model in the Civil War.

REVOLVER DEVELOPMENTS

When the Smith & Wesson patents expired inventors and manufacturers throughout the world proceeded to turn out a tremendous assortment of breech-loading revolvers in calibers from .22 to .50. Many of the revolver developments which today are considered quite modern first appeared at this time. These include features such as the recessed-head cylinder and counter-bored chambers which allow the entire case heads to rest below the outer circumference of the cylinder to protect the shooter in case a weak cartridge head ruptures. The early Italian Glisenti revolvers, several Austrian patterns, and in America the Colt House pistol and the Remington Elliot all utilized such devices. The rimfire was made in a wide variety of calibers, only the .22, .30, .32, .38 and .41 really achieving extensive distribution, however. All used copper cases in standard practice.

SUMMARY OF CHANGES

Center-Fire

The center-fire development followed so hard on the heels of the rimfire, and with their brass cases were so superior for higher-powered cartridges, that the major revolver manufacturers promptly concentrated on weapons to fire them. Only the .22, which is difficult to prime centrally because of its small case head diameter, has survived.

Colt was able to take immediate advantage of the center-fire cartridge because his original basic revolver design had been developed with an idea of mass production. With relatively few alterations the new metallic cartridge design was quickly put into major production, while competitors had to tool up extensively.

In place of the front rammer used to seat the bullets in the chambers from the muzzle end, the altered Colt heavy frame design introduced the sliding rod ejector below the barrel. A hinged loading gate was provided on the right side of the revolver which when swung open gave access to individual chambers. Bringing the hammer to half-cock freed the cylinder from its lock and permitted it to be turned by hand to bring each individual chamber in line with the loading gate for successive loading with center-fire cartridges. For ejecting, with the gate opened, pressure on the ejector rod rammed the rod up in the chamber in line with the loading gate to push out the empty case. Releasing the rod permitted its spring to return into place, whereupon the cylinder could be turned to line up the next chamber for unloading. Slight variations of this ejecting system were immediately applied to revolver designs throughout Europe. However, this efficient but slow method posed an immediate military problem.

Dodge Fast-Loading System

The next advance was that produced by a W. C. Dodge, an American. He projected the idea of a hinged-frame revolver in while the act of bending the barrel down after releasing the locking catch would allow the extractor to be automatically raised out of the center of the cylinder in the barrel assembly to extract and eject all empty cartridge cases simultaneously. At the end of the stroke, an escapement permitted the extractor spring to return the extractor to its seat in the head of the cylinder. The arm was now ready for rapid reloading. This is the fastest revolver loading system ever devised, but hinging the frame results in structural weakness, unless construction is very heavy.

The European Galand, in which the barrel and cylinder were slid forward while the empty cases were retained by an extractor until clear of the chambers, is often cited in European records as the first instance of automatic extraction. Dodge effectively

A Belgian-made hinged-frame revolver, now obsolete.

repudiated this claim in his day. The awkward Galand system has not survived the test of time, incidentally.

Dodge had applied for British, French, and Belgian patents. The boat on which his check was carried in the mails covering his patent fee for the British patent arrived a few days late. On this technicality his British patent was voided. Similar action was taken shortly thereafter in France and Belgium. Dodge always insisted, and the weight of the evidence bears him out, that agents of the British patent office released information on his application before the patent was granted. In substantiation of his claim, it is possible to point out that a very large number of Birmingham manufacturers released revolvers utilizing the Dodge patent within six weeks of the time the British patent was invalidated. The Liege Small Arms Company also produced a revolver utilizing the Dodge principle in a matter of weeks after the invalidation of the Belgian patent. Some revolvers, apparently manufactured by the English firm of Webley although they did not carry a manufacturer's name, were released at this time; these were an exact copy of an original Smith & Wesson .22 which incorporated the Dodge principle, S&W having purchased the Dodge extracting patent.

The Pryse Rebounding Hammer

English revolvers made by Adams and Tranter as well as by Webley were customarily solid frame types provided with hand-operated ejectors for removing cartridge cases individually along the general system of the single-action Colt. Developments on the Continent at this period were also confined to the solid frame construction whether the design was Belgian, German, Austrian, or Italian; only the freak Galand system was different. The Pryse English patent of 1876 covering a revolver of the general Smith & Wesson design was the next true advance. It utilized the now common cam-functioned ejector. Its breech-locking system was operated by dual spring catches, one on either side of the standing breech. Truly monstrous revolvers in this general design were manufactured by Webley in England in calibers as high as .476, while the Belgian manufacturers produced arms of the type of the Montenegrin .45. Occasional freaks were made in calibers approaching .600!

These were the first European arms to incorporate the principle of the rebounding hammer. This is the safety system by which spring tension, applied through various systems direct and indirect, automatically withdraws the hammer from possible contact with the cartridge case as the trigger is released. In some of these systems the hammer rebounds automatically before release of the trigger pressure by transferrence of mainspring thrust from a firing to a withdrawing position on the hammer.

This Pryse system also incorporated the safety device (even then familiar in Europe though not in the U. S.) of a metal block or bar rising automatically between hammer and frame to prevent the hammer from striking the cartridge or firing pin at any time except when a deliberate pull on the trigger caused the hammer action.

"Safety" Revolvers

The most famous publicized example of this safety wedge system is that in our own current so-called "Colt Positive Lock." Varieties of this principle actually appeared in Austria and in Italy as early as 1871. Because of inadequate firearms research by patent attorneys, the Colt's Company in the United States was able to exercise an American monopoly on this system of safety during the life of a group of patents.

Early in the twentieth century American firms such as Hopkins and Allen produced safety revolvers in which the hammer operated on an eccentric. In this system, the hammer could strike the firing pin only when the pull on the trigger dropped the hammer and during the course of the period in which the trigger was held back. Release of the trigger automatically worked the hammer on its eccentric to bring it to a position where it could not accidentally hit the firing pin.

The famous Iver Johnson American system of so-called "hammer safety" was another expedient to get around this Colt positive lock system. In the Iver Johnson system, cocking the hammer raises an independent steel bar between the hammer and the firing pin face. If the trigger is deliberately pulled, the hammer blow can be transmitted through this bar to the firing pin. At all other times the hammer is resting on the face of the standing breech above the line of the firing pin and no external violence to the head of the hammer can be transmitted to the firing pin.

Smith & Wesson achieved the same general idea through use of a recoil block, a spring-actuated wedge system. In recent years, however, all Smith & Wesson revolvers have utilized a form of the locking principle found in the Colt Positive Lock. While the operation is somewhat different, the effect is the same. A steel bar is mechanically interposed between hammer and cartridge access at all times except when the trigger is deliberately pulled.

In an endeavor to evade the Dodge patents, numerous manufacturers such as Thomas in England, Galand in both Belgium and France, and Merwin and Hulbert in the United States all developed freak systems in which the barrel assembly could be unlocked and swung away from the standing breech for cartridge case removal and reloading. These were all too complicated to stand in the face of the superior, simply hinged-frame system of Dodge and the later swing-out cylinder systems.

Passing mention must be made of the fantastic method in which the Government Enfield Small Arms Factory in England attempted to develop a unique service revolver which would bypass any of the patents then existing. This was the 1882 model Enfield in caliber .476. Releasing the lock on the standing breech first permitted the barrel to be tipped down and then the cylinder to slide straight ahead to eject! This impractical design illustrates again the futility of permitting nationalistic pride to determine firearms design for government use. It was rapidly supplanted by the hinged frame Webley of 1887.

Frame Design

In 1887 the British government adopted as the Mark I Webley, a slightly modified version of the original Webley hinged frame. Unlike the American systems in which the breech lock is fastened to the extension of the barrel, Webley introduced the stirrup lock which is pivoted on the standing breech and which locks over the barrel extension. This is the strongest hinged-frame system ever devised. Through the years it has remained the official British service pattern whether used in the original .455 caliber or in the World War II .38 caliber. It has, of course, been supplemented in times of emergency by Colt and Smith & Wesson revolvers purchased in America. It will be replaced in British official service by the Belgian Browning-type 9mm automatic pistol in due course, partly because of ammunition standardization since this automatic pistol cartridge is the one used in the official British submachine guns such as the Sten.

The revolvers of the 1871-1880 period in France, Italy, and the Balkans are still in use in many instances. Such revolvers will be found fully covered in Part II. Mauser designed several solid frame and hinge-up pattern revolvers in Germany in 1878 and 1879.

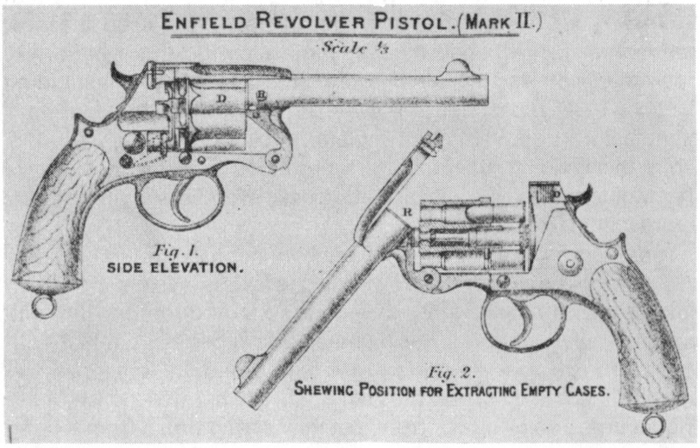

This Enfield saw some service in the early days of the Royal Canadian Northwest Mounted Police. It was replaced by Colts. Today the organization has become the Royal Canadian Mounted Police armed with Smith & Wesson revolvers.

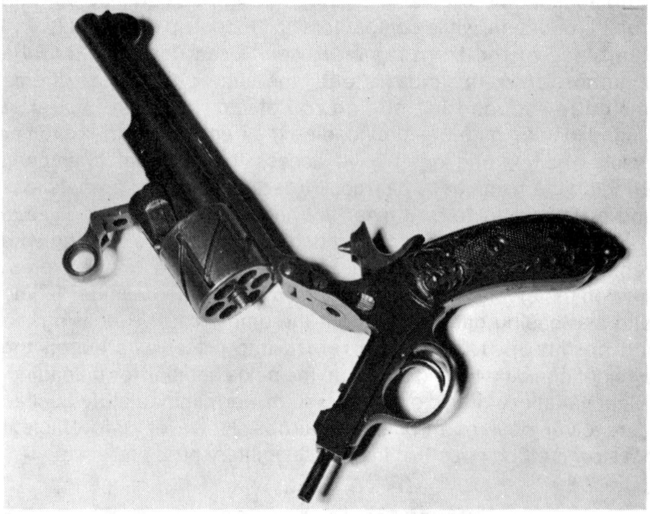

Early Mauser revolver open (above) and closed (below). The cylinder-turning mechanism appeared later on the Webley-Fosbery Automatic Revolver series.

They were noteworthy only for their novel system of cylinder rotation through zig-zag cylinder slot operation. German interest in repeating and auto pistol design nullified most German attempts at revolver design.

Passing notice must be made, however, of one development of the period about 1885. This was a solid-frame revolver of the Nagant pattern originally manufactured in Belgium by Pieper.

This development was a cylinder design which, as the arm was cocked, not only permitted turning the cylinder to line up a chamber with the barrel for firing, but which also thrust the cylinder forward so that the chamber to be fired was slid forward slightly over the mouth of the breech of the barrel. As in the old flintlock Collier and even earlier arms, the idea here, of course, was to attempt to minimize the escape of gas at the juncture of the chamber and the barrel. In actual practice no such system works effectively. However, Pieper incorporated this device into solid frame rod ejecting revolvers in both single-action and double-action designs which were adopted by the Russians as standard equipment in 1895. Revolvers of this type using the 7.62mm Russian cartridge were widely reported by the Germans as used in World War II.

In this same general period, many other solid frame revolvers appeared both here and abroad using various forms of the double-action firing systems. This elementary solid frame pattern continues in manufacture commercially. Its use is confined almost entirely to low-priced protective and plinking arms. It was never utilized in the United States as a military weapon. On the other hand, weapons of this elementary sort were encountered, even during World War II, in use by many of the European powers, notably France and Italy.

The Swing-Out Cylinder System

The next system appeared as early as 1892 in France and the United States. This utilizes a solid frame into which the barrel is screwed. The frame is machined to receive a crane mechanism which carries the cylinder. Releasing a catch, usually on the standing breech (though some European variants have the release at the forward end of the extractor rod) permits the cylinder and crane to be swung out to the side for loading. Pressure on the extractor rod which passes through the center of the cylinder actuates the star-shaped extractor head to withdraw all cases or cartridges from the cylinder simultaneously. The extractor spring returns the extractor to its rest position when rod pressure is released.

This form has a far greater strength than the hinged frame system as a general rule. As a result, for large caliber use it has completely replaced the hinge frame in the United States. While the normal system of swing-out cylinder operation is for the crane to operate from the left side, many European versions such as the French swing out to the right.

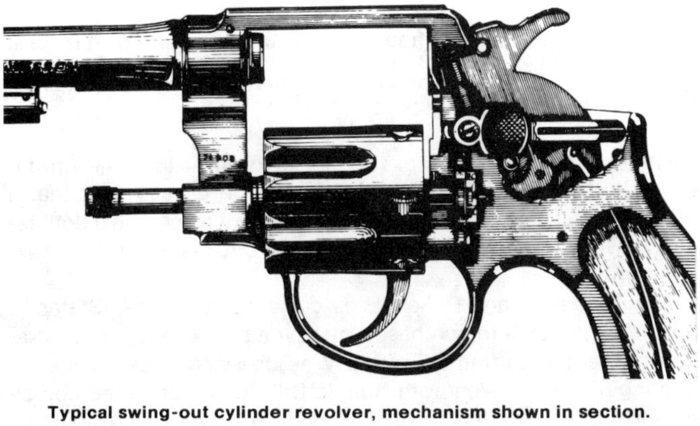

Typical swing-out cylinder revolver, mechanism shown in section.

In all production forms, this revolver incorporates a double action firing system, single action systems of this type being always inventor or freak models. Colt and S&W make only solid frame revolvers of this type. Hi-Standard is currently preparing to manufacture one.

The Dardick Pistol

Although the Dardick pistol is not, in the conventional sense of the term, a revolver it has more similarity to a revolver than it does to an automatic pistol. The Dardick does not have any reciprocating parts such as are used on all automatic pistols and is based on the open chamber system. A three-legged piece somewhat similar to a star wheel serves as the cylinder. This cylinder rotates clockwise and serves three purposes:

1. Indexing of cartridges, as they are fed upwards by the spring-loaded magazine.
2. As two walls of the triangular sided chamber, the receiver serves as the third and top wall.
3. To eject the fired cases from the right of the receiver.

The cartridge used in the Dardick pistol is called a "tround." It has a triangular-shaped outer case of plastic. The bullet, which does not protrude beyond the end of the case, and the propellent are contained in the aluminum sleeve. The primer is seated in a pocket at the end of the plastic case. Plastic cases which act as adaptors for standard .38 caliber revolver cartridges or 9mm Parabellum cartridges were also made for this weapon.

Dardick pistol, Model 1500, caliber .38.

Dardick Model 1500 pistol shown with open loading gate. This model holds 15 cartridges.

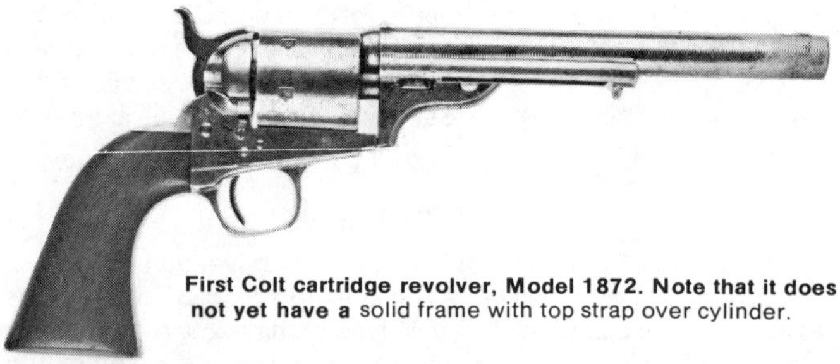

First Colt cartridge revolver, Model 1872. Note that it does not yet have a solid frame with top strap over cylinder.

The Dardick pistol can be loaded either with single rounds or by charger through a loading gate. The 20-shot models have two loading gates. The pistol was made in three basic models: a 20-shot model, 15-shot model, and an 11-shot model. Barrels from 3 to 6 inches in length were available and weight ran from 25 ounces to 39 ounces. The Dardick pistol never caught on and very few were ever sold; so few as a matter of fact, that it is now a collector's item.

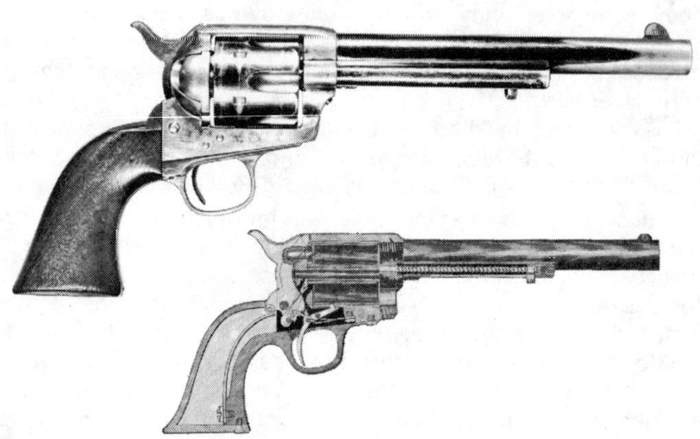

Colt Army revolver, model of 1873. Caliber .45, six-shot, single-action. Manufactured for nearly 70 years with little change. The first solid frame U. S. military cartridge revolver. Note top strap. Insert shows the 1873 sectionalized.

THE REVOLVER IN MILITARY SERVICE

As we have seen, the revolver received its first great impetus as a military weapon during the American Civil War, though attention had been focussed on its possibilities in the earlier Mexican-Texas encounters. While basically this was was fought with muzzle-loading cap-fired revolvers, the earliest Smith & Wesson .22 and .32 caliber rimfire cartridge revolvers were also used, particularly by officers who were fortunate enough to be able to purchase them. They were in extremely short supply at that time.

The S&W .44 American was actually the first breech-loading revolver officially adopted for U.S. Army service. Its service was short and limited. This was a single-action hinged frame design.

Colt's first successful breech-loading revolver was issued after the expiration of the Smith and Wesson patent coverage. This Single-Action Army 45, popularly known as the Colt "Peacemaker," was adopted for Army use in 1873.

Coincident with the adoption of the solid frame Colt 45, a new model Smith & Wesson .45 Schofield was also introduced in 1873. A special barrel catch was furnished on this revolver. Military theory at the time required this on the grounds that the more common latch form permitted an opponent in hand-to-hand fighting to free the formal type of catch and permit him to open the weapon, thereby rendering the revolver useless. While in theory such things can happen, in actual practice such a contingency is almost unthinkable. While wasting time on this sort of technicality, ordnance authorities overlooked the matter of cartridge standardization. The .45 ammunition used in the Colt and the S&W were not interchangeable.

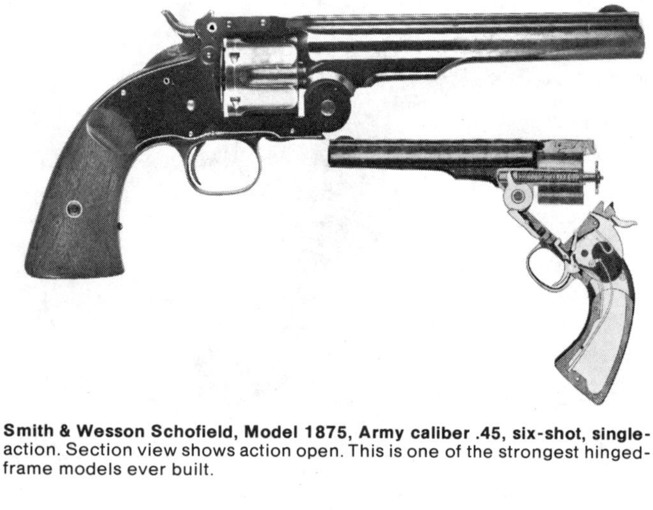

Smith & Wesson Schofield, Model 1875, Army caliber .45, six-shot, single-action. Section view shows action open. This is one of the strongest hinged-frame models ever built.

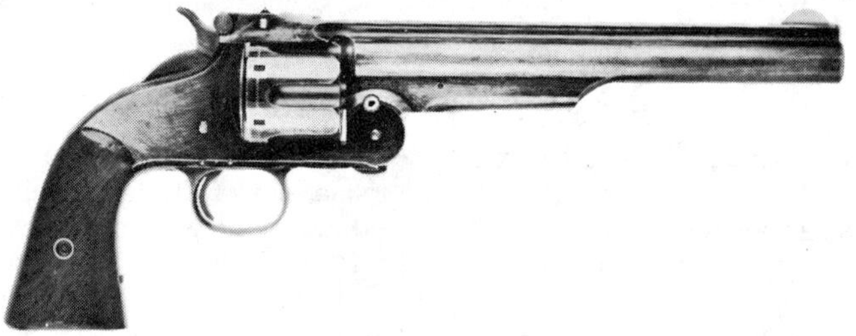

Smith & Wesson Army six-shot, single-action, center-fire revolver. Model 1869. Caliber .44. The first breech-loading revolver adopted by the U. S.

Colt Army revolver, model of 1892, six-shot, double-action. The first solid frame, swing-out cylinder, .38 caliber U. S. service revolver. The modified 1894 version was used in the Spanish-American War, and later in the Philippine Insurrection.

The double-action system was not favored at the time on the grounds that the basic use would be by the cavalry, and that a mounted man during the course of firing might discharge a double-action weapon accidentally if his horse bucked or shied unexpectedly.

The Russian Grand Duke Alexis came to the United States at this time intending to purchase Colt revolvers. After a demonstration of the hinged frame Smith & Wesson system, he went on a hunting trip to the West with Buffalo Bill Cody, during the course of which both used the Smith & Wesson revolver extensively. As a result of the Grand Duke's report, the Russian Government placed an order for 150,000 of the Smith & Wesson .44 Russian Models, hinged frame and single-action. Even in this day, no more accurate heavy revolver or cartridge has yet been designed!

These revolvers were so sturdy that some of them were encountered during World War II by the Germans. When our own troops occupied Japan, quantities of them were also found there, representing revolvers which the Japanese had taken from captured Russian officers at the turn of the century.

Ironically enough, this tremendous order (for those days) occupied the S&W plant to such an extent for export work that Colt was left master of the field of domestic supply of revolvers for the development of our own West. The name and fame of the Colt spread largely because of this monopoly.

When Colt's applied the double-action cocking system to the revolver they called the Double-Action Army Model, a small quantity was purchased for Army experimental use. It was never, however, officially adopted. This pattern is often known as the "Bird's Head Model," the shape of the grip when turned upside down having some resemblance to a bird's head. Varieties of this model with large trigger guards (so a gloved finger could be inserted) were used in the early days of the Klondike, and are popularly known as "Alaskan Models."

The next official United States Army adoption was the 1892 version of the Colt with swing-out cylinder in .38 caliber double action. This used the new .38 Long Colt cartridge, a load not so powerful and far less accurate than the later .38 special.

This .38 Colt saw service in the Philippines campaign. Field tests indicated it did not have sufficient man-stopping power for use against savage groups. On the other hand, the .45 single-action proved an effective man stopper. The British in their African and Indian campaigns had already established this principle to their satisfaction and were using a .455 caliber revolver, the Webley.

In 1909 our Army officially adopted a heavier version of the swing-out cylinder Colt, the New Service .45 caliber. This revolver was also made commercially and for export in various heavy calibers. It was used in both Canada and Great Britain in caliber .455.

Smith & Wesson introduced their version of the swing-out cylinder revolver to compete with Colt's. Both types have seen war service since the time of their adoption in wars throughout the world.

REVOLVERS IN WORLD WAR I

While the .45 Colt Automatic Pistol was the primary sidearm of United States Forces during this war, the shortage of such weapons compelled the adoption of swing-out cylinder double-action designs of both Smith & Wesson and Colt manufacture. These were substitute standard weapons. Under the designation of Model 1917, both patterns were chambered to take the standard .45 automatic pistol cartridge to minimize ammunition supply problems.

Smith & Wesson developed a special half-moon type clip which would hold three of these rimless cartridges. Using these clips the revolvers could be loaded and unloaded very rapidly. The Smith & Wesson chambers permitted firing the cartridges in an emergency without the use of the clip. This was possible because the .45 ACP cartridge is designed to seat in the chamber on the mouth of the case. In such an instance the standard rim-type revolver extractor could not operate. As these cartridges have no rims, it was necessary to punch or pry the empty cases out of the chambers. The Colt, on the other hand, was so chambered that normally the cartridges passed too far into their chambers to be fired at all unless their heads were supported by the half-moon clips.

Colt's and S&W revolvers were also manufactured for use by Canadian and British forces in caliber .455 Eley, the official rimmed cartridge of the British Government during World War I. These required no modification except caliber.

Because of the tremendous number of these .45 A.C.P. caliber revolvers available after the war, the Peters Cartridge Company (now a division of the Remington organization) produced the .45 automatic cartridge with a special rim under the designation of

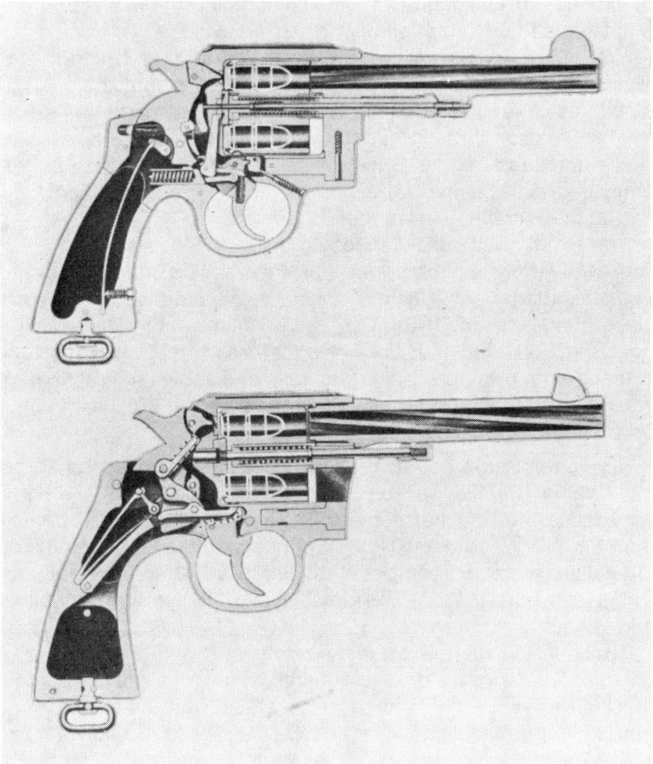

Section views of World War I U. S. revolvers. Upper: Smith & Wesson. Lower: Colt. Caliber .45 (U. S. Gov't).

.45 Auto-Rim. This cartridge of course does not require the use of the half-moon extracting device. Some World War II use was made of this revolver and cartridge combination.

The official revolver of the British forces during World War I, though supplemented by Colts and Smith & Wessons as required, was the Webley hinged-frame in caliber .455.

The other revolvers then in European use were those of France, Italy, Russia, and to a limited extent Austria. All these are covered in detail in Part II.

REVOLVERS IN WORLD WAR II

At the outbreak of World War II, Great Britain again found herself very low on sidearms. While the .455 Webley was still in general use, the official revolver was now a weapon of the same general design made at Enfield in the new .38 British caliber.

This .38, the equivalent of the low-powered American .38 Smith & Wesson cartridge, is not to be confused with the ultra-powerful .38 Smith & Wesson Special types of cartridges used in general by American police.

The original .38 Smith & Wesson cartridge is practically identical with the .38 British service cartridge. The cases are shorter than the American .38 special, while their case diameter is greater than the .38 special. This stems from the fact that when the .38 special powerful cartridge was designed for swing-out cylinder revolvers, it was necessary to develop a form which could not be inserted in the weaker hinged frame .38 models. As a result, the .38 Smith & Wesson cartridge and the British service .38 cartridge are too large in diameter to be inserted in the American .38 special type of revolver. Conversely, the .38 American Special cartridge is too long to permit it to be inserted in the cylinder of the .38 British type of revolver or of S&W or Colt's revolvers chambered for British use during World War II.

Tests led the British to the belief that for close quarters work the .38 was even more efficient in the hands of the average military user than was the larger caliber. The lower recoil, the lighter weight of the weapon, and lighter weight of ammunition were all contributing factors to the adoption of this lesser caliber.

Great Britain turned to Smith & Wesson of Springfield, Massachusetts at the outbreak of the War to manufacture large quantities of revolvers for the British .38 caliber cartridge, since Smith & Wesson in their Military and Police Model had a revolver easily adaptable to the British cartridge. Large numbers of the equivalent Colt's model were also adapted to the British cartridge and sent to England during this period.

As a result, the British service during World War II used a wide variety of revolvers. The primary ones were their Enfield, not only in the standard double-action hinged frame, but also in a special close-quarters model with a special hammer which allowed only double-action firing, plus the Smith & Wesson and Colt's revolvers chambered for their cartridges as indicated above. In addition, of course, considerable stocks of the earlier .455 calibers were pressed into service during this emergency.

During this same period the United States again issued some 1917 Model Smith & Wesson and Colt's .45 caliber ACP revolvers for service, but did not manufacture additional ones. However, both Smith & Wesson and Colt's produced very large quantities of .38 caliber revolvers using the .38 Smith & Wesson special cartridge for special purpose work in our Navy, Marine Corps, Aviation and various guard services.

Russia at this time used some revolvers chambered generally for their 7.62mm cartridge. In general, however, it may be stated that the Russians early came to the conclusion that the submachine gun could be made cheaper and faster than the revolver, and throughout the course of World War II they concentrated on furnishing the former weapons. Revolvers saw minimal service.

French and Italian services at the beginning of the war also used earlier revolvers of solid frame and swing-out cylinder construction as substitute standard arms. These were generally not important, but all are considered in Part II.

Except in the United States, experience with revolvers during the course of World War II convinced military thinkers that this arm should be replaced by automatic pistols utilizing the same caliber ammunition as the submachine gun. This will unquestionably be the foreign military trend of the future.

During World War II the Japanese troops used revolvers only to a very limited extent. Most of these were very crude Japanese-manufactured forms of the early Smith & Wesson hinged-frame type chambered for a freak Japanese rimmed cartridge of approximately 9mm caliber. They fired double-action only. Such other revolvers as were used were generally of American manufacture.

POSTWAR DEVELOPMENTS IN REVOLVERS

In general it may be stated that for all practical military purposes, the revolver design has reached its zenith. The preferred form is the solid frame swing-out cylinder revolver manufactured by Colt's and Smith & Wesson. Mechanically, these revolvers cannot be substantially improved. Changes since the end of the war are basically those which stemmed directly from ammunition improvements and from metallurgical advances; design changes as such are relatively inconsequential.

To be of military value, revolvers must be relatively light in weight and compact in size, and must be effective within the limitations of their sighting equipment. For all practical purposes, a revolver cylinder cannot be made large enough to take more than six cartridges of relatively large caliber. This is an elementary matter of dimension. In this respect, of course, the revolver must always be inferior to the automatic pistol where the cartridges are in a magazine one on top of the other without the necessary metal separation of the revolver chambers. As to weight, research by the aviation industry on aluminum developments has produced qualities of aluminum forgings which are entirely suitable for all practical revolver frame purposes, though in field use it has been determined that high tensile steels are still necessary for ultra-powerful cartridges insofar as cylinder and barrel construction are concerned.

Since the bullet must jump a gap between the chamber and the barrel, there must always be a substantial loss of gas at the juncture. Colt and Smith & Wesson both hold tolerances at this point to an absolute minimum, but any device to give a better seal must inevitably require mechanical complications which would offset its value.

As metallurgical knowledge advances, titanium will unquestionably influence revolver design from the standpoint of producing a lighter weapon with the approximate strength of the standard steel patterns, though such use will probably be confined to

Section view of a revolver. The cartridges are loaded into individual chambers in a revolving cylinder. Design is the modern Smith & Wesson revolver; 95% of American police units use this type side arm. Working mechanism is exposed by removing the sideplate shown below.

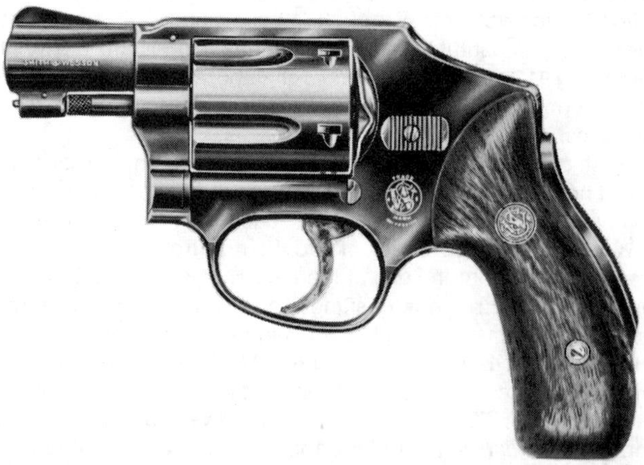

Smith & Wesson Centennial Model Hammerless (enclosed hammer) with grip safety. For the very powerful .38 S&W Special cartridges. A special close-quarters arm.

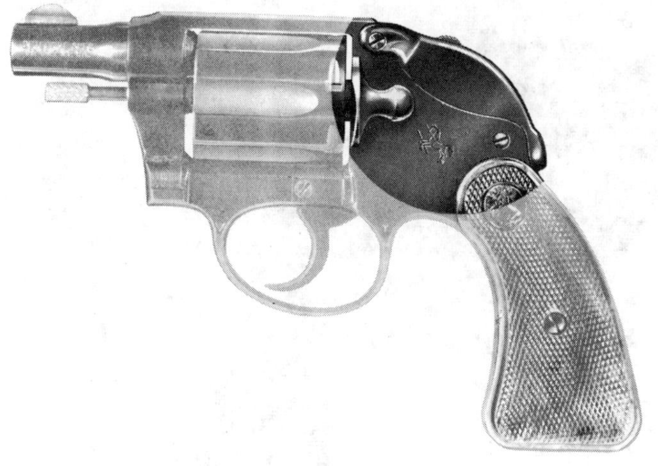

Colt "Hammer Shroud" to cover hammer and permit quick draw from or firing through pocket.

cylinder and barrel production as the lighter aluminum has proven itself sturdy enough for the frame sections. In all practical use, if the cylinder holds during the course of firing, the frame is not subject to any great pressures. Because of this factor, both Smith & Wesson and Colt's, after extensive development of the .38 special revolver using aluminum cylinders, found it advisable to return to the use of high tensile steel in their smaller models. While the forged aluminum cylinders as produced by these firms are adequate for factory loaded ammunition, they will not necessarily hold up under the terrific hand loads introduced by experimenters. In the interests of safety, therefore, the steel cylinders are still recommended.

For practical military use, the power of cartridges also cannot be substantially increased without unreasonably increasing the weight of the revolver. Otherwise the shock to the firer frightens the amateur and interferes with both accuracy and rapidity of operation by the expert. Any major increase in revolver power must depend basically upon ammunition advances, not on any possible improved revolver design.

As to increased accuracy, we have already pointed out that the original .44 Smith & Wesson Russian revolver cartridge of 1873 cannot be substantially improved upon today from the viewpoint of accuracy. For all practical purposes, almost any modern revolver will shoot far more accurately than any human being can hold.

As a military arm the revolver is normally intended for use at ranges under 10 yards. An expert can use it effectively at 50 yards. It seldom has any true value at longer ranges, though of course exceptions do occur.

Smith & Wesson Combat Masterpiece. A favorite U. S. Marine revolver in Korea. Caliber .38 Special.

Smith & Wesson Military and Police Model. Caliber .38 Special. Used by U. S. police and by the Royal Canadian Mounted Police.

Smith & Wesson Chief's Special (steel or airweight). Used in steel by Japanese police units. An Airweight (aluminum frame) Model is used by various U. S. Government services. Handles the powerful .38 Special cartridges.

As to mechanical perfection, the utilization of coil springs and of parts held to absolute minimum tolerances, as now evidenced in the Smith & Wesson and Colt's manufacture, has resulted in the simplest form of mechanism capable of reliable functioning.

Only one slight form of improved revolver for special duty purposes might have military application as distinct from those commonly available, and it is merely a new application of an old principle. This is the form represented by the Smith & Wesson "Centennial Model" revolver. This is their standard solid frame swing-out cylinder revolver for the .38 special cartridge, but the frame is designed to completely enclose the hammer. Colt has produced a "hammer shroud" to enclose the external hammers of some of their small frame revolvers. The intent here is the same as in the Smith & Wesson design, which is basically to furnish an arm which can be fired only double action for close-quarters work. The advantage of this system over the typical exposed hammer revolver might lie in an enlarged trigger guard; troops could carry it in the side pockets of reefers or overcoats in extremely cold climates. Such an arm provides a very efficient close-quarters defense weapon. In case of emergency it can be fired through the pocket of the coat until the cylinder is emptied without any danger of mechanical interference with its operation. The standard type of service revolver given the same usage will normally foul because the exposed hammer will snag in the coat lining. An automatic pistol under similar conditions will not function after the first shot because the reciprocating slide and ejected case will inevitably jam the action.

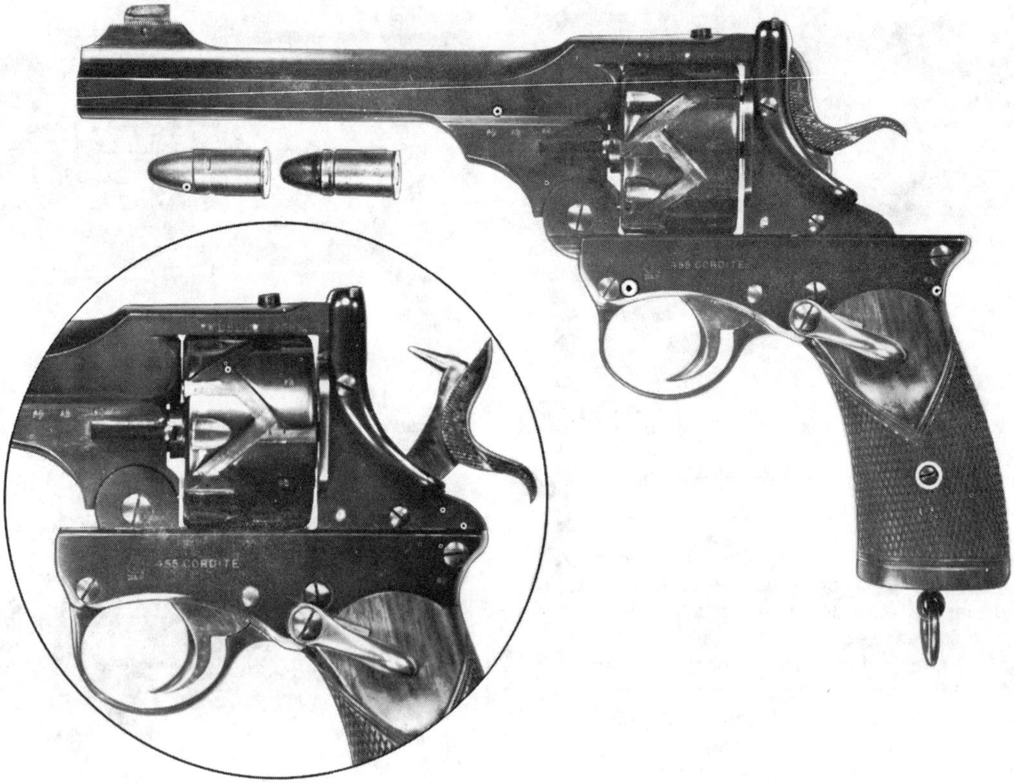

Webley-Fosbery Automatic Revolver. Insert shows barrel and cylinder assembly in full recoil. Cylinder has been revolved halfway to next chamber and hammer cocked. Recoil spring will thrust assembly home and revolve cylinder ready for next shot. Model shown is the .455. This revolver was also made in caliber .38.

Smith & Wesson .32 Hand Ejector. Used by police and military in many South American nations.

The original Smith & Wesson intent in introducing this model (which is equipped with a grip safety to prevent firing except when the weapon is held firmly in the hand), was in the interests of a safe weapon for home use. The general intent was to provide an arm which, because it required a relatively large hand to hold it and simultaneously operate both the safety lever in the back of the frame and the trigger, would preclude children accidentally firing the arm.

Its use for special service purposes in cold areas has not yet been explored. It is worthy of military attention, however.

Based on experience, the United States Navy, Air Force, and to some extent the Marines and Army have found that under some circumstances a revolver utilizing the .38 special cartridge is a more practical close-quarters defense weapon than the bulky .45 automatic pistol. As a result, very large quantities of S. & W. short-barreled lightweight revolvers of this pattern utilizing aluminum frames and cylinders with steel barrels have been furnished to the Government services for special use. Colt's also has developed equivalent arms to handle the .38 special cartridge for similar service use.

In passing it must be noted that since the close of World War II there has been a tremendous resurgence of interest in the use of the revolver as a sidearm in South America and Asia. This is generally in accordance with American police tradition. In Japan, for example, the police and quasi-military bodies are being generally equipped with the short Smith & Wesson .38 caliber special revolver known as the Chief's Special, the name deriving from the interest it excited as an undercover weapon among police chiefs in the United States. Huge quantities of this and other models in both steel and aluminum patterns, and in calibers as small as .32, have been sold to police and military organizations in South and Central America and in some areas in Asia. Calibers above .38 are normally not well received in such areas, due to their excessive recoil and weight. In general, of course, revolvers in use by the military are actually used for police purposes.

The only other areas in which revolver manufacture is proceeding at a rate worthy of attention are Spain and Italy and to a minor degree in Belgium. All these areas are currently manufacturing basic imitations of Smith & Wesson and Colt swing-out cylinder type revolvers for general world export, with particular reference to South American export in competition to the Colt and Smith & Wesson patterns. Such arms from the standpoint of material, manufacture, and tolerances are invariably inferior to their prototypes; the designs range from direct copies to manufacturing modifications.

AUTOMATIC REVOLVERS

The only other type of revolver requiring any consideration is the so-called "automatic revolver" form produced for some time in England as the Webley-Fosbery. It had its counterpart in a small American model manufactured for a time by the Union Arms Company. This was unsuccessful.

This pattern of revolver is basically the Webley hinged frame type, so designed that the barrel and cylinder assembly can recoil within the grip-frame. The revolver must be hand-cocked for the first shot. It is hinged open for loading and extraction as in the standard hinged framed Webley revolver.

As the Webley-Fosbery fires, the barrel and cylinder assembly recoil in the frame. This causes a stud to act through a zig-zag series of cuts in the cylinder to perform the action of turning the cylinder the distance of the next chamber, and to lock the chamber in line with the barrel upon completion of the return action. The return of the assemblies is brought about by a recoil spring mounted in the grip which stores energy during the recoiling operation. The recoil also causes a barrel extension to cock the hammer ready for the next shot. This design was made in England in .38 and in .455 calibers. It was made in the United States in .32 and .38 calibers. While the design is ingenious, it suffers from complication. If the grip is not held quite firmly, the action will not function properly. Designs of this sort have at best only theoretical advantages over the standard double action system. They must be considered as a passing phase, with nothing to recommend them for future experimentation because of their complexity and inherent unreliability.

THE REVOLVER VS. THE AUTOMATIC PISTOL

In England the military trend is away from the revolver and toward the automatic pistol. The British concept is becoming like that of the Continent, strictly that of the practical military approach. There is also, of course, the fact that the utilization of the same ammunition in pistol and submachine gun is a distinct economic and logistical factor.

In this connection it must be pointed out that in Europe generally the revolver has been considered obsolete since the introduction of the successful automatic pistol about the turn of the century. In Germany, for instance, this feeling is so deep-seated that even the police will use revolvers only under direct compulsion.

At the close of the war the United States and Great Britain both equipped German police forces in areas under their control with quantities of Smith & Wesson revolvers. As rapidly as possible, these police organizations found ways to trade these excellent revolvers as part payment on decidedly inferior automatic pistols manufactured in Spain and France! Many of these revolvers were then sold in the world market, some going to the Israel police organizations. Others went to South America. In short, the German military psychology is such that the police felt better with an inferior automatic pistol than they did with a superior revolver.

In the United States, on the other hand, and in areas which have benefited from or been influenced by American police psychology, the revolver is an important defensive arm. In the United States, for example, some 99% of all police organizations are equipped with revolvers to the exclusion of automatic pistols.

In closing, some mention must be made of the superiorities for which the revolver is noted, as well as its inferiorities with reference to automatic pistols. The mechanism is much simpler than that of the automatic pistol. It is easier to train the police officer or soldier in its use. It is inherently safer because when loaded the rims of the cartridges can be clearly seen at all times. If the cylinder is swung out on its crane, a glance will tell if the cartridges are loaded or fired. Firing in single-action fashion from a cocked hammer, the revolver gives a far better pull than can be incorporated in a typical automatic pistol. This, of course, is a great aid to accuracy as it simplifies holding the sights in line during the instant of firing. None of its springs are compressed except when the arm is in the course of being fired or is fully cocked. As a result there is no spring fatigue; this can develop in automatic pistols. No hand-actuated safeties are required, although some are found on occasional European freak designs. As a result, there is nothing to remember or forget when bringing the revolver into action.

Automatic pistols, because of their nature, require in most instances external safeties which must be mechanically thrown off before the arm can be discharged.

Perhaps the most valuable asset of the revolver over the automatic pistol, however, lies in the field of ammunition. If weak or underpowered cartridges are encountered in a revolver, they occasion little or no difficulty since the operation is purely mechanical. If the powder charge is weak or the priming defective, there is nothing to interfere with the firing of the next shot.

With the automatic pistol, on the other hand, when a weak powder load will open the action only part way, the resulting jam can cause a serious tieup. Automatic pistols require a certain minimum amount of blowback action for functioning. And finally there is the matter of the misfire.

A misfire in an automatic pistol is an extremely serious thing in combat. In truly modern arms where the automatic pistol is equipped with a double-action mechanism as in the revolver, pulling the trigger a second time will allow the hammer to drop again on the defective cartridge, which may or may not fire it depending upon the primer and propellant condition. In all automatic pistols, should there be a misfire due to any one of the several failures which may cause it, the slide or bolt mechanism must be withdrawn by hand to eject the defective cartridge from the chamber and to permit the feeding of the next live round for firing. With the revolver, of course, in case of a misfire, another pull of the trigger finger will move the dead cartridge out of line and bring the next one into line for firing. For defensive purposes this is the most serious consideration of all.

European (more specifically German) military psychology accepts the personal danger in the course of operation of an automatic pistol because of defective ammunition and the like. It operates on a basis of the average, not of the individual.

American psychology is more aligned to individual and police thinking that the possible advantage of greater magazine capacity or higher rate of fire is more than offset by the dependability factor. In passing it must be noted that ammunition manufacturing variables, as well as age and storage conditions involved, contribute far more to stoppages and malfunctions in firearms than do the mechanical designs or effectiveness of the weapons themselves.

11 Military Pistols—Historical Outline

A pistol is defined as a short firearm intended to be held normally with one hand. Its chamber is generally formed by reaming and enlarging the breech end of the barrel so it will receive the cartridge case.

The term "pistol" unless otherwise qualified means a single-shot weapon. It is often popularly—but incorrectly—applied to mean a revolver or an autoloading pistol.

DERIVATION OF THE NAME

There are almost as many explanations of the source of the name "pistol" as there are types of pistols. Reference works generally state the name came from the fact that the weapon was "invented" in 1540 at Pistoia, Italy by a gunmaker named Vettelli. This is quite unlikely. In fact the weight of evidence is against it.

Short firearms of the "pistol" type existed a good 200 years before 1540. The Chronicles of Modena, Italy, for the year 1364 A.D., just to cite an Italian instance, list "four little scioppi for the hand" as part of the town's inventory. "Scioppi" evolved from the Latin "sclopetum" which is the authorized Latin word for pistol. Perugia's historical records of about the same period list "500 portable bombards a span's length." The Roman "span" in use at that period was about 7 1/2 inches. Hundreds of similar records in various European centers show that the one-hand small firearm was in use long before Pistoia ever heard if it.

Still another explanation is that the name derived from the fact that the caliber was originally the diameter of a common coin, the "pistole." Historical records and such early samples as are known do not bear this theory out. There was a tremendous divergence in calibers in the early days, and standardization was unknown.

Researchers who combine a knowledge of languages with etymology and also with knowledge of firearms (and these are few indeed) lean to the explanation that the name derived from the fact that the early pistols were used by cavalry; and that "pistol" evolved from the fact that the arms were commonly carried in holsters positioned on the "pistallo" or pommel of the saddle. Henry II of France named such troops "pistoleers."

Whatever the derivation of the name, there is no question that the history of the pistol parallels that of the longer firearms—and may even have preceded them.

EARLY MILITARY PISTOLS

FIRST USE IN EUROPE

In addition to single-shot pistols, there have been multi-barrel pistols, revolving pistols, magazine pistols, and a host of other freaks and legitimate variations in the course of the evolution of the pistol. We can touch here only on those which have had some relation to military usage, or which have some special historical value.

In the days of the Snaphaunce and the true Flintlock, the pistol was known in all these early forms. Books could be written on each type. We can no more than note the existence of hinged frame and other patterns of breechloaders, 3-and 4-barreled pistols, magazine pistols of the "Cookson" type mentioned, over-unders and every other design which had appeared earlier—and was to be repeated later as ignition systems advanced.

In the early period of the Cannon Lock, the German Black Knights terrified the French forces when they introduced the "secret weapon" of that day, the "petronel." This was a one-hand gun commonly hung around the neck on a lanyard. When firing the butt end was rested against the chest (poitrine) for support. A lighted match was applied by the other hand to the touchhole to fire the piece.

The German Ritters about 1520 employed wheel lock one-hand guns. At the battle of Renty (1544) they used them with terrifying effect against the French. This was the first use of the dread "caracole" where a line of horsemen galloped in close to the massed French soldiers, fired their pistols at point-blank range, then wheeled to the rear to reload while a second line of Ritters galloped in to repeat the performance. These Ritters were mostly paid mercenaries, and Henry II finally hired them himself and named them "pistoleers."

Contemporary illustration showing use of the Petronel. From Latin Ms 7239 in the Bibliotheque Richelieu.

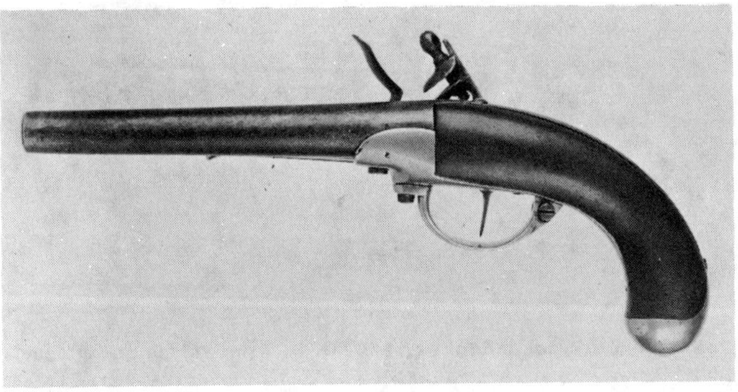

North & Cheney flintlock pistol Model 1799.

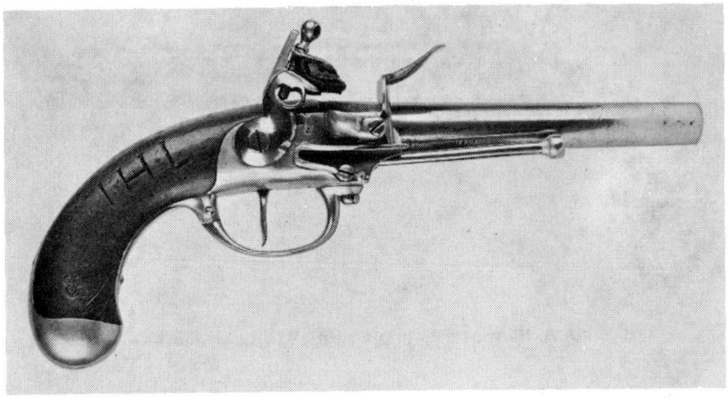

French 1777 pattern flintlock pistol made in Connecticut by North & Cheney for American forces.

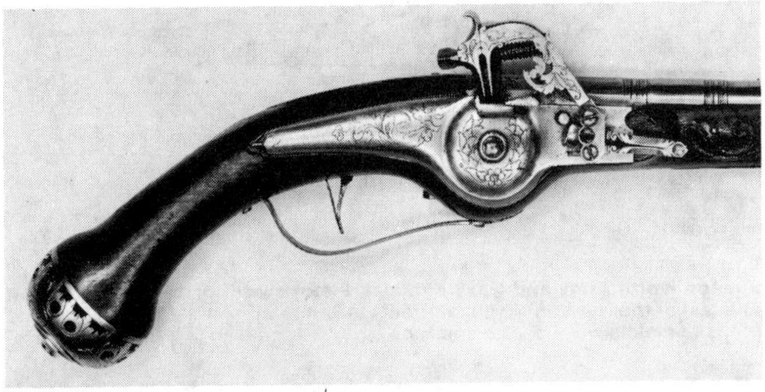

Early German wheel lock pistol. There were thousands of varieties of this type.

A novel called "Jewel House" published in England in 1594 carries an account of a rifled pistol, an interesting item in view of the fact that English patent records disclose no records of rifling before 1635. This is another instance of the difficulty of being definite about dates and names in the history of firearms.

EARLY U. S. MILITARY PISTOLS

The early pistols used in our own country were largely for defense, of course, and authentic records disclose very little about them. Hubbard's HISTORY OF THE PEQUOT INDIAN WARS (published in 1677), for instance, tells of the use of "pieces laden with ten or twelve pistol bullets," but say nothing of the arms themselves except that they were used against the savage Pequots. Another early history, Bodge's SOLDIERS IN KING PHILLIP'S WAR, tells of an attack in which Captain John Gallup was "armed with two guns and two pistols and with buckshot for bullets." Beyond the fact that pistols were used, the historian tells us nothing about them.

John Kim, a gunsmith, advertised in "The Boston News Letter," in the year 1720, that he manufactured pistols to individual order. These, of course, were flintlocks. During the early days of our Revolution when General Gage, British Commander at Boston, called on citizens to surrender their arms, we learn from Frothingham's SIEGE OF BOSTON that 634 pistols were turned in. This is an index of the use of pistols; yet it is strange how little is authentically known of what they were or who made them. The famous "Pitcairn Highlanders," carried by the British commander at the Battle of Lexington, are still to be seen in the little museum in Lexington. One of these captured pistols probably fired "the shot heard round the world" which set off the shooting on Lexington Common. These are flintlocks made in Scotland. They were carried, of course, in holsters across the saddle pommel, which accounts for their capture. Pitcairn's horse was shot and the major was thrown. Minute Men took the pistols from the stricken animal's back.

Rappahannock Forge Pistol

The earliest official American military pistol of which we have record as having been made in America is the "Rappahannock Forge" Flintlock. The locks were imported, but the rest of the pistol was made and the assembly was done in Virginia, where the Legislature set up a forge. It was destroyed early in the Revolution, and the pistols are rare. North & Cheney in 1799 produced flintlock pistols patterned after the French army pistol of the day, the caliber being .69. Deringer's Flintlock made in Philadelphia is representative of the general patterns used by our forces in the early days through the War of 1812.

Government-Made Pistols

The first U. S. Government-made flintlock pistol is the Harper's Ferry 1806. This .54 caliber pistol had a rifled barrel and its general design conformed to that of the rifles produced at Harper's Ferry Armory. Among the best known of these early American pistols for military use are the various models made by Simeon North at Berlin, Connecticut. The 1819 model featured a safety catch

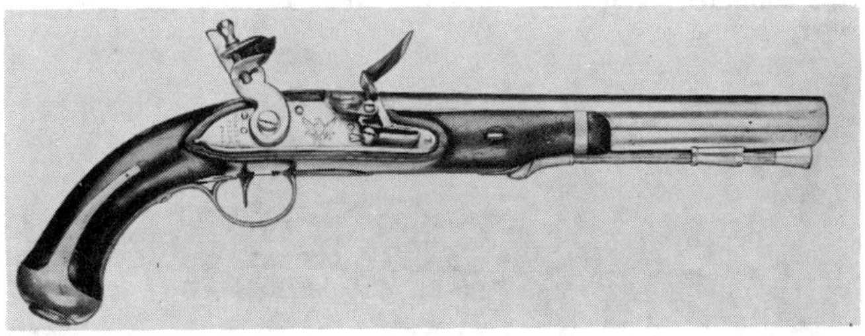

Harper's Ferry. Army and Navy Model 1807 flintlock pistol.

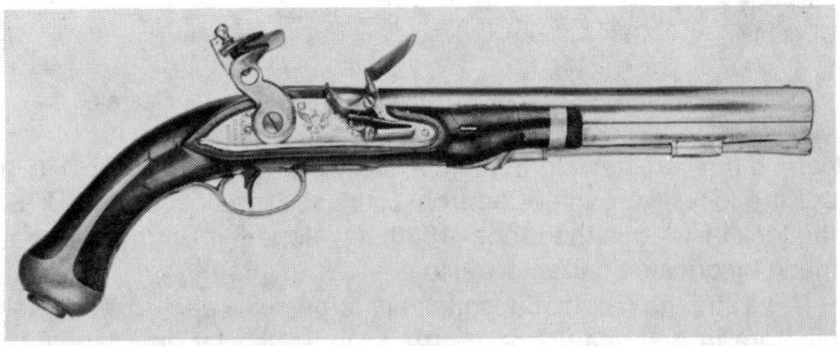

Harper's Ferry. Army and Navy Model 1806 flintlock pistol.

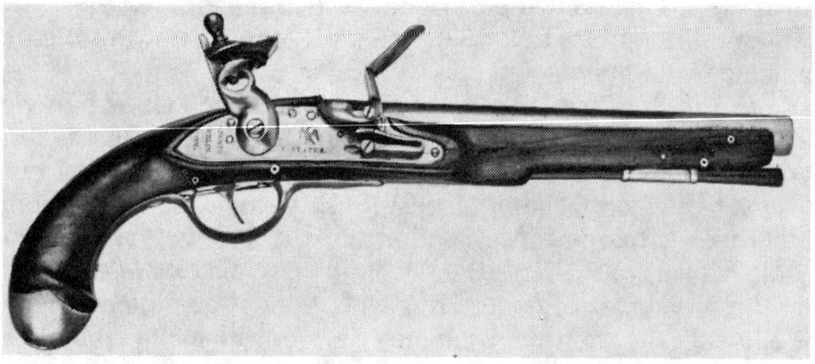

U. S. Pistol, Model 1808 by S. North, Navy caliber .69.

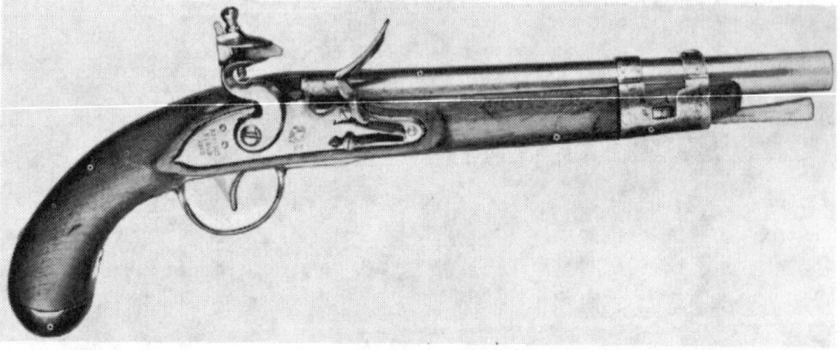

U. S. Flintlock Pistol, model of 1818, U. S. Army. Made at Springfield Armory.

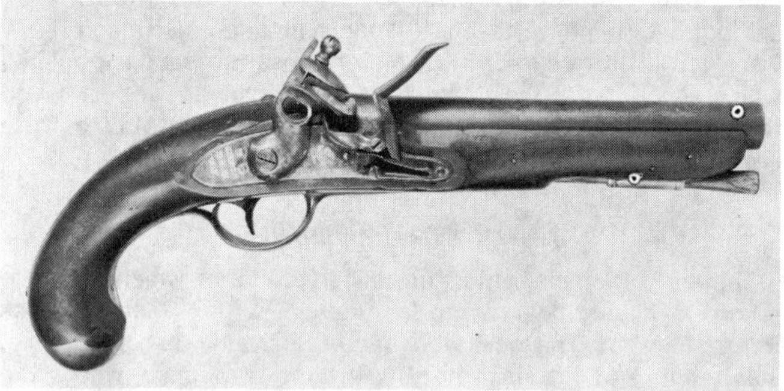

United States Flintlock Pistol model of 1810.

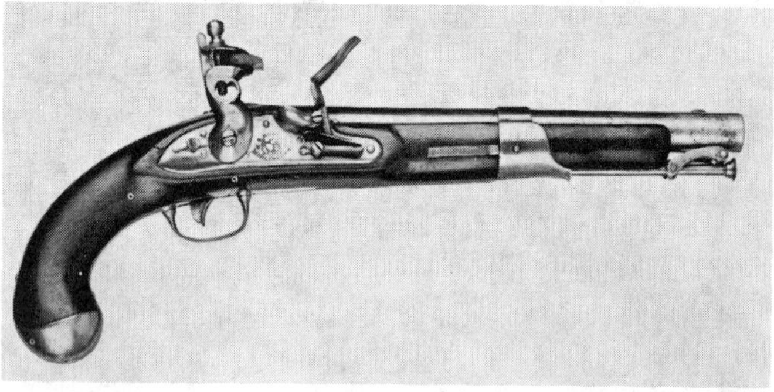

U. S. Flintlock Pistol, Model 1819 by S. North.

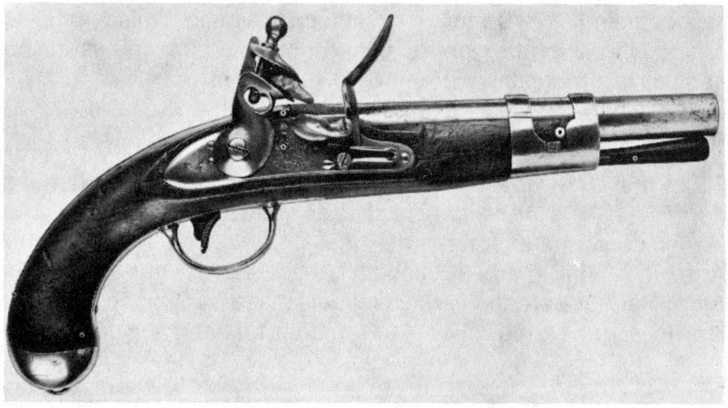

U. S. Flintlock Pistol Model 1813 by Simeon North. Used by Army and Navy.

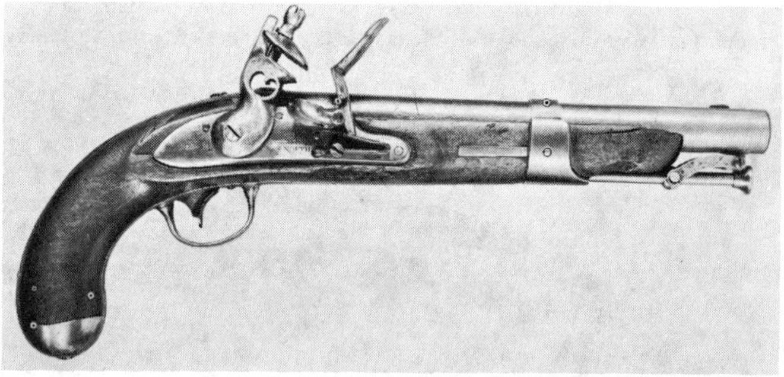

Simeon North Army and Navy Flintlock Pistol model of 1826. Caliber .54. The last of the Simeon North contract U. S. martial pistols. Three thousand of these pistols were contracted for.

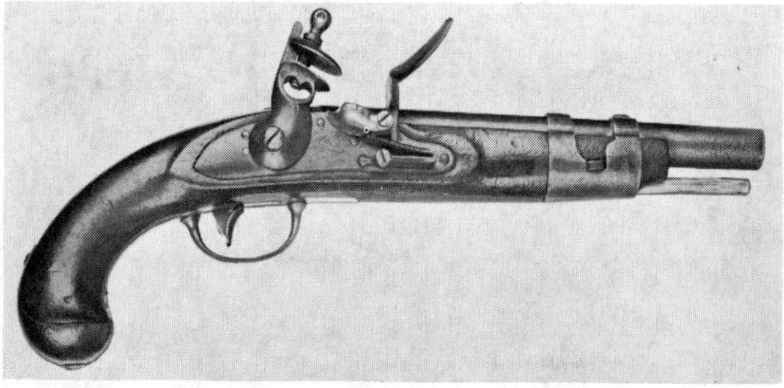

U. S. Flintlock Pistol, model of 1816 by S. North. Caliber .54.

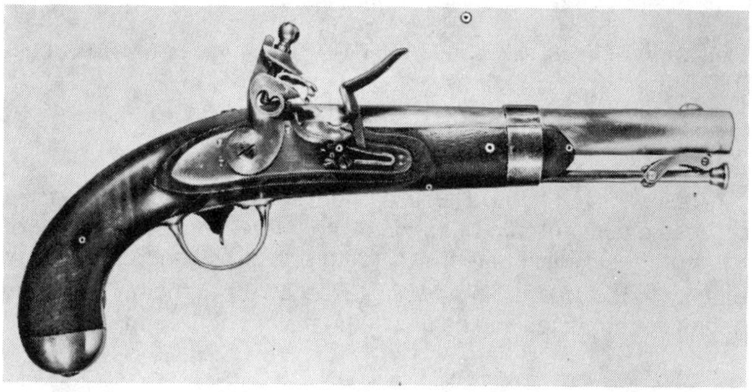

U. S. Flintlock Pistol, Model 1836 by R. Johnson. Caliber .54 taking a one- half ounce spherical ball and a charge of 50 grains of black rifle powder. In 1848 the flintlock arms in Government armories were inspected and the serviceable weapons were ordered to be altered to the percussion system.

behind the hammer. Springfield Armory produced pistols in 1818, but the locks were imported from England. The final official U. S. flintlock pistol was the Model 1836, developed at Springfield, but made by private contractors also until about 1844.

It is a strange fact that despite very large-scale early pistol manufacture in America, there is little to indicate that any particular attention was given to multi-barrel or freak types. Manufacture and design in this period held very closely to single-shot design; while during the same period in Europe every conceivable pattern was attempted.

When the percussion system arrived, many of the early flintlocks were altered to the new ignition pattern. New manufacture varied

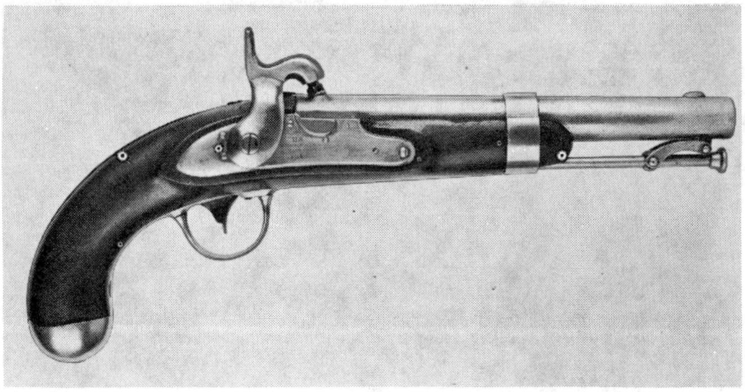

R. Johnson Army Model, caliber .54. Model of 1836 converted to the percussion system.

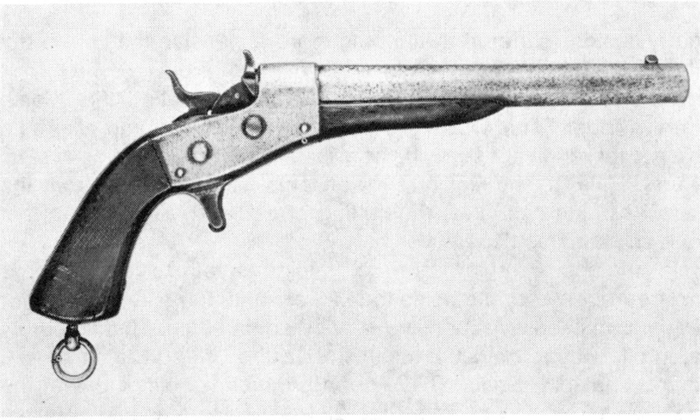

U.S. single-shot Navy Pistol. Early model of 1866 by Remington. Caliber .50 rimfire. This was the first metallic cartridge pistol issued to the U.S. services. Later model Remingtons employ the more secure rolling-block lock and the formal trigger and trigger guard.

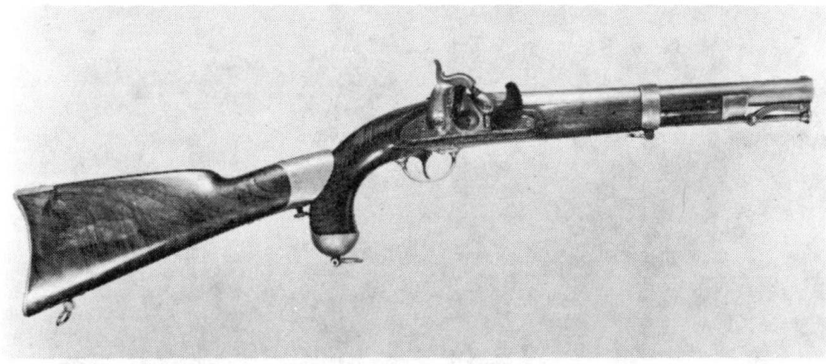

U. S. percussion pistol with detachable carbine stock. Although such stocks are of little practical value, recurrent use is made of the principle. Spanish machine pistols currently made employ it.

very little in overall design except as required for percussion firing. The U. S. Model 1842, made at Springfield and by private firms, also was very much like the 1836 flintlock in its lines, even the caliber of .54 being retained. The 1855 Model U. S. Springfield was issued as a pistol-carbine. It had a pistol form, but was fitted with a detachable shoulder stock for use as a carbine. These arms were equipped with the Maynard primer, the "cap-roll" design we have already dealt with. During the percussion era there was a rash of double-barrel, pepperbox, and similar designs, but none were officially accepted for military use.

With the coming of the metallic cartridge, the military single-shot pistol was doomed. However, even though revolvers were available at the time, Springfield Armory actually designed a .50 caliber pistol with the cam-lock system used in the rifle of the day! This 5-pound wonder was never issued, but models were made. Our Navy did purchase a quantity of Remington Rolling Block pistols Model 1867 to handle a .50 caliber cartridge. This was followed by the Army purchase of the improved Remington Model 1871 in .50 caliber. This pistol is still sought by gun enthusiasts for conversion to handle modern cartridges because of its strong lock and its fine balance.

Pepperbox Pistols

The "pepperbox" design came into its own for a period in percussion days. It is a question just who first introduced it here at this time. Ethan Allen patented his in 1845, and it was a favorite weapon of those who went to California in the gold rush days of 1849. At about the same time, the system was widely manufactured in England and on the Continent.

Original Lefaucheux Pepperbox pinfire shown at London 1851 Exhibition. The pepperbox is usually percussion-fired and there are thousands of variations.

This design had little military significance. It is interesting, however, as an early form of the "double-action" lock mechanism where a pull of the trigger operates both the hammer and the feed mechanism as in the case of modern revolvers. Pepperbox patterns were used in Matchlock days, but the 19th century use of the system is the first known where operation of the mechanism was mechanical by springs and finger pressure. All previous patterns required manual turning of the barrels.

The pepperbox design consists of a series of barrels around a central axis, or alternatively a series of barrels all drilled out of one solid piece of metal much like an elongated revolver cylinder. Usually there were six barrels—sometimes more, sometimes less. Each one was loaded from the muzzle end. Caps were then placed over nipples on a band at the breech. Pulling the trigger raised the hammer, turned a barrel into firing line, and dropped the hammer on the cap to fire the barrel. They were heavy and not very accurate; at best they constituted a transitional design.

Deringer Pistols

Among the earliest developers of small percussion pistols was Deringer in Philadelphia. One of his pistols, as every arms enthusiast knows, was used to assassinate President Lincoln. While soldiers often carried such arms for defensive purposes, they were never official issue.

U. S. Percussion Pistol, model of 1843 by Deringer. Navy Model, caliber .54. The first U. S. martial percussion box-lock pistols actually issued to the service. Although a model 1843, these pistols were made and issued before the Model 1842. This was a matter of manufacturing production.

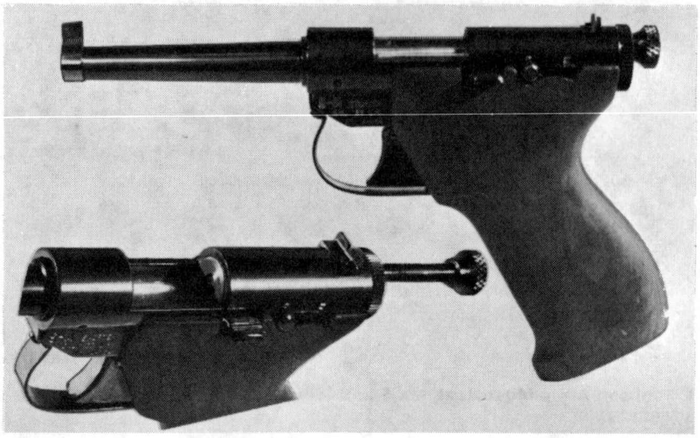

American S-M single-shot pistol closed. Insert shows action open after firing.

SINGLE-SHOTS FOR TRAINING

Today some single-shot pistols are again being introduced for training purposes. One such is the S-M Pistol formerly sold by Sydney Manson of Alexandria, Va. This arm is made in .22 caliber and is intended for training in the handling of automatic pistols. It resembles an automatic pistol in general outline and balance. The bolt is drawn back manually to load the chamber. A release catch is pressed to close the breech as in regular service auto pistol use. When the trigger is pressed the arm fires. The action opens automatically and ejects the empty case. The action stays open ready for reloading. This system simulates quite well the action of a service pistol and affords training of value. At the same time its single shot feature precludes any danger from accidental firing of successive rounds as happens at times with novices unfamiliar with automatic pistol operation. The cost of both pistol and ammunition is low, an important element in a trainer.

Designs somewhat like this have appeared in the past in the form of the English Webley & Scott and the Belgian Pieper. Both these early designs suffered from being more expensive than the pistol they were intended to train for.

Colt's formerly manufactured a single-shot called the "Camp Perry Model." This was intended as a target pistol and also as a trainer for revolver use, since its grip, lockwork and loading system were similar to the Colt's revolver. This pistol suffered from the factor of high cost also, the selling price being higher than that of a standard Colt's revolver.

In England a single barrel conversion is sold to replace large caliber revolver cylinders so that .22 ammunition may be used for training instead of expensive service ammunition. This is merely a unit, however, not an actual pistol. In the United States special barrel and slide devices have been produced to allow use of the service .45 auto as a single-shot .22 trainer. Here again they have not been accepted because of the high cost of such conversion units. Colt's, Springfield Armory, and Sedgley Arms Corp. all introduced such units without success.

WORLD WAR II SPECIAL USES

GUERRILLA USE OF PISTOL

During World War II the United States actually issued a considerable quantity of single-shot pistols for military use, although for all practical purposes the single-shot has been outmoded since 1872. The pistol in question was a single-shot arm very cheaply made from stamping, tubing, and screw machine parts. It fired our standard .45 caliber service auto pistol cartridge. The arm had a trap in the butt where a few extra cartridges could be carried. With each weapon went cartridges, a short stick to be used in pushing the empty case out after firing (no extractor was provided), and a simple set of line illustrations showing without words how to load, fire, and empty the weapon. The single-shot was distributed to guerrilla forces in several theaters of war. It was reported as being used very effectively against the Japanese in particular. The barrel was not rifled. As a close-quarters weapon it was extremely deadly, not only because of the heavy slug it threw, but also because the user knew that he had only one chance. He stopped his enemy the first time or he was in trouble. There is reason to believe, on the basis of data furnished by intelligence groups, that more killings were actually done with this simple, crude pistol than with all the service .45 automatics issued! In general these pistols were issued to natives by the O.S.S. (The Office of Strategic Services, headed by General "Wild Bill" Donovan. This organization was a highly specialized military intelligence organization.)

American-made single-shot pistol, caliber .45 Automatic. Made for our Office of Strategic Services and distributed by them to various underground movements during World War II. The lock lifts up as in the early German Theiss rifle.

In the Balkan areas some use was made of the huge "Bavarian Lightning" single-shot produced at Nuremberg in 1867; but these were merely arms which had been in family or police possession since time of manufacture, not current war period production as was our own pistol mentioned above.

"SECRET" WEAPONS

Passing mention might also be made of two "secret" weapons which were dreamed up during War II, but which were completely impractical. One was a small tube-and-pocket-clip affair resembling a fountain pen. It was to be carried attached to coat or shirt by the clip. When ready for use, the clip was pulled all the way back and let slip. It was under spring pressure, and releasing it fired a .22 short cartridge housed within the barrel section. Cloak-and-dagger operators who carried such items explained that they were to be used as suicide arms in case of capture. Others were supposed to use them in fights for very close-up use, particularly shooting at the exposed area of the medulla oblongata. The only actual use authoritatively reported was when someone carrying such a device shot himself accidentally but not fatally!

Yet another freak was developed in Naval Intelligence circles. This beauty was a flat steel plate device holding one husky .38 S&W special cartridge. It was attached to a heavy leather glove with rivets. A firing release protruded from the front. The idea here was that the wearer of the glove could crack a skull with a blow; and in case of emergency could jam the striker release against the face of an opponent, discharging the .38 into said opponent. Such inane devices crop up in every war period.

At the other extreme were single-shot pistols used by all services, but most highly developed by the British Intelligence, for close quarters use with silencers. Some were useful on Commando raids. Others were frankly weapons for military or political assassination. Revolvers are not efficient with silencers because gas—and hence noise—escapes at the jointure of barrel and cylinder. Auto pistols are not too effective because of noise as the breech opens and gas is blasted out, and because of the mechanical clatter as the slide slams home. The single shot was used because it made a truly silent arm when properly constructed. Arms such as these are intended only for specialist use, hence will not be dealt with in this volume.

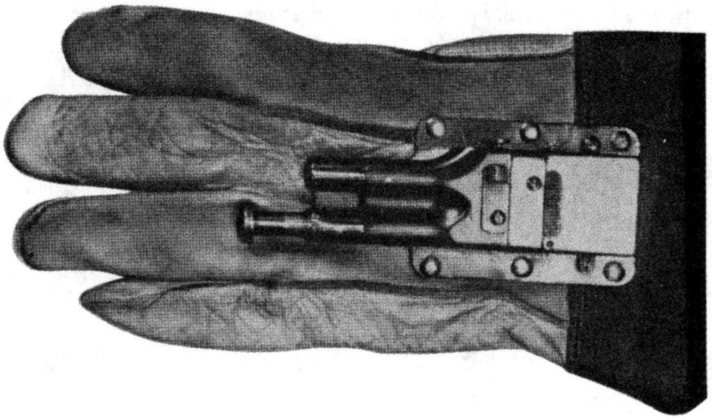

Glove pistol of World War II. When fist is closed, the projecting striker will fire the pistol when struck against the opponent's body. Made in caliber .38 S&W and .38 S&W Special.

DOUBLE AND MULTI-BARREL PISTOLS

DOUBLE-BARREL PISTOLS

All through the course of firearms history double barrel pistols have been known. The common military forms in early ignition periods were side-by-side barrels as in the modern shotgun construction. None are of value to this outline.

In the early percussion era, as we have seen, much attention was given to small designs, a few of which were used as auxiliary side arms. Again none are of particular interest here.

The one truly outstanding double barreled pistol in America was the old Remington Derringer for the copper cased .41 Rimfire cartridge. Note that such arms in varying calibers and barrel systems were made by many other makers. Only the Remington warrants attention here. It is still a weapon carried by many police officers in the West and Southwest as an auxiliary "hideaway" weapon. The two barrels are placed one above the other. They are hinged at the top of the breech. Turning the release catch allows them to be swung up for loading. A manually operated extractor is provided on most of these. The pistol must be thumb-cocked to fire each barrel in succession. The heavy, slow-travelling slug is deadly at close range.

During both World Wars many U. S. officers carried these deadly old Remingtons as hidden arms, much as the German officers tended to carry small and rather ineffective .25 auto pistols concealed for emergency use. Note that such arms are called "Derringers," with two r's. This spelling stems from the fact that the original inventor of the early "Deringer" refused to allow competitors in the short pistol field to capitalize upon his name; so they evaded by doubling the "r."

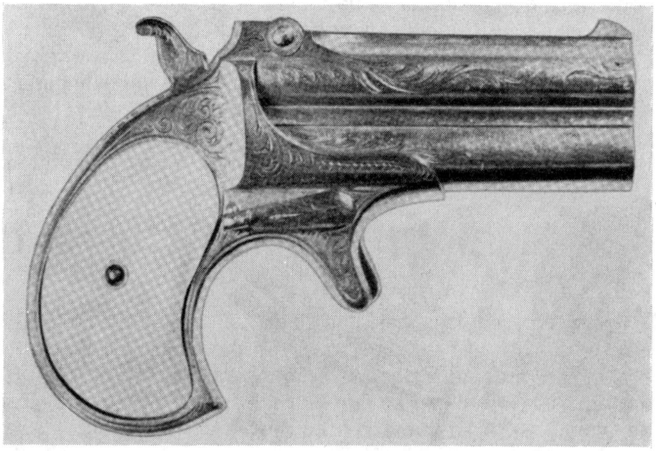

American derringer by Remington. Caliber .41 rimfire. Still highly regarded as a defense weapon by police in some areas, though both gun and cartridge are long obsolete.

MULTI-BARREL PISTOLS

England produced one fantastic multi-barrel type before accepting the revolver design. This was the Lancaster, a 4-barreled pistol to take .476 caliber cartridges. The design was a strong hinge-frame breechloader. Barrels were stationary. Pull-

English Lancaster, caliber .476 center-fire 4-barrel pistol.

Captured German belt-buckle pistol of World War II. Property of the Honorable Gordon Persons, former Governor of Alabama. A very rare specimen.

ing the trigger actuated an internal striker mechanism which acted through individual firing pins to discharge each barrel in succession by double action as the trigger was pulled. This monstrosity saw some little private use by the British in Africa and Asia, but was not officially adopted.

The only military use of pistols with several barrels in recent times was the appearance of freaks such as the early 4- and 6-barreled designs made late in the 19th Century by Braendlin Armory in England and the Lancaster already mentioned. These were hinged-frame types fired by a double-action pull. They could be used with modern .455 British service ammunition, and after Dunkirk even relics such as these appeared, so desperate were the British for arms.

In general, designs of this type have been made with all the barrels drilled in a single steel block; but a few freaks have consisted of several barrels brazed together. They have no historical importance here except as indicated.

Other weird multi-barrel forms such as the European "Regnum" and "Reform" pistols were occasionally found on prisoners during World War II, but did not constitute military issue as such. The former has four barrels drilled one above the other in a solid block. The caliber is usually 6.35mm (.25). Each barrel is fired in succession as the trigger is pulled. This pistol hinges open to load. The "Reform" has a somewhat similar barrel structure. The action is different, however. The first pull fires the top barrel. Each succeeding pull moves the barrel block up and gas from the exploding cartridge ejects the empty case from the barrel above it! The sole value of such pistols is their flat shape allowing easy concealment as an auxiliary arm.

The Germans during War II made a unique belt-buckle pistol. The only samples the author has ever seen were hand-made models. However, former Governor Gordon Persons of Alabama, an ardent collector of unusual arms, possessed a sample which seemed to be a production item. Production was probably halted by events of the war. The pistol is a great rarity, a collector's prize of the first order. Pressing a catch on the buckle allows the face of the buckle to spring away and the barrels to emerge and fire—a dangerous close quarters gadget to be used upon the unsuspecting.

MAGAZINE PISTOLS

None of these are worthy of more than passing notice, since none were ever practical military arms. They have appeared all through firearms history in one form or another, however.

In metallic cartridge days, the early Austrian Schwarzlose magazine pistols were experimented with by the Austro-Hungarian army but were not adopted. (These are not to be confused with later Schwarzlose auto pistols.)

These freak arms differ in operation from the so-called "automatic" pistol where the forces of the explosion function the action. "Magazine pistols"—the Schwarzlose, Bittner, Fiala, and others—looked like auto pistols of their day. However, after firing each shot, the action had to be opened by hand to eject the empty case and feed in the next cartridge in the magazine. Such transition arms and freaks have a single barrel and chamber, but carry a reserve supply of cartridges in the butt to be fed in as the action is manually functioned.

REPEATING PISTOLS

These too are freaks, having a single barrel and chamber and a reserve stock of cartridges usually in the handle section. The actions are mechanically or spring-operated as the trigger is pulled. They are more advanced than the "magazine" types mentioned above in that a pull on the trigger feeds and ejects without use of the other hand. They are inferior to the automatic type in that the ejection and feeding is mechanical rather than by utilizing the explosive forces, hence they are slower and more clumsy. The French and Belgian plants which made these types were never successful in selling them for military use, though again some were taken from prisoners during World War II—"hide-away" arms they had personally bought.

THE SINGLE-SHOT AUTO EJECTOR

The final form of true pistol is that in which the cartridge is manually inserted, but in which the forces of the explosion are utilized as in the automatic pistol to open the action ready for reloading.

We have already mentioned this pattern as having some value as military trainers when, as in the case of the former American S-M .22 pistol, they sell at low price. As they appeared in the discontinued Belgian Pieper and the British Webley & Scott, their training value was offset by their cost.

12

Military Automatic Pistols—Historical Outline

The principle of the so-called automatic pistol design was familiar almost from the inception of gunpowder. We find written references to arms utilizing the principle as far back as Sir Robert Moray's report to the Royal Society in 1664. However, the practical application of the principle was never truly possible until after the development of the successful metallic pistol cartridge and smokeless powder.

As we have seen, American patent office records as early as 1863 showed attempts to develop gas-operated arms. Historically, however, the first successful automatic pistol was that marketed in Austria by the inventor Schönberger.

The successful automatic pistol was also dependent to a large degree upon the development of the rimless and semi-rimmed case. These were at once strong enough to stand the violent jerk of immediate automatic extraction, utilized a metal jacketed bullet which could be securely seated in the case and would feed easily, and permitted cartridges to be placed one on top of the other in the familiar box magazine design.

EARLY AUTOMATIC PISTOLS

The 1893 barrel-recoil-operated pistol of Andrea Schwarzlose of Austria was never put into production, but it paved the way for his later designs.

The original Borchardt pistol.

PISTOL DESIGNS

The Borchardt

The first automatic pistol which was truly successful commercially was of German manufacture but of American design, the 7.63mm Borchardt. Hugo Borchardt of Connecticut early in life worked at the Winchester Repeating Arms Company in Hartford. He developed a very clumsy but nevertheless thoroughly efficient automatic pistol with a toggle-locked breech and detachable box magazine in the stock. For it he developed a special cartridge of bottleneck design with a jacketed bullet. With very little modification, this cartridge is the familiar 7.63mm Mauser cartridge today. An even smaller modification makes it also the Russian 7.62mm automatic pistol and submachine gun cartridge.

Borchardt was unable to interest American manufacturers in producing the design. He went to Germany with it. The Berlin firm of Ludwig Loewe and Company at that time was engaged in production of Mauser rifles. They employed Borchardt as an

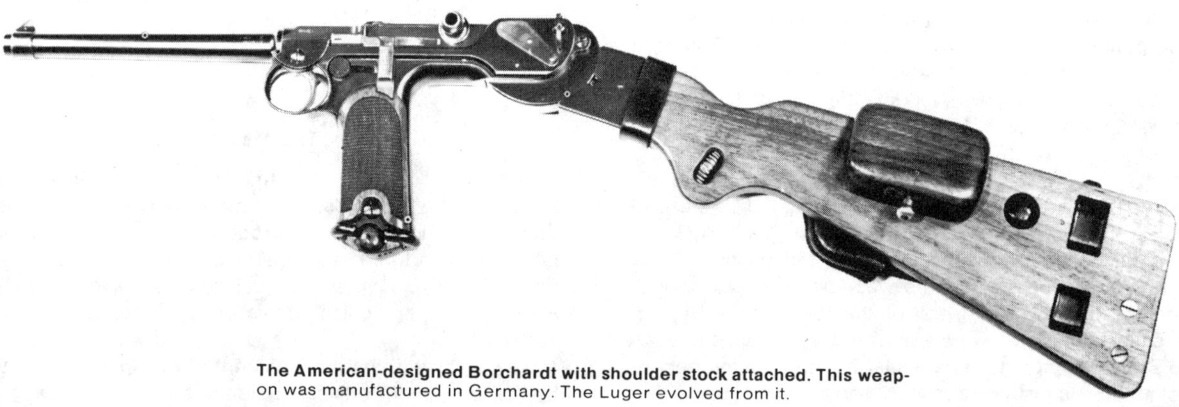

The American-designed Borchardt with shoulder stock attached. This weapon was manufactured in Germany. The Luger evolved from it.

engineer and put his pistol into production. It was marketed in 1893. As originally issued it was accompanied by a detachable rifle stock, the idea being to make it either a one-or a two-hand weapon for short or long-range firing. In practice, shoulder use of pistols is not desirable.

One of Borchardt's assistants at the Loëwe plant was George Luger. He redesigned the mechanism to produce the first model of the world-famous Luger pistol, an arm which from 1908 until 1938 was the official sidearm of the German military forces.

Bergmann Pistols

The long line of German Bergmann pistols began in 1894. These models included locked and unlocked breech types, exposed and internal hammers, in-built and detachable magazines, and many calibers and styles. One of these Bergmann's was manufactured in Belgium under the name "Simplex." It utilized an 8mm cartridge developed specially for it. This cartridge was modified by John Browning and the F. N. Works in Belgium and is familiar throughout the world today as the .32 Colt Automatic Pistol cartridge or the 7.65mm Browning Automatic Pistol cartridge. No other pocket pistol cartridge in history achieved the success this one did.

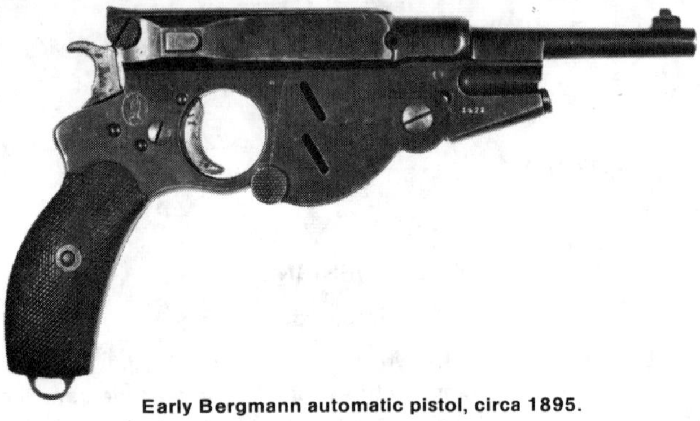

Early Bergmann automatic pistol, circa 1895.

Bergmann-Bayard

The German designer Bergmann negotiated manufacture of a locked-breech powerful design with an external hammer and a magazine forward of the trigger guard which first appeared in Belgium under the name of Bergmann Bayard, and later simply as the Bayard. These were issued in 1908 and 1910 and are still encountered in Scandinavia and in Spain where they were official military arms for a time.

The Mauser Pistol

The automatic pistol as a truly military arm began its history in 1895, when Mauser introduced his 7.63mm Military Model, a pistol still in wide general use throughout the world. Hugo Borchardt was officially credited by German ammunition manufacturers, who first produced this cartridge, with not only the cartridge design but also with having done much of the experimental engineering work on the Mauser pistol itself, although the basic design was that of Mauser.

Winston Churchill as a "Leftenant" in the British Lancers used a Mauser automatic pistol in the cavalry charge at Omdurmann. He recommended the pistol highly, stating that because of its efficiency and magazine capacity he was able to shoot his way out of a native trap, killing several of the fuzzy-wuzzy warriors in the course of saving his own life. However, it was many years before the British service gave particular attention to the general adoption of automatic pistols for military service.

Mauser 9mm Parabellum M1912 "Armeepistole." This pistol, which out wardly resembles the 1910 Mauser Pocket Pistol, was never manufactured in quantity.

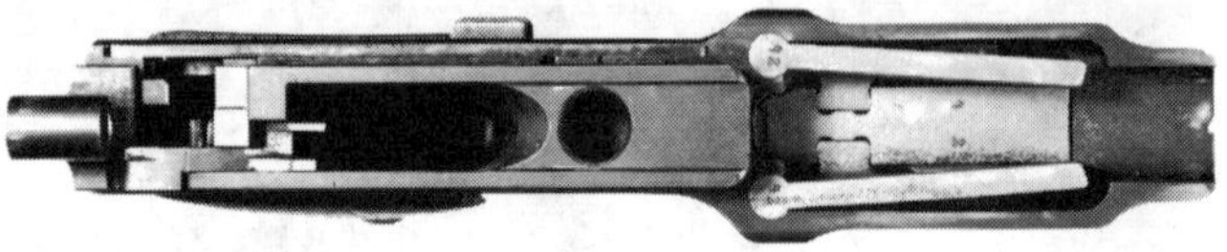

Top view of receiver of Mauser M1912 "Armeepistole." Note locking flaps which are similar to those of the 1898 Mauser semiautomatic rifle.

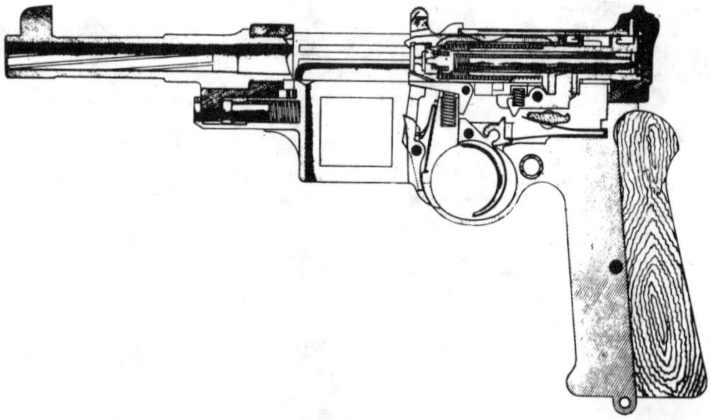

Section drawing of another of Mauser's semiautomatic pistols. This one was not produced commercially. A variant of its locking mechanism is used in some automatic arms today.

The Maxim Pistol

In England in 1896 the American, Hiram Maxim, patented a blowback automatic pistol and made samples in various calibers. These were never put into production, however. The design was noteworthy in that it was one of the first to successfully use the rimmed .455 British service cartridge. An American design, the Reifgraber, later duplicated this feat by using the .32 S & W rimmed cartridge.

However, until the development of the special magazines used by the .22 Colt Woodsman, automatic pistols generally would not operate satisfactorily with rimmed ammunition. Much work is being done today in sporting and quasi-military circles on the development of automatic pistols to shoot the super-accurate .38 Smith & Wesson Special rimmed cartridge.

The Schwarzlose

The first automatic arm to provide a device to hold the action open when the magazine is empty was that used in 1898, again by Andrea Schwarzlose of Austria. The short-recoil locking principle found in this pistol is basically that of the German machine gun of 1934 and the American Johnson light machine rifle. The pistol used 7.63mm Mauser cartridges, but was never generally used for military service.

Browning Pistols

By far the most successful of all automatic pistol designers, of course, was our own John M. Browning. As in the case of so many other inventors, he had to go to Europe to have his original design manufactured. His first pistol was introduced there in 1899. It had been patented in the previous year. This arm was manufactured by F. N. (Fabrique Nationale D'Armes de Guerre, Europe's largest arms manufacturer today, at Herstal, Belgium.)

Browning's first pistol made in the so-called 7.65mm or .32 automatic caliber has been used as a side arm in practically all military groups in the world at one time or another. It has in all instances, however, been a substitute standard weapon or one purchased by officers personally.

In 1900, Colt introduced a new model Browning automatic pistol to handle a newly-developed cartridge, the .38 Colt automatic. The first model of this arm made by Browning was actually a full automatic enclosed hammer model, a single pull on the trigger emptying the magazine.

The arm fired so rapidly and uncontrollably that it was almost impossible to hit anything with it, the rapid successions of recoil throwing the muzzle up so fast that anyone nearby was in danger.

In June, 1900, Browning wrote to the editor of American sporting magazine that he had experimented with various types of safeties, both grip type and thumb actuated. He also explained his reasons for designing the .38 automatic with a rotating hammer, his earlier F. N. pistol having been fired by a conventional rifle-type striker.

From this design he evolved the Colt .38 automatic pistol in Pocket Model with short barrel and Military Model with six-inch barrel. The cartridge developed was, and still is, a remarkable achievement. Our official .38 caliber Army Revolver at the time, using a black powder load, had a muzzle velocity of only 750 feet per second. The new .38 Colt Automatic cartridge developed velocities as high as 1,350 feet per second in unofficial tests! The official velocity was given at 1,260 feet per second. the penetration was tremendous, and even today is practically unsurpassed except by giants such as the S & W .357 Magnum and some special freak loads. The initial penetration was given as 11 inches in pine. The penetration of this pistol resulted in its use throughout the world by big game hunters who carried it as an extra weapon for close quarters use in case of emergency. The pistol itself introduced Browning's first recoil-operated, locked-breech mechanism, varieties of which are today in use in all the locked-breech Colt and Browning pattern pistols as well as in those of many imitators ranging from Poland and Spain to Russia.

The angle of the grip and the size and balance of the arm were all poor. Moreover, the slide was secured to the receiver by a cross wedge, the receiver being so machined as to permit the slide to travel in it. When, as occasionally happened, the wedge crystallized and gave way, the slide could be blown off the back of the pistol.

Browning's next military design in .45 caliber was quite similar to the .38 and was equally a failure.

However, his later .45, as developed for our government tests, was in all except minor details the same superb arm it remains today. This locked-breech short-recoil-operated weapon was so constructed that the slide could be removed only from the forward end of the receiver. The pitch of the grip was improved, as was the balance and weight distribution.

The F. N. Model of 1903 developed by Browning appeared as the Colt Hammerless .32 Automatic Pistol. The design was blow-back, the hammer was enclosed, and a grip safety provided. From this development came a modification in 1910 from the F. N. factory. Hundreds of variants of this pistol have been manufactured in Belgium, Italy, France, Germany, Spain, and Czechoslovakia since the introduction of the arm itself.

In 1908, Browning's small pocket pistols of the so-called vest pocket type (which have seen extensive military use by staff officers in various European armies through both wars) were introduced in the United States as the .25 Colt Vest Pocket Automatic model and in Belgium as the F. N. Baby Browning. This too represents a type which has been tremendously imitated and copied throughout the world. Items resembling it very closely may be encountered with literally hundreds of varying trade names, representing pistols manufactured in Spain, Italy, and Belgium for the most part.

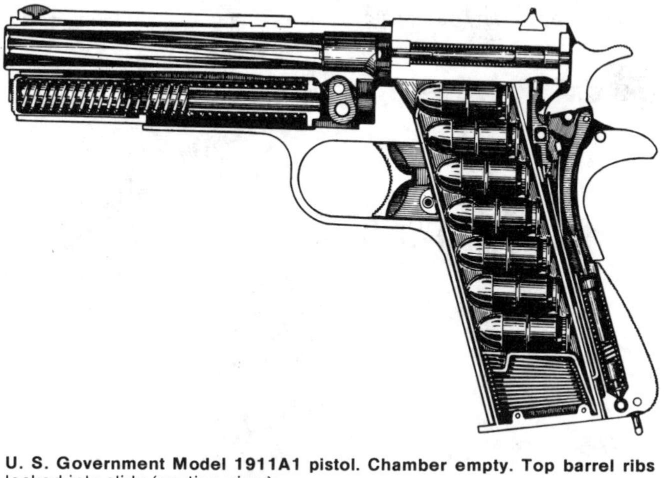

U. S. Government Model 1911A1 pistol. Chamber empty. Top barrel ribs locked into slide (section view).

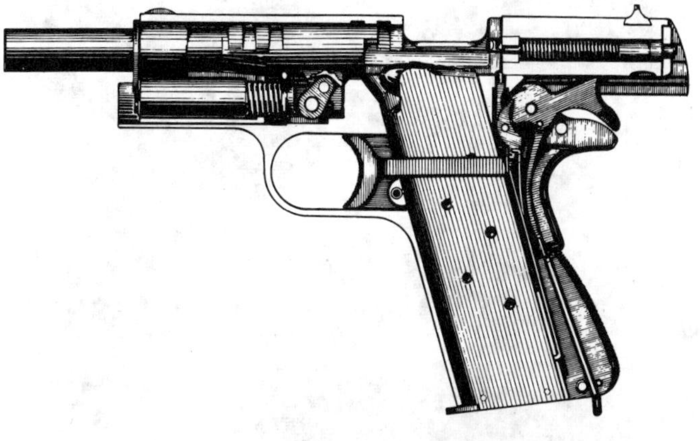

U. S. Government Model 1911A1 pistol. Breech open, barrel down on link and out of slide engagement. Short recoil operation (section view).

The United States .45 Colt Government Model Automatic Pistol designed by John M. Browning was introduced in 1911. With very few modifications, this is still the official pistol of the United States forces. Post-War II commercial models have been altered to utilize aluminum frames and a modified hammer.

Mannlicher

Mannlicher introduced an automatic pistol in 1901. This was not particularly successful, although it is still occasionally encountered. One of its defects was that it used a special cartridge obtainable only from Austria.

Luger

In 1900 the great Deutsche Waffen and Munitions-Fabriken, world renowned before World War I as "DWM," produced the pistol popularly known after its designer as the Luger, which was a refinement of the American Borchardt. This was originally introduced with still another bottlenecked cartridge of considerable velocity and power, the 7.65mm.

In 1908 the German government adopted the Luger as its official service pistol but the caliber was increased to 9mm. However, the 9mm cartridge itself is similar to the earlier one in body diameter. The 9mm case was developed directly from the 7.65mm so that interchangeability of the two cartridges is very simply effected merely by replacement of barrels. Only the chambers and the bores differ. These pistols have never been manufactured commercially in any other calibers than 7.65mm and in 9mm. Early in the century a quantity were hand-manufactured in caliber .45 for United States government tests. These are collector's arms, and although they were advertised commercially they were never manufactured in commercial quantities.

World War I Luger with 32-shot snail magazine. This was a forerunner of the submachine gun.

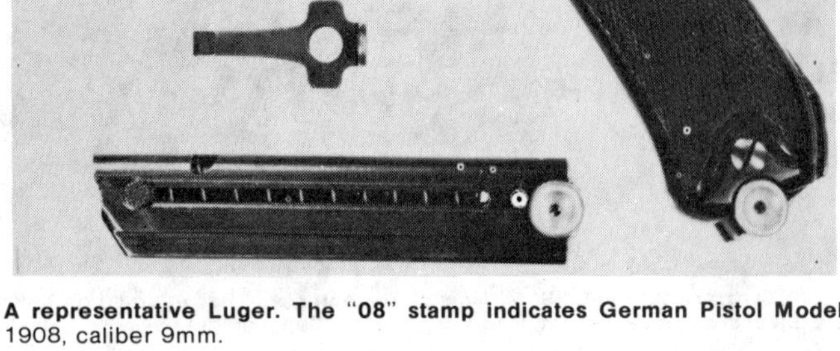

A representative Luger. The "08" stamp indicates German Pistol Model 1908, caliber 9mm.

Webley

The 1904 Webley automatic pistol manufactured in England was produced in .455 caliber and was not a successful design. In 1913 a new and highly improved Webley locked-breech design in .455 caliber was adopted under the designation Mark I by the Royal Navy. It saw some limited use in both World Wars. In general the arm is too heavy and clumsy to be compared with the Colt pattern of automatic pistols. The Webley pocket type automatic pistol, an elementary blowback form of 1906, was used for some time in British police circles and constituted a defense issuance arm to some extent during the war periods.

In 1913 Webley introduced an elementary blowback which they called a "9mm high velocity" but which in effect was an unlocked breech arm for the inferior 9mm Browning long cartridge, a cartridge seldom encountered.

The year 1910 saw the introduction in England of a locked-breech Webley following closely the design of their .455 Mark I Navy pistol but in caliber .38 automatic. This did not meet with general success and was abandoned.

Savage Pistols

In the United States the Savage Arms Company in 1906 produced in limited quantity a .45 caliber automatic pistol of interesting design with a hesitation locked breech in which the barrel rotates slightly to unlock. The angle of rotation is only five degrees, which actually furnishes a delaying or hesitation action rather than a true locked-breech pattern. This pistol was submitted for United States Government tests, but did not stand up in the face of the opposition from the .45 Colt pistol developed by Browning. It is encountered as a collector's item only. It was followed, however, in 1908 by the first of a series of very fine pocket automatic pistols by Savage utilizing the same general principle. Some models of this design are in official military use in caliber .32 in Portugal. They have had little military usage, however, except of course as emergency arms.

American Savage .45 automatic. A collector's item.

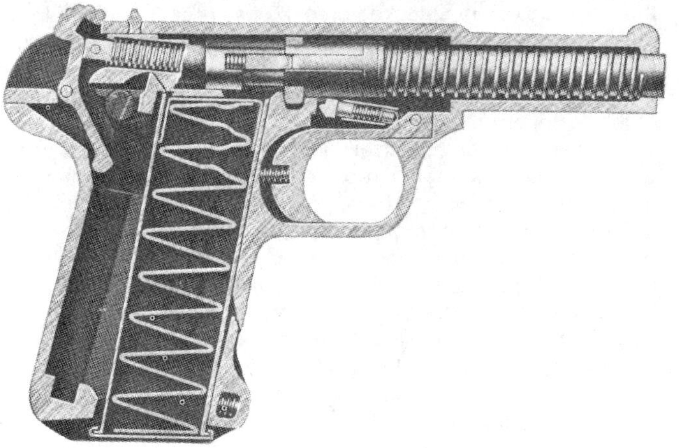

Section view -- Savage pistol.

A representative Italian Beretta pistol stripped to show simple design. The quick-removable barrel is a feature of this type.

Glisenti

The Italian official Glisenti pistol was introduced in 1906, but was not adopted by the Italian Army until 1910. This interesting but weak design is discussed in detail in Part II in the chapter on Italy.

Roth-Steyer

Austro-Hungary in 1907 adopted for cavalry use a freak 8mm pistol by Roth-Steyr. This had been patented several years earlier. It was never a successful design although it saw military usage in both World Wars. This arm used a firing system which was spring operated, though ejecting and loading were produced by recoil forces.

Steyr-Hahn

In 1911, Austria introduced its 9mm Steyr-Hahn, a form converted later in World War II to the German service caliber of 9mm Parabellum or cartridge pistol 08. This interesting rotating barrel design loads with a strip-in clip instead of a detachable box magazine.

Frommer

Hungary introduced the long-recoil pistol made by Frommer in 1912. This is a very unusual form of locked-breech action entirely unnecessary for the type of cartridge for which it was designed, the .32 automatic. This arm saw considerable use in the Balkans as a military arm despite its ineffective cartridge.

Beretta

In 1915, Italy introduced the Beretta, another elementary blowback pistol which in modified form has since received tremendous manufacturing attention, both as a military arm and for export.

DEVELOPMENTS BETWEEN WORLD WARS I AND II

DESIGNS IN VARIOUS COUNTRIES

U. S.—Remington

After World War I, Remington produced a .380 automatic pistol with a hesitation-locked breech of very unusual meritorious design. However, the American field for pocket automatic pistols at that time did not warrant production continuing and the pistol was discontinued. The locking mechanism in this pistol, functioning on the two-part breechblock hesitation mechanism, is susceptible of considerable development for possible application to other automatic arms.

Spain—(Imitations)

During the succeeding years, very extensive manufacture of automatic pistol design was carried on in Spain for export use. In general these were modifications or direct copies of known locked-breech types. Copies and modifications were made of the Colt system also for .38 caliber, 9mm, and .45 automatic calibers.

Spain during this period added to its line of imitations of Colt automatic pistols and Mauser automatic pistols by offering items such as the Star, Astra, and Azul pistols. Except that some of these are fitted with full automatic switches to permit continuous fire on one pull of the trigger (a totally useless device on a small arm), their production was purely that of uninspired patterns. All were inferior in workmanship and materials, though in recent years they have been much improved.

Spanish Star. General design and appearance are Colt-Browning. The design is often fitted with full-auto switch on right side of frame; latest models have improved takedown system. Calibers from 9mm Luger to .45 A.C.P.

Argentina—(Imitations)

In Argentina pistols such as the Criolla appeared; they were merely imitations of the Spanish Star.

Argentine semiauto cal. .45 A.C.P. Made by HAFDASA. This is a slightly modified version of the Colt M1911.

Czechoslovakia—Nickl and Strakonice

In 1922 C. Z. in Czechoslovakia introduced the Nickl pattern of locked-breech pistol in .380 caliber, some parts of the mechanism other than the lock closely resembling that of the German pocket Mauser. This weapon was officially adopted by the Czech Army. Some later patterns involved one of the early forms of double-action trigger mechanism commonly found in revolver construction. Next came a standard 7.65mm or .32 automatic of the blowback pattern marketed in 1927, which also saw use in the Czech military services.

Czechoslovakia improved its Strakonice double-action automatic pistols, introducing a .380 model with a barrel hinged at the muzzle end for ease of cleaning. The design generally, however, was heavy and clumsy for the cartridge used.

France—Le Francais and M1935A

In 1928 a Military Model of the Le Francais blowback pistol was introduced in France. An elementary blowback, the pistol had some unusual design features. Its firing mechanism is not cocked by the opening of the slide as in most pistols of the design. While the slide functions in normal fashion to permit extraction, ejection, and reloading, the firing mechanism is not cocked thereby. A long pull on the trigger is necessary to cock and slip the firing striker for each shot.

The French modification of the Browning pistol locking system, the MAS pistol made at St. Etienne in 1935, is again merely a minor Browning variant utilizing the common principles of the short recoil-action. The cartridge, based on the "mysterious" American Pedersen device cartridge of World War I, is little more efficient than the .32 Colt automatic pistol cartridge, which may make it a desirable lightweight arm for staff officers but labels it impractical for basic military usage.

Belgium—FN Browning High Power

Only minor modifications of design were undertaken in the United States during this period. Great Britain paid no attention whatever to such production. In 1935 the F. N. plant in Belgium introduced its 1935 Browning High Power, a modification of the familiar Browning pistol with several interesting items added. In the original Browning locked-breech design as it appears in the .45 Colt automatic pistol, the movement of the barrel traveling in a vertical plane is utilized to lock the breech. The rear end of the barrel is secured to the receiver by a link, locking ribs on the top of the barrel engaging in mating cuts in the underside of the slide. The front section of the barrel serves as a pivot. The 1935 model, while essentially the same, replaces the swinging link with a solid nose section cam which is shaped to ride up and down in the receiver during the locking and unlocking motions. In the 1935 pattern the barrel is supported by a lug at its lower section which is an integral part of the barrel forging itself. Through a cut in this lug passes a transverse bar in the receiver. As the barrel and slide recoil, a cam-face on the lug strikes the bar forcing the barrel down to disengage the locking lugs on the barrel from their seats in the slide. At this point, the barrel motion being halted, the slide can recoil the remaining distance to extract and eject.

Poland—Radom

Poland shortly thereafter introduced the Radom pistol which is a relatively minor modification of the Belgian Browning High Power.

Italy and Austro-Hungary

In 1923 and later in 1934, several modifications were made in Italy of the basic Beretta pistol, all of which are covered in Part II in the chapter on Italy. None are important particularly from a military standpoint because of the relatively ineffective cartridges of .32 and .380 caliber which they used. The same applies to a considerable degree to Austrian types and other development during this period; except that they produced some forms of the double action automatic system, whereby a pull on the trigger will fire the first shot mechanically without the hammer being cocked. These, however, were all small pocket arms for commercial use. The 1937 Hungarian modification of the Browning system, too, is merely of passing interest.

Norway

Norway introduced in 1914 a .45 caliber automatic pistol which was merely a minor variation of the familiar Colt and Browning system. The arm was not produced in quantity and has nothing original about it worthy of notice here.

Sweden

The Lahti is a short-recoil locked-breech pistol of 9mm caliber originating in Finland. It was later made in Sweden. Externally resembling the familiar Luger, but involving its own locking system, it is a relatively heavy and clumsy arm of its type by today's standards. It is no longer in general production.

Japan—Nambu

In 1925 Japan introduced the Nambu. This locked-breech design is fully described in Part II in the chapter on Japan. Originally introduced in 8mm service caliber, small production was later instituted on a 7mm caliber of reduced size. This is a short-recoil-operated arm with a hinged locking block cammed up from below the receiver to lock the breechblock well to the rear of the firing chamber. It is striker fired. While the design is interesting, it is doubtful that it will ever again be considered as an important design factor in military arms. The cartridge itself is ineffective by modern military standards. The experience of the Japanese military and police groups with small arms of American manufacture since the close of the war will undoubtedly influence Japanese design in the future. We have supplied both .45 and also latest model .38 S & W and Colt's revolvers to Japan recently. The

French 1935 pistol and section view showing mechanism.

latter take the .38 Smith & Wesson special revolver cartridge, a cartridge so much more powerful and efficient than the original Japanese 8mm that its superior characteristics cannot be ignored.

Germany—Walther Pistols

Germany, during the period before World War II, saw the introduction of a wide range of advanced pistols basically of pocket type, but most were later pressed into military service. Perhaps the best known of these were the Walther PP and PPK automatics, the designations indicating Polezei Pistole and Polezei Pistol Kriminal or plainclothes model—from the fact that the latter had a shorter barrel and was intended to be concealed, while the police model was intended basically for holster use on the belt.

The basic function of the double action as provided by Walther was to allow carrying the pistol with complete safety with a cartridge in the chamber yet ready for instant action. While the thumb safety was provided, it was not necessary to apply it as the arm could only be fired by direct pull on the trigger. Carried in this manner, a long pull on the trigger would fire the first shot, the hammer being cocked by the mechanism for succeeding shots in standard automatic pistol fashion. The varieties of these designs will be found covered fully in Part II in the chapter on German World War II Materiel. It may be stated, however, that the Walther

Representative Walther-pattern PPK Model, engraved.

World War I Walther 9mm blowback type pistol.

Hammerless version of the P-38. Very few made. Locked breech.

designs were a radical improvement over any of their predecessors. Mauser later introduced its HSc and Sauer and Sohn also introduced double-action enclosed hammer models with some interesting features in an endeavor to compete with Walther.

None of these initial pistols were in other than the pocket calibers adaptable for blowback action. Walther next turned their design ability to producing the Walther-Heeres Pistole (Army Pistol) in caliber 9mm Luger or PP 08 to use the German military terminology. A slightly modified form of this double-action locked-breech pistol was officially adopted in 1938, the German Government intending to use it to replace the celebrated Luger. As the P-38 it was the basic German sidearm of World War II although it was supplemented, of course, by Lugers, Mausers, and in the final stages by all types of available small arms.

USSR—The Tokarev

In 1930 the Russians introduced their Tokarev automatic pistol. This pistol is credited to their designer Tokarev. It resembles quite closely in most external lines, except hammer, the familiar Colt .32 automatic pistol. However, it is unlike it in that it is a fully locked-breech pistol. The locking system is patterned after that of our .45 Automatic Pistol, also a Browning development, of course. Tokarev's contribution to the Browning design was basically a highly simplified form of manufacture and a compact firing mechanism. While the arm itself is crude in appearance and does not have the balance nor the pointing qualities of the American product, it is cheap to manufacture and completely reliable as a military design. It is chambered for the 7.62mm Russian cartridge which is a bottlenecked rimless cartridge practically identical with the 7.63mm Mauser cartridge.

Remarkable late-World War II stamped pistol intended for ultra-low-cost production. Production did not get under way before the war ended.

Russian Tokarev pistol, caliber 7.62mm. Will handle the Mauser 7.63mm cartridge.

MILITARY USE OF PISTOLS IN WORLD WAR II

Statistics compiled by American military authorities during the course of the war indicate that the actual casualties inflicted upon enemies with the automatic pistol were so few as to render the combat value of the arm questionable. This is, and will continue to be, a subject for military discussion for years to come. The psychological value of the pistol to many military personnel is unquestionable, though its efficiency or its usage in general combat would hardly seem to warrant the expenditure of great sums of money in attempting to further develop the arm.

The standard side arm of the American Forces using a pistol during World War II was, of course, the familiar .45 Colt Government Model. (Officially this is the United States Government Model 1911 or M1911A1, these arms having been made in various plants besides Colt's on government contract. Ithaca, Remington Rand, and Union Switch & Signal all did some production.)

The British used substantial quantities of 9mm Browning High Power auto pistols, most of which were manufactured in Canada by the firm of John Inglis. These pistols proved of considerable

Commercial Colt Model 1911A1.

value in Commando tactics since they had a large magazine capacity (13 shots as against 7 in the Colt .45) and since they used the same ammunition as the Sten and other submachine guns which were the basic weapons of the British specialists. Since these pistols were used by highly trained troops of an "elite guard" nature, their efficiency, which was considerable, is not necessarily a gauge of the value of the arm for general troop use.

The German forces at the outset of the war leaned heavily on automatic pistol distribution to officers and noncoms. As war progressed, with particular reference to the campaigns against the Russians, the German military thinking tended to follow that of the Russian, i.e., to issue submachine guns (Maschinen Pistoles) to special service troops and to noncoms in place of pistols.

The Russians early came to the conclusion that in the hands of their troops offensive arms of the nature of submachine guns were cheaper to make and far more deadly than the familiar pistol. They, therefore, tended to minimize the issuance of the one-hand gun in any form.

Japanese use of the pistol was confined largely to officers. Their Nambus with their relatively inefficient cartridges were not in very wide usage at any time. The widespread issuance of American revolvers to Japanese quasi-military forces may later affect their thinking with reference to pistol issuance for military purposes. Their past experience has tended to minimize both revolver and pistol usage, however.

The Italians issued very large quantities of varying forms of Beretta and lesser known automatic pistols, mostly in .32 and .38 caliber, with of course available supplies of the earlier 9mm calibers such as the Glisenti. The fighting record of the Italian forces, of course, gives no index to the actual value of the pistol as a modern weapon of war any more than does the experience of the French forces. Certainly a study of casualty statistics of forces opposed to these units would not indicate that in their hands the automatic pistol was a particularly efficient arm. Its most noted use occurred during the Italian attack on Ethiopia when official films showed Italian officers using pistols to execute natives, the natives being held by Italian troops while the pistols were presented against their heads for firing. This hardly constitutes combat use.

All later reports from the Korean and Vietnam theaters of activity indicated a concentration by Communist-controlled Chinese and Vietnamese troops on submachine guns, usually of Russian pattern. Pistols as a class have been relegated to use by officers and are not normally encountered when engaging regular troops. Guerrilla use, of course, inevitably produces large quantities of pistols because of the possibility of secreting them on the person. On the other hand, among our own troops on Asiatic duty, there was a persistent demand particularly for revolvers, but the influencing factors appear to be psychological rather than directly military.

Chinese-made Mauser military pattern pistol, cal. .45 A.C.P. Except for cali ber (the Germans made these in production only in calibers 7.63mm, 9mm Mauser, and in 9mm Luger) the Chinese version follows the German design very closely. This is the only commercially successful semiauto pistol to use any magazine except the butt type. Here the magazine is positioned forward of the trigger.

AUTOMATIC PISTOL DEVELOPMENTS SINCE WORLD WAR II

DEVELOPMENTS IN THE UNITED STATES

In the United States since the close of the World War II, Colt has produced a variation of the familiar government Model 1911 .45 caliber pistol using aluminum slide and receiver to reduce weight. The hammer form has been changed to conform with that of the typical H. P. Browning pattern. Other changes have been inconsequential. The basic design has not been altered or appreciably improved.

Smith & Wesson are in small-scale production at this time of an automatic pistol currently designed to handle the 9mm Luger pattern cartridges. This pistol is basically only a modification of the familiar Browning 9mm High Power, whose locking system it utilizes. The locking lug with cam nose on the barrel affords, at least theoretically, better barrel stability when firing than does the familiar link type found in the Colt pattern, but technical matters such as these are of interest only to the ultra precision shooters.

Smith & Wesson has adopted and adapted several of the Browning modifications as found in the Belgian and also in the Polish variant of the 9mm Browning High Power, specifically such items as a magazine disconnector which prevents the firing of a round in the chamber if the magazine is withdrawn. Many accidents occur during the course of cleaning automatic pistols when the soldier withdraws the magazine and fails to recall that there may be a cartridge in the firing chamber. Another adaptation is that of a slide safety device which permits dropping the hammer on a bar which is automatically interposed between the hammer and the firing pin for purposes of safety. Another common source of accidents in military usage of automatic pistols is that encountered when the inexperienced man attempts to lower the hammer while there is a cartridge in the firing chamber and a slipping of the hammer may fire the pistol accidentally.

Smith & Wesson have also made arrangements to produce their new military type pistol with a double-action trigger, under which, as in the case of the earlier Walthers and others, a cartridge may safely be carried in the firing chamber without a manual safety being applied and without any springs being compressed. In a critical moment, it is only necessary to pull the trigger without having to cock the hammer (a very clumsy operation with the typical automatic pistol). The pull on the trigger mechanically actuates the hammer to fire the first shot, the slide automatically cocking the hammer for the next shot.

Other developments in automatic pistols in the United States have been confined, as far as production is concerned, to .22 caliber varieties. The Hi-Standard Company at New Haven, Connecticut has introduced a very fine line of improved versions of its earlier .22 target and plinking pistols, though none are of military design. During the war this firm produced a quantity of .380 caliber blowback automatic pistols which were not suitable for military purposes, being much too heavy for the cartridge they employ. They also did experimental production of both blowback and locked-breech pistols for our Navy to use the standard .38 S & W revolver cartridge. These were discontinued when no order materialized. These designs may, however, be introduced at a later date because of the interest among target shooters; though it must again be noted that, because of the variety of loads required by the target shooter plus the fact that the automatic pistol will operate satisfactorily only on a relatively stable type of loading, the development may not be practical commercially. Certainly it will not be militarily.

The postwar Sturm-Ruger organization developed and produced in considerable quantity a very popular, low-priced automatic pistol in .22 caliber, the external lines resembling very closely those of the familiar Luger. This, however, is a straight blowback weapon intended for plinking and in some models for target use.

Post-World War II "Commander" Model, Colt .45, aluminum frame. Weight reduced to 26.5 ounces. Barrel shortened one inch. Hammer modified.

Smith and Wesson 9mm auto pistol. This is the military model with double-action trigger. Barrel 4 inches. Overall 7.4 inches. Weight with light alloy frame, 28 oz.

The design is ingenious, the low price being made possible by the utilization of stampings, screw machine parts, and furnace brazing techniques. Except as a military trainer, of course, this type of weapon is not of military application, though Ruger's manufacturing methods could be applied to low-cost military designs without loss of reliability.

Several small independent organizations are currently engaged in the development of automatic pistols in the United States. Almost without exception these are of the blowback type intended for small caliber target or low-powered pocket pistol cartridges, the relatively large imports of arms of this sort from abroad having demonstrated the continuing market for them in the United States. These again are of only correlative military interest, being arms which would be usable only in an emergency. They require no particular attention here.

POSTWAR DEVELOPMENTS IN EUROPE

In Switzerland, the Societe Industrielle Suiesse Neuhausen introduced in 1944 their new model pistol based on the Colt-Browning design in accordance with the Charles Petter patent

The Smith and Wesson 9mm auto pistol. Military model with standard single- action trigger. Specimen shown is a factory hand-engraved model from initial production. Pistol is produced in both steel and light alloy frame.

modifications. The French model 1935A automatic pistol which we have mentioned was the first pistol designed by this inventor. The Swiss firm introduced their pistol to handle the familiar 9mm Parabellum (Luger) cartridge. The barrel locking system while still on the familiar Browning pattern utilizes a modified version of the Browning high power 1935 lock. This pistol and its modifications will be found described in Part II in the chapter on Switzerland. The pistol was introduced in two modifications, the SP44/8 and the SP44/16. They differ only insofar as grip and stock modifications required in relation to magazine capacity, the 8 being a straight-line single column magazine holding 8 cartridges, the other being a double line staggered box holding 16 cartridges. Very few of the Model SP44/16 were made. An interesting feature of the Petter design is a hammer assembly which is a modification of the Russian Tokarev incorporating hammer, mainspring, sear, and disconnector assembly in a single unit easily removed from the frame or receiver. Switzerland adopted the 8-round version.

Since the barrel is a floating unit easily removed from the slide and receiver assemblies without tools, as in the case of the Colt .45 and the High Power Browning designs, the Swiss have taken advantage of this feature to furnish the pistol in calibers 7.65mm and 9mm Parabellum. Only a different barrel and recoil spring are needed with the basic mechanism to change the caliber from 9mm to 7.65mm.

This changeover is possible because of the original cartridge design developed by George Luger in Germany at the turn of the century. His original cartridge was a bottleneck 7.65mm which as we have pointed out in discussing German production was later altered when the German high command demanded a pistol of large caliber. Removing the bottleneck in the case and providing a bullet of 9mm caliber to replace the 7.65mm brought about the change. Thus, the only difference in the two barrels is in the forward end of the chamber and in the barrel bore and rifling. In the Luger design the barrel is screwed into the barrel extension and requires tools for removal. It is, therefore, basically a gunsmith's job to change one caliber to another, although only barrel removal and possible recoil spring adjustment is needed to accomplish this since all other parts including the magazines are interchangeable.

In this new Swiss pistol, however, any amateur can take the pistol apart and replace the 9mm barrel and recoil spring with a 7.65mm barrel and recoil spring to convert the arm. The Swiss Army prior to the official adoption of this pistol was equipped with the 7.65mm Luger pistol in the 06-29 variations, which are minor modifications of Luger's original design. The new Swiss Service pistol, by featuring interchangeable barrels and calibers, permits use of ammunition stocks on hand while awaiting complete changeover to the 9mm caliber.

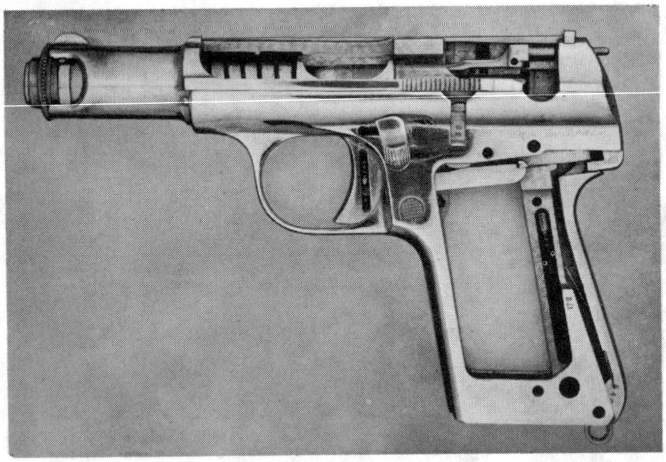

Representative postwar Spanish Astra pistol cutaway to show mechanism closed, hammer down.

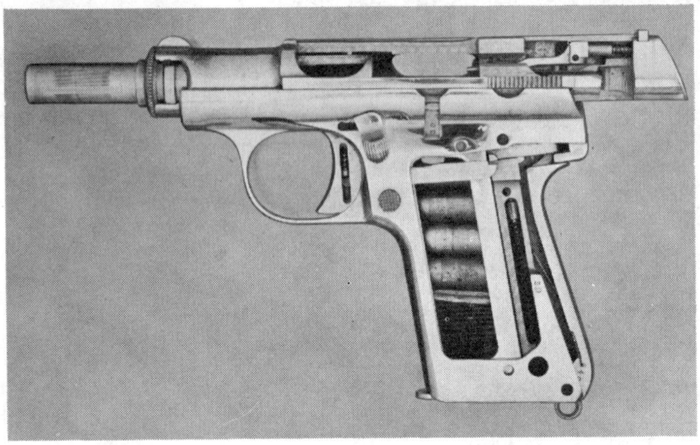

Astra, action open, magazine partly loaded to show feed system.

As an added, but not particularly practical group of accessories, this Swiss pistol can also be furnished with special recoil spring, barrel, magazine, and slide to convert the frame to a .22 caliber pistol for practice work. Theoretically, this is an excellent arrangement. In actual practice, the cost and the difficulty, slight though it is, of the interchange have never permitted this type of element to achieve financial success in the United States.

As evidencing the liaison between Switzerland and Denmark, it is interesting to note that the SP47/8 has been adopted as the official Danish service pistol, also in 9mm caliber.

The Czech national arms factory in 1946 produced under the designation CZ a modification of their wartime pistol. It is a double-action operated weapon in which the hammer is cocked and dropped by mechanical pull through on the trigger, the slide serving only to eject the empty and reload from the magazine. Unlike the Walther type of double action, it must be kept in mind that this Czech design is mechanically actuated at all times, which means that the trigger pull is uniformly bad. The Czech military consider pistols basically defensive weapons only for the closest use and the safety factor inherent in a weapon where the springs are compressed only during the firing cycle, and where a long heavy pull is needed to fire the chamber cartridge, is actually considered an asset. The caliber has been increased to 9mm Parabellum from the previous less powerful .380 cartridge. Where the previous form except for experimental models was blowback operated, the CZ47 is short recoil, still another variation of the familiar Colt-Browning lock, though again the actuation is caused by a barrel lug rather than a swinging link. This weapon is not in common use in Czechoslovakia.

The German Walther PP and PPK designs are being manufactured in small quantity under license by a Turkish factory for government use. These same designs, also from the blueprints and working drawings of the Walther organization, are being used in the manufacture of .22, .32, and .380 caliber pistols identical with the Walther PP and PPK models. These were manufactured in France under Walther license and are now being manufactured in West Germany.

In Spain, Astra pocket pistols are being produced in calibers .22, .25, .380, and 9mm. These are all blowback weapons of elementary design but excellent manufacture. The Model 600 using the powerful 9mm Luger cartridge is equipped with special recoil springs to stand the shock of this very heavy load which in other pistols of lesser weight utilizes locked-breech designs. These designs are refinements only of earlier blowbacks, however.

Also from Spain are coming considerable quantities of elementary blowback pistols in calibers .22, .25, .32, and .380, under the Llama trademark. These for the most part are minor modifications of the Colt-Browning systems.

Echeverria of Eibar in Spain is marketing under their trade name (Star) a wide variety of pistols which are basically modifications of the Colt pattern. These include blowback arms in the .25, .32, and .380 caliber classifications in pocket types, plinking, and target

Postwar Italian Bernardelli .25 (6.35mm) auto. Arms such as these are used at times by European staff officers.

Postwar Italian Bernardelli .32 (7.65mm). A representative modern blow-back auto. Derived generally from an earlier German Walther pattern, the mechanism has been modified in the U.S.A. for production of a low-priced .22 pistol.

pistols, which are basically the same design but with elongated barrels and locked-breech modifications of the Colt-Browning system in calibers 9mm, .38 Colt Automatic, and .45 Automatic.

Czechoslovakia is exporting elementary blowback pistols in .25 caliber in small vest pocket designs, as well as double-action models in .32 caliber, all being blowback design, of course.

A wide range of pocket pistols in calibers .25, .22, and .380 are also being produced and exported from Italy. These include the Bernardelli, the Galesi, and a wide line of Berettas. None of these are in any true sense military pattern arms. None have any essential improvements likely to influence design in any particular way; and currently, at least, none are in competition in the United States because these calibers are not as yet being manufactured here in automatic pistols.

In France the Manufacture d' Armes Bayonne is producing a formal pattern blowback automatic in calibers .22, .32, and .380 principally for export.

Also in France the Manufacture d' Armes Hendaye are producing good blowback pistols in calibers .22, .32, and similar pocket designs. These pistols, as well as some of the Spanish Astra and Star patterns, have been sold to German police agencies. None of them, however, including this French line which is trademarked Unique, have any military features or design features of any importance from an American developmental standpoint. They are elementary patterns of blowback weapons.

SPECIAL FEATURES OF THE AUTOMATIC PISTOL

As we have previously mentioned, the automatic pistol is actually misnamed. It is truly an automatic loading pistol. Since a truly automatic weapon requires a minimum weight of about 8 pounds to permit holding it down during full automatic fire, and since a pistol for personal use cannot be permitted to go much beyond 40 ounces in weight and should preferably be very considerably below that, it is obvious that pistol design must incorporate a device which will allow only one shot to be fired for each individual pull of the trigger.

However, the name "automatic pistol" has come to be generic. In discussing it herein, therefore, we have adhered to that terminology.

The automatic pistol has several advantages over the revolver from a military standpoint, though for sheer dependability and for close quarters defense for all but truly military use, the revolver is unquestionably superior.

The design of the typical automatic pistol in which the magazine is housed in the grip permits a very much larger ammunition supply to be carried in the weapon than does any revolver. The standard number is 7 or 8, but staggered box designs such as the Browning High Power hold 13 cartridges and the Swiss SP44/16 held 16 cartridges. For military pattern cartridges of 9mm or larger, the single line box with 8 cartridges is accepted as the optimum, however. Because a supply of loaded magazines can be carried, the arm can be reloaded so much more rapidly than a revolver that it can pour out a tremendously higher volume of fire than the revolver after the first loading has been exhausted.

It is a much more compact sidearm, caliber for caliber, than the comparable revolver. Since the arm in most designs is cocked by action of the slide during recoil, it can place a larger number of aimed shots in a given period than can the comparable revolver.

It is true that a revolver expert may occasionally fire six shots as fast as from an automatic. But because of the revolver disadvantage of long trigger pull to achieve this rapidity of fire, accuracy suffers correspondingly except at close quarters. Since the cartridge chamber is part of the barrel itself in an automatic pistol, this type of arm is not subject to bullet shaving and gas wastage as occurs in the revolver where the bullet must jump the gap from the chamber in the cylinder to the barrel. In addition, a properly designed automatic pistol is far simpler for the amateur to dismount, maintain, and assemble than any comparable revolver.

For military purposes, even at close quarters where a sustained volume of rapid fire is desirable, the superiority of the automatic pistol cannot be questioned. For all other practical military and police use, however, the additional reliability of the revolver seems more desirable. The automatic pistol when loaded and carried for service use has many springs constantly under compression, the magazine spring at all times, and the hammer and main springs when the arm is carried cocked. The susceptibility to mechanical failure is therefore necessarily higher in the automatic pistol than in the revolver.

However, in all arms as in all things mechanical, there is no all-purpose item. The needs and requirements of the individual services under individual conditions must inevitably lead to compromise. This factor explains and justifies U.S. purchases of lightweight revolvers, for example.

POSSIBLE ADVANCES IN FUTURE MILITARY AUTOMATIC PISTOL DESIGN

Like the revolver, the automatic pistol would appear to have approached the zenith of its design. Further important developments must depend entirely upon advances in ammunition or metallurgical technology.

In face of the world efforts toward standardization of armaments in non-Communist countries, and in the face of the ballistic background of the 9mm Luger cartridge both for the pistol and submachine gun use, there is great likelihood that this cartridge in the future will be the basis for most military design.

On the other hand, the tremendous United States investment in both weapons themselves and manufacturing tools for both .45 caliber automatic pistols and ammunition, has limited the rapid changeover.

Even before the war, Walther experimented extensively with aluminum frames for their PP and PPK designs in automatic pistols. Sauer did likewise. One of the postwar developments now underway with Smith & Wesson at Springfield is the production of their new 9mm automatic pistols in both single- and double-action, made with aluminum frames as alternates for the steel construction. Further advances may be expected in relation to the materials used in automatic pistols with a view to reducing weight or increasing tensile strengths. However, the requirements in the way of weight because of inertia factors, which are a necessary part of any automatic pistol design, must obviously control the minimum weights of the arms with relation to the power of the ammunition used. For all practical purposes blowback pistols must continue to be confined to cartridges with relatively little power. All conceivable forms of springs to compensate for lack of the necessary weight factor in moving breechblocks have been experimented with from the earliest days of the development of this type of weapon.

The long recoil system for pistols has been invalidated by

experience. It is and must continue to be unnecessarily complicated for use in a weapon intended to be fired from one hand. Gas operation in all its forms can be truly successful only with weapons involving considerable weight, hence it is ruled out. The types of mechanisms required for its proper functioning are not applicable to lightweight compact pistol design.

Hesitation, or delaying the blowback, involves the utilization of mechanical factors calling for considerably more cost, skill, and technical engineering than is advisable in a pocket or one-hand type arm.

Soviet .22 caliber Margolin match pistol, Model 1949.

The short-recoil system, therefore, is likely to remain the most practical for relatively high-power automatic cartridge pistols.

Potentialities do exist in connection with the principle of primer projection for unlocking, and for utilization of some of the fields of applied inertia through utilization of permanent magnets. However, for all practical purposes, no major developments are to be expected in the field of automatic pistol mechanisms unless and until a new form of ignition is evolved.

Automatic pistols are normally arms to be used within a distance of 30 feet. In the hands of the target shooter and the expert, given optimum conditions, the pistol is an extremely dangerous weapon at 50 to 75 yards. There are very occasional shooters who can use this one-hand weapon effectively at ranges of 100 yards and even more. But for all practical purposes, the limitations set up by the requirements for light weight and compact form, taken together with the inability of the average shooter to effectively handle an arm with too heavy recoil, plus the sighting deficiencies inherent in a short arm with a minimum distance between sights, all conspire to hold practical pistol development within its present confines.

Of the practical limitations involved in one-hand gun design, it might be pointed out that during World War II the firm of Smith & Wesson conceived the idea of producing a revolver which would use the half-moon clip design they originally developed during World War I for the .45 automatic pistol cartridge, but which would actually chamber the .30 M1 Carbine cartridge. The general theory was that such a revolver issued to paratroopers armed with carbines would give them an additional close-quarters weapon with all the value of interchangeability of ammunition.

When these .30 M1 revolvers were tested by average personnel, it soon developed that the shock of firing the arm more than offset any possible logistical gain in ammunition supply. Although the carbine cartridge was designed to have its best results by having the powder burned in an 18-inch barrel, the resulting ballistics were generally inferior to those of standard handgun cartridges.

It is quite possible with improved metals to produce a Smith & Wesson revolver of relatively light weight to take even their ultra powerful .357 magnum cartridge. However, the shock of successive firing of this cartridge in a lightweight weapon not only destroys accuracy but actually hampers the use of the revolver by all but the most expert shooters.

There will, of course, continue to be modifications in practically all the accepted designs, but the basic physical facts involved in automatic pistol design preclude the possibility of any startling mechanical developments in weapons of this type.

PART 2

Current Weapons

13 Argentina

The Argentine Army is currently equipped with the 7.62mm NATO FN "FAL" rifle which is manufactured at the Fabrica Militar de Armes Portatiles Domingo Matheu at Rosario, Santa Fe. The heavy-barrel version of this weapon is also produced at the Rosario plant. The caliber .45 M1916 and M1927 automatic pistols were the standard pistols; these are copies of the Colt M1911 and M1911A1 respectively. The current standard pistol is the 9mm Browning High Power. The copies of the Colt were made and the 9mm Browning is being made at Rosario as were the 7.65mm Browning light and heavy machine guns which were standard for many years, but are now being replaced by the FN 7.62mm NATO "MAG" machine gun. There are still a few 7.65mm Madsen machine guns in use; these are also being replaced. The caliber .50 Browning machine gun is standard and is used on armored vehicles. The 9mm PAM 1 and PAM 2 are standard submachine guns and are Rosario products.

The Argentine police use 7.65mm M1891 carbines, Star full automatic pistols, the Israeli Uzi submachinegun and the U.S. Caliber .45 M19281 and M1 Thompsons.

ARGENTINE PISTOLS

Argentina has manufactured copies of the caliber .45 Colt M1911 and M1911A1 which are called the M1916 and M1927 respectively.

ARGENTINE BALLESTER MOLINA PISTOL

The firm "HAFDASA" of Buenos Aires manufactured the Ballester Molina and Ballester Rigand pistols. This firm is no longer in business and the pistols are no longer being manufactured. The Ballester Molina caliber .45 pistol is still in wide use as it was manufactured in quantity during World War II.

The Ballester Molina is a slightly modified copy of the Colt .45 M1911A1 pistol. Loading, firing, and functioning of the pistol are the same as for the U.S. pistol.

Characteristics of the Ballester Molina

Caliber: 45 M1911 automatic pistol cartridge.
System of operation: Recoil, semiautomatic fire only.
Weight, empty: 2.25 lb.
Length, overall: 8.5 in.
Barrel length: 5 in.
Feed device: 7-round, single column, detachable box magazine.
Sights: Front, blade; rear, notched bar.
Muzzle velocity: 830 f.p.s.

Major Differences Between the Ballester Molina and the Colt M1911A1

Hammer strut: The hammer strut is much smaller than that of the U. S. M1911A1. It is 0.75 inches in length and 0.158 inches in diameter.

Firing pin stop: The firing pin stop is not recessed on the sides, as it is on the U.S. models.

Safety lock: The safety lock is redesigned and the pin is larger in diameter than the safety lock pin on the U. S. model.

Mainspring housing: The mainspring housing, although arched as in the U. S. Model M1911A1, is an integral part of the receiver.

Trigger: The Argentine pistol has a pivoting trigger. An extension from the trigger, along the right side, cams the disconnector and engages the sear.

Argentine caliber .45 M1927 pistol

Caliber .45 Ballester Molina pistol.

Magazine and magazine catch: The magazines are interchangeable. The magazine catch is located in the same place as the catch on the U. S. M1911A1. The assembly of the catch is somewhat different, but it operates in the same manner as the U.S. model.

Slide: There is no slide stop disassembly notch as on the U. S. models.

Field Stripping the Ballester Molina

Disassembly is the same as for the U. S. M1911A1, except that the pin for the hammer and the sear must be driven out. The trigger is held by a trigger pin. Upon removing the trigger and the trigger extension, the disconnector can be removed downward. After the sear has been removed, the main spring can be removed.

7.65mm M1905 MANNLICHER PISTOL

Argentine 7.65mm M1905 pistol.

This was formerly the Argentine service pistol and quantities of them were sold to surplus arms dealers in the U.S. recently. The pistol was developed at Steyr and introduced commercially in 1901. The weapon is essentially a blowback-operated type and is loaded with a stripper-type clip (charger), from the top. The weapon can be unloaded from the top by pulling the slide to the rear and pulling down the catch on the right side of the pistol.

Characteristics of the 7.65mm M1905 Mannlicher Pistol

Caliber: 7.65mm Mannlicher (called 7.63mm Mannlicher in Austria).
System of operation: Blowback with slight retardation.
Weight: 2 lb.
Length overall: 9.62".
Barrel length: 6.31".
Feed device: 8 round non-detachable box magazine.
Sights: Front, blade; rear, notch.
Muzzle velocity: approx. 1025 f.p.s.

ARGENTINE RIFLES

RIFLES IN USE

The M1891 Argentine Mauser is quite similar to the 7.65mm M1890 Turkish Mauser. The M1909 Mauser is a slight modification of the German Gewehr 98 (rifle 98).

The M1891 Carbine is still used as a police weapon; but the rifles are obsolete. The F.N. 7.62mm NATO "FAL" rifle and the 7.62mm NATO heavy-barrel rifles are now standard.

Characteristics of the M1891 and M1909 Argentine Mausers

	M1891	M1909
Caliber:	7.65mm rimless.	7.65mm rimless.
System of operation:	Manually operated bolt action.	Manually operated bolt action.
Length, overall:	48.6 in.	49 in.
Barrel length:	29.1 in.	29.1 in.
Feed device:	5-round, single column box magazine.	5-round, staggered-row box magazine.
Sights: Front	Barleycorn.	Barleycorn.
Rear	Leaf.	Tangent leaf.
Muzzle velocity: (w/spitzer-pointed ball)	2755 f.p.s.	2755 f.p.s.
Weight (empty):	8.58 lb.	9.2 lb.

7.65mm Model 1891 Argentine Mauser.

7.65mm Model 1909 Argentine Mauser.

ARGENTINE CARBINES

Argentina has had a number of 7.65mm Mauser carbines in service. The Model 1891 carbine is stocked to the muzzle, has a 17.6 inch barrel and is 37 inches overall. There are two versions one with and one without bayonet lug. There are four common versions of the Model 1909 carbine. The cavalry carbine is stocked to the muzzle and is illustrated on page 90. Some cavalry carbines have bayonet lugs and others do not. The Engineer carbine has a typical military stock and a bayonet stud which protrudes horizontally from the upper band like that of the German Mauser 98. A carbine similar to the Engineer carbine, sometimes called the Mountain carbine has a bayonet stud which protrudes vertically in front of and below the upper band.

A large number of these weapons have been converted into sporters in Argentina.

ARGENTINE SUBMACHINE GUNS

NEW MODELS

Several new submachine guns have appeared in Argentina. The Halcon Model ML57 is a 9mm Parabellum weapon capable of selective fire and is considerably lighter and less complicated than the early model Halcon guns. A later model, the Halcon ML 60, is similar but has two triggers, one for automatic and one for semiautomatic fire, rather than the selector lever of the ML 57.

THE "MEMS" SUBMACHINE GUNS

The firm of Armas & Equipos S.R.L. located in Cordoba, Argentina, has developed a series of light, easily manufactured submachine guns. These are the Models 52/58, 52/60, AR 163 and 67. All are generally similar in design being conventional blowback operated guns making extensive use of stamping and fabrications in their design. The Model 67 is advertised as being particularly suitable as a counter-insurgency weapon because of its relative simplicity and ease of manufacture. All parts, excepting the barrel are made with low tolerances. The barrels have microgroove rifling.

OLDER MODELS

The older guns in Argentina are the Ballester Molina caliber .45 Type C3, the Halcon caliber .45 M1946, the Halcon caliber .45 M1943, the P.A.M. 1, a 9mm copy of the U.S. M3A1 made at the government arsenal at Rosario and the 9mm Ballester Rigaud and HAFDASA C-4. The P.A.M. 2 is also made at Rosario. With the exception of the P.A.M. 2, it is believed that all the older Argentine submachine guns are out of production.

The characteristics of Argentine submachine guns follow:

Argentine 9mm Parabellum "MEMS" Model 52/60 Submachine Gun.

Argentine 9mm Parabellum "MEMS" Model AR63 Submachine Gun.

Argentine 9mm Parabellum "MEMS" Model 52/58 Submachine Gun.

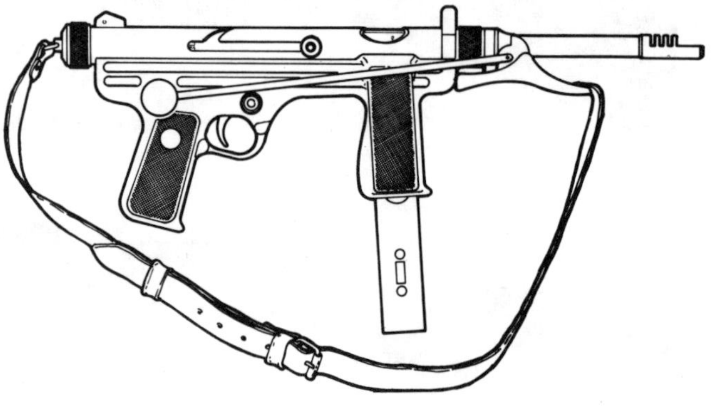

Argentine 9mm Parabellum "MEMS" Model 67 Submachine Gun.

CHARACTERISTICS OF ARGENTINE SUBMACHINE GUNS

Weapon	Caliber	Type of operation	Overall length	Feed device	Barrel length	Cyclic rate	Muzzle velocity	Weight
Ballester Molina C3	.45	Blowback selective fire	33.2 in.	40 round detachable, staggered box magazine	12.62 in.	500-600 rpm	900 fps	8.6 lbs.
Halcon M1943	.45	Blowback selective fire	33.40 in.	17 or 30 round detachable, staggered box magazine	11.5 in. W/ compensator	700 rpm	950 fps	10.45 lbs.
Halcon M1946	.45	Blowback selective fire	Stock folded; 24.4 in. Stock extended: 31.1 in.	17 or 30 round detachable, staggered box magazine	6 in.	700 rpm	920 fps	8.90 lbs.
Hafdasa C-4	9mm Parabellum	Blowback selective fire	Stock folded: 21 in. Stock extended: 31.2 in.	30 round detachable in-line box magazine	7.75 in.	600 rpm	1200 fps	7 lbs.
Halcon M. L. 57	9mm Parabellum	Blowback selective fire	Stock folded: 21 in. Stock extended: 30.7 in.	40 round detachable, staggered box magazine	8.86 in.	520 rpm	1200 fps (approx.)	6.95 lbs.
P.A.M. 1	9mm Parabellum	Blowback full automatic	Stock folded: 21.2 in. Stock extended: 28.6 in.	30 round detachable, in-line box magazine	7.9 in.	450 rpm	1200 fps (approx.)	6.6 lbs.
Model 52/58	9mm Parabellum	Blowback full automatic	Stock extended: 35.2 in.	40 round detachable, staggered box magazine	12 in. (approx.)	750-800 rpm	1200 fps (approx.)	6.8 lbs.
Model 52/60	9mm Parabellum	Blowback selective fire	Stock extended: 35 in.	40 round detachable, staggered box magazine	12 in. (approx.)	750-800 rpm	1200 fps (approx.)	6.8 lbs.
Model AR 163	9mm Parabellum	Blowback selective	Stock extended: 35 in.	40 round detachable, staggered box magazine	12 in. (approx.)	750-800 rpm	1200 fps (approx.)	6.8 lbs.
MEMS M67	9mm Parabellum	Blowback selective	Stock folded: 25.6 in. Stock extended: 34.7 in.	40-round detachable staggered row magazine	7.9 in.	750-800 rpm	1100 fps (approx.)	6.1 lbs.

14

Australia

Australia uses the 7.62mm NATO F.N. "FAL" rifle as modified by the United Kingdom, the L1A1, the 9mm F1A1 (Aust) submachine gun, the 7.62mm NATO M60 (U.S. designed) machine gun and the Caliber .50 Browning machine gun. The Bren gun converted to 7.62mm NATO is also used.

Previously standard weapons such as the caliber .303 Rifle No. 1 Mark III*, the 9mm Austen and Owen submachine guns, and the caliber .303 Bren and Vickers guns are probably still available in varying quantities as reserve weapons.

As with other members of the British Commonwealth, Australia has, on occasion, taken an independent tack in the development of small arms. The development of the X3 (now the F1) submachine gun is an example of this, as was the non-acceptance by Australia of the rifle No. 4. Australia continued to manufacture the Rifle No. 1 Mark 3*, SMLE, through World War II and also developed some prototype lightweight versions of that weapon during World War II.

AUSTRALIAN RIFLES

With the establishment of the Australian government rifle factory at Lithgow in 1912, Australia began manufacture of her own rifles. The caliber .303 Rifle No. 1 Mark III* rifle was made until 1955; Australia never adopted the No. 4 rifle. A lightweight modification of the No. 1 rifle, called the Rifle No. 6 (Aust) was developed at Lithgow during World War II. This rifle was a shortened and lightened version of the No. 1 rifle; it was never produced in quantity. A total of 640,000 Lee Enfield rifles were produced at Lithgow and it's World War II feeder plants (Orange and Bathurst were the principal feeder plants) before the end of manufacture in 1955.

AUSTRALIAN SUBMACHINE GUNS

Australia has produced a number of native submachine gun designs. The Austen and the Owen were developed during World War II and were considered excellent designs for their time. Both guns are out of production and have been replaced in Australia by the 9mm Parabellum F1A1 (Aust) which is also of Australian design.

THE AUSTEN SUBMACHINE GUNS

These guns were made by Diecasters Ltd. and W. T. Carmichael of Melbourne. About 20,000 guns were made during World War II. Although the Austen resembles the British Sten externally, internally it resembles the German MP 38 and MP 40 (Schmeisser). The Austen has the same telescoping type cover over its recoil spring and firing pin assembly as do the MP 38 and MP 40.

Australian 9mm Austen Mark I Submachine Gun, no longer being made.

Characteristics of the Mark I Austen

Caliber: 9mm Parabellum.
System of operation: Blowback, selective fire.
Length overall: Stock fixed: 33.25 in.
Stock folded: 22 in.
Barrel length: 7.8 in.
Weight: 9.2 lb.
Feed device: 32-round, detachable, staggered row box magazine.
Sights; Front: Barley corn.
Rear: Aperture set for 100 yards.
Muzzle velocity: Approx. 1280 f.p.s.
Cyclic rate: 500-550 r.p.m.

Stock folded, magazine in place -- the Austen Mark I 9mm Submachine Gun.

Loading and Firing the Austen

Load magazine exactly as for automatic pistol. Magazine will hold 32 cartridges, but gun will function better if about 28 are used.

Insert magazine in housing on left side of receiver and push forward until it clicks.

Pull the cocking handle back as far as it will go. It will be held back by the sear, and is now ready for firing when the trigger is pressed.

The folding skeleton stock may be opened by pushing down on the release plunger and pulling back the stock.

To put the weapon on safe, pull bolt handle to the rear and lock in slot cut in top of receiver.

How the Austen Gun Works

Starting with the gun loaded and cocked the action is as follows:

When the trigger is pressed it moves the sear down out of its contact with the bent of the bolt. The compressed recoil spring in the telescoping tube is now free to drive the bolt forward. The firing pin is seated inside a hole in the bolt and protrudes through its hole in the front face of the bolt. It is blocked by the heavy extractor from striking the cartridge until the weapon is in full forward position. Feed ribs are cut in the side of the bolt and these strip the top cartridge from between the cutaway lips of the magazine and push it into the firing chamber.

As the cartridge is chambered, the heavy extractor cams up over the base of the cartridge, and its hook snaps, under the tension of the extractor spring, into the cannelure of the cartridge case. At this point the firing pin is free to strike the primer.

As the powder burns, the bullet is driven down the barrel. The rearward action starts at the same time, but since the weight and inertia of the recoiling parts are very much greater than that of the bullet, opening the breech is delayed long enough to assure that the moment of dangerous breech pressure passes before it opens.

As the bolt starts back, the extractor draws the empty cartridge case with it; this case is struck against the ejector and pivoted out the right side of the gun and then the extractor spring snaps the extractor back into its place. As the bolt in its rearward travel passes the mouth of the magazine, the magazine spring forces the next cartridge into line ready to be picked up on forward motion.

The rear buffer end of the recoil spring tube is firmly supported by the cap and retaining ring through which it passes. Hence as the bolt goes back, it telescopes the heavy steel sections, one over another, compressing the spring inside.

If the selector has been set for single-shot fire, the bolt rides over the sear nose which springs up and catches in the bent of the bolt holding it open for the next press of the trigger. If the selector has been set for full automatic fire, it is not retained and the bolt is free to move forward to fire the next cartridge.

THE MARK II AUSTEN SUBMACHINE GUN

A later version of the Austen machine carbine was also introduced. This design, known as the Mark II, resembles the original in outward appearance, but differs radically in construction.

The receiver consists of two pieces of cast aluminum, and

Field Stripping the Mark I Austen

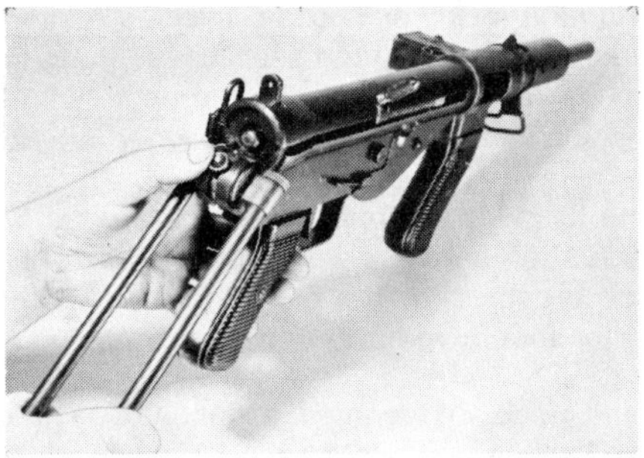

Push down on stock locking plunger and bend the stock down out of line with the rear of the gun.

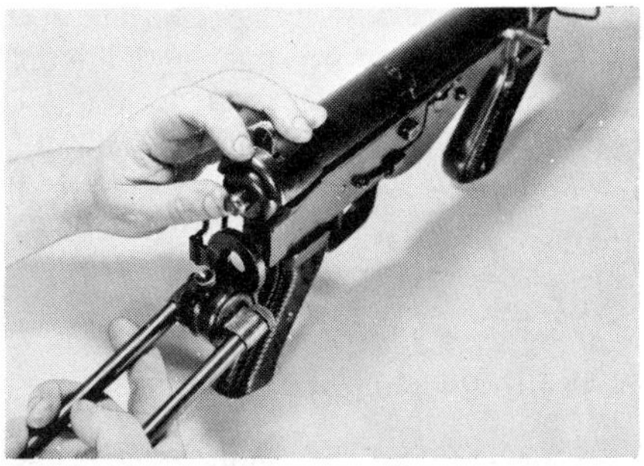

With the left thumb press in on the head of the spring tube protruding through the back of the gun. With the right hand grasp the stock firmly and

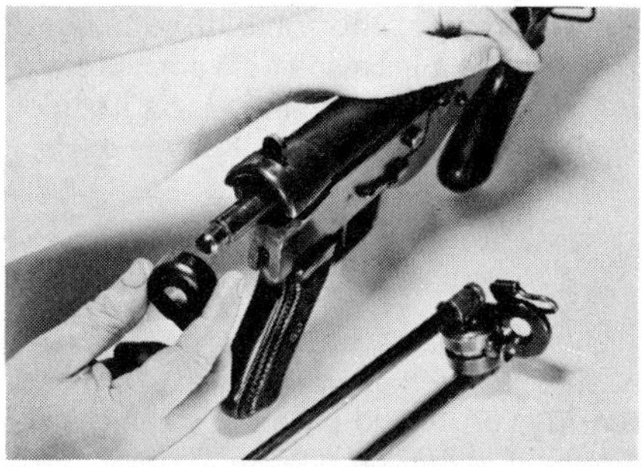

slide it down out of its locking groove in the rear of the receiver. Lift out the buffer cap.

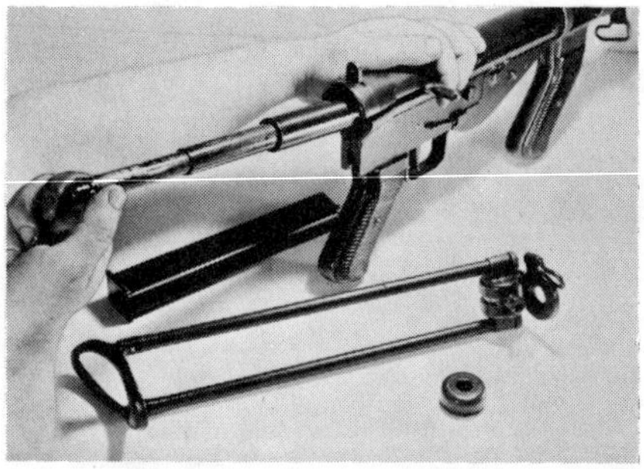

Pull back the cocking handle which will bring the recoil and buffer spring tube back and pull the telescoping tube out of the receiver.

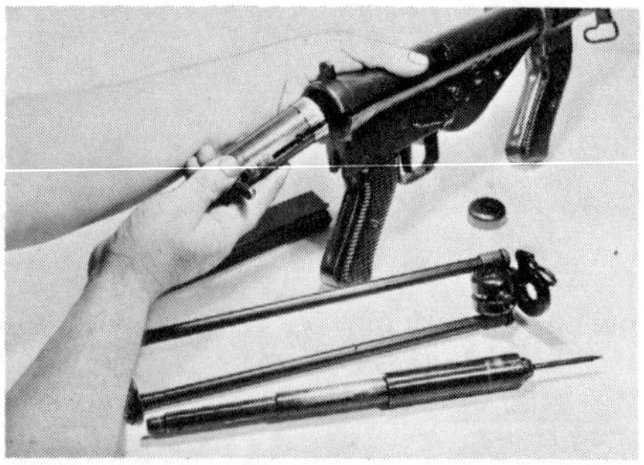

Now pull the bolt back out of the gun. The extractor may be punched out of the bolt if necessary.

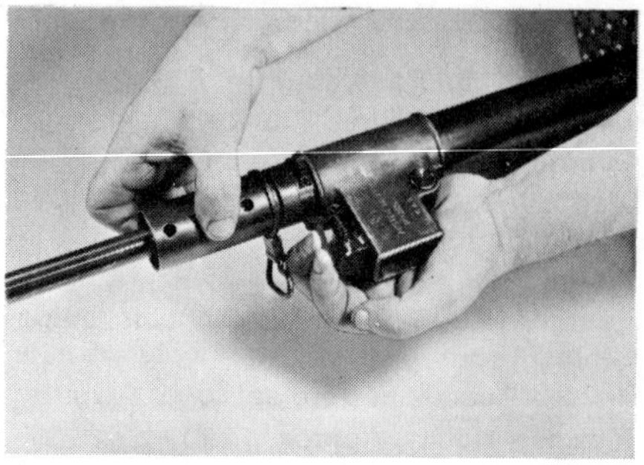

With the fingers of the left hand, pull back the barrel nut catch against its spring tension (it is at the front end of the magazine housing on the left side of the gun) and with the right hand unscrew the barrel casing and barrel nut to the right.

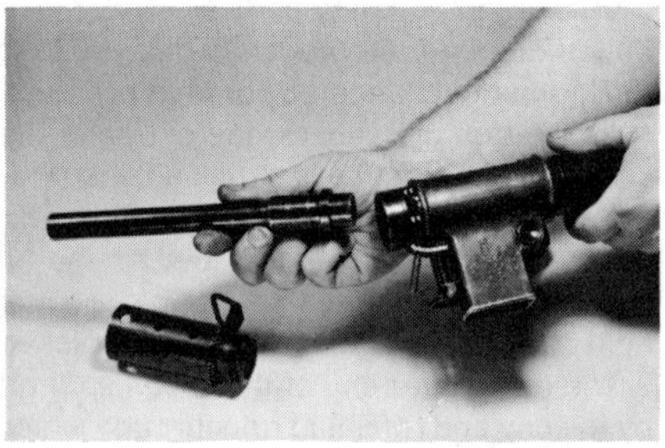

Pull the barrel nut and casing forward out of the receiver. Pull barrel straight forward out of the receiver.

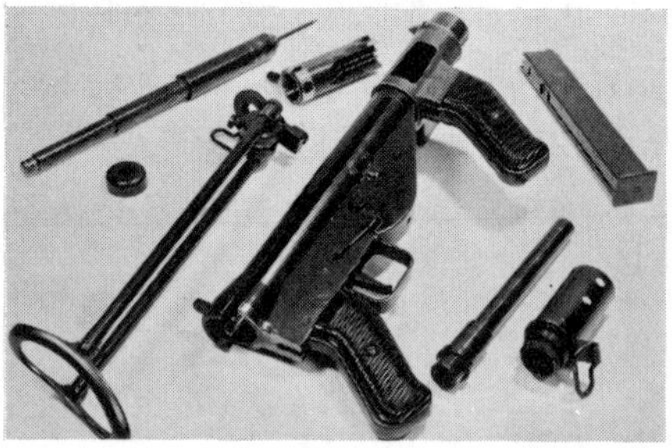

This completes field stripping.

represents an endeavor on the part of the designers to speed up production and at the same time provide a sturdy weapon.

The receiver may be separated when stripping the weapon by pushing a button just ahead of the front hand grip on the left side; then pulling back on the rear grip. The rear section of the receiver containing the trigger mechanism will come straight back out of its joint with the front section.

The barrel rests in the front section of the aluminum receiver, while a cylindrical steel bolt travel-piece acts as a barrel extension. It is attached to the barrel, but extends back into the rear section of the aluminum receiver. Both the barrel extension and the rear of the receiver are cut away to permit travel of the cocking handle to be unimpeded. Note that the bolt travels back and forth in this steel barrel extension—at no time does it come in touch with the aluminum receiver shell. The customary Sten-type push-through button is provided to give single-shot or full automatic fire.

As the tube in which the bolt travels is of heavy steel and is partly enclosed in and cushioned by the aluminum receiver, injury to or deformation of the travel tube is not as likely to occur as in submachine guns made from straight steel stampings. The bolt and recoil spring tube, as in the Mark I Model, follow the Schmeisser design.

THE OWEN SUBMACHINE GUN

Over 40,000 Owen submachine guns were made by Lysaghts Newcastle Works, New Castle, South Wales, Australia during World War II. The Owen is somewhat unusual in having a top-mounted magazine, like the current F1A1 and a quick-change barrel. The ease of barrel removal is of help in maintenance of the weapon, but is not intended for change in battle as are the quick-change barrels of machine guns.

Characteristics of the Owen Mark I

Caliber: 9mm Parabellum.
System of operation: Blowback, selective fire.
Length overall: 31.8 in.
Barrel length: 9.8 in.
Weight: 9.37 lb.
Feed device: 30-round, detachable staggered row, box magazine.
Sights; Front: Off-set barley corn.
Rear: Off-set aperture.
Muzzle velocity: Approx. 1300 f.p.s.
Cyclic rate: 800 r.p.m.

How to Load and Fire the Owen

When a loaded magazine is inserted in the housing on top of the receiver and pushed in until it locks, pulling the cocking handle back until the bolt is caught and held by the sear leaves the gun ready for firing. The gun may be set to fire a shot for each pull of the trigger; or for full automatic fire as long as the trigger is held and cartridges feed.

It will be noted that the barrel is equipped with a compensator at the muzzle to help hold the gun down when firing, with radial cooling surfaces near the breech end. The barrel catch releases the barrel and front grip for speedy removal under favorable conditions.

The operating spring and guide are mounted in standard fashion in the head of the bolt and within the receiver tunnel, where they are retained by the cap at the rear.

The metal skeleton stock is easily removed by pressing the spring held catch mounted within its forward section.

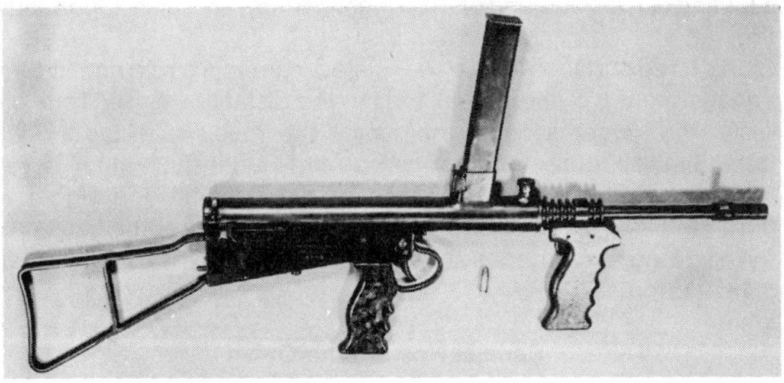

Owen 9mm Submachine Gun, no longer being made.

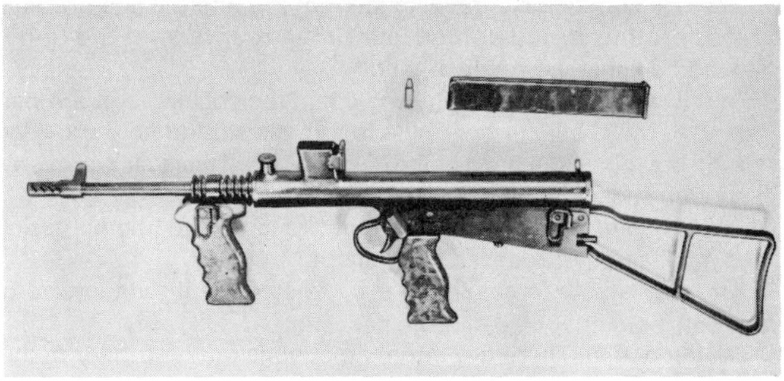

Owen 9mm Submachine Gun, magazine removed.

How the Owen Works

When a loaded magazine is in position and the bolt is withdrawn to its fullest extent, if the change lever is set for single-shot fire, pressure on the trigger forces the rear end of the sear out of contact with the bolt. As the operating spring starts to drive the bolt forward, the bent on the front end of the sear slips over the bent on the trigger permitting the sear to return to its former position under action of the sear spring. This leaves it free to engage and hold the bolt when it returns to the rear.

As the bolt is driven forward by the operating spring, either the right or the left upper feed piece machined in the bolt strikes the base of the first cartridge in the magazine and drives it straight ahead through the opening in the front end of the magazine lips.

The bullet nose is guided by the barrel feed into the chamber as the rear of the cartridge clears the lips of the magazine. As the cartridge enters the chamber, it lines up with the bolt, enabling the head of the cartridge to seat in the base of the bolt-head recess, in which the firing pin is machined.

The cartridge comes to rest when the front end of the case stops against the square shoulder at the front end of the chamber. As the cartridge comes to rest, the bolt continues forward to drive the firing pin against the primer and discharge the cartridge. At this time the extractor is sprung over the groove in the cartridge case.

The gases generated in the cartridge case drive the light bullet forward and exert rearward pressure through the base of the cartridge case to the base of the bolt head. The bolt starts to the rear but in view of the much greater weight of the moving parts and the spring tension in relation to the comparatively light bullet weight, the action does not open appreciably until the bullet has left the muzzle. By this time the breech pressure has dropped to safe limit.

When the bullet emerges from the muzzle, the gas behind it expands in the compensator.

The pressure wave thus created thrusts downward against the solid lower half of the compensator while the gases expanding upward strike against the inclined surfaces cut into the compensator to result in a forward and downward thrust at the muzzle end. This tends to hold the muzzle down during automatic fire.

(Note that the compensator merely acts to stabilize the gun. It has nothing whatever to do with the functioning of the weapon itself.)

As the bolt starts to the rear, the empty cartridge case is held in the bolt face gripped by the extractor. When it clears the chamber far enough, the upper face of the cartridge strikes the ejector (which in the Owen is a part of the rear magazine wall.) The empty shell is hurled out the ejection opening which in this weapon is in the lower part of the receiver tube.

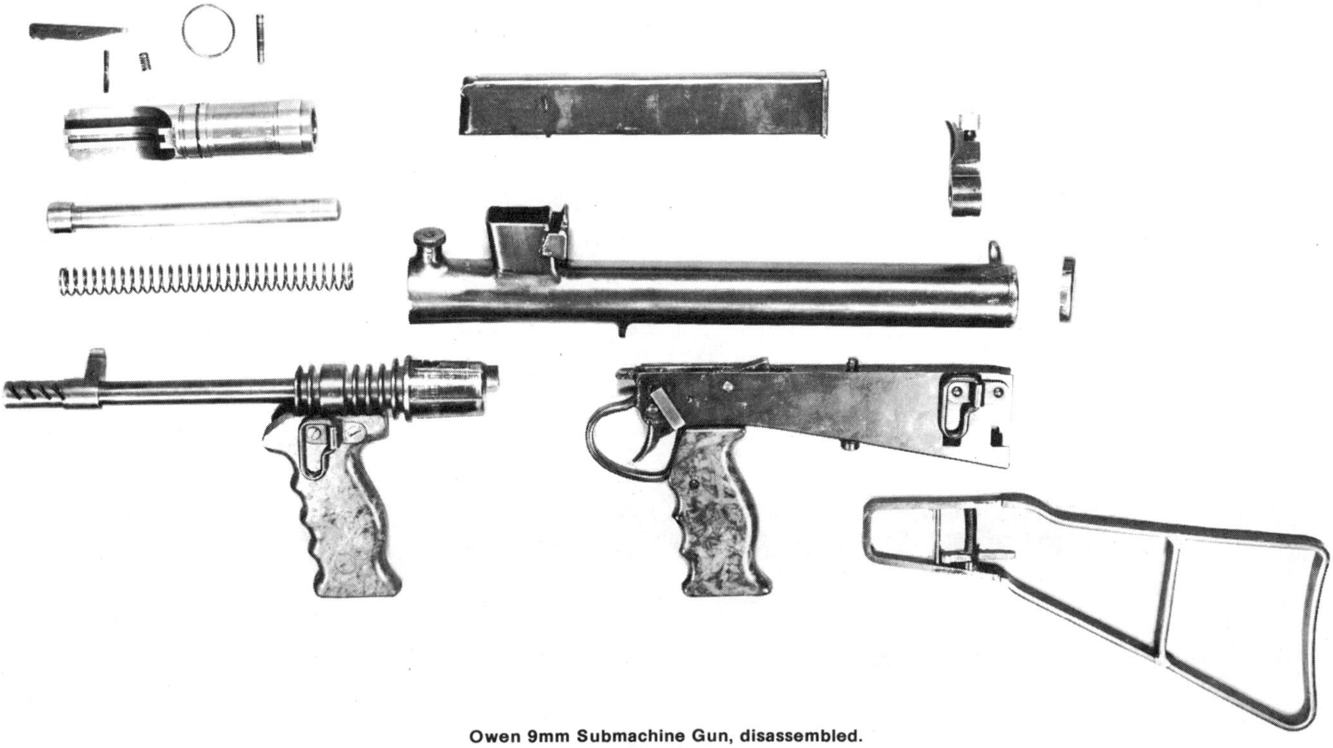

Owen 9mm Submachine Gun, disassembled.

The bolt continues to travel to the rear in a straight line pressing the operating spring behind it until the rear face of the bolt is stopped against the receiver plug.

In this weapon the cocking handle and the cocking bolt are permitted to travel still further to the rear in the slotted hole provided for the bolt pin in the head of the cocking bolt. This action prevents a sudden shock on the bolt pin.

At this point the main spring is at practically full compression and it halts further rearward movement.

The sear spring forces the sear up to catch in the underside of the bolt, holding it ready for the next shot.

Trigger action. There are three projections on the trigger in the Owen which can engage with the sear. The upper projection locks the sear from rising if the change lever is set at the safe position. The central projection forms a bent which is accurately located from the trigger axis. When the change lever is adjusted to the single-shot position, the rise of the trigger is strictly limited assuring that the sear will hold the bolt back on its first rearward movement. The lower projection engages the underface of the sear and carries one end of the trigger spring. It permits continuous fire.

The three surfaces on the change lever are in a circle and in turn engage the top of the trigger to limit the distance of rise. Thus on "safe" the trigger is locked; on "single shot" the rise of the trigger being limited, only one shot can be fired until the trigger is released and pulled back again; and on automatic fire the trigger is permitted full movement which permits the bolt to shuttle back and forth so long as cartridges are fed into the chamber and the trigger is kept depressed.

Special Note on the Owen

Two other varieties of this gun were issued. The first is called the "Mark I Wood Butt" type. This is a lightened version of the Mark I in which some of the metal is cut away from the receiver behind the rear grip and which is provided with a wooden butt.

The second type of Owen is the Mark II. In this type the shape of the receiver to the rear of the rear hand grip is still further modified

Left side of the 9mm Submachine Gun F1 (Aust) with bayonet fixed.

Right side of the F1, bayonet fixed.

resulting in a weight with butt and empty magazine of only eight pounds and three ounces while without the butt it is only eight pounds. The trigger assembly differs from the Mark I.

9mm SUBMACHINE GUN F1 (AUST)

Australia has been the most prolific designer of submachine guns among the Dominions of the British Commonwealth. In a desire to secure a weapon with the reliability of the Owen, but lighter in weight, with lower rate of fire, and easier to produce, the Australians designed the F1. (The weapon was called the X3 while in development.)

Characteristics of the F1 (Aust) Submachine Gun

Caliber: 9mm Parabellum.
System of operation: Blowback, automatic fire only.
Weight, loaded with bayonet: 9.88 lbs.
Length, overall: 28.12 in.
Feed device: 34-round staggered detachable box magazine.
Sights: Front: Blade mounted on the right side of magazine guide.
Rear: Aperture in stamped leaf type sight.
Muzzle velocity: Approx. 1300 f.p.s.
Cyclic rate: 600 r.p.m.

How the F1 Works

The F1 has a separate cocking handle and cover which do not reciprocate with the bolt during firing. The pistol grip and butt plate are the same as those used on the Australian-made F. N. L1A1 rifle. The top loading magazine of the F1 is the same as that used on the British L2A3 and the Canadian C1 (the Sterling or Patchett submachine gun). The weapon has a bayonet lug on the left side.

Loading and Firing the F1

Pull cocking handle to the rear. Insert a loaded magazine in the guide on top of the receiver. Press trigger and the weapon will fire. To put the weapon on safe, push safety catch located on the left side of the pistol grip, to down position, word "Safe" will be exposed.

AUSTRALIAN MACHINE GUNS

The Lithgow plant was expanded and tooled to produce the caliber .303 Vickers Machine Gun Mark I between 1925 and 1930. Tooling for the caliber .303 Bren Gun was done in 1938-39. Production of machine guns at Lithgow during World War II amounted to over 12,000 Vickers and 17,000 Bren guns.

Australia adopted the U. S. 7.62mm NATO M60 general purpose machine gun in the late fifties. The caliber .50 Browning gun is also used.

Austria 15

The Austrian Army uses the FN 7.62mm NATO "FAL" rifle which is made under license at Steyr. They also use the 9mm Walther P 38 pistol and the 7.62mm Model 42/59 machine gun made by Rheinmetall.

After the signing of the Austrian peace treaty, the new Austrian Army was initially equipped with U.S., British and Soviet arms. These weapons have been relegated to a reserve status or disposed of.

AUSTRIAN PISTOLS

Steyr Daimler Puch is again making pocket pistols, however, no pistol of Austrian design is receiving any significant military usage in the world today.

AUSTRIAN ROTH STEYR 8mm PISTOL M1907

This pistol was used by the Austrians in World War I. It was produced by the Oesterreichische Waffenfabrik (Steyr), and by the Fegyvergyr in Budapest. The design of the pistol is based on patents issued to George Roth, G. Krnka, and K. Krnka. Limited quantities of these pistols were apparently used in World War II.

Characteristics of Roth Steyr 8mm Pistol M1907

Caliber: 8mm Roth Steyr, 8mm Steyr M7
System of operation: Recoil.
Feed device: 10-round in line non-removable magazine.
Muzzle velocity: 1045 f.p.s.
Barrel length: 5.1 in.
Overall length: 9.1 in.
Sights: Blade with notch.

Special Feature of Roth Steyr

This weapon was originally designed for use by cavalry. The recoil of the weapon ejects the empty case and strips a new cartridge into the firing chamber as in other automatic pistols. However, it does not cock in the regular fashion. The striker is drawn back and released to fire the cartridge by pulling the trigger, exactly as in hammerless revolvers. This makes the weapon safe to handle but difficult to shoot accurately.

Both pistol and cartridge are generally considered obsolete.

9mm STEYR PISTOL M12

The Steyr Model 12 Pistol was the most widely used of the various pistols used by the Austro-Hungarian forces in World War I. It was also used by Rumania, and by the Germans, (in 9mm Parabellum) to a limited extent. There is considerable confusion over the correct nomenclature for this pistol; many call it the Model 1911 or M 11, others call it the Model 1912 or M 12. Both designations are correct - the commercial designation for the weapon is Model 1911; the official Austrian Army nomenclature for the pistol was Selbstiade Pistol M 12. The pistol is also called the Steyr Hahn. During World War II the Germans rebarreled a number of these weapons for the 9mm Parabellum Cartridge. These weapons can be identified by the "08" stamped on the slide.

Although there were about 250,000 of these pistols made they are no longer used as service pistols anywhere in the world and have not been made since 1919.

Austrian Roth-Steyr 8mm Pistol M1907.

Austrian 9mm Steyr Pistol M12.

Characteristics of 9mm Steyr Pistol M12

Caliber: 9mm Steyr.
Feed device: 8-round. Located in handle; cartridges must be stripped into it from the top of the pistol.
Capacity: 8 cartridges.
Muzzle velocity: 1112 f.p.s.
Barrel length: 5.2 in.
Overall length: 8.5 in.
Weight: 2.12 lb.
Sights: Blade with notch.
Locked: By cam ribs on barrel which lock in cam slots on inside of top of slide. As bullet passes down barrel, barrel tends to twist to the right. As barrel and slide move to the rear under recoil, cam rib twists barrel to the left and opens lock permitting slide to continue backward and function the action.
Type of fire: Single-shot only.
Magazine loading arrangement: Clip guide on top of slide permits insertion of loaded clip when action is opened, and slide is locked back with safety.
Position of slide when last shot is fired: Open.
Safeties: (a) A thumb safety somewhat like that on the Colt .45 Automatic will be found on the left side of the pistol just below the hammer. Turning this up into its notch in the slide makes the pistol safe. (b) An automatic disconnector on the right side of the pistol under the slide prevents this pistol from being fired until the action is wholly closed.

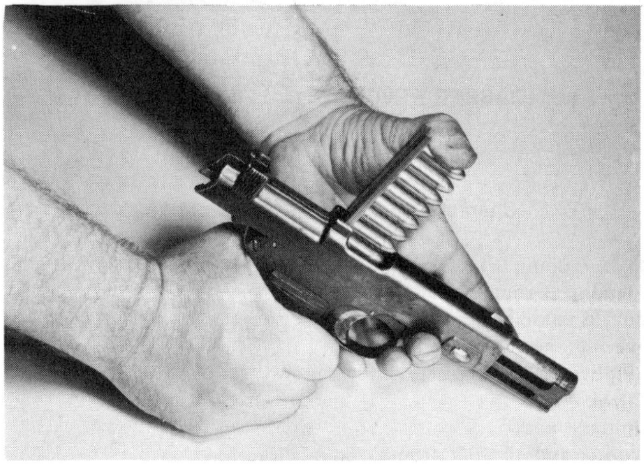

Stripping in a clip of cartridges -- loading the Steyr M12 9mm pistol.

M12 Steyr chambered for 9mm Parabellum cartridge, stamped "08" on slide.

Austrian 9mm Steyr Pistol M12 -- field-stripped.

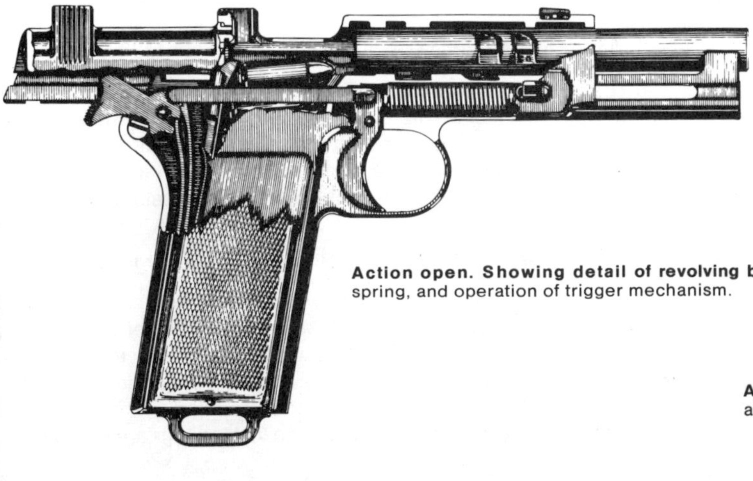

Action open. Showing detail of revolving barrel lock, function of recoil spring, and operation of trigger mechanism.

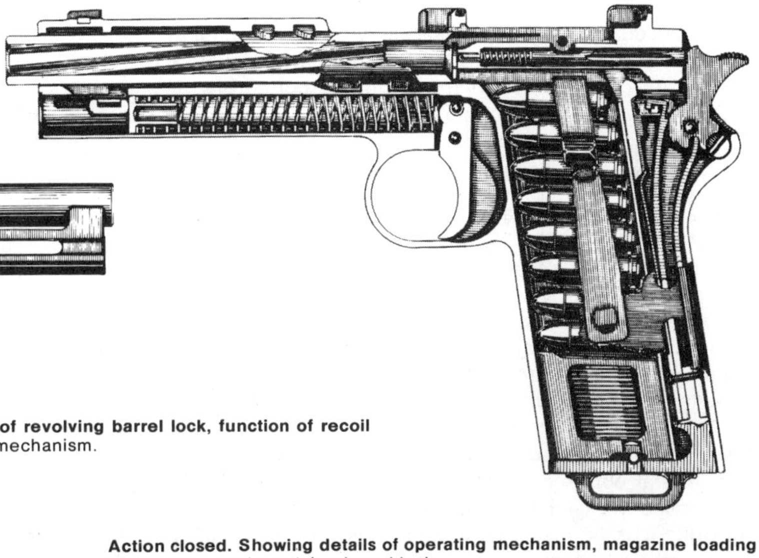

Action closed. Showing details of operating mechanism, magazine loading and release, and revolving barrel lock.

AUSTRIAN REVOLVERS

RAST-GASSER REVOLVER CAL. 8mm, MODEL 1898

Austrian Rast-Gasser revolver, 8mm, Model 1898.

Characteristics of Rast-Gasser, M 1898

Caliber: 8mm, 8.1 x 27mm, 8.2 x 27.5mm Revolver M98.
Cylinder: 8 chambers.
Muzzle velocity: About 787 feet per second.
Overall length: 9 in.
Weight: 2.06 lb.
Barrel: 4.5 in.
Sights: Fixed.
Type of action: Solid frame; rod ejector.
Double action: May be fired by thumb cocking and pulling trigger, or by straight pull through on the trigger.
Safety: Rebounding hammer which strikes the firing pin, which is set in the frame, only when the trigger is pulled through its entire length.

AUSTRIAN RIFLES

As previously noted, the current Austrian Service Rifle is the 7.62mm FN "FAL." There are no military rifles of Austrian design in significant military use anywhere at the present time. There were, however, a large number of Austrian rifles used during World War I and II. These were mainly of the Mannlicher straight-pull design, although there were some Steyr-made Mausers as well.

Austria-Hungary was apparently prepared to drop the straight-pull Mannlicher in favor of the 98 Mauser design in 1914. The 7mm Model 1914 rifle is a rare specimen, having the pointed-type pistol grip and bands as found on the earlier Mannlicher. The Model 29 was a Mauser made by Steyr for export; a quantity were made for the German Air Force during World War II and are marked G29/40 on the left side of the receiver. The receiver ring is marked 660—which was the numerical code for the Steyr plant.

AUSTRIAN MANNLICHER SERVICE RIFLES

None of the Austrian Mannlicher service rifles are used in significant quantity in active military service anywhere today. They were used extensively in World Wars I and II, however, and still exist in large numbers in the hands of collectors. A short description of each model follows.

11mm Rifle Model 1885 (Repetier Gewehr M85)

This was the first magazine rifle to be used by the Austro-Hungarian Empire. The M85 has a straight-pull non-rotating bolt, which is locked by a block pivoted from the underside of the rear section of the bolt. This block abuts against a shoulder in the receiver when in the locked position. The Model 85 introduced the Mannlicher magazine system in which the clip is inserted into the magazine of the rifle and functions as a part of the magazine—as does the clip of the U.S. M1 rifle. The Model 85 was chambered for the 11mm M77 cartridge and was not made in significant quantity. Its design led directly to the next of the Mannlichers, but has several outstanding differences. The principal difference is that, unlike the later Mannlichers in which the clip drops out of the bottom of the magazine when the last cartridge is loaded into the chamber, the M85 has a spring-loaded clip ejector which ejects the clip out of the top of the magazine when the last cartridge case is ejected in a fashion similar to that of the M1 rifle. The clip can be inserted into the rifle, in only one way, i.e., it is not reversible as regards top and bottom as is the Mauser type charger or the M1 clip.

11mm Rifle Model 1886 (Repetier Gewehr M86)

The M86 is similar in most respects to the M85. It is the first of the Austrian service rifles to introduce the feature of the clip dropping out of the bottom of the magazine when the last round is chambered. The 11mm cartridge was improved with the introduction of this rifle and as a result it had better ballistics than the M85. The sights of the M86, as the M85 and all other Austrian weapons until after World War I, are graduated in "Paces" (one pace equals 29.53 inches) a term similar to the "Arshin" formerly used as a Russian standard of measurement. Approximately 90,000 M86 rifles were made by Steyr.

8mm Rifle Model 1888 (Repetier Gewehr M88)

The M88 is chambered for the black-powder M88 cartridge, and its rear sight is graduated for that cartridge; with these exceptions it is the same as the M86.

8mm Rifle Model 1888-90 (Repetier Gewehr M88-90)

In 1890 the Austrian 8x50mm cartridge with smokeless powder charge was introduced. The sights of the M88 rifle were modified for the new and more powerful cartridge by the addition of new graduation scales which were engraved on plates and attached to the sides of the sights. Rifles thus modified are called M88-90.

8mm Carbine Model 1890 (Repetier Carabiner M90)

This weapon introduced the straight-pull bolt with rotating bolt to the Austrian Service. Although Mannlicher had introduced a rotating straight-pull bolt in 1884 it was not very successful and was never made in quantity. The bolt is of two-piece design. The bolt handle and bolt body are one piece; mounted within the bolt body is the bolt shaft or bolt cylinder. The locking lugs are mounted

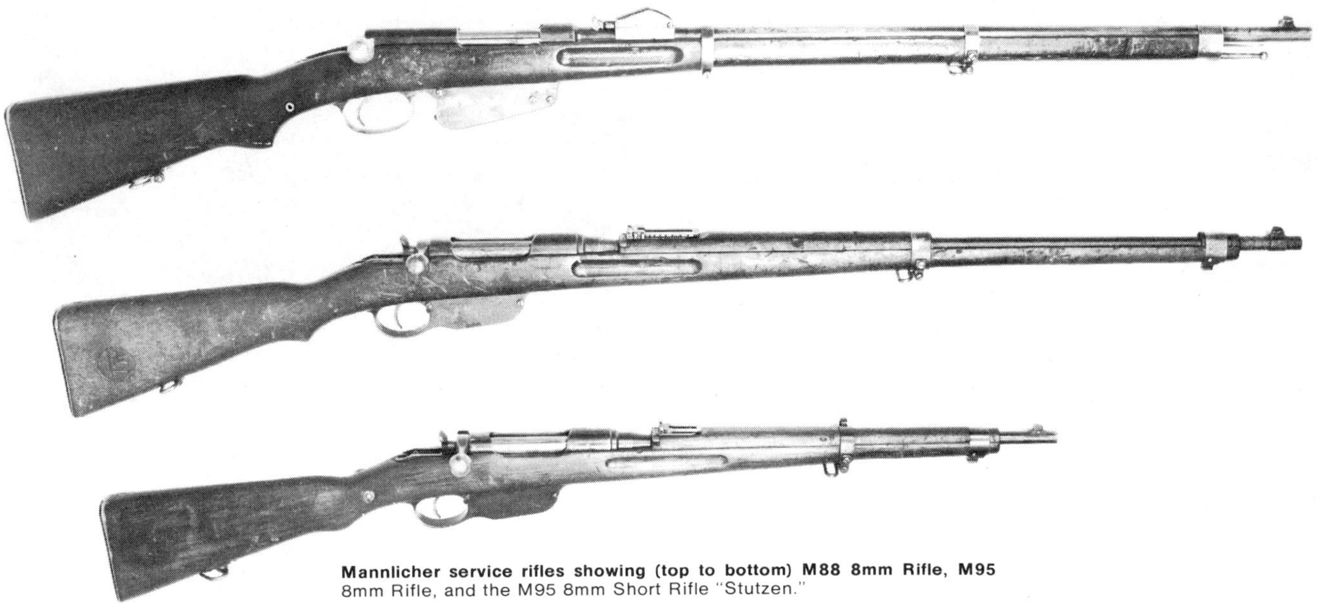

Mannlicher service rifles showing (top to bottom) M88 8mm Rifle, M95 8mm Rifle, and the M95 8mm Short Rifle "Stutzen."

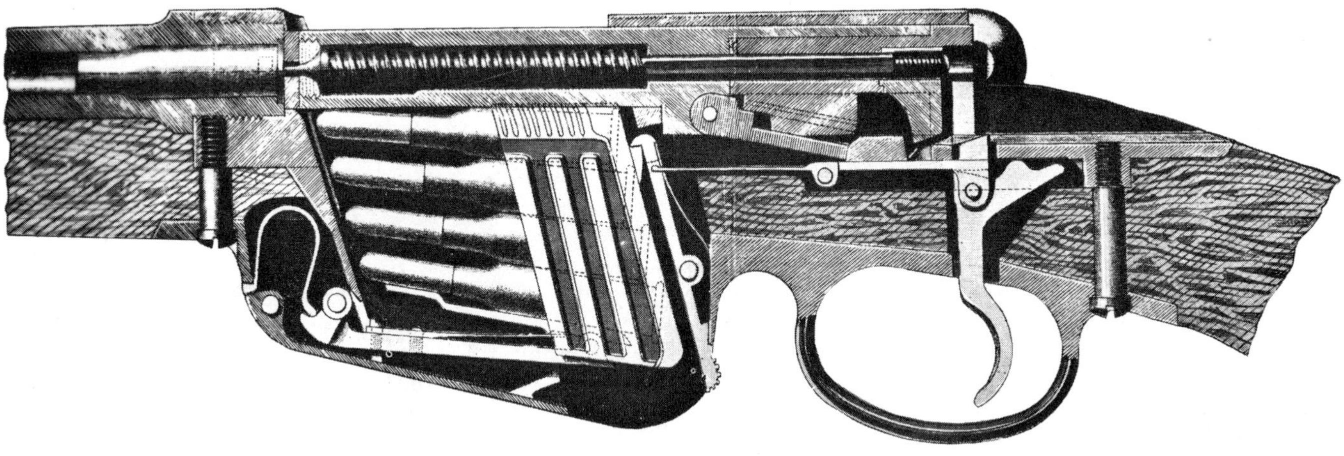

Austrian M1888 and 1888-90. Note engagement of clip catch above clip projection. Locking block is wedged down into locking position in the receiver.

on the head of the bolt cylinder and the bolt cylinder rotates within the bolt body during the locking and unlocking process. This bolt system is used with all the later Austrian straight-pull bolt-action Mannlichers and, since it provides for frontal locking, is considered by many to be a stronger system than that of the Models 84, 86, and 88. The magazine system adopted with the M86 is used in the M90 carbine and the later rifles. The M90 carbine has no hand-guard and the sight swivels are mounted on the left side of the stock; it is not fitted for a bayonet and the cocking piece is round. All later models have a thumb-shaped cocking piece. The M90 carbine is a relatively rare piece these days.

8mm Rifle Model 1895 (Repetier Gewehr M95)

This weapon which was made at Budapest as well as Steyr, as were all of the M95 series of weapons, and was the principal Austro-Hungarian rifle of World War I. It was also used in large quantities by the Italians - who had received them from the Austrians as World War I war reparations—in World War II. It was also used by the Bulgarians, Yugoslavs, and to some extent by the Greeks. The M95 rifle was made in tremendous quantities; large numbers of them are in this country.

8mm Carbine Model 1895 (Repetier Carabiner M95)

The carbine version of the M95 rifle, in addition to its short length, can be distinguished by the following: (1) sling swivels on side of stock only; (2) no provision for bayonet lug; and (3) no stacking hook.

8mm Short Rifle M1895 (Repetier Stutzen M95)

The M95 "Stutzen" is frequently confused with the M95 carbine. It apparently was designed for use by special troops, i.e. Engineer, Signal, etc., and not for Cavalry, since it is fitted with a bayonet stud and has sling swivels fitted to the underside as well as the side. This weapon also has a stacking hook which screws into the upper band. When the rifle is fired with bayonet fixed a blade on top of the bayonet barrel ring is used as the front sight to compensate for changes in center of impact due to the weight of the bayonet on the barrel.

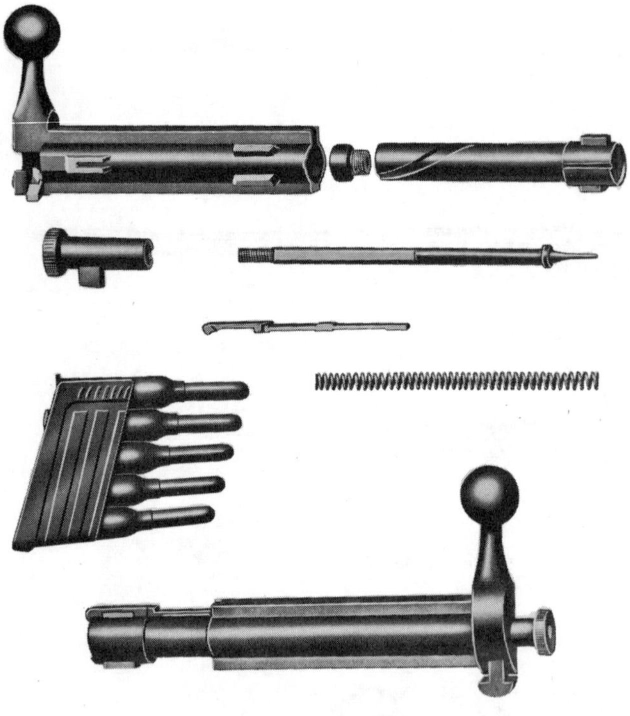

Bolt of the 8mm M90 Carbine.

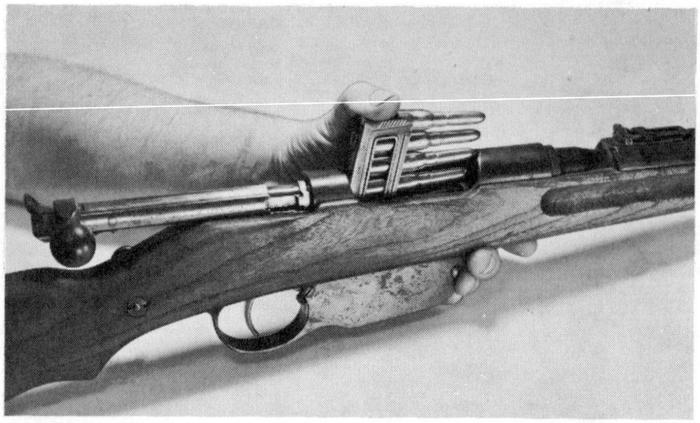

Note that both clip and cartridges are inserted into the magazine when loading the Steyr-Mannlicher Model 95 series 8mm rifles.

MODIFIED AUSTRIAN SERVICE RIFLES

The Austro-Hungarian Empire, as a loser in World War I, had to provide large amounts of war material to the Allies. Among those countries which benefitted from the Austrian war booty were: Italy, Yugoslavia, and Greece. Italy used large quantities of the M88-90 and M95 series weapons in World War I without modification and made large quantities of the 8x50mm cartridges for those rifles.

Yugoslav 7.92mm M95

Yugoslavia converted many of the M95 weapons to 7.92mm. These weapons can be distinguished by the addition of the stamped letter "M" after the "M95" which is on the top of the receiver. These rifles have a clip permanently fixed in their magazines and therefore can be loaded with the standard Mauser five round charger.

7.92mm M95/24

These rifles were apparently used by Bulgaria. They have the markings "/24" stamped on the receiver after the "M95." They are generally similar to the Yugoslav M95M in their magazine arrangements.

Conversion of M88-90 to 7.92mm

This rifle is believed to have been used by Greece. The barrel has been shortened and a wooden handguard added (the Austrians frequently used a laced canvas handguard on the M88 and M88-90 rifles in World War II), and a new rear sight added. This weapon also has a clip permanently attached to the interior of the magazine. The conversion was apparently done in Belgium, as the specimen examined has Belgian proof marks.

8mm M95 Stutzen Converted to 8x56mm

There are two versions of this weapon extant. The Austrian version appeared in 1930 with the adoption of the Model 30 (8x56mm) cartridge. The Model 30 was a large-rimmed cartridge with a pointed bullet—Spitzgeschoss—and therefore, a letter "S" twelve millimeters high was stamped on the receiver to distinguish it from the unconverted weapons. These weapons were used considerably by the German police in World War II and steel-cased Austrian-made ammunition, bearing the date 1938 plus the German Eagle and Swastika marking, has been found in quantity. In 1931 the 8x56mm cartridge was adopted by Hungary, who

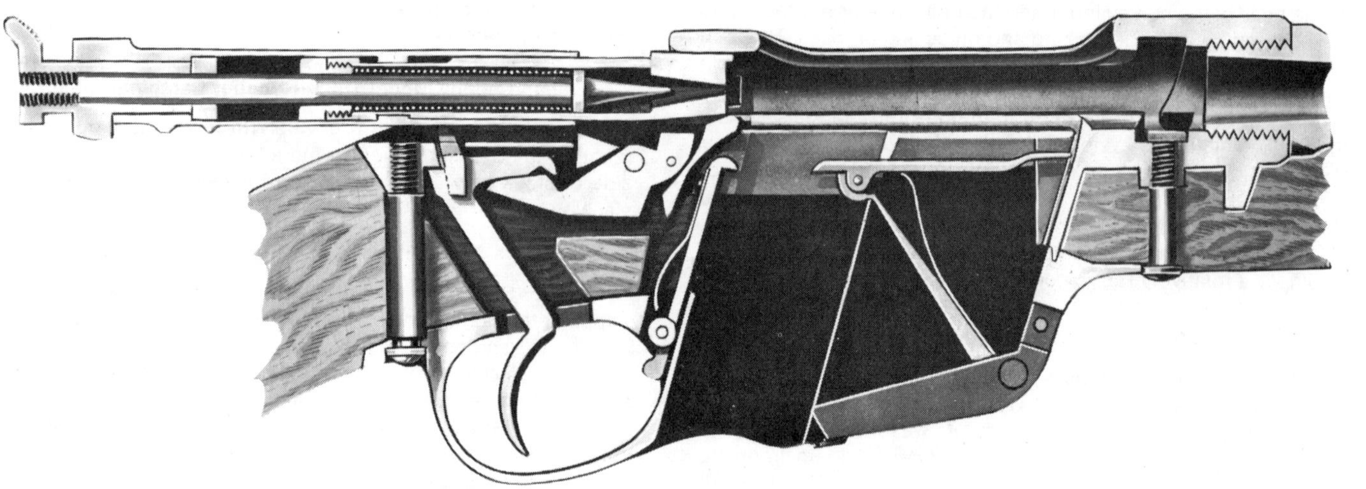

Austrian Model 95 Mannlicher straight-pull section and mechanism.

CHARACTERISTICS OF AUSTRIAN STRAIGHT PULL MANNLICHER SERVICE RIFLES

Weapon	Method of locking	Overall length	Barrel length	Feed device	Sights	Muzzle velocity	Weight
Rifle: 11mm M86	Wedge	52 in.	31.75 in.	5-round, single column, fixed box magazine	Front: Barley corn Rear: V notched tangent with long range side sight	1610 f.p.s	10 lbs.
Rifle: 8mm M88	Wedge	50.38 in.	30.14 in.	5-round, single column, fixed box magazine	Front: Barley corn Rear: V notched tangent with long range side sight	1750 f.p.s.	9.7 lbs.
Rifle: 8mm M88/90	Wedge	50.38 in.	30.14 in.	5-round, single column, fixed box magazine	Front: Barley corn Rear: V notched tangent with long ranged side sight	2115 f.p.s.	9.7 lbs.
Carbine: 8mm M90	Frontal locking lugs	39.5 in.	19.5 in.	5-round, single column, fixed box magazine	Front: Barley corn Rear: V notched tangent	1900 f.p.s.	6.9 lbs. (approx.)
Rifle: 8mm M95	Frontal locking lugs	50 in.	30.12 in.	5-round, single column, fixed box magazine	Front: Barley corn Rear: Leaf	2030 f.p.s.	8.31 lbs.
Short Rifle: 8mm M95	Frontal locking lugs	39.5 in.	19.65 in.	5-round, single column, fixed box magazine	Front: Barley corn Rear: Leaf	1900 f.p.s.	7.5 lbs.
Carbine: 8mm M95	Frontal locking lugs	39.5 in.	19.65 in.	5-round, single column, fixed box magazine	Front: Barley corn Rear: Leaf	1900 f.p.s.	7 lbs. (approx.)

called it the 31M (Model 31). The Hungarians had large quantities of M95 "Stutzen" on hand and converted many of these to the 31M cartridge; these weapons can be distinguished by the letter "H" stamped on the receiver. It should be noted that M95 rifles rebarreled for the 8x56mm cartridge—the M30 or M31—cannot be used with the old conical-nosed 8x50mm cartridge. Although both are rimmed, the 8x56mm is considerably longer and more powerful—these cartridges are definitely NOT interchangeable. Both the Austrian and Hungarian conversions require special clips.

AMMUNITION FOR AUSTRIAN MANNLICHER SERVICE RIFLES

The 11mm M77 Austrian cartridge is a typical black-powder cartridge. Although it was once made in the United States, it is now basically a collector's item, as is the 8mm M88 black-powder loaded cartridge. The 8mm M90 and its slightly improved version, the M93, may be encountered in quantity on occasion but it should be born in mind that all military loads are at least 20 years old and all the above rifles chambered for this cartridge require a special clip to be used as magazine loaders. The situation is the same for the 8x56mm (M30 or M3l) cartridge. This cartridge has not, to the writer's knowledge, been manufactured for 20 years.

The conversions which are chambered for the 7.92mm cartridge present a far simpler problem. They do not require special clips and the 7.92mm (8mm Mauser) cartridge is available in quantity both in military and sporting configurations. A word of warning is in order, however: 7.92mm cartridges can be found with pressures up to 55,000 p.s.i. The Model 88-90 converted to 7.92mm was built to take a maximum pressure of about 40,000 p.s.i., therefore shooting this weapon with some of the military and commercial cartridges currently available could be EXTREMELY HAZARDOUS! Shooting any of these weapons unless previously checked by a reliable gunsmith can be hazardous, especially if the weapon shows any signs of hard usage.

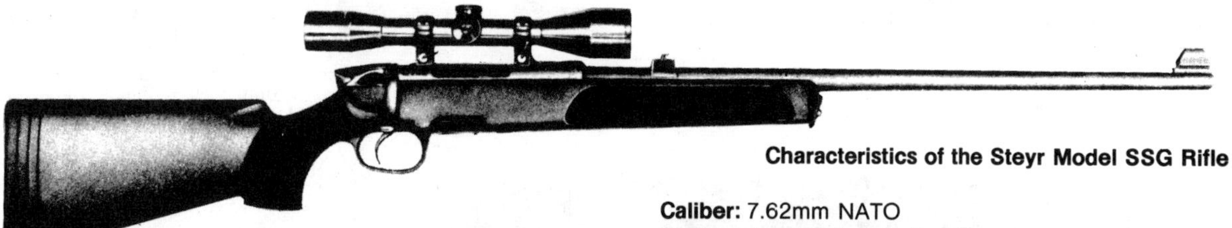

Steyr 7.62mm NATO Model SSG Rifle.

STEYR 7.62mm NATO MODEL SSG RIFLE

Steyr Daimler Puch has developed a rifle based on their post war bolt action system. This rifle has been adopted by the Austrian Army as a sniper rifle. It has a heavy, cold hammered barrel. The bolt has six rear-mounted locking lugs. Steyr claims ten round 2.75 inch groups with this rifle at 300 meters.

Characteristics of the Steyr Model SSG Rifle

Caliber: 7.62mm NATO
System of operation: Bolt action
Weight (w/o telescope): 10.25 pound
Length overall: 45 in.
Barrel length: 25.6 in.
Feed mechanism: 5 round, removable rotary
Sights: Front: (w/telescope) hooded post
Rear: (w/ telescope) notch
Muzzle Velocity: 2800 f.p.s.

May also be fitted with Walther micrometer rear and hooded front sight with changeable inserts.

AUSTRIAN SUBMACHINE GUNS

THE AUSTRIAN MP34 SUBMACHINE GUN

Although Austria did little if anything in the line of submachine guns, they did have one gun which attained fairly wide usage during World War II. The MP34 is commonly known as the Steyr Solothurn and is a product of German design worked out at Waffenfabrik Solothurn A.G. of Solothurn Switzerland, a Swiss plant owned by Rheinmetall of Germany, during the period when German military arms development was restricted by the Versailles Treaty.

The MP34 was taken over by the Germans when they took over Austria in 1938 and was called by the Germans MP34 (Ö) - Maschinen Pistole 34 Österreich-, (Österreich meaning Austrian). The weapon was widely used by German police and rear area units. The weapon in various modifications was offered commercially and was also used by Chile, El Salvador, Bolivia, and Uruguay. It was used, in extremely limited quantities, by the Japanese in 7.63mm Mauser. The commercial designation for the weapon is SI-100. It is probable that all the MP34s used by Austria were made by Steyr as from 1930 on the gun was known as the Steyr Solothurn and the two concerns had a joint marketing arrangement.

It should be noted that the MP34 as used by the Austrian Army was chambered for the 9mm Mauser cartridge. It was used by the Austrian police in 9mm Steyr; both calibers were found in German service or police units. It may also be found chambered for the 9mm Parabellum cartridge.

Characteristics of Austrian MP34 Submachine Gun

Caliber: 9mm Mauser (Army Model).
System of operation: Blowback.
Weight loaded: 9.87 lbs.
Length overall: 33.5 in.
Barrel length: 7.80 in.
Feed mechanism: 32 round detachable, staggered box magazine.
Sights: Front: Barley corn.
Rear: Tangent with "V" notch graduated from 50-500 meters in 50-meter increments.
Muzzle velocity: 1360 f.p.s. (For 9mm Mauser).
Cyclic rate of fire: 500 rounds per minute.

Unusual Features of MP34

The weapon is typical of the period in which it was made in that it is heavy and expensive, being made of heavy forgings. The only unusual feature is the magazine loader which is machined into the magazine housing. The magazine is inserted into the underside of the magazine housing and is then loaded with ten round chargers—stripper clips—through the opening in the top of the magazine housing.

STEYR SUBMACHINE GUN

Steyr Daimler Puch has developed a new submachine gun chambered for the 9mm Parabellum cartridge. This is the first military weapon to be developed in Austria since the 1930s.

The Steyr submachine gun is a compact weapon which resembles the Israeli UZI in several respects. Like the UZI, it has a long barrel and a short overall length; accomplished by having a bolt which telescopes the barrel for about 2/3rds of the barrel length.

Steyr 9mm Parabellum Submachine Gun.

Austrian 9mm Parabellum Steyr Submachine Gun field stripped. 1. Oper-ating spring assembly. 2. Barrel. 3. Bolt. 4. Barrel nut. 5. Receiver. 6. Stock. 7. Trigger housing assembly. 8. Magazine.

Austrian 9mm Model 34 Steyr Solothurn Submachine Gun.

Characteristics of the Steyr Submachine Gun

Caliber: 9mm Parabellum
System of operation: blowback, selective fire
Weight: 6 pounds
Length overall—
stock retracted: 18 in.
stock extended: 25 in.
Barrel length: 10.2 in.
Feed mechanism: 25 or 32-round detachable staggered row box magazine
Sights—front: post with protecting ears
rear: "L" w/apertures set for 100 and 200 meters
Muzzle velocity: approx. 1350 f.p.s.
Cyclic rate: 550 r.p.m.

Description of the mechanism of the Steyr Submachine Gun

The Steyr as most submachine guns, fires from an open bolt. Pressure on the trigger, which pulls the trigger half of its length of travel, produces semi-automatic fire; pulling the trigger all the way to the rear produces automatic fire. As with the UZI, the ejection port is open only when the bolt moves to the rear with the fired case and on its return to the closed position. The push-button type safety is located on the left side of the trigger housing above and to the rear of the trigger. The magazine catch is located at the bottom rear of the pistol grip. The magazine well is in the pistol grip as on the UZI. There is a cocking safety which prevents the bolt from closing and firing if it slips in cocking. The weapon is easily field stripped. The barrel is moved by depressing the barrel catch lock at the right top front of the receiver and unscrewing the barrel nut. The receiver cover is disengaged by pushing in on a catch at the top rear of the receiver. Before attempting disassembly, the stock must be in the fixed position.

AUSTRIAN MACHINE GUNS

The Austro-Hungarian Empire used the Schwarzlose machine gun in several models. The background of this weapon has been covered in the historical section. The Schwarzlose in addition to being used in Austria was also used in Sweden - the 6.5mm Model 14, in the Netherlands, - the 08, 08/13, 08/15 some using the 8x57mm rimmed cartridge, in Czechoslovakia in 7.92mm; and in Italy in the form of Austrian war booty, in 8mm. The Schwarzlose is not in active use in any country, and it is doubtful if this weapon will ever see active service again.

SCHWARZLOSE 8mm M 07/12 MACHINE GUN

Characteristics of the Schwarzlose 8mm M 07/12 Machine Gun

Caliber: 8mm (8x50mm).
System of operation: delayed blowback.
Weight, gun: 44 lbs.
Weight, tripod: 43.75 lbs.
Length overall: 42 in. (approx.)
Barrel length: 20.75 in.
Feed mechanism: canvas belt.
Sights: Front: Barley corn.
Rear: V notched tangent.
Muzzle velocity: 2,000 feet per second (approx.).
Cyclic rate of fire: 400 rounds per minute.

This weapon has an extremely heavy recoil spring which must be very carefully removed.

AUSTRIAN 8mm MODEL 30S MACHINE GUN

This weapon was adopted by the Austrians in 1930 and was among the materiel taken from Austria by the Germans after the annexation of Austria. A similar weapon designated as the 31M (Model 31) was adopted by Hungary in 1931. Both guns were marketed by Waffenfabrik Solothurn A.G., a Swiss plant which was owned by Rheinmetall-Borsig A.G. They were developed from the designs of Louis Stange, a Rheinmetall engineer. Assembling and marketing the gun by Solothurn was a neat dodge by the Germans to avoid Versailles Treaty restrictions. The gun was offered first in 7.92mm as the Solothurn Model 29. In addition to being adopted by the Austrians and Hungarians as ground guns, the weapon, slightly modified, was adopted in 7.92mm by the Germans as a fixed aircraft gun—the MG15, and as a flexible aircraft gun the MG17.

Austrian Schwarzlose 8mm Machine Gun M07/12.

Characteristics of the Austrian 8mm Model 30S Machine Gun

Caliber: 8mm, M30, 8x56R-will not use the 8x50R.
System of operation: Recoil operated, selective fire.
Weight: Approximately 18.5 lbs.
Length overall: 46.25 in.
Barrel length: 23.6 in.
Sights: Front: Blade with guards.
Rear: V notch tangent graduated from 100 to 2,000 meters.
Cyclic rate: 450-500 rpm.
Feed device: 25-round detachable box magazine.
Muzzle velocity: 2,395 f.p.s.

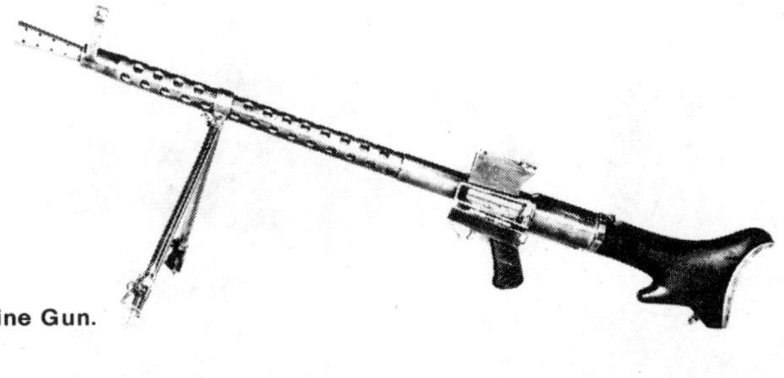

Austrian 8mm Model 30S Machine Gun.

16 Belgium

The Belgian Army is currently equipped with the following small arms: the 9mm Parabellum Browning FN G. P. (High Power) pistol; the 9mm Parabellum Vigneron M2 submachine gun; the 7.62mm NATO FN FAL rifle; the 7.62mm NATO FN MAG; general-purpose machine gun; and the caliber .50 Browning machine gun.

The Belgian Army was stripped of weapons during World War II and for a period after World War II was equipped mainly with British weapons, i.e., Lee Enfields, Vickers, and Bren guns. Re-equipment with FN-made weapons started very rapidly; a caliber .30-06 bolt-action Mauser was among the first of the weapons of native origin to be issued to Belgian troops. The .30-06 FN-made Browning machine guns and the .30-06 FN M1949 self-loading rifle were also issued. The Mausers have been sold and the other weapons are presumably held in reserve.

THE FN PLANT

Fabrique Nationale d'Armes de Guerre (FN) of Herstal lez Liege is the most prolific designer of successful small arms still in service in the Western World today. The FN Browning G.P. (Grande Puissance - High Power) pistol is used in many countries, as is the FN FAL rifle and the MAG machine gun. FN-made Browning pistols of .32 and .380 caliber, as well as FN-made Browning machine guns, automatic rifles, and the FN commercial weapons, are found throughout the Western World.

The FN organization was founded in 1889 by a combine of Liege interests and Ludwig Loewe and Co. of Berlin, to manufacture the Model 1889 Mauser rifle for the Belgian government. The company had a stroke of luck early in the present century when John Browning—upset over his financial arrangements with Winchester—and, according to a many times told story, more than a little bit annoyed at having to cool his heels in the anteroom of a prominent American gun manufacturer waiting to see some now unknown dignitary who had unfortunately died the night before, decided to take the design of his now renowned long recoil operated shotgun to Liege and to FN. Thus began a close relationship with the outstanding genius of American gun designers which ended with his death in 1926 in Belgium. Browning brought more than his automatic shotgun to FN. They also produced his automatic pistols—the M1900, M1903, M1906 (.25 automatic), M1907, M1910, M1922, and the M1935—the High Power. They produced his commercial semiautomatic rifle and after World War I produced the Browning Automatic Rifle and Browning machine gun, which Browning had developed for his native land during the war. Although John M. Browning is generally known in the U.S.A., especially to those who have used his weapons in service, in Europe his name is a household word for automatic pistol. It should be noted that Browning designs similar to those produced and marketed by FN in the Eastern Hemisphere were produced and marketed by American manufacturers in the Western Hemisphere. Colt produced in one form or another successful Browning pistol designs including the still standard U.S. Army caliber .45 Colt automatic pistol. Colt also produced the automatic weapons, Remington and Savage-Stevens produced the shotguns, and of course Winchester continued manufacture of many of the earlier rifles and are still producing the famed .30-30 Model 94 carbine.

FN, in addition to producing the Browning automatic rifle and Browning machine guns (water cooled, air cooled, aircraft and heavy), began producing Mauser bolt-action rifles based on the Model 98 action in 1924. These weapons were quite successful and modified forms of these military Mausers were in production as late as 1964 for small Middle Eastern countries which could not afford the more modern semiautomatic and automatic rifles. FN and the Zbrojovka Brno plant of Czechoslovakia had the world's Mauser military rifle market to themselves to a great extent during the period from 1924 to 1938; the Belgians, being extremely astute businessmen in addition to manufacturing a fine product, got at least their share of the business. (There is actually no real difference in the quality of the products made by the two concerns

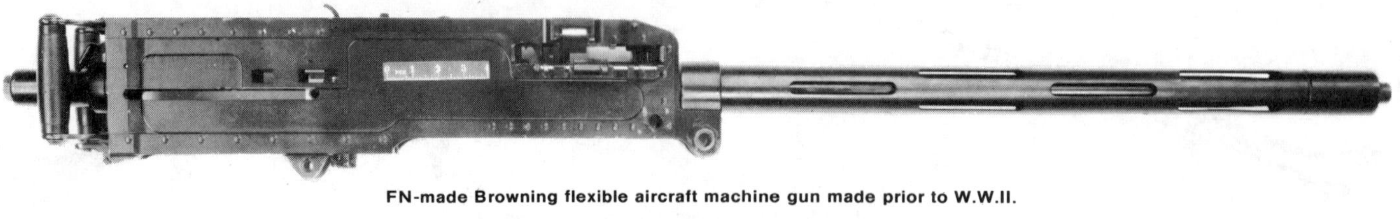

FN-made Browning flexible aircraft machine gun made prior to W.W.II.

during this period; they are among the finest quality military rifles ever built.)

World War II found Belgium again an occupied country. The Germans continued manufacture of weapons suited to their ammunition system, i.e. the 7.92mm Mauser carbine, the 9mm Parabellum High Power pistol (called Pistol 640(b) by the Germans), the 9mm Browning short (.380 ACP) M1922 (called pistol 626 (b) by the Germans), and various weapon parts. The weapons made during World War II are not equal in quality to the pre- or post-war product.

Fortunately the FN plant was not destroyed by Allied air strikes or by the Germans when they departed in late 1944. The plant was able to set up rapidly for the manufacture of Browning automatic rifle and Browning machine-gun parts for the U.S. Army. At the conclusion of hostilities FN got a large contract to rebuild a large number of U.S. weapons in Europe. The first post-war weapon they put on the market was actually designed prior to the war by D. Saive. It was the FN self-loading rifle known variously as the ABL, SAFN, or Model 1949. This rifle was followed by the Type D Browning Auto rifle, the FAL rifle, and its heavy-barreled version, and the MAG machine gun.

The FN organization currently has probably the largest gun plant in the Western world, certainly the largest gun plant in Western Europe. In addition to the manufacture of FAL rifles, MAG machine guns, and a few Mauser military rifles, FN manufactures a full line of sporting rifles, shotguns, and pistols. They also manufacture a full line of military and sporting ammunition, agricultural machinery, trucks, and jet engines.

BELGIAN PISTOLS

Belgium has been a prolific producer of military pistols in the past, but only the FN pistols have any significant military usage today. Among the Belgian-designed pistols and revolvers which had extensive military use but are rarely encountered in service today, are the Bergmann Bayard, the Nagant revolver, and the various Pieper and Clement automatics.

MODELS 1900, 1903, 1910, AND 1922

The M1900 7.65mm (.32 ACP) FN Browning Pistol

This pistol had limited use as a military pistol. It is the first of John Browning's automatic pistols produced in quantity by FN. A blowback-operated pistol, it is of somewhat unusual design in that the recoil spring is mounted in a separate tunnel above the barrel. This pistol and its earlier versions—the Model 1898 and Model 1899, which were made in very limited quantity—introduced the 7.65mm Browning cartridge or .32 ACP cartridge as it is known in the U.S. This cartridge was apparently designed by Winchester with the assistance of Browning. Copies of the M1900 may be encountered in the Orient with various weird markings which are attempts of Chinese counterfeiters of the pistol to reproduce the FN markings.

FN Browning M1900 7.65mm Pistol.

The M1903 9mm FN Browning Pistol

This handgun is a blowback-operated pistol similar in operation and construction to the Colt caliber .32 and .380 pocket model automatics which appeared during the same period. This pistol was quite extensively used as a military automatic, but is no longer

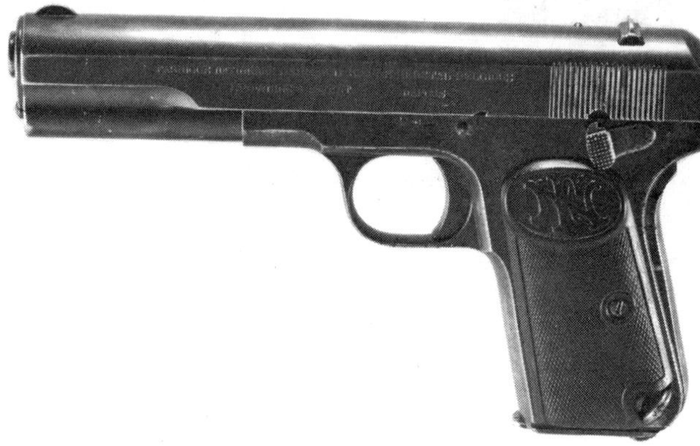

FN Browning M1903 9mm automatic pistol.

used as a service pistol and hasn't been produced for many years. This weapon is chambered for the 9mm Browning long cartridge which has not been manufactured in the U.S. for many years.

The M1903 Browning was a standard service pistol in Sweden (pistol M/07), Belgium, Denmark, the Netherlands, and Turkey.

The M1910 FN Browning Pistol

This weapon may be found chambered for either the 7.65mm (.32 ACP) or 9mm Browning Short (.380 ACP) cartridge. It is still manufactured and is distributed in the U.S. by the Browning Arms Co. of St. Louis. This blowback-operated pistol is basically an improvement on the Model 1903 and has been extensively used abroad as a police pistol. It has also had some use as a limited standard service pistol.

The M1922 FN Browning Pistol

This is basically an enlarged Model 1910. It has been made in caliber .32 ACP and caliber .380 ACP and is blowback operated.

FN Browning M1910 Pistol.

FN Browning M1922 Pistol.

This pistol, frequently called the Model 10/22, was apparently designed with an eye to the military market and was adopted as a service pistol by several countries including the Netherlands, Yugoslavia, and Belgium. It's manufacture in both calibers was continued by the Germans during the occupation of Belgium. Specimens of the pistol bearing the German Waffenamt stamps (eagle over Swastika with letters Waa) are not of good quality and should be avoided except as collectors' items.

THE 9mm FN BROWNING HIGH POWER PISTOL

The FN Browning High Power is one of the most extensively used military pistols in the world today. It is also widely distributed as a commercial weapon. This pistol was the last developed by John Browning and first appeared in prototype form in 1926. It was introduced to the market in 1935 in two forms: an "Ordinary Model" with fixed sights and an "Adjustable Rear Sight Model" which had a tangent-type rear sight graduated to 500 meters and had a slotted grip for attachment of a wooden shoulder stock. This shoulder stock is attached to a leather holster in contrast to the all wooden shoulder-stock holsters made for the Canadian made FN Browning pistols No. 1 mark 1 and No. 1* during World War II. The High Power, which is called the G.P. (Grande Puissance) in Belgium, uses the Browning Colt parallel ruler system of locking but is considerably simplified in many ways in comparison with the U. S. caliber .45 Model 1911.

The High Power has been or is being used as a service pistol by Belgium, Lithuania, Denmark, the Netherlands, Nationalist China, Canada, the United Kingdom, Rumania and other countries. It was manufactured by the John Inglis Company of Toronto, Ontario during World War II. (For further information see under Canada.) Large quantities were manufactured for the Germans during the occupation of Belgium; the High Power was used as a first-line weapon by the Germans because of its caliber, 9mm Parabellum.

FN Browning 9mm High Power pistol, standard model.

FN Browning 9mm High Power, with tangent type rear sight.

The pistol in commercial form is distributed in the U.S. by the Browning Arms Co. of St. Louis.

Listed below are some of the nomenclatures which have been or are being used for the High Power in various countries:

Belgium: Pistolet Automatique Browning Modele a Grande Puissance (GP).
Canada, U.K.: Pistol, Browning, FN, 9mm HP No. 1, Marks 1 and 1.* and No. 2, Marks 1 and 1.*
Denmark: 9mm Pistol M/46.
Germany: Pistole 640 (b).
Netherlands: Pistool, 9mm Browning, FN, GP.

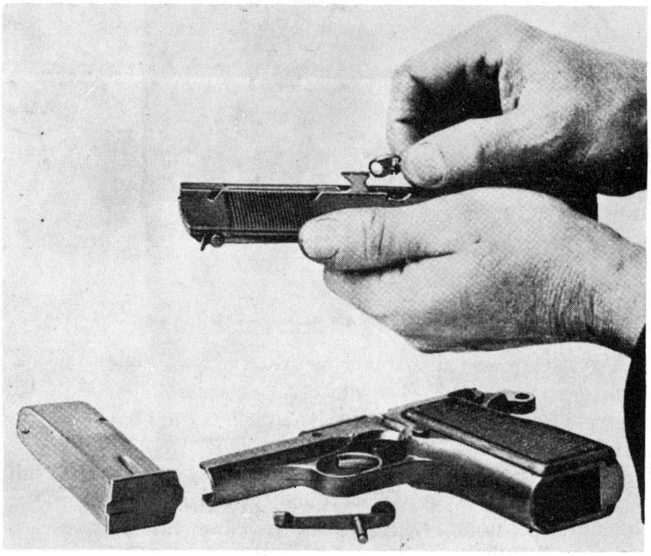

Field stripping the Browning High Power.

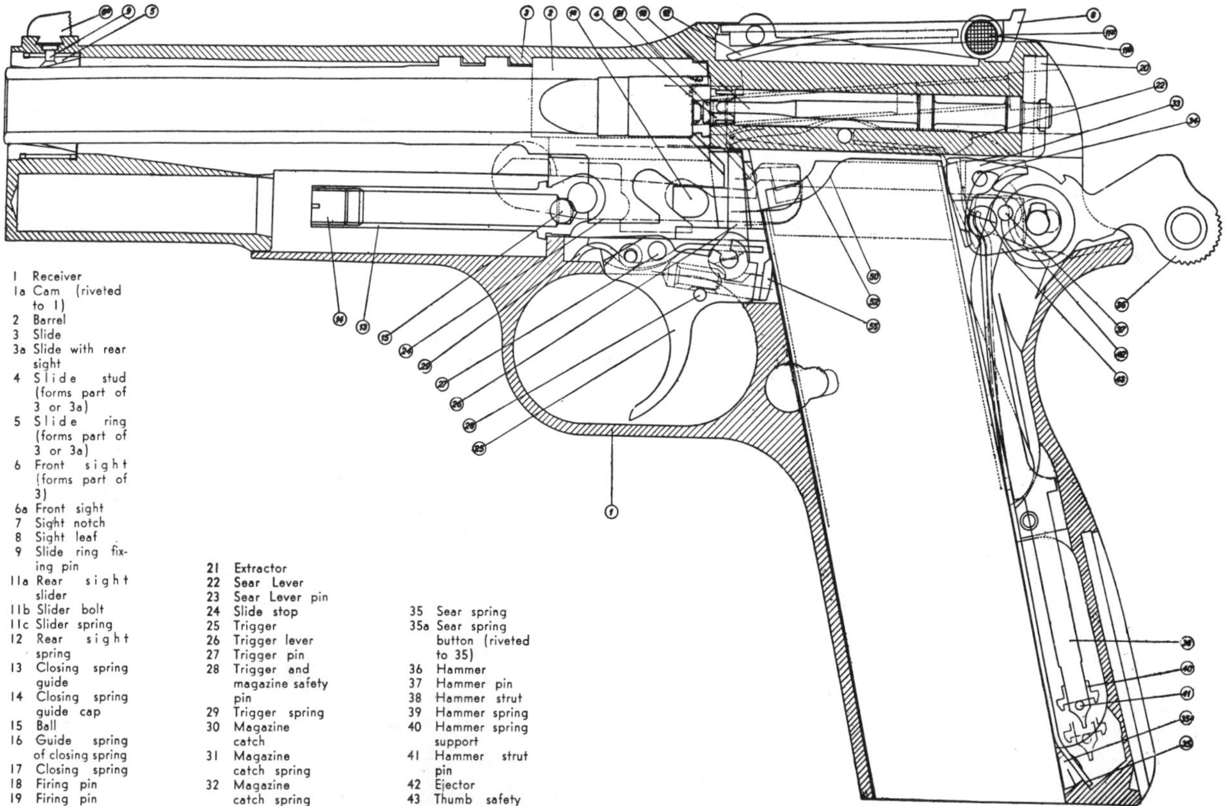

1 Receiver
1a Cam (riveted to 1)
2 Barrel
3 Slide
3a Slide with rear sight
4 Slide stud (forms part of 3 or 3a)
5 Slide ring (forms part of 3 or 3a)
6 Front sight (forms part of 3)
6a Front sight
7 Sight notch
8 Sight leaf
9 Slide ring fixing pin
11a Rear sight slider
11b Slider bolt
11c Slider spring
12 Rear sight spring
13 Closing spring guide
14 Closing spring guide cap
15 Ball
16 Guide spring of closing spring
17 Closing spring
18 Firing pin
19 Firing pin spring
20 Firing pin fixing plate
21 Extractor
22 Sear Lever
23 Sear Lever pin
24 Slide stop
25 Trigger
26 Trigger lever
27 Trigger pin
28 Trigger and magazine safety pin
29 Trigger spring
30 Magazine catch
31 Magazine catch spring
32 Magazine catch spring guide
33 Sear
34 Sear pin
35 Sear spring
35a Sear spring button (riveted to 35)
36 Hammer
37 Hammer pin
38 Hammer strut
39 Hammer spring
40 Hammer spring support
41 Hammer strut pin
42 Ejector
43 Thumb safety
44 Safety stud
45 Safety pin
46 Safety spring

Section view and parts listing — FN Browning High Power pistol.

Field Stripping the High Power Browning

Pull slide back and push thumb safety up into the second notch.

Press magazine catch and withdraw magazine.

Push pin from right side of receiver and lift out the pin and slide stop unit.

Holding firmly to the slide, depress the safety catch and permit the slide assembly to go forward and off the receiver runners.

Holding slide assembly upside down, pull recoil spring toward the muzzle and lift it out of engagement. Remove it and the spring.

Remove barrel from rear of slide.

Characteristics of the FN Browning High Power Pistol

Caliber: 9mm Parabellum.
System of operation: recoil, semiautomatic.
Magazine: Box type, double line staggered. Capacity 13 cartridges.
Muzzle velocity: 1040 to 1500 feet per second depending on type and manufacture of ammunition.
Barrel length: 4.75 in.
Overall length: 8 in.
Weight: 1.9 lb.

Description of the Mechanism of the Browning High Power

The use of a double-row staggered magazine gives greater magazine capacity but at the same time necessarily increases the thickness of the grip. This additional width, together with an arched lower section of the handle section of the receiver, gives the pistol better than usual instinctive pointing qualities.

While in general the design follows that of the Colt .45 Automatic, in detail it is quite different. The positive hammer safety is positioned as in our service pistol. When it is forced up into its notch in the slide, a projection on its lower side fits at the rear of the sear to prevent the release of the hammer.

The slide stop, which is forced up by the magazine follower to hold the pistol open when the last shot has been fired, and the button magazine release both follow the standard Browning form. No grip safety is provided. This is held to be non-essential on a military pistol.

Other new features include a magazine safety. When the magazine is withdrawn the safety spring forces this safety out, swinging the trigger lever forward out of engagement with the sear. Inserting a magazine presses the safety against its spring tension and swings the trigger lever back and under the sear to permit finger pressure to be transmitted.

CAUTION. Many of these pistols made for the Germans in World War II had the magazine safety removed.

In place of the Colt stirrup-type trigger, this weapon is fitted with a comfortable trigger which, when pressed, forces a trigger lever upward. This rotates the sear lever which acts upon the sear arm, causing it in turn to swivel and release the hammer.

Unless the slide is fully forward and the barrel securely locked to it, this sear lever remains at the rear and the trigger lever cannot act upon it. This acts as a positive disconnector to prevent the weapon from being fired. Should the trigger be held back after a shot has been fired, the trigger lever is retained in raised position but is also forced forward by the sear lever on the forward motion of the slide. This prevents the trigger lever from acting upon either the sear or the sear lever, and so the hammer cannot fall until the

Prototype of 9mm Browning High Power.

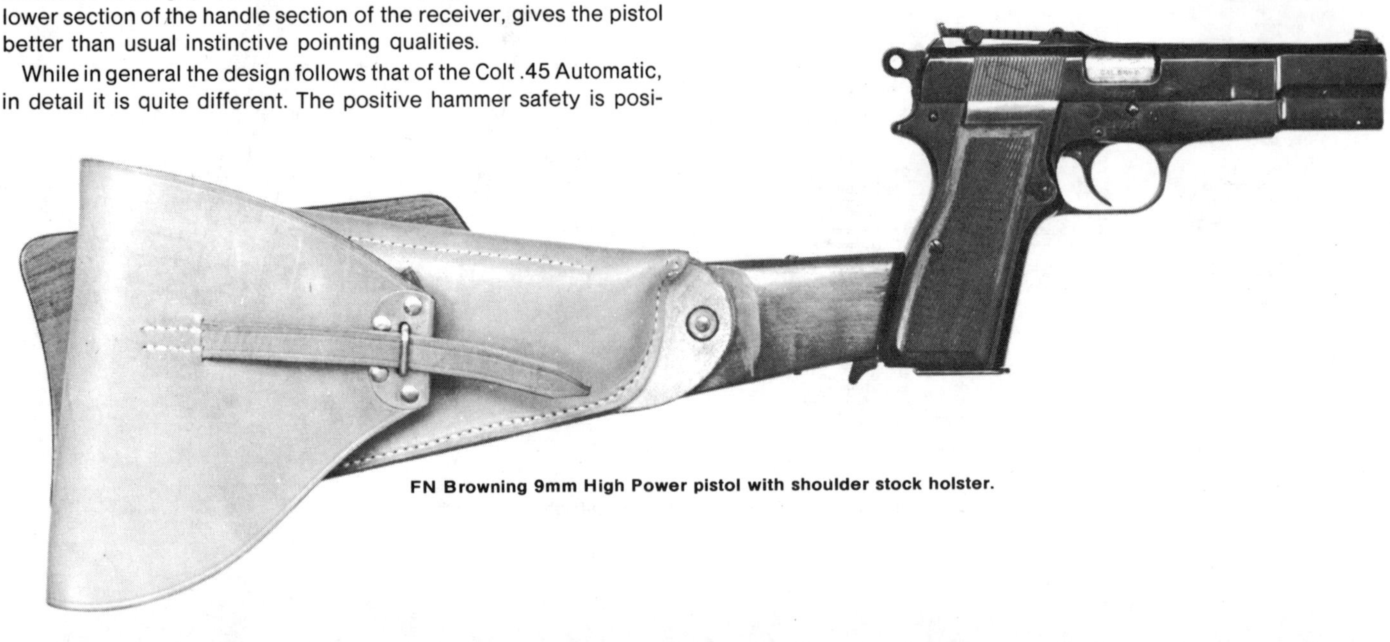

FN Browning 9mm High Power pistol with shoulder stock holster.

trigger is permitted to move to its normal forward position. Thus only one shot is possible on each pull of the trigger.

The slide in particular is an improvement over the original Browning design. Its forward end has only one opening, that for the barrel. The front of the slide below the muzzle opening is solid, doing away with the weak barrel bushing of previous models. The recoil spring and its guide seat in the hollow below the barrel, and the head of the guide sets into the barrel nose (or lock) where it is securely retained by the transverse slide stop pin.

The barrel lock is also improved. While it retains the basic Browning locking idea (that of ribs on top of the barrel fitting into corresponding grooves inside the slide at the moment of firing), this new Browning does away with the swinging link and pin used in our service pistol. It provides instead a "barrel nose" which is part of the barrel forging itself. This barrel nose is placed directly below the heavily reinforced chamber section, and has a guiding slot which is controlled by a cam machined into the receiver. This arrangement gives a much more rigid barrel support and permits simplification of the recoil spring system.

How the Browning High Power Pistol Works

At the moment of discharge the barrel is locked securely to the slide as the top-locking ribs engage in the locking slots in the slide. As the slide starts back in recoil carrying the barrel with it, and the bullet leaves the barrel, the sear lever is disconnected from the trigger lever.

As the barrel pressure drops to safe limits, the lower section of the notch of the barrel nose contacts the cam in the receiver, and the rear end of the barrel is thus drawn down until its ribs are free from the locking slots. At this point the rearward barrel movement is stopped as the barrel nose brings up sharply against its receiver stop.

The slide continues on backwards, riding over the hammer to cock it. The extractor claw carries the empty cartridge case back until it strikes the ejector and is tossed out of the pistol. Meanwhile the recoil spring is compressed around its guide. Rearward motion of the slide stops when its lower part strikes against the forward end of the receiver.

For the return motion, the recoil spring pushes the slide forward to strip the top cartridge from the magazine into the chamber. The breech end of the slide strikes the barrel; under the action of the cam in the receiver acting against the upper part of the barrel nose notch, the rear of the barrel is brought up into locking position and its ribs fit firmly into their slots in the slide.

The trigger, sear, and hammer mechanisms hook up properly and the pistol is ready for the next pull of the trigger.

The FN Hi-Power is now being made with a side-mounted pivoting extractor. The spring loaded extractor lies in a cut in the slide behind the ejection port and is visible from the outside, rather than being mounted in a hole cut through from the rear of the slide to the slide breech face as previously.

BELGIAN RIFLES

The story of modern Belgian rifles begins with the 7.65mm M1889 Mauser which is considered the first of the Modern Mauser rifles and is covered below. FN began the manufacture of Mauser 98-type military rifles and carbines around 1924 and still produces them to a limited extent. The principal military rifle produced in Belgium at the present time is the FN "FAL" rifle. FN is the only current producer of military rifles in Belgium; the government arsenal Fabrique d'Armes de L'Etat no longer manufactures military rifles.

The Browning Automatic Rifle (BAR) has been produced by FN since the twenties. Prior to World War II the Model 30 was manufactured in the greatest quantity. After World War II the type D

Top to bottom: 7.65mm M1889 Rifle, 7.65mm M1889 Carbine, 7.65mm M1936 Rifle, 7.65mm M1935 Rifle.

appeared; this weapon is considerably modified and improved as compared to the Model 30.

BELGIAN SERVICE BOLT-ACTION RIFLES AND CARBINES

Belgium adopted what might be called the first of the modern Mauser rifles in 1889. The 7.65mm M1889 Belgian Mauser was the first Mauser to have a solid bolt body bored from the rear with locking lugs at the head of the bolt. The Model 1889 was also the first rifle to use the Mauser type charger, (stripper clip). This type of rifle was used by Belgium until 1935 when a rifle generally similar to the German Kar 98K was adopted.

Description of Belgian Mauser Bolt-Action Rifles

The various models of Belgian Mauser bolt-action rifles are listed below.

7.65mm Rifle M1889. Long barreled, with metal jacket covering barrel to serve as handguard, straight bolt handle, single line magazine protrudes below stock, magazine not normally detached.

The Model 1889 series were made by FN, Fabrique d'Armes de L'Etat (the former Belgian government arms factory) and also, during World War I, by Hopkins and Allen of Norwich Conn. and an arms plant established by Belgian refugees in Birmingham, England. The 1889 series were made for loads of about 39,000 P.S.I. pressure and should not be used with higher-powered loads.

7.65mm Rifle M1935. This rifle has the 98 Mauser bolt system, and flush magazine. Except for bands, the front sight guard and sling swivels are the same as the German 7.92mm Kar 98K.

7.65mm Rifle M1936. Sometimes called the Model 89/36. This rifle was converted from the M1889 rifle. The barrel is partially covered with a wooden handguard, the tubular metal handguard being removed. An upper band and front sight guard similar to that of the Model 1935 rifle are used. The bolt system has been altered by the fitting of a bolt sleeve similar to that ot the 98 and a new 98 type cocking piece, firing-pin system.

Caliber .30 FN Rifle M1924/30. Used in limited quantity by the Belgian Army after World War II. A standard Model 98 type, it is covered in detail on later pages. This rifle is also known as the M1930; both model designations given are FN model designations.

In every mechanical respect they are duplicates of the German and Czech Service rifles, and the descriptions of mechanisms given for those arms cover the FN line as well. Model modifications are minor and deal with externals.

Most Mauser military actions are based on the same receiver and bolt. The standard 7mm, 7.65mm and 7.9mm Mauser cartridges were developed by Paul Mauser to permit simplicity of conversion, as well as low cost design. Thus the machinery which in time of peace made 7mm rifles for South America could be readily converted to 7.9mm German Service when needed.

Characteristics of M1924 Rifles in Caliber 7mm, 7.65mm, or 7.9mm

Length of rifle without bayonet: 43.3 in.
Length of rifle with bayonet: 58.2 in.
Length of barrel: 23.2 in.
Weight without bayonet: 8.5 lbs.
Number of cartridges in magazine: 5.

7.65mm M1889 Carbines

There are four M1889 carbines; all have the year Model designation 1889. They are as follows:

Carbine M1889 with Bayonet. Differs from the rifle only in dimensions, uses the standard knife type rifle bayonet; length--41 inches.

Carbine 1889 with "Yatagan". A "yatagan" is a curved blade sword without hand guard. The Belgians had a bayonet of this design. This carbine generally resembles the M1889 with bayonet, but has a turned down bolt handle. Used by foot gendarmes and fortress artillery.

Carbine 1889, "Lightened". This arm has a turned down bolt and is shorter--35 inches--than the preceding M1889 carbines. The stock terminates at the lower band and the rear sight is mounted immediately ahead of the receiver. On the left side of the butt, there is a catch with lock which is for mounting on a metal stud worn on straps going over the back of cavalrymen equipped with this arm.

Carbine 1889 "Lightened with Yatagan". Same length as the

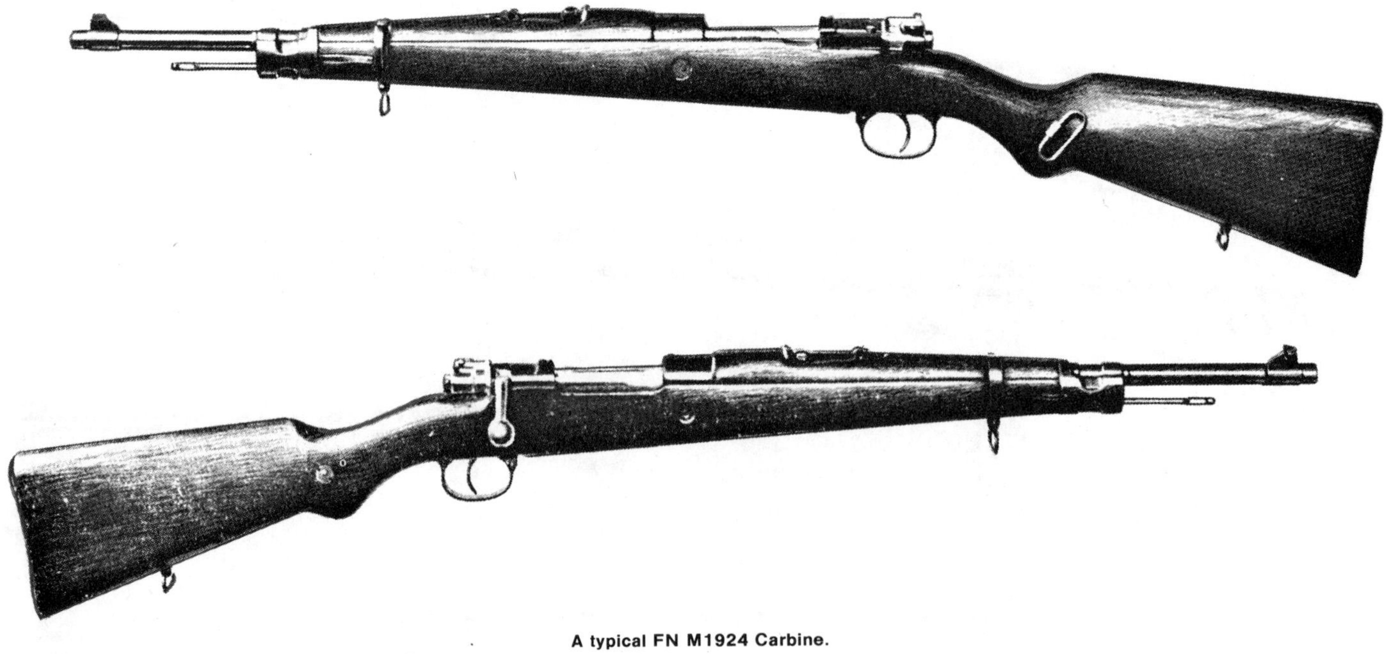

A typical FN M1924 Carbine.

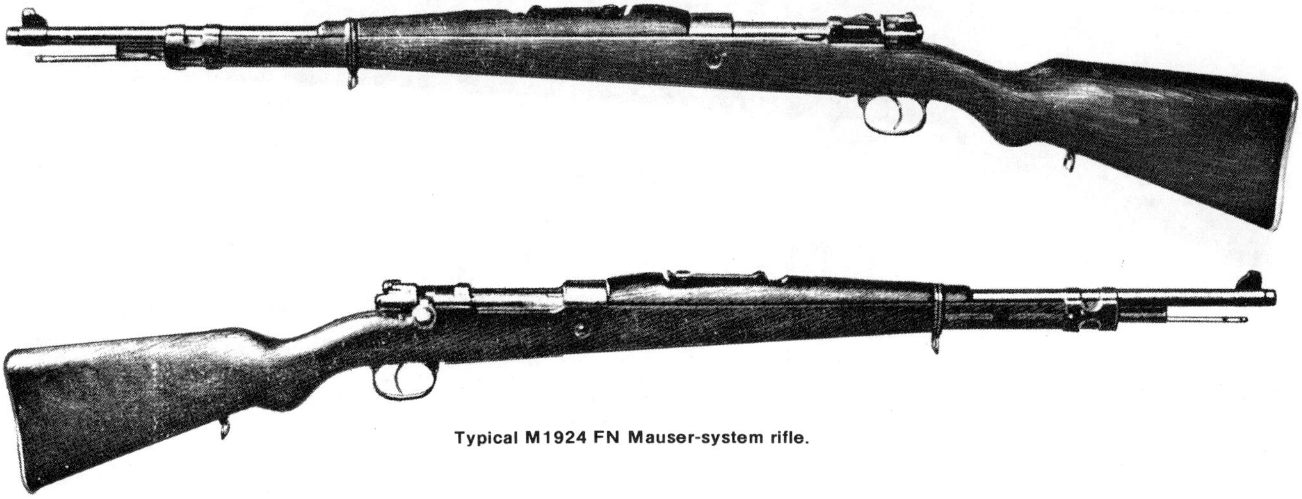

Typical M1924 FN Mauser-system rifle.

1889 lightened carbine, but has a longer stock. The lower band is a very short distance from the upper band. It has swivels mounted on the butt and lower band as do all the other carbines, except the 1889 Lightened, but also has a bracket on the right side of the butt. This model was used by mounted gendarmerie.

Characteristics of M1924 Carbines in Caliber 7mm, 7.65mm, or 7.9mm

Length of carbine: 37.4 in.
Length of barrel: 17.3 in.
Weight of carbine: 7.3 lb.
Number of cartridges in magazine: 5.
Lowest rear sight graduation in meters: 200.
Highest rear sight graduation in meters: 1400.

The Model 1924 and 1924/30 rifles and carbines were made for Argentina, Belgium, Bolivia, Brazil, Chile, China, Columbia, Costa Rica, Ecuador, Iran, Liberia, Lithuania, Luxembourg, Mexico, Paraguay, Peru, Turkey, Uruguay, Venezuela, Yemen and Yugoslavia.

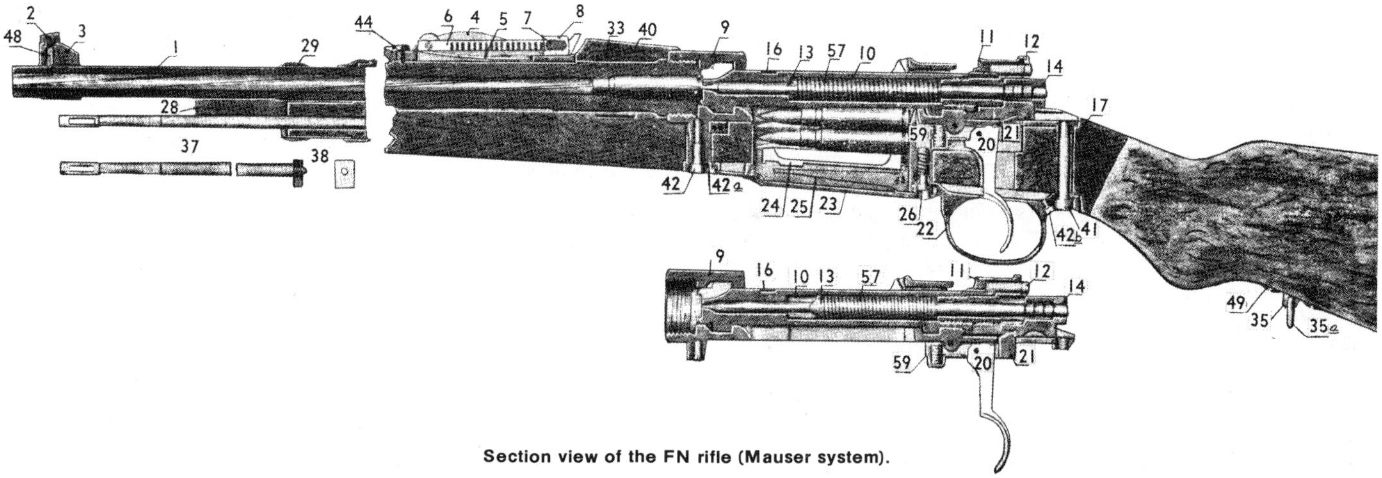

Section view of the FN rifle (Mauser system).

1. Barrel.
2. Front sight.
3. Front sight ring.
4. Rear sight bed ring.
5. Rear sight spring.
6. Rear sight tangent.
7. Spring pawl.
8. Sight slide.
9. Body.
10. Bolt.
11. Bolt plug.
12. Safety wing.
13. Firing pin.
14. Cocking piece.
16. Extractor ring.
17. Bolt stop.
20. Trigger.
21. Trigger bar.
22. Trigger guard.
23. Magazine cover plate.
24. Magazine platform.
25. Magazine spring.
26. Magazine cover plate catch.
28. Bayonet attachment.
29. Upper band.
37. Cleaning rod.
38. Cleaning rod stop.
40. Handguard.
41. Trigger guard screw (rear).
42. Trigger guard screw (front).
42a. Check screw of front trigger guard screw.
42b. Check screw of rear trigger guard screw.
44. Rear sight bed screw.
48. Front sight ring screw.
57. Firing pin spring.
59. Trigger bar spring.

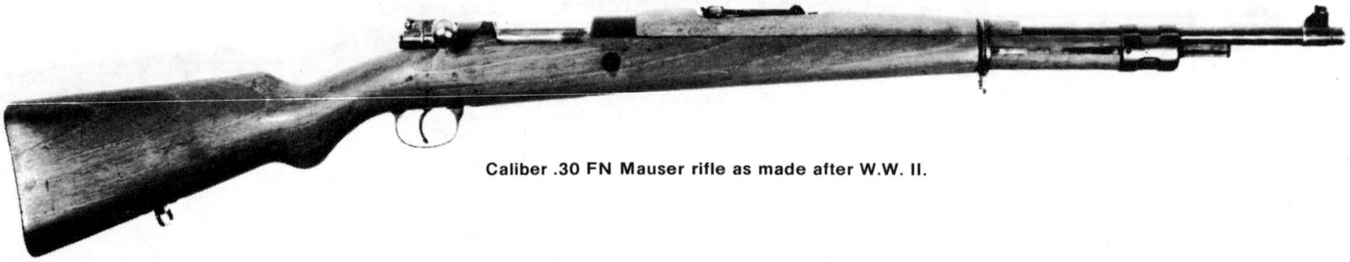

Caliber .30 FN Mauser rifle as made after W.W. II.

CHARACTERISTICS OF PRE-WORLD WAR II BELGIAN SERVICE BOLT-ACTION RIFLES AND CARBINES

	Rifle M1889	Carbine M1889*	Rifle M1935	Rifle M1936
Caliber:	7.65mm Mauser.	7.65mm Mauser.	7.65mm Mauser.	7.65mm Mauser.
Overall length:	50.13 in.	41.16 in.	43.6 in.	Approx. 43 in.
Barrel length:	30.69 in.	21.65 in.	23.5 in.	23.7 in.
Feed device:	5-round in line box magazine.	5-round in line box magazine.	5-round staggered box magazine.	5-round in line box magazine.
Sights: Front:	Barley corn.	Barley corn.	Barley corn.	Barley corn.
Rear:	V notch, leaf sight.	V notch, leaf sight.	V notch, tangent.	V notch, tangent.
Muzzle velocity: (at date of adoption)	2034 f.p.s.	1900 f.p.s. (approx).	2755 f.p.s.	2378 f.p.s.
Weight:	8.88 lbs.	7.75 lbs.	9. lbs.	8.7 lbs. (approx).

*The other three carbines vary in detail.

BELGIAN AUTOMATIC RIFLES

FN SELF-LOADING RIFLE (SAFN)

Prior to World War II Dieudonne Saive developed a gas-operated rifle which was intended to replace the bolt-action Mausers of the Belgian Army and also to be offered to the armies of the world as a replacement for their bolt-action rifles. It should be remembered that only one nation - the U.S.A. - had adopted a semiautomatic rifle as the standard shoulder weapon at that time. The occupation of Belgium by the Germans in 1940 halted work on the self-loading rifle and it did not appear on the world market until the end of World War II. The rifle was offered in caliber .30, 7mm, 7.65mm, and 7.92mm and was adopted by Belgium in caliber .30 model 1949, Egypt in 7.92mm, the former Netherlands East Indies in caliber .30, Brazil in caliber .30, Venezuela in 7mm (model 49), Luxembourg in caliber .30, Argentina in caliber .30, the former Belgian Congo in caliber .30, and Colombia in caliber .30.

The tilting bolt of this rifle locks on a bar set in the bottom of the receiver and is cammed into and out of the locked position by cam slots on the bolt carrier in engagement with lugs on the rear of the bolt. This bolt system is essentially the same as that of the Soviet Tokarev rifles and of the FN "FAL" rifle. There were specimens of this rifle made which had selective fire capability.

This rifle is frequently referred to as the ABL or SAFN. ABL (Armee Belgique), is the marking found on the rifles made for the Belgian government; SAFN stands for semiautomatic FN. The British government tested this weapon in 7.92mm before they came out with the EM-2 design and the U.S. tested the weapon in caliber .30.

Characteristics of FN Self-loading Rifle (SAFN)

Caliber: .30 M2, 7mm, 7,65mm, and 7.92mm Mauser.
System of operation: Gas, semiautomatic.
Feed device: Projecting steel box. Capacity 10 cartridges. Loaded single shot or from 5-shot clip.
Barrel length: 23.2 in.
Overall length: 43.7 in.
Weight: 9.48 lbs.
Sights: Front, shielded post; rear, tangent.

Operation: Tappet driving back through hole in receiver above line of barrel strikes bolt carrier and starts it to the rear until pressure has dropped. At unlocking point, the housing is machined to cam the bolt up a ramp at its rear end, thus allowing the carrier and bolt to travel to the rear with the bolt carrying the cartridge case in its face, held by the extractor, until it strikes the ejector and is tossed out of the weapon. Recoil springs compressed during this motion start the housing and bolt forward at the end of the recoil stroke. Upon closing, as the cartridge is chambered and the bolt face is against the breech face of the barrel, the housing still has continuing forward movement which enables it, through its machined surfaces, to depress the rear of the bolt into its locking recess with the rear locking surface at the top of the ramp.

Loading and Firing the FN Self-Loading Rifle

Magazine is hand- or clip-loaded. The cocking handle on the right side is drawn back to unlock the action and compress the springs. Releasing the cocking handle permits the springs to drive the mechanism forward to chamber the top cartridge from the magazine and lock the bolt. A pull on the trigger fires the cartridge, and thus one shot is fired, the empty case extracted and ejected, the weapon cocked, and a new cartridge loaded into the chamber ready for firing on the next trigger pull. An individual pull is required to fire each shot on most of these weapons.

How the FN Self-Loader Works

The hammer mechanism in the design is an adaptation of the familiar John Browning hammer hook system as used originally in his automatic shotgun. Variations of this design are encountered in most sporting arms of successful types today, and a minor variant of it is used in the United States Rifle M1 and the Carbine M1. It is not unusual that a modification of it, and in many senses an improvement of it, should be encountered in this Belgian rifle, since its designer Monsieur Saive was a very close associate of John Browning throughout the period of that great inventor's years at the FN establishment.

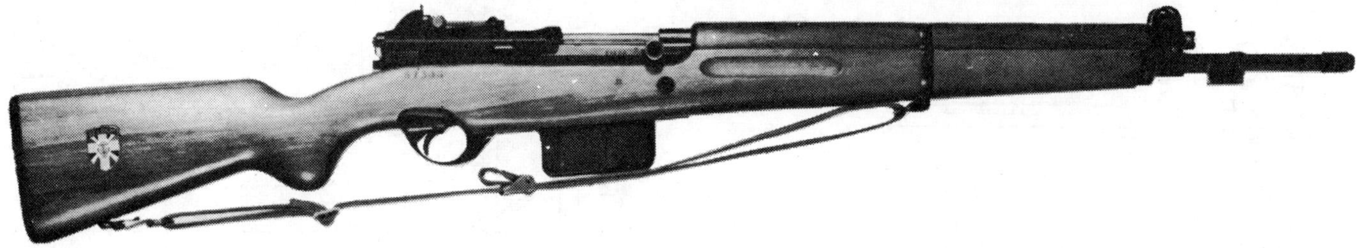

FN Semiautomatic Rifle M1949 as used by Belgian troops with the U.N. in Korea. This rifle is chambered for the U.S. .30 M2 cartridge.

FN semiautomatic rifle (left side).

The hammer hook system is so designed that during recoil, the rear upper hook on the hammer is engaged automatically to prevent full automatic fire. When the trigger is deliberately released, the upper hook (or sear) releases, the hammer spring reacts but the forward holding sear, which is at a lower level than the automatic holding one, grips a lower cut or hook in the hammer. One innovation in the design, (which has, however, appeared on German rifles, notably of Walther commercial design), is that the hammer spring guide is designed to protrude slightly below the line of the trigger guard when the hammer is in cocked position. It is remembered that the hammer is concealed in this type of weapon. The FN design permits the holder of the weapon to tell by sight or touch if the hammer is cocked as evidenced by the protruding nose of the mainspring guide.

The safety system also includes a variation of the one utilized in the Garand in that the bolt housing or carrier is so designed that it interferes with the hammer striking the firing pin until both the bolt is locked and the housing itself is in its full forward position. Thus if the housing is not completely closed, while the hammer may fall it can only strike the rear of the housing and cannot fire the cartridge in the chamber. In such an instance, it is necessary to pull back the cocking handle to recock the arm before it can be fired. The firing pin is automatically retracted by a return spring after being driven forward to fire.

The face of the bolt is slotted to allow it to travel back over the ejector which also helps to serve as a travel guide. As the proper motion nears the end of the recoil stroke, the ejector is far enough out from the face of the bolt so that it can pivot the cartridge out of the gun.

The manual safety is on the right side of the trigger guard and is in the form of a turning lever with a half round block which not only locks the trigger to prevent any movement, but also drops its bar down far enough to interfere with the trigger finger being inserted into the trigger guard as a warning that the safety is applied.

A study of the detail drawings will disclose all the salient features and show the resemblance to the Russian designs in particular. While this is a beautifully constructed rifle, the very nature and quality of the workmanship in it make it a relatively costly one to produce.

Full advantage of all the fine points of construction were incorporated in the later Belgian design.

How to Load and Fire the FN Self-Loading Rifle

Apply safety by pushing down, and pull bolt operating handle completely to the rear. Bolt will remain to the rear and two 5-round chargers can be loaded into the magazine in a manner similar to that used when loading a Mauser or Springfield rifle. Pull bolt slightly to the rear and release; the bolt will run forward and chamber a cartridge. The magazine may also be loaded by inserting the cartridges one by one in the magazine. Disengage safety by pushing it back up away from the rear of the trigger. Pressure on the trigger will fire the rifle and for each individual pull of the trigger a round will be fired until the magazine is exhausted.

How to Field Strip the FN Self-Loading Rifle

At the rear of the receiver is a locking key which seats on the rear end of the operating spring guide which protrudes through the rear end of the receiver. Making sure that the bolt is forward, turn locking key 180° upward, push receiver cover forward against the pressure of the operating spring lifting rear end of cover to release it from the guide track in the receiver. Pull cover to the rear and remove cover and operating spring assembly. Pull bolt operating handle to the rear until the bolt carrier guides are in line with clearance cut in receiver track. Lift front-end of bolt carrier/bolt assembly and remove from receiver. Remove bolt from the receiver. The piston and piston spring are removed by depressing the gas cylinder plug, located at the front of the gas cylinder tube under the front sight, and rotating it 90°. It can then be removed and tilting the rife forward will cause the gas piston and spring to slide forward and out of the gas cylinder tube. The magazine can be removed by pushing up the magazine catch with the point of a bullet or some similar shaped object. The amount of gas let into the rifle can be regulated on this rifle. Remove gas cylinder plug; then remove front end cap screw and front end cap. Remove front hand guard by swinging its front end upwards. The gas adjusting sleeve can be turned by inserting a bullet or pointed instrument in the holes in its body. To increase gas pressure, screw the sleeve forward; to decrease the gas pressure screw the sleeve to the rear.

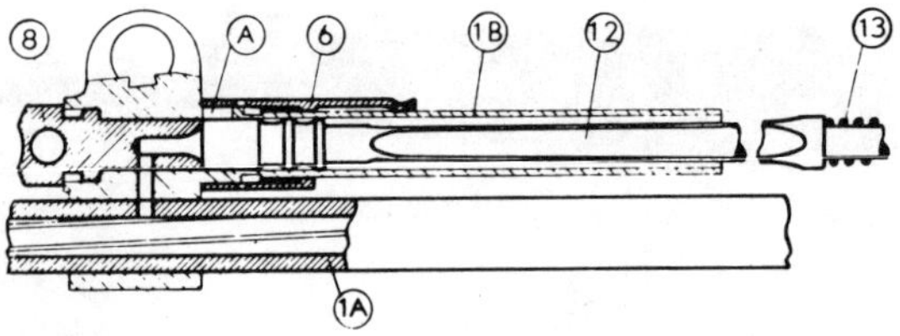

(1) **Gas action. May be sealed off by turning nut. Standard gas port top of** barrel bleeds gas into cylinder to drive piston back on tappet principle as for German Kar. 43.

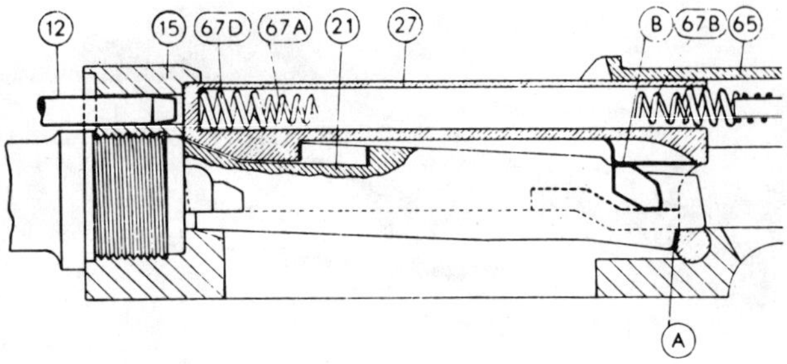

(2) **Unlocking action. Bolt locked. Tappet hits bolt carrier starting it back.** Spring returns piston to battery. Slide has initial free movement before camming action starts.

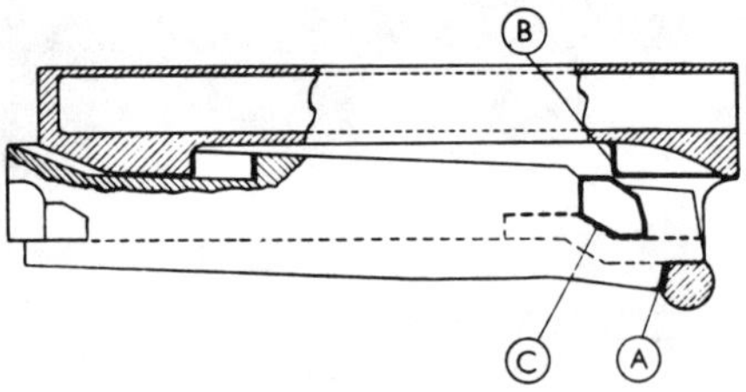

(3) **Cams on carrier raise bolt out of receiver engagement as with Russian** Tokarev rifle.

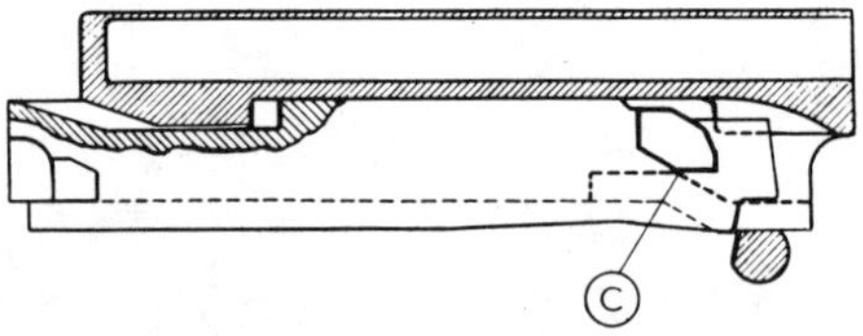

(4) **Bolt and carrier go back. Hammer is ridden down to cock. Recoil spring** compressed in standard fashion.

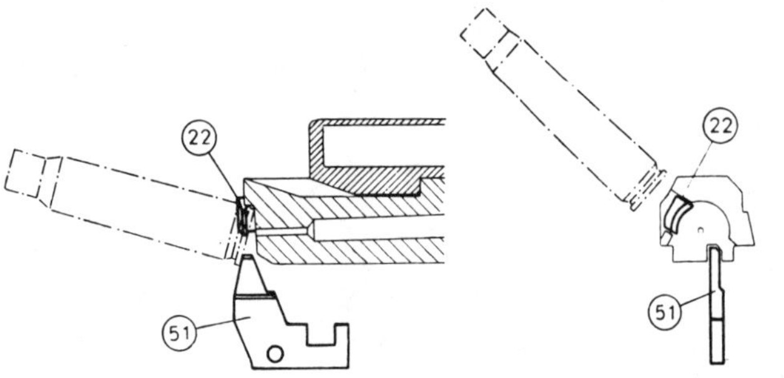

(5) **Near end of recoil stroke the cartridge case held by the extractor hits** the ejector and is tossed out of the rifle.

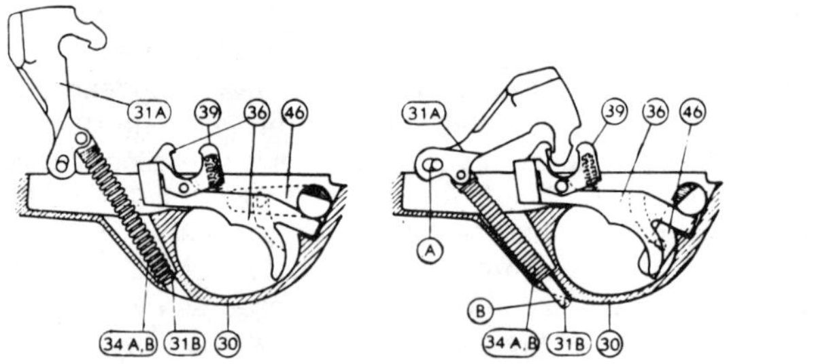

(6) **Left—Hammer and mechanism in fired position. On recoil hammer hook** will be engaged Browning-style on rear sear to prevent firing until trigger is released. Right—Hammer held by forward sear ready for firing. Mainspring guide pin projects as signal that hammer is cocked. Walther system.

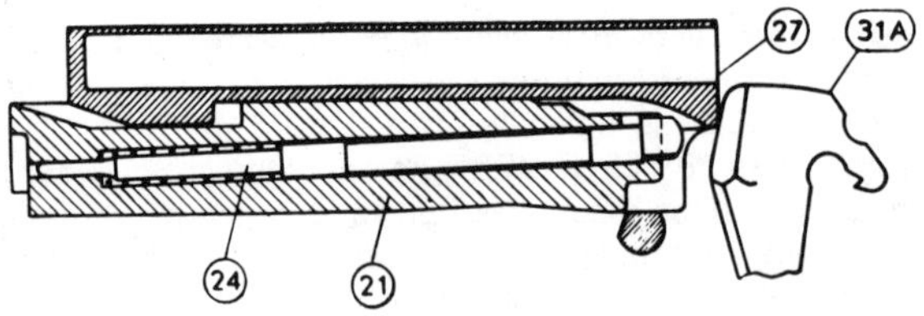

(7) **Safety feature. If action is not locked, hammer will hit carrier instead** of firing pin.

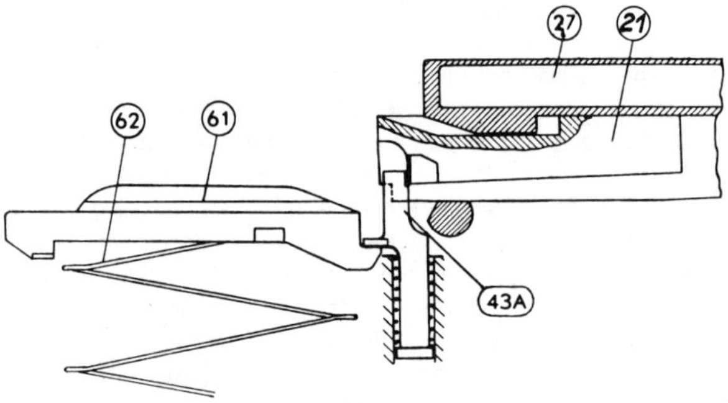

(8) **Magazine follower pushed up plunger to hold bolt open when last shot** is fired.

Operation of the FN Semiautomatic Rifle.

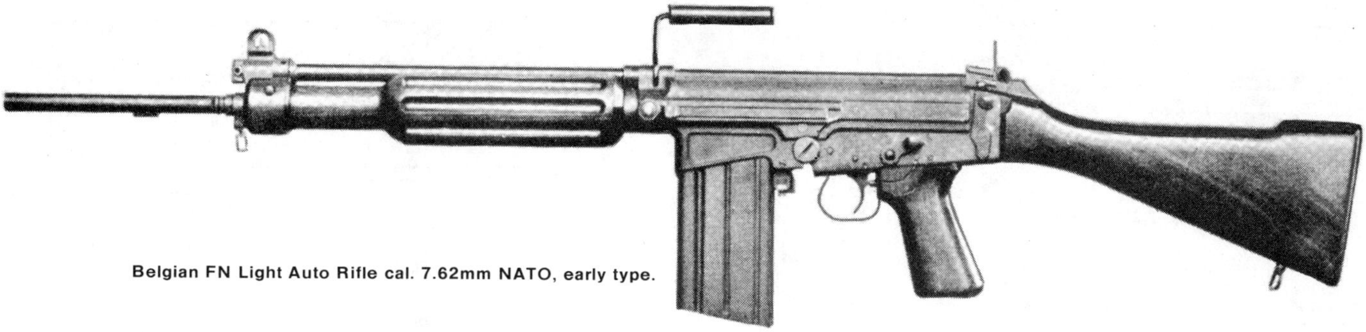

Belgian FN Light Auto Rifle cal. 7.62mm NATO, early type.

THE FN LIGHT AUTOMATIC RIFLE (FAL)

The FAL (Fusil Automatique Legere - light automatic rifle), has the widest distribution of any postwar developed rifle in the non-Communist world. It is currently used in Argentina, Austria, Australia, Belgium, Brazil, Burundi, Cambodia, Canada, Chile, Congo, Cuba, the Dominican Republic, Ecuador, India, Indonesia, Ireland, Israel, Kuwait, Liberia, Libya, Luxembourg, Morocco, Mozambique, Muscat and Oman, New Zealand, Paraguay, Peru, Portugal, Quatar, Ruanda, South Africa, Syria, Thailand, the United Kingdom (Britain) and Venezuela. It was used in quantity by West Germany. In addition to being manufactured at the FN plant it has been or is being manufactured at Rosario in Argentina, Steyr in Austria, Lithgow in Australia, Long Branch in Canada, B.S.A. and Enfield in the United Kingdom, Ishapore in India, and Pretoria in South Africa.

This weapon exists in many variations as can be noted from chart on a later page, but all are chambered for the 7.62mm NATO cartridge. It was extensively tested in the U.S. as the T48 and the Harrington and Richardson Arms Co. converted the drawings from metric to English and made 500 rifles for test by the U.S. In addition the U.S. purchased approximately 3,300 from FN for field tests. It should be noted that those FAL rifles as well as those FN MAG machine guns supplied to Cuba were purchased by the Batista regime and shipped to Cuba prior to 1960.

The FAL was originally made for the German 7.92mm PP43 cartridge, (7.92mm short). It was also made chambered for the developmental British .280 and .280/30 cartridges. After the United Kingdom and Canada adopted the U.S. caliber .30 T65E3 cartridge case (the current 7.62mm NATO), the FAL was made for that cartridge.

Characteristics of FN Light Automatic Rifle (FAL)

Caliber: 7.62mm NATO.
Weight w/empty magazine: 9.06 lbs.

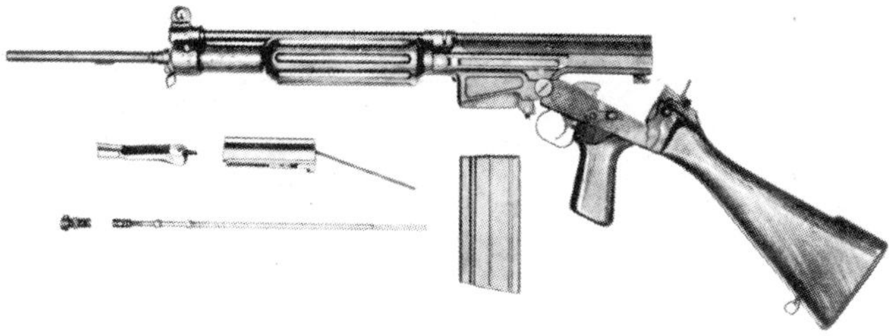

Belgian FN Light Auto Rifle. Operating parts removed. Field-strip without tools.

Overall length: 40 in.
Barrel length: 21 in.
Method of operation: Gas, selective fire.
Feeding: From detachable box magazine, staggered 20 round capacity.
Sights: Front; Hooded post. Rear aperture graduated in 100 meter steps to 600 meters.
Cyclic rate of fire: 650 to 700 per minute. Most of the weapons in service throughout the world are only used as semiautomatic weapons.

How the FAL Works

Pull cocking handle on left side of receiver to rear as in BAR. This leaves the right hand on the pistol grip ready for firing. The cocking handle does not move during firing, again as in the case of the BAR. This removes danger to the firer's face and does not interfere with aiming.

If there is an empty magazine in the rifle at the end of the cocking handle's stroke, the bolt is held open automatically. A loaded magazine is inserted in the magazine housing and pushed in until the retaining catch secures it.

Heavy-barrel version of the FN 7.62mm "FAL." This rifle is in service in Israel and Peru among other countries. It is used as the squad automatic weapon. A modified version of this weapon is manufactured in Australia and Canada.

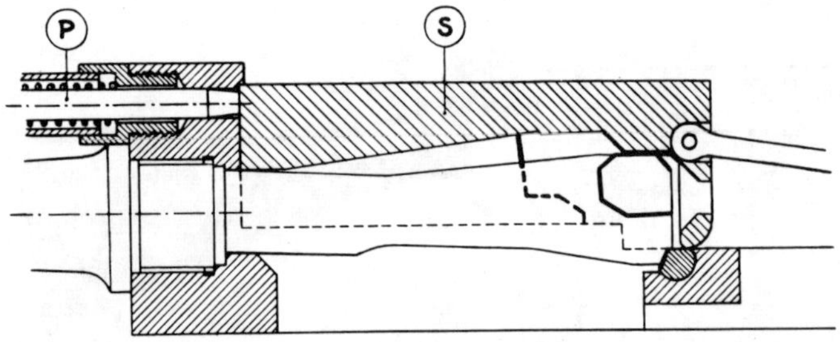

Piston driven back by gas hits movable carrier above locked bolt, driving it back and compressing return spring.

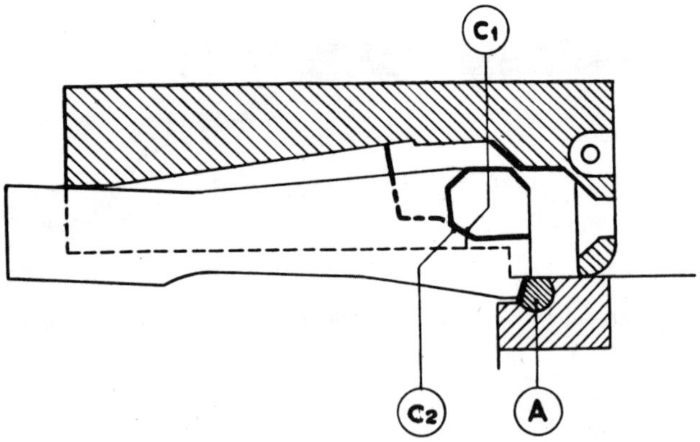

Ramps in the carrier engage cam (Cl) on bolt as bullet leaves barrel, thus raising the rear of bolt out of its locking seat (A) in the receiver.

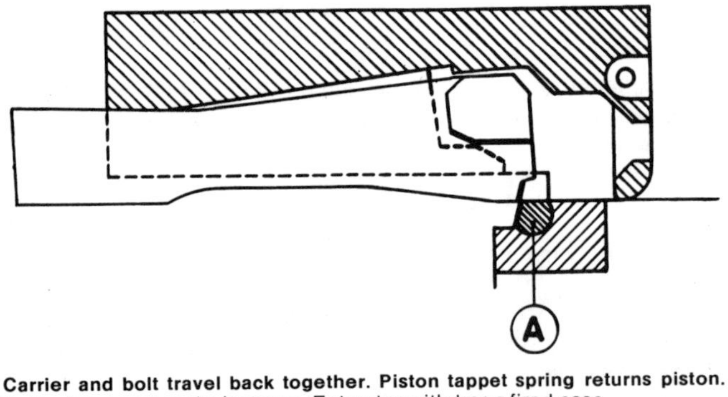

Carrier and bolt travel back together. Piston tappet spring returns piston. Bolt rotates and cocks hammer. Extractor withdraws fired case.

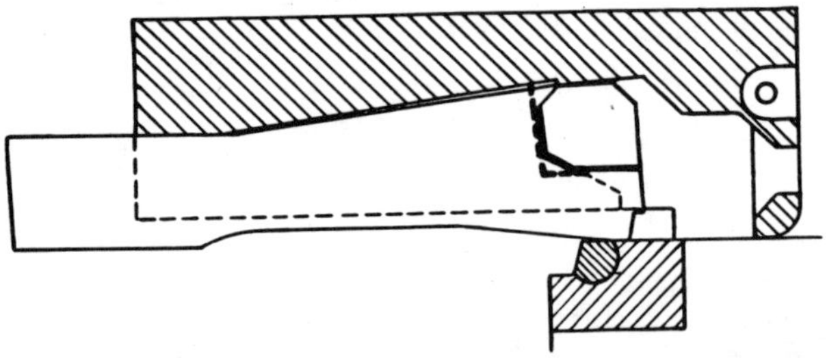

Bolt and carrier reach end of stroke. Case hits ejector. Return spring in butt is fully compressed.

Pressing the release stud on the left side near the magazine catch releases the bolt which is driven forward by its compressed return spring.

This return spring is housed in the butt. It acts on the bolt carrier through a rod pivoted at its rear face. As the bolt moves forward, its feed face strips the top round from the magazine in standard fashion and thrusts it up the ramp into the barrel chamber, where the extractor engages the cannelure in the case. The gun is now ready for firing.

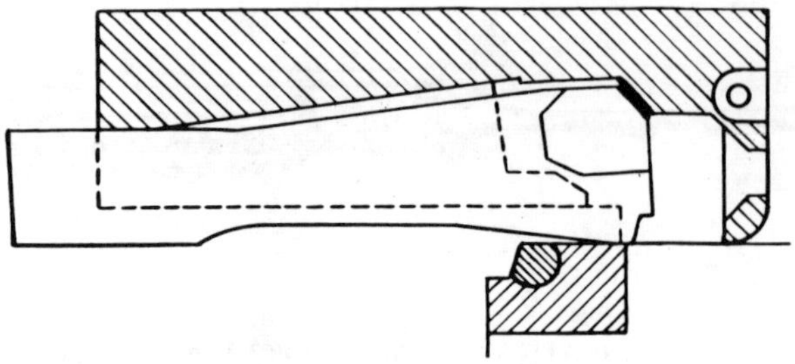

Spring starts carrier and bolt forward. Bolt picks up top cartridge in maga- zine and starts it toward chamber.

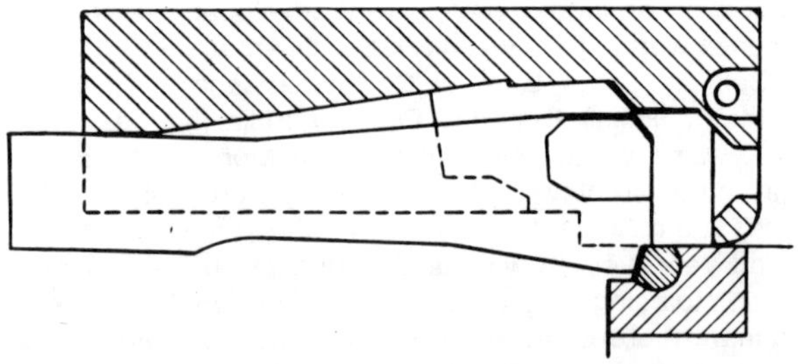

Face of bolt which chambers cartridge is stopped against face of barrel. Carrier is still driven forward, until it cams bolt down into its locking seat in the receiver.

Firing the FAL

The change lever on the left side of the receiver may be set for "safe," "single shot," or "full automatic" fire. The positions are widely spaced deliberately so that in the dark one can tell by sense of touch the position of the change lever.

If the change lever is set for single shot fire, pressure on the trigger releases the hammer to strike the rear face of the firing pin and fire the cartridge. Since the weapon fires from a closed breech, there is no such disturbance of aim as occurs in weapons like the submachine gun and the Browning machine rifle, where the mass of the breech mechanism moves forward with the pull on the trigger.

The operation is the standard gas system whose reliability has long been established for this type of weapon. Part of the gases following the bullet down the barrel pass through the port into the forward section of the gas cylinder. A gas regulator, previously adjusted, provides sufficient gas to satisfactorily operate the piston which is driven back to function the mechanism. Remaining gas passes to the open air through holes in the gas cylinder.

The piston, acting on the tappet principle, is driven to the rear in its tube on top of the barrel. It strikes the bolt carrier and pushes it to the rear.

Ramps machined into the bolt carrier engage a cam on the bolt after the bullet has emerged from the barrel and the pressure has dropped to safe limits. The bolt is cam lifted out of engagement with its locking shoulder in the receiver—this unlocks the action.

The bolt carrier and bolt now travel together to the rear. They ride down the hammer to cock it.

During rearward motion of the bolt, the extractor has carried the empty case out of the chamber with it. The case strikes the ejector and is thrown out the right side of the gun.

After the piston tappet strikes the bolt carrier and imparts the necessary impetus to it, during which travel the spring around the piston has been compressed, the spring operates to return the piston to forward (battery) position. The return spring within the butt is fully compressed during rearward movement of the bolt and carrier in standard fashion. At the end of the recoil stroke, the

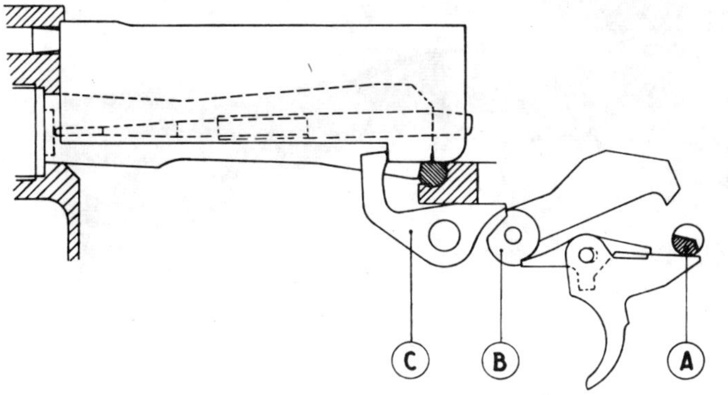

Gun with safety applied. Change-lever stud (A) is locking trigger to prevent movement.

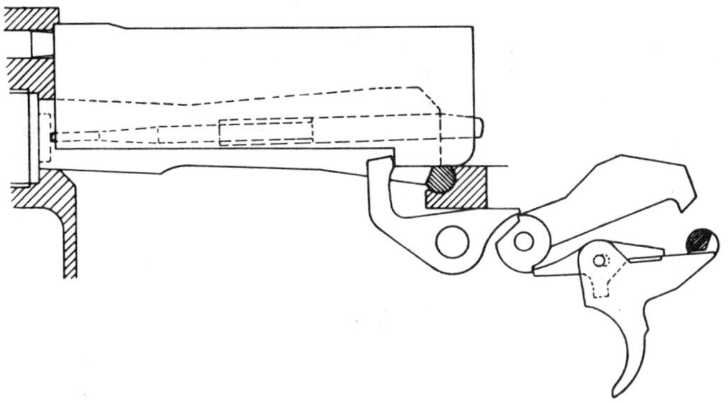

Change-lever stud set for single-shot fire. Pressure on trigger will free sear from hammer contact, but as the rifle action opens, the safety sear will hold the hammer until finger pressure is released.

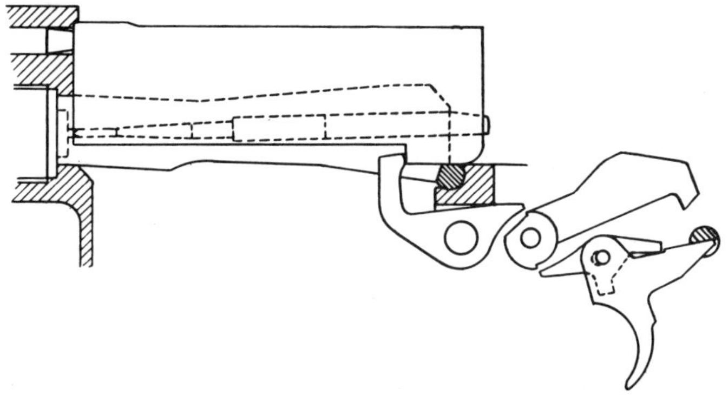

Lever is set for full-auto fire. Sear is out of engagement except when the trigger is released. Safety sear controls firing to prevent premature discharge.

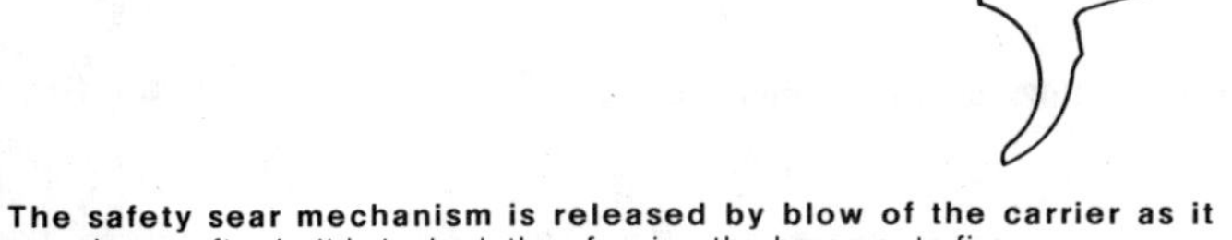

The safety sear mechanism is released by blow of the carrier as it goes home after bolt is locked, thus freeing the hammer to fire.

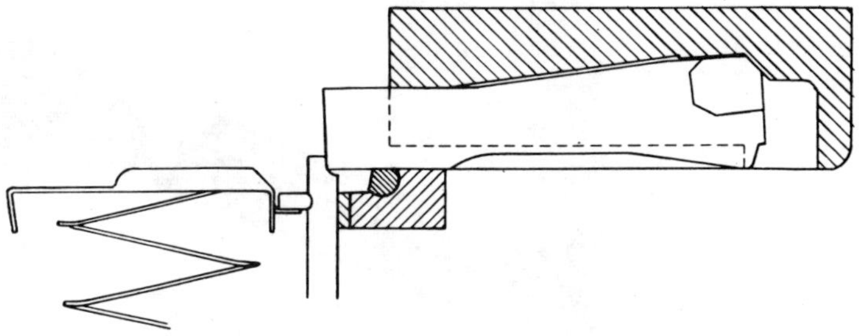

When last shot has been fired, the magazine follower forces a holding plunger up in front of the bolt to hold the action open.

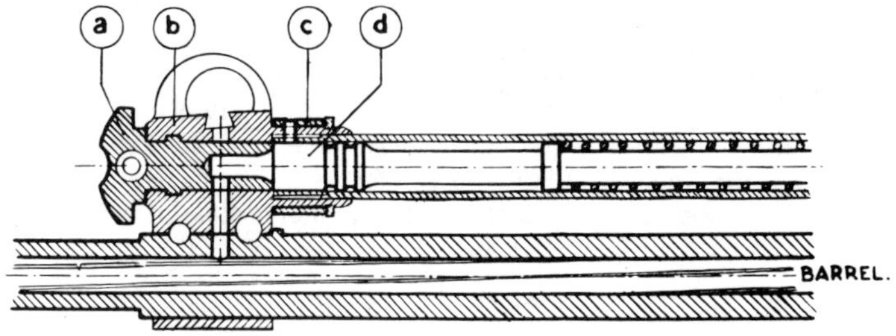

Gas plug (a) is positioned in end of gas cylinder (b). It may be turned to shut off gas when using rifle for grenade launching or as a straight-pull rifle. Gas regulator (c). Head of gas piston (d).

compressed spring reasserts itself and working through its connecting rod thrusts the bolt and carrier forward. The bolt strips a cartridge from the magazine and chambers it and stops against the face of the barrel. The bolt carrier, which still has continuing movement under thrust from the recoil spring guide and spring, works through the cam and ramps to force the rear locking end of the bolt down its locking recess in the receiver. At the end of the movement of the carrier, it is resting against the receiver above the line of the chamber where it is in line with the rear end of the piston tappet. The extractor, of course, snaps over the cannelure of the cartridge as the bolt chambers the cartridge.

When the change lever is set for automatic firing, the operation is identical with that already described except that the hammer is automatically released to continue fire until pressure on the trigger is released.

FAL Trigger Operation

Turning the change lever on the left side of the receiver to the safe position turns a stud into position at the extreme rear of the trigger to lock it mechanically. In this position, trigger movement is impossible.

When the change lever is set for single shot fire, the stud is turned into position to clear the rear of the trigger and to permit a pull on the trigger to draw the tip of the sear attached to the trigger down out of engagement with the hammer notch to release the hammer for firing. There are actually two sears—the rear or firing sear which connects hammer and trigger during single-shot operation; and the front or safety sear. This safety sear performs two functions. It blocks the hammer to prevent it from falling except when the bolt and the bolt carrier are fully home. It is struck by the bolt carrier at the final movement of closing travel when it releases the hammer so the firing sear may be operated on single-shot operation to release the trigger, or so the hammer may fire in full automatic fashion if the firing sear is out of engagement due to the automatic fire control being set on the change lever.

The entire operation is extremely simple mechanically and a

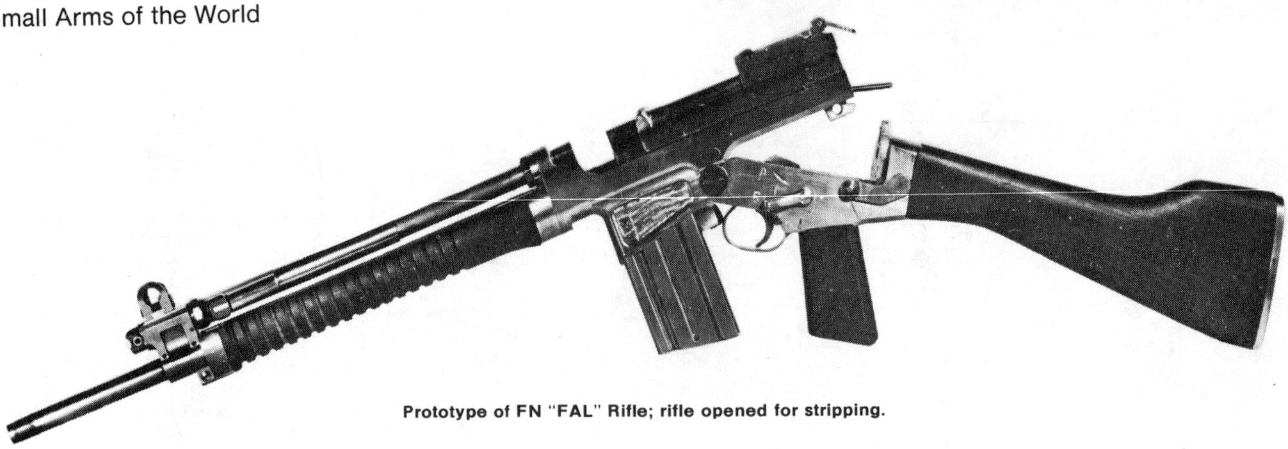

Prototype of FN "FAL" Rifle; rifle opened for stripping.

study of the drawings accompanying the explanation will clearly demonstrate all the mechanical features involved.

The ejector is mounted in the receiver and serves as a guide in its slot in the lower side of the bolt. At the end of the recoil stroke, the cartridge case strikes the ejector as the bolt passes slightly past the nose of the ejector to hurl the empty case out of the gun.

FAL Hold-Open Device

The pin to hold the bolt in rear position when the magazine is empty is mounted in the receiver to the right of the ejector.

It is in the form of a heavy plunger which is elevated by a nib on the rear of the magazine follower through a small pin working in a slot.

In rest position this plunger is held down by its spring. However, when the magazine follower rises as the last cartridge is fired, upward pressure of the heavy magazine spring overcomes the lighter spring attached to the pin and the pin is thrust upwards. The plunger is thereby raised to its complete height and projects into the forward path of the bolt.

When a full magazine is inserted in the gun, drawing the bolt slightly to the rear allows the plunger spring to react and lower the plunger, and the bolt driven by the recoil spring now moves forward to chamber a cartridge.

7.62mm T-48 Rifle, the FN "FAL" rifle as tested by the United States. Note prong-type flash suppressor.

New version of 7.62mm NATO FN Heavy-barrel Rifle with short handguard.

FAL Gas Regulator

The gas regulator and a gas plug control the quantity of gas permitted to reach the piston. The gas plug A is fixed into the end of the gas cylinder B. This plug has two positions. One permits full access of the gas from the barrel directly to the gas cylinder. The other when turned through 180 degrees blocks off all entry of gas. In this condition the arm will not function automatically. In this condition, however, it can be operated manually as a straight-pull bolt-action rifle. In other words, pulling the cocking handle back and releasing it will load and cock the gun ready for firing. After firing, another full pull to the rear and release of the cocking handle will eject the empty case and load a new round ready for another pull.

Many American military experts believe that the full automatic feature is entirely too wasteful of ammunition by the average infantryman. This device permits a commander to control the type of fire developed by his troops if he so desires. By putting the gas plug in its closed position, he can have the equivalent of a manually-operated arm in the hands of his personnel to save ammunition. When desired, opening the gas plug gives him the full advantage of either semi-automatic or full automatic fire.

In this closed position the arm may be used as a grenade launcher, since automatic action is not desired for such a purpose.

A gas regulator C consists of a shroud around the end of the gas cylinder. Unscrewing allows gas to escape.

This system of gas regulation by exhaust keeps fouling of the piston to an absolute minimum, and allows regulation of power.

The gas cylinder is placed above the barrel. The center of gravity of the weapon is in line with the axis of the barrel. As a result, the recoil does not tend to pull the weapon upwards as much as many semiautomatic rifles.

FAL Field Stripping

Stripping and assembling this weapon for normal maintenance and repairs is done without the aid of tools.

With the magazine removed, the release catch at the rear of the top of the receiver is pressed and the body of the gun is hinged exactly as in the case of the familiar shotgun. The entire bolt and gas operating assembly may now be withdrawn as a unit from the open rear of the receiver. The extractor and firing pin (the normal breakage points in any automatic design) may be withdrawn in a matter of seconds and replaced. All parts are self-contained. There are no loose springs, guides, screws, or pins to be removed during this operation. The recoil spring is retained within the butt when the gun is opened for field stripping. When the assembly is returned to the receiver, the guide rod at the end of the bolt, resting against the compression plug in front of the butt, serves to compress the spring housed within when the action is cocked or in recoil motion.

For all practical purposes, no further stripping of this arm is necessary.

FAL Carrying Handle

A special folding carrying handle is provided above and forward of the magazine housing. This handle is placed at the center of gravity to make the arm easy to carry in rapid advances. It may also be used as a carrying handle for marching and general field order. It can be quickly turned down out of the line of sight and out of the way.

The grip of pistol type design behind the triggerguard aids greatly in stabilizing fire and, since all operations of loading and cocking and change lever adjusting are done on the left side of the

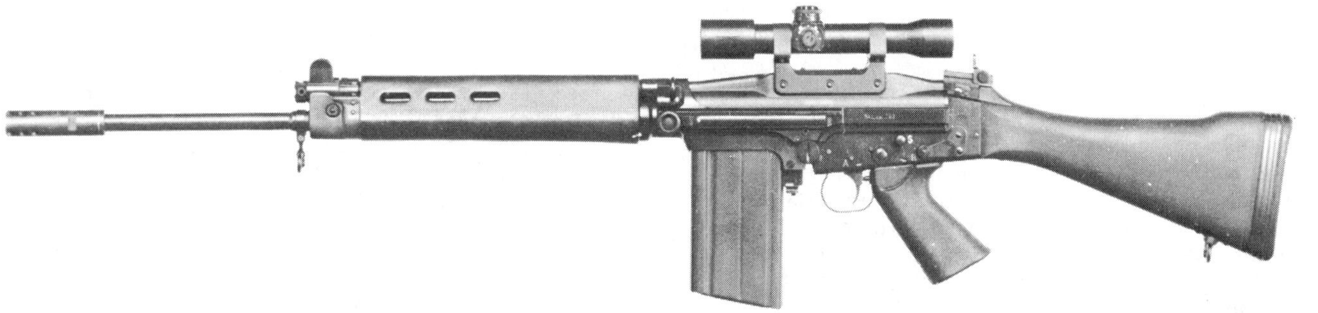

FN 7.62mm NATO "FAL" Rifle with high-mounted scope.

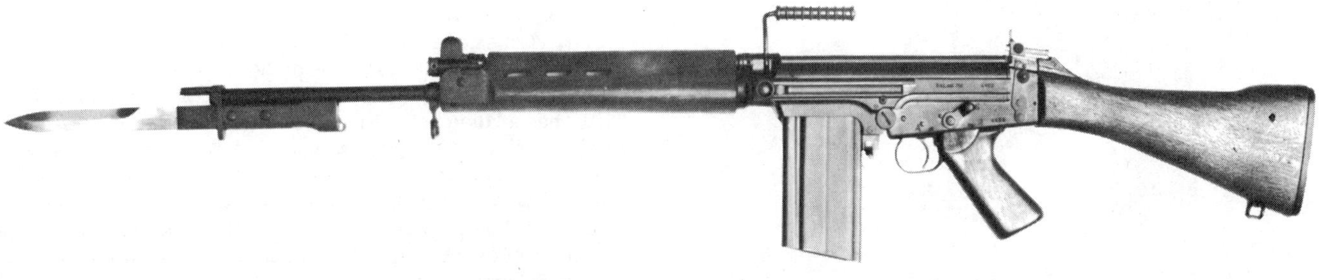

FN 7.62mm "FAL" as made for Peru with bayonet.

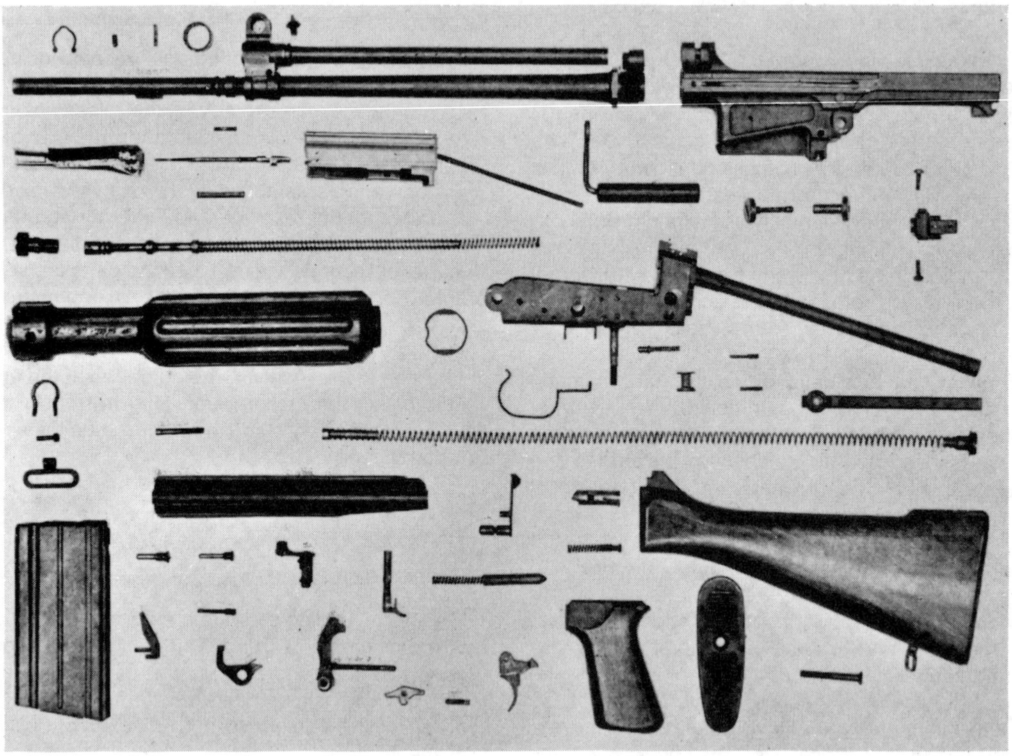

FN Light Auto Rifle, detail stripped.

receiver, full control of the arm may be maintained at all times. In addition, of course, this pistol grip gives the weapon the advantage of ease of operation for firing at waist level under appropriate conditions. The forestock is designed for comfort and secure control as a normal left hand hold for shoulder firing or waist firing.

As special accessories, a muzzle brake and flash hider are available. A grenade launcher is also part of the accessory equipment as is a special bayonet and a detachable folding bipod mount.

FAL Further Dismounting

Remove the magazine. Draw the cocking lever to the rear and check that the chamber is empty. Move the cocking handle back to free the bolt and allow it to go forward to locked closed position.

Push down the butt locking lever on the left rear side of the receiver as far as it will go. Simultaneously push the butt itself downwards. The butt will pivot together with the lower receiver section. This allows the arm to open as in the familiar case of the hinged-frame shotgun.

Pull back the spring rod attached to the bolt; this will pull the bolt and carrier assembly to the rear out of the gun.

Slide the receiver cover to the rear off the receiver.

Lift the front end of the bolt while pressing it forward into the carrier and continue lifting the front end to raise the rear gently out of carrier contact against the pressure of the firing pin spring.

Push out the cross retaining pin while holding onto the rear of the firing pin and the pressure of its spring will force it out of its housing.

Insert the nose of a bullet under the extractor and pry outwards and upwards to withdraw the extractor.

The gas plug can be removed if desired and a rod passed through the gas cylinder to clean fouling.

The gas plug can be turned with the nose of a bullet for removal. This permits removal of the piston and spring from the gas cylinder. While there are very few remaining parts, and additional stripping of the firing mechanism can be done easily, it is not normally done by the average soldier.

FN 7.62mm "FAL" Rifle with low-mounted scope.

FN 5.56mm "CAL" RIFLE

As long ago as 1963, FN was working on a military rifle for the 5.56mm (.223) cartridge. At that time they were considering scaling down the "FAL" from 7.62mm to 5.56mm and did build some "FAL" prototype rifles chambered for the 5.56mm cartridge. The "CAL"-- Carbine Automatique Legere-- or light automatic carbine--does resemble the "FAL" externally in its overall configuration. The "FAL" has a tilting bolt system, but the bolt of the "CAL" rotates to lock and unlock. The bolt head has an interrupted screw-type locking mechanism. There are two sets--top and bottom--of buttress type lugs which are inclined at an angle to the axis of the bolt body. This inclination provides for slow initial extraction, i.e., an initial loosening of the case in the chamber before the case is withdrawn from the chamber.

Variations Among Some FN Light Automatic Rifles Produced at FN.

	Barrel 1. Smooth muzzle 2. Threaded muzzle 3. Threaded muzzle for combined grenade launcher/flash suppressor	**Flash suppressor** 1. Without 2. With 3. Combined with grenade launcher	**Bayonet** 1. Flash suppressor type or normal 2. Tubular type 3. Without	**Extractor in** 1. One piece 2. Two pieces	**Butt stock** 1. Without front socket 2. With front socket	**Butt plate** 1. Without butt trap 2. With butt trap	**Handguard** 1. Wood 2. Metal 3. Molded material	**Bipod** 1. Without 2. With	**Loading** 1. Without charger (Stripper clip) 2. With charger
Austria	3	3	3	2	2	2	2	2	1
Belgium	1	1	1	2	2	1	3	1	2
Cambodia	1	1	1	2	2	1	3	1	1
Chile	2	2	1	2	2	1	3	1	1
Ecuador	2	2	1	2	2	1	3	1	1
Indonesia	1	1	1	2	1	2	1	1	1
Ireland	2	3	2	2	2	2	3	1	1
Israel	1	1	1	1	1	1	1	1	1
Kuwait	1	1	1	2	1	2	1	1	1
Libya	1	1	1	2	2	1	3	1	1
Luxembourg	1	1	1	1	1	2	1	1	2
Netherlands	3	3	2	2	2	1	2	2	1
Paraguay	1	1	1	2	1	2	1	1	1
Peru	1	1	1	2	2	2	1	1	1
Portugal	1	1	1	2	2	1	3	1	1
Qatar	1	1	1	1 & 2	1	2	1	1	1
Santo Domingo	1	1	1	1	1	1	1	1	1
South Africa	1	1	2	2	2	1	3	1	1
Syria	1	1	1	1	1	2	1	1	1
West Germany	1	2	3	2	1	2	2	2	1
Venezuela	2	2	3	1	1	1	1	1	1

There are other variations as for example, the British Produced L1A1 and the Canadian Produced C1A1.

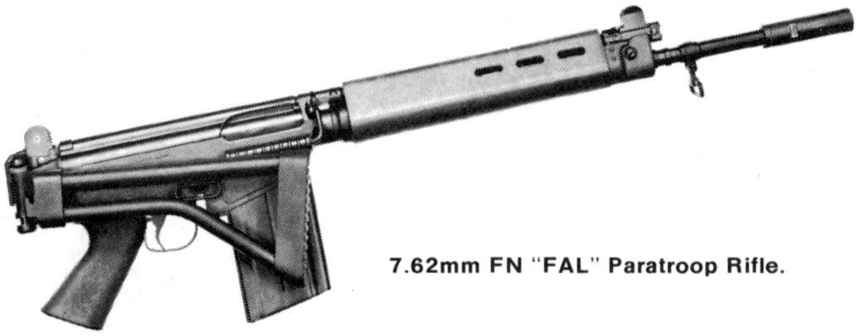

7.62mm FN "FAL" Paratroop Rifle.

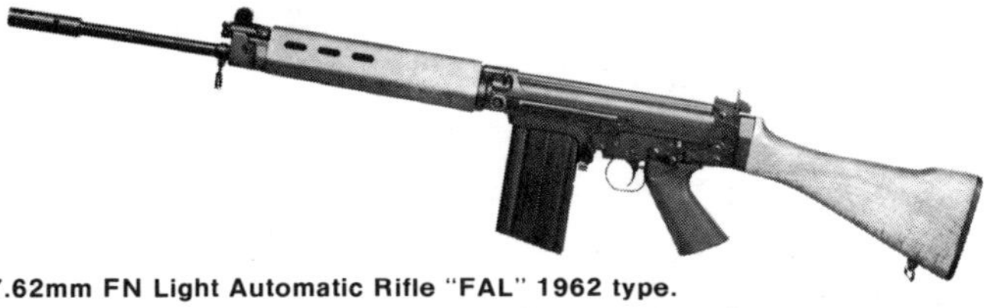

7.62mm FN Light Automatic Rifle "FAL" 1962 type.

7.62mm FN "FAL" Rifle as used by Belgian Army.

Characteristics of the "CAL"

Caliber: 5.56mm
System of operation: gas, selective fire
Length overall: 38.6 in.
Barrel length: 18.4 in.
Feed device: 20-round, detachable, staggered row box magazine
Sights: Front: protected post
Rear: "L" with apertures
Muzzle velocity: 3182 f.p.s.
Weight: 7.3 lbs. loaded w/light alloy magazine
Cyclic rate of fire: approx. 850 r.p.m.

Description of the "CAL"

The "CAL" makes extensive use of steel stampings--the forend, receiver and magazine and trigger housing are all of stamped construction. The rotating bolt, described above, locks into a barrel extension therefore keeping the strain on the receiver to a minimum. The barrel is locked in the receiver by a nut which is slipped over the front end of the barrel and is screwed onto a thread near the breech-end of the barrel, and bears against the conical surface at the front of the receiver. The gas system is basically the same as that of the "FAL", the spring loaded piston strikes the bolt carrier and then returns to its forward position; a cam slot cut in the top and side of the bolt carrier cams a stud on the bolt causing it to rotate into and out of the locked position. The trigger mechanism has a 3-round burst selector in addition to the provisions for full and semi-automatic fire. If the safety/selector lever is set on the figure 3, the weapon will fire three rounds and cease fire until the trigger is pulled again.

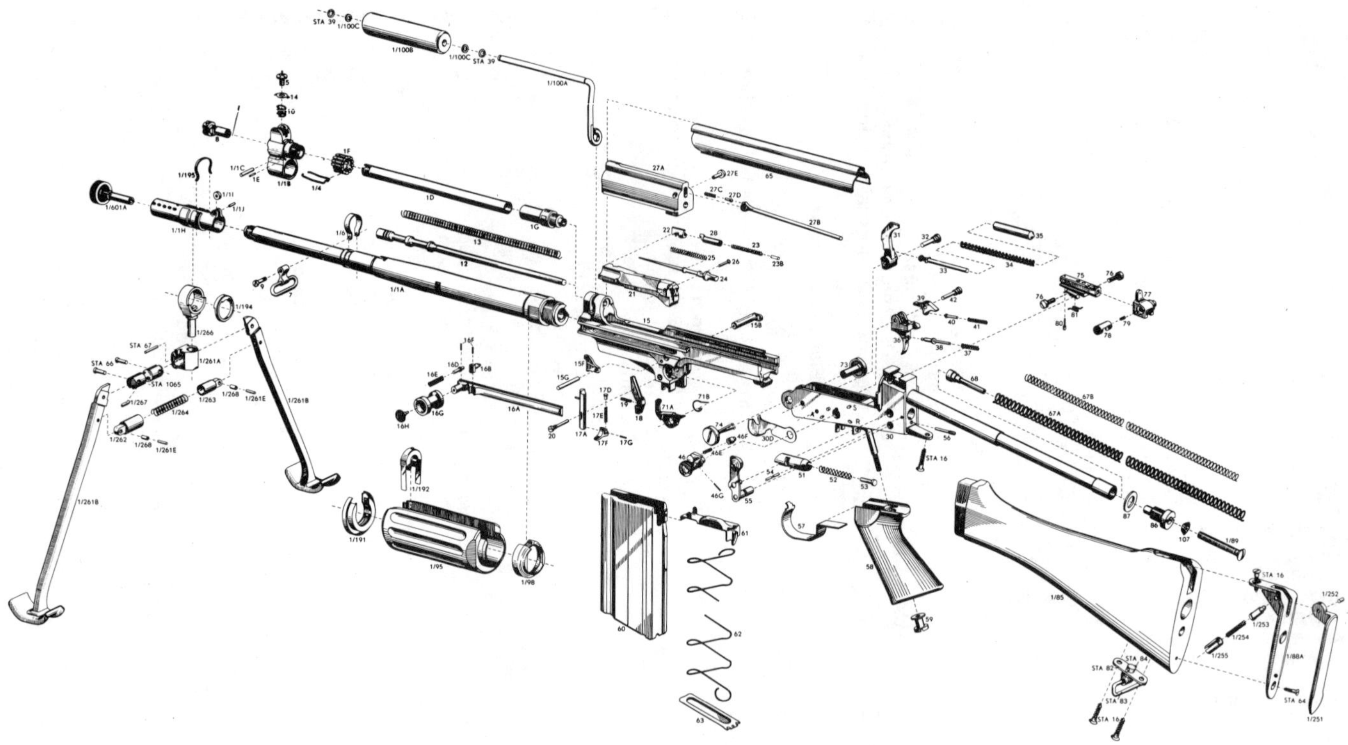

FN Heavy-barrel Automatic Rifle as made for South Africa. Caliber 7.62mm.

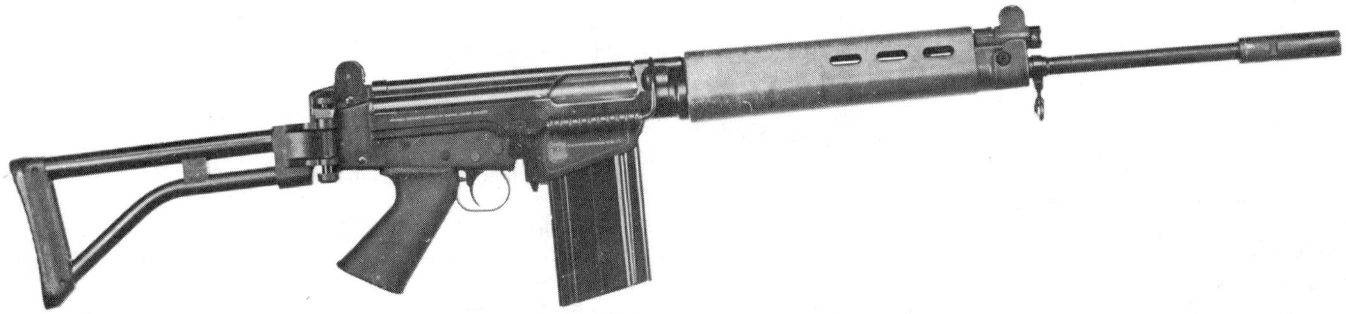

FN 7.62mm NATO "FAL" Paratroop Rifle, stock extended.

FN 7.62mm NATO "FAL" Rifle, 1964 pattern with plastic stock, handguard and pistol grip.

Belgian FN 5.56mm "CAL" Rifle.

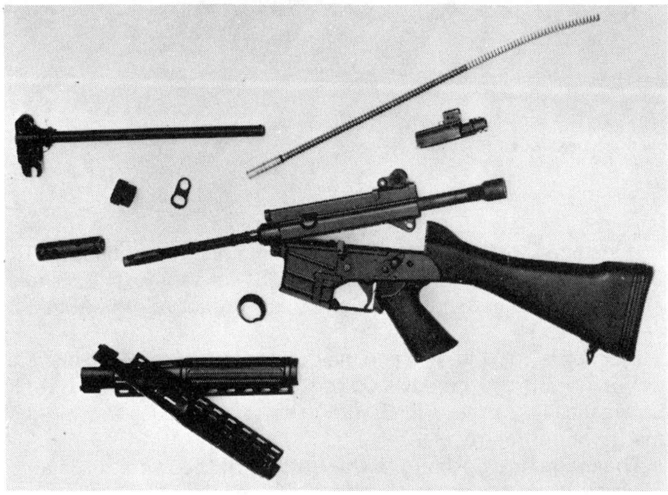

Belgian FN 5.56mm "CAL" Rifle stripped.

Folding Stock Version of the CAL

The 5.56mm CAL is now being made in a folding stock as well as a fixed stock version. The overall length of the folding stock version with stock fixed is 39.3 inches and with stock folded it is 29.6 inches. The weight of the weapon with empty magazine is approximately 7.8 pounds.

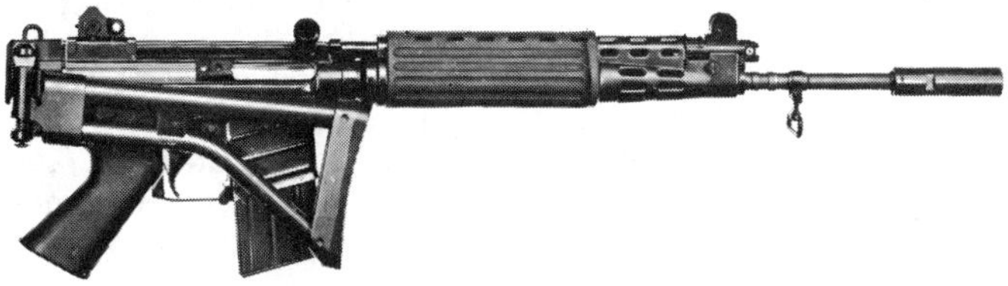

FN 5.56mm CAL rifle with folding stock; stock folded.

Belgian 5.56 "CAL" Rifle with 40mm Grenade Launcher attached; launcher sight in aiming position.

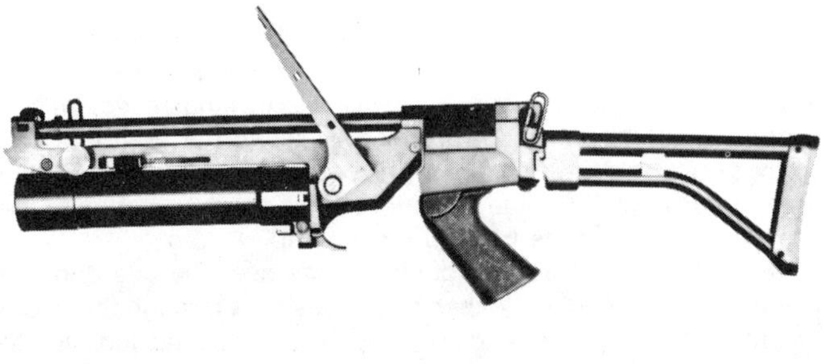

Belgian FN 40mm Grenade Launcher on frame, stock in fixed position.

Field Stripping the "CAL"

The "CAL" breaks open like a shotgun does (and as does the "FAL") for easy field stripping. In addition to the bolt/bolt carrier assembly, and the piston, the barrel can also be removed if the hand guards are detached, but this should not be necessary for normal cleaning.

40mm Grenade Launcher for the "CAL"

FN has developed a 40mm grenade launcher which can use the U.S. 40mm grenade used with the M79 grenade launcher. This launcher can be attached to the "CAL" as shown in the photograph or can be used on a frame with folding stock. The launcher can be reloaded on the "CAL" by tripping the lever at the rear underside of its barrel. An automatic safety keeps the grenade launcher from being fired unless the launcher sight is moved into the aiming position. The trigger is mounted on the left top side of the launcher.

THE MODEL 30 BROWNING AUTOMATIC RIFLE

FN produced the Browning Automatic rifle for Chile, China, Belgium and other countries before World War II. Most of these weapons were variations of the Model 30 and were similar to the U.S. Model 1918A1 BAR. The Model 30 can be distinguished from the U.S. issue "BAR" by the magazine and ejection port covers, the separate pistol grip, the ribbed barrel (the U.S. M1922 BAR had a ribbed barrel, but few of these were made) the shape of the fore-end and the dome-shaped gas regulator. Some Model 30 rifles were made with quick change barrels and all could be mounted on a special tripod made by FN. The Model 30 was made in the following calibers: 7mm, 7.65mm, and 7.92mm. The Model 30 Browning automatic rifle is an obsolescent weapon and is not likely to be found in the hands of troops today.

BROWNING AUTOMATIC RIFLE TYPE D

After World War II, FN introduced the Type D Browning automatic rifle. This weapon features major improvements not found in other versions of the BAR. The Type D has a quick change barrel and a rate-reducing mechanism as do several other versions of the BAR. It is the only version of the BAR, however, to have a rapid method of field stripping. The stock is hinged and by removal of the trigger guard assembly pin and butt access pin, the piston slide and bolt assembly can be removed. The recoil (operating) spring of the Type D is mounted in the butt rather than in the piston slide assembly, as it is in the U.S. BAR. A clockwork type rate reducer is used with the Type D rather than the buffer type found on the U.S. Browning Automatic Rifle M1918A2. The Type D was purchased by Egypt in 7.92mm during the reign of King Farouk and by Belgium in caliber .30-06.

Pre-war FN Browning Automatic Rifle Model 30.

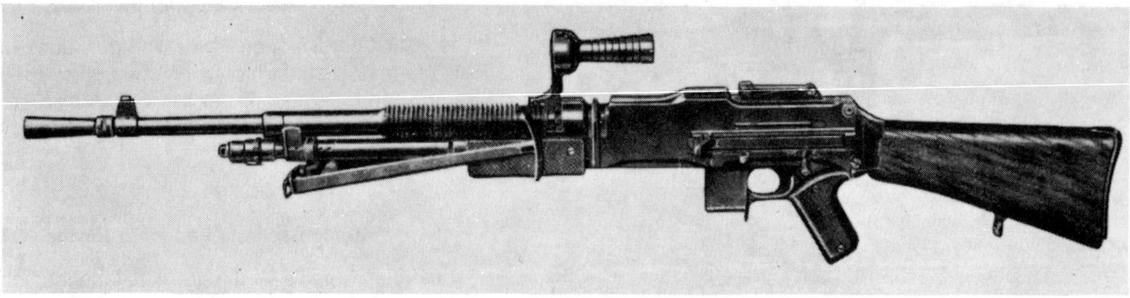

Belgian Type D. Quick-removable barrel version.

Characteristics of Type D

Caliber: U.S. caliber .30 and 7.92mm.
Magazine: Detachable box. Holding 20 cartridges in staggered line. Magazine positioned directly below receiver.
Muzzle velocity of cartridge: As given for specific cartridges.
Barrel length: 19.7 in.
Length with flash hider attached: 45.1 in.
Weight without magazine: 20.3 lbs.
Weight of magazine: .56 lb.
Sights: Front: post type. Rear: tangent.

Description of Type D

Barrel group. The barrel has a bedding point at the front end to receive the front sight. It also carries a gas cylinder bracket which forms the housing for the gas regulator.

A hole in the underside of the barrel passes through the gas cylinder bracket. This allows gas to enter the gas cylinder during firing, and provides the energy for operation.

The carrying handle and barrel removal bushing are mounted at gravity center on the barrel and connected with a locking nut. The nut has an external and an internal differential thread. While the internal thread engages the barrel, the external thread engages the receiver. Pulling the handle loosens it from its notch in the locking nut and allows it to swing freely. In this position it is used as a carrying handle for transporting the weapon.

On the other hand, when it is desired to remove the barrel or to tighten it in the receiver, swinging the handle tightens or releases the differential thread as required in accordance with direction of motion applied.

The barrel muzzle is threaded for a protecting muzzle ring. This may be replaced with a flash hider. A blank firing device also may be threaded on for use in maneuvers.

The rear of the barrel is machined to furnish a cut for the extractor and also a ramp for guiding cartridges into the chamber.

The flanges on the barrel provide a greater area for radiation to increase air cooling.

Regulator. This is the exhaust type. It permits adjusting the amount of gas utilized to function the action to meet varying conditions.

When it is screwed in, most of the gas passes into the cylinder giving maximum thrust. As it is unscrewed, the regulator allows progressively larger amounts of gas to escape into the atmosphere and the smaller quantities used for thrust slow down the action correspondingly.

Receiver. This is the heart of the gun. It receives all the mechanism. The rear sight is mounted conventionally at the upper rear end of the receiver. The cocking handle groove guide is on the left side plate. Above it is a push button to free the bolt for quick dismounting.

The left side plate also carries the trigger guard retaining pin. It is stamped with the letter "M" indicating full automatic fire, "R" for automatic firing at slower rate, and "S" for safety. (N.B. Single shot fire is obtained in this design by fast trigger release from the "R" position.)

Ejection is from the right and the port there is fitted with a moving cover. When the gun is cocked or fired, this cover opens automatically. At other times it serves to keep dust and foreign matter out of the receiver.

The magazine opening in the bottom of the receiver is also closed by a dust cover when the magazine is not inserted. To the rear of this is the trigger guard opening. The butt support at the rear of the receiver is secured by a removable axis and axis screw. Two grooves are machined inside the receiver to guide the reciprocating slide. Also machined are an opening for the bolt guide and its spring, locking recess for the bolt, and recess for the two bolt guides.

An interrupted thread at the forward end of the receiver is provided to lock the barrel in. The gas recess cylinder at the front end carries the fore-arm plate and the magazine opening cover.

Gas cylinder. The cylinder is fastened to the receiver. It is open at its front end to receive the gas regulator. Two wings on the upper front section of the cylinder support and guide the barrel to assist in rapid barrel assembly. There are three gas exhaust ports bored in the front of the gas cylinder to regulate firing rate.

Bipod mount. The mount is secured to the front end of the gas cylinder by an assembly block pinned to a head. The carrying sling swivel is fastened on the bipod mount head.

Operation of Type D

A magazine is inserted in the bottom and pushed in until the catch holds it. The cocking handle is then pulled to the rear to cock the gun and thrust forward to its normal position.

As the trigger is pulled it operates the right sear to release the slide. The recoil springs drive the operating mechanism forward.

The lower feed face on the bolt hits the upper part of the first cartridge in the magazine and drives it up the barrel ramp into the chamber. As the base of the cartridge case slides up the front face of the bolt behind the extractor, the lower cam surface of the bolt lock hits the end of the bolt supports and buckles the bolt lock up.

The bolt stops against the rear face of the barrel, but the bolt lock continues forward driven upward by its link and by the reciprocating slide. The bolt lock is now in its locking recess in the receiver, but the slide continues forward under action of the springs, carrying the hammer with it to strike the firing pin and fire the cartridge. The slide shoulder strikes the rear section of the gas cylinder halting its forward movement.

Recoil Movement. As the bullet passes over the gas opening in the barrel, some of the expanding gases pass through the opening into the gas cylinder and drive the piston and attached slide to the rear.

The slide acting through its link pulls the bolt lock down out of

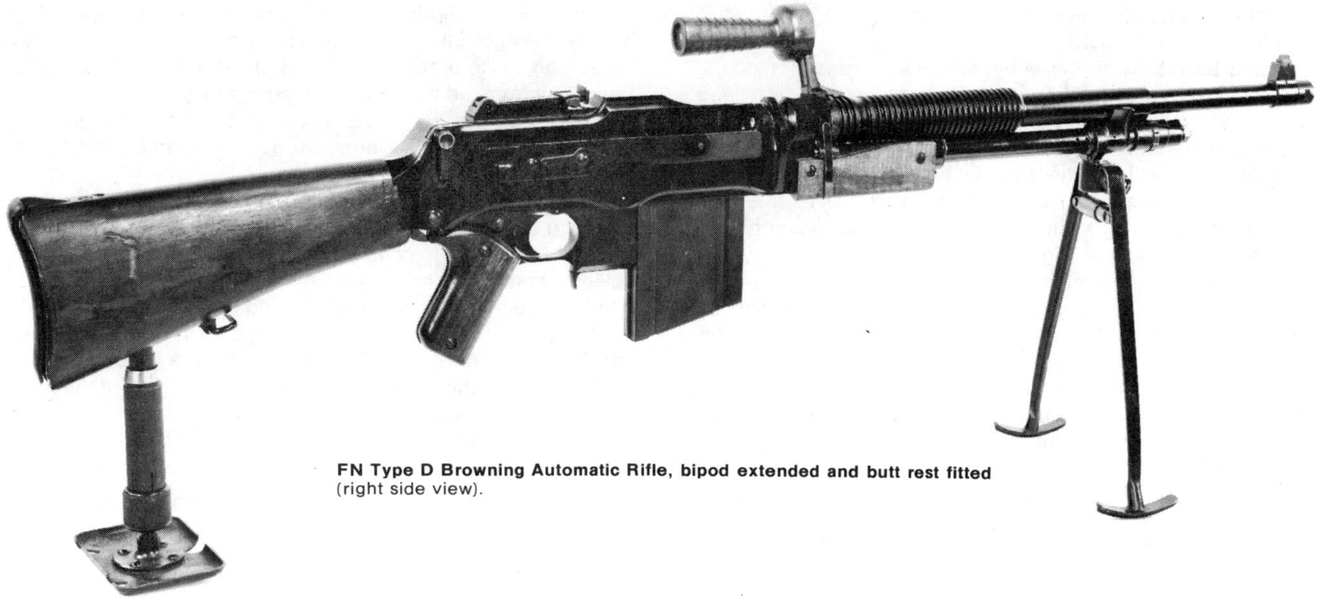

FN Type D Browning Automatic Rifle, bipod extended and butt rest fitted (right side view).

the locking recess and as motion continues backward the bolt (which starts very slowly relatively) gains speed when the bolt lock is near its completely unlocked position. At this point of complete unlocking, the bolt and the bolt lock travel at the same speed as the slide.

The firing pin is withdrawn during this rearward motion by the action of the slope in the bolt lock on the firing pin heel. Meanwhile, the empty case held by the extractor against the front face of the bolt travels back with the bolt until it strikes the ejector and is thrown out the right side through the ejection port.

The rearward movement is limited by the buffer which compresses the return springs. At this point where the recoil movement is complete, the slide and the attached sear catch the slide.

Note: One of the most important features of the Browning action is the two-stage extraction system. As the opening movement begins, the bolt lock is started slightly backwards by a progressive movement as its cam comes in contact with the rear section of the bolt guides. This loosens the case from the chamber in very efficient primary extraction. The loosened cartridge case is completely extracted by direct pull only after the breech pressure has been dissipated and the action is unlocked.

This is one of the basic reasons for the reliability of this type of design.

Automatic Fire. When the gun is cocked the change lever is pushed to position "M." In this position the left sear is cut off and does not interfere with the movement of the slide. As a result, when the trigger is pulled, and the action is operated by the gas piston and slide, at the end of the recoil stroke the sear is not in position to grasp the slide. This results in the slide continuing to go forward to fire automatically until the trigger is released to catch the recoiling slide.

Automatic Fire—Slow Rate. Cocking the gun and pushing the change lever to "R" position serves to slow down the normal high rate of full automatic fire. It is not single-shot fired and should not be so construed.

In position "R," the slide is held at the rear by the right sear which is somewhat longer than the left one. Pulling the trigger lowers the right sear and releases the slide to be driven forward by the recoil springs, but it is momentarily caught by the left sear which moves slightly forward through the action of its pin traveling in an elongated hole. The left sear moves forward a catch which in turn releases the slowing up device lever. The lever is raised by action of its spring and the movement is slowed by interference of a rack, pinion, and ratchet. This effectively slows down the forward travel of the moving mechanism.

As the lever is raised it swings up the forward end of the left sear and pulls down the rear end of the sear. When this raising movement is completed the sear disengages completely from the notch in the slide releasing it for forward travel. The moving slide depresses the lever which is again caught and held by the latch.

Single-Shot Firing. When the gun is cocked and the change lever set at position "R," the rate of fire is slowed down enough that the trigger may be released without difficulty to furnish single shot fire. In effect, the action therefore is one merely of slowing down the automatic operation so that manual tapping of the trigger gives single shot fire, but at the same time the gun may be held in position to allow a burst of automatic fire at a relatively low rate in case of emergency.

Buffer Operation. As the slide is driven back by the attached piston when the gun fires, the opening movement is slowed by the return springs through their spring rods.

When the slide contacts the recoil plug, it pushes back the first friction ring and cone, the first cone pushes the next one, and so on until the final compression of the buffer spring is completed.

These friction rings are split and they open slightly under action of their cones rubbing against the tube containing the buffer. The friction thus engendered, working with the action of the return springs and the buffer spring itself, absorbs the violent recoil of the mechanism.

Gas Adjustment. The gun is first placed in firing position. By use of a special tool provided, the gas shroud is unscrewed until the gas escape holes are very nearly uncovered. The change lever is then set in position "R" and the gun cocked. In this position when the gas adjustment is correct, the mechanism will remain open after each shot and ejection must be normal (meaning an empty case projection of about four to five yards from the gun). If this condition does not ensue, the screw must be further adjusted to achieve it.

To Load. Pull cocking lever as far back as possible until slide is caught and held by the sear. Push cocking lever forward. Open magazine cover and insert a loaded magazine through bottom of gun. Set change lever at position desired. Gun is ready to fire by pulling the trigger.

Barrel Removal. First check to be sure that the end of the handle is inserted in the groove in the locking nut. Press in the locking lever and turn the handle upward to vertical position. Push forward on handle and remove barrel.

Replacing Barrel. Seize barrel by handle with regulator downward. Insert the rear of the barrel into the receiver, its front end resting on its vee support in the front section of the gas cylinder. Push the regulator into the gas cylinder. Draw the barrel to the rear of the receiver and swing to the left as far as possible on the handle. This will lock the barrel securely to the receiver.

Complete Dismounting. With weapon in firing position on bipod, remove magazine and pull trigger and ease mechanism forward.

Withdraw trigger guard retaining pin and trigger guard.

Pull the butt access pin out to the right. This allows butt to be swung completely downward, the gun being held standing by the bipod and the butt.

Using the recoil spring rod, pull the operating mechanism out the rear of the weapon. (This mechanism carries the slide, piston, bolt, bolt lock, link, hammer, and return spring rod.)

Firing pin may be removed from the bolt if desired.

Removing hammer pin allows slide to be separated from hammer and link-bolt bolt lock group. Hammer may now be withdrawn from the slide. Return spring rod may now be removed. Punching out link pin permits link to be separated from bolt and bolt lock group. Use firing pin to press under head of extractor to disengage and remove it from the bolt. Extractor spring may then be removed.

Reassembling. Replace spring in the extractor and insert assembly in the bolt by pressing head of extractor and pushing completely home.

Use link pin to reconnect bolt and bolt lock group to the link. Replace the return spring in the slide followed by the hammer.

Use hammer pin to connect bolt link group to slide and hammer taking care that the head of the pin is on the right side of the slide. The firing pin should now be replaced on the bolt.

The assembled mechanism should then be inserted in the receiver. Care must be taken that the slide is properly started in its housing and the bolt is kept in the upper section of the receiver so it travels in its guide grooves.

The butt is then swung upwards and secured to the receiver by pushing home the axis pin. The trigger guard is then replaced and its retaining pin pushed into position. Replacing the magazine completes the operation.

BELGIAN SUBMACHINE GUN

Belgium used the Schmeisser MP 28 II prior to World War II. This 9mm parabellum weapon is described in some detail under Germany. The postwar Belgian Army uses the Vigneron M2 submachine gun. A number of other submachine guns have been developed and/or produced in Belgium since World War II. Outstanding among these is the FN-produced Uzi which is of Israeli design and is covered in detail under that country. FN designed and produced a 9mm parabellum submachine gun prior to the Uzi, but it was not a very successful design. At the present time FN is working on a new submachine gun which has many interesting features.

Repousemetal of Belgium developed an interesting weapon in 9mm called the RAN submachine gun. It has an internal cooling system which uses the bolt as a pneumatic ram to force air through a system of helical grooves around the barrel. It was produced in limited quantities in several versions including one with a folding bayonet. Another departure from conventional submachine gun design was in the fitting of a bipod to one model. The value of a bipod on a weapon firing 9mm parabellum ammunition is questionable because of the limited accurate range of this, or any other, pistol cartridge.

Several other submachine guns which are basically modified Sten guns have been developed in Belgium since World War II.

THE VIGNERON M2 SUBMACHINE GUN

The Vigneron is a conventional post World War II submachine gun and is mainly of stamped construction. Loading and firing of the weapon are basically the same as those of the British Sten with the exception that it has a grip safety which must be squeezed to fire the gun and a selector on the left side of the weapon which can be set on semiautomatic, automatic fire or safe positions. The grip safety prevents accidental discharge if the gun is dropped with a loaded magazine in place and the bolt in the forward position.

The Belgian 9mm Vigneron M2 Submachine Gun.

FN 9mm Uzi submachine gun.

Characteristics of the Vigneron M2 Submachine Gun

System of operation: Blowback selective fire.
Weight loaded: 8.74 lb.
Length overall: 34.9 ins. w/stock extended 24 ins. w/stock telescoped.
Barrel length: 12 in.
Feed device: 32 round, detachable, staggered box magazine.
Sights: Front: Blade.
Rear: Nonadjustable aperture.
Muzzle velocity: 1224 f.p.s.
Cyclic rate of fire: 600 r.p.m.

How to Field Strip The Vigneron M2

Remove magazine and check chamber for a cartridge. Unscrew the receiver cap at rear of receiver and remove the bolt. Unscrew barrel nut on front of receiver and remove the barrel. Further disassembly is not recommended. Reassembly is performed by reversing the above steps.

BELGIAN MACHINE GUNS

The Lewis gun, which was probably the outstanding light machine gun of World War I, was first manufactured in Belgium circa 1913 by "Armes Automatique Lewis" in Liege. It was first encountered by the Germans in the hands of Belgian troops and they called it "the Belgian rattle snake." The last part may be legend but in any event the Belgians appreciated the value of automatic fire power that was truly mobile.

As has already been mentioned, FN produced Browning machine guns prior to World War II, and although they are no longer in production, has produced the caliber .30 Browning air-cooled for the Belgian Army since World War II. FN produced rifle caliber and heavy caliber air-cooled and water-cooled Browning machine guns. These weapons were sold world-wide and among the purchasers were Argentina, China, Siam (Thailand), the Netherlands, Greece and Belgium.

THE FN MAG MACHINE GUN

The MAG is another development of FN and, like most of the products of that concern, it demonstrates first-class engineering ability. The type MAG has been adopted by Belgium and by Sweden (where it is called the Model 58), and has been adopted by the U.K. and many other nations. The gun combines the operating system of the Browning automatic rifle (BAR) with a belt feed mechanism similar to that of the German MG 42. The bolt mechanism of the BAR has been changed in the MAG so that it locks on the bottom of the receiver, rather than on the top as with the BAR. It has a chrome-plated and stellite-lined bore and chamber in its quick-change barrel. The MAG, like the German World War II guns and the U.S. M60, is designed to be used on a bipod as a light machine gun and on a tripod as a heavy machine gun. Its rate of fire can be adjusted, through the use of its gas regulator, from a low cyclic rate of 700 rounds per minute to a high cyclic rate of 1000 rounds per minute.

Caliber .30 Browning FN machine gun as used by Belgian Army.

The Belgian FN General Purpose Machine Gun, Type MAG.

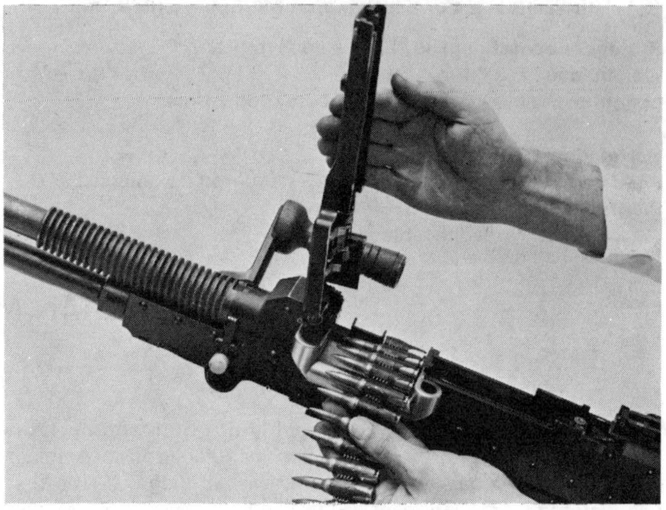

Loading the MAG.

Characteristics of MAG Machine Gun

Caliber: Has been made in 7.62mm NATO and 6.5 Swedish.
System of operation: Gas, automatic only.
Weight, w/butt and bipod: 23.92 lb.
Weight, w/o butt and bipod: 22.22 lb.
Weight of FN tripod: 22 lb. (constructed of aluminum alloy).
Length, overall: 49.21 in. w/flash suppressor.
Barrel length: 21.44 in.
Feed mechanism: Link belt. (Nondisintegrating push-out type links similar to those used on the MG 34 and MG 42 or U.S. M13 disintegrating links may be used.)
Sights: Front: Folding type with blade, or type adjustable for height.
Rear: Combined battle-sight peep and leaf with notch; peep adjustable to 600 meters and leaf adjustable to 1400 meters.
Muzzle velocity: 2800 f.p.s. (approx.) 7.62mm NATO ball cartridge.
Cyclic rate of fire: 700 to 1000 r.p.m.

How to Load and Fire the MAG

Pull the operating handle to the rear; since the MAG fires from an open bolt, the slide and bolt will remain to the rear. Push the button-type safety mounted in the pistol grip from the left side, so that the letter "S" is exposed on the right side. Open cover by pressing cover catch at rear of cover. Lay cartridge belt on feedway so that the first cartridge abuts against the cartridge stop. Close cover securely and push safety button from right to left. Squeeze trigger and the weapon will fire.

How the MAG Works

Essentially, the operation of the MAG is the same as that of the Browning automatic rifle, with the exception of its belt feeding mechanism and bottom receiver locking. A stud mounted on the top of the bolt operates in a track in the belt feed lever, which moves the belt feed slide back and forth, pulling cartridges into position for ramming by the bolt. The trigger mechanism of this weapon appears to be much simpler than that of the BAR. No rate-reducing mechanism is used with this gun.

Field Stripping the MAG

Open the cover and check to insure that the weapon is not loaded. Push in on stock catch located on front underside of butt and slide butt up and off the receiver. Push in on recoil spring rod

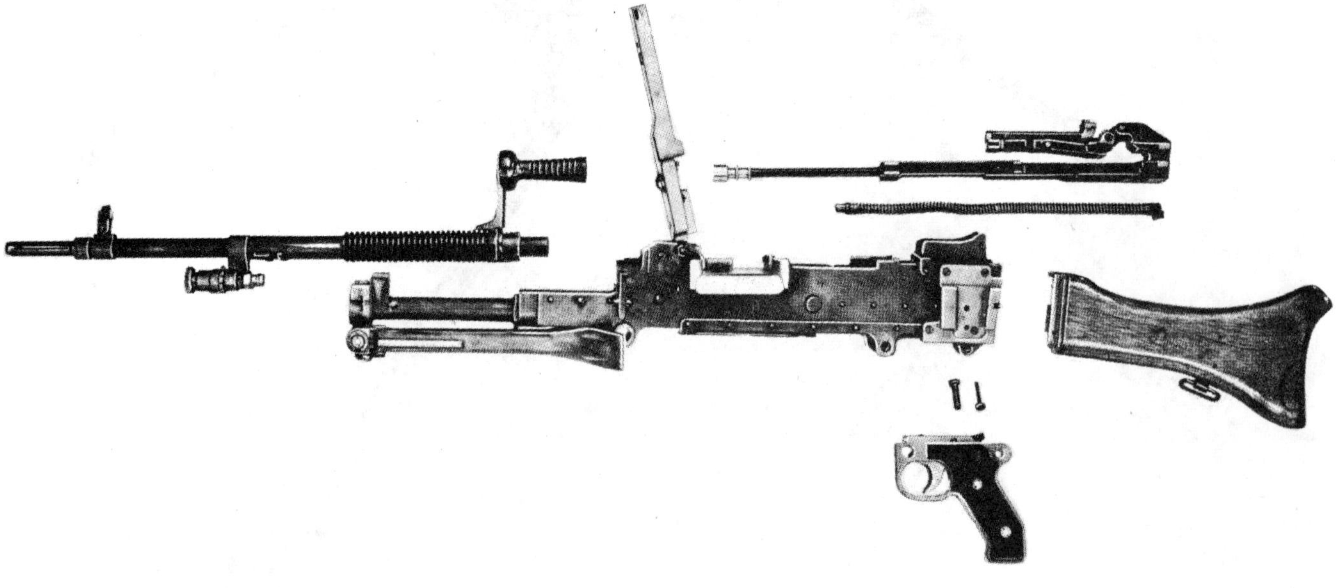

The MAG field-stripped.

(1) Removing butt assembly.

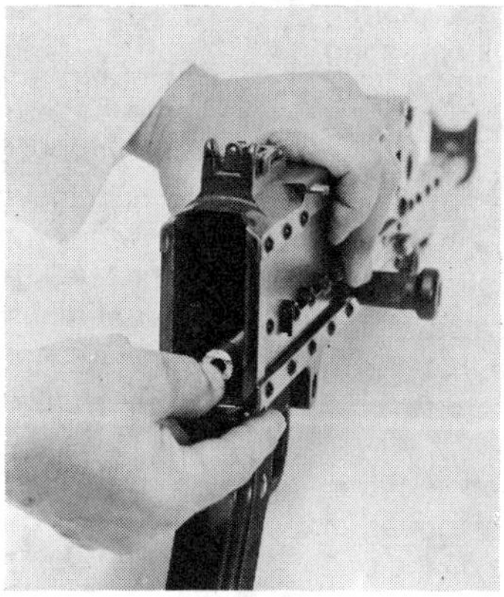

(2) Disengaging recoil spring rod.

(3) Removing recoil spring assembly.

(4) Removing bolt, piston, and slide.

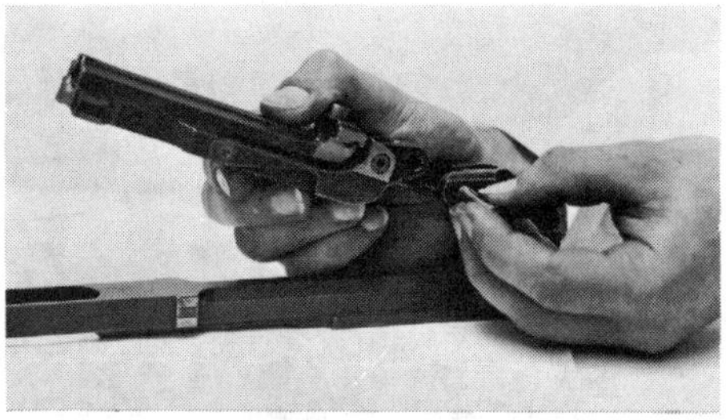

(5) Removing bolt lock and link.

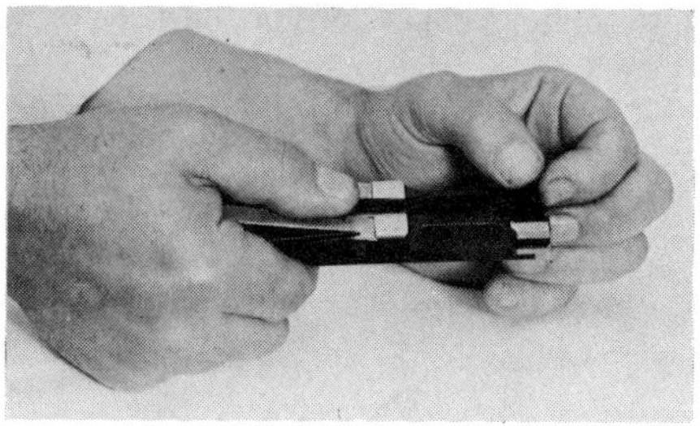

(6) Removing extractor.

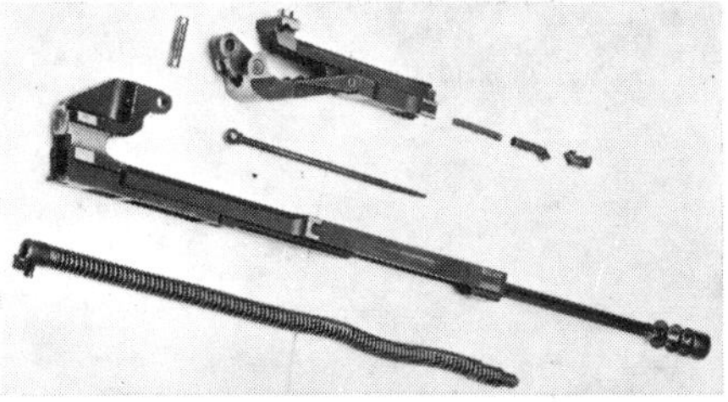

(7) Bolt and slide components.

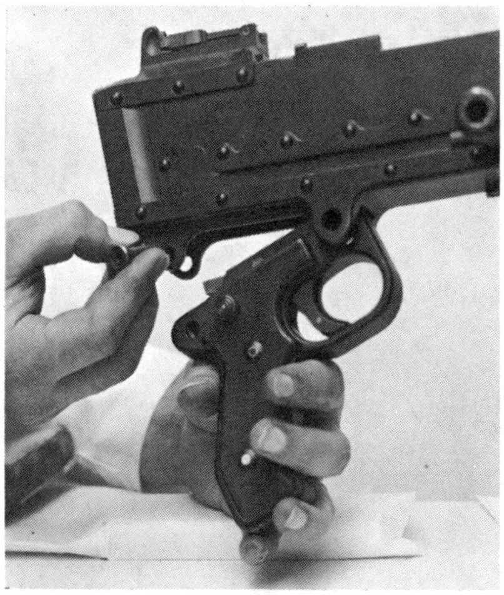

(8) Removing grip trigger assembly.

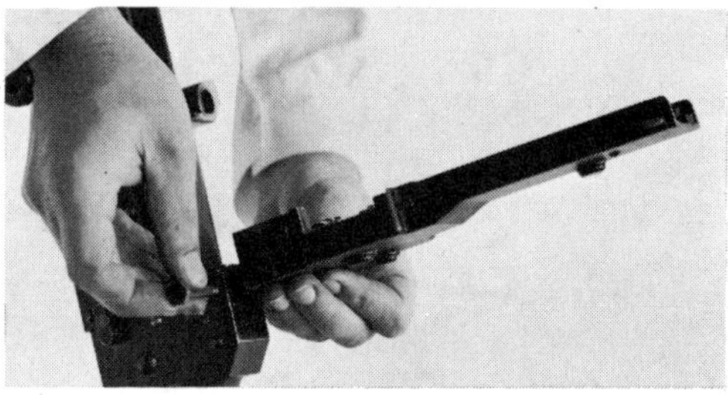

(9) Removing cover.

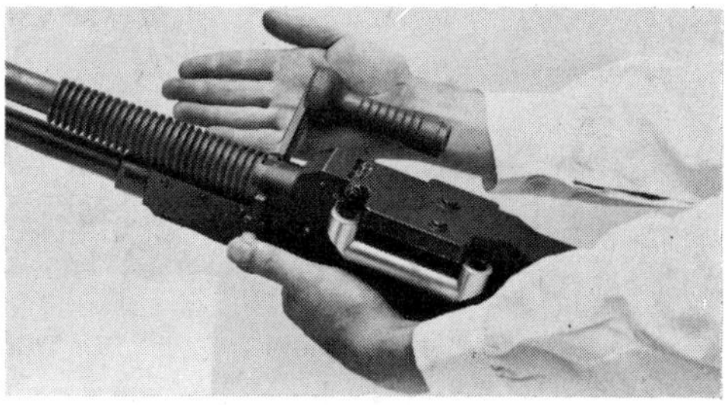

(10) Removing barrel.

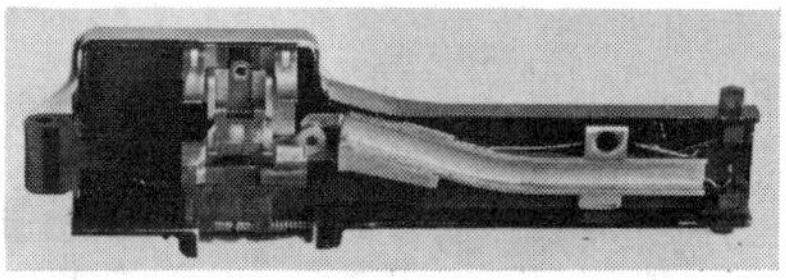

(11) Cover assembly.

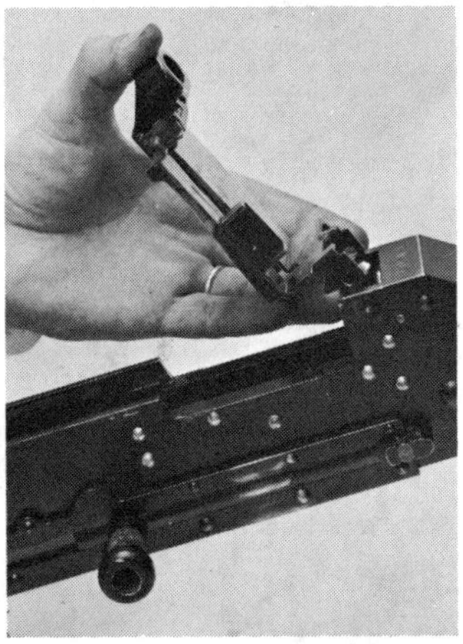

(12) Removing feed plate.

(13) Replacing barrel.

(14) Drawing barrel to rear.

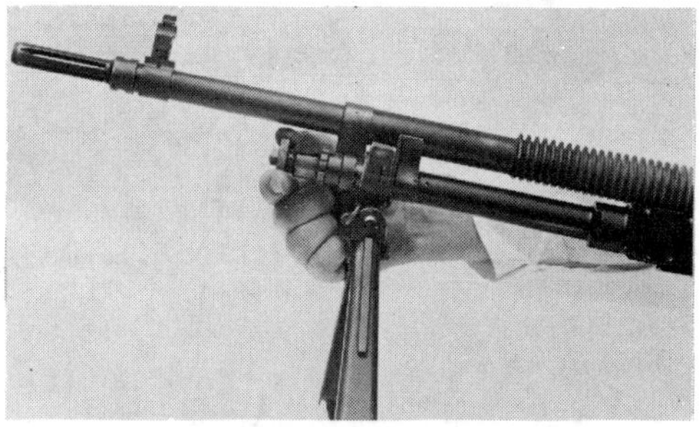

(15) Adjusting gas cylinder aperture.

and disengage it from the bottom of receiver; remove recoil spring assembly (recoil spring and rod are packaged unit). Pull operating handle to the rear and the slide and bolt will move to the rear. Grasp slide and bolt assembly by the slide post and withdraw assembly from the receiver. Remove link pin; the link, bolt lock, and bolt can be removed from the slide. To remove the pistol grip, pull out the retaining pin from the right side. To remove the cover and the feed tray, pull out the cover pin from the right side. To remove barrel, push barrel lock in (barrel lock is located at left front of receiver), move the barrel handle to the left so that it is in the vertical position, and pull barrel straight out. No further disassembly is recommended.

Reverse the above procedure to reassemble the weapon. When reassembling the bolt slide assembly to the receiver, the head of the bolt must be supported so that the forward grooves on the bolt engage the mating ridges on the sides of the receiver.

Special Note on the MAG

The MAG can be used with any rifle cartridge which has the same base dimension as the 7.92mm Mauser; this includes 7.62mm NATO, simply by changing the barrel. The butt and bipod can be removed from the gun for use in transport vehicles or tanks. The weapon can be used on the tripod with the butt removed.

The MAG has proven to be quite a popular machine gun and has been purchased by the following countries; Argentina, Belgium, Cuba, Ecuador, India, Israel, Kuwait, Libya, New Zealand, the Netherlands, Northern Rhodesia, Peru, Quatar, Ruanda, Sierra Leone, Southern Rhodesia, South Africa, Sweden, Tanganyika, Uganda, the United Kingdom (Great Britain), and Venezuela. With the exception of Sweden which chose 6.5mm, all countries adopted the gun in 7.62mm NATO. It is probable that the Swedish guns will also be converted to this caliber in the future. The MAG (machine gun, general purpose L7A1) is being manufactured at the Royal Small Arms Factory at Enfield Lock.

The 7.62mm FN MAG. machine gun as currently made on new model tripod. Note smooth barrel.

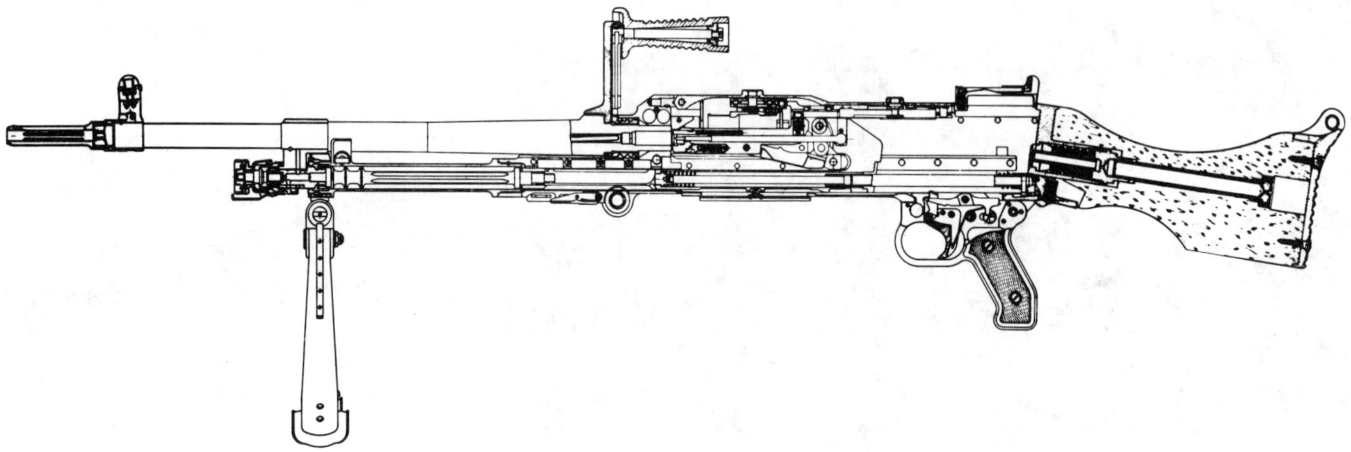

Section view of FN MAG. machine gun.

17 Britain (United Kingdom) and the British Commonwealth

It is no longer realistic to attempt to collect under Britain all weapons in use in the British Commonwealth, since many of the member nations of the Commonwealth already have taken an independent course on weapons, and more and more of these nations will tend to do so. This independence in weaponry should not be too surprising, inasmuch as the United Kingdom does not dictate to the member nations of the Commonwealth what weapons they will or will not adopt. In the past these nations have usually adopted the same weapons as the UK, because in most cases (but not all, by any means) they were members of defense treaties with the UK, and because their political, social, and economic orientation would put them on the same side as the UK in any major war. Since World War II, however, the defense arrangements have tended to be more general and to include nations outside the Commonwealth, notably the United States. It is therefore likely that the future will see many United States weapons added to those that are already standard in the member nations of the British Commonwealth.

The standard small arms in the British Army are the 9mm Browning FN H.P. pistol in its various marks, the 7.62mm rifle L1A1, the 9mm submachine gun L2A3, the 7.62mm L7A1 general purpose machine gun, and the caliber .50 Browning machine gun. Some caliber .30 Browning M1919A5 and M37 machine guns are still in service on various armored vehicles.

The United Kingdom has procured quantities of Colt 5.56mm (.223) AR 15 rifles for use by forces which might have to operate in tropical climates. The 7.62mm NATO L4A2 Bren gun has been continued in service with Infantry in the Far East and is used by other branches of the Army on a world wide basis.

A separate chapter is now given each to Australia, Canada, India, and New Zealand.

BRITISH SMALL ARMS NOMENCLATURE

The model-designation procedures of the UK are somewhat involved and can be confusing. Prior to World War I, British model designations for small arms were comprised of the word Mark followed by a Roman numeral, e.g., Mark I, Mark II. Two exceptions to this form of designation were the Pattern 13, which was an experimental cal. .276 rifle tested in 1913, and the Pattern 14, which was made in cal. .303 as the production version of the Pattern 13.

Between World Wars I and II, rifles and pistols were given number designations in addition to mark designations. Thus the Rifle Mark III SMLE became Rifle No. 1 Mark III SMLE, and the Pattern 14 became the Rifle No. 3 Mark I. Toward the end of World War II, the British began using Arabic numerals for both the number and mark designations. An additional complication was the star symbol (*) which was frequently found tacked to the end of everything else, e.g., Rifle No. 4 Mark 1*. This star indicates a minor modification from the Mark design.

Since the early fifties, weapons entering the British service have received an L designation, and modifications have been indicated by an A, as in L2A3. The L stands for "Land service." Older weapons that have been considerably modified, such as the Brens rebarreled for the 7.62mm NATO cartridge, also receive an L designation, as do (in most cases) United States weapons adopted as standard or limited standard by the UK.

A revolver is usually called a "pistol" by the British.

BRITISH PISTOLS AND REVOLVERS

The British army was the last major army in the world to use the revolver as a standard service weapon. Lieutenant Winston Churchill of the 21st Lancers used a personally procured Mauser Automatic against the Dervishes at the battle of Omdurman in the Sudan in 1898 and carried the same weapon during his stint as a war correspondent in the Boer War. This put Sir Winston Churchill approximately fifty years ahead of his country's army, but he was ahead of most people on many other things as well. After World War II, in which the caliber .38 Enfield revolvers were standard (Webley, Smith and Wesson, and Colt revolvers and Canadian-made Browning, Webley, Colt, Star, and various other automatic pistols were also used), the U.K. ran a series of pistol tests. These tests confirmed that a 9mm Parabellum automatic pistol was the best service arm and the Canadian made Browning FN H.P. pistol was adopted as standard. Troops were equipped with pistols on hand which had been used by British paratroop and commando units during the war.

Recently the United Kingdom purchased a quantity of 9mm Parabellum Browning Hi-Power pistols for the RAF from FN. These are the latest type with extractor mounted on the slide and two piece barrel. They are called "Pistol 9mm L9A1 by the British.

WEBLEY REVOLVERS

The British Government used Webley revolvers as standard or limited standard for 60 years. The Mark I was adopted in November 1887 and the last of the standard Webley revolvers, the No. 1 Mark VI was declared obsolete in 1947. All the standard British issue Webley revolvers were caliber .455 and all were similar in design. The Webley is a top breaking revolver which is locked by a heavy stirrup-type barrel catch. The first five Marks have "birds head" type grips and the Mark VI has a square grip. The Mark VI (called No. 1 Mark VI after 1927), which was adopted in May 1915, was made in the greatest quantity, over 300,000 of these revolvers having been made by Webley & Scott at Birmingham during World War I. A quantity of the Mark VI were also made at Enfield Lock after World War I.

After World War I the British decided that .455 was too heavy for the most effective use and after test decided upon the use of a caliber .38 cartridge based on the .38 Smith & Wesson cartridge case. Webley designed a new pistol using many of the features of their commercial Mark III caliber .38 revolver. The design was taken over by Royal Small Arms Factory and as completed was not compatible with the Webley pistol; parts are not interchangeable. The Webley Mark IV caliber .38 revolver was adopted as limited standard in World War II.

Webley .455 Mark V "pistol," 6-inch barrel pattern introduced in 1915.

The .38 Webley Mark IV. Used during World War II.

THE WEBLEY .455 "PISTOL" NO. 1 MARK VI

Although this revolver has been obsolete in Great Britain since 1947, it is widely distributed throughout the British Commonwealth and former British territories.

Characteristics of Webley .455 "Pistol" No. I Mark VI

Caliber: .455 Webley.
System of operation: Single or double-action, top break revolver.
Weight: 2.37 lb.
Length: overall: 11.25 in.
Barrel: 6 in.
Feed device: Cylinder with 6 chambers.
Sights: Front: Blade.
Rear: Notch.
Muzzle velocity: 620 f.p.s.
NOTE: Some No. 1 Mark VI revolvers have been rechambered for the caliber .45 Colt automatic cartridge. These revolvers use the three-round clip used with the Colt and Smith and Wesson Model 1917 revolvers.

Loading and Firing the Webley Mark VI

Push forward on the curved tail of the pivoted barrel catch which is on the left side of the revolver just below the hammer. As the catch is pushed it pivots on its screw drawing the upper latching end back over the barrel strap, freeing the barrel to be tipped down on its hinge. As the barrel is bent down, the extractor will rise on

Webley .455 "Pistol" No. 1 Mark VI.

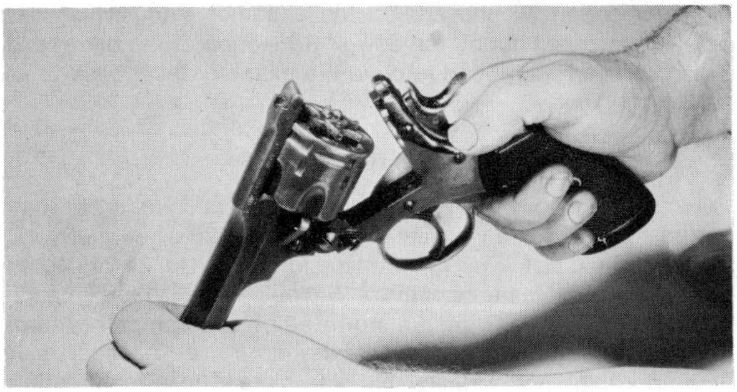

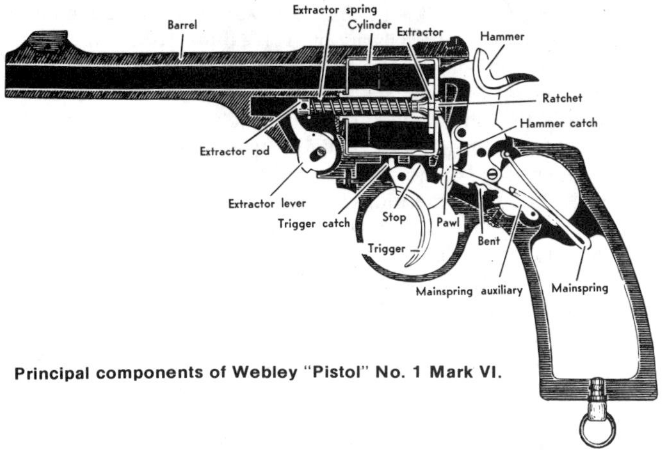

Principal components of Webley "Pistol" No. 1 Mark VI.

Field Stripping the Webley Mark VI

The only stripping necessary and recommended for this revolver is removing the cylinder. The bottom screw at the extreme forward end on the left side of the receiver is the cylinder catch retaining screw. Unscrew this. Now push the bottom of the cylinder catch retainer directly above the screw upwards. This will depress the rear of the catch and permit the cylinder to be lifted out.

How the Revolver No. 1 (Mark VI) Works

Drawing the hammer back compresses the mainspring. The pawl is moved upward against the ratchet and revolves the cylinder. When the chamber containing the cartridge to be fired next is in line with the barrel, the cylinder stop engages in a notch in the cylinder and prevents further rotation. Simultaneously the trigger catch rises to its peak in the cylinder, holding the cylinder securely so that it cannot turn in either direction. The trigger nose drops into the bent holding the hammer at full cock. The mainspring is fully compressed when its lower arm is raised by the mainspring auxiliary. When the trigger is squeezed, the hammer falls, striking the primer of the cartridge exploding the charge.

Extractor: When the barrel catch has been drawn back and the barrel tipped down, the extractor lever tooth catching against the frame is stopped in its movement. The extractor lever arm stops the motion of the extractor rod, and as the barrel and cylinder move down the extractor is forced up out of its seating carrying with it its stem until the revolver is fully opened, at which point the extractor under the influence of its spring will slip back into its place in the cylinder.

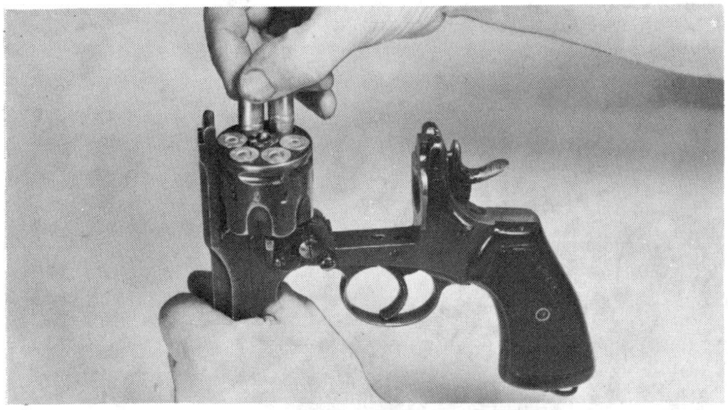

Now load the six chambers. With a little practice this may be done two chambers at a time. If the cylinder is to be only partly loaded, remember that the cylinder revolves clockwise; and that the first cartridge must be to the left of the chamber in direct line with the hammer nose when the weapon is closed. Cocking the hammer automatically turns the cylinder the distance of one chamber.

Now turn barrel and loaded cylinder up to the fullest extent. The heavy catch will automatically be sprung over the barrel strap and lock it securely.

If you have time, and accuracy is desired, always pull back the hammer with the thumb to full cock for each shot. For close quarters or emergency firing, drawing the trigger straight back will raise the hammer to full cock and turn the cylinder and trip the hammer, completing the firing. It is necessary to release the pressure on the trigger after each shot to permit the mechanism to engage for the next shot. Accurate shooting except at close range is difficult when shooting double action.

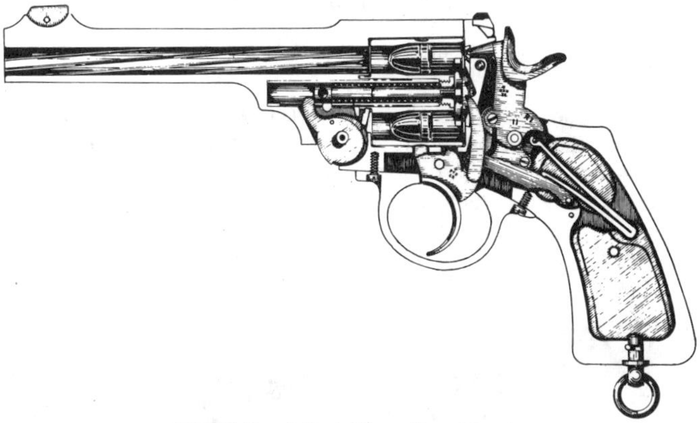

"Pistol" No. 1 Mark VI - action closed.

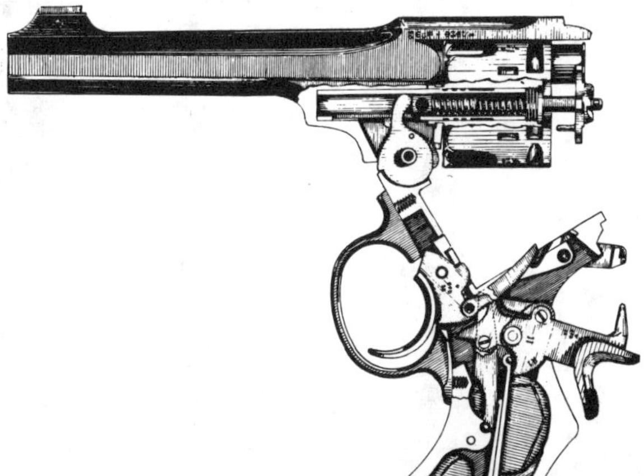

Action open showing details of extraction and locking.

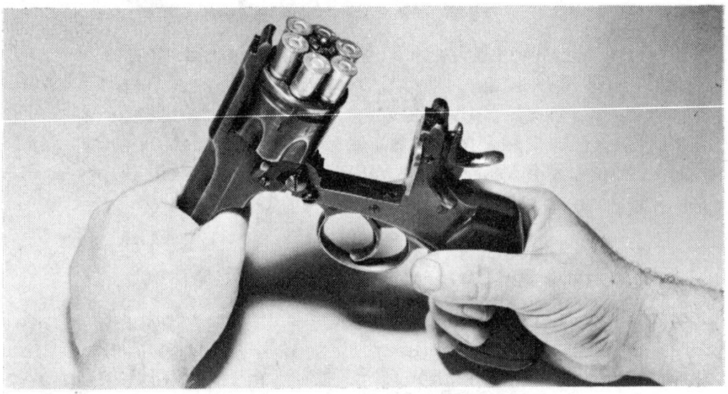

To extract cartridges, break the revolver and push barrel down.

Caliber .22 L.R. version of the Webley "Pistol" No. 1 Mark VI which was used by the British for training.

the cartridges, the rims of which have been resting on the base of the extractor. When the barrel nears its completely open position, a corner of the barrel joint passes over the tooth and presses it to the rear. It is thus forced into the groove in the frame. The extractor spring (wound around the extractor stem) which has been compressed during the downward motion of the barrel and cylinder, is now permitted to drive the extractor lever back to its seat in the cylinder.

ENFIELD REVOLVERS

As noted previously the British government and Webley & Scott parted company on uniformity of design in 1926 when the No. 2 Mark 1 Enfield pistol was in prototype form. The No. 2 Mark 1 had many of the best features of the .455 Webley Mark VI and in addition had a movable firing pin mounted on the hammer (all the earlier Webley government revolvers had a fixed hammer nose type firing pin), and a removable side plate. These features had appeared in the commercial .38 Webley Mark III.

The revolver, called "Pistol" .38 No. 2 Mark 1 was produced from 1927 to 1938; it was officially adopted on 2 June 1932. On 22 June 1938 the first modification of this revolver, - the No. 2 Mark 1* - was introduced. The modification consisted of the removal of the spur and of the single-action cocking notch on the hammer. The No. 2 Mark 1* can therefore only be used double-action. Since this requires lifting of the hammer, firing and rotation of the cylinder by the pulling of the trigger, the trigger pull is very hard. As a result this revolver is of very limited accurate range. In 1942 another Model was introduced, the pistol No. 2 Mark 1**. This model has no hammer safety stop. As originally issued the No. 2 Mark 1 had Mark 1 walnut grips of rather square configuration, at a later date Mark 2 black bakelite or walnut grips with thumb recesses were adopted. These are usually seen on the No. 2 Mark 1* and Mark 1** but may occasionally be seen on the No. 2 Mark 1 as well.

Complete revolvers were made by Enfield, and Albion Motors at Glasgow. Singer Sewing Machine of Great Britain made parts which were assembled into complete revolvers at Enfield. In 1957 the Enfield revolvers were dropped as standard and replaced by the FN Browning Hi-Power automatic. These revolvers are still

British .38 Enfield "Pistol" No. 2 Mark I.

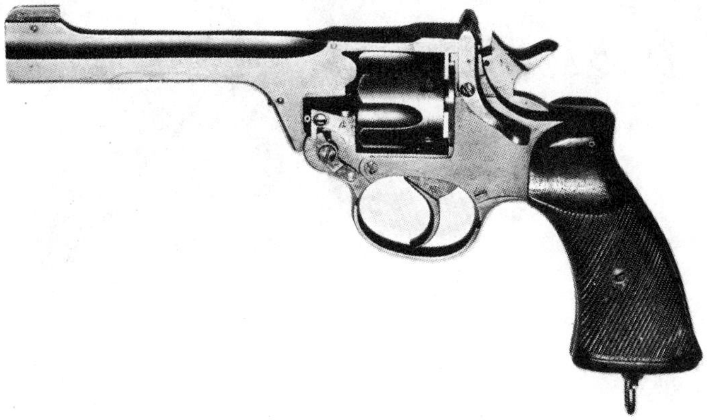

Enfield .38 "Pistol" No. 2 Mark I with Mark II grips.

Cutaway drawing of Enfield "Pistol" No. 2 with action closed.

in extensive use in former British territories and are considered a reserve weapon in the United Kingdom.

The Enfield revolver, as the Mark IV Webley uses the British .380 (or .38), revolver cartridge. It can also be used with commercial U.S. caliber .38 Smith and Wesson ammunition (not S&W special), but will have a tendency to shoot high with this ammunition, since the issue front sight is set for the heavier British Mark 2 caliber .38 bullet. Higher sight blades can be obtained from arms dealers in the U.K. or the blade can be built up by brazing.

Characteristics of Enfield Revolver

Caliber: .38, .380 in. revolver, .38 S&W, .38 Webley.
System of operation: No. 2 Mark 1 single or double action, No. 2 Mark 1* and No. 2 Mark 1** double action only.
Weight: 1.58 lb. (1)
Length, overall: 10.25 in.
Barrel length: 5 in.
Feed device: Cylinder with 6 chambers.
Sights: Front: Blade.
Rear: Square notch.
Muzzle velocity: 600.

(1) No. 2 Mark 1* and No. 2 Mark 1** weight about an ounce less.

How Enfield Revolvers Work

Single-Action. As the hammer is drawn back it rotates on its axis pin and compresses the main spring. While this is happening, the bent of the hammer bears against the nose of the trigger to rotate the trigger. When the hammer is all the way back, the nose of the trigger is forced into the bent of the hammer by the main spring lever. The hammer is now full cocked.

When the trigger is pressed the nose of the trigger is released from the bent of the hammer and the compressed main spring reacts to drive the hammer forward. The hammer nose inset in the hammer strikes the primer and fires the cartridge.

As the trigger pressure is released, the mainspring lever pushed by the mainspring forces the lower part of the hammer forward a short distance; this withdraws the upper part of the hammer and the hammer nose is brought clear of the cartridge primer. At the same time the mainspring lever pushes the trigger forward again.

Double-Action. The action here is the same as for single-action with the exception that rotation of the hammer and compression of the mainspring is brought about by pressure on the trigger instead of by the thumb on the hammer.

The trigger nose bears behind the hammer catch to lift the hammer until the nose of the trigger rises high enough to slip over the end of the hammer catch, and allow the mainspring to drive the hammer forward to fire the cartridge.

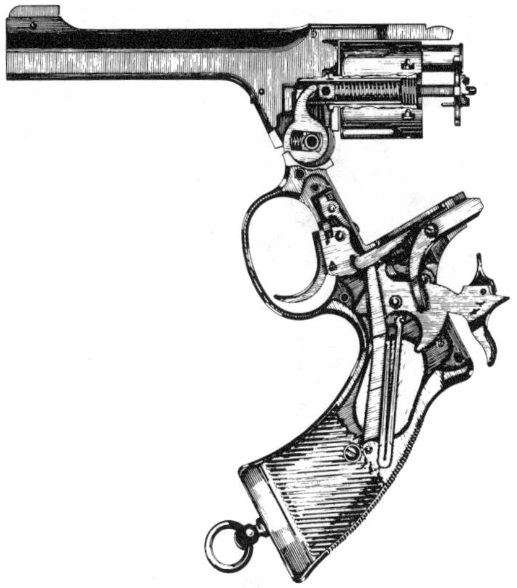

Action open, showing details of extraction and cocking.

Cylinder Rotation. The rotating pawl is pivoted to the trigger. Thus when the trigger moves the pawl correspondingly rises and falls. It engages with the teeth of the ratchet on the cylinder; hence an upward movement of the pawl revolves the cylinder. As soon as the trigger is released, the pawl drops to engage behind the next tooth of the ratchet where it is ready to rotate the cylinder once more when the trigger is pressed. In single-action, cocking the hammer thus operates the pawl.

Cylinder Lock. When the hammer is at full cock, the trigger has lifted the pawl to its highest possible position. It holds it here against the ratchet. This prevents the cylinder from rotating backwards.

Forward rotation of the cylinder is prevented by the cylinder stop which at this point has risen to lock into one of the recesses in the face of the cylinder. These two engagements hold the cylinder in line so that the chamber to be fired is properly alined to the barrel.

Loading and Firing the Enfield

Loading and firing the Enfield Mark 1 are accomplished exactly as in the Webley Mark VI. The Mark 1* is handled in the same manner, except that it can be used only double-action. The barrel latch is the same as on the Mark VI.

Enfield .380 "Pistol" No. 2 Mark I*.

Enfield .38 No. 2 Mark I fitted with Parker-Hale .22 conversion unit.

Loading the Enfield No. 2, Mark I.

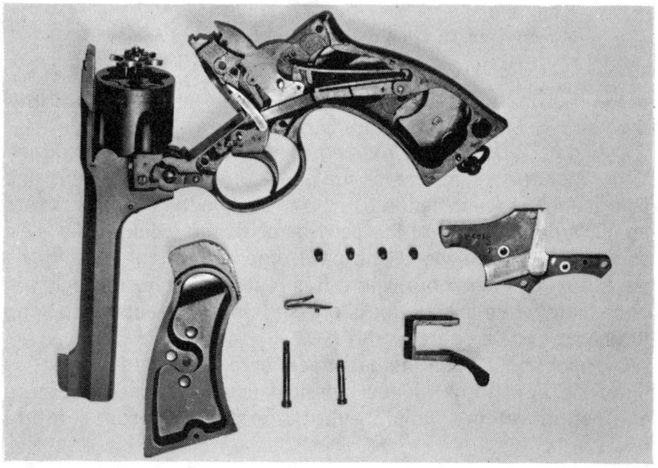

The Enfield No. 2 Mark I* field-stripped.

Field Stripping the Enfield

Unscrew cam lever fixing screw. Push barrel catch to open pistol and remove the cylinder.

Unscrew the stock and side plate screws on the left side of the weapon; and remove the stock and the side plate.

All parts are now exposed and further dismounting is not recommended except by a competent armorer, as springs and parts may be injured unless properly handled.

SUBSTITUTE STANDARD AND NON-STANDARD BRITISH PISTOLS AND REVOLVERS

In both World War I and World War II, Great Britain found it necessary to obtain automatic pistols and revolvers abroad in order to meet their military requirements. Most of these weapons were procured from the U.S. although Spanish and Argentine weapons were used as well.

During World War I, Smith and Wesson and Colt made large quantities of caliber .455 revolvers for Britain. Although there were a number of models supplied, the most common models were the Colt New Service and the Smith and Wesson Mark II Hand Ejector of which 73,650 were supplied to the U.K. and Canada. Colt caliber .32, .38, .45, and .455 automatic pistols were also purchased by the U.K. and many of these weapons have since come back to the U.S. with British proof and broad arrow (signifying government ownership) marks on all major components. Approximately 10,000 .455 M1911 Colt automatics were supplied to the U.K. during World War I. After the war these pistols were issued to the R.A.F. and most bear the markings of that organization.

During World War II the U.S. supplied Great Britain with 20,000 Colt and Smith and Wesson caliber .45 Model 1917 revolvers after Dunkirk. Large quantities of Smith and Wesson .38/200 K200 revolvers and Colt .45 Model M1911A1 automatics were supplied under Lend-Lease. Apparently in 1940 the U.K. purchased every type of pistol that has any military potential, as they also procured quantities of Smith & Wesson K-38 target revolvers. Ballester Molina caliber .45 automatics were purchased from Argentina, and Star (among other) automatics were purchased from Spain. Since World War II the United Kingdom has disposed of all of these non-standard weapons, mainly by sale to surplus arms dealers.

There is considerable significance to the fact that Great Britain, one of the greatest industrial powers in the world and formerly one of the leading exporters of small arms, has had to become an importer of small arms in both world wars in order to meet military requirements. Extremely restrictive gun laws in the U.K. have, to a great extent, killed the initiative of the British manufacturers and of course cut down on production during peace time. The end result is the lack of adequate production facilities and a trained labor force ready to produce military small arms in quantity when desired. Although this shortcoming of the U.K. has resulted in considerable profit to some American manufacturers, including the erection, at British government expense, of some of the largest arms and ammunition plants in the U.S., it is not to the overall advantage of the U.S.A. that its most powerful ally be myopic about the production of basic weapons and their ammunition.

BRITISH SMITH & WESSON .38 "PISTOL"

Characteristics of British S&W .38 "Pistol"

Caliber: British Service .380 inch. Also .38 S & W, 148- or 200-grain bullet. (Note: British Service Ammunition has metal jacketed bullet).
Cylinder: 6 chambers.
Muzzle velocity: 600 feet per second with British Service Ammunition.
Barrel length: 5 inches.
Overall length: 10.2 in.
Weight: 1.81 lb.
Sights: Front, blade. Rear, square notch.
Other data: Essentially the same in operation and stripping as the United States Smith & Wesson .45 1917 Revolver. This revolver is almost identical to the S & W Military and Police Model.
Note: This revolver, which in British terminology is called a pistol, will not handle the .38 Smith & Wesson Special type cartridges. It will handle the shorter and wider .38 S & W type. This cartridge is known in England as the Webley .380 inch. These weapons are considered obsolescent in the U.K. at present.

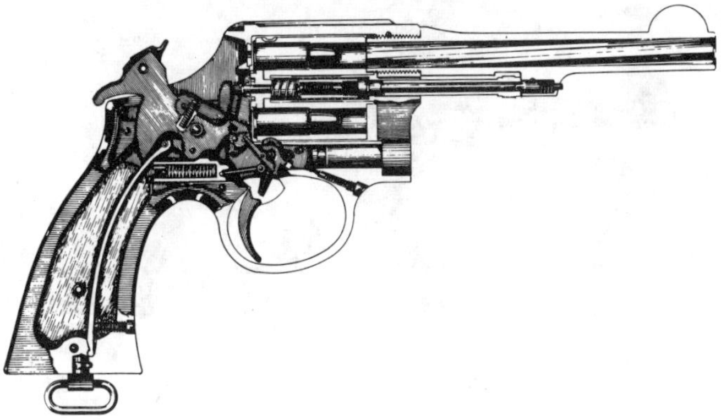

Hammer back, showing details of lockwork and front and rear cylinder locking mechanism.

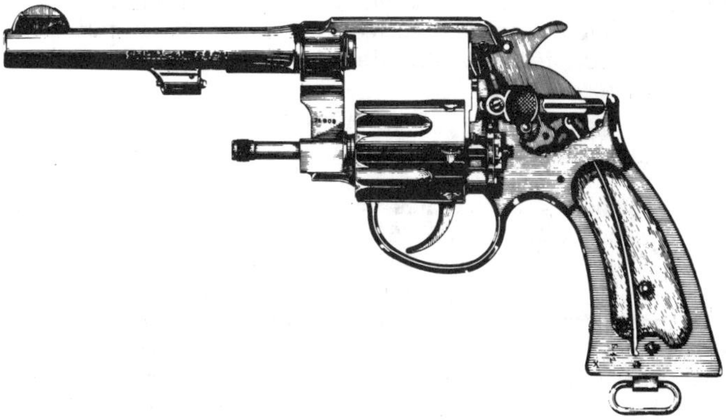

Hammer down, cylinder swung out to show details of extraction. Sideplate cut away to show detail of thumb lock.

British Smith & Wesson .38 "Pistol" (.38/200).

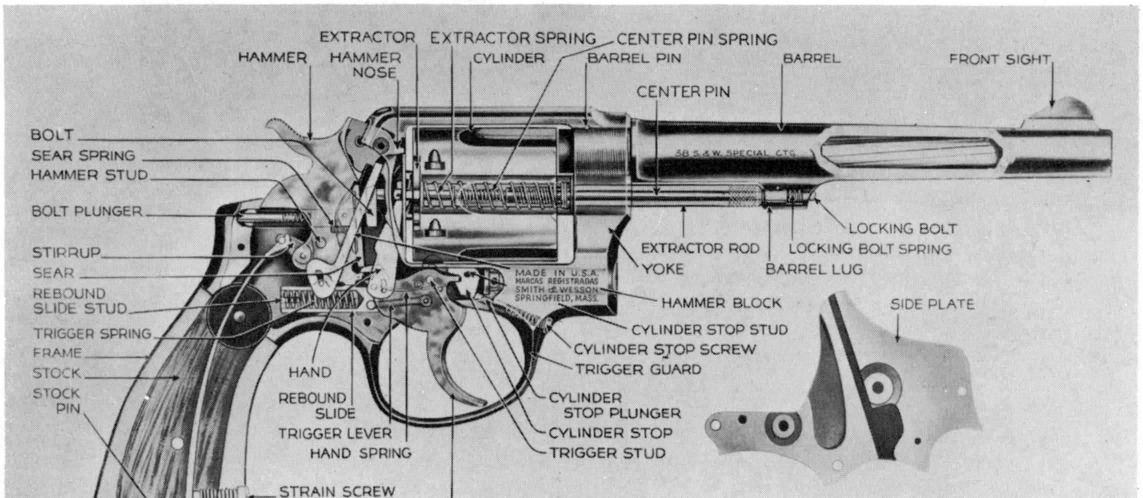

Nomenclature of all S&W swing-out revolvers.

Smith & Wesson .455 Mark II (hand-ejector).

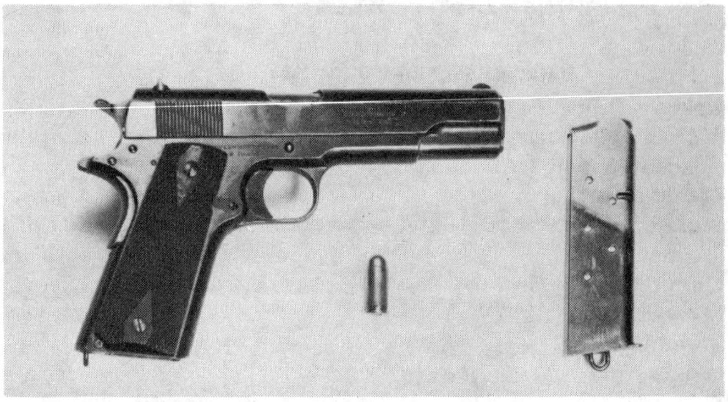

British Colt .455 Automatic Pistol.

WEBLEY FOSBERY AUTOMATIC REVOLVER

Although the Webley-Fosbery was never an item of issue in the British Army it was widely used by British officers in World War I who by regulation, could purchase any pistol chambered for the service cartridge they desired. It was overly sensitive to mud, however, and soon fell out of favor. The Webley-Fosbery is a collector's item at present.

BRITISH COLT .455 AUTOMATIC PISTOL

The standard United States .45 pistol cartridge may be used in this weapon. However the .455 cartridge will not chamber in our .45 service pistol.

This one way interchangeability of ammunition is occasioned by the fact that the actual bullet diameter of the U.S. .45 Auto cartridge is .4515 inch; while that of the .455 Webley S. L. is actually .455 inch. Thus while the smaller diameter .45 will chamber in the .455, the reverse is not true. The caliber .455 will be found stamped on the right side of the receiver in this weapon.

THE .455 WEBLEY AUTOMATIC PISTOL

The Webley Automatic pistol was standard issue in the Royal Navy from 1912 until the end of World War II. There are two basic models, the Mark 1 and the Mark 1 No. 2. The Mark 1 No. 2 has a different type rear sight than the Mark 1 and has a different type manual safety. During World War I some .455 Webley automatics were fitted with shoulder stocks to be used by the Royal Flying Corps.

Characteristics of British Webley Automatic Pistol

Caliber: .455 Webley automatic.
System of operation: Recoil operated.
Weight: 2.43 lb.

Webley-Fosbery Automatic Revolver, caliber .455, 1902 model. The safety lever is on the left side of the frame by the grip.

Webley .455 Automatic Mark I — obsolete in the U.K.

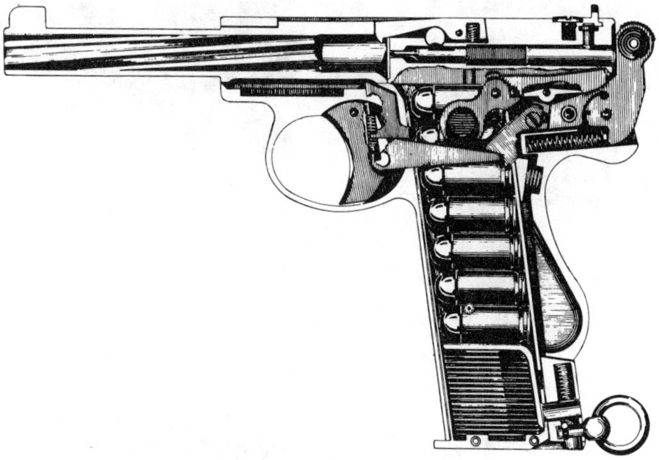

Webley automatic pistol — action closed.

Length, overall: 8.5 in.
Barrel length: 5 in.
Feed device: 7-round, in line detachable box magazine.
Sights: Front: Blade.
Rear: Mark 1 fixed notch, Mark 1 No. 2 adjustable.
Muzzle velocity: 710 f.p.s.
Special feature: Magazine is provided with two catch notches in the magazine, one above the other. Push the magazine all the way in and the catch will lock in the lower notch leaving the pistol ready for magazine fire. If the magazine is pushed only part way in so that the catch locks in the upper notch, the pistol can be loaded with single cartridges inserted through the open breech; and action closed by pressing slide release catch. After each shot thus fired the slide will remain open ready for the next cartridge. Meanwhile the magazine remains loaded in the handle held in reserve. To achieve magazine fire it is only necessary to push the magazine in until it catches in the second lock notch.

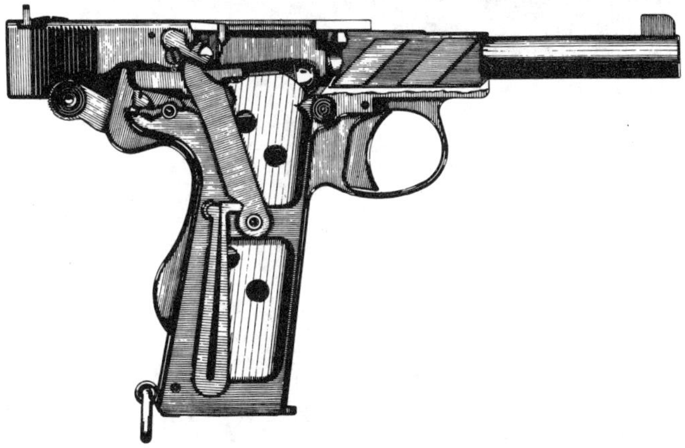

Action open and cutaway to show mainspring and sear.

BRITISH RIFLES AND CARBINES

The current standard British Service rifle is the 7.62mm L1A1. Quantities of the No. 4 Mark 2, Mark 1/2, and Mark 1/3 rifle may be held for Naval use and in reserve.

THE BRITISH BOLT ACTION MILITARY RIFLES FROM 1888 TO 1951

The British and Canadian governments have disposed of most of their bolt action rifles since World War II. The caliber .22 No. 7, No. 8, and No. 9 rifles are still used as training rifles. The No. 1, Mark III and Mark III* and the No. 4 Mark I, I* and II are still extensively used throughout the former British territories and in a few other countries such as Greece. The greater percentage of these rifles have been sold on the U.S. market.

The bolt action rifles which follow are not all of the various models that have been used by Great Britain by any means; they are the most common. As with pistols, Britain has had to import rifles during both world wars to meet military requirements. During World War I contracts were let with Remington Arms, Ilion, N.Y., Remington Arms of Eddystone, Pa., and Winchester for the production of the .303 Pattern 14 (Rifle No. 3 Mark I*). This rifle was continued in manufacture for the United States in caliber .30-06, as the U.S. Rifle M1917. Although large quantities of Pattern 14 rifles were made, they were apparently used on the battle front in very limited quantities and principally as sniper rifles.

During World War II the British were desperately short of rifles, especially in 1940-41, due to losses at Dunkirk and temporary loss of industrial plants due to bombing. In order to make up for these losses Canada gave the U.K. 70,000 Ross rifles and the United States supplied 785,000 caliber .30 M1917 Enfield rifles out of United States war reserve stocks, at a price of $7.50 per rifle. The British government contracted with Remington Arms to produce caliber .30 Springfield M1903 rifles; this contract was later taken over by the U.S. government. Stevens Arms of Chicopee, Massachusetts produced over one million No. 4 Mark I and Mark I* rifles for Great Britain. All contracts let with U.S. manufacturers were supervised by the U.S. government after the introduction of Lend-Lease in 1941. The Ross, M1917 Enfield, and Springfield M1903 rifles were used by Home Guard units and have since been disposed of by the British government.

Safety Measures and Inspection Criteria

British rifles are usually well made and, if in good condition, are safe enough. Rifles of earlier marks than the Long Lee Enfield Mark I should not be used with the Mark 7 or other heavily loaded cartridges. The No. 4 and later rifles, if they are in good condition will safely use any .303 cartridge loaded for rifles. Some United States commercial ammunition is not loaded too heavily in cal. .303, and will not bother any of the Long Lee Enfields or later weapons. Wherever there is any doubt about safety, the weapon should not be fired until checked by a reliable gunsmith.

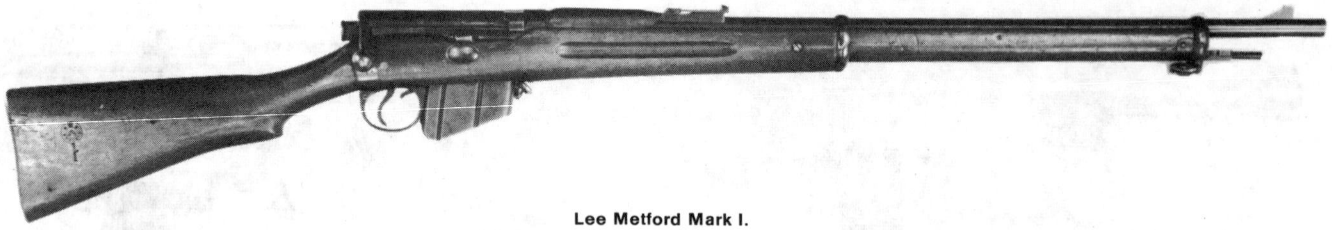

Lee Metford Mark I.

Lee Metford Mark II.

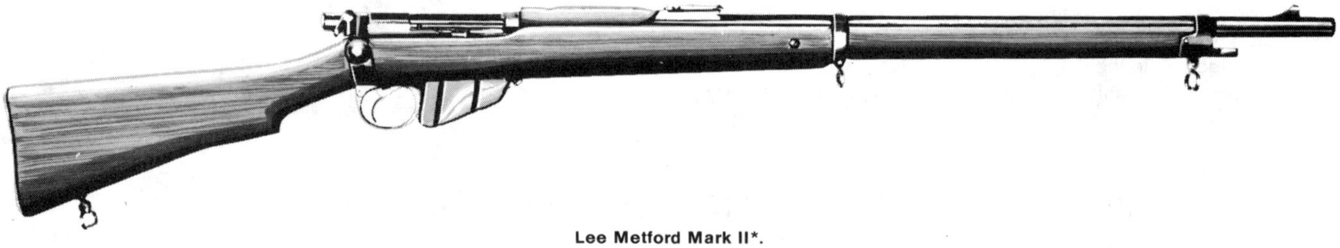

Lee Metford Mark II*.

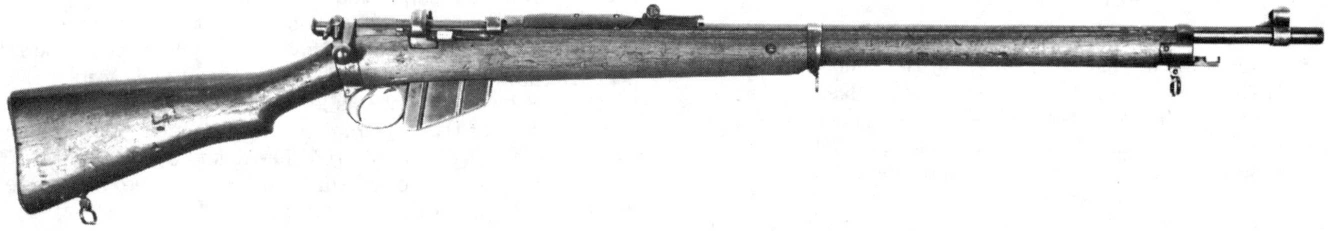

Rifle, Charger Loading, Long Lee Enfield Mark I*.

In passing, it is worth noting that the .303 British and .303 Savage are not the same cartridge, and are not interchangeable. The .303 British, as currently loaded by one American cartridge manufacturer, has a 215-grain bullet with a muzzle velocity of 2,180 feet per second. This is quite close to the loading of the British Mark 6 cal. .303 cartridge, which had a 215-grain bullet with a velocity of 2,060 feet per second. The Mark 6 cartridge was used with the Long Lee Enfields and the early No. 1s.

As an aid to gunsmiths and others who will undoubtedly encounter many British Lee Enfield rifles in the future, some of the inspection criteria for the weapons are listed below.

Headspace. Since the .303 is a rimmed cartridge, headspace is measured from the rear face of the barrel to the face of the bolt. The headspace of the .303 rifle should not exceed .074 inches, although—as a wartime measure—a maximum of 0.08 inches was allowed. Minimum headspace is .064 inch.

CHARACTERISTICS OF BRITISH BOLT ACTION RIFLES AND CARBINES

	Lee Metford Rifle Mark I*	Lee Metford Rifle Mark II	Lee Metford Carbine Mark I	Lee Enfield Rifle Mark I
Caliber	.303	.303	.303	.303
Overall length	49.85 in	49.85 in	40 in	49.5 in
Barrel length	30.19 in	30.19 in	20.75 in	30.19 in
Feed device	8 rd detachable box w/cut-off	10 rd detachable box w/cut-off	6 rd detachable box w/cut-off	10 rd detachable box w/cut-off
Sights: Front	Barley corn	Barley corn	Barley corn w/ protecting ears	Barley corn.
Rear	Vertical leaf and ramp	Vertical leaf and ramp	Vertical leaf and ramp	Vertical leaf and ramp
Muzzle velocity (at date of adoption)	2000 FPS	2000 FPS	1940 FPS	2060 FPS
Weight	10.43 lbs.	10.18 lbs	7.43 lbs	9.25 lbs

	Pattern 14 Rifle (Rifle No. 3 MKI*)	Short Lee Enfield Rifle Mark I (Rifle No 1 SMLE MK 1)	Short Lee Enfield Rifle Mark III (Rifle No 1 SMLE MK 3)	Short Lee Enfield Rifle Mark III* (Rifle No 1 SMLE MK 3*)
Caliber	.303	.303	.303	.303
Overall length	46.25 in	44.5 in	44.5 in	44.5 in
Barrel length	26 in	25.19 in	25.19 in	25.19 in
Feed device	5 rd integral magazine	10 rd detachable box w/cut-off	10 rd detachable box w/cut-off	10 rd detachable box
Sights: Front	Blade w/ protecting ears	Barley Corn w/ protecting ears	Blade w/ protecting ears	Blade w/ protecting ears
Rear	Vertical leaf w/ aperture battle sight, long range side sights	Tangent leaf w/notch long range side sights	Tangent leaf w/notch long range side sights	Tangent leaf w/notch
Muzzle velocity (at date of adoption)	Apprx 2500 FPS	2060 FPS	2060 FPS	2440 FPS
Weight	9.62 lbs	8.12 lbs	8.62 lbs	8.62 lbs

	Rifle No 2 Mark 4	Rifle No 4 Mark 1	Rifle No 5 Mark 1	Rifle No 8 Mark 1
Caliber	.22	.303	.303	.22
Overall length	44.5 in	44.5 in	39.5 in	41.05 in
Barrel length	25.2 in	25.2 in	18.7 in	23.3 in
Feed device	Single shot	10 rd detachable box	10 rd detachable box	Single shot
Sights: Front	Blade w/ protecting ears	Blade w/ protecting ears	Blade w/ protecting ears	Blade w/ protecting ears
Rear	Tangent leaf w/notch	Vertical leaf w/ aperture battle sight or L type	Vertical leaf w/ aperture battle sight	Vertical leaf w/ aperture battle sight
Muzzle velocity (at date of adoption)	1050 FPS	2440 FPS	2400 FPS	1050 FPS
Weight	9.19 lbs	8.8 lbs	7.15 lbs	8.87 lbs

Characteristics are listed only for the principal models.
Lengths are with normal butt.

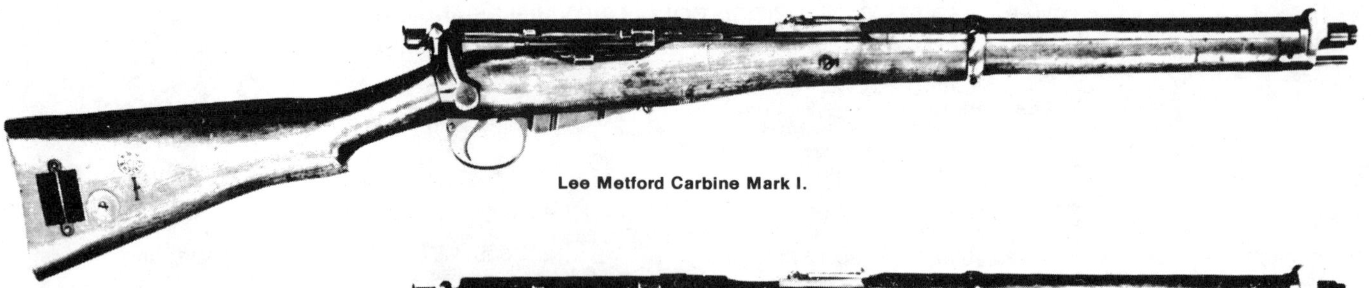

Lee Metford Carbine Mark I.

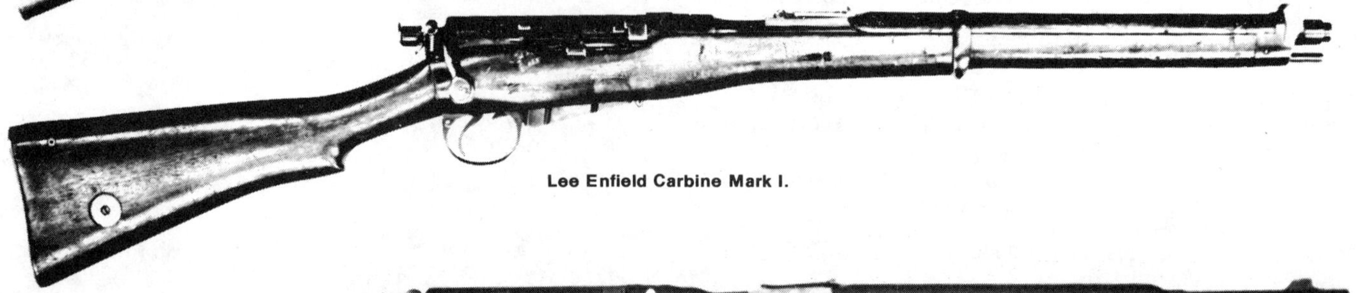

Lee Enfield Carbine Mark I.

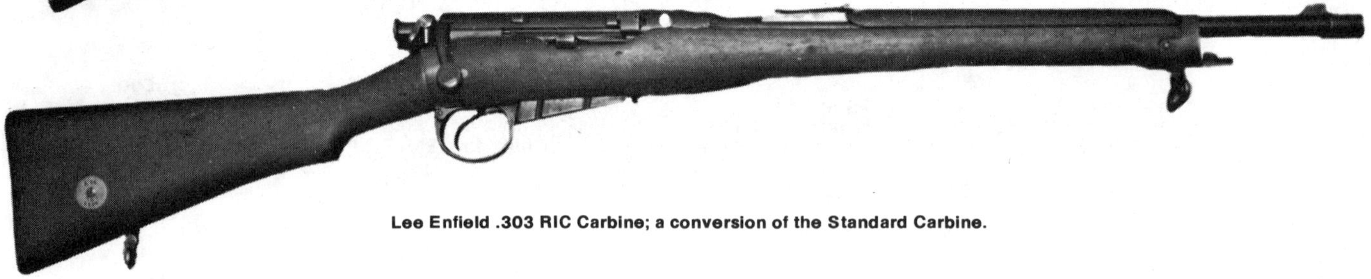

Lee Enfield .303 RIC Carbine; a conversion of the Standard Carbine.

Barrel gaging. The bore diameter should be from .301 to .304 inches, in a new barrel. To gage a used barrel, plug gages from .303 to .310 inches should be used. The .303 gage should run through the barrel; the .307 gage should not run through the barrel. The .308 gage should not enter the muzzle more than .25 inches, and the .310 gage should not enter the breech more than .25 inches.

Firing pin protrusion. The high for firing pin protrusion for the No. 1's is .055 inches, and the low is .050 inches. The high for the No. 4's and No. 5's is .050 inches, and the low is .040 inches.

Trigger pull. The first pull or slack should be from 3 to 4 pounds. The second pull should be from 5 to 6 pounds. To increase or decrease the trigger pull weight, alter the angle of the cocking piece sear notch.

Buttstock lengths. Butts for the No. 1 rifles were made in long and short lengths, and during World War I a special short butt called the Bantam was made. These butts will be marked "L," "S," or "B" on the top of the stock, approximately one inch from the butt plate tang. Butts for the No. 4 and No. 5 rifles come in long, short, and normal lengths.

The Cal. .303 Lee Metford Rifles and Carbines

Rifle, Magazine, Lee Metford Mark I. Adopted December 1888. Was the first British production Lee. Chambered for the cal. .303 black-powder-loaded cartridge. Had an eight-round magazine and a full-length cleaning rod.

Rifle, Magazine, Lee Metford Mark I*. Adopted January 1892. Was a conversion of the Mark I; the sights were changed from "Lewes" and "Welsh" pattern to barleycorn front and V-notch rear sight.

Rifle, Magazine, Lee Metford Mark II. Adopted April 1892. Was the first of the series to be fitted with a 10-round magazine. The bolt was modified and the outside contour of the barrel was changed. A half-length cleaning rod was fitted to the gun, and the brass marking disk on the buttstock was omitted.

Carbine, Magazine, Lee Metford Mark I. Adopted 1894.

Rifle, Magazine, Lee Metford Mark II* . Adopted 1895. Had a safety catch added to the bolt. (Previous marks had a half-cock notch on the cocking piece as their only safety.)

The Cal. .303 Lee Enfield Rifles and Carbines

Rifle, Magazine, Lee Enfield Mark I. Adopted November 1895. Had the deep Enfield rifling, rather than the shallow Metford rifling used on previous marks. The sights were also modified.

Rifle, Magazine, Lee Enfield Mark I* . Adopted 1899. Had no cleaning rod mounted in the stock.

Carbine, Magazine, Lee Enfield Mark I. Adopted 1896. Same as Lee Metford carbine except for rifling.

Carbine, Magazine, Lee Enfield Mark I* . Same as the Mark I carbine but has no cleaning rod and no sling bar in the left side of the butt.

Carbine, Magazine, Lee Enfield, RIC Model. Adopted in 1905 when 10,000 of this model were made up from Lee Enfield Carbines. The carbine nose cap was removed and the stock cut back and slimmed down to take an upper band with bayonet stud to fit the pattern 88 knife bayonet. These carbines were made up for the Royal Irish Constabulary-RIC, which was disbanded in 1922.

Rifle No. 1, Short Magazine, Lee Enfield Mark I. Adopted December 1902. Was the first of the short rifles (SMLE). Was stocked to the muzzle and charger loaded. The right side charger guide is on the bolt head, and the left charger guide is on the receiver. Has a V-notch rear sight with adjustable windage and a barleycorn front sight. Was the first of what later came to be called the No. 1 series of rifles.

Rifle, No. 1, Short Magazine, Lee Enfield Mark II (CNVD). Essentially the same as the SMLE No. 1 Mark I, but was converted from earlier Mark II and Mark II* Lee Metford's and Long Lee Enfield's.

Rifle No. 1, Short Magazine, Lee Enfield Mark I* . A minor variant of the SMLE No. 1 Mark I.

Rifle No. 1, Short Magazine, Lee Enfield Mark II* . A minor variant of the No. 1 Mark II SMLE.

Lee Enfield Mark I* Carbine.

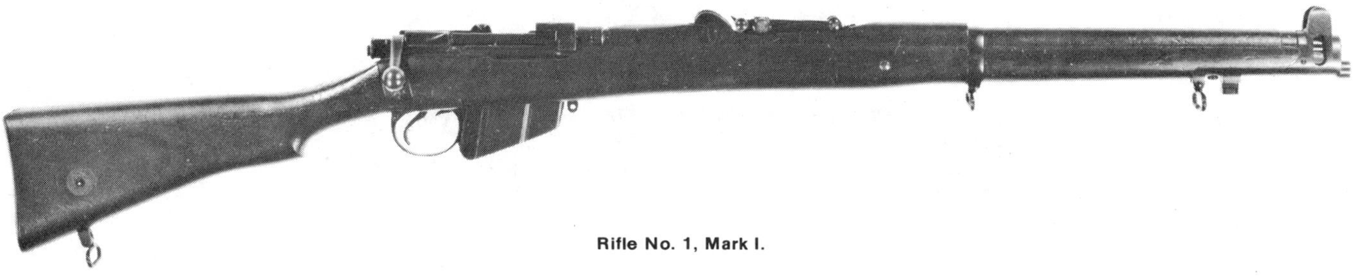

Rifle No. 1, Mark I.

Rifle No. 1, Short Magazine, Lee Enfield Mark III. Adopted January 1907. Was the backbone of the British Army in World War I, and was also used extensively in World War II. Is still in use in many of the areas of the British Commonwealth today.

Rifle No. 1, Short Magazine, Lee Enfield Mark IV (CNVD). Adopted July 1907. Basically the same as the No. 1 Mark III; converted from Long Lee Metford's and Long Lee Enfield's.

Rifle, Charger Loading, Long, Lee Metford Mark II. Was converted to charger loading in 1907 for use of the Territorial Army, and converted to rifle charger loading Lee Enfield Mark I* in 1909. Few of these were made.

Rifle, Charger Loading, Long, Lee Enfield Mark I. A 1907 conversion of early marks of Long Lee Enfield to charger loading. The Mark I* version is more common. A large number of these weapons were used by British forces in the early days of World War I.

Rifle No. 1, Short Magazine, Lee Enfield Mark III* . Was adopted during World War I, and made in very large quantities. Is still in widespread use throughout the world. Does not have the long-range side sights of the Mark III and earlier marks, and does not have a magazine cutoff.

The Royal Ordnance Small Arms Factory at Enfield Lock made over 2 million of this model and the No. 1 Mark III during World War I. During the same period, B.S.A. made 1,601,608 and L.S.A. made several hundred thousand. This rifle was last manufactured in the U.K. by B.S.A. in 1943. The Australian arsenal at Lithgow and the Indian plant at Ishapore manufactured the Mark III* after the adoption of the No. 4. Lithgow produced 415,800 from 1939 to 1955 when production was switched to the FN rifle.

Rifle No. 1, Short Magazine, Lee Enfield Mark V. Appeared around 1922. The rear sight is mounted on the receiver bridge, and an additional stock band is mounted to the rear of the nose cap.

Rifle No. 1, Short Magazine, Lee Enfield Mark VI. Was developed in the period 1924-1930. Was the forerunner of the No. 4 rifles. Had rear sight on the receiver bridge. Had a lighter nose cap, heavier barrel, and smaller bolt head than the earlier marks. Had cut-off and left receiver wall is cut low as the Mark III.

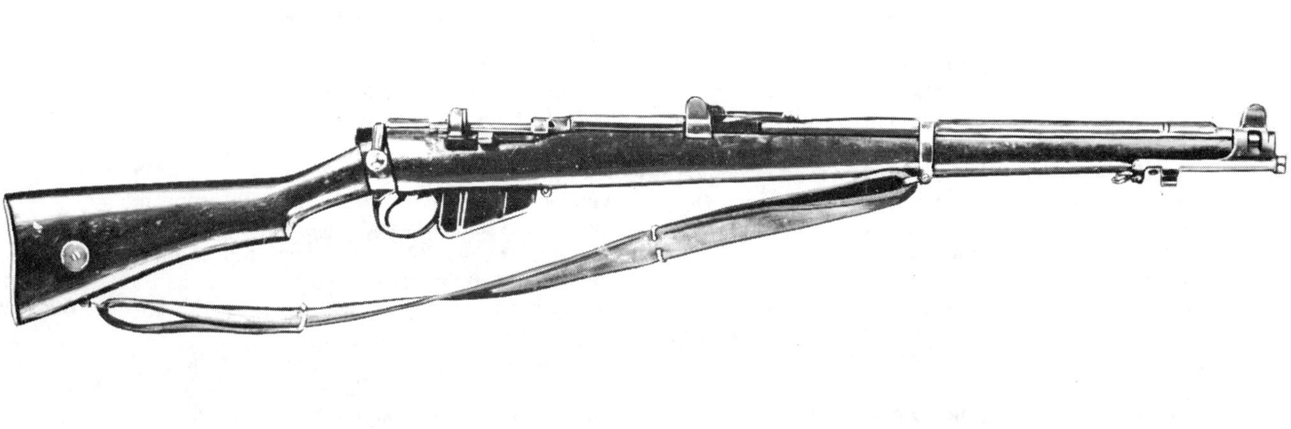

Rifle No. 1, Mark III.

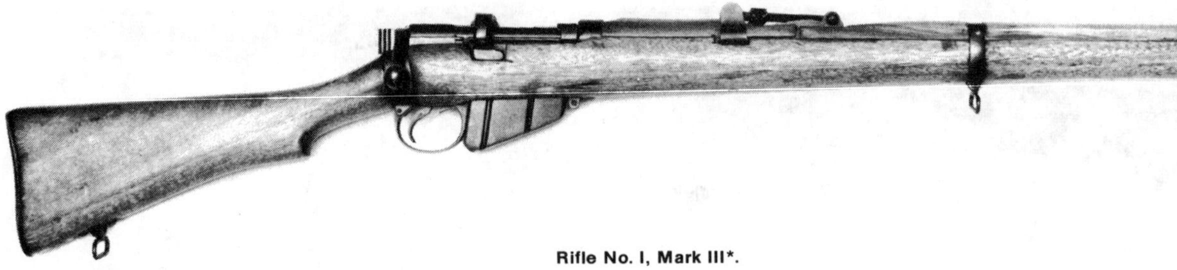

Rifle No. I, Mark III*.

Rifle No. 1, Mark V.

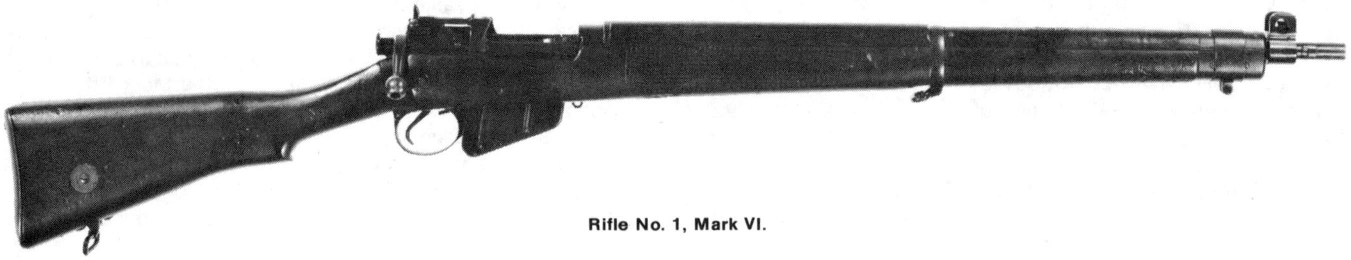

Rifle No. 1, Mark VI.

Rifle No. 4 Mark 1. Originally appeared in 1931. Was finely made, and was generally similar to the No. 1 Mark VI except that it had a heavier receiver. Was redesigned for mass production around 1939 and became, with the No. 4 Mark 1*, the British "work horse" of World War II. Stamped bands were used, and various manufacturing shortcuts were taken to increase production. Three different marks of rear sights may be found on this weapon, ranging from a finely machined adjustable leaf to a simple L-type. Many of these weapons are still in service in the British Commonwealth and in former British territories.

Rifle No. 4 Mark 1* . Was the North American production version of the No. 4 Mark 1. The principal difference was that the bolthead catch, which was situated behind the receiver bridge on the No. 4 Mark 1 (and earlier Marks), was eliminated on the No. 4 Mark 1*, and a cutout on the bolt head track was used for bolt removal. Over five million No. 4 rifles were made during World War II in the UK, Canada, and the United States (Stevens Arms). Australia did not adopt the No. 4, but continued production of the No. 1 Mark III* at Lithgow during World War II.

Canadian Rifle No. 4 Mark I* (light weight). This weapon was produced at the Canadian arsenal at Long Branch in prototype form. It has a one piece stock, and its trigger is pinned to the receiver. Weight about 6 3/4 pounds. Barrel length about 23 inches. Overall length about 42 1/2 inches. One piece stock. Receiver wall cut down and stock inletted to reduce weight. Sporting type Hawkins rubber buttplate. Micrometer sights with peep battle sight. Sight adjustable in clicks and 100 yard steps from 100 to 1300 yards. This arm may be used for grenade launching. Has a Mauser type trigger.

Rifle No. 4 Mark 2. Was developed at the end of World War II. Differed from the earlier marks by having its trigger pinned to the receiver rather than to the trigger guard.

Rifle No. 4 Mark 1(T) and No. 4 Mark 1* (T). Are the sniper versions of the No. 4. Are fitted with scope mounts on the left side of the receiver, and have a wooden cheek rest screwed to the butt. The No. 32 telescope is used on these weapons. There are also sniper versions of the No. 1 and No. 3 rifles (Pattern 14). The Canadians also used the No. 4 Mark 1*(T) with the Telescope C No. 67 Mark 1.

Rifle No. 4 Mark 1/2 and Rifle No. 4 Mark 1/3: These are conversions of the No. 4 Mark I and No. 4 Mark I* respectively to the pattern of the No. 4 Mark II. These rifles, like the No. 4 Mark II, are still in extensive use and are probably held as reserve weapons by the U.K.

Rifle No. 4, Mark I stripped.

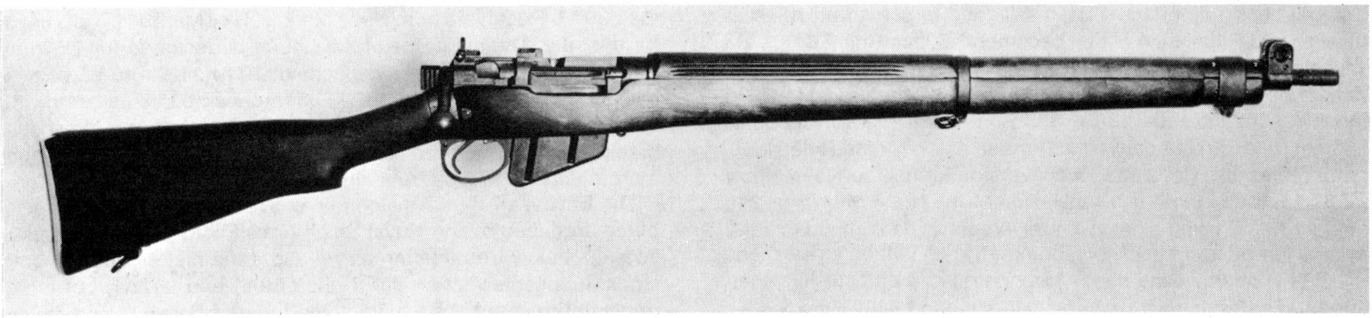

British Rifle No. 4 Mark I*.

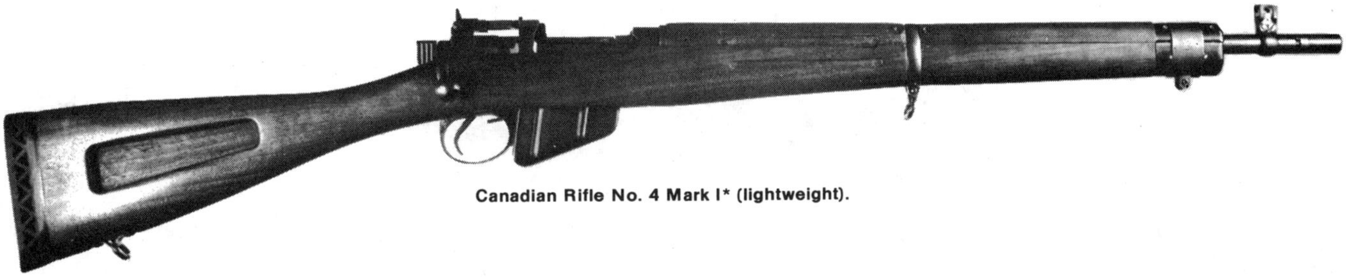

Canadian Rifle No. 4 Mark I* (lightweight).

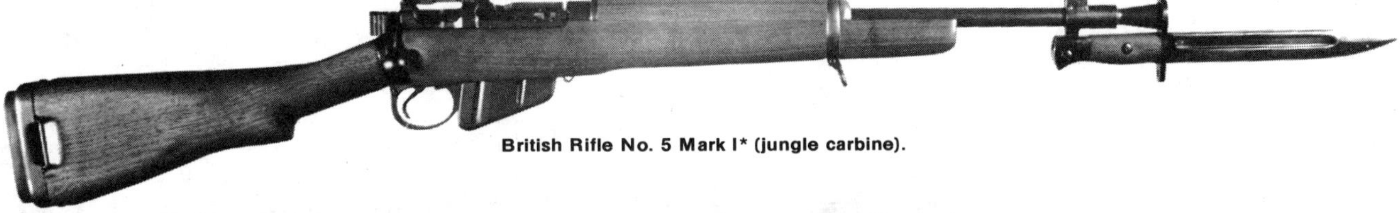

British Rifle No. 5 Mark I* (jungle carbine).

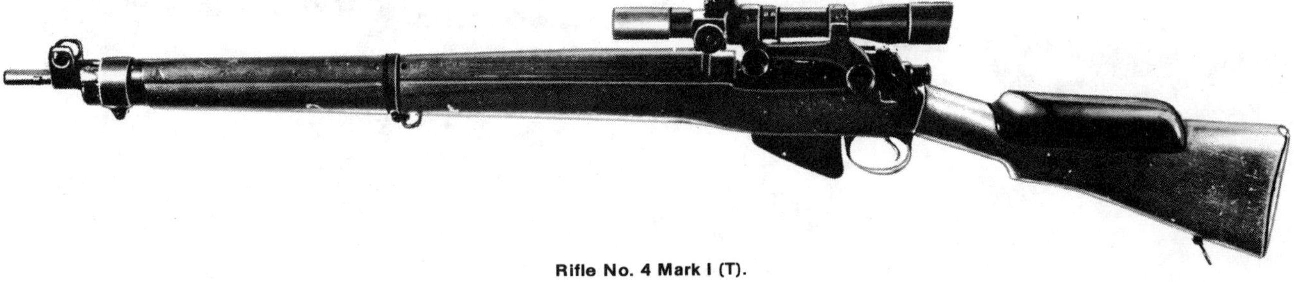

Rifle No. 4 Mark I (T).

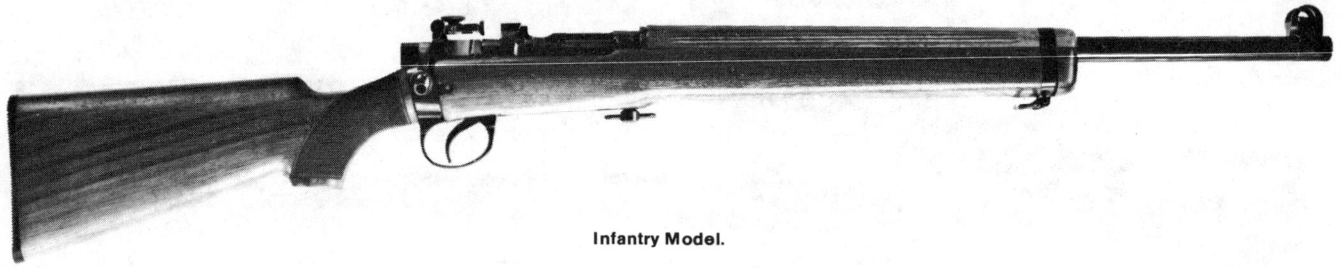

Infantry Model.

Conversion of No. 4 Rifles to 7.62mm NATO. The Royal Small Arms Factory at Enfield Lock has developed a conversion kit for the No. 4 rifles to convert them to use the 7.62mm NATO cartridge. This kit consists of a new barrel, extractor, magazine, charger guide liner, front sight block fixing pin and a barrel breeching washer. This kit can be fitted to an existing No. 4 rifle with normal armorers tools and certain special purpose tools, i.e. a special drift, a taper pin reamer, a breeching gage etc.

Model designations have been assigned to converted rifles as follows: .303 Rifle No. 4 Mark I becomes 7.62mm Rifle L8A4; .303 Rifle No. 4 Mark I* becomes 7.62mm Rifle L8A5; .303 Rifle No. 4 Mark 1/2 becomes 7.62mm Rifle L8A2; .303 Rifle No. 4 Mark 1/3 becomes 7.62mm Rifle L8A3; .303 Rifle No. 4 Mark II becomes 7.62mm Rifle L8A1. A conversion kit also exists for the Rifle No. 5.

B.S.A. has also developed a conversion kit for the No. 4 rifles.

7.62mm Rifle L39A1 is a conversion of the No. 4 MK1/2 or 2 to 7.62mm NATO using a heavy barrel. A special hand guard is fitted and the forend is cut back in sporting style so that its forward end is just beyond the lower band. A standard No. 4 butt stock is used. This rifle is designed for competitive shooting within the British services. A front sight base is provided, but no front sight or rear sight is provided. Units are supposed to supply their own and a variety of sights are available. The L39A1 uses a 10 round box magazine, but the magazine, which is .303, serves only as a loading platform. The barrel is 27.5 inches long and overall length is 46.5 inches. The trigger pull is lighter than that of the No. 4. L39A1 weighs 9.75 pounds.

7.62mm Rifle L42A1 is the current standard sniper rifle. It has a heavy 7.62mm NATO barrel and the forend is cut back the same as the L39A1. Receivers from No. 4 Mark 1 (T) or Mark 1* (T) are used for this rifle. The magazine of the L42A1 is designed for 7.62mm NATO cartridges and has a capacity of 10 rounds. The butt stock has the same type "screw on" wooden check piece as used with the No. 4 Mark 1 (T). The left side of the receiver has a telescope bracket for the telescope No. 32 Mark 3. A leaf type rear sight and a protected blade type front sight are also used.

The Envoy, a 7.62mm match rifle conversion of the No. 4 is also being produced by the Royal Small Arms Factory. This rifle has a swaged, heavy free floating barrel and a full pistol grip type butt stock. It is somewhat similar to the L39A1 and L42A1, but more finely finished and has a match type tunnel front sight and match type aperture rear sight fitted.

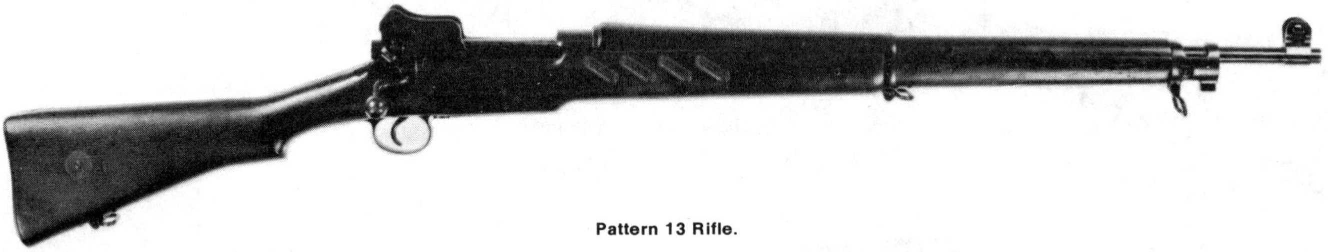

Pattern 13 Rifle.

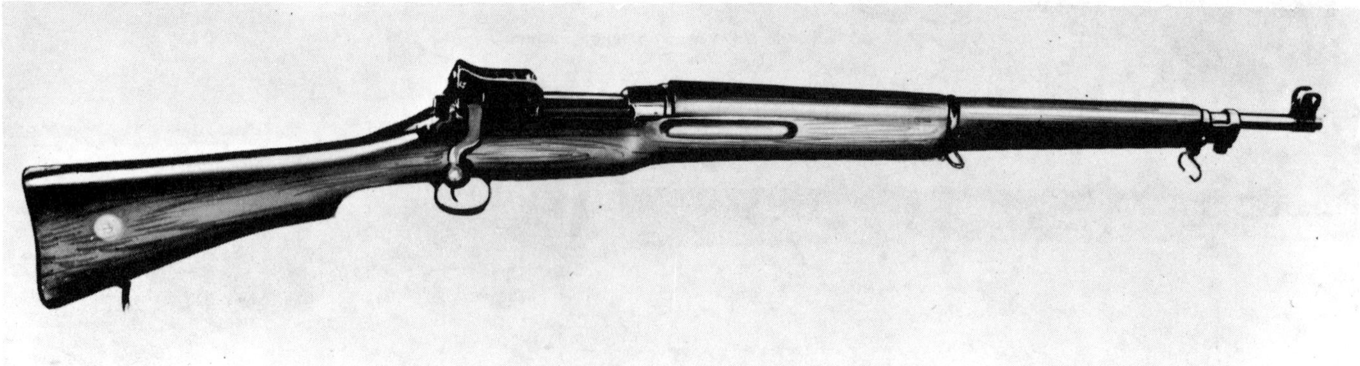

Pattern 14 Rifle.

Rifle No. 5 Mark 1. Appeared toward the end of World War II. Was a lightweight weapon, and was commonly called the jungle carbine. Has a lightened and shortened barrel which is fitted with a flash hider. Fore-end has been cut back and rounded, giving weapon the appearance of a sporting rifle. A rubber recoil pad is fitted to the butt.

Rifle No. 6(Aust). Appeared only as prototype; 18" barrel version of No. 1. Developed at Lithgow.

The Cal. .22 Rifles

Cal. .22 R.F. Short Rifle Mark I. Adopted in 1907. This rifle is a conversion of the Lee Metford Mark I rifle; it is approximately the same length as the SMLE. Sights are an adjustable blade front sight and a tangent type rear sight with adjustment for windage.

Cal. .22 R.F. Long Rifle Mark II. Adopted in 1911. Conversion of Long Lee Enfield to .22 rimfire. The Mark I pattern of .22 R.F. long rifles was converted from Long Lee Metfords.

Cal. .22 R.F. Short Rifle Mark I*. Conversion to .22 rimfire in shortened form of Lee Metford Mark I*. Mark II of this pattern was converted from Lee Metford Mark II.

Cal. .22 R.F. Short Rifle Mark III. Adopted in 1912 - this is a conversion of SMLE Marks II and II* to .22 rimfire. A number of different patterns of Lee were converted during World War I to .22 caliber rimfire. Some were fitted with new barrels and others had their .303 barrels bored out and a caliber .22 liner inserted. The .22 R.F. pattern 1914 had "conveyors" - cartridge adaptors or auxilary cartridges which fed through .303 magazines.

Rifle No. 2 Mark IV. Is a conversion of cal. .303 SMLE's to cal. .22. Some have new .22 barrels, and some were "Parker Rifled," i.e., a .22 liner was placed in a bored-out .303 barrel. A special bolt head was made for these rifles.

Rifle No. 2 Mark IV*. A variant of the No. 2 Mark IV.

Rifle No. 7. Developed at Long Branch; single shot version of No. 4 Mark I*. Called Rifle "C" No. 7, .22 in Mark I. Also has been made by B.S.A. with a 5-shot magazine.

Rifle No. 8 Mark 1. Two variations of this rifle were developed simultaneously. They were called the Infantry Model and the Match Model and differed principally in sights and length of barrel. The Infantry Model has sights similar to the No. 4 rifle and a shorter barrel than the Match Model. The Match Model, in addition to the longer barrel, has match type sights. The Infantry Model was adopted in 1950 as the Rifle No. 8 Mark 1.

Rifle No. 9. Converted to .22 by Parker Hale, from No. 4 rifles, single shot.

Remington-made Springfield .303 Rifle.

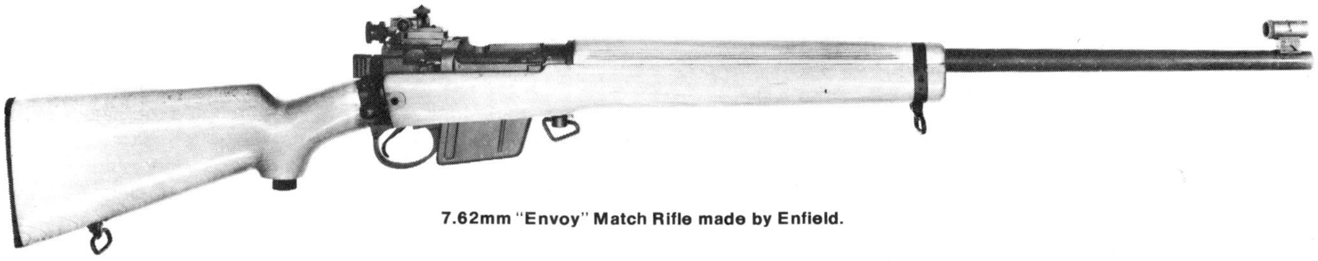

7.62mm "Envoy" Match Rifle made by Enfield.

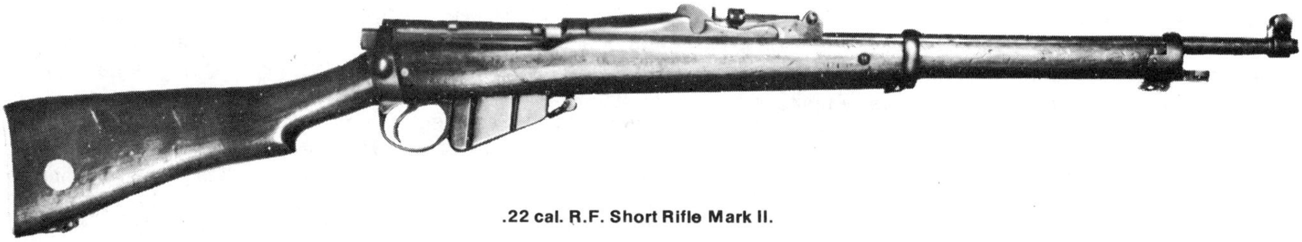

.22 cal. R.F. Short Rifle Mark II.

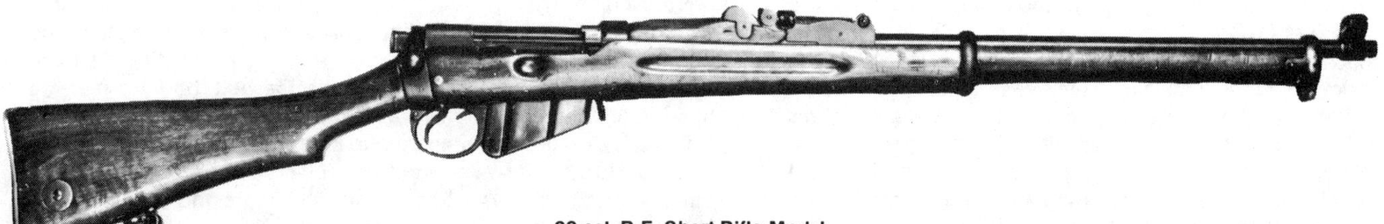

.22 cal. R.F. Short Rifle Mark I.

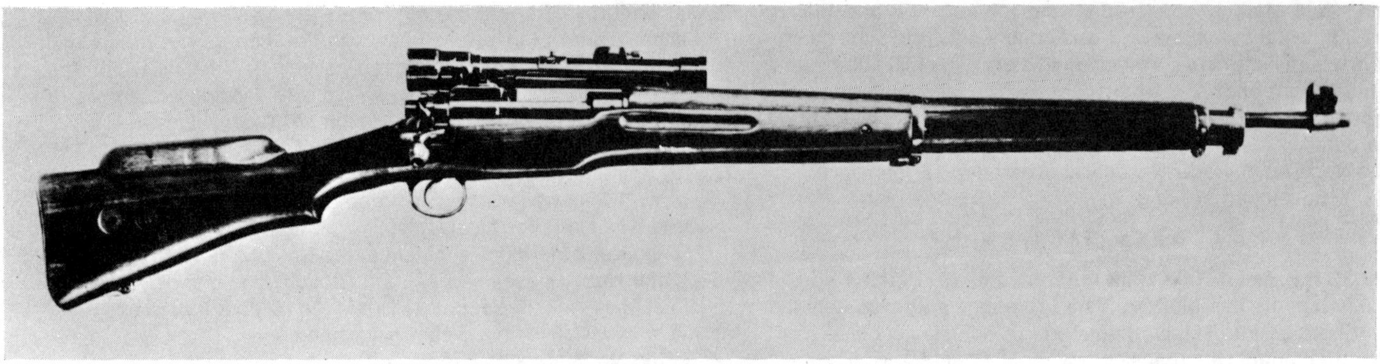

Rifle No. 3, Mark I*(T)A — Pattern 14(T)A.

The Mauser-Type Rifles

Pattern 13 (P-13). Tested in 1913. Was a modified Mauser (it cocked on the forward stroke of the bolt), chambered for a large cal. .276 cartridge. The cartridge was remarkably similar to the Canadian cal. .280 Ross cartridge. The rifle was made in comparatively small numbers for field trials.

Pattern 14 (P-14). Was the production model of the P-13. Was made in the United States in cal. .303 for the UK, during World War I. The weapon was classed as limited standard in the British Army, and, except for sniping, was not too widely used. Upon the entrance of the United States into World War I, the design was changed to U.S. Cal. .30, and the weapon was produced as the U.S. rifle, cal. .30, M1917 and was commonly known as the Enfield. Between World Wars I and II, the British changed the nomenclature of the P-14 to Rifle No. 3 Mark I.

Pattern 14 Sniper Rifles. The P-14 was extensively used as a sniper rifle in World War I. The two basic patterns were the P-14 (T) and the P-14 (T) A. The former has a Pattern 1918 telescope adjustable for range and windage and the latter has an Aldis telescope adjustable for range only. In 1926, when all British small arms were given number designations, these weapons were renamed the Rifle No. 3 Mark I* (T) and Rifle No. 3 Mark I* (T) A, respectively.

Springfield .303. This rifle which apparently has no official nomenclature, is a modification of the U.S. M1903 designed by Remington Arms at the request of Great Britain in 1941. Production of the Springfield rifle by Remington Arms during World War II was initially—until Sept 1941—set up for a British contract. The .303 version was made only as a prototype and development on this rifle stopped in September 1941. The type "C" stock was used with the pistol grip modified to the Enfield type, the bolt face, extractor, follower and magazine were modified to fit the .303 cartridge. The receiver bridge was lengthened. The "L" type rear sight was similar to the Mark 2 sight used on the No. 4's. Note that the bayonet mounting lugs are on the barrel as with the No. 4.

Loading and Firing Lee Enfield Rifles

Turn bolt handle up as far as it will go and pull it straight back to the limit of travel.

Insert loaded clip in the clip guide in the receiver and strip the cartridges down into the magazine. Remove the empty clip. Insert a second clip, push these cartridges down and remove clip. This will leave the magazine fully charged with 10 cartridges.

Pushing bolt handle fully forward and down loads the firing chamber, cocks, and locks ready for firing with a pull of the trigger.

Unless weapon is to be fired immediately, pull the thumb rocker on the left rear of the receiver to "Safe."

Replacing bolt, No. 1 Lee Enfield Rifle.

Field Stripping Lee Enfield Rifles

Remove magazine. This may be done by pushing in or pulling up, as different rifles may require, the magazine catch located in the forward end of the triggerguard. This will release the heavy sheet steel box which may be withdrawn from the bottom of the receiver.

Removing the magazine follower and its spring is simply done. Hold the magazine, open end up, and push the rear of the magazine follower down inside the casing. This will permit you to ease the front end of the follower up and out of the casing and remove it and the spring.

In order to remove the bolt, it is first necessary to rock forward the safety catch just above the rear end of the triggerguard, on the left side of the rifle. Then turn the bolt handle up and turn it back as far as it will go. Catch your right forefinger under the head of the bolt. Pull the bolt head up until it is released from its spring catch. Then withdraw it straight to the rear.

Field Stripping for Rifle No. 4 Mark I* . The No. 4 Mark I* rifle has a different method of removal of the bolt from the rifle than do the other Lee Enfield rifles. On the bolt head track—right side of receiver—there is a cut-out; draw bolt back until bolt head is over this cut-out; then lift bolt head straight up and draw bolt out of rifle.

Note on Replacing Bolt. These bolts are not interchangeable and the number on the bolt should always be checked against the number on the rifle when there has been any possibility of substitution of another bolt. Before inserting the bolt, be sure that the head is fully screwed home, and that the cocking piece lines up with the lug on the underside of the bolt. Insert the bolt in the boltway and thrust it forward, and then pull it back as far as it will go until the head touches the resistance shoulders and force the bolt head down over the spring retaining catch. Then push it forward to the forward position. Turn down bolt handle and press trigger.

How the Lee Enfield Rifle Works

Starting with the rifle loaded and cocked, the action is as follows: When the trigger is pressed, it draws down the sear until the sear nose reaches the bottom of the full bent. (This provides the first pull or slack, which is a feature of the best military rifles.) As the trigger pressure continues, the upper part of the sear is drawn still further down until the sear nose clears the bent allowing the cocking piece on the striker to be driven forward by the compressed mainspring. The striker nose, or firing pin, passes through a hole in the face of the bolt head and discharges the cartridge in the firing chamber.

Special Note on the Lee Enfield System. The locking system on this rifle makes it the fastest operating bolt action rifle in the world. The abrupt turning action of the Mauser system will not permit it to attain a speed of operation possible with the Lee Enfield.

This rifle, since it has no locking recesses cut into the receiver, is much easier to clean than the Mauser type, and functions well under all battle conditions.

BRITISH EXPERIMENTAL SEMI-AUTOMATIC RIFLES

In 1893 the "Automatic Rifle Syndicate" submitted the Griffiths-Woodgate, a short recoil operated rifle, to the British Small Arms Committee for testing. The rifle was not tested, but it was one of the first automatic rifle designs submitted to the British government for test. Other early British designs were the long recoil operated Schlund and Arthur of 1885, Gabbett—Fairfax of 1896, Ross of 1896—both of which had toggle joint actions—and the Clausim of 1895.

In the first decade of the 20th century, Great Britain tested the following semi-automatic rifles: Kjellman, Fidgeland, Rexer (D.R.R.S./Madsen Schoubee), Sjogren, Mondagon, Halle, Frommer, Mannlicher, Tatarek, Mauser, Bang, Farquhar Hill, and various Enfield designs. None of these tests resulted in any real change of attitude toward the semi-automatic rifle. It was considered over-complicated and wasteful of ammunition.

After World War I, additional designs of semi-automatic rifles were tested. Among these were the Thompson, a delayed blow back which was made in a number of models by B.S.A. The Bang, which trapped gas at the muzzle to operate the action was tested after World War I—1927—as well as before the war. It had a sliding muzzle cap that was pulled forward by the gas after the bullet left the bore. This forward movement was translated through a series of rods and levers into a rearward movement of the bolt. In 1913, Bang submitted a rifle chambered for the .276 cartridge which was developed for the P-13 rifle. The White, which used a gas expansion system was tested in the 30's. The Adams, a British design, was also tested during this period.

The production of the Pedersen .276 by Vickers and its test by Britain is an interesting story. The Pedersen was developed at Springfield Armory by Mr. J. D. Pedersen. In 1924, the United States government hired Pedersen on a contract basis to develop a semi-automatic rifle and a cartridge. Pedersen developed a .276 cartridge which originally had a 125-grain bullet, and a muzzle velocity of 2700 feet per second; a later version had a 140 grain bullet and a muzzle velocity of 2600 feet per second. Pedersen's rifle was a delayed blowback with a toggle joint action. The cases were coated with a hard wax to aid extraction and an "en bloc" clip similar to that of the later Garand M1 was used. In 1930 Pedersen made an arrangement with Vickers for them to tool up to produce the rifle. The British government was interested in the rifle and cartridge, because it appeared that the United States was going to adopt the rifle. The Pedersen lost out to the Garand in the United States—partially because the .276 cartridge was rejected by the United States—and the Pedersen ran into troubles in the U.K. because of it's waxed cases which were not popular. The .276 Pedersen cartridge was made in limited quantities by Kynoch and Greenwood and Bately from 1930-34. The Farquhar-Hill in a different model appeared again during the 30's. Most of the Farquhar-Hill prototypes were made by Webley.

In the late thirties the Janecek, a delayed blow back with fluted chamber, was made in .303 caliber for British testing. The Czech ZK 381 in 7.92mm was also tested. The year 1938 was one of the most active in the testing of rifles in the U.K. One of the rifles tested was a conversion of the P-14 into a gas operated semi-automatic by the Italian, Scotti. His own Model X, which was used in limited quantities by Italy during World War II, was tested in Britain in the mid 30's.

The U.S. M1 rifle was also extensively tested by the U.K. and adoption of the Faive designed FN semi-automatic—commonly called ABL or M1949—was seriously considered after World War II. Some tooling up to produce the rifle was done at Enfield in 1946. The 7.92mm rifle was called the SLEM or SLEM I or II—Self Loading Experimental Model. The above listing of rifles is by no means a complete list of semi-automatic and selective fire rifles that have been tested by the United Kingdom since the beginning of the 20th century. Activity increased after World War II since it was obvious to all that the days of the bolt action rifle, as a first line weapon, were over.

E.M.1 AND E.M.2

During the same period Enfield was developing several rifles for the British-developed .280 (7mm) cartridge. This "intermediate" size cartridge had a 140-grain bullet with a muzzle velocity of 2415 feet per second from a 24.5-inch barrel. Two weapons, the E.M. 1 and E.M. 2 were developed by a team of experts under Mr. Noel Kent-Lemon. Both of these weapons are gas operated and are of "Bull Pup" design, i.e., the magazine is to the rear of the pistol grip and the shooter's face is parallel to the receiver. Barrel length of both rifles is about 24.5 inches and overall length is about 36 inches. Both weapons use a 20-round magazine and weigh about 9 pounds.

The E.M. 1 and E.M. 2 rifles are capable of semiautomatic and automatic fire; cyclic rate of fire is about 450 rounds per minute. Both have a telescopic sight built into the carrying handle.

In 1949 the United Kingdom was ready to adopt the .280 E.M. 2 as standard, but the carrying out of NATO talks on standardization of cartridges and possible standardization of rifles delayed it. A series of tests of cartridges and rifles were held in the United Kingdom and the United States. The cartridges were the caliber .280 cartridge and the U.S. caliber .30 T65E3 cartridge. The FN "FAL" was made in .280 for these tests and was one of the leading contenders. The United Kingdom developed a .280/30 cartridge which was basically a caliber .30 bullet in the .280 case and a number of E.M. 2 rifles were made for this cartridge.

An impasse was reached at the technical level, but a decision was made at the highest political level in the United Kingdom to adopt the U.S. cartridge (which is now called the 7.62mm NATO) and the FN rifle. This decision was highly criticized in the House of Commons, but considering the logistical history of World Wars I and II, it made a great deal of sense. Although the United States has not been an importer of military ammunition in any significant quantity from the United Kingdom during the world wars, the U.K. has consistently been an importer of military ammunition from the U.S. during these periods. By adopting the 7.62mm NATO cartridge, Britain thus guaranteed that any future orders for ammu-

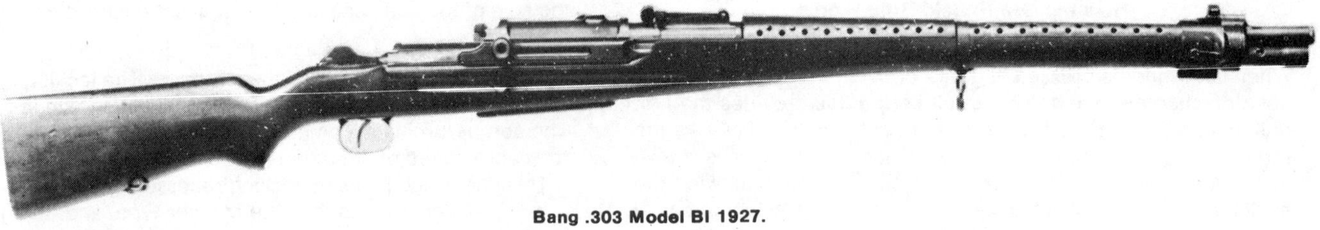

Bang .303 Model Bl 1927.

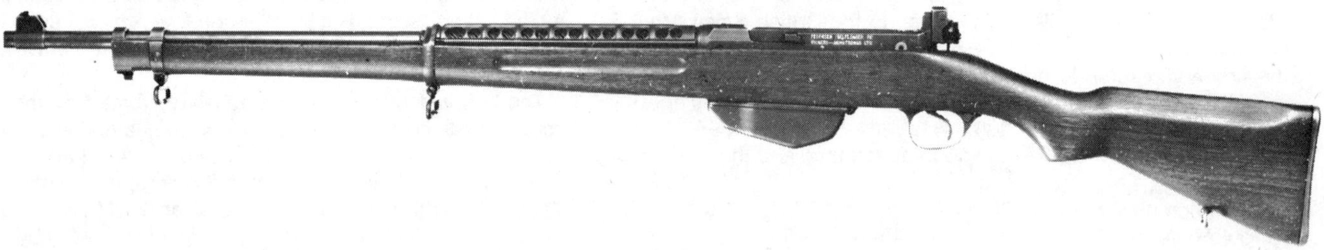

Pedersen .276 Rifle as made by Vickers Armstrong.

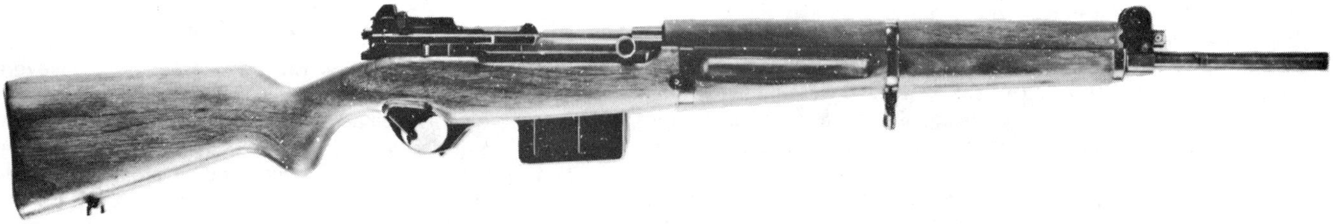

7.92mm self-loading experimental Model FN tested by UK in the late 1940s.

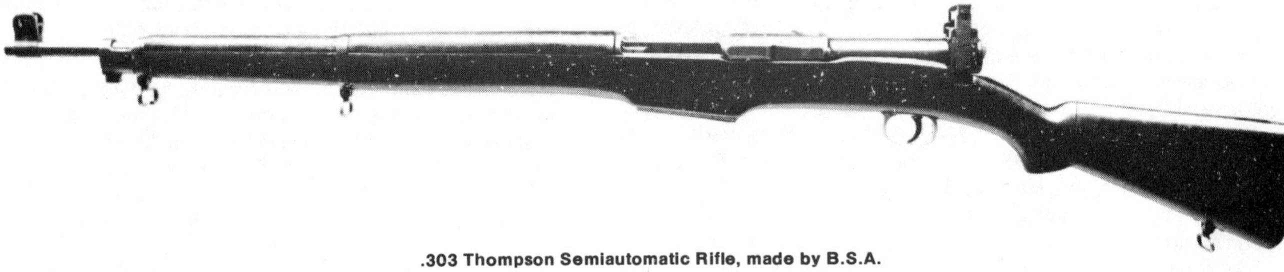

.303 Thompson Semiautomatic Rifle, made by B.S.A.

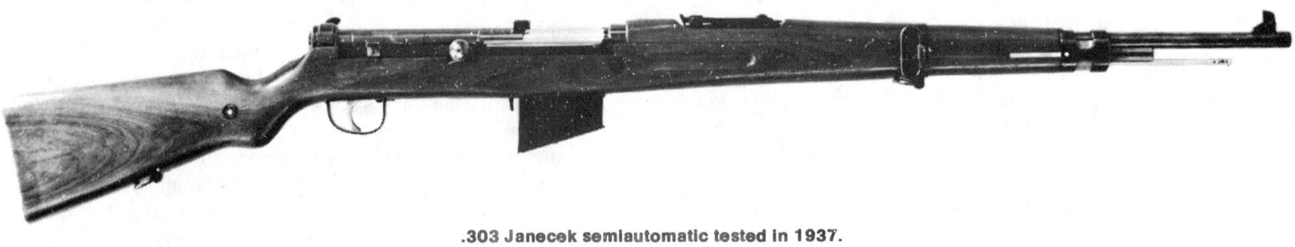

.303 Janecek semiautomatic tested in 1937.

Caliber .280 E.M. 1 Rifle.

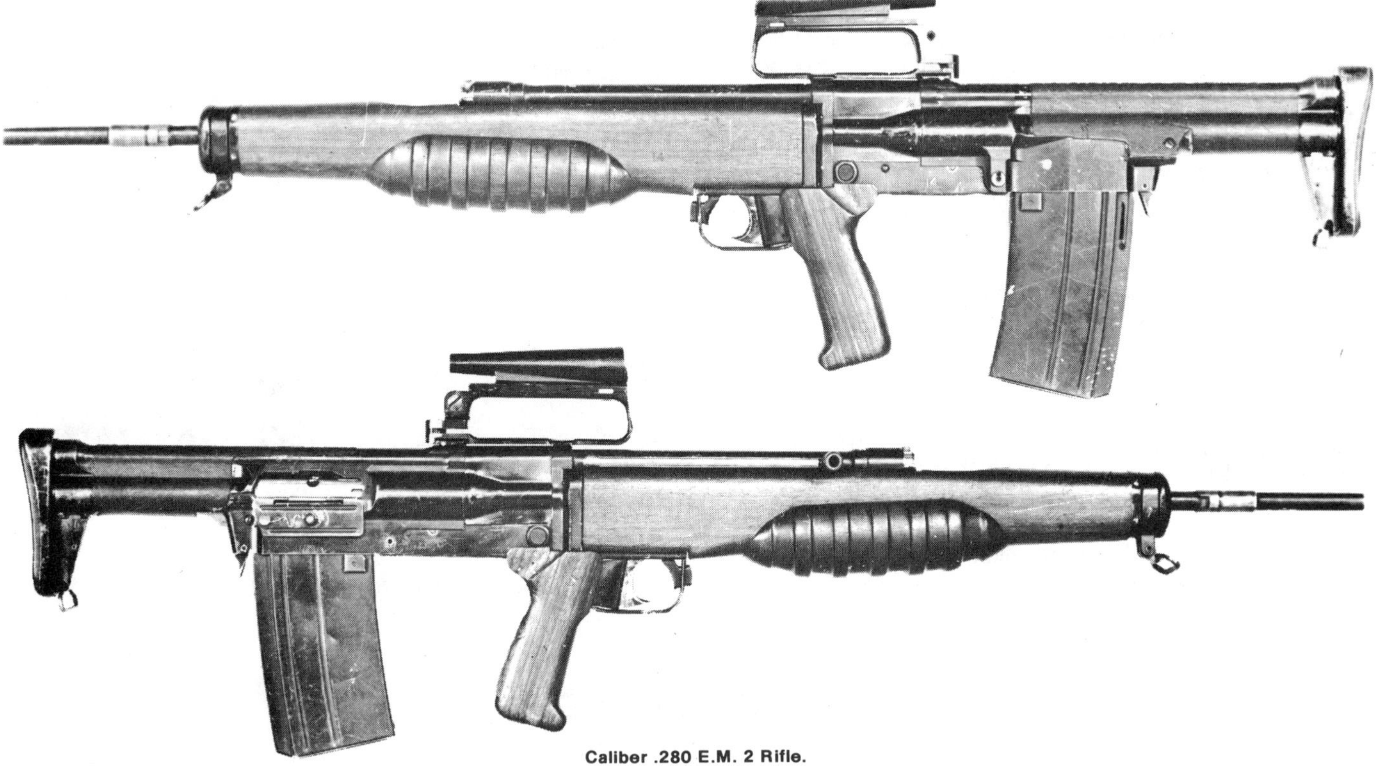

Caliber .280 E.M. 2 Rifle.

nition placed in the U.S.A. during an emergency would be met with a timely and substantial response.

The technical aspects of the controversy over the relative merits of the .280 and 7.62mm NATO cartridges and the rifles concerned are not as easy to explain as the logistical aspects. Basically a number of problems, which bulked much larger 10 or 15 years ago than they do now, had to be solved in a reasonable period of time and could not be solved within the time allotted. The two following factors probably influenced the controversy:

1. The specific requirements laid down by the using arms of the countries concerned did not agree on what was most desirable or necessary in a rifle. This is because the standardization of equipment was begun before standardization of tactics—and tactics dictate weapons requirements.

2. The attempted standardization of rifles was among the first of the NATO standardization projects and occurred in the early days of the alliance. The military and research and development personnel of the countries concerned hadn't developed the close relationships which now exist.

7.62mm RIFLE L1A1

The UK has adopted the 7.62mm NATO FN "light automatic rifle" and produced it locally at BSA and the Royal Ordnance Small Arms Factory. The official British nomenclature for the weapon is: Rifle, 7.62mm, L1A1.

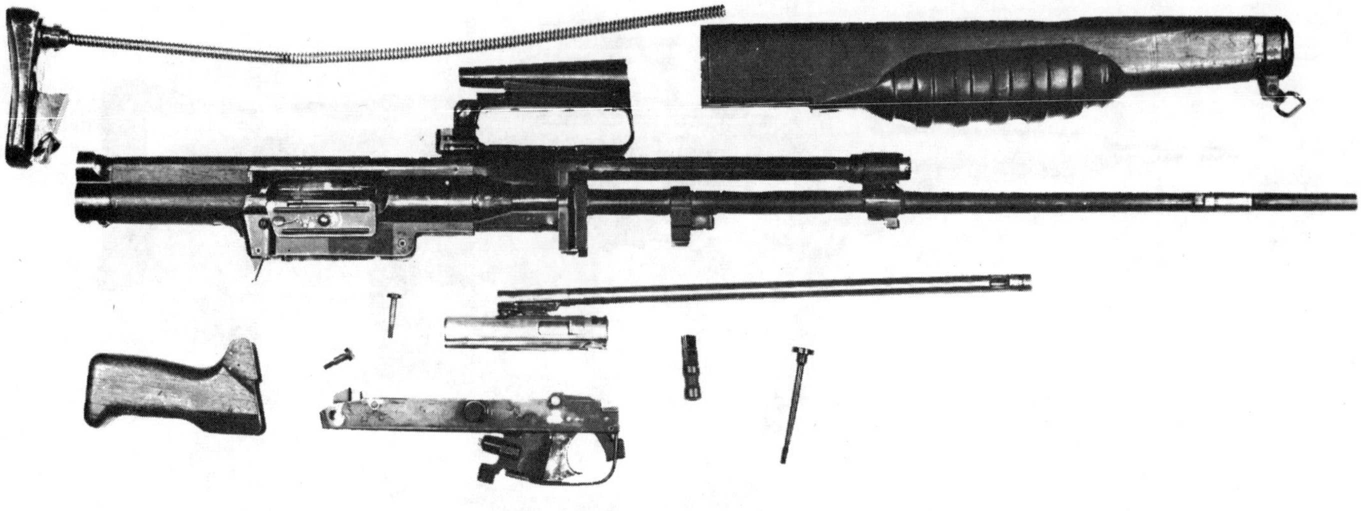

Caliber .280 E.M. 2 Rifle—field-stripped.

The L1A1 is the British-produced version of the FN NATO "light rifle," or "FAL," as it is commonly called in Belgium. The FN design has been modified in a few components, but basically the weapon is the same as that covered in detail in the chapter on Belgium. Cuts have been made in the bolt carrier to serve as gathering places for dirt and dust that may enter the action. These cuts are deep enough so that a good deal of foreign matter can accumulate in them without impairing the normal functioning of the weapon. The L1A1 fires the 7.62mm NATO cartridge.

Smith and Wesson 9mm Semiautomatic Carbine.

Characteristics of the L1A1
(British Version of the FN NATO Light Rifle)

System of operation: Gas, semiautomatic fire only (can be modified to selective fire).
Weight, loaded: 10.48 lb.
Length, overall: 44.5 in.
Barrel length: 21 in.
Feed device: 20-round, detachable, staggered box magazine.
Sights: Front: Post w/protecting ears.
Rear: Aperture, adjustable from 200 to 600 yd.
Muzzle velocity: 2800 f.p.s.

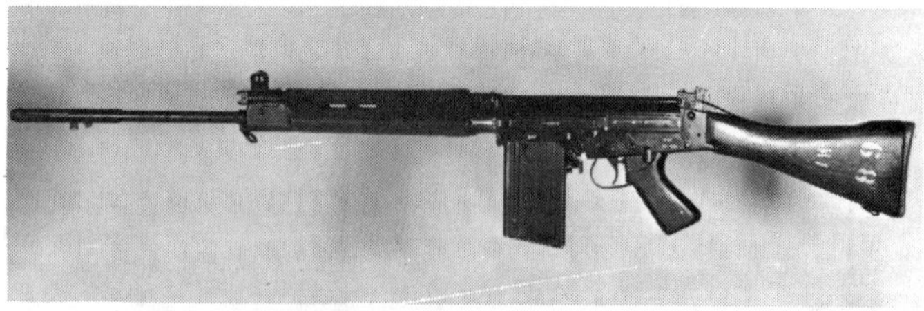

British 7.62mm Rifle L1A1.

BRITISH SUBMACHINE GUNS

Although B.S.A. had developed a number of modifications of the Thompson Submachine Gun during the 1920s, the British Army did not show much interest in submachine guns until after World War II started. In 1940 large contracts were let for the manufacture of the caliber .45 Thompson Submachine gun M1928A1 by the Auto Ordnance Corporation of Bridgeport, Conn.

Britain was desperate for weapons at this time and a limited quantity of the 9mm Parabellum Smith and Wesson semi-automatic carbines were purchased. These guns, which were very finely made and very expensive wound up in the hands of the Royal Navy.

During World War II the Lanchester and Sten guns were designed and produced during the war. These guns are covered in detail later in this chapter. A number of other submachine guns were produced in prototype form. Among these were the Welgun which was designed as a paratroop weapon, the B.S.A. V-42 and V-43, and the Patchett. The Patchett (name of designer) was devel-

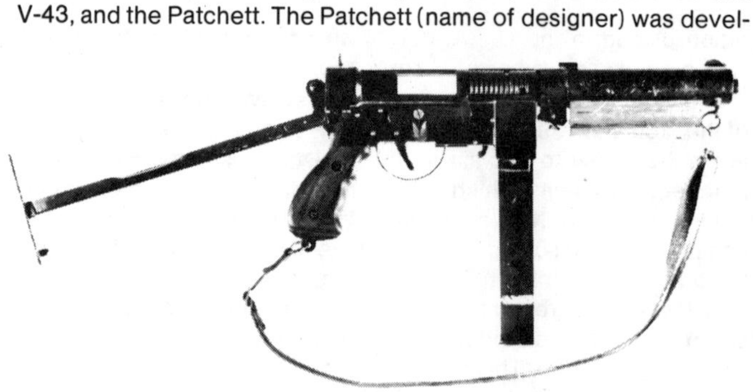

British 9mm Welgun, produced in limited quantities.

oped by the Sterling Engineering Company and in considerably modified form is the L2A3, the current standard British submachine gun.

The EM2 9mm submachine gun shown below was developed in the United Kingdom toward the end of World War II. It has many interesting features, as can be seen in the photograph. Note that it has the telescoping type bolt which is now so popular and is used by the Israeli Uzi, the Beretta M12, and the Czech M23, M24, M25, and M26 submachine guns.

The compactness of this weapon would appear to make it ideal for a guerrilla or underground type weapon. The shoulder stock is a hollow steel frame which also serves as a holster. The barrel is

EM2 British 9mm submachine gun.

EM2 British 9mm submachine gun, stripped.

approximately 8.25 inches long and the weapon, without butt, is about 14 inches long. The magazine capacity is 18 rounds and the magazine catch is somewhat unusual in that it pivots 90° to completely block the magazine well, preventing entrance of dirt, when the magazine is removed. All in all this is a very interesting weapon and has many commendable features. Another version of this weapon called the EM6 takes a bayonet. The EM2 and the EM6 were under development from 1944 to 1946.

Before the Sterling submachine gun was adopted in 1953, B.S.A. submitted a number of prototype 9mm Parabellum submachine guns for test. The weapon illustrated below was tested by the British government in 1949. It has one unusual feature; the ribbed section surrounding the barrel is used to retract the bolt and cock the gun. The ribbed sleeve does not reciprocate with the bolt. The B.S.A. gun is, as are most of the British-developed guns, capable of selective fire and has a folding stock and magazine housing. The folding magazine housing, in addition to being convenient for paratroops or armored troops to carry, acts as a safety feature in this weapon. If the weapon is dropped, the magazine housing with magazine are unlatched and swing out of the way, preventing the bolt, if it rebounds, from picking up a cartridge and firing the weapon.

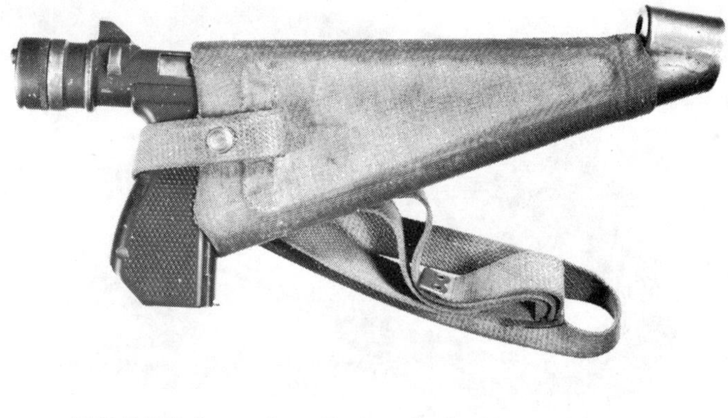

EM2 British 9mm submachine gun in shoulder stock holster.

Prototype B.S.A. 9mm submachine gun.

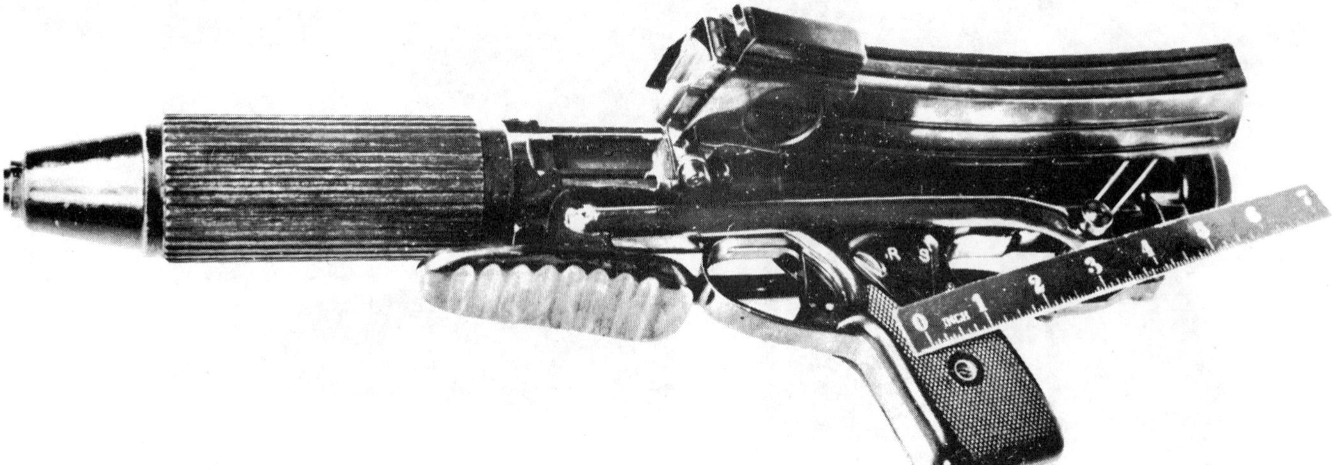

Prototype B.S.A. 9mm submachine gun with magazine housing and butt folded.

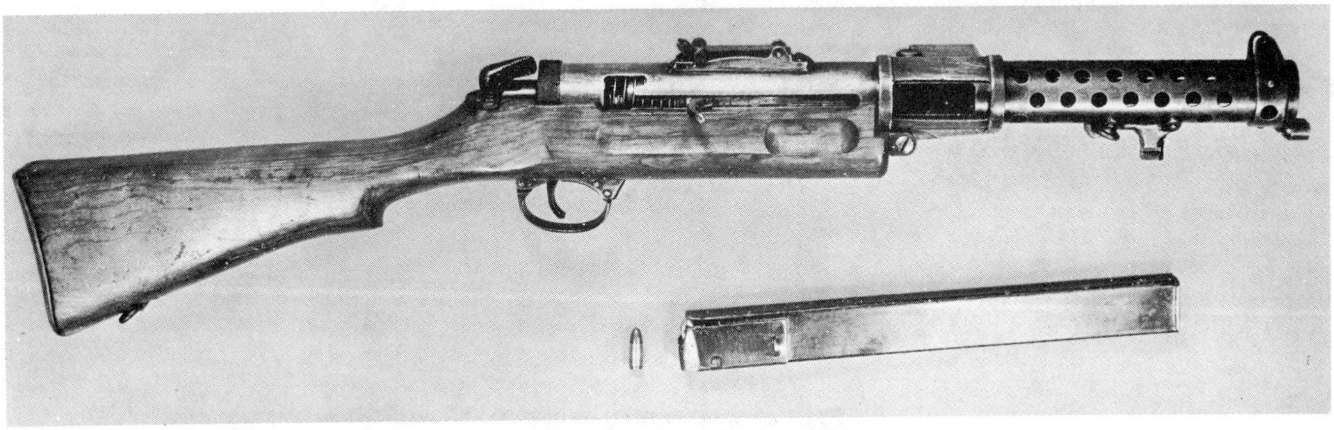

Right side of the Lanchester 9mm Mark I submachine gun, bolt cocked ready for firing. Note recoil spring compressed around end of firing pin unit which protrudes from rear of bolt. Ejection port is exposed.

The Lanchester — field-stripped.

THE LANCHESTER MARK I

This submachine gun was designed by G.H. Lanchester; it was manufactured by the Sterling Engineering Company, the same firm which developed the L2A3. The design of the Lanchester is based on that of the German MP 28 II. The selector lever is positioned differently than that of the MP 28 II and the Lanchester has a bayonet boss and stud for the Mark I (Pattern 1907) bayonet.

The Lanchester is a typical pre-World War II submachine gun in that it is of heavy construction and is relatively expensive and difficult to manufacture. The Mark 1, a selective fire weapon, was introduced in 1941. Later in the war a model appeared capable of automatic fire only—the Mark 1*. The Lanchester was used by the British Navy and is now obsolete.

THE STEN GUNS

The Stens which are variously known as the "plumbers delight," the "Woolworth gun" and sometimes unflatteringly as the "Stench gun," introduced a new era in submachine gun design and manufacture. The Stens filled the need of the United Kingdom for an easily made, cheap weapon which did not require a large usage of scarce machine tools in their manufacture. Although the early Stens had many shortcomings, they were just as effective in killing people as were more expensive weapons. They have been given the greatest flattery by being copied in Germany, China, Argentina, Belgium, and Indonesia.

The Stens were made by the millions by a number of basic manufacturers who in turn were supported by a number of subcontractors. In the United Kingdom, the primary producers were B.S.A. and the Royal Ordnance Factory at Fazakerley. B.S.A. made over 400,000 Stens at a special plant at Tysely; some were made at their Shirley plant prior to September 1941. As subcontractors, B.S.A. had firms that made cheap jewelry, lawn mowers, hardware, children's scooters and the engineering department of a brewery among others. The gun was also extensively made in Canada.

The basic Sten gun was developed at Enfield by R.V. Shepperd

Sten Mark I.

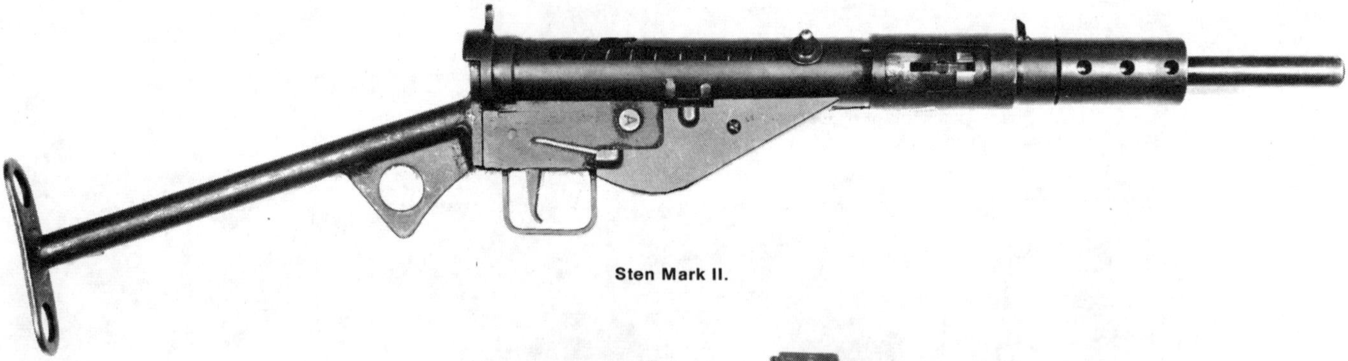

Sten Mark II.

Sten Mark 2S, fitted with silencer.

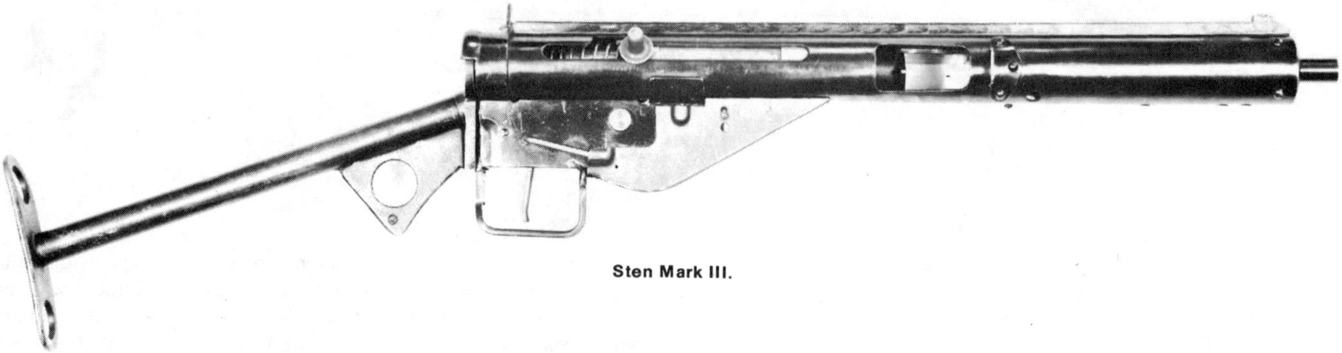

Sten Mark III.

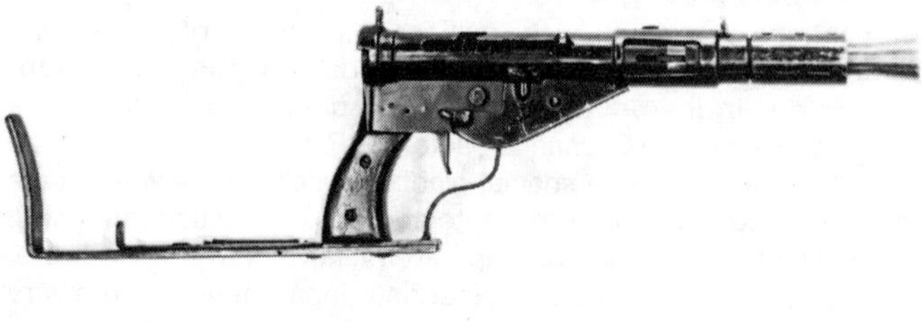

Sten Mark IV, Model A, with stock fixed.

Sten Mark IV, Model B, with stock folded.

and H.J. Turpin and its name is derived from the first letters of their last names and the first two letters of Enfield. In addition to being used by the troops of the British Commonwealth, the Sten was dropped in large numbers into occupied Europe during World War II. The later model Stens are still in extensive use throughout the world, but the Stens are no longer used as standard weapons by the United Kingdom.

Sten Mark I

Adopted in 1941, the Mark I has a complete barrel jacket, a flash hider, a wooden fore end and a vertical fore grip which can be folded up under the barrel jacket. Two basic butt stocks are used with this weapon—the No. 1 Mark I, made of steel with a wooden piece in its forward section. The No. 2 Mark II stock is made of tubular steel and does not have the wooden brace.

Sten Mark I*

A simplification of Mark I without flash hider and wooden fore end. A stamped steel housing replaces the fore end. Most of the Mark I* guns do not have a wooden fore grip.

Sten Mark II

The weapon differs from the Mark I only in externals. The barrel and barrel jacket were shortened, the design of the bolt handle was altered and a simplified buttstock was issued with this gun. The Mark II may be found with a number of different buttstocks as may all of the Sten guns. Butt stocks are interchangeable among the various models. The Mark II Sten magazine housing can be turned on the axis of the receiver so that it acts as a dust cover for the magazine and ejection ports.

Sten Mark II S

This weapon is the Mark II with a shorter barrel, silencer, a lighter bolt, and a shorter recoil spring. The weapon should only be used semiautomatic as automatic fire burns out the silencer very rapidly.

Sten Mark III

The barrel of the Mark III is not detachable as are those of other models. The receiver and barrel jacket are made of one welded steel tube and the magazine housing is welded to the receiver. The Mark III is probably the most cheaply made of the Sten guns.

Sten Mark IV

This weapon was made in two models—A and B—but very few were manufactured—about 2,000 total. The Mark IV was designed for special units and is a very compact weapon. The Model A has a pistol grip and trigger just to the rear of the magazine port whereas the Model B has the pistol grip and trigger at the rear of the receiver as do the other Stens and has the same type trigger assembly cover as does the Mark II. Both weapons have a flash hider and a very short barrel.

Sten Mark V

This is the last basic design of Sten and was the standard Sten until the adoption of the Sterling (Patchett) in 1953. The Mark V has a number of features not found on most of the earlier Stens. These are: a wooden pistol grip, a wooden stock, a front sight with protective ears (same as that of the Rifle No. 4 Mark 1), the barrel has lugs for the No. 7 Mark I and the No. 4 Mark II bayonet. Early specimens had a wooden vertical fore grip.

Sten Mark VI

This weapon is the Mark V fitted with a shortened barrel and a silencer. As with the Mark II S, automatic fire is discouraged.

Loading and Firing Sten Guns

A small special hand loader is provided as part of the equipment of every Sten gun. This is very helpful as compressing cartridges in this magazine is quite difficult due to the cartridge capacity and heavy spring.

The loader is clamped over the mouth of the magazine. The ring is pulled down as illustrated and a cartridge inserted into the mouth of the loader.

The ring is then lifted up to force the cartridge down and back under the magazine lips. It is then brought down to permit insertion of the next cartridge.

Insert loaded magazine, bullets pointing forward, into magazine housing on left side of gun just ahead of forward end of cocking handle slot. Push in until magazine locks with a click.

Pull back cocking handle and turn down into safety slot if the model is Mark I. (If model is Mark II, III, or V, the safety slot is up—so turn cocking handle up into slot.)

When ready to fire, turn cocking handle out of the safety slot.

Directly under the safety slot is a button passing through the gun from side to side. (a) If you wish to fire one shot with each pull of the trigger, push the button from the left side. (It is marked "R," meaning "Repetition.") (b) If you wish to fire full automatic, push the button through on the right side of the gun where it is marked "A," meaning "Automatic."

Note. To remove magazine press down with the left thumb on the magazine catch (which is at the rear of the magazine housing), and at the same time grasping and pulling the magazine out with the fingers of the left hand.

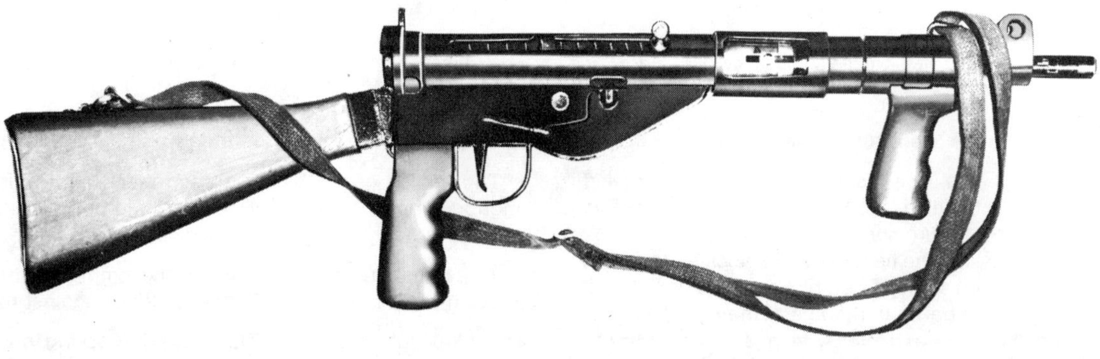

Sten Mark V, early type with fore-grip.

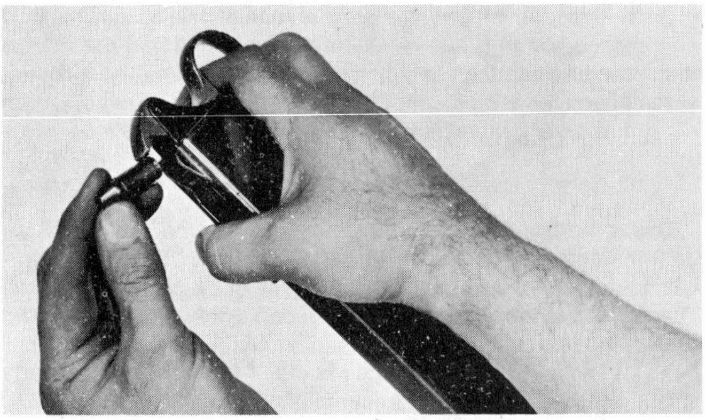

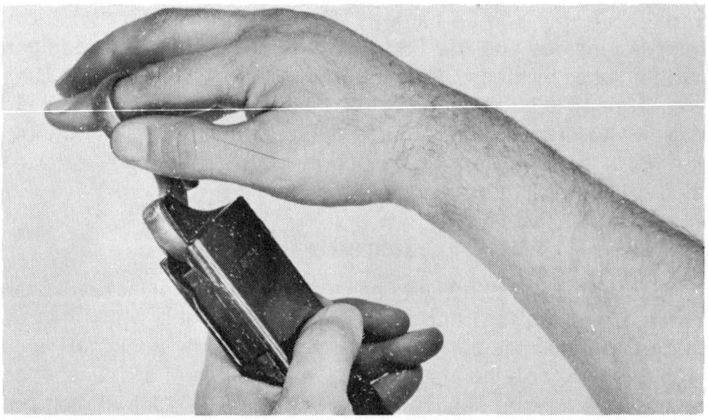

Using special Sten gun magazine loading accessory. Ring down at left to permit cartridge insertion; ring raised up at right to force cartridge down into the magazine.

Field Stripping Sten Guns

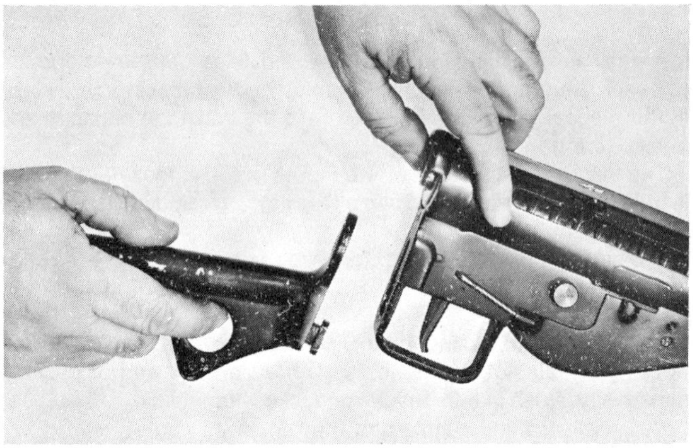

Press in the stud on the return spring housing to clear hole and slide butt down out of its slots.

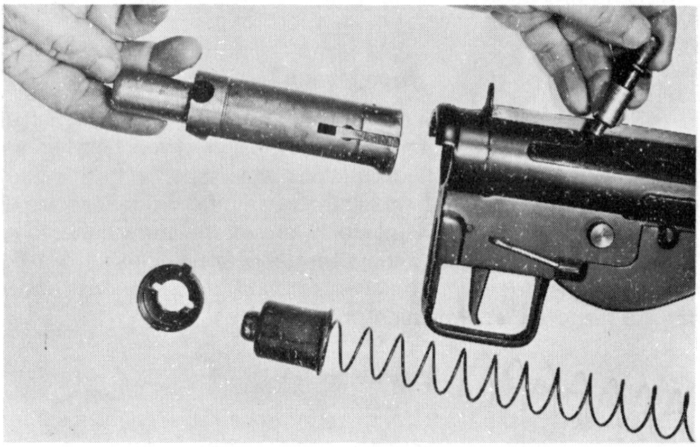

(a) Pull cocking handle back to safety slot. (b) Rotate until it can be pulled out of breechblock. (c) Tip up gun and slide out breechblock.

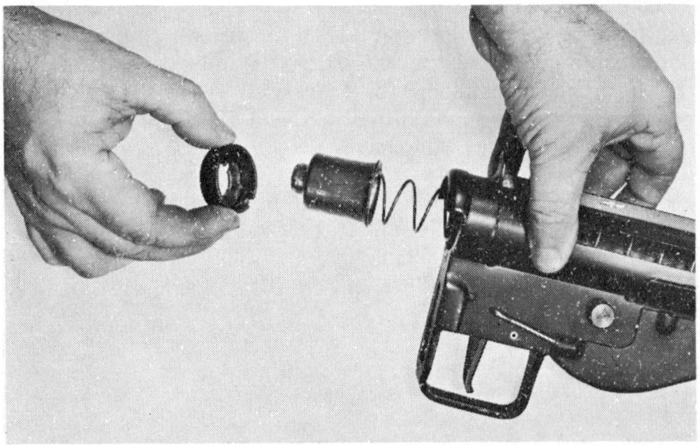

Press in on stud and spring cap and twist to the left to unlock lugs. Ease out spring cap, return spring and return-spring housing and remove.

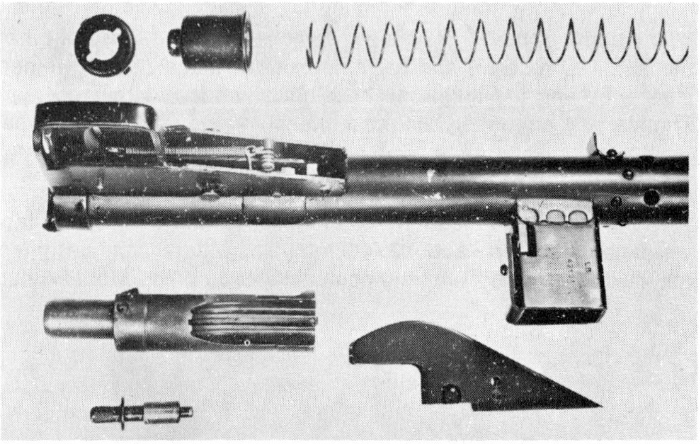

Weapon disassembled showing trigger and feeding mechanism.

How the Sten Gun Works

A loaded magazine being inserted in the magazine housing until it locks, the cocking handle is then pulled back to the cocked position, compressing the return spring.

When the trigger is pressed, the heavy breech block is freed and driven forward by the return spring. Feed ribs on the breech block strip the top cartridge from between the lips of the magazine and drive it into the chamber. The extractor, which is attached to the breechblock, snaps into the cannelure in the cartridge case and the firing pin strikes the cartridge primer exploding the powder.

The inertia of the heavy breech block and spring in forward motion keeps the breech closed until the bullet has left the barrel and the breech pressure has dropped to safe limits.

The remaining pressure drives the empty cartridge case and moving parts to the rear. The case strikes against the ejector and is hurled out of the gun. The magazine spring pushes the next cartridge in line for feeding.

CHARACTERISTICS OF BRITISH WORLD WAR II SUBMACHINE GUNS

	Lanchester Mark I	Sten Mark I	Sten Mark II	Sten Mark II S	Sten Mark III	Sten Mark IV (Model A)	Sten Mark IV (Model B)	Sten Mark V
Caliber:	9mm.	9mm.	9mm.	9mm.	9mm.	9mm.	9mm.	9mm.
System of operation:	Blowback, selective fire.	Blowback, selective fire.	Blowback, selective fire.	Blowback, selective fire.	Blowback, selective fire.	Blowback, selective fire.	Blowback, selective fire.	Blowback, selective fire.
Overall length:	33.5 in.	35.25 in.	30 in.	37 in.	30 in.	Stock extended 27.5" Stock folded 17.5"	24.5" 17.5"	30 in.
Barrel length:	7.9 in.	7.75 in.	7.75 in.	3.61 in.	7.75 in.	3.85 in.	3.85 in.	7.8 in.
Feed device:	50-rd. box magazine.	32-rd. box magazine.	32-rd box magazine.	32-rd. box magazine.	32-rd. box magazine.	32-rd. box magazine.	32-rd. box magazine.	32-rd. box magazine.
Sights: Front:	Barleycorn.	Barleycorn.	Barleycorn.	Barleycorn.	Barleycorn.	Barleycorn.	Barleycorn.	Barleycorn.
Rear:	Tangent, adj. to 600 yds.	Fixed. Aperture.	Fixed. Aperture.	Fixed. Aperture.	Fixed. Aperture.	Fixed. Aperture.	Fixed. Aperture.	Fixed. Aperture.
Muzzle velocity:	1280 f.p.s.	1280 f.p.s.	1280 f.p.s.	1280 f.p.s.	1280 f.p.s.	Approx 1200 f.p.s.	Approx 1200 f.p.s.	1280 f.p.s.
Cyclic rate:	575-600 r.p.m.	540 r.p.m.	540 r.p.m.		540 r.p.m.	575 r.p.m.	5.75 r.p.m.	575 r.p.m.
Weight:	3.62 lb.	7.8 lb.	6.62 lb.	7.48 lb.	7 lb.	7.5 lb.	7.5 lb.	8.5 lb.

Notes:
1. All weapons on this chart, use the 9mm Parabellum cartridge.
2. The fixed aperture sight on all of the Sten guns is set for 100 yds.

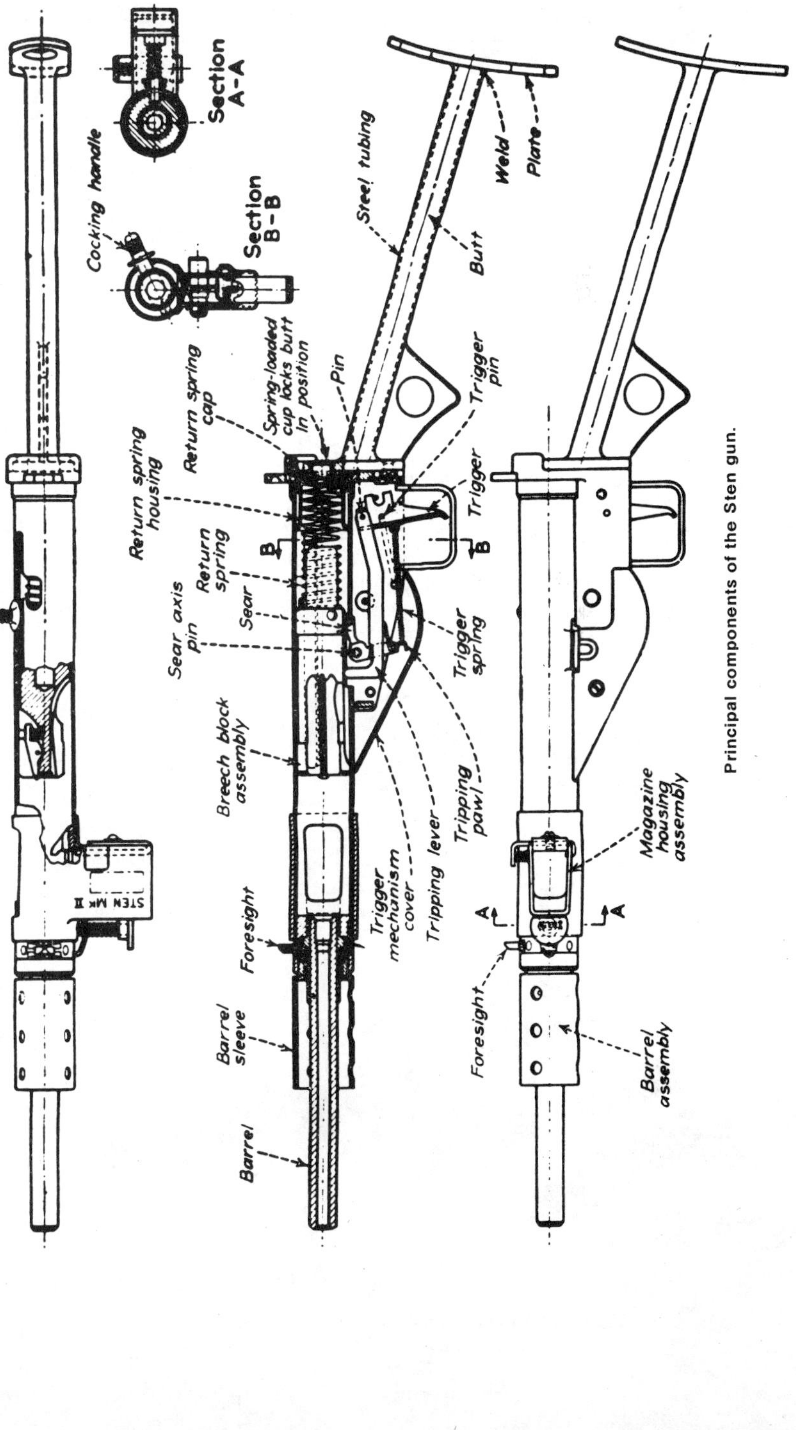

Principal components of the Sten gun.

THE STERLING (PATCHETT) GUN

This weapon was developed by G.W. Patchett, toward the end of World War II, at the Sterling Engineering Co. of Dagenham, Essex. The weapon was tested by the United Kingdom as the "Patchett" in several different forms and was chosen for extensive field test after the competitive trials held around 1949. It was issued in limited quantities in 1951 and in a modified form was issued as Submachine Gun L2A1 in 1953. The current standard model is the L2A3. The Sterling has been adopted by New Zealand, Canada, India, and a number of other countries in addition to the United Kingdom. In addition to the selective fire military version of the Sterling, there is also a semiautomatic version called the Sterling Police Carbine. This weapon was sold quite extensively to planters in Kenya during the Mau Mau uprising.

Patchett Machine Carbine

The original gun was considerably different than the weapons in service today. The buttstock was made of heavy flat steel stripping and was of different design than the later weapons. The fire selector/safety is on the front of the trigger housing as opposed to its position on the left top of the pistol grip on the later guns. The overall impression of the weapon is that it is much heavier than the later guns.

L2A1 Submachine Gun

The basic difference between this weapon and the later L2A2 is that the L2A1 had parts—grip screw and cocking (bolt) handle—which could be used for the removal of the barrel screws and the inner block of the bolt had an extension for removal of the extractor pin.

9mm Sterling Police Carbine Mark 4.

L2A2 Submachine Gun

Parts were not used as stripping tools, a forward finger guard was added, the rear sight was modified by repositioning of the sight flip-over lever and increasing the size of the 100 yard aperture. The butt was strengthened, a fouling plunger added to the bolt to prevent improper assembly and the chamber was modified to feed under adverse conditions.

British 9mm Submachine Gun L2A1.

L2A3 Submachine Gun

This is the standard service gun and differs from the L2A2 as follows: rear sight flip lever has been deleted, the butt has again been redesigned and made as a complete stamping rather than as a fabrication with the butt plate indexing and the position along barrel jacket changed. The chamber was modified to the NATO standard and the trigger guard made removable. The previous model had a special Arctic trigger which was mounted on the trigger guard. An early model of the Sterling Gun had a folding bayonet; all later models have a bayonet boss and stud for a knife type bayonet.

Characteristics of L2A3

System of operation: Blowback, selective fire.
Weight: Unloaded, w/o bayonet: 6 lb.
Loaded, w/bayonet: 8.25 lb.
Overall length: Stock extended: 28 in.
Stock folded: 19 in.

The L2A3 — field-stripped.

L2A3 — stock folded.

Barrel length: 7.8 in.
Feed device: 34-round, detachable, staggered box magazine.
Sights: Front: Blade w/protecting ears.
Rear: Flip-type aperture, graduated for 100 and 200 yd.
Muzzle velocity: 1280 f.p.s. w/British 9mm service ball.
Cyclic rate: 550 r.p.m.

How to Load and Fire the L2A3

Pull the cocking handle to the rear; the bolt will remain to the rear, since the weapon fires from an open bolt. Engage the safety by turning the change lever (located on the left side of the pistol grip) to the letter "S." Insert a loaded magazine in the magazine guide, checking to insure that it locks in place. Move change lever to letter "R" for semiautomatic fire, or letter "A" for automatic fire. Squeeze the trigger and the weapon will fire.

Field Stripping the L2A3

Elementary Stripping. Before stripping, insure that the weapon is not loaded, and remove sling if fitted. Set change lever to "A"; place butt in the folded position and bolt forward.

To Remove Return Spring and Bolt. Press back-cap catch for full depth. Push back-cap forward and rotate counterclockwise until locking lugs disengage from locking recesses. Remove back-cap and draw cocking handle to rear of weapon. Lift cocking handle outward and withdraw return spring assembly from rear of receiver. Remove bolt from rear of receiver. Reassemble in reverse order. The spring-loaded fouling pin will prevent misassembly, since the cocking handle cannot be inserted until this pin is pushed forward by the center pin on the spring assembly. This ensures that the cocking handle must pass through the hole in the center pin.

To Remove Trigger Group. With a small coin or the rim of a cartridge, turn the slot in the head of the trigger group retaining pin until it is in line with the word "FREE" on the right side of the pistol grip. With the nose of a bullet or the blunt end of the cocking handle, push the trigger group retaining pin out and remove. Press the trigger, and pull the trigger group toward rear of weapon, disengaging it from the step in underside of barrel case; then swing front of trigger group out and remove from receiver.

Note: Elementary stripping does not include any further stripping of trigger group.

Assembly

Assemble in the reverse order of stripping.

How the L2A3 Works

The L2A3 is a blowback-operated weapon, and its bolt operation is essentially the same as that of most other blowback-operated guns. The action of the trigger mechanism differs among these weapons, however, and accordingly is described below in detail for the L2A3.

Semiautomatic Fire. When the change lever is set to the single shot position "R," the inner arm of the change lever is located under the tail of the tripping lever.

When the weapon is cocked and the trigger is pressed, the sear cradle is rotated about the sear axis pin, the sear is lowered, and the bolt is carried forward by the pressure of the return spring. During this movement the tail of the tripping lever contacts the inner arm of the change lever, causing the tripping lever to partially rotate. Continued pressure on the trigger causes further rotation of the tripping lever until the upper arm disengages from the step on the sear. At the same time the sear plunger and spring are compressed.

When the round is fired, the sear is held down by contact with the undersurface of the bolt, but as the bolt reaches the end of its rearward movement and is clear of the sear, the sear is forced upward by pressure of the sear plunger and spring. Then, as the bolt is moving forward, the sear engages against the notch on the face of the bolt and holds the bolt in cocked position.

When the pressure on the trigger is released, the rear end of the sear cradle rises, lifting the tripping lever and causing it to rotate about its axis pin until the upper arm of the tripping lever re-engages the step of the sear.

The trigger must be fully released and again pressed for each single shot.

Automatic Fire. With the change lever set at automatic fire position "A" and the weapon cocked, when the trigger is pressed the projection on the upper part of the trigger lifts the end of the sear cradle, rotating it about its axis pin. This action depresses the sear, freeing it from contact with the face of the notch on the bolt, and allowing the bolt to fly forward. The movement of the sear cradle compresses the sear cradle spring.

The weapon will now continue firing until either the trigger is released or the magazine is empty.

When the trigger is released, the sear cradle returns to its former position under the action of the sear cradle spring, and the sear is raised into the boltway where it engages the notch on the bolt, to hold the bolt in the cocked position.

Applied Safety. When the weapon is cocked, and the change lever is set at the safe position "S," the inner arm of the change lever is positioned directly under the short arm of the tripping lever. When the trigger is pressed, the sear cradle and sear cannot be depressed because the short arm of the tripping lever is held immovable by the inner arm of the change lever.

When the bolt is forward, and the change lever is set at the safe position, the weapon cannot be cocked because the sear is engaged in the safety slot at the rear of the bolt, and the sear cannot be depressed because it is held immovable as described in the previous paragraph.

The Butt Mechanism. To open butt, hold the weapon with the left hand near the rear sight, with the barrel pointing toward the ground. Pull the butt plate outward with the right hand to release the butt catch, and swing the butt to the rear of the weapon. With the thumb of the left hand press the back-cap catch and snap the butt into engagement with the lugs on the back-cap. Open the butt frame to form a triangle and the butt catch will engage, to lock.

To close butt, release the butt plate catch and collapse the triangle by pushing the tubular member into the frame. With the thumb of the left hand, press the back-cap catch; at the same time, push the back-cap forward and swing the butt away from the back-cap. Pivot the butt to its folded position, swing the butt plate out to operate the butt catch, to engage in the barrel casing, then fold the butt plate flat to lock in position.

L34A1 Submachine Gun

This weapon is a silenced version of L2A3. The barrel jacket is covered by a silencer casing, which is supported by front and rear supports. The barrel has gas escape holes throughout its length and is threaded at the muzzle. The barrel has a metal wrap and diffuser tube; the extension tube extends beyond the silencer casing and barrel. Beyond the barrel is a spiral diffuser; this is a series of discs and is held in place by tie rods which run from the end cap at the muzzle to the front support. The spiral diffuser has a hole through its center to allow passage of the bullet. L34A1 uses the standard British 9mm Parabellum cartridge.

British 9mm L34A1 Submachine Gun.

Characteristics

Caliber: 9mm Parabellum

System of operation: blowback, selective fire

Weight: approx. 8 lb.

Overall length: Stock extended: 34 in.
Stock folded: 26 in.

Barrel length: 7.8 in.

Feed device: 34-round, detachable, staggered row, box magazine

Sights: Front: blade w/protecting ears
Rear: "L" type w/apertures

Muzzle velocity: approx. 1200 f.p.s.

Cyclic rate: 550 r.p.m.

Special Note on British Submachine Guns

Prior to the mid-fifties the United Kingdom called submachine guns—"Machine Carbines," since that time they have adopted the same terminology as the U.S., but in Britain it is slightly differently arranged - "Sub-machine gun."

BRITISH MACHINE GUNS

Britain adopted her first true machine gun—the .450 Maxim—around 1891 and the Maxim in one form or another was used by British forces until quite recently. During World War I the .303 Vickers, the .303 Lewis, and the .303 Hotchkiss were the main machine guns. Between the wars the British looked for a replacement for the Lewis gun; the replacement was found when the Bren gun was adopted in 1935. Manufacture of the Bren started slowly, however, and did not really gain volume until World War II started.

As with all other small arms, Britain was short of machine guns during World War II and 87,000 machine guns from U.S. war reserves were sold to Britain in 1940. The following weapons—all caliber .30—were in the 1940 shipments:

1,157 M1917 Lewis ground guns
7,071 M1915 Vickers ground guns
2,602 M1918 Marlin tank guns
15,638 M1917 Marlin aircraft guns
5,124 Vickers aircraft guns
38,040 Lewis aircraft guns
10,000 M1917 Browning ground guns

These guns were used by the Home Guard and to some extent by the Merchant Marine, who used the stripped-down Lewis for defense against low level air attack. British forces in the field used

Prototype B.S.A. 7.62mm NATO General Purpose Machine Gun on tripod.

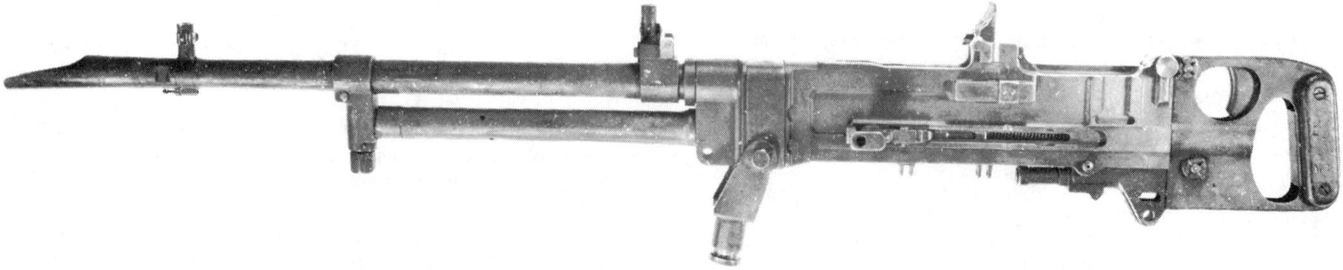

Vickers .303 gas-operated (G.O.) machine gun.

Prototype B.S.A. 7.62mm NATO Machine Gun on bipod.

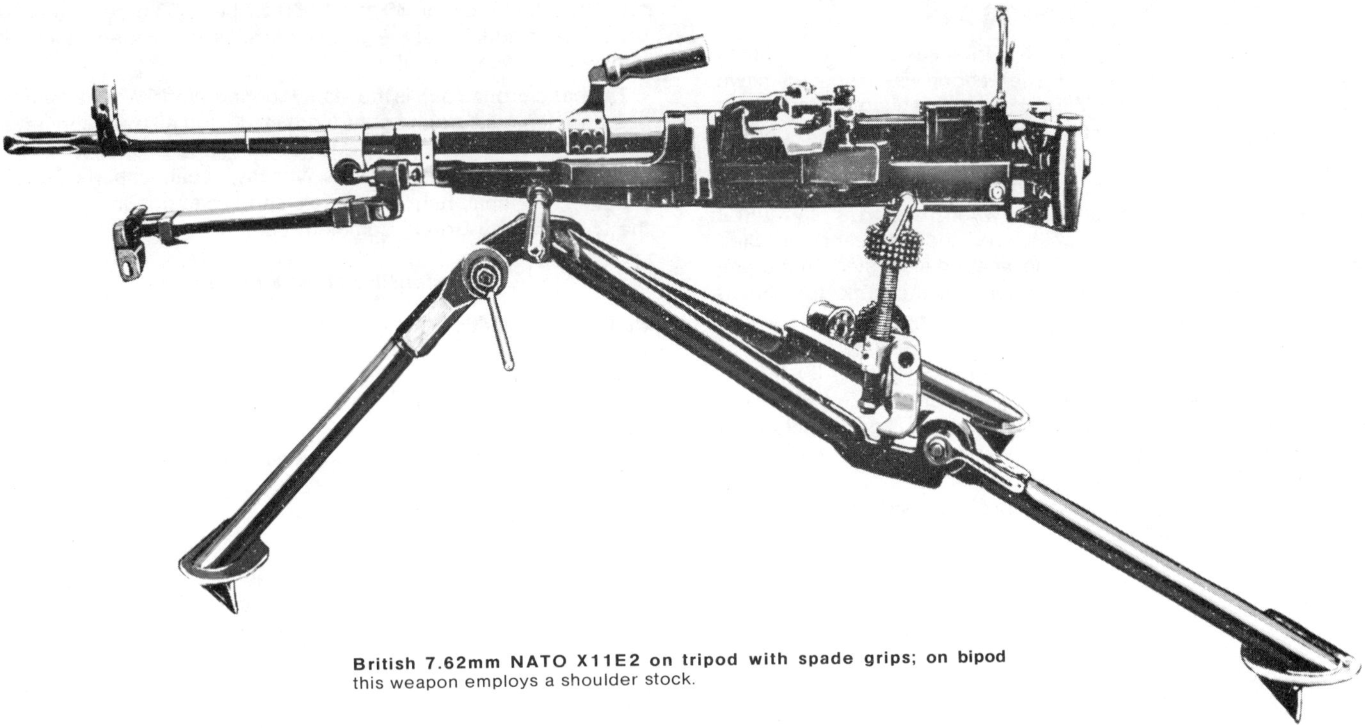

British 7.62mm NATO X11E2 on tripod with spade grips; on bipod this weapon employs a shoulder stock.

the Mark 1 Vickers, the Bren, the Besa and, to a limited extent, the Vickers gas-operated Mark 1 which, although designed as an aircraft gun, was used as a vehicular gun. All the rifle caliber machine guns used in the field were caliber .303 except for the 7.92mm Besas.

After World War II Britain looked for a new machine gun to replace the Vickers and, if possible, the Bren. Although the Vickers was a reliable and proven weapon it was overly heavy and bulky and not as tactically flexible as modern general purpose machine guns. Two of the guns tested by the U.K. were the B.S.A. general purpose machine gun and the Enfield developed X11E2. In basic design both of these guns are quite similar. Their basic design is based on that of the ZB26-Bren family of weapons, excepting the fact that they are belt-fed rather than being magazine-fed. It is interesting to note that the forerunner of the ZB26, the Praga 24, was also belt-fed.

After extensive trials conducted around 1957, the U.K. decided to adopt the FN 7.62mm NATO MAG machine gun. According to accounts appearing in British papers at the time, the MAG was adopted because it was the best available gun then in production. For reasons of economy the British Government did not want to pay, at that time, for the industrial engineering and tooling up necessary with the developmental British weapons. Be that as it may, the MAG which was tested as the X15E2, was adopted as the L7A1 machine gun.

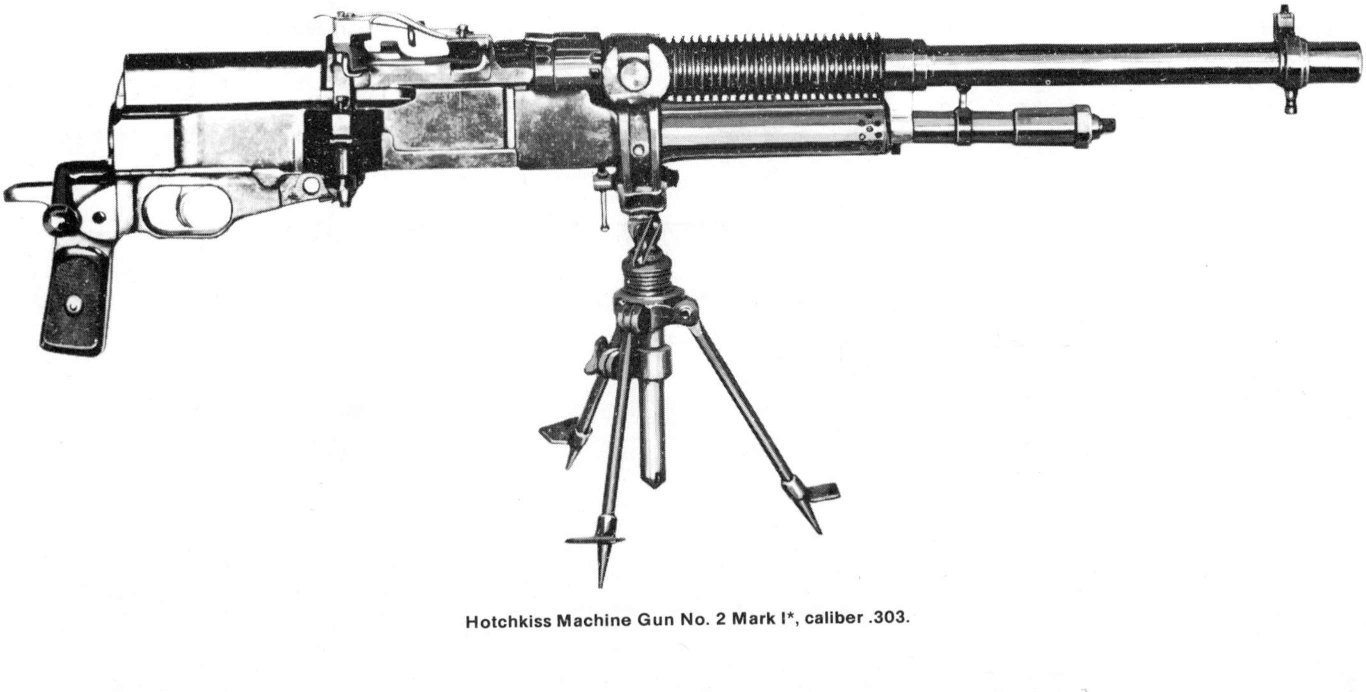

Hotchkiss Machine Gun No. 2 Mark I*, caliber .303.

THE HOTCHKISS GUN

The Hotchkiss was widely used by Britain as a cavalry and tank machine gun during World War I. The weapon was produced in two models, the Mark I and the Mark I* (later called No. 1 Mark I and No. 1 Mark I*). The principal difference between the two guns is that the Mark 1 gun can only use feed strips while the Mark I* can use feed strips or a belt. When used as a cavalry gun the weapon has a wooden butt and small tripod; when used as a tank gun a pistol grip and different rear sight is used and the weapon is called No. 2 Mark I or No. 2 Mark I*. The weapon illustrated on the preceding page has the pistol grip used in the tank gun, but is mounted on the tripod used with the cavalry gun.

The British Hotchkiss is much like the U.S. caliber .30 M1909 Benét Mercié Machine Rifle and the French 8mm M09/13 light machine gun. The belt used by the No. 1 Mark I* gun is actually a series of articulated feed strips; a similar belt was developed for use with the French 8mm M1914 Hotchkiss.

The British Hotchkiss guns are currently obsolete and are unlikely to be encountered in the field.

Characteristics of the Hotchkiss

Caliber: .303 British.
System of operation: Gas, selective fire.
Weight: No. 1 Mark I 27.25 lbs.
No. 2 Mark I* 26 lbs.
Length, overall: 46.75 in.
Barrel length: 35.5 in.
Feed device: Metal strips holding 9, 14, or 30 rounds and on MK I* articulated strip belt holding 50 rounds.
Sights: Front: Barleycorn.
Rear: No. 1 guns-leaf, No. 2 guns-tubular.
Muzzle velocity: Approx. 2,500 f.p.s.
Cyclic rate: 550 rounds per minute.

THE LEWIS MACHINE GUN

First produced in quantity in Belgium, the Lewis gun was the principal light machine gun of the British Army in World War I and was used by the Home Guard and Merchant Marine (for defense against low-level air attack), in World War II. It was also used—in caliber .30—by the U.S. Marine Corps and Navy until World War II. The U.S. gunboat Panay, which was sunk by Japanese bombers in China during 1937, had Lewis guns as part of its antiaircraft armament as did many other U.S. naval vessels.

The Lewis was made in large quantities during World War I and a few were assembled by Savage Arms early in World War II. B.S.A. made 145,397 guns at Small Heath; Savage Arms Corporation of Utica, N.Y., produced Lewis guns for the U.K. and Canada and produced 2,500 caliber .30 and 1,050 caliber .303 ground guns for the U.S.A. In addition to the ground guns, large numbers of aircraft guns were made as well.

The basic ground gun is the Mark I; during World War I a number of different model ground guns were made, but all were converted to Mark I after the war. The Lewis gun, in one form or another, was used by France, the Netherlands, Norway, Japan, Imperial Russia, Belgium, Portugal, Italy, Honduras, and Nicaragua in addition to the U.S. and the British Commonwealth.

Characteristics of Mark I Lewis Gun

Caliber: .303 British.
System of operation: Gas, automatic fire only.
Weight: 27 lbs.
Length, overall: 50.5 in.
Barrel length: 26.04 in.
Feed device: 47 round drum magazine, a 97 round drum designed for aircraft use also exists.
Sights: Front: Barley corn.
Rear: Leaf w/aperture.
Muzzle velocity: 2440 f.p.s.

During World War II many of the caliber .30 aircraft Lewis guns that were sold to the U.K. by the U.S. were converted to ground use for the Home Guard. An aperture sight fixed for 400 yards was mounted on the rear aircraft gun sight base and either the standard ground gun wooden butt or a steel skeleton stock with wooden cheek rest were substituted for the aircraft type spade grip. They initially had no mounts and were to be laid over walls or fired from the hip; a non-telescoping bipod was later issued. During the same time a number of British ground guns were modified for antiaircraft use on ships. The radiator casing and radiator were removed, the butt was shortened by two inches, a forward hand grip was added, and a light steel guard was fitted over the gas cylinder.

Although the Lewis, whose basic design was actually the work of Samuel Neal McClean, is no longer in service in any quantity, its memory lingers on in the operating mechanism of the U.S. M60 machine gun. A conventional operating spring has replaced the clock-type operating spring, but the bolt operation is the same.

THE VICKERS MACHINE GUN

The Vickers, originally called the Vickers Maxim, was adopted by Great Britain in 1912. It was their principal heavy rifle-caliber machine gun in both world wars and was a standard weapon until the adoption of the L7A1 general purpose machine gun in the early sixties. The Vickers, which is a modified Maxim gun, has the reputation of being one of the most reliable and rugged machine guns ever built. The weapons used by the U.K. were made by

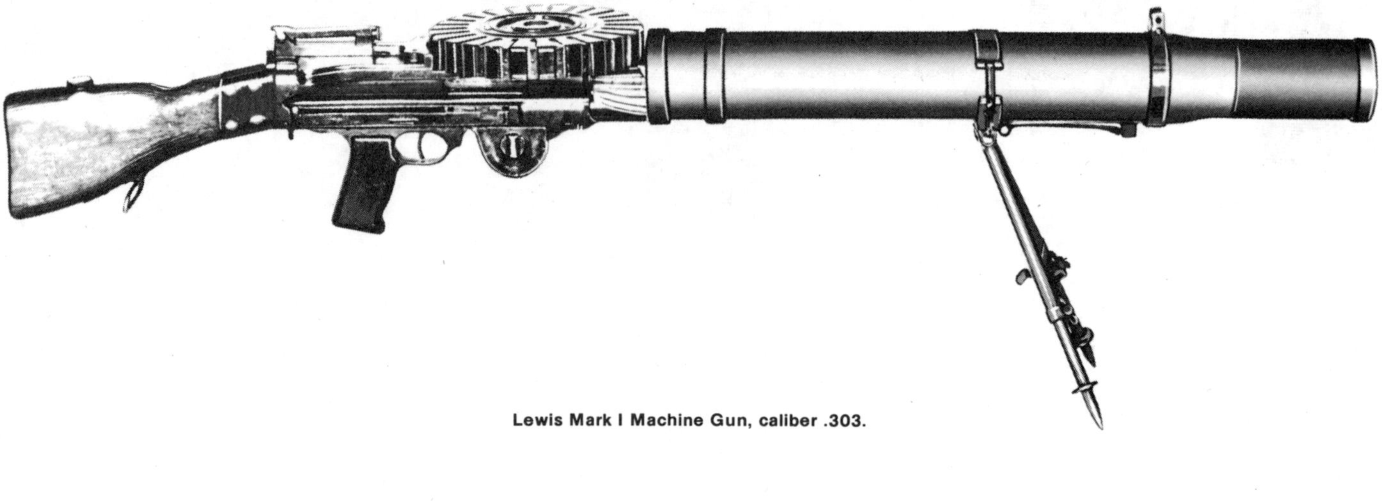

Lewis Mark I Machine Gun, caliber .303.

British .303 Vickers Machine Gun Mark I on tripod, Mark IVB, with ammun-tion box and steam condensing assembly.

Vickers at Crayford, Kent, in both ground and aircraft versions. The U S. had bought some Colt-made Maxims—the Model 1904—in very limited quantities prior to World War I. Colt tooled up to produce the Vickers in World War I; and in caliber .30, it was adopted as the U.S. Machine Gun Model 1915. Due to the emergence of the Browning in 1917 and the limited quantities of Vickers Colt was able to produce (it was a difficult gun to make), U.S. troops received few Vickers ground guns during World War I. Those on hand in 1940 were sold to the U.K. and were used by the Home Guard. These caliber .30 weapons had a red stripe painted on the receiver, the mouth of the feed block and on the side lever, to distinguish them from the .303 Vickers. The U.K. made all the Vickers ground guns needed for their forces in both wars.

The weapon can be found with two different water jackets (barrel casings): a corrugated type and a smooth-surfaced type. The smooth-surfaced type is made of slightly heavier metal than the corrugated type. The feed block bodies may be made of either steel, gun metal (bronze), or gun metal with steel strips.

The Vickers may still be found in use throughout the former British territories and is probably still held in reserve in the United Kingdom. It is likely to be found in service in various odd areas of the world for some years to come.

Characteristics of Vickers Machine Gun

Caliber: .303 British, Mark 8z ball normally used.
System of operation: Recoil with gas boost from muzzle booster, automatic only.
Weight:
gun w/o water---33 lbs.
gun w/ water---Approx. 40 lbs.
tripod---50 lbs.
Overall length: 43 in.
Barrel length: 28.4 in.
Feed device: 250-round canvas belt.
Sights: Front: Hooded blade.
Rear: Leaf with aperture, 400 yard battle sight.
Muzzle velocity: 2440 f.p.s.
Cyclic rate: 450-550 r.p.m.

Because of the differences in the barrel erosion characteristics of the .303 Mark 7 ball (cordite loaded) and .303 Mark 7z or .303 Mark 8z ball (single-base nitrocellulose loaded), barrels should be used with either the Mark 7 round or the others—not both. British experience in World War II indicates barrels which have been used with Mark 7 cartridges should NEVER be used with Mark 7z or Mark 8z cartridges for OVERHEAD fire because of erratic wear pattern.

Loading the Vickers Gun

Under normal firing conditions, the rear leg of the tripod will be aligned with the target. The gunner sits behind the gun to the rear of this leg with his legs on either side of the tripod. The knees are drawn up slightly so that the elbows can rest of the inside of his thighs while his hands grasp the traversing handle.

Proper Hand Position. Both thumbs are rested lightly on the thumb trigger. The forefingers are wrapped around the top of the handle. The second fingers are placed underneath the ring safety catch. The other two fingers of each hand grasp the traversing handles firmly but without strain.

To Load. See that ammunition box is placed on right side of gun directly below the feed block.

If the gun is equipped with a shutter, open the shutter.

Vickers .303 Machine Gun Mark I with smooth water jacket.

Pass the brass tag-end of the belt through the feed block from the right side and grasp it firmly with the left hand.

With the right hand pull the crank handle back on its roller as far as it will go, and while holding it in that position, pull the belt sharply through the feed block with the left hand as far as it will go.

Release the crank handle and let it fly forward under the influence of the spring. This action grips the first cartridge firmly between upper and lower portions of a gib at the top of the extractor. Now pull the crank handle back on its roller once again. Give the belt another sharp tug to the left as far as it will go, and again let the crank handle fly forward under the influence of the spring. This action withdraws the cartridge from the belt, places it in the chamber ready for firing, and grips the next cartridge by the gib in the upper part of the extractor.

The gun is now cocked and ready to fire, whenever the safety catch is lifted and the trigger pushed in.

Note on Unloading. Because of the method of feeding, safely unloading this weapon requires special consideration. Without touching the belt, pull the crank handle back onto the roller as far as it will go and release it.

Again pull the crank handle back as far as it will go and permit it to fly forward. The first motion of the crank handle extracts the cartridge from the firing chamber, and drops or ejects it through the bottom of the gun. The cartridge in the feed block is withdrawn for positioning by this movement, and is fed into the chamber, but as the belt is not moved across, no new cartridge is gripped by the top of the extractor. Thus when the crank handle is run back for the second time, the second live cartridge is dropped through the bottom of the gun, leaving the firing mechanism empty.

With the left hand raise the finger plate of the bottom pawls and simultaneously push down the top pawl by squeezing the pawl grips. Keeping the pawls disengaged, pull the belt out of the block to the right. The pawls hold the belt in position in the feed block, the top pawls being behind the first cartridge and the bottom pawls behind the second. During recoil of the gun the top pawls feed the cartridge into position while the bottom ones prevent any backward lash of the belt; thus it is necessary to release the pawls from their position before the belt can be pulled out of the gun.

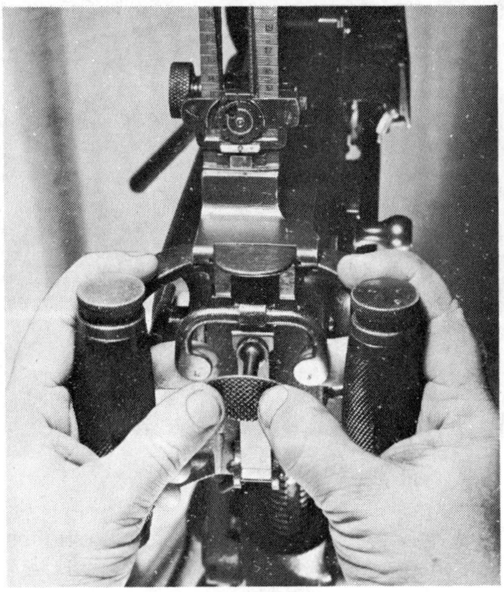

Illustrating correct hand position for firing.

Firing the Vickers Gun

With forefingers wrapped over the top of the traversing handle raise the safety catch with the second fingers of the hands, wrap the other fingers around the traversing handles, and with both thumbs press in on the thumb trigger.

The gun will now fire as long as the trigger is kept pushed in and cartridges are fed into the gun. Releasing either trigger or safety will stop the gun.

Remember that this gun is supported by the tripod, and that the hands are intended only for use in firing. Thus no particular effort is required on the part of the gunner.

How the Vickers Gun Works

Starting with the gun loaded and cocked the action is as follows: The safety catch is normally held down by a spring which also holds the firing lever to the rear. This catch prevents any forward movement of the firing lever while it is in safe position. When this catch is raised by the second finger, it clears the way for the thumb piece of the firing lever to push the lower end of the trigger bar forward. The trigger bar lever is engaged with the trigger bar in the rear cover and as the lever moves back it draws the trigger bar to the rear also.

The forward end of the trigger bar is in engagement with the trigger situated in the lock; and as the trigger moves to the rear, it releases the striker, contained in the lock, from its spring and the striker is driven forward to fire the cartridge in the chamber.

When the cartridge fires, the bullet is driven down the barrel and the locked barrel and locking mechanism start rearward.

As the bullet leaves the barrel, the gases behind it expand in the muzzle attachment. Some of these gases strike against a cone which surrounds the muzzle and rebound to strike the front face of the muzzle cap cup, fastened over the muzzle.

Thus it will be seen that the rearward action of this gun is brought about by two forces: (1) by recoil (the rearward thrust of gases in the barrel against the cartridge case and the lock as the bullet is forced ahead); and (2) by the effect of the rebounding gases after the bullet has left the muzzle, giving an added backward push against the muzzle cup.

On the rear of the left side plate is a protruding metal box. A powerful fusee spring is attached to the front end of this box. The rear end of the box is locked to the body of the gun. At the front end of this box is the fusee which is attached to the fusee spring. The rear of this spring, being attached to the fusee, can be drawn straight backwards, extending the spring and storing up energy to provide the return movement of the action. The recoil forces the tail of the crank handle to roll against its roller and rotate the crank, which is attached to the fusee. This winds the fusee chain and extends the fusee spring while the lock is traveling to the rear. The sharp backward thrust caused by the recoil forces the lock, crank, and crank handle to move back as the crank handle continues to roll against the roller.

While the mechanism is moving backward, a stud on the bottom lever of the feed block located in a recess in a prolongation of the left side plate is forced to the rear, taking with it the bottom lever, which being connected with the top lever carries the bottom one over to the right, thus causing the feed block slide to move over to the right. This movement causes the top pawls in the feed block to drive to the right and slip over and behind the next cartridge in the belt, which is being held in place by the bottom pawl.

Meanwhile the lock had been moving backwards. The extractor attached to it removes the loaded cartridge from the belt at the same time it draws the empty cartridge case from the chamber. Horns on this extractor ride along the top of solid cams in the breech casing sides; and as the cartridge is drawn clear of the belt, the horns clear the ends of these cams. Rims in the rear cover force the extractor downward and bring the live cartridge into line with the chamber. During this downward movement of the extractor the empty cartridge case usually drops out. If it fails to, it will be ejected during the forward movement of the extractor. The loaded cartridge is held firmly in position in the extractor by the gib which has a bottom projection to prevent the cartridge from slipping down out of the extractor face.

During recoil the backward rotation of the crank moves the connecting rod and side lever head upwards. The side lever head bears on the tail of the tumbler and rotates it, thrusting the firing pin to the rear.

Further rotation of the tumbler in the lock completely withdraws the firing pin as the long arm of the lock spring bears against the projection of the pin. Thus the lock spring is compressed until the trigger nose, forced by the short arm of the lock spring, is pushed over the bent of the tumbler. The firing pin is withdrawn still farther until the sear spring forces the bent of the sear into the bent of the firing pin which thus holds it in cocked position.

It should be noted that in this weapon part of the action attributable to backward movement is actually to start some parts forward. The crank handle, continuing to roll against the roller during recoil movement, actually starts the recoiling portions forward while the lock is still moving backwards.

Return Movement of the Action. The force of the recoil having expended itself, the stretched fusee spring now reasserts itself, unwinds the fusee chain, moves the link to rotate the crank in the forward direction and forces the connecting rod and side lever head to drive the lock forward.

The stud of the bottom lever of the feed block is carried forward in its recess in the prolongation of the side plate, moves the bottom lever of the feed block forward, thus causing the top lever and the slide to move over to the left. As the pawls move and are gripping the next cartridge in the belt, the loaded cartridge is moved into position against the cartridge stops, ready to be gripped by the extractor on the next rearward movement.

As the lock is driven forward, the extractor supported by the gib is carrying the cartridge into the firing chamber.

The extractor is now raised and its levers are pushed by the side levers; the gib is depressed against its spring, thus letting go its hold of the cartridge as the round is chambered and the gib is forced back into the face of the extractor.

The upper end of the extractor slips up around the rim of the cartridge in the feed block and the gib is pushed forward by its spring to grip the head of the cartridge to place it in proper position in the extractor when the next rearward motion of the lock will draw it out of the feed block. Springs located in the side plates engage in slots in the side of the extractor to hold it in its highest position to prevent it from falling should there be no cartridges left in the belt.

If the empty cartridge case has not dropped off the extractor face, it is ejected as the extractor rises during the forward movement by striking against the ejection feeding in the barrel casing.

When the lock approaches its fully forward position, its side lever head is forced slightly below the horizontal by the connection rod. It now depresses the sear, disengaging it from the firing pin, and allowing the firing pin to move forward slightly so that the trigger nose engages the bent in the tumbler.

If pressure on the thumb pieces is maintained for automatic fire, the trigger nose is held clear of the bent in the tumbler. The firing pin is free to spring forward under compression of the lock spring when the sear is depressed by the side lever head. However, it should be noted that the depression of the sear is so arranged that the firing pin cannot possibly be released until the lock is fully home and in firing position.

Locking Principle. This gun is locked securely at the moment of firing by a toggle joint. This is the principle developed by Hiram Maxim. The Vickers gun is a modification of the Maxim gun. This locking principle was used in Maxim guns throughout the world, notably in Germany and in Russia.

The simplest way to explain this principle is to compare it with the human knee.

When in firing position, the lock on the Vickers gun fits securely against the firing chamber. Now picture the human foot with the heel held firmly in the position of this lock against the head of the cartridge. Pivoted to the lock is the connection rod, a heavy metal bar, thrust straight forward. This connection rod is like the lower part of the leg but it can buckle at the ankle where it joins the foot. The crank is attached to the connection rod by a hinge pin and extends to the rear. This crank forms a bending knee where it joins the connection rod; it resembles the upper part of the human leg. However, the knee in this mechanical device is actually below the line of the connection rod and crank.

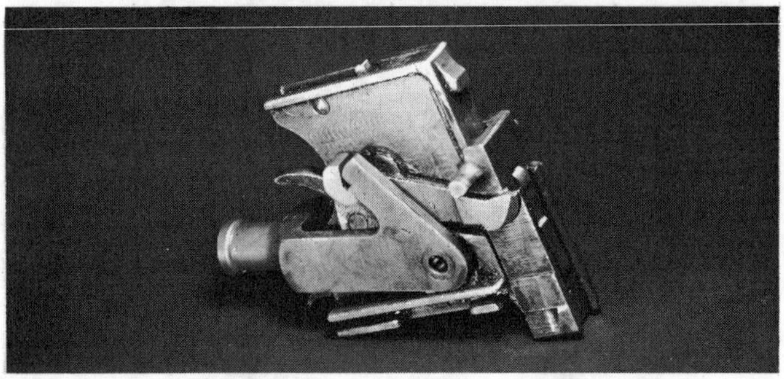

Vickers lock.

This crank is rigidly supported from below by the inside plates of the weapon and pressure applied to it by the side levers, during the opening movement of the recoil, merely presses the crank down harder on the plates.

Attached to the crank is a crank handle which travels back with it and after the gun has recoiled far enough to permit the bullet to leave the barrel, the tail on the lower side of this handle is forced back in contact with a roller which causes the crank handle to rotate upwards. This raises the axis of the crank pin and permits the knee-like joint to buckle. (Thus, as the human foot is driven backwards, pressure applied to the underside of the knee will buckle the knee but draw the foot straight back.)

The connection rod is locked securely by a twisting motion inside the side lever head, which projects beyond the lower rear end of the lock. As the connection rod buckles, it naturally raises the side lever head with it, and this raises a tumbler which cocks the lock.

Field Stripping the Vickers Machine Gun

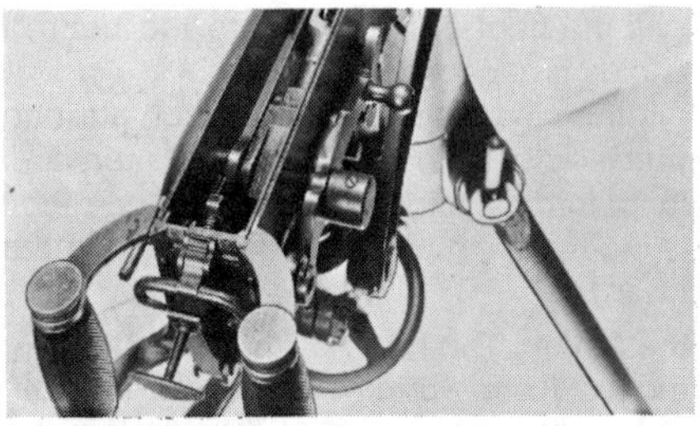

At the rear of the gun above the safety is the rear cover catch. It is held in place by a spring. Push up on the catch and raise the cover up as far as it will go. Now pull back the crank handle

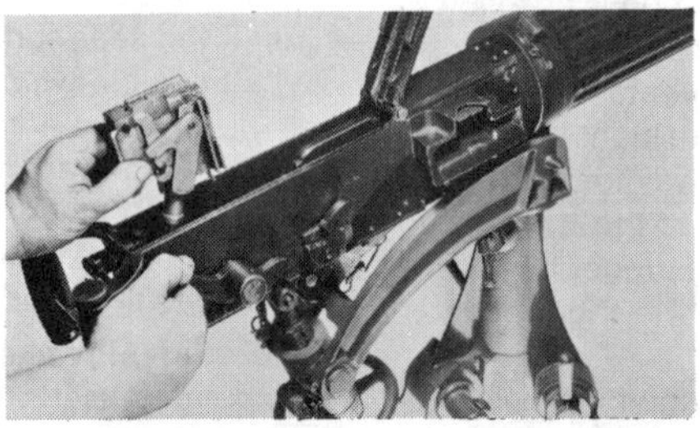

against the tension of the fusee spring. Hold it firmly. Reach inside the gun and lift out the lock which is fastened to the connection rod. Now twist the lock on the connection rod about one-third of a turn to the right, to release it from the

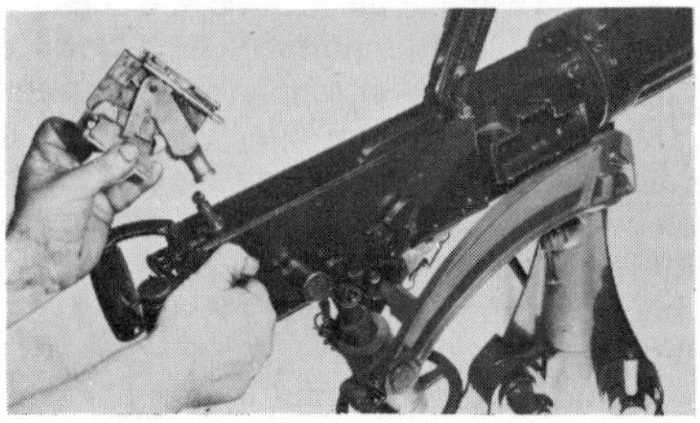

connection rod, which in its turn is connected to the crank. Lift the lock out, ease the crank handle home under the tension of the fusee spring. Then close the cover.

Turn the cover latch (on the forward end of the cover on the left side of the gun) up to the left as far as it will go. This releases the front cover which should now be raised as high as it will go.

Now lift complete feed block directly up and out of the gun. Go to the forward end of the gun. Pull out the split pin and twist the outer casing through about one-sixth of a turn. It can now be pulled off to the front. The muzzle cup and the front cone may also be unscrewed and removed.

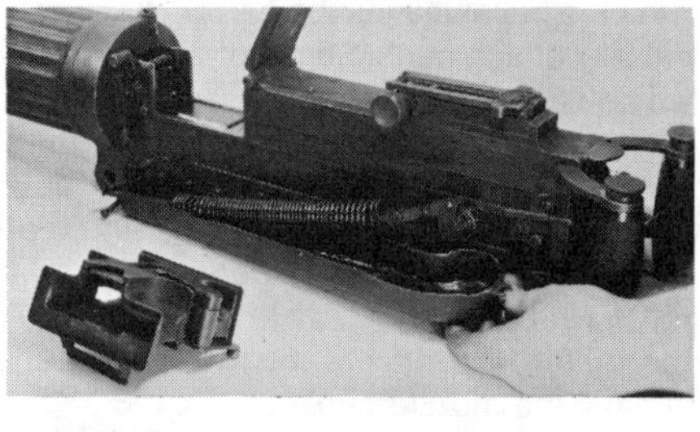

(The gland and packing should be removed only if absolutely necessary.)

Grasping the front end of the spring box with the left hand, push forward on the rear end with the right hand until the hooks which fasten the box at front end and rear can be sprung out of their studs. Disconnect the box from the gun and unhook the fusee spring from the fusee. The fusee may now be turned until its lugs are free and then it can be withdrawn with its chain from the left.

Now lift the rear cover and unscrew the large key pin protruding from the left side at the rear of the gun. Pull this pin out to the left, and it will permit the handles and their enclosed mechanism to be swung down to a horizontal position. Slides which travel in the body at the rear may now be pulled straight out. The right slide carries the roller with it. Now pull the crank handle stem directly to the rear which will withdraw the crank together with the right and left inside plates and the barrel. Disconnect the right and left side plates from the crank and the barrel. This completes field stripping.

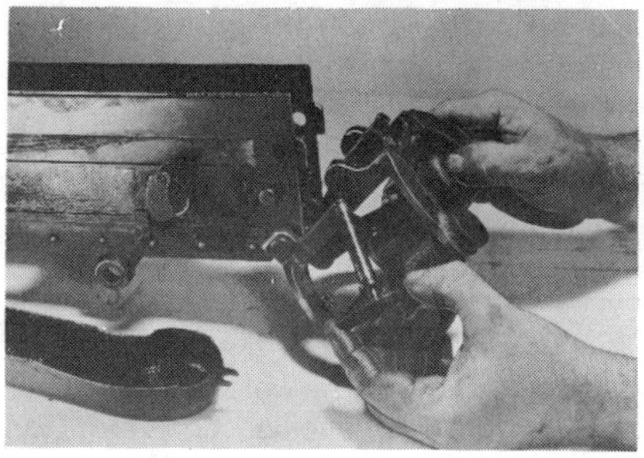

Further Notes on Stripping the Vickers Gun

As this is one of the world's basic machine gun types, a more detailed explanation of stripping should be of value. An understanding of the Vickers is particularly helpful in understanding all the German and Russian type Maxim guns.

Stripping the Lock. The lock is cocked as it comes out of the gun. Should it not be, due to having been snapped when withdrawn, it may be cocked by raising the side lever head.

A split pin with a bushing fastens the combined side lever head and side levers to the lock casing. Force these out. The side levers, and the extractor and extractor levers may now be removed.

The tumbler, the finger-like projection protruding from the locked casing just above the side lever, is fastened by an axis pin. Push this lever out and remove the tumbler. Now push down the tail of the sear which will release the lock spring.

Push out the trigger axis pin (the tip of the trigger protrudes from the top of the lock) and the trigger and lock spring may be removed.

Push downward on the sear and remove the firing pin and the sear and sear pin.

The gib may be removed by pushing out its spring cover and removing spring and gib.

To Assemble the Lock. First insert the sear with spring downward, making sure that the sear jaws engage with the sear pivot.

Next insert the firing pin in its groove. Then replace the tumbler and fasten it with its tumbler axis pin.

Insert the trigger, fasten it by the trigger axis pin.

Replacing the gib and its spring and cover in position on the extractor, slide the assembled extractor from the bottom up in the guides in the locked casing.

Replace the extractor levers and side levers and fasten with the bushing and pin.

With sear held down by side lever head, pull back the trigger and press down the tumbler. Now insert the lock spring with its long arm facing toward the extractor and force it down and home.

The firing pin must be released only when the extractor is up in the casing as far as it will go; as only at this point is it lined up so that it will pass through the hole in the extractor. The pin will be injured if it strikes against a solid steel surface in the extractor.

Stripping the Feed Block. The split pin holds the top and bottom block together and this must be forced out to permit separating the top and the bottom levers.

Pull out the slide with the top pawls and springs. They may be removed from the slide.

Now pull out the bottom pawl axis pin, which will permit removal of the bottom pawl and spring.

Assembling the Feed Block. Reverse the above procedure.

Adjusting the Vickers Gun

The most important adjustment on a machine gun of this type is the head space. This is the correct space between the end of the barrel at the firing chamber and the face of the lock. Should this space be too great, the head of the cartridge will not be held firmly during the moment of high breech pressure. This may bulge the cartridge case so that extraction will be extremely hard, or it may rip the head completely off the cartridge case, resulting in an even more serious jam.

On the other hand, if there is insufficient head space, the lock cannot go forward completely and as a result the side lever head on the lock will not push the sear down far enough to permit the gun to fire.

To Adjust Head Space. Remove the lock and the fusee spring.

Place crank handle in vertical position. Put the No. 1 washer on the outer face of the adjusting nut, making sure that the nut is tight. Now replace the lock in the rear position.

Reach up from below the breech and insert a dummy round or the correct armorer's gauge in the extractor over the firing pin hole, and raise the extractor to its highest point with the fingers. (Use a live cartridge only under suitable range conditions as this is a dangerous operation.)

Make sure that recoiling portions are all locked fully forward and guide the round or gauge into the firing chamber.

Rotate the crank handle forward while guiding the cartridge into the chamber.

Aside from the pressure necessary to compress the sear, a slight check will be felt when the crank handle reaches the check lever if the connecting rod is adjusted to the proper length to give the correct test space.

If no check is left, separate No. 1 or No. 2 washers should be added as required to the outer face of the adjusting nut to provide the correct length.

When the correct length has been ascertained, the washers are assembled permanently on the shoulder of the connecting rod and secured by its nut. (This is done by unscrewing and removing the nut with a combination tool, placing the washers on the connecting rod, and replacing and screwing up the adjusting nut on the washer.)

Tests should be made to be sure that adjustment has been done correctly before completing assembly.

Water Glands. To prevent the cooling water from leaking out of the casing, glands are provided at the muzzle end and oiled asbestos packing wound in a cannelure cut around the breech end of the barrel.

If water leaks at the rear or breech end, empty the casing and then strip the weapon to remove the barrel. A piece of oil soaked asbestos string is then wound into the cannelure of the barrel; and pressed in with the point of a screw driver until the cannelure is full. Now oil this packing and smooth it down until it is flush with the barrel, then reassemble the weapon.

Should water leak at the muzzle, stand the gun up on its traversing handles and remove the muzzle attachment together with the cup and unscrew the muzzle gland.

Remove the asbestos string packing, reoil it, and wind it loosely around the barrel, pushing it in with a punch or piece of wood. Then screw the gland on as tightly as possible by hand.

This should stop the leakage and yet permit the recoiling portions to move freely.

Cooling System. Whenever it is at all possible, the barrel casing should be kept full during the firing period. The water will boil after firing two belts. It evaporates at the rate of one and a half pints for every 500 rounds, or two belts, if fired continuously. If 2000 rounds are fired, casing will require refilling.

Weighing the Recoiling Portions. Remove the fusee spring and put the crank handle in almost vertical position. Now place the loop of the spring balance on the crank shaft and draw it slowly to the rear. The weight should not exceed 4 lbs. If more than 4 lbs. is required to remove the recoiling portions, it indicates that the packing is pressing too hard on the barrel and the gland nut must be removed and one or two strands taken out of the asbestos.

To Check Weight of Fusee Spring. Remove the lock and place the loop of the spring balance over the top of the crank handle. Stand to the left of the gun and press down the check lever with the left hand; and with the right pull the balance vertically upwards. When the crank handle begins to move, the weight should be between 7 and 9 lbs.

If necessary to adjust it, wind the vice pin at the forward end of the fusee spring box on the left side of the gun. Six clicks turning from right to left increases the weight one pound; while six clicks turning from left to right decreases the weight one pound.

Improper adjustment of the fusee spring will jam the weapon.

THE BESA TANK MACHINE GUNS

The Besa guns were developed by B.S.A. from the Czech ZB53 (Model 37) machine gun and were used by the U.K. for tank armament. In 1936 B.S.A. signed an agreement with Zbrojovka Brno allowing them to manufacture the 7.92 ZB53. In April 1938 the War Office placed its first order for the gun and in 1939 production commenced. B.S.A. soon discovered, however, that considerable modification would have to be made to the gun if it was to be capable of mass production; the modified gun was called the Besa. B.S.A. made 59,322 7.92mm Besa guns during World War II.

The Besa was produced in four different models: Mark 1, Mark 2, Mark 3, and Mark 3*. These weapons differ in minor details, but principally in that the Mark 1 and Mark 2 have two rates of automatic fire, which can be selected by moving the selector lever at the left rear of the receiver ("L" is for low rate and "H" is for high rate), and the Mark 3 and 3* guns have only one rate of fire.

The Besa guns have one unusual feature; although they are gas operated they have a recoiling barrel. The cartridge is chambered and fired before the recoiling barrel is completely in the battery position (fully forward). The recoil of the firing cartridge, therefore, must overcome the inertia of the forward-moving barrel. This action helps in buffing the bolt and reduces the shock on the weapon and the mounting when firing.

The Besa is no longer a standard weapon in Britain, having been replaced by the Browning caliber .30. It may still be found, however, on some of the older armored vehicles. The Besa tank gun is still widely used on older British armored vehicles throughout the British Commonwealth and former British Territories.

Characteristics of the Besa Tank Machine Gun

Caliber: 7.92mm.
System of operation: Gas, automatic only.
Weight: Mark 1 47 lbs.
Mark 2 48 lbs.
Mark 3 54 lbs.
Mark 3* 53.5 lbs.
Overall length: 43.5 in.
Barrel length: 29 in.
Feed device: 225 round link or metal and (canvas) belt.
Sights: None fitted to gun, telescopic sights used on vehicles.
Muzzle velocity: 2,700 f.p.s.
Cyclic rate: MK 1 MK 2 - 450-750 r.p.m.
MK 3 MK 3* - 450 r.p.m.

Loading and Firing the Besa

The trigger guard is part of a heavy steel unit called the trigger guard body. Press forward on the cocking catch thumbpiece which is mounted on the left side of the trigger guard body to disengage the cocking catch which is in a recess in the underside of the receiver. Be careful not to touch the trigger.

Push the trigger guard body forward and the sear will click into engagement in a bent in the piston extension.

Still keeping the finger away from the trigger, and holding firmly to the pistol grip, pull the trigger guard body back with a quick motion. This withdraws the working parts and compresses the return spring. The operating parts will be held to the rear by the sear which is now engaged in the bent of the piston extension. The cocking catch engages in a recess in the underside of the receiver and locks the trigger guard in firing position.

Push the tab-end of the belt through the feed block from the right side and pull through to the left as far as possible. This places the first cartridge in line with the chamber, bullet pointing downwards. The weapon is now ready to fire.

The pistol grip of the trigger guard body in this weapon is fitted with a grip safety somewhat like the one on the Colt .45 automatic.

Besa 7.92mm Mark 2 Machine Gun.

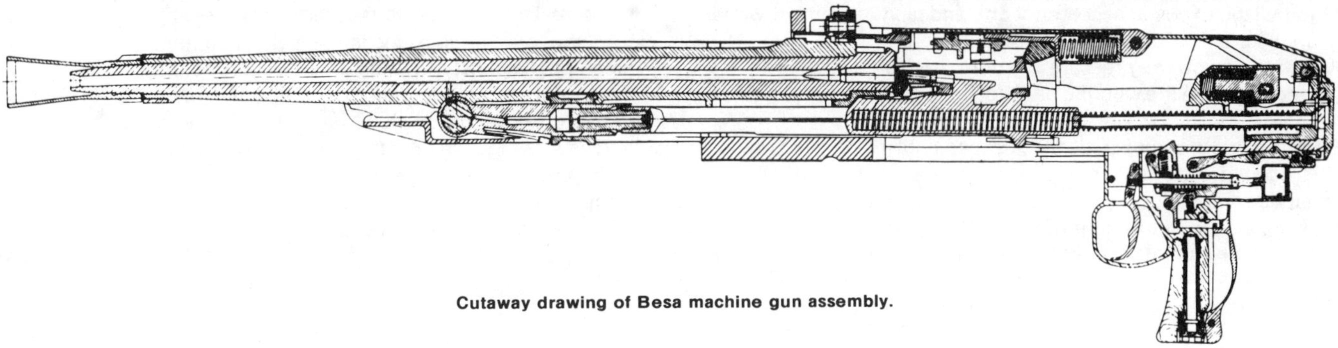

Cutaway drawing of Besa machine gun assembly.

With finger on trigger, compress the hand to push in the safety catch lever. This will rotate the safety catch to the rear until it clears the underside of the sear; as the trigger is pressed, the sear is pulled out of the bent of the piston extension permitting the return spring to force the working parts forward. The gun will fire as long as the trigger is held back and the safety catch lever is kept depressed. The gun will stay open between shots when trigger is released.

Note: When the belt is emptied, it is expelled from the gun; but the action will go forward to close on an empty chamber.

Unloading the Besa

Release feed pawl depressor lever which will free the feed pawl from engagement with the belt. Open cover to release retaining pawl which permits engagement with belt links.

Lift out the belt. Close the cover, lower the depressor lever, and ease the working parts forward.

Field Stripping the Besa

See that the gun is cocked.

Pull the barrel retainer carrying handle up about half an inch and push it forward until it rests on the ramp.

Press in the cover catch, push the carrying handle straight forward, then strike the handle with the palm of the hand to bring it to the upright position.

Lift the rear of the barrel until it clears the barrel extension, then ease it forward. This frees the slides on the barrel sleeve from their guides in the body and permits the barrel to be pulled out.

Remove the body cover. Pull the cover locking pin out as far as it will go, press in on the cover catch, and lift the cover. Raise it until it can be lifted out of the body (receiver).

Press in the catch on the belt guide and lift it out of the receiver. Lift out the feed block and remove the feed slide.

The breech block may now be lifted out.

Besa machine gun, cover open.

Now pull the exposed accelerator arm and plunger cap outwards, pull up the crank arm, and lift out the accelerator.

Maintain downward pressure on the barrel extension with the left hand and ease the working parts forward. Pull the trigger guard to the rear.

Push the return spring guide block ahead until it clears the guides in the receiver, then lift it up. The guide lock and return spring may now all be removed.

Grasp piston with the right hand and the rear end of the barrel extension with the left and lift both pieces out, then slide out the piston.

The feed lever may now be inclined inward and lifted out of the weapon.

At the rear of the receiver is the trigger guard catch. Raise this, and press the trigger guard easily to the rear to its fullest extent. Then release the trigger, jerk back on the trigger guard till it clears the gun, and then lower the catch.

Field Assembly, Besa Machine Gun

Reverse the stripping procedure. Start by replacing the trigger guard, and make sure that it is in its normal position. In replacing feed lever, check that the upper arm is slightly to the right so it will engage correctly its stud with the groove in the piston.

In assembling barrel extension and piston, take care that the upper flange on the piston engages in the lower groove in extension. Work piston backwards and forwards when inserted to be sure that stud on lower arm of feed lever is properly engaged with piston extension groove.

Place return spring in piston, and while holding the spring firmly, place the guide rod in the spring, push forward until the rod enters the piston extension. Then press the guide lock down into the receiver and release the pressure. This will permit the spring to position the guide lock properly.

Hold the barrel extension down with the left hand while cocking the gun with the right and replace accelerator, breech block, feed block body, and feed slide assembly. Be sure that the stud on the feed slide engages in the slot of the upper arm of the feed lever belt guide.

When replacing the body cover, check that the cover bearings engage properly with the body trunnions.

Barrel Replacement

Keeping rear end of the barrel elevated, insert the slides on the barrel sleeve in the guides on the body. Then draw the barrel to the rear until the flanges at the breech end of the barrel are just above the grooves in the barrel extension. Then lower the barrel into the barrel extension, push the carrying handle over to the right until it rests on the ramp, strike it back sharply with the palm of the hand and push the handle down to lock the barrel.

How the Besa Gun Works

Starting with the gun loaded and cocked, the action is as follows. As the safety catch lever is squeezed in, the trigger is free to engage with the sear, drawing it out of the bent in the piston extension and permitting the return spring to start the piston and moving parts forward.

As the piston and piston extension move ahead, the extension acts to move the breechblock ahead, a projection on the front of the block strikes the rim of the cartridge in front of it, strips it out of the belt, and chambers it.

The extractor rides up over the cannelure of the cartridge during the closing movement of the breechblock and snaps into engagement.

As the rear end of the breechblock is lifted, it disengages from the shallow stop in the piston extension by engaging with projections in the barrel extension. It rides up an inclined plane and engages with a resistance face on the barrel extension.

As the breechblock locks, the piston extension carrying the piston post against the firing pins discharges the cartridge.

Also during the forward movement, the piston extension acts on the lower arm of the feed lever moving the upper arm and the feed pawl over to the right permitting the feed pawl to be depressed by the belt and cartridge and engage in the link of the third round in the belt.

The retaining pawl is engaged behind the link of the cartridge just stripped into the firing chamber and holds the belt stationary during the movement.

Return Movement of the Action. As the cartridge in the chamber is exploded, the bullet moves down the barrel and the rearward thrust of the recoil bearing against the cartridge case forces the breechblock, piston, barrel extension, and barrel to the rear as a unit. The barrel extension presses against the recoil spring in the cover easing the rearward motion; and after the initial shock the barrel extension and barrel are forced forward by the action of the recoil spring.

Gas Action. As the bullet passes over the gas vent in the barrel, some gas escapes through and into the gas cylinder, then through the gas regulator to strike a sharp blow against the piston head.

This sudden thrust drives the piston and its extension to the rear, and the inclined ramp on the piston post bearing against the breechblock lowers the rear end of the block out of engagement with the resistance space at the rear of the barrel extension, unlocking the weapon.

Further rearward motion of the piston extension pulls the breechblock clear of the barrel. The extractor draws the fired case out of the chamber until the base of the empty cartridge case strikes the ejector, which is a fixed projection on the belt guide. This snaps the case free from the breechblock, permitting the extractor to snap back into place, and the empty is hurled through a slot in the piston extension and out the ejection opening in the bottom of the receiver. Also during the backward movement of the piston extension, the return spring is compressed between its place in the extension and the return spring guide lock.

If the trigger has been released, the sear spring will reassert itself and push the nose of the sear into engagement with the bent of the piston extension holding it to the rear. If the trigger is kept depressed, on completion of rearward motion the return spring will drive the piston and other moving parts forward and the cycle of firing will be repeated.

The Accelerator (MK 1 and 2 Guns). When the accelerator is set in its high position, the rear of the piston extension will not make contact with the accelerator and the gun will fire at its normal rate of 450 to 550 per minute.

If, however, the accelerator is set in its low position, the complete backward motion of the piston extension is halted by the rear of the extension striking the front face of the accelerator. This forces the accelerator casing to the rear, thereby compressing the heavy accelerator springs. Thus, when the forward movement starts, the operating parts are forced forward not only by the return spring, but also by the accelerator springs. This extra spring action speeds up the forward movement of the working parts and the rate of fire is increased to 750 to 850 shots per minute.

THE 15mm BESA MARK 1 MACHINE GUN

Production was started on the 15mm Besa at B.S.A. in 1939 and the first guns were delivered in May 1940. The 15mm Besa is a modification of the Czech 15mm ZB60 (Model 38). In basic design the 15mm Besa is similar to the 7.92mm Besa guns, but it can be used for semiautomatic as well as automatic fire and has only one cyclic rate.

The 15mm Besa had a more limited usage than the 7.92mm gun; its main use was as primary armament on certain British armored

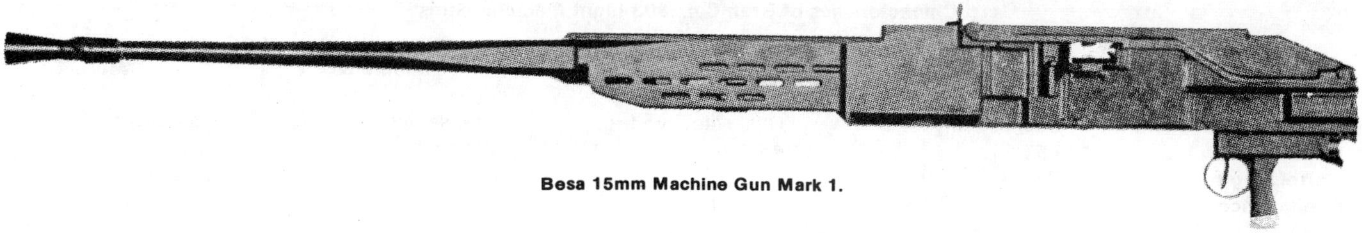

Besa 15mm Machine Gun Mark 1.

cars. B.S.A. made 3,218 15mm Besa guns during World War II. This weapon is obsolete in Britain.

Characteristics of 15mm Besa

Caliber: 15mm.
System of operation: Gas, selective fire.
Weight: 125.5 lbs.
Length overall: 80.75 in.
Length of barrel: 57.6 in.
Feed device: 25-round link belt.
Sights: None mounted on gun, telescope or armored vehicle.
Muzzle velocity: Approx. 2685 f.p.s.
Cyclic rate: 400-500 r.p.m.

Barrel Removal, 15mm Besa

Removing the barrel on this model is a two-man job. While one man holds the barrel, the other raises the carrying handle of the barrel retainer a half-inch and pushes the handle forward until it can be turned up. About 13" from the breech end a special slot is provided to permit this action.

Second man now raises the rear end of the barrel until it clears the barrel extension, then both ease it forward and lift it out. Slides on the barrel are freed from the guides of the body by this forward motion.

THE BREN LIGHT MACHINE GUNS

As previously noted, the Bren was developed from the Czech 7.92mm ZB26 by Enfield and ZB in the mid-thirties. Production of the gun started at Enfield in 1937. Most of the production capabilities of Enfield were used to produce Bren guns during World War II. The Bren was also made by Inglis in Canada, both in .303 for British and Canadian service and in 7.92mm for the Chinese Nationalists. By 1943 Canada was making 60% of the Bren guns.

The Bren was one of the best light machine guns of World War II and is still considered a fine gun. The Bren converted to 7.62mm NATO models L4A2, L4A4 and L4A6 is still in service with all British troops except Infantry throughout the world and is used by British Infantry in Asia. There are still many .303 Brens in use throughout the world. ZB produced the Bren at Brno for commercial sale and the gun was listed in some of their early post World War II catalogs. It was called the ZGB by the Czechs.

Types of Bren Guns

Bren Light Machine Gun Mark 1. The Mark 1 has a radial type sight, and the butt is shaped differently from the later models. Early versions had a wooden handle which was hinged under the butt.

Bren Light Machine Gun Mark 1 (M). This weapon was only manufactured in Canada and differs from the Mark 1 in the following ways: the bipod legs do not telescope, the gas vent in the barrel has been enlarged, and the stock has been simplified by the removal of the shoulder support (butt strap) simplification of the butt plate and removal of the butt plate buffer spring.

Bren Light Machine Gun Mark 2. This weapon was made in both the U.K. and Canada. It has the simplified butt and a leaf-type rear sight.

Bren Light Machine Gun Mark 3. The Mark 3 has been lightened and has a shorter barrel.

Bren .303 Light Machine Gun Mark 1.

Characteristics of Bren Cal. .303 Light Machine Guns

	Mark 1	Mark 2	Mark 3	Mark 4
System of operation	Gas, selective fire.	Gas, selective fire.	Gas, selective fire.	Gas, selective fire.
Overall length	45.5 in.	45.6 in.	42.6 in.	42.9 in.
Barrel length	25 in.	25 in.	22.25 in.	22.25 in.
Feed device	30-round box or 100 round drum.	30-round box or 100-round drum.	30-round box.	30-round box.
Sights:				
Front	Blade w/ears.	Blade w/ears.	Blade w/ears.	Blade w/ears.
Rear	Aperture w/radial drum.	Leaf w/aperture.	Leaf w/aperture.	Leaf w/aperture.
Muzzle velocity w/MK 7 ball	2,440 fps.	2,440 fps.	Approx 2,400 fps.	Approx 2,400 fps.
Cyclic rate	500 rpm.	540 rpm.	480 rpm.	520 rpm.
Weight of barrel	6.28 lbs.	6.46 lbs.	5.09 lbs.	5 lbs.
Weight of gun	22.12 lbs.	23.18 lbs.	19.3 lbs.	19.14 lbs.

Mark 2 Bren.

Bren Light Machine Gun Mark 4. The butt assembly of the Mark 4 differs in minor details from the Mark 2 butt used with the Mark 2 and 3 guns. There have been other minor changes as well.

Bren Light Machine Gun L4A2. This weapon is a conversion of the later model Brens to 7.62mm NATO. A new magazine, a new bolt, and a new barrel are used with this gun. The receiver has been modified to ensure feeding of the 7.62mm cartridge. The barrel has a prong-type flash suppressor.

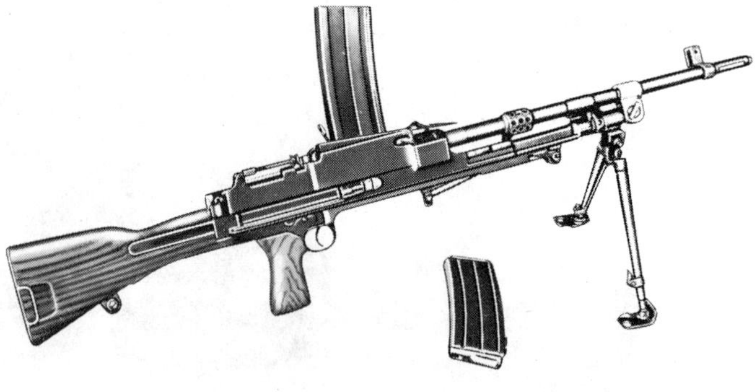
L4A2 Bren 7.62mm NATO.

Mark 3 Bren.

Loading and Firing the Bren

To load the magazine by hand: The magazine is rested on the thigh, or on a solid object, and cartridges placed in the magazine as for ordinary automatic pistol. They should be inserted with the right hand, and pressed down into place with the thumb of the left hand. Unlike our United States cartridge, the British service cartridge had a rim. In inserting cartridges in magazine, therefore, care must be taken to see that the rim of each cartridge is placed in front of the round already in the magazine. If rim gets behind rim, jams will inevitably result.

Magazine Filler. Push the magazine into the mouth of the filler, and swing the filling lever as far as it will go to the left. Fill the hopper and push the filling lever over to the right and back to its limit 6 times; this will put 30 rounds into the magazine. If the filler is the small hand type, push magazine in until the magazine catch engages, and then insert a loaded cartridge charger (or clip as it is called in the United States) into the mouth of the filler over the head of the magazine. See that the tip of the operating lever is against the topmost cartridge and push down slowly and firmly with the operating lever.

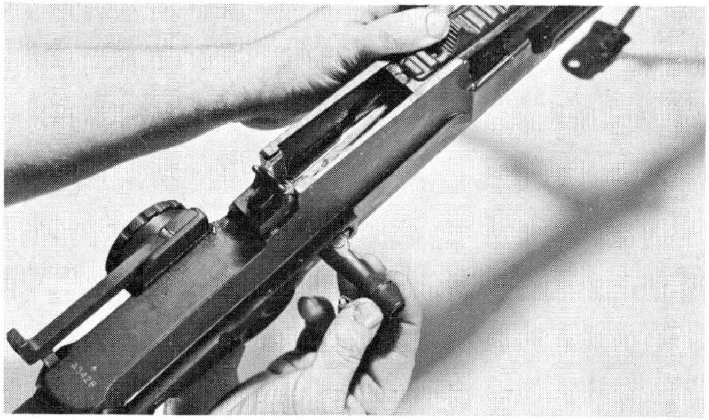

Note on Magazine. While the magazine capacity is 30, it is better practice to use 27 or 28 cartridges so as not to strain the magazine spring.

The magazine opening on top of the receiver is fitted with a sliding cover; push this opening cover forward as far as it will go.

Holding the magazine mouth downward in the right hand, insert the lip at the front end into the magazine opening and hook it there; then press downward the rear of the magazine until the magazine catch engages on the magazine rim.

Draw the cocking handle back as far as it will go to cock the action and push it forward again. If weapon has a folding cocking handle, fold it over.

Set the change lever on the left side of the receiver at the desired position of "Automatic," "Safe," or "R" for single shot.

Note on Ejection. A cover over the ejection opening will automatically spring open when the trigger is pulled to permit ejection of empty cartridge case.

Caution: Always remember that gun fires from an open bolt. The bolt should never be permitted to go forward while there is a magazine in the gun unless you intend it to fire. The magazine must be removed first, and the action eased forward second in unloading the weapon.

Field Stripping the Bren

Be sure there is no magazine in the gun and all moving parts are forward.

The body locking pin passes through the receiver from right to left directly under the aperture of the rear sight. Push it with the point of a bullet from the left side and withdraw it from the right.

Grasp the back sight drum firmly with the left hand, and with the right pull back the butt group as far as possible. The return spring rod, which is housed in the butt, will now protrude from the butt through the buffer.

With the thumb and forefinger of the left hand pull the return spring rod to the left out of line with the piston; and with the right

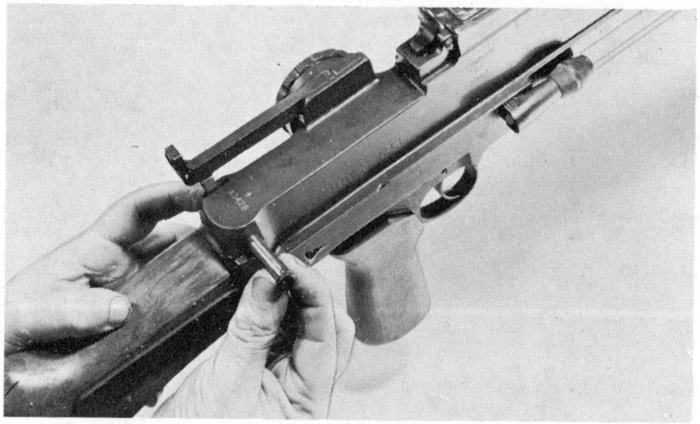

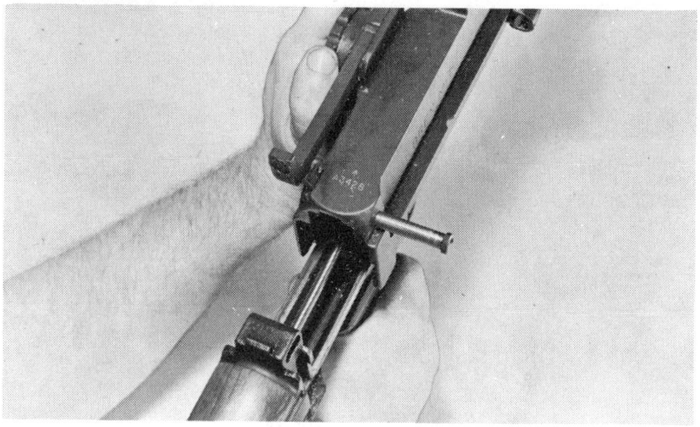

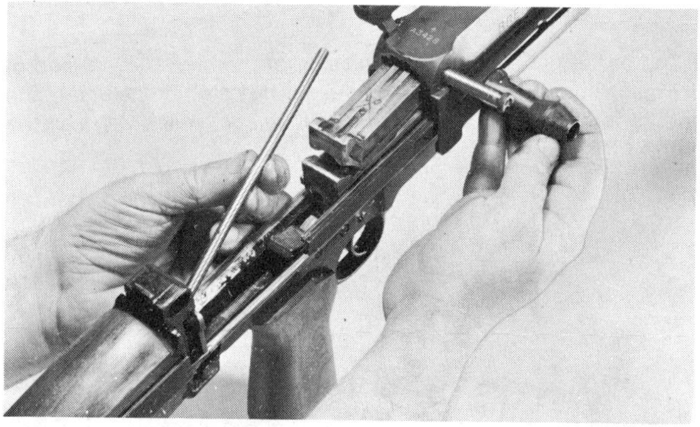

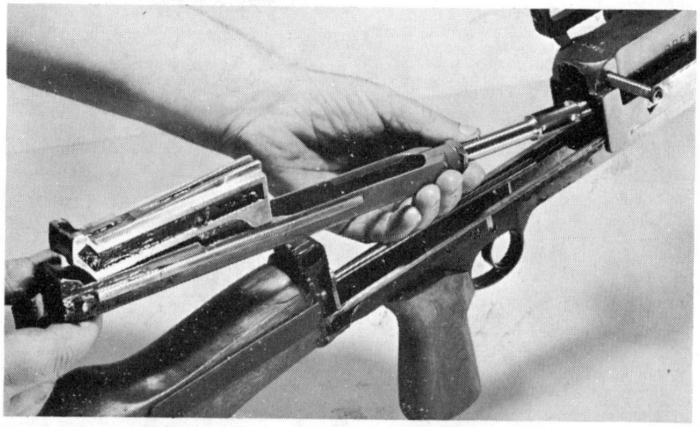

hand pull the cocking handle back with a rapid motion. The piston and breechblock will now come out of the receiver and may be removed from the gun.

The claws at the front end of the breechblock are in engagement with grooves on the piston, and if the breechblock is slid to the rear it can be lifted out of this engagement and removed.

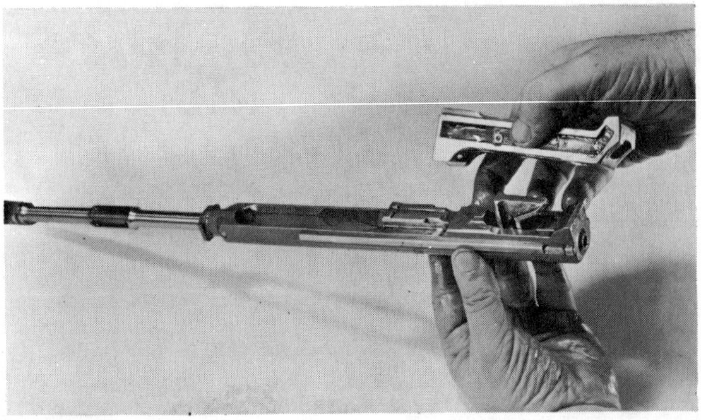

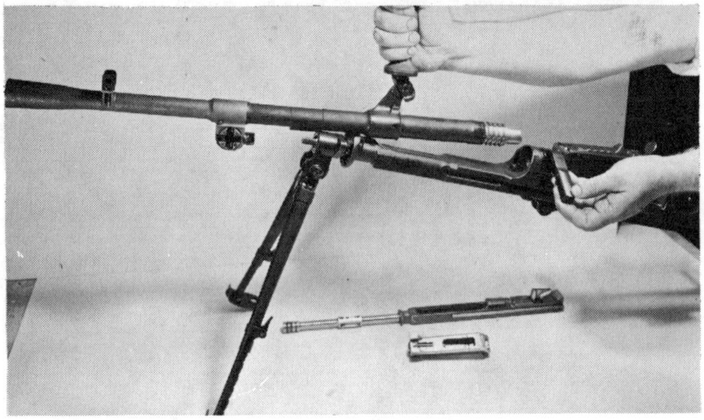

The barrel nut catch lies on the side of the barrel just ahead of the magazine opening. Force in the spring catch on its underside and lift the barrel nut catch as far as it will go; which will free the barrel for removal.

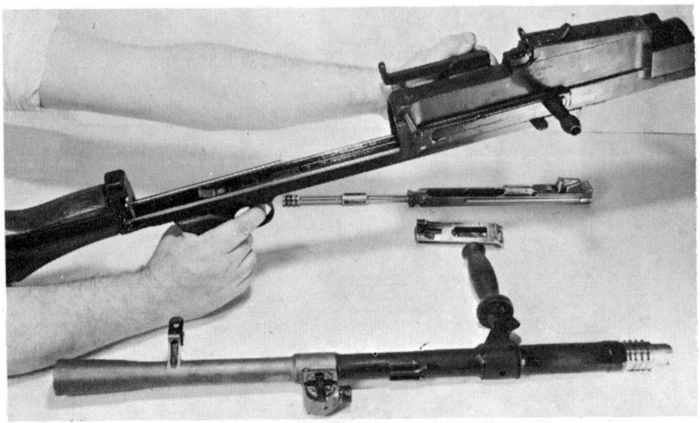

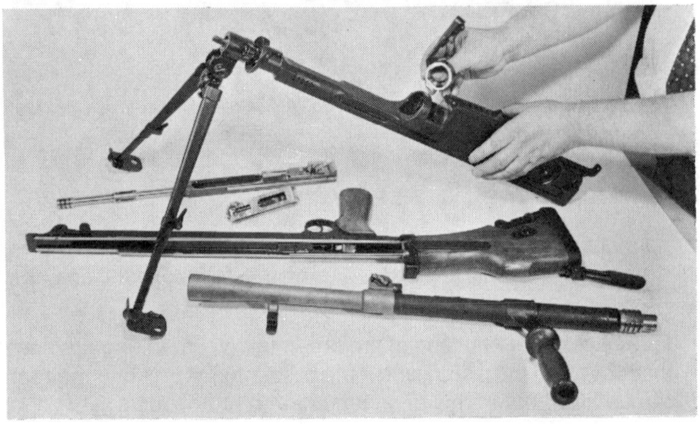

Grasp the rear sight drum firmly with the left hand and with the right hand pull directly back on the butt. The entire butt group may now be removed.

The barrel nut may be removed by lifting the catch as far as it will go and pushing down the small stud in front of the magazine opening cover. The barrel nut is then lifted out vertically.

Now lift the front of the body with the right hand and with the left pull the left leg of the bipod as far forward as possible—slide bipod sleeve off the front end of the gas cylinder.

Notes on Assembling. Reverse the stripping order.

In replacing bipod take care the mount is fully home.

In Mark 1 guns check that the stop on the left of the forward end of the butt group is in front of the barrel nut catch before lowering the catch.

In replacing barrel on Mark 1, make sure the long groove underneath between gas block and carrying handle engages properly with stud on top of receiver.

Be sure the barrel nut catch is fully locked and catch has engaged on rib in the body or receiver.

When replacing breechblock on piston, slide the claws down into the groove as far forward as possible and then let the tail of the breechblock drop.

When inserting the assembled breechblock and piston, make sure that the breechblock is fully forward and that the two are pushed into the receiver before attempting to push forward the butt group.

Be sure that the return spring rod engages in a recess for it in the end of the piston when the butt group is being pushed forward.

Gas Regulator. The gas regulator is mounted on the barrel near the muzzle. It faces to the left. The correct setting is usually the No. 2 size. There are four different ports. Lifting the retainer pin permits the gas regulator to be turned to increase the size of the port. Should the gun become sluggish in action, the gas regulator is altered to the next larger hole to increase the amount of pressure available.

How the Bren Gun Works

Starting with the gun loaded and cocked the action is as follows: If the change lever is set at "R" pressing the trigger pulls a connecting tripping lever, which in its turn draws down the sear out of engagement with the pin on the piston. This action also compresses the coil sear spring. The compressed return spring, situated in the butt, pushes the rod forward, and this in turn pushes against the seat in the piston driving the piston forward, carrying

Showing change-lever for Safe, Automatic, or Single-shot fire settings.

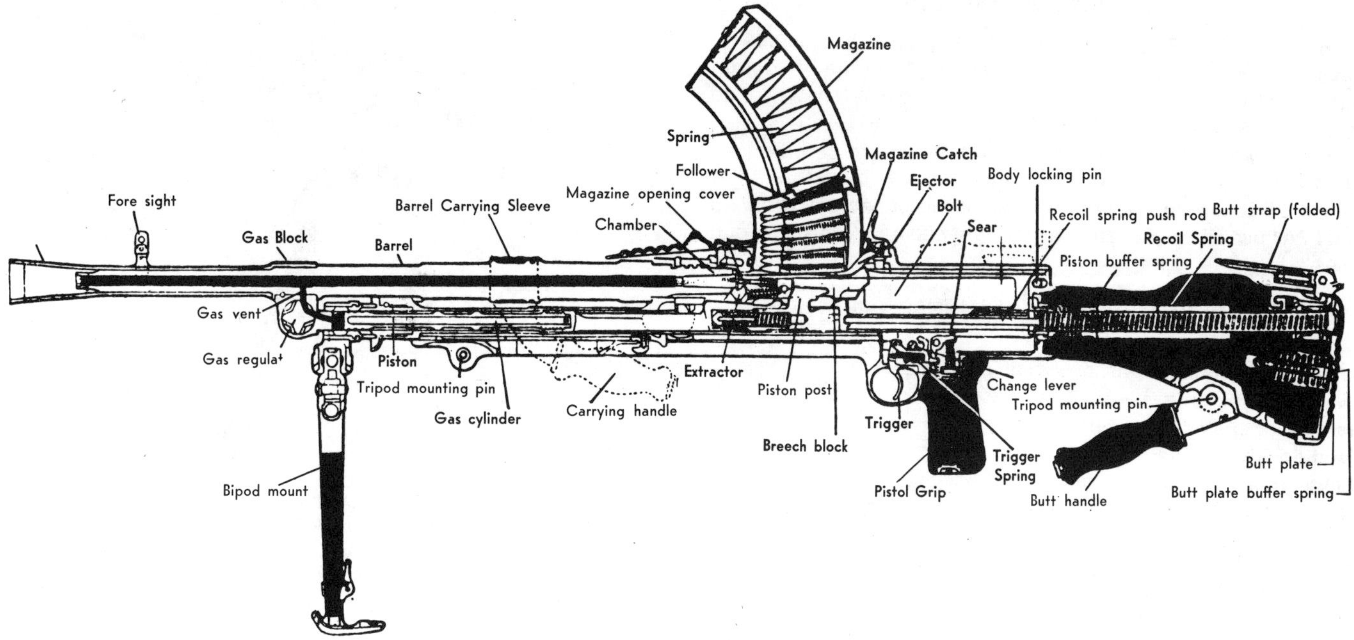

Principal components of the Bren light machine gun.

with it the locking and firing mechanism. Meanwhile the sear spring pushes the sear back into place. The breechblock mounted on the top of the piston is carried forward, and the feed piece strikes the base of the first round in the magazine and forces it forward out of lips of the magazine and into the chamber, with the extractor slipping over the rim. The rear end of the breechblock is cammed up into a locking recess in the top of the receiver as the cartridge is properly chambered; in its final move the piston post drives the firing pin against the primer of the cartridge, exploding it.

As the bullet passes over the small gas vent cut in the barrel, a short distance from the muzzle, a small amount of gas under high pressure passes through the vent and through the gas regulator (where the size of the port selected determines the amount of gas to be let in) and escapes into a well where it expands with a hammerlike thrust against the piston. As the piston is driven back in its cylinder, the gas can now escape through holes provided for it.

Meanwhile the sudden thrust on the piston drives it back and forces the return spring rod back into the butt where the return spring is compressed; this action being finally stopped by the piston buffer.

Details of tripod mount.

Bren gun on antiaircraft mount.

The empty cartridge case, gripped by the extractor and carried to the rear in the face of the breechblock, strikes its face against the base of the ejector and is hurled downward through the ejection slot in the piston and out of the weapon. During this rearward action the upper locking surfaces of the breechblock are forced down into line, so that in its final movement, the piston and breechblock travel together in a straight line.

Note: The buffer spring is in the butt below the line of the return spring.

The Bren Tripod Mark 2 and 2/1

A tripod was issued for use with the Bren light machine gun. Approximately one tripod was issued for every three guns. The tripod is of Czech design and is basically the same as that used with the Czech ZB26 and ZB30 machine guns. A modified form of this tripod is used with the Chinese copy of the U.S. 57mm recoilless rifle.

THE L7A1 MACHINE GUN

The L7A1 is the British version of the FN MAG 7.62mm NATO machine gun. Enfield has made a few changes in the FN design particularly in the barrel. In addition Enfield has developed a tripod for use in the sustained fire role.

Originally a heavy barrel with Stellite liner was to be produced for this gun to use on its role as a tripod mounted sustained fire weapon. Due to manufacturing difficulties this project was dropped and L7A1 uses the light barrel in both light and heavy machine gun roles.

Characteristics of L7A1 Machine Gun

Caliber: 7.62mm NATO.
System of operation: Gas, automatic only.
Overall length: 49.7 in.
Barrel length: W/flash suppressor 24.75 in.
Feed device: Disintegrating link belt.
Sights: Front: Protected blade.
Rear: Peep battle sight of tangent type and leaf.
Weight of gun: 24 lb. with light barrel.
Weight of tripod L4A1: 29 lb.
Muzzle velocity: 2800 f.p.s.
Cyclic rate: 700 to 900 r.p.m.

This weapon has been adopted by the U.K. for use as a tank machine gun to replace the caliber .30 Browning currently being used.

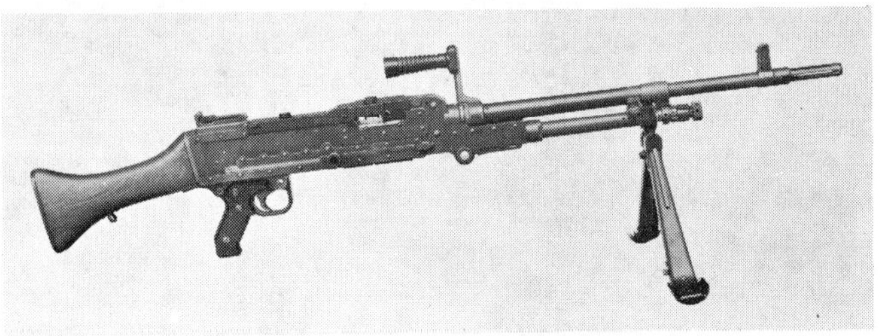

L7A1 7.62mm Machine Gun on bipod.

The L7A1 is being modified so that it will have a sear with double nose with slide notched to match. Other versions of this weapon are as follows:

L7A2: has attachment for 50-round belt box on left side of the receiver, double feed pawls and double bent sear with slide machined to match.

L8A1: This is a tank gun; the barrel has a fume extractor—bore evacuator—incorporated. It has a three position non-venting gas regulator and the trigger is designed for use with a solenoid. It has a folding pistol grip for emergency manual operation and a feed pawl depressor is fitted. It is used on the "Chieftain" tank.

L20: This is an experimental version of the L8 for use in aircraft gun pods. It is capable of left or right feed. It has a hybrid barrel assembly having the L8 gas regulator on the L7 barrel. The front sight and the carrying handle have been removed from the barrel.

L37A1: This gun is a mixture of L7 and L8 components produced to make a gun for armored vehicles other than "Chieftain", which can be removed and used as a normal ground gun. It is basically an L8 with barrel from L7. It may also be found with butt, bipod and trigger group from L7.

7.62mm Machine Gun L7A1 on L4A1 tripod. The gun shown has a grooved barrel which was developed only in prototype form.

L19A1: This weapon has an 8 pound barrel; it has not been issued to troops.

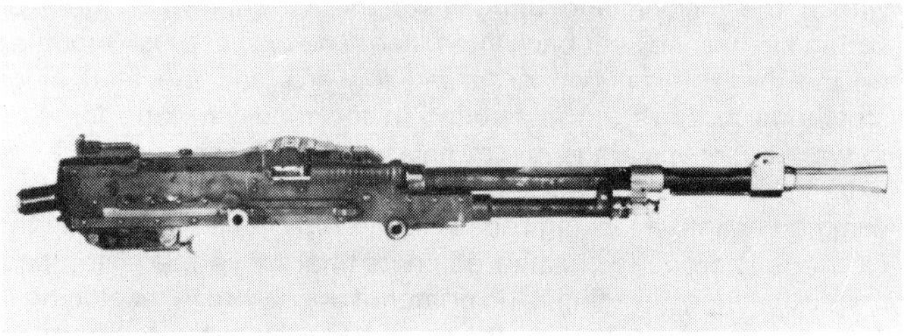

7.62mm Tank Machine Gun L8A1.

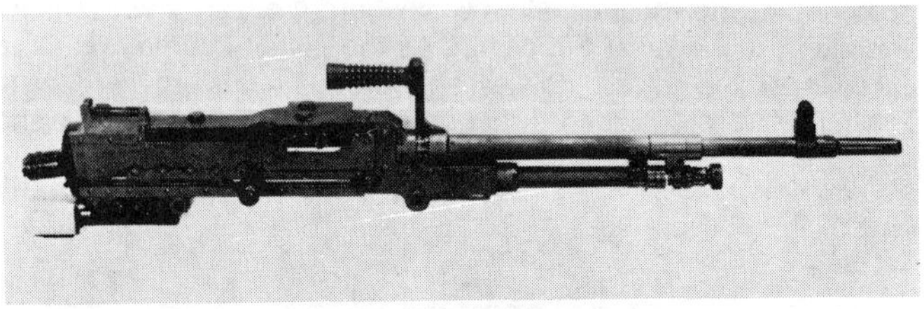

British 7.62mm Machine Gun L20.

U.S. MACHINE GUNS IN BRITISH SERVICE

The United Kingdom has adopted the U.S. Browning caliber .50 M2 heavy barrel machine gun which they call the L1A1. The caliber .30 Browning M1919A4 and 1919A5 machine guns are also used on armored vehicles.

The .30 caliber Brownings in service are called the L3A1 Fixed, L3A2 Flexible, L3A3 and L3A4.

OTHER COMMONWEALTH MEMBERS

South Africa, which is no longer a member of the Commonwealth, is producing the 7.62mm NATO L1A1 (British FN "FAL") rifle at Pretoria. South Africa has also adopted the 7.62mm NATO FN MAG general purpose machine gun, as have Zambia, Southern Rodesia, Tanzania, Uganda, and Sierra Leone.

Canada 18

Canada uses the 9mm FN Hi-Power pistol of native manufacture. The Canadians also use rifles 7.62mm NATO C1A1 and C2A1, the Canadian counterparts of the FN "FAL" and its heavy barrel version, respectively. These rifles were considerably modified from the original FN design by the Canadian government arsenal at Long Branch, Ontario where they are manufactured. Canada was the first country to put the FN in mass production and the first member of NATO to have its regular army completely equipped with 7.62mm weapons.

Canada has adopted the British 9mm Sterling (Patchett) submachine gun in slightly modified form as the submachine gun C4. The standard machine gun in Canada is the 7.62mm NATO C1. This gun is a converted caliber .30 Browning Model 1919A4. Bren guns and other .30 caliber Browning guns may be held in reserve stocks. The caliber .50 Browning M2 H.B. machine gun is also standard in Canada.

CANADIAN PISTOLS

Canada used the British service pistol, the Webley .455 Pistol No. 1, Mark VI to some extent in World War I. However, they extensively used Colt and Smith & Wesson revolvers and Colt automatics. Canada, being the northern and geographically larger half of the North American continent, has always been influenced by gun developments in the U.S. In addition, the pioneer experience in Canada probably gave the Canadians a different appreciation of handguns than would be held in the mother country. In any event, Canada was the first member of the British Commonwealth to adopt and produce a truly modern automatic pistol. The Canadian-made 9mm Parabellum Browning Hi-Power pistol is standard in Canada and in the United Kingdom. This weapon was supplied to the U.K. during World War II to arm Commandos and paratroop divisions. The pistol was originally put into production for the Chinese Nationalist Army.

THE CANADIAN HI-POWER BROWNING PISTOL

The John Inglis Company of Toronto, Ontario produced the 9mm Parabellum Browning Hi-Power pistols in several models for the Canadian and Chinese Nationalist Governments during World War II. The weapons produced were given the following nomenclature and differed as explained below.

Pistol, Browning, FN, 9mm, HP, No. 1 Mark 1

The butt is machined for a shoulder-stock holster, and a tangent-leaf rear sight graduated from 50 to 500 meters is fitted.

Pistol, Browning, FN, 9mm, HP, No. 1 Mark 1*

Also machined for a shoulder-stock holster but the height of the ejector has been increased, and the tangent-type rear sight (similar to the No. I, Mark 1) has been machined to accommodate the increased height of the Mark 2 ejector. The extractor also differs and cannot be interchanged with the Mark 1 extractor.

Pistol, Browning, FN, 9mm, HP, No. 2 Mark 1

Not machined for shoulder stock holster; has the smaller ejector and a notched, fixed rear sight. Uses Mark 1 extractor and Mark 1 ejector.

Canadian Browning FN 9mm HP No. 2 Mark 1.

Right side of the No. 2 Mark 1.

Canadian lightweight version of the 9mm Browning FN Hi-Power pistol.

Canadian caliber .45 NAACO "Brigadier" pistol.

Pistol, Browning, FN, 9mm, HP, No. 2 Mark 1*

Same as No. 2 Mark 1, except for ejector and extractor. Uses extractor Mark 2 and ejector Mark 2 for which slide clearance is machined.

There are two models—Mark 1 and 2— of hammer and link for these pistols; however, these are interchangeable. The wooden shoulder-stock holster is no longer in common use; these pistols are used now as "one-hand" weapons.

The characteristics of the Canadian Browning FN HP pistol are basically the same as those of this pistol as produced in Belgium and given in the chapter on Belgium.

POST-WAR CANADIAN PISTOLS

After World War II a number of pistol tests were conducted in Canada, the U.K. and the U.S. Among the competitors was a lightweight version of the Canadian Hi-Power.

This pistol has lightening cuts on both sides of the slide. The pistol tests turned out to be of little consequence since economics and logistics—the quantity of weapons and ammunition on hand and tooled for—dictated that pistols in hand would be used.

The Brigadier .45 Pistol

NAACO—the North American Arms Corporation of Canada—developed a caliber .45 pistol called the "Brigadier." This pistol is a modified Browning Hi-Power chambered for a new caliber .45 cartridge of considerably more power than the .45 automatic cartridge.

It should be noted that the "Brigadier" has its safety catch on the slide where it blocks the firing pin rather than on the receiver as does the Browning (and the Colt).

CANADIAN RIFLES

BRIEF HISTORICAL SUMMARY

Canada had a rifle of native origin for part of World War I—the Ross in its various models. The Ross was dropped as standard in 1916 and the British Short Magazine Lee Enfield Mark III (Rifle No. I Mark III), was adopted. The U.S. bought 20,000 Ross rifles from Canada in 1917 for training purposes. The Lee Enfield No. 1 rifles were used between World Wars I and II and Canada adopted the No. 4 rifle at about the same time as the U.K. Long Branch tooled up to produce the No. 4 early in World War II and produced a total of 952,000, of which the greater part by far were No. 4 Mark 1*. Canada was among the first to adopt the FN "FAL" rifle and was the first to mass produce the weapon. Prior to the adoption of the production model (the C1) Canada, as well as the U.K. tested a number of experimental models. The Ex 1 model was quite similar to the British-adopted L1A1 in that it could not be fed with chargers and had a rear sight with a fixed-size aperture.

The EX 2 could not be fed with chargers either, but had an optical sight similar to that of the British E.M. 2 rifle. Both rifles had barrels without flash suppressors.

Similar rifles were tested by the U.K. as rifle 7.62mm, FN BR X8E1, Type A (iron sight type), and rifle 7.62mm, FN BR, X8E2, Type B (optical sight type.) Canada made additional modifications and adopted the rifle 7.62mm FN (C1) in June 1955.

7.62mm AUTOMATIC RIFLES FN C1 AND C1A1

The obvious external difference between the C1 and most other versions of the FN "FAL" family are the rear sight and the charger guide which allows feeding of the magazine, when in the rifle, with five-round chargers.

The sight of the C1 is similar to that found on some sporting rifles. A disc containing five differently-sized apertures is held in a frame. The edge of the disc is serrated for ease of turning; a flick of the disc turns up the range aperture desired from 200 to 600 yards. The range for which the aperture is set, is indicated by numbers from 2 to 6 which are visible in the lower part of the sight. The sight can be folded when not in use.

Characteristics of C1 and C1A1 Rifles

Caliber: 7.62mm NATO.
System of operation: Gas, semiautomatic only.
Overall length: 44.75 in.
Barrel length: 21 in.
Weight: 9.4 lb.
Feed device: 20-round, detachable, staggered row. Box, can be fed with 5-round chargers.
Sights: Front: Protected post.
Rear: Revolving disc with apertures.
Muzzle velocity: 2,750 f.p.s.

Experimental 7.62mm FN "FAL" Rifle CDN EX 1.

Experimental 7.62mm FN "FAL" rifle CDN EX 2.

Modification of C1—the C1A1

The C1 was modified slightly around 1959, the modification is called: rifle 7.62mm FN C1A1. The principal modifications were in the firing pin which was altered to two-piece configuration, and a new plastic carrying handle replaced the wooden type.

Presumably C1 rifles in service will be converted to C1A1 rifles. Both C1 and C1A1 have prong-type flash suppressors fitted to the muzzle of the barrel.

7.62mm AUTOMATIC RIFLE FN C2

Characteristics of 7.62mm Automatic Rifle FN C2

Caliber: 7.62mm NATO.
Weight loaded (30-round magazine): 15.25 lbs.
Overall length (normal butt): 44.75 in.
Barrel length: 21 in.
Feed device: 20 or 30 round box magazine.
Sights: Front: Protected post.
Rear: Tangent w/aperture graduated from 200-1000 yds.
Muzzle velocity: 2800 f.p.s.
Cyclic rate of fire: 675-750 r.p.m.

The C2 is the heavy-barreled selective-fire version of the Canadian semiautomatic FN rifle C1. The method of operation and field strip of the C2 are similar to that of the C1. C2 like C1 has a prong-type flash suppressor and can be fitted with a bayonet or grenade launcher. The bipod legs are fitted with wooden strips, allowing the bipod to be used as a fore-end when in the folded position. Differences between the C2 and C1 rifles other than noted above are as follows:

a. The rear sight of C2 is different.
b. C2 has a bipod.
c. C2 has no handguard.
d. C2 change lever has 3 positions: safe, semiautomatic, and automatic fire.
e. C2 gas block assembly includes a mounting for the bipod.

This weapon is the Canadian squad automatic weapon and has the advantage of using most of the components of the basic rifle including magazines.

C2 is, like C1, made at the Canadian government small arms plant at Long Branch, Ontario.

7.62mm AUTOMATIC RIFLE FN C2A1

About 1960, C2A1—a modification of C2—was adopted. C2A1, like C1A1, has a two-piece firing pin and a plastic carrying handle. C2A1 is about one-quarter pound lighter than C2. As with all of the Canadian FN rifles, C2A1 can be fitted with any of three lengths of buttstock; normal, long, and short.

ROSS RIFLES

Description of Weapons

Although Ross rifles have not been used as first-line weapons since mid-World War I, there are still a fairly large number of them in circulation among collectors and sportsmen. There are two basic

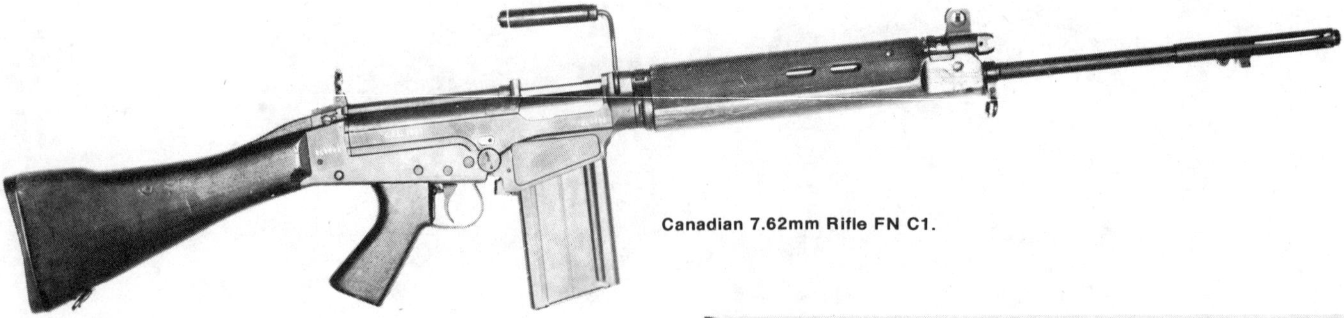

Canadian 7.62mm Rifle FN C1.

Receiver detail of 7.62mm Rifle FN C1.

Rear sight of 7.62mm FN C1 Rifle.

variations of the .303 Ross and a number of minor variations to the basic types; all are straight pull bolt actions. The Mark II, which is frequently called the Model 1905, has solid bolt locking lugs and the Harris type magazine. This magazine, which is flush with the stock, cannot be loaded with chargers. There is a magazine lifter thumb lever on the right side of the rifle ahead of the receiver. The Mark II may be found with tangent type sight or with leaf type sight. A cut-off projects into the forward part of the trigger guard. The Mark III has several variations in rear sights and front sights. The Mark III is frequently called the M1910. It is easily distinguished from the Mark II by its magazine which protrudes below the stock. The Mark III magazine can be loaded with a charger and the Mark III cut-off is mounted at the left rear of the receiver in the same position as that on the U.S. Springfield M1903. It also functions the same as that of the Springfield. The Mark III has locking lugs with interrupted screw thread.

7.62mm Rifle FN C1A1.

7.62mm Automatic Rifle FN C2.

Receiver detail of FN C2 Rifle.

7.62mm Rifle C2 with bipod fixed.

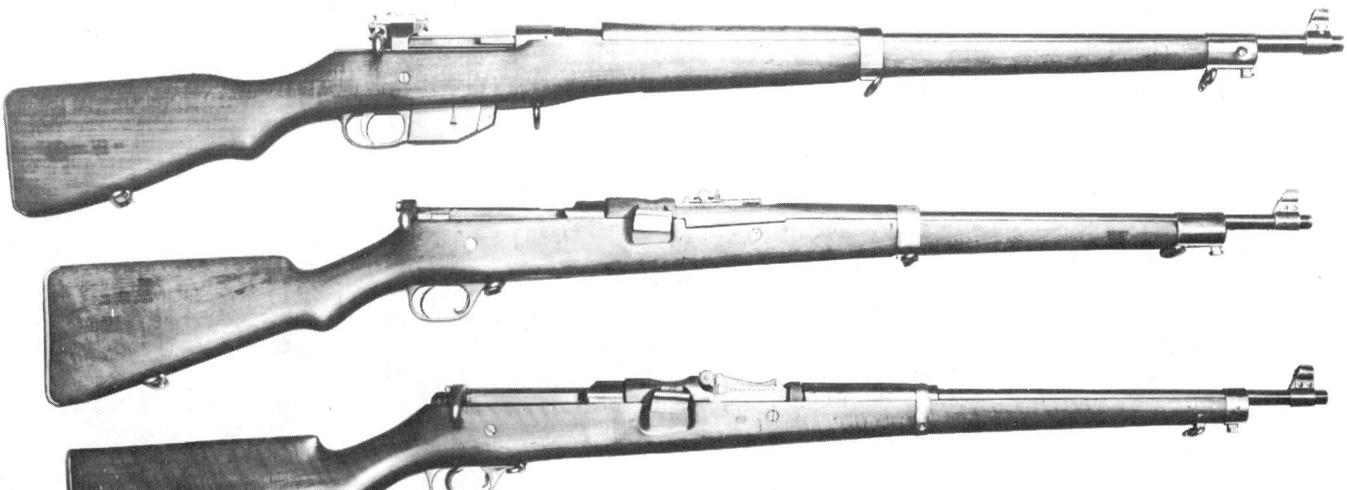

Caliber .303 Canada Ross rifles. From top down: Mark III, Mark II*, Mark II, and the Mark III*.

Special Note on the Ross Rifles

The Ross rifles are well made of good materials, but they had several serious design defects which caused their abandonment as an infantry weapon by Canada in 1916. The action is suitable for sporting rifles, but was found eminently unsuited for the mud of Flanders. In an attempt to make up for the poor extracting qualities of these weapons when dirty, the chamber was relieved slightly. This makes the Ross a rather poor rifle for a hand loader, since cases are badly stretched on firing.

The most serious problem with the Ross for the modern shooter is the fact that on most models the bolt can be reassembled wrong and yet put in the weapon and may fire a cartridge in an unlocked condition with resulting **serious injury, if not death** to the shooter. The bolt is assembled wrong if the distance between the bolt head and the bolt sleeve is less than one inch when the bolt is withdrawn from the rifle (bolt in unlocked posture). The bolt is exceedingly difficult to disassemble and reassemble and it is best to take it to a gunsmith if in doubt.

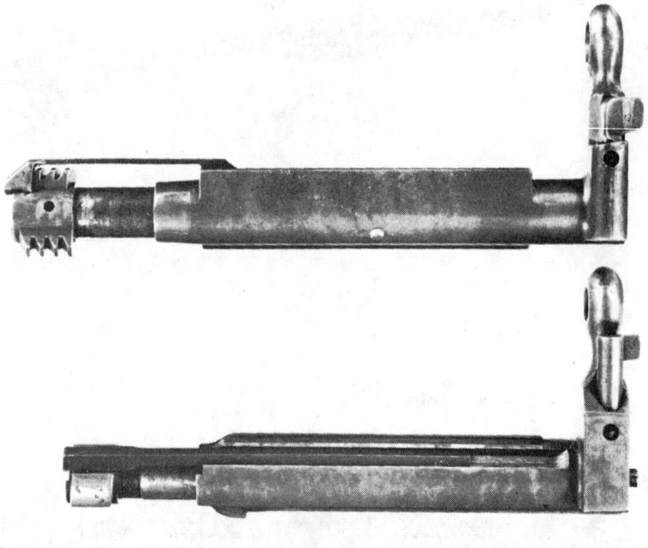

Bolt of the Ross M1910 Mark III Rifle at top and that of the Ross M1905 Mark I Rifle below.

CANADIAN SUBMACHINE GUNS

STEN SUBMACHINE GUNS

The Sten gun was produced in Canada in tremendous quantities during World War II. In addition, a good deal of development work was done on submachine guns at Long Branch during the war.

Modified Mark 2 Sten

A modified Mark 2 Sten was developed by Anton Roscziewski of that arsenal and has several unusual features for a Sten.

The trigger is curved at the top and the bottom; pressure on the top of the trigger produces automatic fire and pressure on the bottom produces semiautomatic fire. This feature is not unique, for it is essentially the same feature used on the German MG 13 and other weapons; it is unusual for a Sten. The other unusual feature is the operation of the magazine. The reciprocal action of the bolt works the magazine through a spring and plunger linkage. This method of feed control ensures a constant feed pressure on the cartridge thereby cutting down feed stoppages resulting from the irregular spring pressure frequently experienced with conventional magazines.

Experimental Rosciszewski Model 2

In 1945 Mr. Rosciszewski developed another submachine gun called the Model 2.

This weapon used the same type trigger mechanism as the modified Sten, but had several different features. The magazine is parallel to the barrel and therefore the cartridges must go through an arc of 90° to feed into the chamber. The breech block activates the cartridge lifter strut which pivots the cartridge lifter through 90°. The bolt is also different in that it has a forward extension, similar to that of the German G3 rifle, which rides in a tunnel over the barrel and to which the bolt handle is attached.

Canada has adopted the Sterling (Patchett) submachine gun as submachine C4. The Canadian Sterling differs in minor details from the L2A3.

Long Branch Arsenal

The Long Branch Arsenal was organized in 1940 at Long Branch Ontario, a suburb of Toronto as "Small Arms Ltd." On 1 January 1946 it was taken over by Canadian Arsenals Ltd., a government-owned corporation, and renamed Small Arms Division. The principal product in World War II was the No. 4 Mark I* rifle, but development work on a number of different type weapons was performed. Messrs. A. Rosciszewski, F. Kearsey, and D. Miller were the principal designers.

CANADIAN MACHINE GUNS

Canada used the .303 Lewis and Vickers guns in World War II. In World War II the John Inglis firm in Toronto manufactured large quantities of .303 Mark 1 (M) and Mark 2 Bren guns. In addition Inglis made substantial quantities of 7.92mm Mark 2 Brens for the Nationalist Chinese. A few .30-06 Brens were made, apparently for experimental work.

The Canadian forces engaged in the Korean conflict were mainly supplied with U.S. weapons in order to simplify the logistics problem. In addition Canada had considerable stocks of caliber .30 Browning guns, mainly from vehicles, left over from World War II. Canada ran tests of several 7.62mm NATO machine guns during the fifties and decided to convert existing stocks of caliber .30 Browning Model 1919A4 machine guns to 7.62mm NATO. The converted gun is the 7.62mm NATO C1 machine gun.

7.62mm NATO C1 Machine Gun

The C1 is supplied in fixed, for vehicular mounting, and flexible, for infantry, versions. In external appearance it differs very little from the U.S. caliber .30 Browning M1919A4. In the ground role it is used on a light tripod. The principal differences between the 1919A4 and the C1 are the barrel and the feed mechanism. The feed mechanism of the Browning was originally designed for a "pull out" type belt or metallic link. The feed mechanism of the C1 has been modified so that a "push through" type link -the U.S. M13- can be used with a "pull out" type mechanism. The front barrel bearing plug has also been modified to give the action more gas boost in order to provide more power to extract cartridges from the links. A link guide has been installed on the feedway and a short round stop has been fitted to the receiver in front of the feedway.

In addition to these and several other minor changes, two spring loaded ball bearings have been added to the T slot of the bolt. They prevent rounds from sliding completely through the T slot. The cartridge extractor has also been modified by removal of the ejector finger, spring and pin; and the rear sight leaf has been engraved with new graduations in meters to match the ballistics of the 7.62 mm NATO round.

7.92mm Mark 2 Bren made by Inglis for Nationalist China during World War II

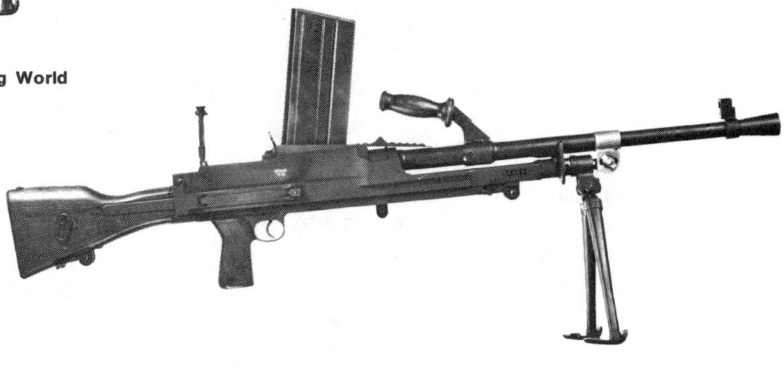

.30-06 Canadian Mark 2 Bren, Serial Number X38

Experimental Rosciszewski Model 2 submachine gun.

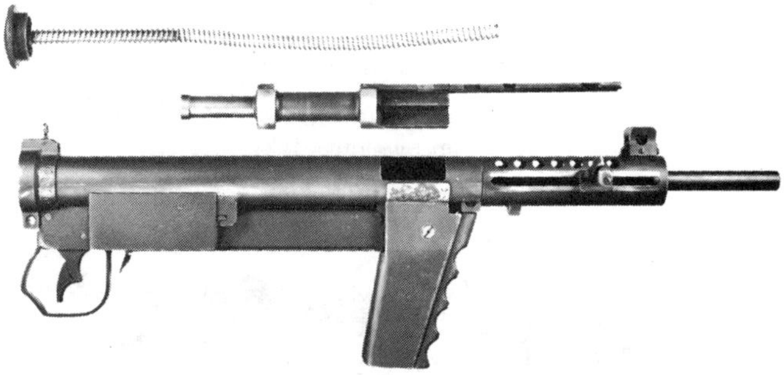

Experimental Model 2 submachine gun—field-stripped.

Canadian 9mm Submachine Gun C4.

Chile

Chile has adopted the FN 7.62mm light and heavy barrel rifles as standard. In addition they have purchased SIG weapons, German 7.62mm MG42/59s and have a quantity of U. S. cal. . 30 M1 rifles and cal. .30 carbines. They also have a number of 7mm Madsen machine guns.

19

CHILEAN RIFLES

As a result of Chile's adoption of the FN rifle, United States arms dealers have purchased quantities of Chilean Mausers which are now in this country in quantity for the first time.

SPECIAL NOTE ON CHILEAN MAUSERS

The Chilean Mausers are generally similar to the Spanish M1893 Mauser. The specimens imported into the United States are all of German manufacture, and are of good finish and materials. Loading, firing, field stripping and functioning is the same as other bolt action Mausers. It should be noted that the Chilean M95, as the Spanish M1893, does not have a bolt sleeve lock. Therefore in field stripping, the bolt sleeve assembly will turn off easily when the bolt is withdrawn with safety in the middle position (straight up).

There is also a Chilean M95 short rifle which falls between the rifle and carbine in length and weight.

Characteristics of Chilean Mauser M1895 Rifle and M1895 Carbine

	M1895 Rifle	M1895 Carbine
Caliber:	7mm	7mm
System of operation:	Manually operated bolt action.	Manually operated bolt action.
Weight:	8.9 lbs.	Approx. 7.5 lbs.
Length, overall:	48.5 in.	37 in.
Barrel length:	29.06 in.	18.25 in.
Feed device:	5-round, non-detachable staggered box magazine.	5-round, non-detachable staggered box magazine.
Sights: Front	Barleycorn.	Barleycorn.
Rear	Leaf.	Leaf.
Muzzle velocity:	Approx. 2700 f.p.s.	Approx. 2600 f.p.s.

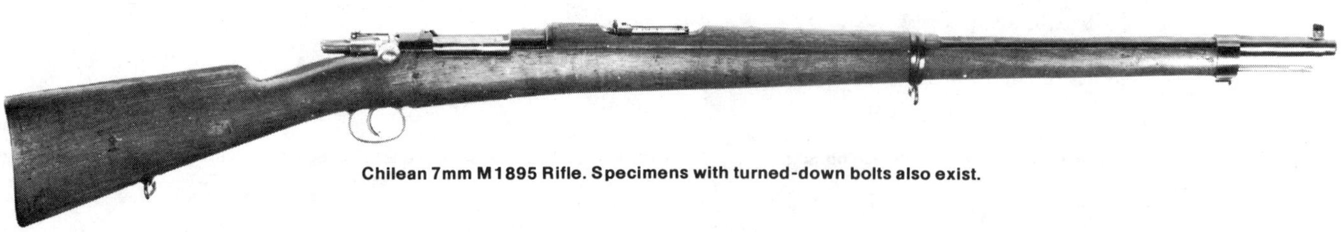

Chilean 7mm M1895 Rifle. Specimens with turned-down bolts also exist.

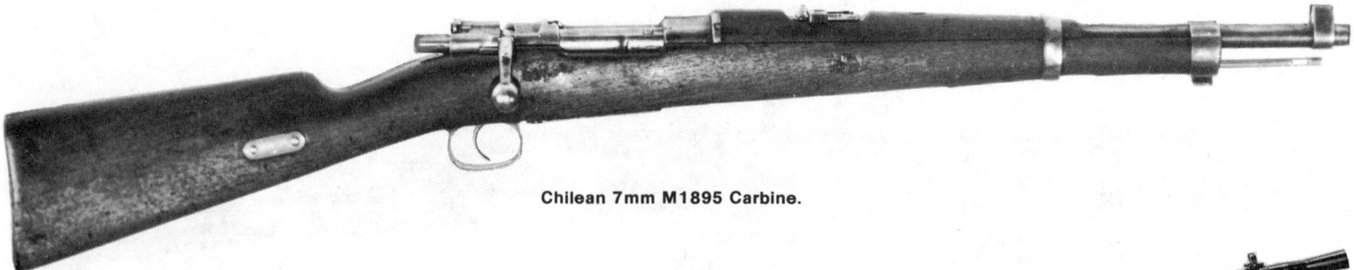

Chilean 7mm M1895 Carbine.

Chilean 7mm M1895 Short Rifle.

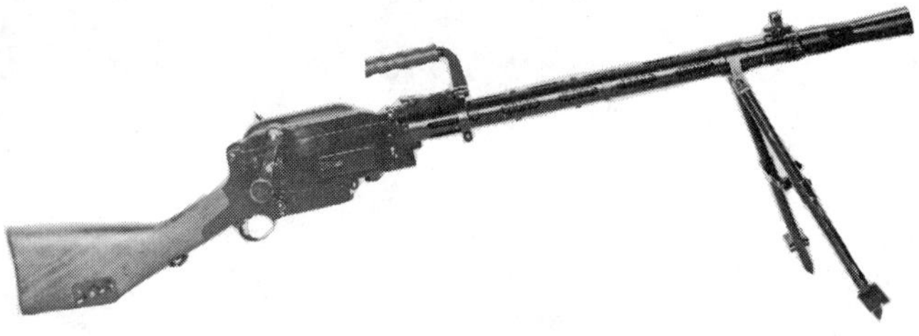

Chilean 7mm Madsen machine gun.

Free China: "The Republic of China" (on Taiwan)

20

The Chinese Nationalist Army usually used the term "Type" rather than "Model" for the nomenclature of weapons. The Chinese character for "Type" can be translated as "SHIH" or "SHIKI" and is the same as that used by the Japanese. Since a type designation may be followed by a model designation to indicate a modification (as with the old Japanese system, for example, the Type 34 Model 1), Chinese markings on weapons must be carefully examined to secure the proper nomenclature. On many (but not all) Chinese Nationalist weapons, the Type designation indicates the date of adoption of the weapon in number of years since the Chinese revolution—1911. Thus the Type 36 submachine gun was adopted in 1947, the Type 41 light machinegun was adopted in 1952. The Chinese Communists, however, use the calendar year designation, i.e., 51 for 1951, etc.

Although for a period of time the Chinese (free Chinese) and the Chinese Communists armies used the same types of small arms, this picture has changed considerably since the Korean war. During World War II, the Chinese Nationalists received large quantities of small arms from the United States. After World War II, the Chinese Nationalists started production of some U. S.-type weapons on the mainland of China, and also took over Japanese ordnance plants in the Manchurian area. The production of Japanese weapons was continued, as was the production of Japanese small arms ammunition. However, in an effort to standardize on ammunition, some of the Japanese weapons were rebarreled or made for 7.92mm cartridges.

Thus, at the time the Chinese Nationalists were forced to leave the mainland, they had some of their earlier weapons chambered for the 7.92mm cartridge, American weapons chambered for U. S. cartridges, and Japanese weapons chambered for 7.92mm and also for Japanese cartridges. Since the Chinese Nationalist Army has been on Taiwan (Formosa), it has been able to standardize its weapons, and is from a technical and logistical point of view, immeasurably better off than it was on the mainland. The Chinese Nationalist Army at the present time uses U. S. cal. .30 M1, M1903, and M1903A4 rifles. It also uses all other U. S. small arms, excepting 7.62mm NATO weapons and those others which are not as yet completely in service in the U. S. Army. A Chinese-modified copy of the Bren gun, the Type 41, is also made and used by the Nationalists. This weapon is also chambered for U. S. cal. .30 cartridges. Since many of the Chinese-made or Chinese-modified weapons are still in being (although not accepted as standard anywhere), a short description of these weapons, and illustrations of some of them, are included in this section.

The Republic of China produces the 7.62mm NATO M14 rifle and M60 machine gun.

CHINESE PISTOLS

As with all other small arms, the Chinese had a wide collection of pistols. The Mauser in 7.63mm was a great favorite, and Chinese-made copies of the Mauser in cal. .45 have been encountered. During World War II, the United States supplied the Chinese with cal. .45 M1911A1 automatics, with Colt and Smith & Wesson M1917 revolvers, as well as some cal. .38 revolvers. The Canadians supplied the Chinese with the 9mm Browning Hi-Power pistol made by John Inglis in Toronto.

Chinese 9mm FN Browning Hi-Power pistol made by Inglis at Toronto.

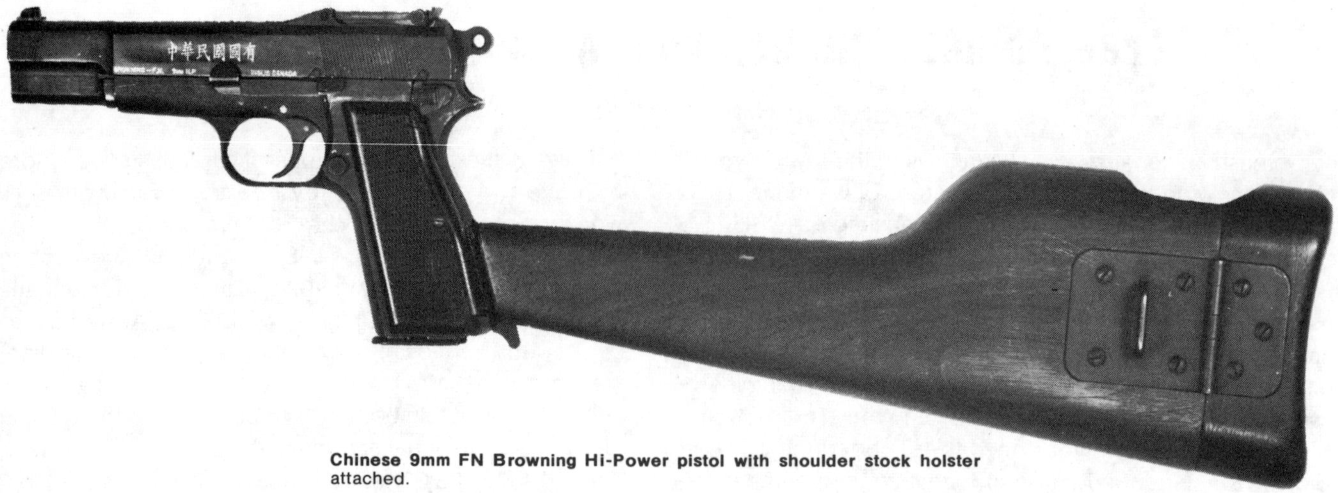

Chinese 9mm FN Browning Hi-Power pistol with shoulder stock holster attached.

CHINESE RIFLES

The pre-World War II Republic of China had many rifles in service. The oldest was probably the Type 88 or Hanyang rifle as it was called from its place of manufacture. This weapon is a copy of the German M1888 rifle, large quantities of which were sold to China by Germany after the Germans adopted the Model 98 rifle. This rifle is chambered for the old 7.92mm x 57mm rimless cartridge, and has a .318 bore rather than the .323 bore of the 98 and later 7.92mm weapons. Therefore, this weapon should not be used with 7.92mm x 57mm IS (sometimes called JS) ammunition. The Chinese issued a special conical-nosed ball cartridge for this weapon. It should also be noted that Chinese weapons of pre-World War II manufacture are widely variable in quality of materials and construction. Chinese weapons manufactured since World War II are, so far as can be determined, made of first-class materials and show fine workmanship. Other Chinese rifles of this period are:

7.92mm Belgian FN M1924 and M1930 rifles, and Chinese copies.

7.92mm Czechoslovak Brno M1924 rifles, and Chinese copies.

The Mauser 7.92mm "Standard Model," a Mauser export model which was copied by the Chinese in 1935 and is called the "Generalissimo" or "Chiang Kai Shek" model. This model is now called the type 79 by the Chinese Communists. All these weapons were used by Chinese Communist troops in Korea.

The 7.62mm NATO Type 57 Rifle

The Chinese Nationalists manufacture a copy of the U.S. M14 rifle called the Type 57 (1968). This rifle varies only in minor details from the U.S. made rifle. The most noticeable difference is the shape of the stock immediately behind the receiver. The Chinese made rifle has a more pronounced flat surface immediately behind the rear of the receiver than does the U.S. made rifle.

Type 57 7.62mm NATO Chinese Nationalist Rifle, copy of U.S. M14.

Chinese copy of the Model 1928 Thompson submachine gun.

Chinese 7.92mm Type 88—Hanyang—Rifle; a modified copy of the German Model 88.

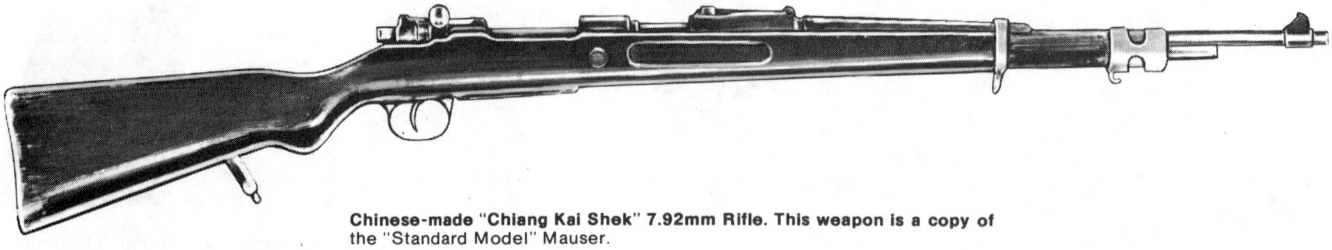

Chinese-made "Chiang Kai Shek" 7.92mm Rifle. This weapon is a copy of the "Standard Model" Mauser.

CHINESE SUBMACHINE GUNS

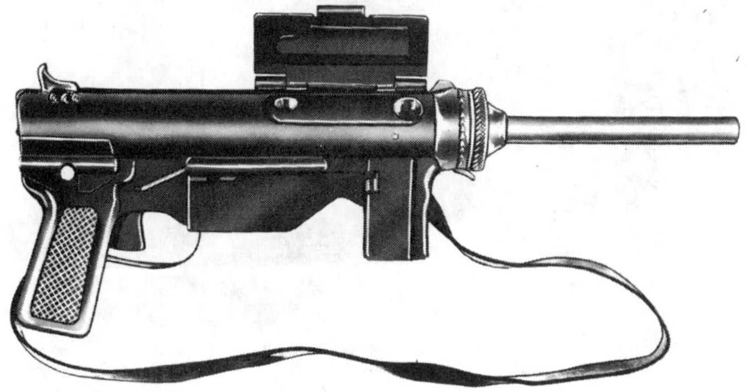

Type 36 submachine gun. A caliber .45 Chinese-made copy of the U. S. M3A1 submachine gun. Similar weapons in 9mm Parabellum were made by the Chinese.

Caliber .45 submachine gun, based on design of Japanese Type 2, believed to be of Chinese manufacture.

The most common submachine guns in the Chinese Army prior to and during World War II were:

All the models of the U.S. Thompson in cal. .45 (frequently called 11mm by the Chinese) plus a Chinese-made copy of the 1928 Thompson.

The U.S. M3 and M3A1, and a Chinese copy of the M3A1 called the Type 36, in cal. .45, and Type 37 in 9mm Parabellum.

British Stens chambered for the 9mm Parabellum cartridge. The Chinese also made some copies of these.

The Thompson, the M3A1, and the Type 36 are still in use in the Chinese Nationalist Army.

CHINESE MACHINE GUNS

The Chinese used a wide variety of machine guns prior to and during World War II. Most of these weapons again appeared in Korea during the early stages of the Chinese Communist commitment. The most common weapons were:

The Chinese-made Type 24 heavy machine gun. This 7.92mm weapon is a modification of the German Maxim Model 08, (MG 08).

Czechoslovak- and Chinese-made 7.92mm ZB 26 and ZB30 light machine guns. These weapons were called Type 26 and Type 30 by the Chinese.

Swiss SIG-made KE 7 light machine guns in 7.92mm caliber.

Danish Madsen-made 7.92mm light machine guns.

Chinese-made 7.92mm Maxim aircooled machine guns.

French-made Hotchkiss M1914 heavy and Model II light machine guns in 7.92mm caliber.

Colt and FN-made Browning automatic rifles, also chambered for the 7.92mm cartridge. There apparently were not too many of these.

Colt and FN-made Browning watercooled machine guns in 7.92mm. These were also apparently in short supply.

Chinese-made Type 24 heavy machine gun. A 7.92mm weapon similar to the German 08 Maxim gun.

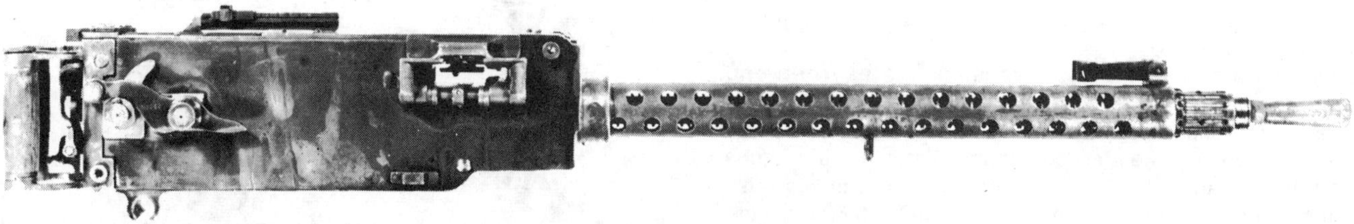

Chinese 7.92mm air-cooled Maxim machine gun.

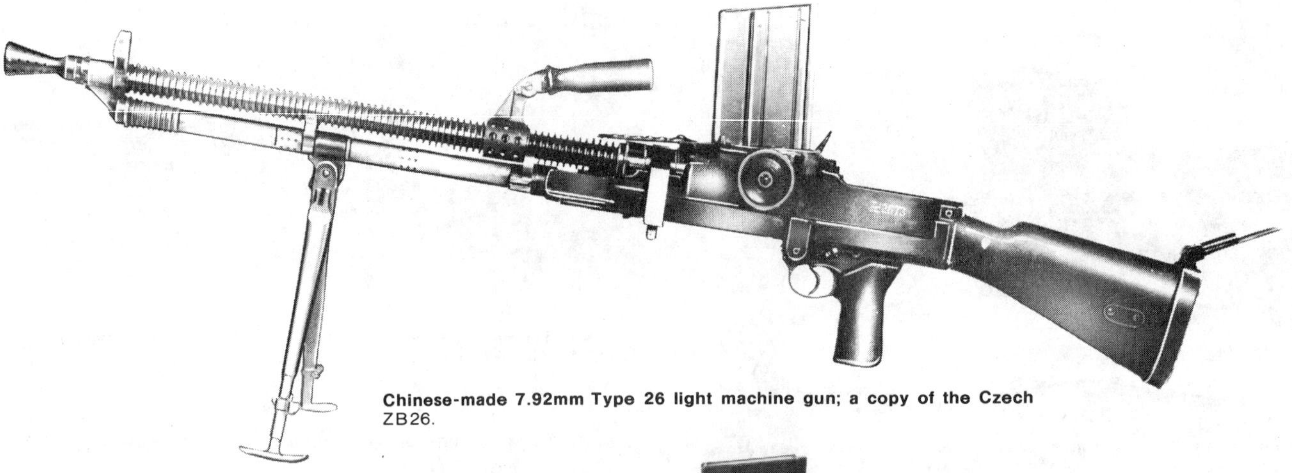

Chinese-made 7.92mm Type 26 light machine gun; a copy of the Czech ZB26.

Chinese-made Type 41 light machine gun. Bren-modified for U. S. caliber .30.

Chinese 7.92mm M1937 Madsen machine gun.

7.92mm Bren guns Mk. 2, made in Canada for the Chinese during World War II.

The list given above is by no means complete. When one considers purchases, Lend-Lease, war booty, etc., the Chinese Army was a gun collector's paradise and an ammunition supply officer's nightmare. All the standard United States machine guns and most of the standard British machine guns were in wide use before the end of World War II. The 7.62mm NATO M60 machine gun, made in Taiwan, in now the standard machine gun of the Nationalist Chinese Army.

Colt Model 38B Machine Gun. Modified Browning M1917A1. Supplied to China in 7.92mm before World War II.

Communist China:

"The People's Republic of China" (on Chinese mainland)

21

At the conclusion of World War II, the Chinese Communists inherited large quantities of Japanese weapons from the Soviets, who had taken them in Manchuria. During the Chinese civil war, they captured respectable quantities of materiel from the Chinese Nationalists, including quantities of U. S. weapons. When the Chinese Communists entered the Korean War, they had far more Chinese, Japanese, and U. S. materiel in the battle area than Soviet materiel. However, battle losses, breakages, and probably the impossibility of fighting a war of attrition like the Korean War with their crazy-quilt weapons system, resulted in the Chinese Communists ending the Korean War with far more Soviet equipment than any other type.

The Chinese Communists used all the Soviet pre-World War II and World War II weapons in Korea. They also put some of these weapons into production, using their own year of adoption as their model designations. The Soviet the Chinese Communists as the Type 50. This weapon is like the Soviet weapon covered in the chapter on the USSR. The major difference between the Chinese-and the Soviet-made guns is that the Chinese-made weapon has a differently shaped buttstock and is usually found only with a box magazine. The Soviet 7.62mm TTM1933 pistol was adopted and put into production as the Type 51. The Soviet M1944 Mosin Nagant carbine was put into production as the Type 53. The Soviet DPM light machine gun was put into production as the Type 53.

The Chinese Communists have come out with a new series of small arms of native design within the last decade. Among these weapons are the 7.62mm Type 68 rifle, the 7.62mm Type 67 light machine gun, and the 7.62mm Type 64 submachine gun, a silenced weapon.

SERVICE WEAPONS OF THE CHINESE COMMUNIST ARMY

The Chinese Communist army is currently equipped with Chinese-made copies of Soviet small arms as follows:

- Soviet 7.62mm TT M 1933 Pistol, **Chinese Communist Type 51 Pistol.**
- Soviet 7.62mm AK-47 assault rifle, **Chinese Communist 7.62mm Type 56 assault rifle.**
- Soviet 7.62mm SKS carbine, **Chinese Communist 7.62mm Type 56 carbine.**
- Soviet 7.62mm RPD light machine gun, **Chinese Communist Type 56 machine gun.**
- Soviet 7.62mm DPM light machine gun, **Chinese Communist Type 53 light machine gun.**
- Soviet 7.62mm RP-46 light machine gun, **Chinese Communist Type 58 light machine gun.**
- Soviet 7.62mm SG-43 heavy machine gun, **Chinese Communist Type 53 heavy machine gun.**
- Soviet 7.62mm SGM heavy machine gun, **Chinese Communist Type 57 heavy machine gun.**
- Soviet 12.7mm DShK M1938/46 heavy machine gun, **Chinese Communist Type 54 heavy machine gun.**

The 14.5mm Soviet heavy machine gun KPV is also used on the ZPU-1, ZPU-2, and ZPU-4 mounts which have one, two, and four guns, respectively.

Many of the older model Soviet weapons covered later in this text are still in use and many of these weapons, as well as some of the smaller, newer weapons listed above, have been shipped to the Viet Cong in South Vietnam.

Chinese Communist 7.62mm Type 57 heavy machine gun; copy of Soviet SGM.

Chinese Communist Silenced Pistol

This weapon is quite clever in design. It may be used as a single shot weapon loading the cartridges by manually working the slide or as a standard semi-automatic pistol by pressing the push-through type bar mounted in the slide from left to right or vice versa. When the weapon is used as a single shot locking lugs on a rotating bolt engage locking surface in the receiver. There is no noise of slide movement and no ejected empty cartridge case when used in this mode. When used as a semi-automatic, the rotating bolt in the slide is locked into position within the slide so that its locking lugs do not engage the locking surfaces in the receiver. There is therefore the noise of the slide reciprocating and the empty cartridge cases are forcefully ejected. The cartridge used by this weapon is similar to the standard 7.65mm (.32 ACP) cartridge but is rimless rather than semi-rimmed.

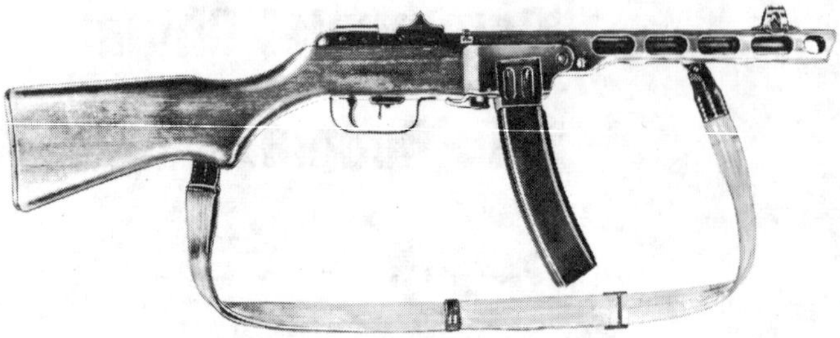

Type 50 7.62mm submachine gun. A copy of the Soviet PPSh M1941.

Communist Chinese 7.65mm silenced pistol, drawn partially to the rear.

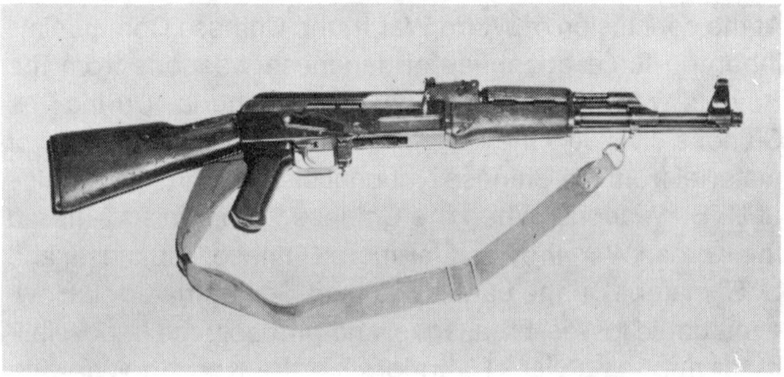

Chinese Communist-made 7.62mm Type 56(AK) assault rifle.

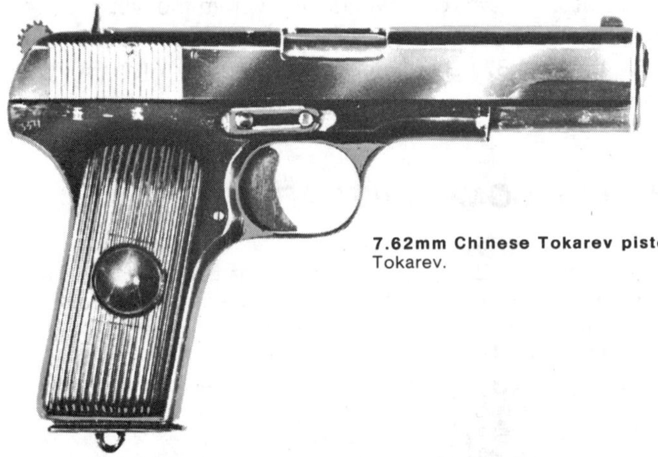

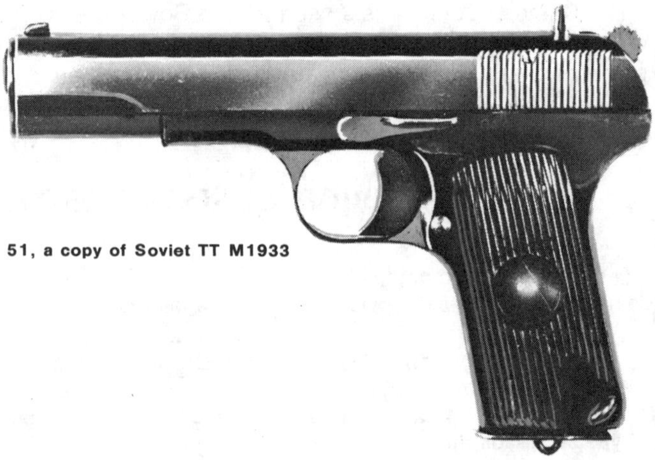

7.62mm Chinese Tokarev pistol Type 51, a copy of Soviet TT M1933 Tokarev.

Chinese Communist 7.62mm Type 56-1 Assault Rifle.

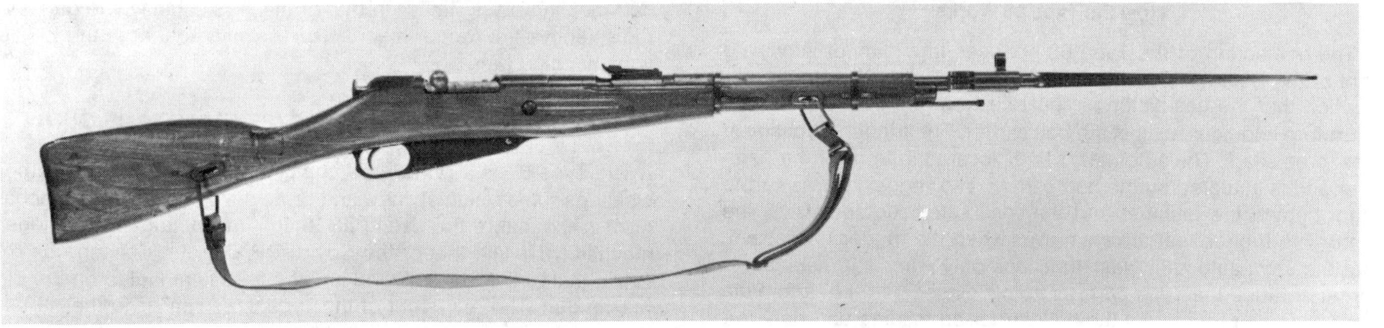

Chinese Communist 7.62mm Type 53 carbine; copy of Soviet M1944 carbine.

Communist Chinese 7.62mm Type 56 Assault Rifle with folded bayonet.

Communist Chinese 7.62mm Type 53 Rifle with grenade launcher.

Communist Chinese Modifications of Soviet Designed Weapons

The Chinese Communists have modified several of the copies of Soviet weapons which they manufacture. The standard Soviet 7.62mm AK assault rifle, manufactured in Communist China as the Type 56, has a removable knife type bayonet. The Communist Chinese still use the Type 56, but are also making a modification which has a folding type, cruciform section, bayonet attached to a stud on the under section of the front sight base. The Chinese Communists are also making the 7.62mm Type 56 Carbine (copy of Soviet SKS) with a cruciform section folding bayonet.

The 7.62mm Type 53, which is a copy of the Soviet 7.62mm M1944 Mosin Nagant carbine has appeared with a rifle grenade launcher. The launcher is of the removable type and has a clamp type lock which engages behind the front sight in a manner similar to the U.S. M7 and M8 rifle grenade launchers. When the rifle grenade launcher is attached, the folding bayonet cannot be fixed.

CHINESE COMMUNIST 7.62mm TYPE 68 RIFLE

The Type 68 is a native Chinese designed selective fire weapon. Externally it resembles the Soviet SKS carbine modified to use a detachable box magazine. Internally it is quite different; the bolt and bolt carrier mechanism are basically the same as that of the AK but the piston is a separate piece. The Type 68 is made in two different versions one having a stamped steel receiver and the other having a machined steel receiver. Unlike the SKS or the AK the Type 68 has a gas regulator whose retainer is mounted on the left of the gas cylinder beyond the hand guard. The amount of gas fed into the gas cylinder tube can be regulated by pressing the retainer in until it disengages from the hand guard and then rotating the retainer down until it can be pulled out of the gas cylinder. The gas regulator has indicator holes which correspond to the size of the gas ports in the regulator and the indicator hole that is closest to the barrel shows how much gas is being fed to the action.

Characteristics of the Type 68 Rifle

Caliber: 7.62mm (7.62 X 39)
System of operation: gas, selective fire
Weight: 7.7 lb.
Length overall:' 40.5 in.
Barrel length: 20.5 in.
Feed device: 15 round, staggered row, detachable box magazine
Sights: Front: hooded post
Rear: tangent with notch
Muzzle velocity: 2395 f.p.s.
Cyclic rate: 750 r.p.m.

How to load and fire the Type 68

The Type 68 rifle can be loaded either with SKS type chargers, (stripper clips), or by removing the magazine and loading the cartridges into it individually. Pulling the operating handle to the rear with a loaded magazine in place will chamber a cartridge. The combination safety/selector lever is on the right side of the rifle in front of the trigger guard. The figure "O" setting is used to put the rifle on safe, the figure 1 setting is used for semi-automatic fire and the figure 2 setting is used for automatic fire. The bolt remains open after the last round is fired.

How to Field Strip the Type 68

Clear the weapon, press in the lock at the left rear of the receiver and pull the cover rearward off the receiver. If the rifle has a machined receiver pull the driving spring guide up and out of its seat in the rear of the receiver and remove the driving spring assembly. If the rifle has a stamped receiver press in lock and push the driving spring and it's guide forward until it is clear of its seat. Ease the spring forward then remove it. Pull the operating handle rearward while keeping downward pressure on the bolt carrier. When the bolt carrier is about one inch from the rear of the receiver it can be lifted up and out of the receiver. The bolt is removed from the carrier by pushing it to the rear, turning the bolt head and pulling it forward and out of the carrier.

The gas regulator has to be removed in order to remove the gas piston and its spring. It can be removed by turning its retainer down as explained earlier. Lower the muzzle and the gas cylinder and its spring will slide out. The hand guard can then be slid forward off the gas cylinder. The heat shield which may be found on the gas cylinder can be slid to the rear and removed.

The trigger guard catch is at the rear of the trigger guard. Rotate it and the trigger group is disengaged; the trigger guard can now be pulled free of the rifle. Swing the bayonet downward and the stock can be disengaged from the barrel and receiver. Reassemble in reverse order.

How the Type 68 Works

The bolt action of the Type 68 operates in a manner similar to that of the Soviet AK assault rifle, (Chicom Type 56), with the exception that the gas piston is separate from the bolt carrier and therefore impinges against the bolt carrier. The trigger mechanism has three sears. The automatic sear is located in front of the hammer and is actuated by the bolt carrier. The trigger sear is to the right behind the hammer and the semi-automatic sear is to the right. The trigger sear always moves when the trigger is pressed, but the semi-automatic sear functions only when the selector is set at 1. Since the Type 68 fires from a closed bolt the automatic sear has been tripped by the bolt carrier on loading and only the trigger sear holds the hammer cocked. When the trigger is pressed the trigger sear releases the hammer firing the weapon. As the bolt carrier moves to the rear it pushes the hammer rearward and releases the automatic sear to hold the hammer in cocked position until the bolt carrier returns fully forward at which time the full automatic sear releases the hammer and the rifle fires again, and will continue to fire until the trigger is released or the magazine emptied. The above happens when the selector lever is set on the figure 2. When the selector lever is set on the figure 1 the selector releases the semi-automatic sear which is then controlled by the trigger. When the trigger is pressed it causes the trigger sear to release the hammer and allows the semi-automatic sear to move to a position where it can catch the hammer; therefore the trigger must be pressed for each shot. If the selector is set at 0 the selector blocks movement of the trigger sear and prevents release of the hammer.

THE 7.62mm TYPE 64 SUBMACHINE GUN

The Type 64 is a blowback operated weapon chambered for the 7.62mm pistol cartridge and fitted with a Maxim type silencer. The bolt mechanism is similar to that of the PPS submachine gun and the selective fire trigger mechanism is similar to that of the ZB 26 light machine gun.

THE CHINESE COMMUNIST TYPE 67 MACHINE GUN

The Type 67 is a rather odd combination of a number of other weapons some relatively new and some quite old. The bolt mechanism is like that of the Czech ZB30 light machine gun, the trigger mechanism is like that of the Soviet DP, the quick change barrel is like that of the Soviet SG43 and the feed is modeled on the old Maxim gun. The rear sight of the Type 67 is similar to that of the Czech Type 59 machine gun, but has a permanently fixed AA rear sight post on its top. A removable speed ring type AA front sight is mounted at the front of the receiver. The Soviet style stock of the Type 67 is made of plastic. The type 67 also has the screw type headspace adjustment feature of the Soviet SGM machine gun.

The Type 67 is chambered for the 7.62mm X 54mm rimmed cartridge and can be used either on a bipod or a tripod. The tripod can be used for AA fire as well as ground fire.

Characteristics of the Type 67 Machine Gun

Caliber: 7.62mm (7.62 X 54)
System of operation: gas, automatic fire only
Weight: 21.8 lb.
Length overall: 45 in.
Barrel length: 23.5 in.
Feed device: 100 round metallic link belt
Sights: Front: protected post
Rear: leaf
Muzzle velocity: 2,650 f.p.s.
Cyclic rate: 600 - 700 r.p.m. est.

CHINESE COMMUNIST MILITIA WEAPONS

The Communist Chinese have a "People's Militia" which numbers in the millions. This militia, which is a part-time local-defense type force, is armed with an amazing variety of small arms. Apparently the Chinese never scrap or throw away a weapon. Among the weapons still used are Japanese 6.5mm Type 38 rifles made at Shansi and Shenyang arsenals and called Type 65, 7.92mm Mauser "Standard" models—same as Chinese Nationalist "Generalissimo" model—called Type 79, the Japanese Type 99 7.7mm rifle—also found in 7.92mm, U.S. caliber .30 rifles and carbines of all models since the M1903, all the pre-World War II and World War II Japanese machine guns, German 7.92mm MG 34s and U.S. British, Chinese, and German submachine guns. The effectiveness of this force from a logistical point of view is, to put it mildly, dubious.

Chinese Communist 7.62mm Type 56 light machine gun, copy of Soviet R.P.D.

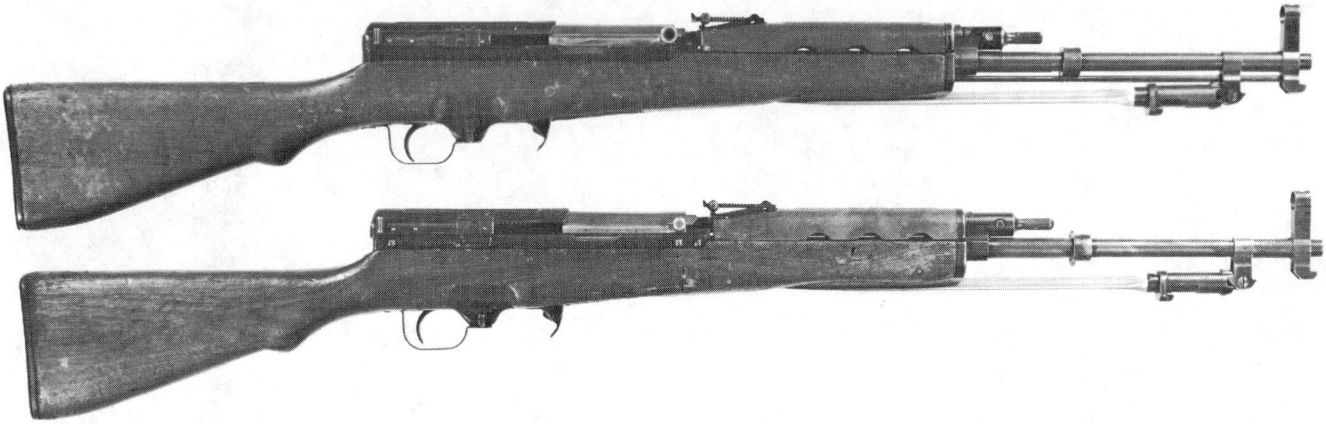

7.62mm Chicom Type 68 Rifles without magazines. Top rifle has machined steel receiver; rifle below has a stamped steel receiver.

Chinese Communist 7.62mm Type 68 Rifle field-stripped.

Close-up of action, Chicom Type 68 Rifle.

Communist Chinese 7.62mm Type 53 Heavy Machine Gun on Chinese tripod.

22

Cuba

Cuba is currently reported to have the largest army in the Western Hemisphere next to the army of the United States.

The Cuban Army, prior to the seizure of power by Fidel Castro, was mainly equipped with U.S. small arms. The U.S. caliber .30 M1903 Springfield rifle and the U.S. caliber .30 M1 rifle were used as were U.S. Browning machine guns.

Purchases were made of Belgian 7.62mm NATO FN "FAL" rifles and 7.62mm NATO MAG machine guns prior to the embargo imposed by the Western nations on arms shipments into Cuba. Quantities of these Cuban "FAL" rifles were found in Venezuela where they had been shipped to arm Communist forces attempting to overthrow the Venezuelan government.

Cuba has been receiving arms shipments from the Soviet bloc on a very substantial scale for the past several years. Soviet bloc small arms displayed in parades and newsreels by the Cubans include the following: Czech 7.92mm Model 37 machine guns, Czech 12.7mm Quad DShK M1938/46 machine guns, Soviet 7.62mm DP, DT, and DTM machine guns, Soviet 7.62mm SG43 Goryunov machine guns, Czech 9mm Model 23 and 25 submachine guns, and Czech 7.62mm Model 52 rifles. The Soviet 7.62mm AK47 assault rifle and the Soviet 14.5mm KPV heavy machine gun on the Quad ZPU4 mount are also currently in service in Cuba.

All the weapons mentioned above are covered in the chapters on the countries of origin of the weapons.

The multiplicity of types and calibers of weapons currently in use in Cuba creates for the Cubans a very large logistical headache from the point of view of weapon maintenance and ammunition supply. Although Cuba has the capability to produce limited amounts of spare parts, it is not economically or industrially feasible for them to produce spares for all the weapon types they have.

23

Czechoslovakia

Czechoslovakia is entering her third post World War II generation in some categories of small arms. The current small arms are the 7.62mm Model 52 pistol, the 7.65mm Model 61 submachine gun, the 7.62mm Model 58 assault rifle, the 7.62mm Model 54 sniper rifle, the 7.62mm Model 52/57 light machine gun, the 7.62mm Model 59 general purpose machine gun and the Czech-made 12.7mm DShK M1938/46 heavy machine gun. All these weapons with the exception of the DShK are of native design. The 9mm Models 23 and 25 and the 7.62mm Models 24 and 26 submachine guns are no longer standard and, like the now also obsolescent 7.62mm Models 52 rifle and machine gun, have been exported in large quantities.

All weapons now in service use standard Soviet type ammunition—the 7.62mm M43 (called M57 in Czechoslovakia) for assault rifle and light machine gun; and the 7.62mm rimmed for the Model 59 machine gun and 12.7mm for the DShK. An exception is the Model 61 submachine gun, which uses the 7.65mm—.32 Colt ACP cartridge. The Czech-designed 7.62mm Model 52 cartridge is no longer standard.

A word on Czech nomenclature: the Czech word for Model is Vzor. This word is usually abbreviated to "VZ" in markings on Czech weapons and has frequently been picked up in English language publications as part of the nomenclature as per example "Light Machine Gun Model VZ 26." This of course is the same as saying "Model Model" and in the following text the English term "Model" is used for "VZ" wherever applicable.

CZECHOSLOVAK PISTOLS

Czechoslovakia has developed a number of automatic pistols of generally good design. The first Czech designed and made service automatic, the 9mm Short (.380 ACP) Model 22 was actually based on the German Nickl design and was produced to a limited extent, by Mauser as well as by Ceska Stan Zbrojovka, Brno. This pistol has a rotating locked barrel somewhat similar to the M12 Austrian Steyr pistol.

An earlier pistol, the Praga, had been produced for police use by Zbrojovka Praga (Prague). The Model 22 was apparently made in very limited quantity and was soon followed by the Model 24 which has a slightly modified locking and firing mechanism. This pistol was made in considerable quantity and production apparently continued until 1937.

The 7.65mm Model 27 was the last of this series of Czech pistols. Originally made in Prague by Ceska Zbrojovka, like the Model 24, manufacture was carried on after World War II at Strakonice until 1950 or '51. During the German occupation of Czechoslovakia, the name of the plant in Prague was changed to Böhmische Waffenfabrik A.G. and pistols made there during the period between the seizure of Czechoslovakia and the beginning of World War II bear that marking. Pistols made during the war are marked "fnh." The Ceska Zbrojovka marking was resumed after the war; after 1948 (the time of the Communist take-over) the words "Narodni Podnik" (People's Factory or Cooperative Enterprise) were added. Although the M24 and the M27 are externally quite similar the M27 is a blowback operated arm, while the M24, like the M22, is recoil operated.

In 1938 Ceska Zbrojovka, commonly known as CZ, introduced a new pistol, the 9mm Short (.380 ACP) Model 38 for the Czech Army. This weapon was designed by Frantisek Myska and was considerably different than the earlier designs. The trigger mechanism is double-action and the barrel is permanently mounted in a collar which is hinged to the front of the receiver. The Germans called this pistol the Model 39(t).

Since World War II the Czechs have produced several pistols which have military or police application. The 9mm Parabellum CZ47 was apparently produced as a prototype only, as recent Czech literature does not even mention the weapon. The 7.65mm M1950 is basically a modified copy of the Walther PP and was designed as a police weapon and for commercial sale. The 7.62mm Model 52 is entirely different from the earlier Czech pistols; it fires a heavily-loaded cartridge and utilizes a unique locking system for a pistol. The Model 52 is the current Czech service pistol. A number of pocket automatics, target pistols, and a target revolver chambered for the .38 special cartridge, are also currently made in Czechoslovakia. The ZKP524 is a recent Czech automatic chambered for the 7.62mm cartridge and copied from the Colt Browning design.

CZECH 9mm SHORT M22 PISTOL

This pistol is the first of the series of Nickl designs made by the Czechs. The pistol was made on license from Mauser. Mauser submitted a modified form of this pistol chambered for the 9mm Parabellum cartridge, to the German Army during the thirties for consideration in the tests which resulted in the Germans adopting the P38. The Model 22 and its successor, the Model 24, are somewhat unusual in that they are locked-breech pistols, using a relatively low-powered cartridge.

How to Load and Fire the Model 22 Pistol

The magazine catch is on the base of the receiver (frame) grip at the rear; pulling this to the rear releases the magazine. The magazine is loaded round by round and then inserted smartly into the grip until the magazine catch snaps into place. Pull the slide to the rear and release; this will chamber a cartridge and the weapon is now loaded. Pressure on the trigger will cause the weapon to fire; pressure must be applied for each shot. On firing the last round the slide will move to the rear and stay open. When the magazine is removed the slide will run forward. The safety is on the left side and is similar to that of the M1910 Mauser pistol (Mauser M1910 9mm short magazines can be used in this pistol). A lever is pushed down to put the pistol on safe, pressing the button below releases the safety.

Field Stripping the Model 22 Pistol

With empty magazine in gun, pull slide to rear. Slide will remain in position; push down on dismounting catch on side of receiver over front of trigger guard. Remove magazine and slide may then be moved forward and lifted off the receiver. The barrel can then be removed by turning barrel bushing 30°, bushing can then be pulled off slide and barrel. Pull barrel out of slide. No further disassembly is recommended. Reassemble by reversing the above procedure.

Czech 9mm Model 22 Pistol.

How the Model 22 Pistol Works

Pressure on the trigger causes the trigger and its spring-loaded bearing piece to operate the extension sear, which in turn operates the hammer. The hammer strikes the inertia-type spring-loaded firing pin, causing it to strike the primer of the cartridge, thereby firing the cartridge. The barrel and slide are rigidly locked by two rectangular lugs on either side of the barrel engaging in recesses in the side walls of the slide. A helical-portion camming lug on the barrel underside rides in a corresponding helical groove in a rigidly pinned block on the upper surface of the receiver. Upon functioning of the cartridge, the barrel is rotating clockwise through 22°, causing disengagement of the lugs and permitting further recoil of the slide. Recoil of the slide compresses against the recoil spring mounted on its guide which passes through the helical camming block below the barrel. During recoil the extractor, mounted in the slide, has withdrawn the fired case which is ejected by the ejector, which is mounted over the rear of the magazine well. The slide returns forward under the pressure of the compressed recoil spring and strips a cartridge from the magazine feeding it into the chamber. Pressure on the trigger again will repeat the above process.

Disconnection—prevention of full automatic fire—is accomplished by the camming down of the trigger-bearing surface so that it cannot contact the trigger extension. The disconnector is mounted in a recess in the slide on the left side.

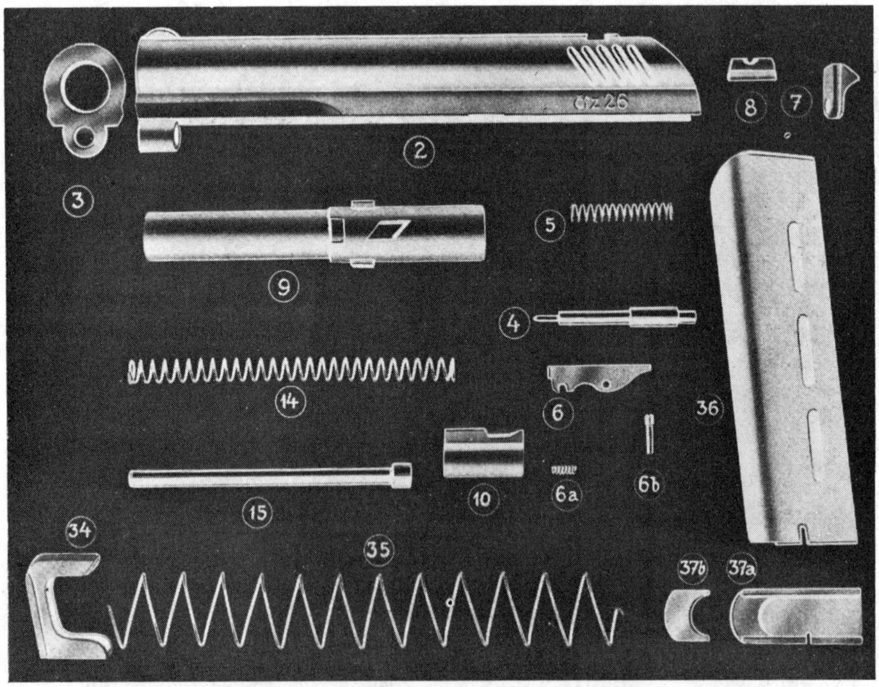

Details of slide and magazine components. No. 10 is a removable block which is placed in the receiver extension above the trigger. The guide lug on the barrel rests in a cam cut on top of this block to compel the barrel to turn.

THE CZECH 9mm SHORT PISTOL MODEL 24

The Model 24 is also a locked-breech pistol and is generally similar to the Model 22.
The pistols vary in the following details:

Locking action: A stopping stud is provided on the under side of the barrel ahead of the helical lug. This stud butts against the helical camming block to act as a stop during recoil. The helical lug is widened and the camming block is machined so that it can be assembled without regard to position. The block of the M1922 has an arrow to indicate assembly position.

Magazine: A magazine safety has been added, which blocks trigger motion and prevents firing the weapon without a magazine in place.

Trigger mechanism: The trigger bearing piece is incorporated into the trigger extension where it serves as a disconnector and the separate disconnector is no longer used. The trigger extension is loaded with a coil spring rather than a wire form spring and the trigger mounting and pin differ.

Other parts: The hammer spring is separate from the magazine spring, the former being retained by a screw.

Earlier types have one-piece unchecked wood grips; later types have one-piece checked molded-plastic grip with CZ insignia.

Side plate on left side is relatively flat, retained by a small screw; 1922 has no screw but has a step projecting from its lower edge to close off the trigger mounting.

The slide rib is matted and the frame and slide show different machining. A variant of this pistol with a 10-round rather than an 8-round magazine, has been reported.

Loading, firing, and field stripping of the Model 24 are essentially the same as that of the Model 22.

Czech 7.65mm Model 27 Pistol with pre-1940 German markings.

THE CZECH 7.65mm M27 PISTOL

The Model 27 was made in the largest quantity of all the pre-World War II Czech automatic pistols. It was extensively used by the Germans, who called it Pistol 27(t), during World War II.

The Model 27 differs from the Models 22 and 24 in caliber and in being blowback operated. The barrel is not rigidly fixed to the receiver as in many blowback operated pistols such as the Walther PP, but is removable in a fashion similar to the Colt .32 and .380 pocket automatics.

During the course of the war various simplifications in manufacture were introduced by the Germans. Pistols which bear the marking

Czech 7.65mm Model 27 Pistol of wartime manufacture; this pistol has special barrel to be used with silencer.

Czech 7.65mm M27 Pistol of post-1948 manufacture.

"fnh" were made with the following modifications: Trigger extension bar, safety, safety release, magazine release, and firing pin retaining plate are stamped; magazine catch can be used as lanyard loop; and the retaining screw on side plate is eliminated.

Manufacture of the pistol was continued by CZ after the war. The modifications made during the war were retained for the most part and initially the old type Czech markings were used. After 1948, however, the "Narodni Podnik" marking was added.

As can be noted from the photograph, finish is extremely rough, especially for a weapon intended for export as indicated by the marking "made in Czechoslovakia."

The Model 27 is loaded, fired, and field stripped in a manner similar to the earlier models with one exception: the barrel is held in place by parallel ribs (similar to Colt pocket pistols), and must be rotated to be removed from the slide.

THE CZECH 9mm SHORT M38 PISTOL

This pistol was not made in the quantity of the earlier models; it was used by the Germans as Pistole Modell 39(t). There are a number of variations of this pistol and some are fitted with safety catch. Essentially the pistol has the Model 22, 24, 27, trigger mechanism modified to double-action and ordinary blowback operation.

Czech 9mm Short Model 38 Pistol.

Czech 9mm Model 47 Pistol.

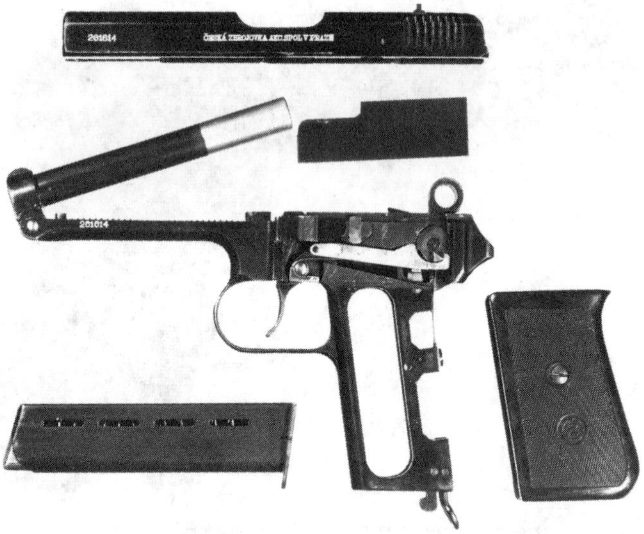

Czech 9mm Short Model 38 Pistol—field-stripped.

Loading and firing of the Model 38 are the same as the earlier Czech pistols with one principal difference. The Model 38 is double-action, i.e., a pull through on the trigger cocks and fires the weapon. The hammer, which is difficult to cock because of its low position, can be set at half cock with a cartridge in the chamber and if the trigger is pulled the hammer will go to the rear and fall firing the weapon.

Field Stripping the Model 38 Pistol

Remove magazine, push dismounting catch on left side of receiver above trigger, pull slide to the rear and up. Barrel may be lifted on hinge for clearing and side plate may be removed for cleaning trigger mechanism. No further disassembly is recommended. To assemble, reverse procedure.

CZECH 9mm PARABELLUM MODEL 47 PISTOL

The Model 47 was apparently made only in prototype form. It combines the double-action trigger mechanism of the Model 38 with the Browning-Petter locking mechanism.

This pistol really has very little to recommend it and could not compete with the Browning and Colt on the world market or the Tokarev within the Communist bloc.

Czech 7.62mm ZKP 524 Automatic Pistol.

The weapon was designed by Frantisek Myska, a designer who has produced much better work than the Model 47 pistol.

7.62mm ZKP524 PISTOL

This pistol chambered for the 7.62mm rimless pistol cartridge was developed by Francis and Joseph Koucky. It is a recoil operated pistol using the basic Browning type action and has an 8-round magazine. It was produced only as a prototype and was probably designed in competition with the Model 52 which was adopted as the Czech service pistol.

CZECHOSLOVAK 7.65mm M1950 PISTOL

Characteristics, 7.65mm M1950 Pistol

System of operation: Blowback, semiautomatic fire only.
Weight, unloaded, w/magazine: 1.5 lb.
Length, overall: 6.8 in.
Barrel length: 3.8 in.
Feed device: 8 round removable box magazine.
Muzzle velocity: 919 f.p.s.
Sights: Front: Blade.
Rear: Round notch.

The M1950 is no longer used as a service pistol in Czechoslovakia, but is probably still used as a police weapon. It is chambered for

CHARACTERISTICS OF PRE-WORLD WAR II CZECH PISTOLS

	Model 22	Model 24	Model 27	Model 38
Caliber:	9mm Short*	9mm Short*	7.65mm †	9mm Short*
System of operation:	Recoil, semi-automatic only.	Recoil, semi-automatic only.	Blowback, semi-automatic only.	Blowback, semi-automatic only.
Weight:	1.37 lb.	1.5 lb.	1.56 lb.	2 lb.
Length overall:	6 in.	6 in.	6.3 in.	7.8 in.
Barrel length:	3.44 in.	3.56 in.	3.9 in.	4.7 in.
Feed device:	8-round in line. Detachable box. magazine.	8-round in line. Detachable box. magazine.	8-round in line. Detachable box. magazine.	8-round in line. Detachable box magazine.
Muzzle velocity:	984 f.p.s.	984 f.p.s.	919 f.p.s.	1,000 f.p.s.
Sights: Front:	Blade.	Blade.	Blade.	Blade.
Rear:	V notch.	V notch.	V notch.	V notch.

*same as .380 A.C.P. † same as .32 A.C.P.

the 7.65mm automatic pistol cartridge (cal. .32 ACP). Specimens have been examined, and appear to be of good quality manufacture, but there is nothing unusual about the design of the weapon.

How to Load and Fire the M1950 Pistol

Remove magazine from pistol grip by depressing the release button on left side of receiver and pulling magazine out. Fill magazine with eight cartridges. Insert magazine in the well in pistol grip. Draw slide smartly to the rear, and release. With the safety off, the trigger may now be squeezed and the weapon will fire; when the trigger is released, it may be squeezed to fire again. This process may be repeated till the weapon is empty.

The safety of the M1950 operates like that of the Walther PP and PPK. The M1950 is a double action pistol; with a cartridge in the chamber and the hammer in the down position, pulling the trigger to the rear will cause the hammer to rise and fall, firing the weapon.

How the M1950 Pistol Works

In its functioning, the M1950 pistol is generally similar to the Walther PP and PPK.

How to Field Strip the M1950 Pistol

The dismounting button is pushed, and the slide is pulled to the rear as far as it will go. The rear end of the slide is then drawn upwards, and the slide is pushed forward and off the weapon. The recoil spring which encircles the barrel can be removed. The barrel is fixed to the receiver and should not be removed. No further disassembly is recommended. Reverse the procedure to reassemble the weapon.

CZECH 7.62mm MODEL 52 PISTOL

The Czechoslovaks have used several types of service pistols since the conclusion of World War II. The 7.62mm Model 52 is the current service weapon.

The Model 52 pistol is a native Czechoslovak design which has borrowed its locking system from the German MG42 machine gun. The pistol is chambered for the Czechoslovak-made version of the Soviet 7.62mm Type P pistol cartridge, which the Czechoslovaks call the Model 48. The Soviet and Czechoslovak cartridges are interchangeable with the 7.63mm Mauser, but are considerably hotter loadings than are the United States commercial loadings of this cartridge. For this reason, the functioning of Soviet and Czechoslovak weapons with commercially loaded 7.63mm cartridges is, at best, marginal. The Czechoslovak cartridge has a particularly heavy loading, being about 20% heavier than the Soviet.

Characteristics of 7.62mm Model 52 Pistol

Caliber: 7.62mm pistol cartridge.
System of operation: Recoil, semiautomatic fire only.
Weight, loaded: 2.31 lb.
Length, overall: 8.25 in.
Barrel length: 4.71 in.
Feed device: 8-round removable box magazine.
Muzzle velocity: 1600 f.p.s.
Sights: Front: Blade.
Rear: Square notch.

Czechoslovak 7.65mm M1950 Pistol.

Czechoslovak 7.62mm Model 52 Pistol.

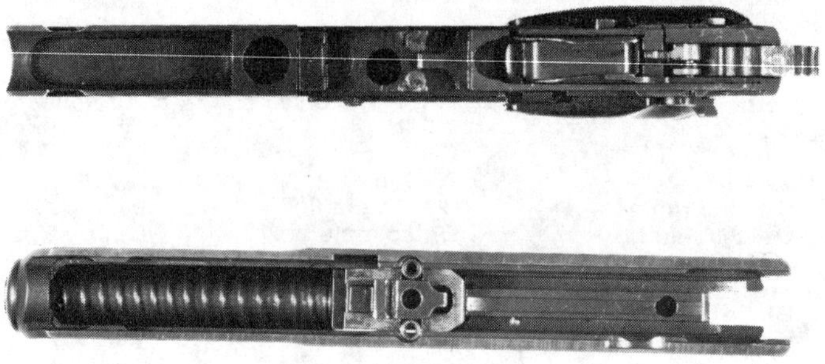

Top view of receiver and bottom view of slide of Model 52 Pistol, barrel locked.

How to Load and Fire the Model 52 Pistol

Remove the magazine by forcing to the rear the magazine catch, which is located at the bottom rear of the grip. Load the magazine with eight cartridges; insert magazine in well in pistol grip. Draw slide smartly to the rear, and release; the weapon is now loaded. With the safety in the "fire" position, squeeze the trigger and the weapon will fire; after the trigger is released, it can be squeezed again to fire the pistol. This sequence can be repeated until the weapon is empty. The slide remains to the rear on the last shot. The slide cannot be released from the outside of the weapon as can that of the United States M1911A1 automatic pistol. The slide must be drawn to the rear before it will release. The Model 52's safety is mounted on the left rear of the receiver and has three positions; the lowest position is "fire," and the middle and topmost positions are "safe." The Model 52 is not a double-action automatic. The hammer, if it is in the down position with a cartridge in the chamber, must be manually cocked before the weapon can be fired.

Field Stripping the Model 52 Pistol

The slide-dismounting catches at the front of the trigger guard are pulled down, and the slide assembly is pushed forward until it can be drawn straight up, to the rear, and off the receiver. With the slide in an upside down position, the barrel can be removed from the slide by pushing the roller cam and barrel forward so that the rollers clear their locking recesses. (To force the barrel forward, a screwdriver or a punch must be inserted in the hole in the roller cam.) Then the breech end of the barrel can be swung upward, clear of the slide, after which the barrel can be removed toward the rear of the slide. This process is very difficult and must be performed carefully, because the barrel, while being withdrawn, is under the pressure of the partly compressed recoil spring.

How the Model 52 Pistol Works

Loading. A loaded magazine is inserted in the magazine well in the grip of the receiver. The slide is drawn to the rear, cocking the hammer, and then released. As it moves forward, the slide forces a cartridge out of the lips of the magazine, up the bullet ramp, and into the chamber. The claw of the extractor, which is mounted on the right side of the slide behind the ejection port, slips over the rim of the cartridge.

Firing. Pulling the trigger to the rear draws forward the trigger bar, which is pinned to the trigger. If the slide is now in battery position (locking the weapon), the trigger bar pulls the sear ahead, allowing the hammer to snap forward under the pressure of the mainspring. The hammer strikes the firing pin and then rebounds to a position out of contact. The firing pin in turn strikes the primer of the cartridge and causes the cartridge to explode.

Operation of the Slide Group. Both the barrel and the slide, which are locked together by the roller bearings, move to the rear for three-sixteenths of an inch. During this period, the roller bearings are locked in their recesses in the slide and in the barrel lug (located under the barrel) by the thick center section of the roller cam. The roller cam, a sleeve-like piece which encircles the barrel ahead of the barrel lug, has a tongue-shaped section which extends through a longitudinal groove cut in the underside of the barrel lug. The roller cam is kept from going to the rear by another lug which is located in the mid-section of the receiver, forward of the bullet ramp. When the narrowed section of the roller cam tongue lines up with the roller bearings and their locking slot in the barrel lug, the roller bearings are cammed completely into the cutout portion of the barrel lug by the edges of the roller bearing recesses in the slide. The gun is now unlocked.

When the barrel and slide have traveled three-sixteenths of an inch to the rear, the barrel is stopped by the abutment of the barrel lug against the forward edge of the bullet ramp of the receiver. The slide continues to the rear and compresses the recoil spring, which is mounted on the barrel between the roller cam and the front of the slide.

The cartridge case, which is held by the extractor, goes rearward with the slide until it hits the ejector. The ejector, mounted on the sear pin on the left rear side of the receiver, then throws the case out the ejection port.

The lower front of the slide abuts against the receiver, and the compressed recoil spring forces the slide back into battery position. As the moving slide picks up the next round from the magazine and pushes it into the chamber, the extractor claw slips over the rim of the cartridge and engages the extractor groove. The roller bearings are cammed out of their position in the cutout of the barrel lug and into the locking recesses of the slide. They are then held in the locked position (i.e., half in the slide and half in the cutout of the barrel lug) by the wide central portion of the roller cam.

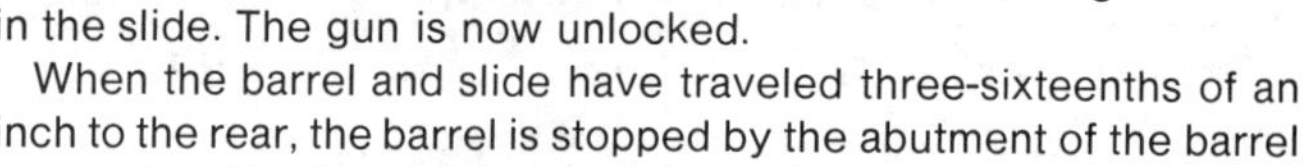

Bottom view of slide of Model 52 Pistol showing barrel unlocked.

If the trigger is pulled again, the process is repeated. After the last shot is fired, the slide is held to the rear by the slide stop, which is mounted on the left side of the receiver, and which engages a cutout on the underside of the slide.

Operation of the Receiver Group. The spur-type trigger is attached to the trigger bar, which disengages the sear from the hammer when the trigger is pulled to the rear. On the top side of the trigger bar, near the center, is a lug which serves as a means of disconnection. When the trigger is pulled, this lug rises into a cutout portion in the right rear underside of the slide. If the weapon is not fully locked, however, the trigger bar lug will strike the underside of the slide rather than the cutout portion; thus prevented from rising, the trigger bar will be unable to contact the sear, and the weapon will not fire.

Pulling the trigger also causes an arm of the sear to rise and force up the spring-loaded firing pin lock, which is mounted in the feed rib of the slide. A portion of the firing pin lock is seated in a semicircular notch cut in the inertia-type firing pin, near its head, and both locks the firing pin and cams it back from the primer when the slide and barrel start to move to the rear. The forcing of the firing pin lock upward by the sear arm allows the firing pin to move forward when struck by the hammer. Rearward movement of the slide disengages the firing pin lock from the sear arm, allowing the firing pin lock to come down, under the pressure of its spring, and to cam the firing pin rearward out of engagement with the cartridge primer.

The hammer is the rebound type. The sear engages the rebound notch of the hammer, which then cannot move forward to strike the firing pin unless the trigger is pulled to the rear.

The safety, which is mounted on the left rear side of the receiver, in a location similar to that of the safety lock of the United States cal. .45 M1911A1 pistol, has three positions. Its lower position is "fire"; in this position, the safety does not engage the hammer or the trigger mechanism. Its central position, indicated by a red dot, is "safe"; when pushed upward to cover the red dot, the safety blocks rearward movement of the trigger bar and engages the rebound notch of the hammer, preventing its forward movement. With the safety in this position, not only can the hammer be locked, but the slide can be drawn to the rear, permitting the unloading of the chamber. With hammer cocked, the safety may be pushed all the way up to its top position, which also is a "safe." The hammer will fall, but, restrained by the safety stud which enters the rebound notch, cannot strike the firing pin.

The mainspring is mounted on the mainspring strut, which is positioned to the rear of the magazine well of the receiver. The mainspring exerts pressure upon the hammer when the hammer is cocked. It also presses against the magazine catch, mounted at the bottom rear side of the receiver, in the rear of the magazine well, thus enabling the catch to hold the magazine firmly in place.

CZECHOSLOVAK RIFLES

When the Republic of Czechoslovakia was founded in 1919 it continued for a short time in the path of the Empire from which it had come—the Austro-Hungarian—insofar as weapons were concerned. Mannlicher M1895 rifles were the first rifles produced in the new republic, but various arrangements with Mauser were made and in 1924 Ceskaslovenske Zobrojovka Brno—"ZB"—began the production of Mauser rifles for the Czech Army and for export. The Czechs, in competition with the Belgians of FN, took over a goodly part of the military rifle arms trade of the world.

All the Mausers produced by ZB were based on the M98 action. The models produced were: the Rifle 98/22, Rifle 98/29, Short Rifle 98/29, Carbine 12/33, and the rifle Model 24. The rifle Model 24 was adopted by Czechoslovakia and is covered in detail later in this chapter; the 98/29 weapons are covered in detail under Iran. The 98/22 differs from the 98/29 mainly in the use of front sight guards--"Sight ears"--on the 98/29. The 98/22 was used by Turkey, among other countries. Rumania, Guatemala, Yugoslavia, and China used the Model 24 rifle in addition to Czechoslovakia. Model 24(t) was the German Army designation for the Czech Mauser and eleven divisions of the German Army were equipped with it and other Czech weapons in September 1939 when World War II began.

Czechoslovakia also adopted the 16/33 Carbine which has a small diameter receiver ring, as the Model 33 rifle. This weapon was kept in production at ZB during World War II with minor modifications as the 33/40 and was issued to German paratroops and mountain troops. During the war the production of the Model 24 was gradually changed so that the rifles which were being produced at the end of the war were actually Kar 98ks. Those rifles marked 24(t) differ from the standard Model 24 in having a cup type butt plate, a firing pin disassembly disk on the buttstock and a slot through the buttstock for the sling. As finally produced at ZB, the Mauser rifle has a stamped oversize trigger guard, firing pin dismounting hole in the butt plate, slotted buttstock, German-type stamped bands and modified gas escape holes in the bolt. This rifle was produced until circa 1950 and was sold to Israel and Pakistan. It is actually the Kar 98k in its last production version.

Prior to World War II, a number of semiautomatic rifle designs were produced in Czechoslovakia and at least one—the ZH 29 rifle designed by E. Holek—was produced in quantity and was tested by the United States in 1929. Although the United States tests indicated that the ZH 29 was not as reliable in function as a military rifle should be, it was sold in limited quantities to a number of countries including Ethiopia. In 1938, another semiautomatic rifle appeared, the Model 38 designed by Vaclav Polanka and Jan Kratochvil. This was a gas-operated weapon chambered for a 7.62mm cartridge. The ZK381 designed by Josef Koucky was a 7.92mm gas-operated weapon which was tested by the U.S.S.R. in 1938 and by France in February 1939. The ZK391 was designed by Josef and Frantisek (Francis) Koucky and was a gas-operated 7.92mm weapon. According to Czech sources, this weapon was adopted by the Italians and manufactured in Cremona, Italy circa 1943. It was tested in Denmark in 1946. Other weapons developed included the ZK425, ZK420, and ZK420-S. The ZK472 was designed for an "intermediate-sized" 7.5mm cartridge with a muzzle velocity of about 2500 feet per second. This cartridge was

Czech copy of German Kar. 98k rifle.

Czech 7.92mm Model 24 (VZ 24).

the direct ancestor of the Czech 7.62mm M52. This list of semiautomatic rifles developed is by no means complete; there were other semiautomatic rifles designed during this period as well as those listed above.

With all the design and prototype production activity, however, the first semiautomatic rifle to be adopted by the Czech Army was the Model 52. This weapon was chambered for the Czech designed 7.62mm Model 52 cartridge. The Model 52 is a surprisingly poor design for a country which has produced so many worthwhile designs; part of it—trigger mechanism and gas system—is pure unadulterated plagiarism and poor plagiarism at that. When the Czechs adopted the Soviet 7.62mm M43 cartridge, which varies more in shape than it does in performance from the Czech cartridge, the Model 52 was modified to chamber this cartridge and became the Model 52/57. In 1958 a new weapon, the 7.62mm Model 58 assault rifle, appeared. This rifle, a gas-operated selective fire weapon, is replacing the Model 52/57 and the 7.62mm Model 24 and 26 submachine guns in the Czech Army.

THE 7.92mm CZECH MODEL 24 RIFLE

The Model 24 Rifle was the standard Czech rifle prior to World War II and as previously stated was widely used by the Germans and by other countries. It uses the basic Mauser 98 action and differs from the German Kar 98k mainly in fittings, having a full length handguard, sling swivels on both the side and underside of the stock, and a straight bolt handle.

Characteristics of the 7.92mm Czech Model 24 Rifle

Caliber: 7.92mm.
System of operation: Bolt action, manually operated.
Weight: 8.98 lb.
Length overall: 43.3 in.
Barrel length: 23.2 in.
Feed mechanism: 5-round staggered row, non-detachable box magazine.
Sights: Front: Barley corn.
Rear: Tangent graduated from 300-2000 meters by 100 meter steps.
Muzzle velocity: 2700 f.p.s.

Operation of the Mechanism, M24

The bolt handle is raised to vertical position, then pulled back to expose the magazine.

The clip, with five or less cartridges, is put in the clip guides in the receiver. On pressing down the top cartridge with the thumb, all the cartridges are forced into the magazine. The magazine platform, being raised by the flat magazine spring, has the top cartridge always ready for loading into the chamber. As the bolt starts ahead the extractor glides underneath the edge of the cartridge case so that its hook engages in the cannelure of the cartridge case. By pushing and turning the handle of the bolt to return it to its horizontal position, the cartridge is loaded into the chamber, and the arm is ready for firing.

The sight is adjusted by pressing on both sides the slide catches, and moving the slide on the range scale.

The trigger can be pulled lightly until certain resistance is felt. As soon as this is overcome and the movement continued, the trigger nose descends, thus releasing the firing pin. The firing pin is thrown forward by the main spring, the striker strikes the primer, firing the cartridge. The empty cartridge case is drawn back out of the chamber by the extractor. During the backward movement of the bolt the ejector strikes the bottom of the cartridge case and ejects it to the right of the receiver. This happens before the bolt is checked by the stop.

Dismounting and Assembling the Bolt, M24

Before removing the bolt from the receiver, turn the safety catch into its central position between its safe and fire position, i.e., with the wing turned upwards. The bolt can be removed from the receiver only when the bolt stop is moved to the side with a slight pressure, whereby the groove in the receiver is (fig. 1) opened to permit bolt passage.

Turn the bolt sleeve to the left, then the latter with the safety catch, the cocking piece and the firing pin may easily be separated from the bolt. While removing these parts the main spring is still in a state of compression (fig. 2).

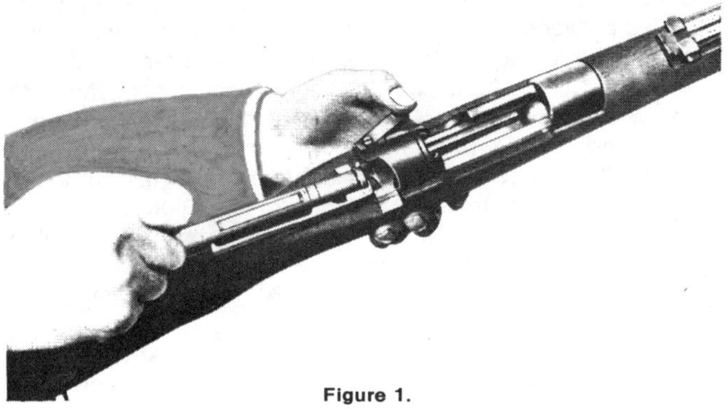

Figure 1.

For further dismounting put the striker with its point turned downward against a convenient bearing (for this purpose a hole is provided in the stock sleeve) (fig. 3). Press home the main spring until the cocking piece can be turned 90°, remove the latter and, in releasing the pressure of the spring, take out the firing pin. The safety catch may be dismounted by turning the wing to the right side.

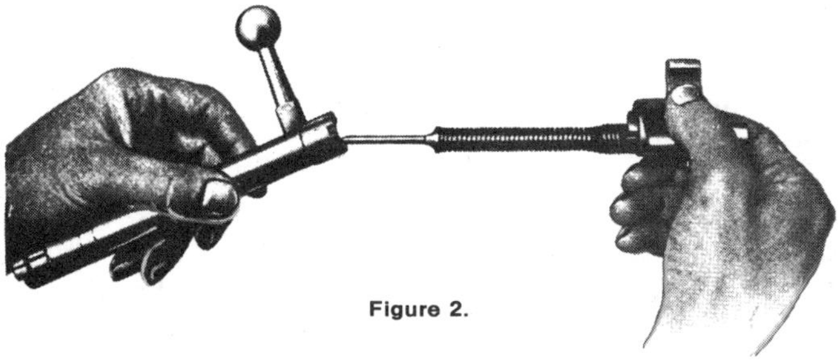

Figure 2.

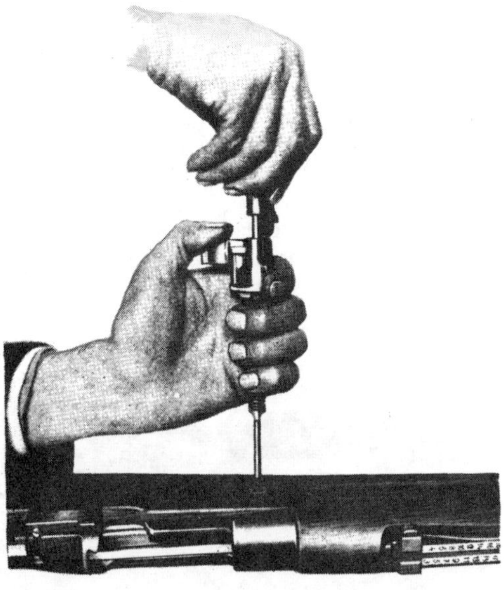

Figure 3.

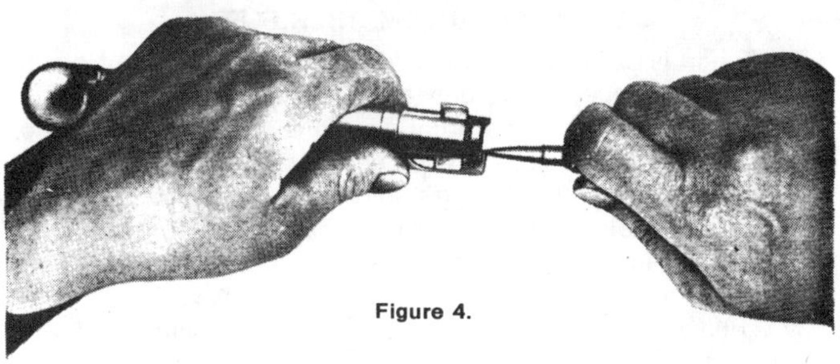

Figure 4.

To remove the bolt sleeve lock, pull it into a position of 90°, where its spindle can be turned in the notch of the bolt lock.

To remove the extractor, pull it to the side with the help of a solid object (for instance with a cartridge), till it disengages the dovetail before the bolt lug (fig.4). Then turn it to an angle of 90°, and, holding it against a table edge, draw it out of the groove of the extractor ring (fig. 5).

The extractor ring of the bolt should not be dismounted. The extractor and the bolt sleeve lock should seldom be removed.

The assembling is done in the opposite order. Mount the extractor on the extractor ring, until it can be turned and engaged in the groove of the bolt. Put into the bolt lock the bolt sleeve lock spring, then the bolt sleeve lock itself, and turn 90°. Place the main spring on the firing pin, place the safety catch into the bolt sleeve, and compress the main spring, just enough to insert the cocking piece, and turn 90°. The bolt sleeve, thus assembled, is screwed on the bolt. Now the bolt can be pushed into the receiver. When doing so, it is not necessary to push aside the bolt stop, for it disengages itself by the pressure on the bolt.

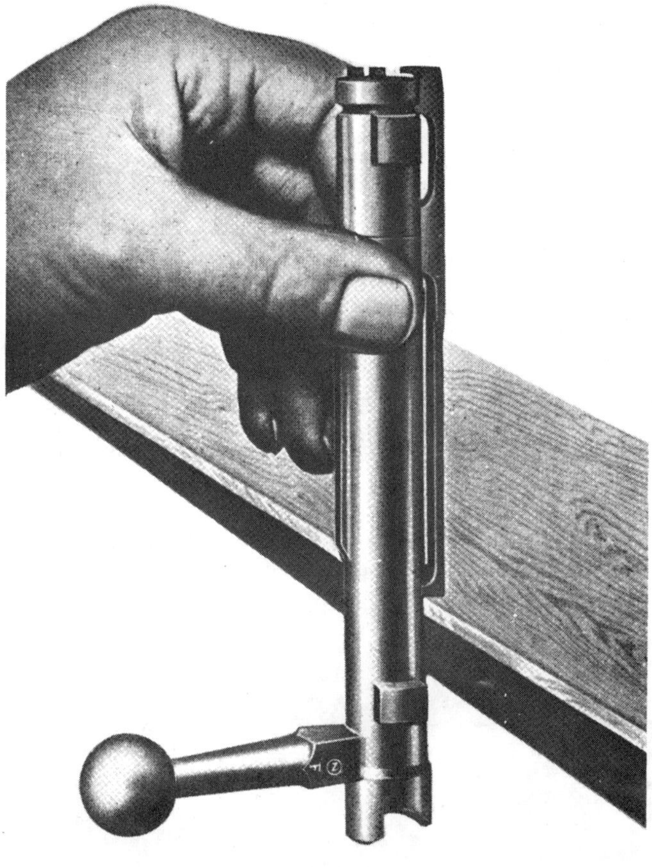

Figure 5.

CZECH SEMIAUTOMATIC RIFLES

The first truly Czech semiautomatic rifle was probably the Netsch which was developed at CZ between 1922-24. This 7.92mm rifle was gas operated and had a drum magazine. The ZH 29 was the first Czech rifle of this category to become well known and to be exported in any quantity. The design is credited by the Czechs to Emmanuel Holek of ZB. This rifle, chambered for the .276 Pedersen cartridge, was tested by the United States at Aberdeen Proving Ground in 1929.

This rifle was used by Ethiopia and Thailand to a limited extent.

Characteristics of Czech ZH29

Caliber: 7.92mm.
System of operation: Gas, semiautomatic fire (specimens capable of full automatic fire have been reported).
Weight: Approx. 10 lb.
Length overall: 45.5 in.
Barrel length: 21.5 in.
Feed mechanism: 10-or 25-round, detachable box magazine.
Sights: Front: Protected blade.
Rear: Tangent leaf adjustable from 100-1400 meters.
Muzzle velocity: Approx. 2650 f.p.s.

An unusual feature of the ZH 29 is the aluminum cooling jacket fitted over the barrel and gas cylinder ahead of the wooden hand guard. This is an interesting commentary on the military concepts of the time since it indicates that the designer thought that in normal battle usage the rifle would be fired frequently enough in one fight to require such a gadget for cooling. The sad and bloody experiences of the last forty years indicate that most fire fights are short and sharp and that heating is only a major problem with automatic support weapons. Far more handguards are burned off in proving ground and user tests than in battle! The main purpose of the handguard over the barrel today is to prevent burning of the hand in the very remote and rare possibility of a bayonet fight after a fire fight.

The CZ Model 38 was designed by Vaclav Polanka and Jan Kratochil. This gas operated rifle was made in 7.62mm as well as 7.92mm and was tested in the Soviet Union. This rifle broke open like a shotgun for disassembly in a manner somewhat similar to the current FN "FAL". The hinge point is beneath the forward portion of the receiver.

The ZK381 was developed by Joseph Koucky and is also a gas operated box magazine fed weapon. This weapon was tested by the U.S.S.R. in August 1938, by France in February 1939 and by Great Britain in approximately the same period.

In 1939 Joseph and Francis Koucky brought out another rifle with the ZK391. The ZK391 more closely resembles the ZK420 series than do any of the earlier Czech semiautomatic rifles. In 1943 a modification of this rifle was manufactured in limited quantities in Cremona, Italy. The ZK391 was tested by Denmark in 1946.

THE CZECH 7.92mm ZK420 RIFLE

The ZK420 design was produced by Josef and Frantisek Koucky in 1942 and went through a number of modifications until the final model—the ZK420S appeared after World War II. As originally made, the ZK420 had Mauser-type bands and bayonet stud and a full-length type stock. The magazine was not removable and was loaded with chargers. In 1946 the first modification of this design appeared; the 1946 version has a sporter-type stock, a removable magazine, and in general resembles the ZK420S. The outstanding external difference between the 1946 Model and the ZK420S is the configuration of the magazine and the prominent safety catch lever—as that of the U.S. M1 rifle—on the ZK420S. The latter rifle appeared between 1947 and 1950 and was the last of the series.

The ZK420 series of rifles received extensive testing abroad. The 1942 and 1946 Models were tested by Denmark and the 1946 Model was tested by the United Kingdom, Sweden, Ethiopia, Egypt, Israel, and Switzerland. A version of the ZK420S with permanently attached bayonet was made for Israel for test purposes. The 1946 Model was made in 7.92mm, 7mm, .30-06, and 7.5mm Swiss. It is somewhat surprising that the Czech Army never adopted this weapon, since it was extensively developed and certainly advertised throughout the world; the ZK420 appears in a post 1948 ZB catalog. The design in certain features, i.e. bolt and gas system appear to be at least equal to, if not superior to that of the rifle finally adopted—the Model 52. The ZK420 series weapons are, however, generally heavier designs and undoubtedly much more expensive to manufacture than is the Model 52.

Characteristics of the ZK420S Rifle

Caliber: 7.92mm.
System of operation: Gas, semiautomatic fire only.
Weight: 10.58 lb.
Length overall: 41.73 in.
Barrel length: 21.65 in.
Feed mechanism: 10-round, double staggered row, detachable box magazine.
Sights: Front: Hooded post.
Rear: Notched tangent with ramp.
Muzzle velocity: Approx. 2700 f.p.s.

The ZK420S has a trigger mechanism similar to the U.S. M1, as do the later Czech M52 and M52/57 rifles. The bolt mechanism of this weapon bears a superficial resemblance to that of the U.S. M1, but it actually more closely resembles that of the Soviet AK47. The body or shaft of the one-piece bolt is enclosed by the bolt carrier and is cammed by a camway within the bolt carrier. The head of the bolt bears the locking lugs and protrudes forward of the bolt carrier. Since the original design of the ZK420 appeared in 1942, it may have had some influence on Mikhail Kalashnikov, the designer of the AK47.

In 1942 the Koucky's brought out another rifle called the ZK425. This gas operated rifle has a very neat outward appearance. It was tested in Denmark in 1946. The last of this series of rifles the ZK472 is also gas operated. It has a 10 or 15 round box magazine and is chambered for an "intermediate" sized 7.5mm cartridge. The ZK472 is 39.4 inches long and the barrel is 20.5 inches long. The muzzle velocity is 2526 feet per second.

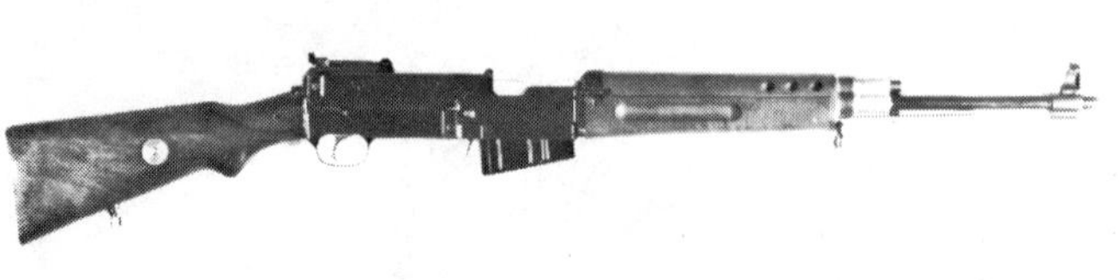

Czech 7.92mm ZK 381 Semiautomatic Rifle

Czech CZ 38 Semiautomatic Rifle

Czech 7.92mm Model ZK-420S

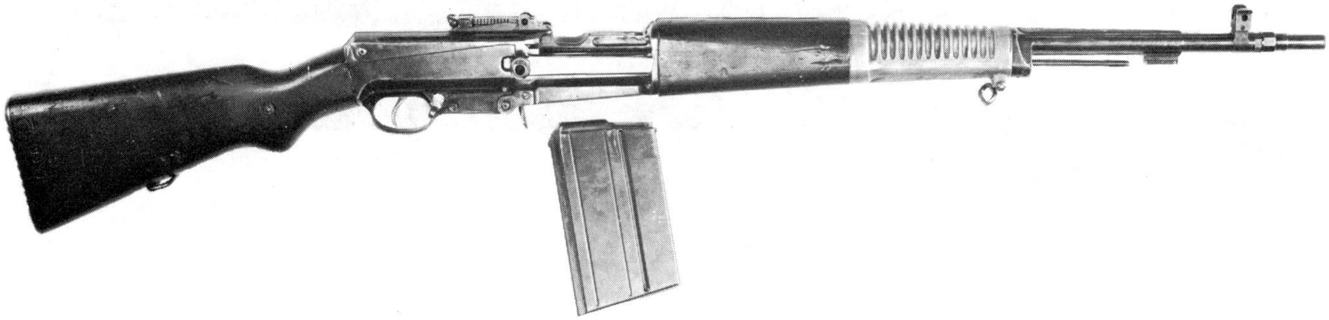

Czech ZH 29 Semiautomatic Rifle.

CZECH 7.62mm MODEL 52 RIFLE

Characteristics of 7.62mm Model 52

Caliber: 7.62mm, Czechoslovak Model 52.
System of operation: Gas, semiautomatic fire only.
Weight, loaded: 9.8 lb.
Length, overall: 39.37 in.
Barrel length: 20.66 in.
Feed mechanism: 10-round, double staggered, detachable box magazine (loaded with 5-round chargers).
Sights: Front: Hooded blade (removable hood).
Rear: Notched tangent with curved ramp.
Muzzle velocity: 2440 f.p.s.

The Model 52 is chambered for the Czechoslovak 7.62mm Model 52 cartridge. Its design is a combination of several older designs added to a few native ideas. The gas system is generally similar to that of the German MKb 42W and some of the earlier Walther commercial semiautomatic rifle designs. The trigger mechanism is similar to that of the United States M1 rifle. The bolt appears to be a native design; it is somewhat unusual in that it is a tipping bolt design with frontal locking lugs. For the first time the Czechoslovaks have put a nondetachable bayonet on one of their rifles, possibly as a result of Soviet influence. The Model 52 makes fairly extensive use of stampings, but, all in all, the design is hardly revolutionary and in some respects is not too remarkable considering the date it appeared—1952.

The Czechs have issued a slightly modified version of the 52 rifle called the Model 52/57 and chambered for the Soviet designed 7.62mm M43 cartridge. The Model 52 and Model 52/57 rifle are no longer standard in the Czech Army.

How to Load and Fire the Model 52 Rifle

Pull the operating handle all the way to the rear; since the weapon has a bolt-holding-open device, the bolt and carrier will remain open. Insert the end of a five-round charger in the charger (clip) guide at the front of the receiver cover and force the cartridges down into the magazine with the thumb of the left hand. Repeat this process with another five-round charger; then draw the operating handle slightly to the rear and release. The bolt and carrier will run home, chambering a cartridge. The weapon is now loaded and, with safety off, the trigger will cause it to fire. The trigger must be released after each shot. The safety is similar to that on the United States M1 rifle. When the safety is forward the rifle will fire; when the safety is drawn to the rear, the weapon is on "safe."

Field Stripping the Model 52 Rifle

The magazine is released by pressing forward on the magazine lever. After clearing the weapon to assure that the chamber is empty, disassemble the receiver cover—carefully, as the driving spring is always compressed. Push the receiver cover forward until the side rails are out of the receiver slots. Lift the cover slightly and slide back along the top of the receiver until the driving spring snaps down. Hold bolt carrier to the right and slide it to the rear until it reaches the disassembly notch. Remove bolt carrier and bolt. To remove bolt from bolt carrier, slide bolt to the rear and press down the front end until it is disengaged from the bolt carrier.

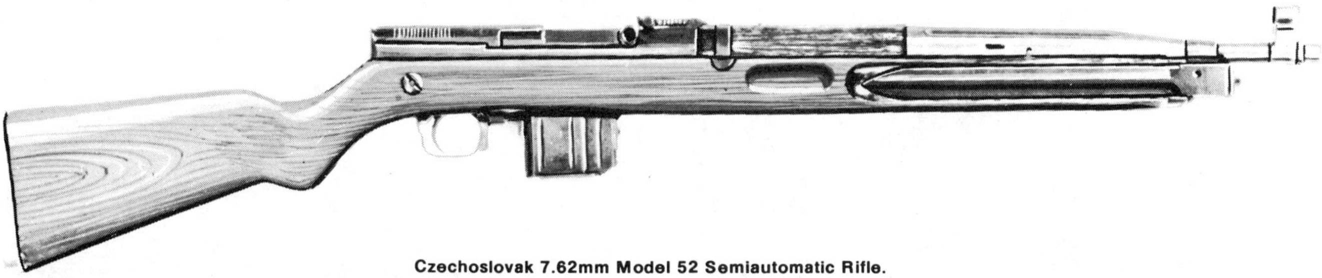

Czechoslovak 7.62mm Model 52 Semiautomatic Rifle.

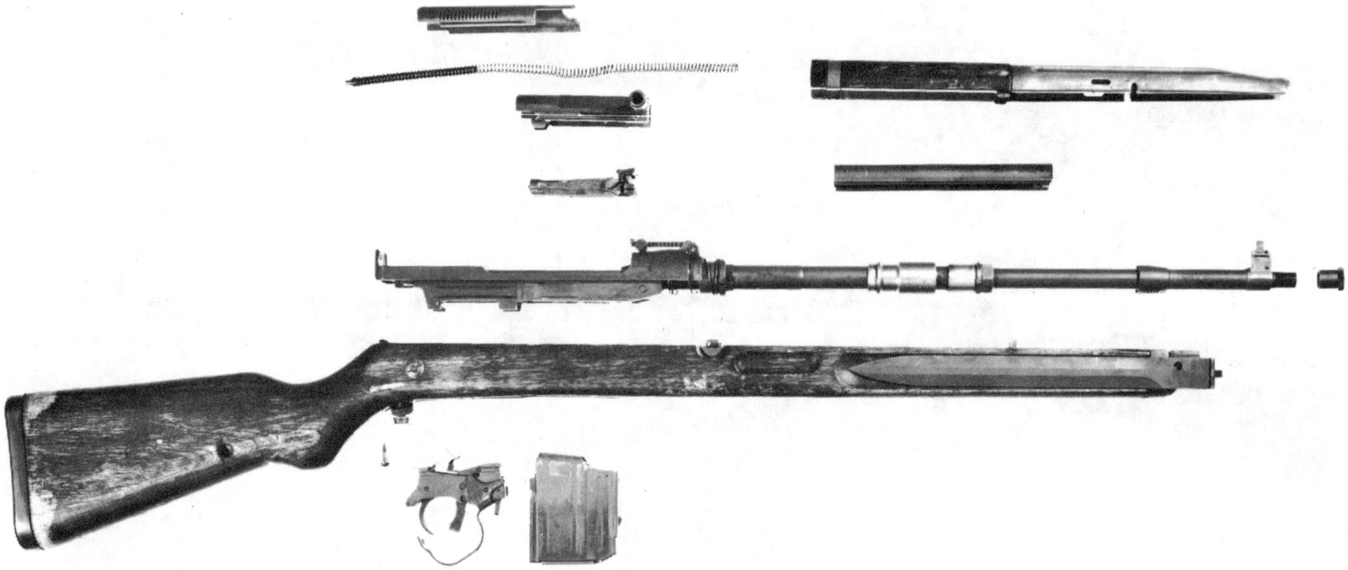

Czech Model 52—field-stripped.

Slide the bolt forward out of the carrier. After applying the safety, remove the trigger housing group by forcing the rear of the trigger guard back and rotating it downward. (Disassembly is difficult, as the rear of the trigger guard is hard to grasp and pull to the rear to disengage guard from housing.) Slide the trigger housing group forward and downward. Remove the handguard over the barrel by depressing ears on each side of the stock, at the rear end of the handguard. If the barrel and receiver assembly is to be removed from the stock, depress plunger visible through hole in bayonet, and slide the barrel sleeve forward. Assemble in the reverse order.

How the Model 52 Rifle Works

The magazine is loaded and the operating handle is pulled to the rearmost position, compressing the driving spring, and released. The driving spring drives the bolt carrier and bolt forward. The top round is stripped from the magazine and chambered during the forward stroke of the bolt. As the bolt chambers the round, the extractor enters the extractor groove of the cartridge case. Locking occurs during the last 0.3 of an inch of bolt carrier travel, when the bolt locking lugs are cammed down into the receiver locks. The weapon can now be fired by pulling the trigger, which releases the hammer. The hammer strikes the firing pin to fire the round. After firing, powder gases pass through the gas port and enter the gas chamber formed by the gas sleeves and barrel. The gases cause the actuator to strike the bolt carrier. Momentum of the bolt carrier unlocks the action and carries the bolt to the rearmost position, compressing the driving spring. During the bolt recoil stroke, the extractor pulls the cartridge case back with the bolt until ejection occurs as the ejector strikes a lug in the rear of the receiver recess. The driving spring returns the bolt carrier and bolt to battery to complete one cycle of fire.

CZECH 7.62mm MODEL 58 ASSAULT RIFLE

Characteristics of Model 58 Assault Rifle

Caliber: 7.62, uses Czech copy of Soviet M1943 rimless cartridge.
System of operation: Gas, selective fire.

Czech 7.62mm Model 58P Assault Rifle with wooden stock.

Czech 7.62mm Model 58 Assault Rifle with folding metal stock.

Czech 7.62mm Model 58V Assault Rifle with metal stock folded.

Czech 7.62mm Model 58 Assault Rifle on bipod; note plastic stock and hand guards.

Weight, loaded:
Wooden stock version: 8.75 lbs.
Metal stock version: 8.75 lbs.
Weight, empty: 7.31 lbs.
Length, overall:
Wooden stock version: 33 in.
Metal stock version, stock folded: 25 in.
Metal stock version, stock fixed: 33 in.
Barrel length: 15.8 in.
Feed mechanism: 30 round, staggered column, detachable box magazine.
Sights:
Front: Protected post.
Rear: Tangent leaf, adjustable in 100-meter increments from 100 to 800 meters.
Cyclic rate: 700-800 r.p.m.
Muzzle velocity: 2300 f.p.s.

The Model 58 assault rifle is the standard shoulder weapon of the Czech Army. It is similar in concept and in external appearance to the Soviet AK47, but is quite different internally. Its outstanding characteristic is its light weight but this is of dubious value considering its use in automatic fire from the shoulder. Lighter arms rise more rapidly in automatic fire than do heavier arms, even if they do have a straight-line stock configuration as does the Model 58.

The Model 58 may be found with wooden stock and handguards, plastic stock and handguards, or with folding steel stock. The version of the Model 58 with fixed stock is called the Model 58P; the version of the Model 58 with folding metal stock is called the Model 58V.

How to Load and Fire the Model 58 Assault Rifle

Push magazine catch in front of trigger guard forward and remove magazine. Fill magazine with cartridges and insert in magazine well. Set safety on safe by turning safety-selector lever to the safe position. Pull operating handle to the rear and release; a cartridge will be chambered. The selector has two fire positions; up for semi-automatic fire and midway for automatic fire; push safety-selector lever to position desired and the weapon will fire. When last round is fired the bolt will remain open.

Field Stripping the Model 58 Assault Rifle

Remove magazine and check chamber to insure that weapon is not loaded. Press the trigger to release hammer if cocked. Pull cover assembly retaining pin, mounted at rear of receiver cover, to the right and remove receiver cover. Move bolt and bolt carrier assembly to the rear and lift it up and out of the receiver. Rotate the hammer-striker approximately 1/8 turn counterclockwise and withdraw to the rear so that the bolt carrier, bolt assembly, and locking lugs are separated. Pull handguard retaining pin, located at front of rear sight base to the right and remove handguard by lifting up its rear portion and withdrawing it to the rear. Pull the piston to the rear and pull the piston head up so that piston and piston return spring can be removed. Reassemble by reversing the above procedure.

How the Model 58 Assault Rifle Works

Unlike most weapons of this type, the Model 58 does not have a rotating hammer—it has a linear travel hammer firing pin. Pressure on the trigger causes the sear (there are two sears—semiautomatic and automatic) to release the hammer and fire the cartridge. Gas from the cartridge is tapped off at the gas port into the gas cylinder and it moves the gas piston to the rear at high speed; the rear end of the spring-loaded piston protrudes through the rear sight base and strikes the top front of the bolt carrier. The bolt carrier moves to the rear, camming the bolt lugs out of their locking recesses in the receiver and compressing the operating spring. The extractor in the face of the bolt withdraws the empty cartridge case which is ejected out of the weapon. The operating spring forces the bolt and bolt carrier forward and the bolt picks up a cartridge from the magazine and chambers it. If the weapon is set on automatic the automatic sear will remain disengaged—assuming pressure is continued on the trigger—and the weapon will fire again. If the weapon is set on semiautomatic fire the semiautomatic sear will rise and hold the hammer to the rear until pressure is released on the trigger and the trigger is pressed again.

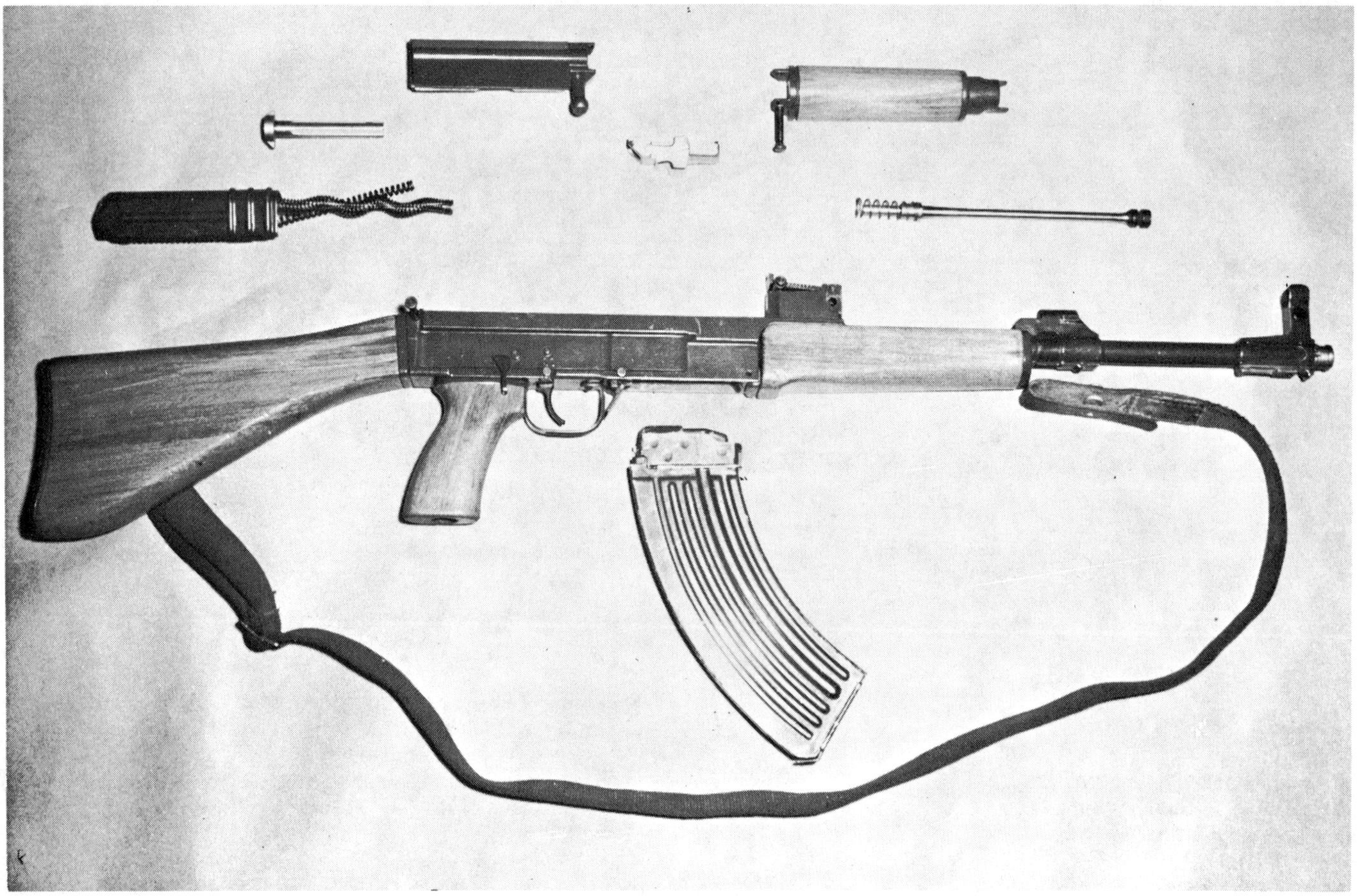

Czech 7.62mm M58P Assault Rifle—field-stripped.

CZECHOSLOVAK SUBMACHINE GUNS

The Czechoslovaks have come out with a large number of submachine gun designs since World War II. Most of these designs appeared only in prototype form and were intended for oversea sales. Among these were the ZK466, CZ247, ZB47, and ZK476. These weapons are not used by the Czechoslovak Army. Before the Communists took over Czechoslovakia, her army used the 9mm Parabellum as the service cartridge for pistols and submachine guns. Two submachine guns chambered for that cartridge were adopted by the Czechoslovak Army: the Model 23 and the Model 25 submachine guns. After the adoption of the Soviet 7.62mm pistol cartridge by the Czechoslovaks, two weapons similar to the Models 23 and 25, but chambered for the 7.62mm pistol cartridge, were produced, the Model 24 and the Model 26. These weapons have been replaced, for most purposes, by the Model 58 assault rifle and in some cases by the 7.65mm Model 61 submachine gun.

ZK466 9mm Submachine Gun.

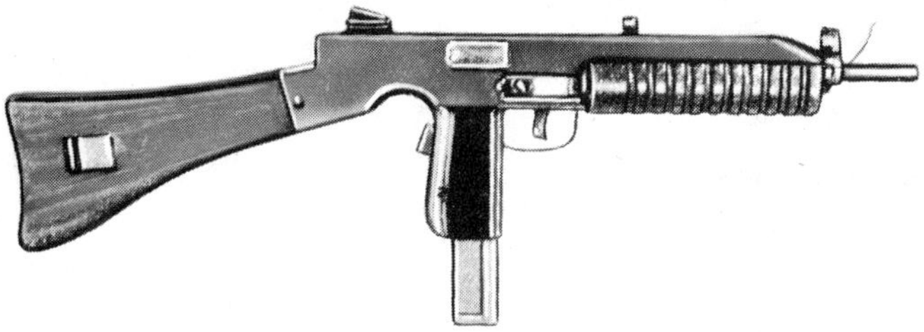

ZK476 9mm Submachine Gun.

Czech 9mm ZK383 Submachine Gun with bipod fixed.

THE CZECH ZK383 SUBMACHINE GUN

The ZK383 was designed and produced in some quantity prior to World War II and was carried in a post-1948 ZB catalog. The weapon was designed by Josef and Frantisek Koucky and was produced in three slightly different designs. The basic ZK383 has a folding bipod attached and was used by the Bulgarians and several South American countries. It was continued in production at ZB (then called Waffenwerke Brunn by the Germans) and was apparently used by the Waffen S.S. to some extent during the war. The ZK383P was developed for police use and does not have a bipod or a removable barrel like the other models.

The ZK383H is the third variation of this weapon and differs mainly in having a folding-type magazine housing which is fitted to the bottom of the weapon rather than to the left side as with the other models.

Characteristics of the ZK383

Caliber: 9mm Parabellum.
System of operation: Blowback, selective fire.
Length overall: 34.4 in.
Barrel length: 12.8 in.
Weight: 9.6 lb.
Sights: Front: Protected blade.
Rear: V-notch tangent graduated from 100-800 meters.
Feed mechanism: 30-round detachable, staggered row box magazine.
Muzzle velocity: approx. 1400 f.p.s.
Cyclic rate: 500 r.p.m., normal; 700 r.p.m., accelerated.

The bolt of this weapon has a six-ounce removable block. Removal of this block provides the higher rate of fire.

THE CZECH CZ 247 SUBMACHINE GUN

The CZ247 gun is a direct descendent of the CZ1938 submachine gun and was designed by Frantisek Myska. The CZ1938 used either a box or drum magazine but the CZ247 uses only a 40-round box magazine. The weapon is capable of selective fire and is blowback operated.

Czech Submachine Gun Model ZK-383-P.

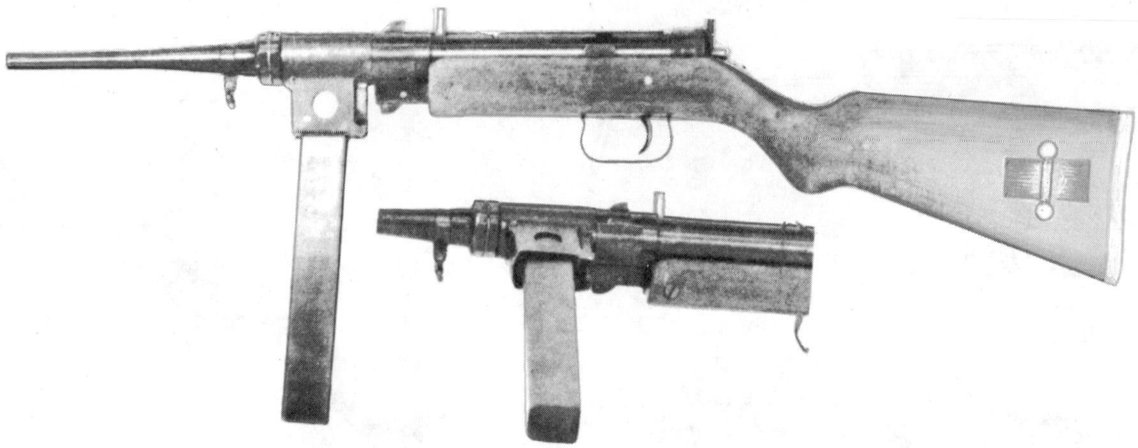

Czech 9mm Parabellum CZ 247 Submachine Gun.

This weapon was never extensively manufactured and has only one unusual feature. The magazine housing may be turned on the barrel so that the magazine will feed from under or from the left side of the barrel.

THE CZECH MODEL 23 AND MODEL 25 SUBMACHINE GUNS

Characteristics of Models 23 and 25

Caliber: 9mm Parabellum.
Weight, M23, loaded: 8 lb., with 24-round magazine. 8.4 lb., with 40-round magazine.
M25, loaded: 8.7 lb., with 24-round magazine. 9.0 lb., with 40-round magazine.
Length, overall: 27 in. for Model 23 and Model 25 with stock extended; 17.5 in. for Model 25 with stock folded.
Barrel length: 11.2 in.
Feed mechanism: 24-round or 40-round detachable, staggered, box magazine.
Sights: Front: Hooded barleycorn.
Rear: V-notch on rotary base; adjustable for 100, 200, 300, and 400 meters.
Muzzle velocity: 1470 f.p.s.
Cyclic rate of fire: 600-650 rounds per minute.

The Model 23 and Model 25 are basically the same weapon, but the Model 23 has a wooden stock and the Model 25 has a folding metal stock. The weapons have several outstanding features and are good examples of modern submachine gun construction. They share most of their unusual features with the Israeli Uzi submachine gun. Among these features are:

The magazine well is in the pistol grip. This gives the magazine far better support than is normally obtained with a submachine gun.

The guns have a very short overall length in comparison to their barrel length. This is made possible by a hollow bolt, which telescopes the rear 6 1/8 inches of the barrel. Only a 1 7/8-inch length of the 8-inch bolt is solid; the remainder is hollowed out to telescope the barrel.

The trigger mechanism on the Czechoslovak weapons is so designed that a short pull on the trigger gives semiautomatic fire, while pulling the trigger all the way to the rear gives full automatic fire.

The ejection port is closed at all times to the entry of dirt, except when the weapon is ejecting.

These weapons also have an unusually simple method of disassembly and assembly.

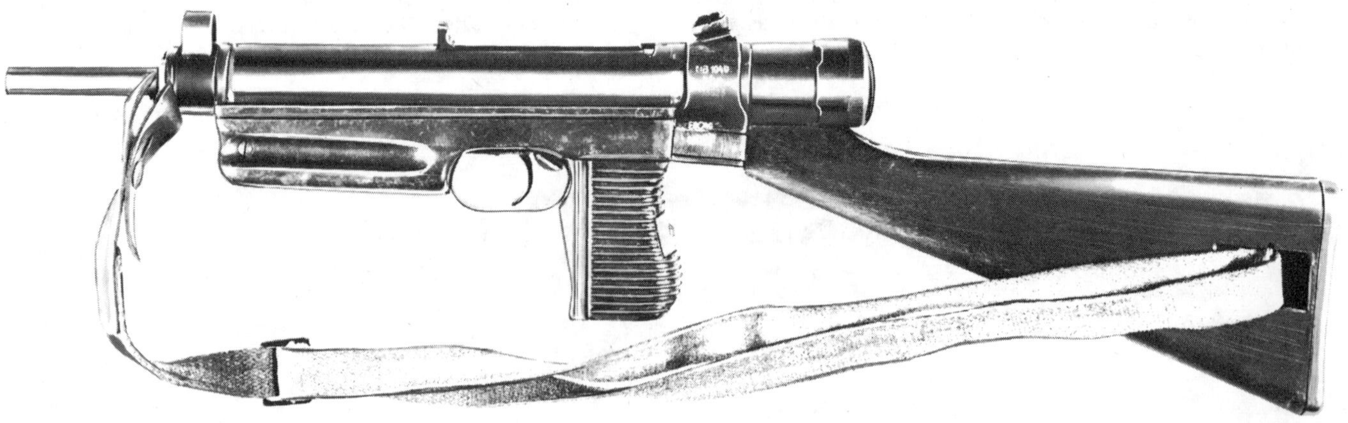

Czech 9mm Model 23 Submachine Gun.

Czech 9mm Model 25 Submachine Gun.

How to Load and Fire the Model 23 and Model 25 Submachine Guns

Draw the operating handle to the rear. The weapon fires from an open bolt, and therefore the bolt will remain to the rear. Insert a loaded magazine in the well of the pistol grip. Squeeze the trigger. If the trigger is pulled about halfway to the rear, the weapon will fire single shots with occasional doubles. If the trigger is pulled all the way to the rear, the weapon will fire automatically until it is empty or until the trigger is released. To remove the magazine, pull the magazine catch (located at the bottom rear of the pistol grip) to the rear, and pull the magazine down and out of the weapon. The safety blocks the rear of the trigger, and also locks the bolt. To put the weapon of SAFE, push the safety lever to the right.

Field Stripping the Model 23 and Model 25 Submachine Guns

Disassembly of Weapon. Remove magazine. Push forward on button in center of receiver barrel jacket cap; at the same time, turn cap 1/8 turn to right or left and remove to the rear.

Slide bolt assembly rearward by means of the operating handle.

Pull trigger, and slide bolt assembly to the rear, out of receiver/ barrel-jacket assembly.

Use bolt assembly as tool to loosen barrel locking nut. Place bolt assembly around barrel, with slots at front of bolt engaging lugs of barrel locking nut. Unscrew locking nut by turning bolt counterclockwise.

Pull barrel forward from receiver/barrel-jacket assembly.

To unfasten stock, remove screw holding the metal neck of stock to the stock support. To remove from receiver/barrel-jacket assembly, pull stock to the rear and left.

Use same procedure for Model 23.

Disassembly of magazine. Through the rectangular opening in magazine base, depress flat metal stop; slide flat metal base cover to the rear, thereby releasing magazine spring. Withdraw spring and cartridge follower through bottom of magazine.

Assembly. Assembly is accomplished by performing the steps for disassembly in reverse order.

How the Model 23 and Model 25 Submachine Guns Work

Chambering. With the bolt in rearward position, the trigger is pulled. This depresses the sear, releasing the bolt. The operating spring drives the bolt forward. The lower edge of the bolt face strips a round from the magazine and forces it forward into the chamber.

Locking. No locking step takes place, since the weapon is of the straight blowback type.

Firing. Firing occurs when the operating spring drives the bolt forward against the chambered round, and the fixed firing pin strikes the primer.

Extraction. The extractor engages in the extraction groove of the round as it is chambered. When blowback drives the bolt rearward, the extractor pulls the cartridge case out of the chamber.

Ejection. When the cartridge case (held by the recoiling bolt) clears the chamber, it strikes the tip of the stationary ejecting rod and pivots to the right. Simultaneously, the ejection ports of the bolt and receiver move into alinement, allowing the cartridge case to be thrown clear of the weapon.

Cocking. Since the gun fires only from an open bolt, it is cocked by moving the bolt rearward until the sear engages the sear notch

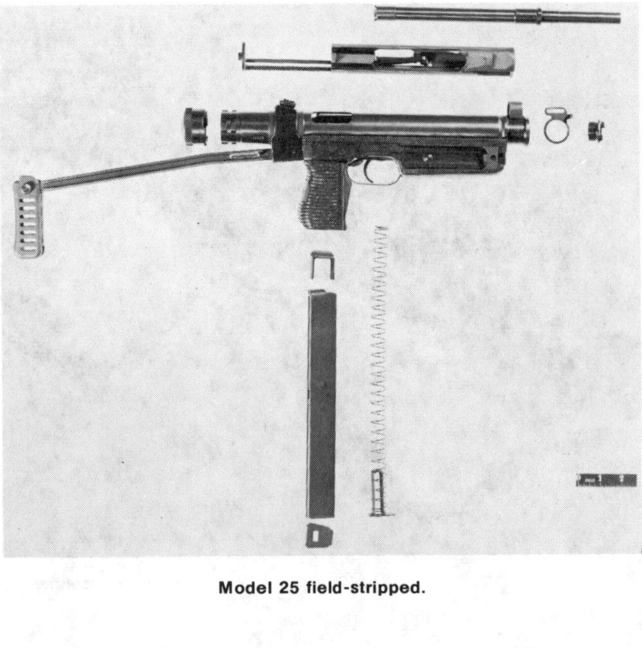

Model 25 field-stripped.

in the bolt, thus holding the bolt to the rear against pressure of the operating spring. Cocking may be accomplished either manually or, when firing, by blowback.

Semiautomatic Fire. The sear assembly is pivoted at the front and contains a spring-loaded plunger extending to the rear. The trigger, which is pivoted at the rear, has a forward projection which rests on the sear plunger. As the trigger is pulled, its projection bears against the sear notch in the bolt. The bolt then moves forward to chamber and fire the round. As the trigger is pulled beyond the point at which the bolt is released, the trigger projection rotates out of engagement with the sear plunger, thus allowing the sear to rise under pressure of the sear spring. Upon firing, the bolt is blown back to the rear until the sear engages the sear notch in the bolt.

NOTE. Since no positive disconnector is provided, a slow squeeze on the trigger can result in short bursts.

Full Automatic Fire. In full automatic fire the trigger is pulled all the way to the rear, beyond the semiautomatic fire position. A projection on the trigger now engages the sear directly, disengaging it from the bolt and allowing full automatic fire until the trigger is released or the magazine emptied.

7.62mm Replacement for 9mm Models 23 and 25

The 7.62mm submachine gun which replaced the 9mm Models 23 and 25 in the Czechoslovak Army are basically the same as the Models 23 and 25. A few minor changes have been made, but function, loading, firing, assembly, and disassembly remain the same. The major change is the chambering of the weapons for the 7.62mm pistol cartridge. The heavy loading of the Czechoslovak-made 7.62mm pistol cartridge gives the weapons a velocity which falls in the class of the United States M1 and M2 carbines.

Characteristics of 7.62mm Submachine Gun Model 24

System of operation: Blowback, selective fire.
Length, overall: 27 in.
Barrel length: 11.2 in.
Feed mechanism: 32-round detachable, staggered-box magazine.
Sights: Front: Hooded blade.
Rear: Square notch on rotary base, adjustable for 100, 200, 300, and 400 meters.
Muzzle velocity: 1800 f.p.s.
Cyclic rate of fire: 600-650 r.p.m.
Weight, loaded: 8.8. lbs.

Note that only the weight and overall length with stock folded will be different for the version with metal stock.

Special Note on Czechoslovak Submachine Guns

The Czechoslovak 9mm Models 23 and 25 submachine guns, as well as the 7.62mm submachine guns, have a magazine filler built into the right side of their plastic fore-ends. The ammunition for these weapons is packed in 8-round chargers (clips). The charger is laid base down in the guide slot of the magazine filler, with the weapon on its left side. The empty magazine is then pushed down the guide, mouth forward (toward the muzzle of the weapon), and the cartridges are stripped into the magazine.

THE CZECH MODEL 61 SUBMACHINE GUN

This weapon might be described as a "machine pistol" since it is designed to be fired from one hand as well as from the shoulder. It could be considered the Czech equivalent of the Soviet Stechkin machine pistol or the M1932 Mauser. It uses a relatively low powered cartridge, the 7.65mm Browning short or the .32 ACP as it is called in the United States, and is probably relatively easy to control in full automatic fire from the shoulder because of that fact. As a matter of interest, the .32 ACP cartridge as loaded in the United States has less muzzle energy than the .22 Long Rifle cartridge in high velocity loads. With a weapon of the M61 type however, there is a possibility of obtaining multiple hits on a target.

Characteristics of Model 61 Submachine Gun

Caliber: 7.65mm (.32 ACP).
System of operation: Blowback, selective fire.
Length overall: With stock fixed: 20.55 in.
With stock folded: 10.62 in.
Barrel length: 4.5 in.
Weight: 2.87 lb.
Sights: Front: Protected post.
Rear: Flip-over notch graduated for 75 and 150 meters.
Feed Device: 10- or 20-round detachable staggered row box magazine.
Muzzle velocity: 1040 f.p.s.
Cyclic rate: 750 r.p.m.

Czech 7.62mm submachine gun.

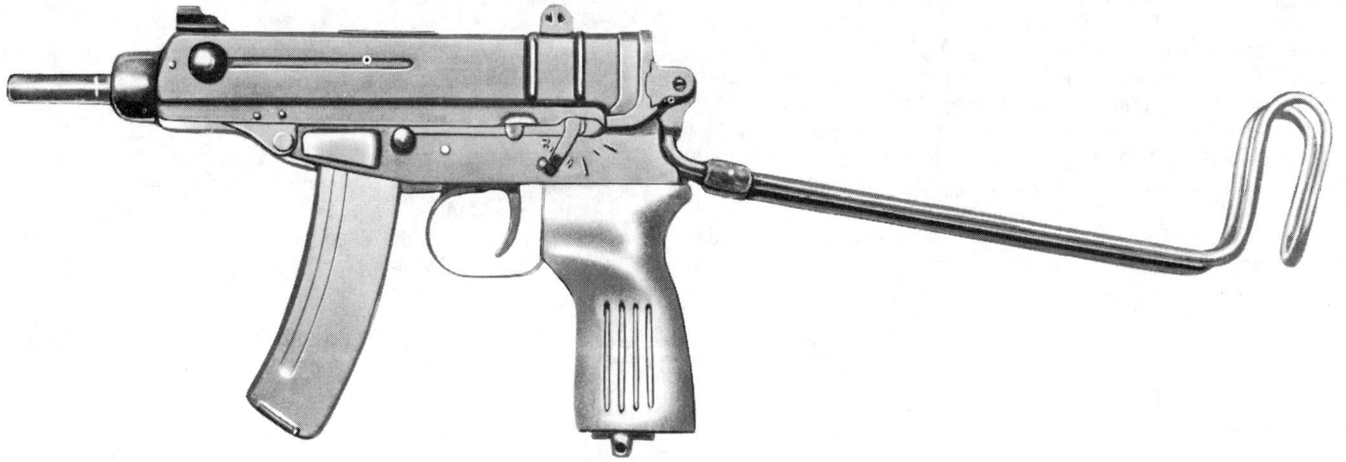

Czech 7.65mm Model 61 Submachine Gun with stock fixed.

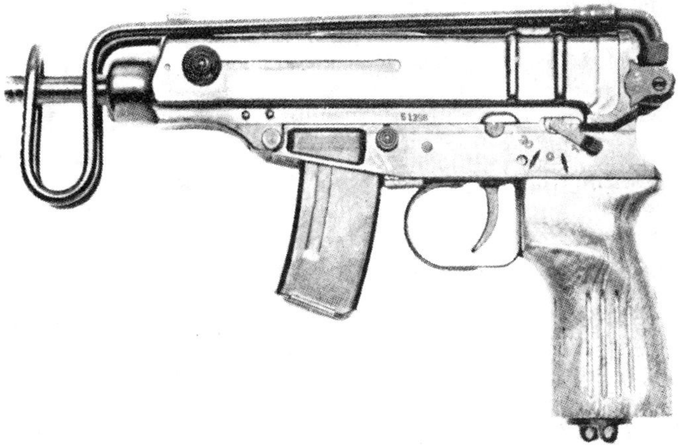

Czech 7.65mm Model 61 Submachine Gun with stock folded.

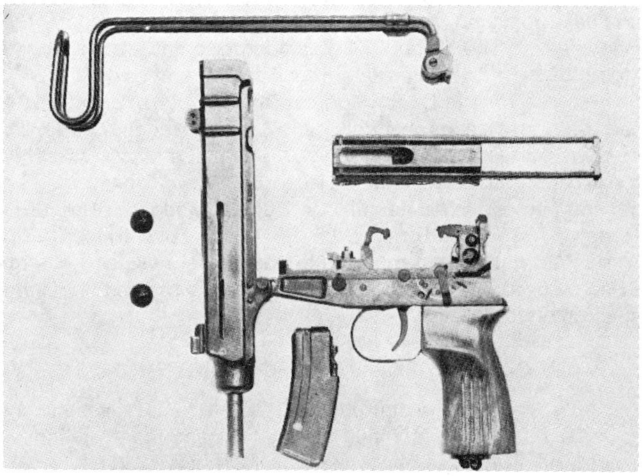

Czech 7.65mm Model 61 Submachine Gun—field-stripped.

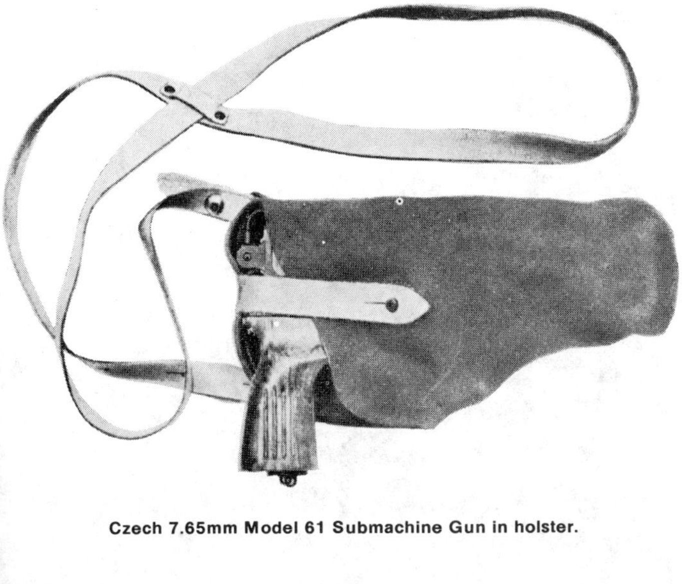

Czech 7.65mm Model 61 Submachine Gun in holster.

How to Load and Fire the Model 61 Submachine Gun

The magazine can be removed by pressing the button located on the left side of the receiver in front of the trigger guard. Pull bolt to rear, bolt operating knobs are on both sides of the receiver. Set the weapon on safe by pushing the safety-selector lever, located over the pistol grip on the left side of the receiver, down to its central position. Insert a loaded magazine and set safety-selector lever for type of fire desired—forward for automatic fire—to "20" and to the rear for semiautomatic fire to "1." Press trigger and weapon will fire; bolt remains open on the last shot.

Field Stripping the Model 61 Submachine Gun

Remove magazine and pull bolt to the rear to clear gun. Fix shoulder stock. Push out pin—at lower front of the receiver—to the left. Pull receiver forward and hinge upward, remove bolt operating knobs. Remove bolt and operating springs assembly. Further disassembly is not recommended. Reassemble by reversing above procedure.

How the Model 61 Submachine Gun Works

This weapon is a pure blowback, i.e., the cartridge is held in place during firing by the weight of the bolt and the recoil spring. The rate reducer, used to lower the cyclic rate, is a hook at the rear of the receiver which holds the bolt to the rear momentarily after each shot. This hook is released by pressure from a spring loaded tripper-plunger mounted in the pistol grip. As the bolt comes to the rear, it cams down this tripper-plunger compressing its spring. The tripper-plunger and spring are mounted in a tube and the tripper-plunger travels a considerable distance down the tube before it is pushed up by the spring causing the hook to tip up and release the bolt. This system, which is quite similar to that used on the Soviet Stechkin machine pistol, has the effect of reducing the cyclic rate of fire. The bolt telescopes the barrel as in the earlier Model 23 series submachine guns, but is square in configuration, like the Israeli Uzi rather than round like the Model 23.

Special Note on Model 61 Submachine Gun

There is a silencer available for this weapon which should be effective because of the relatively low velocity of the bullet fired. Luminescent night sights are also available for the weapon.

CZECHOSLOVAK MACHINE GUNS

Czechoslovakia has been a prolific producer of machine gun designs since the early twenties. The famed ZB26 (Model 26 light machine gun) is still used extensively throughout the world, as is the ZB30 and ZB53 (Model 37 heavy machine gun). The British Bren was developed from the ZB series of light machine guns, and still has an excellent reputation as a light machine gun.

Since World War II the Czechs have come out with several new machine gun designs. The Model 52, a rather versatile weapon, originally replaced the 7.92mm Model 26 as the standard light machine gun. The Model 52 is chambered for the Czech 7.62mm Model 52 cartridge, but a later modification, the Model 52/57, is chambered for the Czech copy of the Soviet 7.62mm M1943 cartridge. The Czechs abandoned the 7.92mm Model 37 as a heavy machine gun and used the Soviet 7.62mm Goryunov for awhile; this weapon has now been replaced by a new "Universal"—i.e., general purpose machine gun, the 7.62mm Model 59. The 15mm Model 38 (ZB60) used prior to the war, and by the Germans during World War II, has been replaced by the Soviet designed 12.7mm DShK M38/46, which is used on vehicles and on a Czech-designed quadruple mount.

CZECH ZB26 and ZB30 MACHINE GUNS

In 1924 Vaclav Holek introduced a belt-fed light machine gun. In the same year he modified the weapon and it was called the Praga Model 24. This weapon is now known as the ZB26 (although it is frequently still called the Model 24 in Czechoslovakia; the gun is stamped VZ26) and was one of the most popular light machine guns in the world. The Model 26, Model 30, and Model 30J were used in 24 countries throughout the world. Models 26 and 30 have been manufactured in China, the Model 30 in Iran and Rumania, and the Model 30J in Yugoslavia. All three models were carried in a post-1948 ZB catalog.

These weapons are the direct ancestors of the Bren Gun, which was manufactured by ZB as well as by the British, Canadians, and Australians. The ZB26 was the standard light machine gun of the Czech Army and many ZB26 and ZB30 machine guns were used by the Germans during World War II.

Characteristics of ZB26 Machine Gun

Caliber: 7.92mm.
System of operation: Gas, selective fire.
Length overall: 45.8 in.
Barrel length: 23.7 in.
Weight loaded: 21.28 lb.
Sights: Front: Protected blade.
Rear: Radial tangent.
Feed device: 20-round, detachable staggered row box magazine.
Muzzle velocity: 2500 f.p.s.
Cyclic rate: 550 r.p.m.

These weapons can be mounted on a tripod similar to that of the Bren Gun. A tripod of this type is also being used by the Chinese Communists to mount a copy of the U.S. 57mm recoilless rifle.

Loading and Firing the ZB26

Pull back the bolt handle to cock the bolt and then push the handle forward. Pull back magazine cover and insert loaded magazine mouth end first. Push down until it locks. For semiautomatic fire move the selector on the left side of the trigger group to the rear; for full auto fire, push it forward. When the selector is in the vertical position, it acts as a safety.

Field Stripping the ZB26

Push out receiver locking pin and withdraw the frame group. The slide, bolt, and gas piston will now come out the rear of the receiver.

Release the barrel nut catch and lift up to the right as far as it will go. This releases the barrel which may now be slid off to the front.

Push the butt plate catch and remove the two buffer springs.

ZB26 (Brno) L.M.G. The magazine is a special oversized one.

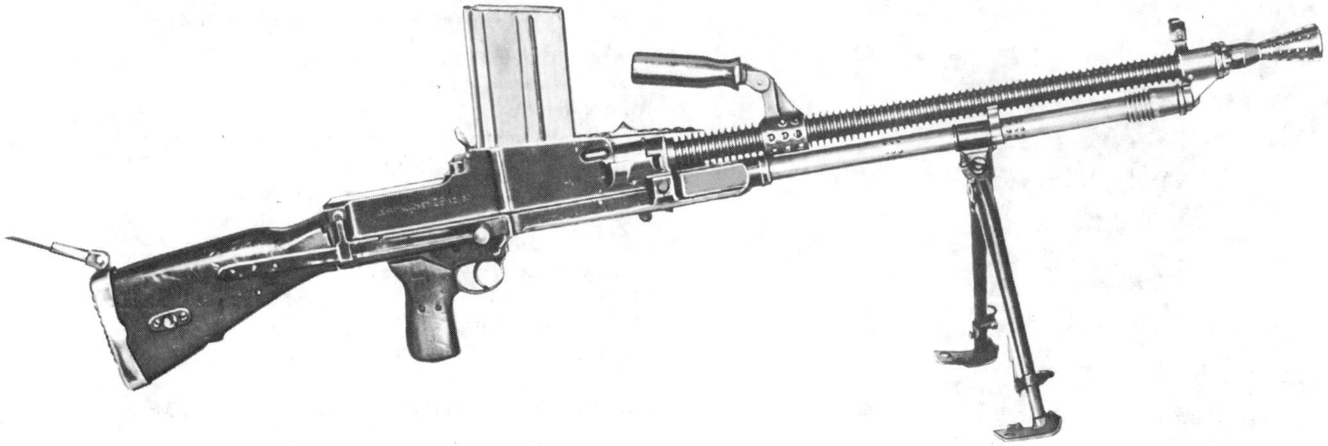

7.92mm ZB30 L.M.G. of Czech manufacture.

Differences Between the ZB26 and ZB30

Outwardly, the two weapons are almost identical in appearance. The bolt of the ZB26 does not ride on the piston post as does that of the ZB30 (and the Bren). It is cammed into the locked position by a built-up rear section of the piston/slide assembly, but does not "sit" on the post. The ZB30J is similar to the ZB30, but has a knurled section on its barrel to the rear and just in front of the carrying handle.

CZECH MACHINE GUN MODEL 37 (ZB53)

The Model 37 heavy machine gun has been extensively manufactured for export. The military designation for the gun is Model 37 (ZB37); the commercial designation is Model 53 (ZB53).

The Model 37 is an air-cooled, gas-operated weapon with selective slow (500 r.p.m.) and fast (700 r.p.m.) rates of fire. It is fed from the right side by a metal belt of either 100- or 200-round capacity. By use of an attachment to the Model 45 tripod, it can be quickly adapted to antiaircraft fire.

This weapon was the forerunner of the British-made Besa tank machine gun. The main functioning features of the Model 37 are the same as those given for the Besa machine gun, which can be found in the chapter on Britain.

Czechoslovakian Heavy Machine Gun Model 37 (ZB 53).

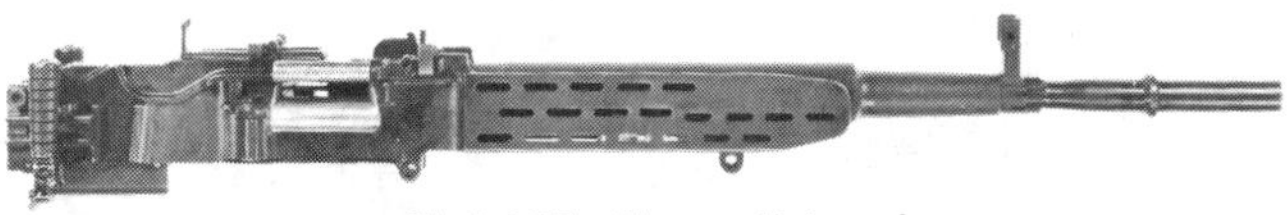

Model 37 with smooth barrel.

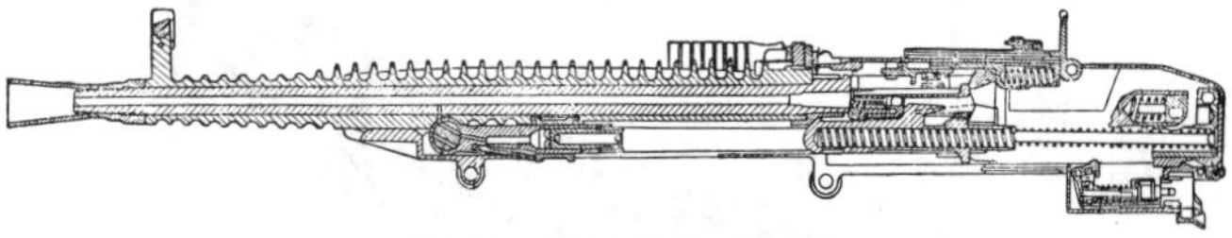

Section view of Model 37.

Characteristics of the Model 37

Caliber: 7.92mm.
System of operation: Gas operated, two cyclic rates of fire.
Weight: 41.8 lb.
Length, overall: 43.5 in.
Barrel length: 26.7 in.
Sights: Front: Blade with guard.
Rear: Folding, leaf, graduated from 300 to 2000 meters in 100-meter increments, fixed 200-meter battle sight.
Cyclic rate: 450-550 r.p.m. or 700 r.p.m.
Feed device: Metallic link belt, 100- or 200-round capacity.
Muzzle velocity: 2600 f.p.s. (approx.).

CZECH 12.7mm QUAD DShK M1938/46 HEAVY MACHINE GUN

This weapon consists of four Czechoslovak-made (but Soviet-designed) DShK M 1938/46 heavy machine guns on a Czechoslovak-designed, two-wheeled antiaircraft mount. The weapon is therefore generally similar in purpose to the United States M45 quad cal. .50. The Czechoslovak mount, however, lacks powered operation and also suffers from having magazines with a very limited capacity (fifty rounds for each gun).

The weapon is the same as the Soviet DShK, which is covered in detail in the chapter on the U.S.S.R. The weight of the complete equipment is 1411 pounds, and the mount is capable of 360-degree traverse and 90-degree elevation.

Czech 12.7mm Quad DShK M1938/46 Heavy Machine Gun.

CZECH MODEL 52 LIGHT MACHINE GUN

The Model 52 light machine gun is also chambered for the Czechoslovak 7.62mm Model 52 "intermediate-sized" cartridge. The weapon can be fed from a belt or a box magazine without changing feed covers. When fed from the belt, the feed is similar to that of the Czechoslovak Model 37 heavy machine gun; when fed from the magazine, it is similar to that of the ZB26. The weapon makes extensive use of stampings and is, all told, a very sophisticated weapon. Possibly its greatest shortcoming is that it may be too sophisticated; in general, it can be said that the simplest weapons give the best performance in the field. Although the Model 52 is a well-designed weapon, it is not particularly simple.

A slightly modified version of the model 52 light machine gun, chambered for the Soviet designed 7.62mm M43 cartridge, was adopted in 1957 as the Model 52/57.

Characteristics of M52 Machine Gun

System of operation: Gas, selective fire.
Length, overall: 41 in.
Barrel length: 21.3 in.
Feed mechanism: 25-round box magazine, or 100-round non-disintegrating link belt (push-out type link).
Sights: Front: Blade with removable hood.
Rear: U-notch, adjustable for elevation and windage, graduated from 200 to 1200 meters.
Muzzle velocity: 2450 f.p.s.
Cyclic rate of fire: 1140 r.p.m. with belt, 900 r.p.m. with box magazine (approx.).
Weight without belt or magazine, with bipod: 17.6 lb.

How to Load and Fire the Model 52 Light Machine Gun

Cock the gun by pressing down on the lug which protrudes from the safety (on top left side of pistol grip) and pull pistol grip to the rear. The bolt will remain to the rear since this weapon fires from an open bolt. If the pistol grip is in the rearward position, press down on the lug and push the pistol grip forward until the slide is engaged by the sear; then pull the pistol grip rearward until the pistol grip catches on the rear lock.

To load the gun with a belt of ammunition, push upward on the feed-way cover. Lay the first cartridge in the belt between the belt holding pawls, and close and lock the cover. The link ejection port cover must be in its raised position or belt movement will be blocked.

To load the gun for magazine fire, press forward on the magazine feed port latch, allowing the cover to spring forward. Invert the magazine and insert it in the magazine opening until the latch engages the magazine. The belt feedway cover and the link ejection port cover should be closed. Pressure on the bottom half moon of the trigger will produce automatic fire; pressure on the top half moon of the trigger will produce semiautomatic fire. To put the weapon on safe, push the safety lever (located on top left side of the pistol grip) down.

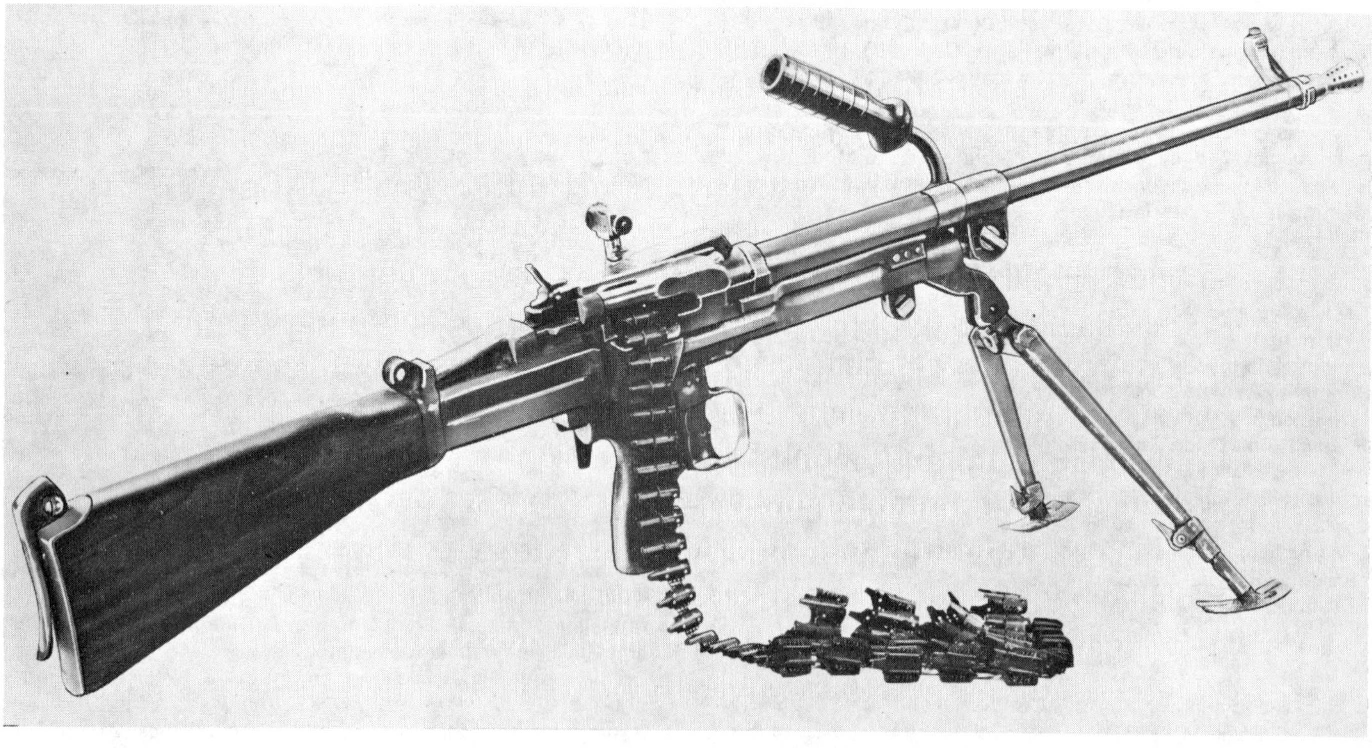

Czechoslovak 7.62mm Model 52 Light Machine Gun.

Field Stripping the Model 52 Light Machine Gun

Driving Spring and Rod. The driving spring is retained by a cap with internal bayonet-type slots that engage projections of the spring tube within the stock. To remove the spring assembly, press in on the cap, turn it counterclockwise to disengage the slots, and withdraw the spring and rod from the stock.

Barrel and Bipod. Press forward on the magazine feed port latch, allowing the cover to spring open. Using the cover as a handle, turn the barrel lock clockwise and draw the barrel forward out of the receiver. Turn the bipod assembly to disengage its key, and remove the bipod from the weapon housing.

Receiver Assembly, Bolt Carrier, and Bolt. The receiver, bolt carrier, and bolt are removed simultaneously. Draw the bolt carrier rearward so that the piston is located slightly to the rear of the gas chamber in the receiver housing. Lift up on the front end of the receiver and remove the receiver, bolt carrier, and bolt from the weapon. Remove the bolt carrier and bolt through the rear of the receiver.

Trigger Mechanism Assembly. Depress the pistol grip lock lever and push forward on the pistol grip. Slide the trigger mechanism assembly out of the front end of its slot in the weapon housing. Unhook the dust cover from the rear of the trigger mechanism housing.

Reassembly. Reassemble the weapon by reversing the procedure described above.

Adjustment of Gas Cylinder. There are four gas port openings that may be selected to vary the power of the weapon. To change the port setting, the barrel must first be removed from the gun as described above. To change port setting, turn the gas regulator until the desired port is aligned with the barrel gas port. The ports are identified by different-size indents in the quadrants formed by the crossed cylinder locking slots on the left side of the regulator. Alignment of one of these indents with the indent in the cylinder body selects the appropriate port size.

How the Model 52 Light Machine Gun Works

When the gun is fired, gas from the barrel is bled through a port into the gas cylinder and into the gas chamber at the front of the receiver housing. The expanding gas operates on the gas piston to

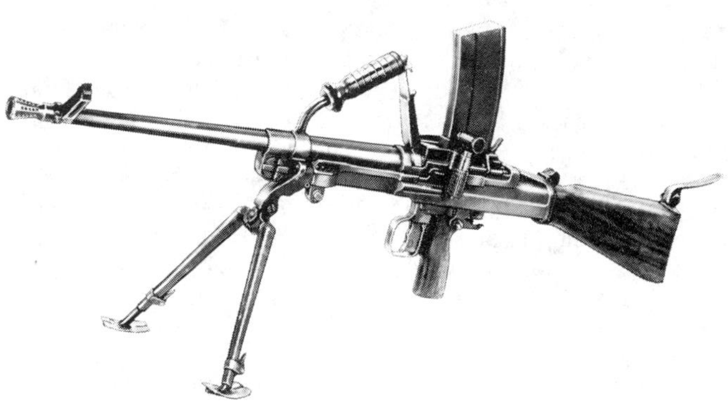

Czech Model 52 with box magazine.

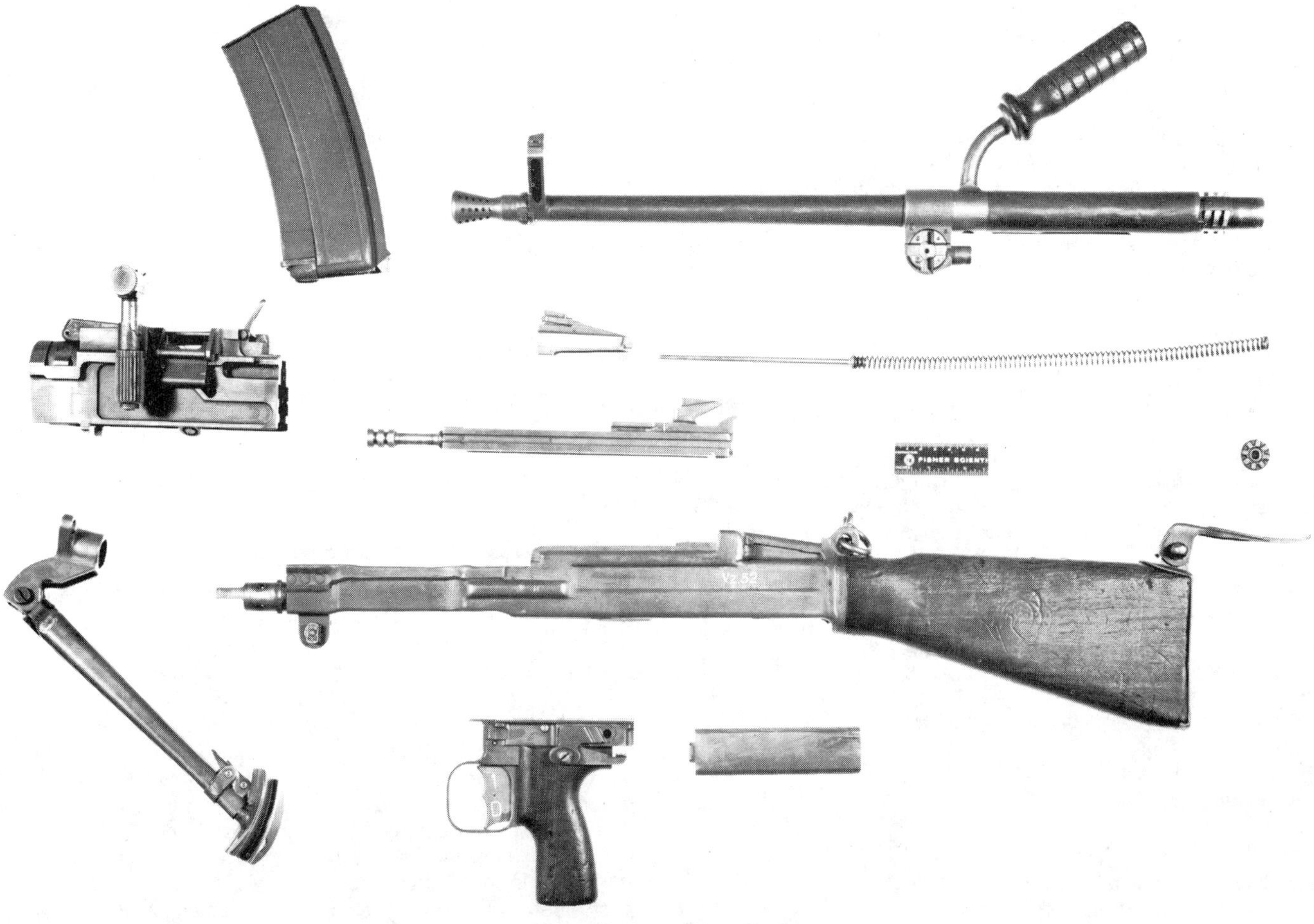

Czech Model 52, field-stripped.

force the bolt carrier rearward in recoil. For a short distance, the bolt carrier travels alone. This short period is followed by an unlocking period, during which the rear of the bolt rotates downward our of engagement with the receiver locking abutments. Upon completion of unlocking, the bolt is carried to the rear by the carrier, the spent cartridge is extracted from the barrel, and the case is ejected forward.

Upon completion of recoil, the residual energy of the bolt and bolt carrier is absorbed by a recoil plate at the rear of the receiver housing, and the bolt carrier moves forward under the force of the compressed driving spring in the stock. If the gun is being fired semiautomatically, the sear will engage the carrier and hold it until the trigger is again pressed. When the trigger is pressed, the carrier will move forward; as the rear of the carrier clears the sear, it will depress the disconnector, releasing the sear so that it can again be in position to engage the carrier.

Feeding takes place on the counterrecoil stroke of the bolt and bolt carrier. The cartridge to be fed is held in the center of the feedway by the belt holding pawl or by the pressure of the magazine spring, depending upon the type of feed being used. The forward-moving bolt strikes the lower edge of the cartridge rim, stripping if from the belt link or magazine lips. The nose of the cartridge is depressed by a ramp in the receiver breech ring, and, as the bolt continues forward, the ramp depresses the rear of the cartridge seating the rim between the extractor claws.

Receiver of Model 52.

Model 52 showing magazine in place.

Special Note on the Czech Model 52 Light Machine Gun

This weapon has many very interesting features. Because of its ability to feed from either a box magazine or a belt, without changing feed components, it has a great deal of tactical flexibility. Its trigger mechanism, while hardly new in concept (the double half moon trigger can be traced back at least as far as World War I), adds considerably to the weapon. The use of a large stamped receiver housing with a much smaller machined receiver within the housing also is advantageous from an industrial point of view. It should be noted, however, that the machined receiver is quite a complex piece from a manufacturing point of view. Dirt may be kept out of the Model 52 by closing the feed and link ejection port covers, and by pulling the pistol grip forward and locking it by engaging the safety. When the weapon is ready to fire, however, the bottom of the receiver is open to dust and dirt.

The bolt is carried on a post on the slide/piston assembly in a fashion similar to that of ZB26 or the Bren gun, but it does not lock like these guns. The two locking lugs on the rear of the bolt ride in cut-outs in the receiver and are locked in the side of the receiver.

THE CZECH MODEL 59 MACHINE GUN

The Model 59 is called a "Universal" machine gun by the Czechs. This basically means that it is a general purpose machine gun, similar to the U.S. M60, which is used on a bipod as a light machine gun and on a tripod as a heavy machine gun. The Model 59 comes equipped with heavy and light barrels which are used in its heavy gun and light gun roles, respectively.

Characteristics of the Model 59

Caliber: 7.62mm rimmed (7.62mm Russian).
System of operation: Gas, automatic only.
Length overall: W/heavy barrel: 47.9 in.
W/light barrel: 43.9 in.
Barrel length:
Heavy: 27.3 in. w/flash hider.
Light: 23.3 in. w/flash hider.
Weight: W/heavy barrel on tripod, 42.4 lb.
W/light barrel on bipod, 19.1 lb.
Feed device: 50-round non-disintegrating metallic link belt.
Muzzle velocity: W/heavy barrel: 2723 f.p.s.
W/light barrel: 2657 f.p.s.
Cyclic rate: 700-800 r.p.m.

When used with the light barrel, this weapon is called the Model 59L.

How to Load and Fire the Model 59 Machine Gun

The linked cartridges come in 50-round, non-disintegrating link belts; five of these belts are usually joined together for use with the weapon when it is used on the tripod. When used as a light machine gun—on bipod—a box with a 50-round belt may be attached to the side of the receiver. The cover is opened and the belt—open side of links down—is laid on the feedway from right to left; close cover. If bolt is forward, push lug which protrudes from the safety mounted on left rear side of the pistol grip/trigger group, down pull the trigger and push pistol grip as far forward as it will go, release the trigger then draw to the rear. The belt may be inserted whether the bolt is in the forward or rearward position. If the trigger is pulled, the weapon will now fire. To put gun on "safe", push safety lever down. The gas cylinder block has four ports and can be adjusted as required for reliable operation.

Czech 7.62mm Model 59 Machine Gun on tripod.

Field Stripping the Model 59 Machine Gun

Check to insure weapon is not loaded and let bolt forward. Remove barrel by opening cover and turning one quarter turn to the right and pulling barrel forward out of the receiver. Push pin at middle rear of receiver from left to right. **Bolt must be forward so that tension on recoil spring is released.** Remove butt; recoil spring and buffer spring are mounted in the butt. Withdraw slide/piston assembly and bolt assembly from the rear of the receiver by pulling pistol grip to the rear. Pistol grip/trigger assembly will also come off the receiver. The feed plate can be lifted up off the receiver. Reassemble in reverse order.

How the Model 59 Machine Gun Works

With a belt in the feedway and the bolt in the rear position, pressure on the trigger causes the sear to disengage from its notch in the rear underside of the slide/piston assembly. The slide/piston assembly with the bolt is forced forward by the compressed recoil spring (operating spring) and strips a cartridge out of the belt forward and downward into the chamber. The bolt lock is cammed into the locked position by cam rails on the receiver and the piston post continuing its travel a short distance further engages the striker causing it to strike the cartridge primer functioning the cartridge. Gas from the cartridge is drawn off through the gas port in the barrel and travels through the gas cylinder impinging on the piston head driving the piston/slide assembly to the rear. Rearward movement of the piston/slide cams the bolt lock up out of the locked position and the bolt starts moving to the rear with the empty cartridge case; the ejector knocks the case downward out of the bottom of the gun.

Feeding is achieved through the operation of a cam surface on the slide against the belt feed pawl which is mounted on the right side of the receiver. At the bottom of the feed pawl, is a roller which contacts the cam surface on the slide causing the pawl to go in and out, i.e., move from right to left in relation to the side of the receiver. Movement of the feed pawl pulls the linked rounds in one at a time. The pawl engages the linked cartridge as the slide/piston moves to the rear and is then moved to the left on the forward run of the bolt. The cartridge is held in place by the spring loaded belt feed pawl.

Special Note on the Czech Model 59 Machine Gun

The Model 59 is one of a series of "universal" or general purpose machine guns which have appeared since World War II. It is used on a tripod as a heavy machine gun or on a bipod as a light machine gun. In Czech service, it is used with a heavy barrel for tripod use and light barrel for bipod use and a medium weight barrel is advertised as well in trade brochures. The weapon has been made in 7.62mm NATO in which caliber it is called the Model 59N, to attract sales to non-Communist nations. Barrels, gas piston, gas cylinder block and bolt are chrome plated. There are spring loaded dust covers over the feed, link ejection and case ejection ports. The case ejection port cover is opened and closed by pulling the trigger.

The Model 59 is an interesting gun in many ways. There is no major feature of the weapon that is new in concept. The feed mechanism is basically the same as that of the pre-World War II ZB37 and is also used for belt feed in the Model 52.

The cocking mechanism is also similar to that of the ZB-37 and exactly the same as that of the Model 52. The bolt mechanism is a modification of the basic ZB-26, ZB-37, Model 52; the separate bolt locking lug piece is similar to that used on the Czech Model 58 Assault rifle. The method of barrel removal is the same as that of the Model 52 machine gun. A notable feature of the Model 59 as opposed to the previous Model 52 is the comparatively small number of stampings or fabrications used in its manufacture. The receiver is a milled forging and most other parts of the weapon except the feed plate and dust covers, seem to be forgings as well. The design is basically good, but as pointed out above, hardly novel.

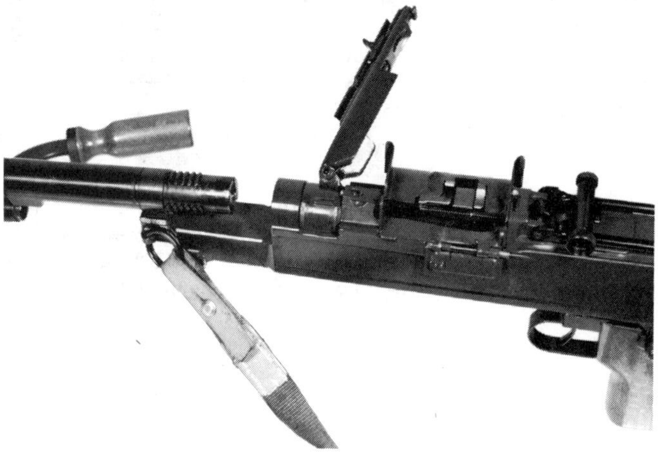

Barrel change on Czech 7.62mm Model 59 Machine Gun.

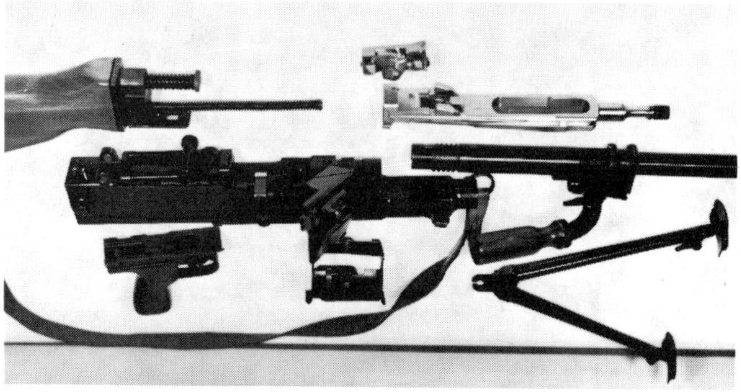

Czech 7.62mm Model 59 Machine Gun, field stripped.

Czech 7.62mm Model 59L Machine Gun on bipod.

THE CZECH "URZ" WEAPONS FAMILY

Czechoslovakia has come out with a new weapons family called the "URZ" - which translates to Universal Small Arms - which, in concept, appears to be modeled on the Stoner 63 system. A common receiver system is used for an automatic rifle (AP), a light machine gun (LK), a heavy machine gun (TK) and a tank machine gun (T).

The automatic rifle (AP) has a lighter barrel than the other models. A grenade launcher is built onto the end of the barrel and a grenade launcher sight is attached to the rear of the front sight. It uses a fifty-round, light alloy drum type magazine. It can be used with the telescopic sight normally used with the heavy machine gun TK. The AP takes a bayonet.

The light machine gun LK has a heavier barrel than the AP rifle, a folding bayonet, and a chrome lined barrel. It also uses the 50-round drum magazine.

The heavy machine gun TK is actually the light machine gun LK with a belt type feeder mounted on a tripod and normally using a telescopic sight. This weapon is fed with a 250-round metallic link belt.

The Tank machine gun T is intended for use on vehicles and aircraft. It is belt fed and fired by the use of a solenoid.

Characteristics of URZ Weapons Family

	Rifle AP	LMG LK	Hvy MG TK	Tank MG T
Caliber:	7.62mm			
System of operation:	gas, selective fire			
Length overall:	39.2 in.	39.2 in	47.2 in.	34.5 in.
Weight: (less magazine)	8.62 lb.	11.5 lb.	24.3 lb. w/tripod mount	12.6 lb.
Feed device:	50 rd drum	50 rd drum	250 rd belt	250 rd belt
Muzzle velocity:	2625 f.p.s.			
Cyclic rate:	800 r.p.m.	800 r.p.m.	800 r.p.m.	1100 r.p.m.

The URZ weapons family as advertised is not chambered for the 7.62mm "intermediate" sized Czech Model 57 (Soviet Model 43) cartridge. It is advertised for cartridges of the size and muzzle energy of the 7.62mm NATO or the Soviet rimmed 7.62 X 54mm cartridges. This does not necessarily mean that the Czechs intend to adopt this type of system themselves. Czechoslovakia is a large exporter of arms to many countries outside of the Warsaw Pact and the URZ family is probably intended for this trade.

24
Denmark

The Danish Army currently uses the small arms listed below. Shown for each weapon is: (1) its common name; (2) its Danish nomenclature; and (3) the chapter in this book where the weapon is covered in detail.

- (1) The 9mm FN Browning Hi Power Pistol; (2) 9mm P M/46; (3) see Belgium.
- (1) 9mm SIG Pistol 47/8; (2) 9mm P M/49; (3) see Switzerland.
- (1) 9mm Suomi Submachine Gun; (2) 9mm Mp M/41; (3) see Finland.
- (1) 9mm 37-39 Suomi Submachine Gun; (2) 9mm Mp M/44 (37); (3) see Sweden.
- (1) 9mm Hovea Submachine Gun M/49; (2) 9mm Mp M/49: (3) See Denmark.
- (1) 7.62mm NATO Rifle G3; (2) 7.62mm G M/66; (3) see West Germany
- (1) .30 cal. Rifle M1; (2) 7.62mm G M/50; (3) see U.S.A.
- (1) .30 cal. Rifle M1917; (2) 7.62mm G M/53 (17); (3) see U.S.A.
- (1) 7.62mm NATO MG 42/59 (MG 1) (2) 7.62mm Mg/62 (3) see West Germany
- (1) .30 cal. Madsen M48 Machine Gun; (2) 7.62mm Mg M/48; (3) see Denmark. (Note. The characteristics for the M48 Madsen machine gun are basically the same as those for the M1950, given later in this chapter. Operating characteristics are the same.)
- (1) .30 cal. SIG M50 Machine Gun; (2) 7.62mm Mg M/51; (3) see Switzerland.
- (1) .30 cal. M1919A4 Machine Gun; (2) 7.62mm Mg M/52-1 (19); (3) see U.S.A.
- (1) .30 cal. M1919A5 Machine Gun; (2) 7.62mm Mg M/52-11 (19); (3) see U.S.A.
- (1) .50 cal. M2 Browning Heavy Barrel; (2) 12.7mm Mg M/50; (3) see U.S.A.
- (1) 7.62mm NATO MG42 (MG42/59) (3) see Germany.

DANISH PISTOLS

The Danish Army, which had previously used the Gasser revolver, adopted the Bergmann Bayard automatic pistol Model 1908 in 1911, calling it the Model 1910. This pistol is a slightly modified version of the Bergmann 1903 which was made at the Bergmann plant in Germany. The Model 1908 was made by Anciens Etablissements Pieper of Herstal, Belgium. The weapon as made either in Germany or Belgium was used by Greece and Spain; pistols used by Denmark were made in Belgium or, after 1922, in Denmark.

In 1922 the pistol was slightly changed and the model designation was changed to Model 1910/21. All original Model 1910/21 pistols—a total of 2204—were made at the Army Arsenal. Between 1922 and 1935, the 4840 Belgian-made pistols were converted to the 1910/21 pattern. These pistols are obsolete in Denmark and many have been sold in the U.S.A.

In 1940 Denmark decided to adopt the 9mm Parabellum FN Hi-Power Browning pistol. A few pistols were delivered to Denmark before it was invaded by the Germans. In 1946 the order was re-instituted and 1577 pistols were purchased. The Hi-Power is called the Model 46 by the Danes. Among the weapons which the Danish Brigade (troops who had been maintained in Sweden during the war) brought back with them, were Swedish Model 40 Lahti pistols. These pistols were called Model 40S by the Danes. In 1948 Denmark adopted the Swiss S.I.G. 9mm Parabellum Model 47/8 which they call the pistol Model 49. This is the current standard Danish service pistol.

Danish 9mm Model 1910 Pistol.

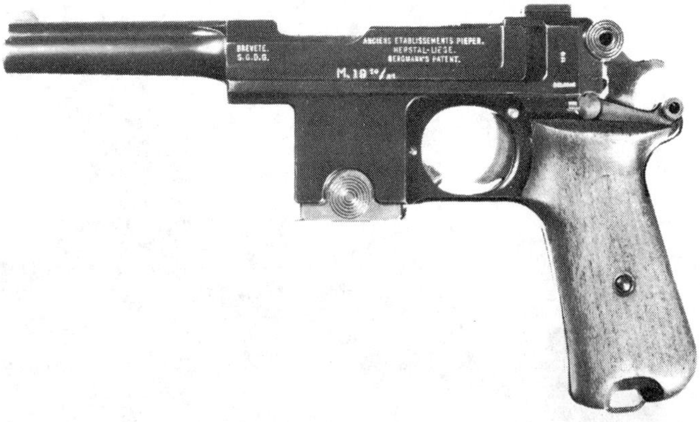

Danish 9mm Model 1910/21 Pistol converted from Model 1910.

DANISH 9mm M1910/21 PISTOL

The Model 1910/21 is a recoil-operated weapon of heavy construction. It is chambered for the 9mm Bergmann-Bayard cartridge (in Spain where it is still used, this cartridge is called the 9mm Largo) which is quite similar to the Austrian 9mm Steyr

Danish 9mm Model 1910/21 Pistol as made in Denmark; magazine has been removed.

cartridge and the U.S. Super .38 automatic pistol cartridge. It is not made in the United States.

These pistols are quite heavy, weighing about 2.25 pounds, and have a relatively short barrel—about four inches—in comparison to their overall length of about 10 inches. There are six-, eight-, and ten-round magazines for these weapons.

One of the strongest and most powerful pistols ever made. Good materials and fine workmanship. Positive lock. However, design is clumsy and bulky and should be considered obsolete.

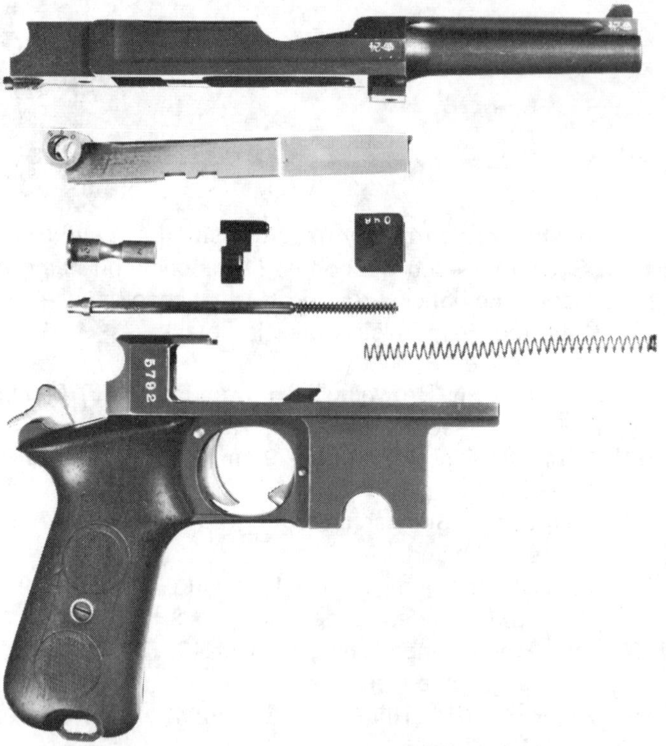

Danish 9mm M1910/21 Pistol, field-stripped.

DANISH RIFLES

Denmark was the first country to adopt the Krag Jorgensen—in 1889—and continued to use this weapon, in a series of models, through World War II. After the war Denmark was supplied with U.S., British, and Swedish rifles and eventually the U.S. caliber .30 M1 was adopted as standard. Recently the German 7.62mm NATO G3 has been adopted and this rifle will eventually replace all others in the Danish service.

The Danish Home Guard was equipped with Swedish 6.5mm M94, M96, and M38 carbines and rifles until 1953, when these weapons were returned to Sweden. The Home Guard was then issued U.S. caliber .30 M1917 (Enfield) rifles, called Model 53 by the Danes, and U.S. M1 rifles. During World War II, the Danish sporting rifle firm of Schultz and Larsen produced a police carbine (the Model 1942) for the 8mm Danish cartridge. This rifle has four locking lugs on the bolt like the current Schultz and Larsen sporters, and a box magazine which protrudes below the level of the stock.

In 1896 the Danish Navy and Coast Guard adopted the Madsen M1896 recoil operated semiautomatic rifle. Denmark therefore was probably the first country to use a semiautomatic service rifle.

DANISH KRAG RIFLES

Denmark was the first country to adopt the rifle developed by Krag and Jorgensen. This weapon in slightly modified form was later adopted by the United States and Norway. The principal point of difference between the Danish and the Norwegian and U.S. Krags are: the loading gate swings out horizontally on the Danish weapon, on the U.S. and Norwegian Krags the loading gate swings down to open and is pushed up to close; the Danish 1889 rifle and several of the carbine models have a metal barrel jacket, the U.S. and Norwegian Krags use wooden handguards. All Danish Krags were chambered for the 8mm rimmed cartridge.

Danish Madsen M1896 Semiautomatic Rifle.

Danish 8mm Rifle Model 1889/10.

Characteristics of M1889 Rifle

Caliber: 8mm Danish Krag rimmed.
System of operation: Manually operated bolt.
Length overall: 52.28 in.
Barrel length: 32.78 in.
Weight: 9.5 lb.
Feed device: 5-round in-line magazine, loaded singly through side gate.
Sights: Front: Barley corn.
Rear: Leaf, can be used as tangent with leaf down.
Muzzle velocity: 1968 f.p.s. w/M1889 ball.
2460 f.p.s. w/M1908 ball.

Danish Krag Models

The models of the Danish Krag are as shown below.

Rifle M1889. This weapon whose characteristics are given above, is typical of its period in having a long barrel and stock without pistol grip. As noted above, a metal handguard encircles the barrel in a fashion similar to that of the Belgian M1889 rifle. As originally issued this rifle had no safety catch; a half-cock notch on the cocking piece—firing pin assembly served this purpose. In 1910 this weapon was modified by the addition of a manual safety, which was placed on the left side of the receiver just behind the closed bolt handle.

8mm Infantry Carbine M1889. Introduced in 1924, this weapon also has a metal barrel jacket and a stud for a bayonet. A tangent-type rear sight is used on this weapon rather than the leaf-type rear sight of the rifle. This carbine has the stamp "F," i.e., Fodfolk or Infantry, before the serial number. Overall length is 43.3 inches with a barrel length of 24 inches and the carbine weighs 8.8 pounds. A horizontal-type bolt handle is used.

8mm Artillery Carbine M1889. This carbine was also introduced in 1924. It is generally similar to the Infantry carbine but it has a turned-down bolt handle, a triangular upper sling swivel, and a stud on the left side of the stock. This stud was used to hang the carbine from a leather hanger which was worn on the gunner's back.

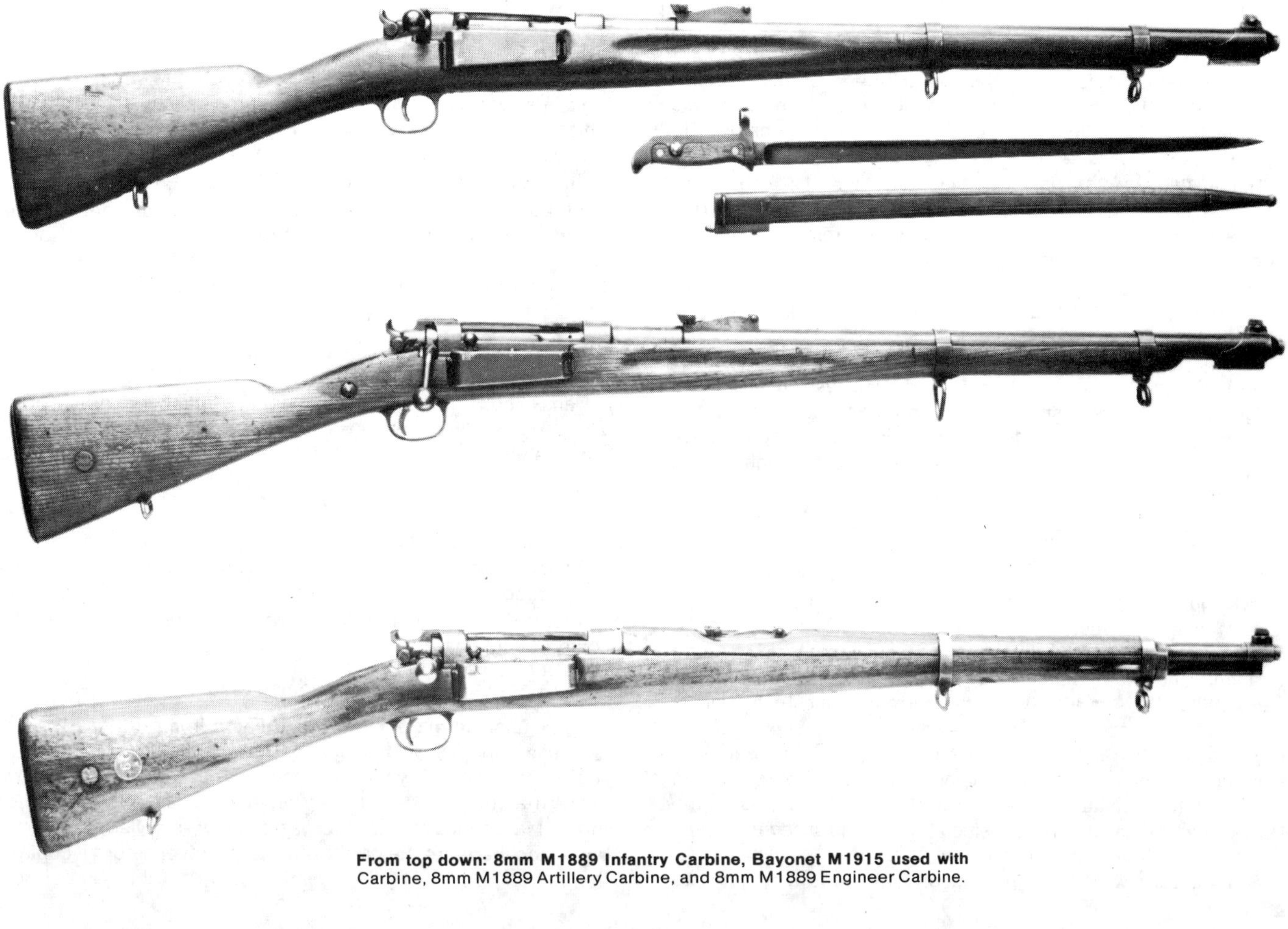

From top down: 8mm M1889 Infantry Carbine, Bayonet M1915 used with Carbine, 8mm M1889 Artillery Carbine, and 8mm M1889 Engineer Carbine.

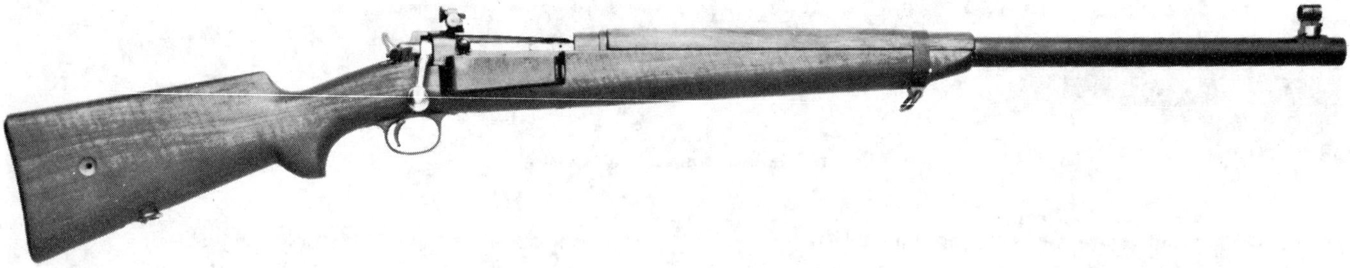

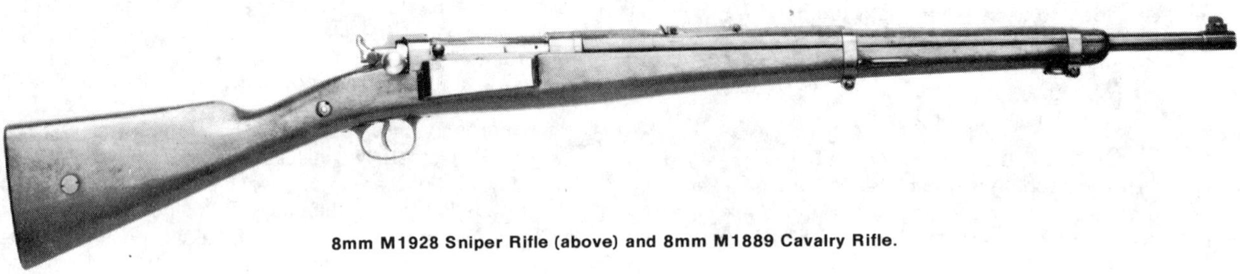

8mm M1928 Sniper Rifle (above) and 8mm M1889 Cavalry Rifle.

8mm Engineer Carbine M1889. The Engineer carbine was also introduced in 1889. It has a wooden handguard and the barrel was shortened to 23.6 inches to accommodate the muzzle cap of the Cavalry Rifle M1889. The letter "I" is found before the serial number.

Sniper Rifle M1928. This is an alteration of the rifle M1889 and has a heavier barrel with wooden handguard, a sporting-type stock with pistol grip, a turned-down bolt handle, micrometer-type rear sight, and a hooded target-type front sight. This rifle, which resembles the U.S. caliber .30 Style "T" rifle in general configuration, weighs 11.7 pounds and is 46 inches in length with a 26.3 inch barrel.

Cavalry Rifle M1889. This weapon was introduced in 1914. The rear sling swivel is mounted on the left side just ahead of the trigger guard. It has a straight-bolt handle and has a mounting stud similar to that of the M1889 cavalry carbine on the left side of the stock. The letter "R" is found before the serial number. This rifle is not fitted for a bayonet.

DANSK INDUSTRI SYNDICAT (MADSEN)

After over 60 years of manufacturing weapons of fine quality, Dansk Industri Syndicat—DISA (or Madsen, as it is commonly known)—the manufacturers of the Madsen gun, are out of the small-arms business. This company has produced many outstanding weapons, and has had much influence on arms design since the early 1900s. DISA is a victim of the current political situation in which weapons are almost given away by the opposing power blocs, basically for political reasons. Only the giant weapons manufacturers who have many other irons in the fire can stay in the international military small-arms business in these days. In the U.S.A., the field has opened considerably in the past several years, but it has not in the international field.

The western world has no reason, nor does Denmark, to congratulate itself about the demise of DISA in the military small-arms business. This is just one less source of imagination and skill which is lost at a time when we need all of the imagination and skill we have.

DISA still makes and develops machine gun mounts, notably the tripod mount for the German MG1, which has been adopted by the West German government, and a vehicle mount for the same weapon. A tripod for the FN "MAG" machine gun has also been developed.

Madsen Post-War Rifles

Dansk Industri Syndicat developed a number of rifles since World War II. None of these developments were commercially successful.

The Danish M47 Rifle

The Model 47 bolt-action rifle was designed primarily for the smaller races of the world. It weighs about 7.5 pounds and is fitted with a rubber recoil pad. Owing to the availability of large stocks of World War II surplus rifles at very low prices, it was sold in limited quantities to Columbia.

The Ljungman Rifle

The Ljungman made by Madsen differs from the Swedish and Egyptian varieties of this weapon in one important respect. In the Swedish and Egyptian versions of this rifle the gas blows through a straight gas cylinder directly against the bolt carrier. In the Madsen-made gun, the gas cylinder is coiled around the barrel and therefore the gas has further to travel before it contacts the bolt carrier. This had the tendency to cool the gas before it

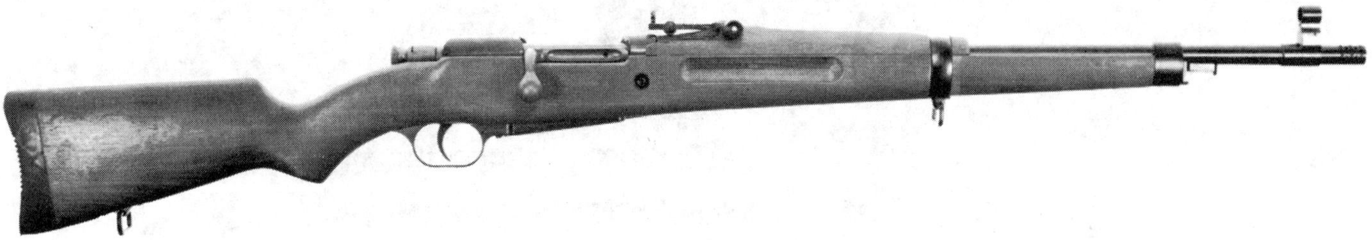

Danish Madsen .30 caliber Model 47 Bolt Action Rifle as supplied to Colombia.

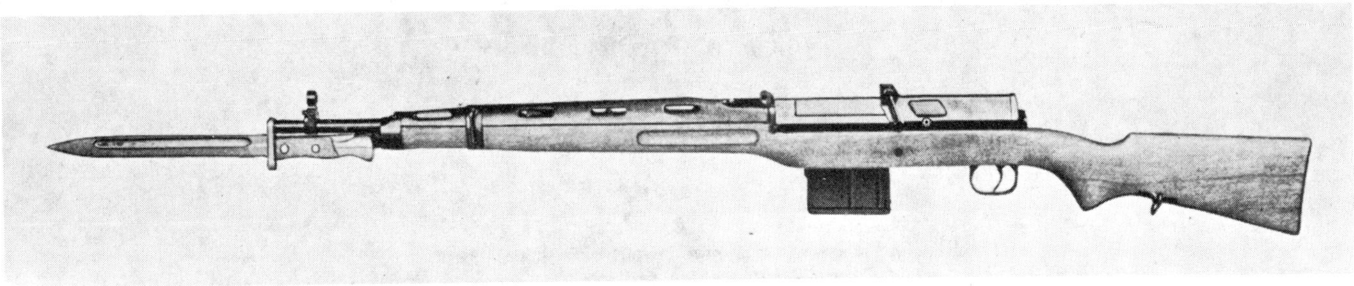

Madsen-made Ljungman semiautomatic rifle.

hits the bolt carrier and to reduce the thrust of the gas, thereby making the action less abrupt. Possibly because the Ljungman is basically an expensive and rather heavy weapon, this rifle was not a commercial success.

MADSEN LIGHT AUTOMATIC RIFLE

Characteristics of Madsen Light Automatic Rifle

Caliber: 7.62mm NATO.
System of operation: Gas, selective fire.
Weight w/loaded magazine: 10.6 lbs.
Length, overall: 42.3 in.
Barrel length: 21.1 in.
Feed device: 20 round detachable, staggered box magazine.
Sights: Front—Hooded blade.
Rear—Aperture, graduated from 100 to 600 meters.
Muzzle velocity: Approx. 2650 f.p.s.
Cyclic rate: 550-600 rounds per minute.

This weapon currently exists in prototype form only. Extensive use is made of light weight materials in its construction. Aluminum alloy is used in the receiver, receiver cover, trigger guard, rear sight, magazine, and bipod.

How to Load and Fire the Madsen Light Automatic Rifle. Insert a loaded magazine into the magazine port and push home until locked by the magazine catch. Cock the weapon by pulling the operating handle to the rear as far as it will go, release the handle, and let it run forward. The weapon is now cocked and ready to fire. The safety selector lever can be set on safe, semiautomatic, or automatic as desired.

Madsen 7.62mm NATO Light Automatic Rifle with wooden stock.

Madsen 7.62mm NATO Light Automatic Rifle with tubular steel stock.

Special Note on the Madsen Automatic Rifle. The Madsen has its return spring mounted above the barrel and circles the piston rod, thus it pulls the bolt forward into battery position rather than pushes it forward as with most weapons. The bolt is similar to that of the Soviet AK-47 in that it rotates to lock and is rotated by means of a lug on the bolt operating in a cam in the bolt carrier. However, the piston rod of the Madsen is not permanently attached to the bolt carrier as with the AK. The piston rod of the Madsen has a ball-shaped end which fits in a cut-out in the bolt carrier. The trigger mechanism of the Madsen is similar to that of the AK. The Madsen has a grenade launcher built into the end of the barrel. It can also be easily fitted with a bipod and a telescopic sight. There are two versions of the weapon: one with a tubular metal stock and one with a wooden stock.

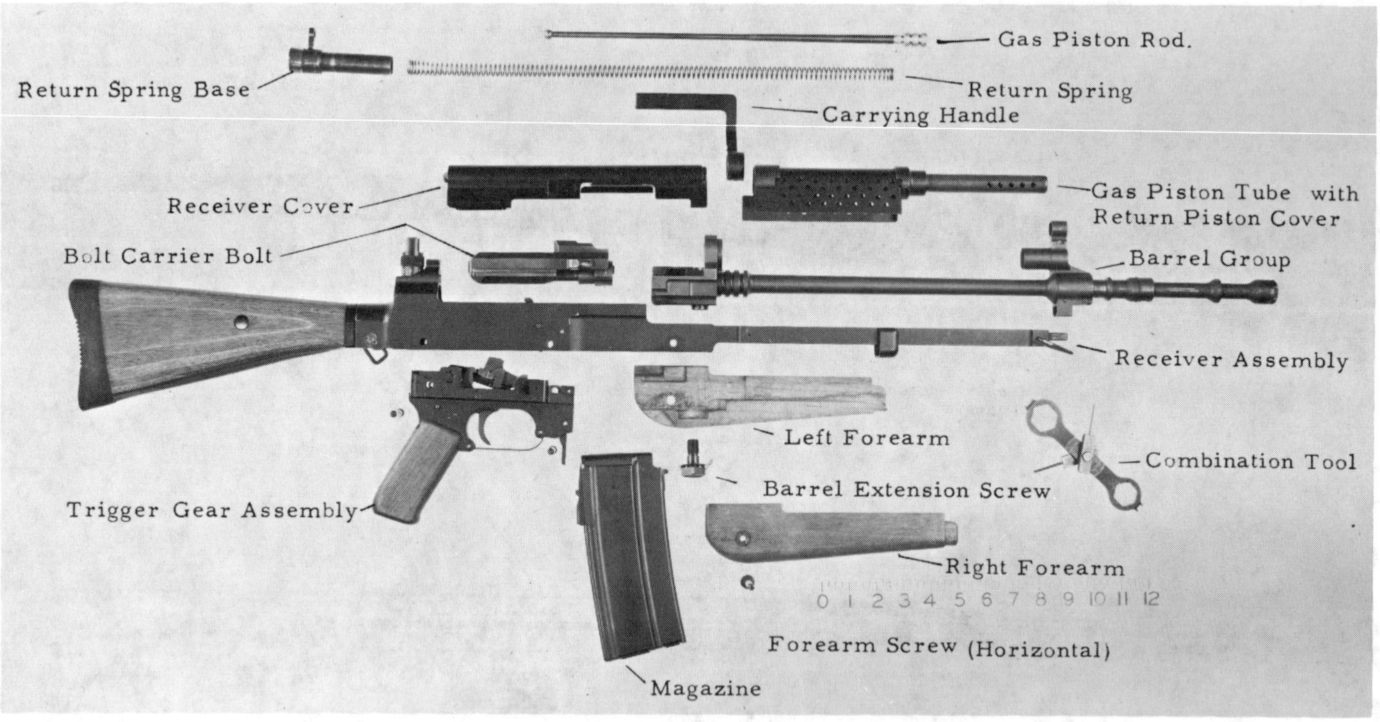

Madsen 7.62mm NATO Light Automatic Rifle, field-stripped.

Folding-stock version of Madsen 7.62mm NATO Light Automatic Rifle.

DANISH SUBMACHINE GUNS

The standard submachine gun of the Danish Army is the 9mm M49 "Hovea." However, several versions—Finnish and Swedish—of the 9mm Suomi are used as well. DISA has developed several submachine guns since World War II. The initial post-war submachine developed by Madsen was the Model 45; although it had several interesting features, it was not a success. The Model 1946 and its successors, the Model 50 and Model 53, have been fairly extensively manufactured and are in use in a number of countries throughout the world.

DANISH MADSEN SUBMACHINE GUN MODEL 1945

While this gun has been rendered obsolete by the improved 1946 and later patterns, it is worthy of some attention because of a few original design factors which may have some future application in other arms.

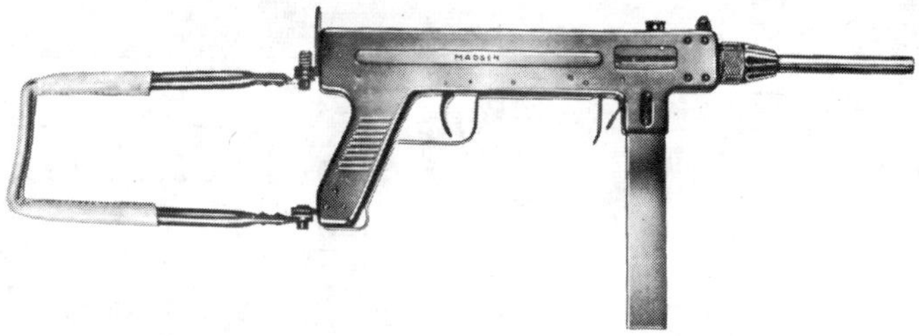

Madsen Model 1950 Submachine Gun. One of the finest examples of modern low cost, fast production stamped designs. Also made in Brazil in caliber .45 A.C.P.

Characteristics of M1945

Caliber: 9mm Parabellum.
Magazine: Standard box. 50 cartridges mounted below receiver.
Overall length: 31.5 inches.
Weight, excluding magazine: 7.1 pounds.
Operation: Blowback. Standard general operation but unusual inertia movements added, selective fire.
Cyclic rate of fire: About 800 per minute.
Special feature: Safety pin lock on firing pin.

Despite the fact that this arm uses a rifle type wooden stock and fore-end, the designers achieved the extremely light weight for a submachine gun of only 7.1 pounds.

A cocking cover over the barrel breech in the form of a slide (not unlike that of the typical automatic pistol design) has serrated sides forward of the magazine. Cocking, instead of by customary handle, is by withdrawing this member, which travels with the true breechblock in recoil. The recoil spring is positioned around the barrel below this sliding cover member. Utilization of this spring position, together with the sliding cover, were used to supplement the inertia and mass of a light breechblock, to thereby achieve minimum weight.

In an endeavor to produce a safety factor to overcome accidental discharge during slamming of the breechblock when the gun is dropped or violently put aside, a sear interrupter was introduced in this model. The striker can move forward through the breechblock to fire only when the trigger is depressing the sear. Dropping the weapon will not cause accidental firing.

The spring position around the barrel, of course, subjects the spring to considerable heat under continuous fire, inevitably producing crystallization and spring breakage.

MADSEN M1950 SUBMACHINE GUN

Characteristics of M1950

Caliber: 9mm Parabellum.
System of operation: blowback, automatic fire only.
Magazine: 32-round detachable straight line box.
Weight: 7.6 pounds, excluding magazine.
Overall length: 30.71 inches with folding butt extended.
Barrel length: 7.87 in.
Muzzle velocity: approx. 1200 f.p.s.
Cyclic rate: 500-550 r.p.m.

Construction and Design, M1950

This is one of the most unusual submachine gun designs ever produced. The gun is designed to lend itself to high-speed production at extremely low cost. This is the type of weapon which could be manufactured in American automotive or similar plants equipped with heavy duty punch presses at a fraction of the cost of a typical submachine gun of the Thompson M1A1 design.

Because of its unusual design, we will consider it here in considerable detail. The construction itself is most ingenious. The receiver (or frame) is flat. It is divided vertically in longitudinal section and is hinged at the rear. The pistol grip and magazine guide are a simple stamping.

The barrel is fastened to the receiver by a locking nut which when unscrewed and thrust forward permits the entire left side of the receiver to be folded back exposing the right side in which all the moving parts are housed. The barrel may be lifted out for immediate replacement or for cooling. While this system of design has been applied in Europe to revolvers in the past, this is the first production example of the application of the design principle to the submachine gun. It permits not only simplified manufacture but extreme ease of fitting and assembly. The photographs and drawings herewith will establish in clear detail all the manufacturing and operational factors of importance which are involved.

The stock is a folding metal skeleton design permitting the gun to be used easily either from the shoulder or from the hip.

The stock when folded does not interfere with access to the trigger. The sling swivel positions on the left side of the gun are designed to permit the weapon to be slung across the chest for immediate use at a moment's notice.

A very unusual factor, yet one of extreme simplicity, is the special automatic safety provided. This is a lever positioned to the rear of the magazine housing. When firing, the normal manner of gripping an arm of this type is with the right hand about the grip and the left hand around the magazine or magazine housing. In this position, the firer's left hand in this new design also embraces the safety lever. Should he release this grip at any time, or should he stumble and lose control of it, the lever automatically blocks the path of the breechblock so that it cannot chamber a cartridge from the magazine for firing. Pulling the cocking handle all the way back will put the gun back in readiness for operation.

The gun consists of the following units: receiver or frame, barrel, breechblock with inserted cocking handle, recoil spring, trigger mechanism, safety device, magazine, and folding shoulder stock.

The receiver is composed of two nearly identical sections of stamped sheet steel. The pistol grip and magazine housing are formed as sections of these individual halves of the receiver. The two frame sections are hinged at the rear. At the front they are secured to the barrel by the barrel bearing nut.

The left side of the receiver serves as an actual cover for the right section in which the moving members are housed. The sling swivels and sights are on the left-hand half. The front sight is positioned at the forward end of the receiver instead of on the barrel, thus eliminating need for accurate barrel positioning, since barrel locking ridges are cylindrical. It is a standard blade design. May be adjusted laterally for windage. The rear sight positioned at the rear of the receiver section is a fixed aperture sight. The gun is sighted in at a practical hundred meters range. This is an entirely sensible sighting range and system for the caliber and design.

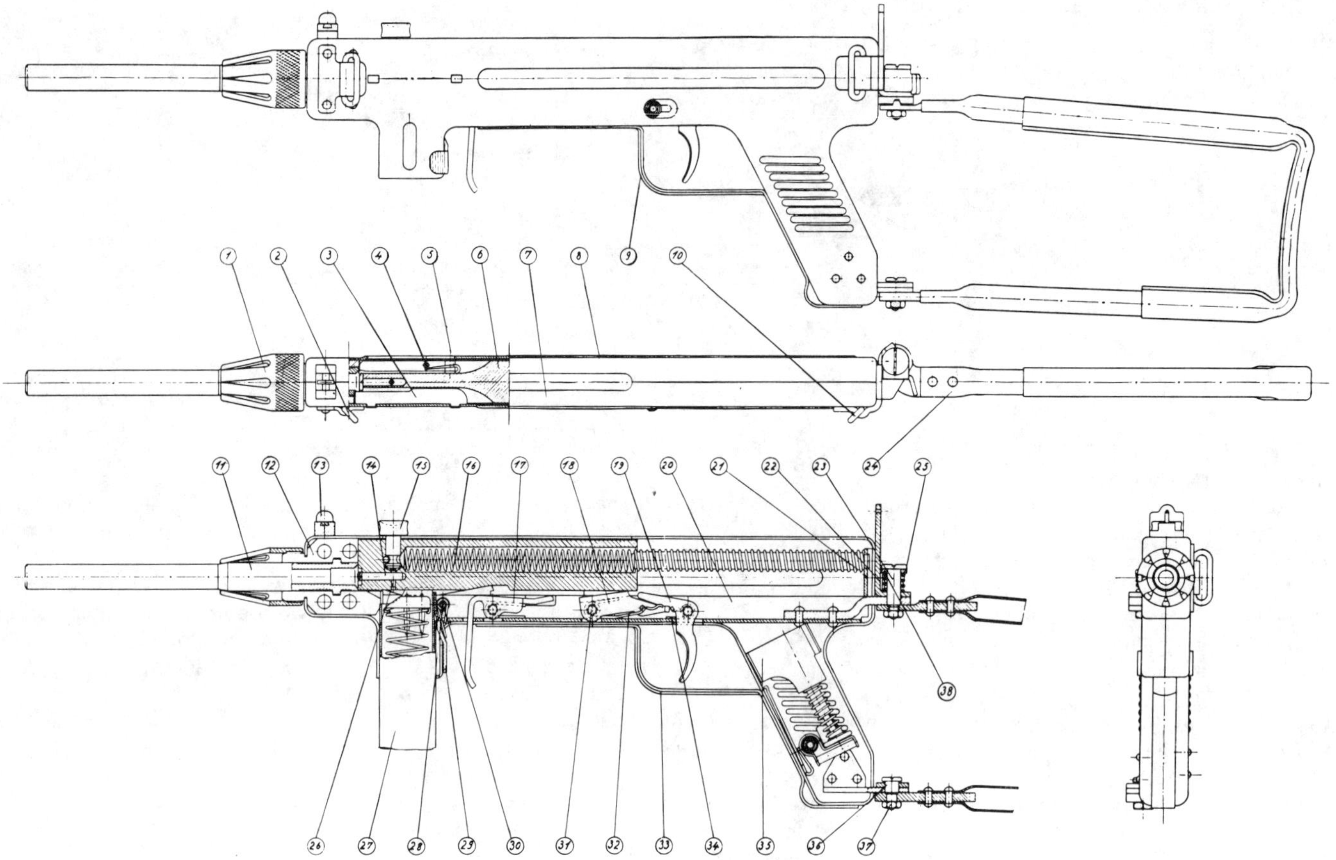

Detail drawings showing construction and parts of the Madsen Model 1950 Submachine Gun.

1. Barrel-bearing nut
2. Sling swivel
3. Ejector
4. Extractor pin
5. Extractor
6. Breechblock
7. Frame half (left)
8. Frame half (right)
9. Safety catch
10. Sling swivel
11. Barrel
12. Barrel bearing
13. Front sight
14. Firing pin
15. Cocking handle
16. Return spring
17. Breech block retainer
18. Trigger rod
19. Trigger
20. Trigger plate
21. Return spring guide (complete)
22. Rear sight plate
23. Shoulder piece bolt (upper)
24. Shoulder piece
25. Shoulder piece spring
26. Firing pin rivet
27. Magazine
28. Magazine catch pin
29. Magazine catch
30. Magazine catch spring
31. Trigger rod pin
32. Trigger rod spring
33. Trigger guard
34. Trigger spring
35. Magazine loading apparatus
36. Shoulder piece bolt (lower)
37. Shoulder piece bolt nut
38. Shoulder piece lock

While the sighting radius is relatively short, it is still adequate. The positioning of the front sight is determined by the fact that if it were placed forward on the barrel muzzle to give maximum sight radius, its height would be too great, resulting in possibility of damage to it. Moreover, barrel positioning would require closer tolerances and increased manufacturing costs.

The receiver is smooth finished on the outside. The jointure of the parts gives a dustproof fit.

A projecting rib at the forward end of the left section of the receiver serves as the ejector. The right side of the receiver is pierced for the ejection port.

Except for the barrel, which is detachable by merely unscrewing its nut, all the other operating parts are housed in the right hand side of the receiver.

The Barrel. No radiating rings or cooling flanges are provided. The barrel is a smooth taper screw machine or lathe-turned unit and is quite light. Because of interchangeability factors, the heavier barrel is not considered essential, though for hard military usage a heavy or flanged barrel could readily be provided.

The breech section of the barrel is housed in the forward breech section of the receiver. It is furnished with an external rib which fits into corresponding grooves in both frame halves to prevent the barrel from moving forward or rearward. A groove cut in the rear of the chamber section of the barrel mates with the ejector rib in the left section of the receiver. This positively prevents any barrel rotation when assembled.

Breechblock. This rectangular unit reciprocates inside the receiver on the flat bottom wall of the right receiver section. The firing pin is an integral part of the breechblock for complete simplicity. The extractor at the front right side section of the breechblock is secured with an elementary vertical pin.

Cocking Handle. This is a separate member which is inserted in the top of the breechblock and is fastened by a cross pin. It travels in a slot in the top of the frame between the two halves. While this may be considered militarily objectionable, as in the case of the original Thompson 1928 model, the system could be subject to side operation as in the case of the later Thompson modification, the M1A1, as it is in the Brazilian model.

Recoil Spring. This is positioned in a pilot hole in the breechblock. The rear compression point is against a spring base loosely inserted in the frame receiver against the right rear wall. A protruding tubular spring guide is furnished.

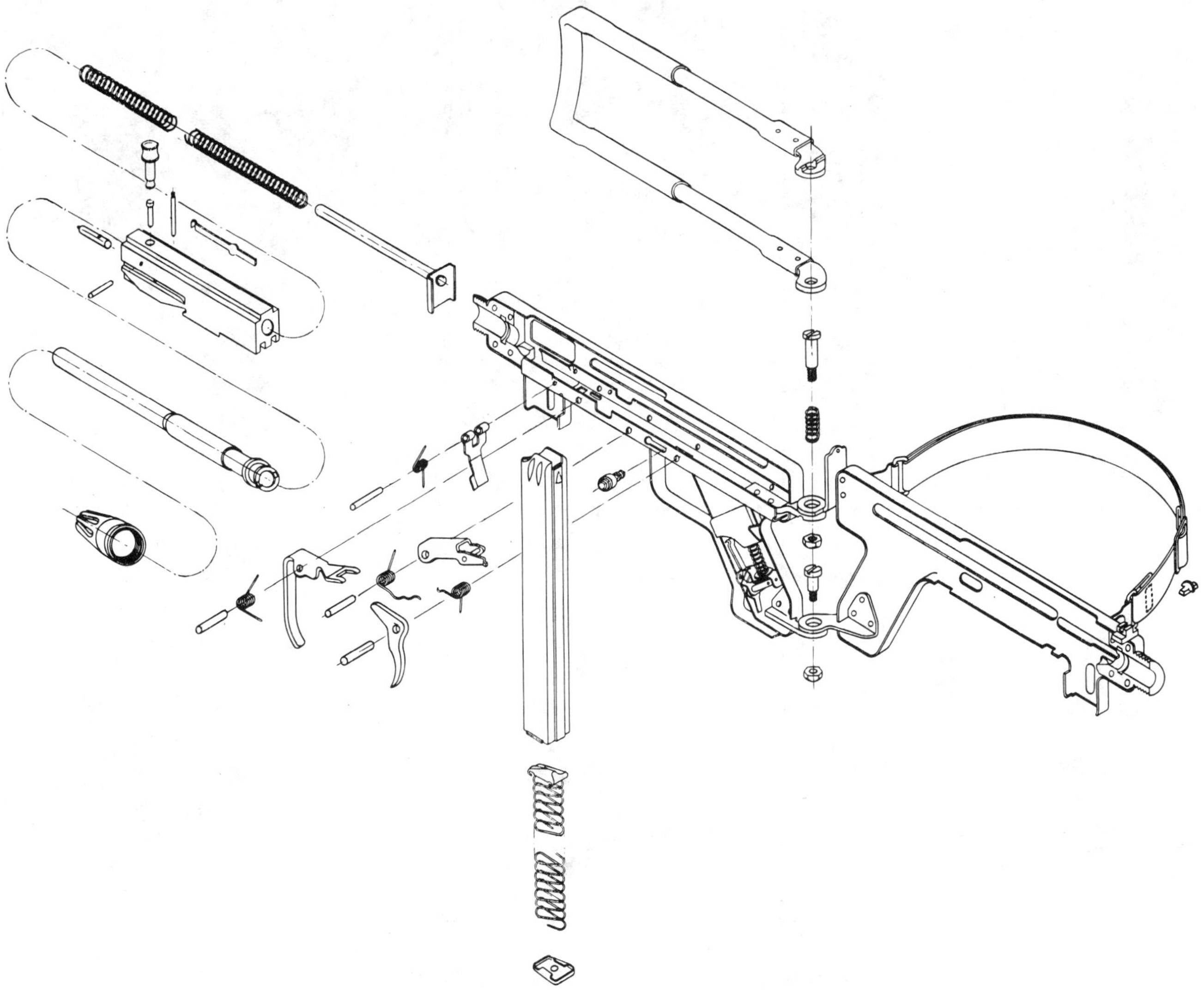

Detail exploded view of the Madsen Model 1950 Submachine Gun as made in Denmark.

Trigger Mechanism. This is positioned at the bottom of the right receiver section. The unit consists simply of the trigger and arm, the trigger rod, and their respective springs. The rod (or sear member) is thrust upwards by its spring to catch in a notch in the bottom of the breechblock when in cocked position. The trigger arm projects down through the opening in the bottom of the right frame section, where it is protected by the trigger guard.

Safety Catch. This is at the bottom of the frame. It can be locked only when the breechblock is cocked. Pushing back the catch in its short travel slot in the left side of the frame effectively locks the trigger rod in position to prevent any movement of the breechblock.

Breechblock Retainer. This is in the bottom of the frame forward of the safety catch. This special safety squeezer member blocks the forward travel of the breechblock, except when its lever is gripped when firing. The pressure pivots the lever on its pin to lower the upper surface out of the travel path of the breechblock to the rear of the magazine.

Magazine. Box design holding 32 cartridges. This is inserted in standard fashion into the magazine housing from below. The magazine retaining catch is in the rear wall of the housing. Pushing the catch backwards releases the magazine. The magazine may be inserted when the breechblock is open or closed. However, if the breechblock is in closed position, more force must be applied to insert the magazine as it must be thrust up enough for the top cartridge (which presses against the underside of the breechblock) to be thrust further down into the magazine itself.

Buttstock. This is a skeleton folding butt of steel tubing partly leather covered for comfort. It is hinged to the rear end of the right section of the receiver, its lower hinge being behind the pistol grip. In both open and closed positions the stock is held by a notch and a lug in the hinge. A slight jerk will free it to be moved to either position. However, a lock on the upper shoulder piece bolt allows it be securely locked into extended position if desired. When folded forward it lies along the right side of the frame to make a very compact fold, without interfering with the use of the arm in any way.

Dismounting and Reassembling the M1950

To Dismount. Press the release catch and remove the magazine. Fold the shoulder piece. Ease the breechblock to forward position if it is not already there. Unscrew the barrel bearing nut and slide forward. Place the gun on its right side. Pull the front wing swivel

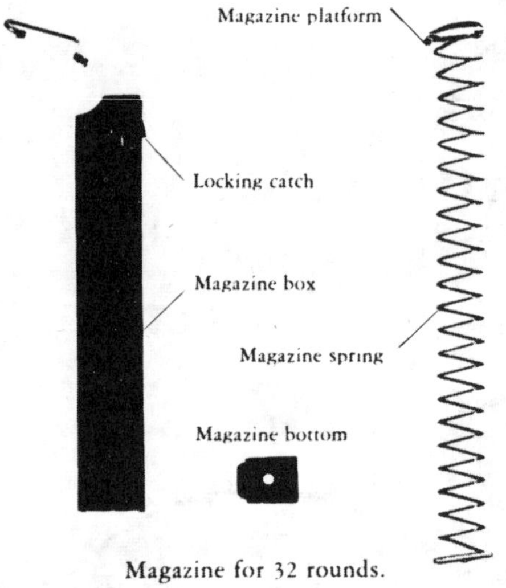

Magazine for 32 rounds.

SAFETY

The breech block retainer serves to prevent a round from being fired accidentally owing to an incomplete cocking motion or as a result of a shock to the gun if dropped or laid down hard.
It does not require any attention, however, to release the breech block retainer, because the correct grip for firing the gun, whether kneeling, standing, sitting or prone, is with left hand firmly round the magazine housing, the thumb in the most natural way pressing against the downward lever of the breech block retainer, and with the right hand round the pistol grip.

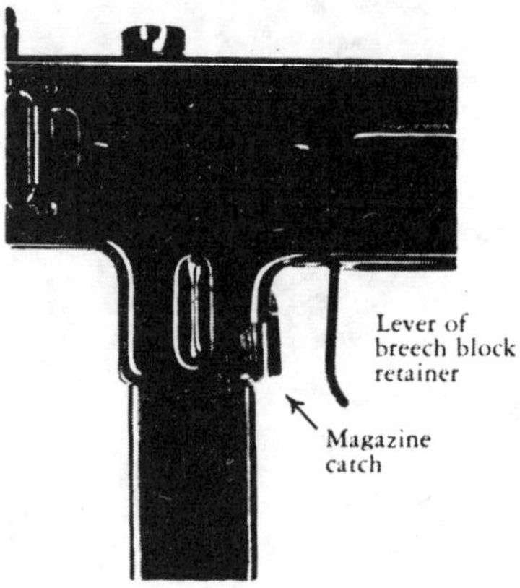

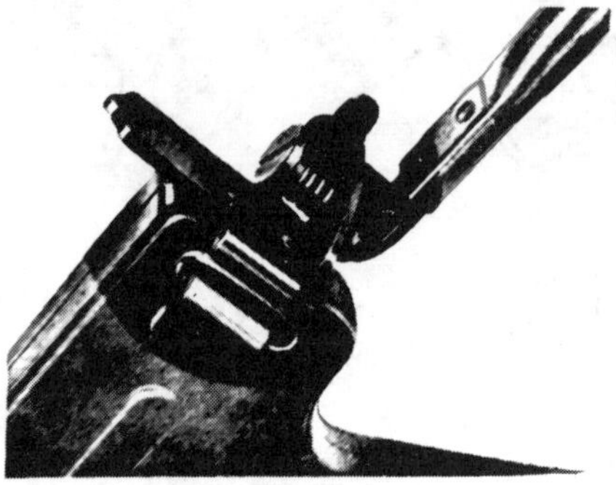

Upper shoulder piece hinge with lock.

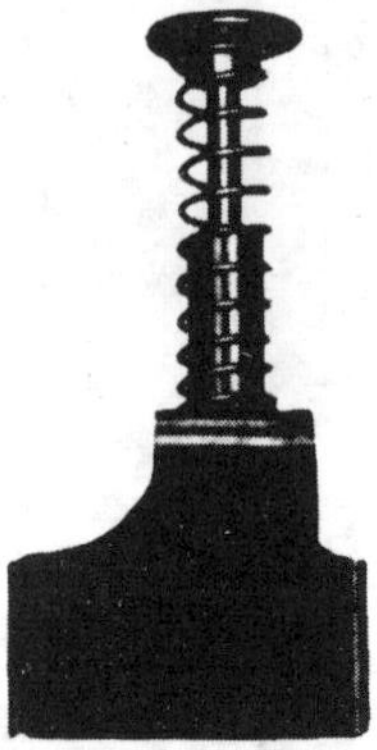

Magazine loader.
Standard equipment. Kept in pistol-grip.

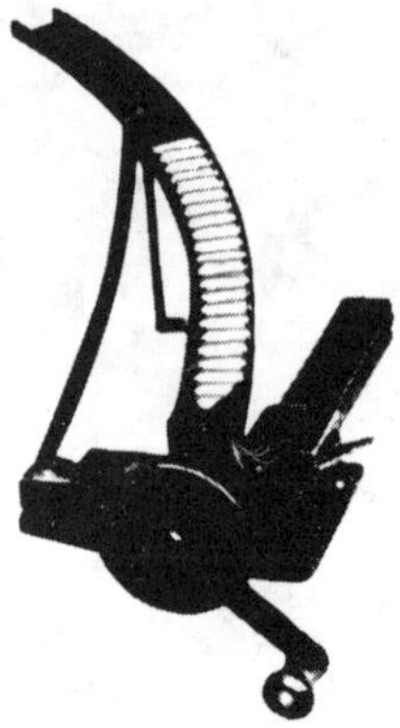

Loader for stationary use.

Details and accessories. Danish Madsen Model 1950 Submachine Gun.

with the right hand while holding the left hand pressing against the barrel. This will raise the left receiver section, freeing the barrel. Withdraw the barrel.

Press the base of the return spring forward and upwards. This will allow it to be withdrawn from the breechblock. Lift the breechblock up and out. No further disassembly is normally required. The magazine bottom may be slid out to remove the spring and follower, although this too is not commonly necessary. Reassembly is merely reversal of this procedure.

Special Note. This gun in Caliber .45 U.S. Govt. is currently being made in Brazil under Madsen license. Many of the design and manufacturing features lend themselves readily to application to other small arms designs.

Operation of M1950

A magazine is inserted in the housing and thrust in until the magazine catch secures it. The cocking handle on top of the breechblock is drawn to the full rear position to compress the recoil spring. At the end of the stroke the trigger rod spring thrusts the rod up to catch in its notch in the under side of the breechblock. The safety catch may be pushed to lock the rod into the breechblock if desired. Otherwise the weapon is now ready to fire on pull of the trigger.

Firing. Pressing the trigger causes it to pivot and its forward arm moves the attached trigger rod down out of contact with the breechblock. The compressed recoil spring drives the breechblock ahead. If the left hand is not supporting both the magazine housing and the safety lever behind it, the breechblock will be halted before striking the cartridge in the magazine lips.

Assuming that the supporting hand is pressing in the safety catch towards the magazine housing, the breechblock impelled by the recoil spring is free to move ahead, since the upper section of the safety lever is not in its path. Its feed face strips the top cartridge from between the lips of the magazine and drives it into the chamber. The nose of the bullet is guided by the barrel feed section of the breech into the chamber as the rear section of the cartridge clears the magazine lips. When the cartridge enters the chamber, it lines up with the firing pin which is fixed in the face of the breechblock. The firing pin strikes the cartridge primer while still moving forward. At the same time the extractor on the right side of the breechblock springs over the cannelure in the cartridge case.

In this quite standard practice for submachine guns, the heavy breechblock and spring members are still moving forward at the actual instant of firing before the chambering is really complete. This serves as an additional inertia factor to offset the recoil of the discharge.

Recoil Movement. The gases developed inside the cartridge case on firing drive the bullet forward and force the cartridge case against the side walls and back against the face of the breechblock in standard blowback practice. The breechblock starts to the rear

Brazilian cal. .45. Similar to the Danish Madsen Model 1946 except for caliber. Presented to Gen. J. Lawton Collins, then Chief of Staff, U. S. Army by the Brazilian Minister of War. This gun is manufactured by INA in Brazil. Its bolt handle is on the right side rather than on the top.

Danish Madsen Model 1953 9mm Submachine Gun.

under this thrust. However, since the breechblock weight and return spring tension are far greater than the comparatively light bullet, the opening movement is barely started when the bullet leaves the muzzle. The breech pressure drops rapidly to safe limits but the residual pressure in the bore continues to exert a backward force against the case head which in turn passes it on the breechblock in a continuing thrust.

The breechblock moving to the rear, its cartridge extractor draws the empty case back with it. When the case clears the chamber, the left side hits the ejector, pivoting the case through the ejection port opening on the right.

The breechblock continues its rearward travel and as it clears the top of the magazine, the magazine spring pushes the next cartridge up against the lips of the magazine ready to be picked up on the counter-recoil motion. The recoil spring is compressed between its seat in the breechblock in the rear of the frame as the breechblock travels in the tracks provided for it in the frame walls. The travel slot for the cocking handle is greater than the distance of travel of the breechblock in order to prevent sudden halting shock. If the trigger is still retained, the breechblock will continue forward to fire in full automatic cycle. This firing continues as long as the trigger is depressed, the safety lever depressed, and there are cartridges in the magazine.

If, however, the trigger is released, the springs will reassert themselves to thrust the trigger rod up into the breechblock notch to hold it back ready for the next forward motion.

The gun is normally cocked with the left hand. Accidental discharge is almost impossible. Should the trigger be accidentally pulled when the safety is not being thrust in for firing, or should the breechblock be jarred loose under any conditions when the left hand is not gripping the safety, the breechblock, as we have pointed out, cannot complete its forward travel. In this position, it is necessary to pull back the cocking handle to recock the arm before it can be fired.

DANISH MADSEN SUBMACHINE GUN M1953

This is a newer model of the submachine gun Model 1946. This weapon was changed in 1950, and again in 1953. This model is called the M53.

The M53 is very similar to the 1950 Model, but it has incorporated several improvements. The most noticeable, at first glance, is the design of the magazine, which is curved instead of being straight as on the 1946 and 1950 models. The curved magazine is considered a better design for feeding purposes.

Another new feature in the M53 submachine gun is that the

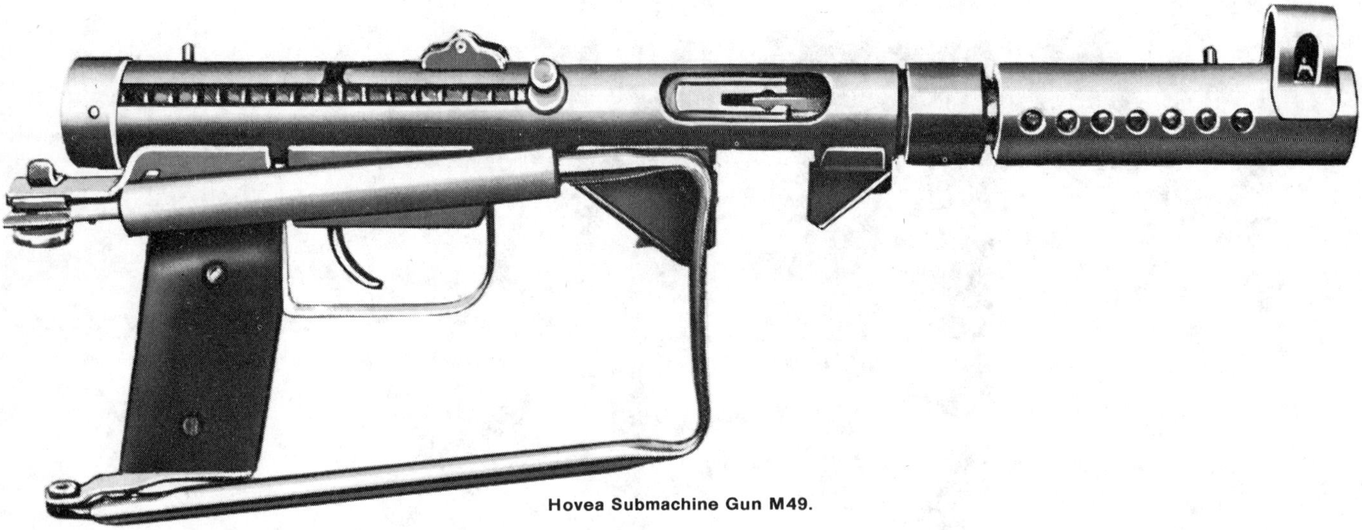

Hovea Submachine Gun M49.

barrel bearing nut now screws onto the barrel, rather than onto the front of the receiver as on the M46 and M50 models. This feature gives the M53 added strength and stability for the barrel. The bolt has also been streamlined to aid in better functioning.

If desired, the weapon can be supplied with a removable barrel jacket to which a special short bayonet can be fixed. All the Madsen models can be made with or without selection-fire features.

The characteristics and field stripping are the same for this weapon as for the 1950 model given earlier in this chapter.

Hovea 9mm Submachine Gun M1949

This weapon, commonly called the Hovea, was developed by Husqvarna in Sweden. It is similar in construction to the Swedish M45 submachine gun. Functioning, disassembly, and assembly are similar to that described under the Swedish 9mm M45 submachine gun.

Characteristics of Hovea 9mm Submachine Gun M1949

Caliber: 9mm Parabellum.
System of operation: Blowback.
Weight: 8.9 lbs. (loaded).
Length, overall: 31.8 in. (w/stock extended)
21.6 in. (w/stock folded)
Barrel length: 8.4 in.
Feed device: 35-round detachable, staggered, box magazine
Sights: Front: Hooded post.
Rear: L-type with setting for 100 and 200 meters.
Muzzle velocity: 1263 f.p.s.
Cyclic rate: 600 r.p.m.

DANISH MACHINE GUNS

Denmark has used various models of the Madsen machine gun since 1904 in the standard light gun versions and in aircraft versions as well as heavy (20mm) guns of this design. After World War II the Danish Army was equipped with British .303 Bren guns, Swedish 6.5mm Model 37 Browning guns, and U.S. caliber .30 M1919A4 and A5 and caliber .50 M2 Browning Heavy Barrel machine guns. In 1948 Denmark adopted the last of the true Madsens in caliber .30 and in 1950 they adopted the SIG 50 as the Model 50 in caliber .30. Recently Denmark adopted the German 7.62mm NATO MG1 (MG42/59) as standard and also adopted the DISA Model F 197 tripod mount for this gun.

THE MADSEN MACHINE GUN

The Madsen machine guns have been among the most popular in the world since their introduction in the early 1900s. They have world-wide distribution and will continue to be encountered in service for many years. There are many variations of this gun in existence, but all operate basically in the same way. The Madsen is an expensive gun to manufacture and requires quality ammunition for reliability of function. These factors limited its use among the major powers during the world wars. In 1926 the Madsen was issued in a water-cooled version; a quantity of these weapons were sold to Chile. The Madsen has been sold to 34 countries in a dozen different calibers.

Loading and Firing the Madsen

Like the Chatellerault and the Bren guns, the Madsen uses a top-loading magazine. This requires the sights to be set off to the side of the gun. The magazine is arc-shaped, for use with rimmed cartridges which cannot lie flat on top of each other.

Pull the cocking handle back as far as it will go and release it.

Put the forward end of the magazine into the forward end of the magazine opening and lower the rear end down into place, snapping it down until it locks. Now set the selector on the left side of the receiver above the trigger in the fire position. **Note:** Remember that this weapon fires as the bolt goes forward and no attempt should ever be made to let the action go forward while there is a magazine mounted on top of the gun.

How the Madsen Gun Works

This gun fits into a subdivision of functioning principles known as the "long barrel recoil type." In the short recoil types, the barrel moves backward a less distance than the length of the case.

In the long recoil type the breech has to move back far enough to permit feeding up the entire cartridge in one operation. This is done by the barrel going forward while the lock is held back until the cartridge has partly entered the chamber. In this type of action, the rate of fire is much lower than in the short recoil.

West German 7.62mm NATO MG1 (MG 42/59) on DISA tripod.

Madsen 8mm light machine gun.

Starting with the gun cocked and in firing position the action is as follows: As the trigger is pressed, the spring below it is compressed while the trigger nose is pulled down out of the bent of the recoil lever. This permits the recoil spring in the butt to force the lever downward, and, as it is engaged in the rear of the breech mechanism, it thrusts the recoiling parts forward. As the recoiling mechanism nears forward position, the recoil arm is still up somewhat. The hump on the recoil lever now bears on the side of the sear and forces the sear downward. The nose of the sear is thus relieved from the bent of the firing lever, and compresses the sear spring. The firing lever is now forced downward by its spring and strikes the tail of the hammer. The front of the hammer drives the firing pin forward to explode the cartridge in the firing chamber. A coiled spring around the firing pin, which is compressed by the hammer movement, pulls the firing pin back into the face of the breech block as the cartridge is fired.

As the recoiling parts are thrust forward by the recoil lever spring, a circular stud in the lower part of the breech block, working in the guide grooves of a switch plate which is fitted to the non-recoiling portion of the receiver, strikes the rear of the center block in the plate and so guides the breech block downward, leaving the chamber ready for the cartridge to be inserted. As the stud continues forward, it strikes the lower cam surface of the switch plate causing the breech block to rise and close the breech. Now the stud is lined up with the horizontal slot in the switch plate, down which it travels during the final half-inch forward motion, securely locking the breech.

Also during the forward thrust of the recoiling parts, an arm is forced up by a cam on the left side of the receiver, forcing outwards the distributor against the tension of its spring and permitting the first cartridge from the magazine which was resting on the distributor to drop into the magazine opening.

Meanwhile the front surface of the rear claw of the feed arm engages with the rear surface of the feed arm actuating block, thus rotating the feed arm forward. The arm strikes the head of the

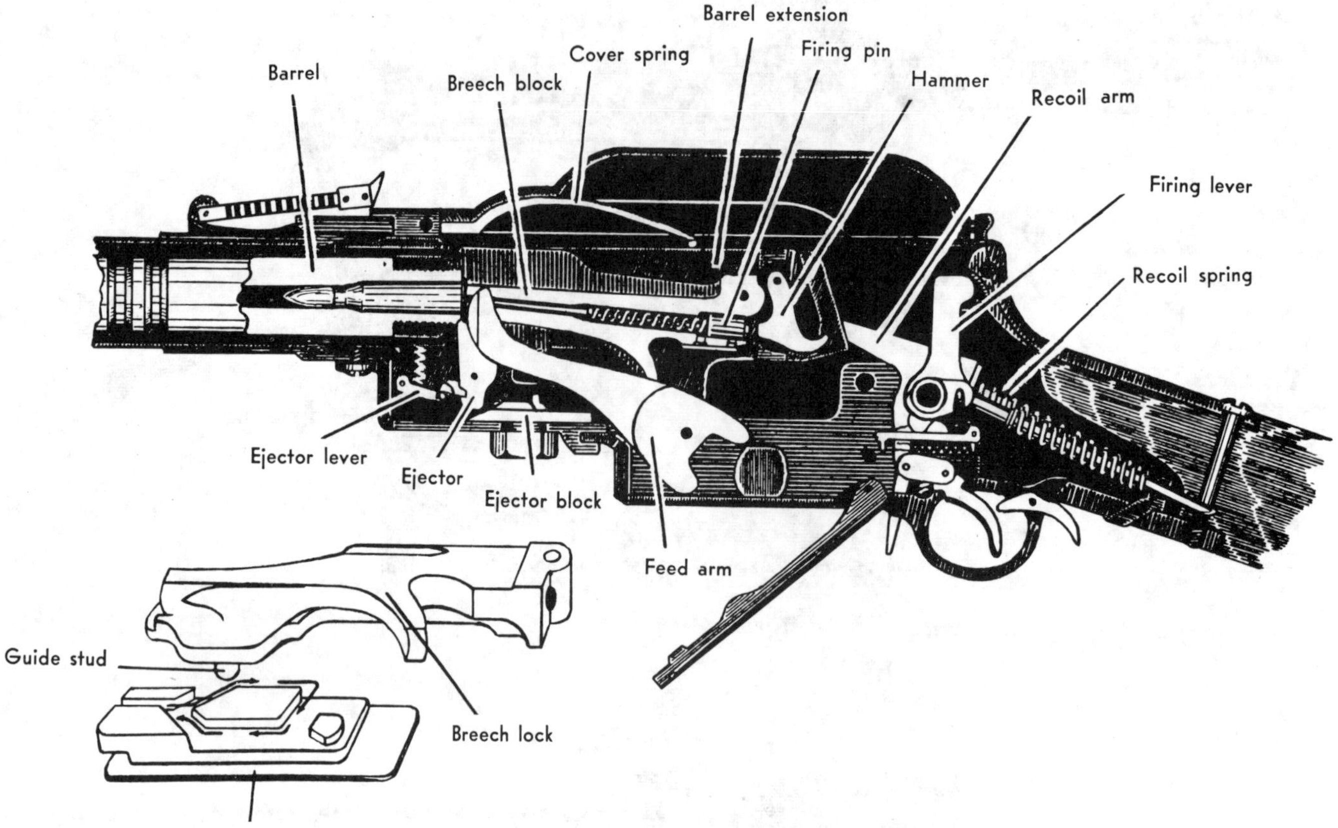

Principal parts of the 8mm Madsen L.M.G.

Madsen M1903/14 Machine Gun.

cartridge in its seat against the left flange of the breech block, pushing it ahead into the chamber. The rear claw of the feed arm rises up to the rear surface of the feed arm actuating block and travels along its upper surface. The bottom of the rear claw, now being above the feed arm actuating block, moves the cartridge in the chamber to allow the breech block to rise. It also prevents any rebound of the feed arm.

As the forward action starts, the ejector is positioned alongside the rear of the ejector block. As the recoiling parts go forward a stud on the ejector lever rides down the sloping cam and forces an ejector downward on its spring, thus bringing the lever in contact with the tail of the ejector. As soon as the breech block starts to rise, the tail of the ejector is clear of the ejector block and it is raised in position by its lever. As the breech closes, the ejector is able to rise to the vertical under the influence of the ejector lever spring and falls in place just below the chamber with its hook below the rim of the cartridge.

Note that in this type of gun, the cartridge is not set into the chamber by the breech block; and until the front of the breech block rises to a complete locking position, the hammer, firing pin, and cartridge are not in alinement and so there can be no accidental discharge during feeding.

Return Movement of the Action. At the breech, the barrel is joined to the breech block casing in which are the hinged breech blocks. These three units recoil together, with the breech remaining closed and locked for about 1/2". This sudden rearward thrust forces the firing lever up as it is struck by the rear of the breech mechanism, and frees it from the hammer, which permits the hammer to pivot back and the firing pin to be withdrawn by the firing pin spring. The guide stud is now passed out of the horizontal groove and travels up the upper cam of the switch plate which pivots the breech block upwards at its nose to permit ejection. The extra stud travels along the top of the cam and the cover spring then forces the front of the breech block downward, compelling the stud to drop out of the rear stud of the switch plate.

Note: This gun has no individual extractor. The ejector pulls the empty cartridge out of the chamber and hurls it from the gun.

With the first movement to the rear, the inclined slope in the front of the ejector block raises the ejector which is held in vertical position by its lever which is engaged in a recess in the bottom of the ejector. From the influence of this separate movement, a hook on the ejector catches the rim of the empty cartridge case and the bottom of the ejector lies on top of the front flats of the ejector block.

Now the stud on the ejector lever runs up the sloping cam on the left of the ejector block to compress the ejector lever spring. This also disengages the ejector lever from the ejector, allowing the ejector tail to be tripped forward by the step on the ejector block, and as the ejector is pivoted about its center, the tripping motion of the tail forces the hook to the rear, pulling the empty cartridge case out and hurling it from the bottom of the gun.

An ejection guide on the breech block guides the empty cartridge case as it is hurled out. The ejector lever stud now rests above the sloping cam holding the ejector levers upwards and free of the ejector, which lies on top of the rear flap of the ejector block beneath the breech block. During the recoil movement, the distributor arm rides down its cam and rotates the distributor inwards and downwards under the influence of the spring. This places the cartridges in the feedway against the left flange of the breech block. Meanwhile, the rear surface of the front feed arm claw engages with the front face of the feed arm actuating block, and rotates the feed arm backwards. The bottom of the front claw now rides along the top of the feed arm block, preventing rebound of the feed arm.

Further Note on Recoil System. Some Madsens are fitted with a so-called "recoil increaser" which forms a choke at the muzzle. By reversing the two parts of the increaser, the rear portion forms a collar which forces the gases escaping at the muzzle to rebound onto the barrel and give an additional thrust to the rearward action. This speeds the gun up greatly.

When the gun is cocked, the claws of the feed arm automatically open the ejector cover. It must be closed by hand on cease fire.

Field Stripping the Madsen

Remove magazine from gun and ease recoiling parts forward.

Lift the locking bolt lever into vertical position and withdraw it to the left.

Push the butt to the right front while gripping the receiver with the left hand; and then remove the butt.

Holding forefinger of right hand ahead of feed arm axis bar, with the hand draw back barrel and breech mechanism.

Now pull out the barrel. Remove barrel very carefully as it is easy to damage the front end ring. This barrel ring is one of the weak points in the weapon. Handle it carefully.

Now remove the breech block bolt.

Pull feed arm to the rear. Lift the front and lower the end of the breech block, and then pivot the front up to vertical position, when the feed arm may be eased forward and lifted out of the block.

Further dismounting need not be attempted.

Madsen M1929 Machine Gun.

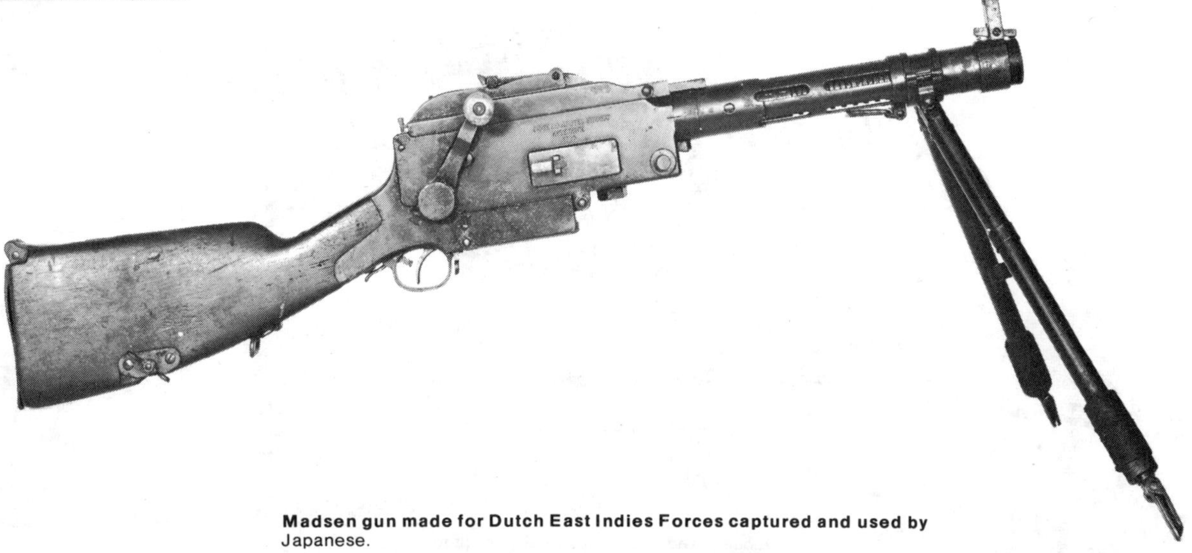

Madsen gun made for Dutch East Indies Forces captured and used by Japanese.

DANISH MADSEN MACHINE GUN M1950

Characteristics of the Madsen Machine Gun Model 1950

Caliber: Advertised for any rifle cartridge, usually found in .30-06.
System of operation: Recoil, selective fire.
Length overall: 45.9 in.
Barrel length: 18.8 in.
Weight: 22 lb.
Feed device: 30-round, detachable box magazine.
Sights: Front: Blade.
Rear: Tangent, graduated from 200-1800 meters.
Muzzle velocity: Approx. 2700 f.p.s. w/ .30-06.
Cyclic rate: 400 r.p.m.

Special Note on Operating Characteristics of M1950

The operating characteristics of the Model 1950 Madsen machine gun, except for minor variations as required by feed alterations in some instances, are the same as described previously.

This model, while recoil operated on the Madsen principle, is a modification of the earlier production. The change is in the barrel removal. This may be removed without the use of tools or without removing any component parts.

This Madsen may be fired from the bipod with or without the shoulder stock. It may also be fired from a light tripod which is convertible to an antiaircraft high-angle fire mount. In addition, it may be fired from the shoulder by the rifleman in kneeling, standing, or prone positions.

Madsen Model 1950 mounted on bipod with flash hider and top mounted magazine.

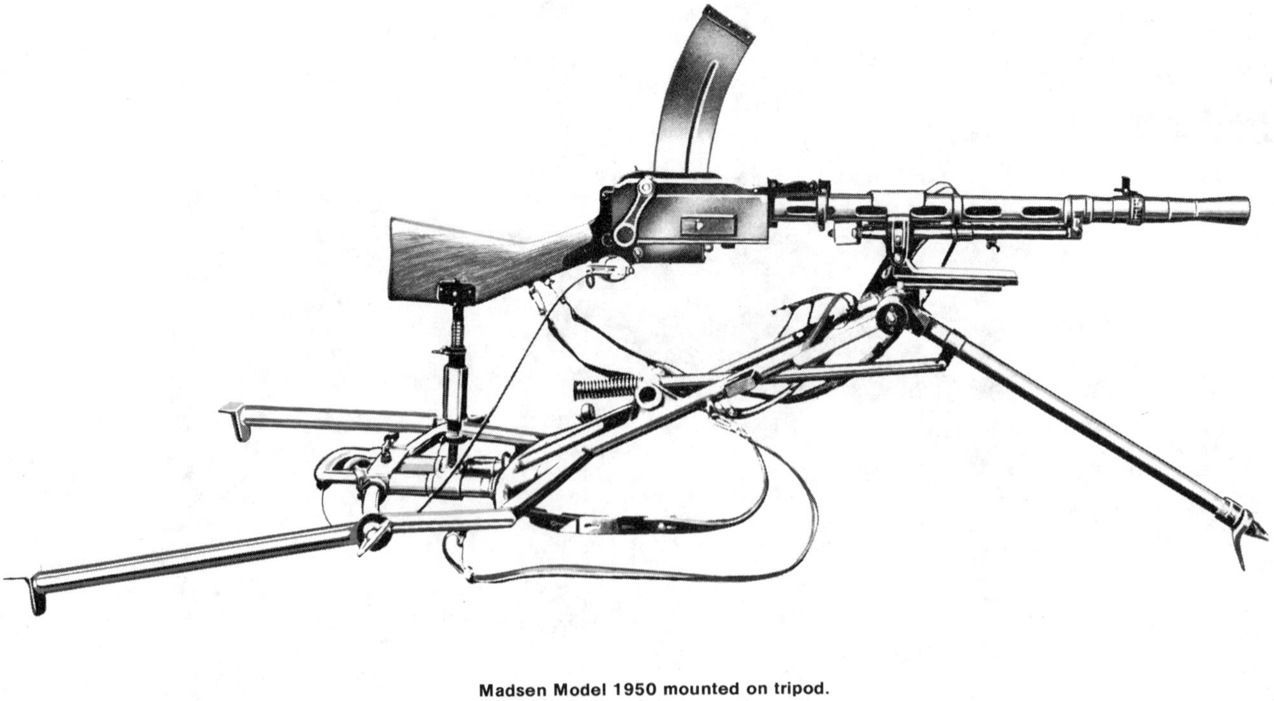

Madsen Model 1950 mounted on tripod.

Distribution of Madsen Machine Guns

As a matter of record, background on the distribution of Madsen machine guns may be of interest. The models are listed by country in which used.

Argentina. Madsen machine guns Models 1910, 1925, 1926, 1928, 1931 and 1935. Most in calibers 7.65mm Mauser.

Bolivia. Model 1925. Caliber 7.65mm Mauser.

Brazil. Models 1908, 1913, 1916, 1925, 1928, 1932, 1934, 1935 and 1936, in calibers 7mm Mauser and 1946 Model machine gun in caliber .30-06.

Bulgaria. Models 1915 and 1927. Caliber 8mm (8 x 50R).

Chile. Models 1923, 1925, 1926 water cooled, and 1928 and 1940. Most in caliber 7mm Mauser. Model 1946 in caliber .30-06.

China. Models 1916, 1930 and 1937. All in the standard caliber 7.92mm Mauser.

Czechoslovakia. Models 1922 and 1923. Caliber 7.92mm German.

Denmark. Models 1904, 1916, 1919, 1924, 1939 in caliber 8mm and Model 1948 in caliber .30-06.

El Salvador. Models 1951 in caliber .30-06 and Model 1934 in caliber 7mm Mauser.

Esthonia. Models 1925 and 1937 in caliber .303 British.

Ethiopia. Models 1907, 1910, 1934 and 1935. All in caliber 7.92mm.

Finland. Models 1910, 1920, 1921 and 1923. Most in caliber 7.62mm, Russian.

France. Models 1915, 1919, 1922 and 1924. All in caliber 8mm Lebel.

Germany. Models 1941, 1942 in caliber 7.92mm.

Great Britain. Models 1915, 1919, 1929, 1931 and 1939 in caliber .303 British.

Holland. Models 1919, 1923, 1926, 1927, 1934, 1938 and 1939. All in caliber 6.5mm Dutch.

Honduras. Models 1937 and 1939 in caliber 7mm Mauser.

Hungary. Models 1925 and 1943 in caliber 7.92mm.

Indonesia. Model 1950 in caliber .30-06.

Italy. Models 1908, 1910, 1925 and 1930 in caliber 6.5mm Italian.

Lithuania. Model 1923. Caliber 7.92mm.

Mexico. Models 1911 and 1934, Caliber 7mm Mauser.

Norway. Models 1914 and 1918. Caliber 6.5mm Mauser.

Pakistan. Model 1947. Caliber .303 British.

Paraguay. Model 1926. Caliber 7.65mm Mauser.

Peru. Model 1929. Caliber 7.65mm Mauser.

Portugal. Models 1930, 1936 and 1952. Caliber 7.7mm. Also in Portugal, Models 1936, 1940 and 1947 in caliber 7.92mm.

Russia. Models 1904 and 1915 in caliber 7.62mm Mosin-Nagant, Russian.

Spain. Model 1907 and 1922 in caliber 7mm Mauser.

Sweden. Models 1906, 1914 and 1921. Caliber 6.5mm Mauser.

Thailand. Models 1925, 1930, 1934, 1939, 1947 and 1949. Caliber 8mm and Model 1951 in caliber .30-06.

Turkey. Models 1925, 1926, 1935 and 1937. Caliber 7.92mm.

Uruguay. Model 1937. Caliber 7mm Mauser.

Yugoslavia. Various in caliber 7.92mm.

MADSEN/SAETTER RIFLE-CALIBER MACHINE GUN

In the new line of Madsen weapons is the Madsen/Saetter machine gun. This weapon is belt fed and gas operated. It is usually mounted on a light field tripod, but can be fired from a bipod or from the hip. The gun is, in appearance and design, a modern weapon, embodying all the experience gained during the last few years in the development of machine guns. The component parts are easily mass produced by punching, turning, and precision casting without detracting from reliability and durability.

The Madsen/Saetter machine gun has gone through three changes. These three models are the Mk I, Mk II, and Mk III. The Mk III was not completely developed until 1959. The Mk III is more reliable and shorter than the Mk I or Mk II. Madsen also made a tank machine gun model which has no buttstock, and no bipod. The return spring is placed round the gas piston. A special "light-weight" tripod is made for this weapon for use outside the tank.

Characteristics of Madsen/Saetter Machine Gun

Caliber: Any military rimless cartridge from 6.5mm to 8mm.
System of operation: Gas operated, full automatic fire only.
Weight: 25.6 lb., w/heavy barrel.
Weight of tripod: 36.2 lb.
Length, overall: 48 in.
Barrel length: 26 in.
Feed device: By non-disintegrating metallic belts of 50 rounds, which can be joined to any desired length. The weapon can also be supplied with two box magazines that fasten to the receiver directly below the action; one magazine holds 50 rounds, and the other holds 100 rounds, in metal link belts. The feed system used on this weapon is a copy of the German MG42.
Sights: Front: Barley corn.
Rear: Open, with graduations up to 1200 meters.
Muzzle velocity: Standard for ammunition employed.
Cyclic rate: 700 to 1000 r.p.m.

Field Stripping the Madsen/Saetter

Insure that the chamber is empty by opening the feed cover, and pull back the cocking handle.

To remove the buttstock, seize the pistol grip with one hand, and, with the thumb, press the trigger gear housing latch. The buttstock, now released, should be turned 90 degrees with the other hand, and pulled to the rear.

To remove the triggerguard housing, the cocking handle should be pulled fully back, and then pushed forward again. This brings the bolt carrier to the rear. The triggerguard housing can now be turned downward and taken out of engagement.

To remove the gas piston: when the bolt assembly is in its rearmost position, the gas piston head is just outside a clearance at the end of the receiver and can be withdrawn.

To remove the bolt, let it slide to the rear, out of the receiver.

To remove the barrel, turn the barrel handle forward until it is free.

To strip the bolt assembly, press out the bolt carrier pin, pull back the action head, and remove components.

To assemble, follow the above instructions in reverse order.

The Madsen/Saetter light machine gun in caliber .30 is used by and manufactured in Indonesia.

Dansk Industri Syndicat developed a new version of the rifle caliber Madsen/Saetter. The Mark IV is shorter and lighter than the earlier Marks and is used with a lighter tripod. Like the earlier Marks, it can be made for any rimless cartridge from 6.5mm to 7.92mm and, of course, for the U.S. cal. .30 rifle and machine gun cartridge. Any of the Madsen/Saetter machine guns can be easily modified to use disintegrating metallic links.

MADSEN/SAETTER 7.62mm TANK MACHINE GUN

This is a version of the rifle caliber Madsen/Saetter made specifically for use on tanks and armored vehicles. It can also be used as a ground gun on a tripod. The arrangement of the gas cylinder and piston is interesting. As with the Soviet RPD light machine gun, there is a definite air gap between the gas cylinder and piston when the gun is cocked. This gap serves a useful purpose during functioning—a good deal of the gas which is bled through the gas port into the gas cylinder is dissipated into the atmosphere after it has served to force the piston to the rear. This has several advantages in that it results in less build-up of carbon

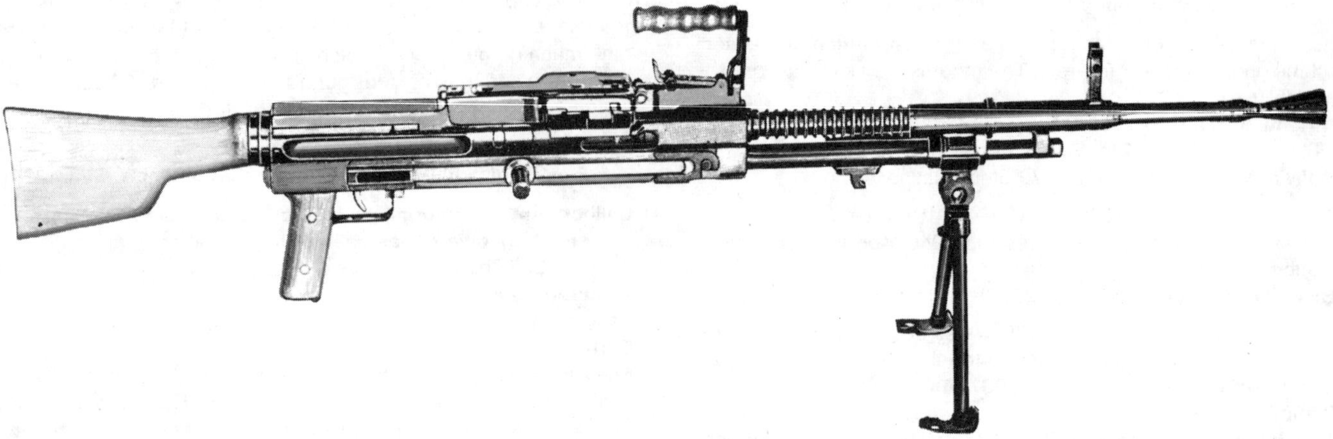

The Danish Madsen/Saetter machine gun (rifle caliber).

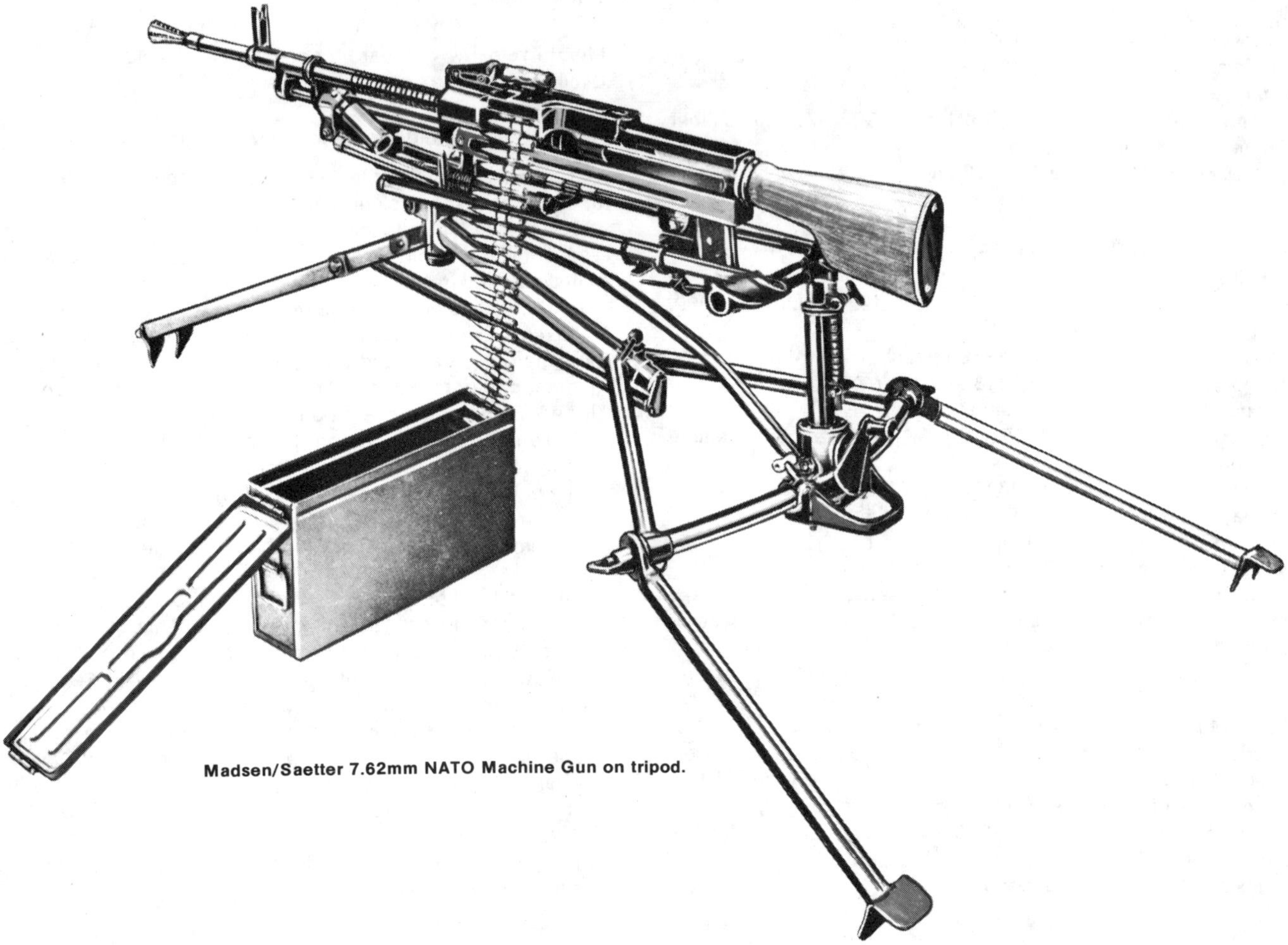

Madsen/Saetter 7.62mm NATO Machine Gun on tripod.

in the gas cylinder and piston tube and, of special importance in a tank machine gun, it cuts down the amount of "operating" gas which filters back into the receiver and thereby into the tank.

This weapon was never manufactured in quantity.

Characteristics of Madsen/Saetter Tank Machine Gun

Caliber: 7.62mm NATO (can be made in other calibers).
System of operation: Gas.
Weight: 22.3 lbs.
Length, overall: 38.2 in.
Barrel length w/flash hider: 22.2 in.
Sights: Front: Protected blade.
Rear: Tangent with V-notch.
Feed device: Non-disintegrating 50 round metallic link belt which can be joined to other belts.

The weapon can be built to use disintegrating links of the M13 (U.S.) type. Belt is normally contained in a box attached to left side of receiver.

Muzzle velocity: 2800 f.p.s.
Cyclic rate: 700-800 r.p.m.

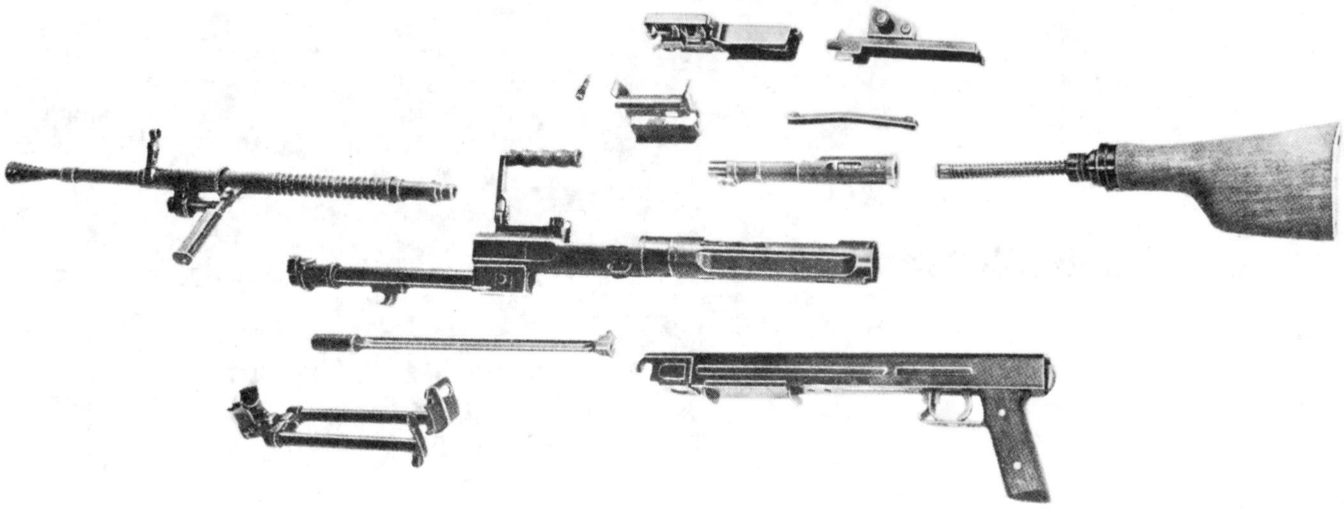

Stripped view of early model 7.62mm Madsen/Saetter.

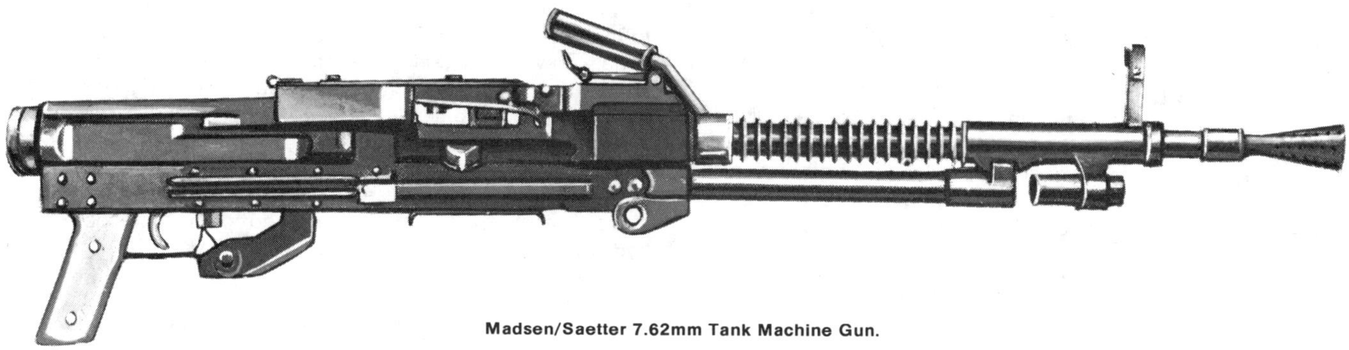

Madsen/Saetter 7.62mm Tank Machine Gun.

DANISH MADSEN/SAETTER CAL. .50 MACHINE GUN

This caliber .50 weapon exists only in prototype form. This machine gun utilizes the same basic system and design features as the Madsen/Saetter rifle-caliber machine gun, with the exception of the mounting. The cal. .50 weapon can be adapted for special mounts, for use in armored cars and tanks, or for antiaircraft or antipersonnel use.

There are two types of mounts for this weapon. One serves a dual purpose, as it can be set up for antiaircraft fire, or, with the addition of rubber wheels, can be towed and used against troops on the ground. This mount is generally similar in principle to that of the Soviet DShK M1938/46 heavy machine gun mount, which serves the same dual role.

The other mount used is a light tripod for use on the ground or on armored vehicles.

Characteristics of Madsen Cal. .50 Machine Gun

Caliber: Cal. .50 (12.7mm).
System of operation: Gas operated, full automatic fire only.
Weight: 61.7 lb.
Length, overall: 64 in.
Barrel length: 39.4 in.
Feed device: 50-round non-disintegrating metallic link belts; box magazine holding 50 rounds may be attached to left side of receiver.
Sights: Open sights for ground use; special antiaircraft sights are used when set up for antiaircraft fire.
Muzzle velocity: Standard for cal. .50 ammunition.
Cyclic rate: 1000 r.p.m.

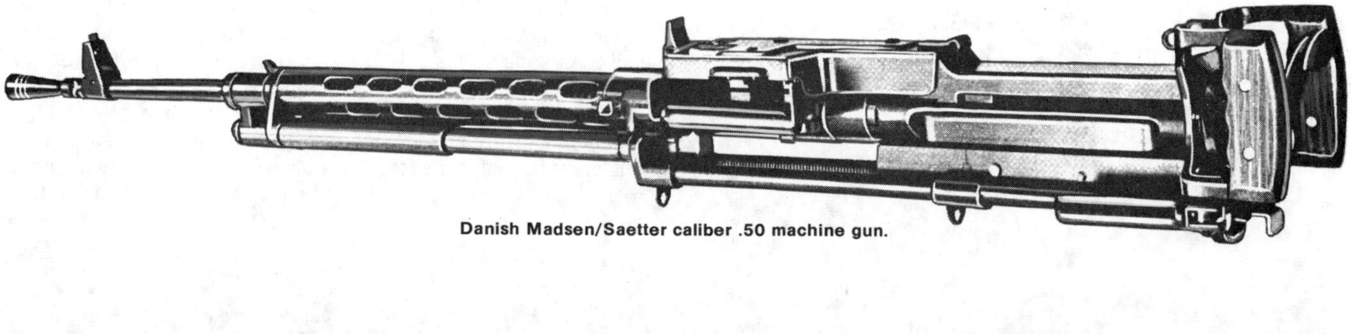

Danish Madsen/Saetter caliber .50 machine gun.

25

Dominican Republic

The Dominican Army has a wide variety of small arms, but within the past several years has attempted to standardize by selling some of the older weapons. Colt caliber .45 M1911 Pistols, FN Browning 9mm Hi-Power Pistols, and Colt and Smith & Wesson caliber .38 revolvers are used. The caliber .30 Cristobal Carbine Model 2 is used in large quantities. The 7mm M1908 Mauser rifle is being replaced by the 7.62mm NATO FN "FAL" and the 7.62mm NATO G3 rifles.

The most common machine guns are the Browning caliber .30 M1919A4 and M1917A1, and the caliber .50 Browning M2 HB and M2 water cooled. There are also some 7mm Madsen guns in service.

DOMINICAN AUTOMATIC CARBINE CRISTOBAL MODEL 2

The Dominican Republic manufactures an automatic carbine chambered for the U.S. cal. .30 carbine cartridge. This weapon is called the automatic carbine Cristobal Model 2 by the Dominicans. The weapon is produced in a plant which has been run with the help of technicians from P. Beretta of Italy and Hungary; it is, therefore, not too surprising that the Model 2 has some striking similarities in outward appearance to the Beretta submachine guns. Internally, however, there are some significant differences. The Beretta Model 38-series submachine guns are blowback operated; the Model 2 Cristobal carbine is a delayed blowback.

Characteristics of the Cristobal Model 2

Caliber: Cal. .30 (U.S. M1 carbine cartridge.)
System of operation: Delayed blowback, selective fire.
Weight, w/o magazine: 7.75 lb.
Length, overall: 37.2 in.
Barrel length: 16.1 in.
Feed device: 25- or 30-round, detachable, staggered-row box magazine.
Sights: Front: Hooded blade.
Rear: Notch with elevator.
Muzzle velocity: 1875 f.p.s.
Cyclic rate: 580 r.p.m.

Caliber .30 Cristobal Model 2 Automatic Carbine.

How to Load and Fire the Cristobal Model 2

Insert a loaded magazine in the magazine well. Pull the handle on the right side of the receiver to the rear. The bolt will remain to the rear, since this weapon fires from an open bolt. Push the handle forward—it does not reciprocate with the forward and rearward movements of the bolt. If the forward trigger is pulled, the weapon will fire single shots; if the rear trigger is pulled, the weapon will produce automatic fire. To put the weapon on safe, pull the lever mounted on the left side of the receiver to the rear. The safety blocks the sear and the triggers, and prevents rearward movement of the bolt. This weapon has no bolt-holding-open device, and therefore the bolt must be pulled to the rear every time a new magazine is loaded.

How to Field Strip the Cristobal Model 2

Remove the magazine by pushing the magazine catch forward. Press in the receiver cap lock and turn the receiver cap, removing it from the rear of the receiver. The recoil spring and the bolt can now be removed from the receiver. The bolt can be disassembled by removing the cross pin at its forward part, and slipping off the inertia lock. No further disassembly is recommended. To reassemble the weapon, perform the above steps in reverse order.

How the Cristobal Model 2 Works

The trigger mechanism of this weapon is similar to that of the Beretta Model 38-series of weapons. The bolt consists of two main parts: the bolt body, and a heavy part called the striker. These parts are joined by a two-armed inertia lever which is seated in the rear end of the bolt proper. The upper long arm of the movable inertia lever engages the striker, and lower short arm of the lever projects down from the bolt. When the bolt is closed, the short arm of the inertia lever stands before a stationary shoulder which is firmly attached to the bottom of the receiver. When a round is fired, the gases thrust rearward on the cartridge case base, which pushes back against the face of the bolt. Before the bolt can open, however, its inertia must be overcome. When the bolt begins to move rearward, the lower arm of the inertia lever (which could be called an inertia lock) bears against the bottom shoulder of the receiver. The rearward movement of the bolt then causes rearward rotation of the inertia lever, swinging the bottom arm up and out of engagement with the receiver, and forcing the upper arm back against the heavy striker. It is claimed that this system of delayed opening offers as great a resistance to the cartridge thrust as that of a bolt three to five times heavier than the one used with this weapon. During the closing of

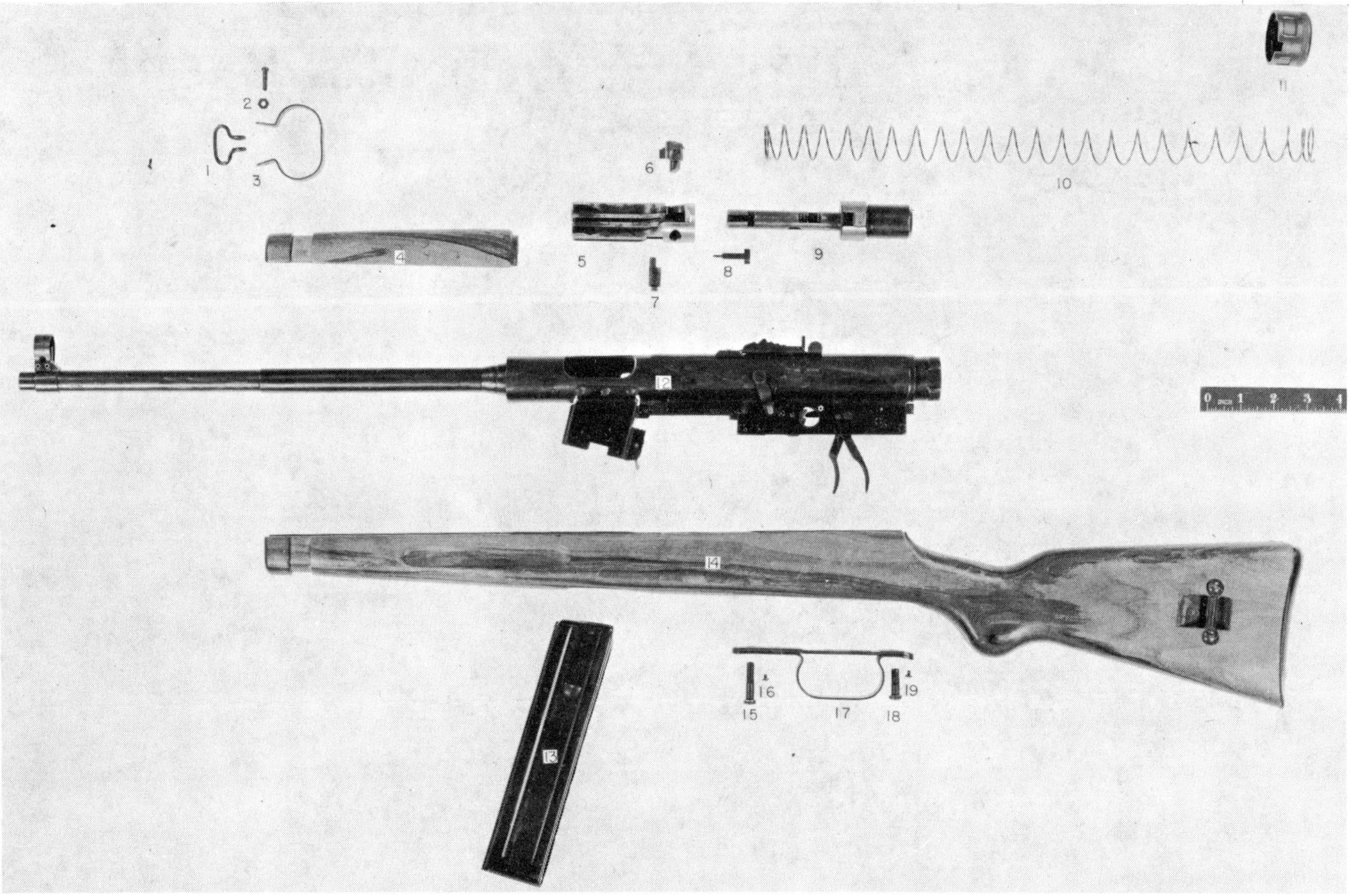

Cristobal carbine, stripped.

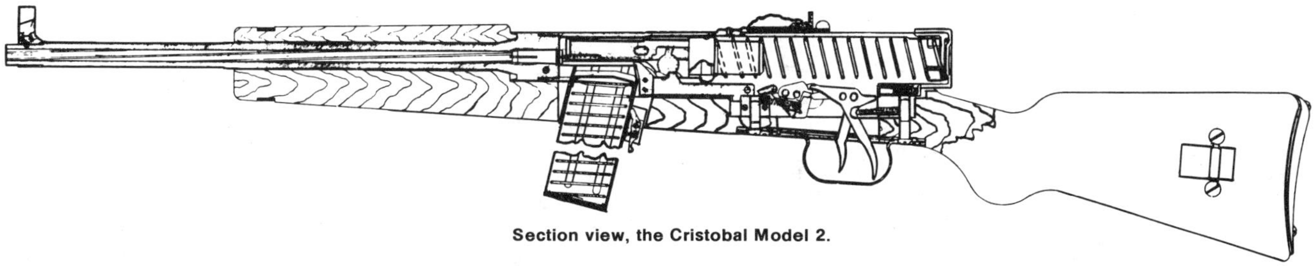

Section view, the Cristobal Model 2.

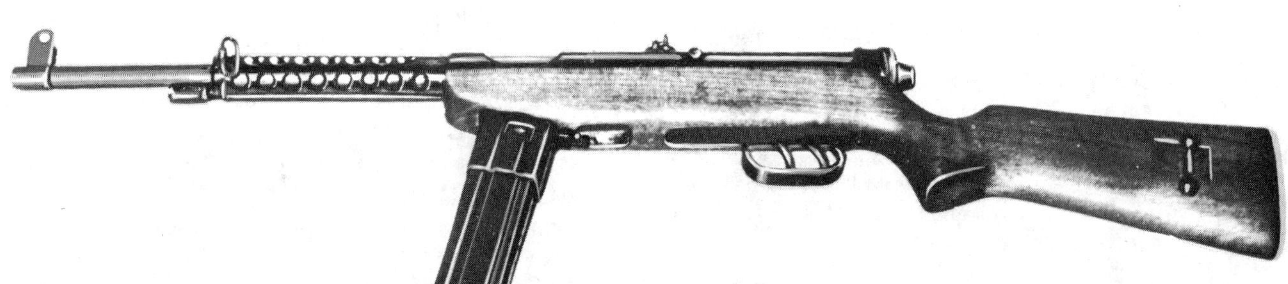

Dominican caliber .30 Automatic Carbine Model 1962.

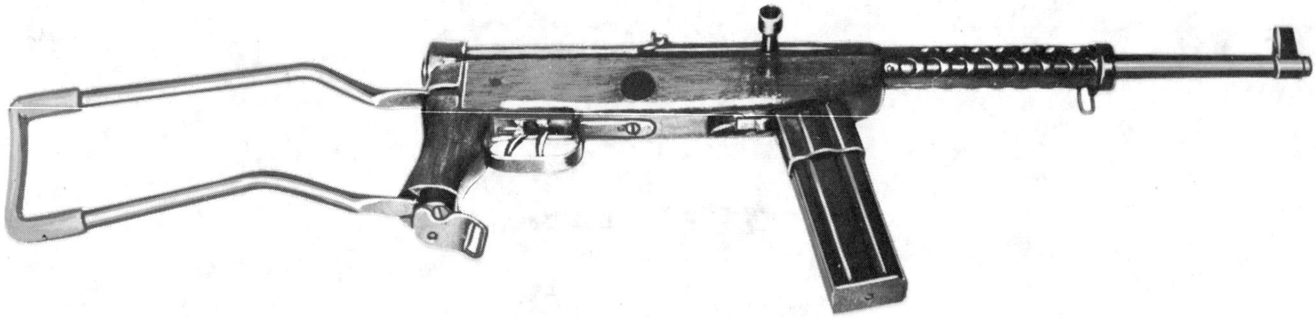

Dominican caliber .30 Automatic Carbine Model 1962 with folding metal stock.

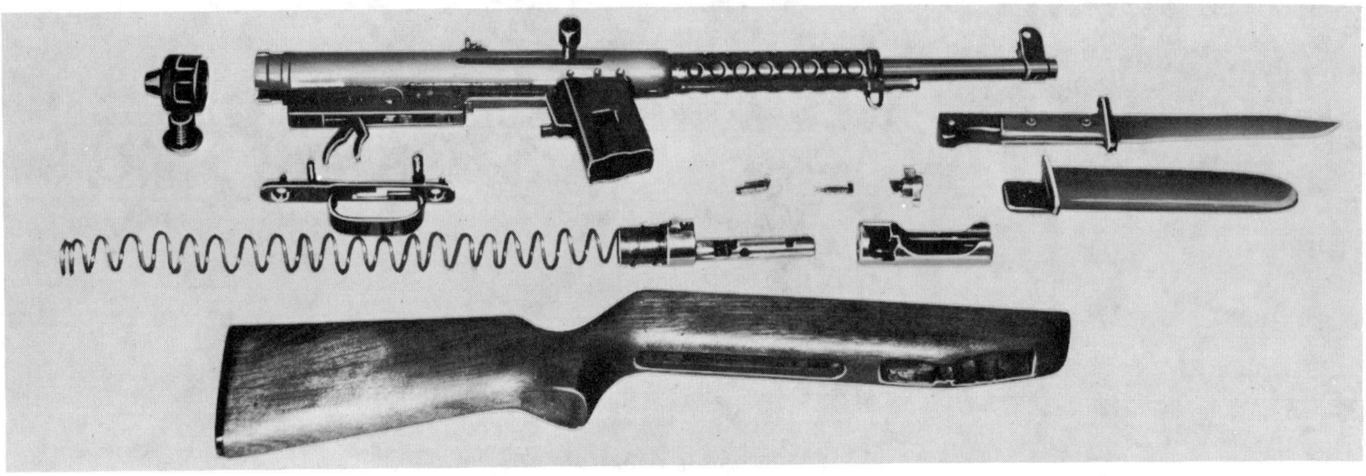

Dominican caliber .30 Automatic Carbine Model 1962, field-stripped.

the bolt, another lateral arm of the inertia lever slides on the receiver wall, and retains the inertia lever and the striker in a cocked position long enough for the bolt to run fully home. When the bolt is fully closed, the lateral arm of the inertia lever moves above a slot in the receiver wall, permitting forward rotation of the lever by the striker and the firing of the round. The inertia lever therefore prevents the weapon from firing before the bolt is fully closed.

DOMINICAN CARTRIDGES

The Dominican Republic also manufactures its own cal. .30 rifle and carbine cartridges. These are loaded with Berdan primers, rather than the Boxer type used in the United States and Canada.

DOMINICAN CAL. .30 AUTOMATIC CARBINE M1962

A new model of the Cristobal carbine has been developed. The Model 1962 Automatic carbine may be found either with a fixed wooden stock or a folding metal stock. Its loading, firing, field stripping, and functioning are the same as those of the Cristobal Model 2. The main difference in construction, other than the folding steel stock, is the use of a perforated metal barrel jacket on the Model 1962 weapons.

Characteristics of the M1962 Automatic Carbine

Caliber: Cal. .30 (U.S. M1 carbine cartridge).
System of operation: Delayed blowback, selective fire.
Weight, loaded:
  W/wooden stock—8.7 lbs.
  W/folding steel stock—8.2 lbs.
Length, overall:
  W/wooden stock—34.1 in.
  W/steel stock extended—37 in.
  W/steel stock folded—25.6 in.
Barrel length: 12.2 in.
Feed device: 30-round, detachable, staggered-row box magazine.
Sights: Front: Protected blade.
  Rear: L type
Muzzle velocity: 1870 f.p.s.

DOMINICAN 7.62mm AUTOMATIC RIFLE MODEL 1962

The Model 1962 automatic rifle combines the gas system of the U.S. M14 rifle with the bolt mechanism of the FN light automatic rifle. This weapon was apparently made only as a prototype.

Characteristics of the M1962 Automatic Rifle

Caliber: 7.62mm NATO.
System of operation: Gas, selective fire.
Weight loaded: 10.4 lbs.
Length, overall: 42.5 in.
Barrel length: 21.3 in.
Feed device: 20-round, detachable, staggered row box magazine.
Muzzle velocity: 2700 f.p.s.

Dominican 7.62mm Automatic Rifle Model 1962.

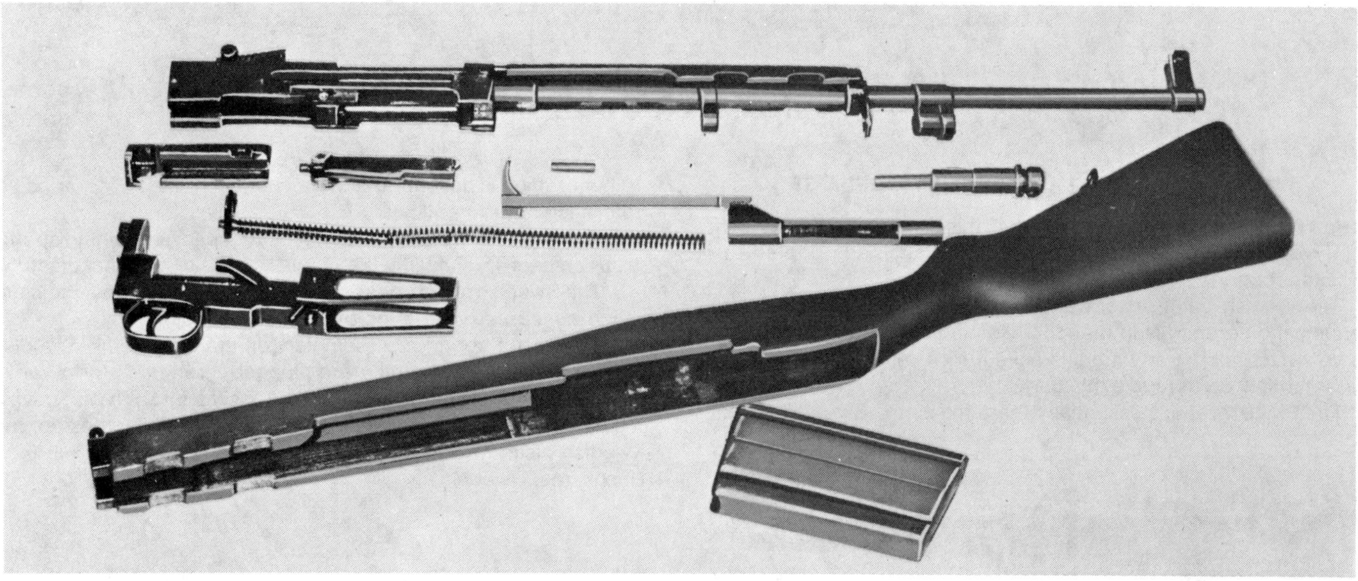

Dominican 7.62mm Automatic Rifle Model 1962, field-stripped.

Finland

26

Finland uses the 9mm Lahti pistol Model L35. The 9mm M31 and Model 44 Suomi submachine gun may still be found, but will probably be replaced by the 7.62mm Model 60 and 62 assault rifles.

The Finns adopted the 7.62mm Soviet M1943 "intermediate"-size cartridge and the Model 60 assault rifle and Model 62 assault rifle, to replace their bolt-action 7.62mm Mosin-Nagant rifles and carbines. These weapons, like the older Finnish machine guns, are chambered for the 7.62mm Russian rimmed cartridge. The 7.62mm Lahti Saloranta light machine gun is being replaced by the Model 62 light machine gun, which is chambered for the Soviet 7.62mm M1943 cartridge.

FINNISH PISTOLS

THE FINNISH 9mm LAHTI PISTOL MODEL L-35

In 1935 the Lahti pistol was adopted; this pistol was made by VKT at Jyvaskyla, Finland. (Earlier, in 1923, Finland had adopted the 7.65mm Luger.)

The Finnish Lahti pistol was issued in several variations. The principal difference is in the lock retaining spring; the Lahti uses a yoke-type lock. Early models have a lock retaining spring, but later models do not have this part.

The Finnish Lahti is essentially the same as the Swedish Model 40 Lahti which is described in detail under Sweden. The principal differences are as follows:

a. The Swedish weapon does not have the lock retaining spring; in this respect it is similar to the Finnish pistols of later manufacture.

b. The Swedish pistol does not have the loaded chamber indicator which is mounted on the top of the Finnish pistol.

c. The recoil spring is assembled differently on the Swedish weapon. It is assembled on a rod plugged through the grip frame, on a projection of the grip frame that passes into the bolt cavity.

d. The grips of the Finnish pistol are marked "VKT"; those of the Swedish pistol have the trademark of Husqvarna Vapenfabrik of Husqvarna, Sweden.

Finnish 7.65mm Luger.

The 9mm Lahti pistol (left side).

The 9mm Lahti pistol (right side).

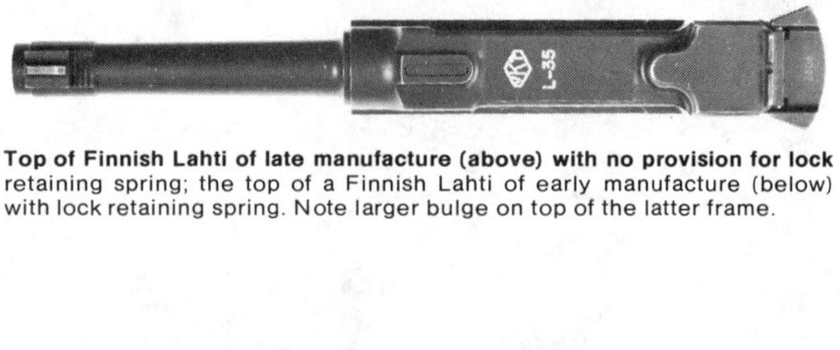

Top of Finnish Lahti of late manufacture (above) with no provision for lock retaining spring; the top of a Finnish Lahti of early manufacture (below) with lock retaining spring. Note larger bulge on top of the latter frame.

FINNISH RIFLES

Former Finnish service rifles were all based on the Russian 1891 Mosin Nagant action. The Finnish models of the Mosin Nagant vary from the Russian mainly in sights, stocks, fittings, etc. In general it can be said that the Finnish-made weapons are of higher quality manufacture than the Russian weapons. Various Finnish service rifles chambered for the 7.62mm rimmed cartridge are the Model 91 Carbine, Model 27 Rifle, Model 28 Rifle, Model 28-30 Rifle, Model 30 Rifle, and the Model 39 Rifle. The Model 62 assault rifle is now standard.

FINNISH 7.62mm M60 ASSAULT RIFLE

Characteristics of M60 Assault Rifle

Caliber: 7.62mm; uses copy of Soviet M43 rimless.
System of operation: Gas, selective fire.
Weight, loaded: 9 lbs.
Length, overall: 36 in.
Barrel length: 16.5 in.
Feed device: 30-round detachable, staggered box magazine.
Sights: Front: Hooded post.
Rear: Tangent w/aperture.
Cyclic rate: 650 r.p.m.
Muzzle velocity: 2362 f.p.s.

Finnish 7.62mm M60 Assault Rifle.

This weapon is a modified copy of the Soviet AK-47. Loading and firing, field stripping, and functioning are the same as described under that weapon in the chapter on the U.S.S.R.

The fore-end and pistol grip of the M60 are made of plastic and the weapon has a tubular steel stock. The rear sight is mounted on the receiver cover rather than on the receiver, as on the AK. An unusual feature of the weapon is that it does not have a full trigger guard; a post protrudes downward from the receiver and blocks the trigger from the front. This allows easy use of the weapon with heavy winter mittens. The bayonet of the M60 and its manner of mounting are different from that of the Soviet AK47.

FINNISH 7.62mm ASSAULT RIFLE M62

The Model 62 Assault Rifle is a more recent version of the M60 Assault Rifle. A conventional trigger guard is used and the plastic stock has been improved. Finland now makes the Model 62 assault rifle with a stamped steel receiver similar to that of the Soviet AKM. It can also be found with a wooden stock.

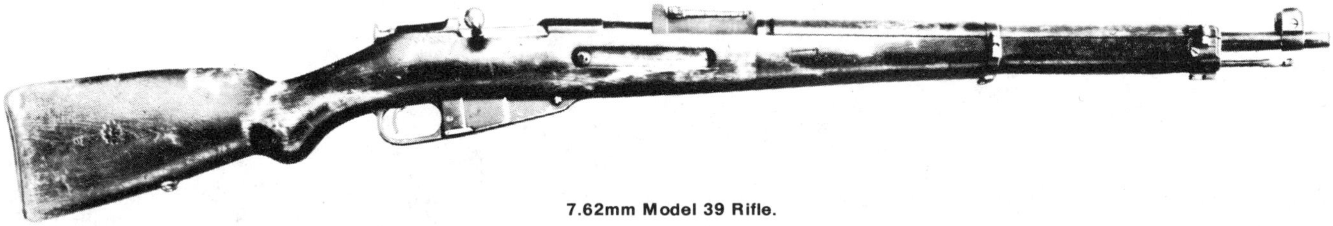

7.62mm Model 39 Rifle.

7.62mm Model 28-30 Rifle.

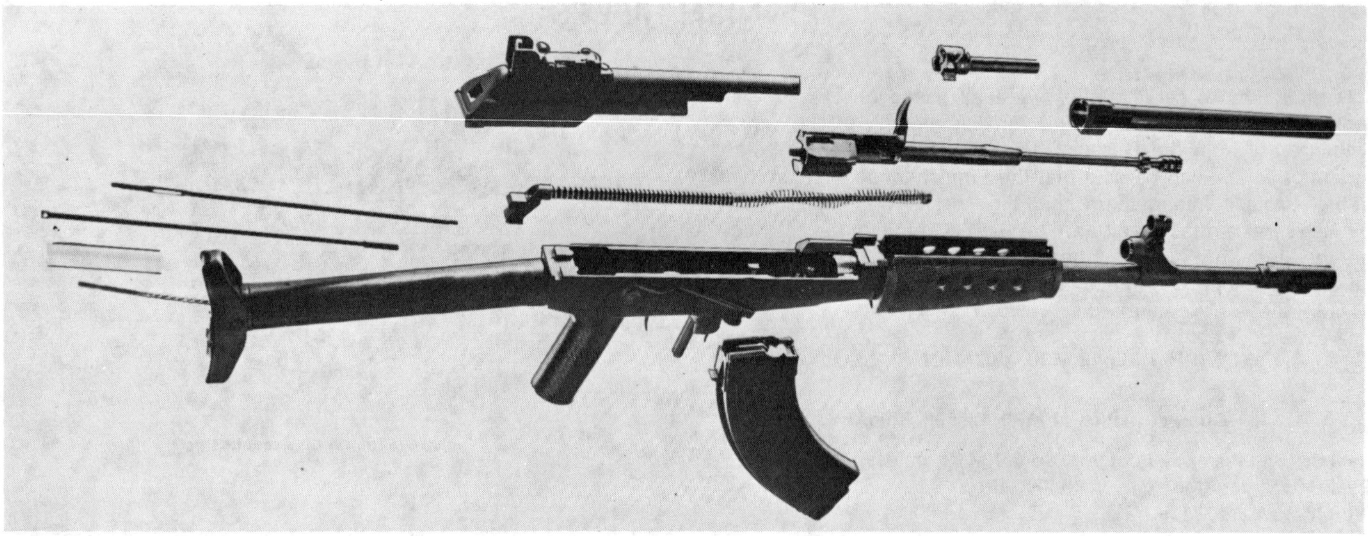

Finnish 7.62mm M60 Assault Rifle, field-stripped.

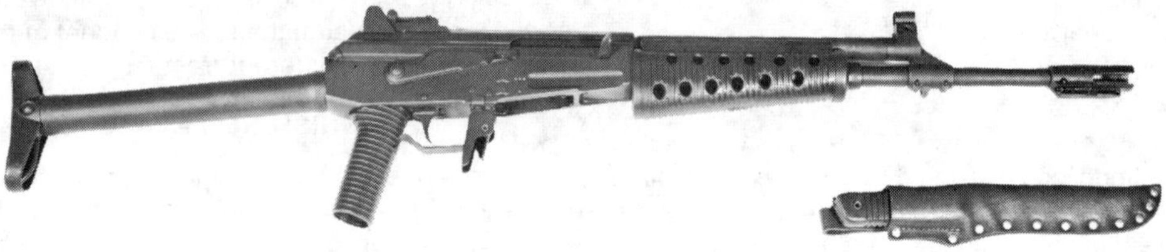

Finnish 7.62mm M62 Assault Rifle without magazine.

Finnish 7.62mm M62 Assault Rifle, bayonet fixed.

FINNISH SUBMACHINE GUNS

Although the Finns have used a number of submachine guns and made a modified copy of the Soviet PPS43, the story of the submachine gun in Finland is really the story of the Suomi. The Suomi was developed at Tikkakosi O/Y, Sakara, Finland.

THE FINNISH SUOMI SUBMACHINE GUN

The Suomi has been made in a number of different models and is still used extensively. In Finland the weapon is used in two models: the Model 31 and Model 44; both are chambered for the 9mm Parabellum cartridge. The Suomi was also used in Sweden where, as made in Sweden, it was known as the Model 37-39 and as imported from Finland, it was called the Model 37-39F. In Switzerland, where the gun was made by Hispano-Suiza and at the Waffenfabrik, Bern, the gun is known as the Model 43/44. In Denmark the Suomi has been made by Madsen and the Swedish Model 37-39 version of the gun is still in limited use as the Model 44 (37).

A delayed blowback version of the Suomi chambered for a rimmed 7.62mm "intermediate-sized" cartridge was made in very limited quantity.

Characteristics of the Model 31 Suomi

Caliber: 9mm Parabellum.
System of operation: Blowback, selective fire.
Length overall: 34 in.
Barrel length: 12.62 in.
Weight: 11.31 lb. w/ empty 50-round box magazine.
Feed device: 70-round drum, 25 or 50-round box magazine.
Sights: Front: Blade.
Rear: Tangent graduated from 100-500 meters.
Cyclic rate: 800-900 r.p.m.
Muzzle velocity: Approx 1300 f.p.s.

Loading and Firing the Suomi

A loaded magazine is inserted below into the magazine housing and pressed up until it locks. The cocking handle protrudes from the rear of the weapon under the milled recoil spring cap. Grip it firmly, pull back to the rear to compress the bolt spring and cock the bolt, and allow to run forward under the influence of its own spring. Pressure on the trigger will now fire the weapon. The bolt will stay back between shots. When a continuous burst of fire is required, maintain a firm stiff pressure.

Finnish Suomi 9mm submachine gun.

Field Stripping the Suomi

At the rear of the receiver is a heavy milled cap. Unscrew this cap and ease out the housing recoil spring guide, and recoil spring.

Drawing back on the cocking handle will now pull the bolt back for removal from the weapon.

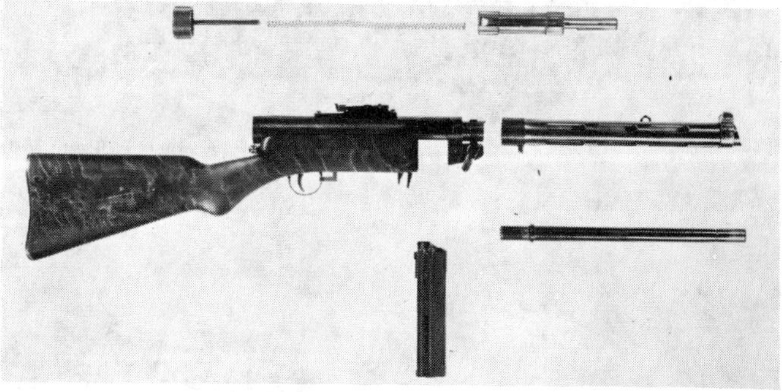

9mm Suomi submachine gun, field-stripped.

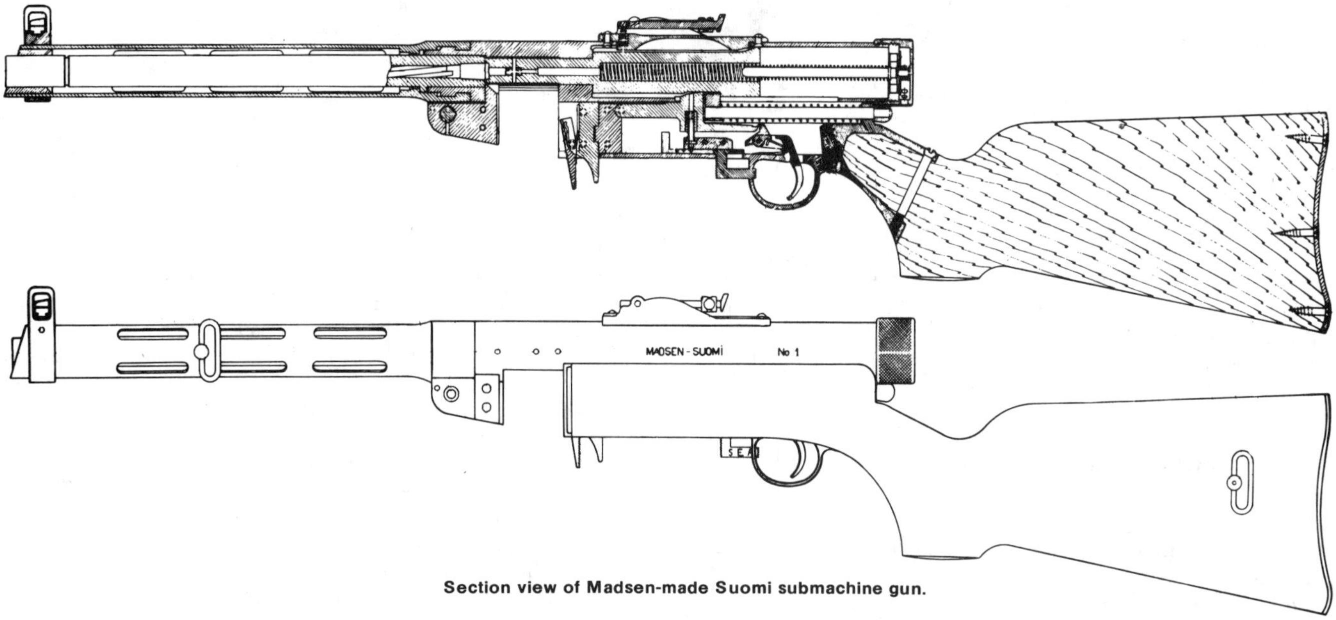

Section view of Madsen-made Suomi submachine gun.

FINNISH MACHINE GUNS

The Finns used three 7.62mm heavy machine guns. Apparently, all were Maxim types; their model designations were Model 09, Model 21, and Model 32.

The light machine gun formerly used was the Model 26 Lahti Saloranta. Although this weapon was developed for international sale and was advertised as being suitable for any service caliber, so far as known it has only been made in 7.62mm caliber for the Finnish Army, and in 7.92mm for the Chinese prior to World War II.

FINNISH LAHTI SALORANTA LIGHT MACHINE GUN

The Model 26 was one of the first of the post-World War I "true" light machine gun types. It was considered a noteworthy gun at its time of development and, although somewhat lacking in adaptability as compared with the post-World War II guns, is still a basically sound weapon.

Characteristics of the Model 26 Lahti Saloranta

Caliber: 7.62mm rimmed.
System of operation: Recoil, selective fire.
Weight, w/loaded 20-rd. magazine: 23 lbs.
Length, overall: 46.5 in.
Feed device: 20-round box, or 75-round drum.
Cyclic rate: 500 r.p.m.
Muzzle velocity: 2625 f.p.s.

How to Load and Fire the Model 26 Lahti Saloranta

Pull the operating handle to the rear, insert a loaded magazine in the magazine port (when using the box magazine) and set the selector on the type of fire desired. Squeeze the trigger and the weapon will fire. The Model 26 fires from an open bolt. To attach

Finnish 7.62mm Model 32 Heavy Machine Gun.

The 7.62mm Model 26 Lahti Saloranta Light Machine Gun.

the 75-round drum magazine to the gun, remove the magazine support and push the magazine up into position until the holding latch clicks. To remove the barrel, turn the lever 180 degrees. This releases the catch holding the butt to the receiver. Lift the receiver cover and the barrel, barrel extension, and bolt can be lifted out as a unit. Usually the complete unit is replaced.

How the Model 26 Lahti Saloranta Works

Two distinct feed systems are available with this gun. A spring loaded clip which holds 25 cartridges may be used. The alternate is a flat drum magazine mounted below the gun with a capacity of 75 rounds.

Maximum rate of fire is normally set at 500 per minute.

In the Lahti the bolt is automatically held to the rear after the trigger has been released. The effect of this is to keep the action open to prevent a round in the chamber cooking off. This also, of course, permits air circulation through the barrel for cooling purposes.

The barrel is removed by turning the dismounting lever 180 degrees to release the catch which holds the buttstock to the receiver. The receiver cover is lifted, and the barrel extension together with the barrel and bolt are then pulled out to the rear in a manner not unlike that of the Swiss Furrer.

When using the flat 75-shot drum, the magazine support is removed and the drum is pushed up until its holding catch snaps into place. The operating handle on the right is pulled all the way to the rear. The springloaded sear rises to catch the notch in the bottom of the bolt and holds it back. The arm is now cocked and ready to fire. Setting the selector from safe to automatic fire prepares it for action.

As the trigger is pulled the sear is freed from the bolt. The bolt is driven ahead by the compressed operating spring. A feed rib machined into the top of the bolt pushes the first round out of the lips of the magazine and toward the chamber.

The bolt seats behind the cartridge when about one-half inch from final position and the claw extractor snaps over the case.

Simultaneously the bolt locking piece is cammed downwards into its locking notch in the bolt. This releases the holding catch which has been preventing the barrel and extension from moving forward.

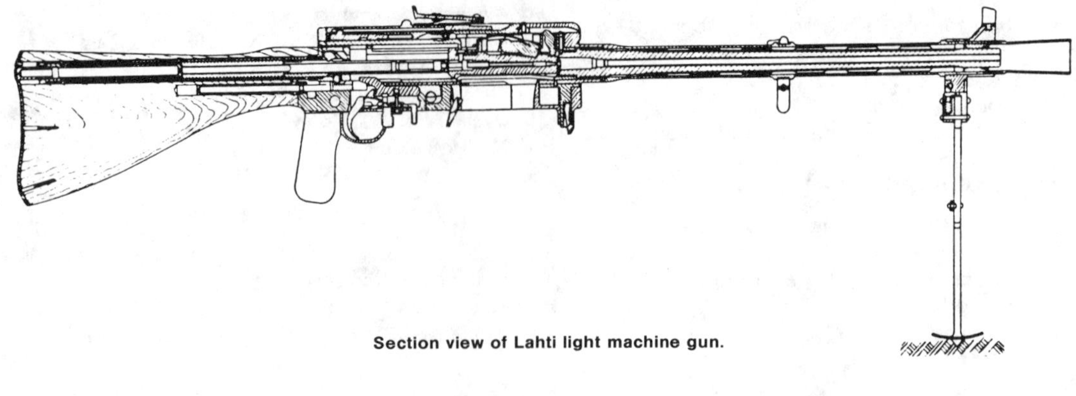

Section view of Lahti light machine gun.

The locked barrel with its extension and the bolt start the final movement ahead. When it nears complete battery position, a pivoting pin in the bolt body which has been blocking the firing pin hits a ramp in the receiver and is pushed up out of the way to permit the firing pin to be released and go forward to fire the cartridge.

Again rather on the Furrer principle, we find the timing such that the recoil starts before the forward moving locked parts strike the receiver. This buffing action eases the recoil and makes for smoother functioning.

During the recoil stroke the parts are locked together for about one-half inch of travel. At that point the stud on the bolt lock engages a cam in the receiver to lift it out of its locking recess there.

The bolt is allowed a rearward motion of a few thousandths of an inch before complete unlocking sets in, thereby permitting the extractor to loosen the cartridge before the gas pressure affects it directly. (This allows the extractor to break the gas seal. The cartridge is thereby entirely loosened in primary extraction at this point.)

As the gun fires, energy is transferred from the recoiling barrel to the bolt through an accelerator which pivots to speed the bolt to the rear while the extractor withdraws the empty cartridge case until its rim hits the solid ejector and is knocked out the ejection port to the right.

The firing pin is withdrawn during the initial recoil movement. The firing pin sear in the top of the bolt drops in front of a circular projection over the body of the pin. As the bolt is released from locking, the barrel and barrel extension are held rearward by the holding catch. The compressed operating spring is ready at the end of the stroke to return the members if the trigger is on full automatic.

FINNISH 7.62mm M60 LIGHT MACHINE GUN

The M60 is used on a bipod or on a light weight tripod. The weapon has a prong-type flash suppressor and a tubular steel stock. Like the M60 Assault rifle it has a bar in front of the trigger in lieu of a conventional trigger guard. The ammunition container is hung on the right side of the receiver, allowing easy movement of the gun with ammunition by one man.

Characteristics of Model 60

Caliber: 7.62mm uses copy of Soviet M43 rimless.
System of operation: Gas.
Weight: 16.8 lbs.
Feed device: 100 round non-disintegrating metallic link belt.
Sights: Front—Hooded post.
Rear—Tangent leaf.
Cyclic rate: 1050 r.p.m.

The M60 7.62mm Light Machine Gun.

27 France

The 9mm Parabellum M1950 pistol is the standard service pistol. The standard service rifle is the 7.5mm M1949/56; M1949 rifles are also in service. The 9mm M1949 (MAT49) submachine gun is standard as is the 7.5mm general purpose machine gun, Model 52 (AAT52). The 7.5mm M1924 M29 light machine gun (called an automatic rifle by the French) is still used extensively as is the U.S. caliber .50 Browning M2 HB machine gun.

Photographs in French magazines indicate that there are still many United States small arms in use in reserve units. These weapons were supplied to France mainly during World War II.

FRENCH SERVICE PISTOLS

The French Army during World War I was equipped with the 8mm Model 1892 revolver, and the 11mm Model 1873 revolver. During this war, extensive purchases were made of 7.65mm (.32 ACP) Star and Ruby automatics from Spain. Numbers of the above weapons were apparently still in service during World War II.

Between the wars, the French developed two automatic pistols that were quite similar, the 7.65mm M1935A and M1935S. These pistols were developed for the 7.65mm long cartridge which is much like the caliber .30 cartridge developed for the U.S. Pedersen device of 1918. One story of this resemblance, which may or may not be true, is that John Browning demonstrated his Model 30-18 automatic rifle, which was chambered for the Pedersen cartridge, for the French authorities after World War I. Although the French were not interested in the rifle, a little-known weapon which was one of the few Browning designs that went nowhere, they were interested in the cartridge. In any event the French used it as their principal pistol and submachine gun cartridge until 1949.

The M1935 pistols are based on the patents of C. Petter of SACM, and are essentially modifications of the Browning recoil-operated pistols. These pistols were used throughout World War II, supplemented by earlier weapons and the U.S. caliber .45 M1911 and M1911A pistols, plus some British pistols. There were sufficient M1935 pistols in stock for them to be used extensively by the French Government Forces during the 1945-54 fighting in Indo-China. They are not likely to be found in quantity with French forces at present.

French 9mm M1950 Pistol.

In addition to the pistols listed above, the French used extensive quantities of Lugers and P38 pistols in the immediate post-war period. The 9mm Parabellum cartridge was adopted as standard for pistol and submachine gun and a pistol designated the SE-MAS 1948 was developed for this cartridge at the Saint Etienne Arsenal. This pistol was improved and finally adopted as the Model 1950, the current standard pistol.

FRENCH 9mm PISTOL M1950

The M1950 pistol is similar in most respects to the U.S. cal. .45 Colt automatic. It incorporates a few improvements over that weapon, but in loading, firing, and functioning is essentially the same as the cal. .45. It is a much better made and more lethal weapon than any of the earlier French pistols.

Characteristics of the M1950 Pistol

Caliber: 9mm Parabellum.
System of operation: Recoil, semi-automatic fire only.
Weight: 1.8 lb.
Length, overall: 7.6 in.
Barrel length: 4.4 in.
Feed device: 9-round, single-column, detachable box magazine.
Sights: Front: Tapered post.
Rear: U-notch; the top of the slide has a matted ramp.
Muzzle velocity: 1156 f.p.s.

How to Load and Fire the M1950 Pistol

The M1950 pistol is loaded and fired the same as the U.S. M1911A1 cal. .45 automatic pistol. However, it cannot be fired with the magazine removed. A loaded-chamber indicator is mounted in the slide. The safety catch is mounted on the left rear of the slide.

How to Field Strip the M1950 Pistol

Field stripping is the same as for the U.S. .45, with the following exceptions:

There is no barrel bushing and recoil spring plug; the recoil spring is dismounted after the slide has been removed from the receiver.

The hammer, its spring and lever, and the sear assembly are contained in a housing which can be lifted out as one piece when the slide is removed.

How the M1950 Pistol Works

The functioning of the M1950 pistol is essentially the same as that of the U.S. service automatic.

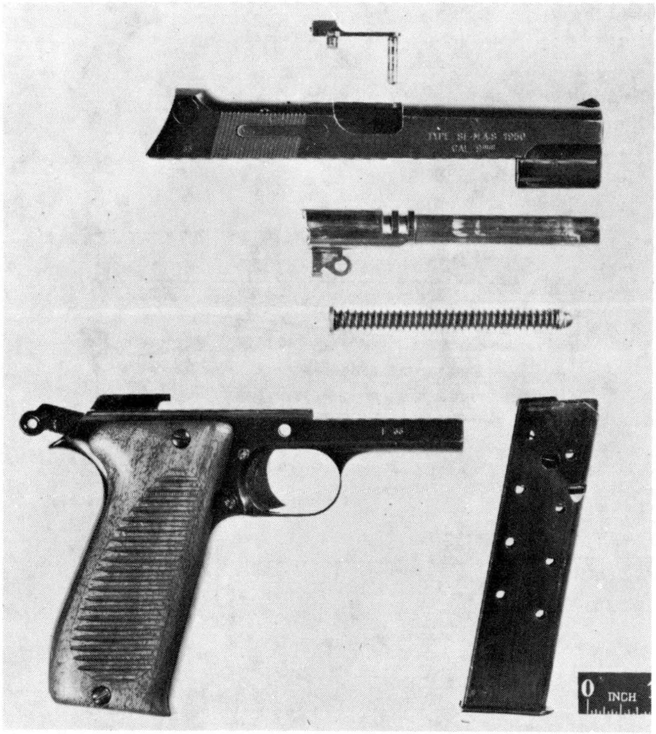

M1950 field-stripped.

Special Note on the M1950 Pistol

The M1950 pistol incorporates most features of the M1935 French pistols, and has a few changes borrowed elsewhere. The safety on the slide was used in the M1935A and M1935S; it blocks the hammer from the firing pin. The internal mounting of the recoil spring is used on practically all the Browning and modified Browning-design pistols in service today except the U.S. cal. .45 and those pistols which are direct copies of the U.S. cal. .45.

9MM PISTOL MODEL F 1

France has adopted for competition and possible service use, the MAB designed and produced Model "P-15 Competition" pistol; official French government nomenclature is 9mm PAP Mle F 1. This pistol is currently available on the U.S. commercial market and is of quality manufacture. Other French service pistols have either a parkerized or enamelled finish; the PA-15 has a blued finish.

French 9mm Parabellum Model F 1.

Characteristics of the Model F 1

Caliber: 9mm Parabellum.
System of operation: Recoil operated, semi-automatic.
Length overall: 9.64 in.
Barrel length: 6 in.
Weight: 2.43 lb.
Feed device: 15-round, staggered row, detachable box magazine.
Sights: Front: Blade.
Rear: Notch adjustable for elevation and windage.
Muzzle velocity: Approx. 1200 f.p.s.

How to Load and Fire the Model F 1 Pistol

The Model F 1 is loaded and fired in the same manner as the U.S. caliber .45 M1911A1 pistol. The pistol has a magazine safety and cannot be fired with the magazine removed.

How to Field Strip the Model F 1 Pistol

The Model F 1 is field stripped in a manner similar to that of the French 9mm M1950 except that it has a barrel bushing. The barrel bushing is removed by pressing in on the barrel bushing catch, a spring loaded detent mounted on the under side of the muzzle end of the slide, and unscrewing the barrel bushing.

How the Model F 1 Works

The F 1, or "-15 Competition", to use both the designations, this pistol has been known by, is the match version of the MAB "P-15" pistol. There is also a version of this pistol known as the Model P-8 which has an 8-shot magazine. All have a rotating barrel type locking mechanism. The barrel has lugs mounted on its top and bottom which engage a cut-out section in the top of the slide and a cam track cut in a locking piece mounted in the receiver respectively. This piece is held in place in the receiver by the pin portion of the slide stop in a manner reminiscent of the Mexican Obregon pistol; the recoil spring guide rod and recoil spring are mounted on the front portion of the piece. Rearward movement of the slide on firing causes the bottom lug to rotate in the camway of the receiver locking piece, thereby rotating the locking lug on top of the barrel out of its locked position allowing the slide to continue its travel to the rear with the empty cartridge case gripped in the slide mounted extractor. The ejector, mounted on the right side of the receiver, engages the base of the case and forces the case out of the ejection port. At the same time the slide is forcing the externally mounted hammer to the rear causing the hammer to engage the sear and remain to the rear in cocked position. Rearward movement of the slide has compressed the recoil spring and the slide reaches its limit of rearward travel; the spring forces the slide forward. The slide, during its forward movement, strips a cartridge from the magazine and feeds it into the chamber. The cam slot in the receiver locking piece begins camming the barrel around so that the locking lug on the barrel is engaged in its recess in the slide when the slide is in its forward position. Pressure on the trigger will fire the pistol again and the whole process is repeated for each separate pull of the trigger until the magazine is empty.

This pistol is also called the "PA-15".

THE FRENCH M1935 PISTOLS

The Model 1935A and 1935S pistols are quite similar in design and differ principally in method of locking. The Model 1935A has two lugs on the upper surface of the barrel which lock into mating grooves in the slide in a fashion similar to the U.S. Colt M1911A1. The M1935S, on the other hand, has a step machined on the top of the barrel at a point slightly ahead of the chamber and this step locks into a cut-out section in the slide.

As previously stated, these pistols are of Petter-Browning design. They were made at Chatellerault (MAC), Saint Etienne (MAS), Tulle (MAT), Société Alsacienne de Construction Mecanique (SACM), and Société d'Applications Génerales Electriqueset Mécaniques (SAGEM).

Characteristics of the Model 1935 Pistols

	Model 1935A	Model 1935S
Caliber:	7.65mm long	
System of operation:	Recoil operated, semiautomatic	
Length overall:	7.6 in.	7.4 in.
Barrel length:	4.3 in.	4.1 in.
Weight:	1.62 lb.	1.75 lb.
Feed device:	8-round, detachable, in-line box magazine	
Sights: Front:	Blade	
Rear:	Rounded notch	
Muzzle velocity:	1132 f.p.s.	

Loading, firing, and field stripping of these pistols is essentially the same as that of the U.S. caliber .45 M1911A1 or the 9mm M1950. These pistols have packaged hammer, main spring, and sear assemblies which can be removed after the slide is dismounted from the receiver.

French 7.65mm Long M1935A Pistol.

French 7.65mm Long M1935S Pistol.

FRENCH REVOLVER MODEL 1892

This weapon is obsolete and is unlikely to be encountered in military use at present. The model 1892 was commonly called the Lebel or Model d'Ordonnance and was the first swing out cylinder revolver adopted in Europe.

Characteristics of the Model 1892 Revolver

Caliber: 8mm ball Model 1892.
System of operation: Double-action revolver.
Length overall: 9.25 in.
Barrel length: 4.5 in.
Weight: 1.8 lb.
Feed device: 6-chambered cylinder.
Sights: Front: Blade.
Rear: V notch.
Muzzle velocity: 738 f.p.s.

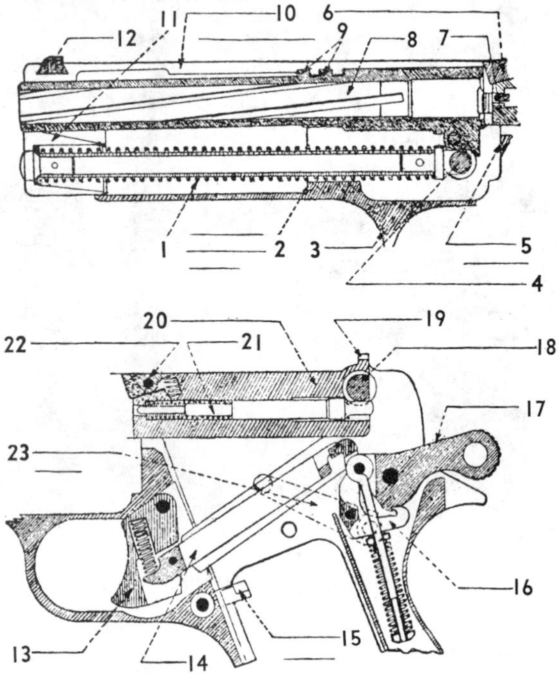

French Pistolet 1935A (Drawing from French Manual).

1. Recoil spring.
2. Receiver stop.
3. Receiver.
4. Barrel pivot pin.
5. Barrel stop.
6. Loading indicator.
7. Face of breechblock.
8. Barrel.
9. Dual barrel locking lugs.
10. Slide-breechblock unit.
11. Front of slide.
12. Front sight.
13. Trigger.
14. Trigger bar.
15. Magazine catch.
16. Sear.
17. Hammer.
18. Manual Safety.
19. Rear sight.
20. Breech section of slide.
21. Firing pin.
22. Loading indicator.
23. Magazine safety.

French 7.65mm Long M1935A Pistol, field-stripped.

French-made Walther PPK, caliber .380. This pistol was made by Manhurin in France before Walther postwar resumption of manufacture. It was not used as a French service pistol and is no longer being made in France. It is currently manufactured in Germany.

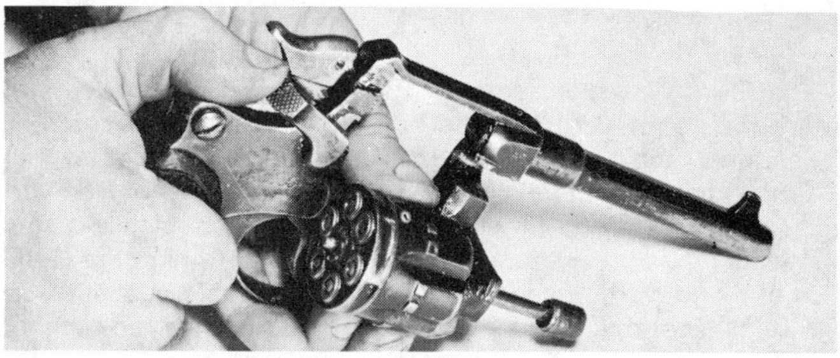

Use thumb lever on right side to swing out the cylinder.

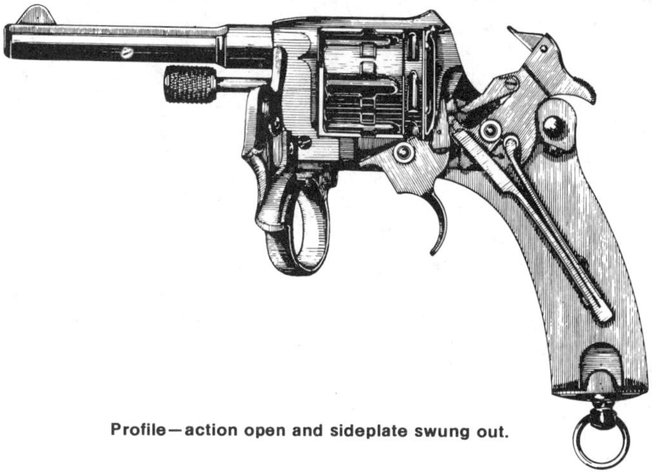

Profile—action open and sideplate swung out.

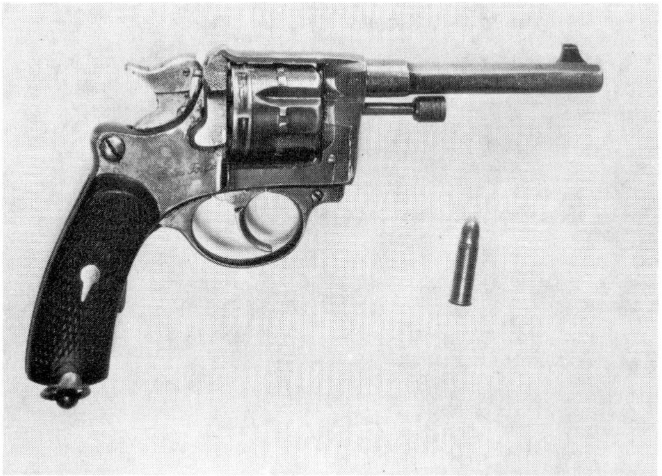

French 8mm Modele d'ordonnance 1892 Revolver.

FRENCH REVOLVER MODEL 1873

This weapon is obsolete, but was a common souvenir among United States troops returning from World War II. It is generally similar in operation to the U.S. caliber .45 M1873 revolver (the Colt Frontier or Peacemaker revolver). The principal difference in operation between these two weapons is the double-action feature of the French revolver as compared to the single-action operation of the U.S. revolver.

Some 1873 revolvers were modified to use the 8mm cartridge. The Model 1873, as originally issued, was designed for a low-pressure, black-powder cartridge which was manufactured only in France. It should NOT be used with any other cartridge.

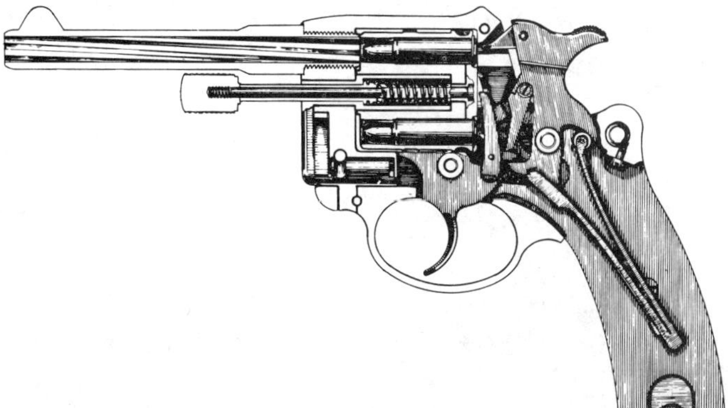

The French Model 1892 Revolver—action closed.

Obsolete 11mm M1873 Revolver.

FRENCH BOLT-ACTION RIFLES AND CARBINES

The French developed the first modern small-bore magazine rifle—the Lebel—in 1886. In 1890, they dropped the tubular magazine of the Lebel for the Mannlicher-type magazine of the Berthier rifles and carbines. The Lebel and the Mannlicher Berthier rifles and carbines were the standard service weapons of the French Army until the thirties. An interesting fact is that, during the early part of World War I, France purchased considerable quantities of the single-shot rolling block Remington rifle, chambered for the 8mm Lebel cartridge. Remington also produced 8mm M1907/15 rifles for the French during World War I.

The French realized that the 8mm Lebel cartridge had several outstanding shortcomings—a peculiar shape and rim which made feeding in automatic weapons difficult—and soon after World War I, developed a new 7.5mm rimless cartridge, the Model 1924. The Model 1924 was quite similar to the 7.92mm Mauser; so similar in fact, that a number of accidents occurred when people used 7.92mm cartridges in 7.5mm weapons. (The Model 1924 light machine gun was in limited service at this time). In 1929 the French shortened the case of the 7.5mm cartridge and produced the M1929 cartridge which is still the principal rifle cartridge of the French Army. A number of prototype bolt-action rifles were made for this cartridge, the principal rifle being the M1932. A modification of this rifle, the Model 1936, was adopted and is still used extensively in France and former French territories. Several modifications of this rifle—the M1936 CR39 and M1936 M51—also exist. At approximately the same time France, having huge stocks of World War 1 materiel on hand, began to modify and convert some of the older bolt-action arms on hand. Many of the 8mm Model 1907/15 rifles were altered and rebarreled with 7.5mm barrels. A modification of the 8mm Lebel rifle M1886 M93 was also produced, the 8mm M1886 M93 R35—during this period.

During World War II, the Free French Army was provided with over 69,000 U.S. caliber .30 M1903 and M1917 bolt-action rifles and Browning Automatic rifles.

FRENCH RIFLES AND CARBINES OF PRE-WORLD WAR II DESIGN

The older French rifles are still commonly encountered in Middle Eastern areas and among collectors in the United States. Therefore, coverage on them has been expanded to include a table of characteristics of the various models, and a few words have been added to explain the major differences among the various categories of these weapons. See table.

BOLT-ACTION CATEGORIES

Basically the bolt-action French rifles and carbines can be broken down into three categories: (1) the M1886; (2) the M1890; and (3) the M1932 and its production model, the M1936.

The M1886 (Commonly Called the Lebel)

The M1886 uses the tubular-type magazine; and its modifications. These weapons are chambered for the 8mm M1886 rimmed cartridge.

FRENCH RIFLES AND CARBINES OF PRE-WORLD WAR II DESIGN

Weapon	System of Operation	Overall Length	Barrel Length	Feed Device	Sights	Muzzle Velocity	Weight
Rifle: 8mm M1886 M93	Manually operated 2-piece bolt	51.3 in.	31.4 in.	8-rd tubular magazine	Front: Notched blade Rear: Leaf	2380 f.p.s.	9.35 lb.
Rifle: 8mm M1886 M93R35	Manually operated 2-piece bolt	37.64 in.	17.7 in.	3-rd tubular magazine	Front: Notched blade Rear: Leaf	2080 f.p.s.	7.84 lb.
Carbine: 8mm M1890	Manually operated 2-piece bolt	37.2 in.	17.7 in.	3-rd Mannlicher-type, integral, in-line magazine	Front: Blade Rear: Leaf	2080 f.p.s.	6.83 lb.
Mousquetoon 8mm M1892.	Manually operated 2-piece bolt	37.2 in.	17.7 in.	3-rd Mannlicher-type, integral, in-line magazine	Front: Blade Rear: Leaf	2080 f.p.s.	6.8 lb.
Rifle: 8mm M1902 "Indo-China Model"	Manually operated 2-piece bolt	38.6 in.	24.8 in.	3-rd Mannlicher-type, integral, in-line magazine	Front: Blade Rear: Leaf	2180 f.p.s.	7.9 lb.
Rifle: 8mm M1907 "Colonial Model"	Manually operated 2-piece bolt	52 in.	31.4 in.	3-rd Mannlicher-type, integral, in-line magazine	Front: Blade Rear: Leaf	2380 f.p.s.	8.6 lb.
Rifle: 8mm M1907/15	Manually operated 2-piece bolt	51.42 in.	31.4 in.	3-rd Mannlicher-type, integral, in-line magazine	Front Notched blade Rear: Leaf	2380 f.p.s.	8.38 lb.
Rifle: 8mm M1916	Manually operated 2-piece bolt	51.42 in.	31.4 in.	5-rd Mannlicher-type, integral, in-line magazine	Front Notched blade Rear: Leaf	2380 f.p.s.	9.25 lb.
Carbine: 8mm M1916	Manually operated 2-piece bolt	37.2 in.	17.7 in.	5-rd Mannlicher-type, integral, in-line magazine	Front: Blade Rear: Leaf	2080 f.p.s.	7.17 lb.
Rifle: 7.5mm M1907/15 M34	Manually operated 2-piece bolt	43.2 in.	22.8 in.	5-rd integral, staggered-row, box magazine	Front: Blade Rear: Leaf	2700 f.p.s.	7.85 lb.
Rifle: 7.5mm M1932	Protype of M1936, made in limited quantities						
Rifle: 7.5mm M1936	Manually operated 1-piece bolt	40.13 in.	22.6 in.	5-rd integral, staggered row, box magazine	Front: Barley-corn w/guards Rear: Ramp w/aperture	2700 f.p.s.	8.29 lb.
Rifle: 7.5mm M1936 CR39	Manually operated 1-piece bolt	Stock extended 34.9 in. Stock folded: 24.3 in.	17.7 in.	5-rd integral, staggered-row, box magazine	Front: Barley-corn w/guards Rear: Ramp w/aperture	2700 f.p.s.	8 lb. (approx.)

The M1890 Carbine (Sometimes Called the Mannlicher Berthier)

This carbine uses the Mannlicher-type feed system (requiring a special clip which functions as part of the magazine); and the many later rifles and carbines which use this system. These weapons also are chambered for the 8mm cartridge.

The M1932 and Its Production Model, the M1936

These use a 5-round Mauser-type magazine and a one-piece bolt. These weapons are chambered for the 7.5mm rimless cartridge.

The 7.5mm M1907/15 M34 is a variant; this is the M1907/15 rifle cut down and rebarreled for the 7.5mm cartridge. This weapon also was fitted with a Mauser-type 5-round magazine.

The early models have so many variants—Gendarmerie models, Artillery models, etc.—that it would be impossible to list them all here. An interesting point about the early rifles is that many of their parts are interchangeable; with a good supply of parts, and a vivid imagination, anyone can make up a weapon that would keep the collectors guessing for a long time.

A postwar version of the M1936, the 7.5mm M1936 M51, has a rifle grenade launcher built into the muzzle end of the barrel.

THE FRENCH MAS 1936 RIFLE

The M1936 is a 7.5mm caliber rifle with modified Mauser magazine. Action is shown ready for trigger pull. Note that dual locking lugs are engaged in recesses within the receiver bridge directly above the trigger. Placing the lugs this far back from the chamber permits a short bolt stroke but results in a rifle locking design weaker than the conventional Mauser front-lug type.

This turn bolt rifle has a modified Mauser magazine with quick removable bottom plate. Primary extraction is by action of bolt handle extension working against a cam in the receiver bridge.

The breech locking system of this rifle is somewhat different than the Mauser. Dual lugs at the rear of the bolt body turn into seatings in the receiver bridge as the handle is turned down. No manual safety.

Locking lugs are on rear of bolt cylinder near handle knob; they lock into receiver bridge to rear of magazine. Note cam shape of left face of receiver bridge which affords leverage for primary extraction cam on modified Mauser system. Rifle is loaded with Mauser-type charger clip. Note cuts in receiver for finger pressing cartridges into magazine. This is necessary because of large bolt diameter.

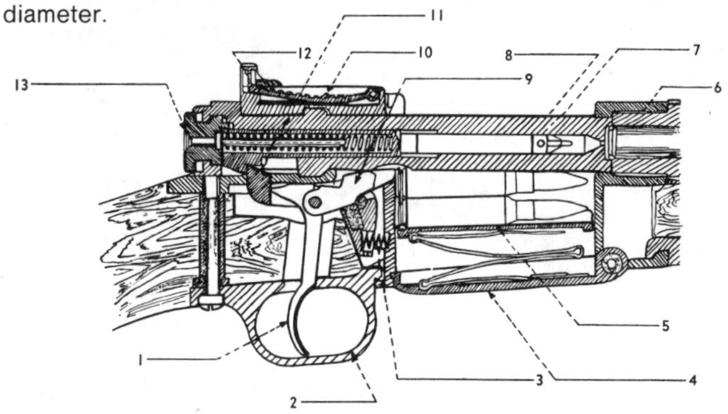

French Fusil 1936 (Drawing from French Manual).

1. Trigger
2. Triggerguard
3. Sear spring
4. Magazine bottom plate
5. Magazine follower
6. Barrel
7. Bolt
8. Firing pin
9. Ejector & Bolt stop unit
10. Receiver bridge
11. Sear
12. Peep sight
13. Cocking piece head.

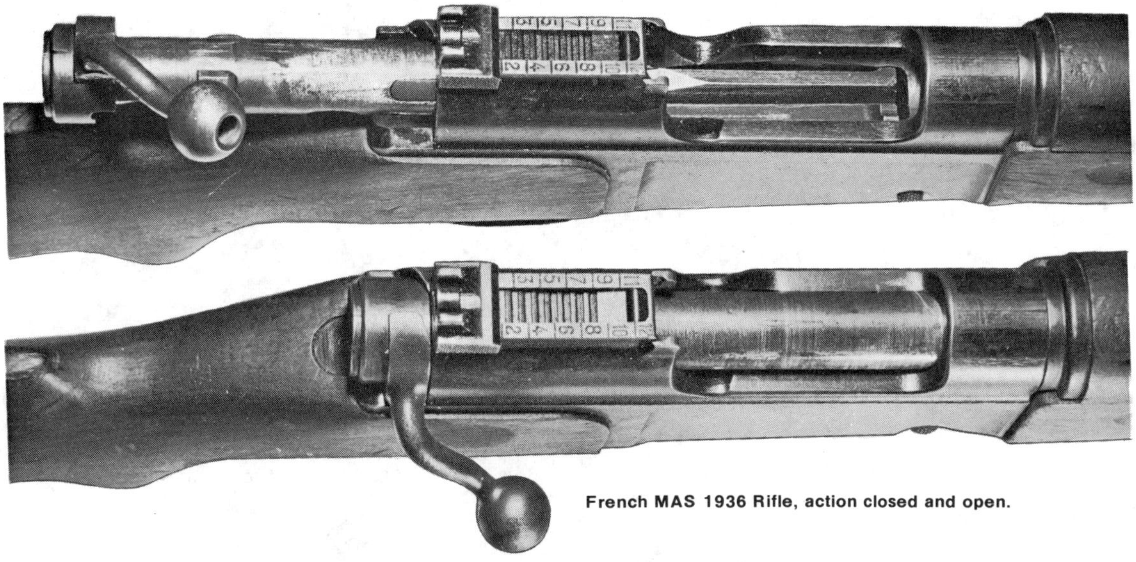

French MAS 1936 Rifle, action closed and open.

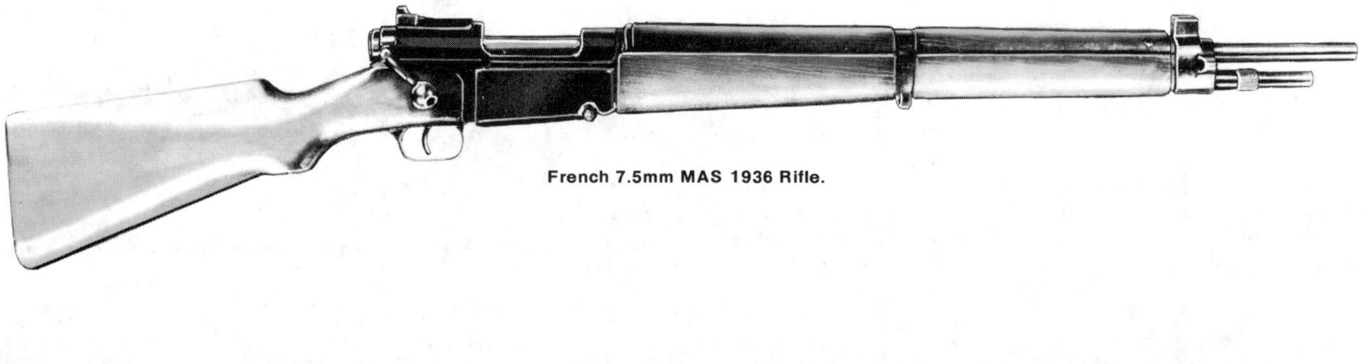

French 7.5mm MAS 1936 Rifle.

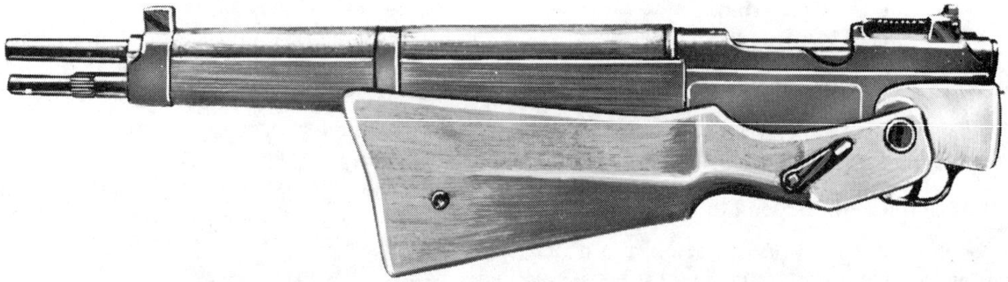

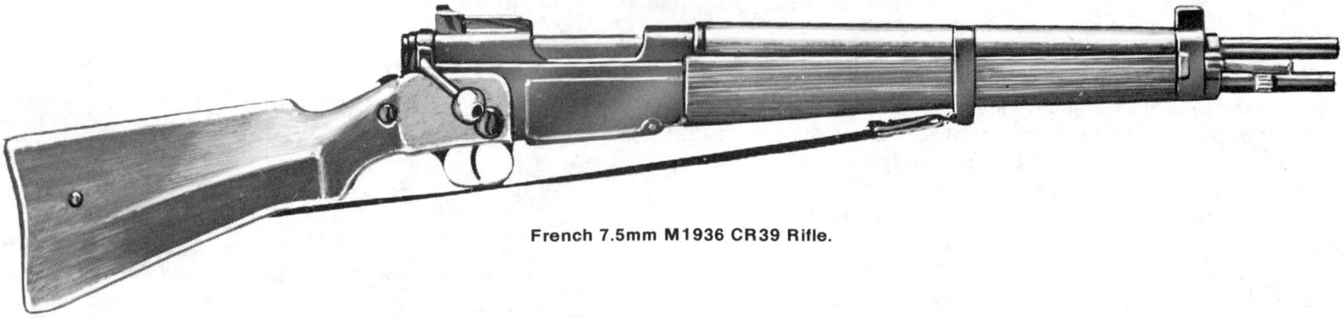

French 7.5mm M1936 CR39 Rifle.

French 8mm Rifle Model 1886 M93.

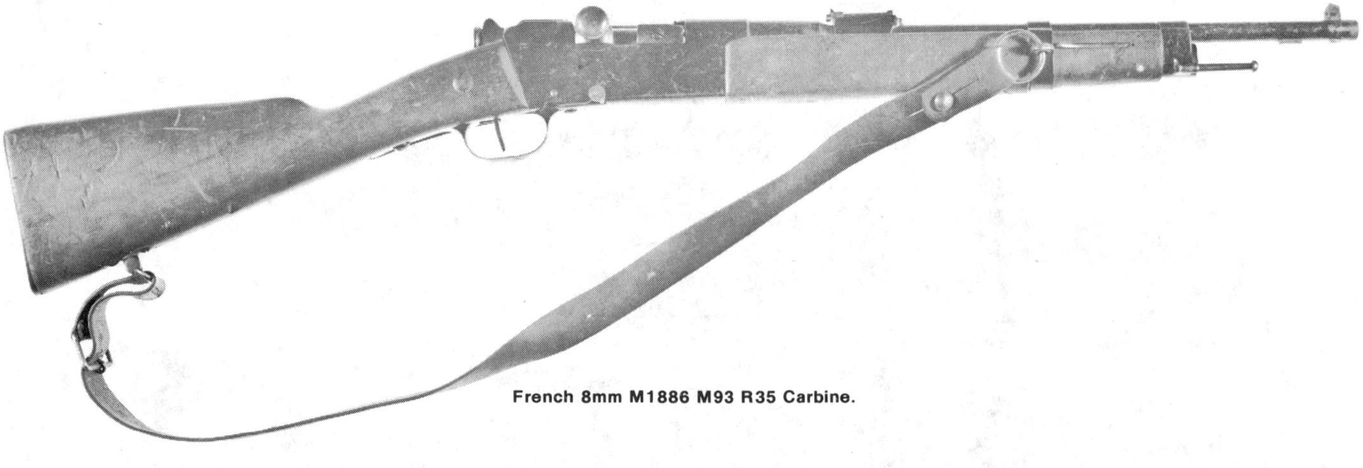

French 8mm M1886 M93 R35 Carbine.

The 7.5mm M1936 CR39 is the same as the Model 1936 but has a folding aluminum buttstock.

RIFLE M1886, M93

This rifle, commonly called the Lebel, was introduced in 1886 to use a new 8mm cartridge loaded with the then revolutionary smokeless powder of Paul Vielle.

The Lebel bolt is quite similar to that of the 11mm Gras rifle Model 1874 and the feed mechanism is similar to the 11mm Model 1878 Kropatschek rifle used by the French Navy.

The M1886 is the parent of all modern small-bore military rifles. It was slightly modified in 1893 by strengthening the receiver, boring a gas-escape hole in the bolt, change of rear sight mounting and leaf, and adding a stacking hook to the upper band. It was still in limited service in World War II, while numerous modified rifles were built around many spare parts of the old 86/93.

This is a turn-bolt action rifle of conventional design. The magazine is a tube in the forestock below the barrel, being loaded through the open action. The bolt is a two-piece design with a long detachable bolt head which carries the dual locking lugs.

Section view of Lebel M1886 action. Completion of bolt pull will eject empty and raise elevator under cartridge.

Typical 8mm M1892 Carbine as modified in 1927; 3-round capacity.

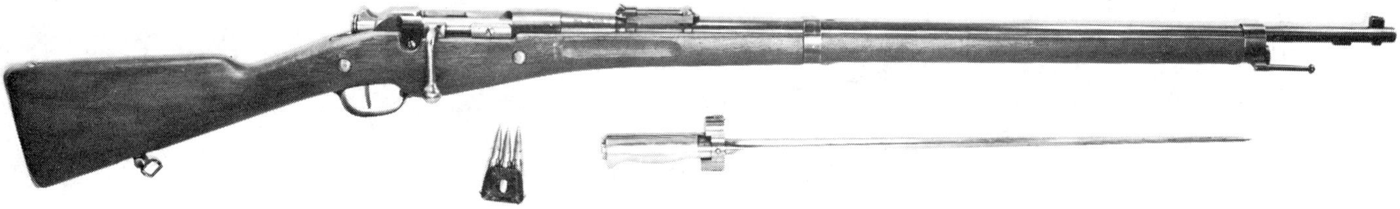

8mm Fusil 1907/15.

8mm Mousqueton (carbine) M1892.

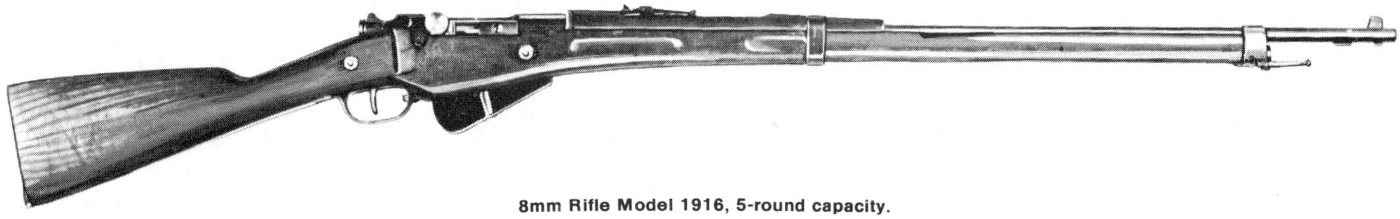

8mm Rifle Model 1916, 5-round capacity.

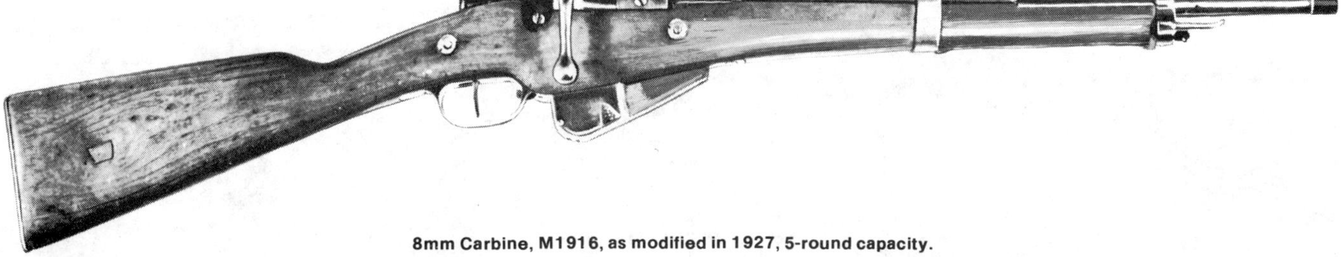

8mm Carbine, M1916, as modified in 1927, 5-round capacity.

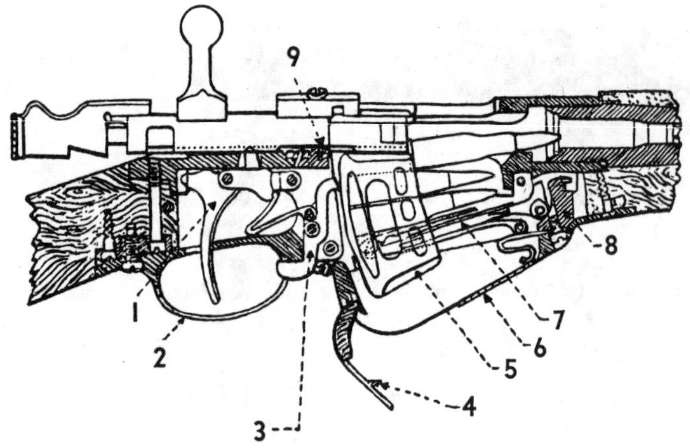

French Fusil 1916 (Drawing from French Manual).

1. Trigger
2. Trigger guard
3. Clip catch
4. Magazine plate
5. Mannlicher-type clip
6. Magazine bottom
7. Follower
8. Follower support
9. Ejector

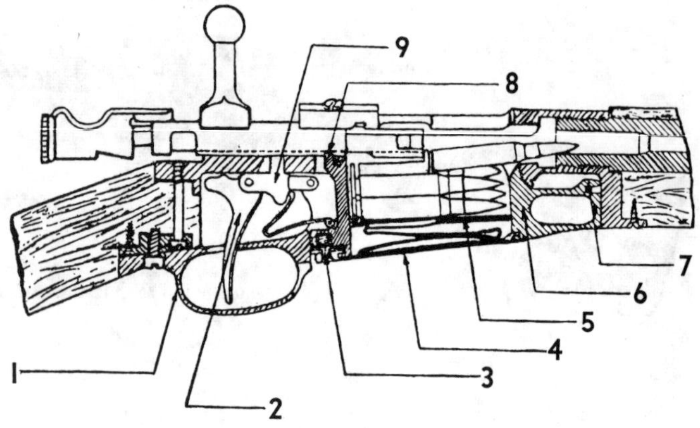

French Fusil 07/15 M34 (Drawing from French Manual).

1. Trigger guard
2. Trigger
3. Magazine plate lock
4. Bottom plate
5. Follower
6. and 7. Forward magazine section
8. Ejector
9. Sear

bolt head which carries the dual locking lugs.

To remove the bolt, the action must be opened. A holding screw is then removed from a projecting strap on top of the bolt body. The bolt body can then be pulled out of the receiver, leaving the bolt head in the boltway ahead of the receiver bridge. The head can then be picked out of the boltway. Primary extraction is given as the bolt handle is lifted by the projecting bolt strap working against a cam face in a long overhang on the top of receiver.

As the bolt is withdrawn it operates an elevator which raises a cartridge in its trough and lines it up with the bolt. A hook on the bottom of the elevator trough blocks the cartridge in the magazine tube at this point. When the cartridge is chambered, the bolt motion lowers the elevator trough, permitting the magazine spring in the tube to force the next cartridge into the trough ready to be raised on the following rearward bolt stroke.

Modification—1886 M93 R35

This carbine modification was issued in 1935. It is merely a shorter form with different furniture and smaller magazine. Model number is on receiver.

MANNLICHER BERTHIER CARBINES AND MODIFICATIONS

French 8mm Carbine M1892

This is one of the first Berthier arms. Bolt is two-piece type with removable head as in the earlier Lebel tube loader (1886). It is modified, however, to feed from a Mannlicher-type fixed box magazine.

Cartridges for these arms come in Mannlicher-style clips which are inserted in the receiver with the cartridges to form a part of the magazine action. Clips fall out bottom of receiver when last cartridge has been chambered. The three-round models use the Model 1892 clip; five round models use the Model 1916 clip.

All the Mannlicher Berthier carbines (1890, 1892, 1916) were modified in 1927. Modification consisted of removing cleaning rods and adding a stacking swivel.

French 8mm Fusil, M1907/15/34 and M1916

The French drawing shows details of Fusil (Rifle) 1916. The hinged magazine plate is opened to show how clip is dropped out of magazine when empty. Bolt has been thrust forward far enough to strip cartridge out of clip and guide bullet into chamber. This is a turn bolt rifle. This design uses the Mannlicher-clip loading system.

The 07/15 design was modified in 1934 to handle the 7.5mm rimless cartridge developed in 1929 for light machine guns. The changeover included replacing the old Mannlicher type magazine with a standard 5-shot Mauser type. Cartridges may be loaded into the magazine at any time, staggering from side to side in regular Mauser fashion. The detachable bolt head locks with two heavy lugs into a solid receiver section behind the cartridge head. The Mauser magazine permitting the use of cartridges without clips is an important feature.

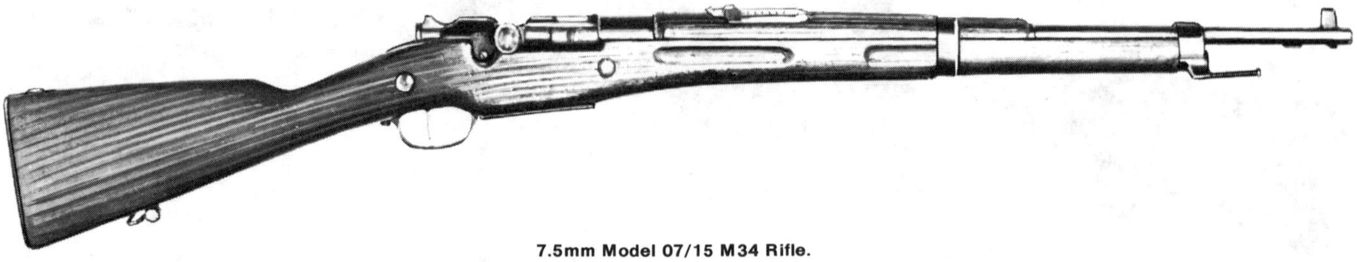

7.5mm Model 07/15 M34 Rifle.

FRENCH SEMIAUTOMATIC RIFLES

France started working on semiautomatic rifles with the Clair brothers developments circa 1888. Serious work on a military rifle began in the nineties. A total of 15 different prototype rifles were made between 1897 and 1911. These weapons were developed by a number of French military establishments in great secrecy and were developed for a number of different cartridges to include 6mm, 6.5mm, 7mm, 8mm, and 8.5mm. Gas-operated, short-recoil, and long-recoil types were represented among these weapons. Seven of the weapons were given extensive test.

In 1910, the STA8 rifle—also known as the A6 rifle—developed by Monsieur Meunier at the Technical Section of Artillery (Section Technique De L'Artillerie) was adopted. Manufacture was started in 1913, but was stopped by World War I. It is interesting, at this point in time, to consider what a nasty jolt the Germans might have gotten if the French had several hundred thousand of these rifles on hand in 1914! The STA8 rifle was long-recoil operated and chambered for a 7mm rimless cartridge.

The adoption of the automatic rifle Model 1915 CSRG slowed down development of semiautomatic rifles, but requirements from the front soon resulted in the adoption of the 8mm M1917 rifle. The 1917 rifle had many problems and was followed by the 8mm Model 1918 which was an improvement on the 1917. During this period about 1000 of the 7mm STA8 rifles were made as a possible replacement for the 1917; manufacturing problems and the logistical problem of the different ammunition stymied this plan.

Little real effort seems to have been expended in France between the wars on semiautomatic rifles, this having been the period of economy on weapons development in France. In 1935, the Model 1917 and 1918 rifles were converted to manual operation. In 1944, a semiautomatic rifle was produced at Saint Etienne—the 7.5mm MAS44. This rifle was further developed into the Model 1949. The current standard rifle, the Model 1949/56, is a modification of the Model 1949.

During the forties, several carbines were developed by Saint Etienne (MAS) and Chatellerault (MAC). The 7.65mm MAS 1949 carbine is chambered for a French developed 7.65mm "intermediate-sized" cartridge. This cartridge appeared to be modeled on the German 7.92mm short (PP43) cartridge. The MAS49 carbine has a folding stock.

The MAC 1949 carbine was chambered for the U.S. caliber .30 carbine cartridge and bears a superficial resemblance to the U.S. carbine. Internally this weapon is quite different from the U.S. carbine, since the MAC 1949 is a delayed blowback operated weapon. Both of these weapons appeared only in prototype form.

France developed a 7.62mm NATO assault rifle called the AP59. This weapon, which makes extensive use of stampings, is selective fire. The gas system has a short piston at the end of a long gas cylinder tube and a bolt system similar to that of the FN "FAL". It is fitted with a grenade launcher similar to that of the French Model 49/56 rifle and the gas cylinder is cut off when the grenade launcher sight is elevated. It is fed with a box magazine and has a horizontal fore grip similar to that of the French Model 1915 Chauchat. It's current status is unknown.

French 7.65mm Carbine MAS 1949.

The United States supplied the Free French Forces with 96,983 U.S. caliber .30 carbines during World War II and the French resistance forces were supplied with approximately 18,000 U.S. carbines. The French Army was also provided with U.S. caliber .30 M1 rifles via Military Assistance Plans after World War II.

THE FRENCH MODEL 1917 and MODEL 1918 SEMIAUTOMATIC RIFLES

The 8mm Model 1917 rifle is also known as the R.S.C. for Ribeyrolle, Sutter and Chauchat who worked on the design. Although the rifle is commonly known as the Saint Etienne in the United States, it is probable that much of the development work was done at Puteaux. The 1917 was adopted in May 1916 and mass production began in April 1917. Components were made at Manufacture d'armes de Paris, Chatellerault, Tulle, and Saint Etienne. The weapons were assembled at Saint Etienne; a total of 86,333 were made.

Service use indicated many deficiencies and broken parts were numerous. There were also complaints about the excessive weight

French .30 cal. MAC 1949 Carbine.

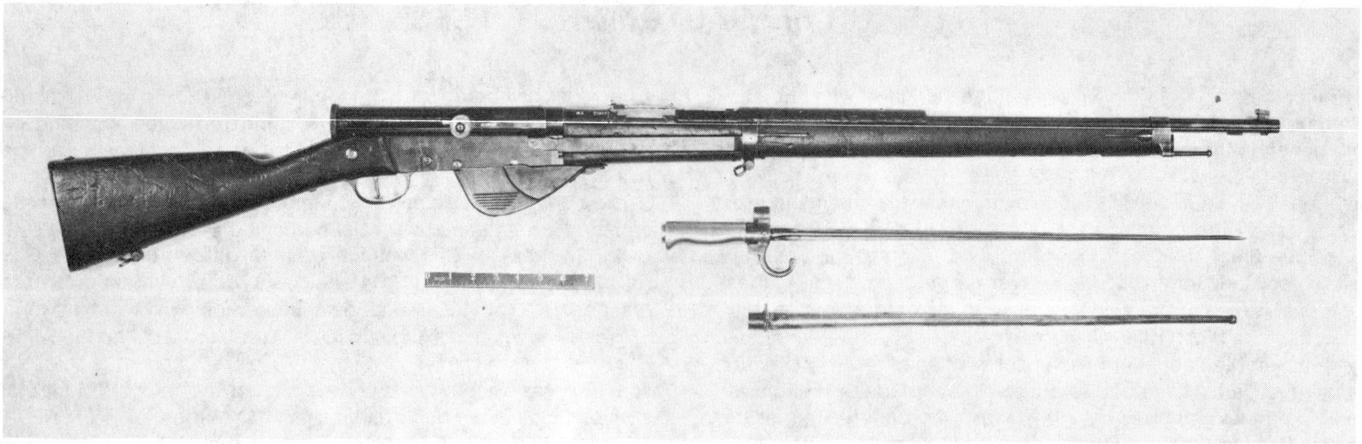

French 8mm Semiautomatic Rifle Model 1917.

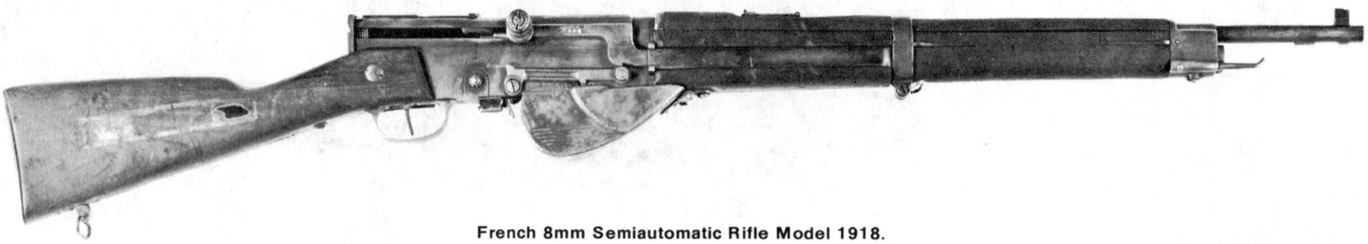

French 8mm Semiautomatic Rifle Model 1918.

of the rifle. As a result the 8mm Model 1918 rifle was developed. The 1918 is a modification of the 1917 and did not get into active service during World War I, since manufacture did not begin until November 1918.

Characteristics of the Model 1917 and Model 1918 Rifles

	Model 1917	Model 1918
Caliber:	8mm Lebel	
System of operation:	gas, semiautomatic	
Overall length:	52.4 in.	44.1 in.
Barrel length:	31.4 in.	23.1 in.
Weight:	11.6 lb.	10.5 lb.
Feed device:	5-round, non-detachable, in-line box magazine	
Sights: Front:	Blade	Blade
Rear:	Leaf	Leaf
Muzzle velocity:	2380 f.p.s.	2180 f.p.s.

Operation of 8mm Semiautomatic Rifles M1917 and M1918

The operation of both of these rifles is basically the same. The piston is attached to an operating rod which is attached in turn to the bolt carrier. The bolt, which has an interrupted thread-type head (multiple) lug, rides in spiral splines in the bolt carrier (the bolt carrier could also be considered the bolt body of a two-piece bolt).

Five-round Mannlicher-type clips are used to feed these weapons. The clips remain in the action (as with the U.S. M1 rifles) and are necessary for semiautomatic fire. The Model 1917 uses a special clip and the Model 1918 uses the standard French 8mm Model 1916 clip, which is also used with the Model 1916 Berthier bolt-action rifles.

In 1935, both of these rifles were converted into manually-operated straight-pull rifles by the welding of a block under the gas port. Their model designations in this form are Rifle Model 1917-1935 and Model 1918-1935. Both rifles are obsolete.

THE MODEL 1949 AND 1949/56 RIFLES

These gas-operated weapons show considerable designing skill. They are relatively simple in operation and are very easy to strip and clean. They use a tilting bolt system, and have no moving parts in the gas system. The forward part of the barrel is used as a grenade launcher on both models and both have built-in

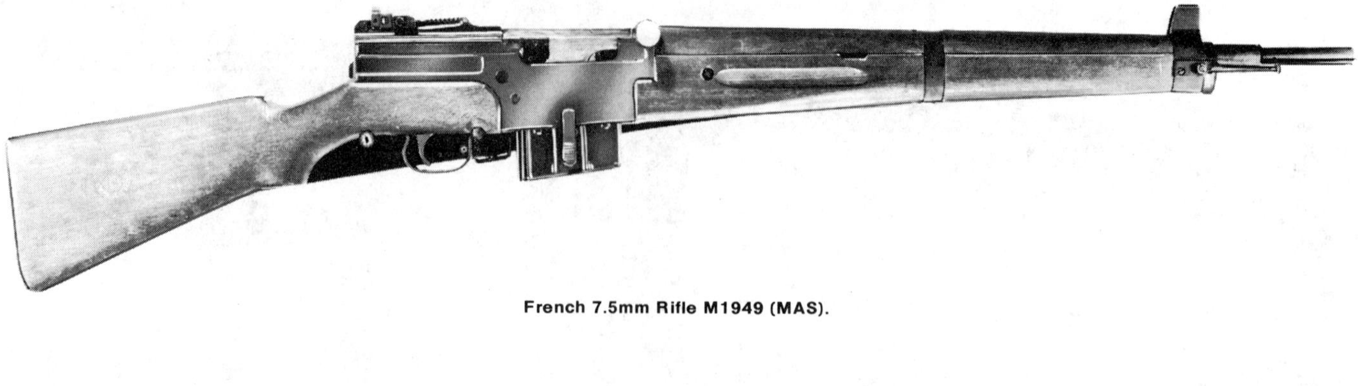

French 7.5mm Rifle M1949 (MAS).

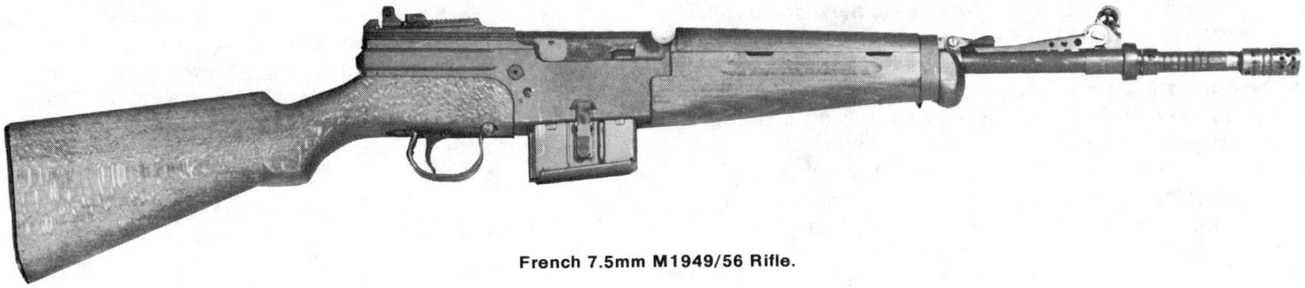

French 7.5mm M1949/56 Rifle.

grenade launcher sights. On the M1949 the grenade launcher sight is on the left front side of the fore-end; on the M1949/56, the grenade launcher sights are on top of the barrel, to the rear of the front rifle sight. Both of these weapons have two-piece stocks, as did the M1936 series of weapons. The M1949 does not use a bayonet, but the M1949/56 does.

Characteristics of M1949 and M1949/56

	M1949	M1949/56
Caliber:	7.5mm French M1929 cartridge, for both models.	
System of operation:	Gas, semiautomatic fire only, for both models.	
Weight:	10.4 lb.	8.6 lb. w/o magazine.
Length, overall:	43.3 in.	43.4 in.
Barrel length:	22.8 in.	20.7 in.
Feed device:	10-round, detachable, staggered-row box magazine, normally loaded with 5-round chargers: for both models.	
Sights: Front:	Blade w/protecting ears, for both models.	
Rear:	Ramp w/aperture, for both models.	
Muzzle velocity:	2705 f.p.s.	2700 f.p.s.

How to Load and Fire the M1949-Series Rifles

The weapon can be loaded by pulling the operating handle to the rear and filling the magazine, using chargers (clips) in a manner similar to that used with the U.S. M1903 rifle; or the magazine may be removed (note that the magazine catch is a part of the magazine, not of the weapon), and then filled by hand. If the bolt is to the rear, it will require a short rearward jerk to release the bolt latch—the bolt remains open after the last round is fired—allowing the bolt to go forward and chamber a cartridge. If the bolt is forward, it will have to be pulled to the rear and released. The safety is located at the right front side of the trigger-guard. Upon firing the last shot, the bolt will remain open.

How to Field Strip the M1949-Series Rifles

Remove magazine by depressing magazine catch on right side of magazine and pulling magazine down and out of the receiver. Release receiver cover by depressing the latch, located at the rear of the receiver; push the cover forward slightly and lift it from its grooves in the receiver. **CAUTION:** The bolt must be in the forward position before attempting to remove the receiver cover. (When the bolt is to the rear, the recoil spring, which is housed in the receiver cover, is compressed.) The bolt and bolt carrier can now be lifted out of the receiver. The trigger group can be removed by removal of one screw. No further disassembly is recommended.

To reassemble the weapon perform the above steps in reverse.

How the M1949-Series Rifles Work

When a cartridge is fired, propellent gas is tapped off through tbe gas port in the barrel. This gas blows back through the gas tube directly into a hole in the top face of the bolt carrier. The gas

M1949 Rifle, field-stripped.

tube protrudes a small distance, so that a portion of it actually enters the bolt carrier. The bolt carrier moves to the rear and, after a slight "dwell" time, the bolt is cammed up and out of its locked position in the bottom of the receiver, and starts to travel back with the bolt carrier. The ejector is a pin type that protrudes from the face of the bolt like that of the U.S. M1 rifle; it ejects the spent case when the case mouth clears the face of the barrel. The trigger mechanism is quite similar to that of the U.S. M1 rifle, and operates in a similar fashion. The bolt and bolt carrier compress the recoil spring, which decompresses at the end of their travel and returns the bolt and carrier to the battery position. On the return of the bolt and its carrier, the bolt strips the top cartridge from the magazine and feeds it into the chamber. The bolt carrier cams the lugs of the bolt down, and the rear end of the bolt is brought down into engagement with the locking bar in the bottom of the receiver. The weapon is ready to fire again.

Special Note on the M1949-Series Rifles

The gas system of the M1949-series rifles is similar to that of the Swedish Ljungman Model 42. Its tilting bolt (or propped breech, as it is sometimes called) has been used in a great number of systems, e.g., the Tokarev, the FN, the Ljungman, and so on. These weapons are an illustration of the fact that a good modern weapon can be produced by selectively choosing from proven past designs and modifying them to suit the need.

Both of these weapons have mounting grooves, cut on the left side of the receivers, for the mounting of telescopic sights.

THE 7.5mm MODEL F 1 RIFLE

This bolt action rifle is made in 3 versions: Version A—a sniper rifle with scope and bipod for use by infantry; Version B—designed for use in competitive shooting limited to iron sights. It is equipped with micrometer type rear sight and a tunnel type front sight with changeable inserts; Version C—a sporting rifle with scope.

The action is basically the same as that of the MAS 36; the trigger mechanism has been modified to allow adjustment and a 10-round, box type magazine is used. All versions have a muzzle brake and a rubber recoil pad. The length of the butt stock can be adjusted by removing or adding pieces.

Characteristics of the 7.5mm Model F 1 Rifle

Caliber: 7.5mm
System of operation: Manually operated bolt action
Length overall: 44.8 in.
Barrel length: 22.8 in.
Feed device: 10-round, staggered row, detachable box magazine.
Sights: Front: Tunnel type.
Rear: Telescope or micrometer.
Weight: 9.9 lb.
Muzzle velocity: Approx. 2700 f.p.s.

This rifle has been made in 7.62mm NATO as well as 7.5mm French. The French claim that this rifle will consistently group 10 rounds into a circle smaller than 7.8 inches at 200 meters when used with good ammunition.

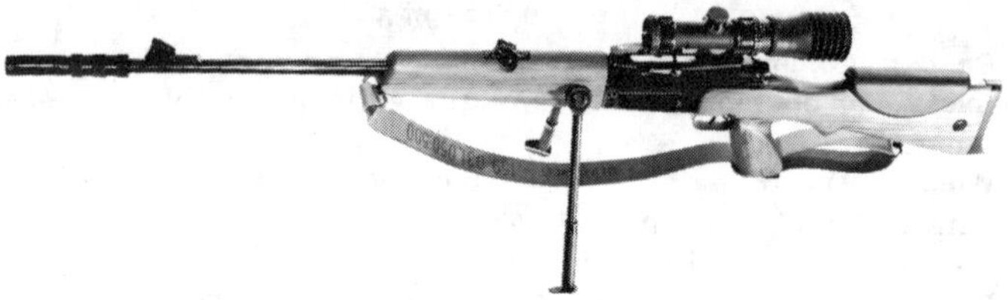

French 7.5mm Model F 1 Rifle.

FRENCH SUBMACHINE GUNS

France used the German 9mm Parabellum Vollmer Erma submachine gun to a limited extent prior to 1941. French development of a native submachine gun began at MAS (Saint Etienne) during the thirties. In 1935 the first of the MAS weapons developed for the 7.65mm long pistol cartridge appeared. This weapon, called the 7.65mm L Type SE-MAS 1935, was quite similar to the later and more common MAS 1938.

The 7.65mm long MAS 1938 was the standard French submachine gun until 1949. It is still in wide use with French police forces. After World War II, submachine gun development was very active in France. The Hotchkiss firm developed a submachine gun which had some overseas sales and limited service use in Indo-China.

The French government arsenals were also active in this line

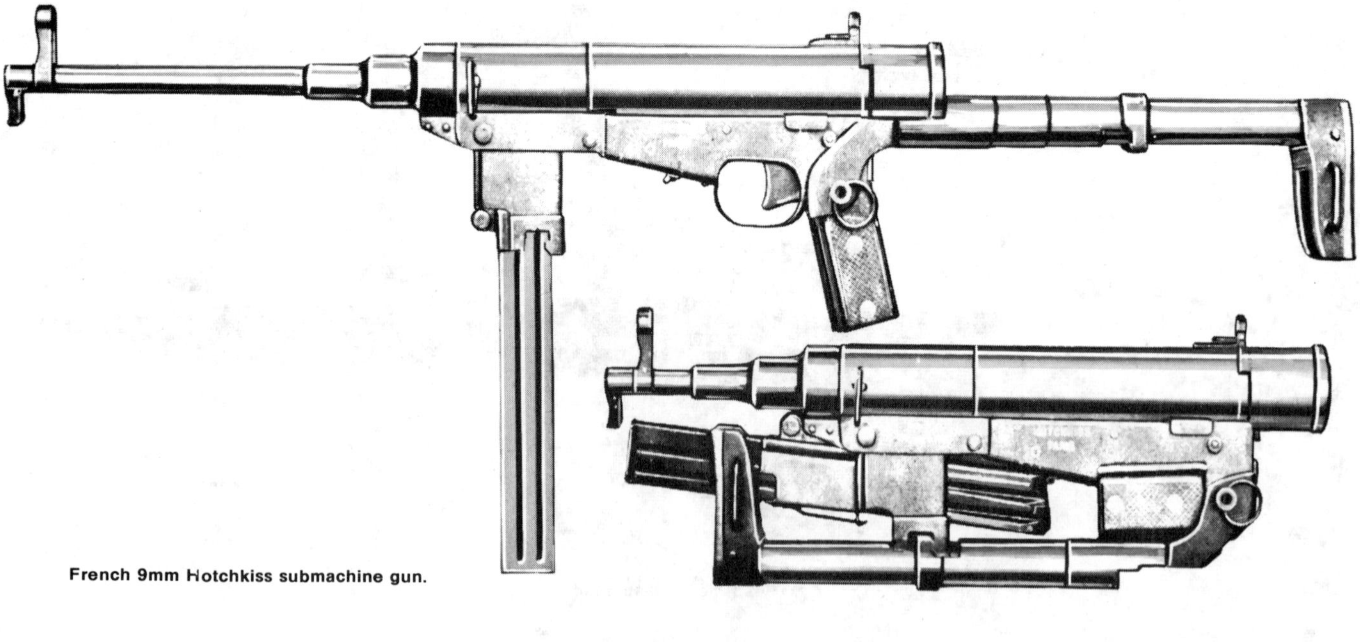

French 9mm Hotchkiss submachine gun.

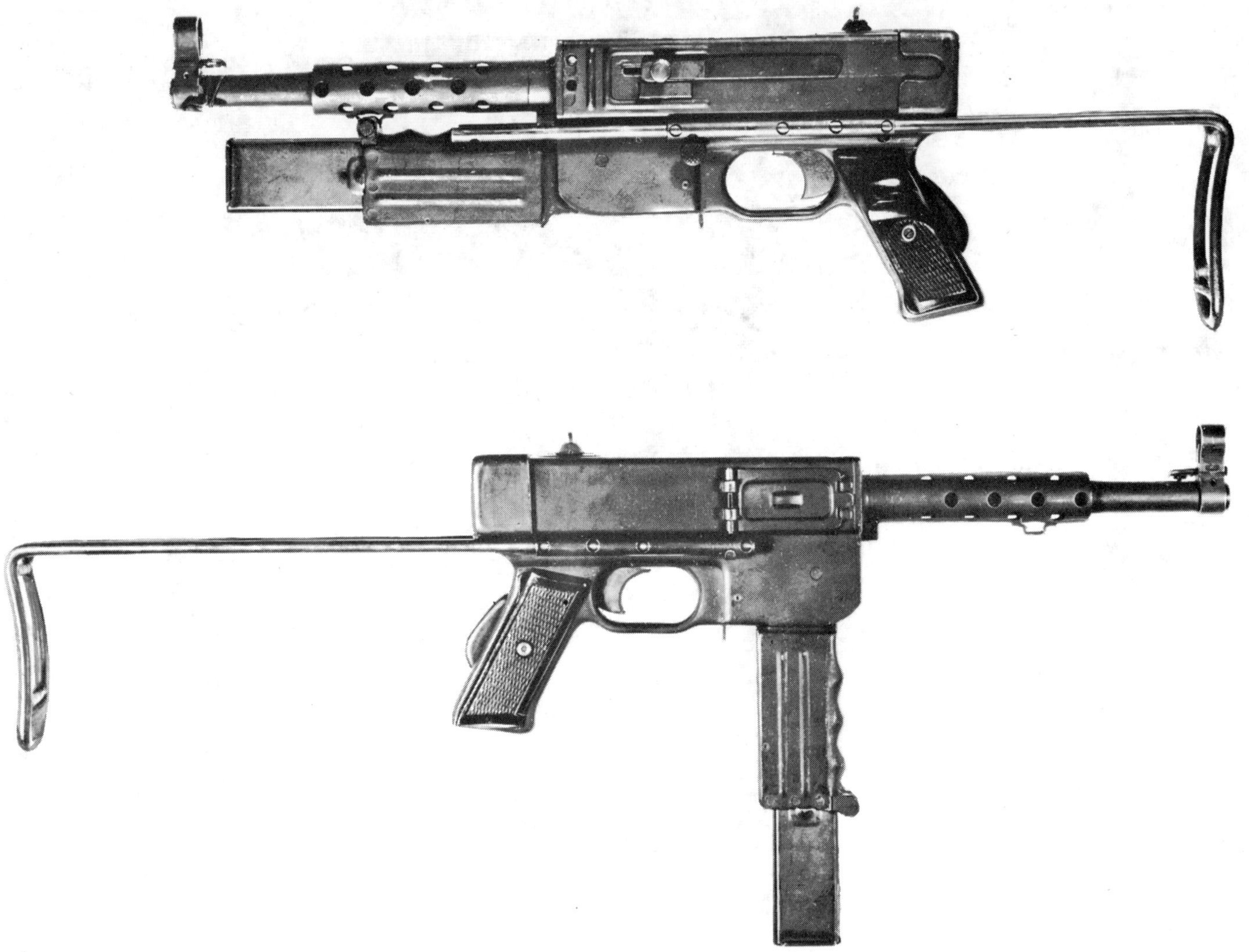

French 9mm Submachine Gun M1949 (MAT 49).

and a number of designs were produced. A 9mm Parabellum weapon was built by Tulle (MAT) in 1948 and the present service weapon the MAT Model 1949 was adopted.

The United States provided the Free French Army with 20,856 Thompson submachine guns during World War II. The French underground was supplied with several thousand 9mm U.S. (Marlin) submachine guns, in addition to British Stens and other types of submachine guns.

THE FRENCH MODEL 1949 SUBMACHINE GUN

The M1949 submachine gun, which has seen extensive service in Indo-China and Algeria, is well made and of excellent design. It has a very good reputation with French troops. It has a telescoping steel stock, and magazine which folds up under the gun when not in use; this makes the weapon very handy for use by armored or airborne troops. The M1949 also has a grip safety and an ejection port cover. The grip safety prevents the gun from firing when accidentally dropped, and the ejection port cover helps to keep dirt out of the internal mechanism of the gun.

Characteristics of the M1949 Submachine Gun

Caliber: 9mm Parabellum.
System of operation: Blowback, automatic fire only.
Length, overall:
  Stock extended: 28 in.
  Stock retracted: 18.3 in.
Barrel length: 9.05 in.
Feed device: 32-round, detachable, staggered-row box magazine.
Sights: Front: Hooded blade.
  Rear: L-Type w/apertures for 100 and 200 meters.
Muzzle velocity: 1237 f.p.s.
Cyclic rate: 600 r.p.m.
Weight, loaded: 9.41 lb.

How to Load and Fire the M1949 (MAT 49) Submachine Gun

Insert a loaded magazine in the well of the magazine housing; if the housing is in the horizontal position, swing it down to the vertical, making sure that the lock located on the underside of the

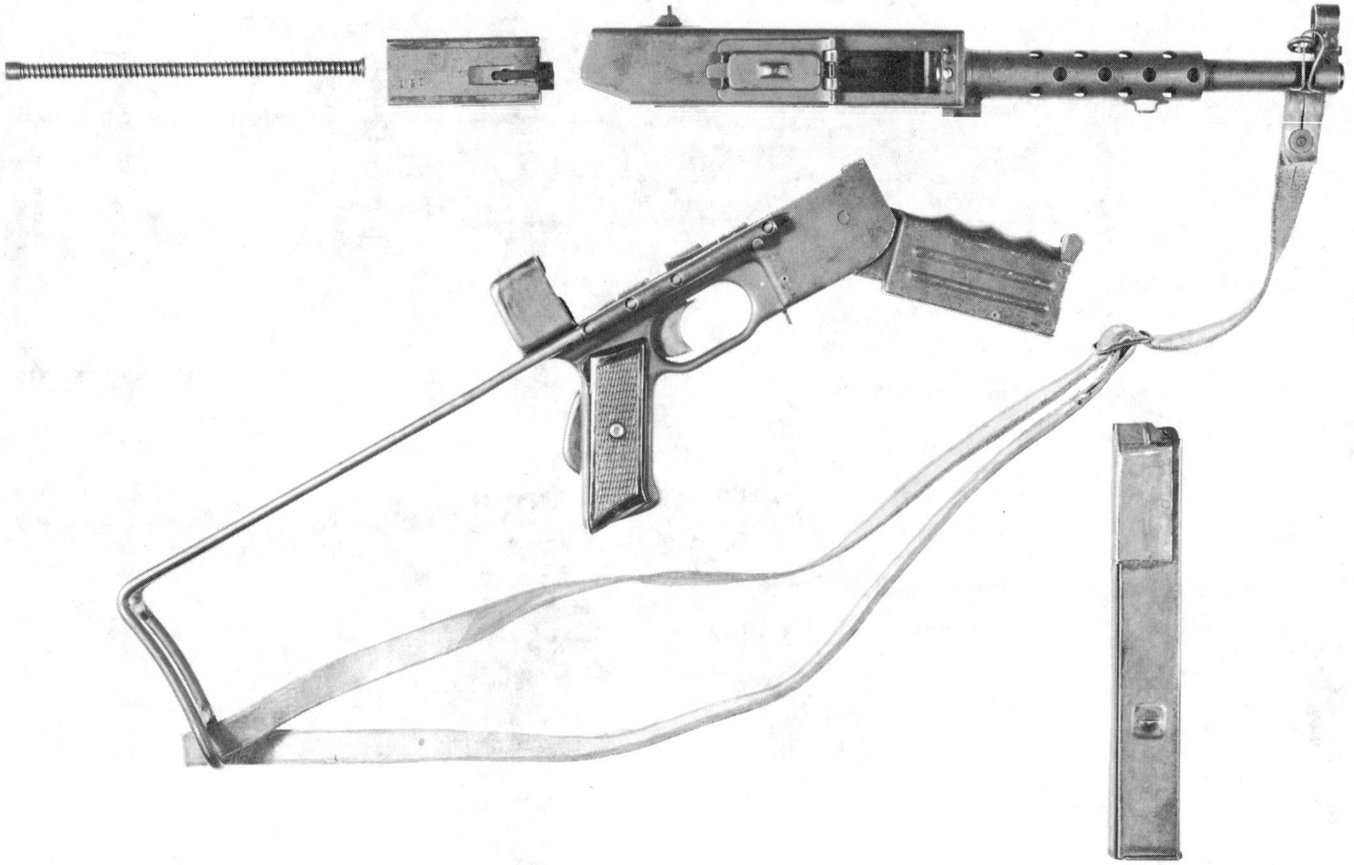

M1949 (MAT 49) Submachine Gun, field-stripped.

trigger housing engages the magazine housing. Pull the cocking handle, located on the left side of the receiver, to the rear. The bolt will remain to the rear, since this weapon fires from an open bolt. Push the cocking handle to its forward position. Squeeze the grip safety with the rear of the hand, pull the trigger, and the weapon will fire. To remove the magazine, engage the magazine catch located at the bottom rear of the magazine housing. To lift the magazine housing and magazine into the horizontal position under the barrel, depress the lock located on the underside of the trigger housing and swing the magazine up till the lug at the forward end of the housing engages the clip on the underside of the barrel jacket. To change the position of—or remove—the stock, depress the catch located on the left side of the trigger housing.

How to Field Strip the M1949 (MAT 49) Submachine Gun

Remove the magazine and clear weapon. Pull trigger and let bolt go forward. Press the knurled bar located on the rear section of the barrel jacket to the rear. The barrel and receiver assembly can now be pulled upward and forward off the frame. The operating spring and bolt can be removed from the rear of the receiver. Further disassembly is not recommended.

Special Note on the M1949 (MAT 49) Submachine Gun)

The M1949 submachine gun is composed mainly of steel stampings. Since its functioning is basically the same as that of any other blowback-operated submachine gun, no extended coverage on functioning is given. The ejection port cover is automatically opened by the forward or rearward movement of the bolt.

THE FRENCH MODEL 1938 SUBMACHINE GUN

The 7.65mm Long Model 1938 submachine gun, or MAS 38 as it is commonly known, is a relatively simple blowback-operated weapon. It does have a few unusual features however: the use of a folding trigger for a safety and the angular travel of the bolt.

Characteristics of the Model 1938 Submachine Gun

Caliber: 7.65mm Long.
System of operation: Blowback, automatic fire only.
Length overall: 24.8 in.
Barrel length: 8.8 in.
Weight: 6.3 lb.
Feed device: 32-round, detachable staggered row, box magazine.
Sights: Front: Block with notch.
Rear: Two folding leaves, 100 and 200 meter.
Muzzle velocity: Approx. 1200 f.p.s.
Cyclic rate: 700 r.p.m.

How the MAS Model 38 Operates

A hinged cover seals the mouth of the magazine housing in the bottom of the receiver in this weapon when the magazine is withdrawn. Pushing this forward on its hinge opens the mouth of the

French 7.65mm Model 1938 Submachine Gun.

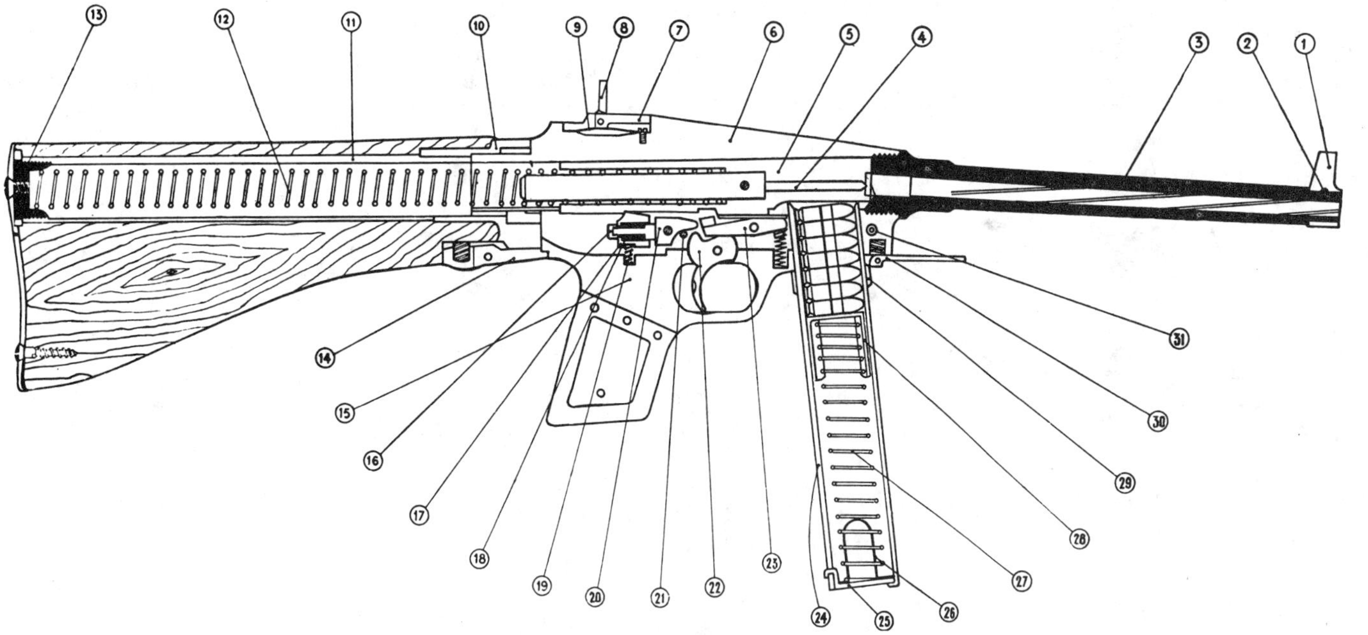

MAS 1938 Submachine Gun (Drawing from French Manual).

1. Front sight.
2. Fixing pin.
3. Barrel.
4. Striker.
5. Breechblock (bolt).
6. Receiver.
7. Rear sight leaf, down position.
8. Rear sight leaf, up.
9. Spring.
10. Recoil spring tube bridge.
11. Recoil spring tube.
12. Recoil (or operating) spring.
13. Tube butt cap.
14. Takedown lever.
15. Grip.
16. Sear nose.
17. Sear.
18. Sear buffer seat.
19. Sear spring.
20. Sear bar.
21. Stop pin.
22. Trigger.
23. Safety lever.
24. Magazine.
25. Floorplate.
26. Spring guide.
27. Spring.
28. Follower.
29. Magazine stop.
30. Opening plate (used to close magazine opening when magazine is withdrawn).
31. Magazine release.

housing and permits insertion of a loaded magazine from the bottom. The magazine is pushed in until it locks in standard fashion.

When the bolt handle on the right side of the receiver is pulled to the rear, it draws the bolt back and presses the recoil spring behind it which extends from the rear of the receiver down into a steel tube inside the shoulder stock. When the bolt has been drawn back far enough, it rides over the sear which is then forced up by its spring to catch in the bent or notch in the underside of the bolt and hold it open ready for firing.

Pressure on the trigger will now withdraw the sear from the bolt and permit the compressed recoil spring to drive the bolt up the slightly inclined surface machined into the receiver, stripping a cartridge from the magazine and thrusting it into the chamber.

Notes on the MAS Model 38

The bolt is not a true cylindrical piece moving in a straight line as in most submachine guns. It has a somewhat tilted face to permit it to bring up flush with the head of the cartridge as it is chambered, while the main body of the bolt is at an angle to the bore; this tends to give the effect of a hesitation action.

Just how practical this system is is open to question, in view of the fact that even with a low-powered cartridge and a comparatively light bolt, the recoil action is somewhat stiff. The bolt not only travels to the rear in the receiver, but actually passes down the tube inside the stock itself. During rearward movement, the extractor withdraws the empty case and strikes it against the ejector which is part of the fastening pin to which the forward sling swivel is attached. The empty case is ejected from the port on the right hand side of the receiver.

The sliding cover over the bolt opening or ejection port should be pulled further to the rear after the weapon has been cocked so that it will not have to be carried forward by the bolt in forward travel. This is a full automatic weapon in which the action will continue to function so long as the trigger is held back. When the trigger is released, a spring will force the sear up to catch and hold the bolt to the rear. The weapon fires from an open bolt.

The sights are rough and are of a rather impractical design. The rear sight consists of two separate leaves which fold down into the top of the receiver. One aperture is for one hundred meters and the other is for two hundred meters. The front sight is offset to the left and is too clumsy to permit of anything but typical submachine gun burst fire.

Note that this weapon cannot be fired when the shoulder stock is removed. Additional travel of the bolt inside the tube in the stock is necessary to function the weapon.

Magazine. The magazine catch which is pressed in to release the magazine for withdrawal from the receiver is on the left side of the housing. The magazine itself may be dismounted by pushing the protruding rib at the lower front end of the box. This will tilt the floor plate so that it can be slid out rear end first. The zigzag spring and the follower may be withdrawn from the magazine box.

Design. An unusual safety is incorporated in this weapon. When the trigger is pushed forward it folds up and hence cannot be pressed. Thus, if the bolt is cocked, the weapon cannot be fired until the trigger is deliberately unfolded. On the other hand, if the bolt is in forward position and the chamber empty, pushing the trigger forward will force a catch up into a hole in the bolt and prevent the bolt handle from being pulled back to open the weapon.

Field Stripping the MAS

Push in the spring buttstock catch at the under side of the forward end of the stock and, as it releases, twist the butt to the right.

Ease the butt and the recoil or operating spring out of the receiver.

Pressing the trigger will lower the sear and permit the trigger-guard unit to be withdrawn from its guide in the receiver.

Pull back the bolt handle which will bring the bolt to the rear when it and the recoil spring may be withdrawn from the receiver.

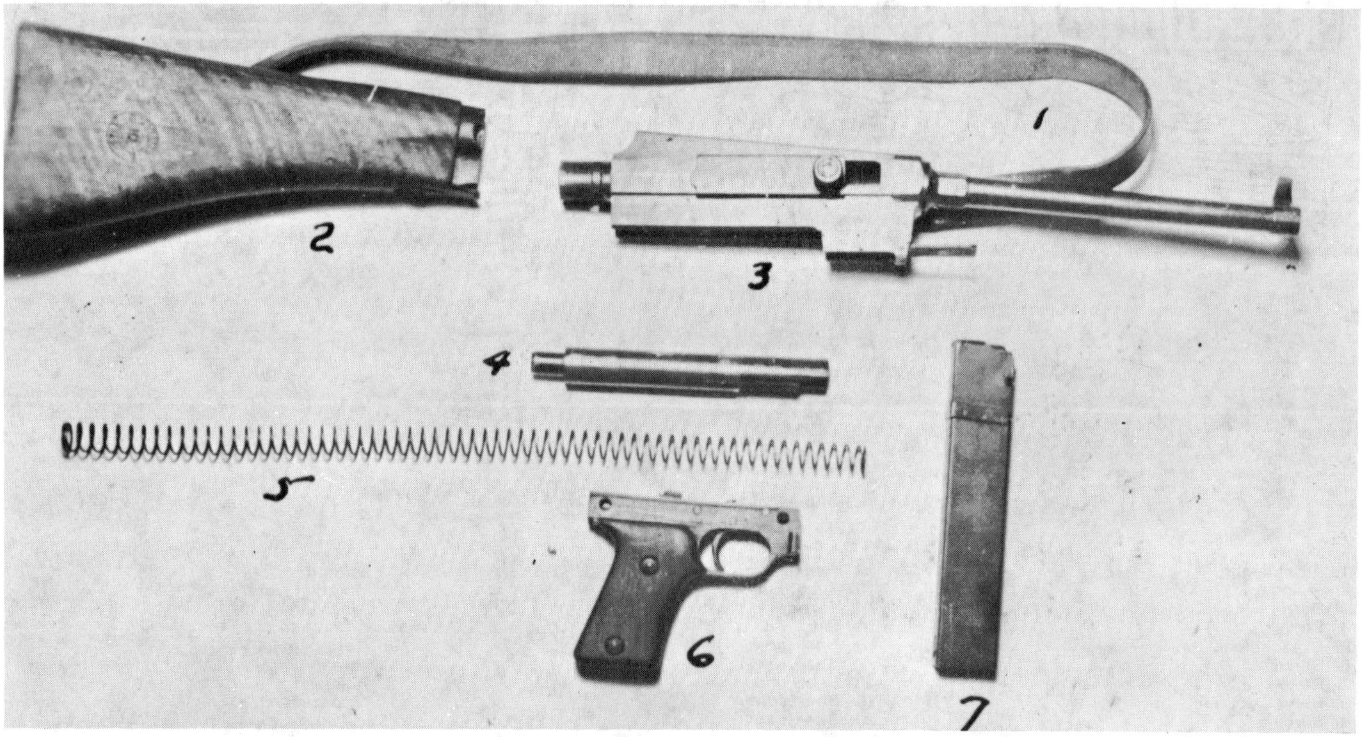

The stripped MAS.

FRENCH MACHINE GUNS

France adopted her first true machine gun in 1897—the Hotchkiss. Other guns soon followed: the Models 1900 and 1914 heavy Hotchkiss, 09/13 light Hotchkiss (Benét-Mercié) the arsenal-designed guns, Puteaux of 1905 and the Saint Etienne of 1907 and its modification, the 1907 T. The Lewis gun was used as an aircraft gun as was the Darne. The Hotchkiss was also made in 11mm during the war.

After World War I, the 7.5mm Chatellerault Model 1924 appeared. This weapon, which was apparently the French reaction to the Model 1918 Browning Automatic rifle, was modified for the

French 7.62mm Type MAS 1950 Machine Gun.

French 7.5mm M1931A Machine Gun with drum magazine.

new 7.5mm cartridge in 1929 to become the Automatic Rifle Model 1924 M29. This weapon is still in wide use. The basic operating system of the Chatellerault is used with the later M1931 tank gun and the Chatellerault aircraft machine gun. The weapons of World War II, prior to the fall of France, were the 8mm Hotchkiss Model 1914 and the 7.5 Chatellerault in its various models. The 13.2mm Hotchkiss was adopted in 1931.

During World War II, the Free French Forces were supplied with British Bren and Vickers guns and 10,731 U.S. Browning guns. At the conclusion of the war, German 7.92mm MG34 and MG42 machine guns were used until adequate quantities of French guns could be provided. France has developed a general purpose machine gun, the 7.5mm Model 52 (AAT) since World War II. A number of other weapons appeared in prototype form, but were not adopted by France.

French 7.5mm Machine Gun M1931A on U.S. M2 tripod.

The 7.62mm Type MAS 1950 was made for the U.S. caliber .30-06 cartridge. It is a gas-operated weapon, which can be fed from either a box magazine or a link belt. It appeared only as a prototype.

FRENCH MODEL 1931A MACHINE GUN

The French 7.5mm tank and fortress machine gun M1931A, was adapted for use on the U.S. cal. 30 M2 tripod because of French post-war shortage of machine guns. This weapon is still in use on French-made armored vehicles. Since it is likely to be encountered by some of the readers of this book, it will be covered in detail even though it is not a post-war weapon.

The M1931A machine gun is the tank and fortress version of the French 7.5mm M1924 M29 light machine gun (commonly known as the Chatellerault). The weapon was adapted to the U.S. M2 mount after World War II, but is only likely to be encountered on French armored vehicles these days. A French tripod mount, the MAS M1945, was also used with the gun. The weapon has a very heavy barrel, since it is designed for a high rate of sustained fire and does not have a quick-change barrel. Two magazines can be used with the gun—a box type or a large drum which fastens to the side of the gun.

Characteristics of M1931A Machine Gun

Caliber: 7.5mm M1929 French rimless.
System of operation: Gas, automatic fire only.
Length, overall: 40.5 in. w/flash hider.
Weight, empty: 27.48 lb.
Weight of Mounts:
Tripod M2: 16.22 lb.
Tripod M1945: 32.15 lb.
Barrel length: 23.5 in.
Feed device: 36-round, detachable, staggered-row box magazine, or 150-round drum magazine.
Sights: Front: Blade.
Rear: Tangent leaf w/open notch.
Muzzle Velocity: 2750 f.p.s.
Cyclic Rate: 750 r.p.m.

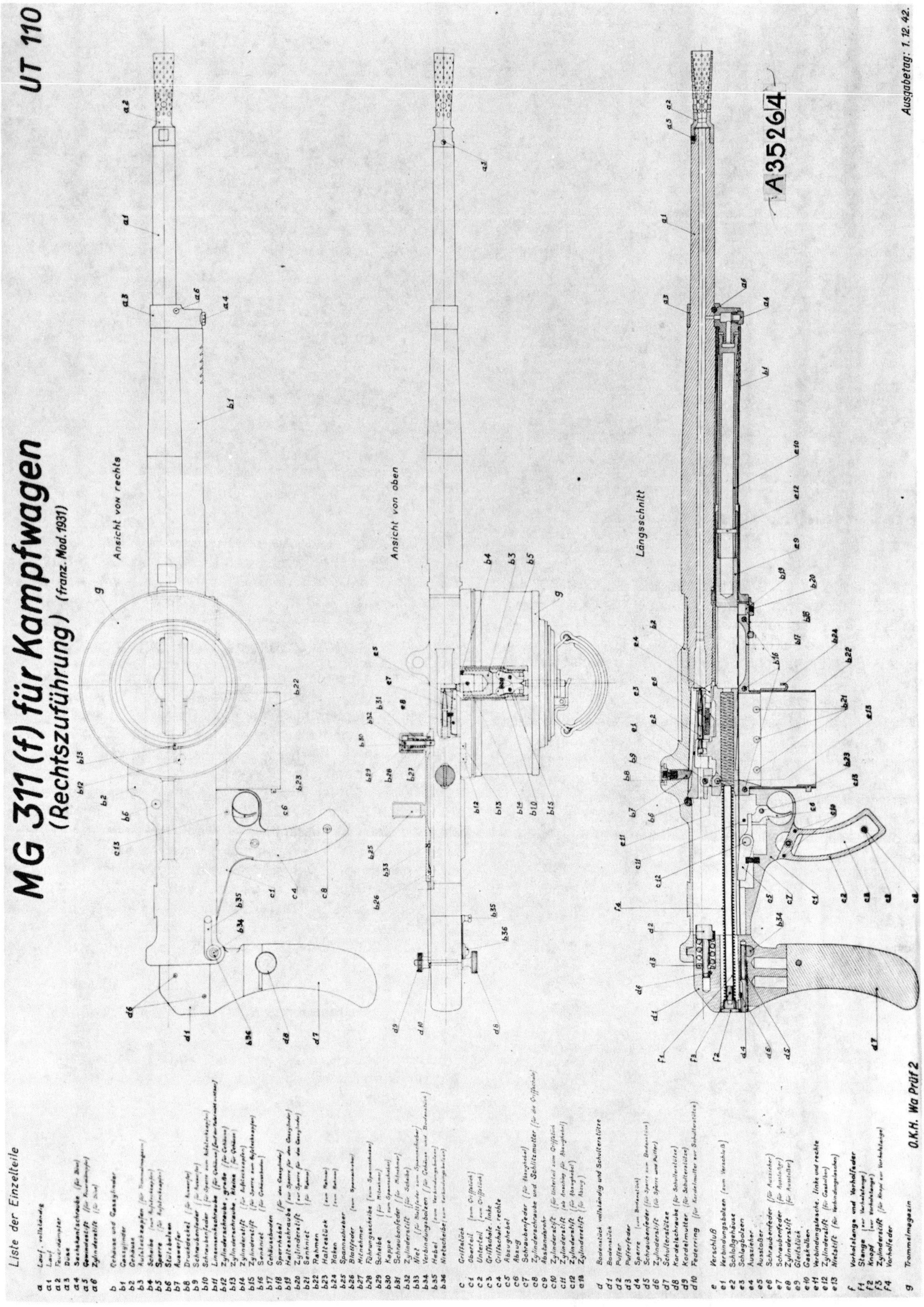

German WW II training chart on M1931A.

How to Load and Fire the M1931A Machine Gun

Cock the gun by pulling the operating handle to the rear. If the box magazine is to be used, rotate it into the feed fixture on the right side of the gun, so that the lug on the magazine engages in the recess in the fixture. Then pull the magazine toward the rear of the gun, so that the rear of it is locked in place in the feedway. Pull the trigger and the weapon will fire. This weapon fires from an open bolt and has no safety. To load the drum-type magazine, fit the magazine on the projection of the receiver which is just ahead of the feedway. A stud to the rear of the projection serves as an index when assembling the magazine to the gun. Firing is carried on in the same way with the drum as with the box magazine. To remove the box magazine, engage the magazine catch and swing the magazine forward and out. To remove the drum magazine, pull the handle on the outside of the drum, and pull the magazine straight off. When this gun is unloaded, using either magazine, there may be a round in the feedway, unless all the rounds have been fired. To clear the gun, the round must be pushed forward so that it falls free inside the receiver and out the bottom of the gun. USE EXTREME CARE IN UNLOADING THE GUN! DO NOT ALLOW ANYONE TO GET IN FRONT OF THE GUN!

How to Field Strip the M1931A Machine Gun

Check to see that the weapon is not loaded. Depress the spring-loaded catch in the rear of the backplate. Rotate the lock 90 degrees, and remove the recoil spring and guide. Remove the screw at the lower rear of the receiver; a lever attached to the screw assists in its removal. Remove the backplate; the trigger housing and recoiling parts can then be removed. No further disassembly is recommended.

To reassemble the weapon, perform the above steps in reverse order.

How the M1931A Machine Gun Works

The piston and gas cylinder are similar to those on the M1924 M29 light machine gun. However, because the M1931A ejects through the bottom of the receiver, the piston assembly is cut away to permit cases or complete rounds to pass through. The operating spring is positioned at the rear of the piston assembly, with only a small portion of it inside the piston. Locking is accomplished in the same manner as on the M1924 M29 light machine gun. A spring-loaded lever, positioned in the right side of the bolt, engages the base of the cartridge case for feeding and firing. The ejector operates in a groove in the top of the bolt. The firing pin operates inside the bolt, but is held rigidly to a projection on the rear of the piston assembly.

Special Note on the M1931A Machine Gun

The M1931A cannot be considered an up-to-date machine gun for use either as a tank gun or as a ground gun. It probably will be replaced by the M1952 machine gun in armored vehicles, as well as in the ground role.

FRENCH MODEL 52 MACHINE GUN

The M1952 is called the Arme Automatique Transformable (AAT) by the French, since it is designed to be used as both a heavy and light machine gun. The weapon can be used as a light machine gun with a light barrel, a bipod, and a butt support. It can also be used as a heavy machine gun on a U.S. M2 tripod with a French adapter. This gun has been made in 7.62mm NATO caliber for test and for export.

Characteristics of the Model 52 Machine Gun

Caliber: 7.5mm French M1929 rimless.
System of operation: Delayed blowback, full automatic fire only.
Length, overall:
Stock extended: 45.9 in.
Stock retracted: 38.6 in.
Barrel length (Quick-change barrel):
Heavy barrel, w/o flash hider: 23.6 in.
Light barrel, w/o flash hider: 19.3 in.
Feed device: 50-round, metallic, nondisintegrating link belt; when used as a light gun, a box containing one link belt section is hung on the side of the receiver.
Sights: Front: Barleycorn.
Rear: Tangent leaf w/U-notch.
Muzzle velocity: 2690 f.p.s.
Cyclic rate: 700 r.p.m. (approx.).
Weight, w/light barrel: 21.7 lb.
w/heavy barrel: 23.28 lb.

How to Load and Fire the M1952 (AAT Mle. 52) Machine Gun

Pull the operating handle to the rear, cocking the weapon. Push the cover catch forward and open the cover; lay the belt on the feedway, so that the first cartridge in the belt is positioned against the cartridge stop. Close the cover and press the trigger; the weapon will fire. An alternate method is to feed the loading tab of the belt into the feed port on the left side of the gun, and pull it through to the right until a cartridge clicks into position. Then pull the charging handle to the rear and fire. The weapon fires from an open bolt.

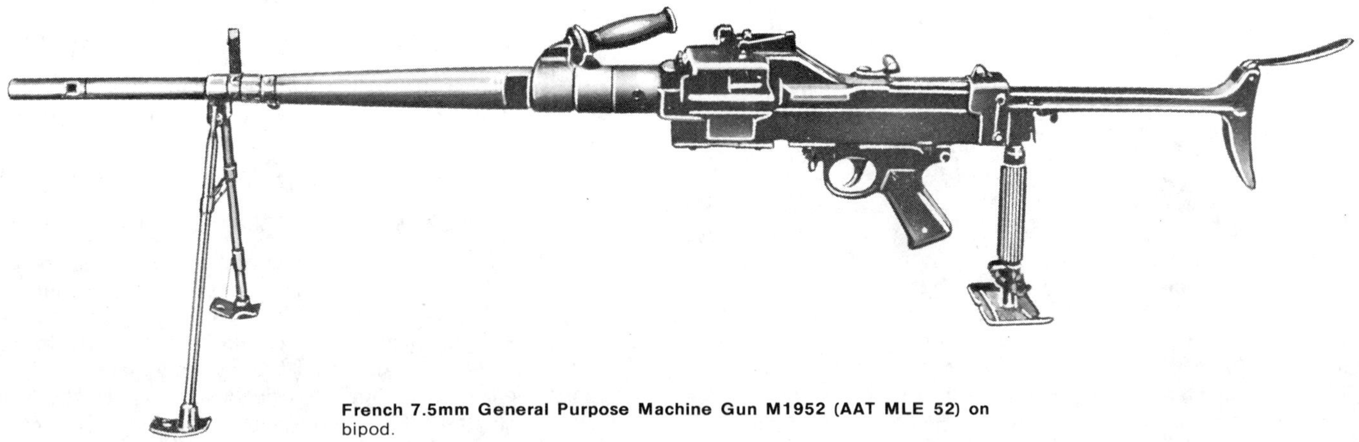

French 7.5mm General Purpose Machine Gun M1952 (AAT MLE 52) on bipod.

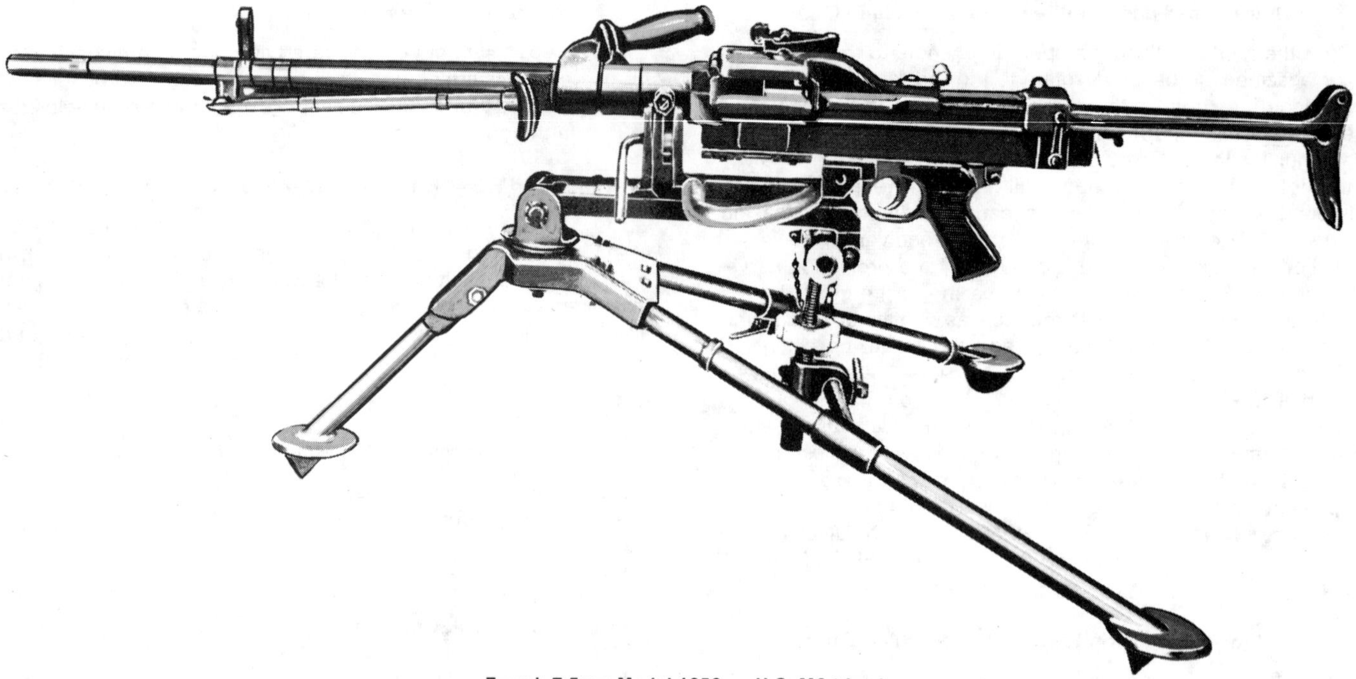

French 7.5mm Model 1952 on U.S. M2 tripod.

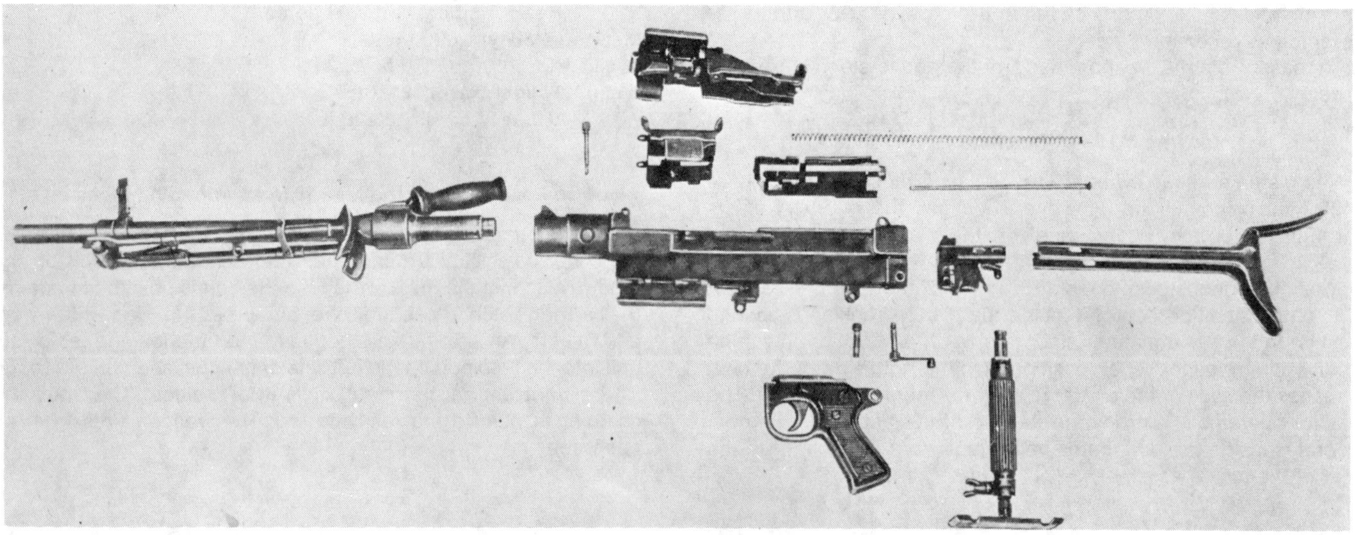

M1952 field-stripped.

How to Field Strip the M1952 (AAT Mle. 52) Machine Gun

With the stock in the extended position, press down on the stock bolt lever to remove the two stock bolts; then pull the butt to the rear and remove. Unscrew and remove the assembly pin. Pull back the bolt assembly by swinging it up. Pull the cocking handle to the rear. Remove the recoil spring and the recoil spring guide; then remove the bolt assembly by sliding it out the rear. Remove the trigger guard retaining pin and disengage the trigger guard from the receiver. Pull back the barrel catch with the left hand and, with the right hand, twist the carrying handle 1/6th turn to the right; then pull the barrel forward. Put the cover and the feed plate at a 90-degree angle from the receiver and remove the cover pin. Separate the bolt head by moving the head forward and up. Remove the firing pin and the bolt lock from the bolt.

To reassemble the weapon, perform the above steps in reverse order.

How the M1952 (AAT Mle. 52) Works

The delayed-blowback bolt system of the M1952 machine gun is similar in principle to that of the Spanish CETME assault rifle and the Swiss Model 57 assault rifle. The bolt is in two pieces—a head and a body; the head is much smaller than the body. Instead of using two roller bearings for locks, as do the Spanish and Swiss weapons, the M1952 employs a locking lever. The bolt head cannot move to the rear until the pressure is high enough to cause

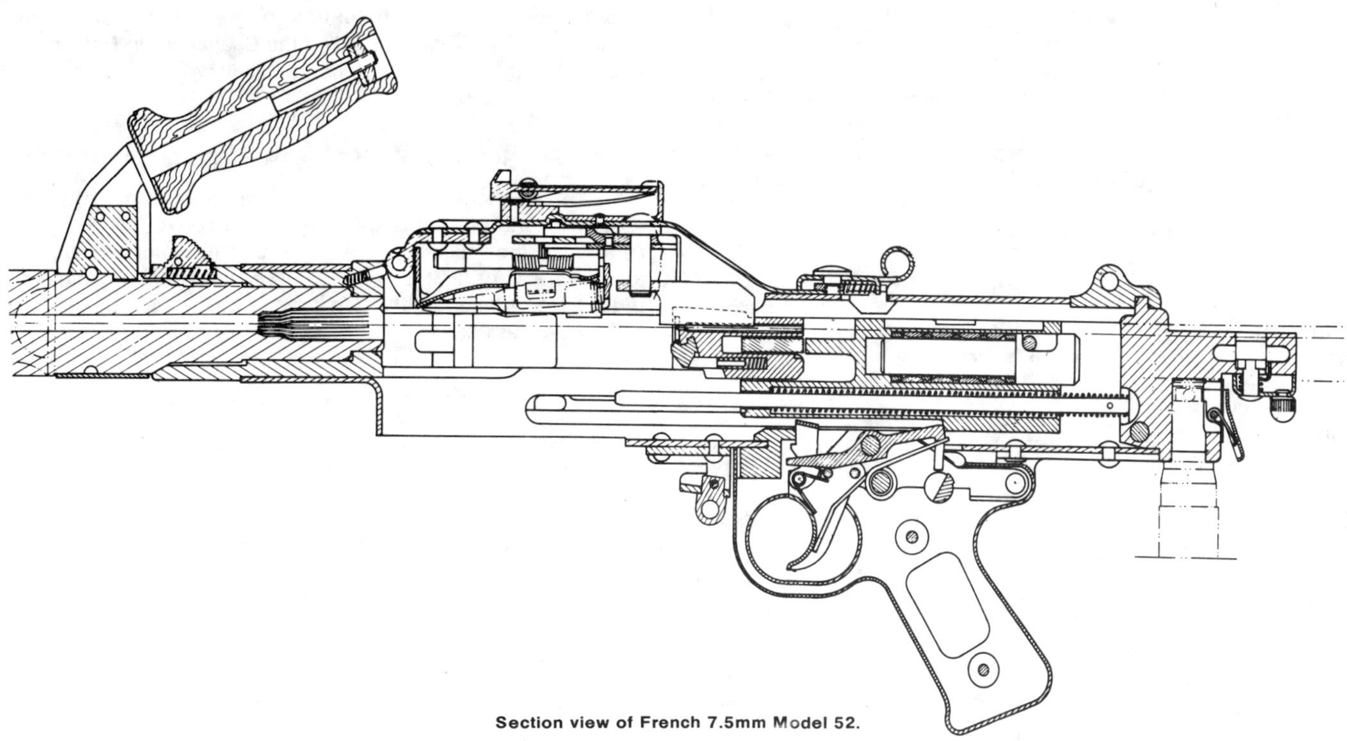

Section view of French 7.5mm Model 52.

the locking lever to start opening and, in turn, force back the bolt body and firing pin. This weapon also uses a partially fluted chamber, to make up for its lack of slow initial extraction. The feed system appears to be quite similar to that of the MG42.

Special Note on the M1952 (AAT Mle. 52)

The M1952 machine gun is quite an interesting design; it represents a great departure from pre-World War II French practice. It is considered likely that this machine gun will eventually replace the M1924 M29 as the squad automatic weapon, since the French refer to the gun as a "fusil Mitrailleur" (automatic rifle) when it is used in the light role on a bipod. On the subject of squad (or "section," as the British refer to squad) automatic weapons, it is interesting to note that the U.S., Belgium, France, and Sweden call their squad-level guns "automatic rifles." Most other countries call them "light machine guns"—e.g., the Bren, the RPD, etc.

THE FRENCH MODEL 1915 LIGHT MACHINE GUN C.S.R.G.

Although this weapon, as the Model 1924 M29, is called an automatic rifle by the French, it is usually considered as a machine gun in the United States. This weapon was developed by Chauchat, Sutter, and Ribeyrolle at Puteaux from the prototype APX 1910 rifle. Some American publications indicate that Chauchat, Ribey-

Chauchat 8mm Light Machine Gun Model 1915, C.S.R.G. (called Fusil Mitraileur Chauchat Sutter Ribeyrolle-Gladiator).

rolle and Sutter were members of a commission which approved the weapon; French publications indicate that they actually worked on the design.

This weapon is long-recoil operated and is of rather poor manufacture; but it was a wartime product and built by a number of subcontractors, who shipped parts to a central plant for assembly. You could consider it a World War I Sten gun insofar as manufacture is concerned. The main problem was that it was not as reliable in performance as the Sten, according to most reports.

Characteristics of the Model 1915 Light Machine Gun

Caliber: 8mm Lebel.
System of operation: Long recoil, selective fire.
Length overall: 45.2 in.
Barrel length: 18.5 in.
Weight: 20 lb. w/bipod.
Feed device: 20-round, detachable, in-line crescent shaped magazine.
Sights: Front: Blade.
Rear: Tangent with notch.
Muzzle velocity: 2300 f.p.s.
Cyclic rate: 240 r.p.m.

FRENCH CAL. .30 M1918 LIGHT MACHINE GUN, C.S.R.G.

United States forces used the C.S.R.G., commonly called the Chauchat (or, by the Doughboys, the "Shosho") during World War I. The United States purchased 15,988 of these weapons in 8mm and 19,241 in caliber .30-06. The .30-06 weapons were made by the French specifically for the United States Army and have a straight 16-round box magazine rather than the crescent-shaped magazine required by the rim and taper of the French 8mm cartridge.

The C.S.R.G. was adopted by Belgium in 7.65mm after World War I, and by Greece, where it was called the Gladiator, in 8mm Lebel. It was used to some extent in World War II as well as in World War I. This weapon is called the Chauchard in England; this is probably the result of a typographical error sometime in the past forty years.

THE FRENCH MODEL 1914 HOTCHKISS MACHINE GUN

The 1914 Hotchkiss was the principal machine gun of the French Army in World War I and was also still in service by the French Army until the fall of France in 1940. The Hotchkiss appeared again, to a very limited extent, during the French campaign in Indo-China (Vietnam). Although a rather heavy and bulky gun, the 1914 Hotchkiss had a good reputation for reliable performance.

Characteristics of the Model 1914 Hotchkiss

Caliber: 8mm Lebel.
System of operation: Gas, automatic fire only.
Length overall: Approx. 51.6 in.
Barrel length: 31 in.
Weight: Gun, 55.7 lb.; tripod, 60 lb.
Feed device: 24- and 30-round strip or 250-round belt consisting of articulated strips.
Sights: Front: Blade.
Rear: V-notch.
Muzzle velocity: 2325 f.p.s. w/ball 1932N.
Cyclic rate: 450-500 r.p.m.

This weapon was used by twelve U.S. divisions in France in 1918. A total of 5,255 of these machine guns were procured from France for the United States Army. Most of the guns used by U.S. troops were converted to caliber .30-06. This model Hotchkiss was also used by Spain in 7mm and by China in 7.92mm. It is unlikely to be found in service in any army today.

Caliber .30 M1918 Light Machine Gun, C.S.R.G.

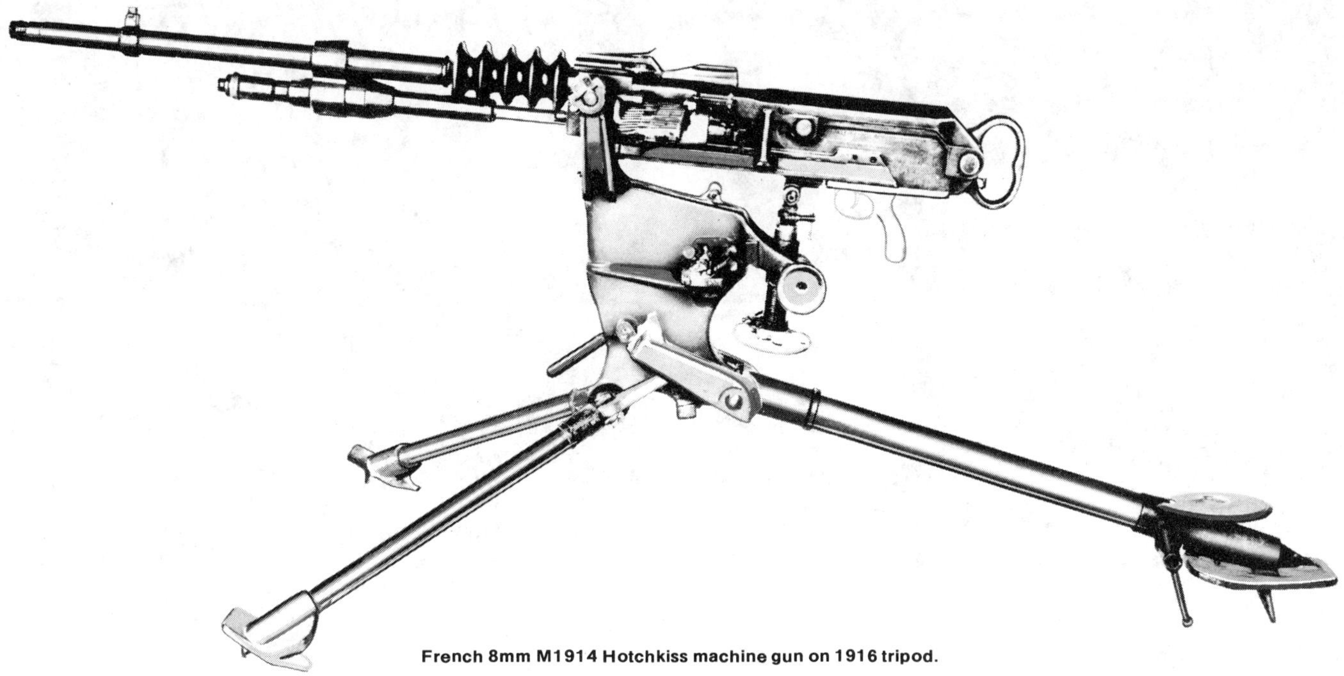

French 8mm M1914 Hotchkiss machine gun on 1916 tripod.

THE FRENCH MODEL 1924 M29 LIGHT MACHINE GUN

This weapon is called an automatic rifle by the French and is used as the squad automatic weapon. It has had extensive combat use and is a very popular weapon with French troops. The basic design, except for the top mounted magazine and the double trigger, is quite similar to the U.S. Browning Automatic rifle Model 1918. The original gun, the Model 1924, was issued chambered for the 7.5mm Model 24 cartridge. When the cartridge was shortened, the machine gun was modified to chamber the new cartridge and designated Model 1924 M29. This weapon may be found in any of the former French colonies or mandates.

Characteristics of the Model 1924 M29

Caliber: 7.5mm French.
System of operation: Gas, selective fire.
Length overall: 42.6 in.
Barrel length: 19.7 in.
Weight: 24.51 lb. w/bipod.
Feed device: 25-round, detachable, staggered row, box magazine.
Sights: Front: Blade.
Rear: Tangent.

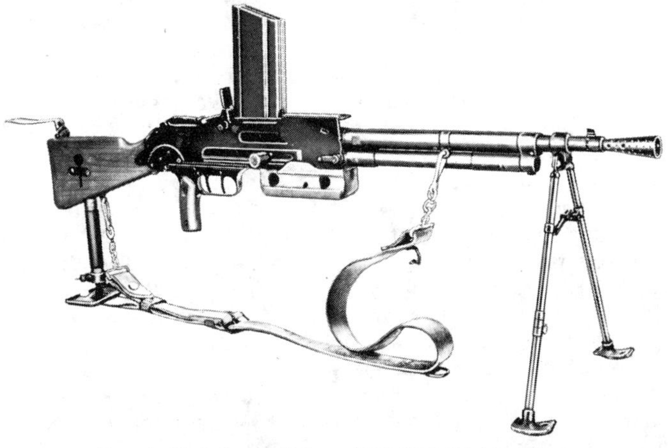

French Chatellerault 7.5mm 1924 M29 Machine Gun.

Muzzle velocity: 2590 f.p.s.
Cyclic rate: 550 r.p.m.

Field Stripping the M1924/29

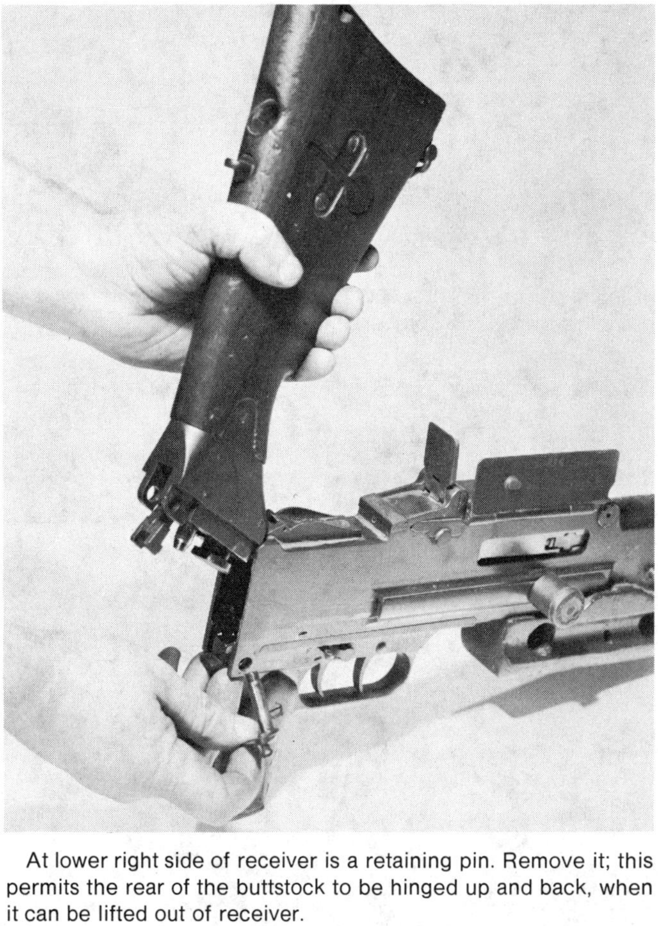

At lower right side of receiver is a retaining pin. Remove it; this permits the rear of the buttstock to be hinged up and back, when it can be lifted out of receiver.

Trigger guard assembly will now swing forward on hinge. Pushing first in, then out, unhook front end and remove.

Pulling head of ejector out of its slot will permit it to be removed from the rear. Remove recoil spring and guide.

Withdraw bolt and piston with slide out of receiver.

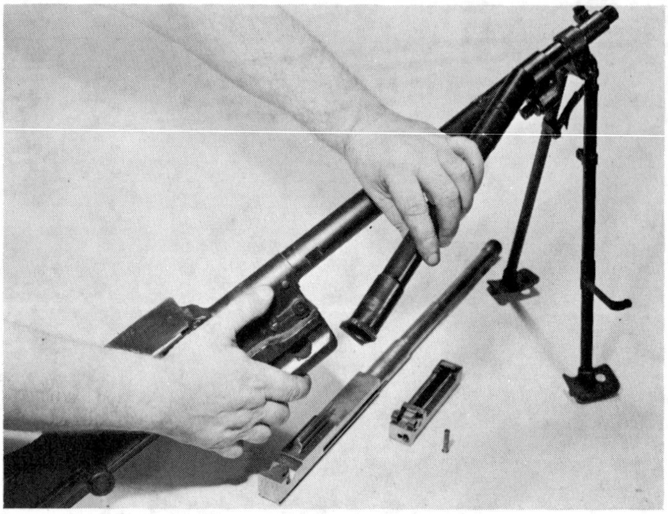

Gas cylinder tube lock is at lower front of receiver. Turn it to "O" mark on receiver, then raise rear end of tube. Tube can now be pulled out of receiver. Barrel lock is on upper front of receiver. Turn it to "O" and unscrew barrel to right.

Loading and Firing the M1924/29

A loaded magazine is inserted vertically in the top of the gun. The magazine opening is fitted with a dust cover which must be hinged up and forward to expose the magazine opening. The magazine release catch is positioned at the rear opening. Pressing it forward will release the magazine to be lifted out with the hand which is operating the catch.

Pull back the cocking handle to cock the weapon and compress the mainspring. Safety lever on the trigger guard may be used to lock the weapon in this position.

If front trigger is pulled, pressure will be exerted on the sear to release the bolt to fly forward and fire the cartridge stripped out of the magazine by the bolt in its forward travel. If the second trigger is pulled, the sear will not engage as long as the trigger is held back, and the weapon will continue to fire full automatic.

How the Model 1924/29 Works

Starting with the gun loaded and cocked the action is as follows: If the first trigger is pulled, it pushes up the forward tip of the sear, bending down the rear end against the tension of the sear spring and pulling it out of the bent in the slide. The slide carries the striker in the front of its face. The slide and piston form a single unit; the bolt is fastened by a swinging link pin to the rear end of the slide. As the slide goes forward, pulled by the compressed recoil spring, carrying the moving components with it, the feed rib on top of the bolt passes between the lips of the magazine and strips a cartridge out. The extractor is forced over the base of the cartridge by the magazine spring, bringing the cartridge to proper feeding position in the bolt face.

As the forward motion nears its completion, the front end of the bolt stops against an abutment in the receiver. The rear end of the bolt, fastened on a rotating link to the slide, is now lifted up as the forward-moving slide rotates the link about the pins, and locks firmly in locking recess in the top of the receiver (somewhat like the Browning Automatic Rifle).

The slide, which carries the striker, continues to move ahead during the period of link rotation and the pin passes through its hole in the face of the bolt and strikes the cartridge now locked in the firing chamber.

Return Movement of the Action, M1924/29

Gas escaping through the port as the bullet passes over it (a short distance from the muzzle) passes through the gas cylinder and expands violently against the cup end of the piston in standard Hotchkiss fashion. As the piston starts back under the impact of the thrust, the gas escapes through slots in the gas cylinder tube. The piston, slide, and link go back while the bolt is still securely locked.

Then as the dangerous period of pressure is passed, the rearward moving slide pulling on the bottom of the swinging link rotates it down, drawing with it the locked rear end of the bolt. As the movement continues, the bolt moves back in straight line with the other parts, carrying the empty cartridge case in the extractor. This case strikes the ejector and is expelled from the gun.

In semiautomatic fire, the rearward movement is completed as the end of the slide strikes the buffer sear release attached to the sear mechanism, causing a rebounding action which catches the sear in the side to hold it back; while the back end of the bolt strikes the buffer which absorbs shock through a coil spring. In automatic fire, of course, the sear cannot engage; and the slide goes forward immediately.

An interesting development in this weapon is an actuator in the buttstock which is used to check the rate of fire.

A plunger rod rides in a tube with a spring behind the rod. The front end of this rod passes through a hole into the receiver in line with the slide, and touches a sear trip attached to the sear mechanism. This device adds resistance to recoil and increases the time between shots, as the sear is held in engagement longer than usual.

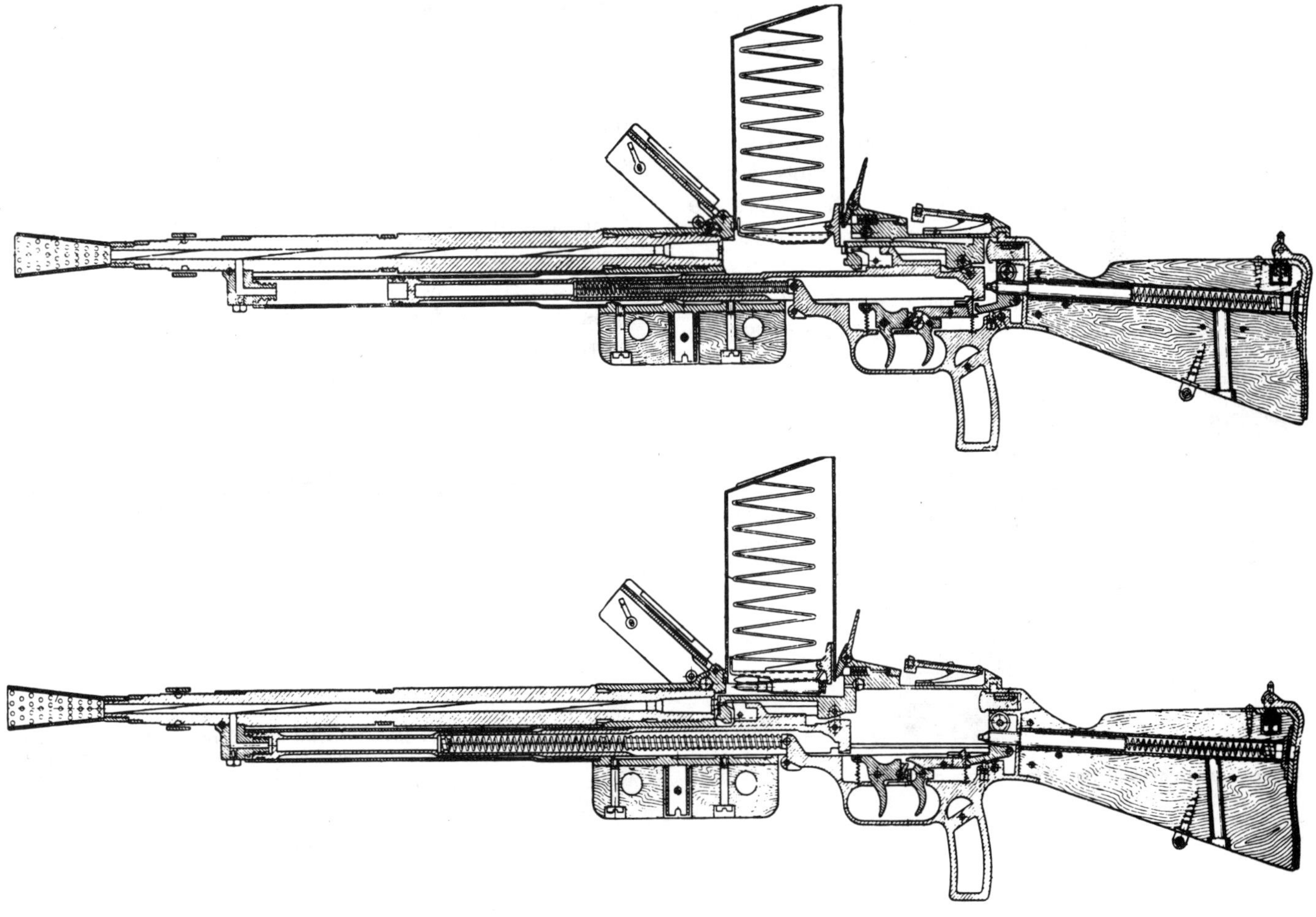

Section view of French 7.5mm M1924/29 Machine Gun; bolt open (top); bolt closed (below).

28

East Germany: "German Democratic Republic"

The East German Army is currently equipped with most of the Soviet post-war small arms. The 9mm Makarov pistol and 9mm Stechkin machine pistol are used as are the 7.62mm AK47 and AKM assault rifles and the 7.62mm RPD and 7.62mm RPK light machine guns. The 7.62mm RP46 light machine gun and 7.62mm SG43 and SGM heavy machine guns are in service as well as the 12.7mm DShK M1938/46 and the 14.5mm ZPU2 and ZPU4 machine guns.

The 7.62mm General purpose PK/PKS machine gun is now in service in the East German Army and will probably eventually replace the 7.62mm RP 46, SG 43 and SGM machine guns in that Army.

Some of the pre-world War II and World War II Soviet weapons such as the 7.62mm PPSh M1941 submachine gun and the the 7.62mm DT and DTM machine guns are still in service with border guards and armored forces respectively (on older armed vehicles).

The workers militia was armed with German World War II weapons such as the 7.92mm Kar 98K and the 7.92mm MP44. They also have older weapons. The East Germans manufacture or have manufactured the 9mm Makarov pistol, the 7.62mm AK 47 and AKM assault rifles, the 7.62mm SKS carbine, and the RPK light machine gun. The arms plant in the Suhl area are back in production and manufacture sporting arms in addition to military arms.

The East Germans use slightly different nomenclature for the Soviet weapons than do the Soviets. The usual procedure is to use the first letter of the designer's name tacked on to the German abbreviation of the weapon type. Thus the AK 47 assault rifle with wooden stock is the MPi K and with folding stock it is the MPi KmS. The AKM is the MPi KM, the RPK is the LMG K, the RPD is the LMG D, etc.

Some of the Soviet designed weapons manufactured in East Germany differ in minor detail from those made in the U.S.S.R. The Soviet made AK47 and SKS have cleaning rods mounted under their barrels; the cleaning rods are not carried on the East German-made weapons.

East German copy of 9mm Soviet Makarov pistol.

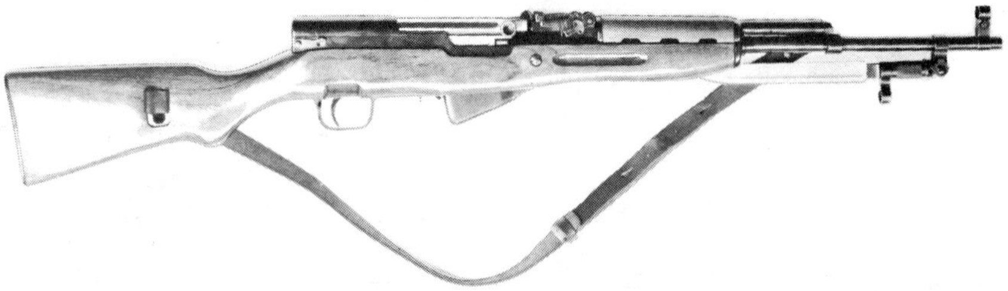

East German 7.62mm copy of Soviet SKS carbine. Note that sling is mounted similar to Kar 98k and that no cleaning rod is carried under barrel.

East German 7.62 MPi KM (AKM).

East German 7.62mm MPi-K (AK-47) Assault Rifle with folding stock.

29 West Germany: "Federal Republic of Germany"

The West German Army uses the following small arms:
9mm Walther P38 pistol called P1;
9mm Israeli Uzi submachine gun;
7.62mm Rifle G3A3, G3A4 and G3A32F;
7.62mm NATO machine gun Model 42/59 MG3.

U.S. caliber .50 Browning guns are also in service, particularly on armored vehicles.

The West German Army when originally organized was armed mainly with U.S. small arms. Most of these weapons are no longer in service.

WEST GERMAN PISTOLS

WALTHER PISTOLS

The P1 (P38), is made at the new Walther plant at Ulm. It is essentially the same as the P38 of World War II, which is covered in detail in the next chapter, German World War II weapons.

Walther is again producing the PP and PPK pistols in three calibers.

West German 9mm P38 of current manufacture. The P-1 of the West German Army.

HECKLER AND KOCH VP70 MACHINE PISTOL

The HK VP70 is a rather unusual weapon in that it is a normal blowback operated pistol when used without its shoulder stock holster and when the shoulder stock holster is fitted it can be used as a full automatic weapon to fire 3 round bursts. The change lever that accomplishes this is mounted on the shoulder stock holster. It operates through a toggle lever, mounted on the top front of the shoulder stock holster, which goes through a slot in the top rear of the pistol receiver. Another somewhat unusual feature is that this pistol fires double action only; for this reason it is not normally fitted with a safety.

Characteristics

Caliber: 9mm Parabellum
System of operation: blowback, selective fire - 3 round bursts.
Weight; w/o stock: 2.5 lb. loaded
w/stock: 3.5 lb. loaded
Length, overall: w/o stock 8 in.
w/stock 21.5 in.
Feed device: 18-round, detachable, staggered row
Sights: Front: blade
Rear: square notch
Muzzle velocity: 1,180 f.p.s.
Cyclic rate: 2,200 r.p.m.

Loading and Firing the VP70

When the VP70 is used without the shoulder stock holster it is loaded and fired much like any other self loading pistol. A loaded magazine is inserted and the slide pulled to the rear and released, chambering a cartridge. Pressure on the trigger will fire the weapon, one round for each separate pull of the trigger. When the VP70

Heckler and Koch 9mm VP 70 Pistol.

is fired with the shoulder stock holster attached the selector switch on the left, top front of the shoulder stock determines the mode of fire. If the selector is put on the figure 1, one round is fired for each squeeze of the trigger; if the selector is put on 3, three rounds are fired for each squeeze of the trigger. As pointed out previously this weapon has no safety catch although Heckler and Koch advise that they can fit it with one if desired. The magazine catch is at the bottom rear of the grip.

Field Stripping of the VP70

Remove magazine and pull slide back to insure that pistol is empty. Release slide and pull the retaining catch, located just to the rear of the trigger guard bow, down and pull the slide to the rear. Lift the slide off upward and to the rear and guide it forward and off the barrel.

Heckler and Koch 9mm Parabellum Model P9 Pistol.

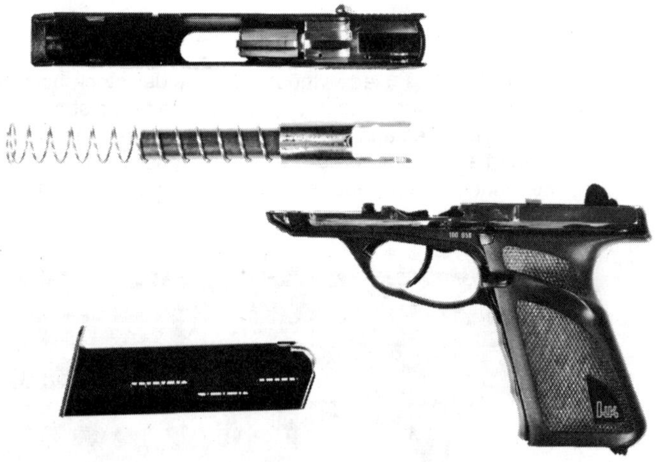

Heckler and Koch Model P9 Pistol field-stripped.

HECKLER AND KOCH 9mm PISTOL MODEL P9

Heckler and Koch have developed a new self loading pistol chambered for the 9mm Parabellum cartridge. The P9 uses the same type roller locking system as the G3 rifle series. This pistol makes extensive use of stampings in its manufacture. It has a pin type indicator which shows that the pistol is cocked. This pin protrudes from the rear of the receiver. The loaded chamber indicator is mounted above and to the rear of the chamber somewhat like that of the Luger. The P9 is double action and has a cocking lever mounted on the left side of the receiver. The safety is mounted on the left side of the slide; it operates by blocking the firing pin. The pistol can be cocked or unloaded with the safety in the "on" position.

The P9 has polygonal rifling i.e. the edges of the rifling are rounded. The plastic grips are of the wrap around type.

Characteristics of the P9

Caliber: 9mm Parabellum
System of operation: delayed blowback
Weight: 1.93 lb.
Length overall: 7.5 in.
Barrel length: 4 in.
Feed device: 9 round in-line detachable box magazine
Sights: Front: blade
Rear: square notch
Muzzle velocity: 1180 f.p.s.

Heckler and Koch 9mm Parabellum Model V70 machine pistol with shoulder-stock-holster attached.

WEST GERMAN RIFLES

The West Germans were initially equipped with U. S. M1 rifles. They then purchased a quantity of FN "7.62mm NATO FAL" rifles, which they called Rifle G1.

The West Germans were interested in the Spanish CETME assault rifle which had been designed by a group headed by L. Vorgrimmler, formerly of Mauser Werke. The design of the CETME was based on the design of the German StG45(M), a prototype assault rifle chambered for the German 7.92mm short cartridge. The West German firm Heckler and Koch assisted in the industrial engineering of the CETME. The design of the CETME was modified somewhat and the weapon was produced by Heckler and Koch and Rheinmettal of Dusseldorf for the West German Government. Heckler and Koch has developed a number of modifications of the G3 design in various calibers.

WEST GERMAN RIFLE 7.62mm G3

Characteristics of the G3

Caliber: 7.62mm NATO.
System of operation: Delayed blowback, selective fire.
Weight (loaded w/o bipod): 9.9 lbs.
Length, overall: 40.2 in.
Barrel length: 17.7 in.
Feed device: 20-round, detachable, staggered-row box magazine.
Sights: Front: Hooded post.
Rear: L type.
Muzzle velocity: 2624 f.p.s.
Cyclic rate: 500-600 r.p.m.

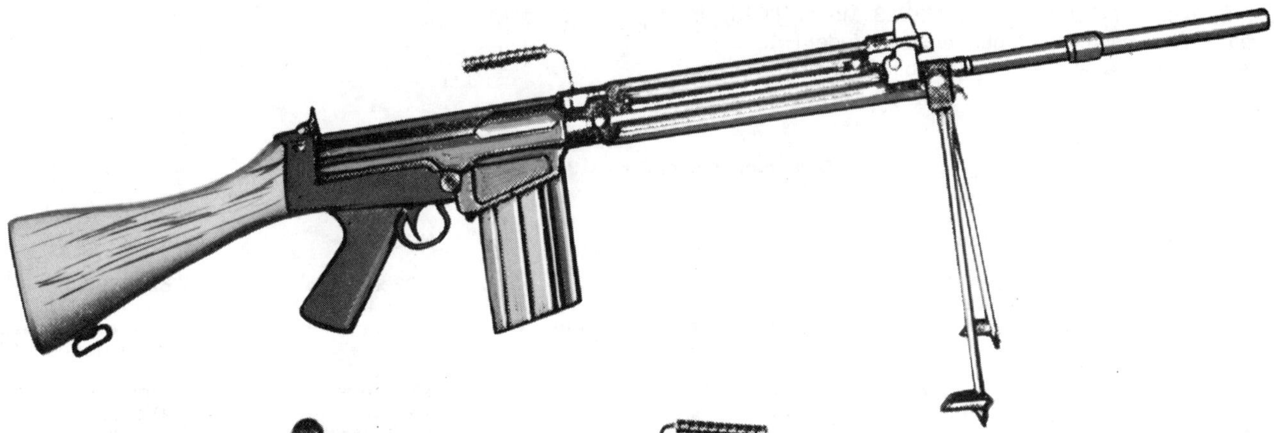

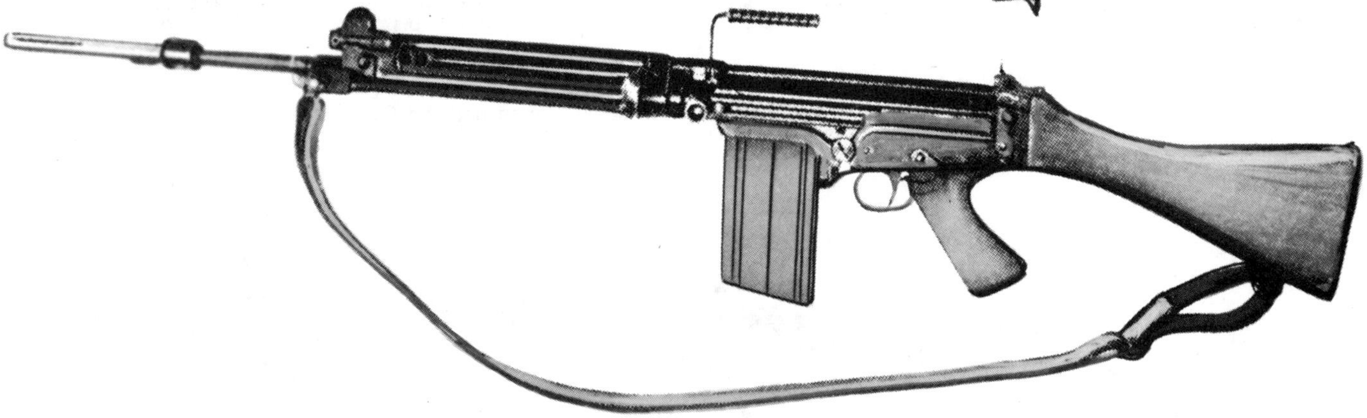

7.62mm FN Rifle G1. Note metal handguards.

Loading, Firing, Field Stripping, and Functioning of the G3

Loading, firing, field stripping, and functioning of the G3 is covered in full in the chapter on Spain. The safety-selector settings on the G3 are: "S" (safe); "E" (semiautomatic) and "F" (automatic).

The G3 can be used with a bipod, a scope sight, and an infrared snooper scope, and has a combination flash hider and grenade launcher built into the barrel. This grenade launcher accommodates the more or less standardized rifle grenades, those with a tail boom internal diameter of 22mm, in use in the free world today.

G3 is mainly made of steel stampings; only the barrel and bolt components are machined. Handguards on first production were plastic; the latest production has wooden handguards. It has been adopted in 7.62mm NATO, by Portugal as Rifle 7.62mm M1961, and has also been adopted by Norway, Sweden, Denmark, Pakistan, and the Dominican Republic.

The 7.62mm NATO G3 rifle as originally made for West Germany has a flip over type sight and wooden butt. The G3A1 has a folding type stock and flip over sight. The G3A2 has a rotating type rear sight; the current weapon being manufactured is the G3A3. It has a rotating type rear sight, a modified front sight guard and a prong type flash suppressor. G3A4 is similar, but has a retractable type stock, and G3A2F is the G3A3 with a scope mounted for sniping.

The 7.62mm NATO G3A3 rifle is being manufactured at the French Arsenal at St. Etienne (MAS) under contract; it is called Rifle Fr 3 by the French.

COMMERCIAL VARIATIONS OF THE G3 RIFLE

Heckler and Koch has produced assault rifle versions of the G3 rifle in caliber .223 (model HK33 and HK33K) 7.62X39 Soviet M1943 cartridge (Model HK32 and 32K), belt-fed light machine

West German 7.62mm NATO Rifle G3A4.

versions in 7.62mm NATO (Model HK21), .223 (Model HK 13), and a 9mm parabellum submachine gun version (Model HK54).

The assault rifles variations have the following characteristics:

Characteristics of Four G3 Variations

	HK33	HK33K	HK32	HK32K
Caliber:	.223	.223	7.62 x 39mm.	7.62 x 39mm.
Barrel length:	15.35 in.	12.4 in.	15.35 in.	13.6 in.
Length overall:	36.1 in.	Stock retracted, 24.4 in. Stock extended, 32.7 in.	36.1 in.	Stock retracted, 25.6 in. Stock extended, 33.9 in.
Weight:	6.6 lb.	6.6 lb.	6.6 lb.	6.6 lb.
Magazine capacity:	20 rounds.	40 rounds.	30 rounds.	30 rounds.
Cyclic rate:	600 r.p.m.	600 r.p.m.	600 r.p.m.	600 r.p.m.
Muzzle velocity:	3182 f.p.s.	3018 f.p.s.	2560 f.p.s.	2493 f.p.s.

All the above rifles operate the same as the G3 and many components are interchangeable among these weapon and the G3 rifle.

West German Rifle 7.62mm G-3.

West German 7.62mm NATO Rifle G3, stripped.

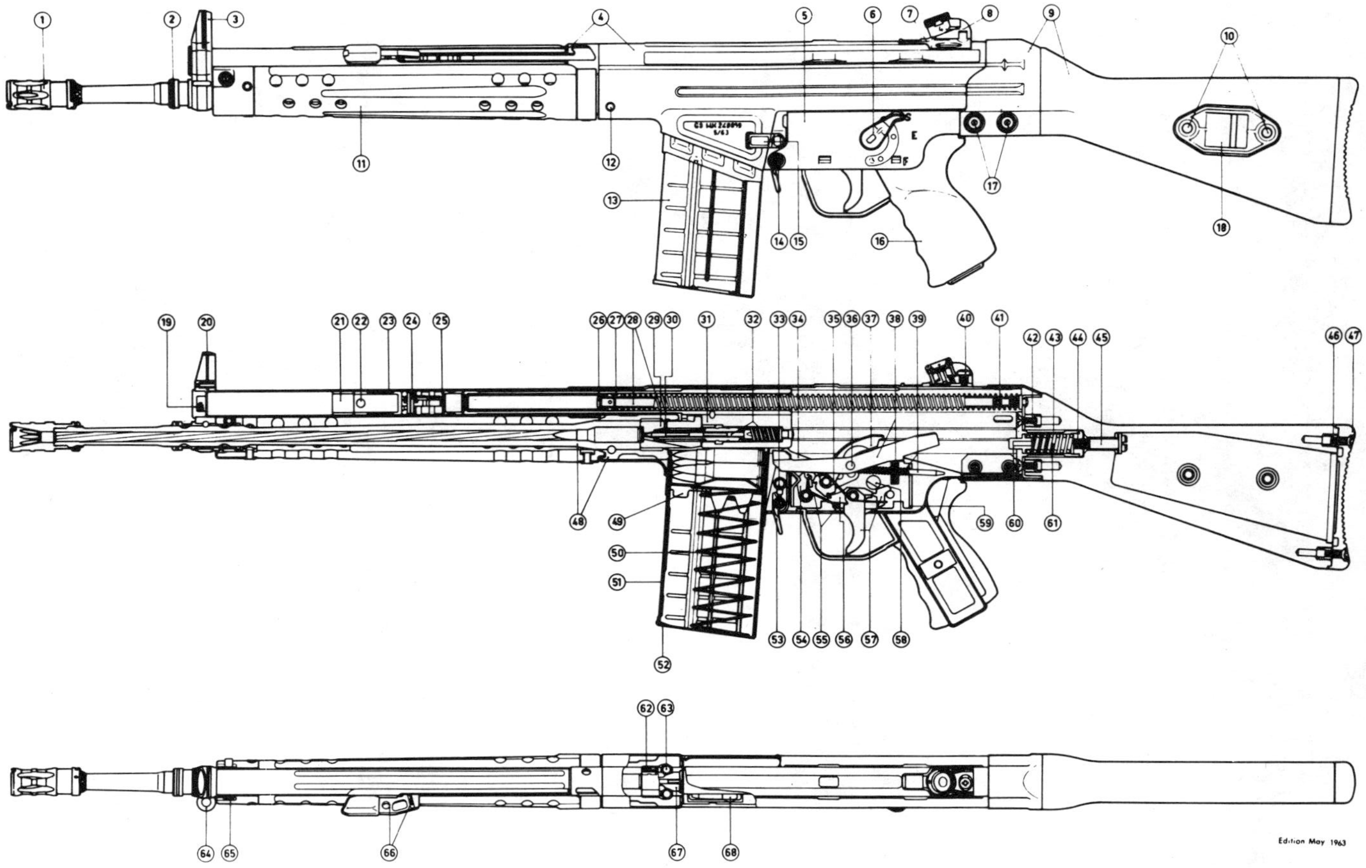

Numerical Table

1 Flash hider (grenade launcher) with retaining spring for flash hider
2 Snap ring to mount grenade
3 Front sight holder
4 Receiver with loading lever housing
5 Grip assembly
6 Safety mechanism
7 Sight base
8 Rotary rear sight
9 Back plate with butt stock
10 Tubular rivet for sling holder
11 Forearm
12 Cylindrical pin
13 Magazine
14 Grip assembly locking pins
15 Magazine catch
16 Grip
17 Butt stock locking pins
18 Sling holder
19 Cap
20 Front sight
21 Stop abutment
22 Stop pin
23 Stop
24 Loading handle spindle
25 Loading handle support
26 Cross pin for recoil spring
27 Recoil spring guide ring
28 Recoil spring tube with recoil spring
29 Bolt head
30 Clamping sleeve and holder for locking roller
31 Bolt head carrier with recoil spring tube
32 Firing pin with firing pin spring
33 Contact piece
34 Release lever
35 Elbow spring for trigger
36 Ejector spindle
37 Hammer
38 Ejector with ejector pressure spring
39 Pressure shank and pressure spring
40 Fixing screw
41 Bolt for spring guide
42 Countersunk screw
43 Buffer housing
44 Buffer bolt
45 Screw for buffer
46 Oval-head countersunk screw
47 Butt stock plate
48 Barrel with barrel extension
49 Follower
50 Follower spring
51 Magazine tube
52 Magazine cover
53 Release lever
54 Catch
55 Elbow spring with roller
56 Trigger lever
57 Trigger
58 Safety pin
59 Trigger assembly
60 Buffer pin
61 Buffer spring
62 Extractor with extractor spring
63 Locking roller
64 Eyebolt
65 Forearm locking pin
66 Loading handle with elbow spring
67 Locking piece
68 Bolt head locking lever

Section view of G3 Rifle.

7.62mm NATO G3 Rifle with 30-round magazine used by Dominican Army.

Heckler & Koch Model HK33 cal. .223 Assault Rifle.

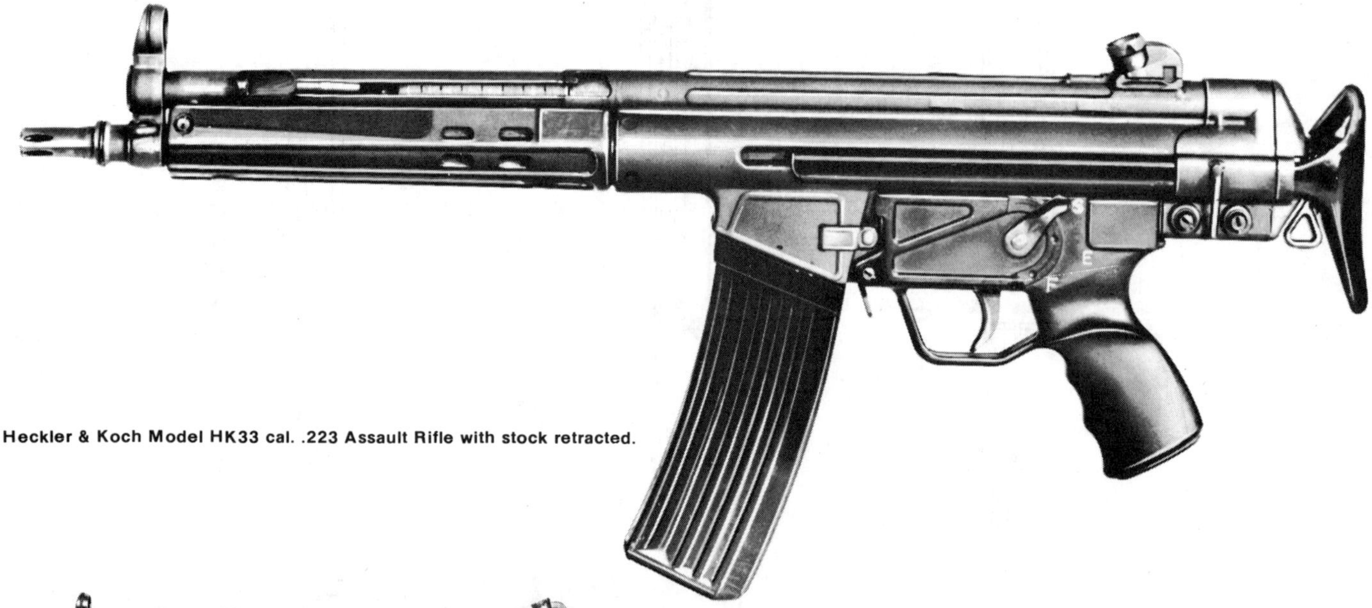

Heckler & Koch Model HK33 cal. .223 Assault Rifle with stock retracted.

Heckler and Koch 5.56mm HK 33Ka1 Rifle.

West German G3A3 Rifle.

RHEINMETALL RH4 ASSAULT RIFLE

The Rheinmetall firm which brought back the MG42 and is very active in the development of 20mm guns developed a gas operated, roller locked, selective fire rifle in the mid sixties. This rifle was chambered for the Soviet 7.62X39mm cartridge. Little has been heard of the weapon since it was developed in two versions: a fixed wood stock version and a retractable metal stock version. The RH4 has a 30 round detachable magazine and is 37 inches long in the fixed stock version and with metal stock out in the retractable stock version. With the stock retracted the overall length is 28.5 inches. The barrel is 16.5 inches long, and cyclic rate of fire is 500-600 rounds per minute.

Heckler and Koch HK33A2 Rifle.

Rheinmetall 7.62mm RH4 Rifle with wood stock.

RH4 Rifle with retractable metal stock.

West German G3A3 ZF Rifle.

Heckler and Koch HK 33A3 Rifle.

WEST GERMAN SUBMACHINE GUNS

The standard submachine gun of the West German Army is the Israeli-designed Uzi, but a number of native West German designs have appeared since the late fifties. Mauser, Walther, Heckler & Koch, Anschütz, and Erma have all prepared designs. These weapons have mainly appeared in prototype form and are usually made of stampings and are of short length.

9mm Parabellum Walther MPL submachine gun with stock extended.

DUX 53

The DUX 53 is a weapon designed at the Oviedo Arsenal in Spain by W. Daugs and L. Vorgrimmler and based on the design of the Finnish 9mm Model 44 submachine gun, which in turned was based on the Soviet PPS M1943. A quantity of these weapons were manufactured for the West German Border Police. This weapon and various modifications made by Mauser and Anschütz were tested by the West German Army during the mid-fifties.

THE WALTHER SUBMACHINE GUN

The Walther submachine gun was introduced in 1963. There are two basic versions of this gun: the MPL (long model), and the MPK (short model). The Walther is made mainly of steel stampings and has a folding steel stock which can be folded to either side. The bolt is guided through the receiver by a guide rod which is mounted in the receiver above the bolt face. The mass of the bolt is above the bolt face and the guide rod, which also guides the recoil spring, goes through a tunnel in this top portion of the bolt.

The safety lever on the Walther gun is fitted on both sides of the receiver.

Walther MPL, field-stripped.

Characteristics of the Walther Submachine Gun

	MPL	MPK
Caliber:	------------9mm Parabellum------------	
System of operation:	Blowback, full automatic only*	
Length overall: w/stock extended	29.42 in.	25.96 in.
w/stock folded	18.1 in.	14.75 in.
Barrel length:	10.25 in.	6.75 in.
Weight:	6.62 lb.	6.27 lb.
Feed device:	--32-round, staggered row, detachable box magazine.---	
Sights: Front:	------------ Protected blade -----------	
Rear:	---Flip over, notch for 75m and aperture for 125m. -----------------	
Muzzle velocity:	Approx. 1370 f.p.s.	Approx. 1250 f.p.s.
Cyclic rate:	----------------550 r.p.m.----------------	

* Can be made selective fire in special order.

9mm Parabellum Walther MPK submachine gun with stock folded.

THE HECKLER AND KOCH HK54 SUBMACHINE GUN

The HK 54 is the submachine-gun version of the G3 rifle. It is, as the G3 rifle, a delayed blowback operated weapon and fires from a closed bolt. Although not unique-there are many delayed blowback submachine guns and some that fire from a closed bolt-it is somewhat unusual for a submachine gun. There are some advantages and disadvantages to this system.

There is a theory that delayed blowback submachine guns have less vibration and rise than blowback-operated submachine guns. On the other hand, they are more complex and usually more expensive.

A finer degree of accuracy can be obtained with a gun that fires from a closed bolt, since the only disturbing influence on "hold" is the forward movement of a light hammer and/or firing pin as opposed to the forward movement of a heavy bolt. The "lock time" (the period from trigger/sear release to ignition of primer) is also less on a weapon which fires from a closed bolt than on a weapon which fires from an open bolt; the other side of the coin in this case, however, is the "cook-off" problem. Automatic fire heats up a weapon rather rapidly and a point is reached when the temperature of the chamber will cause cartridges to function i.e., to "cook-off" spontaneously as it were. Some designs have solved this problem by firing from an open bolt in automatic fire and from a closed bolt on semiautomatic fire, as is done by the 7.92mm

FG42 German World War II paratroop rifle and the U. S. Johnson.

The HK54 comes in two models; one with fixed wooden stock and one with retractable metal stock. Many parts of the HK54 are common to the G3 rifle.

Characteristics of the HK54 Submachine Gun

Caliber: 9mm Parabellum.
System of operation: Delayed blowback, selective fire.
Length overall: w/fixed stock, 26 in.
w/retractable stock fixed, 26 in.
w/retractable stock retracted, 19.3 in.
Barrel length: 8.85 in.
Weight: 5.5 lb.
Feed device: 30-round, detachable box magazine.
Sights: Front: Post.
Rear: Flip-type w/"U" notch.
Muzzle velocity: Approx. 1312 f.p.s.
Cyclic rate: 600 r.p.m.

The West German Border Police have adopted the HK54 which with fixed stock is called the MP5 and with retractable metal stock is called the MP5A1. A later version of this weapon with modified front sight, rotating rear sight and modified front section of barrel is called MP5A2 with fixed stock and MP5A3 with retractable metal stock.

Top: West German 9 mm Parabellum MP5 Submachine Gun.
Bottom: West German 9mm Parabellum MP5A1 Submachine Gun.

9mm Parabellum HK54 submachine gun. (MP5)

HK54 submachine gun with metal stock retracted. (MP5A1)

WEST GERMAN MACHINE GUNS

WEST GERMAN 7.62mm MACHINE GUN MG1

West Germany has adopted the 7.62mm NATO MG 42/59 made by Rheinmettal. This weapon, which is almost identical to the 7.92mm World War II German MG42, is called MG1 by the West Germans. An improved model with a stellite-lined chromed barrel and a modified barrel booster flash hider is now on the market.

The 7.62mm NATO MG1 is loaded, fired, and field stripped in the same manner as the 7.92mm MG42 which is covered in detail under German World War II materiel. This machine gun in 7.62mm NATO is used by Austria, Denmark, Italy, and other countries as well as West Germany. It is manufactured in 7.92mm as the "SARAC" M1953 by Yugoslavia.

Modifications of the MG1

Three basic modifications of the MG1 have been produced by Rheinmettal of Duesseldorf, manufacturers of the weapon. These weapons and their basic modifications are:

MG1A1 hard chrome plated barrel, sight corrected for 7.62mm NATO cartridge, trigger mechanism slightly modified.

MG1A2 receiver modified, and case ejection port widened, heavy bolt. Can be used with the German non-disintegrating link belt or the U. S. M13 disintegrating link belt. The cocking stud on the cocking slide has been shortened, buffer has been modified the recoil booster and flash hider have been modified. The barrel guide sleeve (front barrel bearing) has been modified, barrel similar to MG1.

MG1A3 sight modified, barrel similar to MG1A1, an additional stud has been added to the bolt housing and other minor modifications of the bolt, belt-feed lever and feed mechanism modified, trigger mechanism modified and trigger pull lightened, stock modified in dimension of sleeve which threads on to buffer, butt plate mounting screw modified, bipod modified, recoil booster and flash hider made in one piece, barrel bearing chrome plated, barrel chrome plated.

There are two bolts and two buffers for the MG1 series guns. Use of the light bolt, called V550 by Rheinmettal, and the Type N buffer produces a cyclic rate of 1100-1300 rounds per minute. Use of the heavy bolt, called V950 by Rheinmettal, and the Type R buffer produces a cyclic rate of 700-900 rounds per minute. The 550 and 950 are the weights of the bolts in grams.

The current model of this machine gun used by the West German government is the G3. The G3 differs in minor details from earlier models. The most noticeable differences are the addition of an anti-aircraft sight in front of the rear sight and the shape of the barrel booster flash hider unit.

West German Border Police were equipped with 7.92mm MG42 machine guns in the early fifties. These machine guns were of German World War II manufacture and were purchased from France which had captured them during the war (1944-45). Rheinmettal "reverse engineered" the 7.92mm MG42 to produce the 7.62mm NATO MG42 or MG42/59 as it is now called commercially, since the production drawings of the weapon were "lost, strayed or stolen" at the end of World War II.

Some interest was shown in the re-development of the prototype MG45 by the West German Army, but apparently this interest died. MG45 was essentially the same as MG42, but had a delayed blowback action similar to the G3 rifle, rather than being recoil operated.

There has been some commercial machine gun development by Heckler and Koch. This firm has come out with machine-gun versions of the G3 rifle in 7.62mm NATO and .223. The design is similar to the Mauser—CETME machine gun, which was advertised by Mauser and the Dutch firm NWM during the late fifties.

THE HECKLER AND KOCH MODEL 11, 12, 13, 21, 21A1 AND 23 MACHINE GUNS

These weapons are variations on the basic G3 design guns. The Model 11 can be fed with either a saddle drum magazine or a box magazine and is chambered for the 7.62mm NATO cartridge. The Model's 12 and 13 are the same as the Model 11, but are chambered for the Soviet 7.62 X 39mm and the 5.56mm (.223) cartridges respectively. Models 12 and 13 are also somewhat lighter than the Model 11.

The Model 21 is belt fed, but can also, with a special adaptor, be used with a box magazine. The Model 21 may be chambered for the 7.62mm NATO, 7.62 X 39mm or 5.56mm (.223) cartridges. The Model 21A, which has been adopted by Sweden, differs from the Model 21 in having a hinged rather than a removable feed tray. A reduced weight 5.56mm (.223) version of the Model 21A is the Model 23.

All of the above weapons have quick change barrels. Rear sights are usually of the radial type, but the rotating type aperture, as on the G3A3 rifle may also be found.

All of these weapons fire from a closed bolt and all have a quick change barrel. Many of the components of these weapons are interchangeable with the components of the G3 rifle. The concept of having a weapon system consisting of a number of weapons of basically similar construction with different application, i.e. individual weapon, squad automatic weapon, and (in some cases) support machine gun has become quite popular these days. The Soviets now have a squad-level system in the 7.62mm AKM assault rifle and the 7.62mm RPK light machine gun. The Stoner 63 weapons system is similar in concept and has support and tank machine gun versions as well; all being built around the same basic action. Colt has developed a generally similar system to the G3, using the AR15 rifle as the base.

This type of weapon system design has a definite advantage from a training and logistical point of view, but frequently has some technical drawbacks. The remarks above about "cook off" dangers with automatic rifles apply with more force to belt-fed machine guns with comparatively light weight barrels which fire from a closed bolt. Briefly, gun design like all else is a series of "trade-offs"; what one gains in one area one loses in another, and there is no weapon in any category which is better than all other weapons of its type in all characteristics.

West German 7.62mm NATO MG1 (MG 42/59).

Heckler and Koch 7.62mm Model HK 21 showing belt feed arrangement.

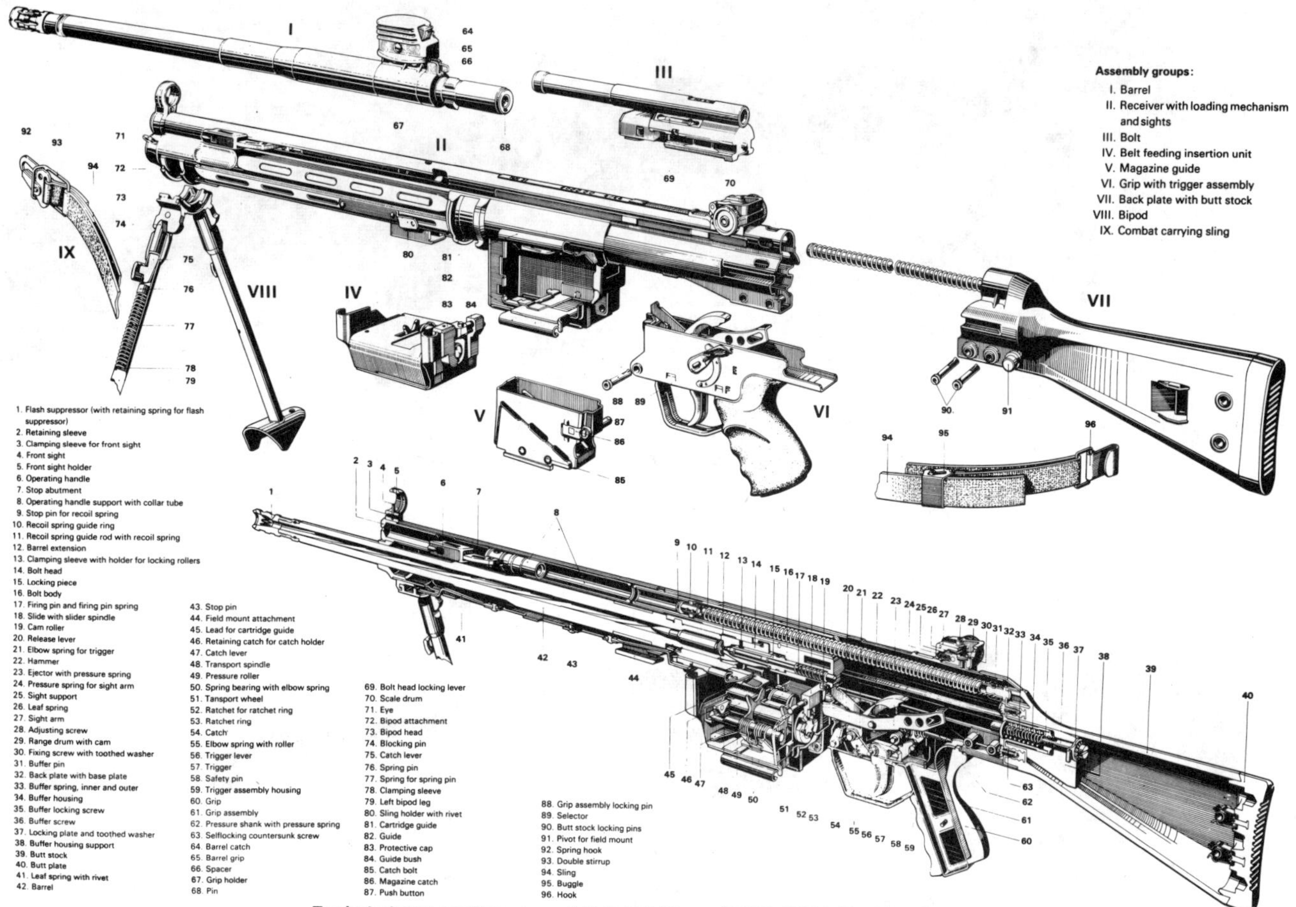

Exploded and section view of H & K 7.62mm NATO HK21 Machine Gun. This weapon has been adopted by Portugal.

Characteristics of the Heckler and Koch Machine Guns

Model:	**HK11**	**HK21**	**HK13**
Caliber:	7.62mm NATO	7.62mm NATO	.223
System of operation:	Delayed blowback, selective fire		
Length overall:	40.15 in.	40.1 in.	38.6 in.
Barrel length:	17.71 in.	22.63 in.	22.13 in.
Weight w/o magazine:	15 lb.	14.7 lb. w/o bipod.	8 lb.
Feed device:	80 round saddle drum or 20 round box	Metallic link belt.	100-round detachable drum or 20-round detachable box magazine
Sights: Front:	Protected post		
Rear:	Rotary rear with V notch and apertures		
Cyclic rate:	850 r.p.m.	750 r.p.m.	600 r.p.m.
Muzzle velocity:	2589 f.p.s.	2625 f.p.s.	3248 f.p.s.

Heckler and Koch .223 HK 13 Machine Gun.

West German 7.62mm MG3 Machine Gun.

View of HK Model 21A Machine Gun showing hinged feed tray.

Heckler and Koch Model 23 Machine Gun.

Heckler and Koch 7.62mm Model II Machine Gun.

German World War II Small Arms

30

Germany manufactured over 13,000,000 small arms during World War II and in addition used tremendous quantities of captured arms and arms made in occupied countries. There were not sufficient standard German small arms on hand at any time during World War II to arm all German forces. Eleven divisions of the German Army were armed with Czech small arms at the beginning of World War II.

As a result of the continual arms shortage, the Germans used a large variety of small arms during the war, but they did standardize to the extent that all front-line units used weapons chambered for the standard German service cartridges and also had the highest priority for the standard service weapons. Service units, police units, and various other German organizations used anything available.

The standard service pistol was the 9mm Walther P38 and the 9mm Luger 08 was substitute standard throughout the war. The 9mm MP38 and MP40 submachine guns were standard weapons which were to be replaced by the 7.92mm MP43, 44 series of weapons. The 7.92mm Mauser Kar98k and the 7.92mm G33/40 were standard as was the semiautomatic Kar43, but these rifles were also scheduled to be replaced by the 7.92mm MP43, 44 weapons. The 7.92mm MG42 was the standard machine gun at the end of the war, but it never completely replaced the 7.92mm MG34.

GERMAN PISTOLS

A total of twenty-seven different models of German designed and made pistols (not including caliber .22 pistols) were approved for service use by the German forces between 1914 and 1945. Only a few models were ever purchased in quantity. These pistols were in 6.35mm (.25ACP), 7.65mm (.32 ACP), 9mm short (.380 ACP), 9mm Parabellum, 7.65mm Luger (caliber .30 Luger), and 7.63mm Mauser.

Toward the end of World War II, Walther and Mauser developed 9mm Parabellum pistols composed mainly of steel stampings which were to be used to arm the Volksturm. This was a Home Guard type organization composed of those males too old or too young or too infirm for the regular armed forces or those having essential jobs in civilian industry. Both designs were disapproved by the German Army Ordnance Office.

Because of the large number of different types of pistols used by the German Army only those which were used in large quantities will be covered in detail in this book. The Germans also used many foreign pistols, some of which they had captured and some made in occupied countries. The most common among these were the Polish 9mm Parabellum, Radom M1935, the FN Hi-Power, the Czech 7.65mm Model 27, the Czech 9mm short Model 38, the Hungarian 7.65mm Model 37, and the Belgian 7.65mm and 9mm short FN Browning Model 1922.

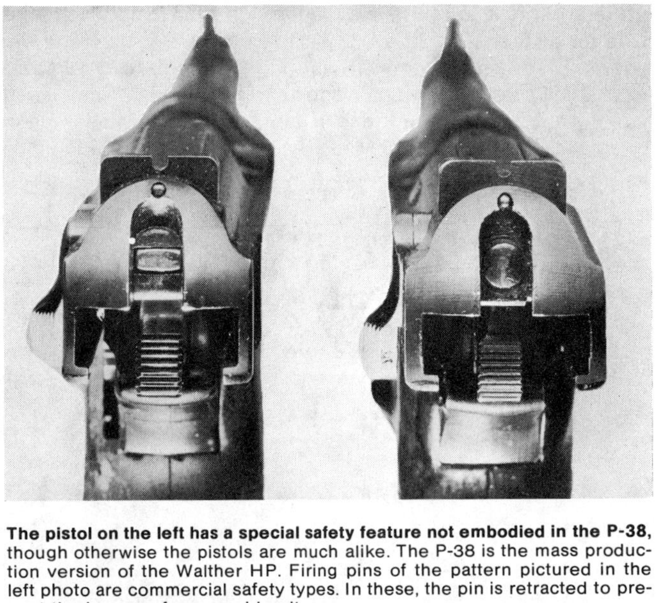

The pistol on the left has a special safety feature not embodied in the P-38, though otherwise the pistols are much alike. The P-38 is the mass production version of the Walther HP. Firing pins of the pattern pictured in the left photo are commercial safety types. In these, the pin is retracted to prevent the hammer from reaching it.

Prototype 9mm Walther P38 marked "M P".

THE 9mm P38 PISTOL

The P38 was developed by the Waffenfabrik, Carl Walther at Zella Mehlis Thuringia. It is descended from a double-action 9mm Parabellum pistol developed around 1935 which used the double-action system introduced with the Walther PP pistol in 1929. In 1937 the Walther HP Heeres (Army) pistol appeared on the commercial market. In addition to 9mm Parabellum, the HP was made in caliber .45 and Super .38. The Heeres pistol has a rectangular section firing pin, a machined slide stop as opposed to the fabricated type used on the P38, and anvil surface hammer and different type grip pieces (usually checkered wood).

The P38 was adopted in 1938. As originally made for the German army it bore the Walther marking. During the war the German code letters were used to identify the manufacturers: "ac" for Walther, "cyq" for Spree werke, "byf" and "svw" for Mauser, "dov" for Brunn (Brno), and "ch" for FN, which manufactured approximately 3500 sets of components. About one million P38s were made during World War II.

A considerable impression was made on U.S. personnel by the double-action feature of the P38. While not a new concept, for Walther introduced it in 1929 with the PP and it was used on earlier pistols, it was apparently almost unknown in the U.S. The value of the double-action feature is somewhat exaggerated. If a man is in battle and is carrying a pistol, he will have it in his hand cocked; the main advantage of the double-action feature is the possibility of striking a defective primer a second blow with a pull of the trigger. Even with wartime ammunition, defective primers are rare and the long trigger pull through with double action, if used for all first shots (the P38 hammer can be cocked and the weapon fired single-action) will have a tendency to impair accuracy. There is a safety advantage to the double action; when carrying a pistol with a cartridge in the chamber, but single-action automatics like the U.S. .45 M1911A1 with inertia-type firing pins can be carried with the hammer down on a loaded chamber with safety if reasonable precautions are taken.

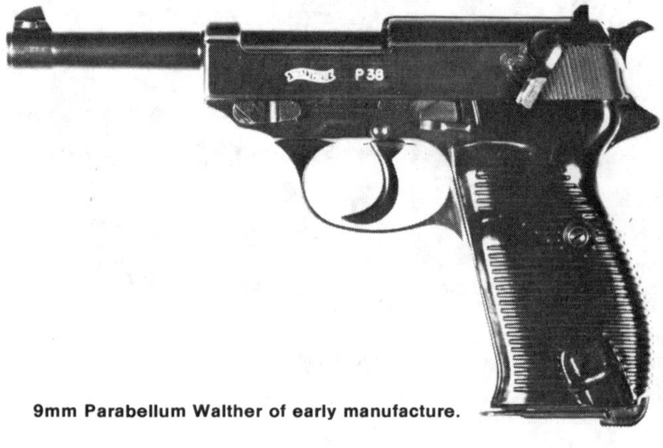

9mm Parabellum Walther of early manufacture.

The P38 was used by Sweden who called it P39. The P38 saw some use in France, East Germany, and in some of the Soviet satellites for a time after World War II.

The Walther "Armee" pistol is a hammerless version of the Walther HP which was made for commercial sale prior to World War II. It was made for the 9mm Parabellum cartridge and a few were made in caliber .45 for sale in the U.S. One variant of the "Armee" pistol has been reported with a duraluminum receiver.

Loading and Firing the P38. Load and insert magazine in handle as is customary in all automatic pistols of this general design.

Draw back the slide, which will ride over the head of the hammer and cock it, meanwhile compressing the recoil springs and permitting a cartridge in the magazine to rise in line with the breechblock. Release the slide and let it go forward under the influence of the recoil springs.

If the safety catch has been in the "Fire" position during the backward movement of the slide, then the hammer will remain at full cock and the pistol is now ready for immediate firing.

If the safety catch was in the "Safe" position when the slide was retracted, then the chamber will load but the hammer will go forward safely to its resting place. The pistol cannot be fired now as the safety is locking the firing pin. However, if you now move the safety catch, the hammer may be cocked by the thumb, or the pistol may be fired by pulling straight back on the trigger bringing the double action factor into play.

Warning. Applying the thumb safety when hammer is cocked automatically locks the firing pin and then drops the hammer on it. In case of a defective or crystallized safety, the pistol may fire accidentally. It is not advisable to use this hammer lowering device on a loaded chamber with pistols of late World War II manufacture.

Action open.

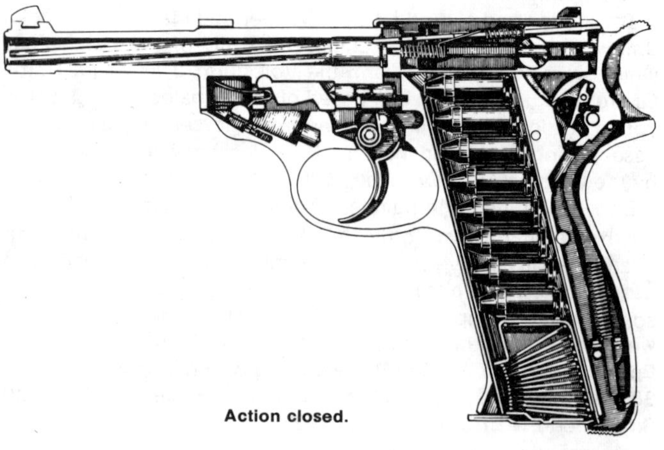

Action closed.

Field Stripping the P38

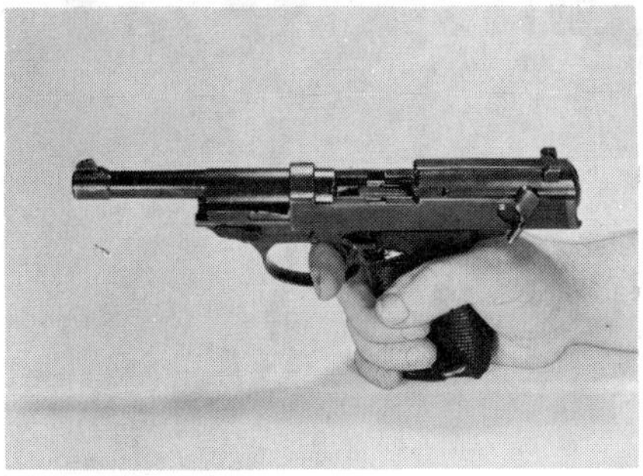

(1) Set the safety catch in the "safe" position and pull the slide back over the empty magazine, so that the inside catch on the slide stop will be forced up by the magazine follower and hold the slide open. Then remove the magazine.

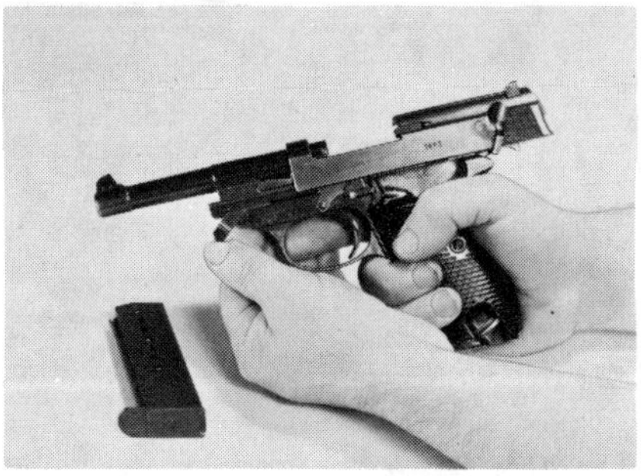

(2) At the front end of the frame below the receiver is a lever-type locking pin. Turn this down and around as far as it will go.

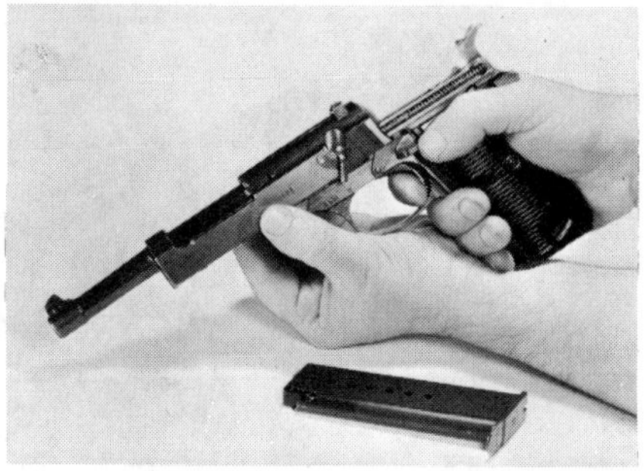

(3) Hold the slide under control with the left hand and with the right thumb push down the slide stop. Now press the trigger and pull the barrel and slide directly forward in their runners on the receiver, sliding them out of the guides.

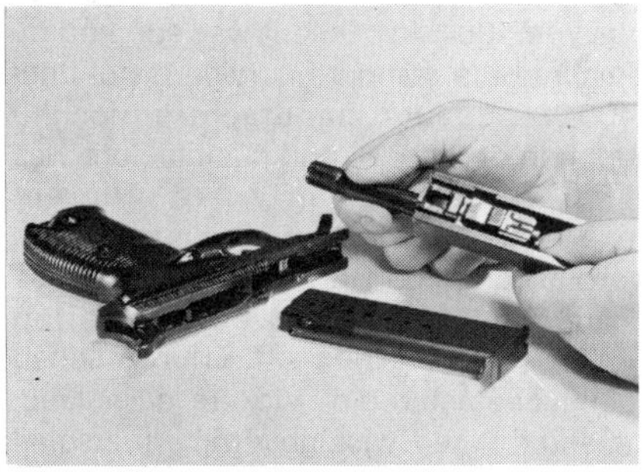

(4) Now turn barrel and slide upside down. A small locking plunger will be seen at the rear of the barrel assembly. Push the plunger and spring out the white metal locking cam block.

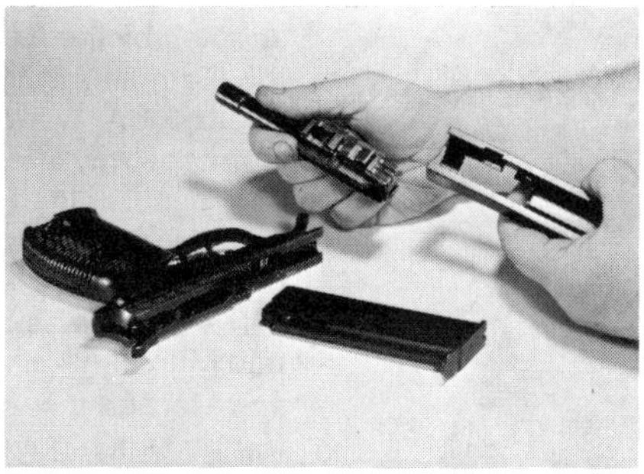

(5) Now slide barrel directly ahead out of slide. Lock will come forward with barrel.

(6) Push forward and up on locking cam block and lift it out of its recess. This completes field stripping.

(7) Top to bottom, left: Slide, barrel, lock, magazine. Top to bottom, right: Receiver, stock.

Note on Assembly of P38

Reverse stripping procedure.

When replacing locking block, be sure that its lugs are in line with the wide ribs on both sides of the barrel.

Insert barrel assembly as far as it will go into slide, then push the locking block into its locked position.

Hammer must be uncocked, and the ejector and the safety mechanism levers pushed down to prevent them from catching on the rear end of the slide.

With safety catch at "safe" position, hold the locking block in the locked barrel position, and push the slide and barrel onto the receiver in the guide. Force the slide all the way back against the tension of the springs and raise the stop to catch and retain the slide in the open position.

Turn the locking lever around on its pin as far as it will go to its original position. Press the slide stop and permit the slide to run home. Insert magazine.

Special Note on the P38 Pistol

When the pistol is loaded with a cartridge in the chamber, a floating pin protrudes from the slide above the hammer. A glance or (in the dark) a touch will always tell whether the chamber is loaded, making it unnecessary to pull back the slide as in other pistols.

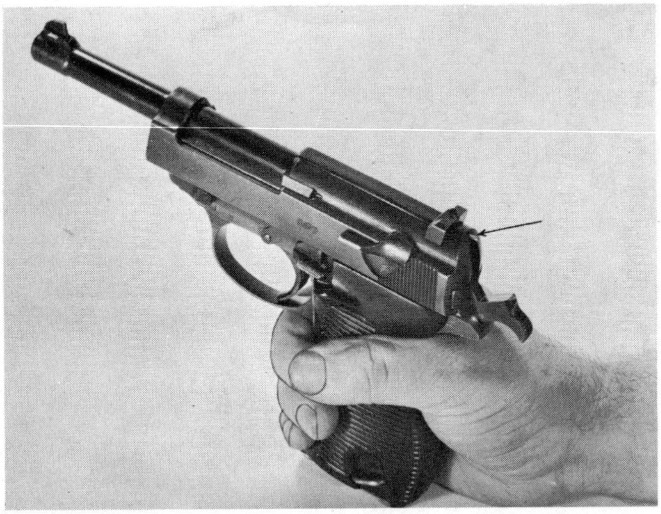

Protruding pin (arrow) on the P-38 indicates a loaded chamber.

Walther Armee Pistol, internal hammer.

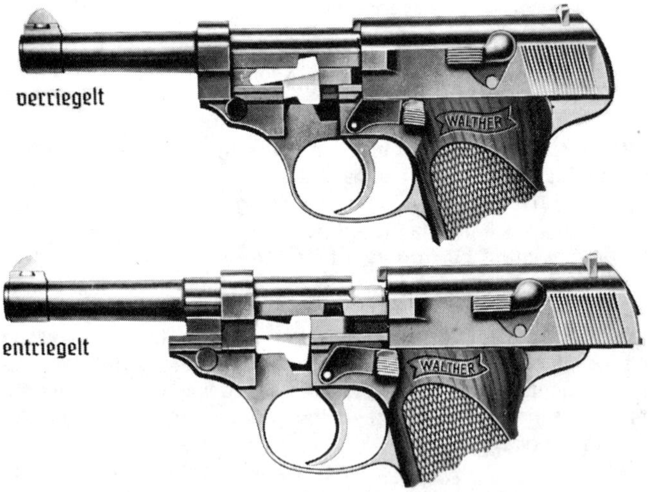

Walther Armee model pistol, caliber 9mm, hammerless; breech locked (above) and unlocked (below).

THE 9mm LUGER (PO8)

The Luger is one of the best known pistols in the world and there are several excellent books in English which deal with this pistol in detail far beyond the scope of this book. The Luger was first adopted as a service pistol by Switzerland in 1900; this model was chambered for the bottle-necked 7.65mm (called caliber .30 Luger in the U.S.) Luger cartridge. The 9mm Parabellum cartridge version, which introduced the most widely-used pistol and submachine gun cartridge in the world, appeared in 1902 and was adopted by the German Navy in 1904. In 1908, it was adopted by the German Army and remained the standard service pistol until 1938. There were over 400,000 Lugers manufactured for the German Army after the adoption of the P38 and manufacture was continued until 1943. There are at least thirty-five different variations of the Luger in existence, including numerous variations of the basic PO8 which were used by the German Army.

German 9mm Parabellum Model 08 Pistol (Luger PO8).

The Luger was manufactured in Germany by Deutsch Waffen und Munitions fabriken (DWM), Simson, Krieghoff, Erfurt Arsenal, and Mauser. It has also been manufactured by Vickers in England, for the Dutch government, and by the Swiss government arsenal at Bern for the Swiss government. Total Lugers manufactured is unknown, but it is probably at least two million and possibly considerably more.

The Luger is no longer in production as a military pistol, but a caliber .22 version of the Luger is now being produced by Erma in West Germany. The Luger is a fine-hanging pistol, very pleasant to shoot and it introduced an exceptionally fine cartridge, the 9mm Parabellum. It is not, however, a standard pistol in any major country today because it is prone to stoppages if mud or sand gets into the action. The Colt Browning type of pistol, which does not have as much of its working mechanism exposed as does the Luger, is a much more reliable pistol in operation under muddy or sandy conditions.

The Navy Luger and long-barreled 1908 type (frequently called the Model 1914, Artillery, or Model 1917) were also used in German service. The Navy Luger has a six-inch barrel, adjustable rear sight and is ridged for a wooden shoulder-stock holster. The long-barreled 1908 has an eight-inch barrel and a tangent-type rear sight and was frequently issued with the 32-round "Snail" magazine. The latter type appeared toward the end of World War I and is made of PO8 components except for barrel and sights.

Many Lugers have their grips ridged for shoulder-stock holsters and the following facts should be brought to the attention of U.S. collectors. Possession of a pistol which is ridged or slotted for a shoulder stock and of the stock which fits the pistol, requires under U.S. firearms laws, that the weapon be registered with the Firearms Branch, Alcohol Tax Unit, U.S. Internal Revenue Service, the branch of the U.S. government which registers submachine guns and machine guns and enforces the National and Federal Firearms Acts. Failure to register this type of weapon with the above agency can result in fine or imprisonment, and at a minimum considerable trouble and embarrassment. It should be noted that in addition to the Federal statutes, most states have similar laws.

Field Stripping the PO8 Luger

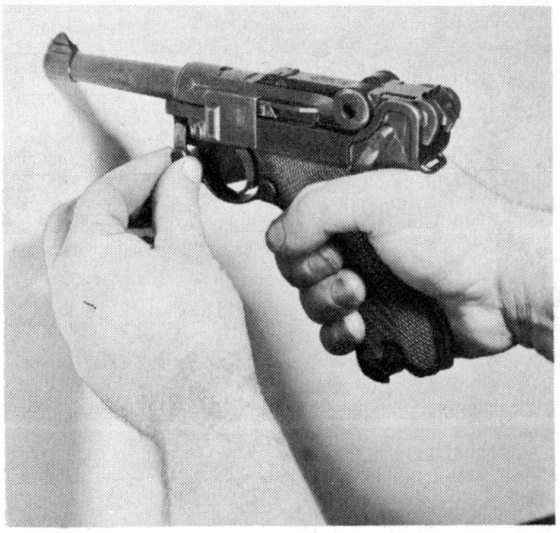

(1) Holding pistol in right hand press muzzle down firmly on a hard surface about 1/2 inch to release tension on the recoil spring. With the tension removed, the thumb catch on the side-plate may now be turned down to a vertical position.

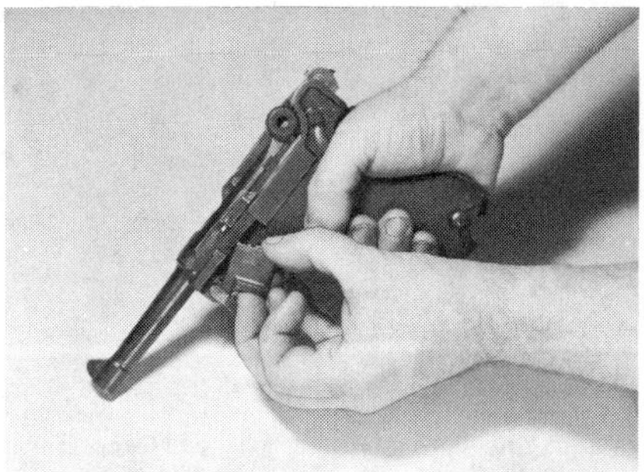

(2) Now lift out the sideplate.

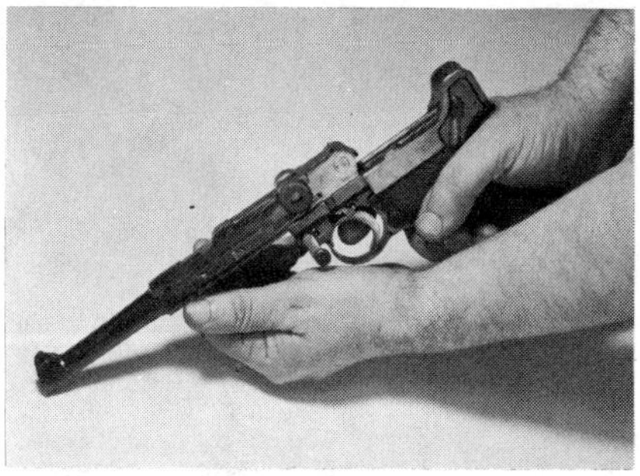

(3) Slide the complete barrel and toggle assembly directly to the front and out of the receiver.

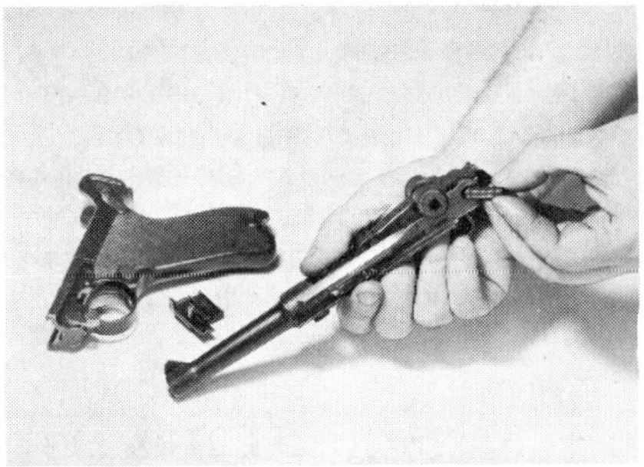

(4) Buckle the toggle slightly to relieve tension and extract retaining pin on the left hand side.

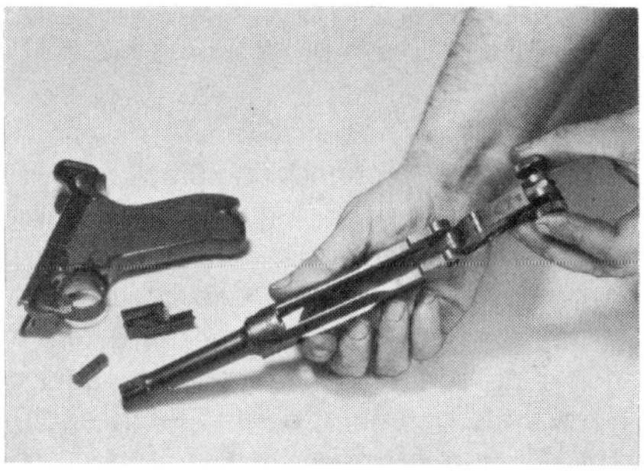

(5) Now pull toggle assembly, breechblock containing firing pin and extractor directly back in their guide and out of the frame. No further dismounting is necessary nor recommended.

Note on Reassembling. Merely reverse stripping procedure. Take care hook suspended from rear of the toggle assembly drops into proper place, which is in front of the inclined ramps. Also note that when replacing the side plate, the tongue on the rear end must be inserted in the recess in the receiver and the projecting section of the trigger bar must fall into the proper slot at the top of the trigger.

Instructions for Loading and Firing the Luger

(1) To extract magazine: Press magazine release stud near trigger on left hand side and withdraw magazine from butt of pistol. To load magazine: Hold magazine firmly in left hand. Pull down stud attached to magazine platform. This will compress spring and permit cartridge to be dropped into the magazine.

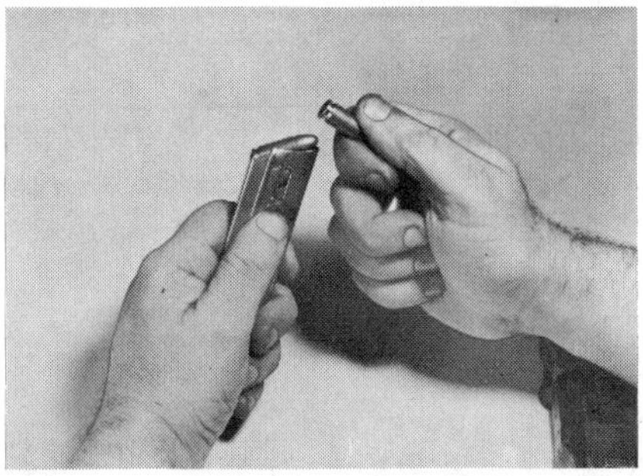

(2) To load chamber: Holding pistol pointed down toward ground with right hand, grip the milled knobs on the toggle and pull up and back as far as the breechblock will go. This compresses the recoil spring in the grip and permits the first cartridge in the magazine to rise in line with the breechblock.

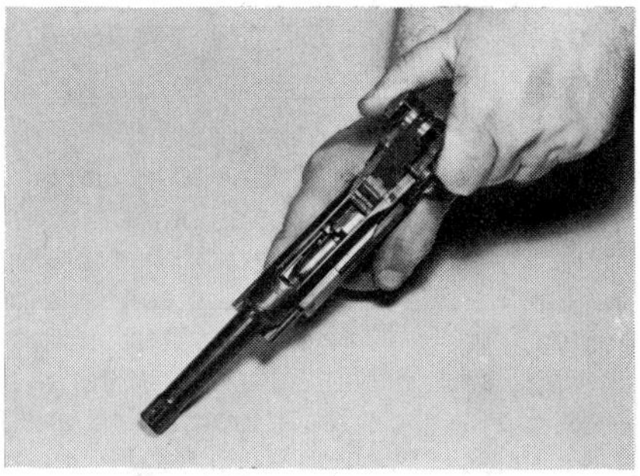

(3) Release grip and spring will force breechblock back into locked position driving a cartridge into the chamber.

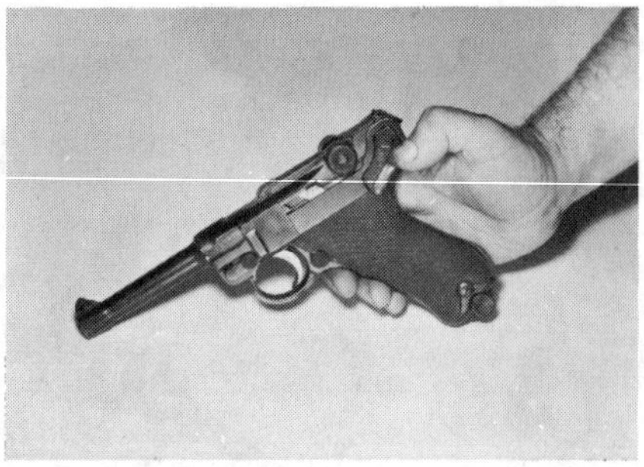

(4) To set thumb safety: Pull thumbpiece back and down. This will expose the German word "Gesichert," "Made safe." At the same time a flat solid steel piece will be seen to rise directly in front of the milled knob on the toggle. This locks the sear so the weapon cannot be fired.

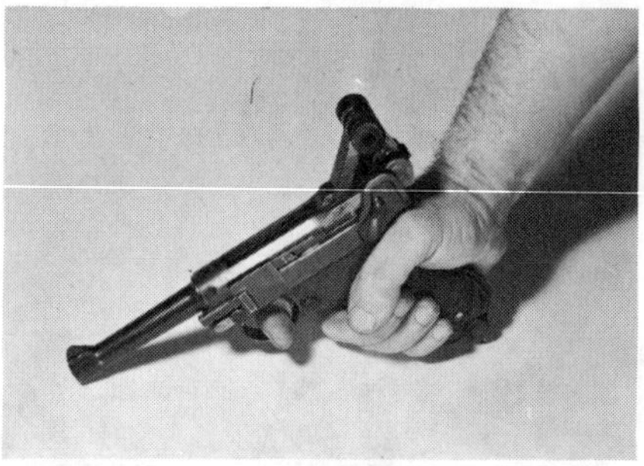

(5) Breechblock stop: When the last cartridge has been fired, the stud of the magazine follower will force the catch up and hold the breech open with toggle joint buckled. Reloading from open breech: (a) Remove empty magazine. (b) Replace with loaded magazine. (c) Pull back on milled surfaces and permit breechblock to drive forward loading chamber.

How the Pistol (Luger) 08 Works

Starting with the pistol loaded and cocked, the action is as follows: The trigger being pressed, a connecting piece forces back a pin which presses out a spring-retaining lock permitting the striker to go forward and fire the cartridge in the chamber. (The involved trigger system of this pistol is one of its greatest weaknesses.) As the bullet goes forward down the barrel, the barrel and the recoiling mechanism locked together move backwards about half an inch. There is a toggle joint behind the breechblock which functions exactly as a human knee. A strong pin at the rear fastens this toggle securely to the barrel extension. At the point where the breech pressure has dropped to safe limits, the center part of this toggle joint strikes against a sloping part of the frame, buckling the toggle exactly as a human knee; but continuing to draw the breech block in a direct line in its guide in the barrel extension. During this opening movement a short coil spring, which drives the firing pin and is situated inside the breech block, is compressed and caught and held by the sear. The extractor, fitted in the top front of the breech block pulls the empty cartridge case back until it strikes an ejector piece and is hurled out of the pistol. A small coil spring snaps the extractor back into place. As the toggle joint buckles upward, a hook lever hanging from its pin and hooked under claws attached to the recoil spring in the grip, compresses the recoil spring, storing up energy for the return movement of the action. The magazine spring forces a cartridge up into line with the breech block. A shoulder below the rear end of the barrel contacts a shoulder in the receiver in front of the magazine space stopping further travel. The rearward motion is now complete.

Return Movement of the Action. The compressed recoil spring pulls down against the hook lever drawing down the bent toggle exactly as a bent knee straightens out when one stands up. This forward action drives the attached breech block straight ahead in its guide and strips the top cartridge out of the magazine and into the chamber. The extractor springs over the head of the cartridge and locks in the cannelure of the cartridge case. This raises the height of the extractor so that it is above the face of the breech block. Looking at the extractor—or touching it if in the dark—tells if the chamber is loaded. The breech block and the two levers of the toggle are now in a straight line and the axis of the toggle is slightly below the other axes. The pistol is thus securely locked. The sear now connects with the trigger mechanism, trigger spring pushes trigger into place; and pistol is ready for the next shot.

How the Thumb Safety Works. While the normal method of applying the thumb safety on this pistol is to push it back and down, there are models in which the procedure is the exact reverse. When the pistol is on safe, a flat steel piece attached to thumb piece is forced up out of the receiver on the left-hand side just above the stock. This prevents the outward expansion of springs which lock the striker back in firing position.

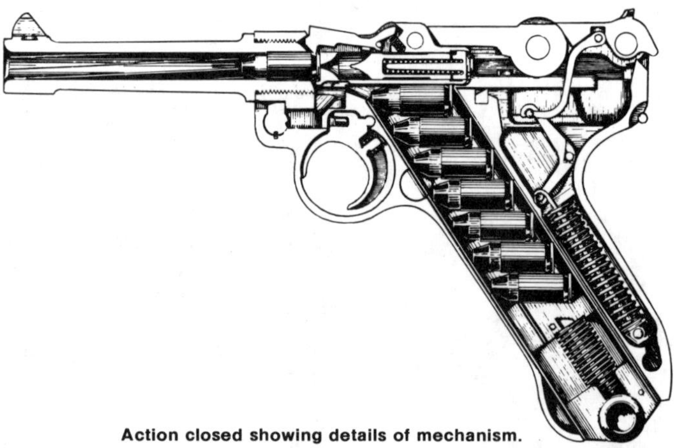

Action closed showing details of mechanism.

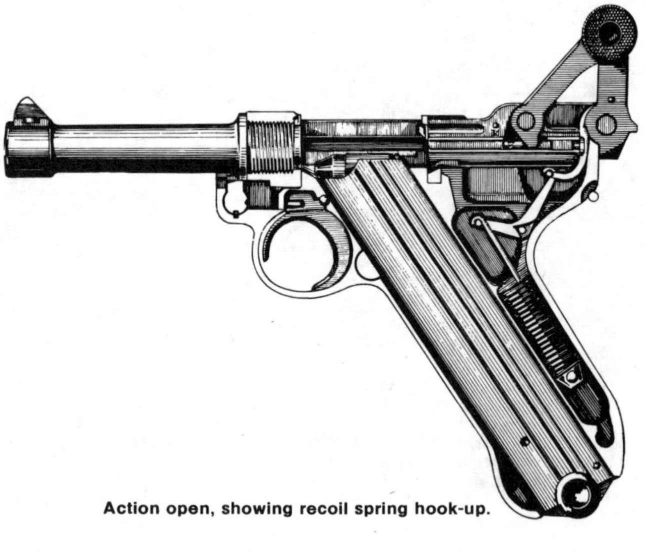

Action open, showing recoil spring hook-up.

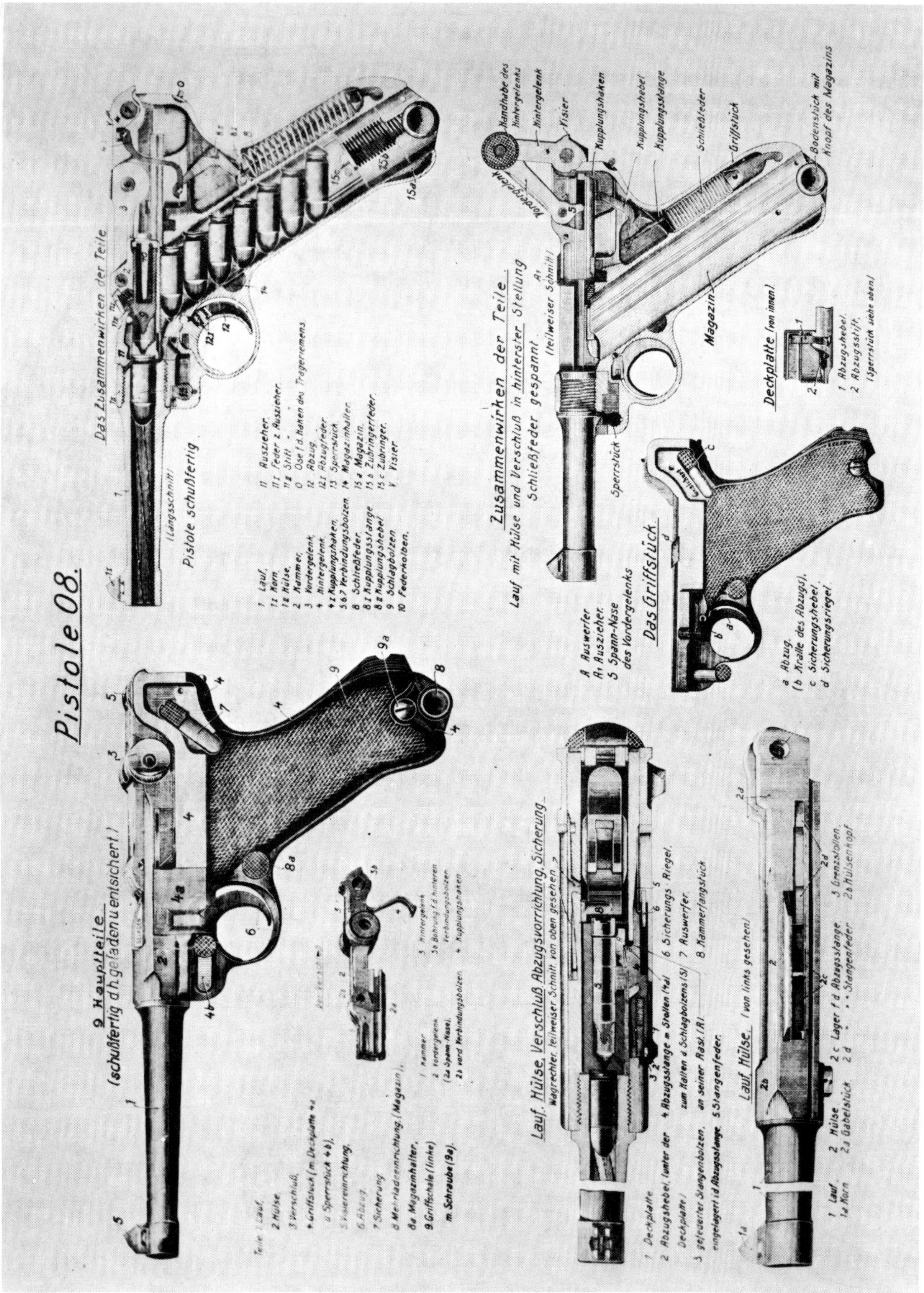

German training chart on the 9mm Parabellum PO8.

Luger Navy Pistol, Model 1904.

Long barrel PO8 Luger with 32-shot magazine.

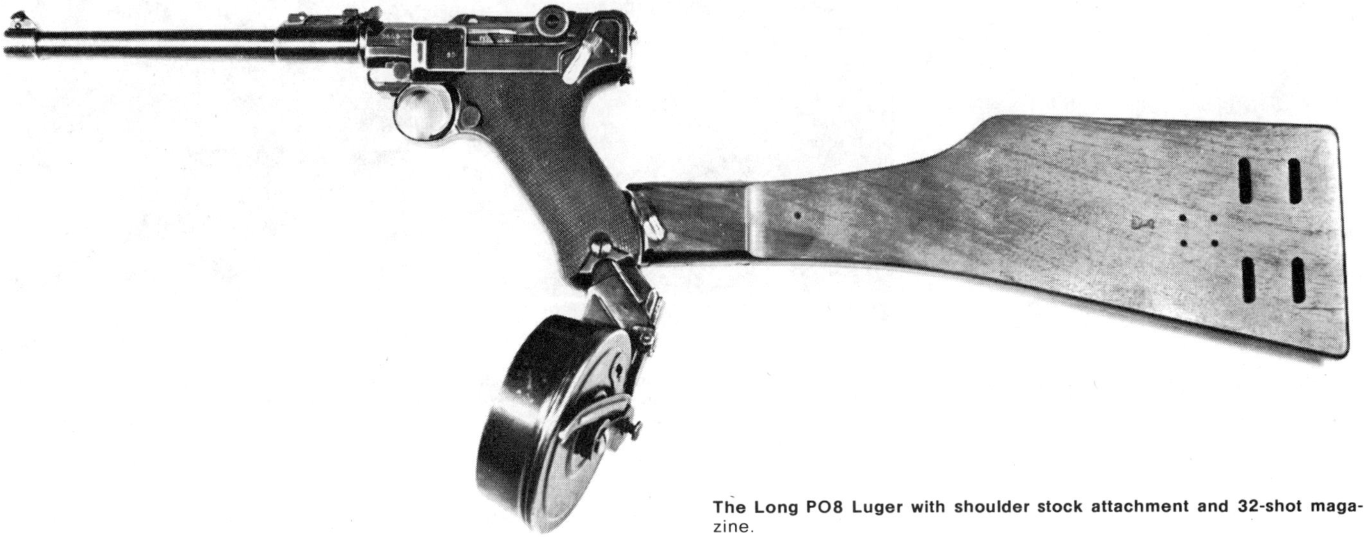

The Long PO8 Luger with shoulder stock attachment and 32-shot magazine.

To Load 32-Shot Magazines

This magazine was issued during World War I. Each magazine comes with a filler. The lever must be wound against the tension of the magazine spring and locks into position by means of a spring loaded catch. The loading tool is slid over the mouth end of the magazine and by a pumping action the cartridges are forced downward into the magazine. The magazine must be loaded cautiously since the lever is under heavy spring tension and if released it can do serious damage to the fingers.

(Note: this magazine is also used in the early model of the MP 18I Submachine Gun.)

THE MAUSER MILITARY MODEL

The Mauser was one of the first successful automatic pistols to appear but was never used by any power in the massive quantities of the Luger, P38, Colt M1911, Tokarev, or Browning Hi-Power. The Mauser has usually been the substitute standard pistol, probably because it is basically a rather expensive and somewhat awkward weapon. It does not have the natural pointing qualities of the Luger or the ruggedness under poor environmental conditions of the Colt Browning pistols. Granting all this, it is an interesting weapon and represents a stage in the development of the modern self-loading pistol.

The Mauser pistol first appeared in 1895 and with it appeared the bottle-necked 7.63mm (caliber .30 Mauser) cartridge. The cartridge became more popular than the pistol, as it was adopted in slightly modified form by the U.S.S.R. as their standard pistol and submachine gun (7.62mm Type P) cartridge from 1930 to the late 1940's and is still extensively used by the Soviet Bloc. At least thirty models of this pistol have been made and it is beyond the scope of the book to cover them all.

In addition to being made in 7.63mm, the Mauser was made in 9mm Mauser and, for the German Army during World War I, in 9mm Parabellum. The 9mm Mauser cartridge is a rather large cartridge which is not interchangeable with any other 9mm cartridge. Copies of the Mauser have been made in Spain and in China. Some of the Chinese copies have been in .45 caliber.

The 1912 Mauser was issued in 9mm Parabellum during World War II, equipped with shoulder stock and was called Model 1916. A large figure "9" is branded on each grip piece of this weapon. The 7.63mm Model 1932, (commercial designation Model 712), was used by the German security units during World War II. This model of the Mauser has a detachable magazine—10- and 20-shot magazines are used—but it can also be loaded with 10-round stripper clips from the top of the receiver. It is a selective-fire weapon. The 7.63mm Mauser cartridge was made for German service use with steel cased cartridges during World War II; this would indicate that there were a considerable quantity of 7.63mm Mauser pistols in service at that time.

MAUSER MODEL 1910 PISTOL

The 1910 Mauser was extensively used as a pistol for service troops during World War I and was used, among others, by SS police units during World War II. The Mauser 1910 was very widely distributed through commercial channels. It is a straight blowback weapon which has only one really unusual feature. When the last shot is fired, the slide remains open and the insertion of a magazine—after removal of the magazine originally in the weapon—whether loaded or empty causes the slide stop to release the slide and lets it return to the closed position chambering a cartridge in the process (if a loaded magazine was inserted). The slide will also remain open if it is drawn manually to the rear with an empty magazine in the gun.

In 1934 Mauser modified the 1910. The modification consisted principally of various changes in components to ease manufacture, substitution of stampings for machined parts etc., and the use of streamlined type grips.

Mauser 7.63mm Model 1932 Selective Fire Pistol.

Loading and Firing the Mauser

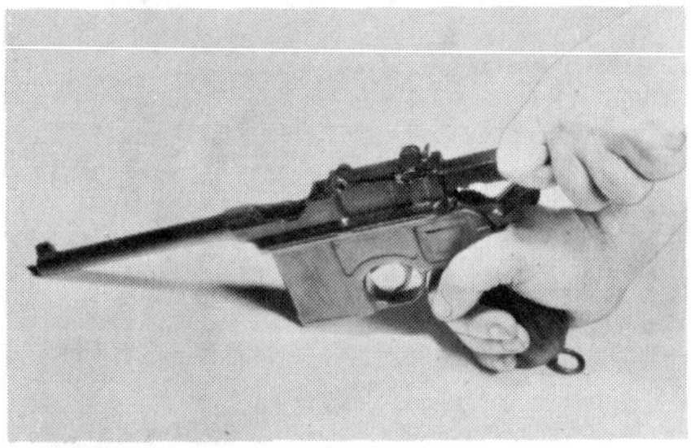

(1) Grasp bolt wings firmly and draw bolt straight to the rear as far as it will go. The magazine follower will rise and hold the bolt open.

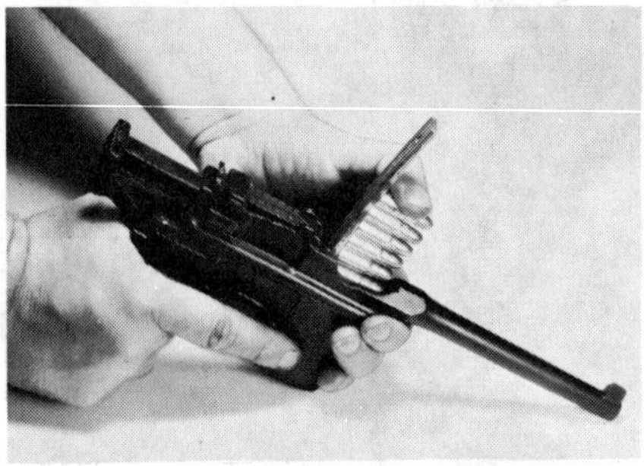

(2) Now insert a loaded clip in the clip guide directly in front of the rear sight. Exert firm even pressure with thumb and strip cartridges down into the magazine.

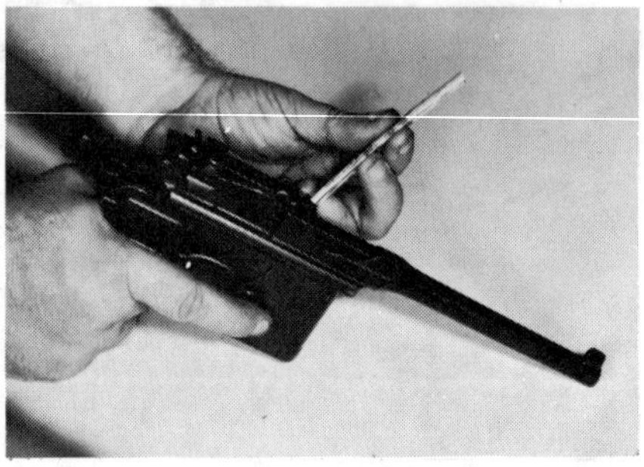

(3) Now pull clip straight up out of pistol. The bolt will run forward and load the chamber the instant the clip is withdrawn.

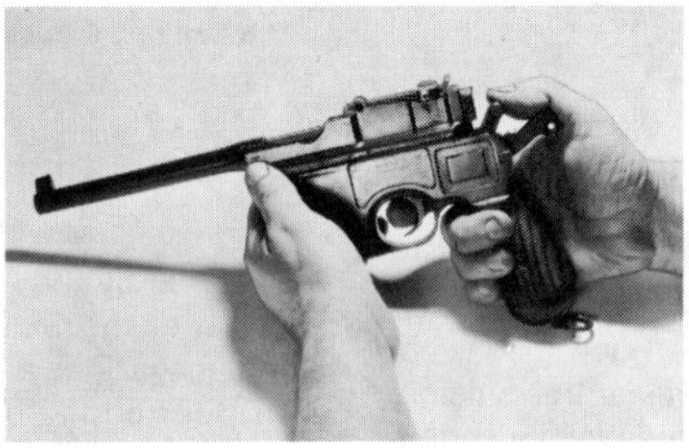

(4) Unless pistol is to be fired immediately, set the thumb safety by rocking it forward as far as it will go. Thumb safety may be easily released by pulling back slightly on hammer and tilting pistol up and back.

Field Stripping the Mauser

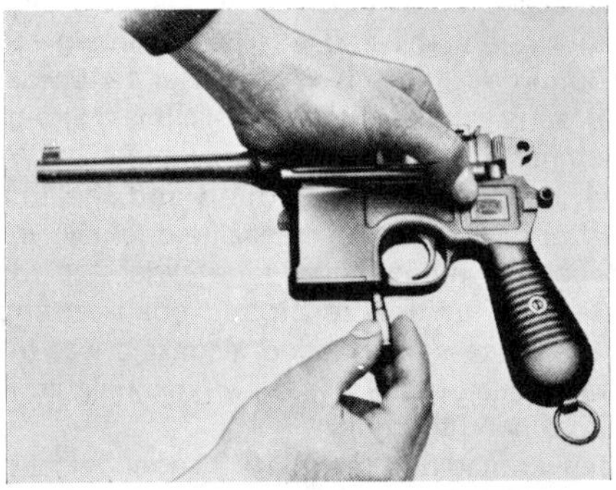

(1) To dismount magazine: Insert nose of a bullet in hole in magazine base plate and push up lock stud. With tension thus removed, slide plate forward slightly. It may now be eased out together with magazine spring and magazine follower.

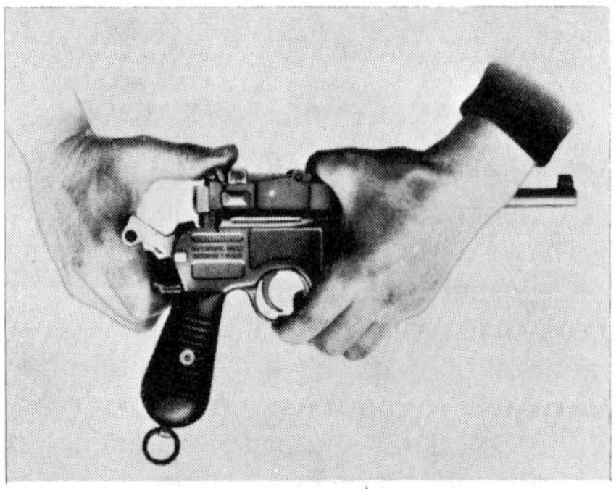

(2) Cock the hammer. Using a cartridge clip, or screwdriver, press up the catch just below the base of the hammer and pull back barrel.

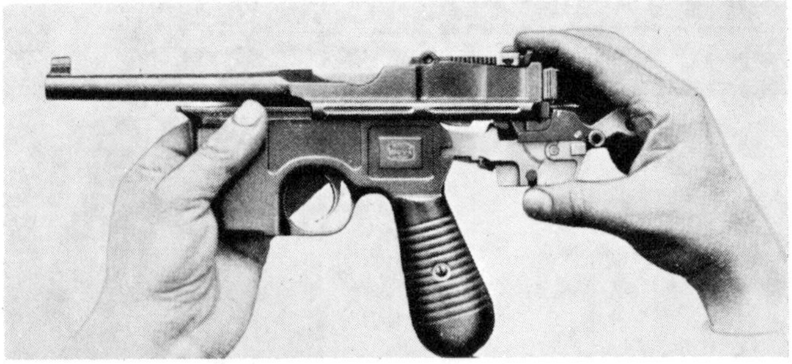

(3) Now withdraw barrel, barrel extension, and hammer mechanism clear of the receiver.

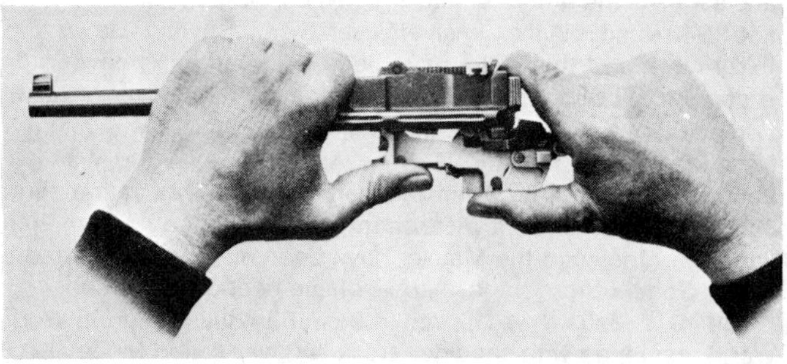

(4) Lift hammer mechanism out of barrel extension.

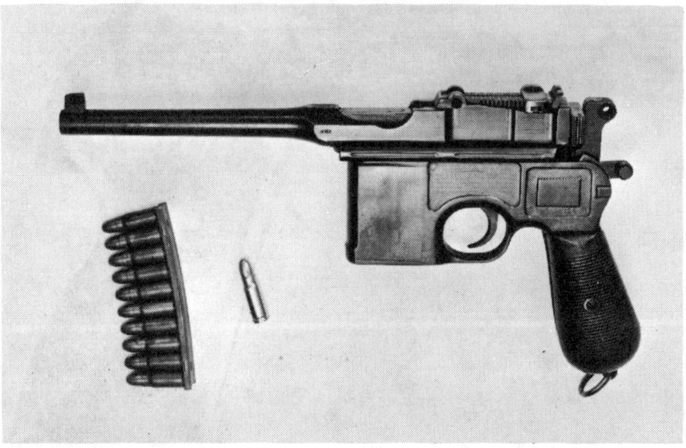

German Mauser 7.63mm Automatic, Model 1912.

Mauser 7.65mm Model 1910 Pistol.

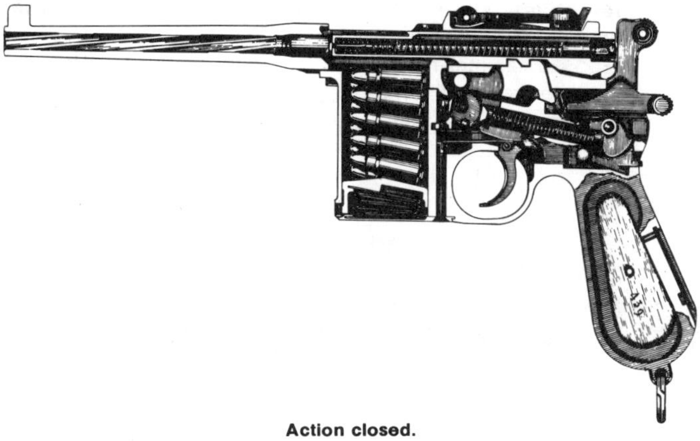

Action closed.

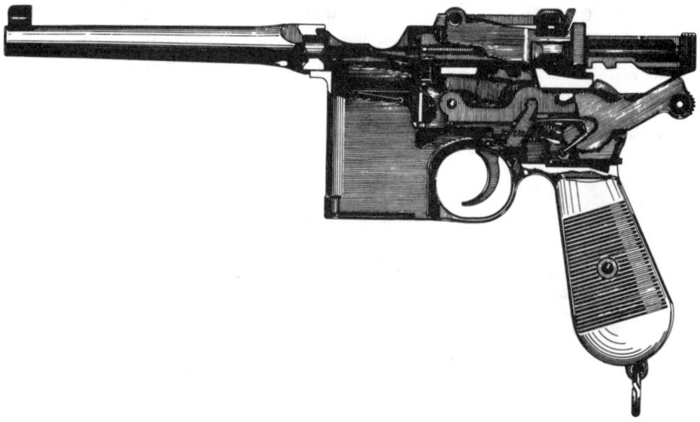

Action open and side cut away to show details of cocking mechanism.

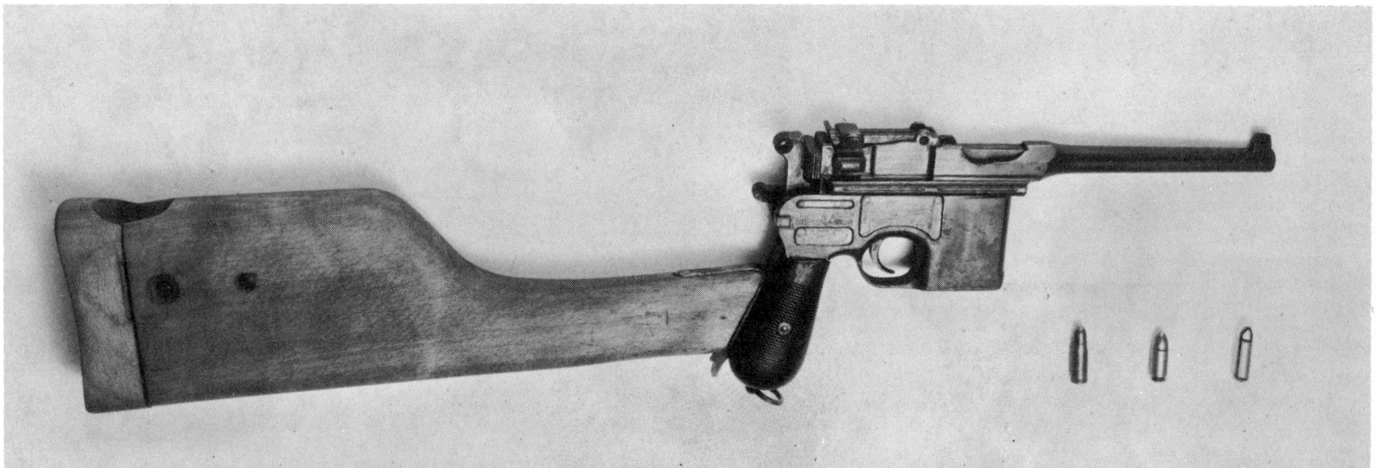

The Mauser 7.63mm automatic pistol with shoulder stock holster.

Magazine catch is in the bottom of the butt and must be pushed back to release the magazine.

When loaded magazine has been inserted in handle, slide is drawn fully to the rear exactly as in the case of the Colt automatic pistol and then permitted to run forward under the influence of the compressed recoil spring. The indicator pin protrudes from the rear of the breech block when the pistol is cocked, giving warning that the weapon is dangerous.

Pressing down on the milled thumb catch on the left side of the pistol, just back of the trigger, sets the pistol at "safe." To release this safety, press in the small button directly below the thumb piece.

MAUSER MODEL HSc PISTOL

This pistol was frequently called the Mauser Pistole neuer Art (M.n.A.), or Mauser new-type pistol by the Germans. The design was produced in the late thirties and was intended for military, police, and service use. The weapon was widely used by German service and police units during World War II and although the finish varies depending on date of manufacture, it is, generally speaking, a well-designed weapon of good manufacture.

The HSc is a double-action pistol; an enlarged version, the 9mm parabellum HSv, was Mauser's entry in the German service pistol tests of the late thirties which resulted in the selection of the Walther P38 as the German service pistol.

Loading and Firing the Mauser HSc Pistol

The slide is pulled back its full length to cock the hammer and permit the magazine spring to force a cartridge up to the feed way. This movement also cocks the hammer. Releasing the slide permits the recoil spring, which is wrapped around the barrel, to pull the slide forward and load the firing chamber.

When the last shot has been fired, the magazine follower holds the slide open as a notice. When the magazine is removed, and then reinserted, the slide goes forward automatically.

A positive magazine safety is incorporated in this weapon. When the magazine is withdrawn the trigger cannot be pulled.

A positive disconnector prevents more than one shot being fired for each pull of the trigger.

The action, being a straight blowback, permits the weapon to be of comparatively simple design, as no locking mechanism is necessary to shoot this low-power cartridge.

The exposed hammer may be lowered with the thumb. When it is necessary to fire, a pull on the trigger will function the hammer in the same general way that it does in a double-action revolver.

Mauser 7.65mm Model HSc Pistol.

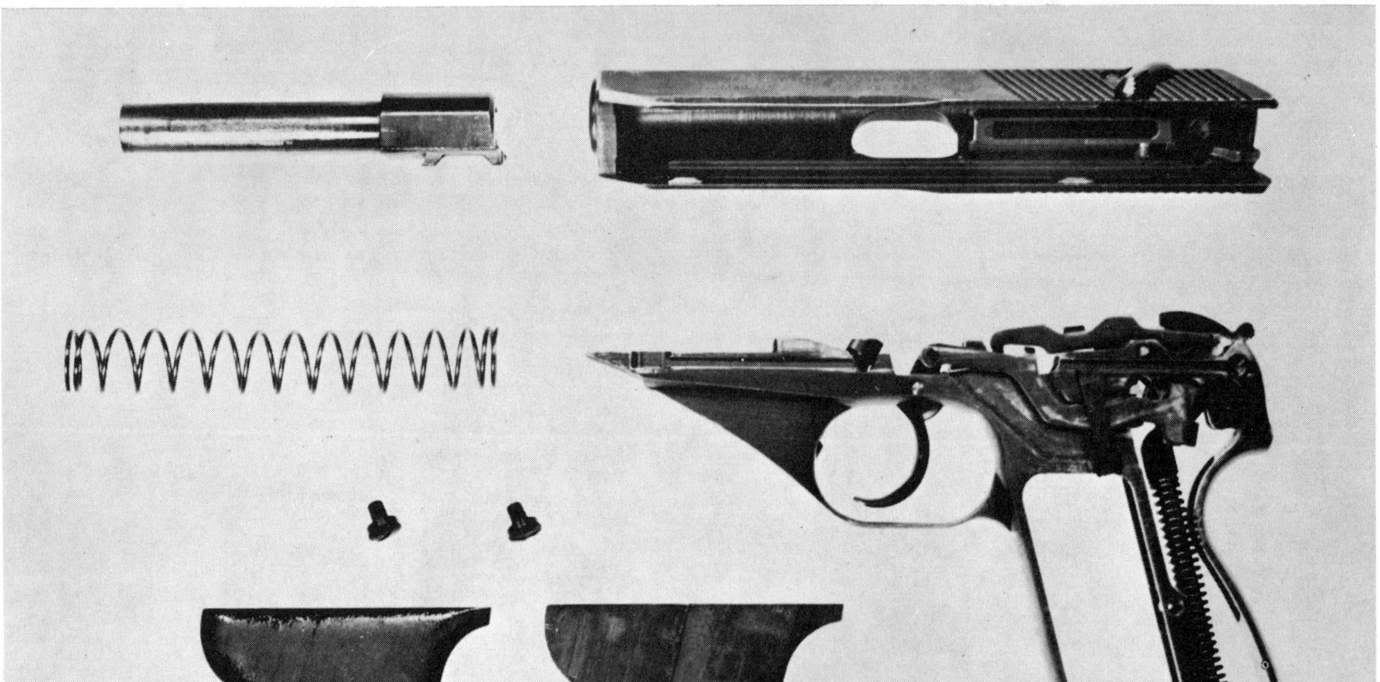

The Model HSc stripped.

Field Stripping the HSc Mauser

Push the small spring supported piece inside the trigger guard directly in front of the trigger. While maintaining this pressure, push the slide forward a short distance.

Move the slide slightly backward. It may then be lifted up and off the receiver.

The barrel may be pushed forward against tension of the recoil spring and lifted up and out of the slide.

Removing the stock screws and lifting off the stocks exposes the firing mechanism.

THE SAUER MODEL 38 PISTOL

The Model 38 is a double-action pistol with internal hammer. It was a very popular weapon with the Germans and has several good features. Like the Walthers, it has a loaded-chamber indicator which projects from the rear of the slide when the cartridge is in the chamber.

Field Stripping the Model 38 Sauer

When the magazine is removed, the weapon cannot be fired. The thumb safety not only blocks the hammer when it is applied, but pushes it back out of engagement with the sear.

The double-action firing system is one of the best developed. It utilizes a minimum of springs and is entirely enclosed.

One unique feature is an exposed cocking lever which does not move with the action. When the weapon has been loaded and cocked by a movement of the slide, pressing down on this lever will safely lower the hammer so that a cartridge may be carried in the firing chamber in complete safety. Pulling straight through on the trigger will cock and trip the hammer to fire the cartridge. However, pressure on the cocking lever will also recock the hammer should it be desirable to take more deliberate aim.

Sauer 7.65mm Pistol Model 38.

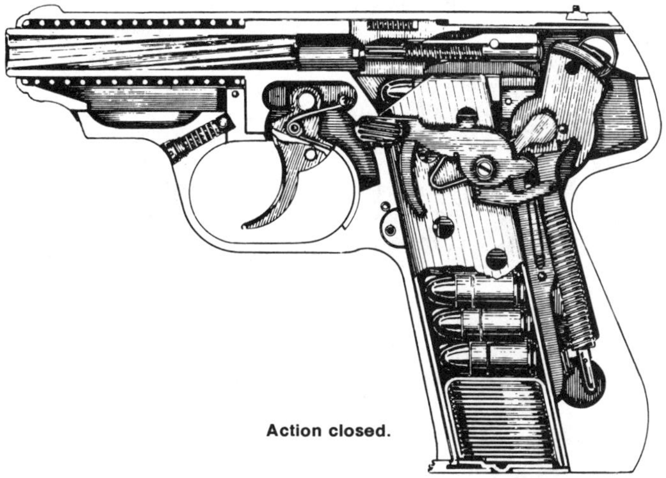

Action closed.

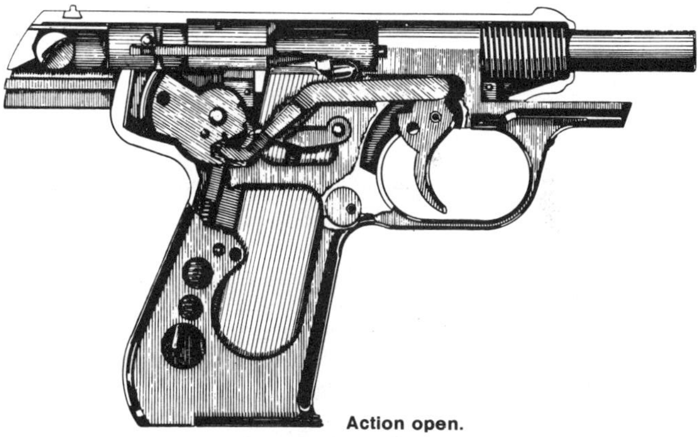

Action open.

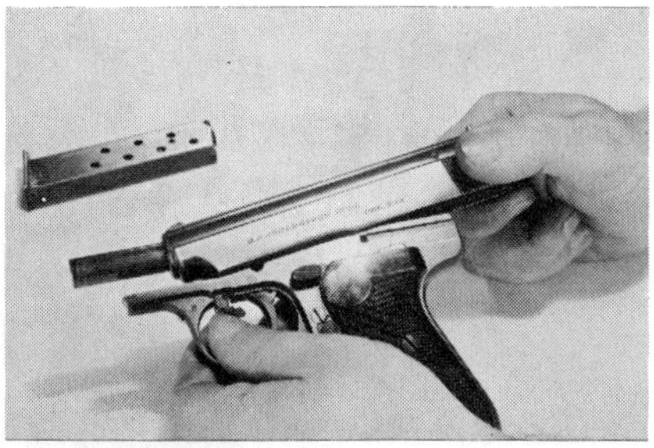

Pull down the locking latch in front and above the trigger. Draw the slide back to its full extent, lifting upward as you draw.

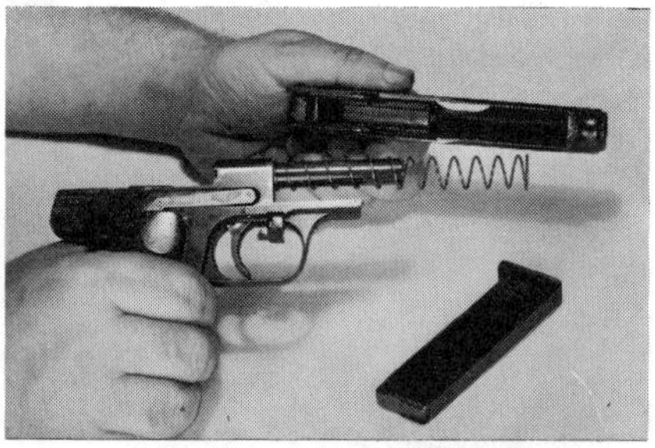

Let the disengaged slide move forward and push it off the barrel.

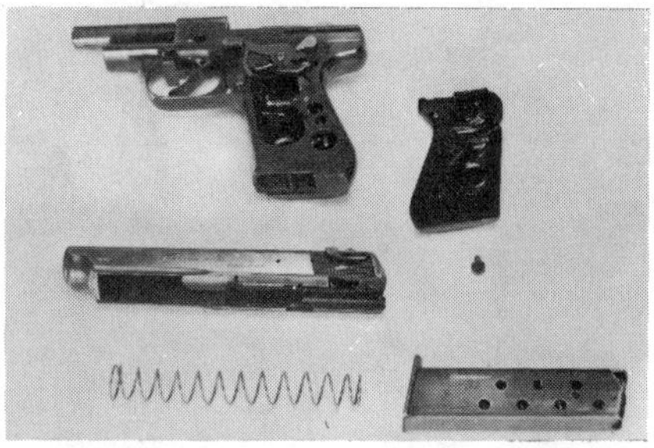

Recoil spring may be drawn off the barrel. Removing the grips exposes all working parts for attention. No further stripping need be done.

THE WALTHER PP AND PPK PISTOLS

These pistols are among the best known in the world and were made in large quantities for German service use and commercial sale. The Model PP (Polizei Pistole) was introduced in 1929 and the PPK (Polizei Pistole), a shorter version of the PP, was introduced in 1931. Both were originally chambered for the 7.65mm (.32 ACP) cartridge, but caliber .22, 6.35mm (.25 ACP), and 9mm short (.380 ACP) versions were also produced.

The PP and PPK had considerable influence on pistol design in Germany prior to World War II and throughout the rest of the world since. Copies of the PP have been made in Turkey (Kirikkale), Hungary (M48), and under license in France (Manurhin), since World War II. The PP and PPK may have been the first production pistols made with lightweight alloy receivers, since they were on the market with light receivers during the thirties. The PP and PPK are currently being made by Carl Walther at Ulm a/d, West Germany.

The Model PP was approved for German service use in 9mm short as well as 7.65mm. Some PP pistols made during World War II do not have the pin-type loaded chamber indicator which normally protrudes from the rear of the slide.

How to Load and Fire the PP and PPK

The magazine is removed by pressing the magazine catch on the left side behind the trigger guard. Fill magazine with cartridges and insert smartly in grip. Push safety catch—on left rear of slide—down; pull slide to the rear and release. Push safety up into the off position and pistol is ready to fire double action by pulling through on trigger. If a lighter trigger pull is desired, cock hammer and then press trigger. The slide will remain open on the last shot. It can be released on a new loaded magazine by pulling it slightly to the rear and releasing. The hammer will be cocked and the pistol can be fired by pressure on the trigger or the safety can be applied, in which case the hammer will fall and remain in the down position.

How the PP and PPK Work

These are blowback pistols of advanced design. External hammers, double-action triggers, positive manual safeties.

The recoil spring is positioned around the barrel. When the slide is drawn back over the top of a loaded magazine in standard automatic pistol fashion, the rear of the slide runs over and cocks the hammer. The recoil spring is compressed between a shoulder in the front end of the shaped slide (which surrounds the barrel muzzle) and the receiver abutment into which the barrel is secured. Releasing the slide permits the spring to pull it forward. The

Walther 7.65mm Model PPK Pistol.

breechblock face of the slide chambers a cartridge from the top of the magazine. The extractor claw in the right side of the breechblock is snapped into cartridge engagement by its spring.

Pulling the trigger will cause an attached trigger bar to draw the sear out of engagement with the hammer to fire the cartridge.

When the chamber is loaded, the front end of a floating pin in the slide is raised. The rear end of the pin projects from the rear of the slide. If this pin can be seen or felt, the chamber is loaded.

Safety Systems. A special hammer block of steel prevents any forward hammer movement until the trigger is deliberately pulled. When the trigger bar pulls the upper section of the rotating sear, a nose on the sear raises the hammer block until it is opposite a cut in the hammer face. Only at this point can the hammer fall far enough to hit the firing pin.

The firing pin is the spring-loaded type which is shorter than the length of its stroke. Its spring pulls it back into the breechblock as soon as its forward drive is halted.

The rearward thrust of the gas within the cartridge case drives the slide to the rear to extract and eject and reload in standard blowback fashion.

Disconnector. The opening movement of the slide runs over and forces down a section of the trigger bar. This disconnects the trigger bar from effective sear contact. Until the trigger is released permitting the spring to force it ahead into firing position, the trigger bar attached to the trigger cannot rise into a slide undercut. Only when it can rise in this undercut can the trigger bar tip draw the sear out of engagement to let the hammer fall.

Thus one pull is necessary for each shot fired.

Walther 7.65mm Model PP Pistol.

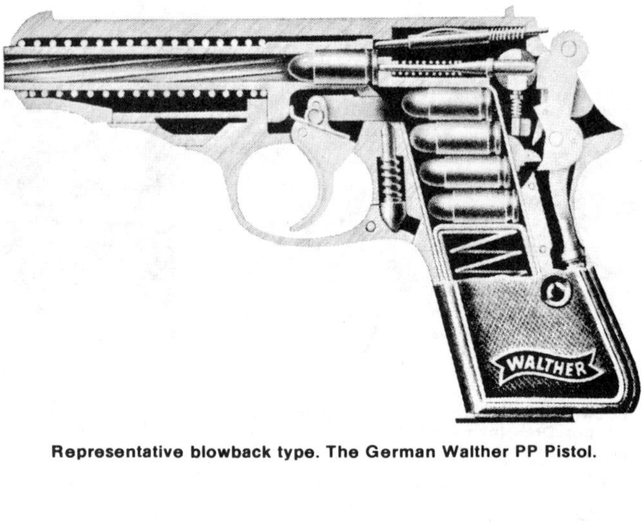

Representative blowback type. The German Walther PP Pistol.

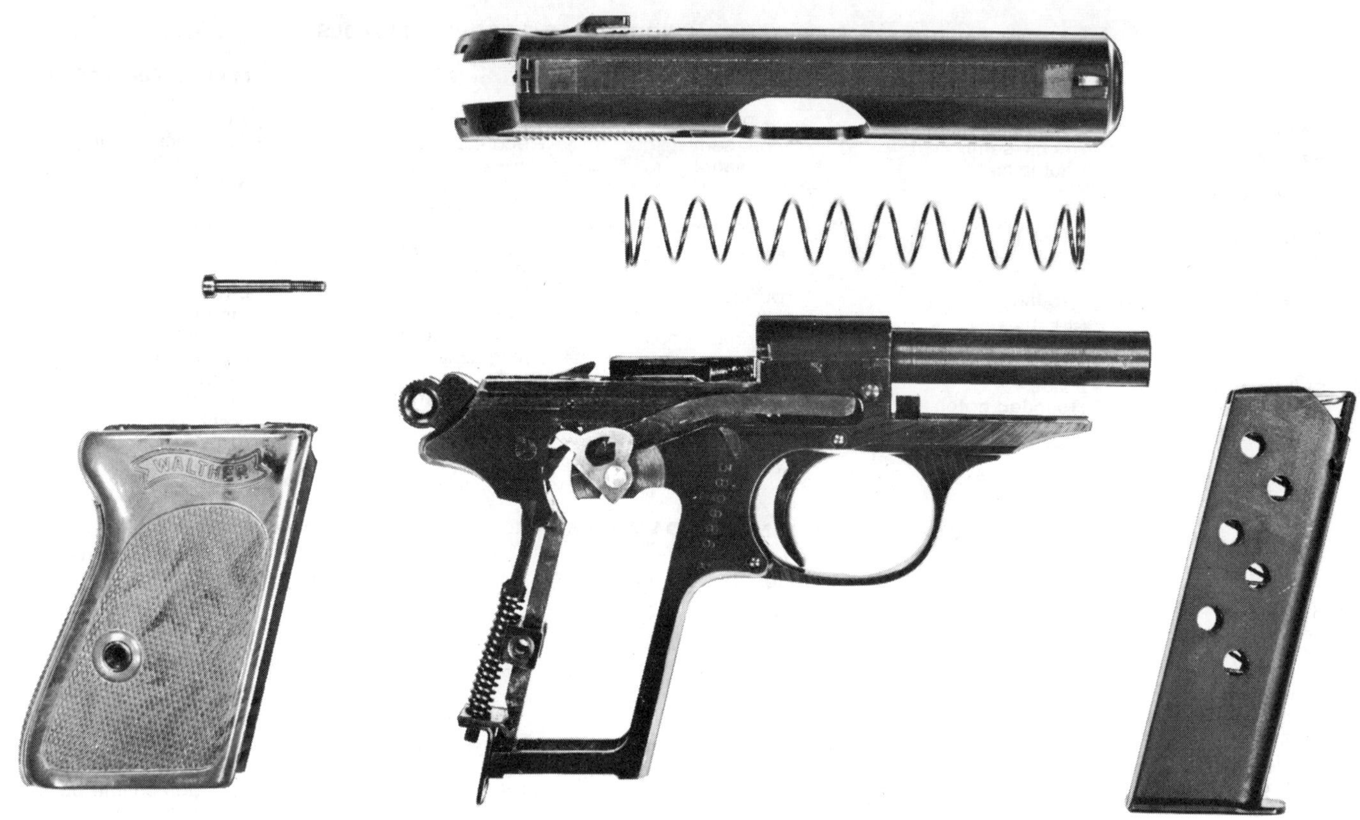

PPK stripped; hammer is cocked. Pull on trigger will draw trigger bar ahead to disengage sear from hammer. Pulling trigger guard down lowers slide locking lug seen under barrel.

Manual Safety. A thumb piece projects from the left side of the slide. When the hammer is down, if this safety is turned it will bring a steel face between the hammer and the head of the striker and will lock the sear so that the trigger cannot be pulled far enough to raise and drop the hammer.

If the thumbpiece is turned while the hammer is cocked, the safety face will come between the hammer and the firing pin, and the safety arm will then trip the sear to let the hammer fall on the steel safety. Note that this differs entirely from the P-38 design, in which the pin is locked, but the hammer hits the pin head.

THE WALTHER MODEL MP

The MP is essentially an oversized Model PP chambered for the 9mm Parabellum cartridge. This weapon was a pre-World War II commercial development and was dropped by Walther with the emergence of the Model HP.

Specimens of this pistol in caliber .45 ACP have been reported.

The Sauer Behorden Modell (S & S BM)

The Behorden modell was introduced in 1930 and was used by German military police units. Like the earlier Model 1914 Sauer, the Behorden modell has a bolt, externally visible, at the rear of the slide, which reciprocates rather than having the slide reciprocate as do most self-loading pistols.

Walther 9mm Parabellum Model MP Pistol.

S&S Behorden modell, 7.65mm.

CHARACTERISTICS OF GERMAN WORLD WAR II SERVICE PISTOLS

	P38	P08	Mauser Model 1932	Mauser Model 1910
Caliber:	9mm Parabellum.	9mm Parabellum.	7.63mm.	7.65mm.
System of operation:	Recoil, semi-automatic.	Recoil, semi-automatic.	Recoil, selective fire.	Blowback, semi-automatic.
Length overall:	8.6 in.	8.75 in.	11.75 in. w/o stock. 25.5 in. w/stock.	6.2 in.
Barrel length:	4.9 in.	4.06 in.	5.63 in.	3.4 in.
Weight:	2.1 lb.	1.93 lb.	2.93 lb. w/o stock. 3.93 lb. w/stock.	1.3 lb.
Feed device:	8-round, in-line, detachable box magazine.	8-round, in-line, detachable box magazine.	10-or 20-round, staggered row, detachable box magazine.	8-round, in-line, detachable box magazine.
Sights: Front:	Blade.	Blade.	Blade.	Blade.
Rear:	Rounded notch.	V notch.	Tangent leaf.	Round notch.
Muzzle velocity·	1115 f.p.s.	1050 f.p.s.	1575 f.p.s.	950 f.p.s.
Status:	Army Standard.	Army Limited Standard.	Police and S.S. use.	Army and Police use.

*Also made in 9mm short (.380 ACP)

CHARACTERISTICS OF GERMAN WORLD WAR II SERVICE PISTOLS, Continued

	Mauser HSc	Sauer Model 38	Walther PP	Walther PPK	Sauer Bohorden Modell
Caliber:	7.65mm.	7.65mm.	7.65mm*.	7.65mm*.	7.65mm.
System of operation:	Blowback, automatic.	Blowback, semiautomatic.	Blowback, semiautomatic.	Blowback, semiautomatic.	Blowback, semiautomatic.
Length overall:	6.5 in.	6.3 in.	6.8 in.	6.1 in.	5.9 in.
Barrel length:	3.4 in.	3.5 in.	3.9 in.	3.4 in.	3.1 in.
Weight:	1.3 lb.	1.56 lb.	1.5 lb.	1.25 lb.	1.37 lb.
Feed device:	8-round, in-line, detachable box magazine.	8-round, in-line, detachable box magazine.	8-round, in-line, detachable box magazine.	7-round, in-line, detachable box magazine.	7-round, in-line, detachable box magazine.
Sights: Front:	Blade.	Blade.	Blade.	Blade.	Blade.
Rear:	Round notch.	Round notch.	Round notch.	Round notch.	Round notch.
Muzzle velocity:	950 f.p.s.	920 f.p.s.	948 f.p.s.	919 f.p.s.	886 f.p.s.
Status:	Army and Police use.	Army and Police use.	Army and Police use.	Army and Police use.	Police use.

GERMAN WORLD WAR II RIFLES

The Germans used a great variety of rifles during World War II, but only a few were of German manufacture and considered standard by the German Army. The 7.92mm (called 7.9mm by the Germans) Mauser Kar 98k was the most widely used standard rifle; none of the semiautomatic rifles were ever made in the quantity of the bolt action Kar 98k. The German plan of 1944/45 was to replace the bolt action and semiautomatic rifles with the selective-fire assault rifles of the MP43/44, StG44 series; the conclusion of the war prevented the fulfillment of this plan.

Many of the older German rifles such as the 11mm Mauser Models 1871, 1871/84, Model 98, 98a, and the Model 88 were used by Volksturm (Home Guard Units) as were large quantities of captured weapons. Some weapons originally of foreign origin were adopted by the Germans, such as the bolt-action Model 33/40, which was basically a slight modification of the Czech Model 33 Carbine, and the Model 98/40 which was an alteration of the Hungarian Model 35 Rifle. Many other foreign rifles were not adopted or standardized in the American sense of the term but were used in quantity such as the Czech and FN Model 24 Mausers, and the numerous varieties of these found in Rumania, Yugoslavia, Greece, and Bulgaria. It is probable that at least one-third of the rifles brought back to the United States by returning soldiers and fondly referred to as German service rifles are not German at all, but in all probability were used by the Germans in one fashion or another. Even the 6.5mm Norwegian Model 1894 Krag was made in limited quantities for the Germans and some Italian Model 38 Mannlicher Carcano rifles were, around 1943, made in 7.92mm for the Germans. The Polish 7.92mm Model 29 Mauser was also used extensively.

A peculiarity of the German rifle program between the wars was that the Germans never seemed, until it was too late, to appreciate the advantage of semiautomatic rifles. The Germans used two semiautomatic rifles early in World War I—the Mexican-designed, Swiss-made 7mm Mondragon which was called the Aircraft self-loading carbine Model 1915 and the Mauser 7.92mm aircraft self-loading carbine, both of which were used by the Germans as aircraft guns before their aircraft were fitted with machine guns. A full-stocked version of the Mauser aircraft rifle, the 7.92mm Model 1916, was issued in limited quantities during the war.

Mauser had developed semiautomatic designs as early as 1898 and in 1900, 1902, 1908, and in 1935, Walther had produced a semiautomatic rifle prior to World War II; but the German Army adopted a modified 98 Mauser bolt-action rifle, the Kar 98k in 1935, one year before the United States adopted the semiautomatic M1 rifle. This dim spot in German weapons technology may have been due to their accent on the machine gun, an area in which they were very advanced indeed.

Desperate measures were taken during the war to make up for the shortage of rifle fire power. A high percentage of the squad was equipped with submachine guns. The 7.92mm Rifle 41(M) and Rifle 41(W) were made in small quantity and were not very successful. The 7.92mm Model 43 Rifle was made in larger quantity but too late to have any real effect on the battlefield.

Mauser 7.92mm Aircraft Self-loading Carbine.

While the Germans did not move out with the speed or energy they usually show in weaponry on the semiautomatic rifle, they were the first to develop the idea of the comparatively light-weight assault rifle which would replace the rifle and submachine gun. Development of an "intermediate" sized cartridge was started in Germany prior to World War II and the conclusion of the war in 1945 prevented the Germans from putting their plan into fruition.

One of the most interesting German World War II rifles from a design point of view—the Paratroop Rifle (FG 42)— was not adopted by the Army. It was adopted by the Air Force (Luffwaffe) who controlled the airborne divisions. The FG42 is a very impressive rifle from many points of view, and lives on to some extent in the operating mechanism of the U. S. M60 machine gun.

THE 11mm MAUSER MODEL 1871 AND 1871/84 RIFLES

These rifles were not used by any German Army units during World War II, but were issued to the Home Guard. The Model 1871 was the first Mauser rifle to be adopted by any country and is a single-shot, black-powder weapon. Many of these rifles were sold as surplus to the smaller states of the world as were the Model 1871/84. A little known fact about the Model 1871 is that it was the principal weapon of the Irish rebels during the 1916 Easter Monday rebellion in Ireland. The rifles were purchased, with DWM cartridges, in Germany prior to the beginning of the war in 1914. They were landed at Howth, near Dublin, in a well-publicized gun-running expedition and are known in the Irish Republic as the Howth rifles.

The Model 1871/84 is basically the same as the Model 1871 but has a nine-round tubular magazine. The Model 1871, 1871/84, and all their variations use a two-piece bolt. The 11mm Model 71 black powder cartridge, which has a round-nosed bullet, and the 11mm Model 71/84 black-powder cartridge, which has a flat-nosed bullet, were developed for use with these weapons. The Model 1871 rifle was widely used in China for many years and was one of the principal weapons of the Chinese during the Boxer rebellion in 1900. They were also still in wide use by German African colonial troops during World War I.

THE 7.92mm MODEL 1888 RIFLES AND CARBINES

The adoption of the 8mm Lebel, using smokeless powder cartridges, by the French in 1886 caused the Germans to search for a suitable counter weapon. The 7.92mm Model 1888 rifle and carbine were the German answer to the Lebel.

This rifle is frequently called a Mauser, and also a Mannlicher; it was actually developed by a German Army Commission and combines the magazine of the Mannlicher with bolt features of the Mauser Model 1871/84. With the introduction of the Model 1888, Germany introduced the 7.92mm cartridge (called 7.9 in the German service). The 7.92 x 57mm Model 88 cartridge had the same case as found in this cartridge today, but had a .318-inch bullet, as opposed to the current .323-inch bullet (the "S" bullet which was introduced circa 1904-05). Many of these rifles were modified later to use the larger-sized bullet, but as a matter of course the 1888 pattern weapons should not be used with current 7.92mm cartridges. In addition to the bore diameter problem, the chamber pressure of currently available 7.92mm cartridges, especially military rounds, far exceed that for which the Model 1888 weapons were made.

The Model 1888 uses a five-round Mannlicher clip which can be loaded with either side down, unlike the Mannlicher 1886 clip which had to be loaded from one end only. The clip functions as part of the magazine and drops out the bottom of the protruding magazine box when the last round has been chambered.

The Model 1888 rifle and carbine and the Model 1891 rifle do

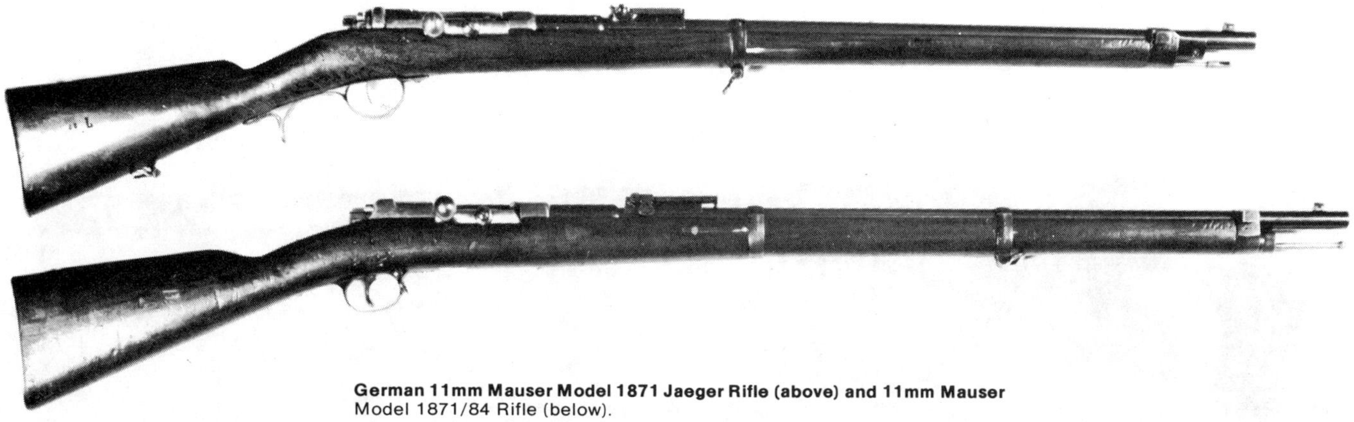

German 11mm Mauser Model 1871 Jaeger Rifle (above) and 11mm Mauser Model 1871/84 Rifle (below).

German Gew 71, Short Rifle.

not have wooden handguards. They have a sheet metal barrel jacket which covers the barrel from the receiver to the muzzle. Theoretically this barrel jacket provided for better accuracy since it prevented the changes in center of impact frequently caused by the change of bearing on barrels of wooden stocks and handguards, as a result of humidity, etc. In actual practice the metal barrel jacket suffered from many shortcomings: it was easily dented, water would seep into the joints and rust both the jacket and the outer portion of the barrel, and it was expensive and difficult to replace.

The Model 1888 was made in a rifle version, a carbine version, and a carbine with stacking hook was introduced in 1891—for some reason this carbine was called Rifle Model 1891.

Numbers of these weapons were modified by Germany at a later date. There were three basic modifications:

(1.) Some were fitted with a plunger and spring to eject the clip out of the top of the magazine after the last round was ejected in a fashion similar to that used by the U. S. M1 rifle. The clip ejection slot in the bottom of the magazine was covered.

(2.) Some were modified to use the charger used with the Model 98 rifle by milling a charger guide on the upper front end of the receiver guide and fitting a spring-loaded cartridge retaining rib on the upper side of the magazine.

(3.) Many were modified by relieving the chamber neck and forcing cone (lead) to use the "S" (.323 inch) bullet. These weapons are stamped "S" on the receiver.

The various modifications are called 88/05, 88/14, and 88S.

The Model 1888 was made at the German arsenals and by Ludwig Loewe, Haenel, Schilling, and Steyr. This rifle was used to some extent by Austria during World War I. China bought many 1888 pattern weapons from Germany and made a modified copy, the Type 88 or "Hanyang," rifle. Yugoslavia and Ethiopia also used the Model 1888 in limited quantities.

THE 7.92mm MODEL 98 AND ITS VARIATIONS

Model 98 Rifle (Gew 98)

The rifle 98 (Gewehr 98) introduced in 1898 is the most successful bolt-action design ever produced. In one form or another, the 98 action has been used by most of the countries of the world since 1898. As originally produced in Germany, the 7.92mm Model 98 rifle had the smaller sized (.318) bore of the Model 88 rifle; in 1903 the rifles were altered to use the larger diameter "S" bullet and bore diameter was set at .323. At the same time the rear sight was modified to match the ballistics of the "S" bullet.

The rifle 98 was the principal rifle of the German Army in World War I and a number of variations of the rifle appeared during that war. One of the first was the 98 rifle with turned-down bolt handle used by bicycle troops. Sights were again modified to reduce the battle sight setting from 400 meters to 150 meters; the marking disc on the left side of the buttstock was replaced with a washer type disc used to assist in disassembly of the firing pin. A variant of the 98 (the Model 18) with a sliding breech cover similar in concept to that of the British Lee Metford and the Japanese Type 38 rifle, and with detachable 5-, 10-, and 25-round box magazines, was developed toward the end of World War I.

The Model 98/17 also had a bolt cover and had a square shoulder on the follower to prevent closing the bolt on an empty magazine in the heat of action. The 98/17 had a 100-meter sight setting.

The Model 98 rifle also appeared in a caliber .22 training version, which was made from the standard 98 by fitting a liner in the barrel. Some Model 98 rifles were fitted with tangent-type rear sights.

Model 98 Carbine (Kar98)

The original Model 98 carbine was apparently never made in quantity since, unlike the Model 98 rifle, it is a rare item these days and photographs of German troops in World War I rarely show this weapon in evidence. The rear sight is similar to that of the Model 98 rifle and it has a peculiar stock and band arrangement. The stock runs to the muzzle as with the Model 88 carbine and 91 rifle, but is reduced in diameter at a point about six inches to the rear where a lower band, similar to the upper band of the 98 rifle, complete with bayonet mounting bar is fitted.

Model 98a Carbine (Kar 98a)

Originally called Kar98, this was the most popular carbine version of the Model 98 in World War I and it had limited usage

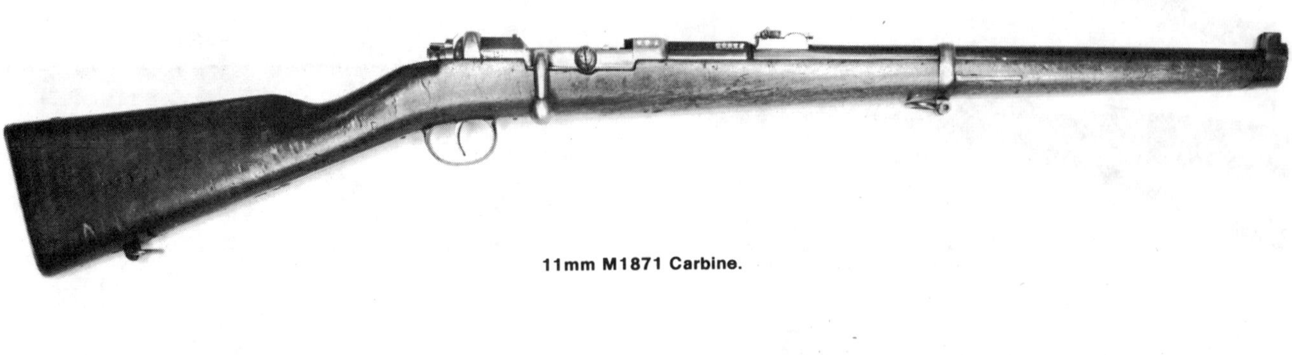

11mm M1871 Carbine.

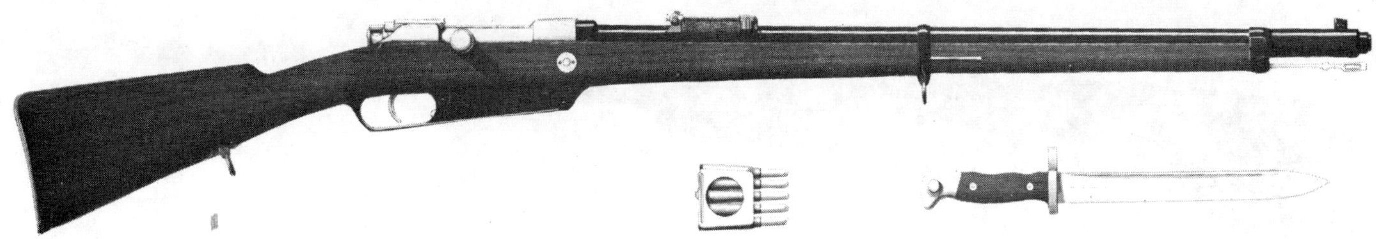

German Gew 88.

7.92mm Model 1888 Carbine.

M1888 sectioned to show locking, firing, and magazine systems. Magazine loaded. Bolt-head is detachable. Note relationship of locking lug seats to the face of the breech. Also see magazine follower driving cartridges up between clip walls. This is the Mannlicher clip system.

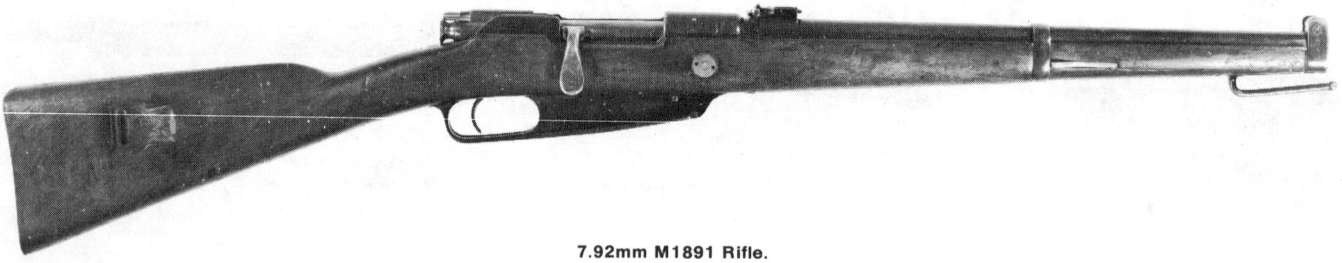

7.92mm M1891 Rifle.

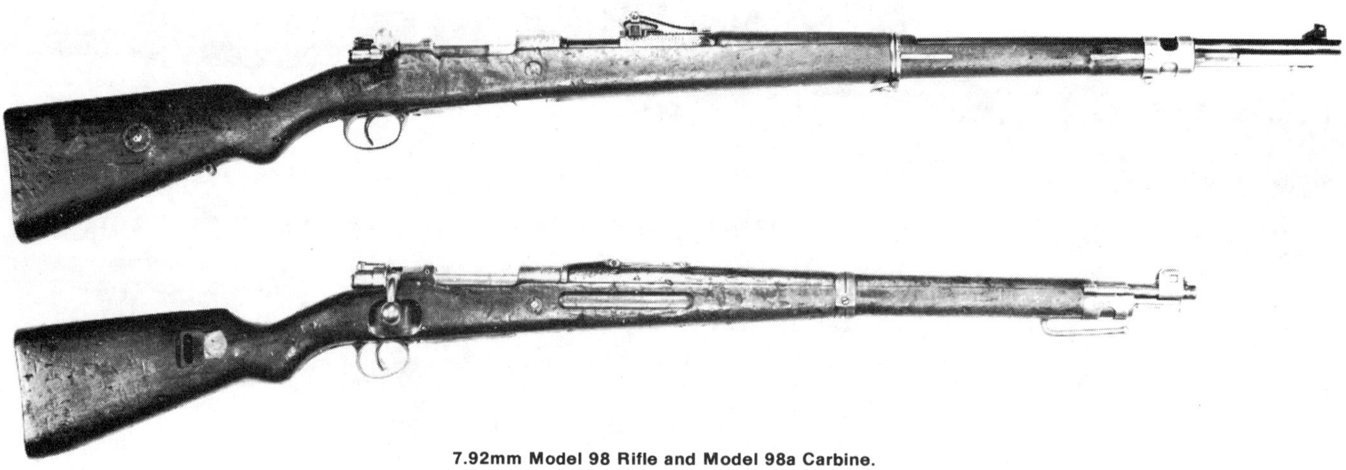

7.92mm Model 98 Rifle and Model 98a Carbine.

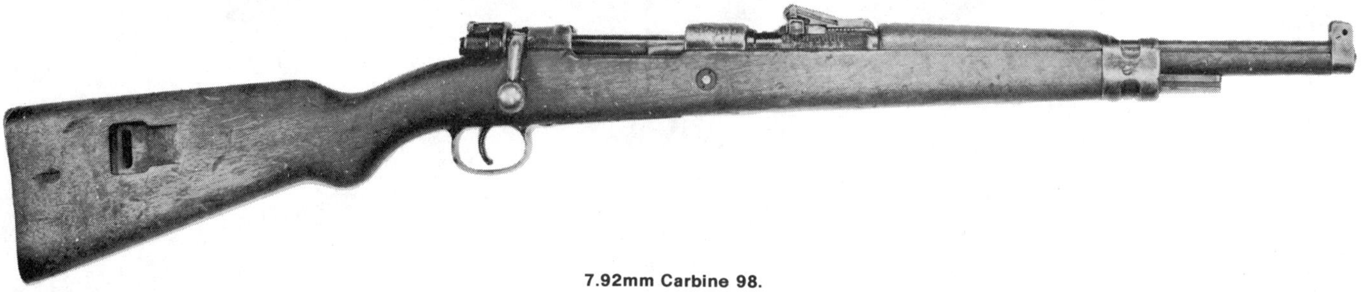

7.92mm Carbine 98.

in World War II. The Carbine 98a appeared in 1904 and was made in tremendous quantities until approximately 1918. It has been claimed that the appearance of the Mark I Short, Magazine, Lee Enfield in 1903, and the Springfield of the same year, influenced the Germans in the adoption of this weapon. In any event it is of handy size and was very popular with German troops. It introduced the tangent-type sight in the 98 series and was sighted for the "S" bullet as first issued. The prominent stacking hook, jointed upper band, front sight guard and full length handguards distinguish this weapon from the other German Mauser Service weapons. This weapon served as the model for the Polish Model 98 carbine.

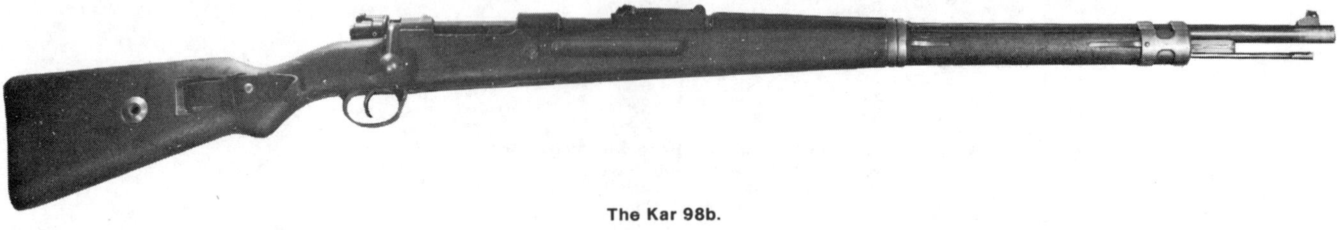

The Kar 98b.

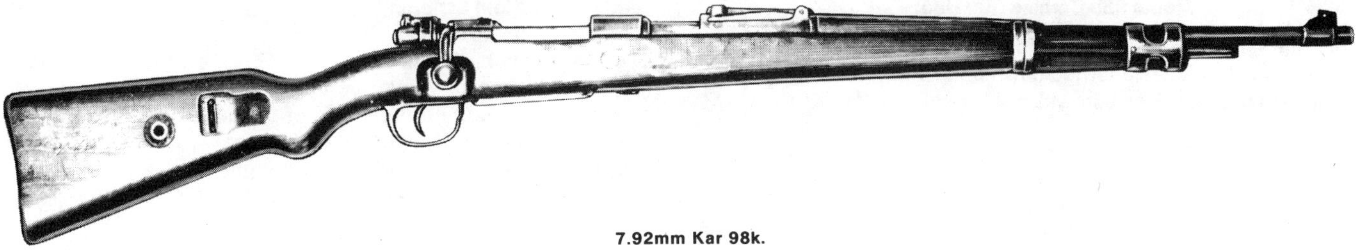

7.92mm Kar 98k.

The Gew 33/40.

Folding stock version of Gew 33/40.

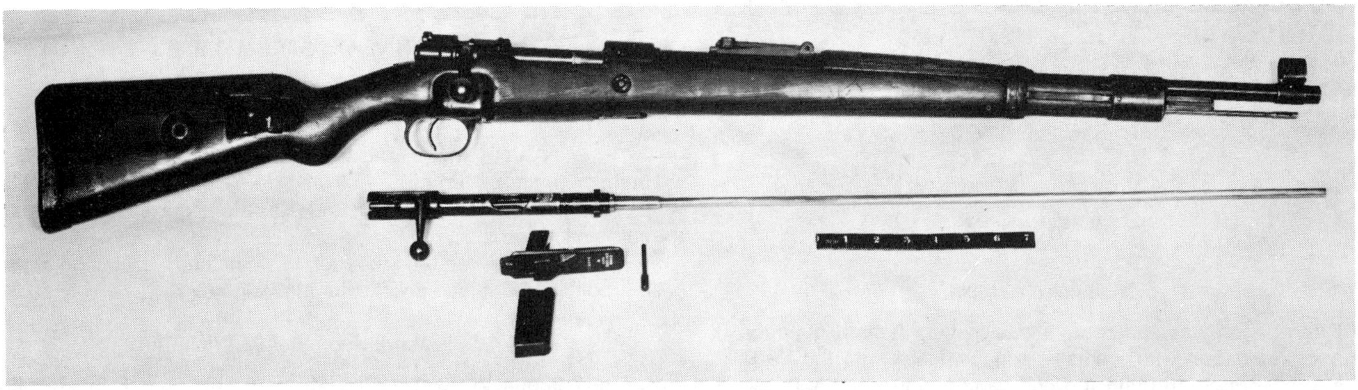

Caliber .22 conversion unit for Kar 98k.

Model 98b Carbine (Kar 98b)

This rifle, although designated a carbine, has the same length as the 98 rifle, but it has a turned-down bolt handle and a tangent sight like that of the Kar 98a which is graduated for SS (heavy ball) bullet.

Model 98k Carbine (Kar 98k)

This weapon was the standard rifle of the German Army in World War II and has been made in tremendous quantities. It was adopted by Germany in 1935 and has many of the features of the commercial Mauser "Standard Model." The "k" in "98k" stand for "kurz," which means "short" in English, and which is somewhat surprising since it is longer than the original Kar98 and about the same length as the Kar98a—even the methodical Germans apparently can be rather disorganized at times! Kar 98k has a half-length handguard, a tangent-type rear sight, turned down bolt handle and a hole bored through the stock in lieu of rear sling swivel (the Kar 98 and Kar 98a also have this feature).

In addition to the normal wooden furniture, Kar 98k has been issued with laminated and wooden stocks. Kar98k may still be found in service in various places in the world and rebuilt specimens of this weapon were taken from the Viet Cong in South Vietnam. The Kar98k was made without bayonet mounting bar and with stamped bands in 1944-45.

Rifle 33/40 (Gew 33/40)

The nomenclature of this weapon is another example of the inconsistency of German nomenclature—this weapon is one of the shortest barreled Mausers used by the Germans yet it is called a rifle! The 33/40 is the German version (made at Brno in occupied Czechoslovakia) of the Czech Model 33 carbine. It is distinguished by its light weight (the receiver has lightening cuts), short length, and the extension of the shoe-type butt plate on the right side of the butt. It was used by German mountain and paratroop divisions, and because of its light weight and short barrel, has a very sharp blast and heavy recoil. A folding-stock version of this rifle was made in limited quantity.

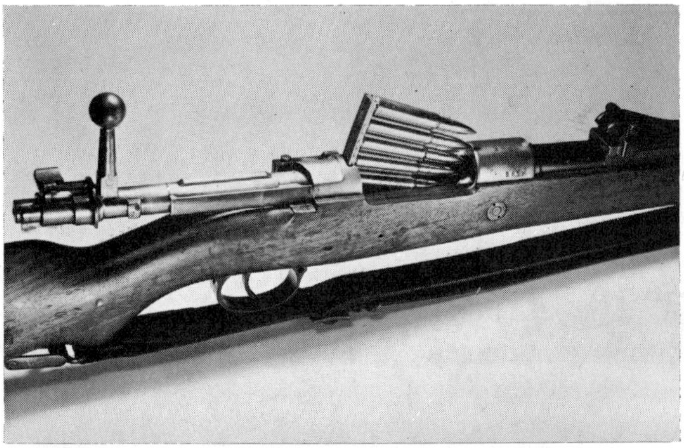

The Kar 98k is clip-loaded.

Rifle 40k (Gew40k)

This weapon was apparently made in very limited quantity. It is a short weapon with a smaller trigger guard than the Kar98k. It also has a hole through the bolt handle, and does not have a lower band.

Field Stripping the Mauser Carbine 98k

To remove bolt. Proceed as for U. S. Rifle Model 1917. With rifle cocked and safety lever vertical, half way between safe and locked position, pull out near end of bolt stop on left side of receiver and draw bolt straight out to the rear.

To dismount bolt. Proceed as for Springfield.

To remove magazine mechanism. Same as for Springfield.

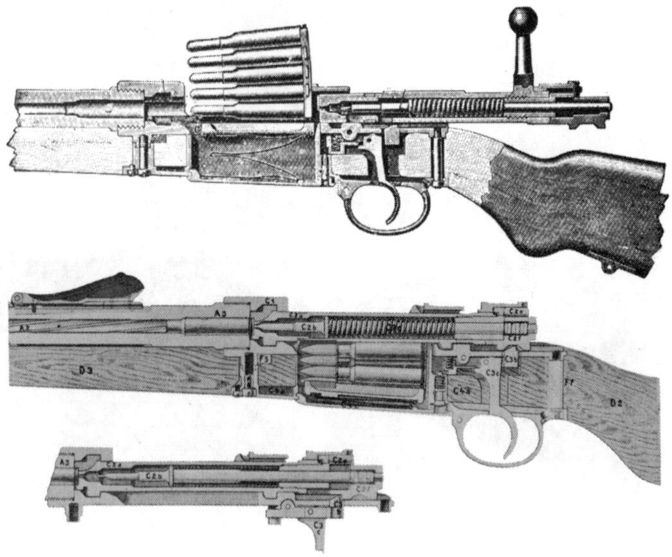

Kar 98k sectioned to show details of locking, firing and magazine systems.

Note position of dual forward locking lugs which assure maximum support to case head at instant of firing.

A 2. Rifling. A 3. Reinforced chamber section of barrel screwed into receiver. C 1. Receiver. C 2a. Bolt body (cylinder). C 2b. Striker. C 2c. Mainspring. C 2e. Manual safety wing. C 2f. Cocking piece. C 3c. Trigger. C 3b. Trigger bar (carries sear nose at its rear). F 7. Trigger guard screw. D 2. Stock. C 4c. Magazine spring.

The 98/40 Rifle (Gew 98/40)

The 7.92mm 98/40 is not a Mauser. It is based on the design of the 8mm Model 1935 Hungarian Rifle. The 98/40 was made only in Hungary by the Danuvia Arms Works. The 98/40 has a two-piece Mannlicher-type bolt, staggered row box magazine which is flush with the bottom of the stock, and a two-piece stock similar to that of the Lee Enfield.

The Hungarian Model 43 rifle is quite similar to the 98/40, but the 98/40 can be distinguished from the Model 43 by the bayonet mounting bar under the barrel and the sling mounting slit drilled through the buttstock.

GERMAN VOLKSSTURM RIFLES

Volkssturm Gewehr 1 (VG1)

The 7.92 VG1 was a last ditch weapon which was made in a number of shops during the closing days of World War II. It has a crudely-made bolt and stock and uses the magazine of the semiautomatic Model 43 rifle.

Firing the VG1 can be a risky affair, since they were made at the low point of German manufacture in World War II.

Volkssturm Karabiner 98 (VK98)

The 7.92mm VK98 uses the Model 98 action combined with miscellaneous barrels from old German and foreign Mausers.

The Gew 98/40.

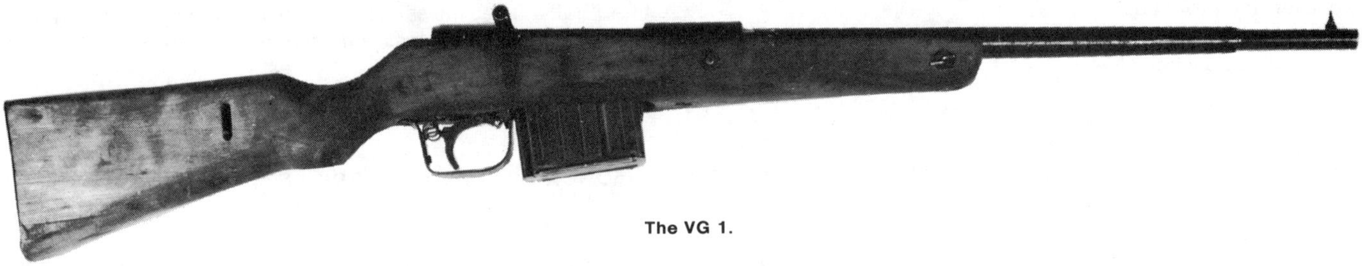

The VG 1.

The VK98.

Caliber .22 DSM34 Rifle.

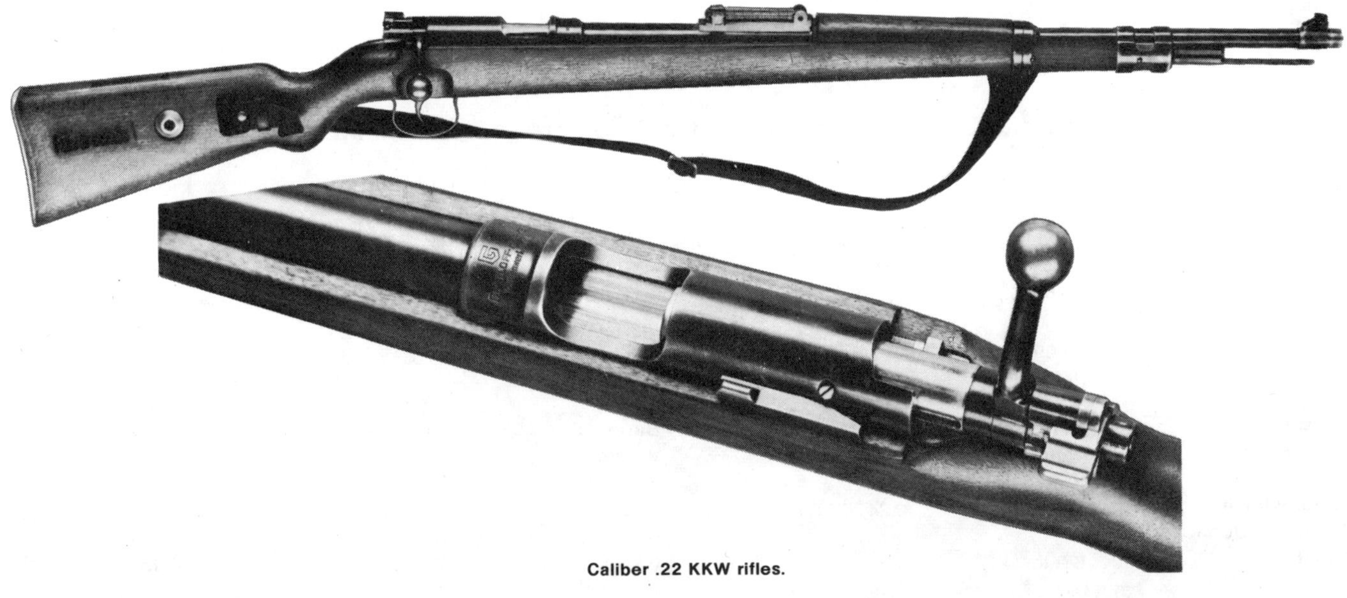

Caliber .22 KKW rifles.

The stock is very crude and is of unfinished unseasoned wood. Most of these weapons are single shot, but some were fitted with the semiautomatic Model 43 rifle magazine.

GERMAN TRAINING RIFLES

The Germans used a number of rifles for training, including service rifles such as the Model 98 rifle converted to caliber .22 and also had a conversion unit which could be easily inserted into a service rifle to convert it to .22 caliber; they had similar devices to convert the pistol 08.

They also had two caliber .22 training rifles: the Sport Model 34 and the Small Caliber KKW.

The German Sport Model 34 (DSM34)

This single-shot rifle, which was sold commercially in addition to its military use, was made by most of the standard German rifle makers, i.e., Mauser, Walther, Simson, etc. It has military-type sights and has a sling which is mounted on the left side as was done on the Kar 98k.

The Small Caliber Rifle (KKW)

This rifle is also single shot, but is about 1/2 pound heavier than the DSM34. It has the same type bands as the Kar 98 k and has a slightly improved action as compared to the DSM34.

CHARACTERISTICS OF GERMAN BOLT-ACTION RIFLES AND CARBINES

	Rifle 1888	Carbine 1888	Rifle 1891	Rifle 98	Carbine 98
Caliber:	7.92mm.	7.92mm.	7.92mm.	7.92mm.	7.92mm.
Length overall:	48.91 in.	37.4 in.	37.3 in.	49.2 in.	37.4 in.
Barrel length:	29.1 in.	17.6 in.	17.6 in.	29.1 in.	16.9 in.
Feed device:	5-round in-line, fixed, box magazine.	5-round, in-line, fixed, box magazine.	5-round, in-line, fixed, box magazine.	5-round, staggered row, fixed, box magazine.	5-round, staggered row, fixed, box magazine.
Sights: Front:	Barley corn.	Barley corn.	Barley corn.	Barley corn.	Barley corn.
Rear:	Leaves with V notch.	Leaves with V notch.	Leaves with V notch.	Bridge type tangent or tangent leaf, V notch.	Bridge type tangent w/V notch.
Muzzle velocity: (at date of adoption)	2099 f.p.s.	1935 f.p.s.	1935 f.p.s.	2099 f.p.s.	590 m/s.
Weight:	8.56 lb.	6.88 lb.	6.8 lb.	8.81 lb.	Approx. 7.5 lb.

CHARACTERISTICS OF GERMAN BOLT-ACTION RIFLES AND CARBINES (Cont'd)

	Carbine 98a	Carbine 98b	Carbine 98k	Rifle 40k	Rifle 33/40
Caliber:	7.92mm	7.92mm	7.92mm	7.92mm	7.92mm
Length overall:	43.3 in.	49.2 in.	43.6 in.	39.1 in.	39.1 in.
Barrel length:	23.6 in.	29.1 in.	23.6 in.	19.2 in.	19.29 in.
Feed device:	5-round, staggered, fixed, box, magazine.	5-round, staggered, fixed, box magazine.	5-round, staggered, fixed, box magazine.	5-round, staggered, fixed, box magazine.	5-round, staggered, fixed, box magazine.
Sights: Front:	Barley corn.	Barley corn.	Barley corn.	Barley corn.	Barley corn.
Rear:	Tangent w/V notch.	Tangent w/V notch.	Tangent w/V notch.	Tangent w/V notch.	Tangent w/V notch.
Muzzle velocity: (at date of adoption)	2853 f.p.s.	2574 f.p.s.	2476 f.p.s.	Approx 2400 f.p.s.	Approx 2400 f.p.s.
Weight:	8 lb.	9 lb.	8.6 lb.	8.3 lb.	7.9 lb.

CHARACTERISTICS OF GERMAN BOLT-ACTION RIFLES AND CARBINES (Cont'd)

	Rifle 98/40	VGI	VK98	DSM34	KKW
Caliber:	7.92mm.	7.92mm.	7.92mm.	Cal. .22 L.R.	Cal. .22 L.R.
Length overall:	43.6 in.	43 in.	40.6 in.	43.3 in.	43.7 in.
Barrel length:	23.6 in.	23.2 in.	20.8 in.	25.98 in.	25.98 in.
Feed device:	5-round staggered row, fixed, box magazine.	10-round staggered row, detachable, box magazine.	Mostly single shot.	Single shot.	Single shot.
Sights: Front:	Barley corn.	Post.	Barley corn.	Barley corn.	Barley corn.
Rear:	Tangent with V notch.	V notch.	V notch.	Tangent with V notch.	Tangent with V notch.
Muzzle velocity: (at date of adoption)	2476 f.p.s.	2476 f.p.s.	2400 f.p.s. (approx)	1500 f.p.s. (approx)	1500 f.p.s. (approx)
Weight:	8.9 lb.	8.3 lb.	6.9 lb.	7.7 lb.	8.6 lb.

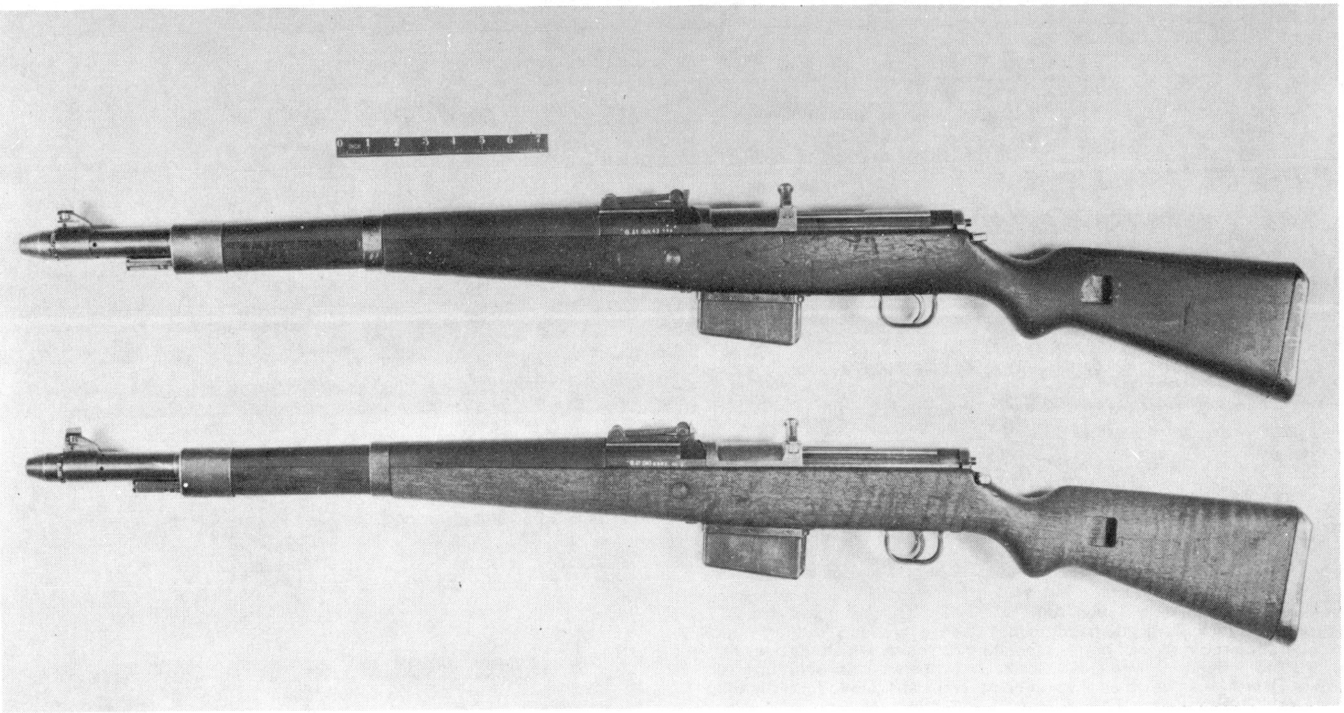

The Gew 41 (above) and the Gew 41(W) (Below). Both are 7.92mm.

THE 7.92mm MODEL 41(W) SEMIAUTOMATIC RIFLE (GEW41(W))

In 1941 Mauser and Walther both introduced semiautomatic rifles which were issued in limited quantities to the German Army and used in what was apparently a competitive combat trial. The Walther Model 41(W) was the most successful of the two designs, since it was developed into the Model 43 rifle, which was made in considerable quantity. The 41(W) is quite similar to the Model 43 except for the gas system. The bolt of the Model 41(W) has locking flaps which are pushed into the locked position by the forward movement of the firing pin and are cammed out of the locked position by the rearward movement of the bolt carrier and the firing pin.

The gas system is a modification of the Bang system. A muzzle cone traps gas which rebounds against a piston, forcing it to the rear. The piston in turn forces an operating rod to the rear and the operating rod forces the bolt carrier to the rear, thereby unlocking the bolt. The 41(W) has a fixed magazine and is loaded with two five-round chargers. This weapon is a finely-machined weapon and is much better made than the Model 43. There are specimens of this rifle which are stamped G41. They do not have a bolt-release catch.

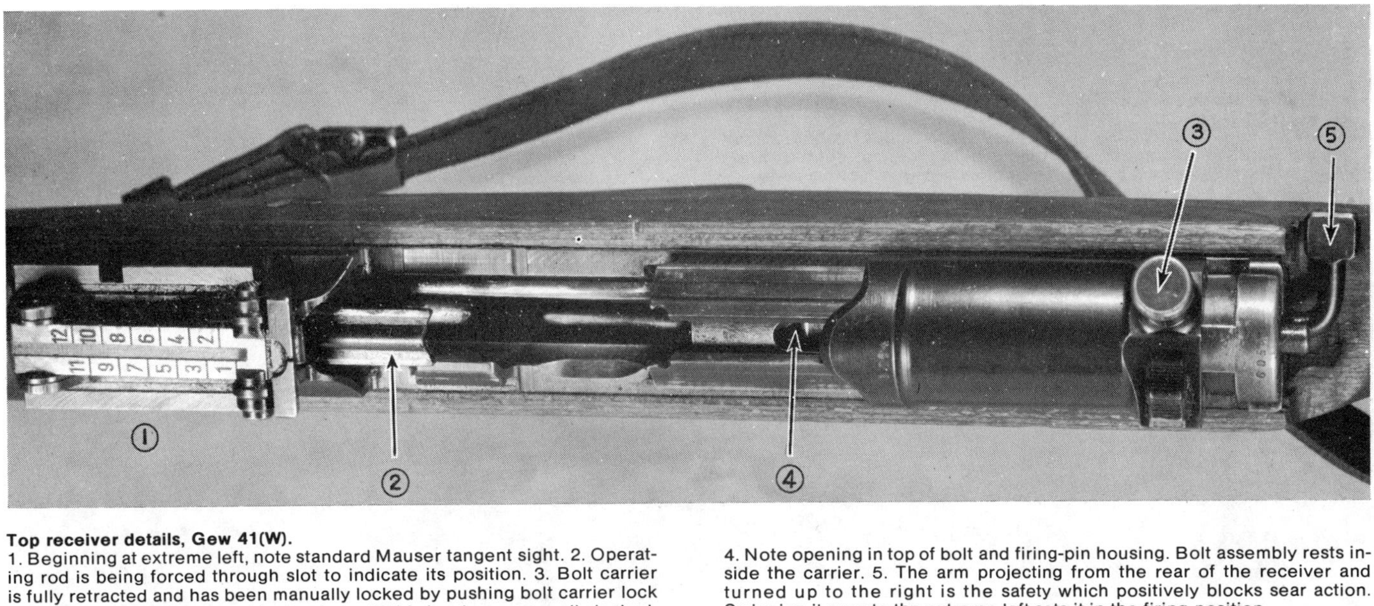

Top receiver details, Gew 41(W).
1. Beginning at extreme left, note standard Mauser tangent sight. 2. Operating rod is being forced through slot to indicate its position. 3. Bolt carrier is fully retracted and has been manually locked by pushing bolt carrier lock to the right. Magazine cannot be loadeduntil this has been manually locked. 4. Note opening in top of bolt and firing-pin housing. Bolt assembly rests inside the carrier. 5. The arm projecting from the rear of the receiver and turned up to the right is the safety which positively blocks sear action. Swinging it over to the extreme left sets it in the firing position.

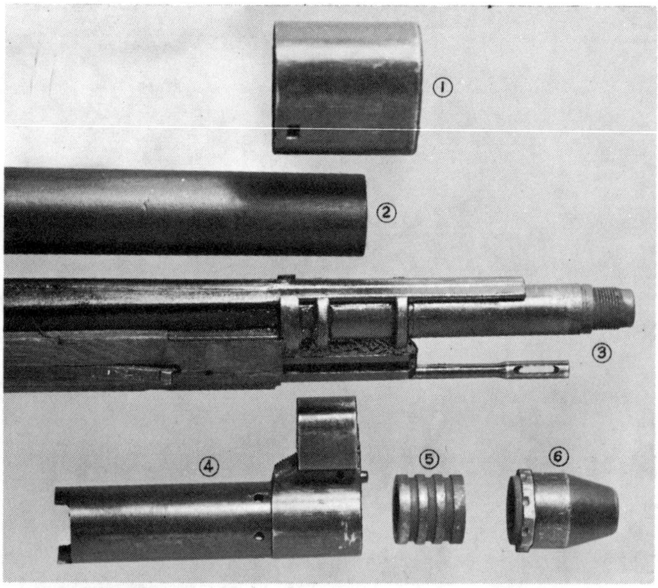

Gas operating assembly, Gew 41(W).
1. Front band lock. 2. Plastic handguard. 3. Barrel threaded for blast cone. Flat operating rod on top of barrel. Operating rod cover which fits over barrel. 5. Piston which mounts around barrel and floats inside operating rod cover. 6. Blast cone which screws on barrel and traps gas to force floating piston back against operating rod.

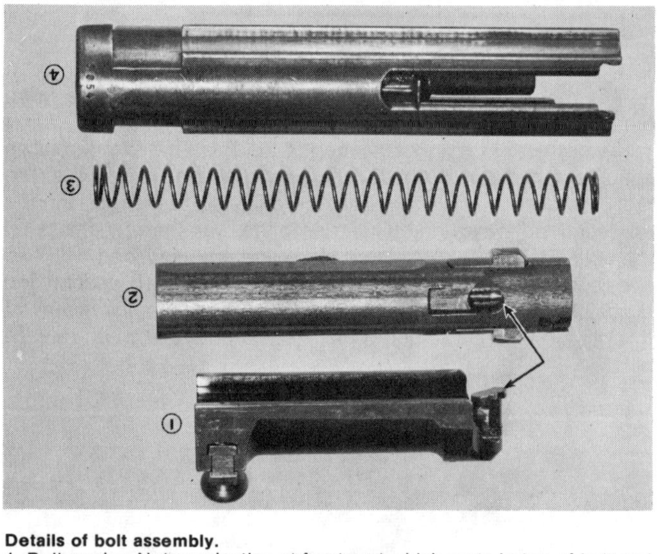

Details of bolt assembly.
1. Bolt carrier. Note projection at front end which seats in top of bolt and when driven back functions the firing-pin carrier to unlock the bolt. 2. Bolt complete with two lugs in locked position. Note firing pin housed inside carrier within the bolt. 3. Operating and recoil spring, which seats inside the hollow bolt. 4. Stamped bolt housing guides the travel of the bolt.

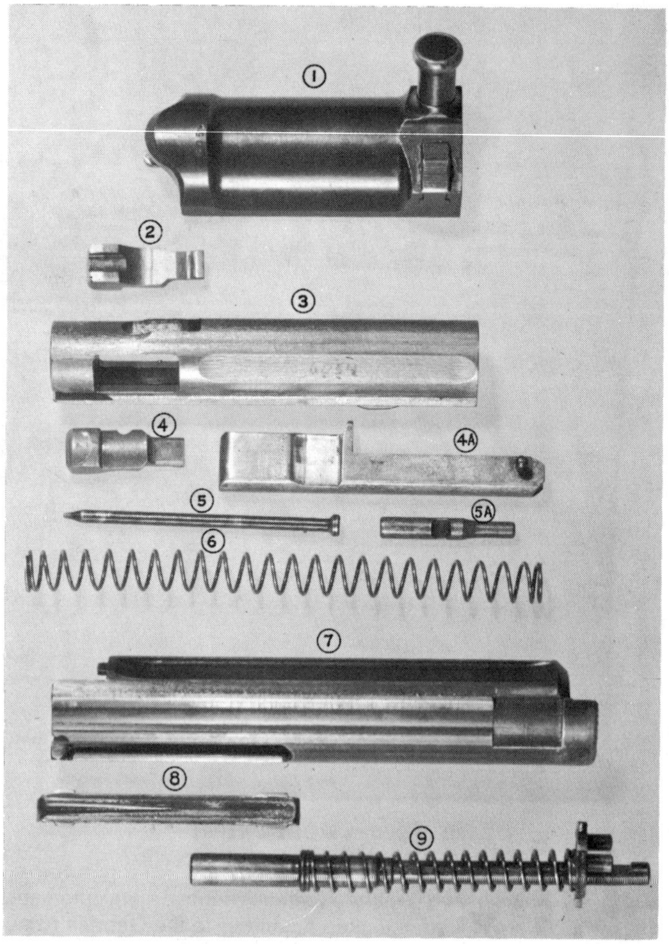

Details of bolt assembly, Gew 41(W).
1. Bolt carrier. Note projection on front end at left for pulling back firing-pin housing. Note bolt handle at upper right and carrier lock in line with it. 2. Bolt lock right side. 3. Detail of bolt. Note slot in left side to receive bolt lock. Also cut in top to receive bolt carrier projection. Also note cam-face at lower right of bolt body which serves to cock the hammer by riding over and depressing it. 4. Left side bolt lock and 4A. Hollow firing-pin housing. 5. Firing-pin and 5A. Firing-pin extension which are inserted in the housing. Extension is retained by pin at rear of housing. 6. Recoil or operating spring. 7. Bolt housing. 8. Sliding cover for housing. 9. Recoil spring guide with secondary or buffer spring affixed. Right end of this rod projects through bolt housing. The two lugs lock in slot in the receiver to hold the bolt assembly securely in place.

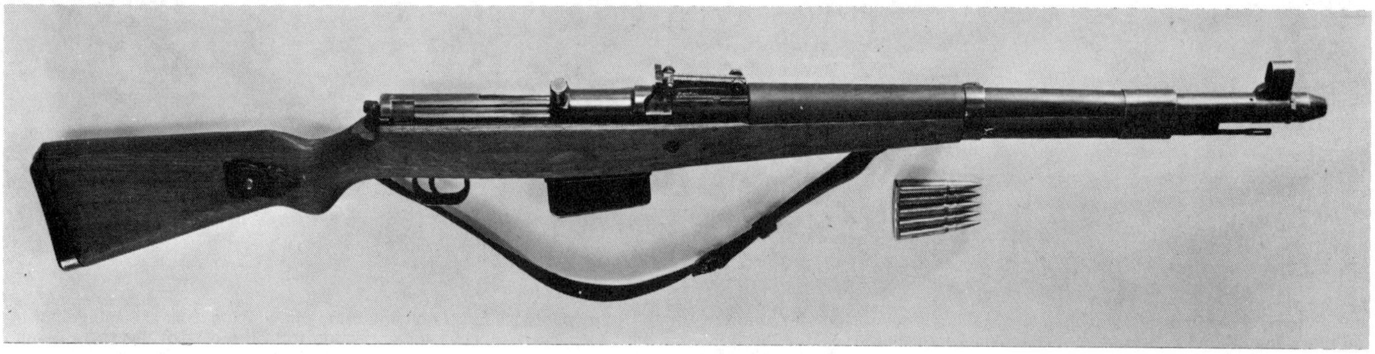

Gew 41(W), right side, bolt closed.

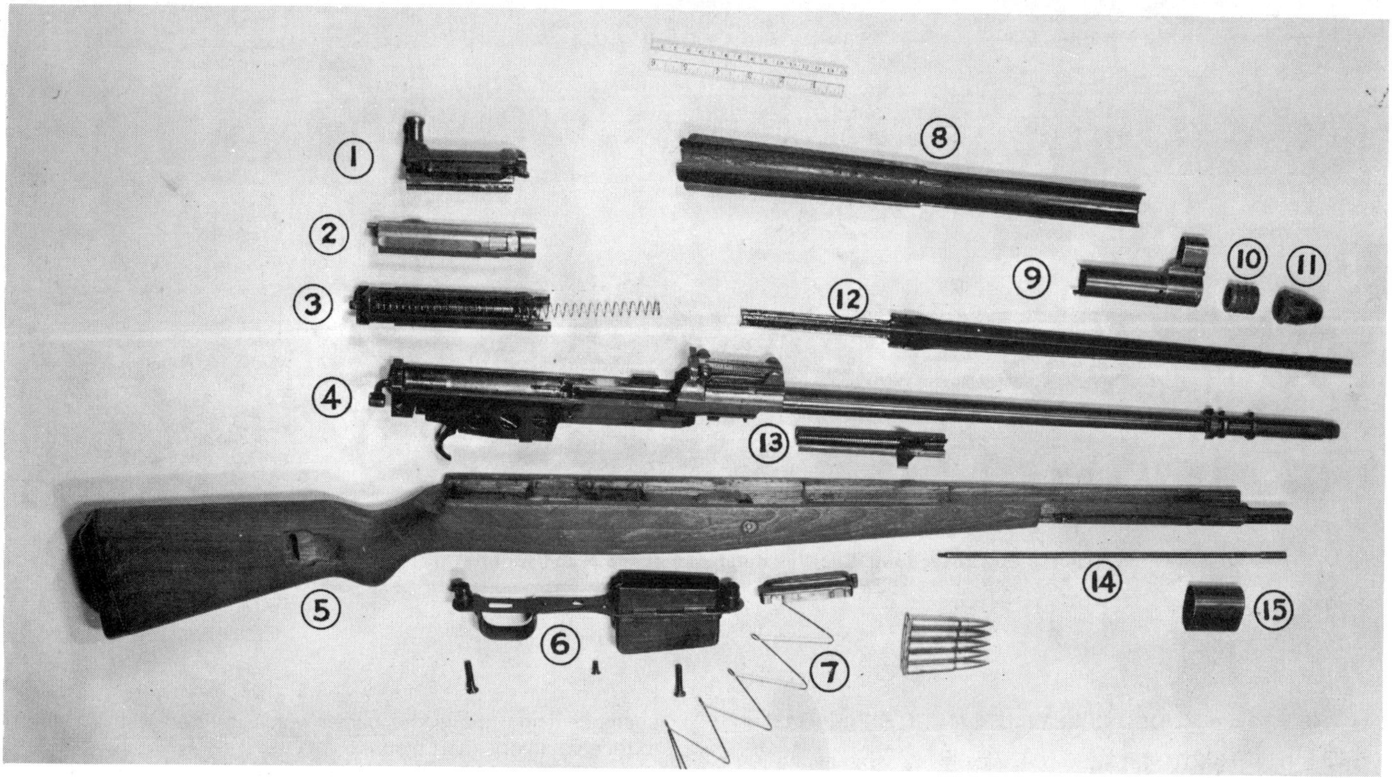

Detailed stripping to show gas and locking systems.
1. Bolt carrier. 2. Bolt. 3. Bolt housing with buffer and recoil springs. 4. Receiver and barrel assembly. 5. Stock. 6. Trigger guard and magazine. 7. Magazine follower and spring. 8. Forearm. 9. Gas cylinder. 10. Gas piston. 11. Gas cone. 12. Operating rod. 13. Operating rod trough and spring. 14. Cleaning rod. 15. Front band.

THE 7.92mm MODEL 41(M) (GEW41(M)) SEMIAUTOMATIC RIFLE

The 41(M), a Mauser development, was not a very succesful design, and was abandoned in 1943. Apparently there were not very many of these rifles made, as they are comparatively rare today. The 41(M), like the 41(W), draws its operating gas at the muzzle. The gas rebounds from a muzzle cone and strikes a piston mounted under the barrel. The piston forces back an operating rod which is connected to the rear section of the two-piece bolt. The rear section of the bolt pulls the front section backward, causing the frontally-mounted bolt locking lugs to be cammed out of their locking recesses in the receiver.

The 41(M) has an operating handle which has the same appearance as a bolt handle on a manually-operated bolt-action rifle and is operated in much the same manner, but it does not reciprocate with the action when the weapon is fired. The magazine of the 41(M) is fixed and is loaded with two five-round chargers.

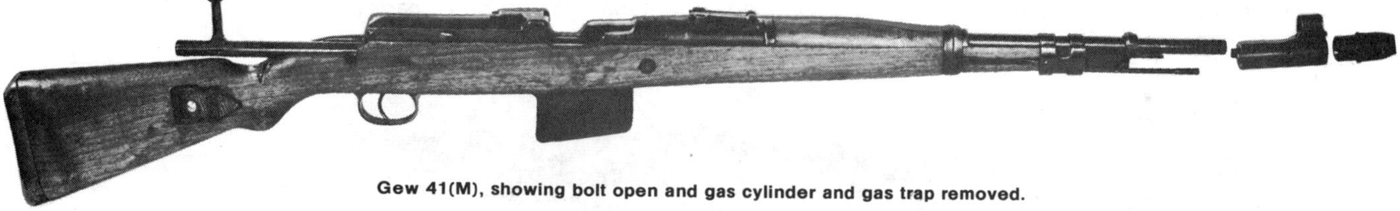

Gew 41(M), showing bolt open and gas cylinder and gas trap removed.

Gew 41(M).

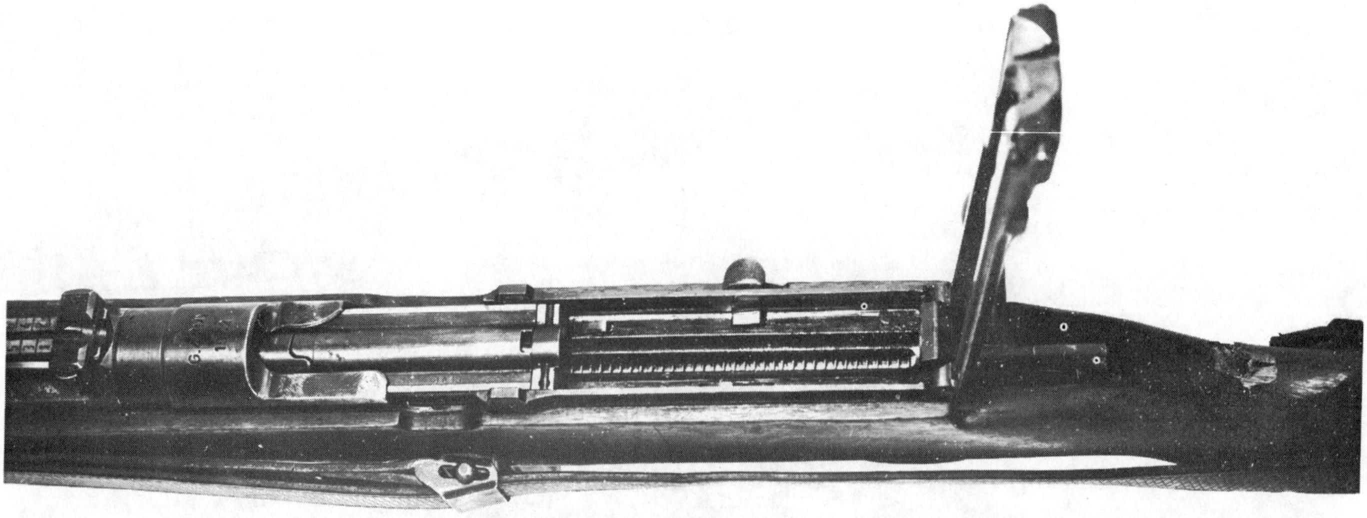

Bolt cover open. Showing detail of bolt and recoil spring. This is the Gew 41(M). It employs a turning bolt-head for locking.

THE 7.92mm MODEL 43 SEMIAUTOMATIC RIFLE (G43)

In the Gewehr 43 (G43), which is basically the same as the Karbiner 43 or Kar43, Walther combined the bolt mechanism of the Model 41(W) with a gas system quite similar to that of the Soviet M1940 Tokarev rifle. G43 was made in large quantities and a number of variations may be found among these rifles. The hand guard may be of wood or plastic and the bolt carrier latch, which locks the bolt carrier and bolt to the rear, may be on the left or right side of the bolt carrier, or may not exist at all.

The G43 is very roughly made of many stampings, castings, and forgings, which are machined only where necessary. All G43 rifles have a scope mounting to be used with the 1 1/2 power ZF41 scope. The G43 has a detachable ten round magazine which can be loaded while in the weapon with two five round chargers.

The G43 was used to a limited extent by the Czech Army for a few years after World War II.

How the Gewehr 43 Works

This rifle has a gas vent drilled in the barrel about 12 1/2 inches back from the muzzle. It is on top of the barrel and leads into a gas chamber rising above the barrel which is fitted with a cylinder to receive a very short piston.

As the bullet passes the gas vent in the barrel, a portion of the gas enters the port and expands in the cylinder driving back the short piston violently.

This piston, acting on the tappet principle, strikes the operating rod which extends backward into the receiver. As the end of this rod passes through a hole drilled in the receiver above the line of the bolt, the thrust imparted to it is transmitted to the bolt carrier on top of the bolt. Meanwhile a spring around the operating rod is compressed to provide energy to return the rod to its forward position.

At the start of its rearward travel the bolt carrier moves independently, leaving the weapon securely locked until the pressure has dropped. The slide carries the firing pin housing back independently of the bolt at this point also.

After a short travel, the slide and firing pin housing attached to the bolt pick up the bolt. The firing pin housing is so constructed that it cams in two locking lugs, drawing them out of their seats in the receiver walls and into the surface of the bolt.

From this point on the parts travel to the rear together extracting

Gew 43 (above) and Kar 43 (below), both 7.92mm.

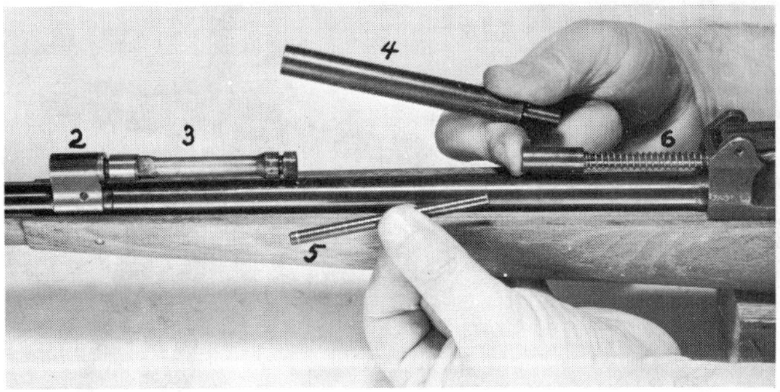

Gas piston operation.
2. Gas chamber above port in barrel into which gas escapes as bullet travels down barrel. 3. Gas cylinder screwed on. 4. Outside gas piston just pulled off the gas cylinder. 5. Connecting tappet piece. When piston is in place over cylinder, the front end of this tappet piece seats in the recess in the front end of the moving piston while its rear seats in the base of the operating rod. 6. Operating rod and spring being held to permit removal of gas piston and tappet.

and ejecting the empty shell and compressing the recoil spring around its guide.

Forward Movement of the G43 Action. The recoil spring compressed around its guide now exerts a forward thrust against the bolt assembly. The bolt strips a cartridge from the magazine and chambers it, the extractor set in the face of the bolt snapping into the extracting groove of the cartridge as it is chambered.

As the bolt reaches its fully forward position, it stops against the face of the chamber. The spring still exerts forward pressure and drives the firing pin housing straight ahead independently of the bolt. This is so constructed that it forces the locking lugs which are loosely set into each side of the bolt out into the receiver walls in a fashion not unlike that of the Russian Degtyarev light machine gun.

When the bolt is fully home in forward position, the carrier mounted on top of it is forced still farther forward into its niche in the receiver, where it rests against the operating rod hole.

When the trigger is pressed, the sear is rotated away from the hammer (which has been ridden over and cocked by the rearward

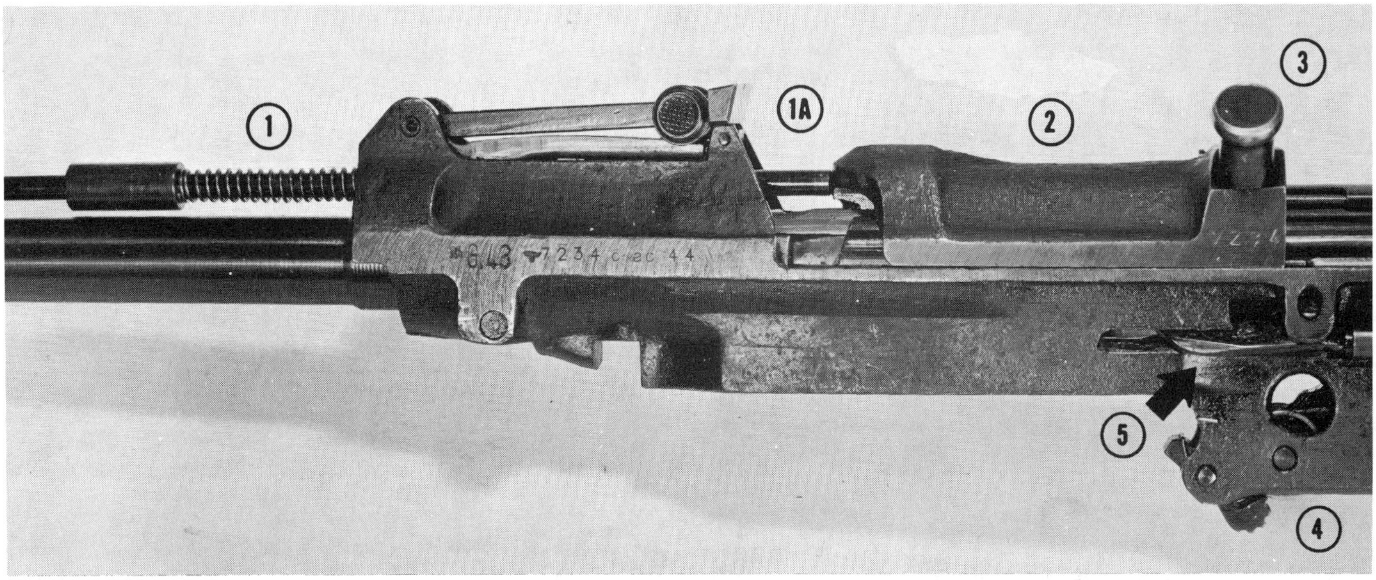

Details of operating rod system, Gew 43.
1. Operating rod being driven to the rear. As it passes th ough the hole in the receiver above the line of the bore, the spring around it is compressed to store up energy for the forward movement. 1-A. Rear end of operating rod moving back to drive the bolt carrier to the rear. 2. Bolt slide moving back in its tracks. Finger on its underside will carry the firing-pin carrier to cam the bolt locks in so that the bolt may be unlocked and travel back to the rear. 3. Bolt handle used for cocking weapon. 4. Magazine release catch. 5. Ejector. (Note that the receiver and the bolt carrier are steel castings.)

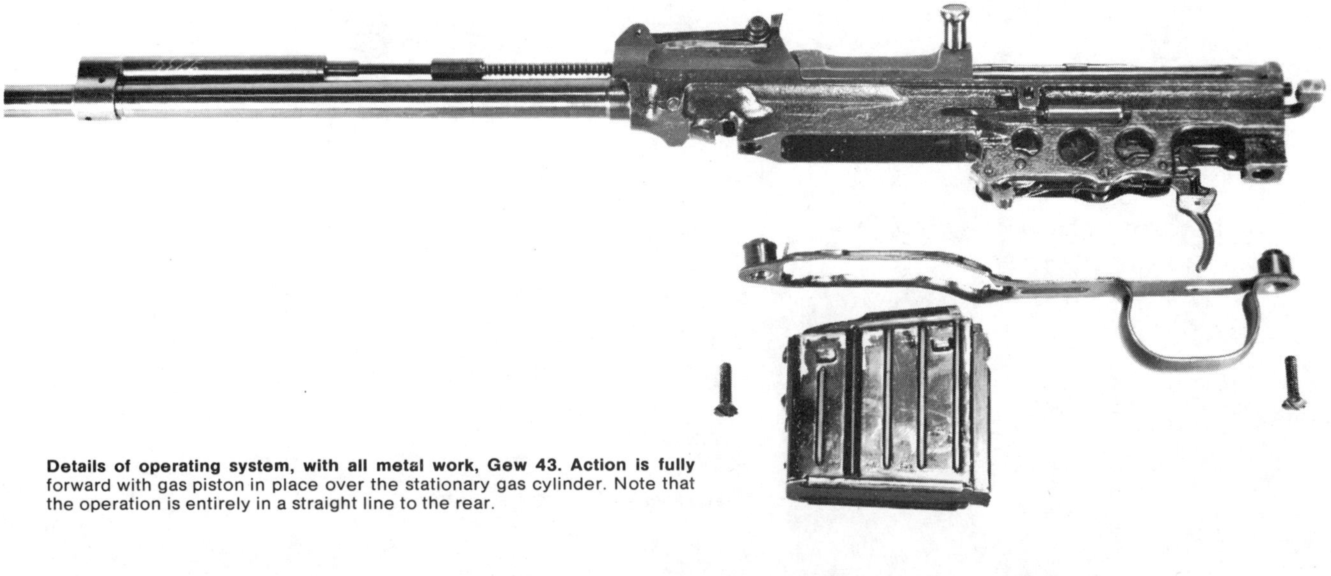

Details of operating system, with all metal work, Gew 43. Action is fully forward with gas piston in place over the stationary gas cylinder. Note that the operation is entirely in a straight line to the rear.

7.92mm FG42 with wooden stock.

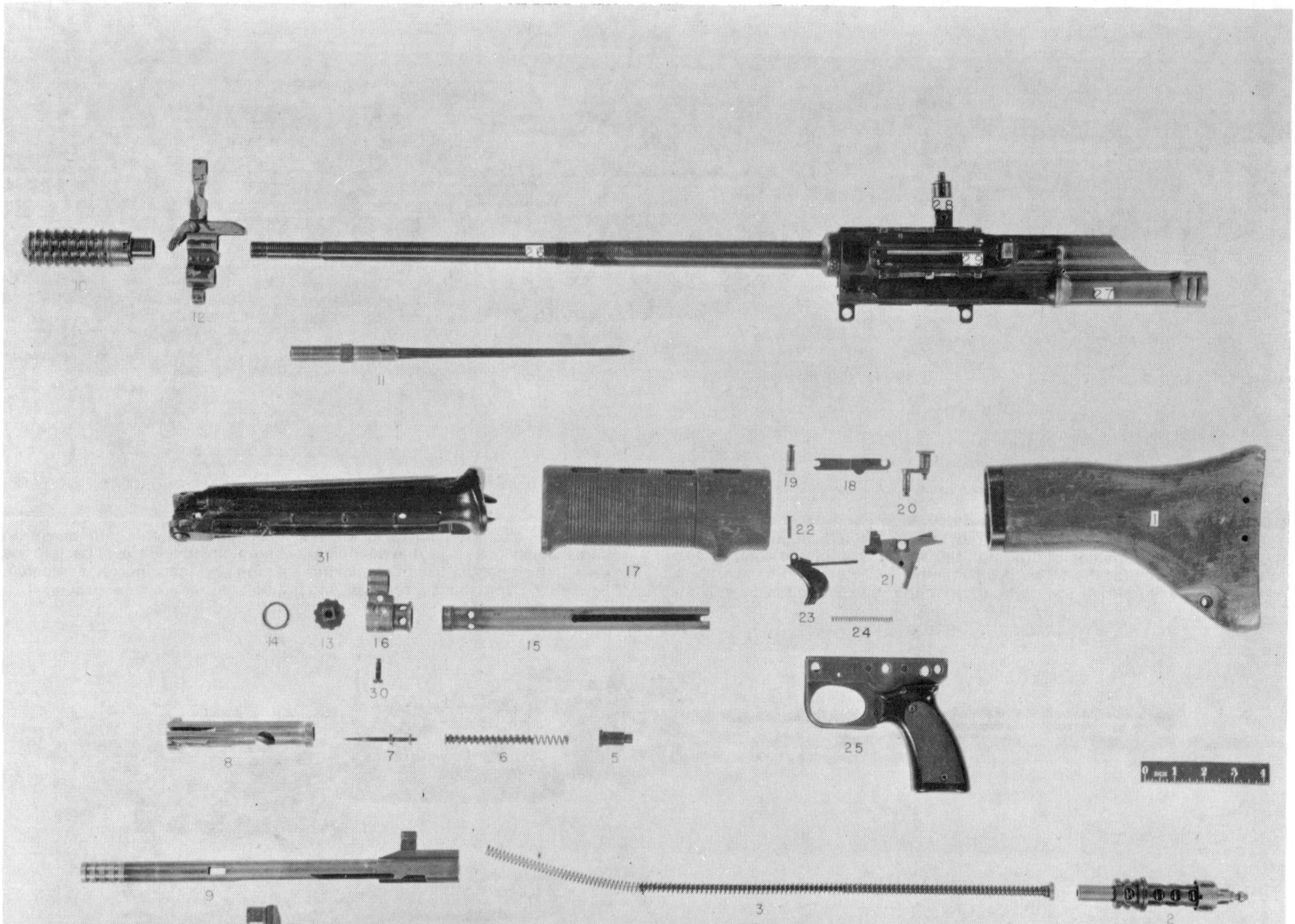

FG42, stripped.

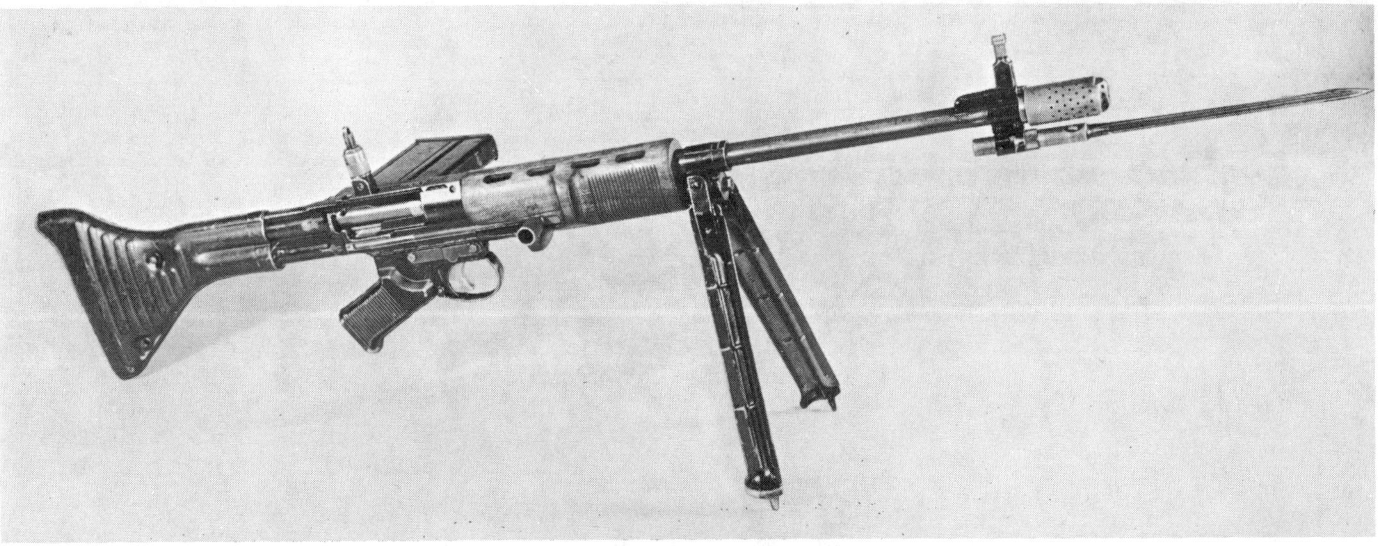

German 7.92mm Fallschirmjaeger Gewehr 42 automatic rifle with steel butt.

motion of the bolt). The hammer spring drives the hammer forward to strike the firing pin extension. This in turn strikes the firing pin and drives it against the primer to fire the cartridge in the chamber.

THE 7.92mm PARATROOP RIFLE MODEL 42 (FG42)

The Fallschirmjäger Gewehr (FG42) was developed by Rheinmettal-Borsig at the request of the German Air Force. Paratroops in the German service were under the Air Force, and therefore the Air Force was responsible for procuring armament for the paratroops.

FG42 was, in many ways, a remarkable weapon and it is somewhat surprising that the Germans did not make more of them. It has been reported that only 5,000 of these rifles were made. FG42 had the following good features:

(1) Straight-line stock configuration and muzzle-brake compensator to assist in holding down the weapon in automatic fire from the shoulder.

(2) Reduction of recoil by use of a recoil-spring sliding shoulder-stock system.

(3) The weapon fires from a closed bolt in semiautomatic fire and from an open bolt in automatic fire—this solves the "cook off" problem.

FG 42 was designed to replace the rifle, machine gun in the light role, and the submachine gun.

The bolt mechanism of the FG42 was copied from that of the Lewis gun, but a standard multiple coil recoil spring is used rather than the clock work type spring used with the Lewis Gun. The FG42-type bolt mechanism is currently used in the U. S. M60 machine gun. The trigger mechanism is cleverly designed and features a swivel mounted sear which can be moved left or right to engage the semiautomatic or automatic sear notch.

A short gas piston rod is used with this weapon; the operating handle is connected to the piston rod, which also has a stud to operate the bolt. The stud operates in a camway in the bolt, rotating it into and out of the locked position. FG42 has a spike-type bayonet which is carried in under the barrel point reversed when not fixed in a manner similar to that of the French MAS36 rifle. A light stamped bipod, which failed in U. S. tests of the rifle at Aberdeen Proving Ground during World War II, is also used with this weapon.

Some of these rifles are fitted with a stamped steel stock and others have a wooden stock. All things considered, FG 42 was one of the most interesting of the German World War II designs. It did not introduce any revolutionary design principles, but it did combine a number of previously uncombined principles to produce an advanced selective-fire weapon for full-size rifle cartridges. The first U. S. weapon patterned on the FG42 was the 7.92mm T44 light machine gun. This weapon was rather unusual; it is belt-fed—it uses the MG42 type belt-feed mechanism—but the feed cover is mounted on the side of the receiver so that the belt feeds in a vertical position rather than horizontally as is usual. This weapon was developed by the Bridge Tool and Die Manufacturing Corp. under contract with the Ordnance Corps.

It should be noted that several authoritative publications credit the design of FG42 to the Heinrich Krieghoff Plant of Suhl, Saxony, but German publications credit it to Rheinmettal.

THE 7.92mm VG1-5 SEMIAUTOMATIC RIFLE

The Volkssturm Gewehr 1-5 is a rather unusual weapon in many ways. The first unusual thing about the rifle is that it was never apparently approved by the Waffenamt (Ordnance Office) in Berlin and does not bear the usual government acceptance stamps. The VG1-5 was put into production at a time when control was crumbling in Germany and the Nazis had given the local Gauleiters authority to draw up contracts for arming the Volkssturm in their own districts with whatever weapons they could beg, borrow, or steal.

Considering this somewhat dubious ancestry, VG1-5 has some rather good design features. The weapon was designed and produced by the Gustloffwerke at Suhl and was apparently made in limited quantity. Stampings are extensively used in this weapon, the wooden furniture is left rough, and machining is of very simple type.

How the VG1-5 Operates

A standard MP 43-44 magazine loaded with 30 cartridges is inserted in the magazine housing from below and pushed in until it locks.

The bolt handle which is a heavy steel piece riveted to the left side of the housing is pulled back. This draws all the moving members to the rear, and the bolt rides over and compresses and cocks the hammer which is held by a simple sear arrangement.

When the bolt handle is released the compressed recoil spring pushes the members forward and strips a cartridge from the magazine into the firing chamber, where the extractor grips it as it seats.

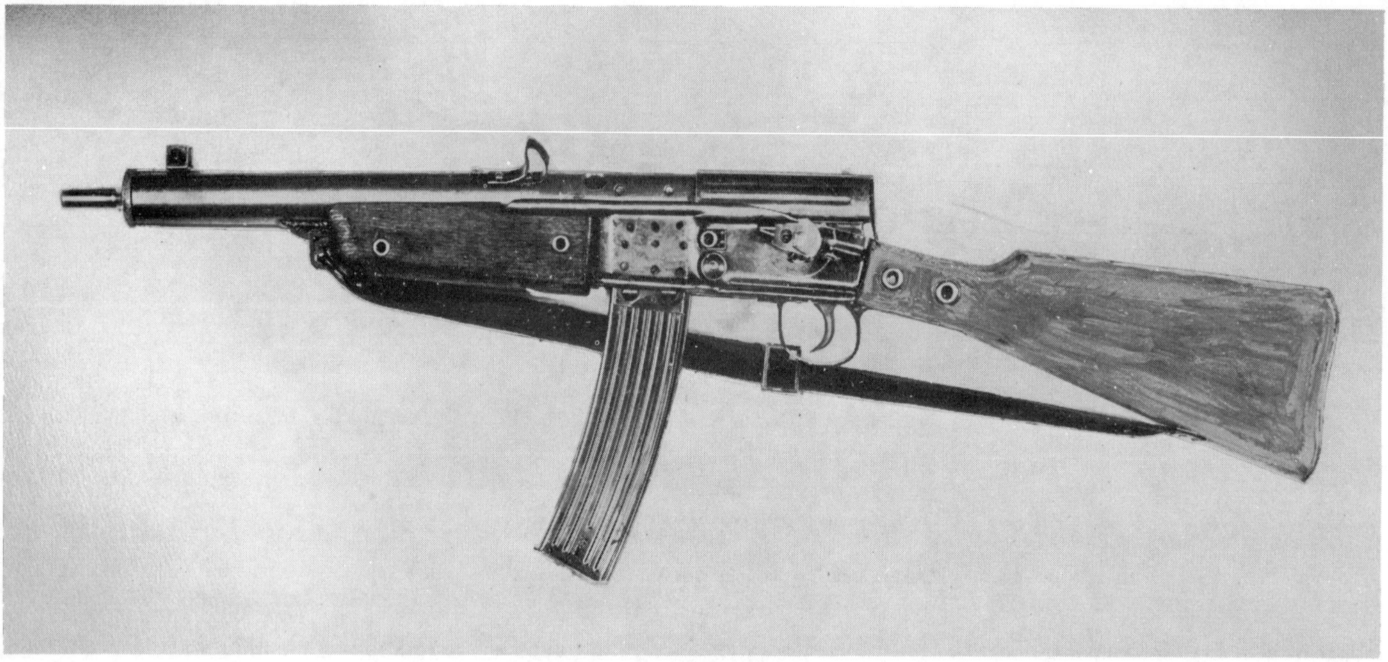

German 7.92mm VG1-5.

Pressure on the trigger is communicated through the sear to release the hammer, which flies forward to strike the firing pin and discharge the cartridge.

As the bullet travels down the barrel it passes over the 4 gas ports about 2 1/2 inches before reaching the muzzle. Thus it will be seen that gas escapes from the barrel into the space between the removable sleeve and the housing to exert a forward thrust as the bullet continues out of the muzzle. The sleeve is shaped to insure that most of the gas thrust will be exerted toward the concave forward end.

As this is a retarded blowback weapon without a lock, the recoil force of the rearward action of gas against the head of the cartridge case drives the action to the rear. However, the thrust has to overcome not only the inertia and weight of moving parts as in the standard blowbacks, but also the forward action of the gases which have expanded in the moving housing. This check system delays the opening of the action long enough to permit the use of a cartridge so powerful that normally a locking device would be required to permit its use.

Rearward action ejects, recocks, and reloads in standard semiautomatic fashion.

THE 7.92mm MP43, MP44, AND StG44 ASSAULT RIFLES

The Germans had decided after World War I that their 7.92 X 57mm cartridge was overly powerful for shoulder weapons. Analysis of the average ranges at which rifles were commonly used and the marksmanship capabilities of the average soldier, especially under the stress of battle, led them to the conclusion that a cartridge with considerably less ballistic potential than the 7.92 X 57mm would be adequate and in addition would result in shorter, generally lighter weapons, allow the soldier to carry more cartridges on his person, cause less fatigue from recoil, and result in a considerable saving of materials in the manufacture of propellents, cartridge cases, and bullets.

It is interesting to note that the U. S. came to the same conclusions at about the same time and but for logistic considerations—time and expense of changing over to a new cartridge—the U.S. might have fought World War II with the .276 Pedersen cartridge, a cartridge of considerably less size and ballistic potential than the caliber .30-06 which we used during World War II and Korea.

The German requirement was solidified in 1934 and prototype cartridges were produced by Gustav Genschow, Rheinisch Wesphalische Sprengstoff (RWS) and Polte. Polte was given the development contract in 1938 and produced the "7.9mm Infanterie Kurz Patrone" by 1941. This cartridge had a case 33mm long with 24.6 grains of propellent as opposed to the 57mm case and 45-50 grains of propellent of the standard full-size 7.92mm rifle cartridge.

To parallel the cartridge development, Haenel was awarded a contract in 1938 for development of a weapon for these cartridges. Hugo Schmeisser of Haenel produced a gas-operated weapon for the 7.92mm Kurz cartridge by 1940 and 50 specimens of the prototype were produced by July 1942. Walther started development of the weapon for the cartridge in 1940, basing their design upon that of an earlier semiautomatic rifle of their conception—the G A115.

Machine Carbine MKb42(H) and Machine Carbine MKb42(W)

Both the Haenel and Walther designs were produced in limited quantity—approximately 7,800 of each—as machine carbines (Maschinen Karabiner) designated MKb42(H) and MKb42(W), respectively. They were extensively used on the Russian front and the Haenel design proved superior to that of Walther.

The Walther design MKb42(W) has a somewhat unusual gas system which was carried on in the design of the Czech Model 52 rifle. Rather than the conventional gas tube, usually found on gas-operated rifles, the gas in this design is confined by a steel jacket around the barrel and drives a piston which encircles the barrel, and an operating sleeve. There are two gas ports in the barrel, and the bolt is operated by the sleeve. The bolt has frontal locking lugs.

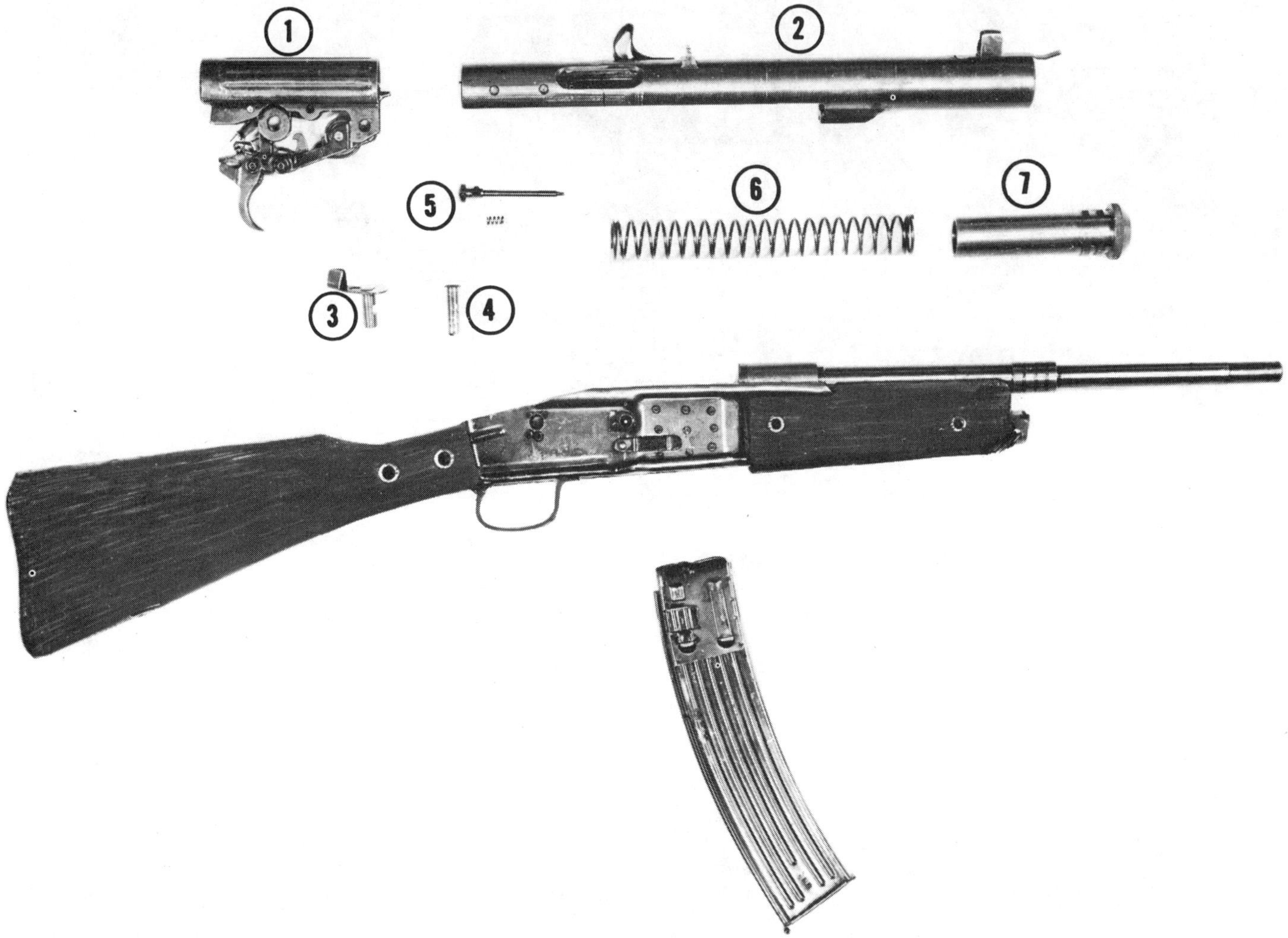

Main components, VG1-5.
1. Bolt housing and firing assembly. The bolt and slide assembly can move back in this housing when it is in place in the receiver. 2. Bolt and slide assembly. The bolt is riveted into the rear of this hollow sliding member. 3. Safety. 4. Retainer pin. 5. Firing pin and spring. 6. Recoil spring. 7. Gas chamber which locks inside front end of slide when the latter is in place around the barrel. Barrel, stock and receiver assemblies; note slide stop abutment on barrel at forward end of the forearm. Standard MP44 magazine for 7.92mm Short cartridge.

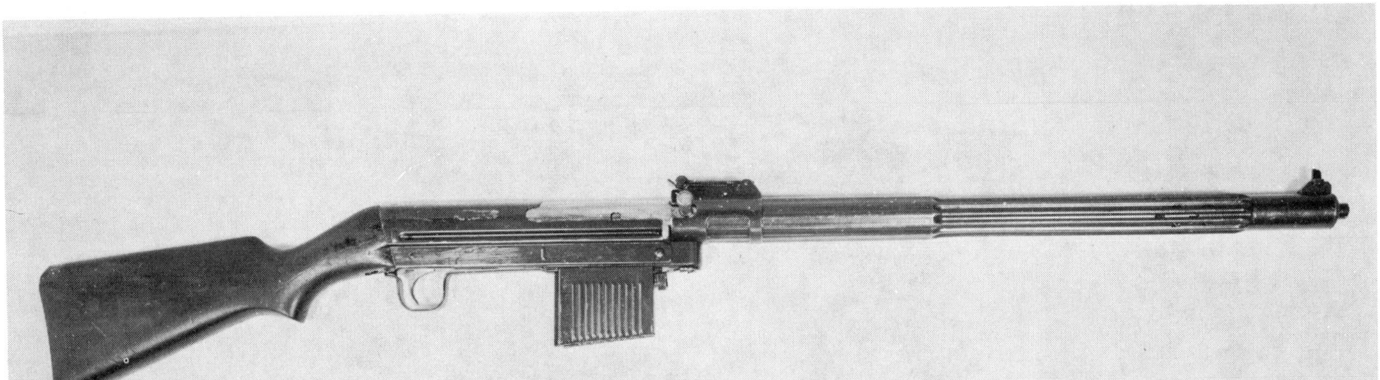

Walther 7.92mm GA115 semiautomatic rifle.

7.92mm MKb42(W).

Differences Between MKb42(H) and MP43

The Haenel MKb42(H) is generally similar in internal design to the MP43 series of weapons which are covered next. The principal differences between the MKb42(H) and the MP43 series of weapons are as follows:

(1) The piston of the MKb42(H) is longer than that of the MP43 and is mounted in a separate tube, divided by a visible air space above the barrel. In the MP43 the piston rides in a tunnel immediately above the barrel.

(2) There is a cut-out for the bolt handle on the receiver of the MKb42(H); this cut-out is not present on the MP43.

(3) MKb42(H) has a bayonet lug (its prototype does not). MP43 does not have a bayonet lug.

There are also differences in the stock, fittings, etc.

7.92mm MKb42(H) without magazine.

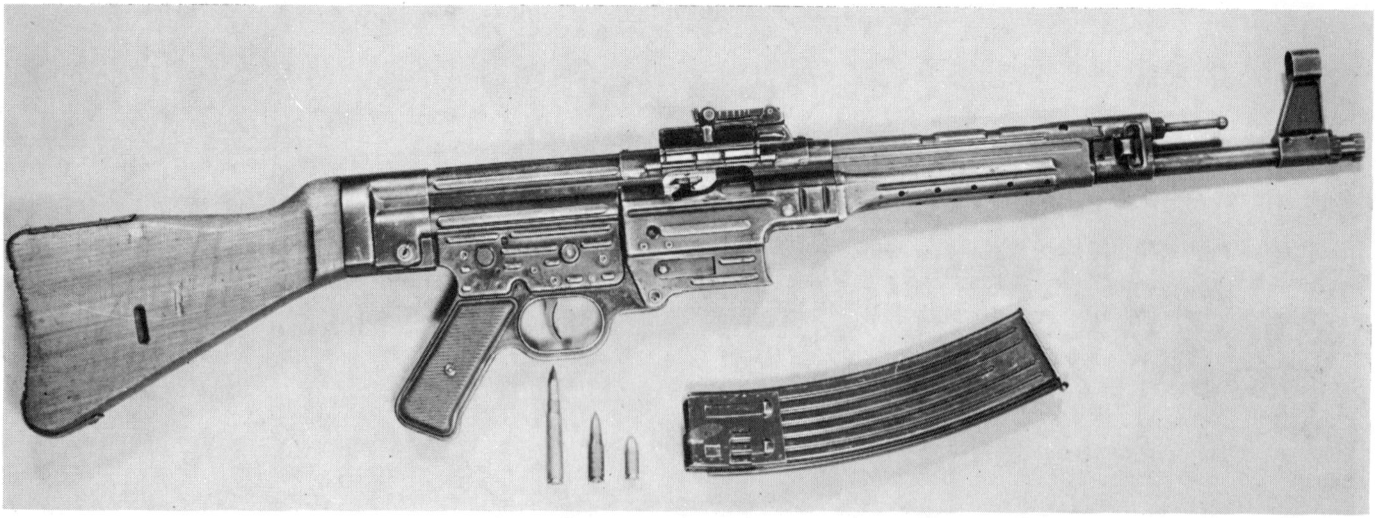

MP43.

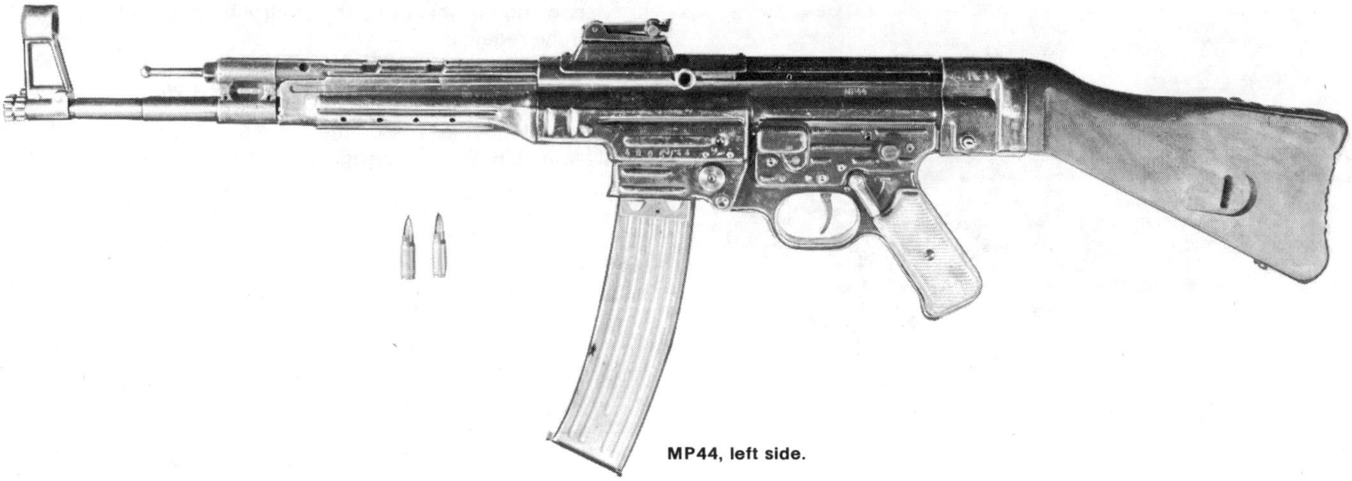

MP44, left side.

The MP43, MP43/1, MP44, and StG44 Assault Rifles

Schmeisser reworked the MKb42(H) in the spring of 1943 and the MP43, 44 series of weapons were born. This was a significant event in current military small-arms history, since it introduced for the first time, in large quantities, the concept of the selective fire assault rifle chambered for the "intermediate"-sized cartridge. It is interesting to note that the Soviet 7.62mm "intermediate"-sized cartridge is Model 1943, although an assault rifle—the AK47—did not appear for this cartridge until 1947.

The MP43 was adopted by the Waffen Amt as a standard weapon, and after 1944 was scheduled to replace the rifle, submachine gun, and light machine gun in the infantry squad. By February 1944 production of this weapon had risen to about 5,000 per month. Producers of this family of weapons were Hanenel, Mauser, and Erma; at least seven subcontractors made components. The term "family" as it refers to these weapons means the MP43, MP43/1, MP44 and StG44. All are essentially the same weapon and minor differences are as follows:

(1) MP43/1 is the same as MP43, but has a screw on type grenade launcher rather than the clamp on type used with MP43.

(2) There is no apparent reason for the change in nomenclature from MP43/1 to MP44. Most MP43/1 rifles have the V-type telescope mounting bracket on the right side of the receiver, some MP44s have this bracket, but no MP43s have been found with the bracket.

The change in nomenclature to StG44 was politically inspired but "Assault Rifle" (StG Sturmgewehr) is more truly descriptive of the role of the weapon than is "Submachine Gun" or "Machine Pistol" (MP-Maschinen Pistole).

The StG44(P) and StG44(V) were experimental versions of StG44, which had a 90° curved barrel and a 40° curved barrel, respectively. Both were rejected by the Waffen Amt.

Field Stripping the MP43—MP44 Series

A spring-held pin passes through the receiver and stock from the right. Pull this out from the right side.

The stock may now be withdrawn exposing the recoil spring.

Press the trigger and swing the trigger guard and all its contained units (these are not dismountable) down on its hinge.

Pull back the bolt handle. This will bring back the recoil spring, bolt, and bolt carrier and the piston for removal.

With the small steel tool found in the butt trap inserted in the hole of the gas cup protruding from the casing at the front end, the gas cup may be unscrewed and withdrawn. The cylinder may

MP44 field-stripped. Note that the gas piston, operating rod, operating handle, spring guide, bolt camming and locking units are actually only two units. A pin secures the rod section to the rear operating section. The bolt is a separate unit.

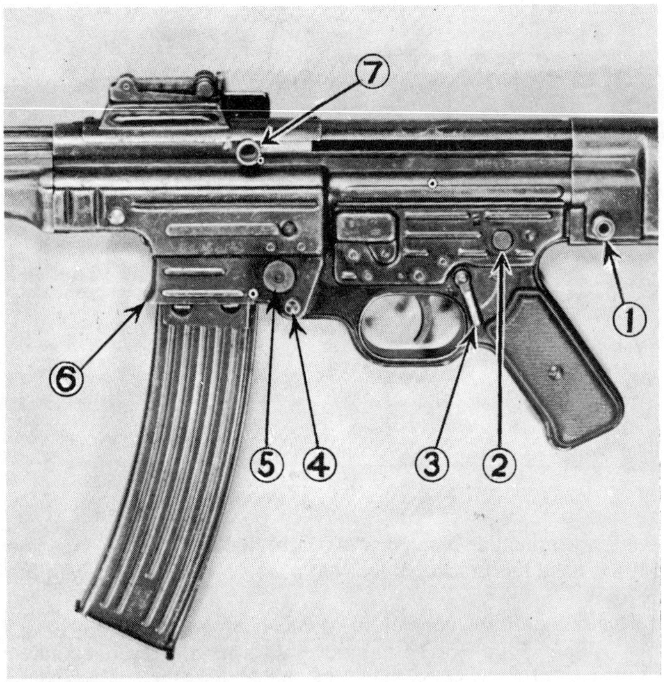

Main Components, MP44.
1. Receiver lock pin. 2. Push-through firing control switch. 3. Thumb safety, in off position. 4. trigger housing pivot pin. 5. Magazine release catch. 6. Magazine housing. 7. Bolt handle.

now be cleaned without difficulty.

The bolt may be lifted off the bolt carrier and the extractor and firing pin removed.

Loading and Firing the MP43-44

There is a trap in the top of the stock on this weapon. A special magazine filler will be found in the hollow.

Place this filler in the mouth of the magazine. Insert cartridges and then force them directly down with thumb pressure. The magazine will hold 30.

Pull bolt handle to compress recoil spring and cock the hammer and let it fly forward. It will pick up the top cartridge from the magazine and load it into the chamber. The dust cover will open automatically.

Push safety up unless weapon is to be fired at once. If it is left down, pressing the trigger will fire a single shot or full automatic fire depending on which way the control button has been pushed through the receiver.

How the MP43-MP44 Weapons Work

When the trigger is pressed the hammer is released to strike the inertia firing pin. This is a wedge-shaped pin which does not have a conventional spring. It is primer retracted.

As the bullet passes the gas port in the barrel it expands into the gas chamber or cup which is screwed into the housing around the barrel and on top of it.

The gas impinges on a piston somewhat resembling the old Lewis piston and drives it to the rear. In the start of its rearward travel, the piston (which has attached to it by a fixed pin the bolt carrier) can move without interfering with the secure locking of the weapon. A gas vent in the top of the casing permits the gas to escape as the gas end of the piston clears it.

After a short rearward travel the bolt carrier hook picks up the separate bolt member, mounted below it, and pulls it down and back to perform the unlocking action.

The recoil spring is mounted behind the bolt extending back to the stock. This spring is compressed as the moving members travel to the rear to extract and eject the empty case in normal fashion and to cock the hammer.

This weapon is fitted with a disconnector.

THE StG45(M) AND OTHER PROTOTYPES

Development of assault rifles by the various arms companies in Germany became very active after adoption of the MP43 series. One disadvantage of the early assault rifles was their weight; they are quite heavy in relation to the muzzle energy of the 7.92mm short cartridge. There was also a continual effort at this time to simplify weapons from the manufacturing point of view and from the point of view of saving on materials.

Gustloff Werke, Haenel, Mauser, and possibly Erma all developed prototypes and although none were accepted, the Mauser weapon had a great influence on future weapon developments. The Mauser development, originally called GerätO6(H), was a delayed blowback weapon weighing only 8.18 pounds as opposed to the over 11 pounds of the StG44. The original GerätO6(H) had a combination of gas and blowback operation; but the gas element of operation was dropped in the final design.

The StG45(M) introduced the delayed blowback with roller bearings now used in the Spanish CETME, West German G3, the Swiss StuG57, and in modified form in the French Model 52 machine gun. The construction of this bolt is explained in detail under the Spanish CETME assault rifle (see chapter on Spain).

7.92mm StG45(M) Assault Rifle.

StG45(M), stripped.

Characteristics of German World War II Semiautomatic and Selective Fire Rifles

	Rifle 41(M)	Rifle 41(W)**	Rifle 43***	FG42	VG1-5
Caliber:	7.92 x 57mm.	7.92 x 57mm.	7.92 x 57mm.	7.92 x 57mm.	7.92mm Kurz (short) (PP 43 m.e.).
System of operation:	Gas, semi-automatic only.	Gas, semi-automatic only.	Gas, semi-automatic only.	Gas, selective fire.	Delayed blowback, semiautomatic only.
Overall length:	46.25 in.	44.25 in.	44 in.	37 in.	35 in.
Barrel length:	21.75 in.	21.5 in.	21.62 in.	19.75 in.	14.75 in.
Feed device:	Fixed, 10-round*, staggered row, box magazine.	Fixed, 10-round, staggered row, box magazine.	Detachable, 10-round, staggered row, box magazine.	Detachable, 20-round, staggered row, box magazine.	Detachable, 30-round, staggered row, box magazine.
Sights: Front:	Barley corn.	Barley corn.	Barley corn.	Barley corn on folding base.	Fixed post.
Rear:	Tangent leaf w/U notch.	Tangent leaf w/U notch.	Tangent leaf.	Aperture on folding base.	Non-adjustable U notch
Muzzle velocity:	Approx. 2550 f.p.s.	Approx 2550 f.p.s.	Approx 2550 f.p.s.	Approx 2500 f.p.s.	2163 f.p.s.
Cyclic rate:	-----	----	-----	750-800 r.p.m.	-----
Weight:	11.25 lb.	11.08 lb.	Approx 9.5 lb.	9.93 lb.	10.18 lb.

*The magazine can be removed, but is not easily removable for reloading.
**This weapon is also called Rifle 41 (SG41).
***This weapon may be marked G43 or K43.

Characteristics of German World War II Semiautomatic and Selective Fire Rifles (Cont'd)

	MKb42(W)	MKb42(H)	StG44	StG45 (M)
Caliber:	7.92mm Kurz.	7.92mm Kurz.	7.92mm Kurz (PP43 m.e.).	7.92mm Kurz (PP43 m.e.).
System of operation:	Gas, selective fire.	Gas, selective fire.	Gas, selective fire.	Delayed blowback, selective fire.
Overall length:	36.75 in.	37 in.	37 in.	35.15 in.
Barrel length:	16.1 in.	14.37 in.	16.5 in.	15.75 in.
Feed device:	Detachable, 30-round, staggered row, box magazine.	Detachable, 30-round, staggered row, box magazine.	Detachable, 30-round, staggered row, box magazine.	Detachable, 30-round, staggered row, box magazine.
Sights: Front:	Hooded barley corn.	Hooded barley corn.	Hooded barley corn.	Hooded barley corn.
Rear:	Tangent w/U notch.	Tangent w/U notch.	Tangent w/U notch.	Tangent w/U notch.
Muzzle velocity:	2132 f.p.s.	Approx 2100 f.p.s.	2132 f.p.s.	Approx 2100 f.p.s.
Cyclic rate:	600 r.p.m.	500 r.p.m.	500 r.p.m.	350-450 r.p.m.
Weight:	9.75 lb.	11.06 lb.	11.5 lb.	8.18 lb.

GERMAN SUBMACHINE GUNS

GERMAN MODELS AND MODIFICATIONS OF FOREIGN MODELS

The first German submachine gun was the Bergmann 9mm MP18I which appeared in 1918. This blowback-operated submachine gun was fed with the 32-round "snail"-type magazine developed for the Luger pistol. The MP18I was designed by Hugo Schmeisser, who was by far the best-known of the German submachine gun designers.

The weapon was introduced into battle during the last part of World War I; about 35,000 were made before the war ended. Many of these weapons were used by the German civil police after the war. The MP18I was modified after the war by removing the magazine housing for the "snail"-type magazine and fitting a magazine housing for a box-type magazine. This modification was done by Haenel.

Further modification resulted in the MP28II, which was also produced by Haenel. The 28II has selective fire and a tangent-type rear sight. This weapon was extensively used by the German police, to include SS Police units, but was never officially adopted by the German Army. This does not mean that army personnel did not use the weapon at one time or another; as with all other small arms, men in the German Army used almost anything they could get. The MP28II was manufactured in Belgium by Pieper and was adopted by the Belgian Army as the "Mitraillete Model 34." It was also used by Bolivia, in addition to several other South American countries.

The next German submachine gun which was produced in quantity was the Bergmann 9mm 34/I. This gun, unlike the earlier Bergmanns, was not designed by Hugo Schmeisser. The prototypes of this gun were made in Denmark circa 1932; production of the weapon in Germany was at the Walther plant in Zella Mehlis, since Bergmann did not have production facilities. This weapon was not adopted by the German Army, but was exported on a limited scale. The Model 34/I can be distinguished by its bolt handle, which resembles, and is operated in a fashion similar to, that of a manually-operated bolt-action rifle, and it's trigger mechanism. The weapon has two triggers; pressure on the outer trigger produces semiautomatic fire until the inner trigger is engaged, at which time the weapon fires automatically. The Model 34/I was produced in long barrel and short barrel versions.

The 9mm Model 35/I is a modified Model 34/I. This weapon was produced during World War II for the "SS" by Junker and Ruh. German police manuals of this period refer to this weapon as the Model 35. The Model 35/I was adopted by Ethiopia and by Sweden which called the weapon the "M/39."

The Erfurter Werkzeug and Maschinenfabrik "Erma Werke" of Erfurt produced a submachine gun designed by Heinrich Vollmer which had fairly wide distribution as the Vollmer Erma. The most common model is called the EMP or MPE. Versions of the Erma were sold to France, Mexico, and Yugoslavia. This weapon, as well as many of the earlier German submachine guns, was extensively used in the Spanish Civil War. The Erma submachine gun MPE was used by the German police and Waffen SS Units; it was never adopted by the German Army.

The first submachine gun to be adopted by the German Army after the MP18I was the 9mm MP38. The MP38 and its successor the MP 40 were developed by the Erma Werke at the request of the German Army. Although the weapon is commonly called a "Schmeisser," Hugo Schmeisser had little if any connection with its design. The Haenel firm, of which Schmeisser was general manager, was among the producers of the weapon during the war and Schmeisser developed the MP41, a modification of the MP40. The MP38 and 40 were the standard submachine guns of the German Army during World War II and over a million of these weapons were made.

During the war several new designs of submachine guns were produced by the Germans in rather limited quantities. In general

9mm MP18I.

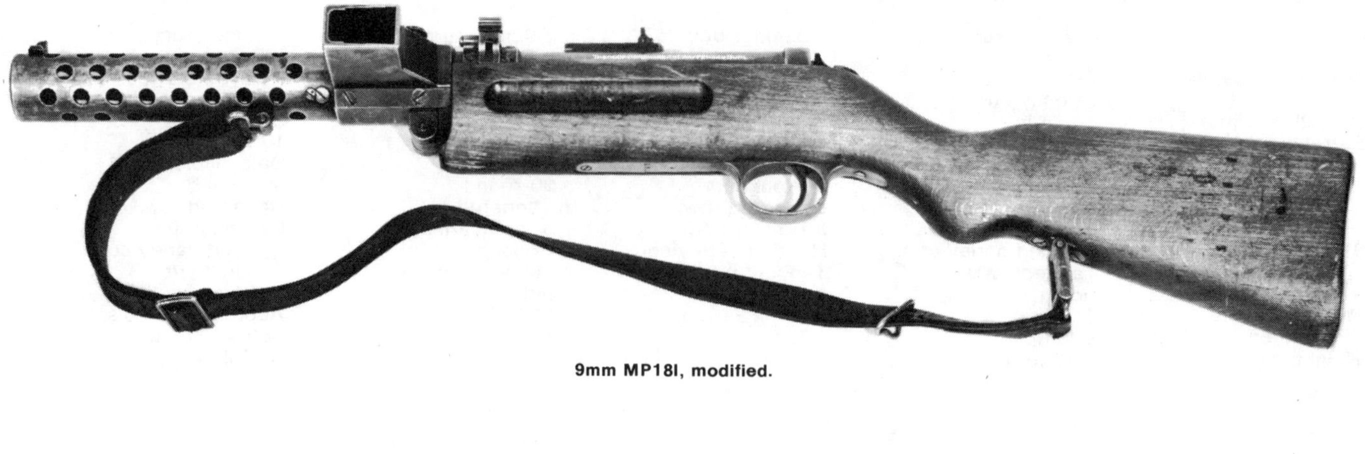

9mm MP18I, modified.

Schmeisser Machine Pistol 28II.

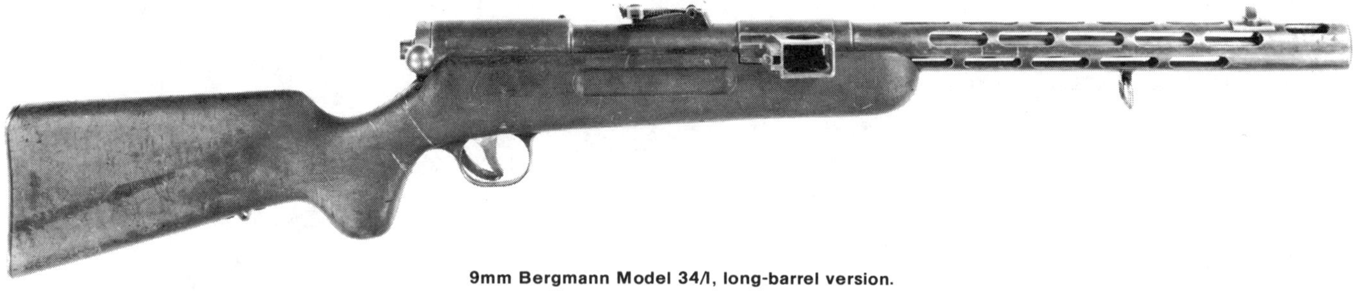

9mm Bergmann Model 34/I, long-barrel version.

their development seemed to be an attempt to produce a cheap, easily-made submachine gun in order to conserve materials and manufacturing facilities.

There are indications, however, that the Mauser-made copy of the British Mark II Sten may have been intended for some clandestine use, since even the British markings were copied. This weapon is called the Gerät Potsdam.

The MP3008 is also a copy of the Sten and was produced by Mauser and six other firms. This weapon which appeared in a number of versions was intended for Volkssturm use. Production did not begin until the closing months of the war.

The EMP44 was developed by Erma Werke; it is a relatively simply made weapon consisting mainly of welded steel tubing. It appears to have been made only as a prototype and was probably designed for special use.

The Germans modified a number of Soviet PPSh M1941 submachine guns for their own use by altering them to use the MP38 and 40 magazine and fitting them with 9mm Parabellum barrels. In addition, the Germans used other foreign-made weapons. The Italian 9mm Beretta Model 38/42 was made for the German Army with German markings and acceptance stamps. The Steyr Solothurn submachine gun adopted by Austria was used by the Germans as the MP34 (Ö). Weapons produced prior to 1939 are chambered for the 9mm Mauser cartridge; those produced during 1939 and 1940 were chambered for the 9mm Parabellum cartridge. The Austrian Police used the MP34, to use the Austrian nomenclature, in a version chambered for the 9mm Steyr cartridge, and these weapons were used to some extent by the German "Ordnungs polizei."

The 9mm Model 38, Model 40 and Model 41 Submachine Guns (MP38, 40 and 41)

The MP38 was the first submachine gun developed for the German Army since the MP18I of World War I. Although the design has been credited to Schmeisser in many publications, it was probably designed by Erma and first production was carried on at that plant. The telescoping, multi-piece recoil spring and firing pin assembly were developed from those used with the Erma submachine gun. The MP38 was made from 1938 to 1940 at the Erma plant.

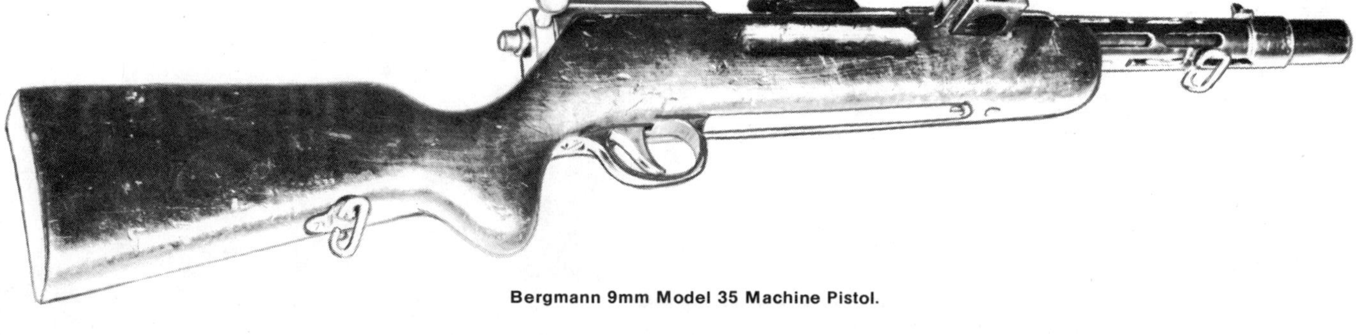

Bergmann 9mm Model 35 Machine Pistol.

Erma 9mm Machine Pistol (EMP).

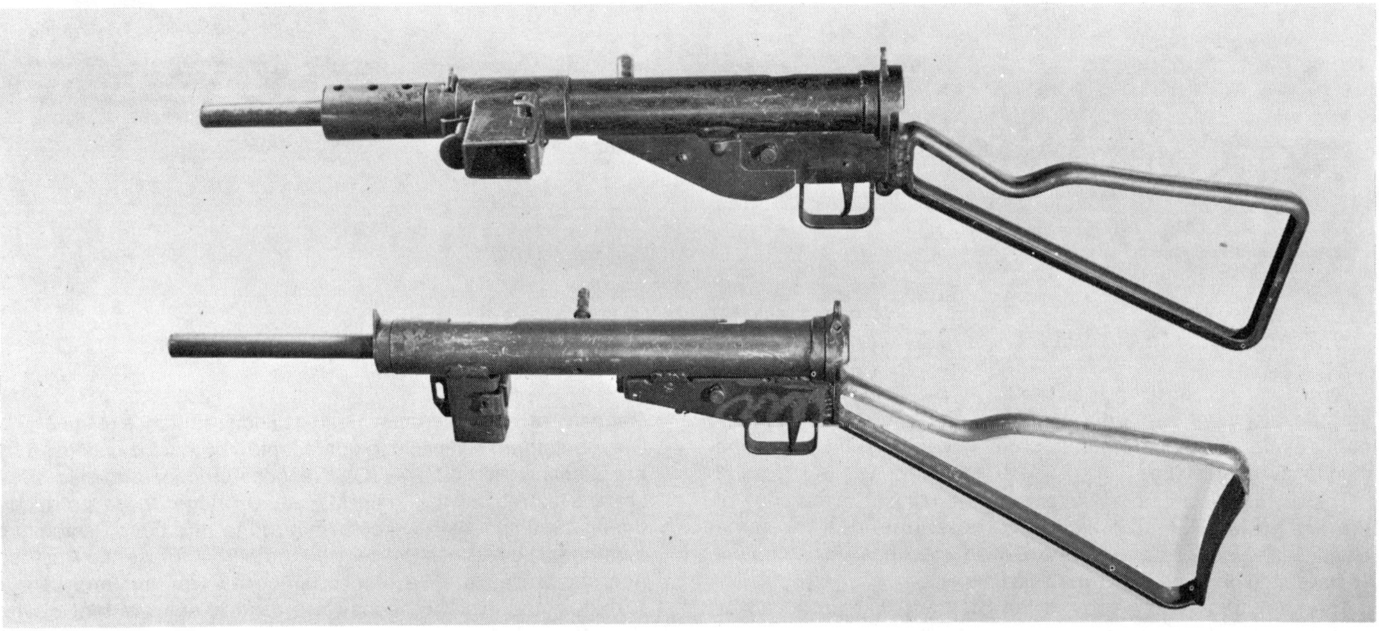

Comparison of British Sten gun (above) and German imitation (MP3008) (below).

9mm EMP44 submachine gun.

The plastic receiver nousing and aluminum frame and folding steel stock of the MP38 were unique and the design of this weapon had considerable influence on later submachine guns. The receiver of the MP38 is made of steel tubing.

The MP38 had one serious deficiency—which is shared by most submachine guns—it was not completely safe to handle. The only safety, a cut-out in the receiver into which the bolt handle locked when the gun was cocked, did not allow the gun to be carried safely with the bolt forward and a loaded magazine in the gun. If the gun received a severe jolt, such as falling on its breech end, the bolt could bounce back far enough to pick a round up from the magazine and fire. The MP38 was modified to remedy this defect by the fitting of a two-piece bolt handle and the cutting of a slot above the front of the bolt receiver track to lock the bolt in the forward position. This modification was called the MP38/40.

The MP38 was somewhat expensive to manufacture and the weapon was re-engineered to cut down the use of expensive tooling. The weapon produced as the result of this redesign was called the MP40, which differs from the MP38 in the following: new ejector, magazine release assembly, receiver (ribbing eliminated), grip frame of the MP38 is cast aluminum while that of the MP40 is formed, and the stamped middle tube of the recoil spring assembly is drawn and pinched on the MP40. There are numerous other minor differences. MP40 was made in much larger quantities than was the MP38 and was manufactured by Steyr, Haenel, and Erma with the assistance of a number of subcontractors. Over 1,000,000 MP40s were made from 1940 to 1944.

There were several modifications of the MP40. The most common modification of the MP40 has a stamped, ribbed magazine housing and uses the two-piece bolt handle. This weapon, which apparently was called MP40/I, is far more common than the MP40 itself. A rarer modification is the MP40/II which is fitted with a magazine housing to accommodate two magazines. The magazines are held in a sliding housing arranged to allow each magazine to feed in turn.

The MP41 was developed at Haenel by Schmeisser. It was made in very limited numbers and it was not used by the German Army or Police; it may have been made for export. MP41 has the receiver and barrel assembly and bolt assembly of the MP40, but the stock and trigger mechanism are modeled on that of the MP28II.

Loading and Firing the MP38

Six spare magazines and a special magazine loader are issued in a web haversack with each one of these guns. The loader is a simple lever device with an attached housing into which the magazine is inserted. Snapping a cartridge into the top of the housing and pushing down firmly on the lever loads the individual cartridge into the magazine. This motion is repeated until the magazine is filled. If no loader is available, cartridges may be inserted by the normal procedure for loading automatic pistol magazines; leverage for inserting the last few cartridges may be exerted by pressure of both thumbs, once the cartridge has been seated.

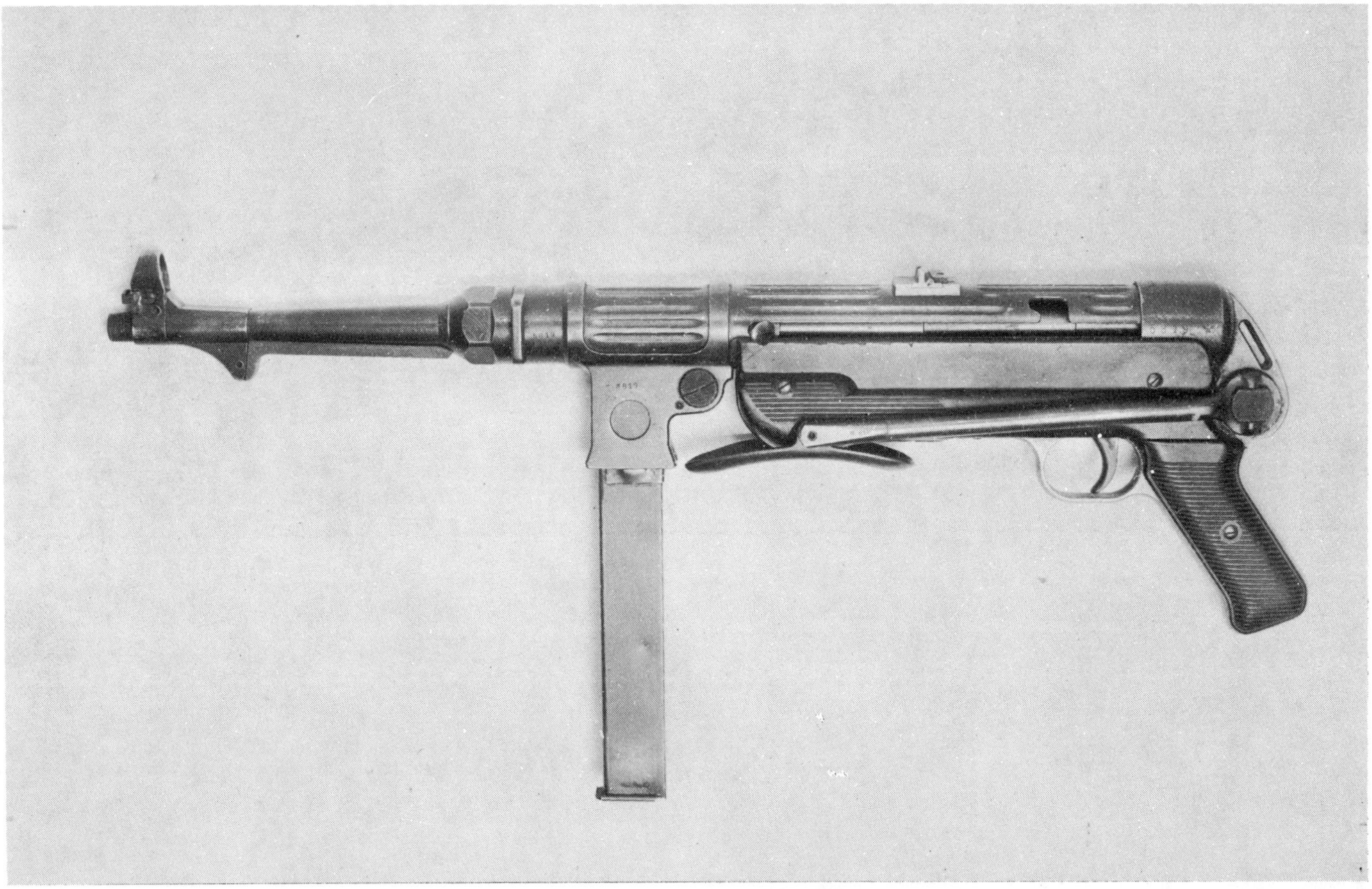

German 9mm Model 38 Machine Pistol.

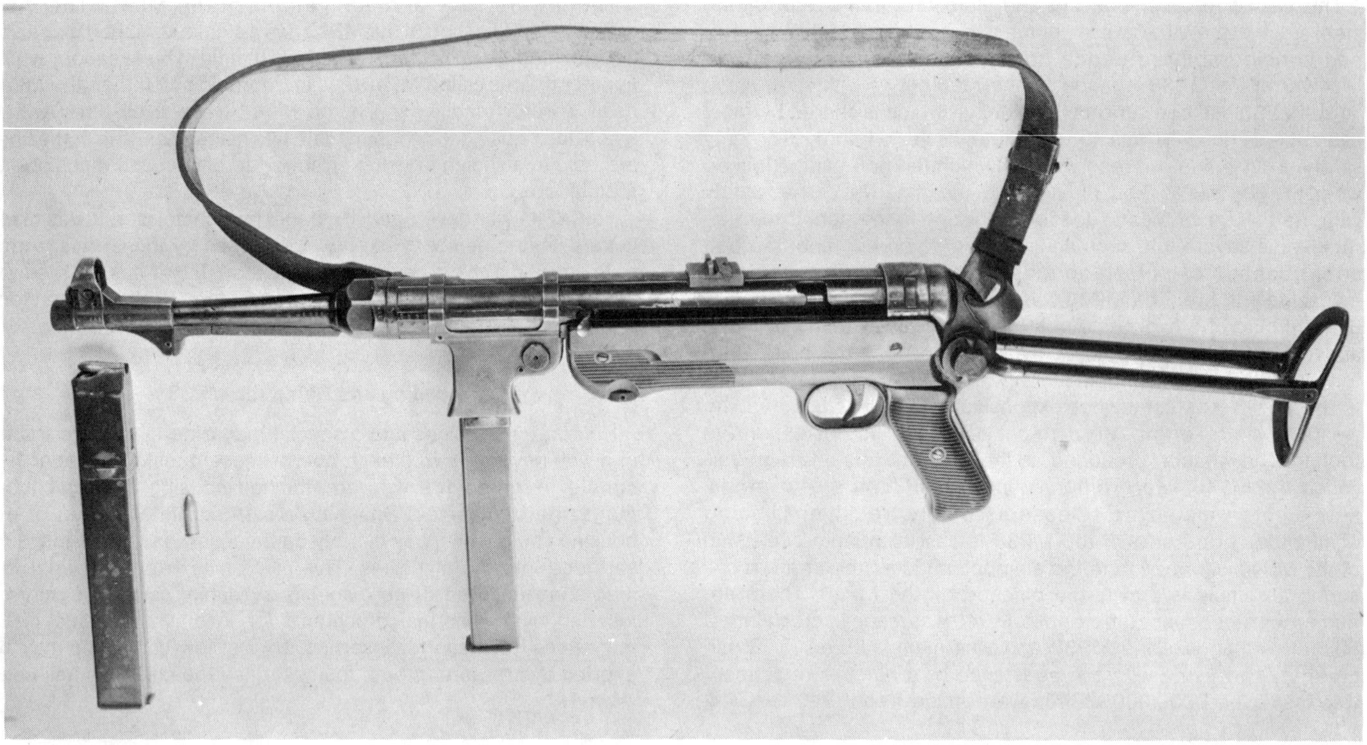

German 9mm Model 40 Machine Pistol.

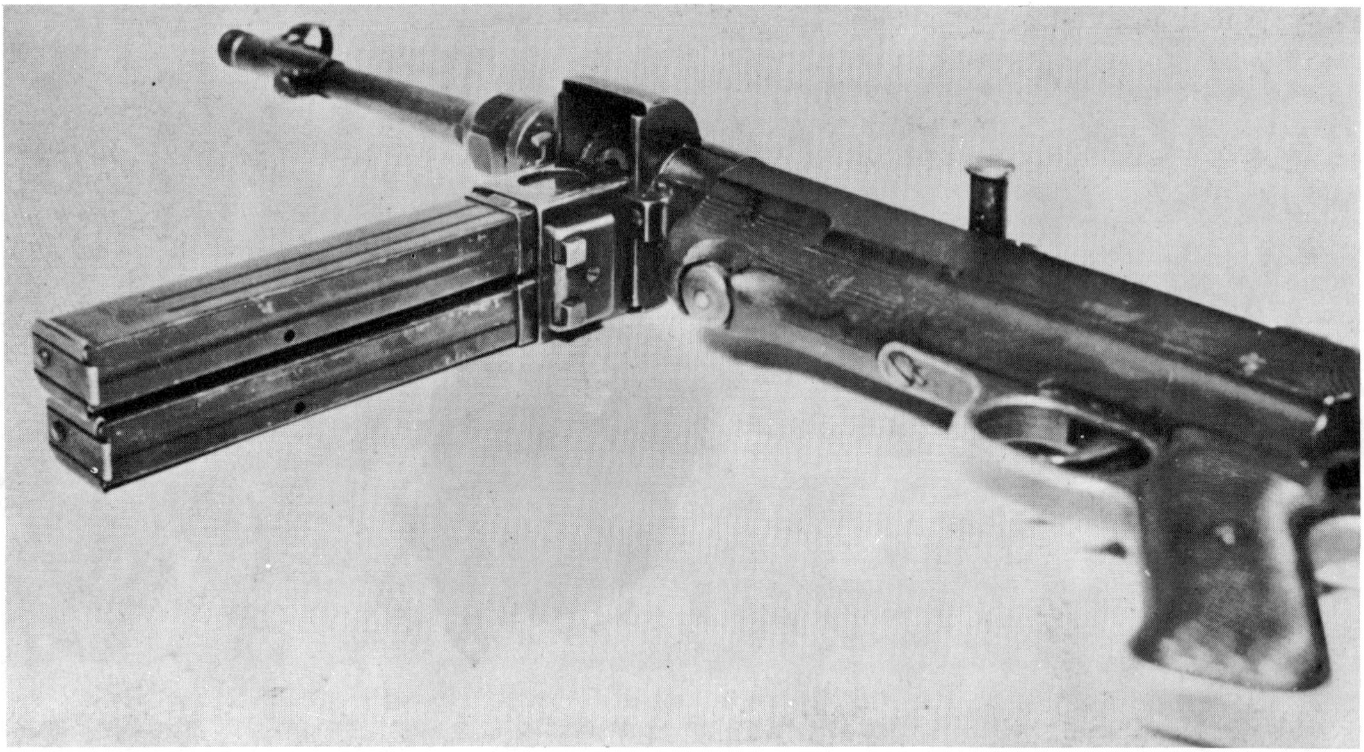

9mm MP40/II.

9mm MP41.

Loading the MP38/40

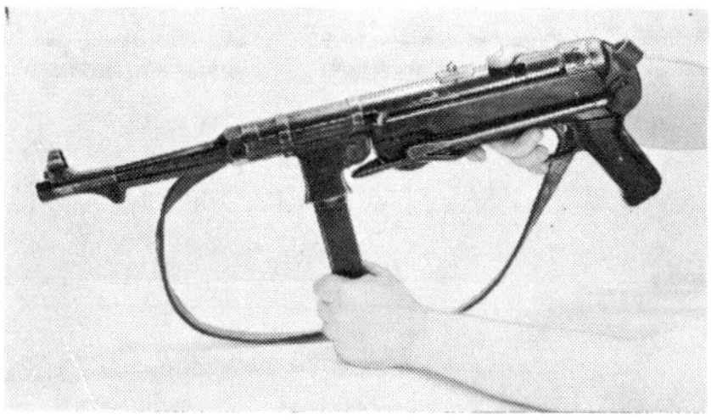

Insert loaded magazine from below into the magazine housing and push up until it locks. Note: a stud on the outside of the magazine will prevent it from going in beyond the proper length.

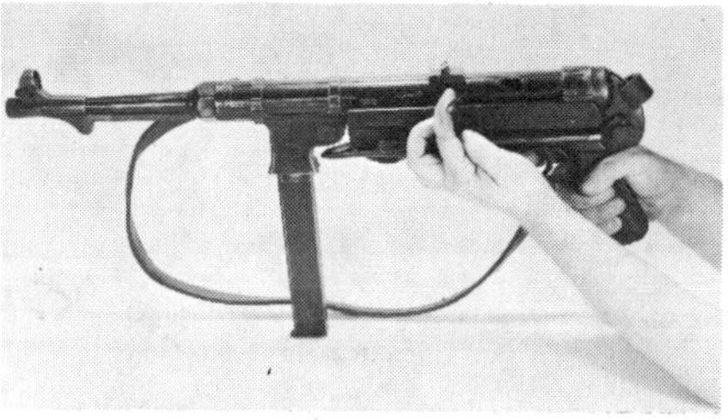

Warning: always remember that this weapon fires when the bolt goes forward! Never, therefore, let the bolt go home while the loaded magazine is in position. Unless you wish to fire the weapon, always remove the magazine before easing the bolt home.

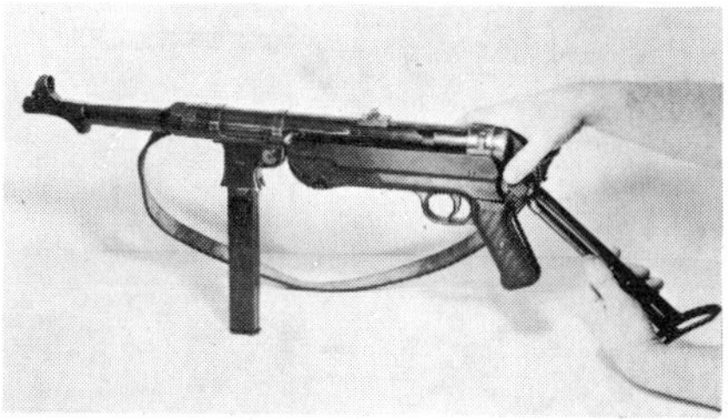

Whenever possible, always use this weapon as a carbine. To do this, press the catch stud as indicated. This will release catch and permit you to unfold the stock and turn the butt piece down into proper place for firing from the shoulder.

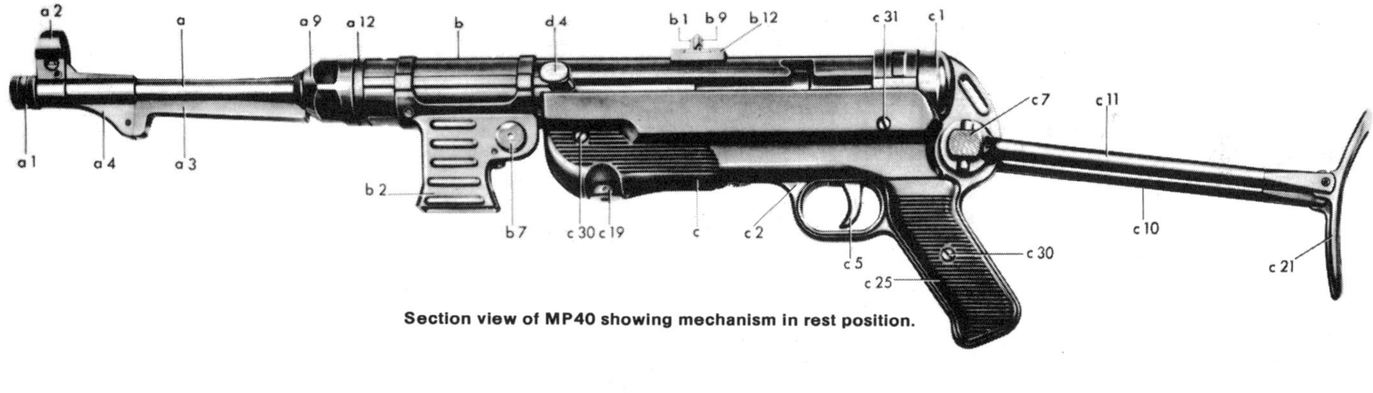

Section view of MP40 showing mechanism in rest position.

CHARACTERISTICS OF GERMAN SERVICE SUBMACHINE GUNS

	MP18I	MP38	MP40	MP40II	MP41
Caliber:	9mm Parabellum.	9mm Parabellum.	9mm Parabellum.	9mm Parabellum.	9mm Parabellum.
System of Operation:	Blowback, full automatic only.	Blowback, full automatic only.	Blowback, full automatic only.	Blowback, full automatic only.	Blowback. Selective fire.
Length:					
Stock extended:	32.1 in.	32.8 in.	32.8 in.	32.8 in.	34 in.
Stock folded:	----	24.8 in.	24.8 in.	24.8 in.	----
Barrel length:	7.88 in.	9.9 in.	9.9 in.	9.9 in.	9.9 in.
Feed device:	32-round, "snail" drum type, detachable, box magazine.	32-round, detachable, staggered row, box magazine.	32-round, detachable, staggered row, box magazine.	64 rounds in 2 detachable staggered box magazines.	32-round, detachable, staggered row, box magazine.
Sights: Front:	Barley corn.	Hooded barley corn.	Hooded barley corn.	Hooded barley corn.	Hooded barley corn.
Rear:	Notched flip-over leaf.	Notched flip-over leaf.	Notched flip-over leaf.	Notched flip-over leaf.	Notched flip-over leaf.
Weight:	9.2 lbs.	9.5 lbs.	8.87 lbs.	10 lb.	8.15 lb.
Cyclic rate:	350-450 r.p.m.	500 r.p.m.	500 r.p.m.	500 r.p.m.	500 r.p.m.
Muzzle velocity:	Approx 1250 f.p.s.	Approx 1300 f.p.s.	Approx 1300 f.p.s.	Approx 1300 f.p.s.	Approx 1300 f.p.s.

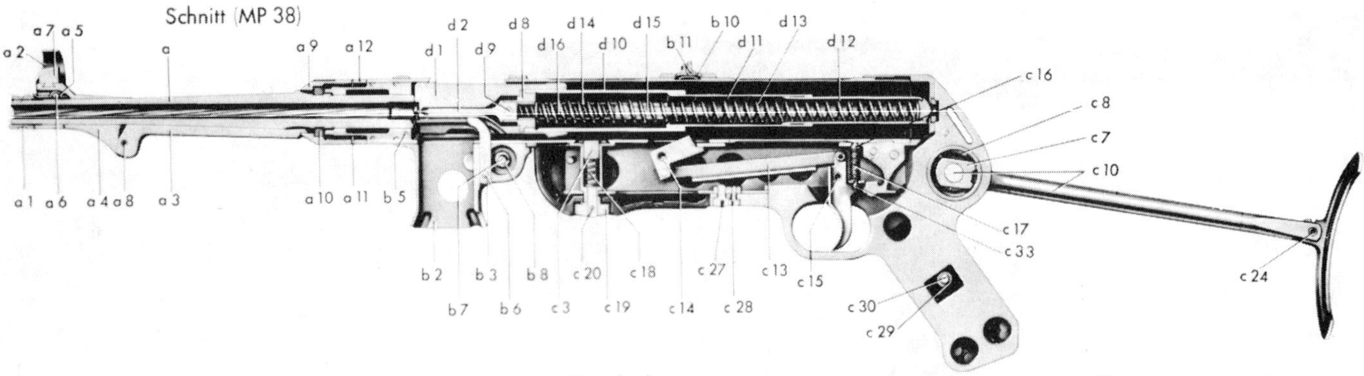

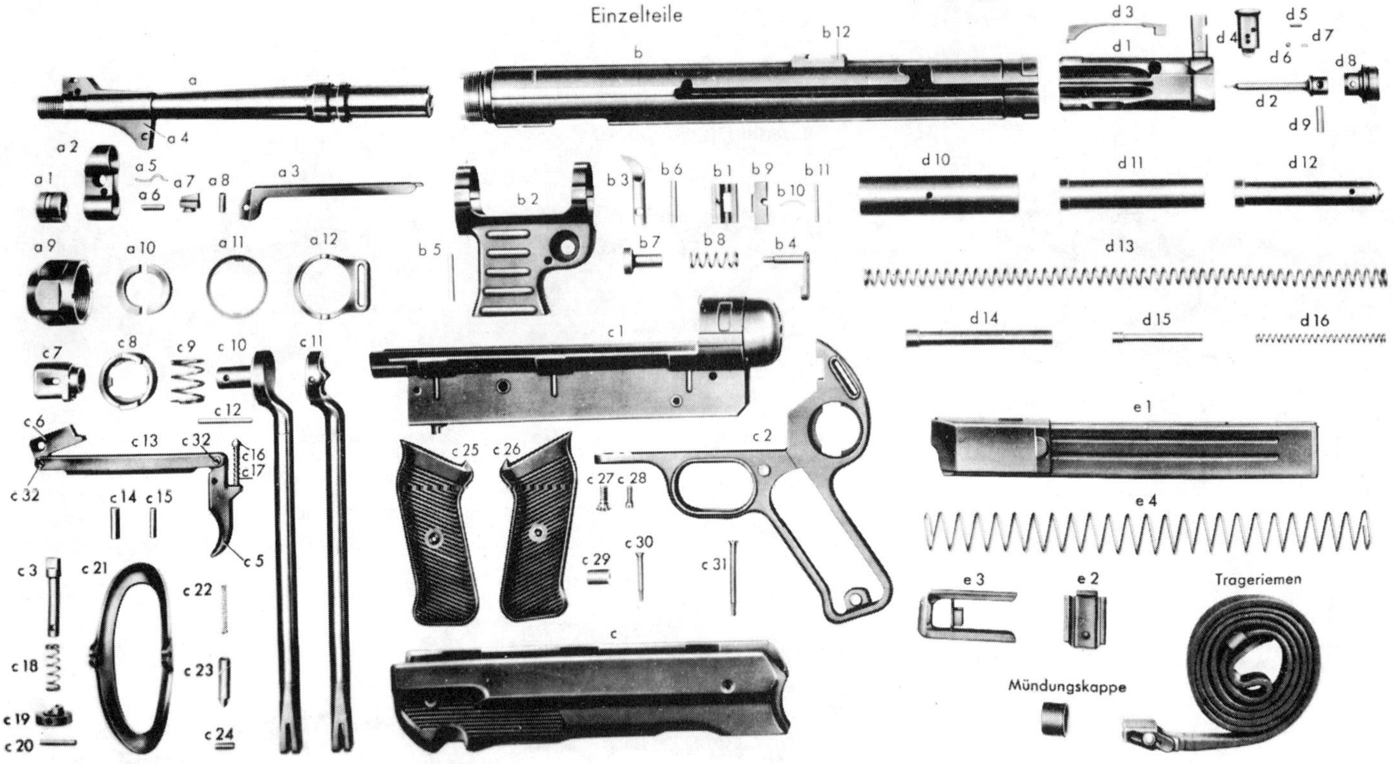

a2 Front cap cover
d10 Recoil spring tube large
a7 Front sight
a5 Front sight retainer
a Barrel
a9 Barrel nut
a12 Collar
d1 Bolt
d2 Firing pin
d9 Firing pin retaining pin
d8 Recoil spring tube end
d14 Recoil spring
d10 Recoil spring tube large
d15 Recoil guide
b10 and b11 Rear sight leaf spring
d11 Recoil spring second tube
d13 Buffer spring
d12 Buffer spring tube
c10 Stock arm
c7 Stock pivot
c8 Stock release
c24 Shoulder piece pivot

a1 Barrel cap
a6 Cover retainer
a4 Barrel jacket
a8 Resting bar pin
a3 Resting and retracting bar
a10 Barrel nut washer
a11 Barrel threads
b2 Magazine guide
b7 Magazine release cap
b3 Magazine release screw
b6 Washer
c3 Receiver lock
c20 Receiver lock screw retainer
c19 Receiver lock screw
c18 Receiver lock spring
c14 Sear
c27 and C28 Frame screws
c13 Sear lever
c15 Trigger axis screw
c17 and c16 Trigger spring
c29 and c30 Grip screws
a Barrel

a1 Barrel cap
a2 Front sight cover
a3 Resting bar
a9 Barrel nut
a12 Collar
b Chamber cover
b2 Magazine guide
b7 Magazine release cap
d4 Bolt handle
b1 and b9 Rear sight leaves
b12 Rear sight base
c31 Lock frame screw
c1 Buffer housing
c7 Stock release button
c10 and c11 Stock arms
c21 Shoulder piece
c30 Fore-end screw
c19 Dismounting screw
c Fore-end
c2 Trigger guard
c5 Trigger
c25 Pistol grip
c30 Grip screw

How the MP38 Works

The loaded magazine inserted from below is held securely in place by the magazine lock. The firing pin is attached to the forward end of the telescoping housing. It passes through the hole in the center of the bolt, while the abutment behind it lodges into the head of the bolt recess. As the bolt is drawn back by its handle, or forced back by the functioning cartridge, it telescopes the three-piece recoil spring housing (which carries the firing pin) and compresses the recoil spring which is inside the telescope. The rear of the recoil spring housing rests against the inside of the rounded buffer end of the frame, which is securely locked to the receiver.

When the bolt is in the fully cocked position, the sear locks into the bottom of the bolt and connects with the trigger. Pressing the trigger depresses the sear and permits the bolt to run forward under the influence of the recoil spring acting through the telescopic section to force the bolt forward.

As the feed ribs on the bottom of the bolt strip the top round from the magazine and push it into the firing chamber, the face of the extractor set in the bolt blocks the base of the cartridge. When the cartridge is fully seated, the further forward movement of the bolt pushes the heavy extractor to snap it into the extracting groove. At the same time the bolt face strikes against the base of the cartridge. The firing pin, a separate unit from bolt, is under the pressure of the recoil spring and functions in the same manner as a fixed firing pin—it protrudes at all times. This pin now strikes the cartridge and discharges it.

During the backward action the extractor hook withdraws the empty cartridge case, carrying it back until it strikes against the ejector and is ejected. This cycle of operation continues as long as the trigger is held back and there are any cartridges left in the magazine.

Field Stripping the MP38/40

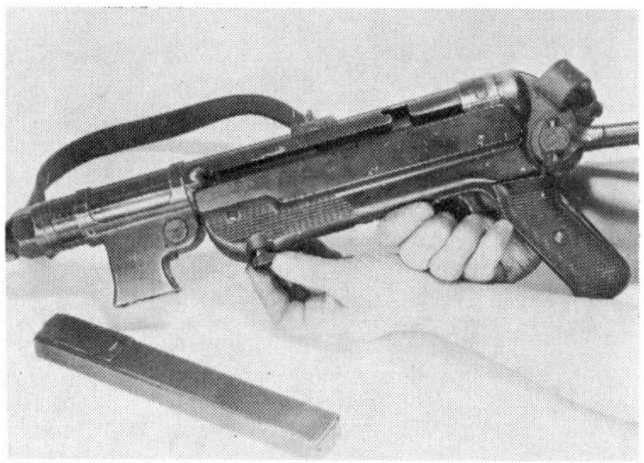

(1) After extracting the magazine, and seeing that the bolt is in its forward position, pull out the receiver lock against the tension of the spring and twist it to keep it locked in the outward position. (This stud is on the bottom of the frame at its forward end.)

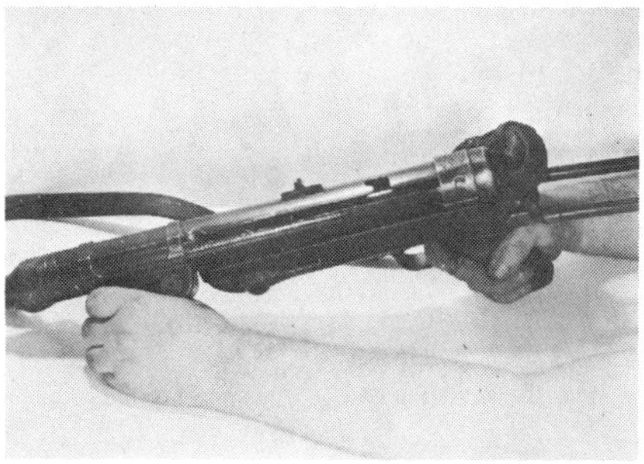

(2) While pressing the trigger with the right forefinger, hold firmly to the magazine housing with the left hand, then twist the pistol grip to the right, about 80°; this will revolve the entire frame assembly and the components.

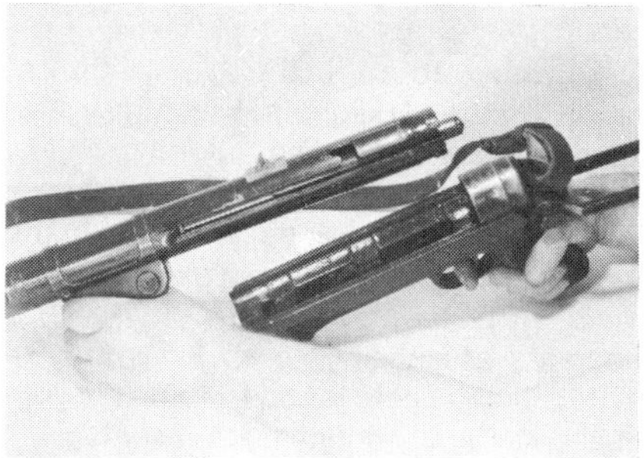

(3) Now draw the frame group back and out of the receiver.

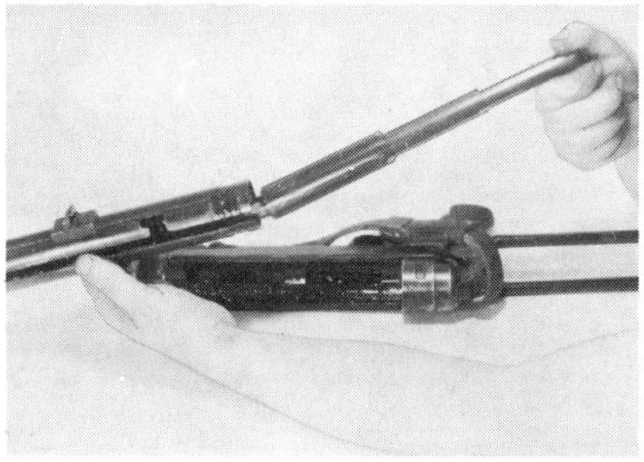

(4) Draw back slightly on cocking handle. This will bring out a telescoping tube inside which is the recoil spring, and at the front of which is the firing pin. Remove this unit.

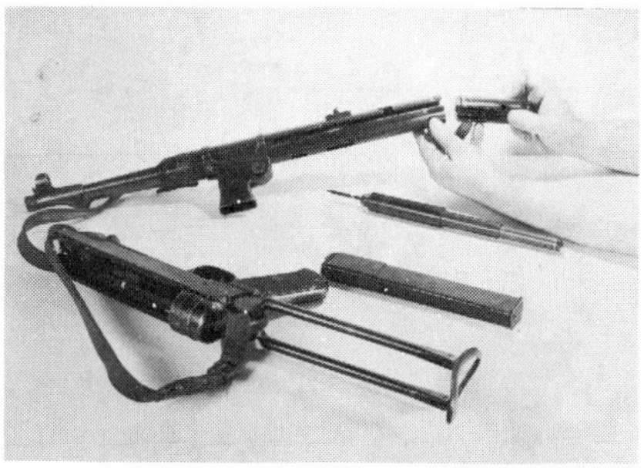

(5) Now draw straight back on the cocking handle which is a part of the bolt and withdraw the bolt from the receiver. No further stripping is required.

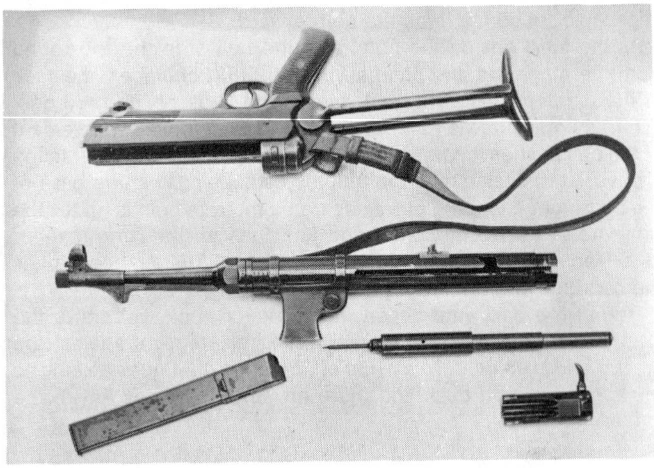

MP40 dismounted, complete field strip.

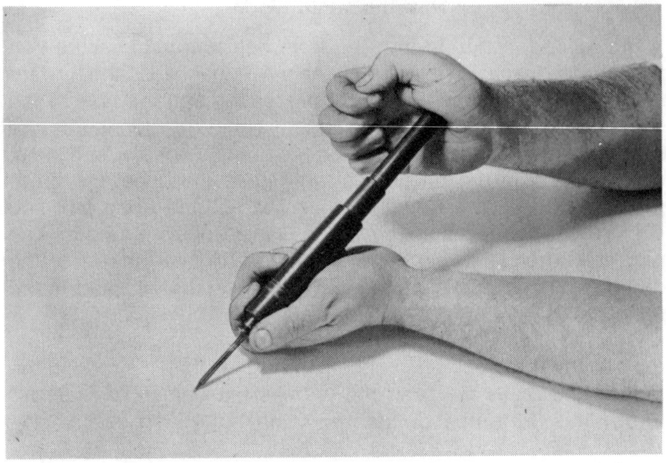

Recoil spring housing, showing firing pin and telescoping of tube.

GERMAN MACHINE GUNS

Germany adopted the Maxim gun about 1899 and in 1908 produced the Maxim gun, which may have the doubtful honor of killing more people than any other military instrument designed by man; it was certainly the most murderous weapon of World War I. The 7.92mm Model 1908 Maxim machine gun (MG08) was the standard German heavy machine gun of World War I and was made in tremendous quantities. A water-cooled weapon, it operates essentially the same as the British Vickers described in detail in the chapter on Britain. It is unlikely that it will be found in service in any country at present.

Although the MG08 was a very effective weapon, it was also quite heavy on the sleigh-type mount used during World War I and a lightened version, the Model 08/15 (MG08/15), was introduced. The 08/15 is fitted with a shoulder stock, bipod, modified receiver, and barrel jacket and its ammunition belt is carried on a reel-type drum magazine which is mounted on the side of the receiver. Its operation is the same as that of the MG08.

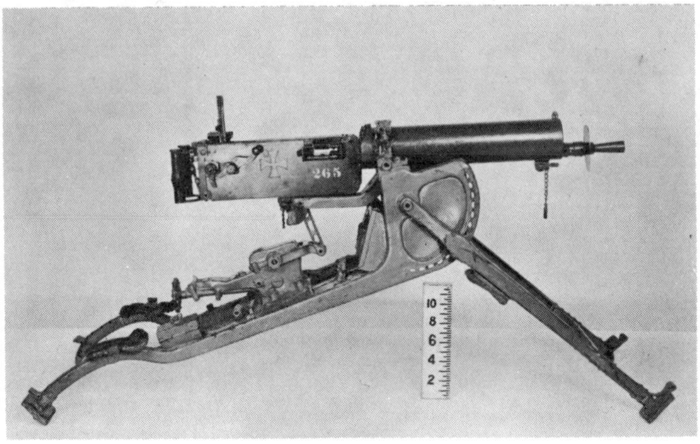

7.92mm MG08.

Maxim 7.92mm MG08/15 (light MG).

Dual 7.92mm MG81.

The MG08/15 was adapted for aircraft use by fitting it with a ventilated-type barrel jacket in place of the water jacket. This gun and the Parabellum, a modified Maxim, were the principal German aircraft guns of World War I. Toward the close of the war the MG08/18 appeared; this was essentially a ground version of the aircraft 08/15 being air cooled with a ventilated barrel jacket.

Machine-gun development and production in Germany after World War I was restrained by the Versailles Treaty, but the Germans managed to "keep their hand in" through development done by German-owned firms in foreign countries. None of the ground gun designs produced by foreign firms—Waffenfabrik Solothurn is the principal example—were adopted by the German Army, but they did add to the German capability in that they gave them an experience factor in translating military requirements into design that might otherwise have been lost.

The 7.92mm MG13 was adopted as a standard machine gun by the German Army about 1932. MG13 was made up from Dreyse M1918 water-cooled light machine guns which had been manufactured in the last year of the war, Simson of Suhl doing the work. These weapons existed in very limited numbers and were apparently all sold to Portugal in 1938. The Mauser-developed MG34 was the first true general purpose machine gun made in quantity, i.e., a gun that is used on a bipod as a light machine gun and on a tripod as a heavy machine gun. MG34 was made in very large quantities and was the standard 7.92mm ground machine gun during World War II until the adoption of the MG42. An aircraft version of MG34, MG81 was also developed and made in quantity. MG81 differed from MG34 principally in its high rate of fire—1000-2000 rounds per minute—and its lack of a semiautomatic capability.

Solothurn had developed a gun called MG29 which was rejected by Germany, but adopted in improved form by Austria as the 8mm Model 30 and by Hungary as the 8mm Model 31. Rheinmetall developed two 7.92mm aircraft guns using the basic operating system of the MG30—the MG17 fixed gun and the MG15 flexible gun. Late in World War II MG15 was fitted with an improvised stock and a bipod and used as a ground gun. One other German aircraft machine gun, the MG151 was modified for use as a ground machine gun and was found mounted on the U.S. caliber .50 machine-gun tripod.

The most famous of all German World War II machine guns is the 7.92mm MG42. This weapon, now chambered for the 7.62mm NATO cartridge, is the current standard machine gun of the West German Army. Like the MG34, MG42 is a dual-purpose machine gun and might be considered something of a pace setter in its method of manufacture. The weapon is composed mainly of stampings and its barrel change system, and feed and locking mechanisms have had considerable influence on post-war machine gun design.

At the close of the war a delayed blowback machine gun—the MG45 or MG42V was under development at Mauser. This weapon utilized the same type of bolt mechanism as the prototype StG45(M) assault rifle; the feed mechanism and barrel change was the same as that of the MG42. The SIG MG710-1 is quite similar to the MG45.

As with all other weapons, Germany used foreign machine guns to some extent. Rear area and police units were likely to be found armed with any type of machine gun for which there was sufficient

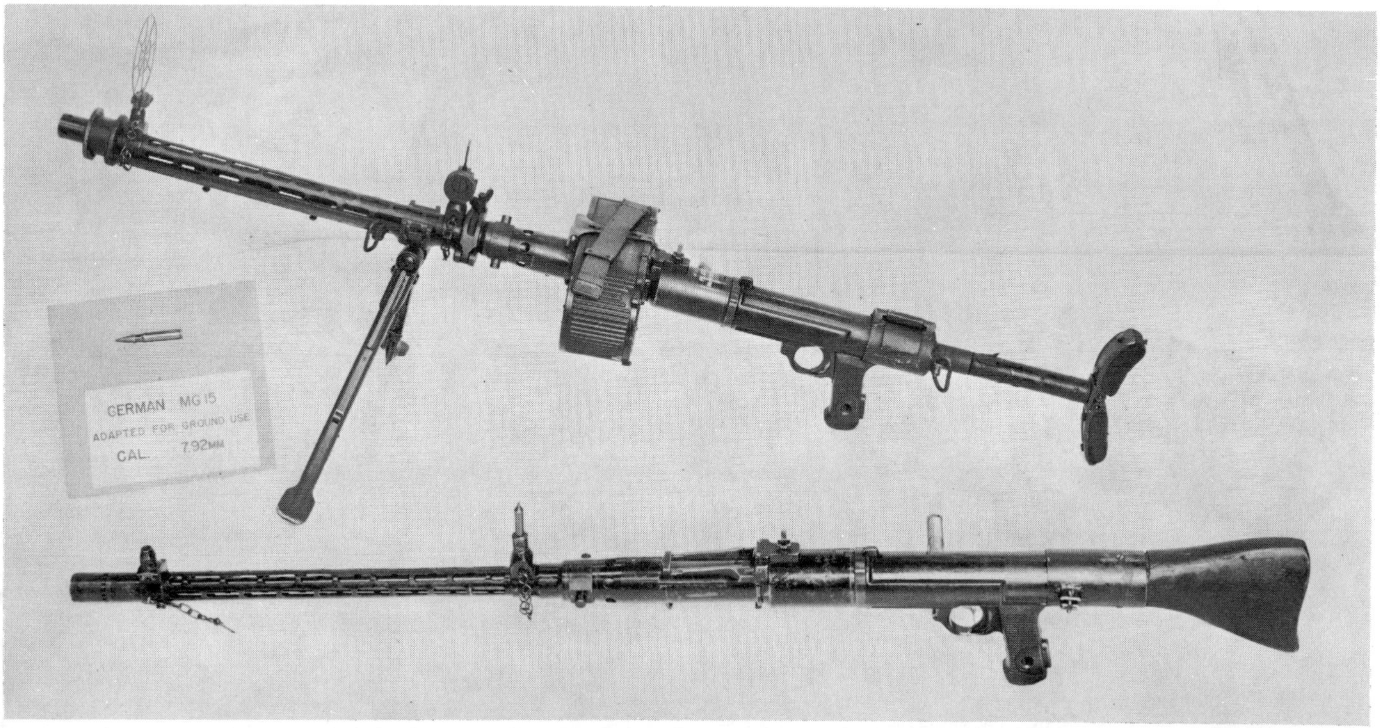

7.92mm MG15 with improvised butt stocks.

ammunition on hand. The Czech ZB26 and ZB30 were continued in production at Brno after the plant was taken over by the Germans until it was tooled up to produce MG34. The ZB53 (Model 37) was also widely used by the Germans, particularly on vehicles. These Czech guns were considered as limited standard by the Germans and manuals concerning their usage and maintenance were issued by the German Army.

THE MG34

This weapon was designed by Mauser at the direction of the Waffen Amt. It is the first modern general-purpose machine gun to be produced in large quantities. The design of MG34 incorporated many of the best features of previously developed weapons and had some outstanding features of its own. Among these were: A good method for changing barrels; a simple method of field stripping, major components being held together with bayonet-type catches; a high impact plastic stock; and a combined recoil booster, flash hider and barrel bearing.

The trigger mechanism of the MG34 is similar to that of the MG13 in that the trigger is pulled at the top for semiautomatic fire and at the bottom for automatic fire. MG34 is frequently confused with the Solothurn MG30. The Solothurn gun is based on patents of Louis Stange, a Rheinmettal engineer. The locking mechanism of the MG30 is a rotating ring which locks the barrel and bolt together, while in the MG34 the bolt head rotates and locks into a barrel extension which is permanently attached to the barrel. Stange's design undoubtedly did influence MG34, but it is different.

Various modifications of the basic MG34 were produced during World War II. These modifications and the ways in which they differ from the basic MG34 are as follows:

ZB26 with German markings.

7.92mm MG34 machine gun.

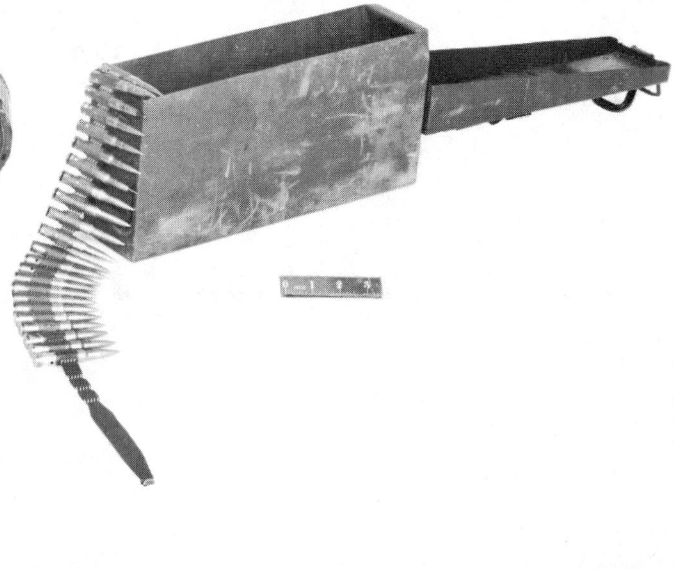

German machine gun belt and box, saddle drum magazine, and carrier with two baskets. Used with MG34.

MG34 (modified) has a heavier barrel jacket than MG34, developed for use in armored vehicles.

MG34S and MG34/41: (1) Are several inches shorter than MG34 and have shorter barrels; fire automatic only and have a simple spur-type trigger and simpler trigger mechanism; have a larger buffer; the diameter of the barrel at the muzzle is increased in order to give more surface for the gas trapped in the recoil booster to bear against; the firing pin nut on the rear of the bolt has been eliminated; and minor changes in the feed system.

MG34 has been used since World War II by the Czechs, the Israelis, and the French; and MG34s have been captured from the Viet Cong in South Vietnam. Of interest is the fact that the United States Army had the MG34 analyzed by the Savage Arms Corporation during World War II. Savage concluded that the weapon would require the use of considerable numbers of machine tools in its manufacture, at a time when machine tools were in short supply; therefore, further investigation was discontinued. MG34 can be cranky on occasion and the Germans have admitted that it took quite a while to work the bugs out of the weapon when it was first issued. One of the basic problems with the weapon is that it is too finely made, with very close fitting parts. Automatic weapons that have operating parts designed to work with plenty of play operate much better under adverse conditions of dust and mud, since they have plenty of space in which the dirt can lie without causing a malfunction.

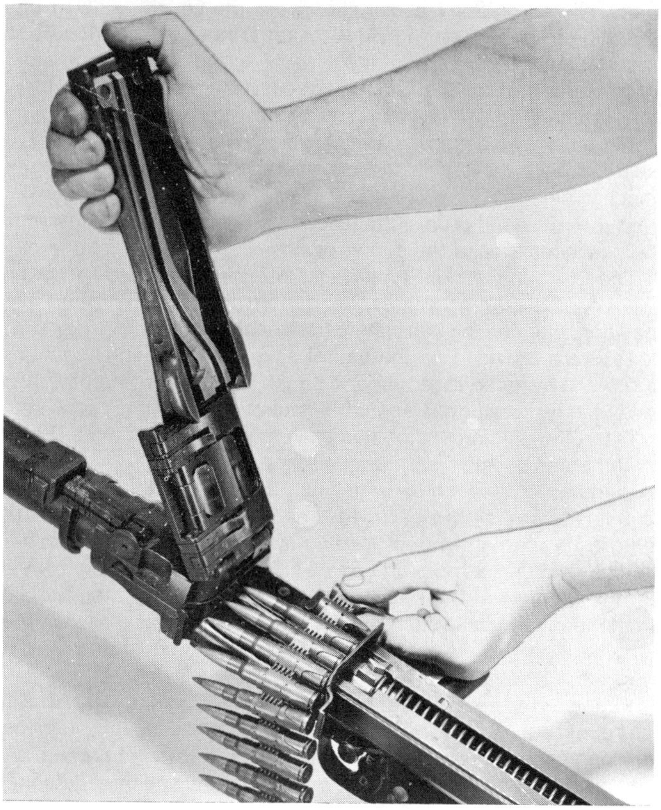

Loading and Firing the MG34

A machine is provided but is not needed to load this form of belt. The belt consists of a series of individual metal links, joined together by small pieces of coiled wire. These links are shaped

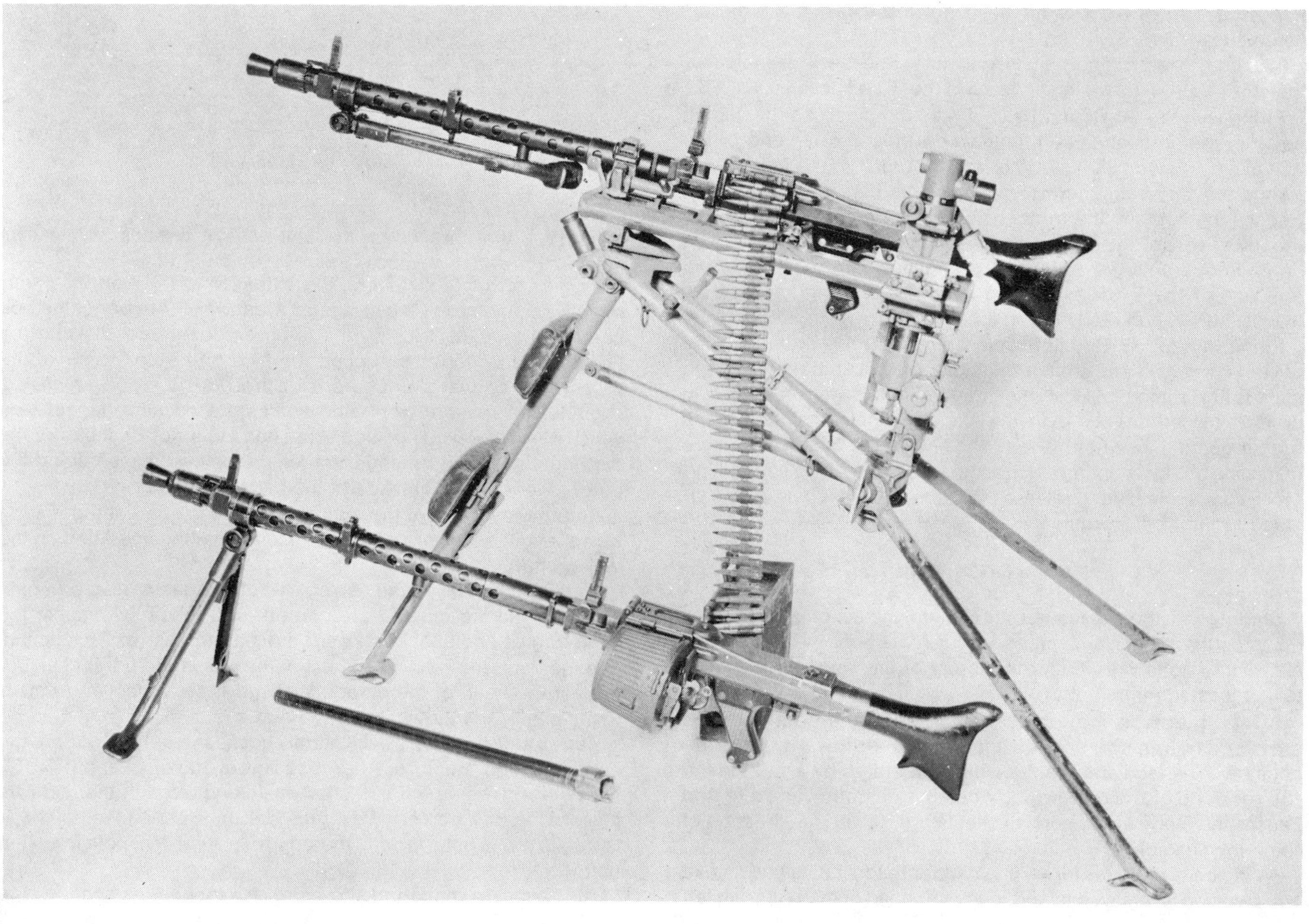

much like an ordinary pencil clip. Press the cartridge down into the clip so that the spring sides fasten around the cartridge and retain it. A nib at the end of the clip will spring into the cannelure of the cartridge and hold it in the correct position. It will be evident that in this form of belt there can be no malfunction of the type common to web belts which may expand when wet and to brass-studded belts which must pass through a complicated feed mechanism.

In the 50-round drums, the loaded belt is inserted in the drum, being wound around the center piece.

The 75-round, saddle-type drums do not use a belt. The drum itself contains the cartridges. The springs force them around into position, one coming alternately from each side.

Tabs are provided on the end of each belt. If several sections are being fastened together, or if no tab is available, then the first two or three cartridges should be removed from the metal belt.

Insert the feeding end of the belt in the feedway on the left side of the receiver, and pull through as far as it will go.

Warning: Unlike the Browning and the Vickers, the belt on this gun lies on top of the cartridges as they pass through the feed block. An alternate way of loading is to push forward the cover catch (which is on top of the receiver at the rear of the gun) and lift the feed cover to vertical position. The belt may then be laid in the feedway; make sure that the first cartridge rests against the stop on the right side of the guide. Close the cover and snap it down in place.

Pull back the cocking handle as far as it will go and the bolt will be caught and held in rearward position by the sear. Now push the cocking handle forward as far as it will go. If this is not done it will be carried forward as the bolt moves to the front, and this additional weight may cause malfunctioning.

Pressing the upper part of the trigger will now fire a single shot. Pressing the lower part of the trigger will fire the weapon automatically.

NOTE: If the cocking handle will not come back, it indicates that the safety is on. Move the lever to the "Fire" position.

Firing with the 50-Round Drum. Press the catch on the sliding cover of the drum and open the cover so that the tag end of the belt can be pulled out. Insert the tag of the belt in the feedway as for the ordinary belt. The narrow end of the belt is the front end. Engage the hook on the front end with the lug on the rear end of the lower part of the feed plate. Now swing the rear end of the drum around until the spring catch engages with the lug on the rear end of the feedway. Pulling back the cocking handle now leaves the weapon ready for firing.

Firing with the 75-Round Saddle Drum. With this drum the feed cover is removed and a magazine holder is substituted. The feed plate is also removed. Belts are not used in this type of feed. The drum is placed directly over the magazine holder ahead of the trigger guard. Its center piece pushes down the dust cover in the magazine holder. A spring catch at the top center of the connecting piece can be pressed to release the drum and a hand-strap is provided to lift it off the gun.

How the MG34 Works

Starting with the gun loaded and cocked, the action is as follows: Pressing the trigger pulls the sear out of its bent in the bolt and allows it to go forward under the thrust of the compressed recoil spring located in the butt.

A feed piece on the top of the bolt strikes the base of the cartridge in line and pushes it from the belt toward the firing chamber. The feed arm is hollow and is operated by a stud on the top rear end of the bolt, which rides in this hollow groove and causes the feed pawl to push the next cartridge in the direction of the firing chamber.

As the bolt continues forward, two inner rollers on its head strike two cams on a cam sleeve and rotate the head of the bolt from left to right so that threads on the bolt engage threads on the cam sleeve; this effectively locks the bolt to the barrel.

MG34 on AA tripod.

As the cartridge chambers, the extractor in the bolt face slips over the cannelure of the cartridge. Meanwhile the rear of the bolt continues forward, tripping the firing pin lever and allowing the firing pin to go forward through the face of the bolt to strike the primer. The forward movement of the bolt is stopped when a shoulder on its right front side strikes the cocking handle stop which is in its forward position at the end of its slot. Just before the cartridge is fired, a locking catch on the bolt engages behind the outer roller on the right side of the head of the bolt.

Return Movement of the Action. This gun is fitted at the muzzle with a recoil increaser somewhat resembling that operating on the Vickers gun.

As the bullet leaves the barrel, part of the gas pressure behind it expands in the muzzle attachment and rebounds against the cone to give additional backward thrust to the barrel. This action, together with the rearward thrust of the gas in the firing chamber against the head of the empty cartridge case, which transmits it to the bolt, starts the action to the rear.

Barrel and bolt start back, firmly locked together during the period of high pressure. After a backward travel of about 3/16", the outer rollers on the bolt head again engage with the two cam faces in the forward end of the receiver, thus forcing the bolt head to rotate from right to left, thereby unlocking the bolt from the barrel.

The rearward motion of the barrel is stopped, as soon as the

Field Stripping the MG34

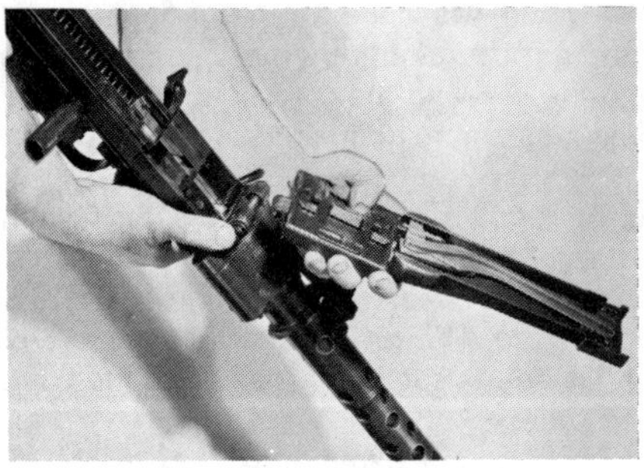

(1) Order of stripping: Push the spring catch at the extreme rear of the cover on top of the receiver and lift the cover to a vertical position. Push the cover hinge pin from the right and lift out the cover. The feed block may be lifted off.

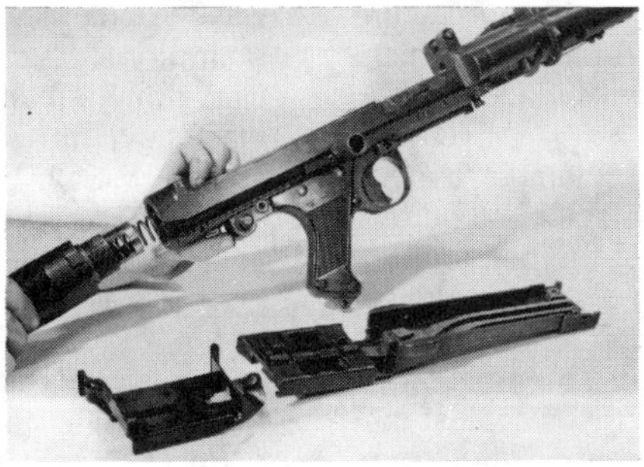

(2) The butt catch is on the underside of the receiver a few inches behind the pistol grip. Press this up with the left thumb. With the right hand, turn the butt a quarter-turn left or right. (Note: the bolt should be in forward position when this stripping motion is being done. Otherwise, the very powerful recoil spring cannot be controlled.) The recoil spring will now force the butt out of the receiver. Now remove the recoil spring.

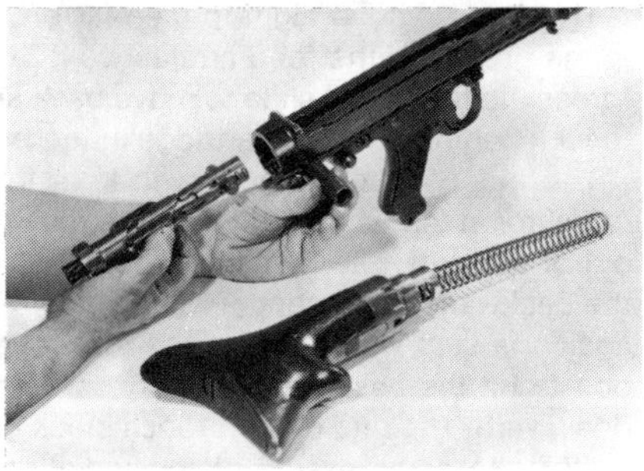

(3) Pull the cocking handle back with a quick motion. (A jerking motion is required here because the action in releasing the bolt twists the barrel extension and the barrel. Watch that the bolt and its carrier do not fly out the back of the receiver.)

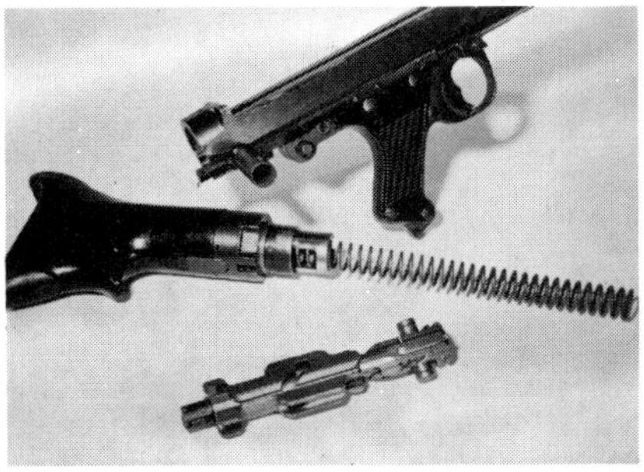

(4) Bolt and carrier may now be removed.

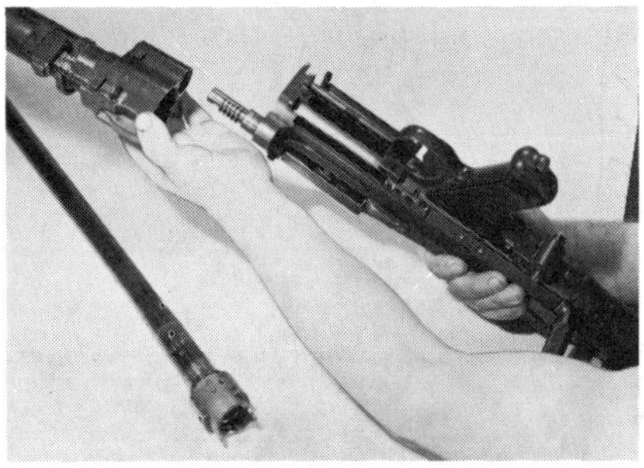

(5) Pressing the locking catch on left of receiver, below and behind rear sight, twist receiver from left to right until it clears the barrel casing. Raise the muzzle and slide the barrel out of the casing. A hinge pin catch will be found on the underside of the barrel casing, near its end and to the right. Press this up and while maintaining pressure twist the receiver, left to right, until it has completed a half-turn. It may now be pulled out to the rear.

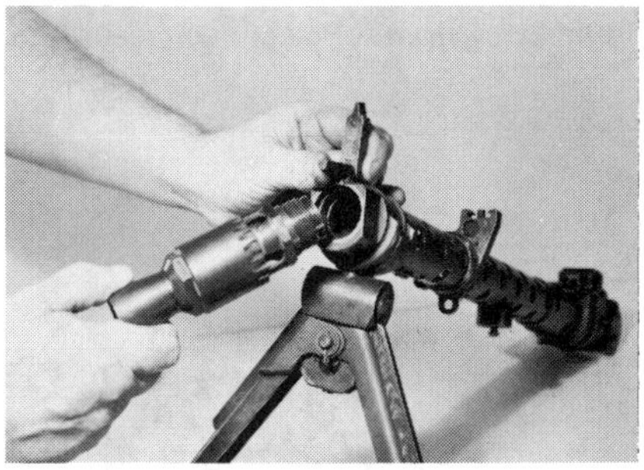

(6) A catch will be found in front of the foresight. Lifting this permits you to unscrew the flash hider over the muzzle. Inside it is a mouthpiece and a recoil cone. Remove them. The trigger assembly is locked to the receiver by two automatic locking pins. Pinching the split ends together permits them to be pulled out. (Removal of this assembly is not recommended without suitable tools.)

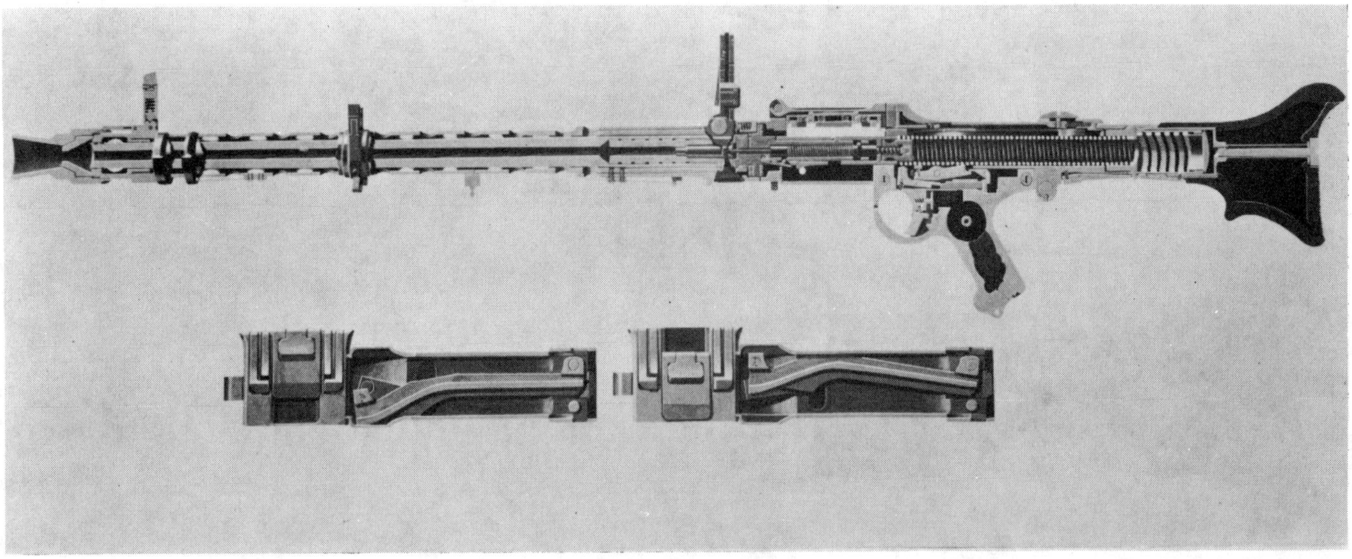

Section view of MG34.

unlocking operation is completed, when its cam sleeve strikes against shoulders in the front end of the receiver.

The stud riding straight to the rear on the bolt, its head caught in the groove in the feed arm above it, twists the feed arm which forces the feed pawl slide to move back and permits the feed pawl to lock behind the next cartridge in the belt.

The empty case, being drawn from the firing chamber by the extractor in the face of the bolt, is struck by the ejector and hurled out of the gun. The ejector is a pin in the top of the bolt; during the backward movement of the bolt the rear end of this pin strikes against a stop which forces the front end through its hole in the bolt to hit the base of the empty cartridge case. The ejection is downward. The end of the breech block carrier strikes against the buffer, the compression of the recoil spring is completed, and, if the semi-automatic portion of the trigger is being pulled, the bolt will stop open, engaged from below by the sear forced up by its spring. If the automatic trigger is being pressed, the firing cycle will be completed and continued as long as there are any cartridges left in the belt.

MG42

MG42 was developed by Dr. Grunow of Grossfuss in Dobeln reportedly from a Polish design seized by the Germans in 1939. The main contribution of Dr. Grunow was to design the weapon to be made principally from stampings and fabrications. The MG42 was the first machine gun which was made for the greater part of stampings. Its design and adoption cut down on machine tool usage in Germany during the war and the use of stampings in machine guns has become much more common since the appearance of MG42.

The MG42 introduced the recoil operated roller locking system and an unusually rapid barrel change system. The rapid barrel change was needed in this weapon because of its high cyclic rate of fire—1200 rounds per minute. The feed mechanism of MG42 was very successful and has been used in the design of the U. S. M60 machine gun among others.

West Germany uses the weapon in 7.62mm NATO; as the MG1, it has also been adopted by Italy where it is known as the 7.62mm machine gun M42/59. The weapon has been produced in 7.92mm by Yugoslavia as the Model 53 or "SARAC." It is being produced and marketed by Rheinmettal of Dusseldorf and may well be adopted by other countries in the future.

The United States produced a caliber .30 prototype version of the MG42 during World War II. This weapon, called the T24, was unsuccessful mainly because of engineering error in making allowances throughout the receiver assembly for the difference in length between the U. S. caliber .30 and the German 7.92mm cartridges.

The MG34 stripped.

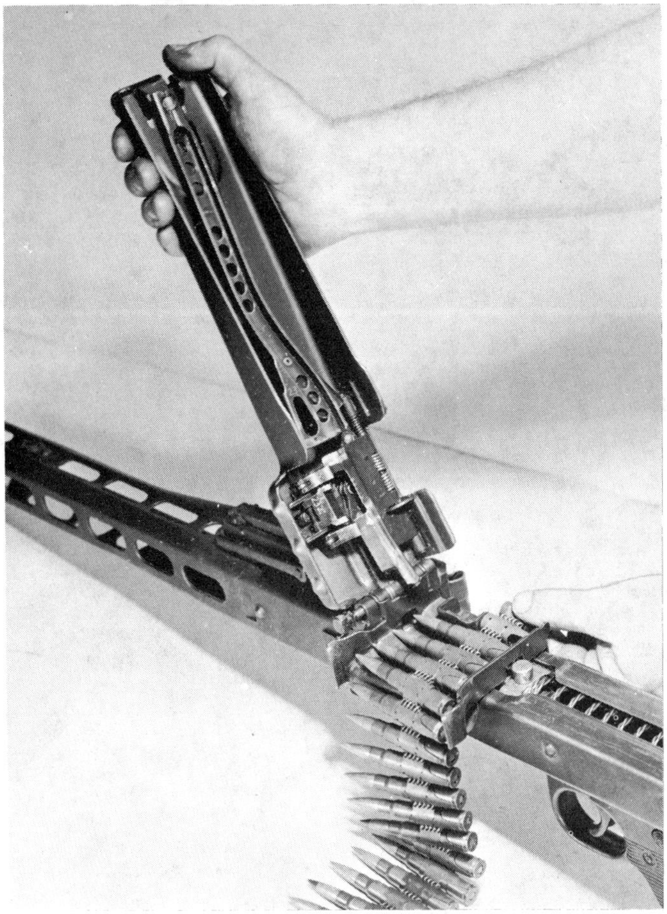

MG42, feed cover open.

Loading and Firing the MG42

To Load: (As for MG34). Feed cover may be open or closed. Be sure that the first cartridge rests against the stop on the right side of the feed guide.

Pull back cocking handle on left side of gun as far as it will go. The bolt will stay open. Then shove the cocking handle fully forward until it clicks.

Unless gun is to be used immediately, set the safety. On this gun the safety is just above the pistol grip. Push the button from the right side and it sets the safety. Push the button from the left as far as it will go and the gun is ready to fire.

NOTES ON UNLOADING. Unloading this gun is a very simple operation. First pull back the cocking handle as far as it will go. Then set the safety. Push forward the cover catch and raise the cover as high as it will go. Lift the belt out of the gun.

It is not good practice to permit the bolt to go home on an empty chamber. Always hold the cocking handle firmly while pressing the trigger and ease the bolt into forward position.

Note that there is a spring cover over the ejection opening in these guns. It flies open when the trigger is pressed. On "Cease Fire" always push it shut. This will keep dirt and dust out of the mechanism.

How the MG42 Works

In general this gun follows the operating detail of the MG34. However, an entirely new design of bolt and locking mechanism is employed. The barrel and bolt in this gun travel back in a straight line during the period of recoil. There is no turning action.

A heavy barrel extension is screwed onto the chamber end of the barrel. In its sides are slots into which cams are machined. As the bolt goes forward, a movable locking stud on each side of the front end of the bolt strikes a corresponding cam in the barrel extension. This forces locking lugs out and into slots in the barrel extension as the face of the bolt comes flush with the base of the cartridge in the firing chamber. The extractor slips over the base of the cartridge. The firing pin, mounted in the rear of the bolt assembly, is driven forward to explode the cartridge. Note that a stud, driving from the top of this rear bolt assembly, travels in a groove in the curved feed arm and shuttles the feed across and back to operate the feed mechanism.

During the recoil movement, barrel extension and bolt are firmly locked together during the moment of high breech pressure. Then as the barrel extension and barrel are stopped in rearward travel, the studs on the bolt head are cammed out by the camming surfaces on the barrel extension and the locking lugs are thus withdrawn from their seats in the barrel extension permitting rearward direct line motion in the action. This action is patterned after a simple pile-driver, the bolt resembling the pile-driver hammer being pulled up (or out) to its full extent, then the gripping surfaces being cammed out to release it.

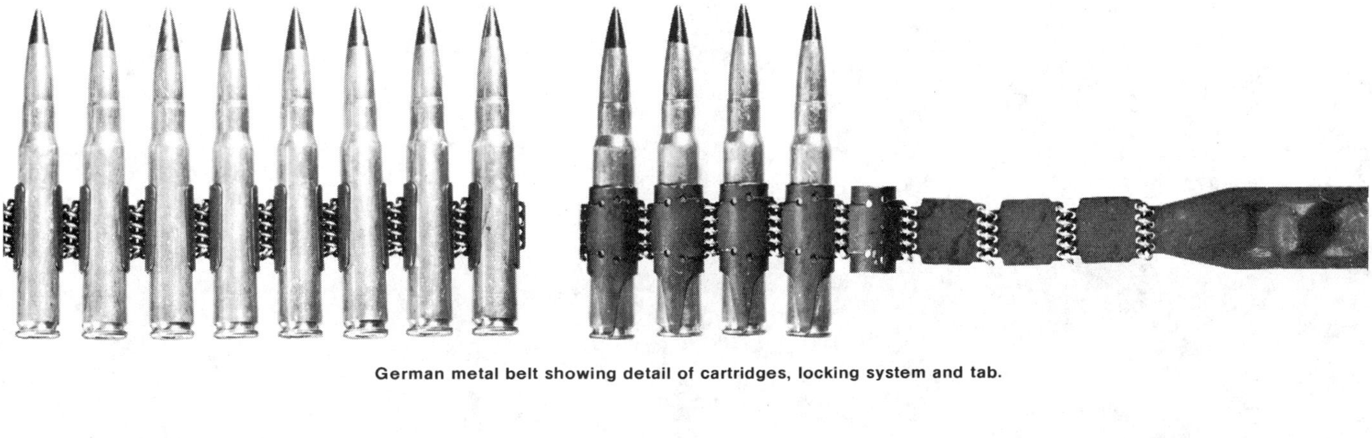

German metal belt showing detail of cartridges, locking system and tab.

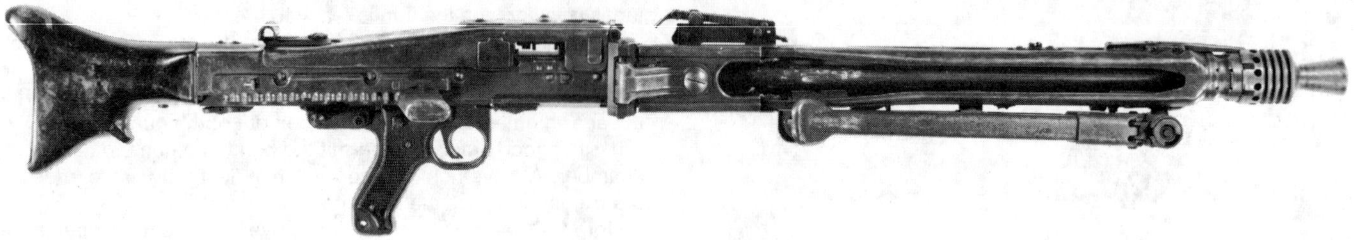

7.92mm MG42 Machine Gun.

MG42 on tripod.

Field Stripping the MG42

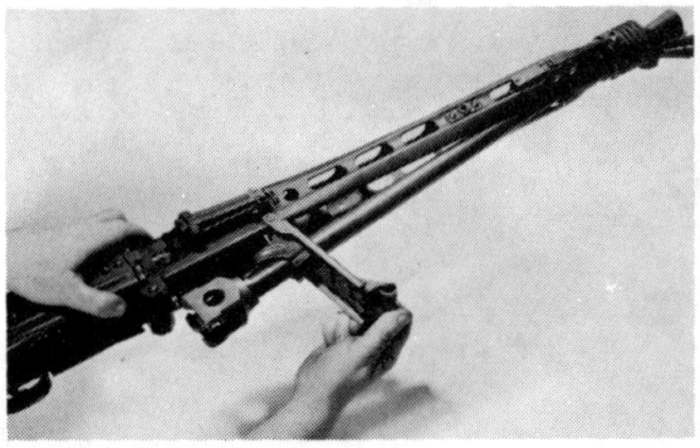

(1) In general field stripping this gun is similar to the MG34. There are, however, some few differences. To remove barrel: It is first necessary to cock the weapon. This is done by pulling back the cocking handle. Then thrust forward and outward on the heavy release catch jutting out from the rear of the barrel extension on the right hand side of the gun, below the feed block. This draws the rear of the barrel out of its seat and permits it to be drawn from the rear of the gun.

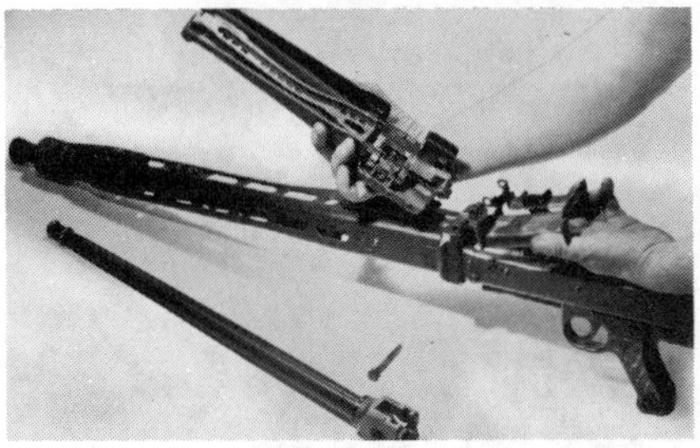

(2) The feed resembles the MG34. Push forward the feed cover catch on top of the gun near the stock and lift the cover. Pull out the feed cover hinge-pin and remove the feed block from the gun. Dismounting this is very simple.

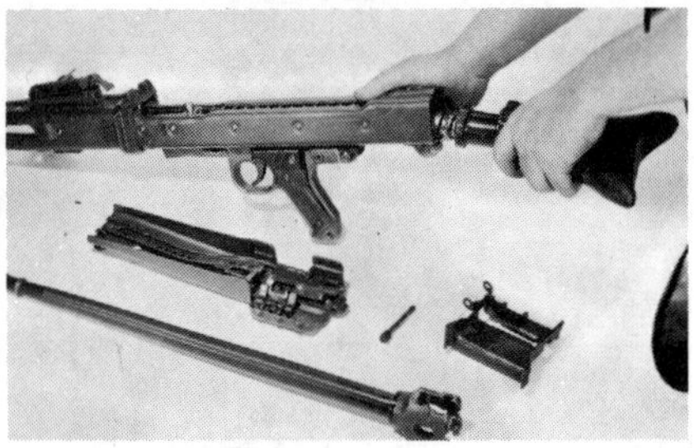

(3) Remove buttstock, same as for MG34. Be sure bolt is in forward position before removing buttstock. Catch is on the underside of the stock. Push it and twist the butt a quarter turn, right or left.

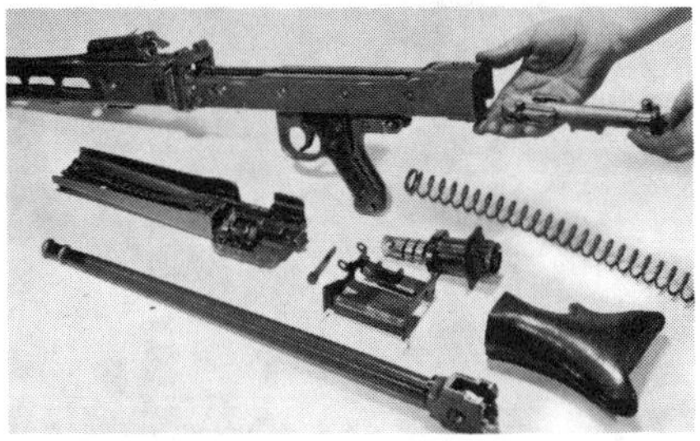

(4) Remove buffer and recoil spring. As in the MG34, the housing catch is on the rear end of the receiver, just back of the pistol grip. Press the catch and control the buffer housing that moves away from the receiver under tension of the powerful spring. Remove the bolt. Press the trigger and strike the cocking handle a sharp rearward blow. This will drive the bolt to the open rear of the receiver where it may be withdrawn. No further stripping is normally necessary.

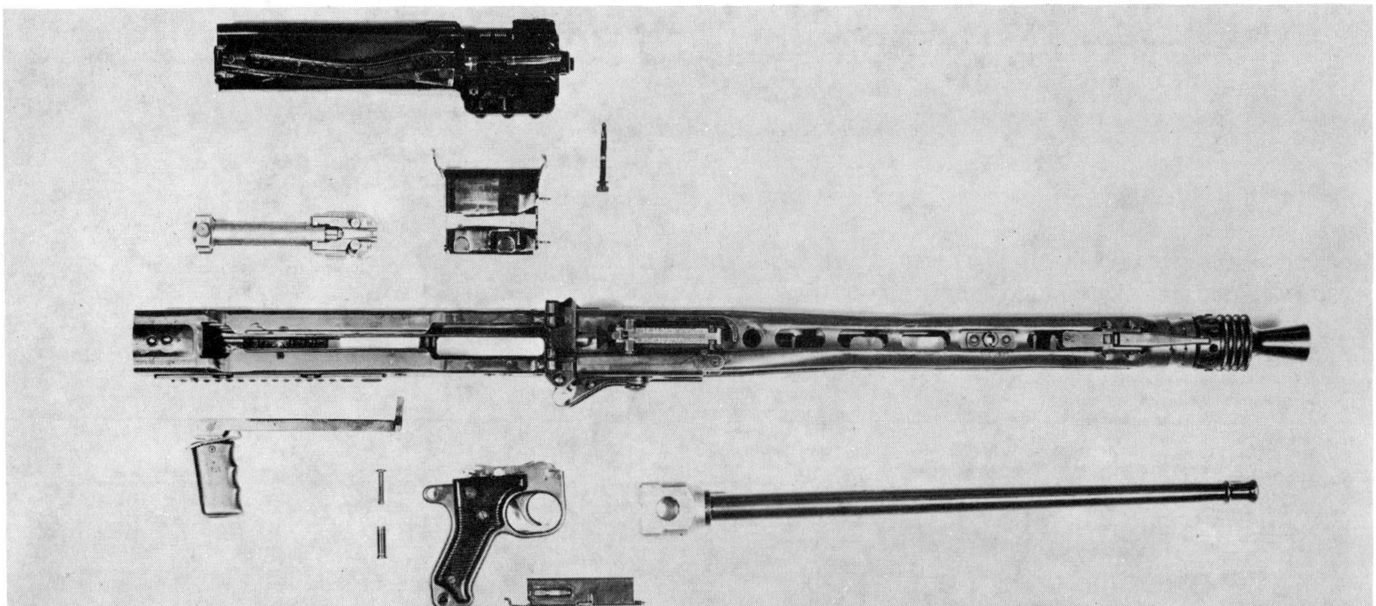

MG42, stripped.

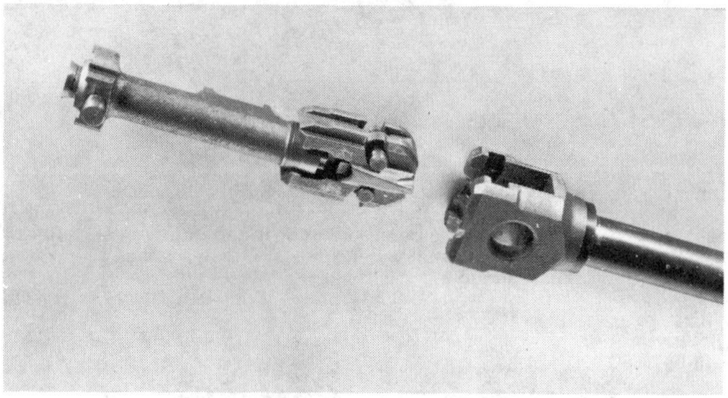

Showing detail of the bolt with feed operating stud, bolt locks, and barrel.

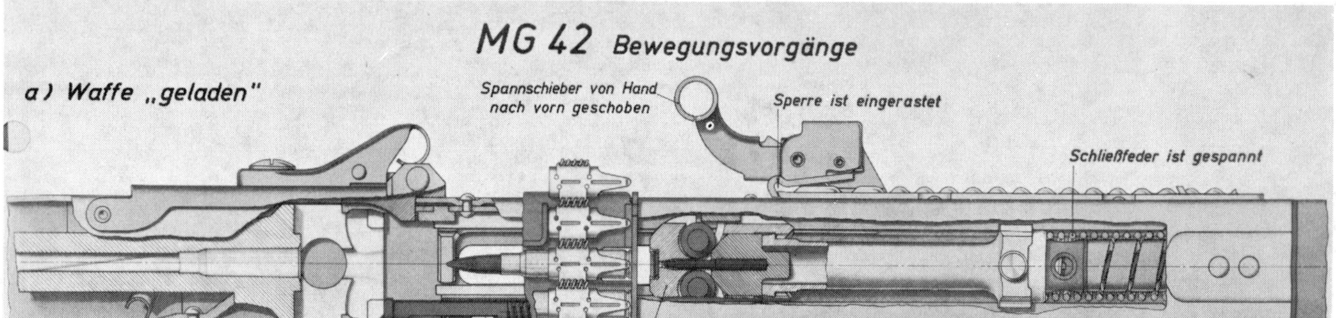

Views of MG42 mechanism showing bolt feeding into chamber, bolt unlocked.

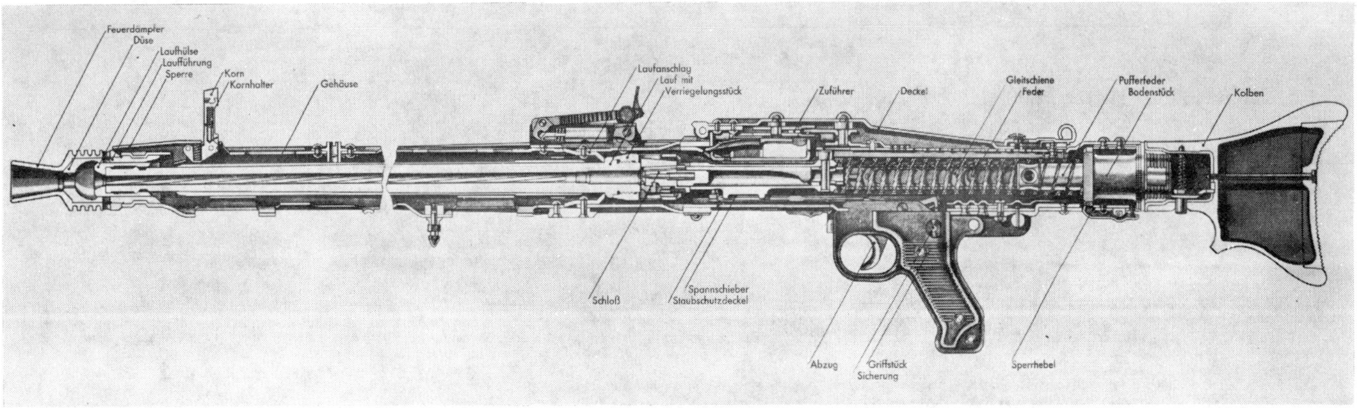

Phantom drawing showing all operating parts in full closed position. View of left side. Nomenclature, left to right, top: 1. Flash hider. 2. Barrel mouth. 3, 4, and 5. Blast cone assembly. 6 and 7. Front sight and spring. 8. Housing. 9. Barrel retainer. 10 and 11. Barrel and barrel locks. 12. Feed. 13. Cover. 14 and 15. Recoil spring and guide. 16. Buffer spring. 17 and 18. Butt assembly and butt. Left to right, bottom: 1. Lock. 2. Bolt. 3. Bolt extension. 4. Trigger. 5. Safety. 6. Grip. 7. Buffer release.

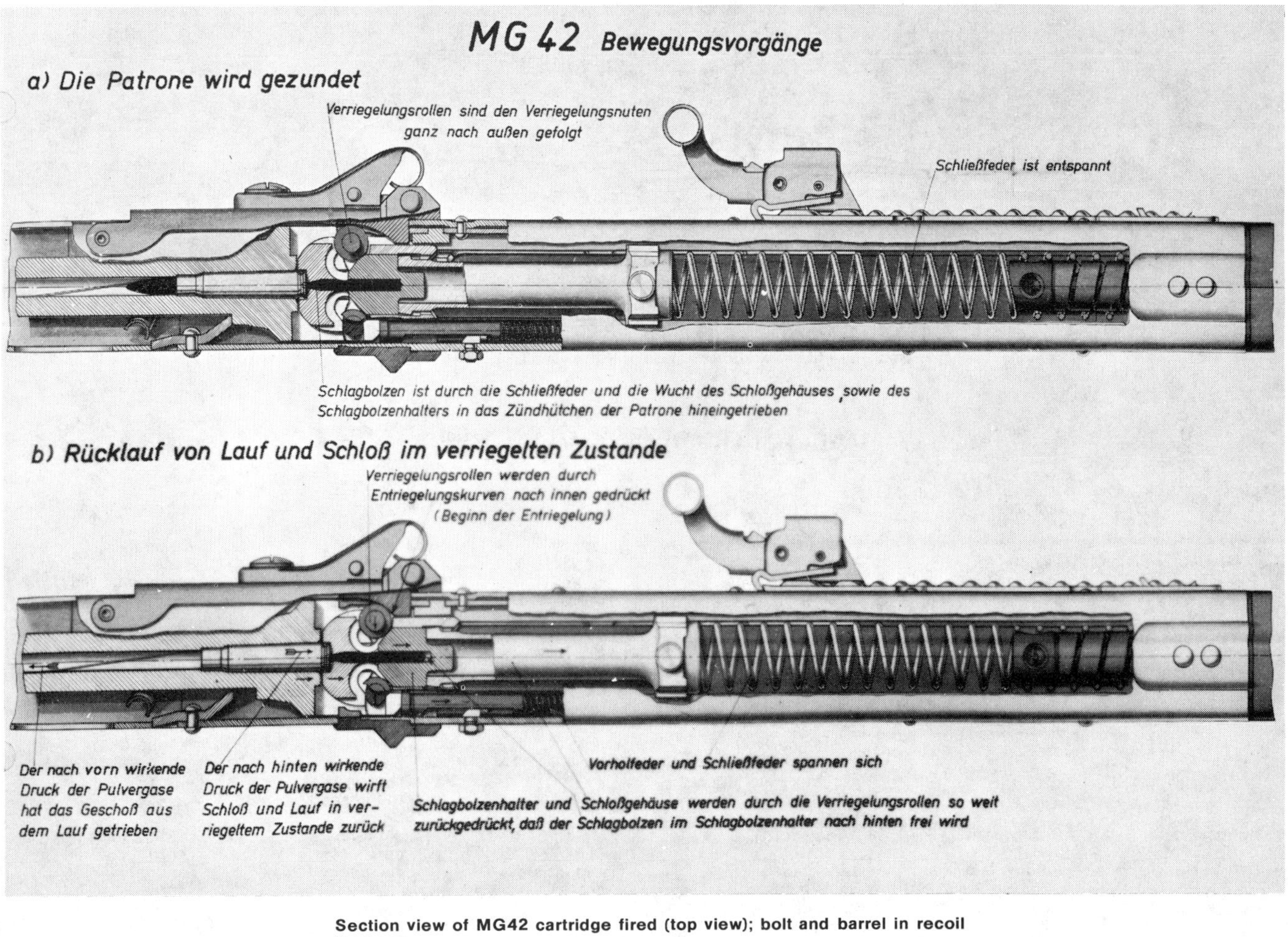

Section view of MG42 cartridge fired (top view); bolt and barrel in recoil (bottom).

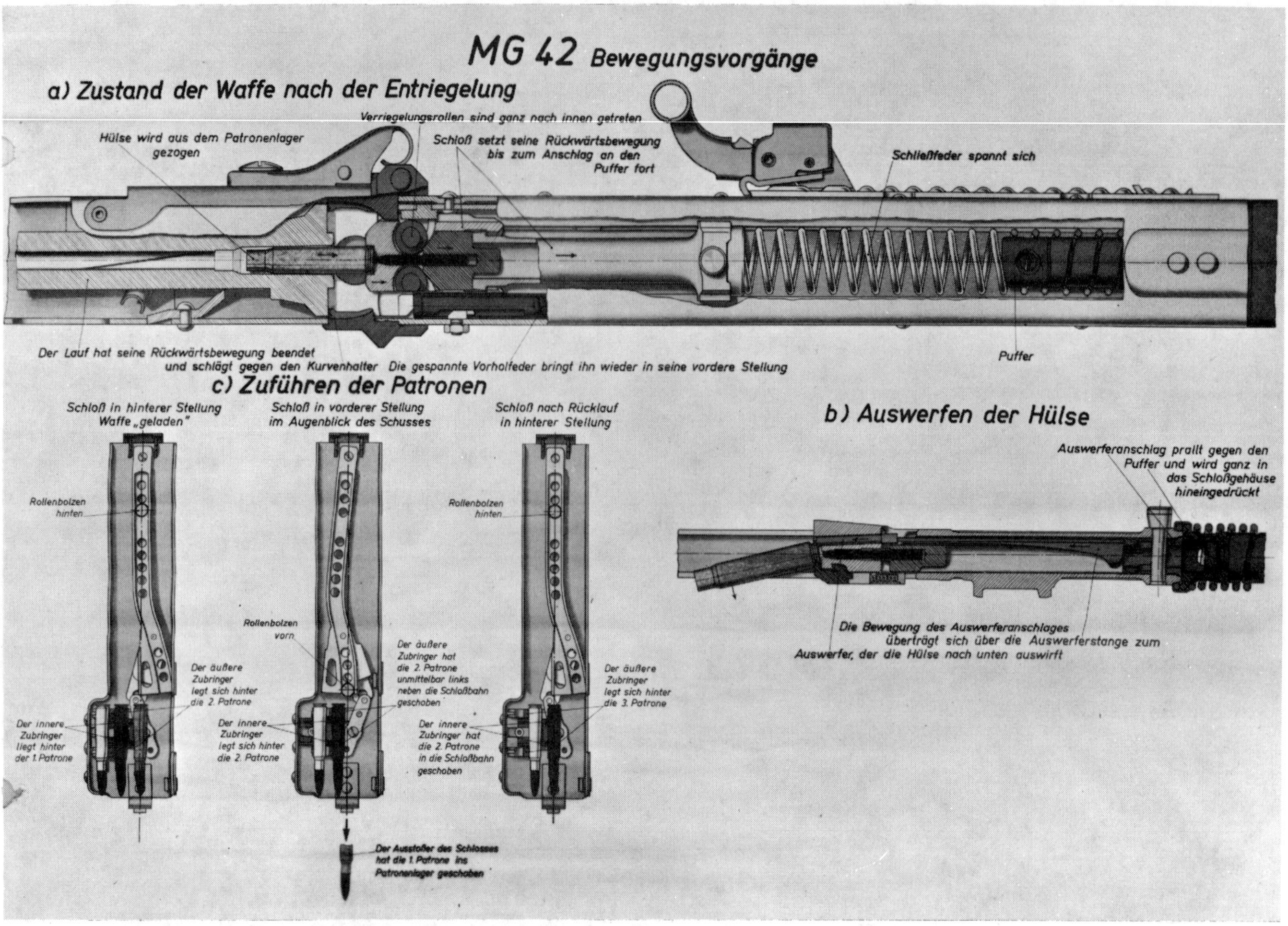

Top: Section view of MG42, gun unlocked, cartridge being withdrawn from chamber by bolt. Bottom left: View showing movement of belt feed mechanism. Bottom right: View shows action of ejector.

STANDARD GERMAN ARMY MACHINE GUNS

	MG08	MG08/15	MG13	MG34*	MG42
Caliber:	7.92mm.	7.92mm.	7.92mm.	7.92mm.	7.92mm.
System of operation:	Recoil, automatic only.	Recoil, automatic only.	Recoil, selective fire.	Recoil, selective fire.	Recoil, automatic only.
Length overall: (gun only)	46.25 in.	Approx 57 in.	Approx 57 in.	48 in.	48 in.
Barrel length:	28.25 in.	28.25 in.	28.25 in.	24.6 in.	21 in.
Cooling:	Water.	Water.	Air.	Air.	Air.
Feed device:	100-and 250-round fabric belt.	50-, 100-, & 250-round fabric belt. Can be used with spool type container with 50-round belt or with ordinary ammo can.	25-round box magazine or 75-round saddle drum.	50-round non-disintegrating belt, linked together to form 250-round belt. 50-round belt drum. 75-round saddle drum.	50-round non-disintegrating belt usually joined into 250-round belt, 50-round drum.
Sights: Front:	Barley corn.	Barley corn.	Folding barley corn.	Folding barley corn.	Folding barley corn.
Rear:	Folding leaf w/v notch.	Tangent leaf w/V notch.	Tangent leaf w/V notch.	Leaf w/V notch.	Tangent w/V notch.
Muzzle velocity:	2750 f.p.s. (S ball)	2750 f.p.s. (S ball)	2750 f.p.s. (S ball)	Approx 2500 f.p.s. (sS ball)	2480 f.p.s. (sS ball)
Cyclic rate:	400-500 r.p.m.	400-500 r.p.m.	750 r.p.m.	800-900 r.p.m.	1100-1200 r.p.m.
Weight of gun:	40.5 lb.	31 lb.	26.4 lb. w/bipod.	26.5 lb. w/bipod	25.5 lb. w/bipod.
Weight of mount:	83 lb. sled mount. 65.5 lb. tripod mount.	51 lb.	Approx 25 lb.	42.3 lb.	42.3 lb.

*Data is given for basic MG34; other versions vary as follows:
MG34S and MG34/41----automatic fire only.
Overall length: approx. 46 in.
Barrel length: approx. 22 in.

Representative German machine guns. 1. MG34. Single-shot and full auto. 2. Modified MG34. Single-shot and full auto. 3. MG34S. Full auto. 4. MG34-41. Full auto. 5. MG42. Differs mechanically from 34's.

31 Greece

The Greek Army is currently equipped with a mixture of U. S. and British service arms. The Greeks lost the greater percentage of their small arms during World War II. The British supplied the Greek Army in their fight against the Communists until the U. S. started its military assistance in 1947. British caliber .38 and .455 revolvers, the No. 2 9mm Browning Hi-Power, caliber .303 No. 4 rifles and Bren and Vickers guns are in service. U. S. weapons used are the caliber .45 M1911A1 pistols, caliber .30 M1 rifles, M1918A2 Browning automatic rifles, and the caliber .30 and .50 Browning machine guns.

GREEK PISTOLS

Greece has developed no native designs in pistols and has used a wide variety of pistols in the past, including the 9mm M1903 Bergmann Bayard, and various models of the Browning.

GREEK RIFLES AND CARBINES

MANNLICHER SCHOENAUER 6.5mm 1903 RIFLE

The Greeks adopted the Mannlicher Schoenauer rifle in 6.5mm in 1903. This rifle combines the Mannlicher two-piece rotating bolt with the rotating spool-type magazine developed by Otto Schoenauer. This system is still used in the sporting-type Steyr rifles made currently and the 6.5 x 54 Greek cartridge is still popular throughout most of the world as a sporting cartridge.

A modification of the 1903 model, the 1903/14, was adopted in 1914. Differences between the two models are relatively minor and are principally in the graduations on the rear sight, shape of grasping grooves on the stock, and in the length and ease of removal of the handguards.

The Greek Army lost many of these weapons during the first World War and as part of the reparations for that war received a mixed batch of Austrian Mannlicher rifles as well as some Turkish Mausers. During that war the Greeks were supplied with a number of French weapons including French 8mm Lebel and Mannlicher Berthier rifles.

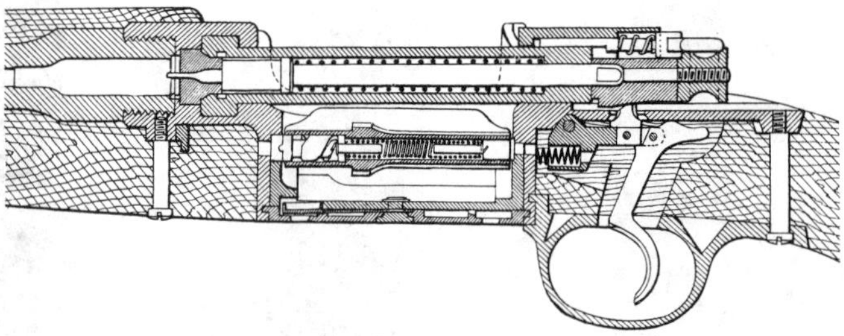

Section of rifle with action closed. Note position of locking lugs in rear of bolt head.

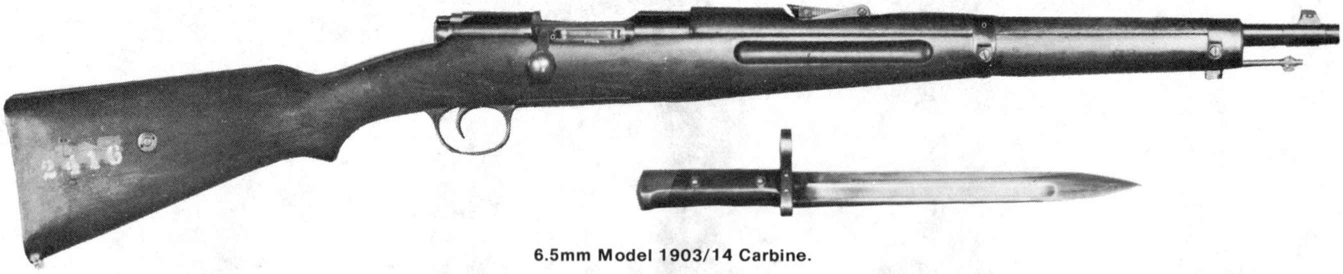

6.5mm Model 1903/14 Carbine.

6.5mm Mannlicher Schoenauer M1903/14 Rifle.

OTHER GREEK SHOULDER ARMS

The Greeks had a number of Austrian 8mm M1888/90 and 8mm M1895 rifles converted to 7.92mm. The Model 1895 conversion was called the model 95/24. In 1930 the Greeks adopted the Model 1924 FN Mauser in 7.92mm which they called the Model 30. This rifle is covered in detail under Belgium.

Shortly before the beginning of World War II, the Greek Powder Company announced that it was developing an automatic rifle. This gas-operated 7.92mm weapon was apparently never produced in quantity. It weighed 9.1 lbs, was 35.4 inches long, with a 15.74-inch barrel and a cyclic rate of 750 rounds per minute.

GREEK SERVICE RIFLES AND CARBINES

	M1903 Rifle	M1903 Carbine	M95/24	M1930
Caliber:	6.5mm.	6.5mm.	7.92mm.	7.92mm.
System of operation:	Turn bolt.	Turn bolt.	Straight-pull bolt.	Turn bolt.
Overall length:	48.3 in.	40.3 in.	39.5 in.	43.3 in.
Barrel length:	28.5 in.	20.5 in.	19.6 in.	23.2 in.
Feed device:	5-round, revolving spool, non-detachable box magazine.	5-round, revolving spool, non-detachable box magazine.	5-round, in line* non-detachable box magazine.	5-round staggered row, non-detachable box magazine.
Sights:				
Front:	Barley corn.	Barley corn.	Barley corn.	Barley corn.
Rear:	Tangent with 'V' notch.	Tangent with "V" notch.	Leaf with 'V' notch.	Tangent with 'V' notch.
Weight:	8.31 lbs.	Approx. 8 lbs.	6.8 lbs.	8.5 lbs.
Muzzle velocity:	2225 f.p.s.	Approx. 2125 f.p.s.	Approx. 2410 f.p.s.	2500 f.p.s.

*Magazine is charger fed; charger does not remain in magazine.

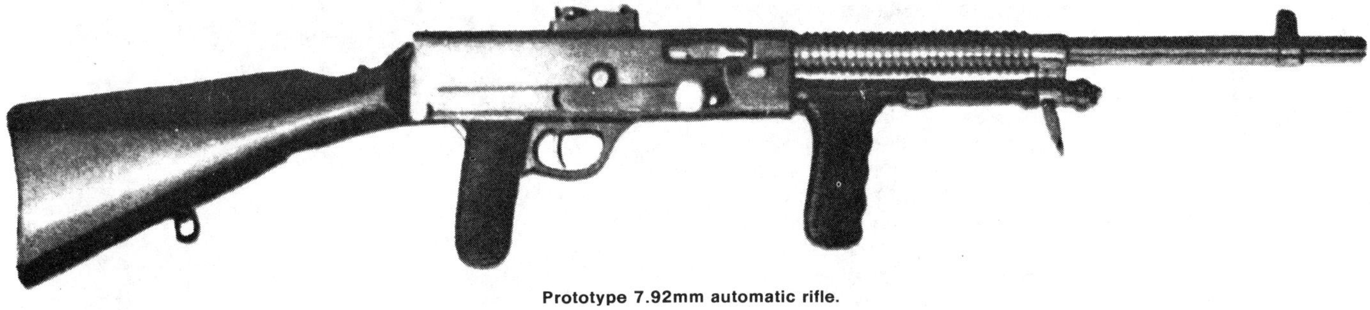

Prototype 7.92mm automatic rifle.

GREEK MACHINE GUNS

The Greek Army was equipped with the French 8mm Saint Etienne M1907 heavy machine gun and the 8mm C.S.R.G. Model 1915 light machine gun during and immediately after World War I. For some peculiar reason, the 8mm Lebel cartridge was called the 7.8mm cartridge in Greece. The German Maxim MG08 in 7.92mm was among the captured weapons used by the Greeks. During World War II the Greeks captured all types of Italian machine guns as well as all other types of Italian small arms.

32 Hungary

The Hungarian Army is currently equipped with copies and/or modified copies of Soviet World War II and post-war small arms. The service pistol is the 7.62mm Model 48, a Hungarian copy of the Soviet 7.62mm TT33 Tokarev. The service rifle is a modified copy of the 7.62mm AK made in Hungary. Service machine guns include the 7.62mm RPD, 7.62mm SGM, the 12.7mm DShK M1938/46 and the 14.5mm ZPU-2 and ZPU-4 heavy antiaircraft machine guns. There are probably still Hungarian-made 7.62mm Model 48 submachine guns (copy of Soviet PPSh M1941) and Hungarian-made 7.62mm Model 48 rifles (copy of the Soviet 7.62mm M1891/30 rifle) and some of the old Degtyarev series machine guns still in service.

HUNGARIAN SERVICE PISTOLS

THE HUNGARIAN 7.65mm FROMMER STOP PISTOL MODEL 19

Hungary was, until 1918, part of the Austro-Hungarian Empire and in general used the standard weapons of that empire. Because of local development capabilities, and probably local politics, the Hungarians used a different pistol during World War I than the rest of the empire forces. This pistol was the 7.65mm (.32 ACP) Model 19 Frommer Stop pistol. This long recoil-operated pistol was developed in 1912 and was used by the Hungarian police and service forces after World War I. A 9mm (.380 ACP) version which, according to some authorities, was never issued as a military weapon, appeared some time between 1916 and 1919. The designation Model 19 for the 7.65mm Frommer is rather unusual since the weapon appeared in 1912 and was adopted by the Hungarian Army (Honved) prior to World War I, but official Hungarian documents indicate that this is the correct model designation. The Frommer Stop was manufactured by Fegyvergyan at Budapest. It is not basically a good design, is somewhat delicate for a service weapon, and was reportedly not popular with Hun-

Frommer M1939. Caliber .380. Obsolete in Hungary.

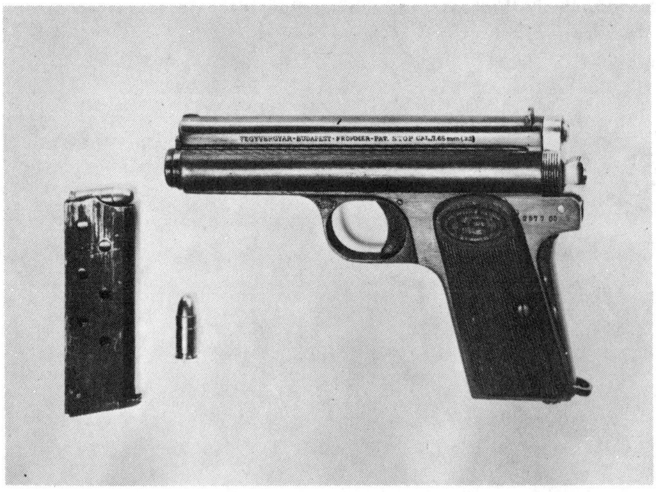

Frommer 7.65mm Automatic, Model 19.

9mm Model 29 Pistol.

garian troops. The complication of a long recoil system is absolutely wasted on relatively low-powered cartridges which can be adequately handled by a simple blowback action. A later 9mm (.380 ACP) version of this pistol is marked "1939M". This was apparently not a service weapon and the purpose for which it was made is not apparent.

7.65mm Model 48 Pistol.

THE HUNGARIAN 7.65mm (AND 9mm) PISTOL MODEL 37

In 1929 the Hungarians adopted a modified Browning blowback design, the 9mm (.380 ACP) Model 29. This weapon has an internal bolt assembly which is fixed to the slide rather than having the bolt (breech) portion machined in the rear of the slide. This weapon was apparently made in very limited quantity and is over complicated, as is the Frommer Stop.

A version of the Model 29 with a conventional slide appeared in 1937. The Model 37 is a conventional blowback pistol chambered for both the 7.65mm (.32 ACP) and the 9mm (.380 ACP). The Model 37 was made in large quantities for the Hungarian Army and is probably the most common Hungarian pistol in the U.S. today.

During World War II, Hungary became—for all practical purposes—a satellite of Germany. Hungarian arms plants produced large quantities of weapons for the Germans. Among these weapons was the Model 37 pistol in 7.65mm. The weapon, as made for the Germans, bears German acceptance markings and the three letter manufacturers code "jhv." The weapons manufactured for the Germans have, in addition to the grip safety fitted to the earlier-made Model 37 pistols, a plate-type safety mounted on the left side of the receiver.

THE HUNGARIAN 7.62mm PISTOL MODEL 48

After World War II a modified copy of the German 7.65mm Walther PP was introduced in Hungary. This pistol, which has a loaded chamber indicator over the chamber rather than at the rear of the slide as does the Walther, is called the Model 48 by the Hungarians. It is not a military weapon but is used by police and has been sold commercially in 9mm (.380 ACP) as the "Walam." These pistols are three quarters of an inch longer than the PP.

Loading, firing, and field stripping of these pistols is the same as that of German Walther PP and PPK.

The current Hungarian service pistol, the 7.62mm Model 48, is a copy of the Soviet 7.62mm TT33 and is loaded, fired, and field stripped as is that pistol.

9mm Model 37 Pistol.

7.65mm Model 37 Pistol made for the Germans.

9mm Walam 48 Pistol.

The Model 48 field-stripped.

Hungarian 7.62mm Model 48, a copy of Soviet TT M33.

9mm Tokagypt 58 Pistol.

CHARACTERISTICS OF HUNGARIAN SERVICE PISTOLS

	Frommer Stop Model 19	7.65 mm Model 48	9mm Model 37	7.62mm Model 48
Caliber:	7.65mm (ACP).	7.65mm (.32 ACP).	9mm (.380 ACP) *	7.62mm.
System of operation:	Long recoil, semiautomatic.	Blowback, semi-automatic.	Blowback, semi-automatic.	Recoil, semi-automatic.
Overall length:	6.5 in.	7 in.	6.8 in.	7.68 in.
Barrel length:	3.8 in.	3.94 in.	3.9 in.	4.57 in.
Feed device:	7 round, in line detachable, box magazine.	8-round, in line detachable, box magazine.	7-round, in line detachable, box magazine.	8-round, in line detachable, box magazine.
Sights:				
Front:	Blade.	Blade.	Blade.	Blade.
Rear:	V notch.	U notch.	V notch.	U notch.
Weight:	1.31 lbs.	1.92 lbs.	1.62 lbs.	1.88 lbs.
Muzzle velocity:	920 f.p.s.	920 f.p.s.	984 f.p.s.	1,378 f.p.s.

*Also made in 7.65mm (.32 ACP)

THE TOKAGYPT 58

The Tokagypt 58 was reportedly developed for the Egyptian police. It is a modified copy of the Tokarev TT33 chambered for the 9mm Parabellum cartridge. A safety catch has been added to the basic Tokarev design, which is fitted on the left top side of the receiver.

In addition, a plastic wrap-around type one-piece grip and a finger support on the magazine are found on this pistol. It is not a Hungarian service pistol.

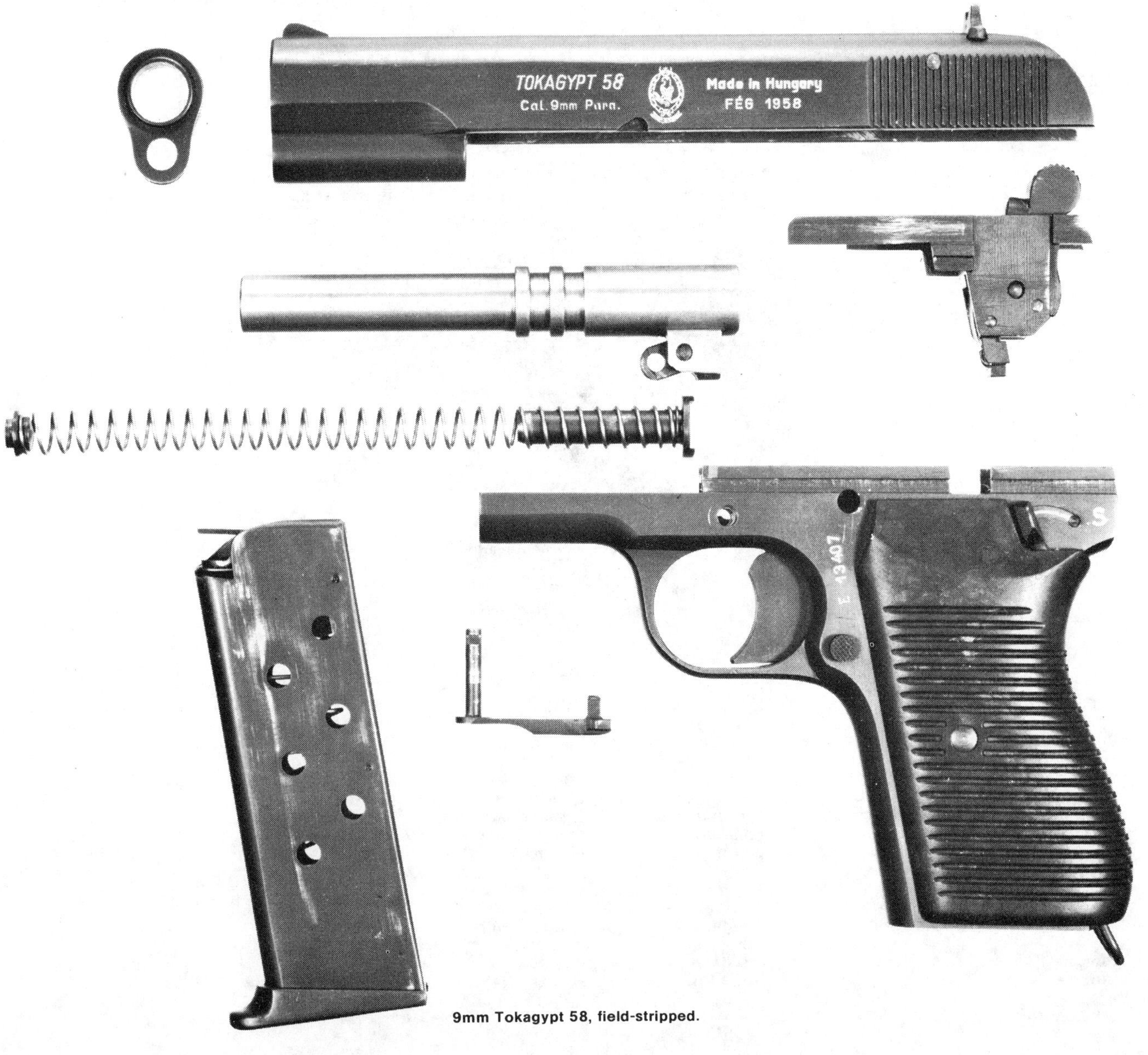

9mm Tokagypt 58, field-stripped.

HUNGARIAN RIFLES

HUNGARIAN 8mm STUTZEN RIFLE MODEL 31

The Hungarians were equipped during World War I with various models of the straight pull 8mm Steyr rifle chambered for the 8mm M93 cartridge. After the adoption of the 8x56mm rimmed M31 cartridge, a number of the 8mm M95 short rifles (the Stutzen) were modified to take this cartridge. This rifle is called the Model 31 and is identified by a letter "H," 5/16 of an inch high, stamped on the barrel or receiver.

HUNGARIAN 8mm RIFLE MODEL 35

In 1935 a turn-bolt Mannlicher-type was adopted, the Model 35, chambered for the 8mm Model 31 cartridge. The Model 35 has a two-piece bolt with the bolt handle positioned ahead of the receiver bridge when the bolt is forward. The magazine is an in-line type which protrudes below the line of the stock. The Model 35 has a two-piece stock similar in concept to the British Lee Enfield.

HUNGARIAN 7.92mm RIFLE MODEL 43

In 1940 the Model 35 was redesigned for German use. The caliber was changed to 7.92mm, a staggered row Mauser-type box magazine flush with the stock was fitted and German-type bands and bayonet lug were used on this rifle, which was called the Model 98/40 by the Germans. In 1943 Hungary adopted the 7.92mm cartridge and started issue of a slightly modified 98/40—the Model 43—to the Hungarian Army. This rifle differs from the 98/40 principally in its bands and fittings, i.e., bayonet lug, sling swivels, etc.

HUNGARIAN 7.62mm RIFLE MODEL 48 (COPY OF SOVIET MOSIN NAGANT)

The conclusion of the war found Hungary disarmed and behind the Iron Curtain. The Hungarian Army was equipped with Soviet 7.62mm Mosin Nagant rifles and carbines. Some copies of these weapons were made in Hungary and called Model 48. Within the past decade the post-war Soviet rifles and carbines have been adopted and both the 7.62mm SKS carbine and 7.62mm AK assault rifle have been made in Hungary.

HUNGARIAN 7.62mm MODIFIED SOVIET AKM RIFLE

Recently the Hungarians have adopted a modification of the 7.62mm AKM. This rifle has a plastic stock and pistol grips and a metal hand guard. The forward pistol grip is grasped during fire rather than the hand guard or the magazine. It has no top hand guard. This weapon is loaded and fired, and field stripped, in a manner similar to that of the Soviet AKM.

HUNGARIAN AMD ASSAULT RIFLE

The Hungarians have come out with another variation of the AKM. The AMD has a short barrel which has a large muzzle brake. It has a single strut steel folding stock. The plastic fore grip and the metal forend are the same as those on the standard Hungarian modification of the AKM. Because of the relatively low energy of the 7.62 X 39mm cartridge recoil with this weapon is probably not excessive, but the muzzle blast is probably very high. The AMD is 33.5 inches long with stock fixed and 23.5 inches long when the stock is folded. The barrel is 12.6 inches long.

Hungarian 7.62mm Model AMD Assault Rifle.

HUNGARIAN SERVICE RIFLES

	Model 31	Model 35	Model 43	Model 48	Modified AKM
Caliber:	8x56mm.	8x56mm.	7.92mm.	7.62 Mosin Nagant.	7.62 M43.
System of operation:	Straight pull bolt.	Turn bolt.	Turn bolt.	Turn bolt.	Gas, selective fire.
Overall length:	39.5 in.	43.7 in.	43 in.	48.5 in.	34.25 in.
Barrel length:	19.65 in.	23.6 in.	23.8 in.	28.7 in.	16.34 in.
Feed device:	5-round single column, fixed box magazine.	5-round single column, fixed box magazine.	5-round staggered row, fixed box magazine.	5-round single column, fixed box magazine.	30-round staggered row, detachable box magazine.
Sights:					
Front:	Barley corn.	Hooded barley corn.	Barley corn.	Hooded post.	Protected post.
Rear:	Leaf with notch.	Tangent with notch.	Tangent with notch.	Tangent with notch.	Tangent with notch.
Weight:	7.5 lbs.	8.9 lbs.	8.6 lbs.	8.7 lbs.	Approx. 10.7 lbs.
Muzzle velocity:	Approx. 2300 f.p.s.	2395 f.p.s.	Approx. 2480 f.p.s. (SS ball)	2800 f.p.s.	2329 f.p.s.
Cyclic rate:					600 r.p.m.

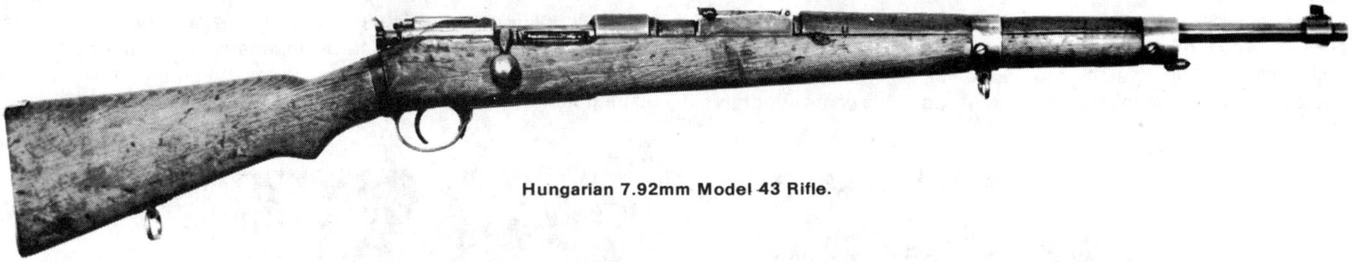

Hungarian 7.92mm Model 43 Rifle.

The 8mm Model 35 Rifle.

Hungarian modification of 7.62mm AKM Assault Rifle.

HUNGARIAN SUBMACHINE GUNS

HUNGARIAN 9mm SUBMACHINE GUN MODEL 39

A native-designed submachine gun was produced in Hungary in the late thirties and adopted in 1939. The design of this weapon, which is chambered for the 9mm Mauser cartridge, is credited to Pal D. Kiraly and resembles in many respects that of the SIG (Swiss) MKMO submachine gun. Kiraly was concerned with the design of the Dominican Cristobal carbine; the bolt design of the Cristobal is similar to that of the Model 39. The folding magazine system of the Model 39 is similar to that of the SIG MKMO.

The standard Model 39 submachine gun has a one-piece stock; a version with a folding wooden butt was produced as the Model 39/A. Both the Model 39 and 39/A were produced in very limited numbers. The fire selector/safety is the circular cap located on the rear of the receiver and is operated by rotating the cap to align with one of the three settings:- 'E' for semiautomatic fire, 'S' for automatic fire and 'Z' for safe.

9mm Mauser Model 39 Submachine Gun.

HUNGARIAN 9mm SUBMACHINE GUN MODEL 43

The Model 43, which was made in much larger quantity than the Model 39, is essentially the same as the Model 39, but has a folding metal stock. The folding stock has wooden strips on the side of the metal stock frame. The magazine of the Model 43 is canted slightly forward when in the fixed position as opposed to the straight vertical position of the Model 39 magazine, and the barrel of the Model 43 is approximately three inches shorter than that of the Model 39. The magazines of the Models 39 and 43 are not interchangeable.

9mm Mauser Model 43 Submachine Gun.

9mm Mauser Model 43 Submachine Gun with stock folded.

HUNGARIAN 7.62mm SUBMACHINE GUN MODEL 48 (COPY OF SOVIET PPSh M1941)

After World War II the Hungarians were supplied with Soviet weapons and manufactured a copy of the Soviet 7.62mm PPSh M1941 as the Model 48. A submachine gun intended basically for police purposes known as the Model 54 was also produced.

Currently the submachine gun is used as a reserve weapon or as a Border Patrol weapon by Hungary. As with other Warsaw Pact nations, the assault rifle is replacing the submachine gun in the Army.

Model 48 7.62mm Submachine Gun.

CHARACTERISTICS OF HUNGARIAN SUBMACHINE GUNS

	Model 39	Model 43	Model 48
Caliber:	9mm Mauser.	9mm Mauser.	7.62mm.
System of operation:	Delayed blowback, selective fire.	Delayed blowback, selective fire.	Blowback, selective fire.
Overall length:	41.25 in.	Stock extended: 37.5 in. Stock folded: 29.5 in.	33.15 in.
Barrel length:	19.65 in.	16.7 inches	10.63 inches
Feed Device:	40-round, staggered row, detachable box magazine.	40-round, staggered row, detachable box magazine.	71-round drum or 35 round detachable box magazine.
Sights:			
Front:	Barley corn.	Barley corn.	Hooded post.
Rear:	Tangent with V notch.	Tangent with V notch.	L type.
Weight:	8.2 lbs.	8 lbs.	11.9 lbs with loaded drum magazine.
Muzzle velocity:	1475 f.p.s.	1450 f.p.s.	1640 f.p.s.
Cyclic rate:	750 r.p.m.	750 r.p.m.	700 - 900 r.p.m.

HUNGARIAN MACHINE GUNS

8mm Model 31 Machine Gun.

The Hungarian Army, after the foundation of a separate Hungary in 1919, was equipped with the 8mm Schwarzlose 07/12 heavy machine gun. Hungary followed the lead of Austria in adopting the Solothurn machine gun developed by Louis Stange. The Hungarian gun was chambered for the 8mm Model 31 cartridge (8 x 56 R) and was called the Model 31. The gun has the same characteristics as that covered under Austria as the Model 30.

Some Schwarzlose machine guns were modified to use the 8mm Model 31 cartridge; they are called the Model 7/31. The Hungarians also produced an aircraft gun chambered for the Model 31 cartridge which was called the Model 34/AM. During World War II the Hungarians changed their machine gun caliber at the same time as they changed caliber with rifles - in 1943. The Hungarian Model 43 machine gun is essentially the same as the Model 31, but is chambered for the 7.92mm cartridge.

As with all their other weapons, the Hungarians were equipped with Soviet machine guns after World War II. Initially they adopted the older Soviet weapons such as the 7.62mm DP, DPM, and DTM machine guns. They currently use the 7.62mm RPD, 7.62mm SGM, 12.7mm DShK 38/46, and the 14.5mm ZPU-2 and ZPU-4 machine guns. All these weapons are covered in detail in the chapter on the Soviet Union.

The Indian Army is mainly armed with British-type service weapons. The 7.62mm FN "FAL" rifle in the L1A1 version (British) is now standard, as is the 9mm Sterling submachine gun. The Sterling is called the SAF in India and is made at Cawnpore. The Indians have Vickers Berthier, Bren, and Browning light machine guns and Vickers heavy machine guns, but these weapons will eventually be replaced by the 7.62mm FN MAG general purpose machine gun.

India

33

INDIAN RIFLES

The government small arms factory at Ishapore has been making rifles for a considerable time and produced the Lee Enfield No. 1 Mark III* through World War II. The Ishapore rifles varied in minor details from the rifle as made in the United Kingdom, frequently having no stacking swivels, the stacking swivel mounting lug being left solid. Ishapore made 692,587 Mark III* rifles during World War II. India is now producing the 7.62mm rifle L1A1 (British FN "FAL").

Marking on Short Lee Enfield No. 1 Mark III* made at Ishapore.

INDIAN MACHINE GUNS

India produced 8,357 machine guns during World War II and was preparing to produce Bren Guns at a new factory at Hyderabad when the war ended. The current standard Indian machine gun is the 7.62mm FN MAG general purpose gun. The standard machine gun of the Indian Army during World War II was the Vickers Berthier.

INDIAN CAL. .303 VICKERS BERTHIER MACHINE GUN

During the period between World Wars I and II, the Indian Army, then a semi-autonomous branch of the British Army, adopted the cal. .303 Vickers Berthier Light Machine Gun Mark III. This weapon is still in limited use in India. It is quite similar in construction and external appearance to the Bren. The main differences between the Bren and the Berthier lie in the breechblock, the feed, the holding-open device, the gas cylinder arrangement, the barrel change, and the sights.

Characteristics of Vickers Berthier Machine Gun

Caliber: .303 British.
Weight, loaded: 24.4 lb.
Overall length: 45.5 in.
Barrel length: 23.9 in.
Feed mechanism: 30-round, detachable, staggered box magazine.
Sights: Front: Hooded blade.
Rear: Leaf.
Muzzle velocity: 2400 f.p.s. w/Mk 7 ball (approx).

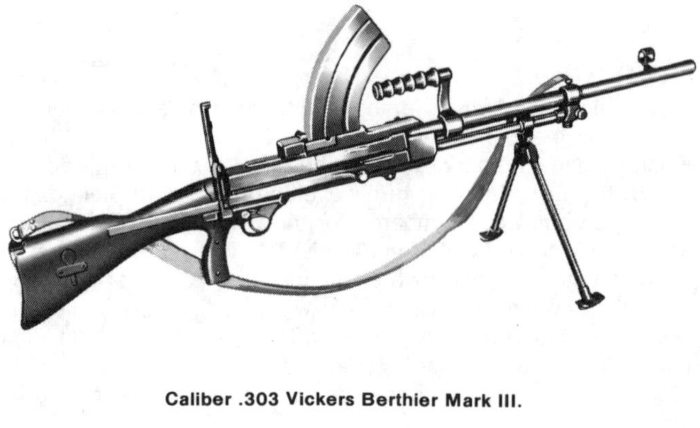

Caliber .303 Vickers Berthier Mark III.

Indonesia has an unusually large variety of service small arms, from a diverse number of sources. The U.S. caliber .45 M1911A1, the 9mm FN Hi-Power Browning, the 7.62mm Tokarev, and various models of the Walther are some of the pistols used by the Indonesians. Rifles and carbines used include the caliber .30 M1 rifle, caliber .30 M1 carbine, the caliber .30 M1918A2 Browning automatic rifle, the caliber .30 FN semiautomatic rifle (the M1949), the 7.62mm G3 rifle, 7.62mm M59 rifle, the Soviet 7.62mm AK47 assault rifle, the Soviet 7.62 SKS carbine, the British caliber .303 No. 1 and No. 4 rifles and the 7.92mm Kar98k. The submachine guns in service include the 9mm Beretta, 9mm Sten, 9mm M1950 Madsen, the Swedish 9mm M45 Carl Gustaf, the caliber .45 Thompson, the 9mm Owen and the Czech 7.62mm M24 and M26. The standard machine gun is the caliber .30 Madsen Mark II, which is made in Indonesia. The caliber .30 Browning guns; the caliber .303 Bren and Vickers guns; the 7.62 Degtyarev and Goryunov and the 12.7mm DShK M1938/46 are also used.

34 Indonesia

INDONESIAN RIFLES

The Indonesians had considerable quantities of Dutch 6.5mm Mannlicher M1895 rifles and carbines, and caliber .303 rifles, at the conclusion of their rebellion against the Dutch. A number of Mannlicher carbines were rebarreled for the .303 cartridge for the Dutch forces during World War II in the then Dutch East Indies. The fall of the Netherlands in 1940 cut off their supply of 6.5mm ammunition. At a later date the Indonesian Army started to standardize on caliber .30 weapons, the principal weapon being the U.S. M1 rifle. The Indonesians purchased some of these from Beretta in Italy and from U.S. sources. Purchases were also made of Soviet 7.62mm AK-47 assault rifles and SKS carbines. The Indonesians have converted some of their M1 rifles to the 7.62mm NATO BM59 (a Beretta design).

INDONESIAN SUBMACHINE GUNS

A number of prototype 9mm Parabellum submachine guns were developed by the Indonesians. None were made in quantity.

Late models of the 9mm Parabellum Beretta submachine gun are currently manufactured in Indonesia.

9mm PM Model VIII, 1957.

35

Iran (Persia)

The Iranian government, as a result of various defense agreements with the United States, has adopted most U.S. small arms. The Iranian Army uses the U.S. cal. .30 M1 rifle and some cal. .30 Browning machine guns, the M3A1 submachine gun, and other U.S. weapons.

IRANIAN RIFLES

Iran adopted the Czech ZB 7.92mm Model 98/29 short rifle and the Model 98/29 carbine as the Model 1309 and Model 1317 (Moslem calendar years), respectively. The Christian calendar years for these are 1930 and 1938. The rifles were originally made in Czechoslovakia, but the Iranian arsenal produced them as well. Production at the Iranian arsenal probably started during World War II, since Czech supplies were shut off during the war.

The Iranians developed a modification of the Model 1930 short rifle, the Model 1949 (Model 1328 by the Moslem calendar); the Model 1949 differs from the Model 1930 only in bands and sling swivels. These rifles are loaded, fired, and field stripped in the same manner as the Czech Model 24 rifle, which is covered in the chapter on Czechoslovakia. They are being replaced in the Iranian Army by the U.S. M1 rifles.

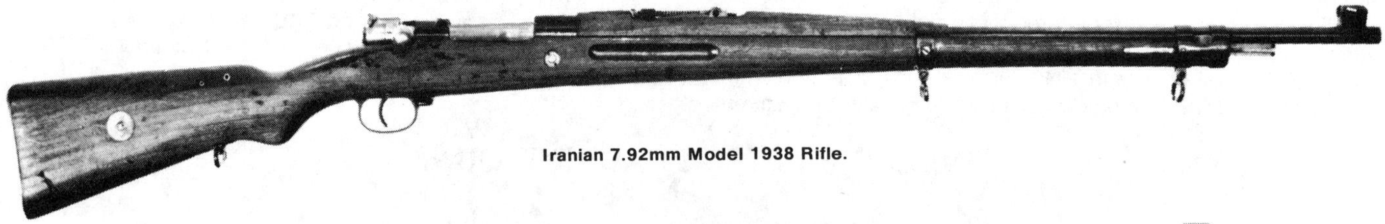

Iranian 7.92mm Model 1938 Rifle.

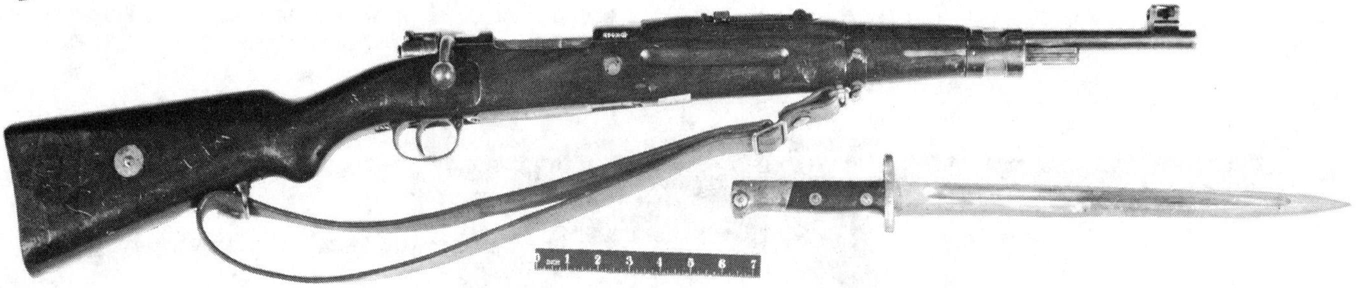

Iranian 7.92mm Model 1930 Short Rifle.

Characteristics of Iranian Rifles

	1938 Rifle	1930 Short Rifle
Caliber:	7.92mm.	7.92mm.
System of operation:	Manually operated turning bolt action.	Manually operated turning bolt action.
Weight:	9.1 lbs.	8.4 lbs.
Length overall:	49.2 in.	38 in.
Barrel length:	29.13 in.	17.91 in.
Feed device:	5-rd, staggered row. non-detachable box magazine.	5-rd, staggered row, non-detachable box magazine.
Sights: Front:	Protected barley corn.	Protected barley corn.
Rear:	Tangent with V-notch, grad. from 100-2000m.	Tangent with V-notch, grad. from 100-2000m.
Muzzle velocity:	Approx. 2800 f.p.s.	Approx. 2600 f.p.s.

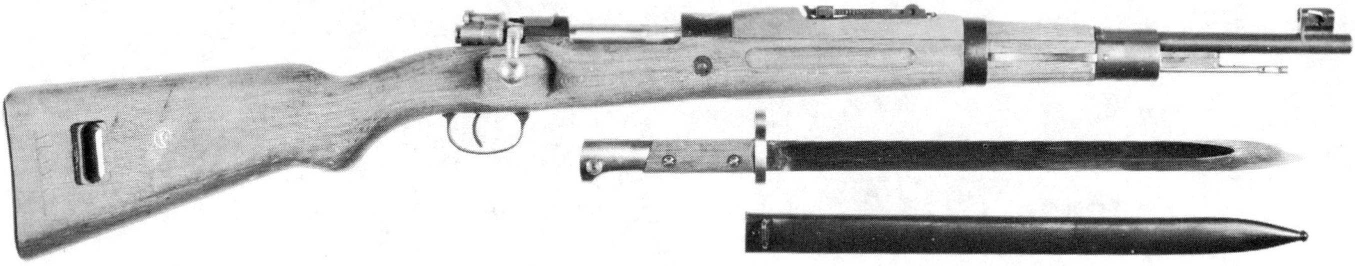

Iranian 7.92mm Model 1949 Short Rifle.

IRANIAN SUBMACHINE GUNS

Iran currently has quantities of U.S. caliber .45 M3A1 submachine guns. Iran has also produced copies of the Soviet 7.62mm PPSh M1941 submachine gun. The Iranians began producing the gun during World War II for the Soviets on a contract basis, and continued to produce the weapon for their own use. Characteristics of the weapon are the same as those given in the chapter on the Soviet Union.

IRANIAN MACHINE GUNS

The Iranians currently have U.S. caliber .30 and .50 Browning machine guns. They also have the Czech ZB30 machine gun which they make in Iran. This weapon was originally manufactured in 7.92mm, but is currently used by the Iranians in cal. .30. Characteristics, loading, firing, and field stripping of this weapon are basically the same as those given for the ZB30 in the chapter on Czechoslovakia.

Iranian copy of Soviet PPSh M1941 Submachine Gun.

Iranian 7.92mm Model 30 Light Machine Gun.

Israel

The current Israeli service weapons are: the Beretta 9mm M1951 pistol, the 7.62mm NATO FN "FAL" rifle and the heavy barrel version of the "FAL," the 9mm Uzi submachine gun, and the 7.62 NATO FN general purpose machine gun. The FN Browning Automatic Rifle Type D is also used. Israel has developed a new 5.56mm Rifle, the Galil, which will replace the FAL's and the Browning automatic rifle. Older weapons that may still be in limited service or reserve are the 7.62mm Kar 98k, the 7.92mm MG34 machine gun, the caliber .303 Bren and Vickers guns, the caliber .303 Lee Enfield rifle, the 9mm Sten gun, and various caliber .38 and .455 British pistols. The U.S. caliber .30 and caliber .50 Browning machine guns are in service on various armored vehicles and the 7.92mm Besa machine gun may also be found on some armored vehicles.

36

ISRAELI PISTOLS

The Israelis had a variety of pistols at the time of the foundation of the State. For the most part these were caliber .38 and caliber .455 Enfield and Webley revolvers, but there were odd quantities of 9mm FN Browning Hi-power, Luger, and P38 automatics as well. The Israelis developed a modified copy of the Smith & Wesson Military and Police pistol, which is unusual in that it is chambered for the 9mm Parabellum cartridge. The use of this rimless cartridge requires the use of two three-shot clips similar to that used with the U.S. caliber .45 Colt and Smith & Wesson service revolvers. This revolver, which bears Israeli Security Forces markings, is apparently basically a police weapon. The Israelis use the 9mm Parabellum M1951 Beretta pistol as their standard military service pistol. This pistol is covered in detail in the chapter on Italy.

ISRAELI RIFLES

The Israelis were originally equipped mainly with British caliber .303 No. 1 and No. 4 rifles. About 1948 purchases were made of 7.92mm Kar 98k rifles of both World War II German and post-war Czech manufacture. Because of the procurement of 7.92mm machine guns at the same time, the 7.92mm cartridge was adopted as standard. Towards the end of the 50s Israel standardized the 7.62mm NATO cartridge and adopted the FN "FAL" rifle and the heavy-barrel version of this weapon. Israel has converted her 7.92mm Kar 98K Mausers to 7.62mm NATO.

THE ISRAELI 5.56mm GALIL RIFLE

Israeli's new rifle the Galil was first issued to troops in May 1973. The Galil is a gas operated weapon which outwardly looks quite a bit like the Finnish 7.62x39mm Model 62 assault rifle. Internally the Finnish Model 62 is basically the same as the Soviet AK assault rifle; the Galil may be quite similar since it does have a selector lever, receiver cover and various other external details that resemble the AK.

The Galil has a folding type steel stock and a bipod that folds between the hand-guards when not in use. It also has a grenade launcher on the end of the barrel. The cyclic rate has been reported as 650 rounds per minute. The magazine probably will take about 50 rounds.

The Israelis have announced that the Galil is to be used to replace the "FAL" and the heavy barrel "FAL"; it therefore has quite a heavy barrel.

ISRAELI SUBMACHINE GUNS

The Israelis used the 9mm Sten in its various configurations in their early fighting with the British and the Arab States; some German 9mm MP40s also procured from Central Europe. With the development of a native submachine gun—the Uzi—Israel made her greatest impression on the small arms world.

THE ISRAELI 9mm UZI SUBMACHINE GUN

There are several different models of the Uzi. The early model guns had a wooden stock and a wooden or fiber fore-end and pistol grip. The model with folding steel stock and plastic fore-end followed soon after the introduction of the gun, circa 1952. About 1960 the Uzi was again modified and a larger bolt-retracting handle was fitted and the fire-selector/safety was modified. The model supplied to West Germany has the bolt-handle track in the receiver cover serrated to prevent forward movement of the bolt if accidentally released during retraction. The bolt-retraction handle of the Uzi does not reciprocate with the bolt.

The Uzi is named for Major Uziel Gal, its developer. It is the standard submachine gun of the Israeli Army, and also has been sold to other countries. In basic operating principles, the Uzi is quite similar to the Czech ZK 476 and its descendants, the Czech Models 23 and 25 submachine guns. As far back as 1945, the UK had an experimental gun which also had a bolt that telescoped the barrel—the telescoping bolt is a feature common to all of these weapons and to the Beretta Model 12 as well.

The Uzi is a very well-made gun, and its performance leaves little to be desired. FN of Belgium is now marketing this weapon in Europe.

Characteristics of the Uzi Submachine Gun

Caliber: 9mm Parabellum.
System of operation: Blowback, selective fire.
Weight: 8.9 lb. w/loaded 25-round magazine and metal butt.
8.8 lb. w/loaded 25-round magazine and wooden butt.
Length, overall: 25.2 in. (wooden-butt model, and metal-butt weapon with stock extended). The length of the metal-butt model with the stock folded is 17.9 in., and the length of the wooden-butt model with the butt removed is 17.3 in.
Barrel length: 10.2 in.
Feed device: 25-round, 32-round or 40-round, detachable, staggered box magazine.
Sights: Front: Truncated cone w/protecting ears.
Rear: L-type w/setting for 100 and 200 yards.
Muzzle velocity: 1310 f.p.s. w/8-gram 9mm Parabellum bullet.
Cyclic rate: 650 r.p.m.

How to Load and Fire the Uzi

Insert a loaded magazine in the magazine well located in the pistol grip. Pull the cocking handle, located on the top of the receiver, to the rear. The Uzi fires from an open bolt, and the bolt will remain to the rear. To fire single shots, set the change lever, located on the top left side of the pistol grip, to the letter R (this is the mid-position). To fire automatic fire, set the change lever at the letter A (this is the forward position). To put the weapon on safe, set the change lever at S (this is the rear position). Pressure on the trigger will fire the weapon, if the change lever is set on semiautomatic or automatic fire positions. The grip safety must be squeezed when firing the weapon.

9mm Parabellum Israeli revolver showing 3-round clip.

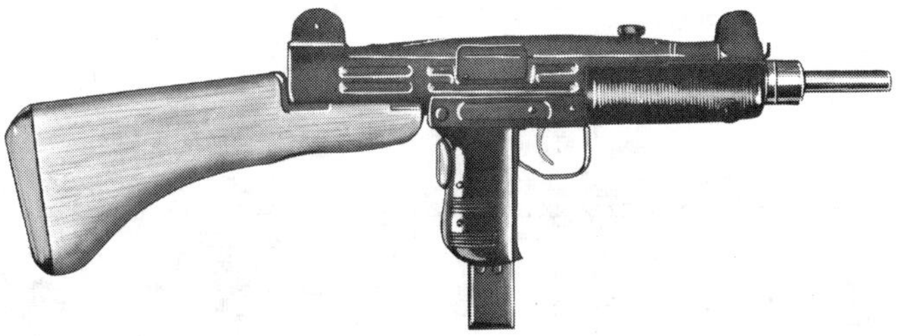

Israeli 9mm Uzi Submachine Gun with wooden stock.

Field Stripping the Uzi

To release the cover, press the catch in front of the rear sight housing to the rear. Raise the cover and remove it from the receiver. Raise the bolt from the front; when its forward part is completely disengaged, withdraw it from the receiver together with the recoil spring, giving the whole assembly a short pull forward. To remove the extractor, take out its retaining pin. Press in the lock of the barrel retaining nut; unscrew the nut and withdraw the barrel. Take out the split pin and the sleeve in which it is housed, and remove the trigger group and pistol grip from the assembly. To strip the folding metal butt, unscrew the hexagonal head of the nut retaining the butt (the nut is located in the rear of the receiver). To remove the wooden butt, press in the butt retaining catch, and pull the butt to the rear. No further disassembly is recommended.

To reassemble the weapon, follow the above procedure in reverse order.

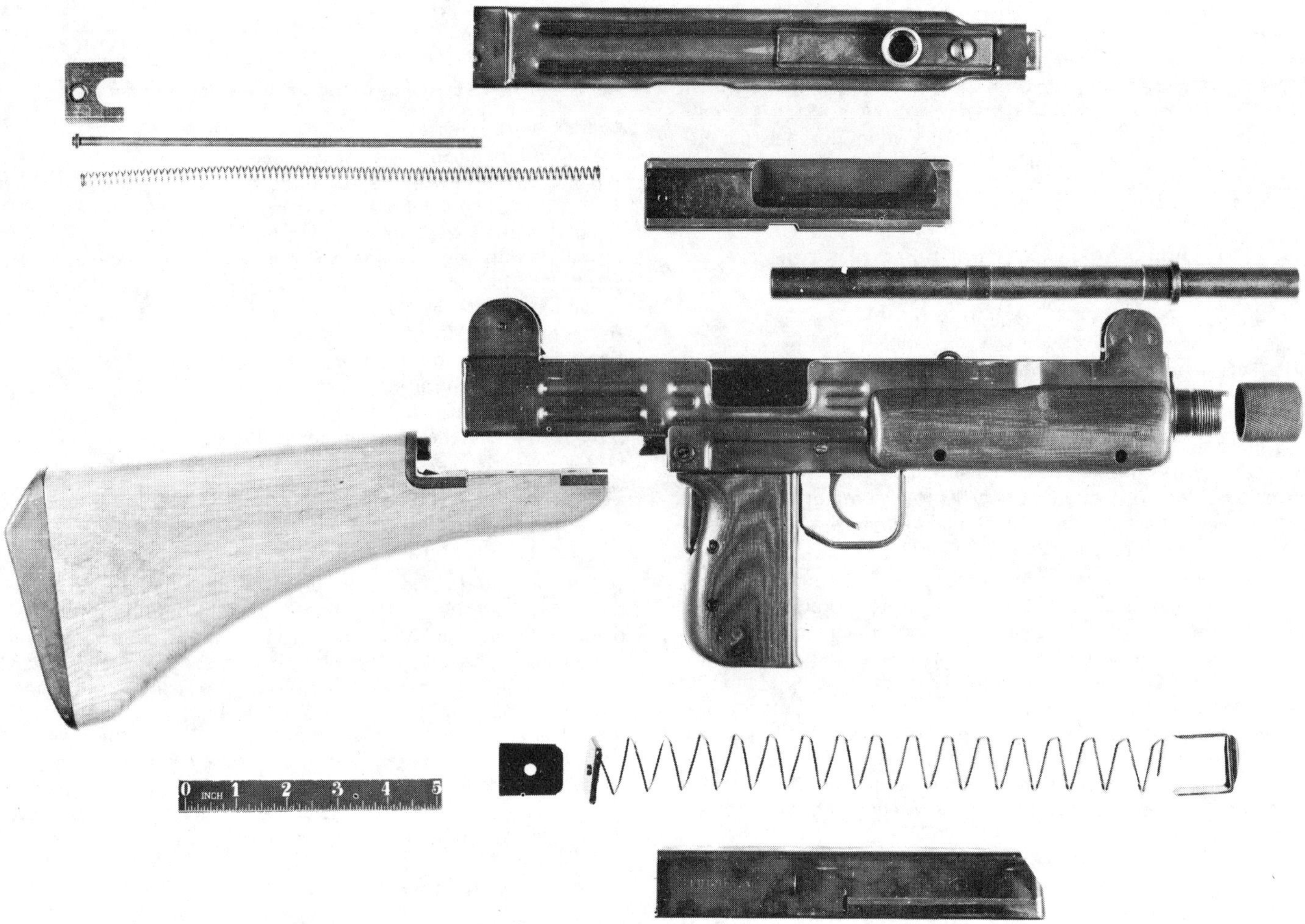

Uzi field-stripped.

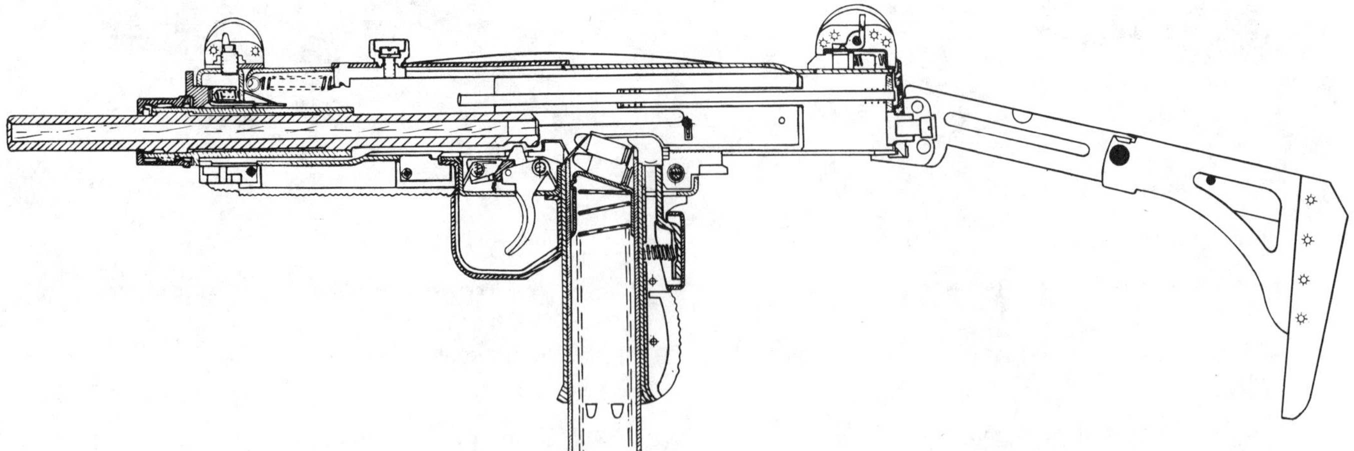

Section view, Uzi 9mm submachine gun.

Israeli 5.56mm Galil Rifle.

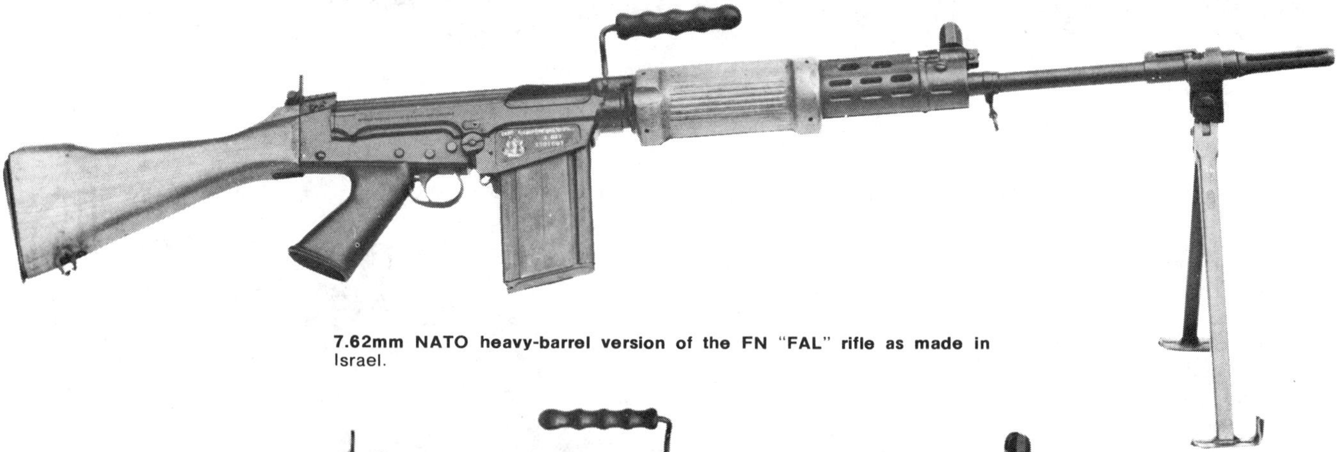

7.62mm NATO heavy-barrel version of the FN "FAL" rifle as made in Israel.

7.62mm NATO FN "FAL" Rifle as made in Israel.

How the Uzi Works

With the exception of the trigger mechanism, the functioning of the Uzi is quite similar to that of the Czech Model 23 submachine gun. (See the chapter on Czechoslovakia.)

Special Note on the Uzi

The current model designation of the Uzi in Israeli service is No. 2 Mark A. The Uzi has a grip safety (as have the Belgian Vigneron and several other submachine guns available today), which makes the weapon much safer to handle than were various World War II submachine guns. As an example, the Uzi cannot be fired accidentally when dropped, since its grip safety locks the sear to the extent that it prevents the rear of the bolt from over-riding, which is necessary if the bolt is to go far enough to the rear with enough kinetic energy to pick up a cartridge and fire it. The Uzi is composed mostly of stampings. It is one of the best examples of a modern post-World War II submachine gun. The West German government has adopted the Uzi, as have the Netherlands, Peru, and several other countries.

ISRAELI MACHINE GUNS

Originally equipped with British caliber .303 Bren and Vickers guns, the Israelis acquired German 7.92mm MG34s and U.S. caliber .30 and .50 Browning guns in the late '40s. Besa 7.92mm machine guns were also used during this period.

The Israelis developed a recoil-operated, magazine-fed, 7.92mm light machine gun called the "Dror." This weapon, which externally closely resembled the Johnson M1944 machine gun, was apparently not a very successful design, since few were made. The Israelis have adopted the FN 7.62mm NATO MAG machine gun as their standard ground machine gun.

37

Italy

Current standard Italian service small arms are the 9mm Parabellum Pistol M1951, the 9mm Parabellum Beretta Model 4 submachine gun, the 7.62mm NATO BM59 Mod I, and the 7.62mm NATO MG42/59 machine gun. All these weapons are currently manufactured in Italy and are of post World War II design or modifications of pre-war or World War II weapons.

Other weapons which are in limited use are as follows:- 9mm (.380 ACP) Beretta M1934 pistol, earlier models of the 9mm Parabellum Beretta submachine gun, the caliber .30 and caliber .50 Browning machine guns (used mainly on armored vehicles), the 8mm Breda Model 37 machine gun, and caliber .30 Browning Automatic rifle M1918A2, and caliber .303 Bren guns.

ITALIAN SERVICE PISTOLS

ITALIAN SERVICE PISTOLS, 1889 to 1951

The Italians have had a number of service pistols since World War I and the long use in service of some of them is rather surprising, considering the generally excellent pistols produced in Italy, particularly by the Beretta firm. The continued manufacture and use of the 10.35mm revolver Model 1889 and its various modifications through World War II is a case in point.

The Model 1889 Glisenti revolver is a double-action Chamelot Delvigne design revolver, loaded through a loading gate on the right side. These weapons were made in small shops and will be found with brass frames, cast steel frames, forged steel frames, and even frames made of copper plates brazed together. The M1889 may be found with a folding trigger and no trigger guard, and is also found with a conventional spur trigger and trigger guard. The manufacture of these revolvers continued through the 1920's and the use of these weapons by the Italians in World War II would be equivalent to the U.S. use of the caliber .45 Model 1873 (Frontier Model) revolver in the same conflict.

The above would suggest that the pistol development in Italy was comatose, but such was not the case. The Brixia automatic pistol appeared in Italy around 1906 and (although it is reported of Swiss origin, developed from patents of Haensler and Roch) it was first manufactured by Metellurgica Bresciana Tempini at

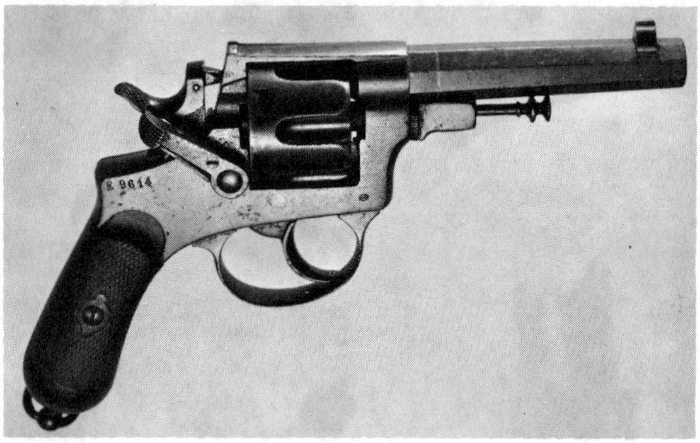

10.35mm Revolver M1889.

9mm Brixia pistol.

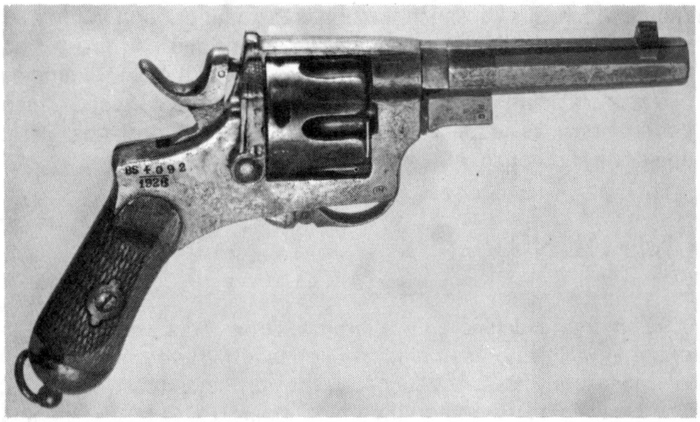

10.35mm Revolver M1889 without trigger guard.

Brescia, Italy. This weapon, which is chambered for a low-powered version of the 9mm Parabellum cartridge, was with slight modification, adopted as the 9mm Model 1910 Glisenti service pistol. The

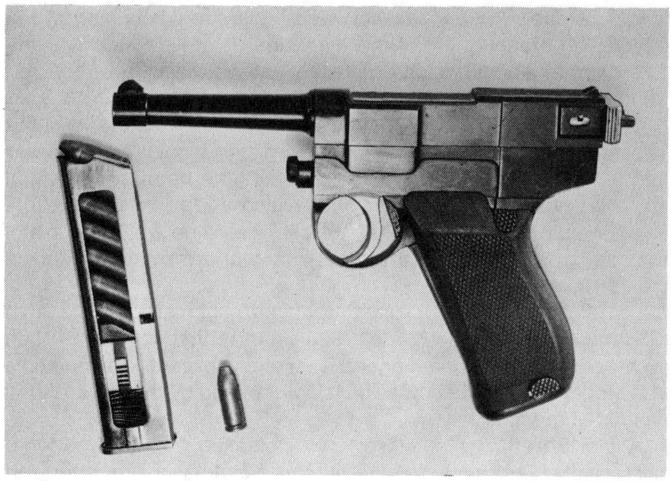

Glisenti 9mm 1910 Automatic.

7.65mm Model 1915 Beretta Pistol.

Glisenti is a retarded blowback pistol and used the 9mm (sometimes called 8.9mm), M1910 cartridge. This cartridge is dimensionally interchangeable with the 9mm Parabellum cartridge. There are differences of opinion as to whether or not the higher standard loadings of the 9mm Parabellum cartridge should be used with this pistol, but the outstanding authorities agree that use of the standard loads of 9mm Parabellum in the Glisenti is not advisable. The Italian Government apparently agreed with this course as well, since their 9mm Parabellum loads for the M1938 Beretta submachine gun are stamped M938, probably to prevent their usage in the Glisenti. The Glisenti cartridge has a truncated conical bullet as opposed to the conical nosed bullet used with the 1938 and later Beretta submachine guns.

In 1915 the first Beretta automatic pistol was introduced into Italian service. As of this date, Beretta designed and made automatic pistols are still the standard service pistol in Italy. The most widely distributed Model 1915 is chambered for the 7.65mm cartridge and is a hammerless blowback operated pistol.

A model of the 1915 chambered for the 9mm Glisenti cartridge was also produced; this weapon differs from the 7.65mm only in

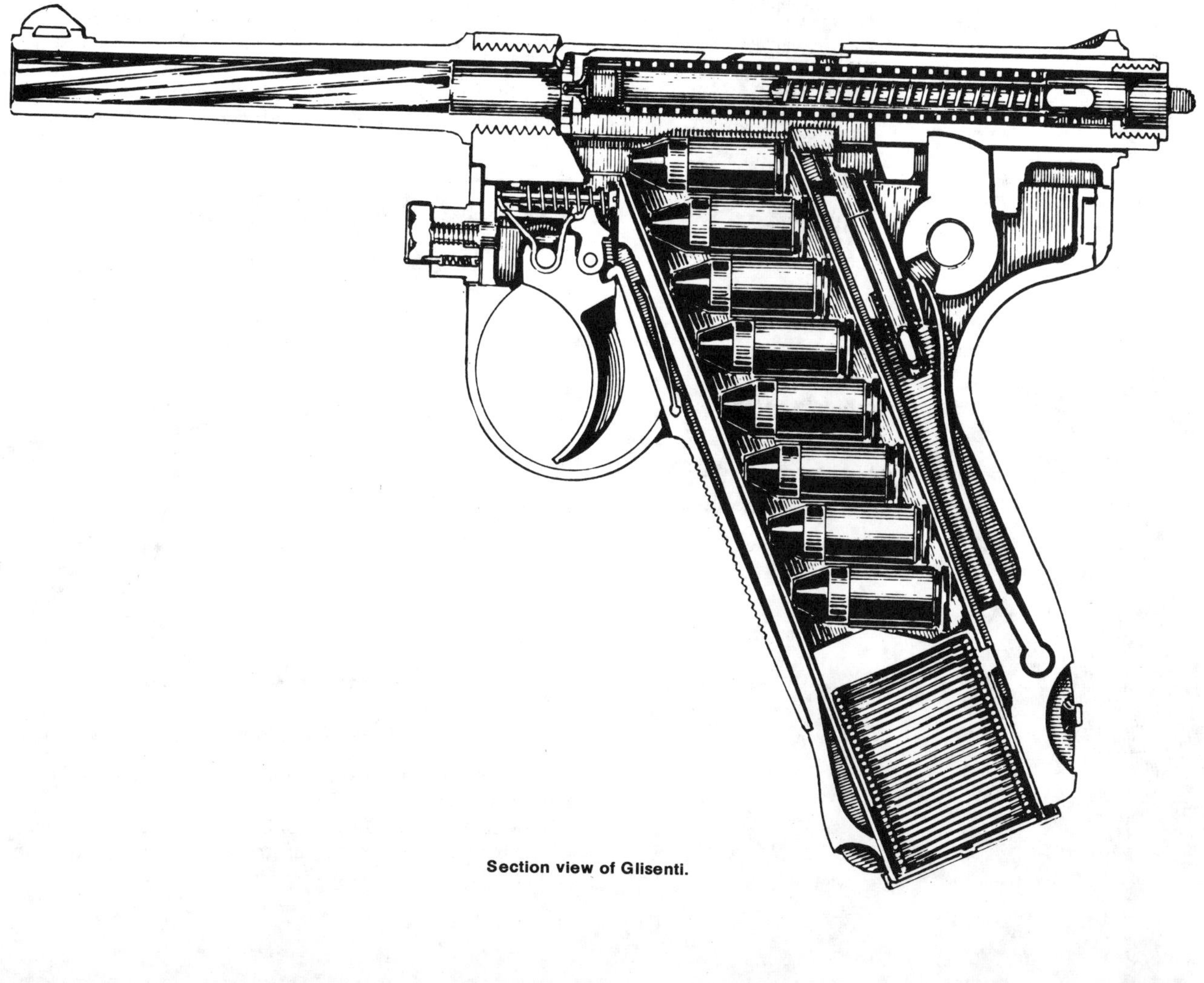

Section view of Glisenti.

Beretta Model 1915 Glisenti 9mm Pistol.

7.65mm Model 1915-1919 Beretta Pistol.

9mm Model 1923 Beretta Pistol.

minor details such as checked wood grips, the positioning of the ejector, the functioning of the magazine as a slide stop and the presence of a small mechanical safety at the rear of the receiver, which can be engaged when the slide is to the rear.

The 7.65mm Model 1915-1919 Beretta is also hammerless, but resembles the later Berettas in many respects, such as the larger portion cut out on the top of the slide and the barrel mounting, a grooved block under the chamber which slides into channels machined in the receiver ahead of the magazine well. This pistol was apparently not issued to the Army, but the specimen shown bears Naval markings.

In 1923 another Beretta was introduced, chambered for the 9mm Glisenti cartridge. The Model 1923 was the first service Beretta to have an external hammer and is, in other respects, the same as the Model 1915-1919. These pistols were used by the Italian Army in limited quantities.

A later Beretta, which closely resembles the well-known Model 1934, is the 7.65mm Model 1931. This pistol has a straighter grip than the Model 34 and issue was limited to the Italian Navy.

The 9mm corto (.380 ACP) Model 34 Beretta is probably the best known of the Italian automatic pistols and was one of the best Italian weapons in World War II. It is still used by Italian forces to some extent and is manufactured for commercial sale as the "Cougar." The Model 1934 has also been made in 7.65mm. The 9mm Parabellum Model 1951 is the current standard service pistol. It is the first Beretta pistol with a positively locked breech and utilizes full-powered 9mm Parabellum cartridges. It is sold commercially as the Beretta Brigadier.

THE 9mm BERETTA PISTOL MODEL 1934

The Model 1934 is a very finely-made weapon and was very popular with the Italian Army during World War II, as well as with U.S. troops who managed to acquire them in one way or another. Specimens marked "RE" (Regio Esercito) or "Royal Army" were Italian Army issue; other specimens may be found with the marking "PS" which indicates police or carabinieri issue. The Italian service designation for the .380 ACP cartridge is 9mm Model 34.

9mm (.380 ACP) Model 1934 Beretta.

Loading and Firing the M1934 Beretta

Load magazine exactly as for Colt automatic pistol. Insert in handle and push in until it locks.

Draw back slide exactly as for Colt automatic. Release slide and let it drive forward pushing cartridge into firing chamber and closing pistol.

Push safety catch around into locking position unless weapon is to be fired.
(Note: The exposed hammer may be let down gently on the firing pin if care is taken and the operation is done with both hands. It may also be set at half-cock.)

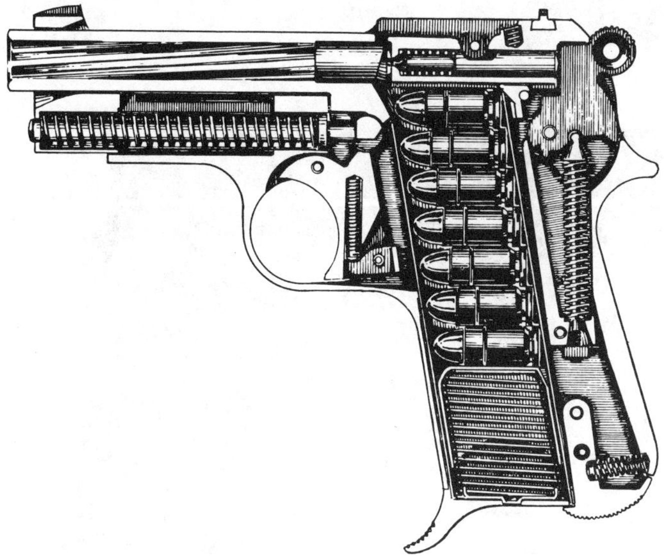

Action closed, showing detail of firing mechanism.

Action open, showing operation of recoil and feeding mechanism.

Field Stripping the M1934 Beretta

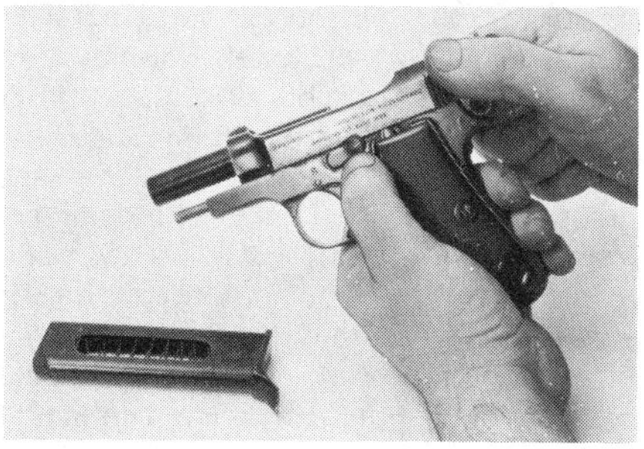

(1) With magazine out of weapon, pull back slide as far as it will go; hold it with right hand, and with left thumb push safety catch up into locking notch in underside of slide.

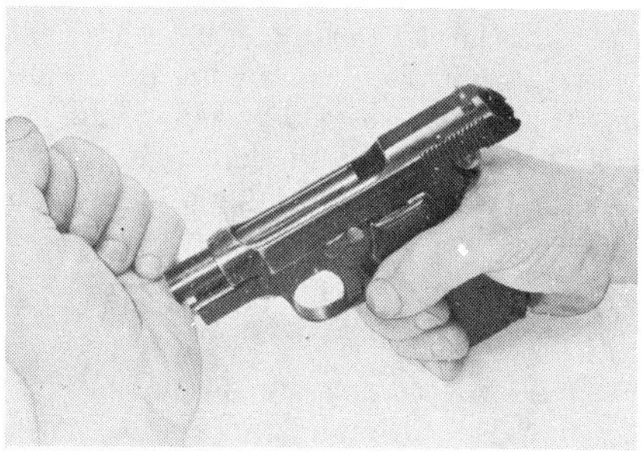

(2) Now push straight back on barrel with palm of hand. This will free barrel from its locking recess.

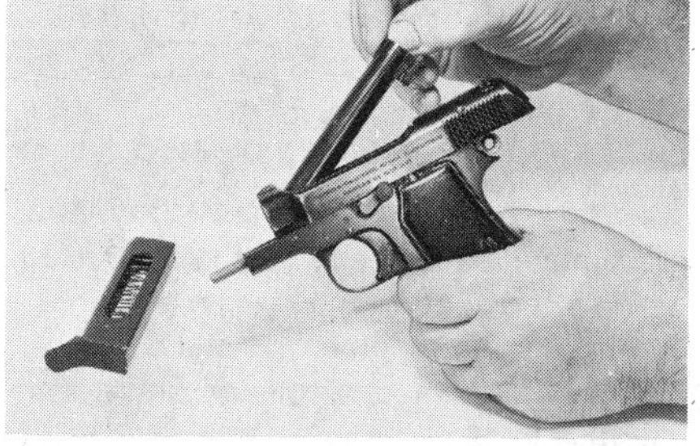

(3) Pull barrel straight back and up, drawing it out of the slide as shown.

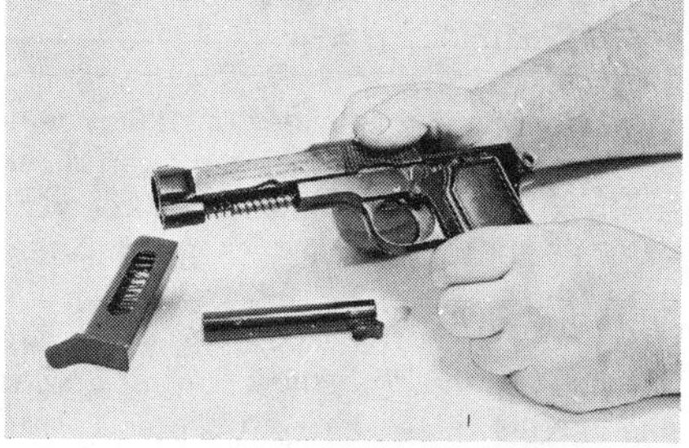

(4) Push slide, recoil spring and recoil spring guide straight forward off the receiver. Spring and guide may now be removed from their seats. Safety locking stud may also be lifted out now. No further dismounting is necessary with this pistol.

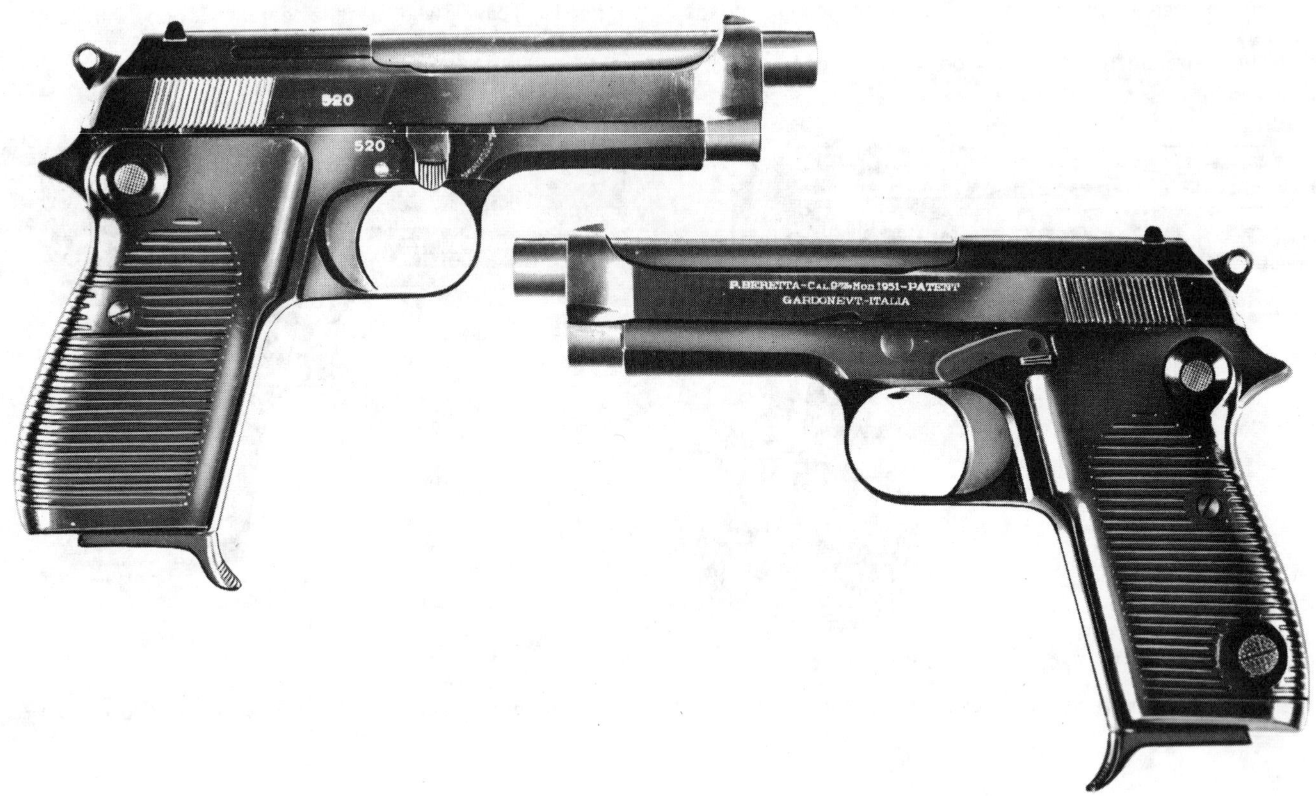

THE 9mm BERETTA M1951 PISTOL

The interesting 9mm M1951 Beretta pistol has been adopted by the Italian Army and Navy, and is also used by the Israeli and Egyptian Defense Forces. It is a very well-made weapon, and has many interesting features. It is made with an aluminum alloy or steel receiver. The weapon, which fires the 9mm Parabellum cartridge, also is known as the Model 951.

How to Load and Fire the M1951 Pistol

The magazine is loaded the same as any single-column pistol magazine. Insert loaded magazine in magazine well which is located in pistol grip. Pull the slide sharply to the rear, and release; the slide will run home and chamber a cartridge. The weapon may now be fired by squeezing the trigger. The safety is a button type, and is mounted at the top rear of the grip; to put the weapon on "safe," push the button from right to left. The M1951 is not a double action weapon. With a cartridge in the chamber and the hammer down, the weapon must be manually cocked before it will fire.

How the M1951 Pistol Works

Loading, Cocking, and Chambering. A magazine containing eight rounds is inserted into the bottom of the grip section of the receiver. The slide is drawn to the rear by hand; it forces the hammer downward and to the rear. The full-cock notch of the hammer engages the nose of the sear and the hammer remains in its rearward position. When the slide is released, it goes forward under the pressure of the recoil spring. The slide feed rib forces a round from the magazine into the chamber of the barrel.

Locking. As the slide goes forward, the locking wedge is cammed into its locked position in the locking grooves of the slide by the cam surfaces on the receiver. The locking mechanism of this pistol is similar to that of the Walther Pistole 38, covered in the chapter on Germany in World War II.

Firing. When the trigger is pulled, the trigger bar is forced to the rear. The trigger bar pushes the disconnector up into the disconnector notch in the slide. If the weapon is not completely locked, the disconnector will be held down by the slide, and will force the trigger bar down and out of contact with the trigger (this arrangement is similar to that of the M1934 Beretta). When the weapon is locked, the trigger bar forces the sear backward against the pressure of the sear spring; the hammer then disengages itself from the sear and moves forward under the pressure of the hammer spring. The hammer strikes the firing pin, forcing it forward against the pressure of the firing pin spring. The firing pin strikes the primer of the cartridge, and immediately rebounds under the pressure of the firing pin spring. The M1951 is not a double-action pistol; if a misfire occurs, the hammer must be recocked manually.

Unlocking. The barrel and slide recoil together for a distance of one-half inch; then the unlocking plunger on the rear barrel lug abuts against the receiver and, moving forward, forces the locking wedge out of the locking grooves in the slide, and down into the cammed surface on the receiver. Having been disconnected from the barrel, the slide (with the empty cartridge case gripped by the extractor) moves to the rear for an additional two inches, cocking the hammer in the process. The cartridge case is knocked out of the grip of the extractor claw, and out of the pistol, by the nose of the ejector. When the last round is fired, the slide stop is forced up by the magazine follower and holds the slide to the rear.

Field Stripping the M1951 Pistol

Remove magazine by pushing the magazine catch inward.

Draw the slide to the rear and align the dismounting latch with the dismounting notch of the slide.

Push the dismounting latch upward, toward the muzzle end of the pistol.

Pull the slide group forward and off the receiver.

Pull the recoil spring and guide assembly slightly to the rear, and remove it from its position in the slide.

Push the barrel backward and withdraw it from the rear of the slide.

No further disassembly is recommended.

Selective Fire Beretta M1951A

Beretta has come out with a selective fire version of the Model 1951. This pistol which is called the Model 951 "A Raffica" has a longer barrel, 4.92 inches, and uses 10 or 15 round magazines. It has a folding, vertical fore grip mounted at the front end of the receiver and a selector switch mounted on the right of the receiver above and behind the trigger. This weapon weighs 4.6 pounds loaded.

M1951 field-stripped.

9mm Parabellum Sosso Pistol Model 1941

This pistol was developed in 1936 by Guilio Sosso. The Sosso is a recoil operated, double action weapon and is unusual in one major respect. It takes nineteen cartridges which are fed by a chain link belt within the magazine. The cartridges are placed into carriers within the magazines and the belt is driven by the action. FNA - Fabrica Nazionale d' Armi - had tooled up to produce these pistols and a few were produced. Most were confiscated by the Germans as was the tooling after the Italian surrender in 1943. This, unfortunately for the Italians, was the fate of many of their World War II arms developments.

Characteristics of Italian Service Pistols

	Model 1889	Model 1910	Model 1915	Model 1915
Caliber:	10.35mm (10.4mm).	9mm Glisenti (M910).	7.65mm.*	9mm Glisenti (M910).
System of operation:	Double-action revolver.	Delayed blowback, semiautomatic.	Blowback, semi-automatic.	Blowback, semi-automatic.
Length overall:	10.25 in.	8.1 in.	6 in.	6.75 in.
Barrel length:	5.25 in.	3.9 in.	3.4 in.	3.75 in.
Feed device:	6-round revolving cylinder.	7-round in-line detachable, box magazine.	7-round in-line detachable, box magazine.	7-round in-line detachable, box magazine.
Sights:				
Front:	Bead.	Barley corn.	Blade.	Blade.
Rear:	V notch.	V notch.	V notch.	V notch.
Weight:	2.2 lb.	1.87 lb.	1.25 lb.	2 lb.
Muzzle velocity:	Approx 840 f.p.s.	1050 f.p.s.	Approx 960 f.p.s.	Approx 1035 f.p.s.

* .32 ACP.

Characteristics of Italian Service Pistols (Cont'd)

	Model 1915-19	Model 1923	Model 1934	Model 1951
Caliber:	7.65mm.*	9mm Glisenti (M910).	9mm Corto.**	9mm Parabellum (M1938).
System of operation:	Blowback, semi-automatic.	Blowback, semi-automatic.	Blowback, semi-automatic.	Recoil, semi-automatic.
Length overall:	6 in.	6.2 in.	6 in.	8 in.
Barrel length:	3.4 in.	3.88 in.	3.5 in.	4.51 in.
Feed device:	8-round in-line detached, box magazine.	8-round in-line detachable, box magazine.	7-round in-line detachable, box magazine.	8-round in-line detachable, box magazine.
Sights: Front:	Blade.	Blade.	Blade.	Blade.
Rear:	V. Notch.	V notch.	V notch.	V notch.
Weight:	1.31 lb.	1.87 lb.	1.25 lb.	1.93 lb. ***
Muzzle velocity:	Approx 960 f.p.s.	Approx 1035 f.p.s.	970 f.p.s.	1182 f.p.s.

* .32 ACP.
** .380 ACP.
*** 1.57 lb with aluminum receiver.

ITALIAN RIFLES

The Italian Army was equipped with a number of different types of rifles prior to and during World War II. Considering everything, it would be true to say that the Italian Army was the poorest armed of the major powers in World War II, insofar as rifles were concerned. The 6.5mm M1891 Mannlicher Carcano and its various 6.5mm and 7.35mm variants were the principal weapons, but numerous 8mm M1889 and 1895 Mannlicher rifles, received as war booty from the first war, and even some ancient Vetterli Vitali rifles which were rebarreled for 6.5mm rifle, were used. This rifle was also used as a caliber .22 training rifle.

THE MANNLICHER CARCANO RIFLE

The Mannlicher Carcano, also known occasionally as the Mauser Paravicino, is a modified Mauser. It has a Mauser-type one-piece bolt with frontal locking lugs but, unlike most Mausers, the bolt handle is in front of the receiver bridge when locked. The magazine is a Mannlicher in-line type, fed with a six-round clip which stays in the magazine until the last round is chambered.

There have been a number of rumors and old wives tales passed around about the safety of the Mannlicher Carcano rifle. When in good condition and used with the proper ammunition, it is as safe as any other military rifle. No nation is going to adopt a weapon which is dangerous to its own troops.

The Carcano was developed at the Italian Government arsenal at Turin, by M. Carcano. Early models have gain twist rifling, i.e., the twist of the rifling gradually increases toward the muzzle. In 1938 the 7.35mm cartridge was introduced and a rifle and two carbines chambered for this cartridge were introduced. The entrance of Italy into World War II in 1940 caused some second thoughts on the use of another cartridge, however, and in 1940 the 6.5mm caliber was re-introduced and all Mannlicher Carcano's manufactured from that date on were chambered for the 6.5mm cartridge. In addition, many of the 7.35mm weapons already in existence were rebarreled for 6.5mm.

The Mannlicher Carcano rifle became a matter of international interest when Lee Harvey Oswald assassinated John F. Kennedy, 35th President of the United States, on November 22, 1963, with a 6.5mm Model 1938 (91/38) rifle, serial number C 2766, made in 1940 at the Italian arsenal at Terni. This rifle was part of the Italian Army surplus sold in this country and was fitted with a four-power Japanese commercial scope by an American dealer.

The Mannlicher Carcano was made in limited numbers in 7.92mm, apparently for the Germans, toward the close of World War II. These rifles were apparently made in very limited quantities.

A somewhat unusual modification of the Mannlicher Carcano has a side-mounted grenade launcher. When the grenade launcher is used, the bolt is removed from the rifle and used in the launcher.

Italian Semi-automatic and Automatic Rifles

Italy's experiences with semi-automatic rifles before World War II were discouraging for several reasons. One of the primary reasons would appear to be the lack of financial means to rearm the Italian Army with promising designs, and another was that just as a decision was made to rearm, the imminence of war made it impracticable. In 1905 a semi-automatic rifle was patented by Filippo Genovesi and the Italian government showed enough interest in it to buy the patent in 1908 and order the production of 6,000 rifles. Production was suspended in 1912 however because the rifle did not function satisfactorily in the Libyan sand. This rifle was chambered for the standard 6.5mm Mannlicher Carcano cartridge. About 1911 Lieutenant Colonel Amasigo Cei-Rigotti designed a gas operated rifle with piston, but the initial design had too violent an action and his later design had the piston acting against the bolt handle rather than the bolt. Nothing of any consequence was done during the first World War, but after the war the Italian Army staff came up with a requirement for a selective fire rifle. The requirement did not specify that the weapon had to use the 6.5mm cartridge. Development began at about this time on the 7.35mm cartridge and the government arms plant at Terni produced an automatic carbine for this cartridge in 1921; the Rome arms plant produced an automatic carbine, chambered for the same cartridge, in 1922. A government commission ran a competition between these weapons and submachine guns produced by Fiat, Ansaldo, (the Crocetti carbine) and a semi-automatic carbine produced by Beretta. The last three weapons were chambered for the 9mm Glisenti cartridge, a low powered 9mm Parabellum.

The government commission rejected the 9mm weapons because the requirement called for accuracy at 600 meters. They determined that the Model 1921 Terni was the best of the 7.35mm rifles. The story of this rifle is somewhat like that of the U.S. .276 Pedersen T2E1 in that Italy determined, as did the U.S. with the Pedersen, that it was not advisable to change calibers at this time. In 1928 Terni produced a modification of the Model 1921 called the Model 1921/28. Other semi-automatic and selective fire rifles developed in Italy around this time were the Revelli of 1928 patented by Bethel Abiel Revelli, the developer of the Fiat machine gun, the Brixa Model 1931, and a Model 91 rifle converted to semi-automatic fire in 1930.

Beretta brought out two recoil operated semi-automatic weapons during this period, the Model 1931 and the Model 1937. Both are recoil operated weapons and differ mainly in caliber, the Model 1931 is 6.5mm and the Model 1937 is loaded with a clip from above. Breda came out with their Model 1935 GP rifle, a selective fire weapon that was made in 7mm for Costa Rica. The Scotti Model X was introduced in 1934 and 250 of these rifles were made for test. More of these rifles were made at a later date and issued to the Italian Army, but the quantity was comparatively small. In 1937 competitive tests were run with the following Italian designed rifles: two semi-automatic designs from Terni, the semi-automatic Breda C.R.5, recoil operated Scotti, the Pavesi - F.N.A. a semi-automatic from the National Arms Plant at Brescia, the S.I./S.T.A.R. and the Revelli Manifatturo Armaguerra recoil operated semi-automatic rifle later called the Model 39. This rifle was developed by the son of the developer of the Fiat machine gun.

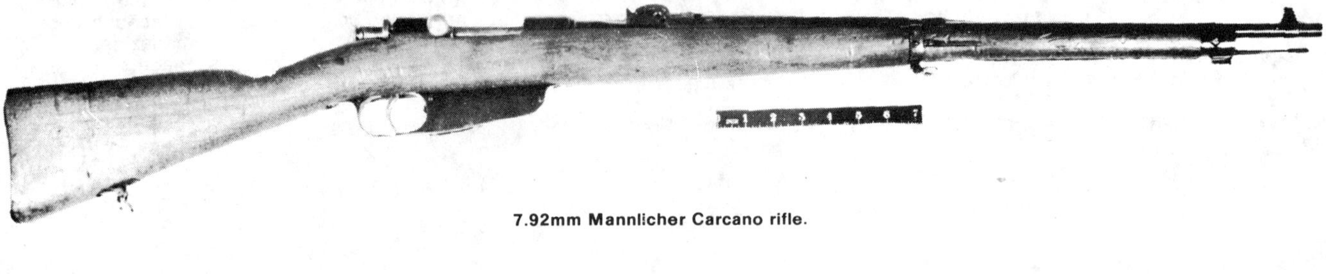

7.92mm Mannlicher Carcano rifle.

Representative Italian rifles. 1. Model 1891 Rifle. 2. Model 91TS Carbine. 3. Model 91 Carbine. 4. Model 1941 Rifle. 5. Model 91/24 Carbine. 6. Model 38 (7.35mm). 7. Model 38TS (7.35mm) Carbine. 8. Model 38 (7.35mm) Carbine. 9. Model 1938 Rifle (caliber 6.5mm). 10. Model Youth Rifle (Moscheto) Ballila, caliber special 6.5mm. Barrel 14.43". (Weapon does not fire ball cartridges.)

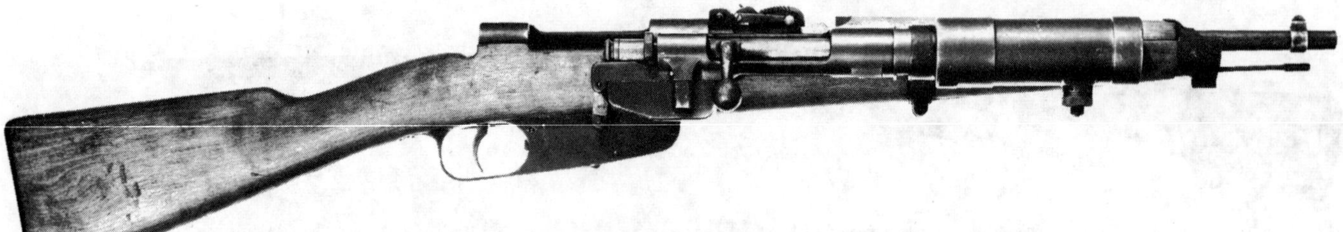

6.5mm Mannlicher Carcano Model 38 Carbine with attached grenade launcher.

The design commission completed the tests in 1939 and decided that although none of the rifles really met all the requirements, the Revelli was the best. An order was placed for 10,000 of these rifles, but the ongoing change in cartridges - 6.5mm to 7.35mm - the higher priority need for the Breda Model 37 machine gun and the onset of World War II kept the rifle from being produced in quantity.

In 1952 the Italian Army was issued the caliber .30 M1 rifle. This rifle was made in Italy by Beretta and Breda. Beretta also manufactured the M1 for Denmark and Indonesia and introduced the 7.62mm BM59 family of rifles. The current standard Italian service rifle is the selective fire 7.62mm NATO BM59 Mark I. Semi-automatic BM59's converted from M1's are also on hand.

THE BERETTA BM59 SERIES OF RIFLES

The basic concept which produced the Beretta Model 59 rifles was the conversion of existing M1 rifles to the 7.62mm NATO cartridge, in most models, to selective fire, and modification of the feed mechanism to use a 20-round detachable magazine rather

Characteristics of Mannlicher Carcano Bolt-Action Rifles and Carbines

	Rifle M1891	**Carbine M1891**	**Carbine M1891 TS**	**Carbine M1891/24**
Caliber:	6.5mm.	6.5mm	6.5mm.	6.5mm.
Overall length:	50.8 in.	36.2 in.	36.2 in.	36.2 in.
Barrel length:	30.7 in.	17.7 in.	17.7 in.	17.7 in.
Feed device:	6-round in line non-detachable box magazine.	6-round in line non-detachable box magazine.	6-round in line non-detachable box magazine.	6-round in line non-detachable box magazine.
Sights: Front:	Barleycorn.	Barleycorn.	Barleycorn.	Barleycorn.
Rear:	Tangent w/V notch graduated from 500-2000 M; leaf turned over for battle sight.	Tangent w/V notch graduated from 500-1500 M; leaf turned over for battle sight.	Tangent w/V notch graduated from 500-1500 M; leaf turned over for battle sight.	Tangent w/V notch graduated from 500-1500 M; leaf turned over for battle sight.
Weight:	8.6 lbs.	6.6 lbs.	6.9 lbs.	6.9 lbs.
Muzzle velocity:	2395 f.p.s.	2297 f.p.s.	2297 f.p.s.	2297 f.p.s.
Remarks:	Basic rifle, straight bolt handle, uses knife type bayonet.	Bayonet permanently attached, bolt handle bent.	Uses knife-type bayonet, bolt handle bent.	Uses knife-type bayonet, bolt handle bent, except for lower band the same as M1891 TS carbine.

	Rifle M1938	**Carbine M1938**	**Carbine M1938 TS**	**Rifle M1941**
Caliber:	7.35mm.	7.35mm.	7.35mm.	6.5mm.
Overall length:	40.2 in.	36.2 in.	36.2 in.	46.1 in.
Barrel length:	20.9 in.	17.7 in.	17.7 in.	27.2 in.
Feed device:	6-round in line non-detachable box magazine.	6-round in line non-detachable box magazine.	6-round in line non-detachable box magazine.	6-round in line non-detachable box magazine.
Sights: Front:	Barleycorn.	Barleycorn.	Barleycorn.	Barleycorn.
Rear:	Fixed.	Fixed.	Fixed.	Tangent w/V notch graduated from 300-1000 M; leaf turned over for battle sight.
Weight:	7.5 lbs.	6.5 lbs.	6.8 lbs.	8.21 lbs.
Muzzle velocity:	2482 f.p.s.	Approx. 2400 f.p.s.	Approx. 2400 f.p.s.	Approx. 2360 f.p.s.
Remarks:	First of Italian rifles chambered for 7.35mm cartridge, bolt handle bent. Some have knife-type folding bayonet which can be carried folded on rifle or removed.	7.35mm version of M1891 carbine, has permanently attached folding bayonet, bolt handle bent.	7.35mm version of M1891/24 carbine, uses knife-type bayonet, bolt handle bent.	Basically the same as the Rifle M1891 except for length and rear sight.

	Rifle M1938	**Carbine M1938**	**Carbine M1938 TS**
Caliber:	6.5mm.	6.5mm.	6.5mm.
Overall length:	40.2 in.	36.2 in.	36.2 in.
Barrel length:	20.9 in.	17.7 in.	17.7 in.
Feed device:	6-round in line non-detachable box magazine.	6-round in line non-detachable box magazine.	6-round in line non-detachable box magazine.
Sights: Front:	Barleycorn.	Barleycorn.	Barleycorn.
Rear:	Fixed.	Fixed.	Fixed.
Weight:	7.6 lbs.	6.6 lbs.	6.9 lbs.
Muzzle velocity:	2320 f.p.s.	2297 f.p.s.	2297 f.p.s.
Remarks:	6.5mm version of 7.35MM Rifle M1938, made after beginning of World War II.	6.5mm version of 7.35mm Carbine M1938, made after beginning of World War II.	6.5mm version of 7.35mm Carbine M1938 TS, made after beginning of World War II.

Notes: (1) All the pre-World War II 6.5mm weapons have right-hand gain twist (progressive) 19.25 to 8.25; 7.35mm weapons have constant right-hand 10-inch twist. (2) All weapons use 6-round Mannlicher type clips. (3) Some of the Model 1938 weapons were made in 7.92x57mm Mauser for the Germans during World War II. (4) Caliber is marked on sight base of all 1938 series weapons.

6.5m. Beretta Model 1931 Semiautomatic Rifle.

6.5mm Beretta Model 1931 Semiautomatic Rifle, field-stripped.

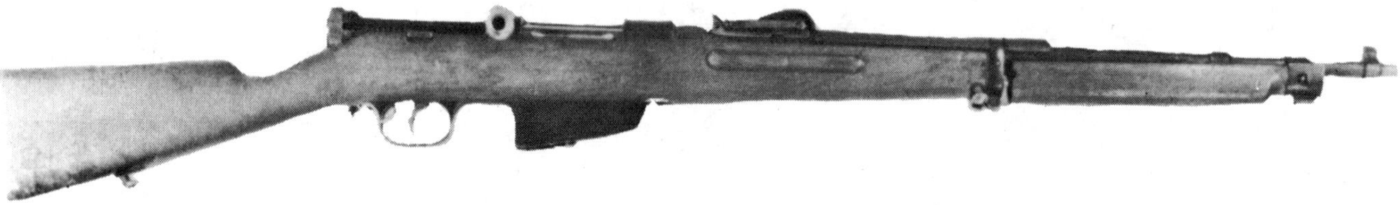

6.5mm Scotti Brescia Model X Rifle.

6.5mm Model 1935 GP Breda Rifle.

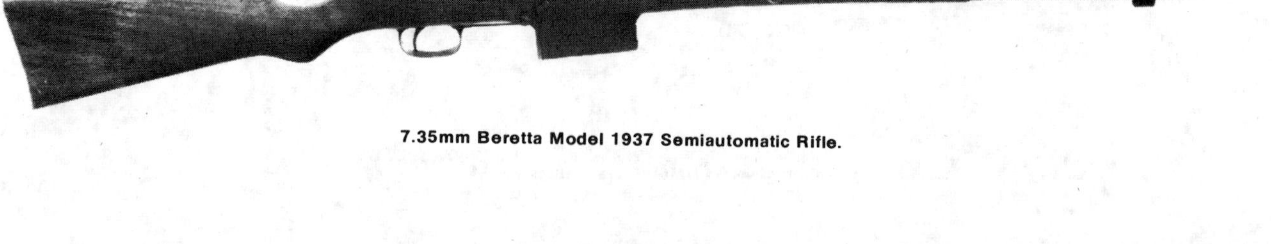

7.35mm Beretta Model 1937 Semiautomatic Rifle.

Beretta 7.62mm NATO BM59 Rifle.

Beretta 7.62mm NATO BM59D Rifle. Stock changed and bipod attached.

Beretta 7.62mm NATO BM59GL Rifle. Permanently attached grenade launcher and grenade sights.

7.62mm NATO BM59 Mark I Rifle.

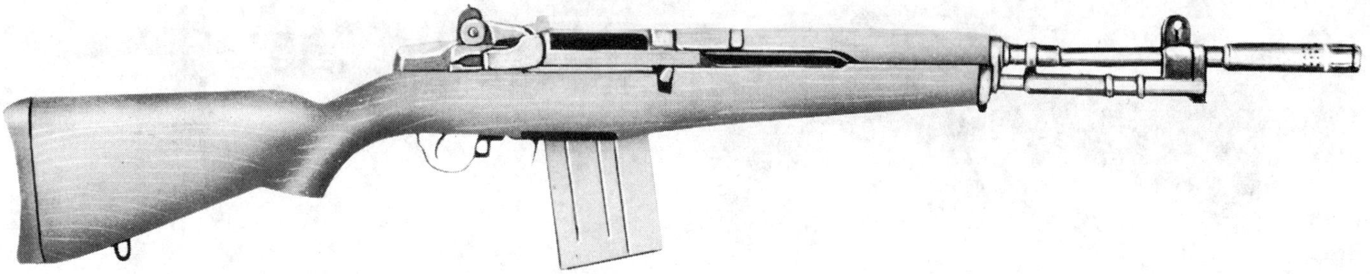

7.62mm NATO BM59 Mark I Rifle (Modified).

7.62mm NATO BM59 Mark II modified with winter trigger.

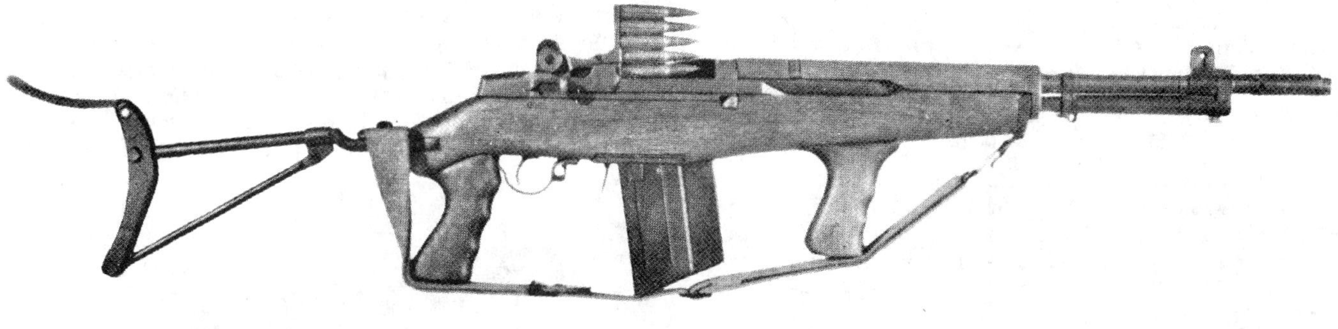

7.62mm NATO BM59 Mark III with folding stock.

7.62mm NATO BM59 Mark III (modified).

7.62mm NATO BM59 Mark IV.

than the 8-round en bloc clip of the M1.

The BM59, the first of the series, uses the M1 stock and appears generally similar to the M1 except for its 20-round magazine, charger guide, and hand guard gas cylinder arrangement.

The BM59D has a straight-line type stock will full pistol grip, a bipod, and a grenade launcher attached to the muzzle of the barrel.

The BM59GL is basically the same as the BM59, but has a grenade launcher and grenade launcher sight attached.

7.62mm NATO BM59 (Mark I).

LATER BM59 SERIES

A later series of BM59 rifles consists of the Mark I, Mark I (Modified), Mark II, Mark II (Modified), Mark III, Mark III (Modified), and the Mark IV. All are pictured and the differences can be noted in the pictures.

The BM59 has been made in a semiautomatic version for commercial sale. This rifle has a prong-type flash suppressor and can easily be distinguished from the full automatic versions of the BM59 by its lack of a selector on the left side of the receiver.

The BM59 Mark "I" and Mark "I" - A

The "I" in these designations stands for "Italian" since these are the weapons which were adopted by the Italian Government. They are also known as the BM59 Mark Ital and Ital-A rifles. These rifles are equipped with bipods, and grenade launchers, and differ from each other mainly in their stocks. The Mark "I"-A is for paratroops and armored troop use.

Loading and Firing the BM59 Mark "I" and Mark "I"-A. Pull operating handle to the rear, locking the weapon. Put safety, located in forward part of trigger guard, on safe by pushing rearwards. Insert a loaded magazine in the magazine well and pull upward and to the rear until the magazine catch engages. If there is already a magazine in the weapon it can be loaded by five-round chargers (stripper clips). Set the selector, located on the far left side of the receiver, on "A" for automatic fire, or "S" for semi-automatic fire. Disengage the safety and press the trigger to fire the weapon. This weapon has a winter trigger which can be used when heavy gloves are required. It can be swung down into position when required.

Field Stripping the BM59 Mark "I" and Mark "I"-A

This rifle is field stripped in a manner similar to the U.S. M1 rifle, which is covered in detail in the chapter on the U.S.

BERETTA .223 MODEL 70/.223 ASSAULT RIFLE

Beretta has developed a selective fire rifle chambered for the 5.56mm cartridge. It resembles the SIG Model 530-1 externally, but the field stripped view indicates that there are significant differences in design. The Model 70/.223 has a combination grenade launcher/flash suppressor built into the barrel. The grenade launcher sight is positioned behind the front sight, pivoting on the front sight base as with those versions of the BM59 that have grenade launchers. There are variations in the weapon: the 70/.223 SC and the 70/.223 LM in addition to the basic 70/.223 AR. They vary in length and weight, the SC having a folding stock and the LM (Light Machine Gun) having a bipod and carrying handle.

7.62mm NATO BM59 for Alpine troops.

Beretta .223 Model 70/.223 Assault Rifle.

7.62mm NATO BM59 Semiautomatic Rifle.

Beretta .223 Model 70/.223LM light machine gun.

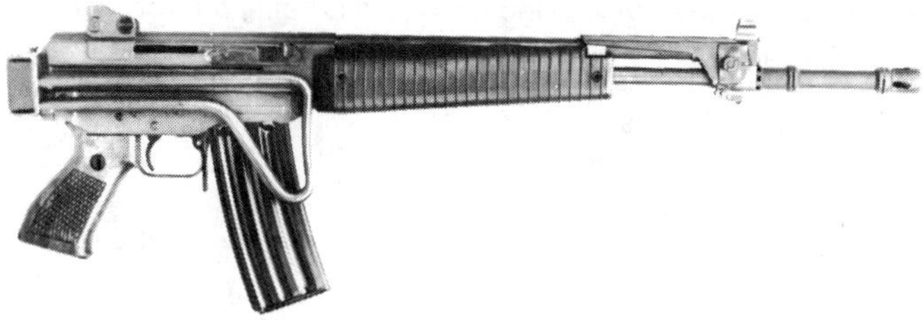

Beretta .223 Model 70/.223SC assault rifle.

Beretta .223 Model 70/.223AR field stripped.

CHARACTERISTICS OF THE BERETTA .223 MODEL 70/.223 RIFLES

	70/.223 AR	70/.223 SC	70/.223 LM
Caliber:	.223 (5.56mm)		
System of operation:	gas, selective fire		
Length overall: w/stock fixed:	37.6"	37.8"	37.6"
w/stock folded:	28.8"		
Barrel length:	17.8"		
Feed device:	30-round, staggered row, detachable box magazine		
Sights: Front:	protected post		
Rear:	aperture adjustable for 100, 250 and 400 meters		
Weight:	8 lb.	8.5 lb.	8.9 lb.
Cyclic rate:	630 r.p.m.	630 r.p.m.	600 r.p.m.
Muzzle velocity:	3,180 f.p.s.		

CHARACTERISTICS OF THE BERETTA BM59 RIFLES

	BM59	BM59D	BM59GL	Mark I	Mark II	Mark III	Mark IV	Mark Ital
Caliber:	7.62mm NATO							
System of operation:	Gas, selective fire.							
Length overall:	37.2 in.	37.2 in.	43 in.	39.4 in.	48.9 in.*	48.9 in.	42.5 in.	43.1 in.
Barrel length:	17.7 in.	17.7 in.	21 in.	17.7 in.	17.7 in.	17.7 in.	21 in.	19.3 in.
Feed device:	20-round, detachable box magazine							
Sights: Front:	Protected blade							
Rear:	Aperture, adjustable for windage & elevation							
Weight:	8.15 lb.	9 lb.	9 lb.	8.9 lb.	8.9 lb.	8.9 lb.	12 lb.	10.4 lb.
Cyclic rate:	750 r.p.m.							
Muzzle velocity:	2620 f.p.s.	2620 f.p.s.	2730 f.p.s.	2620 f.p.s.	2620 f.p.s.	2620 f.p.s.	2730 f.p.s.	2700 f.p.s.

*29.5 in. with stock folded.

Luigi Franchi 7.62mm NATO LF 59 Rifle

The LF 59 is a selective fire weapon which is quite similar to the FN FAL in design. The magazine of the LF 59 is interchangable with that of the FAL and it has a tipping bolt as does the FAL. The bolt carrier of the LF 59 however is attached to the piston and the recoil spring is mounted in a telescoping tube which protrudes from the back of the bolt carrier. The stock of the fixed stock model is made of plastic. The receiver is made principally of stampings. Both fixed and folding stock models were made only as prototypes. The LF 59 has a rate reducer.

Luigi Franchi .caliber .30 LF58 Carbine

The LF 58 is a selective fire weapon chambered for the caliber .30 carbine cartridge. The magazine of this weapon is interchangeable with that of the U.S. caliber .30 carbine. Like the LF 59 it is made principally of stampings. The bolt is different from that of the LF 59 in that it has two plates which cam the bolt in to and out of the locked position by action on lugs mounted on the bolt carrier. The LF 58 has a rate reducer which reduces the cyclic rate from 720-740 rounds per minute to 375 rounds per minute. This weapon has also been made for the Soviet 7.62 X 39mm (M43) cartridge. In all versions it was made only as a prototype.

Beretta Caliber .30 Carbine P30 Model 781

Beretta has produced a gas operated, selective fire carbine chambered for the U.S. caliber .30 carbine. The Beretta weapon is quite similar in outer appearance to the U.S. carbine, but is quite different internally. The bolt is of the tipping type and locks against the top of the receiver. There are two triggers; the front trigger produces semi-automatic fire and the rear trigger provides full automatic fire. In addition to the fixed wooden stock version of this weapon there is a folding steel stock version. The Beretta carbine is in service in Morocco, and is being produced there. This weapon which was designed around 1957 was one of the last designs of Tullio Marangoni, who developed most of Beretta's automatic weapons.

	LF 59	LF 58	P30 M781
Caliber:	7.62mm NATO	cal. .30 carbine	cal. .30 carbine
System of operation:	gas, selective fire	gas, selective fire	gas, selective fire
Weight:	9.5 lb.	7.5 lb.	6.5 lb.
Length overall			
stock fixed:	40.5 in.	36.4 in.	37.3 in.
stock folded:	31.7 in.	27.5 in.	approx. 27 in.
Barrel length:	20.9 in.	16.9 in.	17.7 in.
Feed device:	20-round, detachable staggered box magazine	30-round, detachable staggered box magazine	30-round, detachable staggered box magazine
Sights: Front:	aperture	rotary aperture	"L" adjustable for windage
Rear:	protected blade	hooded blade	protected blade
Muzzle velocity:	2690 f.p.s.	1900 f.p.s.	1950 f.p.s.
Cyclic rate:	610-630 r.p.m.	375 r.p.m.	550 r.p.m.

Luigi Franchi 7.62mm NATO LF59 Rifle, field-stripped.

Luigi Franchi 7.62mm NATO LF59 Rifle with folding stock.

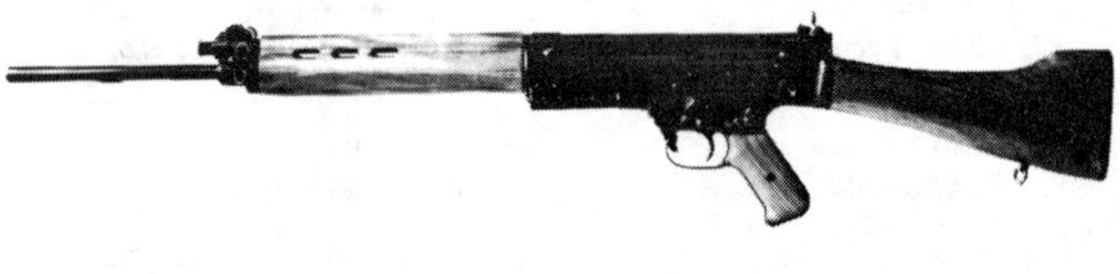

Luigi Franchi 7.62mm NATO LF59 Rifle.

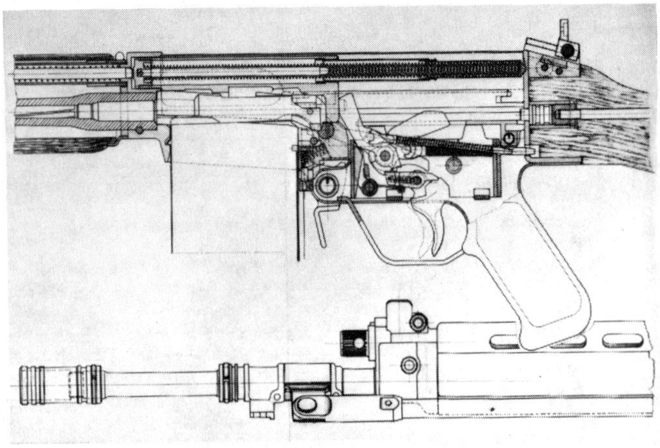

Section view of action of LF58 Rifle.

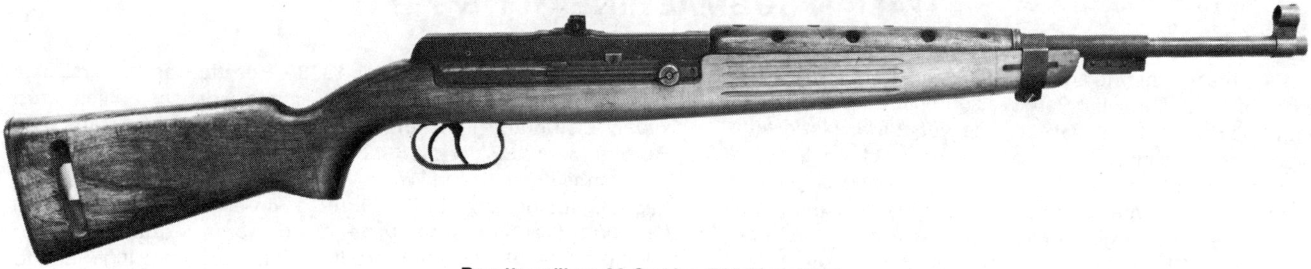

Beretta caliber .30 Carbine P30 Model 781.

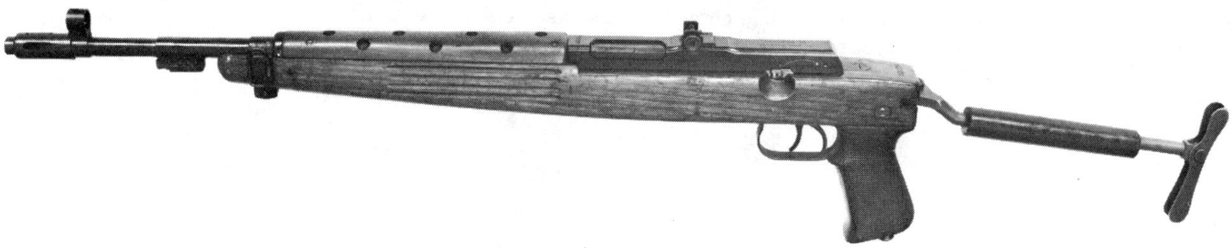

Beretta caliber .30 Carbine P30 Model 781 with folding stock.

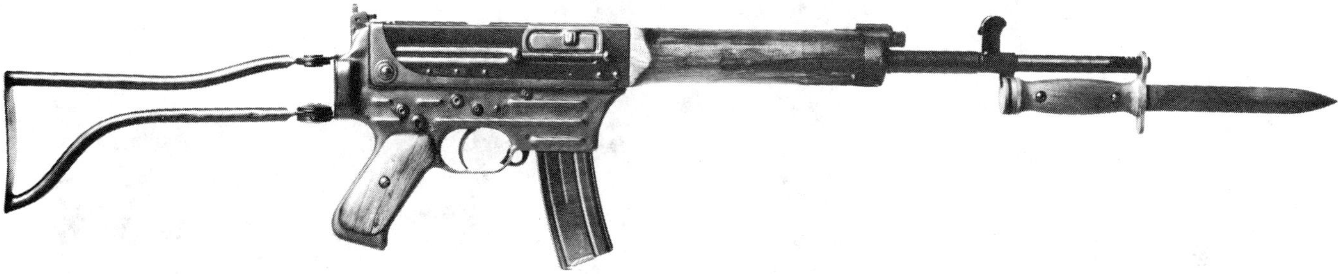

Luigi Franchi cal. .30 Carbine LF58 with folding stock.

Luigi Franchi cal. .30 Carbine LF58 field-stripped.

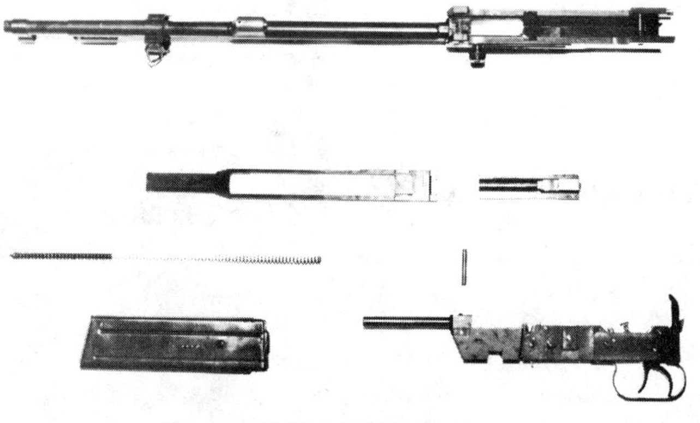

Beretta P30 Model 781 field-stripped.

ITALIAN SUBMACHINE GUNS

Italy was the first country to adopt a submachine gun—the 9mm Villar Perosa of 1915. The Villar Perosa as originally produced had no stock and was mounted in dual sets. The weapon was fired with thumb-type triggers, similar to those used on machine guns. It was mounted, usually with a shield, on motorcycle side cars, aircraft, and on various types of tripods and bipods. At a later date a stock was added to the gun and it became a selective-fire individual weapon. The Villar Perosa and all other early Italian submachine guns were chambered for the 9mm Glisenti cartridge—basically a low-powered 9mm Parabellum cartridge.

Beretta produced a modified copy of the Villar Perosa in 1918 and produced a selective-fire weapon called the Beretta Moschetto Automatico, or the M1918 - 1930.

The Model 1918 Beretta was retarded blowback as was the Villar Perosa and was designed by Tullio Marengoni, who designed all the Beretta submachine guns until the late 1950s.

The 9mm Parabellum Model 1938A Beretta was the first of a series of very well-designed finely-made weapons, which were widely distributed in other countries in addition to Italy. The Beretta Model 38A and 38/42 were considered the finest Italian small arms in service in World War II. The first Model Beretta 38 had longitudinal slots in the barrel jacket as opposed to holes in the barrel jackets of later production. Some Model 38As had folding bayonets and early production had bayonet lugs for the mounting of a removable knife type bayonet.

The early Model 38A did not have the push-through type full automatic safety located behind the full automatic trigger, and had a dual port compensator rather than the multi-slotted compensator found on later models. The most common variation of the Model 38A was without bayonet or bayonet lug, and with a multi-slotted compensator.

This model was sold to Rumania and Argentina, and was made in

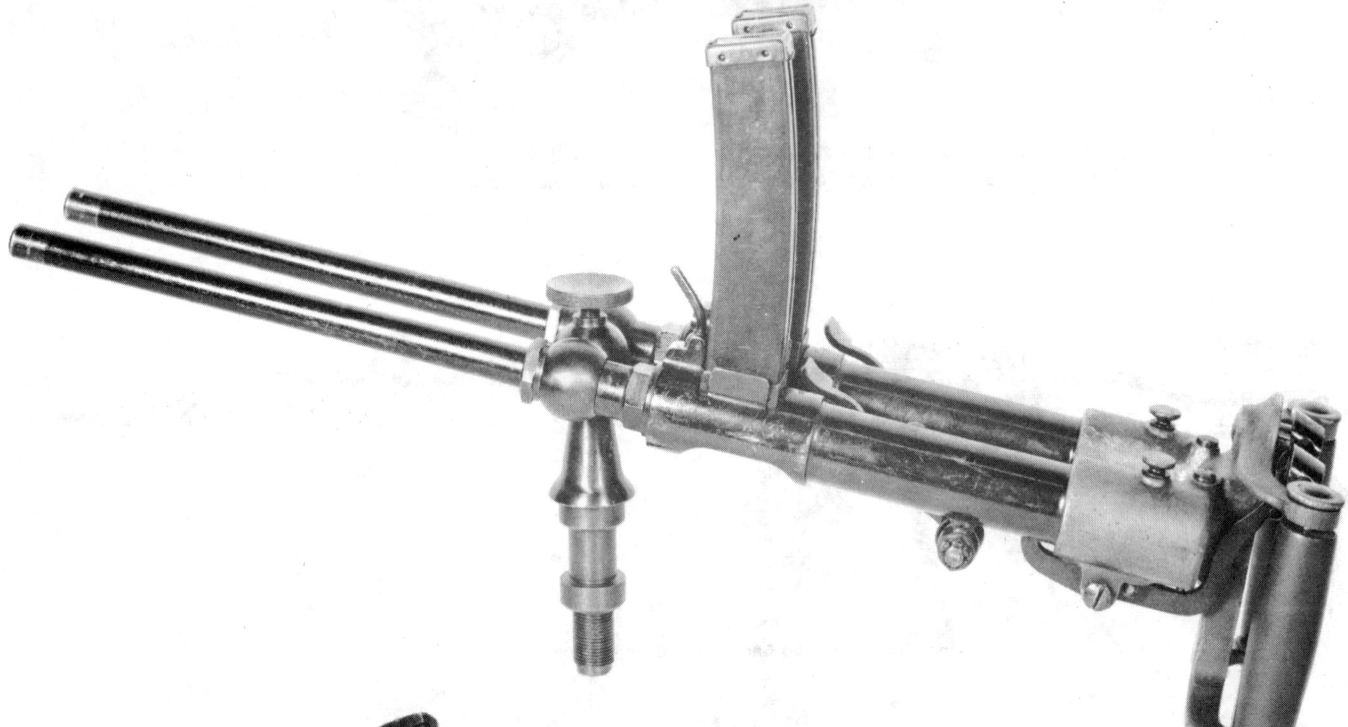

9mm Villar Perosa Submachine Gun.

9mm O.V.P. Submachine Gun.

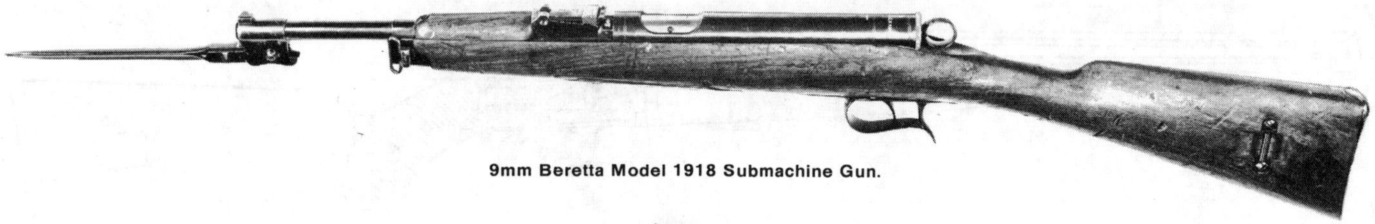

9mm Beretta Model 1918 Submachine Gun.

9mm Beretta Model 1918-30 Submachine Gun.

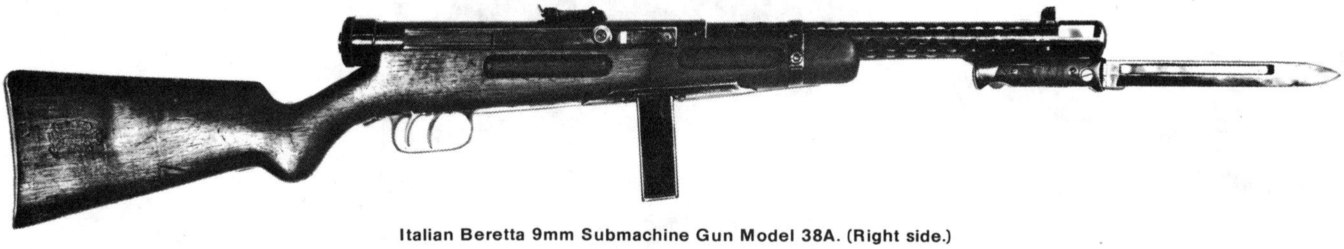

Italian Beretta 9mm Submachine Gun Model 38A. (Right side.)

Italian Beretta 9mm Submachine Gun Model 38A. (Left side.)

tremendous quantities for the Italian Army. The Model 38A was produced to some extent after World War II and in 1949 a modification of the weapon, using the cross bolt safety mounted in the stock as used with the Model 38/49 (Model 4), was produced in limited quantity. Although it has generally been considered that the fixed firing pin was introduced with the Model 38/42, late Model 38As with a fixed firing pin were apparently made. It is quite possible that these weapons were made after the introduction of the Model 38/42, since the Model 38A was made as late as 1950.

The Beretta 9mm Parabellum Model 38/42 is basically a simplified Model 38A; the barrel jacket is not used with the 38/42 and all Model 38/42s have a fixed firing pin. The 38/42 uses a stamped receiver and magazine housing and has a fluted barrel (early production). Models of the 38/42 with smooth barrel are called 38/43 by Beretta. There are three distinctly different models of the 38/42 - 38/43 and an additional similar-appearing weapon, the 38/44, which has a shorter bolt and does not have the operating spring guide found on the earlier Berettas. This can be noted by the absence of the recoil spring guide rod head protruding through the cap on the end of the receiver. The external differences can be noted in the pictures below. Pakistan, Iraq, and Costa Rica purchased Model 38/44 submachine guns.

An unusual Beretta Model 38A submachine gun, which apparently appeared in prototype form, has an aluminum barrel jacket rather than the multi-perforated barrel jacket. Another Beretta which appeared only in prototype form was the Model 1.

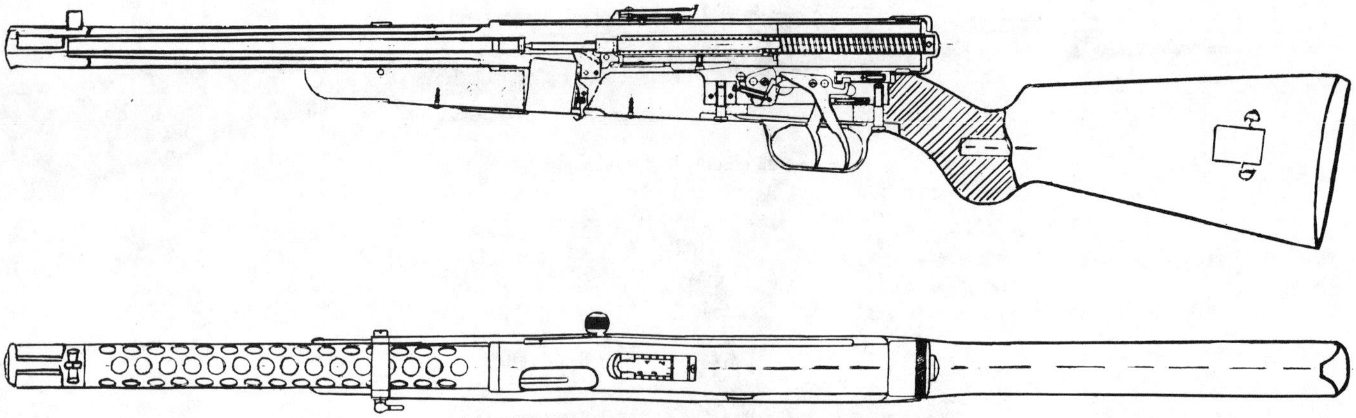

Diagram of the Model 38A from official Italian manual.

9mm Parabellum Beretta Model 38/42, early version.

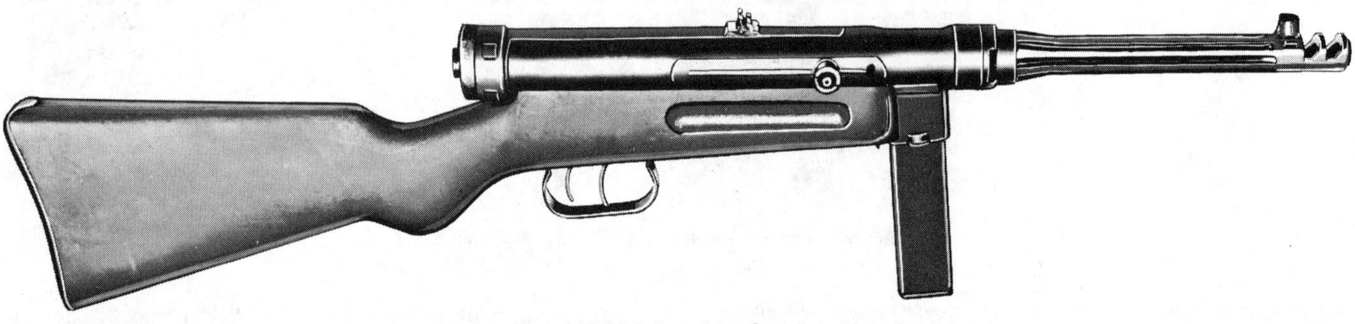

Beretta Model 38/42 Submachine Gun, fluted barrel.

This weapon was developed prior to the Model 38/42 and has a folding stock similar to that of the German MP38; it was apparently an expensive gun to make and was dropped in favor of the Model 38/42. A gun called the 38/44 Special by Beretta closely resembles the Model 1, but does not have the fluted barrel of the Model 1 and does have a cross bolt type safety mounted in the fore-end. This weapon is also called the Model 2.

The Model 38/49, also known as the Model 4, is the current standard submachine gun and is covered in detail separately. The Model 5 Beretta submachine gun is basically the same as the Model 38/49 except that it has a grip safety located in the fore-end.

The Model 5 was introduced in 1957. Another weapon somewhat similar in appearance to the Model 2, and also made only as a prototype, is the Model 4. The Model 4 has a grip safety and a sliding-wire type stock similar to that on the U.S. M3 submachine gun. The weapon illustrated below is similar to the Model 4, but has a wooden fore-end. This weapon is described as a modified 38/44 by Beretta.

Beretta Model 38/44 (above) and Model 38/43 (below), both 9mm Parabellum.

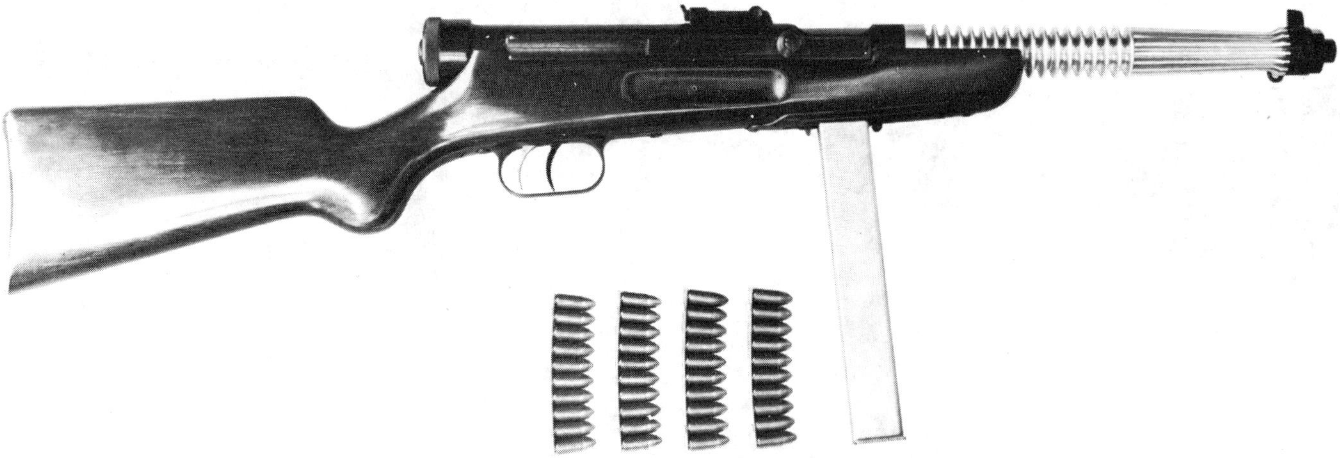

9mm Beretta Model 38A with aluminum barrel jacket.

9mm Beretta Model 38/44 Special (Model 2).

Throughout the 1950s Beretta developed a number of prototype guns, including the Model 6 (1953) and the Model 10 (1957), which were steps in the development of their latest gun, the 9mm Model 12. The Model 12 has a bolt which telescopes the barrel in a fashion similar to the Czech Model 23 and the Israeli Uzi. The Model 12 has been adopted by the Italian Government and is covered in detail later in this chapter.

There were other submachine guns developed in Italy in addition to those developed by Beretta, although Beretta has been the most prolific developer. The Fabbrica Nazionale d'Armi of Brescia developed a submachine gun which was manufactured in limited quantities during the war. This weapon, called the FN A-B Model 1943 is rather unusual for the time in which it was made, since it is machined out of steel forgings. It has a single strut folding steel stock and the magazine can be folded up under the barrel jacket when not in use. The combination muzzle brake compensator resembles that used on the Soviet PPSh41 submachine gun. The Model 1943 is a retarded blowback, selective-fire weapon. Reportedly 7,000 of these weapons were made for the Italian Army.

The TZ45 is another 9mm Parabellum submachine gun which was manufactured in limited quantities toward the end of the war. It was designed by Toni and Zorzoli Giandoso in about 1944. The TZ45 is a relatively crudely-made weapon and has a grip safety which is fitted behind the magazine housing in a fashion similar to the later model Madsen submachine guns. The TZ45 was adopted in a modified form by Burma after the war. It is manufactured in Burma, where it is called the BA52.

The Bernardelli firm produced a modified copy of the Beretta Model 38/42, a few of which were made in 1948 and 1949. This weapon, called the VB, was well made but had no unusual features.

Fabbrica Nazionale d'Armi developed a gun in the 1950's called the X4. The X4 was composed almost entirely of stampings, has a sheet steel retracting slide which partially surrounds the receiver and is used to retract the bolt. A shorter version of the X4 called the X5, which had a barrel a bit over 4.5 inches long, was brought out in the mid 1950s.

The firm of Luigi Franchi has also brought out a family of submachine guns called the LF57. The LF57 is another weapon composed mainly of stampings and is of relatively short length. It is 16.52 inches overall with stock folded, using an 8.1 inch barrel. A semiautomatic only version of this weapon with 16-inch barrel was also offered for sale.

Beretta 9mm Model 5 Submachine Gun.

9mm Beretta prototype, improved 38/44.

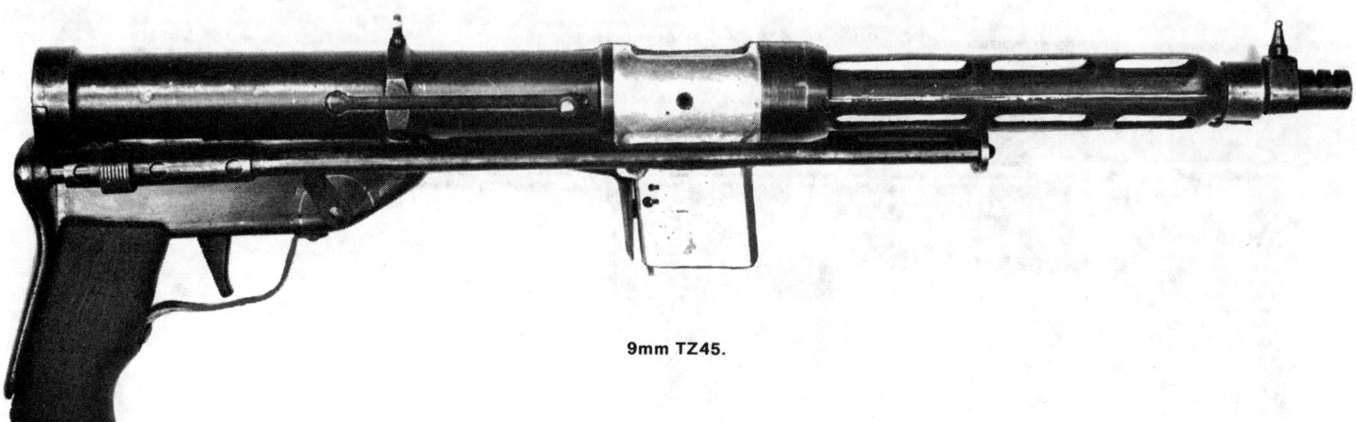

9mm TZ45.

9mm F.N.A.-B Model 1943.

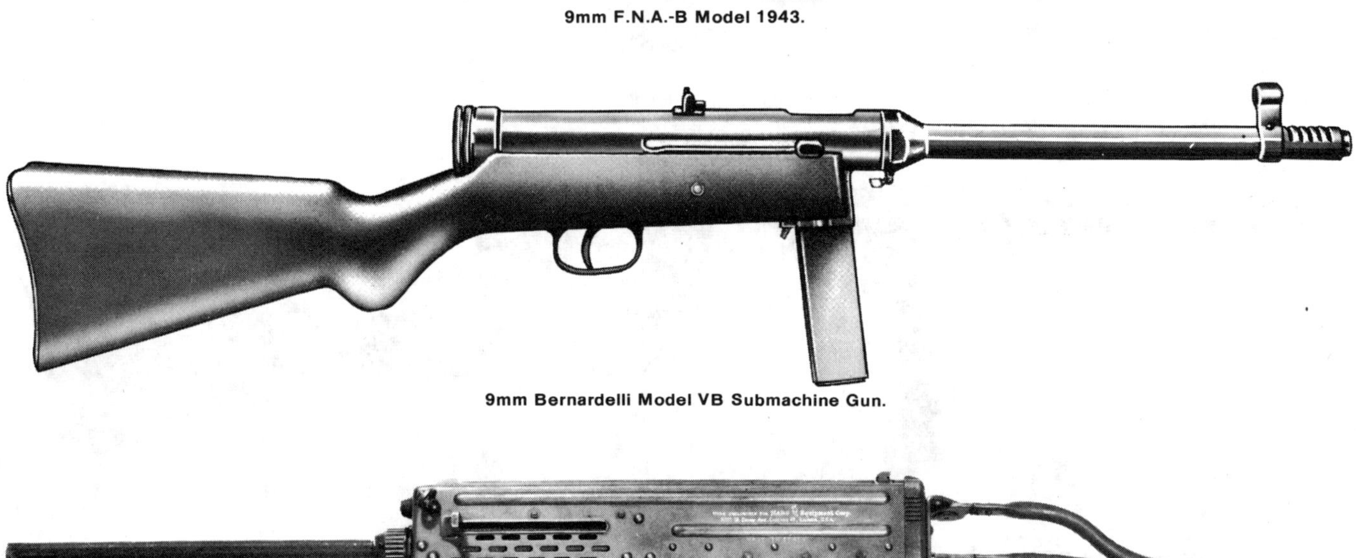

9mm Bernardelli Model VB Submachine Gun.

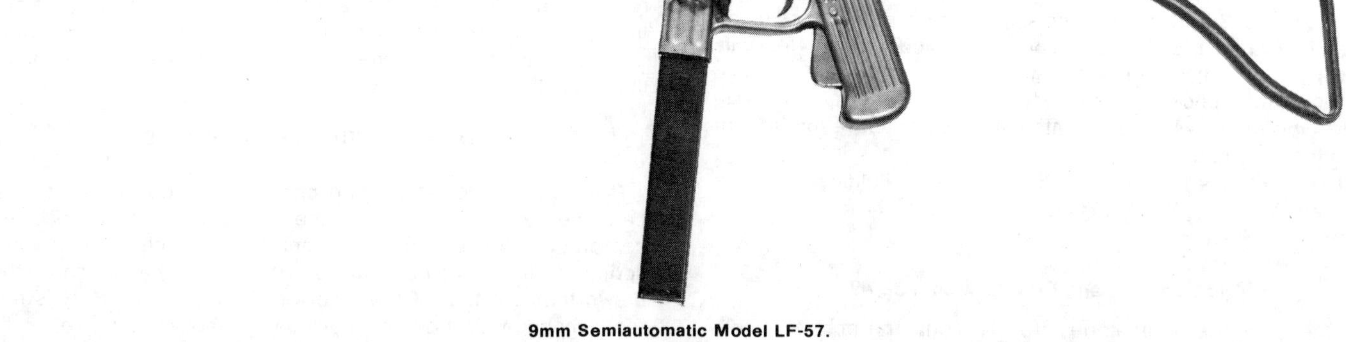

9mm Semiautomatic Model LF-57.

9mm F.N.A. Model X-4.

Beretta 9mm Model 38/49 Submachine Gun.

THE 9mm BERETTA MODEL 38/49 SUBMACHINE GUN MODEL 4

The Model 38/49 is one of the standard submachine guns of the Italian Army. It is a modification of the Model 38/44; like this weapon, it has no recoil spring guide, and a fixed firing pin. The principal difference between the two weapons is the use of a cross-bolt type safety, which is mounted in the stock above the front of the trigger guard.

The 38/49 was sold to Costa Rica, Egypt (with folding bayonet), Yemen, Tunisia, West Germany, Indonesia, Thailand, and the Dominican Republic.

How to Load and Fire the Model 38/49

Insert a loaded magazine into the magazine port. Pull bolt retracting handle (located on the right side of the receiver) to the rear and cock the bolt. Push forward retracting handle (it does not reciprocate with the bolt). The safety is engaged by pushing in from the left side. Pressure on the forward trigger will produce semi-automatic fire and pressure on the rear trigger will produce automatic fire.

How to Field Strip the Model 38/49

Remove the magazine and check weapon to insure that a cartridge has not remained in the chamber. Twist the receiver cap one quarter-turn to the left and remove with the operating spring assembly. The bolt may now be pulled to the rear and withdrawn from the rear of the receiver. Further disassembly is not recommended. To assemble, reverse the above procedure.

THE 9mm BERETTA MODEL 12 SUBMACHINE GUN

The Model 12 comes in both folding metal stock and detachable wooden stock models. As previously pointed out, this weapon has a bolt which telescopes the barrel for about three-quarters of the length of the bolt. The weapon therefore appears to have a much shorter barrel than it really has. An excellent feature of the weapon is that the grooves extend the length of the receiver. The grooves serve as dirt catchers and allow operation even with a considerable amount of dirt in the action. The Model 12 has a grip safety.

How to Load and Fire the Model 12

Remove magazine, if an empty magazine is in the weapon, by pushing magazine catch located at the bottom front of the trigger guard forward and (after pulling magazine down and out of the magazine guide) insert loaded magazine and pull bolt retracting handle, located on the left side of the receiver to the rear. The manual safety is a push button located above the grip; to put the weapon on safe, push the button from left to right. The fire selector is also of the push-button type and is located ahead of the pistol grip. Pushing the button in from the left sets the weapon for semi-automatic; pushing the button from the right produces automatic fire. Push manual safety button from right to left, select type of fire desired, and press trigger.

How to Field Strip the Model 12

Remove the magazine. Pull barrel-locking-nut catch down and unscrew the nut at front of receiver. The barrel and bolt can be removed from the front of the receiver. Push in catch on receiver cap (rear of receiver) and screw off receiver cap. The operating spring can be removed from the rear of the receiver. No further disassembly is recommended.

CHARACTERISTICS OF ITALIAN SERVICE SUBMACHINE GUNS

	Villar Perosa Model 1915	Villar Perosa (O.V.P.)	Beretta Model 1918	Beretta Model 1938A	Beretta Model 1938/42
Caliber:	9mm Glisenti.	9mm Glisenti.	9mm Glisenti.	9mm Parabellum.	9mm Parabellum.
System of operation:	Retarded blowback, full automatic.	Retarded blowback, selective fire.	Retarded blowback.	Blowback, selective fire.	Blowback, selective fire.
Overall length:	21 in.	35.5 in.	33.5 in.	37.25 in.	31.5 in.
Barrel length:	12.56 in.	11 in.	12.5 in.	12.4 in.	8.4 in.
Feed device:	25-round, staggered row, detachable, box magazine.	25-round, staggered row, detachable, box magazine.	25-round, staggered row, detachable, box magazine.	10, 20, 30, or 40-round, staggered column, detachable, box magazine.	20- or 40-round, detachable, staggered row, box magazine.
Sights: Front:	None on some, aperture cut into brace between barrels.	Blade.	Blade.	Blade.	Blade.
Rear:	Y shaped plate with notch.	U notch.	V notch.	Tangent with V notch.	"L" type flip over with U notch.
Weight:	14.30 lb. for twin gun.	Approx 8 lb.	7.2 lb.	9.25 lb.	7.2 lb.
Cyclic rate:	1200 r.p.m.	900 r.p.m.	900 r.p.m.	600 r.p.m.	550 r.p.m.
Muzzle velocity:	1312 f.p.s.	1250 f.p.s.	1275 f.p.s.	1378 f.p.s.	1250 f.p.s.

CHARACTERISTICS OF ITALIAN SERVICE SUBMACHINE GUNS (Cont'd)

	Beretta Model 38/49 (Model 4)	FN A-B Model 1943	TZ 45	Beretta Model 12
Caliber:	9mm Parabellum.	9mm Parabellum.	9mm Parabellum.	9mm Parabellum.
System of operation:	Blowback, selective fire.	Retarded blowback, selective fire.	Blowback, selective fire	Blowback, selective fire.
Overall length:	31.5 in.	Stock fixed: 31.1 in. Stock folded: 20.7 in.	Stock fixed: 33.5 in. Stock folded: 21.5 in.	Stock fixed: 25.39 in. Stock folded: 16.43 in.
Barrel length:	8.4 in.	7.8 in.	9 in.	7.9 in.
Feed device:	20-and 40-round, detachable, staggered row, box magazine.	20-and 40-round, detachable, staggered row, box magazine.	20-and 40 round, detachable, staggered row, box magazine.	20-, 30-, & 40-round staggered row, detachable, box magazine.
Sights: Front:	Blade.	Blade.	Post.	Blade.
Rear:	"L" type flip over with U notch.	V notch.	Aperture.	"L" flip over with U notch.
Weight:	7.2 lb.	7.06 lb.	7.2 lb.	6.6 lb.
Cyclic rate:	550 r.p.m.	400 r.p.m.	550 r.p.m.	550 r.p.m.
Muzzle velocity:	1250 f.p.s.	1225 f.p.s.	1265 f.p.s.	1250 f.p.s.

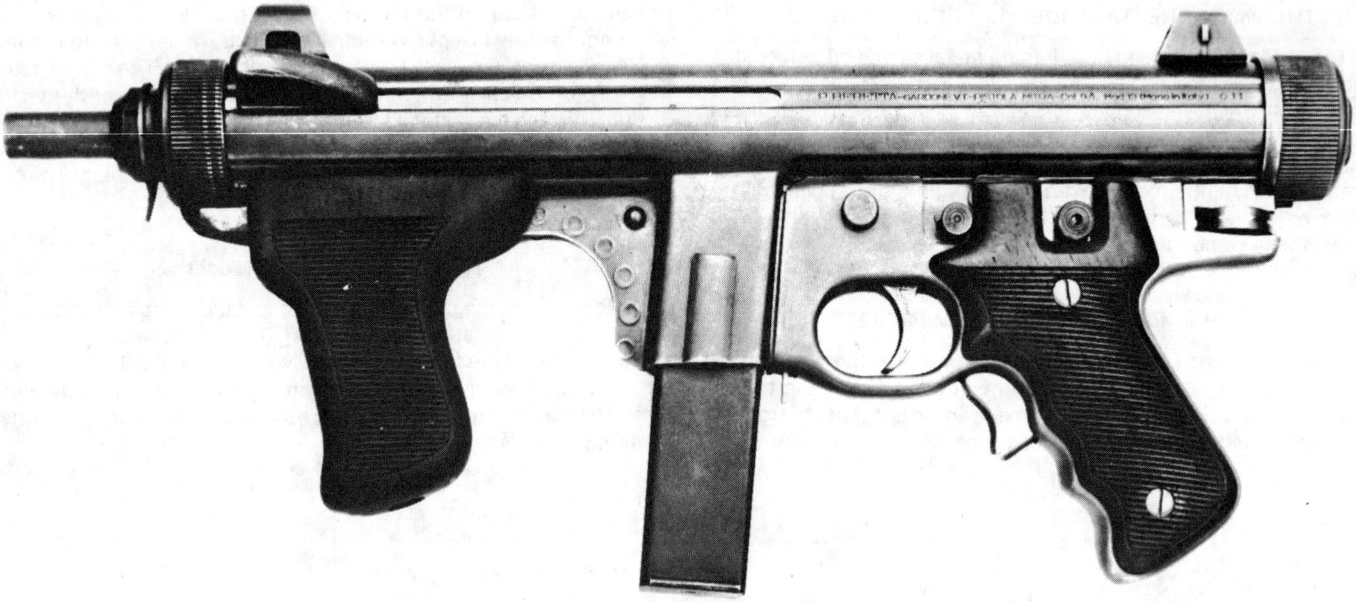

Beretta 9mm Model 12 Submachine Gun.

Beretta Model 12—field-stripped.

ITALIAN MACHINE GUNS

Italy adopted the Maxim gun in 1906 and used two models chambered for the 6.5mm cartridge, the Model 1906 and Model 1911. Apparently few of these guns were made, as they rarely appear in photos of Italian troops in World War I. The 6.5mm water-cooled Revelli Model 1914 was the first Italian-designed gun to appear on the scene in quantity. This retarded blowback gun was unusual in that it was fed with a magazine composed of ten compartments, each holding five rounds. To provide for the lack of slow initial extraction—a slight bolt turning movement to loosen the case before extraction to the rear begins—the cartridge cases were lubricated by an oil pump built into the receiver. The Revelli was manufactured by Fiat and is frequently known by that name.

During World War I, Italy purchased quantities of the Colt Model 1914, Browning "potato digger" in 6.5mm to supplement their machine gun supply. The French supplied the Italians with large quantities of the M1907 F St. Etienne heavy machine gun chambered for the 8mm Lebel cartridge. At the conclusion of World War I, the Italians received tremendous quantities of 8mm (8 x 50R) Schwarzlose Model 1907/12 from the Austrians as part of their share of war reparations. These weapons were still used by the Italians in World War II.

The Breda 6.5mm Model 1924 was one of the first Italian light machine guns. It is, except in details such as the stock and use of a thumb-operated trigger, basically the same as the 6.5mm Model 1930 Breda. The Model 1930 was the standard Italian light machine gun in World War II. A few thousand of the Model 1924 Breda were made, as well as a few thousand 6.5mm Fiat SAFAT light machine guns—a light version of the Revelli Model 1914. A number of machine guns which were never made in quantity (such as the Brixia) were introduced during the twenties and thirties. Part one, the historical section of this book, contains data on these weapons.

The 6.5mm Model 1930 Breda was an unusual weapon in many respects. It is a retarded (delayed) blowback-operated gun with recoiling barrel, as is the Model 1914 Revelli. This weapon has a massive bolt with multiple-locking lugs and an oil pump which lubricates the cartridges prior to chambering. The magazine is permanently attached to the side of the gun and is swung forward to be loaded with a 20-round horse-shoe type charger. The chargers may be brass or cardboard.

This weapon was among the first to be introduced with a quick-change barrel. All in all, the Breda Model 30 was hardly a satisfactory weapon. Any weapon which requires lubricated cases

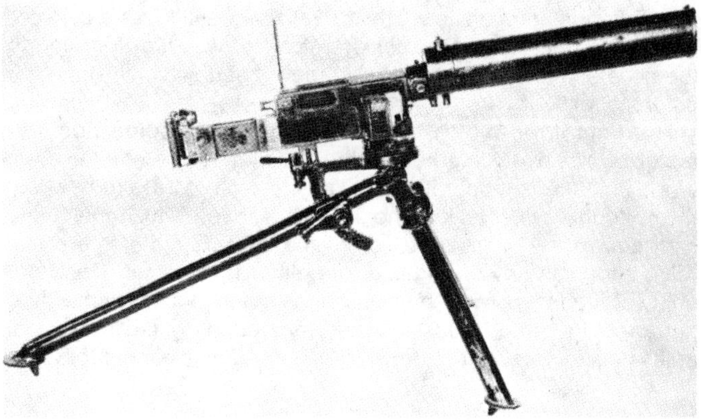

Model 1914 Revelli (Fiat) 6.5mm Machine Gun.

Colt 6.5mm Model 1914.

6.5mm Breda Model 1924 Light Machine Gun.

will be unreliable in sandy or dusty conditions, because the lubricant will pick up the foreign matter and introduce it into the action. This weapon was sold in 7.92mm to Portugal and Lithuania. It was made in limited quantities in 7.35mm as the Model 38 for Italy.

In 1931, the 13.2mm Breda antiaircraft and tank machine gun appeared. This weapon, unlike earlier Italian machine guns, had nothing in its design which would make it "peculiar" from a machine-gun point of view. The 13.2mm Model 31 was a conventional, gas-operated, magazine-fed gun and was the first step in the development of the best Italian developed machine gun—the 8mm Model 37.

The 8mm Model 37 Breda, although it had some peculiar features, was by far the best of the Italian machine guns used in World War II. A gas-operated gun, it is fed by feed trays from which the cartridge is pushed into the chamber and into which the empty cartridge case is reinserted before the feed tray is ejected. The reason for this tidy arrangement is rather mysterious, since the utility of feed trays with empty cartridge cases to a heavy machine gun crew in battle is possibly only exceeded by the proverbial utility of deep freezes to the Eskimos or vacuum cleaners to South Sea Islanders without electricity! Be that as it may, the Breda Model 37 was a very reliable weapon and was used in many varying climatic conditions with good results. The Model 37 was used after World War II and there was some discussion about converting the weapon to caliber .30, but this idea was apparently dropped since there is a Model 55 tracer cartridge for the weapon. Although the weapon is gas operated, the cartridges are lubricated by an oil pump since no provision is made for slow initial extraction. The Breda has been replaced in the Italian service by the 7.62mm NATO MG 42/59. The Breda Model 37 was made for Portugal in 7.92mm and is known by them as the Model 1938.

In 1938, a tank version of the Model 37, chambered for the same Breda 8mm Model 35 cartridge, fed by a top-mounted box magazine, was introduced and adopted.

Aircraft versions of the Breda in 7.7mm (.303 British), 7.92mm, and 12.7mm were also produced. The 12.7mm gun was also produced as an antiaircraft gun. The Italian government used the 7.7mm and 12.7mm versions of this gun. It might be pertinent to point out that the Italians used seven different machine-gun cartridges during World War II in addition to four different pistol and submachine gun cartridges. Pity the poor Italian ammunition

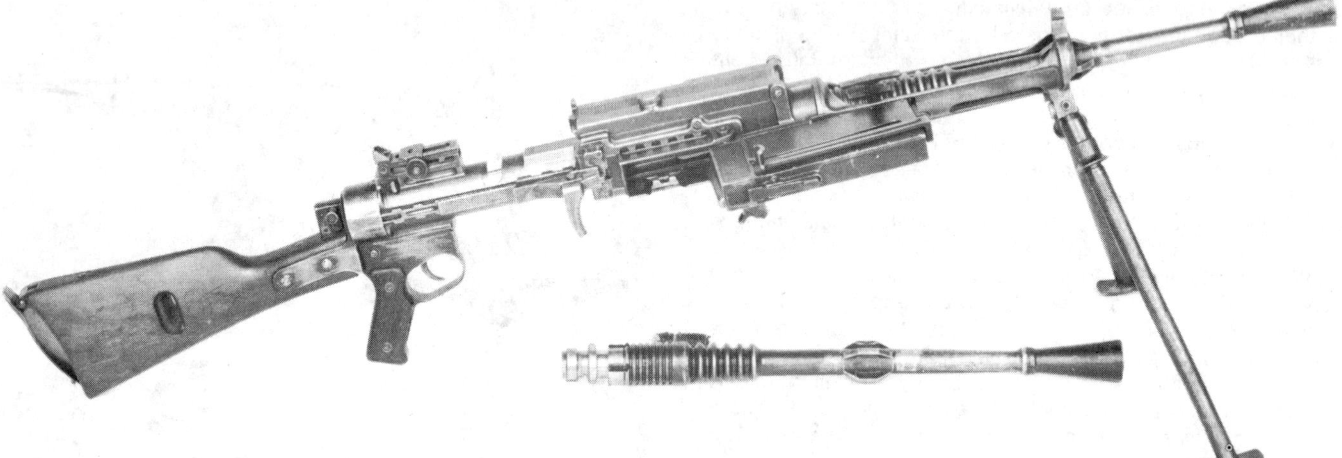

Breda 6.5mm Model 30 Machine Gun.

13.2mm Breda Model 31 Heavy Machine Gun.

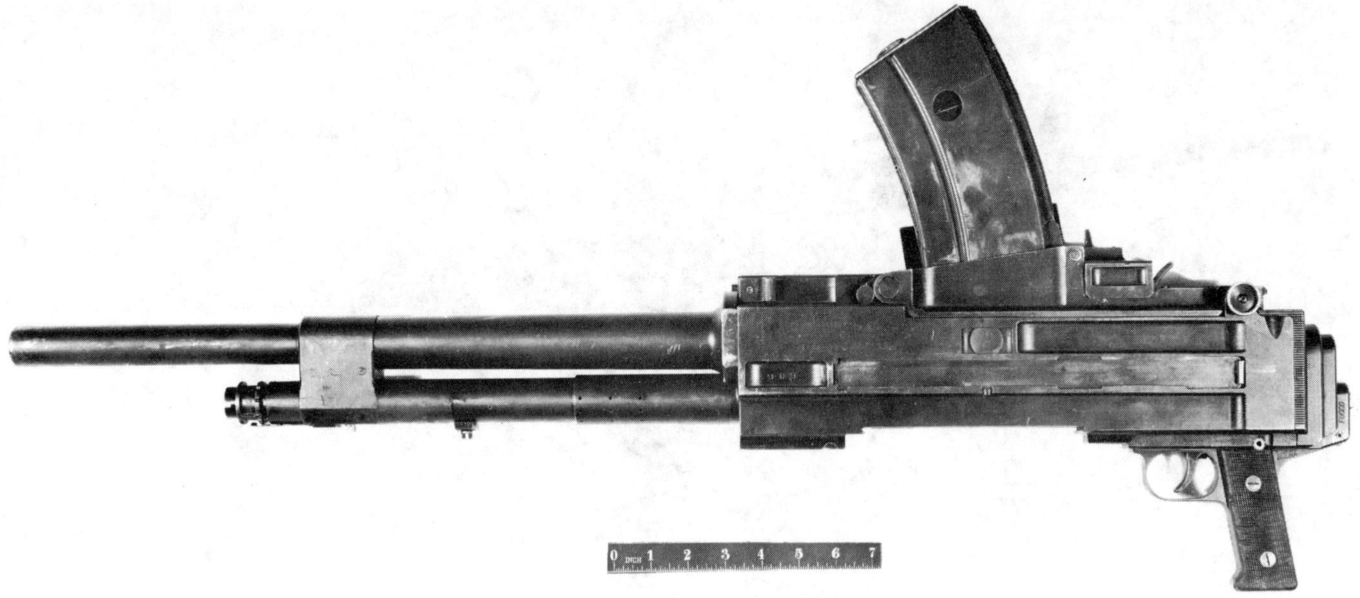

8mm Breda Model 38 Tank Gun.

8mm Breda Model 37 Heavy Machine Gun.

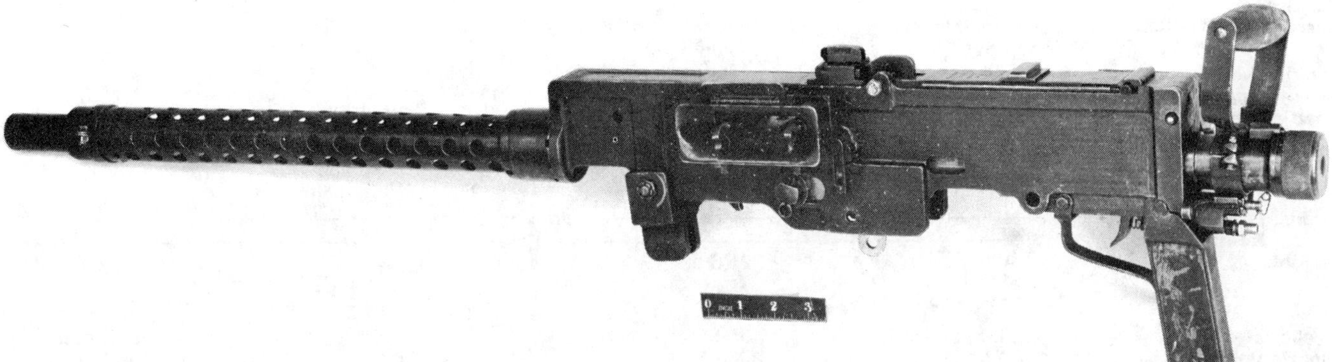

7.7mm Breda Aircraft Gun.

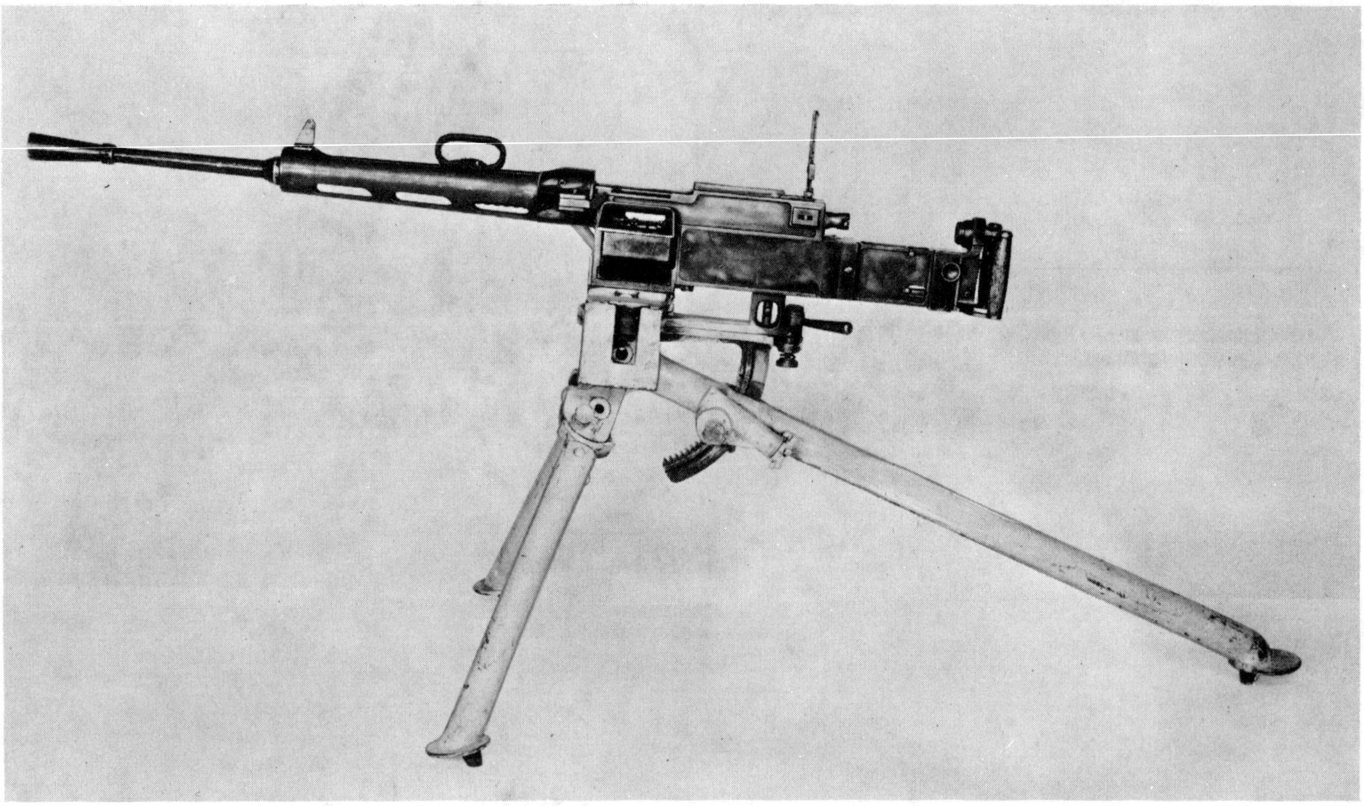

8mm Fiat (Revelli) Model 35.

supply officer of those days!

The Fiat (Revelli) Model 1935 8mm heavy machine gun is essentially a modified Revelli Model 1914; many of the Model 35 Fiats were actually converted Model 1914 weapons. There were some of the weapons which were of new manufacture. The Model 1935 is air cooled and has a fluted chamber rather than an oil pump (it was also a retarded blowback). It was fed by a non-disintegrating belt rather than by the compartmented magazine of the Model 1914. The Model 35 was not a very successful gun; it was actually worse than the Model 1914. The ammunition still required lubrication for proper function and the weapon, unlike the Model 1914, fired from a closed bolt with resultant "cook-offs" after periods of sustained fire. This weapon as the 1914 had an operating rod which reciprocated outside to buff against a pad mounted on the front of the back plate. This made clearing a gun with a cartridge in a hot barrel a hazardous proposition and may account for the nickname it received during World War II—the "knucklebuster." Somewhat needless to say, this weapon is no longer in service.

After World War II, Italy was supplied with United States and British machine guns. The current standard Italian machine gun is the 7.62mm NATO MG42/59 which is called the Mitragliatrice Leggere 42/59 in Italy. This weapon is covered in detail in the chapter on Germany. The MG42/59 is currently being manufactured in Italy.

ITALIAN SERVICE MACHINE GUNS

	Model 1914 Revelli	Model 1930 Breda*	8mm Breda Model 37	8mm Fiat (Revelli) Model 35
Caliber:	6.5mm.	6.5mm.	------------8mm (Italian Model 35 Cartridge)------------	
System of operation:	Delayed blowback, selective fire.	Delayed blowback, automatic only.	Gas, automatic only.	Delayed blowback, selective fire.
Length overall:	46.5 in.	48.5 in.	50 in.	49.75 in.
Barrel length:	25.75 in.	20.5 in.	Approx. 25 in.	25.75 in.
Feed device:	50-round "mouse-trap" type magazine.	20-round, non-detachable magazine, fed by changers.	20-round strip.	300-round non-disintegrating belt.
Sights: Front:	---------- Barley corn ----------			
Rear:	----------Leaf----------			
Weight:	37.5 lb. (less water) tripod—49.5 lb.	22.75 lb.	42.8 lb. Tripod—41.5 lb.	39.75 lb. Tripod—41.5 lb.
Cyclic rate:	450-500 r.p.m.	450-500 r.p.m. .	450 r.p.m.	500 r.p.m.
Muzzle velocity:	2080 r.p.s.	2063 f.p.s.	2600 f.p.s.	2600 f.p.s.

*7.35mm Model 38 has essentially the same characteristics.

38 Japan

The Japanese Ground Self-Defense Forces (JGSDF) were initially equipped completely with United States World War II type small arms and are still, for the most part, equipped with these weapons. U. S. caliber .45 Model 1911A1 pistols and M3A1 submachine guns, U.S. caliber .30 M1 and M2 carbines, caliber .30 M1 rifles, Model 1918A2 Browning Automatic rifles, and U.S. caliber .30 and caliber .50 Browning machine guns are all in service. In addition, however, the caliber 7.62mm NATO Type 64 rifle, and caliber 7.62mm NATO Type 62 light machine gun, all of Japanese manufacture, are being introduced into the JGSDF.

JAPANESE PISTOLS

The firm of Shin Chuo Kogyo K.K. (New Central Industrial Co. Ltd. of Tokyo) has developed a number of pistols since World War II. These are: a modified copy of the U.S. caliber .45 M1911A1, called the Type 57 New Nambu; a smaller caliber .32 automatic, the Type 57B New Nambu; and a caliber .38 special revolver similar to the Smith & Wesson, which is called the New Nambu Type 58.

JAPANESE 9mm TYPE 57 NEW NAMBU AUTOMATIC PISTOL

Characteristics of Type 57 Pistol

Caliber: 9mm Parabellum (also made in .45 cal.).
System of Operation: Recoil operated, semi-automatic only.
Weight: 2.12 lbs.
Length, overall: 7.8 in.
Barrel length: 4.6 in.
Feed device: 8-round, single line, detachable box magazine.
Muzzle Velocity: 1148 f.p.s.

This weapon which is called the "New Nambu" is a modified copy of the U.S. M1911 automatic pistol. It does not have a grip safety and has its magazine catch mounted on the bottom of the grip, but is otherwise much the same as the U.S. pistol. It is manufactured by Shin Chuo Kogyo K.K. (New Central Industrial Company, Ltd., of Tokyo). Loading, firing, field stripping, and functioning are the same as the U.S. M1911 pistol with the exception of the features noted above. The pistol is well made and of very good finish. This pistol was produced only as a prototype.

CAL. .32 TYPE 57B NEW NAMBU AUTOMATIC PISTOL

Characteristics

Caliber: .32 ACP (7.65mm Browning).
System of operation: Blowback, semi-automatic only.
Weight: 1.3 lbs.
Length, overall: 6.3 in.
Feed device: 8 round, detachable in-line box magazine.

This weapon is also called a New Nambu. It is basically a Browning type blowback pistol and loading, firing, field stripping and functioning are the same as the Browning M1910 pistol.

A modification of the Model 57, the Model 57A, has been introduced by Shin Chuo Kogyo. The Model 57A differs from the Model 57 in having a button type magazine catch on the left rear of the grip near the bottom, larger stocks, slightly different shaped receiver and slide stop. Although there were indications quite a while ago that the Japanese Self Defense Forces were going to adopt the Model 57, it was produced only as a prototype and as of mid-1968, the Model 57A also appeared only in prototype form. It is likely that the Model 57A in time will replace the .45 M1911A1 automatic.

9mm Parabellum Model 57A "New Nambu" Pistol.

JAPANESE RIFLES

The JGSDF is mainly equipped with U.S. caliber .30 M1 rifles and carbines. A limited number of 7.7mm Type 99 rifles were in service until recently. The Howa Kogyo K.K. Ltd. of Nagoya started development of a 7.62mm NATO rifle for the JGSDF in 1957. In 1962 the Type R6E (modified) was developed; this weapon was adopted as the Type 64.

THE 7.62mm NATO TYPE 64 RIFLE

The Type 64 is the end result of a number of prototypes developed by Howa Machinery Ltd. Prior to adoption, the rifle was known as R6E (modified). Type 64 was specifically developed with the needs of small-statured Asiatic troops in mind, but is chambered for a high muzzle energy cartridge. The Type 64 normally uses a low-powered Japanese issue 7.62mm NATO cartridge, which has a purple or lavender bullet tip to distinguish it from full powered 7.62mm NATO cartridges. The weapon has an adjustable gas regulator, which is set for a smaller intake of gas if full powered 7.62mm NATO cartridges are to be used, than it is for the Japanese reduced charge cartridges. The gas regulator can be set to shut off the gas completely when launching rifle grenades from the combination muzzle brake grenade launcher attached to the end of the barrel. The Type 64 has a folding-type shoulder rest fitted to the top of the butt, as does the U.S. M14 rifle; but unlike the M14, all Type 64 rifles are equipped with bipods.

Characteristics of the Type 64 Rifle

Caliber: 7.62mm NATO.
System of operation: Gas, selective fire.
Length overall: 38.97 in.
Barrel length: 17.71 in.
Feed device: 20-round, detachable, box magazine.
Sights: Front: Hooded folding blade.
Rear: folding aperture adjustable for windage.
Weight: 9.5 lb. w/o magazine.
Cyclic rate: 450-500 r.p.m.
Muzzle velocity: 2650 f.p.s. with full-charge cartridge.
2347 f.p.s. with reduced-charge cartridge.

7.62mm R6A Rifle; one of the prototypes of the Type 64 Rifle.

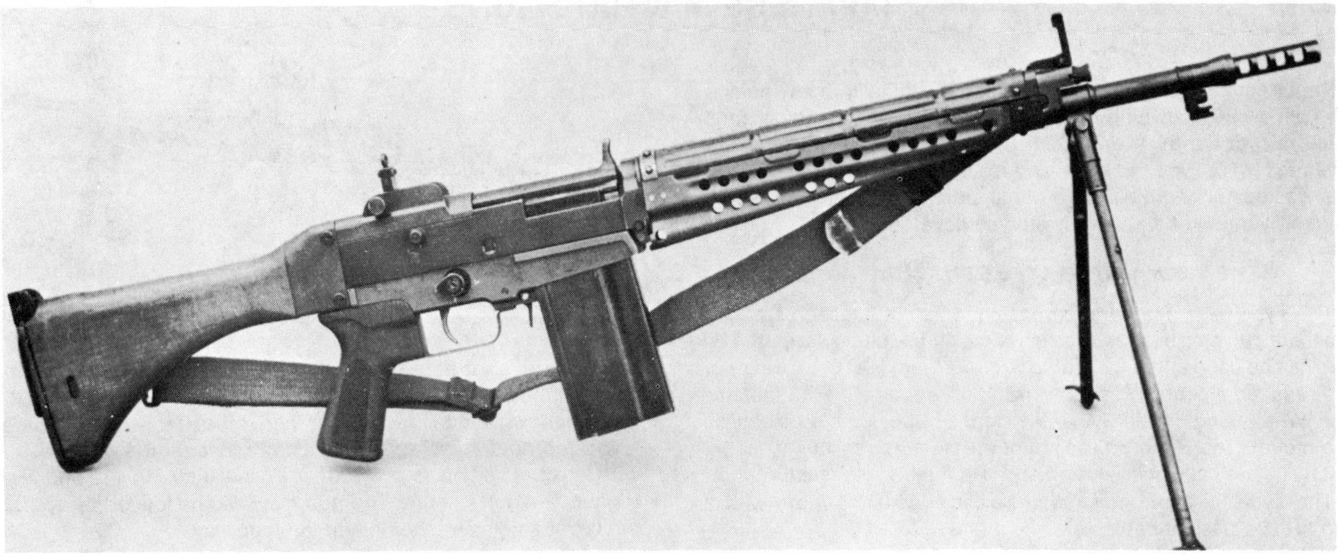

7.62mm Type 64 Rifle.

How to Load and Fire the Type 64 Rifle

The magazine can be removed, if in weapon, by pushing in on the magazine catch, located behind the magazine port, and pulling magazine down and out. The safety/selector switch is located over the trigger on the right-hand side of the receiver. Insert a loaded magazine in the magazine port, pushing up until the magazine catch engages. Select type of fire desired and pull operating handle—top front of receiver—to the rear and release it, chambering a cartridge.

Note on Method of Operation

The bolt of the Type 64 is of the tipping type and is held by a bolt carrier which cams it down into and up out of the locked position in a fashion quite similar to that of the Soviet 7.62mm SKS carbine. The bolt carrier is forced to the rear by a piston rod which is mounted above the barrel; the pistol is spring loaded and returns to the battery position after firing. The operating spring and guide, which return the bolt to the locked position, are mounted in a tunnel in the top of the bolt carrier.

JAPANESE SUBMACHINE GUNS

The Shin Chuo Kogyo Company is currently developing a submachine gun chambered for the 9mm Parabellum cartridge. It is a blowback operated, selective fire weapon with folding steel stock. It is possible that this weapon may be adopted by the Japanese Self Defense Forces.

SHIN CHUO KOGYO SUBMACHINE GUN

This is a conventional blowback operated weapon made mainly of stampings and fabrications.

Note that this gun has a grip safety attached to the magazine housing in a manner similar to that of the Madsen and the Italian TOZ. The ejection port has a cover similar to the U.S. M3A1 submachine gun.

Characteristics of the Shin Chuo Kogyo Submachine Gun

Caliber: 9mm Parabellum
System of operation: Blowback
Length overall: Stock folded--30 inches
Stock fixed--19.75 inches
Barrel length: 6 or 7.3 inches
Feed device: 30-round, detachable box magazine
Sights: Front: Hooded post
Rear: "L" type, graduated for 100 and 200 meters
Weight: 8.8 lb., loaded
Cyclic rate: 600 r.p.m.
Muzzle velocity: 1181 or 1220 f.p.s.

9mm Parabellum Shin Chuo Kogyo Submachine Gun.

JAPANESE MACHINE GUNS

Requirements for a general purpose machine gun to replace the caliber .30 Browning M1919A4 and M1917A1 machine guns were laid down by the JGSDF in 1956. The current gun, the 7.62mm NATO Type 62, is the end result of this requirement. The gun was developed by Nittoku Metal Industry Co. and represents an improvement on the third prototype developed.

THE 7.62mm NATO TYPE 62 MACHINE GUN

The Type 62 is a general purpose gun, i.e., it is used as a light machine gun on a bipod and a heavy machine gun on a tripod. The gun can be used either with full charge 7.62mm NATO cartridge or the reduced charge 7.62mm NATO. It uses the U.S. M13 metallic link. When used on the U.S. M2 tripod, a buffer unit is placed on the mount. There are no really unusual features to this weapon, but it is a basically good design and is of quality manufacture.

The Type 62 has a quick change barrel which is fitted with a prong-type flash suppressor.

7.62mm NATO Type 62 Machine Gun.

Characteristics of the 7.62mm NATO Type 62 Machine Gun

Caliber: 7.62mm NATO.
System of operation: Gas, automatic fire only.
Length overall: 47.3 in.
Barrel length: 23.6 in.
Feed device: Disintegrating metallic link belt.
Sights: Front: Hooded blade.
Rear: Folding leaf with aperture, adjustable for windage.
Weight: 23.6 lb.
Cyclic rate: 650 r.p.m. with full-charge cartridges.
600 r.p.m. with reduced-charge cartridges.
Muzzle velocity: 2800 f.p.s. with full-charge cartridges.
2530 f.p.s. with reduced-charge cartridges.

Loading and Firing the Type 62 Machine Gun

If the bolt is to the rear, pull trigger and allow bolt to run forward. Insert end link of belt into feedway so that feed pawls grasp the link. Pull bolt retracting handle, located on the right side of the receiver down and to the rear and then return it forward. The safety is located on the trigger guard; when forward, it is on "fire" and when to the rear it is on "safe." Push safety forward and press trigger. The gun will fire. The gun may also be loaded by opening the cover and laying the belt on the feed way.

How to Field Strip the Type 62 Machine Gun

Open cover and pull bolt to rear to insure that there are no cartridges in the weapon. Close cover and allow bolt to run forward. Push stock retaining pin out of the receiver to the left. Remove stock assembly with recoil spring guide, and buffer assembly from rear of receiver. Raise the cover, releasing the barrel-locking plunger, draw bolt handle fully to the rear, and remove. Draw slide and bolt assembly from the rear of the receiver. Align carrying handle dismounting stud with dismounting notch in barrel-locking ring; push barrel-locking plunger rearward and turn carrying handle until dismounting guide lines are aligned, then pull barrel forward out of the receiver. Remove gas piston return spring from receiver. Assemble in reverse order.

JAPANESE WORLD WAR II SMALL-ARMS AMMUNITION

As the reader who will read through the chapter which follows on Imperial Japanese World War II small arms must realize, the Japanese small-arms ammunition situation during World War II bordered on anarchy. Although outside the scope of this book, it should be noted that a similar situation existed in the mortar and artillery ammunition areas as well.

In the pistol and submachine gun area, things were simple; there were only two cartridges: the 9mm Type 26 revolver cartridge and the 8mm Type 14 pistol and submachine gun cartridge. But real anarchy reigned in the rifle and machine gun ammunition situation. The rifle and machine gun cartridges used by the Japanese Army, Navy, and Air elements in World War II are listed below, together with the weapons in which they were used.

Cartridge	Type of Rim	Using Weapons
6.5mm Type 38	Semi-rimmed	Type 38 and 97 rifle, Type 38 and 44 carbine, Type 3, 11, 91, 96, and 38 (Hotchkiss) machine guns.
7.7mm Type 92	Semi-rimmed	Type 92 machine gun (Army).
7.7mm Type 99	Rimless	Type 99 and 2 rifle, Type 92, 99, 1 and 97 machine guns.
7.7 Type 92 (Navy only, copy of .303 British)	Rimmed	Type 89, Type 92, and Type 97 Navy machine guns.
7.92mm.	Rimless	Type 98 and 100 machine guns, Type 1 (Navy copy of Type 98).
12.7mm	Semi-rimmed	Type 1 (HO-103) machine gun.
13.2mm Type 93	Rimless	Type 93 and Type 3 machine guns.

39

Imperial Japanese World War II Weapons

The Imperial Japanese Forces had a rather heterogenous assortment of weapons during World War II, especially in their collection of automatic weapons. Japan, like Italy, was caught in the midst of a change of caliber of rifles and machine guns when the war began. This added considerably to the logistic problems of a country which already was using four different rifle caliber cartridges in machine guns. In addition, the Japanese adopted the German 7.92mm MG15 aircraft gun as the Type 98 and had 12.7mm aircraft guns and 13.2mm ground guns as well.

One of the basic reasons for this eminently undesirable situation was the fact that the Army, Navy, and Air Forces proceeded along on their merry way, adopting whatever caliber they wanted without any higher authority forcing standardization of weapon and/or ammunition. In addition, use was made of captured materiel; thus 7.92mm FN-made Browning Automatic Rifles captured from the Chinese were used against United States forces in the Philippines. Other captured weapons which were used were: Czech- and Chinese-made 7.92mm ZB26 and ZB30 light machine guns; Dutch 6.5mm Madsen guns; British caliber .303 Bren and Lewis guns; and various Dutch, British, and Chinese rifles.

Japanese World War II weapons are no longer standard in the armies of any major power, but they will still be found in the hands of the Chinese Communist Militia and have been used by the Viet Cong in South Vietnam. Their future use is doubtful since, unlike the standard German World War II small arms, they are not chambered for cartridges which are still extensively manufactured throughout the world. Limited manufacture of Japanese rifle and machine-gun ammunition was carried on in Communist China as late as the Korean War.

JAPANESE PISTOLS

The Japanese Army adopted a 9mm revolver, the Type 26, in 1893. This weapon is a break-open type similar to the Smith & Wessons of the period, but has lock work similar to the Austrian Rast and Gasser. This revolver is unusual in two respects: it is double-action only (like the later models of the .38 Enfield); and is chambered for a "one-of-a-kind" 9mm rimmed pistol cartridge. Although this revolver was replaced as the standard arm by an automatic in 1914, it was used in quantity during World War II. The only good feature about the weapon is a hinged side plate which allows for easy exposure of the lock work.

The Nambu 8mm automatic pistol which appeared in 1904 introduced the bottle-necked 8mm Japanese pistol cartridge. The Nambu may never have been a standard Japanese service arm, but it was used by Japanese troops during World War II and was also exported. The pistol was developed by Colonel Kijiro Nambu and manufactured by the Kayoba Factory Co. Ltd.

Japanese Revolver 9mm Type 26 (1893).

The 1904 Nambu is a recoil-operated weapon fitted with a grip safety which is mounted on the front of the grip below the trigger guard. This weapon is usually found with a slot cut in the rear of the pistol grip to accommodate a shoulder-stock holster. A smaller version of this weapon chambered for a 7mm bottle-necked cartridge is known as the "Baby Nambu." This was apparently a non-standard weapon although it may have had limited issue in the Air Force. The "Baby Nambu" has a "V"-notch type rear sight, rather than the ramp-type rear sight of the 8mm Nambu.

In 1925, a modified form of the 8mm 1904 Nambu pistol was adopted by the Japanese Army as its standard pistol. The principal differences between the 1925 Nambu, called Type 14 by the Japanese and the 1904 arm are as follows:

(1) The 1904 Model has a grip safety, the Type 14 has a manual safety.
(2) The 1904 Model has no magazine safety, the Type 14 does.
(3) The 1904 Model has a tangent-type sight, the Type 14 has a fixed notch.

A modification of the Type 14 appeared somewhat later. This modified pistol is easily distinguishable from the early Type 14 pistols by its enlarged trigger guard, which allows use with heavily-gloved hands. Additional modifications are: (1) the firing pin spring guide is less elaborate than that of the early Type 14; and (2) the modified pistol has a spring mounted on the lower front of the grip which engages a cut-out in the magazine to hold it more securely in place. Although the magazines of the early Type 25s do not have the cut-out for the spring, they will function in the modified type 14 pistols.

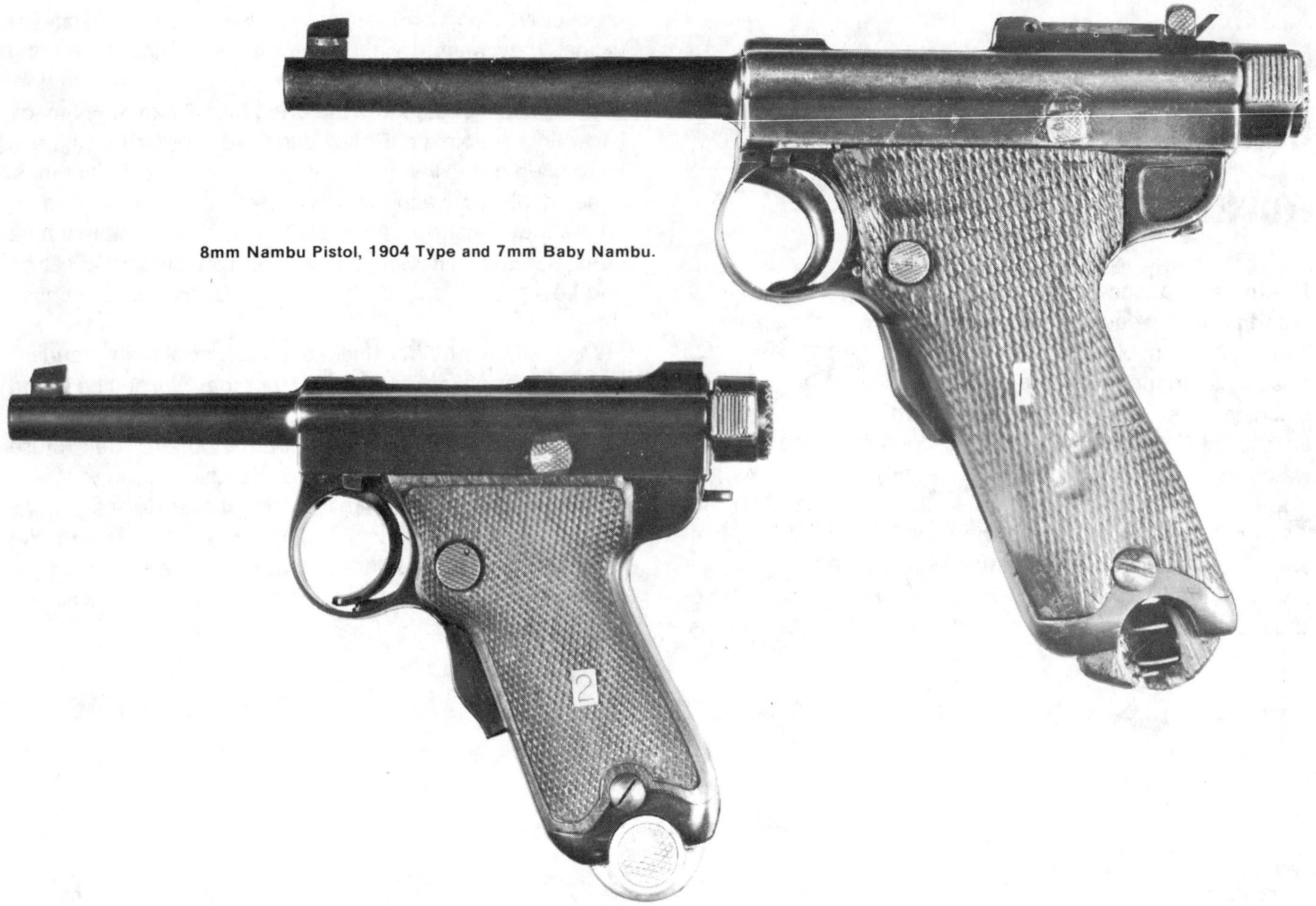

8mm Nambu Pistol, 1904 Type and 7mm Baby Nambu.

8mm Nambu Type 14 Pistol.

Modified 8mm Type 14 Pistol.

In 1934, a new 8mm pistol was introduced in Japan. This weapon was apparently intended principally for export sale, but was used as a service pistol during World War II. This pistol, called the Type 94, is mainly distinguished by having an externally-mounted extension bar sear; it is recoil operated; most specimens show evidence of poor manufacture.

In 1942 work was started on the design of a simplified pistol chambered for the 8mm pistol cartridge. The result of this work was the Type II pistol; approximately 500 being made at Nagoya Arsenal by the end of the war. Although this weapon has been reported as being recoil operated, stripped views indicate that it is blowback operated. In any event, if any are in existence today, they are collectors' items.

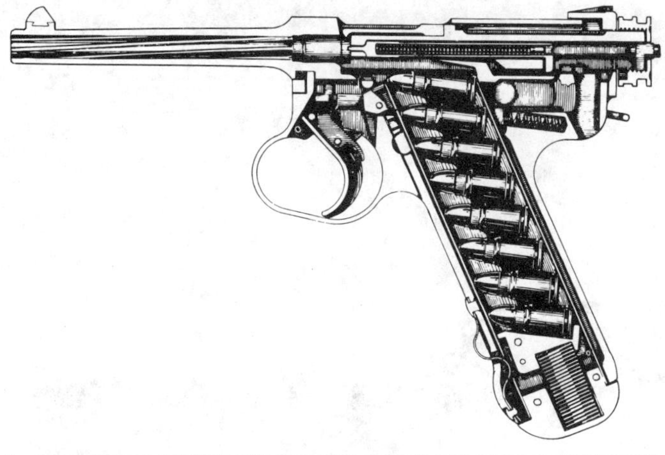

Japanese Type 14 (1925) with magazine loaded and action in full forward position.

8mm Type 94 Pistol.

CHARACTERISTICS OF JAPANESE WORLD WAR II PISTOLS

	Type 26	Type 14	Type 94	1904 Nambu*	"Baby Nambu"
Caliber:	9mm.	8mm.	8mm.	8mm.	7mm.
System of operation:	Double-action only, revolver.	Recoil operated semiautomatic.	Recoil operated semiautomatic.	Recoil, semi-automatic.	Recoil, semi-automatic.
Length overall:	9.4 in.	9 in.	7.2 in.	9 in.	6.75 in.
Barrel length:	4.7 in.	4.7 in.	3.8 in.	4.7 in.	3.25 in.
Feed device:	6-round revolving cylinder.	8-round, in-line detachable box magazine.	6-round in-line detachable box magazine.	8-round in line detachable box magazine.	7-round, in-line detachable box magazine.
Sights: Front:	Rounded inverted V.	Barley corn.	Barley corn.	Barley corn.	Barley corn.
Rear:	"V" notch.	Undercut notch.	Square notch.	Tangent w/notch.	"V" notch.
Weight:	2 lb.	2 lb.	1.68 lb.	1.93 lb.	1.43 lb.
Muzzle velocity:	634 f.p.s.	1065 f.p.s.	1000 f.p.s.	1065 f.p.s.	1050 f.p.s.

*Although these pistols were apparently never service pistols, they were used by Japanese Forces.

JAPANESE RIFLES

The first rifle of Japanese design to appear was the Type 13 11mm Murata, a single-shot bolt-action weapon. The Murata was followed by the 8mm Type 20 (1887) rifle and Type 27 Carbine. These were magazine arms chambered for an 8mm cartridge of Japanese design. In 1897 the first Arisaka rifle, the 6.5mm Type 30, appeared and was the standard Japanese rifle in the Russo-Japanese War of 1904-05. A carbine version of this weapon appeared in the same year. The Type 30 introduced to Japan the Mauser action in a modified form and the 6.5mm semi-rimmed cartridge.

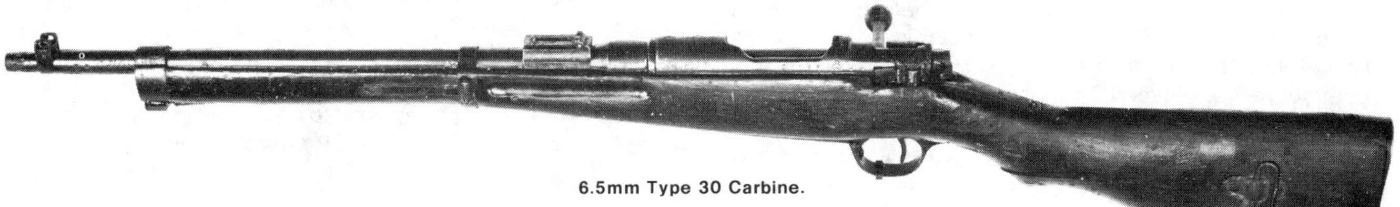

6.5mm Type 30 Carbine.

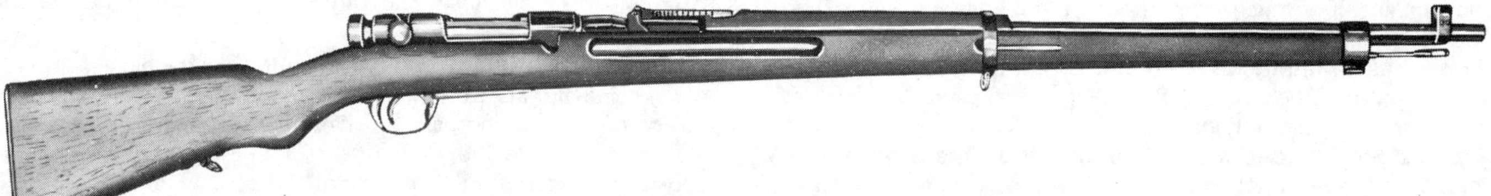

6.5mm Type 38 Rifle.

6.5mm Type 38 Carbine with hinged stock.

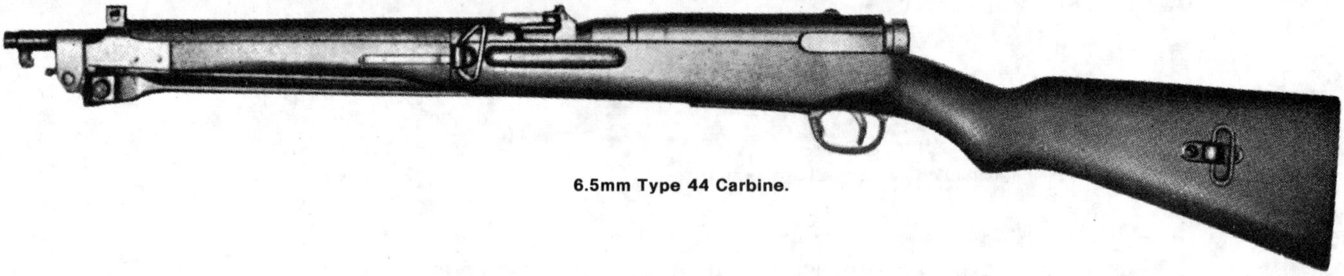

6.5mm Type 44 Carbine.

6.5mm Type 97 Sniper Rifle.

The 6.5mm Type 38 (M1905) was one of the two principal rifles used during World War II. The Type 38 differs from the 98 Mauser action only in the firing pin safety arrangement. The Type 38 has a large knob on the rear of the bolt which is pushed in and rotated to put the weapon on "safe" rather than the flag-type safety mounted on the bolt sleeve as on the Mauser. The Type 38 also has a sliding bolt cover similar to that of the Lee Metford, but this bolt cover was usually removed from rifles by soldiers in the field because of the amount of noise they made in operation.

The Ariska cocks on forward movement of the bolt in a fashion similar to that of the Enfield. The Type 38 carbine is a shortened version of the rifle and has a bayonet lug for the standard knife bayonet. A number of type 38 carbines were converted for paratroop use by the fitting of a hinged buttstock.

In 1911, the Type 44 Carbine was introduced; this weapon differs from the Type 38 Carbine in having a permanently attached folding bayonet.

In 1937 a sniping version of the 6.5mm Type 38 rifle—the Type 97—was adopted.

A rifle which was used to some extent by the Japanese Forces and remains very much a mystery, is the 6.5mm so-called Type "I" rifle. This rifle, which uses the standard Japanese 6.5mm cartridge, has a Mannlicher Carcano action and a Mauser-type magazine. The stock is of typical Japanese two-piece type, i.e., the lower half of the butt and pistol grip is a separate piece which is pinned and glued to the body of the stock. These rifles were apparently made in Italy but why the Japanese purchased them is unknown.

Japanese experiences in China showed the need for a more powerful cartridge than the 6.5mm. A 7.7mm semi-rimmed cartridge was already in service with the Type 92 (1932) heavy machine guns. A rimless version of this cartridge was developed for use in rifles. Four trial rifles were submitted, including one each from Nagoya and Kokura arsenals. Several models were patterned after the Type 38 and Type 44 carbines, but tests indicated that the recoil of these weapons was excessive for the short-statured Japanese soldiers. A decision was made to develop a short rifle for cavalry and special troops and a long rifle for infantry. In 1939 the second series of tests was run at Futsu Proving Ground and the Nagoya designed Rifle "Plan No. 1," which was similar to the Type 38, was adopted. A third series of tests, to dispose of accuracy "bugs" and to check out improved ammunition, was completed and the Type 99 rifles were adopted in mid-1939. In 1942, a Type 99 rifle with four-power scope was introduced for sniper use.

The Type 99 has the sliding bolt cover of the Type 38, but in addition has a folding wire monopod which can be used to support the rifle when firing from the prone position or from a support. The rear sight has folding antiaircraft lead arms which are extended out at 90° each side of the sight leaf when the weapon is used against aircraft.

In 1943, a substitute Type 99 was introduced which was made of inferior materials, without bolt cover and sling swivels, and without chrome-plated bores. The rifles have fixed rear sights. It is inadvisable to fire them, since they can be dangerous. On the subject of materiel and the strength of actions, tests conducted after World War II showed that the 6.5mm Type 38 action was

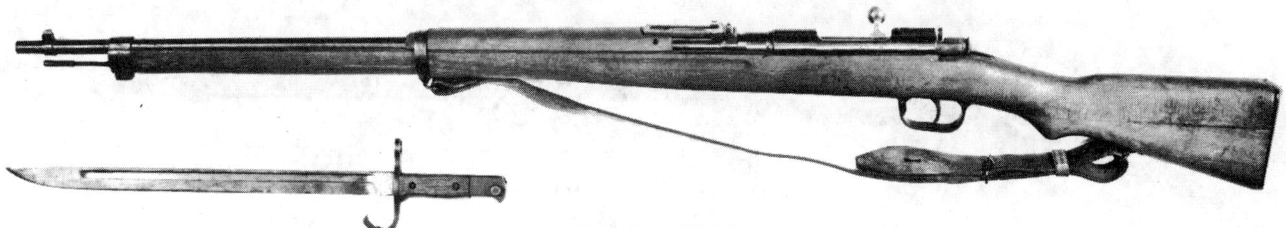

6.5mm Type "I" Rifle.

7.7mm Type 99 Long Rifle.

stronger than the U. S. Springfield, 1917 Enfield, or the German Mauser action.

A take-down version of the Type 99 with interrupted screw-type dismounting was introduced in 1942. The barrel of this rifle had a tendency to loosen in service and further work in this area was done by Japanese Ordnance. The result was the Type 2 take-down rifle which has a barrel-locking mechanism similar to that used on some machine guns. A locking key goes through the receiver and engages a slot in the barrel.

Although Japan started experiments with semiautomatic rifles in 1922, no semiautomatic rifle was produced for use in World War II. At the time the United States was considering adoption of the Pedersen rifle in caliber .276; a few copies of this rifle, fitted with a rotary magazine, were made in Japan.

There were many requests from the Japanese Forces in the field during World War II for a semiautomatic rifle to counter the U.S. M1 rifle. The Japanese Navy produced in 1945 a modified copy of the U.S. M1 rifle chambered for the 7.7mm cartridge. This weapon, called the Type 5, used a 10-round box magazine loaded with two 5-round chargers rather than the 8-round en bloc type clip of the U.S. M1.

Type 99 rear sight, showing AA lead arms.

Numerous difficulties were encountered with the rifle, principally because of the low quality materials then available. Only about twenty of these rifles were made.

The Type 99 Short Rifle (above) and Type 2 Take-down Rifle (below), both 7.7mm.

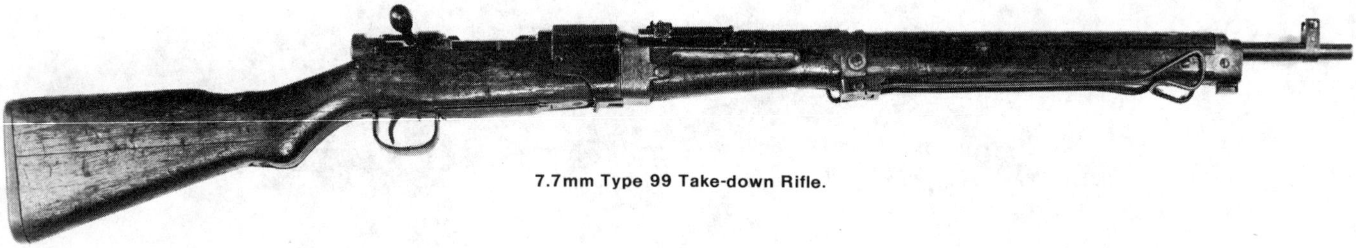

7.7mm Type 99 Take-down Rifle.

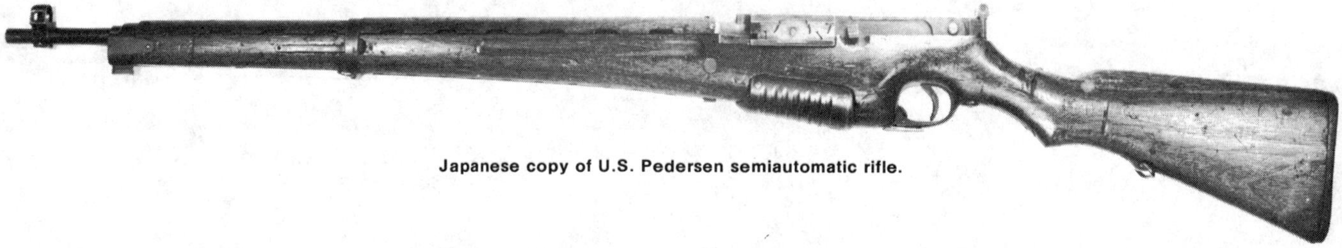

Japanese copy of U.S. Pedersen semiautomatic rifle.

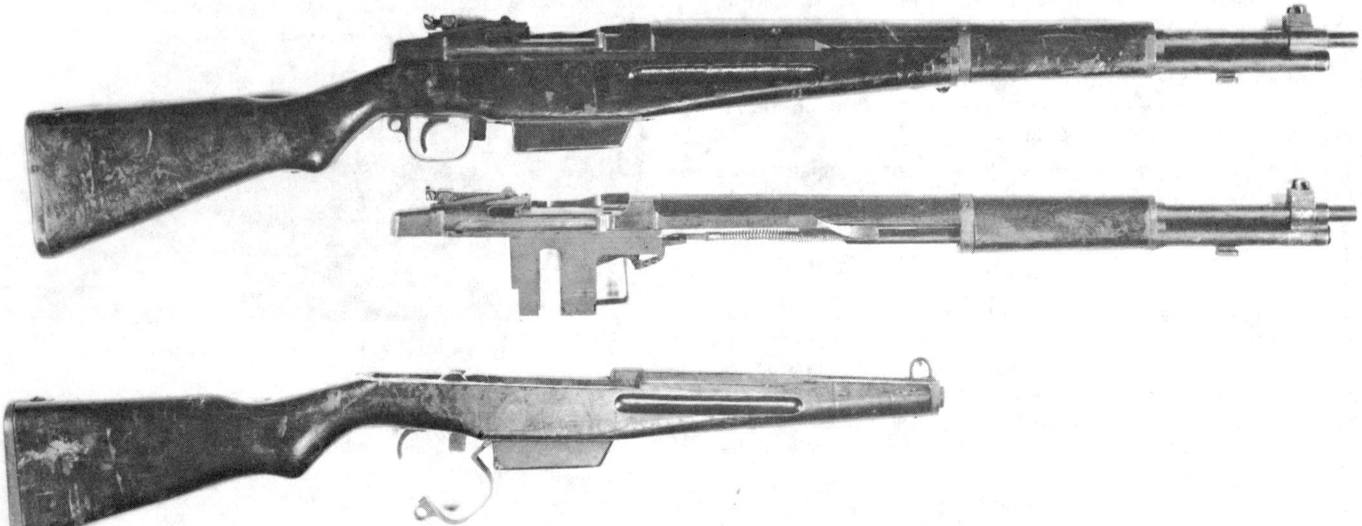

7.7mm Type 5 Semiautomatic Rifle, Japanese copy of the M1.

CHARACTERISTICS OF JAPANESE WORLD WAR II RIFLES

	Type 38 Rifle	Type 38 Carbine	Type 44 Carbine	Type 97 Rifle
Caliber	6.5mm.	6.5mm.	6.5mm.	6.5mm.
System of operation:	Turn bolt.	Turn bolt.	Turn bolt.	Turn bolt.
Length overall:	50.2 in.	34.2 in.	38.5 in.	50.2 in.
Barrel length:	31.4 in.	19.9 in.	19.2 in.	31.4 in.
Feed device:	5-round, nondetachable, staggered row, box magazine.	5-round, nondetachable, staggered row, box magazine.	5-round, nondetachable, staggered row, box magazine.	5-round, nondetachable, staggered row, box magazine.
Sights: Front:	Barley corn with protecting ears.	Barley corn with protecting ears.	Barley corn with protecting ears.	(Same as Type 38 plus 2.5X scope.)
Rear:	Leaf.	Leaf.	Leaf.	
Weight:	9.25 lb.	7.3 lb.	8.9 lb.	11.2 lb. w/scope.
Muzzle velocity:	2400 f.p.s.	Approx 2300 f.p.s.	Approx 2300 f.p.s.	2400 f.p.s.

CHARACTERISTICS OF JAPANESE WORLD WAR II RIFLES (Cont'd)

	Type 99 Long Rifle	Type 99 Short Rifle	Type 2 Rifle
Caliber:	7.7mm Type 99.	7.7mm Type 99.	7.7mm Type 99.
System of operation:	Turn bolt.	Turn bolt.	Turn bolt.
Length overall:	50 in.	43.9 in.	43.9 in.
Barrel length:	31.4 in.	25.8 in.	25.8 in.
Feed device:	5-round nondetachable, staggered row, box magazine.	5-round nondetachable, staggered row, box magazine.	5-round nondetachable staggered row, box magazine.
Sights: Front:	Barley corn with protecting ears.	Barley corn with protecting ears.	Barley corn with protecting ears.
Rear:	Leaf.	Leaf*.	Leaf.
Weight:	9.1 lb.	8.6 lb.	8.9 lb.
Muzzle velocity:	2390 f.p.s.	Approx 2360 f.p.s.	Approx 2360 f.p.s.

*Late production models of this rifle may be found with a fixed notch type rear sight.

JAPANESE SUBMACHINE GUNS

The Japanese did little in the line of submachine gun development until the thirties. They purchased quantities of Bergmann submachine guns manufactured by SIG in Switzerland during the 1920s. These weapons, which were basically the same as the MP 18 I modified to use a box magazine, are chambered for the 7.63mm Mauser cartridge. A bayonet mounting bar was added to the weapon to take the standard rifle bayonet. These weapons were apparently mainly used by the Japanese Special Landing Forces (Marines) and were encountered during the Bataan campaign by United States and Filipino forces.

Japanese Ordnance authorities and the Nambu firm developed two prototype model submachine guns: the Type I which was chambered for the standard Japanese 8mm pistol cartridge using a 50-round magazine and the Type II chambered for 6.5mm with a newly developed bullet using a 30-round magazine. Type I was tested in 1930 and 1937 and the results were promising, but indicated that further development work was required. The Type II design was entirely unsatisfactory and was dropped. In August 1937 an improved Type I was submitted to the Cavalry School for tests which indicated that more development work was required. In April 1939, the Nambu firm submitted a design which was an improved Type I. Testing indicated that with little improvement this weapon, called Type III, was generally satisfactory. The improved weapon, called Type IIIB was approved for issue and adopted as the 8mm Type 100 (1940).

The Type 100 was issued initially to paratroopers, but by 1944 there were demands from the infantry for submachine guns. The Type 100 may be found fitted with a bipod and comes in three basic types:

(1) Type 100 with fixed stock and bayonet lug bar; some of these are fitted with a compensator.
(2) Type 100 with folding stock and bayonet lug bar.
(3) Type 100, circa 1944, with fixed stock, bayonet lug on barrel jacket, compensator, and fixed aperture-type rear sight.

Time caught up with the Japanese on submachine guns. The Type 100 was never really satisfactory, possibly because the Japanese had the concept of an automatic rifle in mind while working on the design rather than the concept of a submachine gun as understood by most other countries.

Prototype development continued during the war and the 8mm Type II was discovered by United States personnel in Japan after the war was over. The 8mm Type II was a step in the right direction insofar as weight and length were concerned. It also had one unusual feature which was borrowed from the 1926 Finnish Suomi, an airlock-type buffer/piston arrangement which can be used to regulate the cyclic rate.

A special air lock is provided to accomplish this. The lock is at the rear of the receiver. As the bolt is blown back it is secured to an extension arm on the piston of the air lock. An escape valve can be set to allow the air compressed within the lock to escape at different rates. By thus speeding or slowing down the travel rate of the bolt, the cyclic rate of fire can be increased or decreased.

CHARACTERISTICS OF JAPANESE WORLD WAR II SUBMACHINE GUNS

	Bergmann 1920	Type 100	Type 100 (1944 version)
Caliber:	7.63mm.	8mm.	8mm.
System of operation:	Blowback, automatic only	Blowback, automatic only	Blowback, automatic only
Length overall:	32 in.	w/stock extended: 34 in.* w/stock folded: 22.2 in.	36 in.
Barrel length:	8 in.	9 in.	9.2 in.
Feed device:	50-round, staggered row, detachable box magazine.	30-round, staggered row, detachable box magazine.	30-round, staggered row, detachable box magazine.
Sights: Front:	Barley corn.	Barley corn with protecting ears.	Barley corn.
Rear:	Tangent with notch.	Tangent with notch.	Fixed aperture set for 100 meters.
Weight:	Approx 9.5 lb.	7.3 lb.	8.5 lb.
Muzzle velocity:	Approx 1350 f.p.s.	Approx 1100 f.p.s.	Approx 1100 f.p.s.
Cyclic rate:	600 r.p.m.	450 r.p.m.	800 r.p.m.

*Also exists in fixed stock version.

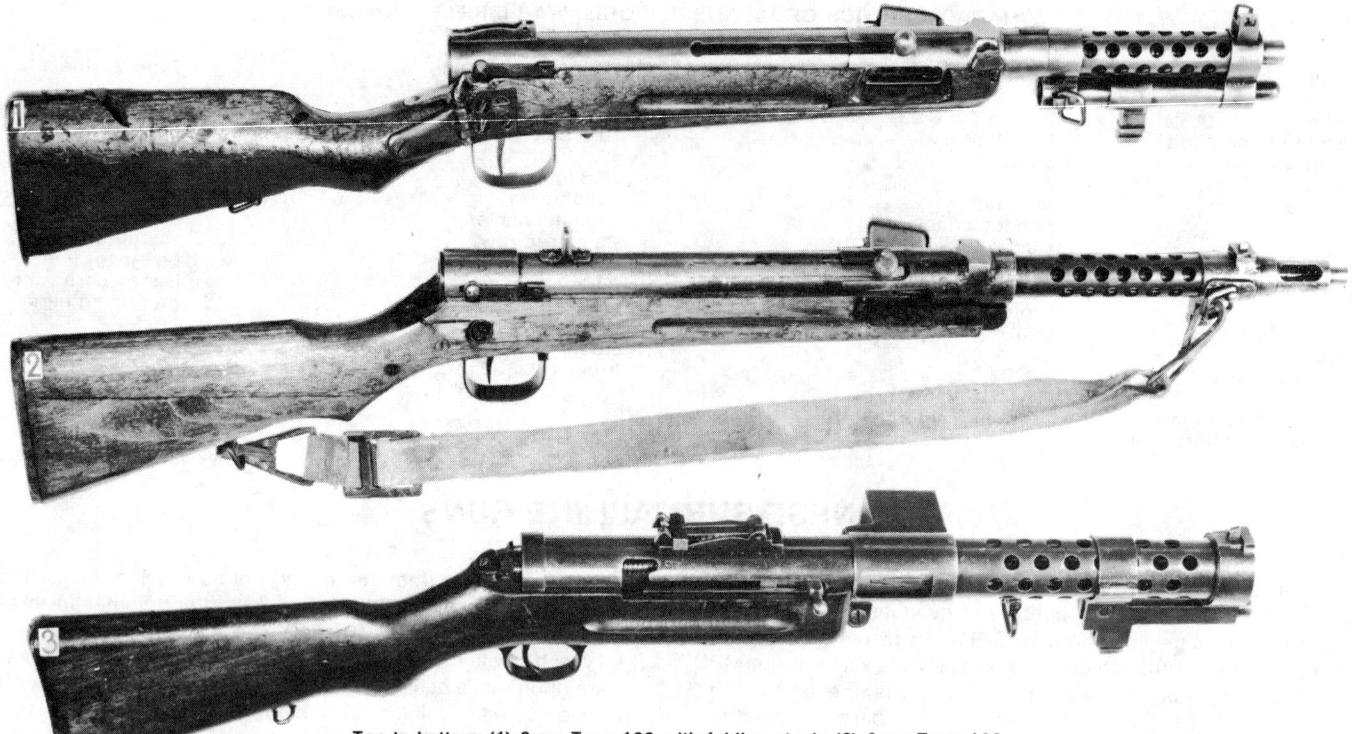

Top to bottom: (1) 8mm Type 100 with folding stock; (2) 8mm Type 100 (1944 model); (3) 7.63mm Bergmann used by Japan.

JAPANESE MACHINE GUNS

The Japanese adopted the Hotchkiss gun at the time of the Russo-Japanese War (1904-05) and also adopted the later 1914 Hotchkiss, in a modified form. Both models were chambered for the 6.5mm cartridge, which left something to be desired insofar as long-range performance was concerned, but the initial adoption of the Hotchkiss set the tone for Japanese machine-gun development for many years to come.

The Type 3 (1914) is a modified Hotchkiss 1914 developed by General Nambu. Like all later Nambu/Hotchkiss machine guns, it has a gravity-fed oil reservoir which, through a spring-loaded lubricator, oiled all cartridges before they were fed into the chamber. The Nambus also differ from the Hotchkiss in the method of ejection. They use the ejection system of the Lewis gun.

During World War II a modification of the Type 3 was encountered; the modified gun was lighter and the barrel could be removed more rapidly than on the original Type 3. The 6.5mm Type 11 was introduced in 1922 and is designed basically for use as a light machine gun, although it was also used on a tripod.

An unusual feature of the Type 11 is the feed hopper, which is fed with six 5-round chargers (stripper clips). The chargers are the same as those used with the 6.5mm Arisaka rifle and it would appear that there would be logistical advantages with this system; but very little computation is necessary to realize the amount of space required for 200 rounds of ammunition on 5-round chargers as opposed to the same amount of ammunition in 30-round staggered row magazines.

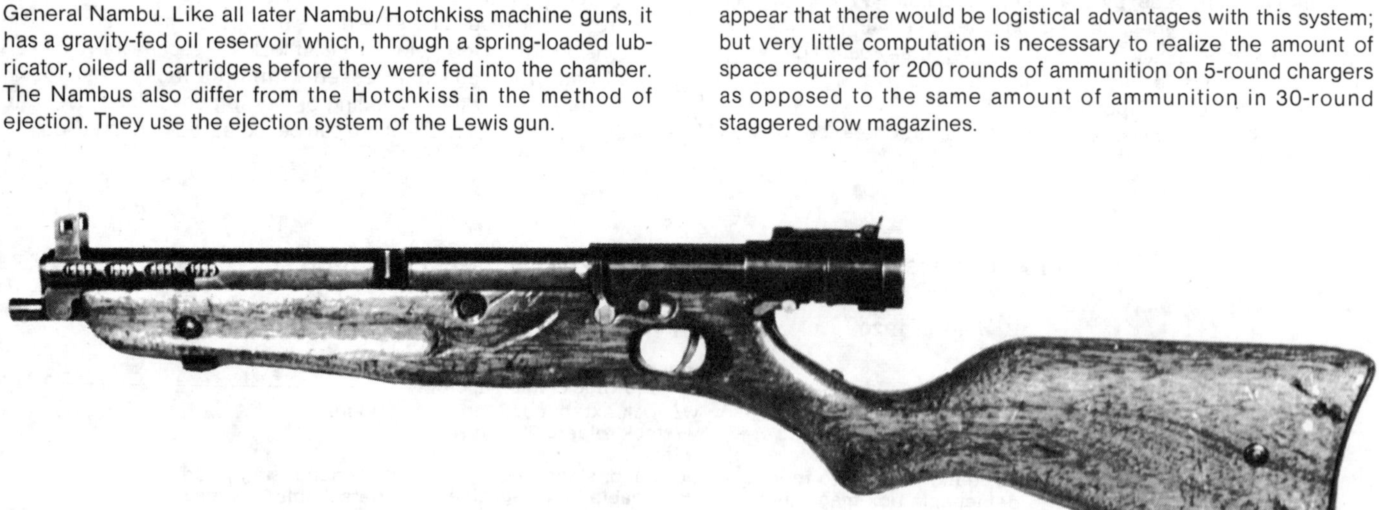

8mm Type II Submachine Gun.

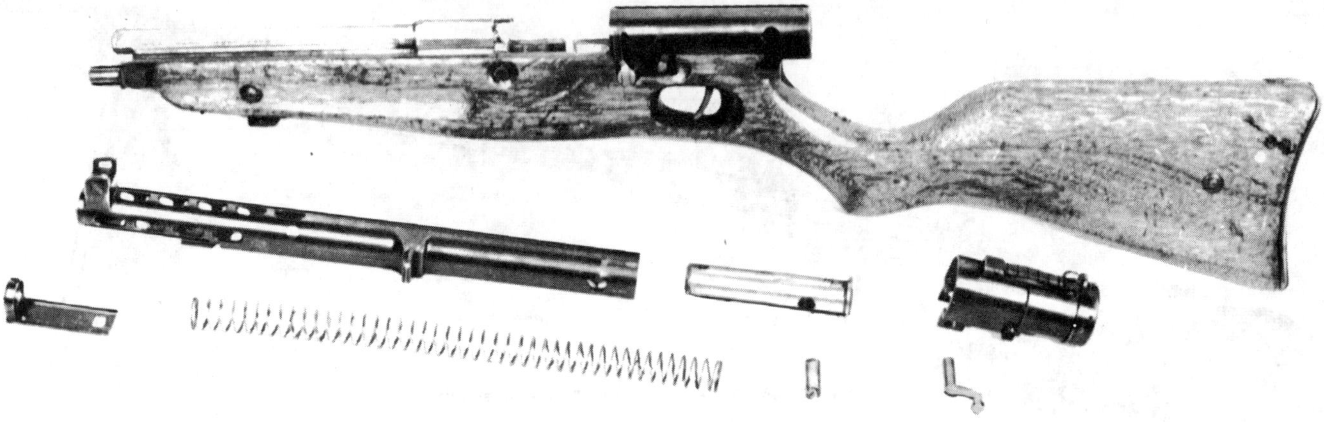

Type II, field-stripped.

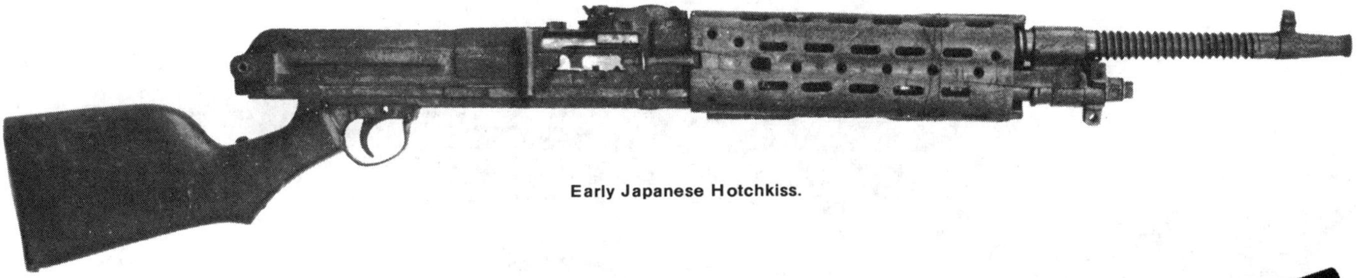

Early Japanese Hotchkiss.

6.5mm Type 3 (1914) Machine Gun.

Of more importance, it has a relatively complicated feed system which accumulates more dirt; this is especially bad with a weapon that requires lubricated cartridges, as compared to a box magazine fed weapon.

The Type 11 also appeared in a tank version called the Type 91 (1931). The Type 91 has a telescope mounted and has a larger feed hopper than the Type 11.

The 7.7mm Type 97 (1937) tank gun was the other standard Japanese tank machine gun of World War II. The Type 97 is a copy of the Czech ZB26 as an examination of the stripped-view below will show.

The Type 97, although a better tank gun than the Type 91, was not completely satisfactory because it was magazine fed—hardly an ideal method of feeding a rifle caliber tank machine gun. Research was started during the war to develop a belt-fed gun with a high rate of fire.

A Browning-type gun, developed from the Japanese aircraft Browning, was to be introduced as the 7.7mm Tank machine gun Type 4, but the conclusion of the war prevented its introduction.

Japanese machine-gun development was a very chaotic type of activity since guns were developed by the Army, Navy, and for aircraft without any apparent coordination. (The reader will have

Nambu 6.5mm Type 11 (1922) Light Machine Gun.

6.5mm Type 11 on tripod.

to forgive the skipping back and forth throughout the text from one type of machine gun to another, but this is apparently what happened in Japanese Ordnance circles at the time. One thing is quite obvious—the Japanese pre-war military had no true appreciation of the logistic difficulties they would encounter in an "all-out" war.)

As pointed out previously, the 6.5mm cartridge was not a good performer at long range and in 1932 the Japanese introduced a new cartridge—the 7.7mm semi-rimmed Type 92 cartridge and a new weapon chambered for this cartridge—the Type 92 Heavy Machine gun. This weapon is essentially an improved Type 3 and was the most widely-used Japanese heavy machine gun in World War II.

There is nothing very unusual about the Type 92. As with all the Japanese heavy-machine guns, the mount is built to be carried by means of pipes fitted to each of the front legs and with a single fork-type pipe fitted to the rear leg. This allows two men to carry the gun and mount in firing position relatively rapidly in moving the gun position over limited distances.

The need of a further lightened modification of the Type 92 was recognized by 1937 and a requirement was laid down for a gun and

Feed mechanism of Type 11 Machine Gun.

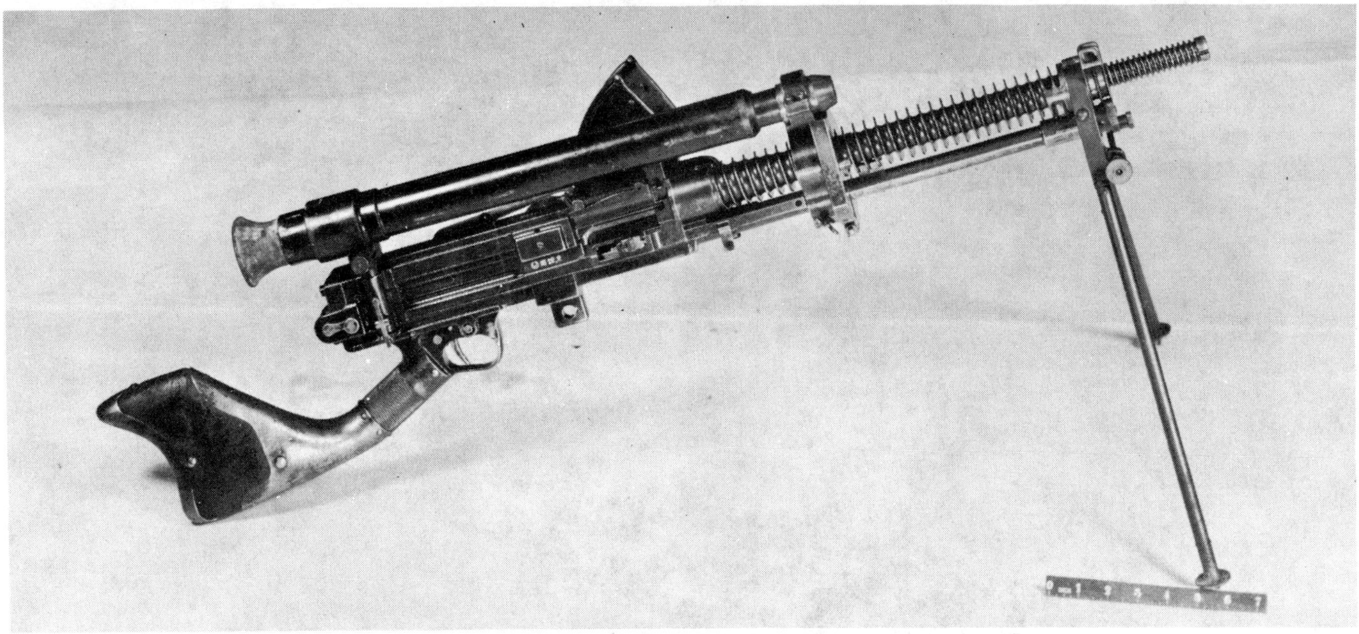

6.5mm Type 91 Tank Gun, fitted with bipod for ground use.

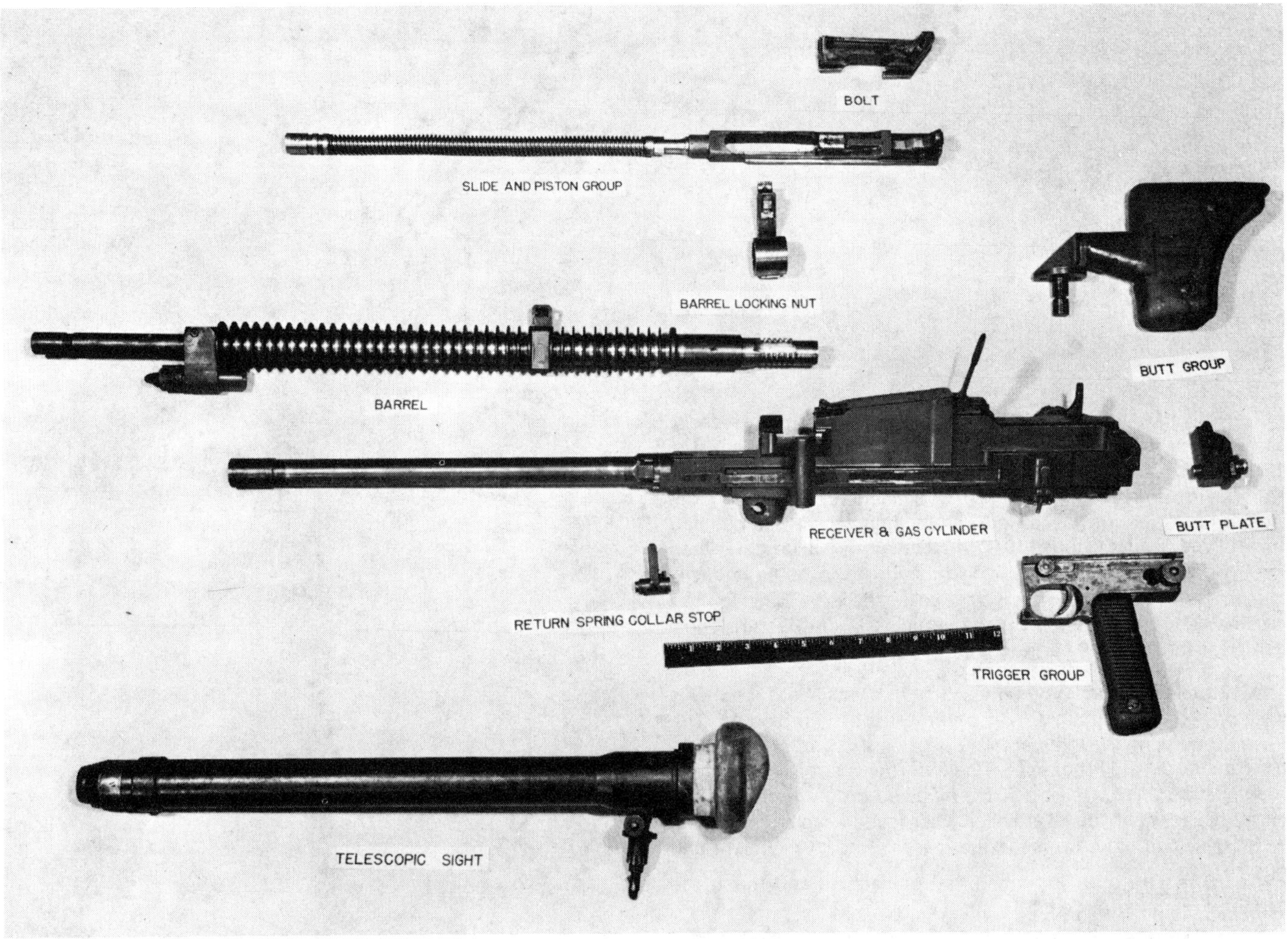

7.7mm Type 97 Tank Gun, stripped.

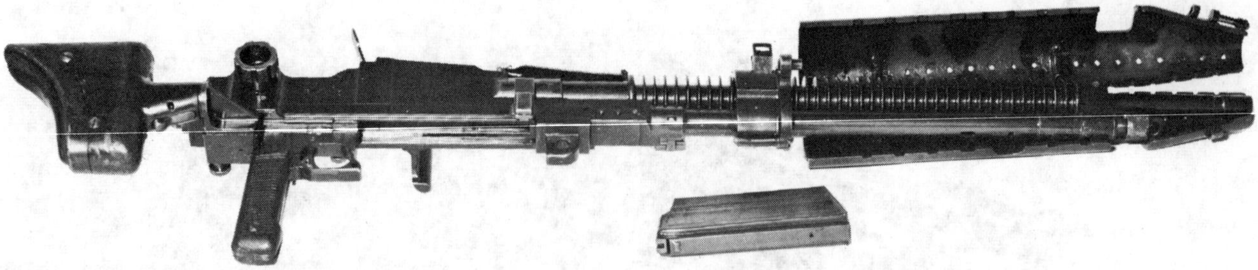

Japanese 7.7mm Type 97 Tank Machine Gun. Right side, magazine with- drawn, and protective shield open to show cooling rings. This weapon is a copy of the Czech ZB26.

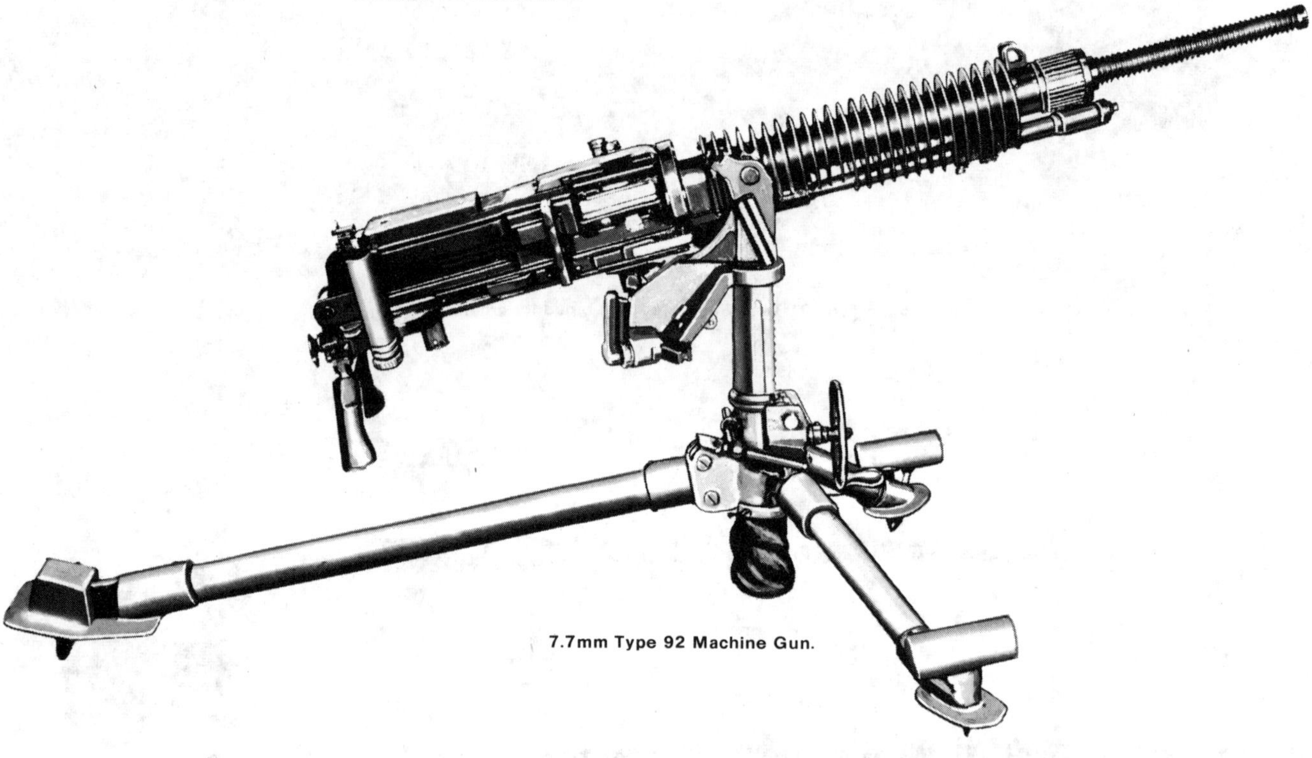

7.7mm Type 92 Machine Gun.

mount weighing less than 88 pounds, with a type of mount easily carried by two men. In March 1940 the first prototype was tested and found unsatisfactory. In June 1940, a second model was tested and with modifications was found to be suitable for service. A modified Type 92 mount was issued with the gun, which was adopted as the Type 1 in November 1942. In addition to being lighter than the Type 92, the barrel of the Type 1 can be removed more easily than that of the Type 92. The Type 1, unlike the Type 92 which can use the semi-rimless Type 92 or the rimless Type 99 cartridges, only uses the Type 99 cartridge.

In 1936, the 6.5mm Type 96 machine gun was introduced. This weapon represented a great improvement over the Type 11. The Type 96 does not have a cartridge oiler on the gun, has a quick-change barrel, and is box-magazine fed. The weapon is frequently found with a 2.5 power telescope mounted on the receiver. The cartridges for the Type 96 are oiled when loaded into the magazine by an oiler built into the magazine loader. Both the Type 11 and Type 96 lacked built-in slow initial extraction and therefore required oiled cartridges.

The Type 99 (1939) light machine gun was designed to obviate the necessity for lubricated cartridges. The barrel lock has a head space adjustment nut and machining of the critical components

7.7mm Type 92 on an AA mount.

was held to closer tolerances than with early guns. The Type 99 came into being because of the need for a light machine gun chambered for the 7.7mm rimless Type 99 cartridge. Four different types of prototype guns were built. Type 1 was modeled on the Type 96 light machine gun. Type 2 was a lightened version of the Type 92 with shoulder stock, bipod, and quick-change barrel. Type 3 was similar to the Type 97 Tank gun and Type 4 was another light variation of the Type 92 heavy machine gun. The requirement was for a weapon weighing not more than 24.7 pounds with sights graduated to 1500 meters. A flash hider was also required since the muzzle flash of the 7.7mm cartridge was much greater than that of the 6.5mm cartridge. The modified Type 96 prototype was chosen and adopted as the 7.7mm Type 99.

A paratrooper version of the Type 99 machine gun was built at the Nagoya Arsenal in 1943. This weapon was easily broken down into barrel and receiver group, buttstock, barrel, and piston and could be rapidly assembled.

The Japanese Navy and Navy Air Corps used the Lewis gun in ground and air versions. These were chambered for 7.7mm rimmed cartridges which were the same as the caliber .303 British cartridge. Both guns are called Type 92; note the complication of having two ground machine guns which use different cartridges called by the same model designation. This situation was due to the almost complete lack of coordination between the Army and Navy Ordnance authorities.

Japanese 7.7mm Heavy Machine Gun, Type I. Standard cartridge strip which is fed into gun from the left side is shown in foreground.

6.5mm Type 96 Light Machine Gun.

Japanese 7.7mm Light Machine Gun Type 99 (1939).

7.7mm Type 99 Paratroop Type Machine Gun, broken down.

The Japanese copied a number of foreign weapons to use as aircraft guns, and in one case (the Browning) produced it in a larger caliber—20mm—than it was produced elsewhere. The Japanese also purchased specimens and the right to manufacture a number of German aircraft machine guns, which they made in 7.92mm including the MG15 called Type 98. The British Vickers aircraft gun was copied as the 7.7mm (.303 British) Type 89.

A Japanese copy of the U.S. caliber .50 Browning chambered for the Japanese 12.7mm cartridge was also quite common. This weapon was called the Type 1.

A great variety of other aircraft machine guns were used as well.

The standard heavy caliber ground and naval machine gun was the 13.2mm Type 93 (1933). This magazine-fed weapon was fitted on a number of different type mounts and could be found in dual as well as single mountings. It is basically a copy of the French 13.2mm M1932 Hotchkiss gun.

The Japanese also produced various types of blowback-operated training machine guns which do not seem to conform to any set pattern. At least five different variations were found in Korea. They are all rather delicate in construction and are probably intended for use with a reduced-charge cartridge.

Type 92 7.7mm Japanese Navy Lewis gun.

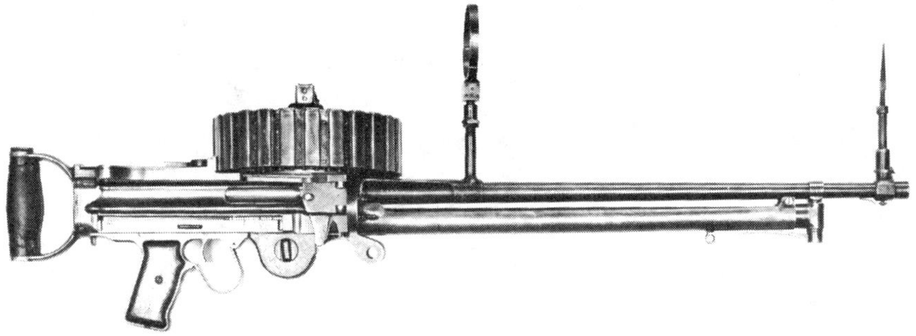

7.7mm Type 92 Aircraft Gun, copy of the Lewis.

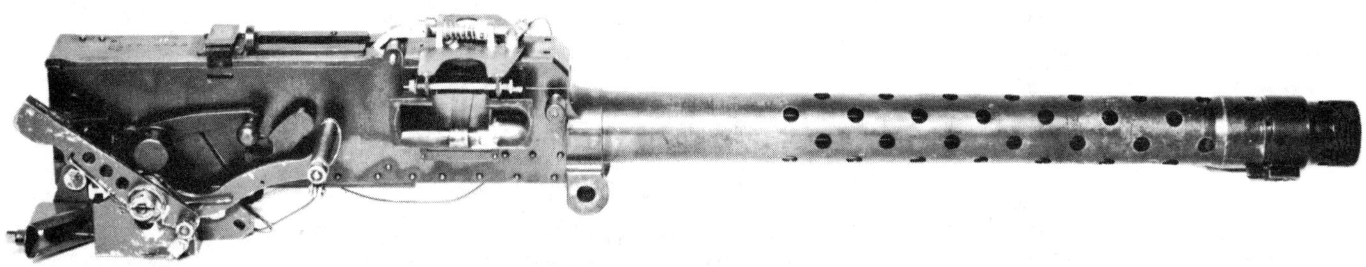

7.7mm Type 89 Aircraft Gun, copy of the Vickers.

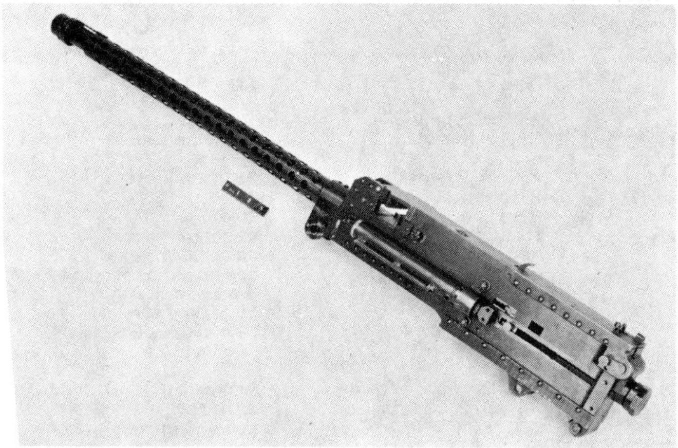

12.7mm Browning, Type 1, copy of U.S. caliber .50 Browning.

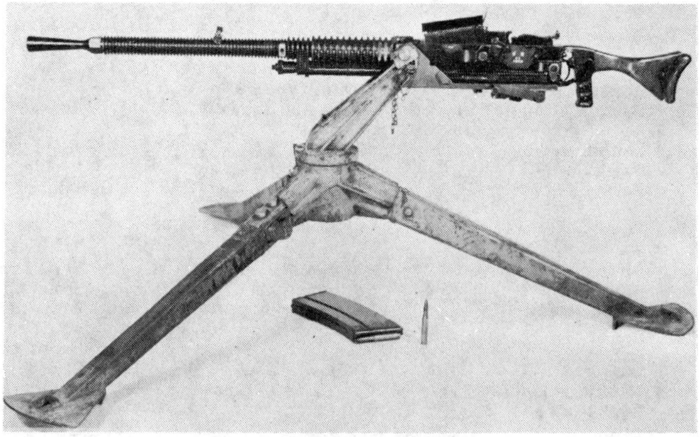

13.2mm Type 93 Heavy Machine Gun.

Training machine gun, caliber 6.5mm.

CHARACTERISTICS OF JAPANESE WORLD WAR II ARMY MACHINE GUNS

	Type 3	Type 11	Type 92	Type 96	Type 99
Caliber:	6.5mm	6.5mm.	7.7mm Type 92 or 99.	6.5mm.	7.7mm Type 99.
System of operation:	Gas, automatic only.	Gas, automatic only.	Gas, automatic only.	Gas, automatic only.	Gas, automatic only.
Length overall:	Approx 47 in.	43.5 in.	45.5 in.	41.5 in.	46.75 in.
Barrel length:	Approx 29 in.	19 in.	Approx 29 in.	21.7 in.	Approx 21.5 in.
Feed device:	30-round strip.	30-round hopper.	30-round strip.	30-round staggered row, detachable, box magazine.	30-round staggered row, detachable, box magazine.
Sights: Front:	Barley corn with protecting ears.	Barley corn with protecting ears.	Barley corn with protecting ears.	Barley corn with protecting ears.	Barley corn with protecting ears.
Rear:	Folding ring A.A. sight and tangent with aperture.	Tangent with "V" notch.	Tangent with aperture or telescope.	Radial wheel tangent with aperture.	Radial wheel tangent with aperture.
Weight:	122lb. w/tripod.	22.5 lb.	122 lb. w/tripod.	20 lb.	23 lb.
Muzzle velocity:	2434 f.p.s.	Approx 2300 f.p.s.	Approx 2400 f.p.s.	Approx 2400 f.p.s.	Approx 2350 f.p.s.
Cyclic rate:	450-500 r.p.m.	500 r.p.m.	450-500 r.p.m.	550 r.p.m.	850 r.p.m.

All barrel lengths include flash hider length.

CHARACTERISTICS OF JAPANESE WORLD WAR II ARMY MACHINE GUNS (Cont'd)

	Type 1	Type 91	Type 97	Type 93
Caliber:	7.7mm Type 99.	6.5mm.	7.7mm Type 99.	13.2mm.
System of operation:	Gas, automatic only.	Gas, automatic only.	Gas, automatic only.	Gas, automatic only.
Length overall:	42.4 in.	33 in.	46.5 in. w/stock.	95 in. (approx).
Barrel length:	23.2 in.	19.2 in.	28 in.	65 in.
Feed device:	30-round strip.	Hopper.	30-round staggered row, detachable, box magazine.	30-round staggered row, detachable, box magazine.
Sights: Front:	Barley corn with protecting ears.	Barley corn with protecting ears.	1.5X scope.	Barley corn with protecting ears.
Rear:	Tangent with aperture.	Tangent with "V" notch and telescope.		Leaf and speed ring type A.A.
Weight:	70 lb. w/tripod.	22.4 lb.	24.5 lb.	Approx 213 lb w/tripod.
Muzzle velocity:	Approx 2400 f.p.s.	Approx 2300 f.p.s.	Approx 2400 f.p.s.	2210 f.p.s.
Cyclic rate:	550 r.p.m.	500 r.p.m.	500 r.p.m.	450-480 r.p.m.

All barrel lengths include flash hider length.

The Mexican Army is currently armed with caliber .45 Colt automatic pistols. The .45 caliber Obregon is now relatively rare. Mexico has purchased 7.62mm NATO FN FAL rifles and 7.62mm NATO FN MAG machine guns. The caliber .30 M1954 and various U.S. caliber .30 rifles and carbines may still be in service as may the caliber .30 Mendoza light machine gun and various models of the Browning machine gun.

Older weapons such as the 7mm Model 1936 rifle and the 7mm Model 1934 light machine gun may be held in reserve.

40 Mexico

MEXICAN SERVICE PISTOLS

Development of pistols in Mexico has been rather limited, although a number of commercial pistols have been developed and sold; only one military pistol of native design has been produced in any significant quantity. This weapon is the caliber .45 Obregon which is covered in detail below. The Obregon is no longer manufactured, mainly because the cost per weapon was high as a result of relatively low total production. Mexico, because of it's proximity to the United States, has closely followed the United States in pistol development. The standard United States military and commercial pistols are well known in Mexico and United States and Mexican made ammunition is readily available for these pistols. The Colt Model 1911A1 automatic pistol in caliber .45 and super .38 are widely used by the Mexican Armed Forces and Federal Police. Earlier United States service pistols such as the cal. .45 Model 1873 Colt single-action and the caliber .38 Model 1892 Colt were also used by Mexican armed forces in the past.

THE CALIBER .45 OBREGON PISTOL

Characteristics of Obregon Pistol

Caliber: .45 Model 1911 (11.43mm).
System of operation: Recoil, semiautomatic.
Length overall: Approx. 8.5 in.
Barrel length: Approx. 5 in.
Feed device: 7-round, in-line, detachable box magazine.
Sights: Front: Blade.
Rear: Dove-tailed bar with notch.
Weight: Approx. 2.5 lb.
Muzzle velocity: 830 f.p.s.

How to Load and Fire the Obregon Pistol

This pistol is loaded and fired in a manner similar to that of the U.S. Caliber .45 M1911A1, which is covered in detail in the chapter on the United States.

How to Field Strip the Obregon

Field stripping is essentially the same as that for the U.S. M1911A1 with the exception of the barrel unlocking cam, which is held in place by the combination slide stop safety. The unlocking cam will be removed during field stripping and must be reinserted with the cam surface up so that it will engage the camming lug of the barrel.

Caliber .45 Obregon pistol.

Caliber .45 Obregon pistol, field-stripped.

How the Obregon Works

The locking design is on the Obregon system, which differs radically from the Browning. As the slide moves to the rear, an unlocking cam on the underside of the barrel moves in the locking sleeve camcut causing the barrel to rotate after initial locked short travel with the slide, until its top lugs pass out of engagement with their locking slots in the underside of slide. There is, of course, no downward hinging movement of the barrel as in the Colt-Browning practice.

The functioning of the mechanism, except for the breech locking, is quite similar to that of the conventional Browning pattern. The locking system is a Steyr-Hahn variant.

This pistol was made for the Mexican Army in caliber 11.43mm (.45 Colt ACP). Its general manufacture and design are a credit to its national origin. The pistol has a slide stop to hold it open on the last shot. This weapon is no longer in manufacture.

MEXICAN RIFLES

Mexico adopted a 7mm Mauser in 1895; this rifle is almost identical to the Spanish 7mm Mauser Model 1893. In 1902 another 7mm Mauser was adopted; this rifle has the Model 98 type action. Except for the action, the M1895 and M1902 are almost identical. In 1912 another 7mm Mauser was purchased by Mexico. The Model 1912 is similar to the German Rifle Model 98, but has a tangent-type rear sight and a longer handguard than does the Model 98.

7mm Model 1902 Mauser Rifle.

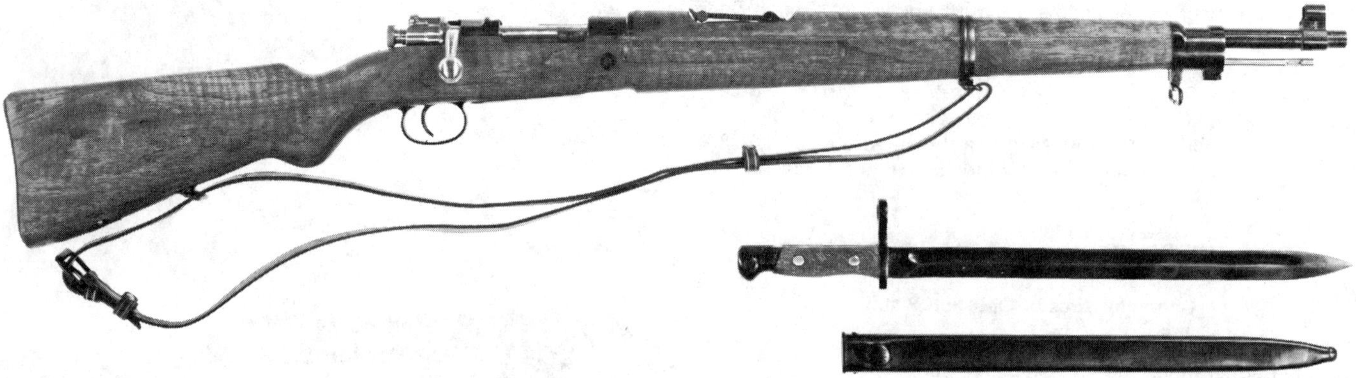

7mm Model 1936 Rifle.

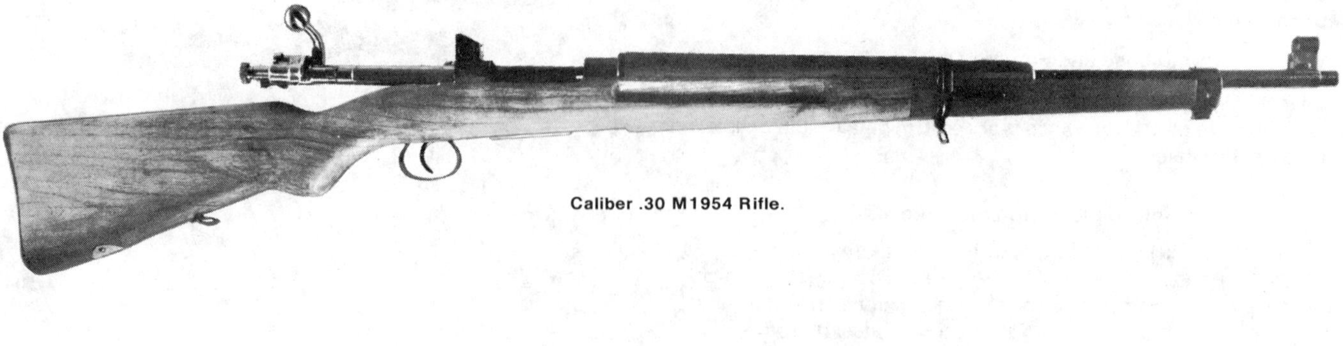

Caliber .30 M1954 Rifle.

During the Mexican revolution, an extended period that began about 1910 and ended about 1920, Mexico secured arms from many sources. At this time, the Mexican Arisaka (the Japanese Type 38 rifle) was purchased from Japan. These are 7mm weapons with the Mexican escutcheon stamped on the receiver. They are relatively rare. In 1936 a Mauser of Mexican design was introduced. The 7mm Model 1936 externally resembles the Springfield Model 1930A1, having a Springfield-type cocking piece, bands and stacking swivel, but the action is of the Mauser "short" type. The Model 1936 is of Mexican manufacture and is very well made.

The Model 1954 rifle is patterned on the Springfield Model 1903A3, but also uses the Mauser-type action, i.e., one-piece firing pin. The stock is made of laminated plywood. A carbine version of this weapon, which is chambered for the .30-06 cartridge, has also been reported.

THE MEXICAN MONDRAGON SEMIAUTOMATIC RIFLE

A number of semiautomatic rifles have also been produced in Mexico, the best known being the Mondragon, one of the earliest reasonably successful semiautomatic rifles of military pattern. It was invented by the Mexican General Mondragon in 1908. Manufacture of the weapon, which was a gas-operated locked breech rifle, was undertaken in Switzerland at Neuhausen by Societe Industrielle Suisse. When World War I broke out, these rifles were shipped to Germany. They did not stand up well in trench service, but were among the earliest rifles issued to observers in aircraft as the Model 1915, before the introduction of the machine gun.

The Mondragon was gas-operated from a port near the muzzle. The mechanism was actuated by the gas piston driven back by the expanding gas in the cylinder. While complicated and heavy by modern standards, the Mondragon was a very serious attempt at early semiautomatic rifle production.

7mm M1908 Mondragon Semiautomatic Rifle. This rifle was produced by SIG in Switzerland.

CHARACTERISTICS OF MEXICAN RIFLES

	Model 1895	Model 1902	Model 1912	Model 1936	Model 1954
Caliber:	7mm.	7mm.	7mm.	7mm.	.30-06.
System of operation:	Turn bolt.	Turn bolt.	Turn bolt.	Turn bolt.	Turn bolt.
Length overall:	48.6 in.	49.2 in.	49.1 in.	42.9 in.	Approx 48 in.
Barrel length:	29.1 in.	29.1 in.	29.1 in.	19.29 in.	24 in.
Feed device:	5-round, non-detachable, staggered row box magazine				
Sights: Front:	Barley corn.	Barley corn.	Barley corn.	Hooded barley corn.	Hooded barley corn.
Rear:	Leaf.	Leaf.	Tangent with "V" notch.	Tangent w/"V" notch.	Ramp type aperture.
Weight:	8.8 lb.	8.8 lb.	9 lb.	8.3 lb.	Approx 9 lb.
Muzzle velocity:	Approx 2300 f.p.s.	Approx 2300 f.p.s.	Approx 2300 f.p.s.	Approx 2600 f.p.s.	Approx 2800 f.p.s.

MEXICAN SUBMACHINE GUNS

Mexico has used various models of the caliber .45 Thompson submachine gun and the noted Mexican small-arms designer Rafael Mendoza developed a series of simple lightweight submachine guns during the 1950s. These weapons have been made in caliber .45, .38 super, and 9mm Parabellum and have limited use in the Mexican service.

MEXICAN MACHINE GUNS

Mexico has used a number of foreign-developed machine guns and developed a number of her own. The Mexicans have used the 7mm Model 1911 and 1934 Madsen guns, the 7mm Model 1896 Hotchkiss gun, the 7mm Browning Model 1919, and 7mm Colt guns.

MEXICAN MENDOZA MACHINE GUNS

In the early 1930s Rafael Mendoza, then working at the Mexican National Arms factory, developed a gas-operated light machine gun, which was standardized in the Mexican Army as the Model C-1934.

While the gun itself utilizes principles already established by the Hotchkiss and Lewis guns, it possesses a number of unique developments of its own. The Lewis-type action is improved with the incorporation of a double cam slot, which helps to equalize the torque. Locking friction is materially reduced thereby.

While the gas cylinder is modeled somewhat after the Hotchkiss, the piston system of the Hotchkiss is not employed. The gas assembly is in the form of a cup shaped projection on the barrel. Within it is housed a short piston. The operating stroke is very short, and at its limit the gases are dissipated openly into the air, thereby reversing the operation of the typical gas mechanism of

7mm Light Machine Gun Model C-1934.

7mm Mendoza machine gun.

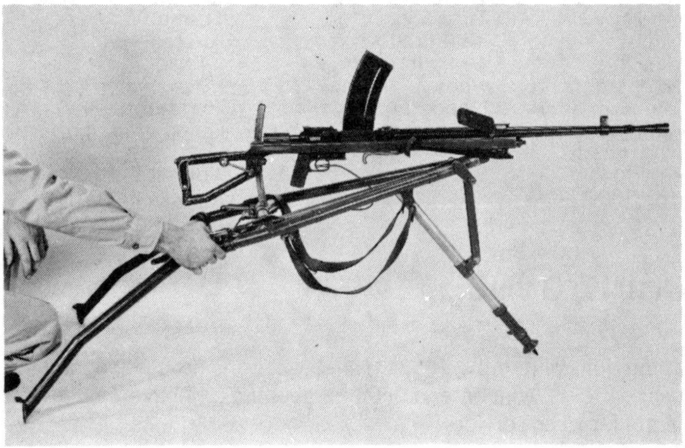

7mm Mendoza machine gun on tripod.

this sort. A firing switch on the left side of the receiver forward of the trigger guard may be set for full or semiautomatic fire in the standard manner.

The cocking handle projects from the left side and is placed well forward. The gun fires from an open bolt position. A special firing pin is provided which carries a firing tip at each end. In the event of firing pin breakage, it is necessary only to reverse the pin to continue firing.

The ejector is an integral part of the hold-back assembly. The quick-change barrel is not threaded. As a result no provision is made for head spacing.

When the gun is cocked, the barrel latch at the forward end of the receiver can be pressed in. A large lug, which is holding the barrel to the receiver, is thereby freed. The barrel can then be pulled forward out of the receiver.

The rear of the barrel is slotted to permit the locking key to pass through. This key, of course, passes through corresponding receiver slots and is retained by the retaining pin.

A three-corner stop on the magazine follower rises above the lips of the magazine into the receiver as the magazine is emptied. This interferes with the movement of the hold-back pawl. Insertion of a loaded magazine releases the rear sear and allows the bolt to move forward enough to engage a notch in the actuator. Pulling the trigger releases the sear. The operating parts are driven ahead by the tension of the operating spring. The bolt face passes over the double column magazine and drives the first cartridge ahead. The extractor cams itself over the cannelure of the cartridge.

As the cylindrical bolt chambers the round and stops, the light locking lugs on the bolt are in line with the fixed lugs. The bolt extension is still three-fourths of an inch from firing and is continuing forward. It rotates the bolt, locking it firmly to the receiver.

The gas port is about 11 inches from the breech end of the barrel. When the bullet passes over it, gas is released into the piston area. The bolt extension starts rearward some three-fourths of an inch before its lug hits the cam in the bolt slot. The firing pin is started backward with the actuator. Continuing rearward movement unlocks the bolt and carries it to the rear.

Mendoza developed a number of other machine guns, some of which have been produced in limited quantity. Early in World War II a modified version of the Model C-1934, chambered for the 7mm cartridge, was produced. This weapon was designed for use as a

Caliber .30 Model 45 Light Machine Gun.

Mexican Mendoza Model RM-2 Light Machine Gun, caliber .30-06.

Characteristics of Mexican Machine Guns

	Model C1934	Model RM2
Caliber:	7mm.	.30-06.
System of operation:	gas, selective fire.	gas, selective fire.
Length overall:	46. in.	43.3 in.
Barrel length:	Approx 25 in.	24 in.
Feed device:	20-round, detachable box magazine	20-round, detachable box magazine
Sights: Front:	Hooded barley corn.	Hooded barley corn.
Rear:	Adjustable aperture.	Adjustable aperture.
Weight:	18.5 lb.	14.1 lb.
Muzzle velocity:	Approx 2700 f.p.s.	Approx 2750 f.p.s.
Cyclic rate:	400-500 r.p.m.	600 r.p.m.

Mendoza L.M.G. Model RM-2. Hinged open for quick withdrawal of all operating parts.

light machine gun on a bipod or as a heavy machine gun on a tripod. This weapon was apparently not introduced into Mexican service.

In 1945, Mendoza introduced another light machine gun chambered for the .30-06 cartridge. This weapon is basically an improved Model C-1934 and has the same type quick-change barrel as the Model C-1934, but has a different receiver configuration and a multi-perforated muzzle brake.

The caliber .30 RM-2 is another Mendoza design. It has been especially designed for easy and low cost manufacture.

While it does not have the quick removal barrel feature of the earlier model, it stresses a very simple, rapid and effective takedown. Pressing the release catch at the rear of the receiver allows the gun to be broken on its hinge much in the manner of the Belgian FN automatic rifle. The gas piston and slide with recoil spring and the few operating components may be withdrawn from the rear of the gun. The unusual Mendoza feature of reversible firing pin again appears in this new design but in improved form. In the event of firing pin breakage in the field (a relatively weak point in any machine gun) merely dropping the rear of the gun will allow the mechanism to be drawn back far enough to lift the bolt off the piston slide. The firing pin has a firing point at each end. If one should be broken, merely withdrawing and reversing the pin and reinstalling it in the bolt prepares the gun for firing again.

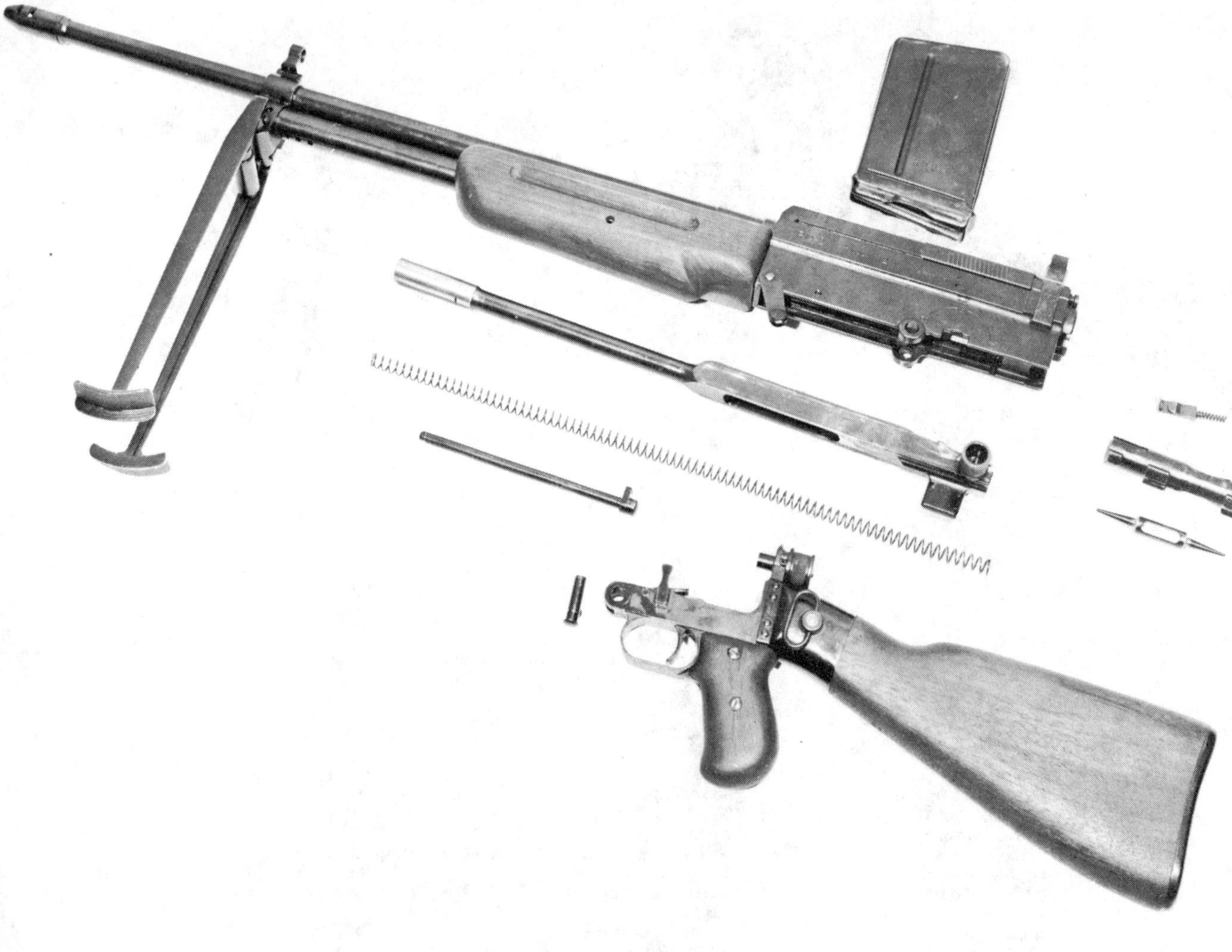

Mexican RM-2, field-stripped.

41
Netherlands

The standard small arms of the Netherlands Army are as follows: the 9mm Browning, FN Hi-Power pistol, the 7.62mm NATO FN "FAL" rifle, the 9mm Uzi M61 submachine gun, and the 7.62mm NATO FN MAG general purpose machine gun. The U.S. caliber .30 and .50 Browning machine guns are still in service on armored vehicles. The Netherlands were furnished with United States and British small arms after World War II.

NETHERLANDS SERVICE PISTOLS

The Netherlands Army adopted a 9.4mm service revolver in 1873 which was, with minor modifications, used as late as World War II, especially in the former Dutch East Indies. Prior to World War I, the Netherlands adopted the M1903 Browning 9mm Long pistol, and in the early 1920s adopted the 9mm (.380 ACP) Browning M1922 pistol which they called Pistool M25 No. 2. The Netherlands also used the 9mm Parabellum Luger, some of German pre-war manufacture similar to the P-08, and approximately 9,000 of Vickers (British) production which are usually called the Model 1920. Whether or not these particular pistols were made by or assembled by Vickers is a controversial question in which the author has no desire to be involved.

The 9mm Parabellum FN Hi-Power pistol was adopted by the Netherlands after World War II.

NETHERLANDS SERVICE RIFLES

The Netherlands adopted the 6.5mm Mannlicher rimmed cartridge and rifle in 1895. The Netherlands Mannlicher is a turnbolt type, using the standard 5-round Mannlicher type clip which is loaded in the weapon and drops out when the last round is chambered. The Dutch Mannlicher is basically the same as the Rumanian Model 1893 Mannlicher.

Seven carbine versions of this weapon exist, including a model with folding bayonet for gendarmerie use. These carbines vary only in minor details, such as position of sling swivels and length of handguards. The Model 95 No. I O.M. carbine has a sporter-type stock and used an old type triangular bayonet. The Model 95 No. 1 N.M., Model 95 No. 3 O.M., Model 95 No. 3 N.M., Model 95 No. 4 O.M., and Model 95 No. 4 N.M., all use knife-type bayonets. An unusual feature of the carbines is that most models have a wooden piece covering the left side of the magazine. This piece is glued and doweled to the stock. The No. 3 O.M. and No. 3 N.M. carbines have handguards which extend beyond the stock, almost to the muzzle. Many of these 6.5mm Mannlicher carbines were

6.5mm Model 95.

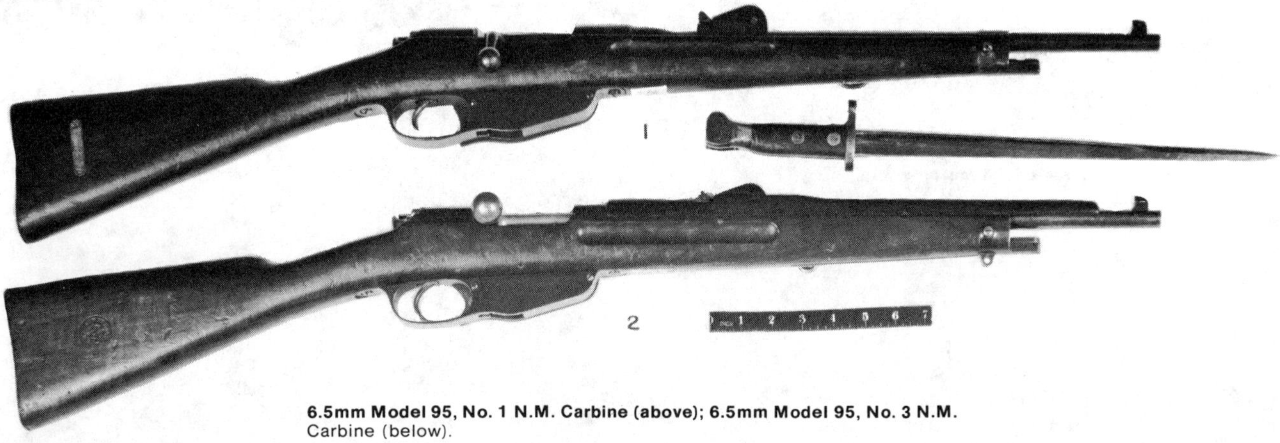

6.5mm Model 95, No. 1 N.M. Carbine (above); 6.5mm Model 95, No. 3 N.M. Carbine (below).

6.5mm Model 95, No. 4 N.M. Carbine.

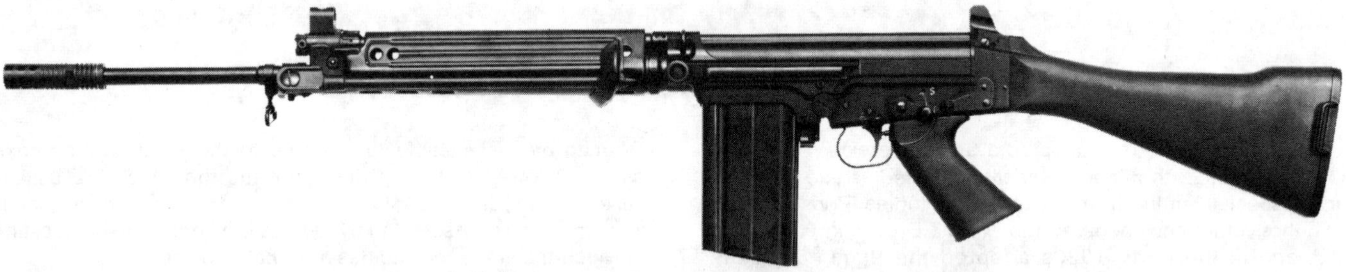

FN 7.62mm NATO Light Automatic Rifle as made for the Netherlands.

converted to caliber .303 British by the Indonesians in the early 1950s.

These weapons are no longer in service in the Netherlands.

In 1917 a version of the Model 95 rifle chambered for the 7.92mm rimmed cartridge was introduced. This weapon was made in very limited quantity.

In 1940, the Netherlands government purchased quantities of the .30-06 Johnson semiautomatic rifle for the East Indies Forces and the Royal Netherlands Navy. These rifles are called the Model 1941. The United States and Great Britain supplied the Netherlands with caliber .30 M1 rifles and caliber .303 Lee Enfield rifles. The Netherlands also purchased a quantity of caliber .30 FN bolt action carbines for police use. About 1960, the 7.62mm NATO FN "FAL" rifle was adopted. The Netherlands version of the "FAL" has metal handguards, a folding bipod, and protecting ears on the rear sight. The Dutch have also made several modifications in the bolt of the "FAL" since adopting the weapon.

The Netherlands government arsenal Artillerie Inrichtingen manufactured the 7.62mm NATO AR-10 rifle in limited quantities on a contract basis. This weapon was never adopted by the Netherlands Army.

NETHERLANDS MACHINE GUNS

The Dutch adopted the Schwarzlose machine gun in 7.92mm rimmed in 1908. This weapon is basically the same as the Austrian 07/12 Schwarzlose, and was used by the Dutch Army during World War II. The Dutch were also one of the largest users of the Madsen, having used the Model 1919, 1923, 1926, 1927, 1934, 1938, and 1939—all chambered for the 6.5mm cartridge. The Dutch also adopted the Lewis gun in 6.5mm in 1920. These weapons are not as frequently encountered as are the Dutch Madsens.

The Vickers Model 1918 was also purchased in 7.92mm.

After World War II, the Dutch received United States Browning guns and British Bren guns. After extensive tests the 7.62mm NATO FN MAG machine gun was adopted in the early sixties.

8mm Model 8 Machine Gun.

6.5mm Model 20, Lewis gun.

42

New Zealand forces have always been equipped with British pattern arms. Some of these have been of British manufacture, some of Australian manufacture (Lithgow) and some of local manufacture. The pistols used have been the caliber .455 Webley Marks V and VI and the Enfield caliber .38 revolvers (No. 2 Marks 1, 1* and 1**). Rifles used in the past have been the No. 1 Mark III and III*. The current rifle is the 7.62mm NATO L1A1 (FN "FAL") as made at Lithgow, Australia. Submachine guns have been produced in New Zealand. Approximately 10,000 Sten and 2000 Charlton machine guns were made in New Zealand during World War II. The Charlton is a native design.

New Zealand

It should be noted that, in the area of submachine guns and machine guns, New Zealand has conformed more closely to the United Kingdom than to Australia. Australia, as previously noted, has taken a rather independent tack, starting with the World War II development and manufacture of the Owen and Austen, which is currently reflected in the use of the F1 submachine gun and the adoption of the U.S. 7.62mm NATO M60 machine gun. New Zealand, on the other hand, has adopted the 9mm Parabellum L2A3 (Sterling submachine gun)—the standard submachine gun in the United Kingdom—and the FN developed 7.62mm NATO MAG general purpose machine gun. The MAG, in modified form, is the L7A1 general purpose machine gun in the United Kingdom and the version of the MAG used in New Zealand is probably the same as the British gun.

THE CHARLTON .303 LIGHT MACHINE GUN AND RIFLE

Philip Charlton, a New Zealander developed a design for converting the Lee Enfield rifle into a selective fire weapon. He was issued a patent in 1941 and, due to the shortage of automatic weapons, work commenced on the conversion of old long Lee Enfields on hand into Charlton machine guns. A plant was established in Hastings, New Zealand to carry out this conversion. Approximately 2000 long Lee Enfields were converted when production was ceased. The Charlton weighed 16.5 pounds and had a cyclic rate of fire at about 700 rounds per minute. It is a gas operated weapon fed with a 30-round magazine, but could use the standard 10-round Lee Enfield magazine as well.

Essentially the conversion consisted of fitting a cylinder on the right side of the weapon originating at a point approximately midpoint of the barrel and extending beyond the rear of the action. This cylinder acted as a gas cylinder in its forward part and a piston tube for the remainder. The piston rod is attached to a slide which has an inclined cam that operates on the bolt handle stub pulling it to the rear and opening the action. A spring attached to the lower portion of the piston is mounted in a cylinder below the gas cylinder and returns the slide and piston forward. The standard Lee trigger mechanism has been modified by the addition of a disconnector system. A selector is mounted behind the trigger and engages the trigger. Setting it on safe prevents rearward movement of the trigger, setting it on "R" allows the trigger to rotate far enough to the rear to allow single shots to be fired and setting it on "A" lets the trigger come fully to the rear producing full automatic fire.

A hinged cover is placed over the action to protect it from dirt and a circular fin assembly, to assist in cooling the barrel, is placed over the rear half of the barrel. The semi-pistol grip of the Lee Enfield is cut away and a full pistol grip is mounted to the rear of the trigger guard. A heavy flash hider with protected front sight is mounted on the front end of the barrel. Australia also converted several thousand Lee Enfields to the Charlton design. The work in Australia was done by the Electrolux Company.

An earlier effort of Charlton's was the conversion of the .303 Rifle No. 1 Mark 3* to a semi-automatic rifle. This design was tested by the United Kingdom but not adopted.

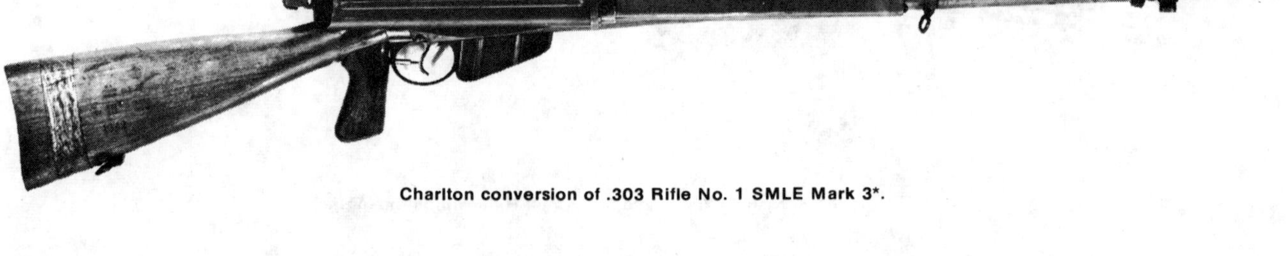

Charlton conversion of .303 Rifle No. 1 SMLE Mark 3*.

42-A

North Korea

The North Korean army was originally equipped with Soviet World War II weapons such as the 7.62mm Tokarev pistols, the 7.62mm M1944 carbine and M1891/30 rifle, the 7.62 mm DP and DPM machine guns, the 7.62mm Maxim machine gun, and the 12.7mm DShK M1938 machine gun. North Korea made the 7.62mm PPSh M1941 Submachine gun during the Korean war.

North Korea however is now equipped with more modern arms most of which are of native manufacture. The 7.62mm Model 68 pistol is standard, and the 7.62mm Model 68 assault rifle is a copy of the Soviet AKM, The 7.62mm SKS carbine, RPD light machine gun and the 7.62mm TUL-1 light machine gun are all used as are the 12.7mm DShK M1938/46 and the 14.5mm ZPU-1, ZPU-2 and ZPU-4 heavy machine guns.

NORTH KOREAN 7.62mm TYPE 68 PISTOL

The Type 68 is a North Korean variation of the Soviet Tokarev TT M1933 pistol. The Type 68 is shorter than the Tokarev and is considerably different in internal construction from the Tokarev. The Tokarev, like the Colt M1911 uses a link to pull down the barrel in unlocking. The Type 68 uses the cam type barrel lug as used on the Hi-Power Browning. The magazine catch on the Type 68 has been relocated to the bottom rear of the grip as opposed to the button type catch on the upper receiver of the Tokarev. The method of holding in the firing pin has also been changed. The same 7.62mm cartridge is used in both pistols.

Characteristics of the Type 68 Pistol

Caliber: 7.62mm
System of operation: recoil operated, semi-automatic only
Weight: 1.75 lb.
Length overall: 7.3 in.
Barrel length: 4.25 in.
Feed device: 8 round, in line, detachable box magazine
Sights: Front: blade
Rear: square notch
Muzzle velocity: 1295 f.p.s.

North Korean 7.62mm Type 58 Assault Rifle without magazine.

NORTH KOREAN 7.65mm TYPE 64 PISTOL

This pistol can be best explained as a mystery, it is a copy of the Model 1900 Browning. The mystery is why the North Koreans brought out this rather ancient piece of ordnance in the past decade. A version of this pistol with silencer has a shortened slide and fine thread on the barrel for the attachment of a silencer.

NORTH KOREAN 7.62mm TYPE 58 ASSAULT RIFLE

The Type 58 is a copy of the AK and comes in both wooden stocked and folding steel stocked versions. It does not appear to be as finely finished as AK's manufactured in other countries.

NORTH KOREAN 7.62mm TYPE 68 ASSAULT RIFLE

This weapon is a copy of the Soviet AKM. It does not have the rate reducer found on the AKM. It uses the standard 7.62 X 39mm "intermediate" sized cartridge.

NORTH KOREAN 7.62mm TUL-1 LIGHT MACHINE GUN

The TUL is somewhat like the Soviet RPK light machine gun, but is built on the AK type receiver rather than the AKM. It is also chambered for the 7.62 X 39mm "intermediate" sized cartridge.

North Korean 7.62mm Type 68 Pistol.

North Korean 7.65mm Type 64 Pistols; top—Silencer type; bottom—Type 64 without silencer.

43 Norway

The Norwegian Army is currently equipped with the following weapons: caliber .45 Model 1914 pistol, and the caliber .45 Model 1911A1 pistol; various submachine guns including the caliber .45 Thompson, the caliber .45 M3A1, and 9mm Parabellum Sten guns. The caliber 7.62mm NATO G3 rifle is standard, but U.S. caliber .30 M1 rifles, 7.92mm Kar98k and British Lee Enfield rifles are probably still held. U.S. caliber .30 Browning guns, Browning automatic rifles and British Bren caliber .303 Bren guns were furnished to Norway after World War II. The U.S. caliber .50 Browning machine gun is also used. Norway has adopted the 7.62mm NATO MG 42/59 machine gun as standard.

NORWEGIAN SERVICE PISTOLS

Norway adopted a single-action Nagant revolver in 1883, but apparently only 794 of the weapons were procured. In 1893, the 7.5mm Model 1893 Nagant revolver was adopted. This weapon, which is basically the same as the Russian 7.62mm Model 1895 Nagant, was the standard pistol until the adoption of the caliber .45 Model 1914 pistol. The Model 1914 is basically the same as the U.S. caliber .45 M1911 Colt pistol except for the external shape of the slide stop.

The first 300 caliber .45 pistols were procured by Norway from Colt and are unmodified Model 1911s. All of the Model 1914 arms were made at the Norwegian government arsenal at Kongsberg. Since World War II, Norway has obtained quantities of caliber .45 Model 1911A1 automatics from the United States.

In the period after World War II, considerable quantities of 9mm P-08 Luger and P-38 pistols were used by the Norwegian forces.

Norwegian caliber .45 Model 1914 Automatic Pistol.

NORWEGIAN RIFLES AND CARBINES

Norway adopted the Krag Jorgensen rifle, chambered for the 6.5 x 55mm rimless Mauser cartridge, in 1894 and this rifle and its carbine versions were used by the Norwegian Army until the majority of the weapons were lost to the Germans in World War II. Some Model 1894 rifles were made at Steyr but the majority were made by Kongsberg. A few were made during the German occupation; these bear German Waffenamt inspection stamps.

THE NORWEGIAN KRAG

The Norwegian Krag is generally similar to the U.S. caliber .30 Krag, but does not have a cutoff. As with the U.S. Krag, there is a single frontally-mounted locking lug. It has been stated many times that the Krag action is not suitable for high-pressure cartridges, but the Norwegian Krag was made after World War II (for a limited period of time) chambered for the 7.92 x 57mm cartridge —hardly a low pressured cartridge in anyone's estimation. This is mainly a matter of metallurgy—Norwegian Krags produced since World War I are obviously made of steel alloys that are better than those used in the United States Krag, which was made from 1892-1902 when metallurgy was not too precise.

Distinctive Details of Norwegian Rifles and Carbines

The various model of rifles and carbines can be distinguished by the following characteristics:

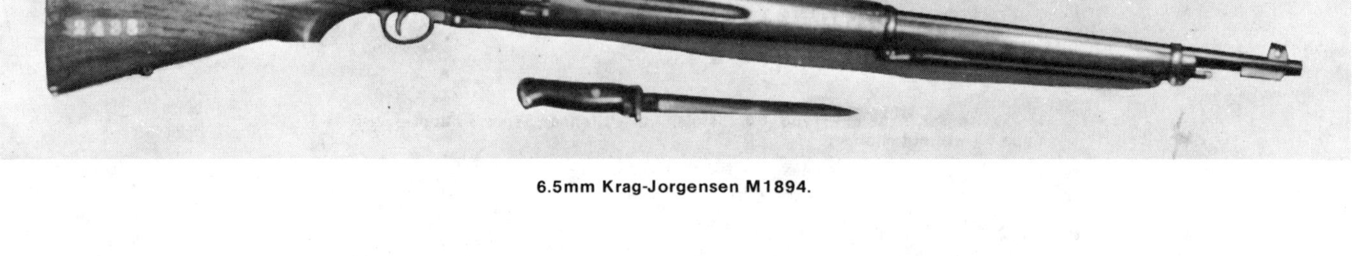

6.5mm Krag-Jorgensen M1894.

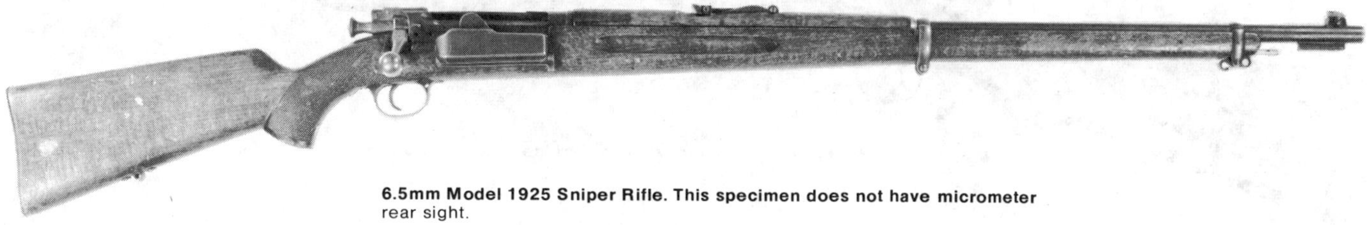

6.5mm Model 1925 Sniper Rifle. This specimen does not have micrometer rear sight.

Rifle Model 1894. Stock with pistol grip, half-length cleaning rod mounted in forward part of stock; bayonet lug mounted under the barrel and a half-length handguard is fitted.

Sniper Rifle Model 1923. Full-length stock with full checked pistol grip; full-length handguard, wide upper band/ nose cap with bayonet lug, micrometer-type rear sight. The rifle is marked M/1894.

Sniper Rifle Model 1925. Basically the same as the Rifle Model 1894, but has checked full pistol grip, micrometer-type rear sight, marked M/25.

Sniper Rifle Model 1930. Has sporter-type stock with checked full pistol grip, heavy barrel, no bayonet lug, micrometer rear sight, marked M/1894/30.

Carbine Model 1895. Sporter-type stock, no bayonet lug, similar in appearance to the United States Krag carbine.

Carbine Model 1897. Same as the Model 1895 carbine, but butt swivel is positioned further toward the rear of the butt.

Carbine Model 1904. Full stock with pistol grip and full length handguard, no bayonet lug.

Carbine Model 1907. Basically the same as the Model 1904, but sling swivels are positioned on rear band and on the butt.

6.5mm Carbine Model 1912. Stocked to the muzzle with full-length handguard, similar to the Sniper Rifle Model 1923. Has bayonet lug mounted on combination upper band nose cap.

After World War II, the Norwegian Army had few Norwegian Krags, but many German 7.92mm Mauser Kar98ks. United States and British rifles were also supplied. In 1959, a caliber .30-06 heavy barrel modification of the Mauser with sporter-type stock and micrometer sight was produced for the Norwegian National Rifle Association. A few Krags were produced after the war, but costs were prohibitively high and there was no intention to equip the armed forces with the Krag, since it can hardly be considered a modern rifle.

In 1964, the Norwegians adopted the West German 7.62mm

Characteristics of Norwegian Rifles and Carbines

	Rifle M1894	Sniper Rifle Model 1923	Sniper Rifle Model 1925	Sniper Rifle Model 1930
Caliber:	6.5 x 55mm.	6.5 x 55mm.	6.5 x 55mm.	6.5 x 55mm.
System of operation:	Turn bolt.	Turn bolt.	Turn bolt.	Turn bolt.
Length overall:	49.9 in.	Approx 44 in.	49.7 in.	48 in.
Barrel length:	29.9 in.	Approx 24 in.	30 in.	29.5 in.
Feed device:	5-round horizontal box.	5-round horizontal box.	5-round horizontal box.	5-round horizontal box.
Sights: Front:	Barley corn.	Barley corn.	Barley corn.	Hooded barley corn.
Rear:	Tangent.	Micrometer with aperture.	Micrometer with aperture.	Micrometer with aperture.
Weight:	9.38 lb.	Approx 9 lb.	9.9 lb.	11.46 lb.
Muzzle velocity:	Approx 2625 f.p.s.	Approx 2600 f.p.s.	Approx 2625 f.p.s.	Approx 2625 f.p.s.
Cyclic rate:	------------	------------	------------	------------

Characteristics of Norwegian Rifles and Carbines (Cont'd)

	Carbine M1895	Carbine M1904	Carbine M1912	G-3
Caliber:	6.5 x 55mm.	6.5 x 55mm.	6.5 x 55mm.	7.62mm NATO.
System of operation:	Turn bolt.	Turn bolt.	Turn bolt.	Gas, selective fire.
Length overall:	40 in.	40 in.	43.6 in.	40.2 in.
Barrel length:	20.5 in.	20.5 in.	24 in.	17.7 in.
Feed device:	5-round horizontal box.	5-round horizontal box.	5-round horizontal box.	20-round staggered row box magazine.
Sights: Front:	Barley corn.	Barley corn.	Barley corn.	Hooded post.
Rear:	Tangent.	Tangent.	Tangent.	Aperture.
Weight:	7.5 lb.	8.4 lb.	8.8 lb.	9.9 lb.
Muzzle velocity:	Approx 2575 f.p.s.	Approx 2575 f.p.s.	2600 f.p.s.	2624 f.p.s.
Cyclic rate:	----------------	---------------	--------------	500-600 r.p.m.

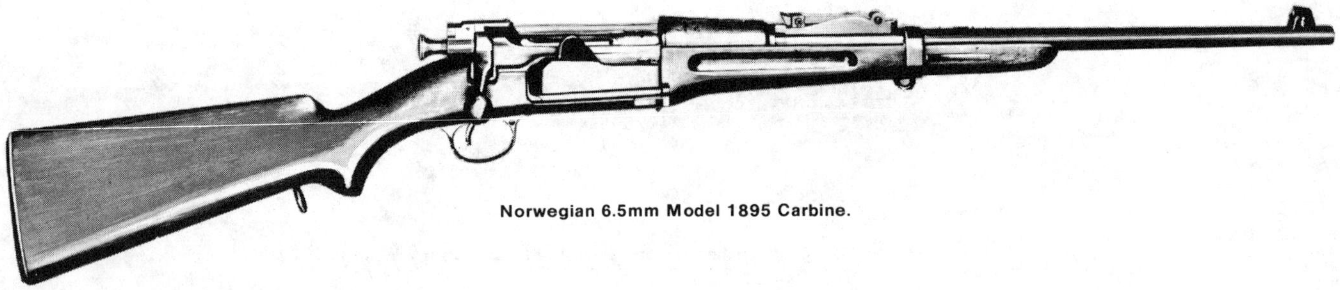

Norwegian 6.5mm Model 1895 Carbine.

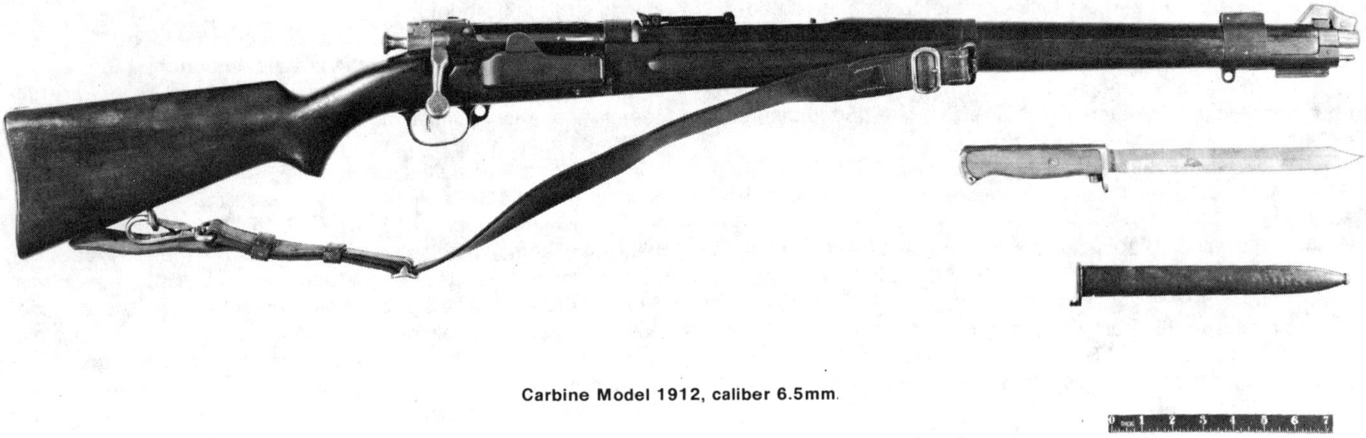

Carbine Model 1912, caliber 6.5mm.

NATO G3 rifle and that rifle is now in production at Kongsberg. The Norwegians have modified the G3 slightly. The bolt operating handle has been removed and the bolt is pulled to the rear by means of a finger hole, like the U.S. M3A1 submachine gun; the sling arrangement has been changed; and two different stock lengths are issued. A semiautomatic sniper model with telescope has also been adopted. The 7.62mm NATO MG 42/59 machine gun is now standard in Norway where it is called the LMG 3.

NORWEGIAN MACHINE GUNS

Norway adopted the Hotchkiss gun chambered for the 6.5mm cartridge in 1911, and also used 6.5mm Model 1914 and 1918 Madsen guns. The Browning water-cooled machine gun chambered for a 7.9 x 61mm cartridge was also used and called the Model 29. This weapon is similar to the Colt M38B machine gun.

After World War II, Norway obtained U.S. caliber .30 Browning guns, British machine guns, and also used 7.92mm MG34 and MG42 machine guns.

44

The Polish Army is currently mainly equipped with Soviet designed weapons. Pistols used include the Polish 9mm Model 64, the Soviet 9mm Makarov and the Polish 9mm Model 63 Machine Pistol. The Soviet designed Polish made 7.62mm AK and AKM assault rifles and 7.62mm SVD sniper rifles are used as are the 7.62mm RPK, RPD, PK/PKS and PKT machine guns.

The 7.62mm SGM machine gun and its tank version, the SGMT are still used as is the 7.62mm RP-46 light machine gun. The 12.7mm DShK M1938/46 is used on vehicles and the 14.5mm ZPU2 and ZPU4 heavy machine guns are also used.

Older Soviet weapons such as the 7.62mm Model 1944 carbine, the 7.62mm PPSh M1941 submachine gun, the 7.62mm DPM and SG43 machine guns, as well as the Polish 7.62mm M43/52 submachine gun, may still be encountered.

Poland

POLISH PISTOLS

THE POLISH 9mm MODEL 35 ("RADOM") PISTOL

A modification of the Colt Browning locked-breech pistol was designed by the Poles at the Fabryka Broni Radom, a government arms plant. The design is credited to P. Wilniewczyc and J. Skrzypinski. This 9mm Parabellum weapon, called the Vis Model 35 or more popularly the Radom, was the Polish service pistol and the Germans continued its manufacture after they occupied Poland. The Model 35 Radom locks and unlocks in a manner quite similar to that of the Browning FN Hi-Power pistol, in that it has a nose forged on the rear underside of the barrel which serves as a cam to pull the barrel down out of the locked position, rather than using the barrel link of the U.S. Colt Model 1911 pistol.

As manufactured for the Polish forces, the weapon has the firing pin retracting device mounted on the left rear of the slide and a slide lock, which appears to be a safety, mounted on the left rear side of the receiver. Most of the pistols manufactured during the German occupation do not have the slide lock. The firing pin retracting device is used to lower the hammer on a loaded chamber. This weapon, although it has a grip safety, does not have the conventional Colt-Browning manual-safety catch. The slide lock is used to lock the slide to the rear to assist in field stripping. Pistols manufactured prior to the German occupation are very well made and are marked on the slide with the Polish eagle and Polish markings. The Model 35s manufactured for the Germans are of much rougher construction and bear the German Waffenamt proof marks. The Radom Model 35 is no longer in service in any Army.

Manufacture of the 7.62mm Tokarev Model TT M1933 was begun in Poland after World War II; this pistol is being replaced by the 9mm Makarov (PM) pistol and the 9mm Stechkin (APS) machine pistol. Characteristics of these weapons will be found in the chapter on the U.S.S.R.

Characteristics of Model 1935 Pistol

Caliber: 9mm Parabellum.
System of operation: Recoil, semiautomatic.
Length overall: 7.8 in.
Barrel length: 4.7 in.
Feed device: 8-round, in line, detachable box magazine.
Sights: Front: Blade.
Rear: "V" notch.
Weight: 2.25 lb.
Muzzle velocity: Approx. 1150 f.p.s.

9mm Model 35 Pistol manufactured for the Polish Army.

9mm Model 35 Pistol manufactured for the Germans with slide lock.

9mm Model 35 Pistol manufactured for the Germans without slide lock.

Polish 9mm Model 64 Pistol.

THE POLISH 9mm MODEL 64 PISTOL

This double action, blowback operated pistol is smaller in size than the Soviet Makarov pistol, but is chambered for the same 9mm cartridge. In some respects it seems to resemble the Walther PPK more than it does the Makarov as for example the trigger bar linkage system and the internally mounted slide stop, but the system is actually different than either the Makarov or the Walther.

Characteristics of 9mm Model 64 Pistol

Caliber: 9mm Makarov
System of operation: Blowback, semi-automatic only
Length overall: 6.1 in.
Barrel length: 3.3 in.
Feed device: 8-round, in-line, detachable box magazine
Sights: Front: Blade
Rear: Notch
Weight: 1.5 lb.
Muzzle velocity: 1017 f.p.s.

The Model 64 is disassembled in a manner similar to that of the Walther PP and PPK. The magazine catch is at the bottom rear of the grip and the safety catch is mounted on the slide and is pushed down to engage. In this position, it blocks the firing pin.

Polish 9mm Model 64 field stripped.

POLISH 9mm MODEL 63 MACHINE PISTOL

This weapon is similar to the Soviet 9mm Stechkin and the Czech 7.65mm M61 "Skorpion" machine pistols in concept. It is a selective fire weapon which can be used as a pistol or as a shoulder weapon. According to Polish magazines, it is intended for issue to junior officers in combat units and to armored troops.

Characteristics 9mm Model 63 Machine Pistol

Caliber: 9mm Makarov
System of operation: Blowback, automatic only
Length overall: w/folded stock--13.1 in.
Feed device: 15- and 25-round, detachable box magazines
Sights: Front: Blade
Rear: "L" type
Weight: 3.96 lb.
Muzzle velocity: 1065 f.p.s.
Cyclic rate: 600 r.p.m.

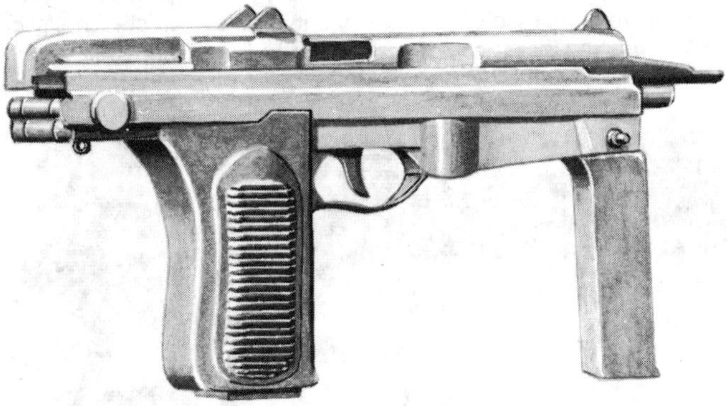

Polish 9mm Model 63 Machine Pistol.

This weapon has a sliding type metal shoulder stock which is used in conjunction with a vertical fore grip. When the stock is in the retracted position and the vertical fore grip is pushed up, it extends beyond the muzzle of the weapon; the weapon can be carried in a holster and fired with one hand. The butt plate portion of the shoulder stock folds up under the rear of the receiver.

POLISH RIFLES AND CARBINES

When Poland gained her freedom after World War I, stocks of Russian 7.62mm M1891 rifles and carbines, and 7.92mm Mauser 98 carbines and rifles were available to them.

POLISH-PRODUCED RIFLES

The first rifle produced by the Poles contained features of both Russian and German design. The Polish 7.92mm Model 91/98/25 rifle has the Russian Mosin Nagant action, but Mauser-type bands and fittings. The bolt head has been modified to use the 7.92mm cartridge.

The manufacture of Mauser rifles and carbines started at the Warsaw Arsenal soon after World War I. The Polish 7.92mm Karabin 98a is basically the same as the later German rifle 98. Similarly the Polish Karabinek 98 is basically the same as the German Kar 98a.

The 7.92mm Karabin 29 is a variant of the Czech Model 24, differing from the Czech weapon mainly in the location of the sling swivels and the front sight.

After World War II, the Poles were initially equipped with various models of the Soviet 7.62mm Mosin-Nagant rifle and carbine. The Poles now produce the Soviet designed 7.62mm AKM assault rifle. The characteristics of these weapons are covered in the chapter on the U.S.S.R.

7.92mm Model 91/98/25 Rifle.

Characteristics of Polish Rifles

	Rifle 91/98/25	Rifle 98a	Carbine 98	Rifle 29
Caliber:	7.92mm			
System of operation:	Turn bolt			
Length overall	43.3 in.	49.2 in.	43.3 in.	43.4 in.
Barrel length:	23.6 in.	29.1 in.	23.6 in.	23.6 in.
Feed device:	5-round, in-line, nondetachable box magazine.	5-round staggered row non-detachable box magazine.	5-round staggered row non-detachable box magazine.	5-round staggered row non-detachable box magazine.
Sights: Front:	Barley corn.	Barley corn.	Barley corn with protecting ears.	Barley corn with protecting ears.
Rear:	Leaf.	Tangent.	Tangent.	Tangent.
Weight:	8.16 lb.	9 lb.	8 lb.	9 lb.
Muzzle velocity:	Approx 2470 f.p.s.	Approx 2570 f.p.s.	Approx 2470 f.p.s.	Approx 2470 f.p.s.

7.92mm Carbine Model 98.

7.92mm Model 29 Rifle.

POLISH 7.62mm AK ASSAULT RIFLE WITH GRENADE LAUNCHER

This weapon, which has been encountered in the hands of the Viet Cong, is a modification of the AK made in Poland fitted with a rifle grenade launcher. It has been called the PMK-DGN and also the KbKg Model 1960.

The grenade launcher, which is not the same size as those used in the NATO countries, is screwed onto the threaded end of the barrel and locked in place by the spring loaded plunger which is found mounted in the front sight base of the AK. A valve is fitted into the right side of the gas cylinder to allow shutting off the gas from the gas port of the barrel when firing grenades. The grenade launcher sight is mounted on the hand-guard retaining pin. A heavy rubber recoil pad is attached to the stock with straps and a special ten-round magazine is used when launching grenades from this rifle. This magazine has a filler block in it that prevents it from being used with bulleted cartridges.

Polish 7.62mm AK Assault Rifle with Grenade launcher (PMK-DGN or KbKg M1960)

POLISH SUBMACHINE GUNS

The Poles were supplied with the Soviet 7.62mm PPSh M1941 and PPS M1943 submachine guns. They developed a variation of the PPS M1943, which was manufactured in Poland.

THE POLISH 7.62mm M1943/52 SUBMACHINE GUN

The M1943/52 submachine gun is a modification of the Soviet PPS M1943. Its characteristics are basically the same as those of the Soviet weapon, except that it has a wooden stock and is reportedly capable of semiautomatic as well as automatic fire.

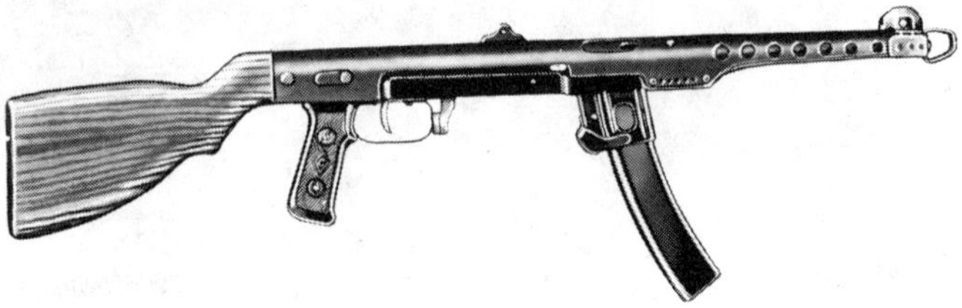

Polish 7.62mm M1943/52 Submachine Gun.

Special Note on the Polish M1943/52

Loading, firing, functioning, and disassembly of this weapon are the same as for the Soviet 7.62mm submachine gun PPS M1943.

Characteristics of M1943/52 Submachine Gun

System of operation: Blowback, selective fire.
Length, overall: 32.72 in.
Barrel length: 9.45 in.
Feed device: 35-round, detachable staggered box magazine.
Sights: Front: Flat-topped post, adjustable for elevation and windage.
Rear: L-type, open V-notch with setting for 100 and 200 meters.
Muzzle velocity: 1640 f.p.s. w/Soviet 7.62mm Type P pistol cartridge.
Cyclic rate: 600 r.p.m.
Weight: 8 lb.

POLISH MACHINE GUNS

The Poles used a 7.92mm water-cooled Browning Machine Gun, which they called the Model 30. They had previously adopted a 7.92mm version of the Browning automatic rifle in 1928. The Polish Browning guns are basically the same as those used by the United States. They are no longer in use in the Polish Army.

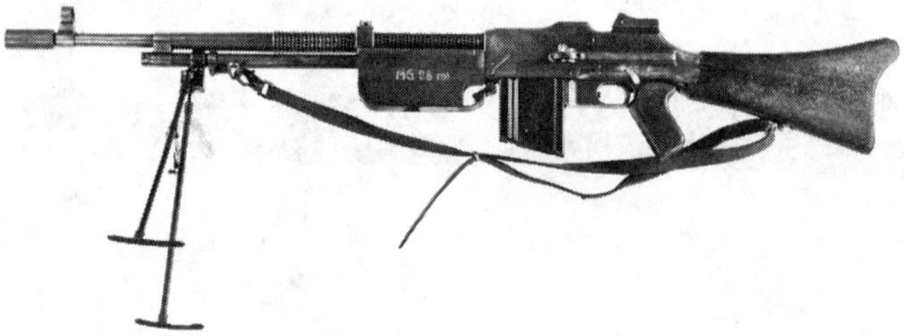

7.92mm Polish Model 28 Browning Automatic Rifle.

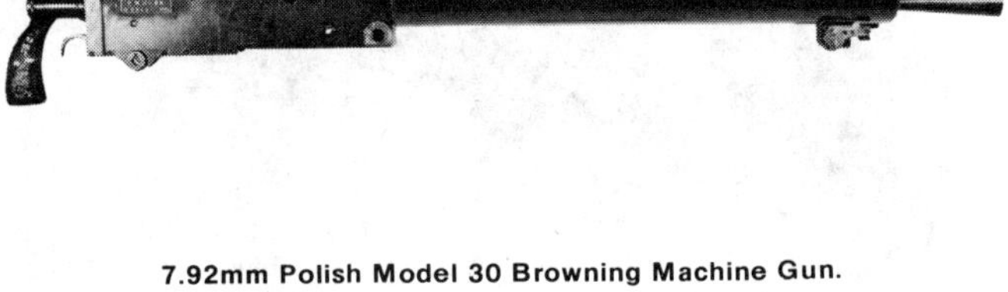

7.92mm Polish Model 30 Browning Machine Gun.

45

Portugal

Portugal has recently been securing odd lots of military small arms from many sources in the West. Most Portuguese service arms are procured from abroad, but the 7.62mm NATO Rifle M961—German G3 rifle—and the 9mm Model 48 F.B.P. submachine gun are made at the Portuguese government arms plant, Fabrica de Braco de Prata. Because of the large variety of weapons in service in Portugal, only those which are unique in Portugal are covered in any detail. The 7.62mm NATO MG42/59 and the HK21 machine guns are among the more recent weapons procured.

PORTUGUESE SMALL ARMS IN WORLD WAR II

During the period preceding World War II, and during the war itself, Portugal procured a large number of different types of small arms from Germany and Great Britain. There were still some quantities of the older Portuguese weapons in service as well. As a result Portugal holds a somewhat varied collection of small arms in various calibers.

Among the weapons used by Portugal are the following:

Pistols: 7.65mm Savage M/908 and M/915, 9mm M/43 Parabellum (Luger) 9mm M/908 Parabellum (Luger).

Rifles: 7.92mm Mauser M/937 (same as Kar. 98k).

Cal. 303 Lee Enfield M/917 (same as SMLE MkIII and III*).

Portugal has adopted the German G3 (CETME) assault rifle in 7.62mm NATO caliber as standard. The Portuguese call this weapon, Rifle 7.62mm M/961. The 7.62mm NATO FN "FAL" rifle is also used.

Submachine Guns: 9mm M1934 Bergmann, 9mm M/43 (Sten), 9mm Model 48 F.B.P., 9mm M/42 (Schmeisser).

Machine Guns: Cal. .303 Lewis M/917 (same as British Mark I Lewis).

Cal. .303 Vickers M/917 (same as British MK 1 Vickers).

7.92mm Breda M/938 (Breda M1937 in 7.92mm).

7.92 Madsen M/940

Cal. .303 M/931 (Vickers Berthier)

7.92mm Dreyse M/938 (same as German MG 13)

Cal. .303 M/43 (Bren)

Cal. .303 M/930 (Madsen).

PORTUGUESE PISTOLS

The only pistol used by the Portuguese which was not used by any other country was the 7.65mm (.32 ACP) Savage M1907 and M1915 pistols, called the M/908 and M/915 pistols by the Portuguese. Although these pistols have been or are being disposed of by the Portuguese government, they are interesting weapons and bear a bit of description.

The Savage Model 1907 is based on the patents of E. H. Searle, first issued in 1904. The weapon is delayed blowback and has an action which is similar in some respects to the Model 12 Steyr. A lug on the top of the barrel is butted against a cam surface on the top underside of the slide. The cam surface is initially angled and then parallel to the axis of the bore, causing the barrel to rotate, delaying the action slightly although the system starts opening from the moment of firing. The M/908 has a rounded and notched cocking lever, which has the appearance of and is mounted in the position of a hammer. The striker of the Savage pistols is spring loaded and is released by the sear. The cocking lever does not strike the firing pin as does the hammer in the Colt

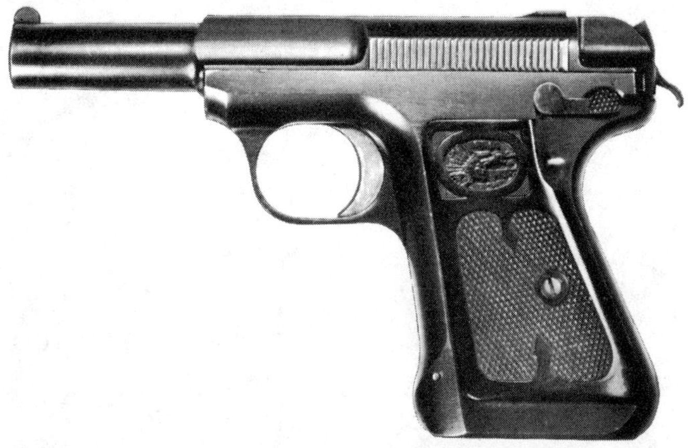

7.65mm Model 915 Pistol.

Browning design. The cocking lever of the M/908 should not be let down with a cartridge in the chamber.

PORTUGUESE 7.65mm M/915 PISTOL

The M/915 differs from the M/908 mainly in that it has a spur-type cocking lever rather than the rounded type used on the M/908. The M/915 has a greater number of grasping grooves to assist in pulling the slide to the rear than does the M/908. The spur-type cocking lever was introduced in the Savage commercial pistols in 1917. The Model 908 is 6.5 inches overall with a 3.8-inch barrel and weighs 1.2 pounds. It has a 10-round, staggered row, box magazine. Characteristics of the Model 915 are generally similar.

PORTUGUESE RIFLES

PORTUGUESE 6.5mm M1904 MAUSER VERGUEIRO RIFLE

The only rifle used by Portugal that is unique to that country is the 6.5mm Model 1904 Mauser Vergueiro. This weapon is somewhat peculiar in that its bolt handle seats ahead of the receiver bridge, like that of the Mannlicher or the Italian Mannlicher Carcano, rather than behind the receiver bridge as with most Mausers. The Model 1904 also has a separate bolt head; the trigger mechanism, bolt stop, and ejector are of Mauser 98 type, but this weapon has no bolt sleeve; therefore the safety is attached to the cocking piece.

These rifles were all made by D. W. M. and originally chambered for the 6.5 x 58mm cartridge which was used only as a military cartridge by Portugal. During the thirties, some of these weapons were converted to 7.92mm. This rifle is not likely to be found in service at present.

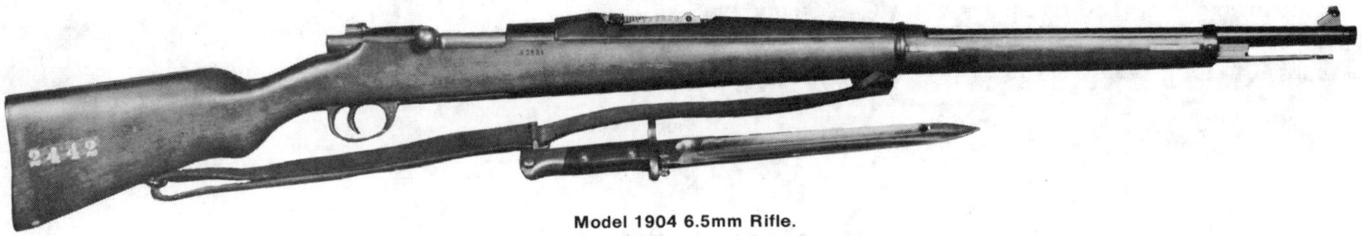

Model 1904 6.5mm Rifle.

PORTUGUESE SUBMACHINE GUN

The 9mm Parabellum Model 48 F.B.P. is the only submachine gun of Portuguese design in service with the Portuguese army. All other weapons in service are covered under their country of origin.

THE 9mm MODEL 48 F.B.P.

There are no really unusual features in the design of this weapon. It has a massive bolt and a telescoping operating spring assembly as originated by Erma and made famous by the MP38 and MP40. This weapon is made in Portugal.

Characteristics of the Model 48 F.B.P.

Caliber: 9mm Parabellum.
System of operation: Blowback, automatic only.
Weight, empty: 8.2 lbs.
Length overall: Stock extended, 32 inches; stock folded: 25 in.
Barrel length: 9.8 ins.
Sights: Front: Blade.
Rear: Fixed aperture graduated for 100 meters.
Feed device: 32-round, detachable staggered row, box magazine.
Muzzle velocity: 1280 f.p.s.
Cyclic rate: 500 r.p.m.

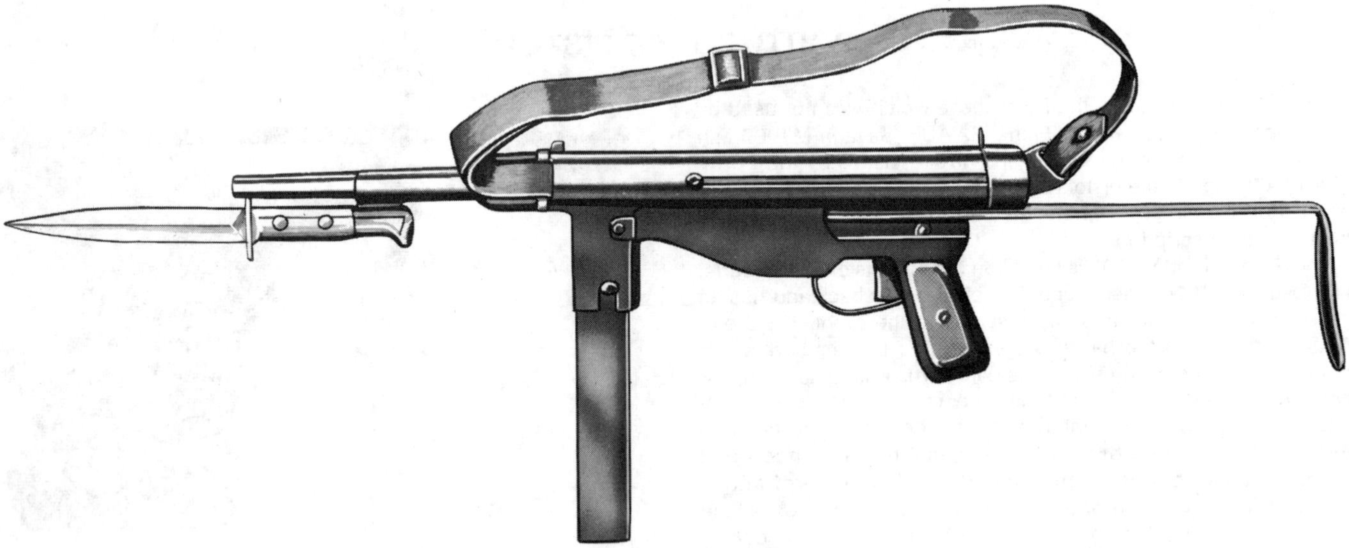

9mm Model 48 F.B.P. Submachine Gun.

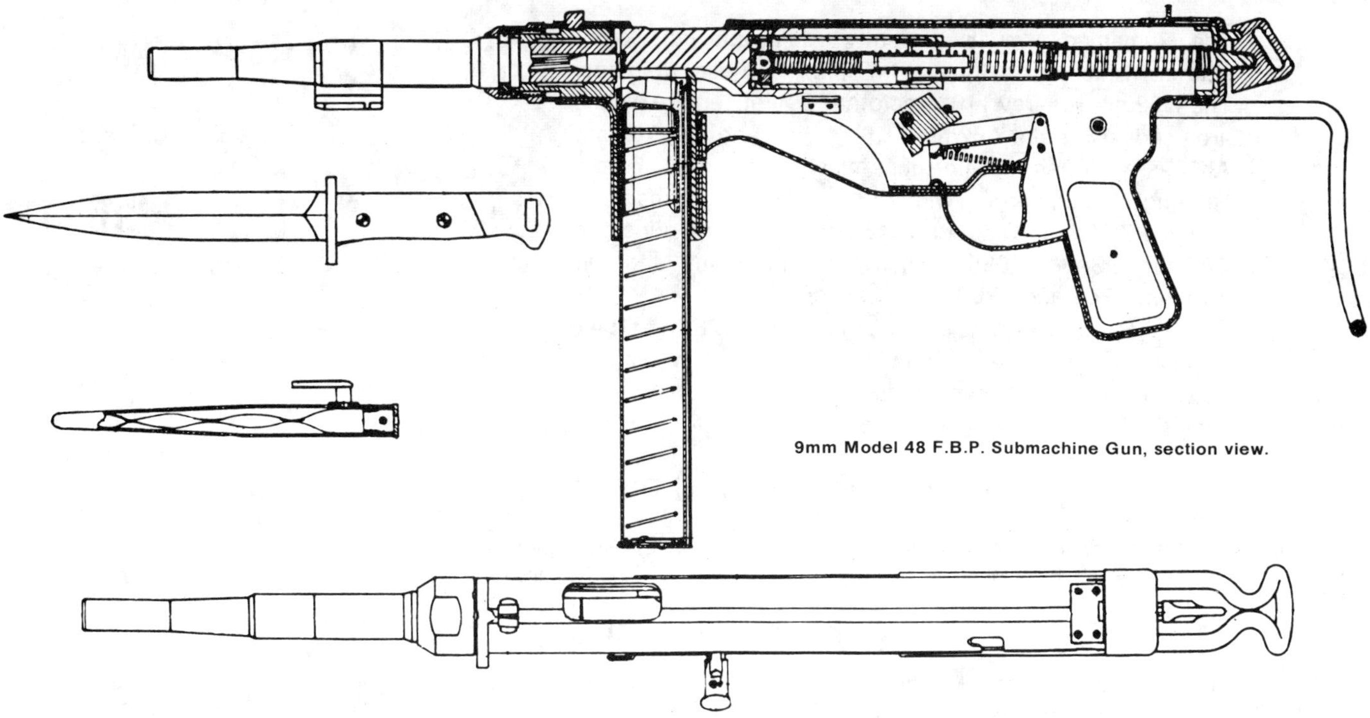

9mm Model 48 F.B.P. Submachine Gun, section view.

PORTUGUESE MACHINE GUNS

There are no machine guns in service in the Portuguese Army which are of Portuguese origin. The 7.92mm M/938 is not covered in detail under Germany, its country of origin; therefore some details on this weapon are given here.

THE 7.92mm M/938 MACHINE GUN

This weapon is the German MG13, which was sold by Germany to Portugal in 1938. It was used as the standard German light machine gun until the adoption of the MG34. The weapon is a modified Dreyse M1918.

Characteristics of M/938 Machine Gun

Caliber: 7.92mm.
System of operation: Recoil, selective fire.
Weight: 26.4 lb.
Length, overall: 57 in. (approx).
Barrel length: 28.25 in.
Sights: Front: Folding blade.
Rear: Tangent leaf.
Cyclic rate: 750 r.p.m.
Feed device: 25-round box magazine or 75-round saddle drum.
Muzzle velocity: 2600 f.p.s. (approx.).

Special Note on the M/938 Machine Gun

It should be noted that this weapon uses two features which were also found on the later MG34: the trigger with two half-moon sections, which are used for either automatic or semiautomatic fire, depending on whether the bottom or the top section of the trigger is squeezed; and the saddle drum magazine. (However, the mounting of the saddle drum on the MG34 is different.)

Portugese 7.92mm M/938 Machine Gun.

46 Rumania

The Rumanian Army is currently equipped with Soviet service weapons or Rumanian copies of these weapons. The 7.62mm Tokarev TT33 and other Soviet service pistols are used, as are the 7.62mm SKS carbine and the 7.62mm AKM assault rifles. Machine guns in use include the 7.62mm RPD light machine gun, the 7.62mm RP46 light machine gun, the 7.62mm SGM and SG43 heavy machine guns, the 12.7mm DShK M1938/46 heavy machine gun and the 14.5mm ZPU2 and ZPU4 heavy machine guns.

Older Soviet weapons such as the 7.62mm PPSh M1941 submachine gun, the 7.62mm Model 1944 Mosin-Nagant carbine, and the 7.62mm DP and DPM light machine guns may still be found.

RUMANIAN PISTOLS

The principal pistols in service in the Rumanian Army prior to World War II were the Austrian 9mm Model 12 Steyr and the 9mm FN Browning Hi-Power. Since World War II, the 7.62mm Tokarev TT M1933 and other Soviet pistols have been used.

RUMANIAN RIFLES

In 1892, Rumania adopted a 6.5mm Mannlicher Rifle. The Rumanian 6.5mm cartridge is the same as that used by the Dutch and is known as the 6.5mm Model 93 or the 6.5 x 53mm R. The Rumanian Mannlicher is a turn-bolt type with typical removable bolt head and clip-loading magazine with a clip that functions as part of the magazine, as with the U.S. M1 rifle.

The 1892 and Model 1893 Mannlicher rifles differ only in sight graduations, position of the ejector (on the bolt on the Model 1892 and on the receiver on the Model 1893), and in the presence of a stacking rod on the left side of the upper band of the Model 1893.

The Model 1893 rifle differs from the Dutch Model 1895 only in minor details. A carbine version of the Mannlicher was also used by the Rumanians.

After World War I, Rumania obtained considerable quantities of French equipment, including 8mm Mannlicher Berthier rifles and carbines. The 7.92mm Czech Model 24 rifle, and Austrian 8mm Model 1895 rifles and carbines, were also in service. The Model 24 was manufactured in Rumania. The Rumanian Army was initially equipped with various models of the Soviet 7.62mm Mosin Nagant rifles. The Soviet 7.62mm SKS carbine and 7.62mm AK assault rifle are the current standard rifles. Characteristics of rifles other than the Rumanian Mannlicher are given under country of origin.

Rumania now manufactures a modified copy of the Soviet 7.62mm AKM. The Rumanian AKM has a vertical fore grip which extends downward from the forend.

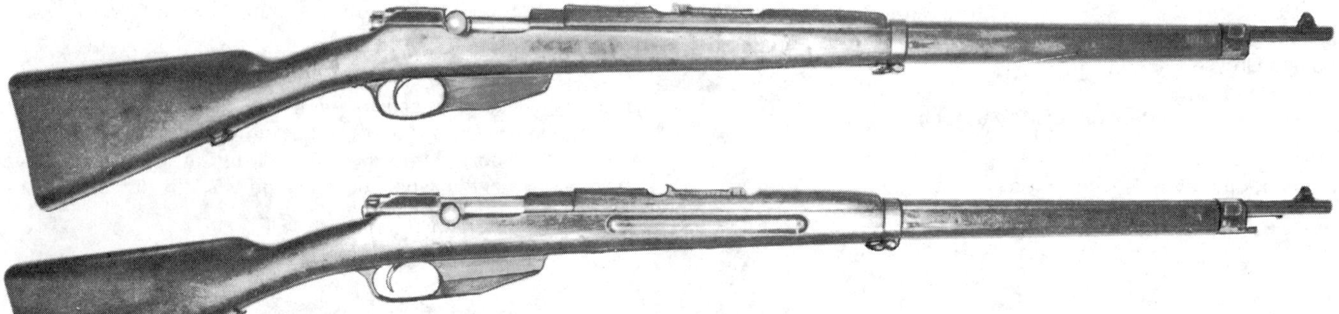

6.5mm Model 1892 Rifle (above); 6.5mm Model 1893 Rifle (below).

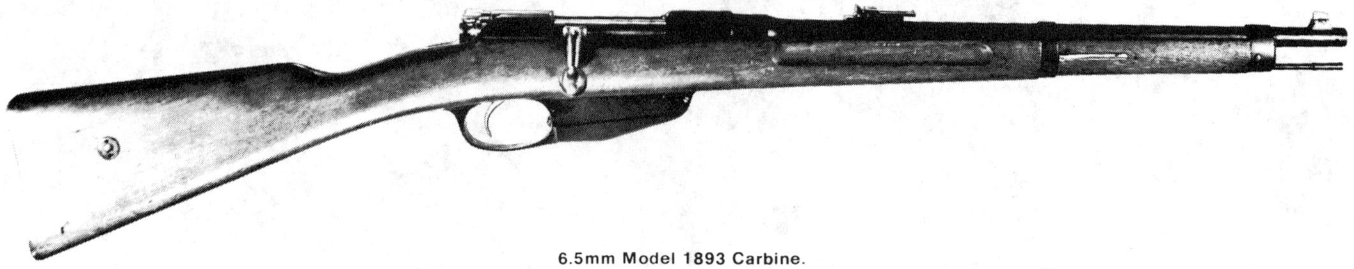

6.5mm Model 1893 Carbine.

CHARACTERISTICS OF RUMANIAN RIFLES

	Rifle M1892	Rifle M1893	Carbine M1893
Caliber:	6.5 x 53mm R.	6.5 x 53mm R.	6.5 x 53mm R.
System of operation:	Turn bolt.	Turn bolt.	Turn bolt.
Length overall:	48.3 in.	48.3 in.	37.5 in.
Barrel length:	28.5 in.	28.6 in.	17.7 in.
Feed device:	5-round, in-line, non-detachable box magazine		
Sights: Front:	Barley corn.	Barley corn.	Barley corn.
Rear:	Leaf.	Leaf.	Leaf.
Weight:	8.9 lb.	9 lb.	7.25 lb.
Muzzle velocity:	Approx. 2400 f.p.s.	Approx. 2400 f.p.s.	Approx. 2325 f.p.s.

RUMANIAN SUBMACHINE GUNS

Prior to World War II, the Rumanian government purchased 9mm Parabellum Model 1938A Beretta submachine guns and during the war purchased 9mm Parabellum Beretta 38/42 submachine guns. In addition, the Rumanians designed and produced a submachine gun of their own, the 9mm Model 1941 Orita.

THE 9mm MODEL 1941 ORITA SUBMACHINE GUN

The only notable weapon of Rumanian design that was used during World War II was the Orita submachine gun. This weapon has been replaced by Soviet weapons in Rumania at the present time, but is covered as a matter of interest.

There is also a version of the Orita submachine gun with a folding metal stock.

Characteristics of Model 1941 Orita Submachine Gun

Caliber: 9mm Parabellum.

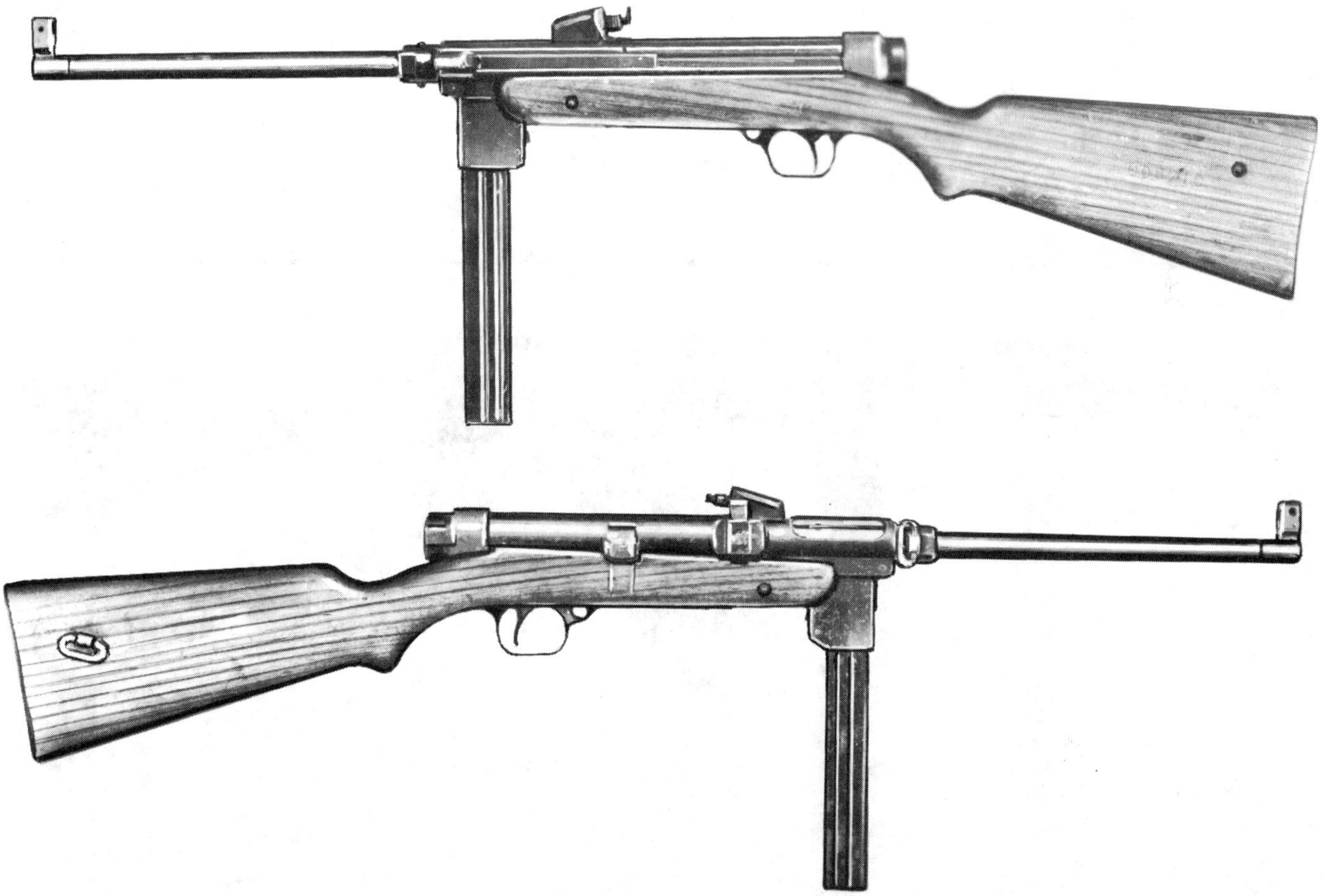

Rumanian 9mm Orita submachine gun.

System of operation: Blowback, selective fire.
Weight, loaded: 8.8 lb.
Length, overall: 35.2 in.
Barrel length: 11.3 in.
Feed device: 32-round, staggered row, detachable box magazine.
Sights: Front: Hooded blade.
Rear: Open V, adjustable, graduated from 100 to 500 meters.
Muzzle velocity: 1280 f.p.s.
Cyclic rate of fire: 400 r.p.m.

RUMANIAN MACHINE GUNS

The Rumanians used a rather mixed collection of machine guns prior to World War II. French 8mm Model 1915 light machine guns (Chauchats) and 8mm Model 1914 Hotchkiss and 13.2mm Model 1932 Hotchkiss guns were used. Schwarzlose 8mm Model 1907/12 and the 7.92mm Czech ZB30 machine guns were also used and the ZB30 was made in Rumania at Cugir.
There were probably other types of machine guns in service as well.

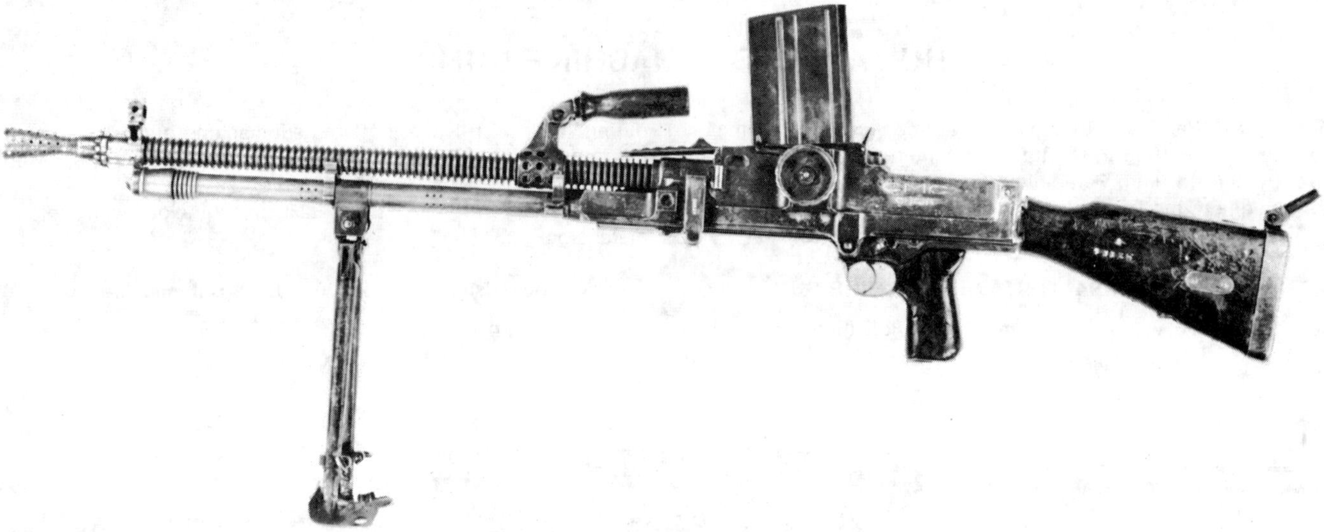

7.92mm ZB30 of Rumanian manufacture.

Rumanian modification of 7.62mm AKM Assault Rifle.

Spain

The Spanish Army is equipped with the following weapons: the 9mm Super Star pistol, the 9mm Star Model Z45 submachine gun, the Star 9mm Model Z-62 submachine gun, the 7.92mm Model 43 rifle, the 7.62mm NATO Model 1916 carbine, 7.62mm NATO Model 58 assault rifle, the 7.62mm NATO FA059 light machine gun, the 7.92mm FAO light machine gun, the 7.92mm ALFA 1944 heavy machine gun. Spain has also adopted the 7.62mm NATO MG 42/59. A quantity has been purchased from Rheinmettal and series manufacture is being carried on at Oviedo. The U.S. caliber .30 and .50 Browning machine guns are in use on U.S. armored vehicles in the Spanish Army.

47

SPANISH PISTOLS

The first automatic pistol adopted by Spain was the 9mm Campo Giro Model 1913. This pistol was chambered for the 9mm Bergmann Bayard cartridge, which is called the 9mm Largo in Spain. The Campo Giro was the basic model of the well-known Astra pistol. It is basically a blowback-operated pistol which has a very heavy recoil spring wrapped around the barrel and is the product of Unceta and Company, originally of Eibar. The pistol was modified slightly in 1916 and issued as the Model 1913-16.

In 1921 the Astra Model 400 (commercial designation), which had the official Spanish government designation of Model 1921, was adopted. The Astra 400 is a most unusual pistol in that it will chamber and fire the following cartridges: 9mm Largo, 9mm Parabellum, 9mm Steyr, 9mm Browning long, and the .38 super automatic cartridge. The military issue pistol has wooden grips and does not have the full commercial markings. The commercial Astra 400 has hard rubber grips, commercial markings and many of the small components are nickeled or chromed. The commercial model was sold widely throughout the world.

A later model, the 9mm Parabellum Astra Model 600, never used as a Spanish service weapon, was made in limited quantities for the Germans during World War II; it differs from the Model 400 only in minor details. A post World War II version of the 9mm Parabellum Model 600—the Condor—has been introduced by Astra. This pistol, although not a standard service pistol, is easily adaptable for such use. Unlike the earlier Astra Model 400 and 600, it has an external hammer.

Unceta and Company produced a copy of the 7.63mm Mauser military model pistol after World War I, several of which were selective-fire weapons. The semiautomatic Astra Model 900 is a copy of the Mauser, with shoulder-stock holster. The Model 902, also known as the Model "F," is a selective-fire weapon, generally similar to the Model 1932 Mauser. These weapons have been used by police and gendarmerie units throughout the world, but they were not used as Spanish service pistols.

9mm Model 1921 Astra Pistol.

9mm Astra Condor Pistol.

9mm Campo Giro Model 1913-1916.

At this point, a bit of history relating to the manufacture of Spanish pistols between World War I and II is in order. Many "job shop" copies of standard Smith & Wesson revolvers and cheap blowback automatics were made in Spain during the twenties and early thirties. These weapons were frequently made of poor materials; they are commonly known as Spanish "booby traps,"

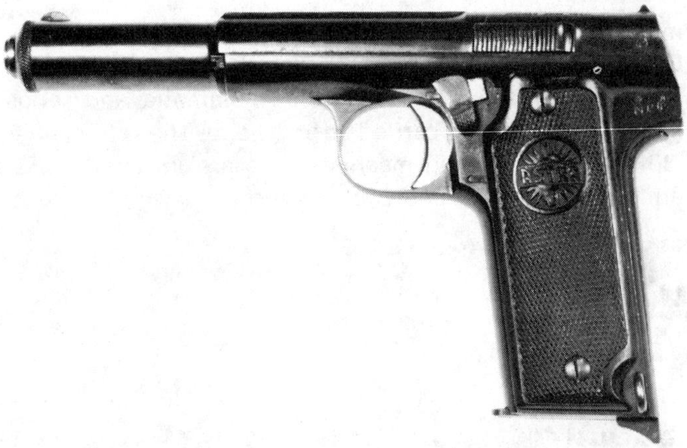

9mm Astra Model 400 Pistol (above); 9mm Parabellum Model 600 Pistol (below).

9mm Largo Llama Pistol Mark IX.

Caliber .32 Llama Pistol.

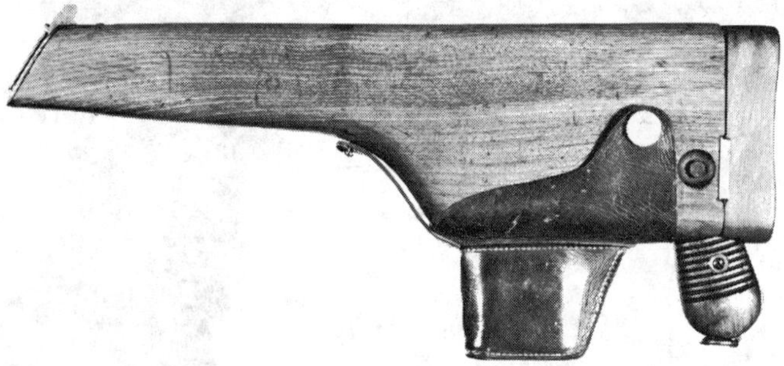

7.63mm Astra Model 902 in shoulder stock holster.

7.63mm Star Pistol Carbine Model "A".

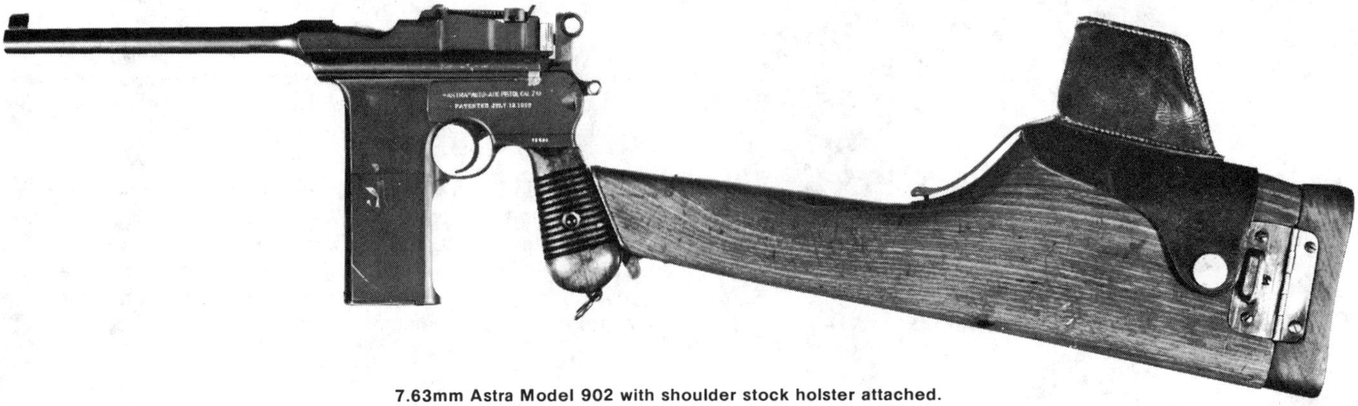

7.63mm Astra Model 902 with shoulder stock holster attached.

9mm Largo Star Model "A" Pistol.

Spanish "Star" Automatic Pistol, caliber .45, Model M.

and gave Spanish-made arms a poor name. Unceta, the manufacturers of Astra pistols; Gabilondo, the manufacturers of Ruby revolvers and Llama pistols and Star; Bonifacio Echeverria S.A.; all made quality weapons throughout this period and still do. World War II and large military export orders brought tighter proof laws to Spain. Only good quality weapons are exported currently.

The "Llama" pistol made by Gabilondo is basically a copy of the U.S. caliber .45 Model 1911 Colt automatic in the heavier calibers and a blowback edition of the Colt Model 1911 in the lighter calibers, i.e., caliber .22, .32, and .380. Differences between these pistols and the Colt are superficial. The Llama has had limited Spanish military use in caliber 9mm Largo, but has been mainly a police and commercial weapon. Like the Astra and the Star, the Llama has been used as a substitute standard wartime pistol by countries other than Spain.

Star Bonifacio Echeverria S.A., of Eibar, has produced a number of military-type pistols, some of which have been used as substitute standard in foreign armies and several of which have been used by the Spanish Army as regulation weapons. The Star, in the higher-powered models, is similar to the U.S. .45 Model 1911, but does not have a grip safety.

During the twenties, Star introduced the caliber 7.63mm Pistol Carbine Model "A." This weapon was adapted for use with a shoulder-stock holster and fitted with a tangent-type rear sight. There are two versions of this pistol: one with a five-inch barrel and one with a 6.5-inch barrel. It was never adopted as a standard service pistol.

The 9mm Largo Star Model "A" pistol was adopted as a Spanish standard pistol prior to World War II. This pistol, chambered for the 9mm Parabellum cartridge, was purchased by a number of governments during World War II. This pistol was made in very large quantities and is quite common throughout the world.

A selective-fire version of the Model "A" was produced during

Caliber .380 Star Pistol.

the thirties as the Model "M." This weapon was made in caliber .45 and possibly in other calibers as well. It was sold to Nicaragua in limited quantities. Cyclic rate is reported to be 800 rounds per minute. The selector switch is mounted on the right side of the slide. The selective-fire feature of this weapon is very impracticable, since the weapon is impossible to control in automatic fire.

As with the Llama, a number of simple blowback versions of the "Star" in calibers .22, .32, and .380 were and are being produced. Many of these models outwardly resemble the U.S. .45 Colt Model 1911.

The current Spanish service pistol is the 9mm Largo Super Star.

SPANISH 9mm SUPER STAR PISTOL

The Super Star is the standard Spanish service pistol. Although it outwardly resembles the U.S. caliber .45 M1911A1 pistol, the locking mechanism resembles that of the FN Browning Hi-Power. The barrel is cammed into and out of its locked position in the receiver by the action of the slide stop pin on a cam-way cut in a lug on the rear underside of the barrel.

The Super Star is field stripped by pushing down and forward on the dismounting lever mounted on the right side of the receiver. The slide, with barrel and recoil spring plug, can then be slid off the receiver from the front. After the slide is removed from the receiver, the recoil spring with guide, recoil spring plug and barrel bushing can be removed from the slide. The Super Star has a pivoting trigger rather than the sliding trigger of the U.S. M1911A1. There is no grip safety on the Super Star. This pistol is of good design and construction and is sold commercially in caliber .380, Super .38 automatic and 9mm Parabellum.

Spanish 9mm Super Star Pistol.

SPANISH SERVICE PISTOLS

	Campo Giro Model 1913-16	Model 1921 Astra	Super Star	Star Model "A"
Caliber:	9mm Largo.	9mm Largo.*	9mm Largo.	9mm Largo.
System of operation:	Blowback, semiautomatic.	Blowback, semiautomatic.	Recoil, semiautomatic	recoil, semiautomatic.
Length overall:	9.7 in.	8.7 in.	8.03 in.	7.95 in.
Barrel length:	6.7 in.	5.9 in.	5.25 in.	5 in.
Feed device:	8-round, in line, detachable box magazine.	8-round, in line, detachable box magazine.	9-round, in line, detachable box magazine.	8-round, in line, detachable box magazine.
Sights: Front:	Blade.	Blade.	Blade.	Blade.
Rear:	"U" notch.	Square notch.	"V" notch.	"V" notch.
Weight:	2.1 lb.	2.1 lb.	2.21 lb.	2.21 lb.
Muzzle velocity:	Approx 1210 f.p.s.	Approx. 1210 f.p.s.	Approx. 1200 f.p.s.	Approx. 1200 f.p.s.

*This weapon will also chamber and fire: the 9mm Steyr, 9mm Parabellum, .38 Super Auto, and the 9mm Browning. Long.

SPANISH RIFLES

Spain adopted its first Mauser in 1891, chambered for the 7.65mm Mauser cartridge. A rifle and carbine version of this weapon, closely resembling the Argentine Model 1891, were each produced. A considerable quantity of these weapons, especially the carbines, were captured by United States troops in Cuba during the Spanish-American War. The 1891 has an in-line type magazine which protrudes below the stock. The 1892 Mauser adopted by Spain introduced the 7mm cartridge and the non-rotating extractor attached to the bolt by a collar, which is found on later Mausers and the U.S. Springfield Model 1903.

The most famous and most significant of the Spanish Mausers was the 7mm Model 1893. The Model 1893 introduced the integral staggered-row box magazine which was used in all future Mausers. The Model 1893 also had a simplified safety lock and an improved bolt stop. A carbine version of this weapon is the 7mm Model 1895 which is stocked to the muzzle as was the fashion with Mauser carbines in those days.

The Model 1893 bolt does not have a third or safety lug or the enlarged shield on the bolt sleeve, as does the Model 98 bolt and bolt sleeve. A number of variations of the Model 1893, in addition to the Model 1895 carbine, were made.

A short rifle version of the Model 1893, and a 1916 short rifle, were also issued. The Model 1916 short rifle was made in very large quantities during the Spanish Civil war. Stocks of 7mm M1893 rifles and 7mm M1916 short rifles on hand are being converted to 7.62mm NATO, probably for issue to reserve forces.

During the Spanish Civil war, large quantities of French, Soviet, Italian, and German rifles came into Spain. The "Standard Model" Mauser was among the weapons procured by Spain at this time and, since it was chambered for the 7.92mm cartridge (the standard rifle and machine-gun service cartridge in Spain from about 1940 to 1957), it was continued in service by the Spanish Army until quite recently.

In 1943, Spain adopted a modified copy of the German 7.92mm Kar 98k which, in the Spanish service, is called the Model 1943. This rifle is currently being replaced by the 7.62mm NATO CETME Model 58 assault rifle, which was adopted in September 1957.

Spanish 7mm Model 1893 Mauser Rifle.

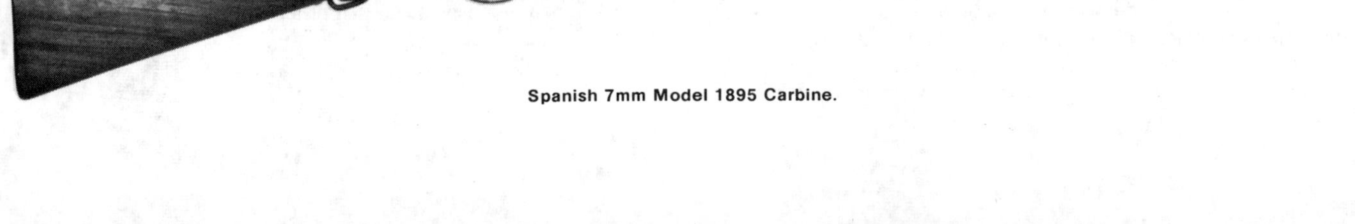

Spanish 7mm Model 1895 Carbine.

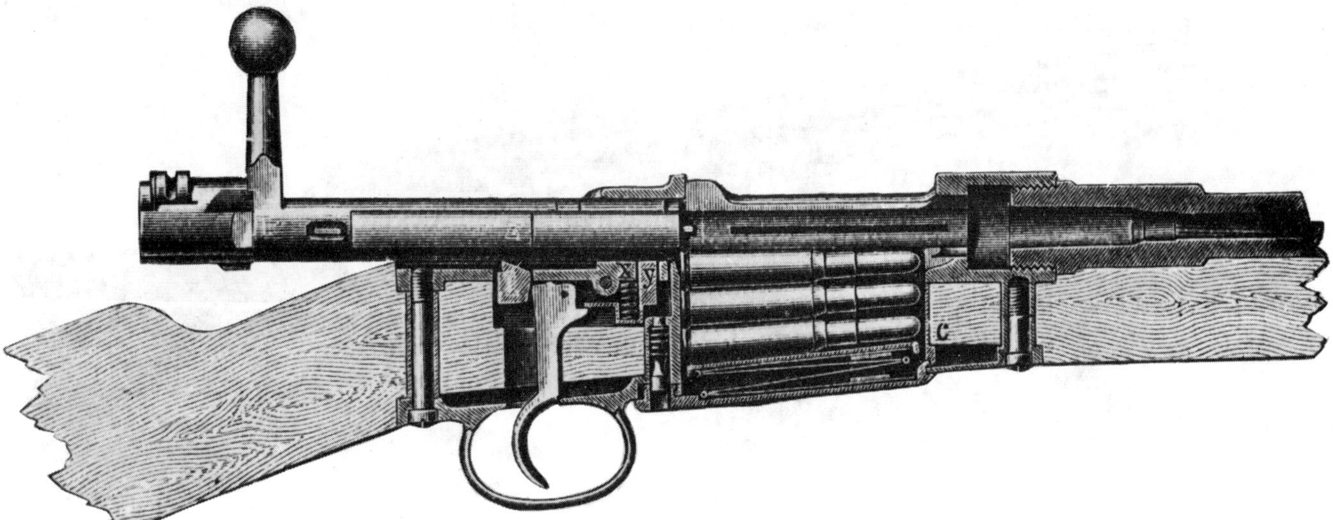

Section view of M1893 Mauser.

7mm M1893 Short Rifle.

7mm Model 1916 Short Rifle.

7.92mm "Standard Model" Mauser.

7.92mm Model 1943 Rifle.

Spanish CETME 7.92mm Assault Rifle, the prototype CETME.

THE 7.62mm NATO CETME MODEL 58 ASSAULT RIFLE

CETME, the Centro de Estudios Tecnicos de Materials Especiales, a Spanish government research establishment, has produced a number of interesting weapon designs, but the CETME assault rifle is by far the most successful of their efforts.

This weapon apparently is mostly the work of Vorgrimmler, a former Mauser Werke engineer. The design is based for the most part on the German prototype StG. 45M. As originally introduced by CETME, the weapon was chambered for an original 7.92mm short cartridge which had a long spire-pointed bullet made of aluminum with a small section of gilding metal jacketing attached to engage the rifling. It has since been made to chamber the 7.62mm NATO cartridge.

The Spanish CETME uses a special 7.62mm NATO load which gives lower pressures than those encountered with the NATO cartridge as loaded in other countries. The CETME can also be used with the NATO cartridges of other countries. The official nomenclature for the CETME rifle in the Spanish Army is: Assault Rifle CETME Model 58.

How to Load and Fire the CETME Assault Rifle

Insert a loaded magazine in the magazine well. Pull the operating handle located on the left side of the operating rod tube (above the barrel) to the rear and release it. If the selector lever has been set on the letter "T," the bolt will run home, chambering a cartridge, and the weapon will fire single rounds when the trigger is pulled. If the selector has been set on the letter "R," the bolt will remain to the rear when cocked. When the trigger is pulled, the weapon will deliver automatic fire. The CETME commences semiautomatic fire from a closed bolt and automatic fire from an open bolt. The operating handle does not reciprocate with the bolt. To put the weapon on SAFE, set the selector lever on the letter "S."

How to Field Strip the CETME Assault Rifle

Remove the magazine by pushing on the magazine catch. Remove the two pins at the rear of the pistol grip-trigger assembly and pull off the stock to the rear. Pull the operating handle to the rear and remove the bolt assembly. No further disassembly is recommended. To reassemble the weapon, perform the above steps in reverse order.

How the CETME Assault Rifle Works

The bolt in this weapon is composed of two major parts: the head and the rear section. The head, which contains the locking rollers, is considerably smaller than the rear section, which contains the firing pin. The two bolt sections are joined together so that they can move horizontally with respect to each other. The large rear section of the bolt has a nose end which is pointed to fit in the head of the bolt. When the bolt runs home, this nose end enters the head of the bolt and cams the locking rollers into their locking recesses in the receiver. When the cartridge is fired, the pressure on the base of the cartridge, after a period of time, forces the locking

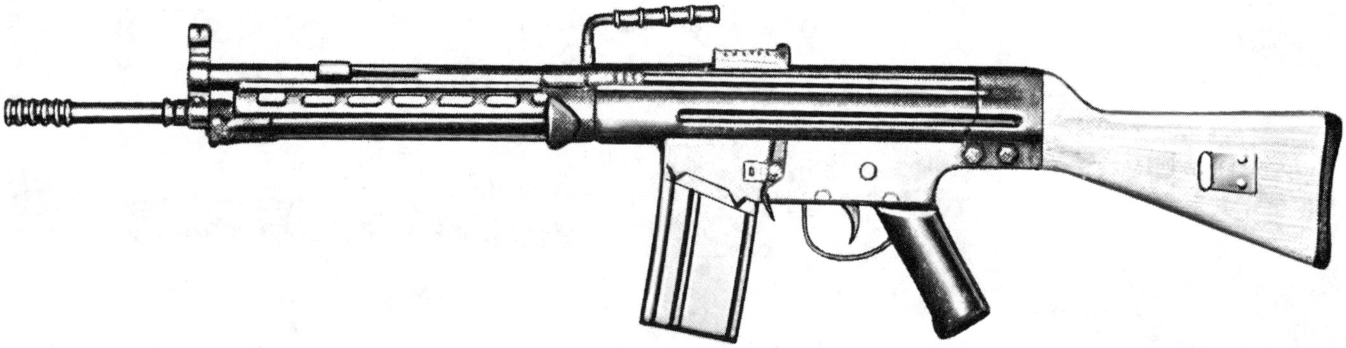

Spanish 7.62mm CETME Model 58 Assault Rifle.

CETME field-stripped.

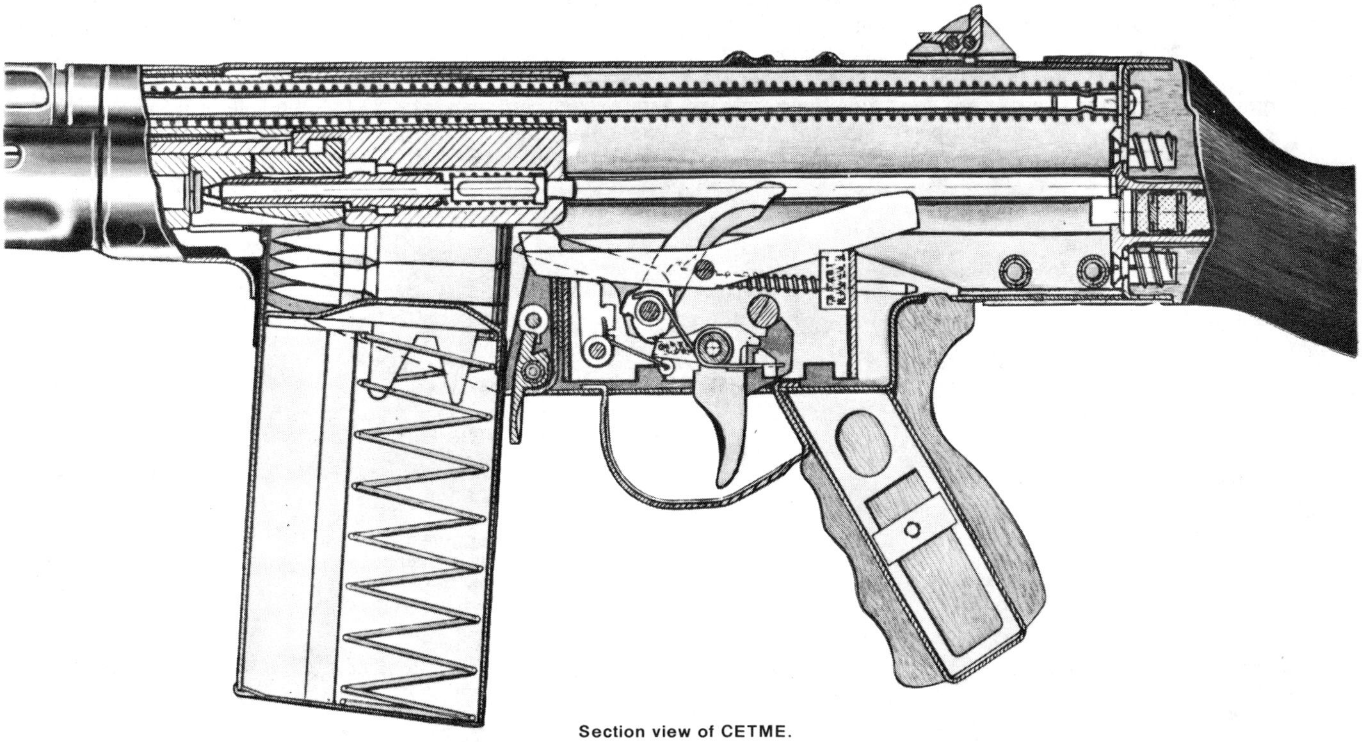

Section view of CETME.

rollers back against the nose of the rear bolt section. The rear bolt section, which is held in position only by the energy of the recoil spring, starts to the rear. The bolt is unlocked and the firing pin drawn to the rear, away from the primer, by the rearward movement of the rear bolt section.

Special Note on the CETME Assault Rifle

The CETME assault rifle is worthy of note in several respects. It is composed mainly of stampings, and CETME claims that the weapon can be made in a total working time of nine hours in series manufacture. There are fewer than twenty parts which require machining. The weapon is fitted with a bipod which, when folded, serves as a fore-end. When supported on the bipod, the gun can be used as a squad automatic weapon. There is, however, one unfortunate aspect to the delayed-blowback system, and this applies to other weapons as well as to the CETME. The system makes no allowance for slow initial extraction, i.e., for a small turning or a short rearward movement of the bolt to loosen up the spent case before it is withdrawn from the receiver. Delayed-blowback action bolts start to the rear with considerable velocity when the pressure arrives at their opening level. Therefore, delayed-blowback weapons depend on lubricated (oiled) cases, as with the Schwarzlose and Breda Model 30, or on fluted chambers, to make up for their lack of slow initial extraction. Lubricating prevents the case from expanding fully against the walls of the chamber and a fluted chamber allows the forward part of the case to float on gas and therefore not expand completely to the walls of the chamber. Without lubricated cases or fluted chambers (the degree of fluting varies from three-quarters the length of the chamber to only the length of the neck and shoulder), a delayed-blowback may pull the base off the case in extraction. These

"Sport" model of 7.6mm NATO CETME Assault Rifle; fires semiautomatic only.

weapons must therefore be designed around brass cartridges of a certain hardness. If a weapon with a fluted chamber is used with brass considerably softer than that for which it was designed, the brass may be engraved by the flutes and the bolt may lose enough kinetic energy in extracting the case to prevent it from going into full recoil, with resultant feed jams. It is probable, however, that a delayed blowback designed around the softest brass likely to be encountered in service will do very well with all sorts of brass, hard or soft. Delayed-blowback weapons do not usually give any trouble with steel cases.

Specimens of the CETME assault rifle have also been made for the Soviet 7.62mm M1943 "intermediate-sized" cartridge. The operating system of the CETME is similar to that of the Swiss SIG Model 57 assault rifle.

CHARACTERISTICS OF SPANISH RIFLES

	Model 1893	**Model 1893 Short Rifle**	**Model 1895 Carbine**	**Model 1916 Short Rifle**
Caliber:	7mm.	7mm.	7mm.	7mm.
System of operation:	Turn bolt.	Turn bolt.	Turn bolt.	Turn bolt.
Length overall:	48.6 in.	41.3 in.	37 in.	40.9 in.
Barrel length:	29.1 in.	21.75 in.	17.56 in.	Approx 23.6 in.
Feed device:	5-round, staggered row, non-detachable, box magazine.	5-round, staggered row, non-detachable, box magazine.	5-round, staggered row, non-detachable, box magazine.	5-round, staggered row, non-detachable, box magazine.
Sights: Front:	Barley corn.	Barley corn.	Barley corn w/ears.	Barley corn w/ears.
Rear:	Leaf.	Leaf.	Leaf.	Tangent.
Weight:	8.8 lb.	8.3 lb.	7.5 lb.	8.4 lb.
Muzzle velocity:	2650 f.p.s.	Approx 2650 f.p.s.	Approx 2575 f.p.s.	Approx 2625 f.p.s.
Cyclic rate:	---------------	---------------	---------------	---------------

CHARACTERISTICS OF SPANISH RIFLES (Cont'd)

	Mauser Standard Rifle	**Model 1943**	**CETME Model 58**
Caliber:	7.92mm.	7.92mm.	7.62mm NATO.
System of operation:	Turn bolt.	Turn bolt.	Delayed blowback selective fire.
Length overall:	43.6 in.	43.7 in.	39.37 in.
Barrel length:	23.62 in.	Approx 23.7 in.	Approx 17 in.
Feed device:	5-round, staggered row, non-detachable, box magazine.	5-round, staggered row, non-detachable, box magazine.	20-round, staggered row, detachable, box magazine.
Sights: Front:	Barley corn.	Barley corn.	Hooded blade.
Rear:	Tangent.	Tangent.	Tangent.
Weight:	8.8 lb.	8.8 lb.	11.32 lb.
Muzzle velocity:	2360 f.p.s.	2360 f.p.s.	2493 f.p.s.
Cyclic rate:	---------------	---------------	600 r.p.m.

SPANISH SUBMACHINE GUNS

A considerable number of submachine guns have been developed in Spain and copies of foreign submachine guns have been made, as well. Prior to World War II a modified copy of the Bergmann MP 28 II, chambered for the 9mm Largo (Bergmann Bayard), was made in Spain in limited numbers. This weapon is distinguishable from the standard MP28II by its oversized bolt handle, sling swivels, and the brass trigger guard and magazine housing.

The Spaniards also manufactured a modified copy of the Vollmer Erma chambered for the 9mm Largo cartridge, which they called the Model 1941/44. The weapon was made at La Coruna until the mid-fifties. The principal difference between the German-produced weapon and the Spanish-produced weapon is that the Spanish weapon has a completely different type of safety from that used on the German weapon.

The Star firm designed a series of submachine guns in the mid-thirties known as the SI35, RU35, and TN35. The SI35 had an adjustable cyclic rate of fire, 300 or 700 rounds per minute. The RU35 has a cyclic rate of 300 rounds per minute and the TN35 has a cyclic rate of 700 rounds per minute. A limited number of the SI35 and RU35 guns were used in the Spanish Civil war, but the weapon was never standard in the Spanish Army.

The Star TN35 was tested by the United States as the "Atlantic" submachine gun in 1942, but was not found acceptable for service.

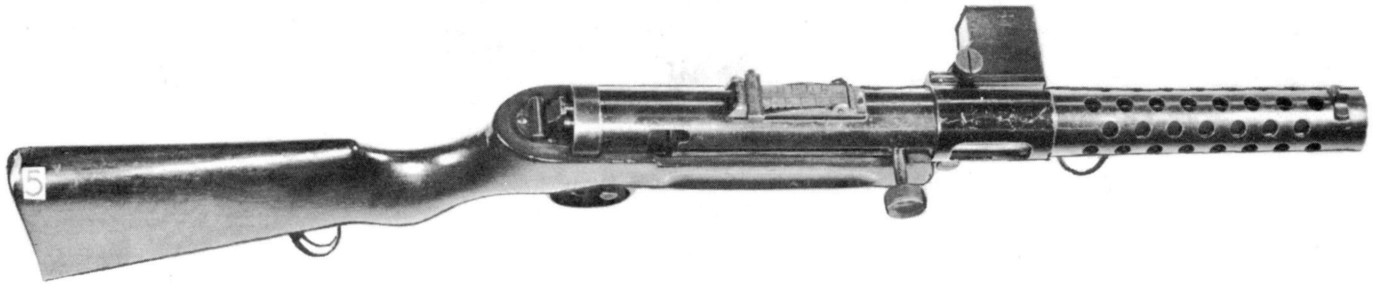

Spanish-made 9mm LARGO MP28 II.

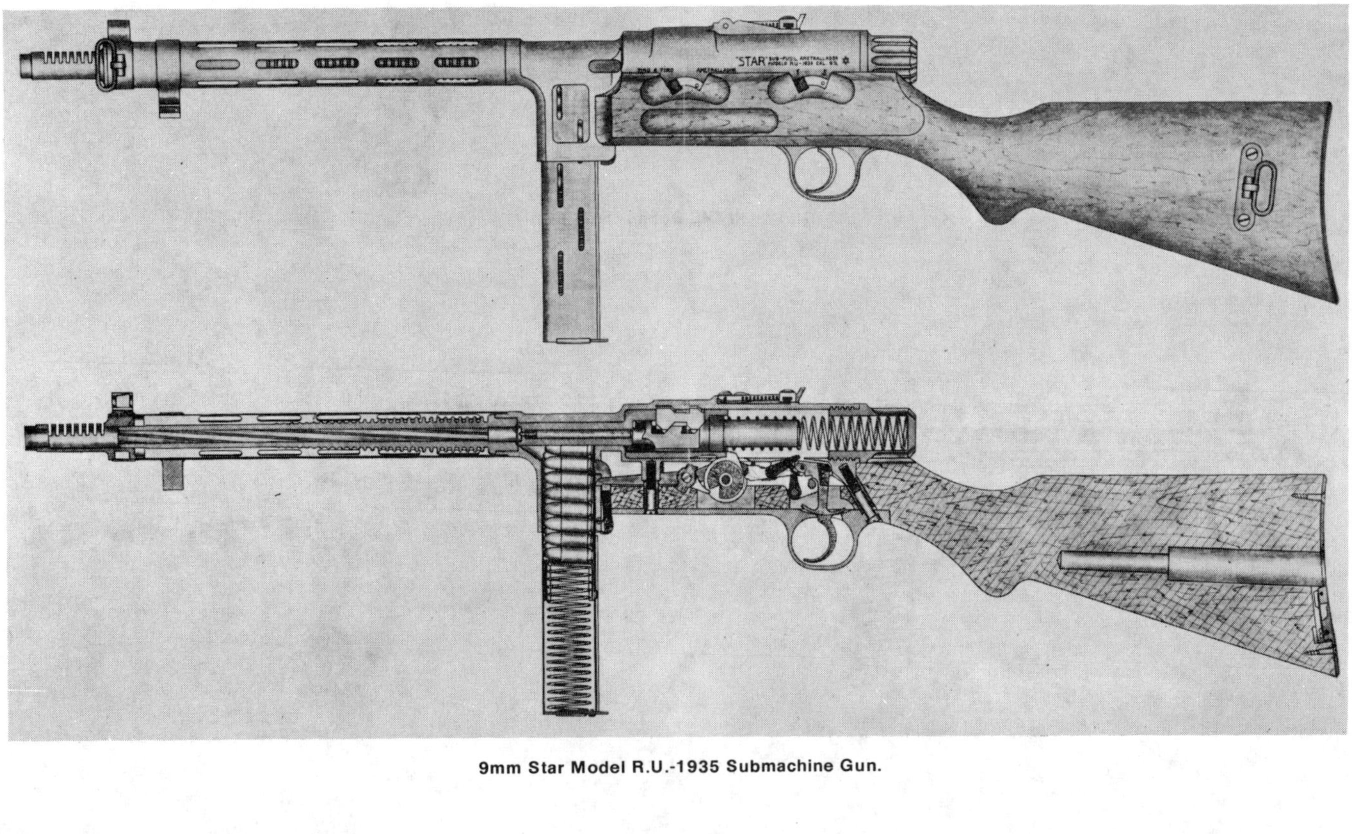

9mm Star Model R.U.-1935 Submachine Gun.

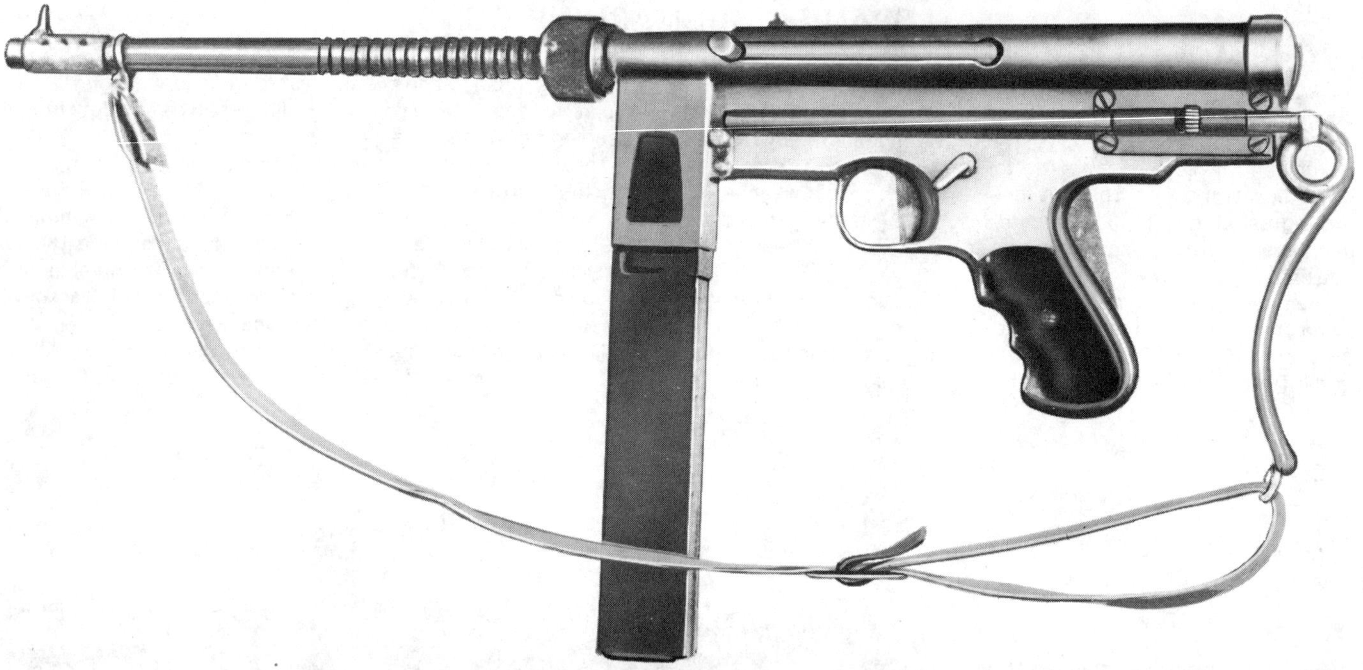

Spanish 9mm Parinco Model 3R Submachine Gun.

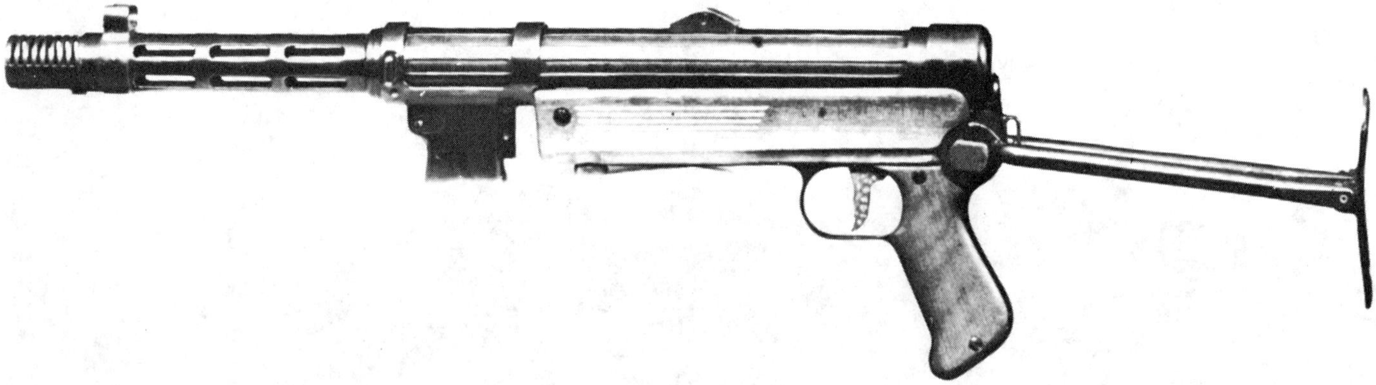

Star 9mm Z45 Submachine Gun.

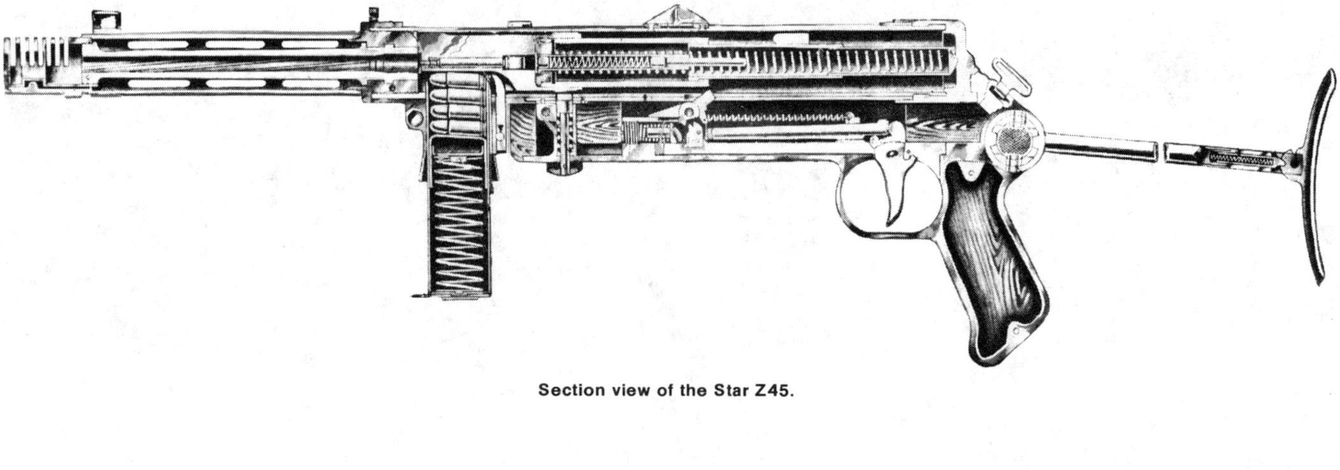

Section view of the Star Z45.

CHARACTERISTICS OF SPANISH SERVICE SUBMACHINE GUNS

	Model Z-45	**Parinco Model 3R**	**A.D.A.S.A. Model 1953**	**Star Model Z-62**
Caliber:	9mm Largo.	9mm Largo.	9mm Largo or 9mm Parabellum.	9mm Largo, also made in 9mm Parabellum
System of operation:	Blowback, selective fire.	Blowback, selective fire.	Blowback, selective fire.	Blowback, selective fire
Length overall:				
Stock extended:	33.1 in.	31.4 in.	31.9 in.	27.6 in.
Stock folded:	23 in.	24.4 in.	24 in.	18.9 in.
Barrel length:	7.9 in.	10 in.	9.85 in.	7.9 in.
Feed device:	30-round, staggered-row detachable box magazine.	32-round, staggered row detachable box magazine.	30-round in-line, detachable box magazine.	20-, 30- or 40-round, detachable box magazine
Sights: Front:	Blade w/protecting ears.	Blade.	Hooded blade.	Blade
Rear:	L-type w/notch.	Notch.	L-type w/notch.	"L" w/apertures for 100 and 200 meters
Weight:	10 lb. (loaded).	6.8 lb.	7 lb.	6.3 lb. w/empty 30-round magazine
Muzzle velocity:	1250 f.p.s.	Approx. 1270 f.p.s.	Approx. 1270 f.p.s.	Approx. 1200-1800 f.p.s.
Cyclic rate:	450 r.p.m.	600 r.p.m.	600 r.p.m.	550 r.p.m.

The British and Germans also apparently tested guns of this series, but were not interested. All activity on these guns ceased about 1942.

The Labora 1938, sometimes known as the Fontbernat, is another Spanish designed submachine gun that was made in limited quantities during the Spanish Civil war. Other than being made out of expensive machined material during a war, and having a push-button type selector, the Labora has no unusual or unique features. In 1953, another submachine gun which was developed by CETME was introduced. This weapon, known as the A.D.A.S.A. Model 1953, will fire either the 9mm Largo cartridge or the 9mm Parabellum. The weapon resembles the Soviet PPS M1943 and was apparently made in very limited quantity for the Spanish Air Force.

In 1959, the Parinco Model 3R was patented. This is another blowback-operated weapon which has appeared in several forms. The Parinco has a plastic frame and trigger housing and a grip safety. The trigger group can be rotated on a pin mounted behind the magazine housing and swung down away from the receiver for maintenance and cleaning. The Parinco has been made in 9mm Parabellum and 9mm Largo.

The standard submachine gun of the Spanish is the Star Model Z45. The Z45 is modeled on the MP40 and differs from that weapon only in the following:

(1) The bolt handle of the Z45 is on the right rather than on the left as on the MP40.

(2) The Z45 has a metal barrel jacket; the MP40 has none.

(3) The barrel and jacket of the Z45 can be removed rapidly by twisting the compensator. The barrel of the MP40 is not easily removable.

(4) The Z45 is a selective-fire weapon; the MP40 has only automatic fire capability.

(5) The Z45 has wooden grips and handguard; the grips and handguard of the MP40 are plastic.

The Z45 has been made with a fixed wooden stock in addition to the more common folding steel stock.

The Z45 has been supplied to Chile, Cuba, Portugal, and Saudi Arabia.

Star has developed a later weapon, the Z62, which is a very compact weapon with a tubular steel combined barrel jacket/receiver similar to the British L2A3 Sterling submachine gun.

Characteristics are limited to current Spanish service submachine guns of Spanish origin; other submachine guns used by Spain are covered under country of origin.

THE STAR Z-62 SUBMACHINE GUN

This weapon has been adopted by the Guardia Civil and the Army of Spain. It has a number of interesting features and is a considerable improvement over the Z-45 in lightness and bulk, safety, maintenance and reliability under rigorous service conditions i.e. sand, mud etc.

How to Load and Fire the Z-62

The magazine is loaded in the normal fashion. If the weapon has a magazine in place, it can be removed by depressing the magazine catch which is on the left side of the magazine housing. Insert loaded magazine in magazine housing as far as it will go. Unlock—pull rearward—the cocking handle located on the left forward side of the barrel jacket. Pull cocking handle completely to the rear and release. The cocking handle will return to the forward position, fold and lock itself to the barrel jacket. The weapon is now cocked and ready to fire; pressure on the lower part of the trigger produces semi-automatic fire and pressure on the top of the trigger produces automatic fire. To put the weapon on "safe", push the safety button, located in the upper section of the pistol grip, from right to left. This blocks the sear with the bolt in either open or closed (battery) position. If the bolt is forward when the safety is engaged, it cannot be drawn to the rear.

9mm Star Model Z-62 Submachine Gun.

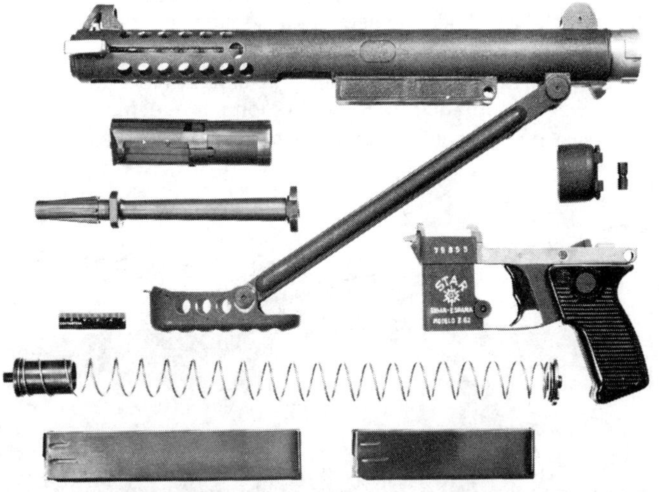

9mm Star Model Z-62 Submarine Gun field stripped.

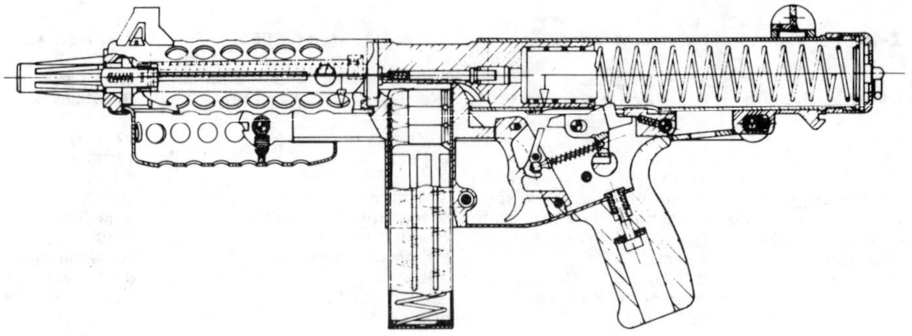

Sectional view of 9mm Star Model Z-62 Submachine Gun.

How to Field Strip the Z-62

Remove the magazine as described above. Push the recoil spring stop, which protrudes through the middle of the receiver cap, in and rotate the receiver cap either way until it disengages from its mounting lugs on the receiver. Carefully remove cap—which is under the tension of the operating spring—and withdraw recoil spring with recoil spring stop and guide. Point the weapon downward and draw cocking handle to the rear; release cocking handle, tilt gun up and the bolt will slide out the rear. The pistol grip/frame can be removed by pushing out the pin mounted at the top rear. The stock must be fixed before the pistol grip/frame can be removed. After the pin, which has a ball type retaining catch, has been removed, the pistol grip/frame can be swung down and disengaged from its mounting in the lower front of the receiver. Grasp barrel at the muzzle and rotate it clockwise a quarter of a turn and tilt the weapon down and slightly to the right. Let the barrel slide back and out of the receiver, ensuring that the "U" shaped notch below the breech is in line with the bolt guide lugs on the inner right face of the receiver.

How the Z-62 Works

The Z-62 is a blowback operated gun and in most respects is conventional enough. It does however have some unusual features which are very interesting. Most submachine guns are potentially dangerous if dropped when the bolt is forward and a loaded magazine is in place. Unless the bolt is locked in place by a grip safety or manual safety, the bolt may go to the rear just far enough to pick up and fire a cartridge. The Z-62 has solved this problem in an unusual way. In the bolt, behind the spring loaded firing pin, are mounted a hammer and a spring loaded inertia bolt safety. The hammer engages the firing pin when a rod-shaped piece, called the tripping lever, engages the hammer. The tripping lever protrudes through the bolt and when the bolt engages the rear of the barrel it is forced rearward striking the hammer which pivots and strikes the firing pin driving it forward to function the cartridge. The bolt inertia safety, which is mounted on the same pin as the hammer, is pushed outward into a recess in the receiver/barrel jacket when the bolt is forward. This prevents the bolt from coming to the rear if the weapon is dropped. The tripping lever disengages this lock at the same time as it strikes the hammer, thereby preventing it from interfering with rearward movement of the bolt after firing.

The spring loaded cocking handle and slide assembly do not reciprocate with the bolt and the automatic folding of the cocking handle makes for ease of handling during firing. Considering everything, the Z-62 is a very well made gun and should be easy to maintain in the field.

SPANISH MACHINE GUNS

The Spaniards adopted the 7mm Hotchkiss machine gun in 1907 and later adopted the Model 1914, similar to the French Model 1914, also in 7mm. The 7mm light Hotchkiss Model II, also called the Model 1922, was adopted by Spain as were the Madsen in 1907 and 1922 models.

During the Spanish Civil war, large quantities of foreign machine guns were shipped into Spain. Among these were French Hotchkiss guns, Soviet 7.62mm Maxim Tokarev, Maxim Koleshnikov, and DP machine guns, Czech 7.92mm ZB26, ZB30, and ZB53 (Model 37) machine guns, and various German and Italian machine guns. The Spaniards started the manufacture of a copy of the Czech ZB26 at the Fabrica de Armas de Oviedo chambered for the 7.92mm cartridge and called it the FAO. This weapon was modified to belt feed and rebarreled for the 7.62mm NATO cartridge and is called the FAO Model 59. It is used on a tripod as well as on a bipod. The weapon uses a drum fitted to the left side, which holds a 50-round belt.

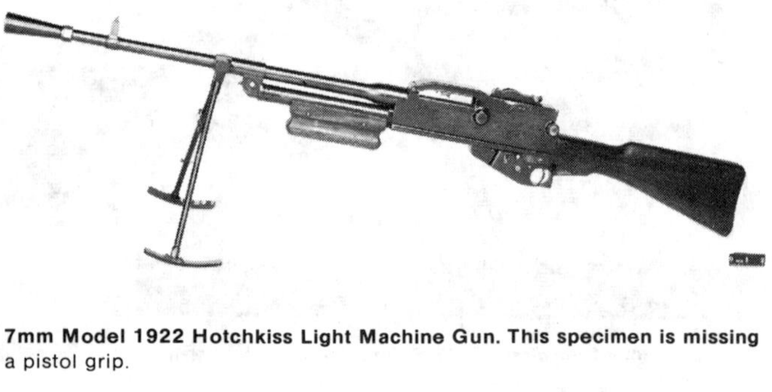

7mm Model 1922 Hotchkiss Light Machine Gun. This specimen is missing a pistol grip.

Spain developed the 7.92mm ALFA Model 1944 heavy machine gun to replace the somewhat motley collection of heavy machine guns she had previously collected. A later version of the ALFA, the Model 55, is chambered for the 7.62mm NATO cartridge. The Model 55 is shorter than the Model 1944, has a ribbed barrel and has a lighter tripod but is otherwise much the same as the Model 1944. It is probable that all these weapons will be replaced by the 7.62mm NATO MG 42/59.

THE 7.92mm ALFA MACHINE GUN MODEL 1944

The ALFA resembels the Breda Model 1937 in some respects. A quantity of these weapons were exported to Egypt prior to 1952.

How to Load and Fire the Alfa Machine Gun

Pull the charging handle to the rear, and release; it will return to its forward position. Insert the end tab of the belt in the feed port on the left side of the receiver, and pull it through the ejection port on the right side of the receiver. Lift the safety lever from the trigger and press the trigger. The weapon will fire; when the trigger is released, the safety lever will again snap into place.

How to Field Strip the Alfa

Cock the gun, to permit the operating rod to clear the gas-port collar. Remove the knurled handle of the barrel latch from the locking recess in the receiver and rotate the latch to lock the handle in the upper locking recess. This operation will permit the barrel to pass through the groove cut in the barrel latch, and will also retract the operating rod housing tube out of the gas chamber.

Grasp the barrel handle and rotate the barrel to the right until it is arrested by the stop in the receiver. At this point the interrupted thread will disengage, and the barrel can be pulled forward and out of the receiver.

To remove the feeder mechanism support, grasp the support retainer key (linchpin) by its handle, and slide it back out of the keyway in the support. Remove the support by pulling it out the left side of the receiver.

To remove the operating spring and guide, release the operating rod by firing the trigger. Press the head of the guide rod and rotate it a quarter-turn to the left, so that the keys on the rod will be in line with the keyways on the backplate. The operating rod spring will force both the rod and spring through the hole, facilitating its removal.

To remove the backplate, remove the backplate retainer pin. If the backplate does not drop out, hit it with a wooden or fiber hammer. The backplate will slide out of the grooves in the receiver.

To remove the sear housing, pull the housing back out of the receiver.

To remove the charging handle, pull the handle back to the end of the slot and then to the right, away from the receiver.

To remove the operating rod and bolt assembly (removal of the charging handle leaves the operating rod partly out of the receiver), a slight pull is necessary. To disassemble the bolt from the operating rod, retract the bolt until the locking block engages the cam on the operating rod, and raise the bolt out of its seat.

To remove the ejector, index the ejector 90 degrees and pull out of the receiver.

To reassemble the weapon, reverse the above procedures.

How the Alfa Works

When the cartridge is fired, the pressure of the propellent gases drives the bullet down the bore. As soon as the projectile passes the gas port in the barrel, a portion of the propellent gases will flow through the opening in the regulator and into the gas cylinder. The gases impinge on the piston, driving the operating rod to the rear and compressing the operating rod spring.

As the operating rod moves to the rear, it withdraws the firing pin from the primer and continues until the ramp raises the locking block. The locking block unlocks, and the operating rod and bolt continue to the rear together. Since the spent case is held by the extractor, it is withdrawn from the chamber and carried backward until the base of the case strikes the ejector blade. The blade deflects the case against the rubber deflection pad on the right side of the receiver, and ejects it through the ejection port.

As the operating rod retracts, the lug on the operating rod frees the lower arm of the belt feed lever which, being spring loaded, drags the belt-feed slide from left to right, placing a new cartridge in feeding position before the chamber. The bolt is locked in position before the chamber. The bolt is locked in position by the action of a lug (at the forward end of the central part of the operating rod) resting against the lower arm of the belt feed lever.

The backward motion of the operating rod is limited by the striking of its rear end against the buffer spring.

Characteristics are limited to weapons of Spanish origin. Other machine guns used by Spain are covered under country of origin.

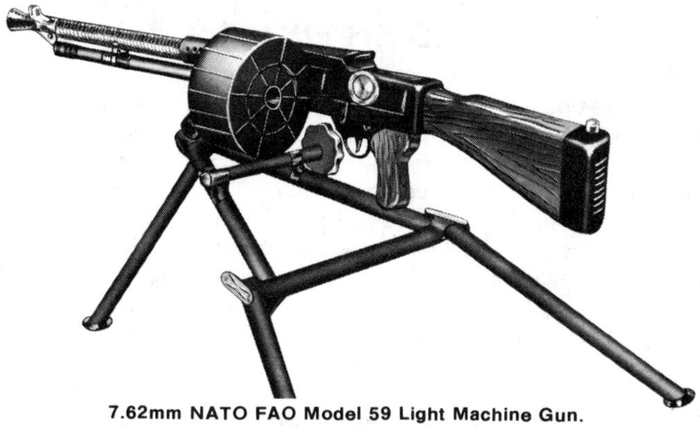

7.62mm NATO FAO Model 59 Light Machine Gun.

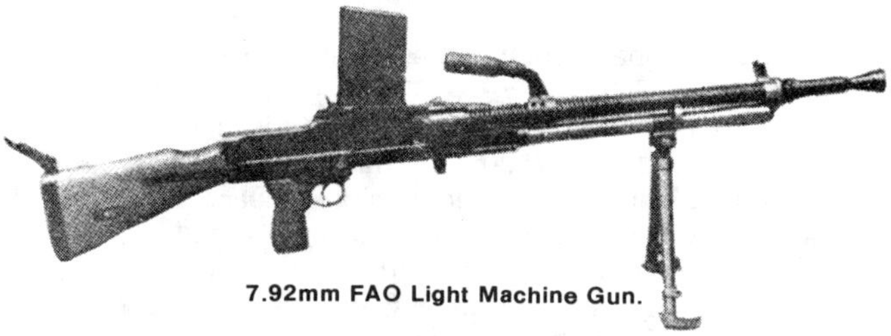

7.92mm FAO Light Machine Gun.

Spanish 7.92mm Alfa Machine Gun M1944.

CHARACTERISTICS OF SPANISH SERVICE MACHINE GUNS

	FAO Model 59	ALFA Model 1944	ALFA MODEL 55
Caliber:	7.62mm NATO.	7.92mm.	7.62mm NATO.
System of operation:	Gas, automatic only.	Gas, selective fire.	Gas, selective fire.
Length overall:	43.3 in.	Approx. 57 in. (gun).	43.4 in. (gun).
Barrel length:	21.8 in.	29.53 in.	Approx. 24 in.
Feed device:	50-round, metallic link belt, loaded in drum.	100-round, metallic link belt, loaded in drum.	100-round metallic link belt, loaded in drum.
Sights: Front:	Hooded blade.	Blade.	Blade.
Rear:	Radial tangent.	Leaf.	Leaf.
Weight:	Approx. 20 lb.	28.66 lb.	28.6 lb.
Tripod weight:	14.3 lb.	59.5 lb.	59.52 lb.
Muzzle velocity:	Approx. 2800 f.p.s.	2493 f.p.s.	2825 f.p.s.
Cyclic rate:	650 r.p.m.	780 r.p.m.	780 r.p.m.

The Swedish Army is currently equipped with the following weapons: 9mm Lahti Model 1940 Pistol, the 9mm Model 45 submachine gun, the 6.5mm Rifle AG42, the 6.5mm Sniper rifle Model 41, the the 7.62mm NATO rifle AK4 (modified copy of West German G-3), the 6.5mm or 8mm Model 36 Browning machine gun, and the 7.62mm NATO FN Model 58 general purpose machine gun. Sweden has recently adopted the Heckler and Koch 7.62mm Model 21A light machine gun.

48 Sweden

SWEDISH PISTOLS

The Swedes adopted a 7.5mm Nagant revolver known as the Model 87. These revolvers remained in limited service until after World War II. Sweden brought the FN 9mm Browning Long M1903 automatic pistol into service as the Model 07. This relatively low-powered, blowback operated pistol was the standard service pistol until the adoption of the 9mm Parabellum Lahti M40 pistol in 1940. A limited number of 9mm Walther P38 pistols were purchased in 1939 and are known in Sweden as the Model 39. The principal Swedish pistol has been the Lahti Model 40.

9mm Parabellum Model 40 Pistol.

9mm SWEDISH GAS OPERATED PISTOL

This pistol which reportedly was developed by Husqvarna is rather unique in operation. It is chambered for the 9mm Parabellum cartridge.

The gas system used to operate the pistol is similar to that of the German World War II MKb42W Walther assault rifle and the Czech Model 52 rifle. Gas is bled from ports near the rear end of the barrel. A gas cylinder piston unit surrounds the barrel at this point and this unit is forced rearward by the gas and impinges upon the front of the slide, driving it to the rear.

The pistol has a compensator slot on the top, front of the barrel. The weapon is striker fired, having no hammer and has a separate bolt within the slide.

SWEDISH 9mm LAHTI MODEL 1940 PISTOL

This pistol is basically the same as the Finnish Model 1935 Lahti, but differs slightly from the later designed Finnish Lahti in the mounting of the recoil spring. The Model 40 has it's grip grooved for a shoulder stock. The Swedish Lahti was manufactured at the Husqvarna Vapenfabrik A.B.

How the Model 40 Works

This pistol is a strange composite of features of the Luger and the Bergmann-Bayard, with several individual characteristics.

As with the Luger, the barrel is screwed into a slide. This slide, however, is enclosed except for an ejection port on the right side and an emergence cut for the breechblock at the rear.

The exterior appearance somewhat resembles the Luger, the pitch of the grip being very much like it.

The takedown also somewhat resembles the Luger. When the breechblock is back, a dismounting lever above the trigger can be turned to release the assemblies. The barrel and breechblock assemblies will then come off the receiver runners to the front.

The recoil spring is positioned in a tunnel in the rear of the separate breechblock. The head of the recoil spring guide rod projects from the rear of the breechblock.

Swedish gas-operated pistol, the recoil spring retainer has been removed from its usual position and the recoil springs are partially withdrawn.

Wings with finger grips are machined at the sides of the rear of the breechblock to permit pulling back for loading.

The breechblock travels in the slide. A surface is machined on its under face to force back and ride over an internal hammer. The firing pin is mounted in a sloped tunnel in the forward section of the breechblock below the recoil spring position.

The breech lock is a separate removable unit. It is housed in the bulge in the extreme rear of the slide. When the pistol is locked, the locking piece is in its slide recesses on right and left, and is also in engagement (its inner surfaces) with cuts in the breechblock. During rearward and forward movement, cam faces on the sides of the locking piece are utilized to raise and lower the lock out of and into breechblock engagement.

Magazine is the familiar box type. It has a follower button as in the Luger to make loading easier. The button actuates a pivoted stop when the magazine is empty, to hold the action open.

The trigger is pivoted. It has a bar crossing the grip which transmits the pull on the trigger to release the sear. The internal hammer has the conventional hammer strut which compresses the coil mainspring below it in the handle. The magazine release catch works off the mainspring.

An unusual device is the accelerator in the forward end of the receiver. It is the Browning MG pivoted type. When barrel travel halts, the barrel flips up the point of the accelerator to deliver an accelerating blow to the breechblock to speed up its rearward movement.

The safety is a positive mechanical block which forces the sear into locked engagement with the hammer notch, making it impossible for the hammer to rotate.

Loading and Firing the Model 40

Grasp wings of breechblock and pull back over loaded magazine. Release breechblock and let recoil spring drive breechblock ahead to chamber a cartridge.

When magazine is empty, breechblock will be held back. Insert a loaded magazine. Pull back slightly on breechblock wings to release the catch and let the breechblock go forward to load.

Field Stripping the Model 40

Pull breechblock back over empty magazine. Remove magazine. Turn down dismounting lever above trigger. Pull back breechblock to free it from stop. Ease assemblies forward off receiver. Remove breechblock from slide.

Note. Do not unscrew barrel unless necessary.

Removing stocks will give access to firing mechanism if necessary.

Characteristics of the Model 40

Caliber: 9mm Parabellum.
System of operation: Recoil operated, semiautomatic.
Length overall: 10.7 in.
Barrel length: 5.5 in.
Feed device: 8-round in-line detachable box magazine.
Sights: Front: Barley corn.
Rear: "U" notch.
Weight: 2.4 lb.
Muzzle velocity: Approx. 1250 f.p.s.

SWEDISH RIFLES AND CARBINES

The Swedes adopted a 6.5mm Mauser carbine in 1894. This short and light weapon, which is illustrated here, is now obsolete in the Swedish Army. Its action is the same as the M96 rifle. The M96 is a very accurate weapon for a military rifle, and very good shooting can be done with it, if it is in good condition and if a micrometer rear sight is fitted to the receiver. The Model 38 is a conversion of the M96, into a shorter and lighter weapon. A sniper version of this weapon, called the Model 41, is fitted with a telescope that is also called the Model 41.

The Swedes also used another shoulder weapon, the 8mm M40 rifle. This weapon is a German-produced Kar. 98k which has been rebarreled for the Swedish 8mm x 63mm Bofors M1932 machine gun cartridge. The weapon is fitted with a muzzle brake and has a four-round magazine. It is no longer in Swedish service.

SWEDISH 6.5mm LJUNGMAN SEMIAUTOMATIC RIFLE

In 1942, Sweden adopted the 6.5mm Ljungman semiautomatic rifle. This weapon is called the AG 42B by the Swedes. As originally issued, it was called AG42, but after some slight modifications was designated the AG42B. The Ljungman has been manufactured in Egypt in caliber 7.92mm and is now manufactured in that country in a modified form for the Soviet 7.62mm Model 43 intermediate-sized cartridge. The Madsen firm in Denmark produced a prototype of a modified form of the Ljungman in which the gas tube encircled the barrel.

The Ljungman gas system is unusual in that it seeks to avoid the use of an intermediary thrusting piston for its actuation.

As the weapon is fired and the bullet passes by the gas port about one-third of the distance from the muzzle, gas is tapped off through a hole into the gas cylinder on top of the barrel in standard fashion. The gas is directed back through the gas cylinder where it impinges upon an extension of the bolt carrier whose forward end passes into a mating receiver cut.

The gas thus delivers its thrust directly to the face of the bolt carrier itself.

The bolt carrier has the customary short free travel in its guides in the receiver, during which time the bolt itself, lying below the carrier, is locked securely to the receiver by lugs at its rear end, being depressed into corresponding receiver cuts.

As the gas pressure drops with the passing of the bullet out of the barrel, cam faces on the bolt carrier engage the bolt and elevate its rear section out of locking engagement. At this point the bolt and carrier with the empty cartridge case gripped in the face of the extractor travel to the rear to compress the recoil springs in the receiver behind the bolt carrier. During the recoil stroke the magazine follower and spring elevate the next cartridge into line for feeding. The bolt and carrier at end of stroke move ahead under the thrust of the counter recoiling spring. The top cartridge is stripped from the magazine and fed up the ramp into the chamber. The extractor snaps into the cannelure of the cartridge case. At this point, where the bolt face is halted against the breech face of the barrel, the bolt carrier still has forward travel under the impetus of the spring behind it. Its cam faces thrust the rear of the bolt down into locking recesses in the receiver. The bolt carrier continues forward as a free member from this point. The front face of the bolt carrier rides into a cup-shaped port in the receiver which forms the emergence port for the gas at the next discharge.

The rifle is equipped with bayonet. The muzzle brake built into the rifle by slotting across the top of the barrel serves to hold the muzzle down to some degree during firing.

The gas system of the Ljungman, in which the gas blows directly back upon the bolt or the bolt carrier, is also used in the French M1949 and M1949/56 and in the United States (Stoner-developed) AR-10 and AR-15 rifles. The Ljungman never replaced the Mauser in Swedish service and was issued on a basis of several per squad.

In 1965 Sweden adopted the 7.62mm NATO G3 rifle with some

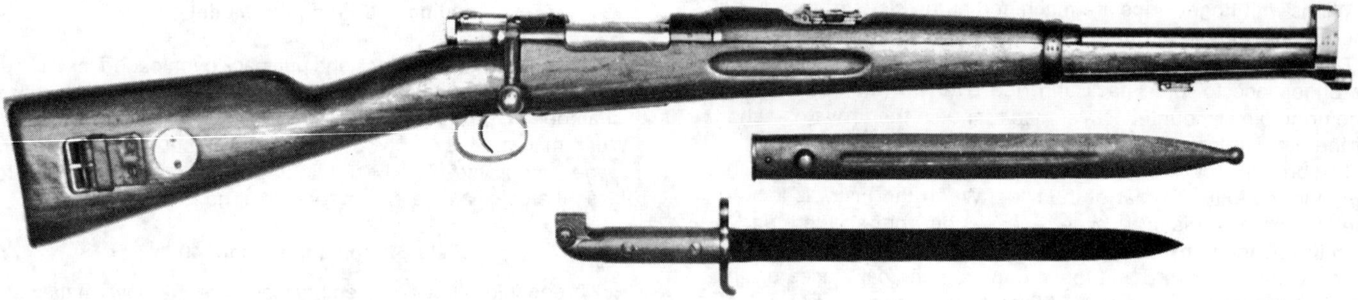

6.5mm Model 94 Carbine.

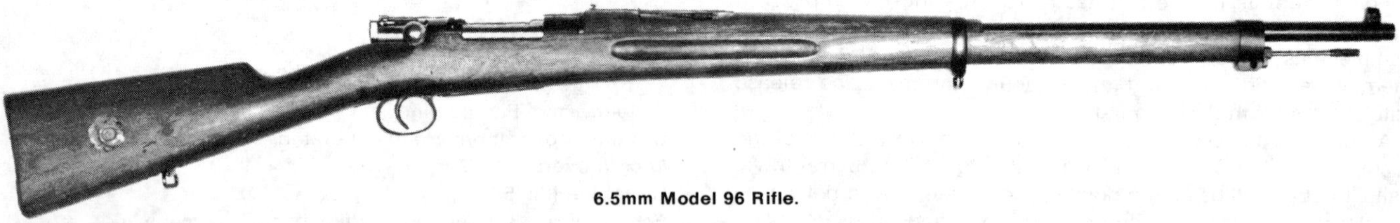

6.5mm Model 96 Rifle.

6.5mm Model 38 Rifle; some had straight bolt handles.

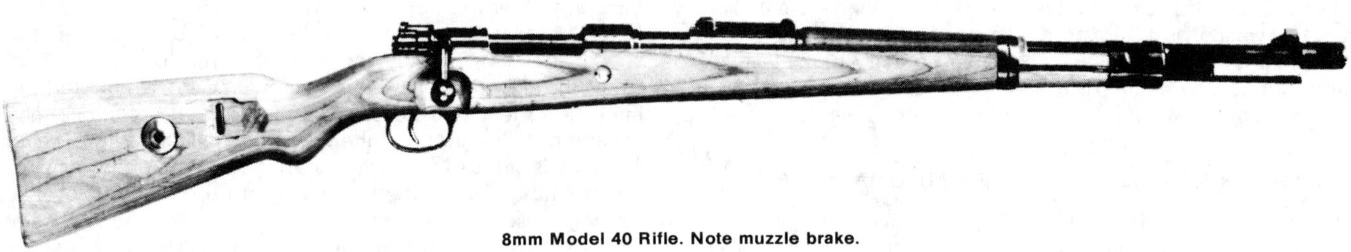

8mm Model 40 Rifle. Note muzzle brake.

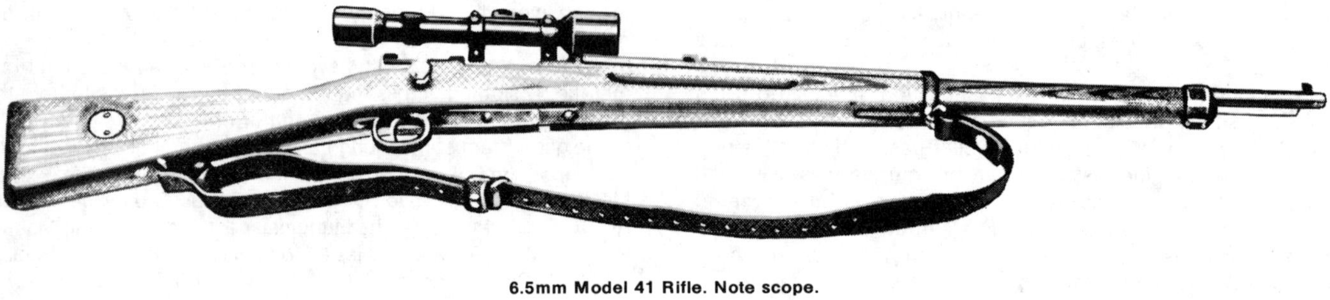

6.5mm Model 41 Rifle. Note scope.

6.5mm Swedish Gevar AG42B.

modifications. The rifle which is to replace the AG42B, the Mausers, and the Model 37 Browning automatic rifle, will be made at the government Carl Gustafs Stads plant at Eskilstuna and at the Husqvarna firm. The Swedish rifle, which is called the AK4, has a plastic handguard and stock, a modified magazine, and has an aperture-type rear sight. The detailed characteristics of the G3 rifle are given in the chapter on West Germany.

CHARACTERISTICS OF SWEDISH RIFLES AND CARBINES

	Model 94 Carbine	Model 96 Rifle	Model 38 Rifle
Caliber:	6.5 x 55mm.	6.5 x 55mm.	6.5 x 55mm.
System of operation:	Turn bolt.	Turn bolt.	Turn bolt.
Length overall:	37.6 in.	49.6 in.	44.1 in.
Barrel length:	17.7 in.	29.1 in.	23.6 in.
Feed device:	5-round, staggered row, non-detachable, box magazine.	5-round, staggered row, non-detachable, box magazine.	5-round, staggered row, non-detachable, box magazine.
Sight: Front:	Barley corn.	Barley corn.	Barley corn.
Rear:	Leaf.	Leaf.	Leaf.
Weight:	7.6 lb.	9.1 lb.	8.5 lb.
Muzzle velocity:	2313 f.p.s.	2625 f.p.s.	2460 f.p.s.

CHARACTERISTICS OF SWEDISH RIFLES AND CARBINES (Cont'd)

	Model 40 Rifle	Model 41 Rifle	Model 42B Rifle
Caliber:	8 x 63mm.	6.5 x 55mm.	6.5 x 55mm.
System of operation:	Turn bolt.	Turn bolt.	Gas-operated, semiautomatic.
Length overall:	Approx 49.2 in.	49.6 in.	47.8 in.
Barrel length:	Approx 29.1 in.	29.1 in.	24.5 in.
Feed device:	4-round, staggered row, non-detachable, box magazine.	5-round, staggered row, non-detachable, box magazine.	10-round, staggered row, non-detachable, box magazine.
Sights: Front:	Barley corn.	3X or 4X telescopic.	Hooded post.
Rear:	Tangent.		Tangent.
Weight:	Approx. 9.5 lb.	11.1 lb.	10.4 lb.
Muzzle velocity:	2428 f.p.s.	2625 f.p.s.	2460 f.p.s.

SWEDISH AUTOMATIC RIFLES

SWEDISH 6.5mm AUTOMATIC RIFLE MODEL 21

Sweden adopted the Browning automatic rifle in 1921, at which time she bought a quantity of this weapon from Colt. The M1921 is now obsolete in Sweden. The weapon is quite similar to the U. S. M1918 BAR, except that it has a separate pistol grip and dust covers for the ejection port and the magazine well.

Characteristics of M21

Caliber: 6.5mm x 55mm Mauser rimless cartridge.
System of operation: Gas, selective fire.
Weight: 19.2 lbs.
Length, overall: 44 in. (approx.).
Barrel length: 26.4 in.
Feed device: 20-round, detachable box magazine.
Sights: Front: Blade.
Rear: Leaf w/notch.
Muzzle velocity: 2460 f.p.s. (approx.).
Cyclic rate: 500 r.p.m.

SWEDISH 6.5mm AUTOMATIC RIFLE MODEL 37

Characteristics of Model 37

Caliber: 6.5mm.
System of operation: Gas, selective fire.
Weight: 20.9 lb.
Length, overall: 46.1 in.
Barrel length: 24 in. (approx.).
Feed device: 20-round, detachable, staggered-row box magazine
Sights: Front: Hooded blade.

Rear: Leaf w/aperture graduated to 1200 meters, aperture battle sight.
Muzzle velocity: 2460 f.p.s. (approx.).
Cyclic rate: 480 r.p.m.

This modification of the Browning Automatic rifle was developed by the Swedish government arsenal, Carl Gustaf State Arms Factory. It will probably be replaced in the Swedish Army by the 6.5mm Model 58 machine gun.

How to Load and Fire the Model 37

Cock the weapon by pulling the operating handle completely to the rear—push operating handle back to the forward position. The bolt will remain open, since the weapon fires from an open bolt. The safety selector lever is on the left side of the receiver at the top of the pistol grip. Settings are marked "P"-semi-automatic fire; "A"—full automatic fire; and "S"—safety. Insert loaded magazine in magazine well in underside of receiver. If the trigger is pulled and safety selector is set on "P" or "A", the weapon will fire.

Barrel Change. Beginning with an empty weapon, cock gun; push barrel latch (located at top left front of receiver) to the rear. Lift carrying handle and turn 1/4 turn to the right. Pull barrel out.

Field Stripping the Model 37

Remove the trigger guard retaining pin, located at left forward end of trigger guard. Pull pistol grip backward and upward and remove. The recoil spring guide bears against the rear of the slide; gently push it back and disengage it from the slide. Line up the hammer pin with the holes in the side of the receiver and with the hole in the operating lever. Push hammer pin out from left to right. Remove operating lever by pulling straight back and remove hammer from the receiver. Push out bolt guide with screwdriver blade and remove bolt link, bolt lock, and bolt. Push barrel latch to rear and remove as described above. Remove gas cylinder retaining pin at forward left side of receiver and slide gas cylinder and bipod straight ahead. Pull slide and gas piston assembly out of receiver.

To reassemble the weapon, reverse the above procedure.

How the Model 37 Works

The Model 37 works basically the same as the U. S. Browning Automatic Rifle M1918 covered in the chapter on the United States. The recoil spring in the Model 37 is located in the butt as with the F.N. Type D Browning Automatic Rifle.

6.5mm Model 21 Automatic Rifle.

6.5mm Model 37 Automatic Rifle.

SWEDISH SUBMACHINE GUNS

The Swedish Army has used a number of submachine guns in the past twenty years. The U. S.-made Thompson was used in limited numbers by the Swedes as the 11mm M40, and the Finnish-designed Suomi was used in two versions, the Model 37-39 and the Model 37-39F. The 9mm Bergmann M34 was used as the Model 39.

THE SWEDISH 9mm CARL GUSTAF SUBMACHINE GUN MODEL 45

The 9mm Model 45 (commonly known as the Carl Gustaf) is the current standard submachine gun in the Swedish Army. This weapon has also been produced in Egypt. It was designed and is produced in Sweden at the Carl Gustafs Stads Gevärsfaktori in Eskilstuna. All submachine guns in service fire the 9mm Parabellum cartridge.

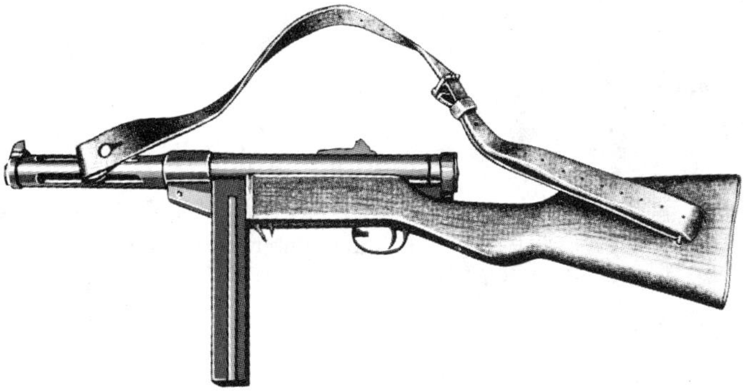

9mm Model 37-39 Submachine Gun.

Characteristics of M45 Submachine Gun

Caliber: 9mm Parabellum
Magazine: Detachable box, staggered dual line.
Magazine capacity: 36 cartridges.
Overall length: w/stock extended: 31.8 in.
w/stock folded: 21.7 in.
Barrel length: 8 in.
Weight: 9.25 pounds (loaded).
Cyclic rate of fire: 550 to 600 per minute.
Type of fire: Automatic.
Operation: Elementary blowback. Actuation of bolt mechanism is by projection of spent case.
Sights: Front: Protected post.
Rear: L type.
Muzzle velocity: 1200 f.p.s.

How to Load and Fire the Model 45

Pull bolt handle to the rear; bolt and bolt handle will remain to the rear since the gun fires from an open bolt. The weapon can be put on safe when the gun is cocked by pulling the bolt handle beyond the sear and engaging it in the cutout section above the bolt handle track of the receiver. Insert magazine, take weapon off safe, pull trigger, and weapon will fire. To put the weapon on safe with the bolt forward, push down on bolt handle.

How to Field Strip the Model 45

Remove magazine; bolt should be forward. Depress catch in center of the receiver cap (rear of receiver), turn receiver cap slightly counter clockwise, and cap will come off. Remove recoil spring and bolt. Push in on barrel jacket nut catch with a drift and unscrew barrel jacket nut. Remove barrel jacket and barrel. To reassemble, carry out in reverse the steps outlined above.

9mm Model 37-39F Submachine Gun.

Swedish 9mm M45 Submachine Gun with fixed magazine guide, stock fixed.

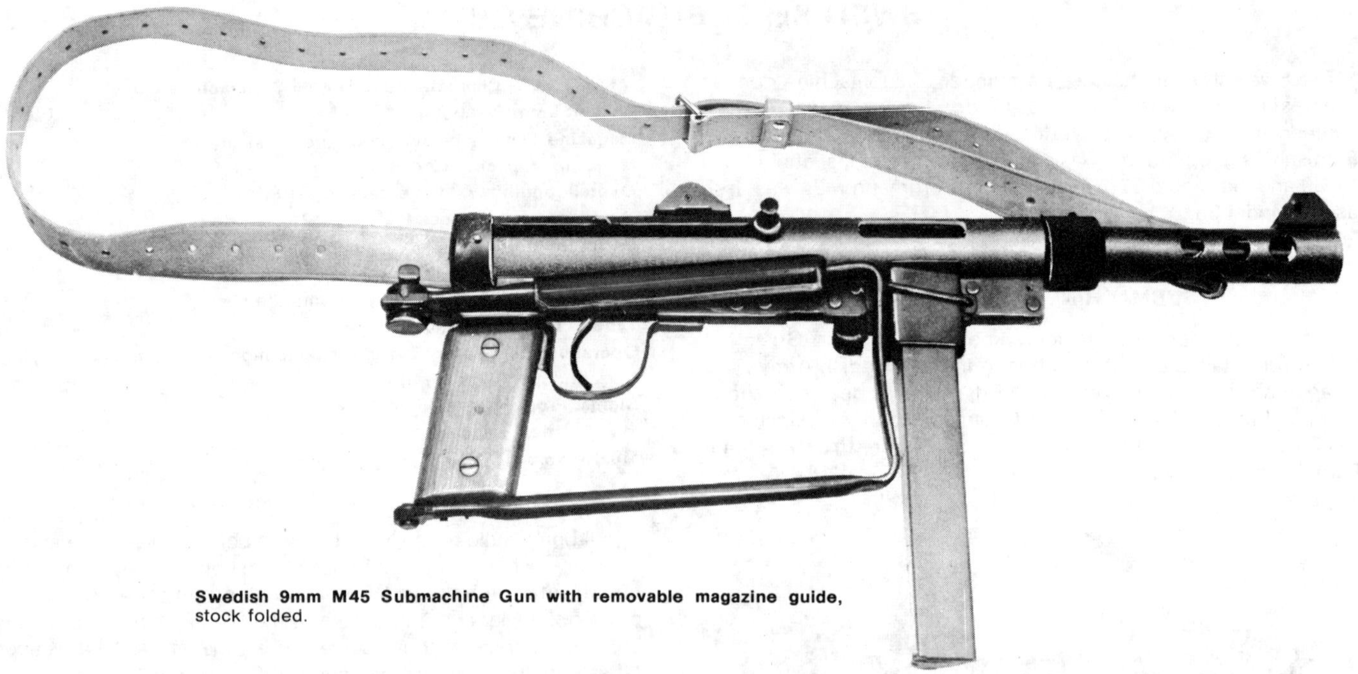

Swedish 9mm M45 Submachine Gun with removable magazine guide, stock folded.

Special Note on the Model 45

There have been at least three different variations of this weapon since it first appeared in 1945. The variations are in the barrel jacket and magazine guide. Some of the M45s have a removable magazine guide. When the guide is removed the weapon can be used with the thicker Suomi submachine gun magazine. The weapon is principally made of stampings and has a folding steel stock.

SWEDISH MACHINE GUNS

The Swedish Army has a collection of various types of machine guns in use at present, but most, if not all, of these weapons will probably be replaced by the Model 58. The Model 58 is the Belgian FN Type MAG, and is covered in detail in the chapter on Belgium. The Schwarzlose in 6.5mm, Model 14, and Model 14-29, were used by Sweden; these weapons are believed to be out of service at present. A series of rifle-caliber Browning guns is used; these weapons are chambered for either the 6.5mm cartridge or the 8mm M1932 cartridge (8mm x 63mm). The Model 36 is a water-cooled gun, and the Model 42 is an air-cooled gun. Both of these guns are similar in loading, firing, functioning, and field stripping to the U. S. cal. .30 Browning guns, although they have a somewhat higher cyclic rate of fire. The Czech-made ZB26 light machine gun in caliber 6.5mm was also used in Sweden as the Model 39. The German-made Knorr Brense in 6.5mm was also used in limited numbers as the Model 40.

6.5mm MODEL 36 AND MODEL 42B MACHINE GUNS

The Swedish Brownings are basically the same as the United States caliber .30 Brownings. The Model 36 is a water-cooled gun similar to the United States Model 1917A1 and the Model 42B is an air-cooled gun similar to the United States Model 1919A6 which can be used on a bipod with a shoulder stock. As originally made, this weapon was called the Model 42. The Swedish Brownings use spade-type grips rather than the pistol-type grip of the United States Brownings. A Model 39 aircraft Browning gun was also used.

The loading, firing, field stripping, and functioning of the Swedish Brownings are essentially the same as those of the United States Brownings.

6.5mm Model 36 Browning Machine Gun.

6.5mm Model 42B Browning Machine Gun on bipod.

Characteristics of Swedish Browning Machine Guns

	Model 36	Model 42
Caliber:	6.5 x 55mm or 8 x 63mm.	6.5 x 55mm or 8 x 63mm.
System of operation:	Recoil, automatic only.	Recoil, automatic only.
Length overall:	43.7 in.	53.19 in.
Barrel length:	23.9 in.	23.9 in.
Feed device:	250-round belt.	250-round belt.
Sights: Front:	Barley corn.	Folding blade.
Rear:	Tangent.	Tangent.
Weight: Gun:	45.4 lb. w/water.	35.2 lb.
Tripod:	Approx 53 lb.	30.14 lb.
Muzzle velocity:	Approx 2460 f.p.s.	Approx 2460 f.p.s.
Cyclic rate:	750 r.p.m.	600-700 r.p.m.

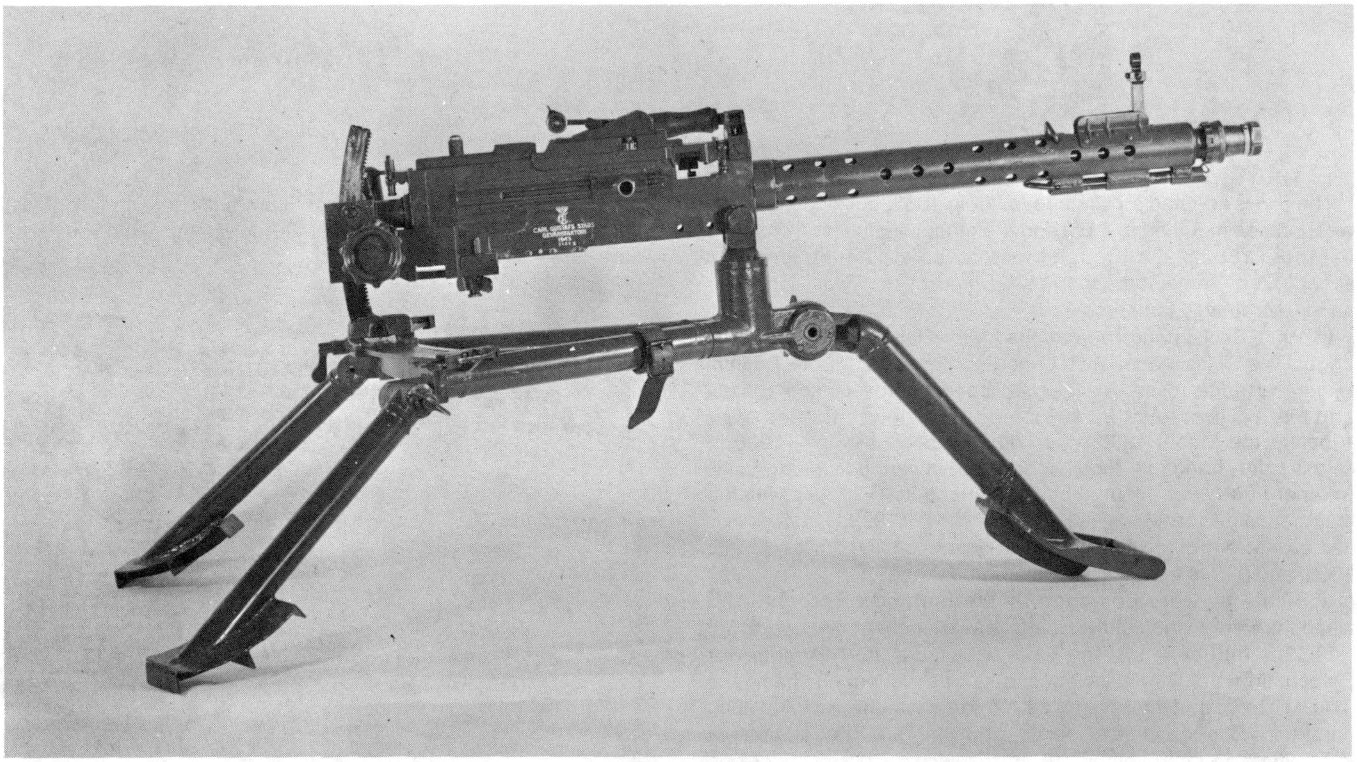

6.5mm Model 42B Browning Machine Gun on tripod.

49

Switzerland

The Swiss Army uses the following small arms: the 9mm Parabellum Model 49 pistol, the 7.5mm Model 31/55 rifle, the 7.5mm Model 57 assault rifle, the 9mm Parabellum Model 43/44 and 41/44 submachine guns, and the 7.5mm Model 51 general purpose machine gun.

Older weapons, such as the 7.5mm Model 31 carbine, may still be found in service in reserve units. The 7.5mm Model 25 light machine gun may also be found in various units.

SWISS PISTOLS

7.65mm Model 1900 Luger.

The Swiss adopted a 7.5mm revolver in 1882; this double-action weapon was modified in 1929 and the model designation changed to Model 1882/29. Although these revolvers are no longer in service, they remained a service weapon of the Swiss Army for an extraordinarily long time.

In 1900, Switzerland adopted the first military production model Luger. The Swiss Model 1900 Luger is chambered for the 7.65mm Luger cartridge (called caliber .30 Luger in the United States) and has a grip safety. In 1906 a modification of this pistol was adopted, the Model 1906 which differs from the Model 1900 in its extractor, firing pin, recoil spring, and in various minor details. Another Luger was adopted by Switzerland in 1929; this weapon may be called the Model 1906/29 or Model 1929. This weapon has a different side plate and take-down catch than the Model 1906, and differs from that Model in other minor points. The 1929 modifications were designed by Waffenfabrik Bern and the weapons were manufactured at that government arsenal.

SIG of Neuhausen Rhine Falls secured a license from the French firm Société Alsacienne de Construction Méchaniques (SACM) to manufacture and further develop the pistol-locking patents of Charles L. Petter of that firm in 1937. Petter's patent is basically an improvement on the Browning Colt recoil-locking system. A few prototypes were made, chambered for the 7.65mm Luger cartridge; about 1937-39, SIG produced a commercial pistol based on the Petter system, chambered for the 9mm Parabellum cartridge.

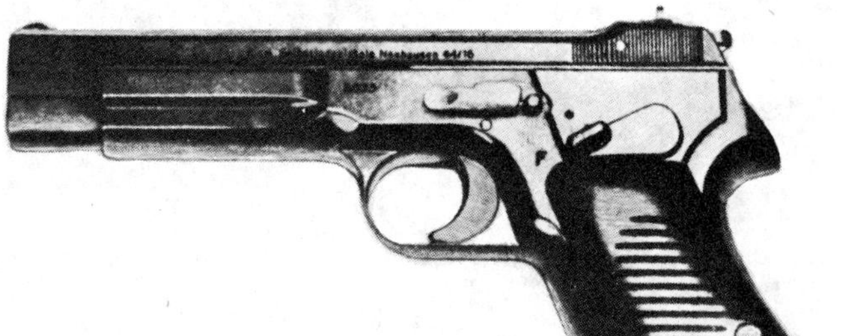

SIG 9mm Model 44/16 Pistol.

SIG Model 210-5 of late manufacture.

In 1942 SIG began modifying the Petter system and after production of a series of prototypes, introduced the 9mm Model 44/16 which has a 16-round magazine and the Model 44/8 which has an eight-round magazine. These pistols were never produced in any quantity. They introduced the internal slide mounting (the slide fits into the receiver rather than over the receiver) found on the current SIG pistol.

Further development of the Model 44 series resulted in the development of the Model SP47/8. This pistol was adopted by Switzerland and Denmark as the 9mm Model 49. A target version of this pistol called the P210-5 is also manufactured by SIG. This pistol can be, like the SP47/8 (now called the P210 by SIG) used with either 9mm Parabellum or 7.65mm Luger barrels.

THE SWISS SIG 9mm SP47/8 PISTOL (MODEL 49)

In general this pistol is merely an adaptation of the familiar Browning system. Instead of the hinged locking link as found in the .45 U. S. Government automatic, the unlocking follows more closely that of the later Browning Hi-Power. A nose on the underside of the barrel is slotted to permit the slide stop pin to pass through, locking it firmly to the receiver. During recoil, the barrel and slide are firmly locked together for approximately 1/4 inch of travel. At this point the cam slot in the locking nose below the barrel, working against the transverse slide stop pin, serves to cam the barrel down to draw its locking lugs out of their engaging seats in the underside of the breech of the slide.

This weapon is beautifully made in the finest Swiss tradition. Because of the care taken in fitting the barrel, it demonstrates considerable accuracy.

The takedown is quite simple. In general it follows the standard Colt-Browning procedure to a marked degree except for the firing mechanism.

How to Field Strip the Model 49

Hold pistol in right hand. Push magazine catch at bottom of butt with left thumb and withdraw magazine.

Grip right hand around rear of receiver and slide and draw slide back about 1/2 inch. With finger of left hand push slide stop from right side. Withdraw slide stop from left. This can be done when the thumb piece is lined up with its dismounting notch in the slide.

Ease slide, barrel and recoil spring forward.

Grip hammer with thumb and forefinger and work free and up out of receiver. This unit carries the hammer, disconnector, sear, and mainspring with its guide compression rod. These may be dismounted if desired, but this is not necessary.

Press recoil spring guide forward from rear and lift out. Spring is captive on its guide rod and will not come loose.

Grip barrel lug and work barrel slightly back and up when it can be lifted out of the slide.

Firing pin may be removed if desired by pushing in on its head and drawing stop down out of slide cut as in Colt Govt. .45.

How the M49 Operates

A loaded magazine is inserted in the butt in standard fashion and pushed in until the catch holds it. The slide is then withdrawn to the full length of its recoil stroke to allow magazine spring to force a cartridge up in line for chambering. Releasing the slide permits the compressed recoil spring to drive the slide forward for chambering and locking in standard Browning pistol fashion.

Applying the thumb safety, which is in a particularly advantageous position for easy operation, locks both the trigger and the trigger bar to prevent firing. This design does not have an auto-

9mm SIG Pistol P210-2, Swiss Army Model 49, this pistol was formerly called the SP47/8 by SIG.

matic grip safety as in the case of the Colt Government Model. It does have, however, the automatic disconnector mechanism standard for the Colt-Browning design. If the slide is not fully forward and the barrel not fully locked into position, the slide is automatically depressing the disconnector which in turn is pushing the trigger bar down out of possible contact with the trigger. When the slide is fully forward and locked, the spring can force the disconnector up into its cut in the underside of the slide. This allows the trigger, the trigger bar and the sear to function for firing. As the action opens, the slide automatically cams the disconnector down, forcing the breakage in the trigger contact so that on forward motion of the slide the chambered cartridge will not fire. Thus the trigger must be definitely released before contact can again be made for firing the next shot.

This design has several features which are superior to the Colt Government Model. The forward bushing in the receiver is eliminated, resulting in more machining difficulties but in an inherently simpler design. The recoil spring is under tension in position around its guide rod and during disassembly and reassembly functions as a unit. The unlocking lug on the bottom of the barrel affords much more rigidity during the operation of the pistol. This tends, theoretically at least, to make for more accurate shooting than with the swinging link design. Combining most of the firing mechanism in a single unit (as in the case of the Russian Tokarev pistol) results in a considerable simplification of both design and machining.

As in all weapons of Neuhausen manufacture, the workmanship and materials are impeccable. However, the pistol is costly by American standards, selling for nearly twice that charged for the equivalent Colt pistol.

Because of the extreme care taken in manufacture, the parts of this pistol are truly interchangeable. This applies not only to parts of the same caliber, but to parts for different calibers except, of course, for barrels and springs.

To convert the pistol from 9mm to 7.65mm Parabellum it is only necessary to replace the recoil spring and the barrel. The magazine and all other elements will function with either cartridge, since the 9mm was developed directly from the bottlenecked 7.65mm.

A special .22 caliber conversion training unit is also available for use with this pistol for practice shooting. It consists of a lighter weight slide, a special barrel with recoil spring and a magazine. This conversion, of course, does not require a locked breech, hence the barrel is not equipped with locking lugs and does not have any movement. Once installed it is locked in place. The operation is otherwise identical with that of the heavier caliber.

An additional 4mm barrel is provided for indoor training if desired. This is a single shot conversion, however.

Considerable attention was given to pitch of grip in the design of this pistol. The center of gravity is at about the trigger point. The locking cam system of barrel operation, as opposed to the swinging link system, prevents movement of the barrel out of the firing axis while the bullet is still in the barrel. This together with the lack of a front bushing makes for a much more solid barrel support.

CHARACTERISTICS OF SWISS SERVICE PISTOLS

	Model 1900	Model 1906	Model 1929	Model 49
Caliber:	7.65 mm Luger.	7.65mm Luger.	7.65mm Luger.	9mm Parabellum.
System of operation:	Recoil, semiautomatic			
Length overall:	9.5 in.	9.5 in.	9.5 in.	8.5 in.
Barrel length:	4.75 in.	4.75 in.	4.75 in.	4.75 in.
Feed device:	8-round, in-line detachable box magazine			
Sights: Front:	Blade.	Blade.	Blade.	Blade.
Rear:	"V" notch.	"V" notch.	"V" notch.	"V" notch.
Weight:	2.25 lb.	Approx. 2 lb.	1.98 lb.	2.14 lb.
Muzzle velocity:	Approx. 1200 f.p.s.	Approx. 1200 f.p.s.	Approx. 1200 f.p.s.	1150 f.p.s.

SWISS RIFLES AND CARBINES

The Swiss adopted the first Schmidt Rubin rifle in 1889; it is covered in detail later in this chapter. The Schmidt Rubins have been replaced in the Swiss Army by the 7.5mm Assault rifle Model 57, developed by SIG as the Model AM55 and now called SG510 by that company.

There were a number of other rifles developed by SIG and the Swiss Government Arms factory at Bern which were not as successful as those listed above. SIG introduced a gas-operated rifle, called the SK46, shortly after World War II. This weapon was semiautomatic and was designed for full-sized rifle cartridges such as the 7.5mm and the 7.92mm. The SK46 taps its gas close to the chamber and the gas cylinder and gas piston are to the side of the receiver. The bolt tips up into a locking surface in the top of the receiver. It shows Schmidt Rubin influence in it's stock, fittings, and the use of a long bolt sleeve.

The AK53, also developed by SIG, is also a gas-operated rifle, but has selective fire capability. It is unusual in that the barrel is pushed forward by the gas piston and an actuator, leaving the fired case in the face of the bolt. The case is ejected by the ejector at the end of the forward movement of the barrel and a fresh cartridge is fed into the face of the bolt. The barrel is forced rearwards over the cartridge by the operating spring and the actuator locks the bolt to the barrel. Because of this system, a low cyclic rate of fire—300 rounds per minute—was obtained with this weapon.

SIG SK46 Rifle.

SIG Model AK43.

Prototype Waffenfabrik Bern 7.5mm Short Rifle.

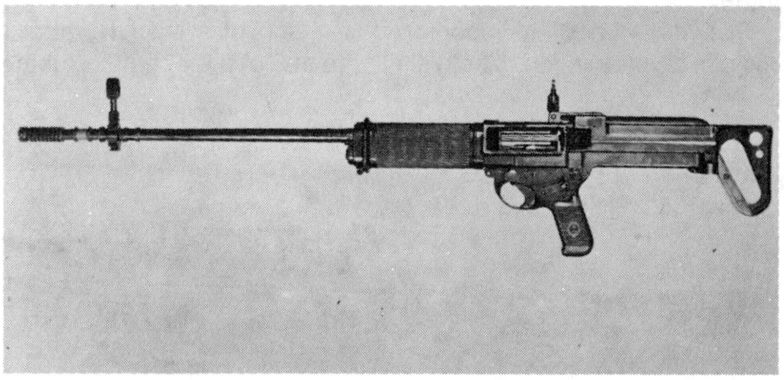

Prototype Waffenfabrik Bern 7.5mm Assault Rifle.

SIG 7.5mm AM55 Rifle.

7.62mm NATO SIG Type SG510-4.

SIG and the Swiss Government Arsenal at Bern both submitted prototypes for test and possible selection as the rifle to replace the Schmidt Rubin in the Swiss service. As early as 1944, a gas-operated 7.5mm semiautomatic rifle was made in limited quantities. In the early fifties the Arsenal at Bern developed a number of selective-fire rifles, including one chambered for a 7.65mm "intermediate" sized cartridge and one chambered for a 7.5mm "intermediate" sized cartridge. The weapon chambered for the 7.5mm "intermediate" sized cartridge is similar in concept to the German FG42 paratroop rifle. At least one prototype weapon was made chambered for the standard Swiss 7.5mm Model 11 cartridge.

SIG submitted a selective-fire rifle chambered for the 7.5mm Model 11 cartridge in the mid-fifties, the AM55. As originally designed the AM55 had a wooden fore-arm and handguard and a metal framed buttstock with wooden inserts. This design was altered and a metal barrel jacket and hard rubber butt stock were fitted.

This rifle with slight modifications, fitting of a stacking hook, use of a bolt retraction handle similar to that of the Schmidt Rubin, and altered so that the magazine for the Swiss Model 25 light machine gun could be used, was adopted by the Swiss government in 1957 as the Sturm Gewehr 57 (StuG 57) or assault rifle model 57.

SIG now calls this weapon the Type 510; and continued producing modifications of the design. The Type SG510-1 is basically the same as the Model 57 assault rifle; the type SG510-2 is a lightweight version of the 510-1; it does not have a bipod and does have a wooden butt. The Type SG510-3 was designed to use "intermediate"-sized cartridges, and is shorter and lighter than the two previous models. The SG510-4 is a carbine version of the 510-2 and has a wooden butt stock, a wooden handguard, and is chambered for the 7.62mm NATO cartridge.

SWISS SCHMIDT RUBIN RIFLES AND CARBINES

The original Schmidt Rubin rifle appeared in 1889. It has a long staggered-row box magazine, and a very long receiver. The bolt of this straight pull rifle had its locking lugs at the rear.

Model 1889 Schmidt Rubin Rifle. Should not be used with latest type Swiss cartridges.

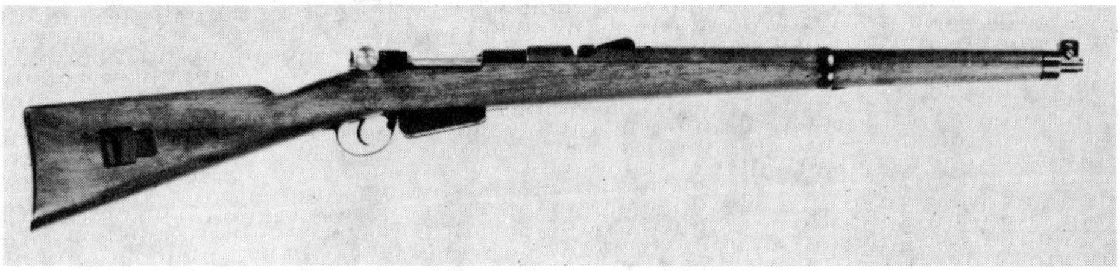

M93 Mannlicher Cavalry Carbine. Few made.

Modifications of the 1889 Schmidt Rubin

M1889/96. The 1889 action was shortened slightly, and the lug system was strengthened.

M97 Cadet Rifle. A single-shot version of the M1889/96 action, made for the training of cadets.

Model 89/00 Short Rifle. A special short rifle designed for the use of Fortress Artillery, Signal, Balloon, and Bicycle troops. This weapon, which weighs approximately seven pounds, has a lower velocity than the M1889 rifle.

Model 1905 Carbine. Weighs approximately 7.5 pounds; uses the same action as the M1889/96 rifle.

M1896/11 Rifle. An alteration of the M1889/96 rifle, and some of the carbines, to make them suitable for use with the Model 11 ammunition.

M1911 Rifle. The first of the Schmidt Rubins made specifically for the 7.5mm Model 11 cartridge. The Model 11 cartridge has a much higher pressure than the earlier Swiss cartridges. Therefore the basic 1889 action design had to be modified to handle the new cartridge safely. The locking lugs were relocated from the rear of the locking sleeve to the front of the locking sleeve. The M1896/11 action is similar. The M1911 rifle also had a bolt-holding-open device.

M1911 Carbine. Differs from the M1911 rifle by having a shorter barrel and stock and being lighter. Its rear sight is graduated to 1500 meters, rather than to the rifle's 2000 meters. The M1911 carbine is still in limited use in Switzerland. It can be identified by the red plastic bolt handle knobs.

M1931 Carbine. Commonly known as the K31. This weapon represents the last basic change in the design of the Schmidt Rubin. The basic difference between this weapon and the earlier Schmidt Rubins is the elimination of the long bolt extension. All earlier Schmidt Rubin bolts had a long bolt extension which actually rammed the cartridge into the chamber, supported its base, and extracted it from the chamber. It could be considered that this piece was the bolt body. This piece did not rotate in unlocking; the bolt sleeve rotated, instead. The bolt sleeve, a piece which supported the rear of the extension, had the locking lugs. (In the rifles before 1911, the lugs were at the rear; in the rifles

Rifle M1911 (above) and Carbine M1911 (below).

Model 31 (K31) Schmidt Rubin. Best of the designs.

after 1911, at the front.) Thus, an abnormally long receiver was needed for these weapons, and the locking lugs were a long way to the rear of the point where they were actually needed. It was not a very satisfactory design from a military point of view, since it meant a great deal of extra length and weight and still did not offer the best type of bolt support. In the K31, the part which was formerly considered the bolt sleeve became the bolt body itself, and the locking lugs were mounted at its head in the position where they were most needed. This shortening of the bolt assembly has resulted in a much better weapon which, although it is about the same overall length as the M1911 carbine, has a barrel 2.5 inches longer (made possible by the shorter receiver). The earlier weapons had their magazines placed a considerable distance in front of the trigger guard, because of the position of the long bolt extension. The magazine of the K31, on the other hand, is immediately ahead of the trigger guard. All in all, the K31 is a vastly superior weapon to the earlier Schmidt Rubins.

Model 31/42 Carbine, K31/42. One of the two sniper versions of the K31. It has a 1.8-power scope permanently built into the left side of the receiver. The head of this scope resembles a periscope, and when in use, is lifted up; when not in use, it is folded down against the side of the carbine. The weapon has a tangent-type metallic rear sight, graduated from 100 to 1000 meters.

Model 31/43 Carbine, K31/43. Equipped with a 2.8-power scope mounted like that of the K31/42. Its metallic rear sight is graduated from 100 to 700 meters.

Model 31/55 Sniper Rifle. Note bipod position and telescopic sight.

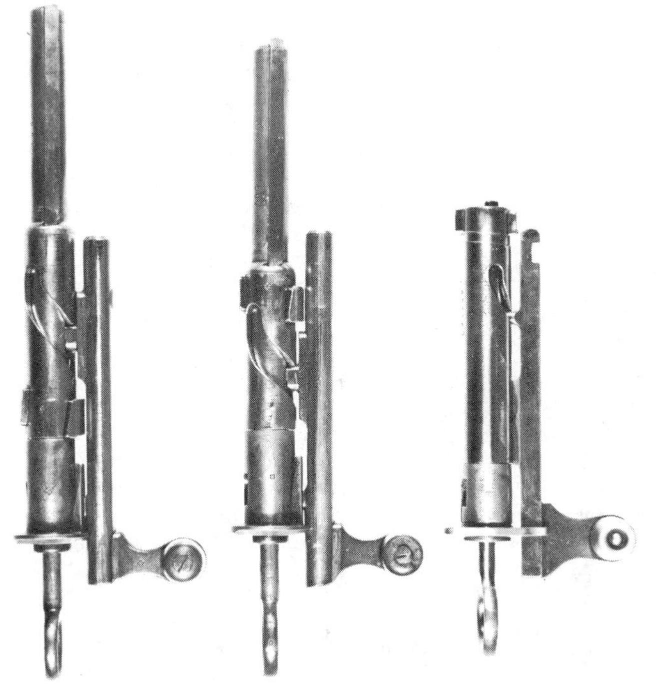

Schmidt Rubin bolts. Left: M1889 with locking lugs at rear of sleeve. Center: M1911 with locking lugs at front of sleeve. Right: M31 with locking lugs at front of bolt.

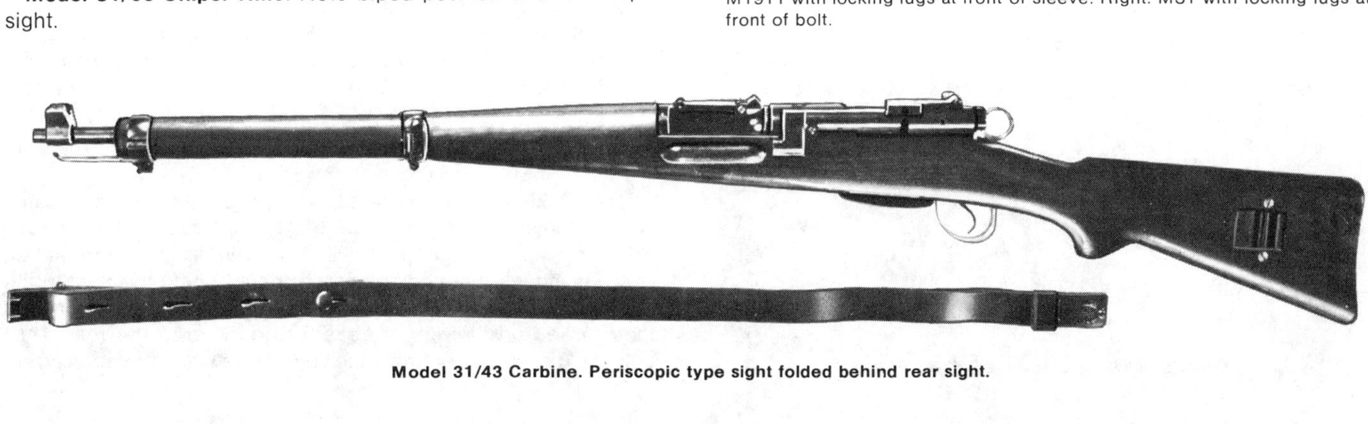

Model 31/43 Carbine. Periscopic type sight folded behind rear sight.

Characteristics of the Significant Schmidt Rubins

	M1889 Rifle	M1911 Rifle	M1911 Carbine	M1931 Carbine
Caliber:	7.5mm.	7.54mm.	7.54mm.	7.51mm.
Land diameter:	.295 in.	.2968 in.	.2968 in.	.2956 in.
Weight:	9.8 lb.	10.15 lb.	8.6 lb.	8.83 lb.
Length, overall:	51.25 in.	51.6 in.	43.4 in.	43.5 in.
Barrel length:	30.7 in.	30.7 in.	23.3 in.	25.67 in.
Magazine:	12-rd. detachable box.	6-rd. detachable box.	6-rd. detachable box.	6-rd detachable box.
Chamber pressure:	38,400 p.s.i.	45,500 p.s.i.	45,500 p.s.i.	45,500 p.s.i.
Muzzle velocity:	2033 f.p.s.	2640 f.p.s.	2490 f.p.s.	2560 f.p.s.

Schmidt Rubin Model 31/55 Sniper Rifle.

Swiss SIG 5.56mm SG 530-1 Rifle with folding butt stock.

Special Note on the Schmidt Rubins

As noted above, there is a considerable difference in the bolt construction of the various models of this weapon. None of the weapons prior to the M1896/11 rifle is considered by the Swiss to be strong enough to use the Swiss Model 11 cartridge. Because of slight differences in bore diameter and rifling, the early rifles will not shoot the Model 11 cartridge very accurately. The early rifles should be used with the Swiss 7.5mm M1890/23 cartridge; this cartridge has a 190-grain, conical-nosed bullet. They should not be used with the Spitzer-pointed Model 11 cartridge which has a 174 grain bullet.

THE .223 SIG MODEL SG 530-1 RIFLE

SIG has developed a new rifle chambered for the .223 (5.56mm) cartridge. This rifle uses a modification of the 7.5mm Model 57 rifle action altered to gas operation. It is made mainly of stampings and fabrications and has a plastic stock, pistol grip and forend. In addition to the fixed plastic stock version, there is a folding metal stock version.

Characteristics of the .223 SIG Model SG530-1 Rifle

Caliber: 5.56mm (.223)
System of operation: Gas operated, selective fire
Length overall: w/fixed stock--39.5 in.
w/folding stock, stock folded--30.8 in.
Barrel length: 18.1 in.
Feed device: 30-round, staggered row, detachable box magazine

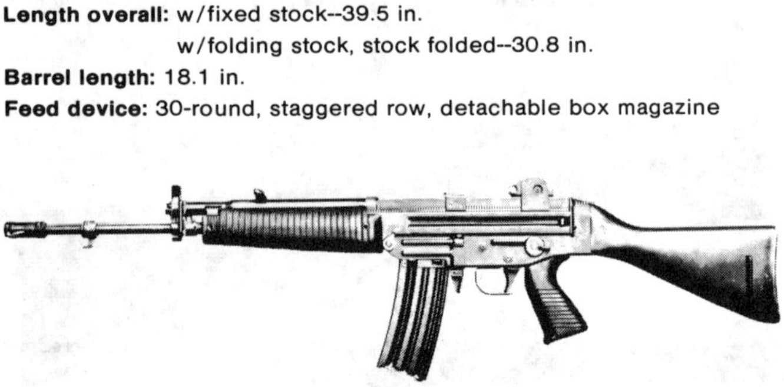

Swiss SIG 5.56mm Model SG 530-1 Rifle.

Sights: Front: Protected post
Rear: Aperture adjustable from 100 to 400 meters
Weight: w/fixed stock--8.35 lb.
w/folding stock--8.44 lb.
Cyclic rate: 550-650 r.p.m.
Muzzle velocity: 3150 f.p.s. with U.S. M193 ball cartridge

How to Load and Fire the SG 530-1

The magazine catch is immediately to the rear of the magazine; push forward to remove magazine. The magazine is loaded in the normal fashion. Set rifle on safe by turning safety/selector lever, located on the left side of the receiver above the pistol grip, to "S". Insert loaded magazine and pull operating handle, located to the rear of the front sight on the left top side of the gas cylinder tube over the barrel, to the rear and release. The cocking handle will return to its forward position and the weapon is loaded. If semi-automatic fire is desired, set the safety/selector lever on the number "1"; if automatic fire is desired, set the safety/selector lever on the number "30". Pressure on the trigger will now fire the rifle.

How to Field Strip the SG 530-1

Remove magazine. Push receiver catch, located at the top rear of the receiver, in and pivot stock, and trigger housing assembly down. The weapon breaks like a shotgun in a fashion similar to the FN "FAL" and the U.S. M16A1. Withdraw operating spring assembly, bolt and operating rod/bolt carrier. Remove bolt from bolt carrier. To assemble, perform the above steps in reverse.

How the SG 530-1 Works

The SG 530-1 is gas operated, but uses a roller bearing locking system somewhat similar to the StG 57. The bolt carrier is forced to the rear by the gas piston. This movement pulls the striker, which is mounted on the bolt carrier right behind the bolt, to the rear and allows the firing pin, which holds the roller bearings in their locked position in the receiver, to move to the rear. Thus unlocking occurs after firing; locking occurs when the bolt carrier, under the pressure of the compressed mainspring, forces the firing pin forward camming the locks into locked position.

The receiver of the rifle is grooved for a scope mount which locks into the front of the rear sight base. A prong-type flash suppressor is used and the fore part of the barrel and the flash suppressor can be used as a rifle grenade launcher.

THE SIG MODEL SG540 AND 542 RIFLES

SIG has produced two new rifles; the SG540 chambered for the 5.56mm cartridge and the SG542 chambered for the 7.62mm NATO cartridge. Outwardly these rifles are generally similar in appearance to the SG530-1, but internally they are quite different. The SG530-1 has a roller locked bolt and the SG540 and 542 rifle has a rotating bolt which is cammed in to and out of the locked position by a cam slot on the bolt carrier. The bolt system is quite similar in concept to that of the AK.

The rear end of the piston fits into a hole in the bolt carrier and is held in place by the operating handle, the end of which fits through a slot cut in the side of the bolt carrier.

The 540 series rifles have flash suppressor/grenade launchers attached to their barrels and can be set to fire three round bursts in addition to full automatic and semi-automatic. They can be supplied with bipod and with fixed or folding butt.

Characteristics of the SIG 540 Series Rifles

	SG540	SG542
Caliber:	5.56mm	7.62mm NATO
System of operation:	----------- gas, selective fire -----------	
Weight loaded (20-round magazine):	7.1 lb.	8 lb.
Length overall w/stock fixed:	37.5 in.	38.9 in.
w/stock folded:	28.7 in.	30.1 in.
Barrel length:	19.3 in.	19.5 in.
Feed device:	20 or 30-round detachable box magazine	20-round detachable box magazine
Sights: Front:	------------- protected post -------------	
Rear:	------rotating drum w/apertures------	
Muzzle velocity:	3215 f.p.s.	2690 f.p.s.
Cyclic rate:	------------- 650-800 r.p.m. -------------	

THE 7.5mm MODEL 57 ASSAULT RIFLE

Characteristics of the Model 57

Caliber: 7.5mm (Swiss Model 11 cartridge).
System of operation: Delayed blowback, selective fire.
Weight, empty: 12.32 lb.
Length, overall: 43.4 in.
Barrel length: 23 in., w/muzzle brake.
Feed device: 24-round, detachable, staggered-row box magazine.
Sights: Front: Folding blade, w/protecting ears.
Rear: Folding aperture, w/micrometer adjustment graduated from 100 to 650 meters.
Muzzle velocity: 2493 f.p.s.
Cyclic rate: 450 to 500 r.p.m.

The selective-fire Model 57 will eventually replace all the Schmidt Rubins in Swiss service. The action of this rifle is similar to that of several weapons which were in prototype stage in Germany at the end of World War II. The developer of this weapon, the Swiss Industrial Company of Neuhausen Am Rheinfalls, Switzerland (commonly known as SIG or Neuhausen) has also developed two machine guns using the rifle's delayed-blowback action—the MG710-1, and the MG710-2; these machine guns, formerly known as the MG 55-1 and MG 55-2, are not used by the Swiss Army.

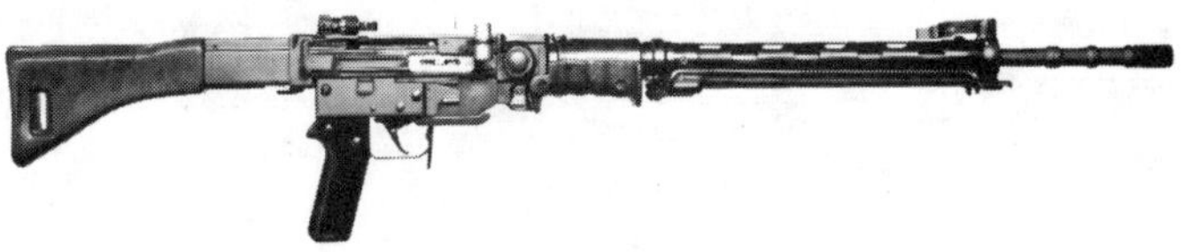

Model 57 Assault Rifle, right side view, current production.

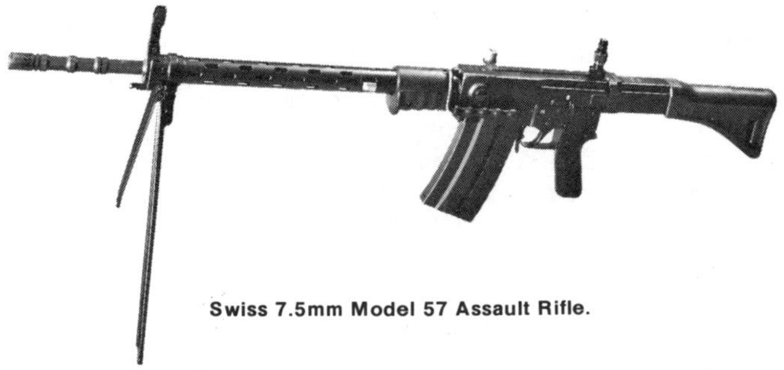

Swiss 7.5mm Model 57 Assault Rifle.

How to Load and Fire the Model 57

Insert a loaded magazine in the magazine well and pull the operating handle to the rear (the operating handle is located on

SIG 7.62mm NATO Model SG 542 Rifle with folding stock.

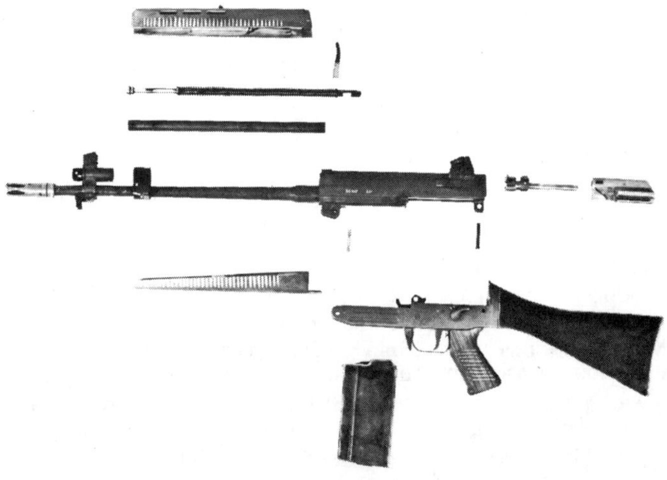

SIG 7.62mm NATO Model SG 542 Rifle field-stripped.

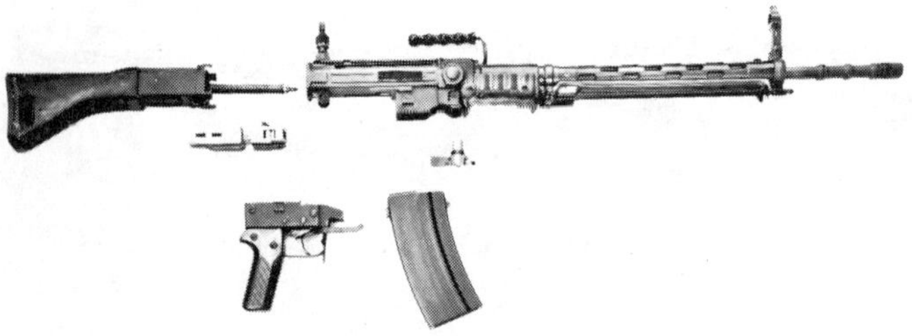

Model 57 field-stripped.

the right side of the receiver). Release the operating handle, and the bolt will go home, chambering a cartridge. Set the safety selector on the letter "E" if semiautomatic fire is desired, or on the letter "M" if full automatic fire is desired. Pressure on the trigger will fire the weapon. To set the weapon on safe, set the safety selector on the letter S.

How to Field Strip the Model 57

Remove the magazine by pushing forward the magazine catch, located at the forward end of the trigger guard. Depress the butt retaining catch, located on the forward underside of the butt; turn the butt approximately one-quarter turn to the right, and remove the butt and recoil spring. Pull the operating handle to the rear, and remove the bolt assembly. No further disassembly is recommended.

To reassemble the weapon, perform the above steps in reverse order.

How the Model 57 Works

When the trigger is pulled, the hammer is released and hits the firing lever mounted on the rear of the bolt. The firing lever strikes the firing pin, which hits the primer of the cartridge. The gas pressure forces the base of the cartridge against the head of the two-piece bolt. When a certain pressure has been reached, the locking rollers begin forcing rearward the nose of the rear section of the bolt. The locking rollers are then free to slip back into their recesses in the bolt head, and the weapon unlocks. The bolt travels to the rear, compressing the recoil spring, extracting and ejecting the spent case, and cocking the hammer. The recoil spring drives the bolt forward again, and the bolt picks up and chambers another round. The nose of the rear section of the bolt cams the locking rollers into their locking recesses in the receiver. If the weapon has been set on semiautomatic fire, the trigger must be pulled again. This weapon fires from a closed bolt on both automatic and semiautomatic fire. If the weapon has been set on automatic fire, it will continue to fire as long as the trigger is held to the rear and there are cartridges remaining in the magazine.

Special Note on the Model 57

The Model 57 has many interesting features. Its delayed-blowback action is worthy of note. The bolt is composed of two principal parts—the head and the rear section. The head is approximately one-fourth as heavy as the rear section. The head section contains two locking rollers which are cammed into recesses in the receiver by the nose section of the rear part of the bolt. The principal difference between this weapon and the Spanish CETME assault rifle, insofar as the bolt is concerned, is that the CETME fires from an open bolt on automatic fire. Since this system has no provision for slow initial extraction, both weapons have fluted chambers to "float" the neck and forward part of the case on gas.

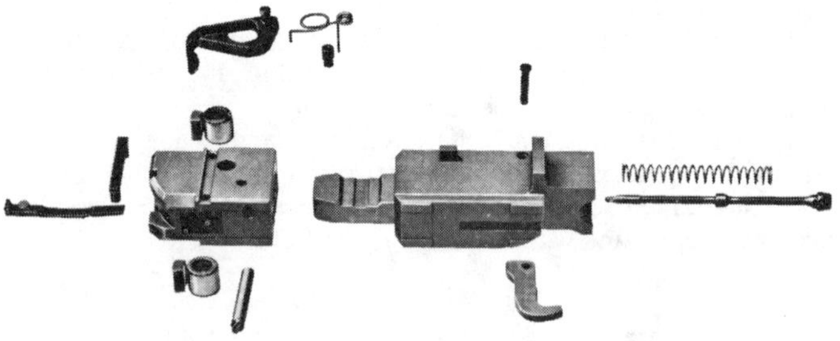

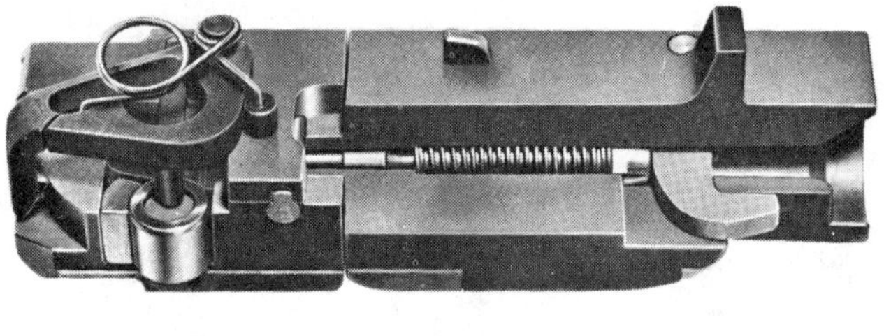

Model 57 bolt.

The Model 57 has a built-in rifle grenade launcher and a rubber buttstock. The rubber butt helps to absorb the shock of firing grenades when a wall, the rear of the trench, etc., is used as a support. A winter trigger is built into the gun. By the use of two spring-loaded catches in the barrel jacket, the bipod can be used to support the weapon either at the rear or at the front of the jacket. Everything considered—method of construction, firepower potential, and ease of maintenance—it is obvious that the Model 57 represents a great advance in the equipment of the Swiss Army.

SWISS SUBMACHINE GUNS

SIG started production of submachine guns in 1920 with the production of a modified Bergmann MP18I. Some of these weapons, which were chambered for the 7.65mm Luger or 7.63mm Mauser cartridge, were sold to Finland, China, and Japan. In 1930 this weapon was modified by the addition of a vertical fore grip but few of these were made, as the SIG designed Model MKMO was introduced in 1933. The MKMO, also known as the MK33, is a delayed blowback weapon which introduced the forward folding magazine on submachine guns, i.e., the magazine can be folded forward and locked into a position in the stock when not in use.

A carbine version of this weapon for police use was produced as the Model MKPO. In 1937 SIG introduced the Model MKMS and MKPS submachine guns. These weapons are generally similar to the MKMO and MKPO, respectively, but use a blowback type action rather than the more expensive retarded blowback of the earlier weapons.

The next submachine gun produced by SIG was the Model 41, a blowback weapon chambered for the 9mm Parabellum cartridge. This weapon was produced to compete in Swiss government submachine trials in 1940-41. A decision was made to adopt the

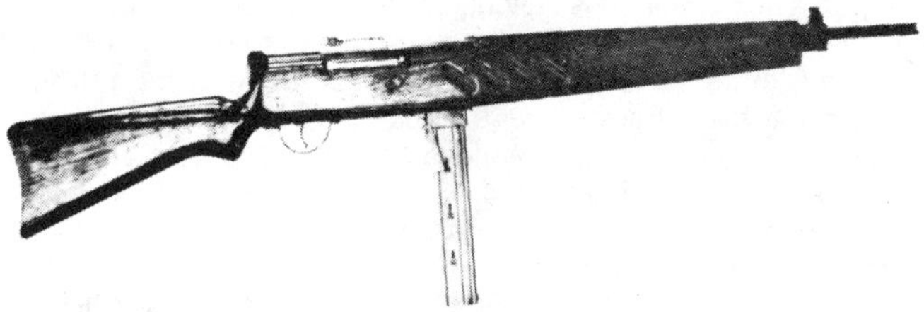

SIG Model MKMO (MK33) Submachine Gun.

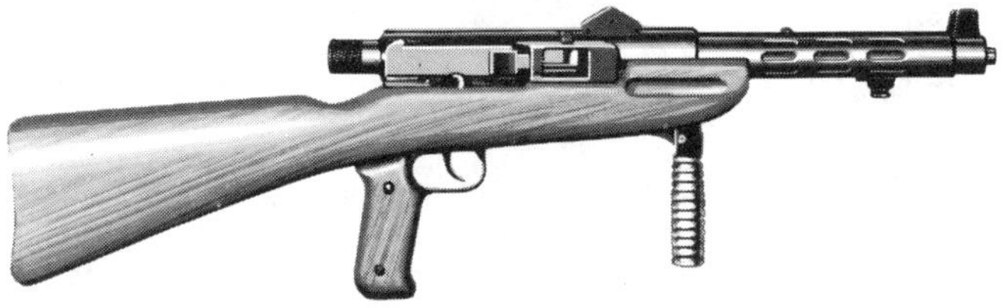

Swiss Model 41/44 Submachine Gun.

Model 43/44 Submachine Gun.

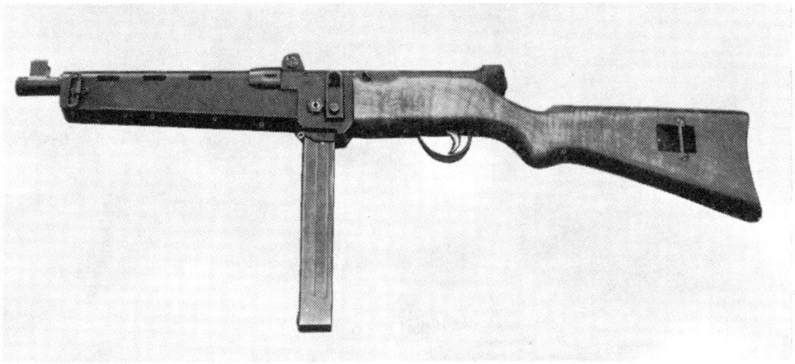

SIG MP46 Submachine Gun.

government-developed Furrer Model 41/44 submachine gun, and the Model 41 was dropped by SIG.

The Swiss 9mm Parabellum Model 41/44 submachine gun was developed by Waffenfabrik Bern and is a most unusual submachine gun in several respects. The weapon is recoil-operated and has a toggle joint action similar to the Model 25 Furrer light machine gun. The vertical fore grip can be folded up under the barrel. It

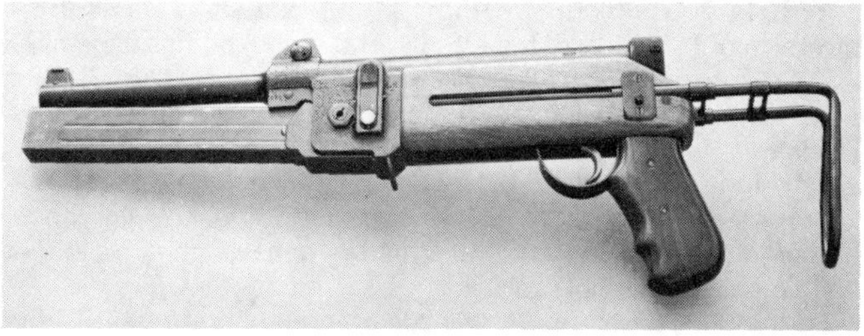

SIG MP48 Submachine Gun.

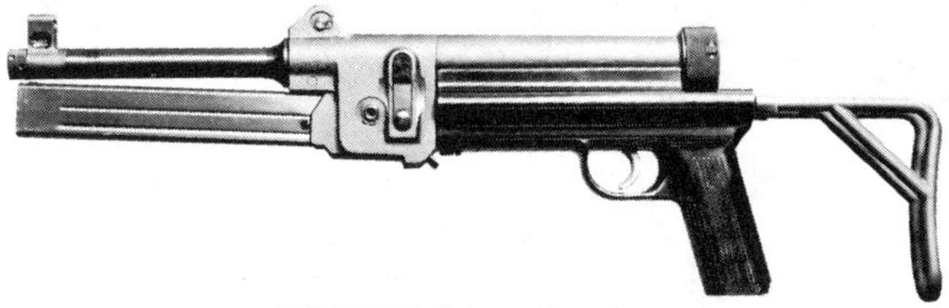

SIG MP310 Submachine Gun.

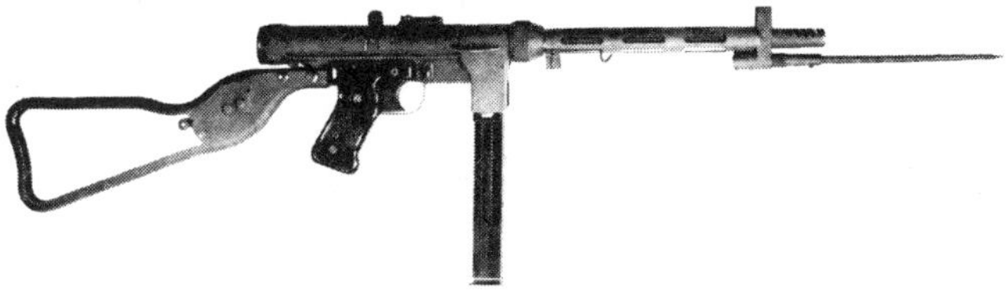

9mm Parabellum Rexim FV4 Submachine Gun.

is an extremely expensive weapon to manufacture and unduly complicated. Unlike the Model 41, the Model 41/44 has a bayonet lug.

The Swiss government also purchased 5,000 Suomi Model 43 submachine guns from Finland. This 9mm Parabellum weapon was the same as that used by the Finnish Army. Hispano Suiza obtained a license to manufacture the weapon in Switzerland. The Suomi as manufactured by Hispano Suiza has a flip over-type rear sight, and a bayonet lug; it is called the MP43/44.

Although the Solothurn (or Steyr Solothurn submachine gun, as it is frequently called) was made in Switzerland, it was of German design. The best known of these weapons was the S1-100. The Steyr Solothurn guns were also made at the Steyr plant in Austria and the Austrian Model 34 submachine gun is a good example of this submachine gun type. The Solothurn was used by Chile, El Salvador, Bolivia, Uruguay, Portugal, and reportedly by China and Japan.

SWISS SERVICE SUBMACHINE GUNS

	Model 41/44	**Model 43/44**
Caliber:	9mm Parabellum.	9mm Parabellum.
System of operation:	Recoil, selective fire.	Blowback, selective fire.
Length overall:	30.5 in.	33.9 in.
Barrel length:	9.8 in.	12.4 in.
Feed device:	40-round, staggered row detachable box magazine.	50-round, in-line, detachable box magazine—has double column divided by central wall.
Sights: Front:	Blade with ears.	Blade.
Rear:	Flip-type w/"U" notch.	Flip-type w/"U" notch.
Weight:	11.4 lb.	10.5 lb.
Muzzle velocity:	1312 f.p.s.	Approx. 1350 f.p.s.
Cyclic rate:	900 r.p.m.	800 r.p.m.

In 1944 SIG introduced their MP44, which is basically a modernized version of the MKMS. The Model 44 has a stamped steel fore-end, and is capable of selective fire. The MP46 is basically the same as the MP44 except for some bolt details. These models were not commercially successful.

The Model 48 was modified by SIG by the fitting of a retractable type steel wire stock, a shorter barrel and in general lightened, and produced as the Model 48. This weapon, which is no longer manufactured, was sold to Chile.

Further improvements were made in the Model 48 by SIG and plastics and castings substituted for machined parts and the weapon which resulted is called MP310 by SIG. The Model 310 is capable of selective fire, as are all other SIG submachine guns starting with the MP44, has the folding magazine of the previous models and the retractable wire stock of the Model 48.

The firm of Rexim S. A. of Geneva introduced a submachine gun known as the Rexim FV Mark 4. This weapon, which was made in limited numbers, with many different barrel lengths and several different stocks, is unusual in that it fires from a closed bolt. The Rexim was also made in Spain at the La Coruna arsenal and was offered for sale as the "La Coruna." Although the Rexim is an interesting design, it has not been a very successful one.

SWISS MACHINE GUNS

Switzerland adopted the Maxim gun in 1894. Later models used were the Model 1900 and Model 1911. The 7.5mm Model 1911 was used as the standard heavy machine gun until 1951.

Colonel Furrer of the Swiss government arms plant at Bern developed the 7.5mm Model 25 light machine gun, also known as the Fusil Furrer. The Model 25 has a toggle-joint action similar to the Luger pistol but which breaks to the side rather than to the top as does the Luger pistol. This weapon was quite impressive in its day, but is being replaced by the 7.5mm Model 57 assault rifle. A link belt-fed 13mm aircraft gun version of the Model 25 was also produced.

The next machine gun to be developed in Switzerland was a commercial development of SIG—the KE7 light machine gun. This weapon had some success in oversea sales (it was sold to China), but was not adopted by the Swiss government. The KE7 is a recoil-operated gun, which fires from an open bolt. It is selective fire and is relatively light, weighing 17.25 lbs. without magazine.

Swiss 7.5mm Model 25 Light Machine Gun.

SIG KE7 Light Machine Gun.

During the post-war period, the Swiss government arms plant at Bern developed a modified copy of the German MG42 which was adopted by the Swiss Army as the MG51. The Swiss seem to have defeated much of the purpose of the MG42 however, by making the MG51 mainly out of heavy milled parts rather than stampings. MG51 has locking flaps in its bolt head, which engage locking recesses in the receiver, rather than locking rollers of the MG42.

Swiss 7.5mm Model 51 Machine Gun.

Hispano Suiza Machine Gun, Type HSS808.

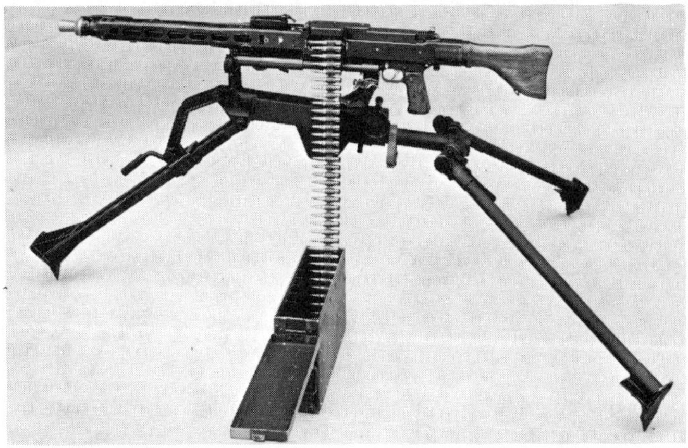

SIG MG710-1 Machine Gun, on tripod.

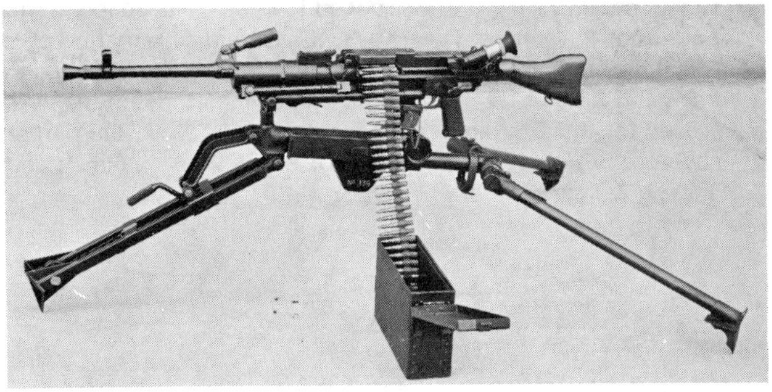

Swiss MG50 on tripod, belt feed, as a medium machine gun.

MG710-1 on bipod with drum magazine.

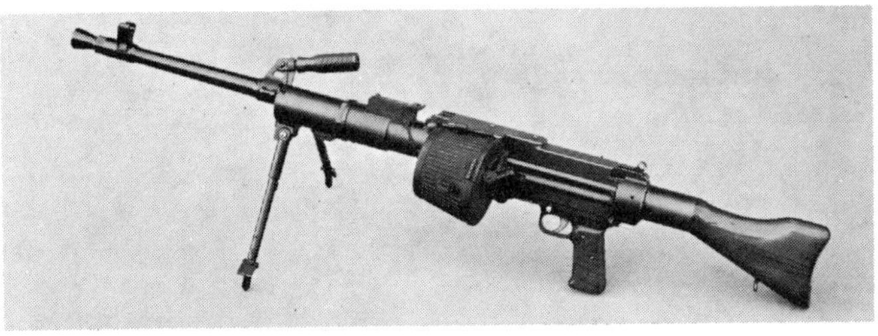

Swiss MG50 on bipod. Drum feed. Used as a L.M.G.

Barrel change and field stripping of the MG51 is generally similar to the MG42.

Hispano Suiza, a well-known manufacturer of aircraft and anti-aircraft cannon, developed a rifle caliber machine gun after World War II. This weapon was not a financial success and was made only as a prototype. It was known as the Type HSS808.

SIG developed a weapon to place in the competition held by the Swiss for a weapon to replace the Model 11 Maxim gun. The SIG Model 50 is a gas-operated weapon designed to be used as a general purpose weapon. It has a quick change barrel and is fed by

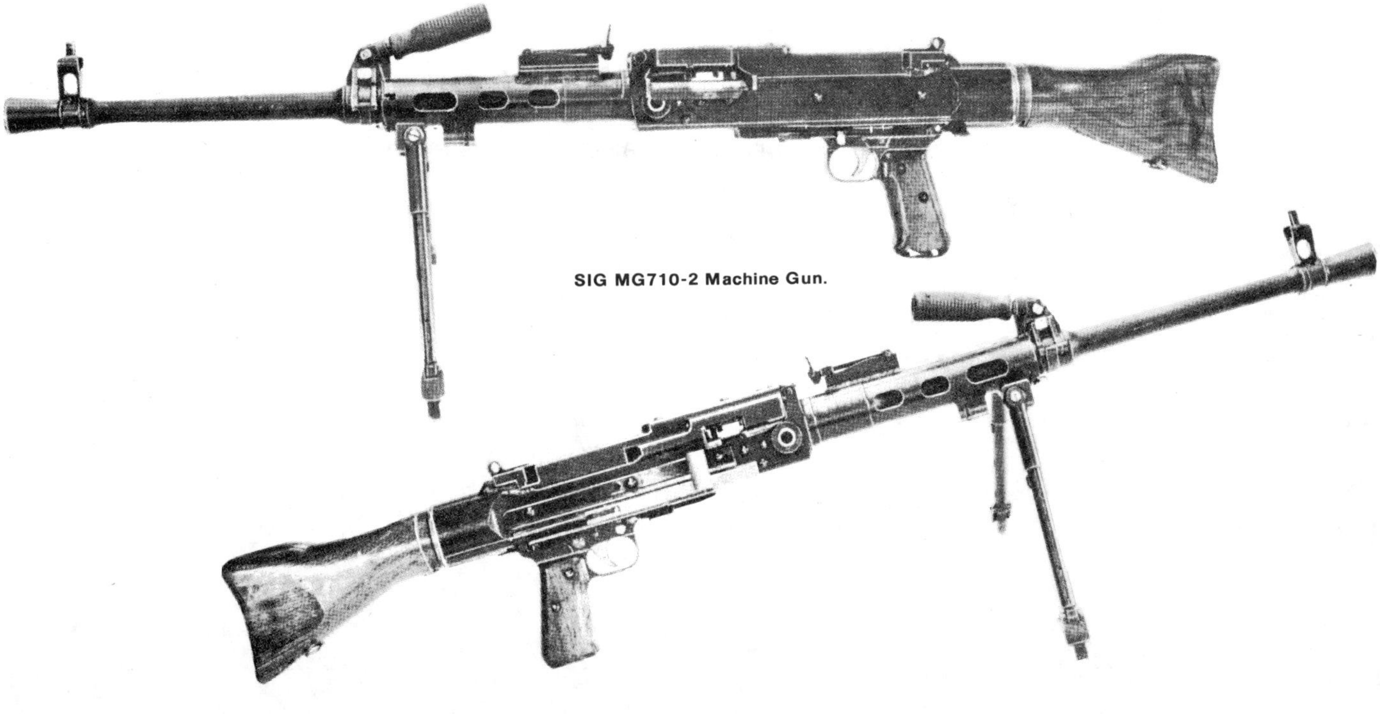

SIG MG710-2 Machine Gun.

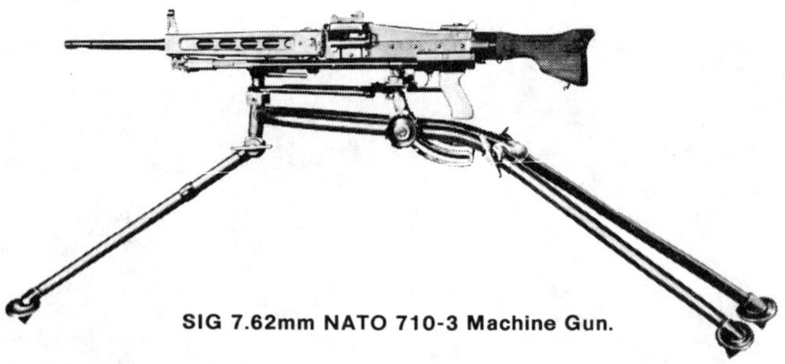
SIG 7.62mm NATO 710-3 Machine Gun.

a non-disintegrating metallic link belt similar to that of the MG42 or by a 50-round drum. This weapon is not used by Switzerland but was adopted by Denmark in caliber .30 as the Model 51 machine gun. It is being replaced in that country by the 7.62mm NATO MG42/59.

After the MG50, SIG developed a series of delayed blowback machine guns using the same locking principles as the Model 57 assault rifle. Originally advertised as the Model 55-1 and 55-2, the first two guns of the series, now called the 710-1 and 710-2, are similar in all respects excepting the barrel, barrel support, and the presence of a barrel jacket on the 710-1. The feed mechanism is similar to that of the MG42 and a link belt or a 50-round drum similar to that of MG42 is used.

The SIG MG710-1 is remarkably similar to the MG42V which was under development in Germany at the close of World War II. The barrel change on this gun is the same as that on MG42.

The SIG 710-2 does not have a barrel jacket and the barrel is held in the receiver by a bayonet-type joint. The barrel handle is used to change the barrel. The 710-2, as the 710-1, is designed to be used as a general purpose machine gun.

The third weapon in this series is the MG710-3. This weapon has a partial barrel jacket and the barrel is released for removal by a catch on the top of the barrel jacket. A large barrel-removal handle is fitted to the right side of the barrel and the barrel is pulled to the right and rear to remove. The 710-3 will use a disintegrating or a non-disintegrating link belt. The 710-3 uses more stampings than do the 710-1 and 710-2, and is a bit lighter than those guns. It is advertised for the 7.62mm NATO cartridge only, while the earlier guns were advertised chambered for the 6.5mm, 7.92mm, and 7.62mm NATO cartridges.

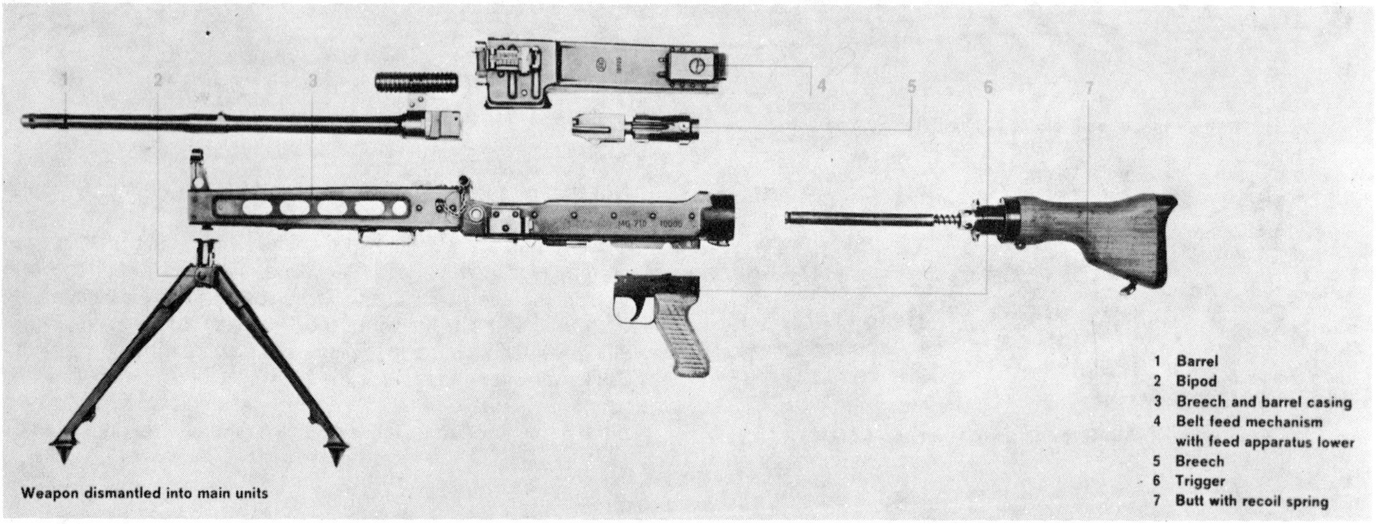

SIG 710-3, field-stripped.

CHARACTERISTICS OF SWISS SERVICE MACHINE GUNS

	Model 11 Heavy	Model 25 Light	Model 51
Caliber:	7.5mm.	7.5mm.	7.5mm.
System of operation:	Recoil, automatic only.	Recoil, selective fire.	Recoil, automatic only.
Length overall:	42.4 in.	45.8 in.	50.1 in.
Barrel length:	28.4 in.	23 in.	22.2 in.
Feed device:	250-round web belt.	30-round, staggered row, detachable box magazine.	Non-disintegrating metallic link belt, in 50 and 250-round segments, 50-round drum.
Sights: Front:	Blade.	Blade w/protecting ears.	Folding blade.
Rear:	Leaf.	Tangent w/notch.	Tangent, optical sight used on tripod.
Weight:	40.8 lb. (gun only).	23.69 lb. w/bipod.	35.3 lb. w/bipod. 57.3 lb. w/tripod.
Muzzle velocity:	2590 f.p.s.	2460 f.p.s.	2460 f.p.s.
Cyclic rate:	Approx. 500 r.p.m.	450 r.p.m.	1000 r.p.m.

50

Turkey

The Turkish Army has a large assortment of service small arms which include most of the standard United States, British and German weapons, of World War II. In addition various French weapons, such as the 7.5mm Model 1924 M29 light machine gun, and the 7.92mm St. Etienne machine gun, Czech 7.92mm Model 1924 Mausers and 7.92mm ZB26 and ZB30 light machine guns are used as well. Many of the older Turkish Mausers such as the Model 1890, Model 1893, and Model 1903 and 1905, now chambered for the 7.92mm cartridge, are still in service.

TURKISH PISTOLS

A native-made pistol is the 9mm Kirikkale. This weapon, which is made at the Kirikkale arms factory in Turkey, is a copy of the Walther PP. It is identical to the late commercial-type Walther, except for variations in machining and the finger-rest extension on the removable magazine floor plate. For data on the Walther PP, see the chapter on German World War II materiel. The Kirikkale is chambered for the 9mm Browning short (.380 ACP) or the 7.65mm (.32 ACP) cartridge.

Turkish 9mm automatic pistol.

TURKISH RIFLES

Turkey responded very rapidly to the advantages of the military small-caliber repeating rifle and adopted a 7.65mm Mauser in 1890—the Model 1890. This rifle, which was chambered for the 7.65mm cartridge introduced with the Belgian Mauser in 1889, has an in-line magazine and bolt similar to the Belgian Model 1889 excepting the buttress threads on the bolt sleeve. The Model 1890 does not have the metal barrel jacket of the Model 1889 and introduced the stepped barrel, which was used in Mausers from that time on.

A number of these rifles were obtained by Yugoslavia after World War I. The Yugoslavs rebarreled with a shorter 7.92mm barrel and fitted a longer handguard. This weapon was called the Model 90T by the Yugoslavs.

The Model 1893 7.65mm Mauser is essentially the same as the

7.65mm Turkish Mauser M1890 Rifle.

7.92mm M90T.

7.65mm Model 1893 Rifle.

7.65mm Model 1903 Rifle.

7.92mm Model 1905 Carbine. This weapon was originally 7.65mm.

7mm Spanish M1893 Mauser, with the addition of a cutoff to the magazine.

The Model 1903 7.65mm rifle is essentially the same as the German Rifle Model 98 except for the rear sight, hand guard, upper band, longer cocking piece and firing pin and modified bolt stop. There is a carbine Model 1905 and Model 1890 which are stocked to the muzzle.

During World War I, the Turks were supplied by Germany with many 7.92mm German Mausers rifles and carbines. After the war, Turkey purchased quantities of Czech 7.92mm Model 1924 Mausers and rebarreled many of the earlier 7.65mm rifles and carbines for 7.92mm.

Only the characteristics of rifles designed specifically for Turkey are given. Characteristics of other rifles used by Turkey will be found under country of origin.

Markings on Turkish ZB30 Machine Gun.

CHARACTERISTICS OF TURKISH RIFLES

	Model 1890 Rifle	**Model 1893 Rifle**	**Model 1903 Rifle**
Caliber:	7.65mm Mauser.	7.65mm Mauser.	7.65mm Mauser.
System of operation:	Turn bolt.	Turn bolt.	Turn bolt.
Length overall:	48.6 in.	48.6 in.	49 in.
Barrel length:	29.1 in.	29.1 in.	29.1 in.
Feed device:	5-round, in line, non-detachable box magazine.	5-round, staggered row, non-detachable box magazine.	5-round, staggered row, non-detachable box magazine.
Sights: Front:	Barley corn.	Barley corn.	Barley corn.
Rear:	Leaf.	Leaf.	Tangent.
Weight:	8.8 lb.	8.8 lb.	9.2 lb.
Muzzle velocity:	2132 f.p.s.	2132 f.p.s.	2132 f.p.s.

Union of Soviet Socialist Republics (USSR) ("Russia")

51

The Soviets use the following small arms: 9mm Makarov (PM) pistol; 9mm Stechkin (APS) machine pistol; the 7.62mm AKM and AK assault rifles; the 7.62mm Model SVD Sniper rifle; the 7.62mm RPK, RPD, and RP-46 light machine guns; the 7.62mm PK/PKS machine gun; the 7.62mm SGM and SGMB heavy machine guns; the 7.62mm DTM, SGMT, and PKT tank machine guns; and the 12.7mm DShK M1938/46 and 14.5mm KPV and KPVT heavy machine guns.

The 7.62mm SKS carbine is no longer a standard shoulder weapon, but may be found in frontier forces as may the SG43 heavy machine gun. It should be noted that all but the DTM, SG43, and DShK M1938/46 machine guns are of post World War II origin and that the feed mechanism of the DShK is of post-war origin.

Most of these weapons are in service in Communist countries throughout the world, and many have been exported to non-aligned countries throughout the world.

SOVIET PISTOLS AND REVOLVERS

The Soviets adopted the 7.62mm Nagant revolver in 1895. This revolver was produced in both single-action and double-action versions and is somewhat unusual in that the cylinder moves forward before the hammer falls and the forward end of the chamber alined for fire telescopes the barrel. The cartridge, which outwardly resembles a blank cartridge, has its bullet seated below the cartridge case mouth. The purpose of these design features is to prevent gas leakage at the joint between the cylinder and barrel. It is doubtful if this complicated system is worth the effort.

Caliber .22 Model R-4 Pistol.

7.62mm Model 1895 Revolver.

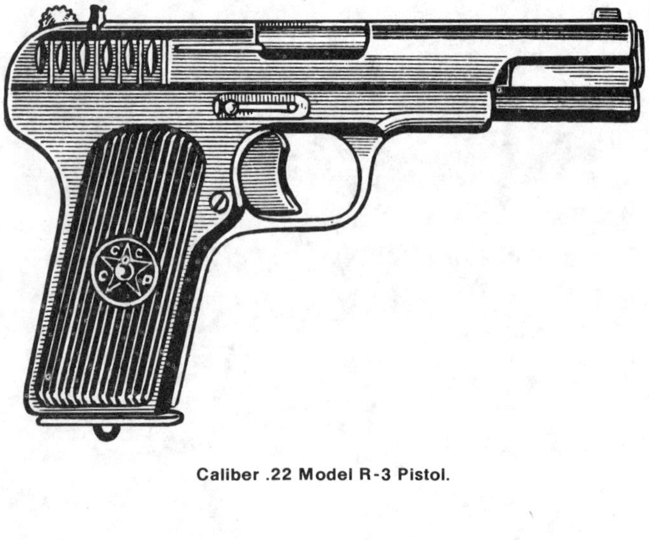

Caliber .22 Model R-3 Pistol.

Although the Model 1895 was manufactured as late as World War II, it is no longer in military service throughout the Communist world. A somewhat smaller version of this revolver was made for police use and a caliber .22 version was made for training purposes.

In 1930 the first model of the 7.62mm TT Tokarev automatic pistol was adopted; a slightly modified version was adopted in 1933. These weapons are no longer in Soviet service, but are widespread throughout the Communist world and are therefore covered in detail later in this chapter. A caliber .22 training version of the Tokarev called the Model R-3 and a caliber .22 target version called the Model R-4 have also been manufactured.

The 7.62mm Tokarev has been replaced in the Soviet service, and in several Soviet satellites as well, by the 9mm Makarov (PM) pistol and the 9mm Stechkin Machine pistol. These weapons are chambered for a new 9mm cartridge which is intermediate in size and power to the 9mm Browning Short (.380 ACP) and the 9mm Parabellum cartridge. It is quite similar to the 9mm Ultra prototype cartridge developed for use in a version of the Walther PP to be used by the Luffwaffe. The 9mm Makarov cartridge has a 94-grain bullet.

The Soviets produced a small 6.35mm (.25 ACP) automatic called the TK (Tula Korovin). This is not a military weapon, but may be used by police and para-military organizations.

6.35mm TK Pistol.

7.62mm TT M1933 Pistol.

THE 7.62mm TOKAREV TT M1930 AND M1933 PISTOLS

The 7.62mm Tokarev or Tula Tokarev-TT—was designed by Fedor V. Tokarev. It is a very slightly modified Colt-Browning design, notably mainly by its "packaged" type sear and hammer assembly, which is removed as a unit. This type of construction is not unique and has been used in a number of foreign pistols. The TT33 has been manufactured in Hungary as the Model 48 and in Communist China as the Type 51. The Soviet 7.62mm Model 1930 Type P cartridge is almost identical to the 7.63mm Mauser cartridge and is interchangeable with that cartridge.

Description of Tokarev Mechanism

The slide closely resembles the Colt Browning type. However, the barrel bushing which is inserted in the front end of the slide to support the barrel and retain the recoil spring, is a heavily forged bushing with one large hole for the barrel and a small hole below it in which the steel disc at the end of the recoil spring seats. This is much stronger than the conventional horseshoe shape bushing and plug.

The firing pin unit is retained in the slide by a simple split pin driven through a hole drilled in the slide, the pin passing through a

Field Stripping the Tokarev

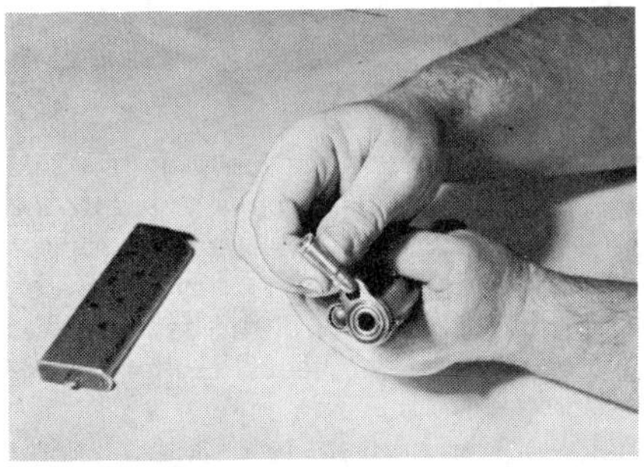

(1) Insert nose of cartridge through hole in bushing below barrel and compress spring until bushing is free to turn.

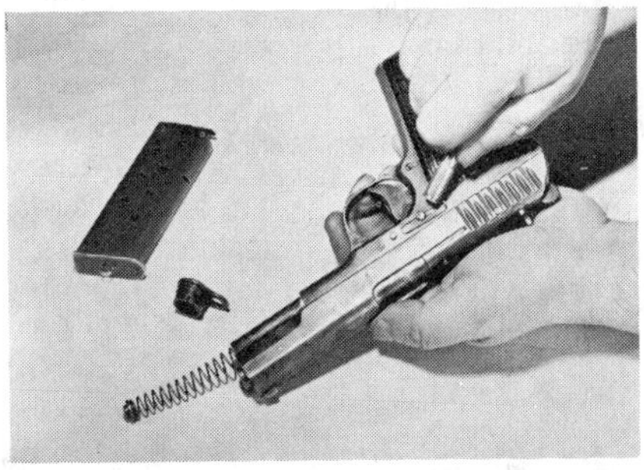

(2) Swing bushing up to the right until its locking lugs disengage. Remove it from the slide. Let the recoil spring protrude.

(3) Using a cartridge or the bottom of the magazine, pull back the spring locking clip on the right side of the pistol. This frees the slide-stop pin.

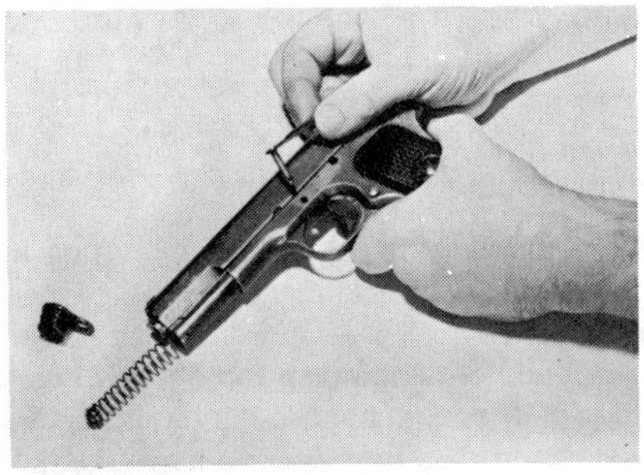

(4) Withdraw the slide-stop barrel-locking pin from the left.

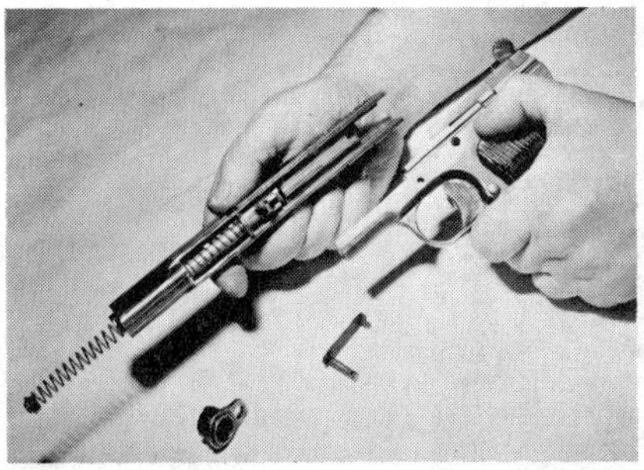

(5) Push the slide and barrel assembly forward out of the guides in the receiver.

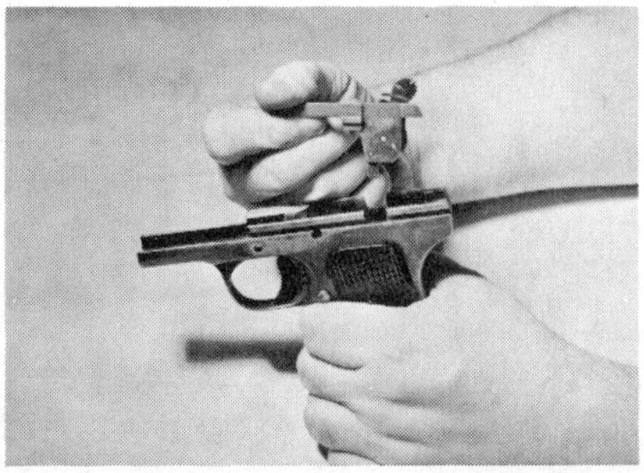

(6) Lift the receiver sub-assembly and hammer mechanism out of the receiver.

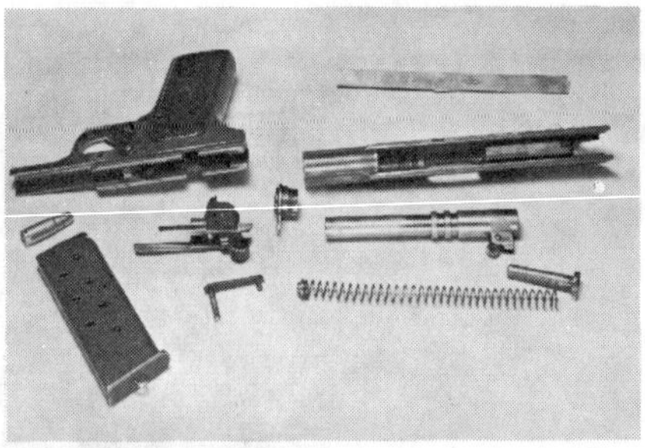

(7) Remove barrel, recoil spring and guide from front of slide.

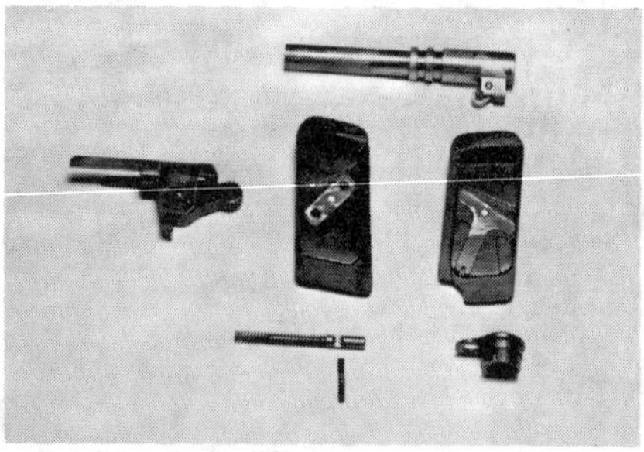

(8) No further dismounting is normally necessary. However stocks may be removed by reaching inside the handle and turning the metal locking buttons. Firing pin may be removed by punching out split pin from right side of slide. Hammer unit may be completely dismounted by pushing out the three retaining pins. The magazine release catch is a split pin which may be driven out from the right side of the receiver. This permits removing trigger and spring to complete entire disassembly.

slotted out section on top of the firing pin. This pin not only retains the floating firing pin but also determines the distance it can fly forward when struck by the hammer.

The magazine is particularly worthy of note in that it can be easily taken apart for cleaning or repairs. The follower, or platform on which the first cartridge rests, is supported from below by a typical magazine spring. The bottom of this spring, however, rests on a flat platform which in turn rests on the magazine bottom. The bottom is tongued into grooves in the magazine case, being slid in from the front. It is retained by a small nib on the platform which protrudes through a hole in the magazine bottom. Pushing this nib in permits the bottom to be pulled forward. The platform, spring, and follower can then be removed.

While the barrel locks in the same manner as the Colt Browning, its manufacture has been greatly simplified. Where the Colt has two locking ribs on the top of the barrel only, the TT33 Tokarev locking ribs run entirely around the barrel circumference. The lower sections have no locking function but permit of simpler barrel manufacture. They also provide, together with a thickened breech end, a strong support for the high powered cartridge used.

The stocks are of black plastic and require neither screws nor machined-in supports. A pivoted flat spring is riveted on the inner face of each stock. When turned crosswise each end locks in the frame. Turning the strip disengages it from the frame and permits the stock to be lifted off.

The magazine release catch is a split pin. When punched out it permits easy removal of the stirrup-type trigger and the flat spring in the handle which acts as the trigger return spring.

The slide stop and barrel locking pin function as in the Colt Browning, but the retaining system has been simplified. Where the pin emerges on the right side of the receiver, a sliding spring clip

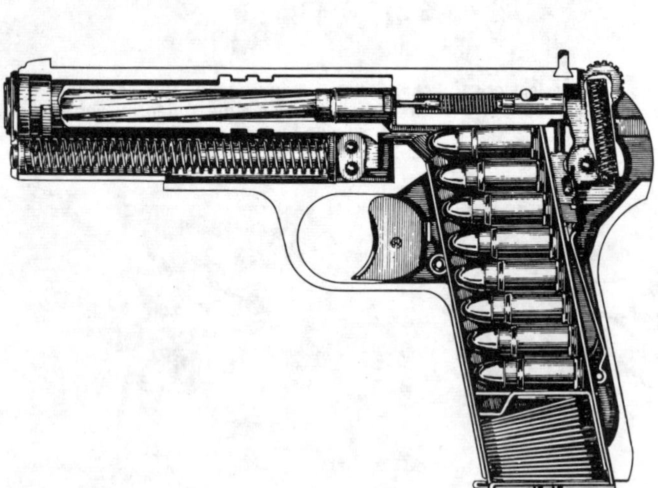

Action closed, showing details of mainspring in hammer and firing mechanism.

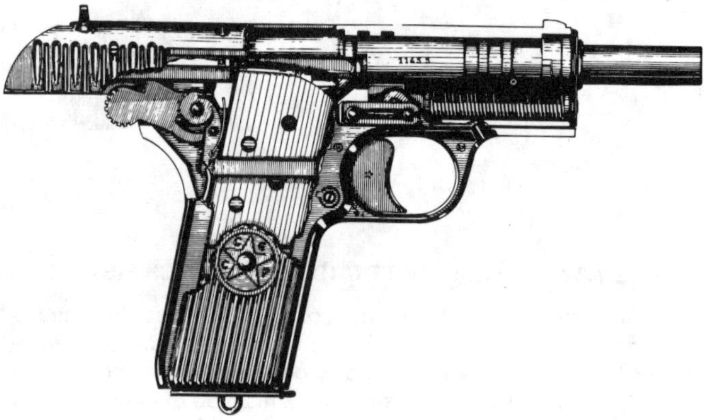

Action open, cut away to show details of recoil system and cocking operation.

is provided. When pushed forward teeth on it engage in notches in the end of the pin. Pulling this back frees the pin to permit removal.

The really startling advance of this pistol over all preceding types, however, is in its hammer mechanism.

This mechanism, consisting of a simple hammer, sear, disconnector, sear spring, and mainspring, is housed in a sub-assembly which forms part of the receiver and carries part of the slide grooves.

This sub-assembly block has two arms of unequal length. While the outside surfaces line up with the regular receiver guides, the inside surfaces are specially grooved to act as cartridge guides to assure proper feeding as each cartridge is stripped from between the very flat lips of the magazine. The effect of this feature is to make the feeding much more positive, as magazine mouths are normally weak points in feeding systems.

The mainspring is a coil which is housed inside the hammer itself. Its lower end rests against a pin driven through the housing in the assembly. As the hammer is rocked back by the slide during recoil, this spring is compressed against the supporting pin to provide the energy source for firing the next shot.

The sear, sear spring, and disconnector are mounted together on a third pin in the housing. When the slide is fully forward, the disconnector rises into a slot machined for it in the slide housing below the line of the firing pin.

Operation of the Tokarev

When the trigger is pressed, its stirrup forces the flat spring in the handle back to provide energy for returning the trigger to firing position when pressure is released.

The lower end of the disconnector is riding on top of the trigger stirrup while its upper end is in its seat in the breech block. In this position, the sear attached to the disconnector is also in contact with the rear of the trigger stirrups. The trigger forces back against the sear and presses the upper end of the sear out of its engage-

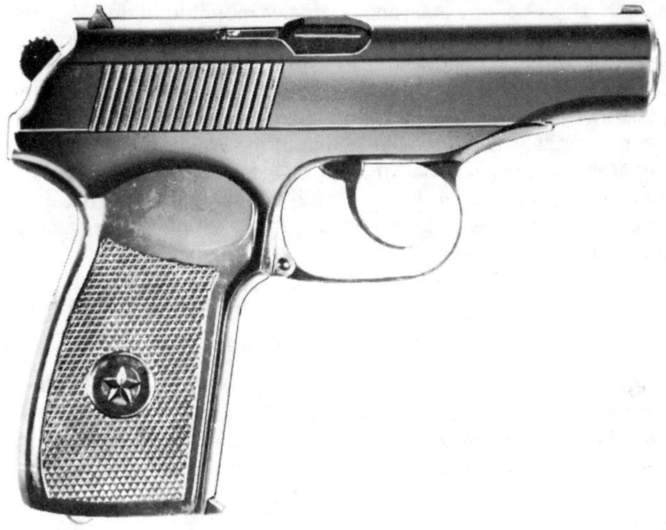

Soviet 9mm Makarov Pistol (PM).

Makarov field-stripped.

ment in the second hammer notch. The hammer under tne pressure of the compressed spring in it rotates on its axis pin to strike the firing pin and drive it forward to fire the cartridge.

As the slide and barrel under force of the recoil start back, they are locked securely together. As the pressure drops, the barrel swings down on its link to unlock from the slide. The slide continues on backwards to extract and eject in normal fashion.

The slide now forces the hammer back to be caught by the sear. When pressure on the trigger is released so that the disconnector can resume its firing position, the weapon is ready for another shot. The slide has moved forward under tension of the recoil spring below the barrel to strip a cartridge from the top of the magazine.

How TT33 Tokarev Differs From TT30

The 1930 type has a dismountable block assembled into the back edge of the grip frame. This block carries the trigger extension bar operating spring and the disconnector spring. The 1933 type has a solid back edge grip frame with the operating spring and disconnector spring assembled directly into the grip frame.

The firing system frame of the 1930 type differs from the 1933 frame in its cutouts and general assembly to the grip frame.

The locking lugs of the 1930 type are machined and ground out only on the upper surface of the barrel ahead of the chamber. The lugs on the 1933 type barrel are machined and ground out around the entire circumference of the barrel ahead of the chamber.

The form of the disconnector differs between the two pistols. The 1930 type has a smaller extension bar contacting surface.

THE 9mm MAKAROV (PM) PISTOL

The Makarov which is called PM (pistol Makarov) by the Soviets, is a double action, blowback operated self-loading pistol which

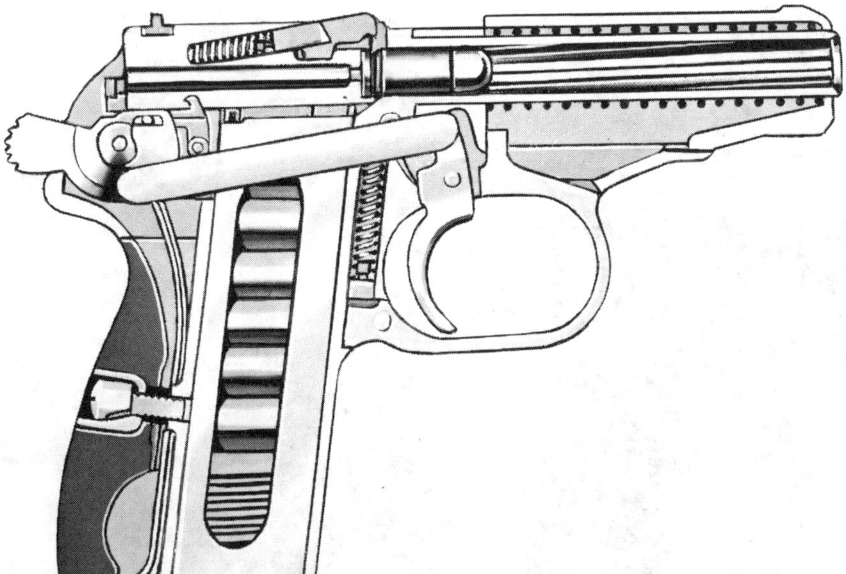

Section view of Makarov pistol.

outwardly looks like a scaled up copy of the German Walther PP. The Makarov is quite different in internal design from the Walther. The principal differences are as follows: the Makarov does not have a loaded chamber indicating pin. The PM uses a leaf type main spring; the Walther used a coil spring. The trigger-sear linkage of the Makarov is considerably different than that of the Walther, as is the disconnector. The magazine catch of the Makarov is of the spring type and is at the bottom of the grip. The magazine catch of the Walther is a button type mounted on the left side of the receiver. The Makarov has an externally mounted slide stop; the Walther slide stop is internally mounted and is released by drawing back the slide and allowing it to run forward. The ejector of the Makarov is the back end of the slide stop bar. The safety of the Makarov is pushed up to put it on "Safe" and pushed down to put it on "Fire", which is opposite to that of the Walther. The safety of the Makarov places a bar in front of the hammer and has a lug which positions itself in front of the rear shoulder of the receiver, thereby locking the slide in position and preventing the hammer from being cocked.

9mm Stechkin Machine Pistol (APS).

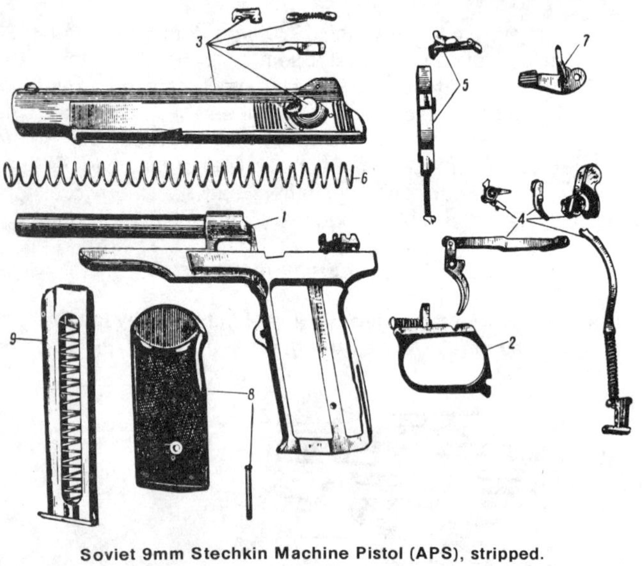

Soviet 9mm Stechkin Machine Pistol (APS), stripped.

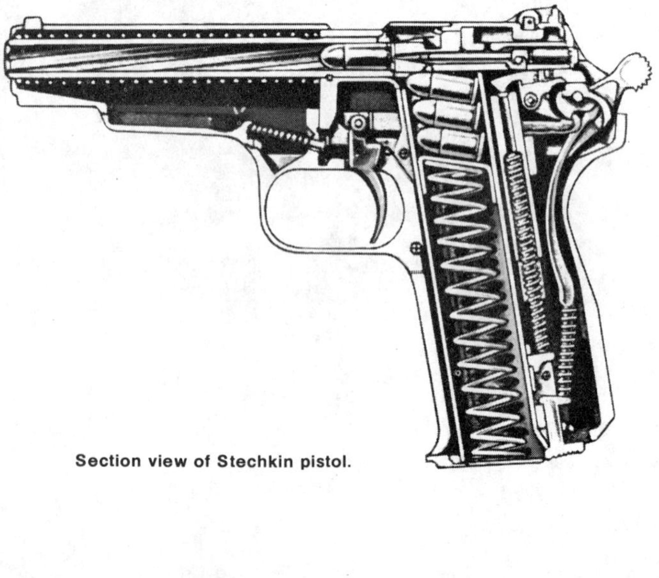

Section view of Stechkin pistol.

Stripping of the Makarov

Pull down the trigger guard; pull the slide to the rear and lift its rear end up; then ease forward the slide and the recoil spring (which encircles the barrel), and remove them from the weapon. Since the Makarov is a blowback-operated weapon, it has a fixed. barrel which should not ordinarily be removed. Further stripping is not recommended.

SOVIET 9mm STECHKIN MACHINE PISTOL (APS)

The Stechkin is a true machine pistol in that it is capable of full automatic and semiautomatic fire. It is equipped with a wooden holster which can be used as a shoulder stock. The Stechkin, like the Makarov, is blowback in operation, but it is a considerably larger weapon than the Makarov and has considerably more potential as a weapon.

Operation of the Stechkin

The Stechkin is loaded by inserting a loaded magazine into the grip (as with the U.S. cal. .45 M1911A1 automatic pistol), then pulling the slide to the rear and releasing it so that it runs home and chambers a cartridge. The safety selector catch is mounted on the left rear side of the slide, as it is on the Makarov and the Walther PP and PPK pistols. The safety selector catch has three positions. The bottom position is safe, the middle position is semiautomatic fire, and the top position is full automatic fire.

The Stechkin with its shoulder-stock holster attached can hit man-sized targets consistently at ranges of 100 to 150 yards, when it is in the hands of a good shot. This, of course, is in semiautomatic fire. The Stechkin's usable range in full automatic fire probably does not exceed twenty-five yards.

Stechkin, with shoulder stock holster.

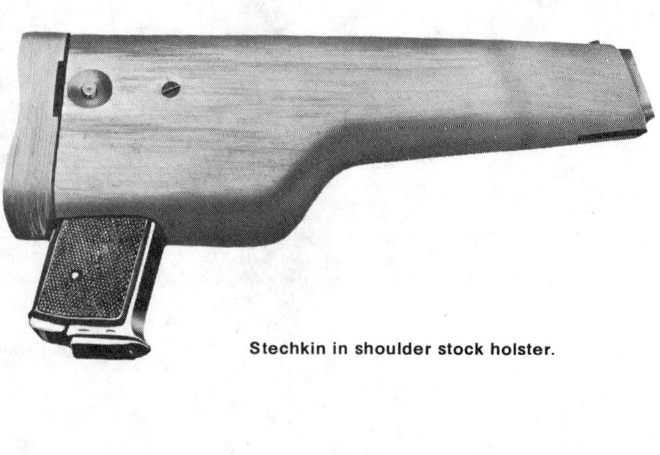

Stechkin in shoulder stock holster.

CHARACTERISTICS OF SOVIET PISTOLS AND REVOLVERS

	Model 1895	TT Model 1933	Makarov (PM)	Stechkin (APS)
Caliber:	7.62mm.	7.62mm.	9mm.	9mm.
System of operation:	Double-action revolver.	Recoil, semiautomatic.	Blowback, semiautomatic.	Blowback, selective fire.
Length overall:	9.06 in.	7.68 in.	6.34 in.	w/shoulder stock—21.25 in. w/o shoulder stock—8.85 in.
Barrel length:	4.33 in.	4.57 in.	3.83 in.	5 in.
Feed device:	7-round cylinder.	8-round, in-line, detachable box magazine.	8-round, in-line, detachable box magazine.	20-round, staggered row, detachable box magazine.
Sights: Front:	Blade.	Blade.	Blade.	Blade.
Rear:	"U" notch.	"U" notch.	Square notch.	Flip over "L" with notches.
Weight:	1.65 lb.	1.88 lb.	1.56 lb.	w/shoulder stock—3.92 lb. w/o shoulder stock—1.7 lb.
Muzzle velocity:	892 f.p.s.	1378 f.p.s.	1070 f.p.s.	1100 f.p.s.
Cyclic rate:	---------------	---------------	---------------	750 r.p.m.

SOVIET BOLT-ACTION RIFLES AND CARBINES

(Through World War II)

THE MOSIN-NAGANT

The Mosin-Nagant rifle was adopted in 1891 by Imperial Russia. The action of the rifle was developed by Colonel S. I. Mosin of the Imperial Russian Army, and the magazine was developed by Nagant, a Belgian. All Soviet bolt-action military rifles and carbines are Mosin-Nagant weapons and all are basically similar to the original Mosin-Nagant rifle adopted by Russia in 1891. These weapons can be considered reasonably effective infantry weapons. Fairly good shooting can be done with them at combat ranges, although their sights do not lend themselves to the finer degrees of accuracy which can be obtained with similar United States weapons. They suffer from an over-complicated bolt, but in other respects are relatively simple to service and maintain. The safety, in that it is extremely hard to engage and disengage, represents a shortcoming of the Mosin-Nagant weapons.

The M1891 Mosin-Nagant

The original rifle M1891 was considerably different than later versions of the same model. The original rifle M1891 had no handguard, was fitted with sling swivels instead of the sling slots used on later versions, and had a leaf rear sight which was designed for the old conical-nosed 7.62mm ball cartridge. In 1908 the Spitzer pointed light ball round (which is still used) was introduced, and the rear sight was changed. About this time handguards were added and the swivels were replaced by sling slots bored in the stock. The original M1891 is now a collector's item. The later versions of the rifle M1891 are obsolete.

The Dragoon Rifle M1891

The Dragoon rifle M1891 was originally developed as a weapon for heavy cavalry. Manufacture of this rifle was discontinued about 1930, when it was replaced by the rifle M1891/30. The Dragoon rifle M1891 is obsolete. Both the M1891 rifle and the Dragoon rifle have hexagonal receivers.

These rifles are all caliber 7.62mm (.30) for rim cartridges. Mechanically they are all basically the same.

The design differs considerably from the Mauser-type on which U.S. bolt-action rifles are all built; and the following descriptive matter is set down to provide an understanding of the Russian design.

Bolt. The bolt is an entirely original design. It has never been imitated. It is a 2-piece design in which a removable bolt head carries the locking lugs. The two locking lugs engage horizontally in the receiver, the opposite of most types. The bolt head turns with the bolt. If this head is accidentally left out when assembling

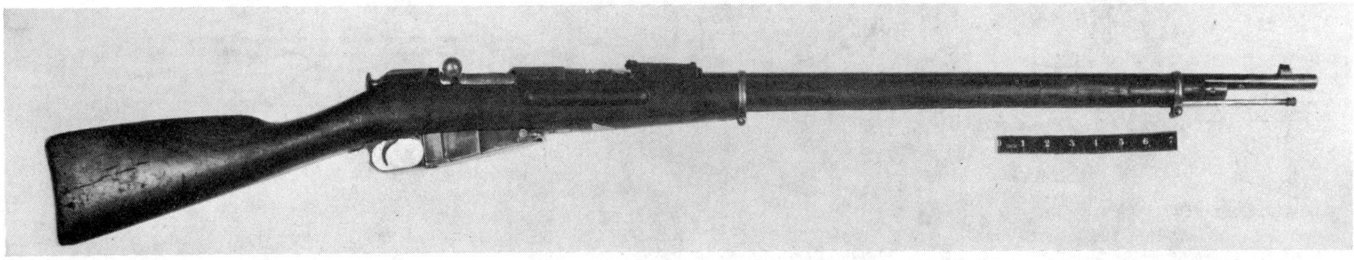

M1891 7.62mm Rifle.

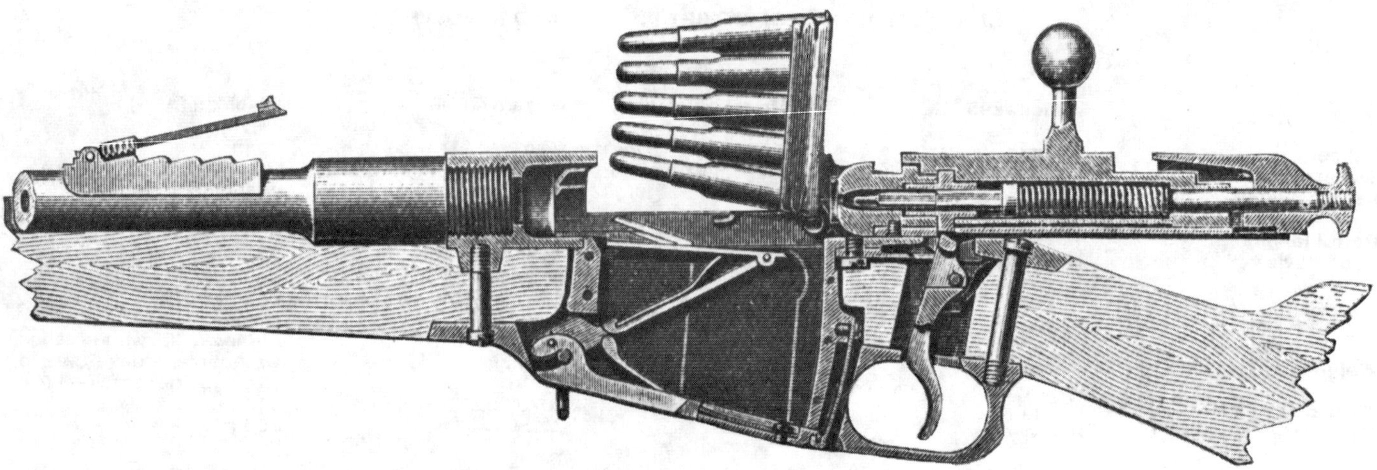

Russian M1891. Details of loading, locking, and firing mechanism.
Cartridges are stripped down off clip, but clip must be manually withdrawn.

the rifle, the weapon cannot be fired. The extractor is carried by the bolt head.

A special connecting bar joins the bolt head to the bolt itself. This bar does not turn with the bolt; it is positioned in the underside of the bolt assembly and lies in a receiver cut in the boltway. This bar also acts as a guide for the cocking piece, and assists in keeping the bolt securely in the receiver. The bar has a projection rising from its forward end. This projection holds a small hollow cylinder. The rear end of this cylinder nests in the hollow front of the bolt, while the forward end of the cylinder nests in the rear of the bolt head.

The handle is part of the bolt forging. A rib extends forward from the hollowed-out bolt. A cut on the underside of this rib takes the projection on the connecting bar. As the bolt handle is lifted, the connecting bar stays fixed in its groove in the receiver below, but its upper projection in the bolt recess holds the bolt assembly securely together.

A small lug on the front end of the connecting bar fits into a groove inside the bolt head; and a small lug on the rear of the bolt head catches in a recess at the front end of the bolt extension rib. Thus as the bolt turns the bolt head is compelled to turn with it.

The striker and mainspring are inserted from the front end of the hollow bolt. The rear of the striker passes through a small rear hole in the bolt, and a cocking piece is screwed onto it. A collar at the front end of the striker serves as forward mainspring compression point; and the firing pin end which is forward of this collar passes through the hollow cylinder on top of the connecting bar. Flat surfaces on the striker at this point and corresponding surfaces within the cylinder prevent the striker from turning.

Cocking Piece. This is a simple and effective design. When it is screwed onto the end of the striker projecting from the bolt, an upper extension of the cocking piece passes along the top of the bolt and through the cut in the top of the receiver bridge.

A small cocking nose is machined on the rear lower surface of the cocking piece.

Safety. When the cocking piece is pulled back slightly it draws the attached striker back far enough for the flat surfaces on the striker to pass out of the connecting bar cylinder; and at this point the striker and cocking piece can be turned to the left. A lug on the underside of the cocking piece extension is thus turned into a locking recess at the rear end of the bolt.

Hence by pulling back and turning the cocking piece the following safety results are obtained: (a) The cocking nose on the cocking piece is turned out of contact with the sear—hence a trigger pull cannot affect it. (b) The striker has been turned so that if it were possible for it to slip it could not hit the cartridge primer, because the flat surfaces at its forward end are turned out of line with the corresponding surfaces inside the connecting bar cylinder through which the striker must pass to hit the primer. (c) The lug on the underside of the cocking piece extension is locked into a cut in the rear of the bolt to prevent any forward movement. (d) The long cocking piece extension is turned to the rear of the cut in the receiver bridge, hence cannot move forward since the receiver bridge is in its path.

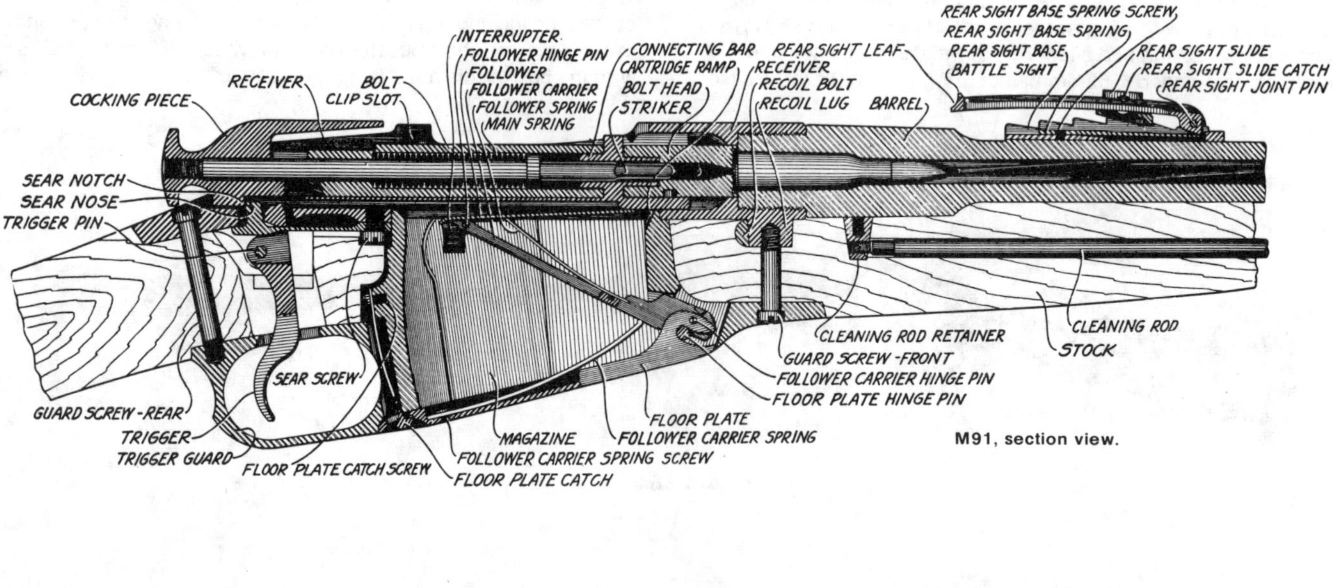

M91, section view.

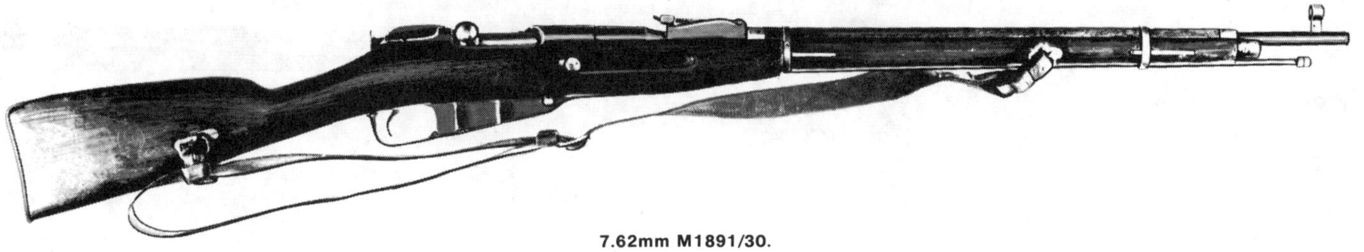

7.62mm M1891/30.

Magazine. The magazine is a projecting single-line box holding five rim cartridges. It forms part of the trigger guard. The magazine floor plate hinges forward and is kept closed by a spring catch. When the catch in the floor plate is drawn back, the magazine follower carrier spring forces the plate down and permits removal of cartridges.

As the accompanying drawing shows, the magazine has not merely a follower on which the first cartridge rests, together with its spring, but also a special "follower carrier" and its spring as part of the cartridge lifting system.

Interrupter-Ejector. This unit is also called "distributor-ejector" and "cartridge valve-ejector." No other rifle needs or uses this device.

The interrupter is a specially shaped plate attached to the left side of the receiver. At its rear is a flat spring held to the receiver by a screw. The thin plate section passes through a cut in the receiver wall above the magazine way. A point on this plate acts as ejector.

The interrupter is operated by action of the bolt. It prevents double feeding. When the top cartridge in the magazine has been chambered, the following cartridge cannot rise because the interrupter is holding it down. Only when the bolt is locked and the extractor engaged around the cartridge case head does a cam-shaped groove in the bolt bear against the interrupter and force it out against the interrupter spring. As the holding projection on the interrupter is forced away from the cartridge, the follower spring can push the top cartridge into line to be ready to feed.

When the bolt handle is raised to open the action, the cam groove in the bolt allows a point on the interrupter to enter the magazine and press against the second cartridge from the top.

An ejecting point on the edge of the interrupter is now in line with a groove in the bolt head. Thus, as the bolt is pulled back, the empty case is struck against this point and ejected from the rifle.

Operation. Lifting bolt handle rotates bolt, bolt head, and extractor as a unit. A cam cut at the rear of the bolt forces against the cocking nose on the underside of the cocking piece. The cocking piece and striker assembly are forced back, thereby pulling the firing pin point back away from the primer. The mainspring is partly compressed.

The rib extending forward from the bolt proper passes along a special cam surface on the top of the overhanging receiver ring. (The bolt head is entirely enclosed by the receiver ring.) This action forces the entire bolt assembly back to give primary extraction, the rearward thrust beginning as soon as the bolt has turned far enough for the locking lugs on the bolt head to clear their seats in the receiver ring.

The turning bolt operates the interrupter as described, permitting the interrupter spring to force the interrupter in to hold the next to the top cartridge in the magazine and to line up the ejector point.

The short extractor in the bolt head carries the empty case back and strikes it against the ejector. At this point a lug on top of the trigger hits against metal at the end of a groove in the connecting bar. (The bar has been traveling back as part of the bolt assembly, once the rearward action started.) This acts as a bolt stop.

Scope on M1891/30 Sniper Rifle.

The upright bolt handle passes back through the cut in the receiver bridge on rearward bolt pull.

As the bolt is pushed forward the bolt head starts the cartridge in its path towards the chamber. At the point where the forward extension rib on the bolt hits the cam face in the receiver ring, the sear nose catches and holds the cocking piece nose. The bolt head locking lugs are at the cam grooves leading to their locking seats in the receiver ring.

As the bolt handle is turned down, the locking lugs are forced along their cam grooves. This draws the bolt forward as the bolt head lugs are locked. Since the sear nose is holding the cocking piece back, this forward pull completes compression of the mainspring. The interrupter works as already described to permit the cartridge to rise below the bolt.

The Rifle M1891/30

The rifle M1891/30 is about the same length as the M1891 Dragoon, but it represents many improvements over the Dragoon. The sights used on the M1891/30 are superior to those of the Dragoon, and, because the metric system of measurement was adopted in Russia during this period, the sights of the M1891/30 are calibrated in meters rather than in arshins. (One arshin equals 0.71 meters or 0.78 yards.) Manufacture of the M1891/30 was initiated in 1930. This model was used in large numbers in the Soviet Army, but was replaced by the carbine M1944 as the standard Soviet infantry shoulder weapon at the end of World War II. It is still in use in some of the satellite countries.

The Sniper Rifle M1891/30

The sniper rifle M1891/30, which is basically the M1891/30 adapted for use with a telescope, is still a standard weapon in some satellite armies. The telescopes employed are somewhat similar to those used on United States hunting rifles.

The Carbine M1910

Although Imperial Russia adopted the Mosin-Nagant rifle in 1891, a true carbine did not appear until 1910. The carbine M1910, with its leaf sight and sling slots, had characteristics of both the

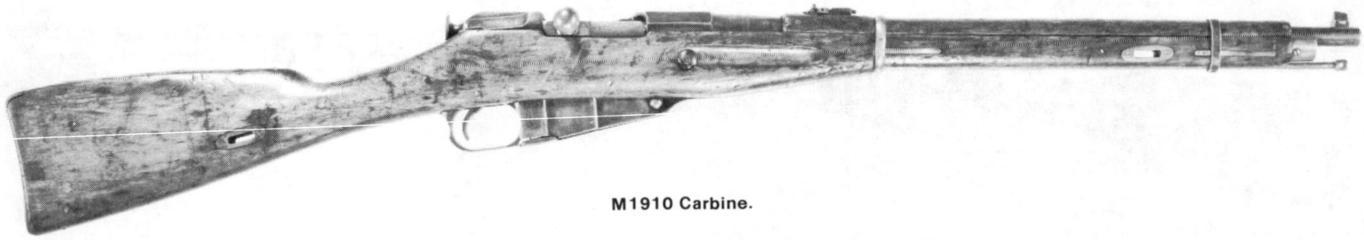

M1910 Carbine.

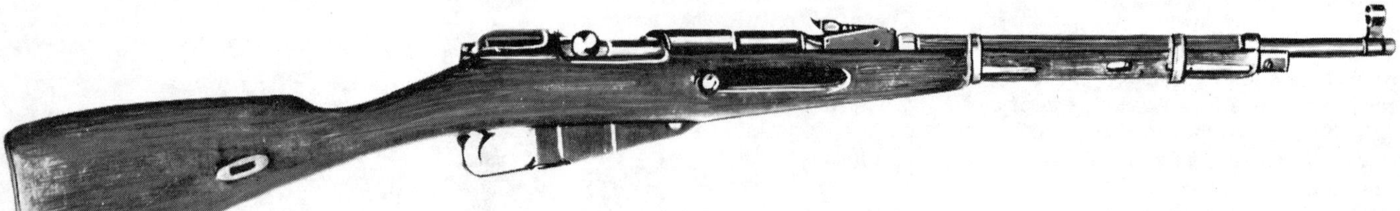

M1938 Carbine.

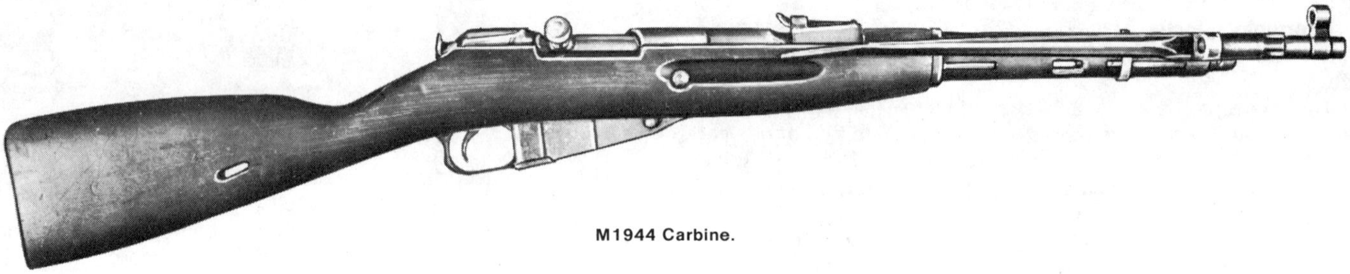

M1944 Carbine.

original and later versions of the rifle M1891. The carbine M1910 has a hexagonal receiver and does not take a bayonet. This model is comparatively rare.

The Carbine M1938

The carbine M1938 replaced the M1910. It is similar in many respects to the rifle M1891/30. It has a tangent-type rear sight, hooded front sight, and rounded receiver. It does not take a bayonet. This model may be encountered in satellite forces, although it is not manufactured at present.

The Carbine M1944

The carbine M1944, introduced during the latter part of World War II, was the last of the Mosin-Nagants. The permanently fixed bayonet folds down along the right side of the carbine stock when not in use. Except for a slightly longer barrel and the addition of the bayonet, the carbine M1944 is identical to the M1938. It is still in use in various Soviet satellites.

CHARACTERISTICS OF SOVIET 7.62MM MOSIN-NAGANT BOLT ACTION RIFLES AND CARBINES

	Rifle M1891	Dragoon Rifle M1891	Rifle M1891/30	Sniper Rifle M1891/30	Carbine M1910	Carbine M1938	Carbine M1944
Weight:							
w/o bayonet & sling:	9.62 lb.	8.75 lb.	8.7 lb.	11.3 lb.	7.5 lb.	7.62 lb.	
w/bayonet & sling:	10.63 lb.	9.7 lb.	9.7 lb.		7.7 lb.		8.9 lb.
Length:							
w/o bayonet:	51.37 in.	48.75 in.	48.5 in.	48.5 in.	40 in.	40 in.	40 in. (folded)
w/bayonet:	68.2 in.	65.5 in.	65.4 in.	65.4 in.			52.25 in. (extended)
Barrel length:	31.6 in.	28.8 in.	28.7 in.	28.7 in.	20 in.	20 in.	20.4 in.
Magazine capacity:	5 rounds.	5 rounds.	5 rounds.	5 rounds.	5 rounds.	5 rounds.	5 rounds.
Instrumental velocity at 78 ft. w/hvy ball:	2660 f.p.s.	2660 f.p.s.	2660 f.p.s.	2660 f.p.s.	2514 f.p.s.	2514 f.p.s.	2514 f.p.s.
Rate of fire:	8-10 r.p.m.	8-10 r.p.m.	8-10 r.p.m.	8-10 r.p.m.	8-10 r.p.m.	8-10 r.p.m.	8-10 r.p.m.
Maximum sighting range:	3200 arshins (2496 yd.)	3200 arshins (2496 yd.)	2000 meters (2200 yd.)	2000 meters* (2200 yd.)	2000 arshins (1560 yd.)	1000 meters (1100 yd.)	1000 meters (1100 yd.)
Front sight:	Unprotected blade.	Unprotected blade.	Hooded post.	Hooded post.	Unprotected blade.	Hooded post.	Hooded post.
Rear sight:	Leaf.	Leaf.	Tangent.	Tangent.	Leaf.	Tangent.	Tangent.
Ammunition**							

*For iron sights when scope is dismounted. Maximum sighting range for the telescopic sight on this weapon is: PE scope—1400m (1540 yd.); PU scope—1300m (1420 yd.).
**Soviet 7.62mm rifle and ground machine gun rimmed ammunition.

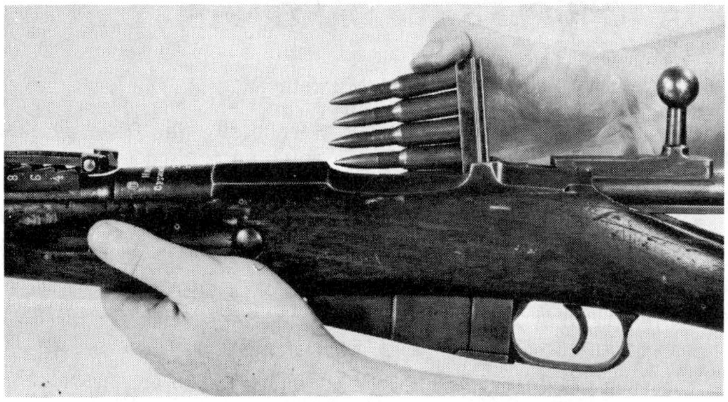

Loading Mosin-Nagant rifle.

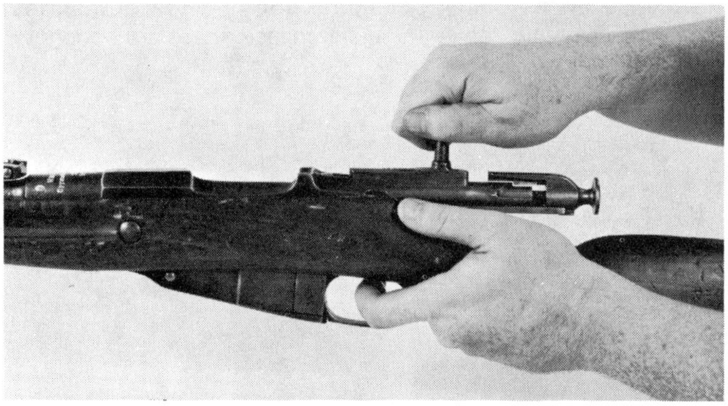

Removing bolt.

How to Load and Fire the Mosin-Nagant Rifles and Carbines

The Rifle M1891. To set the safety, draw back the cocking piece and turn it to the left. This prevents the bolt from opening. To put off safe, pull the cocking piece back, turn it to the right, and allow it to move forward.

The rifle M1891 is loaded in the same manner as the United States Springfield or any Mauser rifle. Open the bolt, place a clip of cartridges in the clip guides, and press the rounds down into the magazine. Close the bolt; the clip will then fall out of the clip guides onto the ground. The weapon is now ready to fire.

To unload the rifle M1891, open the magazine floor plate and remove the cartridges. The magazine floor plate catch is located on the lower rear part of the magazine, forward of the trigger guard. Press the catch rearward; the follower and floor plate will swing down and forward on a pivot pin, and the cartridges will spill out. Open the bolt and extract the round from the chamber.

The M1891 bayonet is attached by a locking ring; if the M1891/30 bayonet is used, a spring-loaded catch holds the bayonet in place.

The Dragoon Rifle M1891. Operating instructions for Dragoon M1891 are the same as for rifle M1891.

The bayonet of the rifle M1891 or M1891/30 is attached to the Dragoon M1891 in the same manner as described for the rifle M1891.

The Rifle M1891/30. Operating instructions for the rifle M1891/30 are the same as those for the rifle M1891.

The bayonet is attached by a spring-loaded catch.

The Sniper rifle M1891/30. Operating instructions for this rifle are the same as those for the rifle M1891. The bayonet for the rifle M1891/30 is attached to the sniper rifle by means of a spring-loaded catch.

The Carbine M1910. Operating instructions for the carbine M1910 are the same as those for the rifle M1891; however,

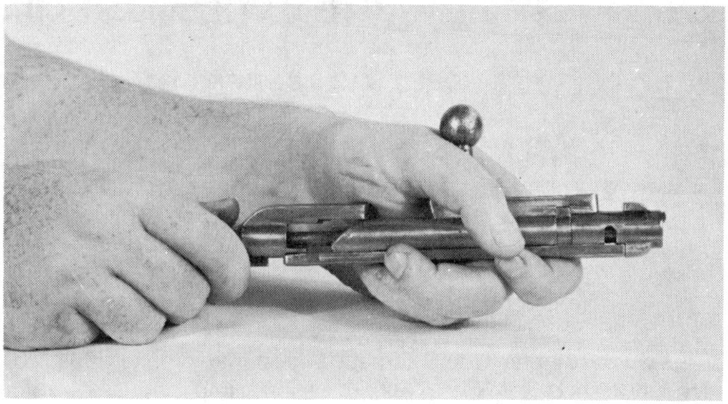

Drawing back cocking piece.

bayonets are not provided for this carbine.

The Carbine M1938. Operating instructions for the carbine M1938 are the same as those for the rifle M1891; bayonets are not provided for this carbine.

The Carbine M1944. Operating instructions for the carbine M1944 are the same as those for the rifle M1891; however, this carbine has a nondetachable bayonet which may be folded or extended by forcing the spring-loaded bayonet tube away from the pivot pin, and then swinging the bayonet to either folded or fixed position.

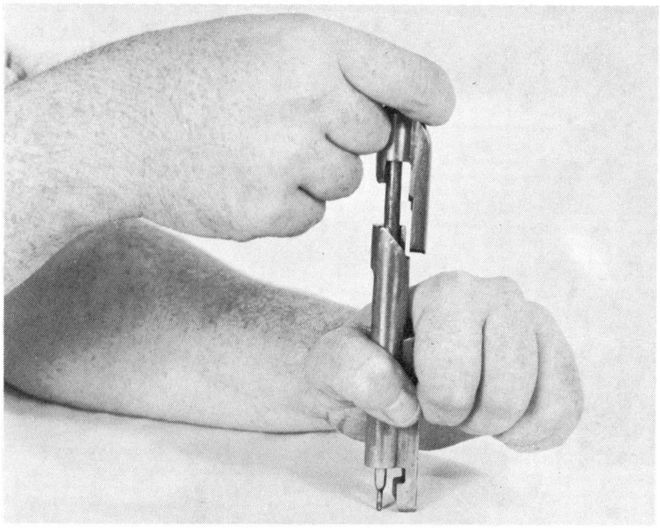

Removal of firing pin.

How to Field Strip the Mosin-Nagant Rifles and Carbines

Open bolt and draw to the rear, pull trigger all the way to the rear and remove bolt from the receiver.

Pull cocking piece to the rear and turn it to the left to relieve the tension of the main spring. Remove bolt head and guide.

Place the firing pin on a solid surface, preferably a block of soft wood, push the bolt body down and unscrew the cocking piece, then remove the firing pin and spring.

The magazine follower can be removed by pushing the magazine floor plate catch, on the bottom of the floor plate just forward of the trigger guard, rearward swinging down the floor plate, follower spring and follower and pressing together, with thumb and forefinger, the follower and floor plate and pulling down to remove them.

Reassembly is in reverse order of disassembly. Care must be exercised when screwing the firing pin into the bolt, that the rear of the firing pin is flush with the cocking piece and that the marks on the rear of the firing pin are alined with those on the cocking piece in order to assure correct firing pin protrusion.

SOVIET AUTOMATIC and SEMIAUTOMATIC RIFLES and CARBINES

PRE-WORLD WAR II DESIGNS

The Federov Model 1916 "Avtomat"

This rifle was made in limited quantity during World War I. It was a selective-fire rifle chambered for the 6.5mm Japanese cartridge. It could be considered one of the first assault rifles.

The Simonov Automatic Rifle M1936 (AVS)

This was one of the first automatic and semiautomatic rifles produced by the U.S.S.R. in 7.62mm caliber. Imperial Russia had previously produced the Federov automatic rifle M1916 in 6.5mm caliber, but the weapon did not prove successful and very few were manufactured. The Simonov rifle M1936 evidently did not meet requirements either.

The Tokarev 7.62mm Semiautomatic Rifle M1938 (SVT)

This was the second of a series of Tokarev rifles. This model has a two-piece stock and is very lightly built.

The Tokarev 7.62mm Semiautomatic Rifle M1940 (SVT) and The 7.62mm Automatic and Semiautomatic Rifle M1940 (AVT)

While considerably sturdier than the M1938, both of these rifles still proved rather flimsy for military use. Considerable difficulty was experienced in repair and maintenance of these weapons during World War II, and they are no longer standard weapons.

The Tokarev Semiautomatic Sniper Rifles M1938 and M1940

Because of their flimsy construction and the difficulties experienced in their repair and maintenance, these are no longer standard weapons. These sniper rifles are merely Tokarev semiautomatic rifles M1938 (SVT) and M1940 (SVT) which have been specially selected for accuracy and adapted for mounting telescopic sights.

Federov Model 1916 "Automat."

Simonov Model 1936 (AVS).

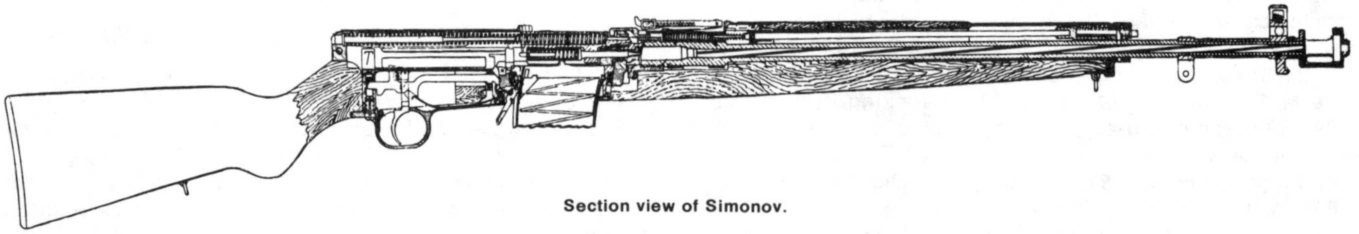

Section view of Simonov.

The Tokarev 7.62mm Semiautomatic Carbine M1932

This was made only in small numbers. Its weaknesses, with regard to durability, repair, and maintenance, were apparently the same as those of the Tokarev rifles. This carbine is not a standard weapon.

How the M1940 Tokarev Rifle Works

A loaded magazine is inserted in the bottom of the receiver until its catch locks it. The bolt handle is drawn back and released to cock the hammer and strip a cartridge into the firing chamber.

The rear end of the bolt is forced down by the bolt carrier into locking position. The carrier proceeds to ride over projections on the bolt to hold it in the locked position.

When the trigger is pressed, it forces a bar forward to rotate the sear. The sear is fitted vertically in the trigger guard. It now engages a bent in the head of the hammer. The hammer is rotated

M1938 7.62mm Tokarev Rifle (SVT), action open.

forward to strike the firing pin and explode the cartridge.

As the bullet passes down the barrel, a portion of the gas escapes through a port drilled in the top of the barrel under the upper section of the wooden fore-end. (The gas regulator has five positions for adjustment.) The expanding gas impinges on the head of the piston rod (which is positioned on top of the barrel) driving it backward for a stroke of about 1 1/2 inches.

The thrust is transmitted to the operating rod which passes it on to the bolt carrier.

The operating rod is mounted above the barrel; its rear end passing through a hole in the receiver directly above the forward face of the locked bolt, and in line with the bolt carrier which rides on top of the bolt. The operating rod return spring is mounted around the rod itself.

The carrier travels back about 1/4 of an inch before unlocking. The bolt travels on back carrying with it the empty cartridge case which is ejected to the right. It also compresses the dual springs mounted in the top of the receiver to store up energy for the return motion. At the same time it forces the hammer to rotate on its axis until it is caught and held by the sear.

When the last shot has been fired, a catch in the bottom center of the bolt way is operated by a projection of the magazine platform and holds the bolt open.

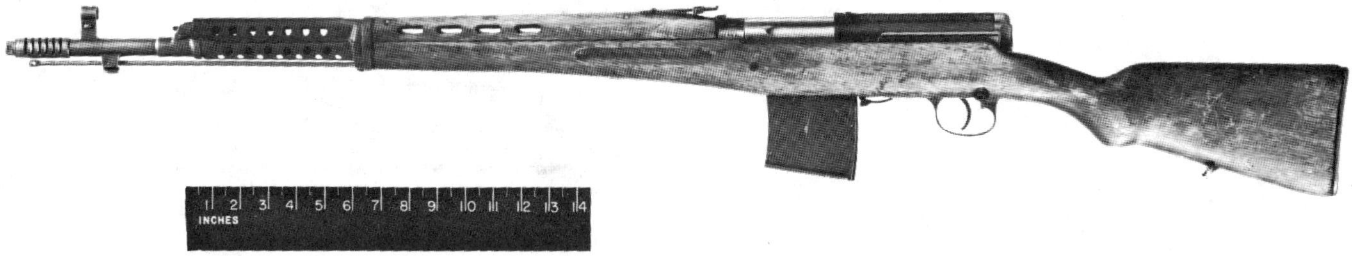

Tokarev 7.62mm M1940 Rifle (SVT).

Field Stripping the Tokarev M1940

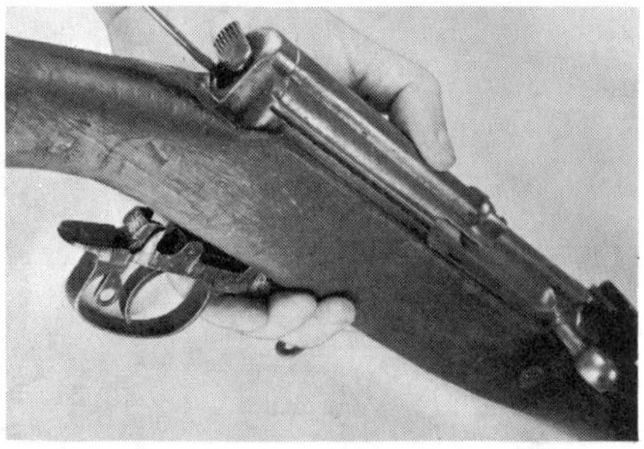

(1) Remove trigger assembly. First turn the thumb latch at the extreme rear of the receiver up to the left as far as it will go. This will expose a cylindrical hole in which the trigger guard locking collar rests. With a screwdriver or a bullet point, push the spring-loaded locking collar in. This will pivot the locking bar in the rear of the receiver and will free the rear end of the trigger guard assembly which may then be removed from the receiver. (Note: When replacing this unit it is necessary that the bolt be back far enough from the chamber so the disconnector is able to rise. Otherwise the front end of the trigger guard unit will not seat in the receiver notches.)

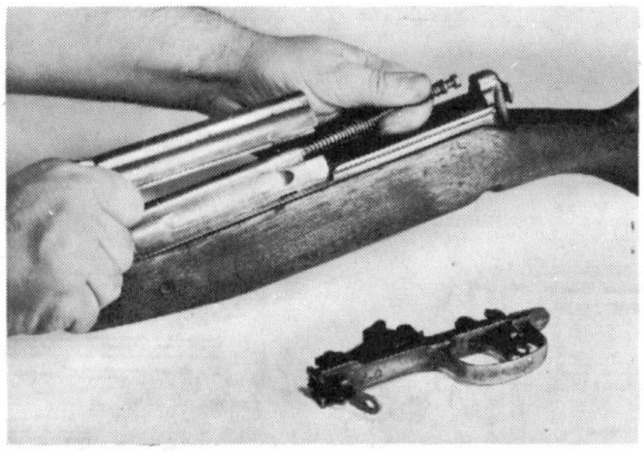

(2) Remove operating spring dust cover. First rest the muzzle of the rifle on the floor, as considerable force must be applied. Then pull the steel cover to the rear of the bolt straight down toward the muzzle. This action will force the buffer and operating springs down inside the bolt, as the buffer guide rod is seated in a groove machined for it at the extreme rear of the cover. When the cover has been pulled down as far as it will go in its guides in the receiver, grasp it firmly with the left hand and squeeze it tightly against the bolt. While retaining this grasp with the left hand, with the thumb of the right hand push the buffer rod down far enough to free its head from the seating in the cover.

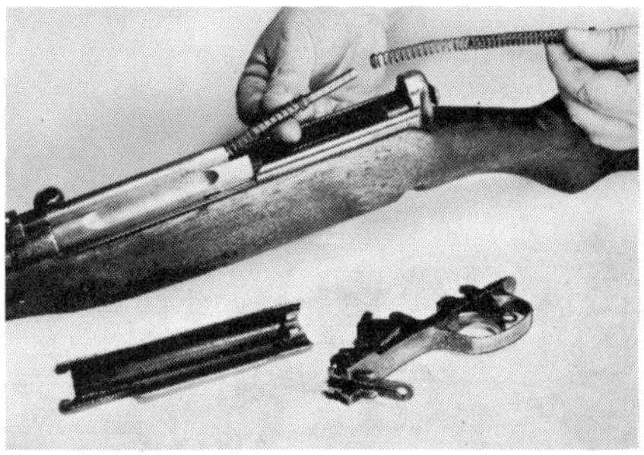

(3) Then push it carefully out away from the cover and ease it and the springs up out of the bolt hole. The cover is free to come off at this point. Remove operating and buffer spring assemblies. Holding the protruding buffer spring and rod close to the bolt hole, ease them up and to your right so the rod will clear the rear of the receiver. When all tension is off the springs, the buffer spring and its guide rod may be lifted out of engagement with the hollow operating spring rod. The operating spring and its rod may now be removed from the bolt hole.

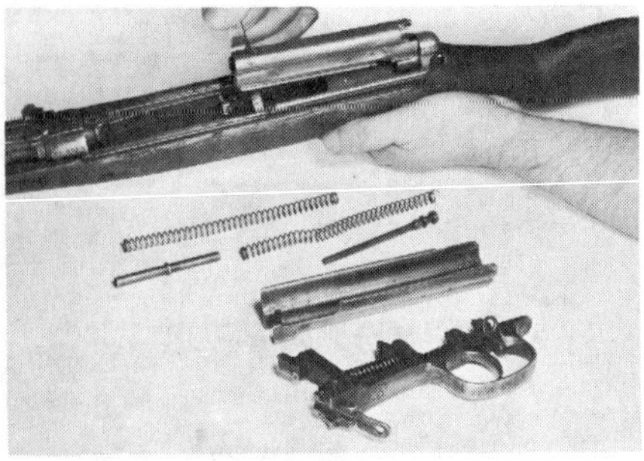

(4) Remove the bolt assembly. Pull the bolt handle back slowly until the bolt carrier and the bolt have cleared the magazine well. Then lift up on the bolt handle and continue to pull back and up. At this point the bolt assembly is free of its receiver guides and may be tilted and lifted up out of the receiver. (Note: When re-assembling this unit, the assembly will drop into its receiver tracks at just the one proper point. Do not exert force, but feel for this point if you cannot locate it immediately.) Remove the bolt from its carrier. Hold the assembly in the palm of the left hand upside down. Place the forefinger over the firing-pin hole and the thumb (of the right hand) over the rear of the firing pin and push the bolt slowly ahead inside the carrier. At the proper point the cam faces on the bolt will mate with the proper surfaces on the bolt carrier and the bolt may then be lifted up out of the carrier. Remove the gas piston. Press in the lock on the front band and push the band forward. This will permit the perforated metal upper cover to be lifted off. The pierced wooden forearm covering the gas piston may then be lifted off. (Note: If it is necessary to remove the stock, unscrew the bolt from the right side of the receiver below the chamber and pull it out. The lower metal fore-end piece may then be removed and the receiver and barrel assemblies lifted out of the stock complete.) Pull the piston back until it clears the face of the gas cylinder over which it is mounted.

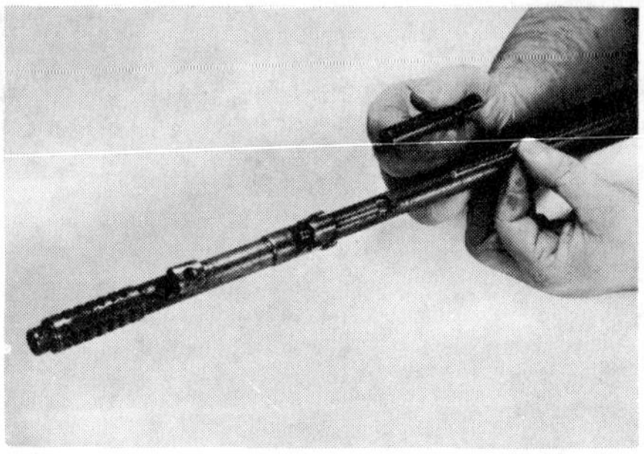

(5) Holding the piston at this point, pull the operating rod in which the piston seats back far enough to permit the piston to be lifted off. Then ease the rod forward against the tension of its spring, and remove the rod and its spring from their seating in the receiver below the line of the rear sight. Remove compensator and gas cylinder assembly. Drift the retainer out preferably from the right. The compensator and assembly may then be unscrewed from the barrel and removed.

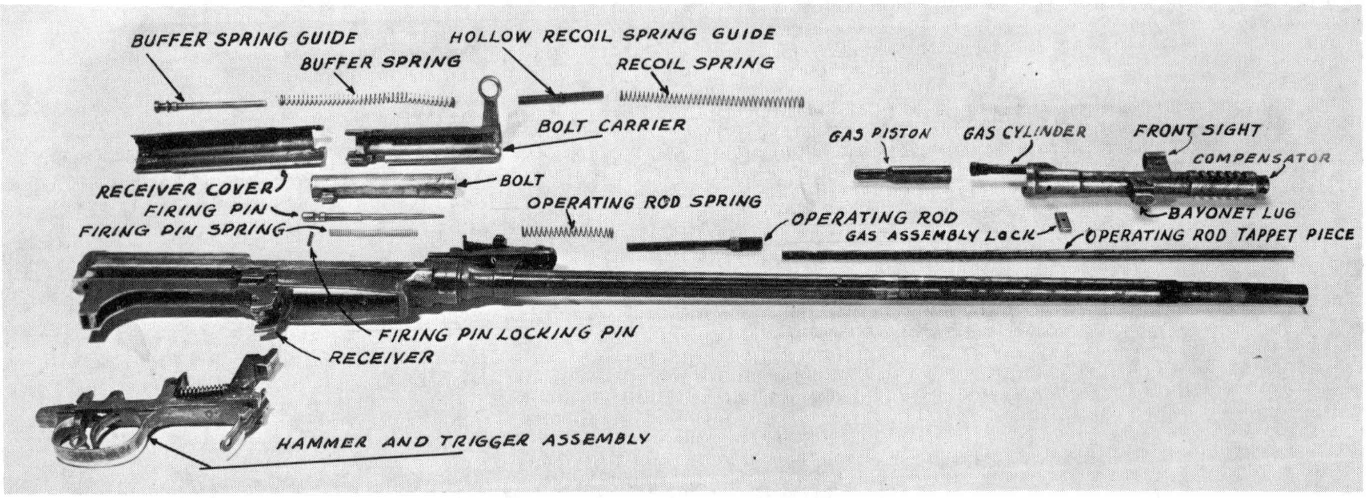

Principal action parts—the Tokarev M1940.

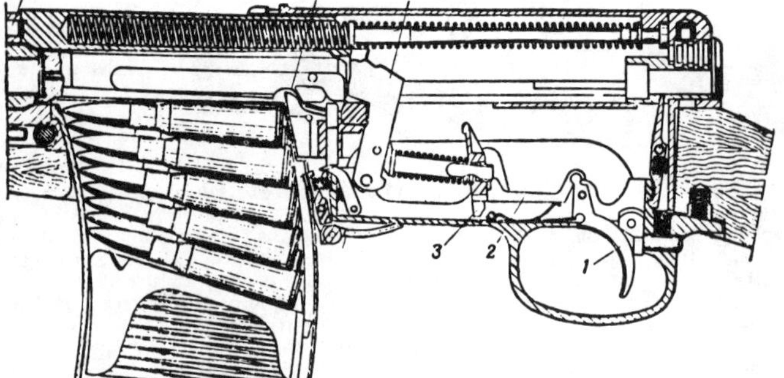

Action closed, magazine loaded, hammer forward. Details of recoil spring shown.

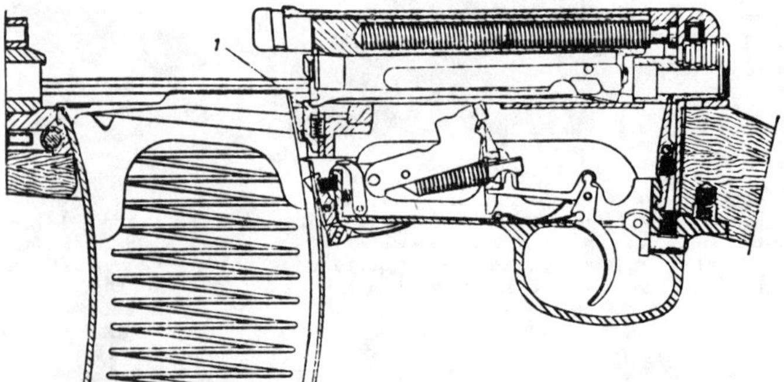

Action open on firing last shot. Recoil spring compressed, hammer cocked, ready for reloading.

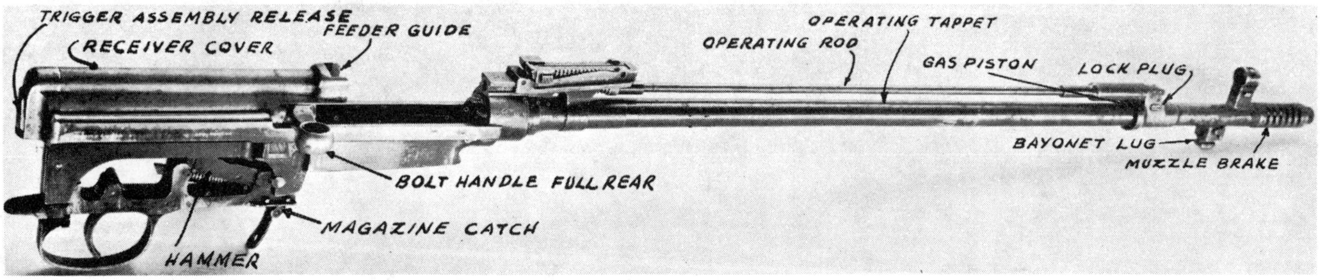

Parts and principal components—the Tokarev M1940.

Metal parts removed from stock to show operation of mechanism and dust cover removed. Operating rod is being driven back through hole in the receiver and has pushed bolt carrier back until carrier has unlocked bolt and is carrying it back with it.

Characteristics of Soviet Pre-World War II 7.62mm Automatic and Semiautomatic Rifles

	Automatic Rifle M1936	Semiautomatic Rifle M1938	Semiautomatic Rifle M1940	Automatic Rifle M1940	Semiautomatic Sniper Rifle M1940	Semiautomatic Sniper Rifle M1938
Weight:						
w/o bayonet & magazine:	8.93 lb.	8.70 lb.	8.59 lb.	8.35 lb.	9.18 lb.	9.52 lb.
w/bayonet & magazine:	. . . .	10.8 lb.	9.48 lb.	9.24 lb.	. . . .	. . . .
Length:						
w/o bayonet:	48.6 in.	48.1 in.	48.1 in.	48.1 in.	48.1 in.	48.1 in.
w/bayonet:	59.3 in.	60.84 in.	57.1 in.	57.1 in.	57.1 in.	60.84 in.
Barrel length:	24.16 in.	25 in.	24.6 in.	24.6 in.	24.6 in.	25 in.
Magazine capacity:	15 rounds.	10 rounds.	10 rounds.	10 rounds.	10 rounds.	10 rounds.
Instrumental velocity at 78 ft. w/hvy ball:	2519 f.p.s.	2519 f.p.s.	2519 f.p.s.	2519 f.p.s.	2519 f.p.s.	2519 f.p.s.
Rate of fire: (semiautomatic)	30-40 r.p.m.	25 r.p.m.	25 r.p.m.	30-40 r.p.m.	25 r.p.m.	25 r.p.m.
Maximum sighting range:	1500 meters.	1500 meters.	1500 meters.	1500 meters	Iron sights: 600 m. (660 yd.) Telescope: 1300 m. (1430 yd.)	Iron sights: 600 m. (660 yd.) Telescope: 1300 m (1430 yd.)
Front sight:	Open guard blade.	Hooded post.	Hooded post.	Hooded post.	Hooded post.	Hooded post.
Rear sight:	Tangent.	Tangent.	Tangent.	Tangent.	Tangent and telescope.	Tangent and telescope.
Principle of operation:	Gas.	Gas.	Gas.	Gas.	Gas.	Gas.
Ammunition:	*	*	*	*	*	*

*7.62mm U.S.S.R. rifle and ground machine gun rimmed ammunition.

Early model 7.62mm AK with stamped receiver.

POST WORLD WAR II SOVIET SHOULDER WEAPONS

The Soviets introduced two new shoulder weapons since World War II and have recently developed an improved version of one of these weapons. The 7.62mm Simonov SKS carbine, a semiautomatic weapon, was introduced around 1946. It is no longer standard with Soviet combat troops and was originally replaced by the 7.62mm Automat Kalashnikov (AK) assault rifle, a selective-fire weapon. The AK, frequently called the AK47, has been modified and a new version called the AKM was introduced several years ago and is replacing the AK.

Soviet Cartridges

The new shoulder weapons are chambered for a new 7.62mm "intermediate-sized" cartridge, i.e., a cartridge intermediate in size between the U.S. cal. .30 carbine cartridge and the U.S. cal. .30 rifle cartridge. This cartridge (which is considerably smaller and has considerably less power than the 7.62mm NATO cartridge or the older Soviet 7.62mm rimmed cartridge) is now used in all Soviet squad-level small arms except the Stechkin pistol. The official Soviet nomenclature for the cartridge is Cartridge, 7.62mm, M1943. There are five types of this cartridge in use:

Ball Type PS (PS indicates the use of a mild steel core)
Tracer Type T-45
Armor-Piercing Incendiary Type BZ
Incendiary Tracer Type Z
Blank (has rosette-type crimp)

The Soviet M1943 cartridge appears to be a further development of the German 7.92mm Kurz, or "short," of World War II. The German cartridge, whose official German nomenclature was Pistolen Patrone 43, was used in the Mkb 42, MP 43, MP 44, and StG 44 series of weapons. Essentially, cartridges of the size of the German 7.92mm short and the Soviet 7.62mm M1943 are cut-down versions of full-size rifle cartridges, and are limited in effective range and penetration; they do, however, have certain advantages over the full-size cartridges.

They are considerably lighter, and therefore the soldier can carry more cartridges on his person. The lightness is also a logistical advantage in shipping and handling.

They require less material in manufacture.

Since they are shorter than full-size cartridges, the weapons made for them can be shorter and lighter than those made for full-size cartridges—although this is not always true in practice.

They have considerably less recoil than do full-size cartridges.

It is significant, however, that the Soviets still use their old full-size 7.62mm rimmed cartridge in their heavy machine guns, which may be used for fire over the heads of their own troops or for long-range interdictory fire. It should also be borne in mind that the Russians who have competed in international rifle matches in recent years do not use the new intermediate-sized M1943 cartridge; they use a 7.62mm Match Cartridge which has the same case as the old 7.62mm rimmed cartridge. In other words, where first-rate accuracy is a requirement, the Soviets still use the full-size rifle cartridge.

THE SOVIET 7.62mm AK ASSAULT RIFLE

The AK has frequently been called a submachine gun. It is true that the AK has replaced the PPSh M1941 and PPS M1943 submachine guns in the Soviet service, but it is not entirely practical to class the AK, which fires rifle-type cartridges, with submachine guns, which fire pistol cartridges. The AK is much more accurate over a longer range than any of the normal run of submachine guns such as the British Sten, U.S. M3A1, or Israeli Uzi, all of which use pistol ammunition. The average submachine gun firing single shots will produce a group from 12 to 18 inches in diameter at 100 yards. The AK will group into 6 inches at 100 yards; this is about average for a military rifle at that range. The AK is outwardly quite similar to the German MP43 and MP44 series of weapons, but internally is quite different. The weapon is made almost completely from milled steel components, few stampings being utilized. Surprisingly enough, this is a characteristic of all of the Soviet postwar small arms, and is in sharp contrast with their extremely large-scale use of stampings during World War II. The AK is made in two versions, one using a wooden stock and the other using a folding metal stock. Originally the AK did not have a bayonet, but now has a knife-type bayonet. The bayonet has a blade of approximately 8 inches. The AK replaced the SKS carbine as the principal shoulder weapon and is being replaced in turn by the AKM. Early versions of the AK had stamped steel receivers. The receiver was stamped out in several sections and held together by being rivetted to machined steel filler blocks. These weapons are relatively rare.

Loading and Firing the AK

The thirty-round magazine is loaded by hand by pressing the cartridges down into the mouth of the magazine with the thumb. Insert the magazine into the underside of the receiver, forward end first; then draw up the rear end of the magazine until a click is heard or until the magazine catch is felt to engage its slot at the

7.62mm Assault Rifle AK (Avtomat Kalashnikov), left side.

Right side, the AK Assault Rifle.

AK Rifle with folding stock.

rear of the magazine. Pull the operating handle (which protrudes from the right side of the receiver) smartly to the rear and release it. The bolt and bolt carrier will go forward under the pressure of the recoil spring and will chamber a cartridge. To put the weapon on safe, push the safety selector catch (mounted on the right side of the receiver) as far up as it will go. When on safe, the safety selector blocks rearward movement of the operating handle and covers a gap to the rear of the bolt carrier. For semiautomatic fire, push the end of the safety selector lever all the way down (so that its forward end is opposite the bottom two Cyrillic letters), aim, and squeeze (do not pull) the trigger. For full automatic fire, push the safety selector lever to the middle position marked by the two Cyrillic letters "AB" (which stand for "automatic"). Although the AK is quite heavy, it climbs rapidly in automatic fire; it is therefore necessary to get a good grip on the weapon before squeezing the trigger for automatic fire.

Field Stripping the AK

Remove the magazine and pull the operating handle to the rear to clear weapon of any live ammunition. At the rear of the stamped receiver cover, a serrated catch protrudes; this is the rear end of the recoil spring guide rod. Push the catch in with the finger and at the same time pull the receiver cover upward and backward from the receiver. After removing the receiver cover, push the recoil spring guide forward, so that the bottom rear lug section of the guide clears the dovetail slot at the rear of the receiver. Remove recoil spring. Note that the recoil spring and its guide rod are a packaged unit; i.e., they are held together as one piece by a collar fitted to the end of the guide rod. This is similar to the recoil spring used on the cal. .50 Browning machine gun. The bolt carrier, piston, and bolt can now be removed as a unit by pulling them back to the cutout section in the receiver track and lifting them upward and rearward. To remove the bolt from the bolt carrier, turn its head so that the guide lug aligns with the cam

AK Assault Rifle with bayonet, Bulgarian manufacture.

North Korean copy of Soviet 7.62mm Assault Rifle, AK.

surface of the bolt carrier, move the bolt as far to rear as possible, rotate the bolt so that the guide lug leaves the camway of the bolt carrier, and pull the bolt forward and out of the bolt carrier. This operation is not as complicated as it seems; the bolt can actually be shaken out of the bolt carrier. Rotate the handguard lock lever (on the forward right side of the rear sight base) upward, and the handguard-piston tube assembly can be lifted out. No further disassembly is recommended.

To assemble the weapon, carry out in reverse the steps outlined above.

How the AK Works

Full Automatic Fire. With the safety selector set on full automatic fire, a cartridge in the chamber (to chamber a cartridge initially, since the AK fires from a closed bolt, it is necessary to pull the operating handle to the rear and release it), and a loaded magazine in the weapon, the following actions occur when the trigger is pulled. The safety selector lug is far enough to the rear to release the rear end of the trigger, but stays directly above the rear end of the disconnector. Thus, while the trigger is free to rotate, the safety selector lug prevents the disconnector from rotating. One large multi-stranded spring is used as both hammer spring and trigger spring in the AK.

Semiautomatic Fire. To fire a single round from the rifle, set the weapon for semiautomatic fire by rotating the safety selector as far downward as possible, and then press the trigger.

When the trigger is pressed, the semiautomatic sear and disconnector rotate. The rear end of the sear (which is actually part of the trigger) raises the ends of the hammer-and-trigger spring. As the trigger rotates, the semiautomatic sear releases the hammer cock notch.

The hammer-and-trigger spring rotates the hammer forward, and the hammer strikes the rear end of the firing pin, pushing it

Loading the AK magazine.

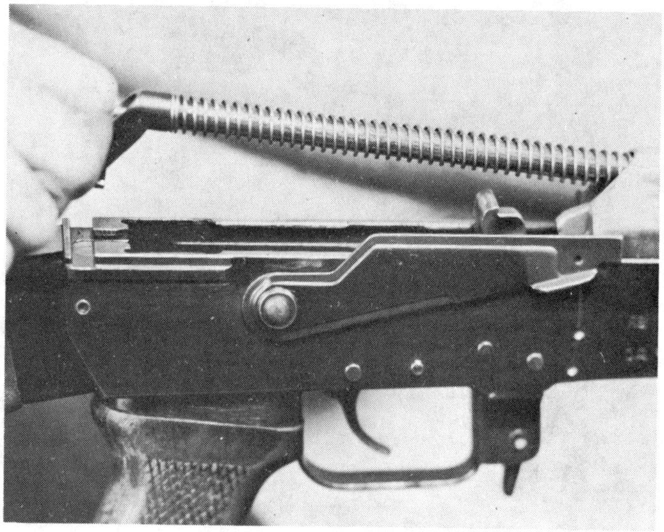

Removing the recoil spring assembly of the AK.

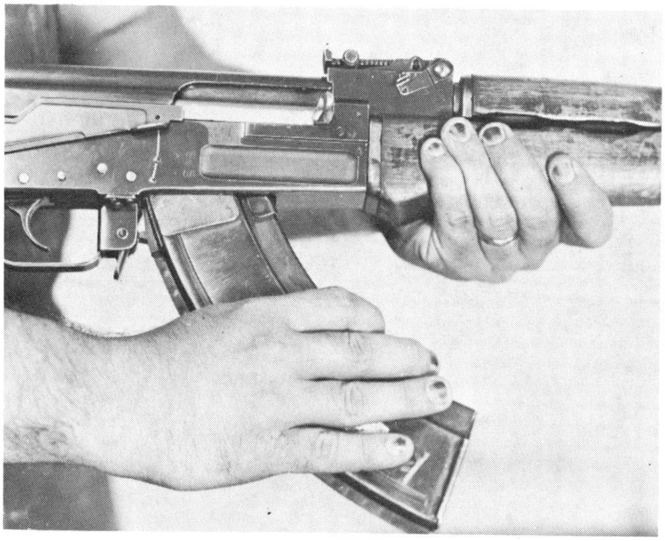

Inserting the AK magazine.

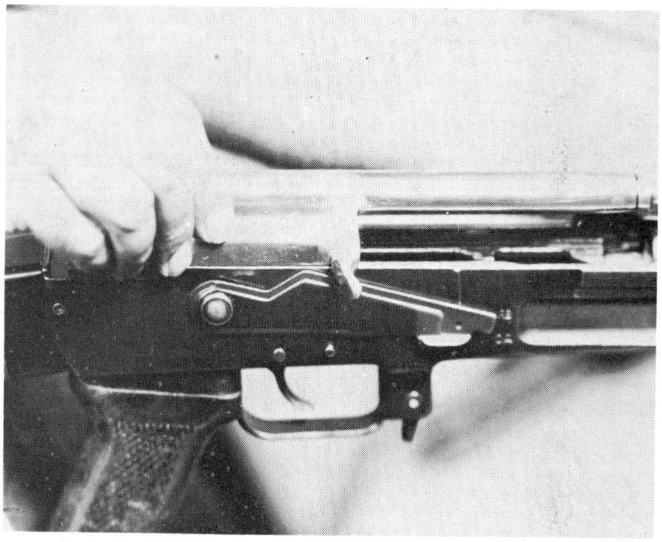

Removing the bolt carrier assembly.

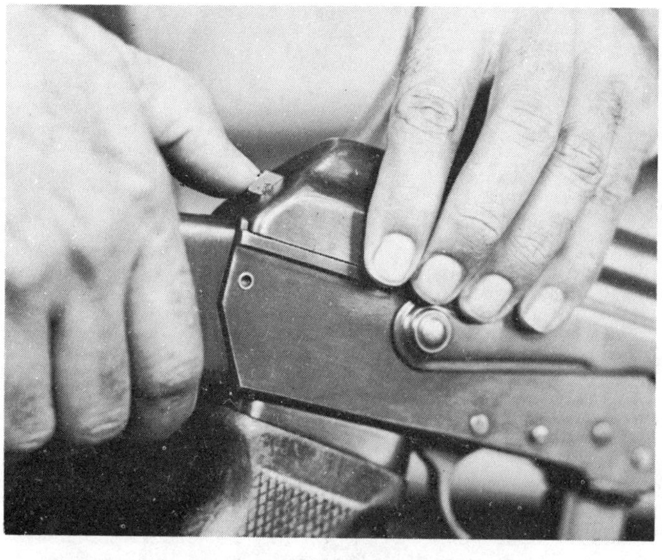

Removing the AK receiver cover.

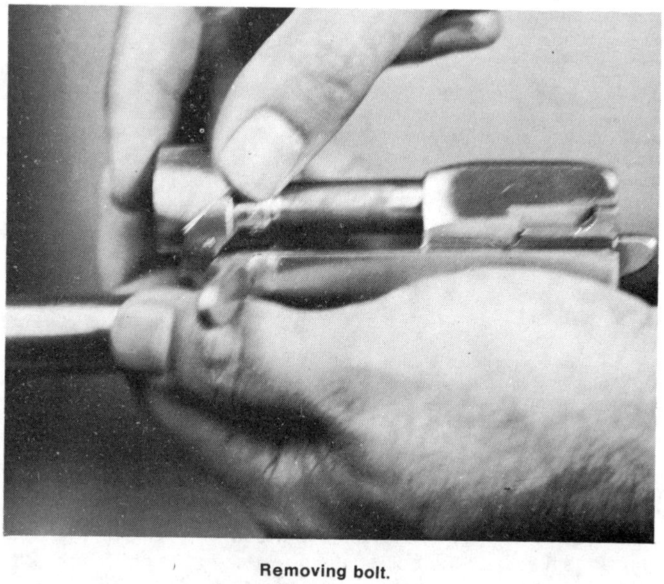

Removing bolt.

The AK, field-stripped.

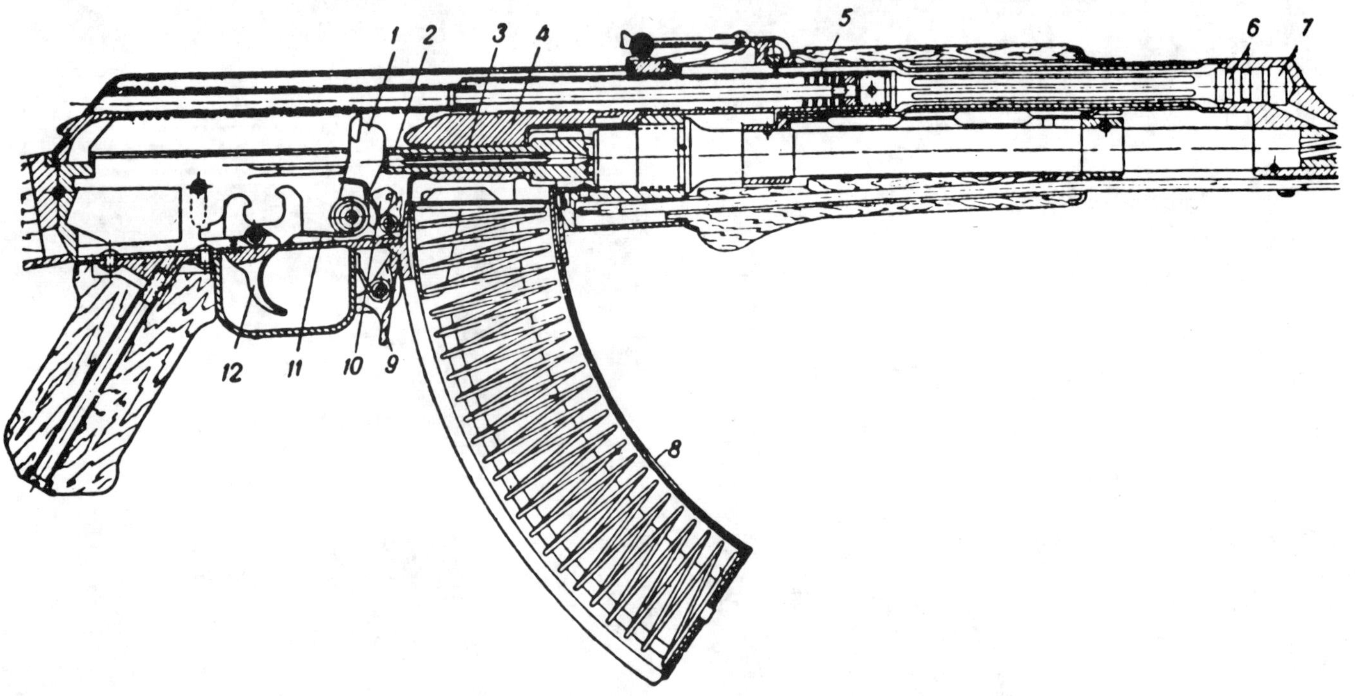

Section view of AK.

1. Hammer
2. Bolt
3. Firing pin
4. Operating rod
5. Recoil spring
6. Gas piston
7. Gas cylinder
8. Magazine
9. Magazine catch
10. Full automatic sear
11. Trigger and hammer spring
12. Trigger

forward so that it strikes the primer of the cartridge. The cartridge fires, and gases from the cartridge flow through the gas port in the barrel into the gas cylinder and force the piston and bolt carrier assembly to the rear. As the piston and bolt carrier assembly move to the rear, the recoil spring is compressed. The bevel in the bolt carrier cam acts on the bolt guide lug, rotating the bolt to the left and thereby unlocking the bolt.

After the bolt is unlocked, the bolt carrier and bolt move to the rear together. The bolt carrier rotates the hammer to the rear, compressing the hammer-and-trigger spring. As the hammer rotates, it rotates the disconnector. When the head of the hammer has passed the notch in the disconnector, the disconnector spring forces the disconnector to engage the disconnector notch in the hammer. This holds the hammer at full cock.

The full automatic sear spring rotates the full automatic sear into engagement with the full automatic sear notch in the hammer. The full automatic sear, however, does not hold the hammer in the cocked position, since the disconnector is already performing this function. As the full automatic sear rotates, the upper end of the sear rises to obstruct the passage of the full automatic disconnector.

As the bolt moves to the rear, the extractor pulls the cartridge case from the chamber. When the case strikes the ejector, it is ejected from the receiver.

The top round in the magazine is forced upward by the follower until it is arrested by the magazine flange.

The rearward movement of the bolt carrier and bolt is arrested by the rear wall of the receiver. Forward movement of these parts is caused by the decompression of the recoil spring.

As the bolt carrier moves forward, the top cartridge in the magazine is stripped from the magazine and forced into the chamber.

As the bolt approaches the barrel face, the first stage in the rotation of the bolt to the right takes place. At the same time, the extractor engages the extractor groove of the cartridge case. As the bolt carrier moves to the extreme forward position, it produces the final rotation of the bolt to the right, locking the bolt.

After the bolt is locked, but while the bolt carrier is still a short distance from the extreme forward position, the full automatic disconnector (which is integral with the bolt carrier) strikes the upper end of the full automatic sear, and rotates the sear forward. This action moves the full automatic sear away from the hammer, so that the hammer will not be prevented from rotating.

The next round is fired by releasing the trigger, and then pressing it again. When the trigger is released, the hammer-and-trigger spring rotates the disconnector and semi-automatic sear to the rear, disengaging the disconnector from the disconnector notch in the hammer. The trigger-and-hammer spring rotates the hammer until the hammer cock notch engages the semi-automatic sear. This is accompanied by an audible click.

When the trigger is again pressed, the semiautomatic sear releases the hammer cock notch. The hammer once again strikes the firing pin, and the entire operating cycle of the automatic mechanism is repeated.

Automatic Fire. To fire the rifle automatically, set the selector at full automatic fire by rotating the indicator until it is opposite the Cyrillic letters AB on the receiver, and press the trigger.

When the trigger is pressed, it rotates on the trigger pin. The disconnector, because it is prevented from rotating by the selector lever, does not engage the disconnector notch in the hammer.

As the trigger rotates, the semiautomatic sear releases the hammer cock notch. The hammer-and-trigger spring rotates the hammer, which strikes the firing pin forcibly. The round is fired. The powder gases act on the gas piston, thrusting the operating rod to the rear, opening the bolt, extracting and ejecting the cartridge case, and cocking the hammer.

The full automatic sear engages the full automatic sear notch in the hammer, holding the hammer at full cock.

The top round in the magazine is raised by the follower.

As the bolt carrier and bolt are moved forward by the recoil spring, the round is fed into the chamber and the bolt is locked.

When the bolt carrier is a short distance from the extreme forward position, the full automatic disconnector strikes the upper end of the full automatic sear and rotates the sear, releasing the hammer. The hammer strikes the firing pin, firing the next round. The entire operating cycle of the automatic mechanism is repeated.

The rifle will continue firing until the last round in the magazine is expended or until the firer releases pressure on the trigger. In the first case, the bolt carrier and bolt will remain in the forward position and the hammer will not be cocked (the AK has no bolt-holding-open device to hold the bolt to the rear after the last round is fired). In the second case, the rifle will be loaded and ready to fire again if the rifleman ceases fire before expending all the rounds in the magazine.

THE SOVIET 7.62mm AKM ASSAULT RIFLE

The AKM is a modification of the AK and probably will eventually replace the AK in Soviet service. The principal ways in which the AKM differ from the AK are:

(1) The AKM has a stamped steel receiver as opposed to the milled receiver of the AK.

(2) The gas relief holes in the AKM gas cylinder tube are semi-circular cutouts at the forward end of the tube which match similar cutouts in the gas cylinder block. The gas relief holes on the AK are cut into the body of the gas cylinder tube—four on each side.

(3) The AKM has a rate reducer attached to the trigger mechanism; the rate of fire is however, the same as that of the AK.

(4) The fore-end of the AKM has a beaver-tail configuration, i.e., it bulges out on both sides.

(5) The rear sight leaf of the AKM is graduated to 1000 meters as opposed to the 800-meter graduation on the AK. The AKM uses the sight leaf of the RPK light machine gun.

7.62mm AKM Assault Rifle.

7.62mm AKM, field-stripped.

(6) The AKM stock and fore-end is made of laminated wood; those of the AK are usually made of ordinary beech or birch.

(7) The bolt and bolt carrier of the AKM are parkerized; those of the AK are bright steel.

THE SOVIET 7.62mm SKS CARBINE

The SKS, like the AK automatic rifle and RPD machine gun, is chambered for the new Soviet "intermediate-sized" M1934 cartridge previously described. The action of the SKS is basically a scaled-down version of the obsolete 14.5mm PTRS antitank rifle. The SKS is a well-designed gun and is very easy to use. Its disassembly is not quite as simple as is the AK's, but is still relatively easy. The SKS is distinguished by its folding bayonet, which—unlike earlier Soviet folding bayonets—is of blade section and folds under the weapon rather than to the side of the weapon. The SKS is no longer used in first-line Soviet units.

NOTE: The Soviets call the SKS a carbine because of its short barrel length. It is not intended to be nor is it used as a replacement for the pistol (as the U. S. carbine was designed to be). The SKS was intended to be used as the tactical equivalent to the U. S. M1 rifle or the new U. S. M14 rifle.

The AKMS is the folding metal stock version of the AKM. The steel struts of the stock are of stamped steel rather than machined steel as used on the folding stock AK. A short compensator is now frequently seen on the muzzle of the AKM and AKMS. The purpose of the compensator is to hold the barrel down in automatic fire. There is now a light alloy magazine weighing .4 pound available for the AK, AKM series weapons.

Field Stripping the SKS

Turn off the safety by rotating it downward. Swing the bayonet upward so that it projects at a 90-degree angle from the muzzle of the rifle. Remove the cleaning rod by disengaging its head from the lugs on the underside of the barrel and pulling the rod forward. Rotate the bayonet downward until it locks in its folded position.

Hold the small of the stock with the left hand and, with the right hand, rotate the receiver cover retaining pin arm (located at the right rear of the receiver) upward; then, pressing on the receiver cover with the thumb of the left hand, move the pin to the right as far as possible, and remove the cover from the receiver. Remove the recoil spring assembly from the bolt carrier. Note

Soviet 7.62mm Simonov Semiautomatic Carbine (SKS).

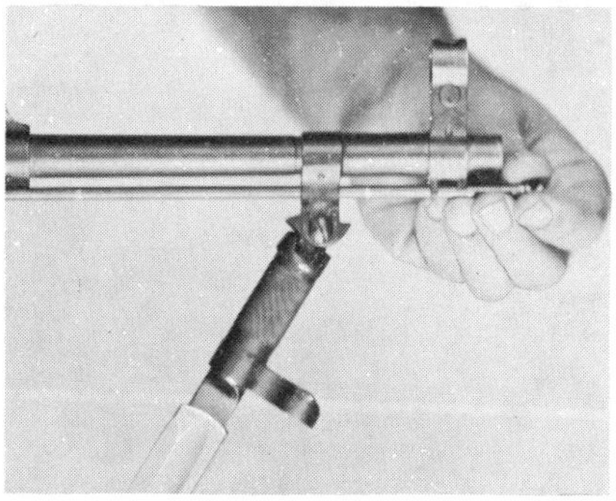

(1) Removing cleaning rod.

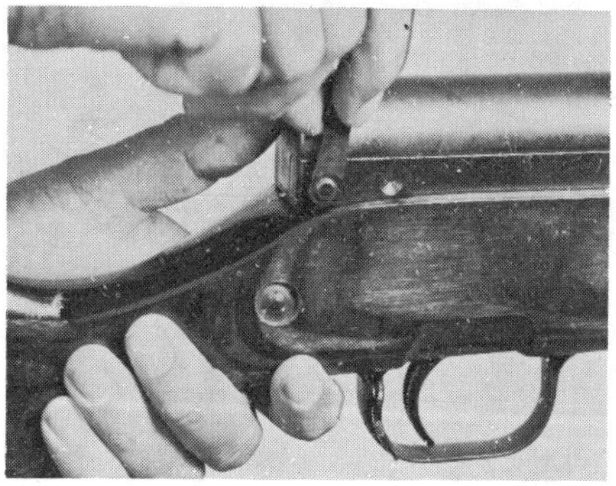

(2) Loosening receiver cover retaining pin.

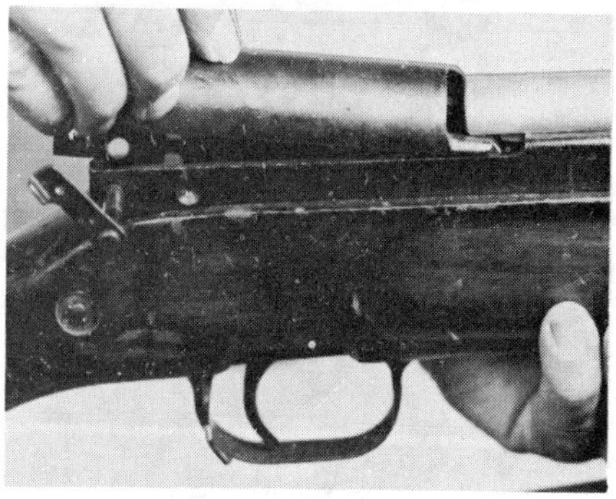

(3) Removing receiver cover.

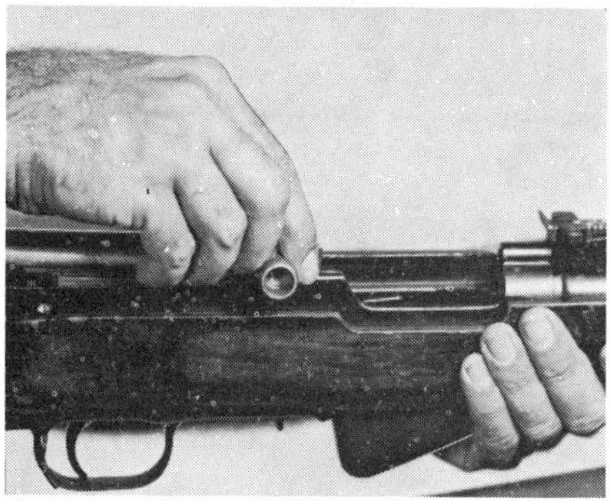

(4) Removing bolt and bolt carrier.

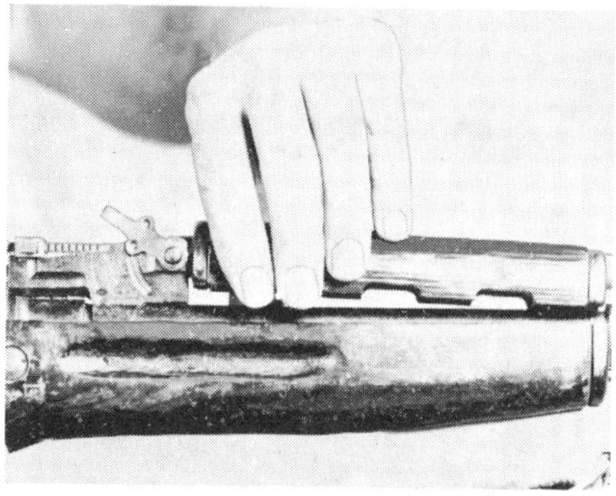

(5) Removing handguard and gas piston assembly.

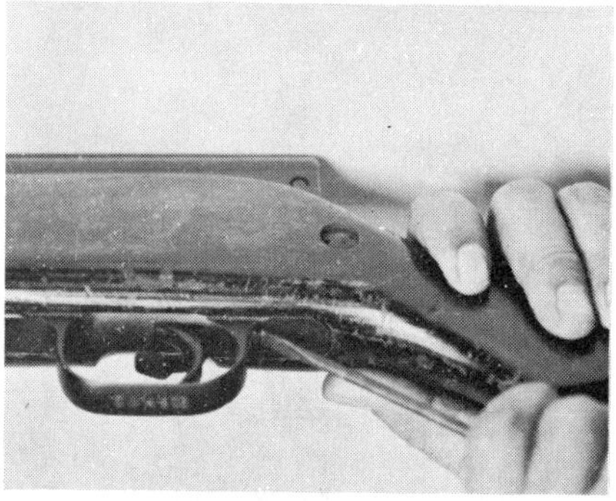

(6) Unlocking trigger guard.

(7) Removing trigger group.

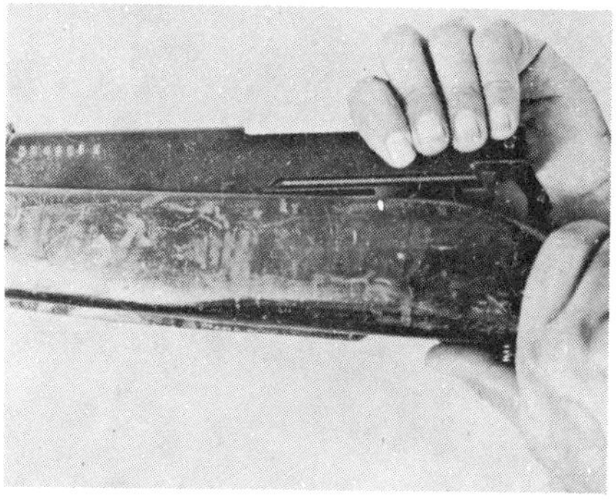

(8) Removing barrel and receiver.

that the recoil spring assembly is, like that of the AK, a packaged unit.

To remove the bolt and bolt carrier from the receiver, pull back on the operating handle; the whole assembly will lift out, and the bolt can be removed from the bolt carrier.

Hold the fore-end of the carbine with the left hand and, with the right hand, rotate the gas cylinder tube lock lever upward (the lock is at the right front side of the rear sight base) until the lock lever lug is stopped by the upper wall of the cutout in the rear sight base. Then the gas cylinder tube and handguard assembly can be slightly raised and pulled rearward off the weapon. The gas piston will slide out of the gas cylinder tube.

To assemble the weapon, follow the above steps in reverse order.

Loading and Firing the SKS

Set the weapon on "Safe" by turning the safety lever up as far as it will go. The safety lever is at the rear of the trigger guard. Pull the operating handle to the rear as far as it will go; the bolt and bolt carrier will be held to the rear by the bolt-holding-open device. If the cartridges are assembled in the ten-round charger normally used, insert one end of the charger in the charger guide which is machined into the top forward end of the bolt carrier. Push down on the cartridges with the thumb until all the cartridges are loaded into the magazine; then remove the empty charger guide. Pull back on the operating handle and release it; the bolt and bolt carrier will now go forward and chamber a cartridge. Disengage the safety by turning it down as far as it will go. The rifle is now ready to fire.

If all ten rounds are fired, the bolt and bolt carrier will stay to the rear, leaving the action open for reloading. If only a few rounds are fired, and it is desired to unload the weapon, the following procedure should be observed. To the rear of the magazine, on the underside of the weapon, is a catch; pull this catch rearward. The magazine body will now swing downward and the rounds in the magazine will spill out. The round in the chamber is removed by pulling the operating handle to the rear.

SKS, field-stripped.

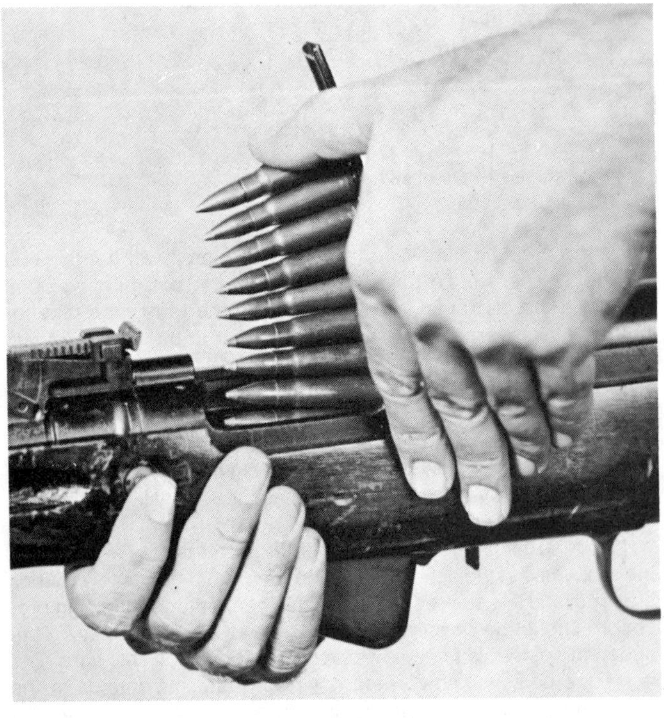

Loading the SKS.

To Fix the Bayonet

To fix the bayonet, the bayonet handle is pulled to the rear, and the bayonet is rotated upward until the bayonet muzzle ring snaps over the muzzle of the weapon. To fold the bayonet, pull the bayonet handle upward until the bayonet muzzle ring clears the muzzle; then swing the bayonet downward and into its groove in the stock.

How the SKS Works

When the trigger is pressed, the trigger rotates and moves the trigger arm forward. The trigger arm moves the sear forward and disengages it from the hammer. The hammer spring, now being free to expand, rotates the hammer, which strikes the firing pin and drives it through the bolt channel. When the hammer is rotated, the hammer heel lowers the forward end of the disconnector and the forward end of the trigger arm. At this time, the trigger arm is disengaged from the sear, which, under the action of the sear spring, returns to the rear position. The firing pin moves through the bolt channel until the firing pin tip emerges from the bolt face and strikes the primer. After the bullet passes the gas port in the wall of the barrel, the gases enter the gas cylinder, exert pressure on the piston, and move the bolt carrier to the rear by means of the piston rod.

The bolt carrier, after traveling a path of eight mm, raises the rear end of the bolt, disengages the bolt locking surface from the receiver lug and brings the bolt to the rear. As the bolt moves to the rear, it extracts the empty cartridge case, which is held in the

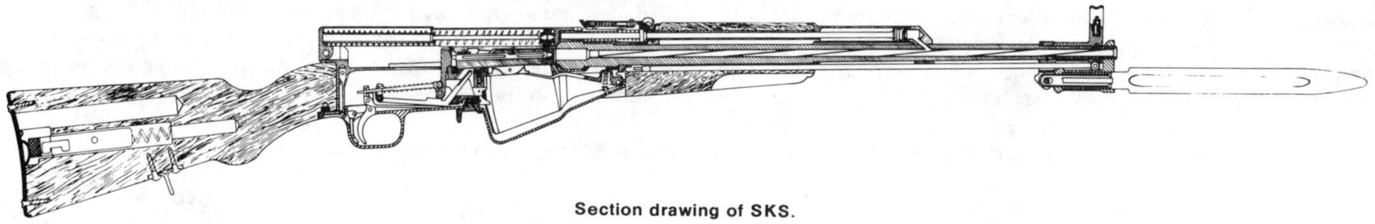

Section drawing of SKS.

bolt face by the extractor lug. After traveling a distance of 70 mm, the base of the cartridge case meets the ejector, and the case is ejected from the weapon. In moving to the rear, the bolt carrier, and then the bolt, cocks the hammer. The forward end of the disconnector, under the action of the hammer spring, is raised upward, and the hammer is positioned on the disconnector notch at the same time. The forward end of the trigger arm is located between the base of the trigger guard and the sear. Under the action of the sear spring, the sear, which is in the extreme rear position, is positioned under the hammer cock notch. Under the action of the follower lever spring, the follower positions the next round for the feed rib. The piston, together with the piston rod, pushes the bolt a distance of 20 mm. The bolt thereafter continues by inertia to the extreme rear position, at the same time compressing the recoil spring, and the piston and piston rod return to the forward position under the action of the piston rod spring.

Under the action of the expanding recoil spring, the bolt moves forward, and the bolt feed rib grasps the next round in the magazine and sends it into the chamber. The bolt carrier lowers the rear end of the bolt upon approaching the barrel face, and the bolt locking surface is positioned in front of the receiver locking lug to seal the bore. The bolt, in lowering, depresses the protruding front end of the disconnector, at which time the hammer is disengaged from the disconnector and is then cocked. As it approaches the barrel face, the extractor, which is now sliding along the base of the cartridge case, is forced to the right to grasp the cartridge case rim. The extractor spring is compressed at this time. Under the action of the follower lever spring, the follower raises the next round until it meets the bolt. Before the next round is fired, it is necessary to release the trigger; under the action of the trigger spring, the trigger returns to the forward position, and the front end of the trigger lever is disengaged from the sear and is positioned opposite the sear shoulders.

To fire the next round, it is necessary to press the trigger. The hammer can be released from the cocked position only when the bolt is in the extreme forward position and the bore is completely sealed. If the bolt is in the rear position (on the stop), or if the bore is not completely sealed (bolt in extreme forward position), the forward end of the disconnector is not depressed, the forward end of the trigger lever presses against the guide lugs for the sear in the trigger guard groove, and the hammer cannot be released. If the bolt travels past the locking lug and depresses the forward end of the disconnector, and the bolt carrier has not attained the extreme forward position, the weapon still cannot be fired. In this event, the hammer, in moving forward will not strike the firing pin, but will strike the vertical surface of the bolt carrier which cocks the hammer.

THE SOVIET 7.62mm SVD SNIPER RIFLE

The SVD or Dragunov (SVD means Self loading Rifle, Dragunov) rifle is the replacement for the M1891/30 sniper rifle. It is chambered for the 7.62mm rimmed cartridge (7.62 x 53 R). The rifle is fitted with a four-power scope Model PSO-1 and has a somewhat unusual stock in that a large section has been cut out of it immediately to the rear of the pistol grip. This lightens the weight of the rifle considerably. The Dragunov uses an action which closely resembles that of the AK. It has a prong type flash suppressor similar to those used on current U.S. small arms.

How To Load and Fire The SVD

The box magazine is loaded in the normal manner; it is removed from the rifle by pushing the magazine catch—located behind the magazine port—forward and pulling down on the magazine. Insert loaded magazine in the magazine port pushing it upward till it securely locks in place. Pull bolt-operating handle which protrudes from the right side of the receiver to the rear and release, thus chambering a cartridge. The weapon is now loaded and will fire one round for each pull of the trigger until the magazine is empty. The safety is similar to that of the AK/AKM assault rifle and is mounted on the right side of the receiver. To put the rifle on safe, push the lever upwards.

How the SVD Works

The bolt operation of the SVD is essentially the same as that of the AK/AKM in semi-automatic fire. The principal difference is that the SVD has a spring-loaded piston rod which is a separate as-

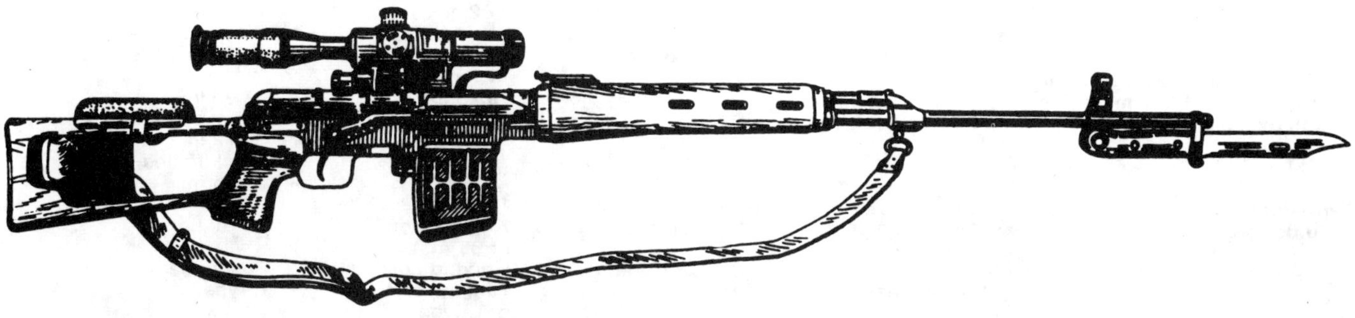

7.62mm SVD (Dragunov) Sniper Rifle.

sembly; it is not attached to the bolt carrier as is that of the AK/AKM. The trigger mechanism is relatively simple consisting of 12 parts including the fabricated trigger housing/trigger guard. It varies from the AK/AKM trigger mechanism in that, among other things, it does not have a full automatic sear or full automatic disconnector and it has a separate trigger spring in addition to the hammer spring as opposed to the one spring which performs both functions in the AK/AKM.

CHARACTERISTICS OF POST WORLD WAR II SOVIET RIFLES AND CARBINES

	SKS Carbine	AK Assault Rifle	AKM Assault Rifle	SVD
Caliber:	7.62mm M43 (7.62 x 39m)	7.62mm M43 (7.62 x 39mm)	7.62mm M43 (7.62 x 39mm)	7.62mm rimmed.
System of Operation:	Gas, semiautomatic.	Gas, selective fire.	Gas, selective fire.	Gas, semi-automatic.
Length overall:	40.16 in.	34.25 in.	34.25 in.	48.2 in.
Barrel length:	20.47 in.	16.34 in.	16.34 in	24 in.
Feed device:	10-round, staggered row non-detachable box magazine.	30-round, staggered row detachable box magazine.	30-round, staggered row detachable box magazine.	10-round, staggered row detachable box magazine.
Sights: Front:	Hooded post.	Post w/ protecting ears.	Post w/ protecting ears.	Hooded post.
Rear:	Tangent, graduated to 1000 meters.	Tangent, graduated to 800 meters.	Tangent, graduated to 1000 meters.	Tangent w/notch*
Weight:	8.8 lb.	10.58 lb.	8.87 lb.	9.5 lb.
Muzzle velocity:	2410 f.p.s.	2330 f.p.s.	2330 f.p.s.	2720 f.p.s.
Cyclic rate:	----	600 r.p.m.	600 r.p.m.	----

*Normally fitted with 4-power scope graduated to 1300 meters.

SOVIET SUBMACHINE GUNS

All pre-World War II and World War II Soviet submachine guns were chambered for the Soviet 7.62mm pistol cartridge Type P. Although the two latest types, the PPSh M1941 and the PPS M1943, are still in use in some of the satellites, all of these guns are obsolete in the U.S.S.R.

THE 7.62mm PPSh SUBMACHINE GUN M1941

Note that the bolt is a very simple machined piece. The recoil spring guide is resting in a hole in the rear of the bolt, with the spring around it; and a plastic buffer at its rear end. This spring arrangement also serves to spring lock the receiver when the weapon is assembled.

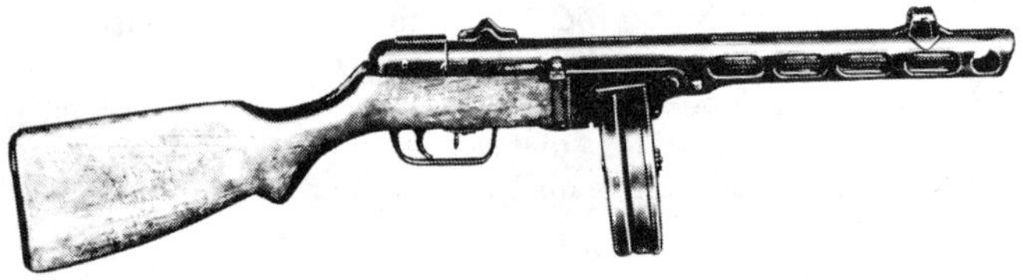

PPSh M1941 with 71-round drum magazine.

Field Stripping The PPSh M1941

Push forward on receiver catch and hinge barrel and barrel casing down.

Draw straight back on bolt handle a short distance, meanwhile exerting an upward pull. The bolt may be lifted out.

Remove recoil spring and buffer.

Barrel is removable from casing. Further dismounting is seldom necessary.

7.62mm PPSh M1941, field-stripped.

CHARACTERISTICS OF PRE-WORLD WAR II AND WORLD WAR II SOVIET SUBMACHINE GUNS

	PPD Model 1934/38	PPD Model 1940	PPSh Model 1941	PPS Model 1943
Caliber:	7.62mm.	7.62mm.	7.62mm.	7.62mm.
System of operation:	Blowback. Selective fire.	Blowback. Selective fire.	Blowback. Selective fire.	Blowback. Automatic fire only.
Weight: w/loaded drum magazine:	11.5 lb.	11.90 lb.	11.99 lb.	
Weight: w/loaded box magazine:	8.25 lb.		9.26 lb.	7.98 lb.
Length overall:	30.63 in.	30.63 in.	33.15 in.	Stock extended—32.72 in. Stock folded—24.25 in.
Barrel length:	10.63 in.	10.63 in.	10.63 in.	9.45 in.
Feed device:	71-rd. drum or 25-rd. box.	71-rd. drum.	71-rd. drum. or 35-rd. box.	35-rd. box.
Sights: Front:	Blade.	Hooded blade.	Hooded post.	Post w/ears.
Rear:	Tangent leaf.	Tangent leaf.	Tangent leaf or L-type.	L-type.
Cyclic rate:	900 r.p.m.	900-1100 r.p.m.	700-900 r.p.m.	650 r.p.m.
Muzzle velocity:	1640 f.p.s.	1640 f.p.s.	1640 f.p.s.	1608 f.p.s.

NOTE: A 1942 version of the PPS M1943 also exists, but differs only in minor details from the Model 1943.

From top down: M34/38, PPD M1940, PPSh M1941, and PPS M1943.

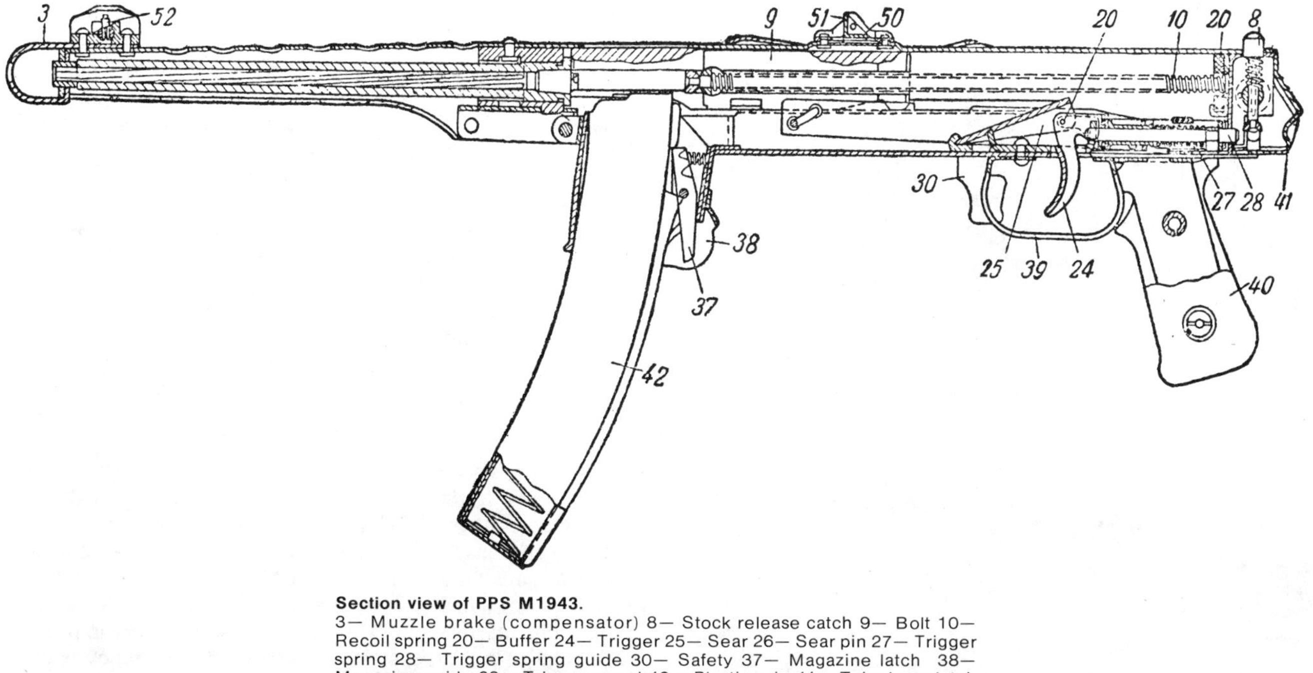

Section view of PPS M1943.
3— Muzzle brake (compensator) 8— Stock release catch 9— Bolt 10— Recoil spring 20— Buffer 24— Trigger 25— Sear 26— Sear pin 27— Trigger spring 28— Trigger spring guide 30— Safety 37— Magazine latch 38— Magazine guide 39— Trigger guard 40— Plastic grip 41— Takedown latch 42— Magazine 50— Rear sight guard 51— "L" type flip sight 52— Front sight post.

SOVIET MACHINE GUNS

7.62mm Model 1905 Maxim.

HISTORICAL SUMMARY

Imperial Russia, the predecessor to the U.S.S.R., adopted the Maxim gun prior to 1900. Russian troops in the Russo-Japanese War (1904-05) used Maxims against the Japanese Hotchkiss guns. The 7.62mm Model 1905 Maxim was the first machine gun manufactured in Russia. This weapon, as all the early Maxims, had numerous brass fittings and was quite heavy; it has a bronze water jacket.

The Imperial government adopted the 7.62mm Model 1902 Madsen as the standard light machine gun. During World War I, the Russians purchased 7.62mm Maxim and Colt Model 1914 (modified Colt Model 1895 "potato digger") machine guns from Colt in the United States. They also purchased a considerable quantity of Lewis guns. Imperial Russia did not have the capability to produce their wartime requirements for rifles or machine guns.

In 1910 the Russians modified the Maxim; the 7.62mm Model 1910 Maxim (SPM) has a ribbed water jacket similar to that of the British Vickers gun. This weapon was used and manufactured through World War II and was modified during World War II by the addition of a tractor-type water entry port.

After World War II, the Soviets started development of their own weapons. The 7.62mm Maxim Tokarev and Maxim Koleshnikov were among the first efforts. These were air-cooled Maxims fitted with bipods and were used in large quantities during the Spanish Civil war. The first originally developed Soviet machine gun was the 7.62mm DP—Degtyarev Infantry—which appeared in 1926. The DP introduced the modified Kjellman Frijberg locking system to the Soviets which is still in service in several machine guns. The DP was the first of a series of Degtyarev machine guns adopted by the Soviet Union. A tank version of the DP called the DT may still be found in service on older Soviet armored vehicles in use in the Soviet satellites. An aircraft version, the 7.62mm DA, was also produced.

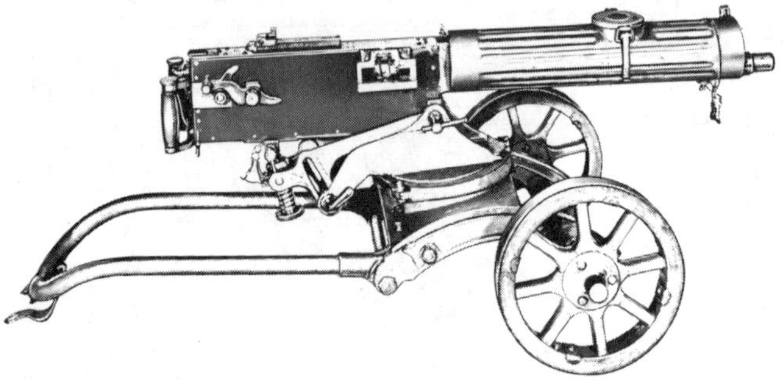

Maxim M1910 Heavy Machine Gun (SPM).

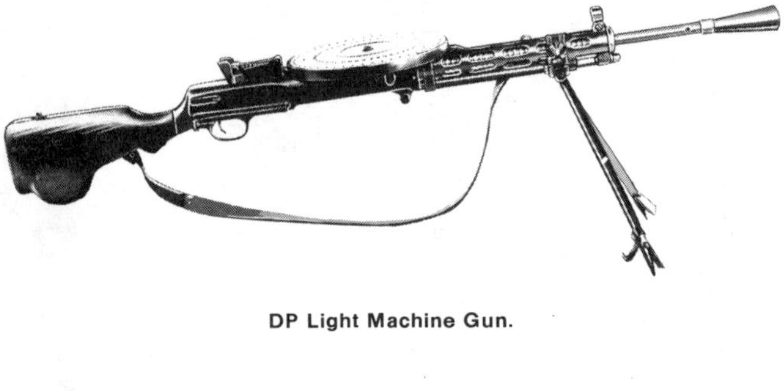

DP Light Machine Gun.

DT, the tank gun version of the DP.

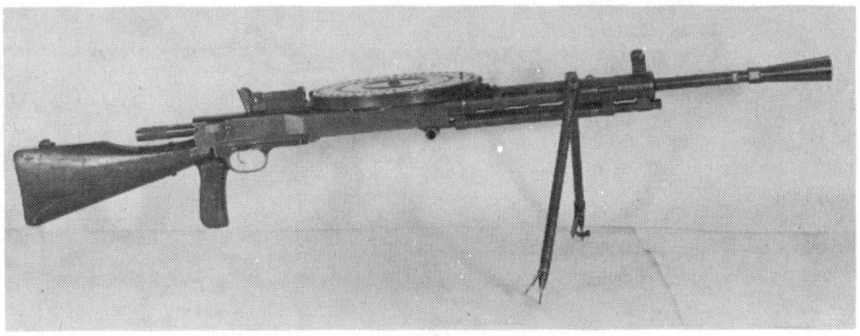

DPM Light Machine Gun, a modified DP.

The DP, DT, and DA have their operating springs coiled on the piston rod which is seated under the barrel. The heating of the barrel caused distortion of the spring with resultant malfunctions. During World War II, a modification of the DP—the DPM—was put into service. The DPM is basically the same as the DP, but the recoil spring is mounted in a tube which projects to the rear of the receiver. A tank version of this weapon, the DTM, was also produced and is still in service on pre-1949 Soviet armored vehicles.

A heavy machine gun version of the DP, the 1939 DS, turned out to be a failure in battle or more properly a failure in the manufacturing plant. The DS was replaced by the Goryunov SG43, a very successful gun which is still in use in Soviet satellites and which in modified form—SGM, SGMT, SGMB—is still in the Soviet Union as a battalion-level machine gun, tank machine gun, and vehicular machine gun. The Goryunov series guns do not use the Degtyarev locking system; the bolt of the Goryunov is cammed to the side to lock in a side wall of the receiver.

The 12.7mm DShK Model 1938 was the first Soviet heavy caliber

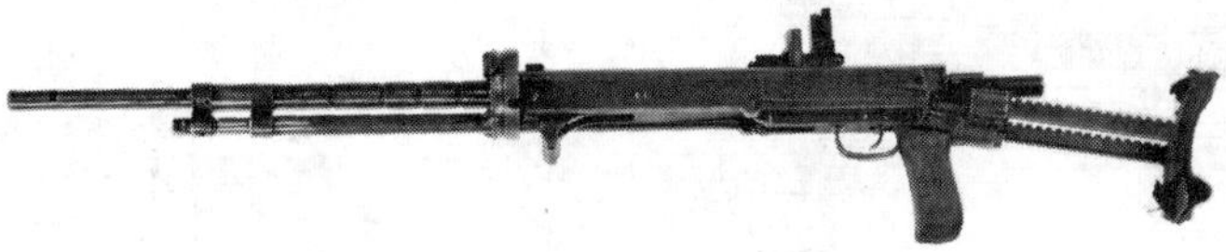

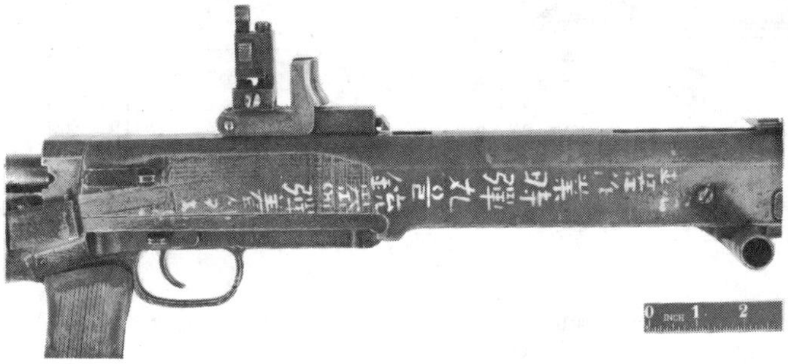

DTM, the tank gun version of the DPM. Instructions in Chinese on receiver.

DS M1939 Heavy Machine Gun.

machine gun to be produced in quantity. It was preceded by the 12.7mm DK which appeared around 1934, but was apparently not a very successful gun. The DShK uses the Degtyarev locking system.

The 7.62mm RPD machine gun is also basically a Degtyarev weapon and has a bolt system generally similar to the earlier Degtyarevs; it is being replaced in Soviet service by the 7.62mm RPK machine gun which is similar to the AK assault rifle in most respects.

Outstanding among the post World War II weapons is the 14.5mm KPV machine gun. This weapon does not resemble any of the earlier Soviet weapons insofar as its operating mechanism is concerned. It bears some resemblance to the weapons of Louis Stange—Austrian MG30, Hungarian Model 31, etc., but has other quite original characteristics.

The weapons listed in the chart below are, with the exception of the DT and DTM, no longer used in the U. S. S. R., although they may be encountered in Soviet satallites. The M1943 Goryunov is covered later in this section, since it is still used in quantity by the Soviets. All these weapons are chambered for the Soviet 7.62mm rimmed cartridge.

CHARACTERISTICS OF PRE-WORLD WAR II AND WORLD WAR II SOVIET 7.62mm GROUND MACHINE GUNS

	Model 1910 Maxim (SPM)	Model 1939 DS	DP	DPM	DT	DTM	SG43
Caliber:	7.62mm.	7.62mm.	7.62mm.	7.62mm.	7.62mm.	7.62mm.	7.62mm.
Type of gun:	Heavy ground gun.	Heavy ground gun.	Squad automatic.	Squad automatic.	Tank gun.	Tank gun.	Heavy ground gun.
System of operation:	Recoil, automatic only.	Gas, automatic only.	Gas, automatic only.	Gas, automatic only.	Gas, automatic only.	Gas, automatic only.	Gas, automatic only.
Weight:	52.47 lb. w/water.	52.47 lb. w/water.	26.23 lb. loaded.	26.9 lb.	27.91 lb.	28.46 lb.	30.42 lb.
Mount weight:	99.71 lb. w/shield.	99.71 lb. w/shield.					59.3 lb.
Length overall:	43.6 in.	46 in.	50 in.	50 in.	Stock extended, 46.46 in. Stock retracted, 39.76 in.	Stock extended, 46.44 in. Stock retracted, 39.76 in.	44.09 in.
Barrel length:	28.4 in.	28.4 in.	23.8 in.	23.8 in.	23.5 in.	23.5 in.	28.3 in.
Feed device:	250 rd.	250-rd. web belt or 50-rd. link belt.	47-rd. drum.	47-rd. drum.	60-rd. drum.	60-rd. drum.	250-round drum metalic link belt.
Sights: Front:	Blade.	Post w/ears.	Post w/ears.	Post w/ears.	None on tanks On bipod: Post w/ears.	None on tanks On bipod: Post w/ears.	Blade.
Rear:	Leaf.	Leaf.	Tangent leaf.	Tangent leaf.	Aperture.	Aperture.	Leaf.
Cycle rate of fire:	500-600.	2 Rates: 500-600, 1000-1200.	500-600.	500-600.	600.	600.	600-700.
Muzzle velocity w/light ball:	2822 f.p.s.	2832 f.p.s.	2756 f.p.s.	2756 f.p.s.	2756 f.p.s.	2756 f.p.s.	2832 f.p.s.
Cooling:	Water.	Air.	Air.	Air.	Air.	Air.	Air.

NOTE: An aircraft version of the Degtyarev, the 7.62mm DA, was also used. It has been obsolete for some time.

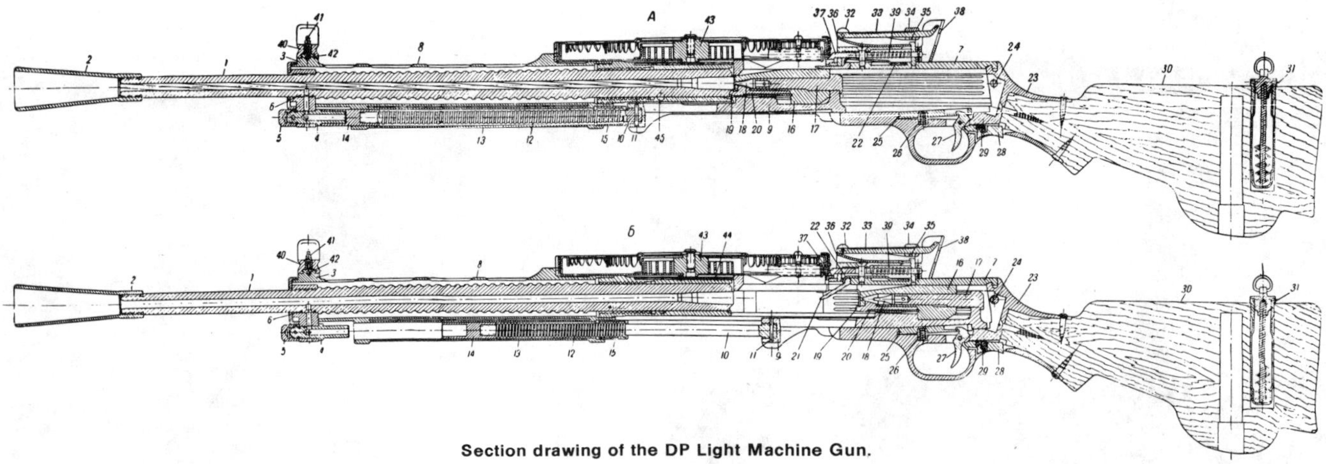

Section drawing of the DP Light Machine Gun.

THE 7.62mm DP AND DPM LIGHT MACHINE GUNS

Field Stripping The DP

To remove the barrel, pull the cocking handle to the rear to cock the weapon. The barrel locking stud is on the left side near the front of the receiver. Press this stud in, which will release the barrel, then twist the barrel up one quarter turn to the right. Now slide the barrel straight forward out of the receiver.

Both sights are carried on casing. Rear sight base serves as magazine catch.

Flash hider is screwed to barrel. Gas cylinder is just forward of barrel cooling rings. Bolt is shown with right and left side locks and firing pin removed.

Gas piston with operating rod and spring shown attached to bolt carrying slide.

Bipod is easily detachable. Bottom view of magazine to show cartridge feed system. A squeezer-type safety is positioned in the rear of the trigger guard; and is automatically pressed in as the fingers tighten around the grip.

Press the trigger and ease the cocking handle forward. Pull out the bolt at the rear of the trigger guard, which leaves the stock and trigger guard free to be turned until the rear of the trigger guard is clear of the receiver. Pull the stock and trigger guard assembly back and out of the receiver.

A small sleeve fits behind the recoil spring at the rear of the gas cylinder tube. Press this forward and twist it to the left; this will free the bolt together with the slide and the gas piston attached to it to be withdrawn at the rear of the receiver.

The bolt may now be lifted from the top of the slide. The firing pin may now be slid out of the rear of the bolt. The bolt locks on each side of the bolt may now be lifted out; and the front of the extractor spring raised and pulled forward to permit removal of it and the extractor.

This completes field stripping. Assembling the gun is equally simple and merely calls for reversing the stripping procedure.

Loading and Firing The DP and DPM

The Magazine. This type drum differs radically from the Lewis type. The inner center rotates, while the outer rim is fastened securely to the gun. The cartridges lie in single line around the inside of the pan.

To Prepare for Firing. Mount a loaded magazine on top of the receiver and press firmly down until it is caught by the magazine catch. (The magazine catch is mounted in the front end of the rear

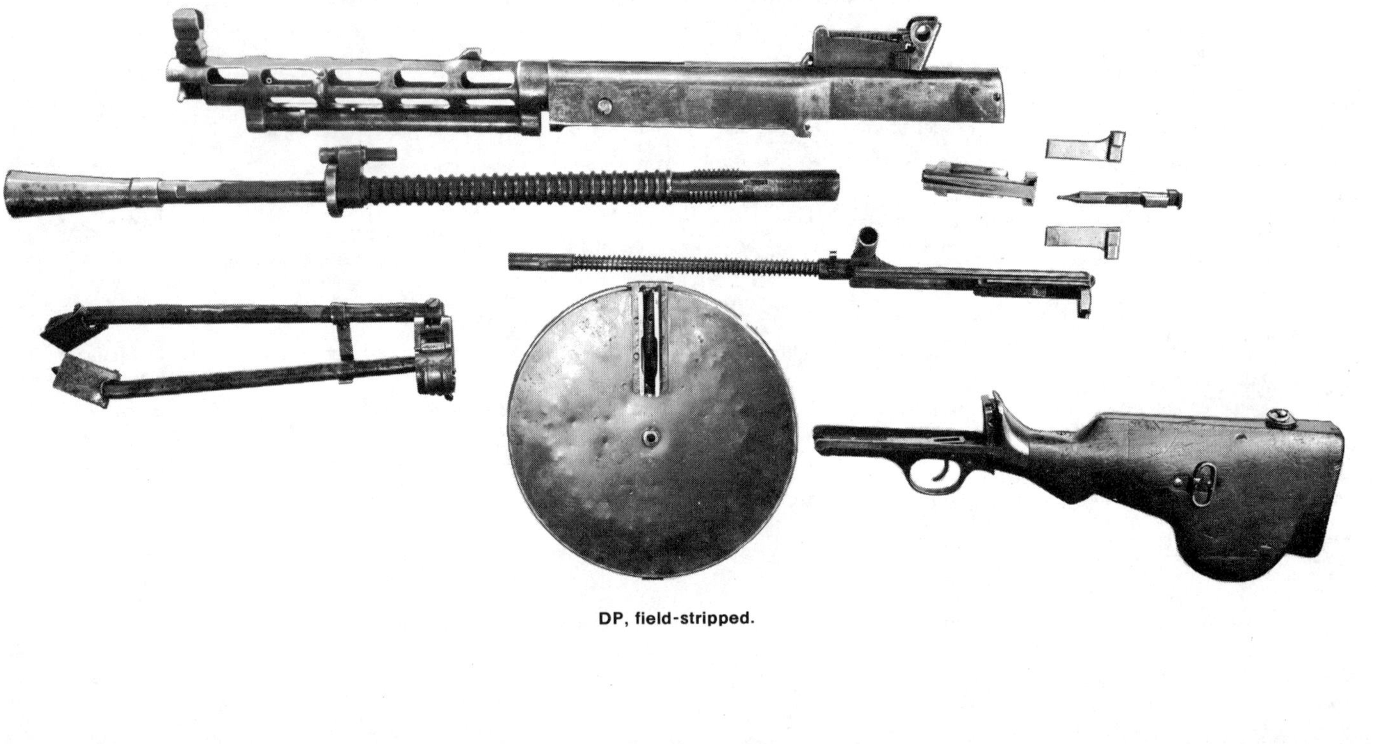

DP, field-stripped.

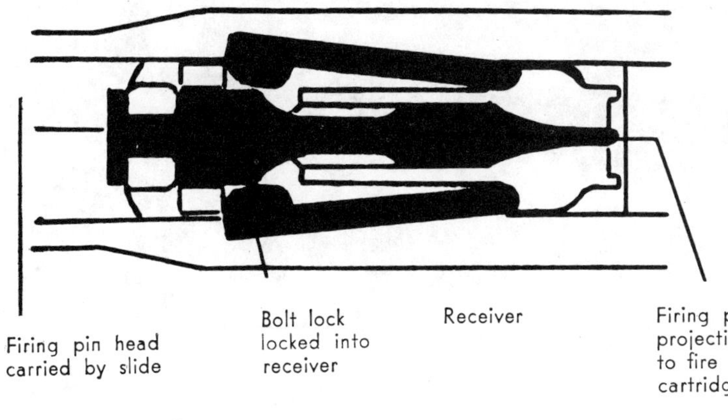

Locking and firing movement of the action.

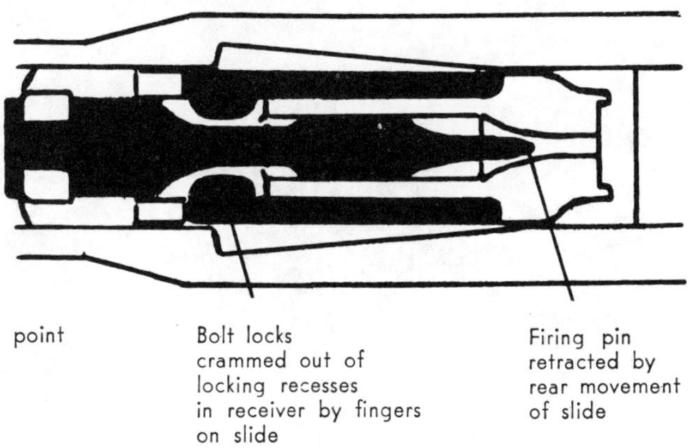

Unlocking action.

sight base.)

Now pull back the cocking handle as far as it will go. It will stay open. Pressing the trigger will now fire the gun. Full automatic fire will ensue as long as the trigger is held down and there are cartridges in the magazine.

Note: A safety catch to the rear of the triggerguard is automatically pressed in as the rifle is gripped ready to fire on the DP; the DPM has a manual safety.

How the DP and DPM Work

A loaded pan magazine is placed on top of the receiver where it engages with a hook on the barrel jacket and is held at the rear by a spring catch, the handle of which forms a guard for the rear sight. The handle on the right side of the gun is drawn back to its full length. This compresses the recoil spring which is positioned below the barrel and which travels with a rod connecting the gas cup to the slide. This spring is compressed against its lock by a gas cup. The rod moves backwards through the center of this spring and lock.

During this opening movement unlocking also takes place and the bolt is held open when it is caught by the sear.

Pressing the trigger rotates the sear down and out of its notch in the bottom of the slide. The bolt is carried on top of the slide. The operating rod is attached to the front end of the slide. The compressed spring is now free to pull the moving members forward.

A moving plate inside the fixed magazine pan has brought a cartridge through the medium of a rotor spring into the feed lips in line with the bolt. The bolt strikes this cartridge and drives it ahead into the firing chamber. On each side of the bolt is a loose plate. These plates are flush with the bolt when it is cocked. As the bolt brings up against the face of the cartridge in the chamber and its forward action is stopped, the firing pin is carried still farther ahead by the operating slide on which it is mounted; and cams on it force the loose locking plates out into recesses machined into the receiver to accommodate them. The firing pin now discharges the cartridge, with the bolt locked securely to the receiver.

Return Movement of the Action. A small quantity of gas passes through the port in the barrel as the bullet goes over it. This passes into a nozzle attached to the barrel which directs the expanding gas against the head of a cup which serves in lieu of a piston. The cup walls extend forward about an inch around this nozzle.

The energy from the expanding gas is transmitted through the gas cup at the end of the rod to the attached slide which it starts backward. The first movement of the slide withdraws the firing pin and from thereon a projection on the bottom of each loose lock plate rides in a cam groove in the slide, camming the projections

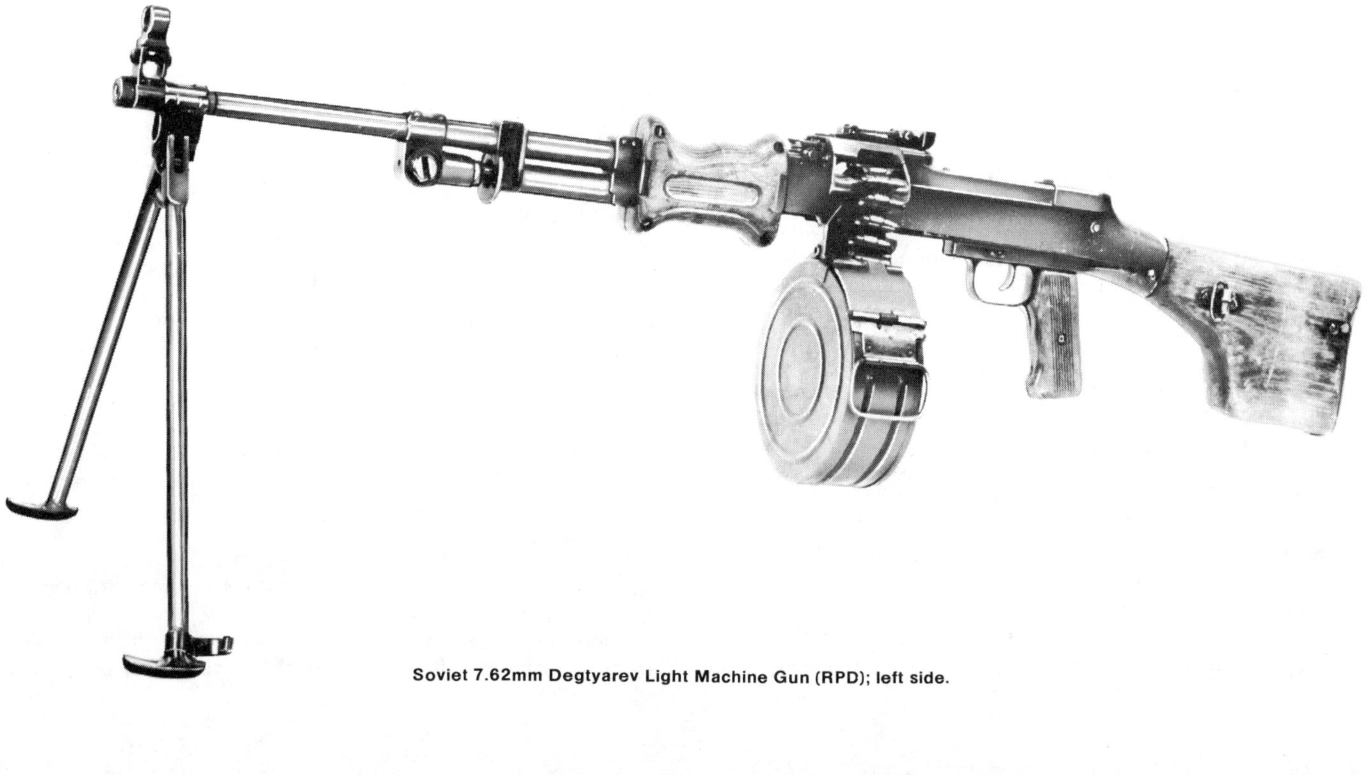

Soviet 7.62mm Degtyarev Light Machine Gun (RPD); left side.

RPD Light Machine Gun, right side.

Late model Soviet 7.62mm RPD Light Machine Gun with dust covers on feed and link ejection ports, and new type operating handle.

toward the center. This pulls the locks out of their recesses in the receiver. The slide is then able to carry the bolt straight back extracting and ejecting the empty cartridge case.

THE 7.62mm RPD LIGHT MACHINE GUN

The Ruchnoi Pulemet Degtyarev, or RPD, as it is commonly called by the Soviets, is the standard squad automatic weapon in the Soviet Army. It has replaced the DP light machine gun and is chambered for the same "intermediate-sized" cartridge as are the AK and the SKS. The RPD is the tactical equivalent to the U.S. Browning automatic rifle or the British Bren gun. It is designed only for use on a bipod and is not designed for, or capable of, long periods of sustained fire. The RPD is a very simple weapon in design and should be relatively easy to service in the field. Its basic operating mechanism is a modification of that of the earlier DP, but the RPD is belt fed, while the DP is fed from a pan-type magazine. The RPD is the lightest belt-fed gun in service in the world today. It should be kept in mind, however, that the RPD is not chambered for a full-size rifle cartridge; if it were, it probably would be several pounds heavier.

Loading and Firing the RPD

The link belt of the RPD is loaded by pushing the individual rounds into the belt so that the bottom hook-like section of the link snaps into place in the cartridge extractor groove. The two fifty-round sections are joined together by slipping the tongue of the end link on one belt section through the slots of the starting link on the other belt section. A cartridge is then inserted, locking the two belt sections together.

The belt is then rolled into a tight circle and fitted into the stamped metal drum. The drum is fitted on the weapon by sliding its top dovetail on the mating surfaces which protrude under the forward part of the receiver. The belt loading tab should be protruding from the spring-loaded trap door of the drum, so that it can be inserted in the receiver. The drum is locked in place by pulling down the lock on the underside of the receiver; this lock keeps the drum from moving backwards off the gun during fire.

Cock the gun by pulling operating handle to the rear. On older guns, which have the non-folding operating handle, the handle will remain to the rear. On the newer type guns, which have the folding type operating handle, the handle should be pushed forward after pulling to the rear.

Open the cover by pushing forward on cover latch and lifting the cover. Lay belt on feedway so that the leading cartridge lies beside the cartridge stop. Close cover.

If the trigger is squeezed, the weapon will now fire. The safety catch is located on the right side of the pistol grip butt group,

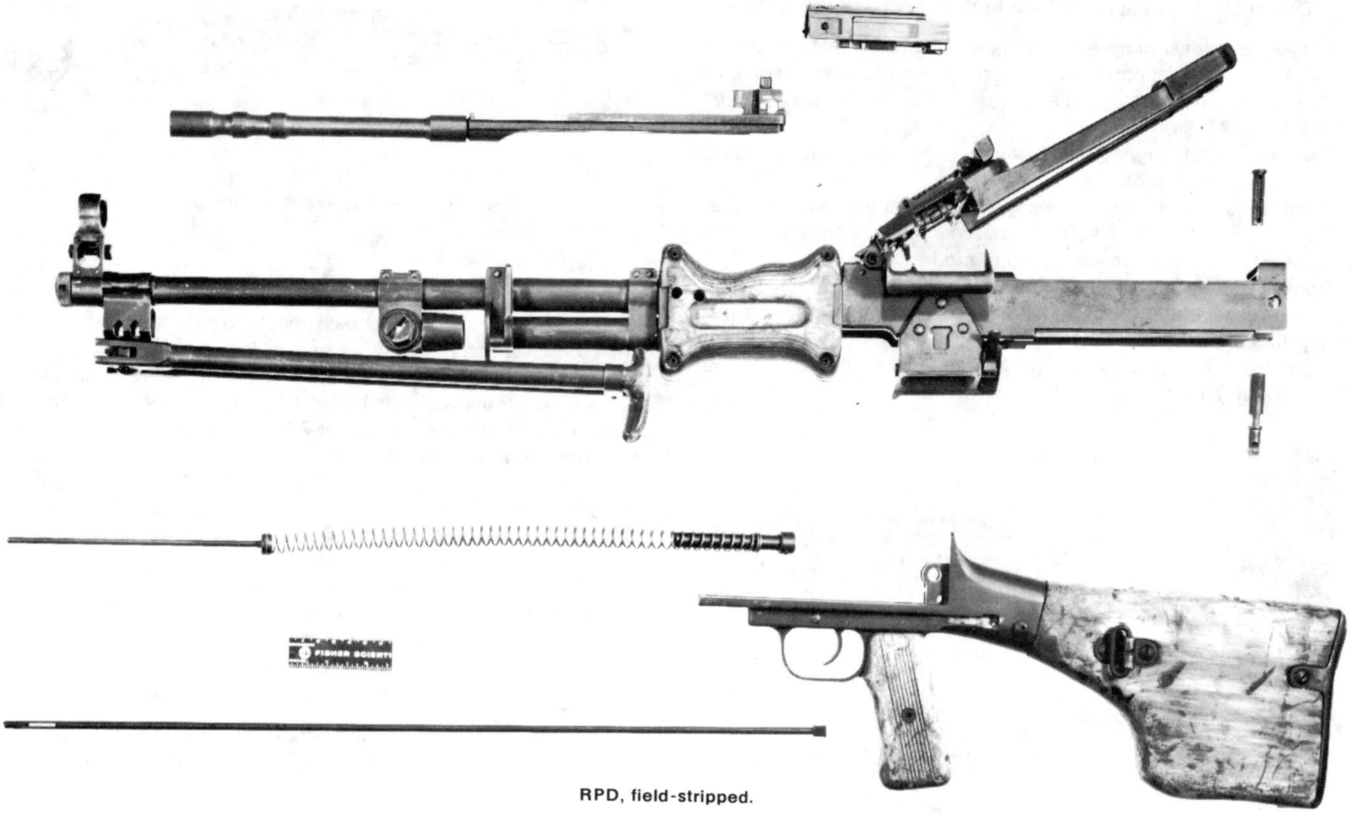

RPD, field-stripped.

immediately above the trigger. When the catch is forward of the trigger the gun is on safe; to put the gun on fire, rotate the catch to the rear position.

To unload the gun, lift the cover by pushing forward on the cover catch located at the rear of the cover. Lift the link belt out and snap the cover shut.

Field Stripping the RPD

Turn the butt trap cover so that it is at right angles to the butt-stock. Place a screwdriver in the top hole in the stock, and turn the recoil spring plug one quarter turn; this will release the plug and the recoil spring and recoil spring guide. Withdraw the recoil spring and its guide from the gun. The entire butt and pistol grip group can then be removed by forcing out the butt retaining pin, which is located in the lower rear section of the receiver. When the retaining pin is pulled out as far as it will go, slide the butt group rearward till it separates from the receiver. Pull the operating handle to the rear until the cutout point on the handle track is reached. Pull out the operating handle. The bolt and the slide and piston assembly can now be withdrawn from the rear of the receiver. The bolt and bolt locks can now be lifted from the slide. Lift the cover; the belt feed lever assembly and the belt feed slide can be removed by pinching together with a pliers the split pin which is located at the right front of the cover. When the end of the pin is compressed, its locking collar can be removed, and all the belt feed components can be removed. Like many other gas-operated weapons, such as the U.S. and FN Browning automatic rifles and the British Bren guns, the gas cylinder of the RPD can be easily adjusted. To adjust the gas cylinder of the RPD to obtain a different size of orifice, a special wrench is used.

To reassemble the weapon, perform the above steps in reverse order.

How the RPD Works

With the belt loaded in the gun and the bolt group to the rear, the trigger is pulled. The trigger pulls the sear down from its engagement in the sear notch on the underside of the slide. The slide piston assembly with the bolt goes forward under the pressure of the decompressing recoil spring. The slide post stud, which operates in the track of the belt feed lever, moves the belt feed lever, which in turn moves the belt feed slide over, indexing the cartridge so that the top of the bolt can engage the cartridge and strip it forward out of the link and downward into the chamber. The locking flaps mounted on the sides of the bolt are cammed out into their locking surfaces in the receiver by the rearward movement of the bolt on the slide, when the bolt abuts the barrel. The outward movement of the locking flaps allows the slide post to strike the rear end of the firing pin, which in turn strikes the primer of the cartridge. Gas from the cartridge enters the gas port in the barrel and forces the piston and slide assembly to the rear. The bolt is also drawn to the rear, and its locking flaps are withdrawn into their unlocked position when the slide post has moved a slight distance to the rear. As the bolt goes to the rear it carries the empty cartridge case in its face until the ejector strikes the upper part of the rim and knocks the case downward and out of the gun. If the trigger is held to the rear, this process will repeat itself until the 100-round belt is expended. The links are forced out the right side of the receiver by the left-to-right movement of the belt feed slide; when the first fifty rounds are fired, the first fifty-round link belt section falls out of the gun.

Special Note on the RPD

Although there is nothing new in the design of the RPD, the weapon is rather remarkable for its simplicity. There are indications that the weapon/cartridge combination has only marginal operating power.

Distinguishing Features of the Four Versions of the RPD

There are at least four versions of the RPD light machine gun in existence. Their distinguishing features are as follows:

First Model: No sight guards for rear sight, a male gas cylinder and a female piston head.

Second Model: Sight guard for rear sight, male piston head, and female gas cylinder.

Third Model: Non-reciprocating type operating handle, dust covers on feed and link ejection ports, rear sight guard and piston and gas cylinder the same as second model.

Fourth Model: Gas cylinder enlarged to cover end of gas piston in cocked as well as in battery (forward) position; a bearing has been added to the rear right of the bolt—apparently to prevent the bolt from rebounding. All other particulars are the same as the third model. This model is called the RPDM.

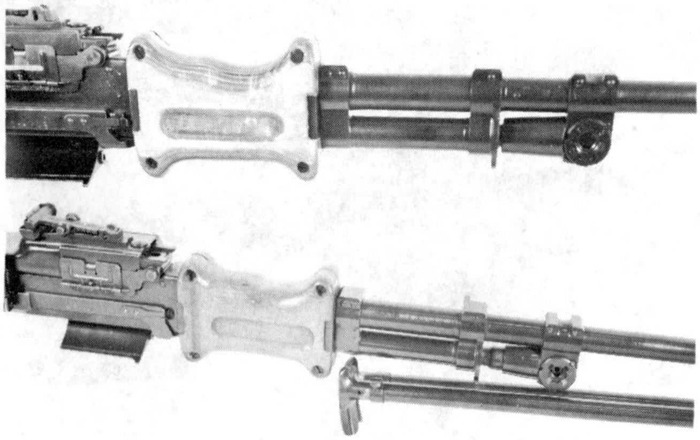

Top: RPDM showing lengthened gas cylinder.

Bottom: Third model RPD.

THE 7.62mm RPK LIGHT MACHINE GUN

The RPK is basically the same as the AKM assault rifle, with longer barrel and a bipod. It uses a 75-round drum magazine, a 40-round box magazine, or the 30-round magazine of the AK and AKM. This weapon is replacing the RPD as the squad automatic weapon (base of fire) of the Soviet Army.

Adoption of the RPK by the Soviets eases their logistical and training problems, since the RPK used for the most part the same parts as the AKM and is operated in the same manner.

RPK does not have a quick-change barrel and as squad automatic weapon it is not designed for long periods of sustained fire.

THE 7.62mm RPKS LIGHT MACHINE GUN

The RPKS is a folding stock version of the RPK light machine gun. The stock folds to the left side of the receiver and although the weapon can be fired with box magazines with the stock folded it is doubtful that the 75-round drum magazine can be used in this model. The RPKS is 40.9 inches long with the stock fixed and approximately 32 inches long with stock folded.

THE 7.62mm LIGHT MACHINE GUN MODEL 1946 (RP46)

The RP46 is basically a modification of the 7.62mm DPM light machine gun, which appeared during World War II. The DPM itself was a modification of the DP. While both the DP and DPM were fed by pan-type magazines, the RP46 can be fed either by a pan

7.62mm RPK Light Machine Gun with drum magazine.

or by the same 250-round nondisintegrating link belt that is used with the 7.62mm Goryunov heavy machine gun. As its name implies, the RP46 is a company-level weapon. It is therefore much heavier than the RPD, and is capable of a higher level of sustained fire since it has a quick-change barrel and the RPD does not. The RP46 is chambered for the old Soviet 7.62mm rimmed cartridge, which is quite similar in performance to the U.S. cal. .30 rifle and machine gun cartridge (the .30-06). It therefore has a much higher muzzle velocity and range than does the RPD, and it can be used for engagement of targets far beyond the range of the Soviet squad-level weapons. The RP46 is the tactical equivalent of the U.S. M1919A4 or A6 machine guns. or to the new U.S. M60 machine gun when it is used as a light gun on a bipod. The RP46 is apparently always used on a bipod.

How to Load and Fire the RP46

Cock gun by pulling operating handle to the rear. Pull rear sight base to the rear and lift cover. Lay belt, open side up, on the feed plate so that the leading round contacts the cartridge stop and close cover. Pull trigger and gun will fire. To unload the gun, pull the rear sight guard to the rear, lift cover, remove belt. If the gun has been fired, there will still be one round in the feed way—remove round. The safety is on when the lever is pointing toward the muzzle.

Field Stripping the RP46

Field stripping of the RP46 is similar to that of the RPD, except

7.62mm Company Light Machine Gun (RP-46).

7.62mm RP-46, field-stripped.

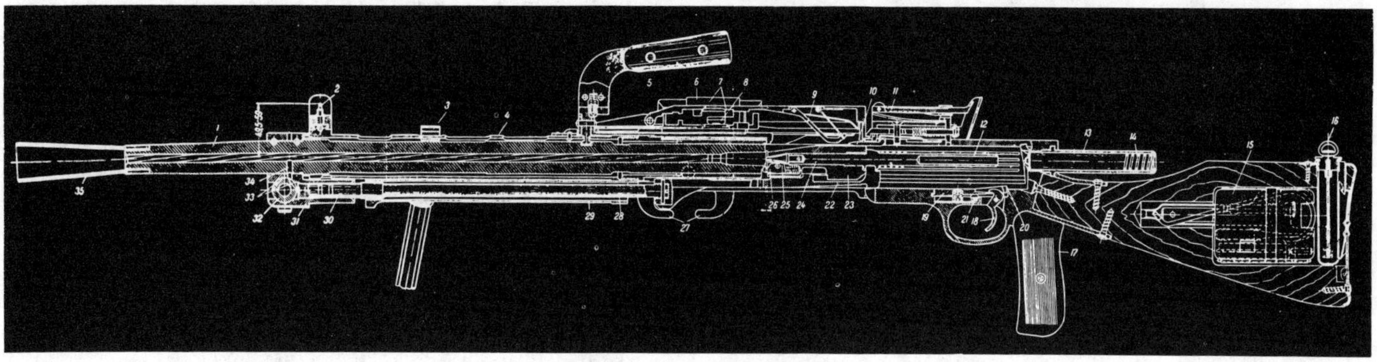

Section drawing of RP-46.

that the recoil spring is mounted in a tube which projects out from the right rear side of the receiver. The tube and spring are dismounted by pushing in on the tube lock and turning it to the right, then withdrawing the tube and spring from the receiver. The barrel can be removed by pushing in the latch on the left forward side of the receiver and pulling the barrel out.

How the RP46 Works

The RP46 has the same basic operating system as the other Degtyarev guns (DP, DPM, and RPD) and operates the same as they do except for its belt feed mechanism. An ingenious system is used on the RP46 to translate the forward and backward movement of the operating handle into a side-to-side movement of the belt-feed lever. The operating handle on its forward and backward travel operates a double-arm type of feed lever which transmits this movement to parts in the feed mechanism; these parts in turn transmit the movement to the belt-feed slide. This system is also used on the Soviet 12.7mm DShK M1938/46 machine gun. Insofar as the bolt parts of the Degtyarev weapons are concerned, there is only one major difference among the lot. In DP, DPM, and RP46 the firing pin is mounted in the bolt, and the slide post itself serves as a hammer.

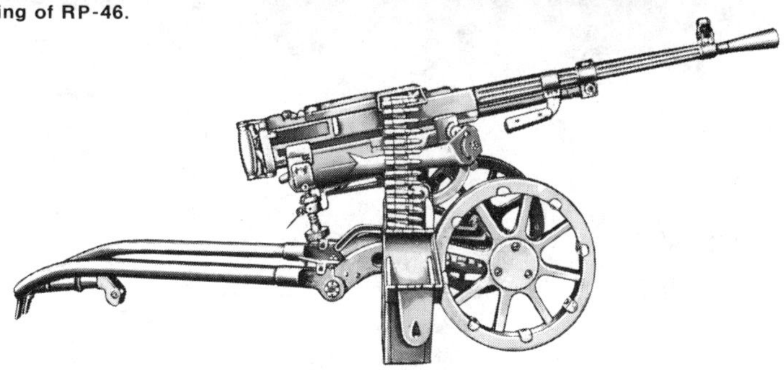

Modernized Goryunov (SGM) on wheeled mount.

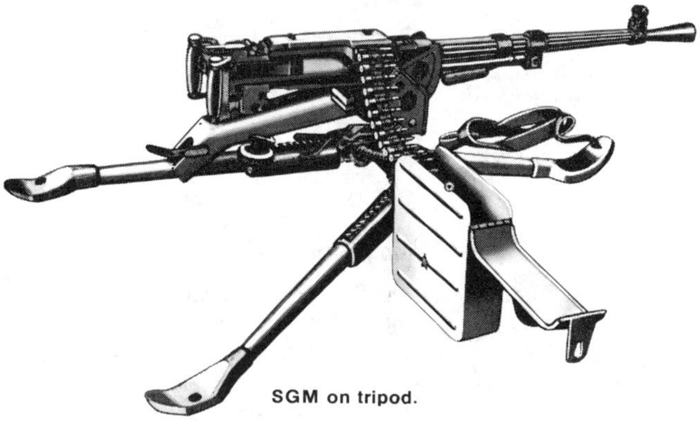

SGM on tripod.

THE 7.62mm SG43 and SGM HEAVY MACHINE GUNS

The Goryunov SG43 heavy machine gun appeared during World War II and has replaced the Maxim M1910 water-cooled gun in the Soviet Army. The Goryunov has itself been modified since World War II, and its modification is called the SGM or Heavy Modernized Goryunov. Both the older and the new Goryunov are apparently in service in the U.S.S.R. The Goryunov is used as a battalion-level gun and is also used on armored personnel carriers. The original Goryunov had a wheeled mount which could be fitted with a shield. The newer gun appeared with two mounts: a wheeled mount somewhat similar to the Sokolov mount used with the Maxim, and a light tripod mount. Both the old and new guns, however, can be used on any of the three mounts. As a matter of passing interest it should be noted that the Soviets use the term "Ruchnoi," which means hand, for light machine guns (i.e., a machine gun which is carried in the hand). They use the term "Stankovy" for heavy rifle-caliber machine guns; "Stankovy" means mounted, i.e., a gun on a mount.

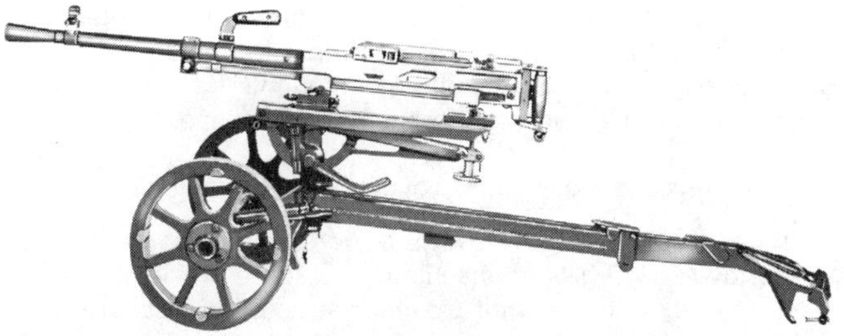

Soviet 7.62mm Goryunov Heavy Machine Gun M1943 (SG43).

How to Load and Fire the Goryunov

To load the Goryunov, open the top cover by pushing forward the cover latch. Insert the link belt on the feedway, placing the rim of the cartridge in the jaws of the feed carrier. Close the cover and pull the operating handle to the rear as far as it will go. The operating handle of the SG43 is at the rear underside of the gun, under the spade grips. After the operating handle has been retracted, it should be pushed forward again; the bolt remains to the rear since the Goryunov fires from an open bolt. Raise the safety lock with the left thumb and press the upper end of the trigger with the right thumb. The weapon will now fire, and will continue to fire until the trigger is released. The weapon can be unloaded by raising the top cover and lifting out the link belt. NOTE: On the SGM, the operating handle projects from the right underside of the gun.

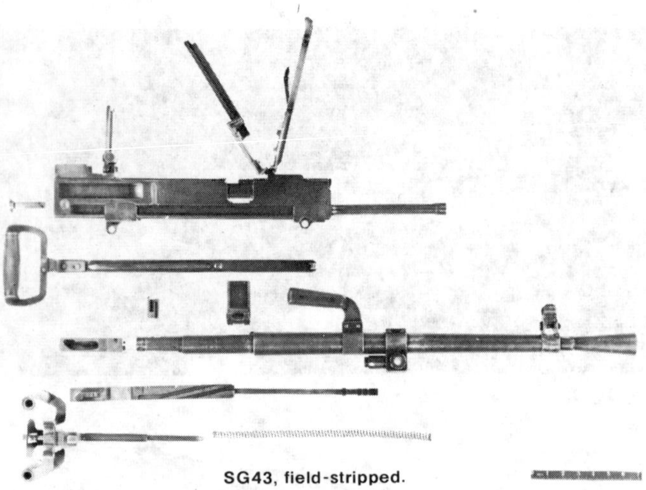

SG43, field-stripped.

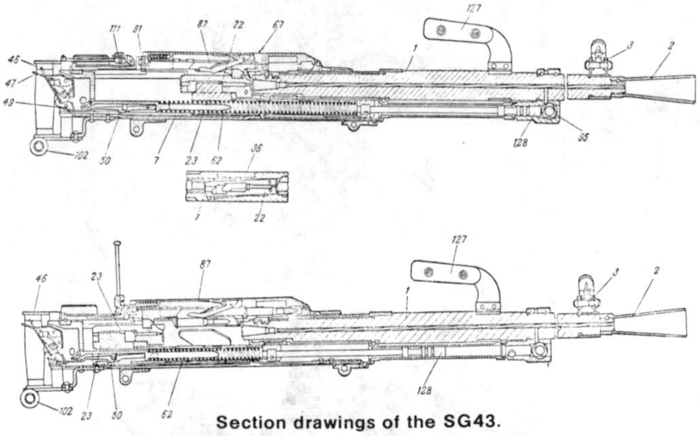

Section drawings of the SG43.

Field Stripping the Goryunov

Open the top cover by pushing the cover catch forward; then lift the feed cover slightly and remove the feed carrier. Move the cover and feed cover to the vertical position. On the SGM, lift the rear sight leaf and with a punch, push out the backplate catch pin and move the catch backward: turn the backplate one quarter-turn to the right, and remove it from the receiver; remove the recoil spring and recoil spring guide. Remove the trigger mechanism from the receiver. Move the slide rearward with the operating handle until the bolt emerges from the receiver; grasp the bolt and slide and remove them from the receiver. Lift the bolt up and out of the slide. Move the operating handle rearward until it rests against the rear clamp of the receiver, and, turning it upward, separate it from the receiver. Move the plate covering the lower opening of the receiver to the rear and remove it from the receiver. Move the belt feed slide to the right and withdraw it from the receiver. To remove the barrel, press on the barrel lock with the thumb, moving the lock as far to the left as it will go; then, grasping the barrel handle, pull the barrel forward and remove it from the receiver.

No further disassembly is recommended. To reassemble the weapon, reverse the above procedure.

The field stripping of the SG43 differs in that the backplate is removed by pulling out its retaining pin, which is located at the bottom rear of the receiver.

How the Goryunov Works

In loading, the belt is positioned in the feedway with the cover raised, so that the spring-loaded holders of the feed carrier engage the rim of the first round. The cover is closed and the operating parts retracted by means of the handle provided. The gun is then ready for firing. The feed carrier, which reciprocates in the feed tray (feed cover), operates in a notch in the top of the bolt. As the bolt is retracted, the feed carrier withdraws the cartridge from the belt. A belt holding pawl is located in the cover. The belt feed slide, reciprocating in a cut in the receiver, operates on a cam of the operating slide.

The gun fires from the open bolt position. The bolt is held to the rear by the sear, which is a part of the backplate assembly. The trigger operates through a connector to disengage the sear from the bolt. When the bolt is released, it is driven forward under the energy of the compressed driving spring which operates against the operating slide. As the round is pushed forward by the bolt, it is forced downward, out of the carrier, and into the feeding tray. From the feeding tray, the round enters the chamber. As the bolt approaches its forward position, a cam on the operating slide forces the rear of the bolt 3/16 of an inch to the right and into a recess in the receiver. The head of the bolt is recessed at an angle with the center of the bolt to give normal support to the base of the round in firing. As the round is chambered, the spring-loaded extractor, which is located in the left side of the bolt and is pivoted on a pin on the right side of its center, is forced over the rim of the round. After the bolt has been forced into the locked position, the slide moves forward to contact the firing pin. The firing pin strikes the primer, causing ignition of the round.

As the bullet is forced from the barrel by the expanding powder gas, some gas passes through the gas port in the barrel and impinges upon the piston attached to the slide, forcing it to the rear. The slide assembly moves independently of the bolt for a fraction of an inch. At this point, the bolt is forced out of engagement in the receiver and is drawn to the rear.

The extractor withdraws the fired case from the chamber. As the operating parts approach the rearward position, the ejector, a pin placed at an angle through the right side of the bolt to contact the base of the cartridge case, contacts the locking recess of the receiver. The ejector and the cartridge case are forced forward. The case, pivoted on the extractor, is ejected through the ejection port in the left side of the receiver.

If the trigger continues to be depressed, the firing cycle will be repeated. However, should the trigger be released, the sear will engage the bolt and hold it to the rear.

Notes on the Goryunov Machine Guns

The idea of a bolt whose rear end locked into the side of the receiver by being turned out of the line of axis of the bore was patented by John Browning many years ago. Browning never did very much with this idea in machine guns, but the Russians did. Goryunov, the Russian developer of this gun, died before his gun was put into service in the Soviet Army.

Since World War II a modified gun, the SGM, has been adopted by the Soviets. The SGM has a longitudinally fluted barrel and the barrel lock has been changed to incorporate a provision for headspace adjustment. The barrel lock has been changed by adding a scale, a slide, and an Allen-screw type of lock and by having multiple grooves and ridges on the barrel lock and barrel. Thus, as the barrel and/or receiver wears, it is possible to unlock the slide, tighten it in the desired position (which can be determined from the scale), and relock it. This, of course, is not done with every barrel change, but only after the components have had enough wear to make a significant difference in headspace.

Variations and Modifications of the Goryunov Machine Guns

As noted above, the SGM is a modernized version of the Goryunov M1943 ground machine gun. There are in Soviet service, and most of the other Communist countries, several variations of the SGM; all have the same basic operating mechanism, are chambered for the 7.62mm rimmed cartridge and have the

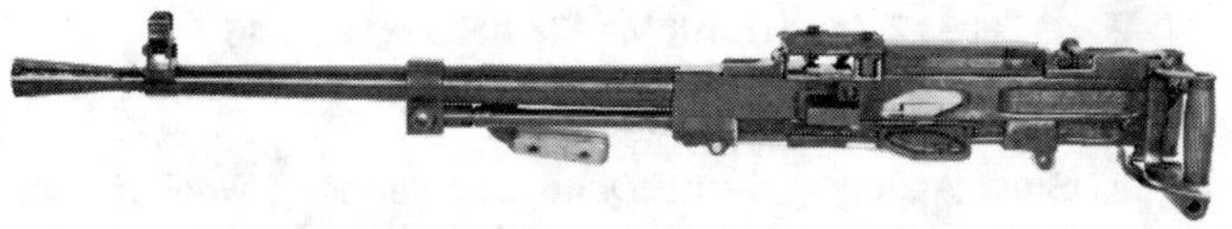

7.62mm SG 43M Machine Gun, dust covers open.

same cyclic rate and muzzle velocity as the SGM. They are as follows:

SGMT: This is the tank version of the SGM. It is fired by a solenoid mounted on the back plate. It is not used on a tripod or wheeled mount as a ground gun.

SGMB: This gun is usually found mounted on armored personnel carriers, but can also be found mounted on ground mounts. It has dust covers over the feedway ports, feed slide and ejection port. Unlike SGM, which has mounting lugs which have holes to accept mounting pins, the SGMB has a semi-circular flange to engage the mount cradle. Spring-loaded clamps hold the gun in position. All versions of the SGM have a separate sear housing.

The original Goryunov (SG 43) has been modified by the addition of dust covers in a manner similar to that of SGMB. This version is called the SG 43M.

THE 7.62mm PK/PKS MACHINE GUN

The Soviets have adopted a general purpose machine gun which will probably eventually replace the 7.62mm RP-46 Company machine gun and the 7.62mm SGM battalion-level machine gun. When used on a bipod, this weapon is called the PK; when used on a tripod, it is called PKS. The PK stands for "Pulomet Kalashnikov"—machinegun Kalashnikov; the S, as in the other Soviet machine guns, stands for "Stankovy"—mounted. With the adoption of the PK, former Soviet Army Master Sergeant Mikhail Kalashnikov, a rather handsome, middle-aged, World War II veteran, now has a near monopoly of small arms designed by him in service at battalion and lower levels in the Soviet and many satellite armies. The PK is a clever combination of the basic operating principles of the AK with some apparently original design on the feed mechanism. Whatever else may be said about the Soviets, it cannot be said that they do not recognize a good mechanical idea when they see one; the PK proves that Mikhail Kalashnikov is no exception. The operating system of the PK is basically that of the AK turned upside down, as was done circa 1955-56 by FN with the mechanism of the Browning Automatic rifle to produce the "MAG". Gene Stoner did the same thing somewhat later to produce the machine gun versions in his Stoner 63 system.

How To Load And Fire The PK/PKS

Open cover by pressing in catch at top rear of cover and lifting cover. Lay cartridge belt in feedway so that first cartridge in belt is flush against the cartridge stop. Close cover and pull operating

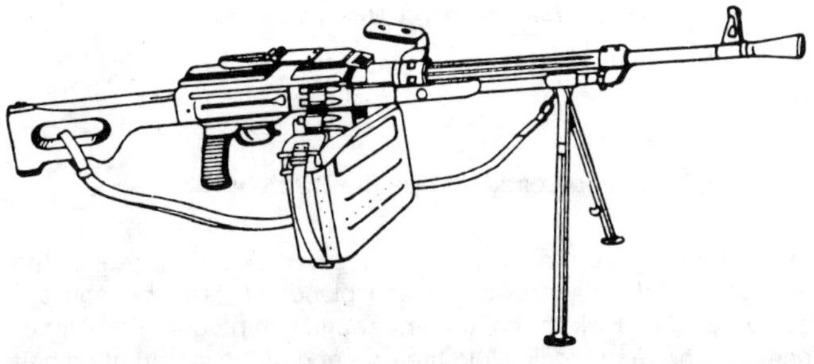

7.62mm PK Machine Gun; although the drawing shows a flash hider, this weapon uses a prong-type flash suppressor.

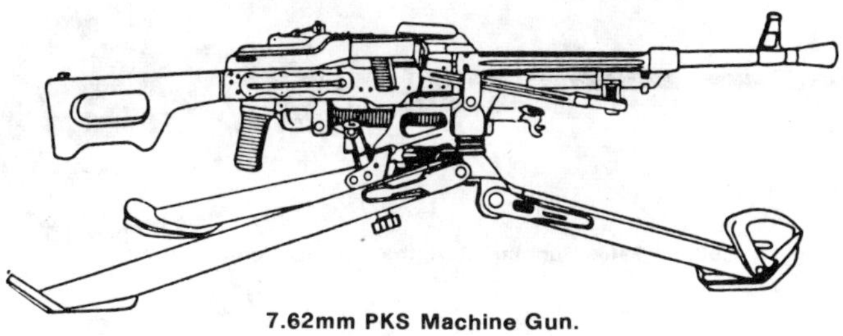

7.62mm PKS Machine Gun.

handle on right side of the receiver all the way to the rear. The weapon is now cocked and pressure on the trigger will produce fire. The safety is located on the receiver at the rear of the trigger. The barrel is removed by sliding out the barrel lock and pulling forward on the carrying handle; it is changed after 500 rounds of sustained fire.

How the PK/PKS Works

The bolt of the PK is similar to that of the AK; it has forward locking lugs and is cammed into and out of its locked position in the barrel by a raised cam lug on the slide. The body of the bolt is mounted in a tunneled-out portion of the slide post at the rear of the slide. The slide is in turn attached to the piston; for ease of disassembly, the slide-piston assembly is articulated. The firing pin is also mounted in the slide post; it rides in a hole in the bolt. The operating spring and operating spring guide, which are seated in the rear of the receiver, go into a hole in the rear of the slide. The underside of the slide has a cut out section to engage the sear nose and hold the bolt-slide-piston assembly in cocked position. With the bolt in the cocked position, the following occurs when the trigger is pulled: the nose of the sear is lowered and the bolt-slide-piston assembly is forced forward by the compressed operating spring and rams a cartridge into the chamber. As the bolt goes forward, it is cammed in a circular motion into the locked position in the barrel by the action of a lug on the bolt body on the cam-way in the raised cam lug section of the slide. After the bolt is locked, the slide-piston continues forward a slight distance causing the firing pin, which is mounted on the slide, to continue through the bolt and strike the cartridge primer functioning the cartridge. At a point about two-thirds down the length of the barrel, gas is tapped off through a gas port and goes through the adjustable gas regulator into the gas cylinder where it impinges upon the gas piston forcing it to the rear. If the trigger is still held to the rear and there are cartridges in the belt, the gun will continue to fire.

Feed Mechanism: Although there are certain features of this feed mechanism which resemble the RP-46 and the Goryunov, for the most part the feed system appears to be original. The feed plate on this weapon is directly over the chamber as in the RP-46 and the cartridge must be withdrawn to the rear, then directed downward in a position to be rammed into the chamber by the bolt. The extractor which pulls the cartridge from the link belt, is mounted on the hook-like piece which protrudes forward from the slide post. This piece, the slide post, and the bottom rear section of the piston resemble a square configured letter "C". When the bolt operating handle is pulled to the rear, the extractor pulls a cartridge from the belt and to the rear. In a fashion similar to the RP-46, it is forced downward by a spring-loaded lever which is mounted on the top cover, and presses the cartridge downward. In addition, the base of the cartridge runs into a cam track—called a cartridge stop—which is curved rearward and downward and causes the cartridge to travel in that position and end in an angular position with the bullet pointing toward the chamber; this piece also keeps the cartridge from travelling all the way to the rear with the slide. Like the SG-43, the PK has two covers—a top cover and a feed cover under

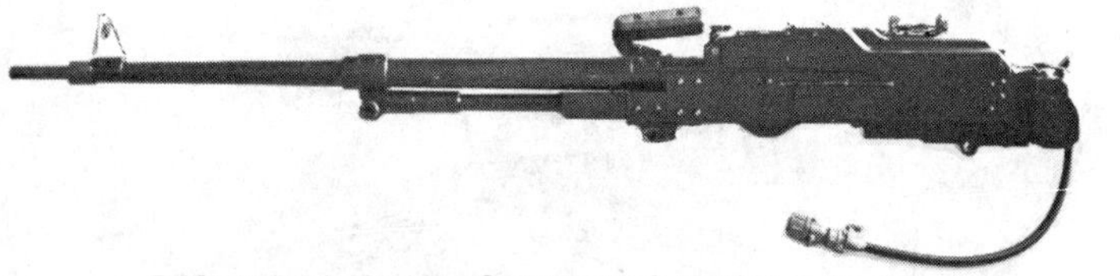
7.62mm Tank Machine Gun; the tank version of the PK.

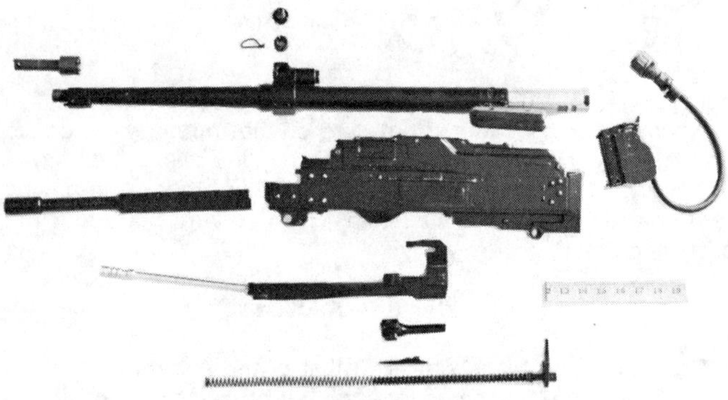
7.62mm PKT Machine Gun stripped.

that. The cartridge is held in the feed lips of this cover until it is engaged by the top of the bolt and rammed into the chamber. This system resembles that of a box magazine if the feed lever pushing down on the cartridge is thought of as a follower and the feed lips of the feed cover are thought of as the feed lips of a magazine. The means of moving the belt across the feedway appears to be original. A bow-shaped feed lever assembly is mounted in the receiver so that it is perpendicular to the receiver in the vertical plane and crosswise to the receiver in the horizontal plane. This feed lever assembly goes completely under the slide-piston-bolt assembly with its left-end engaging the left side of the slide. A roller on this end of the feed lever assembly engages a cammed surface on the slide causing the assembly to move up and down—left to right at the bottom and right to left at the top—in a rocking fashion, so that the left arm of the feed lever is under the slide when the top section—that which operates as a belt feed pawl—is pulling a cartridge on to the feed plate. A short arm of the feed lever—called the feed stud—engages a lip on the right side of the slide and functions as a pivot point for the feed lever. As mentioned the top portion of the feed lever assembly rises and pushes the cartridge belt from right to left. A spring loaded belt holding pawl, mounted in the cover, then engages the belt and holds it in position. Both the feed and link ejection ports have dust covers. The dust cover on the link ejection port is opened when the bolt is pulled to the rear.

Note on the PK/PKS: The tripod of the PKS can be easily adapted for A.A. fire and a 100-round ammunition box can be attached to the side of the receiver allowing for easy movement. Unlike all earlier Soviet machine guns which require a tool, hammer or screwdriver to adjust the gas regulator, the gas regulator of the PK can be adjusted with the rim of an empty cartridge case. PK is, all in all, an impressive weapon. A tank version of this gun, the PKT, is used on coaxial mountings on a number of Soviet Armored vehicles.

SOVIET 7.62mm PKB MACHINE GUN

The PKB is a variant on the PK and like that weapon uses the 7.62X54 rimmed cartridge. The principal difference between the PKB and the PK is that the PKB does not have a fluted barrel and weighs approximately 1.5 pounds less than the PK.

THE 12.7mm DShK M1938 AND M1938/46 HEAVY MACHINE GUNS

The DShK is another of the Degtyarev series of weapons; its basic system of operation is similar to the other Degtyarev guns. The original DShK had a rotating block type of feed; the M1938/46 has a belt-feed lever type similar to the RP46's. These guns are chambered for the Soviet 12.7mm cartridge, which is almost identical in performance to (although not interchangeable with) the U.S. cal. .50 cartridge. The M1938 was the primary Soviet heavy ground gun; in Korea it was frequently used by the North Koreans against aircraft. The weapon has been replaced in the Soviet Army in this role by the 14.5mm ZPU series of weapons. It is still used as antiaircraft armament on tanks and armored personnel carriers, and is beginning to be used as co-axial armament on tanks. The ground mount for the DShK weapons is a wheeled mount, which can be converted into a tripod mount for AA fire.

Soviet 12.7mm DShK M1938 Heavy Machine Gun.

DShK 12.7mm M1938/46 Heavy Machine Gun.

How to Load and Fire the DShK M1938

Push forward the feed cover latch located at the top rear of the feed cover. Lift the feed cover and place the feed belt on the revolving feed block so that the first round can be put in the upper recess of the feed block. Hold the free end of the ammunition belt with the right hand and press the feed belt against the revolving

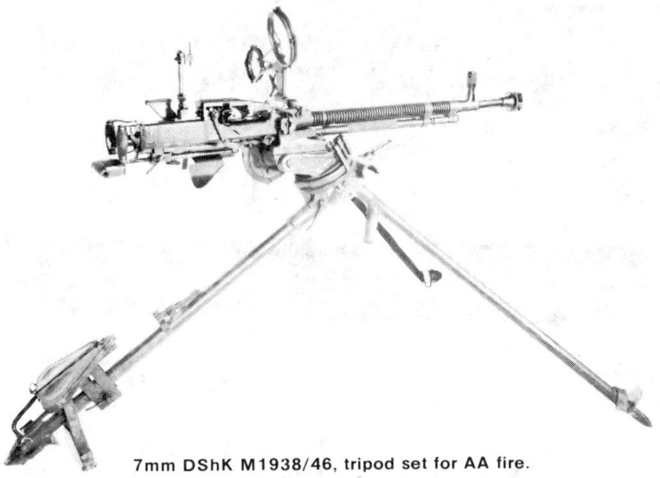

7mm DShK M1938/46, tripod set for AA fire.

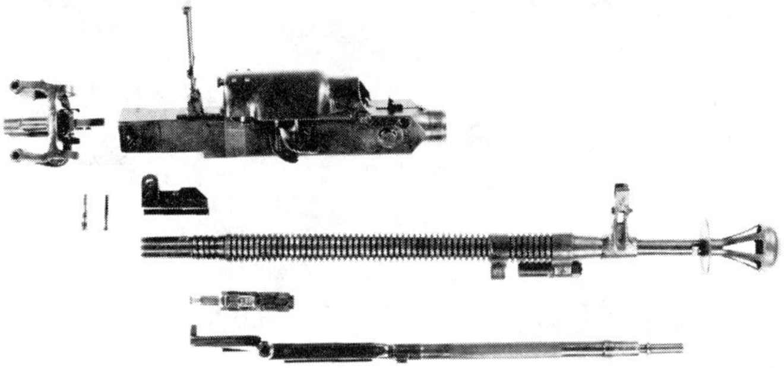

DShK, field-stripped.

block. Rapidly rotate the block with belt as far to the right as it can go (the upper recess should rotate 120°). Close the feed cover. Pull reloading handle to the rear until the slide is engaged by the sear. Hold spade grips with both hands and press the trigger slowly with the index fingers of both hands; the weapon will fire. To put the weapon on safe, rotate the safety catch forward.

How to Load and Fire the DShK M1938/46

Loading and firing of the DShK M1938/46 are the same as for the 7.62mm RP46.

Field Stripping the DShK M1938

Set the safety on fire and loosen (one turn) the latch of the machine gun mounting studs on the gun mount. If the bolt is to the rear, move it forward by pulling the trigger. Stand with back to the muzzle in front of the mount axle with the left foot resting on the axle. Grasp the gas piston tube with both hands and pull it as far forward as it will move. Then turn the tube clockwise with both hands until the support of the tube comes out of its grooves in the barrel. Push all the moving parts (bolt and slide group) to the rear until the recoil stop roller emerges from its recess in the receiver.

Unscrew the connecting screw of the rear machine gun locking bracket. Remove the backplate pin and tap the backplate lightly with a wooden mallet or a copper hammer to separate it from the receiver, meanwhile supporting it with one hand. Remove the trigger housing and then withdraw the bolt and slide group from the rear of the receiver. Remove the firing pin and the bolt and bolt locking flaps. To reassemble the weapon, perform the above steps in reverse order.

How the DShKs Work

In general, both of the 12.7mm Degtyarev guns operate the same as the 7.62mm guns of this series. The M1938 has a circular-type of feed mechanism which operates somewhat like the cylinder of a revolver. The feed lever, which is operated back and forth by the slide, turns the feed drum which carries the cartridges in its recesses. As the cartridges are turned, their links are stripped, and in the final feed step the cartridge is alined with the bolt, which rams it into the chamber. The feed system on the M1938/46 is basically the same as that on the RP46.

THE 14.5mm KPV HEAVY MACHINE GUN

The 14.5mm KPV represents somewhat of a departure in Soviet ground machine gun design. It is the first recoil-operated ground gun used since the 1910 Maxim, and utilizes principles not used in previous Soviet small arms—i.e., its locking system and a quick-change barrel, which is dismounted complete with barrel jacket.

The bolt of the KPV has a two-piece body design. The forward part of the body or bolt head has a "T" slot for the base of the cartridge and has two semicircular sections with clearances top and bottom for the feeding in and ejection out of the cartridge. These semi-circular sections are threaded and engage threads on the outer surface of the rear of the barrel to lock the action. The bolt is turned by a pin which rides in a cam slot cut through the bolt and engages a cam-way in the receiver.

The empty cartridge case is forced out of the cam slot by an ejector which pivots on a pin on the top rear section of the bolt and is cammed down into the "T" slot in the face of the bolt by a cam surface attached to the receiver cover.

The KPV is mounted on a series of towed antiaircraft mounts which are designated as follows: ZPU1, a two-wheeled mount with one gun; the ZPU2, a two-wheeled mount with two guns; and the ZPU4, a four-wheeled mount with four guns. The ZPU2 and ZPU4 are quite common and their use is widespread throughout the Communist world and in some non-alined nations as well. The 14.5mm machine gun is also mounted in double and single mounts on some Soviet armored vehicles.

The 14.5mm cartridge was developed for use with the 14.5mm PTRS and PTRD antitank rifles of World War II. It is comparable to the U.S. caliber .60 cartridge, an experimental round developed for use in United States aircraft guns during World War II. The 14.5mm cartridge exists in armor piercing incendiary, incendiary tracer, and armor-piercing incendiary tracer types.

14.5mm KPV Machine Gun.

Special Note on Soviet Machine Guns

It is of interest to note that all modern Soviet automatic weapons have chromed barrels, and that all current Soviet ground machine guns, with the exception of the RPD, RPK, and the 12.7mm DShK, have quick-change barrels. Soviet machine guns as a group are characterized by ruggedness and simplicity of maintenance. This is not, however, synonymous with simplicity of manufacture. It may be necessary to use very complicated and expensive machine work to manufacture an item which is simple to maintain in the field. The Soviets are apparently moving away from easily made stamped weapons judging by their post-World War II models.

The Soviets have aircraft machine guns in 7.62mm, 12.7mm, 20mm, 23mm, 30mm, and 37mm calibers. The two larger caliber weapons are the only ones used on the latest aircraft. The 7.62mm ShKAS first appeared in service during the Spanish Civil war, at which time it was supplied with Soviet aircraft to the Spanish Loyalist forces. The ShKAS was distinguished by a very high rate of fire for its time (well over 1000 rounds per minute), and a very complicated squirrel-cage type of feed mechanism. The ShKAS is a toolmaker's nightmare, insofar as complexity of manufacture is concerned, and was probably handmade to a large extent. The ShKAS served as the model for the 20mm ShVAKs which saw extensive service in World War II. The 12.7mm UB was also introduced during World War II; it was a much less complicated weapon than the ShKAS and the ShVAK. The 23mm VYa was essentially a scale-up of the UB, and was mounted in the wings of the famed IL-2 Stormovik "tank killer." All the foregoing aircraft guns were gas operated; the later NS, NR, and N series of weapons are recoil operated.

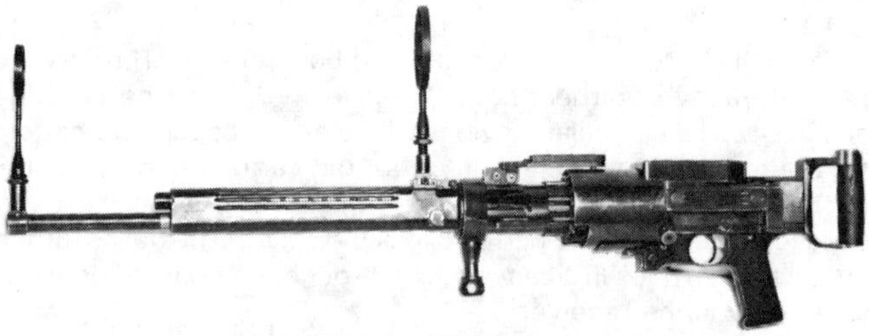

ShKAS 7.62mm Aircraft Machine Gun.

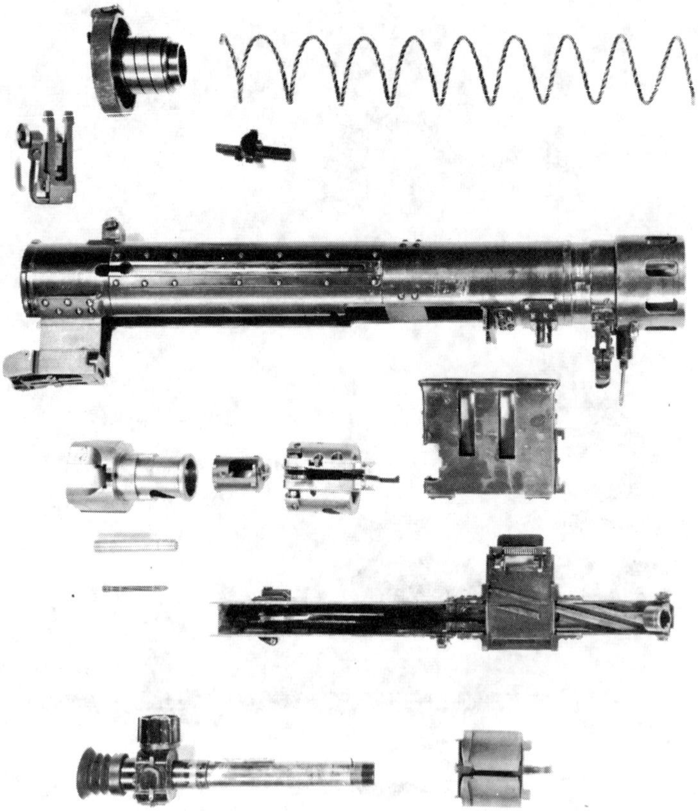

14.5mm KPV Machine Gun, stripped.

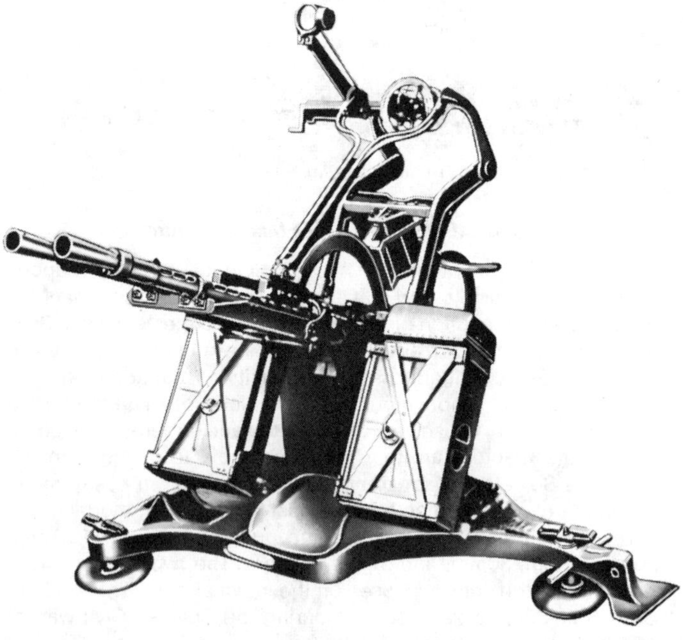

ZPU-2 14.5mm Machine Gun.

ZPU-4 14.5mm Machine Gun.

CHARACTERISTICS OF SOVIET POST-WORLD WAR II MACHINE GUNS

	SGM	RP-46	RPD	RPK	DShK M1938/46	KPV	PK/PKS
Caliber:	7.62mm (7.62 x 53).	7.62mm (7.62 x 53).	7.62mm. (7.62 x 39).	7.62mm. (7.62 x 39).	12.7mm.	14.5mm.	7.62mm rimmed (7.62 x 53)
System of operation:	Gas, automatic.	Gas, automatic.	Gas, automatic.	Gas, automatic.	Gas, automatic.	Recoil, automatic.	Gas, automatic.
Length overall:	44.09 in.	50 in.	40.8 in.	40.9 in.	62.5 in.	78.7 in.	47.2 in.
Barrel length:	28.3 in.	23.8 in.	20.5 in.	23.2 in.	42.1 in.	53.2 in.	25.9 in.
Feed device:	250-round metallic link belt.	250-round metallic link belt.	100-round metallic link belt in drum.	75-round drum and 40-round box magazine.	50-round metallic link belt.	Metallic link belt.	100, 200 or 250-round metallic link belt
Sights: Front:	Blade.	Post w/ears.	Post w/ears.	Post w/ears.	Post w/ears.	Post w/ears.	Protected post.
Rear:	Leaf.	Tangent.	Tangent.	Tangent.	Leaf.	Tangent.*	Tangent w/notch.
Weight, gun:	29.76 lb.	28.7 lb.	15.6 lb.	w/empty drum: 12.3 lb. w/empty box: 11 lb.	78.5 lb.	107.9 lb.	19.8 lb. w/bipod.
Mount:	50.9 lb.**	-----	-----	-----	259 lb.	-----	16.5 lb.
Muzzle velocity:	2870 f.p.s.	2750 f.p.s.	2410 f.p.s.	2410 f.p.s.	2822 f.p.s.	3280 f.p.s.	2700 f.p.s.
Cyclic rate:	600-700 r.p.m.	600-650 r.p.m.	650-750 r.p.m.	600 r.p.m.	540-600 r.p.m.	600 r.p.m.	650-700 r.p.m.

*Weapon is normally used with telescopic or mechanical computing sight.
**A 30.6 lb. tripod mount also exists.

Soviet 14.5mm ZPU-2 AA machine gun on new type carriage. Note that wheels remain on carriage in firing position and the overall weapon has a lower silhouette.

The Egyptian Army uses the following weapons: the 9mm Beretta Model 1951 pistol, the 7.62mm Soviet SKS carbine, the 7.62mm Soviet AK assault rifle, the 7.62mm Czech Model 52 rifle, the 7.92mm Hakim (modified Ljungman rifle), the 7.62mm Rashid rifle, the 7.92mm FN self-loading rifle and 7.92mm Type D automatic rifle. Submachine guns used include the 9mm Port Said (Model 45 Swedish Carl Gustaf), and various models of the 9mm Beretta and 9mm Sten. Machine guns used include the 7.62mm Soviet RPD, the 7.62mm Czech Model 52, the

United Arab Republic (Egypt) 52

7.62mm Soviet SG43 and SGM (Goryunov) machine guns, the 7.92mm Spanish Alfa machine gun, the Soviet 12.7mm DShK Model 1938/46 heavy machine gun, and the Soviet 14.5mm ZPU-2 and ZPU-4 heavy machine gun. British Webley revolvers, caliber .303 No. 1 rifles, Lewis machine guns, Bren machine guns, and Vickers machine guns are probably still used in reserve units.

Generally speaking, Egypt obtained all of her weapons from western nations prior to 1954 and most of the weapons from those nations were obtained before that time. In 1954, for reasons which are beyond the scope of this book, the UAR made the first of a series of extensive arms purchases from the Soviet Bloc. Arms purchases from Western nations since that time have been limited by the embargo on arms shipments to the Middle East observed by most of the Western nations.

Egypt has developed a small-arms industry of its own and is less dependent on imported small arms. This industry, which was established by Swedish technicians prior to the overthrow of the Egyptian monarchy, originally produced copies of foreign weapons. Of late it has started development of native designs, which although they are basically modifications of or combinations of foreign designs, are definitely Egyptian in origin. Cartridges for all UAR service weapons are produced in that country.

EGYPTIAN PISTOLS

Egypt was originally equipped with British caliber .455 No. 1 Mark VI Webley revolvers and Enfield caliber .38 No. 2 revolvers. After World War II, Egypt adopted the 9mm Parabellum as the standard pistol and submachine gun cartridge. A modification of the Tokarev TT M1933 called the Tokagypt Model 58, chambered for the 9mm Parabellum cartridge, was developed in Hungary for Egypt and limited purchases of this weapon were made. The Egyptian authorities were not too pleased with this weapon however, and the usage has been confined to police work. The 9mm Beretta Model 1951—covered in detail in the chapter on Italy—is the standard service pistol.

The Egyptian small arms authorities, in conjunction with Beretta, developed a target version of the Model 1951. This weapon has an adjustable target type rear sight and a ramp mounted front sight, a longer barrel than the standard Model 1951 and target type stocks.

9mm Parabellum target type Beretta, Model 1951; developed for Egypt.

EGYPTIAN RIFLES

The Egyptians used the British caliber .303 rifle No. 1 Mark III and Mark III* rifle from the time of World War I until approximately 1949. At that time, the Royal Egyptian government purchased quantities of the FN self-loading rifle (also known as SAFN) chambered for the 7.92mm cartridge. These rifles can be identified by the Royal Egyptian crest on the receiver.

In this same time frame a small arms manufacturing plant was set up in Egypt with the assistance of Swedish technicians. This plant manufactured the 7.92mm Hakim rifle, a modification of the 6.5mm Ljungman Model 42, and the 9mm Parabellum Port Said submachine gun. The Hakim differs from the Model 42 in having a full length hand guard, tangent type rear sight, enlarged charger guide, modified magazine catch and the shape of the muzzle brake/compensator. A number of training versions of the Hakim rifle are used by Egypt. Beretta made a caliber .22 version of the Hakim and Anschutz made a 4.5mm air rifle version of this rifle for training.

In 1954, the UAR procured significant quantities of the Soviet 7.62mm SKS carbine and the Czech 7.62mm Model 52 rifle. The design of the SKS apparently appealed to the Egyptians insofar as it's shortness, lightness and the permanently attached folding bayonet are concerned. These features of the SKS plus use of the Soviet 7.62mm Model 1943 "intermediate" sized cartridge were incorporated into the design of the Egyptian Rashid rifle.

The Rashid has a modified Ljungman action; the bolt retracting handle on the Rashid is mounted in the right forward section of the action; bolt retraction on the Ljungman is accomplished by pulling the receiver cover forward and to the rear. Very few Rashid rifles apparently were made. The most common rifle in Egyptian service today appears to be the Soviet 7.62mm AK assault rifle.

7.92mm FN self-loading rifle as used by Egypt. Top rifle is sniper version.

7.92mm Hakim rifle.

Caliber .22 training version of Hakim rifle (above); 4.5mm air rifle, training version of Hakim rifle (below).

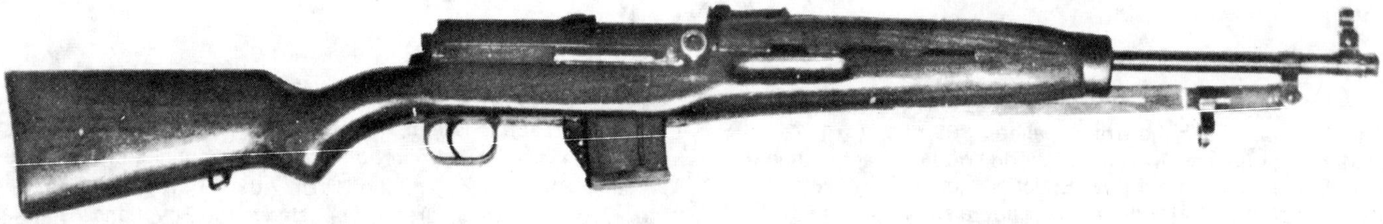

7.62mm Rashid rifle.

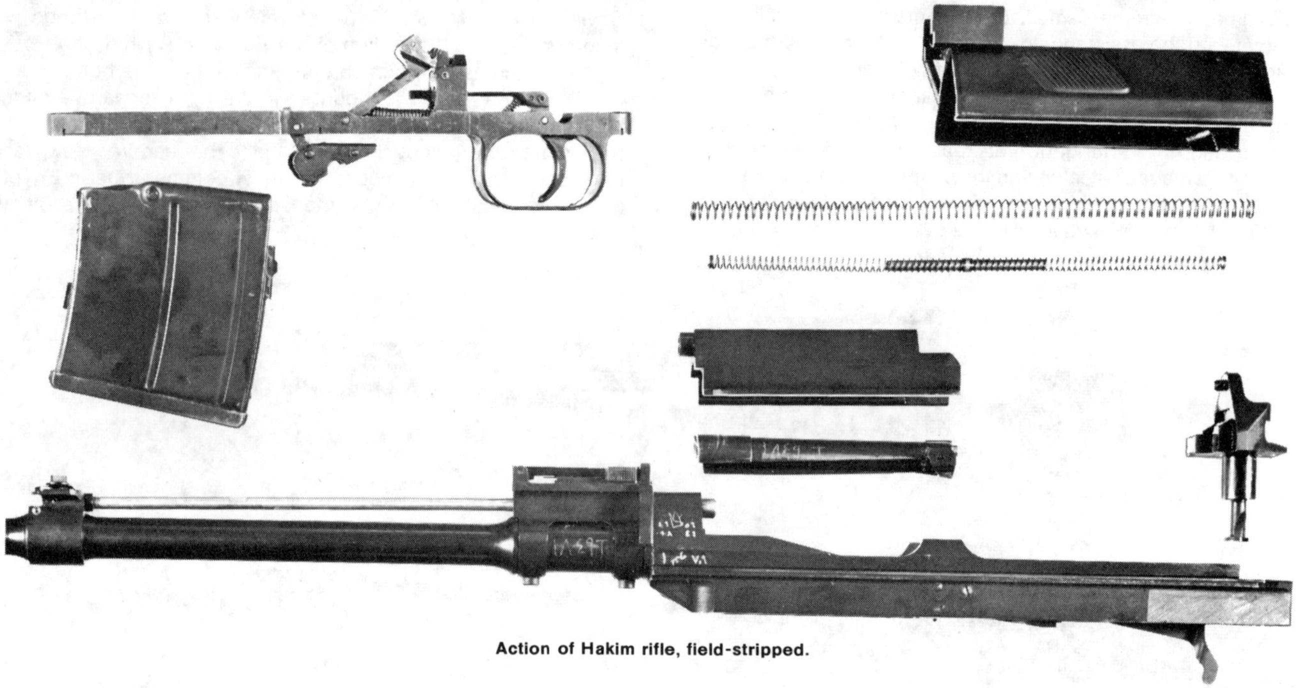

Action of Hakim rifle, field-stripped.

EGYPTIAN SUBMACHINE GUNS

The Egyptians used British 9mm Parabellum Sten guns and also purchased various types of submachine guns in Western Europe to include the 9mm Parabellum Spanish Star Model Z-45, and the 9mm Parabellum Beretta Model 38/42 and 38/49.

As previously noted, Egypt was tooled to produce the 9mm Parabellum Swedish Carl Gustaf (Model 1945) submachine gun, which the Egyptians call the Port Said submachine gun. This weapon is covered in detail under Sweden.

9mm Parabellum Port Said Submachine Gun.

EGYPTIAN MACHINE GUNS

British machine guns formerly predominated in Egypt and prior to the overthrow of the Egyptian monarchy, some Spanish ALFA 7.92mm M1944 machine guns were procured. Soviet machine guns now predominate in the Egyptian forces.

53

United States

The United States forces use the following small arms: caliber .45 Pistol Model 1911A1, caliber .45 submachine gun M3A1, caliber 5.56mm Rifle M16 and M16A1, 7.62mm NATO M14, and M14A1 rifles, the 7.62mm NATO M60 machine gun, the 7.62mm NATO M73 machine gun, the caliber .50 M85 machine gun, the caliber .50 M2 HB Browning Machine Gun, and 40mm grenade launcher M79. Weapons such as the caliber .30 M37 machine gun may be found in use on older armored vehicles and certain handguns are issued on a limited basis as indicated in the text.

The caliber .30 M1 rifle, caliber .30 M2 carbine, caliber .30 Browning Automatic Rifle M1918A2, and the Browning caliber .30 M1919A6 may still be found in use in National Guard and reserve units.

UNITED STATES PISTOLS

The United States adopted the caliber .45 Browning-designed Colt automatic pistol in 1911. All manufacture of this pistol was originally carried on at Colt, but Springfield Armory was tooled to produce the weapon prior to 1914. At the time of United States entry into World War I, 55,553 pistols were on hand. During World War I, Model 1911 pistols were manufactured by Remington Arms and Colt, although eight other firms were given contracts for the weapon. Springfield did not manufacture the pistol, since top priority was given, at that arsenal, to the manufacture of Model 1903 rifles. Approximately 450,000 Model 1911 pistols were made during World War I by Colt and Remington. Colt was by far the largest producer; Remington produced only 13,152 pistols by the close of December 1918.

Caliber .45 Model 1917 Colt Revolver.

Caliber .45 Model 1911 Pistol made by Remington.

Caliber .45 Model 1917 Smith & Wesson Revolver. Issue revolvers did not have checked stocks.

Owing to the shortage of Model 1911 pistols, orders were placed with Colt and Smith & Wesson for a heavy frame revolver chambered for the caliber .45 Model 1911 pistol cartridge. The revolvers chosen were the Colt New Service and the Smith & Wesson Hand Ejector models. Both had been in production, chambered for the .455 Webley cartridge, for the United Kingdom. Modifications were made to accommodate the rimless .45 Model 1911 cartridge. Three round, half-moon clips which fitted in the cartridge case cannelures, were used with these weapons to allow case ejection by use of the ejector rod. They may be loaded and fired without these clips, but the individual cases have to be ejected one by one with a nail or pencil if half-moon clips are not used. These revolvers are conventional swing-out cylinder types; the cylinder latch is pushed forward on the Smith & Wesson to release the cylinder and is pulled back on the Colt. Colt manufactured 151,700 Model 1917 revolvers and Smith & Wesson manufactured 153,311 of their Model 1917 revolver.

The Colt and Smith & Wesson revolvers were still in use by Military Police and security personnel during World War II, but are no longer used in the United States armed forces, or as a standard weapon, in any army.

After World War I, the Colt automatic was modified. The modifications, adopted on 15 June, 1926, caused a change in nomenclature to: Pistol U.S. Caliber .45 Model 1911A1. The changes were as follows:

(1) The main spring housing of the 1911 is flat and smooth; that of the 1911A1 is arched beyond the line of the grip portion of the receiver and is knurled.

(2) The trigger of the 1911 is smooth and is longer than the serrated trigger of the 1911A1.

(3) The tang of the Model 1911A1 is longer than that of the Model 1911.

(4) The 1911A1 front sight is wider than that of the 1911.

(5) Finger clearance cuts were made on the receiver of the 1911A1 immediately behind the trigger; these are not present on the 1911.

(6) Rifling and diameter was reduced and land height was increased.

The Model 1911A1 pistol is covered in detail later in this chapter.

The United States forces have and do use various commercial pistols for specialized purposes, such as air-crew armament, issue to general officers, and issue to security personnel. Among the weapons which have been or are issued for such purposes are the Colt caliber .32 and .380 automatic pistol, the caliber .38 Colt Detective Special Revolver, the caliber .38 Colt Police Positive Revolver, the caliber .38 Colt Special Official Police, and the caliber .38 Smith & Wesson Military and Police Revolver. The above revolvers are all chambered for the .38 special cartridge; Smith & Wesson Military and Police revolvers chambered for the .38 S&W cartridge have also been used. Various commercial target-type caliber .22 pistols and revolvers are also used for training purposes.

THE CALIBER .45 MODEL 1911A1 AUTOMATIC PISTOL

This has been the standard United States military pistol since 1926. During World War II, the Model 1911A1 was manufactured by Colt, Remington Rand, Union Switch and Signal, and the Ithaca Gun Co. Approximately 1,800,000 pistols were made during World War II in addition to the 150,000 or so that were purchased by the United States prior to World War II. Added to the total of caliber .45 Model 1911 pistols made prior to and during World War II, over 2,400,000 of the .45 Colt Automatic have been made for the United States Government. In addition, hundreds of thousands have been made for commercial sale and export to foreign forces.

Caliber .45 Model 1911A1 Automatic Pistol.

While no stripping beyond that illustrated is ever necessary to clean and properly care for this pistol, the following instructions will be helpful to those who wish to master every detail.

To Remove Safety Lock. (1) Cock hammer. (2) Grasp thumbpiece of safety lock between thumb and index finger, pull steadily outward and at same time move back and forth.

To Remove Hammer. (1) Lower hammer—do not snap it. (2) Use safety lock to push out hammer pin, removing from left side. (3) Lift out hammer and hammer strut.

To Remove Mainspring Housing. (1) Using hammer strut, push mainspring housing pin out from right side of receiver. (2) Slide housing and its contained spring down out of its guides. (3) Push in on mainspring cap, at the same time push out mainspring cap pin.

To Remove Sear and Disconnector. Using hammer strut, push out sear pin from left side of receiver and remove sear and disconnector.

To Remove Magazine Catch. Press in checkered left end to permit turning catch lock a quarter turn to left out of its seat in receiver, using long leaf of sear spring. Catch, its lock and spring may now be removed. Be careful not to let spring jump away when released.

To Remove Trigger. Pull straight to the rear.

To Remove Slide Stop Plunger, Safety Lock Plunger, and Plunger Spring. Draw straight to rear.

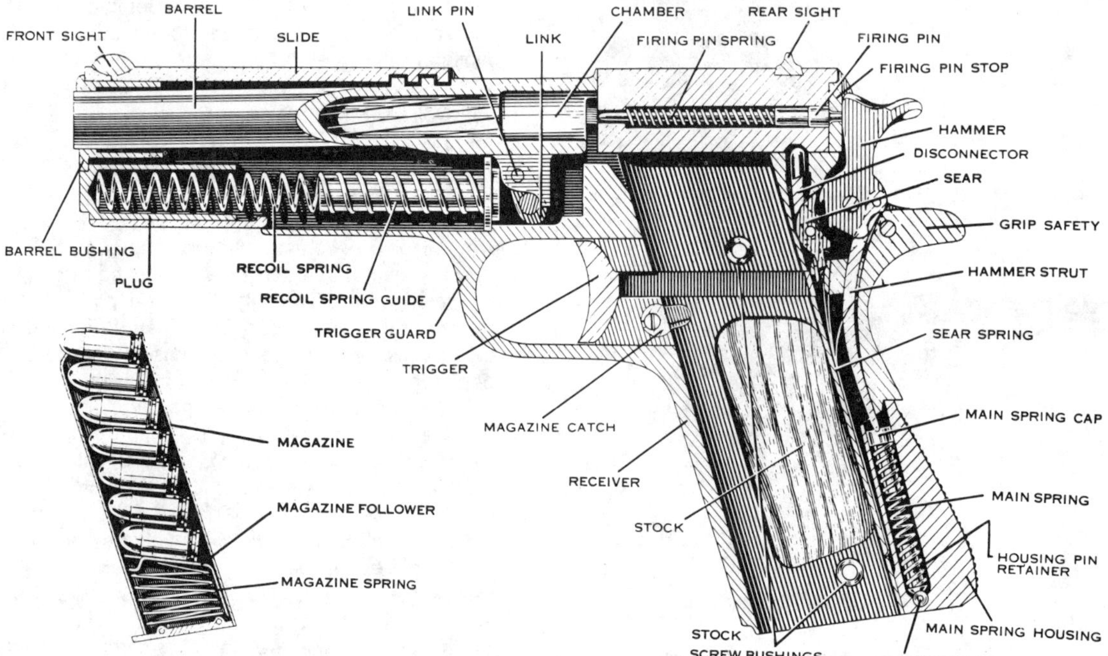

Section drawing, M1911A1.

Instructions for Loading and Firing the M1911A1 Pistol

To remove magazine: Press magazine catch (button). Magazine will normally be ejected and should be caught with left hand. If spring is weak, it may come only part way; withdraw it from handle.

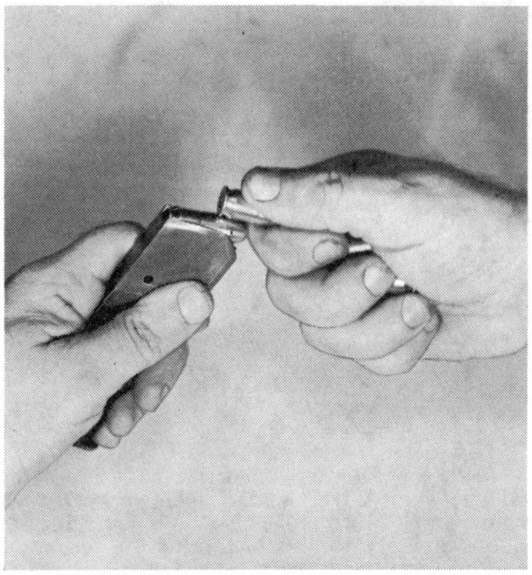

Load magazine: Holding firmly in left hand, press cartridge down in forward end of magazine follower (platform) and slide in under the curved lips of the magazine. Press following cartridges down as illustrated. Any number from 1 to 7 may be inserted.

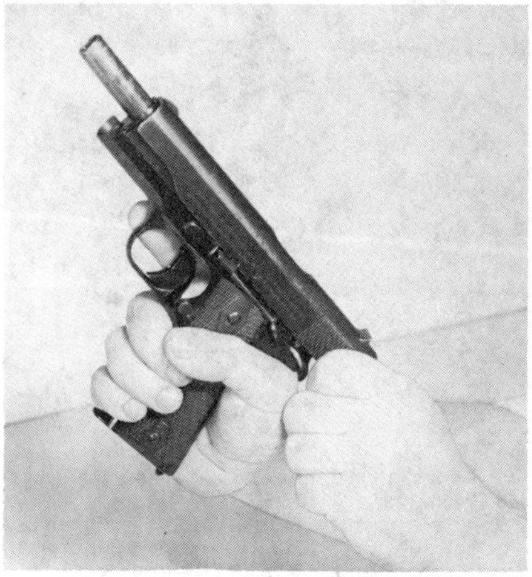

To load chamber: (1) Holding pistol at height of right shoulder and pistol 6 inches from shoulder, insert loaded magazine and press home until it locks with a click. (2) Grasp slide with thumb and fingers of the left hand, thumb on right side of slide pointing upwards and pull back slide as far as it will go. This compresses the recoil spring, cocks the hammer and permits the magazine spring to push the top cartridge into line with the breech block. (3) Release slide. The recoil spring will drive it forward and feed a cartridge into the chamber; barrel will be forced up on its link and will lock into slide; firing mechanism will engage ready for first shot.

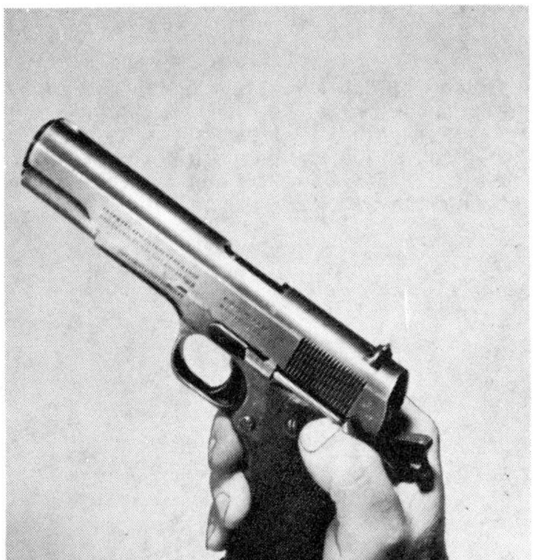

To engage thumb safety: Unless pistol is to be fired at once, always push safety lock up into place as soon as chamber is loaded. A stud on the inner face of the thumb safety locks the hammer and sear when the safety is pushed up into the slide. It can be released by simply pushing down on the thumbpiece.

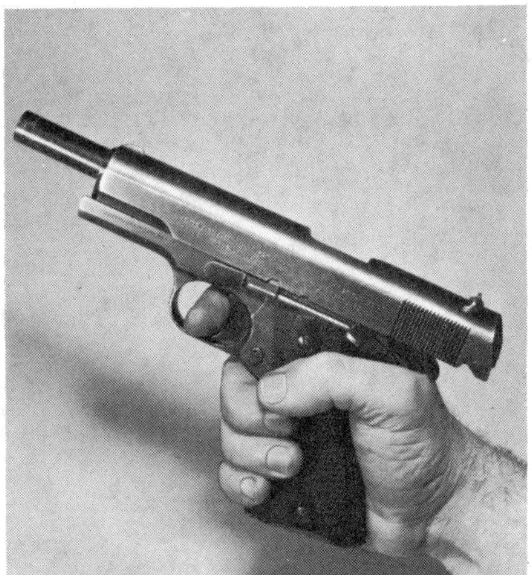

Slide stop: When the last shot has been fired, a section of the front end of the magazine follower, pushed up by the magazine spring, presses against the underside of the slide stop. This forces the stop up into a niche cut in the slide and holds the slide open as an indication that the pistol is empty.

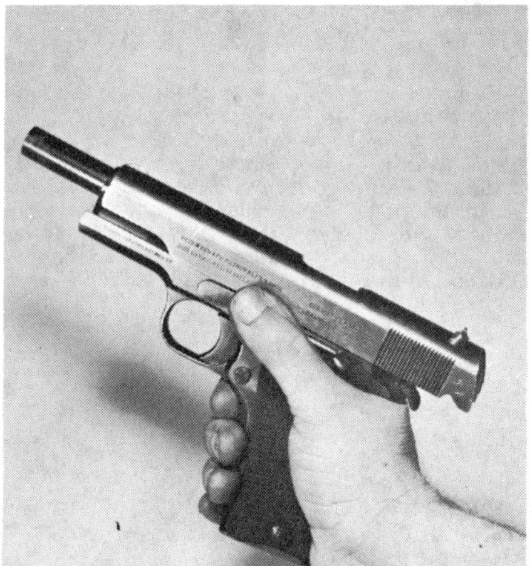

Reloading from open slide: (1) Press magazine catch and extract empty magazine. (2) Insert loaded magazine. (3) Push down on slide stop with right thumb. This will release the slide to drive forward and load the chamber. Note: Slide stop cannot be released while an empty magazine is in the pistol. Slide will go forward only on a loaded magazine or when the magazine has been pulled part way out.

Field Stripping the M1911A1 Pistol

Remove magazine and examine chamber: (1) Press magazine catch and withdraw magazine. (2) Draw back slide and look into chamber through the ejection port to be sure the pistol is empty. Remember that even when the magazine is out, the pistol is still dangerous: there may be a cartridge in the chamber.

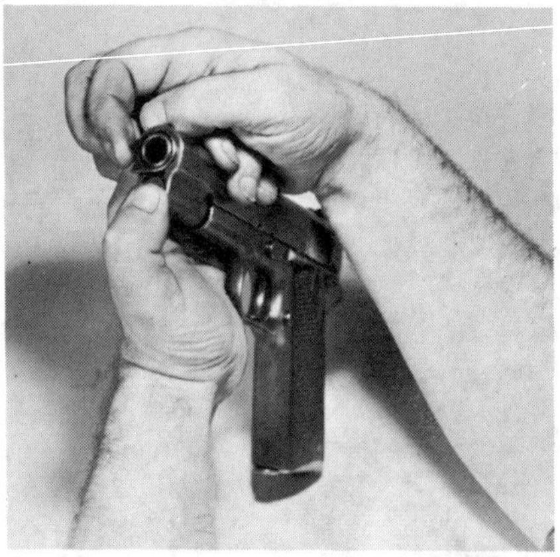

Release tension of recoil spring: (1) Press in on plug which covers end of recoil spring, using thumb or butt of magazine if it is too stiff. (2) Barrel bushing, freed from spring tension, may now be turned to the right side of the pistol.

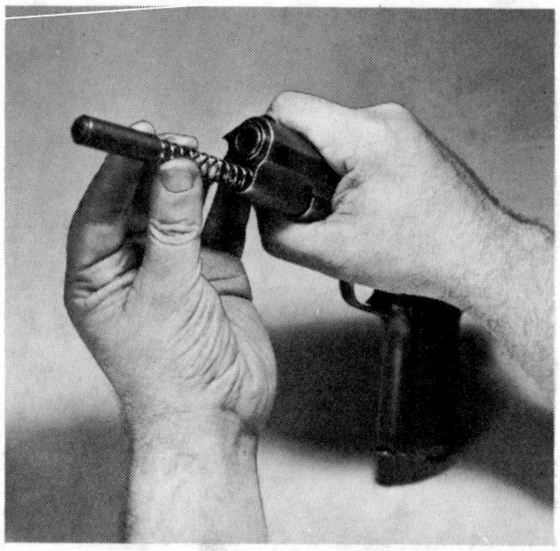

Ease out plug and recoil spring: The spring is very powerful. Take care not to let it fly out of the pistol. Do not withdraw these parts from the pistol yet, as they serve to keep the recoil spring guide in place and make the next step easy.

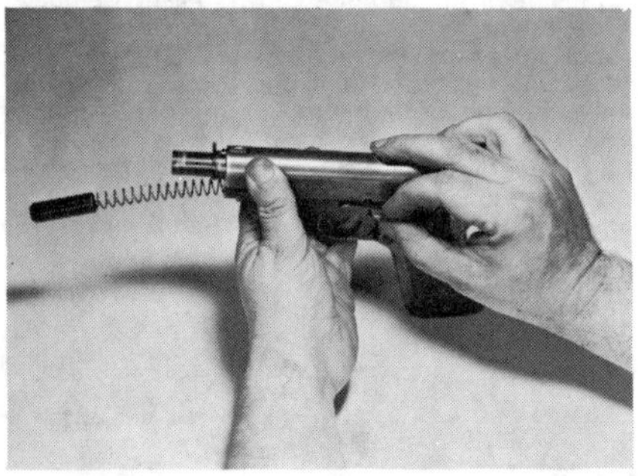

Remove slide stop. (1) Push slide back until the rear edge of the smaller recess in the lower edge of the slide is even with the rear end of the slide stop. (2) Now press from the right side against the protruding pin which is part of the slide stop. This pin passes through the right side of the receiver, then through the barrel link which holds the barrel, then through the left side of the receiver. (3) Now pull slide stop out from left side of pistol.

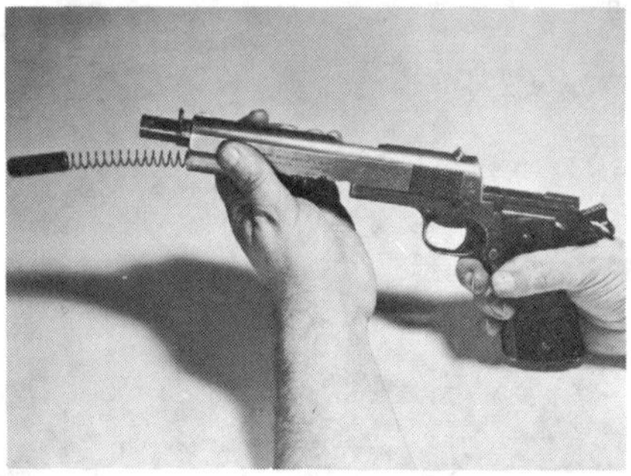

Remove slide and components: Pull slide forward on its guides in the receiver and remove. With the slide will come the barrel, barrel link, barrel bushing, recoil spring and recoil spring guide.

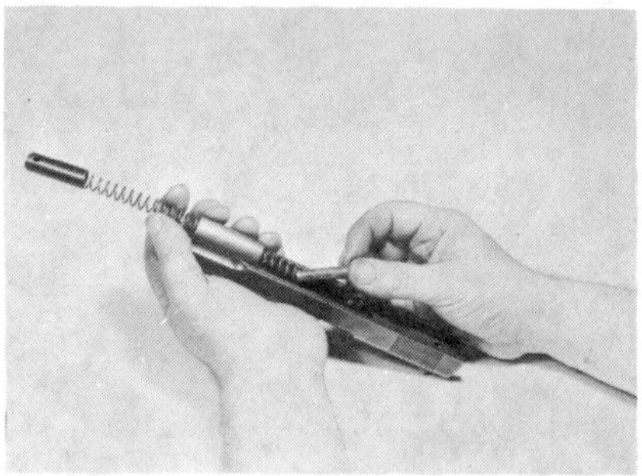

Remove recoil spring guide: (1) The recoil spring guide (on which the recoil spring compresses) may now be lifted out to the rear. (2) The recoil spring and plug are pulled out from the front. (3) The barrel bushing is turned to the left which unlocks it so it can be withdrawn.

Remove barrel: Turn barrel link forward on its pin and withdraw barrel assembly from the front of the slide. Note: Normally no further stripping of this pistol is required.

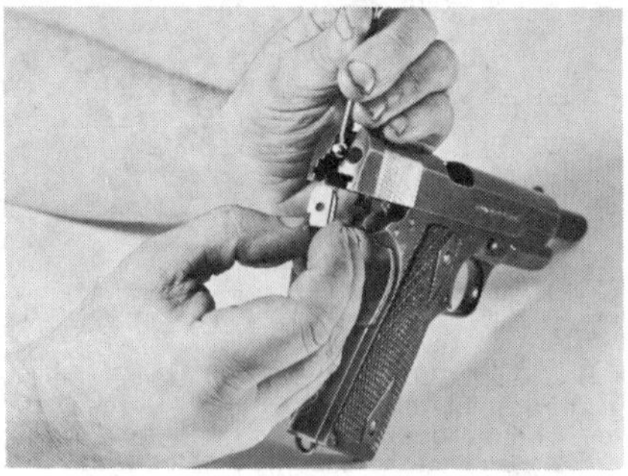

To remove firing pin: (1) Should it be necessary, the firing pin may be easily removed by pressing the pin in against the tension of its spring, at the same time pushing down on the firing pin stop which holds the firing pin in place. This may be done with a nail, match or similar object. (2) Slide the stop down out of its grooves and ease out the firing pin and spring.

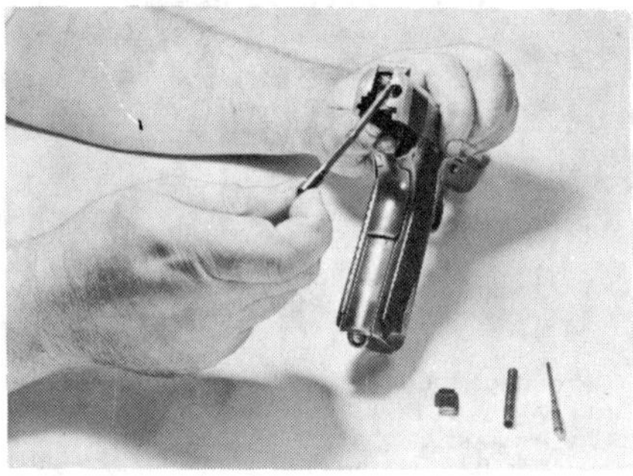

To remove extractor: When the firing pin has been removed the extractor, which is a long piece of spring steel inserted in a hole to the left of the firing pin, may be pried up and pulled out to the rear as illustrated.

Notes on Assembling

Barrel link must be tilted forward and link pin properly in place before it will slide into place in the slide.

Put sear and disconnector together, hold by their lower ends, place them in the receiver, and replace sear pin.

Sear spring should be replaced after sear and disconnector are in place, care being taken that lower end is in its place in the cut in the receiver; upper end of left-hand leaf resting on sear.

Insert mainspring housing until lower end projects about one-eighth inch below frame. Then (1) Replace hammer and pin; (2) Grip safety; (3) Cock hammer and replace safety lock; (4) Lower hammer and push mainspring housing home, and insert pin.

Cock hammer. Insert end of magazine follower to press safety lock plunger home.

When inserting slide stop, make sure that its upper rear end stops on the receiver just below the small slide stop plunger. Then push stop upward and inward with the one motion. This will enable the upper round part of the stop to push the plunger back and let the stop snap into place.

In replacing sear and disconnector, hold the receiver as when firing, then tilt front end down. Insert the sear and the disconnector using the trigger bar as a guide to align the holes in these two parts with the receiver holes. Slight pressure may be applied on the trigger until the holes are properly lined up. When replacing the mainspring housing, it is important that the rear end of the hammer strut be in its place in the mainspring cap.

Description of Mechanism

The pistol has three main parts: receiver, barrel, and slide. The receiver is fitted with guides in which the slide runs. Its handle is hollow to permit insertion of the box magazine which is locked by the magazine catch.

The receiver also holds the trigger whose front end projects through the trigger guard. The firing mechanism, made up of the hammer, sear, automatic disconnector, grip safety, and safety lock, are at the rear of the receiver. Here too are the mainspring and sear spring. The mainspring is a coiled spring seated within the mainspring housing which is held by the mainspring cap pin. The mainspring cap and housing pin retainer are also in this housing.

The sear spring is a flat spring with a rib fitting into a slot in the rear wall of the receiver to prevent the spring from moving vertically. The mainspring housing bears against the rear of the spring and locks it into position to give it the required tension.

A bent metal piece called the hammer strut is fastened to the hammer by a pin in rear of its pivot, while its end rests in the mainspring cap. In a tube above the handle are the slide stop and safety lock plungers whose ends protrude from front and rear of the tube respectively, as well as the spiral spring plunger seated between the two which holds them in position.

The ejector, a solid piece of metal against which the head of the withdrawn cartridge case strikes, is fastened to the top of the receiver near the rear end.

The top of the receiver extending forward above the trigger guard forms a semi-tubular extension to provide a seat for the rear section of the recoil spring.

The barrel has two transverse locking ribs on its rear upper surface. They positively lock into corresponding slots on the inside of the slide when in firing position. The lower rear end of the barrel is attached to the receiver by a swinging link and pin. It can thus move a limited distance lengthwise and downwards.

The heavy slide mounts on the receiver from the front end and the distance of its rearward movement is controlled by a tubular abutment which absolutely prevents it being thrown rearward from the receiver.

In this abutment at the front end of the slide rests the forward portion of the recoil spring and the plug which fits over its end; while the rear end of this spring is fitted over a removable guide which is supported by the shoulder at the front of the receiver.

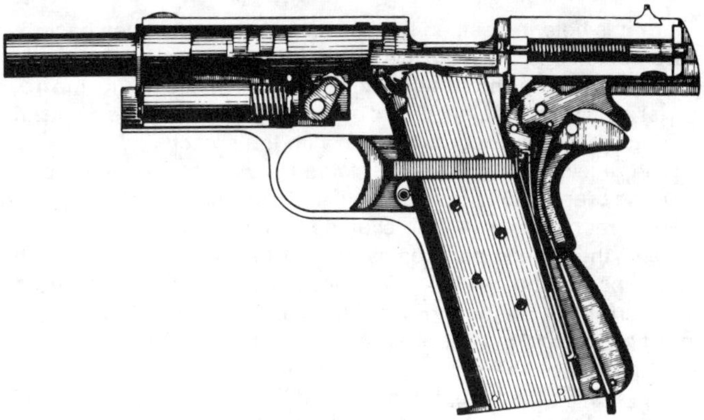

Action open and cutaway to show barrel swung back and down on link for unlocking. Lockwork shows details of cocking and firing mechanism.

A barrel bushing is inserted in the front of the slide over the barrel and locks when turned down into place. It serves to retain the recoil spring and the plug and also supports the muzzle end of the barrel.

When the barrel and slide together are mounted on the receiver, the slide stop is in place, its pin passing through the receiver from side to side and through the barrel link, thereby positively locking slide, barrel, and receiver together.

The exterior of the slide stop is provided with a checkered thumb piece to be pressed for releasing slide from the open position. When the safety lock is pushed into the upward position, it enters a recess on the slide and a stud on the inner face at the same time locks the sear and hammer when in the full cocked position. This safety can only be applied when the pistol is cocked.

The grip safety is a curved metal piece pivoted in the upper part of the receiver. It acts automatically to lock and release the firing mechanism without any attention on the part of the shooter.

The disconnector is a small lever which when pressed down prevents the trigger mechanism from engaging with the sear and hammer in all positions except when breech is fully closed. This also prevents more than one shot following each pull of trigger.

How the M1911A1 Pistol Works

A magazine holding any number of cartridges from 1 to 7 is inserted in the handle and pushed in until it locks. This locking is produced by the thumb-catch lock snapping into a notch cut in the upper front end of the magazine.

As slide is drawn back for the opening movement, it compresses a recoil spring, and forces the hammer back and down to full cock.

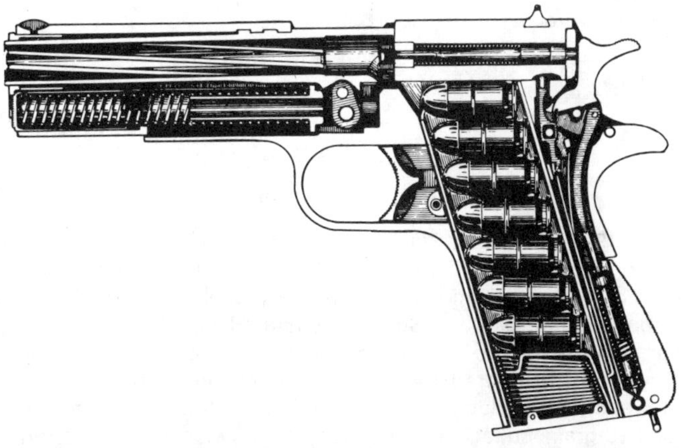

Action cutaway to show details of swinging link, barrel lock and lockwork at rest.

Here it is held in position by the sear. As the slide goes back the disconnector is pressed back, positively preventing the trigger from connecting the firing mechanism. When slide is to the rear as far as it will go, it clears head of the magazine, and the magazine spring feeds a cartridge up in line with the breech.

The slide on being released is forced forward by the recoil spring and the breech bolt face of the slide carries the first cartridge into the barrel chamber. As the slide nears its foremost position, the face of the breech bolt comes against the rear extension of the upper part of the barrel forcing the barrel forward and upward on the barrel link. When the slide and barrel reach the full forward position, the locking ribs are positively locked into the corresponding grooves on the inside of the slide.

The firing-pin spring, firing pin, and the extractor are located in the breech bolt section of the slide. As the cartridge is seated in the barrel chamber, the extractor, which is made of heavy spring steel, springs over the head and into the cannelure of the cartridge, thus preparing it to be drawn back during the rearward motion of the pistol for extraction.

The firing pin is a "flying" one. It is seated in the breechblock surrounded by its spring. It is shorter than the breechblock itself. Thus if the hammer is lowered all the way down on the firing pin, it will push the pin inside the breechblock but cannot push it far enough to make it rest against the primer in the cartridge. The one way in which the firing pin can touch the primer is when it is struck a sharp blow by the hammer itself. As the hammer falls, it drives the firing pin ahead to strike the primer and is stopped itself by the face of the breechblock. The inertia of the firing pin causes it to fly ahead and strike the primer, then its compressed coil spring pulls it back into the breechblock.

The pressure of the powder gases driving the bullet down the barrel also drives the empty case back against the base of the breechblock pushing the slide and barrel together rearward, until the bullet is safely out of the barrel. At that point the barrel swings downward on its pivoted link leaving the slide free to continue direct rearward motion in its guide in the receiver, extracting and ejecting the empty shell, cocking the hammer, and compressing the recoil spring for the next forward movement.

Breech Pressure. When the .45 automatic pistol is fired, a pressure of 16,000 lbs. per sq. inch is generated in the firing chamber. Since the base of the bullet has an area of .159 square inches, a total pressure of 2,225 pounds is exerted against the base of the bullet, driving it forward down the barrel. The same total pressure is exerted back against the face of the breechblock driving it backwards.

The bullet, weighing only a fraction as much as the moving section of the pistol, gets under way much more quickly and utilizes most of the pressure as it travels down the barrel. The pistol's motion as compared to that of the bullet is inversely proportional to the weights of the pistol and the bullet.

Thus the bullet is well out of the barrel before the pistol has recoiled sufficiently to permit the locking mechanism to open. The pressure, of course, lasts only for a very small fraction of a second. However, in high powered military pistols of comparatively low weight, it is necessary that the breech be positively locked until the bullet has left the barrel. The locking principle on the .45 automatic pistol, model 1911, is among the best ever designed.

How the Slide Stop Works. The forward end of the magazine follower is split in two longitudinally for a short length. The left hand section is bent down below the level of the follower.

When the last cartridge has been stripped from the magazine, the follower rises under the influence of the magazine spring, and the bent section presses against the rear of the inside of the slide stop, thus forcing the projection on the top of the stop up into line with the large notch in the base of the slide into which it locks.

The slide will not go forward while the magazine is empty. If the empty magazine is extracted, pressing on the slide stop will permit the slide to run forward by pulling the stop down out of its locking notch. If a loaded magazine is inserted in the handle, then the magazine follower with its bent arm is in position below the cartridges on the inside of the magazine. Hence, if the slide stop is pushed down when a loaded magazine is in the handle, it will release the slide to run forward and load the chamber.

Characteristics of United States Service Pistols

	Colt Model 1917	Smith & Wesson Model 1917	Model 1911A1
Caliber:	.45.	.45.	.45.
System of operation:	Double-action revolver.	Double-action revolver.	Recoil, semiautomatic.
Length overall:	10.8 in.	10.8 in.	8.62 in.
Barrel length:	5.5 in.	5.5 in.	5 in.
Feed device:	6-round, revolving cylinder.	6-round, revolving cylinder.	7-round, in-line, detachable box magazine.
Sights: Front:	Blade.	Blade.	Blade.
Rear:	Square notch.	"V" notch.	Square notch.
Weight:	2.5 lb.	2.25 lb.	2.43 lb.
Muzzle velocity:	830 f.p.s.	830 f.p.s.	830 f.p.s.

UNITED STATES RIFLES AND CARBINES

HISTORICAL SUMMARY

In 1903 the United States adopted the caliber .30 Model 1903 rifle, commonly known as the "03" Springfield. The action is basically a modification of the Mauser Model 98 action. As made from 1903 to 1906 the Springfield had a tangent-type rear sight, a cleaning-rod type bayonet contained in the stock under the barrel and was chambered for the caliber .30 Model 1903 cartridge (.30-03) which is longer than the .30 Model 1906 (.30-06) and used a 220-grain bullet. The adoption of the Model 1906 cartridge resulted in the rebuilding of most of the rifles which had been previously issued, and the Model 1903 appeared in the form which remained standard until the adoption of the Model 1903A1 in 1929.

The Model 1903A1 remained the standard rifle until the adoption of the M1 rifle in 1936. In 1940 the British government contracted with the Remington Arms firm at Ilion, New York, to produce caliber .30 M1903A1 rifles for British forces. Before production was started however, the United States government took over the contract under Lend Lease and plans were made for a simplification of manufacture. A few Model 1903A1s were made by Remington—production figures on the Springfield are given in the detailed section on that rifle—before production of the simplified

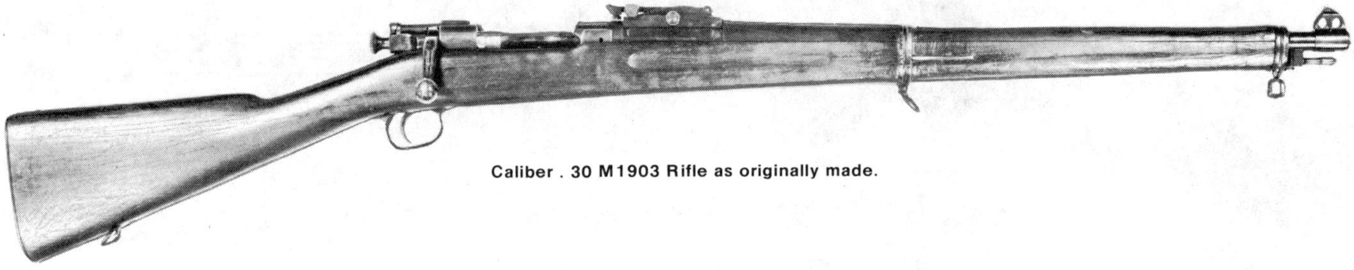

Caliber . 30 M1903 Rifle as originally made.

M1903A3 was begun. A sniper version of the Model 1903A3, the Model 1903A4, was also developed and produced during World War II.

The entry of the United States into World War I found this country with a very limited number of M1903 rifles on hand—approximately 600,000— and only two plants tooled to produce the rifle, Springfield and Rock Island. The government also had 160,000 caliber .30 Krags on hand and purchased 20,000 caliber .303 Canadian Ross rifles and 280,049 Russian 7.62mm Mosin-Nagant rifles from New England Westinghouse and Remington Arms who had been manufacturing this rifle for the Imperial Russian government. The Krag, the Ross, and the Mosin-Nagant were not suitable for service with the American Expeditionary Force in France because of logisistical problems and basic unsuitability for the type of war being conducted.

Three large gun plants—Winchester at New Haven, Conn., Remington Arms at Ilion, N.Y., and Remington Arms Union (later called Midvale Steel and Ordnance Co.) at Eddystone, Pa., were completing large contracts for British-designed caliber .303 P14 (Enfield) rifles. (See the chapter on Great Britain.) The P14 had originally been designed, as the P13, for a rimless caliber .280 cartridge; it was not too difficult to modify the design to use the U.S. caliber .30 M1906 cartridge. Additional modifications were made to simplify production and to standardize parts among the plants and in this manner the U.S. caliber .30 Rifle Model 1917 (or Enfield as it is commonly known) was born. This rifle was used by the United States for training purposes during World War II and large quantities were sold to the United Kingdom in 1940, where for some peculiar reason, it was frequently referred to as "the Springfield."

The Model 1917 is a modified Mauser with frontal locking lugs that cocks, as the British Lee Enfield, on the forward push of the bolt. Bolt removal is by means of a bolt-retaining catch, which also houses the ejector, on the left rear side of the receiver, and is similar to that of the Model 98 Mauser and its many variations. The bolt sleeve is held on the one-piece firing pin by interrupted threads and its removal is rather complicated. The cocking piece is drawn to the rear and turned, by use of a piece of string looped over the sear nose, to allow the firing pin to go forward, thereby relieving the tension on the main spring and allowing the cocking piece to be removed. In the Army, various tricks such as inserting pennies between the cocking piece and bolt body while the bolt was forced forward with the safety in the "on" position were also used to achieve this purpose. Indeed, the possession of a penny was essential to the United States Army user of the Enfield. Since the Enfield did not have a cutoff, as did the Springfield, and the follower of the Enfield held the bolt to the rear of the magazine when empty, a penny was necessary to put in the top of the magazine to hold the follower down so that the bolt could be shoved smartly home during "inspection arms." Woe be unto the penniless private who stood fumbling with his bolt as the rest of the platoon smartly finished their drill—it could cost him a week-end pass or at least a little "extra" marching on the drill field!

Bolt disassembly excepted, the Model 1917 was a fine weapon and had better battle sights than the Model 1903. Total production of the Model 1917 was 2,202,429. Barrels, firing pins, main springs, stocks, and various other parts were made for the rifle during World War II, but the last complete rifles were made after World War I. The Model 1917 is no longer used by any major power and is probably not used in significant quantity by any country.

The search for a suitable semiautomatic rifle began before World War I, but became more urgent at that war's conclusion. After a series of tests in which the St. Etienne, the Liu, the Bang, the Thompson, the Hatcher, the Pedersen, the Garand, and various other rifles were contenders—the Garand was chosen and standardized in 1936 as the United States Rifle caliber .30 M1. The M1 is covered in detail later in this chapter.

Variations of the M1 Rifle

A number of variations of the M1 have been produced, usually only in limited quantity or in prototype form. A number of these weapons are listed below with their outstanding characteristics.

M1E1 Rifle. It differs from the M1 only in having a more gradual cam angle in the operating rod handle—few were made.

M1E3 Rifle. A roller lug was attached to the bolt cam lug and the cam angle of the operating rod was altered to match. Few of these weapons were made, but this system was used in the later selective fire T20E2 rifle and the current M14 rifle.

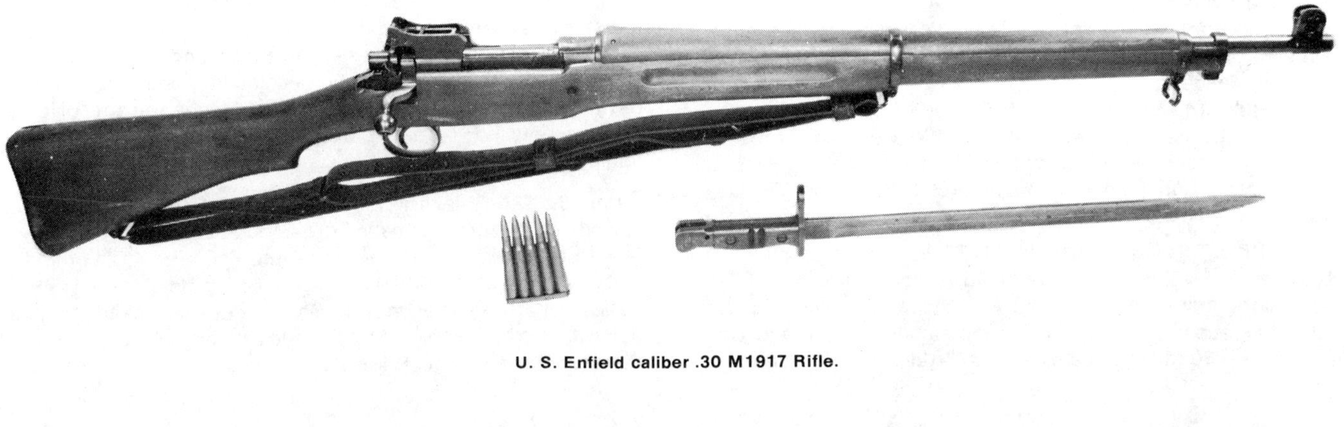

U. S. Enfield caliber .30 M1917 Rifle.

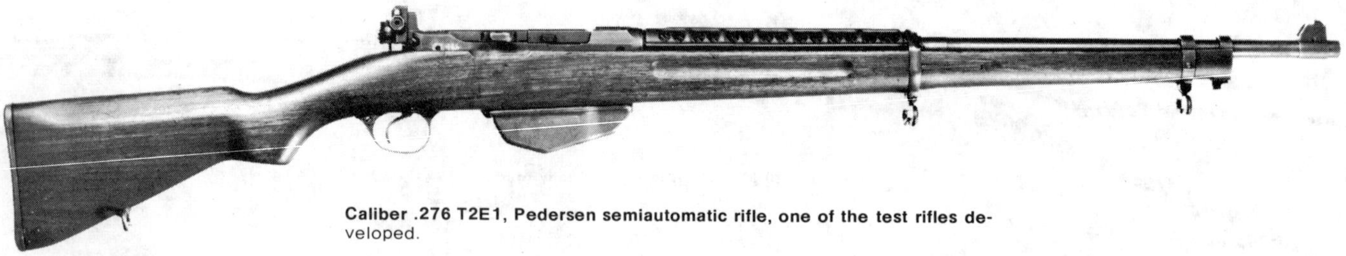

Caliber .276 T2E1, Pedersen semiautomatic rifle, one of the test rifles developed.

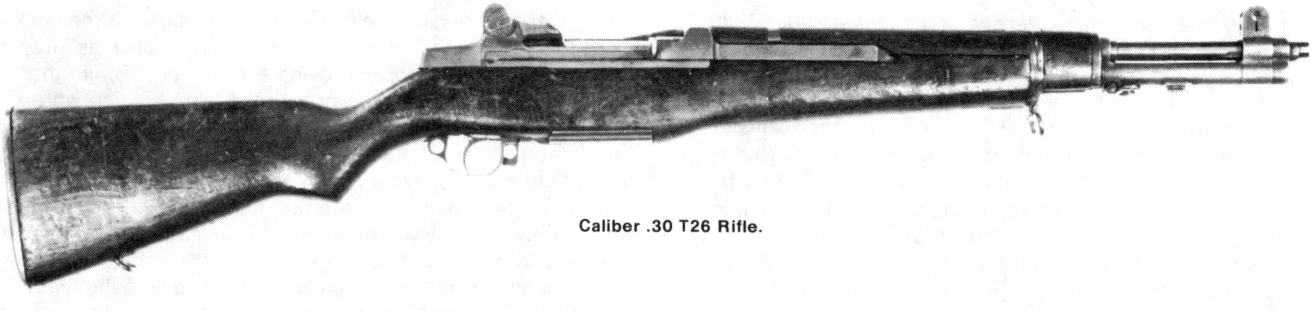

Caliber .30 T26 Rifle.

M1E4 Rifle. In order to provide a "softer action" by increasing action dwell time and to decrease the rearward velocity of the operating rod, the gas system of the M1E4 was altered from that of the M1. The M1 uses an impinging gas system; that of the M1E4 is a gas cutoff expansion system. The gas port is located approximately 8 inches from the muzzle and the gas, after filling the expansion chamber, is cut off from the gas port and is allowed to expand. Immediately after the hammer was cocked by the recoiling bolt, the gas was allowed to escape. This gas system is similar in concept to that used on the current M14 rifle.

M1E9 Rifle. This rifle also has a gas-expansion cutoff system; however, the piston was divided from the operating rod and served as a tappet to the operating rod. This avoided the overheating of the combined piston operating rod experienced with the M1E4. This system is very similar to that used with the current M14 rifle.

M1E5 Rifle. This was a shortened version of the M1 using a pantograph-type folding stock. The barrel was 18 inches long and, although the accuracy was comparable to the M1 at ranges up to 300 yards, blast and muzzle flash were excessive.

T26 Rifle. In July 1945 the Pacific Theater of Operations requested 25,000 shortened M1 rifles. To meet this requirement the M1E5 barreled action was fitted with a standard M1 stock. An order for 15,000 of these rifles was given but was cancelled in August 1945.

M1E2 Rifle. This is the first experimental M1 Sniper rifle. It was equipped with an International Industries telescope and mounts. At the same time an M1 equipped with a Weaver 330 scope and Stith mounts was tested.

M1E6 Rifle. This was an M1 Sniper model with offset telescope designed to allow use of the iron sights in addition to the telescope. None were made.

M1E7 Rifle. This weapon had a Griffin and Howe scope mount and a M73 (Lyman Alaskan) or M73B1 (Weaver 330) scope. A removable leather cheek piece was fitted to the stock. This weapon was standardized in June 1944 as the U.S. Rifle Caliber .30 MIC (Snipers). A removable flash hider was adopted in January, 1945.

M1E8 Rifle. This is similar to the M1E7, but the scope mount is a block-type developed by Springfield Armory. A leather cheek piece on the stock was also used with this rifle. The M1E8 was adopted as substitute standard in September 1944 as the U.S. Rifle Caliber .30 MID (Snipers). The M73 (Lyman) telescope with cross-wire type reticle and rubber eye piece is called the Telescope M81. When equipped with tapered post reticle and rubber eye piece, it is designated as the M82. A later telescope used with the M1C and M1D is the M84.

M1 National Match Model. This is similar to the issue M1 but has a special National Match barrel, sight parts, and a few other special components. Specially assembled for accuracy.

A number of rifles which could be considered predecessors to the M14 rifle were also developed by the United States during the period 1944-1956.

Development Models

In early 1944 Springfield Armory and Remington Arms were directed to develop a selective-fire weapon having the following characteristics:

Weight: 9 pounds less magazine.
Magazine: 20-round.
Bipod equipped.
Semiautomatic fire on closed bolt.
Full automatic fire on open bolt.
Suitable for launching grenades.
Use standard M15 grenade launcher sight.
Weapon to have folding stock.

The development at Springfield was designated the T20 and that at Remington was designated the T22. Winchester developed a weapon of their own. In September 1944, Headquarters Army Ground Forces requested that the M1 be modified to: (1) selective fire, (2) use a bipod, (3) that the mechanism be simple and capable of field stripping, and (4) that a suitable 20-round magazine loaded from the bottom of the piece be developed. Accordingly Springfield and Remington were instructed to change the original requirements they received, dropping those for a short barrel and folding stock.

Cal. .30 T20 Rifle. This rifle had selective fire, originally with closed bolt on semiautomatic fire and open bolt on automatic fire. The first models were delivered to Aberdeen Proving Ground in 1944 for test. A modified 20-round Browning Automatic Rifle (BAR) magazine was used. The gas cylinder was locked into position with a muzzle brake. Aberdeen tests indicated the need for a better magazine and better method of holding it in place.

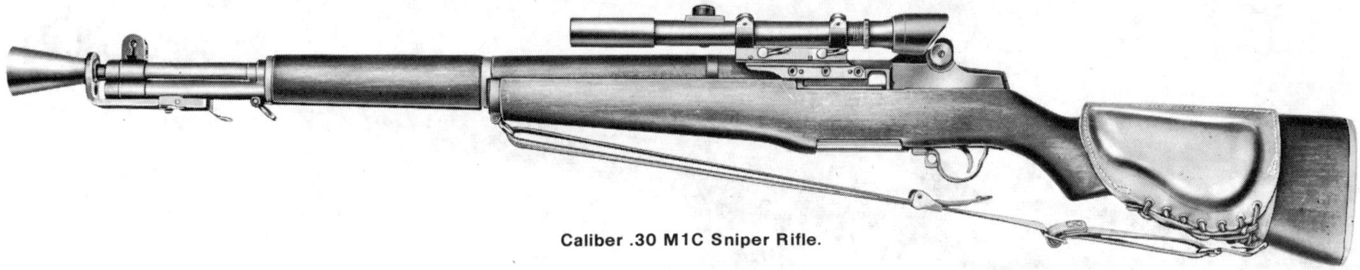

Caliber .30 M1C Sniper Rifle.

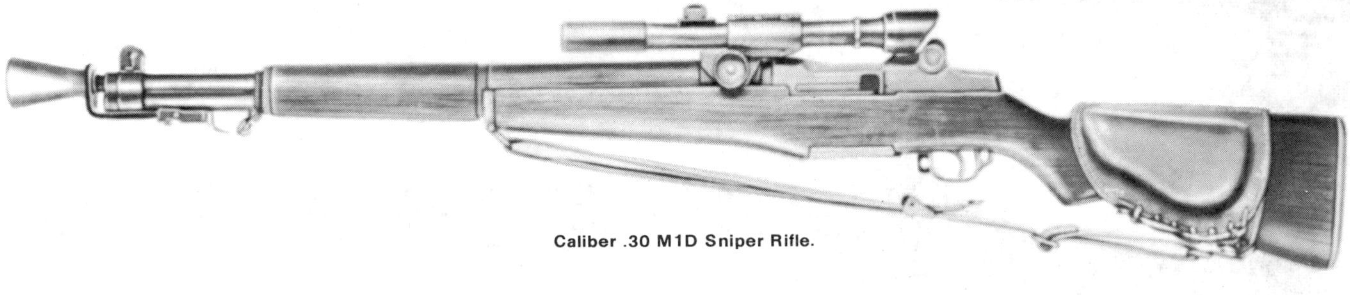

Caliber .30 M1D Sniper Rifle.

Cal. .30 T20E1 Rifle. This model fired from the closed bolt in both semiautomatic and automatic fire. A new magazine catch and new magazine were also used. A new muzzle brake was used; the muzzle brake of the T20 did not allow the use of a bayonet, grenade launcher, or flash hider. Two heat flow arresting grooves were put on the barrel before the chamber. The T20E1 had an adjustable bipod which was not easily removable. The weapon was tested at Aberdeen in January 1945 and the tests were successful. As a result 100 of an improved version were ordered from Springfield Armory.

Cal. .30 T20 E2 Rifle. This rifle differs from the T20E1 in the following: the muzzle brake has been further modified to use a new grenade launcher which, with a new valve, is designed to permit semi or full automatic fire without the removal or manual adjustment of the launcher by the firer. A new bipod with longer legs was adopted, the bolt was slightly modified, and a roller lug added to the bolt lug (similar to the M1E3). The receiver is slightly longer than that of the T20 and the bridge has been modified to mount an operating rod lock to hold the bolt open when desired. The magazine was usable in the BAR, but BAR magazines could not be used in the T20E2. There were numerous minor changes made as well. In May 1945 the Ordnance Committee recommended that 100,000 of these rifles be procured, but no action was taken because of the ending of the war.

Cal. 30 T22 Rifle. This was one of the Remington developments and had a bipod, flash hider, and 20-round magazine. Tests indicated that minor modifications were needed to improve functioning, the magazine release, and retention of the bolt in the open position.

Cal. .30 T22E1 Rifle. The required modifications to the T22 were made in this model and several were tested at Aberdeen.

Cal. .30 T22E2 Rifle. This weapon had a new magazine catch and a slightly changed trigger group. It incorporated the gas cylinder, gas cylinder valve screw, and bipod as used on the T20E2. The T22E2, like all the Remington modifications, used the M1 receiver without lengthening.

Cal. .30 T23 Rifle. This Remington modification of the M1 incorporates selective fire control by means of an independent hammer release. This weapon fired full automatic from an open bolt and semiautomatic from a closed bolt; however the Proving Ground test indicated that it would fire from the open bolt only 20% of the time. More change was required and it was decided to use a closed bolt for both full and semiautomatic fire. T23 used the standard M1 stock and early test models used the M1 8-round "en bloc" clip. Later models used a BAR magazine, but were not too successful and it was decided that a new magazine should be designed.

Cal. .30 T24. The T24 was another Remington modification of the M1 to selective fire in which selective-fire control is obtained by means of an independent sear release. The T24 fired from a closed bolt in both semiautomatic and full automatic fire and used the 8-round "en bloc" clip in test models. It was equipped with a straight high comb stock.

Winchester Automatic Rifle. Winchester developed this weapon on their own initiative. It was based on the carbine gas system and had a bipod and flash hider. Ten specimens were procured in May 1945 for test.

Cal. .30 T25 Rifle. This was a post-war Springfield Armory design using a gas cutoff and expansion system. It is a selective-fire weapon, having a straight line stock. The bolt is a tipping-type which locks on the top of the receiver. The T25 fires on a closed bolt in semiautomatic fire and an open bolt in full automatic. It was later altered to fire selectively from a closed bolt. T25 was chambered for the T65 series caliber .30 case; this case later was developed into the 7.62mm NATO case.

Cal. .30 T27 Rifle. This was an M1 converted to automatic fire by an auxiliary device that could be inserted in the field. It was subsequently redesigned so that it was capable of selective fire. It uses the M1 8-round "en bloc" clip.

Cal. .30 T28 Rifle. The T28 used an adaptation of a German-designed breech and trigger mechanism. Stampings and simplified forgings were used considerably in this design. This rifle was chambered for the T65E3 (NATO) cartridge.

Cal. .30 T31 Rifle. The T31 is an in-line stock type weapon with "bull pup" type stock. The magazine design was incorporated into the T44 Rifle. This rifle was chambered for the T65E3 (NATO) cartridge.

Cal. .30 T33 Rifle. The T33 was a private development chambered for the T65E3 (NATO) cartridge. The design was dropped because it lacked sufficient ruggedness and durability.

Cal. .30 T35 Rifle. This rifle was the M1 modified to use the caliber .30 T65E3 cartridge case (7.62mm NATO). It used the 8-round "en bloc" type clip and feed system similar to that of the M1.

Cal. .30 T36 Rifle. This weapon was a modification of the T20E2 and was chambered for the T65E33 (NATO) cartridge. It used a modified T25 magazine and fired in both automatic and semiautomatic from a closed bolt.

Caliber .30 T20 Rifle.

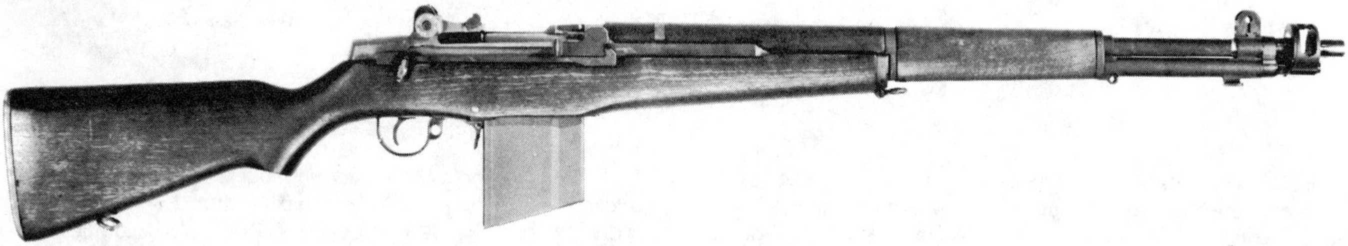

Caliber .30 T20E2 Rifle.

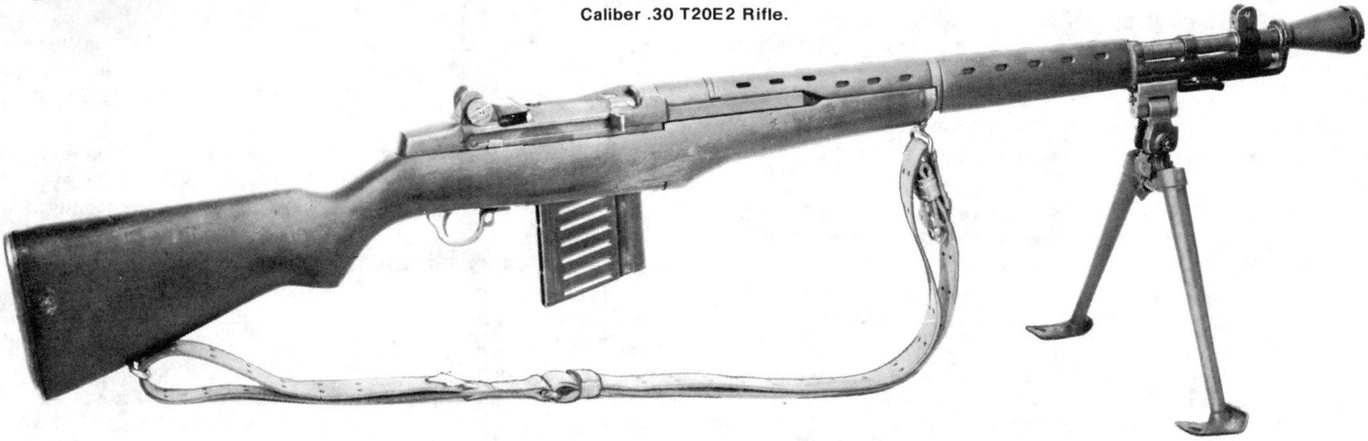

U. S. Rifle caliber .30 T22.

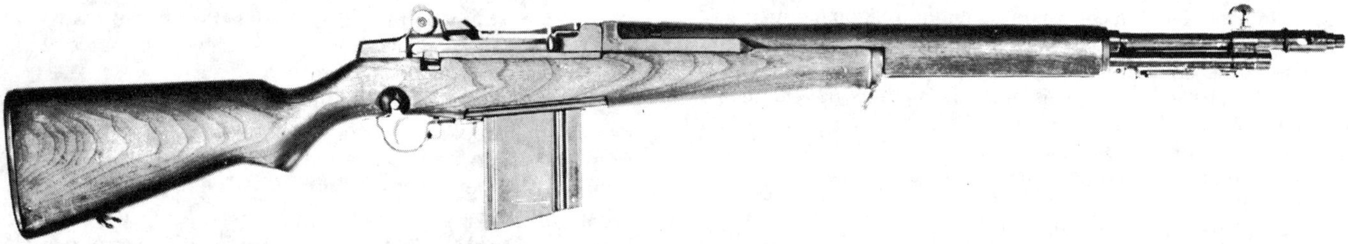

Caliber .30 Rifle T22E2.

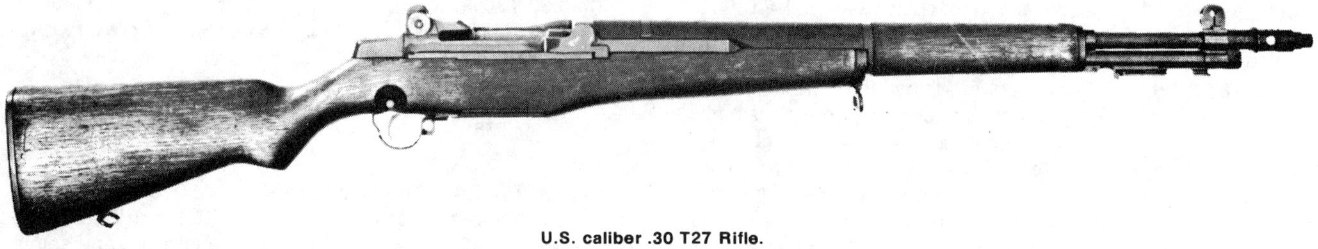

U.S. caliber .30 T27 Rifle.

The T25 Rifle (above) and the T36 (lightened) Rifle (below), both caliber .30.

Cal. .30 T37 Rifle. Chambered for the NATO cartridge, this rifle evolved from the T20E2. A light weight 22-inch barrel with gas port approximately four inches from the muzzle and a shortened stock were used on this weapon.

Cal. .30 T44. Chambered for the NATO cartridge, the T44 is basically the T37 with gas cutoff expansion system. The T44 is the first of the prototypes which directly resulted in the M14 rifle. T44E1 was a heavy-barrel version of the T44. The T44E4 was adopted as the 7.62mm Rifle M14 in June 1957. Two differences of the T44E4 from the production M14 are a plastic ventilated hand guard in the M14 as opposed to a wooden hand guard in the T44E4, and M14 has an aluminum hinged butt plate as opposed to a standard steel butt plate in the T44E4. The T44E5 was adopted as the 7.62mm Rifle M15; a heavy-barrel version of the M14, it was dropped shortly after adoption and never mass produced. It was decided that the selective fire M14 when fitted with bipod and hinged butt would be used to fill the squad automatic role formerly filled by the Browning Automatic Rifle. The T44E6 is a lightweight form of the M14 which was produced only as a prototype. The T44E2 has an impinging type gas system.

Cal. .30 T47 Rifle. Chambered for the NATO cartridge, the T47 is a redesign of the T25 to use a conventional stock.

Cal. .30 T47E1 Rifle. The heavy-barrel version of the T47 rifle, chambered for the NATO cartridge.

Cal. .30 T48 Rifle. Chambered for the 7.62mm NATO cartridge, this is the United States experimental model designation for the Belgian FN "FAL" rifle. In addition to the 3200 purchased from FN, 500 were manufactured by Harrington and Richardson, Worcester, Mass. The T48 competed against the T47, the British EM2, and the T44 during the early 50s.

It should be noted that the 7.62mm NATO cartridge, for which many of these weapons are chambered, was called caliber .30 by the United States before adoption. The cartridge case was the T65E3 and this Model number was frequently, albeit erroneously, referred to as the correct nomenclature for the complete round.

5.56mm Rifle M16. After the 7.62mm NATO M14 rifle was adopted, tests continued to be made of new rifles which appeared. Among those tested were 7.62mm NATO AR10 rifle and the .223 AR15 rifle, both the developments of Mr. Eugene Stoner and both submitted by Armalite Corp. of Costa Mesa, California, at that time a division of Fairchild Engine and Airplane Co. The AR15 was adopted by the United States Air Force and became the 5.56mm Rifle M16. The Army adopted the rifle in slightly modified form as the 5.56mm Rifle M16A1 and the weapon is used by most Army units.

The 5.56mm M16 and M16A1 are covered in detail later in this chapter. Colt obtained the rights for the AR15 rifle and have produced a number of variations of this weapon which they call the Colt 5.56mm CAR-15 weapons system. These are covered in detail later in the chapter.

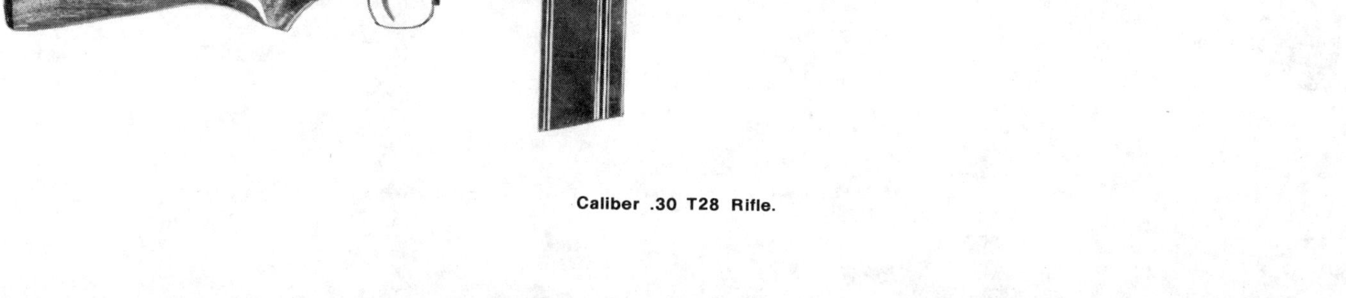

Caliber .30 T28 Rifle.

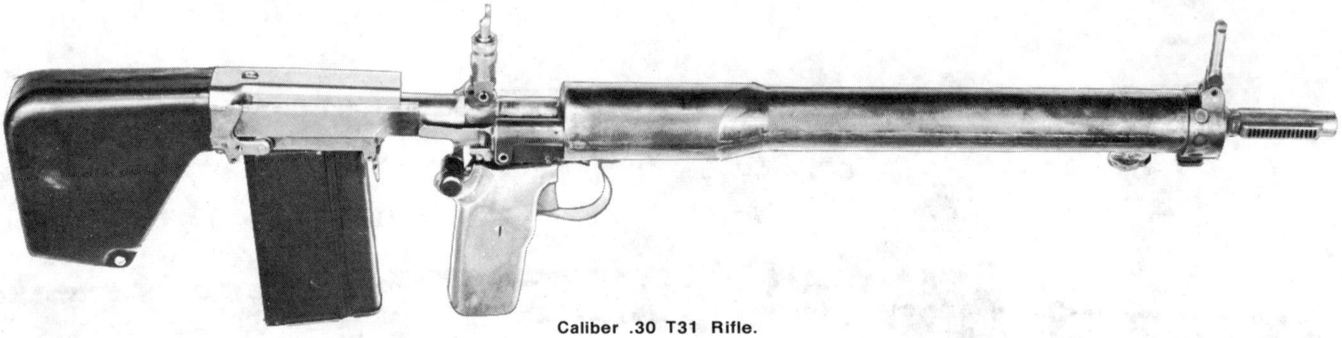

Caliber .30 T31 Rifle.

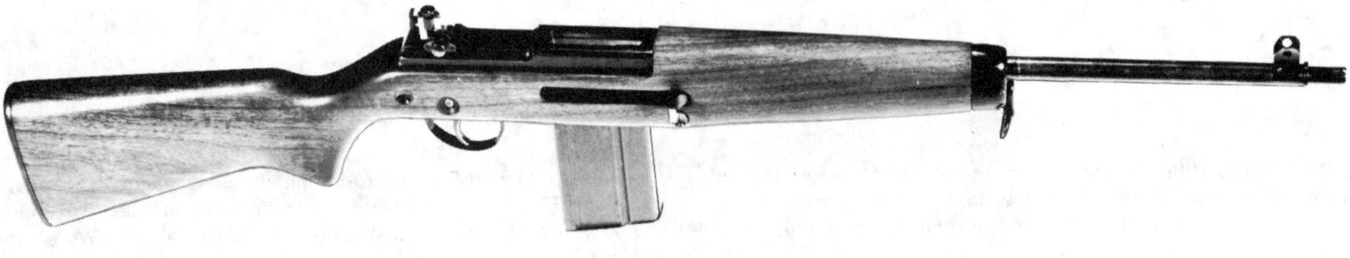

Caliber .30 T33 Rifle.

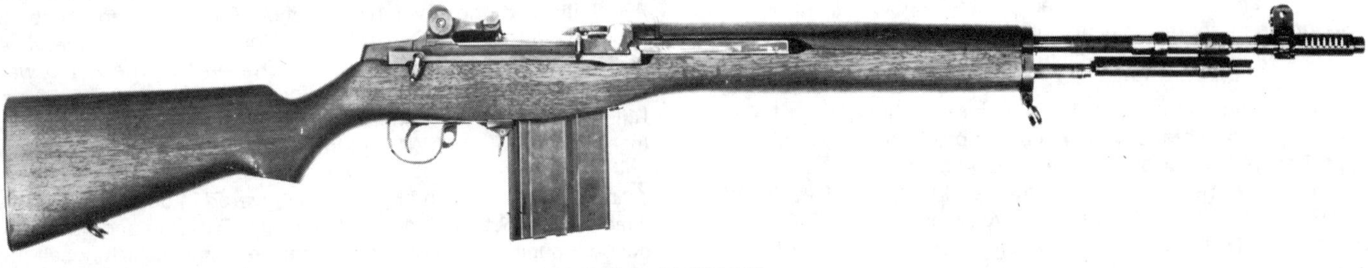

Caliber .30 T37 Rifle.

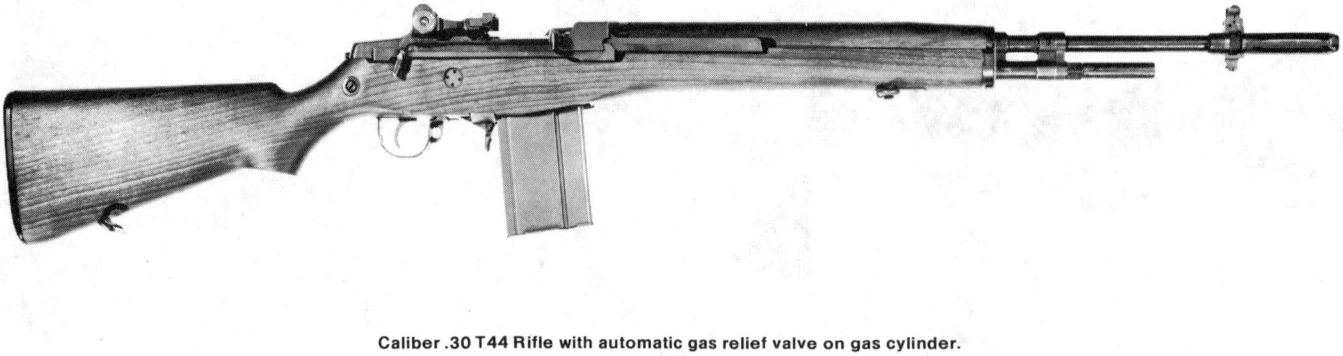

Caliber .30 T44 Rifle with automatic gas relief valve on gas cylinder.

Caliber .30 T44E1 Rifle.

Caliber .30 T44E4 Rifle.

A further development of Mr. Stoner's is the Stoner 63 weapon system. The Stoner 63 system is also chambered for the .223 cartridge and consists of a rifle, carbine, light machine gun belt-fed, light machine gun magazine-fed, and a fixed and flexible medium machine gun, all of which use the same basic components. This system, which has undergone extensive tests by the United States Marine Corps, is being handled by the Cadillac Gage Company. It is covered in detail later in this chapter.

There is a great deal of activity in the development of military rifles in the United States today. A number of versions of the "SPIW"—Special Purpose Individual Weapon—a weapon which will have an area fire as well as a point fire capability, are currently under development.

Springfield Armory and a number of contractors submitted designs for the "SPIW" development. Only one of these designs, that submitted by AAIW Corp., has lasted. All of these designs used flechettes - small arrows - for their point fire capability. The flechette which is smaller than the diameter of the bore is carried through the bore by a plastic sabot which strips off after leaving the muzzle.

There has also been some activity in the area of rifles that are extremely cheap and easy to manufacture and rifles that are easy to maintain. TRW among others submitted prototypes.

TRW prototype, low maintenance rifle.

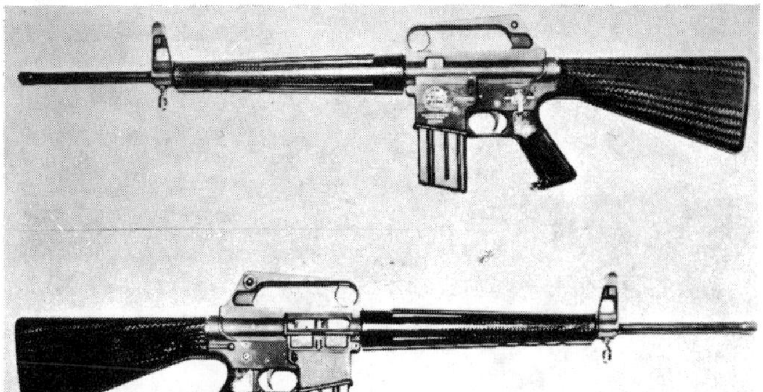

Armalite caliber .222 AR-15 Rifle.

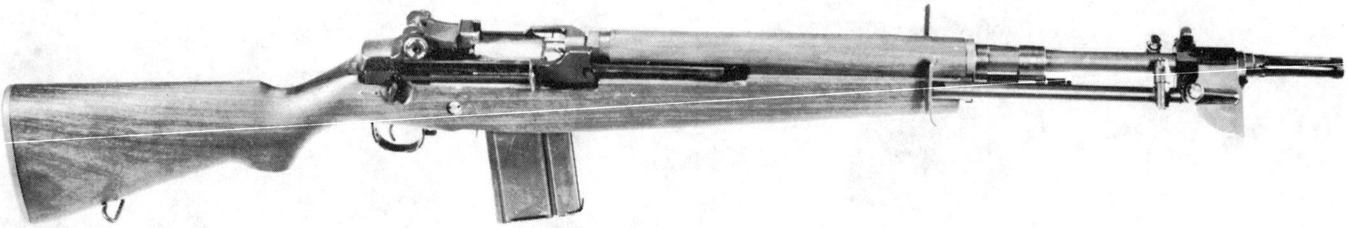

7.62mm NATO Rifle M15 (T44E5) Rifle.

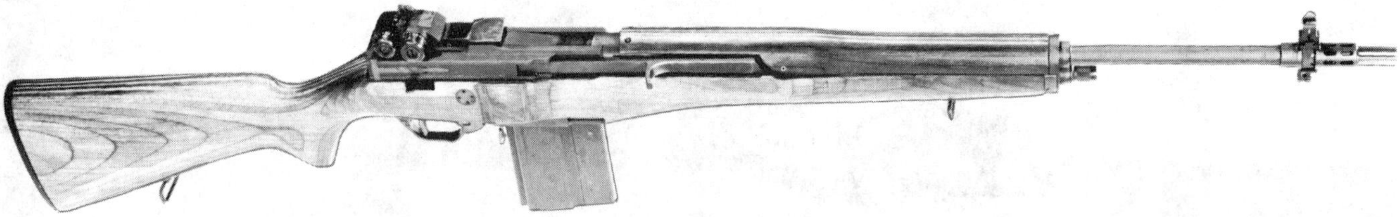

Caliber .30 T47 Rifle.

Caliber .30 T48 Rifle.

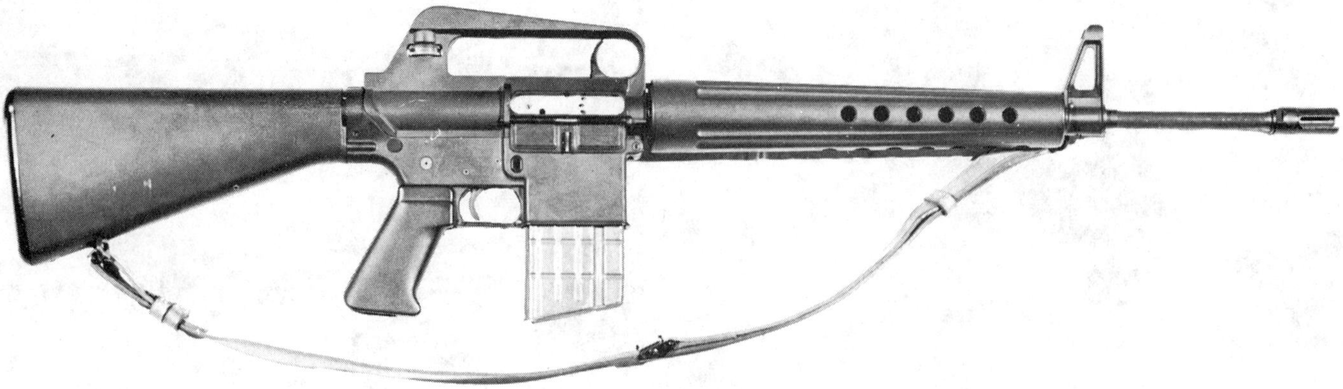

Armalite 7.62mm NATO AR-10 Rifle as made at Hembrug.

THE CALIBER .30 MODEL 1903, 1903A1, 1903A3 and 1903A4 RIFLES

The history of this rifle is given in the introduction to United States Rifles earlier in this chapter.

Differences in 1903 Models

The differences between the various 1903 individual models follow. The basic 1903 with straight stock, has a leaf-type rear sight, and no finger grasping grooves in the fore-end of the stock. The bolt was originally bent straight down, but after 1918 was given a slight bend to the rear. These bolts are stamped "NS" on the handle. These rifles were made by Springfield and Rock Island. Receivers made by Springfield with serial numbers below 800,000 are made of Springfield Class "C" steel and should not be used with stepped-up loads. Later receivers made at Springfield are suitable for use with any factory-loaded .30-06 ammunition, except proof loads. Receivers of Rock Island manufacture up to number 285,506 were also manufactured of Class "C" steel and require similar precautions.

M1903 Mark I. This rifle was altered by the fitting of a new sear mechanism, changes in the cutoff, and a slot was drilled in the left side of the receiver. This was done to accommodate the caliber .30 Pedersen device, which replaced the regular bolt and converted the weapon into a semiautomatic, firing the caliber .30 pistol cartridge. The Pedersen device, or "Pistol caliber .30 Model 1918" as it was known, was never issued and the Mark I rifles were remodified to be used as standard M1903s.

M1903A1. A pistol grip "type C" stock with grasping grooves was fitted, the butt plate was checkered, and the trigger was serrated. This model was adopted in December 1929.

M1903A2. This is not a shoulder rifle; it is a barreled receiver used as a subcaliber rifle in various artillery pieces.

M1903A3. Adopted on 21 May 1942, this rifle was modified to simplify production. Stamped bands, swivels, butt plate, and magazine trigger guard assemblies are used. A one-piece hand guard, simplified front sight, and ramp-type aperture rear sight, mounted on the receiver bridge, are also found on these rifles. Most of these rifles are fitted with straight stocks, however a semi-pistol grip stock was also issued. Four, two, and occasionally six land-and-groove barrels may be found. These rifles were manufactured by Remington Arms and the L.C. Smith Corona Typewriter Co. A total of 945,846 of these rifles were made. Remington also produced approximately 345,000 M1903 and M1903 (modified) rifles before production of the M1903A3 began.

M1903A4. This is the sniper version of the M1903A3; a full pistol grip stock is fitted. The bolt handle is cut away to clear the M73B1 (Weaver 330C) 2.5 power scope. No iron sights are fitted. This rifle was adopted in December 1942. Approximately 26,650 were manufactured.

Model 1942. This is a Marine Corps modification of the M1903A1. The rifle is fitted with a 10x Unertl scope.

Total production of the Model 1903 and 1903A1 rifles was approximately 1,295,000 rifles at Springfield, Rock Island, and Remington Arms.

Loading and Firing the Model 1903

As a Single-Shot Rifle. Check the magazine cutoff on the left side of the receiver, making sure that it is down and the word "Off" can be seen. Turn the bolt handle up as far as it will go. This will release the locking lugs from their recesses in the receiver, and permit the bolt to be drawn straight back. Now place a cartridge in the firing chamber, thrust home the bolt to seat the cartridge properly and permit the extractor to snap over the cannelure of the cartridge case, and turn the bolt handle down as far as it will go to lock the piece. Note: This rifle is cocked as the bolt is rotated clockwise. The knob on the cocking piece will project out of its casing when the weapon is at full cock. If desired, the safety lock may now be applied, by turning the safety lock thumb piece at the rear of the bolt over to the right as far as it will go, when the word "Safe" will be seen on its face.

If safety is not set, pressing the trigger will explode the cartridge in the firing chamber, which then may be extracted and ejected by turning up and pulling back sharply on the bolt handle.

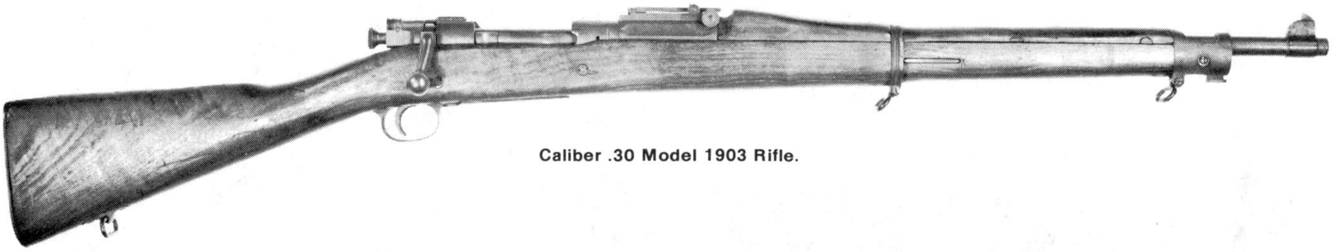

Caliber .30 Model 1903 Rifle.

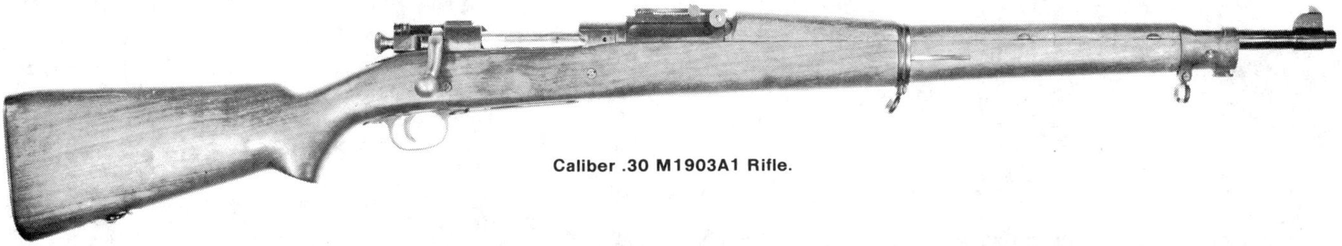

Caliber .30 M1903A1 Rifle.

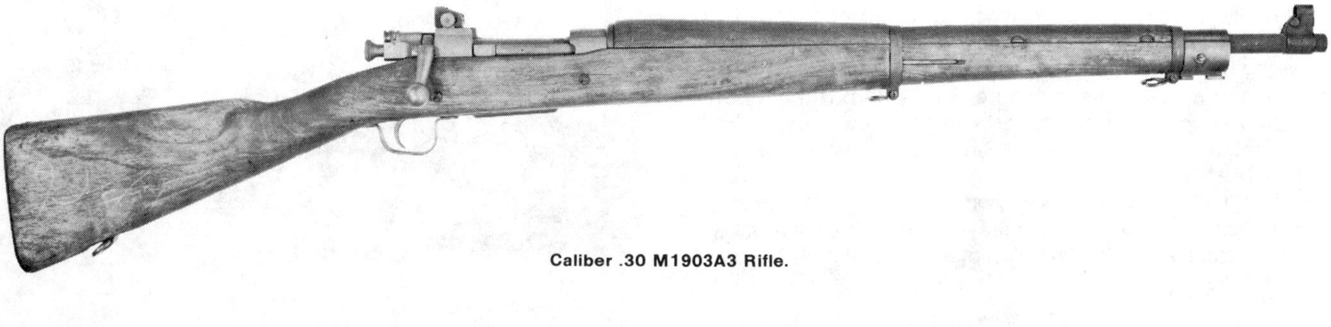

Caliber .30 M1903A3 Rifle.

Caliber .30 M1903A4 Rifle.

To Load Magazine. Turn cutoff up as far as it will go and the word "On" will appear. Now lift up and pull back on the bolt handle as far as it will go, and since the cutoff is in the position of magazine loading, the bolt will be permitted to travel farther back than when single-shot firing is to be done with the cutoff in the "Off" position. When the cutoff is set for magazine loading, the bolt can be drawn back far enough to permit the magazine follower to rise in front of it.

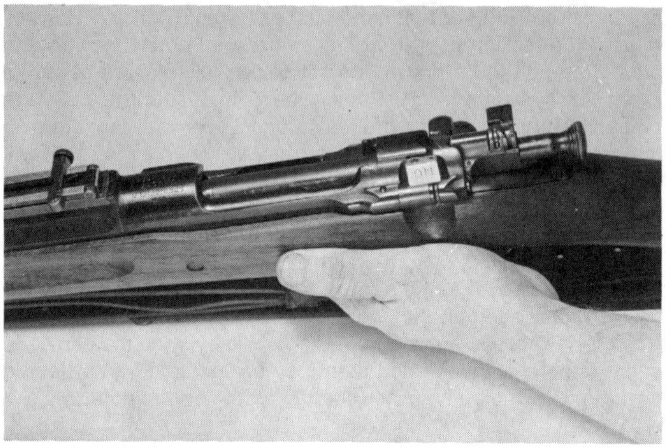

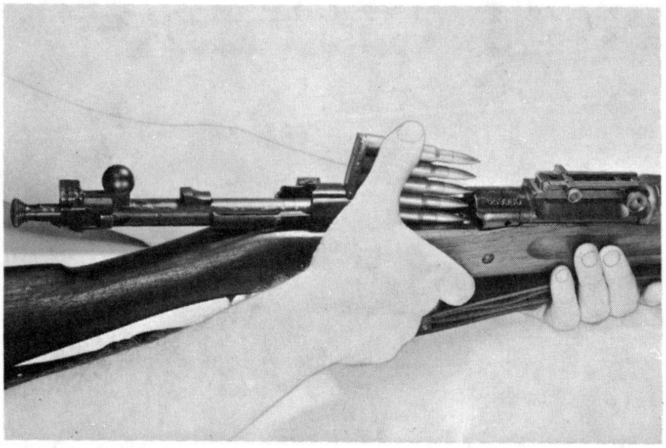

With muzzle pointed up in the air at an angle of 45°, insert a loaded clip in the clip slots; place thumb of right hand over powder space on top cartridge, extend fingers around the rifle with tips resting on magazine floor plate, and with thumb force the five cartridges down into the magazine. Now pull out the empty cartridge clip. Push the bolt home and turn down the locking handle. Set the safety lock on the "Safe" position unless weapon is to be used at once.

To load rifle with six cartridges, proceed as follows: With cutoff set in the "On" position, raise and draw bolt handle back as far as it will go. Load magazine with full clip as above. With left thumb push down on top cartridge, forcing it down below the line of the bolt; and push the bolt slightly over it. Now insert a cartridge in the firing chamber, and push the bolt all the way home, turning and locking it.

Magazine cutoff may now be turned to the "Off" position and the rifle used as a single loader, with the full magazine in reserve. Magazine may be brought into play at any time by merely turning up cutoff.

Setting the Sights on the Model 1903

The elaborate sighting system of the M1903 Springfield rifle has a great deal to do with its reputation as the most accurate military rifle ever developed. To set the sight, the binding screw on the right-hand side of the sight leaf is unscrewed so as to permit the slide to be raised or lowered on the leaf. The index line near the peep is lined up with the range line on the sight leaf which is directly under the number giving the yard range.

Some sights do not have 25- and 50-yard graduation lines, and in such case you must estimate the distance between the 100-yard graduations on the sights. Binding screw must be screwed up tight when the index line is properly aligned for the range at which you wish to shoot.

Wind Gauge. A wind gauge is provided on this sight to compensate for the drift of the bullet caused by wind. The windage screw knob is at the forward end of the sight base on the right-hand side. The graduated windage indicator is at the rear end of the base. Turning the windage screw knob will turn the sighting line to right or left, one point of windage for each calibrated line. If the sight leaf moves to the right, the bullet will strike to the right; if sight leaf moves to the left, bullet will be directed to the left.

The M1903A9 rifle has a ramp-mounted aperture which is slid up and down for range. Windage is adjusted by a knob on the right side of the sight base.

Field Stripping the Model 1903

Turn the magazine cutoff to the center of the position midway between "off" and "on." Draw the cocking piece back to full cock. Turn the safety lock up to the vertical position, midway between "off" and "on." Now raise the bolt handle and draw the bolt straight back to the rear out of the receiver.

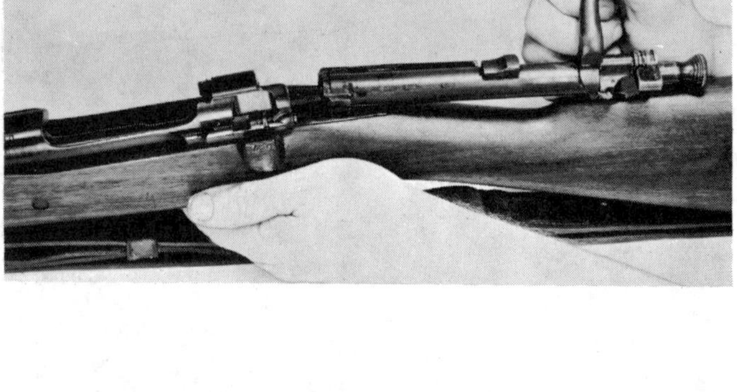

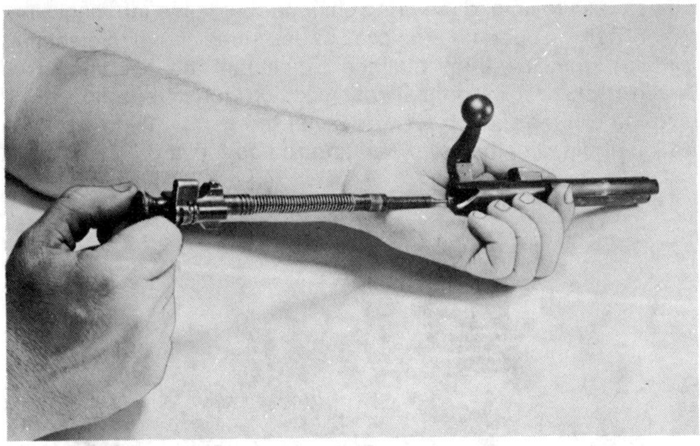

To Dismount Bolt. Holding bolt firmly in left hand, press in bolt sleeve lock with right thumb and unscrew, turning to the left. Bolt sleeve assembly can now be drawn back out of bolt.

Holding firing pin sleeve with left forefinger and thumb, pull back on the cocking piece with the right middle finger and right thumb, and turn the safety lock to the left with the right forefinger to release it. This will relieve part of the tension of the mainspring.

Resting the head of the cocking piece on a firm surface, pull back the firing pin sleeve and remove the striker. Firing pin sleeve, mainspring, and firing pin rod may then be withdrawn.

Extractor is removed by turning it to the right and pushing forward.

Assembling Bolt. Holding bolt handle up in left hand, make sure that the extractor collar lug is in line with the safety lock on the bolt, and insert the extractor collar lug in its undercut in the extractor, then push extractor until tongue comes in contact with bolt face. Now press extractor hook against a rigid surface to spring it into its groove in the bolt.

See that the safety is down and to the left, and assemble firing pin rod and bolt sleeve. Place the cocking piece against a solid surface, draw back the firing pin sleeve and attach the striker. The firing pin must be cocked before the bolt can be screwed on. This is done by pressing the striker point against a wooden surface (which must not be hard enough to injure it). Force the cocking piece back and engage the safety lock.

Assembled bolt sleeve is now replaced in the bolt and screwed until the bolt sleeve lock engages.

With cutoff still turned to center notch, insert bolt in its guide in the receiver, push down the magazine follower and push the bolt home. Now turn safety lock and cutoff down to the left, and press the trigger.

Stripping the Magazine. Turn rifle upside down, insert nose of bullet in the hole in the rear of the floor plate to depress the spring catch.

Retaining pressure, pull back toward the trigger guard. This will release the spring and the magazine follower, and permit them to be removed from the weapon.

Assembling the Magazine. When assembling the magazine, make sure the front end of the floor plate catches on the front end of the magazine opening and push it toward receiver and forward until the spring catch engages.

No further stripping of this weapon is necessary or desirable.

M1903A3, A4. Floor plate cannot be removed, follower is removed from the top of the magazine.

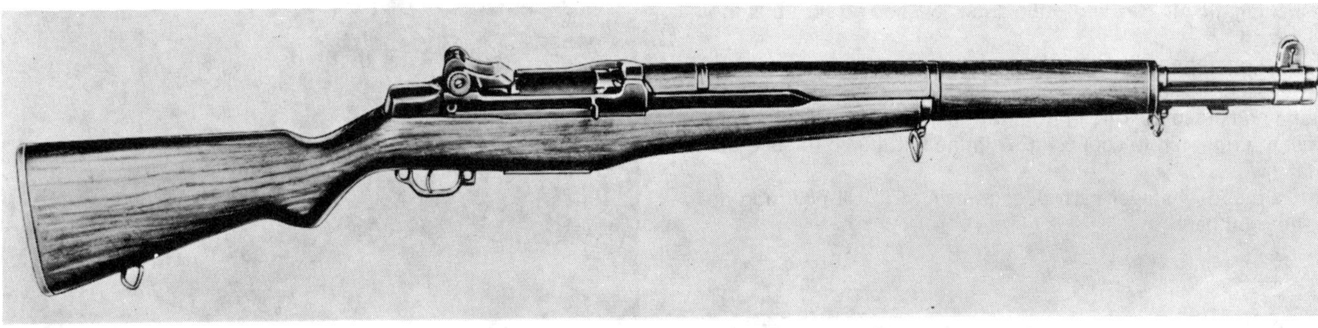

Caliber .30 M1 Rifle (Garand semiautomatic rifle).

Characteristics of United States Bolt-Action Service Rifles

	Model 1903	**Model 1917**	**Model 1903A1**	**Model 1903A3**	**Model 1903A4**
Caliber:	.30 U.S. (.30-06)				
System of operation:	Manually operated turn bolt				
Length overall:	43.2 in.	46.3 in.	43.2 in.	43.5 in.	43.5 in.
Barrel length:	24 in.	26 in.	24 in.	24 in.	24 in.
Feed device:	5-round, staggered row, non removable box magazine				
Sights: Front:	Blade.	Blade with protecting ears.	Blade.	Blade.	2.2X telescopic*
Rear:	Leaf with aperture, notched battle sight.	Leaf with aperture, aperture battle sight.	Leaf with aperture, aperture battle sight.	Aperture on ramp.	
Weight:	8.69 lb. with oiler and thong case.	8.18 lb. w/oiler and thong case.	8.69 lb. w/oiler and thong case.	8 lb.	9.38 lb.
Muzzle velocity: (M2 Ball):	2805 f.p.s.	Approx. 2830 f.p.s.	2805 f.p.s.	2805 f.p.s.	2805 f.p.s.

*Weaver 330 C (M73B1) usually used, some were fitted with the Lymann Alaskan (M73).

RIFLE CALIBER .30 M1

The history of the M1 is given earlier in this chapter. The M1 was produced by Springfield and Winchester during World War II. A total of over 4,040,000 M1s were made; of this total Winchester made 513,582. After the war, an additional 600,000 were made by Springfield, International Harvester, and Harrington and Richardson. The Breda and Beretta firms in Italy also made the M1 for the Italian government. Beretta also made the rifle for Denmark and Indonesia. The U.S. Navy is currently using a number of M1 rifles chambered for the 7.62mm NATO cartridge. These rifles are stamped 7.62mm on the right rear side of the receiver.

Loading and Firing the M1

Pull operating rod straight to the rear. It will be caught and held open by the operating rod catch.

Place the loaded clip on top of the magazine follower; with right side of right hand against the operating rod handle press down with right thumb on the clip until it is caught in the receiver by the clip latch.

Remove the right thumb from the line of the bolt and let go of the operating rod handle which will run forward under the compression of the spring. Push operating rod handle with heel of right hand to be certain that bolt is fully home and locked.

Pressing the trigger will now fire one cartridge. Weapon will be ready for the next pull of the trigger.

If weapon is not to be used at once, set the safety. The safety is in front of the trigger guard. Pulling it back toward the trigger sets it on safe; pushing it forward is the fire position.

Note that the cartridge clip is reversible and may be fed into the rifle from either end.

Unloading the M1 (Garand) Rifle. First check to be sure that the safety is off.

Pull the operating rod back sharply and hold it in rear position. This will eject the cartridge that was in the firing chamber. With the left hand grasp the rifle in front of trigger guard. Hold the butt against the right hip to support it. With the left thumb release the clip latch.

The clip and whatever cartridges remain in it will now pop up into the right hand.

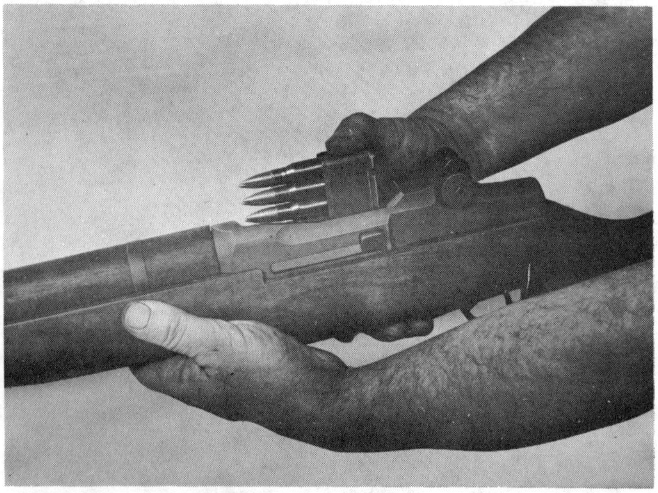

With the right side of the right hand held against the operating rod handle, force the operating rod slightly to the rear. With the right thumb now push down the magazine follower and permit the bolt to move forward about an inch over the end of the follower.

Remove the thumb smartly from the follower and let go of the operating rod. The action will close under the tension of the spring. Now press the trigger.

If you wish to unload the firing chamber but leave the magazine loaded, pull the operating rod back as described above to eject the cartridge from the firing chamber; then pull the operating rod handle back past its normal rear position, force the clip down, ease the rod far enough forward to let the bolt handle ride over the top of the clip, then let operating rod go forward.

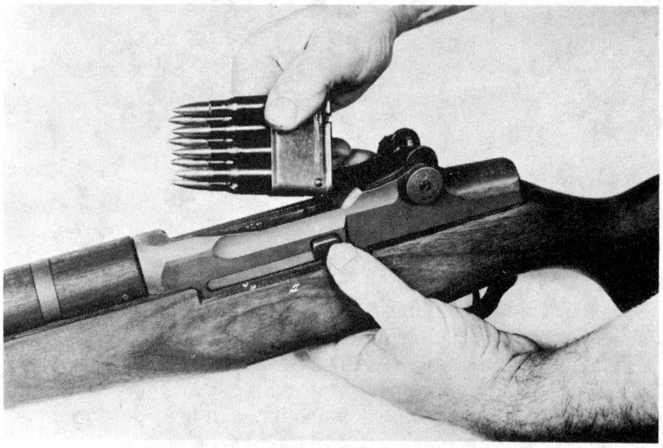

Field Stripping the M1

A thorough knowledge of field stripping is necessary in order to give the rifle the care essential to its correct operation.

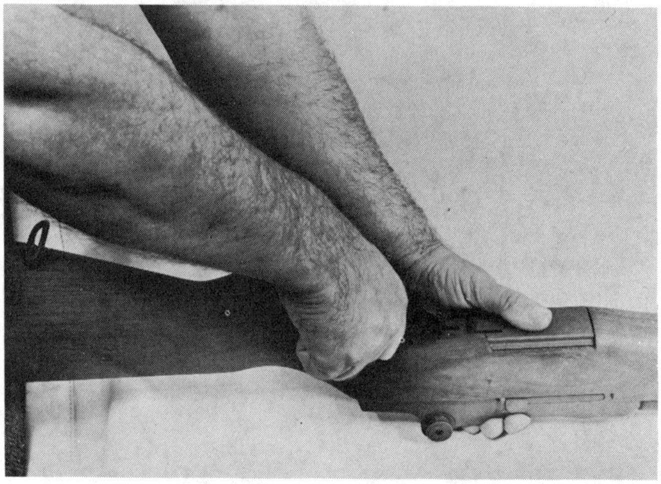

Start by placing the rifle upside down on a firm surface. Holding the rifle with left hand, rest the butt against the left thigh.

With thumb and forefinger unlatch the trigger guard by pulling back on it.

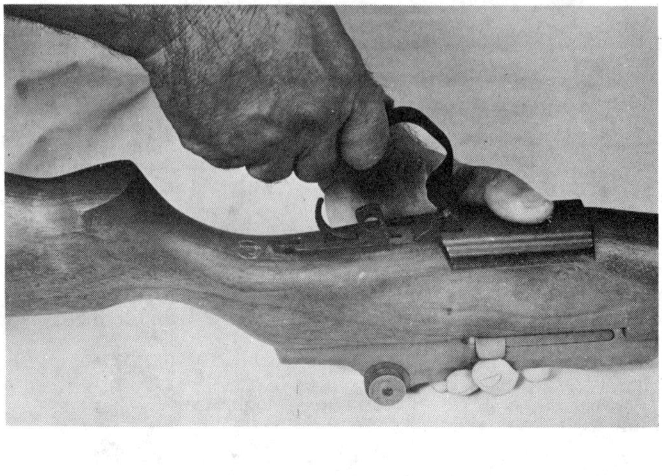

Continue the pressure and pull out the trigger housing group.

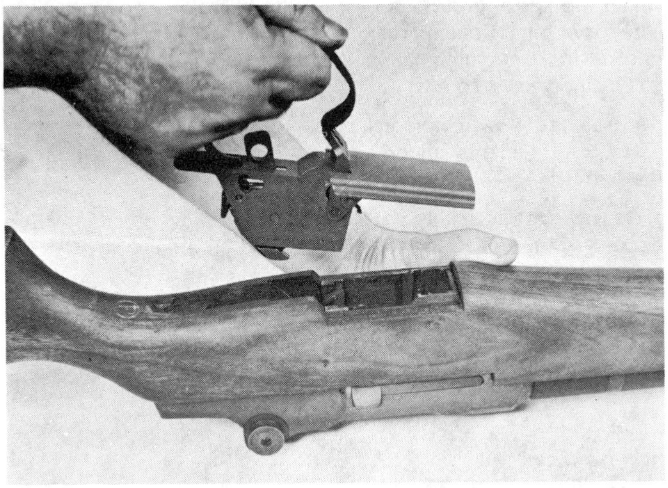

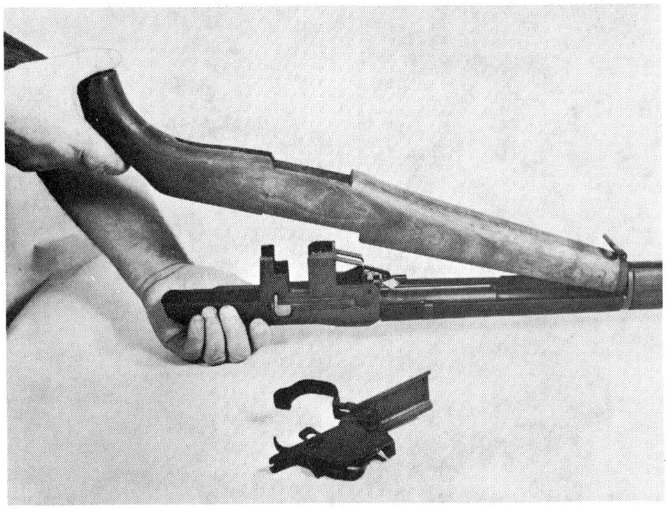

With left hand grasp rifle over rear sight. With right hand strike up against the small of the stock, firmly grasping it at the same time. This will separate the barrel and receiver group from the stock group.

To Dismount Barrel and Receiver Group, M1. With the barrel down, grasp the follower rod at the knurled portion with thumb and forefinger and press it toward the muzzle to free it from the follower arm.

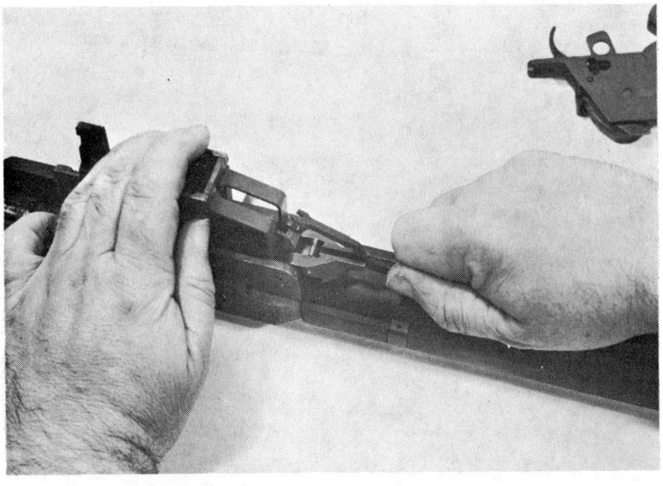

The follower rod and its compensating spring which is attached may now be withdrawn to the right. The compensating spring is removed from the follower rod by holding spring with left hand and twisting rod toward the body with the right hand and pulling slightly to the right.

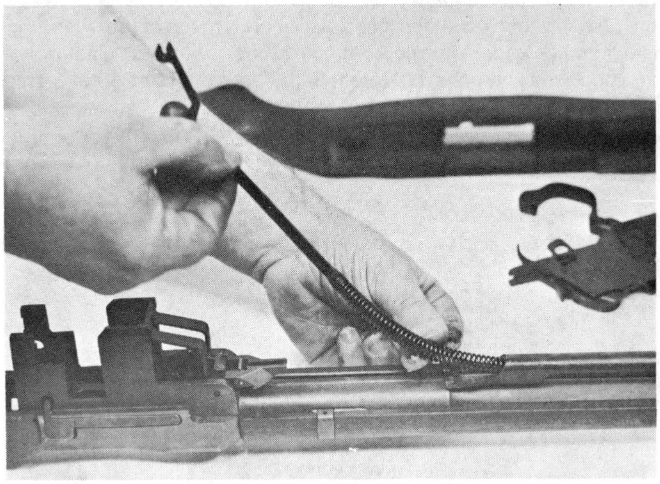

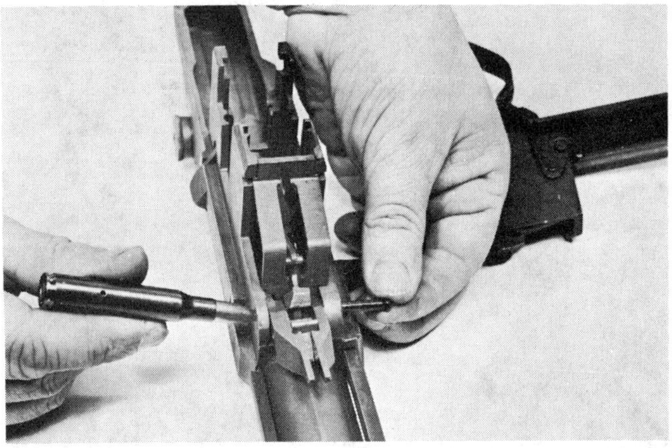

With the point of a bullet, push the follower arm pin from its seat and pull it out with the left hand.

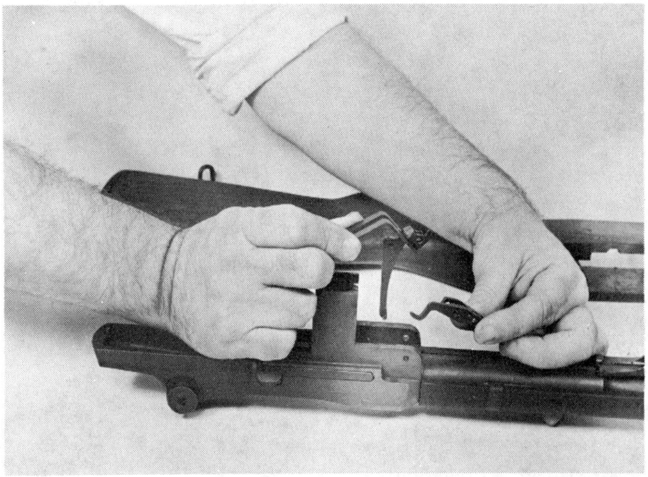

Seize the bullet guide, follower arm, and operating rod catch assembly; draw these to the left until they disengage. The three separate parts may now be lifted out. Accelerator pin is riveted in its seat, so do not attempt to remove accelerator from operating catch assembly.

Lift out follower with its slide attached (do not separate follower from slide).

Holding barrel and receiver assembly with left hand, grasp the operating rod handle with right hand and move it slowly to the rear, meanwhile pulling the rod handle up and away from the receiver. (This disengages operating rod from bolt, when the lug on the operating rod slides into the dismount notch of the operating rod guide groove.) When operating rod is disengaged pull it down and back and withdraw it. (Note that the operating rod is bent. This is intentional. Do not attempt to straighten it.)

Slide the bolt from the rear to the front by pushing the operating lug on it, and lift it out to the right front with a slight twisting motion.

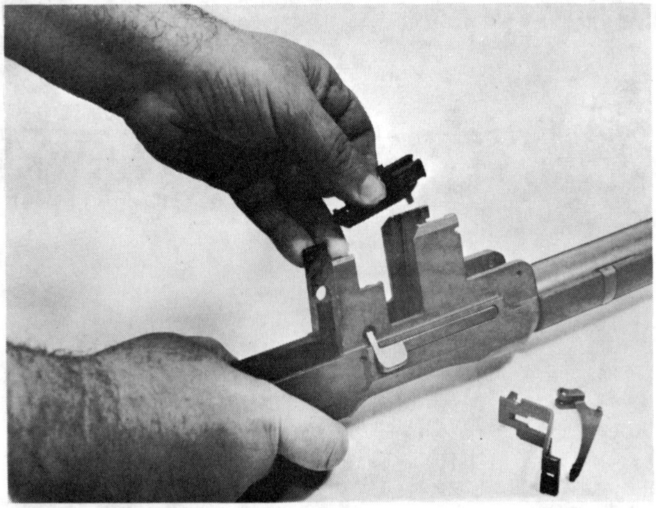

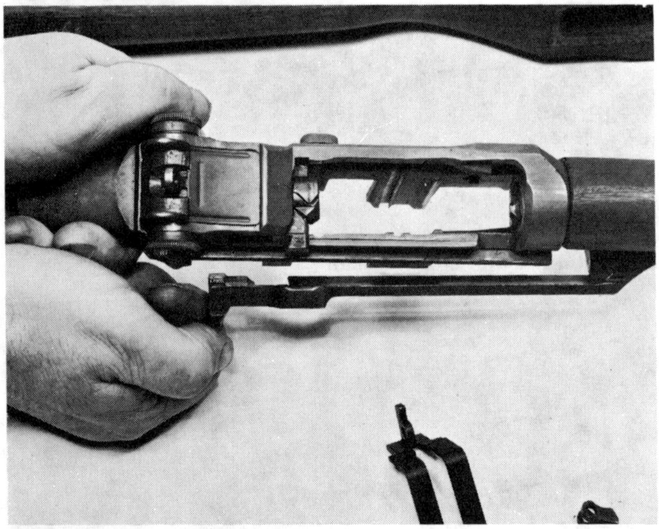

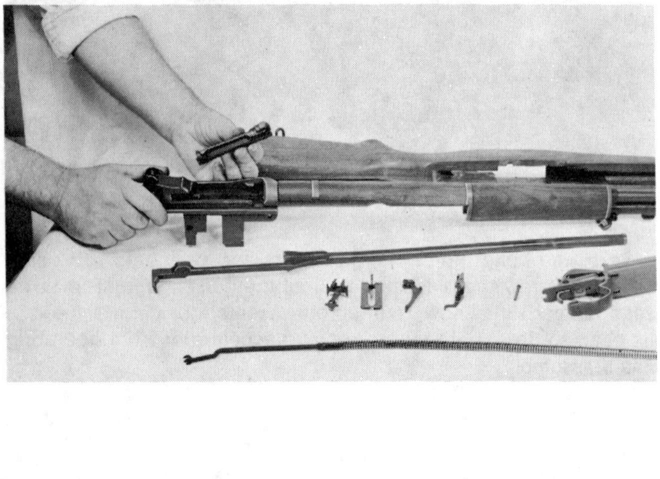

Note on M1 Gas Cylinder. A spline-type gas cylinder is used in which the barrel protrudes beyond the cylinder. The front sight screw is entered from the rear of the sight and is sealed to prevent unscrewing. The combination tool must be used to unscrew the gas cylinder lock screw.

Unscrew the gas cylinder lock.

The gas cylinder is tapped toward the muzzle and removed from the barrel.

Gas cylinder assembly should never be removed except when necessary to replace the front handguard assembly.

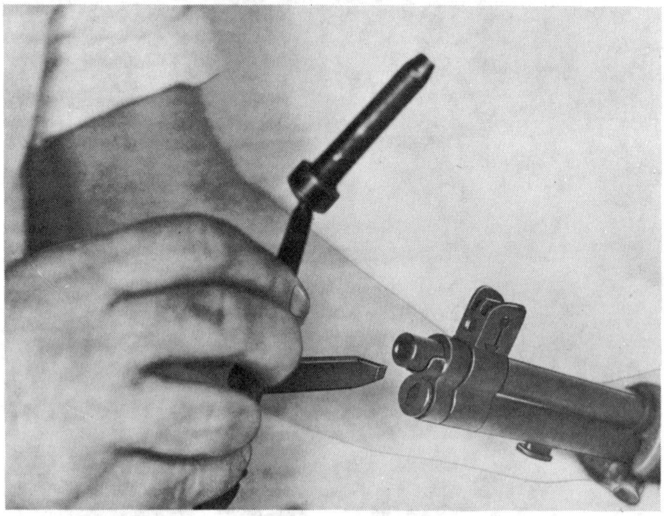

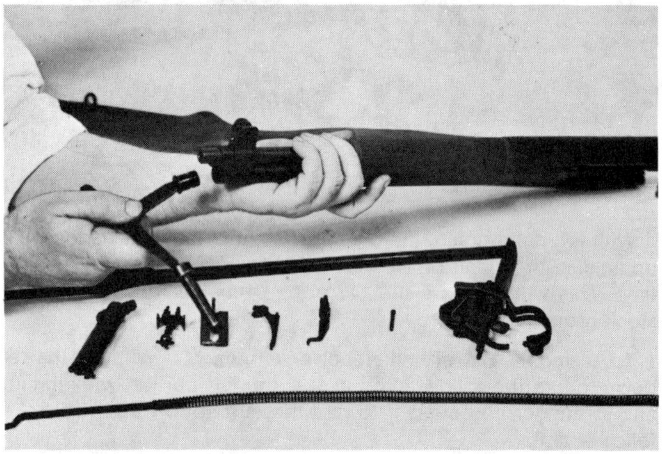

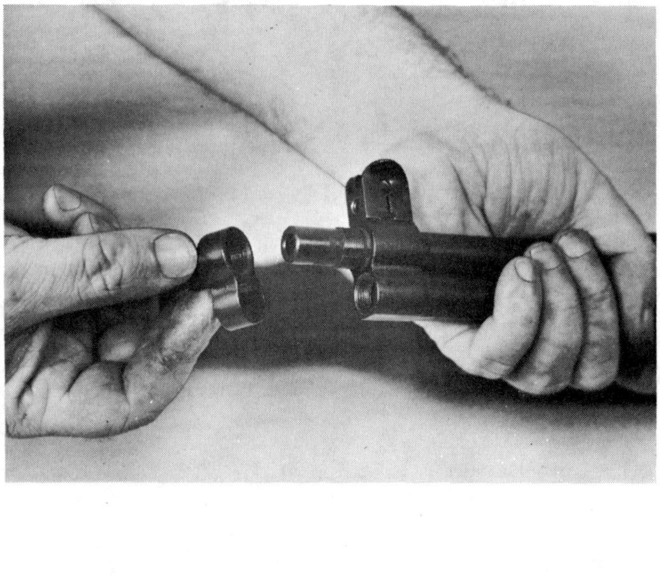

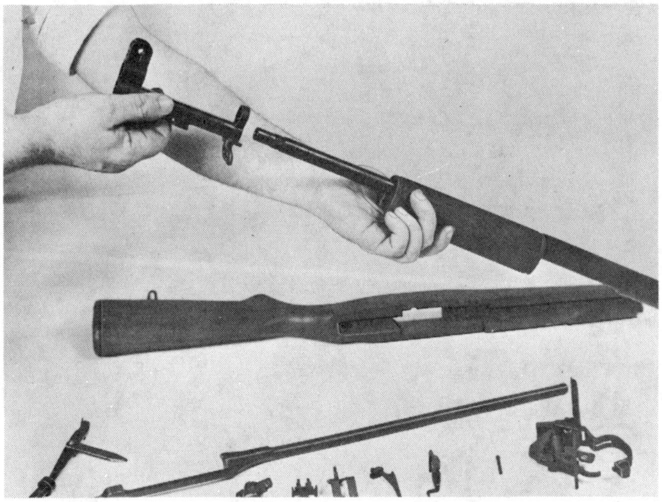

Assembling the M1 Rifle. Replacing gas cylinder, if it has been dismounted, is done by merely reversing the dismounting procedure.

To assemble barrel and receiver group, tilt the barrel and receiver assembly, sight up and muzzle to the front to an angle of about 45°.

Holding the bolt by the right locking lug so the front end of the bolt is somewhat above and to the right of its extreme forward position in the receiver, insert the rear end in its bearing on the bridge of the receiver. Switch it from right to left far enough to let tang of the firing pin clear the top of the bridge. Next guide the left locking lug of the bolt into its groove just to the rear of the lug on the left side of the receiver, and start right locking lug into its bearing in the receiver. Now slide bolt back to its extreme rear position.

Turn barrel and receiver assembly in left hand until barrel is down.

Grasp operating rod at the handle and holding it handle up, insert piston head into gas cylinder about 3/8". Be sure that operating rod handle is to the left of the receiver.

Hold barrel and receiver assembly in left hand and twist to the right until barrel is uppermost.

Adjust operating rod with right hand so that camming recess on its rear end fits over operating lug on bolt. Now press operating rod forward and downward until bolt is seated in its forward position.

With barrel and receiver assembly held barrel down and muzzle to your left, replace the follower with its attached slide so that its guide ribs fit into their grooves in the receiver. (The square hole in the follower must be to the right.) The follower slide rests on the bottom surface of the bolt when the follower is in the correct position.

With left hand replace bullet guide, fitting the shoulders of the guide into their slots in the receiver and the hole in the projecting lug in line with the hole in the receiver.

With left hand replace follower arm passing stud end through bullet guide slot and inserting stud in proper grooves in front end of follower.

Place the forked end of the follower arm in position across the projecting lug on the bullet guide, with pin holes properly alined.

Insert rear arm of operating rod catch into clearance cut in the bullet guide (be sure its rear end is below the forward stud of the clip latch which projects into the receiver mouth). Line up the holes in the operating rod catch, the follower arm, and the bullet guide with those in the receiver; and insert the follower arm pin in the side of the receiver toward your body and press the pin home.

Insert operating rod spring into operating rod; assemble follower rod by grasping the spring in left hand and inserting follower rod with right hand, twisting the two together until the spring is fully seated.

Seize the knurled portion of follower rod with thumb and forefinger of left hand with hump down and forked end to the right.

Place front end of follower rod into operating rod spring and push to the left, seating the forked end against the follower arm.

Insert U-shaped flange of stock ferrule in its seat in the lower band.

Pivoting about this group, guide chamber and receiver group and press into position in the stock.

Replace trigger housing group with trigger guard in open position into the stock opening.

Press into position, close and latch trigger guard. This completes reassembly.

How the M1 Works

Starting with the rifle loaded and cocked the action is as follows: The trigger being pressed, the hammer strikes the firing pin, exploding the cartridge in the chamber. As the bullet passes over the gas port drilled in the under side of the barrel, some of the gas escapes into the cylinder and blasts back against the piston and operating rod with force enough to drive the rod to rear and compress the spring.

During the first 5/16" of rearward travel the operating lug slides in a straight section of the recess on the operating rod; after which the cam surface of this recess is brought in contact with the operating lug which it cams up, thereby rotating the bolt from right to left to unlock its two lugs from their recesses in the receiver.

During the moment of delayed action, the bullet leaves the barrel and the breech pressure drops to a safe point. The further rotation of the bolt then cams the hammer away from the firing pin and pulls the firing pin back from the bolt. The operating rod continues its backward movement carrying the bolt with it as the lug on the bolt has reached the end of its recess.

During this rearward motion of the bolt, the empty case is withdrawn from the chamber by the extractor positioned in the bolt until it is clear of the breech; at which point the ejector, exerting a steady pressure on the base of the cartridge case, throws it to the right front by the action of its compressed spring.

The rear end of the bolt at this point forces the hammer back, rides over it, and compresses the hammer spring; and finally stops in the rear end of the receiver.

As the bolt has now cleared the clip, the follower spring forces the cartridges up until the topmost one is in line with the bolt.

The operating rod spring comes into play at this point to pull the action forward.

Forward Movement of the Action. As the bolt moves forward, its lower front base strikes the base of the cartridge case and pushes it into the firing chamber. The hammer, pressed by its spring, rides on the bottom of the bolt. While it tends to rise, it is caught and held by the trigger lugs engaging the hammer hook, if trigger pressure has not been released. Otherwise the trigger engages the rear hammer hook until letting go the trigger disengages the sear from the hammer. The hammer then slides into engagement with the trigger lugs.

When the bolt nears its forward position, the extractor engages near the rim of the cartridge and the base of the cartridge forces the ejector into the bolt, compressing the ejector spring.

The rear surface of the cam recess in the operating rod now cams the operating lug down and thereby twists the bolt from left to right until the two lugs lock into their places in the receiver.

The operating rod drives ahead for another 5/16". The rear end of the straight section of the operating rod recess reaches the operating lug on the bolt, which completes the forward movement and leaves the rifle ready to fire when the trigger is pressed.

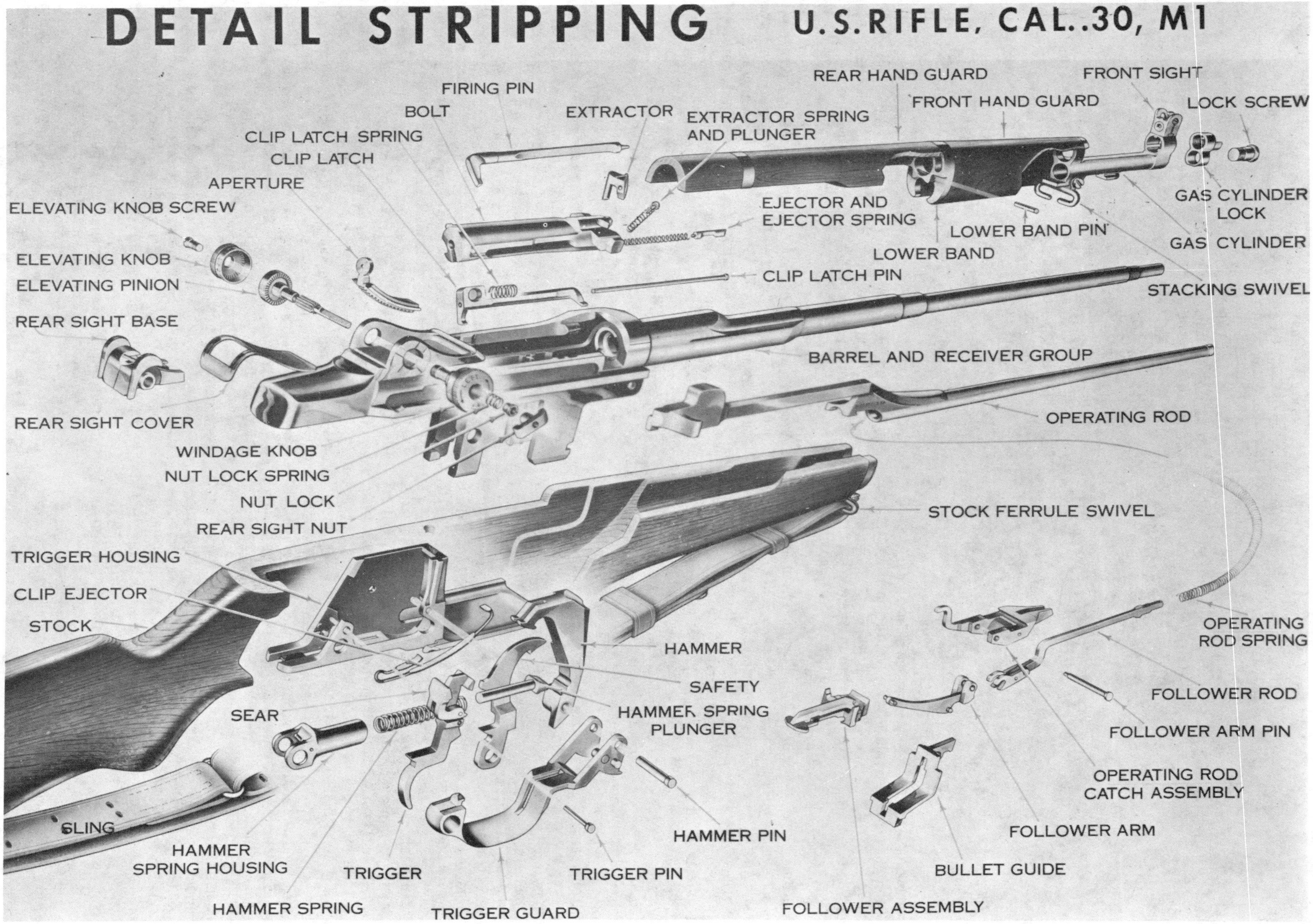
DETAIL STRIPPING
U.S.RIFLE, CAL..30, M1
FIRING PIN
BOLT
EXTRACTOR
EXTRACTOR SPRING
AND PLUNGER
REAR HAND GUARD
FRONT HAND GUARD
FRONT SIGHT
LOCK SCREW
CLIP LATCH SPRING
CLIP LATCH
APERTURE
ELEVATING KNOB SCREW
EJECTOR AND
EJECTOR SPRING
LOWER BAND PIN
LOWER BAND
GAS CYLINDER
LOCK
GAS CYLINDER
ELEVATING KNOB
ELEVATING PINION
CLIP LATCH PIN
STACKING SWIVEL
REAR SIGHT BASE
BARREL AND RECEIVER GROUP
REAR SIGHT COVER
OPERATING ROD
WINDAGE KNOB
NUT LOCK SPRING
NUT LOCK
REAR SIGHT NUT
STOCK FERRULE SWIVEL
TRIGGER HOUSING
CLIP EJECTOR
STOCK
OPERATING
ROD SPRING
HAMMER
SAFETY
FOLLOWER ROD
SEAR
HAMMER SPRING
PLUNGER
FOLLOWER ARM PIN
OPERATING ROD
CATCH ASSEMBLY
SLING
HAMMER PIN
FOLLOWER ARM
HAMMER
SPRING HOUSING
TRIGGER
TRIGGER PIN
BULLET GUIDE
HAMMER SPRING
TRIGGER GUARD
FOLLOWER ASSEMBLY

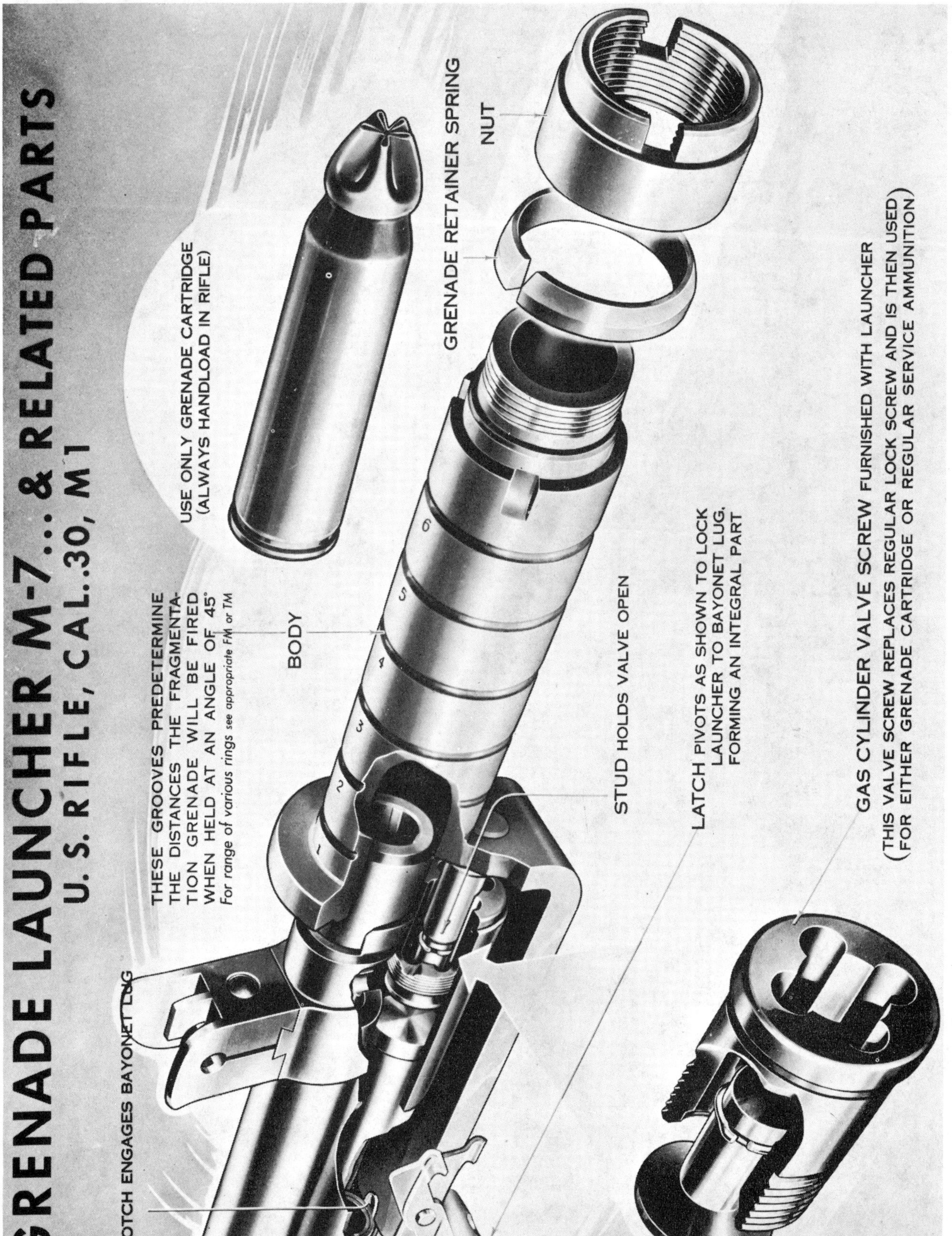
GRENADE LAUNCHER M-7... & RELATED PARTS
U. S. RIFLE, CAL.30, M1
THIS NOTCH ENGAGES BAYONET LUG
THESE GROOVES PREDETERMINE THE DISTANCES THE FRAGMENTATION GRENADE WILL BE FIRED WHEN HELD AT AN ANGLE OF 45°
For range of various rings see appropriate FM or TM
USE ONLY GRENADE CARTRIDGE (ALWAYS HANDLOAD IN RIFLE)
BODY
1
2
3
4
5
6
GRENADE RETAINER SPRING
NUT
STUD HOLDS VALVE OPEN
LATCH PIVOTS AS SHOWN TO LOCK LAUNCHER TO BAYONET LUG, FORMING AN INTEGRAL PART
GAS CYLINDER VALVE SCREW FURNISHED WITH LAUNCHER
(THIS VALVE SCREW REPLACES REGULAR LOCK SCREW AND IS THEN USED FOR EITHER GRENADE CARTRIDGE OR REGULAR SERVICE AMMUNITION)

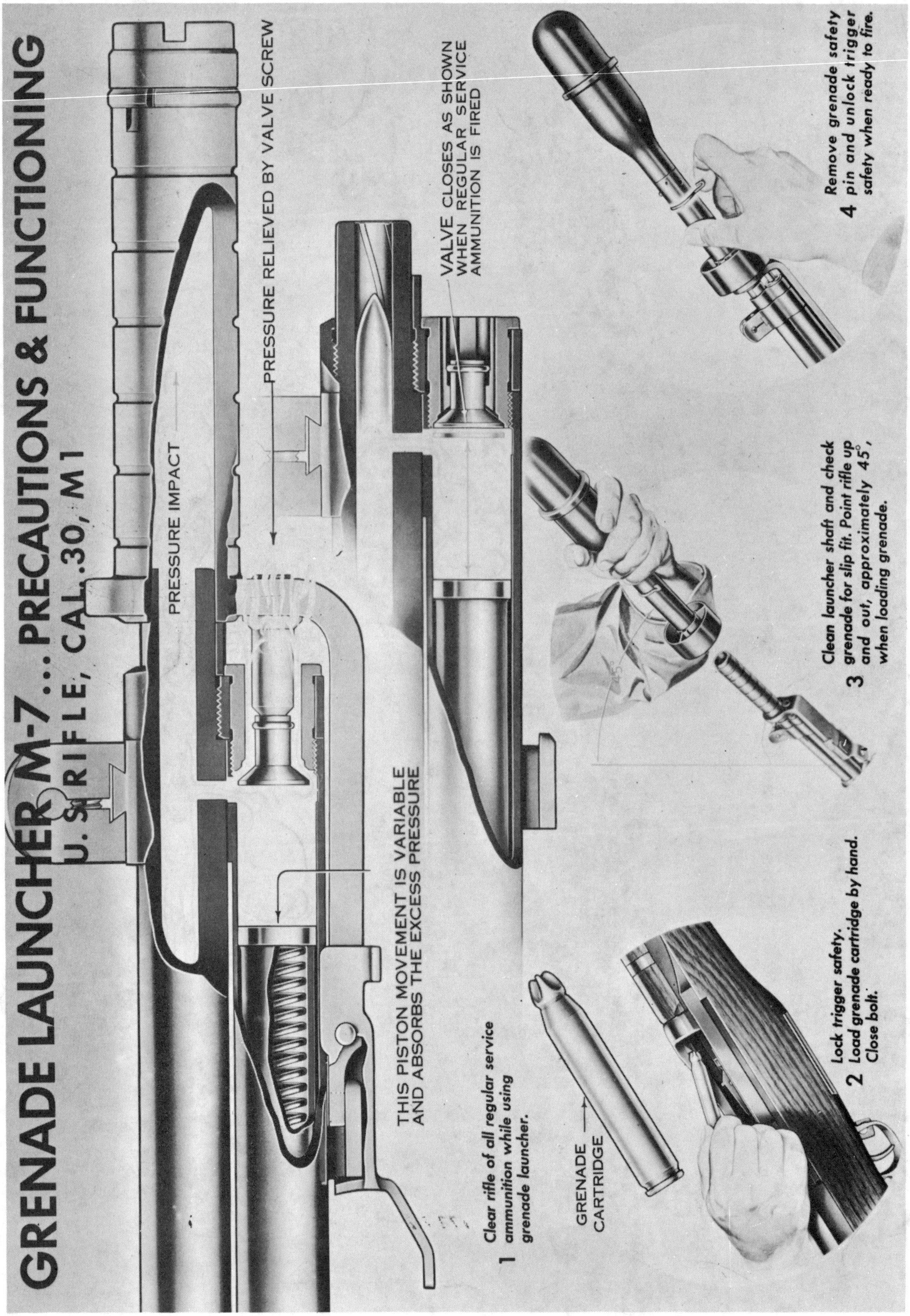
GRENADE LAUNCHER M-7 ... PRECAUTIONS & FUNCTIONING
U. S. RIFLE, CAL. .30, M1
PRESSURE RELIEVED BY VALVE SCREW
PRESSURE IMPACT
VALVE CLOSES AS SHOWN WHEN REGULAR SERVICE AMMUNITION IS FIRED
THIS PISTON MOVEMENT IS VARIABLE AND ABSORBS THE EXCESS PRESSURE
GRENADE CARTRIDGE
1 Clear rifle of all regular service ammunition while using grenade launcher.
2 Lock trigger safety. Load grenade cartridge by hand. Close bolt.
3 Clean launcher shaft and check grenade for slip fit. Point rifle up and out, approximately 45°, when loading grenade.
4 Remove grenade safety pin and unlock trigger safety when ready to fire.

This cycle continues as long as there are cartridges in the magazine and the trigger is squeezed.

Care of the M1 Rifle

The rifle must be kept clean and properly lubricated. Failure to do so may result in stoppages at a critical moment. The rifle should be inspected daily.

To Clean the Bore. A clean patch saturated with bore cleaner should be run through the bore a number of times. Plain water, hot or cold may be used if bore cleaner is lacking. While the bore is still wet, your metal brush should be run through several times to loosen up any material which has not been dissolved by the water. Dry patches should then be pushed through the bore until thoroughly dry. The bore should then be coated with light issue gun oil. Also use the chamber cleaning tool to give the chamber the same attention. Remember that primer fouling in the bore contains a salt which rusts the steel.

To Clean Gas Cylinder. The carbon forming in the gas cylinder varies in amount in different weapons. When the deposit is heavy, the rifle is sluggish in action and may fail to feed. The carbon must be scraped from the exposed surface of the front of the cylinder and the gas cylinder plug and piston head after extensive firing. A knife or similar sharp bladed instrument should be used for this scraping process.

Gas cylinder plugs and grooves in the gas cylinder should be cleaned so they will feed correctly in the plug.

The gas cylinder lock should be removed, and the lock screw inserted in the cylinder far enough to break loose any carbon. Inside the cylinder must be thoroughly wiped clean and oiled at the conclusion of any extensive firing.

When firing is expected to be resumed the next day, tilt the muzzle down and place a few drops of oil into the cylinder between the piston and the walls of the cylinder. Then operate the rod by hand a few times to distribute the oil thoroughly.

Wipe the outside of the gas cylinder and the operating rod and then oil lightly. Should no firing be expected for a week or two, remove the rod and gas cylinder lock screw (or plug) so that the cylinder is open at both ends. Then clean cylinder with rod and patches exactly as the bore of the rifle is cleaned.

Hold the weapon so that no water gets into the gas port. Do not remove the gas cylinder for cleaning.

Piston head and rod should be cleaned with cleaner or with water and dried thoroughly, while the rod and cylinder should be oiled before assembling. Any carbon present should be removed. Do not use abrasive cloth if it is possible to avoid doing so; and should it be used, take proper care that the corners of the plug or lock screw and piston head are not rounded.

Attention to Other Parts of Rifle. Graphite cup grease is used for lubricating bolt lugs, bolt guides, bolt cocking cams, compensating spring, contact surfaces of barrel and operating rod, operating rod cams and springs, and operating rod groove in the receiver.

All other metal parts should be cleaned and covered with a uniform light coat of oil.

Wooden parts must be treated with light coat of raw linseed oil about once a month.

The leather sling should be washed, dried with with a clean rag, and lightly oiled with neatsfoot oil while it is still damp, whenever the sling shows signs of stiffening or drying. Rust should be removed from the metal parts with a piece of soft wood and oil, never with abrasive. Screw heads must be kept clean to prevent rusting.

Be careful not to use too much oil as any heavy coat will collect dirt and interfere with operation.

THE JOHNSON CALIBER .30 SEMIAUTOMATIC RIFLE M1941

How to Load and Fire the Johnson Rifle

Lift magazine cover, on right side of receiver below and parallel to ejection port, insert five rounds either on a Springfield-type charger (stripper clip) or singly. If using a charger, insert horizontally into charger guides in charging port. When last round is stripped or fed singly into magazine, magazine cover will close automatically. Raise the operating handle 20° and pull completely to rear; bolt will run forward, chambering a round. The rifle is now loaded and will fire if the trigger is pulled. When the last round is fired, the bolt will remain to the rear; load as directed above and pull bolt handle slightly to the rear and release it—the bolt will run home chambering a cartridge and the weapon is loaded. The Johnson can be loaded with single rounds or chargers with the bolt home on a loaded chamber. It is therefore easy to replenish the magazine at any time. The magazine can be emptied by depressing the magazine cover with the thumb of the right hand. The safety lock lever is located immediately in front of the trigger guard. When the free end of the lever is at an angle to the right of the axis of the barrel, the weapon is on "safe." When the lever is to the left of the axis of the barrel, the weapon is on "fire."

How to Field Strip the Johnson Rifle

Check rifle to insure that it is empty. With the point of a bulleted round (or a drift), push on the latch plunger of the hinged barrel latch found in the hole in the forward right side of the fore-end and push the barrel rearward. Raise the operating handle with the thumb of the left hand to the unlocked position and withdraw the barrel from the receiver. Disengage the bolt stop plate plunger with the point of a bulleted round and lift out bolt stop plate. Remove bolt stop and disengage the link from the main spring plunger. Raise the operating handle and retract the bolt about two inches. Grasp the knob of the operating handle spindle and pull it outward. Slide the operating handle forward until it is clear of the shoulders in the extractor recess and remove it. Lift out the extractor. Grasp the projecting end of the link and pull it to the rear, withdrawing the bolt through the rear end of the receiver. Rotate the locking cam counterclockwise and remove it from the bolt, remove the firing pin, and push out the link pin and remove the link. Disengage the hammer block pin and push it out with point of bullet; pull off the butt stock. Remove the ejector pin and ejector.

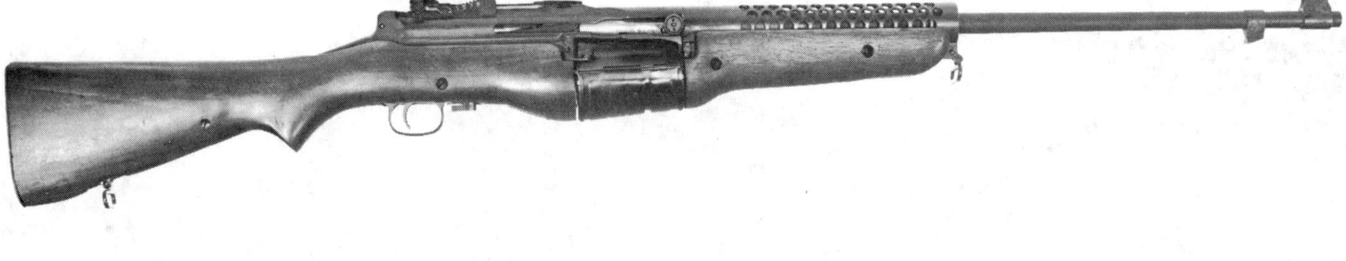

Johnson caliber .30 Semiautomatic Rifle M1941.

Hammer should be cocked before removing the butt stock group. Unscrew the front guard screw and the hammer block screw and lift out the hammer group from the stock. Unscrew the rear trigger guard screw and remove the trigger guard and safety assembly. No further disassembly is recommended. To reassemble, follow the above directions in reverse order.

How the Johnson Rifle Works

When the trigger is pressed, the sear disengages from the hammer. The hammer is driven against the firing pin which protrudes from the rear of the locking cam and the weapon fires. The barrel recoils against the tension of the barrel recoil spring and the main spring (transmitted through the bolt.) When the bullet is at the muzzle the barrel has moved rearward about 1/64 of an inch; when the bullet is about 2 feet from the muzzle the barrel has recoiled about 1/8 of an inch. The camming arm on the bolt engages the camming face in the receiver and unlocking begins. The Johnson has an 8-lug bolt. When the bullet is about 5 feet from the muzzle, the barrel has recoiled its full 5/8 inch and the bolt has rotated 20° and is unlocked. The rearward motion of the barrel is stopped by a shoulder in the receiver. The bolt moves to the rear independently of the barrel due to inertia and residual pressure in the chamber. The extractor gives the empty case a sharp pull and the bolt receives a sharp blow from the locking cam which taps the bolt rearward. The rearward movement of the bolt cocks the hammer and the extracted case is brought into contact with the ejector which throws the case clear of the rifle. The bolt is halted in its rearward travel by the forward end of the link bringing it up against the bolt stop, when the head of the bolt has passed behind the base of the top cartridge in the magazine.

As the bolt moves forward under the pressure of the main spring, the bolt face picks up a cartridge from the magazine and rams it into the chamber. The locking lugs enter the barrel locking bushing and the locking cam rotates the bolt 20° to the locked position. Pressure on the trigger must be relaxed between shots. When the last round has been fired, the bolt remains open.

The Johnson is the only recoil-operated military shoulder rifle which has been manufactured in quantity. It appeared soon after the Army had adopted the M1 and at the time its backers claimed that it was far superior to the M1. A series of tests and demonstrations during the period 1939-40 indicated that the Johnson was not superior to the M1 and Springfield Armory was already tooled up to produce the M1.

Quantities of the 1941 Johnson were used by the U.S. Marines for a limited period of time and significant quantities were made for the Dutch East Indies. The rotary magazine is the common version of the Johnson Rifle, however a vertical feed version was made as well.

CARBINE CALIBER .30 M1, M1A1, M2, AND M3

The carbine was developed to replace the pistols in use by noncommissioned officers, special troops, and company-grade officers. Its historical background is given in Part I of this book.

Manufacturers of the carbine were:

Winchester: 809,451 M1 carbines, 17,500 M2 carbines, and 1,108 M3 carbines.

Inland Manufacturing Div. of General Motors: a total of 2,625,000 carbines including M1s, M1A1s, M2s, and a few M3s.

Underwood Elliot Fisher: 545,616 carbines.

National Postal Meter: 413,017 carbines.

Rock-ola Manufacturing Corp: 228,500 carbines.

Quality Hardware: 359,662 carbines.

Standard Products: 247,155 carbines.

Saginaw: 739,136 carbines.

IBM: 346,500 carbines.

There were more carbines produced than of any other United States weapon.

The variations of the carbines are as follows:

Carbine M1—semiautomatic, originally made with L-type flip over sight which was replaced with a ramp-mounted aperture adjustable for windage, sporter-type stock.

Carbine M1A1—same as M1, but has folding-type metal butt stock.

Carbine M2—selective fire, usually found with fixed wooden stock.

Carbine M3—receiver grooved for Infra Red "Snooper scope," otherwise identical to the M2.

Loading and Firing the M1 Carbine

Load magazine exactly as for automatic pistol with 15 cartridges.

Thrust up into position in the trigger housing until it locks.

Pull back handle of operating slide on right side of gun as far as it will go, opening the action and allowing a cartridge to rise in the magazine in the path of the bolt, and cocking the weapon and compressing the return spring.

Remove hand and permit operating slide to go forward, loading the firing chamber. With heel of hand push operating slide handle forward to be sure it is fully locked. The weapon is now ready to fire.

Push the button safety in the front end of the trigger guard all the way through to the right. This is the safe position. Pushing the button through to the left side as far as it will go releases the safety.

How the M1 Carbine Works

Starting with the gun loaded and cocked, the action is as follows: The trigger being pressed, the hammer is released to strike the firing pin and discharge the cartridge. As the bullet passes down

The Winchester caliber .30 Light Rifle—prototype of the U.S. M1 Carbine.

the barrel, a minute quantity of gas behind it flows down through a very fine hole bored in the under side of the barrel and escapes into a sealed cylinder where it expands against the head of the piston-like operating slide. This operating slide moves back a short distance until a cam recess engages an operating lug on the bolt. During this time the bullet has had sufficient time to leave the barrel and it is safe for the action to open. The extractor fastened in the bolt draws the empty cartridge back to strike the spring-actuated ejector which hurls it out to the right front of the weapon. The bolt is rotated out of its locking recess, simultaneously turning the hammer away from the rear of the firing pin and forcing the firing pin to draw back inside the bolt. This compresses the hammer

Top view of the M2, shown without magazine. Full auto switch is seen at the receiver above the stock. Externally, the M2 resembles the Carbine M1.

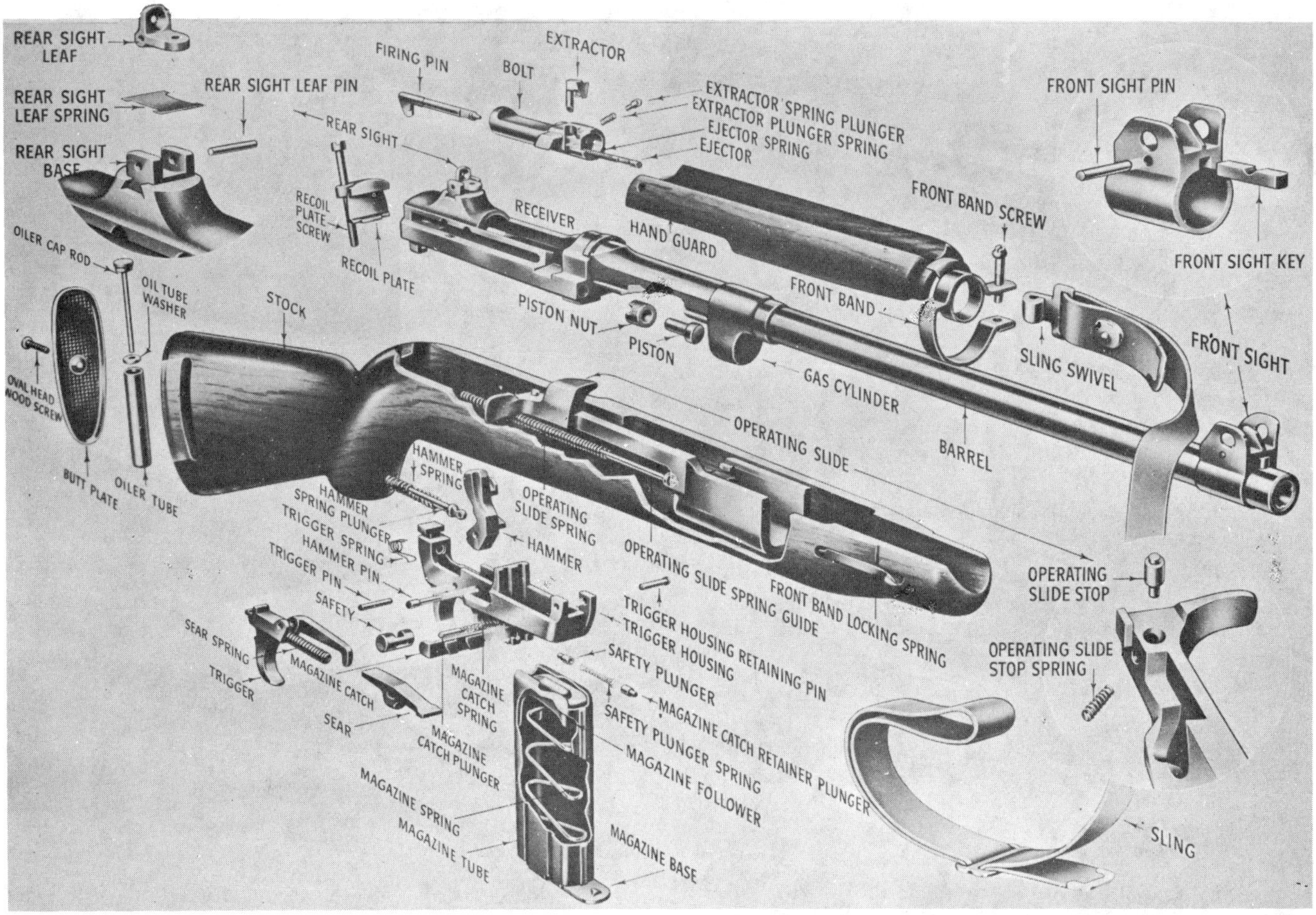

Parts of the M1 Carbine.

spring. The rearward motion of operating slide is completed when rear end of its inertia block strikes against forward end of receiver. Bolt stops when it reaches the end of bolt hole in rear receiver. Boltway is now clear permitting the next round to rise in the magazine in line with the bolt. During this motion the powerful operating slide spring has been compressed. It now drives the bolt forward loading the chamber. The cam recess in the operating slide again comes into play. Pressing against the bolt operating lug, it rotates it from left to right into its locking recess.

Forward movement of operating slide continues until the rear of its inertia block lodges against the piston in the cylinder. This action continues each time the trigger is squeezed until the last cartridge has been fired.

Field Stripping the M1 Carbine

Push the magazine catch to the left (it is positioned just in front

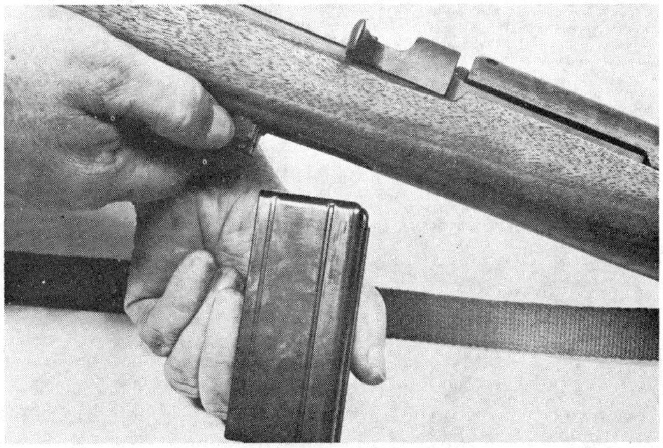

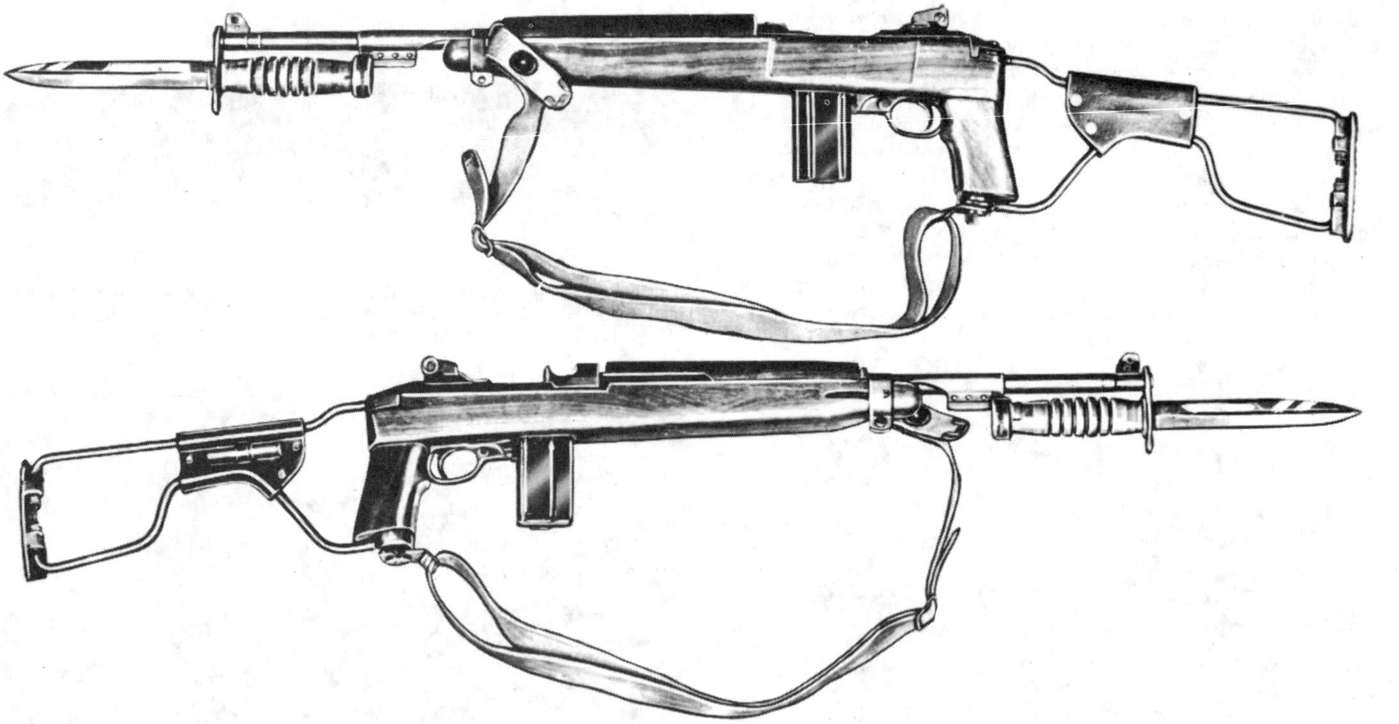

Caliber .30 M1A1 Carbine.

of the trigger guard on the right side), and withdraw magazine from below.

Draw back bolt to examine chamber and make sure that the weapon is unloaded.

At the end of the wooden fore-end is a sling swivel. Push this back against the fore-end and loosen it by unscrewing the front band screw. A cartridge may be used as a screwdriver.

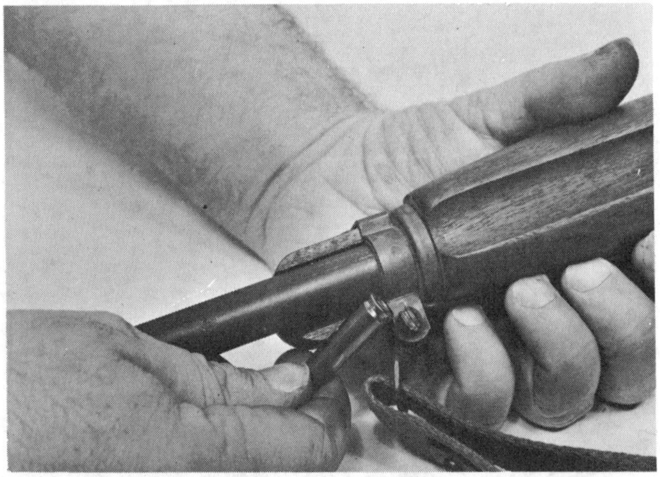

Press the front end of the lock spring toward the rear and slip the front band forward over its locking spring; it will not slip off the barrel unless the front sight is removed.

Now slide the wooden handguard on top of the barrel forward until its liner disengages from the undercut in the forward face of the receiver; it can then be lifted from the barrel.

Holding the stock firmly with the right hand, grasp the barrel near the front end with the left hand and raise it until the lug at the rear of the receiver clears the retaining notch on the face of the recoil plate (the plate just above the pistol grip). The barrel and

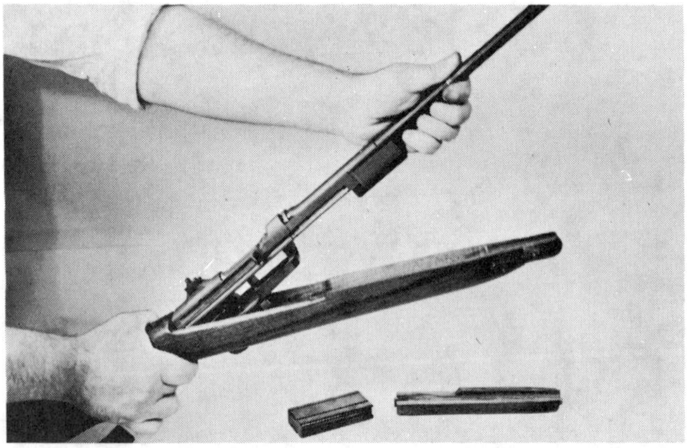

receiver can now be pulled forward and lifted out of the stock; carrying the trigger housing group with them.

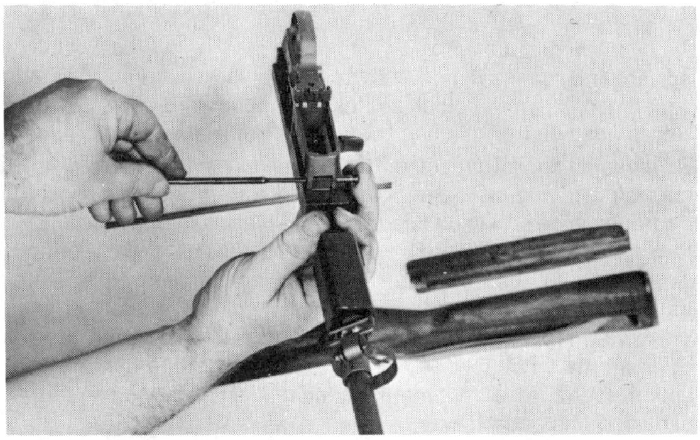

At forward end of trigger guard is the trigger housing retaining pin. Push it out until it clears the lug in the receiver.

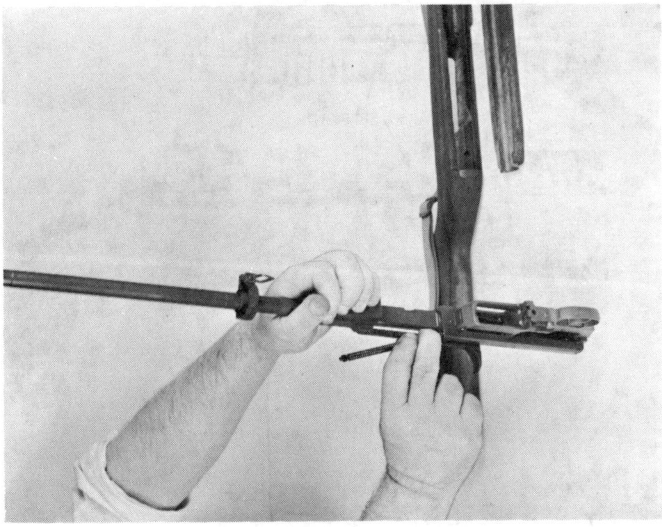

Pull back the operating slide spring guide a short distance until it is free of the operating slide. Pulling it forward and to the right permits it and its spring to be withdrawn.

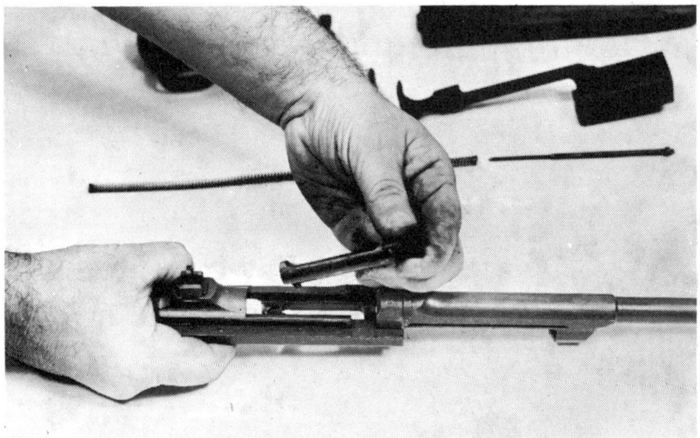

Take hold of the bolt and slide it to the rear until its face is behind the locking shoulder in the receiver. Twist the bolt from right to left, lift it to an angle of 45° and turn it bottom up. It may now be drawn forward and up out of the receiver.

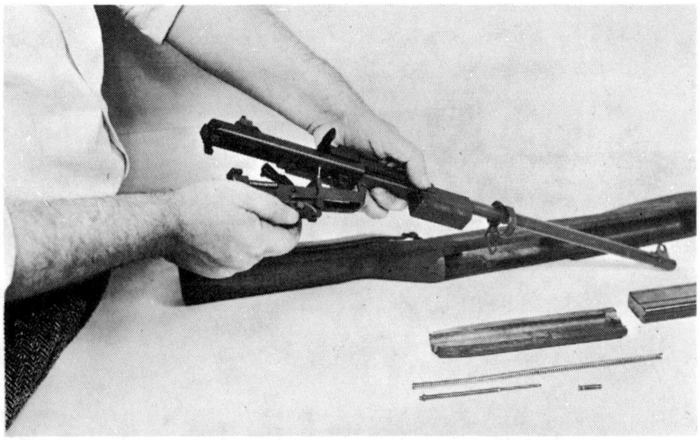

Pull the housing forward until it clears the grooves in the receiver, when it may be lifted out.

Pull the operating slide back until the guide lug at bottom of handle end alines with dismounting cut in receiver. Lift handle up and to right until the guide lug clears the retaining cut in receiver and also disengages from the bolt lug. Then push slide forward until the left barrel guide lip alines with clearance cut on bottom of left barrel guide groove. Rotating the slide body so as to free the left guide lip of the slide from its barrel guide groove will permit removing the slide from barrel.

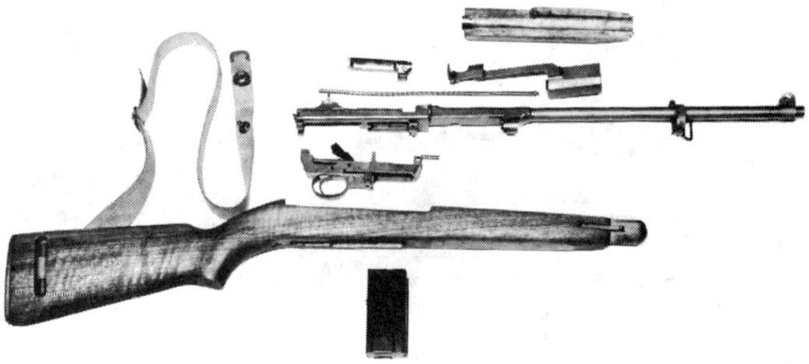

This completes field stripping. To assemble, reverse this procedure.

Caliber .30 M2 and M3 Carbines

The M2 carbine is the selective-fire version of the M1. Parts which differ from those of the M1 carbine are as follows: hammer, sear, trigger housing, operating slide, magazine catch, and stock. Added parts are as follows: disconnector group, disconnector lever assembly, and selector group. All M2 and M3 carbines and many M1 and M1A1 carbines are fitted with a front band assembly which incorporates a bayonet lug. These carbines use the bayonet-knife M4.

The M3 carbine is an M2 with a receiver designed to accommodate an infrared sniperscope. It does not have a conventional rear sight.

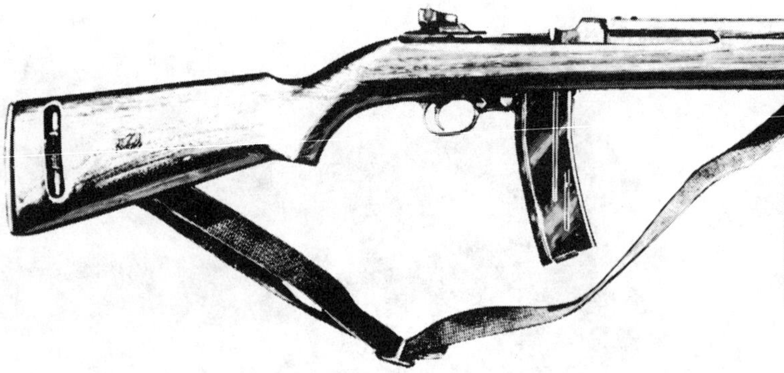

Caliber .30 M2 Carbine.

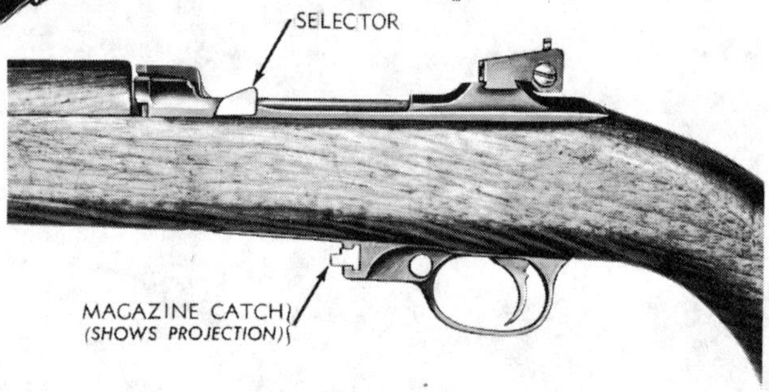

Identifying points of the M2 Carbine.

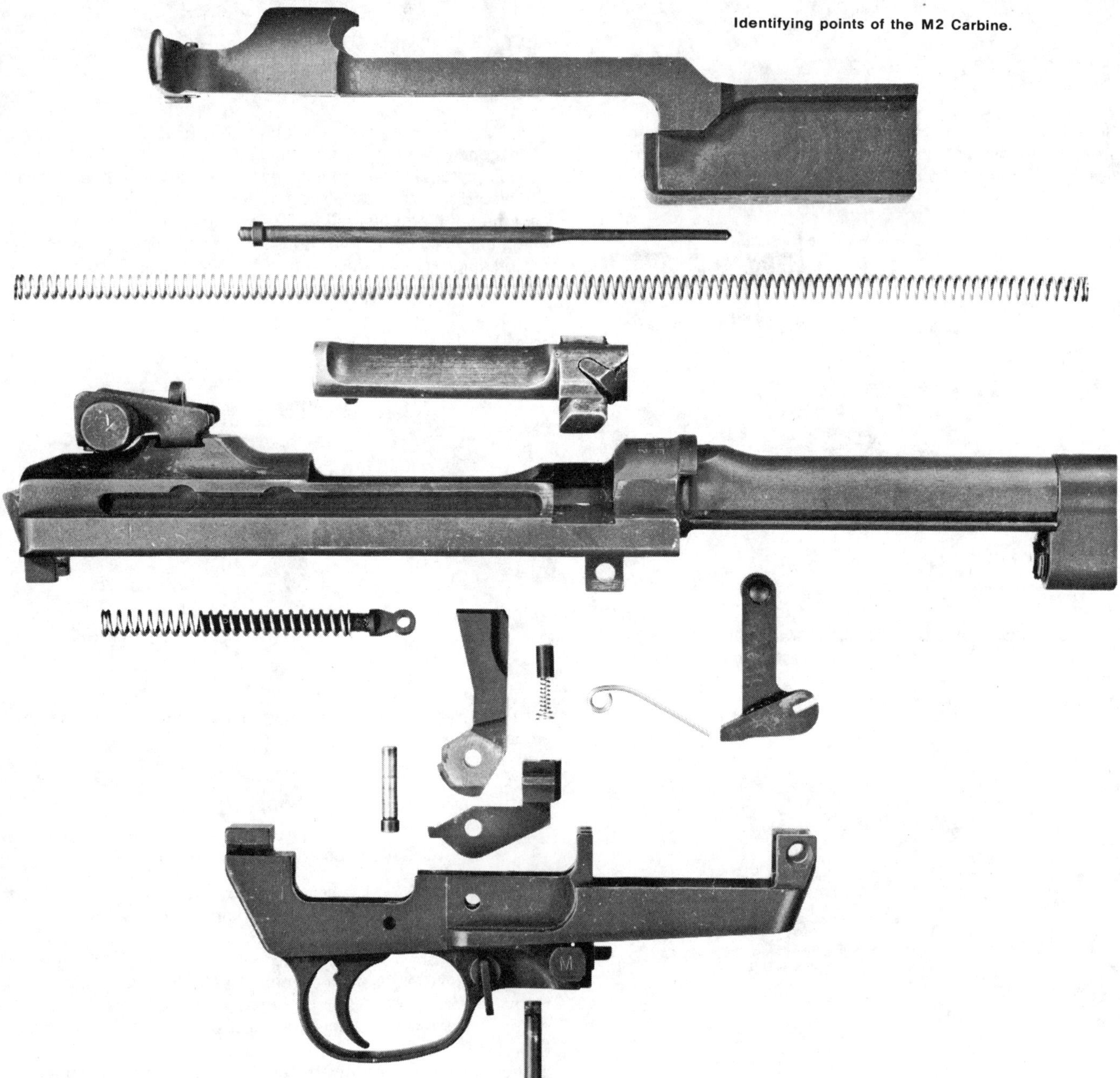

Stripped action of the M2 Carbine.

THE 7.62mm NATO M14 RIFLE

The M14 is a standard rifle of the United States Army. It was produced at Harrington and Richardson Arms Co., Thompson Products (TRW), at the New Haven (Winchester) plant of the Winchester-Western Arms Division of Olin Mathieson Corp. and at Springfield Armory. As stated previously, the M14 will replace the M1 rifle, M3A1 submachine gun, M2 carbine, and the Browning automatic rifle (BAR). A companion piece to the M14—the M15, a heavier barreled weapon—was originally intended to replace the BAR as the squad automatic weapon, but has since been dropped from the program. The M14 is capable of automatic as well as semiautomatic fire, and a certain proportion will be fitted with bipods to serve as squad automatic weapons. The M14 rifle is no longer in production. A total of 1,381,581 M14 rifles were made.

The M14 is an evolution of the M1 rifle; in the design of the M14 many of the shortcomings of the M1 have been eradicated. The basic action of the M1 remains, but the troublesome eight-round en bloc clip has gone. The hanging of the gas cylinder on the end of the M1 rifle's barrel gave some accuracy difficulties; these have been overcome, in the M14, by moving the gas port and gas cylinder back about eight inches from the muzzle. The gas cutoff and expansion system used on the M14 lends itself to better accuracy because its action is not as abrupt as that of the M1. Various other changes have been made to give the Army a basically better weapon than the M1.

How to Load and Fire the M14

Application of Safety. Place the safety in the "safe" position by cocking the hammer and snapping the safety rearward.

Loading of Rifle. Place the safety in the "safe" position.

Insert a loaded magazine into the magazine well, front end leading, until the front catch snaps into engagement; then pull backward and upward until the magazine latch snaps into position.

Pull the operating rod handle to its rearmost position and release; this allows the top round to rise and the bolt to move forward, thus stripping and chambering a round from the magazine.

Semiautomatic Fire with Selector Lock. With the selector lock in the rifle, it cannot be fired automatically. Load the rifle and release the safety. The rifle will now fire one round upon each pull of the trigger.

Semiautomatic Fire with Selector. Press in and turn the selector until it snaps into position with its blank face to the rear and its projection downward. The connector assembly is inoperative in this position since the connector is held forward and out of engagement with the operating rod.

Load the rifle and release the safety. The rifle will now fire one round upon each pull of the trigger.

Full Automatic Fire with Selector. Press in and turn the selector until it snaps into position with the face marked "A" to the rear and the projection upward. This rotation of the eccentric selector shaft moves the sear release to the rear into contact with the sear and moves the connector assembly rearward into contact with the operating rod.

Load the rifle and release the safety.

Pull and hold the trigger. The rifle will fire automatically as long as the trigger is squeezed and there is ammunition in the magazine. To cease firing, release the trigger.

Bolt Lock. When the last round of ammunition is fired, the magazine follower engages the bolt lock and raises it into the path of the retracted bolt; this holds the bolt in the open position.

Unloading the Rifle. Place the safety in the "safe" position.

Grasp the magazine, placing the thumb on the magazine latch, and squeeze the latch. Push the magazine forward and downward to disengage it from the front catch and remove the magazine from the magazine well.

Pull the operating handle rearward to extract and eject a chambered round, and to inspect the chamber. The rifle is now clear.

Gas Shutoff Valve. For semiautomatic and full automatic firing, turn the valve to the open position by pressing in and rotating. The valve is open when the slot in the head of the valve spindle is perpendicular to the barrel. The shutoff valve in the gas cylinder opens and closes the port in the cylinder between the barrel and the gas piston.

Stripping the M14

General Disassembly (Field Stripping). Unload the rifle, remove the magazine, and place the safety in the "safe" position.

Turn the rifle upside down with the muzzle pointing to the left.

Insert the nose of a cartridge into the hole in the trigger guard and pry upward to unlatch the trigger guard.

Swing the trigger guard upward and lift the trigger group from the stock.

Separate the stock from the rifle by cradling the receiver firmly in one hand and by striking upward sharply on the stock butt with the palm of the other hand.

Turn the barrel and receiver group on its side with the connector assembly upward. Press in and turn the selector until the face marked "A" is toward the rear sight knob and the projection forward (this step applies to rifles modified for selective firing). Press forward on the connector with the right thumb until the forward end can be lifted off the connector lock. Rotate the connector clockwise, until the slot at the rear end is aligned with the elongated stud on the rear release. Lower slightly the front end of the connector and lift it from the sear release. **Note:** The connector assembly is a semipermanent assembly and it should not be disassembled.

With the barrel and receiver group upside down, pull forward on the operating rod spring, relieving pressure on the connector lock. Pull the lock outward, disconnect the operating rod spring guide, and remove the spring guide and spring. Turn the barrel and receiver group right side up.

Retract the operating rod until the key on its lower surface coincides with the dismount notch in the receiver. Lift the operating rod free and pull to the rear, disengaging it from the operating rod guide.

Grasp the bolt group by the roller and, while sliding it forward, lift it upward and outward to the right front with a slight rotating motion. The rifle is now field stripped, and basic assemblies such as the bolt and the trigger groups may be disassembled, if required.

Bolt Group. With the bolt in the left hand and the thumb over the ejector, insert the blade of a screwdriver between the extractor and the lower cartridge seat flange. Pry the extractor upward to unseat it. The ejector will snap out against the thumb. Lift out the ejector assembly, and the extractor plunger and spring. Remove the firing pin from the rear of the bolt. **Note:** No attempt should be made to disassemble the roller from the bolt stud.

Barrel and Receiver Group. Disassemble the rear sight as follows: Run the aperture all the way down and record the reading for use in reassembling the sight. Hold the elevating knob and unscrew the nut in the center of the windage knob. Withdraw the elevating knob. Unscrew and remove the windage knob. Pull the aperture

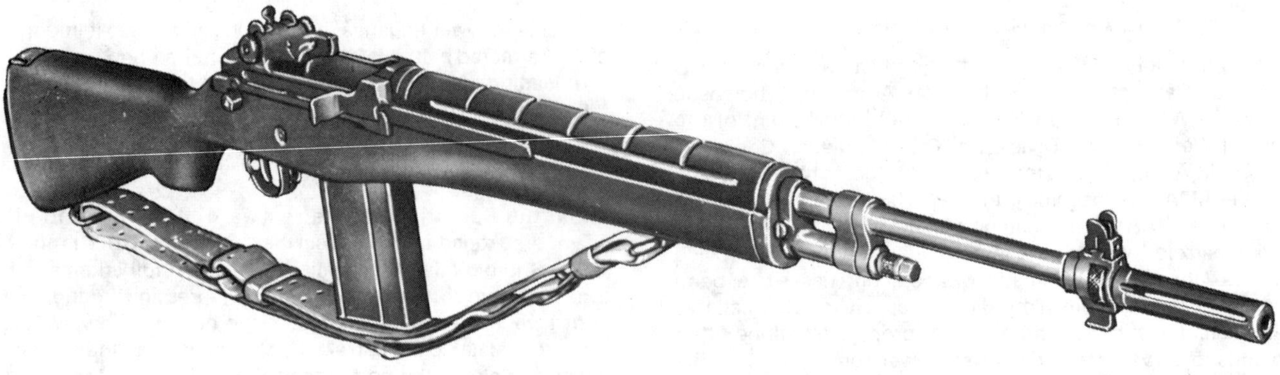

7.62mm M14 Rifle, late production model.

7.62mm Rifle M14. production version.

up about one-half inch. Place the thumb under the aperture and push upward and forward to remove the aperture, cover, and base. Separate the rear sight cover from the rear sight base.

Loosen the setscrew in the base of the front sight lug on the flash suppressor. Unscrew the flash suppressor nut and slide the flash suppressor forward off the barrel.

Loosen and remove the gas plug, using the gas cylinder plug wrench. Tilt the muzzle down and remove the gas piston from the gas cylinder. Unscrew the gas cylinder lock and slide the lock and the gas cylinder off the barrel.

Slip the front band off the barrel. Push the handguard forward and lift it from the barrel.

Trigger Group. To disassemble the trigger group, close and latch the trigger guard. Squeeze the trigger, allowing the hammer to go forward. Hold the trigger housing group with the first finger of the right hand on the trigger and the thumb against the sear. Place the front of the trigger housing against a firm surface. Squeeze the trigger with the finger and push forward on the sear with the thumb. At the same time, using the tip of a cartridge, push out the trigger pin from left to right. Slowly release the pressure with the finger and thumb; this allows the hammer spring to expand.

Lift out the trigger assembly. Remove and separate the hammer spring plunger, hammer spring, and the hammer spring housing.

Push out the hammer pin from left to right, using the tip of a cartridge. Move the hammer slightly to the rear and lift out.

Unlatch the trigger guard. Push out the stud of the safety from its hole. Remove the safety and safety spring. Slide the trigger guard to the rear until the wings of the trigger guard are aligned with the safety stud hole. Rotate the trigger guard to the right and upward until the hammer stop clears the base of the housing. Remove the trigger guard.

Drive out the magazine latch spring pin with a suitable drift to remove the semipermanently assembled magazine latch and spring.

Assembly of M14. To assemble the rifle, reverse the disassembly procedure. However, the following instructions are provided to facilitate and to insure satisfactory assembly:

To assemble the hand guard, when the gas cylinder and related components are in place, position the front end of the guard in the front band and snap the rear band of the handguard assembly into the barrel grooves. **Note:** The handguard need not be reassembled prior to assembly of the gas cylinder and related components.

To assemble the trigger group to the stock and receiver, cock the hammer and swing the trigger guard to the open position. Insert the assembly into the receiver, and close the trigger guard.

To assemble the gas system, replace front band, gas cylinder, and gas cylinder lock. Tighten the lock by hand to its full assembled position and then "back off" until the loop is alined with the gas cylinder. Assemble the piston and the gas cylinder plug.

Special Note on the M14 Rifle

Although the M14 is a selective-fire weapon, most weapons in the hands of troops will have their selectors locked in the semiautomatic position. When desired, these weapons can be made to deliver selective fire by the removal of the selector lock. This feature has been added to the weapon since combat experience with the M2 carbine and troop tests with earlier prototypes of the M14 indicate that troops keep selective-fire weapons set on full automatic as a matter of course. This limits the effectiveness of the weapon at long ranges, since it is effective in off-hand automatic fire only at ranges up to about 100 yards. It also results in a great expenditure of ammunition with little in the way of results to show for this expenditure. Those weapons equipped with bipods for use as squad automatic weapons will not have their selectors locked, and will be capable of selective fire at all times. Production M14s have aluminum butt plates with shoulder support and plastic handguards. Production of the M14 rifle ceased in 1964 at which time 1,380,353 rifles had been produced.

How the M14 Works

Operational System. Operational power is derived from a gas cutoff and expansion system. In this system, gas is bled from the barrel through a port in its underside which registers with a port in the rear loop of the gas cylinder and a port in the wall of the hollow gas piston. Propellent gases released by firing enter these ports and are trapped in the hollow chambers in the gas piston and gas cylinder. The piston, in contact with the operating rod, is driven rearward a complete stroke (one and one-half inches). The resultant force is of sufficient magnitude to drive the operating rod and the

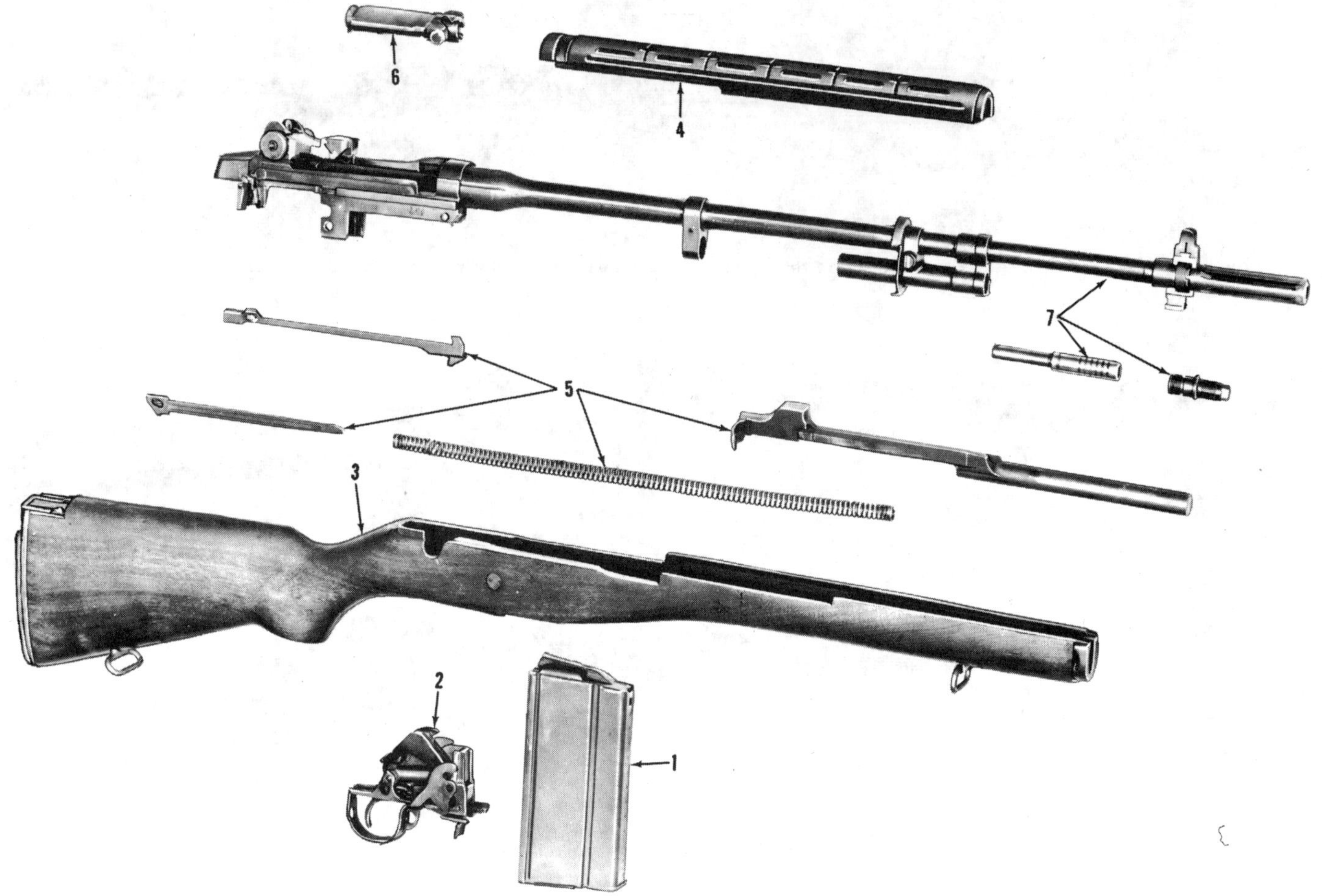

M14 Rifle, field-stripped.
1. Magazine 2. Trigger assembly 3. Stock 4. Hand guard 5. Operating rod group 6. Bolt assembly 7. Gas piston, gas plug.

bolt to their rearmost positions, and to compress the operating rod spring. The exhaust port in the bottom of the gas cylinder is uncovered as the gas piston nears the end of its stroke; this allows the expanded gases to escape.

The reaction of the operating rod spring now returns the bolt and operating rod to their original positions.

The gas system functions in the same manner for all firing cycles unless the port in the gas cylinder is closed by the shutoff valve. When the valve is closed, the gas system becomes inoperative, thus necessitating manual operation of the operating rod.

Semiautomatic Fire. During its rearward travel, the operating rod unlocks and carries the bolt assembly rearward; this action extracts and ejects the fired cartridge case and cocks the hammer. Ammunition is lifted into the path of the retracted bolt by action of the magazine spring and follower.

The operating rod moves forward under spring pressure; this carries the bolt assembly forward, stripping the top round from the magazine, chambering the round, and locking the bolt in battery position.

Full Automatic Fire. After approximately one-eighth inch of rearward travel, the operating rod disengages from the hook on the connector assembly. This allows the spring-loaded connector assembly to move rearward and to rotate the sear release out of engagement with the sear.

With the trigger held back, the sear engages the hammer.

As the operating rod moves forward, the hammer remains in the cocked position until the shoulder on the operating rod engages the hook on the connector assembly. Further travel of the operating rod moves the connector assembly forward, rotates the sear release against the sear, and disengages it from the hammer. This disengagement allows the hammer to fall and to fire the chambered cartridge.

Because the firing pin is prevented from moving forward prematurely by the bridge in the receiver, the weapon cannot be fired until the bolt is locked (or driven into the locked position by action of the nose of the hammer against the cam on the rear end of the bolt).

If the trigger is released at any time prior to the firing of the last round, the hammer will be held in the cocked position by the trigger lugs (secondary sear) and automatic actuation of the sear release by the connector assembly will not release the hammer to fire the chambered cartridge.

Variations of the M14 Rifle

There have been a number of variations of the M14 rifle produced. Two of these variations have steel folding stocks, one of which folds to the side similar to the M1A1 carbine stock—the Type V—and the other folds under the weapon in a manner similar to the stock of the German MP40 submachine gun and the Soviet AK assault rifle—the Type III.

The M14A1. The M14A1 is a variation of the M14 produced for use as a squad automatic weapon. It was originally developed by the United States Army Infantry Board, Fort Benning, Georgia. Springfield Armory made various changes in the design to ease manufacture and maintenance. The M14A1 has a straight-line stock design with full pistol grip and folding forward handgrip.

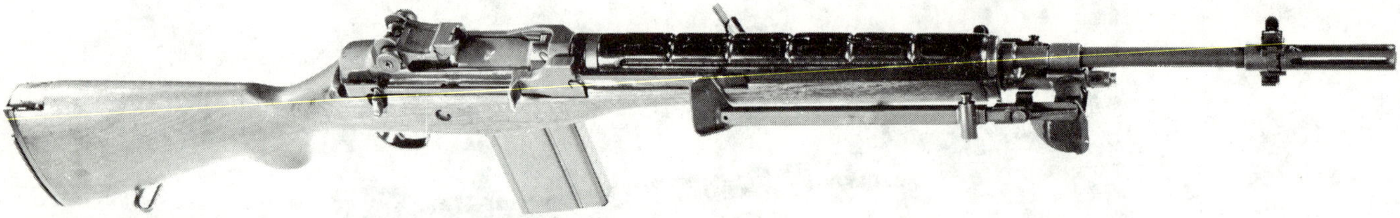

7.62mm Rifle M14, fitted with bipod for use as a squad automatic weapon.

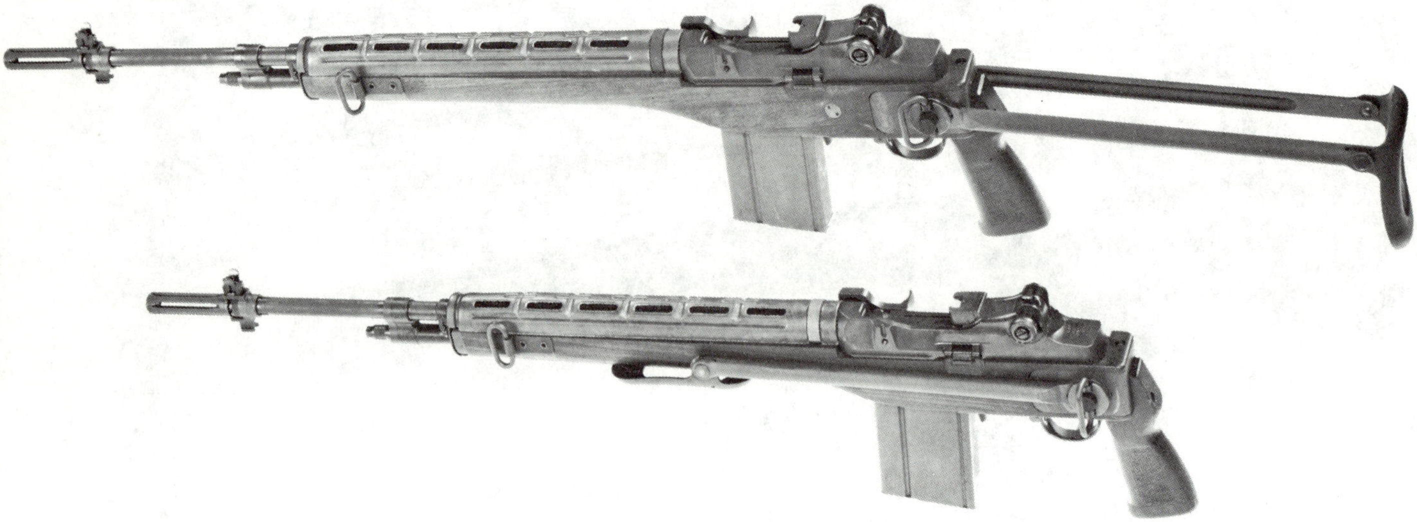

7.62mm NATO M14 Rifle with Type III folding stock.

7.62mm NATO M14 Rifle with Type V folding stock.

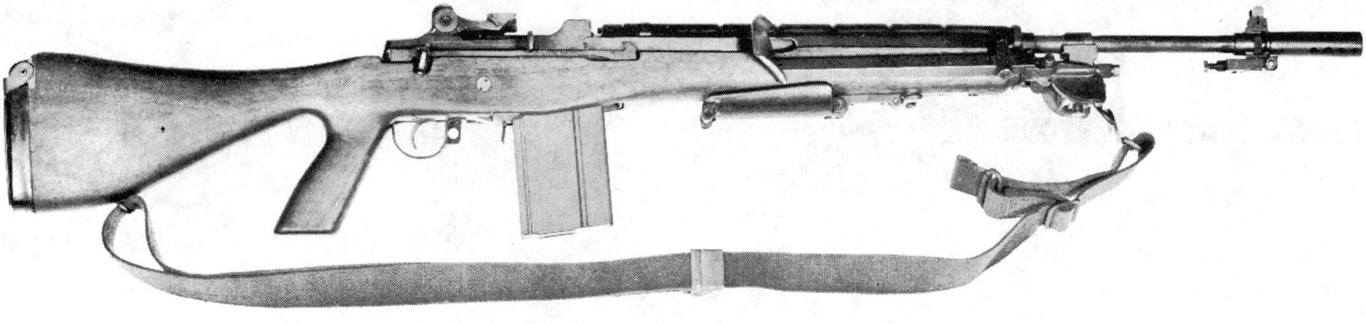

7.62mm NATO M14A1 Rifle. Bottom view shows hand grip in down position.

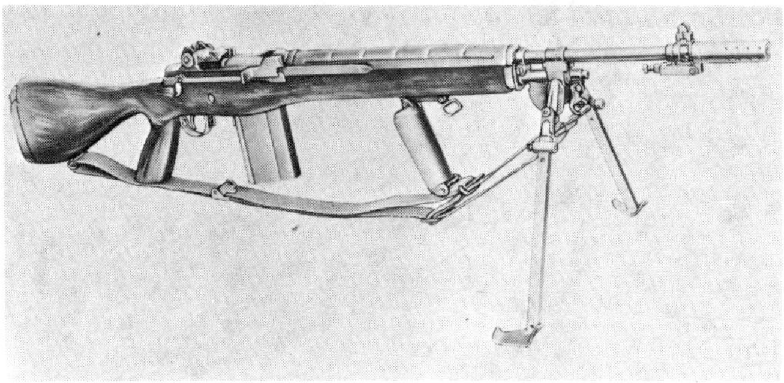

7.62mm NATO M14E2 Rifle.

7.62mm Rifle, M21 (Sniper) with scope sight.

A compensator, which helps to keep the barrel down in automatic fire, is fitted over the flash suppressor. The stock has a rubber recoil pad and folding shoulder rest and the M2 bipod has been modified by the addition of a sling swivel and a longer pivot pin. The Browning Automatic Rifle sling is used on this rifle. The selector lever is found on all M14A1 rifles so that they may be used for automatic or semiautomatic fire.

M14 National Match Rifle. A match version of the M14 rifle for use at the National Matches was developed as the result of a requirement set down in 1959. The M14 National Match Rifle cannot be fired full automatic; it has a hooded aperture rear sight, special sight parts, selected barrel and glass bedded action similar to the National Match Rifle.

The M14M Rifle. The M14M rifle was intended, as of 1963, for issue to NRA affiliated rifle clubs and for sale through the Director of Civilian Marksmanship. This rifle has been modified by welding the selector shaft and lock to eliminate automatic fire capability. The "M" in this rifle's designation stands for "Modified Service."

THE 5.56mm M16 and M16A1 RIFLES

The 5.56mm M16 rifle was designed by Mr. Eugene Stoner while with the Armalite Division of Fairchild Engine and Airplane Co. The rifle was adopted by the United States Air Force as the M16. In the interim Colt's Patent Fire Arms Manufacturing Co. Inc. purchased the manufacturing rights to the weapon and Colt's are currently manufacturing the rifle. The M16 was originally known as the Caliber .223 AR-15 and followed a similar rifle, chambered for the 7.62mm NATO cartridge known as the AR-10. The AR-10 was produced for a time at the Dutch government arsenal—Artillerie Inrichtingen—at Hembrug, Zaandam on a contract basis. It was not used by the Dutch government. The AR-15 was essentially the same as the AR-10 when first made by Armalite, but a number of changes including a bolt retracting handle mounted at the rear of the receiver rather than on top and a different type hand guard were made in the rifle before it was adopted by the Air Force.

The United States Army tested the AR-15 and adopted it as the M16A1. The principal difference between the M16 and M16A1 is the presence of a plunger mounted on the right rear side of the receiver. This plunger is used to push the bolt forward if it becomes stuck due to dirt in the receiver.

The M21 Rifle. The M21 is the sniper version of the M14 rifle. It uses a Leatherwood type variable power scope.

How to Load and Fire the M16 and M16A1 Rifles

Insert loaded magazine. Pull charging handle all the way back, and push. Put selective-fire lever on "Safe," "Semi," or "Auto." The weapon will fire when the trigger is pulled (unless set on "Safe"). Upon firing the last round in the magazine, the bolt remains open; then, after inserting a new loaded magazine, pushing the bolt stop lever allows the bolt to close, feeding a new round into the chamber.

Field Stripping the M16 and M16A1 Rifles

No special tools are required for field stripping and assembling these rifles. Clear the weapon. With the bolt in forward position, press out rear pin, and open the rifle. Withdraw bolt and bolt carrier assembly from upper receiver. Withdraw firing pin retaining pin. Withdraw firing pin. Withdraw cam pin. Withdraw bolt from bolt carrier.

Using firing pin to start it, pull out extractor pin, and remove extractor and extractor spring. Spring will remain with extractor.

After cocking the hammer by hand, use bullet point to depress buffer retainer plunger.

Remove buffer assembly and drive spring.

For assembling the weapon, proceed in the reverse order.

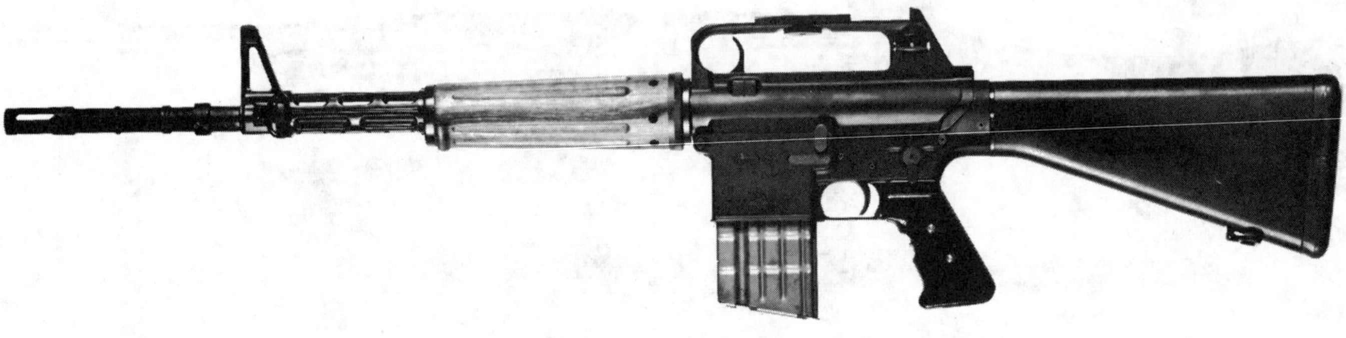

Experimental models of AR-10 Rifle made at Hembrug. Note the built-in grenade launchers.

Further stripping is unnecessary under normal field conditions. If necessary, the trigger group may be stripped for cleaning without the use of tools. The hammer pin should be removed first, using the firing pin to start it. All pins in the trigger group are spring retained, and may be removed and replaced from either side of the receiver. Hammer and trigger pins are interchangeable. Before closing the weapon, insure that the hammer is in cocked position.

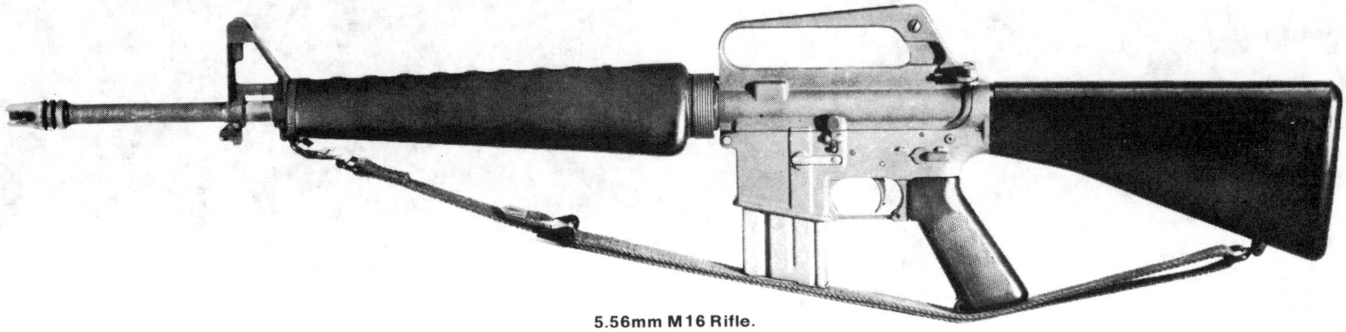

5.56mm M16 Rifle.

U.S. 5.56mm M16A1 Rifle with 40mm M203 grenade launcher attached.

How the M16 and M16A1 Rifles Work

With the weapon loaded and the selective-fire lever on "Semi," pulling the trigger causes the hammer to strike the firing pin, which detonates the primer.

When the bullet passes beyond the gas port, gas passes through the gas tube and into a chamber formed by the bolt carrier and bolt. At this time, the bolt is in the locked position, acting as a stationary piston. The entering gas pressure causes the bolt carrier to move to the rear. In moving rearward, the bolt carrier rotates the bolt, unlocking it and carrying it rearward. As the bolt assembly travels to the rear, the gas is exhausted through a port in the side of the bolt carrier. The cartridge case is then extracted and ejected in the usual manner.

When the selective-fire lever is on "Auto," the automatic sear holds the hammer in cocked position. Upon locking of the bolt, with the trigger held to the rear, the bolt carrier trips the automatic sear, firing the weapon.

When the selective-fire lever is on "Semi," the sear holds the hammer in cocked position. The hammer is released by pulling the trigger.

The firing pin can strike the primer only when the bolt is completely locked.

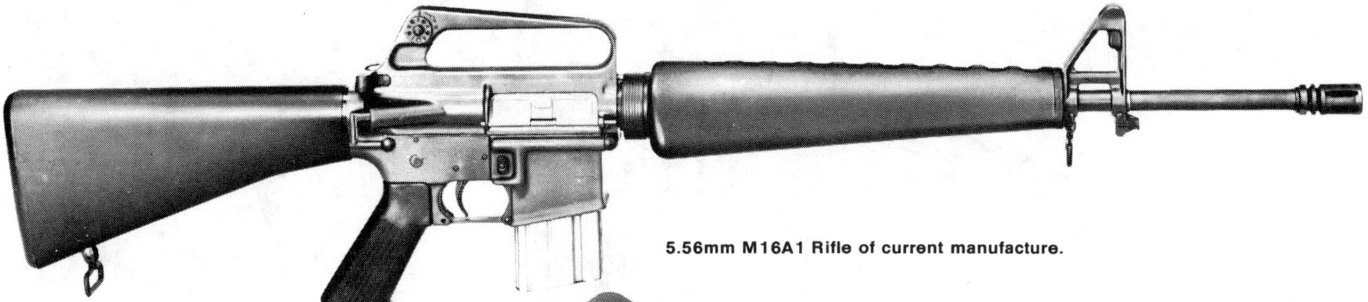

5.56mm M16A1 Rifle of current manufacture.

AR-15 (M16), field-stripped.

Special Note on the AR-10, AR-15, M16, and M16A1 Rifles

The gas system of these weapons is similar to that of the Swedish Ljungman M42 and the French Model 49 and 49/56 rifles, in that no piston rod is used. The gas blows back directly into the bolt (in the other rifles mentioned, it blows back into the bolt carrier). The bolt system, itself, appears to be unique. The use of a fixed carrying handle which acts as a rear sight base and guard is very similar to the combined rear sight/carrying handle of the British cal. .280 EM-2 rifle, which appeared in the late forties. Originally the AR-10 appeared with an aluminum barrel, inside which was a rifled steel insert. Apparently it was soon realized that steel barrels could be made light enough to compare favorably with the composite barrel. The prong-type flash suppressor can be used as a grenade launcher.

It would be less than candid to say that there have been no problems with the M16A1 rifle in service. All new weapons that are introduced into service develop some problems and these are magnified where the weapon is introduced during a period of conflict as was the M16A1. Issue of a new weapon to troops in the battle area without the time or facilities for training with the weapon, which can be provided in training camps in the United States, is always difficult and when the rifle has not had extensive and relatively long troop use in training to show up the bugs which always develop, it is doubly difficult. The M1 rifle was first issued in 1937 and a number of "bugs" showed up in it, most of which were cured prior to World War II. Among these were the "seventh-round" stoppage, to correct this and other problems, a number of changes were made in its construction, some as late as 1950. Similarly the British L1A1 rifle had a number of bugs that had to be worked out.

Although there appear to have been volumes written on the subject, the basic causes of the problem seem to have been:

(1) Small caliber automatic weapons such as the M16A1 need to be cleaned more regularly and thoroughly than the larger caliber weapons. The smaller major parts can be stopped by a smaller quantity of dirt, corrosion, etc., than the larger parts in the larger caliber weapons. The climate and geography of South East Asia undoubtedly aggravated the situation.

(2) A change was made in propellent after the introduction of the weapon. The different burning properties and consequently pressure/time curve, of ball powder as opposed to the tubular IMR originally used in the rifle increased the rate of fire considerably with a resultant higher wear rate of parts.

Major modifications to the rifle include: the adoption of a new buffer system which cuts the rate of fire and chrome plating of the chamber which helps to minimize the corrosion, pitting problem experienced in Vietnam. The prong-type flash suppressor has been changed from an open prong type to a closed prong. A thirty-round magazine has been developed for the M16A1 rifle and its submachine gun version, the XM 177E2.

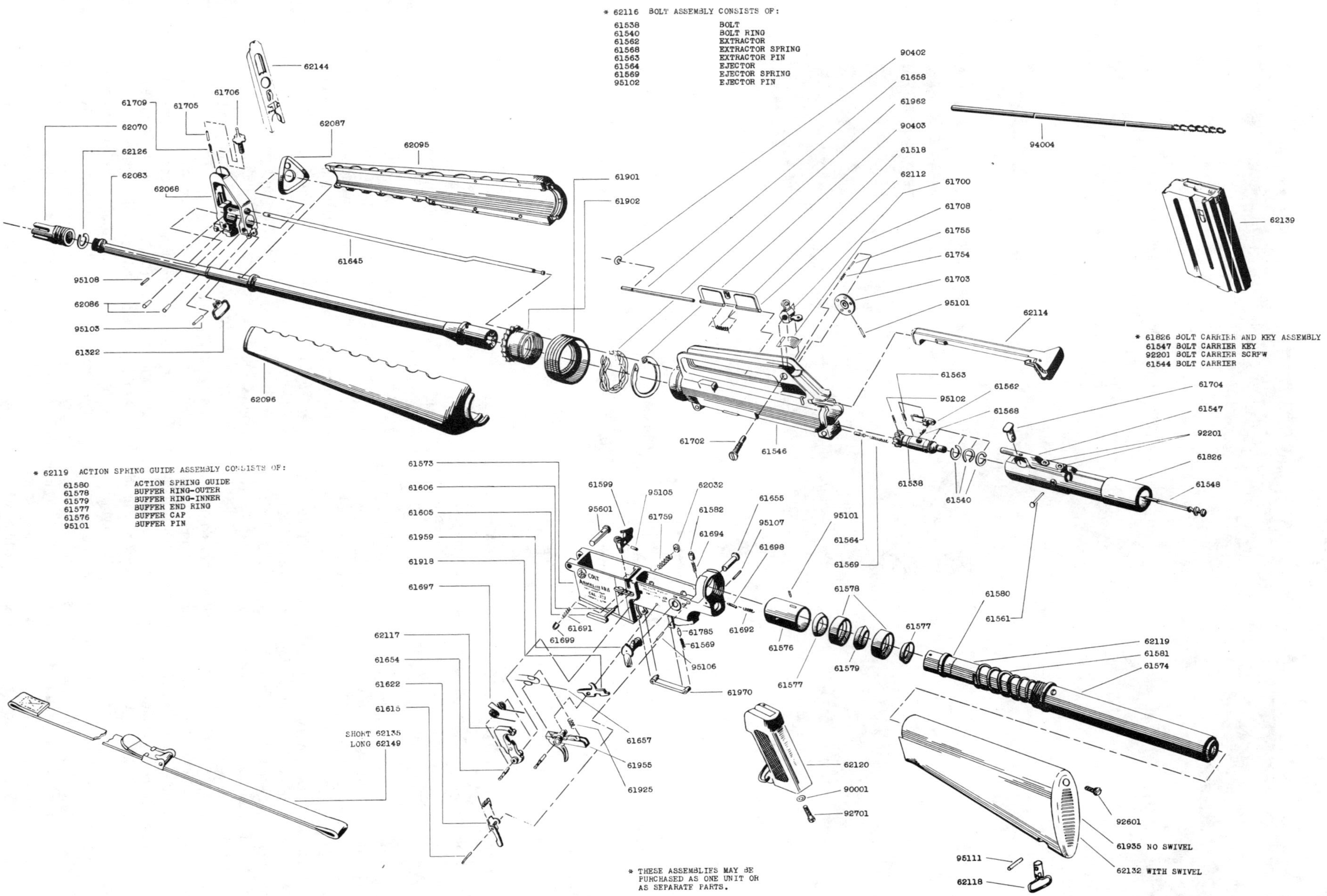

Exploded view of 5.56mm M16 Rifle.

THE COLT CAR-15 WEAPONS SYSTEM

Colt has developed a number of variations of the AR-15 rifle which they call the CAR-15 Weapons System. All operate in the same manner as the AR-15 and are chambered for the caliber .223 (5.56mm) cartridge. They vary only in size and weight and, in a few cases, in configuration. The weapons are the CAR-15 submachine gun, the CAR-15 carbine, the CAR-15 heavy assault rifle, the CAR-15 survival rifle, and the AR-15 with CGL-4 40mm grenade launcher.

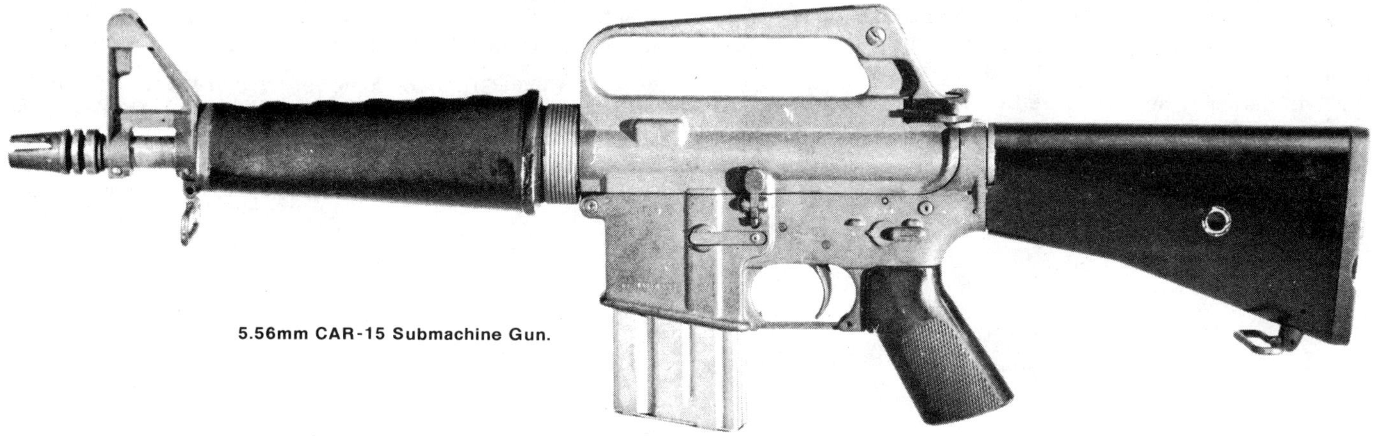

5.56mm CAR-15 Submachine Gun.

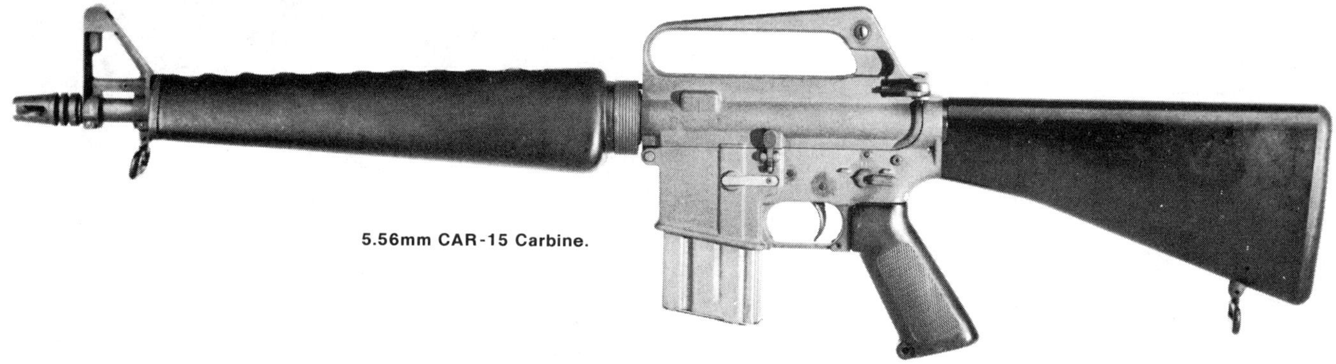

5.56mm CAR-15 Carbine.

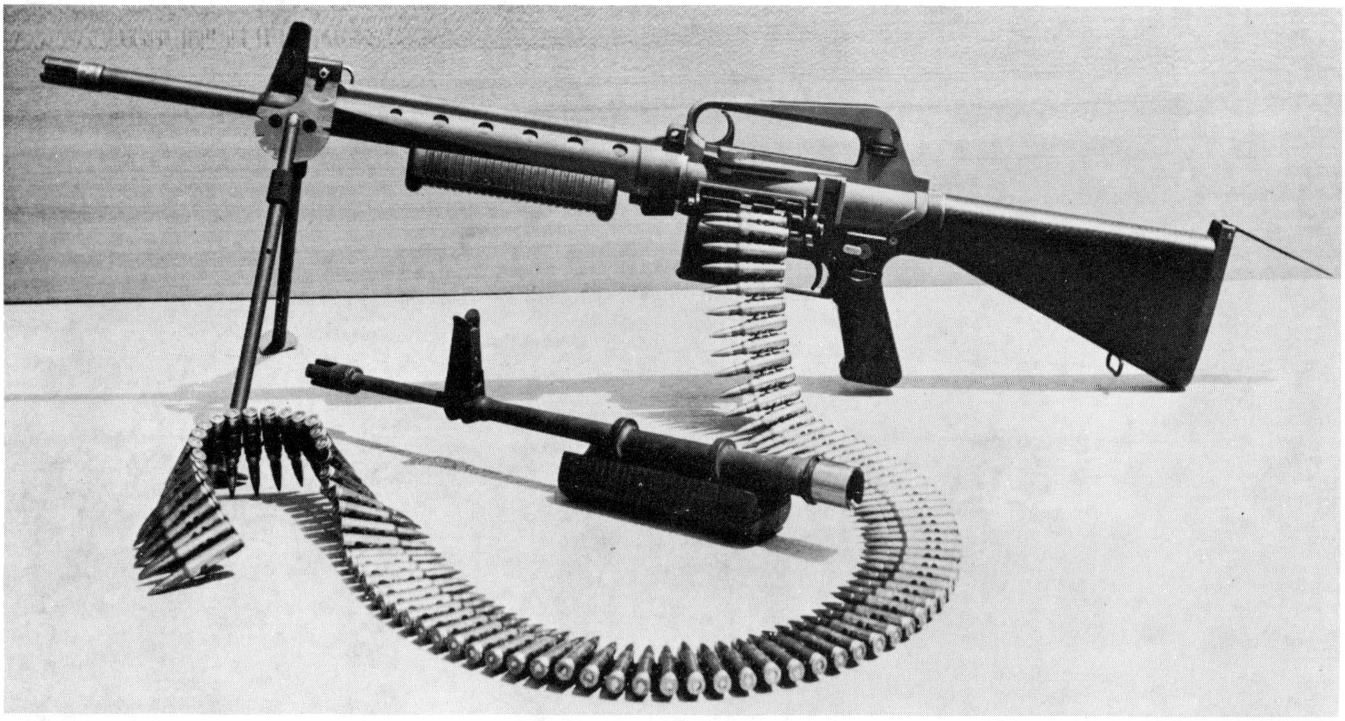

AR-10 7.62mm Light Machine Gun.

Weapons Included in the System

CAR-15 Submachine Gun. This weapon has a 10-inch barrel, is 28.7 inches long overall, weighs 5.9 pounds loaded and has a muzzle velocity of 2750 f.p.s. It has a telescoping type buttstock.

The submachine gun version has been modified and purchased in some quantity by the Army which calls it the 5.56mm submachine gun XM 177E2. It is covered in more detail in the section on submachine guns.

CAR-15 Carbine. The carbine version of the AR-15 has a 15-inch barrel, is 34 inches long overall, weighs 6.8 pounds loaded, and has a muzzle velocity of 3050 f.p.s.

CAR-15 Heavy Assault Rifle M1. This heavy version of the AR-15 has the same barrel length and overall length as the rifle, but has a heavier barrel. It weighs 7.6 pounds empty and 8.4 pounds loaded. Colt has dropped development of the M2 version of the heavy assault rifle. The CAR-15 M2 was similar to the light machine gun version of the AR-10 which could be either belt- or magazine-fed by the change of a few parts.

Car-15 Survival Rifle. This weapon has a tubular stock and a flash hider. It weighs 5.4 pounds loaded, has a 10-inch barrel, and is 29 inches long. Its muzzle velocity is 2750 f.p.s.

CGL-4 40mm Grenade Launcher. This grenade launcher is chambered for the 40mm grenade used with the M79 grenade launcher. The grenade launcher weighs 2.5 pounds and is 13 inches long overall. It is designed to be used with the AR-15 rifle and the total weight of the rifle and launcher, both empty, is 8.8 pounds.

5.56mm CAR-15 Heavy Assault Rifle M1.

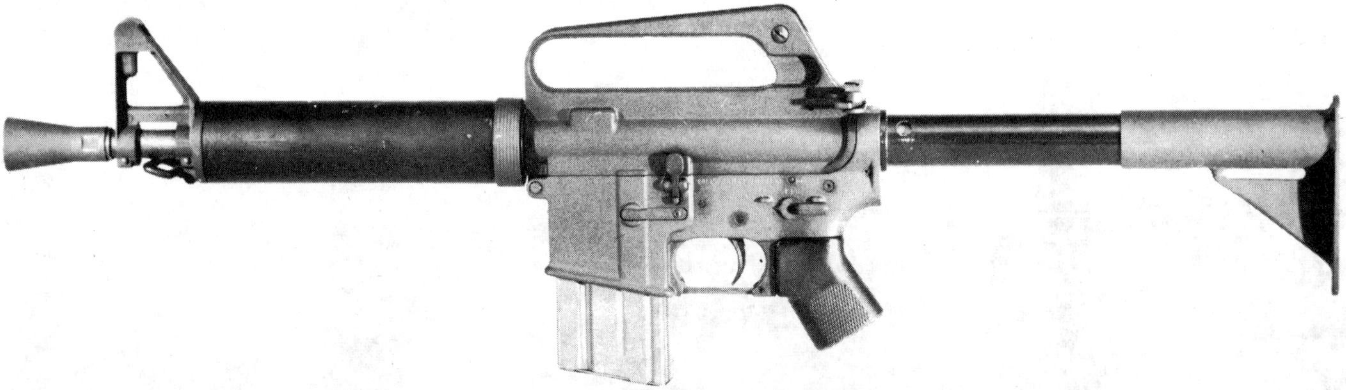

5.56mm CAR-15 Survival Rifle.

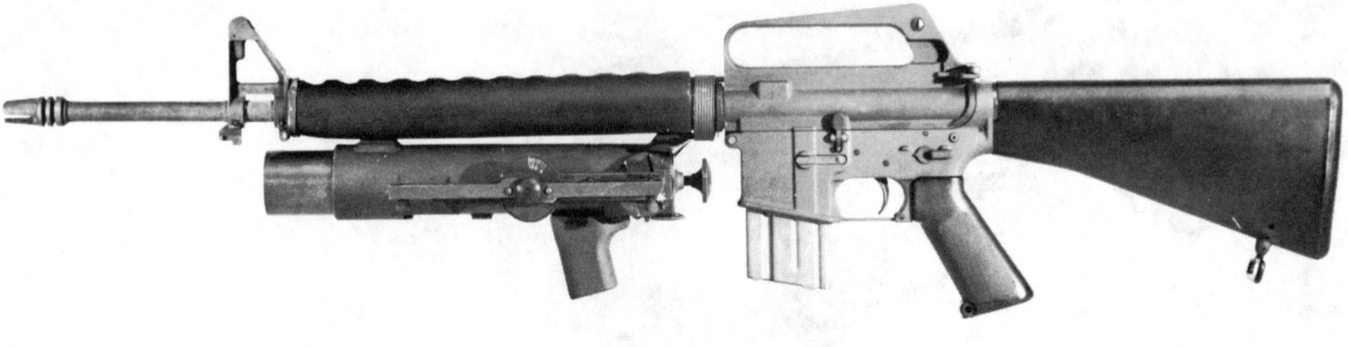

5.56mm AR-15 Rifle with CGL-4 launcher attached.

7.62mm NATO AR-16 Rifle.

THE ARMALITE 7.62mm NATO AR-16 AND 5.56mm AR-18 RIFLES

The 7.62mm NATO AR-16 rifle and 5.56mm (.223) AR-18 rifle was developed at Armalite. Armalite is no longer associated with Fairchild Engine and Airplane Co. and is handling the AR-16 and AR-18 rifles on an independent basis. The AR-16 is not currently being advertised widely, however, and the AR-18 is apparently the weapon which is being pushed by Armalite. The patent on the AR-18 is held by Mr. Arthur Miller, Mr. Charles Dorchester, and Mr. George Sullivan then at Armalite.

The AR-16 has carbine and rifle versions; the carbine is essentially the same as the rifle but has a folding stock.

The AR-16 and AR-18, unlike the Stoner-designed AR-15, has a gas piston rather than the direct gas blowback against the bolt carrier system used in the AR-15. The piston system is similar to that of the German Gewehr 43 and the Soviet Model 40 Tokarev rifle. It impinges against the bolt carrier which holds the frontally locking, multi-lugged bolt. The bolt is cammed into and out of the locked position by a camming pin which is mounted in the bolt and rides in a camway in the bolt carrier.

The AR-16 and AR-18 are field stripped by releasing the receiver from the trigger group and stock and swinging them down on the hinge pin mounted at the forward end of the magazine guide in a manner similar to that of the AR-15 rifle.

These rifles are made mainly of stampings and parts which can be made on automatic screw machines. The AR-18 is 38 inches long with stock fixed and 28.75 inches long with stock folded. The barrel is 18.25 inches long and the weight of the weapon is 6.3 pounds empty. The AR-18 uses a 20-round magazine and has a muzzle velocity of about 3200 feet per second.

Under licensing agreements the Howa Machinery Company of Japan has manufactured some AR 18's and also the semi-automatic sporting version called the AR 180.

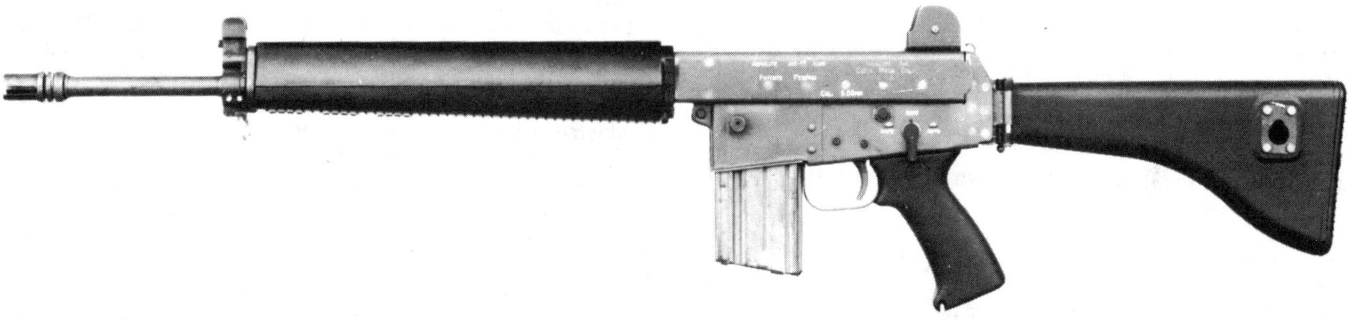

5.56mm AR-18 Rifle.

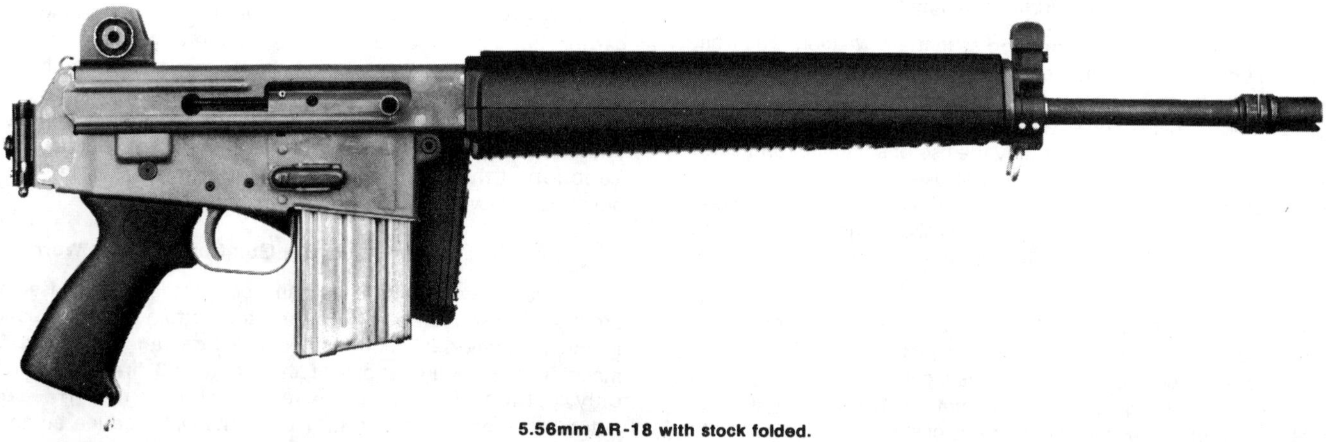

5.56mm AR-18 with stock folded.

5.56mm AR-18 Rifle, stripped.

THE 5.56mm STONER 63 WEAPON SYSTEM

The Stoner 63 system is another development of Mr. Eugene Stoner. It is being handled in the United States by the Cadillac Gage Company of Detroit, Michigan. Essentially the Stoner 63 system consists of fifteen component assemblies plus a machine-gun tripod which can be used in various combinations to make up a rifle, a carbine, a light machine gun magazine-fed, a light machine gun belt-fed, a belt-fed tripod-mounted machine gun, and a fixed belt-fed machine gun developed for tank or helicopter mounting. It is covered here under one heading since it is not practicable or useful to break it down into a separate rifle and machine gun section in different parts of this chapter.

The bolt of the Stoner 63 is similar to that of the AR-18, but the bolt carrier and piston are quite different. The Stoner 63 piston and bolt carrier are assembled together and the bolt carrier has a cap at its rear end which has a roller mounted at its top, which operates the belt-feed lever in the belt-fed configurations of the weapon.

How the Stoner 63 System Works (Rifle and Carbine Configuration)

The Stoner 63 system is arranged so that the weapon fires from a closed bolt when used in the rifle and carbine configuration and from an open bolt when used in the light and medium machine-gun configuration. This is achieved by reversing the vertical position of the bolt carrier and piston and by the use of a hammer and timer in the rifle configuration which is not used in the machine-gun configuration. In the rifle and carbine the bolt carrier piston assembly are placed so that the piston is above the barrel and the bolt closes prior to hammer fall in both automatic and semiautomatic fire.

Semiautomatic Fire. In semiautomatic fire, the hammer is held to the rear by the sear and, if a shot has just been fired, when the trigger is released the disconnector is moved out of engagement with the hammer lug, letting the hammer rotate slightly forward until it is stopped by the engagement of the sear nose with the sear notch of the hammer. Prior to this time the bolt is in the battery (forward) position and the hammer, under pressure of the hammer spring, strikes the spring-loaded firing pin, thus firing the cartridge. The function of the rifle and carbine type weapon from that point on is as follows: gas tapped from the barrel drives the piston and bolt carrier assembly to the rear, unlocking the bolt in basically the same manner as that of the AR-16 and the AR-18. The bolt, bolt carrier, and piston assembly continue to the rear extracting the cartridge and running over the fixed ejector which is mounted so that it extends into the ejector groove in the bolt underside, when in the unlocked position of the bolt, and camming down the hammer. The bolt and bolt carrier gas piston assembly are forced forward by the compressed driving spring, strip a cartridge out of the magazine, and the process is repeated. If the magazine is empty, the magazine follower pushes up the bolt stop and holds the bolt to the rear.

Automatic Fire. In automatic fire, the placing of the selector lever in the "A" (automatic fire) position cams the disconnector back so that it cannot engage the hammer. With the trigger held to the rear and the bolt and bolt carrier piston assembly moving forward, the hammer follows the carrier forward until it is stopped by the timer. The carrier continues to move forward and the tang of the carrier, bottom of the carrier cap, hits the arm of the timer and moves it forward, completing timer disengagement from the hammer when the carrier is still 1/16th-inch out of battery. The carrier continues its movement into battery and the hammer moves up and forward, striking the firing pin and functioning the cartridge. Function of the weapon from that point on is similar to that described above.

How The Stoner 63 Machine-Gun Configuration Works

As noted previously, the bolt and bolt carrier piston assembly are mounted so that the piston is on the bottom and the weapons in the machine-gun configuration fire from an open bolt. The machine-gun configurations of the Stoner 63 fire automatically only; setting of the selector lever in either the "R"—semi-automatic— or "A"—automatic—positions will produce automatic

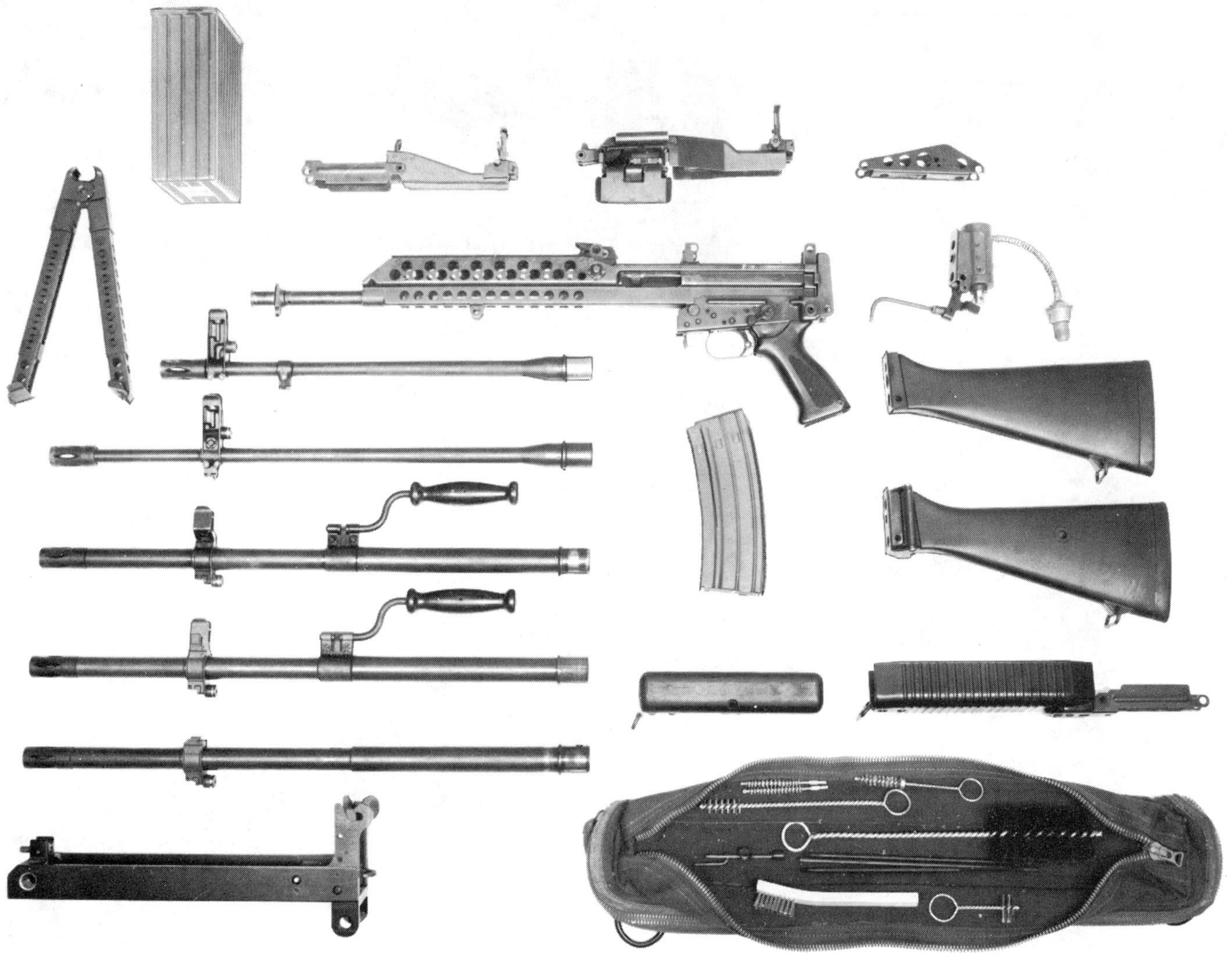

Complete Stoner 63 System.

5.56mm Stoner 63 Carbine version (with folding stock).

fire. In the machine-gun configuration, the hammer and timer are not used with the trigger group. The bolt and bolt carrier piston assembly are held to the rear by the sear engaging a sear notch in the underside of the piston bracket. When the trigger is pulled to the rear, the bolt and bolt carrier piston assembly move forward under the pressure of the expanding driving spring. At this time three phases of the cycle of operation take place before firing is accomplished: feeding, chambering, and locking. The bolt is fully locked when the carrier is 5/32 of an inch from battery. As the carrier continues into battery, the firing pin, which is locked in

5.56mm Stoner Assault Rifle configuration. Note "Flash-suppressor"— grenade launcher combination unit attached to muzzle.

5.56mm Stoner 63 Light Machine Gun—magazine fed. Note offset sights; handle (attached to barrel assembly) is to facilitate rapid change of hot barrels.

5.56mm Stoner 63 Light Machine Gun. Belt fed, with quick attachable ammo box (top cover off) shown as attached to the weapon, normally fitted with bipod.

5.56mm Stoner 63 "Medium" machine gun configuration (mounted on standard U. S. Army M2 tripod.

the carrier, passes through the face of the bolt, strikes the primer and functions the cartridge. All machine-gun configurations of the Stoner 63 have the same type trigger mechanism and therefore the basic functions described above apply to all the machine-gun configurations. The feeding of the machine-gun configurations vary and is explained below.

Feeding of the Stoner 63 Machine-Gun Configurations. The magazine-fed Stoner 63 light machine gun has a top-mounted magazine similar to the Bren, ZB26, Japanese Type 96, etc. It has offset sights—an offset front sight mounted on the barrel and off-set aperture on the rear sight leaf—to compensate for the centrally-mounted magazine. A different receiver cover—housing and sight assembly unit—is used with the magazine-fed weapon than is used with the belt-fed weapon. This housing contains the magazine catch; the magazine adaptor (magazine guide) is a separate component. The cartridge is fed downward into the receiver and the bolt rams the cartridge into the chamber in a downward and forward direction. The barrel is quick change and can be removed by pushing on the barrel latch and pulling forward on the carrying handle. In the magazine-fed light machine-gun configuration, there is no bolt stop to hold the bolt to the rear when the last round is fired as there is in the rifle and carbine configuration.

The belt-fed configurations of the Stoner 63 have a different receiver cover (called a feed cover in this instance) which holds the shuttle-type feed mechanism and a rear centrally mounted leaf-type rear sight, than do the other configurations. They have a quick-change barrel similar to that of the magazine-fed machine gun, but the barrel is heavier in construction. Obviously all Stoner 63 barrels are easily removable; if they were not, the weapon could not be converted into its varied configurations. The feed cover of the belt-fed Stoner 63 operates in a manner quite similar to that of the MG42 and the U.S. M60 machine gun. The bolt carrier cap roller at the top rear of the bolt carrier operates in a track in the belt-feed lever causing it to move laterally—left to right— and move the belt-feed slide which contains the belt-holding panels, belt-feed pawls, and the link retainer. The belt-feed slide is called the pawl carrier in the Stoner 63. The feed pawl moves around into position to be stripped forward and down into the chamber. The link used with the Stoner is similar to the standard M13 push-through type used with the M60 machine gun.

Principal Components. The various Stoner 63 configurations utilize the same principal components to the extent as explained below.

Rifle:
- a. Basic component group, i.e. trigger housing and mechanism and receiver, to include hammer and timer.
- b. Rifle barrel assembly.
- c. Rear sight assembly.
- d. Buttstock.
- e. Magazine adaptor—magazine guide and fore-end assembly.
- f. Magazine.

Carbine:
- a. Basic component group including hammer and timer.
- b. Carbine barrel assembly.
- c. Rear sight assembly.
- d. Folding buttstock.
- e. Magazine adaptor and fore-end assembly.
- f. Magazine.

Light Machine Gun, Magazine fed:
- a. Basic component group—no hammer or timer.
- b. Barrel assembly with off-set front sight and carrying handle.
- c. Machine gun fore-end.
- d. Bipod.
- e. Buttstock.
- f. Magazine adaptor.
- g. Magazine.
- h. Housing (receiver cover) and rear-sight assembly.

Light Machine Gun, belt fed:
- a. Basic component group.
- b. Feed cover including rear sight.
- c. Machine-gun barrel assembly and carrying handle.
- d. Machine-gun fore-end.
- e. Bipod.
- f. Buttstock.

5.56mm Stoner Medium Machine Gun on M2 tripod without butt stock.

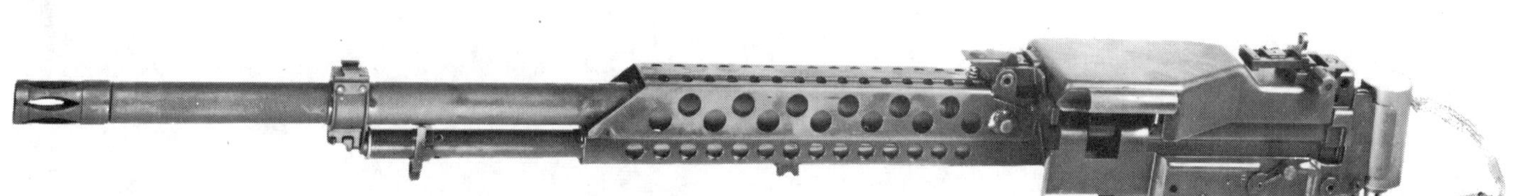

5.56mm Stoner 63 Fixed Machine Gun. Firing solenoid shown as attached to receiver.

Medium Machine Gun, belt fed, tripod mounted:
  a. Basic component group.
  b. Feed cover including rear sight.
  c. Machine-gun barrel assembly and carrying handle.
  d. Stoner or U.S. M2 tripod with adaptor.

Fixed Machine Gun, belt fed:
  a. Basic component group with pistol grip and trigger guard removed.
  b. Solenoid and trigger linkage.
  c. Feed cover without rear sight.
  d. Machine-gun barrel assembly without front sight and carrying handle.

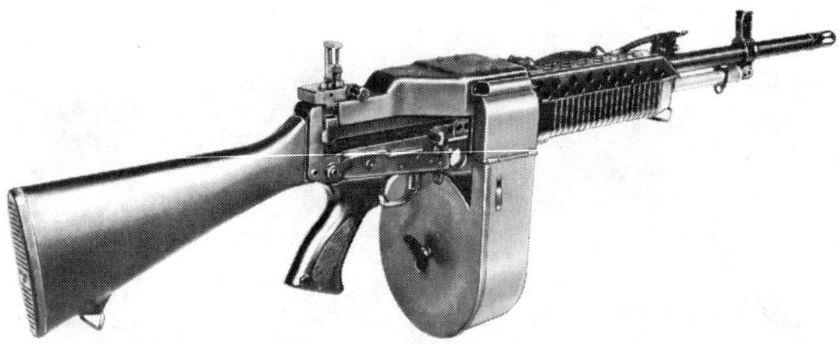

5.56mm Stoner 63A1 Light Machine Gun with 150-round aluminum drum, righthand feed.

Notes on the Stoner 63A System

As a result of engineering and field tests, a number of changes have been made in the Stoner 63 system since it first appeared. The modified system is called the Stoner 63A. The principal changes are as follows: the safety is a separate piece mounted ahead of the trigger guard instead of being a setting on the selector lever. The feed tray is now a post-machined casting rather than the stamped and spot-welded fabrication originally used. The carbine now has a version with a folding steel stock similar to that of the AK in addition to the side-swinging type folding plastic stock. The machine gun versions now have a modified adjustable gas valve. This valve has three positions which are indicated by notches. The widest notch lets the most gas into the system and the narrower settings let in correspondingly smaller amounts of gas. The cyclic rate on the higher setting is about 1000 rounds per minute and about 700 rounds per minute on the lower setting. There have been additional changes made such as dust covers on ejection ports, etc., and it is possible that additional changes will be made in the system. The light machine gun version of the Stoner Model 63A has been assigned the designation XM 207 by the United States Army.

The Navy purchased a limited number of the light machine gun version which they call Light Machine Gun 5.56mm Mark 23. The rifle version tested by the Army was designated the XM22.

Loading and Firing the Stoner System

In general the Stoner 63 is loaded and fired in a manner similar to other weapons of its category. In the rifle and carbine version, a loaded magazine is inserted in the magazine port and the cocking handle, mounted on the top left side of the receiver, is pulled to the rear and then pushed forward—it does not reciprocate with the action. The magazine-fed light machine gun is loaded in a similar manner, but as in the belt-fed machine guns, a setting of either "R" or "A" will produce automatic fire only. With rifle and carbine, set the selector lever, mounted on the left side of the receiver above the pistol grip, on "semi" if semiautomatic fire is desired or "auto" if automatic fire is desired. The machine gun versions fire automatic in either setting of the selector. The safety is engaged by pushing it to the rear. When loading the weapon, the safety should be engaged. It is released by pushing it forward.

The belt-fed guns are loaded by pulling the cocking handle to the rear—cocking the weapon—returning the handle to the original forward position, setting the weapon on "safe," opening feed cover and laying link belt on feed tray—open side of links down— and closing cover. When the weapon is taken off safe and the trigger is pulled, the weapon will fire.

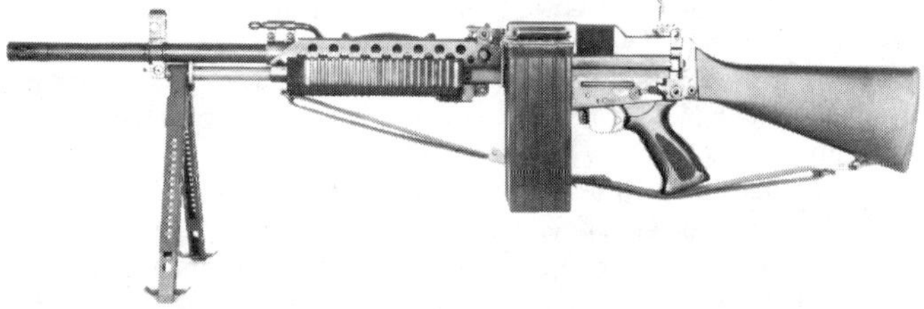

5.56mm 63A belt-fed Light Machine Gun with 100-round plastic ammo box.

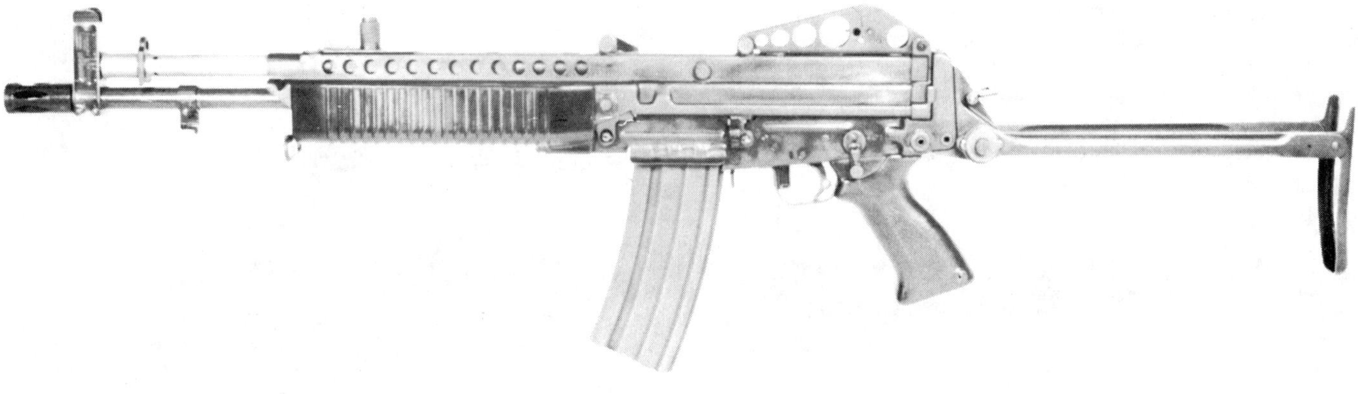

5.56mm Stoner 63A Carbine with folding stock.

Characteristics of the Stoner 63 Weapon System

	Rifle 63A	Carbine	Magazine-Fed Light Machine Gun	Belt-Fed 63A Light Machine Gun	Flexible Medium Machine Gun	Fixed Medium Machine Gun
Caliber:	5.56mm (Caliber .223)	5.56mm (Caliber .223)	5.56mm (Caliber .223)	5.56mm (Caliber .223)	5.56mm (Caliber .223)	5.56mm (Caliber .223)
System of operation:	Gas, selective fire.	Gas, selective fire.	Gas, automatic only.	Gas, automatic only.	Gas, automatic only.	Gas, automatic only.
Length	40.25 in.	Stock extended: 35.87 in. Stock folded: 26.75 in.	40.25 in.	40.25 in.	w/o buttstock: 31 in.	w/o solenoid: 31 in.
Barrel length:	21.67 in*. w/ flash suppressor.	15.7 in.	21.67 in*. w/flash suppressor.	21.67 in*. w/flash suppressor.	21.67 in*. w/flash suppressor.	21.67 in*. w/flash suppressor.
Feed device:	30-round staggered row, detachable, box magazine.	30-round staggered row, detachable, box magazine.	30-round, staggered row, detachable, box magazine.	Disintegrating metallic link belt.	Disintegrating metallic link belt.	Disintegrating metallic link belt.
Sights: Front:	Post with protecting ears.	Post with protecting ears.	Post with protecting ears.	Post with protecting ears.	Post with protecting ears.	None.
Rear:	Aperture.	Aperture.	Leaf w/aperture.	Leaf w/aperture.	Leaf w/aperture.	None.
Weight:	8.20 lb.	8.1 lb.	10.7 lb.	11.68 lb.	Gun: 10.13 lb**.	10.57 lb.
Muzzle velocity:	3250 f.p.s.	3000 f.p.s.	3250 f.p.s.	3250 f.p.s.	3250 f.p.s.	3250 f.p.s.
Cyclic rate:	750-900 r.p.m.	740-800 r.p.m.	Approx 750 r.p.m.	700-1000 r.p.m.	650-850 r.p.m.	650-850 r.p.m.

*True barrel length is 20 inches.
**Adaptor and pintle weight: 3.87 lb.
Tripod (M2) weight: 14 lb.
Elevating and traversing mechanism weight: 3.19 lb.

CHARACTERISTICS OF UNITED STATES SERVICE LIGHT SEMIAUTOMATIC AND AUTOMATIC RIFLES AND CARBINES

	M1 Rifle	M1C Rifle	M1D Rifle	M1 Carbine	M1A1 Carbine	M2 Carbine
Caliber:	.30 (.30-06).	.30 (.30-06)	.30 (.30-06).	.30 Carbine M1.	.30 Carbine M1.	.30 Carbine M1.
System of operation:	Gas, semiautomatic.	Gas, semiautomatic.	Gas, semiautomatic.	Gas, semiautomatic.	Gas, semiautomatic.	Gas, selective fire.
Length overall:	43.6 in.	43.6 in.	43.6 in.	35.6 in.	Stock extended: 35.5 in. Stock folded: 25.4 in.	35.6 in.
Barrel length:	24 in.	24 in.	24 in.	18 in.	18 in.	18 in.
Feed system:	8-round, staggered row, non-detachable, box magazine.	8-round, staggered row, non-detachable, box magazine.	8-round, staggered row, non-detachable, box magazine.	15- or 30-round, staggered row, detachable, box magazine.	15- or 30-round, staggered row, detachable, box magazine.	15- or 30-round, staggered row, detachable, box magazine.
Sights: Front:	Blade with protecting ears.	2.2X telescopic.	2.2X telescopic.	Blade with protecting ears.	Blade with protecting ears.	Blade with protecting ears.
Rear:	Aperture.			Aperture on ramp.	Aperture on ramp.	Aperture on ramp.
Weight:	9.5 lb.	11.75 lb.*	11.75 lb.*	5.5 lb.	6.19 lb.	5.5 lb.
Muzzle velocity:	2805 f.p.s. (M2 ball)	2805 f.p.s. (M2 ball)	2805 f.p.s. (M2 ball)	1970 f.p.s.	1970 f.p.s.	1970 f.p.s.
Cyclic rate:	-----	-----	-----	-----	-----	750-775 r.p.m.

*Includes telescope, flash hider, gun sling, and cheek pad.

CHARACTERISTICS OF UNITED STATES SERVICE LIGHT SEMIAUTOMATIC AND AUTOMATIC RIFLES AND CARBINES (Cont'd)

	M14 Rifle	M14A1 Rifle	M16 Rifle	Johnson M1941 Rifle
Caliber:	7.62mm NATO.	7.62mm NATO.	5.56mm (.223).	.30 (.30-06).
System of operation:	Gas, selective fire.	Gas, selective fire.	Gas, selective fire.	Recoil, semiautomatic.
Length overall:	44.14 in.	44.3 in.	39 in.	45.87 in.
Barrel length:	22 in.	22 in.	20 in.	22 in.
Feed system:	20-round, staggered row, detachable, box magazine.	20-round, staggered row, detachable, box magazine.	20-round, staggered row, detachable, box magazine.	10-round, rotary type, non-detachable magazine.
Sights: Front:	Blade with protecting ears.	Blade with protecting ears.	Post with protecting ears.	Post with protecting ears.
Rear:	Aperture.	Aperture.	Aperture.	Aperture.
Weight:	8.7 lb.	12.75 lb.	6.3 lb. w/o magazine.	9.5 lb.
Muzzle velocity:	2800 f.p.s.	2800 f.p.s.	3250 f.p.s.	Approx. 2770 f.p.s. (M2 ball)
Cyclic rate:	750 r.p.m.	750 r.p.m.	700-900 r.p.m.	-----

THE BROWNING CALIBER .30 AUTOMATIC RIFLE MODEL 1918, 1918A1, and 1918A2

This weapon was developed by John Browning in 1917 to meet the United States requirement for an automatic rifle for service in World War I. The M1918 was made during World War I by Colt, Winchester, and Marlin Rockwell. These concerns made 85,000 of these weapons before the armistice which concluded World War I.

Basic BARs

There are actually four basic Browning Automatic rifles which have been officially adopted by the United States. These weapons and their descriptions follow.

Model 1918. This weapon has no bipod, is capable of selective fire, and is relatively light (16 pounds) compared with the later models. A simple tube-type flash hider is used. There is no shoulder support plate hinged to the butt plate. The rear sight and butt plate are similar to those of the Model 1917 (Enfield) rifle.

Model 1918A1. This model has a shoulder support plate hinged to the butt plate and a bipod attached to the gas cylinder just forward of the forearm. It fires selective fire, has a tube-type flash hider and uses the same sights as the Model 1918. This modification was adopted in 1937.

Model 1918A2. Adopted shortly before World War II, this weapon, as the Model 1918A1, was originally made up from Model 1918s and 1918A1s of World War I manufacture. The bipod, which has skid-type feet as opposed to the spike-type feet of the Model 1918A1, is fitted to the tube-type flash hider. The forearm has been cut down in height around the barrel and shortened. As originally made a removable stock rest, which fitted in a hole in the buttstock, was used with this weapon. It has a hinged shoulder rest attached to the butt plate as does the M1918A1, but the shoulder rest plate is shorter. There is a metal shield inserted horizontally in the forearm to protect the recoil spring guide from barrel heat and right and left magazine guards have been attached to the front of the trigger guard body. The rear sight of the M1918A2 is similar to that of the M1918A4 machine gun and the M1922 Browning Automatic

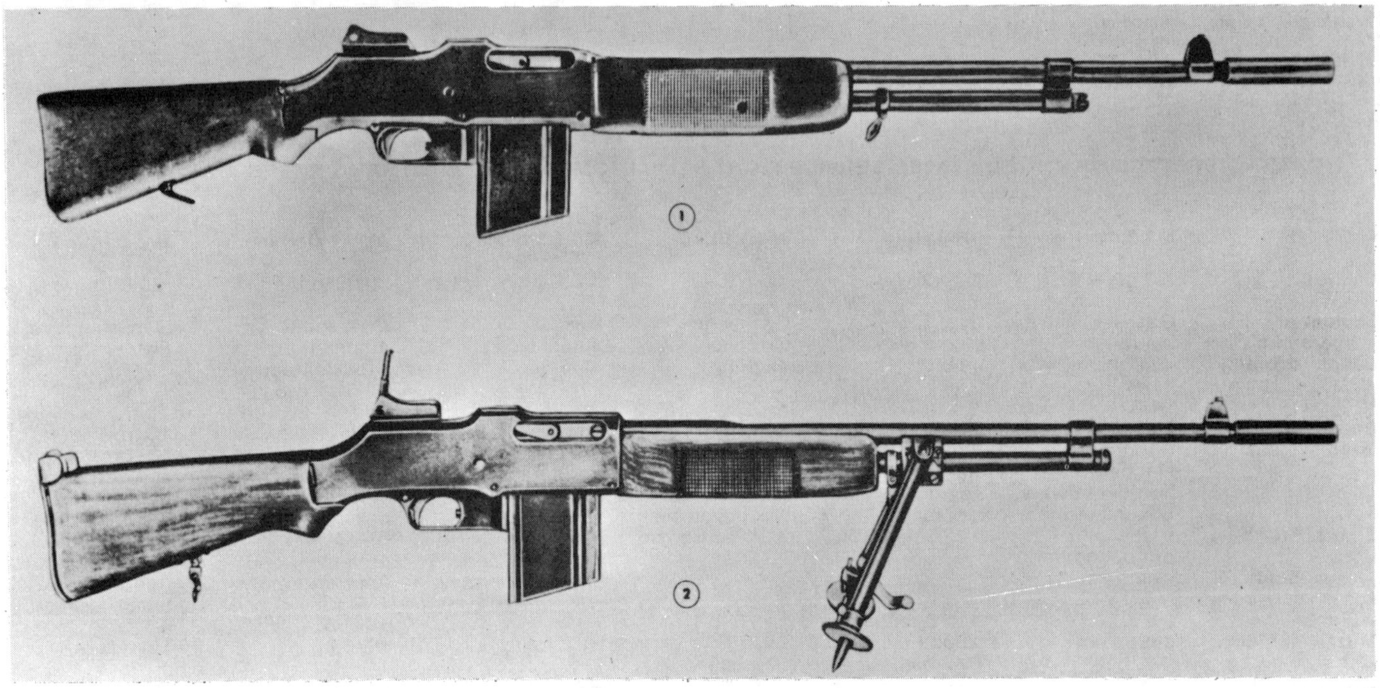

Caliber .30 Browning Automatic Rifles. (1) Model 1918, (2) Model 1918A1.

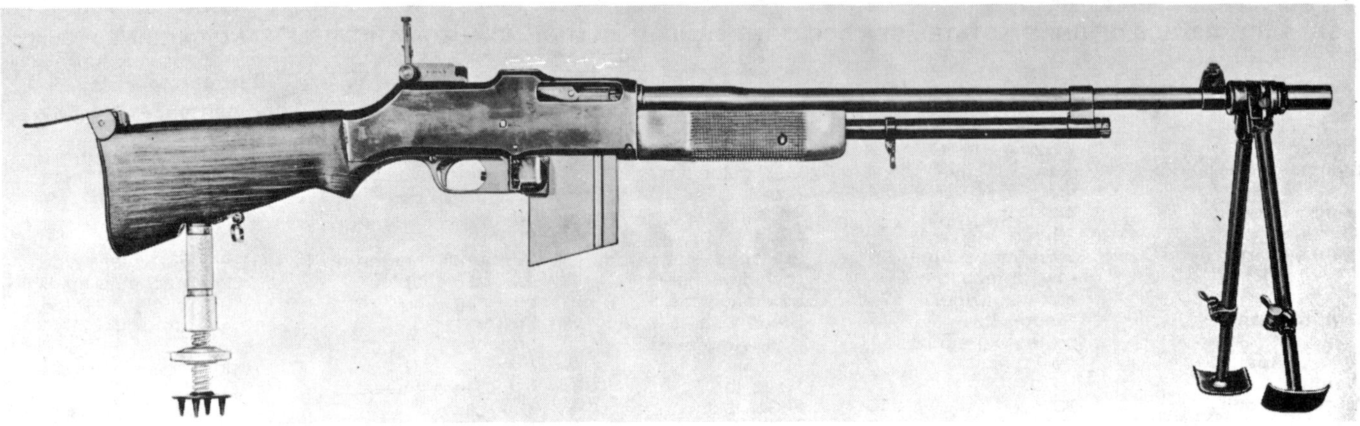
Caliber .30 Browning Automatic Rifle, Model 1918A2 as originally issued.

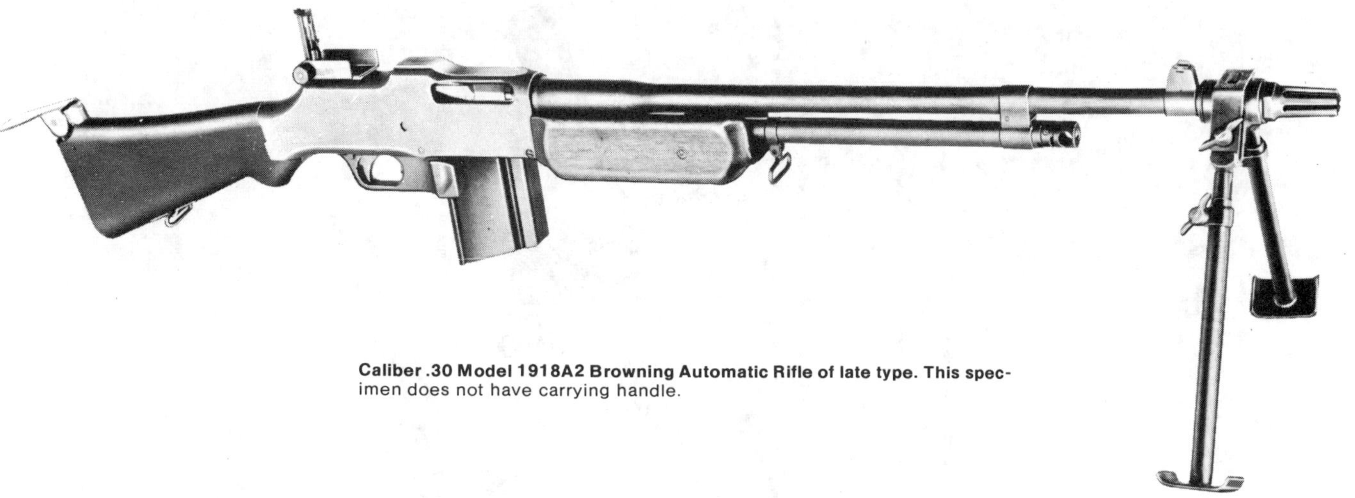

Caliber .30 Model 1918A2 Browning Automatic Rifle of late type. This specimen does not have carrying handle.

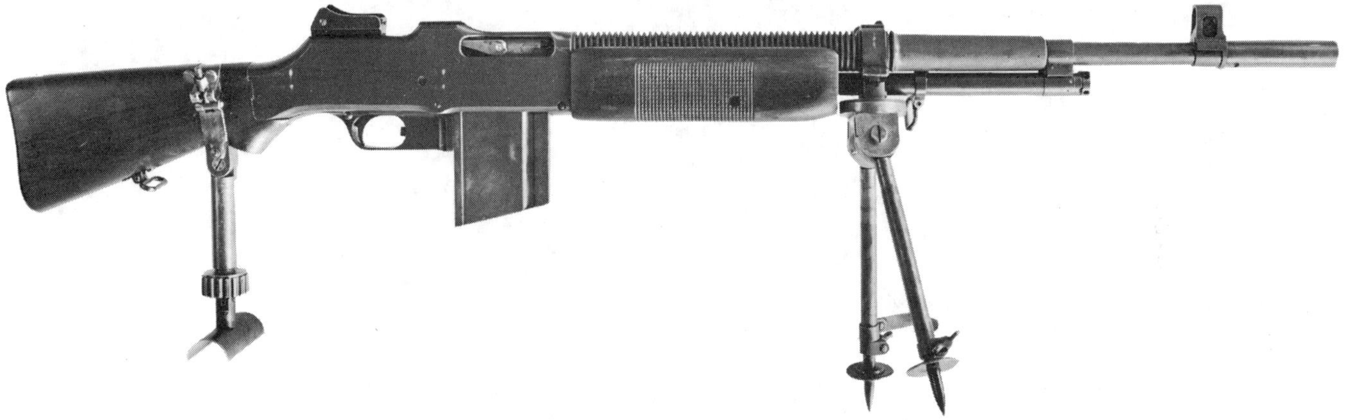

Browning Machine Rifle, caliber .30 M1922 (old type rear sight base) mounted on tripod M1924 and stock rest M1924.

7.62mm NATO T34 Browning Automatic Rifle.

rifle. It is adjustable for windage as well as elevation and uses micrometer screws for adjustment. The 1918A2 is not capable of semiautomatic fire. It has two rates of automatic fire and a rate-reducing mechanism which is explained in detail later in the text.

The Model 1918A2 went through a number of modifications during World War II. Among these were the use of a shortened fore-end with grasping grooves, the abandonment of the stock rest, the use of plastic buttstocks and the development of a carrying handle for the weapon which, due to the conclusion of World War II, did not see much service until the Korean war. I.B.M. and New England Small Arms Corporation manufactured Browning Automatic rifles during World War II. During the Korean war a prong-type flash suppressor was adopted. Royal McBee Typewriter Corp. manufactured 61,000 M1918A2 BARs during this period. A gas cylinder regulator which can be easily turned by hand was also introduced during this period.

Browning Machine Rifle Model 1922. This weapon, which appeared in very limited numbers, was developed to give the horse cavalry of the twenties a light weight sustained-fire capability. It has a heavy barrel with radial cooling fins, butt swivel attached to the left side of the stock, a wide groove around the buttstock

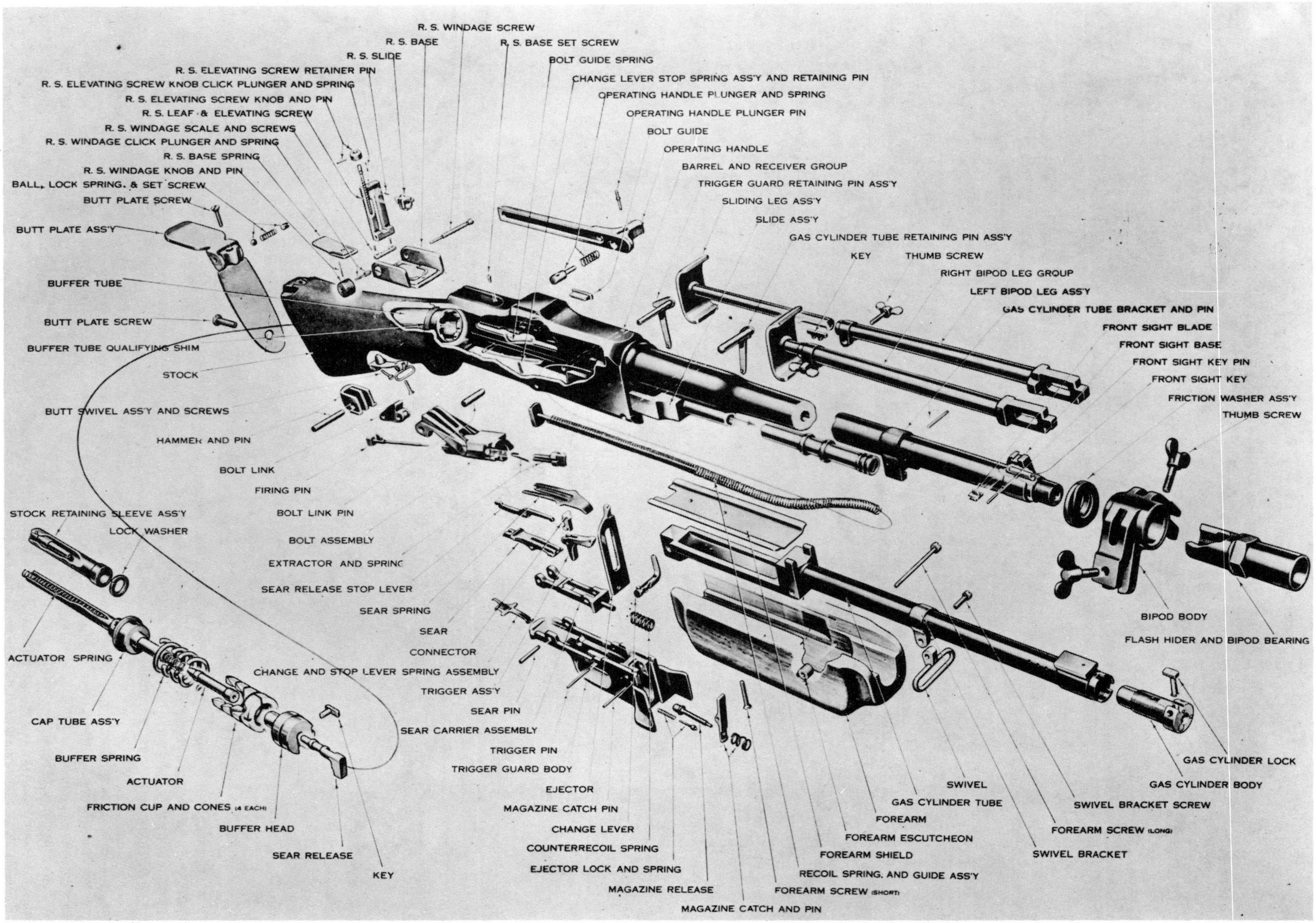

Model 1918A2, stripped.

Characteristics of United States Service Browning Automatic Rifles

	Model 1918	Model 1918A1	Model 1918A2	Model 1922
Caliber:	.30 (.30-06).	.30 (.30-06).	.30 (.30-06).	.30 (.30-06).
System of operation:	Gas, selective fire.	Gas, selective fire.	Gas, automatic only.	Gas, selective fire.
Length overall:	47 in.	47 in.	47.8 in.	Approx. 41 in.
Barrel length:	24 in.	24 in.	24 in.	18 in.
Feed device:	20-round, staggered row, detachable, box magazine.	20-round, staggered row, detachable, box magazine.	20-round, staggered row, detachable, box magazine.	20-round, staggered row, detachable, box magazine.
Sights: Front:	Blade.	Blade.	Blade.	Hooded blade.
Rear:	Leaf w/aperture battle sight w/aperture.	Leaf w/aperture battle sight w/aperture.	Leaf w/aperture adjustable for windage.	Leaf w/aperture adjustable for windage.
Weight:	16 lb.	18.5 lb.	19.4 lb.	19.2 lb.
Muzzle velocity:	2805 f.p.s.	2805 f.p.s.	2805 f.p.s.	Approx. 2700 f.p.s.
Cyclic rate:	550 r.p.m.	550 r.p.m.	Slow: 300-450 r.p.m. Fast: 500-650 r.p.m.	550 r.p.m.

for the butt rest clamp, a bipod which clamps around the barrel, and a rear sight adjustable for windage and elevation similar to that used on the Model 1919 machine guns and the M1918A2 BAR. This weapon was declared obsolete about 1940.

T34 Automatic Rifle. This is a modification of the Browning Automatic Rifle for the caliber .30 T65E3 cartridge case (7.62mm NATO) initiated in June 1949. It is the last United States military Browning Automatic rifle.

The Browning Automatic rifle was made with cast steel receivers during World War II and was the subject of much experimentation in materials and methods of manufacture by the United States, since it is basically a difficult weapon to make and requires a great deal of material and machine time. It is, like so many of the weapons designed during its period, built to last a lifetime with commensurate disabilities in cost and tool expenditure.

Colt's Patent Firearms Manufacturing Company manufactured versions of the weapon for sale to police and foreign governments. The Colt Monitor had a shortened barrel and was widely used as a police weapon; another Colt model was the R75A which had a quick-change barrel somewhat similar to that of the Swedish Model 37 BAR. FN of Belgium also has produced a number of models of the Browning Automatic rifle which are covered in the chapter on Belgium.

How to Load and Fire the Browning Automatic Rifle

Cock weapon by pulling operating handle, located on the left side of the receiver, to the rear. Push the handle back to its forward position—it will now reciprocate with the action. Put change lever on "S"—safe—marking. Insert magazine, rear end cocked up slightly, and slap smartly in with heel of hand. If using M1918 or M1918A1 setting the change lever on the letter "F" will give semi-automatic fire for each pull of the trigger. If using the M1918A2, setting the change lever on "F" will give slow rate—approximately 350 rounds per minute—automatic fire. If the change lever is set on "A", all models will fire automatic fire. To remove the magazine, push magazine release located on the interior front surface of the trigger guard.

Loading and Firing the BAR

If a magazine filler is available, place the wide end over the top of the magazine so the grooves fit over the magazine catch rib. Insert a clip of cartridges in the filler and with the right thumb strip the cartridges into the magazine exactly as though they were being fed into a bolt action rifle. Single cartridges may be so loaded. If no filler is available, cartridges may be loaded singly into the magazine as for automatic pistol.

Insert magazine between sides of receiver in front of the trigger guard and push home until it locks. This will normally be done with the right hand. While the magazine may be inserted with the weapon uncocked, it will ordinarily be done after the rifle has been cocked and set on "safe."

Field Stripping the BAR

Pull the operating handle back to cock the weapon. Then thrust it fully forward.

Rotate the gas cylinder retaining pin (at forward left end of the receiver) and withdraw it from its socket.

Now pull forward the forearm and gas cylinder tube and remove from the rifle. Ease the mechanism forward.

Rotate the retaining pin at forward end of trigger guard and withdraw it. The entire trigger mechanism may now be withdrawn from the bottom of the rifle.

Remove the recoil spring guide. Press in the checkered surface on its head and turn it until the ends clear the retaining shoulders; ease out the guide and the recoil spring and withdraw. Withdraw the handle by lining up the hammer pin holes on the side of the receiver and on the right side of the operating handle. Insert the point of a bullet in the hole in the operating handle with the right hand. Press back against the hammer pin while pushing the slide backward with the left hand.

As the two holes register, the pressure of the bullet will force the hammer pin out of the large hole on the left side of the receiver and it may be withdrawn. This will permit the operating handle to be pulled straight to the rear and out of its guide.

Push the hammer forward out of its seat in the slide and lift it out of the weapon.

Pull the slide directly forward out of the receiver, taking care that the link is pushed well down so that slide can clear it. Remove the slide carefully to avoid striking the gas piston or its rings against the gas cylinder tube bracket female.

With the point of a bullet force out the spring bolt guide from inside the receiver, then lift the bolt, bolt lock, and link by pulling slowly to the rear end of the receiver and then lifting them out. The firing pin may now be lifted out of the bolt and the extractor removed by pressing the small end of the cartridge against the claw and exerting upward and frontal pressure. No further stripping is normally necessary or recommended.

How the Basic Browning Automatic Rifle Action Works

Starting with the gun loaded and cocked, the action is as follows: Pressing the trigger pulls down the nose of the sear, disengaging it and permitting the slide to move forward under the action of the recoil spring. The rear end of the slide contains the hammer which is connected by a link to the bolt. The slide is pulled forward by the compressed recoil spring. During the first quarter inch of travel of the slide, the front end of its feed rib strikes the base of the top cartridge in the magazine, driving it ahead toward the firing chamber.

When the cartridge has traveled about a quarter of an inch, the bullet strikes the bullet guide on the breech and is deflected upward toward the chamber. This action also guides the front end of the cartridge from under the magazine lip. When the head of the cartridge reaches the part of the magazine where the locking lips are cut away and the opening enlarged, the magazine spring forces it out of the magazine. The base of the cartridge now slides across the face of the bolt and under the extractor; and if it fails to position correctly the extractor will still snap over its head as the bolt reaches its forward position. At the time the cartridge leaves the magazine, the bullet nose is so far in the chamber that it is guided from that point on.

When the slide is within two inches of its complete forward position, a circular cam surface on the bottom of the bolt lock starts to ride over the rear shoulders of the bolt support, camming up the rear end of the bolt lock. The link pin rises above the line joining the bolt pin and the hammer pin, so that its joint has a tendency to buckle upwards. As the attached bolt is now opposite its locking recess in the receiver it pivots upward about the bolt lock pin. The link whose lower edge is attached to the hammer pin revolves upward and forces the bolt lock up; the rounded surface on the bolt lock, just above this locking face, slips over the locking shoulder in the hump of the receiver and provides a lever thrust which forces the attached bolt home into final position.

The bolt lock is now above the position of the bolt, and locks firmly in the hump in the receiver as the hammer pin passes beneath the link pin. The firing pin is in the bolt, with a lug on its rear end buried in the slot at the other side of the bolt lock, making it impossible for the firing pin to be struck by the hammer at any time except when the bolt lock is in its recesses in the hump of the receiver. Thus when the hammer pin passes under the link pin, the head of the firing pin is exposed to the center rib of the hammer; and as the slide still continues forward, the hammer drives the firing pin ahead and explodes the cartridge in the chamber.

The forward motion is now halted when the front end of the slide strikes against a shoulder at the rear end of the gas cylinder tube.

Return Movement of the Action. About 6" from the muzzle, a small port is bored in the bottom of the barrel. As the bullet passes over this port, a small amount of gas, still under high pressure, escapes through it and passes through similar ports in the gas cylinder tube bracket, the gas cylinder tube, and the gas cylinder. The gas cylinder port is the smallest of these and acts as a throttle on the barrel pressure. The ports in the gas cylinder lead radially into a small well situated in the head of the gas cylinder. Through this well the pressure is conducted to the gas system plug, through which it acts on the piston for the length of time the bullet is traveling the 6" distance from barrel port to muzzle. This results in a sudden, hard blow, backward against the piston plug.

The gas piston is assembled to the slide, and the sudden blow as it drives the gas piston back also forces back the slide and the parts attached to it and compresses the recoil spring which is seated in the slide.

After the piston has traveled back a little over 1/2", bearing rings on its rear and corresponding ones in the gas piston plug pass out of the gas cylinder. The gas now expands around the gas piston head into the gas cylinder tube where it is exhausted into six portholes in the tube placed just at the rear of the gas cylinder tube brackets.

Two rings on the piston about 1-1/4" from the head prevent most of the gas from traveling back through the gas cylinder tube; and also act as bearings to maintain the front end of the piston in the

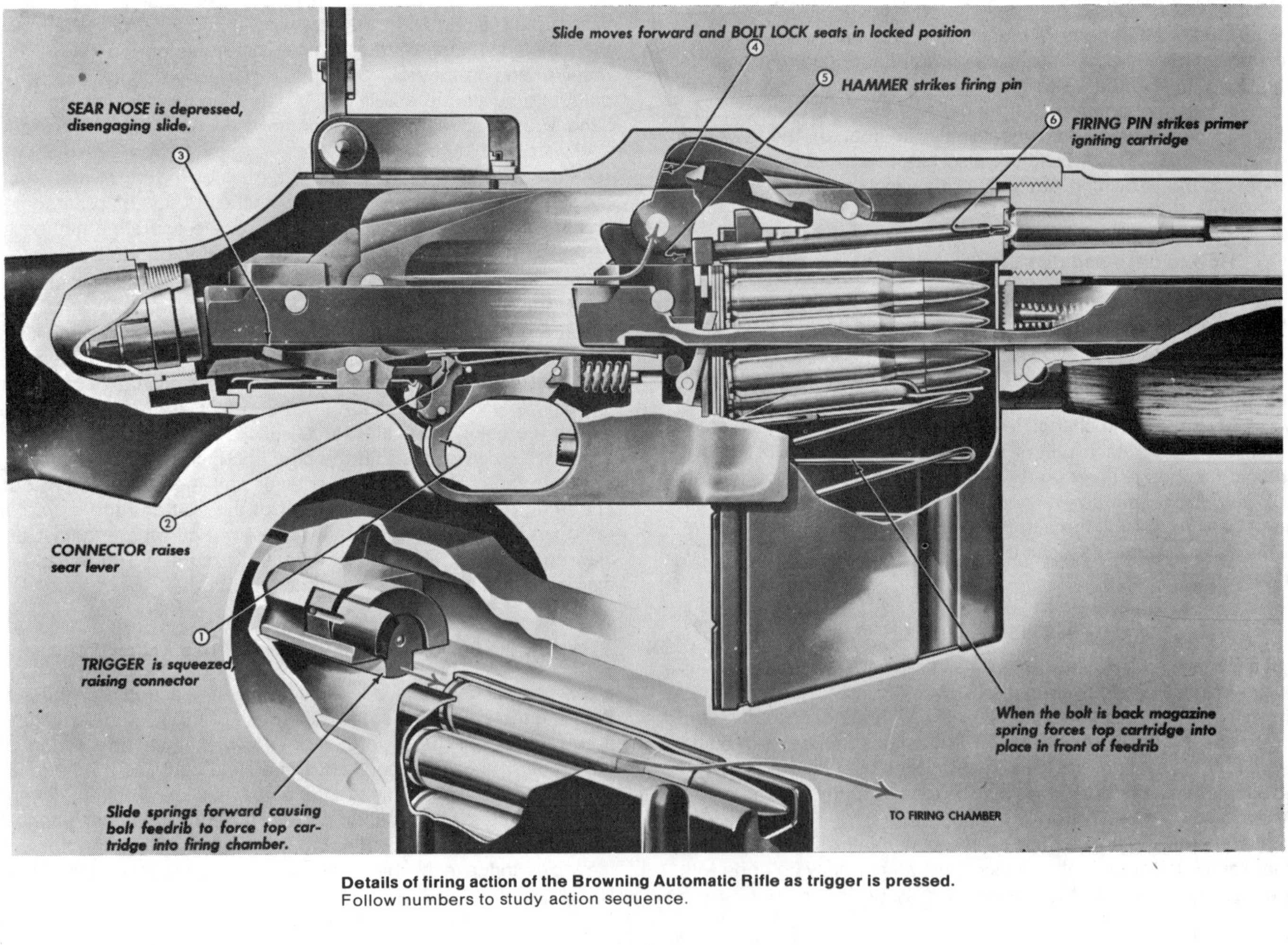

Details of firing action of the Browning Automatic Rifle as trigger is pressed.
Follow numbers to study action sequence.

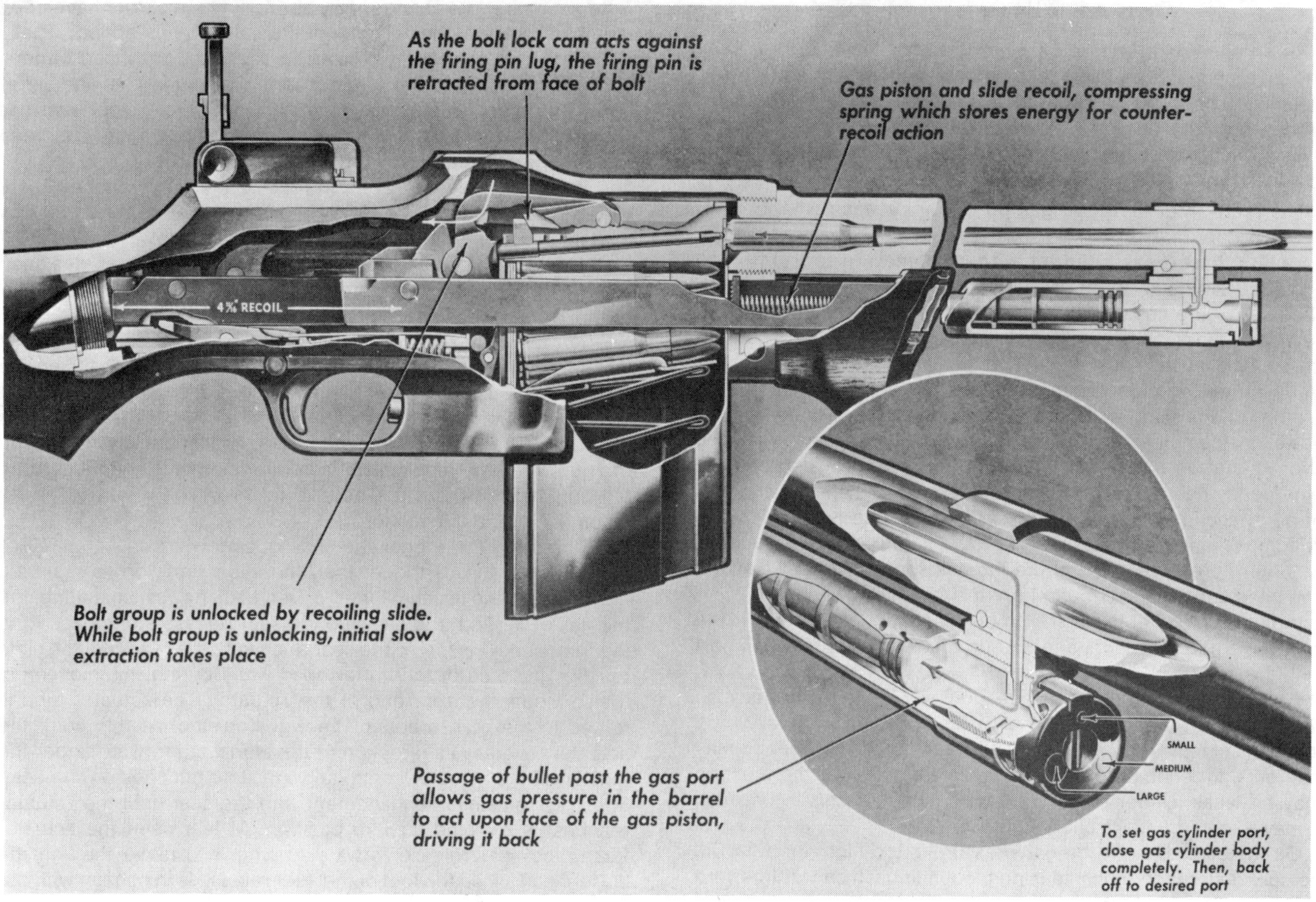

Recoil and unlocking action of the Browning Automatic Rifle M1918A2 showing action beginning as bullet passes over gas port and gas escaping into cylinder which starts the rearward action.

center of the gas cylinder tube after the piston has passed out of the cylinder.

Unlocking Action. As the hammer pin is slightly in advance of the connecting link pin, the initial backward movement of the slide carries the hammer back without moving either the attached bolt lock or bolt; and when the movement has progressed far enough (about 1/5") and the high breech pressure has dropped to safe limits the unlocking action starts. The link is compelled to revolve forward about the hammer pin, and so to draw the bolt lock down out of a hump in the receiver and start it to the rear. The motion of the bolt and bolt lock is now accelerated as the lock is drawn completely out of its locking recess, locking the shoulders in the receiver.

As the bolt lock is prevented from revolving from below the line of backward travel of the bolt, further rearward travel of all moving parts is in a straight line. Meanwhile, however, during the unlocking motion, a cam surface on the slot in the bottom side of the bolt lock has come in contact with a cam surface on the firing pin lug, and has drawn the firing pin away from the base of the bolt.

Also during the backward action, the circular cam surface on the lower part of the bolt lock, operating on the rear shoulders of the bolt support, has produced a lever action tending to loosen the cartridge case in the firing chamber. From that point, the slide and all its moving parts are traveling to the rear at the same speed, carrying along the empty cartridge case held in its seat in the face of the bolt by the extractor (the extractor is positioned in the upper right side of the bolt near the ejection port). Thus as the slide nears the end of its travel and the base of the empty cartridge case strikes the ejector on the left side of the bolt feed rib, the empty case is pivoted about the extractor and through the ejection port. As the front end of the cartridge case passes out of the receiver, it is so pivoted that it strikes the outside of the receiver about an inch to the rear of the ejection port; and hence rebounds toward the right front.

The rearward motion is now completed as the end of the slide strikes against the end of the buffer, and the sear nose catches in the notch at the underside of the slide and holds the weapon open and ready for the next pull of the trigger. (If the weapon is set for full automatic fire, the sear nose is held depressed, so that it does not stop the slide which continues forward, firing the weapon in the full automatic cycle.)

The buffer is a tube in the butt of the rifle in which are placed a buffer head against which the slide stops, a friction cup slit to allow for expansion, a steel cone to fit into the cup, and four more cups and cones in series. Behind these is the coil buffer spring and the buffer nut which is screwed into the end of the tube to form a seat for the spring.

As the rear end of the slide strikes the buffer head, it moves it to the rear, forcing the cups over the cones causing them to expand tightly against the tube, thus producing friction as the cups move back and the buffer spring is compressed. The rearward motion of the slide is therefore checked gradually, and practically no unpleasant rebound occurs. The friction mechanism is returned to its original place by the compression of its spring.

Notes on the BAR

A very important feature of this rifle is that the bolt, bolt lock, and link mechanism start back comparatively slowly and do not attain the speed of the slide itself until after the period of high breech pressure passes. This feature is also important in that it does not subject the mechanism to undue strain as the gas pushes the piston back.

There are three different gas ports. The weapon will normally be set to operate on the smallest port. It is properly alined by screwing in the gas cylinder with combination tool until the shoulder of the cylinder is one turn from the corresponding shoulder of the gas cylinder tube, and the smallest circle on the cylinder head is toward the barrel. (To permit setting the regulator, the split pin must be pushed out sufficiently to permit the regulator to be turned on older weapons.)

If the rifle is sluggish from insufficient gas, the cylinder should be set one complete turn on each side of the original setting. However, it is to be noted that the larger ports are provided only for emergency use. They should be utilized only when through lack of oil or accumulation of dirt or carbon, the rifle is sluggish and conditions make it impossible to properly correct these troubles. It is therefore essential that the threads should be kept cleaned and oiled and cylinder free to turn at all times.

In field service, at the first sign of insufficient gas, unscrew the cylinder a third of a turn, and line up the medium circle and port with the gas opening.

When gas is insufficient, the weapon may fail to recoil because the port is not properly alined or is unusually dirty. A very dirty mechanism may also cause such a stoppage. Or the weapon may not recoil far enough to permit complete ejection or the ejection may be weak. Under some conditions, although this is unusual, it may result in uncontrolled automatic fire.

On the other hand if the gas pressure is too high, the rifle will be speeded up too much causing a pounding which will interfere with accuracy. This may also generate excessive heat in the gas operating mechanism.

How the Browning Automatic Rifle Model 1918A2 Works

The Model 1918A2 is a modification of the M1918A1 Browning Automatic rifle. The change lever spring and the carrier have been modified. Several new components have been added. These include a sear release stop lever, a sear, key and head buffers, sear release actuator and actuator spring, actuator stop, and buffer head. The bipod is attached to the flash hider. A stock rest has been added and the forearm made lighter. A trap plate has been added between the barrel and the gas cylinder tube.

This gun has been modified to replace the single-shot mechanism. Single shots can be fired in this modified version only by pressing and releasing the trigger rapidly.

However, as a compensating factor, the gun has been designed to fire at a low and a high rate of speed somewhat in the manner of the British Besa Machine Guns.

When the change lever is at "F," the rifle is cocked in normal fashion and the sear engages in a notch in the slide. Pressing the trigger in the usual manner results in controlled automatic fire at a reduced rate which will be delivered as long as the trigger is held back. There is a distinct difference noticeable when handling the gun when the mechanism is in operation and when it is firing full automatic without it. There is no provision made for semi-automatic fire in this model.

On pressing the trigger the slide goes forward in normal BAR fashion firing the cartridge; then the slide starts on its rearward movement in the usual manner, driven back by gas expanding into the gas cylinder, but it picks up the sear release and strikes it on the front end. This forces the sear release to the rear until the slide reaches the face of the buffer head while during this movement it also meets the front end of the actuator. The actuator tube is forced to the rear, meeting the actuator stop. At this point the actuator reverses its direction of travel moving forward under the tension of the actuator spring and the slide engages on the sear. The slide remains in engagement with the sear until the actuator reaches its extreme forward position. At that point the actuator forces the sear release forward forcing it to move through the buffer head, while the foot of the sear release is in contact with the angle surface of the rear of the sear. This cams the sear out of engagement with the slide forcing the slide to go forward at this point to fire the cartridge.

The M1918A2 was the standard squad automatic weapon of World War II and Korea.

UNITED STATES SUBMACHINE GUNS

Although the United States was the third country in the world to develop a submachine gun, this type weapon was not adopted by the United States until about 1928 when it was first used by the Marines in Nicaragua and by the Coast Guard in their war with the rum runners of the prohibition period. The weapon used was the caliber .45 Thompson Model 1928, which development of General John T. Thompson and the Auto-Ordnance Corp, had made its first public appearance in earlier form in 1919. The first production Thompson was the Model 1921 and the "Tommy Gun" earned a reputation, probably unfairly and mainly due to lurid movies, as a gangster weapon during the age of the "big gang wars" in the United States. The weapon was widely used by police forces, and the attitude seems to have been adopted both in the United States and the United Kingdom that the submachine gun was basically a police weapon. Be that as it may, the first submachine gun purchased by the United States Army was the caliber .45 Thompson Model 1928A1 and this was purchased in limited quantities, principally for use by armored and reconnaissance units. The Thompson was being produced by Colt at this time; the patent owner—Auto Ordnance Corp.—did not have any manufacturing capability at the time and developed only a limited manufacturing capability during World War II. Colt produced approximately 15,000 Thompsons. In 1940 the British government gave large contracts to Auto Ordnance Corp. for the Model 1928A1 Thompson. Auto Ordnance subcontracted most of this order to the Savage Arms Corp., then at Utica, New York. When Lend-Lease came into effect—1941—the United States government took over these contracts and the Thompson remained in production until 1943. During the course of this contract, modifications were made in the Model 1928A1 and the M1 and M1A1 models were produced. Savage made 1,501,000 Thompson M1928A1, M1, and M1A1 submachine guns. Many of these weapons made on the British contract did not make it across the Atlantic because of the German U-boat campaign which was at its height when shipments of the Thompson to the United Kingdom were at their greatest. The Thompson, while a reliable weapon, has several outstanding shortcomings. It is overly heavy in relation to its muzzle energy and more importantly it is expensive in the use of materials, machine time, and machine tools, all items which are in short supply during a large war. The Army therefore decided to find or develop a weapon to replace the Thompson. In 1941 a requirement was generated for a new weapon and a number of guns were submitted in competitive tests to meet this requirement. Among these were the Hyde 109, first tested in 1939 by the Army and Hyde Inland, the ATMED—also designed by George Hyde—the Star, the Atlantic (also a Star design), the United Defense, the Reising 1 and

Hyde caliber .45 Model 35 Submachine Gun. Shown with 20 and 40-
round magazines.

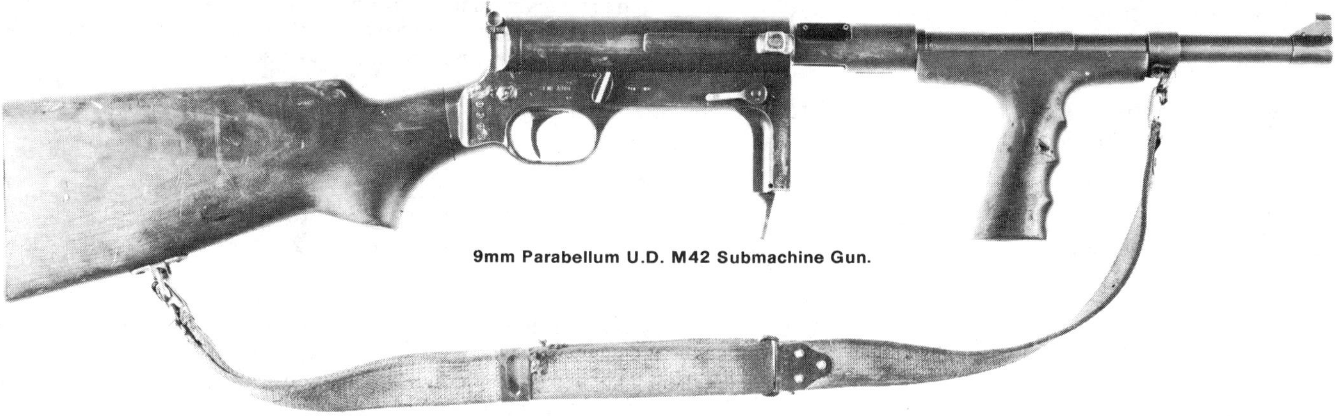

9mm Parabellum U.D. M42 Submachine Gun.

Caliber .45 Reising Model 50 Submachine Gun.

2, the Olsen, the Owen, the Sten Mark III, the Austen, the Woodhull, the Suomi, the Turner, the Smith & Wesson semiautomatic carbine, the standard Thompsons, the Thompson T2, the German MP38 and Bergmann. The MP38 and the Bergmann were not official competitors but were under study.

The Hyde Inland was adopted as the "substitute standard" caliber .45 submachine gun M2 in April 1942. The M2 was never put in mass production and the first production models reached Aberdeen Proving Ground for test in May 1943, five months after the M3 had been standardized.

The Reising Gun, in a fixed-stock version, the Model 50, and a folding stock version, the Model 55, was manufactured by Harrington and Richardson for the Marine Corps, the Home Guard, and the British Purchasing Commission. Approximately 100,000 of these weapons were made. Many wound up in the armories of local police departments throughout the United States. It is no longer in military service.

The United Defense gun, known as the U.D. Model 42, was designed at High Standard by Carl Swebilius and approximately 15,000 were made by Marlin for the United Defense Supply Corporation. This weapon, which was chambered for the 9mm Parabellum cartridge, was used by the OSS and air dropped to various underground organizations. The Dutch government also purchased a few. It was known in the U.K. as the Marlin.

The weapons which eventually developed into the M3 submachine gun were the T15, a selective-fire weapon, and the T20, an automatic only weapon which could be converted from caliber .45 to 9mm Parabellum. This weapon was made of stampings with a minimum of machined steel parts and was at least partially the result of studies made of the British Sten. Mr. George Hyde did the basic design work and the industrial engineering was handled by Mr. Frederick Sampson of the Inland Manufacturing Div. of General Motors Corp. The M3 was adopted in December 1942. In December 1944, the M3A1, a modification of the M3, was adopted as standard.

There have been several prototype submachine guns developed in the United States recently. One, the Colt 5.56mm CAR-15 is, in modified form, being tested by the Army as the XM 177E2. The others are the Smith and Wesson Model 76 and the Ingram Model 10. The Smith and Wesson Model 76 is a selective fire, blowback-operated weapon, chambered for the 9mm Parabellum cartridge. The cyclic rate is 720 rounds per minute and the weapon weighs 8.75 pounds with loaded 36-round magazine. The weapon, which is manufactured mainly of stampings and fabrications, has a folding steel stock.

The Ingram Model 10 is also a blowback-operated weapon chambered for the 9mm Parabellum cartridge. It is similar in general design and appearance to the UZI, in that the top of the bolt telescopes the barrel and the magazine well is located in the pistol

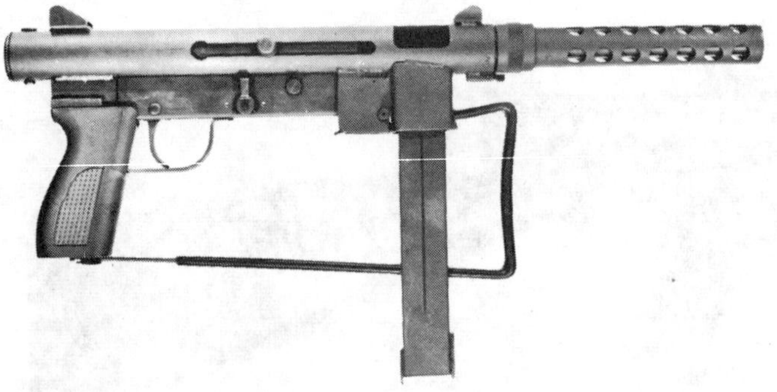

Smith & Wesson 9mm M76 Submachine Gun.

grip. The Ingram is a very short weapon—11.5 inches long with stock retracted and 17.5 inches long with stock fixed. It weighs 7.1 pounds loaded with a 36-round magazine. The cyclic rate of the full automatic Ingram is approximately 700 rounds per minute.

The Thompsons are still widely used by various allied countries, but are no longer in United States military service.

THE THOMPSON CALIBER .45 MODEL 1928A1, M1, and M1A1 SUBMACHINE GUNS

The Model 1928A1 Thompson as originally manufactured has a Cutts compensator mounted to the muzzle and a leaf-type rear sight adjustable for windage and a barrel with radial cooling fins. Before the end of production of the Model 1928A1, specimens had been produced without compensators, with a simple non-movable "L" type rear sight, and a smooth barrel. All Model 1928A1 Thompsons, however, have a top-mounted actuator, the brass

Caliber .45 Model 1928A1 Thompson Submachine Gun.

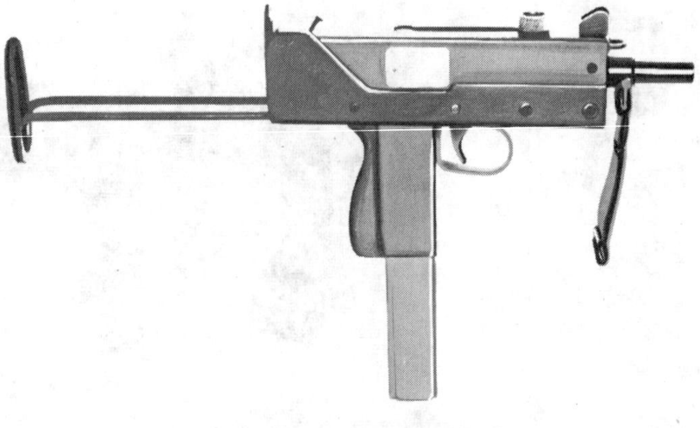

9mm Ingram Model 10 Submachine Gun.

"H" type lock, which operates on the theory of adhesion of different types of metal under pressure (the Blish principle) and breech oiler pads mounted on the receiver. The Model 1928A1 has a removable buttstock and, although usually found with a horizontal fore grip, may be found with a vertical fore grip.

M1 Thompson

The M1 Thompson designs were prepared for the simplification of the M1928A1 and in April 1942 the M1 Thompson was adopted. The Breech lock, actuator, breech oiler, compensator, radial cooling fins on the barrel, and buttstock catch of the M1928A1 were all dropped in the M1 design. The bolt was made a bit heavier, as a result of not being hollowed out for the actuator, and a bolt-retracting handle was fitted into the right side of the bolt where it rode in a track in the receiver. A simple fixed aperture rear sight with sight guards was initially issued; at a later date, the sight guards were dropped. The M1 will not accept the drum-type magazine which was used with the Model 1928A1. The buttstock of the M1 is permanently attached to the trigger housing (frame), and the design of receiver and frame are modified so that the frame slides over a protruding track located on both sides of the receiver and makes a noticeable bulge at the bottom of the receiver which is not present on the M1928A1.

The M1, with the exception of the locking and oiling processes, functions the same as does the M1928A1. It has a spring-loaded firing pin and a hammer the same as the M1928A1.

Caliber .45 Submachine Gun M1.

M1A1 Thompson

The M1A1 differs from the M1 only in having the firing pin machined in the face of the bolt, thereby doing away with the firing-pin assembly and hammers. The thirty-round box magazine was introduced at the same time as the M1-type guns.

To Load Box Magazine. Load as for automatic pistol magazine, but support base of magazine against body or a solid surface if heavy spring tension makes it difficult to force cartridges down.

To Insert Box Magazine. Cock the gun, set the fire control lever for the type of fire desired. Put the safety on "safe." Insert rib at back of magazine in its recess at the front of the trigger guard and push in until the magazine catch engages with a click.

Warning: Remember that when the bolt goes forward in this weapon a cartridge is fired. Hence, if the weapon is not to be fired and you wish to move the bolt forward to prevent straining the recoil spring, first press down the magazine catch and remove the magazine from the gun. **Note:** While it is possible to insert the box type magazine in this gun with the action forward (that is, uncocked) this procedure is not recommended. In so inserting the magazine, make sure that the magazine catch is fully engaged because the overhang of the magazine spring must be taken up before the engagement is securely locked.

Inserting a Drum Magazine. Cock the gun. Set fire control lever for "single" or "full auto" fire. Put safety on "Safe." Hold magazine so that key spring is facing forward. Now insert the two ribs on the magazine into their horizontal grooves in the receiver and slide the magazine into the gun from the left side. Push in until the magazine catch clicks into place.

Warning: While this magazine may be inserted from the right side, it is unwise to do so as this may injure the magazine catch. Also, do not try to insert the magazine when the bolt is in forward position. The bottom of the bolt will strike against the mouth of the magazine and may injure it.

Loading and Firing the Thompson

To load the drum magazine raise the flat magazine key spring to disengage its stud and slide the key off via its slot.

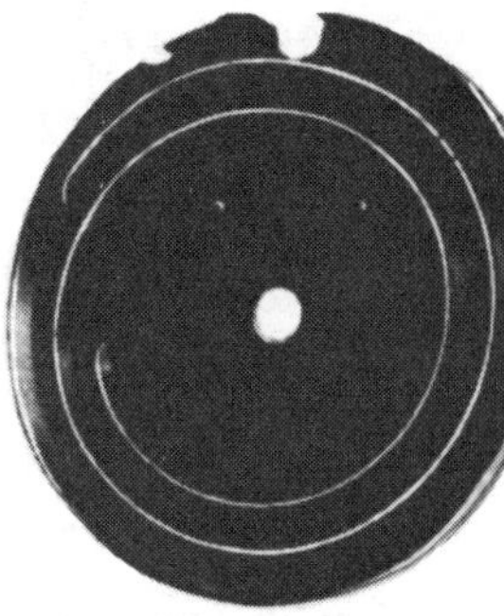

Lift off the magazine cover. Insert 5 cartridges base down in that section of the rotor in which the magazine feed opening is cut.

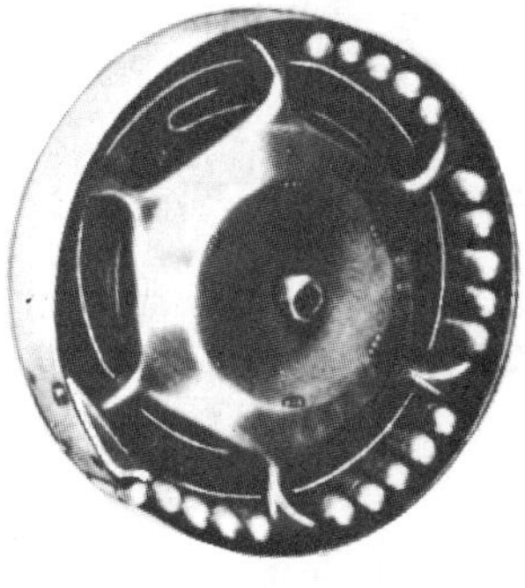

Loading from right to left, place 5 cartridges in each section of the spiral track, taking care to load all outer spirals first.

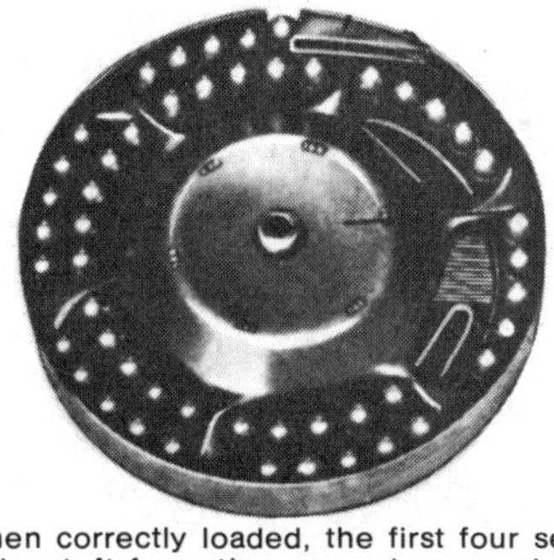

When correctly loaded, the first four sectors starting left from the magazine opening will contain 10 cartridges each, while the last two will have 5 each. Warning: Be careful not to insert any cartridges near the loops opposite the two sectors which hold five cartridges each; any cartridges so placed will jam the magazine when the rotor revolves. Now replace the magazine cover. Make sure that the large slot cut in it engages properly with the cover positioning stud. Slide the magazine key into place. Check to be sure that the stud on the spring correctly engages the center piece. Now wind the key from left to right. As it turns you will hear a distinct click. Count the number of clicks. Stamped on the magazine cover you will find the correct number of clicks necessary to indicate sufficient spring tension to work the magazine properly. (The normal number is 9 or 10 for a 50-shot magazine.) Note: If magazine is not to be used at once, wind up only two clicks to assure proper locking of the magazine and prevent straining

spring. Caution: Never rewind a partially empty magazine. This is unnecessary and may break the magazine mainspring.

Field Stripping the Model 1928A1 Thompson

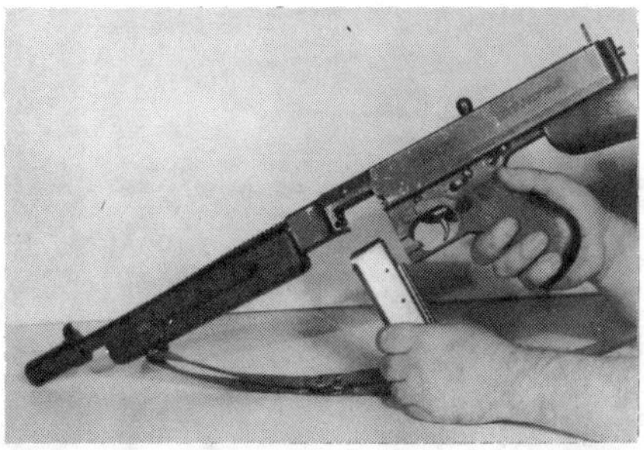

Remove magazine by pressing magazine catch up with the thumb and pulling 20-shot magazine straight down; or sliding 50-shot magazine out to the left.

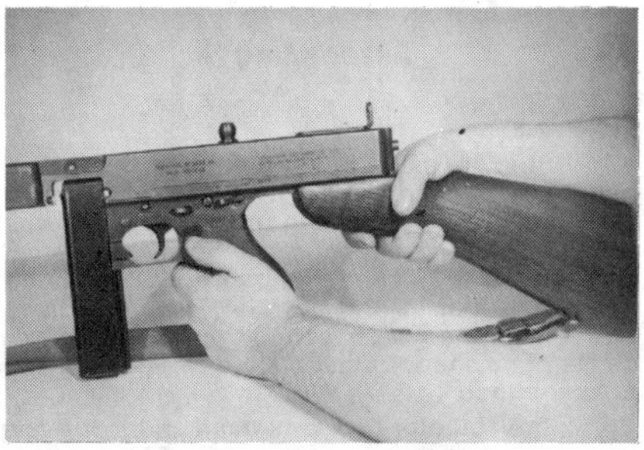

Set safety on "Fire" and set fire control lever on "Full Auto." Remember this can only be done when the weapon is cocked.

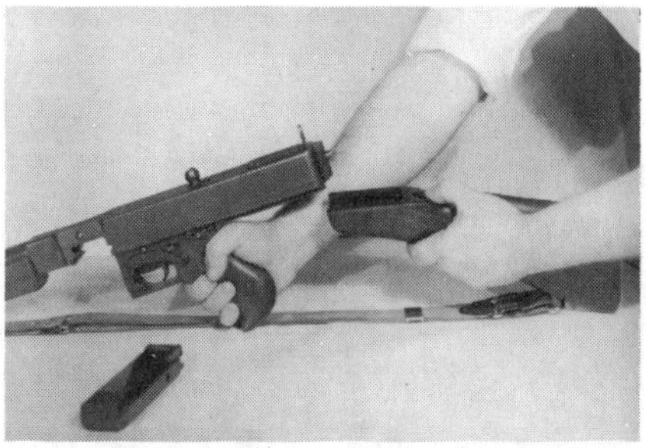

Remove buttstock by pressing its slide catch down and pulling stock straight to the rear out of its guide.

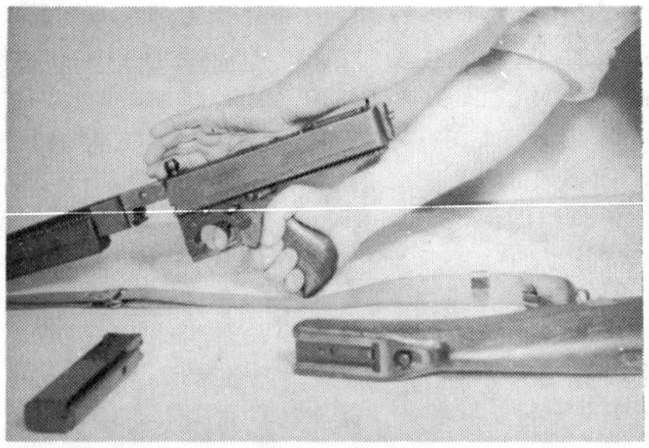

Hold firmly to actuator knob with left hand, pull trigger with right forefinger and ease bolt forward.

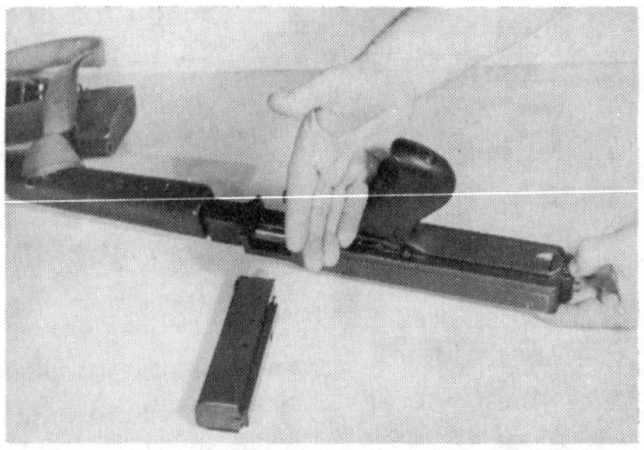

Turn gun upside down on table or knee. Push in the frame latch (the spring plunger on under side of frame behind pistol grip), and tap frame with right hand until it slides back a short distance.

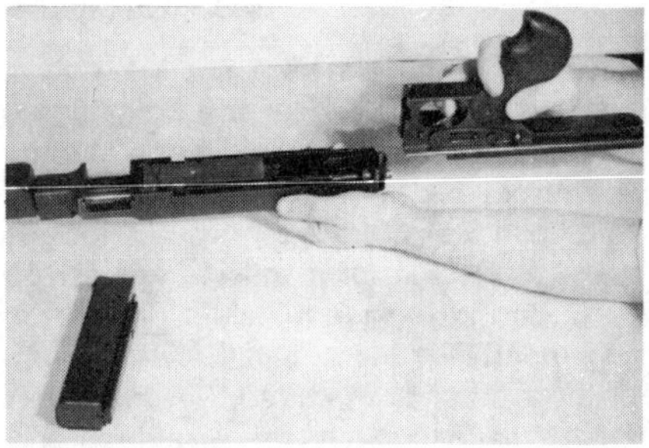

Grasp the rear grip with the right hand and pull the trigger holding the receiver firmly in the left hand, and slide the pistol grip group off out of its grooves.

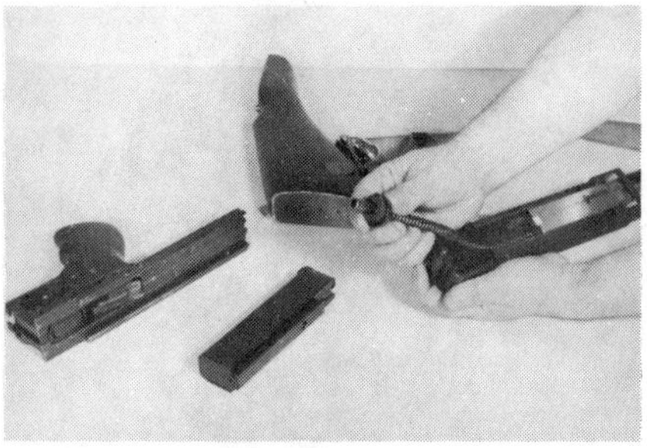

Remove recoil spring, as follows: With gun held firmly, turned upside down, grasp buffer flange with first and second fingers of right hand and pull out with upward and forward motion.

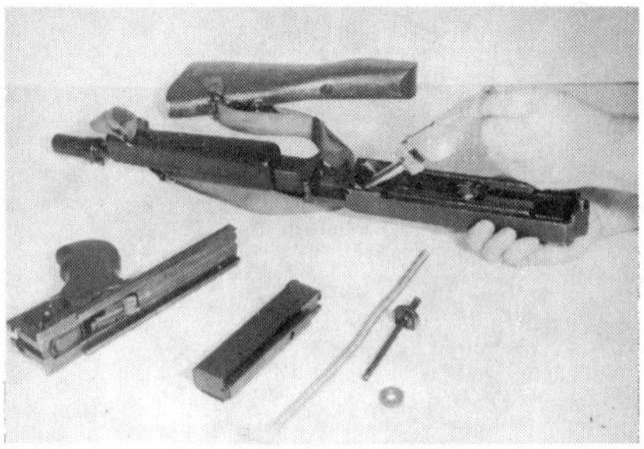

By pulling back on actuator knob, bolt is drawn back and can be removed to the rear.

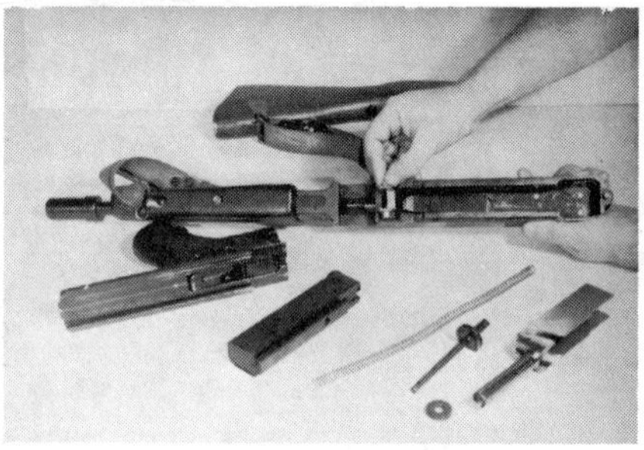

Actuator is then slipped forward with lock, and lock removed through its grooves in the receiver.

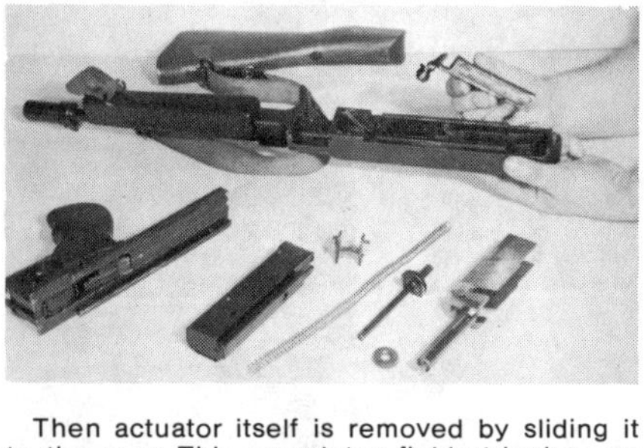

Then actuator itself is removed by sliding it to the rear. This completes field stripping, no further dismounting is necessary.

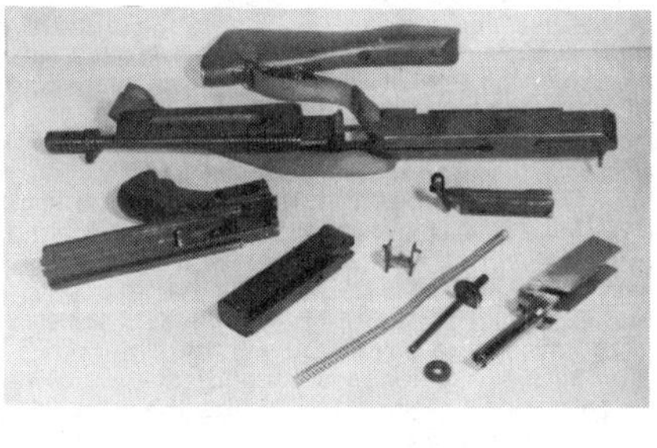

Note: On earlier models of this gun a special tool is required to remove the recoil spring. In this type, the stripping tool is inserted into its hole in the front end of the buffer rod. Then it is pushed in as far as it will go in the direction of the bolt. The rear end of the buffer rod is thus withdrawn from its hole at the rear end of the receiver. By tilting this stripping tool, the buffer may be grasped by the hand, and the recoil spring, fiber buffer disc, and rod will come out with the stripping tool. Buffer rod and spring are to be securely held so they do not fly apart.

Assembling

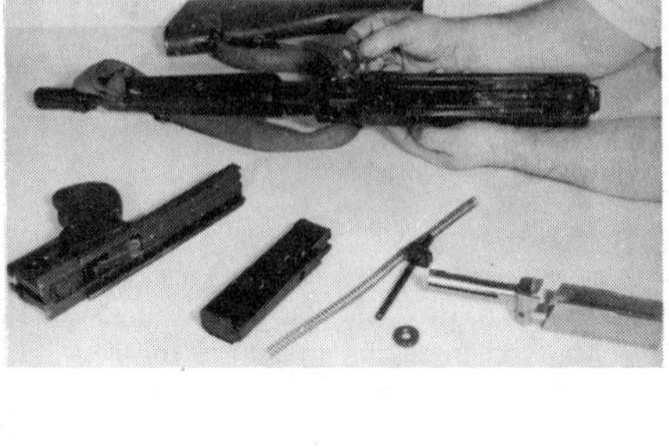

Inserting the actuator knob at its rear position, pull it forward and replace the lock which must be placed in its recesses in the receiver, so that the word "up" stamped on it is in uppermost position and the arrow stamped on it is pointing in the direction of the muzzle. The crosspiece of the lock (the lock is called the "H-Piece" because of its shape) must fit into the jaws of the actuator knob.

Pull the actuator and lock back and insert the bolt. Be sure and insert its bolt-end first so that the inclined cuts line up with the side members of the lock. (Now push the assembly forward as far it will go.)

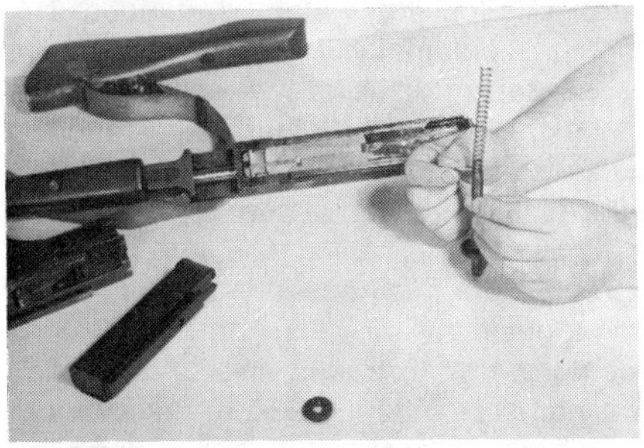

Compress the recoil spring over its rod, push it forward and down and push a nail or clip between the coils and through the hole in the rod.

Insert end of spring in hole in bolt; press rod forward until head of rod will slip into receiver and protrude through its hole to the rear; then withdraw nail.

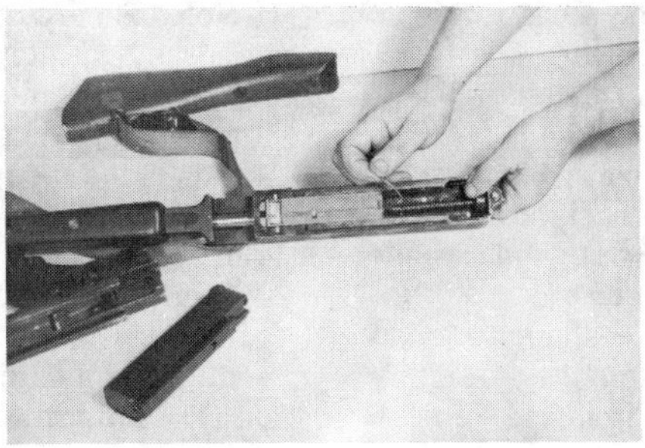

Note: If the gun is the earlier model, put the recoil spring over its rod and push front end of spring into housing and rear of breech block. Compress recoil spring on buffer rod a little at a time. While partly compressed, hold spring on rod with left hand and insert stripping tool into hole and buffer rod to retain spring in position. Replace fiber buffer disc and insert loose end of recoil spring in its hole in the actuator knob. Now place rear end of the buffer rod in its hole at the rear end under receiver. Draw back actuator knob until rear of bolt touches stripping tool. Recoil spring will now enter proper holes. Withdraw stripping tool.

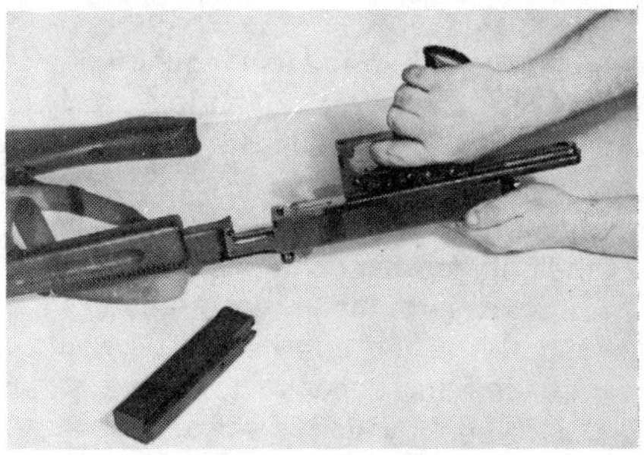

Holding the frame by rear grip pull the trigger and slide the frame forward in its guide in the receiver. Remember that safety must be at "Fire" position and fire control lever at "Full Auto."

Insert undercut of the frame in the buttstock and slide the butt forward until it locks in place.

How the Thompson Gun Works

Starting with the gun loaded and cocked the action is as follows: When the trigger is pressed, it moves the disconnector up to lift the sear lever. The sear lever raises the forward portion of sear, thus depressing the rear section and disengaging it from the notch on the bolt. The bolt is now free to be driven forward by the coiled recoil spring. If the fire control lever is set for semiautomatic fire, the rocker will act on the disconnector and sear lever to leave the sear free to lodge in the bolt on backward motion.

In its forward motion the bolt strips the top cartridge from the magazine, forces it into the chamber and drives the lock downward into locked position. The forward end of the bolt is round to fit in the bolt wall of the receiver, and the rear portion is rectangular to fit into the receiver cavity.

The forward motion of the bolt is halted by the rectangular end abutting against the receiver. The lock is an H-shaped piece of steel with lugs on each side whose center is engaged by the actuator.

The hammer is pivoted in the bolt between the H-piece and the receiver bottom, and as the action closes the lower end of this hammer strikes the abutment somewhat in advance of the bolt so that as the cartridge is seated, the upper end of the hammer strikes the firing pin. (The hammer is made so it can strike the firing pin only when the bolt is completely closed.) The extractor snaps over the cannelure of the cartridge case, and the firing pin strikes the primer.

Return Movement of the Action. The residual breech pressure forces the empty cartridge case back against the bolt, which in

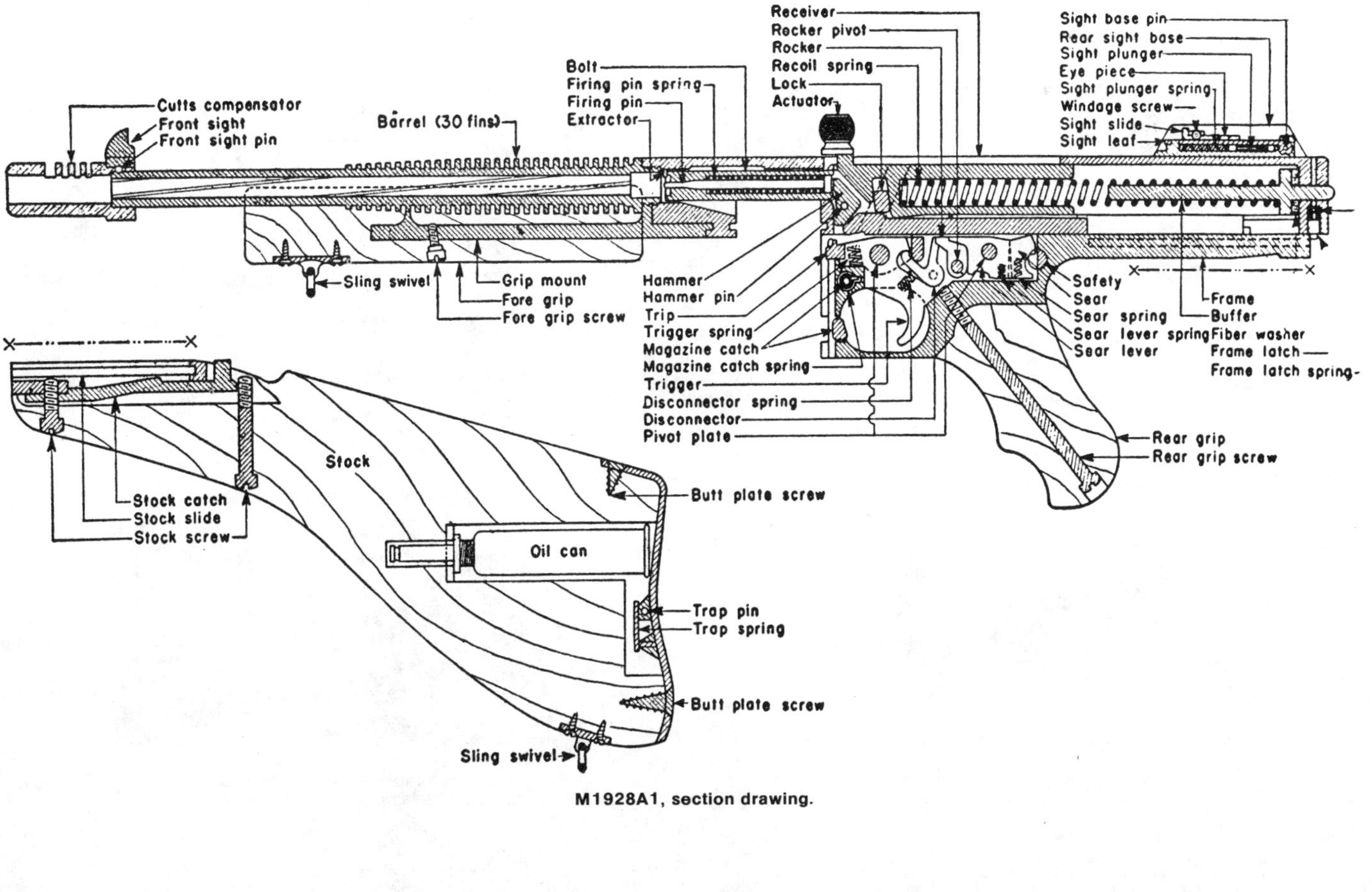

M1928A1, section drawing.

turn transmits the pressure to the H-piece locking device. This in turn transmits it to the locking surface of the receiver.

How the Lock Works. The lock, or H-piece, is situated in a 70° inclined slot in the bolt, with its lugs engaged in short 45° grooves recessed in the receiver. When engaged in the short 45° inclined slot with the H-piece offering resistance to the backward motion of the bolt, resistance to this lifting action is offered by the forward inclined base of the H-piece meeting the rear face of the 70° inclined slot in the bolt itself. The rising of this H-shaped lock is further resisted because its bridge is thrust up into the slot in the actuator knob, which is set at an angle of 10° from the vertical, pointed to the rear.

The general direction of movement of the locking piece, as a result of the movement of these several components, is upward and backward. Thus the bolt is prevented from moving to the rear while the chamber pressure is dangerous.

Because of the rapidity with which the pressure in the bore rises to its maximum on firing, the bolt is said to be supported by adhesion of the inclined surfaces until the pressure has again dropped materially, which acts as a breech locking factor.

For this adhesion lock to work, it is essential that the engaging surfaces remain constantly lubricated by the oil pads in the receiver.

Note: The actual value of this locking system is dubious. The necessity for constant oiling, incidentally, is a source of jams. Regardless of the real or theoretical value of the locking device, the fact remains that the weight of the parts themselves and the inertia of the recoil spring are sufficient to work the weapon safely and satisfactorily when the locking device is removed from the gun.

The forward end of the recoil spring is housed in a cavity in the actuator. The buffer forms a guide for the rear end of the recoil spring, permitting it to compress in a straight line as the action goes backward. A fiber washer is provided to absorb the shock of recoil, and oil pads in the receiver lubricate the locking lugs and bolt sides during the passage of those pieces.

The extractor, which is positioned on the right forward end of the bolt, draws the empty case out of the firing chamber until it strikes the ejector which moves into a clearance cut in the bolt path and hurls it out of the ejection port.

If a box type magazine is in the weapon, the magazine spring forces cartridges up in line bringing the next cartridge into position for the forward movement of the bolt. If the drum magazine is being used, springs inside the drum twist the spiral and feed a cartridge into line.

Field Stripping the M1 and M1A1 Thompson Submachine Guns

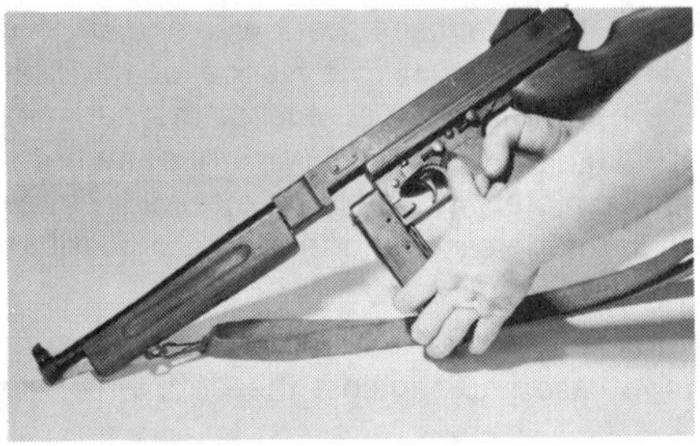

(1) Remove magazine by pressing magazine catch up with the thumb and pulling the box magazine straight down and out of its guide.

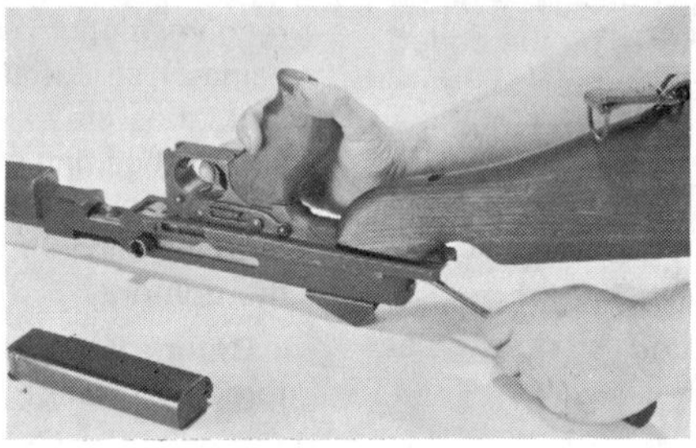

(2) With bolt forward, push in on receiver locking catch.

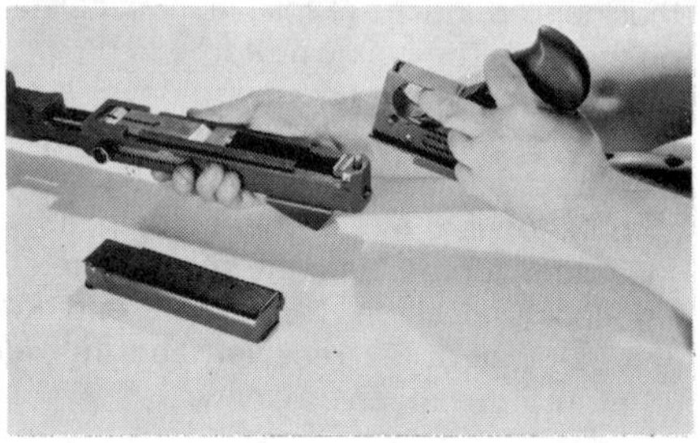

(3) Pressing trigger, draw frame straight back out of its guides in the receiver.

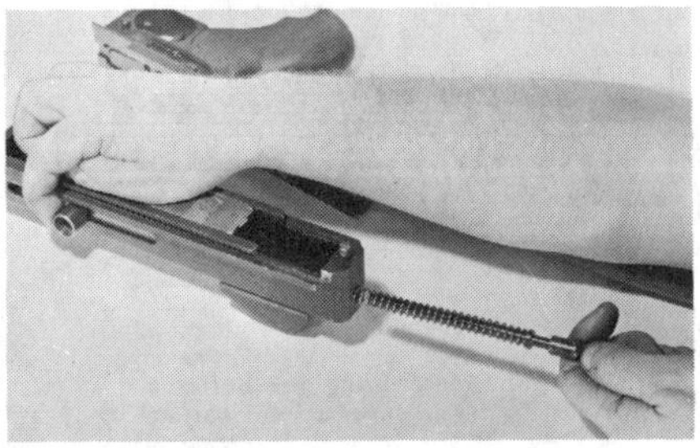

(4) Push the plug protruding from the rear of the receiver (the buffer pilot) in and draw the buffer pad up out of its seat.

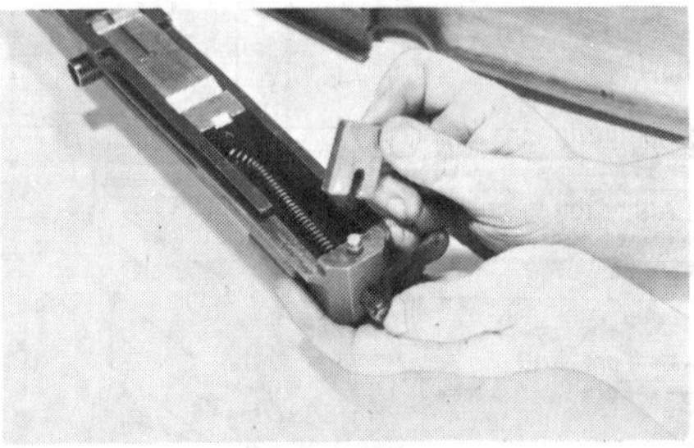

(5) Holding the buffer pilot, pull the bolt back about half way. This will permit removing the recoil spring and the pilot from the rear of the receiver.

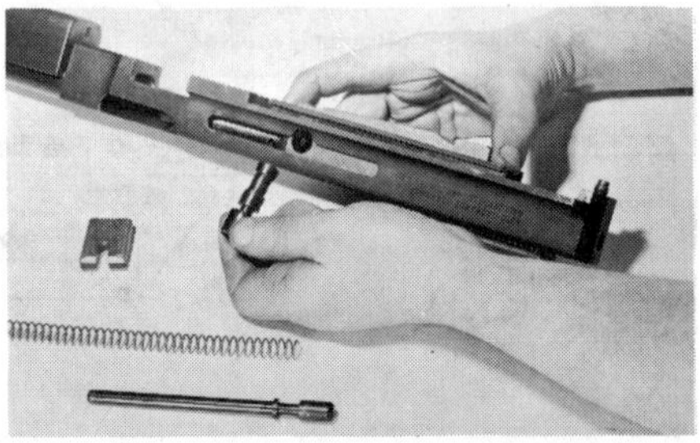

(6) Still holding the weapon upside down, lift the rear end of the bolt when the bolt handle is opposite the low cut in the center of its slot in the receiver.

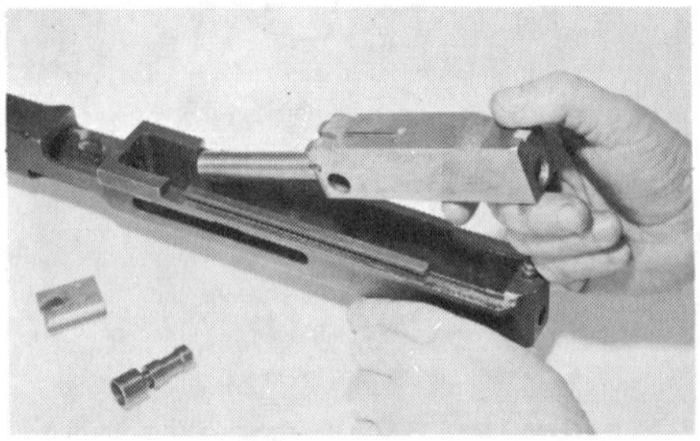

(7) Lift the bolt back and up out of the receiver.

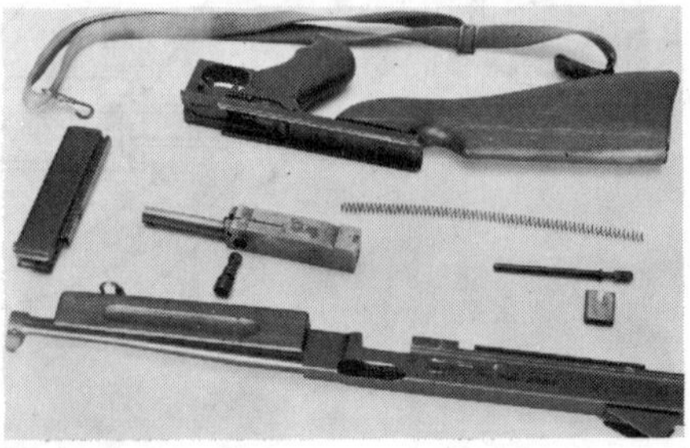

(8) This completes field stripping. No further dismounting is normally necessary.

Field Assembly. Assemble the buffer pilot to the recoil spring and holding the gun on the knee insert the recoil spring through the buffer pilot hole in the receiver from the outside, permitting the recoil spring to slide into the hole in the bolt.

Replace the bolt in the receiver so the bolt handle hole is in the center of the half circle in the receiver slot. Push the back end of the bolt up slightly out of the receiver. Push the hammer back against the shoulder of the bolt and the bolt handle can then be inserted. Then slide the bolt forward.

Now move the bolt halfway in the middle of the receiver, and assemble the buffer pilot with the recoil spring into the receiver and bolt.

When the end of the buffer pilot is flush with the outside of the receiver, the buffer pad can be placed over the buffer.

THE CALIBER .45 M3 and M3A1 SUBMACHINE GUNS

The M3 submachine gun was adopted in December 1942. It had a number of deficiencies which showed up in field service and these were corrected in the M3A1 submachine gun, standardized in December 1944.

The M3 submachine gun was designed so that by changing the barrel and bolt and adding an adaptor to the magazine, it could be used with 9mm Parabellum cartridges. There is a version of the M3 with silencer built into the barrel. Approximately 1,000 of these were made for the OSS during World War II. Guide Lamp Division of General Motors Corp. produced approximately 646,000 M3 and M3A1 submachine guns during World War II.

A curved barrel was made for use with the M3A1 submachine gun after World War II and a flash hider was developed for use with both the M3 and M3A1 submachine guns. Approximately 33,200 M3A1s were made by the Ithaca Gun Co. of Ithaca, New York during the Korean War.

Differences Between M3 and M3A1

The principal differences between these weapons are as follows.

M3. The bolt is pulled to the rear by means of a spring loaded retracting lever assembly.

M3A1. The bolt has a finger hole in its right front side for cocking the gun. The magazine catch has a guard to prevent it being accidentally depressed. The ejection port and its cover are longer and the safety lock on the ejection port cover is placed further to the rear. Disassembly grooves were added so that the bolt can be removed without removing the housing assembly. A stock plate and magazine filler were added to the stock and a larger oil can is fitted inside the grip. The retracting pawl notch is eliminated and a clearance slot for the cover hinge rivets was added. The barrel ratchet was redesigned to provide a longer contact surface for easier disengagement from the barrel collar.

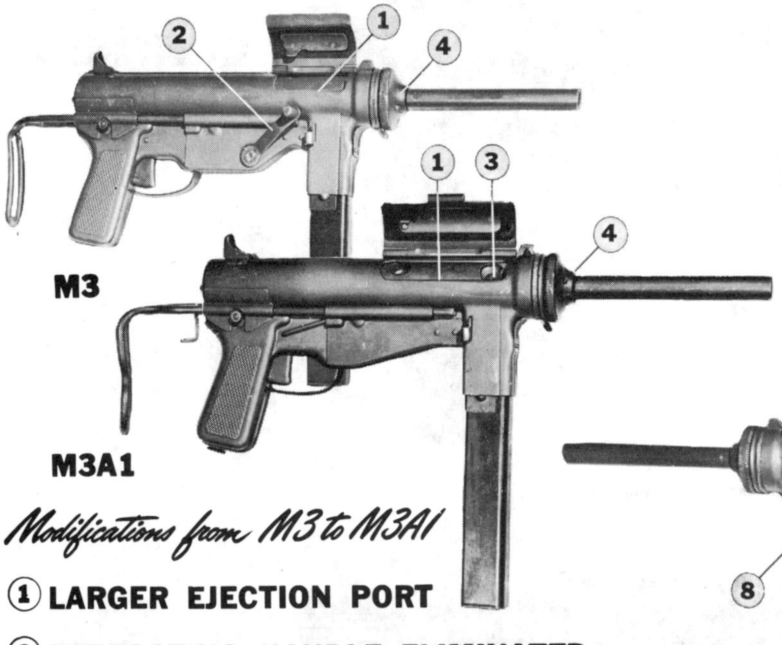

COMPARISON

GUN, SUBMACHINE, CAL. .45

M3 AND M3A1

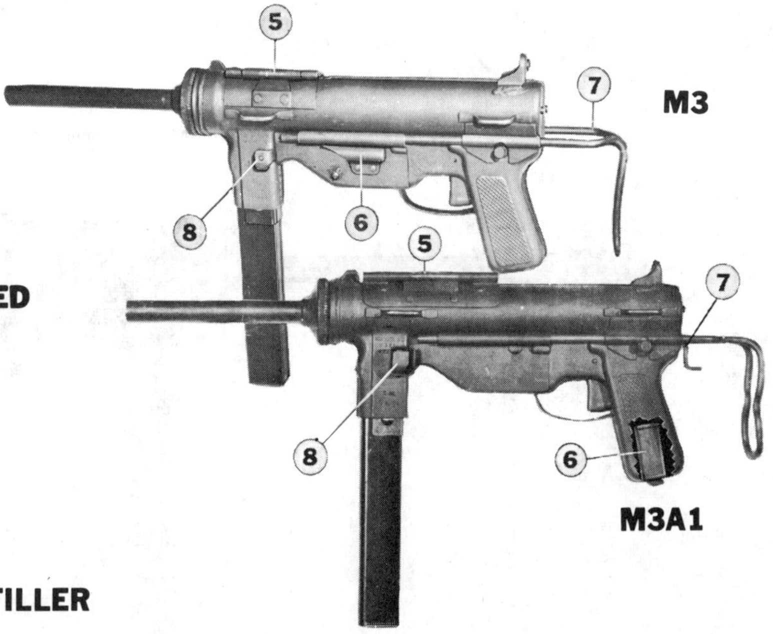

Modifications from M3 to M3A1

1. **LARGER EJECTION PORT**
2. **RETRACTING HANDLE ELIMINATED**
3. **FINGER HOLE FOR COCKING**
4. **DISASSEMBLY GROOVES ADDED**
5. **STRONGER COVER SPRING**
6. **LARGER OIL CAN INSIDE GRIP**
7. **STOCK PLATE AND MAGAZINE FILLER ADDED TO STOCK**
8. **GUARD ADDED FOR MAGAZINE CATCH**

Sectioned caliber .45 M3A1 Submachine Gun with flash hider.

Caliber .45 M3 Submachine Gun.

Caliber .45 M3A1 Submachine Gun with flash-hider.

Loading, Firing, and Field Stripping the M3 and M3A1

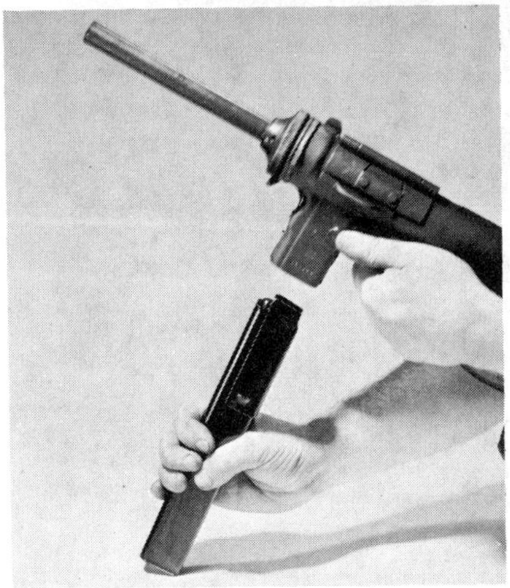

(1) A loaded magazine is inserted in the magazine housing from below until it locks. Pull the cocking handle back as far as it will go to open the ejection port cover to its full extent and draw the bolt back until it is caught and held by the sear.

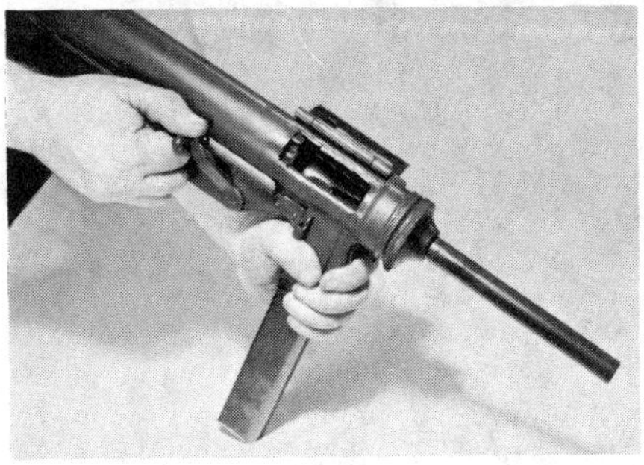

(2) Release the handle and let it fly forward. If it is desired to carry the gun ready for use but on safety, push the cover down into place. Otherwise the gun is ready to fire on a pull of the trigger.

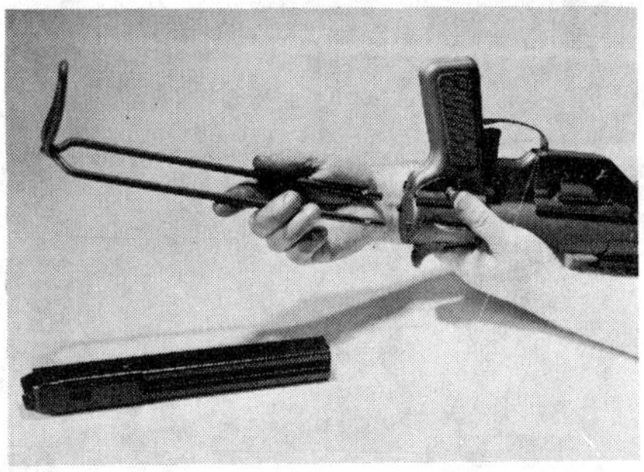

(3) With the magazine out of the weapon, push in the stock catch on the left side of the receiver above the pistol grip with the left thumb and pull the wire stock out of its grooves.

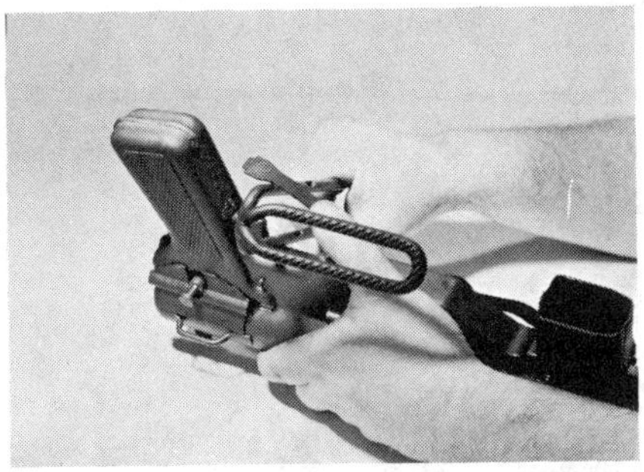

(4) Insert the shoulder end of the wire stock inside the trigger guard to provide a pressure point. Then press down on the lower end of the trigger guard until it springs out of its slot in the pistol grip. Handle this trigger guard carefully as it is of light gauge spring steel. Rotate it toward.the muzzle of the gun. This will unhook it so it can be lifted out.

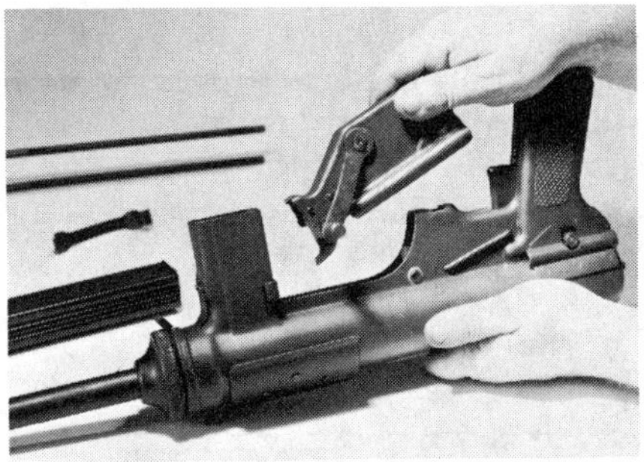

(5) Push down on the housing assembly unit a short distance, and then lift it to the rear until it can be lifted off. This should be done with care to prevent injury to the metal.

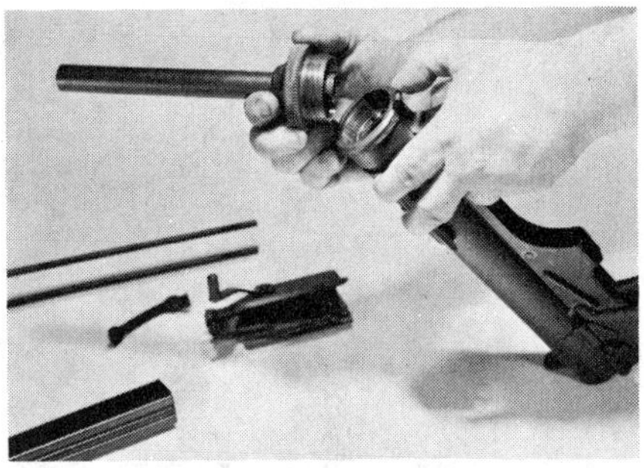

(6) Pull the ratchet catch back with the left thumb and unscrew the barrel assembly from left to right. The barrel and its collar will come out.

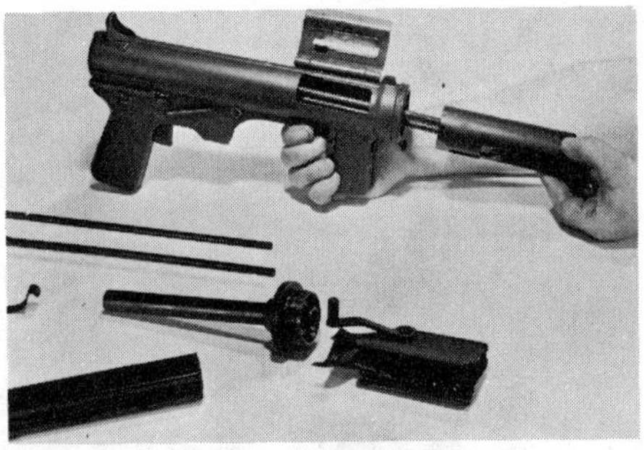

(7) Open the bolt cover and tilt the gun forward. The bolt and the two guide rods and their springs will come forward out the front of the receiver.

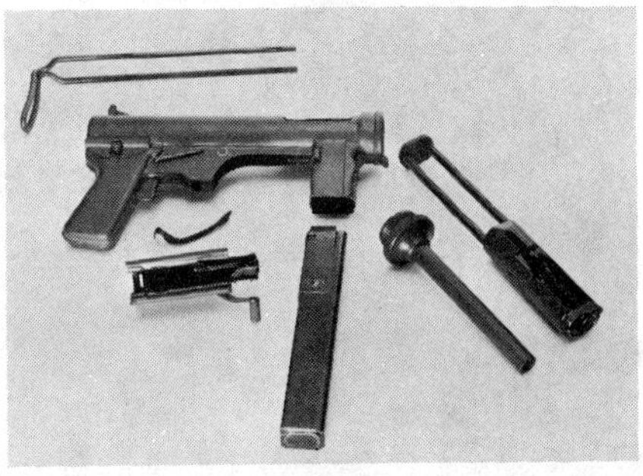

(8) No further stripping is normally necessary.

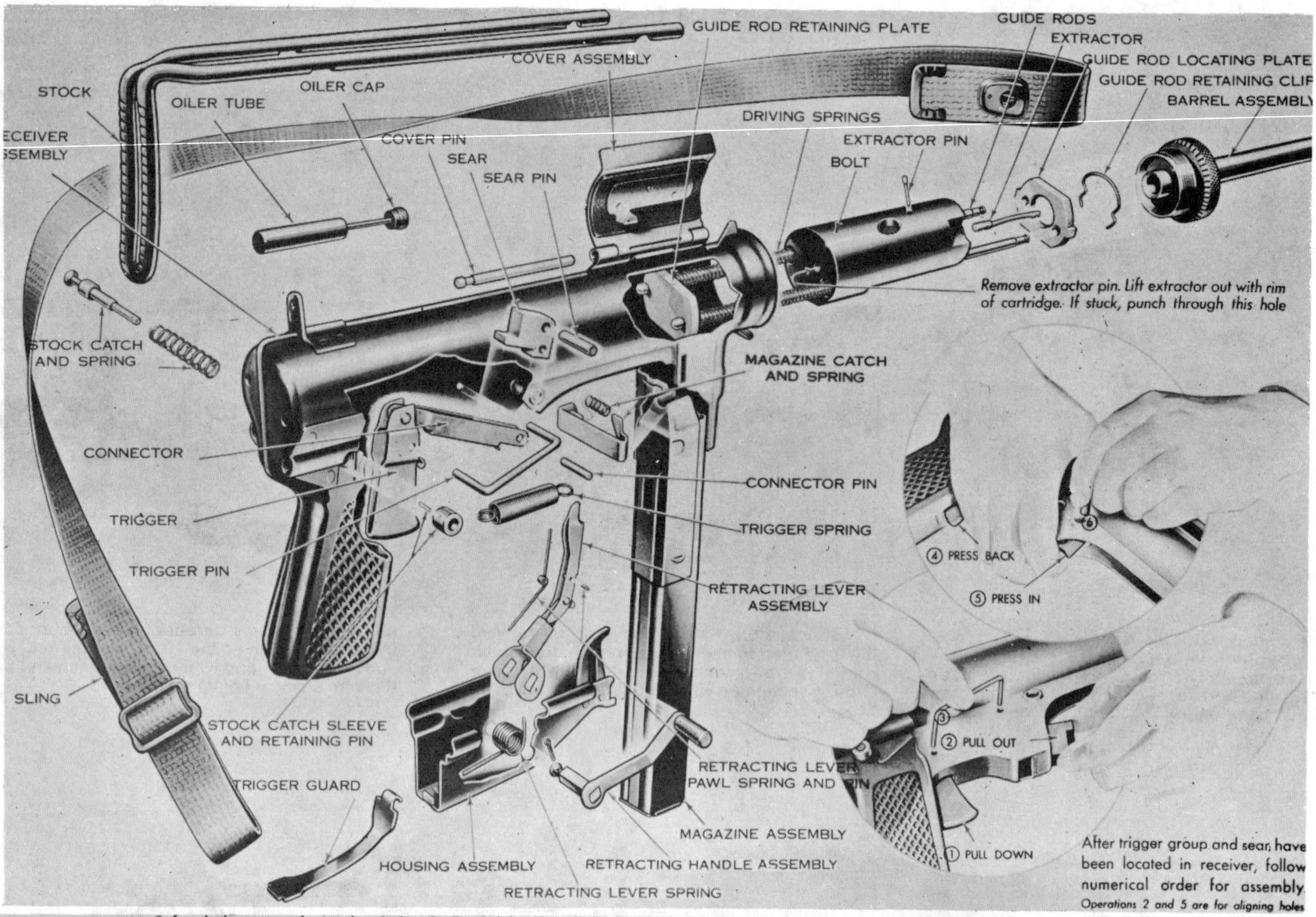
GUIDE ROD RETAINING PLATE
GUIDE RODS
EXTRACTOR
GUIDE ROD LOCATING PLATE
GUIDE ROD RETAINING CLIP
BARREL ASSEMBLY
COVER ASSEMBLY
STOCK
OILER CAP
OILER TUBE
RECEIVER ASSEMBLY
COVER PIN
SEAR
SEAR PIN
DRIVING SPRINGS
EXTRACTOR PIN
BOLT
Remove extractor pin. Lift extractor out with rim of cartridge. If stuck, punch through this hole
STOCK CATCH AND SPRING
MAGAZINE CATCH AND SPRING
CONNECTOR
CONNECTOR PIN
TRIGGER
TRIGGER SPRING
TRIGGER PIN
RETRACTING LEVER ASSEMBLY
④ PRESS BACK
⑤ PRESS IN
SLING
STOCK CATCH SLEEVE AND RETAINING PIN
② PULL OUT
RETRACTING LEVER PAWL SPRING AND PIN
TRIGGER GUARD
MAGAZINE ASSEMBLY
① PULL DOWN
HOUSING ASSEMBLY
RETRACTING HANDLE ASSEMBLY
RETRACTING LEVER SPRING
After trigger group and sear have been located in receiver, follow numerical order for assembly.
Operations 2 and 5 are for aligning holes

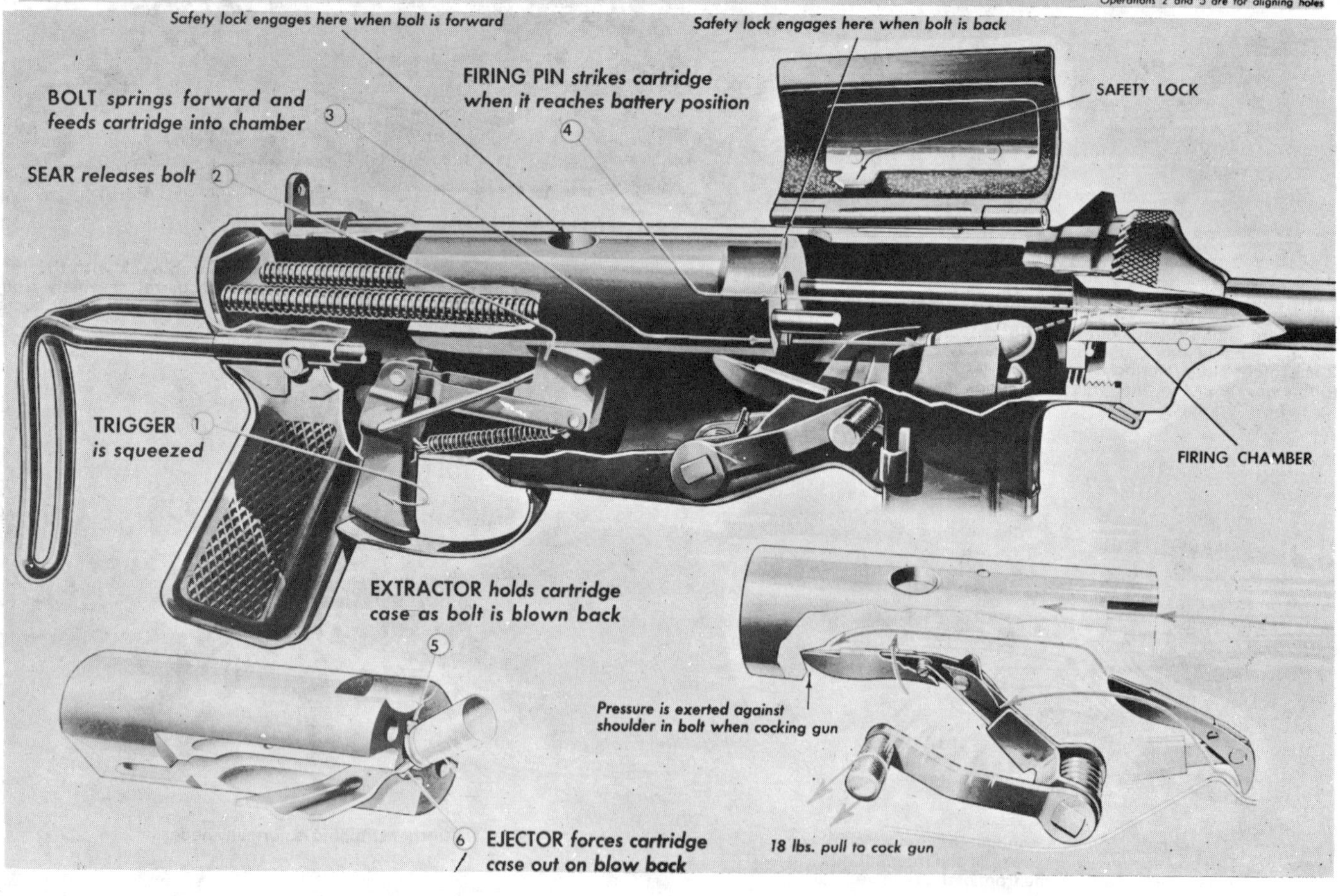
Safety lock engages here when bolt is forward
Safety lock engages here when bolt is back
BOLT springs forward and feeds cartridge into chamber 3
FIRING PIN strikes cartridge when it reaches battery position 4
SAFETY LOCK
SEAR releases bolt 2
TRIGGER 1 is squeezed
FIRING CHAMBER
EXTRACTOR holds cartridge case as bolt is blown back 5
Pressure is exerted against shoulder in bolt when cocking gun
6 EJECTOR forces cartridge case out on blow back
18 lbs. pull to cock gun

How the M3 Submachine Gun Works

When a loaded magazine is inserted into the magazine housing and pushed upward until it locks, the cocking handle on the right side of the gun is drawn back to its full extent. This movement raises the bolt cover (which must be opened full to permit proper ejection) and also cocks the bolt against the tension of two recoil springs mounted in the receiver and extending one on each side into the bolt itself.

When the trigger is pressed the sear releases the bolt. The driving springs force the bolt forward and the feed guides machined into the bolt strike the rear of the topmost cartridge in the magazine and drive it ahead into the firing chamber.

As the cartridge enters the chamber the bolt continues forward and its extractor snaps over the head of the cartridge to fasten in the extracting groove.

The firing pin can now strike the primer and discharge the cartridge.

As the bullet moves out the barrel, the back pressure pushes the empty cartridge case out of the chamber and transmits its force to the bolt face pushing it to the rear.

The bolt assembly as it goes back takes the empty cartridge case with it. The case strikes against the ejector and is thrown out the ejection port.

The driving springs are compressed around their guide rods and the bolt travels back on tracks in the receiver until the energy is absorbed and the bolt is caught by the sear.

The gun will fire as long as the trigger is held back and there are any cartridges in the magazine. If the bolt cover is pushed down in place when the bolt is back, it acts as a safety by holding the bolt off the sear and also interfering with forward travel. It is also a dust cover.

Special Note on the M3A1 Submachine Gun

The M3A1 is loaded and fired in the same manner as the M3 with the exception that the bolt is retracted by placing the right forefinger in the bolt finger hole and drawing the bolt back until it is engaged by the sear. Disassembly differs from that of the M3 in that the bolt and operating springs can be removed after the barrel is removed without removing the housing.

5.56mm SUBMACHINE GUN XM 177E2

This weapon is a submachine gun version of the M16A1 rifle. It is loaded, fired, and functions in the same manner as that weapon. It has a long muzzle attachment which includes a flash suppressor. The stock is of the retractable type. This weapon was developed from the Colt CAR-15 submachine gun.

The barrel of the XM 177E2 is 11.5 inches long; the muzzle velocity is approximately 2750 feet per second. This weapon normally uses a thirty-round magazine.

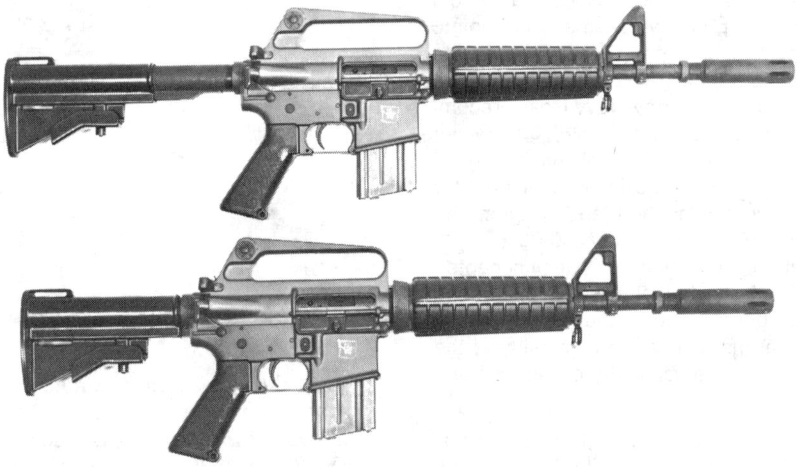

5.56mm XM 177E2 Submachine Gun. Top: stock extended; bottom: stock retracted.

CHARACTERISTICS OF UNITED STATES SUBMACHINE GUNS

	Thompson M1928A1	Thompson M1 and M1A1	M3
Caliber:	.45.	.45.	.45.
System of operation:	Delayed blowback, selective fire.	Blowback, selective fire.	Blowback, automatic.
Length overall:	w/buttstock: 33.75 in. w/o buttstock: 25 in.	32 in.	Stock extended: 29.8 in. Stock retracted: 22.8 in.
Barrel length:	10.5 in.	10.5 in.	8 in.
Feed device:	20- or 30-round, staggered row, detachable, box magazine; 50-round drum.	20- or 30-round staggered row, detachable, box magazine.	30-round, in-line detachable, box magazine.
Sights: Front:	Blade.	Blade.	Blade.
Rear:	Leaf w/ aperture notched battle sight.	Fixed aperture.	Fixed aperture.
Weight:	10.75 lb.	10.45 lb.	8.15 lb.
Muzzle velocity:	920 f.p.s.	920 f.p.s.	Approx. 920 f.p.s.
Cyclic rate:	600-725 r.p.m.	700 r.p.m.	350-450 r.p.m.

CHARACTERISTICS OF UNITED STATES SUBMACHINE GUNS (Cont'd)

	M3A1	Reising M50	Reising M55	U.D. M42
Caliber:	.45.	.45.	.45.	9mm Parabellum.
System of operation:	Blowback, automatic.	Delayed blowback, selective fire.	Delayed blowback, selective fire.	Blowback, selective fire.
Length overall:	Stock extended: 29.8 in. Stock retracted: 22.8 in.	35.75 in.	Stock extended: 31.25 in. Stock retracted: 22.5 in.	32.2 in.
Barrel length:	8 in.	11 in.	10.5 in.	11 in.
Feed device:	30-round, in-line detachable, box magazine.	12- or 20-round, in-line detachable, box magazine.	12- or 20-round, in-line detachable, box magazine.	20-round, staggered row, detachable, box magazine.
Sights: Front:	Blade.	Blade.	Blade.	Blade.
Rear:	Fixed aperture.	Adjustable aperture.	adjustable aperture.	Adjustable aperture.
Weight:	8 lb.	6.75 lb.	6.25 lb.	9.12 lb.
Muzzle velocity:	Approx. 920 f.p.s.	Approx. 920 f.p.s.	Approx. 920 f.p.s.	1312 f.p.s.
Cyclic rate:	350-450 r.p.m.	550 r.p.m.	450-550 r.p.m.	700 r.p.m.

UNITED STATES MACHINE GUNS

HISTORICAL SUMMARY

The United States adopted its first true machine gun, as opposed to the hand-operated Gatling gun, in 1896. The Navy adopted the 6mm, chambered for the 6mm M95 Lee cartridge, Browning-designed, gas-operated machine gun manufactured by Colt. This weapon, because of its jointed gas lever which swings in a vertical plane below the gun, was known popularly as the Colt "potato digger." The Army used the manually-operated Gatling in the Spanish American War—1898—but saw the advantages of a true machine gun. In December 1898 a joint Army-Navy Board recommended adoption of a common caliber—the caliber .30 Krag or .30-40 as it is more popularly known. The Navy Colt machine guns were rebarreled for the .30-40 cartridge and later rebarreled for the .30-06 cartridge. They were known as the Mark I in 6mm, the Mark I Modification I in caliber .30-40 and caliber .30-06.

The Army did not officially adopt a machine gun until it accepted the caliber .30 Maxim Model 1904 manufactured by Vickers and Colt. Few of these guns were bought due to the paucity of the United States defense budget at that time and the inability of the United States military to recognize the capability of the machine gun.

In 1909 the United States Army adopted the light Hotchkiss as the caliber .30 M1909 Benét-Mercié Machine Rifle. This strip-fed weapon is essentially the same as the 8mm French M1908 light Hotchkiss and the caliber .303 British Mark I Hotchkiss. The weapon was manufactured by Colt, but only 670 Benét Mercié machine rifles were on hand at the beginning of World War I. There were also 282 Model 1904 Maxim guns, 353 Lewis guns, and 148 Colt "potato diggers" on hand.

Caliber .30 Model 1904 Maxim Machine Gun.

The Lewis guns were a "windfall" from a British contract with the Savage Arms Corporation for the manufacture of ground and aircraft type Lewis guns. The United States government bought 2,500 Lewis Caliber .30 ground guns and 39,200 caliber .30 flexible Lewis aircraft guns from Savage by December 1918. Only the aircraft guns were used in action; the ground guns were used for training and were later used by the United States Marine Corps and Navy.

There were also 1,050 caliber .303 Lewis ground guns procured for training purposes.

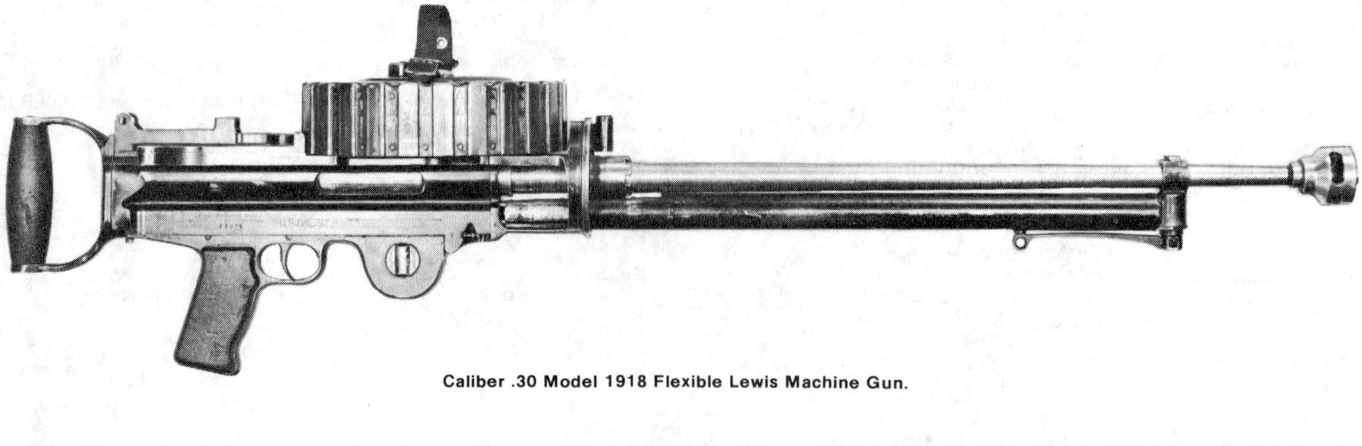

Caliber .30 Model 1918 Flexible Lewis Machine Gun.

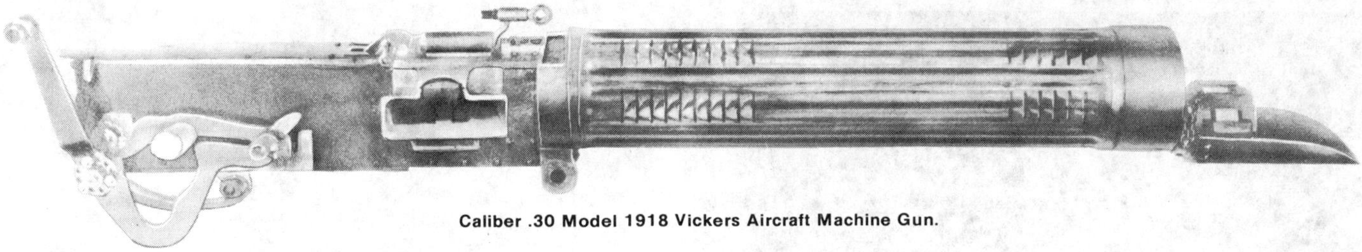

Caliber .30 Model 1918 Vickers Aircraft Machine Gun.

Caliber .30 Browning Model 1917 Machine Gun.

The Army also purchased the caliber .30 Colt "potato-digger," mainly because it was in production for the Italian and Russian governments, and 2,810 of these guns were procured. This venerable weapon, known as the caliber .30 Model 1917 machine gun in its World War I guise, was modified by Carl Swebelius to produce the Marlin caliber .30 aircraft and tank machine guns. The Marlin basically is a Colt Model 1917 (the Model 1895-6 with a lighter barrel) with a one-piece piston rather than the knuckle-jointed piston gas lever arrangement. Marlin-Rockwell Corporation produced 38,000 Marlin aircraft guns and 1,470 Marlin tank machine guns through December 1918. The Army had adopted the Vickers gun in 1915. The Vickers, as adopted by the United States, was the same as the British Mark I Vickers—chambered for the .30-06 cartridge—and Colt manufactured approximately 12,125 Vickers ground guns plus 2,476 Vickers caliber .30 aircraft guns. Vickers guns which were under construction for the Russian government, chambered for the 7.62mm cartridge, were converted to 11mm for use as aircraft anti-balloon guns when the United States ceased deliveries to Russia upon the occasion of the Bolshevik government signing a separate peace with the Germans.

With all these miscellaneous weapons available or on order, the United States had to purchase 5,255 8mm Hotchkiss Model 1914 heavy machine guns, 15,918 8mm Model 1915 (Chauchat) light machine guns, and 19,241 caliber .30 Model 1918 (Chauchat) light machine guns, from the French government. The first 12 divisions of the United States Army which went to France were equipped with French automatic weapons. The next 11 divisions to go overseas were equipped with the caliber .30 Model 1915 Vickers guns, but used the Chauchat as a light machine gun.

John Browning had come to the foreground to help his country in its time of need and produced the Browning Automatic rifle, covered in detail earlier in this chapter, and the Browning caliber .30 Model 1917 machine gun. The Model 1917 was an improvement on Browning's M1901 recoil-operated machine gun design and received acceptance soon after it was introduced. The Model 1917 was produced by Remington, Colt, and New England Westinghouse. A total of 56,608 caliber .30 Model 1917 Browning Machine guns were made by the end of 1918. The last twelve divisions that went to France used the Browning machine gun. In addition to the water-cooled Browning, 580 air-cooled Browning aircraft machine guns were made by Marlin-Rockwell.

At the conclusion of the war, the United States continued development work on the Browning guns. The aircraft Browning served as the basis for the Browning light machine gun which was originally developed for cavalry use. The caliber .30 Model 1919A2 and its various modifications was developed as a cavalry weapon; it was the predecessor of The Model 1919A4 and A6 which were the standard light machine guns of World War II and the Korean war.

A version of the Model 1919A2 was fitted with a shoulder stock but appeared only as a prototype.

At the same time as development of the light Browning was

Caliber .30 Browning Model 1919A2 Cavalry Machine Gun.

Caliber .30 Browning Model 1919A2 with butt stock. Note similarity to Model 1919A6.

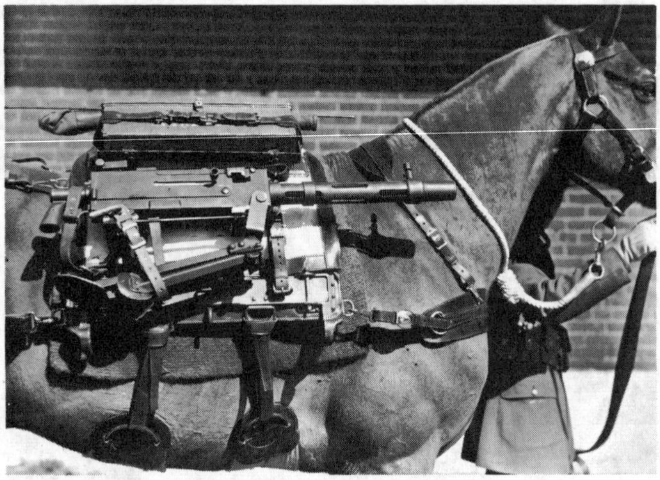

Caliber .30 Browning Model 1919A2, packed on horse.

Caliber .30 Browning Model 1917 on T4 mount in AA position.

going on, there was development work on the mounts for the water-cooled gun and some modification on the weapon itself, which resulted in the rebuilding of all water-cooled guns on hand into the Model 1917A1— the standard battalion-level rifle caliber machine gun of the United States in World War II.

During the early thirties a private contractor developed the caliber .30 T10 machine gun for the Ordnance Corps. The T10 and its successors—the T10E1 (tested in January 1942), the T10E2 (tested in March 1942), the T23, and T23E2 used an operating mechanism similar to the Browning automatic rifle but were belt fed. None of these weapons were found suitable for service use. In April 1940 the Army released to inventors and arms manufacturers of the United States, the characteristics for a new light machine gun to replace the Brownings and invited them to submit weapons for competitive trial.

Springfield Armory submitted a modification of the Browning caliber .30 aircraft gun M2 which had a controllable rate of fire. It weighed just a bit over 20 pounds. Rock Island Arsenal submitted the T13E1, a modification of the Browning Model 1919A4. Colt submitted a modified caliber .30 M2 aircraft machine gun with a rate of fire lowered to 300-350 rounds per minute; the Colt weapon weighed 21.75 pounds. Sedgley submitted a gas-operated selective-fire gun which weighed 23.5 pounds. This weapon, which replaced the empty cartridge case in the link after firing, did not perform satisfactorily under test conditions and was dropped from

Caliber .30 Browning Model 1917 Machine Gun on wheeled mount M1.

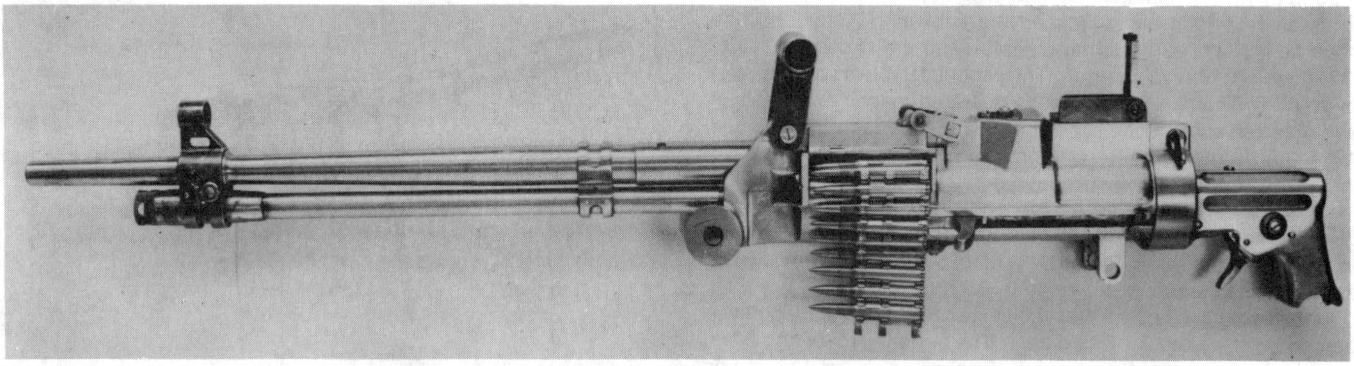

Caliber .30 Sedgley light machine gun.

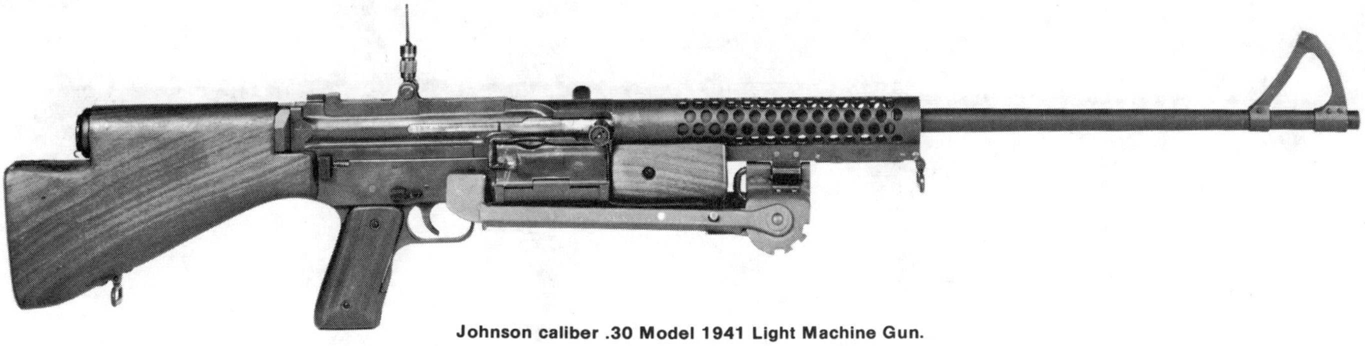

Johnson caliber .30 Model 1941 Light Machine Gun.

Johnson caliber .30 Model 1944 Light Machine Gun.

the tests. The Auto-Ordnance gun was a selective-fire gas-operated weapon designed by Bill Ruger—current manager of the Sturm Ruger Co. It weighed twenty pounds. The Schirgun, designed by Henry B. Schirokauer, was a gas-operated gun and had a cyclic rate of 200 to 600 rounds per minute, controlled by variation in trigger pressure. It weighed twenty pounds. None of the weapons submitted at this time were satisfactory and as too frequently occurred, events caught up with the United States, forcing mass production of existing weapons.

The development of weapons continued during the war and the caliber .30 Browning Model 1919A6 was adopted as a substitute standard weapon on 10 April 1943; this weapon is described in detail later in this chapter. In addition, the German 7.92mm MG34 and MG42 were thoroughly tested. The MG34 was not found to be suitable, but Saginaw Steering Gear made up caliber .30 specimens of the MG42 as the T24. As noted in the historical section of this book, this weapon failed, but it had quite a bit of influence on post-war developments.

Melvin Johnson, who developed the Johnson recoil-operated rifle, developed two light machine guns during the war. The Johnson Model 1941 was manufactured in limited quantity for the Marine Corps and the Army First Special Service Force—a Ranger type force. The Model 1941 is a recoil-operated gun which fires from a closed bolt in semiautomatic and an open bolt in automatic fire.

The Johnson Model 1944 was basically an improvement on the Model 1941. It weighs 14.7 pounds with monopod, is fed with a twenty-round magazine, and is 42.5 inches long with a 22-inch barrel. It has a cyclic rate of 450 to 750 rounds per minute. The Model 1944 was not adopted by the United States but was extensively tested in modified form after World War II as the T48 light machine gun. Russell J. Turner of Butler, Pa., submitted a light machine gun for test in September 1942. It weighed 18.83 pounds; problems developed early in the tests and the weapon was withdrawn from test by the inventor.

At this point, it is appropriate to say a few words about commercial machine-gun manufacture in the United States between the two world wars. Colt manufactured the Browning caliber .30 water-cooled gun and the caliber .30 aircraft guns as the MG38, MG38B, and MG40, respectively. They also manufactured a caliber .30 tank machine gun, the MG38BT, and a caliber .50 Browning machine gun, the MG52. Many of these weapons were sold to South American and Asiatic nations.

After World War II new requirements were prepared for machine guns. The Johnson Model 1944 had been converted to belt feed for test purposes prior to the conclusion of the war (the T30) and was further modified to be used with push-through type links and called the T40. Through 1947-48 a further modified Johnson, called the T48, was tested. Development was terminated in 1948.

The T37, which was a belt-fed Browning Automatic rifle, had begun to take shape in late 1944; work on this weapon continued for a time after World War II and was then terminated.

Work was begun on the caliber .30 T43 and T44 machine guns in February 1946. The T43 disappeared from the scene rapidly, but the T44 incorporated many of the features which would be continued in later guns including the current standard gun—the M60. The T44 incorporated the belt-feed mechanism of the German MG42 and the bolt and piston mechanism of the FG42 rifle in a rather unusual configuration. The feed cover was mounted on the left side of the gun and the link belt fed in from the side.

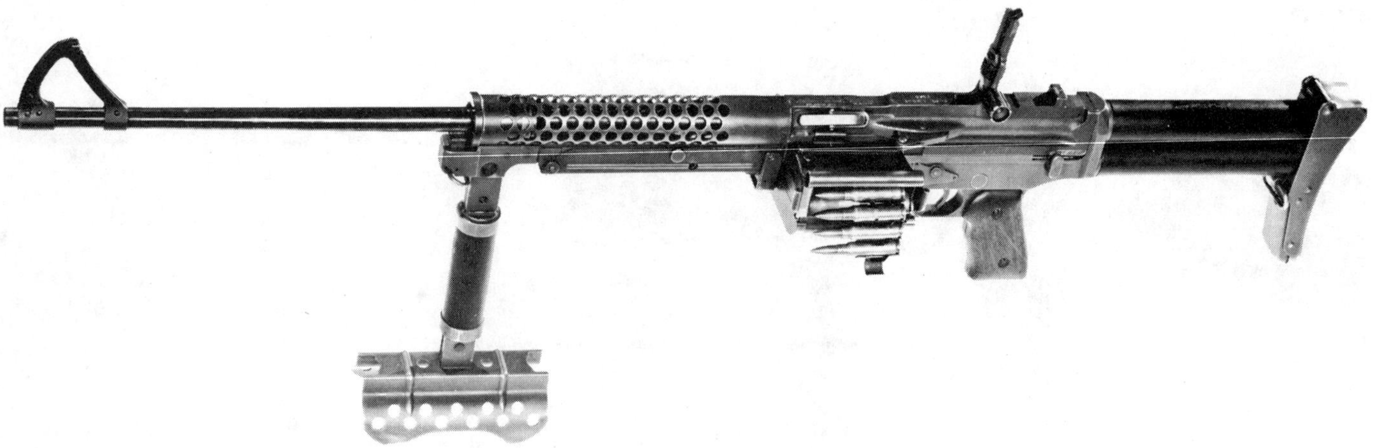

Caliber .30 T48 Light Machine Gun.

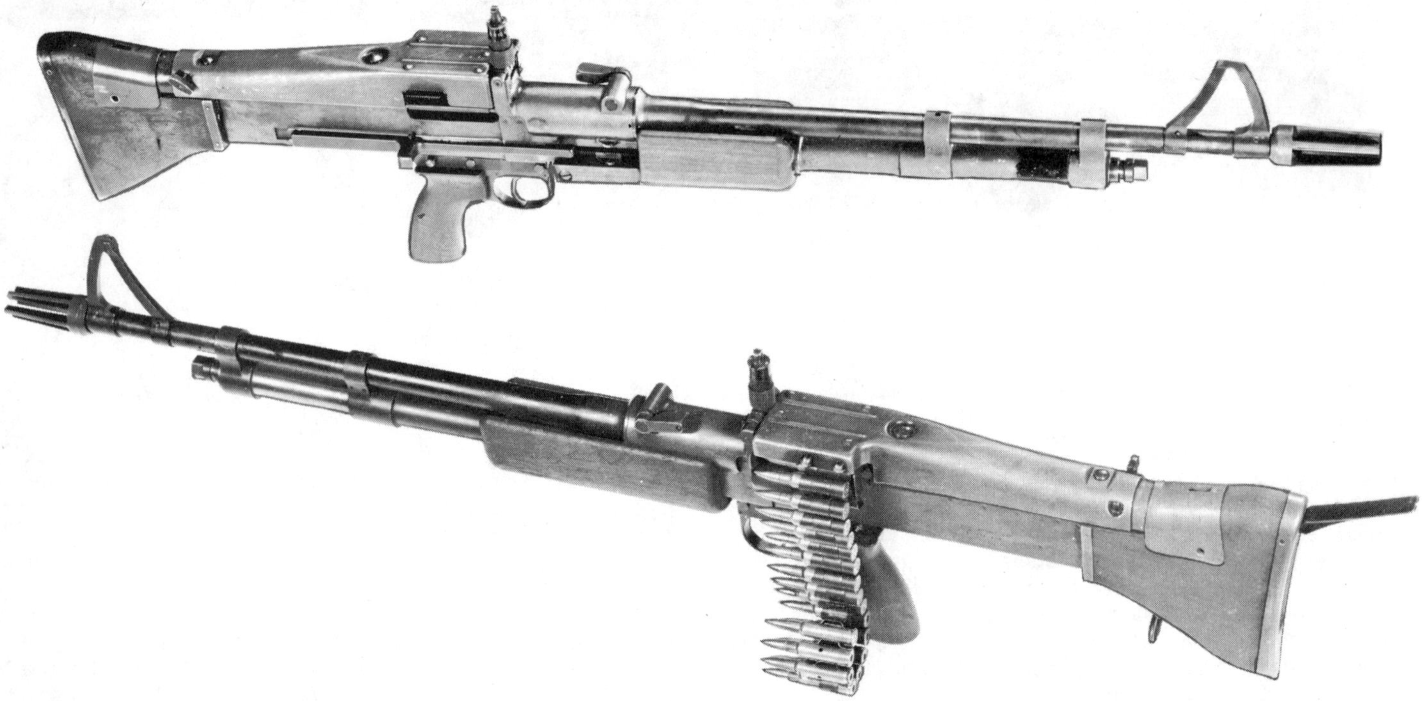

Caliber .30 (T65E3) T52E1 Machine Gun.

Development of the T44 was terminated in 1948.

The Madsen M1946 was tested from 1947 to 1950 as were the T53 and T61 guns. The development of the T52, a modification of the T44 design, was begun in 1947. This design, incorporated a quick-change barrel, and the gas cutoff expansion system currently used with the M60 machine gun was eventually adopted in this design. The T52 was chambered for the then developmental 7.62mm NATO cartridge as were the Madsen, the T53, and the T61. Development of the T52 series of guns was terminated in 1952, but in 1951 development was begun on the T161 series of guns, which were originally designed as caliber .30-06 versions of the T52E3 to provide a back-up for the T51E3 in the event that the caliber .30 T65E3 cartridge was not adopted. In the development of the T161 series, accent was placed on features that would ease manufacture. By the time that the T161E1 was under development, the prospect of adopting the T65E3 cartridge was so definite that the T161E1 was chambered for this cartridge and became in effect the production engineering model of the T52E3.

Additional changes were made in the T161E1 as a result of tests at Aberdeen Proving Ground in March and April 1953 and the modified gun was called the T161E2. There were both light and heavy barrels under consideration for these weapons. Further testing indicated that the heavy barrel was more suitable. Additional minor modifications were made and the T161E3, as the modified gun was called, was adopted as the 7.62mm NATO M60 machine gun in February 1957.

Work was also done from 1949 to 1956 on modification of the Browning M1919A4, 1919A6, and 1917A1 guns to use the T65E3 cartridge (7.62mm NATO). These developments were terminated when it became apparent that the T161E3 would be adopted to replace these guns. These weapons were all fitted with prong-type flash suppressors.

The desirability of a new tank machine gun was recognized during World War II. The receiver of the 1919 Browning gun was too long and fed only from the left side, limiting their usability in armored vehicles. As an interim measure the Browning 1919 was

Caliber .30 T161 Machine Gun.

Caliber .30 (T65E3) T161E2 Machine Gun.

Caliber .30 (T65E3) T214 Machine Gun.

modified by the fitting of an alternate feed and various other features. The flexible gun prototype was the caliber .30 T152 and the fixed gun was called the T153.

The T153 was classified standard in May 1953 as the M37 caliber .30 machine gun. The M37 was modified to use 7.62mm NATO ammunition as the M37E1; this development was suspended in 1957.

Development of tank guns chambered for the 7.62mm NATO cartridge with short receivers began in 1951. The prototypes developed were the T197, T198, T199, and T200. The T197E2 was classified standard as the 7.62mm NATO tank machine gun in May 1959. The other developments were dropped in 1952.

Development of caliber .50 machine guns in the United States began in 1918 at the request of the Army Expeditionary Forces. John Browning sealed up the Model 1917 water-cooled gun and Winchester initially developed the cartridge. An air-cooled gun for use in aircraft was developed almost simultaneously and the first caliber .50 aircraft gun was assembled at Winchester on November 12, 1918. Further development work was carried out on contract by Colt and guns were to be developed for aircraft, antiaircraft, or tank and ground use. In practice this initially narrowed down to two guns: the Model 1921 water-cooled antiaircraft gun and the Model 1921 aircraft gun.

The .50 caliber guns in Army service were modified as the result of extensive test firings held at Aberdeen from 1926-1930 and the modified weapon was called the Model 1921A1. This weapon was declared obsolete on Feb. 15, 1944. In 1933 an improved version of the 1921A1 called the caliber .50 M2 water-cooled antiaircraft machine gun was adopted. This weapon, which was used through World War II, was used on various antiaircraft single and double mounts. It has provisions for continuous pumping of water through the jacket and uses the same receiver as the caliber .50 M2 aircraft gun and the caliber .50 M2 heavy barrel gun.

The M2 aircraft gun was basically an improvement on the alternate feed Model 1923 aircraft gun. By the use of a switch in the belt-feed lever track in the bolt and changeable feed components, any gun could be used to feed from the side desired. The M2 was standardized in October 1933. The cavalry developed a

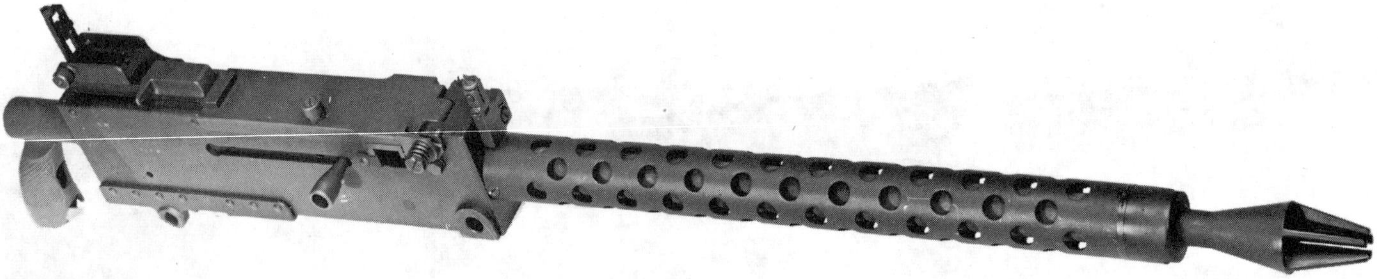

Caliber .30 (T65E3) T66 Machine Gun.

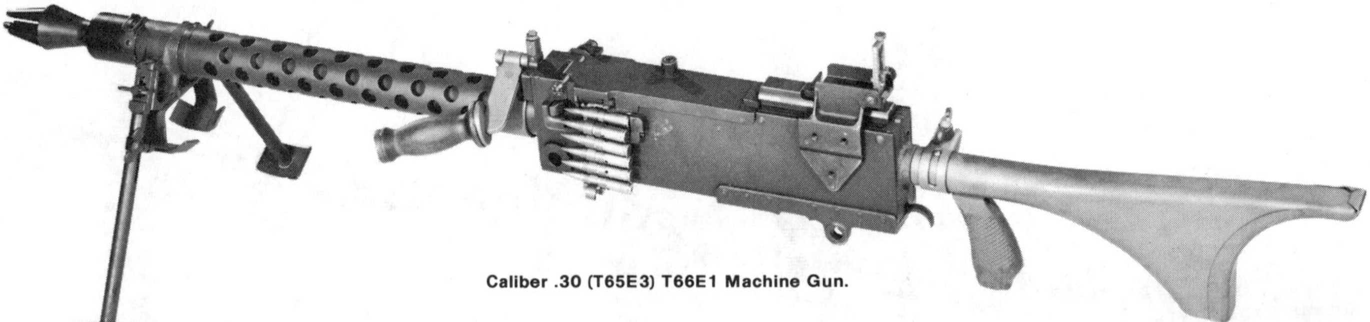

Caliber .30 (T65E3) T66E1 Machine Gun.

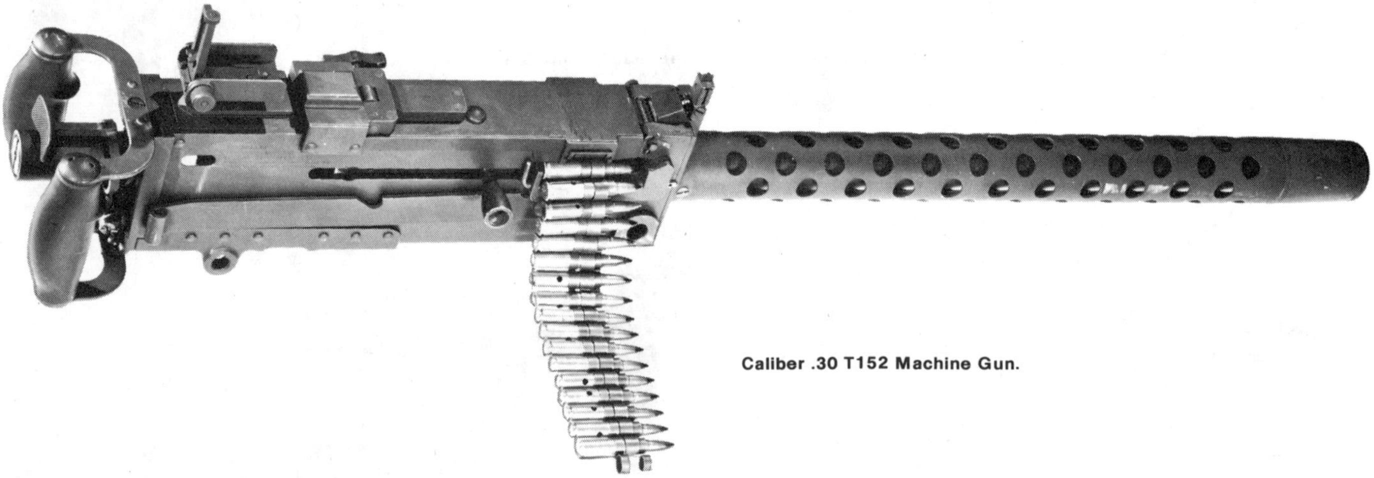

Caliber .30 T152 Machine Gun.

Caliber .50 M2 Machine Gun, water-cooled.

requirement for a heavy-barreled caliber .50 gun and the caliber .50 T2 machine gun was designed to fit this requirement. The T2 differs from the M2 heavy barrel standardized in November 1933 in the length and weight of the barrel, the barrel support, and the oil buffer. The M2 caliber .50 was, and is, one of the best machine guns in its class. No foreign machine gun, except for the various foreign copies of the caliber .50, had the basic reliability and simplicity of the M2. Nearly two million caliber .50 M2 guns of all types, i.e., aircraft, water-cooled antiaircraft, and heavy barrel were manufactured by Frigidaire, AC Spark Plug, Saginaw Steering Gear, Brown-Lipe-Chapman, the Savage Arms Corp., Colt, and the Buffalo Arms Co. Many of these guns are still in service throughout the United States Army and in allied and neutral countries in the world. Unfortunately, many are also being used against United States forces in Vietnam.

An aircraft gun with a higher rate of fire was desired by the Army Air Corps. The M2 design was modified, and the M3 aircraft machine gun, covered in detail later in this chapter, was adopted.

Caliber .50 Machine Gun T2.

Although the Browning caliber .50 is a fine weapon, it has some shortcomings, especially for mounting in armored vehicles. The Army lost interest in the caliber .50 as an antiaircraft gun after World War II. As a matter of interest, the air-cooled caliber .50 M2 heavy barrel and its turret-type variations were much more extensively used as antiaircraft machine guns by the Army than was the caliber .50 M2 water-cooled gun. The quad mount, the M45, used the M2 heavy-barrel turret-type gun as did several other multiple gun mounts. The principal interest in the caliber .50 gun toward the close of World War II was as a tank machine gun. The Browning has an overly long receiver and the rate of fire was considered to be undesirable for tank use.

In September 1945, development was begun on a modified M2 Browning with an increased rate of fire, a flash hider, and a good sustained fire capability. This weapon, called the T42, was under development until 1955. The last model of the T42, the T42E1, was not sufficiently better than the Browning M2 to warrant adoption. In the meantime a project for the development of a new caliber .50 machine gun was initiated in 1951. The requirement stated that the new gun had to have a short receiver with its feed located near the rear of the receiver and have a very durable barrel to allow a high sustained rate of fire. Development was started on three different guns: the T164, T175, and the T176.

The T164 was a caliber .50 version of the M39 20mm revolver type aircraft gun, which is an improvement of the gas-operated German MK213C aircraft gun. Development of the T164 was terminated in 1955. The T175 was a recoil-operated gun with a hydraulically-actuated accelerating mechanism. Oil leakage from the cylinder of the hydraulic system caused modification and a new model, the T175E1, was placed in development in 1955. The T175E1 had a mechanically accelerated mechanism. The T176 was also a recoil-operated gun, in which the bolt locked to the barrel extension by rotary bolt locks.

In August 1954, development was begun on two additional caliber .50 designs. One of these—the T217—was essentially a modified T42 (Browning type) equipped with a gas booster to increase the rate of fire to approximately 1000 rounds per minute. The other, the T28, was a .50 caliber version of the T197 machine gun. In January 1955 it was decided that there was no requirement for a ground gun and the only requirement was for a tank gun. This in effect was a reaffirmation of an earlier decision. It was further determined that the development of the T42 and T164 series and the T217 would be terminated. The T218 was also dropped and all effort was concentrated on the T175E1. Work on the T176E1 was suspended pending determination of the suitability of the T175E1.

Tests of the T175E1 held in 1958 indicated that the links used with the gun were not satisfactory and some parts needed beefing up. A scaled-up version of the 7.62mm NATO M13 link, called the XM15, was developed for use in the weapon. This link required changes in the feed mechanism and the gun which resulted was called the T175E2.

In June 1959 the T175E2 was standardized as the M85 caliber .50 Tank Machine gun.

All the post-war machine guns have stellite-lined and chrome-plated barrels as did the Browning M1919A4 and 1919A6 machine guns after June 1952.

During World War II a requirement was laid down for a .60 caliber machine gun in order to obtain a weapon which would produce a greater effect on a target aircraft. As the result of this requirement a large number of caliber .60 prototype guns were produced, but the front runner was the T17 series guns. The T17 is based on the design of the German 15mm MG151 aircraft gun. The caliber .60 is a very "hot" round, indeed its cartridge case necked out to 20mm is still used in some current U.S. 20mm guns. Because of this fact parts breakages were frequent with early models.

The T17E5 was the last of the weapons in this series. It is a recoil operated belt fed gun weighing 127 pounds. It is 91.7 inches long with a barrel 60 inches long. The cyclic rate is 700-750 rounds per minute and the muzzle velocity 3500 f.p.s. with an 1150 grain projectile. Original work on the T17 series was done by Colt, Springfield Armory, and Frankford Arsenal, GE and the Frigidaire Division of General Motors did most of the later work. The T17E5 was last heard of in the late forties when it was proposed to put four of these guns on a self propelled mount for anti-aircraft work, under the designation T173.

Other caliber .60 weapons were developed by Bendix, who modified the Hispano Suiza 20mm aircraft gun. The caliber .60 T18 was also a Hispano Suiza modification, the T19 was a High Standard project, the T31 was a modification of the Oerlikon by Gazda Engineering Corp., the T39 is an electrically fired T17E3, and the T41 is a further modification of the T17E3 aimed at giving a gas assist to it's basic recoil operation. The T45 was a Gatling type developed by GE. Numerous other caliber .60 guns were developed as well.

THE CALIBER .30 BROWNING MODEL 1917 AND 1917A1 MACHINE GUNS

All weapons of this type used during World War I were of the Model 1917 type. Wartime service indicated that the bottom plate was weak. This was modified by addition of a reinforcing stirrup and 25,000 of these stirrups were mounted on weapons being placed in storage in 1920-21. In 1936 it was determined that further changes were necessary in the weapon and a remanufacturing program was set up at Rock Island Arsenal. The modified weapon was designated the Model 1917A1.

The principal changes made at that time were:

(1). Fitting of a new, bottom plate.

(2). Fitting of a new belt-feed lever.

(3). Fitting of an improved cover-latch assembly.

(4). Fitting of a sight leaf graduated for the caliber .30 M1 and M1906 ball ammunition.

(5). Cover catch assembly which would hold the cover in a fixed position when opened.

(6). The tripod was modified to produce the M1917A1 tripod.

Weapons of new manufacture made during World War II had additional modifications as follows:

(1). A steel end cap was fitted in place of the bronze end caps used with the earlier guns.

(2). A steel trunnion block was fitted in place of the bronze trunnion block used on earlier guns.

(3). An improved steam tube assembly was used.

(4). A new type bunter plate, similar to that of the Model 1919A4 was used.

(5). An improved barrel gland assembly made of non-corrosive material was used.

(6). A recoil plate was fitted in the face of the bolt, i.e., separate ring of steel around the firing pin which can be replaced as firing pin hole wears, thus obviating the necessity of replacing the complete bolt body.

Caliber .60 T41 Machine Gun.

Browning caliber .30 M1917A1 Machine Gun on M1917A1 tripod.

(7). The sight leaf was graduated for caliber .30 M2 ball ammunition.

The Browning M1917A1 is not used in United States service, but can be found in the armies of many allied countries.

Caliber .30 Browning M1917A1 Machine Gun.

Loading the M1917A1 Machine Gun

Check the tripod to see that there is no unusual play; that it is firmly seated, that all its jamming handles are tight, and that the splayed feet of the legs are pushed securely into the ground. Check the water jacket and condenser. Make sure that both are full. Then check to see there is no water leakage at the muzzle glands. See that rear barrel packing is water tight and oil or grease it heavily if it is not. Check headspace adjustment. Remember that this is a most important adjustment on this gun. If it is too tight the gun will refuse to fire; too loose headspace will result in bulged or ruptured cartridge cases which will jam the gun badly. Check that ammunition belts are loaded uniformly and correctly and that they are clean and dry. Check rear sight for vision and working order. See that all moving parts are lightly oiled and work smoothly by drawing the bolt handle back and permitting it to go forward. All mechanism should work smoothly and no unusual effort be required to withdraw barrel and recoiling mechanism to the rear. When bolt handle is released it should position in its fully forward place and bolt should lock home properly. Bore should be inspected to be sure that it is clean.

With ammunition box securely locked in place on the left side of the gun and cover closed, insert the tag of the belt through the feed block as far as it will go to the right and pull the belt sharply to the right.

Pull the bolt handle to the rear as far as it will go. This will compress the barrel plunger spring and the driving rod spring, and when the bolt handle is released, these two springs will drive the action forward, moving the recoiling parts to their forward locked position, and half load the gun.

Now pull the bolt handle back a second time as far as it will go.

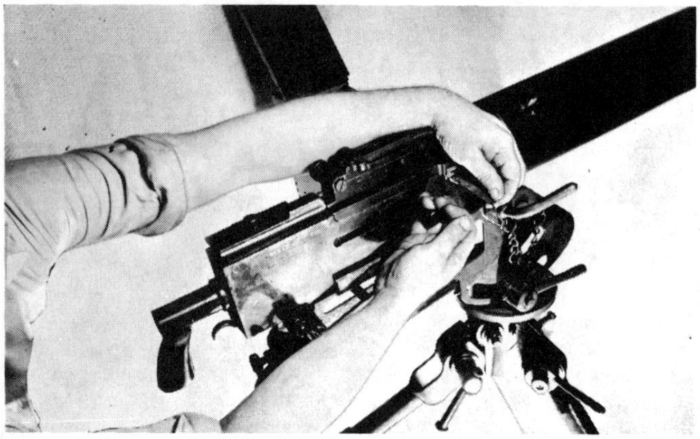

Release it and let it fly forward. This completes the loading, leaving a cartridge properly positioned in the firing chamber and the weapon cocked and ready for firing.

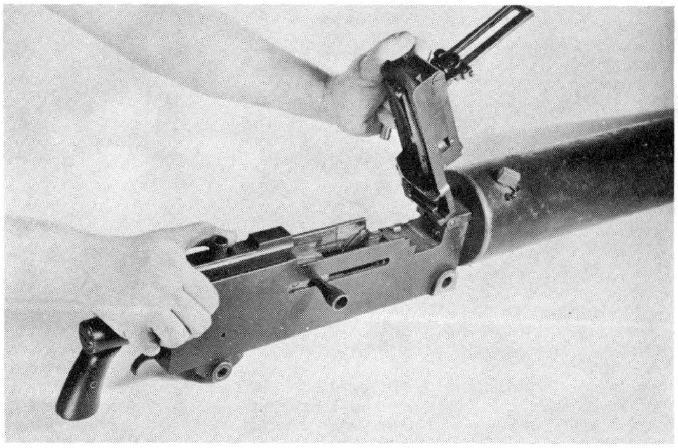

Unloading the M1917A1 Machine Gun

Pull back the cover latch (this is the milled knob on top of the receiver to the rear of the rear sight). The spring controlled section will move backwards and permit you to raise the feed cover.

Lift the belt out to the left and replace it in its box.

Pull the bolt handle to the rear and look inside the bolt way to make sure there is no cartridge in the firing chamber or in the face of the breechblock.

Now push the extractor down to its seat in the front of the breechblock and let go of the cocking handle, permitting the breechblock to go forward.

Snap down the cover and press the trigger.

Stripping the 1917A1 Machine Gun

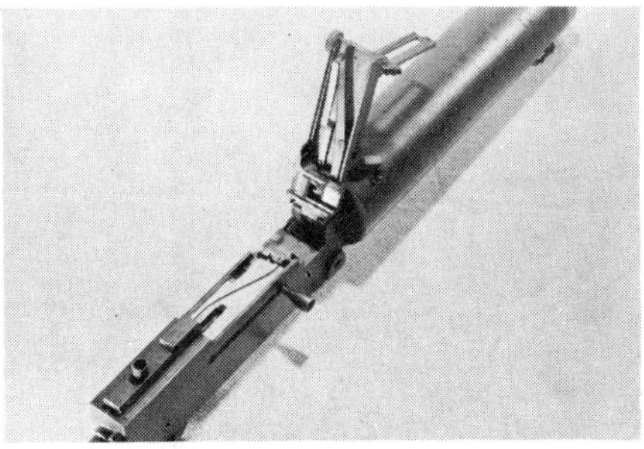

Raise the rear sight. Pull back the cover latch (the knob on top of the receiver behind the rear sight base) and raise the cover.

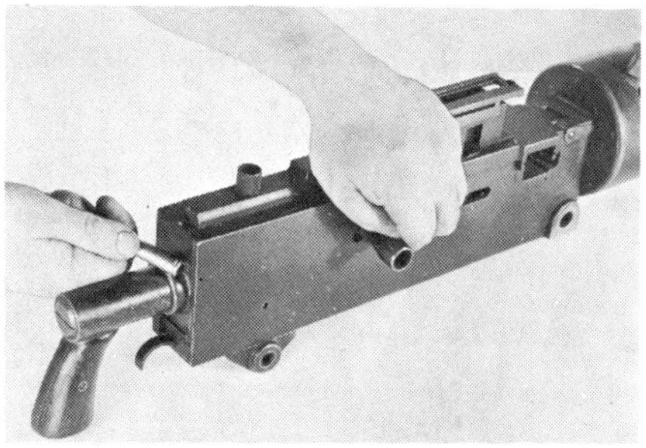

Draw bolt handle back as far as it will go and hold it firmly with the left hand. The driving spring rod protrudes through the back plate of the gun. Insert the base of a cartridge in the slot in the head of the rod. Push the rod in to compress the spring, and turn the rod to the right. This will lock the driving rod and its spring under compression inside the bolt.

Push the bolt handle forward a few inches to draw the driving rod out of the locking hole in the back plate. Then pushing the cover latch forward with the left thumb, raise the pistol grip up and out of its retaining slots in the rear of the receiver.

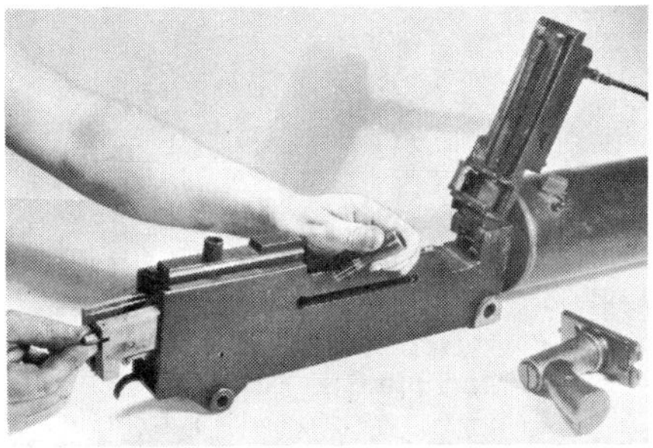

Pull the bolt handle as far as it will go to the rear, at which point it may be pulled to the right out of the bolt and receiver. Reach inside the receiver and grasp the driving rod; then pull the bolt directly to the rear and out of the receiver.

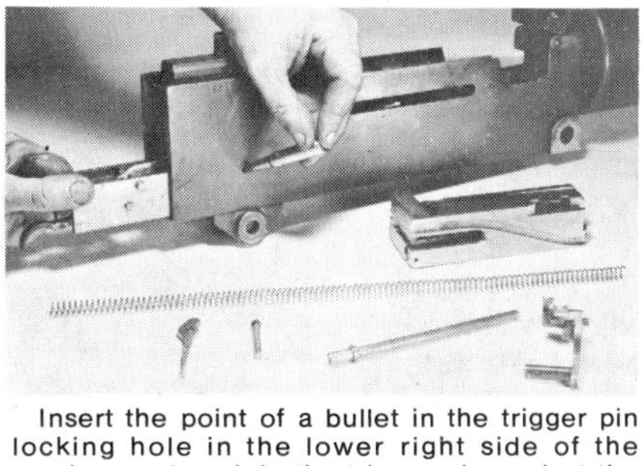

Insert the point of a bullet in the trigger pin locking hole in the lower right side of the receiver and push in the trigger pin against the tension of its spring. This frees the lock frame spacer and other recoiling parts and permits pulling them directly to the rear.

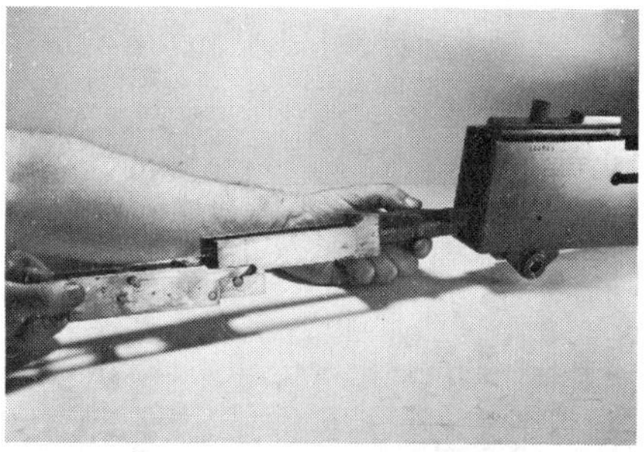

When the bottom projection at the rear of the barrel extension (the part screwed onto the barrel) drops below the bottom of the receiver, pull the combined lock frame spacer, barrel extension, and barrel directly to the rear out of the gun.

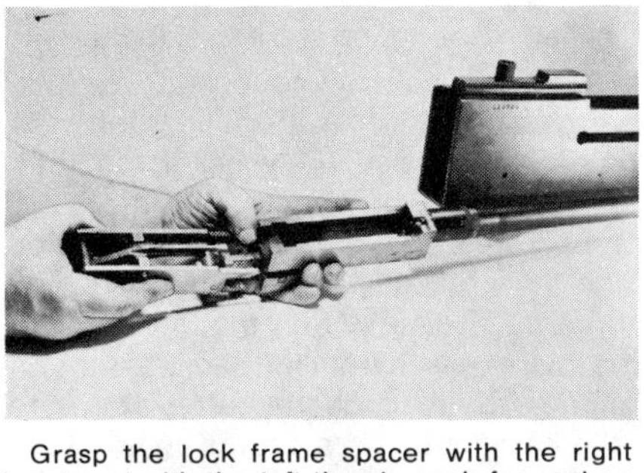

Grasp the lock frame spacer with the right hand, and with the left thumb, push forward on the turned up tips of the accelerator.

This will spring down and forward, separating the lock frame spacer (which holds the trigger and accelerator mechanism) from the barrel extension.

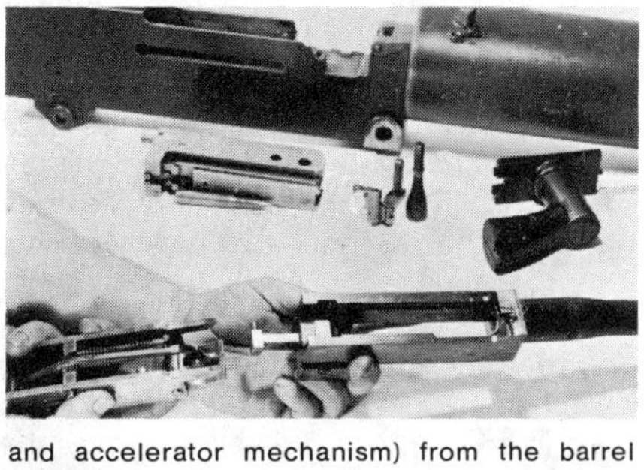

Push the accelerator pin out of the lock frame spacer and remove the accelerator (this is the curved piece of metal with two claws).

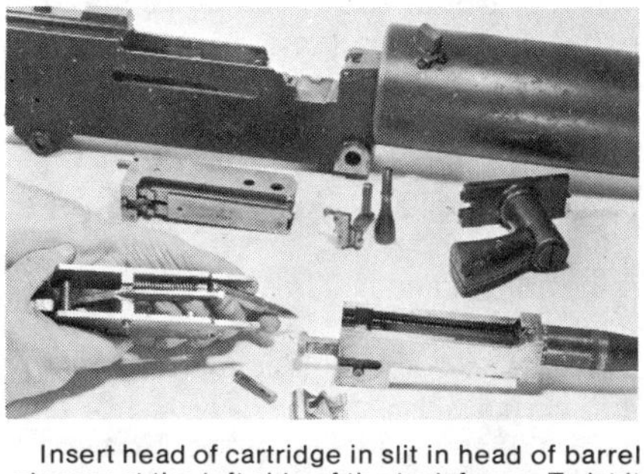

Insert head of cartridge in slit in head of barrel plunger at the left side of the lock frame. Twist it and ease it out. Remember that it is under strong spring tension (if necessary trigger pin may now be pushed out and spring and trigger removed).

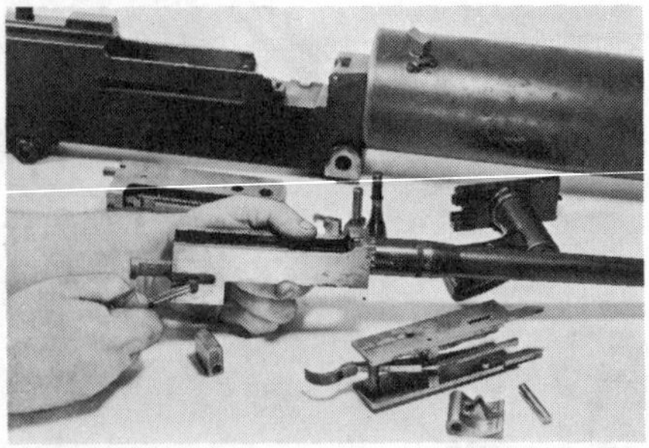

Holding the barrel extension with the left hand, use the point of a bullet to start the breechblock pin from left side and remove it from the right, permitting the breechblock (the heavy wedge whose lower front end is beveled) to drop down out of the barrel extension. (Barrel extension may now be unscrewed from the barrel if necessary.) Turn the extractor (the swinging hook-shaped piece on the left forward end of the bolt) up as far as it will go and pull it out of its hole in the bolt.

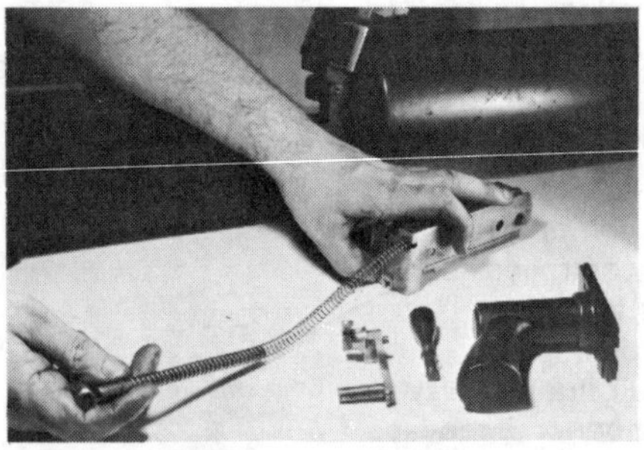

With the base of a cartridge, push in the head of the driving spring rod in the bolt and twist it to the left. Be careful of the very powerful driving spring which will now be free to fly out the back carrying the driving spring rod with it. Grip this firmly and ease it out gradually. Push out the cocking lever pin from the upper front left side of the breechblock and lift out the cocking lever.

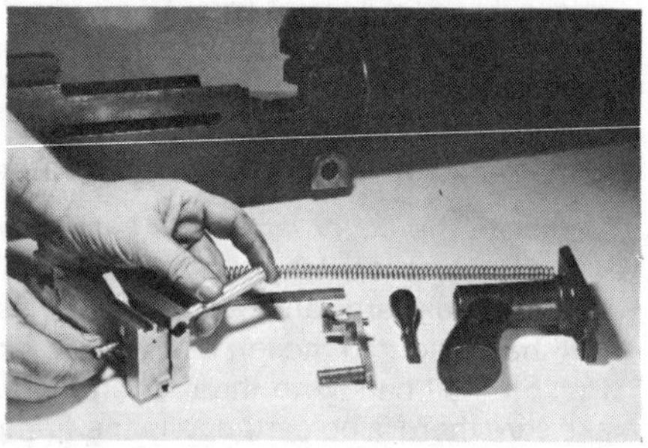

Turning the breechblock upside down, push the sear up with the bullet to release the firing pin spring; turn the breechblock the right way up again and insert the bullet into the slot of the sear spring. Push over to the left, pry the sear spring into a locking recess and the sear drops out. When sear spring is pushed back into normal position its pin may be pushed up and it and the spring removed. The firing pin and its spring may now be dropped out of the back of the breechblock.

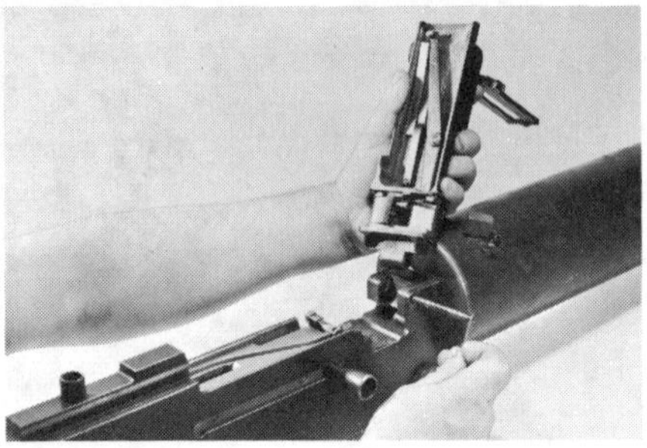

Use the point of a bullet to turn up the cover pin spring at the right side of the receiver just behind the water jacket. Pull the pin out to the right and lift the cover up out of the receiver. Pushing back on the nose of the split pin on the left side of the feed block, control the feed pawl

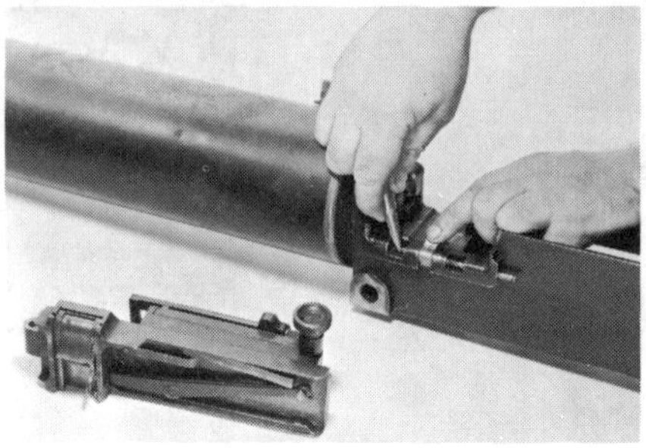

with the left thumb and ease the pawl and its spring up out of the receiver as the locking pin is pulled out. Insert point of bullet between the extractor cam and the long flat piece of metal on the side of the cover opposite the belt feed lever (this is the extractor spring) and pry the

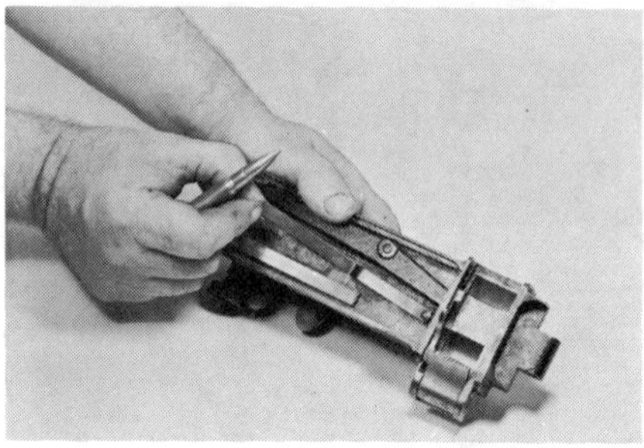

extractor spring out of its seat in the cam, then lift it up and out. The feed lever and slide may be removed if necessary by turning feed lever pivot pin spring outwards. This completes stripping the Browning machine gun.

Assembling the M1917A1 Machine Gun

Start by screwing the barrel into the barrel extension, then insert the barrel and barrel extension into receiver. Slide slowly forward until the lower projection of the barrel extension (which holds the bolt back) is against the bottom plate of the receiver. Holding the lock frame in the right hand, place the accelerator claws between the rear face of the barrel extension and the forward faces of the T-lug extension and at the same time insert the forward projections of the lock frame into their grooves in the barrel extension. Give a quick thrust forward to the lock frame to tip back the accelerator claws and compress the barrel plunger spring; this locks the lock frame to the barrel extension.

Now push the lock frame attached to the barrel extension and barrel farther ahead in the receiver until the trigger pin on the lower right side comes against the side of the receiver. Push in the trigger pin against its spring tension and the whole assembly can now be pushed fully home while the trigger pin will be forced by its spring into its slot in the right side of the receiver.

Now replace the bolt in the receiver being sure the extractor is in position and that the cocking lever projecting through the top is fully forward.

Insert the bolt handle in the end of the slot in the receiver and into its hole in the bolt, then push it forward far enough so that when the cover latch is pushed forward, the pistol grip and back plate may be slid down in their grooves in the receiver. Pull the cover latch back to lock the back plate in position; holding bolt handle back as far as it will go, insert the base of the cartridge into the slit in the driving rod, push in and turn to the left to release the driving rod spring. Now permit the bolt to run forward under the thrust of the driving rod spring.

How the M1917A1 Machine Gun Works

Starting with the weapon loaded and cocked the action is as follows: The trigger being pressed, the trigger bar disengages from the sear block and allows the striker to be driven forward by the spring in the bolt to fire the cartridge.

As the bullet goes down the barrel the recoil drives back against the base of the cartridge base which transmits the blow to the bolt face thereby starting the locked recoil action and barrel to the rear.

The barrel, barrel extension, and bolt recoil, locked together, about 5/8 of an inch. For the first half of this travel, during the period of high breech pressure, they are securely locked together, and then the front projections of the lock frame (which are set against the sides of the pin passing through the barrel extension and the breechblock) force the breechblock pin down, drawing the breechblock down out of its locking slot on the underside of the breechblock. The bolt is thus released from the barrel extension and so can continue straight to the rear. As the barrel extension itself travels to the rear, the barrel plunger spring is compressed and the rear of the barrel extension drives the claws of the accelerator back sharply, flipping the accelerator up and backwards on its pin.

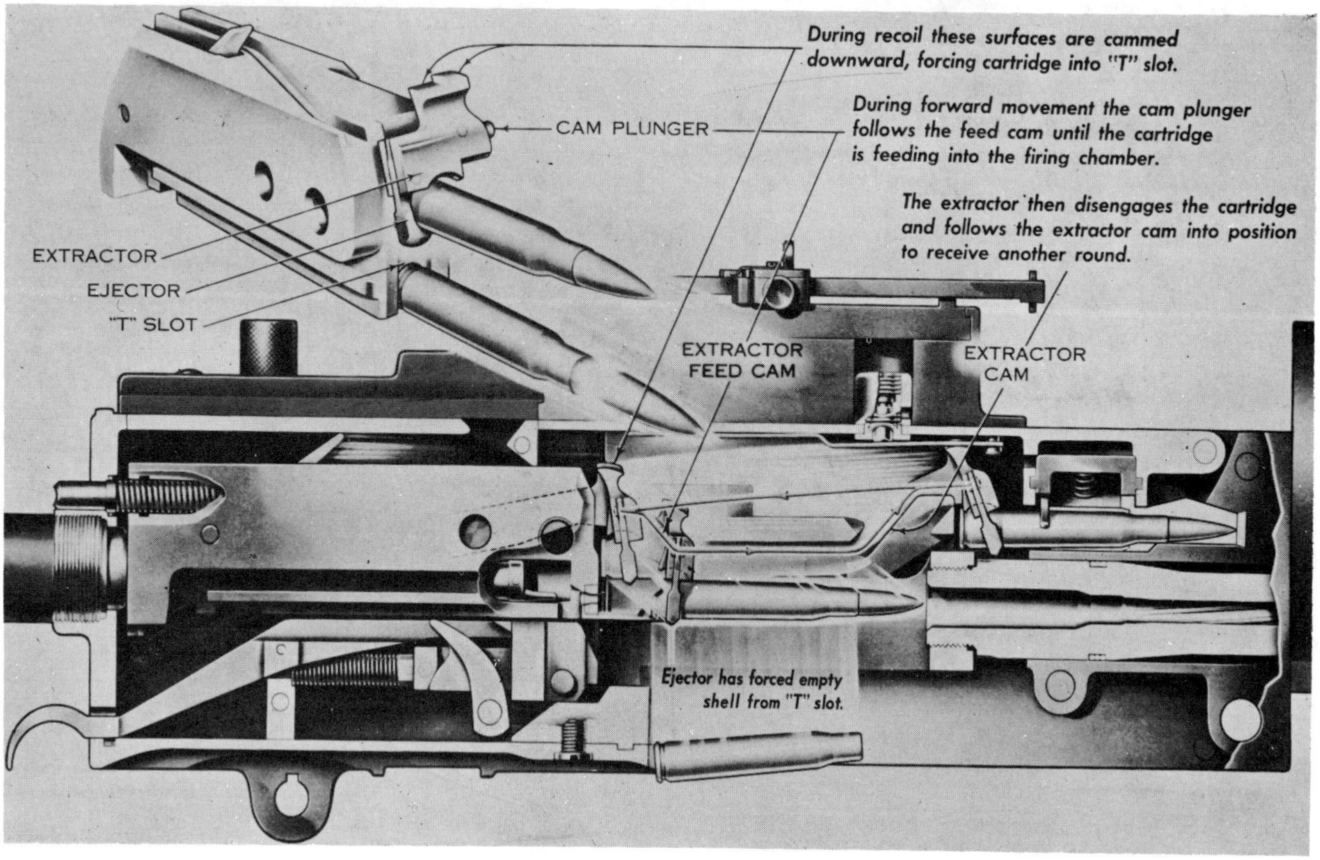

Cross section of parts in firing position.

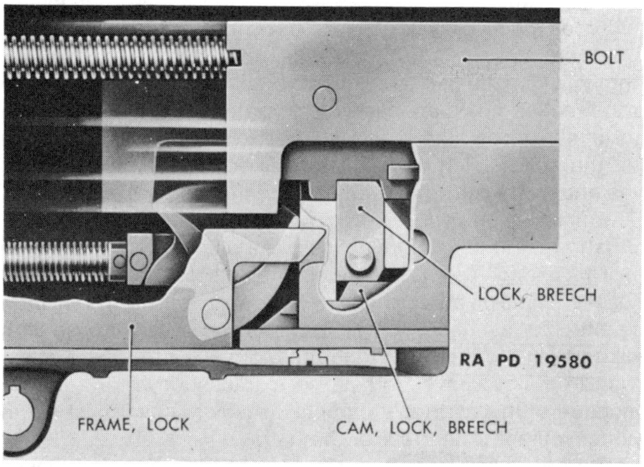

Breech locked.

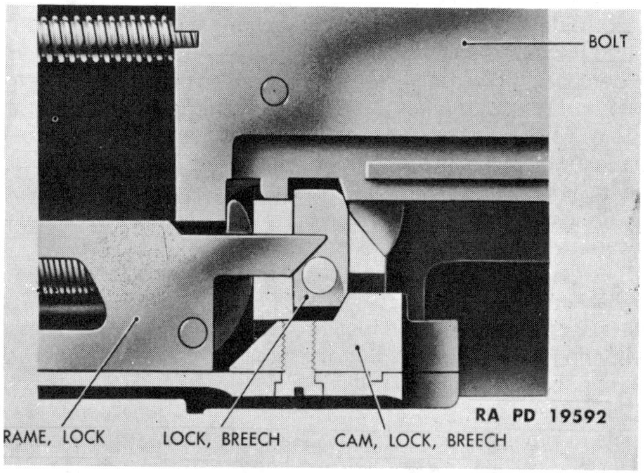

Breech lock beginning to open.

As the accelerator turns, the tips of its claws strike bottom projections on the bolt and thus accelerate the rearward motion of the bolt by transmitting to it the thrust absorbed from the barrel extension, which is now held in rearward position locked to the frame spacer slots. Speeding the rearward motion of the bolt, at the same time that the barrel is slowing up, permits the empty case to be extracted from the chamber without the sudden tug that would normally occur. (This makes special lubrication unnecessary.) The accelerator claws while engaging the shoulder of the T-lug firmly lock the barrel extension in the rearward position to the lock frame. A stop prevents the accelerator from going backwards too far and the barrel plunger spring is held compressed.

During backward motion of the bolt, the driving spring is compressed over the driving spring rod whose head is held securely in the back plate of the receiver. The extractor fitting over the top front of the bolt draws a loaded cartridge from the belt at the same time that the T-slot machined into the face of the bolt draws the empty cartridge case from the firing chamber. The extractor cam plunger (which rides along the top of the extractor cam and the extractor feed cam) is finally forced in by the beveled section of the extractor feed cam. The cover extractor cam thus forces the extractor down and the plunger spring out behind the extractor feed cam.

During the backward movement of the bolt, a stud on the belt feed lever (which is mounted in the front of the cover on a pivot) moves to the right in the cam groove cut in the top of the bolt. Thus the belt feed slide which is attached to the lever is moved to the left. The belt feed pawl (located on the under side of the cover

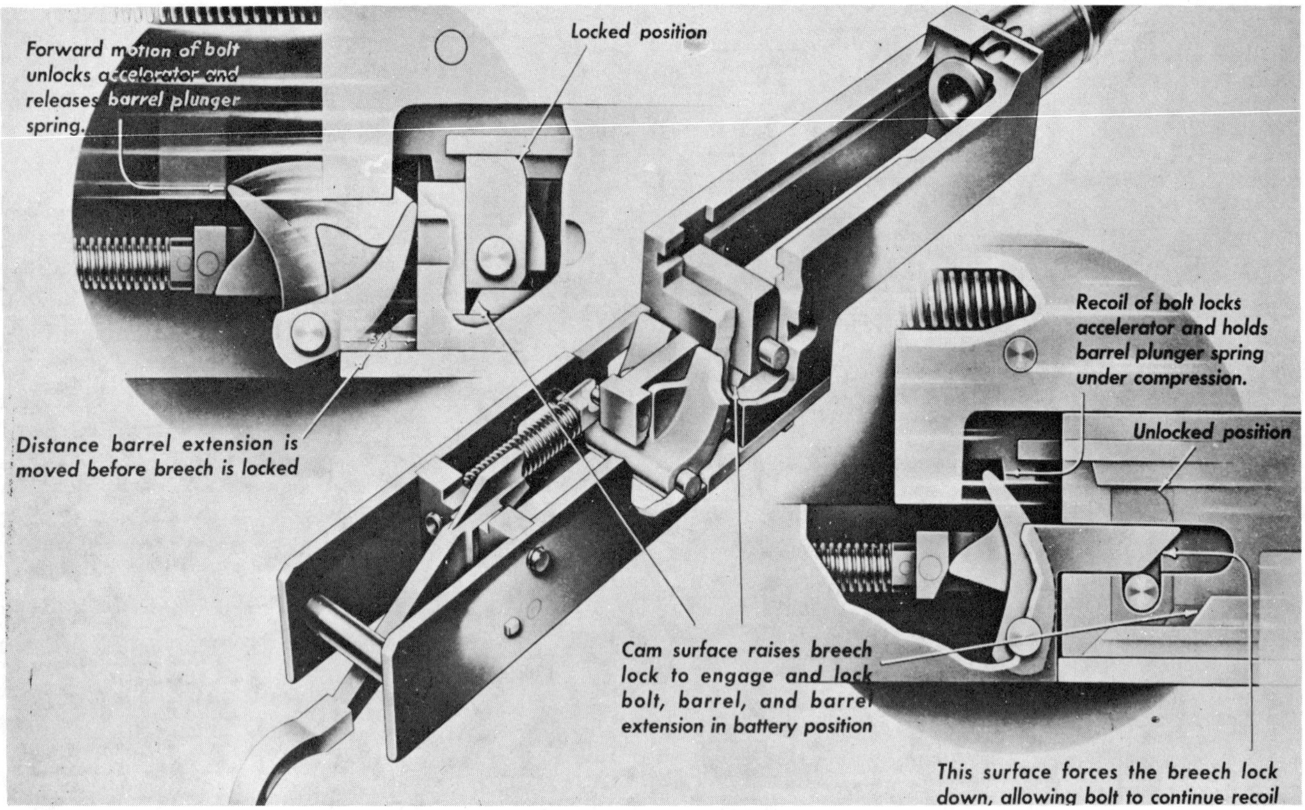

Operation of accelerator mechanism.

above the belt) springs over the left of the first cartridge (which is being held in position by the belt holding pawl below the belt on the left side of the feed block) and supports the cartridge and the belt to prevent feeding trouble.

The cocking lever is fastened by a pivot pin inside the bolt and its top protrudes through the top of the bolt and rests inside the top cover of the receiver. Thus as the bolt starts backwards, pressure on the lower part of this lever drives it back revolving it on its axis and the firing pin spring is compressed drawing back the firing pin. The firing pin engages a notch in the sear (which has been pulled upward by the action of the sear spring).

The rear of the bolt strikes against the buffer plate mounted in the back plate at the upper part of the pistol grip and the remainder of its energy is absorbed in friction and by the buffer disc. (A brass buffer ring is forced over a plug and expands against the inner wall of the grip.)

The driving spring in the bolt which is now fully compressed reacts to drive the bolt forward. During this forward motion, the upper end of the cocking lever is forced to the rear, thus pulling the lower end away from the rear of the firing pin. The extractor feed cam acts on the extractor cam plunger, forcing the extractor down so that the cartridge it is holding drops down the T-slot in the face of the bolt until it reaches a direct line with the firing chamber. The ejector strikes the empty case in the T-slot expelling it through the bottom of the gun and stopping the live cartridge when properly positioned.

Also during the forward motion of the bolt, the bottom projection strikes the top of the accelerator to swing it forward on its axis pin. This unlocks the barrel extension from the lock frame, permitting those two units to move forward as the bolt acts through the accelerator against the rear end of the barrel extension. The forward motion of the barrel extension and barrel is further assisted by a thrust from the barrel plunger spring as it uncoils. The force passed on by the accelerator from the bolt to the barrel extension is sufficient to guarantee proper timing of the locking action.

The actuating stud on the feed lever fits down in a cam groove in the bolt. Thus as the bolt goes forward, the stud rides in the cam and forces the lever on its pivot to the left; the forward end of the lever which carried the feed slide is thus pivoted to the right, bringing with it the belt feed pawl, the belt and the next cartridge. As the motion ends, the cartridge to be fed is held between the cartridge stops and the feedway. The next round to load is pulled over the belt holding pawl which rises behind it. It is in position to be engaged by the belt feed pawl on the next movement.

The final action of the extractor during forward motion of the bolt is to rise under the influence of its plunger riding along the top of the extractor cam. As the ejector pivots forward, the extractor releases its hold on the cartridge which is now well into the chamber. The extractor continues to ride upwards and over the base of the next cartridge in the belt in the feedway. Then the flat extractor spring in the top cover forces the extractor down and into the cannelure of this cartridge gripping it ready to pull it back on the next movement to the rear of the bolt.

The breechblock, mounted in the rear of the barrel extension, strikes a cam as the recoiling parts near the firing position and is forced up this cam and into a recess cut in the bottom of the bolt. Thus as the action comes to a complete close, the breechblock firmly locks the barrel extension (into which is screwed the barrel) to the breechblock.

Special Note on M1917A1 Adjustments

It is extremely important that the headspace on this gun be correctly adjusted before firing. To test the adjustment, pull the bolt handle back and let it run forward several times.

If the bolt does not go home fully and smoothly, it indicates too tight space between the face of the bolt and the face of the firing chamber. If the gun is put into use in this condition, it will fire sluggishly or it may refuse to fire at all.

To correct this condition, it will be necessary to strip the weapon and unscrew the barrel one notch, then assemble and test again.

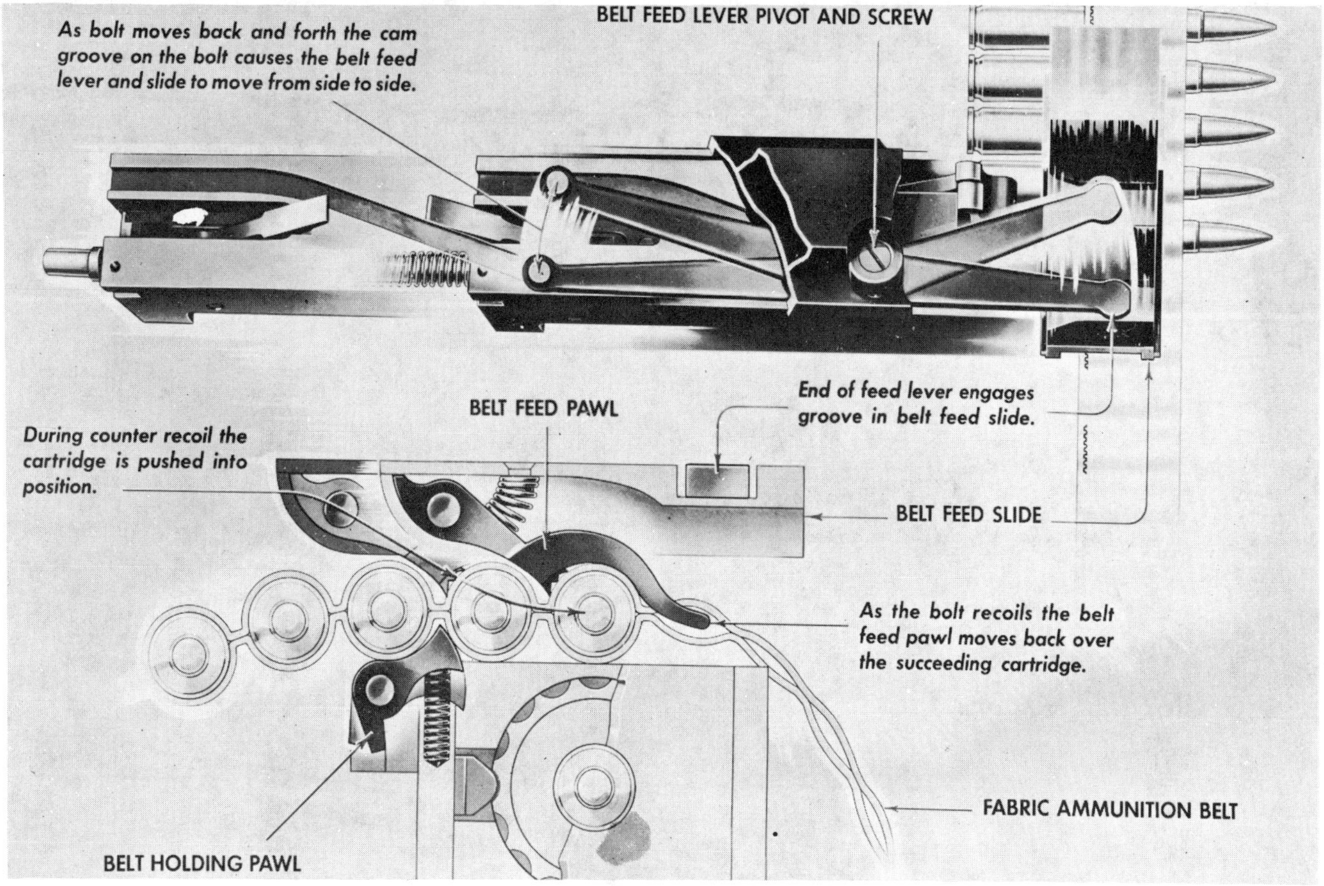

Belt-feed mechanism.

To Test for Loose Headspace. Lift the cover and raise the extractor; then pull the bolt slightly to the rear. If the bolt moves back at all without carrying the barrel and barrel extension with it, then gun is too loose. Fired in this condition, the pressure of the gas in the firing chamber will bulge the head of the empty cartridge case (since it is not fully supported by the bolt), or may rip it off entirely, causing a serious jam.

To correct this condition, screw up the barrel one notch, then retest.

Adjusting Headspace. Screw the barrel into its extension and stop when the first clicking sound is heard (this click is caused by the barrel-locking spring).

Push the breechblock (minus extractor) fully forward on barrel extension.

Push the lock piece up from below to lock the breechblock to the barrel extension; and while holding it firmly, screw up the barrel until resistance is encountered.

Now check to see that barrel locking spring is in a notch and that the lock piece is solidly seated.

Now let go of the lock. If it drops freely, the adjustment is correct. This adjustment should be punched on the barrel, to save time when assembling in future.

Water Leakage. If water leaks from muzzle end, remove the muzzle gland. Wind packing around the barrel. Press it together with combination tool. Then push back on barrel and guide the packing into its seat. Screw the muzzle gland back on. Test by working bolt handle. If there is friction, packing is too tight. This will make a sluggish gun.

If water leaks from breech end, remove barrel and work oiled packing down into barrel cannelure with combination tool. Test as before for undue friction and headspace.

CALIBER .30 BROWNING MODEL 1919A4 AND 1919A6 MACHINE GUNS

Although these weapons are now limited standard in the United States Army and are scheduled to be replaced by the 7.62mm M60 General Purpose Machine gun, they are still in wide use in this country and abroad. The M1919A4 Browning was evolved from the 1919A2 which, through successive modifications, had been developed from the Browning Tank Gun M1919.

The M1919A4

Two types of the M1919A4 have been issued—the fixed gun and the flexible gun. The fixed gun was widely used on World War II armored vehicles; the flexible gun was mainly used as the infantry company-level machine gun. The M1919A4 is used on the caliber .30 M2 mount as a ground gun. The mechanism of the M1919A4 is identical with that of the caliber .30 M1917A1 Browning gun which has been covered in detail. The principal differences between the M1919A4 and the M1917A1 are as follows:

(1) The M1919A4 has a ventilated barrel jacket rather than a water jacket.

(2) The M1919A4 is normally used on a light mount with limited terrain command that is classified as a light machine gun by the United States. The M1917A1 is normally used on a heavier mount with much greater terrain command and is classed as a heavy machine gun; it was the battalion-level gun.

(3) The M1919A4 has a much heavier barrel than the M1917A1, but has a lower sustained rate of fire than the water-cooled gun.

(4) The sights of the M1919A4 and the M1917A1 are different. There are other various minor differences between these guns. The M1919A5 is the same as the M1919A4, except that it has a

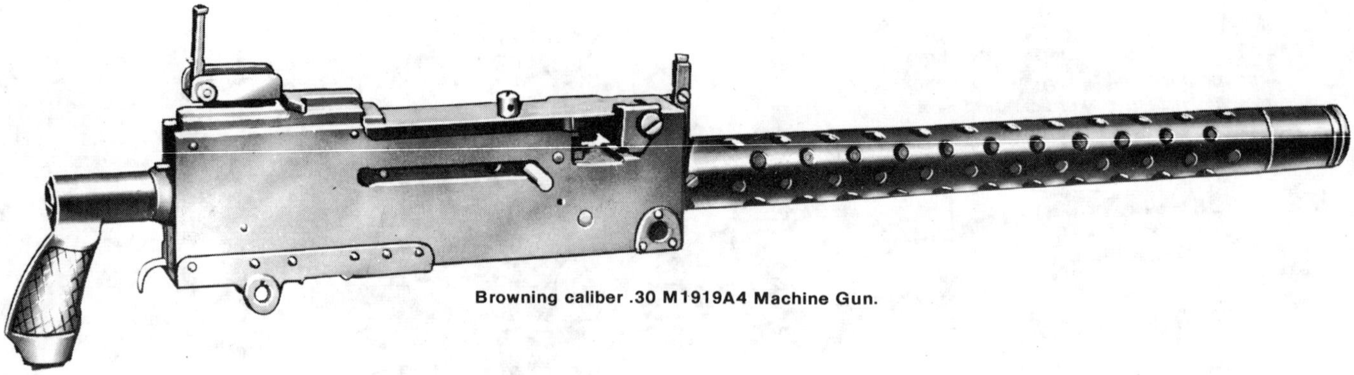

Browning caliber .30 M1919A4 Machine Gun.

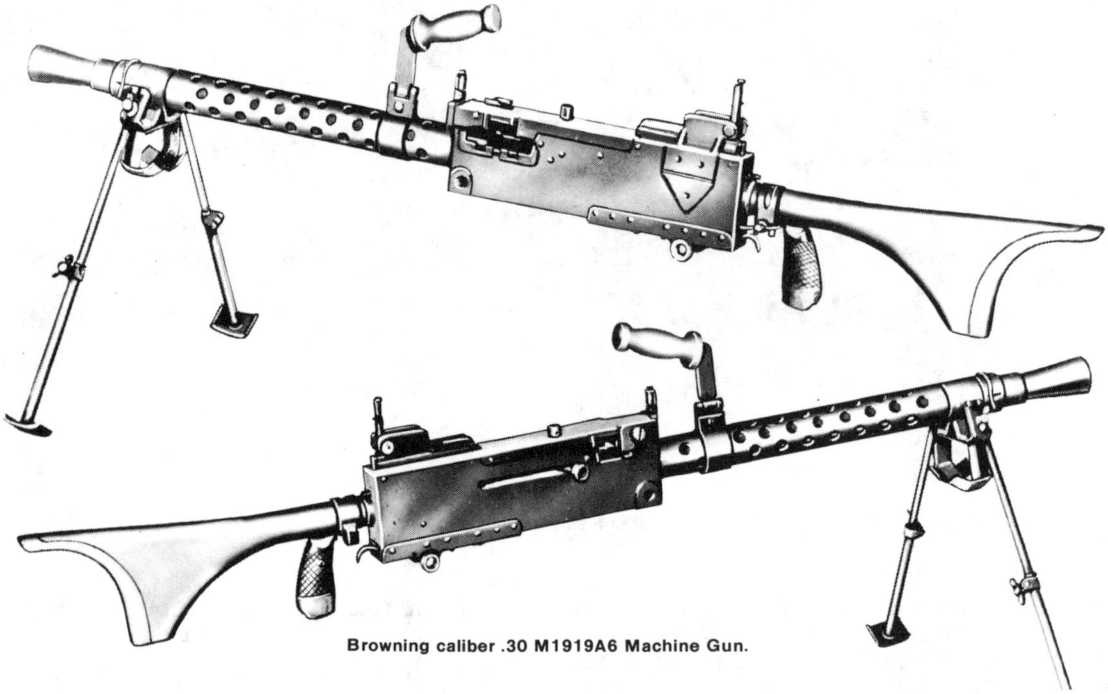

Browning caliber .30 M1919A6 Machine Gun.

bolt retracting slide and a different cover detent. These changes on the weapon were necessary to mount it in the World War II M3 light tank. The U.S. Navy has a number of Browning Model 1919A4 machine guns which have been converted to 7.62mm NATO. These weapons which are fitted with a closed-prong-type flash suppressor are called Machine Gun 7.62mm NATO Mark 21 Mod. O by the Navy. A similar weapon is standard in the Canadian Army as the C1 7.62mm Machine Gun.

The M1919A6

The M1919A6 was a wartime modification of the M1919A4 to give the weapon more tactical flexibility. A bipod, shoulder stock, and carrying handle were added to the basic Browning machine gun action to create a weapon which was easier to move and to get into action than the tripod-mounted M1919A4. The M1919A6 also has a lighter barrel and a different front barrel bearing than the M1919A4. The lighter barrel of the M1919A6 gives it a higher cyclic rate of fire than the M1919A4.

A post-World War II version of the M1919 Brownings (in addition to the M37) also exists; this is the M1919A4E1. This weapon also has a retracting slide similar to the M1919A5.

Loading and Firing; Field Stripping the M1919 Series

The method of loading and firing and the method of field stripping the M1919 series of guns is, in all essentials, the same as that of the caliber .30 M1917A1 water-cooled Browning gun.

The M1919A4 and the M1919A6 may be used with the M1917A1, M2, or M74 tripods, but are normally used with the M2 tripod. The M1919A6 is also used on the bipod shown in the photo.

THE 7.62mm NATO M60 MACHINE GUN

The M60 is called a general purpose gun because it replaces the cal. .30 Browning light and heavy machine guns. The M60 is used on a bipod as a light machine gun, and on a tripod as a heavy machine gun. The M60 (T-161E3) is the result of a series of designs started at approximately the end of World War II. The first of these was called the T44 and was essentially a combination of the belt feed mechanism of the German MG42 with the operating mechanism of the German automatic rifle FG42. A later design, which was considerably modified, was the T52; and from the T52 evolved the T161 series of guns.

The M60 therefore has, in a considerably modified form, the belt feed mechanism of the MG42 and the operating mechanism of the FG42. The FG42 was capable of selective fire but the M60 is designed to give only full automatic fire, although it does have a low cyclic rate which allows the gunner to get off single rounds without too much difficulty. An interesting fact is that the basic FG42 bolt system was itself a further development of the U.S.-designed Lewis gun of World War I fame, which in turn was a modification of a design by McLean, also an American.

Loading and Firing the M60 Machine Gun

Cocking. For all normal purposes this weapon, being fully automatic, should remain cocked at all times. The gun can be cocked only with the safety off or in "F" position. Pull the cocking lever handle, extending from right front of receiver, to its maximum rearward position, where the engagement click of the sear is heard. Return the cocking lever forward to engage its retaining latch. NOTE: Cocking lever will return to its latched position on the first shot, but this practice is not recommended, although it is not harmful if done occasionally. Return the safety lever to the "S" or safe position.

Loading. Rounds should be firmly assembled and positioned in their push-through links. Raise feed cover by lifting cover latch at right rear. Feed cover is retained vertically by a torsion spring and detent. The feed plate should remain in place on the receiver rails. Place the linked belt on the feed plate, links up, with the first round to be fired in the feed plate groove and held by the feed plate retaining pawl engaging the second round. An empty link ahead of the first round will help to position the belt, if desired. Close cover firmly, making sure the cover is latched securely.

Firing. With weapon positioned and aimed, push safety lever forward and up, out of the way, with right thumb. Visual examination will show the lever is directed toward the "F" symbol. When down, near the right thumb, the safety lever is directed toward the "S" symbol. Because of the low cyclic rate (550 r.p.m.) of this weapon, single rounds or short bursts can easily be fired. It is important that the trigger be completely released after each shot, to fire single rounds, or to interrupt firing at any time.

7.62mm Machine Gun M60 on M122 mount. This is the caliber .30 M2 mount with an adaptor for the M60.

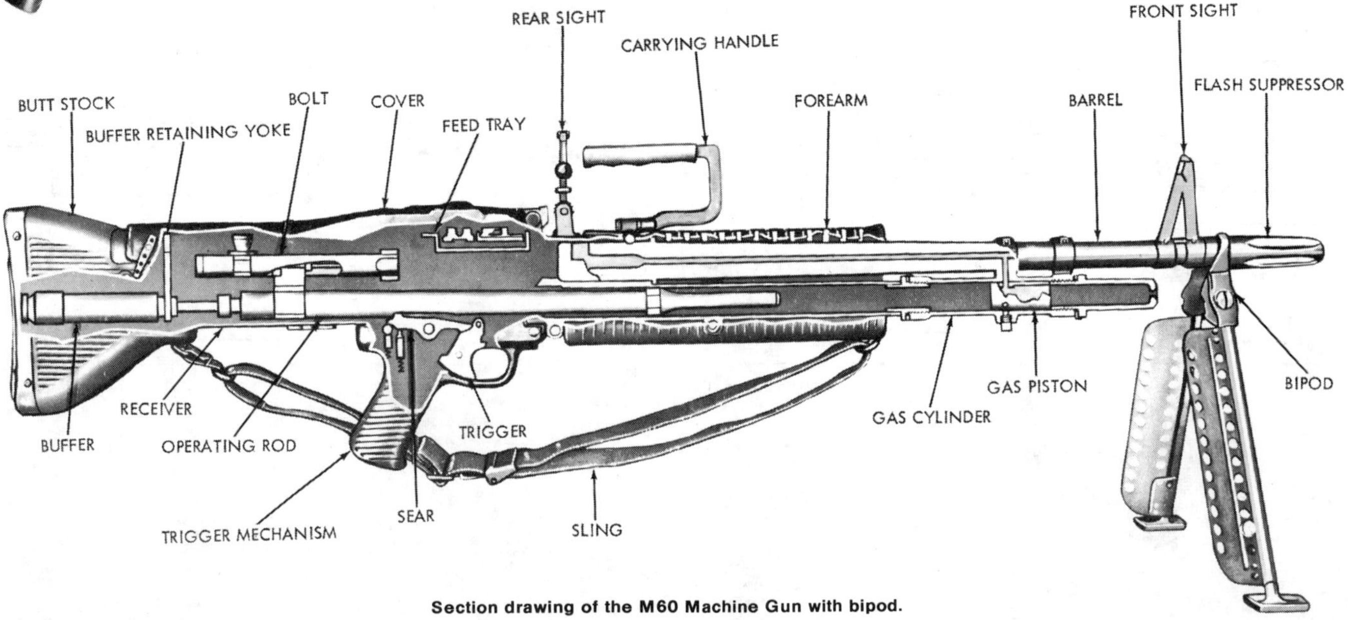

Section drawing of the M60 Machine Gun with bipod.

Reloading. After firing, if the belt has been exhausted, the gun will be closed and must be cocked by hand to reload. The last link of the belt will remain in the feed plate. This link can be removed by the alternate method of loading described above; or, if a leading-link tab is provided, the link can be pushed through by utilizing the tab; or, finally, it can be removed when the cover is open for reloading.

Sights. The weapon as furnished is sighted in at 100 yards. A "quick adjustment" type of rear sight is provided. Barrels are zeroed-in by utilizing the lateral and elevating adjustments on the rear sight. For lateral adjustment, turn the knob at the lower left side of the rear sight. If additional lateral adjustment is required, the spring dove-tail (base) can be moved in its dove-tail on the receiver. The receiver should be restaked to retain the sight in the position permanently. For normal use, the spring action of the sight base is sufficient to hold the sight is place.

For elevating adjustment, a scale slide is provided on the rear sight. Zero the barrel to the aperture; then, after this zeroing is accomplished, the scale can be lined up with the aperture by loosening the lower screw, sliding the scale to its desired location, and tightening the screw securely.

Barrel Changes. Barrels are changed by raising the lever found at the extreme right front of the receiver to the vertical position and withdrawing the barrel. Use the bipod as a handle. Assemble the barrel by inserting from the front, aligning the gas cylinder pilot nut with the receiver extension tube. Press the lock lever down and rearward to the full limit of travel.

Stripping the M60

For ease in cleaning, oiling, and inspection, this weapon can be taken down as follows, using the cartridge as a tool. With gun in closed or fired position:

Unlatch and raise feed cover.

Remove butt stock by lifting shoulder rest and insert cartridge in exposed latch hole. Withdraw butt rearward.

Lift lock plate vertically for removal, holding the buffer to prevent drive spring from ejecting it. Withdraw buffer.

Pull operating rod drive spring guide and spring out of the rear opening. Withdraw operating rod and bolt assembly rearward with the cocking lever, pulling the bolt out of the receiver by hand the remaining distance. As the bolt rotating cam is exposed, insert lock plate in front of cam opening to hold firing pin back. The operating rod can then be withdrawn from the bolt.

CAUTION: Roller bearing and pin retainer are staked in place and should not be removed. These parts are normally also held in place by the firing pin; therefore, care should be exercised to prevent their loss.

Remove trigger housing assembly by pressing spring lock flat against the trigger housing at extreme front pin to unlock it from the pin. Rotate the spring lock down and away from the receiver to disengage it from the pin, and withdraw it. Remove the front pin from the left. Remove trigger housing by sliding it forward to disengage it from the receiver.

Feed cover removal is not necessary for field stripping but can be done if desired. With feed cover raised vertically, and feed plate in place on the receiver (at approximately 90° with cover), use bullet nose to remove cover pin spring, and withdraw cover pin spring and cover pin by hand, rocking cover slightly to aid disengagement. Cover and plate are free for removal. A torsional counterweight spring is also retained by the pin, and is removable when pin is removed.

Remove barrel by raising lock lever to its vertical position and pulling the barrel out in a forward direction.

Handguard removal is not necessary for field stripping, but can be done if desired. Grasp the handguard firmly. Insert a bullet nose in the hole provided at the bottom rear of the handguard.

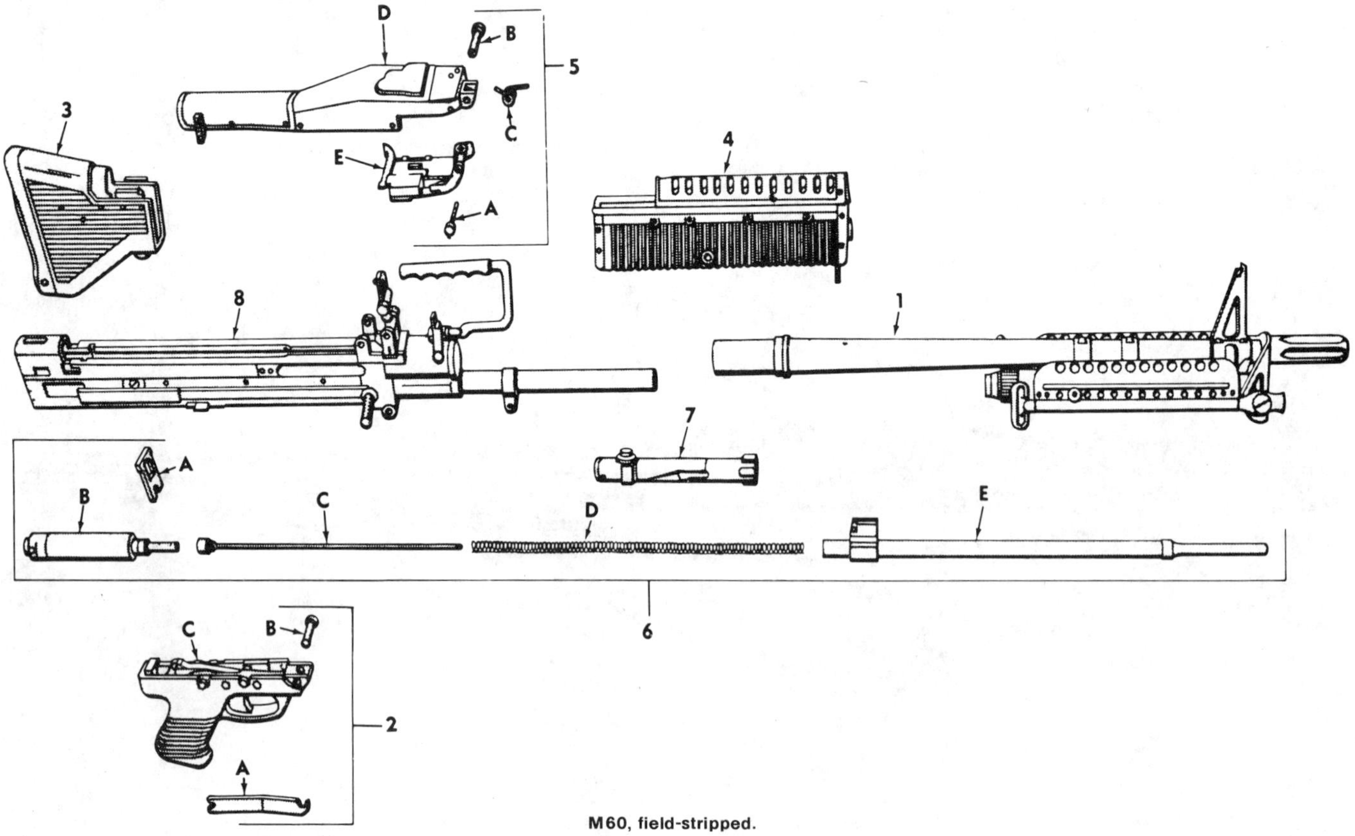

M60, field-stripped.

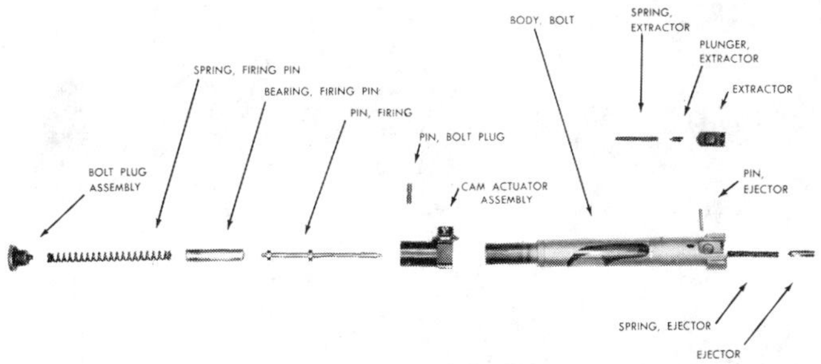

Bolt assembly of the M60.

Push cartridge to disengage spring lock. Slide handguard off to front. It is helpful to position the receiver vertically, with the front extension tube resting on a flat hard surface when disengaging the lock. After unlocking and sliding the handguard forward to disengage it from the receiver, tip the handguard to clear the tripod clevis and remove.

The weapon can be reassembled by reversing the order of the above steps.

How the M60 Works

With the gun loaded, in the cocked position, the following is the sequence of operation:

Release the safety, and pull the trigger. The trigger raises the front of the sear, dropping the rear of the sear out of its engagement notch in the operating rod.

The operating rod, released by the sear, is propelled forward by the energy stored in the drive spring by the cocking operation. The bolt, engaged by the operating rod cam yoke is carried forward with the rod, as is the gas piston, should it be positioned to the rear.

The bolt top locking lug strikes the rear edge of cartridge in the feed plate groove, where the cartridge is positioned by the belt link and the feed cover guides and springs. The bolt strips the cartridge from its link, carrying the cartridge forward out of the feed plate. The empty link is retained in the feed plate by the cartridge guide and the next round.

As the bolt enters the barrel socket, the round has been deflected downward and into the chamber by the front cartridge guide, the receiver feed ramps, and the barrel socket feed ramp. The barrel socket lead cams impart clockwise rotation to the bolt sufficient to enter the locking lugs into the socket locking cam. At this point the round is seated in the chamber, its base contacting the extractor and ejector. The forward motion of the operating rod and bolt compresses the ejector spring, and moves the extractor out to snap over the rim of the cartridge.

The firing pin, held back by the bolt cam and its engagement with the operating rod, has also compressed its drive spring during cocking. As the bolt rotates, its rotating cam releases the firing pin. The firing pin spring now contributes its energy to the operating rod through the bolt rotating cam and aids the drive spring to complete the bolt locking rotation. The operating rod continues forward, carrying the firing pin with it until the pin strikes the cartridge primer and ignites it. The firing pin is stopped by its seat in the bolt at the pin front bearing spool, stopping the operating rod. The operating rod has also positioned the piston in its forward position.

After ignition of the powder charge by the primer, the bullet is forced down the bore by the gases released. At a point about 8 inches from the muzzle, the bullet passes the barrel gas port, allowing part of the gases under pressure to enter the gas cylinder and hollow piston through their respective ports, which are aligned with the barrel port.

Gas under pressure, bled from the bore by the gas port, fills the hollow piston until enough pressure is built up to overcome the mass of the piston and operating rod and the load of the operating rod spring. The piston begins its travel rearward, propelled by gas pressure against the forward end of the cylinder. After a very short travel rearward, the piston ports and collector ring are out of line with the barrel gas port, sealing off a measured charge of gas under pressure in the cylinder and hollow piston. By this time, the bullet is well out of the bore, and gas pressure in the bore rapidly falls to help provide easy extraction of the spent round.

The piston, propelled rearward by the trapped expanding hot gases, forces the operating rod rearward to contact the firing pin rear spool. This compresses the firing pin spring withdrawing the firing pin from the cartridge primer as well as compressing the operating rod spring. Rearward motion of the operating rod continues in the "dwell" slot of the bolt rotating cam, until the rod yoke rollers contact the spiral, or counterrotating part of the bolt cam.

The bolt now begins its counterclockwise rotation and unlocking, urged by the energy of the operating rod. At full unlock, the firing pin spring is fully cocked, the roller of the rod cam yoke contacting the rear of the bolt rotating cam.

Bolt and rod together continue to the rear, propelled by the energy of the expanding gas, withdrawing the spent cartridge from the barrel chamber. As the spent cartridge leaves the chamber it is moved sideward in the direction of the ejection port, the ejector spins the case sideward to pivot about the extractor lip and disengage, allowing the case to be ejected with force out of the port, and against the ejection deflector which propels it downward.

In this interval the piston has reached the limit of its travel. Spent gases are exhausted through ports provided in the gas cylinder, uncovered by the piston in its rearward travel.

Air behind the piston is exhausted through a set of ports near the rear of the gas cylinder, allowing escape of dirt and powder residue.

The bolt and operating rod continue rearward, propelled by stored energy and inertia, compressing the operating rod drive spring. As the drive spring is fully compressed, the operating rod contacts the buffer through the drive spring guide rod collar. Energy remaining in the rod and bolt is now transferred to the buffer. As the buffer plunger is forced rearward against its pads and preloaded springs, the rubber pads are compressed, acting as a high-rate spring. The rubber is forced to flow radially, to expand frictional surfaces tight against the buffer tube wall. Further rearward motion compresses the buffer return spring and the sliding friction of the pads on the tube wall absorbs the remaining energy of the rod and bolt. The moving parts now come to rest, and the counter-recoil cycle begins.

The energy stored in the drive spring now forces the operating rod and bolt forward. The frictionally damped buffer springs return the buffer plunger to its initial position. The firing pin spring also contributes some force until it is halted by the front of the rod yoke, which stops the firing pin in the "cocked" position in the bolt rotating cam. As the rod continues forward with the bolt, if the trigger has been released, the sear is held up against the bottom of the rod by the sear spring. As the sear engages the sear notch in the rod, forward motion stops, and the weapon is held in cocked position, ready for fire. If the trigger has not been released, the weapon continues to fire until interrupted by the sear, or until the ammunition is exhausted. If the last round is fired and the trigger held down, the gun will close on an empty chamber, and must be recocked manually.

Feed Cycle. For simplification of description, this portion of the operation is considered separately. As the bolt travels forward for firing, the feed cam mounted in the feed cover is engaged by the feed cam actuator roller attached to the rear of the bolt. Forward motion of the bolt and actuator causes the feed cam to swing to the right, forcing the front of the feed cam lever to the left. This carries the cartridge feed pawl plate assembly to the left where the pawls drop over and engage the next or second round for transport, and remain there until the round in the plate is stripped and fired.

As the bolt recoils, the actuator forces the feed cam to the left, and the feed pawl assembly transports the round to the right into the feed plate groove. The round is forced down into the groove by the cartridge guides and their springs. The empty link left on the plate is pushed out the port in the feed plate by the new round as it is fed. The feed plate contains a spring-loaded retaining pawl that retains the belt, holding the second round when the first round is in the plate groove. Two anti-friction rollers help guide the belt in place, and support the hanging belt. The weapon is provided with sufficient reserve power to lift a 100-round belt vertically under all normal operating conditions.

Special Note on the M60 Machine Gun

The M60 is the first United States machine gun to have a true quick-change barrel. It was specifically designed for light weight, and components design was simplified for manufacture. Stampings or fabrications were used wherever possible. The quick-change barrel, light weight, and adaptability for use as either a heavy or light machine gun, as well as relative ease of manufacture, make the M60 superior to the Browning guns it is replacing. The performance of the lined and plated barrel of the M60 in sustained fire is really exceptional. United States experience in the Korean war with the Chinese Communist "human wave" type of tactics indicated the need for a weapon capable of long periods of sustained fire. The old water-cooled Browning M1917A1 stood up fairly well, but it is far too heavy and bulky, and it lacks tactical flexibility. Moreover, it is not inconceivable that, at some time in the future, United States troops might be forced to fight in an area where water would be worth its weight in gold. An added factor of advantage in the case of the M60 versus the Brownings is that the M60, like most other weapons with a quick-change barrel, has no headspace adjustment problem.

The M60 has been modified somewhat since original manufacture. The receiver has been strengthened by the addition of several pins and the feed tray now has a hanger assembly pinned to it. The hanger assembly is used with a 100 round ammunition box, for some peculiar reason called a bandoleer in some publications. The box or bandoleer is merely slipped down over the hanger and makes it relatively easy to move the gun around with ammunition in place ready to fire.

7.62mm M60E1 MACHINE GUN

A modified version of the M60 machine gun, the M60E1, has been developed. The principal reason for the development is to simplify barrel change and decrease the number of parts. This weapon differs from the M60 as follows: (1) The barrel does not have the bipod or gas cylinder attached to it. They are attached to the gas cylinder tube; (2) The bipod is attached semi-permanently to the rear of the gas cylinder; (3) The gas cylinder has been simplified and has no threads. It has a "U" shaped key to retain the gas cylinder extension; (4) The operating rod guide tube has a lug which retains the gas cylinder and bipod on the weapon and eliminates the gas cylinder nut; (5) The modified spool type gas piston has no holes; (6) The modified rear sight has the lateral adjustment increased by 20 mils; (7) The modified die-cast feed cover eliminates parts and allows the cover to be closed whether the bolt is in the forward or cocked position; (8) The modified feed tray eliminates parts; (9) The magazine hanger fitted to the left side of the weapon, eliminates parts and can be used either with a modified magazine or a modified bandolier; (10) The new die-cast forearm eliminates parts and eases changing the barrel because of the absence of the forearm cover allows the carrying handle to be fitted to the barrel; (11) The sling swivels have been relocated to the left side of the forearm and the top rear of the buttstock easing the carrying of the weapon; (12) The carrying handle has been increased in diameter.

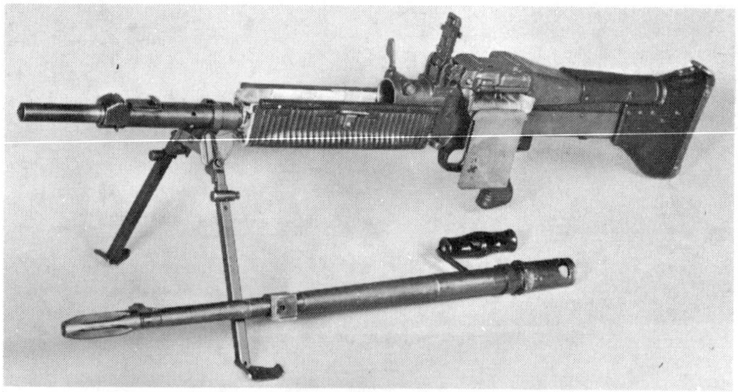

7.62mm M60E1 Machine Gun with barrel removed.

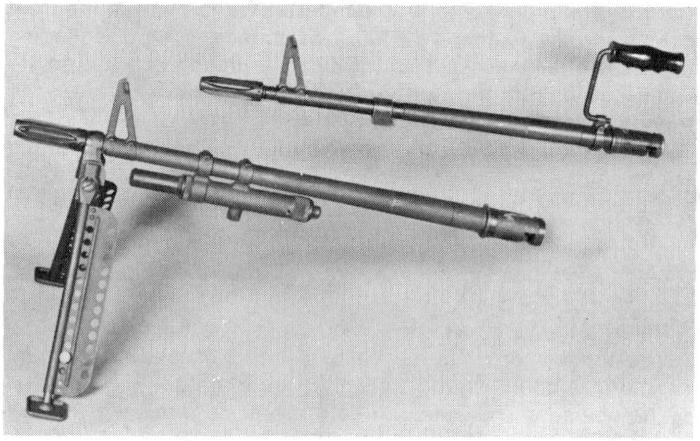

Top: barrel for M60E1; bottom: barrel for M60.

M60C AND M60D

The M60C is a modification of the M60 machine gun for use on helicopter armament kits. The weapon has the stock removed and is remotely charged and fired. The M60D also has the stock removed and the pistol grip as well. The trigger has been relocated to the rear and spade type grips have been fitted. This gun is used on flexible pedestal type and other mounts on helicopters and gunships.

The Maremont Corporation, manufacturers of the M60, has also developed modifications to the weapon. A tank version has been developed with a barrel extension and a gas evacuator tube which protrudes beyond the gas cylinder and taps off the gas from the gas cylinder, preventing most of it from going rearward in the gun. The purpose of these two parts is to cut down on the quantity of gas brought into the interior of the armored vehicle.

THE COLT 5.56mm (cal. .223) CMG-1 MACHINE GUN

The CMG-1 is a gas-operated weapon which can be used on a tripod as a medium machine gun, on a bipod with a stock attached

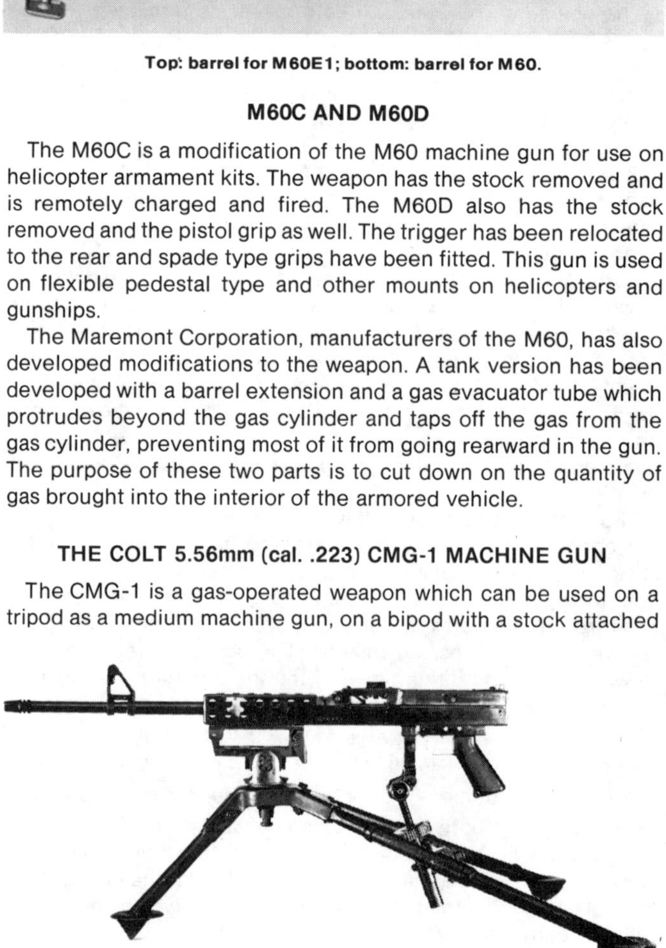

CMG-1 5.56mm Machine Gun; weapon shown is a prototype.

as a light machine gun, without sights, pistol grip, or stock—to be fired by a solenoid—for fixed positions in vehicles or aircraft, and with pistol grip for flexible vehicle applications.

The CMG-1 gun weighs 12.5 pounds in its medium machine gun configuration, 15.4 pounds in the light machine gun configuration, 14.7 pounds in its fixed gun (solenoid fired) configuration, and 13.6 pounds in its flexible vehicle configuration. The overall length of the gun varies from 40.7 inches for the light machine gun to 32 inches for the medium and flexible vehicle configurations. Barrel length is 20 inches in all cases and the muzzle velocity is 3250 feet per second. Cyclic rate of fire for all versions except the fixed gun version, is 550-650 rounds per minute. The cyclic rate for the fixed gun is 650-850 rounds per minute. The medium gun is used on the M2 tripod.

COLT 5.56mm CMG-2

Colt has developed a new light machine gun. This gas operated weapon with a quick change barrel is fed by a disintegrating link belt. It has a quick change barrel and a vertical fore-grip. The butt stock can be removed and the weapon fitted with a short barrel and the weapon can be used as a submachine gun. It can be fitted with a magazine that holds 150 linked rounds.

The CMG-2 with bipod weighs 15.1 pounds and is 41.9 inches long. This weapon delivers automatic fire only at a cyclic rate of 650 rounds per minute. The design of the CMG-2 is similar in some respects to the Czech ZB 53, Model 52 and Model 59. As with these weapons, the weapon is cocked by pushing the pistol grip forward and pulling it, and the slide which is engaged by the sear, to the rear. The feed mechanism is also similar to the Czech guns in that the feed pawls work off cams on the bolt carrier. The multi lugged bolt is cammed into and out of the locked position by a cam follower working on a camway in the bolt. The CMG-2 has two driving springs.

CALIBER .30 TANK MACHINE GUN M37

The M37 is basically a modification of the Browning M1919 series of weapons. It was designed to produce a weapon which would have more usefulness in tank mountings than did the Browning M1919A4 and A5.

Notes on the M37 Tank Machine Gun

The principal differences between the M37 and the Browning M1919 series of guns are given below.

The M37 can be fed from either the right or left side. The bolt has a dual track for the belt feed lever stud. By positioning two switches, the ejector, and various feed components, the weapon can be easily changed to feed from either side. This arrangement is generally similar to that found on the Browning cal. .30 M2 aircraft gun and the cal. 50 M2 and M3 aircraft guns.

The cover catch has been changed so that it can be opened easily from either side.

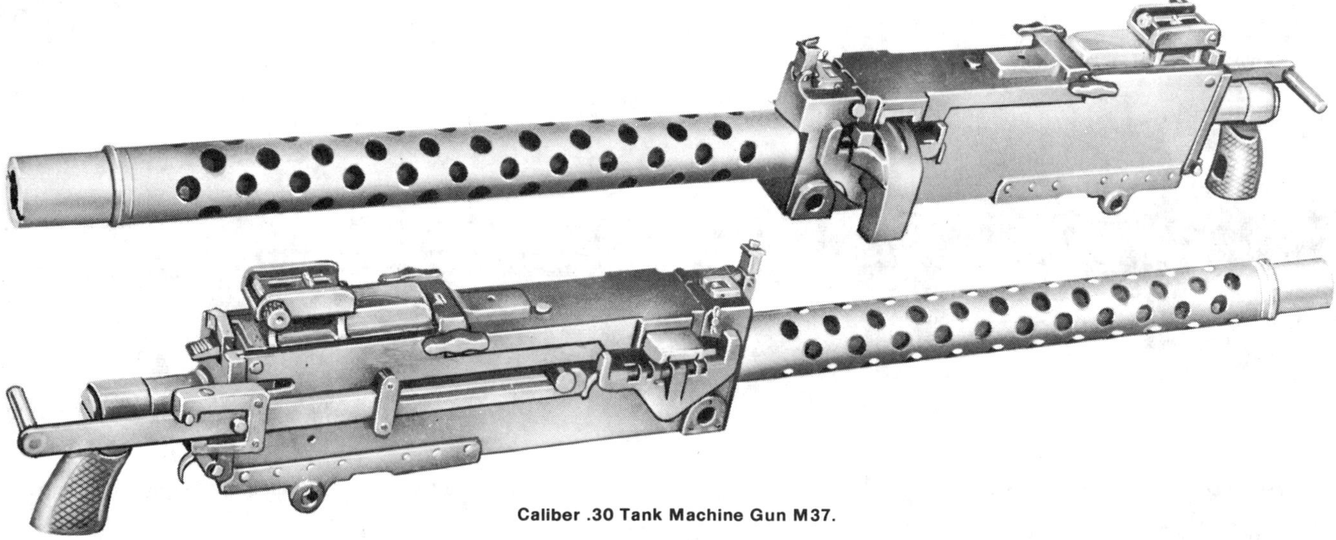

Caliber .30 Tank Machine Gun M37.

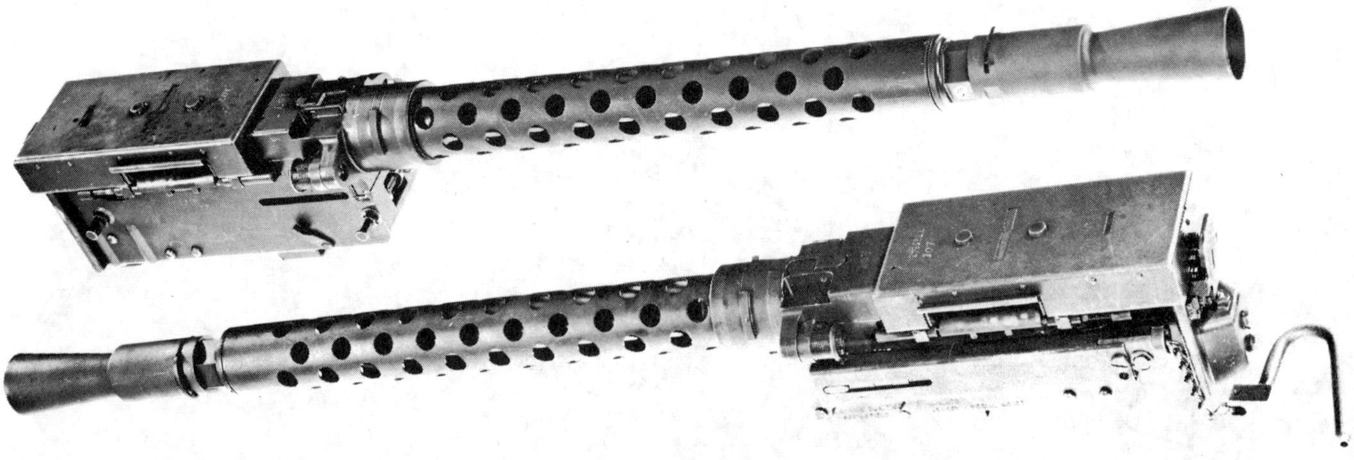

7.62mm Tank Machine Gun M73.

Backplate removal has been simplified by the addition of a new type of backplate latch.

The driving spring assembly has been made into a packaged unit similar to that used on the cal. .50 Browning gun, and a stranded wire spring is used.

A metal link ejection chute has been added; this component is changed from side to side when the feed components are changed.

The functioning of the M37, as well as its field stripping, loading, and firing, is essentially the same as for the earlier M1919-series guns. An earlier modification of the Browning M1919 series, the cal. .30 M1919A4E1, fed from only one side. It was fitted with a retracting bar, as is the M37, for ease of charging the gun in a tank coaxial mount. Both the M1919A4E1 and the M37 can be used as ground guns on the M1917A1 or M2 mount.

THE 7.62mm NATO M73 TANK MACHINE GUN

The development of the M73 (formerly known as the T197E2) was the result of the need for a rifle-caliber machine gun designed specifically for use in tanks. It had become evident that the use

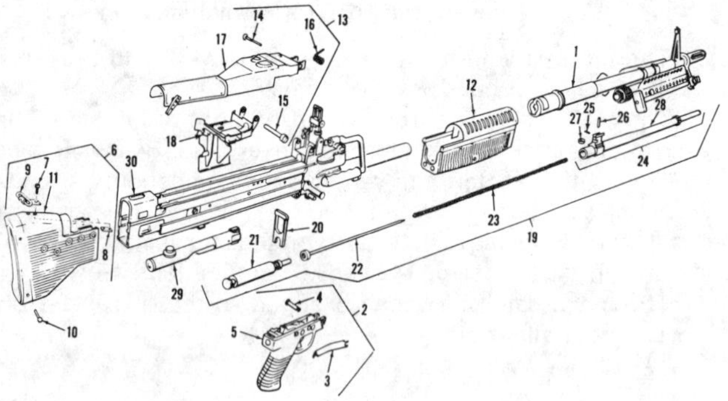

Exploded view of 7.62mm M60 Machine Gun of late manufacture.

Colt CMG-2, field-stripped.

Colt 5.56mm CMG-2 Light Machine Gun.

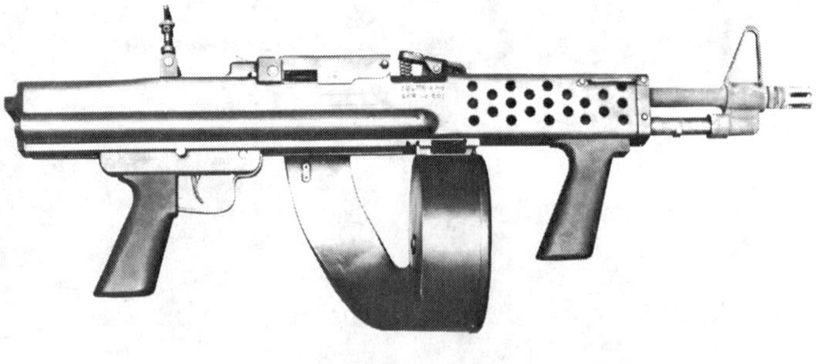

The Colt CMG-2 without butt stock and with short barrel.

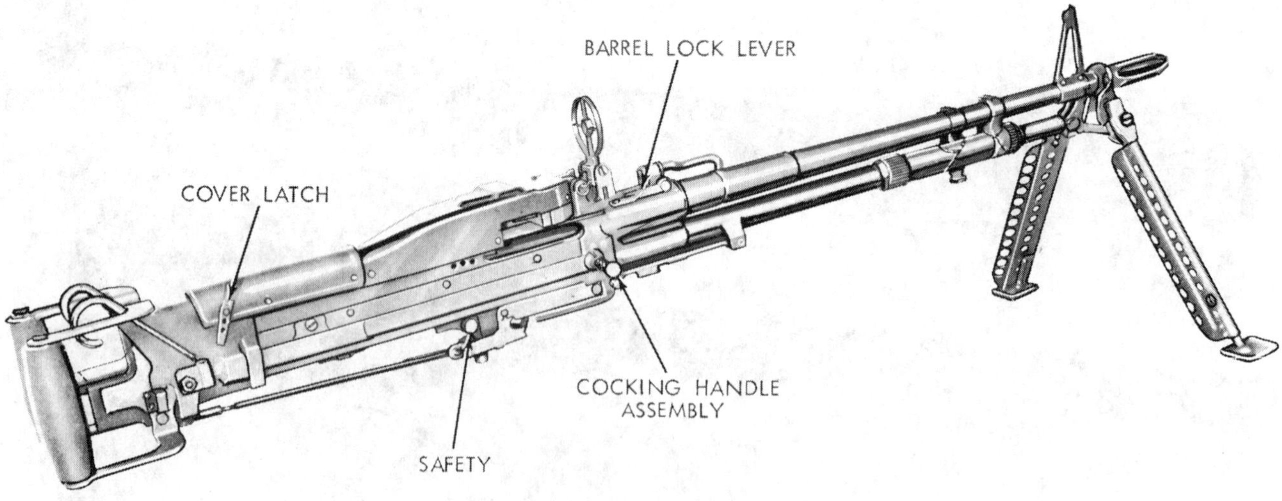

7.62mm M60D Machine Gun.

of the Browning-type guns was not as satisfactory as it could be. The outstanding shortcoming of the Brownings for use in tanks was their long receivers. Space is at a premium inside the turret and hull of a tank. There was a need for a much shorter gun which would at least equal the performance of the Brownings. The M73 has a quick-change barrel, and can be fed from either side. The M73 uses the same link, the M13, that is used with the M60 general purpose machine gun. It can be charged and fired either by manual operation or solenoid. The barrel jacket of the M73 is to be attached to the tank in a semipermanent manner, and all the working components of the weapon, plus the barrel, can be removed from inside the tank for cleaning and repair.

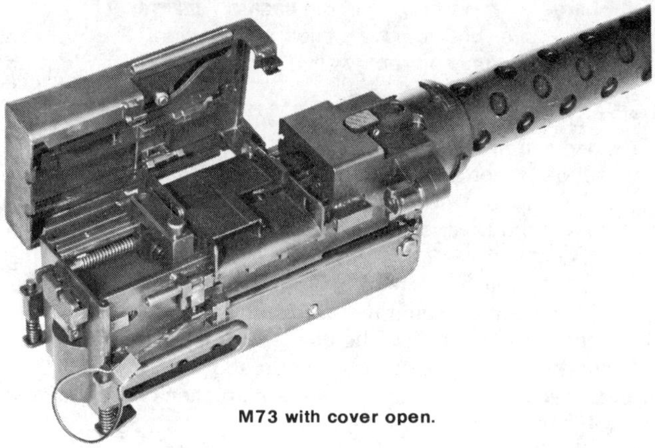

M73 with cover open.

How to Load and Fire the M73

Preparation for Firing. Open cover by pressing the cover latching rod on the side of the receiver from which the belt will be fed. Pivot the cover to an open position.

Retract the action to seared position, using the charging mechanism; make sure that the safety is at the "F" position. NOTE: Mount the charger on either left or right side of the receiver as determined by the mounting conditions of the weapon. Slide the charger connector to the proper side by pressing the retainer and sliding the connector manually.

Quick-change barrel of M73.

Slide the safety to the "safe" position, exposing the letter "S" on the back plate.

Visually inspect the barrel chamber and inside of receiver for assurance against possible obstructions.

To load the weapon: Insert the cartridge belt with the first round in the slot of the feed tray (open side of link loops facing downward).

Press the cover latch rod, close the cover, and release the cover latch rod to the lock position.

Firing. Slide the safety in the back plate to the "fire" position, exposing the letter "F" on the back plate ("S" for safe).

Actuate the solenoid for remote firing.

Press the trigger on the back plate for manual firing.

Immediate Action Procedure (to resume fire after a stoppage). When a stoppage occurs BEFORE COMPLETING A 200-ROUND SERIES (starting from a cool gun), perform the operations listed below in the given order. (If weapon starts after the first operation, do not perform the next, etc.)

(1) Charge weapon fully to sear position and fire.

(2) Repeat above twice if weapon does not respond.

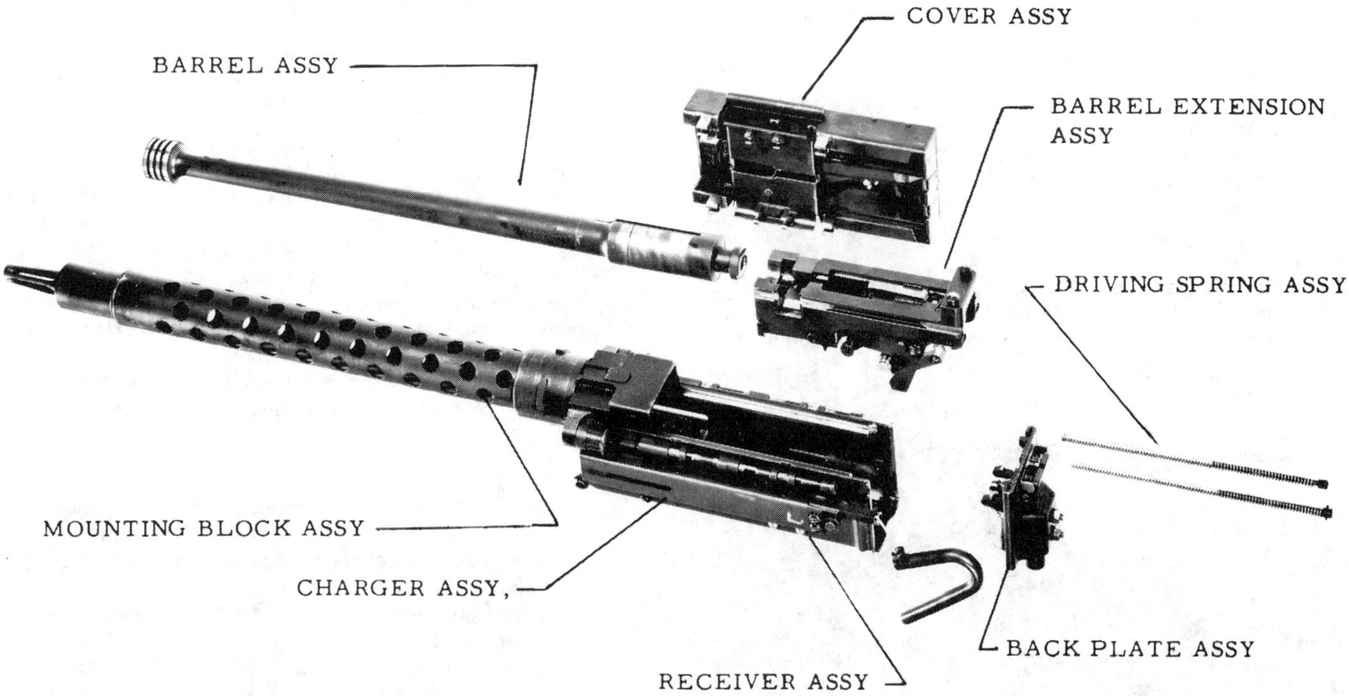

M73, field stripped.

(3) Charge weapon fully and hold back on charger. Then—
(4) Open cover and remove belted ammunition.
(5) Open feed tray and remove live or spent cartridges from the action.
(6) Charge and hand-function weapon.
(7) Load and fire.
(8) If the weapon still does not fire, inspect for broken parts. NOTE: To prevent the possibility of an open-breech cook-off with the cover open, proceed as follows: When a stoppage occurs AFTER FIRING MORE THAN A 200-ROUND SERIES (starting from a cool gun), charge and fire three times. If weapon does not resume fire at this point, charge fully (do not open cover) and allow the gun to cool to near-ambient temperature before opening cover as above.

Feed Change. Observe the following instructions when changing from left-hand to right-hand feed:

Remove the cover assembly from the receiver.

Swing the feed tray to a vertical position relative to the receiver by disengaging it from one of the cover latching rods. Press the plunger of the round stop, and slide the round stop to engage the locating hole on the left side of the feed tray. In this position, the "R" on the round stop will be adjacent to the "R" on the feed tray ("L" and "L" for left-hand feed).

Follow operations listed below to convert the cover assembly from left to right.

Stripping the M73

Actuate the trigger to insure that the weapon is in the forward position.

Press the cover latch rod at either side of the receiver, and raise the cover and feed tray assemblies to clear the latch rod. Release the cover latch rod, and allow the cover and feed tray to rest on the rod. Press the second rod, and remove the cover and feed tray from the receiver.

With action forward, press and rotate counterclockwise the driving spring rods. Withdraw the driving spring and guide rod assemblies through the holes in the backplate.

Slide the backplate assembly vertically from the receiver housing.

Depress the buffer support lever at either side of the receiver, and slide the barrel extension with the barrel to the rear, by pulling the charger handle. Remove the barrel extension and the barrel from the rear of the receiver. NOTE: An alternate method of field stripping is to pull on either receiver disconnector and pivot the receiver down and about the opposite receiver disconnector. Remove the receiver from the mounting block by withdrawing the second receiver disconnector. Proceed as above, and withdraw barrel from mounting block assembly.

Assembly. Assemble the weapon by reversing the procedure given above.

Slide the feed support retainer away from the feed cam.

Slide the feed cam fully forward and lift it out, retaining it in hand.

Lift out the feed support and retain this in the same hand.

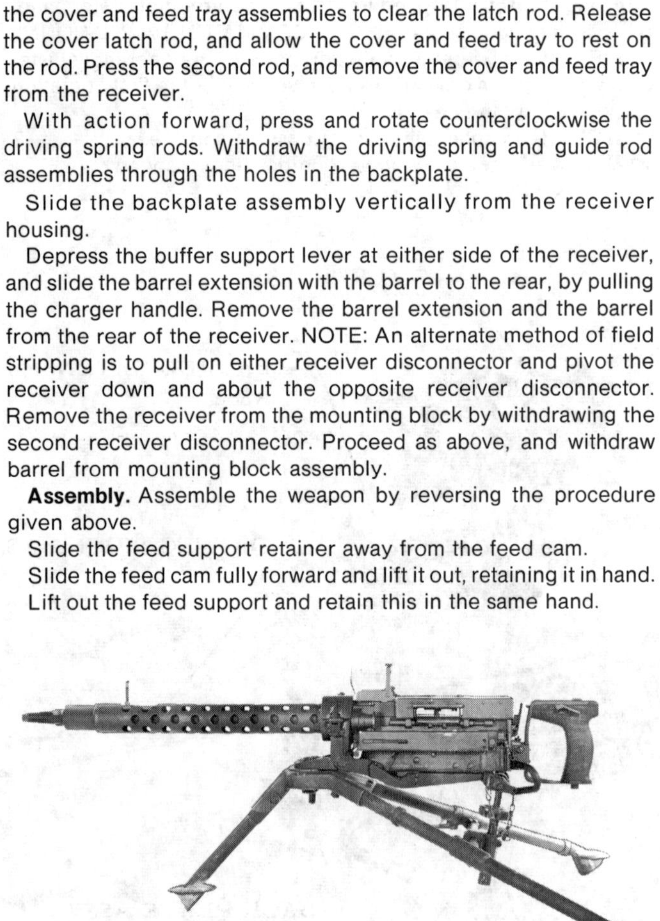

M73C Machine Gun on XM132 Tripod.

Remove the feed track and slide as a unit. Reassemble this unit to the cover after turning it end for end (180°). This will align the "R" of the feed track with an "R" in the cover ("L" and "L" for left-hand feed).

Slide the feed support retainer fully to the opposite side of the cover.

Replace the feed support.

With the "R" side of the feed cam facing up, assemble to the cover in the fully forward position ("R" on cam adjacent to "R" on cover), picking up the feed slide roller during the operation. (Assembly notches prevent improper assembly.) Slide feed cam to rear of cover.

Slide the feed support retainer to the central lock position where it will secure the assembly. NOTE: In changing from right-hand to left-hand feed, reverse the foregoing procedure.

Reassemble the cover to the receiver by placing it in proper position and pressing the two cover latching rods.

How the M73 Works

Operational Power. The energy of recoil is supplied by the momentum of the recoiling parts and a muzzle booster driving the action rearward to buffer contact. Counterrecoil energy is supplied by the driving springs and the buffer spring that were compressed during the recoil cycle. The open-bolt action employs the driving springs alone and counterrecoil for the first shot from the sear position.

Recoil Movement. During the rearward movement of the barrel extension, the attached lever linkage actuates the rammer assembly by means of opening closing cams located in the sides of the receiver. The rearward movement of the barrel extension enables the sliding breechblock to move transversely to the right and away from the base of the chambered cartridge case. The extractor with the rammer grips the rim of the case and removes the spent round from the chamber, carrying it rearward to engage the round carrier grips. At this point, the spent case is transferred from the extractor to the round carrier grips. In the interim, the hammer which is assembled to the barrel extension is cocked by the cocking cam and is secured in the seared position by the sear assembly. At this stage of the recoil cycle, the ammunition will have been fed into the path of the retracted rammer assembly by the barrel extension and the connected feed cam in the cover assembly. The buffer assembly that is attached to the receiver trunnion block limits the rearward travel of the recoil mechanism.

Counterrecoil Movement. Energy of the driving springs and buffer return spring forces the barrel extension forward. During this movement, the rammer assembly strips and chambers the next round; the round carrier transports the empty case downward where it is dislodged by a fixed ejector; the breech assembly locks the next round in position; the rate control slide is released to actuate the hammer sear and allow the hammer to fall on the firing pin extension. The forward motion of the barrel extension is limited by the trunnion block in the receiver.

Firing will cease when the trigger is released and the barrel extension is engaged in the open-bolt position by the sear.

Special Note on the M73

As the result of field experience a number of changes have been made to the M73 machine gun. They are as follows:

(1) A new front barrel bearing, jacket booster assembly and flash hider assembly have been fitted.
(2) New right-hand and left-hand round (cartridge) carriers have been fitted.
(3) A new case carrier link assembly has been fitted.
(4) A new retainer lock has been fitted.
(5) A new sear hammer has been fitted.

(6) Modified springs have been fitted to the barrel extension, the case carrier assembly, and the trigger.
(7) The barrel has been modified.
(8) The firing solenoid lever has been modified.
(9) The trigger sear has been modified.
(10) The feed pawl has been modified.
(11) Rings and a washer have been added to the barrel disconnector to ease barrel removal.

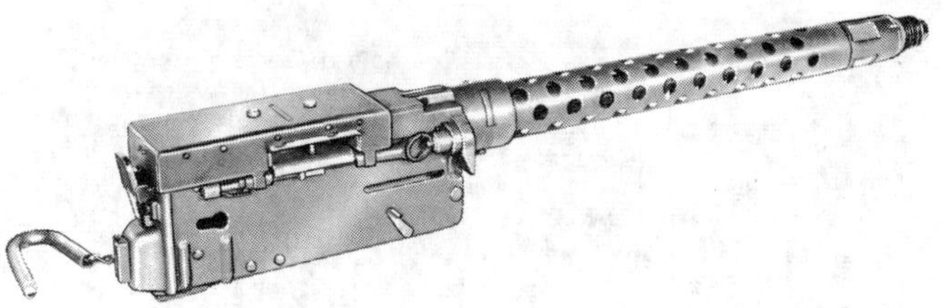

7.62mm M73E1 Machine Gun.

7.62mm Machine Gun M73C on XM132 Tripod

This is the flexible version of the fixed M73 tank machine gun. It is basically the same gun as the M73, with sights and a pistol grip trigger added. The solenoid, which is normally used to fire the gun in a tank, is integral with the back plate and remains on the M73C. The weapon is loaded as the M73 and is fired as follows:

Slide the safety in the back plate to the fire position—letter "F" exposed. Push down on trigger.

The XM132 tripod mount is the caliber .30 M2 mount with a special adaptor for the M73C. This weapon exists only in prototype form.

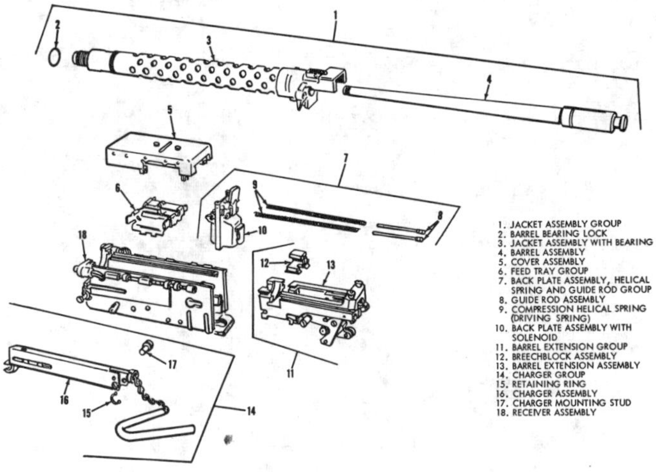

Exploded view 7.62mm M73E1 Machine Gun.

7.62mm MACHINE GUN M73E1

The M73E1 is a simplification and product improvement of the M73. The basic difference in design of the two weapons is in the ejection system. The ejection system has been simplified considerably in that the cartridge case carrier mechanism used in the case ejection cycle of the M73 has been replaced by fixed ejectors located on the underside of the feed tray. The rammer assembly, buffer rod, buffer support tension spring and receiver are modified to be compatible with the fixed ejectors. It is anticipated that the M73E1 will replace the M73.

BROWNING CAL. .50 HEAVY BARREL M2 MACHINE GUN

The caliber .50 M2 Heavy Barrel is the ground gun of the M2 Browning series, which also include the aircraft and water-cooled antiaircraft guns. A turret-type version of the M2 heavy barrel also exists; it was mainly used on multiple A.A. mounts. The aircraft and water-cooled antiaircraft guns are not too frequently encountered at present, but the heavy barrel is still in wide use throughout the world.

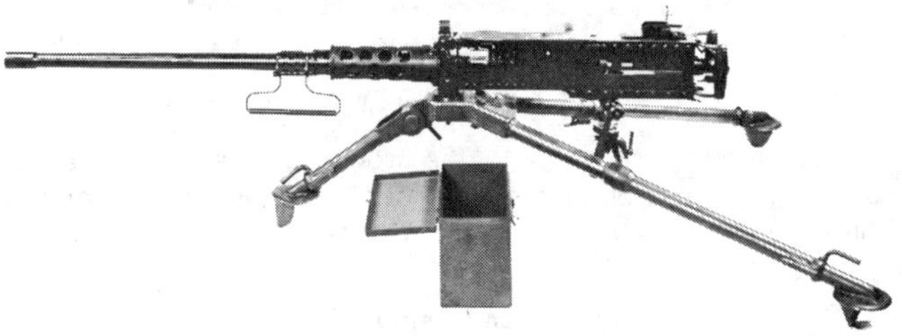

Caliber .50 Browning M2 Heavy Barrel Machine Gun.

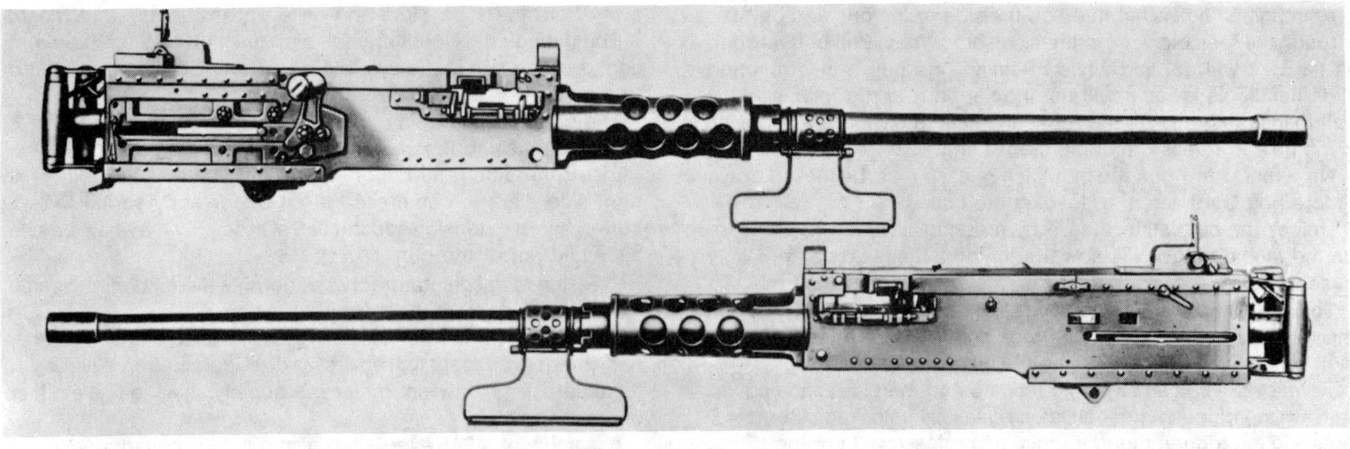

The M2, Heavy Barrel, Flexible.

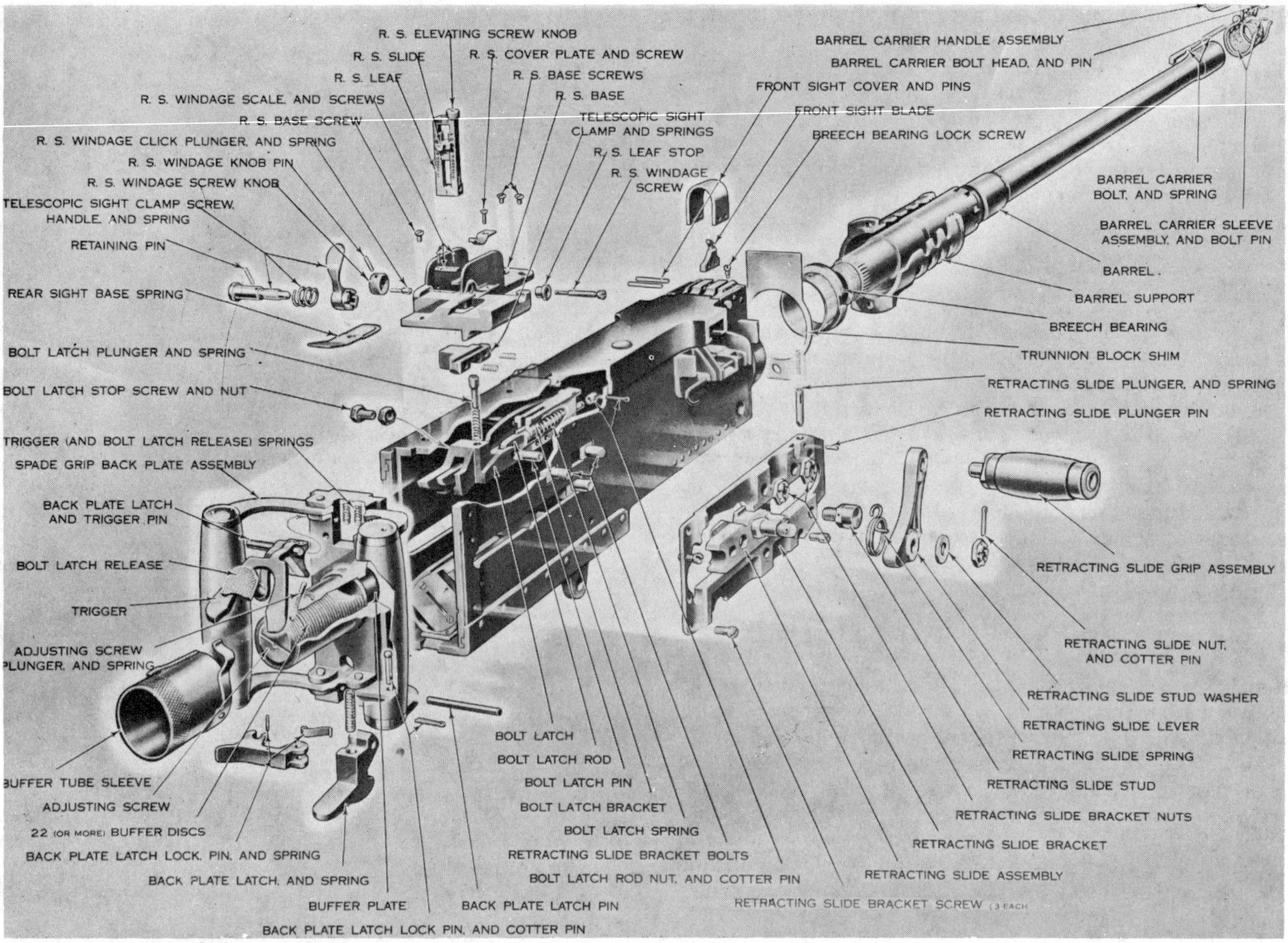

Heavy barrel adaption.

Tripod Mount

A tripod mount is provided for the Browning machine gun caliber .50 HB M2. The tripod assembly weighs about 40 1/2 pounds, while the pintle and elevating mechanism assemblies weigh another 4 pounds.

Loading and Firing the M2

To provide for mounting in aircraft or in vehicles where position or space available require a right-hand feed, this gun is fitted with a bolt and feed mechanism which is inter-changeable for right or left hand feed.

Ammunition box is mounted on the side of the gun and belt fed through the feed block from the side set for feed. Pull belt through as far as it will go, and while retaining grip pull back retracting slide handle as far as it will go and permit it to run forward. This half-loads the gun. Pull the retracting handle back again as far as it will go and release to complete loading.

Note that in this gun the bolt latch release must be locked down before bolt is retracted for loading the gun.

Unlock the bolt latch release by pressing down on it. Pressure on the thumb trigger will now fire the gun. It should be fired only in short bursts.

To Unload the M2. Lift the cover and remove the belt. Pull back the retracting slide handle and look and feel in the feedway, the slot, and the chamber to be sure the gun is unloaded.

Release the bolt and let it go forward and then lower the cover.

Press the trigger. If the bolt latch release is unlocked, alternately pressing the trigger, then the bolt latch release will fire the single shots.

If the bolt latch is locked down and the trigger pressed and held, the gun fires until the trigger pressure is released.

Field Stripping the M2

Grasp the barrel handle firmly and unscrew until the barrel is free from the barrel extension, then withdraw it to the front.

Release cover latch and raise cover as far as it will go.

Release back plate latch lock and also the back plate latch; this will permit the back plate to be lifted up out of the top of the receiver.

Push the protruding end of the driving spring rod forward and away from the slide plate and ease out the spring and the rod.

Pull the bolt back until the bolt stop lines up with the hole in the center of the slot in the side plate and then pull the bolt stud out to the right.

The complete bolt may now be removed from the rear of the casing. Driving spring unit need not be removed.

Insert the point of a bullet in the small hole at the rear of the right side plate to compress the oil buffer body spring lock. Oil buffer, barrel extension and the barrel assembly may now be taken back and out of the gun.

Pressing the accelerator forward permits the oil buffer assembly to be detached from the barrel extension.

This completes the field stripping. Cover should not be removed or dismounted except for repairs, as considerable force is required to compress the pawl spring for reassembly, making this a difficult operation.

Assembly of M2. Reverse the dismounting procedure.

Insert oil buffer into the oil buffer body from the rear, making sure that the cross groove in the piston is on the upper side where it can engage the shank of the barrel extension.

Assemble the buffer and buffer body to the extension. Holding the accelerator up under the barrel extension shank, start the breech lock depressors into their guideway in the barrel extension, and press forward permitting the shank of the barrel extension to engage the cross groove in the piston rod. Thrust sharply forward as far as oil buffer will go. The parts will now lock together and may be assembled into the receiver as a single unit. Press forward until the oil buffer spring locks in its recess in the right side plate.

Insert extractor in bolt and check that cocking lever is fully forward. Then insert bolt into rear of receiver. Press the rear end of the bolt down to elevate the front end just enough to clear the accelerator, otherwise the accelerator will be tripped and will not permit the bolt to be moved forward. When the accelerator has been cleared, raise the rear of the bolt to clear the buffer body. To do this it will be necessary to raise the bolt latch by reaching under the rear of the top plate with thumb or finger of one hand, while the other hand pushes the bolt forward. Bolt latch must be kept in raised position until rear of bolt passes in front of it. If this is not done, its spring will force it downward and engage the notch in the rear of the bolt preventing the bolt from going forward.

Push the bolt forward until the bolt stud hole lines up with the hole in the slot in the side plate; then insert the bolt stud until its shoulders are inside the side plate.

Insert the driving spring rod assembly and push the bolt all the way forward and keep the stud at the rear end of the driving spring rod at the recess in the right side plate.

Holding out the back plate latch lock, insert the back plate from the top. Press the trigger.

Make sure that the bolt is fully home, then close the cover. If the bolt is not fully forward, the feed lever will be forced down in front of the bolt which may result in malfunctioning.

Holding barrel by barrel handle, use both hands to insert it carefully in the front end of the barrel support. Guide the rear end over the breech bearing until it contacts the threads of the barrel extension, then screw it in until definite resistance is met. Now back off two notches to make headspace adjustment.

Headspace Adjustment of M2

As in the case of the Browning cal. .30 machine gun, the headspace adjustment is the most important adjustment on this gun. It is not necessary to remove the barrel to make this adjustment on the caliber .50.

Remember that headspace means that space between the rear end of the barrel and the front face of the bolt; and that if this space is too wide the gun will function sluggishly or not at all and may pull the head away from the cartridge case causing serious jams; while if it is too tight, the recoiling parts will not go fully home and the gun may refuse to fire.

Screw the barrel up tight into the barrel extension, then pull back the slide and let bolt go forward to test the action.

If the action does not close fully, unscrew the barrel one notch. Then the test should be made again by pulling the bolt back and then letting it go forward.

The barrel may be unscrewed a notch at a time by pushing with the point of a bullet to rotate the barrel when the cover is raised and the bolt is in rearward position.

Work the bolt by hand several times and if the breech does not close without effort, unscrew the barrel one notch.

Raise the cover and lift the extractor, then pull the bolt slightly to the rear. If it moves independently of the barrel extension, the adjustment is too loose. Screw the bolt up one notch and then repeat the test. When a dummy cartridge is in the chamber, there must be no rearward motion of the bolt independent of that of the barrel extension before the unlocking action takes place.

Headspace test should be made whenever gun is prepared for firing.

How the M2 Works

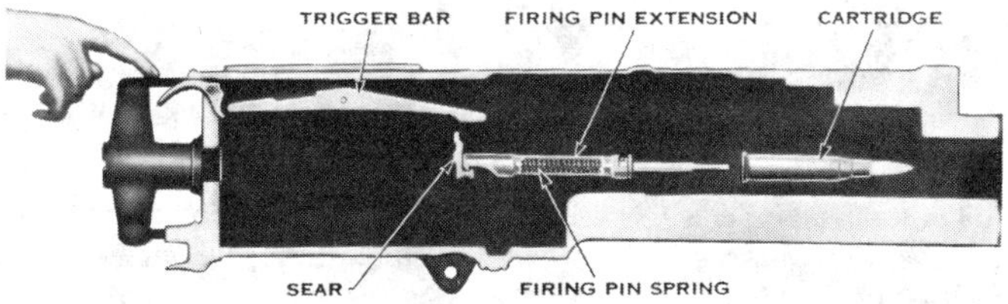

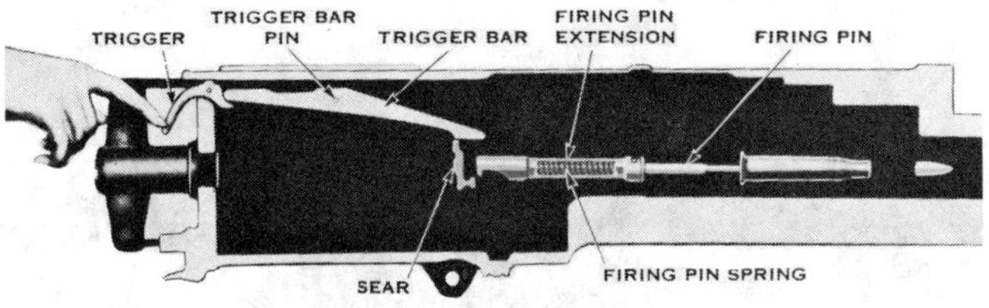

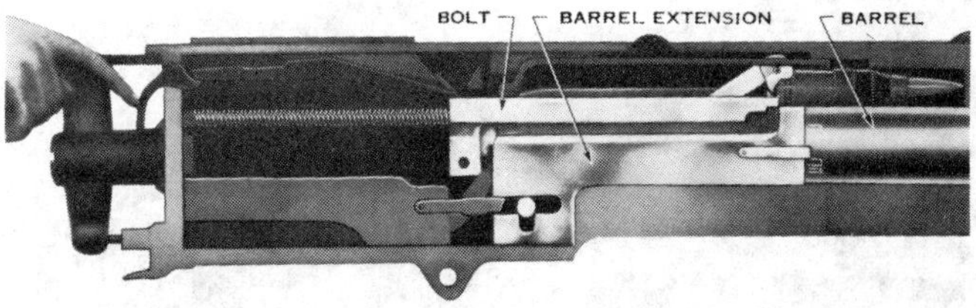

Starting with the gun loaded and cocked, the action is as follows: Pressing the trigger raises the back end of the trigger bar which pivots on the trigger bar pin and presses its front end down on the top of the sear. The sear is forced down until its notch disengages from the shoulder of the firing pin extension. This permits the firing pin extension and the firing pin to be driven forward by the coiled firing pin spring. The firing pin strikes the primer of the cartridge and explodes the powder.

Recoil Action. As the bullet starts down the barrel, the rearward force of the recoil drives the securely locked recoiling mechanism directly to the rear. During this initial motion, the bolt is supported securely against the base of the cartridge by the breech lock which rides up from the barrel extension into a notch in the underside of the bolt.

After a travel of about 3/4", during which time the bullet has left the barrel, the breech lock is pushed back off its cam. It is forced down out of its locking notch on the underside of the bolt by the breech lock depressors riding up and over the lock pin which passes through the breech lock and protrudes on either side. This action unlocks the bolt.

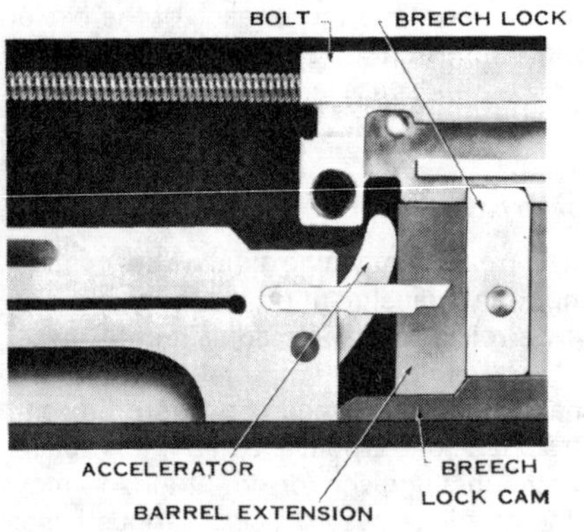

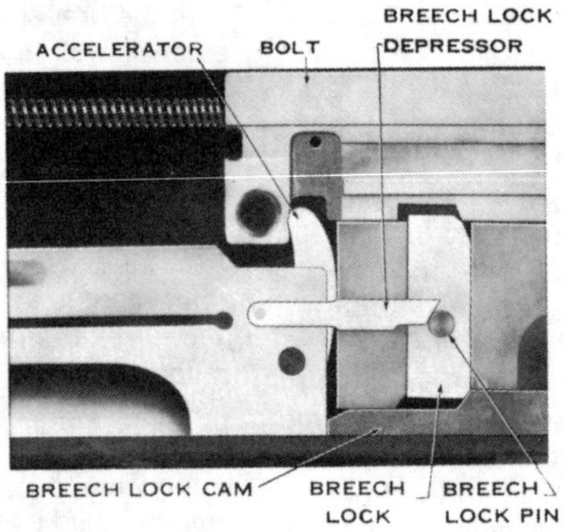

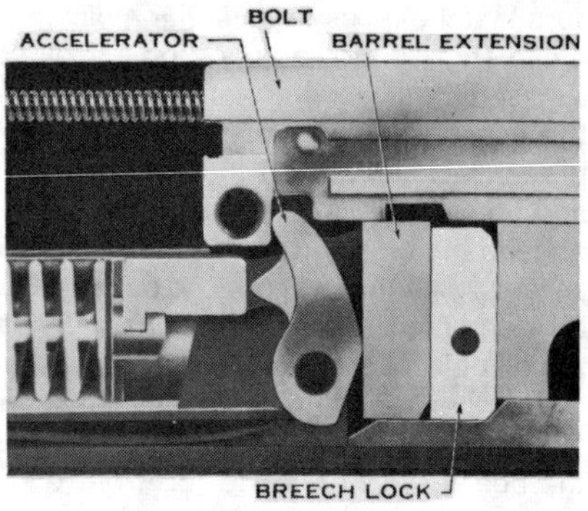

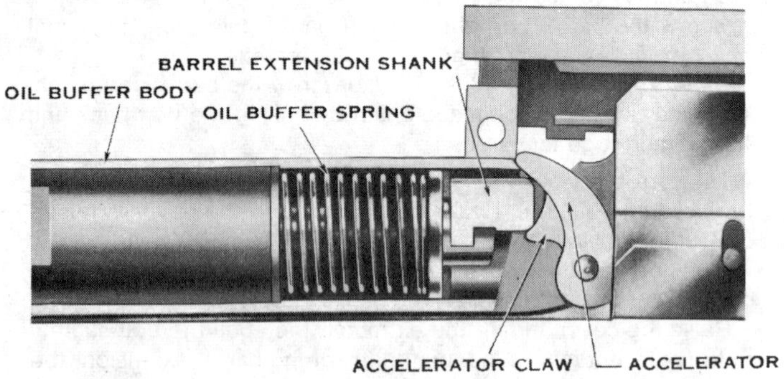

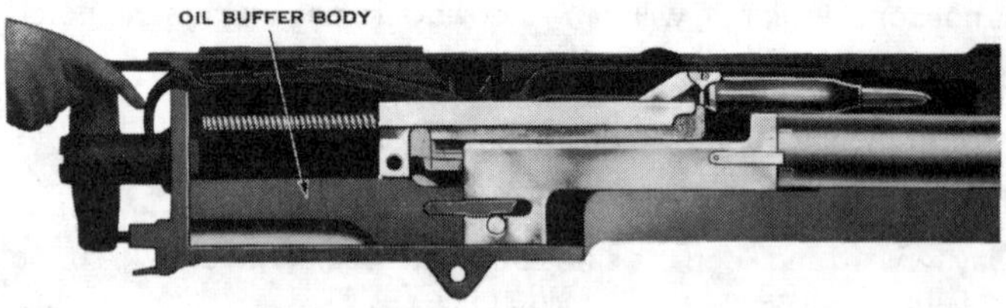

The barrel extension trips the accelerator up and to the rear of its pin. The tips of the accelerator striking the lower projection on the rear of the bolt accelerate its rearward travel. After a travel of about 1 1/8", the barrel and barrel extension have completed their rearward travel. They are stopped by the oil buffer body assembly, whose oil buffer spring has been compressed in the oil buffer body by the shank in the barrel extension. The flipped-up claws of the accelerator lock the spring in compressed position as they are moved against the shoulders of the barrel extension shank.

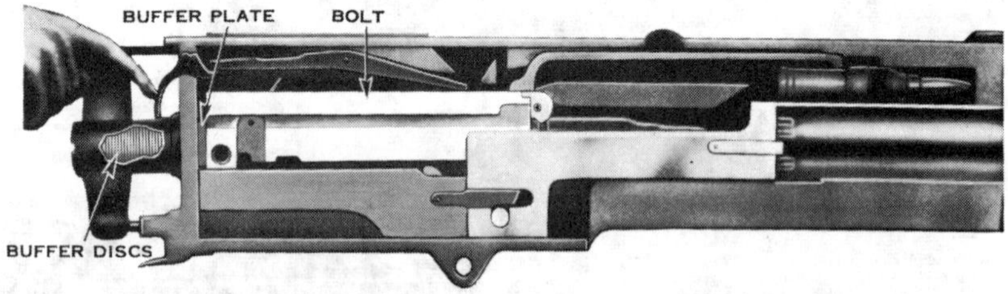

Meanwhile a piston rod head in the oil buffer assembly is forced from front to rear end of the oil buffer tube, and presses against oil in the tube to absorb the rearward shock of recoil until the oil escapes through the front side of the piston. This oil flow is through notches between the edge of the piston rod head and the oil buffer tube. This cushions the recoil and brings the rearward motion to a complete stop when the recoiling functions have been completed.

As the bolt travels to the rear, the driving springs inside it are compressed and its rearward motion is stopped when the rear of the bolt strikes the buffer plate.

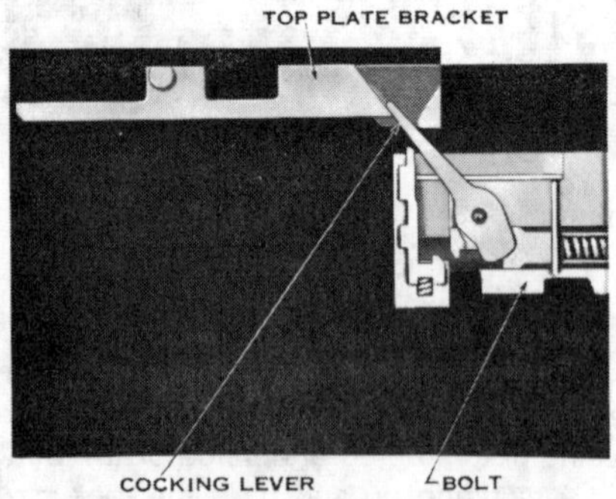

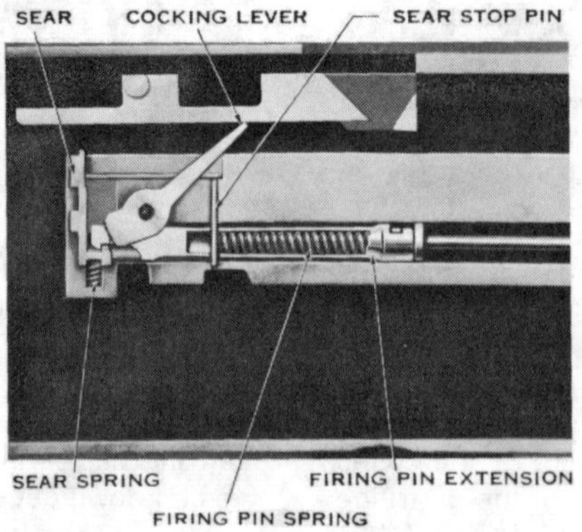

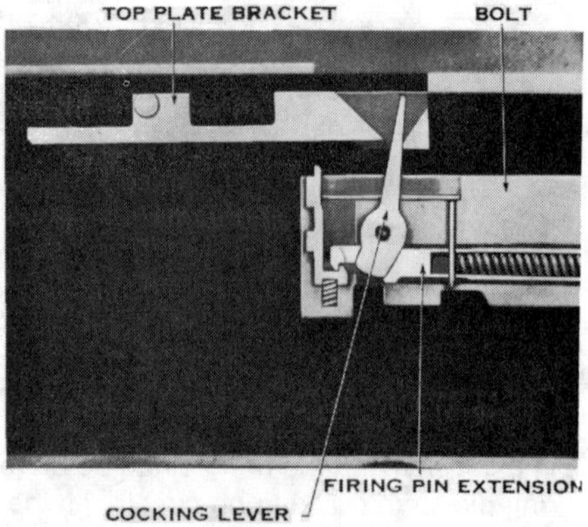

Cocking Action. The tip of the cocking lever protrudes through the top of the bolt where it lies in a V-slot in the top plate bracket. As the bolt starts to recoil, the tip of this cocking lever is pushed forward and its lower end is pivoted to force the firing extension rearward. This compresses the firing pin spring against the sear stop pin until the shoulder at the rear end on the firing pin extension hooks over the notch in the bottom of the sear under pressure of the sear spring.

When the bolt goes forward after the completion of the rearward motion, the tip of the cocking lever enters the V-slot in the top plate bracket, thus pivoting the bottom of the cocking lever out of the path of the firing pin extension to release the firing pin.

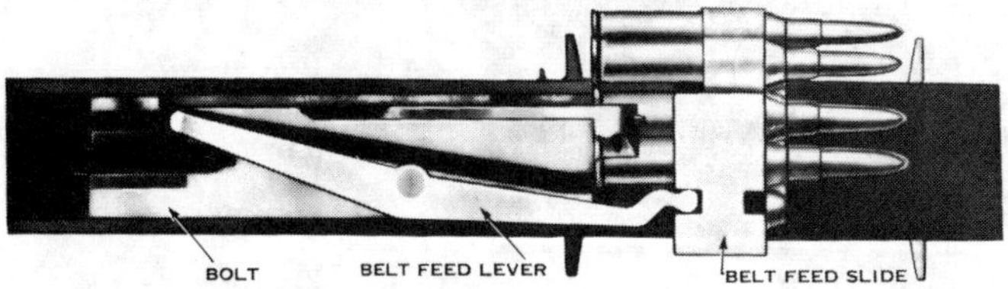

Feeding During the Rearward Motion. As the bolt moves to the rear, the stud at the rear of the belt feed lever is engaged in the diagonal groove on top of the bolt. This bolt stud thus serves to move the feed lever, which is pivoted near its center, and carry the belt feed slide at the front end of the lever out of the side of the gun where its spring snaps it down over the next cartridge in the ammunition belt.

The belt is pulled into the gun by the belt feed pawl attached to

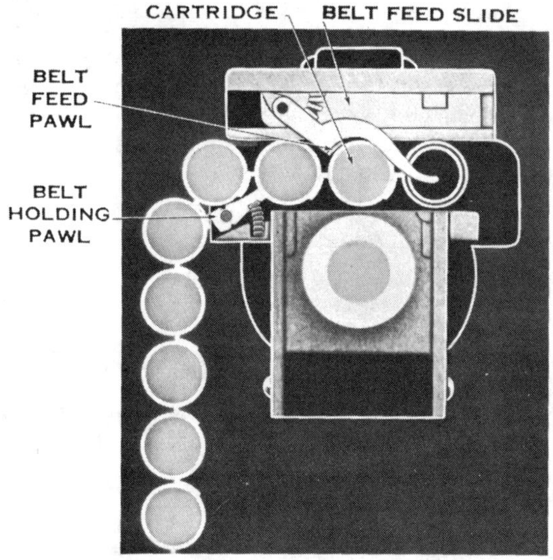

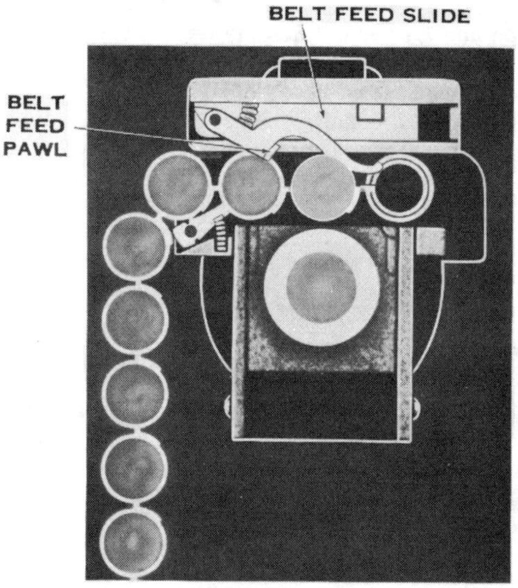

the belt-feed slide and ridges over the next cartridge.

As the recoiling motion is completed, the belt-feed slide has traveled far enough to permit the belt-feed pawl to be snapped down by its spring behind the next cartridge, ready to pull the belt forward into the gun on the next motion.

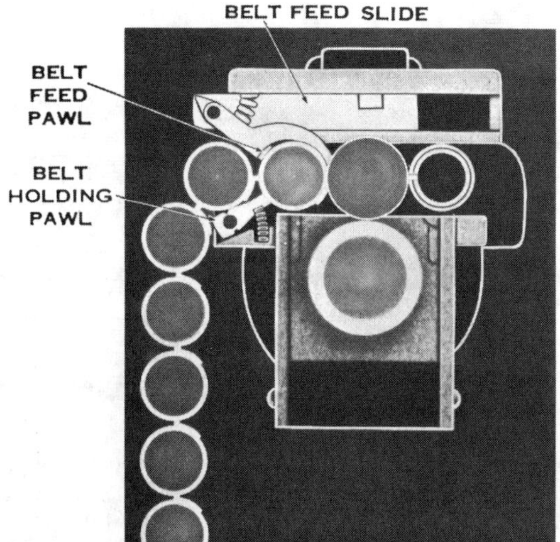

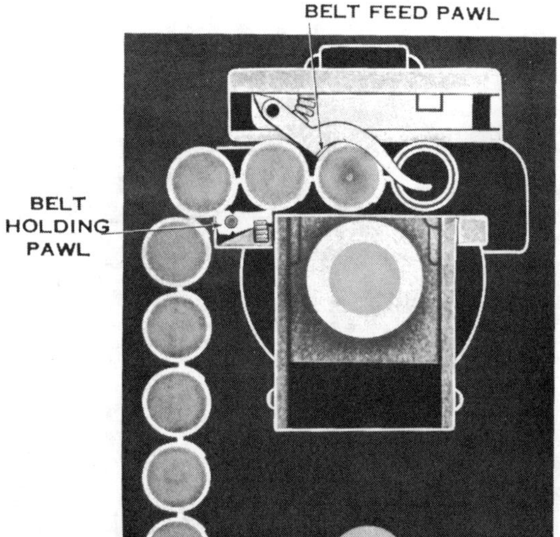

Forward Feeding Motion. As the bolt moves forward, the stud riding in its top pulls on the pivotal belt feed lever. The belt holding pawl is forced downwards as the cartridge is pulled over it and the belt holding pawl snaps up behind the next cartridge.

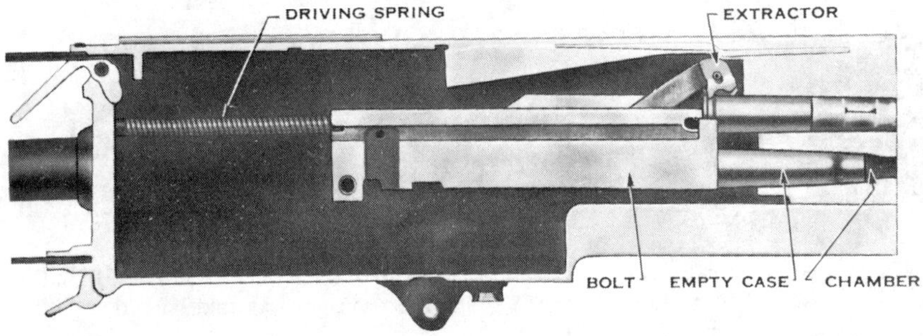

Extraction and Ejection. When the rearward motion starts, the extractor mounted in the side of the bolt and with its head above the bolt level snaps down into the cannelure of the cartridge in the belt, then draws the cartridge back out of the ammunition belt. The empty cartridge case is held in the T-slot in the front face of the bolt and the bolt withdraws it from the chamber.

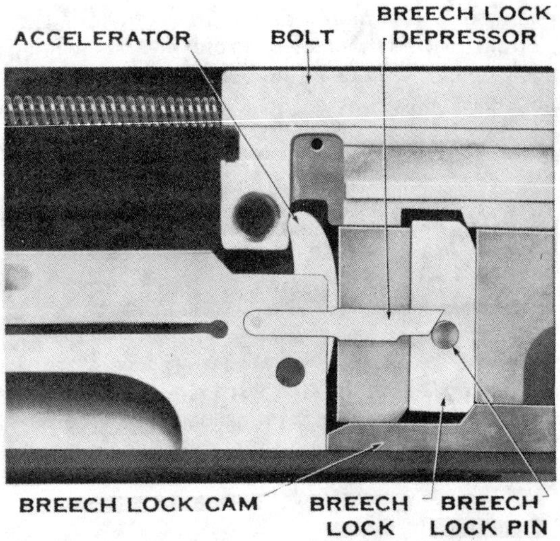

The top front edge of the breech lock and the front side of the notch in the bolt are beveled to start withdrawals of the empty cartridge case slowly to prevent the case from being torn apart by the sudden jerking motion. As the breech lock is unlatched, the bolt pulls away from the barrel and barrel extension easily enough to prevent rupturing the cartridge case.

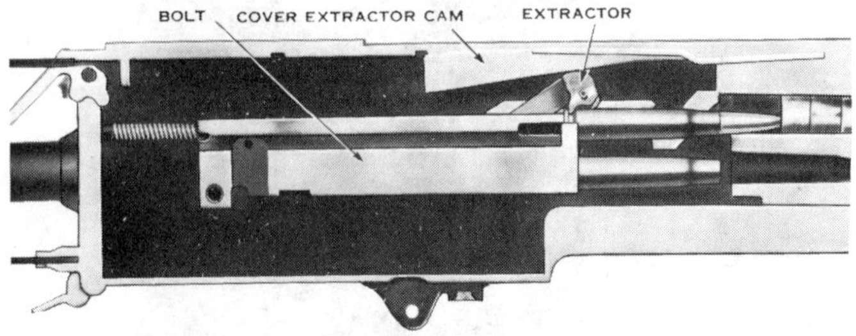

The cam on the inside of the cover forces the head of the extractor down, pushing the loaded cartridge into the mouth of the T-slot in the bolt. A lug on the side of the extractor rides against the top of the switch causing it to pivot downward at the rear; and as the recoiling motion comes to an end the lug on the extractor

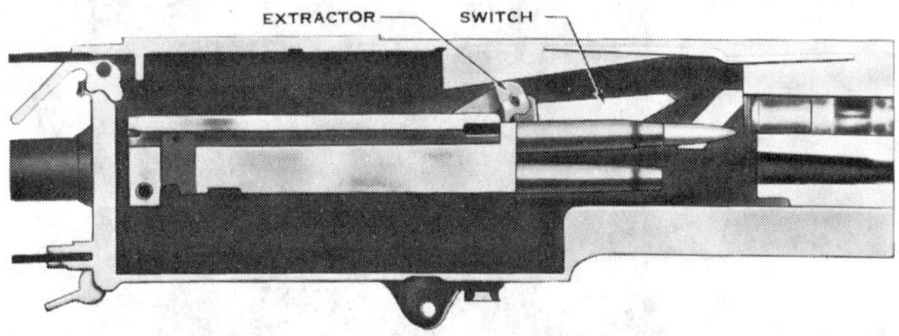

overrides the end of this switch, permitting it to snap up into normal position.

During this movement the empty cartridge case drops down out of the T-slot and is expelled through the bottom of the gun.

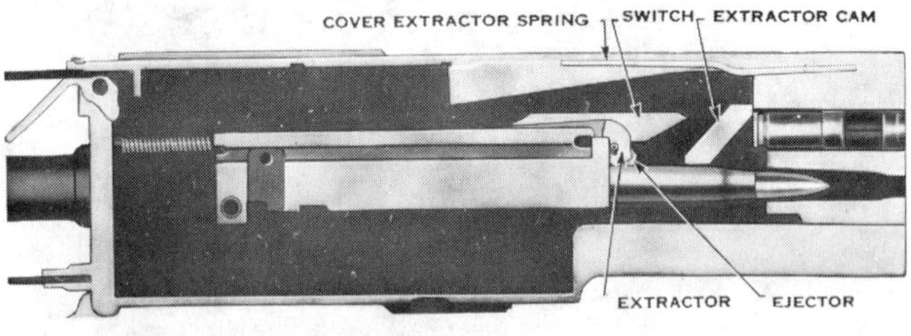

Forward Motion of the Extractor. As the bolt goes forward, the extractor lug riding under the switch forces the extractor farther down, thus forcing out the empty cartridge case if it has not already dropped out of the gun. A pin in the bolt limits the travel of the extractor and the cartridge, assisted by the ejector, is fed directly into the firing chamber.

When the cartridge is nearly chambered, the extractor rides up

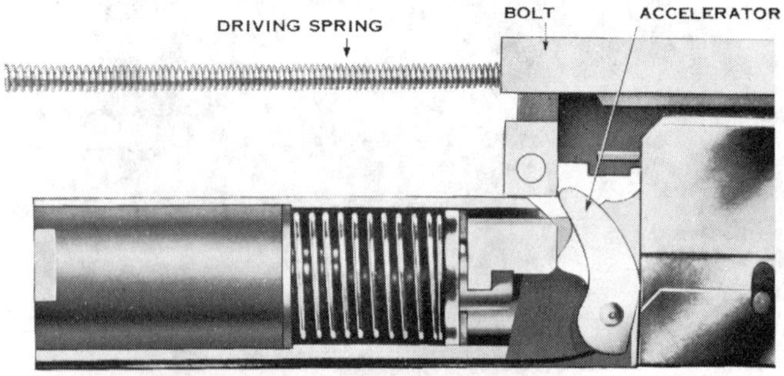

its cam compressing the cover extractor spring and is snapped into the cannelure of the next cartridge.

Further Action During Forward Movement. After the recoiling motion has been completed, the compressed driving spring and the compressed buffer disc force the bolt forward. The bolt travels about 5", when the projection on its bottom strikes the tips of the accelerator, rolling the accelerator forward on its pin.

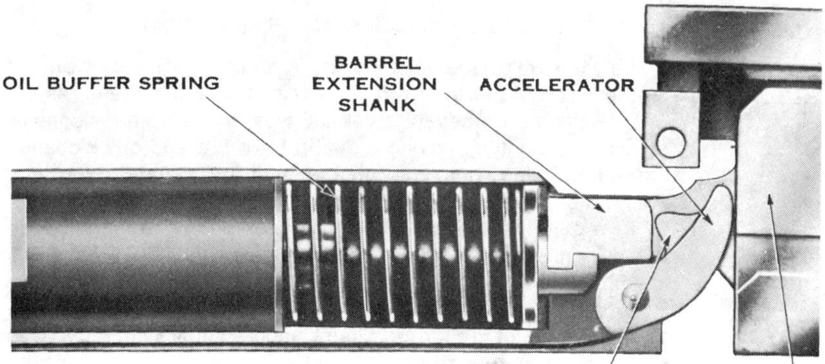

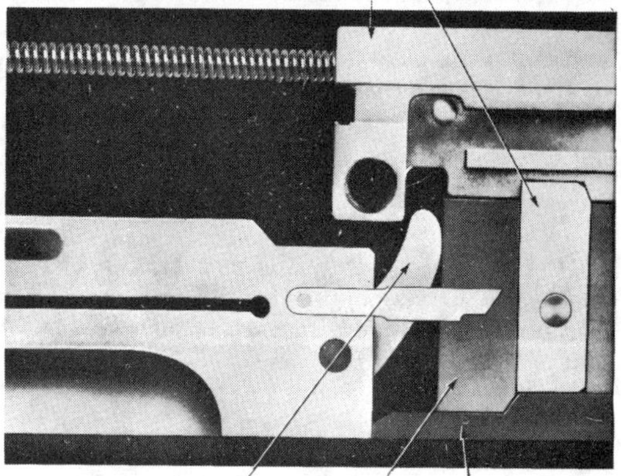

The accelerator claws are pulled away from the shoulder of the barrel extension shank, releasing the oil buffer spring. This spring now shoves the barrel extension on the barrel forward.

As the barrel extension goes forward, the breech lock strikes its cam and is forced upward on its pin. At that moment the bolt has reached the position where the notch on its underside is directly above the breech lock; and the breech lock rides up its cam and engages in this slot in the underside of the bolt. The bolt is locked to the breech end of the barrel just before the recoiling section reaches firing position.

Automatic Fire. If the trigger is pressed and held down, the sear is depressed as its tip is pressed against the cam surface of the trigger bar by the forward motion of the bolt just before it completes its forward motion. The notch in the bottom of the sear releases the firing pin extension and firing pin, automatically firing the cartridge as the forward motion is completed and continuing the action as long as the trigger is held and cartridges are fed into the gun.

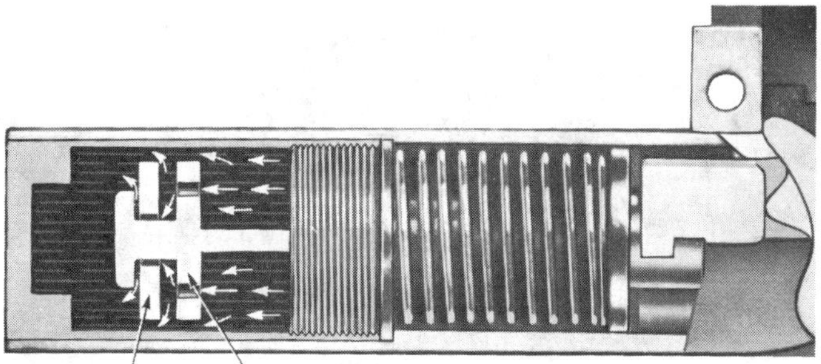

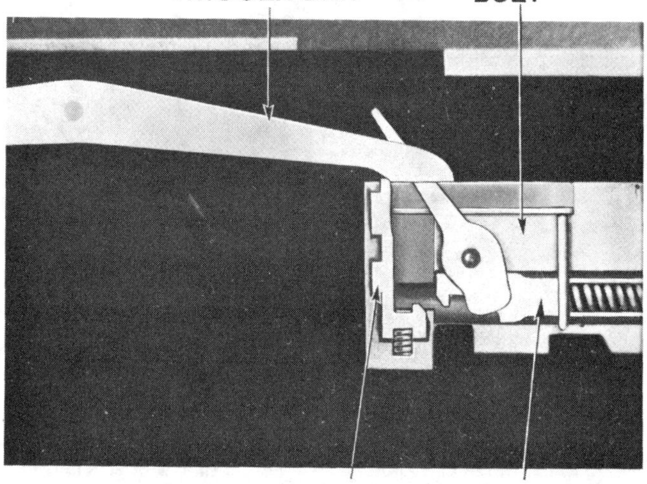

Oil Buffer. As the action moves forward additional openings for oil flow are provided in the piston rod head of the oil buffer assembly. The piston valve is forced away from the rod head as the parts move forward to uncover these openings. Thus the oil is permitted to escape freely from the opening in the center of the piston valve as well as at the edge of the valve near the tube wall, and so prepare it for the rearward motion.

Oil Buffer Adjustment. The oil buffer provides a method of regulating the speed of fire of this gun. Fire rate may be regulated by turning the oil buffer tube the required number of clicks. Turning the buffer tube to the left opens the oil buffer and permits oil to pass through the large ports, increasing the rate of fire. Turning the buffer tube to the right tightens up the oil buffer allowing it to absorb more recoil and reduce the rate of fire. This tube may be turned by inserting a screwdriver in the slot in the rear of the buffer tube.

CALIBER .50 AIRCRAFT MACHINE GUN AN-M3

The AN-M3 was developed toward the end of World War II and during the period immediately afterward. It is a modification of the famed cal. .50 M2 Browning aircraft gun of World War II fame. The M3 gun saw extensive service in Korea, since most of the United States F-86 fighters were equipped with it.

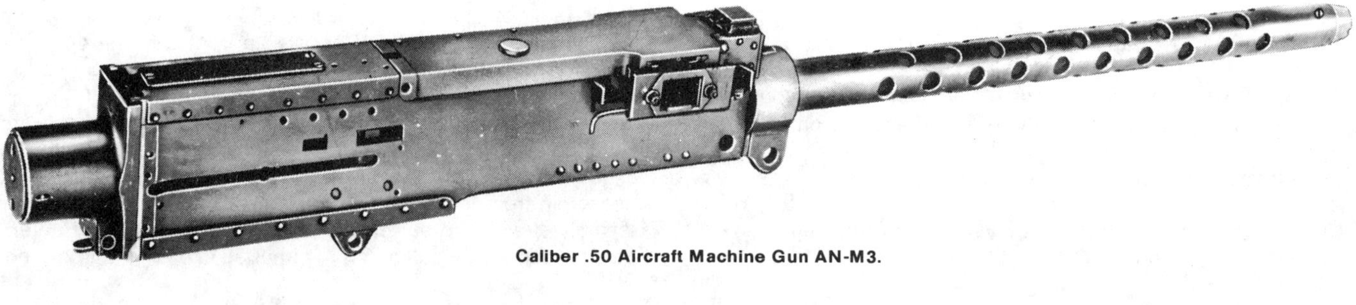

Caliber .50 Aircraft Machine Gun AN-M3.

The main modifications on the M2 were made to produce a higher rate of fire. The basic system of operation is the same as for the other Browning cal. .50 guns, and many, but not all, of the components of the M3 are inter-changeable with those of the M2.

Notes on the AN-M3 Machine Gun

No backplates with spade-type grips are used with this gun, since it is always used in fixed or turret-type aircraft mounts. It can be used in the cal. .50 AA machine gun mounts M3 or M63, but will rarely ever be found so mounted, since it is basically an aircraft weapon.

CALIBER .50 TANK MACHINE GUN M85

The M85 was developed by the Aircraft Armaments Corp. of Cockeysville, Md. as the T175E2 machine gun. It fills a requirement of the Armored Forces for a cal. .50 weapon suitable for co-axial or cupola mounting, having a dual rate of fire and quick-change barrel and being shorter and lighter than the cal. .50 Browning. This weapon is not yet in service.

The requirement for a dual rate of fire originated from the desire to have a high rate of fire for use against low-flying aircraft, and a low rate of fire for use against ground targets. While the cal. .50 Browning M2, HB, has an almost ideal rate for use against ground targets, its rate of fire for use against modern attack aircraft leaves something to be desired.

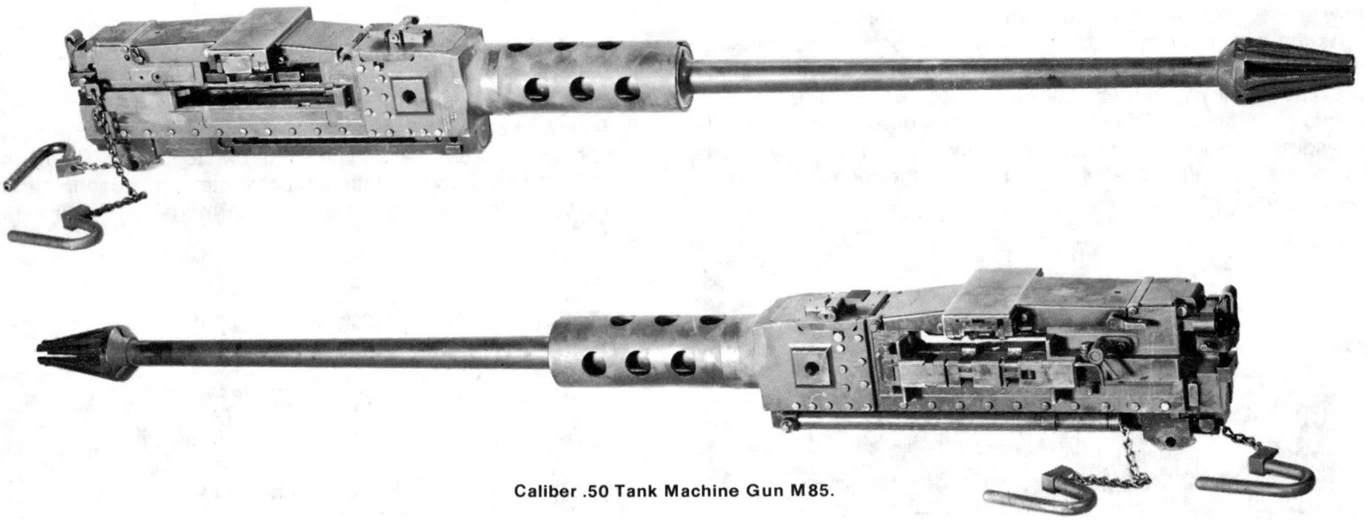

Caliber .50 Tank Machine Gun M85.

How to Load and Fire the M85

Open the cover and visually inspect the chamber to assure that chamber is clear. The bolt must be in the battery (closed) position before the ammunition belt is set in the feedway. Place the first round in the belt inboard of the two belt holding pawls. The belted ammunition must be placed in the gun with the open side of the links downward. Close the cover; the gun can be loaded without opening the cover if necessary. Pull bolt to the rear with the hand charger. Place safety on "safe" position. Set the rate selector lever to the proper position for the desired rate of fire. For high rate of fire, turn the lever completely to the left. For low rate of fire, turn the lever completely to the right. Do not change rate selector while firing a burst. Release the safety and fire the gun electrically by depressing the electrical trigger switch or manually by pushing forward on the manual trigger.

How to Field Strip the M85

Check to see that gun is not loaded. With hand charger control, allow the driving spring to slowly return the bolt to battery as follows:

(1) Pull back and hold hand charger.

(2) Depress trigger.

(3) Allow the driving spring to slowly return the charging handle to the original position.

Remove the barrel. Transversely depress the lock on the barrel latch, full depress the barrel latch, rotate the barrel 90° until the head of the "unlock" arrow is in line with the head of the arrow on the barrel support, and pull the barrel forward out of the barrel support. Remove the cover and tray assemblies. Open the cover, withdraw the quick release pins by inserting the rim of a cartridge case in the annular groove at the top of the pin; using the cartridge case as a lever, pry out the pin; withdraw the tray. Depress the latch lock on the lower left side of the back plate, then depress the back plate latch and lift the back plate straight up off the receiver. Remove back plate slowly for the first inch as the preload on the driving spring may cause the buffer assembly to jump out. CAUTION: Do NOT attempt to disassemble gun with bolt in the REAR position. The driving spring is then heavily loaded and may cause injuries. Remove the bolt buffer assembly from the rear of the receiver. Depress the sear block detent, visible through the right side of the receiver, with the nose of a cartridge and withdraw the sear assembly out of the rear of the receiver. Remove the feed and ejector assembly; withdraw the front and rear quick release pins which fasten the feed and ejector assembly to the receiver. Remove the feed assembly by first pulling its front end out from the side of the receiver, then pull the assembly forward out of the receiver. Pull the barrel extension assembly, with the bolt in it, out through the rear of the receiver. Withdraw the quick release pin from the side of the receiver and lift the accelerator assembly straight up out of the receiver. Disengage the detent from the receiver by pulling on the knurled knob at the front end,

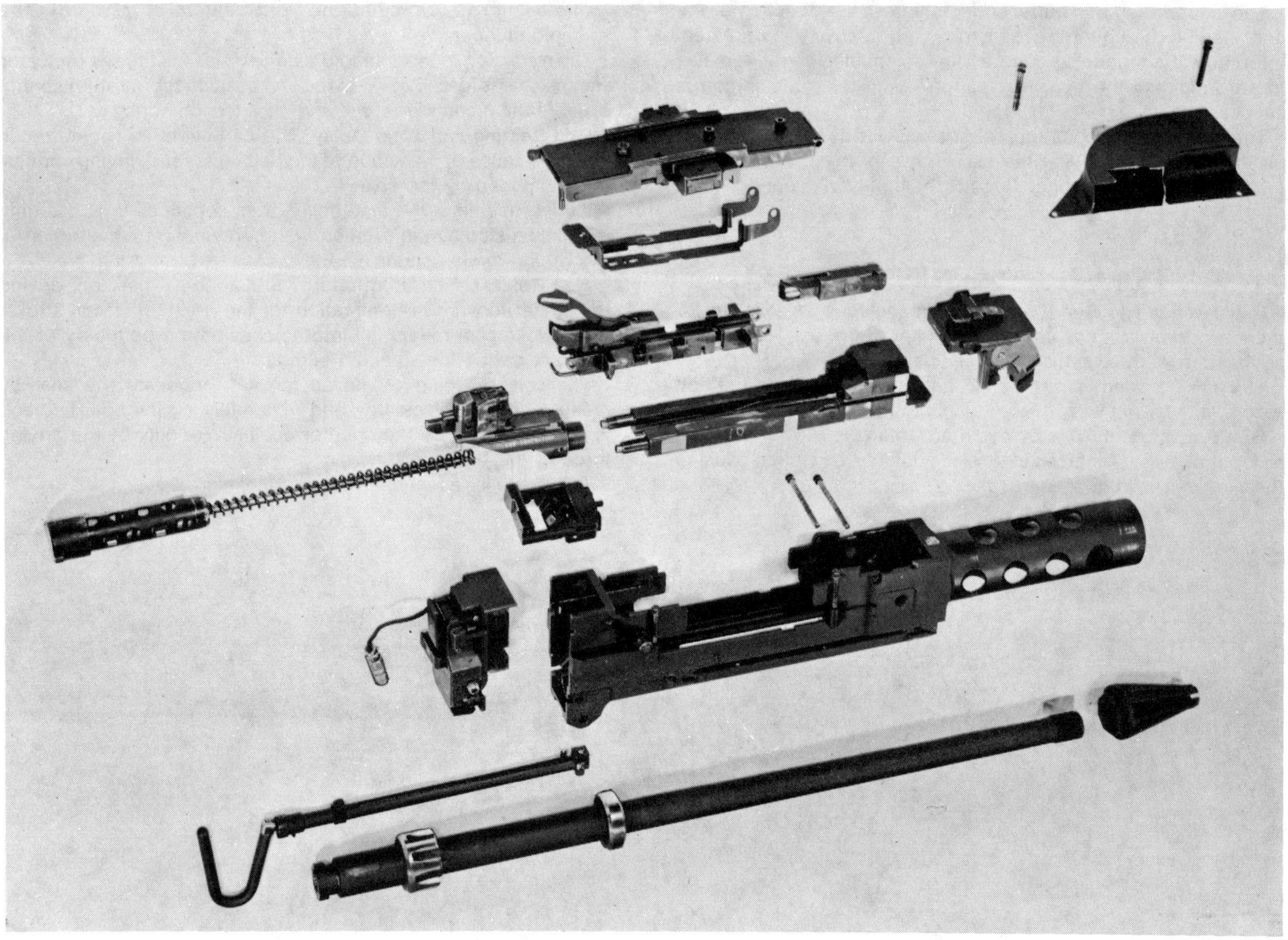

M85, field-stripped.

and slide the hand charger forward. To reassemble, reverse the procedure given above.

Cyclic Operation—High Rate of Fire

Charging the Gun. To begin the cycle, the loaded gun must be hand charged. During the charging stroke, the feed mechanism is actuated by the bolt, positioning a round on the center line of the gun, in the forward path of the bolt.

The bolt is held in the open-bolt position by the sear, with the bolt-driving spring compressed.

Stripping and Chambering of Round. When the firing switch is depressed, or the hand trigger is pressed, the solenoid plunger pushes the sear actuator forward about its pivot, camming the sear up and out of the bolt notch.

The bolt is driven toward battery position by the compressed bolt-driving spring.

The round in the feedway is picked up by the stripper on the bolt, is stripped through the belt link by the action of the bolt, and is deflected downward into the chamber by the chambering ramp.

Locking and Firing. The forward motion of the bolt block is arrested by the base of the chambered round, and the bottom extractor engages the extraction groove at the rim of the cartridge.

The bolt slide continues forward, wedging the bolt locks in place.

As the bolt slide completes its forward motion, the firing pin protrudes through the bolt block and strikes the cartridge primer, firing the round and initiating the recoil stroke.

Recoil Stroke. The feed mechanism, operated by the recoiling barrel extension through a feed spring and a feed cam, is actuated at the beginning of the recoil stroke.

As the barrel and the barrel extension begin to recoil, the barrel actuates the accelerator, thereby driving the bolt slide rearward at an accelerated speed.

As the bolt slide starts to move rearward, it cams the bolt locks inward, releasing the bolt block which moves toward the rear, pulling the spent cartridge case out of the chamber and compressing the bolt-driving spring.

Ejection. The ejector mechanism is actuated by the action of the recoiling bolt on the ejector lever.

Just before cartridge ejection, the bottom extractor is disengaged from the extraction groove of the cartridge by the cam grooves in the barrel extension.

As the bolt continues rearward motion, the springloaded stripper is depressed by the income round.

The barrel group and the accelerator now start their return to battery position, driven by the barrel return springs and accelerator return springs.

As the ejector strikes the cartridge case, the case rotates about the side extractor and is thrown clear of the gun.

Bolt Buffing and Counterrecoil. As the bolt nears the end of its recoil stroke, it compresses the bolt buffer spring, which stops the rearward motion of the bolt.

By this time, the barrel group and the accelerator have returned to their battery position.

The compressed bolt buffer spring reverses the direction of the bolt, starting the counterrecoil stroke, and the cycle is repeated.

Firing will continue as long as the solenoid plunger remains forward, holding the bolt sear up, providing that ammunition is supplied.

The velocity of the counterrecoil stroke of the bolt is much higher on those cycles following the initial cycle because the force of the buffer spring adds to the force of the bolt-driving spring.

Cyclic Operation—Low Rate of Fire

Initial Action and Bolt Recoil. The rate selector lever is set for the low rate of fire, positioning the striker in the path of the bolt so that it is ready to actuate the time-delay drum.

The gun is loaded, hand charged, and initially fired, and the bolt recoils at the same velocity as during the high rate of fire.

Bolt Actuation of Time-Delay Drum. Near the end of the bolt recoil stroke, the bolt block extension contacts the striker, causing the time-delay drum to start rotating.

The drum rotation cams the yoke rearward, retracting the solenoid plunger.

The retraction of the solenoid plunger releases the sear actuator and the sear is forced down to locking position by its return spring. Drum rotation continues, winding up a torsion spring.

Bolt Searing and Time Delay. The bolt starts its counterrecoil stroke because of the action of the bolt buffer and driving springs, but is stopped by the sear.

During this time, the time-delay drum continues to rotate until it strikes a stop. Drum rotation is then reversed by the action of the wound torsion spring.

Bolt Release. The drum returns to its original position, driving the striker forward in preparation for the next bolt recoil stroke.

The yoke cam rollers fall into notches on the periphery of the drum, releasing the solenoid plunger.

The solenoid plunger moves forward, operating the sear by means of the sear actuator, and thus releasing the bolt. The bolt is driven forward on the counterrecoil stroke only by the driving force of the bolt-return spring.

This cycle is repeated as each round is fired.

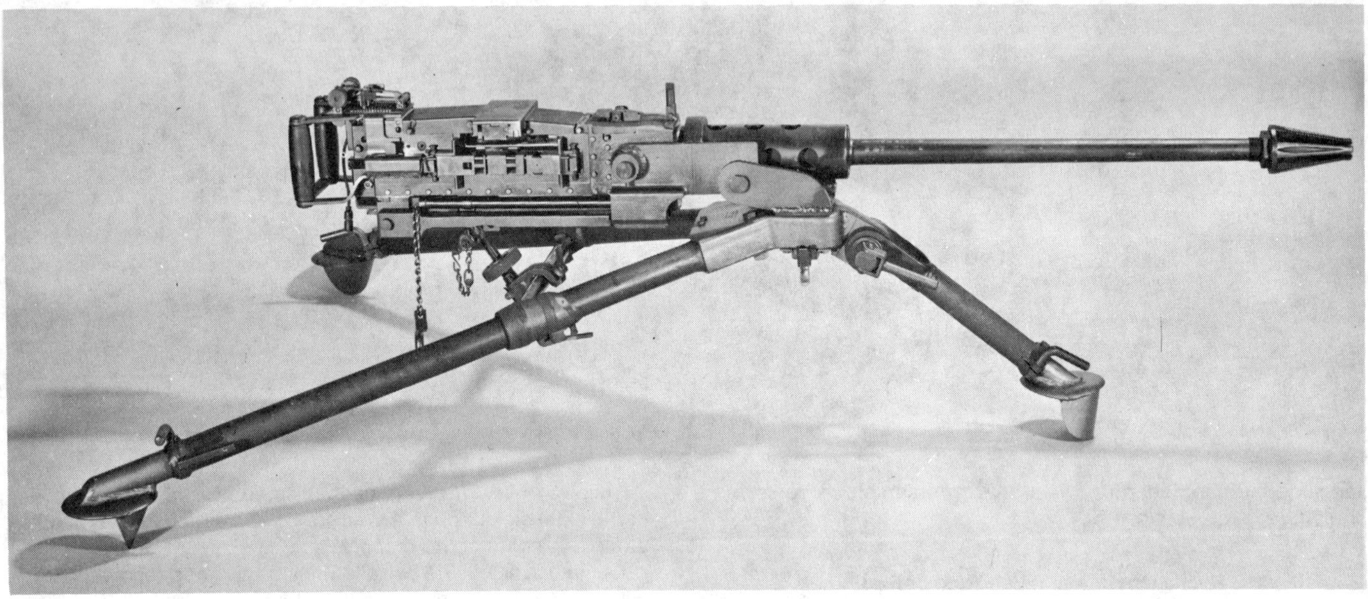

Caliber .50 Machine Gun M85C.

CHARACTERISTICS OF UNITED STATES SERVICE MACHINE GUNS

	Colt M1917	Browning M1917A1	Vickers M1918 Aircraft	M1919A4
Caliber:	.30.	.30.	.30.	.30.
System of operation:	Gas, automatic.	Recoil, automatic.	Recoil with gas assist, automatic.	Recoil automatic.
Length overall:	40.8 in.	38.5 in.	44.19 in.	41 in.
Barrel length:	28 in.	24 in.	28.4 in.	24 in.
Feed device:	250-round, fabric belt.	250-round, fabric belt or disintegrating link belt.	250-round, fabric belt or disintegrating link belt.	250-round, fabric belt or disintegrating link belt.
Sights: Front:	Blade.	Blade.	none—aircraft type fitted to the weapons as required.	Blade.
Rear:	Leaf.	Leaf.		Leaf.
Weight: gun:	35 lb.	41 lb. w/water.	25 lb.	31 lb.
Mount:	61.25 lb.	53.15 lb. (M1917A1).		14 lb. (M2).
Muzzle velocity:	Approx. 2800 f.p.s.	2800 f.p.s.	Approx. 2800 f.p.s.	2800 f.p.s.
Cyclic rate:	480 r.p.m.	450-600 r.p.m.	800-900 r.p.m.	400-550 r.p.m.

CHARACTERISTICS OF UNITED STATES SERVICE MACHINE GUNS (Cont'd)

	M1919A6	**M2 Aircraft**	**M37 Tank**	**M60**	**M73 Tank**
Caliber:	.30.	.30.	.30.	7.62mm NATO.	7.62mm NATO.
System of operation:	Recoil, automatic.	Recoil, automatic.	Recoil, automatic.	Gas, automatic.	Recoil, w/gas assist, automatic.
Length overall:	53 in.	39.9 in.	41.75 in.	43.75 in.	34.75 in.
Barrel length:	24 in.	23.9 in.	24 in.	25.6 in.	22 in.
Feed device:	250-round, fabric belt or disintegrating link belt.	Disintegrating link belt.	Disintegrating link belt.	Disintegrating link belt.	Disintegrating link belt.
Sights: Front:	Blade.	None permanently attached to gun.	Blade.	Blade.	None permanently attached to gun.
Rear:	Leaf.		Leaf.	Leaf.	
Weight: Gun:	32.5 lb.	21.5 lb. (fixed gun). 23 lb. (flexible gun).	31 lb.	23.05 lb.	28 lb.
Mount:	14 lb. (M2)			15 lb. (M122)	
Muzzle velocity:	2800 f.p.s.	2800 f.p.s.	2800 f.p.s.	2800 f.p.s.	2800 f.p.s.
Cyclic rate:	400-500 r.p.m.	1000-1350 f.p.s.	400-550 r.p.m.	600 r.p.m.	450-500 r.p.m.

CHARACTERISTICS OF UNITED STATES SERVICE MACHINE GUNS (Cont'd)

	Johnson M1941	**M1921A1 Antiaircraft**	**M2 Aircraft, Basic**	**M2 Heavy Barrel**
Caliber:	.30	.50.	.50.	.50.
System of operation:	Recoil, selective fire.	Recoil, automatic.	Recoil, automatic.	Recoil, selective fire.
Length overall:	42 in.	56 in.	56.25 in.	65.1 in.
Barrel length:	22 in.	36 in.	36 in.	45 in.
Feed device:	20-round, in line, detachable box magazine.	250-round, fabric belt or disintegrating link belt.	Disintegrating link belt.	disintegrating link belt.
Sights: Front:	Blade.	Hooded blade.	None permanently attached to gun.	Hooded blade.
Rear:	Adjustable folding aperture.	Leaf.		Leaf.
Weight: Gun:	Approx. 13 lb.	79 lb. w/o water.	61 lb.	84 lb.
Mount:				44 lb. (M3)
Muzzle velocity:	2800 f.p.s.	2840 f.p.s.	2840 f.p.s. (M2 ball).	2930 f.p.s. (M2 ball).
Cyclic rate:	400-450 r.p.m.	Approx. 500 r.p.m.	750-850 r.p.m.	450-550 r.p.m.

CHARACTERISTICS OF UNITED STATES SERVICE MACHINE GUNS (Cont'd)

	M2 Antiaircraft	**M3 Aircraft**	**M85**
Caliber:	.50.	.50	.50
System of operation:	Recoil, automatic.	Recoil, automatic.	Recoil, automatic.
Length overall:	66 in.	57.25 in.	54.5 in.
Barrel length:	45 in.	36 in.	36 in.
Feed device:	Disintegrating link belt.	Disintegrating link belt.	Disintegrating link belt.
Sights: Front:	Hooded blade.	None permanently attached to gun.	None permanently attached to gun.
Rear:	Leaf (may be found without rear sight).		
Weight: Gun:	121 lb. w/water.	68.75 lb. w/recoil adaptor.	61.5 lb.
Mount:	401 lb. (A.A. M3).		
Muzzle velocity:	2930 f.p.s. (M2 ball).	2840 f.p.s. (M2 ball).	2840 f.p.s. (M2 ball).
Cyclic rate:	500-650 r.p.m.	1150-1250 r.p.m.	Low rate: 400 ± 50 r.p.m. High rate: 1050 ± 50 r.p.m.

MISCELLANEOUS

CALIBER .50 M8C SPOTTING RIFLE

The M8C spotting rifle is a special purpose type of weapon; it is not an antipersonnel weapon like most other small arms. It is mounted on the 106mm BAT (battalion antitank) recoilless rifle, and serves as a ranging rifle for that weapon. The cal. .50 cartridge used with the M8C is a considerably smaller cartridge than the cal. .50 used with the Browning guns. Its ballistics are matched with those of the 106mm high-explosive antitank (HEAT) projectile and the bullet contains a spotting charge which on impact gives off a bright flash and a cloud of smoke. By using the spotting rifle to get on target, and then firing the recoilless rifle at the same range and azimuth setting, the first-round hit probability with the recoilless weapon is considerably increased. A high first-round hit probability with recoilless weapons is of very great importance, because recoilless weapons are comparatively easy to spot during firing. Their great back-blast throws up clouds of dust and causes considerable movement in trees, brush, etc.

Characteristics of M8C Spotting Rifle

Caliber: .50.
System of operation: Gas, semi-automatic only.
Weight w/20-rd. magazine (empty): 26.10 lbs.
Overall length of weapon: 44.9 in.
Length of barrel: 32 in.
Feed device: 20-rd. detachable box magazine.
Sights: None on spotting rifle (optical sights are mounted on the major weapon).
Muzzle velocity: 1760 f.p.s. with cal. .50 M48 spotting cartridge.

Caliber .50 Spotting Rifle M8C (Modified).

How to Load and Fire the M8C (Modified)

Loading. Ammunition for the spotting rifle is contained in a box-type (10-round or 20-round) magazine which fits into a well at the front of the firing mechanism housing and is retained by a pair of catches. The catches support the magazine at front and rear. The front catch is brazed to a rectangular yoke that surrounds the magazine well in the housing and is linked to and operated by the rear catch assembly. This arrangement is similar to that used in the M8 spotting rifle, except that the rear catch for the M8C (Modified) can be folded against the underside of the firing mechanism housing to prevent accidental release of the magazine or damage to the catch.

Charging. With a full magazine inserted, the spotting rifle is placed in readiness to fire by pulling the charging handle all the way to the rear and releasing. This automatically cocks the weapon and chambers the top cartridge in the magazine. The charging for the M8C (Modified) rifle is offset to clear the ejection path and provides a better grip than that of the M8 rifle. A plunger is incorporated in the handle to engage it with the slide and prevent accidental disassembly. The M8C (Modified) rifle is also provided with a bolt and slide retainer assembly. This assembly consists of a spring-loaded plunger operated by a knob on the top rear of the receiver. The retainer relieves the gunner of the necessity of holding the charging handle to the rear to inspect the chamber or clear the bore. When the charging handle is pulled to the rear and the retainer knob depressed, the plunger seats in a slot in the rear of the slide and prevents the weapon from going into battery when the charging handle is released. The slide is released by pulling the charging handle to the rear until the retainer plunger disengages. The weapon will then go into battery when the charging handle is released.

Firing. The weapon is normally used with a firing cable or lanyard arrangement which is attached to a trigger-type mechanism on the major weapon mount or inside the Ontos vehicle.

Stripping the M8C (Modified)

Magazine Assembly. Push forward on the rear magazine catch assembly and pull the magazine straight out of housing.

Disassemble magazine by sliding base off the tube. The magazine spring and follower can then be removed from the tube.

Guide Rods and Driving Springs. Press forward on rear of guide rods and turn them counterclockwise to disengage from bayonet slots in buffer bushings.

Draw rods and springs from the receiver.

Buffer Assembly. Press the buffer catch away from the buffer. Rotate buffer counterclockwise and remove from receiver.

Charging Handle Assembly. Lift the charging handle plunger knob and slide the handle assembly forward out of its slot in the slide.

Slide and Bolt Assembly. Push the bolt and slide to the rear by pressing at the front of the slide.

Remove bolt and slide from rear of receiver.

Remove bolt assembly from slide by pushing bolt assembly rearward and pushing rear end of bolt away from the slide.

Removal of Firing Pin and Retractor from Bolt. Turn the bolt so the wide end of the retractor is down. The retractor will fall out, and the firing pin can then be removed from the end of the bolt.

Removal of Extractor from Bolt. Drive out the extractor plunger stop pin with a 1/8-inch drift.

With the drift, slide the extractor out of its slot. Use caution to avoid losing extractor plunger and spring.

Remove the plunger and spring.

Removal of Firing Mechanism Housing Assembly. Using a 1/8-inch drift, drive out the front and rear mounting pins.

Remove the buffer catch assembly and the buffer catch spring by pulling the assembly rearward.

Invert the weapon, lift the firing mechanism housing at the rear, and disengage it from the receiver.

Firing Mechanism Assembly. Release the hammer by pressing back on the connector block.

Remove the hammer pin cotter pin and drive out the hammer pin. Remove the hammer and hammer spring.

Remove cotter pins from the front retaining pin, firing cable housing retaining pin, and safety shaft. Drive out the front retaining pin, the safety shaft, and the firing cable housing retaining pin.

Lift the safety from the top of the channel and slide the channel out of the rear of the housing.

Push out the front stop pin, and slide the firing cable housing, connector block, and sear block out of the rear of the channel.

Release the ball of the firing cable from the connector block, and separate the sear connector from the sear block, releasing the connector spring.

Push out the sear pin to release the sear from its block. Push out the connector pin to release the connector from the connector block.

Stripping of Gas Operating System. Unscrew the gas regulator lock ring bushing, turning it counterclockwise until the gas regulator components can be removed as a unit.

Using a spanner wrench, rotate the gas cylinder 180 degrees counterclockwise and pull it forward out of the gas cylinder body.

Pull the operating rod assembly forward and disengage its rear end from the hole in the receiver. The rod can then be pulled rearward and outward, away from the gas cylinder body and the barrel.

How the M8C (Modified) Works

Bolt and Slide Action. The reciprocating bolt and slide action of the spotting rifle is contained in the cylindrical receiver to which the barrel is threaded. The slide moves in keyways in the sides of the receiver, and the bolt is suspended in the underside of the slide by a pair of bolt guides. These guides are rectangular at one end and cylindrical at the other. The rectangular ends fit into keyways at the front end of the bolt, and the cylindrical ends fit into transverse holes at the front of the slide. The rectangular ends permit a relative sliding motion between the bolt and the slide, and the cylindrical ends permit a small amount of bolt rotation relative to the slide. Locking action of the bolt is based on this relative bolt and slide motion. As the slide moves forward in the receiver, it cams the rear of the bolt down into a locking recess (bolt lock) in the bottom of the receiver; after firing, the slide cams the bolt up out of the recess and carries the bolt rearward in the recoil.

Gas Cylinder and Operating Rod Action. The slide is driven in recoil by an operating rod assembly, which derives its energy from a gas cylinder group located on the barrel. When a cartridge is fired, a small amount of the propellent gas is drawn through a port in the side of the barrel and into the gas cylinder. Gas pressure in the cylinder drives the piston to the rear, transmitting energy through a spring-loaded operating rod which drives the slide.

Driving Springs and Buffer Assembly. The rear of the receiver is enclosed by a buffer assembly. On the inner face of the buffer is a plastic disc that absorbs residual slide recoil energy. The main force of slide recoil is stored in a pair of driving springs. These springs are seated over a pair of guide rods supported at the rear in a pair of buffer bushings and at the front in a pair of longitudinal holes through the slide. The springs bottom in the slide holes and are compressed when the slide recoils, storing energy for counterrecoil. When the bolt and slide stop against the buffer disc, the energy stored in the compressed drive springs forces the slide and bolt forward into battery, chambering a fresh cartridge.

Feeding. Feeding takes place as the bolt approaches the magazine. The lower front end of the bolt contacts the rear (or base) of the top cartridge, driving it forward out of the magazine and into the barrel chamber.

Firing mechanism. The firing mechanism and its housing are pinned to the underside of the receiver. The mechanism is controlled by a cable connected to a firing knob on the elevation handwheel of the 106mm rifle system. When the knob is pulled, the following action takes place:

The cable draws the connector block, connector, sear block, and sear to the rear as a unit, releasing the hammer.

As the hammer rotates upward and forward to strike the firing pin, it first contacts the long leg of the connector, rotating the connector rearward and disengaging it from the sear block.

The sear block, released by the connector, is pushed forward by a spring, placing the sear into position to re-engage the hammer during recoil.

The hammer rotates the connector to the rear, insuring that the short leg of the connector is free of the sear block.

The hammer contacts and rotates the sear rearward, compressing the connector spring slightly; then, as the hammer rotates below the sear, the spring snaps the sear back into position, blocking forward movement of the hammer. When the firing knob is released, the sear block retainer spring moves the connector block forward and the sear connector reengages the sear block. The firing knob may now be pulled again to fire the next round.

Safety Action. The safety is operated by a lever on the side of the housing and consists of a block which is rotated to permit or prevent movement of the connector block.

Firing Pin and Retractor Action. The firing pin is contained in the bolt and is actuated by the hammer, as described above, and also by the firing pin retractor. The retractor is a flat, key-like component with rounded ends. The retractor fits through a slot in the firing pin and extends through the sides of the bolt. Its rounded ends are acted upon by cam surfaces on the inner sides of the slide. On recoil, the retractor is cammed to withdraw the firing pin from the cartridge primer; on counter-recoil, the retractor is held to the side to prevent forward movement of the firing pin until the bolt enters the battery position. This delayed camming prevents firing unless the bolt is safely locked in battery position.

Extraction. The extractor is seated in a T-slot in the top front of the bolt. A small amount of outward travel in the slot is opposed by a spring-loaded plunger arrangement. This outward travel permits the extractor claw to cam outward and past the rim of a chambered cartridge, whereupon the spring-loaded plunger draws the extractor inward to seat on the cartridge rim.

Ejection. The ejector is pinned to the front of the bolt lock in the bottom of the receiver. As the bolt travels rearward in recoil, the extractor draws the empty cartridge case against the ejector, which rotates the case free of the extractor claw and flips it through the opening in the top of the receiver.

Special Note on the M8 Series of Spotting Rifles

An earlier weapon, the M8 is also in extensive use. It differs from the M8C in the following:

It has a ten-round magazine rather than a twenty-round magazine.

Its firing mechanism housing cannot be used with the remote-controlled safety mechanism as can that of the M8C.

The M8C (Modified) rifle differs from the M8C in the following:

(1) It has a large number of new components which are not interchangeable with those of the M8 and M8C.

(2) It has an entirely new firing mechanism.

The M8C and M8C (Modified) were developed for use in the Ontos vehicle, which is used by the Marine Corps.

40mm GRENADE LAUNCHER M79

The M79 grenade launcher is a shotgun-type weapon designed to fire a high explosive grenade considerably more accurately than a grenade can be fired from a rifle grenade launcher.

Characteristics of Grenade Launcher

Caliber: 40mm.
System of operation: Single-shot, break-open type.
Weight of launcher (loaded): 6.45 lbs.
Length of launcher: 28.78 in.
Length of barrel: 14 in.
Muzzle velocity: 250 f.p.s.
Sights: Front: Protected blade.
Rear: Leaf, adjustable for windage.

How to Load and Fire the M79

Move barrel locking latch FULLY to the right and break open breech. Moving latch fully to right automatically puts the weapon on "safe." Insert cartridge in chamber until the extractor contacts the rim of the cartridge case. Close the breech, push safety to forward position exposing the letter "F." Pressure on the trigger will now fire the weapon.

How to Field Strip the M79

Under normal conditions it should not be necessary (for maintenance purposes) to do more than break the weapon using the barrel locking latch. The firing pin retainer in the face of the standing breech can be tightened up from time to time by use of the lugs on the combination wrench supplied with the weapon. If the weapon has been immersed in water or snow the following procedure should be followed: Remove the fore-end assembly

40mm Grenade launcher M79.

by taking out screw which passes through the rear mounting hole of the front sling swivel. Pull front end of fore-end away from barrel until lug on the rear sight base is clear of the hole in upper surface of fore-end bracket. Keeping lug clear of hole, pull forward on fore-end assembly until it is free of receiver assembly. Operate barrel locking latch and open breech, holding the stock and receiver stationary, move the barrel rearward in the receiver until it is disengaged from the fulcrum pin. Separate barrel from receiver group. From bottom of stock, near front end, remove machine screw, lock washer, and flat washer which secure stock to receiver. Separate stock from receiver. To reassemble, perform the steps listed above in reverse.

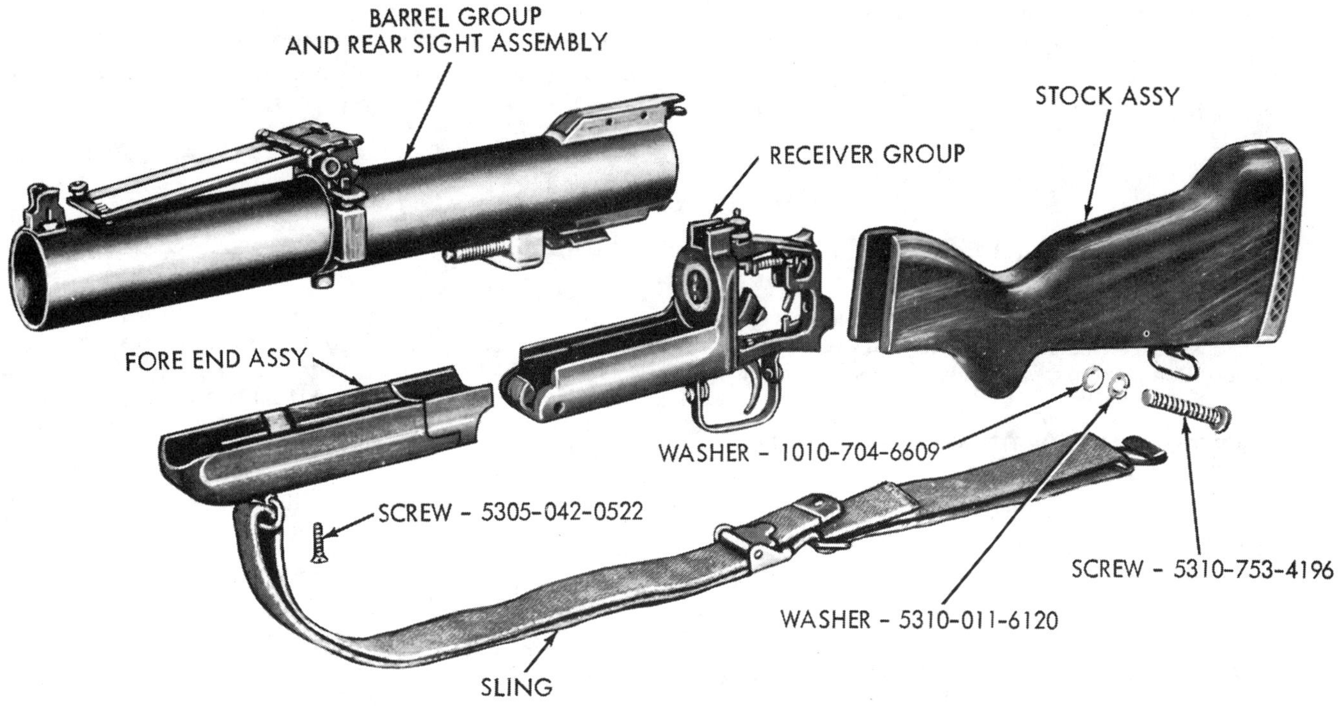

40mm Grenade launcher, field-stripped.

North Vietnam ("Democratic Republic of Vietnam")

54

The army of North Vietnam is equipped with the following weapons: 7.62mm Soviet.TT Model 1933 pistol, the Soviet 7.62mm M1944 carbine, the Soviet 7.62mm AK assault rifle, the Soviet 7.62mm SKS carbine, the Soviet 7.62mm DPM light machine gun, the Soviet RPD light machine gun, the Soviet 7.62mm RP46 light machine gun, the 7.62mm SG43 and SGM heavy machine guns, the 12.7mm DShK M1938/46 heavy machine gun and the Soviet 14.5mm ZPU4 heavy machine guns. The Soviet 7.62mm PPS M1943 and PPSh M1941 submachine guns are also used.

A large number of the weapons used, although of Soviet design, are not Soviet-made; they are of Chinese Communist manufacture and bear the Chinese Communist type designations in their markings. Many of these weapons have been shipped to the Viet Cong.

NORTH VIETNAMESE SUBMACHINE GUNS

A number of submachine guns which are basically modifications of other weapons have appeared in Vietnam—in the South, in the hands of the Viet Cong, and in the North, in the hands of the North Vietnamese army and militia.

7.62mm MODIFICATION OF THE FRENCH MODEL 49 SUBMACHINE GUN

The French apparently lost considerable quantities of the 9mm Parabellum Model 49 (MAT49) submachine guns. The Communist forces used this weapon as issued until their stocks of captured ammunition ran low. They then rebarreled the weapon for the Soviet 7.62mm Type P pistol cartridge (Chinese Communist Type 50 pistol cartridge). The rebarreled MAT49 has a noticeably longer barrel than the original weapon, but in other respects appears the same as the MAT49.

This weapon is loaded, fired, and field stripped in the same manner as the MAT49, which is covered in detail in the chapter on France.

7.62mm Modified MAT 49 Submachine Gun.

MODIFICATION OF THE 7.62mm TYPE 50 SUBMACHINE GUN

This weapon, which has appeared in the hands of the Viet Cong as well as the North Vietnamese, is a modification of the 7.62mm Chinese Communist Type 50, a copy of the Soviet Model PPSh M1941. The weapon is made up of components of the Type 50, some of the Type 56 (Chinese Communist copy of the AK), and the wire stock of the French MAT49.

The pistol grip and front sight are the same as those of the 7.62mm Type 56 assault rifle; the receiver and magazine are the same as those of the Type 50 submachine gun. The barrel jacket of the Type 50, reduced in length, is also used.

Characteristics of the Modified Type 50 Submachine Gun

Caliber: 7.62mm (Type 50 cartridge).
System of operation: Blowback, selective fire.
Length overall: Stock extended: 29.75 in.
Stock retracted: 22.56 in.
Barrel length: Approx. 10.5 in.
Feed device: 35-round, box magazine.
Sights: Front: Hooded post.
Rear: "L" type aperture.
Weight: 9 lb.
Muzzle velocity: 1640 f.p.s.
Cyclic rate: Approx. 900 r.p.m.

7.62mm Modified Type 50 Submachine Gun.

WEAPONS OF THE VIET CONG

In the days before the fighting in Vietnam developed into the large scale operations which are now being carried on, the Viet Cong—as the Viet Minh who were the Vietnamese north and south who forced the French to leave Indo-China—were equipped with a motley collection of homemade weapons. These were captured French weapons, United States weapons either captured from the French or supplied by Communist China, and a relatively few Soviet-designed Chinese Communist-made weapons. Before Diem took over control of South Vietnam, a number of dissident groups maintained private armies and took over some of the jungle workshops which had been established by the Viet Minh. Copies of the U.S. caliber .45 Model 1911A1 pistol and the 9mm Browning Hi-Power were made in these shops as were copies of the U.S. M1 rifle.

Within the past four years, the amount and types of Chinese Communist copies of Soviet small arms in the hands of the Viet Cong has increased. Appreciable quantities of French and United States small arms are still used by the Viet Cong, including the U.S. caliber .50 Browning M2 HB machine gun, modified to fit on the mount for the 12.7mm DShK machine gun and to use the Soviet-type antiaircraft sights.

All the small arms in use in North Vietnam except the 14.5mm machine gun, plus Chinese 7.92mm Mausers, rebuilt 7.92mm German Kar 98k's, Chinese-made 7.92mm ZB26 and Type 24 machine guns, Japanese 7.7mm Type 99 machine guns, German 7.92mm MG34 machine guns, and German 9mm MP40 submachine guns, are used by the Viet Cong. The tendency appears to be to standardize on the later model Chinese Communist weapons, but the older weapons will be encountered for some time to come.

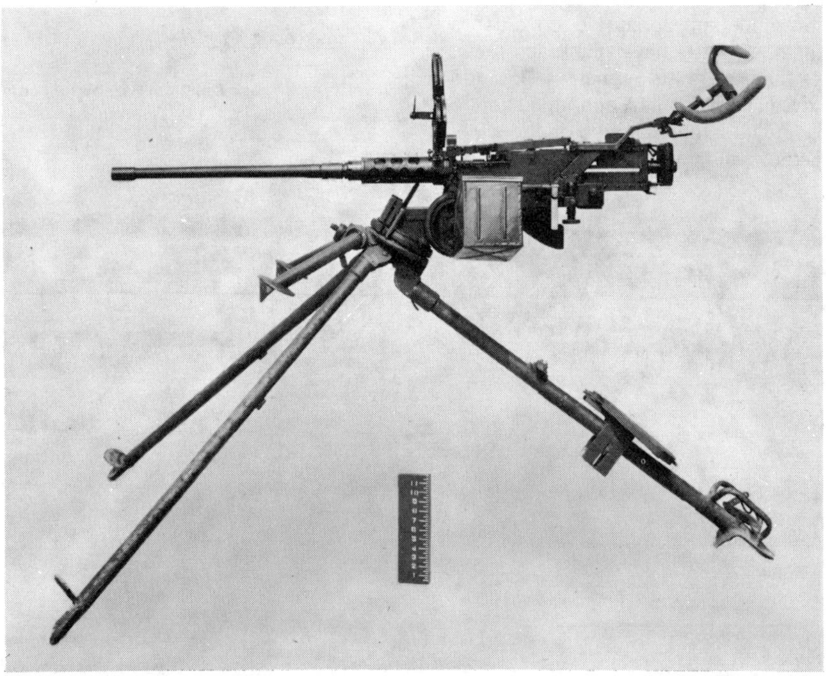

U.S. caliber .50 M2HB on DShK mount.

South Vietnam ("Republic of Vietnam")

The Army of the Republic of Vietnam is equipped mainly with United States Army weapons to include: the caliber .45 Model 1911A1 pistol, the caliber .30 M1 and M2 carbines, the caliber .30 M1 rifle, the caliber .30 Model 1918A2 Browning automatic rifle, the 5.56mm M16A1 rifle, the 7.62mm M60 machine gun, the caliber .30 Browning Model 1919A4 and 1919A6 machine guns, and the caliber .50 Browning machine gun. The caliber .45 Thompson M1 and M1A1 and the M3A1 submachine guns are also used. For details on these weapons, see the chapter on the United States.

French weapons such as the 9mm Parabellum Model 1950 pistol, the 9mm Model 1949 submachine gun (MAT 49) and the 7.5mm Model 36 and Model 1924A29 light machine gun may still be found in use by local militia units.

55

Prior to the establishment of the Diem government in South Vietnam, there were a number of dissident groups which had, in effect, private armies. The Cao Dai were one of these groups and manufactured arms in rather primitive workshops. Two of these weapons, a copy of the FN Browning Hi-Power and the U.S. Colt .45 Model 1911A1—both chambered for the 9mm Parabellum cartridge—are shown here. The finish on these pistols is surprisingly good considering the circumstances under which they were made. The quality of the metallurgy is questionable however.

Copies of the U.S. caliber .30 M1 rifle were also made by dissident groups. It is not certain that all the parts for these weapons were made in Vietnam, some parts may have been of U.S. manufacture and obtained from stocks of parts originally held by the French in that area.

Although Cambodia is not part of South Vietnam, it is geographically contiguous and it is therefore appropriate to consider a weapon of Cambodian manufacture. They have manufactured a 9mm weapon which combines features of the French M1950 and the U.S. M1911A1 pistols. The Cambodian pistol, although shaped like the French M1950 and having its safety mounted on the slide, has a barrel bushing, recoil-spring-plug-arrangement similar to that of the U.S. M1911A1.

Cao Dai 9mm Parabellum copy of U.S. M1911A1.

Cao Dai 9mm Parabellum copy of FN Browning Hi-Power.

Cambodian 9mm Pistol.

56

Yugoslavia

Yugoslavia uses the following weapons: The 7.62mm Pistol Model 57, the 7.62mm Rifle Model 59/66, Model 59, Model 64, Model 64A and Model 64B, the 7.62mm submachine guns Model 49 and Model 56, the 7.62mm light machine guns Model 65A and Model 65B, and the 7.92mm general purpose machine gun Model 53. U.S. caliber .50 Browning machine guns are also used as is the Soviet 7.62mm SGMT.

Older weapons such as the 7.92mm rifle Model 48 and various World War II Soviet and German weapons may still be held in reserve.

YUGOSLAV PISTOLS

Yugoslavia inherited quantities of Austrian 9mm Model 12 Steyr pistols when the State was established at the end of World War I. The new state adopted the FN Browning 9mm short (.380 ACP) Model 1922 pistol prior to World War II.

After World War II, Yugoslavia had quantities of 9mm Parabellum Luger and P38 pistols taken from German forces, and Beretta pistols captured from the Italians. The Tokarev 7.62mm TT M1933 pistol was adopted in the late forties. Yugoslavia manufactures a copy of the Tokarev which they call the Model 57.

The Yugoslav Model 57 differs from the Tokarev in that it has a nine round magazine rather than an eight round magazine. The Yugoslavs also manufacture a copy of the Tokarev in 9mm Parabellum which has been called the Model 65 and the Model 70. It also has a nine round magazine.

FN Browning M1922 caliber .380 (9mm Short) made for Yugoslavia prior to World War II.

Yugoslav 7.62mm Model 57 Pistol.

YUGOSLAV RIFLES

The newly created Kingdom of the Croats and Slovenes—Yugoslavia—obtained many Austrian Mannlicher rifles and Turkish Mausers at the time of its foundation. Few of the 7mm Serbian Mausers (Serbia became part of Yugoslavia), the Model 1889, 99/07, 99/08, or 1910 survived World War II and they were never a major factor in Yugoslav armament. Austrian Model 1895 Mannlichers were converted to 7.92mm and called Model 95M and Turkish Model 1890 and 1893 were also converted to 7.92mm, the conversion being called the Model 90T. French 8mm M1886M93, M1907/15, and Model 1916 rifles and carbines were also procured.

The Czech ZB 7.92mm Model 24 rifle was adopted as standard and weapons were manufactured at Kragujevac in Yugoslavia, in addition to purchases made from Czechoslovakia. This rifle is the same as the Czech Model 24 rifle, which is covered in detail in the chapter on Czechoslovakia.

After World War II, the Yugoslavs had quantities of 7.92mm Kar 98k's, and some Italian rifles. They also were furnished with 7.92mm Model 1944 carbines and Model 1891/30 rifles by the Soviets.

7.92mm Yugoslav Model 24 Rifle.

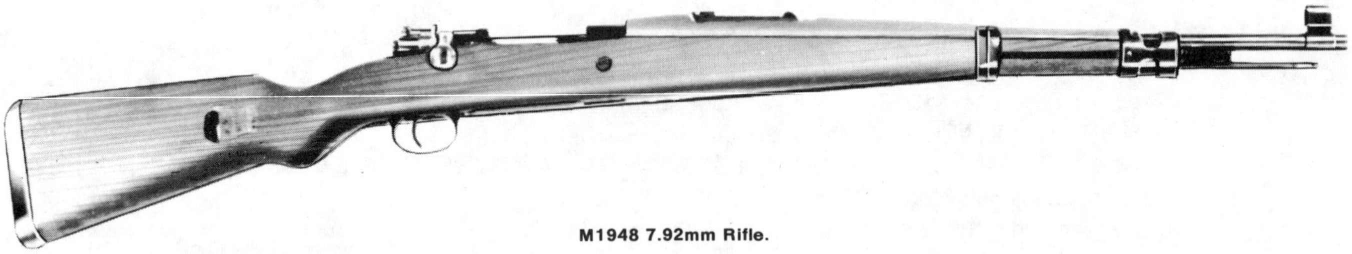

M1948 7.92mm Rifle.

THE YUGOSLAV 7.92mm RIFLE MODEL 1948

The Yugoslavs adopted a slightly modified copy of the Kar 98k as the Model 1948. This rifle is manufactured at Kragujevac.

The Model 1948 is loaded, fired, and field stripped in the same manner as the Kar 98k.

YUGOSLAV 7.62mm MODEL 59 RIFLE

The Model 59 is a copy of the Soviet 7.62mm SKS carbine and uses the same 7.62 X 39mm cartridges that the SKS does. Loading, firing, functioning and field stripping are identical with the SKS.

Characteristics of the Model 48 Rifle

Caliber: 7.92mm.
System of operation: Manually operated bolt.
Weight: 8.62 lb.
Length, overall: 42.9 in.
Barrel length: 23.3 in.
Feed device: 5-round, integral, staggered-row box magazine.
Sights: Front: Hooded blade.
Rear: Tangent w/ramp.
Muzzle velocity: 2600 f.p.s. (approx.).

YUGOSLAV SUBMACHINE GUNS

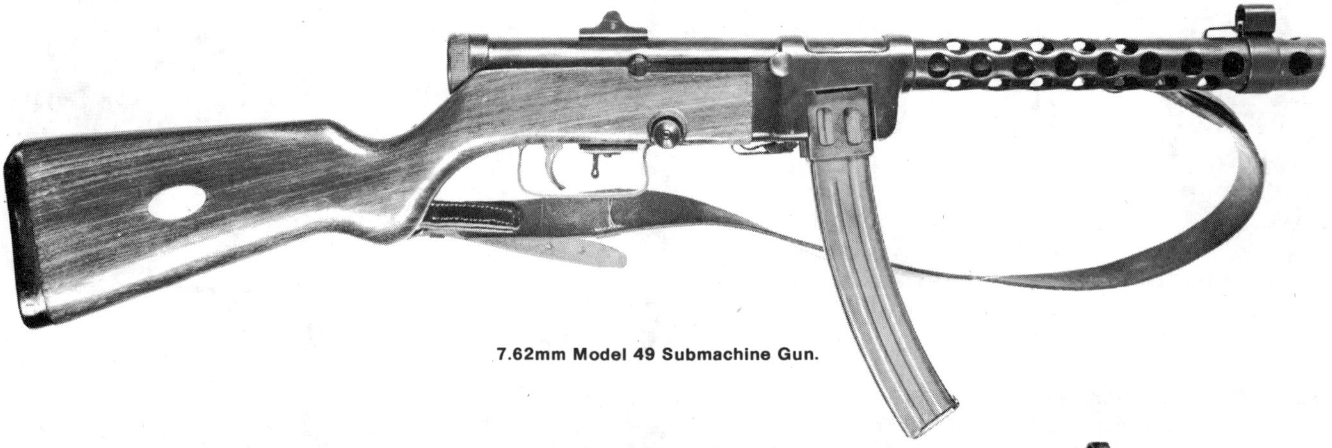

7.62mm Model 49 Submachine Gun.

7.62mm Model 56 Submachine Gun.

YUGOSLAV 7.62mm MODEL 59/66 RIFLE

The Model 59/66 is a modified copy of the Soviet SKS. Like the SKS it is chambered for the 7.62 X 39mm intermediate sized cartridge. The major differences from the SKS are the presence of a gas shut-off valve on the gas cylinder, and integral grenade launcher fitted to the barrel. A grenade launcher sight is mounted on the front top of the barrel. The grenade launcher sight pivots upward from the front sight base. As a matter of interest the diameter of the grenade launcher is 22mm which is the same as that of the grenade launchers in service in the Western World, but is not the same diameter as the grenade launchers in service in Poland, and some of the other Communist countries. The Model 59/66 has a rubber recoil pad fitted to the butt and its blade type bayonet is slightly longer than that of the SKS. It also has night sighting aids fitted to the front and rear sights. Yugoslavia advertises this weapon for export. With the exception of the gas shut-off valve which in the on position - turned up - is used for the launching of grenades, loading firing and functioning of the Model 59/66 is the same as that of the SKS.

YUGOSLAV 7.62mm MODEL 64, MODEL 70 (64A) AND MODEL 70A (64B) ASSAULT RIFLES

These weapons are all essentially modifications of the Soviet AK assault rifle. They are in the Yugoslav "FAZ" family of weapons which includes two light machine guns, the Models 65A and 65B. The Model 64 is essentially the same as the AK but has a longer barrel. As with all the rifles of this series it has a built in grenade launcher sight that pivots on the gas cylinder. The Model 64 uses a 20-round magazine as opposed to the 30-round magazine normally utilized in AK type weapons. Since the "FAZ" family of weapons has a bolt stop—bolt holding open device—which works on a notch cut in the magazine the standard AK magazines cannot be used with these guns.

The Model 70, which was previously called the Model 64A, is essentially the same as the standard wooden stocked AK except for grenade launcher sight, and magazine. The Model 70A, which was previously called the Model 64B, is the same as the Model 70 but has a folding steel stock. All of these weapons can be fitted with a 22mm grenade launcher or a short compensator/muzzle brake.

Characteristics of the Model 59/66

Caliber: 7.62mm (7.62 X 39mm)
System of operation: gas, semi-automatic only
Weight: 9.36 lb.
Length overall: 43.9 in.
Barrel length: 19.7 in.
Feed device: 10 round, staggered row, integral box magazine
Sights: Front: hooded post
Rear: notched tangent
Muzzle velocity: 2411 f.p.s.

Characteristics of the Model 64, Model 70 and M70A Rifles

	Model 64	Model 70	Model 70A
Caliber:	7.62mm (7.62X39)		
System of operation:	gas operated selective fire		
Weight:	8.6 lb.	8.3 lb.	8.2 lb.
Length overall: stock fixed:	40.9 in.	37.7 in.	37.7 in.
stock folded:			27.2 in.
Barrel length:	19.7 in	16.3 in.	16.3 in.
Feed device:	20 round detachable box	30 round detachable box	30 round detachable box
Sights: Front:	hooded post		
Rear:	tangent with notch		
Muzzle velocity:	2395 f.p.s.	2297 f.p.s.	2297 f.p.s.
Cyclic rates:	600-650 r.p.m.		

Yugoslav Model 70A (64B) Assault Rifle with grenade launcher fitted.

Yugoslav 7.62mm Model 59/66 Rifle.

Yugoslav 7.62mm Model 70A (64B) Assault Rifle with compensator/muzzle brake.

Yugoslavia adopted the 9mm Parabellum Vollmer Erma in the mid-thirties. After World War II, quantities of the German 9mm Parabellum MP38 and MP40 were available, as were Italian Beretta submachine guns and British Sten guns, as well as 7.62mm PPD 7.62mm Soviet pistol cartridge, which is called the Model 49. The and PPSh M1941 submachine guns furnished by the Soviet Union.

YUGOSLAV 7.62mm SUBMACHINE GUN MODEL 49

Yugoslavia developed a submachine gun chambered for the Model 49 is similar in appearance to the PPSh M1941, but differs from it internally. The bolt is similar to the Beretta Model 38A and the buffer is considerably more complicated than the plastic or rubber piece found on the PPSh M1941. The buffer of the Model 49 has a spring, separate from the operating spring, and split ring assembly which is retained by a collar on the end of the operating spring tube guide. The Model 49 is field stripped by twisting the receiver cap a quarter turn and removing the buffer, operating spring, and bolt assembly.

YUGOSLAV 7.62mm SUBMACHINE GUN MODEL 56

The Yugoslavs have developed a new and somewhat less complicated gun called the 7.62mm Model 56. The Model 56 has a folding stock similar to the German MP40 and uses a knife-type bayonet. The Model 56 submachine gun is now being made in 9mm Parabellum as the Model 65. It is essentially, except for caliber, identical to the Model 56.

YUGOSLAV MACHINE GUNS

7.92mm ZB30J Machine Gun.

Yugoslavia has used the 8mm 07/12 Schwarzlose, the 7.92mm Maxime Model 8M- a conversion of Serbian 7mm and Bulgarian 8 x 50mm Maxim guns to 7.92mm—the 8mm St. Etienne, and various models of the Madsen gun. The principal light machine gun, prior to World War II, was the 7.92mm ZB30J, a slightly modified version of the ZB30. The principal noticeable difference between the ZB30 and ZB30J is the presence of a knurled ring on the barrel just ahead of the receiver on the ZB30J. This shows up clearly in the photograph above. This weapon was made in Yugoslavia as well as in Czechoslovakia.

After World War II, Yugoslavia had 7.92mm MG34 and MG42 machineguns and various Italian machine guns. They were supplied with some 7.62mm DPs and Model 1910 Maxims by the Soviets. They tooled up for the manufacture of the 7.92mm MG42, which is called the Model53 or "Sarac" in Yugoslavia. This is the current standard machine gun.

THE YUGOSLAV 7.62mm MODEL 65A AND 65B LIGHT MACHINE GUNS

These guns are part of the Yugoslav "FAZ" weapons family. They, as the rifles covered earlier, are based on the Soviet AK in so far as basic design is concerned. The Model 65A and 65B are essentially the same in all basic respects except one: The 65A has a finned, quick change barrel. Both guns have bipods and flash hiders.

Characteristics of Yugoslav Submachine Guns

	Model 49	Model 56
Caliber:	7.62mm.	7.62mm.
System of operation:	Blowback, selective fire.	Blowback, selective fire.
Length overall:	34.4 in.	Stock extended: 34.25 in. Stock folded: 23.25 in.
Barrel length:	10.5 in.	9.84 in.
Feed device:	32-round, detachable staggered row magazine	32-round, detachable staggered row magazine
Sights: Front:	Hooded blade.	Hooded blade.
Rear:	L-type with "U" notch.	L-type with "U" notch.
Weight:	9.44 lb.	6.61 lb.
Muzzle velocity:	Approx. 1700 f.p.s.	Approx. 1700 f.p.s.
Cyclic rate:	Approx. 700 r.p.m.	570-620 r.p.m.

Characteristics of the 7.62mm Model 65A and 65B Light Machine Guns

	Model 65A	Model 65B
Caliber:	7.62mm (7.62X39)	7.62mm (7.62X39)
System of operation:	gas, selective fire	gas, selective fire
Weight:	12.3 lb.	11.4 lb.
Length overall:	41.5 in.	41.5 in.
Barrel length:	approx. 20 in.	approx. 20 in.
Feed device:	30 round, staggered row detachable box	30 round, staggered row detachable box
Sights: Front:	hooded post	hooded post
Rear:	notched tangent	notched tangent
Muzzle velocity:	2411 f.p.s.	2411 f.p.s.
Cyclic rate:	600-650 r.p.m.	600-650 r.p.m.

Small Arms Ammunition

57

For every military small arm in existence, there is at least one and, usually, a large number of cartridges. To cover in detail all of the cartridges used in the weapons covered in this book would require another book several hundred pages in length. The purpose of this chapter on small arms ammunition is to provide the reader with some appreciation of the differences between the various cartridges and thereby to give him an elementary course in cartridge identification.

TABULAR AMMUNITION DATA

Shown below are charts of pistol cartridges, rifle cartridges, and heavy machine-gun cartridges. In each of the three charts there is shown, opposite each cartridge (and case type) listed, the following data: complete round weight, bullet weight, and propellant weight; and complete round length, case length, bullet diameter, and case diameter. Also shown for each cartridge are some common synonyms and the best known weapons in which each cartridge is used.

The dimensions and weights given are approximate. There are many variations insofar as weights are concerned due to the different bullets used with type variations within a given caliber and the differing propellant weights among different types. Weights given are for the basic ball round in each type, except the Soviet 14.5mm, in which data on the API cartridge is given. Diameters given are for the point of maximum diameter of bullet and cartridge case.

The data given in the charts should NOT be used as a guide in the reloading of cartridges. There is considerable variation in the performance of the various propellants in use throughout the world. Therefore no presentation such as that given here can serve as a guide to handloading unless the specific military or commercial model designation for the propellant is given.

PISTOL CARTRIDGES

Cartridge and Case Type	Complete Round Weight	Bullet Weight	Propellant Weight	Complete Round Length	Case Length	Bullet Diameter	Case Diameter	Some Common Synonyms	Best Known Weapons in Which Used
7.62mm Pistol Rimless Bottlenecked	167 gr.	87 gr.	7.71 gr.	1.36 in.	.97 in.	.307 in.	.39 in.	7.62mm Type P (Soviet) 7.62mm M48 (Czech)	TT M1933 Pistol, M34/38 SMG, PPD 40 SMG, PPSh 41 SMG, PPS 43 SMG (All Soviet) SMG and M52 Pistol (Czech) and various satellite copies of Soviet pistol and SMGs, can also be used in weapons chambered for 7.63mm Mauser cartridge.—Note that Czech cartridge has a heavier charge.
7.65mm Browning Straight Semi-rim	121.0 gr.	73 gr.	2.47 gr.	.98 in.	.68 in.	.308-.311 in.	.354 in.	.32 ACP, 7.65mm Browning short	Colt Automatic, Walther PP & PPK, (German & various copies), Browning M1910 & M1922, large numbers of pocket automatic pistols throughout the world.
7.65mm Long Rimless Straight	132 gr.	85-88 gr.-	5.0 gr.	1.19 in.	.78 in.	.309 in.	.335 in.	7.65 Longue Pour Pistole et Pistolet Mitrailleur .30 cal. Pistol Cartridge M1918 (Pedersen Device Cartridge)	French M1935A and M1935S Pistols and M1938 SMG.
7.65 Luger Rimless Bottlenecked	160.0 gr.	93 gr.	5.2 gr.	1.17 in.	.85 in.	.308 in.	.392 in.	Cal. .30 Luger	Swiss M 06/29 Pistol, some Pre-W.W. II Bergmann and SIG SMGs.
8mm Jap Rimless Bottlenecked	177.5 gr.	102 gr.	5.0 gr.	1.23 in.	.83 in.	.320 in.	.411 in.	8mm Type 14	Japanese Type 14 and Type 94 pistols and Type 100 SMG.
8mm Lebel Rimmed Straight	194.5 gr.	120 gr.	14.0 gr. black powder 11.5 gr. smokeless	1.44 in.	1.07 in.	.323 in.	.408 in. (Rim)	8mm M1892	French M1892 Revolver.
9mm Short Rimless Straight	148.8 gr.	95 gr.	3.5 gr.	.98 in.	.68 in.	.356 in.	.372 in.	9mm Browning Short, 9mm Kurz, 9mm Corto, .380 A.C.P., 9mm Court	Colt automatic, Walther PP and PPK and various copies, Hungarian M37, Italian M34 Beretta.
9mm Parabellum Rimless Straight	164. gr.	115 gr.	5.6 gr.	1.17 in.	.76 in.	.356 in.	.457 in.	Pistolen Patrone 08, 9mm M1, 9mm Mk1, 9mm M38, 9mm M39, 9mm Luger	German P08, P38, MP38, MP40, British Browning No. 2 Pistol, Sten SMGs and L2A3 SMG, French M50 Pistol and M49 SMG. Belgian Browning HP Pistol and Vigneron SMG, Swedish and Finnish Lahti Pistol and Suomi SMGs. Swedish M45 SMG, Israeli Uzi SMG, Madsen SMGs. Beretta M38, 38/42, 38/49, 5 & 12 SMG, Beretta M951 Pistol.
9mm Largo Rimless Straight	192 gr.	124 gr.	6.6 gr.	1.32 in.	.91 in.	.354 in.	.385 in.	9mm Bergmann Bayard	Spanish Super Star Pistol and Star Model Z45 SMG.
9mm Makarov Rimless Straight	156 gr.	94 gr.	3.7 gr.	.97 in.	.71 in.	.363 in.	.389 in.		Soviet Makarov and Stechkin pistols.
.380 MK II Rimmed Straight	246.5 gr.	181 gr.	5.0 gr.	1.22 in.	.76 in.	.357 in.	.435 in.	.38 S&W and .38 Webley Scott	British Pistol No. 2 (Revolvers) in all Marks, Webley Mark IV Pistol, S & W .38/200.
.45 M1911 Rimless Straight	327. gr.	230 gr.	5 gr.	1.275 in.	.91 in.	.452 in.	.480 in.	.45 ACP, 11.43mm, 11.25mm, 11mm M40	U. S. Pistol M1911 and M1911A1, M1928 A1, M1, M1A1, M3, and M3A1 SMGs. Norway—pistol M1914. Argentina-Ballester.
.455 Webley Rimless Straight	324.0 gr.	224 gr.	7 gr.	1.22 in.	.91 in.	.455 in.	.502 in.	.455 Webley auto.	Webley Automatic Pistol.
.455 Revolver Rimmed Straight	350 gr.	265 gr.	7.5 gr.	1.22 in. to 1.26 in.	.76 in.	.455 in.	.532 in.	.455 Webley revolver .455 in. Mark II	Pistol No. 1 Marks 4, 5, & 6.

RIFLE CARTRIDGES

Cartridge and Case Type	Complete Round Weight	Bullet Weight	Propellant Weight	Complete Round Length	Case Length	Bullet Diameter	Case Diameter	Some Common Synonyms	Best Known Weapons in Which Used
5.56mm Rimless Bottlenecked	182 gr.	55 gr.	25 gr.	2.26 in.	1.76 in.	.224 in.	.378 in.	.223 5.56mm M193	U.S. M16, M16A1, and AR-18 rifles, Stoner 63 System.
6.5mm Dutch Rimmed Bottlenecked	356 gr.	156 gr.	37 gr.	3.03 in.	2.11 in.	.263 in.	.526 in.	6.5mm Rumanian 6.5mm M93 6.5 x 53mm R	Dutch M92, M95 rifles and M95 carbine. Rumanian M92, M93 rifle & M93 carbine.
6.5mm Italian Rimless Bottlenecked	350 gr.	162 gr.	35 gr.	3.01 in.	2.06 in.	.2655 in.	.447 in.	6.5mm Mannlicher Carcano, 6.5mm M91-95	Italian M91 rifles and carbines Breda M30 MG.
6.5mm Jap Semi-rim Bottlenecked	326 gr.	139 gr.	33 gr.	2.98 in.	2.00 in.	.262 in.	.476 in.	6.5mm Type 38	Japanese Type 38 rifle, Type 38 and Type 44 carbine, Type 11 and Type 96 MG.
6.5mm Mauser Rimless Bottlenecked	363 gr.	139 gr.	36 gr.	3.15 in.	2.17 in.	.264 in.	.478 in.	6.5mm M94, 6.5mm Norwegian, 6.5mm Swedish, 6.5 x 55mm	Norwegian M94 rifle and M12 carbine. Swedish 94 carbine. M96, M38 and M42 rifles, M21 and M37 automatic rifles (used in some Swedish MGs as well).
7mm Mauser Rimless Bottlenecked	377 gr.	172 gr.	38 gr.	3.00 in.	2.24 in.	.284 in.	.474 in.	7 x 57mm	Spanish M93 rifle and M95 carbine.
7.5mm French Rimless Bottlenecked	363 gr.	139 gr.	44 gr.	2.99 in.	2.13 in.	.307 in.	.484 in.	7.5mm M1929	French M 07/15, M34, M36, M49, M49/56 rifles and M1924M29 automatic rifle. M31 and M52 MG.
7.5mm Swiss Rimless Bottlenecked	404 gr.	174 gr.	49.35 gr.	3.05 in.	2.18 in.	.308 in.	.496 in.	7.45mm Swiss 7.5mm M11 7.5mm Schmidt Rubin	Swiss M11 rifle and M11, M31, M31/42 and M31/43 carbines, M57 assault rifle. M11, M25, and M51 machine guns.
7.62mm NATO Rimless Bottlenecked	375 gr.	150 gr.	48 gr.	2.80 in.	2.01 in.	.308 in.	.496 in.	7.62mm Ball M59 7.62 x 51mm 7.62mm M1954 7.62mm OTAN .308 Winchester 7.62mm Ball M85	U. S. M14 rifle, M60 and M73 machine guns. British L2A1 rifle, Belgian FN FAL rifle and FN MAG. M.G. Canadian C1 rifle and C2 LMG. W. German FN and G3 rifles.
7.62mm Russian Rimmed Bottlenecked	348 gr.	148 gr.	50 gr.	3.03 in.	2.11 in.	.311 in.	.564 in.	7.62mm M1908 type L 7.62 x 54mm	Soviet Mosin-Nagant rifles and carbines, Simonov M36 and Tokarev M32, M38, and M40 rifles and carbines. Maxim M1910, DP, DPM, DT, DTM, DA, ShKAS, SG-43, SGM, and RP-46 MGs. Soviet Satellite copies of these weapons.
7.62mm M43 Rimless Bottlenecked	253 gr.	122 gr.	25 gr.	2.20 in.	1.52 in.	.311 in.	.45 in.	7.62mm Russian Short 7.62mm M1934 Type PS 7.62 x 39mm	Soviet AK assault rifle, SKS carbine, RPD LMG.
7.62mm M52 Rimless Bottlenecked	293.7 gr.	132 gr.	27 gr.	2.35 in.	1.77 in.	.310 in.	.442 in.	7.62mm Czech Short	Czech M52 Rifle and LMG.
7.65mm Mauser Rimless Bottlenecked	390 gr.	174 gr.	38.6 gr.	2.95 in.	2.11 in.	310 in.	.472 in.	7.65mm M30 7.65 x 54mm	Belgian M89 rifles and carbines, M35 and 36 rifles. Argentine M91 and M09 rifles. Turkish M91 and M03 rifles.
7.7mm Rimless Jap Rimless Bottlenecked	415 gr.	182 gr.	43.13 gr.	3.14 in.	2.25 in.	.310 in.	.47 in.	7.7mm Type 99	Japanese Type 99 rifle and LMG.
7.7mm Semi-rim Jap, Semi-rim Bottlenecked	429 gr.	200 gr.	44.18 gr.	3.14 in.	2.25 in.	.310 in.	.49 in.	7.7mm Type 92	Japanese Type 92 MG.
7.92mm Mauser Rimless Bottlenecked	408. gr.	198 gr.	47 gr.	3.15 in.	2.24 in.	.323 in.	.468 in.	7.9mm SS 7.9 x 57mm 7.9 IS 7.9 JS 8mm Mauser .315 Mauser rimless 8 x 57mm M98.	German M98 rifles and carbines, MO8, 08/15, 08/18, 15, 17, 34 and 42 MGs. FG-42. Czech M24, M33 rifles and carbines, M26, M30, M30J and M37 MGs. British BESA MGs, Chinese-Brens MK 2.
8nn Austrian Rimmed Bottlenecked	437 gr.	244 gr.	42 gr.	3.00 in.	1.99 in.	.323 in.	.554 in.	8mm M1893 8 x 50mm R	Austrian M90 rifles and carbines, M07/12 MG, M95 rifles and carbines.
8mm M31 Rimmed Bottlenecked	441 gr.	208 gr.	55 gr.	3.02 in.	2.21 in.	.330 in.	.55 in.	8mm M30S 8 x 56mm R	Hungarian M35 rifle and M31 MG. Austrian M30 MG, rebarrelled M95 Mannlichers.
8mm Lebel Rimmed Bottlenecked	429 gr.	198 gr.	43 gr.	2.95 in.	1.98 in.	.328 in.	.629 in.	8mm M1886 D(AM)	French M86M93, M1890, M92, M07, M07/15, M16, M17, M18 rifles and carbines. M14 and M15 MGs.
Cal. .30 Carbine Rimless Straight	195 gr.	110 gr.	14.5 gr.	1.67 in.	1.28 in.	.308 in.	.356 in.	Cal. .30 M1 Carbine	U.S. M1 and M2 carbines.
Cal. .30-06 Rimless Bottlenecked	396 gr.	150 gr.	50 gr.	3.33 in.	2.49 in.	.308 in.	.469 in.	Cal. .30 M2 Ball Cal. .300 Browning 7.62 x 63mm	U. S. M03, 03A1, 03A2, 03A3, 03A4 Springfield rifles, M17, M1, M1C, M1D rifles. Browning automatic rifle M18, M18A1, 18A2. Browning M17, 17A1, 19A4, 19A5, 19A6, M37 MGs.
Cal. .303 Rimmed Bottlenecked	384 gr.	174 gr.	37.5 gr.	3.04 in.	2.21 in.	.311 in.	.53 in.	.303 in. Mark 7 .303 British	British No. 1, MKs 1, 2, 3, 3*, 4, 5, 6, No. 3, MK 1, 1*, No. 4 MK 1, 1*, 2, No. 5 MK 1, 1* rifles. Lewis MG, Hotchkiss MG, Bren MKs 1, 2, 3, and 4 MGs. Vickers MK 1 MG. Canadian Ross M05 and M10 rifles.

HEAVY MACHINE GUN CARTRIDGES

Cartridge and Case Type	Complete Round Weight	Bullet Weight	Propellant Weight	Complete Round Length	Case Length	Bullet Diameter	Case Diameter	Some Common Synonyms	Best Known Weapons in Which Used
Cal. .50 Rimless Bottlenecked	1800 gr.	709 gr.	240 gr.	5.45 in.	3.90 in.	.511 in.	.804 in.	Cal. .50 Ball M2 Cal. .50 Browning	U. S. M2, M2HB, M3 MGs.
12.7mm Soviet Rimless Bottlenecked	2160 gr.	788 gr.	271 gr.	5.76 in.	4.25 in.	.511 in.	.85 in.	12.7mm	Soviet DShK M38, DShk M38/46 and UB MGs.
13.2mm Rimless Bottlenecked	1836 gr.	805 gr.	232 gr.	5.35 in.	3.90 in.	.531 in.	.897 in.	13.2mm Hotchkiss 13.2mm Type 93	French M30 MG. Jap Type 93 MG.
14.5mm Rimless Bottlenecked	3089 gr.	979 gr.	.445 gr.	6.130 in.	4.48 in.	.585 in.	1.06 in.	14.5mm B-32	Soviet KPV MGs.

58 Sporting Arms

by George C. Nonte, Jr.

For all practical purposes, sporting arms development has paralleled that of military arms. As can be seen in Part I of this book, virtually all forms of rifles, pistols, and revolvers were originally developed primarily for military use. This is quite understandable since a far greater market lies in the military field. Single-shot, lever and bolt-action, self-loading rifles, and both revolving and self-loading pistols were all adapted to sporting use from military developments. Even today's most modern target arms are primarily improvements upon basic designs developed for military use. However, in many instances, the designs either failed to be accepted or have since been discarded by military establishments. An excellent example of this is the M98 Mauser, long discarded as a major military rifle, but still the basis for dozens of currently successful hunting and target rifle models. Nearly all double-action revolvers produced for sporting use—well over a million per year—are simply modernized versions of original military designs. However, in recent years, sporting rifle designs have diverged from the military. The vast firepower requirements of military use have resulted in selective-fire and full-automatic designs that have no real sporting application and that are far too costly for the sporting market. Consequently, recent sporting designs bear no resemblance to contemporary military types; though, they do use the same basic mechanisms and operating principles.

Of rifles, only one design form has a purely sporting heritage and that is the slide or trombone action. It is manually-actuated by a sliding fore-end. Aside from that, basic principles found in other types are utilized. In fact, the slide-action rifles (and shotguns) offered today generally have actions virtually identical to existing semiautomatic models. Only the method of applying actuating power is different.

One may note, too, as a generality applicable to this chapter, that a rather narrow construction necessarily has been attached to the term "Sporting Arms," for coverage of a number of rifles—principally many .22s, and some of the big-bores at the very opposite end of the caliber spectrum, of interest to hunters of very large game—could not be attempted owing to space limitations.

Shotguns, since the beginning of the cartridge period, have developed completely apart from military arms. While they do have limited military uses, such requirements are normally met by procurement of standard or slightly modified off-the-shelf sporting guns. Repeating shotgun mechanisms are generally quite similar to those used in rifles of the same basic type—modified only as necessary to accommodate the lower chamber pressures and larger size of shotshells.

The traditional double-barrel shotgun exists in such widely diverse forms and designs that it is impossible to be specific. The basic design types solidified well before the turn of the present century and are produced in countless makes and models all over the world. Because of the diversity, it isn't possible, at this time, to cover them in this book.

BOLT-ACTION RIFLES

BSA (BIRMINGHAM SMALL ARMS CO.), BRITAIN

The BSA Monarch rifle is distributed in the U.S. by J. L. Galef & Sons in standard and varmint styles. The BSA Monarch action is essentially a modified Mauser type with dual opposed locking lugs at the bolt head. It utilizes a counterbored bolt face and a short spring-loaded hook extractor let into the counterbore. Firing pin and cocking system are typically Mauser; however, the bolt sleeve completely encloses the head of the firing pin, and the safety has been transferred to the trigger assembly and protrudes alongside the rear receiver tang. A fully-adjustable single-stage trigger is fitted, as is a hinged magazine floor plate retained by a pivoted latch in the front of the guard bow. Functioning and manipulation is identical with that of other Mauser-type rifles.

Characteristics of BSA Monarch Rifle

Caliber: .222 and 7mm Remington Magnum; .243, .270, .308 Winchester; .30-06.
Barrel length: 22 in.
Overall length: 41 in.
Weight: 7 lbs. 2 oz.
Magazine capacity: 5 rounds; 3 in belted magnum calibers.

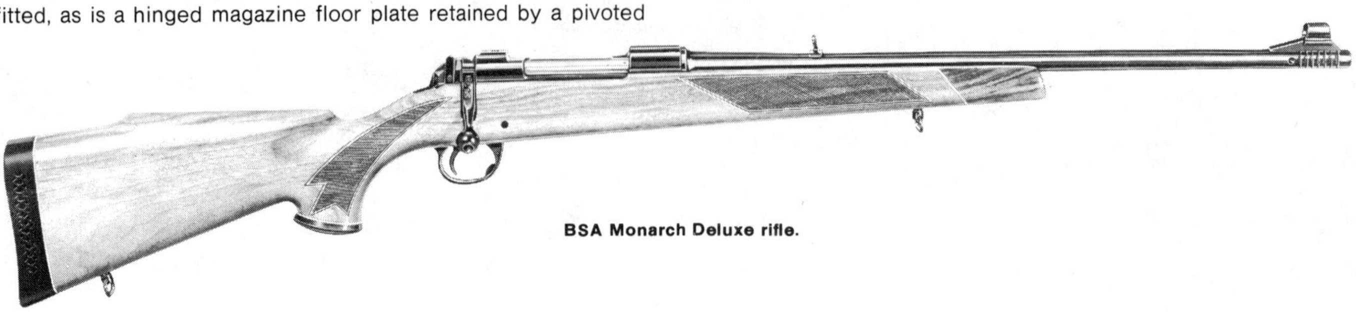

BSA Monarch Deluxe rifle.

BROWNING, U.S.A.

Browning distributes an extensive line of high-power, bolt-action sporting rifles in the U.S., though it does not produce those guns in this country. The bulk of the Browning models are simply the FN rifle manufactured to Browning's own particular specifications of finish, weight, and stock. Browning rifles for cartridges of the .30-06 and larger class are the FN. The so-called "short-action" Brownings are manufactured by SAKO of Finland to Browning specifications.

In both instances, the detailed information and characteristics listed elsewhere in this volume under the names of the manufacturers indicated above apply.

CHAMPLIN FIREARMS INC., U.S.A.

Champlin produces several variations of a single basic model built upon a massive action of original design. The large-diameter bolt is characterized by three equally-spaced lengthwise ribs running its full length. The ribs are interrupted to form six locking lugs—three front, three rear. The ribs between the lugs function as bolt guides, sliding in corresponding slots in the receiver. The bolt face is deeply counter-bored to shroud the case head, and is fitted with a hook-type extractor. At the rear, the bolt is closed by a streamlined, Weatherby-style bolt sleeve completely enclosing the cocking piece. The receiver is octagon in section through its upper half, the bottom being square and fitted with a conventional Mauser-type recoil lug. The manual safety is located shotgun-style on the rear receiver tang. The trigger assembly is characteristic of the modern, single-stage, fully-adjustable type with pivoted sear. The trigger guard/magazine assembly is typical of the Mauser M98, fitted with a hinged floor plate secured by a pivoted latch in the front of the guard bow.

Champlin rifles are essentially custom-built with many optional features, finishes, and dimensional and weight variations offered. In view of this, there are no "standard" specifications, and prices begin in the $700 range. Loading, firing, and manipulation are as for Mauser-type actions.

FN, BELGIUM

Fabrique Nationale has manufactured Mauser rifles of various types continuously since 1889. Following WW II, this firm began producing M98 Mauser-type sporting rifles for sale throughout the world. The first such rifles were identical in all respects to the M98 rifles produced by FN during the war except for stock and sights. However, a distinct sporting variation soon developed. It consisted of the basic M98 barrel, bolt, and trigger guard fitted with a single-stage fully-adjustable sporting trigger containing a manual safety which protruded at the right side of the rear receiver tang. The bolt sleeve was streamlined and stripped of the original military-style safety. The magazine is fitted with a hinged floor plate retained by a catch in the front of the trigger guard bow. Functioning and manipulation of FN rifles is identical to the standard M98 Mauser, which see.

The FN action is widely used by other makers to produce Mauser-type sporting rifles under a variety of names. At the present time, Firearms International is the exclusive U.S. distributor of FN rifles and actions and utilizes the latter in its own style rifle called the "Musketeer," which is assembled in this country.

Characteristics of the FN Rifle

Caliber: Most popular U.S. sporting from .222 to .458.
Operation by: Manual; Mauser-type rotating bolt.
Barrel length: 22 in. to 24 in.
Overall length: 42-1/2 in. to 44-1/2 in.
Weight: 7-3/4 lbs.
Sights: Open, adjustable.
Magazine capacity: 5 rounds; 3 in belted magnum calibers.

Browning Safari Grade, standard model rifle.

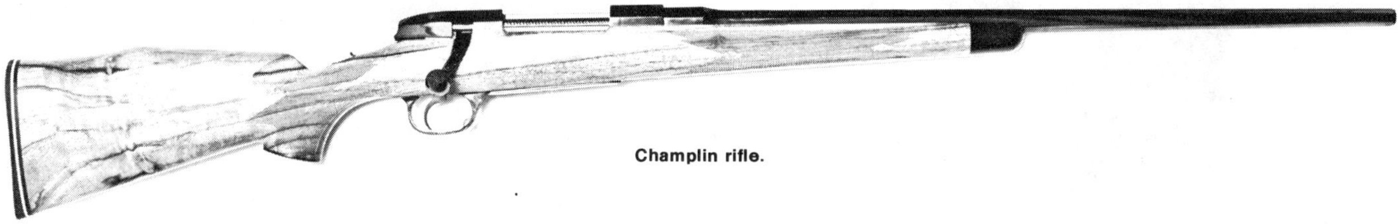

Champlin rifle.

Fabrique Nationale Musketeer rifle.

HARRINGTON & RICHARDSON, U.S.A.

During the early 1960's, Harrington & Richardson added a high-power, bolt-action sporting rifle to its extensive line. This rifle is designated as Model 300, 301, or 330, depending upon finish, barrel, and stock. All are built upon the FN "Supreme" Mauser-type action by H&R. Particulars given under "FN" are applicable. In addition, H&R builds a .17 caliber wildcat rifle on the SAKO L-461 action for which particulars will be found under "SAKO" elsewhere in this volume.

COLT SAUER, GERMANY

In collaboration with Sauer & Sohn, W. Germany, Colt introduced the rear-locking Colt Sauer rifle in late 1972. It is of unusual design, having a non-rotating bolt locked at the rear by three retractable locking lugs actuated by a rotating cam on the bolt handle. It also features unique barrel attachment, a chamber indicator, and roller-bearing trigger assembly. The entire rifle is of fine quality and excellent finish.

Characteristics of Colt Sauer Rifle

Caliber: .243, .30-06 and similar.
Operation by: Manual, rotating bolt handle.
Barrel length: 22".
Overall length: 42".
Weight: 6-3/4 lbs.
Sights: None, fitted for scope.
Magazine capacity: 4.

ITHACA GUN COMPANY, U.S.A.

Ithaca first entered the center-fire rifle field in early 1969. At present, a single model, the LSA-55, is available in Standard and Deluxe versions. This is essentially a modified Mauser-type, bolt-action rifle of conventional style and configuration. It utilizes a recessed hook-type extractor rather than the long leaf type of the M98. A manual safety is pivoted on the right side behind the bolt handle, and the hinged floor plate is secured by a latch inside the trigger guard bow. Integral male scope-mounting dovetails are machined in the receiver ring and bridge. The LSA-55 series is manufactured in Scandinavia.

Characteristics of the Ithaca Rifle

Caliber: .243 and .308 Winchester; .22-250 and 6mm Remington.
Operation by: Manual, rotating bolt.
Barrel length: 23 in.
Overall length: 42-1/2 in.
Weight: 6-1/2 lbs.
Sights: Open, adjustable.
Magazine capacity: 4 rounds.

MANNLICHER SCHOENAUER, AUSTRIA

Mannlicher Schoenauer sporting rifles have been produced at Steyr Daimler Puch in Austria since the early 1900's. The basic sporting rifle is based upon the military M1903 model described in detail under "Greece." Numerous improvements have been made on the original action, and the current production model is

Harrington & Richardson M330 "Ultra" rifle.

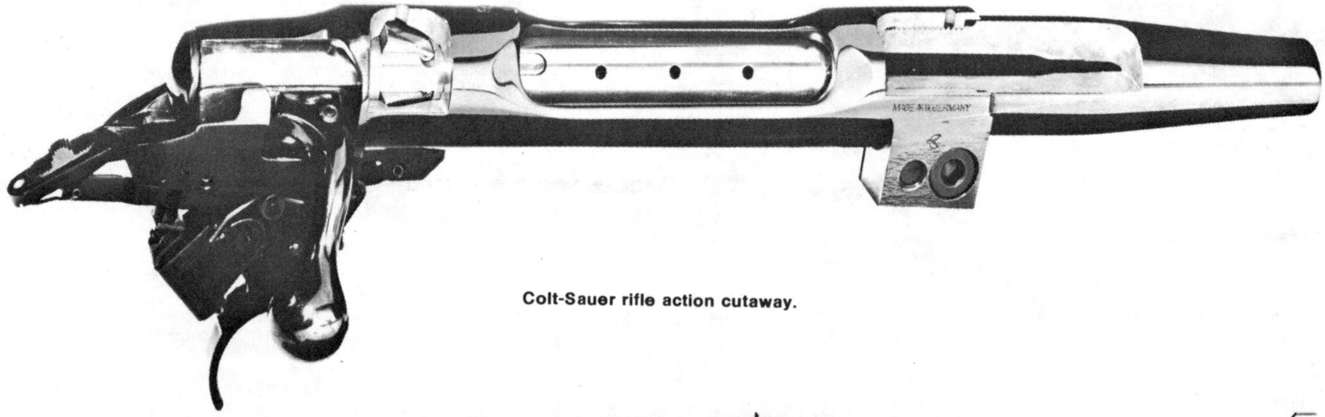

Colt-Sauer rifle action cutaway.

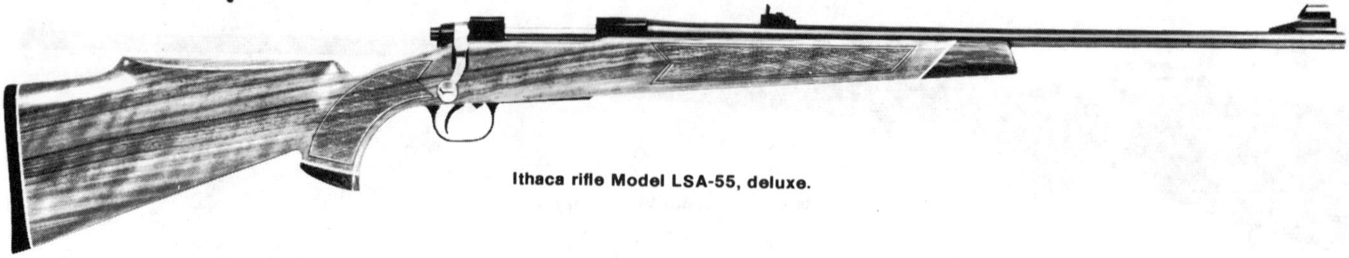

Ithaca rifle Model LSA-55, deluxe.

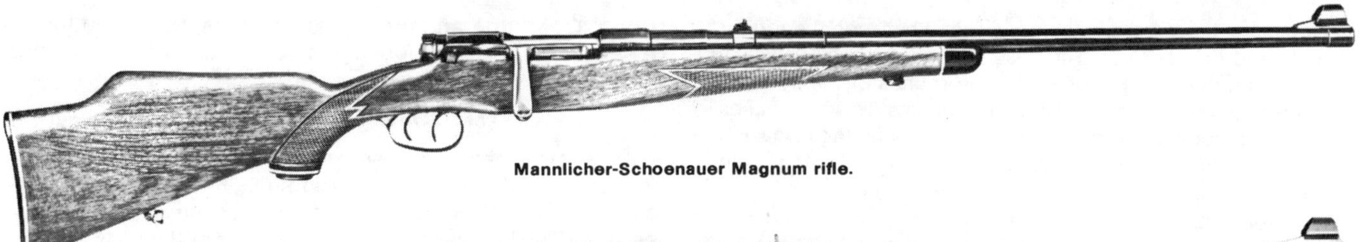

Mannlicher-Schoenauer Magnum rifle.

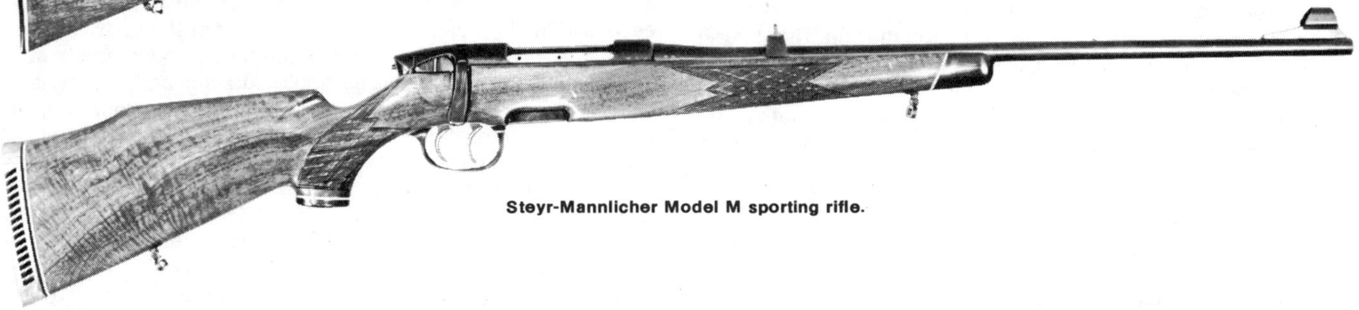

Steyr-Mannlicher Model M sporting rifle.

known as the "1961-MCA." It is distributed in the U.S. by Stoeger Arms Corp. This is a top-quality sporting arm demonstrating a very high degree of workmanship and reliability. Mechanically, the current version differs from the M1903 primarily in that the safety has been moved from the bolt sleeve to the right side of the receiver tang and a sophisticated fully-adjustable, single-stage or optional double-set trigger mechanism is offered. It is also larger in all respects than the M1903 in order to accommodate today's larger cartridges. Functioning and manipulation are identical with those for the M1903, which see.

Characteristics, Mannlicher Schoenauer Rifle

Caliber: 6.5x54mm M-S; .243, .270, and .308 Winchester; .30-06.
Operation by: Manual; Mannlicher-type rotating bolt; rotary magazine.
Barrel length: 18-1/4 in. to 22 in.
Overall length: 42 in. to 44 in.
Weight: 7-1/4 to 8 lbs.
Sights: Open; hooded front.
Magazine capacity: 5 rounds.

STEYR-MANNLICHER, AUSTRIA

In 1968, the Steyr-Mannlicher design was introduced by the U.S. distributor, Stoeger Arms Corp. This is a lightweight design intended to follow in the tradition of the Mannlicher-Schoenauer carbine, both in styling and in the more desirable mechanical features. The locking system consists of six rearward lugs locking into the receiver bridge. Cocking is accomplished in the Mauser manner. Only minimum cutouts are made in the receiver to permit ejection and feeding; and the rear of the bolt is closed by a streamlined sleeve which blends into the lines of the receiver. A detachable rotary-type magazine is employed. A fully-adjustable single-stage trigger is offered, with a European-styled double-set trigger optional at no extra cost. Particularly unusual is the appearance of the barrel. The barrel is hammer-forged about a rifling mandrel, and the forging marks on the exterior are retained—resulting in a many-faceted surface rather than the usual smooth polished barrel finish. The present models SL and L are based on a short action suitable for cartridges from the .222 Remington to the .308 Winchester. Two other lengths are being prepared—the Model M for .30-06 class cartridges and the Model S for the larger cartridges up through the .375 and .458 Magnums.

Characteristics of the Steyr-Mannlicher Rifle

Caliber: .222, .222 Magnum, and .223 Remington; .22-250; .225, .243, and .308 Winchester.
Operation by: Manual; Mannlicher-style bolt.
Barrel length: 23-5/8 in.
Overall length: 42-1/2 in.
Weight: 6 lbs. 4 Oz.
Sights: Open; fixed.
Magazine capacity: 5 rounds.

MAUSER, WEST GERMANY

In 1966, Mauser Werke introduced a radically new rotating-bolt, manually-operated, bolt-action magazine rifle. The entire action design is a complete departure from previous Mauser practice. That the design is not as new as the date of introduction might indicate is evident by the fact that this writer has examined a quite similar prototype acquired from the Mauser plant late in WW II. The action is unusually short; measuring only 4-1/2 in. in length. The receiver, as we know it, does not exist. It is replaced by a simple frame containing guide tracks for the bolt and a mortised seat for the quick-detachable barrel. The short bolt body is fitted with a conventional handle near its front end and carries dual, opposed locking lugs. The lugs engage abutments inside a barrel extension, thus locking the bolt to the barrel proper instead of joining the two by a receiver ring as in past practice. The bolt body is largely enclosed by a sliding, non-rotating sleeve engaging tracks on the receiver.

In unlocking, the bolt is first rotated independent of the sleeve; then the two are drawn rearward and thrust forward as a single unit to accomplish extraction, ejection, feeding, chambering, locking, and unlocking. This extremely short action requires that the

Mauser Model 660 rifle.

double-column box magazine be sandwiched between the trigger assembly and the receiver. This results in an unusually shallow magazine with the capacity of only three rounds of .30-06 class cartridges and two rounds in the belted magnums. This also makes it necessary to route the trigger-sear linkage around the magazine box. Single-stage and double-set triggers, both fully-adjustable are offered. The barrel attachment method is unique.

The barrel proper is fitted with a barrel extension which contains locking abutments to engage the locking lugs of the bolt. The lower surface of this extension carries ribs fitting very closely in corresponding grooves in the receiver. And, the barrel assembly is secured by a clamp screw. Since the bolt locks into the extension which is rigidly attached to the barrel, there is no longitudinal stress upon the receiver, so little strength in that direction is necessary. However, accurate alignment is quite important. This method of barrel attachments makes possible quick interchange of barrels in different weights and/or calibers. In fact, the "Diplomat" Model is supplied cased with barrels of three calibers. Interchangeability of calibers is limited only by magazine shape and size, and bolt face dimensions. The Mauser 660 is distributed in the U.S. by Mauser-Bauer.

Characteristics of the Mauser 660 Rifle

Caliber: Most popular U.S. from .243 Winchester through .458; most popular metric.
Operation by: Manual; rotating telescoping bolt.
Barrel length: 20 in. to 25-1/2 in.
Overall length: 38 in. to 43-1/2 in.
Weight: 6 lbs. 10 ozs. to 9 lbs. 14 ozs.
Sights: Open, adjustable.
Magazine capacity: 3 rounds; 2 in belted magnums.

MODEL 3000

This model represents what might be called the ultimate Mauser development of the original M98. It utilizes the same general style of bolt, but with a recessed hook-type extractor and plunger-type ejector recessed into the bolt head. The firing-pin is of two-piece construction, and cocking and primary extraction are accomplished by cams in the same manner as the M98. A fully-adjustable single-stage trigger is fitted; and the trigger guard and magazine are of typical Mauser construction. The bolt sleeve is streamlined in Weatherby fashion, and the large, bolt stop has been replaced by a simple plunger. The manual safety fitted to the right side of the bolt sleeve is pressed upward to engage, down to prepare for firing. Operation and manipulation of the Mauser M3000 is essentially the same as for other modern Mauser-type bolt rifles.

O.F. MOSSBERG & SONS, U.S.A.

After having been in business since 1919, Mossberg introduced its first center-fire bolt-action rifle in 1965-66—the M800. It utilizes a rotating bolt with six locking lugs positioned in three rows of two each, and the lugs are of the same diameter as the body. The bolt face is deeply counterbored to shroud the cartridge head and is fitted with a narrow, recessed hook extractor and plunger-type ejector. The firing-pin is unusually slender in its front half, the larger rear portion being counterbored after the fashion of the Japanese Arisaka to accept a relatively short mainspring. The rear of the bolt is closed by a threaded bolt sleeve which sweeps downward to join the curve of stock. The sliding manual safety is installed on the rear slope of the sleeve; it moves rearward to engage and forward to make ready for firing. The receiver is tubular, and utilizes the Remington-type recoil lug clamped in place by the barrel. The trigger mechanism is housed within an upward extension of the plastic trigger guard. The double-column box magazine is fitted with a hinged floor plate, the latter secured by a sliding latch ahead of the guard bow. Several variations of the M800 are offered, ranging from a full-stocked carbine to a heavy-barrel varmint rifle.

Characteristics of the Mossberg M800 Rifle

Caliber: .243 and .308 Winchester; .22-250.
Operation by: Manual; rotating bolt.
Barrel length: 20 in. to 24 in.
Overall length: 40 in. to 44 in.
Weight: 6-1/2 to 9-1/2 lbs.
Sights: Open.
Magazine capacity: 4 rounds; 3 in .22-250 caliber.

PARKER-HALE, BRITAIN

This is a straight-forward design consisting of a Spanish-made Mauser M98-type receiver and bolt to which Parker-Hale adds the remaining parts of its own manufacture. It differs from the basic military M98 mechanically only in that an adjustable, single-stage trigger assembly incorporating a manual safety lying alongside the receiver tang is used. The magazine and trigger guard are typically Mauser, and the floor plate is hinged, secured with a catch inside the trigger guard bow. Functioning and manipulation is as for other Mauser-type rifles.

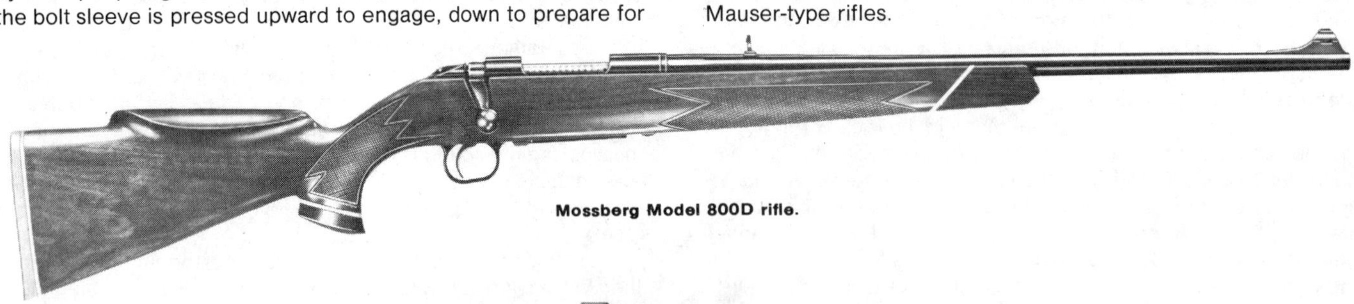

Mossberg Model 800D rifle.

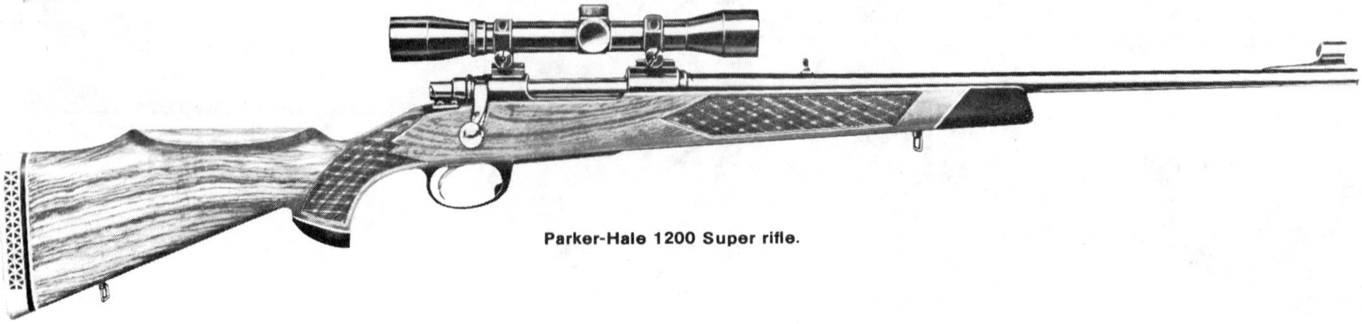

Parker-Hale 1200 Super rifle.

Characteristics of the Parker-Hale Rifle

Caliber: .243, .270, and .308 Winchester; .30-06; .308 Norma Magnum; 7mm Remington Magnum; .300 Winchester Magnum.
Operation by: Manual; rotating Mauser-type bolt.
Barrel length: 22 in. to 24 in.
Overall length: 42-13/16 in. to 44-13/16 in.
Sights: Open, adjustable; folding rear.
Weight: 7-1/2 lbs.
Magazine capacity: 5 rounds; 3 in belted magnum calibers.

RANGER ARMS, INC., U.S.A.

Ranger Arms is a relatively new firm producing several styles of one basic model rifle called the "Texas Magnum." The bolt is of unusually large diameter, being the same size as the circle swept by the three equally-spaced locking lugs protruding from the reduced-diameter head. The bolt face is deeply counterbored and is fitted with a plunger-type ejector and a recessed, pivoted hook extractor. The handle portion is separate and screws into the body after being assembled to the firing-pin, mainspring, and cocking piece. The cocking piece is completely enclosed in a large streamlined bolt sleeve which screws on the handle portion. Cocking and primary extraction are supplied by Mauser-type cams in the bolt handle and receiver ring. The receiver is tubular, machined from bar stock, threaded at the front for barrel attachment, and cut away for feeding and ejection. A Remington-type separate recoil lug is sandwiched between barrel and receiver. A fully-adjustable, single-stage trigger of the pivoting sear type is fitted. This unit is unusual in that it is attached to the trigger guard rather than the receiver and may be removed or interchanged by removing the trigger guard from the stock. A cross-bolt safety is fitted in the front of the trigger guard bow. The magazine box and hinged floor plate are of the type employed in the Winchester Model 70, and the latter is retained by a latch in the front of the guard bow. A total of three screws turned into the receiver holds the guard and floor plate in the stock.

Functioning and manipulation of the Texas Magnum are essentially the same as for modern Mauser-type rifles. Specifications vary considerably, but those listed below are representative.

Characteristics of the Texas Magnum Rifle

Caliber: Most popular U.S.
Operation by: Manual, rotating bolt.
Barrel length: 24 in.
Overall length: 44 in.
Weight: 7-1/2 lbs.
Sights: None.
Magazine capacity: 4 rounds.

REMINGTON ARMS COMPANY, U.S.A.

REMINGTON M700

The Remington M700 rifle is simply a logical improvement of the M721-722 series introduced in 1948. The M700 makes full use of design features and production techniques that were considered quite advanced, if not almost revoluntionary in sporting arms in '48. Other manufacturers who first decried such things have since adopted many of the features and techniques introduced at that time by Remington. The bolt is essentially Mauser in character with dual opposed locking lugs at the head and helical cocking cam at the rear. However, it utilizes a spring-clip type extractor fully enclosed in a recess inside the bolt face. The extractor design permits complete enclosure of the head of the cartridge case within a counterbore within the face of the bolt. The ejector is a Garand-type, spring-loaded plunger in the bolt face. The receiver is round in section; the traditional integral recoil lug being replaced by a barrel bracket protruding downward, sandwiched between barrel and front face of receiver ring.

The magazine is of Mauser type, though simplified by utilization of a separate sheet-metal box sandwiched between floor plate and receiver. The fully-adjustable single-stage trigger assembly of the M700 is considered by many to be the best offered on any commercial sporting rifle, and the assembly incorporates a manual safety that lies closely alongside the rear receiver tang and does not interfere with scope mounting.

Characteristics of the Remington M700 Rifle

Caliber: All popular U.S.
Operation by: Manual; rotating bolt.
Barrel length: 22 in. to 26 in.
Overall length: 42-1/2 in. to 46-1/2 in.
Sights: Open, adjustable.
Weight: 7 to 7-1/2 lbs.
Magazine capacity: 3 to 5 rounds.

The M40X series of bench-rest and target rifles are built upon the basic M700 action in both single-shot and magazine form.

There is also the M660 Carbine which possesses all of the basic design features of the M700, but is made shorter to accommodate smaller cartridges and to achieve maximum weight reduction. This model also utilizes a one-piece plastic magazine box/trigger guard.

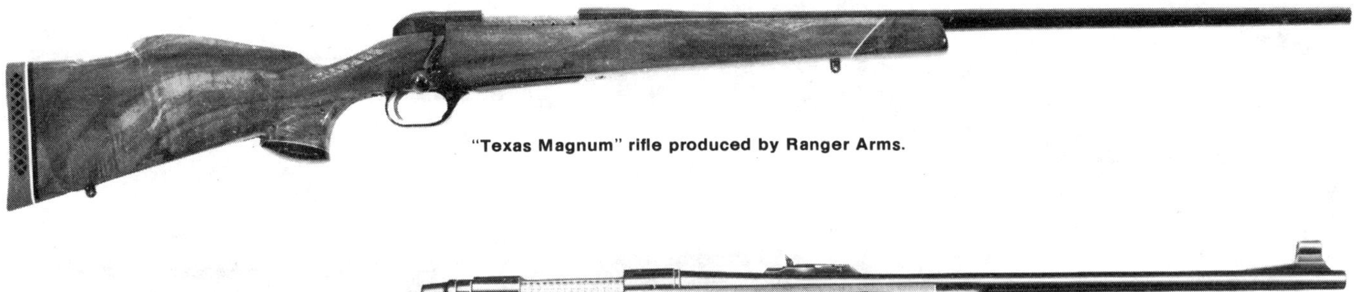

"Texas Magnum" rifle produced by Ranger Arms.

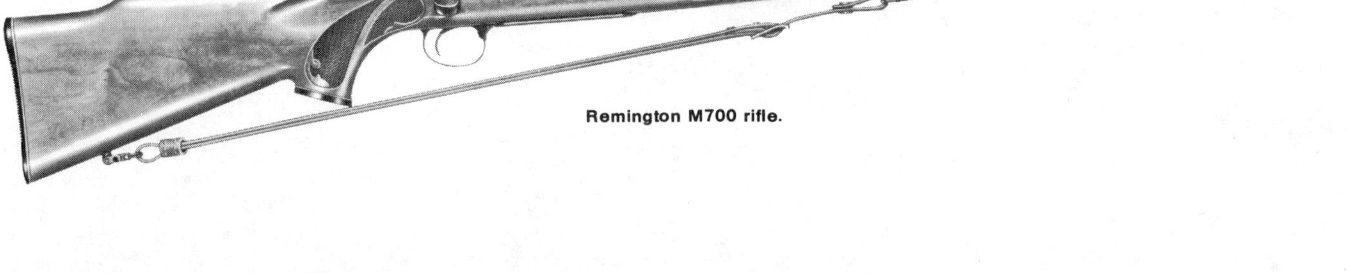

Remington M700 rifle.

Remington M788

In 1967, Remington introduced the M788 rifle built upon an entirely new action designed for maximum rigidity and accuracy and for the most economical production of advanced techniques. The bolt is designed for screw-machine production and carries nine small locking lugs in three rows of three just forward of the bolt handle. The balance of the bolt design is mechanically identical with that of the M700. The receiver is simply a thick-wall steel tube containing ejection and feeding cutouts and abutments against which the locking lugs bear. The trigger and safety mechanism are virtually identical to that of the M700. However, a detachable sheet-metal box magazine and stamped trigger guard are utilized. Aside from the differences in location of locking lugs, functioning and manipulation are the same as in the M700. In 1969, Remington introduced a left-hand version of the M788.

Characteristics of the Remington M788 Rifles

Caliber: .222, .22-250, 6mm Rem.; .243, .308, .30-30 Win.; .44 Magnum.
Operation by: Manual; rear-locking bolt.
Barrel length: 22 in. to 24 in.
Overall length: 41 in. to 43-5/8 in.
Sights: Open, adjustable.
Weight: 7 to 7-1/2 lbs.
Magazine capacity: 4 rounds; 5 in .222 caliber.

SAKO, FINLAND

Sako rifles are produced in Finland and distributed exclusively in the U.S. by the Garcia Corporation. The action is of slightly modified Mauser-type, utilizing dual opposed locking lugs at the bolt head, combined with a recessed spring-loaded hook extractor on the right side of the bolt. The balance of the bolt is of typical Mauser design; the manual safety is incorporated in the adjustable, single-stage trigger assembly. The Sako receiver is tubular in section, and carries integral tapered scope-mounting dovetails on its upper surface. The bolt stop is original in design and functions quite well. Magazine and trigger guard are of typical Mauser construction except that the box proper is a separate sheet-metal part clamped in place between the guard and receiver. The floor plate is hinged, secured by a latch inside the forward portion of the guard bow.

A wide variety of Sako rifles are produced on the above described action; however, the action is made in three lengths. All major parts of the three different actions are dimensionally different so that there is actually a single action design produced in three different sizes—not simply minor variations of a single size to accommodate cartridges of different lengths. Shortest of the series is the 6-1/2 in. long L-461 (Vixen) action weighing 2-1/8 lbs; the L579 (Forester), 7-3/8 in. long, weighing 2 lbs; and, the L61 (Finnbear) which is 8-3/8 in. long and weighs 2-3/4 lbs. Sako is the only firm producing so complete a line of actions tailored specifically to different cartridge lengths. Essentially the same style rifles are offered built on each of the actions.

Characteristics of the SAKO Rifles

Caliber: Most popular U.S. from .222 upward.
Operation by: Manual; rotating bolt.
Barrel length: 20 in. to 24 in.
Overall length: 40 in. to 44 in.
Sights: Open.
Weight: 6-1/2 to 7-1/2 lbs.
Magazine capacity: 4 rounds; 3 in belted magnum calibers.

SAVAGE ARMS CO., U.S.A.

Savage Model 110

In 1958, Savage introduced this highly-modified, Mauser-type bolt-action rifle. The bolt contains an unusual number of parts; the body alone consists of the head, front and rear baffles, friction washer, retaining-pin and handle—all separate pieces. The firing mechanism consists of 10 parts, exclusive of the sear. All parts are, however, designed for maximum production economy, and result in low overall cost. The design is noteworthy in that there are two baffles to obstruct the rearward flow of gas should a case or primer rupture. The Savage extractor is similar in principle to the Remington enclosed type, but is fitted in the outside of the bolt head, with the claw entering the bolt face counterbore through a slot. The cocking piece is housed completely within the bolt, and a pin protrudes from it to engage a Mauser-type cocking cam in the side of

Remington M788 rifle.

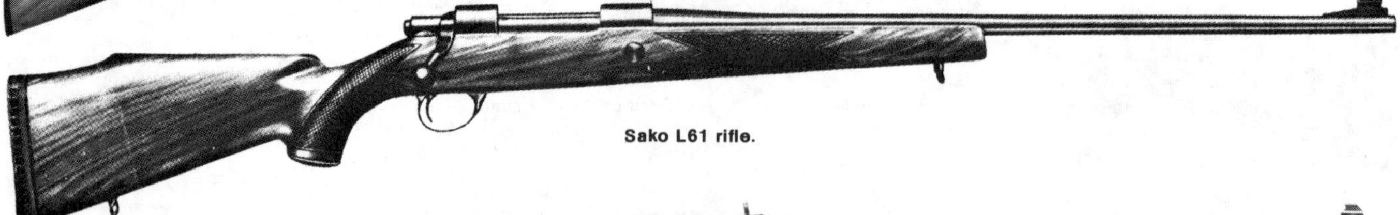

Sako L61 rifle.

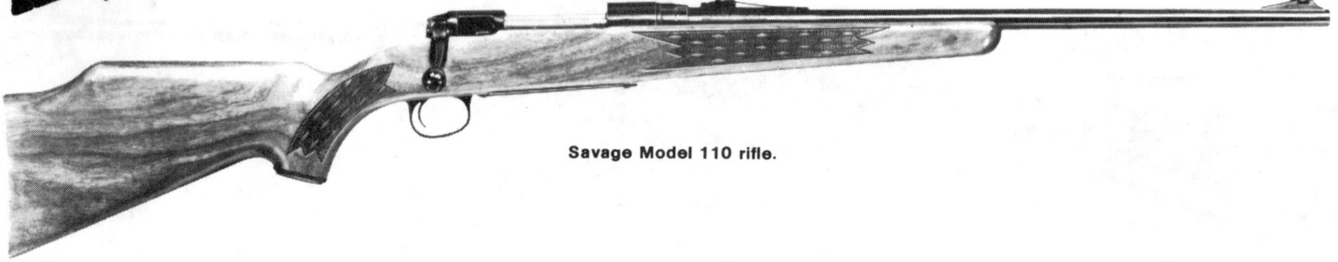

Savage Model 110 rifle.

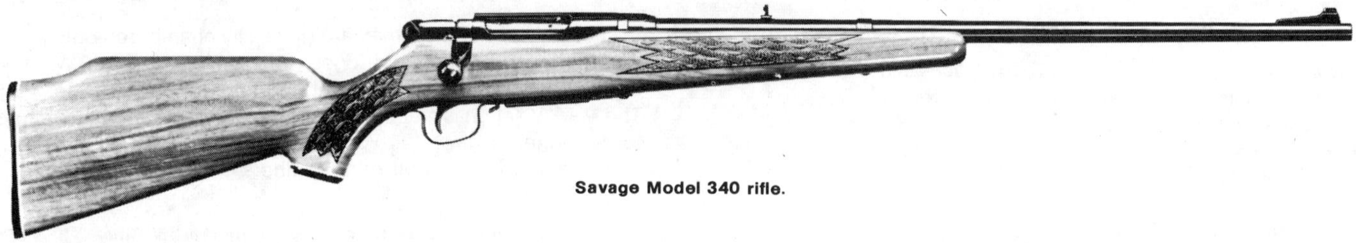

Savage Model 340 rifle.

the bolt body, ahead of the handle. The unusual pivoted sear protrudes upward on the right side of the bolt to engage the cocking-piece pin and has an external projection which serves as a cocking indicator and also as a bolt release.

The receiver is tubular and of essentially Mauser form, fitted with a separate recoil lug in Remington fashion. The barrel screws into the receiver, sandwiching the recoil lug between the two, but it is also fitted with a threaded barrel lock nut. The barrel proper does not screw up tightly in assembly as in other makes; final clamping of all the parts is accomplished by drawing the barrel lock nut up tightly. The trigger assembly is of single-stage type, and the manual safety is situated on the rear receiver tang in shotgun fashion. The trigger guard and magazine floor plate are of multiple-piece construction while the double-column magazine box is attached directly to the receiver.

Characteristics of the Savage Model 110 Rifle

Caliber: .243, .270, and .308 Winchester; .30-06; 7mm Remington Magnum.
Operation by: Manual, rotating reciprocating bolt.
Barrel length: 20 in. to 22 in.
Overall length: 40-1/2 in. to 45 in.
Weight: 6-3/4 to 8 lbs.
Sights: Open, folding rear.
Magazine capacity: 4 rounds; 3 in belted magnum calibers.

Savage Model 340

Savage also produces the Model 340 which is an economy-priced center-fire rifle of distinctive design. The bolt utilizes a single, large locking lug at its head, and the bolt handle functions as a safety lug by virtue of being seated deep into a recess in the receiver. Cocking and extraction are accomplished in Mauser fashion; and the receiver is tubular, requiring a minimum of machining. A pivoted manual safety is fitted to the right rear of the receiver. The trigger is a simple, single-stage design. A detachable, single-column box magazine is installed ahead of the trigger guard.

Characteristics of the Savage Model 340 Rifle

Caliber: .222 Remington; .30-30; .225 Winchester.
Operation by: Manual, rotating single-lug bolt.
Barrel length: 20-22 in.
Weight: 6-1/2 lbs.
Sights: Open; folding rear.
Magazine capacity: 3 rounds; 4 in .222 caliber.

SCHULTZ-LARSEN, DENMARK

The Schultz-Larsen design is one of the few not based upon the Mauser M98. It has been produced in Denmark for many years, and the action has long been highly popular among Europeans for building best-quality target rifles, including those for international competition. The bolt consists of a massive cylinder with four large locking lugs equally spaced around its circumference several inches back from the bolt face. It is thus what we commonly call a "rear-locking" bolt. The lugs engage corresponding abutments in the ring of the massive tubular receiver and a thick bolt-handle root also doubles as a safety lug by turning down into a notch in the receiver. Multiple gas-escape ports are drilled in the bolt body, exposed in the ejection port, and the bolt face is deeply counterbored to surround the cartridge-case head. The firing mechanism and cocking method follows Mauser practice closely. A fully-adjustable single-stage trigger is fitted, as is a hinged floor plate on the magazine. Aside from the unusually short bolt lift occasioned by the four locking lugs; functioning and manipulation are identical with Mauser-type rifles.

Characteristics of the Schultz-Larsen Rifle

Caliber: Most popular U.S. and foreign.
Operation by: Manual; rotating reciprocating rear-lock bolt.
Barrel length: 24 in.
Overall length: 44-1/2 in.
Weight: 7-1/2 lbs.
Magazine capacity: 4 rounds; 3 in belted magnum calibers.

STURM-RUGER, U.S.A.

Ruger M77

Introduced in 1968, this was Ruger's first entry into the bolt-action field. On the surface, it looks more like a cleaned-up M98 Mauser then anything else—but several advance design and production features are incorporated. The bolt is typical Mauser, including the long non-rotating extractor and bolt-guide rib. The firing pin and cocking system is also typically Mauser. The safety, however, is eliminated from the bolt sleeve and placed upon the rear receiver tang in shotgun fashion. The receiver is somewhat slab-sided and intricate in contour and carries a Mauser-type bolt stop at the left rear.

Scope-mounting dovetails are formed integrally with the top of the receiver and Ruger supplies rings to match. The single-stage trigger is quite crisp and is fully-adjustable. The trigger guard/magazine assembly resembles the Winchester M70 and is held in place by three guard screws. An unusual feature is that the head of the forward guard screw is angled rearward so that tightening it not only pulls the receiver down into the stock, but rearward, seating the recoil lug solidly in the wood. The magazine floor plate is hinged and held in place by a quick-release latch in the front of the trigger guard bow.

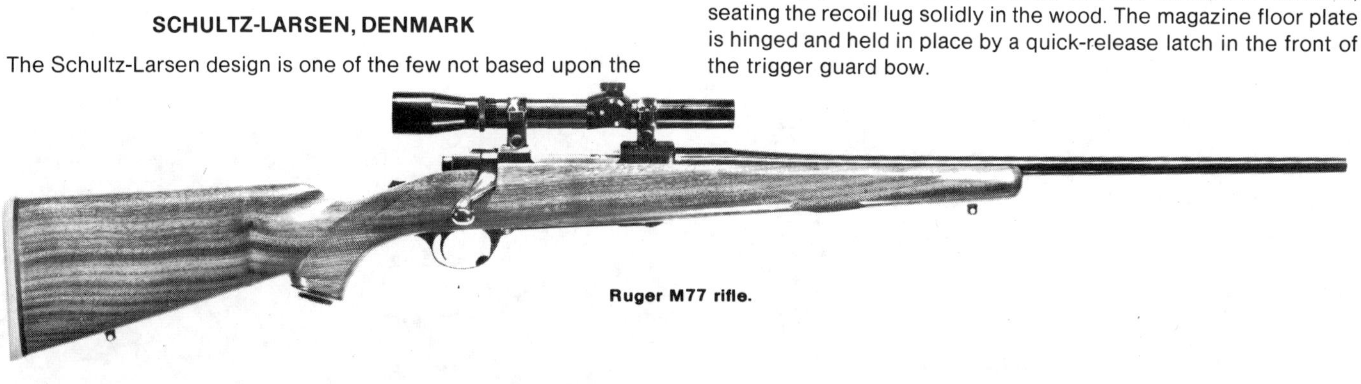

Ruger M77 rifle.

Characteristics of the Ruger M77 Rifle

Caliber: .22-250; 6mm Remington; .243, .308, and .284 Winchester; 6.5mm and .350 Remington Magnum.
Operation by: Manual; rotating reciprocating bolt.
Barrel length: 22 in.
Overall length: 42 in.
Weight: 6-1/2 lbs.
Sights: None; furnished with scope bases.
Magazine capacity: 5 rounds; 3 in magnum calibers.

WEATHERBY INC., U.S.A.

The first proprietary Weatherby rifle design, the MARK V, was introduced in 1958. At that time, it represented a radical departure from current action design. The bolt body is a massive cylinder, carrying three equally-spaced rows of three small locking lugs, each on a reduced diameter portion at the head. Lugs and bolt body are both of the same major diameter. The body contains longitudinal flutes which reduce weight and friction, and provide escape areas for dust and dirt. The bolt face is deeply counterbored to completely surround the cartridge-case head. The only opening into the counterbore is a slot for the recessed hook extractor. The rear of the bolt is closed by a streamlined sleeve which completely encloses the cocking piece, except for an indicator tongue which protrudes from underneath the sleeve when the gun is cocked.

A manual safety is pivoted to the right side of the bolt sleeve. A sheet-metal box magazine is held between the receiver and Mauser-style trigger guard, to which is fitted a hinged floor plate. The trigger is of single-stage design and is fully-adjustable. Weatherby rifles are characterized by high-gloss finish and unusual stock wood. Also, they are generally chambered only for the Weatherby proprietary cartridges which include the .460 Weatherby Magnum—the most powerful sporting cartridge in the world. The Mark V action is made in two sizes—the smaller being adapted to cartridges of the .224 Weatherby and .22-250 class, and the larger action handles all other calibers.

Characteristics of the Weatherby Mark V Rifle

Caliber: The full line of special Weatherby cartridges; any other on special order.
Operation by: Manual; Weatherby rotating bolt.
Barrel length: 24 in. to 26 in.
Overall length: 42-1/4 in. to 46-1/2 in.
Sights: None.
Weight: 6-1/2 to 10-1/2 lbs.
Magazine capacity: 4 rounds; 3 in .378 and .460 calibers.

WINCHESTER REPEATING ARMS COMPANY, U. S. A.

Winchester Model 70

First introduced in the middle 1930's, the M70 underwent minor evolutionary improvements, until in 1964, the completely redesigned, currently-produced version was introduced. It is considered by some to have been the most popular American bolt-action rifle. The design is relatively straight-forward, utilizing Mauser-type cocking system. A plunger-type ejector and sliding/hook-type extractor are let into the bolt face. The receiver is typically Mauser, containing locking abutments at its front and threaded for barrel attachment. The magazine is of basic Mauser type and is considerably modified for economy of fabrication. The trigger mechanism is adjustable.

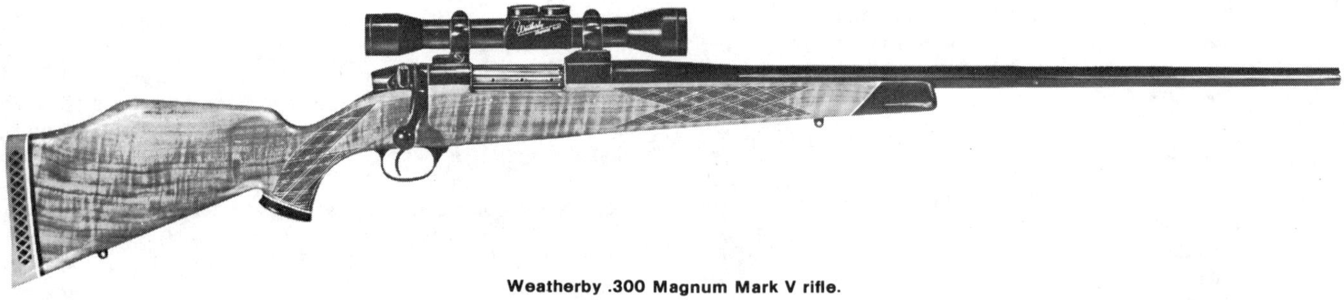

Weatherby .300 Magnum Mark V rifle.

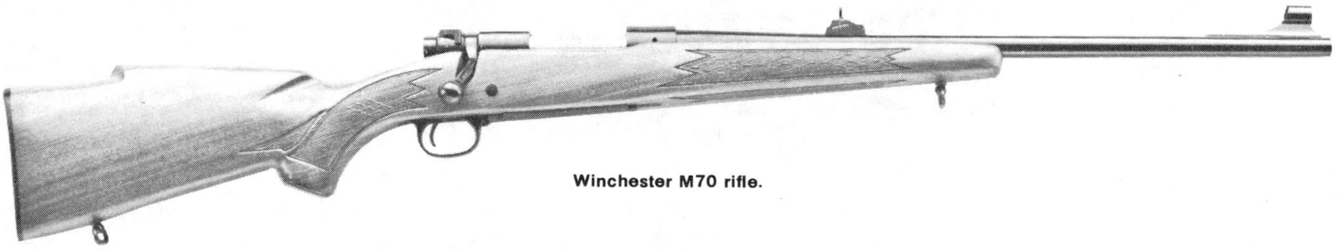

Winchester M70 rifle.

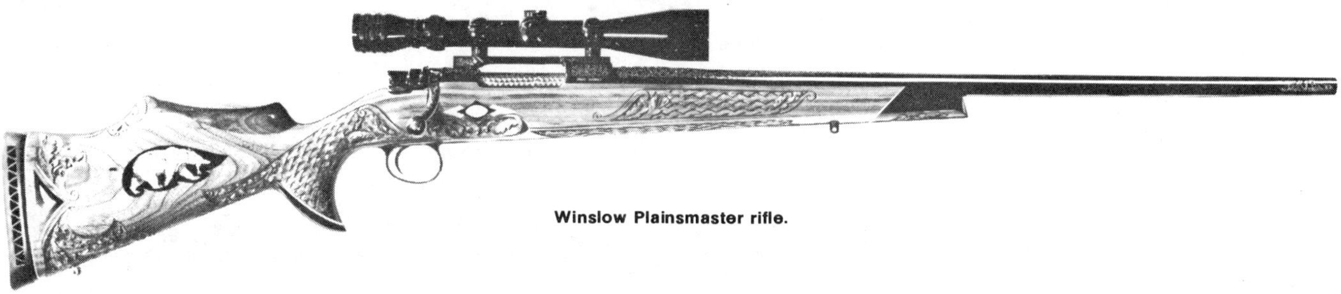

Winslow Plainsmaster rifle.

Characteristics of the Winchester M70 Rifle

Caliber: All popular U.S.
Operation by: Manual; rotating bolt.
Barrel length: 19 in. to 24 in.
Overall length: 39-1/2 in. to 44-1/2 in.
Sights: Open, adjustable.
Weight: 7-1/2 to 8-1/2 lbs.
Magazine capacity: 5 rounds; 3 in belted magnum calibers.

Winchester Models 770 and 670 are simply less-costly versions of the basic M70. Functioning and manipulations are the same.

WINSLOW ARMS, U.S.A.

Winslow Arms takes existing actions, domestic barrels, and assorted woods and accessories, and assembles them into unusual and often highly-ornate sporting rifles on a semi-production basis. The FN action is used for the larger calibers, the short Sako L-461 for the shorter and smaller calibers. Full particulars on the actions will be found under their own names.

It is primarily in the matter of styling and finish that Winslow rifles differ from those assembled by many other makers on the same actions. Stock styling, for example, is "way-out" and incorporates such features as extra-long, hooked pistol grip; fluted fore-end; extra-large roll-over comb; and sweeping curves not encountered in other makes. In addition, exotic inlays, checkering patterns and carvings are found on the more costly models. Considering the fact that some models cost several thousand dollars, the result is often quite striking. Due to the semi-production basis upon which Winslow rifles are built, specifications vary a great deal, so no precise values can be given. Virtually any caliber, barrel length, or weight may be ordered.

GENERAL

In addition to the individual makes and models of bolt-action rifles already listed and described, a number of firms import rifles marked with their own name, but which are manufactured abroad as only minor variations of existing models. These rifles are normally based upon one of the basic Mauser M98-type actions. In some instances actual assembly takes place in the U.S. The actions may be of new European or Scandinavian manufacture, or may be, in some instances, refurbished military actions or commercial actions purchased in partly-finished condition and made into complete rifles by any one of a number of manufacturers. Generally speaking, these rifles are identical in characteristics, operation, and loading and firing procedures to the M98 or FN Mauser or their variations already described in detail. Following are some of the names under which these rifles will be encountered: Colt's; Reinhart Fajen; Golden State; Herter's; L. A. Distributors; Santa Barbara; Smith & Wesson; Dan Wesson Arms.

LEVER-ACTION RIFLES

MARLIN FIREARMS, U.S.A.

Marlin M336

This design originated in 1889, making it the oldest still produced in the U.S. It has undergone considerable improvement but has remained little changed since 1948 when the breech bolt was changed from square to round section and the M36 became the M336. It utilizes a solid receiver, vertical sliding lock and reciprocating bolt, side ejection, tubular under-barrel magazine, and a two-piece stock. Operation is by a swinging finger lever forming the trigger guard. An unusual feature is the two-piece firing pin, which is aligned by the locking block to prevent firing when the breech is not fully locked. Also made since 1972 in .45-70 as the M1895.

Characteristics of the Marlin M336 Rifle

Caliber: .30-30 Win.; .35 Rem.; .444 Marlin; .45-70 (M1895)
Operation by: Swinging finger lever; reciprocating, rear-locked bolt.
Barrel length: 20 in. to 24 in.
Overall length: 38-1/2 in. to 42-1/4 in.
Sights: Adjustable, open.
Weight: 7 to 7-1/2 lbs.
Magazine capacity: 5-7 rounds.

Marlin 1894

This is actually a new (1969 introduction), modernized copy of the original M1894 once made for the short .44/40 cartridge.

Characteristics of the Marlin 1894 Rifle

Caliber: .44 Magnum.
Operation by: Swinging finger lever; reciprocating bolt.
Barrel length: 20 in.
Overall length: 37-1/2 in.
Sights: Adjustable, open.
Weight: 6 lbs.
Magazine capacity: 10 rounds.

Marlin M336 rifle.

Marlin M1894 rifle.

Mechanically, physically, and functionally, it differs from the M336 only in bolt section (square) and its full-length bolt slot in the receiver wall. Operation and handling are as described for the M336.

O.F. MOSSBERG U.S.A.

In 1973 Mossberg introduced a traditional Western-style lever-action carbine designed M472. In both appearance and design it greatly resembles the Marlin M336. It is the only exposed-hammer, lever-action rifle which utilizes a separate manual safety instead of a safety position of the hammer. It utilizes a vertical rising-block locking system similar to the Marlin & Winchester and functions generally the same.

Characteristics of Mossberg M472

Caliber: .30-30 Winchester.
Operation by: Manual, finger lever.
Barrel length: 20".
Weight: 7-1/4 lbs.
Sights: Open.
Magazine capacity: 6 rounds.

SAKO, FINLAND

Sako Finnwolf

The only European high-power lever-action rifle manufactured; first introduced in 1962-63 and produced only in relatively small quantities, primarily for sale in the U.S.A. It is of solid-frame, hammerless construction, using an unusual system of rack and pinion gears to translate lever movement into bolt movement. A curved rack inside the receiver rotates a gear segment on the lever connected by a crank and links to the bolt carrier. Bolt is of reciprocating type with a separate rotating head. Stock is one-piece; a detachable box magazine is fitted.

Characteristics of the SAKO Finnwolf Rifle

Caliber: .243; .308 Win.
Operation by: Swinging finger lever, reciprocating rotating bolt.
Barrel length: 22 in.
Overall length: 42-1/4 in.
Sights: Hooded front only; integral scope-mount bases on receiver.
Weight: 7 lbs.
Magazine capacity: 4 rounds.

SAVAGE ARMS, U.S.A.

Savage M99

This model has been in continuous production since its introduction in 1899, though many minor improvements have been made during that time. It has a solid receiver, tipping, rear-locked reciprocating bolt, rotary magazine, and is actuated by a swinging finger lever. Many minor variations have been produced, and the current M99C utilizes a detachable double-column box magazine.

Characteristics of the Savage M99 Rifle

Caliber: .243; .308; .284; .358 Win.; .300 Sav.
Operation by: Swinging finger lever; tipping reciprocating bolt.
Barrel length: 20 in. to 22 in.
Overall length: 39-3/4 in. to 41-3/4 in.
Sights: Adjustable, open.
Weight: 6-1/4 to 6-3/4 lbs.
Magazine capacity: Rotary, 5; box, 4; .284 caliber, 3.

Savage M99 rifle.

WINCHESTER REPEATING ARMS, U.S.A.

Winchester M88

In 1955, Winchester introduced the M88 as the first really new lever gun since the 1890's. The action is quite similar to that of the M100 semi-automatic. It utilizes an entirely different firing mechanism combined with a finger lever. Bolt movement is obtained by the lever, rather than an operating slide as in the M100. Considerable parts interchangeability exists between the two.

Characteristics of the Winchester M88 Rifle

Caliber: .243; .284; .308 Win.
Operation by: Swinging finger lever, reciprocating rotating bolt.
Barrel length: 19 in. to 22 in.
Overall length: 39-1/2 in. to 42-1/2 in.
Sights: Adjustable, open.
Weight: 7 to 7-1/2 lbs.
Magazine capacity: 4 rounds; 3 in .284.

Winchester M94

Introduced in 1894 and produced virtually without change until 1964, at which time the design was revised, it continues in production. In the minds of most, it typifies the "American-style" lever-action arms developed only in this country during the last half of the 19th Century. It is built around a solid receiver, two-piece stock, and tubular magazine. The reciprocating bolt is locked at the rear and it ejects fired cases directly upward. Operation is by swinging the finger lever.

Characteristics of the Winchester M94 Rifle

Caliber: .30-30 Win.; .44 Magnum.
Operation by: Swinging finger lever; reciprocating bolt.
Barrel length: 20 in. to 26 in.
Overall length: 37-3/4 in. to 43-3/4 in.
Sights: Adjustable, open.
Weight: 6-1/8 to 8 lbs.
Magazine capacity: 6 to 10 rounds.

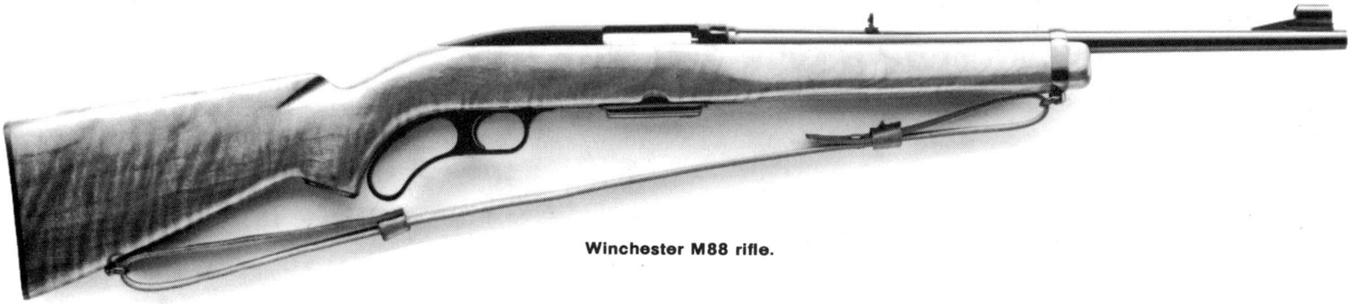

Winchester M88 rifle.

PUMP-ACTION RIFLES

REMINGTON ARMS, U.S.A.

Remington M760

Introduced in 1952 to replace the earlier M14 and M141 pump rifles, this was the first such action capable of handling full-power cartridges in the .30-06 class. It remains so. Except that power to operate the action is supplied by manual reciprocation of the forestock, it is identical mechanically, physically, and functionally to the Remington M742, which see. Most internal parts are interchangeable between the two.

Characteristics of the Remington M760 Rifle

Caliber: .223 Rem.; .243 Win.; .270 Win.; 30-06; .308 Win.
Operation by: Reciprocating forestock, rotating bolt.
Barrel length: 20 in. to 22 in.
Overall length: 40 in. to 42 in.
Sights: Adjustable, open.
Weight: 7-1/2 lbs.
Magazine capacity: 4 rounds.

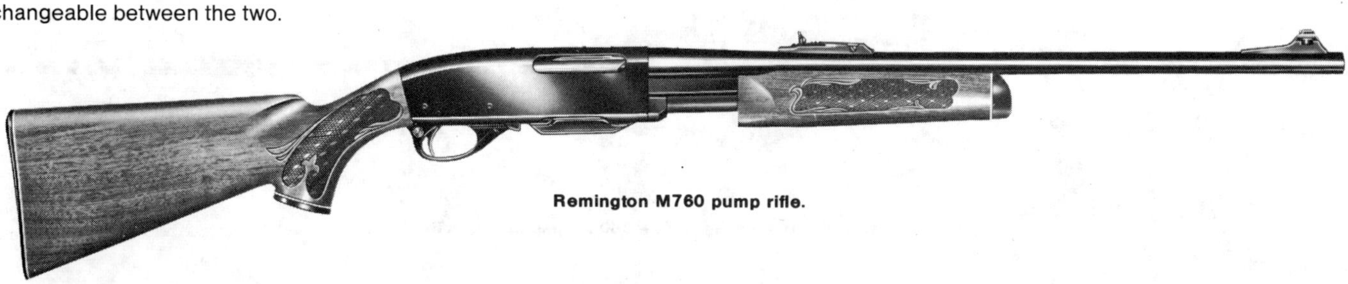

Remington M760 pump rifle.

SAVAGE ARMS U.S.A.

Introduced in 1971 the Model 170 is a lightweight, slide-action hunting rifle offered only in .30-30 Winchester caliber. It is of very simple hammerless locked-breech design and construction and priced quite low.

Characteristics of Savage M170

Caliber: .30-30 Winchester
Barrel length: 22"
Weight: 6-3/4 lbs.
Overall length: 41-1/2"
Sights: Open
Magazine capacity: 3 rounds

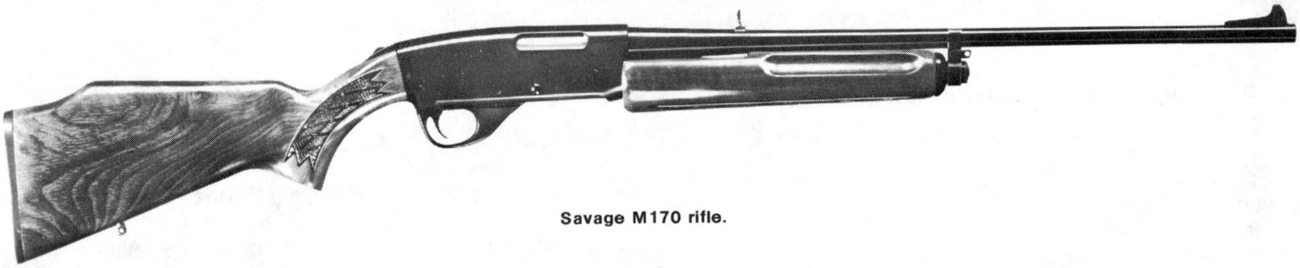

Savage M170 rifle.

SEMIAUTOMATIC RIFLES

BROWNING ARMS, U.S.A.

Browning BAR

In 1967, the U.S. Browning firm introduced its first center-fire self-loading rifle in this country. It is produced by Fabrique Nationale. The design utilizes a short-stroke gas system with a two-piece reciprocating bolt and characteristically Browning double-hook firing mechanism. It uses a solid receiver, two-piece stock, and an unusual hinged box magazine. Present models have the distinction of being the only self-loading design chambered for belted magnum cartridges.

Characteristics of the Browning BAR

Caliber: .243 Win.; .308 Win.; .30-06; 7mm Rem. Mag.
Operation by: Gas; with reciprocating, rotating bolt.
Barrel length: 22 in.
Overall length: 42-7/8 in.
Sights: Adjustable, open.
Weight: 7-3/8 lbs.
Magazine capacity: 4 rounds; 3 in belted magnum.

HARRINGTON & RICHARDSON U.S.A.

H&R Ultra M360

About 1966, H&R announced its new M360 full-caliber, self-loading rifle. It is modern in style; operated by a short-stroke gas piston; uses a rear-locked, tipping, reciprocating bolt; one-piece stock; and detachable box magazine. The design is relatively simple for a self-loader and the bolt resembles somewhat that of the Reising SMG once made by H&R.

Characteristics of the H&R Ultra M360

Caliber: .243 and .308 Winchester.
Operation by: Short-stroke gas; tipping bolt.
Barrel length: 22 in.
Overall length: 43-1/2 in.
Sights: Open adjustable.
Weight: 7-1/2 lbs.
Magazine capacity: 4 rounds.

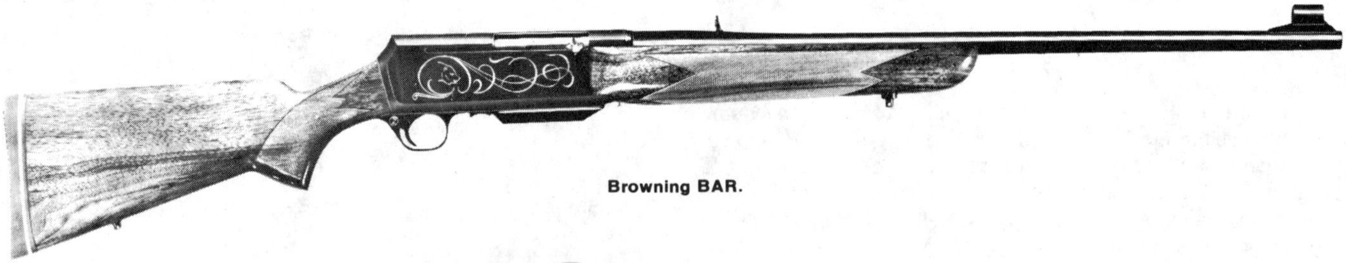

Browning BAR.

Harrington & Richardson Ultra M360 semiautomatic rifle.

REMINGTON, U.S.A.

Remington M742

This was the first successful full-power and the first gas-operated semi-auto to be offered on the American market. It was introduced in 1955 as the M740, and subsequently redesignated the M742 in 1960 as improvements were added. It utilizes a typical under-barrel gas cylinder and piston; the latter connected to twin operating rods which carry a two-piece, reciprocating, multiple-lug bolt at their rear. It utilizes a solid receiver with side ejection and two-piece stock.

Characteristics of the Remington M742 Rifle

Caliber: 6mm Rem.; .243 Win.; .308 Win.; .280 Rem.; .30-06.
Operation by: Gas; reciprocating, rotating bolt.
Barrel length: 20 in. to 22 in.
Overall length: 40 in. to 42 in.
Sights: Adjustable, open.
Weight: 7-1/2 lbs.
Magazine capacity: 4 rounds.

STURM-RUGER, U.S.A.

Ruger .44 Carbine

This is the only semi-auto rifle chambered for the .44 Magnum revolver cartridge. Introduced in 1961 as a short-range deer rifle, it is gas-operated, utilizing a short-stroke floating piston, heavy operating slide, and short one-piece rotating reciprocating bolt. Uniquely, it is the only center-fire self-loader using a tubular, under-barrel magazine. It is styled after the U.S. M1 Carbine, capitalizing on that gun's popularity.

Characteristics of the Ruger .44 Carbine

Caliber: .44 Remington Magnum.
Operation by: Short-stroke gas; rotating bolt.
Barrel length: 18-1/2 in.
Overall length: 37 in.
Sights: Open, adjustable.
Weight: 5-2/3 lbs.
Magazine capacity: 4 rounds.

WINCHESTER, U.S.A.

Model 100

Winchester introduced this gas-operated auto in 1960. It has unusually clean, smooth lines and is a thoroughly modern design exploiting mass-production techniques, including stampings and weldments. It utilizes an under-barrel gas cylinder; twin-armed operating slide/gas piston; two-part rotating, reciprocating bolt; and pivoted-hammer firing mechanism. Stock is one-piece and the detachable box magazine holds 4 rounds. When assembled, the rear of the receiver is closed by a solid steel recoil block attached permanently to the stock in M1 Carbine fashion.

Characteristics of the Winchester Model 100

Caliber: .243; .284; .308 Winchester.
Operation by: Gas; reciprocating, rotating bolt.
Barrel length: 19 in. to 22 in.
Overall length: 39-1/2 in. to 42-1/2 in.
Sights: Adjustable, open.
Weight: 7 lbs.
Magazine capacity: 4 rounds, 3 in .284.

MISCELLANEOUS MAKES AND MODELS

ARMALITE AR-180, U.S.A: Semi-automatic version of Armalite AR-18 .223 military rifle, which see.
COLT AR-15 Sporter, U.S.A: Semi-automatic version of U.S. M16 .223 rifle, which see.

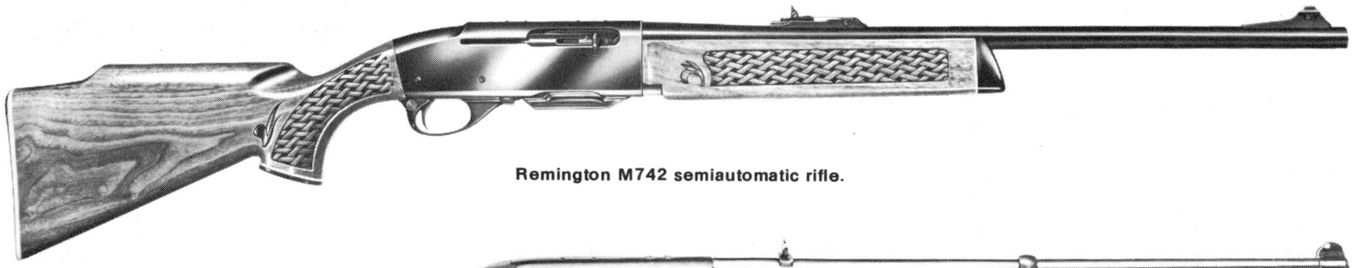

Remington M742 semiautomatic rifle.

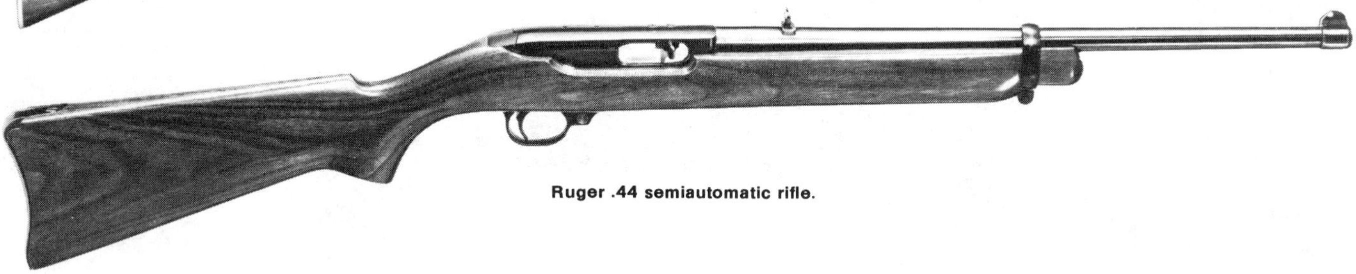

Ruger .44 semiautomatic rifle.

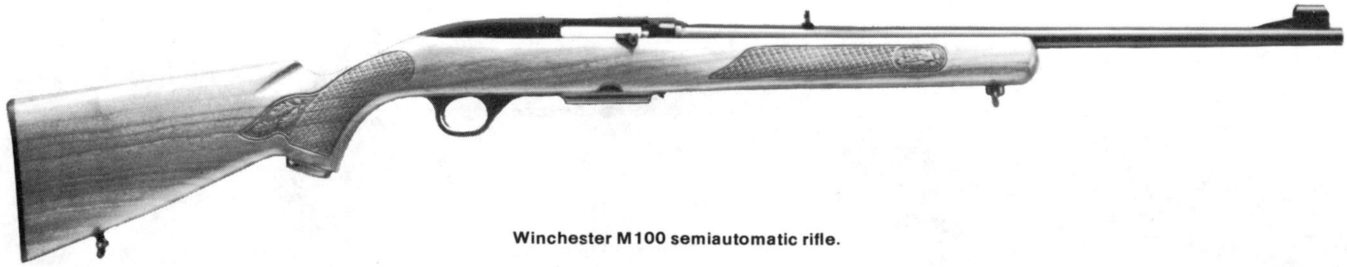

Winchester M100 semiautomatic rifle.

PLAINFIELD M1 Carbine, U.S.A: Identical to the U.S. 30 M1 carbine, which see.
UNIVERSAL M1 and FERRET Carbines, U.S.A: Identical to U.S. M1 Carbine, which see.
CETME Sport: A very slightly modified semi-automatic version of the CETME rifle adopted by the Spanish Army, which see.
SIG/ATM: Commercial semi-automatic version of the Swiss Army STuG57 Service rifle, which see.
BERETTA BM59: Commercial semi-automatic version of the Italian Army 7.62mm BM59 Service rifle, which see.

SINGLE-SHOT RIFLES

In addition to the rifles shown and described, a wide variety of obsolete single-shots are now reproduced by both major makers and small shops. Of the former, Harrington & Richardson is making tens of thousands of Trap-Door Springfield reproductions; Navy Arms offers both the Remington Rolling Block and the Martini. Of the latter both the Sharps-Borchardt and Winchester M1885 High Wall are being made by different small shops. A variety of other modern, semi-modern, and semi-copies of old actions are being built in small quantities by various small shops.

HARRINGTON & RICHARDSON, U.S.A.

Topper Model 158

A very basic single-shot caliber .30-30 rifle based upon the standard H&R break-open, exposed-hammer, single-shot shotgun. It has no distinguishing or unique characteristics other than its economy and utility. Weight, 5-1/2 lbs.; barrel length, 24 in.; open sights. Available with optional 20-gauge barrel. A .45-70 version is available.

STURM-RUGER, U.S.A.

Ruger No. 1

A completely modern, single-shot rifle of top quality, designed in the classic style, but utilizing modern manufacturing techniques. Even the receiver is produced in virtually finished form by investment casting. The action combines features of many others, but most resembles the British Farquharson. It features a separate bar to support the forestock; two-piece stock; through-bolted buttstock; 1/4-length barrel rib containing telescope bases; internal hammer firing mechanism; cocking indicator; selective ejector-extractor; powerful extraction; and adaptability to any center-fire sporting cartridge. Introduced in 1968. A plain carbine version is now available as the No. 3, in .45-70 only.

Characteristics of the Ruger No. 1 Rifle

Caliber: Any standard U. S. center fire on special order.
Operation by: Falling-block, under-lever actuated.
Barrel length: 22 in. to 26 in.
Overall length: 42 in. (26 in. bbl.).
Sights: None; Fitted for telescope bases.
Weight: 6-3/4 to 8-1/2 lbs.

Harrington & Richardson M158 single-shot rifle.

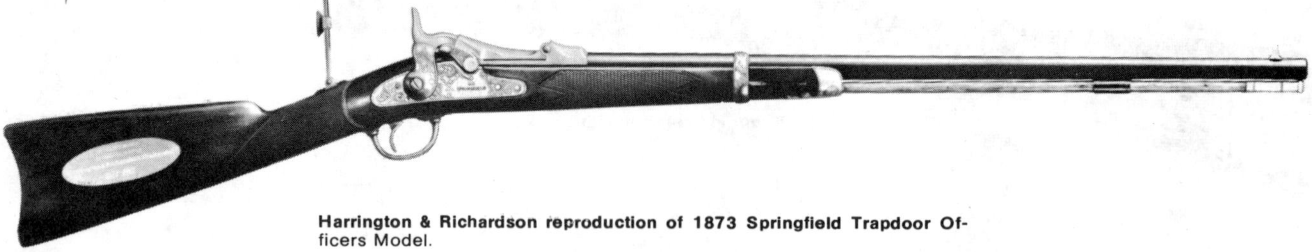

Harrington & Richardson reproduction of 1873 Springfield Trapdoor Officers Model.

Ruger No. 1 single-shot rifle.

COMBINATION GUNS

SAVAGE ARMS, U.S.A.

Savage produces several different model combination guns based on essentially the same action. All are in the M24 series, and most are chambered for the .22 Long Rifle or .22 MRF in the top barrel, and .410 or 20 gauge in the bottom barrel. The most recent development, the Model 24V series, is chambered for the .222 Remington and the 20-gauge shotshell. All are of external hammer, break-open, hinge-barreled design and utilize pivoted locking blocks seating in a wedge-shaped recess in the rear face of the barrel under lug. All utilize a pivoted top lever for unlocking and opening, except the 24S and 24SM series in which the design is somewhat modified to utilize a pivoted side-lever for this purpose. The M24V is generally of more robust construction and has barrels of larger diameter to accommodate the more powerful cartridges. Barrels of the M24V are separate and are inserted into a single barrel block at the chamber; however, the less powerful models have both barrels machined from a single block of steel. Barrel lengths in all models are 24 in., weight 6-3/4 lbs.

TIKKAKOSKI OY, FINLAND

Tikkakoski produces a rather unusual shotgun/rifle combination sold in this country by Ithaca as the "Turkey Gun." One unique feature of this gun is the muzzle attachment which is brazed securely to the shot upper barrel and surrounds—but is not attached to—the muzzle of the rifle barrel. This permits the rifled tube to elongate freely as it heats. The latter is shorter than the shot barrel, permitting the attachment to function as a muzzle brake by means of the eight gas-escape slots machined in its sides. The chamber ends of both barrels are silver-brazed into a solid steel block which contains a recess into which a conventional double-bite locking block seats when the action is closed. Unlocking is accomplished by a typical shotgun-type pivoted top lever. A conventional exposed hammer is utilized, and its impact is transmitted to the upper or lower firing-pin as desired by means of a sliding block controlled by a button on the left side of the receiver. The receiver itself is also quite unusual in that instead of being machined or cast as a single unit, it is formed by brazing together pre-cut laminations of steel to form the necessary internal cavities. After brazing, the exterior of the receiver is shaped.

Characteristics of the "Turkey Gun"

Caliber: 12 gauge and .222 (5.7x43mm) Remington.
Operation by: Hinged-frame, top lever; manual cocking.
Barrel length: Shot, 26 in.; rifle, 24-1/2 in.
Weight: 7-1/4 lbs.
Sights: Open, adjustable.
Magazine capacity: One round, each barrel.

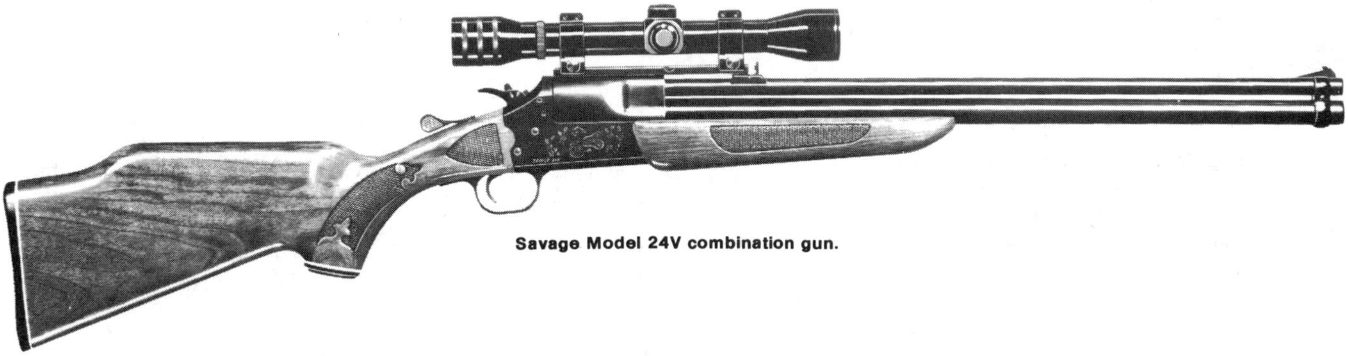

Savage Model 24V combination gun.

Ithaca Turkey combination gun.

PUMP-ACTION SHOTGUNS

BERETTA, ITALY

Beretta SL-2

Externally this gun is virtually identical to the Beretta AL-2 semiautomatic, and is, in fact, built upon the same action with maximum parts interchangeability between the two. Only such modifications as are necessary to permit manual power to be substituted for the gas piston have been made. This consists primarily of substituting for the gas piston and cylinder a reciprocating fore-end attached to the action bar.

Characteristics, functioning, manipulation, loading, and firing are essentially identical to the AL-2, which see.

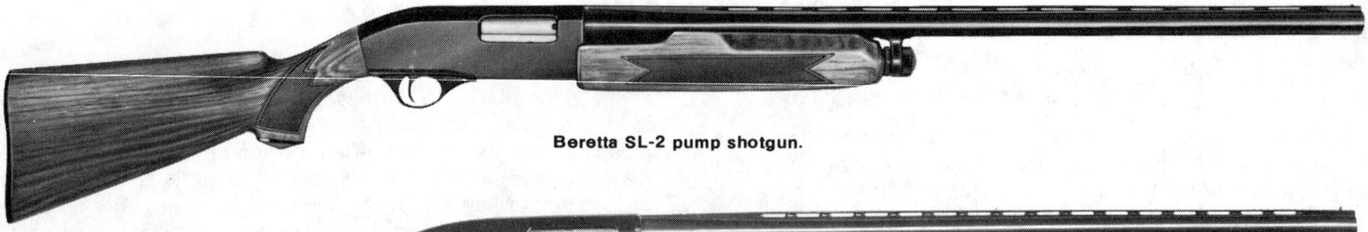

Beretta SL-2 pump shotgun.

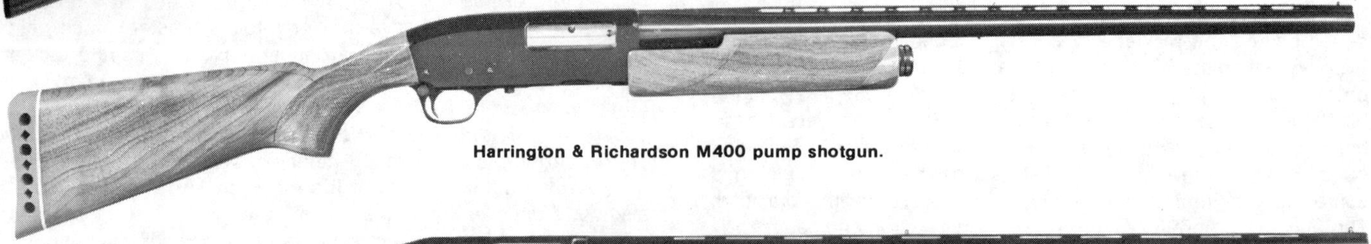

Harrington & Richardson M400 pump shotgun.

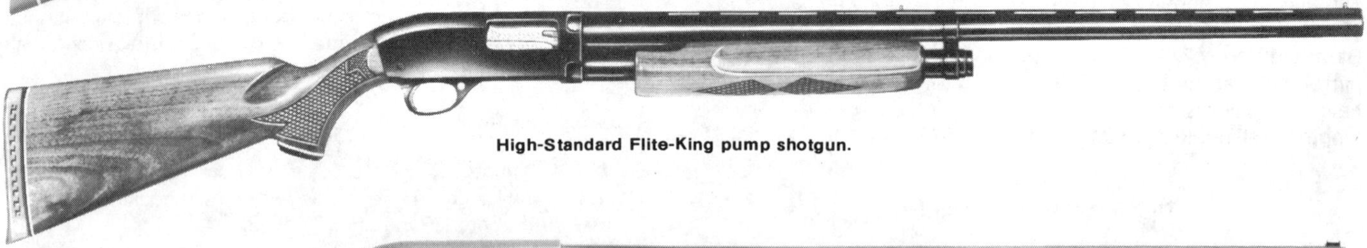

High-Standard Flite-King pump shotgun.

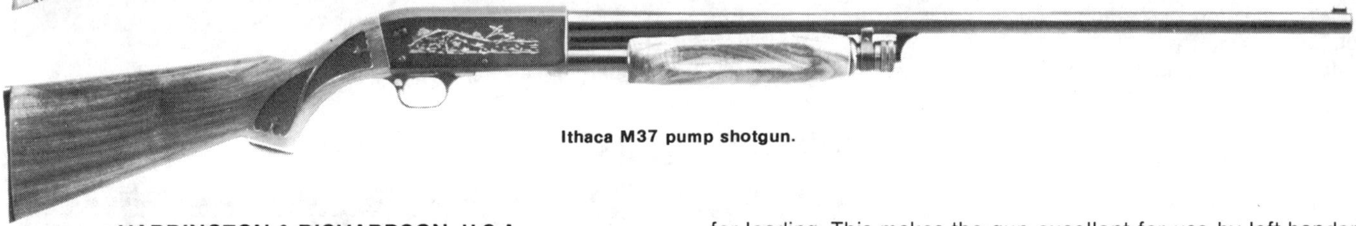

Ithaca M37 pump shotgun.

HARRINGTON & RICHARDSON, U.S.A.

Harrington & Richardson produces a relatively conventional pump-action shotgun, the basic model of which is known as the M400. It utilizes a typical rear-locked, tipping-bolt locking system and pivoted-hammer firing mechanism, combined with an ordinary pivoted carrier and side ejection.

Characteristics of the H&R M400 Shotgun

Gauges: 12, 16, 20, and .410 gauge.
Operation by: Manual; reciprocating fore-end, tipping-bolt lock.
Barrel length: 26 in. to 28 in.
Overall length: 47 in. to 49 in.
Weight: 6-1/4 lbs.
Sights: Bead.
Magazine capacity: 4 rounds.

HIGH STANDARD, U.S.A.

The High-Standard "Flite King" pump shotgun was introduced simultaneously with the Supermatic auto-loader. In keeping with modern practice, the action of this gun is identical to that of the automatic with the exception of those changes necessary to adapt it to manual power applied through a reciprocating fore-end. Most parts within the actions interchange. Characteristics, functioning, loading, and firing are identical to the automatic, which see.

ITHACA GUN COMPANY, U.S.A.

In 1937, Ithaca introduced its M37 "Featherlight" shotgun. At that time, it was the lightest 12-gauge repeating gun available—and at 6-1/2 lbs., still compares very favorably with the most modern lightweight developments. The design is unique in that fired cases are ejected downward through the same port utilized for loading. This makes the gun excellent for use by left-handers and also avoids having ejected cases strike a shooter nearby. A tipping-bolt locking system is utilized, and the firing mechanism is not fitted with a disconnector—the latter meaning that the gun will fire if the trigger is held down as the action is closed on a loaded cartridge. Overall, the design is simple, straightforward, and contains a relatively small number of parts.

Characteristics of the Ithaca M37 Shotgun

Gauges: 12, 16, and 20 gauge.
Operation by: Manual; reciprocating fore-end, tipping bolt.
Barrel length: 20 in. to 30 in.
Weight: 5-3/4 to 6-1/2 lbs.
Sights: Bead; open adjustable on slug-gun models.
Magazine capacity: 5 rounds.

LASALLE, FRANCE

The LaSalle pump-action shotgun is imported by Firearms International for distribution in the U.S. This gun, in keeping with present-day practice, is built upon virtually the same action as the LaSalle Automatic mentioned elsewhere in this volume. The pump-action differs only in that it is slightly modified to permit operating power to be applied by means of a manually-actuated reciprocating fore-end sliding on the magazine tube. Characteristics, manipulation, functioning, loading, and firing are essentially identical to that of the LaSalle Automatic, which see.

O. F. MOSSBERG & SONS, U.S.A.

Mossberg produces a number of variations on its basic, economy-priced M500 pump gun. This gun is conventional in design, construction, and appearance. It utilizes a non-rotating bolt locked to the receiver by a pivoted locking block passing through a recess in the center of the bolt. A single action bar connects the sliding fore-end to a bolt slide which functions in a conventional manner.

Characteristics of the Mossberg M500 Shotgun

Gauges: 12, 16, 20, and .410 gauge.
Operation by: Manual; reciprocating fore-end, non-rotating bolt.
Barrel length: 24 in. to 30 in.
Overall length: 43-1/4 in. to 49-1/4 in.
Weight: 5-3/4 to 6-3/4 lbs.
Sights: Bead; open adjustable on slug models.
Magazine capacity: 5 rounds; 4 in 3 in. Magnums.

REMINGTON ARMS, U.S.A.

In 1950, Remington introduced the M870 pump gun to replace the Model 31 discontinued in 1949. The new design bears considerable resemblance to the old, but it incorporates many improvements and more modern production methods. It utilizes a non-rotating bolt locked into the barrel extension by a pivoted block swinging vertically through the center of the bolt. Dual action bars connect the reciprocating fore-end to an action slide which contains cam and engaging surfaces to actuate the bolt and bolt lock. The firing mechanism is conventional in design, and with only minor exceptions, it is virtually identical to that of the M742/M760 rifles. Most parts will interchange.

Characteristics of the Remington M870 Shotgun

Gauges: 12, 16, and 20-gauge (small-frame version also in .410 and 28-gauge).
Operation by: Manual; reciprocating fore-end.
Barrel length: 20 in. to 30 in.
Overall length: 40-1/2 in. to 50-1/2 in.
Weight: 6-1/2 to 7 lbs.
Sights: Bead; adjustable open on slug guns.
Magazine capacity: 5 rounds.

SAVAGE ARMS, U.S.A.

The Savage Model 30 series pump shotguns represent the current development of a long line of similar designs. The design is relatively conventional and straight forward, and utilizes a non-rotating tipping-bolt locking system which incorporates a number of well-proven principles. A single action bar connects the reciprocating fore-end (which slides on the magazine tube) to a slide which transmits movement to the bolt and performs locking and unlocking functions.

Characteristics of the Savage Model 30 Shotgun

Gauges: 12, 16, and 20 gauges; .410 in M77 version.
Operation by: Manual; reciprocating fore-end, tipping bolt lock.
Barrel length: 26 in. to 30 in.
Overall length: 45-1/2 in. to 50 in.
Weight: 6-1/4 lbs. (.410) to 8 lbs.
Magazine capacity: 4 rounds; 3 in 3 in. length.

Mossberg M500 pump shotgun.

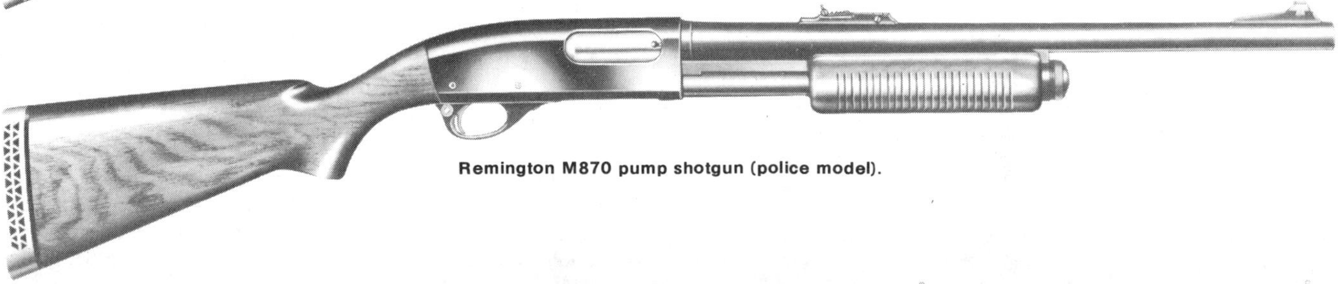

Remington M870 pump shotgun (police model).

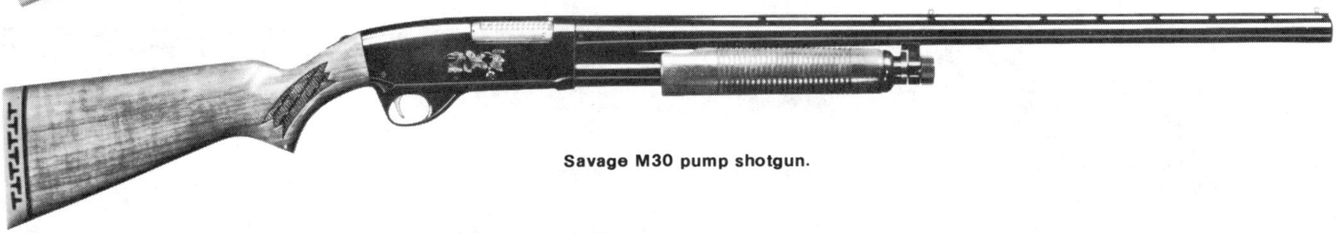

Savage M30 pump shotgun.

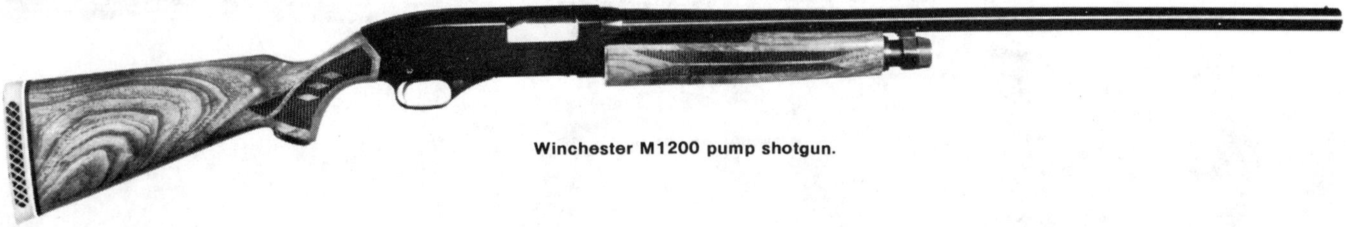

Winchester M1200 pump shotgun.

WINCHESTER, U.S.A.

Winchester M1200

In 1963, Winchester introduced the successor to its justly-famed M12 pump gun—the Model 1200. In external appearance, the new gun is virtually identical to the old. However, internally, it differs vastly. The bolt is a two-part unit, consisting of a heavy rectangular carrier and a separate rotating bolt head carrying four short locking lugs spaced in pairs around its perimeter. An extension screwed solidly to the barrel contains abutments behind which these lugs lock. This arrangement, whereby all the longitudinal stress of firing is taken up within the barrel extension, allows simple takedown and/or barrel interchangeability. A ring on the barrel slips over the magazine tube and a nut is screwed on after it to hold the barrel in place. All major structural parts are made of machined forgings and modern low-cost production techniques are used in the balance of the gun. Some parts previously produced by costly machining are replaced by sophisticated stampings and precision castings. The rotating-bolt locking system is much stronger than that of its predecessor, and the dual action bars make for very smooth operation. The M1200 cannot be fired by slamming the fore-end forward with the trigger depressed as could occur with its predecessor. The M1200 (and companion) is the only production gun available optionally with a recoil-absorbing mechanism built into the butt stock. This is called the "Winchester Recoil Reduction System," and is offered only on 12-gauge guns.

In 1972 Winchester reintroduced its famous M12 pump shotgun.

Characteristics of the Winchester Model 1200 Shotgun

Gauges: 12, 16, and 20-gauge (including 3" Magnum).
Operation by: Manual; rotating bolt; sliding fore-end.
Barrel length: 20 in. to 32 in.
Overall length: 40-5/8 in. to 51 in.
Weight: 6-1/2 to 8-1/4 lbs.
Sights: Bead; open-adjustable on slug-gun models.
Magazine capacity: 2 rounds.

SEMIAUTOMATIC SHOTGUNS

BERETTA, ITALY

Armi Beretta is currently producing its Model AL-2 semiautomatic shotgun which is distributed in the U.S.A. by Garcia Sporting Arms Corp. This is a relatively conventional gas-operated design combining several time-tested principles of operation into an integrated functional mechanism. It utilizes an annular gas piston sliding over the magazine tube and a non-rotating bolt fitted with a separate tipping locking block engaging the receiver. Disassembly and barrel removal or exchange are quite simple and accomplished without tools.

Characteristics of the Beretta Model AL-2 Shotgun

Gauges: 12 gauge, 2-3/4" only.
Operation by: Gas; non-rotating bolt, tipping lock block.
Barrel length: 26 in. to 30 in.
Weight: 7 to 7-1/4 lbs.
Sights: Bead.

BREDA, ITALY

The Breda Mark II semiautomatic shotgun is imported and distributed in this country by Continental Arms Corp. The present series was introduced in 1957, though several variations were produced previously, beginning in 1948. The manufacturer is Breda Meccanica Bresciana, Brescia, Italy. The design is essentially a highly-modified form of the original Browning long-recoil system which has been widely copied since the early 1900's. While functioning remains essentially the same, the Breda utilizes a two-part bolt and a vertically-sliding, rather than pivoted, locking block. The Browning friction-ring system is utilized in conjunction with the recoil spring, and requires adjustment when switching between light and heavy loads. In outline, the Breda is quite similar to the Browning except for a rounding-off of the upper rear corner of the receiver. None of the Breda parts will interchange with the Browning. Functioning, manipulation, loading, and firing are essentially the same as for the Browning, which see.

Characteristics of the Breda Mark II Shotgun

Gauges: 12 gauge, 2-3/4" only.
Operation by: Long recoil; Browning system.
Barrel length: 26 in. to 28 in.
Weight: 7-1/4 lbs.
Sights: Bead.
Magazine capacity: 3 rounds.

BROWNING ARMS CO., U.S.A.

The "Browning Automatic" was introduced in Europe shortly after the turn of the century. In both that form and as the Remington Model 11, it became the first commercially successful self-loading shotgun in the world. The Browning Automatic has been produced continuously since, thus making it the oldest repeating shotgun still in production. It is a long-recoil type; the barrel, barrel-extension and bolt recoiling several inches as a unit first, then unlocking to allow the routine functioning cycle. This design

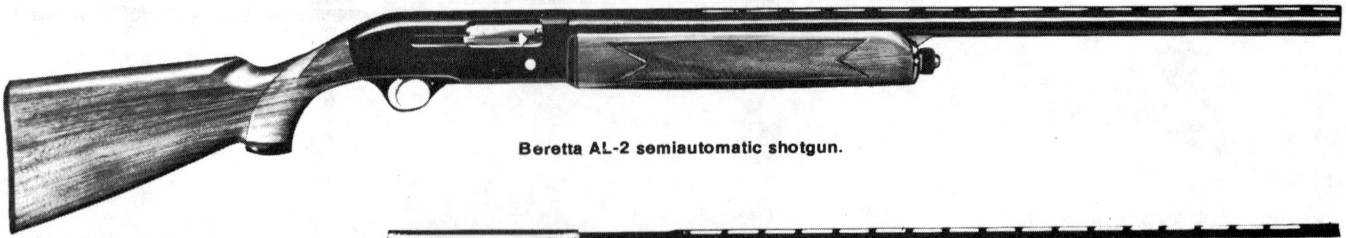

Beretta AL-2 semiautomatic shotgun.

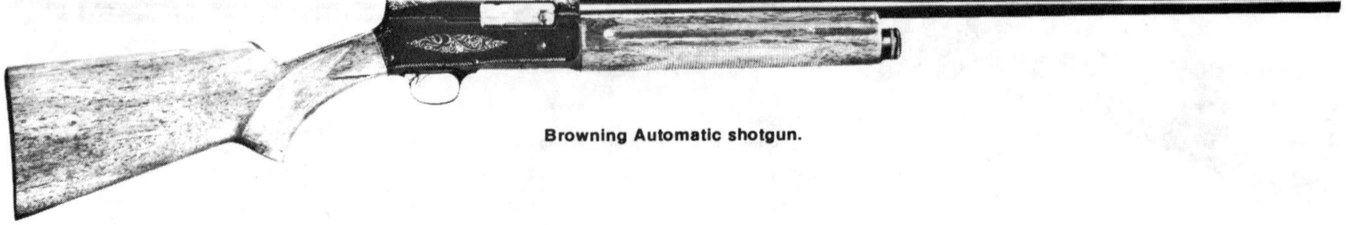

Browning Automatic shotgun.

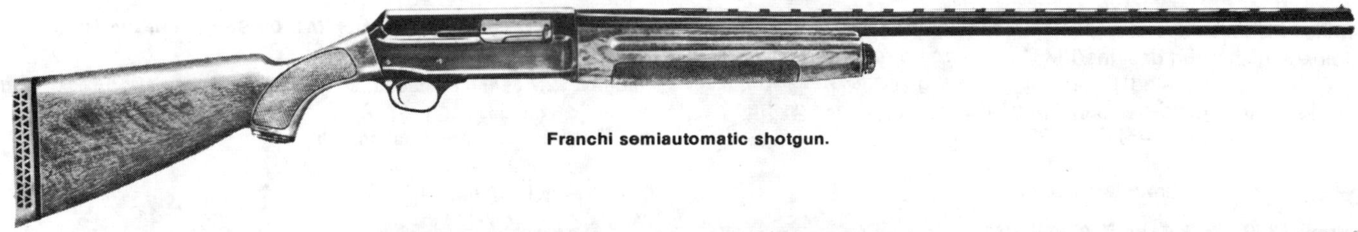

Franchi semiautomatic shotgun.

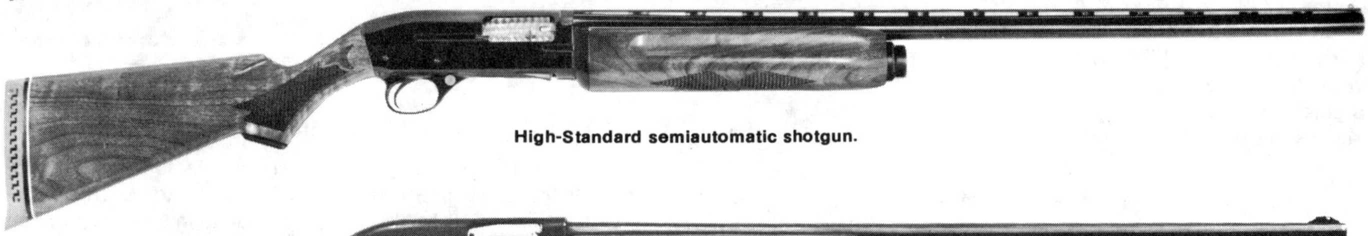

High-Standard semiautomatic shotgun.

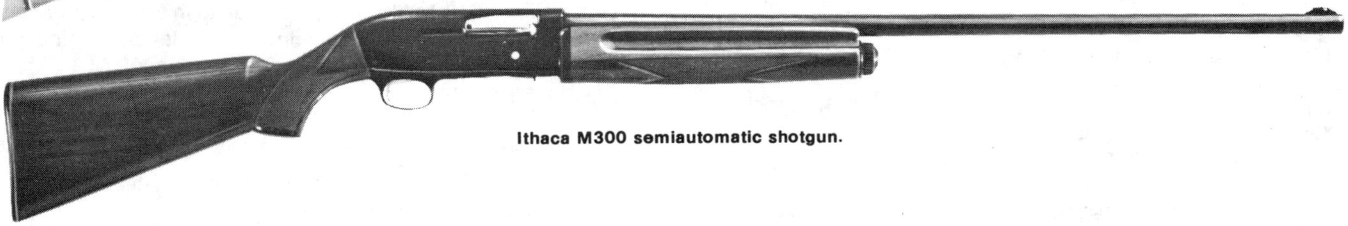

Ithaca M300 semiautomatic shotgun.

results in a rather massive receiver of rectangular outline which has earned it the nickname "Humpback." As a result of its success and durability, this design has been widely copied—both line-for-line, and in varying modified forms by numerous makers.

Characteristics of the Browning Automatic Shotgun

Gauges: 12, 16, and 20-gauge.
Operation by: Long recoil; rising block lock.
Barrel length: 24 in. to 32 in.
Weight: 7 to 8-1/4 lbs.; lightweight models from 6 lbs-2 oz. upward.
Sights: Bead; open-adjustable on slug-gun models.
Magazine capacity: 3 to 5 rounds.

Other Producers

As mentioned above, this design has been widely copied in various forms. Among the names that will be encountered on Browning-type designs of this sort are: Breda, Franchi, Savage Arms, Sears Roebuck, Stevens Arms Co., Tradewinds, and Universal.

FRANCHI, ITALY

The Franchi semi-automatic shotgun is imported and distributed in the U.S. by Stoeger Arms Corp., and is manufactured by Luigi Franchi, Brescia, Italy. Essentially, this design may be considered a highly-modified form of the Browning, which see. It utilizes the long-recoil system, and the bolt and locking mechanisms are identical in principle and extremely similar in outline to the Browning. None of the parts will interchange with the Browning; however, all perform the same functions and are similar in outline. The interior of Franchi barrels are chrome-plated to resist corrosion and leading. The outstanding feature of this design is its unusually light weight—6 lbs. 4 ozs. in 12-gauge and 5 lbs. and 2 ozs. in 20-gauge. Magazine capacity is 4 rounds. Numerous variations are available in 12, 16, and 20-gauge for 2-3/4 in. shells only. Functioning, operation, loading, and firing are essentially identical to those of the Browning.

HIGH-STANDARD, U.S.A.

In 1960, the High-Standard Company introduced under its own name a gas-operated semiautomatic shotgun it had been manufacturing under contract for Sears, Roebuck and Co. It is of generally conventional design and construction, utilizing an annular gas piston sliding on the magazine tube and connected to the two-part bolt with rising-block lock by dual operating rods. It was one of the first gas-operated automatics to incorporate a reliable automatic "gas-bleed" valve which eliminated necessity for any adjustments between light and heavy loads. The gas mechanism does make this gun unusually thick through the fore-end; however, the balance of its lines are quite smooth and modern. This gun, the "Supermatic," has an excellent reputation for durability and reliability. This is one of the few models available in the smaller .410 and 28 gauges.

Characteristics of the High-Standard Semiautomatic Shotgun

Gauges: 12, 16, 20, and .410 gauges; also 3 in. Magnums.
Operation by: Gas; annular piston and 2-part bolt.
Barrel length: 22 in. to 30 in.
Overall length: 41-3/4 in. to 49-3/4 in.
Sights: Bead; open-adjustable on slug guns.
Magazine capacity: 5 in 2-3/4 in.; 4 in 3 in.

ITHACA GUN COMPANY, U.S.A.

In 1968, Ithaca introduced its M300 semiautomatic shotgun, climaxing a development program which expanded the line to include shotguns of all basic types. The M300 is also offered in a deluxe version, the M900. This is essentially the original Browning long-recoil design streamlined in appearance, simplified somewhat, and adapted to more modern production techniques. The squarish profile of the Browning receiver has been changed to a pleasing curve that blends with the stock; but aside from that, the appearance is essentially the same. Functioning, manipulation, loading, and firing are essentially identical to that of the Browning, which see.

Characteristics of the Ithaca Semiautomatic Shotgun

Gauges: 12 and 20-gauge, 2-3/4 in. and 3 in. respectively.
Operation by: Recoil; Browning-type.
Barrel length: 26 in. to 30 in.
Weight: 6-1/2 to 7 lbs.
Sights: Bead.
Magazine capacity: 4 rounds.

Ithaca also introduced in 1972 an excellent new gas-operated autoloading shotgun designed M51. It utilizes a rotating bolt with 3 locking lugs actuated by a self-regulating gas system which handles light and heavy loads without adjustment. It has been rated highly by reviewers.

Characteristics of Ithaca M51

Gauge: 12, 20 (2-3/4" or 3" chambers).
Operation by: Gas.
Barrel length: 26", 30".
Weight: 7-1/2 lbs.
Sights: Bead.
Magazine capacity: 3 rounds.

LA SALLE, FRANCE

The LaSalle semiautomatic shotgun is produced by Manufrance, and is distributed in this country by Firearms International. This gun is of fairly conventional design and manufacture, and utilizes a non-rotating bolt fitted with a rising locking block which locks it to the receiver roof. Styling is conventional, and the gun is unusually lightweight—6-1/2 lbs. in 12-gauge.

REMINGTON ARMS, U.S.A.

Remington Model 1100

In 1962, Remington introduced its first gas-operated semiautomatic shotgun—the M1100. This design carried forward the original concept of maximum interchangeability of parts initiated with the M870 shotgun, and continued through the M740-42 and M760 rifles. Essentially, the action of the M1100 is the same as that of the M870 (which see), modified only as necessary to permit a gas piston and cylinder to supply the power for operation. Bolt and locking mechanism remain the same; however, the dual action bars are combined into a single unit with the action slide. A separate action-bar sleeve is attached to the action bar and encircles the magazine tube. The forward end of the sleeve enters the gas cylinder (attached to the barrel) and functions as an annular piston.

Characteristics of the Remington M1100 Semiautomatic Shotguns

Gauges: 12, 16, and 20-gauge; small-frame versions also in .410 and 28-gauge.
Operation by: Gas; non-rotating bolt.
Barrel length: 22 in. to 30 in.
Overall length: 42 in. to 50 in.
Weight: 6-1/4 to 7-1/2 lbs.
Sights: Bead; open-adjustable on slug models.
Magazine capacity: 5 rounds.

SAVAGE ARMS, U.S.A.

Since the 1930's, Savage Arms has manufactured a very close copy of the original Browning Automatic shotgun. The current model designation is M750 and it is produced in several variations. For all practical purposes, functioning, manipulation, loading, and firing are as for the Browning Automatic, which see.

Characteristics of the LaSalle Semiautomatic Shotguns

Gauges: 12 and 20-gauge; standard or magnum length.
Operation by: Gas; non-rotating bolt.
Barrel length: 26 in. to 30 in.
Weight: 6-1/2 lbs.
Sights: Bead.
Magazine capacity: 3 rounds.

Characteristics of the Savage M750 Shotgun

Gauges: 12-gauge only.
Operation by: Recoil; long-recoil, Browning system.
Barrel length: 26 in. to 28 in.
Overall length: 45-1/2 in. to 47-1/2 in.
Weight: 7-1/4 lbs.
Sights: Bead.
Magazine capacity: 4 rounds.

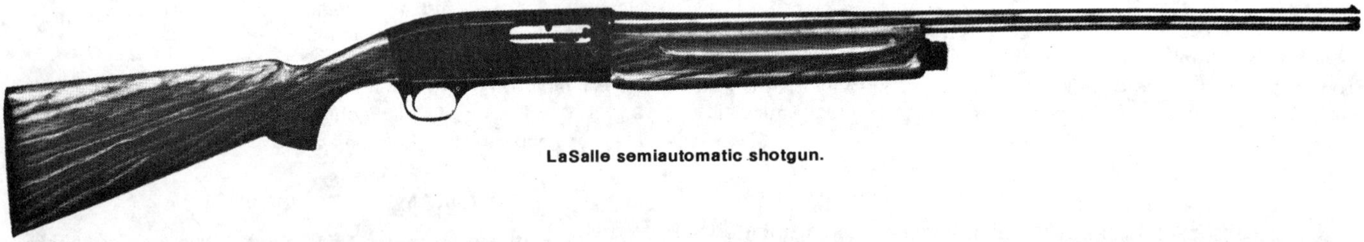

LaSalle semiautomatic shotgun.

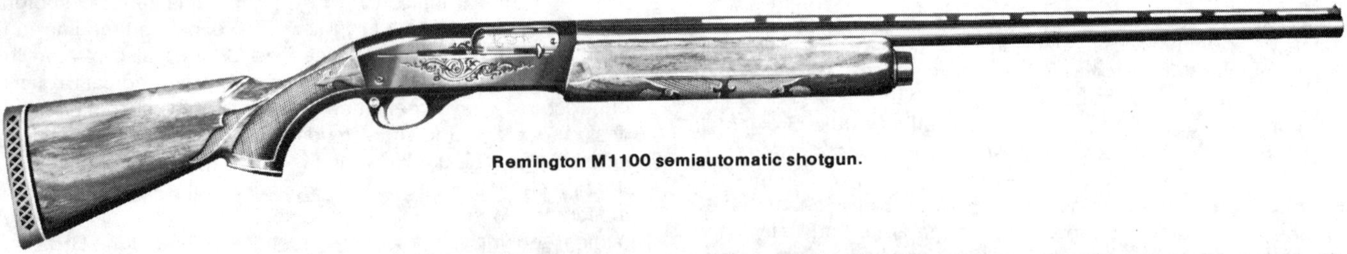

Remington M1100 semiautomatic shotgun.

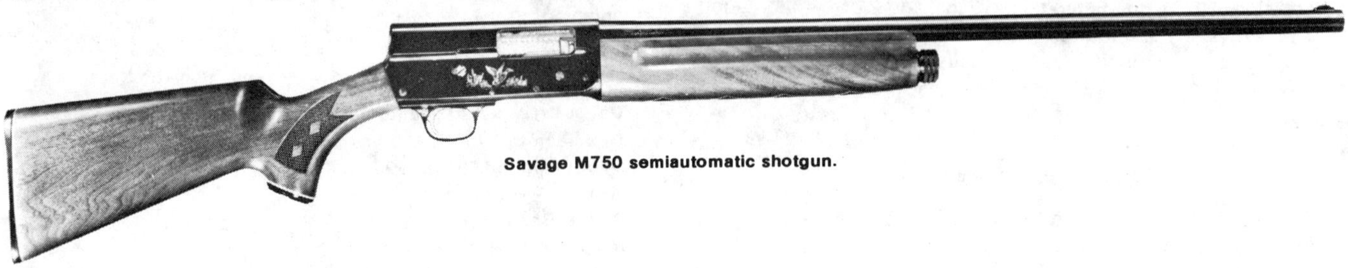

Savage M750 semiautomatic shotgun.

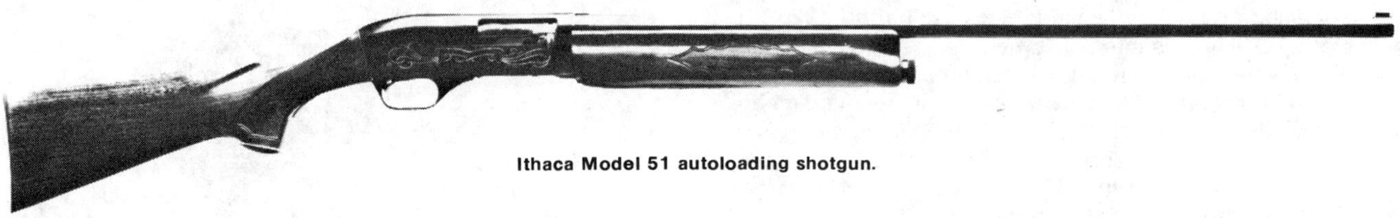

Ithaca Model 51 autoloading shotgun.

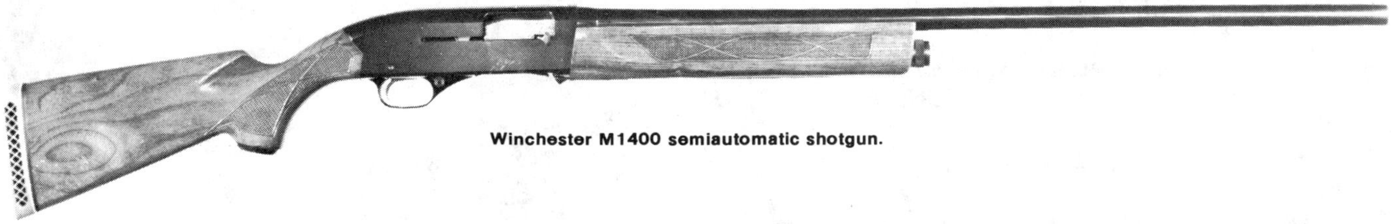

Winchester M1400 semiautomatic shotgun.

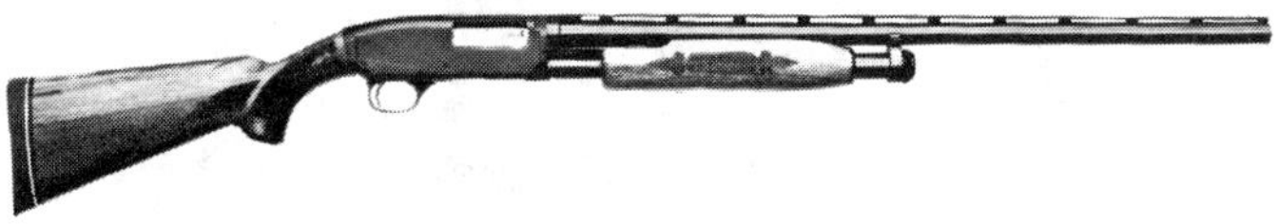

Marlin Model 120 pump shotgun.

WINCHESTER, U.S.A.

Winchester M1400

In 1964, Winchester introduced its new gas-operated M1400 semiautomatic shotgun. In external appearance, the M1400 is virtually identical to the M1200 pump gun. The action is also only slightly changed. A very high degree of parts interchangeability exists between models. The difference lies only in that the M1400 has been modified so that a conventional gas piston and cylinder provides the power for operating the action. Dual action bars are attached to the gas piston, and at their rear extremities, they are identical to the action bars of the M1200. The forward portion of the magazine tube functions as the gas cylinder and the action bars are attached by means of a pin protruding through a slot in the tube. The magazine cap which, as in the M1200, holds the barrel in position, contains a gas relief valve to bleed off gas over and above the designed operating level. This permits use of light, heavy, and magnum loads through the same mechanism without adjustment. The action release of the M1200 is replaced by a bolt latch activated by the magazine follower. After the last round is fired, the bolt is locked to the rear. Pressure on the magazine follower disengages the latch, allowing the recoil spring to drive the bolt forward. This makes it necessary for some slight change in loading procedures. With the action open, drop a round in the ejection port. Then either depress the follower by hand, or with the cartridge being inserted in the magazine. All other functioning, manipulation, loading, and firing procedures are essentially identical to those of the M1200 pump gun, which see. Specifications and characteristics are identical.

REVOLVERS

CHARTER ARMS, U.S.A.

Undercover Model

Having been introduced around 1965, this is the most recent, completely-new, U.S. designed revolver ever to achieve full production status. It is Charter Arms' only product as this is written. It is a solid-frame, swing-out cylinder, double/single-action revolver of minimum weight and size adaptable to the .38 Special cartridge. The design dispenses with a frame side plate, access being provided by removing the trigger guard. It is unique in that the cylinder may be unlatched by either pressing a thumb piece forward, or by pulling forward on the extractor rod. In 1973 a .44 Special version was introduced.

Characteristics of the Charter Arms Revolver

Caliber: 38 Special; 32 S&W Long; .44 Special
Operation by: Manual; double- or single-action.
Barrel length: 2 in. to 3 in.
Overall length: 6-1/4 in. to 7-1/4 in.
Weight: 16 to 17 oz.
Sights: Open, fixed.
Cylinder capacity: 5 rounds; 6 in. 32.

COLT, U.S.A.

Colt solid-frame, swing-out cylinder, double/single-action revolvers all utilize the same basic design, with the exception of the Trooper Mark III introduced in mid-1969. The design dates to before 1900. Though it has undergone improvement from time to time, it remains essentially the same. The most significant change

Charter Arms revolver.

was the addition of a safety hammer block during the 1920's. It consists of a bar which is interposed between hammer face and frame except when the trigger is held fully rearward. It prevents the firing pin from reaching a cartridge or, in rebounding-pin models, prevents the hammer from reaching the firing pin.

The various models are based on variations of 2 basic frame sizes, the old .38 Long and the .41 Colt. In 1946, a third larger frame, the New Service .45 produced since 1898, was discontinued. It also used the same basic action. Current models differ only in minor detail, sights, etc. Lightweight versions use the same frames made of aluminum alloy. All have 6-shot cylinders.

Characteristics of the Colt Revolvers

	.38 Frame (1)	.41 Frame (2)
Caliber:	.22; .32; .38; .357.	.22; .38; .357.
Operation by:	Manual, double, or single-action.	------------------------
Barrel length:	2 in. to 5 in.	2-1/2 in. to 6 in.
Overall length:	6-3/4 in. to 9-3/4 in.	7-1/2 in. to 11-1/4 in.
Weight:	21 to 24 oz.	33 to 44 oz.
Sights:	Open, fixed; target-type same models	------------------

(1) Cobra, Agent, Detective Special, Diamondback, Police Positive Special.

(2) Official Police, Trooper, Officer's Model, Python.

In 1972 Colt revised its small-frame Detective Special and Cobra guns by changing to a heavy Python-style barrel with ejector-rod enclosure and ventilated rib.

Colt Trooper Mark III

The Trooper Mark III contains completely redesigned lock work which in detail functions differently from other standard models. It differs as follows: The bolt is of the S&W type actuated by a forward projection on the trigger; rebound action and trigger return are furnished by separate springs instead of the old-style lever. All springs except that of the hand are the coil-type.

Loading and firing procedures are identical to other Colt models. Various hammer, trigger, and grip options are offered.

Characteristics of the Colt Trooper Mark III Revolver

Caliber: .357 Magnum; .38 Special.
Operation by: Manual, double- or single-action.
Barrel length: 4 in. to 6 in.
Overall length: 9-1/4 in. to 11-1/4 in.; with target grips 1/8 in. longer.
Height: 4-8/10 in.; 5-1/4 in. with target grips.
Weight: 40 to 42 oz.
Sights: Open, micrometer-adjustable target type.
Cylinder capacity: 6 rounds.

Colt Single-Action Army

This revolver has been in production since 1873 in its present form, though the basic action design actually dates before 1850. It was at one time the standard U. S. Army side arm in .45 caliber and is known to have been produced in 36 calibers for sale throughout the world. The action is quite simple mechanically, and capable of hand manipulation with various parts broken or removed, though not as durable as often claimed. The SAA Colt has long been considered the typical American handgun and has

Characteristics of the Colt Single-Action Army Revolver

Caliber: .357 Magnum; .45 Colt.
Operation by: Manual, single-action only.
Barrel length: 4-3/4 in. to 5-1/2 in.; 7-1/2 in. to 12 in.
Overall length: 10-3/4 in. to 18 in.
Weight: 40 to 43 oz.
Sights: Open, fixed.
Cylinder capacity: 6 rounds.

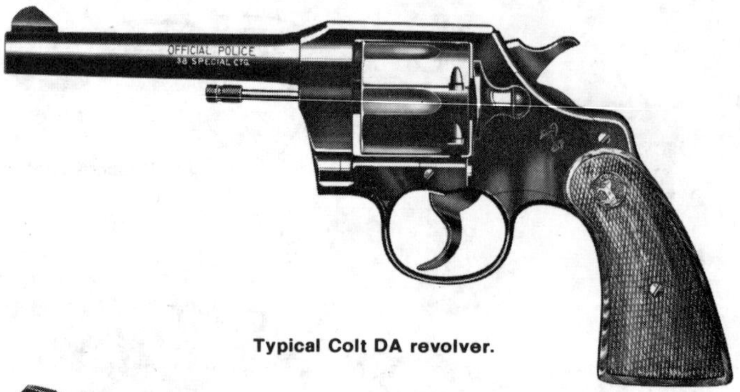

Typical Colt DA revolver.

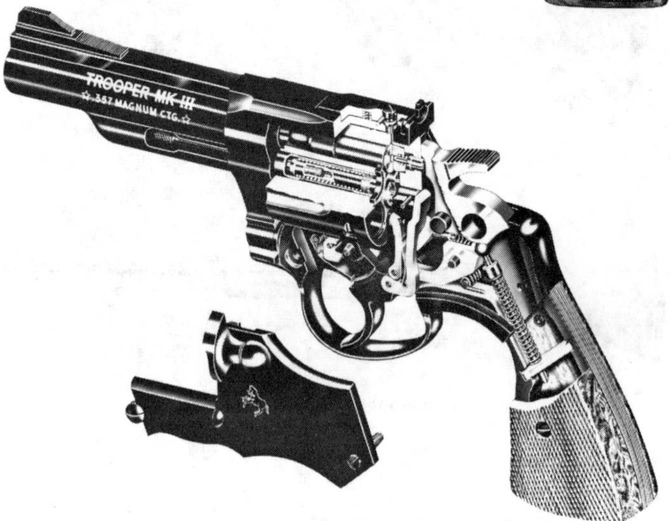

Colt Trooper Mark III revolver.

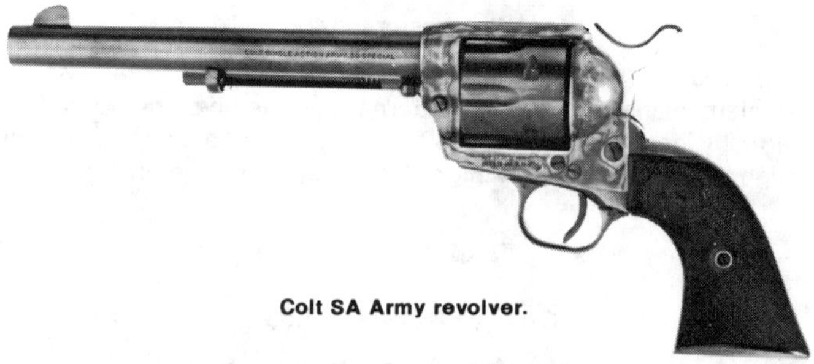

Colt SA Army revolver.

been widely copied, even today, as will be seen elsewhere in this volume. An updated version fitted with target sights is designated "New Frontier."

Colt Frontier Scout

This is simply a .22 RF caliber, reduced-scale version of the SAA. The same basic action, with minor changes and reduction in size is used. Also used is a frame-mounted rebounding firing pin. The grip frame is a single part, rather than two as in the SAA. Operation, loading, and firing are as in the SAA.

Characteristics of the Colt Frontier Scout Revolver

Caliber: .22 RF; .22 MRF.
Operation by: Manual, single-action only.
Barrel length: 4-3/4 in. to 9-1/2 in.
Weight: 24 to 34 oz.
Sights: Open, fixed.
Cylinder capacity: 6 rounds.

CATTLEMAN, WEST GERMANY

Manufactured in West Germany for L. A. Distributors, this is a single-action revolver of solid-frame, rod-ejection, "Frontier" style identical to Colt and Ruger models. Operation and loading-firing instructions for those guns apply.

Characteristics of the Deputy Revolver

Caliber: .22 RF; .22 MRF; .357 Magnum; .44 Magnum.
Operation by: Manual; single-action only.
Barrel length: 4-3/4 in. to 7-1/2 in.
Overall length: 10-3/4 in. to 13-1/2 in.
Weight: 30 to 42 oz.
Sights: Open, fixed. Target-type on some versions.
Cylinder capacity: 6 rounds.

Deputy single-action revolver.

HARRINGTON & RICHARDSON

Harrington and Richardson produces a number of low-to-medium price rimfire and center-fire revolvers. Internally, all are essentially the same, but three different frame designs are used. The top of the line is represented by the "Sportsman" which is of traditional hinged-frame, top-break, automatic-ejection type. The "Sidekick" utilizes a solid frame and swing-out type cylinder, while the "Snap-Out" has merely a quick-removable cylinder. Bottom of the line is the Model 622 which requires removal of the base pin before the cylinder can be taken out for loading or ejection of empties. All models are made without side plate, with access to the interior being obtained by removal of the trigger guard. The basic action design is somewhat unusual in that a separate lifter pivoted to the trigger transmits trigger motion to the hammer in double-action fire, and the hammer motion to the trigger in single cocking. In addition, a separate pivoted sear holds the hammer at full cock until rotated by the trigger to disengage the hammer, which then

Harrington & Richardson M929 revolver.

falls to fire the cartridge. Hammer rebound action is obtained through differential action of a cam-shaped head on the mainspring guide. Another unusual feature supplied only on the more costly of the Sidekick series is a key-operated lock in the butt which prevents firing when engaged.

Characteristics of the H&R Revolvers

Caliber: .22 L.R.; .38 S&W; .32 S&W.
Operated by: Manual; double or single-action.
Barrel length: 2-1/2 in. to 6 in.
Overall length: 6-1/4 in. to 11 in.
Sights: Open, fixed; adjustable on some models.
Weight: 20 to 33 oz.
Magazine capacity: 5 to 9 rounds.

HIGH-STANDARD MFG. CORP., U.S.A.

Sentinel

This modern, cast-frame double/single-action revolver of solid-frame, swing-out design was introduced in 1955. It represented the first domestic departure from traditional solid-frame and hinged-frame .22 rimfire revolvers at a practical price. The frame is aluminum-alloy; all other parts are of steel. It is unique in that the cylinder is released to swing outward only by pulling forward on the extractor rod.

Characteristics of the Sentinel Revolver

Caliber: .22 RF.
Operation by: Manual, double or single-action.
Barrel length: 2-3/8 in. to 6 in.
Overall length: 7-1/4 in. to 11 in.
Weight: 15 to 24-1/2 oz.
Sights: Open, fixed.
Cylinder capacity: 9 rounds.

In 1973 High Standard introduced its Sentinel MKII and MKIII .357 Magnum revolvers. Both are the basic D. B. Wesson (which see) very slightly modified and produced under contract for High Standard with appropriate markings.

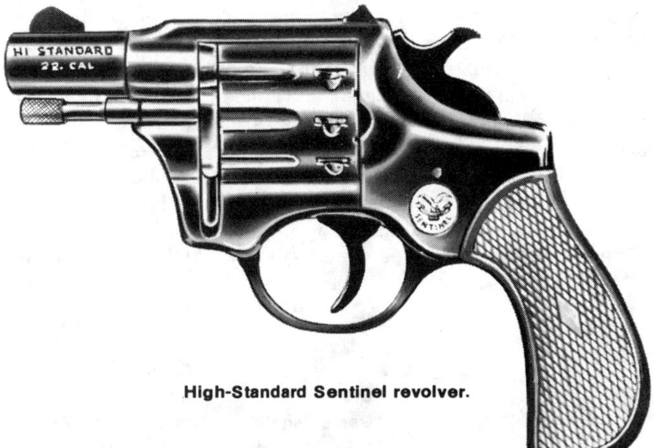

High-Standard Sentinel revolver.

IVER JOHNSON, U.S.A.

The Iver Johnson Arms and Cycle Works has produced reliable, economy-priced rimfire revolvers since well before the turn of the Century. A number of models are offered, all utilizing essentially the same internal mechanism, but differing in frame design. At the top of the line is the Model 67 "Viking," of hinged-frame, top-break design and it possesses the "hammer-the-hammer" internal safety greatly publicized many years ago. The other models are of solid-frame design. The "Sidewinder" series features Frontier styling and Colt-type rod ejection. The bottom of the line is represented by the Model 55 series which has a solid frame and requires the base pin to be withdrawn to remove the cylinder for loading and ejection.

The Iver Johnson lock work utilizes a separate pivoted lifter acting between the trigger and hammer to transfer trigger movement to the hammer in double-action fire, and hammer movement to the trigger when cocked single-action. This lifter also functions as the "hammer-the-hammer" safety in that an upward projection on it must be in position between the hammer and the firing pin in order for a cartridge to be fired when the hammer falls. The lifter cannot be in this position unless the trigger is pulled fully rearward. The action also utilizes a separate pivoted sear which must be acted on by the trigger to fire the gun single-action.

Characteristics of the Iver Johnson Revolvers

Caliber: .22 RF; .32 & .38 S&W.
Operation by: Manual; double or single-action.
Barrel length: 2-1/2 in. to 6 in.
Overall length: 7-1/4 in. to 10-3/4 in.
Sights: Open, fixed; adjustable rear on top models.
Weight: 25 to 34 oz.
Magazine capacity: 5 rounds (CF), 8 (RF).

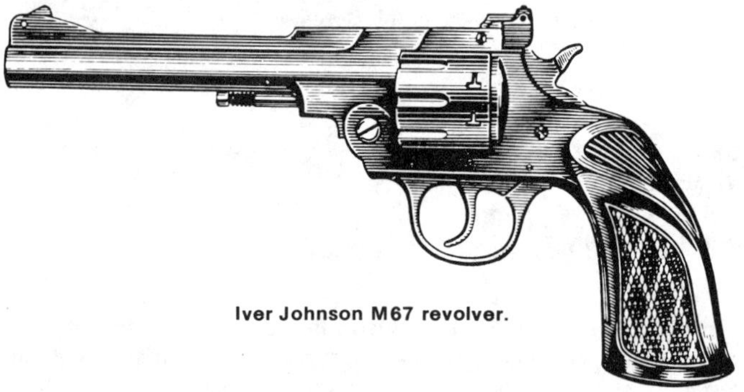

Iver Johnson M67 revolver.

SMITH & WESSON, U.S.A.

Smith & Wesson solid-frame, side-swing, exposed-hammer, double/single-action revolvers are known and respected throughout the world. The basic design is identical in virtually all models except the Bodyguard and Centennial. The design dates to the turn of the Century in the 1902 Military and Police Model with relatively minor improvements in the late 1920's and early 1950's. The various models are produced on three different size frames: the .32 size of the 5-shot guns; the .38 size of the 6-shot guns through .38-.357 caliber; and the .44 size Magnum frame of the 6-shot .41, .44, .45 and heavy .357 guns.

The basic design as represented by the U.S. M1917 .45 revolver under "U.S.A." and the .38/200 K200 "Victory Model" under "Britain" is covered in those sections of this book. Present commercial models differ only in minor areas such as the shortened hammer fall introduced in 1950, micrometer-adjustable sights on target models, and frame-mounted rebounding firing pins in the rimfire models. Most of the .32 and .38 frame models are available as "Airweight" with aluminum-alloy frames.

Smith & Wesson has pioneered in development of stainless-steel handguns. It introduced the M60 "Stainless Chief" in 1965 and in 1972 began delivery of the M64, M66, and M67 revolvers, stainless-steel versions of the M10, M15, and M19 respectively. Specifications are identical with their plain-steel counterparts.

Smith & Wesson Stainless-Steel M67.

Ruger Security-Six revolver.

S&W Bodyguard, Model 38

Built on the .32 frame and differs from the basic model only in that frame side-walls are extended to surround the hammer. The hammer is altered in profile and the checked spur protrudes through a slot just enough to permit single-action cocking, if desired. It is intended primarily for double-action use.

Characteristics of the Smith & Wesson Revolvers

	.32 Frame (1)	.38 Frame (K) (2)	.44 Frame (M) (3)
Caliber:	.22; .32; .38.	.22; .32; .38; .357.	.357; .44; .45; .41.
Barrel length:	2 in. to 6 in.	2 in. to 8-3/8 in.	3-1/2 in. to 8-3/8 in.
Overall length:	6-1/4 in. to 10-1/2 in.	6-1/8 in. to 12-7/8 in.	9-3/8 in. to 13-3/4 in.
Weight:	*17 to 20 oz.	*28 to 42-1/2 oz.	42 to 51-1/2 oz.
Cylinder capacity:	5 rounds.	6 rounds.	6 rounds.

*Airweight models weigh approximately 1/3 less in 2-in. barrel length.
(1) Terrier, Chief Special, Kit Gun.
(2) M&P, K22-32-38, Combat Masterpiece, Combat Magnum.
(3) .357, .41, and .44 Magnums; .44 Mil; .45 Target; .38/44 Outdoorsman.

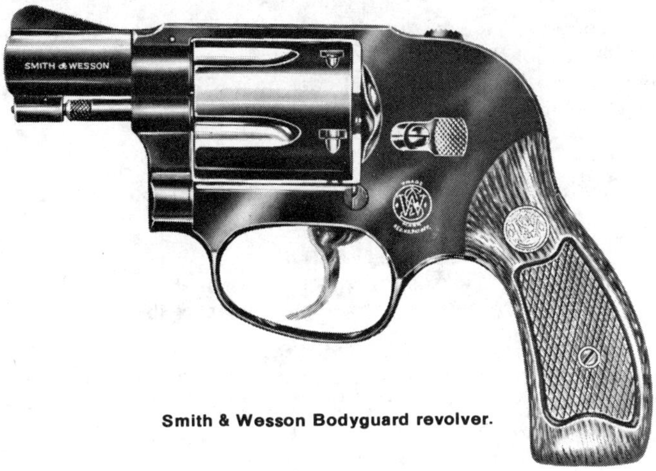

Smith & Wesson Bodyguard revolver.

Characteristics of the S&W Model 38 Pistol

Caliber: .38 Special.
Operation by: Manual; double- or single-action.
Barrel length: 2 in.
Overall length: 6-3/8 in.
Sights: Open, fixed.
Weight: 20-1/2 oz.; Airweight 14-1/2 oz.
Cylinder capacity: 5 rounds.

STURM RUGER, U.S.A.

Ruger Blackhawk

This is a modern center-fire revolver of traditional American single-action "Frontier" styling. In appearance and manipulation, it is identical to the Colt SAA. Mechanically, however, it differs considerably. The grip frame is a one-piece casting rather than a two-screw assembly of parts. The lock work, while quite similar and functioning the same as in the Colt, is redesigned to use coil and torsion springs throughout. The entire gun in all versions is designed for modern production techniques, especially precision-casting and screw-machine fabrication.

In addition to the Blackhawk, there is a smaller-frame version made for rimfire cartridges and designated "Single Six."

Characteristics of the Ruger Revolvers

	Single Six	**Blackhawk**
Caliber:	.22 RF; .22 MRF.	.357; .41; .44 Magnum.
Operation by:	Manual, single-action only.	----------------------------------
Barrel length:	5-1/2 in. to 9-1/2 in.	4-5/8 in. to 7-1/2 in.
Overall length:	10-7/8 in. to 14-7/8 in.	10-1/4 in. to 13-3/8 in.
Weight:	36 to 38 oz.	37 to 48 oz.

Ruger Blackhawk revolver.

Sights:	Open, fixed*	Open, adjustable target type.
Cylinder capacity:	6 rounds.	6 rounds.

*Super version has target-type adjustable sights.

DAN WESSON ARMS, U.S.A.

Wesson M12

This is a most recent and modern U.S. double-action revolver development, introduced in 1969, by Dan Wesson, direct descendant of Daniel B. Wesson of the original Smith & Wesson firm. Conventional in appearance, the M12 differs considerably from contemporary revolvers. Most unusual is its interchangeable barrel system, never before offered in any high-power revolver. The barrel proper is a simple, thin-wall rifled tube threaded on both ends. It is screwed into the receiver against a feeler gage laid across the cylinder face. The gage insures proper cylinder/barrel gap. A separate housing is then slipped over the barrel to butt against the receiver. A muzzle nut is then turned tightly on the barrel to lock both barrel and housing securely in place. Barrels of different lengths, each with its own matched housing, may be freely exchanged on the same gun. The housing also surrounds the extractor rod, protecting it from impact damage.

The M12 also differs in having its cylinder latch located at the front of the frame, moving vertically to act directly on the crane. Being designed primarily for double-action work (though capable of single-action), the hammer fall is 25% less than contemporary designs.

Characteristics of the Wesson M12 Revolver

Caliber: .357 Magnum.
Type: Solid-frame, swing-out cylinder, double- and single-action, manually-operated revolver.
Barrel length: 2-1/2 in. to 6 in.
Overall length: 9 in. with 4 in. barrel.
Height: 5-1/2 in.
Sights: Open; rear adjustable for windage, front for elevation.
Cylinder capacity: 6 rounds.

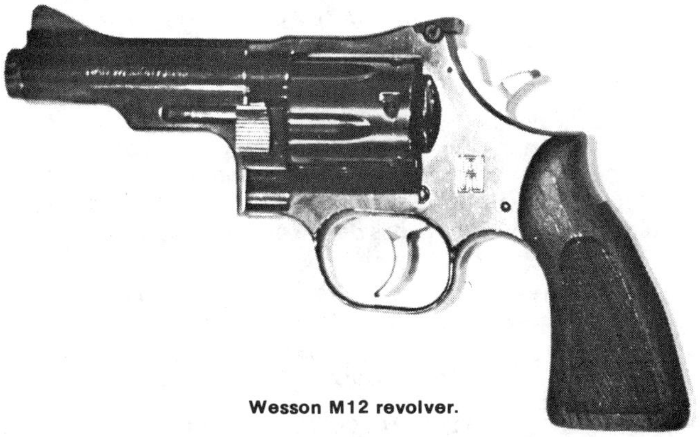

Wesson M12 revolver.

FOREIGN COPIES OF DOMESTIC MODELS

There are numerous foreign-manufactured revolvers which are either outright copies or only slightly-modified versions of basic domestic models. They include copies of both Colt and Ruger single-action models as well as S&W and Colt double-action guns. Some are encountered regularly in this country, others not often. Generally speaking, there is sufficient resemblance to domestic models that they are readily recognized. Some are of good quality, others are not, and each should be examined and judged on its own merits. The more common are listed here.

Stallion, Italy

Imported by J. Galef & Son. A relatively good modified copy of the Colt Frontier Scout offered in .22 RF and .22 MRF caliber; 6-shot.

Dokata, Italy

Imported by Intercontinental Arms. A good copy of the Colt SAA. Offered in .22, .357, .41 and .44 Magnum, .45 Colt. Fixed and target-sighted models available; 6-shot.

Herters, Belgium

Imported by Herters, Inc. Modified copy of Ruger Blackhawk series in .22 RF, .22 MRF, .357, .401, and .44 Magnum. Target sights; 6-shot.

Deputy, Germany

Imported L. A. Distributors. Modified copy of Colt SAA; 6-shot.

Target Master, Germany

Imported by L. A. Distributors. A modified copy of the basic S&W design. Target sights; 6-shot.

Hawes, Germany

Imported by Hawes Firearms. Modified copy of Colt SAA; 6-shot.

Llama Ruby, Spain

Imported by Stoeger Arms in .32 and .38 frame sizes, both service and target styles and sights. They are excellent slightly-modified copies of the basic S&W DA design in .32 S&W Long and .38 Special calibers; 6-shot.

Taurus, Argentina

Imported by Firearms International. A modified copy of the basic S&W .38 frame size in .38 Special caliber; 6-shot.

I.N.A. Tiger

Imported by Universal Firearms. A considerably modified copy of the basic S&W .38 frame size in .38 Special caliber; 5-shot.

Astra Cadix, Spain

Imported by Firearms International. A highly-modified version of the basic S&W design in small frame size; .22 RF (6-shot) and .38 Special (5-shot).

ROSSI, BRAZIL

Imported by Firearms International. A slightly-modified copy of the basic Smith & Wesson .38 Special. A well finished and durable gun.

SEMIAUTOMATIC PISTOLS

BERETTA, ITALY

With the exception of the Brigadier and M90, all currently-available Beretta pistols imported by J. L. Galef & Son are based on the M1934 design. The M1934 is described and pictured in detail in the "Italian" section. Manipulation and functioning of the Jaguar, Puma, and Sable are essentially the same as for the Beretta 9mm Model 1934.

Characteristics of the Beretta Pistol

	Jaguar	Puma	Sable
Caliber:	.22 L.R.	.32 ACP.	.22 L.R.
Operation by:		Blowback.	
Barrel length:	6 in.	6 in.	6 in.
Overall length:	8-3/4 in.	9-1/2 in.	8-3/4 in.
Sights:		Open, adjustable rear.	
Weight:	19 oz.	19 oz.	26 oz.
Magazine capacity:	10 rounds.	10 rounds.	10 rounds.

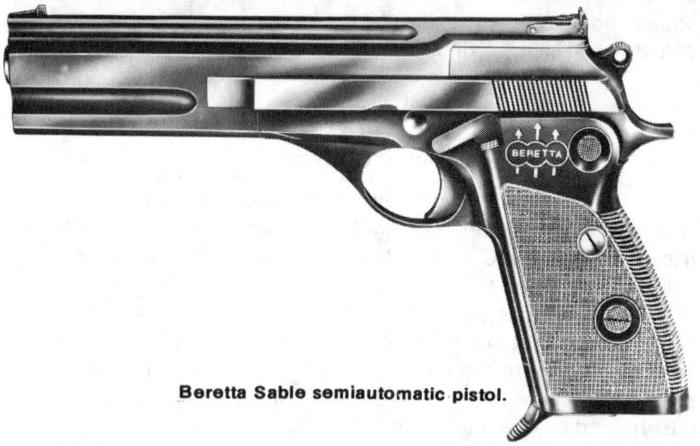

Beretta Sable semiautomatic pistol.

Beretta Brigadier

This is simply a commercial version of the 9mm M1951 pistol adopted by the Italian Army. It is chambered for the 9mm Parabellum (Luger) cartridge and is described and pictured in detail in the "Italian" section of this book.

In 1971 Beretta introduced a very modern double-action .32 caliber pocket-size autoloader designated M90. Of blowback design, it features a stainless-steel removable barrel, concentric recoil spring, and exposed hammer.

BROWNING ARMS CO. U.S.A. & BELGIUM

Browning High-Power

Probably the most popular Browning handgun in the U.S.A., this model is identical in all respects save finish to the standard Belgian 9mm FN High-Power described in Part Two of this book. It is chambered only for the 9mm Parabellum cartridge and is produced for Browning by Fabrique Nationale.

Browning .32 or .380 Automatic

This is the basic M1910 FN/Browning manufactured by FN since that year for both military and commercial sales. This design introduced the barrel-mounted recoil spring and bayonet-type removable barrel bushing or "slide ring," and copied the rotating-barrel takedown system of the 1905 Browning .25 caliber pistol. The entire mechanism is at least as simple and trouble-free as any to come since. It is striker-fired and fitted with a grip safety.

Characteristics of the M1910 FN/Browning Pistol

Caliber: .32 ACP, .380 ACP.
Operation by: Blowback.
Barrel length: 3-7/16 in.
Overall length: 6 in.
Height: 3-7/8 in.
Weight: 20 oz.
Sights: Open, fixed, recessed.
Magazine capacity: 7-6 rounds.

Browning Medalist

The Medalist was introduced in 1962 as a top-grade .22 rimfire target pistol. Essentially, it is a blowback-operated, concealed-hammer, detachable-barrel semiautomatic pistol of conventional construction and is manufactured for Browning by Fabrique Nationale. It is fitted with a ventilated sighting rib and finely adjustable target-type sights. The rear sight is mounted on an integral extension of the rib rearward of the barrel in order to place both

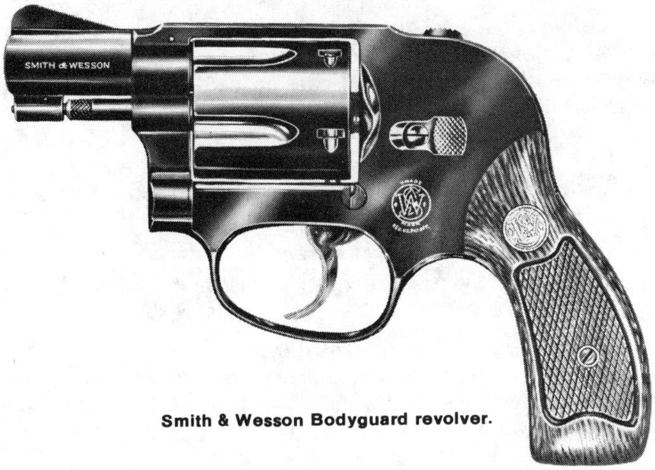

Smith & Wesson Bodyguard revolver.

Characteristics of the S&W Model 38 Pistol

Caliber: .38 Special.
Operation by: Manual; double- or single-action.
Barrel length: 2 in.
Overall length: 6-3/8 in.
Sights: Open, fixed.
Weight: 20-1/2 oz.; Airweight 14-1/2 oz.
Cylinder capacity: 5 rounds.

STURM RUGER, U.S.A.

Ruger Blackhawk

This is a modern center-fire revolver of traditional American single-action "Frontier" styling. In appearance and manipulation, it is identical to the Colt SAA. Mechanically, however, it differs considerably. The grip frame is a one-piece casting rather than a two-screw assembly of parts. The lock work, while quite similar and functioning the same as in the Colt, is redesigned to use coil and torsion springs throughout. The entire gun in all versions is designed for modern production techniques, especially precision-casting and screw-machine fabrication.

In addition to the Blackhawk, there is a smaller-frame version made for rimfire cartridges and designated "Single Six."

Characteristics of the Ruger Revolvers

	Single Six	Blackhawk
Caliber:	.22 RF; .22 MRF.	.357; .41; .44 Magnum.
Operation by:	Manual, single-action only.	----------------------------------
Barrel length:	5-1/2 in. to 9-1/2 in.	4-5/8 in. to 7-1/2 in.
Overall length:	10-7/8 in. to 14-7/8 in.	10-1/4 in. to 13-3/8 in.
Weight:	36 to 38 oz.	37 to 48 oz.

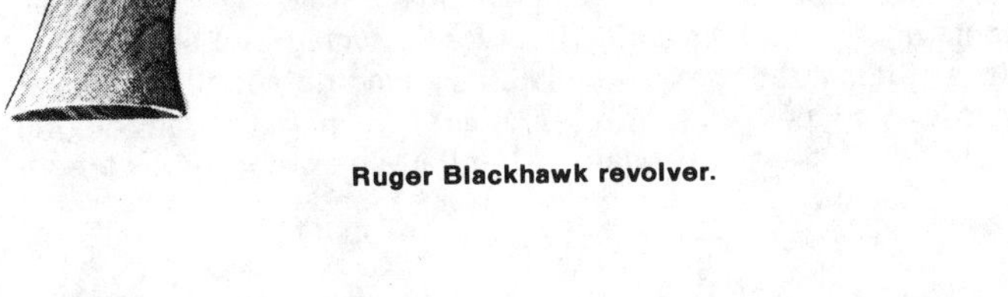

Ruger Blackhawk revolver.

Sights:	Open, fixed*	Open, adjustable target type.
Cylinder capacity:	6 rounds.	6 rounds.

*Super version has target-type adjustable sights.

DAN WESSON ARMS, U.S.A.

Wesson M12

This is a most recent and modern U.S. double-action revolver development, introduced in 1969, by Dan Wesson, direct descendant of Daniel B. Wesson of the original Smith & Wesson firm. Conventional in appearance, the M12 differs considerably from contemporary revolvers. Most unusual is its interchangeable barrel system, never before offered in any high-power revolver. The barrel proper is a simple, thin-wall rifled tube threaded on both ends. It is screwed into the receiver against a feeler gage laid across the cylinder face. The gage insures proper cylinder/barrel gap. A separate housing is then slipped over the barrel to butt against the receiver. A muzzle nut is then turned tightly on the barrel to lock both barrel and housing securely in place. Barrels of different lengths, each with its own matched housing, may be freely exchanged on the same gun. The housing also surrounds the extractor rod, protecting it from impact damage.

The M12 also differs in having its cylinder latch located at the front of the frame, moving vertically to act directly on the crane. Being designed primarily for double-action work (though capable of single-action), the hammer fall is 25% less than contemporary designs.

Characteristics of the Wesson M12 Revolver

Caliber: .357 Magnum.
Type: Solid-frame, swing-out cylinder, double- and single-action, manually-operated revolver.
Barrel length: 2-1/2 in. to 6 in.
Overall length: 9 in. with 4 in. barrel.
Height: 5-1/2 in.
Sights: Open; rear adjustable for windage, front for elevation.
Cylinder capacity: 6 rounds.

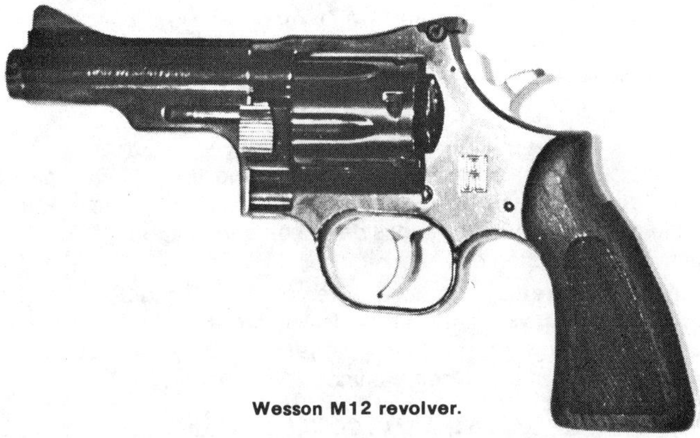

Wesson M12 revolver.

FOREIGN COPIES OF DOMESTIC MODELS

There are numerous foreign-manufactured revolvers which are either outright copies or only slightly-modified versions of basic domestic models. They include copies of both Colt and Ruger single-action models as well as S&W and Colt double-action guns. Some are encountered regularly in this country, others not often. Generally speaking, there is sufficient resemblance to domestic models that they are readily recognized. Some are of good quality, others are not, and each should be examined and judged on its own merits. The more common are listed here.

Stallion, Italy

Imported by J. Galef & Son. A relatively good modified copy of the Colt Frontier Scout offered in .22 RF and .22 MRF caliber; 6-shot.

Dokata, Italy

Imported by Intercontinental Arms. A good copy of the Colt SAA. Offered in .22, .357, .41 and .44 Magnum, .45 Colt. Fixed and target-sighted models available; 6-shot.

Herters, Belgium

Imported by Herters, Inc. Modified copy of Ruger Blackhawk series in .22 RF, .22 MRF, .357, .401, and .44 Magnum. Target sights; 6-shot.

Deputy, Germany

Imported L. A. Distributors. Modified copy of Colt SAA; 6-shot.

Target Master, Germany

Imported by L. A. Distributors. A modified copy of the basic S&W design. Target sights; 6-shot.

Hawes, Germany

Imported by Hawes Firearms. Modified copy of Colt SAA; 6-shot.

Llama Ruby, Spain

Imported by Stoeger Arms in .32 and .38 frame sizes, both service and target styles and sights. They are excellent slightly-modified copies of the basic S&W DA design in .32 S&W Long and .38 Special calibers; 6-shot.

Taurus, Argentina

Imported by Firearms International. A modified copy of the basic S&W .38 frame size in .38 Special caliber; 6-shot.

I.N.A. Tiger

Imported by Universal Firearms. A considerably modified copy of the basic S&W .38 frame size in .38 Special caliber; 5-shot.

Astra Cadix, Spain

Imported by Firearms International. A highly-modified version of the basic S&W design in small frame size; .22 RF (6-shot) and .38 Special (5-shot).

ROSSI, BRAZIL

Imported by Firearms International. A slightly-modified copy of the basic Smith & Wesson .38 Special. A well finished and durable gun.

SEMIAUTOMATIC PISTOLS

BERETTA, ITALY

With the exception of the Brigadier and M90, all currently-available Beretta pistols imported by J. L. Galef & Son are based on the M1934 design. The M1934 is described and pictured in detail in the "Italian" section. Manipulation and functioning of the Jaguar, Puma, and Sable are essentially the same as for the Beretta 9mm Model 1934.

Characteristics of the Beretta Pistol

	Jaguar	Puma	Sable
Caliber:	.22 L.R.	.32 ACP.	.22 L.R.
Operation by:		Blowback.	
Barrel length:	6 in.	6 in.	6 in.
Overall length:	8-3/4 in.	9-1/2 in.	8-3/4 in.
Sights:		Open, adjustable rear.	
Weight:	19 oz.	19 oz.	26 oz.
Magazine capacity:	10 rounds.	10 rounds.	10 rounds.

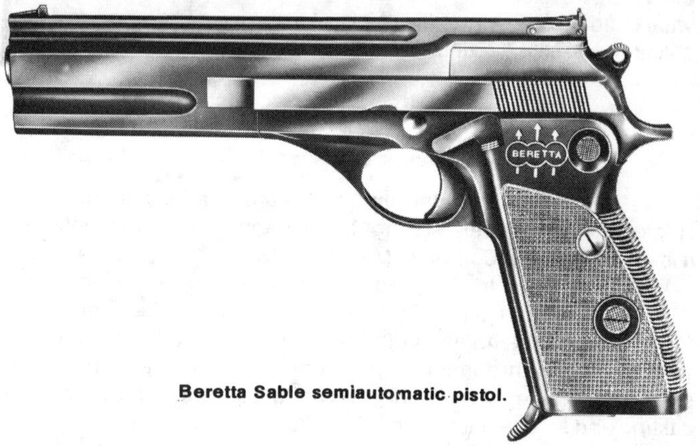

Beretta Sable semiautomatic pistol.

Beretta Brigadier

This is simply a commercial version of the 9mm M1951 pistol adopted by the Italian Army. It is chambered for the 9mm Parabellum (Luger) cartridge and is described and pictured in detail in the "Italian" section of this book.

In 1971 Beretta introduced a very modern double-action .32 caliber pocket-size autoloader designated M90. Of blowback design, it features a stainless-steel removable barrel, concentric recoil spring, and exposed hammer.

BROWNING ARMS CO. U.S.A. & BELGIUM

Browning High-Power

Probably the most popular Browning handgun in the U.S.A., this model is identical in all respects save finish to the standard Belgian 9mm FN High-Power described in Part Two of this book. It is chambered only for the 9mm Parabellum cartridge and is produced for Browning by Fabrique Nationale.

Browning .32 or .380 Automatic

This is the basic M1910 FN/Browning manufactured by FN since that year for both military and commercial sales. This design introduced the barrel-mounted recoil spring and bayonet-type removable barrel bushing or "slide ring," and copied the rotating-barrel takedown system of the 1905 Browning .25 caliber pistol. The entire mechanism is at least as simple and trouble-free as any to come since. It is striker-fired and fitted with a grip safety.

Characteristics of the M1910 FN/Browning Pistol

Caliber: .32 ACP, .380 ACP.
Operation by: Blowback.
Barrel length: 3-7/16 in.
Overall length: 6 in.
Height: 3-7/8 in.
Weight: 20 oz.
Sights: Open, fixed, recessed.
Magazine capacity: 7-6 rounds.

Browning Medalist

The Medalist was introduced in 1962 as a top-grade .22 rimfire target pistol. Essentially, it is a blowback-operated, concealed-hammer, detachable-barrel semiautomatic pistol of conventional construction and is manufactured for Browning by Fabrique Nationale. It is fitted with a ventilated sighting rib and finely adjustable target-type sights. The rear sight is mounted on an integral extension of the rib rearward of the barrel in order to place both front and rear sight on the same fixed base. An unusual external feature is the short wood fore-end which normally covers weight-attaching points. Balance weights supplied with the pistol may be attached beneath the barrel when the fore-end is removed. A most unusual feature is the "dry-fire" device. It permits cocking the lock mechanism without extracting the slide by simply pressing downward on the manual safety. The trigger may then be pressed, duplicating the gun's normal trigger pull and hammer release. Another feature not found in other rimfire pistols is a shell deflector pin set in the top of the right grip. Fired cases strike this pin on being ejected and are deflected forward, thus avoiding disturbing any shooter who might be on the right of the gun. The trigger is adjustable for over-travel and weight of pull, and the breech remains open after firing the last shot and exhausting the magazine.

Characteristics of the Browning Medalist Pistol

Caliber: .22 L.R.
Operation by: Blowback.
Barrel length: 6 3/4 in.
Overall length: 11 1/8 in.
Sights: Open, target-type.
Weight: 46 oz.
Magazine capacity: 10 rounds.

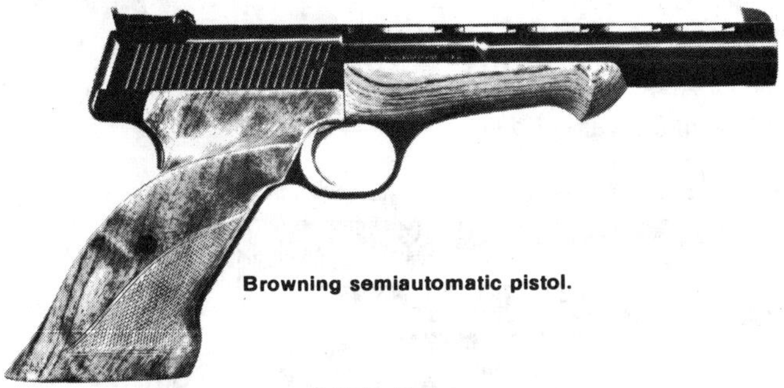

Browning semiautomatic pistol.

COLT, U.S.A.

Colt Commander

Essentially, this is simply a shortened version of the .45 Government Model (U.S. M1911/A1) utilizing an aluminum-alloy receiver. Parts except slide, barrel, and recoil spring interchange with those of the older models. See U.S. M1911 for mechanical and functioning details. Also now offered with steel frame at 33 oz. weight.

Characteristics of the Colt Commander Pistol

Caliber: 9mm Parabellum, .38 Super Auto, .45 ACP.
Operation by: Recoil, Browning system.
Barrel length: 4-1/4 in.
Overall length: 8 in.
Height: 5-1/2 in.
Weight: 26-1/2 oz.
Sights: Open, fixed.
Magazine capacity: 9; 7 in .45.

Colt Super .38 Automatic

This is simply the .45 Government Model chambered for the .38 Colt Super Automatic cartridge—first offered in 1927.

Colt Gold Cup .45 National Match

A highly-refined target version of the original .45 Government Model. It features newly-designed internal parts and adjustable trigger; careful hand-fitting for accuracy; micrometer-adjustable target sights. Mechanically, it is identical with the Government Model, which see.

Colt Gold Cup .38 National Match

Identical to the .45 Gold Cup, except that the locking system has

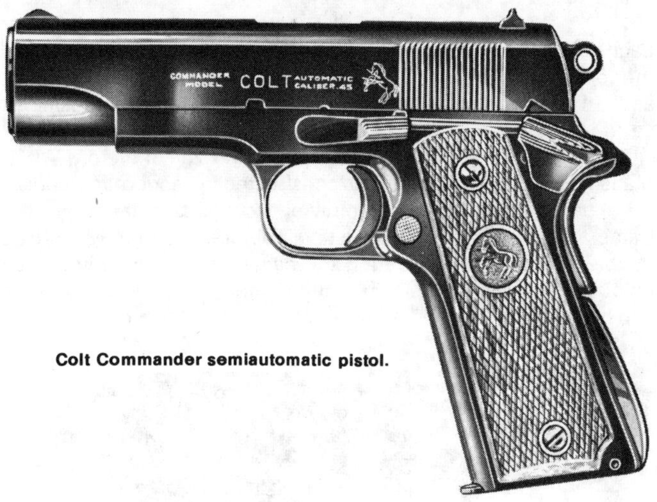

Colt Commander semiautomatic pistol.

been eliminated, converting it to delayed-blowback functioning. Slide is not locked to barrel; barrel is pinned to receiver, but allowed slight free travel in recoil to permit delay in breech opening. In addition, annular grooves in the chamber grip the fired case to further delay opening. Magazine is modified to feed rimmed .38 Special Wadcutters. All other characteristics identical to .45 Gold Cup.

In 1970 Colt modified the entire Government Model series, including the Gold Cups, to utilize a new collet-type barrel bushing and tapered-muzzle barrel. This version is now standard and is marked "Series 70 Mk.IV." Commander models do not have this modification.

Colt Woodsman

First introduced about 1915, continuously produced and improved to date—the first commercially-successful .22 rimfire self-loading pistol. The design is basic blowback and is relatively simple. Several of its innovations have found their way in modified form into later designs. It established the basic style and layout of most .22 caliber pistols to follow. Today, it is produced in several variations.

Colt Match Target Woodsman

This is the basic design fitted with fine target sights and a heavy, slab-sided target barrel. It is more carefully fitted and finished to produce maximum accuracy.

ERMA, WEST GERMANY

ERMA Pistolen

This gun looks like a Luger, but is not. Offered in .22 L.R. caliber only, its design differs from the Luger in that it is a straight blowback and utilizes a conventional striker-type firing mechanism.

Externally, it duplicates the 1908 Luger, but it is entirely different inside. Modern production and design techniques, such as investment castings, are used extensively.

HECKLER & KOCH, WEST GERMANY

HK-4

An unusual design imported by Harrington & Richardson. Its most distinguishing characteristic is that it is sold as a complete 4-caliber set with barrels, springs, and magazines for firing .22 L.R.;

.25 ACP; .32 ACP; or .380 ACP. The need for separate slides for each caliber is avoided by a dual-purpose firing pin and extractor; the former must be switched from center-fire to rimfire, and back again. The extractor automatically adjusts to the different diameter cartridge cases.

Essentially, the HK-4 is a highly-refined version of the pre-WW II Mauser HSc 7.65mm pistol. It is constructed by modern methods—the slide being a welded assembly of castings and stampings; the frame a light-alloy precision casting. The first round may be fired double-action by first lowering the hammer. Disassembly is also unique—depressing the latch in the trigger guard allows the slide and barrel to be lifted off after moving it 1/8 in. forward.

Heckler & Koch HK-4 semiautomatic pistol.

H&K also produce a large locked-breech 9mm P. military and police type autoloading pistol designated P9 (single-action) and P9S (double-action) similar in appearance and construction to the HK-4. Frame and slide are stampings and weldments and part of frame and overlaid with plastic. Locking is by the cam and roller system of the G-3 rifle, which see.

HIGH STANDARD MANUFACTURING CORP., U.S.A.

Since the early 1930's High Standard has been known for its extensive line of top-quality .22 rimfire self-loading pistols. While a large variety of models is cataloged, all target guns are built on the same action in two minor variations. The Military Model series has the rear sight mounted on a vertical yoke attached to the receiver with the slide passing through the yoke. The Military also has a squarish grip frame copying the shape of the U.S. M1911. Standard models have the rear sight mounted conventionally on the slide, and a different grip-frame profile. Within these two series, models differ only in barrel, weight, and accessories. An Olympic model in each series is chambered only for the .22 Short cartridge. All others are .22 L.R. All models are highly-refined target guns with fully-adjustable triggers and carefully-fitted mechanisms. Barrels are readily interchangeable.

Characteristics of the High Standard Pistols

	Military	Standard
Caliber:	.22 L.R.	.22 L.R.
Operation by:	Blowback.	Blowback.
Barrel length:	5-1/2 in. to 7-1/4 in.	5-1/2 in.
Overall length:	9-3/4 in. to 11-3/4 in.	10 in.
Weight:	44-1/2 oz.	42 oz.
Sights:	Open, adjustable, micrometer, target-type.	
Magazine capacity:	10 rounds.	10 rounds.

Two economy-priced, much simplified models called "Duramatic" and "Sport-King" are offered as plinking guns.

LLAMA (GABILONDO & CIA), SPAIN

Llama large-frame pistols in .45 ACP and .38 Colt Super Automatic calibers are simply extremely close copies of the Colt Government Model described in detail under "U.S.A." The more recent models have ventilated sighting ribs and other minor variations, but even so, the resemblance to the Colt is so great that many parts will interchange. A reduced-size version weighing only 20 ozs. is produced in .380 ACP caliber, but is mechanically and functionally identical to the big guns. An apparently identical Llama in .32 ACP has the locking system deleted and functions as a blowback. It is otherwise identical to the .380 and most parts will interchange.

Characteristics of the Llama .380 and .32 ACP Pistols

Caliber: .32 ACP; .380 ACP.
Operation by: Blowback, recoil, Browning system.
Barrel length: 3-11/16 in.
Overall length: 6-1/4 in.
Height: 4-3/8 in.
Sights: Open, fixed.
Weight: 21 oz.; 20 oz.
Magazine capacity: 8; 7 rounds.

In addition, premium-priced, adjustable-sight versions of the .45 and .38 models are offered.

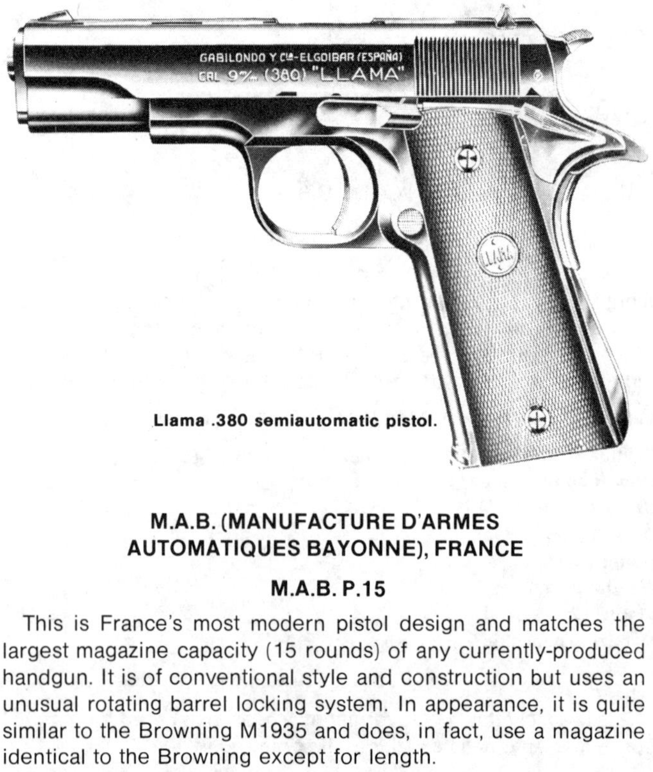

Llama .380 semiautomatic pistol.

M.A.B. (MANUFACTURE D'ARMES AUTOMATIQUES BAYONNE), FRANCE

M.A.B. P.15

This is France's most modern pistol design and matches the largest magazine capacity (15 rounds) of any currently-produced handgun. It is of conventional style and construction but uses an unusual rotating barrel locking system. In appearance, it is quite similar to the Browning M1935 and does, in fact, use a magazine identical to the Browning except for length.

Characteristics of the M.A.B. Pistol

Caliber: 9mm Parabellum.
Operation by: Recoil, rotating barrel system.
Barrel length: 4-6/10 in.
Overall length: 8-1/10 in.
Height: 5-1/4 in.
Sights: Open, fixed.
Weight: 38-1/10 oz.
Magazine capacity: 15 rounds.

M.A.B. P.15 semiautomatic pistol.

MAUSER, WEST GERMANY

In 1968, the famous Mauserwerke again entered the commercial handgun market with a newly-produced version of the original HSc pistol. The highly advanced double-action, pocket-size autoloader made its debut during the late 1930's and hundreds of thousands of 7.65mm specimens saw service in WW II. The new version is offered in .22 L.R., 7.65mm (.32 ACP), and 9mmK (.380 ACP) calibers. Except for the slight differences in the added calibers, the new HSc is as described elsewhere in this volume.

Parabellum

Also in 1968, Mauser introduced newly manufactured Parabellum pistols in 7.65mm (.30 Luger) caliber. This model is identical to pre-WW II "Luger" pistols, but that name cannot be applied to it simply because Stoeger Arms owns the name-copyright. The new gun is actually made on original Swiss-arsenal tooling utilized to produce the M1929 7.65mm pistol for the Swiss Government before WW II. Available now only in 7.65mm caliber, the new Parabellum will be offered in 9mm Parabellum, and perhaps others later. The original design is pictured and described in detail elsewhere in this book. All Mauser pistols are distributed in the U.S. by Interarms.

SIG (SWISS INDUSTRIAL CO.), SWITZERLAND

SIG P210

A series of commercial pistols based on the Swiss SP47/8 (M49) service pistol, which see. A blowback .22 L.R. conversion unit and glossy-finished target models are offered.

SMITH & WESSON, U.S.A.

Smith & Wesson M39

Designed in the early 1950's for U.S.Army lightweight-pistol tests, the M39 was introduced commercially after those tests terminated. It is a modern, locked-breech, double-action, self-loading pistol built upon an aluminum-alloy frame. It contains a magazine safety and a means of dropping the hammer from full cock without firing. It utilizes a single-column box magazine in the butt and is offered only in 9mm Parabellum caliber.

Characteristics of the S&W M39 Pistol

Caliber: 9mm Parabellum.
Operation by: Recoil, modified Browning system.
Barrel length: 4 in.
Overall length: 7-7/16 in.
Weight: 26-1/2 oz.
Sights: Open, rear adjustable for windage only.
Magazine capacity: 8 rounds.

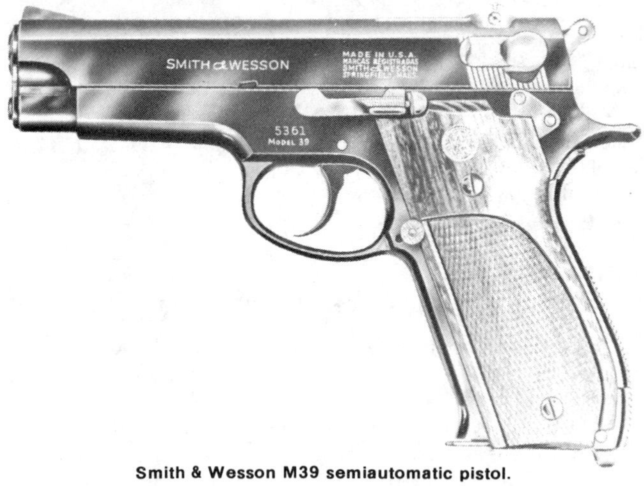

Smith & Wesson M39 semiautomatic pistol.

In 1973 S&W introduced an improved version of the M39 pistol, designated M59. Otherwise the same, it is modified to accept a 15-round, double-column magazine in the butt, making its fully-loaded capacity 16 rounds. Empty it weights 30 oz., loaded 37 oz.

Smith & Wesson M52

This is a longer, heavier version of the M39 intended purely for target shooting. Mechanically and functionally, it is identical to the M39, though much more closely fitted and adjusted. It carries micrometer-adjustable target sights and is chambered only for the .38 Special Wadcutter cartridge.

Characteristics of the S&W M52 Pistol

Caliber: .38 Special Wadcutter.
Operation by: Recoil, modified Browning system.
Barrel length: 5 in.
Overall length: 8-5/8 in.
Weight: 41 oz.
Sights: Micrometer-adjustable, target-type.
Magazine capacity: 5 rounds.

Smith & Wesson M41

In 1957, S&W introduced the M41—its first .22 rimfire self-loader, and its first target pistol. It had produced high-grade target revolvers for many years. The M41 is basically a blowback design bearing some resemblance to old Clement patents. It utilizes a highly sophisticated lock work with fully-adjustable trigger, cocking indicator, enclosed hammer, a rearward barrel extension for sight installation, and micrometer-adjustable target rear sight. It

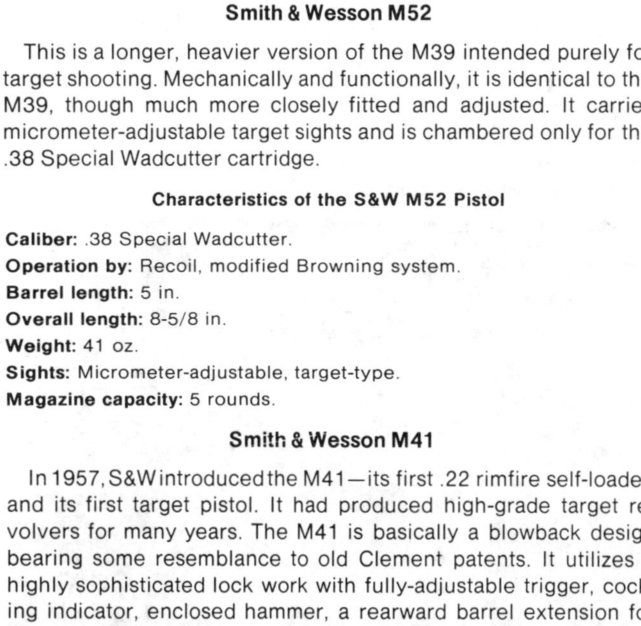

Smith & Wesson M41 semiautomatic pistol.

is equipped with a muzzle brake, barrel weights, and target-type grips. Slide and barrel extension are fitted with grooves designed to expel the greasy fouling of .22 rimfire ammunition. Disassembly is controlled by a pivoted trigger guard which locks the detachable barrel in place.

Characteristics of the S&W M41 Pistol

Caliber: .22 L.R.; .22 Short.
Operation by: Recoil, blowback.
Barrel length: 5 in. to 7-3/8 in.
Overall length: 12 in. (7-3/8 in. bbl. and muzzle brake).
Weight: 43-1/2 oz.; up to 59 oz. with weights.
Sights: Micrometer-adjustable target.
Magazine capacity: 10 rounds.

STAR (BONIFACIO ECHEVERRIA & CIA.), SPAIN

Large-caliber Star pistols are described in detail under "Spain." Commercial models of the "A," "B," and "P," are sold widely, especially in the Americas. In addition, a reduced-scale version in .380 ACP and .32 ACP caliber is available commercially. Called the "S" and "SI", this model duplicates all features, including the locking system of the larger guns. Smaller Star blowback-type pistols are offered in .22, .25, .32, and .380 caliber, but are no longer available in the U.S. because of governmental restrictions.

Characteristics of the Star Pistols

Caliber: .32 ACP; .380 ACP.
Operation by: Recoil, Browning system.
Barrel length: 3-3/4 in.
Overall length: 6-1/2 in.
Height: 4-3/4 in.
Sights: Open, fixed.
Weight: 22 oz.
Magazine capacity: 8; 7 rounds.

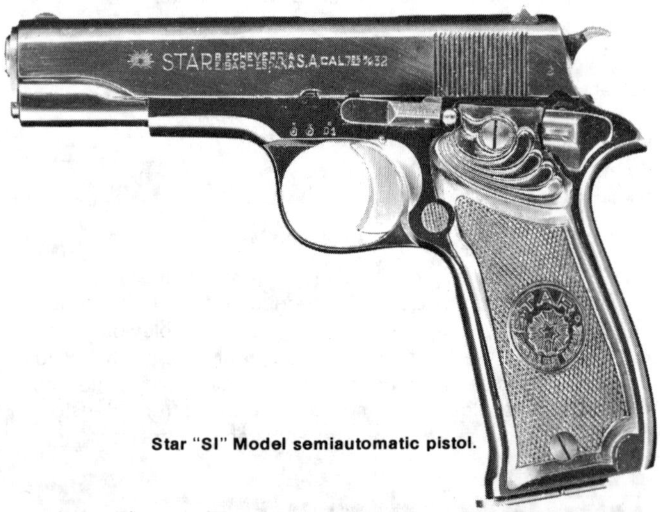

Star "SI" Model semiautomatic pistol.

STOEGER ARMS, U.S.A.

Stoeger Arms owns the "Luger" name copyright and has, therefore, introduced a .22 L.R. "Stoeger Luger." This design copies very closely the style, weight, and dimensions of original P.08 "Luger" pistols. Internally, however, it is entirely different. Though utilizing a toggle-type breech, it operates purely on the unlocked blowback system and incorporates a non-recoiling barrel. This is accomplished by placing the main toggle joint well above the bore center-line where case projection will cause it to open, but slowly. The gun has been extensively tested, and runs of 30,000 consecutive rounds without malfunction have been reported. Production is by subcontractor, utilizing modern technology and materials. A left-hand model is available.

Characteristics of the Stoeger Luger Pistol

Caliber: .22 L.R.
Operation by: Recoil; toggle breech.
Barrel length: 4-1/2 in.
Overall length: 9 in.
Height: 6 in.
Weight: 30 oz.
Sights: Open, fixed.
Magazine capacity: 11 rounds.

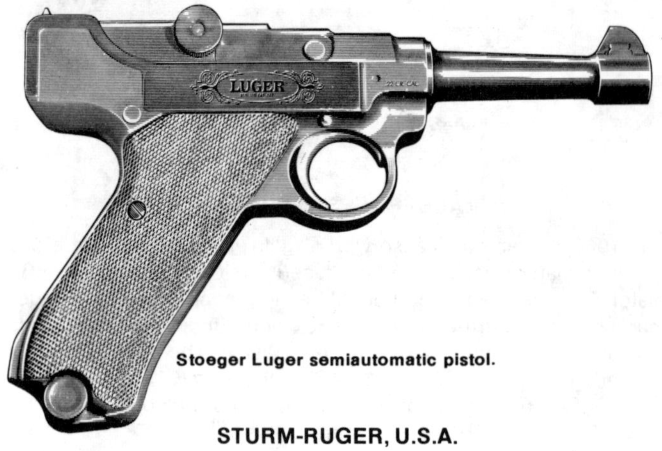

Stoeger Luger semiautomatic pistol.

STURM-RUGER, U.S.A.

Ruger Standard Model

The Sturm-Ruger firm entered the U.S. gun-making field with this new pistol in 1949 and has prospered ever since. While of basic blowback form, this design differs from any other commercially-successful domestic model in that the breech bolt is fully enclosed by a tubular "receiver" which is actually a barrel extension. Its outline greatly resembles the Luger and was no doubt deliberately made so in order to enhance its sales appeal. It was also the first successful domestic design to make use of a stamped and welded frame-assembly and brazed, laminated sheet-steel parts.

Characteristics of the Ruger Standard Model Pistol

Caliber: .22 L.R.
Operation by: Simple blowback.
Barrel length: 4-3/4 in. to 6 in.
Overall length: 8-3/4 in. to 10 in.
Weight: 36 oz. to 37 oz.
Sights: Open, fixed.
Magazine capacity: 9 rounds.

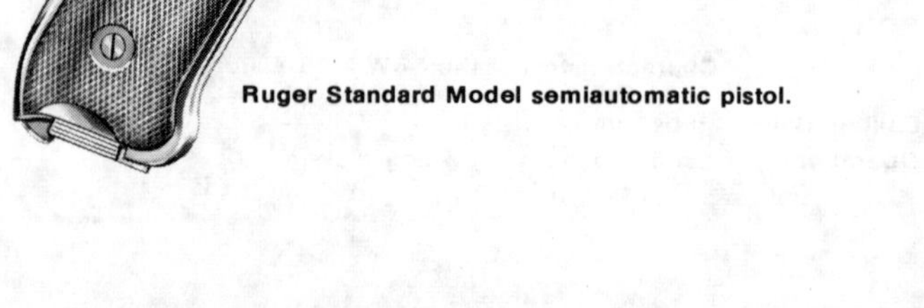

Ruger Standard Model semiautomatic pistol.

As were all subsequent Ruger models, the Standard Model was designed for maximum production ease and economy.

The Ruger Mark I is a target version of the Standard Model. It is fitted with Micro target sights and heavier barrels, and the action is more carefully finished. It weighs 42 oz. with either the 4-3/4 in. "Bull" or 6-7/8 in. tapered barrel.

Smith & Wesson Model 59.

WALTHER, GERMANY

Walther pistols are distributed in the U.S. by Interarms. Naturally, the P-38 is offered, both in its military configuration and finish, as well as in a glossy-finish commercial model. Both are identical to

Walther PP Sport Model semiautomatic pistol.

the aluminum-frame military model described under "West Germany". The P-38 is also furnished commercially in 7.65mm Parabellum (.30 Luger) caliber.

More recent is the .22 rimfire P-38. It is identical to the others except for a light-alloy slide and elimination of the locking system, which converts it to blowback operation.

Walther also now offers the PP Sport Model in .22 L.R. caliber. It is simply the basic PP fitted with a 6 in. barrel protruding from the slide and carrying its own removable front sight. An adjustable target-type rear sight is fitted, and the hammer has a spur added for easier thumb-cocking. See Walther PP elsewhere in this volume.

The highly-specialized Walther OSP International Rapid Fire .22 Short pistol has been revised into the GSP chambered for the .22 L.R. This is a blowback design with bolt fully-housed in the receiver and the magazine located ahead of the trigger guard. It carries finely-adjustable target sights and barrel balance weights.

MISCELLANEOUS PISTOLS

HIGH-STANDARD MFG. CORP., U.S.A.

Model D-100

This is a unique hinged-frame, double-barrel, derringer-type .22 caliber pistol of minimum size. It was introduced in 1962 as a purely defensive arm.

Characteristics of the High Standard Model D-100 Pistol

Caliber: .22 L.R.; .22 MRF.
Operation: Manual, double-action only.
Barrel length: 3-1/2 in.
Overall length: 5 in.
Weight: 11 oz.
Sights: Open, fixed.
Capacity: 2 rounds.

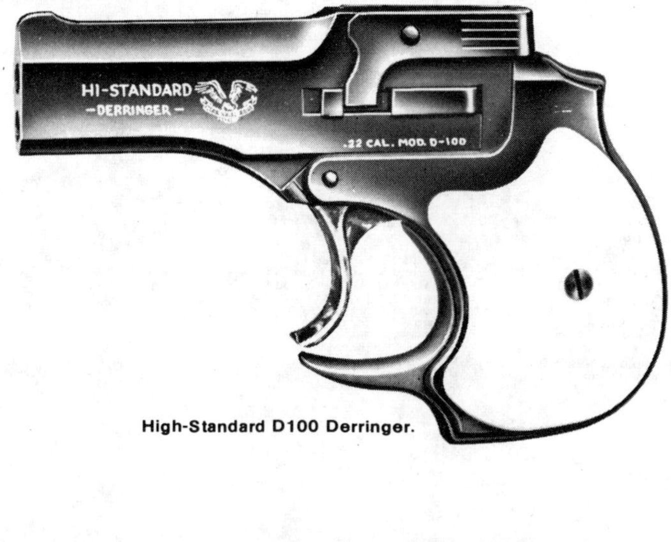

High-Standard D100 Derringer.

Remington XP-100 pistol.

REMINGTON ARMS CO.

Remington XP-100

The XP-100 is quite unusual in that it is the only bolt-action center-fire pistol produced commercially. It also has the distinction of being chambered for a special high-intensity bottleneck cartridge, the .221 Fireball, producing the highest velocity of any known pistol cartridge. The XP-100 was actually designed as a rifle-type bolt-action fitted with a short ribbed barrel and nylon pistol-style stock. Mechanically and functionally, this design is identical to the Remington M600 and M660 carbines; and, in fact, most action parts will interchange. For mechanical and operational details, see Remington M700.

Characteristics of the Remington XP-100 Pistol

Caliber: .221 Remington Fireball.
Operation by: Manual, rotating bolt.
Barrel length: 10-1/2 in.
Overall length: 16-3/4 in.
Sights: Open, adjustable.
Weight: 60 oz.
Magazine capacity: Single-shot.

Index

I

J

K

L

M

S